Trigonometric Formulas

$$\sin\theta = \frac{y}{r} \qquad \cos\theta = \frac{x}{r} \qquad \tan\theta = \frac{y}{x}$$

$$\csc\theta = \frac{r}{y} \qquad \sec\theta = \frac{r}{x} \qquad \cot\theta = \frac{x}{y}$$

$$\sin^2\theta + \cos^2\theta = 1$$
$$1 + \tan^2\theta = \sec^2\theta$$
$$1 + \cot^2\theta = \csc^2\theta$$

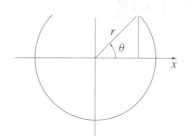

Cosine law- $c^2 = a^2 + b^2 - 2ab\cos C$

Sine law- $\dfrac{\sin A}{a} = \dfrac{\sin B}{b} = \dfrac{\sin C}{c}$

$\sin(A + B) = \sin A \cos B + \cos A \sin B$
$\sin(A - B) = \sin A \cos B - \cos A \sin B$
$\cos(A + B) = \cos A \cos B - \sin A \sin B$
$\cos(A - B) = \cos A \cos B + \sin A \sin B$

$$\tan(A + B) = \frac{\tan A + \tan B}{1 - \tan A \tan B} \qquad \tan(A - B) = \frac{\tan A - \tan B}{1 + \tan A \tan B}$$

$\sin 2A = 2\sin A \cos A$
$\cos 2A = \cos^2 A - \sin^2 A = 1 - 2\sin^2 A = 2\cos^2 A - 1$
$$\tan 2A = \frac{2\tan A}{1 - \tan^2 A}$$

$$\sin A \sin B = \frac{1}{2}[-\cos(A + B) + \cos(A - B)] \qquad \sin A - \sin B = 2\cos\left(\frac{A + B}{2}\right)\sin\left(\frac{A - B}{2}\right)$$

$$\sin A \cos B = \frac{1}{2}[\sin(A + B) + \sin(A - B)] \qquad \cos A + \cos B = 2\cos\left(\frac{A + B}{2}\right)\cos\left(\frac{A - B}{2}\right)$$

$$\cos A \cos B = \frac{1}{2}[\cos(A + B) + \cos(A - B)] \qquad \cos A - \cos B = -2\sin\left(\frac{A + B}{2}\right)\sin\left(\frac{A - B}{2}\right)$$

$$\sin A + \sin B = 2\sin\left(\frac{A + B}{2}\right)\cos\left(\frac{A - B}{2}\right)$$

Inverse Trigonometric Function	**Principal Values**	**Inverse Trigonometric Function**	**Principal Values**
$\mathrm{Sin}^{-1} x$	$-\dfrac{\pi}{2} \leq y \leq \dfrac{\pi}{2}$	$\mathrm{Cot}^{-1} x$	$0 < y < \pi$
$\mathrm{Tan}^{-1} x$	$-\dfrac{\pi}{2} < y < \dfrac{\pi}{2}$	$\mathrm{Csc}^{-1} x$	$-\pi < y \leq -\dfrac{\pi}{2}, \quad 0 < y \leq \dfrac{\pi}{2}$
$\mathrm{Cos}^{-1} x$	$0 \leq y \leq \pi$	$\mathrm{Sec}^{-1} x$	$-\pi \leq y < -\dfrac{\pi}{2}, \quad 0 \leq y < \dfrac{\pi}{2}$

Calculus for Engineers

Fourth
Edition

Calculus for Engineers

Fourth
Edition

Donald Trim

UNIVERSITY OF MANITOBA

PEARSON
Prentice
Hall

Toronto

Library and Archives Canada Cataloguing in Publication

Trim, Donald W.
    Calculus for engineers/Donald Trim.—4th ed.

Includes index.
ISBN-13   978-0-13-157713-8
ISBN-10   0-13-157713-1

1. Calculus—Textbooks. 2. Engineering mathematics—Textbooks. I. Title.

QA303.2.T75 2008   515   C2006-905116-X

ISBN-13   978-0-13-157713-8
ISBN-10   0-13-157713-1

Editor-in-Chief: Gary Bennett
Executive Marketing Manager: Marlene Olsavsky
Acquisitions Editor: Michelle Sartor
Developmental Editor: Kimberley Hermans
Production Editor: Marisa D'Andrea
Copy Editor: Gail Marsden
Proofreader: Margaret Bukta
Production Coordinator: Avinash Chandra
Page Layout: PreTeX, Inc.
Art Director: Julia Hall
Interior Design: Geoff Agnew
Cover Design: Maki Ikushima
Cover Image: First Light

8 9 10   V011   16 15 14 13

Printed and bound in United States.

In memory of my father, who,
like his son, was a man of few words.
Our love was expressed in other ways.

Brief Contents

Contents

About the Author

Donald W. Trim received his honours degree in mathematics and physics in 1965. After proceeding to his masters degree, he obtained his doctorate degree with a dissertation in general relativity. As a graduate student at the University of Waterloo, he discovered that teaching was his life's ambition. He quickly became well known for his teaching. In 1971, he was invited to become the first member of the Department of Applied Mathematics in the Faculty of Science at the University of Manitoba. In 1976, after only five years at the university, he received one of the university's highest awards for teaching excellence based on submissions by faculty and graduating students. At the University of Manitoba, no individual can receive this award twice within a ten-year span. Professor Trim won the award in 1988 and again in 2006. In 1997, he was awarded a 3M teaching fellowship. In addition, the Faculty of Engineering presented him with a gold replica of the engineer's iron ring in appreciation for his service.

Professor Trim's skills in teaching are reflected in the notes and books he has written. These include *Introduction to Applied Mathematics*, *Introduction to Complex Analysis and Its Applications*, *Applied Partial Differential Equations*, and this calculus text. All have received most favourable reviews from students. To bridge the gap between calculus and partial differential equations, he is currently writing a book on ordinary differential equations.

Preface

A Special Focus

This text has been written especially for students in engineering and the physical sciences, but students from other disciplines might well find this book appealing. Like every instructor, I have developed certain views on the best pedagogical approach to calculus. These views are often independent of the area of study of the student.

It is the choice of examples and exercises that help to focus a text for students in a specific discipline. Throughout this book, I have chosen to emphasize the physical applications of calculus. To this end, I have consulted engineers and physicists, and I have searched the literature carefully for physically meaningful examples and exercises. At the same time, there are many examples and exercises from the social sciences, business, medicine, and others.

Three categories of applications recur throughout the book to demonstrate the indispensibility of calculus in engineering and the physical sciences:

- Velocity, speed, and acceleration play a major role in many physical systems. In Chapter 3, velocity and acceleration are introduced as derivatives, and then given a fuller discussion in Chapter 4. They are treated again from an integration point of view in Chapter 5. Instead of expecting the use of standard physics formulas, I encourage students to state time conventions, choose coordinate systems, and develop whatever equations are necessary to solve a particular problem.

- Resistors, capacitors, and inductors and how they relate to charge and current are introduced through derivatives in Chapter 3. As calculus unfolds, they are further developed through antiderivatives.

- Differential equations provide mathematical models for a wide variety of applications. They are introduced briefly in Chapter 3 so that students become familiar with the ideas surrounding this important branch of calculus. Separable differential equations are treated fully in Chapter 5 along with a variety of applications. Other types of differential equations are discussed in detail in Chapter 15.

Other Features

A careful effort has been made to incorporate features that will enhance applications and facilitate learning.

- Chapter 1 provides a review of precalculus material (in addition to some non-review topics that are indispensable in some applications). Each review section begins with a diagnostic test for students to test their familiarity with the material. Should their test score be less than adequate, they are advised to study the material in the section, work the exercises, and retake the test. Even with acceptable test scores, students are advised to at least read all review sections.

- Electronic calculators or software packages such as Maple, Mathematica, MathCad, and MatLab are now commonplace in many engineering and science courses. And the use of these devices is becoming more and more prevalent in the workplace. Accordingly, I have taken advantage of the plotting facilities of calculators and computers in this book. At the same time, a great disparity exists among calculators and computers with respect to taking limits, differentiating, and integrating, and I have not therefore written a totally integrated text that uses these devices to do these operations.

- Numerous Examples with worked-out solutions are given throughout the book.

- Key terms are boldfaced where they are defined in the text, and they are listed near the end of each chapter.

- Important Definitions, Theorems, and Corollaries are highlighted where they are introduced in the text.

- A special icon, consisting of an exclamation mark in a triangular sign, is placed in the margin to alert the reader to pitfalls that commonly entrap students. Such pitfalls may be inappropriate calculations or misinterpretations of ideas. ▽

- A concise Summary is given near the end of each chapter.

- A list of the main Derivative and Integration Formulas is given on the inside of the back cover for easy reference. Some Geometric Formulas and Trigonometric Identities can be found inside the front cover.

- The Exercises at the ends of the sections and the Review Exercises at the ends of the chapters have been graded into three difficulty levels. Exercises with no asterisk are the easiest. They are routine problems designed to reinforce fundamentals. Exercises with one asterisk (*) demand more thought. This grade of exercise contains most of the applications. Fundamentals should be mastered before tackling applications to problems from other fields. Exercises with two asterisks (**) are the most challenging. They should be attempted only after most of the single-asterisk problems have been successfully solved.

- Some Exercises require the use of a calculator or computer. This may be due to intensive calculations with unwieldy numbers or because a detailed plot of a function is needed. These Exercises are marked with a calculator icon, ▤ . I stress the need for students to be able to draw graphs without the need of technology, but at the same time, I suggest they check their work with a computer plot. These Exercises are not given a calculator icon designation.

- Answers to Even-Numbered Exercises, except the challenging two-asterisk ones, are provided after the Appendices.

- Three Appendices are included near the back of the book: Appendix A on Mathematical Induction, Appendix B on Determinants, and Appendix C on Complex Numbers.

Approach

For the most part, the approach in this book is intuitive, making free use of geometry and familiar physical settings to motivate and illustrate concepts. For instance, limits are introduced intuitively, but an optional section is included to allow exposure to the mathematical definition of a limit. Derivatives are introduced and defined algebraically as instantaneous rates of change, and interpreted geometrically as slopes of tangent lines. The definite integral is introduced through two geometric and two physical problems, but its definition is algebraic.

In Sections 1.3 and 1.4, fundamentals of plane analytic geometry are reviewed in order to provide a way of visualizing problems. The most difficult part of a solution to a problem is frequently the initial step. Once started, the solution often unfolds smoothly and easily, but that first step sometimes seems impossible. One of the best ways to start a problem is with a diagram. A picture, no matter how rough, is invaluable in giving a "feeling" for what is going on. It displays the known facts surrounding the problem; it facilitates "seeing" what the problem really is and how it relates to the known facts; and it often suggests that all-important first step. Students should be encouraged to develop the habit of making diagrams at every opportunity — not just to solve problems, but to understand what calculus is all about.

The main goal of this book is to help students think logically. I do not provide an exhaustive list of formulas and procedures to memorize so that students can solve problems by rote. Instead, I provide a few formulas and, I hope, a great deal of insight into the ideas surrounding these formulas. By working on the examples and problems in this book, students learn how to organize facts and interpret them mathematically. They learn how to decide exactly what a particular problem is, and how to produce a step-by-step procedure by which to solve it.

Order of Topics

The order of topics in this book is fairly standard. The main exceptions are the following:

- Conic sections are introduced in Chapter 1 as illustrations of how the form of the equation of a curve dictates its shape and conversely how the shape of a curve influences its equation. Detailed discussions of properties of conic sections are delayed until Chapter 9.

- From the outset, we address what it means for x_0 to approximate a solution of the equation $f(x) = 0$ to a given number of decimal places or with error less than some specified amount. Approximations are introduced in a geometric way in Chapter 1 through the Zero Intermediate Value Theorem. Newton's iterative procedure for approximating solutions to equations is the first application of differentiation in Chapter 4. Numbers in subsequent applications need not then be contrived in order to lead to equations with elementary solutions. Throughout, I demand that the accuracy of approximations be clearly identified.

- Trigonometric, exponential, and logarithm functions are reviewed in Chapter 1. Inverse trigonometric and hyperbolic functions are also introduced in this chapter. Derivatives of these functions are developed in Chapter 3, making them available for applications of differentiation in Chapter 4 and integration in Chapters 5–7.

- Limits at infinity and infinite limits are presented in Chapter 2, along with general discussions on limits.

- Differential equations are so important to students in the physical sciences that separable equations are treated at the earliest opportunity, immediately after the introduction of indefinite integrals in Chapter 5.

- Applications of the definite integral in Chapter 7 are presented before techniques of integration in Chapter 8. In this way, students see the diversity of applications of integration and can appreciate why it is necessary to antidifferentiate more difficult functions. Applications are covered again in the examples and exercises of Chapter 8.

- It is customary to discuss series of numbers in order to familiarize students with the idea of convergence before treating Taylor and power series. Series of functions are treated first to accommodate schools that need Taylor series, but have little use for series of constants. Such would be the case if students required infinite series, say, only to study series solutions of differential equations. They would have little use for series of constants. My experience has shown that because students can so easily visualize partial sums of power series on graphing calculators and computers, they do not have difficulty beginning with series of functions. They can see convergence happening.

Treatment of Particular Topics

A subtle difference in how a topic is treated can significantly affect whether a student fully understands that topic. From my experiences in teaching calculus, I have developed certain preferences which I have incorporated in this book.

- In the review of logarithm functions in Chapter 1, logarithms are treated as powers and as the inverse of exponentials. When we study definite integrals with variable upper limits in Section 6.5, the integral definition of the natural logarithm function is given.

- Relative extrema are discussed in Section 4.3, whereas absolute extrema are discussed in Section 4.7. Because they are different and are used differently, they are treated in separate sections. Curve sketching uses relative extrema, while most applied extrema problems involve absolute extrema.

- The topic of curve sketching has been divided into two separate sections. Section 4.5 contains the standard approach to curve sketching using local extrema, concavity, points of inflection, limits, etc., but not technology. Section 4.6 begins with a computer or calculator plot of a function and uses calculus to analyze what we see (or do not see).

- Antiderivatives or indefinite integrals are treated in a chapter unto themselves, namely, Chapter 5. Relegating them to a section in a chapter on differentiation is insufficient. Conceptually, integration is simple — backwards differentiation. But this operation is so important, and it is so much more difficult to perform, only a separate chapter does it justice. Integration techniques are further explored in Chapter 8.

- Antidifferentiation is approached from an organizational viewpoint. The emphasis is on learning to organize an integrand into a form in which integration is obvious. Needless memorization of formulas is discouraged.

- Antiderivatives are used to evaluate definite integrals. This is the first fundamental theorem of integral calculus in Section 6.4. The fact that a definite integral with a variable upper limit (Section 6.5) can be regarded as an antiderivative is given secondary importance. At first, the importance of the antiderivative as a calculational tool for definite integrals is emphasized.

- In Chapter 6, the definite integral is motivated through four types of problems: area, volume, work, and fluid flow. At the same time, definite integrals are defined algebraically as limits of Riemann sums. This approach permits us to use the definition of a definite integral in a wide variety of applications without fear that it has been associated with any one application in particular. Definitions of double, triple, line, and surface integrals as limit-summations in multivariable calculus follow quite naturally.

- I prefer the notation Sin^{-1} for the inverse sine function. It is so natural once the notation $f^{-1}(x)$ has been introduced for inverse functions in general. The first letter is capitalized as a reminder that this is indeed an inverse trigonometric function and that the -1 superscript should not be interpreted as a power. In Section 1.8, where inverse trigonometric functions are introduced for the first time, the arcsin notation is also presented so that students become familiar with it.

- There is no agreement in the mathematical community on principal values for the inverse secant and inverse cosecant functions. My choice is based on practical reasons for engineers and physical scientists. For values outside the first quadrant, third-quadrant angles stated between $-\pi$ and $-\pi/2$ are used. Third-quadrant angles simplify derivatives of these functions. In addition, choosing angles between $-\pi$ and $-\pi/2$ matches the usual methods when working in polar coordinates and with complex variables. When polar angle θ is restricted to an interval of length 2π, it is most often chosen to be $-\pi < \theta < \pi$. This is especially true in complex analysis where principal values of the argument of a complex number are always chosen in this interval. Principal values of all multivalued functions (logarithm, inverse trigonometric, and inverse hyperbolic) use these principal values.

- I have chosen to keep the polar coordinate r nonnegative. Nothing is lost by this convention. Curves may have slightly different equations depending on which convention is used. But by choosing $r > 0$, considerable simplification is achieved when equations of curves in polar coordinates are rewritten in Cartesian coordinates.

- In Chapter 10, the limit ratio test for convergence of infinite series is applied only to series with positive terms. It is not used directly as a test for convergence of series with positive and negative terms. Convergence is discussed one step at a time. First, convergence of series of positive terms using the comparison, limit comparison, integral, limit ratio, and limit root tests are discussed. Next, series with positive and negative terms are introduced, and the above tests as tests for absolute convergence are used.

- Three-dimensional analytic geometry and vectors in Chapter 11 provide the tools for multivariable calculus. I stress the value of drawing curves and surfaces in space, using the curve sketching tools learned in Chapters 1–4. Such diagrams are essential to the evaluation of double, triple, line, and surface integrals, and to an appreciation of many of the ideas of differential calculus. Vectors are handled algebraically and geometrically. Every algebraic definition is interpreted geometrically, and every geometric definition is followed by an algebraic equivalent. Differentiation and integration of vectors dependent on a single parameter lead to discussions on tangent and normal vectors to curves, curvature and arc length, and three-dimensional kinematics.

- Gradients are useful in many areas of applied mathematics. I introduce gradients in Section 12.4, and apply them to directional derivatives and normal vectors to curves and surfaces in Sections 12.8 and 12.9. In this way, gradients are not associated with any particular application.

- Chain rules for composite multivariable functions are endless in variations and applications. In Chapter 12, I show how to appreciate each term in a chain rule as a contribution of particular variables to the overall rate of change of a function, and then to create a schematic diagram to handle the most complicated functional situations.

- In Chapter 13, definitions of double and triple integrals as limit-summations are analogous to the definition of the definite integral presented earlier. The evaluation by double and triple iterated integrals is geometric. Through representative boxes, rectangles, strips, and columns, I show how to visualize the summation process and affix appropriate limits to integrals. There is no algebraic manipulation of inequalities; a thoughtfully prepared diagram does it all. The geometric approach also helps us visualize integrations in polar, cylindrical, and spherical coordinates. I demonstrate that many of the applications of the definite integral in Chapter 7 are handled much more easily with double integrals.

- For simplicity, definitions are kept to a minimum in Chapter 14. One kind of line integral $f(x, y, z)ds$ is presented and then integrals of the form $Pdx + Qdy + Rdz$ are treated as a special case when $f(x, y, z)$ is the tangential component of some vector function defined along the curve. They can all be evaluated by substituting from parametric equations of the curve. Alternatively, it may be expedient to use Green's or Stokes's theorems or determine whether the integral is independent of path. Likewise, one surface integral $f(x, y, z)dS$ is presented, and then in many applications, $f(x, y, z)$ is treated as the normal component of some vector function defined on the surface.

- Chapter 15 provides a sound introduction to differential equations. It contains sections on separable equations, linear first-order equations, and second-order equations easily reduced to first-order equations. The exercises introduce (first-order) homogeneous and Bernoulli equations. Considerable space (four sections) is devoted to the important topic of linear differential equations. Applications to Newtonian mechanics, vibrating mass-spring systems, and electric circuits are discussed in detail. Other applications are introduced through examples and exercises.

New in This Edition

In writing this edition I have reviewed my approach to each topic, discussions pertaining to it, and choice of examples and exercises. The following changes are the most substantial:

- Each chapter contains "Consulting Projects"— problems that might typically arise in the physical sciences. Solutions require calculus, and are often somewhat complex. They give us the opportunity to illustrate how to think through a multistage problem; to organize its many facets; and provide a step-by-step, logical solution.

- Each section of Chapter 1 that contains review material contains a diagnostic test so that students can judge the preparation they need for calculus.

- Inverse trigonometric and hyperbolic functions have been moved to Chapter 1.

- It has always been my practice in teaching calculus to tell students that they should attempt to estimate an answer to a problem so that they'll know whether their solution to the problem is reasonable. I have incorporated this idea in this edition of the text (I often call it "ballparking" the answer).

- I have tried to emphasize the need to "see" and "feel" calculus — its principles, rules, and applications. This can often be done with a simple picture.

- Many new physical applications are included in the examples and exercises.

Supplements

A supplements package has been carefully prepared to aid instructors and students:

- An Instructor's Solutions Manual, written by the author, providing complete solutions to all the exercises in the text.

- An Instructor's Resource CD-ROM, including the Instructor's Solutions Manual (in PDF format), PowerPoints, Image Library, and TestGen. TestGen is a computerized test bank that allows instructors to edit existing questions, add new questions, and generate tests.

- A Student's Solutions Manual, providing complete solutions to all the even-numbered exercises (except those challenging questions marked with two asterisks), is available to students.

- A Text-Enrichment Website has been created for this text, containing additional material related to topics covered in the text. Visit this site at *www.pearsoned.ca/text/trim* to find material on vector analysis and flux and circulation, as well as translation and rotation of axes.

Acknowledgments

I am grateful to the many people who offered helpful suggestions for the preparation of this edition. I would particularly like to thank the following instructors who provided formal reviews:

Dejan Delic, Ryerson University
Steven J. Desjardins, University of Ottawa
Samuel Dube, Carleton University
David L. Harmsworth, University of Waterloo
Kahina Sid Idris, University of Windsor
Rachel Kuske, University of British Columbia
Patricia Nieva, University of Waterloo

Any errors in the text or in the supplements are the responsibility of the author. I would appreciate having them brought to my attention. Please send them to me care of Acquisitions Editor, Mathematics, Higher Education Division, Pearson Education Canada, 26 Prince Andrew Place, Toronto, Ontario, M3C 2T8.

1 | Calculus Preparation

1.1 Introduction

Over the years, I have learned that the most frequent reason for students failing to achieve optimum results in calculus courses is inadequate preparation. Students may understand calculus concepts completely, they may have every formula and every rule memorized perfectly, they may even have a clear idea of the procedures required to solve problems, but because they lack skills in algebra, geometry, and trigonometry, they cannot put their knowledge to work. It is unfortunate that they never get to demonstrate their calculus knowledge because of poor mathematical preparation. To obtain the best possible grades in calculus, it is essential to have excellent algebraic skills and a good grasp of the elements of analytic geometry and trigonometry. In this chapter we give you the opportunity to test your skills and knowledge in these three areas, and, should it be necessary, the means by which to make improvements. There are also sections that may contain material unfamiliar to many readers. Topics in these sections are essential to some of the applications of calculus; your instructor will indicate whether they are required for your course.

The calculus course at your institution may, or may not, spend time on the review sections in this chapter. If it does not, your instructor may advise you to review certain sections on your own. **DO IT!** You could regret ignoring review material in this chapter in order to get to calculus in Chapter 2 more quickly. Each review section opens with a diagnostic test to determine your knowledge of the material in the section. Give yourself the suggested time to take the test, no longer. You must not only be able to solve the test questions, but you must also be able to do so reasonably quickly. Do **NOT** use a calculator unless specifically instructed to do so. Answers are provided at the end of the section along with marks for each question. Assign yourself partial marks for partially correct answers, but try to be objective in doing so. It is difficult to be specific as to what constitutes an acceptable score on the diagnostic tests. Certainly a score of less than 50% indicates that detailed study of the section is required. A score of more than 50%, but not much more, would also suggest the need for careful review. Marks in the 80%–100% range indicate a good working knowledge of material in the section, but it could be beneficial to give the section a quick reading, paying particular attention to parts corresponding to incorrectly answered test questions. In addition, not only is it helpful to refresh your memory on concepts learned some time ago, it is also wise to become familiar with the terminology, notation, and conventions set forth in these sections. To improve your skills on, and knowledge of, the material in a review section, read the discussions and examples thoroughly, try as many of the exercises as you can, and then retake the diagnostic test. If you are conscientious in your work, we are confident that you will do much better the second time. You will be well rewarded for taking the time to do this; your calculus studies will be so much easier. In fact, the time you spend on calculus preparation now will more than compensate for extra time that you would spend on solving calculus problems later. Believe me; I have taught thousands of students just like you.

1.2 Solving Polynomial Equations

Here is the diagnostic test for this section; give yourself 40 minutes to do it. When you have marked your test using the answers at the end of the section, decide whether a brief reading or a thorough treatment is needed for material in this section.

DIAGNOSTIC TEST FOR SECTION 1.2

Find all real solutions for each of the following polynomial equations. Give multiplicities of any repeated roots.

1. $9x + 5 = 0$

2. $x^2 - 4x - 5 = 0$

3. $6x^2 + 7x - 20 = 0$

4. $x^2 + 3x + 10 = 0$

5. $x^2 + 6x - 4 = 0$

6. $36x^2 - 108x + 81 = 0$

7. $x^3 - 8 = 0$

8. $x^3 - 4x^2 + 5x - 2 = 0$

9. $x^4 - 3x^2 - 4 = 0$

10. $x^4 + x^3 + 3x^2 - x - 4 = 0$

11. $3x^3 + x^2 + x - 2 = 0$

12. $24x^3 + 2x^2 - 27x + 10 = 0$

In questions 13 and 14 you are given the zeros of the polynomial. Write each polynomial in factored form.

13. Zeros of $x^3 + 6x^2 - x - 30$ are $x = -5, -3, 2$.

14. Zeros of $2x^2 + 4x - 10$ are $x = -1 \pm \sqrt{6}$.

A real **polynomial** in x of degree n, where $n \geq 0$ is an integer, is an expression of the form

$$P_n(x) = a_n x^n + a_{n-1}x^{n-1} + \cdots + a_1 x + a_0 \tag{1.1}$$

where $a_n, a_{n-1}, \ldots, a_0$ are real constants with $a_n \neq 0$. (If a_n were equal to zero, the polynomial would not be of degree n.) When $P_n(x)$ is set equal to zero, the resulting equation

$$P_n(x) = a_n x^n + a_{n-1}x^{n-1} + \cdots + a_1 x + a_0 = 0 \tag{1.2}$$

is called a **polynomial equation** of degree n. Examples are $3x^3 - 2x + 5 = 0$ and $2x^2 + 5x - 9 = 0$. In this section, we are concerned with the number of solutions of a polynomial equation, whether solutions are real or complex, and techniques for finding the real solutions. (Complex solutions are dealt with in Appendix C.) Values of x that satisfy equation 1.2 are called **roots** or **solutions** of the equation. They are also called **zeros** of the polynomial $P_n(x)$.

When $n = 1$, equation 1.2 is called a **linear equation** (or equation of degree 1),

$$a_1 x + a_0 = 0. \tag{1.3}$$

Its only solution is $x = -a_0/a_1$. For example, the solution of $2x + 5 = 0$ is $x = -5/2$.

Quadratic equations (equations of degree 2) are obtained when $n = 2$. It is customary in this case to denote coefficients as follows

$$ax^2 + bx + c = 0. \tag{1.4}$$

Quadratic equations can sometimes be solved by factoring, and can always be solved with the quadratic formula. It is preferable to initially attempt to factor. For the example $6x^2 - 5x - 4 = 0$, we write $(3x - 4)(2x + 1) = 0$, from which the two solutions are $x = 4/3$ and $x = -1/2$. Likewise, for $x^2 - 6x + 9 = 0$, we factor in the form $(x - 3)^2 = 0$, and the only solution is $x = 3$. When the quadratic does not factor easily, we use the **quadratic formula**

$$x = \frac{-b \pm \sqrt{b^2 - 4ac}}{2a}. \tag{1.5}$$

Application of this formula to the equation $2x^2 + 4x - 5 = 0$ gives

$$x = \frac{-4 \pm \sqrt{4^2 - 4(2)(-5)}}{2(2)} = \frac{-4 \pm \sqrt{56}}{4} = \frac{-2 \pm \sqrt{14}}{2}.$$

Quadratic formula 1.5 indicates that quadratic equation 1.4 has two distinct real solutions when the **discriminant** $b^2 - 4ac > 0$, two real solutions that are equal when $b^2 - 4ac = 0$, and no real solutions when $b^2 - 4ac < 0$. (Solutions are complex numbers in the last case.)

The next simplest polynomial equation (after the linear and quadratic) is the cubic equation,

$$ax^3 + bx^2 + cx + d = 0, \qquad (1.6)$$

and after that the quartic,

$$ax^4 + bx^3 + cx^2 + dx + e = 0. \qquad (1.7)$$

There are procedures that give exact roots for both of these equations, but they are of so little practical use in this day of the electronic calculator and personal computer, we omit their discussions. For such equations, it is often sufficient to use a numerical procedure to approximate roots to some degree of accuracy (Sections 1.11 and 4.1). On the other hand, when an exact solution of a polynomial equation can be found, it can be removed from the equation, yielding a simpler equation to solve for the remaining roots. The process by which this is done is a result of the factor theorem.

THEOREM 1.1 (Factor Theorem)

$x - a$ is a factor of $P_n(x)$ if and only if $P_n(a) = 0$.

The factor theorem is very useful in solving polynomial equations. It does not find solutions, however. What it does is simplify the problem each time a solution is found. To illustrate, consider the quartic equation

$$P(x) = x^4 + 2x^3 + x^2 - 2x - 2 = 0. \qquad (1.8a)$$

A moment's reflection indicates that $x = 1$ satisfies the equation. The factor theorem then implies that $x - 1$ is a factor of the quartic. The remaining cubic factor can be obtained by long division, synthetic division, or mental long division. The result is

$$P(x) = x^4 + 2x^3 + x^2 - 2x - 2 = (x - 1)(x^3 + 3x^2 + 4x + 2).$$

What this means is that equation 1.8a can be replaced by

$$P(x) = (x - 1)(x^3 + 3x^2 + 4x + 2) = 0. \qquad (1.8b)$$

To find further solutions of quartic equation 1.8a, we need only examine the cubic $x^3 + 3x^2 + 4x + 2$ in 1.8b for its zeros. Once we notice that a zero is $x = -1$, we may factor $x + 1$ from the cubic and replace 1.8b with

$$P(x) = (x - 1)(x + 1)(x^2 + 2x + 2) = 0. \qquad (1.8c)$$

The remaining two solutions are complex numbers. Thus quartic equation 1.8a has two real solutions $x = \pm 1$ and two complex solutions.

Once a solution a of a polynomial equation has been found, $x - a$ can be factored from the polynomial. When a is a fraction, it is recommended that this procedure be modified slightly. To illustrate, consider the equation

$$2x^3 - x^2 - 9x + 9 = 0.$$

A solution of this equation is $x = 3/2$ (we show how we found this solution shortly). When we factor $x - 3/2$ from the cubic, the result is

$$(x - 3/2)(2x^2 + 2x - 6) = 0,$$

but the work involves fractions. They can be avoided by factoring $2(x - 3/2) = 2x - 3$ from the cubic instead. Calculations are simpler, and the result is

$$(2x - 3)(x^2 + x - 3) = 0.$$

The remaining two solutions can be obtained with the quadratic formula

$$x = \frac{-1 \pm \sqrt{1^2 - 4(1)(-3)}}{2} = \frac{-1 \pm \sqrt{13}}{2}.$$

What we are suggesting is that when a rational number $r = p/q$ is a solution of a polynomial equation (with integer coefficients), it is simpler to factor $q(x - p/q) = qx - p$ from the polynomial (and in so doing only integer arithmetic is involved).

All real polynomials can be factored into real linear factors and irreducible real quadratic factors. An **irreducible real quadratic factor** is one that has complex zeros, such as $x^2 + 1$ and $2x^2 + x + 6$; they are characterized by negative discriminants. Finding the linear factors of a polynomial goes hand-in-hand with finding real zeros of the polynomial. We have shown the factorization of three polynomials below.

$$x^4 + 2x^3 + x^2 - 2x - 2 = (x - 1)(x + 1)(x^2 + 2x + 2), \quad \text{(1.9a)}$$

$$x^3 - 3x^2 + 3x - 1 = (x - 1)^3, \quad \text{(1.9b)}$$

$$x^8 + 7x^7 - 86x^5 - 95x^4 + 363x^3 + 486x^2 - 540x - 648 = (x + 3)^4(x - 2)^3(x + 1). \quad \text{(1.9c)}$$

Each linear factor in these polynomials leads to a real zero of the polynomial. For polynomial 1.9a, the real zeros are ± 1, and it also has two complex zeros; for polynomial 1.9b, the zeros are 1, 1, 1; and for 1.9c, the zeros are $-3, -3, -3, -3, 2, 2, 2, -1$. In the case of 1.9b and 1.9c, there are repetitions. We say that $x = 1$ is a zero of **multiplicity** 3 for the polynomial in 1.9b; the multiplicity corresponds to the number of times the factor $x - 1$ appears in the factorization. Each of the zeros in 1.9a is of multiplicity 1. In 1.9c, $x = -3$ has multiplicity 4, the zero $x = 2$ has multiplicity 3, and $x = -1$ has multiplicity 1. These examples suggest that the number of zeros of a polynomial, taking multiplicities into account, is equal to the degree of the polynomial. This is confirmed in the following theorem.

THEOREM 1.2

Every polynomial of degree $n \geq 1$ has exactly n zeros (counting multiplicities).

EXAMPLE 1.1

Find all roots and their multiplicities for the polynomial equation

$$x^3 + x^2 - 16x + 20 = 0.$$

SOLUTION   With a little experimentation, we find that $x = 2$ is a solution of the equation. It follows that $x - 2$ can be factored from the cubic polynomial, and the equation can be written in the form

$$0 = (x - 2)(x^2 + 3x - 10) = (x - 2)(x + 5)(x - 2) = (x - 2)^2(x + 5).$$

Thus, $x = 2$ is a solution with multiplicity 2, and $x = -5$ is a solution with multiplicity 1.

The following theorem provides a quick way to locate all rational roots of a polynomial equation when its coefficients are rational numbers.

> **THEOREM 1.3 (Rational Root Theorem)**
> Suppose that $r = p/q$ is a rational root (in lowest terms) of a polynomial equation $a_n x^n + \cdots + a_1 x + a_0 = 0$ with integer coefficients, and $a_0 \neq 0$. Then, p divides a_0 and q divides a_n.

This is a powerful result. It narrows the field of possible rational solutions of polynomial equations with integer coefficients to a finite set. We illustrate with two examples.

EXAMPLE 1.2

Find all real solutions of the cubic equation $5x^3 + x^2 + x - 4 = 0$.

SOLUTION Since divisors of -4 are ± 1, ± 2, and ± 4, and those of 5 are ± 1 and ± 5, the only possible rational solutions are

$$\pm 1, \ \pm 2, \ \pm 4, \ \pm 1/5, \ \pm 2/5, \ \pm 4/5.$$

Trial and error leads to the fact that $x = 4/5$ satisfies the equation. We now factor $5(x - 4/5) = 5x - 4$ from the cubic,

$$0 = (5x - 4)(x^2 + x + 1).$$

The remaining two solutions are complex numbers.

EXAMPLE 1.3

Find all real solutions of the quartic equation $x^4 - x^3 - 3x^2 + 5x - 2 = 0$.

SOLUTION Since divisors of -2 are ± 1 and ± 2, and those of 1 are ± 1, the only possible rational solutions are the integers ± 1, ± 2. Since $x = 1$ is a solution, we factor $x - 1$ from the polynomial,

$$(x - 1)(x^3 - 3x + 2) = 0.$$

Once again, the only possible rational zeros of the cubic are ± 1, ± 2. We find that $x = 1$ is a zero, and when $x - 1$ is factored from the cubic,

$$0 = (x - 1)(x - 1)(x^2 + x - 2) = (x - 1)^2 (x + 2)(x - 1) = (x - 1)^3 (x + 2).$$

The only roots of the equation are therefore $x = 1$ (multiplicity 3) and $x = -2$.

The list of possible rational roots, as predicted by the rational root theorem, is shortest when any common factors in coefficients of the equation have been removed. For example, $3x^3 + 2x^2 - 2 = 0$ and $9x^3 + 6x^2 - 6 = 0$ have the same solutions. The rational root theorem yields ± 1, ± 2, $\pm 1/3$, and $\pm 2/3$ as possible solutions of the first equation, and ± 1, ± 2, ± 3, ± 6, $\pm 1/3$, $\pm 2/3$, $\pm 1/9$, and $\pm 2/9$ as possible solutions of the second equation. Clearly then it is advantageous to remove the common factor 3 from the second equation.

Given the zeros of a polynomial, it is easy to write the polynomial in factored form. For example, if the zeros of $x^3 + 3x^2 - 6x - 8$ are known to be $x = -4, -1, 2$, the factored form of the polynomial is $(x + 4)(x + 1)(x - 2)$. Zeros of $2x^2 + 9x - 5$ are $x = -5, 1/2$; its factored form is $2(x - 1/2)(x + 5)$ or $(2x - 1)(x + 5)$. Finally, the zeros of $2x^2 + 2x - 4$ are $x = -2, 1$; its factored form is $2(x - 1)(x + 2)$. Watch for the number preceding the product of factors. It can always be determined by examining the coefficient of the highest power of x.

EXERCISES 1.2

In Exercises 1–30 find all real solutions of the polynomial equation. Include multiplicities when they are greater than one.

1. $3x - 2 = 0$

2. $14x + 5 = 0$

3. $x^2 + 2x - 3 = 0$

4. $12x^2 + 11x - 5 = 0$

5. $2x^2 + 5x + 10 = 0$

6. $-4x^2 + 10x + 9 = 0$

7. $x^2 + 8x + 16 = 0$

8. $4x^2 - 36x + 81 = 0$

9. $2x^2 + 5x - 10 = 0$

10. $4x^2 - 8x + 9 = 0$

11. $x^3 - 3x^2 + 3x - 1 = 0$

12. $8x^3 + 12x^2 + 6x + 1 = 0$

13. $x^3 - 2x^2 + 5x - 10 = 0$

14. $x^3 + 4x^2 + 12x + 9 = 0$

* **15.** $x^3 + 12x^2 + 48x + 64 = 0$

* **16.** $x^4 + 7x^3 + 9x^2 - 21x - 36 = 0$

* **17.** $x^4 - 16 = 0$

* **18.** $2x^4 + 9x^3 - 6x^2 - 8x - 15 = 0$

* **19.** $6x^4 + x^3 + 53x^2 + 9x - 9 = 0$

* **20.** $12x^4 + 19x^2 + 5 = 0$

* **21.** $x^3 - 23x^2 - 21x - 72 = 0$

* **22.** $x^4 - 4x^3 - 44x^2 + 96x + 576 = 0$

* **23.** $3x^4 + x^3 + 5x^2 = 0$

* **24.** $6x^3 + x^2 + 19x - 20 = 0$

* **25.** $x^5 + 5x^4 - 9x - 45 = 0$

* **26.** $x^5 - 15x^4 + 85x^3 - 225x^2 + 274x - 120 = 0$

* **27.** $4x^4 + 4x^3 + 17x^2 + 16x + 4 = 0$

* **28.** $25x^4 - 120x^3 + 109x^2 - 36x + 4 = 0$

* **29.** $x^5 + 9x^4 + 47x^3 + 125x^2 + 18x - 200 = 0$

* **30.** $x^6 + 16x^4 - 81x^2 - 1296 = 0$

In Exercises 31–36 you are given a polynomial and its zeros. Write the polynomial in factored form.

31. $2x^2 + 8x - 10$, $x = -5, 1$

32. $2x^2 - 3x - 7$, $x = (3 \pm \sqrt{65})/4$

33. $x^3 - 11x^2 + 4x + 60$, $x = -2, 3, 10$

34. $24x^3 + 22x^2 - 27x + 5$, $x = -5/3, 1/4, 1/2$

35. $x^4 - 2x^3 + 2x - 1$, $x = -1, x = 1$ (muliplicity 3)

36. $16x^4 - 8x^2 + 1$, $x = \pm 1/2$ each of multiplicity 2

37. Find a polynomial that has only zeros $x = -1/3, 4/5, 3$, each with multiplicity 1 and $x = 4$ with multiplicity 3. Is it unique?

38. Prove the following corollary to the rational root theorem, called the "integer root theorem": If r is a rational zero of a polynomial $P_n(x) = x^n + a_{n-1}x^{n-1} + \cdots + a_0$ with integer coefficients, where $a_0 \neq 0$, then r is an integer that divides a_0.

ANSWERS TO DIAGNOSTIC TEST FOR SECTION 1.2

1. $-5/9$ (1 mark)

2. $-1, 5$ (2 marks)

3. $-5/2, 4/3$ (2 marks)

4. No real solutions (2 marks)

5. $-3 \pm \sqrt{13}$ (2 marks)

6. $3/2$ with multiplicity 2 (3 marks)

7. 2 (2 marks)

8. 1 with multiplicity 2, 2 (3 marks)

9. ± 2 (3 marks)

10. ± 1, (3 marks)

11. $2/3$ (3 marks)

12. $-5/4, 1/2, 2/3$ (4 marks)

13. $(x - 2)(x + 3)(x + 5)$ (2 marks)

14. $2(x + 1 - \sqrt{6})(x + 1 + \sqrt{6})$ (3 marks)

1.3 Plane Analytic Geometry and Straight Lines

Here is the diagnostic test for this section; give yourself 30 minutes to do it. When you have marked your test using the answers at the end of the section, decide whether a brief reading or a thorough treatment is needed for material in this section.

1. Find the distance between the points $(-1, 3)$ and $(2, -6)$ and the coordinates of the midpoint of the line segment that joins them.

2. Find the slope of the line $3x + 10y = 14$.

3. Find the x- and y-intercepts of the line $5x - 2y = 11$.

4. Find, in general form, the equation of the line through the point $(3, -5)$ with slope -2.

5. Determine whether the lines $2x - 3y = 4$ and $5x - 8y = 13$ are parallel, perpendicular, or neither.

6. Find the equation of the line through the point $(1, 0)$ parallel to the line $2x - 4y = 5$.

7. Find the equation of the line through the point $(4, -2)$ perpendicular to the line $3x + 2y = 9$.

8. Find the point of intersection of the lines $2x - 3y = 10$ and $x + 5y = 6$.

9. Find the distance from the point $(1, 2)$ to the line $3x - 4y = 5$.

10. Draw the three lines $x + 2y = 4$, $x = 2$, and $x = 3y - 5$ on the same set of axes.

The coordinates of a point

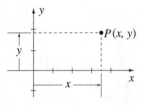

A sound knowledge of analytic geometry, a union of algebra and geometry, is essential to the study of calculus. On the one hand, it provides a way to describe geometric objects algebraically; on the other hand, it permits a geometric visualization of algebraic statements. Our approach to calculus is visual; we draw pictures at every possible opportunity, by hand, by graphing calculator, and by computer. We draw pictures to introduce ideas, to illustrate concepts, to reinforce principles, and to solve problems. We want you to *see* and *feel* calculus in all its aspects. Analytic geometry is the basis for many of these pictures. In this section, we review the fundamentals of plane analytic geometry, paying particular attention to straight lines. In Section 1.4 we review circles, parabolas, ellipses, and hyperbolas.

Points in the plane are identified by an ordered pair (x, y) of real numbers called their **Cartesian coordinates** (Figure 1.1). The x-coordinate of a point P is its perpendicular distance from the y-axis, and the y-coordinate is its perpendicular distance from the x-axis.

Signs of coordinates in the four quadrants

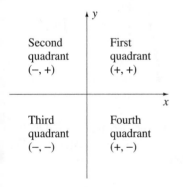

The axes divide the plane into four parts which, beginning at the upper right and proceeding counterclockwise, are called, respectively, the first, second, third, and fourth **quadrants**. The axes themselves are not considered part of any quadrant. Points in the first quadrant have x- and y-coordinates that are both positive; points in the second quadrant have a negative x-coordinate and a positive y-coordinate; and so on (Figure 1.2).

If P and Q are any two points with coordinates (x_1, y_1) and (x_2, y_2), respectively (Figure 1.3), then by the Pythagorean relation for the right-angled triangle PQS,

$$\|PQ\|^2 = \|PS\|^2 + \|QS\|^2,$$

where $\|PQ\|$ denotes the **length of** the **line segment** joining P and Q, and $\|PS\|$ and $\|QS\|$ denote the lengths of the related line segments. But[†]

Length of the line segment joining two arbitrary points

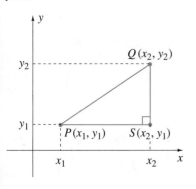

$$\|PS\| = |x_2 - x_1| \qquad \text{and} \qquad \|QS\| = |y_2 - y_1|,$$

and therefore[‡]

$$\|PQ\|^2 = |x_2 - x_1|^2 + |y_2 - y_1|^2 \qquad \text{or}$$

$$\|PQ\| = \sqrt{(x_2 - x_1)^2 + (y_2 - y_1)^2}. \tag{1.10}$$

[†]   The vertical lines around $|x_2 - x_1|$ and $|y_2 - y_1|$ denote **absolute values**. They operate on whatever is between them to produce a nonnegative result according to $|x| = x$ if $x \geq 0$ and $|x| = -x$ if $x < 0$. For example, $|4| = 4$ and $|-3| = 3$. We discuss absolute values from a functional point of view in Section 1.5.

[‡] $\sqrt{x}$ always denotes the *positive* square root of x, so that $\sqrt{4} = 2$.

Length of line joining
two specific points

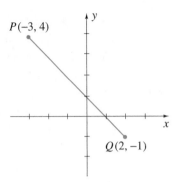

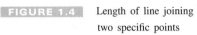

Midpoint of line segment

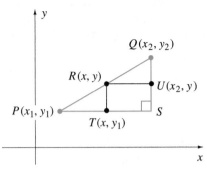

Equation 1.10 expresses the length of a line segment joining points P and Q in terms of their coordinates (x_1, y_1) and (x_2, y_2). For example, in Figure 1.4

$$\|PQ\| = \sqrt{(2+3)^2 + (-1-4)^2} = 5\sqrt{2}.$$

Suppose that $R(x, y)$ in Figure 1.5 is the **midpoint of** the **line segment** joining $P(x_1, y_1)$ and $Q(x_2, y_2)$. Since triangles PTR and RUQ are congruent, it follows that $\|RU\| = \|PT\|$, and for P and Q as shown in Figure 1.5, this condition becomes $x_2 - x = x - x_1$. Consequently,

$$x = \frac{x_1 + x_2}{2}. \tag{1.11a}$$

Similarly, equality of $\|RT\|$ and $\|QU\|$ gives

$$y = \frac{y_1 + y_2}{2}. \tag{1.11b}$$

Derivation of these results assumed that $x_2 > x_1$ and $y_2 > y_1$, but the same formulas are valid for any x_1, x_2, y_1, and y_2 whatsoever. Coordinates of the midpoint of a line segment, therefore, are averages of coordinates of the ends of the line segment.

An equation such as $y = x^2$ specifies a relationship between numbers represented by the letter x and those represented by the letter y: y must always be the square of x. Algebraically, we speak of pairs of values x and y that satisfy this equation. It is customary to write these pairs in the form (x, y), a few simple ones being $(0, 0)$, $(1, 1)$, $(-1, 1)$, $(2, 4)$, and $(-2, 4)$. If we interpret each pair of solutions (x, y) as **coordinates** of a point, we find that all such points lie on the curve in Figure 1.6. This curve, then, is a geometric visualization of solution pairs of the equation $y = x^2$. Every pair of values x and y that satisfies $y = x^2$ is represented by a point on the curve. Conversely, the coordinates (x, y) of every point on the curve provide a pair of numbers that satisfies $y = x^2$. Algebraic solutions, then, are represented geometrically as points on a curve; points on the curve provide algebraic solutions.

If x and y are the coordinates of a point in the plane, then the point is on the curve in Figure 1.6 if and only if x and y satisfy the equation $y = x^2$. In other words, this equation completely characterizes all points on the curve. We therefore call $y = x^2$ the equation of the curve. Thus the **equation of a curve** is an equation that the coordinates of every point on the curve satisfy and at the same time is an equation that no point off the curve satisfies. In the remainder of this section we discuss straight lines and their equations. In Section 1.4, we show that parabolas, circles, ellipses, and hyperbolas can be recognized by the distinctive forms of their equations. Once again note the algebraic-geometric interplay; the form of an equation dictates the shape of the curve, and, conversely, the shape of a curve determines the form of its equation.

Graph of
$y = x^2$

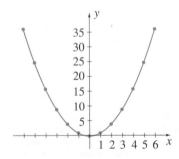

The Straight Line

The **slope of** the **line** through the points (x_1, y_1) and (x_2, y_2) in Figure 1.7a is defined as the quotient

$$m = \frac{y_2 - y_1}{x_2 - x_1}. \tag{1.12}$$

The difference $y_2 - y_1$ is called the **rise** because it represents the vertical distance between the points, and $x_2 - x_1$ the **run**, the horizontal distance between the points. It is easy to show using similar triangles that m is independent of the two points chosen on the line; that is, no matter which two points we choose on the line to evaluate m, the result is always the same. The four numbers x_1, x_2, y_1, and y_2 vary, but the ratio $(y_2 - y_1)/(x_2 - x_1)$ remains unchanged. A horizontal line (Figure 1.7b) has slope zero (since $y_2 - y_1 = 0$), whereas the slope of a vertical line is undefined (since $x_2 - x_1 = 0$). In Figure 1.7c, line l_1, which leans to the right, has positive slope, and line l_2, which leans to the left, has negative slope.

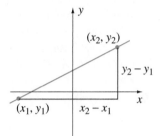

FIGURE 1.7a Slope of a line through two points

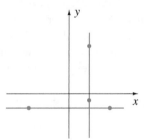

FIGURE 1.7b Slopes of horizontal and vertical lines

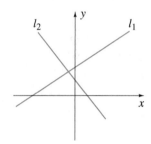

FIGURE 1.7c Lines with positive and negative slopes

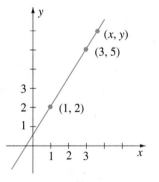

FIGURE 1.8 Equation of line through two given points

In Figure 1.8 we have shown the line through the points $(1, 2)$ and $(3, 5)$. Its slope is $(5 - 2)/(3 - 1) = 3/2$. To find the equation for this line, we let (x, y) be the coordinates of any other point on the line. Since the slope of the line must also be given by $(y - 2)/(x - 1)$, it follows that

$$\frac{y - 2}{x - 1} = \frac{3}{2},$$

and when this equation is simplified,

$$2y = 3x + 1.$$

If (x, y) are the coordinates of any point not on this line, then they do not satisfy this equation because the slope $(y-2)/(x-1)$ joining (x, y) to $(1, 2)$ is not equal to 3/2. Thus, $2y = 3x+1$ is the equation for the straight line through $(1, 2)$ and $(3, 5)$.

We can use this procedure to find the equation for the straight line through any point (x_1, y_1) with any slope m (Figure 1.9). If (x, y) are the coordinates of any other point on the line, then the slope of the line is $(y - y_1)/(x - x_1)$; therefore,

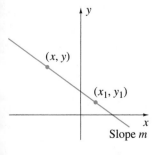

FIGURE 1.9 Equation of a line through given point with given slope

$$\frac{y - y_1}{x - x_1} = m \qquad \text{or} \qquad y - y_1 = m(x - x_1). \tag{1.13}$$

This is called the **point-slope formula** for the equation of a straight line; it uses the slope m of the line and a point (x_1, y_1) on the line to determine the equation of the line.

Other formulas for the equation of a line are available when different characteristics of the line are given. They are listed below. The point-slope formula is included for completeness.

Form of Equation	Name of Equation	Characteristics Determining the Line
$y - y_1 = m(x - x_1)$	Point-slope formula	m is the slope; (x_1, y_1) is a point on the line
$\dfrac{y - y_1}{y_2 - y_1} = \dfrac{x - x_1}{x_2 - x_1}$	Two-point formula	Two points (x_1, y_1) and (x_2, y_2) on the line
$y = mx + b$	Slope y-intercept formula	m is the slope; b is the y-intercept
$y = m(x - a)$	Slope x-intercept formula	m is the slope; a is the x-intercept
$\dfrac{x}{a} + \dfrac{y}{b} = 1$	Two-intercept formula	a and b are x- and y-intercepts

Point-slope formula 1.13 encompasses all of these. The characteristics in the right column determine the slope of a line and a point on it; therefore, with minimal calculations, the point-slope formula can be used in all situations. For example, if we know that the slope of a line is m and the y-intercept is b, then a point on the line is $(0, b)$. Hence, equation 1.13 gives $y - b = m(x - 0)$, or $y = mx + b$, the slope y-intercept formula.

There is one other equation that is worth mentioning. Vertical lines, which do not have slopes, cannot be represented in form 1.13; they have form $x = k$, for some constant k. However, all lines can, for various values of the constants A, B, and C, be represented in the form

$$Ax + By + C = 0. \tag{1.14}$$

This is often called the **general equation of a line**.

EXAMPLE 1.4

Find the general equation of the line through the points $(-1, 1)$ and $(2, 3)$.

SOLUTION   Since the slope of the line is $(3 - 1)/(2 + 1) = 2/3$, we can use the point $(-1, 1)$ and the slope $2/3$ in point-slope formula 1.13 to give

$$y - 1 = \frac{2}{3}(x + 1) \qquad \text{or} \qquad 2x - 3y + 5 = 0.$$

The same result is obtained if the point $(2, 3)$ is used in place of $(-1, 1)$. The two-point formula could also be used.

EXAMPLE 1.5

FIGURE 1.10   Graphs of $4y = 20 - 5x$ and $x = -1$

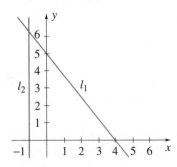

Find equations for the lines in Figure 1.10.

SOLUTION   Since the slope of l_1 is $-5/4$ and its y-intercept is 5, we use formula 1.13 with the point $(0, 5)$,

$$y - 5 = -\frac{5}{4}(x - 0) \qquad \text{or} \qquad 4y = 20 - 5x.$$

The equation of l_2 is clearly $x = -1$.

We do not recommend the use of the slope y-intercept formula $y = mx + b$ for finding the equation of a line; the point-slope formula $y - y_1 = m(x - x_1)$ is more versatile. The slope y-intercept formula is, however, useful in finding the slope of a line with given equation. For instance, to find the slope of the line $2x + 3y = 9$, we express it in slope y-intercept form $y = -2x/3 + 3$. It follows that the coefficient of x, namely, $-2/3$, is the slope of the line.

To find the point of intersection of two straight lines — say, $3y = 2x + 5$ and $4y + x = 14$ (Figure 1.11) — we find the point whose coordinates (x, y) satisfy both equations. If we solve each equation for x and equate the expressions, we obtain $(3y - 5)/2 = 14 - 4y$, which immediately yields $y = 3$. Either of the original equations then gives $x = 2$, and the point of intersection of the lines is $(2, 3)$. Note once again the algebraic–geometric interplay. Geometrically, $(2, 3)$ is the point of intersection of two straight lines. Algebraically, the two numbers constitute the solution of the equations $3y = 2x + 5$ and $4y + x = 14$.

Of particular importance to the study of lines are the concepts of parallelism and perpendicularity.

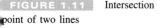
FIGURE 1.11 Intersection point of two lines

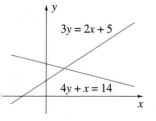

> **DEFINITION 1.1**
>
> Two distinct lines are said to be **parallel** if they have no point of intersection.

EXAMPLE 1.6

Verify that the lines with equations $2x - y = 4$ and $4x - 2y = 7$ are parallel.

SOLUTION   If we solve each of these equations for y and equate the resulting expressions, we obtain

$$2x - 4 = 2x - \frac{7}{2},$$

an obvious impossibility. Consequently, the lines do not intersect.

Geometrically, the following is clear.

> **THEOREM 1.4**
>
> Two distinct lines are parallel if and only if they have the same slope.

For example, if we write the equations of the lines in Example 1.6 in the form $y = 2x - 4$ and $y = 2x - 7/2$, coefficients of the x-terms identify the slopes of the lines. Since each has slope 2, the lines are parallel.

> **DEFINITION 1.2**
>
> Two lines are said to be **perpendicular** if they intersect at right angles.

The following theorem gives a test for perpendicularity of straight lines in terms of slopes.

> **THEOREM 1.5**
>
> Two lines with nonzero slopes m_1 and m_2 are perpendicular if and only if
>
> $$m_1 = -\frac{1}{m_2} \quad \text{or} \quad m_1 m_2 = -1. \tag{1.15}$$

EXAMPLE 1.7

Find the equation of the straight line that passes through the point $(2, 4)$ and is perpendicular to the line $3x + y = 5$.

SOLUTION   Since the slope of the given line $y = -3x + 5$ is -3, the required line has slope $1/3$. Using point-slope formula 1.13, we find that the equation of the required line is

$$y - 4 = \frac{1}{3}(x - 2) \qquad \text{or} \qquad x - 3y = -10.$$

There is a useful formula for finding the (shortest) distance, d, from a point (x_1, y_1) to a line with equation $Ax + By + C = 0$ (Figure 1.12). It is

$$d = \frac{|Ax_1 + By_1 + C|}{\sqrt{A^2 + B^2}}. \qquad (1.16)$$

For example, the distance from the point $(-2, -1)$ to the line $3x + 2y = 2$ in Figure 1.13 is

$$\frac{|3(-2) + 2(-1) - 2|}{\sqrt{3^2 + 2^2}} = \frac{10}{\sqrt{13}}.$$

FIGURE 1.12   Shortest distance from a point to a line

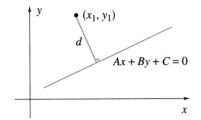

FIGURE 1.13   Distance from $(-2, -1)$ to $3x + 2y = 2$

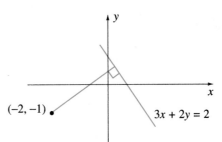

EXERCISES 1.3

In Exercises 1–4 find the distance between the points.

1. $(1, 3)$, $(3, 4)$

2. $(-2, 1)$, $(4, -2)$

3. $(-1, -2)$, $(-3, -8)$

4. $(3, 2)$, $(-4, -1)$

5.–8.   Find the midpoint of the line segment joining the points in Exercises 1–4.

In Exercises 9–18 draw the line described and find its equation.

9. Through the points $(1, 2)$ and $(-3, 4)$

10. Through the points $(3, -6)$ and $(5, -6)$

11. Through the point $(-2, -3)$ with slope 3

12. Through the point $(1, -3/2)$ with slope $-1/2$

13. The y-axis

14. The x-axis

15. Through the point $(4, 3)$ and crossing the x-axis at -2

16. Through the point $(-1, -2)$ and crossing the y-axis at 4

17. Crossing the x- and y-axes at 1 and -3, respectively

18. Through the origin and the midpoint of the line segment joining $(3, 4)$ and $(-7, 8)$

In Exercises 19–26 determine whether the lines are perpendicular, parallel, or neither.

19. $y = -x + 4, y = x + 6$

20. $x + 3y = 4, 2x + 6y = 7$

21. $x = 3y + 4, y = x/3 - 2$

22. $2x + 3y = 1, 3x - 2y = 5$

23. $y = 3x + 2, y = -x/2 + 1$

24. $x - y = 5, 2x + 3y = 4$

25. $x = 0, y = 5$

26. $x + y + 2 = 0, 3x - y = 4$

In Exercises 27–32 find the point of intersection of the lines.

27. $x + y = 0, x - 2y = -3$

28. $x = 1, y = 2$

29. $3x + 4y = 6, x - 6y = 3$

30. $y = 2x + 6, x = y + 4$

31. $x/2 + y/3 = 1, 2x - y/4 = 15$

32. $14x - 2y = 5, 3x + 10y = 12$

In Exercises 33–38 find the (shortest) distance from the point to the line.

33. $(3, 4)$ to $x + y = 1$ **34.** $(1, -3)$ to $x + 2y = 3$

35. $(5, 1)$ to $x - y = 4$ **36.** $(3, 1)$ to $y = -x$

37. $(6, -2)$ to $x = -1$ **38.** $(4, -2)$ to $15x - 2y + 3 = 0$

In Exercises 39–46 find the equation of the line described.

39. Parallel to $x + 2y = 15$, and through the point of intersection of $2x - y = 5$ and $x + y = 4$

40. Perpendicular to $x - y = 4$, and through the point of intersection of $2x + 3y = 3$ and $x - y = 4$

41. Parallel to the line through $(1, 2)$ and $(-3, 0)$, and through the point $(5, 6)$

42. Perpendicular to the line through $(-3, 4)$ and $(1, -2)$, and through the point $(-3, -2)$

43. Crossing through the first quadrant to form an isosceles triangle with area 8 square units

44. Through $(3, 5)$ and crossing through the first quadrant to form a triangle with area 30 square units

45. Has slope 2, and that part in the second quadrant has length 3

46. Parallel to the x-axis, below the point of intersection of the lines $x = y$ and $x + y = 4$, and forms with these lines a triangle with area 9 square units

47.   (a) What is the formula by which we convert temperature T_F in degrees Fahrenheit to temperature T_C in degrees Celsius?

(b) What is the formula by which we convert temperature T_C in degrees Celsius to temperature T_F in degrees Fahrenheit?

(c) Can we interpret the formulas in parts (a) and (b) as equations for straight lines in a $T_F T_C$-plane? Draw both lines.

(d) At what temperature are T_F and T_C numerically equal?

∗ **48.** Repeat Exercise 47 for the relationship between temperatures in degrees Fahrenheit, T_F, and in kelvin, T_K.

∗ **49.** Consider a metal bar of length L_0 at temperature T_0. If the bar is heated or cooled, its length changes. The amount is described by the coefficient of linear expansion α. It is the change in length per unit length per degree Celsius.

(a) Show that the length L of the bar at temperature T is given by the formula $L = L_0[1 + \alpha(T - T_0)]$. Draw its graph.

(b) If steel railroad rails 10 m long are laid with their adjacent ends 3 mm apart at a temperature of $20°C$, at what temperature will their ends be in contact? For steel, $\alpha = 1.17 \times 10^{-5}/°C$.

∗ **50.** A *median* of a triangle is a line segment drawn from a vertex to the midpoint of the opposite side. Find equations for the three medians of the triangle with vertices $(1, 1)$, $(3, 5)$, and $(0, 4)$. Show that all three medians intersect in a point called the *centroid* of the triangle.

∗ **51.** If (x_1, y_1), (x_2, y_2), (x_3, y_3), and (x_4, y_4) are vertices of any quadrilateral, show that the line segments joining the midpoints of adjacent sides form a parallelogram.

∗ **52.** Find the equation of the perpendicular bisector of the line segment joining $(-1, 2)$ and $(3, -4)$. (The perpendicular bisector is the line that cuts the line segment in half and is perpendicular to it.)

∗ **53.** Find coordinates of the point that is equidistant from the three points $(1, 2)$, $(-1, 4)$, and $(-3, 1)$.

∗ **54.** Prove Theorem 1.5.

∗ **55.** Generalize the result of equations 1.11a and b to prove that if a point R divides the length PQ so that

$$\frac{\text{length } PR}{\text{length } RQ} = \frac{r_1}{r_2},$$

where r_1 and r_2 are positive integers, then the coordinates of R are

$$x = \frac{r_1 x_2 + r_2 x_1}{r_1 + r_2} \quad \text{and} \quad y = \frac{r_1 y_2 + r_2 y_1}{r_1 + r_2}.$$

∗ **56.** Is a line parallel to itself?

∗∗ **57.** Prove that in any triangle the sum of the squares of the lengths of the medians is equal to three-fourths of the sum of the squares of the lengths of the sides. (A *median* of a triangle is a line segment drawn from one vertex to the midpoint of the opposite side.)

∗∗ **58.** Let P be any point inside an equilateral triangle (figure below). Show that the sum of the distances of P from the three sides is always equal to the height h of the triangle.

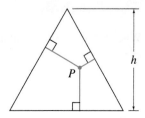

---

1. $3\sqrt{10}$, $(1/2, -3/2)$ (2 marks)

2. $-3/10$ (1 mark)

3. $11/5$, $-11/2$ (2 marks)

4. $2x + y - 1 = 0$ (2 marks)

5. Neither (2 marks)

6. $x - 2y = 1$ (3 marks)

7. $2x - 3y = 14$ (3 marks)

8. $(68/13, 2/13)$ (3 marks)

9. 2 (2 marks)

10. (5 marks)

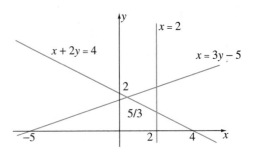

1.4 Conic Sections

Here is the diagnostic test for this section. Give yourself 60 minutes to do it.

In questions 1–10 identify the curve as a straight line, parabola, circle, ellipse, hyperbola, or none of these.

1. $3x - y = 4$

2. $x = y^2 - 2y + 3$

3. $y = x^4 + 3$

4. $3x^2 + y^2 = 4$

5. $y^2 - 2x^2 = x$

6. $x^2 + 4y^2 + 5 = 0$

7. $x^2 + y^2 - 2y = 16$

8. $3x + y^2 = 3$

9. $x^2 + 2y^2 + x = 2y$

10. $x^2 - x + y^2 + y = 0$

11. Which, if any, of the points $(1, 0)$, $(0, 1)$, and $(2, -1)$ are on the circle $3x^2 + 3y^2 + 2y = 6x + 5$?

12. Find the centre and radius of the circle $x^2 + y^2 + 2x - 4y = 25$.

13. Find the centre of the ellipse $2x^2 + 4x + 6y^2 + 9y = 36$. What are its axes of symmetry?

14. Find equations for the asymptotes of the hyperbola $x^2 - 3y^2 = 4$.

15. Find equations for the asymptotes of the hyperbola $4x^2 - 2y^2 = 4x - 10y$.

16. Find the point(s) of intersection of the curves $x + 2y = 5$ and $x^2 + 2y^2 = 9$.

17. Find the highest point on the parabola $y = -x^2 + 6x + 4$.

18. Find the points where the parabola $x = y^2 - 5y - 6$ crosses the x-axis and the y-axis.

19. Find the equation of a parabola that passes through the points $(-1, 1)$, $(0, 3)$, and $(1, 1)$.

20. Find the equation of the ellipse with centre at the origin that passes through the points $(-4, 0)$ and $(3, 4)$.

Parabolas, circles, ellipses, and hyperbolas arise in a multitude of applications. We do not give a complete development of these curves together with their many properties here; we show only how the form of the equation for each conic section relates to its shape. Detailed discussions of conic sections in terms of focus and directrix are given in Sections 9.5 and 9.6.

The Parabola

When the y-coordinate of a point (x, y) on a curve is related to its x-coordinate by an equation of the form

$$y = ax^2 + bx + c, \qquad (1.17)$$

where a, b, and c are constants (with $a \neq 0$), the curve is called a **parabola**. The simplest of all parabolas is $y = x^2$ (Figure 1.6). For every point (x, y) to the right of the y-axis on this curve, there is a point equidistant to the left of the y-axis which has the same y-coordinate; that is, the point $(-x, y)$ is also on the curve. Putting it another way, that part of the parabola to the left of the y-axis is the image in the y-axis (thought of as a mirror) of that part to the right of the y-axis. Such a curve is said to be **symmetric about the y-axis**. It happens whenever the equation of a curve is unchanged when each x therein is replaced by $-x$. In other words, we have the test: *A curve is symmetric about the y-axis if its equation remains unchanged when x is replaced by $-x$.*

The parabola $y = 9 - x^2$ is shown in Figure 1.14. It is symmetric about the y-axis; replacing x by $-x$ leaves the equation unchanged. The parabola is said to *open downward*, whereas the parabola $y = x^2$ in Figure 1.6 *opens upward*. The sign of the coefficient of x^2 dictates which way a parabola opens (positive for upward and negative for downward).

The parabola $y = 2x^2 + 4x - 6$ is plotted in Figure 1.15. It is not symmetric about the y-axis; its equation changes when x is replaced by $-x$. The parabola appears to be symmetric about the line $x = -1$ with lowest point $(-1, -8)$. To prove that this is indeed the case, we rewrite the equation as

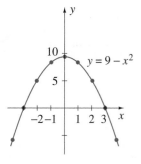

FIGURE 1.14 Parabola opening downward

$$y = 2(x^2 + 2x) - 6 = 2(x + 1)^2 - 8.$$

This form clearly indicates that the smallest value for y is -8, and it occurs when $x = -1$. It also indicates that the parabola is symmetric about $x = -1$. To see this we replace x with $-1 \pm a$, where $a > 0$ is any number whatsoever. The result is $y = 2(-1 \pm a + 1)^2 - 8 = 2a^2 - 8$. This shows that points on the parabola with x-coordinates smaller than -1 and larger than -1 by the same amount a have the same y-coordinate, namely $2a^2 - 8$.

The technique used above to rewrite $y = 2x^2 + 4x - 6$ as $2(x + 1)^2 - 8$ is called *completing the square*. If we apply the same technique to the general parabola $y = ax^2 + bx + c$, we obtain

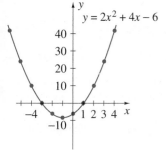

FIGURE 1.15 Parabola opening upward

$$y = a\left(x^2 + \frac{b}{a}x\right) + c = a\left(x + \frac{b}{2a}\right)^2 + \left(c - \frac{b^2}{4a}\right). \qquad (1.18)$$

This form for the equation of the parabola shows the following:

1. When $a > 0$, the parabola opens upward and has a minimum at the point

$$\left(-\frac{b}{2a}, c - \frac{b^2}{4a}\right).$$

2. When $a < 0$, the parabola opens downward and has a maximum at the point

$$\left(-\frac{b}{2a}, c - \frac{b^2}{4a}\right).$$

3. The parabola is symmetric about the line $x = -b/(2a)$ (Figure 1.16); that is, for every point P on one side of the line, there is a point Q on the other side that is the mirror image of P in the line.

4. The parabola crosses the x-axis when

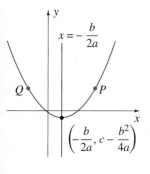

FIGURE 1.16 Symmetry of general parabola

$$0 = y = a\left(x + \frac{b}{2a}\right)^2 + \left(c - \frac{b^2}{4a}\right).$$

To solve this equation for x, we first write

$$\left(x + \frac{b}{2a}\right)^2 = \frac{b^2}{4a^2} - \frac{c}{a},$$

and then take square roots of each side,

$$x + \frac{b}{2a} = \pm\sqrt{\frac{b^2}{4a^2} - \frac{c}{a}}.$$

FIGURE 1.17a Parabola that intersects x-axis in two points

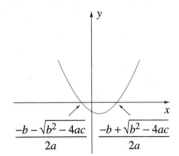

FIGURE 1.17b Parabola that touches x-axis at one point

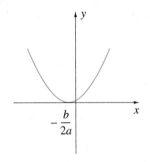

FIGURE 1.17c Parabola that does not intersect x-axis

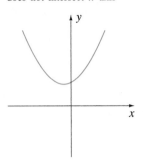

Finally, then,

$$x = -\frac{b}{2a} \pm \sqrt{\frac{b^2}{4a^2} - \frac{c}{a}} = \frac{-b \pm \sqrt{b^2 - 4ac}}{2a}. \qquad (1.19)$$

This is quadratic formula 1.5 that we encountered in Section 1.2, but now in a geometric setting. It determines points where the parabola $y = ax^2 + bx + c$ crosses the x-axis. When $b^2 - 4ac > 0$, the parabola crosses the x-axis twice (Figure 1.17a); when $b^2 - 4ac = 0$, it touches the x-axis at one point (Figure 1.17b); and when $b^2 - 4ac < 0$, the parabola has no points in common with the x-axis (Figure 1.17c).

When x and y in equation 1.17 are interchanged, the resulting parabola,

$$x = ay^2 + by + c, \qquad (1.20)$$

opens to the right or left rather than up or down. Figures 1.18 and 1.19 show the parabolas $x = y^2 + 1$ and $x = -y^2 + 4y - 4$, respectively. The parabola $x = y^2 + 1$ is symmetric about the x-axis; any point (x, y) on the parabola is symmetric with the corresponding point $(x, -y)$. In general, *a curve is symmetric about the x-axis if its equation remains unchanged when y is replaced by* $-y$.

FIGURE 1.18 Parabola opening to the right

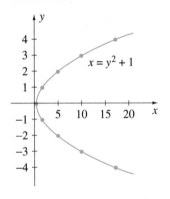

FIGURE 1.19 Parabola opening to the left

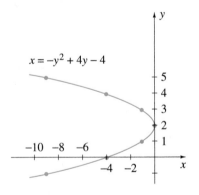

EXAMPLE 1.8

Find equations for the parabolas in Figures 1.20 and 1.21.

SOLUTION The fact that the parabola in Figure 1.20 is symmetric about the x-axis means that b in equation 1.20 must vanish; that is, its equation must be of the form $x = ay^2 + c$. Since the points $(2, 0)$ and $(0, 3)$ are on the parabola, their coordinates must satisfy the equation of the parabola,

$$2 = a(0)^2 + c, \qquad 0 = a(3)^2 + c.$$

FIGURE 1.20   Parabola through three special points

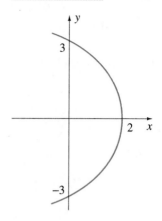

FIGURE 1.21   Parabola through any three points

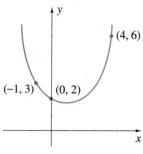

These imply that $c = 2$ and $a = -c/9 = -2/9$. Thus, the equation of the parabola is $x = -2y^2/9 + 2$.

The parabola in Figure 1.21 has no special attributes that we can utilize (such as the position of the line of symmetry in Figure 1.20). We therefore use the facts that its equation must be of the form $y = ax^2 + bx + c$, and the three points $(-1, 3)$, $(0, 2)$, and $(4, 6)$ are on the parabola. Substitution of these coordinates into the equation gives

$$3 = a(-1)^2 + b(-1) + c, \quad 2 = a(0)^2 + b(0) + c, \quad 6 = a(4)^2 + b(4) + c.$$

The second equation yields $c = 2$, and when this is substituted into the other two equations,

$$a - b = 1, \quad 16a + 4b = 4.$$

These can be solved for $a = 2/5$ and $b = -3/5$; therefore, the required equation is $y = 2x^2/5 - 3x/5 + 2$.

Suppose we had not given you the pictures in this example, but only asked for the equation of a parabola passing through the points. There would have been an additional parabola through the points $(-1, 3)$, $(0, 2)$, and $(4, 6)$, one opening to the right; its equation is $x = (2y^2 - 13y + 18)/3$. Only a parabola opening to the left can be found through the points $(0, 3)$, $(0, -3)$, and $(2, 0)$.

EXAMPLE 1.9

When a shell is fired from the artillery gun in Figure 1.22, it follows a parabolic path

$$y = -\frac{4.905}{v^2 \cos^2 \theta} x^2 + x \tan \theta,$$

where v is the muzzle velocity of the shell and θ is the angle at which the shell is fired. Find the range R of the shell and the maximum height attained by the shell.

FIGURE 1.22   Trajectory of an artillery shell

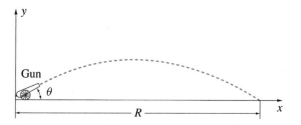

SOLUTION   We can find R by setting $y = 0$ and solving for x:

$$0 = \left(\frac{-4.905}{v^2 \cos^2 \theta} x + \tan \theta \right) x.$$

One solution is $x = 0$, corresponding to the firing position of the shell. The other solution gives the range of the shell,

$$x = R = \frac{\tan \theta \; v^2 \cos^2 \theta}{4.905} = \frac{v^2 \sin \theta \; \cos \theta}{4.905}.$$

Maximum height of the shell is attained when $x = R/2$, in which case

$$y = \frac{-4.905}{v^2 \cos^2 \theta} \left[\frac{v^2 \sin \theta \; \cos \theta}{2(4.905)} \right]^2 + \tan \theta \left[\frac{v^2 \sin \theta \; \cos \theta}{2(4.905)} \right]$$

$$= \frac{-v^2 \sin^2 \theta}{19.62} + \frac{v^2 \sin^2 \theta}{9.81} = \frac{v^2 \sin^2 \theta}{19.62}.$$

In many of the examples and exercises of the book, we ask you to draw and/or plot curves. To *plot* a curve, you are to use a graphing calculator or computer. To *draw* a curve, you are to do so without these devices. Sometimes, as we shall see, a drawing is more informative than a plot.

In Exercises 1–12 draw the parabola. Use a calculator or computer to plot the parabola as a check.

1. $y = 2x^2 - 1$

2. $y = -x^2 + 4x - 3$

3. $y = x^2 - 2x + 1$

4. $3x = 4y^2 - 1$

5. $x = y^2 + 2y$

6. $2y = -x^2 + 3x + 4$

7. $x + y^2 = 1$

8. $2y^2 + x = 3y + 5$

9. $y = 4x^2 + 5x + 10$

10. $x = 10y^2$

11. $y = -x^2 + 6x - 9$

12. $x = -(4 + y)^2$

13. Find x- and y-intercepts for the parabolas (a) $y = x^2 - 2x - 5$ and (b) $x = 4y^2 - 8y + 4$.

In Exercises 14–17 determine the equation for each parabola shown.

14.

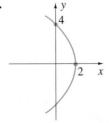

15.

16.

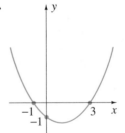

17.

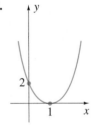

In Exercises 18–23 find all points of intersection for the curves. In each case draw or plot the curves.

18. $y = 1 - x^2$, $y = x + 1$

19. $y + 2x = 0$, $y = 1 + x^2$

20. $y = 2x - x^2 - 6$, $25 + x = 5y$

21. $x = y(y - 1)$, $2y = 2x + 1$

22. $x = -y^2 + 1$, $x = y^2 + 2y - 3$

23. $y = 6x^2 - 2$, $y = x^2 + x + 1$

* **24.** For what angle θ is the range of the artillery shell in Example 1.9 largest?

* **25.** The cable of the suspension bridge in the following figure hangs in the shape of a parabola. The towers are 200 m apart and extend 50 m above the roadway. If the cable is 10 m above the roadway at its lowest point, find the length of the supporting rods 30 m from the towers.

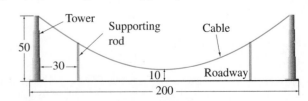

26. Find points of intersection for the parabolas $y = (x - 2)^2$ and $5x = y^2 + 4$.

27. Find the height of the parabolic arch in the figure below.

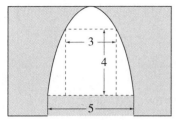

28. Determine the equation of a parabola of type 1.17 passing through the points $(1, 2)$, $(-3, 10)$, and $(3, 4)$.

29. Resistance R in ohms in a platinum resistance thermometer is related to temperature T in degrees Celsius by the equation

$$R = R_0(1 + aT + bT^2),$$

where R_0, a, and b are constants.

(a) Use the facts that resistances at temperatures $0°C$, $100°C$, and $700°C$ are, respectively, $10.000 \ \Omega$, $13.946 \ \Omega$, and $24.172 \ \Omega$ to determine R_0, a, and b.

(b) Plot a graph of the function on the interval $0 \le T \le 700$.

(c) At what temperature is the resistance $20 \ \Omega$?

∗ **30.** The parabola $y = x^2 - 1$ in the following figure represents the base of a wall perpendicular to the xy-plane (x and y measured in metres). A rope is attached to a stake at position $(3, 4)$, pulled tight, wrapped around that part of the base of the wall containing the vertex of the parabola, and tied to the vertex. At what point does it meet the wall?

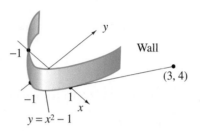

The Circle

When the x- and y-coordinates of points on a curve are related by an equation of the form

$$(x - h)^2 + (y - k)^2 = r^2, \qquad (1.21)$$

where h, k, and $r > 0$ are constants, the curve is called a **circle**. It takes but a quick recollection of distance formula 1.10 to convince ourselves that this definition of a circle coincides with our intuitive idea of a circle. If we write equation 1.21 in the form

$$\sqrt{(x - h)^2 + (y - k)^2} = r,$$

the left side is the distance from the point (x, y) to the point (h, k). Equation 1.21 therefore describes all points (x, y) at a fixed distance r from (h, k), a circle centred at (h, k) with radius r (Figure 1.23). For example, the radius of the circle in Figure 1.24 is equal to 2, and its equation is therefore

$$(x + 1)^2 + (y - 2)^2 = 4.$$

When equation 1.21 is expanded, we have

$$x^2 - 2hx + h^2 + y^2 - 2ky + k^2 = r^2 \qquad \text{or}$$

$$x^2 + y^2 - 2hx - 2ky + h^2 + k^2 - r^2 = 0.$$

This shows that the equation of a circle may be given in another form, namely,

$$x^2 + y^2 + fx + gy + e = 0, \qquad (1.22)$$

where e, f, and g are constants. Given this equation, the centre and the radius can be identified by reversing the expansion and completing the squares of $x^2 + fx$ and $y^2 + gy$. For instance, if $x^2 + y^2 + 2x - 3y - 5 = 0$, then

$$0 = (x + 1)^2 + (y - 3/2)^2 - 5 - 1 - 9/4 = (x + 1)^2 + (y - 3/2)^2 - 33/4.$$

The centre of the circle is therefore $(-1, 3/2)$ and its radius is $\sqrt{33}/2$.

FIGURE 1.23 Circle with centre (h, k) and radius r

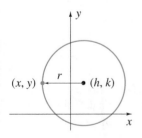

FIGURE 1.24 Circle with centre $(-1, 2)$ and radius 2

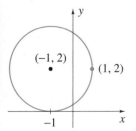

When the centre of a circle is the origin $(0, 0)$, equation 1.21 simplifies to

$$x^2 + y^2 = r^2. \tag{1.23}$$

 Be careful to use equation 1.21, not equation 1.23, when the centre of the circle is not the origin. It is a common error to use equation 1.23.

EXAMPLE 1.10

Figure 1.25 shows an arc of a circle. Find the equation for the circle.

FIGURE 1.25 Equation of a circle passing through 3 points

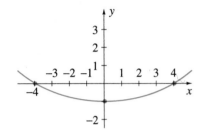

SOLUTION From the symmetry of the figure, we see that the centre of the circle is on the y-axis. Its equation must be of the form

$$x^2 + (y - k)^2 = r^2.$$

Because $(4, 0)$ is a point on the circle, these coordinates must satisfy its equation; that is,

$$16 + (0 - k)^2 = r^2 \quad \text{or} \quad 16 + k^2 = r^2.$$

Similarly, since $(0, -1)$ is on the circle,

$$(-1 - k)^2 = r^2 \quad \text{or} \quad 1 + 2k + k^2 = r^2.$$

If we subtract these two equations, we obtain $2k - 15 = 0$, from which we see that $k = 15/2$. Consequently, $r^2 = 16 + k^2 = 16 + 225/4 = 289/4$, and the equation of the circle is

$$x^2 + (y - 15/2)^2 = 289/4.$$

EXAMPLE 1.11

The straight lines in Figure 1.26a represent a welded frame where AOB is a right angle. The circle represents a wheel rotating on a pin through O perpendicular to the plane of the frame. If the minimum clearance between circle and side AB must be 10 cm, find the maximum radius of the wheel. Locate the point on the circle closest to AB.

FIGURE 1.26a Wheel spinning on welded frame

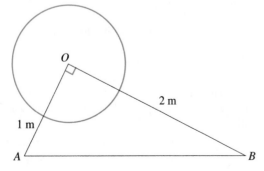

FIGURE 1.26b Reorientation of wheel on frame

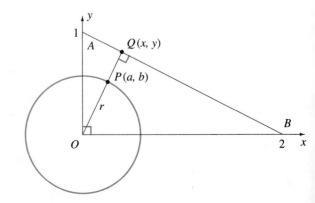

SOLUTION Suppose we rotate and flip the figure and establish the coordinate system in Figure 1.26b. Let the maximum radius of the circle be r and let $P(a, b)$ be the point on the circle closest to AB. Intuitively, the shortest distance $\|PQ\|$ between wheel and AB occurs when line OPQ is perpendicular to AB. Since the slope of the line AB is $-1/2$, its equation is

$$y - 1 = -\frac{1}{2}(x - 0) \qquad \Longrightarrow \qquad x + 2y - 2 = 0.$$

According to formula 1.16, the distance from the origin to line AB is

$$\|OQ\| = \frac{|(0) + 2(0) - 2|}{\sqrt{1^2 + 2^2}} = \frac{2}{\sqrt{5}}.$$

For the length of PQ to be $1/10$ m, it follows that $r + 1/10 = 2/\sqrt{5}$, and the radius of the wheel is $r = (2/\sqrt{5} - 1/10)$ m. To locate P, we note that the slope of OP is 2 (the negative of the reciprocal of the slope of AB). It follows that $b/a = 2$. Furthermore, because P is on the circle (whose equation is $x^2 + y^2 = r^2$), we must have $a^2 + b^2 = r^2$. When we substitute $b = 2a$ into this equation, we obtain

$$a^2 + 4a^2 = r^2 \qquad \Longrightarrow \qquad a = \frac{r}{\sqrt{5}} \quad \text{and} \quad b = \frac{2r}{\sqrt{5}}.$$

EXERCISES 1.4B

In Exercises 1–10 draw the circle. Use a calculator or computer to plot the circle as a check.

1. $x^2 + y^2 = 50$

2. $(x + 5)^2 + (y - 2)^2 = 6$

3. $x^2 + 2x + y^2 = 15$

4. $x^2 + y^2 - 4y + 1 = 0$

5. $x^2 - 2x + y^2 - 2y + 1 = 0$

6. $2x^2 + 2y^2 + 6x = 25$

7. $3x^2 + 3y^2 + 4x - 2y = 6$

8. $x^2 + 4x + y^2 - 2y = 5$

9. $x^2 + y^2 - 2x - 4y + 5 = 0$

10. $x^2 + y^2 + 6x + 3y + 20 = 0$

In Exercises 11–16 find an equation for the circle.

11.

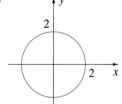

12.

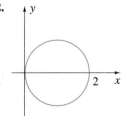

13.

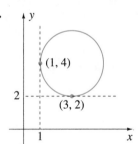

14.

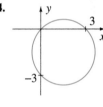

15.

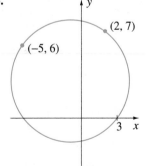

16.

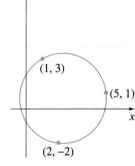

* **17.** A ladder of length L rests vertically against a wall. If the lower end of the ladder is moved along level ground away from the wall while the top of the ladder remains in contact with the wall, find an equation for the curve followed by the midpoint of the ladder.

* **18.** Find the equation of a circle that passes through the points $(3, 4)$ and $(1, -10)$, and has its centre (a) on the line $2x + 3y + 16 = 0$ and (b) on the line $x + 7y + 19 = 0$.

In Exercises 19–22 find points of intersection for the curves.

* **19.** $x^2 + 2x + y^2 = 4$, $y = 3x + 2$

* **20.** $x^2 + y^2 - 4y + 1 = 0$, $2x + y = 1$

* **21.** $x^2 + y^2 = 9$, $y = 3x^2 + 4$

* **22.** $(x + 3)^2 + y^2 = 25$, $y^2 = 16(x + 1)$

* **23.** Show that every equation of the form 1.22 represents a circle, a point, or nothing at all.

* **24.** Prove that the perpendicular bisector of a chord of a circle always passes through the centre of the circle.

* **25.** Two lights are 100 m apart, one at the origin, and the other at point $(100, 0)$ in the xy-plane. The light at the origin is 10 times as bright as the other light. Find, and draw, all points in the xy-plane at which the amount of light received from both sources is the same. Assume that the amount of light received at a point is directly proportional to the brightness of the source and inversely proportional to the square of the distance from the source.

* **26.** Two loudspeakers are 20 m apart. One is at the origin, and the other is at the point $(0, 20)$ in the xy-plane. The speaker at the origin is only 70% as loud as the other. Find, and draw, all points in the xy-plane at which the amount of sound received from both speakers is the same. Assume that the amount of sound received at a point is directly proportional to the loudness of the speaker and inversely proportional to the square of the distance from the speaker.

* **27.** The *circumcircle* for a triangle is that circle which passes through all three of its vertices. Find the circumcircle for the triangle with vertices $A(1, 1)$, $B(-3, 3)$, and $C(2, 4)$ by (a) finding its centre and

radius (see Exercise 24) and, (b) taking the equation of the circle in form 1.21 and requiring A, B, and C to be on the circle.

* **28.** Prove that the three altitudes of the triangle in Exercise 27 intersect in a point called the *orthocentre* of the triangle.

** **29.** Show that if a line $Ax + By + C = 0$ and a circle $(x - h)^2 + (y - k)^2 = r^2$ do not intersect, then the shortest distance between them is the smaller of the two numbers

$$\frac{|(Ah + Bk + C) \pm r\sqrt{A^2 + B^2}|}{\sqrt{A^2 + B^2}}.$$

** **30.** The *incircle* of a triangle is that circle which lies interior to the triangle but touches all three sides. The centre of the incircle is called the *incentre*. Show that the incentre (x, y) of the triangle with vertices $(0, 0)$, $(2, 0)$, and $(0, 1)$ must satisfy the equations

$$|x| = |y| = \frac{|x + 2y - 2|}{\sqrt{5}}.$$

Solve these equations for the incentre, and explain why there are four points that satisfy these equations.

** **31.** Loudspeakers at points (x_1, y_1) and (x_2, y_2) emit sounds with intensities I_1 and I_2, respectively. The amount of sound received at a point (x, y) from either speaker is directly proportional to the intensity of the sound at the speaker and inversely proportional to the square of the distance from the point to the source. Show that all points in the xy-plane at which the amount of sound received from both sources is equal lie on a circle with centre on the line through (x_1, y_1) and (x_2, y_2).

The Ellipse

The set of points whose coordinates (x, y) satisfy an equation of the form

$$\frac{x^2}{a^2} + \frac{y^2}{b^2} = 1, \tag{1.24}$$

where a and b are positive constants, is said to constitute an **ellipse**. Since this equation is so similar to equation 1.23, and is exactly the same when $a = b = r$, it is not unreasonable to expect that the shape of this curve might be similar to a circle, especially when values of a and b are close together. This is indeed the case, as Figure 1.27a and b illustrate. The ellipse is symmetric about the x- and y-axes, and this is consistent with the fact that equation 1.24 remains unchanged when x and y are replaced by $-x$ and $-y$. The ellipse is elongated in the x-direction when $a > b$ (Figure 1.27a), and when $b > a$ it is elongated in the y-direction (Figure 1.27b). The point of intersection of the lines of symmetry of an ellipse is called the centre of the ellipse. For equation 1.24 the centre is the origin since the x- and y-axes are the lines of symmetry.

FIGURE 1.27a   Ellipse elongated in x-direction

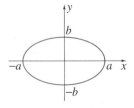

FIGURE 1.27b   Ellipse elongated in y-direction

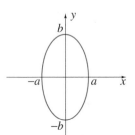

EXAMPLE 1.12

Find the equation of the ellipse that has its centre at the origin, the x- and y-axes as axes of symmetry, and passes through the points $(4, 1)$ and $(-2, 3)$.

SOLUTION   If we substitute coordinates of the points into equation 1.24 (since they are both on the ellipse),

$$\frac{4^2}{a^2} + \frac{1^2}{b^2} = 1 \quad \Longrightarrow \quad \frac{16}{a^2} + \frac{1}{b^2} = 1,$$

$$\frac{(-2)^2}{a^2} + \frac{3^2}{b^2} = 1 \quad \Longrightarrow \quad \frac{4}{a^2} + \frac{9}{b^2} = 1.$$

When the first equation is multiplied by -9 and added to the second, the result is

$$-\frac{140}{a^2} = -8.$$

Thus, $a^2 = 35/2$, and when this is substituted into the first equation,

$$\frac{32}{35} + \frac{1}{b^2} = 1 \quad \Longrightarrow \quad b^2 = \frac{35}{3}.$$

The equation required is therefore $2x^2/35 + 3y^2/35 = 1$, or $2x^2 + 3y^2 = 35$.

When equation 1.24 is changed to

$$\frac{(x - h)^2}{a^2} + \frac{(y - k)^2}{b^2} = 1, \tag{1.25}$$

where h and k are constants, the curve is still an ellipse; its shape remains the same. Just as a change from equation 1.23 to 1.21 for a circle moves the centre of the circle from $(0, 0)$ to (h, k), equation 1.25 moves the centre of the ellipse to (h, k). Lines $x = h$ and $y = k$ are the new lines of symmetry (Figure 1.28), and a and b are the distances between the centre and where the ellipse crosses the lines of symmetry.

FIGURE 1.28   Ellipse with centre (h, k)

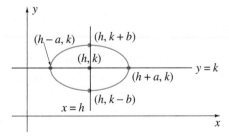

EXAMPLE 1.13

Find the centre of the ellipse $16x^2 + 25y^2 - 160x + 50y = 1175$, and draw the ellipse.

SOLUTION   When we complete squares on the x- and y-terms,

$$16(x - 5)^2 + 25(y + 1)^2 = 1600 \quad \text{or} \quad \frac{(x - 5)^2}{100} + \frac{(y + 1)^2}{64} = 1.$$

The centre of the ellipse is $(5, -1)$; it cuts the lines $y = -1$ and $x = 5$ at distances of 10 and 8 units from the centre, respectively (Figure 1.29).

FIGURE 1.29   An ellipse graphed from its equation

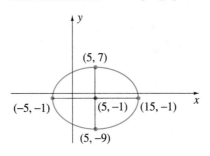

In Exercises 1–8 draw the ellipse. Use a calculator or computer to plot the ellipse as a check.

1. $x^2/25 + y^2/36 = 1$

2. $7x^2 + 3y^2 = 16$

3. $9x^2 + 289y^2 = 2601$

4. $3x^2 + 6y^2 = 21$

5. $x^2 + 16y^2 = 2$

6. $x^2 + 2x + 4y^2 - 16y + 13 = 0$

7. $9x^2 + y^2 - 18x - 6y = 26$

8. $x^2 + 4x + 2y^2 + 16y + 32 = 0$

9. Find the equation of an ellipse that passes through the points $(-2, 4)$ and $(3, 1)$.

∗ **10.** Find the width of the elliptic arch in the figure below.

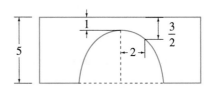

In Exercises 11–16 find all points of intersection for the curves. In each case draw the curves.

11. $x^2 + 4y^2 = 4$, $y = x$

12. $16x^2 + 9y^2 = 144$, $y = x + 3$

∗ **13.** $9x^2 - 18x + 4y^2 = 27$, $2y = \sqrt{3}x + \sqrt{3}$

∗ **14.** $9x^2 - 18x + 4y^2 = 27$, $2y = -\sqrt{3}x + 5\sqrt{3}$

∗ **15.** $x^2 + 4y^2 - 8y = 0$, $y = x^2$

∗ **16.** $x^2 + 4y^2 = 4$, $y = x^2 - 4$

∗∗ **17.** Show that every point on the ladder in Exercise 17 of Section 1.4B follows an ellipse.

The Hyperbola

Changing one sign in the equation of an ellipse leads to a curve with totally different characteristics. The set of points whose coordinates (x, y) satisfy an equation of the form

$$\frac{x^2}{a^2} - \frac{y^2}{b^2} = 1 \qquad \text{or} \qquad (1.26a)$$

$$\frac{y^2}{b^2} - \frac{x^2}{a^2} = 1, \qquad (1.26b)$$

where a and b are positive constants, is called a **hyperbola**. Hyperbola 1.26a crosses the x-axis at $x = \pm a$, but does not cross the y-axis. Since the equation remains unchanged when x is replaced by $-x$ and y is replaced by $-y$, the hyperbola is symmetric about both the x-axis and the y-axis. It follows that if we draw that part of the hyperbola in the first quadrant, we can obtain its second, third, and fourth quadrant points by reflection. By taking positive square roots of

$$\frac{y^2}{b^2} = \frac{x^2}{a^2} - 1,$$

we obtain

FIGURE 1.30a   One-
quarter of the hyperbola
$x^2/a^2 - y^2/b^2 = 1$

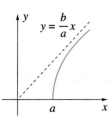

$$\frac{y}{b} = \sqrt{\frac{x^2}{a^2} - 1},$$

from which

$$y = \frac{b}{a}\sqrt{x^2 - a^2}.$$

This equation describes the top half of the hyperbola. Because a is a fixed constant, we can say that for large values of x, values of a^2 are insignificant compared to values of x^2, and values of y are approximately equal to bx/a. This means that for large values of x, the hyperbola is very close to the line $y = bx/a$. We have shown these facts in Figure 1.30a. The complete hyperbola, obtained by reflecting Figure 1.30a in the axes, is shown in Figure 1.30b. The lines $y = \pm bx/a$ that the hyperbola approaches for large positive and negative values of x are called **asymptotes** of the hyperbola.

FIGURE 1.30b   Entire
hyperbola using its symmetry

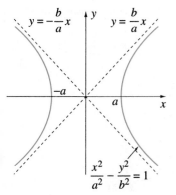

Hyperbola 1.26b is shown in Figure 1.31.

FIGURE 1.31   The hyperbola $\dfrac{y^2}{b^2} - \dfrac{x^2}{a^2} = 1$

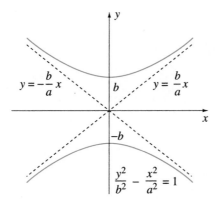

EXAMPLE 1.14

Find the equation of a hyperbola that cuts the y-axis at $y = \pm 5$ and has lines $y = \pm x/\sqrt{3}$ as asymptotes.

SOLUTION   Since the hyperbola crosses the y-axis at $y = \pm 5$, we write its equation in form 1.26b with $b = 5$,

$$\frac{y^2}{25} - \frac{x^2}{a^2} = 1.$$

When we solve this equation for y in terms of x, the result is

$$y = \pm\frac{5}{a}\sqrt{x^2 + a^2}.$$

Since asymptotes of this hyperbola are $\pm 5x/a$, it follows that $5/a = 1/\sqrt{3}$, or $a = 5\sqrt{3}$. The equation of the hyperbola is therefore

$$\frac{y^2}{25} - \frac{x^2}{75} = 1.$$

When x and y in equations 1.26 are replaced by $x - h$ and $y - k$, the resulting equations

$$\frac{(x - h)^2}{a^2} - \frac{(y - k)^2}{b^2} = 1 \quad \text{and} \tag{1.27a}$$

$$\frac{(y - k)^2}{b^2} - \frac{(x - h)^2}{a^2} = 1 \tag{1.27b}$$

still describe hyperbolas. They are shown in Figures 1.32 and 1.33. These are the hyperbolas of Figures 1.30 and 1.31 shifted so that the asymptotes intersect at the point (h, k). Equations of the asymptotes are $y = k \pm b(x - h)/a$, and the lines $x = h$ and $y = k$ are now axes of symmetry.

FIGURE 1.32   Hyperbola with centre at (h, k)

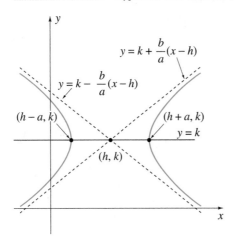

FIGURE 1.33   Hyperbola with centre at (h, k)

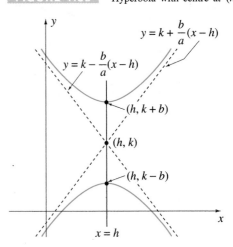

EXAMPLE 1.15

Find asymptotes for the hyperbola $x^2 - y^2 + 4x + 10y = 5$.

SOLUTION   If we complete squares on x- and y-terms, we obtain

FIGURE 1.34   A hyperbola graphed from its equation

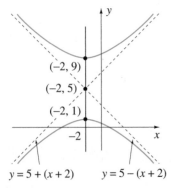

$$(x + 2)^2 - (y - 5)^2 = -16 \quad \text{or} \quad \frac{(y - 5)^2}{16} - \frac{(x + 2)^2}{16} = 1.$$

The axes of symmetry of the hyperbola are $x = -2$ and $y = 5$, intersecting at the point $(-2, 5)$. When we solve for y in terms of x, the result is

$$y = 5 \pm \sqrt{(x + 2)^2 + 16}.$$

Therefore, the asymptotes are $y = 5 \pm (x + 2)$. The hyperbola is shown in Figure 1.34.

Hyperbolas are but one of many curves that have asymptotes. For instance, in Figure 1.35, the curve $y = (x^3 - 3x^2 + 1)/(x^2 + 1)$ is asymptotic to the line $y = x - 3$. The curve $y = 2e^{-x^2/100}$ is asymptotic to the x-axis (Figure 1.36). Limits (discussed in Chapter 2) provide a unifying structure for all types of asymptotes.

FIGURE 1.35 Asymptote of nonhyperbolic curve

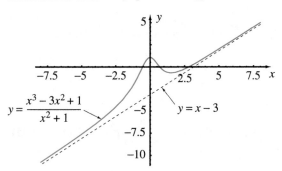

FIGURE 1.36   Illustration of horizontal asymptote

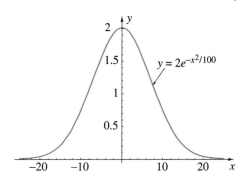

EXAMPLE 1.16

To the right of the right branch of the hyperbola $x^2 - 4y^2 = 5$ in Figure 1.37 is a swamp. A pipeline is to originate from point $(15, 10)$ to meet with a pipeline running north from point $(-1, -100)$ along the line $x = -1$. The pipeline must meet the north-south line as far down the line $x = -1$ as possible. Given that the pipeline from $(15, 10)$ should be straight, determine where it should meet the north-south pipeline.

FIGURE 1.37   Best line along which to build a pipeline

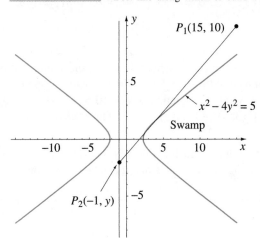

SOLUTION   The pipeline from $P_1(15, 10)$ should meet the line $x = -1$ at $P_2(-1, y)$ so that line $P_1 P_2$ just touches the right half of the hyperbola. If we let m be the slope of $P_1 P_2$, then using point-slope formula 1.13, the equation of line $P_1 P_2$ is

$$y - 10 = m(x - 15).$$

To find the required position of P_2, line $P_1 P_2$ must intersect the hyperbola in exactly one point (most lines intersect in two points, or not at all). Points of intersection are found by solving

$$y - 10 = m(x - 15) \quad \text{and} \quad x^2 - 4y^2 = 5.$$

Substituting from the first equation into the second gives

$$x^2 - 4[m(x - 15) + 10]^2 = 5.$$

If we expand the second term on the left,

$$x^2 - 4[m^2(x - 15)^2 + 20m(x - 15) + 100] = 5.$$

This can be rearranged into the form

$$(1 - 4m^2)x^2 + (120m^2 - 80m)x + (-900m^2 + 1200m - 405) = 0.$$

Given a value for m, solutions for x of this quadratic equation are x-coordinates of points of intersection of the line with slope m and the hyperbola. We want only one solution. Consequently, the discriminant must be equal to zero,

$$(120m^2 - 80m)^2 - 4(1 - 4m^2)(-900m^2 + 1200m - 405) = 0.$$

When terms are expanded, this equation reduces to

$$0 = 176m^2 - 240m + 81 = (4m - 3)(44m - 27),$$

and therefore solutions are $m = 3/4$ and $m = 27/44$. Thus, lines through $(15, 10)$ with slopes $3/4$ and $27/44$ touch the hyperbola at only one point. The line with smaller slope $27/44$ touches the left branch of the hyperbola. The line with larger slope $3/4$ touches the right branch; this is the line we want. Its equation is $y - 10 = (3/4)(x - 15)$, and it cuts the line $x = -1$ when $y = 10 + (3/4)(-1 - 15) = -2$. Thus, the pipeline should meet the north-south pipeline at the point $(-1, -2)$.

We shall find a much easier solution to this problem when we have studied some calculus, but the solution above shows that it can be done without calculus, albeit not easily.

EXERCISES 1.4D

In Exercises 1–10 draw the hyperbola. Use a calculator or computer to plot the hyperbola as a check.

1. $y^2 - x^2 = 1$

2. $x^2 - y^2 = 1$

3. $x^2 - y^2/16 = 1$

4. $25y^2 - 4x^2 = 100$

5. $y^2 = 10(2 + x^2)$

6. $3x^2 - 4y^2 = 25$

7. $x^2 - 6x - 4y^2 - 24y = 11$

8. $9x^2 - 16y^2 - 18x - 64y = 91$

9. $4y^2 - 5x^2 + 8y - 10x = 21$

10. $x^2 + 2x - 16y^2 + 64y = 79$

* **11.** Find the equation of a hyperbola that passes through the point $(1, 2)$ and has asymptotes $y = \pm 4x$.

In Exercises 12–18 find all points of intersection for the curves. In each case draw the curves.

12. $x^2 - 2y^2 = 1, x = 2y$

13. $9y^2 - 4x^2 = 36, y = x$

14. $9y^2 - 4x^2 = 36, x = 3y$

15. $3x^2 - y^2 = 3, 2x + y = 1$

* **16.** $x^2 - 2x - y^2 = 0, x = y^2$

* **17.** $x^2 - 2x - y^2 = 0, x = -y^2$

* **18.** $9(x - 1)^2 - 4(y - 1)^2 = 36, 27x = 5(y - 1)^2$

1. Straight line (1 mark)
2. Parabola (1 mark)
3. None of these (1 mark)
4. Ellipse (1 mark)
5. Hyperbola (1 mark)
6. None of these (1 mark)
7. Circle (1 mark)
8. Parabola (1 mark)
9. Ellipse (1 mark)
10. Circle (1 mark)
11. $(0, 1)$ (3 marks)

12. $(-1, 2)$, $\sqrt{30}$ (3 marks)
13. $(-1, -3/4)$, $x = -1$, $y = -3/4$ (4 marks)
14. $y = \pm x/\sqrt{3}$ (2 marks)
15. $5/2 \pm \sqrt{2}(x - 1/2)$ (3 marks)
16. $(4, -3)$, $(19/3, -2/3)$ (4 marks)
17. $(3, 13)$ (3 marks)
18. $(-6, 0)$, $(0, 6)$, $(0, -1)$ (3 marks)
19. $y = 3 - 2x^2$ (3 marks)
20. $16x^2 + 7y^2 = 256$ (4 marks)

1.5 Functions and Their Graphs

Here is the diagnostic test for this section. Give yourself 60 minutes to do it.

In questions 1 and 2 find the largest possible domain for the function.

1. $f(x) = \sqrt{4 - x^2}$
2. $f(x) = (x + 3)/(x^2 - 2x - 4)$

In questions 3 and 4 find the range of the function.

3. $f(x) = -\sqrt{4 + x^2}$
4. $f(x) = 3 + 2|x|$

In questions 5–8 determine whether the function is even, odd, or neither even nor odd.

5. $f(x) = x^4 - 2x^2 - 5$
6. $f(x) = x^3 + 5x^5 - x$
7. $f(x) = x^4 - 2x$
8. $f(x) = (x + 1)/(x^2 - 3)$
9. Find the even and odd parts of the function $f(x) = x/(x + 1)$.
10. Does the equation $x = y^2 - 2y + 1$ define y as a function of x?
11. What is a rational function?

In questions 12–17 draw the graph of the function.

12. $f(x) = 8 - |x|$
13. $f(x) = \sqrt{9 - x^2}$
14. $f(x) = -2\sqrt{x}$
15. $f(x) = -|1 - x^2| - (x^2 - 1)$

16. $f(x) = |x^3 - 8|$
17. $f(x) = x^3 - 4x$

18. The graph of a function $f(x)$ is shown to the left below. Draw a graph of the function $f(-x)$ on the axes to the right.

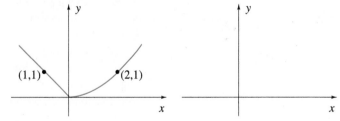

19. If the graph of $f(x)$ is that in question 18, draw a graph of $f(x+2)$ on the left set of axes below.

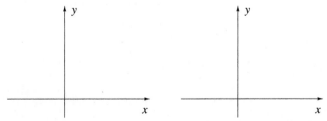

20. If the graph of $f(x)$ is that in question 18, draw a graph of $f(2x)$ on the right set of axes above.

Most quantities that we encounter in everyday life are dependent on many, many other quantities. For example, think of what might be affecting room temperature as you read this sentence — thermostat setting; outside temperature; wind conditions; insulation of the walls, ceiling, and floors; and perhaps other factors that you can think of. Functional notation allows interdependences of such quantities to be represented in a very simple way.

When one quantity depends on a second quantity, we say that the first quantity is a function of the second. For example, the volume V of a sphere depends on its radius r; in particular, $V = 4\pi r^3/3$. We say that V is a function of r. When an object is dropped, the distance d (metres) that it falls in time t (seconds) is given by the formula $d = 4.905t^2$. We say that d is a function of t. Mathematically, we have the following definition.

> ### DEFINITION 1.3
>
> A quantity y is said to be a **function** of a quantity x if there exists a rule by which we can associate exactly one value of y with each value of x. The rule that associates the value of y with each value of x is called the function.

If we denote the rule or function in this definition by the letter f, then the value that f assigns to x is denoted by $f(x)$, and we write

$$y = f(x). \tag{1.28}$$

In our first example above, we write $V = f(r) = 4\pi r^3/3$, and the function f is the operation of cubing a number and then multiplying the result by $4\pi/3$. For $d = f(t) = 4.905t^2$, the function f is the operation of squaring a number and multiplying the result by 4.905.

We call x in equation 1.28 the **independent variable** because values of x are substituted into the function, and y the **dependent variable** because its values depend on the assigned values of x. The **domain** of a function is the set of all specified (real) values for the independent variable. It is an essential part of a function and should always be specified. Whenever the domain of a function is not mentioned, we assume that it consists of all possible values for which $f(x)$ is a real number.

As the independent variable x takes on values in the domain, a set of values of the dependent variable is obtained. This set is called the **range** of the function. For the function $V = f(r) = 4\pi r^3/3$, which represents the volume of a sphere, the largest possible domain is $r > 0$, and the corresponding range is $V > 0$. Note that mathematically the function $f(r)$ is defined for negative as well as positive values of r, and $r = 0$; it is because of our interpretation of r as the radius of a sphere that we restrict $r > 0$. The function $d = f(t) = 4.905t^2$ represents the distance fallen in time t by an object that is dropped at time $t = 0$. If it is dropped from a height of 20 m, then it is clear that the range of this function is $0 \le d \le 20$. The domain that gives rise to this range is $0 \le t \le \sqrt{20/4.905}$.

EXAMPLE 1.17

▼

Find the largest possible domain for the function

$$f(x) = \sqrt{\frac{8 + 2x - x^2}{x + 1}}.$$

SOLUTION   We begin by factoring the quadratic in the numerator

$$f(x) = \sqrt{\frac{(2 + x)(4 - x)}{x + 1}},$$

and note that $f(x)$ is defined whenever $(2+x)(4-x)/(x+1) \ge 0$. To determine when this is true, we examine, in tabular form, signs of the individual factors. The first line in Table 1.1 indicates that $x + 2$ is positive for $x > -2$, and is negative for $x < -2$. The second and third lines show similar results for $4 - x$ and $x + 1$. The last line of the table counts the number of negative signs in the lines above it for the intervals $x < -2$, $-2 < x < -1$, $-1 < x < 4$, and $x > 4$ to arrive at the sign of $(2+x)(4-x)/(x+1)$. When we note that division points

$x = -2$ and $x = 4$ are acceptable, but $x = -1$ is not (it gives division by 0), the largest domain of the function consists of all values of x in the intervals $x \leq -2$ and $-1 < x \leq 4$.

TABLE 1.1

	−3	−2	−1	0	1	2	3	4	5	x
$x + 2$	−		+							
$4 - x$			+					−		
$x + 1$	−			+						
$(2 + x)(4 - x)/(x + 1)$	+		−	+				−		

EXAMPLE 1.18

Does the equation $x - 4 - y^2 = 0$ define y as a function of x for $x \geq 4$?

SOLUTION   For any $x > 4$, the equation has two solutions for y:

$$y = \pm\sqrt{x - 4}.$$

Since the equation does not define exactly one value of y for each value of x, it does not define y as a function of x.

If we add to the equation in Example 1.18 an additional restriction such as $y \geq 0$, then y is defined as a function of x, namely,

$$y = \sqrt{x - 4}.$$

Note that $x - 4 - y^2 = 0$ does, however, define x as a function of y:

$$x = y^2 + 4.$$

In this case, y is the independent variable and x is the dependent variable. In other words, whenever an equation (such as $x - 4 - y^2 = 0$) is to be regarded as defining a function, it must be made clear which variable is to be considered as independent and which as dependent. For example, the equation $x - 4 - y = 0$ defines y as a function of x, and x as a function of y.

In the study of calculus and its applications we are interested in the behaviour of functions; that is, for certain values of the independent variable, what can we say about the dependent variable? The simplest and most revealing method for displaying characteristics of a function is a graph. To obtain the graph of a function $f(x)$ we use a plane coordinatized with Cartesian coordinates x and y (in short, the Cartesian xy-plane). The **graph of the function** $f(x)$ is defined to be the curve with equation $y = f(x)$, where x is limited to the domain of the function. For example, the graph of the function $f(x) = x^3 - 27x + 1$ is shown in Figure 1.38.

FIGURE 1.38   Graph of a function

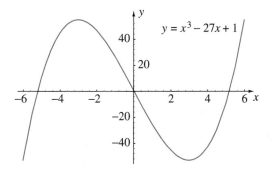

This curve is a pictorial or geometric representation of the function $f(x) = x^3 - 27x + 1$. The value of the function for a given x is visually displayed as the y-coordinate of that point on the curve with x-coordinate equal to the given x. This graph clearly illustrates properties of the function that may not be so readily obvious from the algebraic definition $f(x) = x^3 - 27x + 1$. For instance,

1. For negative values of x, the largest value of $f(x)$ is $f(-3) = 55$; for positive values of x, the smallest value of $f(x)$ is $f(3) = -53$.

2. As x increases, values of $f(x)$ increase for $x < -3$ and $x > 3$, and values of $f(x)$ decrease for $-3 < x < 3$.

3. $f(x)$ is equal to zero for three values of x — one a little less than -5, one a little larger than 0, and one a little larger than 5.

Graphing calculators and computers have become indispensable tools for modern scientists; one reason for this is the ability of these electronic devices to graph functions so quickly. We hasten to point out, however, that machine-generated graphs may sometimes be misleading; care must be taken not to make rash assumptions based on machine output. We illustrate with some examples below. In addition, because calculators and computers have finite screen resolution, it may not always be perfectly clear how to interpret every aspect of a machine-generated graph. Mathematical analysis may be required to confirm or deny what is being suggested by the graph. This is very important! We use machine-generated graphs extensively, but we are always prepared to corroborate what we see, or don't see, with rigorous mathematical analysis. Recall that when we ask you to *plot* a graph, we intend for you to use a graphing calculator or computer. When we ask you to *draw* a graph, we expect you to do so without these devices. What follows are some examples of computer-generated graphs that are misleading.

EXAMPLE 1.19

A very powerful software package once gave the graph for the function $f(x) = x + \sin(2\pi x)$ on the interval $0 \le x \le 24$ in Figure 1.39. (It doesn't in newer versions of the package.) How do you feel about what you see?

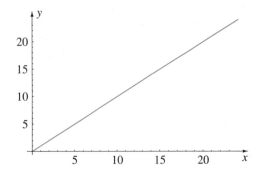

FIGURE 1.39   Incorrect graph produced by computer program

SOLUTION   The graph appears to be the straight line $y = x$, implying that therefore $\sin(2\pi x)$ is always zero. This is ridiculous; it is a fluke, resulting from the choice of sampling points taken by the software package in plotting the graph. Choosing the plot interval to be $0 \le x \le 25$ gave Figure 1.40. This is still misleading. The plot in Figure 1.41 gives a true indication of the nature of the function. Examples like this are rare, but they can occur. Use your common sense.

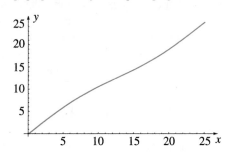

FIGURE 1.40   Another incorrect
graph produced by a computer program

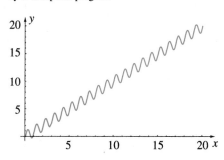

FIGURE 1.41   Correct graph produced
by a computer program

We encountered absolute values in Section 1.3. The **absolute value function**, denoted by $|x|$, is defined as

$$|x| = \begin{cases} -x, & x < 0, \\ x, & x \geq 0. \end{cases} \tag{1.29}$$

It defines the size or *magnitude* of its argument without regard for sign. The graph of the function is composed of two straight lines with slopes ± 1 meeting at right angles at the origin. This does not appear to be the case in the computer-generated plot of Figure 1.42. Why?

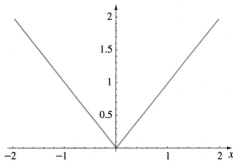

FIGURE 1.42   Graph of $|x|$

EXAMPLE 1.20

The most up-to-date version of the software package that produced the graph in Figure 1.39 yields Figure 1.43a when asked to plot the function $f(x) = x^{1/3}$ on the interval $-8 \leq x \leq 8$. It is also accompanied by warning messages to the effect that values of $f(x)$ are not real for negative values of x. Define input for the computer so that a proper graph is generated.

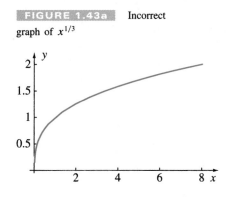

FIGURE 1.43a   Incorrect
graph of $x^{1/3}$

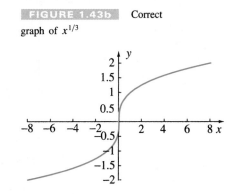

FIGURE 1.43b   Correct
graph of $x^{1/3}$

SOLUTION   To fully explain what the computer is doing, we need complex numbers from Appendix C. To be brief, every nonzero real number has three cube roots, one of which is real and two of which are complex. The computer is programmed to yield the real cube root of positive real numbers ($8^{1/3} = 2$, for instance). Unless directed otherwise, however, the computer, and maybe your calculator, produces a complex number when asked for the cube root of a negative real number such as -8. It does not yield -2, which is also a cube root of -8, unless specifically told to do so. We can instruct the computer to do this by redefining $f(x) = x^{1/3}$ as

$$f(x) = \begin{cases} -|x|^{1/3}, & x < 0, \\ x^{1/3}, & x \geq 0. \end{cases}$$

A plot of this function is shown in Figure 1.43b.

Quite often the argument of the absolute value function is a function of x rather than x itself. Such is the case in the following example.

EXAMPLE 1.21

A computer-generated graph of the function $f(x) = |x^3 + 5|$ is shown in Figure 1.44. It appears to touch the x-axis somewhere between $x = -2$ and $x = -1$, but we cannot be positive of this graphically. In addition, we cannot be sure whether the graph is rounded or whether it comes to a sharp point at this same value of x. Perform some elementary graphing to answer these questions.

FIGURE 1.44   Graph of $|x^3 + 5|$

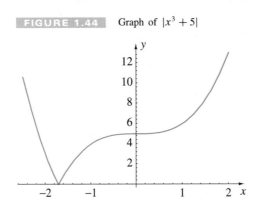

SOLUTION   To draw the curve $y = |x^3 + 5|$, we begin with $y = x^3$ in Figure 1.45a. Next we draw the curve $y = x^3 + 5$ by adding 5 to every ordinate in Figure 1.45a. The result in Figure 1.45b is the curve in Figure 1.45a shifted upward 5 units. It crosses the x-axis at $-5^{1/3}$. The last step is to take the absolute value of every ordinate on the curve $y = x^3 + 5$. This

FIGURE 1.45a

Building the graph of $|x^3 + 5|$

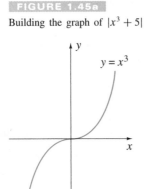

FIGURE 1.45b

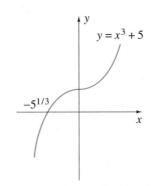

FIGURE 1.45c

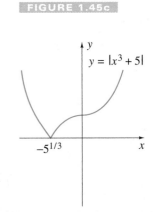

changes no ordinate that is already positive, but changes the sign of any ordinate that is negative. The result and final drawing is shown in Figure 1.45c. Clearly then, the graph touches the x-axis at $x = -5^{1/3}$, and there is a sharp point on the graph at $(-5^{1/3}, 0)$.

We might get the impression from Figures 1.42 and 1.45c that absolute values lead to curves with sharp points. This is not always the case. When absolute values are placed around the function x^3 in Figure 1.46a, the resulting function $|x^3|$ is shown in Figure 1.46b. There is no sharp point at the origin.

FIGURE 1.46a

FIGURE 1.46b

Taking absolute values does not always lead to sharp points on a curve

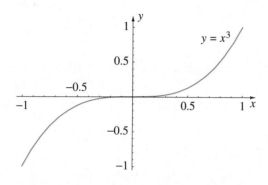

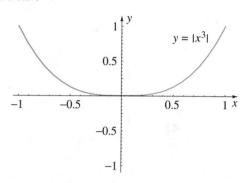

FIGURE 1.47a

FIGURE 1.47b

Graphs to illustrate that taking absolute values sometimes leads to sharp points on a curve

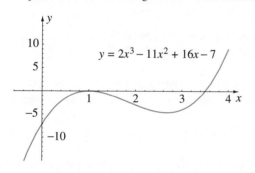

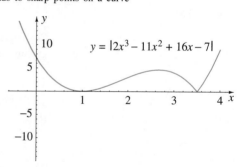

The absolute value of the function $2x^3 - 11x^2 + 16x - 7$ in Figure 1.47a leads to the graph in Figure 1.47b for $|2x^3 - 11x^2 + 16x - 7|$. It has a sharp point at the position between $x = 3$ and $x = 4$ where it touches the x-axis, but not at $x = 1$.

EXAMPLE 1.22

FIGURE 1.48 The top half of an ellipse

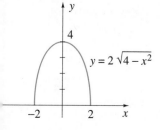

Plot a graph of the function $f(x) = 2\sqrt{4 - x^2}$ (using a calculator or computer). Does the graph look familiar? Confirm what it suggests.

SOLUTION The domain of the function is $-2 \leq x \leq 2$. Its graph in Figure 1.48 appears to be the top half of an ellipse. To confirm this, we square $y = 2\sqrt{4 - x^2}$, obtaining $y^2 = 4(4 - x^2)$, or $4x^2 + y^2 = 16$. This is indeed the equation of an ellipse. Only the top half of the ellipse is defined by $y = 2\sqrt{4 - x^2}$ since the square root requires y to be nonnegative.

EXAMPLE 1.23

Repeat Example 1.22 with $f(x) = \sqrt{4 - x^2}$.

SOLUTION   The graph in Figure 1.49a once again suggests the top half of an ellipse. Squaring $y = \sqrt{4 - x^2}$ leads to $x^2 + y^2 = 4$, a circle, not an ellipse. The reason the graph does not appear to be semicircular is that the computer has chosen different scales for the x- and y-axes. If we instruct the computer to use equal scales, the graph in Figure 1.49b does indeed look like a semicircle.

FIGURE 1.49a   A semicircle that does not look like a semicircle

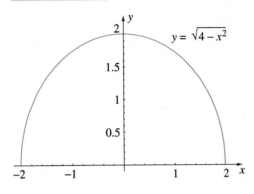

FIGURE 1.49b   A semicircle that does look like a semicircle

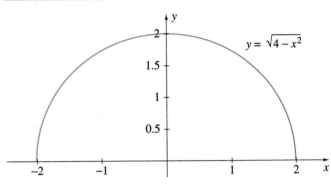

EXAMPLE 1.24

Plot a graph of the function $f(x) = x + \sqrt{3x^2 - 4}$, $x \geq 2/\sqrt{3}$. Does it appear to have an asymptote? Confirm this mathematically.

SOLUTION   The plot in Figure 1.50a appears to be fairly straight, suggesting an asymptote. To confirm this, we note that for large x, the 4 is insignificant compared to $3x^2$, and therefore

$$f(x) \approx x + \sqrt{3x^2} = x + \sqrt{3}x = (1 + \sqrt{3})x.$$

Thus, the line $y = (1 + \sqrt{3})x$ is an asymptote for the graph. It is shown along with the graph of the function in Figure 1.50b.

FIGURE 1.50a   Graph that suggests an asymptote

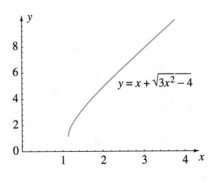

FIGURE 1.50b   Graph showing asymptote

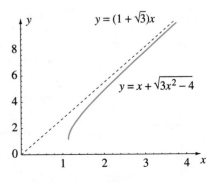

EXAMPLE 1.25

When a ball is dropped from the top of a building 20 m high at time $t = 0$, the distance d (in metres) that it falls in time t is given by the function $d = f(t) = 4.905t^2$. Draw its graph.

SOLUTION   In this example independent and dependent variables are denoted by letters t and d, which suggest their physical meaning — time and distance — rather than the generic labels x and y. With the axes labelled correspondingly as the t-axis and d-axis, we draw that part of the parabola $d = 4.905t^2$ in Figure 1.51. The remainder of the parabola has no physical significance in the context of this problem.

FIGURE 1.51   Distance fallen by a ball dropped from a height of 20 m

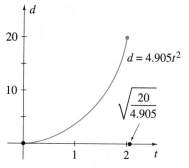

Even and Odd Functions

The parabola $y = x^2 + 1$ in Figure 1.52 is the graph of the function $x^2 + 1$; it is symmetric about the y-axis. The curve in Figure 1.53 is the graph of the function $(x^4 - 2x^2 + 5)/(x^2 + 6)$; it is also symmetric about the y-axis. These are examples of a special class of functions identified in the following definition.

DEFINITION 1.4

A function $f(x)$ is said to be an **even function** if for each x in its domain

$$f(-x) = f(x);\qquad\text{(1.30a)}$$

it is said to be an **odd function** if for each x in its domain

$$f(-x) = -f(x).\qquad\text{(1.30b)}$$

The first of these implies that the equation $y = f(x)$ for the graph of an even function is unchanged when x is replaced by $-x$; therefore, the graph of an even function is symmetric about the y-axis. As a result, $f(x) = x^2 + 1$ and $f(x) = (x^4 - 2x^2 + 5)/(x^2 + 6)$ are even functions.

Equation 1.30b implies that if (x, y) is any point on the graph of an odd function, so too is the point $(-x, -y)$. This is illustrated by the graph of the odd function $f(x) = 2x^5 - 3x^3 + x$

FIGURE 1.52   An even function

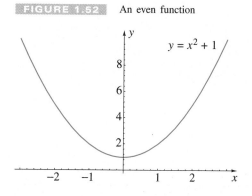

FIGURE 1.53   An even function

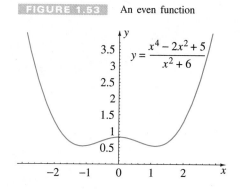

in Figure 1.54. Another way to describe the graph of an odd function is to note that either half ($x < 0$ or $x > 0$) is the result of two reflections of the other half, first in the y-axis and then in the x-axis. Alternatively, either half of the graph is a result of rotating the other half by π radians (one-half a revolution) around the origin.

FIGURE 1.54   An odd function

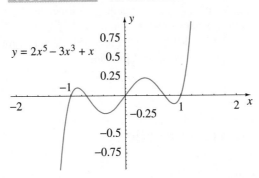

$y = 2x^5 - 3x^3 + x$

EXAMPLE 1.26

Which of the following functions are even, odd, or neither even nor odd?

$$\text{(a) } f(x) = \sqrt{|x|} \qquad \text{(b) } f(x) = x^5 - x \qquad \text{(c) } f(x) = x^2 + x$$

FIGURE 1.55a   Half the graph of an even function

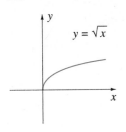

$y = \sqrt{x}$

FIGURE 1.55b   The full graph from symmetry

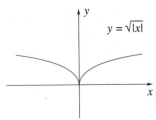

$y = \sqrt{|x|}$

Draw a graph of each function.

SOLUTION

(a) Since

$$f(-x) = \sqrt{|-x|} = \sqrt{|x|} = f(x),$$

this function is even. Its graph, the curve $y = \sqrt{|x|}$, is symmetric about the y-axis. When $x > 0$, this equation becomes $y = \sqrt{x}$, the half-parabola in Figure 1.55a. The complete graph of the function is shown in Figure 1.55b.

(b) Since

$$f(-x) = (-x)^5 - (-x) = -x^5 + x = -(x^5 - x) = -f(x),$$

this function is odd. Its graph is shown in Figure 1.56. The right half of the graph in Figure 1.56 was drawn by analyzing signs of $f(x)$ in the factored form $f(x) = x(x^2 + 1)(x - 1)(x + 1)$. The left half is the result of a half revolution of the right half about the origin.

(c) The function $f(x) = x^2 + x$ is neither even nor odd. Its graph is a parabola that opens upward, crossing the x-axis at $x = 0$ and $x = -1$ (Figure 1.57).

FIGURE 1.56   An odd function $x^5 - x$

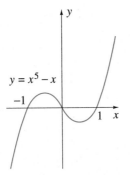

$y = x^5 - x$

FIGURE 1.57   A function that is neither even nor odd

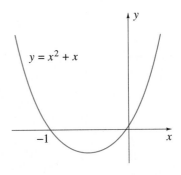

$y = x^2 + x$

We have seen even functions, odd functions, and functions that are neither even nor odd. Can functions be both even and odd? Such a function would have to satisfy both 1.30a and 1.30b, and hence

$$f(x) = -f(x).$$

But this implies that $f(x) = 0$, and this therefore is the only even and odd function.

It is often useful to divide a function into what are called its even and odd parts. For example, if $f(x)$ is defined for all real x, we may write that

$$f(x) = \left[\frac{f(x) + f(-x)}{2}\right] + \left[\frac{f(x) - f(-x)}{2}\right]. \tag{1.31}$$

It is straightforward to show that $[f(x)+f(-x)]/2$ is an even function, and $[f(x)-f(-x)]/2$ is an odd function. In other words, we have written $f(x)$ as the sum of an even function and an odd function; they are called the **even** and **odd parts** of $f(x)$. If we denote them by

$$f_e(x) = \frac{f(x) + f(-x)}{2} \quad \text{and} \quad f_o(x) = \frac{f(x) - f(-x)}{2}, \tag{1.32a}$$

then we have

$$f(x) = f_e(x) + f_o(x). \tag{1.32b}$$

EXAMPLE 1.27

Write the functions (a) $f(x) = x^3 - 2x^2 + x + 5$ and (b) $g(x) = \dfrac{x+1}{x-2}$ in terms of their even and odd parts.

SOLUTION **Solution**

(a) Clearly, the even and odd parts of $f(x)$ are $f_e(x) = -2x^2 + 5$ and $f_o(x) = x^3 + x$, respectively. Thus,

$$f(x) = (-2x^2 + 5) + (x^3 + x).$$

(b) The even and odd parts of $g(x)$ are

$$g_e(x) = \frac{1}{2}\left[\frac{x+1}{x-2} + \frac{-x+1}{-x-2}\right] = \frac{x^2+2}{x^2-4}, \qquad g_o(x) = \frac{1}{2}\left[\frac{x+1}{x-2} - \frac{-x+1}{-x-2}\right] = \frac{3x}{x^2-4}.$$

Thus,

$$g(x) = \frac{x^2+2}{x^2-4} + \frac{3x}{x^2-4}.$$

Polynomials and Rational Functions

Two of the most important classes of functions are polynomials and rational functions. A polynomial of degree n is a function of the form

$$f(x) = a_n x^n + a_{n-1} x^{n-1} + \cdots + a_2 x^2 + a_1 x + a_0, \tag{1.33}$$

where n is a nonnegative integer, and $a_0, a_1, \ldots, a_n$ are real numbers ($a_n \neq 0$). They are defined for all values of x. Graphs of linear and quadratic polynomials are straight lines and parabolas. The cubic polynomial $x^3 + 12x^2 - 40x + 6$ and the quartic $x^4 + 10x^3 + 6x^2 - 64x + 5$ are plotted in Figures 1.58 and 1.59. Figure 1.58 has a low point just to the right of $x = 1$ and a high point just to the left of $x = -9$. We could approximate these points more and more closely by reducing the plot interval. Calculus provides an analytic way to determine them.

FIGURE 1.58   A cubic polynomial

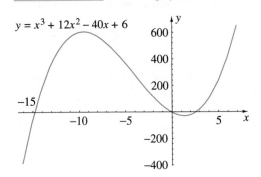

$y = x^3 + 12x^2 - 40x + 6$

FIGURE 1.59   A quartic polynomial

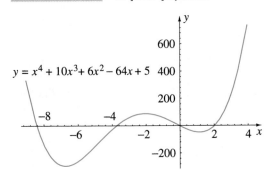

$y = x^4 + 10x^3 + 6x^2 - 64x + 5$

A **rational function** $R(x)$ is defined as the quotient of two polynomials $P(x)$ and $Q(x)$:

$$R(x) = \frac{P(x)}{Q(x)}. \tag{1.34}$$

Rational functions are undefined at points where $Q(x) = 0$. For example, the rational function

$$f(x) = \frac{x^3 + 3x^2 - 12x + 1}{x^2 - 2x - 3}$$

is undefined at $x = -1$ and $x = 3$. The computer-generated graph on the interval $-2 \le x \le 4$ in Figure 1.60a indicates that function values become very large positively and negatively near $x = -1$ and $x = 3$. We investigate this behaviour in detail in Chapter 2. The graph may or may not be accompanied by messages indicating that division by 0 is encountered at $x = -1$ and $x = 3$, depending on the plot interval specified to the plotting device. A plot on the larger interval $-10 \le x \le 10$ in Figure 1.60b suggests that the graph has an asymptote. We shall see why in Chapter 2.

FIGURE 1.60a   Unbounded behaviour of a
rational function

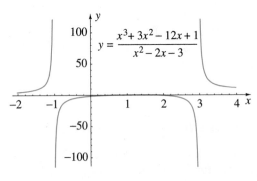

$$y = \frac{x^3 + 3x^2 - 12x + 1}{x^2 - 2x - 3}$$

FIGURE 1.60b   Suggested asymptote of a
rational function

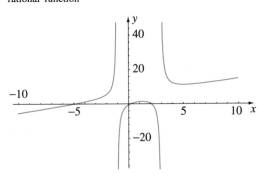

Every function $f(x)$ can be represented pictorially by its graph, the curve with equation $y = f(x)$. But what about the reverse situation? Does every curve in the xy-plane represent a function $f(x)$? The curves in Figure 1.61, which both extend between $x = a$ and $x = b$, illustrate that the answer is no. The curve in Figure 1.61a represents a function, whereas the curve in Figure 1.61b does not, because for values of x between a and c there are two possible values of y. In other words, *a curve represents a function f(x) if every vertical line that intersects the curve does so at exactly one point.*

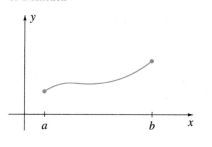

FIGURE 1.61a   Curve that is the graph of a function

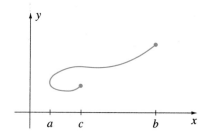

FIGURE 1.61b   Curve that is not the graph of a function

FIGURE 1.62   Translated circles

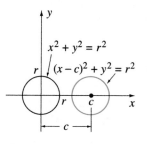

Translation of Curves

In Section 1.4 we saw how graphs of conic sections can be shifted in the xy-plane, and how these shifts, or **translations** as they are called, are reflected in the equations of the curves. This principle applies to all curves, not just conic sections. When every x in the equation of a curve is replaced by $x - c$, where c is a constant, the curve is shifted c units to the right. When every x is replaced by $x + c$, the curve is shifted c units to the left. For example, when x in the equation $x^2 + y^2 = r^2$ is replaced by $x - c$, the centre of the resulting circle is shifted from the origin to the point $(c, 0)$ (Figure 1.62). When each x in the parabola $y = x^2 + x$ is replaced by $x + c$,

FIGURE 1.63   Translated parabolas

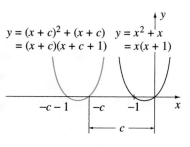

$$y = (x + c)^2 + (x + c) = (x + c)(x + c + 1),$$

the parabola is shifted c units to the left (Figure 1.63).

Vertical shifts result when y is replaced by $y \pm c$ in the equation of a curve. For instance, the curve $x^2 - (y + 1)^2 = 1$ is the hyperbola $x^2 - y^2 = 1$ in Figure 1.64a shifted downward 1 unit (Figure 1.64b).

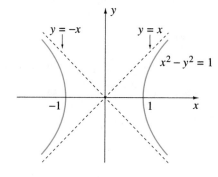

FIGURE 1.64a   The hyperbola $x^2 - y^2 = 1$

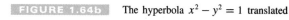
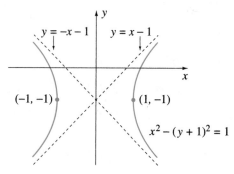

FIGURE 1.64b   The hyperbola $x^2 - y^2 = 1$ translated

EXAMPLE 1.28

Describe the relationship between the curves $|x| + |y| = 1$ and $|x| + |y - a| = 1$, where $a > 0$ is a constant. Draw both curves.

SOLUTION   Since $|x| + |y - a| = 1$ can be obtained from $|x| + |y| = 1$ by replacing y by $y - a$, the first curve is the second shifted a units upward. The curve $|x| + |y| = 1$ is easily drawn without an electronic device. Since this equation remains unchanged when x is replaced by $-x$ and y is replaced by $-y$, the curve $|x| + |y| = 1$ is symmetric about both the x-axis and the y-axis. This means that we can concentrate our efforts on drawing the graph in the first quadrant, where the equation reduces to $x + y = 1$. The segment of this straight line in the first quadrant is shown in Figure 1.65a. To obtain $|x| + |y| = 1$ (Figure 1.65b), we reflect this curve in the axes. Finally, $|x| + |y - a| = 1$ may be obtained by shifting $|x| + |y| = 1$ upward a units (Figure 1.65c).

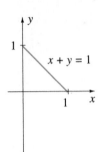

FIGURE 1.65a   One-quarter of
$|x| + |y| = 1$

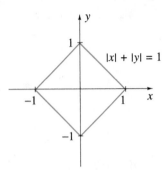

FIGURE 1.65b $|x| + |y| = 1$
from symmetry

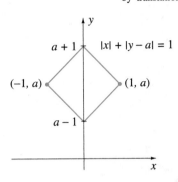

FIGURE 1.65c $|x| + |y - a| = 1$
by translation

EXERCISES 1.5

In Exercises 1–8 find the largest possible domain of the function.

1. $f(x) = \sqrt{9 - x^2}$

2. $f(x) = \dfrac{1}{x - 2}$

3. $f(x) = \dfrac{1}{x\sqrt{x^2 + 4}}$

4. $f(x) = \dfrac{1}{x\sqrt{x^2 - 4}}$

* **5.** $f(x) = \dfrac{1}{x\sqrt{4 - x^2}}$

* **6.** $f(x) = \sqrt{\dfrac{x^2 - 4}{9 - x^2}}$

* **7.** $f(x) = \sqrt[4]{-9 + 6x - x^2}$ * **8.** $f(x) = \sqrt{x^3 - x^2}$

In Exercises 9–18 determine algebraically whether the function is even, odd, or neither even nor odd. Plot each function to confirm your conclusion geometrically.

9. $f(x) = 1 + x^2 + 2x^4$

10. $f(x) = x^5 - x$

11. $f(x) = 12x^2 + 2x$

12. $f(x) = x^{1/5}$

13. $f(x) = \dfrac{x - 1}{x + 1}$

14. $f(x) = x(x^2 + x)$

15. $f(x) = \dfrac{x|x|}{3 + x^2}$

16. $f(x) = x^3 + \sqrt{x}$

* **17.** $f(x) = \dfrac{x^2 + 1}{1 - 2x^4}$

* **18.** $f(x) = \dfrac{x^2 + \sqrt{x^4 + 1}}{x^6 + 3x^2}$

In Exercises 19–24 find the even and odd parts of the function, if they exist.

19. $f(x) = x^3 + 3x^2 - 2x$

20. $f(x) = \dfrac{x - 2}{x + 5}$

21. $f(x) = |x|$

22. $f(x) = \dfrac{x^3}{x^2 + 3}$

23. $f(x) = \dfrac{2x}{3 + 5x}$

24. $f(x) = \sqrt{x - 1}$

25. Verify that when an odd function $f(x)$ is defined at $x = 0$, its value must be $f(0) = 0$.

26. Prove each of the following:

(a) The product of two even functions or two odd functions is an even function.

(b) The product of an even and an odd function is an odd function.

In Exercises 27–56 first draw, then plot, a graph of the function.

27. $f(x) = \sqrt{5 - x^2}$

28. $f(x) = -\sqrt{5 - x^2}$

29. $f(x) = -\sqrt{x^2 - 5}$

30. $f(x) = \sqrt{x^2 - 5}$

31. $f(x) = \sqrt{5 - 4x^2}$

32. $f(x) = -\sqrt{5 - 4x^2}$

33. $f(x) = |x| + 2x$

34. $f(x) = |x| - 2x$

35. $f(x) = -\sqrt{-x}$

36. $f(x) = x^{1/3}$

37. $f(x) = x^{2/3}$

38. $f(x) = 3x^{3/2}$

* **39.** $f(x) = 3|x|^{3/2}$

* **40.** $f(x) = x\sqrt{x + 1}$

* **41.** $f(x) = x\sqrt{x^2 - 1}$

* **42.** $f(x) = -x\sqrt{4 - 9x^2}$

* **43.** $f(x) = x^2\sqrt{4 - x}$

* **44.** $f(x) = x^2\sqrt{x^2 - 4}$

* **45.** $f(x) = x^2\sqrt{4 - 9x^2}$

* **46.** $f(x) = |x^2 - x - 12|$

* **47.** $f(x) = \sqrt{2x - x^2}$

* **48.** $f(x) = \sqrt{2x - 4x^2}$

* **49.** $f(x) = \sqrt{4x^2 - 2x}$

* **50.** $f(x) = \sqrt{2x - x^2 - 4}$

* **51.** $f(x) = x + 2 + \sqrt{x}$

52. $f(x) = \sqrt{9 - 4x^2} + \sqrt{4x^2 - 9}$

53. $f(x) = x^2 + |x| - 2$

54. $f(x) = \sqrt{(x^2 - 4)^2}$

55. $f(x) = \sqrt{2 - \sqrt{1 + x}}$

56. $f(x) = \sqrt{(x^2 - 1)^2 - (x^2 - 1)}$

57. What condition ensures that a curve in the xy-plane represents a function $x = f(y)$?

A curve is translated vertically when each y in its equation is replaced by $y + c$; it is shifted horizontally when each x is replaced by $x + c$. If each x in the equation of a curve is replaced by cx, where $c > 1$ is a constant, the curve is compressed by a factor of c in the x-direction. When $0 < c < 1$, the curve is stretched by a factor of $1/c$ in the x-direction. When each y is replaced by cy, the curve is compressed or stretched in the y-direction. Illustrate this by drawing and plotting the pair of curves in Exercises 58–63 .

58. $y = x^2, \ y = x^2/9 = (x/3)^2$

59. $y = x, \ 5y = x$

60. $x^2 + y^2 = 1, \ x^2 + 16y^2 = 1$

61. $x^2 - y^2 = 1, \ x^2 - 4y^2 = 1$

62. $|x| + |y| = 1, \ |x| + |y/2| = 1$

63. $x^2 - (y + 5)^2 = 1, \ x^2 - (y/2 + 5)^2 = 1$

When each x is replaced by $-x$ in the equation of a curve, the curve is reflected in the y-axis. When each y is replaced by $-y$, the curve is reflected in the x-axis. Illustrate this by drawing and plotting the pair of curves in Exercises 64–67 .

64. $y = x^3 - 3x^2, \ y = -x^3 - 3x^2$

65. $(x - 2)^2 + y^2 = 4, \ (x + 2)^2 + y^2 = 4$

66. $x = y^2 - 2y, \ x = y^2 + 2y$

67. $x = \sqrt{y}, \ x = \sqrt{-y}$

68. The *floor function* is defined by $f(x) = \lfloor x \rfloor = $ greatest integer that does not exceed x.

(a) Draw a graph of the floor function.

(b) If first-class postage is 51 cents for each 50 g, or fraction thereof, up to and including 500 g, draw a graph of this cost function.

(c) Express the cost function in part (b) in terms of the floor function.

In Exercises 69–80 draw the curve. Indicate whether the curve defines y as a function of x.

69. $y = x^2 - x^6$ * **70.** $y = (x + 1)^2 - 2(x + 1)$

71. $(x - 2)^2 + y^2 = 4$ * **72.** $y = \sqrt{4 - (x - 2)^2}$

* **73.** $x^2 - (y + 5)^2 = 1$ * **74.** $x = -\sqrt{64 + 9(y + 5)^2}$

* **75.** $y = (x + 1)^4 - (x + 1)^2$ * **76.** $|x| + |y + 2| = 1$

* **77.** $y = |3 - |x + 2|| - 1$ * **78.** $|x| - |y| = 1$

* **79.** $x = 4 - |y|$

* **80.** $(y - 1)^2 = (x - 1)^2[1 - (x - 1)^2]$

* **81.** An electronic signal is defined by

$$s(t) = \begin{cases} 0, & t < 0, \\ t, & 0 \le t \le 1, \\ (3 - t)/2, & 1 \le t \le 3, \\ 0, & t > 3, \end{cases}$$

where t is time. Draw graphs of the time-shifted signals $s(t - 2)$ and $s(t + 1)$.

* **82.** The graph of an electronic signal $s(t)$ is shown in the figure below, where t is time.

(a) Find an algebraic representation for $s(t)$.

(b) Draw graphs of the time-shifted signals $s(t + 1/2)$ and $s(t - 3)$.

(c) What are algebraic representations for the graphs in part (b)?

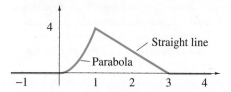

* **83.** A cherry orchard has 255 trees, each of which produces an average of 25 baskets of cherries. For each additional tree planted, the yield per tree decreases by one-twelfth of a basket. If x represents the number of extra trees (beyond 255) and Y the total yield, find Y as a function of x, and draw its graph. How many more trees should be planted for maximum yield?

* **84.** A rectangle with sides parallel to the axes is inscribed inside the ellipse $b^2x^2 + a^2y^2 = a^2b^2$ (figure below). Find a formula for the area A of the rectangle in terms of x. Draw a graph of this function.

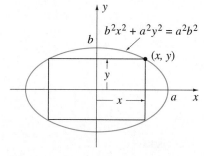

* **85.** A man 2 m tall walks along the edge of a straight road that is 5 m wide. On the other edge of the road stands a street light 10 m high. Find a functional relationship for the length of the man's shadow in terms of his distance from the point on his side of the road directly across from the light. Plot a graph of this function.

∗ **86.** When two substances, A and B, are brought together, a chemical reaction takes place to form a new substance, C. It requires 2 L of A for each litre of B to produce 3 L of C. The rate R at which A and B react to form C is proportional to the product of the amounts of A and B present at that instant. If the original amounts of A and B are 20 L and 40 L, respectively, and if x represents the amount of C present in the reaction at any given time, find a formula for R as a function of x. Draw a graph of this function, and determine when the reaction rate is highest.

∗ **87.** In classical physics and engineering, the mass m of an object is constant, independent of how fast it is moving. In special relativity, however, m is given by the formula

$$m = \frac{m_0}{\sqrt{1 - (v^2/c^2)}},$$

where m_0 is the mass of the object when it is not moving, v is the speed of the mass, and c is a constant (the speed of light). Draw a graph of this function, and draw any conclusions that you feel are suggested.

∗ **88.** A square plate 4 m on each side is slowly submerged in a large tank of water. One diagonal is kept vertical and lowered at a rate of 0.5 m/s, entering the water at time $t = 0$. If A is the area of the submerged portion of the surface (one side only) at time t until complete submersion occurs, find A as a function of t and draw its graph.

∗ **89.** Because of construction, no passing is permitted on a 10-km stretch of highway. If cars travel at v km/h along this stretch, a safe distance between them must be maintained, and this distance increases as v increases. In particular, the highway traffic commission has determined that for speeds over 50 km/h, the distance in metres between cars should be at least

$$d = \frac{3v^2}{500}.$$

If it is supposed that everyone maintains the safe distance and the same constant speed v through the stretch, find the number q of cars leaving the "bottleneck" per hour as a function of speed v. Draw a graph of this function for $50 \leq v \leq 100$, and determine the speed that maximizes q.

∗ **90.** A box measuring 1 m on each side is attached to a rope as shown in the figure below. The rope passes over a pulley 10 m from the ground and a truck pulls on the other end in a horizontal direction along a line 1 m above the ground. We denote positions of the truck and the bottom of the box by x and y, respectively. Find y as a function of x if the truck starts at position $x = 5$ m and stops when the top of the box touches the pulley. Assume that the length of rope between truck and box is 25 m. Draw a graph of the function.

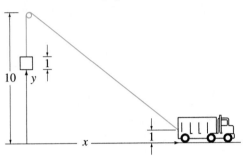

In Exercises 91–94 draw a graph of the function where $\lfloor x \rfloor$ is the floor function of Exercise 68.

∗ **91.** $f(x) = \lfloor 2x \rfloor$ ∗ **92.** $f(x) = x + \lfloor x \rfloor$

∗ **93.** $f(x) = x\lfloor x \rfloor$ ∗ **94.** $f(x) = \lfloor x + \lfloor x \rfloor \rfloor$

∗ **95.** Is $\lfloor f(x) + g(x) \rfloor = \lfloor f(x) \rfloor + \lfloor g(x) \rfloor$?

∗ **96.** If a thermal nuclear reactor is built in the shape of a right circular cylinder of radius r and height h, then neutron diffusion theory requires r and h to satisfy an equation of the form

$$\frac{a^2}{r^2} + \frac{b^2}{h^2} = 1,$$

where a and b are positive constants. Draw a graph of the function $r = f(h)$ defined by this equation for $h \geq 2b$.

ANSWERS TO DIAGNOSTIC TEST FOR SECTION 1.5

1. $-2 \leq x \leq 2$ (1 mark)

2. All reals except $x = 1 \pm \sqrt{5}$ (2 marks)

3. All reals ≤ -4 (2 marks)

4. All reals ≥ 3 (2 marks)

5. Even (1 mark)

6. Odd (1 mark)

7. Neither (1 mark)

8. Neither (1 mark)

9. $-x^2/(1 - x^2)$, $x/(1 - x^2)$ (3 marks)

10. No (1 mark)

11. The quotient of two polynomials (1 mark)

12. (2 marks)

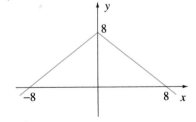

13. (2 marks)

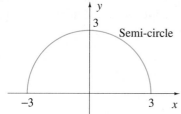

Semi-circle

14. (2 marks)

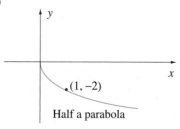

$(1, -2)$

Half a parabola

15. (3 marks)

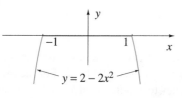

$y = 2 - 2x^2$

16. (3 marks)

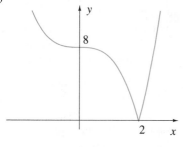

17. (3 marks)

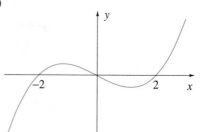

18. (2 marks)

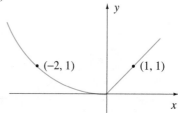

$(-2, 1)$ $(1, 1)$

19. (2 marks)

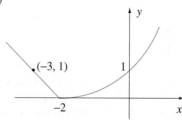

$(-3, 1)$

20. (2 marks)

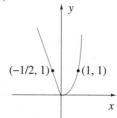

$(-1/2, 1)$ $(1, 1)$

1.6 Inverse Functions

This material is likely to be new for many students. It does not therefore have a diagnostic test associated with it. If you have previously studied inverse functions, it would still be advisable to at least read the section to ensure that you are familiar with the vocabulary, notation, and theory.

When we speak of a function $y = f(x)$, there is a unique y associated with each x; that is, given a value of x in the domain of $f(x)$, the function associates one — and only one — value of y in the range. However, it may happen, and quite often does, that a value of y in the range of the function may be associated with more than one value of x. For example, each $y > 0$ in the range of the function $y = f(x) = x^2$ (Figure 1.66) is associated with two values of x, namely, $x = \pm\sqrt{y}$.

Some functions have the property that each value of y in the range arises from only one x in the domain. For instance, given any value y in the range of the function $y = f(x) = x^3$ in Figure 1.67, there is a unique x such that $y = x^3$, namely, $x = y^{1/3}$. Such a function is said to be one-to-one. Formally, we say that a function $f(x)$ is **one-to-one** if for any two distinct values x_1 and x_2 in the domain of $f(x)$, it follows that $f(x_1) \neq f(x_2)$. To repeat, a one-to-one function $f(x)$ has the property that given any y in its range, there is one — and only one — x in its domain for which $y = f(x)$. We can therefore define a function that maps values in the range of $f(x)$ onto values in the domain, a function that maps y onto x if x is mapped by $f(x)$ onto y. We call this function the **inverse function** of $f(x)$. It reverses the action of $f(x)$. For the function $f(x) = x^3$ in Figure 1.67, the inverse function is the function that takes cube roots of real numbers.

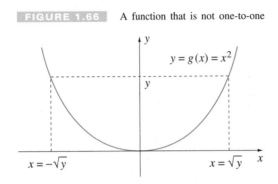

FIGURE 1.66   A function that is not one-to-one

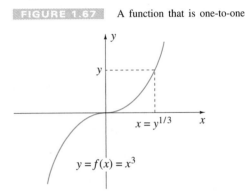

FIGURE 1.67   A function that is one-to-one

Unless there is a good reason to do otherwise, it is our custom to denote the independent variable of a function by the letter x and the dependent variable by y. For the inverse function of a function $y = f(x)$, there is a natural tendency to denote the independent variable by y and the dependent variable by x, since the inverse function maps the range of $f(x)$ onto its domain. When discussing general properties of inverse functions, as we do in this section, it is usually better to maintain our usual practice of using x as independent variable and y as dependent variable even for inverse functions. This may not be advisable in applications when letters for independent and dependent variables represent physical or geometric quantities.

When a function $f(x)$ has an inverse function, we adopt the notation $f^{-1}(x)$ to represent the inverse function. For example, when $f(x) = x^3$, the inverse function is $f^{-1}(x) = x^{1/3}$. Be careful with this notation. Do not interpret the "-1" as a power and write $f^{-1}(x)$ as $1/f(x)$. This is not correct. The notation f^{-1} represents a function, just as tan represents the tangent function and $\sqrt{\ }$ represents the positive square root function.

The inverse function $f^{-1}(x)$ of a function $f(x)$ "undoes" what $f(x)$ "does"; it reverses the effect of $f(x)$. For example, the function $f(x) = x^2$, $x \geq 0$ in Figure 1.68 is one-to-one; it squares nonnegative numbers. Its inverse is $f^{-1}(x) = \sqrt{x}$, the positive square root function (Figure 1.69). Squaring a positive number x and then taking the positive square root of the result returns the original x. Similarly, the inverse of $g(x) = x^2$, $x \leq 0$ (Figure 1.70), is the negative square root function $g^{-1}(x) = -\sqrt{x}$ (Figure 1.71).

FIGURE 1.68   The function x^2, $x \geq 0$

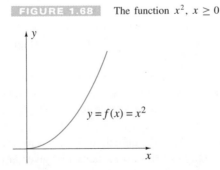

FIGURE 1.69   The inverse function $\sqrt{x}$

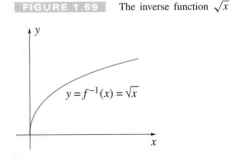

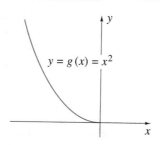

FIGURE 1.70   The function x^2, $x \leq 0$

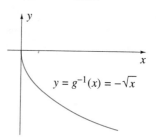

FIGURE 1.71   The inverse function $-\sqrt{x}$

The fact that $f^{-1}(x)$ undoes $f(x)$ can be stated algebraically as follows. For each x in the domain of $f(x)$,

$$f^{-1}\big(f(x)\big) = x. \tag{1.35}$$

This is the defining relation for an inverse function. We understand that the domain of $f^{-1}(x)$ is the same as the range of $f(x)$, so that $f^{-1}(x)$ operates on all outputs of $f(x)$.

EXAMPLE 1.29

What is the inverse function of $f(x) = (x+1)/(x-2)$?

SOLUTION   To find the algebraic definition of $f^{-1}(x)$, we solve $y = (x+1)/(x-2)$ for x in terms of y (you will see why in a moment). First we cross-multiply,

$$x + 1 = y(x - 2);$$

then group terms in x,

$$x(y - 1) = 2y + 1;$$

and finally, divide by $y - 1$,

$$x = \frac{2y + 1}{y - 1}.$$

What have we accomplished by solving for x in terms of y? If we take an x in the domain of $f(x)$, then $f(x)$ produces $y = (x+1)/(x-2)$. Now this same x and y satisfy the equation $x = (2y+1)/(y-1)$ because this equation is simply a rearrangement of $y = (x+1)/(x-2)$. Consequently, if we substitute y into the right side of the equation,

$$x = \frac{2y + 1}{y - 1},$$

we obtain the original x. This equation must therefore define the inverse function of $f(x)$. In other words, if we denote the independent variable by x, the inverse function is

$$f^{-1}(x) = \frac{2x + 1}{x - 1}.$$

Example 1.29 has illustrated that to find the inverse of a function $y = f(x)$, the equation should be solved for x in terms of y, and then variables should be renamed. If the equation does not have a unique solution for x in terms of y, then $f(x)$ does not have an inverse function. Such is the case for the function $g(x) = x^2$ in Figure 1.66. Solving $y = g(x) = x^2$ for x gives two solutions $x = \pm\sqrt{y}$.

So far our discussion of inverse functions has been algebraic. The geometry of inverse functions is most revealing. You may have noticed a relationship between the graphs of $f(x)$

FIGURE 1.72 Graph of x^2 and its mirror image in $y = x$

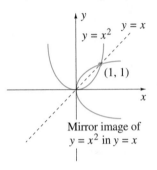

Mirror image of $y = x^2$ in $y = x$

FIGURE 1.73 An increasing function

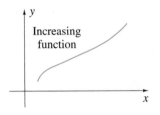

FIGURE 1.74 A decreasing function

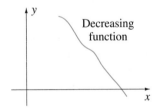

and $f^{-1}(x)$ in Figures 1.68 and 1.69, and of $g(x)$ and $g^{-1}(x)$ in Figures 1.70 and 1.71. The inverse function is the mirror image of the function in the line $y = x$; that is, graphs of inverse pairs are symmetric about the line $y = x$. These two examples are not mere coincidence, graphs of inverse functions are always mirror images of each other in the line $y = x$.

This suggests that a very simple way to graphically determine the inverse of a function $f(x)$ is to take its mirror image in the line $y = x$. Note, too, that if the mirror image does not represent a function, then no inverse function for $f(x)$ exists. For example, the mirror image of $y = x^2$, $-\infty < x < \infty$, in the line $y = x$ is shown in Figure 1.72 and does not represent a function. We conclude as before that this function does not have an inverse.

If $y = f^{-1}(x)$ is the mirror image of $y = f(x)$ in the line $y = x$, then $y = f(x)$ is the mirror image of $y = f^{-1}(x)$. This means that $f(x)$ is the inverse of $f^{-1}(x)$, and that $f(x)$ undoes what $f^{-1}(x)$ does:

$$f\left(f^{-1}(x)\right) = x. \tag{1.36}$$

Geometrically, a function $f(x)$ has an inverse if the reflection of its graph $y = f(x)$ in the line $y = x$ represents a function. But we know that a curve represents a function if every vertical line that intersects it does so at exactly one point. Furthermore, a vertical line intersects the reflected curve at exactly one point only if its horizontal reflection intersects $y = f(x)$ at exactly one point. These two facts enable us to state that a function $f(x)$ has an inverse function if — and only if — every horizontal line that intersects it does so at exactly one point. This is a geometric interpretation of a function being one-to-one. See, for example, the functions in Figures 1.68 and 1.70. Horizontal lines that intersect the curves do so at exactly one point. The function in Figure 1.66, which has no inverse, is intersected in two points by every horizontal line $y = c > 0$.

When the graph of a function always moves upward and to the right on an interval I (Figure 1.73), the function is said to be increasing on the interval. We shall discuss increasing functions in Section 4.2. What is clear is that such a function passes the horizontal line test (and is therefore one-to-one). Likewise, a function whose graph moves downward and to the right on an interval I (Figure 1.74) is said to be decreasing on I; it is one-to-one. A function that is either increasing on an interval I or decreasing on I is said to be a **strictly monotonic function** on I. What we have shown is the following result.

THEOREM 1.6

A function that is strictly monotonic on an interval has an inverse function on that interval.

It is important to realize that being strictly monotonic is a sufficient condition for existence of an inverse function; that is, if a function is strictly monotonic, then it has an inverse. It is not, however, a necessary condition. The function in Figure 1.75a is not strictly monotonic on the interval $-1 \leq x \leq 1$, but it does have the inverse function in Figure 1.75b.

FIGURE 1.75a   **FIGURE 1.75b**

A function that is not strictly monotonic, and its inverse

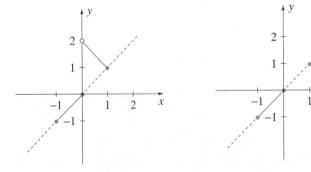

EXAMPLE 1.30

Does the function $f(x) = x^2 - 2x$ have an inverse? Does it have an inverse when its domain is restricted to the interval $x \geq 1$, and to the interval $x \leq 1$?

SOLUTION   The graph of $f(x)$ in Figure 1.76 indicates that $f(x)$ does not have an inverse. When restricted to the interval $x \geq 1$, the function is one-to-one, and does have an inverse. To find it, we solve $y = x^2 - 2x$ for x in terms of y. The quadratic formula applied to

$$x^2 - 2x - y = 0$$

gives

$$x = \frac{2 \pm \sqrt{4 + 4y}}{2} = 1 \pm \sqrt{y + 1}.$$

Since x must be greater than or equal to 1, we must choose $x = 1 + \sqrt{y + 1}$. Hence, the inverse of $f(x) = x^2 - 2x, \ x \geq 1$ is

$$f^{-1}(x) = 1 + \sqrt{x + 1}.$$

FIGURE 1.76   A function that does not have an inverse

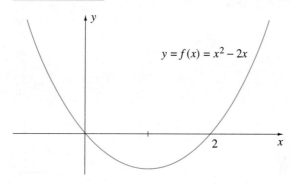

Similarly, when restricted to the interval $x \leq 1$, $f(x)$ has the inverse

$$f^{-1}(x) = 1 - \sqrt{x + 1}.$$

This example illustrates that when a function is not strictly monotonic on an interval I, the interval can usually be subdivided into subintervals on which the function is strictly monotonic, and on each such subinterval the function has an inverse.

EXERCISES 1.6

In Exercises 1–14 determine graphically whether the function has an inverse. Find each inverse function.

1. $f(x) = 2x + 3$

2. $f(x) = \sqrt{x + 1}$

3. $f(x) = x^2 + x$

4. $f(x) = \dfrac{x + 5}{2x + 4}$

5. $f(x) = 1/x$

6. $f(x) = 3x^3 + 2$

7. $f(x) = \sqrt{4 - x^2}, \quad 0 \leq x \leq 2$

8. $f(x) = 2x + |x|$

9. $f(x) = x + |x|$

10. $f(x) = \sqrt{1 - x^2}$

11. $f(x) = x^4 + 2x^2 + 2, \quad x \leq 0$

12. $f(x) = x^2 - 2x + 4, \quad x \geq 1$

13. $f(x) = \left(\dfrac{x + 2}{x - 2}\right)^3$

14. $f(x) = \dfrac{x}{3 + x^2}$

In Exercises 15–20 show that the function does not have an inverse. Subdivide its domain of definition into subintervals on which the function has an inverse, and find the inverse function on each subinterval.

* **15.** $f(x) = x^4$

* **16.** $f(x) = 1/x^4$

* **17.** $f(x) = x^2 + 2x + 3$

* **18.** $f(x) = x^4 + 4x^2 + 2$

* **19.** $f(x) = \dfrac{x^2}{x^2 + 4}$

* **20.** $f(x) = \dfrac{x^4}{x^2 + 4}$

* **21.** Give an example of a function that is defined for all x, but does not have an inverse on any interval whatsoever.

* **22.** Give an example of a function that is defined for all x, has an inverse on the interval $0 \le x \le 1$, but does not have an inverse on any other interval.

* **23.** If a manufacturing firm sells x objects of a certain commodity per week, it sells them at a price of r per object, and r depends on x, $r = f(x)$. In economic theory this function is usually considered decreasing, as shown in the left figure below; hence it has an inverse

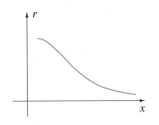

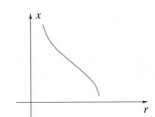

function $x = f^{-1}(r)$. In this function x depends on r, indicating that if the price of the object is set at r, then the market will demand x of them per week. This function is therefore called the *demand function* (right figure). Find demand functions if (a) $r = \dfrac{a}{x+b} + c$, $a, b,$ and c positive constants; (b) $r = \dfrac{a}{x^2 + b} + c$, $a, b,$ and c positive constants. Draw the demand function $x = f^{-1}(r)$ and given function $r = f(x)$ in each case.

* **24.** Show that the demand function

$$x = f(r) = 4a^3 - 3ar^2 + r^3, \quad 0 < r < 2a$$

has an inverse function $r = f^{-1}(x)$. What is the domain of $f^{-1}(x)$? Draw its graph.

* **25.** Find the inverse function for

$$f(x) = \frac{x^2}{(1 + x)^2}, \quad -1 < x \le 0.$$

1.7 Trigonometry Review

Here is the diagnostic test for this section; give yourself 60 minutes to do it. Use a calculator only in problems that specify their use.

In all questions, angles are in radian measure unless indicated otherwise.

1. Express the angles $135°$ and $-270°$ in radian measure.

2. Express the angles $2\pi/3$ and $-9\pi/4$ in degree measure.

3. Evaluate the following quantities:

    (a) $\sin(\pi/3)$

    (b) $\cot(-\pi/4)$

    (c) $\cos(5\pi/3)$

    (d) $\csc(-3\pi/4)$.

4. Find the unspecified angles and the length of the third side of the triangle below. You will need a calculator for this question.

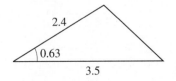

In questions 5–8 draw a graph of the function.

5. $f(x) = 3\sin 2x$

6. $f(x) = 2\cos(x - \pi/3)$

7. $f(x) = \tan(x + \pi/2)$

8. $f(x) = 4\cos(2x + \pi/2)$

In questions 9–11 find all solutions to the equation.

9. $\sin x = \sqrt{3}/2$

10. $\cos x = -1/2$

11. $\tan 2x = 1$

12. Express the function $f(x) = 2\sin 2x - 3\cos 2x$ in the form $A\sin(2x + \phi)$ where A is positive and ϕ is an angle in the interval $0 < \phi < 2\pi$. You may use a calculator.

Many physical systems exhibit an oscillatory nature: vibrations of a plucked guitar string, motion of a pendulum, alternating electric currents, fluctuations in room temperature as the thermostat continually engages and disengages the furnace on a cold day, and the rise and fall of tides and waves. Magnitudes of these oscillations are best described by the sine and cosine functions of trigonometry. In addition, rates at which these oscillations occur can be represented by *derivatives* of these functions from calculus. In this section, we briefly review the trigonometric functions and their properties, placing special emphasis on those aspects that are most useful in calculus.

FIGURE 1.77   Definition of a radian

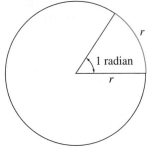

In trigonometry, angles are usually measured in degrees, radians, or mils; in calculus, angles are always measured in radians. In preparation, then, for the calculus of trigonometric functions, we work completely in radian measure. By definition, a **radian** is that angle subtended at the centre of a circle of radius r by an arc of equal length r (Figure 1.77). When an arc has length s (Figure 1.78), the number of *units of radius* in this arc is s/r; this is called the radian measure of the angle θ subtended at the centre of the circle

$$\theta = \frac{s}{r}. \tag{1.37}$$

In particular, if s contains π units of radius ($s = \pi r$), then s represents one-half the circumference of the complete circle, and $\theta = \pi r/r = \pi$ radians. In degree measure, this angle is $180°$; hence we can state that π radians is equivalent to $180°$. This statement enables us to convert angles expressed in degrees to radian measure, and vice versa. If the degree measure of an angle is ϕ, then it is $\pi\phi/180$ radians; conversely, if an angle measures θ radians, then its degree measure is $180\theta/\pi$. For example, the radian measure for $\phi = 45°$ is $45\pi/180 = \pi/4$ radians.

FIGURE 1.78   Relationship between arc length and angle at centre of circle

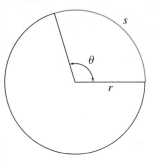

Elementary trigonometry is concerned with relationships among angles and lengths and, in particular, angles and sides of triangles. This naturally restricts angles to the range $0 \leq \theta \leq \pi$. For the purposes of calculus, however, we need to talk about angle θ, where θ is any real number — positive, negative, or zero. To do this we first define what we mean by the **standard position of an angle**. If $\theta > 0$, we draw a line segment OP through the origin O of the xy-plane (Figure 1.79) in such a way that the positive x-axis must rotate counterclockwise through an angle θ to coincide with OP. In other words, we now regard an angle as rotation, rotation of the positive x-axis to some terminal position. If $0 < \theta < \pi/2$, then OP lies in the first quadrant; and if $\pi/2 < \theta < \pi$, then OP is in the second quadrant. But if $\theta > \pi$, we have a geometric representation of θ also. For example, angles $7\pi/4$ and $9\pi/4$ are shown in Figure 1.80. When $\theta < 0$, we regard θ as a clockwise rotation (Figure 1.81).

FIGURE 1.79   Standard position of an angle

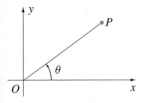

FIGURE 1.80   Positive angles as counterclockwise rotations

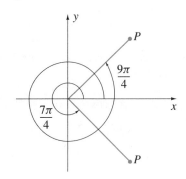

FIGURE 1.81   Negative angle as clockwise rotations

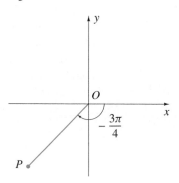

For any angle θ, the angles $\theta + 2n\pi$, where n is an integer, have the same terminal position of OP as θ. They are different angles, however, because the positive x-axis must encircle the origin one or more times before reaching the terminal position in the case of $\theta + 2n\pi$.

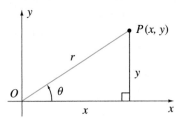

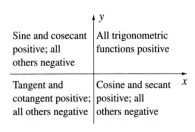

With angles represented as rotations, it is easy to define the six trigonometric functions. If θ is an angle in standard position (Figure 1.82) and (x, y) are the coordinates of P, we define

$$\sin\theta = \frac{y}{r}, \quad \cos\theta = \frac{x}{r}, \quad \tan\theta = \frac{y}{x},$$
$$\csc\theta = \frac{r}{y}, \quad \sec\theta = \frac{r}{x}, \quad \cot\theta = \frac{x}{y}, \tag{1.38}$$

wherever these ratios are defined, and where $r = \sqrt{x^2 + y^2}$ is always assumed positive. Since x is positive in the first and fourth quadrants, and y is positive in the first and second, signs of the trigonometric functions in the various quadrants are as shown in Figure 1.83. Furthermore, since $r \neq 0$, $\sin\theta$ and $\cos\theta$ are defined for all θ, whereas $\tan\theta$ and $\sec\theta$ are not defined for $x = 0$, and $\csc\theta$ and $\cot\theta$ do not exist when $y = 0$.

Definitions 1.38 indicate that

$$\csc\theta = \frac{1}{\sin\theta}, \quad \sec\theta = \frac{1}{\cos\theta}, \quad \cot\theta = \frac{1}{\tan\theta}. \tag{1.39}$$

In addition, the fact that $r^2 = x^2 + y^2$ leads to the identity

$$\sin^2\theta + \cos^2\theta = 1, \tag{1.40a}$$

which, in turn, implies that

$$1 + \tan^2\theta = \sec^2\theta, \tag{1.40b}$$

$$1 + \cot^2\theta = \csc^2\theta. \tag{1.40c}$$

Table 1.2 contains values of the trigonometric functions for the most commonly used angles. You should commit at least values for sine, cosine, and tangent to memory.

	$\sin x$	$\cos x$	$\tan x$	$\csc x$	$\sec x$	$\cot x$
0	0	1	0	undefined	1	undefined
$\pi/6$	$1/2$	$\sqrt{3}/2$	$1/\sqrt{3}$	2	$2/\sqrt{3}$	$\sqrt{3}$
$\pi/4$	$1/\sqrt{2}$	$1/\sqrt{2}$	1	$\sqrt{2}$	$\sqrt{2}$	1
$\pi/3$	$\sqrt{3}/2$	$1/2$	$\sqrt{3}$	$2/\sqrt{3}$	2	$1/\sqrt{3}$
$\pi/2$	1	0	undefined	1	undefined	0

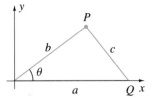

Identity 1.40a is simply a restatement of the Pythagorean relation ($r^2 = x^2 + y^2$), which allows us to express the hypotenuse of a right-angled triangle in terms of the other two sides. If the triangle is not right-angled (Figure 1.84), it is not possible to express one side, c, in terms of the other two sides, a and b alone, but it is possible to express c in terms of a and b and the angle θ between them. The coordinates of P and Q in Figure 1.84 are $(b\cos\theta, b\sin\theta)$ and $(a, 0)$; therefore, by distance formula 1.10,

$$c^2 = \|PQ\|^2 = (b\cos\theta - a)^2 + (b\sin\theta)^2$$
$$= b^2(\cos^2\theta + \sin^2\theta) + a^2 - 2ab\cos\theta.$$

Since $\sin^2\theta + \cos^2\theta = 1$, we have

$$c^2 = a^2 + b^2 - 2ab\cos\theta. \tag{1.41}$$

This result, called the **cosine law**, generalizes the Pythagorean relation to triangles that are not right-angled. It reduces to the Pythagorean relation $c^2 = a^2 + b^2$ when $\theta = \pi/2$.

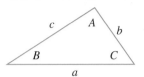

FIGURE 1.85   Sine law

If A, B, and C are the angles in the triangle of Figure 1.85, and a, b, and c are the lengths of the opposite sides, then by drawing altitudes of the triangle we can show that

$$\frac{\sin A}{a} = \frac{\sin B}{b} = \frac{\sin C}{c}. \tag{1.42}$$

This result, known as the **sine law**, is also useful in many problems. The sine and cosine laws can be used to find angles and lengths of sides of triangles as illustrated in the following examples.

EXAMPLE 1.31

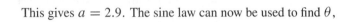

Find length a and angles θ and ϕ in the triangle of Figure 1.86.

FIGURE 1.86   Identifying angles and sides of a triangle

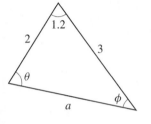

SOLUTION   We can calculate a with the cosine law,

$$a^2 = 2^2 + 3^2 - 2(2)(3)\cos(1.2).$$

This gives $a = 2.9$. The sine law can now be used to find θ,

$$\frac{\sin\theta}{3} = \frac{\sin 1.2}{2.9} \quad\Longrightarrow\quad \sin\theta = \frac{3\sin 1.2}{2.9}.$$

The inverse sine button on a calculator gives $\theta = 1.3$. (We will deal with inverse trigonometric functions in detail in Section 1.8.) Finally, $\phi = \pi - 1.2 - 1.30 = 0.64$.

EXAMPLE 1.32

Distances across water are often much greater than they appear from land. We would like to calculate the distance from the straight shoreline to the island in Figure 1.87a. What we could do is take two points B and C on the shore some distance apart, 1 km say, and measure the angles ABC and ACB as shown in Figure 1.87b. Use this information to calculate how far the island is from shore.

FIGURE 1.87a

Distance from shore to an island

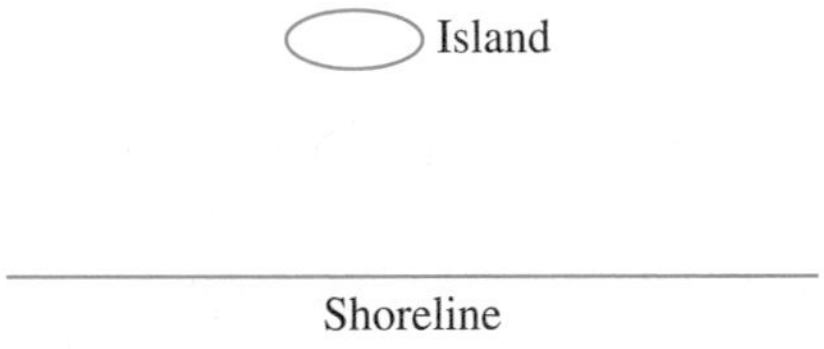

FIGURE 1.87b

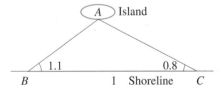

FIGURE 1.87c

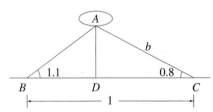

SOLUTION   We require the length of AD in Figure 1.87c. Angle $BAC = \pi - 0.8 - 1.1 = 1.242$. We use the sine law to find b,

$$\frac{\sin 1.1}{b} = \frac{\sin 1.242}{1} \quad\Longrightarrow\quad b = \frac{\sin 1.1}{\sin 1.242} = 0.942.$$

We can now use triangle ADC to calculate that the length of AD is $b\sin 0.8 = (0.942)(0.8) = 0.75$ km.

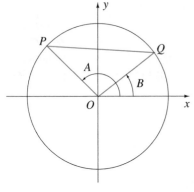

A large number of identities are satisfied by the trigonometric functions. They can all be derived from the following **compound-angle formulas** for sines and cosines:

$$\sin(A + B) = \sin A \cos B + \cos A \sin B, \tag{1.43a}$$

$$\sin(A - B) = \sin A \cos B - \cos A \sin B, \tag{1.43b}$$

$$\cos(A + B) = \cos A \cos B - \sin A \sin B, \tag{1.43c}$$

$$\cos(A - B) = \cos A \cos B + \sin A \sin B. \tag{1.43d}$$

Let us prove one of these, say identity 1.43d, where A and B are the angles in Figure 1.88. If P and Q lie on the circle $x^2 + y^2 = r^2$, then their coordinates are $(r\cos A, r\sin A)$ and $(r\cos B, r\sin B)$, respectively. According to formula 1.10, the length of PQ is given by

$$\begin{aligned}
\|PQ\|^2 &= (r\cos A - r\cos B)^2 + (r\sin A - r\sin B)^2 \\
&= r^2(\cos^2 A + \sin^2 A) + r^2(\cos^2 B + \sin^2 B) \\
&\quad - 2r^2(\cos A \cos B + \sin A \sin B) \\
&= 2r^2 - 2r^2(\cos A \cos B + \sin A \sin B).
\end{aligned}$$

But, according to cosine law 1.41,

$$\begin{aligned}
\|PQ\|^2 &= \|OP\|^2 + \|OQ\|^2 - 2\|OP\|\|OQ\|\cos(A - B) \\
&= r^2 + r^2 - 2r^2\cos(A - B) \\
&= 2r^2 - 2r^2\cos(A - B).
\end{aligned}$$

Comparison of these two expressions for $\|PQ\|^2$ immediately implies identity 1.43d.

By expressing $\tan(A + B)$ as $\sin(A + B)/\cos(A + B)$ and using identities 1.43a and 1.43c, we find a compound-angle formula for the tangent function,

$$\tan(A + B) = \frac{\tan A + \tan B}{1 - \tan A \tan B}, \tag{1.44a}$$

and similarly,

$$\tan(A - B) = \frac{\tan A - \tan B}{1 + \tan A \tan B}. \tag{1.44b}$$

By setting $A = B$ in 1.43a,c and 1.44a, we obtain the **double-angle formulas**,

$$\sin 2A = 2\sin A \cos A, \tag{1.45}$$

$$\cos 2A = \cos^2 A - \sin^2 A, \tag{1.46a}$$

$$= 2\cos^2 A - 1, \tag{1.46b}$$

$$= 1 - 2\sin^2 A, \tag{1.46c}$$

$$\tan 2A = \frac{2\tan A}{1 - \tan^2 A}. \tag{1.47}$$

EXAMPLE 1.33

Use a trigonometric identity to find the cosine of the angle $\theta/2$ if it is known that the cosine of θ is 0.3.

SOLUTION   If we set $A = \theta/2$ in double-angle formula 1.46b, and solve for $\cos^2(\theta/2)$, we obtain

$$\cos\theta = 2\cos^2(\theta/2) - 1 \quad \Longrightarrow \quad \cos^2(\theta/2) = \frac{1 + \cos\theta}{2} = \frac{1 + 0.3}{2} = \frac{1.3}{2}.$$

Hence,

$$\cos(\theta/2) = \pm\sqrt{\frac{1.3}{2}} = \pm 0.81.$$

Without further information about θ or $\theta/2$, we cannot decide which sign to choose. For instance, $\theta = 1.266$ is an angle whose cosine is 0.3. The cosine of half this angle is 0.81. On the other hand, $\theta = 5.0$ is also an angle whose cosine is 0.3, but the cosine of half this angle is -0.81.

When pairs of compound-angle formulas 1.43 are added or subtracted, the **product formulas** result. For example, subtracting 1.43c from 1.43d gives

$$\sin A \sin B = \frac{1}{2}[-\cos(A + B) + \cos(A - B)]. \tag{1.48a}$$

The other product formulas are

$$\sin A \cos B = \frac{1}{2}[\sin(A + B) + \sin(A - B)], \tag{1.48b}$$

$$\cos A \cos B = \frac{1}{2}[\cos(A + B) + \cos(A - B)]. \tag{1.48c}$$

By setting $X = A + B$ and $Y = A - B$ in 1.48, we obtain the **sum and difference formulas**,

$$\sin X + \sin Y = 2\sin\left(\frac{X + Y}{2}\right)\cos\left(\frac{X - Y}{2}\right), \tag{1.49a}$$

$$\sin X - \sin Y = 2\cos\left(\frac{X + Y}{2}\right)\sin\left(\frac{X - Y}{2}\right), \tag{1.49b}$$

$$\cos X + \cos Y = 2\cos\left(\frac{X + Y}{2}\right)\cos\left(\frac{X - Y}{2}\right), \tag{1.49c}$$

$$\cos X - \cos Y = -2\sin\left(\frac{X + Y}{2}\right)\sin\left(\frac{X - Y}{2}\right). \tag{1.49d}$$

The following example is typical of problems in calculus when trigonometric functions are involved.

EXAMPLE 1.34

Write the expression $\cos^4\theta$ in terms of $\cos 2\theta$ and $\cos 4\theta$.

SOLUTION   If we replace A by θ in double-angle formula 1.46b,

$$\cos 2\theta = 2\cos^2\theta - 1.$$

It follows that

$$\cos^2\theta = \frac{1 + \cos 2\theta}{2}.$$

Consequently,

$$\cos^4\theta = (\cos^2\theta)^2 = \frac{1}{4}(1 + \cos 2\theta)^2 = \frac{1}{4}(1 + 2\cos 2\theta + \cos^2 2\theta).$$

But if we now replace θ by 2θ in the identity $\cos^2\theta = (1 + \cos 2\theta)/2$, we obtain

$$\cos^2 2\theta = \frac{1 + \cos 4\theta}{2}.$$

Thus,

$$\cos^4\theta = \frac{1}{4}\left(1 + 2\cos 2\theta + \frac{1 + \cos 4\theta}{2}\right) = \frac{1}{8}(3 + 4\cos 2\theta + \cos 4\theta).$$

By regarding arguments of the trigonometric functions as angles, we have been stressing geometric properties of these functions. In particular, identities 1.43–1.49 have been based on definitions of the trigonometric functions as functions of angles. What is important about a function — be it a trigonometric function or any other kind of function — is that there is a number associated with each value of the independent variable. How we arrive at this number is irrelevant. As far as properties of the function are concerned, only its values are taken into account. Thus, when we discuss properties of a trigonometric function, what is important is not that its argument can be regarded as an angle or that its values can be defined as ratios of sides of a triangle, but that we know its values. Indeed, it is sometimes unwise to regard arguments of trigonometric functions as angles. Consider, for example, the motion of a mass m suspended from a spring with spring constant k (Figure 1.89). If $x = 0$ is the position at which the mass would hang motionless, then when vertical oscillations are initiated, the position of m as a function of time t is always of the form

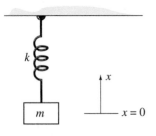

FIGURE 1.89   Displacement of mass in vibrating system

$$x = A\cos\left(\sqrt{\frac{k}{m}}t\right) + B\sin\left(\sqrt{\frac{k}{m}}t\right),$$

where A and B are constants. Clearly, there are no angles associated with the motion of m, and it is therefore unnatural to attempt to interpret the argument $(\sqrt{k/m})t$ as a physical angle.

Henceforth, we consider the trigonometric functions as those of a real variable. If it is convenient to regard the argument as an angle, then we do so, but only if it is convenient. To emphasize this, we replace θ with our usual generic label for the independent variable of a function, namely, x. With this change, the trigonometric functions are $\sin x$, $\cos x$, $\tan x$, $\csc x$, $\sec x$, and $\cot x$. Their graphs are shown in Figures 1.90.

These graphs illustrate that $\sin x$, $\csc x$, $\tan x$, and $\cot x$ are odd functions, and $\cos x$ and $\sec x$ are even. Trigonometric functions are **periodic**. A function $f(x)$ is said to be periodic if there exists a number T such that for all x in its domain of definition,

$$f(x + T) = f(x). \tag{1.50}$$

The smallest such positive number T is called the **period** of $f(x)$. Clearly, then, $\sin x$, $\cos x$, $\csc x$, and $\sec x$ are periodic with period 2π, whereas $\tan x$ and $\cot x$ have period π.

FIGURE 1.90a Sine function

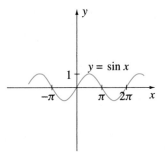

FIGURE 1.90b Cosine function

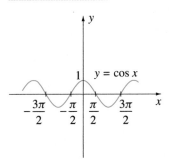

FIGURE 1.90c Tangent function

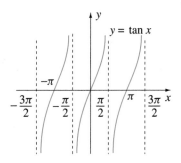

FIGURE 1.90d Cotangent function

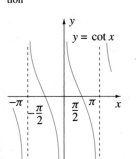

FIGURE 1.90e Secant function

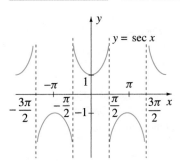

FIGURE 1.90f Cosecant function

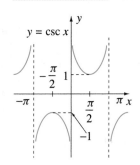

When two periodic functions are added together, the resulting function may, or may not, be periodic. For example, the function $f(x) = 3 \sin 2x + 3\sqrt{3} \cos 2x$ is the addition of two π-periodic functions $3 \sin 2x$ and $3\sqrt{3} \cos 2x$. It also has period π. Function $f(x) = 2 \sin 3x$ has period $2\pi/3$; function $g(x) = -3 \cos 2x$ has period π. Their sum $2 \sin 3x - 3 \cos 2x$ in Figure 1.91 has period 2π; it is the smallest interval in which both functions reproduce themselves. The function $f(x) = \sin x + 2 \sin \sqrt{2}x$ in Figure 1.92 is not periodic even though $\sin x$ and $2 \sin \sqrt{2}x$ are both periodic.

FIGURE 1.91 Graph of $2 \sin 3x - 3 \cos 2x$

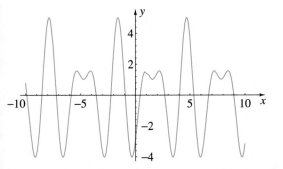

FIGURE 1.92 Graph of a nonperiodic function

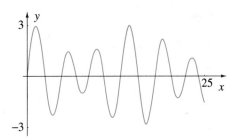

Seldom is the argument of a trigonometric function just x in applications; it is usually a function of x.

EXAMPLE 1.35

Draw the graph of the function $f(x) = \cos(x - \pi/3)$.

SOLUTION   The graph of this function is that of $\cos x$ translated $\pi/3$ units to the right (Figure 1.93).

FIGURE 1.93 Translated graph of cosine function

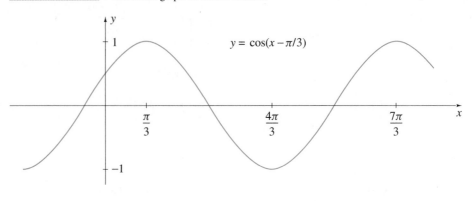

$y = \cos(x - \pi/3)$

EXAMPLE 1.36

When a mass vibrates on the end of a spring as in Figure 1.94, and there is no friction with the surface, or air resistance, the position of the mass (in centimetres) relative to its position when the spring is unstretched takes the form

$$x(t) = 3 \sin 2t + 3\sqrt{3} \cos 2t.$$

Numbers would be different, but $x(t)$ would be a combination of a sine function and a cosine function with the same argument. We have chosen simple numbers here so that unnecessarily complicated calculations do not obscure the significance of the discussion. Plot a graph of $x(t)$ for $t \geq 0$, and then draw a graph. What advantages are derived from the drawing as opposed to the plot?

FIGURE 1.94 Vibrating mass-spring system

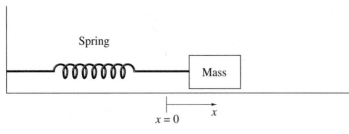

Spring

Mass

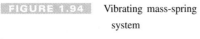

$x = 0$

x

FIGURE 1.95 Computer plot for displacement of mass in vibrating system

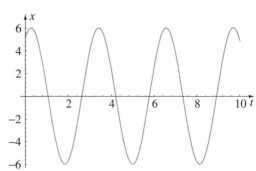

SOLUTION   A plot of the function is shown in Figure 1.95.

To draw a graph of $x(t)$, we first write it in the form $x(t) = A \sin(2t + \phi)$, where A and ϕ are constants. When we equate this to the given expression for $x(t)$, and expand $\sin(2t + \phi)$ with compound-angle formula 1.43a, we obtain

$$3 \sin 2t + 3\sqrt{3} \cos 2t = A \sin(2t + \phi) = A[\sin 2t \cos \phi + \cos 2t \sin \phi].$$

This equation will be true for all t if we choose A and ϕ to satisfy

$$3\sqrt{3} = A \sin \phi, \qquad 3 = A \cos \phi.$$

To solve these for A and ϕ, we square each equation and add the results,

$$27 + 9 = A^2 \sin^2 \phi + A^2 \cos^2 \phi = A^2.$$

This implies that $A = \pm 6$. If we choose $A = 6$ ($A = -6$ works equally well), then

$$3\sqrt{3} = 6 \sin \phi, \qquad 3 = 6 \cos \phi.$$

These equations are satisfied by $\phi = \pi/3$ (there are other angles also), and therefore $x(t)$ can be expressed in the form

$$x(t) = 6 \sin (2t + \pi/3) = 6 \sin [2(t + \pi/6)].$$

The function is most easily graphed by shifting the graph of $f(t) = \sin 2t$ in Figure 1.96a to the left by $\pi/6$ units and changing the scale on the x-axis. The result is shown in Figure 1.96b.

With $x(t)$ expressed in the form $6 \sin (2t + \pi/3)$, it is clear that the function has period π and that oscillations take place between $x = \pm 6$. These are facts that we could surmise from

FIGURE 1.96a Sine function with period π

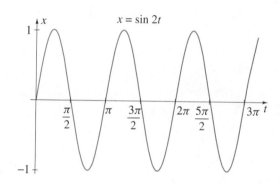

FIGURE 1.96b Hand-drawn graph of displacement of mass in vibrating system

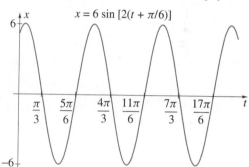

the plot in Figure 1.95, but evidence would not be conclusive. Were we to need values of t at which the function has value 3, say, it is definitely advantageous to have $x(t)$ in the form $6 \sin (2t + \pi/3)$. We can visualize possibilities as t-coordinates of points of intersection of the curve in Figure 1.96b with the horizontal line $x = 3$. There is an infinity of values, and to find them algebraically, we must solve $3 = 6 \sin (2t + \pi/3)$, or $\sin (2t + \pi/3) = 1/2$. We can think of $2t + \pi/3$ as an angle whose sine is 1/2, one possibility being $\pi/6$. But there are many other angles with a sine equal to 1/2. Because the sine function is 2π-periodic, each of the angles $\pi/6 + 2n\pi$, where n is an integer, also has sine equal to 1/2. Furthermore, the sine of $5\pi/6$ is also equal to 1/2, and when multiplies of 2π are added, each of the angles $5\pi/6 + 2n\pi$ has sine equal to 1/2. In other words, all angles that have a sine equal to 1/2 are

$$\frac{\pi}{6} + 2n\pi \qquad \text{and} \qquad \frac{5\pi}{6} + 2n\pi,$$

where n is an integer. In other words, we can set

$$2t + \frac{\pi}{3} = \begin{cases} \dfrac{\pi}{6} + 2n\pi \\[2mm] \dfrac{5\pi}{6} + 2n\pi. \end{cases}$$

Consequently,

$$2t = \begin{cases} -\dfrac{\pi}{6} + 2n\pi \\[2mm] \dfrac{\pi}{2} + 2n\pi \end{cases} \qquad \Longrightarrow \qquad t = \begin{cases} -\dfrac{\pi}{12} + n\pi \\[2mm] \dfrac{\pi}{4} + n\pi. \end{cases}$$

Because we are only interested in positive values of t, we must choose $n \geq 1$ when combined with $-\pi/12$, and $n \geq 0$ when combined with $\pi/4$. The smallest positive value of t is $\pi/4$; the second smallest is $11\pi/12$.

You begin to appreciate the elegance of this solution and the simplicity of the result when you compare the magnitude of the problem were you to attempt to find, using a calculator or computer, all solutions of $3 \sin 2t + 3\sqrt{3} \cos 2t = 3$. Think about it.

The curve in Figure 1.96b is an example of a **general sine function**,

$$f(x) = A \sin(\omega x + \phi), \tag{1.51}$$

where A and ω are positive constants and ϕ is also constant. The graph of the general sine function is shown in Figure 1.97. The number A, which represents half the range of the function, is called the **amplitude** of the oscillations. The period is $2\pi/\omega$, and $-\phi/\omega$ is called the **phase shift**.

FIGURE 1.97   General sine function

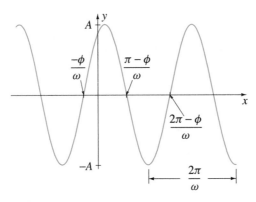

Example 1.36 is an illustration of the following very useful result. A function of the form $f(x) = B \sin \omega x + C \cos \omega x$ can always be expressed in form 1.51 for a general sine function. The amplitude is given by the formula

$$A = \sqrt{B^2 + C^2}. \tag{1.52}$$

Function $B \sin \omega x + C \cos \omega x$ can also be expressed in the form $A \cos(\omega x + \psi)$, and A is once again given by the formula in equation 1.52.

EXAMPLE 1.37

FIGURE 1.98   Schematic for LC-circuit

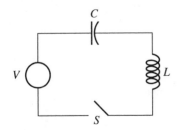

The emf device in the LC-circuit of Figure 1.98 produces a constant voltage of V volts. If the switch is closed at time $t = 0$, and there is no initial charge on the capacitor, the charge thereafter is given by

$$Q(t) = \frac{V}{C}\left(1 - \cos\frac{t}{\sqrt{LC}}\right), \quad t \geq 0.$$

Draw a graph of this function.

SOLUTION   We begin by drawing a graph of $\cos(t/\sqrt{LC})$. It is a standard cosine function with period $2\pi\sqrt{LC}$ (Figure 1.99a). The graph of $Q(t)$ in Figure 1.99b is then obtained by turning Figure 1.99a upside down, shifting it upward 1 unit, and changing the scale on the Q-axis.

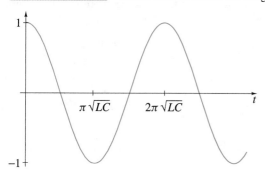

FIGURE 1.99a Cosine function needed for charge on capacitor

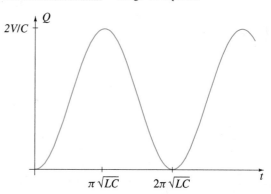

FIGURE 1.99b Charge on capacitor

It is worthwhile pointing out here, as we are sure you realize, that electronic devices cannot plot curves containing unspecified parameters. This is another reason why we must develop our graphing skills and not rely totally on graphing calculators and computers.

EXAMPLE 1.38

Find all solutions for each of the following equations:

(a) $\sin x = \dfrac{1}{\sqrt{2}}$

(b) $\cos 2x = -\dfrac{\sqrt{3}}{2}$

(c) $\tan (3x + 1) = -\sqrt{3}$

SOLUTION

(a) One solution of the equation $\sin x = 1/\sqrt{2}$ is $\pi/4$. This is not the only solution, however; there are many angles with a sine equal to $1/\sqrt{2}$. Since $\sin x$ is 2π-periodic, the angles $2n\pi + \pi/4$, for n any integer, all have sine equal to $1/\sqrt{2}$. Because $\sin (\pi - x) = \sin x$, it follows that $\sin (3\pi/4) = 1/\sqrt{2}$, and therefore $3\pi/4$ is another solution. When multiples of 2π are added to this angle, $2n\pi + 3\pi/4$ are also solutions. Thus, the complete set of solutions is

$$2n\pi + \frac{\pi}{4}, \quad 2n\pi + \frac{3\pi}{4},$$

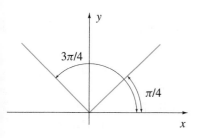

FIGURE 1.100 Simplified representation of solutions to a trigonometric equation

where n is an integer. Figure 1.100 suggests that this set of numbers can be represented more compactly as an initial rotation of $\pi/2$, plus or minus $\pi/4$, and possible multiples of 2π; that is,

$$x = \frac{\pi}{2} \pm \frac{\pi}{4} + 2n\pi = \left(\frac{4n + 1}{2}\right)\pi \pm \frac{\pi}{4}.$$

(b) One solution of the equation $\cos 2x = -\sqrt{3}/2$ for $2x$ is $5\pi/6$. But there are others. Since the cosine function is even, all solutions are given by

$$2x = \pm\frac{5\pi}{6} + 2n\pi \qquad \text{(where } n \text{ is an integer).}$$

Consequently,

$$x = \pm\frac{5\pi}{12} + n\pi.$$

(c) One solution of the equation $\tan(3x+1) = -\sqrt{3}$ for $3x+1$ is $3x+1 = 2\pi/3$. Since the tangent function is π-periodic, all solutions can be expressed in the form

$$3x + 1 = \frac{2\pi}{3} + n\pi = \frac{(3n+2)\pi}{3}, \qquad \text{(where } n \text{ is an integer)}.$$

Consequently,

$$x = \frac{(3n+2)\pi}{9} - \frac{1}{3}.$$

EXERCISES 1.7

In Exercises 1–10 express the angle in radians.

1. $30°$

2. $60°$

3. $135°$

4. $-90°$

5. $-300°$

6. $765°$

7. $72°$

8. $-128°$

9. $321°$

10. $-213°$

In Exercises 11–20 express the angle in degrees.

11. $\pi/3$

12. $-5\pi/4$

13. $3\pi/2$

14. 8π

15. $-5\pi/6$

16. 1

17. -3

18. 2.5

19. -3.6

20. 11

21. What angle is subtended at the centre of a circle of radius 4 by an arc of length (a) 2, (b) 7, and (c) 3.2?

22. The angle of elevation from a transit to the top of a building is 1.30 radians (left figure below). If the transit is 2 metres above the ground, and the distance from the building to the transit is 30 metres, how high is the building?

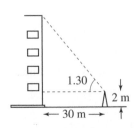

23. Two buildings are 100 metres apart (right figure above). The angle of elevation from the top of the smaller building to the top of the taller building is 11/10 radians. The angle of depression from the top of

the smaller building to the bottom of the taller building is 3/5 radians. How tall are the buildings?

24. If the angle of elevation of the sun is 0.80 radian, and the length of the shadow of a flagpole is 20 metres, how high is the pole?

For each of the triangles in Exercises 25–28 use the cosine law and/or the sine law to find the lengths of all three sides and the measures of all the interior angles.

25.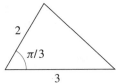

26.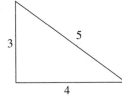

27.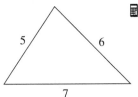

28.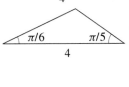

29. Use compound-angle formulas 1.43 to prove identities 1.44.

30. Verify double-angle formulas 1.45, 1.46, and 1.47.

31. Use compound-angle formulas 1.43 to prove product formulas 1.48.

32. Show that the sum and difference formulas 1.49 can be obtained from 1.48.

In Exercises 33–54 draw a graph of the function.

33. $f(x) = 3\sin x$

34. $f(x) = \sin 2x$

35. $f(x) = 3\sin 2x$

36. $f(x) = \sin(x+\pi/4)$

37. $f(x) = 3\sin(x+\pi/4)$

38. $f(x) = \sin(2x+\pi/4)$

39. $f(x) = 3\sin(2x+\pi/4)$

40. $f(x) = 4\cos(x/3)$

41. $f(x) = 2\sin(x/2-\pi)$

42. $f(x) = 5\cos(\pi/2-3x)$

43. $f(x) = \sec 2x$

44. $f(x) = \tan 3x$

45. $f(x) = \csc(x-\pi/3)$

46. $f(x) = \cot(x+\pi/4)$

47. $f(x) = \tan^2 x$

48. $f(x) = \sqrt{1-\cos^2 x}$

49. $f(x) = \sqrt{1 + \tan^2 x}$ * **50.** $f(x) = 5 - 2\sec x$

51. $f(x) = 4 + 2\tan x$ * **52.** $f(x) = \tan|x|$

53. $f(x) = -|\cot 2x|$ * **54.** $f(x) = 3\csc(x/2)$

55. When a javelin is released from height h above the ground with speed v at angle θ with the horizontal (figure below), the horizontal distance R that it travels is given by the formula

$$R = \frac{v^2 \cos\theta}{g}\left(\sin\theta + \sqrt{\sin^2\theta + \frac{2gh}{v^2}}\right),$$

where $g = 9.81$ is the acceleration due to gravity.

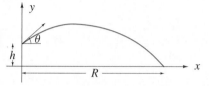

(a) What is R for thrower A, who releases the javelin with speed $v = 20$ m/s at angle $\theta = \pi/4$ and height 2 m?

(b) What is R for thrower B, who, being slightly taller, releases the javelin with the same speed and angle, but at height 2.1 m?

(c) With what speed must thrower A release the javelin if he is to achieve the same R as thrower B, assuming that $\theta = \pi/4$ and $h = 2$ m?

In Exercises 56–69 find all solutions of the equation. All solutions involve the standard angles in Table 1.2. For equations with solutions involving other angles, we require the inverse trigonometric functions of Section 1.8.

56. $\sin x = \sqrt{3}/2$ **57.** $\cos x = 2$

58. $\cos x = -1/2$ **59.** $\cot x = \sqrt{3}$

60. $\sin x = \cos x$ **61.** $2\cos^2 x = 1$

62. $\cos 2x = -1/\sqrt{2}$ **63.** $\tan 3x = -\sqrt{3}$

64. $2\sin 3x - 2 = -1$ **65.** $\sec 4x = -\sqrt{2}$

66. $\sin 2x = \sin x$ * **67.** $\sin^2 x - \sin x - 2 = 0$

68. $3\cot^2 x - 1 = 0$ * **69.** $\sin x + \cos x = 1$

70. When two musical instruments play notes with nearly identical frequencies, *beat notes* result. Suppose, for example, that the signals produced by the instruments are at 180 Hz and 220 Hz, say $\cos(360\pi t)$ and $\cos(440\pi t)$.

(a) Show that the combined signal can be expressed in the form $x(t) = 2\cos(40\pi t)\cos(400\pi t)$.

(b) Plot, on the same axes, graphs of $\pm 2\cos(40\pi t)$ and $x(t)$ for $0 \le t \le 0.8$. The amplitude of $\cos(400\pi t)$ is modulated by $2\cos(40\pi t)$. You would hear the signal $\cos(400\pi t)$ fade in and out as its amplitude rises and falls. This phenomenon is called *beating* of tones in music. Musicians use it to tune two instruments to the same pitch. (See also Exercise 71.)

* **71.** Amplitude modulation is the process of multiplying a low-frequency signal by a high-frequency sinusoid (as occurred in Exercise 70). It is the technique used to broadcast AM radio signals. The AM signal is a product of the form $x(t) = v(t)\cos(2\pi f t)$, where the frequency f is much higher than any frequency in $v(t)$. The cosine term is the *carrier signal* and $v(t)$ is the voice or music signal to be transmitted.

(a) Plot a graph of $x(t)$ when $f = 700$ Hz and $v(t) = 5 + 2\cos(40\pi t)$.

(b) What are minimum and maximum amplitudes of the modulated signal?

In Exercises 72–77 express each function as a general sine function, identifying its amplitude, period, and phase shift. Draw a graph of each function.

* **72.** $f(x) = 3\sin 3x + 3\cos 3x$

* **73.** $f(x) = 2\sin 4x - 2\cos 4x$

* **74.** $f(x) = -2\sin x + 2\sqrt{3}\cos x$

* **75.** $f(x) = -2\sin 5x - 2\sqrt{3}\cos 5x$

* **76.** $f(x) = \sin x\cos x$ * **77.** $f(x) = \sin^2 2x - \cos^2 2x$

In Exercises 78–82 verify the identity.

* **78.** $\cos 3x = 4\cos^3 x - 3\cos x$

* **79.** $\sin 4x = 8\cos^3 x\sin x - 4\cos x\sin x$

* **80.** $\tan 3x = \dfrac{3\tan x - \tan^3 x}{1 - 3\tan^2 x}$

* **81.** $\tan\left(\dfrac{x}{2}\right) = \dfrac{\sin x}{1 + \cos x}$

* **82.** $\dfrac{1 + \tan x}{1 - \tan x} = \tan\left(x + \dfrac{\pi}{4}\right)$

* **83.** Show that a function $f(x) = A\cos\omega x + B\sin\omega x$ can always be written in the form

$$f(x) = \sqrt{A^2 + B^2}\sin(\omega x + \phi),$$

where ϕ is defined by the equations

$$\sin\phi = \frac{A}{\sqrt{A^2 + B^2}} \quad \text{and} \quad \cos\phi = \frac{B}{\sqrt{A^2 + B^2}}.$$

* **84.** In Exercise 83 can we replace the two equations defining ϕ with the single equation $\tan\phi = A/B$?

In Exercises 85–88 find all solutions of the equation in the interval $0 \le x < 2$.

* **85.** $\sin 4x = \cos 2x$

* **86.** $\cos x + \cos 3x = 0$

* **87.** $\sin 2x + \cos 3x = \sin 4x$

* **88.** $\sin x + \cos x = \sqrt{3}\sin x\cos x$

* **89.** Verify that if A, B, and C are the angles of a triangle, then

$$\tan A + \tan B + \tan C = \tan A\tan B\tan C.$$

Hint: Expand $\tan(A + B + C)$ in terms of $\tan A$, $\tan B$, and $\tan C$.

ANSWERS TO DIAGNOSTIC TEST FOR SECTION 1.7

1. $3\pi/4$, $-3\pi/2$ (2 marks)

2. $120°$, $-405°$ (2 marks)

3. (a) $\sqrt{3}/2$ (b) -1 (c) $1/2$ (d) $-\sqrt{2}$ (4 marks)

4. 0.74 radians, 1.8 radians, 2.1 (6 marks)

5. (2 marks)

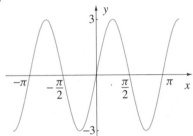

6. (3 marks)

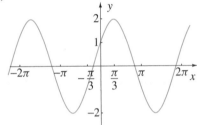

7. (2 marks)

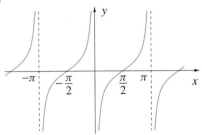

8. (3 marks)

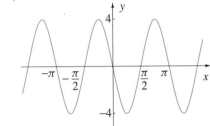

9. $\pi/3 + 2n\pi$, $2\pi/3 + 2n\pi$, n an integer (3 marks)

10. $2\pi/3 + 2n\pi$, $-2\pi/3 + 2n\pi$, n an integer (3 marks)

11. $\pi/8 + n\pi$, n an integer (3 marks)

12. $\sqrt{13}\sin(2x + 5.3)$ (3 marks)

1.8 The Inverse Trigonometric Functions

This material is likely new for many students. It does not therefore have a diagnostic test associated with it.

Functions that involve a finite number of additions, subtractions, multiplications, divisions, and roots are called **algebraic functions**. More specifically, a function $y = f(x)$ is said to be algebraic if, for all x in its domain, it satisfies an equation of the form

$$P_0(x)y^n + P_1(x)y^{n-1} + \cdots + P_{n-1}(x)y + P_n(x) = 0, \tag{1.53}$$

where $P_0(x), \ldots, P_n(x)$ are polynomials in x, and n is a positive integer. A polynomial $P(x)$ is itself algebraic since it satisfies $y - P(x) = 0$; that is, it satisfies 1.53 with $n = 1$, $P_0(x) = 1$, and $P_1(x) = -P(x)$. Rational functions $P(x)/Q(x)$ are also algebraic [$n = 1$, $P_0(x) = Q(x)$, and $P_1(x) = -P(x)$]. The function $f(x) = x^{1/3}$ is algebraic since it satisfies $y^3 - x = 0$. The equation $y^3 + y = x$ defines y as a function of x (a graph of the curve would illustrate that it satisfies the vertical line test). We cannot find the function in the form $y = f(x)$, but according to equation 1.53, the function so defined is algebraic.

A function that is not algebraic is called a **transcendental function**. The trigonometric functions and the exponential and logarithm functions (to be reviewed in Section 1.9) are transcendental. In this section, we consider the inverse trigonometric functions; they are also transcendental.

In Section 1.6 we discussed inverse functions, what it means for one function to be the inverse of another. We learned that, algebraically, a function has an inverse only if it is one-to-one, or geometrically, if its graph passes the horizontal line test. Since the trigonometric functions do not satisfy these conditions, they do not have inverses. But we also learned that it is usually possible to restrict the domain of a function that is not one-to-one in such a way that an inverse function can be defined. We do this for the trigonometric functions in this section.

The function $f(x) = \sin x$, defined for all real x, does not have an inverse; it is not one-to-one; its graph fails the horizontal line test. By restricting the domain of $\sin x$, however, the function can be made one-to-one, and this can be done in many ways. In particular, that part of $\sin x$ on the interval $-\pi/2 \leq x \leq \pi/2$ is one-to-one, and therefore has an inverse function. This function, denoted by

$$y = \text{Sin}^{-1} x, \tag{1.54}$$

and called the **inverse sine function**, is shown in Figure 1.101. The range of the function is

$$-\frac{\pi}{2} \leq \text{Sin}^{-1} x \leq \frac{\pi}{2}. \tag{1.55}$$

The values that fall in this range are called **principal values** of the inverse sine function. They have resulted from our restriction of the domain of $\sin x$ to this same interval.

An equivalent way to derive the inverse sine function is as follows. The reflection of the graph of $y = \sin x$ in the line $y = x$ does not represent a function, but by restricting the range of values of the reflected curve, we can produce a single-valued function (Figure 1.102). Once again this can be done in many ways, and when we do so by restricting the y-values to the interval $[-\pi/2, \pi/2]$, the resulting function is called the inverse sine function $\text{Sin}^{-1} x$. Note very carefully that $\text{Sin}^{-1} x$ is not the inverse function of $f(x) = \sin x$, because the sine function has no inverse. It is, however, the inverse of the sine function restricted to the domain $-\pi/2 \leq x \leq \pi/2$. It is perhaps then a misnomer to call the function $\text{Sin}^{-1} x$ the inverse sine function, but this has become the accepted terminology.

We now know what the inverse sine function looks like graphically, but what does it mean to say that $y = \text{Sin}^{-1} x$? Certainly, given any value of x, we can push a few buttons on an electronic calculator and find $\text{Sin}^{-1} x$ for that x. To use inverse trigonometric functions in practice, we must have a feeling for what they do. To obtain this insight, we note that if (x, y) is a point on the curve $y = \text{Sin}^{-1} x$, then (y, x) is a point on the sine curve; that is,

$$y = \text{Sin}^{-1} x \quad \text{only if} \quad x = \sin y. \tag{1.56}$$

FIGURE 1.101 Inverse of $\sin x$ found by restricting the domain of the function

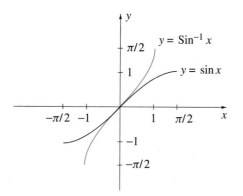

FIGURE 1.102 Inverse of $\sin x$ found by reflecting its graph in line $y = x$ and restricting the range of the reflection

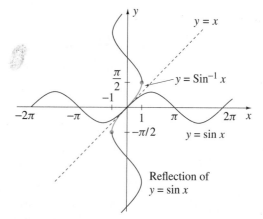

In the latter equation y is an angle and x is the sine of that angle. Thus, when we see $y = \text{Sin}^{-1} x$, we may read "y equals inverse sine x," but we should think "y is an angle whose sine is x." For instance, if $x = 1/2$, then $y = \text{Sin}^{-1}(1/2)$ means "y is an angle whose sine is $1/2$." Clearly, the angle whose sine is $1/2$ is $\pi/6$, and we write $y = \text{Sin}^{-1}(1/2) = \pi/6$. Remember that we must choose the principal value $\pi/6$; there are many angles that have a sine equal to $1/2$, but the inverse sine function demands that we choose that angle in the range $-\pi/2 \leq y \leq \pi/2$, and state it with a number in this interval. For example, $2\pi + \pi/6$ may represent the same angle with the positive x-axis as $\pi/6$, but the number $2\pi + \pi/6$ does not lie between $-\pi/2$ and $\pi/2$. In summary, the function $\sin x$ regards x as an angle and assigns to x the sine of the angle; the inverse sine function $\text{Sin}^{-1} x$ regards x as the sine of an angle and assigns to x the angle with that sine.

Another notation that is commonly used for the inverse sine function is $\arcsin x$. One reason the notation $\arcsin x$ is preferable to $\text{Sin}^{-1} x$ is the possible misinterpretation of $\text{Sin}^{-1} x$. Sometimes students regard the "-1" as a power and write $\text{Sin}^{-1} x$ as $1/\sin x$. This is *not* correct, a fact that you were warned about in Section 1.6. Sin^{-1} is the name of a function, just as $\sin$ is the name of the sine function, and $\sqrt{}$ is the notation for the positive square root function. The capital S in Sin^{-1} should also warn you that this is the inverse sine function.

EXAMPLE 1.39

Simplify each of the following expressions:

(a) $\text{Sin}^{-1}(-\sqrt{3}/2)$ (b) $\text{Sin}^{-1}(1)$ (c) $\text{Sin}^{-1}(3)$
(d) $\text{Sin}^{-1}(3/5) + \text{Sin}^{-1}(4/5)$ (e) $\sin\left[\text{Sin}^{-1}(\sqrt{3}/2)\right]$

SOLUTION

(a) $\text{Sin}^{-1}(-\sqrt{3}/2)$ asks for the angle whose sine is equal to $-\sqrt{3}/2$. Clearly,

$$\text{Sin}^{-1}(-\sqrt{3}/2) = -\pi/3.$$

(b) $\text{Sin}^{-1}(1) = \pi/2$.

(c) $\text{Sin}^{-1}(3)$ is not defined since the domain of $\text{Sin}^{-1} x$ is $-1 \le x \le 1$.

(d) If $\phi = \text{Sin}^{-1}(3/5)$, then ϕ is illustrated in the triangle in Figure 1.103. Since the third side must have length 4, it follows that $\text{Sin}^{-1}(4/5) = \pi/2 - \phi$, and

$$\text{Sin}^{-1}\left(\frac{3}{5}\right) + \text{Sin}^{-1}\left(\frac{4}{5}\right) = \phi + \left(\frac{\pi}{2} - \phi\right) = \frac{\pi}{2}.$$

(e) $\sin\left[\text{Sin}^{-1}(\sqrt{3}/2)\right] = \sin(\pi/3) = \sqrt{3}/2$.

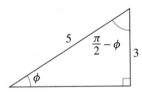

FIGURE 1.103   Triangle to fit the statement $\phi = \text{Sin}^{-1}(3/5)$

EXAMPLE 1.40

Find the values of x for which the following are valid:

(a) $\sin\left(\text{Sin}^{-1} x\right) = x$ (b) $\text{Sin}^{-1}(\sin x) = x$

SOLUTION   These two equations express the fact that the sine function and the inverse sine function are inverses, provided that we are careful about domains:

(a) The equation $\sin\left(\text{Sin}^{-1} x\right) = x$ is valid for $-1 \le x \le 1$. Given an x in this interval, $\text{Sin}^{-1} x$ finds that angle (in the principal value range) which has x as its sine. Then $\sin\left(\text{Sin}^{-1} x\right)$ takes the sine of this angle. Naturally, it returns the original number x.

(b) The function $\text{Sin}^{-1}(\sin x)$ is defined for all x, but only on the domain $-\pi/2 \le x \le \pi/2$ is it equal to x.

Our analysis of the inverse sine function is now complete. We could give a similar discussion for each of the other five trigonometric functions. Instead, we give an abbreviated version for the inverse cosine function and tabulate results for the remaining four functions.

The reflection of the graph of the function $f(x) = \cos x$ is shown in Figure 1.104, and it does not represent a function. If we restrict the y-values on the reflected curve to

$$0 \leq y \leq \pi, \tag{1.57}$$

then we do obtain a function called the **inverse cosine function**, denoted by

$$y = \text{Cos}^{-1} x. \tag{1.58}$$

The values in 1.57 are the principal values of the inverse cosine function. Note again that $\text{Cos}^{-1} x$ is not the inverse function of $\cos x$, but of $f(x) = \cos x$, $0 \leq x \leq \pi$.

 **FIGURE 1.104**   Inverse cosine function from reflection of cosine graph in line $y = x$

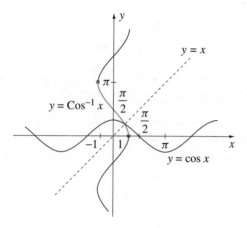

When we write $y = \text{Cos}^{-1} x$, we read this as "y equals inverse cosine x" but it means that "y is an angle whose cosine is x" for if $y = \text{Cos}^{-1} x$, then $x = \cos y$.

The remaining four inverse trigonometric functions, along with those of inverse sine and cosine, are shown in Figure 1.105. Black and blue curves represent reflections in the line $y = x$ of the trigonometric functions $\tan x$, $\cot x$, $\csc x$, and $\sec x$. Blue curves represent principal values of the inverse trigonometric functions $\text{Tan}^{-1} x$, $\text{Cot}^{-1} x$, $\text{Csc}^{-1} x$, and $\text{Sec}^{-1} x$. Principal values of the six inverse trigonometric functions are listed in Table 1.3.

FIGURE 1.105a   Graph of $\text{Sin}^{-1} x$

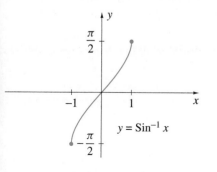

FIGURE 1.105b   Graph of $\text{Cos}^{-1} x$

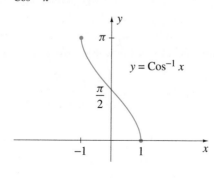

FIGURE 1.105c   Graph of $\text{Tan}^{-1} x$

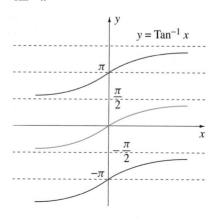

FIGURE 1.105d Graph of $\text{Cot}^{-1} x$

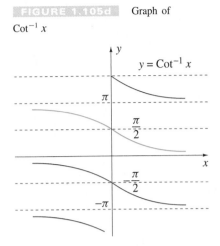

FIGURE 1.105e Graph of $\text{Csc}^{-1} x$

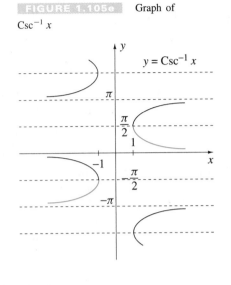

FIGURE 1.105f Graph of $\text{Sec}^{-1} x$

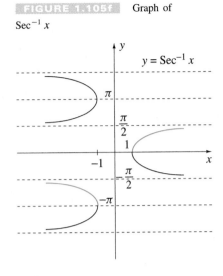

TABLE 1.3

Inverse Trigonometric Function	Principal Values
$\text{Sin}^{-1} x$	$-\dfrac{\pi}{2} \leq y \leq \dfrac{\pi}{2}$
$\text{Tan}^{-1} x$	$-\dfrac{\pi}{2} < y < \dfrac{\pi}{2}$
$\text{Cos}^{-1} x$	$0 \leq y \leq \pi$
$\text{Cot}^{-1} x$	$0 < y < \pi$
$\text{Csc}^{-1} x$	$-\pi < y \leq -\dfrac{\pi}{2}, \ 0 < y \leq \dfrac{\pi}{2}$
$\text{Sec}^{-1} x$	$-\pi \leq y < -\dfrac{\pi}{2}, \ 0 \leq y < \dfrac{\pi}{2}$

It would be reasonable to ask why principal values of $\text{Sec}^{-1} x$ were not chosen as $0 \leq y < \pi/2$, $\pi/2 < y \leq \pi$. Had they been chosen in this way, they would have been very similar to those of $\text{Cos}^{-1} x$ and $\text{Cot}^{-1} x$. Likewise, why are the principal values of $\text{Csc}^{-1} x$ not $-\pi/2 \leq y < 0$, $0 < y \leq \pi/2$? The answer is that they could have been selected in this way, and some authors do indeed make this choice; it is simply a matter of preference. Each choice does, however, create corresponding changes in later work. Specifically, in the exercises of Section 3.10, if principal values of $\text{Sec}^{-1} x$ and $\text{Csc}^{-1} x$ are selected in this alternative way, then derivatives of these functions are modified correspondingly.

In the remainder of this section we solve problems that make use of inverse trigonometric functions.

EXAMPLE 1.41

Find all solutions of the following equations:

(a) $\sin x = 0.4$

(b) $3 \cos 2x = -0.21$

(c) $5 \tan (3x - 1) = 4$

SOLUTION

(a) One solution of this equation is

$$x = \arcsin(0.4) = 0.412 \text{ radians}\quad \text{(to three decimal places).}$$

A second solution is $\pi - 0.412$, and when we add multiples of 2π, we obtain the complete set of solutions

$$2n\pi + 0.412,\quad 2n\pi + (\pi - 0.412),$$

where n is an integer. Following the lead of Example 1.38, Figure 1.106 suggests that this set of numbers can be represented more compactly as an initial rotation of $\pi/2$, plus or minus $\pi/2 - 0.412 = 1.159$, and possible multiples of 2π; that is,

$$\frac{\pi}{2} \pm 1.159 + 2n\pi = \left(\frac{4n+1}{2}\right)\pi \pm 1.159.$$

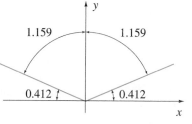

Simplified representation of solutions to a trigonometric equation

Often in problems like this, $x = \arcsin(0.4) = 0.412$ is the only solution given. We can see the reasoning behind this conclusion. There is only one principal value of the inverse sine function, and therefore one solution to the equation. But the equation $\sin x = 0.4$ says nothing about the inverse sine function. We have introduced it simply as a convenience. When we obtain the solution 0.412, we must ask whether there are other solutions to the original equation. For the equation $\sin x = 0.4$, there are other solutions. In some problems, the principal value is the only acceptable solution. Remember, then, if we introduce an inverse trigonometric function into a problem along with its corresponding principal values, we must ask whether there are other possibilities besides the principal values.

(b) Since $\cos 2x = -0.07$, one solution of this equation for $2x$ is

$$2x = \text{Cos}^{-1}(-0.07) = 1.6409 \text{ radians}\quad \text{(to four decimal places).}$$

But there are others. Since the cosine function is even, all solutions are given by

$$2x = \pm 1.6409 + 2n\pi \quad \text{(where } n \text{ is an integer).}$$

Consequently,

$$x = \frac{\pm 1.6409 + 2n\pi}{2} = n\pi \pm 0.820 \quad \text{(to three decimal places).}$$

(c) Since $\tan(3x - 1) = 0.8$, one solution of this equation for $3x - 1$ is

$$3x - 1 = \text{Tan}^{-1}(0.8) = 0.6747 \text{ radians}\quad \text{(to four decimal places).}$$

Since the tangent function is π-periodic, all solutions can be expressed in the form

$$3x - 1 = 0.6747 + n\pi \quad \text{(where } n \text{ is an integer).}$$

Consequently,

$$x = \frac{1 + 0.6747 + n\pi}{3} = 0.558 + \frac{n\pi}{3} \quad \text{(to three decimal places).}$$

EXAMPLE 1.42

Find all solutions of the equation

$$\cos^2 x + 3\cos x = 2.$$

SOLUTION   The quadratic formula applied to the equation

$$(\cos x)^2 + 3(\cos x) - 2 = 0$$

yields

$$\cos x = \frac{-3 \pm \sqrt{9+8}}{2} = \frac{-3 \pm \sqrt{17}}{2}.$$

Since $\cos x$ can only take on values in the interval $-1 \le \cos x \le 1$, the possibility that $\cos x = (-3 - \sqrt{17})/2$ must be rejected, leaving

$$\cos x = \frac{-3 + \sqrt{17}}{2}.$$

From the inverse cosine solution $x = \text{Cos}^{-1}[(\sqrt{17} - 3)/2)] = 0.975$ radians, we obtain all solutions

$$2n\pi \pm 0.975,$$

where n is an integer.

EXAMPLE 1.43

Find all times when the mass in Example 1.36 is 2.5 centimetres to the right of its equilibrium position.

SOLUTION   Since the position of the mass is given by $x(t) = 6\sin[2(t + \pi/6)]$, it is 2.5 centimetres to the right of its equilibrium position when

$$2.5 = 6\sin[2(t + \pi/6)].$$

One solution of this equation for $2(t + \pi/6)$ is $\text{Sin}^{-1}(2.5/6) = \text{Sin}^{-1}(5/12)$. All solutions of the equation are given by

$$2\left(t + \frac{\pi}{6}\right) = \frac{\pi}{2} \pm \left[\frac{\pi}{2} - \text{Sin}^{-1}(5/12)\right] + 2n\pi$$

where n is an integer. We can now solve this equation for t:

$$t + \frac{\pi}{6} = \frac{\pi}{4} \pm \left[\frac{\pi}{4} - \frac{1}{2}\text{Sin}^{-1}(5/12)\right] + n\pi$$

$$t = \frac{\pi}{12} \pm \left[\frac{\pi}{4} - \frac{1}{2}\text{Sin}^{-1}(5/12)\right] + n\pi$$

$$t = \left(\frac{12n + 1}{12}\right)\pi \pm 0.571$$

Since t must be positive, acceptable solutions are

$$t = \begin{cases} \left(\dfrac{12n + 1}{12}\right)\pi + 0.571, & \text{where } n \ge 0 \\ \left(\dfrac{12n + 1}{12}\right)\pi - 0.571, & \text{where } n > 0. \end{cases}$$

EXAMPLE 1.44

Find all solutions of the equation

$$\tan(\cos x) = \frac{1}{\sqrt{3}}.$$

1.02

1.02

SOLUTION   If we set $y = \cos x$, then $\tan y = 1/\sqrt{3}$, and one solution of this equation is $y = \operatorname{Tan}^{-1}(1/\sqrt{3}) = \pi/6$. But there are many other solutions for y, namely

$$y = \frac{\pi}{6} + n\pi,$$

where n is an integer. But $y = \cos x$, and $\cos x$ must take on values in the interval $-1 \le \cos x \le 1$. There is only one possibility for n, namely $n = 0$; hence,

$$\cos x = \frac{\pi}{6}.$$

From the solution $x = \operatorname{Cos}^{-1}(\pi/6) = 1.02$ and Figure 1.107, we obtain

$$x = \pm 1.02 + 2n\pi,$$

where n is an integer, as the complete set of solutions.

EXAMPLE 1.45

Find constants $A > 0$ and $0 < \phi < \pi$ so that the function $f(x) = 3\cos\omega x - 4\sin\omega x$, where $\omega > 0$ is a constant, can be expressed in the form $A\sin(\omega x + \phi)$ for all real x.

SOLUTION   If we expand $A\sin(\omega x + \phi)$ by means of compound-angle formula 1.43a, and equate it to $f(x)$, we have

$$A[\sin\omega x \cos\phi + \cos\omega x \sin\phi] = 3\cos\omega x - 4\sin\omega x.$$

This equation will be valid for all x if we can find values of A and ϕ so that

$$A\cos\phi = -4 \qquad \text{and} \qquad A\sin\phi = 3.$$

When we square and add these equations, the result is

$$A^2\cos^2\phi + A^2\sin^2\phi = A^2 = (-4)^2 + (3)^2 = 25.$$

Consequently, $A = 5$, and

$$\cos\phi = -\frac{4}{5} \qquad \text{and} \qquad \sin\phi = \frac{3}{5}.$$

The only angle in the range $0 < \phi < \pi$ satisfying these equations is $\phi = \arccos(-4/5) = 2.50$ radians. Notice that $\arcsin(3/5)$ does not give this angle. Thus, $f(x)$ can be expressed in the form

$$f(x) = 5\sin(\omega x + 2.50).$$

Perpendicularity and parallelism deal with lines that make a right angle at their point of intersection or that make no angle since parallel lines do not intersect. Lines that intersect usually do so at angles other than $\pi/2$ radians. In order to determine the angle at which two lines intersect, we first define the inclination of a line.

DEFINITION 1.5

The **inclination of a line** l is the angle of rotation ϕ $(0 \le \phi < \pi)$ from the positive x-direction to the line.

Line $y = x + 3$ in Figure 1.108 has inclination $\pi/4$ radians, and line $y = -x + 4$ has inclination $3\pi/4$ radians.

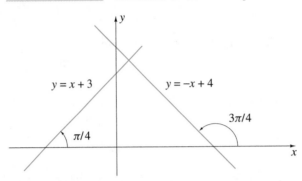

FIGURE 1.108   Inclinations of lines with slopes ± 1

When the slope m of a line is positive, as shown in Figure 1.109a, it is clear that ϕ must be in the interval $0 < \phi < \pi/2$, and m and ϕ are related by $\tan \phi = m$. When the slope of l is negative as in Figure 1.109b, we use identity 1.44b to write

$$m = -\tan(\pi - \phi) = -\frac{\tan \pi - \tan \phi}{1 + \tan \pi \tan \phi} = \tan \phi.$$

FIGURE 1.109a   Inclination of line with positive slope

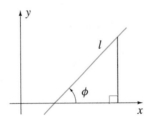

FIGURE 1.109b   Inclination of line with negative slope

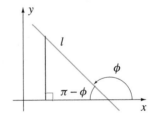

Thus, whenever the slope m of a line is defined, the inclination is related to m by the equation

$$\tan \phi = m. \tag{1.59}$$

In some sense this equation is true even when m is not defined. This occurs for vertical lines, which have no slope. For a vertical line, $\phi = \pi/2$ and $\tan \phi$ is undefined. Thus, equation 1.59 is also valid for vertical lines from the point of view that neither side of the equation is defined. Notice that it is not correct for us to write $\phi = \text{Tan}^{-1} m$, since the principal values of the inverse tangent function $(-\pi/2 < \text{Tan}^{-1} m < \pi/2)$ do not coincide with the specified values for inclination $(0 \leq \phi < \pi)$.

EXAMPLE 1.46

What are the inclinations of the lines $2x - 3y = 4$ and $2x + 3y = 4$?

SOLUTION   From $y = 2x/3 - 4/3$, the slope of the first line is $2/3$. The inclination of this line is $\phi = \text{Tan}^{-1}(2/3) = 0.588$ radians. From $y = -2x/3 + 4/3$, the slope of the second line is $-2/3$. The inclination of this line is $\phi = \pi + \text{Tan}^{-1}(-2/3) = 2.55$ radians.

When two lines l_1 and l_2 with nonzero slopes m_1 and m_2 intersect (Figure 1.110), the angle θ $(0 < \theta < \pi)$ between the lines is given by the equation

$$\phi_1 = \theta + \phi_2 \qquad \text{or} \qquad \theta = \phi_1 - \phi_2.$$

By applying the tangent function to both sides of this equation and using identity 1.44b, we can express θ in terms of m_1 and m_2,

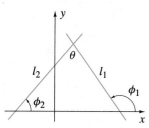

FIGURE 1.110   Angle between intersecting lines

$$\tan \theta = \tan (\phi_1 - \phi_2) = \frac{\tan \phi_1 - \tan \phi_2}{1 + \tan \phi_1 \tan \phi_2} = \frac{m_1 - m_2}{1 + m_1 m_2}.$$

This equation determines θ when $\phi_1 > \phi_2$. When $\phi_1 < \phi_2$, the equation is replaced by

$$\tan \theta = \frac{m_2 - m_1}{1 + m_1 m_2}.$$

In both cases we may write

$$\tan \theta = \left| \frac{m_1 - m_2}{1 + m_1 m_2} \right|,$$

for the acute angle between the lines. It follows then that the **acute angle between two lines with slopes m_1 and m_2** is

$$\theta = \text{Tan}^{-1} \left| \frac{m_1 - m_2}{1 + m_1 m_2} \right|. \tag{1.60}$$

EXAMPLE 1.47

Find the angle between the lines $2x - 3y = 4$ and $x + 4y = 6$.

SOLUTION   Since slopes of these lines are $2/3$ and $-1/4$, it follows that

$$\theta = \text{Tan}^{-1} \left| \frac{2/3 - (-1/4)}{1 + (2/3)(-1/4)} \right| = 0.833 \text{ radians.}$$

EXERCISES 1.8

In Exercises 1–16 evaluate the expression (if it has a value).

1. $\text{Tan}^{-1}(-1/3)$

2. $\text{Sin}^{-1}(1/4)$

3. $\text{Sec}^{-1}(\sqrt{3})$

4. $\text{Csc}^{-1}(-2/\sqrt{3})$

5. $\text{Cot}^{-1}(1)$

6. $\text{Cos}^{-1}(3/2)$

7. $\text{Sin}^{-1}(\pi/2)$

8. $\text{Tan}^{-1}(-1)$

9. $\sin\left(\text{Tan}^{-1}\sqrt{3}\right)$

10. $\tan\left(\text{Sin}^{-1}3\right)$

11. $\text{Sin}^{-1}[\tan(1/6)]$

12. $\text{Tan}^{-1}[\sin(1/6)]$

13. $\sec\left[\text{Cos}^{-1}(1/2)\right]$

14. $\text{Sin}^{-1}[\sin(3\pi/4)]$

15. $\sin\left[\text{Sin}^{-1}(1/\sqrt{2})\right]$

16. $\text{Sin}^{-1}\left[\cos\left(\text{Sec}^{-1}(-\sqrt{2})\right)\right]$

In Exercises 17–26 find all solutions of the equation.

17. $\sin x = 1/3$

18. $\tan x = -1.2$

19. $\cos 2x = 1/3$

20. $\cot 4x + 1 = -1.2$

21. $2 \sin(1 - x) = 1.4$

22. $3 \tan 3x + 2 = -1.2$

23. $4 \sin^2 x - 2 \cos^2 x = 1$

24. $4 \sin^2 x + 2 \cos^2 x = 1$

25. $\cos^2 x - 3 \cos x + 1 = 0$

26. $\sin^2 x - 3 \sin x - 5 = 0$

In Exercises 27–30 draw a graph of the function.

27. $f(x) = 2 + \left(\text{Csc}^{-1}x\right)^2$

28. $f(x) = \sqrt{\text{Tan}^{-1}x} + \sqrt{\text{Sec}^{-1}x}$

29. $f(x) = \text{Sin}^{-1}(x - 3)$

30. $f(x) = \text{Sin}^{-1}x + \text{Csc}^{-1}x$

In Exercises 31–36 find the inclination of the line.

31. $x - y + 1 = 0$

32. $x + 2y = 3$

33. $3x - 2y = 1$

34. $y - 3x = 4$

35. $x = 4$

36. $y = 2$

In Exercises 37–44 determine whether the lines are perpendicular, parallel, or neither. In the last case determine the angle between the lines.

37. $y = -x + 4$, $y = x + 6$

38. $x + 3y = 4$, $2x + 6y = 7$

39. $x = 3y + 4$, $y = x/3 - 2$

40. $2x + 3y = 1$, $3x - 2y = 5$

41. $y = 3x + 2$, $y = -x/2 + 1$

42. $x - y = 5$, $2x + 3y = 4$

43. $x = 0$, $y = 5$

44. $x + y + 2 = 0$, $3x - y = 4$

45. If ϕ is the angle formed by AB and AO in the figure below, find ϕ as a function of θ.

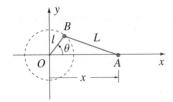

In Exercises 46–51 find all solutions of the equation.

46. $\sin x \tan^2 x - 3 + \tan^2 x - 3 \sin x = 0$

47. $\sin x + \cos x = 1$

48. $\sec (\sin x) = -\sqrt{2}$ **49.** $\cos \left(\mathrm{Sin}^{-1} x \right) = 1/2$

50. $\sec (\tan x) = -\sqrt{2}$

51. $\mathrm{Cos}^{-1} [\tan (x^2 + 4)] = 2\pi - 5$

52. Draw graphs of the following functions: (a) $f(x) = \sin (\mathrm{Sin}^{-1} x)$; (b) $f(x) = \mathrm{Sin}^{-1} (\sin x)$.

53. Draw graphs of the following functions: (a) $f(x) = \cos (\mathrm{Cos}^{-1} x)$; (b) $f(x) = \mathrm{Cos}^{-1} (\cos x)$.

54. Express the function $f(x) = 4 \sin 2x + \cos 2x$ in the form $R \sin (2x + \phi)$, where $R > 0$ and $0 < \phi < \pi$.

55. Express the function $f(x) = -2 \sin 3x + 4 \cos 3x$ in the form $R \cos (3x + \phi)$, where $R > 0$ and $0 < \phi < \pi$.

56. Repeat Exercise 54 for $f(x) = -2 \sin 2x + 4 \cos 2x$.

57. Repeat Exercise 55 for $f(x) = -4 \sin 3x + 5 \cos 3x$.

58. Two electric signals $f(t) = 4 \cos (\omega t + 2\pi/3)$ and $g(t) = 3 \sin (\omega t + \pi/3)$ are fed into the same line so that the resulting signal is $x(t) = f(t) + g(t)$. Express $x(t)$ in the form $A \sin (\omega t + \phi)$ for appropriate values of $A > 0$ and $-\pi < \phi < \pi$.

59. Repeat Exercise 58 but express $x(t)$ in the form $A \cos (\omega t + \phi)$.

60. Two electric signals $f(t) = 2 \sin (\omega t + 4)$ and $g(t) = 3 \sin (\omega t + 1)$ are fed into the same line so that the resulting signal is $x(t) = f(t) + g(t)$. Express $x(t)$ in the form $A \cos (\omega t + \phi)$ for appropriate values of $A > 0$ and $-\pi < \phi < \pi$.

61. Repeat Exercise 60 but express $x(t)$ in the form $A \sin (\omega t + \phi)$.

62. Three electric signals $f(t) = 5 \cos (\omega t + 3\pi/2)$, $g(t) = 4 \cos (\omega t + \pi/3)$, and $h(t) = 2 \sin (\omega t + \pi/4)$ are fed into the same

line so that the resulting signal is $x(t) = f(t) + g(t) + h(t)$. Express $x(t)$ in the form $A \sin (\omega t + \phi)$ for appropriate values of $A > 0$ and $-\pi < \phi < \pi$.

63. Repeat Exercise 62 but express $x(t)$ in the form $A \cos (\omega t + \phi)$.

64. A crank of length R with slider C is rotating clockwise about O as shown below. The slider moves in a slotted lever hinged at A at a distance L from O. Find angle θ as a function of angle ϕ.

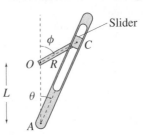

65. The angle of elevation of the top of a tower from A is ϕ, and the angle from B at a distance d from A is θ (figure below). Find a formula for θ in terms of ϕ.

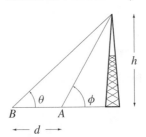

66. A pendulum consists of a mass m suspended from a string of length L (figure below). At time $t = 0$, the mass is pulled through a small angle θ_0 — to the right when $\theta_0 > 0$, and to the left when $\theta_0 < 0$ — and given an initial speed v_0 to the right. Its subsequent angular displacements are given by

$$\theta = \theta(t) = \theta_0 \cos \omega t + \frac{v_0}{\omega L} \sin \omega t, \quad t \geq 0,$$

where $\omega = \sqrt{9.81/L}$, provided that any resistance due to the air is neglected. Show that $\theta(t)$ can be expressed in the form

$$\theta(t) = \sqrt{\theta_0^2 + \frac{v_0^2}{\omega^2 L^2}} \sin (\omega t + \phi),$$

where $\phi = \mathrm{Tan}^{-1} (\omega L \theta_0 / v_0)$.

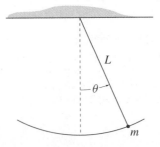

67. What changes, if any, must be made in Exercise 66 if the initial speed v_0 is to the left rather than the right?

68. A mass m is suspended from a spring with constant k (figure below). If at time $t = 0$, the mass is given an initial displacement y_0 and an upward speed v_0, its subsequent displacements are given by

$$y = y(t) = y_0 \cos \omega t + \frac{v_0}{\omega} \sin \omega t, \quad t \geq 0,$$

where $\omega = \sqrt{k/m}$, provided that any resistance due to the air is neglected. Show that $y(t)$ can be expressed in the form

$$y(t) = \sqrt{y_0^2 + \frac{v_0^2}{\omega^2}} \sin (\omega t + \phi),$$

where $\phi = \mathrm{Tan}^{-1} (\omega y_0 / v_0)$.

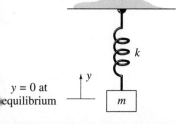

$y = 0$ at
equilibrium

69. What changes, if any, must be made in Exercise 68 if the initial speed v_0 is downward rather than upward?

70. An inductance L, resistance R, and capacitance C are connected in series with a generator producing an oscillatory voltage $E = E_0 \cos \omega t, t \geq 0$ (figure below). If $L, C, R, E_0 > 0$, and $\omega > 0$ are all constants, the steady-state current I in the circuit is

$$I = \frac{E_0}{\sqrt{R^2 + \left(\omega L - \dfrac{1}{\omega C} \right)^2}}$$

$$\bullet \left[R \cos \omega t + \left(\omega L - \frac{1}{\omega C} \right) \sin \omega t \right].$$

Express I in the form

$$I = A \cos (\omega t - \phi),$$

where $A > 0$ and $-\pi/2 \leq \phi \leq \pi/2$.

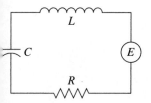

71. It is known that $5 \cos \omega t = A \cos (\omega t - \pi/6) + 5 \cos (\omega t + \phi)$ for all t, where ω is a fixed constant. Find, exactly, all possible values for $A > 0$ and ϕ.

* **72.** Repeat Exercise 71 if $5 \cos \omega t = A \cos(\omega t + 1) + 5 \sin(\omega t + \phi)$.

* **73.** Show that

$$\mathrm{Csc}^{-1} x = \begin{cases} \mathrm{Sin}^{-1}\left(\dfrac{1}{x}\right), & x \geq 1, \\[4mm] -\pi - \mathrm{Sin}^{-1}\left(\dfrac{1}{x}\right), & x \leq -1. \end{cases}$$

* **74.** Show that

$$\mathrm{Sin}^{-1} x = \begin{cases} -\mathrm{Cos}^{-1} \sqrt{1 - x^2}, & -1 \leq x < 0, \\[2mm] \mathrm{Cos}^{-1} \sqrt{1 - x^2}, & 0 \leq x \leq 1. \end{cases}$$

* **75.** Prove that

$$\mathrm{Sec}^{-1} x = \begin{cases} \mathrm{Cos}^{-1}\left(\dfrac{1}{x}\right), & x \geq 1, \\[4mm] -\mathrm{Cos}^{-1}\left(\dfrac{1}{x}\right), & x \leq -1. \end{cases}$$

* **76.** Prove that

$$\mathrm{Cot}^{-1} x = \begin{cases} \mathrm{Tan}^{-1}\left(\dfrac{1}{x}\right), & x > 0, \\[4mm] \pi + \mathrm{Tan}^{-1}\left(\dfrac{1}{x}\right), & x < 0. \end{cases}$$

* **77.** Verify that

$$\mathrm{Sec}^{-1} x = \begin{cases} \mathrm{Tan}^{-1} \sqrt{x^2 - 1}, & x \geq 1, \\[2mm] -\pi + \mathrm{Tan}^{-1} \sqrt{x^2 - 1}, & x \leq -1. \end{cases}$$

* **78.** Verify that

$$\mathrm{Csc}^{-1} x = \begin{cases} \mathrm{Cot}^{-1} \sqrt{x^2 - 1}, & x \geq 1, \\[2mm] -\pi + \mathrm{Cot}^{-1} \sqrt{x^2 - 1}, & x \leq -1. \end{cases}$$

* **79.** Verify that if $0 \leq x < 1$, then

$$2 \, \mathrm{Tan}^{-1} \sqrt{\frac{1 + x}{1 - x}} = \pi - \mathrm{Cos}^{-1} x.$$

1.9 Exponential and Logarithm Review

Here is the diagnostic test for this section. Give yourself 30 minutes to do it.

1. Evaluate the following quantities: (a) $\log_2 32$ (b) $\log_3 (1/27)$
(c) $10^{\log_{10} 2x}$ (d) $e^{2\ln 4}$

In questions 2–4 find all solutions of the equation.

2. $\log_{10} 3x = 2$ **3.** $e^{2x+1} = 4$

4. $\ln x - \ln (x - 1) = 1$

In questions 5–8 draw a graph of the function.

5. $f(x) = e^{3x}$ **6.** $f(x) = \log_{10} (x - 1)$

7. $f(x) = 4^{-x}$ **8.** $f(x) = e^x + e^{-x}$

9. If $y = 4e^{2x-1}$, find x in terms of y.

In elementary algebra we learned the basic rules for products and quotients of powers:

$$a^b a^c = a^{b+c}; \tag{1.61a}$$

$$\frac{a^b}{a^c} = a^{b-c}; \tag{1.61b}$$

$$\left(a^b\right)^c = a^{bc}; \tag{1.61c}$$

where $a > 0$, and b and c are real constants. These rules are used to develop the **exponential function**

$$f(x) = a^x, \tag{1.62}$$

that is, a raised to the exponent x for variable x.

We concentrate on the case when $a > 1$. The meaning of a^x when x is an integer is clear, and when $x = 1/n$, where $n > 0$ is an integer, $a^x = a^{1/n}$ is the nth root of a. When x is a positive rational number n/m (n and m positive integers), 1.61c implies that

$$a^x = a^{n/m} = \left(a^n\right)^{1/m} \quad \text{or} \quad a^x = a^{n/m} = \left(a^{1/m}\right)^n;$$

that is, $a^{n/m}$ is the mth root of a to the integer power n, or $a^{n/m}$ is the integer power n of the mth root of a. When x is a negative rational, we write $a^x = 1/a^{-x}$, where $-x$ is positive. These results lead to the points and the graph of $y = a^x$ in Figure 1.111.

FIGURE 1.111   The exponential function

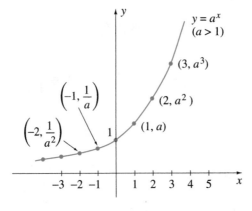

There is a difficulty with the definition of a^x and its graph in Figure 1.111. We are not prepared to resolve the problem now, but we would be remiss in not pointing it out. How do we define a^x when x is an irrational number? For instance, what is the value of $a^{\sqrt{2}}$ or a^{π}? If a^x is undefined whenever x is irrational, the graph in Figure 1.111 is misleading. Although we have joined the points with the smoothest possible curve, there is actually an infinite number of values of x (the irrational numbers) at which there is, as yet, no dot on the curve. We require *limits* from Chapter 2 to deal with this problem, and we therefore set it aside until Section 2.4.

EXAMPLE 1.48

Plot graphs of the exponential functions 2^x and 3^x on the same axes.

SOLUTION   Graphs of these functions are shown in Figure 1.112. Notice that both curves pass through the point $(0, 1)$. More generally, $a^0 = 1$ for any a. When $x > 0$, the graph of 3^x is higher than that of 2^x, whereas the opposite is true for $x < 0$.

FIGURE 1.112   Two exponential functions

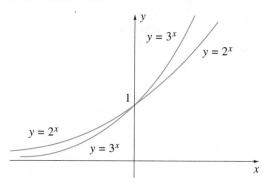

EXAMPLE 1.49

Draw graphs of the functions x^4 (a power function) and 2^x (an exponential function) on the same axes.

SOLUTION   Graphs of these functions are shown in Figure 1.113, but no attempt has been made to use a scale on either the x- or y-axis. Notice here that for $x > 16$, $2^x > x^4$. This is always the situation for power and exponential functions. Given any exponential function a^x $(a > 1)$, and any power function x^n $(n > 1)$, there always exists a value of x, say X, such that when $x > X$, we have $a^x > x^n$. In short, exponential functions grow more rapidly than power functions for large values of x.

FIGURE 1.113   Comparison of graphs of power and exponential functions

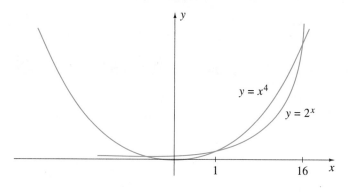

In terms of the exponential function, rules 1.61 take the form

$$a^{x_1} a^{x_2} = a^{x_1 + x_2}, \tag{1.63a}$$

$$\frac{a^{x_1}}{a^{x_2}} = a^{x_1 - x_2}, \tag{1.63b}$$

$$\left(a^{x_1}\right)^{x_2} = a^{x_1 x_2}. \tag{1.63c}$$

The exponential function is one-to-one; its graph passes the horizontal line test. As a result, it has an inverse; we call it the **logarithm function** to base a, denoted by $\log_a x$. As an inverse, the logarithm function reverses the action of the exponential function. For instance, working with base 10, since $10^3 = 1000$, it follows that $\log_{10} 1000 = 3$; since $10^{-2} = 0.01$, we have $\log_{10} 0.01 = -2$. We can work with logarithms to any base, but it is customary to use bases greater than 1. With base 3, say, we can write that $\log_3 81 = 4$ since $3^4 = 81$, and that $\log_3 (1/9) = -2$ since $3^{-2} = 1/9$. These examples demonstrate that the logarithm of a number is a power. To say that y is the logarithm of x to base a is to say that $x = a^y$. In general, *the logarithm of x to base a is the power to which a must be raised in order to produce x*. Algebraically, we write this as

$$y = \log_a x \qquad \text{only if} \qquad x = a^y. \tag{1.64}$$

When the first of these is substituted into the second, the result is

$$a^{\log_a x} = x. \tag{1.65a}$$

When the second is substituted into the first,

$$y = \log_a (a^y),$$

or since y is arbitrary, we may replace it with x,

$$\log_a a^x = x. \tag{1.65b}$$

The second of these is valid for all x, but the first is true only for $x > 0$. These equations simply express the fact that exponential and logarithm functions are inverses of each other; do one, then the other, and you are back where you started (see equations 1.35 and 1.36).

A graph of the logarithm function $f(x) = \log_a x$ can be obtained by reflecting the graph of $y = a^x$ in the line $y = x$ (Figure 1.114). It passes through the point $(1, 0)$ for any a. In other words, $\log_a 1 = 0$. As values of x get closer and closer to 0, their logarithms become very large negative numbers. We cannot take the logarithms of 0 or negative numbers.

FIGURE 1.114   Logarithm function

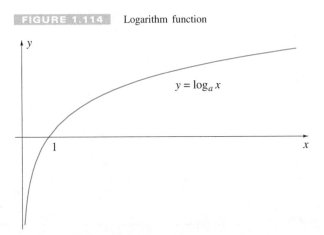

$y = \log_a x$

Corresponding to rules 1.63 for the exponential function are the following rules for logarithms:

$$\log_a(x_1 x_2) = \log_a x_1 + \log_a x_2, \tag{1.66a}$$

$$\log_a\left(\frac{x_1}{x_2}\right) = \log_a x_1 - \log_a x_2, \tag{1.66b}$$

$$\log_a\left(x_1^{x_2}\right) = x_2 \log_a x_1. \tag{1.66c}$$

To prove 1.66a, say, we set $z = \log_a(x_1 x_2)$, in which case

$$
\begin{aligned}
a^z &= x_1 x_2 \\
&= \left(a^{\log_a x_1}\right)\left(a^{\log_a x_2}\right) \qquad \text{(using 1.65a)} \\
&= a^{\log_a x_1 + \log_a x_2}. \qquad \text{(using 1.63a)}
\end{aligned}
$$

Thus,

$$\log_a x_1 + \log_a x_2 = z = \log_a(x_1 x_2).$$

We leave proofs of 1.66b and 1.66c to the exercises.

EXAMPLE 1.50

Simplify the following expressions:

(a) $3^{\log_3(x^2)}$ (b) $10^{-4\log_{10} x}$ (c) $\log_a\left(a^{-x+3}\right)$ (d) $\log_2 8 + \log_3(1/27)$

SOLUTION

(a) Identity 1.65a implies that

$$3^{\log_3(x^2)} = x^2.$$

(b) Since -4 intervenes between the logarithm and exponential operations, we cannot use 1.65a immediately. The -4 can be relocated, however, with 1.66c:

$$10^{-4\log_{10} x} = 10^{\log_{10}(x^{-4})} = \frac{1}{x^4} \quad \text{(if } x > 0\text{)}.$$

(c) Identity 1.65b gives

$$\log_a\left(a^{-x+3}\right) = -x + 3.$$

(d) Since $\log_2 8 = 3$ and $\log_3(1/27) = -3$,

$$\log_2 8 + \log_3(1/27) = 3 - 3 = 0.$$

EXAMPLE 1.51

Solve the following equations:

(a) $\log_5 x = -3$ (b) $\log_{10} x + \log_{10}(x + 1) = 0$ (c) $10^x - 12 + 10^{-x} = 0$

SOLUTION

(a) By means of equation 1.64,

$$x = 5^{-3} = \frac{1}{125}.$$

(b) Since $\log_{10} x + \log_{10}(x + 1) = \log_{10}[x(x + 1)]$, we can write

$$0 = \log_{10}[x(x + 1)].$$

If we now take exponentials to base 10,

$$10^0 = 10^{\log_{10}[x(x+1)]} \quad \text{or} \quad 1 = x(x + 1).$$

This quadratic equation has solutions

$$x = \frac{-1 \pm \sqrt{1 + 4}}{2} = \frac{-1 \pm \sqrt{5}}{2}.$$

Since x must be positive (the original equation demands this), the only solution is $x = (\sqrt{5} - 1)/2$.

(c) If we multiply the equation by 10^x, the result is

$$0 = 10^{2x} - 12(10^x) + 1 = (10^x)^2 - 12(10^x) + 1.$$

But this is a quadratic equation in 10^x, so that

$$10^x = \frac{12 \pm \sqrt{144 - 4}}{2} = 6 \pm \sqrt{35}.$$

Finally, we have

$$x = \log_{10}\left(6 \pm \sqrt{35}\right).$$

EXAMPLE 1.52

The ear hears by detecting pressure variations of impinging sound waves. The loudness of the sound is related to the intensity of the sound wave, which is measured in watts per square metre (energy transmitted by the sound wave per unit time per unit area). The lowest intensity detectable by the ear is normally taken as $I_0 = 10^{-12}$ W/m² at a frequency of 1000 Hz; it is called the *audible sound threshold*. By comparison, the intensity of sound from a jet engine is about 10^4 W/m²; it is 10^{16} times that of the audible sound threshold. Because the range of intensities to which the ear is sensitive is so large, dealing directly with intensities is cumbersome. Logarithms provide a way to reduce this enormous range to a manageable size. As a number increases by a factor of 10, its logarithm increases by 1. For example, the difference between the logarithms of 10 and 100 is $2 - 1 = 1$. If a number increases by a factor of 10^{14}, its logarithm increases by 14. This range is deemed to be a little too compact; it is expanded by a factor of 10 in the following definition. The loudness of a sound is said to be L decibels if

$$L = 10 \log_{10}\left(\frac{I}{I_0}\right),$$

where I is the intensity of the sound and I_0 is the intensity of sound at the audible threshold. Use this definition to answer the following questions.

(a) What is the loudness of sound at the audible sound threshold?

(b) Express the intensity I of a sound in terms of I_0 and its decibel reading L.

(c) If decibel readings for a voice, a car, and a jet engine are 70, 100, and 160, respectively, what are the corresponding intensities of the sound waves relative to I_0?

(d) If the pain threshold for sound has an intensity 10^{14} times I_0, what is its decibel reading?

(e) If the intensity I_1 of one sound is 10 times the intensity I_2 of a second sound, how do their decibel readings compare?

SOLUTION

(a) The decibel reading for the audible sound threshold is $L = 10\log_{10}(I_0/I_0) = 0$.

(b) When we take both sides as exponents of powers of 10, and use properties 1.66c and 1.65a, we obtain

$$10^L = 10^{10\log_{10}(I/I_0)} = 10^{\log_{10}(I/I_0)^{10}} = \left(\frac{I}{I_0}\right)^{10}.$$

If we take 10th roots of both sides, we have

$$\left(10^L\right)^{1/10} = \frac{I}{I_0} \qquad \text{or} \qquad I = I_0 10^{L/10}.$$

(c) Since the decibel level of the normal voice is 70, its intensity is

$$I = I_0 10^{70/10} = 10^7 I_0.$$

Similarly, the intensities of a car and a jet are 10^{10} and 10^{16} times I_0.

(d) For an intensity of $10^{14} I_0$, the decibel reading is

$$L = 10\log_{10}(10^{14}) = 10(14) = 140.$$

(e) If L_1 and L_2 are the decibel readings for sounds with intensities I_1 and I_2, then

$$L_1 = 10\log_{10}(I_1/I_0) \qquad \text{and} \qquad L_2 = 10\log_{10}(I_2/I_0).$$

When we subtract these readings, we obtain

$$
\begin{aligned}
L_1 - L_2 &= 10\log_{10}(I_1/I_0) - 10\log_{10}(I_2/I_0) \\
&= 10[\log_{10}(I_1/I_0) - \log_{10}(I_2/I_0)] \\
&= 10\log_{10}\left(\frac{I_1/I_0}{I_2/I_0}\right) \qquad \text{(using 1.66b)} \\
&= 10\log_{10}\left(\frac{I_1}{I_2}\right) \\
&= 10\log_{10}\left(\frac{10I_2}{I_2}\right) \qquad \text{(since } I_1 = 10I_2) \\
&= 10\log_{10} 10 \\
&= 10.
\end{aligned}
$$

Thus, when the intensity of one sound is 10 times that of another, their decibel readings differ by 10.

It is sometimes necessary to change from one base of logarithms to another. If we take logarithms to base b on both sides of identity 1.65a, we obtain immediately

$$\log_b x = (\log_a x)(\log_b a). \tag{1.67}$$

This equation defines $\log_b a$ as the conversion factor from logarithms to base a to logarithms to base b.

Before the discovery of calculus, the base of logarithms was invariably chosen to be 10. Such logarithms are called *common logarithms*; they correspond to the exponential function 10^x. Another base for exponentials and logarithms, however, that is much more convenient in most applications is a number, denoted by the letter e, and defined in a variety of ways. One way is to consider the numbers

$$\left(1 + \frac{1}{n}\right)^n \tag{1.68}$$

TABLE 1.4

n	$\left(1 + \frac{1}{n}\right)^n$
1	2.000 000
3	2.370 370
5	2.488 320
10	2.593 742
100	2.704 814
1 000	2.716 924
10 000	2.718 146
100 000	2.718 255
1 000 000	2.718 282

for ever-increasing values of n. The numbers in Table 1.4 are steadily increasing but getting closer together. They suggest that for larger and larger values of n, the function $(1 + 1/n)^n$ is indeed getting closer to some number — to 12 decimal places this number is

$$e \approx 2.718281828459.$$

This number e is irrational, with a nonterminating, nonrepeating decimal expansion. Why this number is so convenient as a base for logarithms and exponentials is shown in Section 3.11. For now, let us rewrite some of the more important formulas of this section with a set equal to e. The exponential function to base e is e^x, and equations 1.63 in terms of e^x read

$$e^{x_1} e^{x_2} = e^{x_1 + x_2}, \tag{1.69a}$$

$$\frac{e^{x_1}}{e^{x_2}} = e^{x_1 - x_2}, \tag{1.69b}$$

$$\left(e^{x_1}\right)^{x_2} = e^{x_1 x_2}. \tag{1.69c}$$

Logarithms to base e are usually given the notation $\ln x$ rather than $\log_e x$, and are called **natural logarithms**:

$$\ln x = \log_e x. \tag{1.70}$$

In terms of $\ln x$, rules 1.66 are

$$\ln (x_1 x_2) = \ln x_1 + \ln x_2, \tag{1.71a}$$

$$\ln \left(\frac{x_1}{x_2}\right) = \ln x_1 - \ln x_2, \tag{1.71b}$$

$$\ln \left(x_1^{x_2}\right) = x_2 \ln x_1. \tag{1.71c}$$

Identities 1.65 become

$$x = e^{\ln x}, \qquad x > 0, \tag{1.72a}$$

$$x = \ln (e^x). \tag{1.72b}$$

Graphs of e^x and $\ln x$ in Figures 1.115 and 1.116 have the same shape as those of a^x in Figure 1.111 and $\log_a x$ in Figure 1.114. Only the "steepness" of the curves is affected by a change of base.

FIGURE 1.115   Graph of e^x

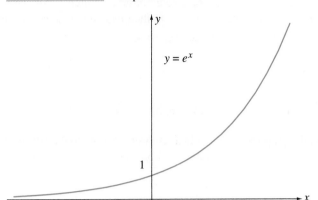

FIGURE 1.116   Graph of $\ln x$

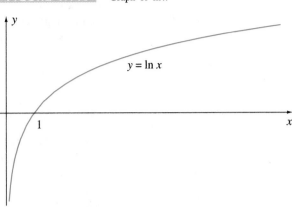

Neither of the functions e^x nor $\ln x$ is even or odd, but functions derived from them may be even or odd. This is illustrated in the following example.

EXAMPLE 1.53

Is the function $f(x) = e^{-ax^2}$, where $a > 0$ is a constant, even or odd? Draw its graph.

SOLUTION   Since

$$f(-x) = e^{-a(-x)^2} = e^{-ax^2} = f(x),$$

the function e^{-ax^2} is even. Graphs for $a = 1, 2,$ and 3 are plotted in Figures 1.117a–c. The value of a controls the spread of the curve. For any value of a, the graph passes through the point $(0, 1)$ and is asymptotic to the x-axis (Figure 1.117d). This curve is very important in statistics. It is called the *bell curve* or *normal distribution*.

FIGURE 1.117a   Graph of e^{-x^2}

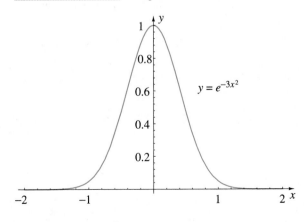

FIGURE 1.117b   Graph of e^{-2x^2}

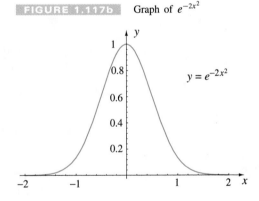

FIGURE 1.117c   Graph of e^{-3x^2}

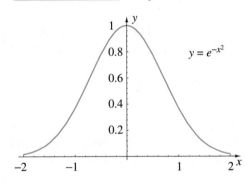

FIGURE 1.117d   Graph of e^{-ax^2}

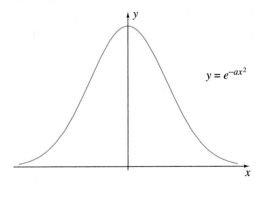

You will be making graphs of functions at every turn in this book. We remind you that when we ask you to plot a graph, you are to use a graphing calculator or computer. When we ask you to draw a graph, you are to do so without these devices. In the preceding example, we plotted graphs of $f(x) = e^{-ax^2}$ for $a = 1, 2,$ and 3, and then drew $y = e^{-ax^2}$ by hand.

EXAMPLE 1.54

Draw a graph of the function $f(x) = \ln(x + a)$, where $a > 0$ is a constant.

SOLUTION   By translating the graph of $f(x) = \ln x$ in Figure 1.118a to the left by a units, we obtain the graph of $f(x) = \ln(x + a)$ in Figure 1.118b.

FIGURE 1.118a   Graph of $\ln x$

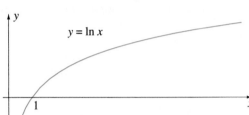

FIGURE 1.118b   Graph of $\ln x$ translated a units to the left

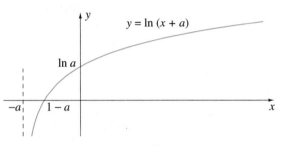

EXAMPLE 1.55

The emf device in the RC-circuit of Figure 1.119 produces a constant voltage of V volts. If the switch is closed at time $t = 0$, and there is no initial charge on the capacitor, the charge thereafter is given by

$$Q(t) = CV[1 - e^{-t/(RC)}], \quad t \geq 0.$$

Draw a graph of this function.

FIGURE 1.119   Schematic for RC-circuit

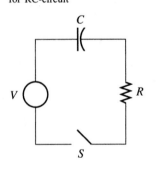

SOLUTION   A graph of $e^{t/(RC)}$ (Figure 1.120a) has the same shape as that in Figure 1.115; the constant RC affects only the steepness of the curve. The graph of $e^{-t/(RC)}$ in Figure 1.120b is that of Figure 1.120a reflected in the vertical axis. The graph of $Q(t)$ in Figure 1.120c is then obtained by turning Figure 1.120b upside down, shifting it upward 1 unit, changing the scale on the vertical axis by a factor CV, and retaining only that part of the curve $t \geq 0$. It is asymptotic to the line $Q = CV$.

FIGURE 1.120a   Graph of $e^{t/(RC)}$

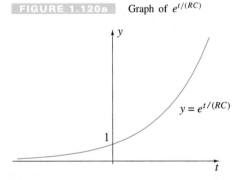

FIGURE 1.120b   Graph of $e^{-t/(RC)}$

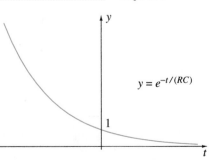

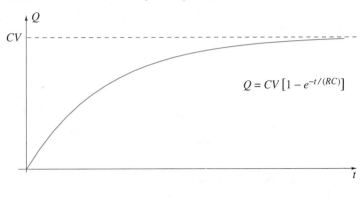

FIGURE 1.120c   Charge on capacitor

$$Q = CV\left[1 - e^{-t/(RC)}\right]$$

EXERCISES 1.9

In Exercises 1–13 find all values of x satisfying the equation.

1. $\log_{10}(2 + x) = -1$

2. $10^{3x} = 5$

3. $\log_{10}(x^2 + 2x + 1) = 1$

4. $\ln(x^2 + 2x + 10) = 1$

5. $10^{5-x^2} = 100$

6. $10^{1-x^2} = 100$

7. $\log_{10}(x - 3) + \log_{10} x = 1$

8. $\log_{10}(3 - x) + \log_{10} x = 1$

9. $\log_{10}[x(x - 3)] = 1$

10. $2\log_{10} x + \log_{10}(x - 1) = 2$

11. $\log_a x + \log_a(x + 2) = 2$

12. $\log_a[x(x + 2)] = 2$

13. $\log_{10}\left[\log_{10}\left(\dfrac{x + 3}{200x}\right) + 4\right] = -1$

In Exercises 14–27 draw a graph of the function. As a check, use a calculator or computer to plot a graph.

14. $f(x) = e^{-x^2}$

* 15. $f(x) = \ln(\cos x)$

16. $f(x) = \log_a |x^2 - 1|$

* 17. $f(x) = a^{\log_a(2x+1)}$

18. $f(x) = \log_{10}(4x)$

* 19. $f(x) = \ln(1 - x)$

20. $f(x) = \ln(1 - x^2)$

* 21. $f(x) = \ln(x^2 - 1)$

22. $f(x) = 10^{x+2}$

* 23. $f(x) = e^{2-x}$

24. $f(x) = e^{x^2}$

* 25. $f(x) = e^{-x^2}\sin x$

26. $f(x) = e^{-x^2}\cos x$

* 27. $f(x) = x^2 e^x, \quad -1 \le x \le 1$

28. A large number N of people are to have their blood tested to determine whether they have been infected by a virus. One way is to test all N people individually, resulting in N tests. An alternative is to divide the N people into groups of x people, pool their blood, and test for the virus. If the blood is disease-free for a group, the individuals need not be tested separately. If the blood test is positive, each of the x people in the group is tested separately. It is shown in probability theory that when the group size is x, then on the average, the expected total number of tests needed to test the N people completely is

$$y = N\left[1 - (0.99)^x + \frac{1}{x}\right].$$

Plot this function for $N = 100$ to determine the group size that minimizes y.

* 29. Show that $f(x) = (1 - e^{-1/x})/(1 + e^{-1/x})$ is an odd function.

* 30. Is there a difference between the graphs of the functions $f(x) = \log_a(x^2)$ and $g(x) = 2\log_a x$?

* 31. (a) In the early period of reforestation, the percentage increase of timber each year is almost constant. If the original amount A_0 of a certain timber increases 3.5% the first year and 3.5% each year thereafter, find an expression for the amount of timber after t years.

   (b) How long does it take for timber of this type to double?

* 32. A new car costs \$20 000. In any year it depreciates to 75% of its value at the beginning of that year. What is the value of the car after t years?

* 33. If the effective height of the earth's atmosphere (in metres) is the solution of the equation

$$10^{-6} = (1 - 2.08 \times 10^{-6}y)^{56},$$

find y.

* 34. Show that if y is the logarithm of x to base a, then $-y$ is the logarithm of x to base $1/a$.

* 35. Prove 1.66b and 1.66c.

* 36. Is identity 1.66a valid for all x_1 and x_2?

* **37.** The magnitude of an earthquake is measured in much the same way as noise level. An earthquake of minimal size is taken as having value 0 on the Richter scale. Any other earthquake of intensity I is said to have magnitude R on the Richter scale if

$$R = \log_{10}\left(\frac{I}{I_0}\right),$$

where I_0 is the intensity of the minimal earthquake being used as reference.

(a) Express the intensity of an earthquake in terms of I_0 and its reading on the Richter scale.

(b) What are readings on the Richter scale of earthquakes that have intensities 1.20×10^6 and 6.20×10^4 times I_0?

* **38.** (a) If P dollars is invested at $i\%$ compounded n times per year, show that the accumulated value after t years is

$$A = P\left(1 + \frac{i}{100n}\right)^{nt}.$$

(b) How long does it take to double an investment if interest is 8% compounded semiannually?

(c) Calculate the maximum possible value of A if i is fixed but the number of times that interest is compounded is unlimited; that is, calculate what happens to A as n gets larger and larger and larger. This method of calculating interest is called *continuously compounded interest*.

(d) What is the accumulated value of a $1000 investment after 10 years at 6% compounded continuously? Compare this to the accumulated value if interest is calculated only once each year.

* **39.** If, in the circuit shown below, V_0 is the voltage across the capacitor at time $t = 0$ when the switch is closed, the voltage thereafter is $V = V_0 e^{-t/(RC)}$.

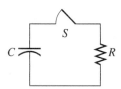

(a) The time constant τ for the circuit is the length of time for the voltage on the capacitor to become V_0/e. Find τ.

(b) Show that if V is the voltage at any time t, then the voltage at time $t + \tau$ is V/e.

* **40.** If, in the circuit shown below, i_0 is the current in the inductor at time $t = 0$ when the switch is closed, the current thereafter is $i = i_0 e^{-Rt/L}$.

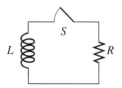

(a) The time constant τ for the circuit is the length of time for the current to become i_0/e. Find τ.

(b) Show that if i is the current at any time t, then the current at time $t + \tau$ is i/e.

In Exercises 41–44 find all values of x satisfying the equation.

* **41.** $3a^{2x} + 3a^{-2x} = 10$

* **42.** $2^x + 4^x = 8^x$

* **43.** $3^{x+4} = 7^{x-1}$

* **44.** $\log_x 2 = \log_{2x} 8$

* **45.** Show that $f(x) = \ln\left(x + \sqrt{x^2 + 1}\right)$ is an odd function. Plot its graph to confirm this geometrically.

46. Repair costs on the car in Exercise 32 are estimated at $50 the first year, increasing by 20% each year thereafter. Set up a function $C(t)$ that represents the average yearly cost of repairs associated with owning the car for t years.

* **47.** A straight-wire conductor has length $2L$ and circular cross-section of radius R. If the wire carries current $i > 0$, then the magnitude of the vector potential at a distance r from the centre of the wire is

$$f(r) = \begin{cases} \dfrac{\mu_0 i}{4\pi}\left[\ln\left(1 + \dfrac{4L^2}{R^2}\right) - 1 + \dfrac{r^2}{R^2}\right], & 0 \le r \le R \\[3mm] \dfrac{\mu_0 i}{4\pi}\ln\left(1 + \dfrac{4L^2}{r^2}\right), & r > R, \end{cases}$$

where μ_0 is a positive constant.

(a) Draw a graph of this function.

(b) Find the radius $r > R$ for which $f(r) = f(0)$.

In Exercises 48–50 solve the given equation for y in terms of x.

* **48.** $y = \dfrac{e^{2x} - e^{-2x}}{2}$

* **49.** $y = \dfrac{e^x + e^{-x}}{2}$

* **50.** $y = \dfrac{e^x - e^{-x}}{e^x + e^{-x}}$

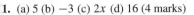
1. (a) 5 (b) -3 (c) $2x$ (d) 16 (4 marks)

2. $e^2/3$ (2 marks)

3. $(\ln 4 - 1)/2$ (2 marks)

4. $e/(e - 1)$ (3 marks)

5. (2 marks)

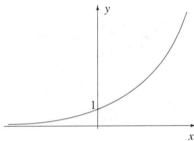

6. (2 marks)

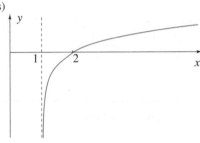

7. (2 marks)

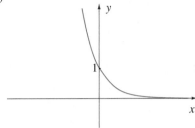

8. (3 marks)

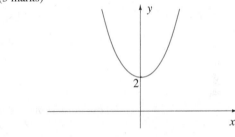

9. $(1/2)[1 + \ln (y/4)]$ (3 marks)

1.10 Hyperbolic Functions

This material is likely new for most students. It does not therefore have a diagnostic test associated with it.

Certain combinations of the exponential function occur so often in physical applications that they are given special names. Specifically, half the difference of e^x and e^{-x} is defined as the **hyperbolic sine function** and half their sum is the **hyperbolic cosine function**. These functions are denoted as follows:

$$\sinh x = \frac{e^x - e^{-x}}{2} \qquad \text{and} \qquad \cosh x = \frac{e^x + e^{-x}}{2}. \qquad (1.73)$$

According to equation 1.32, they are the odd and even parts of e^x.

The names of these hyperbolic functions and their notations bear a striking resemblance to those for the trigonometric functions, and there are reasons for this. First, the hyperbolic functions $\sinh x$ and $\cosh x$ are related to the curve $x^2 - y^2 = 1$ (see Figure 1.123), called the *unit hyperbola*, in much the same way as the trigonometric functions $\sin x$ and $\cos x$ are related to the unit circle $x^2 + y^2 = 1$. We will point out one of these similarities in Example 1.56. Second, for each identity satisfied by the trigonometric functions, there is a corresponding identity satisfied by the hyperbolic functions — not the same identity, but one very similar. For example, using equations 1.73, we have

$$(\cosh x)^2 - (\sinh x)^2 = \left(\frac{e^x + e^{-x}}{2}\right)^2 - \left(\frac{e^x - e^{-x}}{2}\right)^2$$

$$= \frac{1}{4}\left[\left(e^{2x} + 2 + e^{-2x}\right) - \left(e^{2x} - 2 + e^{-2x}\right)\right]$$

$$= 1.$$

Thus the hyperbolic sine and cosine functions satisfy the identity

$$\cosh^2 x - \sinh^2 x = 1, \tag{1.74}$$

which is reminiscent of the identity $\cos^2 x + \sin^2 x = 1$ for the trigonometric functions.

Just as four other trigonometric functions are defined in terms of $\sin x$ and $\cos x$, four corresponding hyperbolic functions are defined as follows:

$$\tanh x = \frac{\sinh x}{\cosh x} = \frac{e^x - e^{-x}}{e^x + e^{-x}}, \qquad \coth x = \frac{\cosh x}{\sinh x} = \frac{e^x + e^{-x}}{e^x - e^{-x}},$$

$$\operatorname{sech} x = \frac{1}{\cosh x} = \frac{2}{e^x + e^{-x}}, \qquad \operatorname{csch} x = \frac{1}{\sinh x} = \frac{2}{e^x - e^{-x}}. \tag{1.75}$$

These definitions and 1.74 immediately imply that

$$1 - \tanh^2 x = \operatorname{sech}^2 x, \tag{1.76a}$$

$$\coth^2 x - 1 = \operatorname{csch}^2 x, \tag{1.76b}$$

analogous to $1 + \tan^2 x = \sec^2 x$ and $1 + \cot^2 x = \csc^2 x$, respectively.

FIGURE 1.121a　Graph of cosh x

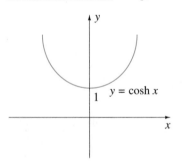

FIGURE 1.121b　Graph of sinh x

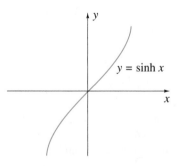

FIGURE 1.121c　Graph of tanh x

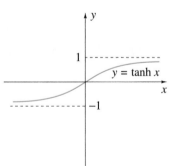

FIGURE 1.121d　Graph of coth x

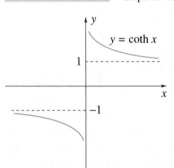

FIGURE 1.121e　Graph of sech x

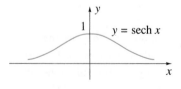

FIGURE 1.121f　Graph of csch x

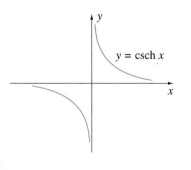

Graphs of the six hyperbolic functions are shown in Figures 1.121. The functions $\cosh x$ and $\operatorname{sech} x$ are even; the other four are odd.

Most trigonometric identities can be derived from the compound-angle formulas for $\sin (A \pm B)$ and $\cos (A \pm B)$. It is easy to verify similar formulas for the hyperbolic functions:

$$\sinh (A \pm B) = \sinh A \cosh B \pm \cosh A \sinh B, \tag{1.77a}$$

$$\cosh (A \pm B) = \cosh A \cosh B \pm \sinh A \sinh B. \tag{1.77b}$$

For example, equations 1.73 give

$$
\begin{aligned}
\cosh A \cosh B - \sinh A \sinh B &= \left(\frac{e^A + e^{-A}}{2} \right) \left(\frac{e^B + e^{-B}}{2} \right) \\
&\quad - \left(\frac{e^A - e^{-A}}{2} \right) \left(\frac{e^B - e^{-B}}{2} \right) \\
&= \frac{1}{4} \big[\left(e^{A+B} + e^{A-B} + e^{B-A} + e^{-A-B} \right) \\
&\quad - \left(e^{A+B} - e^{A-B} - e^{B-A} + e^{-A-B} \right) \big] \\
&= \frac{1}{2} \left[e^{A-B} + e^{-(A-B)} \right] \\
&= \cosh (A - B).
\end{aligned}
$$

With these formulas, we can derive hyperbolic identities analogous to trigonometric identities 1.44–1.49:

$$\tanh (A \pm B) = \frac{\tanh A \pm \tanh B}{1 \pm \tanh A \tanh B}, \tag{1.77c}$$

$$\sinh 2A = 2 \sinh A \cosh A, \tag{1.77d}$$

$$\cosh 2A = \cosh^2 A + \sinh^2 A \tag{1.77e}$$

$$= 2 \cosh^2 A - 1 \tag{1.77f}$$

$$= 1 + 2 \sinh^2 A, \tag{1.77g}$$

$$\tanh 2A = \frac{2 \tanh A}{1 + \tanh^2 A}, \tag{1.77h}$$

$$\sinh A \sinh B = \frac{1}{2} \cosh (A + B) - \frac{1}{2} \cosh (A - B), \tag{1.77i}$$

$$\sinh A \cosh B = \frac{1}{2} \sinh (A + B) + \frac{1}{2} \sinh (A - B), \tag{1.77j}$$

$$\cosh A \cosh B = \frac{1}{2} \cosh (A + B) + \frac{1}{2} \cosh (A - B), \tag{1.77k}$$

$$\sinh A + \sinh B = 2 \sinh \left(\frac{A + B}{2} \right) \cosh \left(\frac{A - B}{2} \right), \tag{1.77l}$$

$$\sinh A - \sinh B = 2 \cosh \left(\frac{A + B}{2} \right) \sinh \left(\frac{A - B}{2} \right), \tag{1.77m}$$

$$\cosh A + \cosh B = 2 \cosh \left(\frac{A + B}{2} \right) \cosh \left(\frac{A - B}{2} \right), \tag{1.77n}$$

$$\cosh A - \cosh B = 2 \sinh \left(\frac{A + B}{2} \right) \sinh \left(\frac{A - B}{2} \right). \tag{1.77o}$$

In Example 1.56, we illustrate a geometric parallel between the trigonometric sine and cosine functions and the hyperbolic sine and cosine functions.

EXAMPLE 1.56

Show that:

(a) every point (x, y) on the unit circle $x^2 + y^2 = 1$ can be expressed in the form $x = \cos t$, $y = \sin t$ for some real number t in the interval $0 \leq t < 2\pi$;

(b) every point (x, y) on the right half of the unit hyperbola $x^2 - y^2 = 1$ can be expressed in the form $x = \cosh t$, $y = \sinh t$ for some real number t.

SOLUTION

(a) If t is the angle in Figure 1.122, then clearly the coordinates of P are $x = \cos t$ and $y = \sin t$. As angle t ranges from 0 to 2π, P traces the circle exactly once.

(b) A sketch of the unit hyperbola $x^2 - y^2 = 1$ is shown in Figure 1.123. If $x = \cosh t$ and $y = \sinh t$ are coordinates of a point P in the plane, where t is some real number, then identity 1.74 implies that $x^2 - y^2 = \cosh^2 t - \sinh^2 t = 1$. In other words, P is on the unit hyperbola. Furthermore, since $x = \cosh t$ is always positive, P must be on the right half of the hyperbola. Finally, because the range of $x = \cosh t$ is $x \geq 1$ in Figure 1.121a, and the range of $y = \sinh t$ is $-\infty < y < \infty$ in Figure 1.121b, it follows that every point on the right half of the hyperbola can be obtained from some value of t. Note that t is *not* the angle formed by the positive x-axis and the line joining the origin to (x, y).

FIGURE 1.122   Relationship between unit circle and trigonometric functions

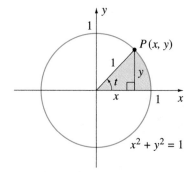

FIGURE 1.123   Relationship between unit hyperbola and hyperbolic functions

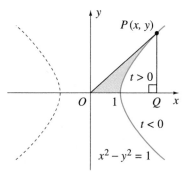

EXERCISES 1.10

In Exercises 1–10 evaluate the expression (if it has a value).

1. $3 \cosh 1$

2. $\sinh (\pi/2)$

3. $\tanh \sqrt{1 - \sin 3}$

4. $\text{Sin}^{-1} (\text{sech } 10)$

5. $\text{Cos}^{-1} (2 \operatorname{csch} 1)$

6. $\coth (\sinh 5)$

7. $\sqrt{\ln |\sinh (-3)|}$

8. $\operatorname{sech} [\sec (\pi/3)]$

9. $e^{-2 \cosh e}$

10. $\sinh [\text{Cot}^{-1} (-3\pi/10)]$

11. Verify the results in identities 1.77c–o.

* **12.** Vertical vibrations of the beam in the figure below involve the function

$$y = f(x) = A \cos kx + B \sin kx + C \cosh kx + D \sinh kx$$

where A, B, C, and D are constants such that $C = -A$, $D = -B$, and A and B must satisfy the equations

$$A(\cos kL - \cosh kL) + B(\sin kL - \sinh kL) = 0,$$

$$A(\cos kL + \cosh kL) + B(\sin kL + \sinh kL) = 0.$$

Eliminate A and B between these equations to show that k must satisfy the condition

$$\tan kL = \tanh kL.$$

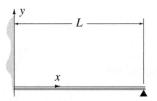

13. Each hyperbolic function has associated with it an inverse hyperbolic function.

(a) Draw the inverse functions $\mathrm{Sinh}^{-1}\, x,\, \mathrm{Tanh}^{-1}\, x,\, \mathrm{Coth}^{-1}\, x,$ and $\mathrm{Csch}^{-1}\, x$ for $\sinh x,\, \tanh x,\, \coth x,$ and $\mathrm{csch}\, x$.

(b) Why do $\cosh x$ and $\mathrm{sech}\, x$ not have inverse functions? It is customary to associate functions $\mathrm{Cosh}^{-1}\, x$ and $\mathrm{Sech}^{-1}\, x$ with $\cosh x$ and $\mathrm{sech}\, x$ by restricting their domains to nonnegative numbers. Draw graphs of $\mathrm{Cosh}^{-1}\, x$ and

$\mathrm{Sech}^{-1}\, x$.

(c) Show that

$$\mathrm{Sinh}^{-1} x = \ln\left(x + \sqrt{x^2 + 1}\right);$$

$$\mathrm{Cosh}^{-1} x = \ln\left(x + \sqrt{x^2 - 1}\right);$$

$$\mathrm{Tanh}^{-1} x = \frac{1}{2}\ln\left(\frac{1 + x}{1 - x}\right), \quad |x| < 1.$$

∗∗ **14.** When an object of mass m falls from rest at time $t = 0$ under the influence of gravity and an air resistance proportional to the square of velocity, its velocity v as a function of time is defined by the equation

$$\frac{1}{2}\sqrt{\frac{\beta}{mg}}\ln\left(\frac{\sqrt{mg/\beta} - v}{\sqrt{mg/\beta} + v}\right) = -\frac{\beta t}{m},$$

where $\beta > 0$ is a constant and $g > 0$ is the acceleration due to gravity.

(a) Show that when this equation is solved for v in terms of t, the result is

$$v(t) = \sqrt{\frac{mg}{\beta}}\tanh\left(\sqrt{\frac{\beta g}{m}}\, t\right).$$

(b) Determine the limit of v for large t, called the *limiting velocity*.

1.11 Approximating Solutions to Equations

This material is likely new for most students. It does not therefore have a diagnostic test associated with it.

In writing this book we have assumed that you have access to a graphing calculator and/or a computer with a mathematical software package. Every software package has one or more equation-solving commands; most graphing calculators have an equation-solving routine. In other words, you have a device that solves equations either exactly or approximately. In this section we discuss some useful concepts related to accuracies of approximations. Suppose, for example, that we are to find all solutions of the equation

$$2x - 4 = \cos(x^2 - 7x + 10). \tag{1.78a}$$

We can express the equation in the equivalent form

$$f(x) = 2x - 4 - \cos(x^2 - 7x + 10) = 0, \tag{1.78b}$$

where solutions are now visualized as x-intercepts of the graph of the function $f(x)$. The graph in Figure 1.124 indicates that the only solution of the equation is between $x = 2$ and $x = 3$.

FIGURE 1.124   Graphical solutions of $2x - 4 - \cos(x^2 - 7x + 10) = 0$

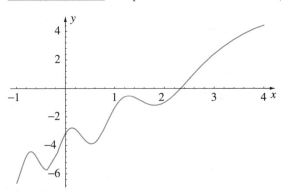

Suppose we use our calculator or computer to find this root and the result is $x = 2.323\,762\,277$. The number of digits depends on the device used and/or how it is programmed. For the purposes of our discussion here, the number of digits is immaterial. Naturally, if $2.323\,762\,277$ constitutes the full display of a calculator, we might be skeptical of the accuracy of the last digit, perhaps even the second-to-last digit. If this is computer output, we might be unsure of how the machine arrived at the last digit. Did it round, and if so, what are its rules for rounding, or did it simply truncate after the 10th digit? There is a very simple way to verify the accuracy of this, or any other, approximate root of an equation. The following theorem provides it.

> **THEOREM 1.7 (Intermediate Value Theorem)**
>
> If a function $f(x)$ is continuous on the closed interval $a \le x \le b$, and k is any number between $f(a)$ and $f(b)$, then there exists at least one value of c between a and b such that $f(c) = k$.

A function $f(x)$ is a **continuous function** on an interval $a \le x \le b$ if its graph can be traced between a and b without lifting pencil from paper. A more mathematical definition of continuous will be given in Section 2.4. For our present purposes, the geometric version suffices.

FIGURE 1.125a   The intermediate value theorem

FIGURE 1.125b   The zero intermediate value theorem

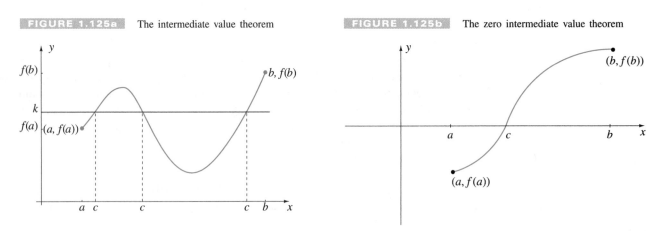

Figure 1.125a illustrates Theorem 1.7 geometrically. Because the graph of $f(x)$ can be traced from the point $(a, f(a))$ to the point $(b, f(b))$ without lifting pencil from paper, it follows that the graph must cross the horizontal line $y = k$ at least once, and this is a value for c. This particular figure shows three such values. Theorem 1.7 is what mathematicians call an *existence theorem*. It states that a number c exists that satisfies $f(c) = k$ for given k, but does not provide a way to find c. We shall encounter other existence theorems in calculus.

What is important for our present purposes is the following corollary.

COROLLARY 1.7.1 (The Zero Intermediate Value Theorem)

If $f(a)f(b) < 0$ for a function $f(x)$ that is continuous on $a \leq x \leq b$, then there exists at least one number c between a and b for which $f(c) = 0$.

The condition $f(a)f(b) < 0$ requires that one of $f(a)$ and $f(b)$ be positive and the other be negative. [We have shown $f(a) < 0$ and $f(b) > 0$ in Figure 1.125b.] The choice of $k = 0$ in Theorem 1.7 gives this corollary.

Without continuity of $f(x)$, we cannot be sure, in general, whether there are solutions to the equation $f(x) = 0$ when $f(a)f(b) < 0$. The function in Figure 1.126a has what is called a *discontinuity* at $x = d$, and there are no solutions of $f(x) = 0$. The function $f(x)$ in Figure 1.126b also has a discontinuity at $x = d$, but there are two solutions of $f(x) = 0$ between a and b.

There are two common ways to discuss the accuracy of an approximation to the solution of an equation, and simple as the zero intermediate value theorem is, it handles both situations.

FIGURE 1.126a   FIGURE 1.126b

Functions not satisfying the conditions of the zero intermediate value theorem

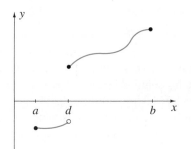
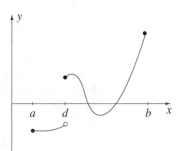

Approximations Rounded to a Specified Number of Decimal Places

We say that x is an approximation to the root α of an equation

$$f(x) = 0, \tag{1.79}$$

correctly rounded to k decimal places, if x has k decimal places, and α rounds to the same k decimal places. For example, the approximation $x = 2.323\,762\,277$ rounded to four decimal places is $x = 2.3238$. How can we verify that the root of equation 1.78b rounds to these same four decimal places? We evaluate $f(x)$ at $x = 2.323\,75$ and $x = 2.323\,85$,

$$f(2.323\,75) = -0.000\,047, \qquad f(2.323\,85) = 0.000\,33.$$

Because one of these values is positive and the other is negative, the zero intermediate value theorem implies that the solution of equation 1.78b must lie between 2.323 75 and 2.323 85. But every number between 2.323 75 and 2.323 85 rounds to 2.3238. In other words, $x = 2.3238$ is a solution of equation 1.78b correctly rounded to four decimal places.

In general, we can say that x is an approximation to a root of the equation $f(x) = 0$, **correctly rounded to k decimal places**, if x has exactly k digits after the decimal, and

$$f\left(x - \frac{10^{-k}}{2}\right) f\left(x + \frac{10^k}{2}\right) < 0. \tag{1.80}$$

Maximum Possible Error

We are often asked to find an approximation to the solution of an equation such as 1.78b, and be sure that the error is less than some given value ϵ, say $\epsilon = 0.0001$ or $\epsilon = 0.000\,000\,1$. The smaller the value of ϵ, the more accurate must be the approximation. To illustrate, suppose an approximation to the solution of 1.78b is required with error less than $\epsilon = 0.000\,01$. We could verify that $x = 2.323\,762\,277$ has error less than $\epsilon = 0.000\,01$, but there is little point in carrying an approximation with nine decimal places when an error of $0.000\,01$ is concerned with the fifth decimal place. We suspect that if we round the approximation to five decimal places, the result $x = 2.323\,76$ has error less than $0.000\,01$. To verify this, we evaluate

$$f(2.323\,75) = -0.000\,047 \quad \text{and} \quad f(2.323\,77) = 0.000\,029.$$

The fact that these values have opposite signs guarantees that the root is between $2.323\,75$ and $2.323\,77$. Since the difference between these numbers is $0.000\,02$ and our approximation $2.323\,76$ is halfway between them, it follows that the error in $2.323\,76$ must be less than $0.000\,01$.

In general, we can say that x is an **approximation to a root of the equation $f(x) = 0$, with error less than ϵ**, if

$$f(x - \epsilon)f(x + \epsilon) < 0. \tag{1.81}$$

There is the potential to use calculators or computers unwisely here. Avoid operating calculators and computers at or near their limits. For example, suppose that the solution $x = 2.323\,762\,277$ of equation 1.78b constitutes the full display of a calculator. It would be unwise to attempt to verify that this solution has error less than 10^{-9} using the same calculator. To do so would require $f(2.323\,762\,276)$ and $f(2.323\,762\,278)$. These values are very, very close to zero. How could we be certain of their positivity and negativity when we are asking the calculator to perform very sensitive calculations with numbers at the limits of its capabilities?

EXAMPLE 1.57

Find an approximation to the smallest root of the equation

$$2x^3 e^{-x} + 5x^2 - 1 = 0$$

correctly rounded to six decimal places.

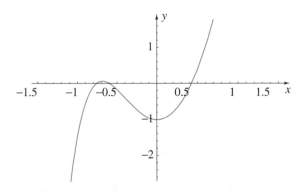

FIGURE 1.127   Graphical solution of $2x^3 e^{-x} + 5x^2 - 1 = 0$

SOLUTION   The graph of $f(x) = 2x^3 e^{-x} + 5x^2 - 1$ in Figure 1.127 shows three solutions. Our computer gives $x = -0.766\,051\,059$ as an approximation to the smallest root. To verify that $x = -0.766\,051$ is an approximation, correctly rounded to six decimal places, we calculate

$$f(-0.766\,051\,5) = -8.1 \times 10^{-7}, \qquad f(-0.766\,050\,5) = 1.0 \times 10^{-6}.$$

The fact that these values have opposite signs confirms the six-decimal-place accuracy of $x = -0.766\,051$.

EXAMPLE 1.58

Find an approximation to the largest root of the equation

$$x + 6\sin x = 0$$

with error less than 10^{-8}.

FIGURE 1.128   Graphical solutions of $x + 6\sin x = 0$

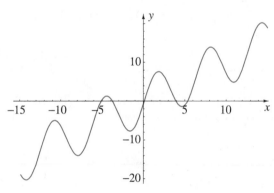

SOLUTION   The graph of $f(x) = x + 6\sin x$ in Figure 1.128 shows five solutions. Our computer yields $x = 5.225\,963\,530$ as an approximation to the largest root. For an approximation with error less than 10^{-8}, we take $x = 5.225\,963\,53$. For verification, we evaluate

$$f(5.225\,963\,52) = -4.1 \times 10^{-8} \quad \text{and} \quad f(5.225\,963\,54) = 3.8 \times 10^{-8}.$$

When we have an approximation to the solution of an equation correctly rounded to k decimal places, we can say that we have an approximation with error no greater than $10^{-k}/2$ (compare equations 1.80 and 1.81).

Knowing an approximation with maximum possible error, however, does not guarantee a predictable number of correctly rounded decimal places. Let us illustrate. Suppose that we have used our calculator or computer to approximate the root of an equation $f(x) = 0$ and the result is 3.115 00. Suppose further that we know that the error is less than 10^{-5}. Can we give an approximation correctly rounded to two decimal places? No! The fact that the error is less than 10^{-5} allows us to say that the solution satisfies

$$3.115\,00 - 10^{-5} < x < 3.115\,00 + 10^{-5} \qquad \Longrightarrow \qquad 3.114\,99 < x < 3.115\,01.$$

Since left and right sides of the latter inequality round to 3.11 and 3.12, we do not know the approximation correctly rounded to two decimal places. However, we do know an approximation correctly rounded to three and four decimal places, namely, 3.115 and 3.1150.

As a second example, suppose we know that an approximation to the root of $f(x) = 0$ is 3.435 with error less than 10^{-3}. How many decimal places can we guarantee? We can say that the solution satisfies

$$3.435 - 0.001 < x < 3.435 + 0.001 \qquad \Longrightarrow \qquad 3.434 < x < 3.436.$$

Since the numbers in the right inequality do not both round to the same two decimal places, we can guarantee only one correctly rounded decimal, namely 3.4.

    To emphasize this point again, we cannot make generalizations to the effect that a maximum possible error of 10^{-k}, say, guarantees any number of decimal places. In every specific example, we will be able to determine how many decimal places are possible, but general statements covering all situations are not possible.

EXAMPLE 1.59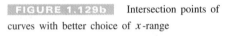

Find points of intersection of the curves

$$y = x^3 - 3x^2 + 2x + 5, \quad y = 6 - 5x^2 - 3x^4.$$

Give coordinates correctly rounded to four decimal places.

FIGURE 1.129a   Intersection points of two curves with poor choice of x-range

FIGURE 1.129b   Intersection points of curves with better choice of x-range

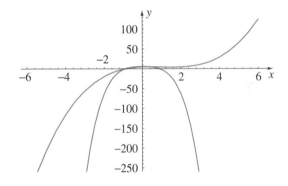
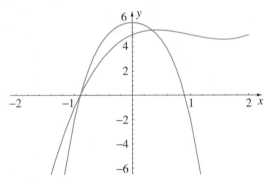

SOLUTION   Graphs of the curves in Figure 1.129a indicate that whatever points of intersection there are, they are in the interval $-2 \le x \le 2$. Plots on this interval in Figure 1.129b indicate two points of intersection. To find x-coordinates of the points of intersection, we solve

$$x^3 - 3x^2 + 2x + 5 = 6 - 5x^2 - 3x^4 \qquad \Longrightarrow \qquad 3x^4 + x^3 + 2x^2 + 2x - 1 = 0.$$

Solutions with four decimal places are -0.8924 and 0.3422. To confirm that all four decimal places are correct, we calculate the following values of $f(x) = 3x^4 + x^3 + 2x^2 + 2x - 1$:

$$f(-0.892\,45) = 3.1 \times 10^{-4}, \qquad f(-0.892\,35) = -4.6 \times 10^{-4},$$

$$f(0.342\,15) = -4.0 \times 10^{-4}, \qquad f(0.342\,25) = 2.1 \times 10^{-5}.$$

Intersection points on the curves corresponding to $x = -0.8924$ and $x = 0.3422$ can be found by substituting these values into the equations for the curves. If we substitute $x = -0.8924$ into $y = x^3 - 3x^2 + 2x + 5$, we obtain $y = 0.115\,379$, whereas in $y = 6 - 5x^2 - 3x^4$, we get $y = 0.115\,459$. These numbers do not agree to four decimal places. What this points out is that the result of calculations with numbers accurate to four decimal places is unlikely to yield numbers accurate to four decimal places; accuracy will be lost. How much depends on the nature and the number of calculations. A similar situation arises with $x = 0.3422$; the equation of one curve gives $y = 5.373\,17$ and the other gives $y = 5.373\,36$. They do not agree to four decimal places. What we should have done is carry more decimal places in intermediate calculations. For instance, we could carry six decimal places with $x = -0.892\,410$ and $x = 0.342\,245$. Verification that $x = -0.8924$ and $x = 0.3422$ are correct to four decimal places is the same. But using $x = -0.892\,410$ and $x = 0.342\,245$ leads to corresponding y-values that agree to four decimal places no matter which equation $y = x^3 - 3x^2 + 2x + 5$ or $y = 6 - 5x^2 - 3x^4$ is used, namely $y = 0.1153$ and $y = 5.3732$.

EXAMPLE 1.60

At time $t = 0$, a 5 Ω resistor, a 2 H inductor, and a 0.01 F capacitor are connected with a generator producing an alternating voltage of $10 \sin 5t$, where $t \geq 0$ is time in seconds (Figure 1.130). The current i in the circuit thereafter is $i(t) = f(t) + g(t)$, where

$$f(t) = -e^{-5t/4}\left[\frac{18}{5}\cos\left(\frac{5\sqrt{31}t}{4}\right) + \frac{12\sqrt{31}}{165}\sin\left(\frac{5\sqrt{31}t}{4}\right)\right]$$

and

$$g(t) = \frac{4}{5}\cos 5t + \frac{2}{5}\sin 5t$$

FIGURE 1.130 Schematic for LCR-circuit

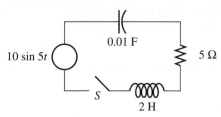

are called the **transient** and **steady-state** parts of the current, respectively. Plot graphs of $f(t)$, $g(t)$, and $i(t)$, and explain why the names for $f(t)$ and $g(t)$ are appropriate. Determine the smallest time (correctly rounded to three decimal places) at which the magnitude of the current in the circuit is 1 A.

SOLUTION Plots of $f(t)$ and $g(t)$ are shown in Figures 1.131a and b; their sum is plotted in Figure 1.131c. The exponential factor $e^{-5t/4}$ causes values of $f(t)$ to approach zero within a few seconds. In other words, $f(t)$ is significant only for small t; it dies off quickly, and therefore the adjective *transient* is appropriate. The function $g(t)$ is periodic with amplitude $2/\sqrt{5}$. It remains for all time, and once $f(t)$ becomes insignificant, current $i(t)$ essentially becomes $g(t)$. In other words, the current eventually settles down to $g(t)$, and hence the terminology *steady state* for $g(t)$ is appropriate.

FIGURE 1.131a Transient part of current

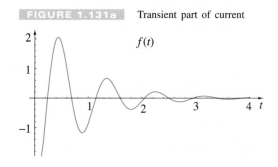

FIGURE 1.131b Steady-state part of current

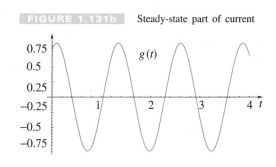

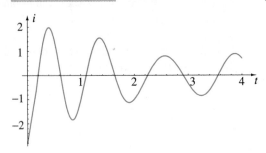

FIGURE 1.131c   Addition of transient and steady-state parts of current

Figure 1.131c makes it clear that there are nine times at which the magnitude of the current is 1 A, four times when the current is positive and five times when it is negative. The smallest is near $t = 0.15$ s. It satisfies the equation $i(t) = -1$, or, writing the equation in our standard form 1.79, we have

$$0 = h(t) = -e^{-5t/4}\left[\frac{18}{5}\cos\left(\frac{5\sqrt{31}t}{4}\right) + \frac{12\sqrt{31}}{165}\sin\left(\frac{5\sqrt{31}t}{4}\right)\right]$$

$$+ \frac{4}{5}\cos 5t + \frac{2}{5}\sin 5t + 1.$$

When solving equations of this complexity for one of several roots, calculators and computers usually require the equation and a reasonable approximation to the root, the closer the better. Our computer returned the root $t = 0.146\,206$ when given $t = 0.15$ as initial approximation. To verify that $t = 0.146$ is correctly rounded to three decimal places, we calculate that

$$h(0.1455) = -0.012 \quad \text{and} \quad h(0.1465) = 0.0052.$$

In Exercises 1–16 use a calculator or computer to find approximations to all roots of the equation accurate to six decimal places. Verify the accuracy of each root. Make a plot in order to determine the number of roots.

1. $x^2 + 3x + 1 = 0$

2. $x^2 - x - 4 = 0$

3. $x^3 + x - 3 = 0$

4. $x^3 - x^2 + x - 22 = 0$

5. $x^3 - 5x^2 - x + 4 = 0$

6. $x^5 + x - 1 = 0$

7. $x^4 + 3x^2 - 7 = 0$

8. $\dfrac{x+1}{x-2} = x^2 + 1$

9. $x - 10\sin x = 0$

10. $\sec x = \dfrac{2}{1+x^4}$

11. $(x+1)^2 = \sin 4x$

12. $(x+1)^2 = 5\sin 4x$

13. $x + 4\ln x = 0$

14. $x\ln x = 6$

15. $e^x + e^{-x} = 10x$

16. $x^2 - 4e^{-2x} = 0$

In Exercises 17–24 use a calculator or computer to find approximations to all roots of the equation with error no greater than that specified.

Verify the accuracy of each root. Make a plot in order to determine the number of roots.

17. $x^3 - 5x - 1 = 0, \quad 10^{-3}$

18. $x^4 - x^3 + 2x^2 + 6x = 0, \quad 10^{-4}$

19. $\dfrac{x}{x+1} = x^2 + 2, \quad 10^{-5}$

20. $(x+1)^2 = x^3 - 4x, \quad 10^{-3}$

21. $(x+1)^2 = 5\sin 4x, \quad 10^{-3}$

22. $\cos^2 x = x^2 - 1, \quad 10^{-4}$

23. $x + (\ln x)^2 = 0, \quad 10^{-3}$

24. $e^{3x} + e^x = 4, \quad 10^{-4}$

In Exercises 25–28 find all points of intersection for the curves accurate to four decimal places.

25. $y = x^3, \quad y = x + 5$

26. $y = (x+1)^2, \quad y = x^3 - 4x$

27. $y = x^4 - 20, \quad y = x^3 - 2x^2$

28. $y = \dfrac{x}{x+1}, \quad y = x^2 + 2$

29. When the beam in the figure below vibrates vertically, there are certain frequencies of vibration, called *natural frequencies*. They are solutions of the equation

$$\tan x = \frac{e^x - e^{-x}}{e^x + e^{-x}}$$

divided by 20π. Find the two smallest frequencies correct to four decimal places.

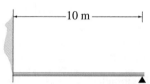

---10 m---

30. A stone of mass 100 g is thrown vertically upward with speed 20 m/s. Air exerts a resistive force on the stone proportional to its speed, and has magnitude 0.1 N when the speed of the stone is 10 m/s. It can be shown that the height y above the projection point attained by the stone is given by

$$y = -98.1t + 1181 \left(1 - e^{-t/10}\right) \text{ m},$$

where t is time (measured in seconds with $t = 0$ at the instant of projection).

    (a) The time taken for the stone to return to its projection point can be obtained by setting $y = 0$ and solving the equation for t. Do so (correct to two decimal places).

    (b) When air resistance is neglected, the formula for y is

$$y = 20t - 4.905t^2 \text{ m}.$$

    What is the elapsed time in this case from the instant the stone is projected until it returns to the projection point?

31. A uniform hydro cable $P = 80$ m long with mass per unit length $\rho = 0.5$ kg/m is hung from two supports at the same level $L = 70$ m apart (figure below). The tension T in the cable at its lowest point must satisfy the equation

$$\frac{\rho g P}{T} = e^{\rho g L/(2T)} - e^{-\rho g L/(2T)},$$

where $g = 9.81$. If we set $z = \rho g/(2T)$, then z must satisfy

$$2Pz = e^{Lz} - e^{-Lz}.$$

Solve this equation for z and hence find T correct to one decimal place.

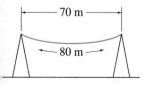

    ---70 m---

    ---80 m---

32. *Planck's law* for the energy density E of blackbody radiation at $1000°$K states that

$$E = E(\lambda) = \frac{k\lambda^{-5}}{e^{c/\lambda} - 1},$$

where $k > 0$ is a constant and $c = 0.000\,143\,86$. This function is shown in the figure below. The value of λ at which E is a maximum must satisfy the equation

$$(5\lambda - c)e^{c/\lambda} - 5\lambda = 0.$$

Find this value of λ correct to seven decimal places.

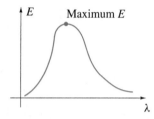

33. Let $f(x)$ be a continuous function with domain and range both equal to the interval $[a, b]$. Show that there is at least one value of x in $a \le x \le b$ for which $f(x) = x$.

34. Use the zero intermediate value theorem to prove that at any given time there is a pair of points directly opposite each other on the equator of the earth that have exactly the same temperature. *Hint:* Take the equator to be the circle $x^2 + y^2 = r^2$. Let $f(x)$ be the temperature on the upper semicircle and $g(x)$ be the temperature on the lower semicircle. Consider the function $F(x) = f(x) - g(-x)$.

35. A marathoner runs the 26-odd miles from point A to point B starting at 7:00 a.m. Saturday morning. Starting at 7:00 a.m. Sunday morning she runs the course again, but this time from point B to point A. Prove that there is a point on the course that she passed at exactly the same time on both days.

36.
    (a) Use the zero intermediate value theorem to prove that when the domain of a continuous function is an interval, so also is its range. *Hint:* Use the idea that a set S of points on the y-axis constitutes an interval if for any two points c and d in S, the points $c < y < d$ are all in S.

    (b) If the domain is an open interval, is the range an open interval?

SUMMARY

In this chapter we have reviewed basic concepts from algebra, analytic geometry, and trigonometry, and introduced material that is essential to many of the applications of calculus. To find real solutions of polynomial equations with integer coefficients, we use the rational root theorem to narrow the field of possibilities and the factor theorem to remove roots from the equation as they are found.

Analytic geometry is a combination of geometry and algebra. Algebraic equations are used to describe geometric curves and curves are the geometric representation of equations. The form of an equation dictates the shape of the curve and, conversely, the shape of a curve influences its equation. To illustrate this fact, we discussed straight lines, circles, parabolas, ellipses, and hyperbolas. The most common forms for equations of these curves are as follows:

Straight line
$$
\begin{cases}
y - y_1 = m(x - x_1) & \text{Point-slope} \\[4pt]
\dfrac{x - x_1}{x_2 - x_1} = \dfrac{y - y_1}{y_2 - y_1} & \text{Two-point} \\[4pt]
y = mx + b & \text{Slope } y\text{-intercept} \\[2pt]
y = m(x - a) & \text{Slope } x\text{-intercept} \\[2pt]
\dfrac{x}{a} + \dfrac{y}{b} = 1 & \text{Two-intercept} \\[4pt]
Ax + By + C = 0 & \text{General}
\end{cases}
$$

Parabola
$$
\begin{cases}
y = ax^2 + bx + c & \text{Vertical axis of symmetry} \\[4pt]
x = ay^2 + by + c & \text{Horizontal axis of symmetry}
\end{cases}
$$

Circle
$$
\begin{cases}
x^2 + y^2 + fx + gy + e = 0 \\[4pt]
(x - h)^2 + (y - k)^2 = r^2
\end{cases}
$$

Ellipse
$$
\begin{cases}
\dfrac{x^2}{a^2} + \dfrac{y^2}{b^2} = 1 \\[8pt]
\dfrac{(x - h)^2}{a^2} + \dfrac{(y - k)^2}{b^2} = 1
\end{cases}
$$

Hyperbola
$$
\begin{cases}
\dfrac{x^2}{a^2} - \dfrac{y^2}{b^2} = 1 \\[8pt]
\dfrac{y^2}{b^2} - \dfrac{x^2}{a^2} = 1 \\[8pt]
\dfrac{(x - h)^2}{a^2} - \dfrac{(y - k)^2}{b^2} = 1 \\[8pt]
\dfrac{(y - k)^2}{b^2} - \dfrac{(x - h)^2}{a^2} = 1
\end{cases}
$$

Basic to all mathematics is the concept of a function, a rule that assigns to each number x in a domain, a unique number y in the range. A function is simply another way of saying "a quantity y depends on x." The notation $y = f(x)$ for a function immediately suggests that a function can be represented geometrically by a curve — the curve with equation $y = f(x)$ — and we call this curve the graph of the function. Polynomials are functions of the form $a_n x^n + \cdots + a_1 x + a_0$, where n is a nonnegative integer and coefficients $a_n, \ldots, a_0$ are constants. Rational functions are quotients of polynomials. Some functions are even, some are odd, most are neither even nor odd, and only $f(x) \equiv 0$ is both even and odd.

Some of the most important functions in mathematics have inverses, in particular, the exponential and logarithmic functions, the trigonometric functions, and the hyperbolic functions. A function $f^{-1}(x)$ is the inverse of $f(x)$ if $f^{-1}\big(f(x)\big) = x$ for each x in the domain of $f(x)$. The inverse function $f^{-1}(x)$ reverses the action of $f(x)$. The graph of $f^{-1}(x)$ is the mirror image of the graph of $f(x)$ in the line $y = x$. Increasing functions have inverses, as do decreasing functions. The domain of a function that does not have an inverse can usually be subdivided into subdomains on which the function does have inverses.

Trigonometric functions play a prominent role in many areas of applied mathematics. Particularly important are descriptions of oscillatory systems by the sine and cosine functions. The sine, cosecant, tangent, and cotangent are odd functions, while cosine and secant are even. All are periodic; sine, cosecant, cosine, and secant have period 2π, and tangent and cotangent have period π. Trigonometric functions satisfy many identities. Recognizing when these identities can be used to advantage to simplify expressions, or write them in alternative forms, is a huge asset.

The trigonometric functions do not have inverses, but their domains can be restricted so as to create inverse functions; these domains turn out to be the principal values of the associated inverse function. The inverse trigonometric functions reverse the roles of trigonometric functions. A trigonometric function such as the sine function associates a value called $\sin x$ with a real number (angle) x. The corresponding inverse function, $\mathrm{Sin}^{-1} x$, regards x as the sine of an angle, and yields the angle in the principal value range with sine equal to x.

Exponential and logarithmic functions are also important in applications; they are inverses of each other. These functions are not periodic, nor are they even or odd. The exponential function a^x raises a to power x. As its inverse, the logarithm function $\log_a x$ does the reverse. It determines the power that a must be raised to produce x.

Hyperbolic functions are special combinations of exponential functions that arise sufficiently often in applications to warrant special consideration. They satisfy identities very similar to the trigonometric functions. Each hyperbolic function has an associated inverse hyperbolic function.

In approximating solutions to equations, it is always necessary to indicate the accuracy of the approximation. This can be done by correctly rounding the approximation to a specified number of decimal places, or determining an approximation with maximum error. In both cases, the zero intermediate value theorem is instrumental in verifying the accuracy.

KEY TERMS

In reviewing this chapter, you should be able to define or discuss the following key terms:

Polynomial	Polynomial equation
Roots	Solutions (zeros)
Linear equation	Quadratic equation
Quadratic formula	Discriminant
Irreducible real quadratic factor	Multiplicity
Cartesian coordinates	Quadrants
Length of a line segment	Absolute values
Midpoint of a line segment	Coordinates
Equation of a curve	Slope of a line
Rise	Run
Point-slope formula	General equation of a line
Parallel lines	Perpendicular lines
Parabola	Symmetric about the x-axis and y-axis
Circle	Ellipse
Hyperbola	Asymptotes
Function	Independent variable
Dependent variable	Domain
Range	Graph of a function
Absolute value function	Even function
Odd function	Even and odd parts of a function
Rational function	Translation
One-to-one	Inverse function
Strictly monotonic function	Radian
Standard position of an angle	Cosine law
Sine law	Compound-angle formulas
Double-angle formulas	Product formulas
Sum and difference formulas	Periodic
Period	General sine function
Amplitude	Phase shift
Inclination of a line	Algebraic function

Transcendental function

Principal values

Exponential functions

Natural logarithms

Intermediate value theorem

Correctly rounded to k decimal places

Approximation with maximum error

Steady state

Inverse trigonometric functions

Acute angle between two lines with slopes

Logarithm functions

Hyperbolic functions

Continuous function

Zero intermediate value theorem

Transient

REVIEW EXERCISES

In Exercises 1–4 find all real solutions of the polynomial equation, giving multiplicities for any repeated roots.

1. $x^3 - x^2 - 4 = 0$ **2.** $2x^3 - 9x^2 + 27 = 0$

3. $2x^4 - x^3 - 9x^2 + 13x - 5 = 0$

∗ **4.** $36x^4 + 12x^3 - 179x^2 - 30x + 225 = 0$

In Exercises 5–6 find the distance between the points and also the midpoint of the line segment joining the points.

5. $(-1, 3)$, $(4, 2)$ **6.** $(2, 1)$, $(-3, -4)$

In Exercises 7–10 find the equation for the line described.

7. Parallel to the line $x - 2y = 4$ and through the point $(2, 3)$

8. Perpendicular to the line joining $(-2, 1)$ to the origin and through the midpoint of the line segment joining $(1, 3)$ and $(-1, 5)$

∗ **9.** Perpendicular to the line $x = 4y - 11$ and through the point of intersection of this line and $x = \sqrt{y^2 + 9}$

∗ **10.** Joining the points of intersection of the curves $y = x^2$ and $5x = 6 - y^2$

In Exercises 11–20 find the largest possible domain for the function.

11. $f(x) = \sqrt{x^2 + 5}$ **12.** $f(x) = \sqrt{x^2 - 5}$

13. $f(x) = \dfrac{1}{x^2 + 3x + 2}$ **14.** $f(x) = \dfrac{x + 4}{x^3 + 2x^2 + x}$

15. $f(x) = (x^3 - 8)^{1/3}$ **16.** $f(x) = x^{3/2}$

∗ **17.** $f(x) = \sqrt{x^2 + 4x - 6}$. ∗ **18.** $f(x) = \dfrac{1}{\sqrt{2x^2 + 4x - 5}}$

∗ **19.** $f(x) = \sqrt{\dfrac{2x + 1}{x - 3} + 2}$ ∗ **20.** $f(x) = \sqrt{x - \dfrac{1}{x}}$

In Exercises 21–32 identify the curve as a straight line, parabola, circle, ellipse, hyperbola, or none of these.

21. $x + 2y = 4$ **22.** $x = y^2 - 2y + 3$

23. $y = x^3 + 3$ **24.** $x^2 + 2y^2 = 4$

25. $y^2 - x^2 = x$ **26.** $x^2 + y^2 + 5 = 0$

27. $x^2 - 2x + y^2 = 16$ **28.** $x + y^2 = 3$

29. $x^2 + 2y^2 + y = 2x$ **30.** $x^2 - x + y^2 + y = 0$

31. $2x^2 + 20x + 38 = 3y^2 + 12y$

32. $2y^2 - x = 3x^2 - y$

In Exercises 33–34 find the distance from the point to the line.

33. $(1, -3)$, $y = 2x + 3$ **34.** $(-2, -5)$, $x = 4 - 3y$

In Exercises 35–58 draw the curve. Then use a calculator or computer to plot the curve as a check.

35. $y = 2x^2 + 3$ **36.** $x^2 = 4 - y^2$

37. $y = x^3 - 1$ **38.** $|y| = |x|$

39. $4x^2 + y^2 = 0$ **40.** $x^2 + 3y^2 = 6$

41. $2y^2 - x^2 = 3$ **42.** $x^2 - 2x - y^2 + 4y = 1$

43. $y = \sin 3x$ **44.** $y = \cos(2x + \pi/2)$

45. $y = \cos(2x - \pi/4)$ **46.** $y = 2\sin(3x + \pi/2)$

∗ **47.** $x^2 - 4y + 2 = 4x - 2y^2$ ∗ **48.** $y = \sqrt{-x^2 + 4x + 4}$

∗ **49.** $y = |x| + |x - 1|$ ∗ **50.** $y = \sqrt{|x - 1| - 1}$

∗ **51.** $x = \tan y$ ∗ **52.** $y = 2\ln(3x + 4)$

∗ **53.** $x = e^{-y}$ ∗ **54.** $y = \sin|x|$

∗ **55.** $|y| = |\sin x|$ ∗ **56.** $y = \sqrt{\sin 2x}$

∗ **57.** $y = \sinh(2x - 1)$ ∗ **58.** $y = 4\tanh 3x$

In Exercises 59–60 find the angle between the lines.

▦∗ **59.** $x + 2y = 4$, $y = 3x - 2$

▦∗ **60.** $x = 4y + 2$, $2x + 3y = 5$

In Exercises 61–64 give an example of a function $y = f(x)$ with the indicated properties.

61. The range of the function consists of one number only.

62. The algebraic formula defining the function cannot be extended beyond $-1 \le x \le 2$.

63. The domain of the function consists of all reals except $x = \pm 1$.

64. The domain of the function is $x \le 0$ and the range is $y \ge 1$.

In Exercises 65–68 show that the function does not have an inverse. Subdivide its domain of definition into subintervals on which the function has an inverse, and find the inverse function on each subinterval.

65. $f(x) = x^2 - 4x + 3$ * **66.** $f(x) = x^4 - 8x^2$

67. $f(x) = \dfrac{x^2 + 2}{x^2 + 3}$ * **68.** $f(x) = \sqrt{\dfrac{x^2}{x + 1}}$

69. Express $f(x) = \cos 2x - \sin 2x$ in the form $f(x) = A\sin(2x + \phi)$. Use this to draw a graph of the function and find the second smallest positive value of x for which $f(x) = 0$.

70. Is the function $f(x) = 2\sin 2x - 3\cos 3x$ periodic? If so, what is its period?

In Exercises 71–80 find all solutions of the equation.

71. $\cos^2 x + 5\cos x - 6 = 0$

72. $4\sin 2x = 1$

73. $\csc^2(x + 1) = 3$

74. $\mathrm{Tan}^{-1}(3x + 2) = 5 - 2\pi$

75. $\cos 2x = \sin x$

76. $\ln(\sin x) + \ln(1 + \sin x) = \ln 3 - \ln 2$

77. $3\,\mathrm{Sin}^{-1}\left(e^{x+2}\right) = 2$

78. $3\sin\left(e^{x+2}\right) = 2$

79. $\tan(x\cosh 2) = 1/4$

80. $\sinh x = 4$

81. Draw graphs of the following functions:
   (a) $f(x) = \tan\left(\mathrm{Tan}^{-1} x\right)$
   (b) $f(x) = \mathrm{Tan}^{-1}(\tan x)$

82. When the mass in the figure below is pulled 5 cm to the right of the position ($x = 0$) it would occupy were the spring unstretched, and given speed 2 m/s to the right, its position thereafter is given by

$$x(t) = \frac{1}{20}\cos 4t + \frac{1}{2}\sin 4t.$$

Find when the mass passes through $x = 0$ for the first time.

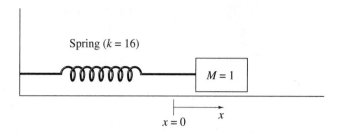

83. A lighthouse is 6 km offshore and a cabin on the straight shoreline is 9 km from the point on the shore nearest the lighthouse (figure below).

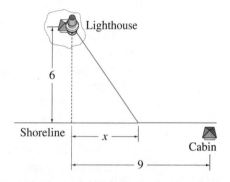

Show that if a man rows at 3 km/h and walks at 5 km/h, and he beaches the boat at distance x from the near point on the shore, then the total travel time from lighthouse to cabin is

$$t = f(x) = \frac{\sqrt{x^2 + 36}}{3} + \frac{9 - x}{5}, \quad 0 \le x \le 9.$$

Plot a graph of this function.

In Exercises 84–85 find all solutions of the equation correctly rounded to three decimal places.

84. $x^3 - 2x^2 + 4x - 5 = 0$

* **85.** $x^2 - 1 = \sin x$

In Exercises 86–87 find all solutions of the equation with error less than 10^{-4}.

86. $x^3 + 12x^2 + 4x - 5 = 0$

* **87.** $x^2 - 1 = 24\sin x$

* **88.** Prove that the diagonals of a rhombus intersect at right angles. (A rhombus is a parallelogram with all sides of equal length.)

CHAPTER 2 | Limits and Continuity

With a solid foundation of fundamentals in Chapter 1, you are well-prepared to study calculus. We hope that you have been conscientious in your review. The better your algebraic skills and the more familiar you are with analytic geometry and trigonometry, the easier calculus will be.

In Chapter 1, we placed tremendous emphasis on graphing. This was by design. The most difficult part of the solution to many problems is frequently the initial step. Once started, the solution often unfolds smoothly and easily, but that first step sometimes seems impossible. One of the best ways to start a problem is with a diagram. A picture, no matter how rough, is invaluable in giving you a "feeling" for what is going on. It displays the known facts surrounding the problem; it permits you to "see" what the problem really is and how it relates to the known facts; and it often suggests that all-important first step. We encourage you to develop the habit of making diagrams at every opportunity — not just to solve problems, but to understand what calculus is all about. We want you to *see and feel* calculus in all its aspects.

We introduce each remaining chapter of the book with an Application Preview, a problem from one of the engineering disciplines, the solution of which requires material to be introduced in that chapter. The solution of the problem is identified as the Application Review Revisited at the appropriate place in the chapter. Here is the Application Preview for this chapter.

Application Preview

The figure on the left below shows a complicated electrical network containing capacitors, inductors, resistors, and a source of electric voltage E. Electrical engineers are interested in the induced current in various parts of the network when the source is turned on and off very quickly.

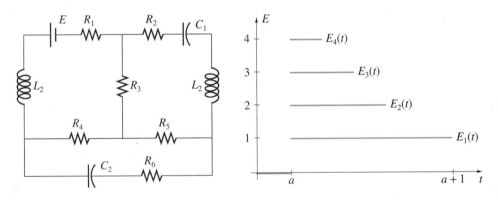

Function $E_1(t)$ in the figure on the right represents 1 V of potential being turned on at time $t = a$, and turned off again one second later. This is not a very short period of time. Graph $E_2(t)$ in the same figure represents 2 V turned on for one-half of a second; $E_3(t)$ is 3 V for one-third of a second; and $E_4(t)$ is 4 V for one-quarter of a second. Were we to continue this process indefinitely, the source would apply ever-increasing voltages over ever-decreasing time intervals, but the product of the voltage and the length of the time interval is always unity. The ultimate result of this process is what is called an instantaneous application of 1 V to the circuit.

THE PROBLEM   How do we represent the result of this process as a mathematical *function*, and how do we operate with this function in equations? (For the answer, see Dirac-delta functions in Section 2.5 on page 141.)

Can you *see* the problem? As *a* gets closer and closer to zero, the graph of *E* is zero everywhere, except at $t = a$, where it becomes "infinite." In what sense is this a function? Get the *feeling* that this cannot be a function as we now understand functions.

The concept of a *limit* is crucial to calculus, for the two basic operations in calculus are differentiation and integration, each of which is defined in terms of a limit. For this reason you must have a clear understanding of limits from the beginning. In Sections 2.1–2.4 we give an intuitive discussion of limits of functions; in Section 2.6 we show how these ideas can be formalized mathematically.

2.1 Limits

The functions in Figure 2.1 all have value 3 at $x = 1$, but behaviours of the functions close to $x = 1$ are totally different. As *x* gets closer and closer to 1 in Figure 2.1a, function values get closer and closer to 3. In Figure 2.1b, function values get closer and closer to 3 if *x* approaches 1 through numbers smaller than 1, but they get closer and closer to 2 if *x* approaches 1 through numbers larger than 1. Function values approach 2 as *x* gets closer and closer to 1 in Figure 2.1c whether *x* approaches 1 through numbers smaller than 1 or larger than 1. In this section we emphasize the distinction between the value of a function at a point, and the values of the function as we approach the point. We can *see* the distinction graphically; we now want to express it algebraically.

FIGURE 2.1a   FIGURE 2.1b   FIGURE 2.1c

The value of a function at $x = 1$ need not equal its limit as *x* approaches 1

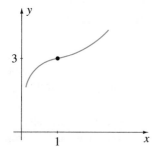 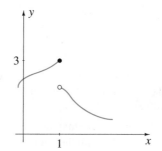 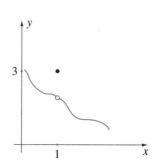

The value of the function $f(x) = x^2 - 4x + 5$ at $x = 2$ is $f(2) = 1$. A completely different consideration is contained in the question,"What number do values of $f(x) = x^2 - 4x + 5$ get closer and closer to as *x* gets closer and closer to 2?" Table 2.1 shows that *as x gets closer and closer to 2, values of $x^2 - 4x + 5$ get closer and closer to 1.*

TABLE 2.1

x	$f(x) = x^2 - 4x + 5$	x
1.9	1.01	2.1
1.99	1.000 1	2.01
1.999	1.000 001	2.001
1.9999	1.000 000 01	2.0001

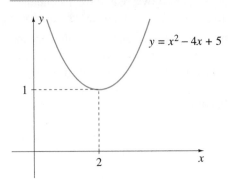

FIGURE 2.2 Limit of $x^2 - 4x + 5$ is 1 as x approaches 2

Likewise, the graph of the function (Figure 2.2) clearly shows that values of $f(x)$ approach 1 as x approaches 2. This statement is not precise enough for our purposes. For instance, the graph also indicates that as x gets closer and closer to 2, values of $f(x)$ get closer and closer to 0. They do not get very close to 0, but nonetheless, values of $f(x)$ do get closer and closer to 0 as x gets closer and closer to 2. In fact, we can make this statement for any number less than 1. To distinguish 1 from all numbers less than 1, we say that $x^2 - 4x + 5$ *can be made arbitrarily close to* 1 *by choosing x sufficiently close to* 2. We can make values of $x^2 - 4x + 5$ within 0.1 of 1 by choosing values of x sufficiently close to 2; we can make values of $x^2 - 4x + 5$ within 0.01 of 1 by choosing values of x even closer to 2 ; we can make values of $x^2 - 4x + 5$ within 0.001 of 1 by choosing values of x yet even closer to 2, and so on. For numbers less than 1, we cannot do this. For instance, it is not true that $x^2 - 4x + 5$ can be made arbitrarily close to 0. The closest the function gets to 0 is 1 unit when $x = 2$. In calculus we say that the **limit** of $x^2 - 4x + 5$ as x approaches 2 is 1 to represent the more lengthy statement "$x^2 - 4x + 5$ can be made arbitrarily close to 1 by choosing x sufficiently close to 2." In addition, we have a notation to represent both statements:

$$\lim_{x \to 2} (x^2 - 4x + 5) = 1. \tag{2.1}$$

This notation is read "the limit of (the function) $x^2 - 4x + 5$ as x approaches 2 is equal to 1," and this stands for the statement "$x^2 - 4x + 5$ can be made arbitrarily close to 1 by choosing x sufficiently close to 2."

We emphasize that the limit in 2.1 is not concerned with the value of $x^2 - 4x + 5$ at $x = 2$. It is concerned with the number that $x^2 - 4x + 5$ approaches as x approaches 2. These numbers are not always the same. For example, the limit as x approaches 1 of the function in Figure 2.1c is 2, whereas the value of the function at $x = 1$ is 3.

Generally, we say that a function $f(x)$ has limit L as x approaches a, and write

$$\lim_{x \to a} f(x) = L \tag{2.2}$$

if $f(x)$ can be made arbitrarily close to L by choosing x sufficiently close to a. Sometimes it is more convenient to write $f(x) \to L$ as $x \to a$ to mean that $f(x)$ approaches L as x approaches a. This is especially so in the middle of a paragraph, as opposed to a displayed equation such as 2.2.

The value of the function $f(x)$ at $x = a$ is irrelevant to the limit of $f(x)$ as x approaches a. In Figure 2.3a, they are the same; the value of the function at $x = a$ is L, the same as the limit as $x \to a$. In Figure 2.3b, the value of the function $f(a)$ at $x = a$ is different from the limit L as $x \to a$. Finally, in Figure 2.3c, the function has no value at $x = a$, but the limit as $x \to a$ is L.

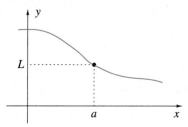

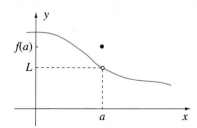

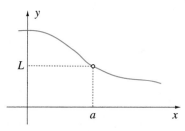

Figures to illustrate that the value of a function at $x = a$ may be different from its limit as $x \to a$.

EXAMPLE 2.1

Evaluate $\lim_{x \to 1} (x^2 + 2x + 5)$.

SOLUTION   As x gets closer and closer to 1, values of $x^2 + 2x + 5$ get arbitrarily close to 8; therefore, we write

$$\lim_{x \to 1} (x^2 + 2x + 5) = 8.$$

This is corroborated by the graph of the function in Figure 2.4.

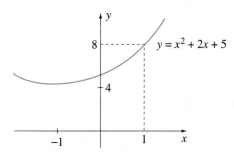

FIGURE 2.4   Limit of $x^2 + 2x + 5$ is 8 as x approaches 1

To calculate the limit of a function $f(x)$ as x approaches a, we evaluate $f(x)$ at values of x that get closer and closer to a. For limits of complicated functions such as

$$\lim_{x \to 5} \frac{x^2(3 - x)}{x^3 + x},$$

it would be tedious to evaluate $x^2(3 - x)/(x^3 + x)$ at many values of x approaching 5. The following theorem provides a much easier method.

THEOREM 2.1

If $\lim_{x \to a} f(x) = F$ and $\lim_{x \to a} g(x) = G$, then

(i) $\displaystyle\lim_{x \to a} [f(x) + g(x)] = F + G.$ (2.3a)

(ii) $\displaystyle\lim_{x \to a} [f(x) - g(x)] = F - G.$ (2.3b)

(iii) $\displaystyle\lim_{x \to a} [cf(x)] = cF,$ when c is a constant. (2.3c)

(iv) $\displaystyle\lim_{x \to a} [f(x)g(x)] = FG.$ (2.3d)

(v) $\displaystyle\lim_{x \to a} \frac{f(x)}{g(x)} = \frac{F}{G},$ provided that $G \neq 0.$ (2.3e)

What this theorem says is that a limit such as $\lim_{x \to 5} [x^2(3 - x)/(x^3 + x)]$ can be broken down into smaller problems and reassembled later. For instance, since

$$\lim_{x \to 5} x^2 = 25, \qquad \lim_{x \to 5} (3 - x) = -2, \qquad \lim_{x \to 5} x^3 = 125, \qquad \lim_{x \to 5} x = 5,$$

we may write

$$\lim_{x \to 5} \frac{x^2(3 - x)}{x^3 + x} = \frac{25(-2)}{125 + 5} = \frac{-50}{130} = -\frac{5}{13}.$$

Although the results of Theorem 2.1 may seem evident, to prove them mathematically is not a simple task. In fact, because we have not yet given a precise definition for limits, a proof is impossible at this time. When we give definitions for limits in Section 2.6, it will then be possible to prove the theorem (see Exercises 31–35 in Section 2.6).

EXAMPLE 2.2

Evaluate $\displaystyle\lim_{x \to -2} \frac{x + 2}{x^2 + 9}.$

SOLUTION   Since $\lim_{x \to -2} (x + 2) = 0$ and $\lim_{x \to -2} (x^2 + 9) = 13$, part (v) of Theorem 2.1 gives

$$\lim_{x \to -2} \frac{x + 2}{x^2 + 9} = \frac{0}{13} = 0.$$

EXAMPLE 2.3

Evaluate $\displaystyle\lim_{x \to -1} \frac{x^2(1 - x^3)}{2x^2 + x + 1}.$

SOLUTION   Using Theorem 2.1, we can write

$$\lim_{x \to -1} \frac{x^2(1 - x^3)}{2x^2 + x + 1} = \frac{(1)(2)}{2 + (-1) + 1} = 1.$$

Be sure that you understand how we obtained the expression

$$\frac{(1)(2)}{2 + (-1) + 1}.$$

In particular, we *did not* set $x = -1$ in $x^2(1-x^3)/(2x^2+x+1)$. Indeed, this is not permitted because to evaluate a limit as x approaches -1, we are not to set $x = -1$; we are to let x get closer and closer to -1. What we did do is take limits of x^2, $1 - x^3$, $2x^2$, and x as x approaches -1, and then use Theorem 2.1.

The following example illustrates what can happen if we substitute $x = a$ into $f(x)$ in the evaluation of $\lim_{x \to a} f(x)$.

EXAMPLE 2.4

Evaluate $\lim_{x \to 3} \dfrac{x^2 - 9}{x - 3}$.

SOLUTION   Because $\lim_{x \to 3} (x - 3) = 0$, we cannot use Theorem 2.1. Nor can we set $x = 3$ in $(x^2 - 9)/(x - 3)$ because it is inherent in the limiting procedure that we do not put $x = 3$. Besides, if we did, we would obtain the meaningless expression $0/0$. Figure 2.5a contains a typical graph of the function $(x^2 - 9)/(x - 3)$ on the interval $-3 \leq x \leq 6$, using a calculator or computer. It shows no anomaly in the behaviour of the function at $x = 3$. The graph may, however, be accompanied by a message indicating that the function is undefined at $x = 3$, as indeed it is. But the fact that the function is undefined at $x = 3$ does not concern us here; we are interested in values of the function near $x = 3$, not at $x = 3$, and the graph indicates that as $x \to 3$, values of the function approach 6. To verify this algebraically, we factor $x^2 - 9$ into $(x - 3)(x + 3)$ and divide out a factor of $x - 3$ from numerator and denominator:

$$\lim_{x \to 3} \frac{x^2 - 9}{x - 3} = \lim_{x \to 3} \frac{(x - 3)(x + 3)}{x - 3} = \lim_{x \to 3} (x + 3) = 6.$$

Dividing by the factor $x - 3$ would not be permissible if $x - 3$ were equal to 0, that is, if x were equal to 3. But once again this cannot happen, because in the limiting operation we let x get closer and closer to 3, but do not set $x = 3$.

FIGURE 2.5a   Computer graph of $(x^2 - 9)/(x - 3)$

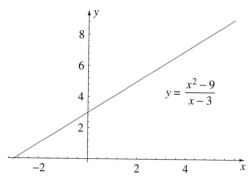

FIGURE 2.5b   Hand-drawn graph of $(x^2 - 9)/(x - 3)$

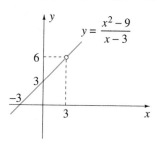

Note in this example that although the limit is 6, there is no value of x for which the function $(x^2 - 9)/(x - 3)$ is ever equal to 6. The graph of the function is a straight line with the point at $x = 3$ removed. We have shown this with an open circle in Figure 2.5b.

EXAMPLE 2.5

Evaluate $\lim\limits_{x \to 0} \dfrac{-2x + 3x^2}{4x - x^2}$.

SOLUTION   The function is undefined at $x = 0$, but

$$\lim_{x \to 0} \frac{-2x + 3x^2}{4x - x^2} = \lim_{x \to 0} \frac{x(-2 + 3x)}{x(4 - x)} = \lim_{x \to 0} \frac{3x - 2}{4 - x} = \frac{-2}{4} = -\frac{1}{2}.$$

Figure 2.6a shows a computer-generated graph of the function on the interval $-2 \le x \le 2$. We have redrawn the graph in Figure 2.6b with a hole at $x = 0$.

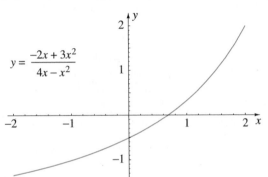

FIGURE 2.6a   Computer
graph of $(-2x + 3x^2)/(4x - x^2)$

$$y = \frac{-2x + 3x^2}{4x - x^2}$$

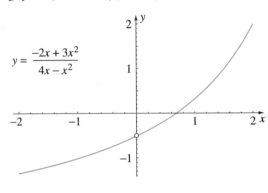

FIGURE 2.6b   Redrawn
graph of $(-2x + 3x^2)/(4x - x^2)$

$$y = \frac{-2x + 3x^2}{4x - x^2}$$

EXAMPLE 2.6

Evaluate $\lim\limits_{x \to 0} \dfrac{\sqrt{1 + x} - 1}{\sqrt{x}}$.

SOLUTION   Since the limit of the denominator as x approaches 0 is 0, we cannot immediately use Theorem 2.1. The function is undefined for $x \le 0$; its graph on the interval $0.001 \le x \le 1$ in Figure 2.7 suggests, although not conclusively, that the limit is 0. To verify this we rationalize the numerator, that is, rid the numerator of the square root by multiplying numerator and denominator by $\sqrt{1 + x} + 1$:

$$\lim_{x \to 0} \frac{\sqrt{1 + x} - 1}{\sqrt{x}} = \lim_{x \to 0} \left(\frac{\sqrt{1 + x} - 1}{\sqrt{x}} \frac{\sqrt{1 + x} + 1}{\sqrt{1 + x} + 1} \right)$$

$$= \lim_{x \to 0} \frac{x}{\sqrt{x}\left(\sqrt{1 + x} + 1\right)}$$

$$= \lim_{x \to 0} \frac{\sqrt{x}}{\sqrt{1 + x} + 1}$$

$$= 0.$$

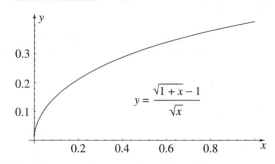

FIGURE 2.7   Suggested limit of $(\sqrt{1+x}-1)/\sqrt{x}$ as x approaches 0

$$y = \frac{\sqrt{1+x}-1}{\sqrt{x}}$$

The following is a second example of this type.

EXAMPLE 2.7

Evaluate $\lim\limits_{x \to 1} \dfrac{(\sqrt{x+3}-2)(\sqrt{x-1}+3)}{x-1}$.

SOLUTION   Once again we cannot use Theorem 2.1; the denominator has limit 0 as $x \to 1$. The graph of the function for $1.001 \le x \le 2$ in Figure 2.8 may suggest a limit, but certainly the evidence is far from conclusive. Following the lead of Example 2.6, we rationalize the term $\sqrt{x+3}-2$ in the numerator by multiplying numerator and denominator by $\sqrt{x+3}+2$,

$$\lim_{x \to 1} \frac{(\sqrt{x+3}-2)(\sqrt{x-1}+3)}{x-1} = \lim_{x \to 1} \frac{(\sqrt{x+3}-2)(\sqrt{x-1}+3)(\sqrt{x+3}+2)}{(x-1)(\sqrt{x+3}+2)}$$

$$= \lim_{x \to 1} \frac{(x+3-4)(\sqrt{x-1}+3)}{(x-1)(\sqrt{x+3}+2)}$$

$$= \lim_{x \to 1} \frac{\sqrt{x-1}+3}{\sqrt{x+3}+2}$$

$$= \frac{3}{4}.$$

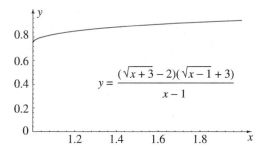

FIGURE 2.8   Suggested limit of $(\sqrt{x+3}-2)(\sqrt{x-1}+3)/(x-1)$ as x approaches 1

$$y = \frac{(\sqrt{x+3}-2)(\sqrt{x-1}+3)}{x-1}$$

EXAMPLE 2.8

Evaluate $\lim\limits_{x \to 0} \dfrac{\sin 2x}{\sin x}$.

SOLUTION   Once again we cannot immediately use Theorem 2.1 since the limit of the denominator is zero. But using the double-angle formula $\sin 2x = 2 \sin x \cos x$, we find that

$$\lim_{x \to 0} \frac{\sin 2x}{\sin x} = \lim_{x \to 0} \frac{2 \sin x \cos x}{\sin x} = \lim_{x \to 0} (2 \cos x) = 2.$$

You may feel that we are overemphasizing limits in which both the numerator and denominator are approaching zero (Examples 2.4–2.8). We stress this type of limit because when we use the definition of a *derivative* in the next chapter, we always encounter this situation.

EXAMPLE 2.9

Do the functions

$$f(x) = \sin\left(\frac{1}{x}\right) \qquad \text{and} \qquad g(x) = x^2 \sin\left(\frac{1}{x}\right)$$

have limits as x approaches 0?

SOLUTION   These limits are more difficult to find. To get a *feeling* for the behaviour of the function $\sin(1/x)$ near $x = 0$, we plot its graph on the interval $-0.01 \le x \le 0.01$ (Figure 2.9a); it is a washout. The graph for $-0.1 \le x \le 0.1$ in Figure 2.9b is more instructive. It shows that the function oscillates back and forth between ± 1 more and more rapidly as x approaches 0. As a result, the limit of $\sin(1/x)$ does not exist as x approaches 0.

FIGURE 2.9a $\sin(1/x)$ for $-0.01 \le x \le 0.01$

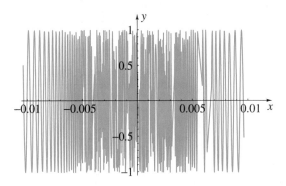

FIGURE 2.9b $\sin(1/x)$ for $-0.1 \le x \le 0.1$

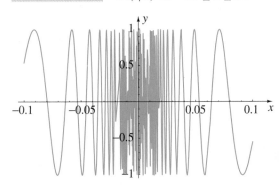

The function $g(x) = x^2 \sin(1/x)$ has exactly the same number of oscillations as $f(x) = \sin(1/x)$, but the oscillations become smaller and smaller as x approaches 0 (Figure 2.10a). In other words, $\lim_{x \to 0} g(x) = 0$.

FIGURE 2.10a   Suggested limit of $x^2 \sin(1/x)$ as x approaches 0

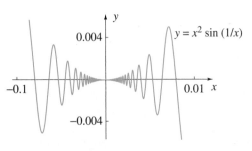

FIGURE 2.10b $x^2 \sin(1/x)$ and $\pm x^2$ for squeeze theorem

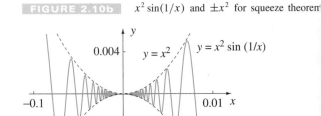

We can confirm the geometric conclusion that $\lim_{x \to 0} g(x) = 0$ in this example using the following theorem.

> ### THEOREM 2.2 (Squeeze or Sandwich Theorem)
>
> Suppose that functions $f(x)$, $g(x)$, and $h(x)$ satisfy the following two properties:
>
> 1. $f(x) \leq g(x) \leq h(x)$ for all x in some open interval containing $x = a$;
> 2. $\lim_{x \to a} f(x) = L = \lim_{x \to a} h(x)$.
>
> Then $\lim_{x \to a} g(x) = L$ also.

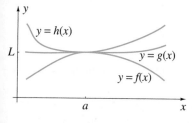

FIGURE 2.11 Illustration of the squeeze theorem

With Figure 2.11 we can *see* this result geometrically. The graph of $g(x)$ is always between those of $f(x)$ and $h(x)$ in an open interval around $x = a$ (condition 1). Since graphs of $f(x)$ and $h(x)$ come together at $x = a$ (condition 2), so also must the graph of $g(x)$.

For $g(x) = x^2 \sin(1/x)$ in Example 2.9, we know that $-1 \leq \sin(1/x) \leq 1$ for all x. If we multiply all terms by x^2, we obtain

$$-x^2 \leq x^2 \sin\left(\frac{1}{x}\right) \leq x^2.$$

This shows that the graph of $x^2 \sin(1/x)$ is between those of $-x^2$ and x^2 (Figure 2.10b). Since $\lim_{x \to 0} (-x^2) = \lim_{x \to 0} x^2 = 0$, the squeeze theorem gives $\lim_{x \to 0} x^2 \sin(1/x) = 0$.

EXAMPLE 2.10

Use the squeeze theorem to evaluate $\lim_{x \to 0} x \cos\left(\dfrac{3}{x}\right)$, if it exists.

SOLUTION The function $\cos(3/x)$, like $\sin(1/x)$, oscillates violently as x approaches zero. But we know that $-1 \leq \cos(3/x) \leq 1$, and multiplication by $|x|$ gives

$$-|x| \leq |x| \cos\left(\frac{3}{x}\right) \leq |x|.$$

Since $-|x|$ is always less than or equal to zero and $|x|$ is always greater than or equal to zero, we can write that

$$-|x| \leq x \cos\left(\frac{3}{x}\right) \leq |x|.$$

Since $\lim_{x \to 0} (-|x|) = \lim_{x \to 0} |x| = 0$, the squeeze theorem requires that $\lim_{x \to 0} x \cos(3/x) = 0$ also.

Here is a good question for you. Why in the last example did we multiply all parts of the inequality $-1 \leq \cos(3/x) \leq 1$ by $|x|$ rather than x?

One-Sided Limits

When we write $L = \lim_{x \to a} f(x)$, we mean that $f(x)$ gets arbitrarily close to L as x gets closer and closer to a. But how is x to approach a? Does x approach a through numbers larger than a, or does it approach a through numbers smaller than a? Or does x jump back and forth between numbers larger than a and numbers smaller than a, gradually getting closer and closer to a? We have not previously mentioned "mode" of approach simply because it would have made no difference to our discussion. In each of the preceding examples, all possible modes of approach lead to the same limit. In particular, Table 2.1 and Figure 2.2 illustrate that 1 is the limit of $f(x) = x^2 - 4x + 5$ as x approaches 2 whether x approaches 2 through numbers larger than 2 or through numbers smaller than 2.

Approaching a number a either through numbers larger than a or through numbers smaller than a are two modes of approach that will be very important; therefore, we give them special notations:

$$\lim_{x \to a^-} f(x) \qquad\qquad\qquad \lim_{x \to a^+} f(x)$$

indicates that x approaches a through numbers smaller than a (often called a **left-hand limit** since x approaches a along the x-axis from the left of a)

indicates that x approaches a through numbers larger than a (often called a **right-hand limit** since x approaches a along the x-axis from the right of a)

Example 2.6 should, in fact, be designated a right-hand limit,

$$\lim_{x \to 0^+} \frac{\sqrt{1+x} - 1}{\sqrt{x}} = 0,$$

since the presence of $\sqrt{x}$ in the denominator demands that x be positive. Similarly, the limit of Example 2.7 should only be right-handed.

 Do not interpret the $-$ and $+$ in a^- and a^+ as approaching a through negative and positive numbers. This is the case only when $a = 0$. For instance, when $a = 5$, 5 is approached through positive numbers whether it is approached from the left or from the right.

A natural question to ask is: What should we conclude if for a function $f(x)$

$$\lim_{x \to a^+} f(x) \neq \lim_{x \to a^-} f(x)?$$

Our entire discussion has suggested (and indeed it can be proved; see Exercise 20 in Section 2.6) that if a function has a limit as x approaches a, then it has only one such limit; that is, the limit must be the same for every possible method of approach. Consequently, if we arrive at two different results depending on the mode of approach, then we conclude that the function does not have a limit. This situation is illustrated in Figure 2.1b and again in the following example.

EXAMPLE 2.11

Evaluate, if possible, $\displaystyle\lim_{x \to 0} \frac{|x|}{x}$.

SOLUTION   If $x < 0$, then $|x| = -x$, and

$$\lim_{x \to 0^-} \frac{|x|}{x} = \lim_{x \to 0^-} \frac{-x}{x} = -1.$$

If $x > 0$, then $|x| = x$, and

$$\lim_{x \to 0^+} \frac{|x|}{x} = \lim_{x \to 0^+} \frac{x}{x} = 1.$$

The function has a right-hand limit and a left-hand limit at $x = 0$, but because they are not the same, $\lim_{x \to 0} (|x|/x)$ does not exist. The graph of $f(x) = |x|/x$ in Figure 2.12 clearly illustrates the situation.

FIGURE 2.12   Illustration of no limit as x approaches 0

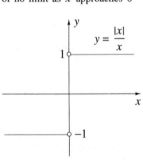

EXAMPLE 2.12

The weight W of an object depends on its distance d from the centre of the earth. If d is less than the radius R of the earth, then W is directly proportional to d; and if d is greater than or equal to R, then W is inversely proportional to d^2. If the weight of the object on the earth's surface is W_0, find a formula for W as a function for d and draw its graph.

SOLUTION When $d < R$, we may write that $W = kd$; and when $d \geq R$, $W = \ell/d^2$, where k and ℓ are constants of proportionality; that is,

$$W = \begin{cases} kd, & 0 \leq d < R \\ \dfrac{\ell}{d^2}, & R \geq d. \end{cases}$$

Since the weight of the object is W_0 on the surface of the earth when $d = R$, it follows that $W_0 = \ell/R^2$, from which $\ell = W_0 R^2$. It now remains to find k. If we physically moved the object from below the surface of the earth to the surface, it would slowly gain weight. Its weight would approach W_0, that on the surface of the earth. In other words, the limit of W as $r \to R^-$ must be W_0; that is,

$$W_0 = \lim_{r \to R^-} kd = kR \qquad \Longrightarrow \qquad k = \frac{W_0}{R}.$$

Thus,

$$W = \begin{cases} \dfrac{W_0 d}{R}, & 0 < d < R \\ \dfrac{W_0 R^2}{d^2}, & d \geq R \end{cases}.$$

A graph is shown in Figure 2.13.

FIGURE 2.13 Weight of an object as a function of distance from centre of earth

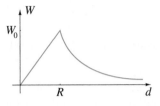

EXERCISES 2.1

In Exercises 1–41 find the indicated limit, if it exists.

1. $\displaystyle\lim_{x \to 7} \frac{x^2 - 5}{x + 2}$

2. $\displaystyle\lim_{x \to -2} \frac{x^3 + 8}{x + 5}$

3. $\displaystyle\lim_{x \to -5} \frac{x^2 + 3x + 2}{x^2 + 25}$

4. $\displaystyle\lim_{x \to 0} \frac{x^2 + 3x}{3x^2 - 2x}$

5. $\displaystyle\lim_{x \to 3^+} \frac{2x - 3}{x^2 - 5}$

6. $\displaystyle\lim_{x \to 2^-} \frac{2x - 4}{3x + 2}$

7. $\displaystyle\lim_{x \to 0^-} \frac{x^4 + 5x^3}{3x^4 - x^3}$

8. $\displaystyle\lim_{x \to 2^+} \frac{x^2 + 2x + 4}{x - 3}$

9. $\displaystyle\lim_{x \to 2} \frac{x^2 - 4}{x - 2}$

10. $\displaystyle\lim_{x \to 3^+} \frac{x^2 - 9}{x - 3}$

11. $\displaystyle\lim_{x \to 5^-} \frac{x^2 - 25}{x - 5}$

12. $\displaystyle\lim_{x \to 3} \frac{x^2 - 2x - 3}{3 - x}$

13. $\displaystyle\lim_{x \to 2} \frac{x^2 - 4x + 4}{x - 2}$

14. $\displaystyle\lim_{x \to 2} \frac{x^3 - 6x^2 + 12x - 8}{x^2 - 4x + 4}$

15. $\displaystyle\lim_{x \to 1} \frac{x^3 - 6x^2 + 11x - 6}{x^2 - 3x + 2}$

16. $\displaystyle\lim_{x \to 2} \frac{x^3 - 6x^2 + 11x - 6}{x^2 - 3x + 2}$

17. $\displaystyle\lim_{x \to 3^+} \frac{x^3 - 6x^2 + 11x - 6}{x^2 - 3x + 2}$

18. $\displaystyle\lim_{x \to 3^-} \frac{x^3 - 6x^2 + 11x - 6}{x^2 - 3x + 2}$

19. $\displaystyle\lim_{x \to 0} \frac{x^3 - 6x^2 + 11x - 6}{x^2 - 3x + 2}$

20. $\displaystyle\lim_{x \to -1} \frac{12x + 5}{x^2 - 2x + 1}$

21. $\displaystyle\lim_{x \to 1} \sqrt{\frac{2 - x}{2 + x}}$

22. $\displaystyle\lim_{x \to 5} \frac{\sqrt{1 - x^2}}{3x + 2}$

23. $\displaystyle\lim_{x \to 0} \frac{\tan x}{\sin x}$

24. $\displaystyle\lim_{x \to \pi/4} \frac{\sin x}{\tan x}$

* 25. $\displaystyle\lim_{x \to 0} \frac{\sin 4x}{\sin 2x}$

* 26. $\displaystyle\lim_{x \to 0^+} \frac{\sin 6x}{\sin 3x}$

* 27. $\displaystyle\lim_{x \to 0^+} \frac{\sin 2x}{\tan x}$

* 28. $\displaystyle\lim_{x \to 2} \frac{x - 2}{\sqrt{x} - \sqrt{2}}$

* 29. $\displaystyle\lim_{x \to 0} \frac{\sqrt{1 - x} - \sqrt{1 + x}}{x}$

* 30. $\displaystyle\lim_{x \to 5^+} \frac{|x^2 - 25|}{x^2 - 25}$

* 31. $\displaystyle\lim_{x \to 5^-} \frac{|x^2 - 25|}{x^2 - 25}$

* 32. $\displaystyle\lim_{x \to 5} \frac{|x^2 - 25|}{x^2 - 25}$

* 33. $\displaystyle\lim_{x \to 0^+} \frac{\sqrt{x + 2} - \sqrt{2}}{\sqrt{x}}$

* 34. $\displaystyle\lim_{x \to 0} \frac{1 - \sqrt{x^2 + 1}}{2x^2}$

* 35. $\lim\limits_{x \to -2} \dfrac{x+2}{\sqrt{-x} - \sqrt{2}}$

* 36. $\lim\limits_{x \to 0} \dfrac{\sqrt{1+x} - \sqrt{1-x}}{x}$

* 37. $\lim\limits_{x \to -2^+} \dfrac{\sqrt{x+3} - \sqrt{-x-1}}{\sqrt{x+2}}$

* 38. $\lim\limits_{x \to 0} \dfrac{x}{\sqrt{x+4} - 2}$

* 39. $\lim\limits_{x \to 0} \dfrac{\sqrt{1+x} - \sqrt{1-x}}{\sqrt{2+x} - \sqrt{2-x}}$

* 40. $\lim\limits_{x \to 0} \dfrac{\sqrt{x+1} - \sqrt{2x+1}}{\sqrt{3x+4} - \sqrt{2x+4}}$

* 41. $\lim\limits_{x \to 1} \dfrac{\sqrt{x+3} - 2}{x - 1}$

In Exercises 42–49 assume that $a > 0$ is a constant and calculate the limit, if it exists.

42. $\lim\limits_{x \to a} \dfrac{x^2 - a^2}{x - a}$

43. $\lim\limits_{x \to a} \dfrac{x^3 - a^3}{x - a}$

44. $\lim\limits_{x \to -a} \dfrac{x+a}{x^2 + ax - x - a}$

* 45. $\lim\limits_{x \to 0} \dfrac{\sin 2ax}{\sin ax}$

* 46. $\lim\limits_{x \to a} \dfrac{\sqrt{x} - \sqrt{a}}{x - a}$

* 47. $\lim\limits_{x \to 0^+} \dfrac{\sqrt{x+a} - \sqrt{a}}{\sqrt{x}}$

* 48. $\lim\limits_{x \to 0} \dfrac{\sqrt{a+x} - \sqrt{a-x}}{x}$

* 49. $\lim\limits_{x \to 0} \dfrac{\sqrt{x^2 + a^2} - \sqrt{2x^2 + a^2}}{\sqrt{3x^2 + 4} - \sqrt{2x^2 + 4}}$

* 50. Plot a graph of $f(x) = (1/x) \sin x$ on the interval $-\pi \le x \le \pi$. What does it suggest for the limit of the function as $x \to 0$? (This will be confirmed in Section 3.9.)

In Exercises 51 and 52 use the squeeze theorem to discuss the limits.

51. $\lim\limits_{x \to 0} x \sin\left(\dfrac{1}{x}\right)$

52. $\lim\limits_{x \to 0} x^4 \cos\left(\dfrac{3}{x}\right)$

* 53. Does the floor function $\lfloor x \rfloor$ of Exercise 68 in Section 1.5 have a limit, a right-hand limit, or a left-hand limit as x approaches integer values?

* 54. Prove or disprove the following statement: If $f(x) < g(x)$ for all $x \ne a$, then $\lim\limits_{x \to a} f(x) < \lim\limits_{x \to a} g(x)$.

* 55. If n is a positive integer, evaluate $\lim\limits_{h \to 0} \dfrac{(x+h)^n - x^n}{h}$. *Hint:* Use either the binomial theorem or the result that

$$a^n - b^n = (a - b)(a^{n-1} + a^{n-2}b + \cdots + ab^{n-2} + b^{n-1}).$$

* 56. Evaluate $\lim\limits_{h \to 0} \dfrac{\sqrt{x+h} - \sqrt{x}}{h}$.

* 57. Plot a graph of the function $f(x) = \dfrac{1 - e^{-1/x}}{1 + e^{-1/x}}$. What does the plot suggest for right-hand and left-hand limits of $f(x)$ as x approaches 0? What is $f(0)$?

* 58. At the present time it is impossible for us to calculate algebraically $\lim\limits_{x \to 0} \dfrac{\sin x - x}{x^3}$. What can we do?

(a) One suggestion might be to use a calculator or computer to evaluate the function $(\sin x - x)/x^3$ for various values of x that approach 0. Try this with $x = 10^{-n}$, $n = 1, \ldots, 7$, and make any conclusion that you feel is justified. How would you feel about the use of the calculator if you knew that the value of the limit were $-1/6$?

(b) Another suggestion might be to plot the function on smaller and smaller intervals around $x = 0$. Do this on the intervals $-0.1 \le x \le 0.1$, $-0.01 \le x \le 0.01$, $-0.001 \le x \le 0.001$, and $-0.0001 \le x \le 0.0001$. What happens?

* 59. Assume the existence of $\lim\limits_{x \to a} \dfrac{f(x) - g(x)}{x - a}$. If $\lim\limits_{x \to a} f(x) = L$, find $\lim\limits_{x \to a} g(x)$.

* 60. If $f(x)$ is an even function and $\lim\limits_{x \to a} f(x) = L$, find, if possible, $\lim\limits_{x \to -a} f(x)$.

* 61. If $f(x)$ is an even function and $\lim\limits_{x \to a^+} f(x) = L$, find, if possible, $\lim\limits_{x \to -a^-} f(x)$.

* 62. If $f(x)$ is an even function and $\lim\limits_{x \to a^+} f(x) = L$, find, if possible, $\lim\limits_{x \to -a^+} f(x)$.

* 63.–65. Repeat Exercises 60–62 for an odd function $f(x)$.

* 66. If $\lim\limits_{x \to 0} f(x) = F$, what do you conclude about $\lim\limits_{x \to 0} f(x) \sin(1/x)$? *Hint:* See Example 2.9.

* 67. If a, b, c, and d are constants, find the following limits, if they exist:

(a) $\lim\limits_{x \to 0^+} \dfrac{a + ce^{-1/x}}{b + de^{-1/x}}$

(b) $\lim\limits_{x \to 0^-} \dfrac{a + ce^{-1/x}}{b + de^{-1/x}}$

(c) $\lim\limits_{x \to 0} \dfrac{a + ce^{-1/x}}{b + de^{-1/x}}$

2.2 Infinite Limits

Functions do not always have limits. Sometimes this is due to the fact that right-hand and left-hand limits are not identical; sometimes it is a result of erratic oscillations. Examples of both of these situations were discussed in Section 2.1 (see Examples 2.9–2.11). Nonexistence of a limit may also be due to excessively large values of the function. For instance, consider the function $f(x) = 1/(x - 2)^2$, which is not defined at $x = 2$. Does it have a limit as x approaches 2? Table 2.2 indicates that as x approaches 2, values of $1/(x - 2)^2$ become very large; in fact, values of the function can be made arbitrarily large by choosing x sufficiently close to 2. Thus, the function does not have a limit as x approaches 2. We express this symbolically in the form

$$\lim_{x \to 2} \frac{1}{(x - 2)^2} = \infty.$$

The symbol ∞ represents what mathematicians call **infinity**. Infinity is not a number; it is simply a symbol that we find convenient to represent various ideas. In the equation above it states that the *limit does not exist*, and indicates that the reason it does not exist is that values of the function become arbitrarily large as x approaches 2.

The graph of $f(x)$ in Figure 2.14 further illustrates this point. We say that the line $x = 2$ is a **vertical asymptote** for the graph.

TABLE 2.2

x	$1/(x - 2)^2$	x
1.9	10^2	2.1
1.99	10^4	2.01
1.999	10^6	2.001
1.9999	10^8	2.0001
1.99999	10^{10}	2.00001

FIGURE 2.14 Unbounded behaviour of $1/(x - 2)^2$ near $x = 2$

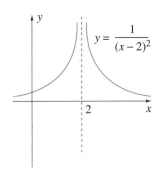

EXAMPLE 2.13

FIGURE 2.15 Unbounded behaviour of $1/(x - 1)$ near $x = 1$

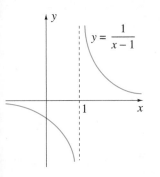

Evaluate $\lim\limits_{x \to 1} \dfrac{1}{x - 1}$, if it exists.

SOLUTION   In this example we consider right- and left-hand limits as x approaches 1. We find that

$$\lim_{x \to 1^+} \frac{1}{x - 1} = \infty \quad \text{and} \quad \lim_{x \to 1^-} \frac{1}{x - 1} = -\infty,$$

the latter meaning that as x approaches 1 from the left, the function takes on arbitrarily "large" negative values. Either one of these expressions is sufficient to conclude that the function $1/(x - 1)$ does not have a limit as x approaches 1. The function is shown in Figure 2.15; the line $x = 1$ is a vertical asymptote.

EXAMPLE 2.14

Discuss left- and right-hand limits of the function

$$f(x) = \frac{x^2 - 9}{x^2 + x - 2}$$

as $x \to 1$ and $x \to -2$.

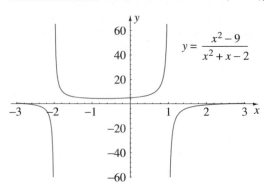

FIGURE 2.16   Unbounded behaviour of $(x^2 - 9)/(x^2 + x - 2)$ near $x = -2$ and $x = 1$

SOLUTION   The function is undefined at $x = 1$ and $x = -2$. (Why?) Its graph is shown in Figure 2.16. To confirm algebraically what we see geometrically, it is advantageous to factor numerator and denominator of $f(x)$ as much as possible,

$$f(x) = \frac{(x + 3)(x - 3)}{(x + 2)(x - 1)}.$$

Consider now the (right-hand) limit as $x \to 1^+$. Each of the four factors has a limit as $x \to 1^+$:

$$x + 3 \to 4, \qquad x - 3 \to -2, \qquad x + 2 \to 3, \qquad x - 1 \to 0.$$

Were we to combine them according to Theorem 2.1, we would write

$$\frac{4(-2)}{3(0)}.$$

In spite of the fact that this is not correct — part (v) of the theorem does not allow a 0 in the denominator — this expression lets us *see* what is happening to $f(x)$ as $x \to 1^+$. It states that as $x \to 1^+$, the numerator of $f(x)$ approaches -8 and the denominator approaches 0. But this means that $f(x)$ must be taking on larger and larger values as $x \to 1^+$. Are these values positive or negative? The numerator is clearly negative, but the sign of the denominator depends on whether the 0 is approached through positive or negative numbers. Recalling that 0 arose from the fact that $x - 1 \to 0$ as $x \to 1^+$, we can be more specific; $x - 1$ must approach 0 through positive numbers since $x > 1$ for $x \to 1^+$. We indicate this by writing $x - 1 \to 0^+$ as $x \to 1^+$. The fraction displayed above is therefore replaced by

$$\frac{4(-2)}{3(0^+)},$$

and it is clearly negative. In other words,

$$\lim_{x \to 1^+} \frac{x^2 - 9}{x^2 + x - 2} = \lim_{x \to 1^+} \frac{(x + 3)(x - 3)}{(x + 2)(x - 1)} = -\infty,$$

as indeed Figure 2.16 indicates. Similarly, as $x \to 1^-$, the fraction

$$\frac{4(-2)}{3(0^-)},$$

which shows limits of the four factors of $f(x)$, indicates that

$$\lim_{x \to 1^-} \frac{(x + 3)(x - 3)}{(x + 2)(x - 1)} = \infty.$$

In the following, the fractions on the right yield results as $x \to -2^+$ and $x \to -2^-$:

$$\lim_{x \to -2^+} \frac{(x+3)(x-3)}{(x+2)(x-1)} = \infty, \qquad\qquad \frac{(1)(-5)}{(0^+)(-3)}$$

$$\lim_{x \to -2^-} \frac{(x+3)(x-3)}{(x+2)(x-1)} = -\infty. \qquad\qquad \frac{(1)(-5)}{(0^-)(-3)}$$

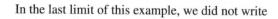

In the last limit of this example, we did not write

$$\lim_{x \to 2^-} \frac{(x+3)(x-3)}{(x+2)(x-1)} = \frac{(1)(-5)}{(0^-)(-3)} = \infty.$$

 To include the fraction $\dfrac{(1)(-5)}{(0^-)(-3)}$ as part of a mathematical equation is not acceptable; division by 0 is impossible. Place the fraction to the right of the limit to aid in its evaluation, but do not include it in the equation.

EXERCISES 2.2

In Exercises 1–24 evaluate the limit, if it exists.

1. $\displaystyle\lim_{x \to 2^+} \frac{1}{x-2}$

2. $\displaystyle\lim_{x \to 2^-} \frac{1}{x-2}$

3. $\displaystyle\lim_{x \to 2} \frac{1}{x-2}$

4. $\displaystyle\lim_{x \to 2^+} \frac{1}{(x-2)^2}$

5. $\displaystyle\lim_{x \to 2^-} \frac{1}{(x-2)^2}$

6. $\displaystyle\lim_{x \to 2} \frac{1}{(x-2)^2}$

7. $\displaystyle\lim_{x \to 1} \frac{5x}{(x-1)^3}$

8. $\displaystyle\lim_{x \to 1/2} \frac{6x^2 + 7x - 5}{2x - 1}$

9. $\displaystyle\lim_{x \to 1} \frac{2x+3}{x^2 - 2x + 1}$

10. $\displaystyle\lim_{x \to 2} \frac{x-2}{x^2 - 4x + 4}$

11. $\displaystyle\lim_{x \to 0} \csc x$

12. $\displaystyle\lim_{x \to \pi/4} \sec(x - \pi/4)$

13. $\displaystyle\lim_{x \to 3\pi/4} \sec(x - \pi/4)$

14. $\displaystyle\lim_{x \to 0^+} \cot x$

15. $\displaystyle\lim_{x \to \pi/2^+} \tan x$

16. $\displaystyle\lim_{x \to \pi/2^-} \tan x$

17. $\displaystyle\lim_{x \to 1} \frac{x^2 - 2x + 1}{x^3 - 3x^2 + 3x - 1}$

18. $\displaystyle\lim_{x \to 0} \frac{\sqrt{1+x} - 1}{x^2}$

19. $\displaystyle\lim_{x \to 0} \frac{2x}{1 - \sqrt{x^2 + 1}}$

20. $\displaystyle\lim_{x \to 4} \frac{|4-x|}{x^2 - 8x + 16}$

21. $\displaystyle\lim_{x \to 0^+} \ln(4x)$

22. $\displaystyle\lim_{x \to 1} \frac{1}{\ln|x-1|}$

23. $\displaystyle\lim_{x \to 0} e^{1/x}$

24. $\displaystyle\lim_{x \to 0} e^{1/|x|}$

In Exercises 25–28 assume that $a > 0$ is a constant and calculate the limit, if it exists.

25. $\displaystyle\lim_{x \to a^+} \frac{x-a}{x^2 - 2ax + a^2}$

26. $\displaystyle\lim_{x \to a} \frac{|x-a|}{x^2 - 2ax + a^2}$

27. $\displaystyle\lim_{x \to 0^-} \frac{\sqrt{a+x} - \sqrt{a}}{x^2}$

28. $\displaystyle\lim_{x \to -a} e^{1/(|x|-a)}$

29. It is not clear whether the limit $\displaystyle\lim_{x \to 0^+} x^2 \ln x$ exists due to the fact that $\lim_{x \to 0^+} x^2 = 0$ and $\lim_{x \to 0^+} \ln x = -\infty$. It depends on which term is more dominant in the product, x^2 or $\ln x$.

   (a) Calculate x^2, $\ln x$, and $x^2 \ln x$ for $x = 10^{-n}$, $n = 1, \ldots, 10$, and use this information to decide on a value for the limit.

   (b) Plot graphs of $x^2 \ln x$ near $x = 0$ to confirm your calculation in part (a).

30. Repeat Exercise 29 for the limit $\displaystyle\lim_{x \to 0^+} x^{10} e^{1/x}$, but pick your own values of x at which to evaluate x^{10} and $e^{1/x}$.

2.3 Limits at Infinity

In many applications we are concerned with the behaviour of a function as its independent variable takes on very large values, positively or negatively. For instance, consider finding, if possible, a number that the function $f(x) = (2x^2 + 3)/(x^2 + 4)$ gets closer and closer to as x becomes larger and larger and larger. The graph of $f(x)$ on the interval $0 \le x \le 100$ in Figure

2.17 suggests that function values are approaching 2 for large x. To confirm this algebraically, we divide the numerator and denominator of $f(x)$ by x^2,

$$f(x) = \frac{2x^2 + 3}{x^2 + 4} = \frac{2 + \dfrac{3}{x^2}}{1 + \dfrac{4}{x^2}}.$$

For very large x, the terms $3/x^2$ and $4/x^2$ are very close to zero, and therefore $f(x)$ is approximately equal to 2. Indeed, $f(x)$ can be made arbitrarily close to 2 by choosing x sufficiently large. In calculus we express this fact by saying that **the limit** of $(2x^2+3)/(x^2+4)$ **as x approaches infinity** is 2, and we write

$$\lim_{x \to \infty} \frac{2x^2 + 3}{x^2 + 4} = 2.$$

Once again we stress that ∞ is not a number. The notation $x \to \infty$ simply means "as x gets larger and larger and larger." We say that the line $y = 2$ is a **horizontal asymptote** for the graph of the function $f(x) = (2x^2 + 3)/(x^2 + 4)$.

We can also find limits of functions as x takes on arbitrarily large negative numbers, denoted by $x \to -\infty$.

FIGURE 2.17   Limit of $(2x^2 + 3)/(x^2 + 4)$ for large x

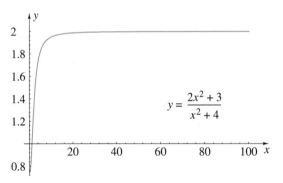

We can also find limits of functions as x takes on arbitrarily "large" negative numbers, denoted by $x \to -\infty$.

EXAMPLE 2.15

Evaluate $\displaystyle\lim_{x \to -\infty} \frac{5x^4 - 3x + 5}{x^4 - 2x^2 + 5}$, if it exists.

SOLUTION   Division of numerator and denominator by x^4 leads to

$$\lim_{x \to -\infty} \frac{5 - \dfrac{3}{x^3} + \dfrac{5}{x^4}}{1 - \dfrac{2}{x^2} + \dfrac{5}{x^4}} = 5.$$

FIGURE 2.18   Limits of $(5x^4 - 3x + 5)/(x^4 - 2x^2 + 5)$ for large positive and negative x

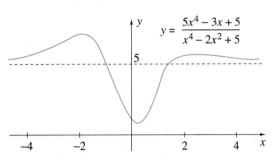

If the limit in Example 2.15 is taken as $x \to \infty$, the same result is obtained. The graph of this function is shown in Figure 2.18. The line $y = 5$ is a horizontal asymptote.

In general, we say that a line $y = L$ is a horizontal asymptote for the graph of a function $f(x)$ if either, or both, of the following situations exist:

$$\lim_{x \to -\infty} f(x) = L \quad \text{or} \quad \lim_{x \to \infty} f(x) = L.$$

In Example 2.15, both of these conditions are satisfied for $L = 5$.

EXAMPLE 2.16

Evaluate the following limits, if they exist:

(a) $\displaystyle\lim_{x \to \infty} \frac{2x^3 - 4}{x^3 + x^2 + 2}$ (b) $\displaystyle\lim_{x \to -\infty} \frac{2x^2 - 14}{3x^3 + 5x}$ (c) $\displaystyle\lim_{x \to \infty} \frac{2x^4 - 14}{3x^3 + 5x}$

SOLUTION

(a) To obtain this limit we divide numerator and denominator of the fraction by x^3:

$$\lim_{x \to \infty} \frac{2x^3 - 4}{x^3 + x^2 + 2} = \lim_{x \to \infty} \frac{2 - \dfrac{4}{x^3}}{1 + \dfrac{1}{x} + \dfrac{2}{x^3}} = 2.$$

(b) Once again we divide numerator and denominator by x^3:

$$\lim_{x \to -\infty} \frac{2x^2 - 14}{3x^3 + 5x} = \lim_{x \to -\infty} \frac{\dfrac{2}{x} - \dfrac{14}{x^3}}{3 + \dfrac{5}{x^2}}.$$

Since the numerator approaches 0 and the denominator approaches 3, we conclude that

$$\lim_{x \to -\infty} \frac{2x^2 - 14}{3x^3 + 5x} = 0.$$

We could also have obtained this limit by dividing numerator and denominator by x^2 instead of x^3,

$$\lim_{x \to -\infty} \frac{2x^2 - 14}{3x^3 + 5x} = \lim_{x \to -\infty} \frac{2 - \dfrac{14}{x^2}}{3x + \dfrac{5}{x}}.$$

Now the numerator approaches 2, but since the denominator becomes very large, the fraction once again approaches zero.

(c) Division by x^3 in this case gives

$$\lim_{x \to \infty} \frac{2x^4 - 14}{3x^3 + 5x} = \lim_{x \to \infty} \frac{2x - \dfrac{14}{x^3}}{3 + \dfrac{5}{x^2}}.$$

Since the numerator becomes arbitrarily large as $x \to \infty$ and the denominator approaches 3, it follows that

$$\lim_{x \to \infty} \frac{2x^4 - 14}{3x^3 + 5x} = \infty.$$

EXAMPLE 2.17

Draw a graph of the function

$$f(x) = \frac{1}{x} \sin x, \quad x \geq \pi.$$

Does it have a horizontal asymptote?

SOLUTION   We draw the graph in Figure 2.19 by making the oscillations of $\sin x$ become smaller and smaller as x gets larger and larger. The graph indicates that the positive x-axis is a horizontal asymptote. This is confirmed by the fact that

$$\lim_{x \to \infty} \frac{1}{x} \sin x = 0.$$

We could have reasoned this out as follows. The function $\sin x$ does not have a limit as $x \to \infty$, but as $x \to \infty$ its values oscillate back and forth between ± 1. Since these values are multiplied by $1/x$, which is getting smaller and smaller, the product $(1/x) \sin x$ must be getting closer and closer to 0. We could also use the squeeze theorem to arrive at the same limit.

Notice that the graph actually crosses the asymptote an infinite number of times.

FIGURE 2.19   Limit of $x^{-1} \sin x$ for large x

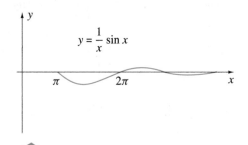

Figures 2.17, 2.18, and 2.19 indicate that the graph of a function $y = f(x)$ can approach an asymptote $y = L$ in three ways: from above (Figure 2.18), from below (Figure 2.17), and oscillating about the asymptote, gradually getting closer and closer to it (Figure 2.19). A computer-generated graph may or may not always make it clear which situation prevails.

EXAMPLE 2.18

Plot a graph of the function $f(x) = (3x - 6)/(x^2 + 5)$. Indicate any horizontal asymptotes and determine how the graph approaches these asymptotes.

FIGURE 2.20   Horizontal asymptote of $(3x - 6)/(x^2 + 5)$

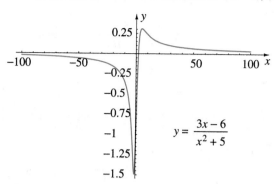

SOLUTION   We begin by plotting a graph of the function on the domain $-100 \le x \le 100$ (Figure 2.20). It suggests that $y = 0$ is a horizontal asymptote, and that the graph approaches $y = 0$ from above when $x \to \infty$ and from below when $x \to -\infty$. We can confirm this algebraically in various ways. To verify the situation as $x \to \infty$, we calculate

$$\lim_{x \to \infty} \frac{3x - 6}{x^2 + 5} = \lim_{x \to \infty} \frac{3 - \dfrac{6}{x}}{x + \dfrac{5}{x}} = 0.$$

This confirms that $y = 0$ is indeed a horizontal asymptote as $x \to \infty$. To decide how the asymptote is approached, we note that for large positive x, both $3 - 6/x$ and $x + 5/x$ are positive, and therefore $f(x)$ must approach 0 through positive numbers. We indicate this by writing

$$\lim_{x \to \infty} \frac{3x - 6}{x^2 + 5} = 0^+.$$

We also could have reasoned as follows. The graph of the function crosses the asymptote $y = 0$ when

$$\frac{3x - 6}{x^2 + 5} = 0,$$

and the only solution of this equation is $x = 2$. Combine this with the graph in Figure 2.20, and we conclude that $f(x) > 0$ for all $x > 2$, and hence the graph must approach $y = 0$ from above as $x \to \infty$.

Similar reasoning shows that the graph approaches $y = 0$ from below as $x \to -\infty$.

In the following example, the function is more complicated. It discusses an alternative way for determining how graphs of rational functions with horizontal asymptotes approach these asymptotes.

EXAMPLE 2.19

Find vertical and horizontal asymptotes for the graph of the function $f(x) = (x^2 - 16)/(x^2 + x - 6)$. Determine how the graph approaches horizontal asymptotes.

SOLUTION   The plot on the interval $-10 \le x \le 10$ in Figure 2.21a suggests vertical asymptotes at $x = 2$ and $x = -3$. These are confirmed with the following limits:

$$\lim_{x \to 2^+} \frac{(x + 4)(x - 4)}{(x + 3)(x - 2)} = -\infty, \qquad \frac{6(-2)}{(5)(0^+)}$$

$$\lim_{x \to 2^-} \frac{(x + 4)(x - 4)}{(x + 3)(x - 2)} = \infty, \qquad \frac{6(-2)}{(5)(0^-)}$$

$$\lim_{x \to -3^+} \frac{(x + 4)(x - 4)}{(x + 3)(x - 2)} = \infty, \qquad \frac{(1)(-7)}{(0^+)(-5)}$$

$$\lim_{x \to -3^-} \frac{(x + 4)(x - 4)}{(x + 3)(x - 2)} = -\infty. \qquad \frac{(1)(-7)}{(0^-)(-5)}$$

The plot in Figure 2.21b suggests that $y = 1$ is a horizontal asymptote as $x \to \infty$. To verify this, we calculate

$$\lim_{x \to \infty} \frac{x^2 - 16}{x^2 + x - 6} = \lim_{x \to \infty} \frac{1 - \dfrac{16}{x^2}}{1 + \dfrac{1}{x} - \dfrac{6}{x^2}} = 1.$$

FIGURE 2.21a   Vertical asymptotes of $(x^2 - 16)/(x^2 + x - 6)$

FIGURE 2.21b   Suggested horizontal asymptote of $(x^2 - 16)/(x^2 + x - 6)$ as $x \to \infty$

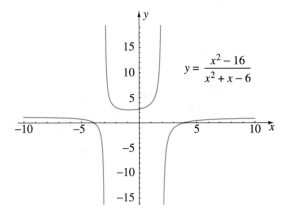

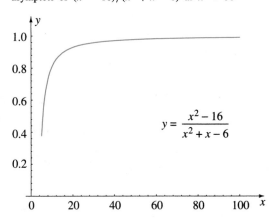

FIGURE 2.21c   Suggested horizontal asymptote of $(x^2 - 16)/(x^2 + x - 6)$ as $x \to -\infty$

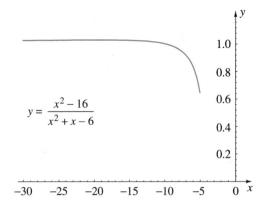

To show whether the graph approaches this horizontal asymptote from above or below, we use long division to express $f(x)$ in the form

$$f(x) = \frac{x^2 - 16}{x^2 + x - 6} = 1 - \frac{x + 10}{x^2 + x - 6}.$$

Because $(x + 10)/(x^2 + x - 6)$ is positive for large x, it follows that $f(x)$ is less than 1 for large x, and the graph approaches $y = 1$ from below as $x \to \infty$.

The plot in Figure 2.21c indicates that $y = 1$ is also a horizontal asymptote as $x \to -\infty$. Since $(x + 10)/(x^2 + x - 6)$ is negative for "large" negative x, it follows that $f(x)$ is greater than 1 for such x, and the graph approaches $y = 1$ from above.

Previous graphs in this section have had at most one horizontal asymptote. The following example has two.

EXAMPLE 2.20

Evaluate $\displaystyle\lim_{x \to \infty} \frac{\sqrt{2x^2 + 4}}{x + 5}$, if it exists. What is the limit as $x \to -\infty$?

SOLUTION   When we divide numerator and denominator by x, and take the x inside the square root as x^2,

$$\lim_{x \to \infty} \frac{\sqrt{2x^2 + 4}}{x + 5} = \lim_{x \to \infty} \frac{\dfrac{1}{x}\sqrt{2x^2 + 4}}{\dfrac{1}{x}(x + 5)} = \lim_{x \to \infty} \frac{\sqrt{\dfrac{2x^2 + 4}{x^2}}}{1 + \dfrac{5}{x}} = \lim_{x \to \infty} \frac{\sqrt{2 + \dfrac{4}{x^2}}}{1 + \dfrac{5}{x}} = \sqrt{2}.$$

In evaluating the limit as $x \to -\infty$, we must be extra careful:

$$\lim_{x \to -\infty} \frac{\sqrt{2x^2 + 4}}{x + 5} = \lim_{x \to -\infty} \frac{\dfrac{\sqrt{2x^2 + 4}}{x}}{1 + \dfrac{5}{x}}.$$

It is not correct in this case to take x inside the square root as

$$\frac{\sqrt{2x^2 + 4}}{x} = \sqrt{\frac{2x^2 + 4}{x^2}}$$

since for negative x, the expression on the left is negative and that on the right is positive. In this case, we should replace x by $-\sqrt{x^2}$, and write

$$\frac{\sqrt{2x^2 + 4}}{x} = \frac{\sqrt{2x^2 + 4}}{-\sqrt{x^2}} = -\sqrt{\frac{2x^2 + 4}{x^2}}.$$

Hence,

$$\lim_{x \to -\infty} \frac{\sqrt{2x^2 + 4}}{x + 5} = \lim_{x \to -\infty} \frac{-\sqrt{\dfrac{2x^2 + 4}{x^2}}}{1 + \dfrac{5}{x}} = \lim_{x \to -\infty} \frac{-\sqrt{2 + \dfrac{4}{x^2}}}{1 + \dfrac{5}{x}} = -\sqrt{2}.$$

The graph of this function (Figure 2.22) confirms these limits; there are two horizontal asymptotes, $y = \sqrt{2}$ and $y = -\sqrt{2}$.

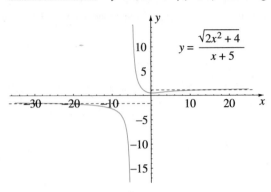

FIGURE 2.22 $y = \sqrt{2x^2 + 4}/(x + 5)$ illustrating two horizontal asymptotes

Graphs of functions have horizontal asymptotes if either, or both, of the conditions $\lim_{x \to \pm\infty} f(x) = L$ hold. What this means is that for large positive or negative values of x, the function is approximately equal to L. The larger the value of x, the better the approximation. Hyperbolas also have asymptotes, but they are not usually horizontal. For example, asymptotes of the hyperbola $x^2 - y^2/9 = 1$ in Figure 2.23 are $y = \pm 3x$. These are often called **oblique** (or slanted) **asymptotes**.

FIGURE 2.23 Oblique asymptotes of hyperbola $x^2 - \dfrac{y^2}{9} = 1$

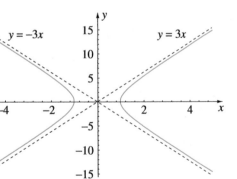

FIGURE 2.24 Vertical and oblique asymptotes of a rational function

$$y = \frac{2x^3 + 4x^2 - x + 1}{2 - x - x^2}$$

Graphs of functions can also have oblique asymptotes. In Figure 1.35, $y = x - 3$ is an oblique asymptote for the graph of $f(x) = (x^3 - 3x^2 + 1)/(x^2 + 1)$. We can confirm this algebraically if we divide $x^2 + 1$ into $x^3 - 3x^2 + 1$. The result is

$$f(x) = \frac{x^3 - 3x^2 + 1}{x^2 + 1} = x - 3 + \frac{4 - x}{x^2 + 1}.$$

As $x \to \pm\infty$, the term $(4 - x)/(x^2 + 1) \to 0$, and therefore $f(x) \to x - 3$. For large values of x, the function $f(x)$ can be approximated more and more closely by $x - 3$. This representation for $f(x)$ can be used to show that the depiction of the asymptote in Figure 1.35 is incorrect. For large positive x, the term $(4 - x)/(x^2 + 1) < 0$, so that $f(x) < x - 3$ and the graph should approach the asymptote from below, not above.

Rational functions $P(x)/Q(x)$ always have oblique asymptotes when the degree of polynomial $P(x)$ is exactly one more than the degree of polynomial $Q(x)$. A second example is the function $f(x) = (2x^3 + 4x^2 - x + 1)/(2 - x - x^2)$ in Figure 2.24. It has an oblique asymptote $y = -2x - 2$ identified from

$$f(x) = \frac{2x^3 + 4x^2 - x + 1}{2 - x - x^2} = -2x - 2 + \frac{x + 5}{2 - x - x^2}.$$

It also has two vertical asymptotes, one at $x = -2$ and the other at $x = 1$.

Functions other than rational functions can also have oblique asymptotes. The graph of $f(x) = x + 2 + 5e^{-x}$ in Figure 2.25 is asymptotic to $y = x + 2$. The straightness of the graph of $f(x) = \sqrt{(2x^4 + 4x^2 - x + 1)/(x^2 + x - 2)}$ for large positive x in Figure 2.26 suggests an asymptote. It is $\sqrt{2}(x - 1/2)$, but to show this algebraically is difficult.

FIGURE 2.25   Oblique asymptote of $x + 2 + 5e^{-x}$

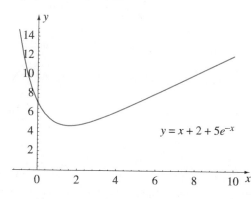

FIGURE 2.26   Oblique asymptote of a root function

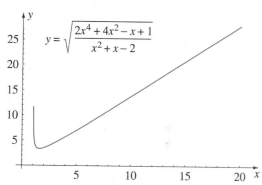

EXERCISES 2.3

In Exercises 1–38 evaluate the limit, if it exists.

1. $\lim\limits_{x \to \infty} \dfrac{x + 1}{2x - 1}$

2. $\lim\limits_{x \to \infty} \dfrac{1 - x}{3 + 2x}$

3. $\lim\limits_{x \to \infty} \dfrac{x^2 + 1}{2x^3 + 5}$

4. $\lim\limits_{x \to \infty} \dfrac{1 - 4x^3}{3 + 2x - x^2}$

5. $\lim\limits_{x \to -\infty} \dfrac{2 + x - x^2}{3 + 4x^2}$

6. $\lim\limits_{x \to -\infty} \dfrac{x^3 - 2x^2}{3x^3 + 4x^2}$

7. $\lim\limits_{x \to -\infty} \dfrac{x^3 - 2x^2 + x + 1}{x^4 + 3x}$

8. $\lim\limits_{x \to -\infty} \dfrac{x^3 - 2x^2 + x + 1}{x^2 - x + 1}$

9. $\lim\limits_{x \to \infty} \dfrac{\sqrt{x^2 + 1}}{2x + 1}$

* 10. $\lim\limits_{x \to \infty} \dfrac{3x - 1}{\sqrt{5 + 4x^2}}$

11. $\lim\limits_{x \to -\infty} \dfrac{\sqrt{1 - 2x^2}}{x + 2}$

* 12. $\lim\limits_{x \to -\infty} \dfrac{\sqrt{1 - 2x}}{x + 2}$

13. $\lim\limits_{x \to \infty} \sqrt{\dfrac{2 + x}{x - 2}}$

* 14. $\lim\limits_{x \to \infty} \dfrac{\sqrt{3 + x}}{\sqrt{x}}$

15. $\lim\limits_{x \to \infty} (x^2 - x^3)$

* 16. $\lim\limits_{x \to \infty} \left(x + \dfrac{1}{x} \right)$

17. $\lim\limits_{x \to \infty} \dfrac{x}{\sqrt{x + 5}}$

* 18. $\lim\limits_{x \to -\infty} \dfrac{x^2}{\sqrt{3 - x}}$

19. $\lim\limits_{x \to -\infty} \dfrac{x}{\sqrt[3]{4 + x^3}}$

* 20. $\lim\limits_{x \to \infty} \dfrac{3x}{\sqrt[3]{2 + 4x^3}}$

* 21. $\lim\limits_{x \to \infty} \dfrac{1}{2x} \cos x$

* 22. $\lim\limits_{x \to -\infty} \dfrac{1}{2x} \cos x$

* 23. $\lim\limits_{x \to \infty} \dfrac{\sin 4x}{x^2}$

* 24. $\lim\limits_{x \to \infty} \dfrac{\sin^2 x}{x}$

* 25. $\lim\limits_{x \to -\infty} \tan x$

* 26. $\lim\limits_{x \to \infty} \dfrac{1}{x} \tan x$

* 27. $\lim\limits_{x \to \infty} \left(\sqrt{x^2 + 1} - x \right)$

* 28. $\lim\limits_{x \to \infty} \left(\sqrt{x^2 + 4} - x \right)$

* 29. $\lim\limits_{x \to \infty} \left(\sqrt{2x^2 + 1} - x \right)$

* 30. $\lim\limits_{x \to -\infty} \left(\sqrt{2x^2 + 1} - x \right)$

* 31. $\lim\limits_{x \to \infty} \dfrac{\sqrt{3x^2 + 2}}{x + 4}$

* 32. $\lim\limits_{x \to \infty} \dfrac{\sqrt{4x^2 + 7}}{2x + 3}$

* 33. $\lim\limits_{x \to -\infty} \dfrac{\sqrt{3x^2 + 2}}{x + 4}$

* 34. $\lim\limits_{x \to -\infty} \dfrac{\sqrt{4x^2 + 7}}{2x + 3}$

* 35. $\lim\limits_{x \to \infty} \left(\sqrt{x^2 + 4} - \sqrt{x^2 - 1} \right)$

* 36. $\lim\limits_{x \to \infty} \left(\sqrt[3]{1 + x} - \sqrt[3]{x} \right)$

* 37. $\lim\limits_{x \to \infty} \left(\sqrt{x^2 + x} - x \right)$

* 38. $\lim\limits_{x \to -\infty} \left(\sqrt{x^2 + x} - x \right)$

In Exercises 39–42 assume that $a > 0$ is a constant and calculate the limit, if it exists.

* **39.** $\displaystyle \lim_{x \to \infty} \frac{x^2 + ax - 2}{ax^2 + 5}$

* **40.** $\displaystyle \lim_{x \to \infty} \frac{x}{\sqrt{ax^2 + 3x + 2}}$

* **41.** $\displaystyle \lim_{x \to \infty} \left(\sqrt{x^2 + ax} - x \right)$

* **42.** $\displaystyle \lim_{x \to -\infty} \frac{\sqrt{ax^2 + 7}}{x - 3a}$

In Exercises 43–48 identify all horizontal, vertical, and oblique asymptotes for the graph of the function. Determine whether the graph approaches horizontal and oblique asymptotes from above or below.

* **43.** $\displaystyle f(x) = \frac{2 - x}{3 + 4x}$

* **44.** $\displaystyle f(x) = \frac{x + 3}{2x - 5}$

* **45.** $\displaystyle f(x) = \frac{3x - 1}{\sqrt{5 + 2x^2}}$

* **46.** $\displaystyle f(x) = \frac{\sqrt{5x^2 + 7}}{2x + 3}$

* **47.** $\displaystyle f(x) = \frac{1 - 4x^3}{3 + 2x - x^2}$

* **48.** $\displaystyle f(x) = \frac{3x^3 + 2x - 1}{1 - 3x + x^2}$

* **49.** What is the value of $\displaystyle \lim_{x \to -\infty} \frac{\sqrt{ax^2 + bx + c}}{dx + e}$, where $a > 0$, and b, c, d, and e are constants?

** **50.** What conditions on the constants a, b, c, d, e, and f will ensure that

$$\lim_{x \to \infty} \left(\sqrt{ax^2 + bx + c} - \sqrt{dx^2 + ex + f} \right)$$

exists? What is the value of the limit in this case?

2.4 Continuity

We have noticed that sometimes the limit of a function $f(x)$ as $x \to a$ is the same as the value $f(a)$ of the function at $x = a$. This property is described in the following definition. This is the situation at $x = 1$ in Figure 2.1a, but not the case in Figure 2.1b or 2.1c.

DEFINITION 2.1

A function $f(x)$ is said to be **continuous** at $x = a$ if it satisfies three conditions:

1. $f(x)$ is defined at $x = a$;
2. $\lim_{x \to a} f(x)$ exists;
3. The value of the function in 1 and its limit in 2 are the same.

All three conditions can be combined by writing the single equation

$$\lim_{x \to a} f(x) = f(a). \tag{2.4a}$$

Equivalent to this, and sometimes more useful, is the equation

$$\lim_{h \to 0} f(a + h) = f(a). \tag{2.4b}$$

(See Exercise 46.)

The graph in Figure 2.27 illustrates a function that is discontinuous (i.e., not continuous) at $x = a, b, c, d$, and e. For instance, at $x = a$, conditions 1 and 2 are satisfied but condition 3 is violated; at $x = b$, condition 2 is satisfied but conditions 1 and 3 are not. Figure 2.27 suggests that discontinuities of a function are characterized geometrically by separations in its graph. This is indeed true, and it is often a very informative way to illustrate the nature of a discontinuity.

FIGURE 2.27 Various types of discontinuities for a function

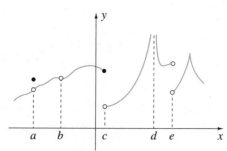

When a function is defined on a closed interval $a \le x \le b$, Definition 2.1 must be rephrased in terms of right- and left-hand limits for continuity at $x = a$ and $x = b$. Specifically, we say that $f(x)$ is **continuous from the right** at $x = a$ if $\lim_{x \to a^+} f(x) = f(a)$, and that $f(x)$ is **continuous from the left** at $x = b$ if $\lim_{x \to b^-} f(x) = f(b)$.

A function is said to be **continuous on an interval** if it is continuous at each point of that interval. In the event that the interval is closed, $a \le x \le b$, continuity at $x = a$ and $x = b$ is interpreted as continuity from the right and left, respectively. Geometrically, a function is continuous on an interval if a pencil can trace its graph completely without being lifted from the page.

EXAMPLE 2.21

Draw graphs of the following functions, indicating any discontinuities:

$$(a) \quad f(x) = \frac{x^2 - 16}{x - 4} \qquad (b) \quad f(x) = \frac{1}{(x - 4)^2} \qquad (c) \quad f(x) = \frac{|x^2 - 25|}{x^2 - 25}$$

FIGURE 2.28a A function with a *hole*

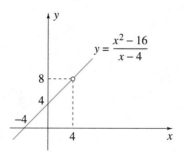

FIGURE 2.28b Computer plots do not show *holes*

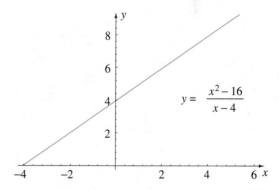

SOLUTION

(a) Since $f(x)$ is undefined at $x = 4$, the function is discontinuous there. For $x \ne 4$,

$$f(x) = \frac{(x + 4)(x - 4)}{x - 4} = x + 4.$$

Consequently, the graph of $f(x)$ is a straight line with a hole at $x = 4$ (Figure 2.28a). The computer-generated graph in Figure 2.28b does not display the discontinuity at $x = 4$; depending on the plot interval, the graph may, or may not, be accompanied by an error message about division by 0 at $x = 4$. Note that $\lim_{x \to 4} [(x^2 - 16)/(x - 4)]$ exists and is equal to 8.

FIGURE 2.29   A function may
be unbounded near a discontinuity

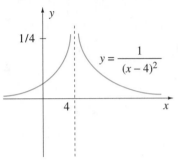

(b) Since $1/(x-4)^2$ is undefined at $x = 4$, the function is discontinuous there. The limits

$$\lim_{x \to 4^+} \frac{1}{(x-4)^2} = \infty, \quad \lim_{x \to 4^-} \frac{1}{(x-4)^2} = \infty, \quad \lim_{x \to \infty} \frac{1}{(x-4)^2} = 0, \quad \lim_{x \to -\infty} \frac{1}{(x-4)^2} = 0$$

are displayed in Figure 2.29.

(c) The function $f(x) = |x^2 - 25|/(x^2 - 25)$ is undefined at $x = \pm 5$, and is therefore discontinuous at these values. When $-5 < x < 5$,

$$f(x) = \frac{25 - x^2}{x^2 - 25} = -1,$$

and when $|x| > 5$,

$$f(x) = \frac{x^2 - 25}{x^2 - 25} = 1.$$

The graph is shown in Figure 2.30a; Figure 2.30b is a computer version. Neither of the following limits exists:

$$\lim_{x \to 5} \frac{|x^2 - 25|}{x^2 - 25} \qquad \text{or} \qquad \lim_{x \to -5} \frac{|x^2 - 25|}{x^2 - 25},$$

although right- and left-hand limits exist as $x \to 5$ and $x \to -5$.

FIGURE 2.30a   A function with different left-hand and
right-hand limits

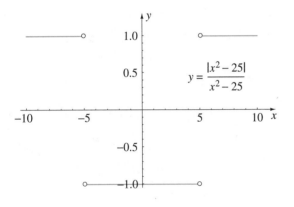

FIGURE 2.30b   Computer plots do not show that a function
may be undefined

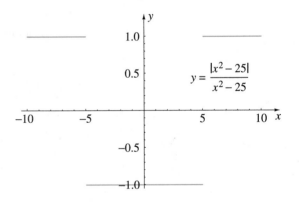

We now mention some general properties of continuous functions. According to the following theorem, when continuous functions are added, subtracted, multiplied, and divided, the result is a continuous function.

THEOREM 2.3

If functions $f(x)$ and $g(x)$ are continuous at $x = a$, then so also are the functions $f(x) \pm g(x)$, $f(x)g(x)$, and $f(x)/g(x)$ [provided that $g(a) \neq 0$ in the case of division].

This is easily established with Theorem 2.1. For instance, to verify that $f(x) + g(x)$ is continuous, we note that since $\lim_{x \to a} f(x) = f(a)$ and $\lim_{x \to a} g(x) = g(a)$,

$$\lim_{x \to a} [f(x) + g(x)] = \lim_{x \to a} f(x) + \lim_{x \to a} g(x) \qquad \text{(by Theorem 2.1)}$$

$$= f(a) + g(a) \qquad \text{[by continuity of } f(x) \text{ and } g(x)\text{].}$$

COROLLARY 2.3.1

If functions $f(x)$ and $g(x)$ are continuous on an interval, then so are the functions $f(x) \pm g(x)$, $f(x)g(x)$, and $f(x)/g(x)$ [provided that $g(x)$ never vanishes in the interval in the case of division].

It is an immediate consequence of this corollary that polynomials are continuous for all real numbers. Rational functions $P(x)/Q(x)$, where P and Q are polynomials, are continuous on intervals in which $Q(x) \neq 0$. The trigonometric functions $\sin x$ and $\cos x$ are continuous for all x; $\sec x$ and $\tan x$ have discontinuities at $x = (2n+1)\pi/2$, where n is an integer; and $\csc x$ and $\cot x$ have discontinuities at $x = n\pi$ (see Figures 1.90). The inverse trigonometric functions are continuous wherever they are defined.

From our graphs of exponential and logarithm functions in Section 1.9, we might also be led to conclude that these functions are continuous. But recall that there is a problem defining exponential functions for irrational exponents. We know what 10^x means when x is a rational number, say, $x = n/m$. It means the mth root of 10^n when n is an integer, or it means the nth power of the mth root $10^{1/m}$. But what does $10^{\sqrt{2}}$ mean? We define $10^{\sqrt{2}}$ as $\lim_{x \to \sqrt{2}} 10^x$, where x approaches $\sqrt{2}$, but does so only through rational numbers. For instance, from the infinite decimal expansion for $\sqrt{2} = 1.414\,213\,562\ldots$, we could create the following sequence of rational numbers that approach $\sqrt{2}$ ever more closely as we include more digits after the decimal:

$$1.4, \quad 1.41, \quad 1.414, \quad 1.414\,2, \quad 1.414\,21, \quad 1.414\,213, \quad 1.414\,213\,5, \quad \ldots.$$

We define $10^{\sqrt{2}}$ to be that number approached by the following sequence of numbers:

$$10^{1.4}, \quad 10^{1.41}, \quad 10^{1.414}, \quad 10^{1.414\,2}, \quad 10^{1.414\,21}, \quad 10^{1.414\,213}, \quad 10^{1.414\,213\,5}, \quad \ldots.$$

Notice that each term in this sequence is 10 raised to a rational number. It can be shown that these numbers do indeed approach a limit, and that this limit is independent of the particular sequence of rationals used to approach $\sqrt{2}$. With this value for $10^{\sqrt{2}}$, the function 10^x is continuous at $x = \sqrt{2}$. In this way, exponential functions a^x can be made continuous for all reals. According to Theorem 2.3, the hyperbolic functions are all continuous except for the hyperbolic cotangent and cosecant, which are discontinuous when their arguments are zero. The following theorem implies that logarithm functions $\log_a x$, which are inverses of exponential functions, are continuous for $x > 0$.

THEOREM 2.4

When a function $f(x)$ is continuous on an interval I, and has an inverse function on I, then $f^{-1}(x)$ is continuous on the range of $f(x)$.

Discontinuities are often characterized according to "size." A discontinuity of a function $f(x)$ at $x = a$ is said to be a **removable discontinuity** if $\lim_{x \to a} f(x)$ exists, and either $f(a)$ is undefined or $f(a)$ is defined and not equal to the limit. The former is illustrated at $x = b$ in Figure 2.27 and the latter at $x = a$ in the same figure. A removable discontinuity can be *removed* from a function by defining or redefining the function at the discontinuity as its limiting value as the discontinuity is approached. For example in Figure 2.28a of Example 2.21, we would define the value of $f(x)$ at the removable discontinuity $x = 4$ as $f(4) = \lim_{x \to 4} f(x) = 8$. This new function, which differs from $f(x)$ only at $x = 4$, is now continuous for all x.

A function $f(x)$ is said to have a (**finite**) **jump discontinuity** at $x = a$ if right-hand and left-hand limits exist as x approaches a, but these limits are different,

$$\lim_{x \to a^+} f(x) - \lim_{x \to a^-} f(x) = \text{finite, nonzero number.}$$

Such is the case at the discontinuities in Figures 2.12 and 2.30a. Jump discontinuities are not removable; the function cannot be defined or redefined only at the discontinuity to create a continuous function.

A function $f(x)$ is said to have an **infinite discontinuity** at $x = a$ if $\lim_{x \to a^+} f(x) = \pm \infty$ or $\lim_{x \to a^-} f(x) = \pm \infty$ or both. Examples can be found in Figures 2.14, 2.15, and 2.16.

Not all discontinuities can be classified according to size. The function $f(x) = \sin(1/x)$ in Figure 2.9 is discontinuous at $x = 0$ due to violent oscillations. This discontinuity is not removable, nor is it a jump discontinuity or an infinite discontinuity.

EXAMPLE 2.22

The floor function (also called the greatest integer function) $f(x) = \lfloor x \rfloor$ of Exercise 68 in Section 1.5 has many applications. Characterize its discontinuities.

SOLUTION   The graph of the function in Figure 2.31 indicates that it has jump discontinuities at integer values of x.

FIGURE 2.31   The floor function

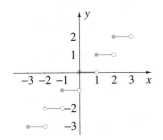

Besides adding, subtracting, multiplying, and dividing functions, functions can also be *composed*, or substituted one into another. For instance, when $f(x) = \sqrt{x^2 + x}$ and $g(x) = e^x$, the function obtained by replacing each x in $f(x)$ by $g(x)$ is called the **composition** of f and g,

$$f\big(g(x)\big) = \sqrt{e^{2x} + e^x}.$$

Composition is often denoted by

$$(f \circ g)(x) = f\big(g(x)\big).$$

Were we required to find the limit of this function as $x \to 1$, say, we would probably write nonchalantly that

$$\lim_{x \to 1} \sqrt{e^{2x} + e^x} = \sqrt{e^2 + e}.$$

Effectively, we have interchanged the operation of taking square roots and that of taking limits; that is, without thinking we have written

$$\lim_{x \to 1} \sqrt{e^{2x} + e^x} = \sqrt{\lim_{x \to 1} \left(e^{2x} + e^x\right)} = \sqrt{e^2 + e}.$$

This is correct according to the following theorem, because the square root function is continuous.

THEOREM 2.5

If $\lim_{x \to a} g(x) = L$, and $f(y)$ is a function that is continuous at $y = L$, then

$$\lim_{x \to a} f\big(g(x)\big) = f\left(\lim_{x \to a} g(x)\right) = f(L). \tag{2.5}$$

An immediate consequence of this result is that the composition of a continuous function with a continuous function yields a continuous function. We state this as a corollary.

COROLLARY 2.5.1

If $g(x)$ is continuous at $x = a$, and $f(y)$ is continuous at $g(a)$, then $f\big(g(x)\big)$ is continuous at $x = a$.

EXAMPLE 2.23

Evaluate $\displaystyle\lim_{x\to 2}\sin\left(\frac{x^2 - 2}{2x - 3}\right)$.

SOLUTION   Since the sine function is continuous for all values of its argument, we may take the limit operation inside the function and write

$$\lim_{x\to 2}\sin\left(\frac{x^2 - 2}{2x - 3}\right) = \sin\left(\lim_{x\to 2}\frac{x^2 - 2}{2x - 3}\right) = \sin 2.$$

We have used computer plots in this section to illustrate various kinds of limits. We wanted you to *see* limits graphically and get a visual *feeling* for them. In the next two examples we illustrate how limits can be used to draw graphs of rational functions in the absence of technology.

EXAMPLE 2.24

Use limits to draw a graph of the rational function $f(x) = \dfrac{x^2 - x + 12}{x^2 + 4x - 5}$.

SOLUTION   Because $x^2 + 4x - 5 = (x - 1)(x + 5)$, the function is undefined at $x = 1$ and $x = -5$. In order to discover the nature of the graph near these discontinuities, we calculate left- and right-hand limits for each value of x,

$$\lim_{x\to -5^-}\frac{x^2 - x + 12}{(x - 1)(x + 5)} = \infty \quad \left[\frac{42}{(-6)(0^-)}\right], \qquad \lim_{x\to -5^+}\frac{x^2 - x + 12}{(x - 1)(x + 5)} = -\infty \quad \left[\frac{42}{(-6)(0^+)}\right],$$

$$\lim_{x\to 1^-}\frac{x^2 - x + 12}{(x - 1)(x + 5)} = -\infty \quad \left[\frac{12}{(0^-)(6)}\right], \qquad \lim_{x\to 1^+}\frac{x^2 - x + 12}{(x - 1)(x + 5)} = \infty \quad \left[\frac{12}{(0^+)(6)}\right].$$

We use long division and limits as $x \to \pm\infty$ to find horizontal asymptotes,

$$\lim_{x\to -\infty}\frac{x^2 - x + 12}{(x - 1)(x + 5)} = \lim_{x\to -\infty}\left(1 + \frac{17 - 5x}{x^2 + 4x - 5}\right) = 1^+,$$

$$\lim_{x\to \infty}\frac{x^2 - x + 12}{(x - 1)(x + 5)} = \lim_{x\to \infty}\left(1 + \frac{17 - 5x}{x^2 + 4x - 5}\right) = 1^-.$$

These limits are shown in Figure 2.32a. To finish the graph as in Figure 2.32b, we join the parts smoothly, add the y-intercept at $-12/5$, and note that the graph has no x-intercepts. When we have learned how to take derivatives in Chapter 3, we will be able to locate the precise positions of the high point of the graph between the vertical asymptotes and the low point to the right of $x = 1$. Notice that the graph crosses the horizontal asymptote once. We can locate this position by solving

$$\frac{x^2 - x + 12}{x^2 + 4x - 5} = 1.$$

The solution is $x = 17/5$.

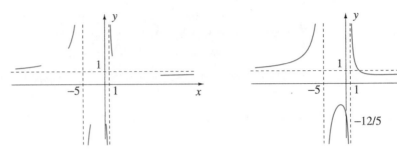

FIGURE 2.32a   FIGURE 2.32b

Using limits to draw the graph of $f(x) = \dfrac{x^2 - x + 12}{x^2 + 4x - 5}$

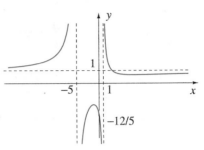

EXAMPLE 2.25

Use limits to draw a graph of the rational function $f(x) = \dfrac{x^4 - 3x^3}{x^3 + 2x^2 + x + 2}$.

SOLUTION   To discover points of discontinuity of the graph we factor the denominator, and for limit calculations, we also factor the numerator,

$$f(x) = \frac{x^3(x - 3)}{(x + 2)(x^2 + 1)}.$$

The function is undefined at $x = -2$, and we therefore calculate

$$\lim_{x \to -2^-} \frac{x^3(x - 3)}{(x + 2)(x^2 + 1)} = -\infty \quad \left[\frac{(-8)(-5)}{(0^-)(5)}\right]$$

and

$$\lim_{x \to -2^+} \frac{x^3(x - 3) \cdot}{(x + 2)(x^2 + 1)} = \infty \quad \left[\frac{(-8)(-5)}{(0^+)(5)}\right]$$

Since the degree of the numerator is one more than that of the denominator, we have a slanted asymptote that can be obtained by long division,

$$f(x) = x - 5 + \frac{9x^2 + 3x + 10}{x^3 + 2x^2 + x + 2}.$$

Because $(9x^2 + 3x + 10)/(x^3 + 2x^2 + x + 2)$ is positive for large x, the graph approaches the slanted asymptote $y = x - 5$ from above as $x \to \infty$. It approaches the asymptote from below as $x \to -\infty$. These facts are shown in Figure 2.33a along with a y-intercept of zero, which is also an x-intercept, and an additional x-intercept of 3. We join these parts of the graph smoothly as shown in Figure 2.33b. We have flattened the graph at the origin because of the x^3-factor in the numerator (just as the curve $y = x^3$ is flat at the origin).

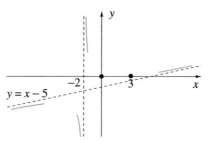

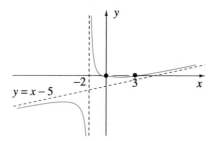

Using limits to draw graph of $f(x) = \dfrac{x^4 - 3x^3}{x^3 + 2x^2 + x + 2}$

The following table indicates when rational functions have horizontal or oblique asymptotes.

TABLE 2.3

Horizontal and Oblique Asymptotes for Rational Functions $\dfrac{P(x)}{Qx}$	
If degree $P <$ degree Q,	graph has horizontal asymptote $y = 0$.
If degree $P =$ degree Q,	graph has horizontal asymptote $y = \dfrac{\text{coefficient of highest power of } P(x)}{\text{coefficient of highest power of } Q(x)}$
If degree $P =$ degree $Q + 1$,	graph has oblique asymptote.
If degree $P >$ degree $Q + 1$,	graph has neither horizontal nor oblique asymptote.

EXERCISES 2.4

In Exercises 1–30 plot a graph of the function indicating any discontinuities. Classify each discontinuity as a removable discontinuity, a jump discontinuity, or an infinite discontinuity, if possible.

1. $f(x) = \dfrac{1}{x + 2}$

2. $f(x) = \dfrac{16 - x^2}{x + 4}$

3. $f(x) = |x^2 - 5|$

4. $f(x) = \dfrac{12}{x^2 + 2}$

5. $f(x) = \dfrac{12}{x^2 + 2x}$

6. $f(x) = \dfrac{12}{x^2 + 2x + 2}$

7. $f(x) = \dfrac{3 + 2x - x^2}{x + 1}$

8. $f(x) = \dfrac{x^3 + x^2 - 2x}{x^2 - x}$

9. $f(x) = \dfrac{x^3 - 2x^2 + 5x - 10}{x - 2}$

10. $f(x) = \tan x$

11. $f(x) = \sec 2x$

12. $f(x) = \sin(1/x)$

13. $f(x) = x^2 \sin(1/x)$

14. $f(x) = \dfrac{x + 12}{x^2 - 9}, \quad -5 \le x \le 4$

15. $f(x) = \dfrac{x^2 + 2x}{x^2 - 9}, \quad -3 \le x \le 6$

16. $f(x) = \dfrac{x^3 - 27}{|x - 3|}$

17. $f(x) = \dfrac{x}{x^2 - 1}$

18. $f(x) = \dfrac{3x + 2}{x^2 - x - 2}$

19. $f(x) = \dfrac{x^2 - 3x + 2}{x^2 + 4x - 5}$

20. $f(x) = \dfrac{3x^2 - 6x}{x^2 - 6x - 7}$

21. $f(x) = \dfrac{x^2 + 3x + 2}{x + 4}$

22. $f(x) = \dfrac{x^2 - 2x + 4}{x - 1}$

23. $f(x) = \dfrac{1}{x^3 - 4x}$

24. $f(x) = \dfrac{2x^3 - 2}{x^3 + 5x^2}$

25. $f(x) = \dfrac{1}{x^4 + 3x^2}$

26. $f(x) = \dfrac{|3x + 1|}{x + 5}$

27. $f(x) = \dfrac{1}{\sqrt{x - 1}}$

28. $f(x) = \dfrac{1}{\sqrt{5 + x}}$

29. $f(x) = \dfrac{1}{\sqrt{5 - x}}$

30. $f(x) = \sqrt{\dfrac{x - 3}{x + 2}}$

In Exercises 31–42 use limits to draw a graph of the function. Use a plot to check the accuracy of your graph.

31. $f(x) = \dfrac{x - 2}{x + 3}$

32. $f(x) = \dfrac{2x - 1}{3 - 4x}$

33. $f(x) = \dfrac{x + 2}{x^2 - 4x - 5}$

34. $f(x) = \dfrac{1 - x}{2x^2 + 5x - 3}$

35. $f(x) = \dfrac{x^2 + x + 2}{x^2 - 6x + 9}$

36. $f(x) = \dfrac{x^2 - 3x - 4}{3x^2 - 14x - 5}$

37. $f(x) = \dfrac{x^2 - x - 2}{3x + 1}$

38. $f(x) = \dfrac{3x^2 - 2x + 5}{1 - x}$

39. $f(x) = \dfrac{x^3}{x^3 - 1}$

40. $f(x) = \dfrac{x^3 - x^2 + 2x - 8}{x^3 - 3x^2 + 3x - 1}$

41. $f(x) = \dfrac{x^2 - 5x + 6}{x^3 - 64}$

42. $f(x) = \dfrac{x^3 - 64}{x^2 - 5x + 6}$

43. Is the function

$$f(x) = \begin{cases} x \sin(1/x), & x \neq 0 \\ 0, & x = 0 \end{cases}$$

continuous at $x = 0$?

44. If $x = f(t)$ represents the displacement function of a particle moving along the x-axis, can $f(t)$ have discontinuities? Explain with graphs.

45. Illustrate graphically that the function $f(x) = x^{-1} \sin x$ has a removable discontinuity at $x = 0$.

46. Verify that the condition in equation 2.4b is equivalent to that in equation 2.4a.

47. (a) The *signum function* (or *sign function*), denoted by sgn, is defined by

$$\text{sgn } x = \begin{cases} -1, & x < 0 \\ 0, & x = 0 \\ 1, & x > 0. \end{cases}$$

Draw its graph, indicating any discontinuities.

(b) Draw a graph of the function $f(x) = \text{sgn}(x + 1) - \text{sgn}(x - 1)$, indicating any discontinuities.

48. (a) Draw a graph of the function $f(x) = \lfloor 10x \rfloor / 10$. Where is the function discontinuous?

(b) Prove that $f(x)$ truncates positive numbers after the first decimal.

49. What function truncates negative numbers after two decimals?

50. (a) Draw a graph of the function $f(x) = \lfloor x + 1/2 \rfloor$. Where is the function discontinuous?

(b) Prove that $f(x)$ rounds positive numbers to the nearest integer.

51. What function rounds positive numbers to:

(a) the nearest tenth;

(b) the nearest hundredth;

(c) 10^{-n}, where n is a positive integer?

52. Determine points of continuity, if any, for the function

$$f(x) = \begin{cases} 1, & x \text{ is a rational number,} \\ 0, & x \text{ is an irrational number.} \end{cases}$$

53. Determine points of continuity, if any, for the function

$$f(x) = \begin{cases} x, & x \text{ is a rational number,} \\ 0, & x \text{ is an irrational number.} \end{cases}$$

2.5 Heaviside and Dirac-Delta Functions

One of the simplest, but at the same time most useful, functions in engineering and physics is the **Heaviside unit step function**. It is defined as follows with the graph in Figure 2.34:

$$h(x) = \begin{cases} 0, & x < 0 \\ 1, & x > 0. \end{cases} \tag{2.6}$$

We have drawn small circles at $x = 0$ to indicate that the function does not have a value there; it has a jump discontinuity. Closely related to this function is that in Figure 2.35; the jump from value 0 to value 1 takes place at $x = a$ rather than at $x = 0$. Since this simply shifts the graph in Figure 2.34 by a units to the right, it is customary to denote this function by $h(x - a)$.

Algebraically, we have

$$h(x - a) = \begin{cases} 0, & x < a \\ 1, & x > a. \end{cases} \tag{2.7}$$

For example, consider the Heaviside function $h(x - 4)$. To evaluate this function at $x = 6$, say, we have two choices. Because $6 > 4$, equation 2.7 implies that the value of the function at $x = 6$ is 1. Alternatively, we can substitute $x = 6$ into $h(x - 4)$ to get $h(6 - 4) = h(2)$, and this is equal to 1 by equation 2.6.

FIGURE 2.34 Heaviside unit step function $h(x)$

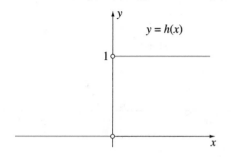

FIGURE 2.35 Heaviside unit step function $h(x - a)$

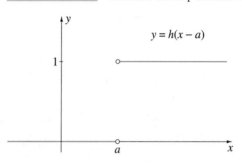

The product of $h(x - a)$ with any other function $f(x)$ results in a function $f(x)h(x - a)$ whose values are equal to those of $f(x)$ for $x > a$, but whose values are 0 for $x < a$. For example, if $f(x) = x^2 + 2$ (Figure 2.36a), a graph of $f(x)h(x + 1) = (x^2 + 2)h(x + 1)$ is as shown in Figure 2.36b.

Think of $h(x - a)$ as a switch that turns values of a function $f(x)$ on for $x > a$. For the function $f(x)$ in Figure 2.37a, $f(x)h(x - a)$ is shown in Figure 2.37b. Values are 0 for $x < a$ and those of $f(x)$ for $x > a$.

FIGURE 2.36a The parabola $x^2 + 2$

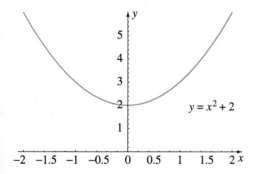

FIGURE 2.36b The parabola $x^2 + 2$ multiplied by $h(x + 1)$

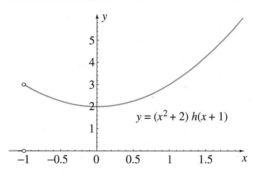

FIGURE 2.37a Any function $f(x)$

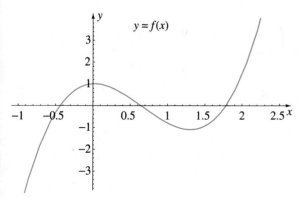

FIGURE 2.37b Function $f(x)$ multiplied by $h(x - a)$

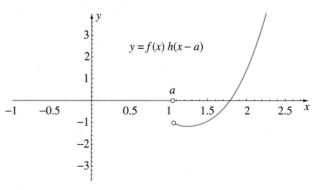

The function $h(x-a) - h(x-b)$ for $b > a$ is shown in Figure 2.38. Algebraically, we have

$$h(x-a) - h(x-b) = \begin{cases} 0, & x < a \\ 1, & a < x < b \\ 0, & x > b. \end{cases} \tag{2.8}$$

When multiplying the function $f(x)$ in Figure 2.37a, it yields the function $f(x)[h(x-a) - h(x-b)]$ with graph in Figure 2.39. Think of $h(x-a) - h(x-b)$ as an on–off switch. It turns values of other functions on for $x > a$ and off again for $x > b$.

FIGURE 2.38   Graph of $h(x-a) - h(x-b)$

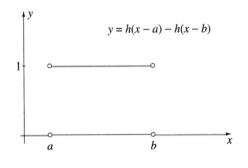

FIGURE 2.39   Graph of $f(x)$ multiplied by $h(x-a)-h(x-b)$

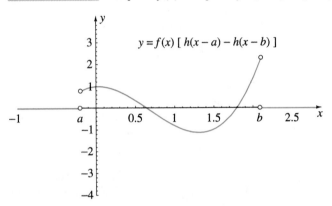

FIGURE 2.40   A function with different definitions on various intervals

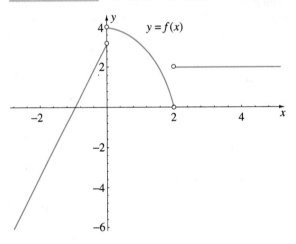

Heaviside functions provide a convenient representation for functions that have different definitions on different intervals. Such functions are said to be **piecewise defined**. For instance, the function in Figure 2.40 is defined as follows:

$$f(x) = \begin{cases} 3x + 3, & x < 0 \\ 4 - x^2, & 0 < x < 2 \\ 2, & x > 2. \end{cases}$$

It can be represented algebraically in the form

$$f(x) = (3x + 3)[1 - h(x)] + (4 - x^2)[h(x) - h(x-2)] + 2h(x-2)$$

$$= 3x + 3 + (-3x - 3 + 4 - x^2)h(x) + (x^2 - 4 + 2)h(x-2)$$

$$= 3x + 3 + (1 - 3x - x^2)h(x) + (x^2 - 2)h(x-2).$$

There are many physical examples in which Heaviside functions are very useful. Consider a mass m attached to a spring as shown in Figure 2.41. Motion is initiated at time $t = 0$ by pulling the mass away from its equilibrium position and releasing it. During the subsequent motion, various forces could act on the mass, including the spring, friction with the surface on which it slides, air resistance, and others. Suppose that among the others, a force to the left with magnitude 5 N is applied for 3 s beginning at time $t = 1$ s. This force can be represented algebraically as

$$F(t) = -5[h(t-1) - h(t-4)].$$

The generator in Figure 2.42, were it operational from time $t = 0$, would produce an oscillating voltage $A \sin \omega t$, where A and ω are positive constants. If the generator is indeed turned on at time $t = 0$, by closing the switch, and off again after 10 s by opening the switch, the voltage applied to the circuit can be expressed in the form

$$E(t) = A \sin \omega t \, [h(t) - h(t-10)].$$

The beam in Figure 2.43 is made more rigid by attaching a second beam over the middle half. This extra support creates additional loading on the original beam. If the support has mass per unit length m, then its weight per unit length is $-9.81m$, and the extra loading per unit length can be expressed algebraically as

$$F(x) = -9.81m[h(x - L/4) - h(x - 3L/4)].$$

FIGURE 2.41   Schematic for vibrating mass-spring system

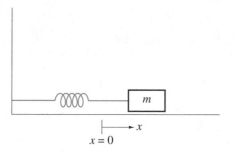

FIGURE 2.42   Schematic for LCR circuit

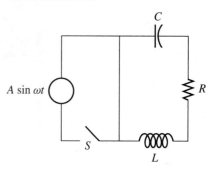

FIGURE 2.43   Schematic for loaded beam

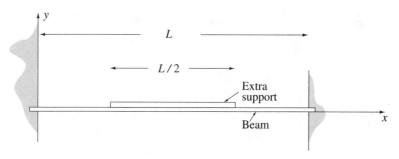

Throughout the text, we will be hired as consulting engineers to tackle projects. Information will be given to us that may not always be sufficient to finish our work. We may have to make justifiable assumptions. Here is Project 1.

Consulting Project 1

The national space association is building a two-stage rocket wherein the rocket burns fuel during the first stage, jettisons the thrusters, burns fuel in the second stage, achieves orbit, and turns the remaining engines off. The initial mass of the rocket is M_0 kilograms, of which M_1 kilograms is the first stage. This consists of the mass of the thrusters and fuel. Thrusters burn fuel at a constant rate of r_1 kilograms per second for t_1 seconds. After the t_1 seconds, the thrusters are jettisoned, and the remaining engines ignite and burn fuel at a constant rate of r_2 kilograms per second for the next t_2 minutes. At this time the rocket achieves orbit and engines are shut down. For enormous numbers of calculations, the association wants a single formula for the mass of the rocket at all times during take off and after reaching orbit. Our task is to find one.

SOLUTION   We quickly realize that there will be different formulas for the mass of the rocket during the various stages. Heaviside functions are ideal for representing piecewise defined functions in a single formula. The mass of the rocket during lift-off is

$$M_0 - r_1 t, \qquad 0 < t < t_1.$$

After t_1 seconds, the thrusters are jettisoned, and the remaining engines ignite and burn fuel at a constant rate of r_2 kilograms per second for the next $60t_2$ seconds. The mass of the rocket during this time interval is

$$M_0 - M_1 - r_2(t - t_1), \qquad t_1 < t < t_1 + 60t_2.$$

Finally, when the rocket achieves orbit and the engines are shut down, the mass of the rocket is

$$M_0 - M_1 - 60t_2 r_2, \qquad t > t_1 + 60t_2.$$

We now use Heaviside functions to bring these expressions into a single formula for $M(t)$, the mass of the rocket at time t:

$$
\begin{aligned}
M(t) = {} & (M_0 - r_1 t)[h(t) - h(t - t_1)] \\
& + [M_0 - M_1 - r_2(t - t_1)][h(t - t_1) - h(t - t_1 - 60t_2)] \\
& + (M_0 - M_1 - 60t_2 r_2)h(t - t_1 - 60t_2).
\end{aligned}
$$

By recombining this expression into terms involving $h(t)$, $h(t - t_1)$, and $h(t - t_1 - 60t_2)$, we can also write

$$
\begin{aligned}
M(t) = {} & (M_0 - r_1 t)h(t) - [M_1 - r_2 t_1 + (r_2 - r_1)t]h(t - t_1) \\
& + r_2(t - t_1 - 60t_2)h(t - t_1 - 60t_2).
\end{aligned}
$$

A graph of this function is shown in Figure 2.44. It has a jump discontinuity at $t = t_1$ when thrusters are jettisoned, and a removable discontinuity at $t = t_1 + 60t_2$ when engines are turned off. We could remove the discontinuity at $t = t_1 + 60t_2$ by defining $M(t_1 + 60t_2) = M_0 - M_1 - 60t_2 r_2$.

FIGURE 2.44   Function representing mass of multi-stage rocket

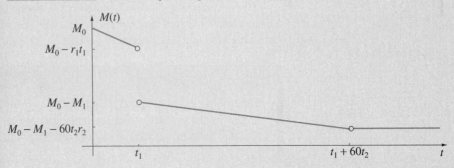

**Application Preview
Revisited**

In the Application Preview we introduced the problem of describing an instantaneously applied voltage to an electrical network. The same situation would arise were we to consider the mathematical representation of striking the mass in Figure 2.41 with a hammer. We would also find the same situation if a *point-load* were to replace the extra support on the beam in Figure 2.43, or be in addition to the extra support.

To find a mathematical representation for the instantaneously applied voltage, the force due to an impinging hammer, and a point-load on a beam, we begin with the function $(b - a)^{-1}[h(x - a) - h(x - b)]$ in Figure 2.45a. What is important to notice here is that the area of the rectangle formed by drawing vertical lines at $x = a$ and $x = b$ is 1. If b is replaced by $a + \epsilon$ so that ϵ is the width of the nonzero portion of the graph, then the area of the rectangle is also 1 if the height of the function is $1/\epsilon$ (Figure 2.45b). The function describing this graph is

$$\frac{1}{\epsilon}[h(x - a) - h(x - a - \epsilon)].$$

This is often called a **unit pulse function**.

FIGURE 2.45a   Unit pulse function $(b - a)^{-1}[h(x - a) - h(x - b)]$ FIGURE 2.45b   Unit pulse function $\frac{1}{\epsilon}[h(x - a) - h(x - a - \epsilon)]$

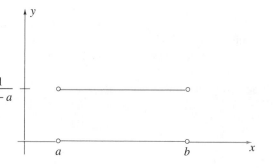
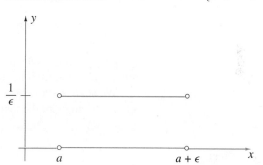

Suppose we make ϵ in Figure 2.45b smaller and smaller, so that $a + \epsilon \to a$ and the height of the horizontal line moves upward (in order that the area under the curve will always be 1). A few smaller values of ϵ are shown in Figure 2.46. If we take the limit as $\epsilon \to 0$, we obtain what is called the **Dirac-delta function**. It is denoted by

$$\delta(x - a) = \lim_{\epsilon \to 0} \frac{1}{\epsilon}[h(x - a) - h(x - a - \epsilon)]. \tag{2.9}$$

FIGURE 2.46   Unit pulse functions leading to Dirac-delta function

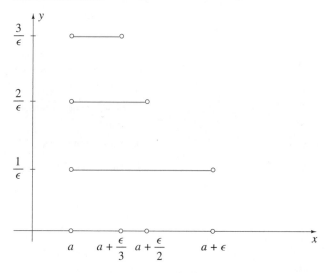

Its value is zero for every $x \neq a$, and somehow its value at $x = a$ is "equal to infinity." Clearly, this is not a function as we understand functions, since functions cannot have infinite values. It is called a *generalized function*. Generalized functions have operational properties that make them very useful in handling point sources in engineering and physics. We will introduce these properties at appropriate places in the text.

Point sources arise in most areas of engineering and physics. For instance, suppose we were to strike the left side of the vibrating mass in Figure 2.41 with a hammer at time $t = t_0$. This would be equivalent to applying a force over a very short time interval. If it were deemed that striking the mass is equivalent not to a 10 N force for 1 s, nor a 20 N force for 1/2 s, nor a 40 N force for 1/4 s, but to the limit of this sequence of forces, we would say that the mass has been struck with a force of 10 N at $t = t_0$, and express the force in the form

$$F(t) = 10\,\delta(t - t_0).$$

If we were to take a sharp object, place it at the centre of the beam in Figure 2.43, and push downward with a force of 200 N, we would represent this force as

$$F(x) = -200\,\delta(x - L/2).$$

It is a point force of 200 N at $x = L/2$.

A unit voltage at time $t = t_0$ for the circuit in Figure 2.42 is $\delta(t - t_0)$. It is equivalent to applying a voltage not equivalent to 1 for 1 s, nor 2 V for 1/2 s, nor 4 V for 1/4 s, but a voltage equivalent to the limit of such voltages.

We shall learn how to manipulate Dirac-delta functions in these applications as calculus unfolds throughout the text.

EXERCISES 2.5

In Exercises 1–6 express the piecewise defined function in terms of Heaviside functions. Draw a graph of each function.

* **1.** $f(x) = \begin{cases} 1, & x < 0 \\ 2 - x, & 0 < x < 2 \\ 2, & x > 2 \end{cases}$

* **2.** $f(x) = \begin{cases} 1 - x^2, & x < 0 \\ x^2, & x > 0 \end{cases}$

* **3.** $f(x) = \begin{cases} \sin x, & x < 0 \\ 2 + 2\cos x, & x > 0 \end{cases}$

* **4.** $f(x) = \begin{cases} 0, & x < -1 \\ x - 2, & -1 < x < 3 \\ x - 4, & 3 < x < 5 \\ 0, & x > 5 \end{cases}$

* **5.** $f(x) = \begin{cases} 0, & x < 0 \\ x, & 0 < x < 1 \\ 1 - x, & 1 < x < 2 \\ 0, & x > 2 \end{cases}$

* **6.** $f(x) = \begin{cases} 0, & x < 0 \\ \sin x, & 0 < x < \pi \\ 2\sin(x - \pi), & \pi < x < 2\pi \\ 3\sin(x - 2\pi), & 2\pi < x < 3\pi \\ 4\sin(x - 3\pi), & 3\pi < x < 4\pi \\ 0, & x > 4\pi \end{cases}$

In Exercises 7–11 the force described acts on the mass on the end of the spring in Figure 2.41. Represent the force in terms of Heaviside functions or Dirac-delta functions, as appropriate.

7. A force of F N to the left for the first T seconds after $t = 0$

8. A periodic force $F(t) = 10\sin 4t$ that is turned on for two periods beginning at time $t = 1$

9. An instantaneous force of 50 N to the left at time $t = 4$

* **10.** A force of 100 N that begins at time $t = 10$ and increases linearly in size to 200 N over the next 50 seconds

* **11.** Every 10 seconds beginning at $t = 0$ s and ending at $t = 60$ s, the mass is stuck with a hammer with a force of 60 N to the right

In Exercises 12–15 the load described is applied to the beam in Figure 2.43 without the extra support. Represent the force in terms of Heaviside functions or Dirac-delta functions, as appropriate.

12. An extra support of length $L/2$ and mass m over the left half of the beam

13. A downward force of $F > 0$ N concentrated at $x = L/3$

14. An upward force of $F_1 > 0$ N concentrated at $x = x_1$ and a downward force of $F_2 > 0$ N concentrated at $x = x_2$

15. A total mass m evenly distributed over the last two-thirds of the beam

16. The function $h(x-a) - h(x-b)$ is an on-off switch. What function would be an on-off-on switch?

17. How would you represent a switch that turns on at time $t = 0$, off at $t = 1$, on again at $t = 2$, off again at $t = 3$, and so on?

18. Is $h(x-a)h(x-b) = h(x-b)$, when $b > a$?

In Exercises 19–24 draw a graph of the function.

19. $f(x) = h(2x - 1)$ * **20.** $f(x) = h(3 - 2x)$

21. $f(x) = h(x^2 - 1)$ * **22.** $f(x) = 5h(4 - x^2)$

23. $f(x) = x^2 h(x^2 - 2x - 3)$

24. $f(x) = (5 - x) h(2 - x^3)$

* **25.** Express the *step function* shown below in terms of Heaviside functions.

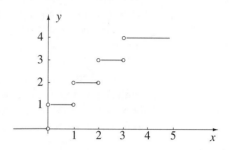

2.6 A Mathematical Definition of Limits

Our work on limits has been intuitive, but some later topics require a precise definition of limits. To obtain such a definition, we begin with our intuitive statement of a limit and make a succession of paraphrases, each of which is one step closer to the definition. We do this because the definition of a limit is at first sight quite overwhelming, and we wish to show that it can be obtained by a fairly straightforward sequence of steps. We hope that this will be a convincing argument that the definition of a limit does indeed describe in mathematical terms our intuitive concept of a limit. Our intuitive statement is:

- A function $f(x)$ has limit L as x approaches a if $f(x)$ can be made arbitrarily close to L by choosing x sufficiently close to a.

  Next we paraphrase "$f(x)$ can be made arbitrarily close to L."

- A function $f(x)$ has limit L as x approaches a if the difference $|f(x) - L|$ can be made arbitrarily close to zero by choosing x sufficiently close to a.

  Now we take the important step — make "$|f(x) - L|$ can be made arbitrarily close to zero" mathematical.

- A function $f(x)$ has limit L as x approaches a if given any real number $\epsilon > 0$, no matter how small, we can make the difference $|f(x) - L|$ less than ϵ by choosing x sufficiently close to a.

  Penultimately, we paraphrase "by choosing x sufficiently close to a."

- A function $f(x)$ has limit L as x approaches a if given any $\epsilon > 0$, we can make $|f(x) - L| < \epsilon$ by choosing $|x - a|$ sufficiently close to zero.

Finally, by making "choosing $|x-a|$ sufficiently close to zero" precise, we arrive at the definition of a limit.

> **DEFINITION 2.2**
>
> A function $f(x)$ has **limit** L as x approaches a if given any $\epsilon > 0$, we can find a $\delta > 0$ such that
> $$|f(x) - L| < \epsilon$$
> whenever $0 < |x - a| < \delta$.

Notice that by requiring $0 < |x - a|$, this definition states explicitly that as far as limits are concerned, the value of $f(x)$ at $x = a$ is irrelevant. In taking limits we consider values of x closer and closer to a, but we do not consider the value of $f(x)$ at $x = a$. This definition states in precise terms our intuitive idea of a limit: that $f(x)$ can be made arbitrarily close to L

by choosing x sufficiently close to a. Perhaps you will get a better *feeling* for this definition if we interpret it graphically. Figure 2.47a indicates a function that has limit L as x approaches a. Let us illustrate what must be done to verify algebraically that $\lim_{x \to a} f(x) = L$. We suppose that we are given a value $\epsilon > 0$, which we should envisage as being very small, although we are never told exactly what it is. We must show that x can be restricted to $0 < |x - a| < \delta$ so that $|f(x) - L| < \epsilon$. The latter inequality is equivalent to $-\epsilon < f(x) - L < \epsilon$ or $L - \epsilon < f(x) < L + \epsilon$, and this describes a horizontal band of width 2ϵ around the line $y = L$ (shaded in Figure 2.47b). What Definition 2.2 requires is that we find an interval of width 2δ around $x = a$, as $|x - a| < \delta$ is equivalent to $a - \delta < x < a + \delta$, such that whenever x is in this interval, the values of $f(x)$ are all within the shaded horizontal band around $y = L$. Such an interval is shown in Figure 2.47c for the given ϵ. Now Definition 2.2 requires us to verify that the δ-interval can always be found no matter how small ϵ is chosen to be. This is always possible for the function illustrated in Figure 2.47, and it is clear that the smaller the given value of ϵ, the smaller δ will have to be chosen. For instance, for the value of ϵ in Figure 2.47d, δ is smaller than that in Figure 2.47c. In other words, the value of δ depends on the value of ϵ. Herein lies the difficulty in using Definition 2.2. In order to ensure that δ can be found for any given value of ϵ, we usually determine precisely how δ depends functionally on ϵ. We illustrate with two examples.

FIGURE 2.47a   Graphs to illustrate the mathematical definition of limit   **FIGURE 2.47b**

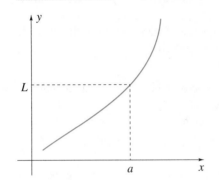

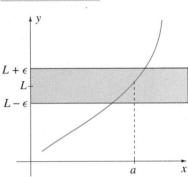

FIGURE 2.47c   **FIGURE 2.47d**

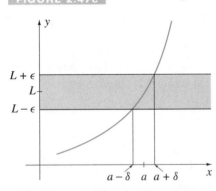

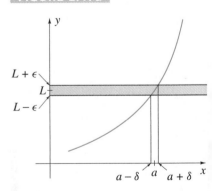

EXAMPLE 2.26

Use Definition 2.2 to prove that $\lim_{x \to 3} (2x + 4) = 10$.

SOLUTION   It is true that based on Theorem 2.1, this is quite obvious, but we are required to use Definition 2.2. We must show that given any $\epsilon > 0$, we can choose x sufficiently close to 3 that

$$|(2x + 4) - 10| < \epsilon.$$

To do so, we rewrite the left side of the inequality with all x's in the combination $x - 3$,

$$|(2x + 4) - 10| = |2x - 6| = 2|x - 3|.$$

We must now choose x so that

$$2|x - 3| < \epsilon.$$

But this will be true if $|x - 3| < \epsilon/2$. In other words, if we choose x to satisfy $0 < |x - 3| < \epsilon/2$, then

$$|(2x + 4) - 10| = 2|x - 3| < 2\left(\frac{\epsilon}{2}\right) = \epsilon.$$

The verification is now complete. We have shown that we can make $2x + 4$ as close to 10 as we want (within ϵ) by choosing x sufficiently close to 3 (within $\delta = \epsilon/2$).

The following example is more complicated, but a manageable one.

EXAMPLE 2.27

Use Definition 2.2 to prove that $\lim_{x \to 2} (x^2 + 5) = 9$.

SOLUTION   We must show that given any $\epsilon > 0$, we can choose x sufficiently close to 2 that

$$|(x^2 + 5) - 9| < \epsilon \qquad \text{or} \qquad |x^2 - 4| < \epsilon.$$

Once again we rewrite the left side of the inequality with all x's in the combination $x - 2$:

$$|x^2 - 4| = |(x - 2)^2 + 4x - 8| = |(x - 2)^2 + 4(x - 2)|.$$

We must now choose x so that

$$|(x - 2)^2 + 4(x - 2)| < \epsilon.$$

Now, all real numbers a and b satisfy the inequality

$$|a + b| \le |a| + |b|. \tag{2.10}$$

With a replaced by $(x - 2)^2$ and b replaced by $4(x - 2)$, we can say that

$$|(x - 2)^2 + 4(x - 2)| \le |x - 2|^2 + 4|x - 2|.$$

As a result, we consider finding x so that

$$|x - 2|^2 + 4|x - 2| < \epsilon.$$

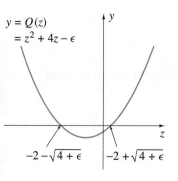

FIGURE 2.48   The parabola $z^2 + 4z - \epsilon$

$y = Q(z)$
$= z^2 + 4z - \epsilon$

$-2 - \sqrt{4 + \epsilon}$ $-2 + \sqrt{4 + \epsilon}$

If we set $z = |x - 2|$ and consider the parabola $Q(z) = z^2 + 4z - \epsilon$ in Figure 2.48, we will be able to *see* what to do. The parabola crosses the z-axis when

$$z^2 + 4z - \epsilon = 0,$$

a quadratic with solutions

$$z = \frac{-4 \pm \sqrt{16 + 4\epsilon}}{2} = -2 \pm \sqrt{4 + \epsilon}.$$

The graph shows that whenever $0 < z < -2 + \sqrt{4 + \epsilon}$, then

$$z^2 + 4z - \epsilon < 0.$$

Since $z = |x - 2|$, we can say that if $0 < |x - 2| < \sqrt{4 + \epsilon} - 2$, then

$$|x - 2|^2 + 4|x - 2| < \epsilon,$$

and therefore

$$|(x - 2)^2 + 4(x - 2)| < \epsilon.$$

Verification is now complete. We have shown that we can make $x^2 + 5$ as close to 9 as we want (within ϵ) by choosing x sufficiently close to 2 (within $\delta = \sqrt{4 + \epsilon} - 2$).

These examples have indicated the difference between "evaluation of" and "verification of" a limit. Evaluation is calculational; it comes first. Verification is much more difficult; it requires a clear understanding of Definition 2.2. Results in succeeding chapters rely heavily on our intuitive understanding of limits and the ability to calculate limits, but use of Definition 2.2 is kept to a minimum.

EXERCISES 2.6

In Exercises 1–9 use Definition 2.2 to verify the limit.

* **1.** $\lim\limits_{x \to 1} (x + 5) = 6$

* **2.** $\lim\limits_{x \to 2} (2x - 3) = 1$

* **3.** $\lim\limits_{x \to 0} (x^2 + 3) = 3$

* **4.** $\lim\limits_{x \to 1} (x^2 + 4) = 5$

* **5.** $\lim\limits_{x \to -2} (3 - x^2) = -1$

* **6.** $\lim\limits_{x \to 3} (x^2 - 7x) = -12$

* **7.** $\lim\limits_{x \to -1} (x^2 - 3x + 4) = 8$

* **8.** $\lim\limits_{x \to 1} (x^2 + 3x + 5) = 9$

* **9.** $\lim\limits_{x \to 2} \dfrac{x + 2}{x - 1} = 4$

In Exercises 10–19 give a mathematical definition for each statement.

* **10.** $\lim\limits_{x \to a^+} f(x) = L$

* **11.** $\lim\limits_{x \to a^-} f(x) = L$

* **12.** $\lim\limits_{x \to \infty} f(x) = L$

* **13.** $\lim\limits_{x \to -\infty} f(x) = L$

* **14.** $\lim\limits_{x \to a} f(x) = \infty$

* **15.** $\lim\limits_{x \to a} f(x) = -\infty$

* **16.** $\lim\limits_{x \to \infty} f(x) = \infty$

* **17.** $\lim\limits_{x \to \infty} f(x) = -\infty$

* **18.** $\lim\limits_{x \to -\infty} f(x) = \infty$

* **19.** $\lim\limits_{x \to -\infty} f(x) = -\infty$

* **20.** Use Definition 2.2 to prove that a function $f(x)$ cannot have two limits as x approaches a.

In Exercises 21–28 use the appropriate definition from Exercises 10–19 to verify the limit.

* **21.** $\lim\limits_{x \to 1} \dfrac{1}{(x - 1)^2} = \infty$

* **22.** $\lim\limits_{x \to -2} \dfrac{-1}{(x + 2)^2} = -\infty$

* **23.** $\lim\limits_{x \to \infty} (x + 5) = \infty$

* **24.** $\lim\limits_{x \to \infty} (5 - x^2) = -\infty$

* **25.** $\lim\limits_{x \to \infty} \dfrac{x + 2}{x - 1} = 1$

* **26.** $\lim\limits_{x \to -\infty} \dfrac{x + 2}{x - 1} = 1$

* **27.** $\lim\limits_{x \to -\infty} (5 - x) = \infty$

* **28.** $\lim\limits_{x \to -\infty} (3 + x - x^2) = -\infty$

* **29.** Prove that if $\lim_{x \to a} f(x) = L > 0$, then there exists an open interval I containing a in which $f(x) > 0$ except possibly at $x = a$.

* **30.** Does the function

$$f(x) = \begin{cases} x \sin(1/x), & x \neq 1/(n\pi) \\ 1, & x = 1/(n\pi), \end{cases}$$

where n is an integer, have a limit as $x \to 0$?

In Exercises 31–35 we use Definition 2.2 to prove Theorem 2.1. In each exercise assume that $\lim_{x \to a} f(x) = F$ and $\lim_{x \to a} g(x) = G$.

* **31.** Show that given any $\epsilon > 0$, there exist numbers $\delta_1 > 0$ and $\delta_2 > 0$ such that

$$|f(x) - F| < \epsilon/2 \quad \text{whenever } 0 < |x - a| < \delta_1 \quad \text{and}$$

$$|g(x) - G| < \epsilon/2 \quad \text{whenever } 0 < |x - a| < \delta_2.$$

Use these results along with identity 2.10 to prove part (i) of Theorem 2.1.

* **32.** Use a proof similar to that in Exercise 31 to verify part (ii) of Theorem 2.1.

* **33.** Verify part (iii) of Theorem 2.1.

34.   (a) Verify that

$$|f(x)g(x) - FG| \le |f(x)||g(x) - G| + |G||f(x) - F|.$$

(b) Show that given any $\epsilon > 0$, there exist numbers $\delta_1 > 0$, $\delta_2 > 0$, and $\delta_3 > 0$ such that

$$|f(x)| < |F| + 1 \qquad \text{whenever } 0 < |x - a| < \delta_1,$$

$$|g(x) - G| < \frac{\epsilon}{2(|F| + 1)} \qquad \text{whenever } 0 < |x - a| < \delta_2,$$

and

$$|f(x) - F| < \frac{\epsilon}{2|G| + 1} \qquad \text{whenever } 0 < |x - a| < \delta_3.$$

(c) Use these results to prove part (iv) of Theorem 2.1.

** **35.**   (a) Verify that when $G \ne 0$,

$$\left| \frac{f(x)}{g(x)} - \frac{F}{G} \right| \le \frac{|f(x) - F|}{|g(x)|} + \frac{|F||G - g(x)|}{|G||g(x)|}.$$

(b) Show that given any $\epsilon > 0$, there exist numbers $\delta_1 > 0$, $\delta_2 > 0$, and $\delta_3 > 0$ such that

$$|g(x)| > \frac{|G|}{2} \qquad \text{whenever } 0 < |x - a| < \delta_1,$$

$$|f(x) - F| < \frac{\epsilon|G|}{4} \qquad \text{whenever } 0 < |x - a| < \delta_2, \qquad \text{and}$$

$$|g(x) - G| < \frac{\epsilon|G|^2}{4(|F| + 1)} \qquad \text{whenever } 0 < |x - a| < \delta_3.$$

(c) Now prove part (v) of Theorem 2.1.

SUMMARY

In Section 2.1 we introduced limits of functions. For the most part our discussion was intuitive, beginning with the statement "$\lim_{x \to a} f(x) = L$ if $f(x)$ can be made arbitrarily close to L by choosing x sufficiently close to a." This idea was then extended to include the following:

Right-hand limits: $\lim\limits_{x \to a^+} f(x)$;

Left-hand limits: $\lim\limits_{x \to a^-} f(x)$;

Limits at infinity: $\lim\limits_{x \to \infty} f(x), \qquad \lim\limits_{x \to -\infty} f(x)$;

Infinite limits: $\lim\limits_{x \to a} f(x) = \infty, \qquad \lim\limits_{x \to a} f(x) = -\infty.$

Keep in mind that the term *infinite limits* is somewhat of a misnomer since in both situations the limit does not exist.

A function $f(x)$ is continuous at a point $x = a$ if

$$\lim_{x \to a} f(x) = f(a).$$

To be continuous at $x = a$, a function must be defined at $x = a$ and have a limit as $x \to a$, and these numbers must be the same. The function is continuous on an interval if it is continuous at each point of that interval. Geometrically, this means that one must be able to trace its graph completely without lifting pencil from page. Most discontinuities can be characterized as removable, finite, or infinite.

We illustrated that discontinuous functions such as the Heaviside and Dirac-delta functions model many physical situations.

In Section 2.6 we developed the mathematical definition of a limit.

KEY TERMS

In reviewing this chapter, you should be able to define or discuss the following key terms:

Limit   Squeeze or sandwich theorem
Left-hand limit   Right-hand limit
Infinity   Vertical asymptote

Limits at infinity   Horizontal asymptote

Oblique asymptotes   Continuous function

Continuity from the right   Continuity from the left

Continuity on an interval   Removable discontinuity

(Finite) jump discontinuity   Infinite discontinuity

Composition   Heaviside unit step function

Piecewise defined function   Unit pulse function

Dirac-delta function

REVIEW EXERCISES

In Exercises 1–20 evaluate the limit, if it exists.

1. $\displaystyle\lim_{x \to 1} \frac{x^2 - 2x}{x + 5}$

2. $\displaystyle\lim_{x \to -1} \frac{x^2 - 1}{x + 1}$

3. $\displaystyle\lim_{x \to -2} \frac{x^2 + 4x + 4}{x + 3}$

4. $\displaystyle\lim_{x \to \infty} \frac{x + 5}{x - 3}$

5. $\displaystyle\lim_{x \to -\infty} \frac{x^2 + 3x + 2}{2x^2 - 5}$

6. $\displaystyle\lim_{x \to -\infty} \frac{5 - x^3}{3 + 4x^3}$

7. $\displaystyle\lim_{x \to \infty} \frac{3x^3 + 2x - 5}{x^2 + 5x}$

8. $\displaystyle\lim_{x \to \infty} \frac{4 - 3x + x^2}{3 + 5x^3}$

9. $\displaystyle\lim_{x \to 2^+} \frac{x^2 - 2x}{x^2 + 2x}$

10. $\displaystyle\lim_{x \to 2^-} \frac{x^2 - 4x + 4}{x - 2}$

11. $\displaystyle\lim_{x \to 0} \frac{x^2 + 2x}{3x - 2x^2}$

12. $\displaystyle\lim_{x \to 1} \frac{x^2 + 5x}{(x - 1)^2}$

13. $\displaystyle\lim_{x \to 1} \frac{\sqrt{x} - 1}{x}$

14. $\displaystyle\lim_{x \to 1} \frac{\sqrt{x} - 1}{x - 1}$

15. $\displaystyle\lim_{x \to 1/2} \frac{(2 - 4x)^3}{x(2x - 1)^2}$

∗ 16. $\displaystyle\lim_{x \to \infty} \frac{\cos 5x}{x}$

∗ 17. $\displaystyle\lim_{x \to -\infty} x \sin x$

∗ 18. $\displaystyle\lim_{x \to -\infty} \frac{\sqrt{3x^2 + 4}}{2x + 5}$

∗ 19. $\displaystyle\lim_{x \to \infty} \frac{\sqrt{3x^2 + 4}}{2x + 5}$

∗ 20. $\displaystyle\lim_{x \to \infty} \left(\sqrt{2x + 1} - \sqrt{3x - 1}\right)$

In Exercises 21–32 draw a graph of the function, indicating any discontinuities. Determine whether discontinuities are removable, jump, or infinite.

21. $f(x) = \dfrac{1}{x - 2}$

22. $f(x) = \dfrac{x}{x - 2}$

23. $f(x) = \dfrac{x^2}{x - 2}$

24. $f(x) = \dfrac{x^2 - 36}{x - 6}$

25. $f(x) = \dfrac{x + 1}{x - 1}$

26. $f(x) = \left|\dfrac{x + 1}{x - 1}\right|$

∗ 27. $f(x) = \dfrac{|x + 1|}{x - 1}$

∗ 28. $f(x) = \dfrac{x + 1}{|x - 1|}$

∗ 29. $f(x) = \dfrac{2x}{x^2 - 3x - 4}$

∗ 30. $f(x) = \dfrac{2x^2}{x^2 - 3x - 4}$

∗ 31. $f(x) = \dfrac{x^3 - 3x^2 + 3x - 1}{x - 1}$

∗ 32. $f(x) = \dfrac{x^3 - 3x^2 + 3x - 1}{x^2 - 2x + 1}$

In Exercises 33–34 express the piecewise defined function in terms of Heaviside functions.

33. $f(x) = \begin{cases} x^2, & x < 0 \\ x, & 0 < x < 4 \\ 5 - 2x, & x > 4 \end{cases}$

34. $f(x) = \begin{cases} 3 + x^3, & x < -1 \\ x^2 + 2, & -1 < x < 2 \\ 4, & x > 2 \end{cases}$

∗ 35. For what values of x is the function $f(x) = \lfloor x^2 \rfloor$ discontinuous? Draw its graph.

3 | Differentiation

Application Preview

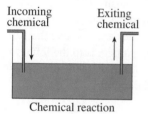

Incoming chemical   Exiting chemical

Chemical reaction

The container in the figure below represents a chemical reactor in which a chemical is either created or broken down. The chemical enters in the form of a solution at one concentration and leaves the reactor at a different concentration.

THE PROBLEM   Given the flow rates at which solution enters and leaves the reactor, and the concentration of chemical entering the reactor, find the concentration at which chemical leaves the reactor. Can you *see* that there is not enough information to solve the problem? Some information must be given, or assumed, about the rate at which chemical is formed or broken down.

Chapters 1 and 2 have prepared the way for calculus. The functions and curves in these chapters yield a wealth of examples for our discussions, and limits from Chapter 2 provide the tool by which calculus is developed. Calculus has two major components, *differentiation* and *integration*. In this chapter we study differentiation. We learn what a *derivative* is algebraically and geometrically, and develop some of its properties; we learn how to differentiate polynomials and rational, trigonometric, exponential, logarithm, and hyperbolic functions; and we see glimpses of the applications that are to follow in Chapter 4.

3.1 The Derivative

Very few quantities in real life remain constant; most are in a state of change. For example, room temperature, the speed of a car, and the angle of elevation of the sun are three commonplace quantities that are constantly changing. A few more technical ones are current in a transmission line, barometric pressure, moisture content of the soil, and stress in vertical members of tall buildings during hurricanes. Rates at which these quantities change are called *derivatives*. To study the concept of a rate of change more thoroughly, we consider two commonplace physical situations.

Displacement and Speed

Calculus plays a central role in discussions that trace the motion of particles and objects moving under the influence of forces. Without calculus, analysis of a simple system such as a mass on the end of a spring would be difficult; planning lift-off, space travel, and re-entry of spacecraft would be impossible. We introduce basic ideas connected with motion here, and with each new development of calculus, we take the analysis a little further.

Suppose a particle is at some position on the x-axis at time $t = 0$, and for $t > 0$, various forces act on it resulting in motion that is confined to the x-axis (Figure 3.1).

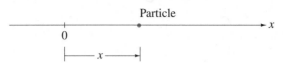

FIGURE 3.1   Displacement of particle moving along the x-axis

The position of the particle can be identified by its x-coordinate, and x can always be regarded as a function of time t. Suppose, for example, that

$$x = f(t) = t^3 - 27t^2 + 168t + 20, \quad t \geq 0,$$

where x is measured in metres and t in seconds. This is called the **displacement function**; it gives a complete history of the particle's motion. It tells us not only where the particle is at any given time, it also contains information about the velocity, speed, and acceleration of the particle. We begin by plotting a graph of the displacement function (Figure 3.2). Ordinates of this curve represent horizontal displacements of the particle relative to $x = 0$. For example, the height of the displacement graph at time $t = 0$ is 20. This means that the particle begins motion 20 m to the right of the origin. At $t = 4$ s, it is 324 m to the right of the origin, and at $t = 14$ s, it is 176 m to the left of the origin. Whenever ordinates are positive, the particle is to the right of the origin. Whenever ordinates are negative, the particle is to the left of $x = 0$. Figure 3.2 indicates that the particle is to the right of $x = 0$ for the first ten seconds and then again after $t \approx 17$ s. Between these times, it is to the left of the origin.

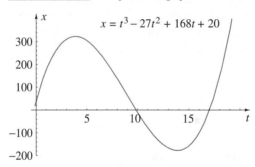

FIGURE 3.2   Displacement graph for motion along the x-axis

For $0 < t < 4$, values of x are getting larger. This means that the particle is moving to the right. For $4 < t < 14$, the particle moves to the left (values of x are getting smaller), and for $t > 14$, it moves to the right once again. This means that the particle stops and reverses direction at $t = 4$ s and at $t = 14$ s.

Everyone has an intuitive idea about speed, when something moves quickly and when it moves slowly. To discuss speed of the above particle quantitatively, we need derivatives, but even now we can derive qualitative information about how fast the particle is moving. The more distance the particle covers in a given time interval, the faster it is moving. In Table 3.1, we have tabulated the distance that the particle travels each second for the first twenty seconds. The first two entries indicate that the particle travels 142 m in the first second, but only 94 m from $t = 1$ s to $t = 2$ s. The particle is therefore travelling faster during the first second of its motion than during the next second. The remainder of the table suggests that the particle slows down until, as we know, it comes to a stop at $t = 4$ s. It then speeds up until somewhere around $t = 9$ s when it starts to slow down again, and comes to a stop at $t = 14$ s. It then picks up speed thereafter.

TABLE 3.1

time (s)	1	2	3	4	5	6	7	8	9	10
distance (m)	142	94	52	16	14	38	56	68	74	74

time (s)	11	12	13	14	15	16	17	18	19	20
distance (m)	68	56	38	32	30	52	94	142	196	256

We can see this graphically as well as in tabular form. The steeper the displacement graph, the larger the speed, the flatter the graph, the smaller the speed. This means that speed is greatest when the graph rises or falls quickly. At the beginning of the particle's motion (for small $t > 0$), the graph is relatively steep, and therefore the particle is moving quickly. Steepness decreases as we head toward $t = 4$ s, and therefore the particle is slowing down. At $t = 4$ s, the particle's speed is zero (it reverses direction). The particle then picks up speed, moving now to the left until steepness is greatest somewhere around $t = 9$ s. It then starts to slow down, coming to a stop at $t = 14$ s. It then moves to the right thereafter picking up speed (as steepness increases). Again, all of this is quite qualitative, but when we learn about derivatives, we can be much more specific and we can add information about the velocity and acceleration of the particle. We begin with velocity.

Figure 3.3 is an enlargement of that part of the graph in Figure 3.2 for the first four seconds. At time $t = 1$ s, the particle is 162 m to the right of the origin. At $t = 4$ s, it is at position $x = 324$ m. During these three seconds, the particle has moved $324 - 162$ or 162 m; its displacement at $t = 4$ s relative to its position at $t = 1$ s is 162 m to the right. When we divide the displacement by time taken to travel it, we obtain $162/3 = 54$. This is called the **average velocity** of the particle during the time interval and has units of metres per second. During the time interval $1 \le t \le 3$, the displacement of the particle is 146 m; therefore, its average velocity in this time interval is $146/2$ or 73 m/s. The average velocity during $1 \le t \le 2$ is 94 m/s. What we are doing is taking average velocities over shorter and shorter time intervals, all beginning at $t = 1$ s. If we continue the process indefinitely, we are in effect taking the limit of average velocities as the length of the time interval starting at $t = 1$ approaches 0. This will be called the *instantaneous velocity* of the particle at $t = 1$ s. Let us introduce notation to describe the limiting process, and do so at an arbitrary time t_0 rather than at $t = 1$.

FIGURE 3.3   Enlargement of graph in Figure 3.2

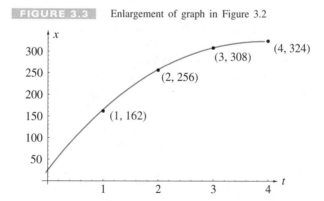

We let Δt represent a small interval of time, often called an **increment** of time. The function value $f(t_0)$ is the position (or displacement) of the particle at time t_0, $f(t_0 + \Delta t)$ is its position at time $t_0 + \Delta t$, and $f(t_0 + \Delta t) - f(t_0)$ is the difference in these displacements. It represents the displacement of the particle at time $t_0 + \Delta t$ relative to its position at time t_0. It may or may not represent the distance travelled by the particle during the time interval Δt. (Can you explain why?) The quotient

$$\frac{f(t_0 + \Delta t) - f(t_0)}{\Delta t}$$

is the average velocity of the particle during the time interval between t_0 and $t_0 + \Delta t$. The limit of this quotient as $\Delta t \to 0$ is called the **instantaneous velocity** of the particle at time t_0; it is denoted by

$$v(t_0) = \lim_{\Delta t \to 0} \frac{f(t_0 + \Delta t) - f(t_0)}{\Delta t}. \tag{3.1}$$

It represents how fast the particle is moving at time t_0; it is the instantaneous rate of change of displacement with respect to time. For example, the instantaneous velocity at $t = 1$ is

$$v(1) = \lim_{\Delta t \to 0} \frac{f(1 + \Delta t) - f(1)}{\Delta t}$$

$$= \lim_{\Delta t \to 0} \frac{(1 + \Delta t)^3 - 27(1 + \Delta t)^2 + 168(1 + \Delta t) + 20 - (1 - 27 + 168 + 20)}{\Delta t}.$$

Simplification of the numerator gives

$$v(1) = \lim_{\Delta t \to 0} \frac{117(\Delta t) - 24(\Delta t)^2 + (\Delta t)^3}{\Delta t},$$

and when we divide numerator and denominator by Δt,

$$v(1) = \lim_{\Delta t \to 0} [117 - 24(\Delta t) + (\Delta t)^2] = 117.$$

The particle is travelling 117 m/s at time $t = 1$ s.

A similar calculation at $t = 5$ s leads to an instantaneous velocity $v(5) = -27$ m/s. The negative sign indicates that the particle is moving to the left. The positive velocity $v(1) = 117$ means that the particle is moving to the right at $t = 1$ s. We shall have much more to say about velocity in Section 3.6 when we relate velocity and speed, and we also introduce acceleration. For now, we simply want you to appreciate that instantaneous velocity is a rate of change; it is the instantaneous rate of change of displacement with respect to time.

Rate of Rainfall

FIGURE 3.4   Cylinder catching rainfall

For our second situation in which to introduce rates of change, consider an open cylindrical container placed outside a house. The depth D (measured in millimetres) of water in the container during a rainstorm (Figure 3.4) is a function of time t (in hours), say $D = f(t)$. A graph of this function might look like that in Figure 3.5. There is no water in the container at time $t = 0$ when the storm begins. As the rain falls, the depth of water increases, and finally, when the rain stops at time $t = \bar{t}$, the depth of water remains at a constant level $\bar{D}$ thereafter.

Intuitively, rain is falling fastest when this curve is steepest, somewhere around the time indicated as $\tilde{t}$. When the curve is flat (just after $t = 0$ and just before $t = \bar{t}$), very little rain is falling. We want to be more specific; we want to be able to say exactly how fast the rain is falling at any

FIGURE 3.5   Depth of rainwater in cylinder as a function of time

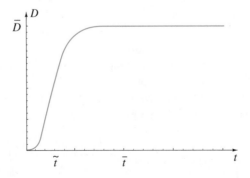

given time t_0. We proceed as we did in defining velocity. If we let Δt be a small increment of time, then the difference $f(t_0 + \Delta t) - f(t_0)$ is the number of millimetres of rain that falls during the time interval Δt after t_0. The quotient

$$\frac{f(t_0 + \Delta t) - f(t_0)}{\Delta t}$$

with units of millimetres per hour is called the *average rate* of rainfall during the time interval. The limit

$$\lim_{\Delta t \to 0} \frac{f(t_0 + \Delta t) - f(t_0)}{\Delta t}$$

is called the *instantaneous rate* of rainfall at time t_0. It is the instantaneous rate of change of depth of water in the cylinder with respect to time, at time t_0. If its value were 5, say, then were rain to fall at this rate over an extended period of time, the depth of water in the container would increase by 5 mm every hour.

The rates of change in these two situations are typical of rates of change that occur in a multitude of applications — applications from such diverse fields as engineering, physics, economics, psychology, and medicine, to name a few. We now reformulate them in a mathematical framework that allows us to introduce them in every area of applied mathematics. Suppose a function $f(x)$ is defined for all x, and $x = a$ and $x = a + h$ are two values of x (a is taking the place of t_0 and h is replacing Δt in the previous situations). The quotient

$$\frac{f(a + h) - f(a)}{h}$$

is called the **average rate of change** of $f(x)$ with respect to x in the interval between a and $a + h$. The limit of the quotient as h approaches zero,

$$\lim_{h \to 0} \frac{f(a + h) - f(a)}{h},$$

is called the **instantaneous rate of change** of $f(x)$ with respect to x at $x = a$. It is also called the derivative of $f(x)$ at $x = a$.

DEFINITION 3.1

The **derivative** of a function $f(x)$ with respect to x at a point $x = a$, denoted by $f'(a)$, is defined as

$$f'(a) = \lim_{h \to 0} \frac{f(a + h) - f(a)}{h}, \tag{3.2a}$$

provided that the limit exists.

The following limit is equivalent to that in equation 3.2a but it avoids the introduction of h:

$$f'(a) = \lim_{x \to a} \frac{f(x) - f(a)}{x - a}. \tag{3.2b}$$

The operation of taking the derivative of a function is called **differentiation**. We say that we differentiate the function when we find its derivative. It would now be appropriate to use Definition 3.1 to calculate derivatives of various functions at various points. Before doing so, however, we feel that it is important to discuss the geometric interpretation of the derivative. It will be prevalent in many applications.

Tangent Lines and the Geometric Interpretation of the Derivative

FIGURE 3.6   Tangent line at one point may intersect curve at another point

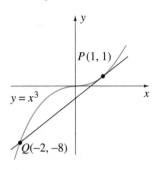

FIGURE 3.7   Tangent line to $y = x^3$ at $(0,0)$ crosses curve

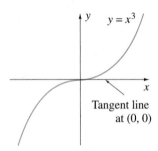

Algebraically, the derivative of a function is its instantaneous rate of change. Geometrically, derivatives are intimately connected to tangent lines to curves, and most students of calculus have an intuitive idea of what it means for a line to be tangent to a curve. Often, it is the idea of *touching*. A line is tangent to a curve if it touches the curve at exactly one point. For curves such as circles, ellipses, and parabolas, this notion is adequate, but in general it is unsatisfactory. For instance, in Figure 3.6 we have drawn what would look like the tangent line to the curve $y = x^3$ at the point $P(1, 1)$. But this line intersects the curve again at the point $Q(-2, -8)$. The tangent line at P does not touch the curve at precisely one point; it intersects the curve at a second point.

You might reply that tangency is a local concept; whether the tangent line at $(1, 1)$ in Figure 3.6 intersects the curve at another point some distance from $(1, 1)$ is irrelevant. True, and perhaps we could remedy the situation by defining the tangent line at $(1, 1)$ as the line that touches the curve at $(1, 1)$ and does not cross it there. Unfortunately, this definition does not always work either. For instance, what is the tangent line to $y = x^3$ at $(0, 0)$ (Figure 3.7)? The only reasonable line is the x-axis, but notice that the x-axis crosses the curve at $(0, 0)$. To the right of $(0, 0)$, the curve is above the tangent line, and to the left of $(0, 0)$, it is below the tangent line. This happens quite often, as we shall see later.

How then are we to define the tangent line to a curve? The idea that a tangent line touches a curve is good, but it needs to be phrased properly. Consider defining what is meant by the tangent line at the point P on the curve $y = f(x)$ in Figure 3.8. When P is joined to another point Q_1 on the curve by a straight line l_1, certainly l_1 is not the tangent line to $y = f(x)$ at P. If we join P to a point Q_2 on $y = f(x)$ closer to P than Q_1, then l_2 is not the tangent line at P either, but it is closer to it than l_1. A point Q_3 even closer to P yields a line l_3 that is even closer to the tangent line than l_2. Repeating this process over and over again leads to a set of lines $l_1, l_2, l_3, \ldots$, which get closer and closer to what we feel is the tangent line to $y = f(x)$ at P. We therefore define the **tangent line** to $y = f(x)$ at P as the limiting position of these lines as points $Q_1, Q_2, Q_3, \ldots$ get arbitrarily close to P.

FIGURE 3.8   Lines approaching tangent line to a curve

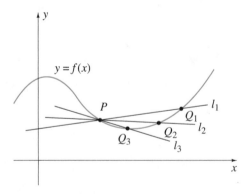

The line in Figure 3.6 satisfies this definition; it is the limiting position of lines joining P to other points on the curve which approach P. It is irrelevant whether this line intersects the curve again at some distance from P. What is the tangent line to the curve $y = x^3$ at the origin $(0, 0)$? According to the above definition, the limiting position of lines joining $(0, 0)$ to other points on the curve is the x-axis; that is, the x-axis is tangent to $y = x^3$ at $(0, 0)$ (Figure 3.7). As mentioned earlier, the tangent line actually crosses from one side of the curve to the other at $(0, 0)$. To the left of $x = 0$, the tangent line is above the curve, whereas to the right of $x = 0$, it is below the curve.

Now that we understand what it means for a line to be tangent to a curve, let us make the connection with derivatives. Suppose $P(a, f(a))$ is a point on the curve $y = f(x)$ in

Figure 3.9. If $Q(a + h, f(a + h))$ is another point on the curve, then the quotient

$$\frac{f(a + h) - f(a)}{h}$$

in equation 3.2a is the slope of the line joining P and Q. As $h \to 0$, point Q moves along the curve toward P, and the line joining P and Q moves toward the tangent line at P. It follows that the limit in equation 3.2a, the derivative $f'(a)$, is the slope of the tangent line to the curve $y = f(x)$ at $x = a$ (Figure 3.10).

FIGURE 3.9   Lines used to find tangent line to a curve

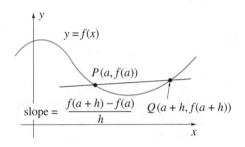

FIGURE 3.10   Tangent line to a curve has slope $f'(a)$

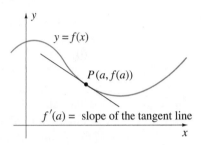

To summarize, algebraically the derivative $f'(a)$ of a function $f(x)$ at $x = a$ is its instantaneous rate of change; geometrically, it is the slope of the tangent line to the graph of $f(x)$ at the point $(a, f(a))$.

Let us now use equation 3.2a to calculate derivatives of some simple functions.

EXAMPLE 3.1

FIGURE 3.11   Tangent line to parabola $y = x^2$ at $(1, 1)$

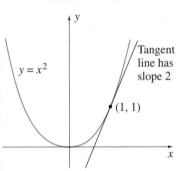

Find $f'(1)$ if $f(x) = x^2$.

SOLUTION   According to equation 3.2a,

$$f'(1) = \lim_{h \to 0} \frac{f(1 + h) - f(1)}{h} = \lim_{h \to 0} \frac{(1 + h)^2 - 1}{h}$$

$$= \lim_{h \to 0} \frac{(1 + 2h + h^2) - 1}{h} = \lim_{h \to 0} \frac{2h + h^2}{h} = \lim_{h \to 0} (2 + h) = 2.$$

Algebraically, the instantaneous rate of change of $f(x) = x^2$ when $x = 1$ is equal to 2. Geometrically, the slope of the tangent line to the curve $y = x^2$ at the point $(1, 1)$ is 2 (Figure 3.11).

EXAMPLE 3.2

FIGURE 3.12   Tangent line may have slope equal to 0

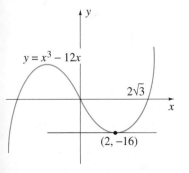

Find $f'(2)$ if $f(x) = x^3 - 12x$.

SOLUTION   Using equation 3.2a,

$$f'(2) = \lim_{h \to 0} \frac{f(2 + h) - f(2)}{h} = \lim_{h \to 0} \frac{[(2 + h)^3 - 12(2 + h)] - (8 - 24)}{h}$$

$$= \lim_{h \to 0} \frac{6h^2 + h^3}{h} = \lim_{h \to 0} (6h + h^2) = 0.$$

This result is substantiated in Figure 3.12, where we see that the tangent line to $y = x^3 - 12x$ is horizontal (has zero slope) at $x = 2$.

In Examples 3.1 and 3.2 we required the derivative for the function at only one value of a, and therefore set a in equation 3.2a equal to this value. An alternative, which is far more advantageous, especially in an example where the derivative is required at a number of points, is to evaluate $f'(a)$ and then set a to its desired value, or values, later. For instance, in Example 3.1 we calculate that

$$f'(a) = \lim_{h \to 0} \frac{f(a+h) - f(a)}{h} = \lim_{h \to 0} \frac{(a+h)^2 - a^2}{h}$$

$$= \lim_{h \to 0} \frac{2ah + h^2}{h} = \lim_{h \to 0} (2a + h) = 2a.$$

This is the derivative of $f(x) = x^2$ at any value $x = a$. For $a = 1$, we obtain $f'(1) = 2(1) = 2$. But it is also easy to calculate $f'(a)$ at other values of a. For example, $f'(0) = 0$, $f'(-1) = -2$, and $f'(4) = 8$.

We now carry this idea to its logical conclusion. The derivative of $f(x)$ at a is denoted by $f'(a)$. But what is a? It is a specific value of x at which to calculate the derivative, and it can be any value of x. Why not simply drop references to a, and talk about the derivative of $f(x)$ at values of x? Following this suggestion, we denote by $f'(x)$ the derivative of the function $f(x)$ at any value of x. With this notation, equation 3.2a is replaced by

$$f'(x) = \lim_{h \to 0} \frac{f(x+h) - f(x)}{h} \tag{3.3}$$

for the derivative of $f(x)$ at x. We call $f'(x)$ the **derivative function**.

When a function is represented by the letter y, as in $y = f(x)$, another common notation for the derivative is $\dfrac{dy}{dx}$. Be careful when using this notation. Do not interpret it as a quotient: dy and dx do not have separate meanings; it is one symbol representing an accumulation of all the operations in equation 3.3. It is not therefore to be read as "dy divided by dx." Typographically, it is easier to print dy/dx rather than $\dfrac{dy}{dx}$, and we will take this liberty whenever it is convenient to do so. But remember, dy/dx for the moment is not a quotient, it is one symbol representing the limit operation in equation 3.3. We will change this in Section 4.12.

Sometimes it is more convenient to use parts of each of these notations and write

$$\frac{d}{dx} f(x).$$

In this form we understand that d/dx means to differentiate with respect to x whatever follows it, in this case $f(x)$. Let us use these new notations in calculating two more derivatives.

EXAMPLE 3.3

Find dy/dx if $y = f(x) = (x-1)/(x+2)$.

SOLUTION   Using equation 3.3,

$$\frac{dy}{dx} = \lim_{h \to 0} \frac{f(x+h) - f(x)}{h} = \lim_{h \to 0} \frac{1}{h} \left(\frac{x+h-1}{x+h+2} - \frac{x-1}{x+2} \right).$$

If we bring the terms in parentheses to a common denominator, the result is

$$\frac{dy}{dx} = \lim_{h \to 0} \frac{1}{h} \left[\frac{(x+h-1)(x+2) - (x+h+2)(x-1)}{(x+h+2)(x+2)} \right].$$

When we simplify the numerator, we find

$$\frac{dy}{dx} = \lim_{h \to 0} \frac{1}{h} \left[\frac{3h}{(x+h+2)(x+2)} \right]$$

$$= \lim_{h \to 0} \frac{3}{(x+h+2)(x+2)} = \frac{3}{(x+2)^2}.$$

EXAMPLE 3.4

Find dv/dt if $v = f(t) = 1/t$.

SOLUTION   In terms of variables v and t, equation 3.3 takes the form

$$\frac{dv}{dt} = \lim_{h \to 0} \frac{f(t+h) - f(t)}{h} = \lim_{h \to 0} \frac{1}{h} \left(\frac{1}{t+h} - \frac{1}{t} \right)$$

$$= \lim_{h \to 0} \frac{t - (t+h)}{t(t+h)h} = \lim_{h \to 0} \frac{-1}{t(t+h)} = -\frac{1}{t^2}.$$

In Section 1.8 we proved that the inclination ϕ of a line l, where $0 \le \phi < \pi$, is related to its slope m by the equation $\tan \phi = m$ (equation 1.59). When l is the tangent line to a curve $y = f(x)$ at point (x_0, y_0) (Figure 3.13), its slope is $f'(x_0)$. Hence, the inclination ϕ of the tangent line to a curve at a point (x_0, y_0) is given by the equation

$$\tan \phi = f'(x_0). \tag{3.4}$$

FIGURE 3.13   Relating slope and inclination of a tangent line to a curve

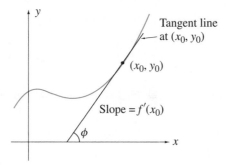

EXAMPLE 3.5

Find the inclination of the tangent line to the curve $y = (x-1)/(x+2)$ at the point $(4, 1/2)$.

SOLUTION   In Example 3.3, we calculated the slope of the tangent line to this curve at any point as

$$\frac{dy}{dx} = \frac{3}{(x+2)^2}.$$

According to equation 3.4, the inclination ϕ of the tangent line at $(4, 1/2)$ is given by

$$\tan \phi = \frac{3}{(4+2)^2} \Rightarrow \phi = 0.083 \text{ radian.}$$

In applications, independent and dependent variables of functions $y = f(x)$ represent physical quantities and have units associated with them. Units for the derivative $f'(x)$ are units of y divided by units of x. We see this from equation 3.3, where units of the quotient $[f(x + h) - f(x)]/h$ are clearly units of y divided by units of x, and the limit of this quotient as $h \to 0$ does not alter these units. For example, if x measures length in metres and y measures mass in kilograms, the units of dy/dx are kilograms per metre.

EXERCISES 3.1

In Exercises 1–10 use equation 3.3 to find $f'(x)$.

1. $f(x) = x + 2$

2. $f(x) = 3x^2 + 5$

3. $f(x) = 1 + 2x - x^2$

4. $f(x) = x^3 + 2x^2$

5. $f(x) = x^4 + 4x - 12$

6. $f(x) = \dfrac{x + 4}{x - 5}$

∗ 7. $f(x) = \dfrac{x^2 + 2}{x + 3}$

∗ 8. $f(x) = x^2(x + 2)$

∗ 9. $f(x) = \dfrac{3x - 2}{4 - x}$

∗ 10. $f(x) = \dfrac{x^2 - x + 1}{x^2 + x + 1}$

In Exercises 11–14 find the specified rate of change.

11. The rate of change of the circumference C of a circle with respect to its radius r

12. The rate of change of the area A of a circle with respect to its radius r

13. The rate of change of the area A of a sphere with respect to its radius r

14. The rate of change of the volume V of a sphere with respect to its radius r

Answer Exercises 15–19 by drawing a graph of the function. Do not calculate the required derivative.

15. What is $f'(x)$ if $f(x) = 2x - 4$?

16. What is $f'(x)$ if $f(x) = mx + b$, where m and b are constants?

17. What is $f'(0)$ if $f(x) = x^2$?

18. What is $f'(1)$ if $f(x) = (x - 1)^2$?

19. What is $f'(0)$ if $f(x) = x^{1/3}$?

In Exercises 20–23 find the equation of the tangent line to the curve at the given point.

20. $y = x^2 + 3$ at $(1, 4)$

21. $y = 3 - 2x - x^2$ at $(4, -21)$

∗ 22. $y = 1/x^2$ at $(2, 1/4)$

∗ 23. $y = (x + 1)/(x + 2)$ at $(0, 1/2)$

In Exercises 24–27 find the inclination of the tangent line to the curve at the given point.

⊞ 24. $y = x^2$ at $(1, 1)$

⊞ 25. $y = x^3 - 6x$ at $(2, -4)$

⊞∗ 26. $y = 1/x^2$ at $(2, 1/4)$

∗ 27. $y = 1/(x + 1)$ at $(0, 1)$

In Exercises 28–32 find $f'(x)$.

∗ 28. $f(x) = x^8$

∗ 29. $f(x) = \sqrt{x + 1}$

∗ 30. $f(x) = \dfrac{1}{(x - 2)^4}$

∗ 31. $f(x) = \dfrac{1}{\sqrt{x - 3}}$

32. $f(x) = x\sqrt{x + 1}$

In Exercises 33–34 find the specified rate of change.

∗ 33. The rate of change of the radius r of a circle with respect to its area A

∗ 34. The rate of change of the volume V of a sphere with respect to its area A

⊞ 35. Find the angle between the tangent lines to the curves $y = x^2$ and $x = y^2$ at their point of intersection in the first quadrant.

∗ 36. Find $f'(x)$ if $f(x) = |x|$.

∗ 37. A sphere of radius R has a uniform charge distribution of ρ coulombs per cubic metre. The electrostatic potential V at a distance r from the centre of the sphere is defined by

$$V = f(r) = \begin{cases} \dfrac{\rho}{6\epsilon_0}(3R^2 - r^2), & 0 \leq r \leq R, \\[2mm] \dfrac{R^3\rho}{3\epsilon_0 r}, & r > R, \end{cases}$$

where ϵ_0 is a constant. Draw a graph of this function. Does $f(r)$ appear to have a derivative at the surface of the sphere? That is, does $f'(R)$ exist? Prove your conjecture using Definition 3.1.

38. Repeat Exercise 37 for the magnitude E of the electrostatic field:

$$E = f(r) = \begin{cases} \dfrac{\rho r}{3\epsilon_0}, & 0 \le r \le R, \\[3mm] \dfrac{\rho R^3}{3\epsilon_0 r^2}, & r > R. \end{cases}$$

** **39.** Find $f'(x)$ if $f(x) = x^{1/3}$.

** **40.** Let $f(x)$ be a function with the property that $f(x + z) = f(x)f(z)$ for all x and z, and be such that $f(0) = f'(0) = 1$. Prove that $f'(x) = f(x)$ for all x.

3.2 Rules for Differentiation

Since calculus plays a key role in many branches of applied science, we need to differentiate many types of functions: polynomial, rational, trigonometric, exponential, and logarithm functions, to name a few. To use equation 3.3 each time would be extremely laborious. Fortunately, however, we can develop a number of rules for taking derivatives that eliminate the necessity of using the definition each time. We state each of these formulas as a theorem.

THEOREM 3.1

If $f(x) = c$, where c is a constant, then $f'(x) = 0$.

PROOF By equation 3.3,

$$f'(x) = \lim_{h \to 0} \frac{f(x + h) - f(x)}{h} = \lim_{h \to 0} \frac{c - c}{h} = 0. \qquad \blacksquare$$

In short, the derivative of a constant function is zero;

$$\frac{d}{dx}(c) = 0. \tag{3.5}$$

THEOREM 3.2

If $f(x) = x$, then $f'(x) = 1$.

PROOF With equation 3.3,

$$f'(x) = \lim_{h \to 0} \frac{f(x + h) - f(x)}{h} = \lim_{h \to 0} \frac{(x + h) - x}{h} = \lim_{h \to 0} \frac{h}{h} = 1. \qquad \blacksquare$$

In short,

$$\frac{d}{dx}(x) = 1. \tag{3.6}$$

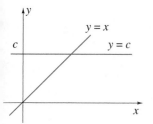

FIGURE 3.14 Derivatives of functions $y = c$ and $y = x$

With the graphs of the functions $f(x) = c$ and $f(x) = x$ in Figure 3.14, we can *see* the results of equations 3.5 and 3.6. The tangent line to $y = c$ always has slope zero, whereas the tangent line to $y = x$ always has slope equal to 1.

THEOREM 3.3

If $f(x) = x^n$, where n is a positive integer, then $f'(x) = nx^{n-1}$.

PROOF Equation 3.3 gives

$$f'(x) = \lim_{h \to 0} \frac{f(x+h) - f(x)}{h} = \lim_{h \to 0} \frac{(x+h)^n - x^n}{h}.$$

If we expand $(x+h)^n$ by means of the binomial theorem (an alternative proof is given in Exercise 35), we have

$$f'(x) = \lim_{h \to 0} \frac{1}{h} \left[x^n + nx^{n-1}h + \frac{n(n-1)}{2}x^{n-2}h^2 + \cdots + nxh^{n-1} + h^n - x^n \right].$$

The first and last terms in brackets cancel, and dividing h into the remaining terms gives

$$f'(x) = \lim_{h \to 0} \left[nx^{n-1} + \frac{n(n-1)}{2}x^{n-2}h + \cdots + h^{n-1} \right] = nx^{n-1}. \qquad \blacksquare$$

In short,

$$\frac{d}{dx}(x^n) = nx^{n-1}. \tag{3.7}$$

This is called the **power rule** for differentiation. Although we have proved the power rule only for n a positive integer, it is in fact true for every real number n. We will assume that equation 3.7 can be used for any real number n, and will prove this more general result in Section 3.11. For example,

$$\frac{d}{dx}(x^3) = 3x^2 \qquad \text{and} \qquad \frac{d}{dx}\left(x^{1/3}\right) = \frac{1}{3}x^{-2/3}.$$

THEOREM 3.4

If $g(x) = cf(x)$, where c is a constant, and $f(x)$ has a derivative, then

$$g'(x) = cf'(x). \tag{3.8a}$$

PROOF By equation 3.3 and Theorem 2.1,

$$g'(x) = \lim_{h \to 0} \frac{g(x+h) - g(x)}{h} = \lim_{h \to 0} \frac{cf(x+h) - cf(x)}{h}$$

$$= c \lim_{h \to 0} \frac{f(x+h) - f(x)}{h} = cf'(x). \qquad \blacksquare$$

Thus, for $y = f(x)$, we may write that

$$\frac{d}{dx}(cy) = c\frac{dy}{dx}. \tag{3.8b}$$

THEOREM 3.5

If $p(x) = f(x) + g(x)$, where $f(x)$ and $g(x)$ have derivatives, then

$$p'(x) = f'(x) + g'(x). \tag{3.9a}$$

PROOF   Equation 3.3 gives

$$p'(x) = \lim_{h \to 0} \frac{p(x+h) - p(x)}{h}$$

$$= \lim_{h \to 0} \frac{[f(x+h) + g(x+h)] - [f(x) + g(x)]}{h}$$

$$= \lim_{h \to 0} \left[\frac{f(x+h) - f(x)}{h} + \frac{g(x+h) - g(x)}{h} \right]$$

$$= f'(x) + g'(x).$$ ■

In short, if we set $u = f(x)$ and $v = g(x)$,

$$\frac{d}{dx}(u + v) = \frac{du}{dx} + \frac{dv}{dx}, \tag{3.9b}$$

or, in words, the derivative of a sum is the sum of the derivatives.

We now use these formulas to calculate derivatives in the following examples.

EXAMPLE 3.6

Find dy/dx if

(a) $y = x^4$ (b) $y = 3x^6 - x^{-2}$ (c) $y = \dfrac{x^4 - 6x^2}{3x^3}$

SOLUTION

(a) By power rule 3.7,

$$\frac{dy}{dx} = 4x^3.$$

(b) Equation 3.9 allows us to differentiate each term separately; by equations 3.8 and 3.7 it follows that

$$\frac{dy}{dx} = 3(6x^5) - (-2x^{-3}) = 18x^5 + 2x^{-3}.$$

(c) If we write y in the form $y = (x/3) - 2x^{-1}$, we can proceed as in part (b):

$$\frac{dy}{dx} = \frac{1}{3}(1) - 2(-x^{-2}) = \frac{1}{3} + \frac{2}{x^2}.$$

EXAMPLE 3.7

If $f(x) = 3x^4 - 2$, evaluate $f'(1)$.

SOLUTION   Since $f'(x) = 12x^3$, it follows that $f'(1) = 12$. Geometrically, 12 is the slope of the tangent line to the curve $y = 3x^4 - 2$ at the point $(1, 1)$ in Figure 3.15. Algebraically, the result implies that at $x = 1$, y changes 12 times as fast as x.

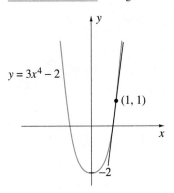

FIGURE 3.15 Tangent line to $y = 3x^4 - 2$ at $(1, 1)$

EXAMPLE 3.8

Find the equation of the tangent line to the curve $y = f(x) = x^3 + 5x$ at the point $(1, 6)$.

SOLUTION Since $f'(x) = 3x^2 + 5$, the slope of the tangent line to the curve at $(1, 6)$ is

$$f'(1) = 3(1)^2 + 5 = 8.$$

Using point-slope formula 1.13 for a straight line, we obtain for the equation of the tangent line at $(1, 6)$

$$y - 6 = 8(x - 1) \qquad \text{or} \qquad 8x - y = 2.$$

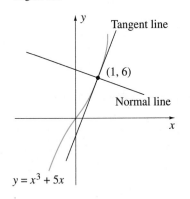

FIGURE 3.16 Normal line to a curve is perpendicular to tangent line

The line through $(1, 6)$ perpendicular to the tangent line in Figure 3.16 is called the **normal line** to the curve at $(1, 6)$. Since two lines are perpendicular only if their slopes are negative reciprocals (see equation 1.15), the normal line at $(1, 6)$ must have slope $-1/8$. The equation of the normal line to $y = x^3 + 5x$ at $(1, 6)$ is therefore

$$y - 6 = -\frac{1}{8}(x - 1) \qquad \text{or} \qquad x + 8y = 49.$$

EXAMPLE 3.9

Find, accurate to four decimal places, points on the curve $y = x^4 - 4x^3 - x^2 + x$ at which the slope of the tangent line is -1.

SOLUTION It is always wise to have an idea about how many solutions to expect for a problem and approximate values for them. (We might call this "ball-parking" the answer.) In this problem, a plot of the curve in Figure 3.17 should provide this information. However, we must be careful in trying to estimate where the slope of the graph is -1. Because scales are different on the axes, a line with slope -1 is not inclined at $\pi/4$ radians with respect to the negative x-axis. The line in the figure has slope -1 for reference. The graph suggests three points at which the tangent line is parallel to this line: one just to the left of $x = 0$, one just to the right of $x = 0$, and one near $x = 3$.

FIGURE 3.17   Finding points on a curve where tangent line has given slope

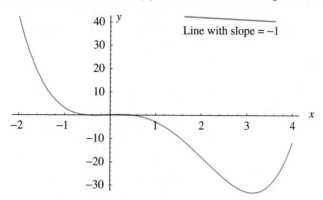

To confirm this, we set the slope of the tangent line to the curve equal to -1:

$$-1 = \frac{dy}{dx} = 4x^3 - 12x^2 - 2x + 1 \quad \Longrightarrow \quad 2(2x^3 - 6x^2 - x + 1) = 0.$$

Six-digit solutions of this equation are $-0.459\,261$, $0.350\,168$, and $3.109\,09$. To confirm -0.4593, 0.3502, and 3.1091 as solutions correct to four decimal places, we calculate $g(x) = 2x^3 - 6x^2 - x + 1$ at the following values:

$$g(-0.459\,35) \;=\; -5.1 \times 10^{-4}, \quad g(-0.459\,25) \;=\; 6.5 \times 10^{-5},$$

$$g(0.350\,15) \;=\; 8.0 \times 10^{-5}, \quad g(0.350\,25) \;=\; -3.7 \times 10^{-4},$$

$$g(3.109\,05) \;=\; -9 \times 10^{-4}, \quad g(3.109\,15) \;=\; 1.1 \times 10^{-3}.$$

The zero intermediate value theorem confirms the four-decimal-place approximations. Using the six-decimal-place approximations, and rounding results to four decimal places, corresponding points on the curve are $(-0.4593, -0.2382)$, $(0.3502, 0.0708)$, and $(3.1091, -33.3326)$.

Increment Notation

In Section 3.1, we have used the letter h to represent a small change in x when calculating derivatives. An alternative notation that is sometimes more suggestive was used in the introductory examples of that section. A small change in x, called an **increment** in x, is often denoted by Δx. It is pronounced "delta x" since Δ is the capital letter delta in the Greek alphabet. When x is given an increment Δx in the function $y = f(x)$, the corresponding change or increment in y is denoted by Δy. It is equal to

$$\Delta y = f(x + \Delta x) - f(x). \tag{3.10}$$

With this notation, equation 3.3 takes the form

$$f'(x) = \lim_{\Delta x \to 0} \frac{\Delta y}{\Delta x}. \tag{3.11a}$$

The notation dy/dx for the derivative fits very nicely with increment notation,

$$\frac{dy}{dx} = \lim_{\Delta x \to 0} \frac{\Delta y}{\Delta x}; \tag{3.11b}$$

the derivative of y with respect to x is the limit of the change in y divided by the change in x as the change in x approaches zero. We use this notation in the following example.

EXAMPLE 3.10

Use equation 3.11 to calculate the derivative of $y = f(x) = 3x^2 - 2x$, and check your answer using the differentiation rules discussed earlier in this section.

SOLUTION   Since

$$
\begin{aligned}
\Delta y &= f(x + \Delta x) - f(x) \\
&= [3(x + \Delta x)^2 - 2(x + \Delta x)] - [3x^2 - 2x] \\
&= 3[x^2 + 2x\,\Delta x + (\Delta x)^2] - 2x - 2\Delta x - 3x^2 + 2x \\
&= \Delta x(6x - 2 + 3\Delta x),
\end{aligned}
$$

equation 3.11 gives

$$
\frac{dy}{dx} = \lim_{\Delta x \to 0} \frac{\Delta y}{\Delta x} = \lim_{\Delta x \to 0} \frac{\Delta x(6x - 2 + 3\Delta x)}{\Delta x} = \lim_{\Delta x \to 0} (6x - 2 + 3\Delta x) = 6x - 2.
$$

Power rule 3.7, and rules 3.8 and 3.9 for differentiation of $3x^2 - 2x$, also yield $6x - 2$.

Consulting Project 2

Figure 3.18a shows parts of two transmission lines, one straight, and the other in the shape of a parabola as it circumnavigates a lake. Numbers represent lengths in kilometres but no attempt has been made to adhere to a scale. The two transmission lines are to be joined by a third that should be as short as possible. We are to find its length and where it should join the existing lines.

FIGURE 3.18a   Schematic for finding shortest distance between two transmission lines

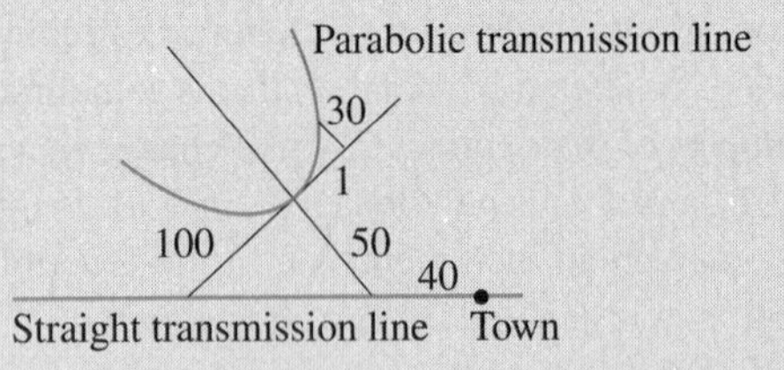

FIGURE 3.18b

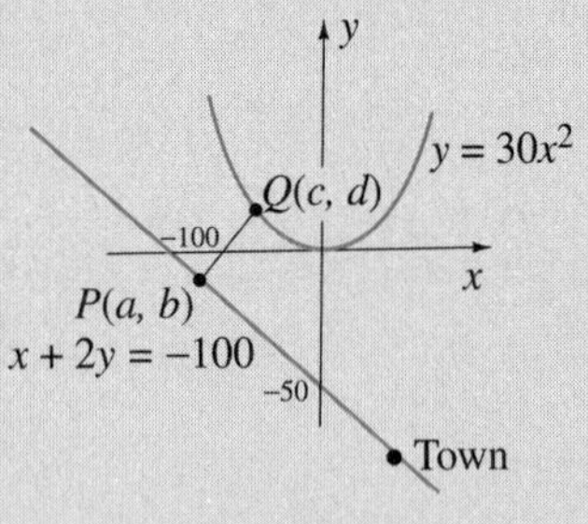

SOLUTION   We begin by choosing a coordinate system to facilitate analysis. Since equations for straight lines are straightforward in any coordinate system, we choose axes to simplify the equation of the parabola (Figure 3.18b). In this coordinate system, the equation of the parabola is $y = 30x^2$ and the equation of the line is $x + 2y + 100 = 0$. Our problem now is to find the shortest distance between the line and the parabola (which must be a straight line distance) and the points on these curves at which the line segment should be drawn. Let the required points on the straight line and parabola have coordinates $P(a, b)$ and $Q(c, d)$, respectively. This is an important step in our analysis. When unknowns are required, identify them, in this case give names to the coordinates of the unknown points. In addition, do not designate either one of the points by coordinates

(x, y). Letters x and y already identify variable points on the straight line and parabola; to identify fixed points with the same letters would lead to confusion. Now the big step. Of the infinite number of line segments joining points P on the straight transmission line to points Q on the parabolic transmission line, which is shortest? Our intuition tells us that the shortest line segment is the one that is simultaneously perpendicular to both transmission lines. Recognizing this, we could attempt to ballpark positions of P and Q by plotting the existing transmission lines with equal scales on the axes (otherwise, lengths and angles are not what they seem). We have done this in Figure 3.19 and drawn a line that seems perpendicular to both transmission lines. It intersects the parabola very close to the origin so that coordinates of Q should both be close to zero. Coordinates of P would appear to be close to $(-20, -40)$.

Using slopes we can now set up equations for the unknown coordinates. Since the slope of line segment PQ is $(d - b)/(c - a)$, and that of $x + 2y + 100 = 0$ is $-1/2$, perpendicularity requires

$$\frac{d - b}{c - a} = 2.$$

The slope of the tangent line to the parabola at any value of x is $60x$, so that the slope of the tangent line at Q is $60c$. Since the slope of the normal line is $-1/(60c)$, and this must be parallel to line segment PQ, we must have

$$\frac{d - b}{c - a} = -\frac{1}{60c}.$$

We now have two equations in the four unknown coordinates. Two more equations can be obtained by using the fact that P is on the line $x + 2y + 100 = 0$ and Q is the on the parabola $y = 30x^2$,

$$a + 2b + 100 = 0, \qquad d = 30c^2.$$

When we equate right sides of the first two equations, we obtain

$$2 = -\frac{1}{60c} \quad \Longrightarrow \quad c = -\frac{1}{120}.$$

This now implies that

$$d = 30\left(-\frac{1}{120}\right)^2 = \frac{1}{480}.$$

We substitute these values for c and d into the first equation,

$$\frac{1}{480} - b = 2\left(-\frac{1}{120}\right) - 2a \quad \Longrightarrow \quad 2a - b = \frac{3}{160}.$$

If we double this equation and add it to $a + 2b = -100$, we obtain

$$5a = \frac{3}{80} - 100 \quad \Longrightarrow \quad a = -\frac{8003}{400} \quad \Longrightarrow \quad b = -\frac{8003}{200} - \frac{3}{160} = -\frac{31\,997}{800}.$$

Thus, the required point on the straight transmission line is $(-8003/400, -31\,997/800) \approx (-20, -40)$ and the point on the parabolic line is $(-1/120, 1/480)$. These agree with our predictions for positions of P and Q. The length of the line segment joining these points is

$$\|PQ\| = \sqrt{\left(-\frac{1}{120} + \frac{8003}{400}\right)^2 + \left(\frac{1}{480} + \frac{31\,997}{800}\right)^2} \approx 44.7.$$

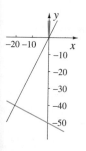

FIGURE 3.19   Scale diagram of transmission lines

The shortest transmission line joining the existing lines is approximately 44.7 km long. Since the length of the line segment joining P to the y-intercept of the straight transmission line is $\sqrt{(8003/400)^2 + (-50 + 31\,997/800)^2} \approx 22.4$, the new transmission line should begin 62.4 km west of the town and be perpendicular to the existing straight transmission line.

EXERCISES 3.2

In Exercises 1–20 find $f'(x)$.

1. $f(x) = 2x^2 - 3$

2. $f(x) = 3x^3 + 4x + 5$

3. $f(x) = 10x^2 - 3x$

4. $f(x) = 4x^5 - 10x^3 + 3x$

5. $f(x) = 1/x^2$

6. $f(x) = 2/x^3$

7. $f(x) = 5x^4 - 3x^3 + 1/x$

8. $f(x) = -\dfrac{1}{2x^2} + \dfrac{3}{x^4}$

9. $f(x) = x^{10} - \dfrac{1}{x^{10}}$

10. $f(x) = 5x^4 + \dfrac{1}{4x^5}$

11. $f(x) = 5x^{-4} + \dfrac{1}{4x^{-5}}$

12. $f(x) = \sqrt{x}$

13. $f(x) = \dfrac{3}{x^2} + \dfrac{2}{\sqrt{x}}$

14. $f(x) = \dfrac{1}{x^{3/2}} + x^{3/2}$

15. $f(x) = 2x^{1/3} - 3x^{2/3}$

16. $f(x) = \pi x^{\pi}$

17. $f(x) = (x^2 + 2)^2$

18. $f(x) = (4x^6 - x^2)/x^5$

19. $f(x) = x^{5/3} - x^{2/3} + 3$

20. $f(x) = (2x + 5)^3$

In Exercises 21–24 find equations for the tangent and normal lines to the curve at the point indicated. In each case, draw the curve and lines.

21. $y = x^2 - 2x + 5$ at $(2, 5)$

22. $y = \sqrt{x} + 5$ at $(4, 7)$

23. $y = 2x^3 - 3x^2 - 12x$ at $(2, -20)$

∗ 24. $x = \sqrt{y + 1}$ at $(3, 8)$

∗ 25. Find the points on the curve $y = x^4/4 - 2x^3/3 - 19x^2/2 + 22x$ at which the slope of the tangent line is 2.

∗ 26. Show that the x-intercept of the tangent line at any point (x_0, y_0) on the parabola $y = ax^2$ bisects that part of the x-axis between $x = 0$ and $x = x_0$.

∗ 27. Draw a graph of the function $x = f(t) = t^3 - 8t^2$. Find the value(s) of t at which the tangent line to this curve is parallel to the line $x = 6t - 3$.

▦ ∗ 28. A *chirp signal* in acoustics is a signal whose frequency changes linearly from a low value to a high value. For example, consider the

signal $x(t) = \cos(1000\pi t^2 + 100\pi t)$, where t is time in seconds.

(a) Plot $x(t)$ for $0 \le t \le 0.1$. Notice how the frequency increases as t increases.

(b) The frequency in hertz of the signal, at any given time, is defined as the derivative of the *phase* $1000\pi t^2 + 100\pi t$ at that time, divided by 2π. Find the frequencies of the signal at $t = 0$ and $t = 0.1$.

∗ 29. The general formula for a chirp signal (see Exercise 28) is $x(t) = \cos(\alpha t^2 + \beta t + \phi)$, where $\alpha > 0$, $\beta > 0$, and ϕ are constants. The derivative of $\alpha t^2 + \beta t + \phi$, divided by 2π, is the instantaneous frequency of the signal. If the signal begins at time t_1 and ends at time t_2, what is the difference in initial and final frequencies of the signal?

∗ 30. At what point(s) on the curve $y = x^3 + x^2 - 22x + 20$ does the tangent line pass through the origin?

∗ 31. At what point(s) on the parabola $y = x^2$ does the normal line pass through the point $(2, 5)$? Can you suggest an application of this result?

∗ 32. Show that the sum of the x- and y-intercepts of the tangent line to the curve $\sqrt{x} + \sqrt{y} = \sqrt{a}$ is always equal to a.

∗ 33. Show that the line segment cut from the tangent line at a point P on the curve $y = 1/x$ by the coordinate axes is bisected by P.

∗ 34. A hill is best described by a parabola containing the three points in the figure below, all measurements in metres. A transmitter 30 m high stands at the point $(-120, 0)$. What is the closest point to the base of the hill on the positive x-axis that a receiver can detect the signal unobstructed by the hill?

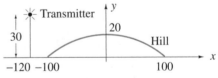

∗ 35. Give an alternative derivation of power rule 3.7 based on the identity

$$a^n - b^n = (a - b)(a^{n-1} + a^{n-2}b + \cdots + ab^{n-2} + b^{n-1}).$$

∗ 36. Find all pairs of points on the curves $y = x^2$ and $y = -x^2 + 2x - 2$ that share a common tangent line.

∗ 37. Show that the tangent lines at any two points P and Q on the parabola $y = ax^2 + bx + c$ intersect at a point that is on the vertical line halfway between P and Q (provided that neither P nor Q is at the vertex of the parabola).

38. Find a formula for $\dfrac{d}{dx}|x|^n$ when $n > 1$ is an integer.

39. Prove that

$$\frac{d}{dx}(ax + b)^n = an(ax + b)^{n-1}$$

when a and b are constants, and $n > 1$ is an integer.

** **40.** Find the two points on the curve $y = x(1 + 2x - x^3)$ that share a common tangent line.

3.3 Differentiability and Continuity

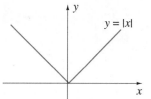

FIGURE 3.20 Function $f(x) = |x|$ has no derivative at $x = 0$

Many functions fail to have a derivative at isolated points. For example, consider the function $f(x) = |x|$ in Figure 3.20. It is clear that for $x > 0$, $f'(x) = 1$, and for $x < 0$, $f'(x) = -1$. At $x = 0$, however, there is a problem. If $f(x)$ is to have a derivative at $x = 0$, it must be given by

$$\lim_{h \to 0} \frac{f(0 + h) - f(0)}{h} = \lim_{h \to 0} \frac{|h|}{h}.$$

But this limit does not exist since

$$\lim_{h \to 0^-} \frac{|h|}{h} = -1 \quad \text{and} \quad \lim_{h \to 0^+} \frac{|h|}{h} = 1.$$

Consequently, $f(x) = |x|$ does not have a derivative at $x = 0$.

The same conclusion can be drawn at any point at which the graph of a function takes an abrupt change in direction. Such a point is often called a *corner*. As a result, the function in Figure 3.21 does not have a derivative at $x = a, b, c$, or d. It is also true that a function cannot have a derivative at a point where the function is discontinuous (see, e.g., the discontinuities in Figure 2.27). This result is an immediate consequence of the following theorem.

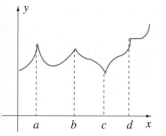

FIGURE 3.21 Points where a continuous function has no derivative

---

THEOREM 3.6

If a function has a derivative at $x = a$, then the function is continuous at $x = a$.

PROOF   To prove this theorem we show that existence of $f'(a)$ implies that $\lim_{x \to a} f(x) = f(a)$, the condition that defines continuity of $f(x)$ at $x = a$ (Definition 2.1 in Section 2.4). We show that $\lim_{x \to a} [f(x) - f(a)] = 0$:

$$\lim_{x \to a} [f(x) - f(a)] = \lim_{x \to a} \left[\frac{f(x) - f(a)}{x - a} \cdot (x - a) \right] = \left[\lim_{x \to a} \frac{f(x) - f(a)}{x - a} \right] \left[\lim_{x \to a} (x - a) \right],$$

provided both limits on the right exist. Obviously, the second limit has value zero, and the first is definition 3.2b for $f'(a)$, which we have assumed exists. Thus,

$$\lim_{x \to a} [f(x) - f(a)] = f'(a) \cdot 0 = 0.$$

The following result is logically equivalent to Theorem 3.6; it is called the *contrapositive* of the theorem. Although equivalent to Theorem 3.6, we shall designate it as a corollary.

COROLLARY 3.6.1

If $f(x)$ is discontinuous at $x = a$, then $f'(a)$ does not exist.

Students are often heard to say that having a derivative is equivalent to having a tangent line (because derivatives are slopes of tangent lines); that is, a function $f(x)$ has derivative $f'(a)$ at $x = a$ if and only if the graph of $f(x)$ has a tangent line at $x = a$. This is not quite true. For example, consider the function $f(x) = x^{1/3}$ (Figure 3.22). At $(0, 0)$, the tangent line to the graph is the y-axis (use the definition of *tangent line* in Section 3.1 to convince yourself of this). The derivative of $f(x)$ is $f'(x) = (1/3)x^{-2/3}$; it does not exist at $x = 0$. Thus, we have a tangent line, but no derivative. The reason is that the tangent line is a vertical line, and vertical lines do not have slopes. What can we say, then? Two things:

1. If $f'(a)$ exists, then $y = f(x)$ has a tangent line at $\left(a, f(a)\right)$ with slope $f'(a)$.

2. If $f'(a)$ does not exist, then $y = f(x)$ either has a vertical tangent line at $\left(a, f(a)\right)$ or does not have a tangent line when $x = a$.

FIGURE 3.22   Tangent line to $y = x^{1/3}$ at $(0, 0)$ is the y-axis

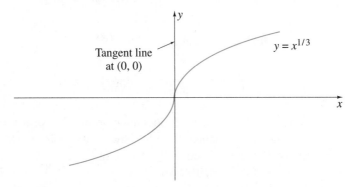

We introduced this section by showing that the derivative of $|x|$ is 1 when $x > 0$; it is -1 when $x < 0$; and it does not exist at $x = 0$. These can be combined into the simple formula

$$\frac{d}{dx}|x| = \frac{|x|}{x}. \tag{3.12}$$

It is straightforward to generalize equation 3.12 and obtain the derivative of $|f(x)|$ at any point at which $f(x) \neq 0$ and $f'(x)$ exists. When $f(x) > 0$, we may write

$$\frac{d}{dx}|f(x)| = \frac{d}{dx}f(x) = f'(x).$$

On the other hand, when $f(x) < 0$, we have

$$\frac{d}{dx}|f(x)| = \frac{d}{dx}[-f(x)] = -f'(x).$$

Both of these results are contained in the one equation

$$\frac{d}{dx}|f(x)| = \frac{|f(x)|}{f(x)}f'(x). \tag{3.13}$$

This is the derivative of $|f(x)|$ at any point at which $f(x) \neq 0$. The exceptional case when $f(x) = 0$ is discussed in Exercise 28.

Right- and Left-Hand Derivatives

When a function $f(x)$ is defined on a closed interval $b \leq x \leq c$, equation 3.3 can be used to calculate $f'(x)$ only at points in the open interval $b < x < c$. For instance, it is impossible to evaluate

$$f'(b) = \lim_{h \to 0} \frac{f(b + h) - f(b)}{h}$$

since $f(x)$ is not defined for $x < b$ and therefore $f(b + h)$ is not defined for $h < 0$. When a function $f(x)$ is defined only to the right of a point, we define a right-hand derivative at the point; and when $f(x)$ is defined only to the left of a point, we define its left-hand derivative.

> **DEFINITION 3.2**
>
> The **right-hand derivative** of $f(x)$ with respect to x is defined as
>
> $$f'_+(x) = \lim_{h \to 0^+} \frac{f(x + h) - f(x)}{h}, \qquad (3.14a)$$
>
> provided that the limit exists. The **left-hand derivative** of $f(x)$ is
>
> $$f'_-(x) = \lim_{h \to 0^-} \frac{f(x + h) - f(x)}{h}, \qquad (3.14b)$$
>
> if the limit exists.

The left-hand derivative at a point is not confined to the situation where a function is defined only to the left of the point; nor is the right-hand derivative restricted to the situation where the function is defined only to the right of the point. We may consider left- and right-hand derivatives at any point, as well as a "full" derivative. Obviously, when a function has a derivative at a point x, its right- and left-hand derivatives both exist at x and are equal to $f'(x)$. It is possible, however, for a function to have both a left- and a right-hand derivative at a point but not a derivative. An example of this is the absolute value function $f(x) = |x|$ at $x = 0$ (Figure 3.20). Its right-hand derivative at $x = 0$ is equal to 1 and its left-hand derivative there is -1.

When a function has a derivative at $x = a$, we say that it is **differentiable** at $x = a$. When it has a derivative at every point in some interval, we say that it is differentiable on that interval. In the event that the interval is closed, $b \leq x \leq c$, we understand that derivatives at $x = b$ and $x = c$ mean right- and left-hand derivatives, respectively.

| EXAMPLE 3.11

What is the derivative of the Heaviside function $h(x - a)$ introduced in Section 2.5?

SOLUTION   The tangent line to the graph of the function (Figure 3.23) is horizontal at every point except $x = a$, where the function is undefined. In other words, $h'(x - a) = 0$ except at the discontinuity $x = a$. Right- and left-hand derivatives do not exist at $x = a$ either.

FIGURE 3.23   Tangent line to $y = h(x - a)$ is horizontal except at $x = a$

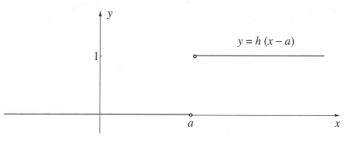

By examining the definition of the derivative of $h(x-a)$, a very useful result for applications emerges. According to equation 3.3,

$$h'(x - a) = \lim_{\Delta x \to 0} \frac{h(x + \Delta x - a) - h(x - a)}{\Delta x}.$$

If we change variables in this limit by setting $\epsilon = -\Delta x$,

$$h'(x - a) = \lim_{-\epsilon \to 0} \frac{h(x - \epsilon - a) - h(x - a)}{-\epsilon} = \lim_{\epsilon \to 0} \frac{h(x - a) - h(x - a - \epsilon)}{\epsilon}.$$

But according to equation 2.9, this is the definition of the Dirac-delta function $\delta(x - a)$. In other words, we may write that $h'(x - a) = \delta(x - a)$.

EXAMPLE 3.12

Is the function

$$g(x) = \begin{cases} x^2 \sin(1/x), & x \neq 0 \\ 0, & x = 0 \end{cases}$$

differentiable at $x = 0$?

SOLUTION   We encountered the function $x^2 \sin(1/x)$ in Example 2.9 of Section 2.1 and drew its graph (except for the point at $x = 0$) in Figure 2.10a. According to equation 3.3, the derivative of $g(x)$ at $x = 0$ is

$$g'(0) = \lim_{h \to 0} \frac{g(0 + h) - g(0)}{h},$$

provided that the limit exists. When we substitute from the definition of $g(x)$ for $g(h)$ and $g(0)$, this limit takes the form

$$g'(0) = \lim_{h \to 0} \frac{h^2 \sin(1/h) - 0}{h} = \lim_{h \to 0} h \sin\left(\frac{1}{h}\right) = 0$$

(see Exercise 51 in Section 2.1).

Angle Between Intersecting Curves

The **angle θ between two curves that intersect** at a point (x_0, y_0) (Figure 3.24) is defined as the angle between the tangent lines to the curves at (x_0, y_0). This can be calculated using formula 1.60 once slopes of the tangent lines are known.

In the event that $\theta = \pi/2$ (Figure 3.25), the curves are said to be **orthogonal** or perpendicular at (x_0, y_0). Should $\theta = 0$, the curves are said to be tangent at (x_0, y_0) (Figure 3.26).

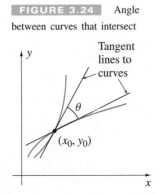

FIGURE 3.24   Angle between curves that intersect

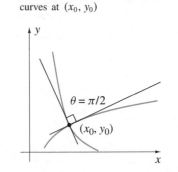

FIGURE 3.25   Orthogonal curves at (x_0, y_0)

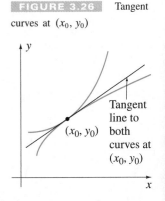

FIGURE 3.26   Tangent curves at (x_0, y_0)

EXAMPLE 3.13

Find the angle between the line $x + 2y = 5$ and the curve $y = x^3 + 31$ at their point of intersection.

FIGURE 3.27 Finding angle between curves at their point of intersection

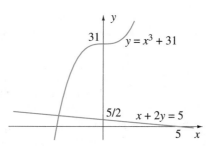

SOLUTION   A quick diagram (Figure 3.27) allows us to ballpark the location of the point of intersection of the curves. To find it, we solve the equation of the line for $y = (5 - x)/2$ and equate it to $x^3 + 31$,

$$x^3 + 31 = \frac{5 - x}{2} \quad \Longrightarrow \quad 2x^3 + x + 57 = 0.$$

According to the rational root theorem of Section 1.2, the only possible rational solutions of this equation are

$$\pm 1, \ \pm 3, \ \pm 19, \ \pm 57, \ \pm \frac{1}{2}, \ \pm \frac{3}{2}, \ \pm \frac{19}{2}, \ \pm \frac{57}{2}.$$

Figure 3.27 makes it clear that we should only try -3. It is indeed a solution, and the only one. The point of intersection of the curves is therefore $(-3, 4)$. The slope of the line is $-1/2$, and to find the slope of the cubic, we calculate that $dy/dx = 3x^2$. The slope at $(-3, 4)$ is 27. Using formula 1.60, the acute angle between the curves at their point of intersection is

$$\theta = \operatorname{Tan}^{-1} \left| \frac{27 - (-1/2)}{1 + 27(-1/2)} \right| = 1.1 \text{ radians.}$$

It looks much larger than this. Why?

EXERCISES 3.3

In Exercises 1–6 determine whether the function has a right-hand derivative, a left-hand derivative, and a derivative at the given value of x.

1. $f(x) = |x - 5|$ at $x = 5$

2. $f(x) = x^{3/2}$ at $x = 0$

3. $f(x) = |x - 5|^3$ at $x = 5$

4. $f(x) = \operatorname{sgn} x$ at $x = 0$ (See Exercise 47 in Section 2.4.)

5. $f(x) = (x^2 - 1)/(x - 1)$ at $x = 1$

6. $f(x) = \lfloor x \rfloor$ at $x = 1$ (See Exercise 68 in Section 1.5.)

7. Does it make any difference in Example 3.11 if we define $h(a) = 0$?

8. Does it make any difference in Example 3.11 if we define $h(a) = 1$?

9. Does it make any difference in Exercise 4 if $\operatorname{sgn} x$ does not have a value at $x = 0$?

In Exercises 10–13 determine whether the statement is true or false.

10. If a function $f(x)$ has a derivative at $x = a$, then its graph has a tangent line at the point $(a, f(a))$.

11. If a function $f(x)$ has a tangent line at a point $(a, f(a))$, then it has a derivative at $x = a$.

12. If a function does not have a tangent line at $(a, f(a))$, then it does not have a derivative at $x = a$.

13. If a function $f(x)$ does not have a derivative at $x = a$, then it does not have a tangent line at $(a, f(a))$.

In Exercises 14–16 show algebraically that $f'(0)$ does not exist. Draw a graph of the function.

∗ **14.** $f(x) = x^{1/3}$

∗ **15.** $f(x) = x^{2/3}$

∗ **16.** $f(x) = x^{1/4}$

In Exercises 17–18 find the angle (or angles) between the curves at their point (or points) of intersection.

17. $y = x^2$, $x + y = 2$ **18.** $y = x^2$, $y = 1 - x^2$

In Exercises 19–20 determine whether the curves are orthogonal.

19. $x - 2y + 1 = 0$, $y = 2 - x^2$

20. $y = 3 - x^2$, $4y - 7 = x^2$

In Exercises 21–22 show that the curves are tangent at the indicated point.

21. $y = x - 2x^2$, $y = x^3 + 2x$ at $(-1, -3)$

22. $y = x^3$, $y = x^2 + x - 1$ at $(1, 1)$

* **23.**   (a) The equation $x + 2y = C_1$, where C_1 is a constant, represents a *one-parameter family of curves*. For each value of C_1, the parameter, a different curve in the family is obtained. Draw curves corresponding to $C_1 = -2, -1, 0, 1, 2$.

    (b) Draw a few curves from the one-parameter family $2x - y = C_2$ on the graph in part (a).

    (c) Two families of curves are said to be *orthogonal trajectories* if every curve from one family intersects every curve from the other family orthogonally. Are the families in parts (a) and (b) orthogonal trajectories?

* **24.** For what value of k are the one-parameter families $2x - 3y = C_1$ and $x + ky = C_2$ orthogonal trajectories?

* **25.** Draw a graph of the function $f'(x)$ when $f(x) = \lfloor x \rfloor$ is the floor function (Example 2.22 in Section 2.4).

* **26.** Is the function $f(x) = x|x|$ differentiable at $x = 0$?

* **27.** Find $f'(x)$ if $f(x) = |x| + |x - 1|$. Draw graphs of $f(x)$ and $f'(x)$.

* **28.** If $f(x)$ is a differentiable function, does $|f(x)|$ have a derivative at points where $f(x) = 0$? *Hint:* Draw some pictures.

* **29.** If $\lim_{x \to \infty} f(x) = L$ (so that $y = L$ is a horizontal asymptote for the graph of the function), is it necessary that $\lim_{x \to \infty} f'(x) = 0$?

* **30.** The Green's function for displacements of a taut string with constant tension τ and length L, and ends fixed at $x = 0$ and $x = L$ on the x-axis is

$$G(x; X) = \frac{1}{L\tau}[x(L - X)h(X - x) + X(L - x)h(x - X)],$$

where $h(x - X)$ is the Heaviside function in Section 2.5. Think of $G(x; X)$ as a function of x that also depends on X where X can have any value between 0 and L. It is the displacement at position x in the string if a unit force in the positive y-direction is applied to the string at position X.

(a) Draw a graph of $G(x; L/2)$. Is it symmetric about $x = L/2$?

(b) Draw a graph of $G(x; X)$ when $L/2 < X < L$.

(c) Show algebraically that $G(x; X)$ is continuous for $0 \leq x \leq L$, except for a removable discontinuity at $x = X$.

(d) Show that dG/dx is continuous for all $x \neq X$, and has a jump discontinuity at $x = X$ of $-1/\tau$.

* **31.** Are $f'_+(a)$ and $\lim_{x \to a^+} f'(x)$ the same? Illustrate with $h(x - a)$.

* **32.** Is the function

$$f(x) = \begin{cases} x \sin(1/x), & x \neq 0 \\ 0, & x = 0 \end{cases}$$

differentiable at $x = 0$?

* **33.** For what values of the real number n is the function

$$f(x) = \begin{cases} x^n \sin(1/x), & x \neq 0 \\ 0, & x = 0 \end{cases}$$

differentiable at $x = 0$?

** **34.** Is the function

$$f(x) = \begin{cases} x^2, & x \text{ a rational number} \\ 0, & x \text{ an irrational number} \end{cases}$$

differentiable at $x = 0$?

3.4 Product and Quotient Rules

In this section we add two more formulas to those of Section 3.2 for calculating derivatives. The first is a rule for differentiating a function that is the product of two other functions.

THEOREM 3.7

If $p(x) = f(x)g(x)$, where $f(x)$ and $g(x)$ are differentiable, then

$$p'(x) = f(x)g'(x) + f'(x)g(x). \tag{3.15a}$$

PROOF By equation 3.3,

$$p'(x) = \lim_{h \to 0} \frac{p(x+h) - p(x)}{h}$$

$$= \lim_{h \to 0} \frac{f(x+h)g(x+h) - f(x)g(x)}{h}.$$

To organize this quotient further, we add and subtract the quantity $f(x+h)g(x)$ in the numerator:

$$p'(x) = \lim_{h \to 0} \left\{ \frac{[f(x+h)g(x+h) - f(x+h)g(x)] + [f(x+h)g(x) - f(x)g(x)]}{h} \right\}$$

$$= \lim_{h \to 0} \left[f(x+h)\frac{g(x+h) - g(x)}{h} + g(x)\frac{f(x+h) - f(x)}{h} \right]$$

$$= f(x)g'(x) + g(x)f'(x).$$

In taking the limit of the first term, we have used the fact that $\lim_{h \to 0} f(x+h) = f(x)$, which follows from continuity of $f(x)$ (see Theorem 3.6 and Exercise 46 in Section 2.4). ∎

This result is called the **product rule** for differentiation. If we set $u = f(x)$ and $v = g(x)$, then the product rule may also be expressed in the form

$$\frac{d}{dx}(uv) = u\frac{dv}{dx} + v\frac{du}{dx}. \tag{3.15b}$$

For a change, we use increment notation to prove the **quotient rule**. Use of h in place of Δx works equally well.

THEOREM 3.8

If $p(x) = f(x)/g(x)$, where $f(x)$ and $g(x)$ are differentiable, then

$$p'(x) = \frac{g(x)f'(x) - f(x)g'(x)}{[g(x)]^2}. \tag{3.16a}$$

PROOF Using equation 3.11 yields

$$p'(x) = \lim_{\Delta x \to 0} \frac{p(x+\Delta x) - p(x)}{\Delta x}$$

$$= \lim_{\Delta x \to 0} \frac{1}{\Delta x} \left[\frac{f(x+\Delta x)}{g(x+\Delta x)} - \frac{f(x)}{g(x)} \right]$$

$$= \lim_{\Delta x \to 0} \frac{f(x+\Delta x)g(x) - g(x+\Delta x)f(x)}{\Delta x g(x)g(x+\Delta x)}.$$

To simplify this limit, we add and subtract $f(x)g(x)$ in the numerator:

$$p'(x) = \lim_{\Delta x \to 0} \frac{[f(x+\Delta x)g(x) - f(x)g(x)] - [g(x+\Delta x)f(x) - f(x)g(x)]}{\Delta x g(x)g(x+\Delta x)}$$

$$= \lim_{\Delta x \to 0} \frac{1}{g(x)g(x+\Delta x)} \left\{ g(x)\left[\frac{f(x+\Delta x) - f(x)}{\Delta x}\right] - f(x)\left[\frac{g(x+\Delta x) - g(x)}{\Delta x}\right] \right\}$$

$$= \frac{1}{[g(x)]^2} \left\{ g(x) \lim_{\Delta x \to 0} \frac{f(x + \Delta x) - f(x)}{\Delta x} - f(x) \lim_{\Delta x \to 0} \frac{g(x + \Delta x) - g(x)}{\Delta x} \right\}$$

$$= \frac{g(x) f'(x) - f(x) g'(x)}{[g(x)]^2}.$$

If we set $u = f(x)$ and $v = g(x)$, then quotient rule 3.16a can be expressed in the form

$$\frac{d}{dx} \left(\frac{u}{v} \right) = \frac{v \dfrac{du}{dx} - u \dfrac{dv}{dx}}{v^2}. \qquad (3.16b)$$

It makes no difference which term in product rule 3.15 is written first; it does make a difference in the quotient rule. Do not interchange the terms in 3.16.

EXAMPLE 3.14

For the following two functions, find $f'(x)$ in simplified form:

(a) $f(x) = (x^2 + 2)(x^4 + 5x^2 + 1)$ (b) $f(x) = \dfrac{\sqrt{x}}{3x^2 - 2}$

SOLUTION

(a) With product rule 3.15,

$$f'(x) = (x^2 + 2) \frac{d}{dx}(x^4 + 5x^2 + 1) + (x^4 + 5x^2 + 1) \frac{d}{dx}(x^2 + 2)$$

$$= (x^2 + 2)(4x^3 + 10x) + (x^4 + 5x^2 + 1)(2x)$$

$$= 6x^5 + 28x^3 + 22x.$$

(b) With quotient rule 3.16,

$$f'(x) = \frac{(3x^2 - 2) \left(\dfrac{1}{2\sqrt{x}} \right) - \sqrt{x}(6x)}{(3x^2 - 2)^2}$$

$$= \frac{\dfrac{3x^2 - 2 - 12x^2}{2\sqrt{x}}}{(3x^2 - 2)^2}$$

$$= -\frac{9x^2 + 2}{2\sqrt{x}(3x^2 - 2)^2}.$$

EXAMPLE 3.15

Draw a graph of the function $y = f(x) = (x - 1)/(x + 2)$. Calculate dy/dx and show qualitatively that the graph agrees with your calculation.

FIGURE 3.28 Relationship between the derivative of a function and its graph's shape

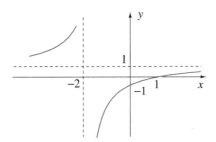

SOLUTION Limits were used to draw the graph in Figure 3.28. Using quotient rule 3.16, we find that

$$\frac{dy}{dx} = \frac{(x+2)(1) - (x-1)(1)}{(x+2)^2} = \frac{3}{(x+2)^2}.$$

The sketch in Figure 3.28 and dy/dx agree that:

(a) The slope of the curve is always positive.

(b) The slope becomes larger and larger as x approaches -2, either from the left or the right; that is,

$$\lim_{x \to -2} f'(x) = \infty.$$

(c) The slope approaches zero as x approaches $\pm\infty$; that is,

$$\lim_{x \to \pm\infty} f'(x) = 0.$$

Statements (a), (b), and (c) in Example 3.15 were arrived at by examining $dy/dx = 3/(x+2)^2$. They could also have been realized by drawing a graph of the derivative function (Figure 3.29). Ordinates of this graph are slopes in Figure 3.28. For example, at $x = 1$, the height of the curve in Figure 3.29 is $1/3$. At $x = 1$ in Figure 3.28, the slope of the tangent line is $1/3$. Notice how clear statements (a), (b), and (c) are from the graph of $f'(x)$ in Figure 3.29. The slope is always positive, $\lim_{x \to -2} f'(x) = \infty$, and $\lim_{x \to \pm\infty} f'(x) = 0$.

FIGURE 3.29 The derivative function

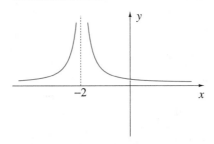

   In Chapter 4 when we apply derivatives to geometric and physical problems, we need to know when the derivative exists and when it does not exist, and when it is positive, negative, and zero. A graph of the derivative function is an excellent way to discover and visualize these properties.

EXAMPLE 3.16

Plot a graph of the function $f(x) = (x^2 + 4x - 1)/(x^3 + 2)$ on the interval $-10 \le x \le 10$. Find, to five decimal places, points where the tangent line to the graph is horizontal.

FIGURE 3.30   Plot to indicate where the tangent line is horizontal

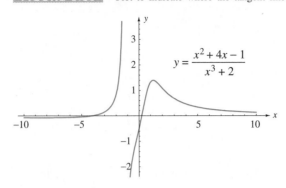

SOLUTION   A graph is shown in Figure 3.30. The graph has a vertical asymptote at $x = -2^{1/3}$. Obviously, there is a point near $x = 1$ at which the tangent line is horizontal. In addition, because the graph crosses the x-axis near $x = -5$, and $y = 0$ is a horizontal asymptote as $x \to -\infty$, there must be at least one point to the left of $x = -5$ at which the tangent line is horizontal. Whether there is more than one such point is not clear in Figure 3.30. The plot of

$$f'(x) = \frac{(x^3 + 2)(2x + 4) - (x^2 + 4x - 1)(3x^2)}{(x^3 + 2)^2} = \frac{-x^4 - 8x^3 + 3x^2 + 4x + 8}{(x^3 + 2)^2}$$

in Figure 3.31a crosses the x-axis near $x = 1$, thus confirming the point in Figure 3.30 where the tangent line is horizontal. It does not make clear, however, the number of points to the left of $x = -5$ at which the graph of $f'(x)$ crosses the x-axis. The graph of $f'(x)$ in Figure 3.31b clearly indicates one, and only one, point to the left of $x = -5$ at which $f'(x) = 0$. To find the two points then where $f'(x) = 0$, we set

$$-x^4 - 8x^3 + 3x^2 + 4x + 8 = 0.$$

Our computer gives two (real) solutions of this equation, $-8.316\,793$ and $1.238\,656\,0$. The zero intermediate value theorem of Section 1.11 guarantees that $-8.316\,79$ and $1.238\,66$ are solutions, correct to five decimal places, when we calculate

$$f'(-8.316\,795) = -3.1 \times 10^{-9}, \qquad f'(-8.316\,785) = 1.5 \times 10^{-8},$$

$$f'(1.238\,655) = 2.2 \times 10^{-6}, \qquad f'(1.238\,665) = -1.9 \times 10^{-5}.$$

FIGURE 3.31a   **FIGURE 3.31b**

Derivative function plots to determine where the derivative is zero

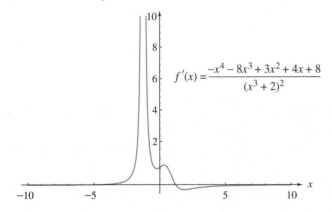

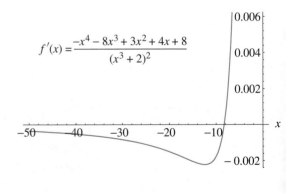

EXAMPLE 3.17

Show that when $f(x)$ is a function, differentiable for all x, and $h(x-a)$ is the Heaviside unit step function of Section 2.5,

$$\frac{d}{dx}[f(x)h(x-a)] = f'(x)h(x-a) \quad \text{when } x \neq a.$$

SOLUTION   The product rule gives

$$\frac{d}{dx}[f(x)h(x-a)] = f'(x)h(x-a) + f(x)h'(x-a).$$

Because $h'(x-a) = 0$ at all x, except $x = a$ where the derivative does not exist, the required result now follows. If we wish a result that includes $x = a$, we could use Example 3.11 to write

$$\frac{d}{dx}[f(x)h(x-a)] = f'(x)h(x-a) + f(x)\delta(x-a).$$

EXERCISES 3.4

In Exercises 1–16 find $f'(x)$ in simplified form.

1. $f(x) = (x^2 + 2)(x + 3)$

2. $f(x) = (2 - x^2)(x^2 + 4x + 2)$

3. $f(x) = \dfrac{x}{3x + 2}$

4. $f(x) = \dfrac{x^2}{4x^2 - 5}$

5. $f(x) = \dfrac{x^2}{2x - 1}$

6. $f(x) = \dfrac{x^3}{4x^2 + 1}$

7. $f(x) = \sqrt{x}(x + 1)$

8. $f(x) = \dfrac{\sqrt{x}}{3x + 2}$

9. $f(x) = \dfrac{2x^2 - 5}{3x + 4}$

10. $f(x) = \dfrac{x + 5}{2x^2 - 1}$

11. $f(x) = \dfrac{x(x + 1)}{1 - 3x}$

12. $f(x) = \dfrac{x^2 + 2x + 3}{x^2 - 5x + 1}$

13. $f(x) = \dfrac{1}{x^3 - 3x^2 + 2x + 5}$

14. $f(x) = \dfrac{x^3 + 3x^2 + 3x + 10}{(x + 1)^3}$

15. $f(x) = \dfrac{x^{1/3}}{1 - \sqrt{x}}$

* **16.** $f(x) = \dfrac{\sqrt{x} + 2x}{\sqrt{x} - 4}$

17. Find equations for the tangent and normal lines to the curve $y = (x + 3)/(x - 4)$ at the point $(1, -4/3)$. Plot the curve and lines.

18.   (a) Plot a graph of the function $f(x) = x^2/(x^2 + x - 2)$.
   (b) Find $f'(x)$ and show that it agrees with the plot.

19. If the total cost of producing x items of a commodity is given by the equation

$$C(x) = ax\left(\frac{x + b}{x + c}\right),$$

where a, b, and c are constants, show that the marginal cost $C'(x)$ is

$$a\left[1 + \frac{c(b - c)}{(x + c)^2}\right].$$

* **20.** Find a rule for the derivative of the product of three functions $f(x)$, $g(x)$, and $h(x)$.

In Exercises 21–23 find the angle (or angles) between the curves at their point (or points) of intersection.

* **21.** $y = x^3$, $y = 2/(1 + x^2)$

* **22.** $y = 2x + 2$, $y = x^2/(x - 1)$

* **23.** $y = 5 - x^2$, $y = 3x/(x - 1)$

* **24.** Find all points on the curve $y = (5 - x)/(6 + x)$ at which the tangent line passes through the origin.

* **25.**   (a) A manufacturer's profit from the sale of x kilograms of a commodity per week is given by

$$P(x) = \frac{3x - 200}{x + 400}.$$

Plot a graph of this function.

   (b) The average profit per kilogram when x kilograms are sold is given by

$$p(x) = \frac{P(x)}{x}.$$

Plot a graph of this function.

   (c) If a point (x, P) on the total profit curve is joined to the origin, the slope of this line is the average profit $p(x)$ for that x. Use this idea to find the sales level for highest average profit.

* **26.**   (a) Find $f'(x)$ if $f(x) = h(x + 1)(x^2 + x)$. Draw graphs of $f(x)$ and $f'(x)$.
   (b) Does the derivative exist at $x = -1$ if we define $f(-1) = 0$?

* **27.** Repeat Exercise 26 if $f(x) = h(x + 1)(x + 1)^2$.

3.5 Higher-Order Derivatives

When $y = f(x)$ is a function of x, its derivative $f'(x)$ is also a function of x. We can therefore take the derivative of the derivative to get what is called the **second derivative** of the function. This can be repeated over and over again. For instance, if $y = f(x) = x^3 + 1/x$, then

$$f'(x) = \frac{dy}{dx} = 3x^2 - \frac{1}{x^2}.$$

We denote the second derivative of y with respect to x by d^2y/dx^2 or $f''(x)$:

$$f''(x) = \frac{d^2y}{dx^2} = \frac{d}{dx}\left(\frac{dy}{dx}\right) = 6x + \frac{2}{x^3}.$$

Similarly,

$$f'''(x) = \frac{d^3y}{dx^3} = 6 - \frac{6}{x^4}.$$

This is called the **third derivative** or the derivative of order three. Clearly, we can continue the differentiation process indefinitely to produce derivatives of any positive integer order whatsoever.

EXAMPLE 3.18

How many derivatives does $f(x) = x^{8/3}$ have at $x = 0$?

SOLUTION   Since

$$f'(x) = \frac{8}{3}x^{5/3}, \qquad f''(x) = \left(\frac{8}{3}\right)\left(\frac{5}{3}\right)x^{2/3}, \qquad f'''(x) = \left(\frac{8}{3}\right)\left(\frac{5}{3}\right)\left(\frac{2}{3}\right)x^{-1/3},$$

and $f'''(0)$ is not defined, $f(x)$ has only a first and a second derivative at $x = 0$.

EXAMPLE 3.19

Find a formula for the second derivative of a product, $d^2(uv)/dx^2$ if $u = f(x)$ and $v = g(x)$.

SOLUTION   Since

$$\frac{d}{dx}(uv) = v\frac{du}{dx} + u\frac{dv}{dx},$$

then

$$\frac{d^2}{dx^2}(uv) = \frac{d}{dx}\left(v\frac{du}{dx} + u\frac{dv}{dx}\right)$$

$$= v\frac{d^2u}{dx^2} + \frac{dv}{dx}\frac{du}{dx} + u\frac{d^2v}{dx^2} + \frac{du}{dx}\frac{dv}{dx}$$

$$= v\frac{d^2u}{dx^2} + 2\frac{du}{dx}\frac{dv}{dx} + u\frac{d^2v}{dx^2}.$$

This formula is very handy. It is worth memorizing. See Exercise 14 for extensions to higher order derivatives of products.

EXERCISES 3.5

In Exercises 1–10 find the derivative indicated.

1. $f''(x)$ if $f(x) = x^3 + 5x^4$

2. $f'''(x)$ if $f(x) = x^3 - 3x^2 + 2x + 1$

3. $f''(2)$ if $f(x) = (x + 1)(x^3 + 3x + 2)$

4. $f'''(1)$ if $f(x) = x^4 - 3x^2 + 1/x$

5. $f''(x)$ if $f(x) = (x + 1)/\sqrt{x}$

6. $f'''(t)$ if $f(t) = t^3 - 1/t^3$

7. $d^9 y/dx^9$ if $y = x^{10}$

8. $f''(u)$ if $f(u) = \sqrt{u}/(u + 1)$

9. $d^2 t/dx^2$ if $t = x/(2x - 6)$

10. $f''(x)$ if $f(x) = x/(\sqrt{x} + 1)$

11. Steady-state temperature T in a region bounded by two concentric spheres of radii a and b (where $a < b$) must satisfy the equation

$$\frac{d^2 T}{dr^2} + \frac{2}{r}\frac{dT}{dr} = 0,$$

where r is the radial distance from the common centre of the spheres.

   (a) Verify that for any constants c and d, the function

$$T = f(r) = c + \frac{d}{r}$$

   satisfies the equation.

   (b) If temperatures on the two spheres are maintained at constant values T_a and T_b, calculate c and d in terms of a, b, T_a, and T_b.

12. If heat is generated at the centre of the sphere in Exercise 11 at a constant rate, the equation satisfied by $T(r)$ is

$$\frac{d}{dr}\left(r^2\frac{dT}{dr}\right) = kr^2,$$

where k is a constant.

   (a) Verify that

$$T(r) = \frac{kr^2}{6} + \frac{c}{r} + d$$

   satisfies the equation for any constants c and d.

   (b) If temperatures on the two spheres are maintained at constant values T_a and T_b, calculate c and d in terms of a, b, T_a, T_b, and k.

13. Find constants a, b, c, and d in order that the function $y = f(x) = ax^3 + bx^2 + cx + d$ has its first derivative equal to 4 when $x = 1$ and $y = 0$, and its second derivative equal to 5 when $x = 2$ and $y = 4$.

14. (a) Find a formula for $d^3(uv)/dx^3$ if $u = f(x)$ and $v = g(x)$.

   (b) On the basis of part (a) and Example 3.19, could you hazard a guess at a formula for $d^4(uv)/dx^4$ (i.e., do you see the pattern emerging)? Exercise 18 asks you to verify the correct formula.

∗ **15.** Evaluate $\dfrac{d^{2n}}{dx^{2n}}(x^2 - 1)^n$ for n a positive integer.

∗ **16.** The Green's function for the deflection of a diving board of length L as it bends under its own weight and any other loading is

$$G(x; X) = \frac{1}{6EI}(x - X)^3 h(x - X) - \frac{x^3}{6EI} + \frac{Xx^2}{2EI}$$

where $h(x - X)$ is the Heaviside function of Section 2.5, and E and I are constants depending on the material of the board and its cross-section. Think of $G(x; X)$ as a function of x that also depends on X where X can take on any value between 0 and L. It is the deflection at position x if the only force acting on the diving board is a unit force in the positive y-direction at position X. It also assumes that the board is massless.

   (a) Draw a graph of $G(x; X)$ when $X = L/2$. Is a part of it straight?

   (b) Verify that the board is straight when $x > X$ for any X.

   (c) Verify that $G(x; X)$, dG/dx, and $d^2 G/dx^2$ are all continuous for $0 \le x \le L$, except for a removable discontinuity at $x = X$.

   (d) Verify that $d^3 G/dx^3$ is continuous for all $x \ne X$, and has a jump discontinuity at $x = X$ of $(EI)^{-1}$.

∗ **17.** The Green's function for a beam of length L that has its ends clamped horizontally at $x = 0$ and $x = L$ is

$$G(x; X) = \frac{1}{6EI}(x - X)^3 h(x - X)$$

$$+ \frac{x^3}{6EIL^3}(-L^3 + 3LX^2 - 2X^3)$$

$$+ \frac{x^2}{2EIL^2}(X^3 - 2LX^2 + L^2 X).$$

(See Exercise 16 for a description of E and I and the interpretation of $G(x; X)$.)

   (a) Draw a graph of $G(x; X)$ when $X = L/2$. Is its tangent line horizontal at $x = 0$ and $x = L$? Is it symmetric about $x = L/2$? Does it have a removable discontinuity at $x = L/2$?

   (b) Verify that $G(x; X)$, dG/dx, and $d^2 G/dx^2$ are all continuous for $0 \le x \le L$, except for a removable discontinuity at $x = X$.

   (c) Verify that $d^3 G/dx^3$ is continuous for all $x \ne X$, and has a jump discontinuity at $x = X$ of $(EI)^{-1}$.

** **18.** If n is a positive integer, and u and v are functions of x, show by mathematical induction that

$$\frac{d^n}{dx^n}(uv) = \sum_{r=0}^{n} \binom{n}{r} \left(\frac{d^r u}{dx^r}\right) \left(\frac{d^{n-r} v}{dx^{n-r}}\right),$$

where $d^0 u/dx^0 = u$, and $\binom{n}{r}$ are the binomial coefficients

$$\binom{n}{r} = \frac{n!}{r!\,(n-r)!}.$$

(If you are not familiar with sigma notation, see Section 6.1. Mathematical induction is discussed in Appendix A.)

3.6 Velocity, Speed, and Acceleration

Most applications of differentiation are dealt with in Chapter 4, but velocity, speed, and acceleration are so important in engineering and the physical sciences that it is important to discuss them as soon as possible. In Section 3.1, we used the position function

$$x(t) = t^3 - 27t^2 + 168t + 20, \quad t \geq 0,$$

to define velocity as the instantaneous rate of change of displacement with respect to time. (Unless we indicate otherwise, *velocity* means instantaneous velocity, rather than average velocity.) We can now say that **velocity** is the derivative of displacement,

$$v(t) = \frac{dx}{dt} = 3t^2 - 54t + 168 \text{ m/s}.$$

In terms of the graph of the displacement function (Figure 3.32), velocity is the slope of the tangent line. For instance, $v(3) = 33$ m/s. The particle is moving to the right at 33 m/s; the slope of the tangent line at $(3, 308)$ is 33. (See the triangle on the tangent line and remember that the axes have different scales.) After 5 s, $v(5) = -27$ m/s. The particle is moving to the left at 27 m/s; the slope of the tangent line is -27 at $(5, 310)$. By factoring $v(t)$ in the form

$$v(t) = 3(t - 4)(t - 14),$$

we see that velocity is zero at $t = 4$ and $t = 14$. This is consistent with horizontal tangent lines at $(4, 324)$ and $14, -176)$.

FIGURE 3.32 Displacement function illustrates the velocity of object

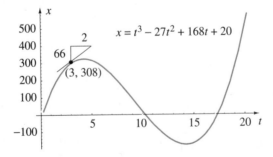

The particle is moving to the right whenever its velocity is positive, to the left when its velocity is negative. Graphically, it moves to the right when the slope of the tangent line is positive $(0 \leq t < 4$ and $t > 14)$, and to the left when the slope of the tangent line is negative $(4 < t < 14)$.

Speed is defined to be the magnitude of velocity,

$$\text{speed} = |v(t)|. \tag{3.17}$$

It represents how fast the particle is moving without regard for direction. For instance, at time $t = 0$, velocity and speed are both 168 m/s, but at $t = 5$, velocity is -33 m/s and speed is 33 m/s. Do not confuse velocity and speed; they are not interchangeable. Velocity can be positive or negative depending on direction of motion; speed is never negative.

EXAMPLE 3.20

Figure 3.33 shows the graph of the displacement function $x(t)$ for a particle moving along the x-axis during the time interval $0 \leq t \leq 10$. Use the graph to answer the following questions:

    (a) Is the particle moving to the left or right at $t = 7$ s?

    (b) How many times does the particle stop moving?

    (c) Is the velocity at $t = 0.5$ s greater or smaller than at $t = 3.5$ s?

    (d) Is the speed at $t = 0.5$ s greater or smaller than at $t = 3.5$ s?

    (e) Estimate the average velocity of the particle over the interval.

    (f) At what times would you say that the velocity is greatest and smallest?

    (g) At what times would you say that the speed is greatest and smallest?

FIGURE 3.33   Displacement function for $x(t)$

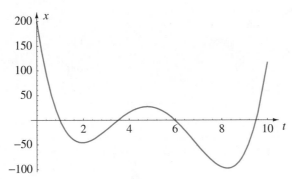

SOLUTION

    (a) Since the slope of the tangent line at $t = 7$ s is negative, the velocity is negative, and the particle is moving to the left.

    (b) The particle stops moving when its velocity is zero, and this occurs when the tangent line to the graph is horizontal. This happens three times.

    (c) At $t = 0.5$ s, the velocity is negative; at $t = 3.5$ s, it is positive. Thus, it is smaller at $t = 0.5$ s.

    (d) The tangent line at $t = 0.5$ s is much steeper than it is at $t = 3.5$ s. Since speed measures steepness (slope without regard for sign), speed is greater at 0.5 s.

    (e) Average velocity is the difference between initial and final displacements divided by the length of the time interval. If we estimate $x(0) = 200$ and $x(10) = 120$,

$$\text{average velocity} \approx \frac{120 - 200}{10} = -8 \text{ m/s}.$$

    (f) Since the slope of the graph appears to be greatest at $t = 10$ s, the velocity must be greatest then. The slope and velocity appear to be smallest at $t = 0$ s.

    (g) Because speed is the magnitude of velocity, it is never negative. Speed is zero if velocity is zero. Since this happens three times, speed is smallest (and has value 0) at the three times when the tangent line is horizontal. Speed is greatest when steepness of the graph is greatest. This is either at $t = 0$ s or $t = 10$ s. It is difficult to tell from the graph, but we favour a steeper tangent line at $t = 0$ s.

EXAMPLE 3.21

You are now told that the displacement function for the graph in Figure 3.33 is

$$x(t) = t^4 - 20t^3 + \frac{521t^2}{4} - \frac{1243t}{4} + \frac{399}{2}.$$

Answer the following questions:

(a) What are the speed and velocity at $t = 1$ s?

(b) Is the velocity equal to zero at $t = 2$ s?

(c) What is the average velocity of the particle for $0 \le t \le 10$?

(d) At what time is speed greatest?

(e) To three decimal places, at what times is the speed of the particle equal to 20 m/s?

FIGURE 3.34a   Graph of velocity

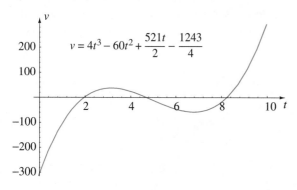

FIGURE 3.34b   Graph of speed

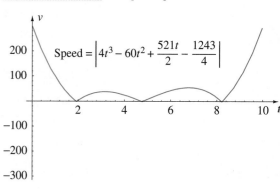

SOLUTION   The velocity of the particle is

$$v(t) = \frac{dx}{dt} = 4t^3 - 60t^2 + \frac{521t}{2} - \frac{1243}{4} \text{ m/s}.$$

The above questions will be answered algebraically, but it is helpful to visualize responses by plotting the velocity and speed functions. We have done this in Figure 3.34. Figure 3.34b reflects that part of the graph in Figure 3.34a below the t-axis, in the t-axis.

(a) At $t = 1$ s, velocity is

$$v(1) = 4(1)^3 - 60(1)^2 + \frac{521}{2} - \frac{1243}{4} = -\frac{425}{4} \text{ m/s}.$$

The speed is 425/4 m/s.

(b) Geometrically, the velocity is zero at $t = 2$ s if the tangent line in Figure 3.33 is horizontal at $t = 2$. It looks close. Equivalently, does the graph in Figure 3.34a cross the t-axis at $t = 2$? Again, the decision is not clear. Since $v(2) = 4(2)^3 - 60(2)^2 + (521/2)(2) - 1243/4 = 2.25$ m/s, the velocity is not zero at $t = 2$ s.

(c) Since displacements at $t = 0$ s and $t = 10$ s are 399/2 m and 117 m,

$$\text{average velocity} = \frac{117 - 399/2}{10} = -8.25 \text{ m/s}.$$

(d) Figure 3.34b indicates that speed is greatest at $t = 0$ s or $t = 10$ s, favouring $t = 0$ s. Because $v(0) = -1243/4$ m/s and $v(10) = 1117/4$ m/s, it follows that speed is greatest at $t = 0$ s [see also part (g) in Example 3.20].

(e) Solutions can be visualized as times when the height of the graph in Figure 3.34b is 20. Clearly, there are six such times. Alternatively, solutions are times when the ordinate in Figure 3.34a is ± 20, the same six times. Solutions of

$$20 = v(t) = 4t^3 - 60t^2 + \frac{521t}{2} - \frac{1243}{4}$$

are 2.306 95, 4.240 43, and 8.452 62. The zero intermediate value theorem can be used to confirm three-decimal-place solutions 2.307 s, 4.240 s, and 8.453 s. Three-decimal-place solutions of $v(t) = -20$ are 1.718 s, 5.300 s, and 7.982 s.

The velocity of a particle moving along the x-axis is the derivative of its displacement function $x(t)$ with respect to t,

$$v(t) = \frac{dx}{dt}. \qquad (3.18)$$

The (**instantaneous**) **acceleration** of the particle is defined as the rate of change of velocity with respect to time:

$$a(t) = \frac{dv}{dt} = \frac{d^2x}{dt^2}. \qquad (3.19)$$

In actual fact, $x(t)$, $v(t)$, and $a(t)$ are the x-components of the displacement, velocity, and acceleration vectors, respectively. In the absence of a complete discussion of vectors, we omit the terms *vector* and *component*, and simply call $x(t)$, $v(t)$, and $a(t)$ displacement, velocity, and acceleration. When distance is measured in metres and time in seconds, velocity is measured in metres per second (m/s). Since acceleration is the time derivative of velocity, its units must be units of velocity divided by units of time [i.e., metres per second per second (m/s/s)]. Usually, we shorten this by saying "metres per second squared" and write m/s^2.

For example, in Figure 3.2 the acceleration of the particle with displacement function $x(t) = t^3 - 27t^2 + 168t + 20$ is

$$a(t) = \frac{d}{dt}(3t^2 - 54t + 168) = 6t - 54 \text{ m/s}^2.$$

The graphical interpretation of acceleration requires the concept of "concavity." This is discussed in Section 4.4.

Consulting Project 3

An industrialist is having a problem with the design and manufacture of a cam. A cam is a machine part that rotates about an axis to cause periodic movement in another part, called a follower. The plate cam in Figure 3.35a rotates about an axis through the origin O and perpendicular to the plate. The follower moves up and down along the y-axis as point A on its end remains in contact with the cam. Suppose Figure 3.35b represents the displacement $y = f(\theta)$ of the follower, from its lowest position (1 cm above O), as a function of angle θ through which the cam has rotated.

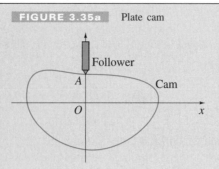

FIGURE 3.35a   Plate cam

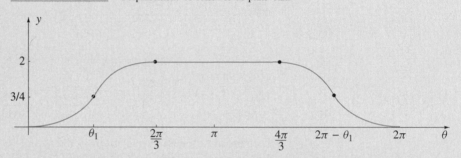

FIGURE 3.35b   Displacement of follower in plate cam

The follower rises 2 cm during a rotation of $2\pi/3$ radians. Because velocity of the follower is proportional to the slope of the graph, velocity increases in the interval $0 \leq \theta \leq \theta_1$, when the follower rises 3/4 cm, and decreases in the interval $\theta_1 \leq \theta \leq 2\pi/3$ when the follower rises the final 5/4 cm. The follower is stationary in the interval $2\pi/3 \leq \theta \leq \pi$. It then retraces its path back to zero displacement above its minimum position in a similar fashion. All curves are parabolas; the first has a minimum at $(0, 0)$ and the second has a maximum at $(2\pi/3, 2)$. The industrialist has two problems for us. One is to ensure that the follower moves smoothly; that is, its motion is not "jerky." Secondly, he wants us to give him a scale diagram of the cam itself. He has not specified angle θ_1.

SOLUTION   We begin be using our knowledge of equations for parabolas. Equations of the first two parabolas must be of the form

$$f(\theta) = \begin{cases} a\theta^2, & 0 \leq \theta \leq \theta_1, \\ A(\theta - 2\pi/3)^2 + 2, & \theta_1 \leq \theta \leq 2\pi/3, \end{cases}$$

as shown here where a and A are constants. For $(\theta_1, 0.75)$ to be a point on both parabolas,

$$0.75 = a\theta_1^2, \qquad 0.75 = A(\theta_1 - 2\pi/3)^2 + 2.$$

Now comes the most difficult consideration of the project, interpretation of the follower moving smoothly. This means that there can be no sudden changes in its velocity. Since velocity is related to slope of the curve, we ask where sudden changes in slope could occur. Slope changes gradually along parabolas, and there is no difficulty in the transition from parabola to horizontal straight line at $(0, 0)$ and $(2\pi/3, 2)$. The only questionable point is where the above two parabolas join. We must ensure that slopes of the two parabolas match at the point $(\theta_1, 0.75)$. Otherwise, there will be a jump in the velocity there. For the left-hand derivative of $a\theta^2$ and the right-hand derivative of $A(\theta - 2\pi/3)^2 + 2$ to be equal at $(\theta_1, 0.75)$, we must have

$$2a\theta_1 = 2A(\theta_1 - 2\pi/3).$$

We now have three equations in three unknowns, a, A, and θ_1. If we solve the first two equations for a and A in terms of θ_1, and substitute into the last equation,

$$2\left(\frac{0.75}{\theta_1^2}\right)\theta_1 = 2\left[\frac{0.75 - 2}{(\theta_1 - 2\pi/3)^2}\right](\theta_1 - 2\pi/3).$$

The solution of this equation is $\theta_1 = \pi/4$. This gives $a = 12/\pi^2$ and $A = -36/(5\pi^2)$. The industrialist therefore has no choice for angle θ_1. Smooth motion of the follower requires $\theta_1 = \pi/4$. The equation of the displacement curve for the follower above its minimum position is therefore

$$y = f(\theta) = \begin{cases} 12\theta^2/\pi^2, & 0 \le \theta \le \pi/4 \\ -36(\theta - 2\pi/3)^2/(5\pi^2) + 2, & \pi/4 \le \theta \le 2\pi/3 \\ 2, & 2\pi/3 < \theta \le \pi. \end{cases}$$

Now consider the shape of the cam itself. When we add unity to $f(\theta)$, we obtain the distance of the end A of the follower above O. This represents the distance from the centre of the cam to its outer edge as a function of angle through which it has rotated, call it $g(\theta) = 1 + f(\theta)$ (Figure 3.36a). To obtain a scale diagram of the cam, we need the equation of this curve. If (x, y) is a point on the curve, then trigonometry indicates that $x = g(\theta)\cos\theta$ and $y = g(\theta)\sin\theta$. Unfortunately, it is not possible for us to eliminate θ between these equations and find an equation in x and y only. On the other hand, electronic devices have programs to plot curves given in this form, and we will deal with them at length in Chapter 9 as parametrically defined curves. The resulting plot and required cam shape is shown in Figure 3.36b.

FIGURE 3.36a   Schematic for shape of cam

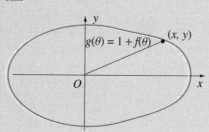

FIGURE 3.36b   Actual shape of cam

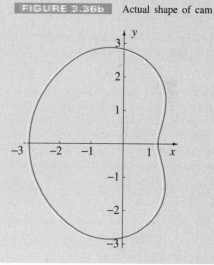

EXERCISES 3.6

1. The figure at right shows the graph of the displacement function $x(t)$ of a particle moving along the x-axis during the time interval $0 \le t \le 6$. Answer the following questions.

    (a) Is the particle to the left or right of the origin at times $t = 1$ and $t = 4$?

    (b) Is the particle moving to the right or to the left at times $t = 1/2$ and $t = 3$?

    (c) How many times does the particle change direction?

    (d) Is the velocity greater at $t = 7/2$ or at $t = 9/2$?

    (e) Is the speed greater at $t = 7/2$ or at $t = 9/2$?

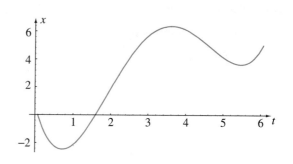

2. You are given that the displacement function in Exercise 1 is

$$x(t) = \frac{t^4}{6} - \frac{32t^3}{15} + \frac{25t^2}{3} - \frac{251t}{30}.$$

Confirm each answer in Exercise 1 algebraically.

3. Repeat Exercise 1 for the graph below.

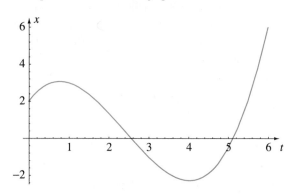

4. You are given that the displacement function in Exercise 3 is

$$x(t) = \frac{14t^3}{45} - \frac{101t^2}{45} + \frac{132t}{45} + 2.$$

Confirm each answer in Exercise 3 algebraically.

5. When $x(t)$ is the displacement function for a particle moving along the x-axis, the third derivative $x'''(t)$ is called *jerk*. Why is this name appropriate? Find jerk for the displacement function in Exercise 2.

6. Find jerk for the displacement function in Exercise 4. (See Exercise 5 for the definition of jerk.)

In Exercises 7–11 find the velocity and acceleration of an object that moves along the x-axis with the given position function. Assume that x is in metres and t is in seconds. Draw graphs of $x(t)$, $v(t)$, and $a(t)$, and examine the graphs from the point of view that ordinates on $v(t)$ represent slopes on $x(t)$, and ordinates on $a(t)$ are slopes on $v(t)$. Draw a graph of speed as a function of t also.

7. $x(t) = 2t + 5, \quad t \geq 5$ **8.** $x(t) = t^2 - 7t + 6, \quad t \geq 0$

9. $x(t) = t^2 + 5t + 10, \quad t \geq 1$

10. $x(t) = 4t - t^3, \quad t \geq 0$

11. $x(t) = 1/t, \quad t \geq 1$

Repeat the calculations for the above five exercises for Exercises 12–16 but plot graphs instead of drawing them.

12. $x(t) = -2t^3 + 2t^2 + 16t - 1, \quad t \geq 0$

13. $x(t) = t^3 - 9t^2 + 15t + 3, \quad t \geq 2$

14. $x(t) = t + 4/t, \quad t \geq 1$

15. $x(t) = (t - 4)/t^2, \quad t \geq 2$

16. $x(t) = (t - 1)^2 \sqrt{t}, \quad t \geq 1$

* 17. An object moving along the x-axis has position given by

$$x(t) = t^3 - 9t^2 + 24t + 1,$$

where x is measured in metres and $t \geq 0$ is time in seconds. Determine algebraically: (a) position, velocity, speed, and acceleration at $t = 3$ s; (b) when the object is instantaneously at rest; (c) when acceleration vanishes; (d) times when the object is moving to the right and left; (e) if and when velocity is 1 m/s; (f) if and when speed is 1 m/s; (g) if and when velocity is 20 m/s; and (h) if and when speed is 20 m/s.

* 18. Repeat Exercise 17 if

$$x(t) = t^3 - 9t^2 + 15t - 2.$$

* 19. Can the position curve $x = x(t)$ of a realistic particle moving along the x-axis be represented by a function $x(t)$ that has a discontinuity? Explain.

* 20. When an object moves with constant acceleration a along the x-axis, its position as a function of time t must be of the form

$$x = x(t) = \frac{1}{2}at^2 + bt + c,$$

where b and c are constants.

(a) If the object is at positions x_1 and x_2 at times t_1 and t_2, what is its average velocity over the time interval $t_1 \leq t \leq t_2$?

(b) At what time in this time interval is the instantaneous velocity equal to the average velocity? Where is the object at this time? Is it at the midpoint of the interval between x_1 and x_2, is it closer to x_1, or is it closer to x_2? (Assume in this part of the problem that $a > 0$ and that the velocity of the object is positive at time t_1.)

* 21. Find $f(\theta)$ in the consulting project of this section when rotation $2\pi/3$ is replaced by θ_2, and displacements $3/4$ and 2 are replaced by y_1 and y_2, respectively. In the process, show that $y_1/y_2 = \theta_1/\theta_2$, and that this implies that the point (θ_1, y_1) is on the line joining $(0, 0)$ and (θ_2, y_2).

* 22. The figure below represents a portion of the displacement $f(\theta)$ of a follower in a cam mechanism. It consists of a straight line portion between A and B and two parabolic portions OA and BC that are horizontal at O and C. Given are the rises y_1, y_2, and y_3, and angle θ_3. It is required that the slope of the curve be continuous at θ_1 and θ_2. By taking $f(\theta)$ in the form

$$y(\theta) = \begin{cases} a\theta^2, & 0 \leq \theta \leq \theta_1 \\ m\theta + b, & \theta_1 < \theta < \theta_2 \\ A(\theta - \theta_3)^2 + y_3, & \theta_2 \leq \theta \leq \theta_3 \end{cases}$$

show that

$$\theta_1 = \frac{2y_1\theta_3}{y_1 - y_2 + 2y_3}, \qquad \theta_2 = \frac{(y_1 + y_2)\theta_3}{y_1 - y_2 + 2y_3},$$

and find a, m, b, and A.

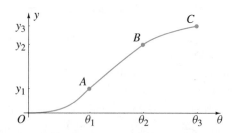

3.7 The Chain Rule and the Extended Power Rule

With the differentiation rules in Sections 3.2 and 3.4, we can differentiate any polynomial whatsoever. For instance, if $f(x) = x^3 - 3x^2 + 2x + 1$, then $f'(x) = 3x^2 - 6x + 2$. But consider the polynomial $f(x) = (2x^2 - 3)^8$, which has been conveniently factored for us. To find its derivative, we could expand $(2x^2 - 3)^8$ by the binomial theorem, say, then differentiate, and finally, simplify. We could also consider using the product rule over and over and over again. Thus, in spite of the fact that the rules of Sections 3.2 and 3.4 permit differentiation of $(2x^2 - 3)^8$, they are not convenient to use. Even more unpleasant would be differentiation of the rational function $g(x) = 1/(3x^2 + 8)^{12}$. In this section, we obtain results that enable us to differentiate quickly much wider classes of functions, which include $(2x^2 - 3)^8$ and $1/(3x^2 + 8)^{12}$.

Suppose that y is defined as a function of u by

$$y = f(u) = \frac{u}{u + 1},$$

and u, in turn, is defined as a function of x by

$$u = g(x) = \frac{\sqrt{x}}{x + 2}.$$

These equations imply that y is a function of x; indeed, y is the composition of f and g,

$$y = f\big(g(x)\big) = \frac{\dfrac{\sqrt{x}}{x + 2}}{\dfrac{\sqrt{x}}{x + 2} + 1}.$$

After some algebraic simplification we find that

$$y = \frac{\sqrt{x}}{\sqrt{x} + x + 2},$$

and we can therefore calculate

$$\frac{dy}{dx} = \frac{(\sqrt{x} + x + 2)\left(\dfrac{1}{2\sqrt{x}}\right) - \sqrt{x}\left(\dfrac{1}{2\sqrt{x}} + 1\right)}{(\sqrt{x} + x + 2)^2}.$$

This can be reduced to

$$\frac{dy}{dx} = \frac{2 - x}{2\sqrt{x}(\sqrt{x} + x + 2)^2}.$$

But notice that if we differentiate the original functions, we obtain

$$\frac{dy}{du} = \frac{(u + 1)(1) - u(1)}{(u + 1)^2} = \frac{1}{(u + 1)^2} \quad \text{and}$$

$$\frac{du}{dx} = \frac{(x + 2)\left(\dfrac{1}{2\sqrt{x}}\right) - \sqrt{x}(1)}{(x + 2)^2} = \frac{2 - x}{2\sqrt{x}(x + 2)^2}.$$

The product of these derivatives is

$$\frac{dy}{du}\frac{du}{dx} = \frac{2 - x}{2\sqrt{x}(x + 2)^2(u + 1)^2}$$

$$= \frac{2 - x}{2\sqrt{x}(x + 2)^2\left(\dfrac{\sqrt{x}}{x + 2} + 1\right)^2} = \frac{2 - x}{2\sqrt{x}(\sqrt{x} + x + 2)^2}.$$

Consequently, for this example, we can write

$$\frac{dy}{dx} = \frac{dy}{du}\frac{du}{dx}.$$

According to the following theorem, the derivative of a composite function $y = f(g(x))$ can always be calculated by the above formula.

THEOREM 3.9 (Chain Rule)

If $y = f(u)$ and $u = g(x)$ are differentiable functions, then the derivative of the composite function $y = f(g(x))$ is

$$\frac{dy}{dx} = \frac{dy}{du}\frac{du}{dx}. \tag{3.20a}$$

Increment notation is particularly useful in proving the chain rule.

PROOF   By equation 3.11, the derivative of the composite function is

$$\frac{dy}{dx} = \lim_{\Delta x \to 0}\frac{\Delta y}{\Delta x} = \lim_{\Delta x \to 0}\frac{f(g(x + \Delta x)) - f(g(x))}{\Delta x}.$$

Now a change Δx in x produces a change $g(x + \Delta x) - g(x)$ in u. If we denote this change by Δu, then it in turn produces the change $f(u + \Delta u) - f(u)$ in y. We may write

$$\frac{dy}{dx} = \lim_{\Delta x \to 0}\frac{f(u + \Delta u) - f(u)}{\Delta x}.$$

Since $f(u)$ is differentiable, its derivative exists and is defined by

$$f'(u) = \lim_{\Delta u \to 0}\frac{f(u + \Delta u) - f(u)}{\Delta u}.$$

An equivalent way to express the fact that this limit is $f'(u)$ is to say that

$$\frac{f(u + \Delta u) - f(u)}{\Delta u} = f'(u) + \epsilon,$$

where ϵ must satisfy the condition $\lim_{\Delta u \to 0}\epsilon = 0$. We may write

$$f(u + \Delta u) - f(u) = [f'(u) + \epsilon]\Delta u,$$

and if we substitute this into the second expression for dy/dx above, we obtain

$$\frac{dy}{dx} = \lim_{\Delta x \to 0}\frac{[f'(u) + \epsilon]\Delta u}{\Delta x} = \lim_{\Delta x \to 0}\left\{[f'(u) + \epsilon]\frac{\Delta u}{\Delta x}\right\}.$$

But

$$\lim_{\Delta x \to 0}\frac{\Delta u}{\Delta x} = \lim_{\Delta x \to 0}\frac{g(x + \Delta x) - g(x)}{\Delta x} = \frac{du}{dx}.$$

Furthermore, since $g(x)$ is differentiable, it is continuous (Theorem 3.6), and this implies that $\Delta u \to 0$ as $\Delta x \to 0$. Consequently, $\lim_{\Delta x \to 0}\epsilon = \lim_{\Delta u \to 0}\epsilon = 0$, and these results give

$$\frac{dy}{dx} = f'(u)\frac{du}{dx} = \frac{dy}{du}\frac{du}{dx}. \qquad \blacksquare$$

This result is called the chain rule for the derivative of a composite function. It expresses the derivative of a composite function as the product of the derivatives of the functions in the composition. From the point of view of rates of change, the chain rule seems quite reasonable. It states that if a variable is defined in terms of a second variable, which is in turn defined in terms of a third variable, then the rate of change of the first variable with respect to the third is the rate of change of the first with respect to the second multiplied by the rate of change of the second with respect to the third. For example, if car A travels twice as fast as car B, and car B travels three times as fast as car C, then car A travels six times as fast as car C.

It is essential to understand the difference between the derivatives dy/dx and dy/du in equation 3.20a. The second, dy/du, is the derivative of y regarded as a function of u, the given function $y = f(u)$. On the other hand, dy/dx is the derivative of the composite function $f(g(x))$.

The chain rule can also be expressed in terms of the circle notation for composite functions. With $f(g(x))$ denoted by $(f \circ g)(x)$, equation 3.20a takes the form

$$(f \circ g)'(x) = f'(g(x)) g'(x), \tag{3.20b}$$

or

$$(f \circ g)'(x) = (f' \circ g)(x) g'(x). \tag{3.20c}$$

EXAMPLE 3.22

Find dy/dt at $t = 4$ when $y = x^2 - x$ and $x = \sqrt{t}/(t + 1)$.

SOLUTION   By the chain rule

$$\frac{dy}{dt} = \frac{dy}{dx}\frac{dx}{dt} = (2x - 1)\left[\frac{(t + 1)(1/2)t^{-1/2} - \sqrt{t}(1)}{(t + 1)^2}\right].$$

When $t = 4$, we find $x = \sqrt{4}/(4 + 1) = 2/5$. We use the notation $\left.\dfrac{dy}{dt}\right|_{t=4}$ to represent dy/dt evaluated at $t = 4$,

$$\left.\frac{dy}{dt}\right|_{t=4} = [2(2/5) - 1]\left[\frac{(4 + 1)(1/2)(4)^{-1/2} - \sqrt{4}}{(4 + 1)^2}\right] = \frac{3}{500}.$$

The Extended Power Rule

When the function $f(u)$ in Theorem 3.9 is a power function, the chain rule gives what we call the **extended power rule**, often considered the most important differentiation formula of calculus.

COROLLARY 3.9.1

When $u = g(x)$ is differentiable,

$$\frac{d}{dx}u^n = nu^{n-1}\frac{du}{dx}. \tag{3.21}$$

It is essentially power rule 3.7 in Section 3.2 with an extra factor du/dx to account for the fact that what is under the power (u) is not just x; it is a function of x. In the special case that u is equal to x, equation 3.21 reduces to 3.7. Power rule 3.7 was verified only for n a nonnegative integer, but we have been using it for any real number n. We shall use its generalization 3.21 for any real number n also in spite of the fact that, in effect, it has only been verified for nonnegative integers. In Section 3.11 we provide the justification.

It is important not to confuse rules 3.7 and 3.21. Rule 3.7 can be used only for x^n; if anything other than x is raised to a power, formula 3.21 should be used. The most common error in using equation 3.21 is to forget the du/dx. With equation 3.21 it is a simple matter to differentiate the functions in the first paragraph of this section:

$$\frac{d}{dx}(2x^2 - 3)^8 = 8(2x^2 - 3)^7 \frac{d}{dx}(2x^2 - 3) = 8(2x^2 - 3)^7(4x) = 32x(2x^2 - 3)^7$$

and

$$\frac{d}{dx}\left[\frac{1}{(3x^2 + 8)^{12}}\right] = \frac{d}{dx}(3x^2 + 8)^{-12} = -12(3x^2 + 8)^{-13}\frac{d}{dx}(3x^2 + 8)$$

$$= \frac{-12}{(3x^2 + 8)^{13}}(6x) = \frac{-72x}{(3x^2 + 8)^{13}}.$$

Notice that in neither of these examples do you see the letter u, although rule 3.21 is stated in terms of u's. We could have introduced u, in the first example, say, by setting $u = 2x^2 - 3$, and proceeded as follows: With $u = 2x^2 - 3$,

$$\frac{d}{dx}(2x^2 - 3)^8 = \frac{d}{dx}u^8 = \frac{d}{du}(u^8)\frac{du}{dx} = 8u^7(4x) = 32xu^7 = 32x(2x^2 - 3)^7.$$

But this is unnecessary; with an understanding of equation 3.21, we should proceed directly to the derivatives without defining an intermediate variable. With a little practice, the writing should be shortened even more. For example, calculation of the derivative of $(2x^2 - 3)^8$ should appear as

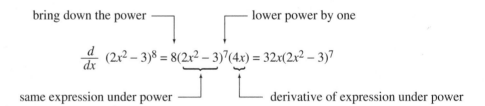

EXAMPLE 3.23

A balloon always remains spherical when it is being filled. If the radius of the balloon is increasing at a rate of 2 mm/s when the radius is 10 cm, how fast is the volume changing at this instant?

SOLUTION   Because the rate of change of the radius r of the balloon is 2 mm/s, we can say that the derivative

$$\frac{dr}{dt} = 0.2 \text{ cm/s},$$

when $r = 10$ cm. Since the volume of the balloon is $V = (4/3)\pi r^3$, and r is a function of time t, we may differentiate this equation with respect to t,

$$\frac{dV}{dt} = \frac{4}{3}\pi\left(3r^2\frac{dr}{dt}\right) = 4\pi r^2\frac{dr}{dt}.$$

When $r = 10$,

$$\frac{dV}{dt} = 4\pi(10)^2(0.2) = 80\pi.$$

Thus, the volume is increasing at 80π cm^3/s.

EXERCISES 3.7

In Exercises 1–8 use the chain rule to find dy/dx.

1. $y = t^2 + \dfrac{1}{t}, \quad t = x^2 + 1$

2. $y = \dfrac{u}{u+1}, \quad u = \sqrt{x} + 1$

3. $y = (u^2 + 1)(u + 1), \quad u = \dfrac{1}{\sqrt{x} - 4}$

4. $y = \dfrac{s}{s^2 - 2}, \quad s = x^2 - 2x + 3$

5. $y = (v^2 + v)(\sqrt{v} + 1), \quad v = \dfrac{x}{x^2 - 1}$

6. $y = \dfrac{t+3}{t-4}, \quad t = \dfrac{x-2}{x+1}$

7. $y = \dfrac{t^2 + 3}{t - 4}, \quad t = (3x + 2)(x^2 + 4x)$

8. $y = u^2(1 + \sqrt{u}), \quad u = \dfrac{x+1}{x - x^2}$

In Exercises 9–36 find dy/dx, where $y = f(x)$.

9. $f(x) = x(x^3 + 3)^4$

10. $f(x) = x\sqrt{x+1}$

11. $f(x) = x^2(2x + 1)^2$

12. $f(x) = \dfrac{x}{\sqrt{2x+1}}$

13. $f(x) = (x+2)^2(x^2 + 3)$

14. $f(x) = \dfrac{(2x-1)^2}{3x+5}$

15. $f(x) = \dfrac{3x+5}{(2x-1)^2}$

16. $f(x) = x^3(2 - 5x^2)^{1/3}$

17. $f(x) = \dfrac{x^3}{(2 - 5x^2)^{1/3}}$

18. $f(x) = (x+1)^2(3x+1)^3$

19. $f(x) = \dfrac{x^{1/3}}{1 - \sqrt{x}}$

20. $f(x) = \dfrac{\sqrt{2-3x}}{x^2}$

21. $f(x) = \left(\dfrac{x^3 - 1}{2x^3 + 1}\right)^4$

22. $f(x) = \sqrt[4]{\dfrac{2-x}{2+x}}$

23. $f(x) = (x^3 - 2x^2)^3(x^4 - 2x)^5$

24. $f(x) = (x+5)^4\sqrt{1 + x^3}$

25. $f(x) = \dfrac{x\sqrt{1 - x^2}}{(3+x)^{1/3}}$

26. $f(x) = x(x+5)^4\sqrt{1 + x^3}$

* 27. $f(x) = \dfrac{x^2(x^3 + 3)^2}{(x - 2)(x + 5)^2}$

* 28. $f(x) = x\sqrt{1 + x\sqrt{1 + x}}$

* 29. $y = t^2 + \dfrac{1}{t^3}, \quad t = \sqrt{4 - x^2}$

* 30. $y = (2s - s^2)^{1/3}, \quad s = \dfrac{1}{x^2 + 5}$

* 31. $y = \dfrac{v^2}{v^3 - 1}, \quad v = x\sqrt{x^2 - 1}$

* 32. $y = \dfrac{u}{u + 5}, \quad u = \dfrac{\sqrt{x} - 1}{x}$

* 33. $y = u^4(u^3 - 2u)^2, \quad u = \sqrt{x - 2x^2}$

* 34. $y = t + \sqrt{t + \sqrt{t}}, \quad t = \dfrac{x^2 + 1}{x^2 - 1}$

* 35. $y = \left(\dfrac{v^2 + 1}{1 - v^3}\right)^3, \quad v = \dfrac{1}{x^3 + 3x^2 + 2}$

* 36. $y = \dfrac{\sqrt{k}}{1 + k + k^2}, \quad k = x(x^2 + 5)^5$

* 37. If $y = f(u)$, $u = g(s)$, and $s = h(x)$, show that

$$\frac{dy}{dx} = \frac{dy}{du}\frac{du}{ds}\frac{ds}{dx}.$$

* 38. When an electrostatic charge $q = 5 \times 10^{-6}$ C is at a distance r metres from a stationary charge $Q = 3 \times 10^{-6}$ C, the magnitude of the force of repulsion of Q on q is

$$F = \frac{Qq}{4\pi\epsilon_0 r^2} \text{ newtons,}$$

where $\epsilon_0 = 8.85 \times 10^{-12}$. If q is moved directly away from Q at 2 m/s, how fast is F changing when $r = 2$ m?

* 39. When a mass m of 5 kg is r metres from the centre of the earth, the magnitude of the force of attraction of the earth on m is

$$F = \frac{GmM}{r^2} \text{ newtons,}$$

where M is the mass of the earth in kilograms and $G = 6.67 \times 10^{-11}$. If m is falling at 100 km/h when it is 5 km above the surface of the earth, how fast is F changing? The mean density of the earth is 5.52×10^3 kg/m^3, and its mean radius is 6370 km.

∗ **40.** Show that if an equation can be solved for both y in terms of x $[y = f(x)]$ and x in terms of y $[x = g(y)]$, then

$$\frac{dy}{dx} = \frac{1}{\frac{dx}{dy}}.$$

∗ **41.** Prove that the derivative of an odd function is an even function, and that the derivative of an even function is an odd function. Recall Definition 1.4 in Section 1.5.

∗ **42.** Determine whether the following very simple proof of the chain rule has a flaw. If Δu denotes the change in $u = g(x)$ resulting from a change Δx in x, and Δy is the change in $y = f(u)$ resulting from Δu, then

$$\frac{dy}{dx} = \lim_{\Delta x \to 0} \frac{\Delta y}{\Delta x} = \lim_{\Delta x \to 0} \frac{\Delta y}{\Delta u}\frac{\Delta u}{\Delta x} = \left(\lim_{\Delta x \to 0} \frac{\Delta y}{\Delta u}\right)\left(\lim_{\Delta x \to 0} \frac{\Delta u}{\Delta x}\right)$$

$$= \left(\lim_{\Delta u \to 0} \frac{\Delta y}{\Delta u}\right)\left(\lim_{\Delta x \to 0} \frac{\Delta u}{\Delta x}\right)$$

$$= \frac{dy}{du}\frac{du}{dx}.$$

In Exercises 43–46 find d^2y/dx^2.

∗ **43.** $y = v^2 + v, \quad v = \dfrac{x}{x+1}$

∗ **44.** $y = (u+1)^3 - \dfrac{1}{u}, \quad u = x + \sqrt{x+1}$

∗ **45.** $y = \sqrt{t-1}, \quad t = (x + x^2)^2$

∗ **46.** $y = \dfrac{s}{s+6}, \quad s = \dfrac{\sqrt{x}}{1+\sqrt{x}}$

In Exercises 47–54 assume that $f(u)$ is a differentiable function of u. Find the derivative of the given function with respect to x in as simplified a form as possible. Then set $f(u) = u^3 - 2u$ and simplify further.

∗ **47.** $f(2x + 3)$

∗ **48.** $[f(3 - 4x)]^2$

∗ **49.** $f(1 - x^2)$

∗ **50.** $f(x + 1/x)$

∗ **51.** $f\big(f(x)\big)$

∗ **52.** $\sqrt{3 - 4[f(1 - 3x)]^2}$

∗ **53.** $\dfrac{f(-x)}{3 + 2f(x^2)}$

∗ **54.** $f\big(x - f(x)\big)$

In Exercises 55–58 show that the families of curves are orthogonal trajectories (see Exercise 23 in Section 3.3). Draw both families of curves.

∗ **55.** $y = mx, \quad x^2 + y^2 = r^2$

∗ **56.** $y = ax^2, \quad x^2 + 2y^2 = c^2$

∗ **57.** $x^2 - y^2 = C_1, \quad xy = C_2$

∗ **58.** $2x^2 + 3y^2 = C^2, \quad y^2 = ax^3$

∗ **59.** If $y = f(u)$ and $u = g(x)$, use the chain rule to show that

$$\frac{d^2y}{dx^2} = \frac{d^2u}{dx^2}\frac{dy}{du} + \frac{d^2y}{du^2}\left(\frac{du}{dx}\right)^2.$$

∗ **60.** Use the result of Exercise 59 to find d^2y/dx^2 at $x = 1$ when

$$y = (u+1)^3, \quad u = 3x - \frac{2}{x^2}.$$

∗ **61.** Find the rate of change of $y = f(x) = x^9 + x^6$ with respect to x^3.

∗ **62.** Find the rate of change of $y = f(x) = \sqrt{1 - x^2}$ with respect to $x/(x+1)$.

∗ **63.** If $y = f(u)$ and $u = g(x)$, show that

$$\frac{d^3y}{dx^3} = \frac{d^3u}{dx^3}\frac{dy}{du} + 3\frac{d^2y}{du^2}\frac{d^2u}{dx^2}\frac{du}{dx} + \frac{d^3y}{du^3}\left(\frac{du}{dx}\right)^3.$$

∗ **64.** At what point(s) on the hyperbola $x^2 - 16y^2 = 16$ do tangent lines pass through the point $(2, 3)$?

∗ **65.** Generalize the result of equation 3.13 to find a formula for

$$\frac{d}{dx}|f(x)|^n \qquad \text{for } n > 1 \text{ an integer.}$$

∗ **66.** The curve $x^2y^2 = (x+1)^2(4 - x^2)$ is called a *conchoid of Nicomedes*. Plot the curve and find points where its tangent line is horizontal.

∗∗ **67.** Repeat Exercise 34 in Section 3.2 if the hill is the arc of a circle.

3.8 Implicit Differentiation

We say that y is defined **explicitly as a function** of x if the dependence of y on x is given in the form

$$y = f(x). \tag{3.22}$$

Examples are $y = x^2$, $y = 3x + \sin x$, and $y = 1/(x+1)$. In each case, the dependent variable stands alone on the left side of the equation. The differentiation rules in Theorems 3.1 to 3.5, 3.7, and 3.8 are applicable to explicitly defined functions.

An equation in x and y may define y as a function of x even when it is not in form 3.22. Equations for which y has not been separated out are often written in the generic form

$$F(x, y) = 0. \tag{3.23}$$

The notation $F(x, y)$ is used to denote an expression that depends on two variables x and y. For example, the volume of a right circular cylinder depends on its radius r and height h; in particular, $V = \pi r^2 h$. In such a case we write that $V = F(r, h) = \pi r^2 h$. Examples of equations of form 3.23 are $y - x^3 = 0$ and $y^3 - 3y - 2x = 0$. Equation 3.23 is said to define y **implicitly as a function** of x in some domain D, if for each x in D, there is one, and only one, value of y for which (x, y) satisfies the equation. The equation $y - x^3 = 0$ defines y implicitly as a function of x for all (real) x. The explicit definition of the function is $y = x^3$. The equation $y^3 - 3y - 2x = 0$ does not define y as a function of x. For each x in the interval $-1 < x < 1$, there are three solutions of the equation for y. For $x < -1$ and $x > 1$, the equation has only one value of y corresponding to each x. This is most easily seen from the plot of the curve in Figure 3.37. We shall have more to say about the equation $y^3 - 3y - 2x = 0$ later in this section.

Each of the following equations defines y as a function of x, and does so implicitly:

$$y + x^2 - 2x + 5 = 0;$$

$$2y + x^2 + y^2 - 25 = 0, \quad y \geq -1;$$

$$x + y^5 + x^2 + y = 0.$$

It is easy to solve the first equation for the explicit definition of the function,

$$y = -x^2 + 2x - 5,$$

and find its derivative,

$$\frac{dy}{dx} = -2x + 2.$$

We can also obtain an explicit definition of the function defined by the second equation, but not so simply. If we write

$$y^2 + 2y + (x^2 - 25) = 0,$$

and use quadratic formula 1.19, we obtain

$$y = \frac{-2 \pm \sqrt{4 - 4(x^2 - 25)}}{2} = -1 \pm \sqrt{26 - x^2}.$$

Since y is required to be greater than or equal to -1, we must choose

$$y = -1 + \sqrt{26 - x^2}.$$

This is the explicit definition of the function, and it is now easy to find the derivative of y with respect to x:

$$\frac{dy}{dx} = \frac{1}{2\sqrt{26 - x^2}}(-2x) = \frac{-x}{\sqrt{26 - x^2}}.$$

The third equation is quite different, for it is impossible to solve this equation for an explicit definition of the function. Does this mean that it is also impossible to obtain dy/dx? The answer is no! To see this, we differentiate both sides of the equation with respect to x, keeping in mind that y is a function of x. Only the term in y^5 presents any difficulty, but its derivative can be calculated using extended power rule 3.21,

$$1 + 5y^4 \frac{dy}{dx} + 2x + \frac{dy}{dx} = 0.$$

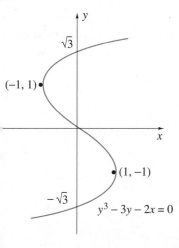

FIGURE 3.37 For $y^3 - 3y - 2x = 0$, y is not a function of x

We can now solve this equation for dy/dx by grouping the two terms in dy/dx on one side of the equation, and transposing the remaining two terms:

$$(5y^4 + 1)\frac{dy}{dx} = -1 - 2x.$$

Division by $5y^4 + 1$ gives the required derivative,

$$\frac{dy}{dx} = -\frac{2x + 1}{5y^4 + 1}.$$

This process of differentiating an equation that implicitly defines a function is called **implicit differentiation**. It could also have been used in each of the first two examples. If $y + x^2 - 2x + 5 = 0$ is differentiated with respect to x, we have

$$\frac{dy}{dx} + 2x - 2 = 0,$$

from which, as before,

$$\frac{dy}{dx} = -2x + 2.$$

When $2y + x^2 + y^2 - 25 = 0$ is differentiated with respect to x, we find that

$$2\frac{dy}{dx} + 2x + 2y\frac{dy}{dx} = 0,$$

from which

$$(2 + 2y)\frac{dy}{dx} = -2x,$$

and therefore

$$\frac{dy}{dx} = \frac{-x}{y + 1}.$$

Although this result appears different from the previous expression for dy/dx, when we recall that $y = -1 + \sqrt{26 - x^2}$, we find that

$$\frac{dy}{dx} = \frac{-x}{-1 + \sqrt{26 - x^2} + 1} = \frac{-x}{\sqrt{26 - x^2}}.$$

These examples illustrate that if implicit differentiation is used to obtain a derivative, then the result may depend on y as well as x. Naturally, if we require the derivative at a certain value of x, then the y-value used is determined by the original equation defining y implicitly as a function of x.

EXAMPLE 3.24   ▼

Assuming that y is defined implicitly as a function of x by the equation

$$x^3y^3 + x^2y + 2x = 12,$$

find dy/dx when $x = 1$.

SOLUTION   When we differentiate both sides of the equation with respect to x, using the product rule on the first two terms, we find

$$3x^2y^3 + 3x^3y^2\frac{dy}{dx} + 2xy + x^2\frac{dy}{dx} + 2 = 0.$$

Thus,

$$(3x^3y^2 + x^2)\frac{dy}{dx} = -(2 + 2xy + 3x^2y^3) \quad \text{and}$$

$$\frac{dy}{dx} = -\frac{2 + 2xy + 3x^2y^3}{3x^3y^2 + x^2}.$$

When $x = 1$ is substituted into the given equation defining y as a function of x, the result is

$$0 = y^3 + y - 10 = (y - 2)(y^2 + 2y + 5),$$

and the only solution of this equation is $y = 2$. We now substitute $x = 1$ and $y = 2$ into the formula for dy/dx to calculate the derivative at $x = 1$,

$$\frac{dy}{dx}\bigg|_{x=1} = -\frac{2 + 2(1)(2) + 3(1)^2(2)^3}{3(1)^3(2)^2 + (1)^2} = -\frac{30}{13}.$$

Implicit differentiation can also be used to find second- and higher-order derivatives of functions that are defined implicitly. Calculations can be messy, but the principles are the same. For example, when the equation $x^5 + y^3 + y^2 = 1$ defines y implicitly as a function of x, it is straightforward to calculate that

$$\frac{dy}{dx} = \frac{-5x^4}{3y^2 + 2y}.$$

To find the second derivative d^2y/dx^2, we differentiate both sides of this equation with respect to x. We use the quotient rule on the right, and when differentiating the denominator we once again keep in mind that y is a function of x. The result is

$$\frac{d^2y}{dx^2} = \frac{(3y^2 + 2y)(-20x^3) + 5x^4\left(6y\dfrac{dy}{dx} + 2\dfrac{dy}{dx}\right)}{(3y^2 + 2y)^2}.$$

We now replace dy/dx by its expression in terms of x and y,

$$\frac{d^2y}{dx^2} = \frac{-20x^3(3y^2 + 2y) + 5x^4(6y + 2)\left(\dfrac{-5x^4}{3y^2 + 2y}\right)}{(3y^2 + 2y)^2},$$

and bring the two terms in the numerator to a common denominator:

$$\frac{d^2y}{dx^2} = \frac{\dfrac{-20x^3(3y^2 + 2y)^2 - 25x^8(6y + 2)}{(3y^2 + 2y)}}{(3y^2 + 2y)^2}$$

$$= -\frac{20x^3(3y^2 + 2y)^2 + 25x^8(6y + 2)}{(3y^2 + 2y)^3}.$$

It would not be a pleasant task to proceed to the third derivative of y with respect to x.

In the above examples we seem to have adopted the principle that both sides of an equation can be differentiated with respect to the same variable. This is not always true. For example, if we differentiate both sides of the equation $4x = 2x$, we obtain the ludicrous result that $4 = 2$. Obviously, then, this equation cannot be differentiated with respect to x.

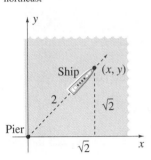

FIGURE 3.38 Ship sailing northeast

Possibly the reason that differentiation fails in the above example is that the equation contains only one variable, namely, x. Perhaps a more reasonable question might be: Can every equation containing two variables be differentiated? To answer this question we consider the situation of a ship heading northeast from a pier. Let us choose east as the positive x-direction and north as the positive y-direction, both originating from the pier (Figure 3.38). Since the x- and y-coordinates of the ship change at the same rate, it follows that the rate of change of the y-coordinate of the ship with respect to its x-coordinate when the ship is 2 km from the pier is equal to 1. We can get this result by noting that the path of the ship is the line $y = x$, and the derivative of this gives $dy/dx = 1$. But consider the following argument.

When the ship is 2 km from the pier, $x = y = \sqrt{2}$, and therefore

$$x^2 + y^2 = 4.$$

If we differentiate this equation with respect to x, we find that

$$2x + 2y\frac{dy}{dx} = 0 \qquad \text{or} \qquad \frac{dy}{dx} = -\frac{x}{y},$$

and with $x = y = \sqrt{2}$,

$$\frac{dy}{dx} = -1.$$

Differentiation of the equation $x^2 + y^2 = 4$, which contains two variables, has led to an erroneous result.

These two examples have certainly illustrated that not all equations can be differentiated. What then distinguishes an equation that can be differentiated from one that cannot? Recall that differentiation is a limiting process. Evaluate whatever is to be differentiated, say $f(x)$, at $x + h$, subtract its value at x, divide by h, and take the limit as $h \to 0$. If both sides of an equation are to be differentiated with respect to a variable, then both sides must be equal for a continuous range of values of that variable. The equation $4x = 2x$ cannot be differentiated because it is true only for $x = 0$. The equation $x^2 + y^2 = 4$ cannot be differentiated in the ship example because *in that example* it is valid only when the ship is 2 km from the pier (i.e., only when $x = y = \sqrt{2}$). The equation $y = x$ can be differentiated because it is valid at every point along the path of the ship. This is an extremely important principle, and we will return to it many times. To emphasize it once again, *we may differentiate an equation with respect to a variable only if the equation is valid for a continuous range of values of that variable.*

Each of the three equations at the beginning of this section was given as defining y as a function of x, and as such defines the function for some domain of values for x. Differentiation of these equations was therefore acceptable according to the principle stated above.

When we are told that equation 3.23 defines y as a function of x, implicit differentiation leads to the derivative dy/dx. But how can we tell whether an equation $F(x, y) = 0$ defines y implicitly as a function of x? Additionally, given that $F(x, y) = 0$ does define y as a function of x, how do we know that the expression for dy/dx obtained by implicit differentiation is a valid representation of the derivative of the function? After all, functions do not always have derivatives at all points in their domains.

Answers to these questions are intimately related. To see how, suppose that an equation $F(x, y) = 0$ in x and y is satisfied by a point (x_0, y_0), and when the equation is differentiated implicitly it leads to a quotient for dy/dx,

$$\frac{dy}{dx} = \frac{P(x, y)}{Q(x, y)}.$$

It can be shown that if $Q(x_0, y_0) \neq 0$, then the equation $F(x, y) = 0$ defines y implicitly as a function of x for some open interval containing x_0, and the derivative of this function at x_0 is $P(x_0, y_0)/Q(x_0, y_0)$. Proofs are usually given in advanced books on mathematical analysis. When $Q(x_0, y_0) = 0$, two possibilities exist. First, the equation $F(x, y) = 0$ might not define y as a function of x in an interval around x_0. Second, the equation might define a function, but the function does not have a derivative at x_0. To illustrate, consider first the equation $y^3 - 3y - 2x = 0$, which we introduced earlier in this section. The curve defined by this equation is shown in Figure 3.37, and it clearly illustrates that the equation does not define y as a function of x. It does define the three functions of x in Figure 3.39. Suppose we differentiate the equation with respect to x,

$$3y^2 \frac{dy}{dx} - 3\frac{dy}{dx} - 2 = 0,$$

and solve for dy/dx,

$$\frac{dy}{dx} = \frac{2}{3(y^2 - 1)}.$$

This derivative is obviously undefined at the points $(-1, 1)$ and $(1, -1)$, and these are precisely the points that separate the original curve into three parts. It is impossible to find a portion of the curve around either of these points that defines y as a function of x. At any other point on the curve, the formula $dy/dx = (2/3)(y^2 - 1)^{-1}$ is a valid representation for the derivative for whichever function of Figure 3.39 contains the point.

FIGURE 3.39   Division of a curve into parts each of which represents a function

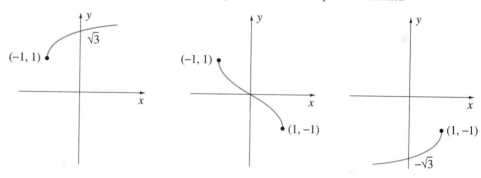

Consider now the equation $x = y^3$, shown graphically in Figure 3.40. Implicit differentiation leads to

$$\frac{dy}{dx} = \frac{1}{3y^2}.$$

Clearly, y is a function of x for all x, but dy/dx is undefined at $(0, 0)$ since the tangent line is vertical at this point.

FIGURE 3.40   Graph with vertical tangent line at $(0, 0)$

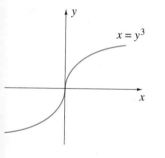

EXAMPLE 3.25

FIGURE 3.41   Curve with no tangent line at $(0, 0)$

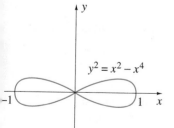

The curve $y^2 = x^2 - x^4$ is called a *lemniscate*. It is plotted in Figure 3.41. Use implicit differentiation to find dy/dx and discuss what happens when the point $(0, 0)$ is substituted into the result.

SOLUTION   Implicit differentiation of $y^2 = x^2 - x^4$ gives

$$2y\frac{dy}{dx} = 2x - 4x^3,$$

from which

$$\frac{dy}{dx} = \frac{x(1 - 2x^2)}{y}.$$

This result is undefined at $(0, 0)$, and Figure 3.41 indicates why. The equation does not define y as a function of x around $x = 0$.

EXAMPLE 3.26

Solve Example 1.16 using derivatives.

SOLUTION   To the right of the right branch of the hyperbola $x^2 - 4y^2 = 5$ in Figure 3.42 is a swamp. We are required to find the point P_2 at which a straight pipeline from $P_1(15, 10)$ meets a pipeline along the line $x = -1$ as far down the line as possible. The required point P_2 occurs when line $P_1 P_2$ is tangent to the hyperbola. Let the point of tangency be $Q(a, b)$. Differentiation of $x^2 - 4y^2 = 5$ with respect to x gives

$$2x - 8y\frac{dy}{dx} = 0 \quad \Longrightarrow \quad \frac{dy}{dx} = \frac{x}{4y}.$$

The slope of the tangent line at $Q(a, b)$ is therefore $a/(4b)$. Since the slope of $P_1 Q$ is $(b - 10)/(a - 15)$, it follows that

$$\frac{b - 10}{a - 15} = \frac{a}{4b}.$$

FIGURE 3.42   Best line along which to build a pipeline

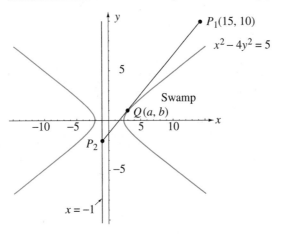

Cross-multiplication leads to

$$a^2 - 4b^2 = 15a - 40b.$$

Since $Q(a, b)$ is on the hyperbola, it also follows that

$$a^2 - 4b^2 = 5.$$

When we subtract these equations,

$$0 = 15a - 40b - 5 \quad \Longrightarrow \quad a = \frac{1}{3}(1 + 8b).$$

Substitution of this into $a^2 - 4b^2 = 5$ gives

$$\frac{1}{9}(1 + 8b)^2 - 4b^2 = 5 \quad \Longrightarrow \quad 1 + 16b + 64b^2 - 36b^2 = 45.$$

Thus,

$$0 = 28b^2 + 16b - 44 = 4(7b + 11)(b - 1),$$

and solutions are $b = 1$ and $b = -11/7$. The second solution gives a point on the left branch of the hyperbola at which a tangent line passes through $(15, 10)$. The solution $b = 1$ gives the point $(3, 1)$ on the right branch of the hyperbola. The slope of the tangent line at $(3, 1)$ is $3/4$, and therefore its equation is $y - 1 = (3/4)(x - 3)$. It cuts the line $x = -1$ when $y = 1 + (3/4)(-1 - 3) = -2$. Thus, P_2 has coordinates $(-1, -2)$.

EXERCISES 3.8

In Exercises 1–10 find dy/dx wherever y is defined as a function of x.

1. $y^4 + y = 4x^3$

2. $x^4 + y^2 + y^3 = 1$

3. $xy + 2x = 4y^2 + 2$

4. $2x^3 - 3xy^4 + 5xy - 10 = 0$

5. $x + xy^5 + x^2 y^3 = 3$

6. $(x + y)^2 = 2x$

7. $x(x - y) - 4y^3 = 2x + 5$

8. $\sqrt{x + y} + y^2 = 12x^2 + y$

9. $\sqrt{1 + xy} - xy = 15$

10. $\dfrac{x}{x + y} - \dfrac{y}{x} = 4$

11. Find equations for the tangent and normal lines to the curve $xy^2 + y^3 = 2$ at the point $(1, 1)$.

In Exercises 12–15 find d^2y/dx^2 wherever y is defined as a function of x.

12. $x^2 + y^3 + y = 1$ * **13.** $2x^2 - y^3 = 4 - xy$

14. $y^2 + 2y = 5x$ * **15.** $(x + y)^2 = x$

16. Find dy/dx at $x = 1$ when $x^3 y + xy^3 = 2$.

In Exercises 17–20 find dy/dx and d^2y/dx^2.

17. $(x + y)^2 = x^2 + y^2$ * **18.** $x^2 y^3 + 2x + 4y = 5$

19. $xy^2 - 3x^2 y = x + 1$ * **20.** $x = y\sqrt{1 - y^2}$

21. Find dy/dx when $x = 0$ if $\sqrt{1 - xy} + 3y = 4$.

22. Find dy/dx when $x = 1$ if $x^2 y^3 + xy = 2$.

23. Find dy/dx and d^2y/dx^2 when $x = 2$ if $y^5 + (x - 2)y = 1$.

24. Find dy/dx and d^2y/dx^2 when $y = 1$ if $x^2 + 2xy + 3y^2 = 2$.

25. Find point(s) on the curve $xy^2 + x^2 y = 16$ at which the slope of the tangent line is equal to zero.

26. Find point(s) on the curve $x^2 + y^{2/3} = 2$ at which the second derivative is equal to zero.

27. If a thermal nuclear reactor is built in the shape of a right circular cylinder of radius r and height h, then, according to neutron diffusion theory, r and h must satisfy the equation

$$\left(\frac{2.4048}{r}\right)^2 + \left(\frac{\pi}{h}\right)^2 = k = \text{constant}.$$

Find dr/dh.

28. The elasticity of a function $y = f(x)$ is defined as

$$\frac{Ey}{Ex} = \frac{x}{y}\frac{dy}{dx}.$$

Calculate elasticity for the function defined by each of the following equations:

(a) $f(x) = x\left(\dfrac{x + 1}{x + 2}\right)$ (b) $x = \dfrac{400y + 200}{3 - y}$

* **29.** Show that the elasticity of a function is equal to 1 if and only if the tangent line to its graph passes through the origin.

* **30.** Find that point $P(a, b)$ on the first-quadrant part of the ellipse $2x^2 + 3y^2 = 14$ at which the tangent line at P is perpendicular to the line joining P and $(2, 5)$.

* **31.** Show that the equation of the tangent line to the hyperbola $b^2 x^2 - a^2 y^2 = a^2 b^2$ at the point (x_0, y_0) is $b^2 xx_0 - a^2 yy_0 = a^2 b^2$.

* **32.** Prove that for any circle $(x - h)^2 + (y - k)^2 = r^2$,

$$\left|\frac{d^2y/dx^2}{[1 + (dy/dx)^2]^{3/2}}\right| = \frac{1}{r}.$$

* **33.** A solution passes through a conical filter 24 cm deep and 16 cm across the top into a cylindrical vessel of diameter 12 cm. Find an equation relating the depth h of solution in the filter and depth H of solution in the cylinder. What is the rate of change of h with respect to H?

* **34.** If x objects are sold at a price of $r(x)$ per object, the total revenue is $R(x) = xr(x)$. Find the marginal revenue $R'(x)$ if price is defined implicitly by the equation

$$x = 4a^3 - 3ar^2 + r^3,$$

where $a > 0$ is a constant, and $0 < x < 4a^3$.

* **35.** The general polynomial of degree n is

$$a_0 + a_1 x + a_2 x^2 + \cdots + a_n x^n,$$

where $a_0, a_1, \ldots, a_n$ are constants. Show that two polynomials of degree n,

$$a_0 + a_1 x + a_2 x^2 + \cdots + a_n x^n = b_0 + b_1 x + b_2 x^2 + \cdots + b_n x^n,$$

can be equal for all x if and only if $a_0 = b_0, a_1 = b_1, \ldots, a_n = b_n$.

* **36.** (a) Find $f'(0)$ if $y = f(x)$ is defined implicitly as a function of x by

$$x\sqrt{1 + 2y} = x^2 - y.$$

(b) Show that by squaring the equation in part (a), we obtain

$$x^2 + 4x^2 y = x^4 + y^2.$$

Differentiate this equation with respect to x to find $f'(0)$. Do you have any difficulties? Explain.

37.–40. Use implicit differentiation to redo Exercises 55–58 in Section 3.7.

41. (a) Use implicit differentiation to find dy/dx if $y^2 = x^2 - 4x^4$.

   (b) Can you calculate dy/dx at $x = 0$ using the result of part (a)? Draw the curve $y^2 = x^2 - 4x^4$ in order to explain this difficulty.

42. (a) Find dy/dx if $\sqrt{x} + \sqrt{y} = 1$, where $0 \le x \le 1$, defines y implicitly as a function of x.

   (b) Draw the curve $\sqrt{|x|} + \sqrt{|y|} = 1$.

43. What is dy/dx if $x^2 + 4x + y^2 + 6y + 15 = 0$?

44. Find dy/dx if $y = u/\sqrt{u^2 - 1}$, and $x^2 u^2 + \sqrt{u^2 - 1} = 4$ defines u implicitly as a function of x.

45. Given that the equations

$$y^4 + yv^3 = 3 \quad \text{and} \quad x^2 v + 3xv^2 = 2x^3 y + 1$$

define y as a function of v, and v as a function of x, find dy/dx in terms of x, y, and v.

46. Show that any function defined implicitly by the equation

$$\frac{x^2}{y^3} - x = C,$$

where C is a constant, satisfies the equation

$$3x^2 \frac{dy}{dx} + y^4 = 2xy.$$

47. Show that any function defined implicitly by the equation

$$2x^2 - 3y = Cx^2 y^3,$$

where C is a constant, satisfies the equation

$$(xy - x^3)\frac{dy}{dx} + y^2 = 0.$$

48. Verify power rule 3.7 in the case that n is a rational number.

49. Find points on the curve $xy^2 + x^2 y = 2$ at which the slope of the tangent line is equal to 1.

50. Show that the families of curves $y^2 = x^3/(a - x)$ and $(x^2 + y^2)^2 = b(2x^2 + y^2)$ are orthogonal trajectories.

51. Let (x, y) be any point on the curve $x^{2/3} + y^{2/3} = a^{2/3}$, where $a > 0$ is a constant. Find a formula for the length of that part of the tangent line at (x, y) between the coordinate axes.

52. The curve described by the equation $x^3 + y^3 = 3axy$, where $a > 0$ is a constant, is called the *folium of Descartes* (figure below). Find points where the slope of the tangent line is equal to -1.

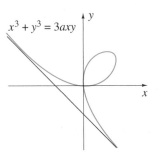

$x^3 + y^3 = 3axy$

53. Use implicit differentiation to solve Exercise 67 in Section 3.7.

54. (a) When a point (x, y) moves in the xy-plane so that the product of its distances from the points $(a, 0)$ and $(-a, 0)$ is always equal to a constant which we denote by c^2 ($c > a$), the curve that it follows is called the *ovals of Cassini*. Verify that the equation for these ovals is $(x^2 + y^2 + a^2)^2 = c^4 + 4a^2 x^2$.

   (b) Show that when $c < \sqrt{2}a$, there are six points on the ovals at which the tangent line is horizontal, but when $c \ge \sqrt{2}a$, there are only two such points.

55. Show that the ellipse $b^2 x^2 + a^2 y^2 = a^2 b^2$ and the hyperbola $d^2 x^2 - c^2 y^2 = c^2 d^2$ intersect orthogonally if $a^2 - b^2 = c^2 + d^2$.

3.9 Derivatives of the Trigonometric Functions

The trigonometric functions, their properties, and the identities that they satisfy were discussed in Section 1.7. We emphasize an important convention in calculus. Arguments of trigonometric functions are always measured in radians, never degrees. With this in mind, we prove the following theorem.

THEOREM 3.10

$$\lim_{\theta \to 0} \frac{\sin \theta}{\theta} = 1 \tag{3.24}$$

This is strongly suggested by the machine-generated plot of $(\sin \theta)/\theta$ in Figure 3.43. There should be an open circle at the point $(0, 1)$ since the function is undefined at $\theta = 0$.

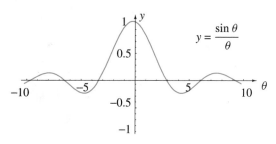

FIGURE 3.43 Plot suggesting that $\displaystyle\lim_{\theta\to 0}\frac{\sin\theta}{\theta}=1$

PROOF   If θ is a positive acute angle as shown in Figure 3.44, then

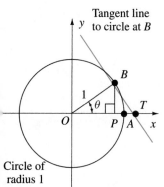

FIGURE 3.44   Proof that

$$\lim_{\theta\to 0}\frac{\sin\theta}{\theta}=1$$

$$\text{area of triangle } BOP < \text{area of sector } BOA < \text{area of triangle } OBT.$$

With θ expressed in radians, the area of a sector of a circle is $(r^2\theta)/2$. Consequently,

$$\frac{1}{2}\|BP\|\|OP\| < \frac{1}{2}(1)^2\theta < \frac{1}{2}\|OB\|\|BT\|.$$

When we multiply each term in these inequalities by 2 and express lengths of the line segments in terms of θ,

$$(\sin\theta)(\cos\theta) < \theta < (1)(\tan\theta).$$

Division by $\sin\theta$ gives

$$\cos\theta < \frac{\theta}{\sin\theta} < \frac{1}{\cos\theta},$$

and when each term is inverted, the inequality signs are reversed:

$$\frac{1}{\cos\theta} > \frac{\sin\theta}{\theta} > \cos\theta.$$

We now take limits as $\theta\to 0^+$. Since $\cos\theta\to 1$ and $1/\cos\theta\to 1$, the squeeze theorem of Section 2.1 implies that

$$\lim_{\theta\to 0^+}\frac{\sin\theta}{\theta}=1.$$

Since $(\sin\theta)/\theta$ is an even function (Figure 3.43), it follows that

$$\lim_{\theta\to 0^-}\frac{\sin\theta}{\theta}=1.$$

Since left- and right-hand limits of $(\sin\theta)/\theta$ are both equal to 1, the full limit as $\theta\to 0$ is also 1,

$$\lim_{\theta\to 0}\frac{\sin\theta}{\theta}=1. \qquad\blacksquare$$

With this result we can now find the derivative of the sine function. Derivatives of the other five trigonometric functions follow easily.

THEOREM 3.11

$$\frac{d}{dx}\sin x = \cos x \tag{3.25}$$

PROOF If we set $f(x) = \sin x$ in equation 3.3 and use the trigonometric identity

$$\sin A - \sin B = 2 \cos\left(\frac{A + B}{2}\right) \sin\left(\frac{A - B}{2}\right),$$

we obtain

$$
\begin{aligned}
f'(x) &= \lim_{h \to 0} \frac{f(x + h) - f(x)}{h} \\
&= \lim_{h \to 0} \frac{\sin(x + h) - \sin x}{h} \\
&= \lim_{h \to 0} \frac{1}{h}\left[2 \cos\left(\frac{2x + h}{2}\right) \sin\left(\frac{h}{2}\right)\right] \\
&= \lim_{h \to 0}\left[\cos\left(x + \frac{h}{2}\right) \frac{\sin(h/2)}{h/2}\right] \\
&= \lim_{h \to 0} \cos\left(x + \frac{h}{2}\right) \lim_{h/2 \to 0} \frac{\sin(h/2)}{h/2} \\
&= \cos x.
\end{aligned}
$$

It is a simple application of the chain rule to prove the following corollary.

COROLLARY 3.11.1

If $u = f(x)$ is a differentiable function,

$$\frac{d}{dx} \sin u = \cos u \frac{du}{dx}. \tag{3.26}$$

EXAMPLE 3.27

Find derivatives for the following functions:

(a) $f(x) = \sin 3x$ \qquad (b) $f(x) = \sin^3 4x$

SOLUTION

(a) According to formula 3.26,

$$f'(x) = \cos 3x \frac{d}{dx}(3x) = 3 \cos 3x.$$

(b) For this function we must use extended power rule 3.21 before equation 3.26,

$$f'(x) = 3 \sin^2 4x \frac{d}{dx} \sin 4x = 3 \sin^2 4x \cos 4x \frac{d}{dx}(4x) = 12 \sin^2 4x \cos 4x.$$

Since the cosine function can be expressed in terms of the sine function, it is straightforward to find its derivative.

THEOREM 3.12

$$\frac{d}{dx}\cos x = -\sin x \qquad (3.27)$$

PROOF   Since $\cos x$ can always be expressed in the form $\cos x = \sin (\pi/2 - x)$, it follows that

$$\frac{d}{dx}\cos x = \frac{d}{dx}\sin \left(\frac{\pi}{2} - x\right) = -\cos \left(\frac{\pi}{2} - x\right) = -\sin x. \qquad \blacksquare$$

COROLLARY 3.12.1
If $u = f(x)$ is a differentiable function,

$$\frac{d}{dx}\cos u = -\sin u \frac{du}{dx}. \qquad (3.28)$$

Derivatives of the other trigonometric functions are obtained by expressing them in terms of the sine and cosine functions. For the tangent function, we use the chain rule and the quotient rule to calculate

$$\frac{d}{dx}\tan u = \frac{d}{du}\left(\frac{\sin u}{\cos u}\right)\frac{du}{dx} = \frac{\cos u(\cos u) - \sin u(-\sin u)}{\cos^2 u}\frac{du}{dx} = \frac{1}{\cos^2 u}\frac{du}{dx}.$$

Consequently,

$$\frac{d}{dx}\tan u = \sec^2 u \frac{du}{dx}. \qquad (3.29)$$

A similar calculation gives

$$\frac{d}{dx}\cot u = -\csc^2 u \frac{du}{dx}. \qquad (3.30)$$

For the secant function, we obtain

$$\frac{d}{dx}\sec u = \frac{d}{du}\left(\frac{1}{\cos u}\right)\frac{du}{dx} = (-1)(\cos u)^{-2}(-\sin u)\frac{du}{dx} = \frac{1}{\cos u}\frac{\sin u}{\cos u}\frac{du}{dx}$$

or

$$\frac{d}{dx}\sec u = \sec u \tan u \frac{du}{dx}. \qquad (3.31)$$

Similarly,

$$\frac{d}{dx}\csc u = -\csc u \cot u \frac{du}{dx}. \qquad (3.32)$$

Notice the relationship between derivatives of the *cofunctions* — cosine is the cofunction of sine, cotangent of tangent, and cosecant of secant — and corresponding derivatives for the functions. Each function is replaced by its cofunction and a negative sign is added.

EXAMPLE 3.28

Find dy/dx if y is defined as a function of x in each of the following:
(a) $y = \sin 2x$
(b) $y = 4\sec (2x^3 + 5)$
(c) $y = \tan^2 4x$
(d) $y = 2\csc 3x^2 + 5x \sin x$
(e) $x^2 \tan y + y \sin x = 5$

SOLUTION

(a) $\dfrac{dy}{dx} = (\cos 2x)\dfrac{d}{dx}(2x) = 2\cos 2x$

(b) $\dfrac{dy}{dx} = 4\sec(2x^3 + 5)\tan(2x^3 + 5)\dfrac{d}{dx}(2x^3 + 5)$

$= 24x^2\sec(2x^3 + 5)\tan(2x^3 + 5)$

(c) $\dfrac{dy}{dx} = 2\tan 4x\dfrac{d}{dx}\tan 4x = 2\tan 4x(4\sec^2 4x) = 8\tan 4x\sec^2 4x$

(d) $\dfrac{dy}{dx} = -2\csc 3x^2\cot 3x^2\dfrac{d}{dx}(3x^2) + 5\sin x + 5x\cos x$

$= -12x\csc 3x^2\cot 3x^2 + 5\sin x + 5x\cos x$

(e) If we differentiate the equation (implicitly) with respect to x, we obtain

$$2x\tan y + x^2\sec^2 y\dfrac{dy}{dx} + \dfrac{dy}{dx}\sin x + y\cos x = 0.$$

When we solve this equation for dy/dx, the result is

$$\dfrac{dy}{dx} = -\dfrac{2x\tan y + y\cos x}{x^2\sec^2 y + \sin x}.$$

EXAMPLE 3.29

In the mechanism of Figure 3.45, crank OA rotates at one thousand revolutions per minute (rpm). Find an expression for the velocity of the piston in terms of R and θ.

FIGURE 3.45   Velocity of piston in crank mechanism

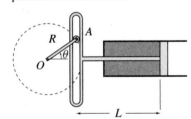

SOLUTION   If the crank starts with $\theta = 0$ at time $t = 0$, then x and y coordinates of the moving end of the crank are $x = R\cos\theta$ and $y = R\sin\theta$. The position of the piston relative to O is $x + L = R\cos\theta + L$. The velocity of the piston is therefore

$$v = -R\sin\theta\dfrac{d\theta}{dt}.$$

Since the crank rotates at 1000 rpm,

$$\dfrac{d\theta}{dt} = \dfrac{2\pi(1000)}{60} = \dfrac{100\pi}{3} \quad \text{radians per second.}$$

Thus,

$$v = -\dfrac{100\pi}{3}R\sin\theta \text{ m/s,}$$

provided R is in metres.

EXAMPLE 3.30

The 1 kg mass in Figure 3.46 is pulled 10 cm to the right of its position when the spring is unstretched (called the *equilibrium position*), and given velocity 3 m/s to the left. Its displacement from the equilibrium position thereafter is given by

$$x(t) = \dfrac{1}{10}\cos 20t - \dfrac{3}{20}\sin 20t \text{ m,} \quad t \geq 0.$$

Find all times when the velocity is zero and all times when the acceleration is zero.

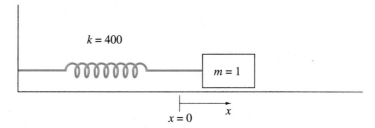

FIGURE 3.46   Velocity and acceleration of oscillating mass

FIGURE 3.47   Simple harmonic motion of oscillating mass

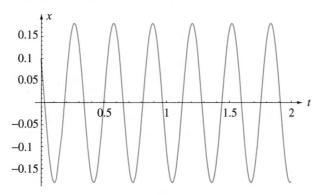

SOLUTION   The displacement function in Figure 3.47 indicates that the mass oscillates back and forth about the equilibrium position. This is called **simple harmonic motion**. It is a direct result of the fact that the mass moves along a frictionless surface and no account has been taken of air resistance.

To determine times at which velocity and acceleration vanish, it is advantageous to follow the lead of Example 1.36 in Section 1.7 and express $x(t)$ in the form

$$\frac{1}{10}\cos 20t - \frac{3}{20}\sin 20t = A\sin(20t + \phi).$$

When we expand the right side with a compound-angle formula, we obtain

$$\frac{1}{10}\cos 20t - \frac{3}{20}\sin 20t = A(\sin 20t\,\cos\phi + \cos 20t\,\sin\phi).$$

This equation will be valid for all $t \geq 0$ if we choose A and ϕ to satisfy

$$\frac{1}{10} = A\,\sin\phi, \qquad -\frac{3}{20} = A\,\cos\phi.$$

Squaring and adding these equations eliminates ϕ:

$$\frac{1}{100} + \frac{9}{400} = A^2 \qquad \Longrightarrow \qquad A = \pm\frac{\sqrt{13}}{20}.$$

If we choose to use the positive value, then

$$\sin\phi = \frac{2}{\sqrt{13}} \qquad \text{and} \qquad \cos\phi = -\frac{3}{\sqrt{13}}.$$

When we let $\phi = 2.55$ be the smallest positive solution of these equations, we express $x(t)$ in the form

$$x(t) = \frac{\sqrt{13}}{20}\sin(20t + \phi) \quad \text{m}.$$

Velocity and acceleration are

$$v(t) = \sqrt{13}\cos(20t + \phi) \text{ m/s}, \qquad a(t) = -20\sqrt{13}\sin(20t + \phi) \text{ m/s}^2.$$

The velocity of the mass is zero when the mass is at the points farthest from the equilibrium position. It occurs when

$$0 = \sqrt{13}\cos(20t + \phi) \qquad \Longrightarrow \qquad 20t + \phi = \frac{\pi}{2} + n\pi,$$

where n is an integer. Thus,

$$t = \frac{\pi}{40} + \frac{n\pi}{20} - \frac{\phi}{20},$$

and these are positive for $n \geq 1$. The acceleration vanishes when

$$0 = -20\sqrt{13}\sin(20t + \phi) \qquad \Longrightarrow \qquad 20t + \phi = n\pi,$$

where n is an integer. Thus,

$$t = \frac{n\pi}{20} - \frac{\phi}{20},$$

and this is positive for $n \geq 1$. Notice that acceleration is zero when the mass passes through the equilibrium position. This is to be expected because at $x = 0$, the spring has no stretch and therefore exerts no force on the mass.

EXERCISES 3.9

In Exercises 1–30 find dy/dx.

1. $y = 2\sin 3x$

2. $y = \cos x - 4\sin 5x$

3. $y = \sin^2 x$

4. $y = \tan^{-3} 3x$

5. $y = \sec^4 10x$

6. $y = \csc(4 - 2x)$

7. $y = \sin^2(3 - 2x^2)$

8. $y = x\cot x^2$

9. $y = \dfrac{\sin 2x}{\cos 5x}$

10. $y = \dfrac{x\sin x}{x + 1}$

11. $y = \sin^3 x + \cos x$

12. $y = \sin 2x \cos 2x$

13. $y = \sqrt{\sin 3x}$

* **14.** $y = \left(1 + \tan^3 x\right)^{1/4}$

* **15.** $2\sin y + 3\cos x = 1$

* **16.** $x\cos y - y\cos x = 3$

* **17.** $4\sin^2 x - 3\cos^3 y = 1$

* **18.** $\tan(x + y) = y$

* **19.** $x + \sec xy = 5$

* **20.** $x^3 y + \tan^2 y = 3x$

* **21.** $x = y^3 \csc^3 y$

* **22.** $y = \cos(\tan x)$

* **23.** $y = x^3 - x^2\cos x + 2x\sin x + 2\cos x$

* **24.** $y = \sin^4 x^2 - \cos^4 x^2$

* **25.** $y = u^3 \sec u, \quad u = x\tan(x + 1)$

* **26.** $y = \sqrt{3 - \sec v}, \quad v = \tan\sqrt{x}$

* **27.** $y = \sqrt{t^2 + 1}, \quad t = \sin(\sin x)$

* **28.** $y = (1 + \sec^3 u)^{1/3}, \quad u = \sqrt{1 + \cos x^2}$

* **29.** $y = \dfrac{\sin^2 x}{1 + \cos^3 x}$

* **30.** $y = \dfrac{1 + \tan^3(3x^2 - 4)}{x^2\sin x}$

In Exercises 31–32 find d^2y/dx^2.

* **31.** $\sin y = x^2 + y$

* **32.** $\tan y = x + xy$

In Exercises 33–38 evaluate the limit, if it exists.

33. $\displaystyle\lim_{x \to 0} \frac{\tan x}{x}$

* **34.** $\displaystyle\lim_{x \to 0} \frac{1 - \cos x}{x}$

* **35.** $\displaystyle\lim_{x \to 0} \frac{\sin 2x}{x}$

* **36.** $\displaystyle\lim_{x \to \infty} \frac{\sin(2/x)}{\sin(1/x)}$

* **37.** $\displaystyle\lim_{x \to 0} \frac{(x + 1)^2\sin x}{3x^3}$

* **38.** $\displaystyle\lim_{x \to \pi/2} \frac{\cos x}{(x - \pi/2)^2}$

39. Draw graphs of the functions $f(x) = |\sin x|$ and $g(x) = \sin |x|$. Where do these functions fail to be differentiable?

40. Does $\lim_{x \to 0} g'(x)$ exist for the function $g(x)$ in Exercise 32 of Section 3.3?

41. The angular displacement of the pendulum in the following figure at time t is given by

$$\theta = f(t) = A \cos(\omega t + \phi), \quad t \geq 0,$$

where A, ω, and ϕ are constants. Verify that $\theta = f(t)$ satisfies the equation

$$\frac{d^2\theta}{dt^2} + \omega^2\theta = 0.$$

42. When the mass m in the figure below moves vertically on the end of the spring, its displacement y must satisfy the equation

$$m\frac{d^2y}{dt^2} + ky = 0,$$

where $k > 0$ is the constant of elasticity for the spring. Verify that the function

$$y = f(t) = A \sin\left(\sqrt{\frac{k}{m}}t\right) + B \cos\left(\sqrt{\frac{k}{m}}t\right)$$

satisfies this equation for any constants A and B.

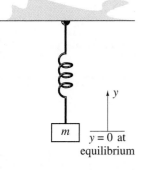

$y = 0$ at equilibrium

* **43.** The two identical cranks in the figure below rotate at constant angular speeds ω_1 and ω_2 radians per second, driving the attachment vertically. Find the angular speed $d\theta/dt$ of the slotted bar in terms of α, β, ω_1, and ω_2.

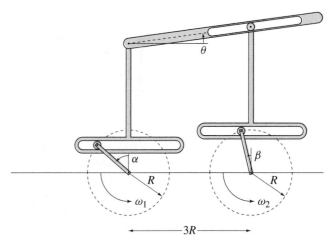

* **44.** Find all positive values of x for which the derivative of the function $f(x) = \cos(x + 1/x)$ is equal to zero.

* **45.** An elliptic cam rotates counterclockwise at 600 rpm (figure below). Find a formula for the velocity of the follower if the ellipse has major and minor axes of lengths $2a$ and $2b$.

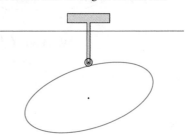

** **46.** Find a formula for the following derivative, simplified as much as possible,

$$\frac{d}{dx}|\sin x|^n, \quad \text{where } n \geq 1 \text{ is an integer.}$$

** **47.** Show that the function

$$f(x) = \begin{cases} x^2 \sin(1/x), & x \neq 0 \\ 0, & x = 0 \end{cases}$$

has a derivative at $x = 0$, but $f'(x)$ is not continuous at $x = 0$.

** **48.** Any cross-section of the reflector in a car headlight is in the form of a parabola, $y^2 = 4x$, $0 \leq x \leq 2$, with the bulb at the point $(1, 0)$ called the *focus* of the parabola. It is a principle of optics that all light rays from the bulb are reflected by the mirror so that the angle between the incident ray and the normal to the mirror is equal to the angle between the reflected ray and the normal. Show that all rays are reflected parallel to the x-axis.

3.10 Derivatives of the Inverse Trigonometric Functions

Derivatives of inverse trigonometric functions are most easily obtained with implicit differentiation. We begin with the inverse sine function.

THEOREM 3.13

$$\frac{d}{dx}\operatorname{Sin}^{-1} x = \frac{1}{\sqrt{1-x^2}}.$$ (3.33)

PROOF   If we set $y = \operatorname{Sin}^{-1} x$, then $x = \sin y$, and we can differentiate implicitly with respect to x:

$$1 = \cos y \frac{dy}{dx}.$$

Thus,

$$\frac{dy}{dx} = \frac{1}{\cos y}.$$

To express $\cos y$ in terms of x we can proceed in two ways:

(i)   From the trigonometric identity $\sin^2 y + \cos^2 y = 1$, we obtain

$$\cos y = \pm\sqrt{1 - \sin^2 y} = \pm\sqrt{1 - x^2}.$$

We know that $y = \operatorname{Sin}^{-1} x$ and the principal values of $\operatorname{Sin}^{-1} x$ are $-\pi/2 \le y \le \pi/2$. Therefore, y is an angle in either the first or fourth quadrant. It follows that $\cos y \ge 0$, and we must choose

$$\cos y = \sqrt{1 - x^2}.$$

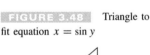
FIGURE 3.48   Triangle to fit equation $x = \sin y$

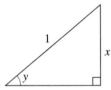

(ii)   We use the triangle in Figure 3.48 to replace the trigonometric identity. The triangle is designed to fit the equation $x = \sin y$. The third side is then $\sqrt{1 - x^2}$. It follows that

$$\cos y = \sqrt{1 - x^2}.$$

To ensure that $\cos y$ is indeed positive, we resort once again to the fact that $y = \operatorname{Sin}^{-1} x$ and principal values are $-\pi/2 \le y \le \pi/2$. Finally, then, we have

$$\frac{dy}{dx} = \frac{1}{\sqrt{1 - x^2}}.$$ ∎

It is a straightforward application of the chain rule to obtain the following corollary.

COROLLARY 3.13.1

If $u(x)$ is a differentiable function, then

$$\frac{d}{dx}\operatorname{Sin}^{-1} u = \frac{1}{\sqrt{1 - u^2}}\frac{du}{dx}.$$ (3.34)

Next we obtain the derivative of the inverse cosine function.

THEOREM 3.14

$$\frac{d}{dx}\text{Cos}^{-1} x = \frac{-1}{\sqrt{1-x^2}}$$
(3.35)

PROOF   If we set $y = \text{Cos}^{-1} x$, then $x = \cos y$, and we can differentiate with respect to x:

$$1 = -\sin y \frac{dy}{dx},$$

and solve for

$$\frac{dy}{dx} = \frac{-1}{\sin y}.$$

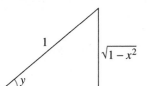

FIGURE 3.49   Triangle to fit equation $x = \cos y$

The triangle in Figure 3.49, obtained from $x = \cos y$, yields $\sin y = \sqrt{1-x^2}$. Since the principal values of $y = \text{Cos}^{-1} x$ are $0 \le y \le \pi$, it follows that $\sin y$ is indeed nonnegative, and therefore

$$\frac{dy}{dx} = \frac{-1}{\sqrt{1-x^2}}.$$

COROLLARY 3.14.1
If $u(x)$ is differentiable, then

$$\frac{d}{dx}\text{Cos}^{-1} u = \frac{-1}{\sqrt{1-u^2}}\frac{du}{dx}.$$
(3.36)

Derivatives for the other inverse trigonometric functions can be derived in a similar way. We list them below and include derivatives of $\text{Sin}^{-1} u$ and $\text{Cos}^{-1} u$ for completeness:

$$\frac{d}{dx}\text{Sin}^{-1} u = \frac{1}{\sqrt{1-u^2}}\frac{du}{dx};$$
(3.37a)

$$\frac{d}{dx}\text{Cos}^{-1} u = \frac{-1}{\sqrt{1-u^2}}\frac{du}{dx};$$
(3.37b)

$$\frac{d}{dx}\text{Tan}^{-1} u = \frac{1}{1+u^2}\frac{du}{dx};$$
(3.37c)

$$\frac{d}{dx}\text{Cot}^{-1} u = \frac{-1}{1+u^2}\frac{du}{dx};$$
(3.37d)

$$\frac{d}{dx}\text{Sec}^{-1} u = \frac{1}{u\sqrt{u^2-1}}\frac{du}{dx};$$
(3.37e)

$$\frac{d}{dx}\text{Csc}^{-1} u = \frac{-1}{u\sqrt{u^2-1}}\frac{du}{dx}.$$
(3.37f)

Note that the derivative of an inverse cofunction is the negative of the derivative of the corresponding inverse function; that is, derivatives of $\text{Cos}^{-1} u$, $\text{Cot}^{-1} u$, and $\text{Csc}^{-1} u$ are the negatives of the derivatives of $\text{Sin}^{-1} u$, $\text{Tan}^{-1} u$, and $\text{Sec}^{-1} u$.

EXAMPLE 3.31

Find dy/dx if y is defined as a function of x in each of the following:

(a) $y = \mathrm{Sec}^{-1}(x^2)$ (b) $y = 2x\,\mathrm{Cot}^{-1}(3x)$ (c) $y = \left[\mathrm{Sin}^{-1}(x^3)\right]^2$

(d) $y = \dfrac{\mathrm{Sin}^{-1}x}{\mathrm{Cos}^{-1}x}$ (e) $\mathrm{Cos}^{-1}(xy) + x^2y + 5 = 0$

SOLUTION

(a) $\dfrac{dy}{dx} = \dfrac{1}{x^2\sqrt{(x^2)^2 - 1}}\dfrac{d}{dx}(x^2) = \dfrac{2x}{x^2\sqrt{x^4 - 1}} = \dfrac{2}{x\sqrt{x^4 - 1}}$

(b) $\dfrac{dy}{dx} = 2\,\mathrm{Cot}^{-1}(3x) + 2x\left(\dfrac{-1}{1 + 9x^2}\right)(3) = 2\,\mathrm{Cot}^{-1}(3x) - \dfrac{6x}{1 + 9x^2}$

(c) $\dfrac{dy}{dx} = 2\,\mathrm{Sin}^{-1}(x^3)\dfrac{d}{dx}\mathrm{Sin}^{-1}(x^3) = 2\,\mathrm{Sin}^{-1}(x^3)\dfrac{1}{\sqrt{1 - x^6}}(3x^2) = \dfrac{6x^2\,\mathrm{Sin}^{-1}(x}{\sqrt{1 - x^6}}$

(d) $\dfrac{dy}{dx} = \dfrac{\mathrm{Cos}^{-1}x\left(\dfrac{1}{\sqrt{1 - x^2}}\right) - \mathrm{Sin}^{-1}x\left(\dfrac{-1}{\sqrt{1 - x^2}}\right)}{\left(\mathrm{Cos}^{-1}x\right)^2} = \dfrac{\mathrm{Cos}^{-1}x + \mathrm{Sin}^{-1}x}{\sqrt{1 - x^2}\left(\mathrm{Cos}^{-1}x\right)^2}$

(e) Differentiation with respect to x gives

$$\dfrac{-1}{\sqrt{1 - x^2y^2}}\left(y + x\dfrac{dy}{dx}\right) + 2xy + x^2\dfrac{dy}{dx} = 0.$$

Thus,

$$\dfrac{dy}{dx}\left(\dfrac{-x}{\sqrt{1 - x^2y^2}} + x^2\right) = \dfrac{y}{\sqrt{1 - x^2y^2}} - 2xy,$$

from which

$$\dfrac{dy}{dx} = \dfrac{y - 2xy\sqrt{1 - x^2y^2}}{\sqrt{1 - x^2y^2}}\dfrac{\sqrt{1 - x^2y^2}}{-x + x^2\sqrt{1 - x^2y^2}} = \dfrac{y - 2xy\sqrt{1 - x^2y^2}}{-x + x^2\sqrt{1 - x^2y^2}}.$$

EXAMPLE 3.32

Find the derivative of the function $f(x) = \mathrm{Sin}^{-1}x + \mathrm{Cos}^{-1}x$. What is your conclusion?

SOLUTION The derivative of the function is

$$f'(x) = \dfrac{1}{\sqrt{1 - x^2}} + \dfrac{-1}{\sqrt{1 - x^2}} = 0.$$

Since this result is valid for the entire domain $-1 \le x \le 1$ of $f(x)$, it follows that the function $f(x)$ must be a constant,

$$f(x) = C = \text{ a constant.}$$

Since $f(0) = \mathrm{Sin}^{-1}(0) + \mathrm{Cos}^{-1}(0) = 0 + \pi/2 = \pi/2$, it follows that $C = \pi/2$. Consequently,

$$\mathrm{Sin}^{-1}x + \mathrm{Cos}^{-1}x = \dfrac{\pi}{2}.$$

EXERCISES 3.10

In Exercises 1–30 y is defined as a function of x. Find dy/dx in as simplified a form as possible.

1. $y = \text{Cos}^{-1}(2x + 3)$

2. $y = \text{Cot}^{-1}(x^2 + 2)$

3. $y = \text{Csc}^{-1}(3 - 4x)$

4. $y = \text{Tan}^{-1}(2 - x^2)$

5. $y = \text{Sec}^{-1}(3 - 2x^2)$

6. $y = x\,\text{Csc}^{-1}(x^2 + 5)$

7. $y = (x^2 + 2)\,\text{Sin}^{-1}(2x)$

8. $y = \text{Tan}^{-1}\sqrt{x + 2}$

9. $y = \text{Sin}^{-1}\sqrt{1 - x^2}$

10. $y = \text{Cot}^{-1}\sqrt{x^2 - 1}$

11. $y = \left[\text{Tan}^{-1}(x^2)\right]^2$

12. $y = x^2\text{Sec}^{-1}x$

13. $y = \tan\left(3\,\text{Sin}^{-1}x\right)$

14. $y = \text{Cot}^{-1}[(1 + x)/(1 - x)]$

15. $y = \text{Csc}^{-1}(1/x)$

16. $y = \text{Sin}^{-1}[(1 - x)/(1 + x)]$

17. $y = \text{Tan}^{-1}(u^2 + 1/u)$, $u = \tan(x^2 + 4)$

18. $y = t\,\text{Cos}^{-1}t$, $t = \sqrt{1 - x^2}$

19. $y^2 \sin x + y = \text{Tan}^{-1}x$

20. $\text{Sin}^{-1}(xy) = 5x + 2y$

21. $y = \sqrt{x^2 - 1} - \text{Sec}^{-1}x$

22. $y = \dfrac{1}{3}\text{Csc}^{-1}\left(\dfrac{x}{3}\right) - \dfrac{\sqrt{x^2 - 9}}{x^2}$

23. $y = x\,\text{Cos}^{-1}\left(\dfrac{x}{2}\right) - \sqrt{4 - x^2}$

24. $y = \dfrac{\text{Csc}^{-1}(3x)}{x} - \dfrac{\sqrt{9x^2 - 1}}{x}$

25. $y = x^2\,\text{Sec}^{-1}x - \sqrt{x^2 - 1}$

26. $y = x\left(\text{Cos}^{-1}x\right)^2 - 2x - 2\sqrt{1 - x^2}$

27. $y = (1 + 9x^2)\,\text{Cot}^{-1}(3x) + 3x$

$*$ **28.** $y = (x - 2)\sqrt{4x - x^2} + 4\,\text{Sin}^{-1}\left(\dfrac{x - 2}{2}\right)$

$*$ **29.** $y = (2x^2 - 1)\,\text{Sin}^{-1}x + x\sqrt{1 - x^2}$

$*$ **30.** $y = \text{Tan}^{-1}\left(\dfrac{\sqrt{2}x}{\sqrt{1 + x^4}}\right)$

$*$ **31.** Evaluate the derivative of

$$f(x) = \text{Sec}^{-1}x + \text{Cot}^{-1}\sqrt{x^2 - 1}.$$

What is your conclusion?

$*$ **32.** (a) Show that if the principal values of $f(x) = \text{Sec}^{-1}x$ are chosen as $0 \le y \le \pi$, $y \ne \pi/2$, then the derivative of the function is

$$\frac{d}{dx}\text{Sec}^{-1}x = \frac{1}{|x|\sqrt{x^2 - 1}}.$$

(b) What is the derivative of $f(x) = \text{Csc}^{-1}x$ if its principal values are chosen as $-\pi/2 \le y \le \pi/2$, $y \ne 0$?

$*$ **33.** Find the angle between the curves $y = \text{Sin}^{-1}x$ and $y = \text{Cos}^{-1}x$ at their point of intersection.

$*$ **34.** Verify the results in equations 3.37c–f.

$*$ **35.** If the crank in Exercise 64 of Section 1.8 rotates at 300 rpm, find an expression for $d\theta/dt$.

3.11 Derivatives of the Exponential and Logarithm Functions

Exponential functions abound in applied mathematics, and much of this is due to the fact that the derivative of an exponential function always returns the exponential function. We obtain derivatives of exponential and logarithm functions in this section, but many of the applications that rely on these derivatives are delayed until Section 5.5. Exponential and logarithm functions and their properties were discussed in Section 1.9. We assume that the reader is familiar with this material. To differentiate logarithm and exponential functions, we must return to the definition of the derivative (see equation 3.3 in Section 3.1). We begin with the natural logarithm function $\ln x$.

THEOREM 3.15

The derivative of the logarithm function $\ln x$ is

$$\frac{d}{dx} \ln x = \frac{1}{x}. \tag{3.38}$$

PROOF   By equation 3.3,

$$\frac{d}{dx} \ln x = \lim_{h \to 0} \frac{\ln(x+h) - \ln x}{h}$$

$$= \lim_{h \to 0} \frac{1}{h} \left[\ln \left(\frac{x+h}{x} \right) \right] \qquad [\text{since } \ln b - \ln c = \ln(b/c)]$$

$$= \lim_{h \to 0} \frac{1}{h} \left[\ln \left(1 + \frac{h}{x} \right) \right]$$

$$= \frac{1}{x} \lim_{h \to 0} \frac{x}{h} \left[\ln \left(1 + \frac{h}{x} \right) \right]$$

$$= \frac{1}{x} \lim_{h \to 0} \ln \left(1 + \frac{h}{x} \right)^{x/h} \qquad (\text{since } c \ln b = \ln b^c).$$

Since the logarithm function is continuous, we may interchange the limit and logarithm operations (see Theorem 2.5 in Section 2.4),

$$\frac{d}{dx} \ln x = \frac{1}{x} \ln \left[\lim_{h \to 0} \left(1 + \frac{h}{x} \right)^{x/h} \right].$$

But if we set $v = x/h$, then

$$\lim_{h \to 0} \left(1 + \frac{h}{x} \right)^{x/h} = \lim_{h/x \to 0} \left(1 + \frac{h}{x} \right)^{x/h} = \lim_{1/v \to 0} \left(1 + \frac{1}{v} \right)^{v} = \lim_{v \to \infty} \left(1 + \frac{1}{v} \right)^{v} = e$$

(see expression 1.68), and therefore

$$\frac{d}{dx} \ln x = \frac{1}{x}.$$

The chain rule gives the following corollary.

COROLLARY 3.15.1

If $u(x)$ is a differentiable function, then

$$\frac{d}{dx} \ln u = \frac{1}{u} \frac{du}{dx}. \tag{3.39}$$

In the rare instance that we might need to differentiate the logarithm function to base $a \neq e$, namely, $\log_a x$, we rewrite the change of base formula 1.67 with a replacing b and e replacing x,

$$\log_a x = (\ln x)(\log_a e).$$

Differentiation with respect to x gives the next corollary.

COROLLARY 3.15.2

$$\frac{d}{dx} \log_a x = \frac{1}{x} \log_a e \qquad (3.40)$$

The chain rule gives the next result.

COROLLARY 3.15.3

If $u(x)$ is a differentiable function of x, then

$$\frac{d}{dx} \log_a u = \frac{1}{u} \frac{du}{dx} \log_a e. \qquad (3.41)$$

We now use implicit differentiation to find the derivatives of exponential functions.

THEOREM 3.16

$$\frac{d}{dx} e^x = e^x \qquad (3.42)$$

PROOF   If we set $y = e^x$, then $x = \ln y$, and we can differentiate (implicitly) with respect to x using equation 3.39:

$$1 = \frac{1}{y} \frac{dy}{dx}.$$

Thus,

$$\frac{dy}{dx} = y = e^x. \qquad \blacksquare$$

The exponential function e^x is therefore its own derivative. In fact, for any constant C whatsoever, the function Ce^x differentiates to give itself. In Chapter 5 we shall see that Ce^x is the only function that is its own derivative. The chain rule gives the derivative of e^u when u is a function of x.

COROLLARY 3.16.1

For a differentiable function $u(x)$,

$$\frac{d}{dx} e^u = e^u \frac{du}{dx}. \qquad (3.43)$$

We can find the derivative of exponential functions a^x with bases other then e by writing

$$\frac{d}{dx}a^x = \frac{d}{dx}\left[e^{\ln(a^x)}\right] = \frac{d}{dx}e^{x\ln a} = e^{x\ln a}(\ln a) = a^x\ln a.$$

We state this as a corollary.

COROLLARY 3.16.2

$$\frac{d}{dx}a^x = a^x\ln a \qquad\qquad (3.44)$$

The chain rule now gives

COROLLARY 3.16.3

For a differentiable function $u(x)$,

$$\frac{d}{dx}a^u = a^u\frac{du}{dx}\ln a. \qquad\qquad (3.45)$$

EXAMPLE 3.33

Find dy/dx if y is defined as a function of x in each of the following:

(a) $y = 2^{3x}$ (b) $y = \log_{10}(3x^2 + 4)$ (c) $y = x^2e^{-2x}$

(d) $y = (\ln x)/x$ (e) $y = \sqrt{1 + e^{2x}}$

SOLUTION

(a) Using equation 3.45 yields

$$\frac{dy}{dx} = 2^{3x}\frac{d}{dx}(3x)\ln 2 = (3\ln 2)2^{3x}.$$

(b) With equation 3.41, we have

$$\frac{dy}{dx} = \frac{1}{3x^2 + 4}\frac{d}{dx}(3x^2 + 4)\log_{10} e = \frac{6x}{3x^2 + 4}\log_{10} e.$$

(c) The product rule and equation 3.43 give

$$\frac{dy}{dx} = 2xe^{-2x} + x^2e^{-2x}(-2) = 2x(1 - x)e^{-2x}.$$

(d) The quotient rule and equation 3.38 yield

$$\frac{dy}{dx} = \frac{x(1/x) - \ln x}{x^2} = \frac{1 - \ln x}{x^2}.$$

(e) With extended power rule 3.21 and 3.43, we obtain

$$\frac{dy}{dx} = \frac{1}{2\sqrt{1 + e^{2x}}}\frac{d}{dx}(1 + e^{2x}) = \frac{2e^{2x}}{2\sqrt{1 + e^{2x}}} = \frac{e^{2x}}{\sqrt{1 + e^{2x}}}.$$

EXAMPLE 3.34

Find values of x for which the first derivative of the function $f(x) = x^2 \ln x$ is equal to zero, and values of x for which the second derivative is equal to zero.

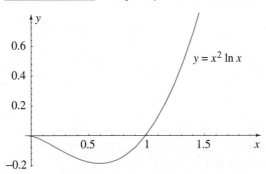

FIGURE 3.50   Graph of $y = x^2 \ln x$ to illustrate where tangent line is horizontal

SOLUTION   A graph of the function is shown in Figure 3.50. It suggests that the tangent line is horizontal at a value of x near $1/2$. To find it we set

$$0 = f'(x) = 2x \ln x + x^2 \left(\frac{1}{x}\right) = x(2 \ln x + 1).$$

Since the function is undefined at $x = 0$, we must have

$$2 \ln x + 1 = 0, \quad \text{or} \quad \ln x = -\frac{1}{2}.$$

Thus, $x = e^{-1/2} = 1/\sqrt{e}$.

The second derivative is equal to zero when

$$0 = f''(x) = (2 \ln x + 1) + x \left(\frac{2}{x}\right) = 2 \ln x + 3.$$

The only solution of this equation is $x = e^{-3/2}$. The geometric significance of the point on the graph at which $f''(x) = 0$ will be discussed in Section 4.4. Can you see it?

In Section 3.2 we proved power rule 3.7 only in the case that n is a nonnegative integer. It is now easy to prove it for all real n, at least when $x > 0$. To do this we write x^n in the form $x^n = e^{n \ln x}$ (see equation 1.72a). Formulas 3.43 and 3.38 give

$$\frac{d}{dx} x^n = \frac{d}{dx} e^{n \ln x} = e^{n \ln x} \frac{d}{dx} (n \ln x) = x^n \left(\frac{n}{x}\right) = nx^{n-1}.$$

A discussion of the power rule in the case that $x \leq 0$ depends on the value of n (see Exercise 49).

EXAMPLE 3.35

If air resistance proportional to velocity and friction with the surface on which the mass slides are taken into account in the mass-spring system of Example 3.30, the displacement function is of the form

$$x(t) = e^{-t/2} \left[\frac{1}{10} \cos \frac{\sqrt{1599}t}{2} - \frac{1}{8} \sin \frac{\sqrt{1599}t}{2} \right] + \frac{1}{4000},$$

from time $t = 0$ when motion begins until the mass comes to rest for the first time. Find this time.

Displacement of mass taking friction and air resistance into account

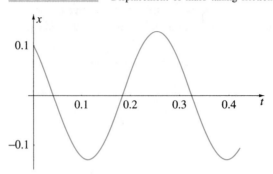

SOLUTION   A plot of $x(t)$ is shown in Figure 3.51, but it can only be used until the tangent line is horizontal for the first time. This is for $t \approx 0.1$. To find it more accurately, we set the velocity equal to 0,

$$
0 = -\frac{1}{2}e^{-t/2}\left[\frac{1}{10}\cos\frac{\sqrt{1599}t}{2} - \frac{1}{8}\sin\frac{\sqrt{1599}t}{2}\right]
$$
$$
+ e^{-t/2}\left[-\frac{\sqrt{1599}}{20}\sin\frac{\sqrt{1599}t}{2} - \frac{\sqrt{1599}}{16}\cos\frac{\sqrt{1599}t}{2}\right].
$$

Of the infinity of possible solutions to this equation, only the smallest positive one is acceptable here. To find it we simplify the equation to

$$
0 = -\frac{1}{80}e^{-t/2}\left[(4 + 5\sqrt{1599})\cos\frac{\sqrt{1599}t}{2} + (4\sqrt{1599} - 5)\sin\frac{\sqrt{1599}t}{2}\right].
$$

This equation implies that

$$
\tan\frac{\sqrt{1599}t}{2} = \frac{4 + 5\sqrt{1599}}{5 - 4\sqrt{1599}}.
$$

The smallest positive angle with tangent equal to $(4 + 5\sqrt{1599})/(5 - 4\sqrt{1599})$ is

$$
\frac{\sqrt{1599}\,t}{2} = \text{Tan}^{-1}\left(\frac{4 + 5\sqrt{1599}}{5 - 4\sqrt{1599}}\right) = 2.2205 \text{ radians.}
$$

Consequently, the time when the mass stops moving to the left is

$$
t = \frac{2}{\sqrt{1599}}(2.2205) = 0.111 \text{ s.}
$$

Application Preview Revisited

The container in Figure 3.52 represents a chemical reactor in which a chemical is either created or broken down. The chemical enters in the form of a solution at one concentration and leaves the reactor at a different concentration. We assume that flow rates in and out are the same, say q cubic metres per second, and therefore the volume V of solution in the reactor remains constant. Let us assume that C_i is the concentration in kilograms per cubic metre of the chemical entering the reactor beginning at time $t = 0$, and there is initially no chemical in the reactor. As the chemical is created (or broken down) in the reactor, its concentration varies throughout the reactor so that concentration depends not only on time but also on position. This makes the problem of predicting concentration of chemical leaving the reactor far more difficult than we can handle at this time. Suppose we add a mixer that is so efficient that concentration of chemical is the same at every point in the reactor. Concentration then depends only on time t, denoted by $C(t)$, and this is also the concentration at which chemical leaves the reactor. Our problem then is to find $C(t)$. Now, concentration of the chemical in the reactor is the amount $A(t)$ of chemical divided by V, the volume of solution. We find it easier to work with $A(t)$ than $C(t)$.

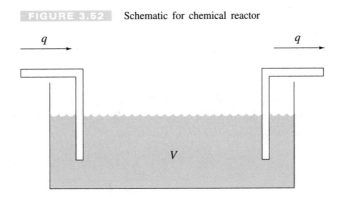

FIGURE 3.52 Schematic for chemical reactor

The derivative dA/dt is the rate of change of the amount of chemical in the reactor at any given time. This must be equal to the rate at which chemical enters less the rate at which it leaves, plus (or minus) the rate at which it is created (broken down),

$$\frac{dA}{dt} = \left\{ \begin{array}{c} \text{rate at which} \\ \text{chemical enters} \end{array} \right\} - \left\{ \begin{array}{c} \text{rate at which} \\ \text{chemical leaves} \end{array} \right\} + \left\{ \begin{array}{c} \text{rate at which chemical} \\ \text{is created or broken down} \end{array} \right\}.$$

The rate at which chemical enters is qC_i. The rate at which it leaves is $q(A/V)$. For many chemical reactions it has been shown experimentally that the rate at which chemical is being created or broken down at any given time is proportional to how much chemical is present at that time. Thus, the rate at which it is created or broken down is kA, where k is a constant ($k > 0$ for chemical formation, $k < 0$ for chemical breakdown). When we substitute these expressions into the above equation, we obtain

$$\frac{dA}{dt} = qC_i - \frac{qA}{V} + kA = qC_i + \left(k - \frac{q}{V} \right) A.$$

This is the equation that the amount of chemical $A(t)$ must satisfy at any given time; it is known as a **differential equation**. A differential equation is an equation that contains one or more derivatives of a function, and the objective is to solve the equation for functions that satisfy the equation. For the equation above, we must find functions $A(t)$ whose first derivatives are equal to $k - q/V$ times themselves plus a constant qC_i. Differential equations are discussed in detail in Section 5.5 and Chapter 15.

But we also know that the initial amount of chemical in the reactor is zero so that the function $A(t)$ must also satisfy the condition $A(0) = 0$. In other words, $A(t)$ must satisfy

$$\frac{dA}{dt} = \left(k - \frac{q}{V} \right) A + qC_i, \qquad A(0) = 0.$$

This is known as an **initial-value problem**. It consists of the differential equation and an initial condition for the unknown function $A(t)$. We shall find that there are many functions satisfying the differential equation, but only one of them also satisfies the initial condition.

Since $k - q/V$ is a constant, equation 3.43 indicates that the derivative of the function $e^{(k-q/V)t}$ is $k - q/V$ times itself. Furthermore, if we multiply $e^{(k-q/V)t}$ by any constant D whatsoever, derivatives of the functions $De^{(k-q/V)t}$ are $k - q/V$ times themselves. This must be a part of $A(t)$. The other part is simple in form but harder to discover. If we express the differential equation in the form

$$\frac{dA}{dt} - (k - q/V)A = qC_i,$$

we need a function that yields qC_i when substituted into the left side of the equation. The constant function $-qC_i/(k - q/V)$ does it. Putting these together, we suspect that $A(t)$ must be of the form

$$A(t) = De^{(k-q/V)t} - \frac{qC_i}{k - q/V}.$$

It is straightforward to check that the derivative of this function does indeed satisfy the required differential equation for any constant D. It remains only to choose D so that $A(t)$ satisfies the initial condition $A(0) = 0$. This requires

$$0 = D - \frac{qC_i}{k - q/V} \quad \Longrightarrow \quad D = \frac{qC_i}{k - q/V}.$$

Consequently, the amount of chemical exiting the reactor at any given time is

$$A(t) = \frac{qC_i}{k - q/V}e^{(k-q/V)t} - \frac{qC_i}{k - q/V} = \frac{qC_iV}{q - kV}\left[1 - e^{(k-q/V)t}\right].$$

The concentration of chemical exiting the reactor is

$$C(t) = \frac{qC_i}{q - kV}\left[1 - e^{(k-q/V)t}\right].$$

The analysis in the above Application Peview Revisited will help us in our next, more challenging consultation.

Consulting Project 4

Figure 3.53 is a schematic for a heat tank. Water enters at constant temperature $10°C$, the water is heated, and it then leaves at higher temperature. The tank is perfectly insulated so that no heat can escape from its sides, and therefore all heat supplied by the heater raises the temperature of the water. When the tank is full, and this is always the case, the mass of water is 100 kg. Water enters the tank at a rate of 3/100 kg/s, and leaves at the same rate. The heater adds energy to the water at the rate of 2000 joules per second (J/s). We are asked to determine whether the temperature of the water in the tank just keeps rising and rising, or whether it somehow levels off after a long time.

SOLUTION   Temperature of the water in the tank depends on both time and position in the tank. To remove spatial dependence (otherwise the problem will be impossible to solve), we add a mixer assumed so efficient that temperature of the water is the same at every point in the tank. Temperature then depends only on time t, denoted by $T(t)$, and this is also the temperature at which water leaves the tank. If we can find a formula for $T(t)$, we will know what happens to the system.

FIGURE 3.53   Schematic for heat tank

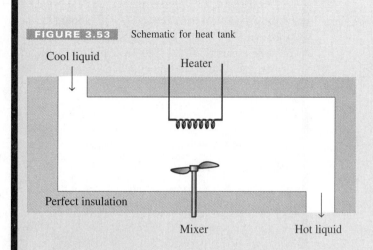

Cool liquid

Heater

Perfect insulation

Mixer   Hot liquid

Where do we start? We want temperature, temperature is due to heat, and heat is a form of energy. We have energy entering the tank in water at temperature $10°$, energy being added by the heater, energy raising the temperature of the water, and energy leaving in water at temperature $T(t)$. (For simplicity, we will ignore energy associated with the mixer. It will have minimal effect anyway.) Since we can calculate rates for these energies, and the tank is perfectly insulated, we must be able to establish an **energy balance equation** for the system. We feel that it should take the form

$$\left\{\begin{array}{c} \text{rate at which} \\ \text{energy enters tank} \\ \text{in cool water} \end{array}\right\} + \left\{\begin{array}{c} \text{rate at which} \\ \text{energy is added} \\ \text{by heater} \end{array}\right\} = \left\{\begin{array}{c} \text{rate at which} \\ \text{energy leaves tank} \\ \text{in warmer water} \end{array}\right\} + \left\{\begin{array}{c} \text{rate at which} \\ \text{energy is used to raise} \\ \text{temperature of} \\ \text{water in tank} \end{array}\right\}.$$

The first three rates are easily calculated; the fourth is more difficult. The second term on the left is 2000. For the remaining terms, we must introduce the *specific heat* $c_p = 4190$ J/kg·C of water. It is the energy required to raise the temperature of 1 kg of water by $1°$C. (To raise 5 kg of water $10°$C, say, requires $(5)(10)(4190) = 209\,500$ J.) Since 3/100 kg of water at temperature $10°$C enter the tank each second, the rate at which energy enters the tank due to this water is $(3/100)(10)(4190) = 1257$ J/s. (This is the energy required to raise 3/100 kg of water from $0°$C to $10°$C.) In a similar way, the rate at which energy leaves the tank in the water at temperature T is $(3/100)(T)(4190) = 1257T/10$ J/s. This leaves only the last term on the right side of the energy balance equation. The rate of change of temperature of the 100 kg of water in the tank is dT/dt; that is, the temperature changes dT/dt degrees each second. It follows that the rate at which energy is used to raise the temperature of this water is $(100)(4190)(dT/dt)$.

When these rates are substituted into the energy balance equation, the result is

$$1257 + 2000 = \frac{1257T}{10} + 419\,000\frac{dT}{dt}.$$

This is another example of a differential equation; find the function $T(t)$ that satisfies the equation. We also need to specify the initial temperature of the water in the tank. We are told that when the heater is turned on, temperature of the water in the tank is the same as the incoming water, namely $10°$C. Let us choose this to be time $t = 0$. The initial-value problem for $T(t)$ is then

$$\frac{dT}{dt} = -\frac{3T}{10\,000} + \frac{3257}{419\,000}, \qquad T(0) = 10.$$

Equation 3.43 suggests that to account for the term $-3T/10\,000$ on the right side of the differential equation, $e^{-3t/10\,000}$ should be involved. Indeed, the derivative of the function $De^{-3t/10\,000}$ is $-3/10\,000$ times itself for any constant D whatsoever. This must be a part of $T(t)$. The other part is simple in form but harder to discover. If we express the differential equation in the form

$$\frac{dT}{dt} + \frac{3T}{10\,000} = \frac{3257}{419\,000},$$

we need a function that yields $3257/419\,000$ when substituted into the left side of the equation. The constant function $\dfrac{3257}{419\,000} \cdot \dfrac{10\,000}{3} = \dfrac{32\,570}{1257}$ does it. Putting these together, we suspect that temperature of the water in the tank must be of the form

$$T(t) = De^{-3t/10\,000} + \frac{32\,570}{1257}.$$

It is straightforward to check that the derivative of this function does indeed satisfy the required differential equation for any constant D. It remains only to choose D so that $T(t)$ satisfies the initial condition $T(0) = 10$. This requires

$$10 = D + \frac{32\,570}{1257} \qquad \Longrightarrow \qquad D = \frac{-20\,000}{1257}.$$

Consequently, temperature of the water at any time t is

$$D(t) = \frac{32\,570}{1257} - \frac{20\,000}{1257}e^{-3t/10\,000}.$$

The graph of $T(t)$ is shown in Figure 3.54. It begins at $T = 10$ and has a horizontal asymptote. This is so because

$$\lim_{t \to \infty} T(t) = \frac{32\,570}{1257} \approx 26.$$

In other words, the temperature of the water does not rise indefinitely; it levels off at $26°C$.

FIGURE 3.54   Temperature of water leaving heat tank

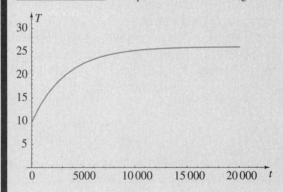

EXERCISES 3.11

In Exercises 1–30 y is defined as a function of x. Find dy/dx in as simplified a form as possible.

1. $y = 3^{2x}$

2. $y = \ln(3x^2 + 1)$

3. $y = \log_{10}(2x + 1)$

4. $y = e^{1-2x}$

5. $y = xe^{2x}$

6. $y = x \ln x$

7. $y = e^{2\ln x}$

8. $y = \log_{10}(3 - 4x)$

9. $y = \ln(\sin x)$

10. $y = \ln(3\cos x)$

11. $y = x\ln(x+1)$

12. $y = x^2 + x^3 e^{4x}$

13. $y = \dfrac{e^{1-x}}{1-x}$

14. $y = \sin(e^{2x})$

15. $y = \ln(\ln x)$

16. $y = e^{-2x}\sin 3x$

17. $y = \ln(x^2 e^{4x})$

18. $y = \dfrac{e^x - e^{-x}}{e^x + e^{-x}}$

19. $y = e^{-x}\ln x$

20. $y = x[\sin(\ln x) - \cos(\ln x)]$

21. $y = e^{\sin u}, \quad u = e^{1/x}$

22. $y = \ln(\cos v), \quad v = \sin^2 x$

23. $\ln(x+y) = x^2 y$

24. $xe^y + x^2\ln y + y\sin x = 0$

25. $y = \ln(\sec x + \tan x)$

26. $y = x\sqrt{x^2+1} - \ln\left(x + \sqrt{x^2+1}\right)$

27. $y = \ln\left(x + 4 + \sqrt{8x + x^2}\right)$

28. $y = x - \dfrac{1}{4}\ln(1 + 5e^{4x})$

29. $e^{xy} = (x+y)^2$

30. $e^{1/x} + e^{1/y} = \dfrac{1}{x} + \dfrac{1}{y}$

31. The figure below shows a long cylindrical cable. Copper wire runs down the centre of the cable and insulation covers the wire. If r measures radial distance from the centre of the cable, then steady-state temperature T in the insulation is a function of r that must satisfy the differential equation

$$\frac{d}{dr}\left(r\frac{dT}{dr}\right) = 0.$$

(a) Verify that the function

$$T(r) = c\ln r + d$$

satisfies the differential equation for any constants c and d.

(b) If temperature on inner and outer edges of the insulation $r = a$ and $r = b$ are constants T_a and T_b, find c and d in terms of a, b, T_a, and T_b.

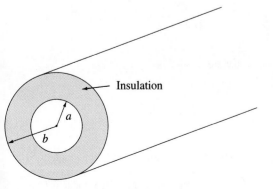

Insulation

* **32.** If heat is generated at a constant rate in the copper wire of Exercise 31 (perhaps because of electric current), the differential equation is replaced by

$$\frac{d}{dr}\left(r\frac{dT}{dr}\right) = k,$$

where k is a constant.

(a) Verify that the function

$$T(r) = kr + c\ln r + d$$

satisfies the differential equation for any constants c and d.

(b) If temperature on inner and outer edges of the insulation $r = a$ and $r = b$ are constants T_a and T_b, find c and d in terms of a, b, T_a, T_b, and k.

* **33.** (a) Two substances A and B react to form a third substance C in such a way that 1 g of A reacts with 1 g of B to produce 2 g of C. If 10 g of A and 15 g of B are brought together at time $t = 0$, the number of grams of C in the mixture as a function of time t is

$$x(t) = \frac{60(1 - e^{-10t})}{3 - 2e^{-10t}}.$$

Verify that $x(t)$ satisfies the (differential) equation

$$\frac{dx}{dt} = (20 - x)(30 - x).$$

(b) Plot $x(t)$. What is the limit of $x(t)$ for large t? Is this reasonable?

* **34.** (a) Two substances A and B react to form a third substance C in such a way that 1 g of A reacts with 2 g of B to produce 3 g of C. If 10 g of A and 30 g of B are brought together at time $t = 0$, the number of grams of C in the mixture as a function of time t is

$$x(t) = \frac{90(1 - e^{-15t})}{3 - 2e^{-15t}}.$$

Verify that $x(t)$ satisfies the (differential) equation

$$\frac{dx}{dt} = (30 - x)(45 - x).$$

(b) Plot $x(t)$. What is the limit of $x(t)$ for large t? Is this reasonable?

* **35.** (a) What is the differential equation satisfied by temperature of the water in the tank of Figure 3.53 if the rate at which water enters and leaves the tank is $100/(t+1)$ kg/s?

(b) Assuming that temperature of the water in the tank is $10°$C at time $t = 0$, verify that

$$T(t) = \frac{4190t + 4189}{419(t+1)} + \frac{t+1}{419}.$$

Plot a graph of $T(t)$.

36. (a) What is the differential equation satisfied by temperature of the water in the tank of Figure 3.53 if the amount of energy supplied by the heater for the first 10 min is $q = 20t$ J/s, $0 \le t \le 600$?

(b) Assuming that temperature of the water in the tank is $10°C$ at time $t = 0$, verify that

$$T(t) = \frac{200t}{1257} - \frac{1\,962\,290}{3771} + \frac{2\,000\,000}{3771}e^{-3t/10\,000}, \quad 0 \le t \le 600.$$

Plot a graph of $T(t)$.

37. (a) What is the differential equation satisfied by temperature of the water in the tank of Figure 3.53 if the temperature of the incoming water is a function of time, $T_0 = 10e^{-t}$, $t \ge 0$?

(b) Assuming that temperature of the water in the tank is $10°C$ at time $t = 0$, verify that

$$T(t) = \frac{20\,000}{1257} - \frac{74\,240\,000}{12\,566\,229}e^{-3t/10\,000} - \frac{30}{9997}e^{-t}.$$

Plot a graph of $T(t)$.

* **38.** A spring with constant 5 N/m is attached to a fixed wall on one end and a 1 kg mass m on the other (figure below). We choose a coordinate system with x positive to the right and with $x = 0$ at the centre of mass of m when the spring is in the unstretched position. At time $t = 0$, m is pulled 1 m to the left of $x = 0$ and given a speed of 3 m/s to the right. During the subsequent motion, a frictional force equal in newtons to twice the velocity of the mass acts on m. It can be shown that if $x(t)$ is the position of m, then $x(t)$ must satisfy the differential equation

$$\frac{d^2x}{dt^2} + 2\frac{dx}{dt} + 5x = 0$$

and the initial conditions $x(0) = -1$ and $x'(0) = 3$. Verify that $x(t) = e^{-t}(\sin 2t - \cos 2t)$ satisfies the equation and initial conditions.

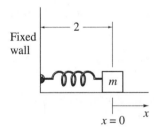

39. The figure below shows a tank containing $10\,000$ L of water in which is dissolved 100 kg of salt.

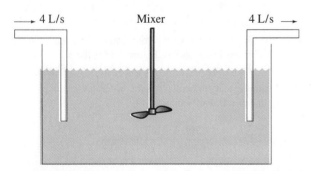

Starting at time $t = 0$, a solution containing 1 kg of salt for each 20 L of solution is added to the tank at 4 L/s. Simultaneously, 4 L

of solution is removed from the tank each second. It is assumed that the mixer is so efficient that the concentration of salt is the same at all points in the tank, and this is the concentration at which solution leaves the tank. If $S(t)$ represents the number of kilograms of salt in the tank at any given time, then dS/dt is the rate of change of the amount of salt in the tank. It must be equal to the rate at which salt enters less the rate at which salt leaves:

$$\frac{dS}{dt} = \left\{ \begin{array}{c} \text{rate at which} \\ \text{salt enters tank} \end{array} \right\} - \left\{ \begin{array}{c} \text{rate at which} \\ \text{salt leaves tank} \end{array} \right\}.$$

The rate at which salt enters the tank is $(1/20)(4) = 1/5$ kg/s. The rate at which salt leaves is $(S/10\,000)(4) = S/2500$ kg/s. Thus,

$$\frac{dS}{dt} = \frac{1}{5} - \frac{S}{2500}, \quad S(0) = 100.$$

Verify that the solution of this differential equation and initial condition is

$$S(t) = 100(5 - 4e^{-t/2500}) \quad \text{kg}.$$

Plot its graph. Is it what you would expect?

40. (a) Show that if solution in Exercise 39 is removed at 2 L/s rather than 4 L/s, the differential equation for $S(t)$ is

$$\frac{dS}{dt} = \frac{1}{5} - \frac{2S}{10\,000 + 2t}, \quad S(0) = 100.$$

(b) Verify that the solution is

$$S(t) = 500 + \frac{t}{10} - \frac{2 \times 10^6}{5000 + t}.$$

Plot a graph of $S(t)$. Does it have an asymptote?

41. (a) Show that if solution in Exercise 39 is removed at 8 L/s rather than 4 L/s, the differential equation for $S(t)$ is

$$\frac{dS}{dt} = \frac{1}{5} - \frac{8S}{10\,000 - 4t}, \quad S(0) = 100.$$

(b) Verify that the solution is

$$S(t) = 100 + \frac{3t}{25} - \frac{t^2}{15\,625}.$$

For how long is the solution valid? Plot a graph of $S(t)$.

(c) What is the maximum amount of salt in the tank?

42. (a) A beer vat contains 2000 L of beer, 4% of which is alcohol. Beer with 8% alcohol is added to the vat at 2 L/s, and well-stirred mixture is removed at the same rate. Show that if $A(t)$ represents the number of litres of alcohol in the vat, then

$$\frac{dA}{dt} = \frac{4}{25} - \frac{A}{1000}, \qquad A(0) = 80.$$

(b) Verify that the solution of this differential equation is

$$A(t) = 160 - 80e^{-t/1000}.$$

Plot a graph of $A(t)$.

(c) When is the beer in the vat 5% alcohol?

43. Two long, parallel rectangular loops lying in the same plane have lengths l and L and widths w and W, respectively (figure below). If the loops do not overlap, and the distance between the near sides is s, the mutual inductance between the loops is

$$M = \frac{\mu_0 l}{2\pi} \ln \left[\frac{s+W}{s\left(1 + \dfrac{W}{s+w}\right)} \right],$$

where $\mu_0 > 0$ is a constant. Show that the derivative of M as a function of s is negative for $s > 0$.

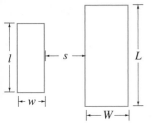

44. Two parallel wires carrying current i (figure below) create a magnetic field. The flux Φ of this field through the loop of dimensions h and w is

$$\Phi = \frac{\mu_0 h i}{2\pi} \ln \left[\frac{R(r+w)}{r(R+w)} \right],$$

where $\mu_0 > 0$ is a constant. Draw a graph of Φ, first as a function of h and then as a function of w.

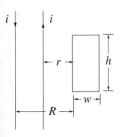

* **45.** Verify that the function

$$y = f(x) = Ax + B\left[\frac{x}{2} \ln \left(\frac{x-1}{x+1} \right) + 1 \right]$$

satisfies the differential equation

$$(1-x^2)\frac{d^2y}{dx^2} - 2x\frac{dy}{dx} + 2y = 0$$

for any constants A and B.

* **46.** The equation $x = e^y - e^{-y}$ defines y implicitly as a function of x. Find dy/dx in terms of x in two ways:

(a) Solve first for an explicit definition for $y = f(x)$.

(b) Differentiate implicitly with respect to x.

* **47.** (a) Show that when a liquid enters the tank in Figure 3.53 at $\dot{m}$ kilograms per second, the mass of liquid in the tank is M kilograms, the temperature of liquid entering the tank is $T_0 °\text{C}$, the specific heat of the liquid is c_p, and the heater adds energy at the rate of q joules per second, the differential equation governing temperature $T(t)$ of liquid in the tank is

$$\dot{m}c_p T_0 + q = \dot{m}c_p T + Mc_p \frac{dT}{dt}.$$

(b) Verify that the solution of this differential equation is

$$T(t) = T_0 e^{-\dot{m}t/M} + \left(T_0 + \frac{q}{c_p \dot{m}} \right)\left(1 - e^{-\dot{m}t/M} \right)$$

when the temperature of the liquid in the tank is T_0 at time $t = 0$, and $\dot{m}$, T_0, and q are all constants.

48. Find d^2y/dx^2 if y is defined implicitly as a function of x by

$$\ln(x^2 + y^2) = 2 \,\text{Tan}^{-1}\left(\frac{y}{x} \right).$$

** **49.** (a) The function $f(x) = x^n$ is not defined for $x < 0$ if n is irrational, that is, is not rational. Verify power rule 3.7 for $x < 0$ if n is rational and x^n is defined.

(b) Discuss the derivative of $f(x) = x^n$ at $x = 0$.

3.12 Logarithmic Differentiation

Compare the following three functions: x^n, a^x, and x^x. The first, in which the exponent is constant and the base is variable, is called a **power function**. The second, in which the base is constant and the exponent is variable, is called an **exponential function**. We have differentiation rules for power functions and exponential functions, namely

$$\frac{d}{dx}x^n = nx^{n-1} \qquad \text{and} \qquad \frac{d}{dx}a^x = a^x \ln a,$$

respectively. The third function x^x, which we consider only for $x > 0$, is neither a power nor an exponential function; therefore, we cannot use either of the formulas above to find its derivative. Instead, we set $y = f(x) = x^x$, and take natural logarithms,

$$\ln y = x \ln x.$$

Implicit differentiation with respect to x now gives

$$\frac{1}{y}\frac{dy}{dx} = \ln x + \frac{x}{x} = \ln x + 1.$$

Consequently,

$$\frac{dy}{dx} = y(\ln x + 1) \qquad \text{or} \qquad \frac{d}{dx}x^x = x^x(\ln x + 1).$$

This process of taking logarithms and then differentiating is called **logarithmic differentiation**.

EXAMPLE 3.36

Find the derivative of the function $f(x) = x^{\sin x}$ when $x > 0$.

SOLUTION   If we set $y = x^{\sin x}$, then

$$\ln y = \sin x \ln x.$$

Differentiation with respect to x gives

$$\frac{1}{y}\frac{dy}{dx} = \cos x \ln x + \frac{1}{x}\sin x,$$

and this can be solved for dy/dx,

$$\frac{dy}{dx} = x^{\sin x}\left(\cos x \ln x + \frac{1}{x}\sin x\right).$$

Logarithmic differentiation can also be used to differentiate complicated products or quotients, as in the following example.

EXAMPLE 3.37

Find the derivative of the function

$$y = f(x) = \frac{x^3(x^2 + 1)^{2/3}}{\sin^3 x}$$

on the interval $0 < x < \pi$. Extend the result to other values of x.

SOLUTION When $0 < x < \pi$, we take natural logarithms of both sides of the definition for y:

$$\ln y = 3 \ln x + \frac{2}{3} \ln (x^2 + 1) - 3 \ln (\sin x).$$

Differentiation with respect to x gives

$$\frac{1}{y}\frac{dy}{dx} = \frac{3}{x} + \frac{2}{3}\frac{2x}{x^2 + 1} - \frac{3}{\sin x}\cos x,$$

and therefore

$$\frac{dy}{dx} = y\left[\frac{3}{x} + \frac{4x}{3(x^2 + 1)} - 3\cot x\right] = \frac{x^3(x^2 + 1)^{2/3}}{\sin^3 x}\left[\frac{3}{x} + \frac{4x}{3(x^2 + 1)} - 3\cot x\right].$$

When x is not in the interval $0 < x < \pi$, this derivation may not be valid. For instance, when $x < 0$, it is not acceptable to write $\ln x$, and when x is in the interval $\pi < x < 2\pi$, the term $\ln (\sin x)$ is not defined. These difficulties are easily overcome by first taking absolute values,

$$|y| = \frac{|x|^3(x^2 + 1)^{2/3}}{|\sin x|^3}.$$

Logarithms now give

$$\ln |y| = 3 \ln |x| + \frac{2}{3} \ln (x^2 + 1) - 3 \ln | \sin x|.$$

To differentiate this equation, we use equation 3.13, which states that when $f(x)$ is differentiable, and $f(x) \neq 0$,

$$\frac{d}{dx}|f(x)| = \frac{|f(x)|}{f(x)} f'(x).$$

When this is combined with formula 3.39, we obtain

$$\frac{d}{dx} \ln |f(x)| = \frac{1}{|f(x)|}\frac{d}{dx}|f(x)| = \frac{1}{|f(x)|}\frac{|f(x)|}{f(x)} f'(x) = \frac{f'(x)}{f(x)}. \tag{3.46}$$

Application of this result to the equation in $\ln |y|$ gives

$$\frac{1}{y}\frac{dy}{dx} = \frac{3}{x} + \frac{2}{3}\frac{2x}{x^2 + 1} - 3\frac{\cos x}{\sin x}.$$

This equation is identical to that obtained for dy/dx when $0 < x < \pi$, but its derivation here shows that it is valid even when x is not in the interval $0 < x < \pi$.

In Exercises 1–24 use logarithmic differentiation to find $f'(x)$.

* **1.** $f(x) = x^{-x}$, $\quad x > 0$

* **2.** $f(x) = x^{4\cos x}$, $\quad x > 0$

* **3.** $f(x) = x^{4x}$, $\quad x > 0$

* **4.** $f(x) = (\sin x)^x$, $\quad 0 < x < \pi$

* **5.** $f(x) = \left(1 + \dfrac{1}{x}\right)^x$, $\quad x > 0$

* **6.** $f(x) = \left(1 + \dfrac{1}{x}\right)^{x^2}$, $\quad x > 0$

* **7.** $f(x) = \left(\dfrac{1}{x}\right)^{1/x}$, $\quad x > 0$

* **8.** $f(x) = \left(\dfrac{2}{x}\right)^{3/x}$, $\quad x > 0$

* **9.** $f(x) = (\sin x)^{\sin x}$, $\quad 0 < x < \pi$

* **10.** $f(x) = (\ln x)^{\ln x}$, $\quad x > 1$

* **11.** $f(x) = (x^2 + 3x^4)^3 (x^2 + 5)^4$

* **12.** $f(x) = \dfrac{\sqrt{x}(1 + 2x^2)}{\sqrt{1 + x^2}}$

* **13.** $f(x) = x\sqrt[3]{1 - \sin x}$
* **14.** $f(x) = (x^2 + 3x)^3 (x^2 + 5)^4$

* **15.** $f(x) = x^2 e^{4x}$
* **16.** $f(x) = x^{3/2} e^{-2x}$

* **17.** $f(x) = x^2 \ln x$
* **18.** $f(x) = \dfrac{e^x}{\ln(x - 1)}$

* **19.** $f(x) = (x^3 + 3)^3 (x^2 - 2x)$

* **20.** $f(x) = \dfrac{\sqrt{x}(1 - x^2)}{\sqrt{1 + x^2}}$

* **21.** $f(x) = \dfrac{x^2 - 1}{x\sqrt{1 - 4\tan^2 x}}$

* **22.** $f(x) = x^3(x^2 - 4x)\sqrt{1 + x^3}$

* **23.** $f(x) = \dfrac{\sin^3 3x}{\tan^5 2x}$

* **24.** $f(x) = \dfrac{\sin 2x \sec 5x}{(1 - 2\cot x)^3}$

* **25.** If $u(x)$ is positive for all x, find a formula for the derivative of u^u with respect to x.

* **26.** If a company sells a certain commodity at price r, the market demands

$$x = r^a e^{-b(r+c)}$$

items per week, where $r > a/b$, and a, b, and c are positive constants.

  (a) Show that the demand increases as the price decreases.

  (b) Calculate the elasticity of demand defined by

$$\frac{Er}{Ex} = \frac{x}{r}\frac{dr}{dx}.$$

3.13 Derivatives of the Hyperbolic Functions

Since the hyperbolic sine and cosine functions are defined in terms of the exponential function, for which we know the derivative, and the remaining hyperbolic functions are defined in terms of the hyperbolic sine and cosine, it follows that calculation of the derivatives of the hyperbolic functions should be straightforward. Indeed, if $u(x)$ is a differentiable function of x, then

$$\frac{d}{dx}\sinh u = \cosh u \, \frac{du}{dx}, \tag{3.47a}$$

$$\frac{d}{dx}\cosh u = \sinh u \, \frac{du}{dx}, \tag{3.47b}$$

$$\frac{d}{dx}\tanh u = \operatorname{sech}^2 u \, \frac{du}{dx}, \tag{3.47c}$$

$$\frac{d}{dx}\coth u = -\operatorname{csch}^2 u \, \frac{du}{dx}, \tag{3.47d}$$

$$\frac{d}{dx}\operatorname{sech} u = -\operatorname{sech} u \tanh u \, \frac{du}{dx}, \tag{3.47e}$$

$$\frac{d}{dx}\operatorname{csch} u = -\operatorname{csch} u \coth u \, \frac{du}{dx}. \tag{3.47f}$$

EXAMPLE 3.38

Find dy/dx if y is defined as a function of x by:

(a) $y = \operatorname{sech}(3x^2)$ (b) $y = \tanh(1 - 4x)$

(c) $y = \cos 2x \sinh 2x$ (d) $y = \cosh\left(\operatorname{Tan}^{-1} x^2\right)$

SOLUTION

$$\text{(a)}\quad \frac{dy}{dx} = -\operatorname{sech}(3x^2)\tanh(3x^2)\frac{d}{dx}(3x^2) = -6x\operatorname{sech}(3x^2)\tanh(3x^2)$$

$$\text{(b)}\quad \frac{dy}{dx} = \operatorname{sech}^2(1 - 4x)\frac{d}{dx}(1 - 4x) = -4\operatorname{sech}^2(1 - 4x)$$

$$\text{(c)}\quad \frac{dy}{dx} = \cos 2x(2\cosh 2x) + (-2\sin 2x)\sinh 2x$$

$$= 2(\cos 2x \cosh 2x - \sin 2x \sinh 2x)$$

$$\text{(d)}\quad \frac{dy}{dx} = \sinh\left(\operatorname{Tan}^{-1} x^2\right)\frac{d}{dx}\operatorname{Tan}^{-1} x^2 = \sinh\left(\operatorname{Tan}^{-1} x^2\right)\frac{2x}{1 + x^4}$$

EXAMPLE 3.39

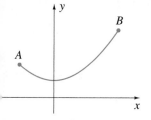

FIGURE 3.55　Shape of hanging cable

When a uniform cable is hung between two fixed supports (Figure 3.55), the shape of the curve $y = f(x)$ must satisfy the differential equation

$$\frac{d^2 y}{dx^2} = \frac{\rho g}{H}\sqrt{1 + \left(\frac{dy}{dx}\right)^2},$$

where ρ is the mass per unit length of the cable, $g > 0$ is the acceleration due to gravity, and $H > 0$ is the tension in the cable at its lowest point. Verify that a solution of the equation is

$$y = f(x) = \frac{H}{\rho g}\cosh\left(\frac{\rho g x}{H}\right) + C,$$

where C is a constant.

SOLUTION   In Exercises 42 of Section 8.4 we derive this solution. For now we simply wish to verify that the hyperbolic cosine is indeed a solution. The first derivative of the function is

$$\frac{dy}{dx} = \sinh\left(\frac{\rho g x}{H}\right),$$

and therefore its second derivative is

$$\frac{d^2 y}{dx^2} = \frac{\rho g}{H}\cosh\left(\frac{\rho g x}{H}\right).$$

On the other hand,

$$\frac{\rho g}{H}\sqrt{1 + \left(\frac{dy}{dx}\right)^2} = \frac{\rho g}{H}\sqrt{1 + \sinh^2\left(\frac{\rho g x}{H}\right)} = \frac{\rho g}{H}\sqrt{\cosh^2\left(\frac{\rho g x}{H}\right)} = \frac{\rho g}{H}\cosh\left(\frac{\rho g x}{H}\right).$$

Thus, the function $y = \left[H/(\rho g)\right]\cosh(\rho g x/H) + C$ does indeed satisfy the given differential equation.

Example 3.39 shows that the many telephone and hydro wires crisscrossing the country hang in the form of hyperbolic cosines. Engineers often call this curve a *catenary*.

EXAMPLE 3.40

In studying wave guides, the electrical engineer often encounters the differential equation

$$\frac{d^2y}{dx^2} - ky = 0,$$

where $k > 0$ is a constant. Verify that $y = f(x) = A \cosh \sqrt{k}x + B \sinh \sqrt{k}x$ is a solution for any constants A and B.

SOLUTION   The first derivative of $y = f(x)$ is

$$\frac{dy}{dx} = \sqrt{k}\, A \sinh \sqrt{k}\, x + \sqrt{k}\, B \cosh \sqrt{k}\, x,$$

and therefore

$$\frac{d^2y}{dx^2} = kA \cosh \sqrt{k}\, x + kB \sinh \sqrt{k}\, x = k(A \cosh \sqrt{k}\, x + B \sinh \sqrt{k}\, x).$$

Thus,

$$\frac{d^2y}{dx^2} = ky.$$

EXERCISES 3.13

In Exercises 1–10 y is defined as a function of x. Find dy/dx in as simplified a form as possible.

1. $y = \operatorname{csch}(2x + 3)$

2. $y = x \sinh(x/2)$

3. $y = \sqrt{1 - \operatorname{sech} x}$

4. $y = \tanh(\ln x)$

5. $\cosh(x + y) = 2x$

6. $y + \coth x = \sqrt{1 + y}$

7. $y = u \cosh u, \ \ u = e^x + e^{-x}$

8. $y = \tan(\cosh t), \ \ t = \cos(\tanh x)$

9. $y = \operatorname{Tan}^{-1}(\sinh x)$

10. $y = \ln \sqrt{\tanh 2x}$

11. Verify the differentiation formulas 3.47.

* **12.** To analyze vertical vibrations of the beam in the figure below, we must solve the differential equation

$$\frac{d^4y}{dx^4} - k^4 y = 0,$$

where $k > 0$ is a constant.

(a) Verify that a solution is

$$y = f(x) = A \cos kx + B \sin kx + C \cosh kx + D \sinh kx$$

for any constants A, B, C, and D.

(b) If the left end ($x = 0$) is fastened horizontally and the right end ($x = L$) is pinned, then $f(x)$ must satisfy the conditions

$$f(0) = f'(0) = f(L) = f''(L) = 0.$$

Show that these restrictions imply that $C = -A$, $D = -B$, and A and B must satisfy the equations

$$A(\cos kL - \cosh kL) + B(\sin kL - \sinh kL) = 0,$$

$$A(\cos kL + \cosh kL) + B(\sin kL + \sinh kL) = 0.$$

(c) Eliminate A and B between these equations to show that k must satisfy the condition

$$\tan kL = \tanh kL.$$

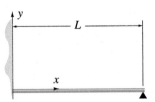

13. Each hyperbolic function has associated with it an inverse hyperbolic function. (See Exercise 13 in Section 1.10.) Obtain the following derivatives of these functions:

$$\frac{d}{dx}\text{Sinh}^{-1}x = \frac{1}{\sqrt{x^2+1}};$$

$$\frac{d}{dx}\text{Cosh}^{-1}x = \frac{1}{\sqrt{x^2-1}};$$

$$\frac{d}{dx}\text{Tanh}^{-1}x = \frac{1}{1-x^2}, \quad |x| < 1;$$

$$\frac{d}{dx}\text{Coth}^{-1}x = \frac{1}{1-x^2}, \quad |x| > 1;$$

$$\frac{d}{dx}\text{Sech}^{-1}x = \frac{-1}{x\sqrt{1-x^2}};$$

$$\frac{d}{dx}\text{Csch}^{-1}x = \frac{-1}{|x|\sqrt{1+x^2}}.$$

3.14 Rolle's Theorem and the Mean Value Theorems

Certain results in calculus are immediately seen to be important. For example, the power, product, and quotient rules that eliminate the necessity of using equation 3.3 to calculate derivatives are clearly indispensable. Even the algebraic and geometric interpretations of the derivative itself are recognized as useful. Through various examples and exercises of this chapter, we have hinted at the variety and quantity of applications of the derivative. These will be dealt with at length in Chapter 4. Other results in calculus, especially those of a theoretical nature, are regarded as less important, or even unimportant, often because it is not obvious how they will be used. In this section we consider three very important theorems, without which we would encounter serious difficulty in treating many of the topics in the remainder of this book. The first theorem is needed to prove the second, and the second leads immediately to the third.

> **THEOREM 3.17 (Rolle's Theorem)**
> Suppose a function $f(x)$ satisfies the following three properties:
>
> **1.** $f(x)$ is continuous for $a \le x \le b$;
> **2.** $f'(x)$ exists for $a < x < b$;
> **3.** $f(a) = f(b)$.
>
> Then, there exists at least one point c in the open interval $a < c < b$ at which $f'(c) = 0$.

FIGURE 3.56 Rolle's theorem

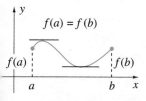

FIGURE 3.57 Two level points for Rolle's theorem

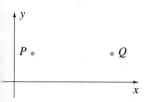

For the function in Figure 3.56 there are two possible choices for c. Geometrically, Rolle's theorem seems quite evident. Begin with two points, say P and Q in Figure 3.57, which have the same y-coordinate. Now try to join these points by a curve that never has a horizontal tangent line while satisfying the following two conditions:

(a) Do not lift the pencil from the page — continuity of $f(x)$;

(b) The curve must have a tangent line at all points, and this tangent line must not be vertical — $f'(x)$ exists at all points.

It is impossible; therefore, a point c where $f'(c) = 0$ must exist.

To verify this theorem directly requires Theorem 4.2 in Section 4.7. Since the latter result is quoted without proof, it seems as reasonable to accept Rolle's theorem on obvious geometric grounds as to base a proof on a theorem which itself is stated without proof. But for those who would like to see a proof based on Theorem 4.2 in Section 4.7, see Exercise 10 in that section.

Like the intermediate value theorem in Section 1.11, Rolle's theorem is an existence theorem; it stipulates the existence of c, but does not provide a way to find it.

Rolle's theorem can be used to prove the following result.

230 Chapter 3   Differentiation

THEOREM 3.18   (Cauchy's Generalized Mean Value Theorem)

Suppose functions $f(x)$ and $g(x)$ satisfy the following three properties:

1. $f(x)$ and $g(x)$ are continuous for $a \leq x \leq b$;
2. $f'(x)$ and $g'(x)$ exist for $a < x < b$;
3. $g'(x) \neq 0$ for $a < x < b$.

Then, there exists at least one point c in the open interval $a < c < b$ for which

$$\frac{f(b) - f(a)}{g(b) - g(a)} = \frac{f'(c)}{g'(c)}. \tag{3.48}$$

PROOF   First note that $g(b) - g(a)$ cannot equal zero. If it did, then $g(a)$ would equal $g(b)$, and Rolle's theorem applied to $g(x)$ on the interval $a \leq x \leq b$ would imply the existence of a point c at which $g'(c) = 0$, contrary to the given assumption. To prove the theorem, we construct a function $h(x)$ to satisfy the conditions of Rolle's theorem. Specifically, we consider

$$h(x) = f(x) - f(a) - \frac{f(b) - f(a)}{g(b) - g(a)}[g(x) - g(a)].$$

Since $f(x)$ and $g(x)$ are continuous for $a \leq x \leq b$, so too is $h(x)$. In addition,

$$h'(x) = f'(x) - \frac{f(b) - f(a)}{g(b) - g(a)} g'(x);$$

therefore, $h'(x)$ exists for $a < x < b$. Finally, since $h(a) = h(b) = 0$, we may conclude from Rolle's theorem that there exists a number c such that $a < c < b$, and

$$0 = h'(c) = f'(c) - \frac{f(b) - f(a)}{g(b) - g(a)} g'(c);$$

that is,

$$\frac{f'(c)}{g'(c)} = \frac{f(b) - f(a)}{g(b) - g(a)}.$$

The next theorem states an important special case of this result that occurs when $g(x) = x$.

THEOREM 3.19   (Mean Value Theorem)

Suppose a function $f(x)$ satisfies the following two properties:

1. $f(x)$ is continuous for $a \leq x \leq b$;
2. $f'(x)$ exists for $a < x < b$;

Then, there exists at least one point c in the open interval $a < c < b$ for which

$$f'(c) = \frac{f(b) - f(a)}{b - a}. \tag{3.49}$$

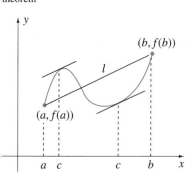

FIGURE 3.58   Mean value theorem

From a geometric point of view equation 3.49 seems as obvious as Rolle's theorem. Figure 3.58 illustrates that the quotient $[f(b) - f(a)]/(b - a)$ is the slope of the line l joining the points $\big(a, f(a)\big)$ and $\big(b, f(b)\big)$ on the graph $y = f(x)$. The mean value theorem states that there is at least one point c between a and b at which the tangent line is parallel to l. In Figure 3.58 there are clearly two such points. Algebraically, equation 3.49 states that at some point between a and b, the instantaneous rate of change of the function $f(x)$ is equal to its average rate of change over the interval $a \leq x \leq b$. Similar interpretations of Theorem 3.18 are given in Section 9.1.

EXAMPLE 3.41

Find all values of c satisfying the mean value theorem for the function $f(x) = x^3 - 4x$ on the interval $-1 \leq x \leq 3$.

SOLUTION   Since $f(x)$ is differentiable, and therefore continuous, at each point in $-1 \leq x \leq 3$, we can indeed apply the mean value theorem and claim the existence of at least one number c in $-1 < x < 3$ such that

$$f'(c) = \frac{f(3) - f(-1)}{3 - (-1)};$$

that is,

$$3c^2 - 4 = \frac{15 - 3}{4} = 3.$$

Consequently, $c = \pm\sqrt{7/3}$. Since $-\sqrt{7/3} < -1$, $c = \sqrt{7/3}$ is the only value of c in the interval $-1 < x < 3$ (see Figure 3.59).

FIGURE 3.59   Mean value theorem for $x^3 - 4x$ on $-1 \leq x \leq 3$

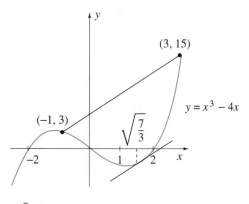

EXAMPLE 3.42

A traffic plane (Figure 3.60) measures the time that it takes a car to travel between points A and B as 15 s, and radios this information to a patrol car. What is the maximum speed at which the police officer can claim that the car was travelling between A and B?

SOLUTION   The average speed of the car between A and B is $500/15 = 100/3$ m/s. According to the mean value theorem, the instantaneous speed of the car must also have been $100/3$ m/s at least once. This is the maximum speed attributable to the car between A and B. It may have been travelling much faster at some points, but from the information given, no speed greater than $100/3$ m/s can be claimed by the officer.

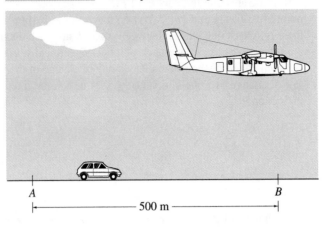

FIGURE 3.60   Traffic plane measuring speed of car

A   B

|←   500 m   →|

EXERCISES 3.14

In Exercises 1–14 decide whether the mean value theorem can be applied to the function on the interval. If it cannot, explain why not. If it can, find all values of c in the interval that satisfy equation 3.49.

* **1.** $f(x) = x^2 + 2x,$ $-3 \le x \le 2$

* **2.** $f(x) = 4 + 3x - 2x^2,$ $1 \le x \le 3$

* **3.** $f(x) = x + 5,$ $2 \le x \le 3$

* **4.** $f(x) = |x|,$ $-1 \le x \le 1$

* **5.** $f(x) = |x|,$ $0 \le x \le 1$

* **6.** $f(x) = x^3 + 2x^2 - x - 2,$ $-3 \le x \le 2$

* **7.** $f(x) = x^3 + 2x^2 - x - 2,$ $-1 \le x \le 2$

* **8.** $f(x) = (x + 2)/(x - 1),$ $2 \le x \le 4$

* **9.** $f(x) = (x + 1)/(x + 2),$ $-3 \le x \le 2$

* **10.** $f(x) = x^2/(x + 3),$ $-2 \le x \le 3$

* **11.** $f(x) = \sin x,$ $0 \le x \le 2\pi$

* **12.** $f(x) = \ln(2x + 1),$ $0 \le x \le 2$

* **13.** $f(x) = e^{-x},$ $-1 \le x \le 1$

* **14.** $f(x) = \sec x,$ $0 \le x \le \pi$

In Exercises 15–18 decide whether Cauchy's generalized mean value theorem can be applied to the functions on the interval. If it cannot, explain why not. If it can, find all values of c in the interval that satisfy equation 3.48.

* **15.** $f(x) = x^2,$ $g(x) = x,$ $1 \le x \le 2$

* **16.** $f(x) = x + 1,$ $g(x) = |x|^{3/2},$ $-1 \le x \le 1$

* **17.** $f(x) = x^2 + 3x - 1,$ $g(x) = x^3 + 5x + 4,$ $0 \le x \le 2$

* **18.** $f(x) = x/(x + 1),$ $g(x) = x/(x - 1),$ $-3 \le x \le -2$

* **19.** Show that if $|f'(x)| \le M$ on $a \le x \le b$, then

$$|f(b) - f(a)| \le M(b - a).$$

* **20.** Show that the value c that satisfies the mean value theorem for any quadratic function $f(x) = dx^2 + ex + g$ on any interval $a \le x \le b$ whatsoever is $c = (a + b)/2$.

* **21.** Use the mean value theorem to show that

$$|\sin a - \sin b| \le |a - b|$$

for all real a and b. Is the same inequality valid for the cosine function?

* **22.** Let $f(x)$ and $g(x)$ be two functions that are differentiable at each point of the interval $a \le x \le b$. Prove that if $f(a) = g(a)$ and $f(b) = g(b)$, then there exists c in the open interval $a < x < b$ for which $f'(c) = g'(c)$.

* **23.** Verify that for a cubic polynomial $f(x) = dx^3 + ex^2 + gx + h$ defined on any interval $a \le x \le b$, the values of c that satisfy equation 3.49 are equidistant from $x = -e/(3d)$.

SUMMARY

In this chapter we defined the derivative of a function $y = f(x)$ as

$$\frac{dy}{dx} = f'(x) = \lim_{h \to 0} \frac{f(x + h) - f(x)}{h}.$$

Algebraically, it is the instantaneous rate of change of y with respect to x; geometrically, it is the slope of the tangent line to the graph of $f(x)$. To eliminate the necessity of using this definition over and over again, we derived the sum, product, quotient, and power rules:

$$\frac{d}{dx}(u + v) = \frac{du}{dx} + \frac{dv}{dx},$$

$$\frac{d}{dx}(uv) = u\frac{dv}{dx} + v\frac{du}{dx},$$

$$\frac{d}{dx}\left(\frac{u}{v}\right) = \frac{v\dfrac{du}{dx} - u\dfrac{dv}{dx}}{v^2},$$

$$\frac{d}{dx}(u^n) = nu^{n-1}\frac{du}{dx}.$$

These four simple rules are fundamental to all calculus.

When a function $y = f(x)$ is defined implicitly by some equation $F(x, y) = 0$, we use implicit differentiation to find its derivative. We differentiate each term in the equation with respect to x, and then solve the resulting equation for dy/dx. We pointed out that care must be taken in differentiating equations. An equation can be differentiated with respect to a variable only if it is valid for a continuous range of values of that variable.

The chain rule defines the derivative of a composite function $y = f(g(x))$ as the product of the derivatives of $y = f(u)$ and $u = g(x)$:

$$\frac{dy}{dx} = \frac{dy}{du}\frac{du}{dx}.$$

These rules and techniques form the basis for the rest of differential calculus. When they are combined with the derivatives of the trigonometric, inverse trigonometric, exponential, logarithm, and hyperbolic functions, we are well prepared to handle those applications of calculus that involve differentiation. Derivative formulas for these trancendental functions are listed below:

$$\frac{d}{dx}\sin u = \cos u\,\frac{du}{dx}, \qquad\qquad \frac{d}{dx}\cos u = -\sin u\,\frac{du}{dx},$$

$$\frac{d}{dx}\tan u = \sec^2 u\,\frac{du}{dx}, \qquad\qquad \frac{d}{dx}\cot u = -\csc^2 u\,\frac{du}{dx},$$

$$\frac{d}{dx}\sec u = \sec u \tan u\,\frac{du}{dx}, \qquad\qquad \frac{d}{dx}\csc u = -\csc u \cot u\,\frac{du}{dx},$$

$$\frac{d}{dx}\log_a u = \frac{1}{u}\frac{du}{dx}\log_a e, \qquad\qquad \frac{d}{dx}\ln u = \frac{1}{u}\frac{du}{dx},$$

$$\frac{d}{dx}a^u = a^u\frac{du}{dx}\ln a, \qquad\qquad \frac{d}{dx}e^u = e^u\frac{du}{dx}.$$

$$\frac{d}{dx}\text{Sin}^{-1} u = \frac{1}{\sqrt{1-u^2}}\frac{du}{dx}, \qquad\qquad \frac{d}{dx}\text{Cos}^{-1} u = \frac{-1}{\sqrt{1-u^2}}\frac{du}{dx},$$

$$\frac{d}{dx}\text{Tan}^{-1} u = \frac{1}{1+u^2}\frac{du}{dx}, \qquad\qquad \frac{d}{dx}\text{Cot}^{-1} u = \frac{-1}{1+u^2}\frac{du}{dx},$$

$$\frac{d}{dx}\text{Sec}^{-1} u = \frac{1}{u\sqrt{u^2-1}}\frac{du}{dx}, \qquad\qquad \frac{d}{dx}\text{Csc}^{-1} u = \frac{-1}{u\sqrt{u^2-1}}\frac{du}{dx},$$

$$\frac{d}{dx}\sinh u = \cosh u\,\frac{du}{dx}, \qquad\qquad \frac{d}{dx}\cosh u = \sinh u\,\frac{du}{dx},$$

$$\frac{d}{dx}\tanh u = \mathrm{sech}^2 u \,\frac{du}{dx}, \qquad\qquad \frac{d}{dx}\coth u = -\mathrm{csch}^2 u \,\frac{du}{dx},$$

$$\frac{d}{dx}\mathrm{sech}\, u = -\mathrm{sech}\, u \tanh u \,\frac{du}{dx}, \qquad\qquad \frac{d}{dx}\mathrm{csch}\, u = -\mathrm{csch}\, u \coth u \,\frac{du}{dx}.$$

Velocity and acceleration have finally been given formal definitions. Velocity is the derivative of displacement, and acceleration is the derivative of velocity, or the second derivative of displacement:

$$v(t) = \frac{dx}{dt}, \qquad a(t) = \frac{dv}{dt} = \frac{d^2 x}{dt^2}.$$

We completed the chapter by using Rolle's theorem to prove two mean value theorems. When $f(x)$ and $g(x)$ are continuous for $a \le x \le b$ and differentiable for $a < x < b$, Cauchy's generalized mean value theorem guarantees the existence of at least one point c between a and b such that

$$\frac{f(b) - f(a)}{g(b) - g(a)} = \frac{f'(c)}{g'(c)},$$

provided also that $g'(x) \neq 0$ for $a < x < b$. When $g(x) = x$, we obtain as a special case the mean value theorem

$$f'(c) = \frac{f(b) - f(a)}{b - a}.$$

KEY TERMS

In reviewing this chapter, you should be able to define or discuss the following key terms:

Displacement function	Average velocity
Increment	Instantaneous velocity
Average rate of change	Instantaneous rate of change
Derivative	Differentiation
Tangent line	Derivative function
Power rule	Normal line
Right-hand derivative	Left-hand derivative
Differentiable	Angle between intersecting curves
Orthogonal curves	Product rule
Quotient rule	Second derivative
Third derivative	Velocity
Speed	Instantaneous acceleratio
Chain rule	Extended power rule
Explicit definition of a function	Implicit definition of a function
Implicit differentiation	Simple harmonic motion
Differential equation	Initial-value problem
Energy balance equation	Power function
Exponential function	Logarithmic differentiation
Rolle's theorem	Chaucy's generalized mean value theorem
Mean value theorem	

REVIEW EXERCISES

In Exercises 1–58 assume that y is defined as a function of x and find dy/dx in as simplified a form as possible.

1. $y = x^3 + \dfrac{1}{x^2}$

2. $y = 3x^2 + 2x + \dfrac{1}{x}$

3. $y = 2x - \dfrac{1}{3x^2} + \dfrac{1}{\sqrt{x}}$

4. $y = x^{1/3} - \dfrac{2}{3}x^{5/3}$

5. $y = x(x^2 + 5)^4$

6. $y = (x^2 + 2)^2 (x^3 - 3)^3$

7. $y = \dfrac{3x^2}{x^3 - 5}$

8. $y = \dfrac{3x - 2}{x + 5}$

9. $y = \dfrac{x^2 + 2x + 2}{x^2 + 2x - 1}$

10. $y = \dfrac{4x}{x^2 + 5x - 2}$

11. $xy + 3y^3 = x + 1$

12. $\dfrac{x}{y} + \dfrac{y}{x} = x$

13. $x^2 y^2 - 3y \sin x = 14$

14. $x^2 y + y\sqrt{1 + x} = 3$

15. $y = \tan^3 (3x + 2)$

16. $y = \sec^2 (1 - 4x)$

17. $y = \dfrac{\sin 2x}{\cos 3x}$

18. $y = \sec (\tan 2x)$

19. $y = x^2 \cos x^2$

20. $y = \sin^2 x \cos^2 x$

21. $y = u^2 - 2u, \quad u = (1 + 2x)^{5/3}$

22. $y = t + \cos 2t, \quad t = x - \cos 2x$

23. $y = \sqrt{1 - t^3}, \quad t = \sqrt{1 + x^2}$

24. $y = v \cos^2 v, \quad v = \sqrt{1 - x^2}$

25. $y = \sqrt{1 + \sqrt{1 + x}}$

26. $x = e^{2y}$

27. $\dfrac{2xy}{3x + 4} = x^2 + 2$

28. $y = (x^2 + 1) \ln (x^2 - 1)$

29. $x \sin y + 2xy = 4$

30. $5 \cos (x - y) = 1$

31. $y = \mathrm{Sin}^{-1} (2 - 3x)$

32. $y = 3 \sinh (x^2)$

33. $y = \dfrac{\mathrm{Cos}^{-1} x}{\mathrm{Sin}^{-1} x}$

34. $y = \mathrm{Tan}^{-1}\left(\dfrac{1}{x} + x\right)$

35. $y = e^{\cosh x}$

36. $\sinh y = \sin x$

37. $\mathrm{Sec}^{-1} (x + y) = xy$

38. $y = x \, \mathrm{Csc}^{-1}\left(\dfrac{1}{x^2}\right)$

39. $y = e^{2x} \cosh 2x$

40. $\ln \left[\mathrm{Tan}^{-1} (x + y)\right] = 1/10$

* 41. $x = \sqrt{1 + x \cot y^2}$

* 42. $x = \sqrt{\dfrac{4 + y}{4 - y}}$

* 43. $y = \sqrt{\dfrac{4 + x^2}{4 - x^2}}$

* 44. $y = \dfrac{x^2 \sqrt{1 - x}}{x + 5}$

* 45. $y = \sqrt{7 - \sqrt{7 - \sqrt{x}}}$

* 46. $\dfrac{x}{x + y} = \dfrac{y}{x - y}$

* 47. $y = \sqrt{\dfrac{4 + t}{4 - t}}, \quad t = \tan x$

* 48. $x = \dfrac{y^2 - 2y}{y^3 + 4y + 6}$

* 49. $y = \dfrac{x^3 - 6x^2 + 12x - 8}{x^2 - 4x + 4}$

* 50. $y = \dfrac{x - 2}{\sqrt{x} - \sqrt{2}}$

* 51. $y = x^{2x}$

* 52. $y = (\cos x)^x, \ 0 < x < \pi/2$

* 53. $y = \dfrac{e^x}{e^x + 1}$

* 54. $y = \log_{10} (\log_{10} x)$

* 55. $y = e^x \ln x$

* 56. $x = e^y + e^{-y}$

* 57. $xy e^{xy} = 1$

* 58. $x^2 y + \ln (x + y) = x + 2$

In Exercises 59–62 find equations for the tangent and normal lines to the curve at the point indicated.

59. $y = x^3 + 3x - 2$ at $(1, 2)$

60. $y = \dfrac{1}{x + 5}$ at $(0, 1/5)$

61. $y = \cos 2x$ at $(\pi/2, -1)$

62. $y = \dfrac{x^2 + 3x}{2x - 5}$ at $(1, -4/3)$

In Exercises 63–65 find $d^2 y/dx^2$ assuming that y is defined implicitly as a function of x.

* 63. $x^2 - y^2 + 2(x - y) = y^3$

* 64. $(x - y)^2 = 3xy$

* 65. $\sin (x + y) = x$

* 66. Draw a graph of $f(x) = \sin x^2$. Is it periodic?

* 67. Find all points on the curve $y = x^3 + x^2$ at which the tangent line passes through the origin.

* **68.** Find that point on the curve $x = y^2 - 4$ at which the normal line passes through the point $(-6, 7)$. What application could be made of this result?

* **69.**   (a) How many functions with domain $-1 \leq x \leq 1$ are defined implicitly by the equation $x^2 + y^2 = 1$?

      (b) How many continuous functions with domain $-1 \leq x \leq 1$ are defined implicitly by the equation?

* **70.** What is the rate of change of the area A of an equilateral triangle with respect to its side length L?

* **71.** In a heated house, the temperature varies as the thermostat continually engages and disengages the furnace. Suppose that at the thermostat, the temperature T in degrees Celsius over a four-hour time interval $0 \leq t \leq 4$ is given by

$$T = f(t) = 20 + 3 \sin (4\pi t - \pi/2).$$

    (a) Draw a graph of $f(t)$.

    (b) How many times is the furnace on during the four-hour period?

    (c) What is the maximum time rate of change of temperature?

* **72.** The curve defined by the equation $(x^2 + y^2)^2 = x^2 - y^2$ and shown below is called a *lemniscate*. Find the four points at which the tangent line is horizontal.

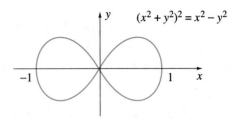

* **73.** Find all points c in the interval $3 \leq x \leq 6$ that satisfy equation 3.49 when $f(x) = x^3 + 3x - 2$.

* **74.** Find all points c in the interval $-1 \leq x \leq 1$ that satisfy equation 3.48 when $f(x) = 3x^2 - 2x + 4$ and $g(x) = x^3 + 2x$.

CHAPTER 4

Applications of Differentiation

Slider-cranks, such as that shown below, transform rotary motion to back-and-forth motion along a straight line, and vice versa. Rod AB, of length r, is pinned at A, and rod BC, of length L, is pinned to rod AB at B. As AB rotates around A, C is confined to move along a horizontal line segment between points D and E, called the *stroke* of the mechanism. Rotary motion of B is transformed to straight-line motion of C along DE. This crank is said to be offset because the extension of line segment DE does not pass through the centre of the circle A; it is offset by an amount e.

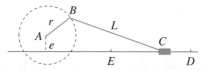

THE PROBLEM   Find the offset that maximizes the stroke. (For the solution, see Example 4.28 on page 284.)

Our discussions in Chapter 3 hinted at some of the applications of the derivative; in this chapter we deal with them in detail. In Section 4.1 we show that derivatives provide one of the most powerful methods for approximating solutions to equations. Sections 4.2–4.7 are devoted to the topic of optimization, its theory and applications. Sections 4.7–4.10, with their wealth of applied problems, show the power of calculus in applied mathematics. In Section 4.7 we illustrate the simplicity that calculus brings to solving applied maxima-minima problems; in Section 4.8 we develop a deeper understanding for the already familiar notions of velocity and acceleration; in Section 4.9 we use the interpretation of the derivative as a rate of change to investigate interdependences of related quantities in a wide variety of applications; and in Section 4.10 we apply derivatives to LCR-circuits. The calculation of many otherwise intractable limits becomes relatively straightforward with L'Hôpital's rule in Section 4.11. In Section 4.12 we discuss differentials, quantities essential to the topic of integration, which begins in Chapter 5.

4.1 Newton's Iterative Procedure for Solving Equations

In almost every area of applied mathematics, it is necessary to solve equations. When the problem is to solve one equation in one unknown, the equation can be expressed in the form

$$f(x) = 0, \tag{4.1}$$

where $f(x)$ is usually a differentiable function of x. It might be a polynomial, a trigonometric function, an exponential or a logarithm function, or a complicated combination of these. The equation may have one solution or many solutions; these solutions are also called *roots* of the equation or *zeros* of the function. Few equations can be solved by formula. Even when $f(x)$ is a polynomial, the only simple formula is quadratic formula 1.5, which solves the equation for second-degree polynomials. Functions other than polynomials rarely have formulas for their zeros.

Since we can seldom find exact solutions of 4.1, we consider approximating the solutions. Sophisticated calculators have routines for approximating solutions to equations; computer software packages have one or more commands for doing this. Unfortunately, with calculators and computers, you really have no idea what is going on inside them. You supply an equation and out comes a number. In this section we develop one of the most powerful techniques for approximating solutions to equations. The method is calculus-based and it, or a modification of it, is used by many calculators and computers to solve equations. If your machine uses this technique you will now understand what it is doing.

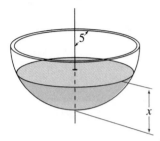

FIGURE 4.1 Volume of a half-filled hemispherical tank

To introduce the technique, consider the problem of finding the depth of water that half fills (by volume) the hemispherical tank in Figure 4.1. It can be shown (see Exercise 21 in Section 7.4) that when the water is x metres deep, the volume of water in the tank is given by the formula

$$\frac{\pi}{3}(15x^2 - x^3).$$

Since the volume of one-half the tank is one-fourth that of a sphere of radius 5, it follows that the tank is half full when x satisfies the equation

$$\frac{1}{4}\left[\frac{4}{3}\pi(5)^3\right] = \frac{\pi}{3}(15x^2 - x^3),$$

and this equation simplifies to

$$f(x) = x^3 - 15x^2 + 125 = 0.$$

How do we solve this equation when we know intuitively that the solution we want is somewhere around $x = 3$? A sketch of $f(x)$ between $x = 3$ and $x = 3.5$ is shown in Figure 4.2. The curvature has been exaggerated in order to more clearly depict the following geometric construction. If we set $x_1 = 3$, then x_1 is an approximation to the solution of the equation — not a good approximation, but an approximation nonetheless. Suppose we draw the tangent line to $y = f(x)$ at $(x_1, f(x_1))$. If x_2 is the point of intersection of this tangent line with the x-axis, it is clear that x_2 is a better approximation than x_1 to the solution of the equation. If we draw the tangent line to $y = f(x)$ at $(x_2, f(x_2))$, its intersection point x_3 with the x-axis is an even better approximation. Continuation of this process leads to a succession of numbers $x_1, x_2, x_3, \ldots$, each of which is closer to the solution of the equation $f(x) = 0$ than the preceding numbers. This procedure for finding better and better approximations to the solution of an equation is called **Newton's iterative procedure** (or the *Newton–Raphson iterative procedure*). We say that the numbers $x_1, x_2, x_3, \ldots$ converge to the root of the equation.

FIGURE 4.2 Geometric interpretation of Newton's iterative procedure

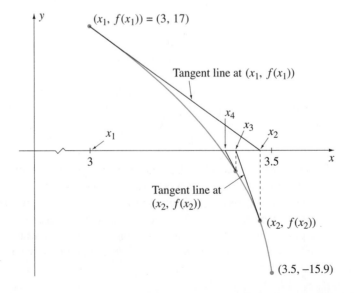

What we need now is an algebraic formula by which to calculate the approximations $x_2, x_3, x_4, \ldots$. The equation of the tangent line to $y = f(x)$ at $\left(x_1, f(x_1)\right)$ is

$$y - f(x_1) = f'(x_1)(x - x_1).$$

To find the point of intersection of this tangent line with the x-axis, we set $y = 0$,

$$-f(x_1) = f'(x_1)(x - x_1),$$

and solve for x,

$$x = x_1 - \frac{f(x_1)}{f'(x_1)}.$$

But the point of intersection of the tangent line at $\left(x_1, f(x_1)\right)$ with the x-axis is the second approximation x_2; that is,

$$x_2 = x_1 - \frac{f(x_1)}{f'(x_1)}.$$

To find x_3 we repeat this procedure with x_2 replacing x_1; the result is

$$x_3 = x_2 - \frac{f(x_2)}{f'(x_2)}.$$

As we repeat this process over and over again, the following formula for the $(n+1)^{\text{th}}$ approximation x_{n+1} in terms of the nth approximation x_n emerges:

$$x_{n+1} = x_n - \frac{f(x_n)}{f'(x_n)}. \tag{4.2}$$

This formula defines each approximation in Newton's iterative procedure in terms of its predecessor.

Let us use this procedure to approximate the solution of $f(x) = x^3 - 15x^2 + 125 = 0$ in the example above. The derivative of $f(x)$ is $f'(x) = 3x^2 - 30x$, and formula 4.2 becomes

$$x_{n+1} = x_n - \frac{x_n^3 - 15x_n^2 + 125}{3x_n^2 - 30x_n}.$$

Calculation of the next four approximations beginning with $x_1 = 3$ gives

$$x_2 = x_1 - \frac{x_1^3 - 15x_1^2 + 125}{3x_1^2 - 30x_1} = 3 - \frac{3^3 - 15(3)^2 + 125}{3(3)^2 - 30(3)} = 3.269\,84;$$

$$x_3 = x_2 - \frac{x_2^3 - 15x_2^2 + 125}{3x_2^2 - 30x_2} = 3.263\,52;$$

$$x_4 = x_3 - \frac{x_3^3 - 15x_3^2 + 125}{3x_3^2 - 30x_3} = 3.263\,518\,2;$$

$$x_5 = x_4 - \frac{x_4^3 - 15x_4^2 + 125}{3x_4^2 - 30x_4} = 3.263\,518\,2.$$

Newton's iterative procedure has therefore produced $3.263\,518$ as an approximate solution to the equation $x^3 - 15x^2 + 125 = 0$. In spite of the fact that we have written six decimals in this final answer, our analysis in no way guarantees this degree of accuracy; we have simply judged on the basis of $x_4 = x_5$ that $3.263\,518$ might be accurate to six decimal places. The zero intermediate value theorem (of Section 1.11) confirms this when we calculate

$$f(3.263\,517\,5) = 4.8 \times 10^{-5} \quad \text{and} \quad f(3.263\,518\,5) = -1.8 \times 10^{-5}.$$

EXAMPLE 4.1

Use Newton's iterative procedure to find the only positive root of the equation

$$3x^4 + 15x^3 - 125x - 1500 = 0$$

accurate to five decimal places. (This equation will be encountered in Exercise 48 of Section 4.7.)

FIGURE 4.3 Initial approximation for Newton's iterative procedure

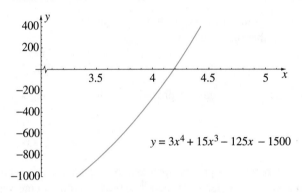

$$y = 3x^4 + 15x^3 - 125x - 1500$$

SOLUTION The plot of $f(x) = 3x^4 + 15x^3 - 125x - 1500$ in Figure 4.3 shows that the solution is between $x = 4$ and $x = 5$. The approximations predicted by Newton's method are defined by

$$x_{n+1} = x_n - \frac{f(x_n)}{f'(x_n)} = x_n - \frac{3x_n^4 + 15x_n^3 - 125x_n - 1500}{12x_n^3 + 45x_n^2 - 125}.$$

Suppose we choose $x_1 = 4$ as the initial approximation; this is closer to the root than $x = 5$. We find that

$$x_2 = 4.199\,56, \qquad x_3 = 4.187\,268, \qquad x_4 = 4.187\,218\,7, \qquad x_5 = 4.187\,218\,7.$$

Since

$$f(4.187\,215) = -5.7 \times 10^{-3} \quad \text{and} \quad f(4.187\,225) = 9.7 \times 10^{-3},$$

it follows that the root is $x = 4.187\,22$, accurate to five decimal places.

In the following example we again use Newton's iterative procedure to approximate the solution of an equation but specify the required accuracy as maximum error.

EXAMPLE 4.2

Use Newton's method to find the smallest root of the cubic equation $x^3 - 3x + 1 = 0$ with error less than 10^{-5}.

FIGURE 4.4 Initial approximation for Newton's iterative procedure

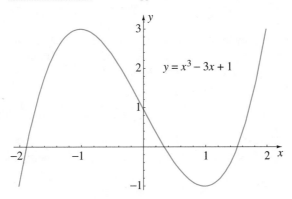

SOLUTION   The graph in Figure 4.4 indicates that the smallest root is just to the right of $x = -2$. We take $x_1 = -2$ and use Newton's iterative procedure to define approximations:

$$x_{n+1} = x_n - \frac{f(x_n)}{f'(x_n)} = x_n - \frac{x_n^3 - 3x_n + 1}{3x_n^2 - 3}.$$

The next four approximations are

$$x_2 = -1.888\,89, \qquad x_3 = -1.879\,45, \qquad x_4 = -1.879\,385, \qquad x_5 = -1.879\,385\,2.$$

For an approximation with error less than 10^{-5}, we suggest $-1.879\,39$. The zero intermediate value theorem confirms this when we calculate

$$f(-1.879\,40) = -1.1 \times 10^{-4} \qquad \text{and} \qquad f(-1.879\,38) = 4.0 \times 10^{-5}.$$

The value of a technique for approximating solutions of an equation depends on two factors, applicability and rate of convergence. Newton's method scores well in both categories. The method can be applied to any equation 4.1, provided that $f(x)$ is differentiable in an interval containing the root. It converges to a solution $x = a$, provided that $f'(a) \neq 0$ and x_1 is chosen sufficiently close to a (see Exercise 68 in Section 10.1). The condition $f'(a) \neq 0$ is included so that the denominator in equation 4.2 does not approach 0 as $x_n \to a$. However, $f(x_n)$ usually approaches 0 faster than $f'(x_n)$ so that even when $f'(a) = 0$, Newton's method may be successful. It is impossible to indicate how close x_1 must be to a in order to guarantee convergence to a. In some examples x_1 can be any number whatsoever, but we are also aware of examples where $|x_1 - a|$ must be less than 0.01 (see Example 4.3). Provided that x_1 is sufficiently close to the root, convergence of the approximations to the root is rapid. This makes Newton's method valuable from the point of view of the second criterion, rate of convergence.

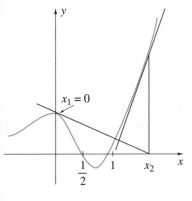

FIGURE 4.5   Initial approximation for Newton's iterative procedure must be sufficiently close to root

When x_1 is not sufficiently close to a, Newton's iterative procedure may not converge, or may converge to a solution other than expected. For instance, if we attempt to approximate the largest root in Example 4.2, and inadvertently choose $x_1 = 1$, we cannot find the second approximation: algebraically because $3x_1^2 - 3 = 0$, and geometrically because at $x_1 = 1$, the tangent line is horizontal, and does not intersect the x-axis. The function $f(x)$ in Figure 4.5 has zeros near $x = 1/2$ and $x = 1$. If we attempt to find the smaller zero using an initial approximation $x_1 = 0$, we find the second approximation x_2 to be larger than 1. Further iterations then converge to the zero near $x = 1$, not the zero near $x = 1/2$. We conclude, therefore, that the initial approximation in Newton's iterative procedure is most important. A poor choice for x_1 may lead to numbers that either converge to the wrong root or do not converge at all.

EXAMPLE 4.3

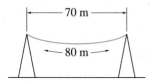

FIGURE 4.6   Finding
tension in hydro cable at its lowest
point

A uniform hydro cable $P = 80$ m long with mass per unit length $\rho = 0.5$ kg/m is hung from two supports at the same level $L = 70$ m apart (Figure 4.6). The tension T in the cable at its lowest point must satisfy the equation

$$\frac{\rho g P}{T} = e^{\rho g L/(2T)} - e^{-\rho g L/(2T)},$$

where $g = 9.81$ m/s^2. If we set $z = \rho g/(2T)$, then z must satisfy

$$2Pz = e^{Lz} - e^{-Lz}.$$

Find an approximation to the solution of this equation for z that yields T correct to one decimal place.

SOLUTION   When we substitute $P = 80$ and $L = 70$, the equation for z becomes

$$160z = e^{70z} - e^{-70z}.$$

The exponential function e^{70z} grows very rapidly for $z > 0$, and e^{-70z} approaches zero very quickly. It follows that the solution of the equation must be quite small. If we plot $f(z) = e^{70z} - e^{-70z} - 160z$ on the interval $0 \le z \le 0.02$ (Figure 4.7), we capture the solution. With $z_1 = 0.013$, Newton's iterative procedure defines further approximations by

$$z_1 = 0.013, \qquad z_{n+1} = z_n - \frac{e^{70z_n} - e^{-70z_n} - 160z_n}{70e^{70z_n} + 70e^{-70z_n} - 160}.$$

To get an idea of how accurate z should be in order to give T correct to one decimal place, we note that for $z = 0.013$, tension is $T \approx 189$. Thus, we should determine z to around five or six figures. Iteration of Newton's procedure gives

$$z_2 = 0.012\,957\,3, \qquad z_3 = 0.012\,957\,0, \qquad z_4 = 0.012\,957\,0.$$

With z approximated by $0.012\,957\,0$, tension is $T = 189.3$ N. We could verify that this is accurate to one decimal place by defining $g(T) = e^{\rho g L/(2T)} - e^{-\rho g L/(2T)} - \rho g P/T$ and evaluating $g(189.25)$ and $g(189.35)$. The first is positive and the second is negative.

FIGURE 4.7   Initial approximation for Newton's iterative procedure

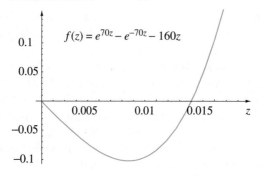

EXERCISES 4.1

In Exercises 1–16 use Newton's iterative procedure to find approximations to all roots of the equation accurate to six decimal places. In each case, make a plot in order to obtain an initial approximation to each root.

1. $x^2 + 3x + 1 = 0$ **2.** $x^2 - x - 4 = 0$

3. $x^3 + x - 3 = 0$ **4.** $x^3 - x^2 + x - 22 = 0$

5. $x^3 - 5x^2 - x + 4 = 0$ **6.** $x^5 + x - 1 = 0$

7. $x^4 + 3x^2 - 7 = 0$ **8.** $\dfrac{x+1}{x-2} = x^2 + 1$

9. $x - 10 \sin x = 0$ **10.** $\sec x = \dfrac{2}{1 + x^4}$

11. $(x + 1)^2 = \sin 4x$ **12.** $(x + 1)^2 = 5 \sin 4x$

13. $x + 4 \ln x = 0$ **14.** $x \ln x = 6$

15. $e^x + e^{-x} = 10x$ **16.** $x^2 - 4e^{-2x} = 0$

In Exercises 17–24 use Newton's iterative procedure to find approximations to all roots of the equation with error no greater than that specified.

17. $x^3 - 5x - 1 = 0, \quad 10^{-3}$

18. $x^4 - x^3 + 2x^2 + 6x = 0, \quad 10^{-4}$

19. $\dfrac{x}{x+1} = x^2 + 2, \quad 10^{-5}$

20. $(x + 1)^2 = x^3 - 4x, \quad 10^{-3}$

21. $(x + 1)^2 = 5 \sin 4x, \quad 10^{-3}$

22. $\cos^2 x = x^2 - 1, \quad 10^{-4}$

23. $x + (\ln x)^2 = 0, \quad 10^{-3}$

24. $e^{3x} + e^x = 4, \quad 10^{-4}$

In Exercises 25–28 find all points of intersection for the curves accurate to four decimal places.

25. $y = x^3, \quad y = x + 5$

26. $y = (x + 1)^2, \quad y = x^3 - 4x$

27. $y = x^4 - 20, \quad y = x^3 - 2x^2$

28. $y = \dfrac{x}{x+1}, \quad y = x^2 + 2$

29. Show algebraically and geometrically that Newton's method never gives the solution of the equation $f(x) = x^{1/3} = 0$, for any initial approximation whatsoever.

30. Show algebraically and geometrically that Newton's method always gives the solution of the equation $f(x) = x^{7/5} = 0$, for any initial approximation whatsoever.

31. Suppose you mortgage your house for P dollars. To repay the loan at an interest rate of $i\%$, amortized over n years, payments made

m times per year are given by the formula

$$M = P \left[\frac{i/(100m)}{1 - \left(1 + \dfrac{i}{100m}\right)^{-mn}} \right].$$

(a) What are monthly payments if $P = 100\,000$, $i = 5$, and $n = 25$?

(b) What interest rate would yield monthly payments of $500 for $P = 100\,000$ and $n = 25$?

32. When the beam in the figure below vibrates vertically, there are certain frequencies of vibration called *natural frequencies*. They are solutions of the equation

$$\tan x = \frac{e^x - e^{-x}}{e^x + e^{-x}}$$

divided by 20π. Find the two smallest frequencies correct to four decimal places.

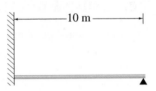

33. A stone of mass 100 g is thrown vertically upward with speed 20 m/s. Air exerts a resistive force on the stone proportional to its speed, and has magnitude 0.1 N when the speed of the stone is 10 m/s. It can be shown that the height y above the projection point attained by the stone is given by

$$y = -98.1t + 1181\left(1 - e^{-t/10}\right) \text{ m},$$

where t is time (measured in seconds with $t = 0$ at the instant of projection).

(a) The time taken for the stone to return to its projection point can be obtained by setting $y = 0$ and solving the equation for t. Do so (correct to two decimal places).

(b) When air resistance is neglected, the formula for y is

$$y = 20t - 4.905t^2 \text{ m}.$$

What is the elapsed time in this case from the instant the stone is projected until it returns to the projection point?

34. *Planck's law* for the energy density E of blackbody radiation at $1000°$ K states that

$$E = E(\lambda) = \frac{k\lambda^{-5}}{e^{c/\lambda} - 1},$$

where $k > 0$ is a constant and $c = 0.000\,143\,86$. This function is shown in the figure below. The value of λ at which E is a maximum must satisfy the equation

$$(5\lambda - c)e^{c/\lambda} - 5\lambda = 0.$$

Find this value of λ correct to seven decimal places.

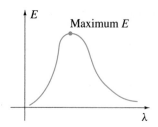

* **35.** The speed of response of an oscillatory system whose graph $y = f(t)$ is shown below is often determined by one of the following three time constants:

    (i) *Delay time*, T_d, defined as the time required for the graph to reach 0.5;

    (ii) *Rise time*, T_r, defined as the time for the graph to rise from 0.1 to 0.9;

    (iii) *Settling time*, T_s, defined as the time to reach and remain within the interval $0.95 \le f(t) \le 1.05$.

Find these times, correct to two decimal places, if $f(t) = 1 + e^{-2.5\sqrt{11}t} \sin(20t - \pi/2)$.

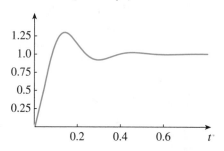

* **36.** Suppose that a cubic polynomial $P(x)$ has three distinct real zeros. Show that when Newton's method uses an initial approximation that is equal to the average of two of the zeros, then the first iteration always yields the third zero.

4.2 Increasing and Decreasing Functions

Many mathematical concepts have their origin in intuitive ideas. In this section we analyze the intuitive idea of one quantity "getting larger" and another "getting smaller." To describe what it means for a quantity to be increasing (getting larger) or decreasing (getting smaller), we first suppose that the quantity is represented by some function $f(x)$ of a variable x. The mathematical definition for $f(x)$ to be increasing or decreasing is as follows.

DEFINITION 4.1

A function $f(x)$ is said to be **increasing** on an interval I if for all $x_1 > x_2$ in I,

$$f(x_1) > f(x_2). \tag{4.3a}$$

A function $f(x)$ is said to be **decreasing** on I if for all $x_1 > x_2$ in I,

$$f(x_1) < f(x_2). \tag{4.3b}$$

The continuous function in Figure 4.8 is increasing on the intervals

$$a \le x \le b, \qquad c \le x \le d, \qquad e \le x \le f,$$

and decreasing on the intervals

$$b \le x \le c, \qquad d \le x \le e.$$

When a function has points of discontinuity, such as in Figure 4.9, the situation is somewhat more complicated. This function is increasing on the intervals

$$a \le x < b, \qquad d \le x < e, \qquad e < x \le f,$$

and decreasing for

$$b < x < d.$$

Pay special attention to whether endpoints of each interval are included.

Figures 4.8 and 4.9 indicate that the sign of $f'(x)$ determines whether a function is increasing or decreasing on an interval. The following test describes the situation.

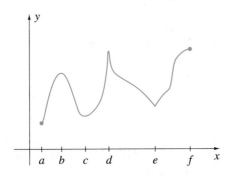

FIGURE 4.8   Intervals on which a continuous function is increasing and decreasing

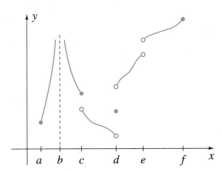

FIGURE 4.9   Intervals on which a discontinuous function is increasing and decreasing

Increasing and Decreasing Test

(i) A function $f(x)$ is increasing on an interval I if on I,

$$f'(x) \geq 0 \qquad (4.4a)$$

and is equal to zero at only a finite number of points.

(ii) A function $f(x)$ is decreasing on an interval I if on I,

$$f'(x) \leq 0 \qquad (4.4b)$$

and is equal to zero at only a finite number of points.

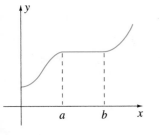

FIGURE 4.10 $f(x)$ is not increasing on $a \leq x \leq b$ if $f'(x) = 0$

Figures 4.10 and 4.11 illustrate why we permit $f'(x)$ to vanish at only a finite number of points and do not, therefore, allow it to vanish on an interval. In Figure 4.10, $f'(x)$ is equal to zero on the interval $a \leq x \leq b$, and certainly, $f(x)$ is not increasing on any interval that contains these points. The function $f(x)$ in Figure 4.11 has $f'(0) = 0 = f'(1)$, and yet $f(x)$ is increasing on the interval $a \leq x \leq b$.

Conditions 4.4 are sufficient to guarantee that a function is increasing or decreasing on an interval; they are not necessary. For example, the function $f(x) = x - \sin x$ in Figure 4.12 has derivative equal to 0 at the infinity of values $x = 2n\pi$, where n is an integer, yet the function is increasing on the interval $-\infty < x < \infty$. What we are saying is that tests more general than 4.4 can be formulated, but we feel that the extra complexity is not worth the gain. For a proof of 4.4, see Exercise 51.

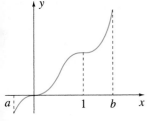

FIGURE 4.11 $f(x)$ is increasing on $a \leq x \leq b$ even if $f'(x) = 0$ at a finite number of points

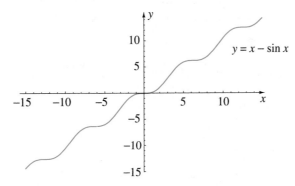

FIGURE 4.12   Even if $f'(x) = 0$ at an infinity of points, $f(x)$ may be increasing

Intervals on which a function is increasing and decreasing are separated by points where the derivative is equal to zero or does not exist. Keep this in mind in the following example.

EXAMPLE 4.4

Find intervals on which the following functions are increasing and decreasing:

(a) $f(x) = 2x^3 + 3x^2 - 5x + 4$ (b) $f(x) = \dfrac{x^2 - 9}{x^2 + x - 2}$

SOLUTION

(a) As mentioned above, the key to finding intervals on which the derivative of a function is positive and negative are points where the derivative is either zero or nonexistent. Since polynomials have derivatives everywhere, we investigate where the derivative is equal to zero,

$$0 = f'(x) = 6x^2 + 6x - 5 \quad \Longrightarrow \quad x = \frac{-6 \pm \sqrt{36 + 120}}{12} = \frac{-3 \pm \sqrt{39}}{6}.$$

To determine intervals on which the function is increasing and decreasing we can proceed in a number of ways. Firstly, if we have a plot of the function (Figure 4.13), it is clear that $f(x)$ is increasing on the intervals $x \le (-3 - \sqrt{39})/6$ and $x \ge (-3 + \sqrt{39})/6$, and decreasing on $(-3 - \sqrt{39})/6 \le x \le (-3 + \sqrt{39})/6$. Secondly, if we do not have a graph of the function, we could visualize that a graph of $f'(x)$ (Figure 4.14) is a parabola that crosses the x-axis at the points $(-3 \pm \sqrt{39})/6$. Notice that we said visualize this graph; we have drawn it, but it would be necessary only to mentally visualize it. Since the parabola opens upward, $f'(x) \ge 0$ on the intervals $x \le (-3 - \sqrt{39})/6$ and $x \ge (-3 + \sqrt{39})/6$, and $f'(x) \le 0$ on $(-3 - \sqrt{39})/6 \le x \le (-3 + \sqrt{39})/6$.

FIGURE 4.13   Graph to determine when function is increasing and decreasing

FIGURE 4.14   Graph to determine the sign of $f'(x)$

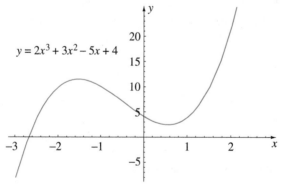

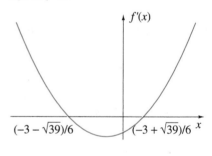

Finally, we could construct a sign table for $f'(x)$ as introduced in Table 1.1 of Section 1.5 (see figure 4.15 below). It also shows that $f'(x) \ge 0$ on the intervals $x \le (-3 - \sqrt{39})/6$ and $x \ge (-3 + \sqrt{39})/6$, and $f'(x) \le 0$ on $(-3 - \sqrt{39})/6 \le x \le (-3 + \sqrt{39})/6$.

FIGURE 4.15

	$(-3 - \sqrt{39})/6$	$(-3 + \sqrt{39})/6$	
$y - (-3 - \sqrt{39})/6$	$-$	$+$	x
$y - (-3 + \sqrt{39})/6$		$-$	$+$
$f'(x)$	$+$	$-$	$+$

(b) The graph of $f(x)$ in Figure 4.16a indicates discontinuities at $x = -2$ and $x = 1$ and the point between $x = -2$ and $x = 0$ where $f'(x) = 0$ separates intervals on which the function is increasing and decreasing. Whether there are other points far out on the x-axis where $f'(x) = 0$ is not clear. To find all points where $f'(x) = 0$, we set

$$0 = f'(x) = \frac{(x^2 + x - 2)(2x) - (x^2 - 9)(2x + 1)}{(x^2 + x - 2)^2} = \frac{x^2 + 14x + 9}{(x^2 + x - 2)^2}.$$

This implies that $x^2 + 14x + 9 = 0$, and there are two points where $f'(x) = 0$, namely $x = -7 \pm 2\sqrt{10}$. Figure 4.16b is an exaggerated version of the graph of $f(x)$ to the left of $x = -2$; it is not a computer plot. It illustrates that $f'(x) = 0$ at $x = -7 - 2\sqrt{10}$ and that the graph is asymptotic to the line $y = 1$ (from above). We can now say that the function is increasing on the intervals

$$-\infty < x \leq -7 - 2\sqrt{10}, \qquad -7 + 2\sqrt{10} \leq x < 1, \qquad 1 < x < \infty,$$

and decreasing for

$$-7 - 2\sqrt{10} \leq x < -2, \qquad -2 < x \leq -7 + 2\sqrt{10}.$$

FIGURE 4.16a	FIGURE 4.16b

Points of discontinuity may also separate intervals on which a function is increasing and decreasing

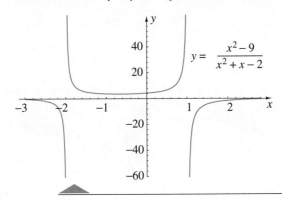

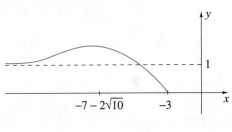

EXERCISES 4.2

In Exercises 1–26 determine intervals on which the function is increasing and decreasing.

1. $f(x) = 2x - 3$

2. $f(x) = 4 - 5x$

3. $f(x) = x^2 - 3x + 4$

4. $f(x) = -2x^2 + 5x$

5. $f(x) = 3x^2 + 6x - 2$

6. $f(x) = 5 + 2x - 4x^2$

7. $f(x) = 2x^3 - 18x^2 + 48x + 1$

8. $f(x) = x^3 + 6x^2 + 12x + 5$

9. $f(x) = 4x^3 - 18x^2 + 1$

10. $f(x) = 4 - 18x - 9x^2 - 2x^3$

11. $f(x) = 3x^4 + 4x^3 - 24x + 2$

12. $f(x) = 3x^4 - 4x^3 + 24x^2 - 48x$

13. $f(x) = x^4 - 4x^3 - 8x^2 + 48x + 24$

14. $f(x) = x^5 - 5x + 2$

15. $f(x) = x + \dfrac{1}{x}$

* 16. $f(x) = x^2 + \dfrac{1}{x^2}$

* 17. $f(x) = \dfrac{x}{2 - x}$

* 18. $f(x) = \dfrac{x^2 + 4}{x^2 - 1}$

* 19. $f(x) = \dfrac{x^3}{x + 1}$

* 20. $f(x) = |x^2 - 1| + 1$

* 21. $f(x) = xe^{-x}$

* 22. $f(x) = x^2 e^{-x}$

* 23. $f(x) = \ln(x^2 + 5)$

* 24. $f(x) = x \ln x$

* 25. $f(x) = \dfrac{x^2 - 9}{x - 3}$

* 26. $f(x) = \dfrac{x^3 + 2x^2 - x - 2}{2 - x - x^2}$

* 27. If the price of a certain car is set at r, then the market demands x cars per year, where

$$x = 4a^3 - 3ar^2 + r^3,$$

where $a > 0$ is a constant, and $0 < r < 2a$. Show that the price function $r = f(x)$ defined implicitly by this equation is a decreasing function.

Figure (a) below contains the graph of a function $f(x)$ and figure (b), the graph of $f'(x)$. Note that $f'(x) \geq 0$ when $f(x)$ is increasing and $f'(x) \leq 0$ when $f(x)$ is decreasing. In addition, the corner in $f(x)$ at $x = 2$ is reflected in the discontinuity in $f'(x)$. In Exercises 28–35 draw similar graphs for the function and its derivative.

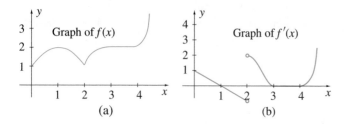

(a)   (b)

* **28.** $f(x) = x^2 + 2x$

* **29.** $f(x) = x^4 - x^2$

* **30.** $f(x) = \dfrac{x-1}{x+2}$

* **31.** $f(x) = \dfrac{x-4}{x^2}$

* **32.** $f(x) = \dfrac{|x|}{x}$

* **33.** $f(x) = |x^2 - 4|$

* **34.** $f(x) = \dfrac{|x^2 - 4|}{x - 2}$

* **35.** $f(x) = \sin 3x$

In Exercises 36–39 find (accurate to four decimal places) intervals on which the function is increasing and decreasing.

* **36.** $f(x) = x^4 + 2x^2 - 6x + 5$

* **37.** $f(x) = 3x^4 - 20x^3 - 24x^2 + 48x$

* **38.** $f(x) = x^2 \sin x,\quad -\pi \leq x \leq \pi$

* **39.** $f(x) = \tan x - x(x + 2),\quad -\pi/2 < x < \pi/2$

* **40.** Show algebraically that the equation $x^{23} + 3x^{15} + 4x + 1 = 0$ has exactly one (real) solution.

* **41.** Show that the equation $ax^5 + bx^3 + c = 0$, where a, b, and c are constants such that $ab > 0$, has exactly one (real) solution.

* **42.** Show that the equation $x^n + ax - 1 = 0$, where $a > 0$ and $n \geq 2$ are constants, has exactly one positive root.

* **43.** Repeat Exercise 42 for the equation $x^n + x^{n-1} - a = 0$.

* **44.** Show that $\sin x < x$ for all $x > 0$. *Hint:* Calculate $f(0)$ and $f'(x)$ for $f(x) = x - \sin x$.

* **45.** Show that $\cos x > 1 - x^2/2$ for all $x > 0$. *Hint:* See the technique of Exercise 44.

* **46.** Use the result of Exercise 45 to prove that for $x > 0$,

$$\sin x > x - \frac{x^3}{6}.$$

* **47.** Use the result of Exercise 46 to prove that for $x > 0$,

$$\cos x < 1 - \frac{x^2}{2} + \frac{x^4}{24}.$$

* **48.** Use the technique of Exercise 44 to verify that for $x > 0$,

$$\frac{1}{\sqrt{1 + 3x}} > 1 - \frac{3x}{2}.$$

* **49.** If $f(x)$ and $g(x)$ are differentiable and increasing on an interval I, is $f(x)g(x)$ increasing on I?

* **50.** If positive functions $f(x)$ and $g(x)$ are differentiable and increasing on an interval I, is $f(x)g(x)$ increasing on I?

* **51.** Verify test 4.4.

* **52.** A number x_0 is called a *fixed point* of a function $f(x)$ if $f(x_0) = x_0$. How many fixed points can a function have if $f'(x) < 1$ for all x?

** **53.** Prove that when $0 < a < b < \pi/2$,

$$\frac{\tan b}{\tan a} > \frac{b}{a}.$$

4.3 Relative Maxima and Minima

One of the most important applications of calculus is in the field of optimization, the study of maxima and minima. In this section we begin discussions of this topic, which continue through to Section 4.7. Fundamental to discussions on optimization are critical points.

> **DEFINITION 4.2**
>
> A **critical point** of a function is a point in the domain of the function at which the first derivative either is equal to zero or does not exist.

Specifically, $x = c$ is a critical point for $f(x)$ if $f'(c) = 0$ or $f'(c)$ does not exist, but in the latter case, $f(c)$ must exist. With the interpretation of the derivative as the slope of a tangent line, we can state that corresponding to a critical point of a function, the graph of the function has a horizontal tangent line, a vertical tangent line, or no tangent line at all. For example, the eight points a through h on the x-axis in Figure 4.17 are all critical. At a, b, c, and d, the tangent line is horizontal; at e and f, the tangent line is vertical; and at g and h, there is no tangent line.

Often overlooked by students, but very important in this definition, is that a function must be defined at a critical point; it must have a value, and therefore, there must be a point on the graph of the function at a critical point. For instance, if the dot in Figure 4.17 at the discontinuity $x = h$ is replaced by an open circle, then $x = h$ is no longer a critical point.

FIGURE 4.17   The critical points of a function

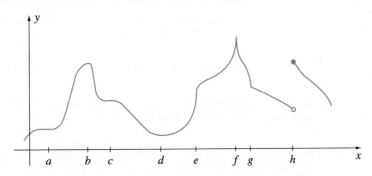

When the domain of a function $f(x)$ is a closed interval $a \leq x \leq b$, the endpoint $x = a$ is said to be critical if the right-hand derivative of $f(x)$ at $x = a$ is equal to zero or does not exist. Likewise, $x = b$ is critical if the left-hand derivative vanishes or does not exist there. This is consistent with the definition of differentiability in Section 3.3. The function in Figure 4.18 has domain $1 \leq x \leq 3$. Both endpoints are critical — $x = 1$ because $f'_+(1)$ does not exist, and $x = 3$ because $f'_-(3) = 0$.

FIGURE 4.18   Critical points at end points when function defined on closed interval

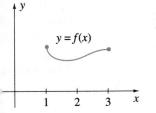

The function in Figure 4.17 is discontinuous at $x = h$, and we know that a function cannot have a derivative at a point of discontinuity (Theorem 3.6). This does not mean that every point of discontinuity of a function is critical. Remember, a function must be defined at a point for that point to be critical. Thus, points of discontinuity are critical only if the function is defined at the point. In the remainder of this section we consider only critical points at which a function is continuous and which are not endpoints of its domain of definition.

EXAMPLE 4.5

Find critical points for the following functions:

(a) $f(x) = x^3 - 7x^2 + 11x + 6$

(b) $f(x) = \dfrac{x^2}{x^3 - 1}$

(c) $f(x) = x \ln x$

(d) $f(x) = |x|$

SOLUTION

(a) For critical points we first solve

$$0 = f'(x) = 3x^2 - 14x + 11 = (3x - 11)(x - 1),$$

and obtain $x = 11/3$ and $x = 1$. These are the only critical points since there are no points where $f'(x)$ does not exist.

(b) For critical points we calculate

$$f'(x) = \frac{(x^3 - 1)(2x) - x^2(3x^2)}{(x^3 - 1)^2} = \frac{-x(x^3 + 2)}{(x^3 - 1)^2}.$$

Clearly, $f'(x) = 0$ when $x = 0$ and when $x^3 + 2 = 0$, which implies that $x = -2^{1/3}$. The derivative does not exist when $x^3 - 1 = 0$ (i.e., when $x = 1$). But $f(x)$ is not defined at $x = 1$ either, and therefore $x = 1$ is not a critical point. There are only two critical points: $x = 0$ and $x = -2^{1/3}$.

(c) For critical points we first solve

$$0 = f'(x) = \ln x + \frac{x}{x} = \ln x + 1.$$

The only solution of this equation is $x = 1/e$. Since $f'(x)$ exists for all x in the domain of $f(x)$, namely, $x > 0$, the function has no other critical points.

(d) The only critical point of $f(x) = |x|$ is $x = 0$. Its derivative is equal to 1 when $x > 0$, and to -1 when $x < 0$, but does not exist at $x = 0$.

At the critical points b and f in Figure 4.19, the graph of the function has "high" points. They are described in the following definition.

DEFINITION 4.3

A function $f(x)$ is said to have a **relative** (or **local**) **maximum** $f(x_0)$ at $x = x_0$ if there exists an open interval I containing x_0 such that for all x in I,

$$f(x) \le f(x_0). \tag{4.5}$$

Since such intervals can be drawn around $x = b$ and $x = f$, relative maxima occur at these points. At a relative or local maximum, the graph of the function is highest relative to nearby points.

Critical points d and h in Figure 4.19, where the graph has "low" points, are described in a similar definition.

FIGURE 4.19   Nature of a graph at critical points

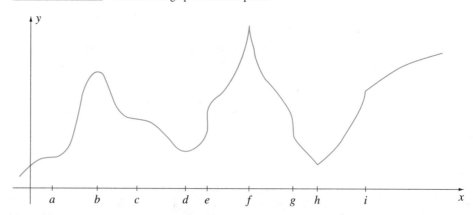

DEFINITION 4.4

A function $f(x)$ is said to have a **relative** (or **local**) **minimum** $f(x_0)$ at $x = x_0$ if there exists an open interval I containing x_0 such that for all x in I,

$$f(x) \ge f(x_0). \tag{4.6}$$

Relative minima therefore occur at $x = d$ and $x = h$ in Figure 4.19.

The critical points $x = a$ and $x = c$, where $f'(x) = 0$ and $x = e$ and $x = g$, where $f'(x)$ does not exist, will be discussed in Section 4.4. At $x = i$, the graph takes an abrupt change in direction. The function has a left- and a right-hand derivative, but $f'(i)$ does not exist. In Section 3.3 we called the corresponding point on the curve a *corner*. Corners can sometimes be relative extrema ($x = h$ yields a corner and a relative minimum).

Relative maxima and minima represent high and low points on the graph of a function relative to points near them. It is not coincidence in Figure 4.19 that the two relative maxima and the two relative minima occur at critical points. According to the following theorem, this is always the case.

THEOREM 4.1

Relative maxima and relative minima of a function must occur at critical points of the function.

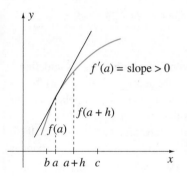

FIGURE 4.20 Proof that relative extrema occur at critical points

PROOF To verify this we prove that at any point at which the derivative $f'(x)$ of a function $f(x)$ exists and is not zero, it is impossible for $f(x)$ to have a relative extremum, that is, a relative maximum or a relative minimum. Suppose that at some point $x = a$, the derivative $f'(a)$ exists and is positive (Figure 4.20):

$$0 < f'(a) = \lim_{h \to 0} \frac{f(a + h) - f(a)}{h}.$$

According to the proof developed in Exercise 29 of Section 2.6, there exists an open interval $I : b < x < c$ around $x = a$ in which

$$\frac{f(a + h) - f(a)}{h} > 0.$$

This implies that when $h > 0$ (and $a + h$ is in I), $f(a + h) - f(a)$ must also be positive, and therefore $f(a + h) > f(a)$. But when $h < 0$ (and $a + h$ is in I), $f(a + h) - f(a)$ must be negative, and therefore $f(a + h) < f(a)$. There is an interval $b < x < a$ in which $f(x) < f(a)$, and an interval $a < x < c$ in which $f(x) > f(a)$. Thus, $x = a$ cannot yield a relative extremum.

A similar proof holds when $f'(a) < 0$. ∎

Although every relative extremum of a function must occur at a critical point, not all critical points give relative extrema. For continuous functions, there is a simple test to determine whether a critical point gives a relative maximum or a relative minimum. To understand this test, consider the critical points in Figure 4.19 as you read the statements.

First-Derivative Test for Relative Extrema of Continuous Functions

(i) If $f'(x)$ (slope of a graph) changes from a positive quantity to a negative quantity as x increases through a critical point at which $f(x)$ is continuous, the critical point yields a relative maximum for $f(x)$.

(ii) If $f'(x)$ changes from a negative quantity to a positive quantity as x increases through a critical point at which $f(x)$ is continuous, the critical point yields a relative minimum for $f(x)$.

Various possibilities can occur if $f'(x)$ does not change sign as x increases through a critical point. For instance, if $f'(x)$ is positive (or negative) on both sides of a critical point [such as is the case for the critical point $x = 0$ of $f(x) = x^3$], then the critical point cannot yield a relative maximum or minimum. It might also happen that $f'(x)$ is both positive and negative in every interval around a critical point, no matter how small the interval. This is investigated in Exercises 76–78.

EXAMPLE 4.6

Find relative maxima and relative minima for the following functions:

(a) $f(x) = 2x^3 - 9x^2 - 23x + 6$

(b) $f(x) = 3.01x - \sin(3x + 1), \quad 0 \le x \le 5$

(c) $f(x) = x^{5/3} - x^{2/3}$

(d) $f(x) = 5$

SOLUTION

(a) The graph of the function in Figure 4.21a indicates a relative maximum near $x = -1$ and a relative minimum near $x = 4$. To locate them precisely, we find critical points by solving

$$0 = f'(x) = 6x^2 - 18x - 23 \implies x = \frac{18 \pm \sqrt{324 + 24 \cdot 23}}{12} = \frac{9 \pm \sqrt{219}}{6}.$$

These are the only critical points, as there are no points at which $f'(x)$ is undefined. Figure 4.21a indicates that a relative maximum occurs at $x = (9 - \sqrt{219})/6$ and a relative minimum at $x = (9 + \sqrt{219})/6$. This can be confirmed algebraically or with the graph of $f'(x)$ in Figure 4.21b. It shows that as x increases through $(9 - \sqrt{219})/6$, $f'(x)$ changes from positive to negative; therefore, $x = (9 - \sqrt{219})/6$ yields a relative maximum of $f\big((9 - \sqrt{219})/6\big) \approx 18.02$. Since $f'(x)$ changes from negative to positive as x increases through $(9 + \sqrt{219})/6$, this value of x yields a relative minimum of $f\big((9 + \sqrt{219})/6\big) \approx -102.0$.

(b) The graph in Figure 4.22 does not make it clear whether the function has relative extrema; the tangent line may become horizontal at or near $x = 2$ and $x = 4$, but we cannot be sure. Critical points of the function are given by

$$0 = f'(x) = 3.01 - 3\cos(3x + 1) \implies \cos(3x + 1) = \frac{3.01}{3}.$$

Since $3.01/3 > 1$, this equation has no solutions. Hence, $f(x)$ has no critical points and there can be no relative maxima or minima.

FIGURE 4.21a   Relative extrema

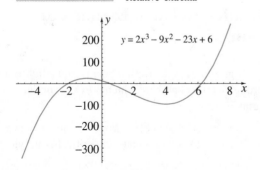

FIGURE 4.21b   Sign changes of $f'(x)$ at critical points

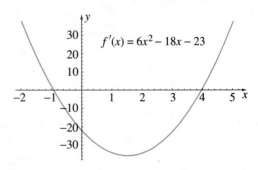

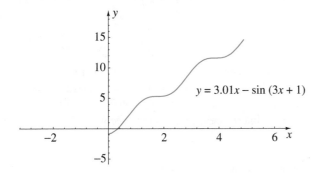

FIGURE 4.22 Plot may not be conclusive evidence of existence of relative extrema

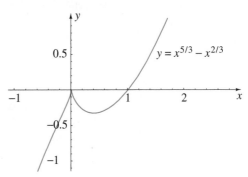

FIGURE 4.23 Graph showing a relative minimum and suggesting a relative maximum

(c) The graph of the function in Figure 4.23 makes it clear that a relative minimum occurs just to the left of $x = 1/2$. It also suggests that $f(0) = 0$ is a relative maximum. Confirmation is provided by critical points, found by first solving

$$0 = f'(x) = \frac{5}{3}x^{2/3} - \frac{2}{3}x^{-1/3} = \frac{5x - 2}{3x^{1/3}}.$$

Clearly, $x = 2/5$ is a critical point. The relative minimum at this point is $f(2/5) \approx -0.33$. In addition, because $f'(0)$ does not exist, but $f(0) = 0$, it follows that $x = 0$ is also a critical point. There are two reasons, each sufficient by itself to conclude that $f(0) = 0$ is a relative maximum. First, $f(x)$ is continuous at $x = 0$, and $f'(x)$ changes from a positive quantity to a negative quantity as x increases through 0. Second, since $f(x) = x^{2/3}(x - 1)$ is negative for all $x < 1$, except at $x = 0$ where $f(0) = 0$, there must be a relative maximum at $x = 0$. It is also interesting to note that

$$\lim_{x \to 0^-} f'(x) = \infty \qquad \text{and} \qquad \lim_{x \to 0^+} f'(x) = -\infty.$$

This means that the graph has a very sharp point at $(0, 0)$.

(d) The graph of this function is a horizontal straight line. Every value of x is critical and at each value of x, the function has a relative maximum and a relative minimum of 5.

EXAMPLE 4.7

The equation $x^2 y + y^3 = 8$ defines y implicitly as a function of x. Find all critical points of the function and determine whether they yield relative maxima or minima for the function.

SOLUTION   When we differentiate the equation implicitly with respect to x, we obtain

$$2xy + x^2 \frac{dy}{dx} + 3y^2 \frac{dy}{dx} = 0 \qquad \Longrightarrow \qquad \frac{dy}{dx} = -\frac{2xy}{x^2 + 3y^2}.$$

The derivative is defined for all x and y except when both are simultaneously zero, a point that does not satisfy the original equation. Consequently we consider when the derivative vanishes. For this to happen, either x or y must be zero. The original equation does not permit y to be zero, and when $x = 0$, the only solution of the equation is $y = 2$. Hence, $x = 0$ is the only critical point of the implicitly-defined function. Since the function must be continuous at $x = 0$ (Theorem 3.6), values of y are close to 2 when values of x are close to zero. It follows that the derivative changes from a positive quantity to a negative quantity as x increases through zero, and the function has a relative maximum at $x = 2$.

EXAMPLE 4.8

The Beattie-Bridgeman equation of state for an ideal gas relates pressure P and volume V according to

$$P = \frac{RT}{V^2}\left(1 - \frac{CT}{V^3}\right)\left[V + B\left(1 - \frac{b}{V}\right)\right] - \frac{A}{V^2}\left(1 - \frac{a}{V}\right),$$

where $R = 0.082\,06$ is the universal gas constant, T is absolute temperature of the gas, and a, A, b, B, and C are constants. For air, $a = 0.019\,31$, $A = 1.3012$, $b = -0.0011$, $B = 0.046\,11$, and $C = 4.34 \times 10^4$. For $T = 300$, plot a graph of P as a function of V on the interval $300 \le V \le 400$. Verify that to one-decimal-place accuracy, $V = 373.5$ gives a relative maximum.

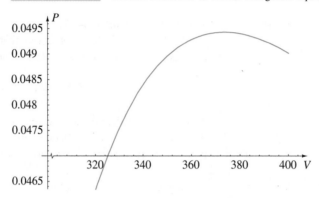

FIGURE 4.24   Relative maximum of Beattie–Bridgeman equation of state

SOLUTION   The plot of the function in Figure 4.24 confirms a relative maximum near $V = 370$. To verify that 373.5 is the critical point yielding this maximum, we calculate

$$P'(V) = RT\left(-\frac{2}{V^3} + \frac{5CT}{V^6}\right)\left[V + B\left(1 - \frac{b}{V}\right)\right]$$

$$+ RT\left(\frac{1}{V^2} - \frac{CT}{V^5}\right)\left[1 + \frac{Bb}{V^2}\right] - A\left(-\frac{2}{V^3} + \frac{3a}{V^4}\right).$$

Since $P'(373.45) = 2.29 \times 10^{-8}$ and $P'(373.55) = -1.19 \times 10^{-7}$, it follows (by the zero intermediate value theorem) that, to one decimal place, the critical point is 373.5.

EXERCISES 4.3

In Exercises 1–44 find all critical points of the function and determine algebraically with the first-derivative test which critical points give relative maxima and relative minima. Use a plot to confirm your findings.

1. $f(x) = x^2 - 2x + 6$

2. $f(x) = 2x^3 + 15x^2 + 24x + 1$

3. $f(x) = x^4 - 4x^3/3 + 2x^2 - 24x$

4. $f(x) = (x - 1)^5$

5. $f(x) = \dfrac{x + 1}{x^2 + 8}$

6. $f(x) = \dfrac{x^2 + 1}{x - 1}$

7. $f(x) = \dfrac{x}{\sqrt{1 - x}}$

8. $f(x) = x^3 - 6x^2 + 12x + 9$

9. $f(x) = x^3/3 - x^2/2 - 2x$

10. $f(x) = x^{1/3}$

* 11. $f(x) = \sin^2 x$

* 12. $f(x) = \dfrac{x^3}{x^4 + 1}$

* 13. $f(x) = 3x^4 - 16x^3 + 18x^2 + 2$

* 14. $f(x) = x + \dfrac{1}{x}$

* 15. $f(x) = 2x^3 - 15x^2 + 6x + 4$

* 16. $f(x) = |x| + x$

* 17. $f(x) = (x - 1)^{2/3}$

* 18. $f(x) = \dfrac{(x - 1)^3}{(x + 1)^4}$

19. $f(x) = (x+2)^3(x-4)^3$

20. $f(x) = x + 2\sin x$

21. $f(x) = x^{1/5} + x$

22. $f(x) = x^2 + \dfrac{25x^2}{(x-2)^2}$

23. $f(x) = \left(\dfrac{x+8}{x}\right)\sqrt{x^2+100}$

24. $f(x) = \dfrac{1+x+x^2+x^3}{1+x^3}$

25. $f(x) = x^{5/4} - x^{1/4}$ * **26.** $f(x) = \dfrac{x^2}{x^2-4}$

27. $f(x) = \dfrac{x^2}{x^3-1}$ * **28.** $f(x) = \dfrac{(2x-1)(x-8)}{(x-1)(x-4)}$

29. $f(x) = \sin^2 x \cos x, \quad 0 \le x \le 2\pi$

30. $f(x) = \sin x + \cos x$

31. $f(x) = 2\csc x - \cot x, \quad 0 < x < \pi/2$

32. $f(x) = \csc x + 8\sec x, \quad 0 < x < \pi/2$

33. $f(x) = \dfrac{\tan x}{x}$

34. $f(x) = x + \sin^2 x, \quad 0 < x < 2\pi$

35. $f(x) = e^{1/x}$ * **36.** $f(x) = x\ln x$

37. $f(x) = x^2\ln x$ * **38.** $f(x) = xe^{-2x}$

39. $f(x) = xe^{-x^2}$ * **40.** $f(x) = x^2 e^{3x}$

41. $f(x) = x^3 - \text{Tan}^{-1} x$ * **42.** $f(x) = \text{Cos}^{-1}(2x) - 5x^2$

43. $f(x) = h(x-a)$ (See equation 2.7.)

44. $f(x) = \lfloor x \rfloor$ (See Exercise 68 in Section 1.5.)

In Exercises 45–50 y is defined implicitly as a function of x. Find all critical points of the function at which its derivative is equal to zero, and classify each as yielding a relative maximum or minimum.

45. $x^4 + y^3 + y^5 = 1$ **46.** $x^2 + y^3 + y = 4$

47. $x^3 y + xy^3 = 2$ * **48.** $y^4 + xy^3 = 1$

49. $x^4 y + y^5 = 32$ * **50.** $x^2 y^4 + y^3 = 1, \quad y \ge 0$

In Exercises 51–52 find all critical points at which the first derivative vanishes for any function, with y as dependent variable, defined implicitly by the equation.

51. $x^2 + 2xy + 3y^2 = 2$ * **52.** $x^4 y + y^5 = 4x$

53. Find, accurate to four decimals, all critical points at which the first derivative vanishes for any function, with y as dependent variable,

defined implicitly by the equation $2x^2 - y^3 + xy = 4$.

* **54.** The equation $y = x^2\sqrt{1-y^2}$ defines y implicitly as a function of x. Find critical points of the function at which the first derivative vanishes by (a) using implicit differentiation, and (b) finding the explicit definition of the function.

* **55.** An N-wave solution of the *Burgers equation* in fluid dynamics is of the form
$$f(x) = \dfrac{x}{1 + be^{x^2/a}},$$
where $a > 0$ and $b > 0$ are constants. Prove that $f(x)$ has exactly two critical points, one the negative of the other.

* **56.** (a) If $f(x) = x^2$ is defined only for $0 \le x \le 1$, are $x = 0$ and $x = 1$ critical points?

   (b) Does the function have a relative minimum of $f(0) = 0$ and a relative maximum of $f(1) = 1$?

In Exercises 57–64 determine whether the statement is true or false.

* **57.** Points of discontinuity of a function are critical if, and only if, the function is defined at the point of discontinuity.

* **58.** When a function is defined only on the interval $a \le x \le b$, the ends $x = a$ and $x = b$ must yield relative maxima or minima for the function.

* **59.** A function is discontinuous at a point if, and only if, it has no derivative at the point.

* **60.** A function can have a relative maximum and a relative minimum at the same point.

* **61.** If the derivative of a function changes sign when passing through a point, the point must yield a relative extremum for the function.

* **62.** If a function has two relative maxima, it must have a relative minimum between them.

* **63.** It is possible for every point in the domain of a function to be critical.

* **64.** On an interval of finite length, a nonconstant function can have only a finite number of critical points at which its derivative is equal to zero.

* **65.** Verify that the function $f(x) = x^3 + \cos x$ has two critical points, one of which is $x = 0$. Find the other critical point correct to three decimal places.

* **66.** The gas equation of Dieterici relating pressure P and volume V is
$$P(V-b) = RTe^{-a/(RTV)},$$
where R is the universal gas constant, T is absolute temperature, and $a > 0$ and $b > 0$ are constants.

   (a) Verify that the function $P(V)$ has two critical points provided $T < a/(4bR)$.

   (b) Verify that when $T = T_c = a/(4bR)$, there is one critical point at which $P = a/(4b^2 e^2)$.

* **67.**   (a) The *Dirichlet function* in digital signal processing is

$$f_L(\omega) = \frac{\sin(\omega L/2)}{L \sin(\omega/2)},$$

   where L is a constant. Plot it for $L = 10$ on the interval $-10 \le \omega \le 10$. Use the plot to aid in answering the following questions.

   (b) Is $f_{10}(0)$ defined? What does the graph suggest for the limit of $f_L(\omega)$ as $\omega \to 0$?

   (c) Is $f_L(\omega)$ even, odd, or neither even nor odd?

   (d) Is $f_{10}(\omega)$ periodic? What is its period?

   (e) For what smallest positive value of ω does $f_{10}(\omega)$ have its smallest positive relative maximum?

   (f) What are the zeros of $f_L(\omega)$?

▦* **68.**   (a) Plot a graph of the function $f(x) = e^{-x} \sin x$, $x \ge 0$.

   (b) For what values of x does $f(x)$ have relative extrema?

In Exercises 69–72 find all critical points of the function correct to four decimal places. Determine whether each critical point yields a relative maximum or a relative minimum.

▦* **69.** $f(x) = x^4 + 6x^2 + 4x + 1$

▦* **70.** $f(x) = x^4 - 10x^2 - 4x + 5$

▦* **71.** $f(x) = x^3 - 2 \cos x$

▦* **72.** $f(x) = \dfrac{x^2 - 4}{(x^2 - 5x + 4)^2}$

* **73.** The distance from the origin to any point (x, y) on the curve $C : y = f(x)$ is given by $D = \sqrt{x^2 + y^2}$. Show that if $P(x_0, y_0)$ is a point on C for which D has a relative extremum, then the line OP is perpendicular to the tangent line to C at P. Assume that $f(x)$ is differentiable, and that the curve does not pass through the origin.

* **74.** The equation $(x^2 + y^2 + x)^2 = x^2 + y^2$ describes a *cardioid* (shown in the following figure). Find the maximum y-coordinate for points on the curve.

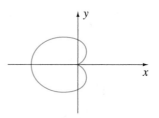

* **75.** The equation $(x^2 + y^2)^2 = x^2 y$ describes a *bifolium* (figure below). Find the points on the curve farthest from the origin.

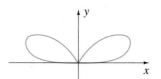

▦* **76.**   (a) Verify that $x = 0$ is a critical point for the function

$$f(x) = \begin{cases} x \sin(1/x), & x \ne 0 \\ 0, & x = 0 \end{cases}$$

   because $f'(0)$ does not exist.

   (b) Use a plot to show that $f'(x)$ does not change sign as x increases through $x = 0$.

   (c) Does $x = 0$ yield a relative maximum or minimum for $f(x)$?

▦* **77.** Repeat Exercise 76 for the function

$$f(x) = \begin{cases} |x \sin(1/x)|, & x \ne 0 \\ 0, & x = 0. \end{cases}$$

▦* **78.** Repeat Exercise 76 for the function

$$f(x) = \begin{cases} -|x \sin(1/x)|, & x \ne 0 \\ 0, & x = 0. \end{cases}$$

** **79.** Show that the function $f(x) = x^3 + px + q$ has three distinct zeros if and only if $4p^3 + 27q^2 < 0$.

4.4 Concavity and Points of Inflection

In the first three sections of this chapter we concentrated on the first derivative of a function. In this section we turn our attention to the second derivative.

Consider the function in Figure 4.25. If we draw tangent lines to the graph at the five points $c_1, c_2, c_3, c_4,$ and c_5, it is clear that the slope is greater at c_5 than it is at c_4, greater at c_4 than at c_3, and so on. In fact, given any two points x_1 and x_2 in the interval $a < x < b$, where $x_2 > x_1$, the slope at x_2 is greater than at x_1,

$$f'(x_2) > f'(x_1).$$

What we are saying is that the function $f'(x)$ is increasing on the interval $a < x < b$.

FIGURE 4.25 $f'(x)$ increasing on $a < x < b$

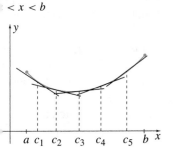

FIGURE 4.26 $f'(x)$ decreasing on $b < x < c$

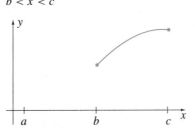

FIGURE 4.27 $f'(x)$ changes from increasing to decreasing at $x = b$

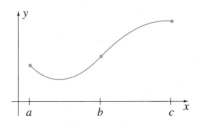

The first derivative of the function $f(x)$ in Figure 4.26 is decreasing on the interval $b < x < c$. In Figure 4.27, we have pieced together the functions in Figures 4.25 and 4.26 to form a function that has $f'(x)$ increasing on $a < x \leq b$ and decreasing on $b \leq x < c$. We assume that the left-hand derivative $f'_-(b)$ in Figure 4.25 and the right-hand derivative $f'_+(b)$ in Figure 4.26 are the same, in which case $f'(b)$ exists in Figure 4.27. We give names to intervals on which the first derivative of a function is increasing and decreasing, and points that separate such intervals in the following definition.

DEFINITION 4.5

The graph of a function $f(x)$ is said to be **concave upward** on an interval I if $f'(x)$ is increasing on I, and **concave downward** on I if $f'(x)$ is decreasing on I. Points on a graph that separate intervals of opposite concavity are called **points of inflection**.

The graph in Figure 4.28 is concave upward on the intervals

$$b \leq x \leq c, \qquad d \leq x < e, \qquad e < x \leq f,$$

and concave downward for

$$a \leq x \leq b, \qquad c \leq x \leq d, \qquad f \leq x < g, \qquad g < x \leq h.$$

The points $x = e$ and $x = g$ are not included in these intervals because $f'(x)$ is not defined at these points. Points on the curve corresponding to $x = b, c, d$, and f are points of inflection; those at $x = e$ and g are not. They do not separate intervals of opposite concavity.

The tangent line has a peculiar property at a point of inflection. To see it, draw the tangent line at the four points of inflection in Figure 4.28. The tangent line actually crosses from one side of the curve to the other (something many readers think a tangent line should not do).

FIGURE 4.28 Intervals on which graph is concave upward and concave downward

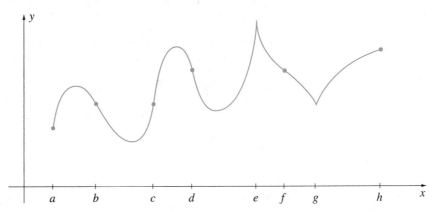

FIGURE 4.29   Horizontal
point of inflection

FIGURE 4.29   Horizontal
point of inflection

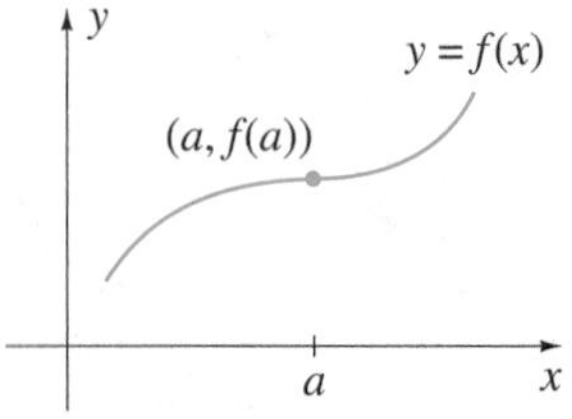

FIGURE 4.30   Vertical
point of inflection

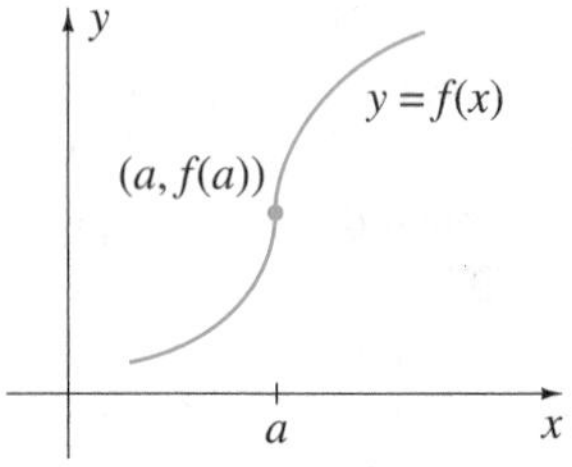

The critical points $x = a$ and $x = c$ in Figure 4.19, where $f'(x) = 0$, and the critical points $x = e$ and $x = g$, where $f'(x)$ does not exist, do not yield relative extrema for $f(x)$. Critical points of these types are illustrated again in Figures 4.29 and 4.30. The tangent line at $x = a$ is horizontal in Figure 4.29 [$f'(a) = 0$]. Since the graph is concave downward to the left of $x = a$ and concave upward to the right, the point $(a, f(a))$ is a point of inflection. We call it a **horizontal point of inflection**. The point $(a, f(a))$ in Figure 4.30 is also a point of inflection. As the tangent line at this point is vertical, we call it a **vertical point of inflection**.

In Section 4.2 we stated that a function is increasing on an interval if its first derivative is greater than or equal to zero on that interval, and equal to zero at only a finite number of points; it is decreasing if its first derivative is less than or equal to zero. Furthermore, points that separate intervals on which the function is increasing and decreasing are points at which the first derivative is either equal to zero (and changes sign) or does not exist. We can use these ideas to locate analogously points of inflection and intervals on which the graph of a function is concave upward and concave downward.

Test for Concavity

(i) The graph of a function $f(x)$ is concave upward on an interval I if on I,

$$f''(x) \geq 0 \tag{4.7a}$$

and is equal to zero at only a finite number of points.

(ii) The graph of a function $f(x)$ is concave downward on an interval I if on I,

$$f''(x) \leq 0 \tag{4.7b}$$

and is equal to zero at only a finite number of points.

(iii) Points of inflection occur where

$$f''(x) = 0 \tag{4.7c}$$

and $f''(x)$ changes sign, or perhaps also where $f''(x)$ does not exist.

Consequently, just as the first derivative of a function is used to determine its relative extrema and intervals on which it is increasing and decreasing, the second derivative is used to find points of inflection and intervals on which the graph of the function is concave upward and concave downward. This parallelism between the first and second derivatives is reiterated in the following table.

First Derivative	Second Derivative
1. A relative extremum occurs at a point where $f'(x) = 0$ and $f'(x)$ changes sign as x increases through the point.	1. A point of inflection occurs at a point where $f''(x) = 0$ and $f''(x)$ changes sign as x increases through the point.
2. A relative extremum may also occur at a point where $f'(x)$ does not exist.	2. A point of inflection may also occur at a point where $f''(x)$ does not exist.
3. $f(x)$ is increasing on an interval if $f'(x) \geq 0$; $f(x)$ is decreasing on an interval if $f'(x) \leq 0$. $f'(x)$ may be equal to zero at only a finite number of points in either case.	3. $f(x)$ is concave upward on an interval if $f''(x) \geq 0$; $f(x)$ is concave downward on an interval if $f''(x) \leq 0$. $f''(x)$ may be equal to zero at only a finite number of points in either case.

These discussions lead to what is often called the *second-derivative test* for determining whether a critical point at which $f'(x) = 0$ yields a relative maximum or a relative minimum.

Second-Derivative Test for Relative Extrema

Suppose $f''(x)$ is continuous on an open interval containing a critical point x_0 of a function $f(x)$ at which $f'(x_0) = 0$. Then:

(i) If $f''(x_0) > 0$, then $x = x_0$ yields a relative minimum for $f(x)$.

(ii) If $f''(x_0) < 0$, then $x = x_0$ yields a relative maximum for $f(x)$.

(iii) If $f''(x_0) = 0$, then no conclusion can be made.

In (i), $f'(x_0) = 0$ implies that $x = x_0$ is a critical point with a horizontal tangent line; $f''(x_0) > 0$ implies that around $x = x_0$ the curve is concave upward, and $x = x_0$ therefore yields a relative minimum. Similarly, $f''(x_0) < 0$ at a critical point in (ii) implies that the curve is concave downward and therefore has a relative maximum.

To show that the test fails in case (iii), consider three functions $f(x) = x^4$, $f(x) = -x^4$, and $f(x) = x^3$ in Figures 4.31, 4.32, and 4.33 respectively. For each, $f'(0) = f''(0) = 0$; yet the first has a relative minimum, the second a relative maximum, and the third a horizontal point of inflection at $x = 0$.

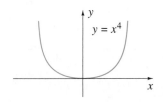

FIGURE 4.31 A relative minimum at $x = 0$

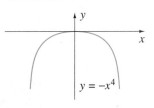

FIGURE 4.32 A relative maximum at $x = 0$

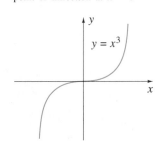

FIGURE 4.33 Horizontal point of inflection at $x = 0$

The second-derivative test very quickly determines the nature of a critical point x_0 when $f'(x_0) = 0$ and $f''(x_0) \neq 0$. For instance, the second derivative of the function $f(x) = 2x^3 - 9x^2 - 23x + 6$ in Example 4.6(a) of Section 4.3 is $f''(x) = 12x - 18$. Since $f''\big((9 - \sqrt{219})/6\big) < 0$, the critical point $x = (9 - \sqrt{219})/6$ yields a relative maximum, and because $f''\big((9 + \sqrt{219})/6\big) > 0$, $x = (9 + \sqrt{219})/6$ gives a relative minimum.

A more refined test to determine the nature of a critical point x_0 when $f'(x_0) = f''(x_0) = 0$ is discussed in Exercise 35.

EXAMPLE 4.9

For each of the following functions, find all points of inflection and intervals on which the graph of the function is concave upward and concave downward.

$$(a) \quad f(x) = \frac{2}{3}x^3 - 6x^2 + 16x + 1 \qquad (b) \quad f(x) = x^{6/5} + x^{1/5}$$

SOLUTION

(a) The graph of the function in Figure 4.34 indicates that there is one point of inflection. To find it, we solve

$$0 = f''(x) = \frac{d}{dx}(2x^2 - 12x + 16) = 4x - 12,$$

and obtain $x = 3$. Since $f''(x)$ changes sign as x passes through 3, $x = 3$ gives the point of inflection $(3, 13)$. The graph $y = f(x)$ is concave downward for $x \leq 3$ [since $f''(x) \leq 0$] and concave upward for $x \geq 3$ [$f''(x) \geq 0$]. Since $f''(x)$ is defined for all x, there can be no other points of inflection.

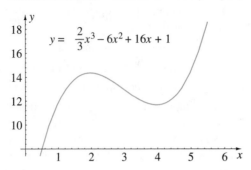

FIGURE 4.34   Intervals on which $y = f(x)$ is concave upward and downward

FIGURE 4.35a   FIGURE 4.35b

Intervals on which $y = f(x)$ is concave upward and downward

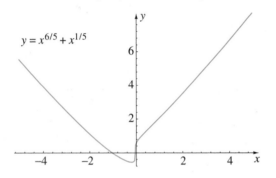

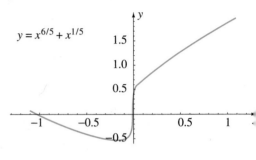

(b) The plot of the function in Figure 4.35a suggests a point of inflection between $x = 0$
and $x = 1$; the curve seems to be concave downward for small positive x and concave
upward for larger x. The graph on the smaller interval in Figure 4.35b does not make
this any clearer, but it suggests that $(0, 0)$ is a point of inflection. To confirm these
suspicions, we first solve

$$0 = f''(x) = \frac{d}{dx}\left(\frac{6}{5}x^{1/5} + \frac{1}{5}x^{-4/5}\right) = \frac{6}{25}x^{-4/5} - \frac{4}{25}x^{-9/5} = \frac{6x - 4}{25x^{9/5}}$$

and find $x = 2/3$. Since $f''(x)$ changes sign as x passes through $2/3$, $x = 2/3$
gives a point of inflection with $f(2/3) \approx 1.5$.

We also note that $f'(x)$ and $f''(x)$ do not exist at $x = 0$, although $f(0)$ does ($= 0$). Since
$f''(x)$ changes sign as x passes through 0, $(0, 0)$ must also be a point of inflection. The fact
that

$$\lim_{x \to 0} f'(x) = \lim_{x \to 0} \left(\frac{6}{5}x^{1/5} + \frac{1}{5}x^{-4/5}\right) = \lim_{x \to 0} \frac{6x + 1}{5x^{4/5}} = \infty$$

indicates that $(0, 0)$ is a vertical point of inflection. With the points of inflection now in place,
we can say that the graph is concave upward for $x < 0$ and $x \geq 2/3$, and concave downward
for $0 < x \leq 2/3$.

EXAMPLE 4.10

The Van der Waals equation for an ideal gas relates pressure P and specific volume V by

$$P = \frac{RT}{V - b} - \frac{a}{V^2},$$

where T is absolute temperature of the gas, R is the universal gas constant, and a and b are positive constants that depend on the gas. Suppose pressure and volume measurements are taken for a specific gas at various temperatures, and the results are plotted as in Figure 4.36. Each curve corresponds to a fixed value of T. One of the curves has a horizontal point of inflection. It can be used to determine values of a and b for this gas. Suppose the temperature of the gas for this curve is T_c, and V_c is the volume, which gives the horizontal point of inflection. Find expressions for a and b in terms of T_c and $P_c = P(V_c)$, called critical temperature and pressure.

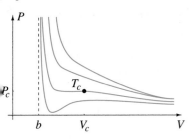

FIGURE 4.36   Plots of Van der Waals equation for various temperatures

SOLUTION   Since $\dfrac{dP}{dV} = -\dfrac{RT}{(V-b)^2} + \dfrac{2a}{V^3}$, and $\dfrac{d^2P}{dV^2} = \dfrac{2RT}{(V-b)^3} - \dfrac{6a}{V^4}$, a horizontal point of inflection occurs for $T = T_c$ and $V = V_c$ if

$$0 = -\frac{RT_c}{(V_c-b)^2} + \frac{2a}{V_c^3}, \qquad 0 = \frac{2RT_c}{(V_c-b)^3} - \frac{6a}{V_c^4}.$$

To find a and b, we solve each equation for a and equate results:

$$\frac{RT_cV_c^3}{2(V_c-b)^2} = \frac{RT_cV_c^4}{3(V_c-b)^3} \quad\Longrightarrow\quad 3(V_c-b) = 2V_c \quad\Longrightarrow\quad b = \frac{V_c}{3}.$$

Substitution of this into the first equation gives

$$\frac{2a}{V_c^3} = \frac{RT_c}{(V_c-V_c/3)^2} = \frac{9RT_c}{4V_c^2} \quad\Longrightarrow\quad a = \frac{9RT_cV_c}{8}.$$

We now have a and b in terms of T_c and V_c. To replace V_c with P_c, we use Van der Waals equation to write

$$P_c = \frac{RT_c}{V_c-b} - \frac{a}{V_c^2}.$$

When we substitute the expressions for a and b into this equation we obtain

$$P_c = \frac{RT_c}{V_c-V_c/3} - \frac{9RT_c}{8V_c} = \frac{3RT_c}{8V_c} \quad\Longrightarrow\quad V_c = \frac{3RT_c}{8P_c}.$$

This then gives $a = \dfrac{9RT_c}{8}\left(\dfrac{3RT_c}{8P_c}\right) = \dfrac{27R^2T_c^2}{64P_c}$ and $b = \dfrac{RT_c}{8P_c}$.

EXERCISES 4.4

In Exercises 1–14 determine where the graph of the function is concave upward, is concave downward, and has points of inflection.

1. $f(x) = x^3 - 3x^2 - 3x + 5$

2. $f(x) = 3x^4 + 4x^3 - 24x + 2$

∗ 3. $f(x) = x^2 + \dfrac{1}{x^2}$

∗ 4. $f(x) = \dfrac{x^2 + 4}{x^2 - 1}$

∗ 5. $f(x) = x + \cos x, \quad |x| < 2\pi$

∗ 6. $f(x) = x^2 - \sin x$

∗ 7. $f(x) = x^2 - 2\sin x$

∗ 8. $f(x) = x^2 - 4\sin x, \quad |x| < 2\pi$

∗ 9. $f(x) = x \ln x$ **∗ 10.** $f(x) = x^2 \ln x$

∗ 11. $f(x) = e^{1/x}$ **∗ 12.** $f(x) = xe^{-2x}$

∗ 13. $f(x) = x^2 e^{3x}$ **∗ 14.** $f(x) = x^2 - e^{-x}$

In Exercises 15–22 use the second-derivative test to determine whether critical points where $f'(x) = 0$ yield relative maxima or relative minima.

15. $f(x) = x^3 - 3x^2 - 3x + 5$

16. $f(x) = x + \dfrac{1}{x}$

17. $f(x) = 3x^4 - 16x^3 + 18x^2 + 2$

18. $f(x) = x^{5/4} - x^{1/4}$

∗ 19. $f(x) = x \ln x$ **∗ 20.** $f(x) = x^2 \ln x$

∗ 21. $f(x) = xe^{2x}$ **∗ 22.** $f(x) = x^2 e^{-2x}$

∗ 23. Is the graph of a function concave upward, concave downward, both, or neither on an interval I if on I its second derivative is always equal to zero?

∗ 24. Prove that the curve $y = 2\cos x$ passes through all points of inflection of the curve $y = x \sin x$.

∗ 25. Show that every cubic polynomial has exactly one point of inflection on its graph.

Draw the graph of a function $f(x)$ that is defined everywhere on the interval $I : a \le x \le b$, and that possesses the properties in Exercises 26–31. In each case assume that $a < x_0 < b$.

∗ 26. $f(x)$ is increasing on I and discontinuous at x_0.

∗ 27. $f''(x) \ge 0$ on I and $f(x)$ is not concave upward on I.

∗ 28. $f(x)$ is decreasing on I, $f(x)$ is continuous at x_0, but $f'(x)$ is discontinuous at x_0.

∗ 29. $f(x)$ is increasing on $a \le x \le x_0$, increasing on $x_0 < x \le b$, but not increasing on I.

∗ 30. $f(x)$ is concave downward on $a \le x < x_0$, concave downward on $x_0 < x \le b$, but not concave downward on I.

∗ 31. $f(x)$ is increasing on $a \le x \le x_0$, concave downward on $a \le x < x_0$, decreasing on $x_0 \le x \le b$, and concave upward on $x_0 < x \le b$.

∗ 32. If $f(x)$ and $g(x)$ are twice differentiable and concave upward on an interval I, is $f(x)g(x)$ concave upward on I?

∗ 33. If the functions in Exercise 32 are also increasing on I, is $f(x)g(x)$ concave upward on I?

∗ 34. Repeat Example 4.10 for Dieterici's equation in Exercise 66 of Section 4.3

∗ 35. The second-derivative test fails to classify a critical point x_0 of a function $f(x)$ as yielding a relative maximum, a relative minimum, or a horizontal point of inflection if $f'(x_0) = f''(x_0) = 0$. The following test can be used in such cases. Suppose $f(x)$ has derivatives of all orders in an open interval around x_0, and the first n derivatives all vanish at x_0, but the $(n + 1)^{\text{th}}$ derivative at x_0 is not zero. If we denote the nth derivative of $f(x)$ at x_0 by $f^{(n)}(x_0)$, these conditions are

$$0 = f'(x_0) = f''(x_0) = \cdots = f^{(n)}(x_0), \qquad f^{(n+1)}(x_0) \ne 0.$$

Then:

(i) If n is even, $f(x)$ has a horizontal point of inflection at x_0.

(ii) If n is odd and $f^{(n+1)}(x_0) > 0$, $f(x)$ has a relative minimum at x_0.

(iii) If n is odd and $f^{(n+1)}(x_0) < 0$, $f(x)$ has a relative maximum at x_0.

A proof of this result requires the use of material from Chapter 10 and is therefore delayed until that time (see Exercise 16 in Section 10.3). Note that the second-derivative test is the special case when $n = 1$. Use this test to determine whether critical points of the following functions yield relative maxima, relative minima, or horizontal points of inflection:

(a) $f(x) = (x^2 - 1)^3$;   (b) $f(x) = x^2\sqrt{1 - x}$.

∗ 36. Show that the points of inflection of $f(x) = (k - x)/(x^2 + k^2)$, where k is a constant, all lie on a straight line.

∗∗ 37. Prove that if a cubic polynomial has both a relative maximum and a relative minimum, then the point of inflection between these extrema is the midpoint of the line segment joining them.

∗∗ 38. Show that the graph of the function $f(x) = \sin(x - \sin x)$ has an infinite number of horizontal points of inflection.

39. Show that the equation $f(x) = x^{n+1} - b^n x + ab^n = 0$, where $a > 0$, $b > 0$, and $n \geq 1$ are constants, has exactly two distinct positive solutions if and only if $a < \dfrac{nb}{(n+1)^{(n+1)/n}}$.

40. (a) In Example 3.12 of Section 3.3, the derivative of the function

$$f(x) = \begin{cases} x^2 \sin\left(\dfrac{1}{x}\right), & x \neq 0 \\ 0, & x = 0 \end{cases}$$

was shown to be zero at $x = 0$ [i.e., $f'(0) = 0$]. It follows that the graph of the function has a horizontal tangent line at $(0, 0)$. Draw the graph and the tangent line.

(b) Show that $f''(0)$ does not exist. *Hint:* See Exercise 47 in Section 3.9.

(c) Is the point $(0, 0)$ a relative maximum, a relative minimum, a horizontal point of inflection, or none of these?

** **41.** Show analytically that the equation

$$2 \sin\theta = \theta(1 + \cos\theta)$$

has no solution in the interval $0 < \theta < \pi$.

4.5 Drawing Graphs with Calculus

In Sections 4.3 and 4.4, we used plots of functions to help you *see* the geometric interpretations of critical points, relative extrema, concavity, and points of inflection. In this section we use critical points, relative extrema, increasing and decreasing functions, concavity, and points of inflection to draw (or sketch) graphs of functions. We assume no access to graphing calculators or computers for the first four examples.

EXAMPLE 4.11

Use calculus to draw a graph of the function $f(x) = (x - 4)/x^2$.

SOLUTION With an x-intercept equal to 4, and the limits

$$\lim_{x \to 0^-} f(x) = -\infty, \qquad \lim_{x \to 0^+} f(x) = -\infty, \qquad \lim_{x \to -\infty} f(x) = 0^-, \qquad \lim_{x \to \infty} f(x) = 0^+,$$

we begin our sketch as shown in Figure 4.37a. This information would lead us to suspect that the graph should be completed as shown in Figure 4.37b. To verify this we find critical points for $f(x)$:

$$0 = f'(x) = \frac{x^2(1) - (x-4)(2x)}{x^4} = \frac{x(8-x)}{x^4} = \frac{8-x}{x^3}.$$

Clearly, $x = 8$ is the only critical point of the function, and since the first derivative changes from a positive quantity to a negative quantity as x increases through $x = 8$, there is a relative maximum of $f(8) = 1/16$. Our sketch is now as shown in Figure 4.37c.

Figure 4.37c makes it clear that there is a point of inflection to the right of $x = 8$, which we could pinpoint with $f''(x)$:

$$f''(x) = \frac{x^3(-1) - (8-x)(3x^2)}{x^6} = \frac{2(x-12)}{x^4}.$$

Since $f''(12) = 0$ and $f''(x)$ changes sign as x passes through 12, $(12, 1/18)$ is the point of inflection. Figure 4.37d contains the final graph.

FIGURE 4.37a Steps in drawing graph of
$$f(x) = \frac{x - 4}{x^2}$$

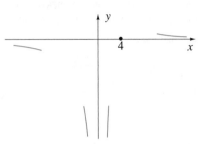

FIGURE 4.37b

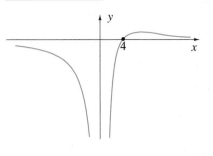

FIGURE 4.37c

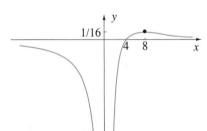

FIGURE 4.37d

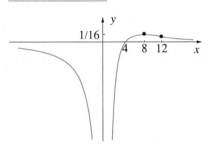

EXAMPLE 4.12

Use calculus to draw a graph of the function $f(x) = x^{5/3} - x^{2/3}$.

SOLUTION   We first note that the three points $(0, 0)$, $(1, 0)$, and $(-1, -2)$ are on the graph. Next we add the facts that

$$\lim_{x \to -\infty} f(x) = -\infty \quad \text{and} \quad \lim_{x \to \infty} f(x) = \infty,$$

as shown in Figure 4.38a. We cannot be sure that the concavity is as indicated, but this will be verified shortly. To find critical points of $f(x)$, we first solve

$$0 = f'(x) = \frac{5}{3} x^{2/3} - \frac{2}{3} x^{-1/3} = \frac{5x - 2}{3x^{1/3}}.$$

The only solution is $x = 2/5$, and because $f'(x)$ changes from negative to positive as x increases through $2/5$, we have a relative minimum of $f(2/5) = (2/5)^{5/3} - (2/5)^{2/3} \approx -0.33$. Since $f'(x)$ does not exist at $x = 0$, this is also a critical point, and we calculate that

$$\lim_{x \to 0^-} f'(x) = \infty \quad \text{and} \quad \lim_{x \to 0^+} f'(x) = -\infty.$$

This information, along with the relative minimum, is shown in Figure 4.38b. We now join these parts smoothly to produce the sketch in Figure 4.38c. To verify that the concavity is as indicated we calculate

$$f''(x) = \left(\frac{5}{3}\right)\left(\frac{2}{3}\right) x^{-1/3} - \left(\frac{2}{3}\right)\left(-\frac{1}{3}\right) x^{-4/3} = \frac{2(5x + 1)}{9x^{4/3}}.$$

Since $f''(x) \leq 0$ for $x \leq -1/5$, the graph is concave downward on this interval. It is concave upward for $-1/5 \leq x < 0$ and $x > 0$ since $f''(x) \geq 0$ on these intervals. A point of inflection occurs at $x = -1/5$ where $f''(x) = 0$ and changes sign, but not at $x = 0$ where $f''(x)$ does not exist. The final sketch is in Figure 4.38d.

FIGURE 4.38a   Steps in drawing graph of
$f(x) = x^{5/3} - x^{2/3}$

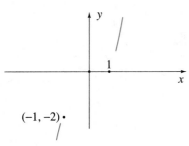

FIGURE 4.38b

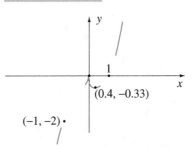

FIGURE 4.38c

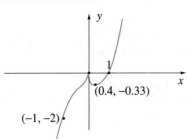

FIGURE 4.38d

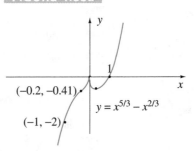

EXAMPLE 4.13

Use calculus to draw a graph of the function $f(x) = x + \sin x$.

SOLUTION   We can add ordinates of the graphs of $y = x$ and $y = \sin x$ in Figure 4.39a. This creates oscillations around $y = x$. They might look like Figure 4.39b, c, or d. To decide which, we find critical points,

$$0 = f'(x) = 1 + \cos x.$$

The solutions of this equation are $x = (2n + 1)\pi$, where n is an integer. Since $f'(x)$ does not change sign as x passes through these points, they cannot yield relative maxima or minima. They must give horizontal points of inflection. The correct graph is Figure 4.39c.

FIGURE 4.39a   Possibilities for graph of
$f(x) = x + \sin x$

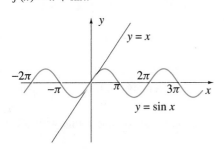

FIGURE 4.39b

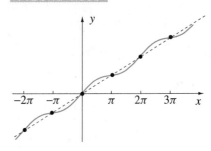

FIGURE 4.39c

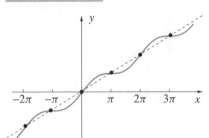

FIGURE 4.39d

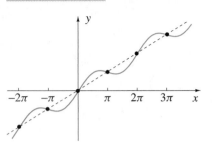

EXAMPLE 4.14

Use calculus to draw a graph of the function $f(x) = (x^2 + 1)/(x - 1)$.

SOLUTION   The function is discontinuous at $x = 1$. We calculate that

$$\lim_{x \to 1^+} \frac{x^2 + 1}{x - 1} = \infty \qquad \text{and} \qquad \lim_{x \to 1^-} \frac{x^2 + 1}{x - 1} = -\infty.$$

If we divide $x^2 + 1$ by $x - 1$, we write $f(x)$ in the form

$$f(x) = x + 1 + \frac{2}{x - 1}.$$

This shows that $y = x + 1$ is an oblique asymptote for the graph (Figure 4.40a). Clearly we should find critical points by solving

$$0 = f'(x) = 1 - \frac{2}{(x - 1)^2} = \frac{(x - 1)^2 - 2}{(x - 1)^2}.$$

This implies that $(x - 1)^2 = 2$, from which $x = 1 \pm \sqrt{2}$. Because $f'(x)$ changes from a positive quantity to a negative quantity as x increases through $1 - \sqrt{2}$, there is a relative maximum at $x = 1 - \sqrt{2}$ of $f(1 - \sqrt{2}) = 2 - 2\sqrt{2}$. Similarly, a relative minimum of $f(1 + \sqrt{2}) = 2 + 2\sqrt{2}$ occurs at $x = 1 + \sqrt{2}$. The final graph is shown in Figure 4.40b.

FIGURE 4.40a   Steps in drawing graph of $f(x) = \dfrac{x^2 + 1}{x - 1}$

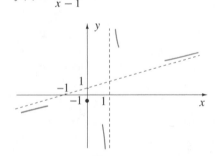

FIGURE 4.40b

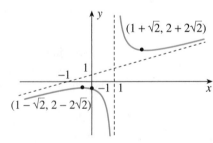

Electronic devices cannot plot curves that contain unspecified parameters. For example, we can use a graphing calculator or computer to plot $y = e^{-ax^2}$ for any given value of a, but we cannot plot the curve without specifying a value for a. What is appropriate is to conjecture the shape of $y = e^{-ax^2}$ from plots with various values of a, and verify the conjecture with calculus. Let us illustrate with two examples.

EXAMPLE 4.15

When two substances A and B are brought together at time $t = 0$, they react to form substance C in such a way that 1 g of A reacts with 1 g of B to form 2 g of C. If initial amounts of A and B are A_0 and B_0 (where $A_0 > B_0$), then the number of grams of C at any time t is given by

$$C(t) = \frac{2A_0 B_0 (1 - e^{-kt})}{A_0 - B_0 e^{-kt}},$$

where $k > 0$ is a constant. Draw a graph of $C(t)$.

SOLUTION Let us get an idea of the shape of the graph by plotting the function with values $A_0 = 20$, $B_0 = 15$, and $k = 0.1$. After experimenting with various domains, we arrive at Figure 4.41 as the most informative.

FIGURE 4.41 Plot of $C(t) = \dfrac{600(1 - e^{-0.1t})}{20 - 15e^{-0.1t}}$

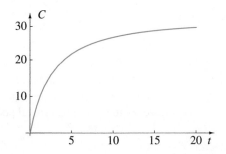

FIGURE 4.42 Graph of $C(t) = \dfrac{2A_0 B_0 (1 - e^{-kt})}{A_0 - B_0 e^{-kt}}$

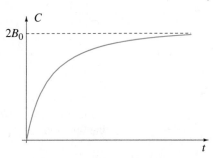

The graph has no relative extrema, no points of inflection, and is asymptotic to the line $C = 30$. We now determine whether these features remain true for all values of $A_0 > B_0$ and k. Critical points of $C(t)$ are given by

$$0 = C'(t) = 2A_0 B_0 \left[\frac{(A_0 - B_0 e^{-kt})(ke^{-kt}) - (1 - e^{-kt})(B_0 ke^{-kt})}{(A_0 - B_0 e^{-kt})^2} \right]$$

$$= \frac{2A_0 B_0 (A_0 - B_0) ke^{-kt}}{(A_0 - B_0 e^{-kt})^2}.$$

There are no solutions and $C'(t)$ always exists, so that $C(t)$ has no relative maxima and minima. For points of inflection, we consider

$$0 = C''(t) = 2A_0 B_0 (A_0 - B_0) k \left[\frac{(A_0 - B_0 e^{-kt})^2 (-ke^{-kt}) - e^{-kt}(2)(A_0 - B_0 e^{-kt})(B_0 ke^{-kt})}{(A_0 - B_0 e^{-kt})^4} \right]$$

$$= \frac{2A_0 B_0 (A_0 - B_0) k^2 e^{-kt} (B_0 + A_0 e^{-kt})}{(A_0 - B_0 e^{-kt})^3}.$$

Once again there are no solutions, and $C''(t)$ always exists. Hence the graph of $C(t)$ has no points of inflection.

The graph begins at $(0, 0)$ with slope $2A_0 B_0 k/(A_0 - B_0)$, and is asymptotic to the line

$$C = \lim_{t \to \infty} \frac{2A_0 B_0 (1 - e^{-kt})}{A_0 - B_0 e^{-kt}} = 2B_0.$$

The graph is shown in Figure 4.42.

EXAMPLE 4.16

When a drug is injected into the blood at time $t = 0$, it is sometimes assumed in biomedical engineering that the concentration of the drug in a nearby organ is given by a function of the form

$$f(t) = k(e^{-at} - e^{-bt}),$$

where k, a, and b are positive constants with $b > a$. Draw a graph of this function indicating any relative extrema and points of inflection.

FIGURE 4.43   Plot of $f(t) = e^{-t} - e^{-2t}$

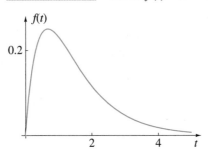

SOLUTION   To get an idea of the shape of the graph we plot the function with values $a = 1$, $b = 2$, and $k = 1$ (Figure 4.43). To determine whether this shape remains constant for all values of a and b (k has no effect on shape of the graph), we begin by finding critical points of the function,

$$0 = f'(t) = k(-ae^{-at} + be^{-bt}).$$

Thus,

$$ae^{-at} = be^{-bt} \implies e^{(b-a)t} = \frac{b}{a},$$

the solution of which is $t = (b - a)^{-1} \ln (b/a)$. To test this for maximum or minimum, and find points of inflection, we use

$$f''(t) = k(a^2 e^{-at} - b^2 e^{-bt}).$$

It is not obvious whether

$$f''[(b - a)^{-1} \ln (b/a)] = k\left[a^2 e^{-\frac{a}{b-a} \ln (b/a)} - b^2 e^{-\frac{b}{b-a} \ln (b/a)}\right]$$

is positive or negative. If we note, however, that at the critical point, $ae^{-at} = be^{-bt}$, and substitute this into $f''(t)$, we can see that

$$f''[(b - a)^{-1} \ln (b/a)] = k[a^2 e^{-at} - b(ae^{-at})] = kae^{-at}(a - b).$$

This is clearly negative (since $b > a$), and therefore $f(t)$ has a relative maximum at $t = (b - a)^{-1} \ln (b/a)$. For points of inflection we set

$$0 = a^2 e^{-at} - b^2 e^{-bt} \implies t = \frac{2}{b - a} \ln \left(\frac{b}{a}\right).$$

To confirm that this value of t gives a point of inflection we should check that $f''(t)$ changes sign as t passes through this value. This is not obvious. We can, however, conclude that a point of inflection must occur here if we examine the information in Figure 4.44a. The graph begins at the origin and achieves a relative maximum at $t = (b - a)^{-1} \ln (b/a)$ where the graph is concave downward. It is also asymptotic to the positive t-axis since $\lim_{t \to \infty} f(t) = 0^+$. Hence, there must be a change in concavity to the right of the maximum, and this can only occur at $t = 2(b - a)^{-1} \ln (b/a)$. The final graph is shown in Figure 4.44b.

FIGURE 4.44a Partial graph of $k(e^{-at} - e^{-bt})$

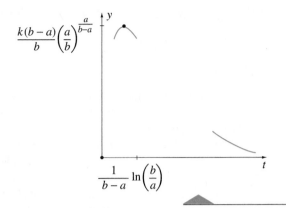

$$\frac{k(b-a)}{b}\left(\frac{a}{b}\right)^{\frac{a}{b-a}}$$

$$\frac{1}{b-a}\ln\left(\frac{b}{a}\right)$$

FIGURE 4.44b   Full graph of $k(e^{-at} - e^{-bt})$

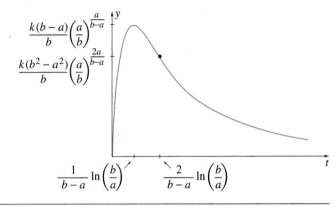

$$\frac{k(b-a)}{b}\left(\frac{a}{b}\right)^{\frac{a}{b-a}}$$

$$\frac{k(b^2-a^2)}{b}\left(\frac{a}{b}\right)^{\frac{2a}{b-a}}$$

$$\frac{1}{b-a}\ln\left(\frac{b}{a}\right) \qquad \frac{2}{b-a}\ln\left(\frac{b}{a}\right)$$

EXERCISES 4.5

In Exercises 1–22 find all relative maxima and minima for the function and points of inflection on its graph. Use this information, and whatever else is appropriate, to draw a graph of the function.

1. $f(x) = \dfrac{x^3}{3} - \dfrac{x^2}{2} - 2x$

2. $f(x) = x^3 - 6x^2 + 12x + 9$

3. $f(x) = 3x^4 - 16x^3 + 18x^2 + 2$

4. $f(x) = 2x^3 - 15x^2 + 6x + 4$

5. $f(x) = x - 3x^{1/3}$ * **6.** $f(x) = \dfrac{x}{x^2 + 4}$

7. $f(x) = (x - 2)^3(x + 2)$ * **8.** $f(x) = x^{2/3}(8 - x)$

9. $f(x) = \dfrac{(x + 2)^2}{x^3}$ * **10.** $f(x) = 2x^{3/2} - 9x + 12x^{1/2}$

11. $f(x) = \dfrac{x^3 + 16}{x}$ * **12.** $f(x) = \dfrac{x^2 + x + 1}{x}$

13. $f(x) = x + \dfrac{1}{x}$ * **14.** $f(x) = \dfrac{x^3}{x^2 - 4}$

15. $f(x) = \dfrac{2x^2}{x^2 - 8x + 12}$ * **16.** $f(x) = \dfrac{x^2 + 1}{x^2 - 1}$

17. $f(x) = \dfrac{(x - 1)^3}{(x + 1)^4}$ * **18.** $f(x) = (x - 1)^{2/3}$

19. $f(x) = (x + 2)^3(x - 4)^3$

20. $f(x) = x + 2\sin x$

21. $f(x) = \text{Sin}^{-1} x + \text{Cos}^{-1} x$

22. $f(x) = x \,\text{Sin}^{-1} x + \sqrt{1 - x^2}$

In Exercises 23–34 find all relative maxima and minima for the function. Use this information, and whatever else is appropriate, to draw a graph of the function.

* **23.** $f(x) = x/(x^2 + 3)$

* **24.** $f(x) = x^4 + 10x^3 + 6x^2 - 64x + 5$

* **25.** $f(x) = x^4 - 2x^3 + 2x$ * **26.** $f(x) = \dfrac{x^2 - 8}{x - 5}$

* **27.** $f(x) = \dfrac{x^2}{x^2 - 4}$ * **28.** $f(x) = x^{5/4} - x^{1/4}$

* **29.** $f(x) = |x^2 - 9| + 2$ * **30.** $f(x) = \dfrac{(2x - 1)(x - 8)}{(x - 1)(x - 4)}$

* **31.** $f(x) = \sin^2 x \cos x, \quad 0 \le x \le 2\pi$

* **32.** $f(x) = \sin x + \cos x$

* **33.** $f(x) = \dfrac{x}{\sqrt{1 - x}}$

* **34.** $f(x) = \left(\dfrac{x + 8}{x}\right)\sqrt{x^2 + 100}$

* **35.** An emf device producing constant voltage V and with constant internal resistance r maintains current i through a circuit with resistance R in the figure below, where

$$V = i(r + R).$$

The power (work per unit time) necessary to maintain this current in r and R is given by

$$P = i^2(r + R) = i^2 r + i^2 R.$$

(a) If we define $P_R = i^2 R$ and $P_r = i^2 r$ as the power dissipated in R and r, respectively, draw P_R and P_r as functions of R.

(b) Draw a graph of $P(R) = P_R(R) + P_r(R)$.

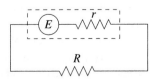

* **36.** The rate of photosynthesis P in a leaf depends on the intensity of light I on the leaf according to

$$P = \frac{MI}{I + K} - R,$$

where $M > R$ and K are all positive constants. Draw a graph of this function.

* **37.** The radial probability density function for the ground state of the hydrogen atom is

$$P(r) = \left(\frac{4r^2}{a^2}\right) e^{-2r/a}, \quad r \geq 0,$$

where $a > 0$ is a constant. Draw a graph of the function identifying its relative maximum and points of inflection.

* **38.** The radial probability density function for the second state of the hydrogen atom is

$$P(r) = \left(\frac{r^2}{8a^3}\right) \left(2 - \frac{r}{a}\right)^2 e^{-r/a}, \quad r \geq 0,$$

where $a > 0$ is a constant. Draw a graph of the function identifying its relative extrema.

* **39.** The *Weibull distribution* for the probability that a certain wind speed v occurs is

$$P(v) = \frac{k}{a} \left(\frac{v}{a}\right)^{k-1} e^{-(v/a)^k},$$

where $a > 1$ and $1 < k < 2$ are constants. Draw a graph of this function identifying its relative extrema.

* **40.** In quantum mechanics, the probability that an energy level E will be occupied is

$$P(E) = \frac{1}{e^{(E-E_f)/(kT)} + 1},$$

where E_f is the Fermi energy, k is the Boltzmann constant, and T is temperature. Draw a graph of the function identifying relative extrema and points of inflection.

* **41.** In Example 3.35 in Section 3.11 we discussed chemical formation in a chemical reactor. What we ignored there was the fact that temperature usually varies in the reactor, and the rate at which the chemical is formed or broken down depends on temperature. In other words, the situation is much more complicated than we presented. When a chemical reactor reaches a stable steady-state condition, a function that is encountered is

$$f(x) = \frac{1}{b + e^{a/x}}, \quad x > 0,$$

where a and b are positive constants.

(a) To get an idea of what this function looks like, plot a graph when $b = 4$ and $a = 10$ on the interval $0 < x \leq 20$. Find its point of inflection.

(b) Now show that $f(x) = (b + e^{a/x})^{-1}$, $x > 0$ has the same shape. Verify that it has no relative extrema, but it has exactly one point of inflection.

* **42.** In the kinetic theory of gases, *Maxwell's speed distribution law* defines the probability P that a molecule of gas moves with speed v as

$$P(v) = 4\pi \left(\frac{M}{2\pi RT}\right)^{3/2} v^2 e^{-Mv^2/(2RT)}, \quad v \geq 0,$$

where T, the temperature of the gas (in kelvin), M, the molar mass of the gas, and R, the gas constant, are all constants. We shall have more to say about this function when we know how to integrate. For now, we simply wish to graph the function.

(a) Plot $P(v)$ for oxygen at 50 K and 300 K using $M = 0.0320$ and $R = 8.31$. Do they have the same shape? The fact that most of the 300 K curve is higher than the 50 K curve means that oxygen molecules are more likely to move faster at higher temperatures than at lower temperatures.

(b) For any gas at any temperature, find the value of v that gives the relative maximum and the values of v that give points of inflection.

* **43.** The function

$$f(x) = \frac{1}{\sqrt{2\pi}\,\sigma} e^{-(x-\mu)^2/(2\sigma^2)},$$

where $\sigma > 0$ and μ are constants is called the *normal probability density function*. Draw its graph.

* **44.** Draw a graph of the function

$$f(x) = \frac{1 + x + x^2 + x^3}{1 + x^3}.$$

4.6 Analyzing Graphs with Calculus

In Section 4.5 we drew graphs of functions without the use of calculators or computers. In this section we take a different approach; we use technology to plot the graph of a function and then use calculus to analyze what we see or do not see on the plot. Remember that we have illustrated on several occasions that electronic output can sometimes be inconclusive and even misleading (see for instance Example 1.19). Electronic devices can never replace sound mathematical analysis. Whenever there is a question about electronic output, we use calculus to find the answer.

EXAMPLE 4.17

Plot a graph of the function $f(x) = (x - 4)/x^2$, and then use calculus to pinpoint significant information concerning the graph.

SOLUTION   A plot of the function on the interval $-10 \leq x \leq 10$ in Figure 4.45a does not show a lot. It certainly indicates that the y-axis is a vertical asymptote, and this is confirmed by the fact that $\lim_{x \to 0} f(x) = -\infty$. The plot also suggests that the x-axis is a horizontal asymptote, confirmed by

$$\lim_{x \to -\infty} f(x) = 0^-, \qquad \lim_{x \to \infty} f(x) = 0^+.$$

The fact that the graph approaches $y = 0$ from above as $x \to \infty$ means that the graph must cross the x-axis at least once. It does so exactly once since $f(x) = 0$ only when $x = 4$. To get a better visualization of the graph for $x > 0$, we now plot it on the interval $1 \leq x \leq 10$, at the same time restricting the y-values to $-1 \leq y \leq 1$ (Figure 4.45b). There must be a relative maximum to the right of $x = 4$ (perhaps more than one). To find it (or them), we solve

$$0 = f'(x) = \frac{x^2(1) - (x - 4)(2x)}{x^4} = \frac{8 - x}{x^3}.$$

FIGURE 4.45a $f(x) = (x - 4)/x^2$ on $-10 \leq x \leq 10$

FIGURE 4.45b $f(x) = (x - 4)/x^2$ on $1 \leq x \leq 10$

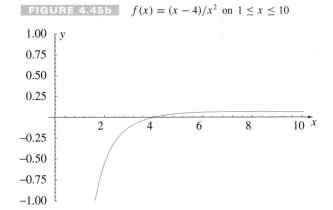

FIGURE 4.45c $f(x) = (x - 4)/x^2$ on $-20 \leq x \leq 20$

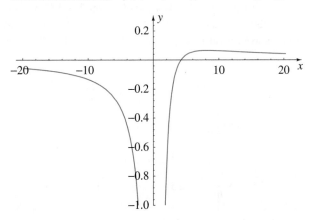

The only solution is $x = 8$. This yields a relative maximum. With the graph concave downward at $x = 8$, and concave upward for large x (recall that $y = 0$ is a horizontal asymptote), there is a point of inflection to the right of $x = 8$. To find it we solve

$$0 = f''(x) = \frac{x^3(-1) - (8 - x)(3x^2)}{x^6} = \frac{2(x - 12)}{x^4}.$$

Since $f''(12) = 0$ and $f''(x)$ changes sign as x passes through 12, there is a point of inflection at $(12, 1/18)$. By restricting y-values even more and plotting on the interval $-20 \leq x \leq 20$, we obtain the plot in Figure 4.45c; it shows all of the foregoing information.

EXAMPLE 4.18

Repeat Example 4.17 for the function $f(x) = x^{5/3} - x^{2/3}$.

SOLUTION   The software package in our computer does not automatically plot a graph of the function $f(x) = x^{5/3} - x^{2/3}$ for negative values of x. (It interprets $x^{5/3}$ and $x^{2/3}$ for negative x as complex numbers.) Your calculator or computer may do the same. To rectify this, define $f(x) = -(-x)^{5/3} - (-x)^{2/3}$ for $x < 0$. A plot then looks like that in Figure 4.46 on the interval $-1 \leq x \leq 2$.

A relative minimum is indicated just to the left of $x = 1/2$. To locate it, we solve

$$0 = f'(x) = \frac{5}{3}x^{2/3} - \frac{2}{3}x^{-1/3} = \frac{5x - 2}{3x^{1/3}}.$$

The only solution is $x = 2/5$, and because $f'(x)$ changes from negative to positive as x increases through $2/5$, we do indeed have a relative minimum of $f(2/5) = (2/5)^{5/3} - (2/5)^{2/3} = -0.33$. The derivative $f'(x)$ does not exist at $x = 0$, but $f(0) = 0$. We calculate that

$$\lim_{x \to 0^-} f'(x) = \lim_{x \to 0^-} \frac{5x - 2}{3x^{1/3}} = \infty \quad \text{and} \quad \lim_{x \to 0^+} f'(x) = -\infty.$$

This means that there is a very sharp point at $(0, 0)$ where the function has a relative maximum. Its tangent line is vertical at this point. It is difficult to get a sense of the concavity of the graph for $x < 0$, it appears so straight. To assess this we consider

$$f''(x) = \left(\frac{5}{3}\right)\left(\frac{2}{3}\right)x^{-1/3} - \left(\frac{2}{3}\right)\left(-\frac{1}{3}\right)x^{-4/3} = \frac{2(5x + 1)}{9x^{4/3}}.$$

Since $f''(x)$ changes sign as x passes through $-1/5$, we have a point of inflection $(-1/5, -0.41)$. The graph is concave downward for $x \leq -1/5$ and concave upward for $-1/5 \leq x < 0$. It is also concave upward for $0 < x < \infty$, so that $(0, 0)$ is not a point of inflection.

FIGURE 4.46   Plot of $f(x) = x^{5/3} - x^{2/3}$

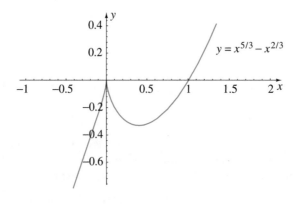

EXAMPLE 4.19

Repeat Example 4.17 for the function $f(x) = x + \sin x$.

SOLUTION   A plot of $f(x) = x + \sin x$ is shown in Figure 4.47. It is not clear whether it has relative extrema. Critical points will provide the answer. We solve $0 = f'(x) = 1 + \cos x$ for $x = (2n + 1)\pi$, where n is an integer. Because $f'(x)$ does not change sign as x passes through these points, they cannot yield relative maxima or minima. They must give horizontal points of inflection.

FIGURE 4.47   Plot of $f(x) = x + \sin x$

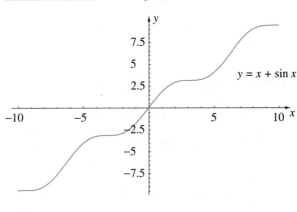

EXERCISES 4.6

In Exercises 1–20 plot a graph of the function. Use derivatives to discuss any significant features of the plot.

1. $f(x) = \dfrac{x^3}{3} - \dfrac{x^2}{2} - 2x$

2. $f(x) = x^3 - 6x^2 + 12x + 9$

3. $f(x) = 3x^4 - 16x^3 + 18x^2 + 2$

4. $f(x) = 2x^3 - 15x^2 + 6x + 4$

5. $f(x) = x - 3x^{1/3}$

6. $f(x) = \dfrac{x}{x^2 + 4}$ * **7.** $f(x) = (x - 2)^3(x + 2)$

8. $f(x) = x^{2/3}(8 - x)$ * **9.** $f(x) = \dfrac{(x + 2)^2}{x^3}$

10. $f(x) = 2x^{3/2} - 9x + 12x^{1/2}$

11. $f(x) = \dfrac{x^3 + 16}{x}$ * **12.** $f(x) = \dfrac{x^2 + x + 1}{x}$

13. $f(x) = x + \dfrac{1}{x}$ * **14.** $f(x) = \dfrac{x^3}{x^2 - 4}$

15. $f(x) = \dfrac{2x^2}{x^2 - 8x + 12}$ * **16.** $f(x) = \dfrac{x^2 + 1}{x^2 - 1}$

* **17.** $f(x) = \dfrac{(x - 1)^3}{(x + 1)^4}$ * **18.** $f(x) = (x - 1)^{2/3}$

* **19.** $f(x) = (x + 2)^3(x - 4)^3$

* **20.** $f(x) = x + 2\sin x$

In Exercises 21–32 plot a graph of the function. Identify all relative extrema.

* **21.** $f(x) = x/(x^2 + 3)$

* **22.** $f(x) = x^4 + 10x^3 + 6x^2 - 64x + 5$

* **23.** $f(x) = x^4 - 2x^3 + 2x$

* **24.** $f(x) = \dfrac{x^2 - 8}{x - 5}$

* **25.** $f(x) = \dfrac{x^2}{x^2 - 4}$ * **26.** $f(x) = x^{5/4} - x^{1/4}$

* **27.** $f(x) = |x^2 - 9| + 2$ * **28.** $f(x) = \dfrac{(2x - 1)(x - 8)}{(x - 1)(x - 4)}$

* **29.** $f(x) = \sin^2 x \cos x, \quad 0 \le x \le 2\pi$

* **30.** $f(x) = \sin x + \cos x$ * **31.** $f(x) = \dfrac{x}{\sqrt{1 - x}}$

32. $f(x) = \left(\dfrac{x+8}{x}\right)\sqrt{x^2 + 100}$

33.   (a) Plot a graph of the function $f(x) = x^8 - 4x^6 - 8x^5 + 40x^3$.

(b) Find all critical points and classify them as yielding relative maxima, relative minima, or horizontal points of inflection.

34.   (a) A company produces x kilograms of a commodity per day at a total cost of

$$C(x) = \dfrac{x^2}{300}\left(\dfrac{x+100}{x+300}\right) + 60, \quad 1 \le x \le 200.$$

Plot a graph of the function. Show that it is always concave upward.

(b) The average production cost per kilogram when x kilograms are produced is given by $c(x) = C(x)/x$. Plot a graph of this function.

(c) Show that the output at which the average cost is least satisfies the equation

$$(x + 300)^2(x^2 - 18\,000) - 60\,000x^2 = 0.$$

Do this in two ways: (i) by finding critical points for $c(x)$; (ii) by noting that the average cost is the slope of the line joining a point $(x, C(x))$ to the origin; hence minimum average cost occurs when the tangent line to the $C(x)$ graph passes through the origin.

4.7 Absolute Maxima and Minima

Seldom in applications of maxima and minima theory do we hear questions such as: What are the relative maxima of this quantity, or what are the relative minima of that quantity? More likely it is: What is the biggest of these, the smallest of those, the best way to do this, the cheapest way to do that? A different kind of extremum is involved — considering all points in the domain of the function, what is the largest (or smallest) value of the function? We define this type of extremum as follows.

DEFINITION 4.6

The **absolute maximum (or global maximum)** of a function $f(x)$ on an interval I is $f(x_0)$ if x_0 is in I and if for all x in I,

$$f(x) \le f(x_0); \tag{4.8a}$$

$f(x_0)$ is said to be the **absolute minimum (or global minimum)** of $f(x)$ on I if for all x in I,

$$f(x) \ge f(x_0). \tag{4.8b}$$

For the function in Figure 4.48, the absolute maximum of $f(x)$ on the interval $a \le x \le b$ is $f(c)$, and the absolute minimum is $f(d)$. For the function in Figure 4.49, the absolute maximum on $a \le x \le b$ is $f(b)$, and the absolute minimum is $f(a)$. For the function in Figure 4.50, the absolute maximum on $a \le x \le b$ is $f(b)$, and the absolute minimum is $f(c)$. Note that we speak of absolute maxima and minima (absolute extrema) of a function only on some specified interval; that is, we do not ask for the absolute maximum or minimum of a function $f(x)$ without specifying the interval I.

Each of the functions in Figures 4.48–4.50 is continuous on a closed interval $a \le x \le b$. The following theorem asserts that every such function has absolute extrema. For a proof of this result, the interested reader should consult books on advanced analysis.

FIGURE 4.48   FIGURE 4.49   FIGURE 4.50

Absolute maxima and minima of continuous functions on closed intervals

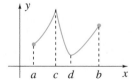
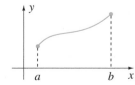
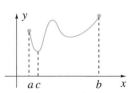

> **THEOREM 4.2**
>
> A function that is continuous on a closed interval must attain an absolute maximum and an absolute minimum on that interval.

The conditions of this theorem are sufficient to guarantee existence of absolute extrema; that is, if a function is continuous on a closed interval, then it *must* have absolute extrema on that interval. However, the conditions are not necessary. If they are not met, the function may or may not have absolute extrema. For instance, if the function in Figure 4.48 is confined to the open interval $a < x < b$, it still attains its absolute extrema at $x = c$ and $x = d$. On the other hand, the function in Figure 4.49 does not have absolute extrema on the open interval $a < x < b$. The function in Figure 4.51 is not continuous on the closed interval $a \leq x \leq d$; it has no absolute maximum on this interval, but it does have absolute minimum $f(a)$. This function is not continuous on $d \leq x \leq b$, but it has absolute maximum $f(d)$ and absolute minimum $f(b)$.

Absolute extrema of the functions in Figures 4.48–4.50 always occur either at a critical point of the function or at an end of the interval $a \leq x \leq b$. This result is always true whether the function is continuous or not, whether the interval is closed or not. If a function has absolute extrema, they occur at critical points or the ends of the interval (if there are ends).

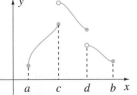

FIGURE 4.51 Absolute maximum and minimum of a discontinuous function

> **THEOREM 4.3**
>
> If a function has absolute extrema on an interval, then they occur either at critical points or at the ends of the interval.

A plot of a function on an interval normally makes it clear where absolute extrema occur, provided of course that they exist. It is then a matter of evaluating the function at the appropriate critical point(s) or endpoint(s). When a plot of the function is unavailable (as would be the case if the function contained unspecified parameters), the following procedure yields its absolute extrema.

Finding Absolute Extrema for a Continuous Function $f(x)$ on a Closed Interval $a \leq x \leq b$

(i) Find all critical points $x_1, x_2, \ldots, x_n$ of $f(x)$ in $a < x < b$.

(ii) Evaluate

$$f(a), \quad f(x_1), \quad f(x_2), \quad \ldots, \quad f(x_n), \quad f(b).$$

(iii) The absolute maximum of $f(x)$ on $a \leq x \leq b$ is the largest of the numbers in (ii); the absolute minimum is the smallest of these numbers.

Note that it is not necessary to classify the critical points of $f(x)$ as yielding relative maxima, relative minima, horizontal points of inflection, vertical points of inflection, corners, or none of these. We need only evaluate $f(x)$ at $x = a$, $x = b$ and at its critical points. The largest and smallest of these numbers must be the absolute extrema.

When the function has discontinuities or the interval is not of finite length, Theorem 4.2 cannot be used. Careful consideration of the function at discontinuities and limits as $x \to \pm\infty$ may be required.

EXAMPLE 4.20

Find the absolute maximum and minimum of the function

$$f(x) = \frac{x^3 - 2x^2 + x + 20}{x^2 + 5}$$

on the interval $0 \leq x \leq 6$, if they exist.

FIGURE 4.52 Graph to illustrate absolute extrema of a function

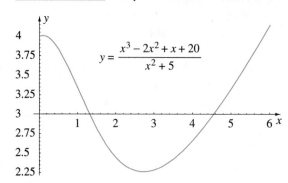

$$y = \frac{x^3 - 2x^2 + x + 20}{x^2 + 5}$$

SOLUTION Since $f(x)$ is continuous on the closed interval $0 \le x \le 6$, the function must have absolute extrema. The plot in Figure 4.52 makes it clear that the absolute minimum is at the critical point to the left of $x = 3$. The absolute maximum appears to be at $x = 6$, but there is a relative maximum near $x = 0$ that should be investigated. For critical points we solve

$$0 = f'(x) = \frac{(x^2 + 5)(3x^2 - 4x + 1) - (x^3 - 2x^2 + x + 20)(2x)}{(x^2 + 5)^2}$$

$$= \frac{x^4 + 14x^2 - 60x + 5}{(x^2 + 5)^2}.$$

To four decimal places, solutions of this equation are $x = 0.0850$ and $x = 2.7188$. The absolute minimum of the function is $f(2.7188) = 2.262$, and because $f(0.0850) = 4.008$ and $f(6) = 4.146$, the absolute maximum is 4.146.

Without the graph in Figure 4.52, we would evaluate

$$f(0) = 4, \qquad f(0.0850) = 4.008, \qquad f(2.7188) = 2.262, \qquad f(6) = 4.146.$$

The absolute minimum of the function is 2.262 and the absolute maximum is 4.146.

EXAMPLE 4.21

Find the absolute maximum and minimum for the function

$$f(x) = \frac{2x^2 + 3x}{x^2 + 4}$$

on the intervals (a) $5 \le x < \infty$ and (b) $1 \le x < \infty$, if they exist.

SOLUTION

(a) The graph in Figure 4.53a appears to be asymptotic to the line $y = 2$ and have a relative maximum at or near $x = 6$. The asymptote is confirmed by

$$\lim_{x \to \infty} \frac{2x^2 + 3x}{x^2 + 4} = \lim_{x \to \infty} \frac{2 + \dfrac{3}{x}}{1 + \dfrac{4}{x^2}} = 2.$$

For critical points, we solve

$$0 = f'(x) = \frac{(x^2 + 4)(4x + 3) - (2x^2 + 3x)(2x)}{(x^2 + 4)^2} = \frac{-(x - 6)(3x + 2)}{(x^2 + 4)^2}.$$

The only critical point in $5 \leq x < \infty$ is indeed $x = 6$. Consequently, the absolute maximum of the function is $f(6) = 9/4$. Because $f(5) = 65/29$ and $\lim_{x \to \infty} f(x) = 2^+$, the function does not have an absolute minimum on the interval $5 \leq x < \infty$.

(b) The graph on the interval $1 \leq x < \infty$ in Figure 4.53b indicates that the absolute maximum is still $f(6) = 9/4$, and the function now has an absolute minimum of $f(1) = 1$.

FIGURE 4.53a $f(x) = (2x^2 + 3x)/(x^2 + 4)$ on $5 \leq x < \infty$

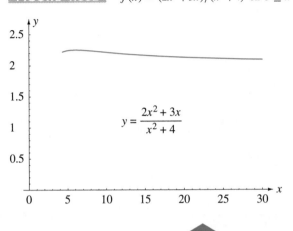

FIGURE 4.53b $f(x) = (2x^2 + 3x)/(x^2 + 4)$ on $1 \leq x < \infty$

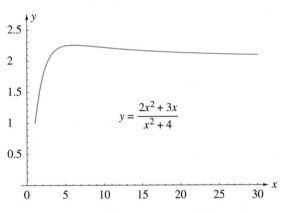

We now consider applied maxima–minima problems.

EXAMPLE 4.22

A rectangular field is to be fenced on three sides with 1000 m of fencing (the fourth side being a straight river's edge). Find the dimensions of the field in order that the area be as large as possible.

FIGURE 4.54   Fencing a rectangular field

SOLUTION   Since the area of the field is to be maximized, we first define a function representing this area. The area of a field of width w and length l (Figure 4.54) is

$$A = lw.$$

This represents the area of a field with arbitrary length l and arbitrary width w. But there is only 1000 m of fencing available for the three sides; therefore, l and w must satisfy the equation

$$2w + l = 1000.$$

With this equation we can express A completely in terms of w (or l):

$$A(w) = w(1000 - 2w) = 1000w - 2w^2.$$

To maximize the area of the field, we must therefore maximize the function $A(w)$. But what are the values of w under consideration? Clearly, w cannot be negative, and in order to satisfy the restriction $2w + l = 1000$, w cannot exceed 500. The physical problem has now been modelled mathematically. Find the absolute maximum of the (continuous) function $A(w)$ on the (closed) interval $0 \leq w \leq 500$.

The graph of this function in Figure 4.55 clearly indicates that area is maximized by the critical point at or near $x = 250$. To find the critical point, we solve

$$0 = A'(w) = 1000 - 4w;$$

the only solution is $w = 250$. The largest possible area is obtained when the width of the field is 250 m and its length is 500 m.

Without the graph in Figure 4.55, we would evaluate

$$A(0) = 0, \qquad A(250) = 125\,000, \qquad A(500) = 0,$$

and draw the same conclusion.

FIGURE 4.55   Area function for rectangular field

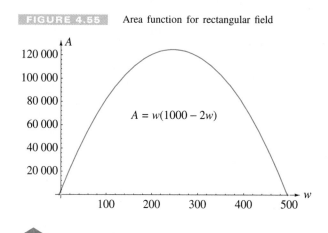

$$A = w(1000 - 2w)$$

EXAMPLE 4.23

Pop cans that must hold 300 mL are made in the shape of right circular cylinders. Find the dimensions of the can that minimize its surface area.

FIGURE 4.56a   Diagram of pop can

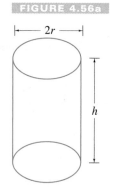

FIGURE 4.56b   Parts of pop can

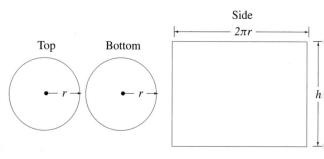

SOLUTION   The surface area of the can (Figure 4.56a) consists of a circular top, an identical bottom, and a rectangular piece formed into the side of the can. The total area of these three pieces is

$$A = 2\pi r^2 + 2\pi rh,$$

where we measure r and h in centimetres (see Figure 4.56b). Since each can holds 300 mL of pop, it follows that

$$\pi r^2 h = 300 \qquad \text{or} \qquad h = \frac{300}{\pi r^2}.$$

This equation can be used to express A completely in terms of r:

$$A(r) = 2\pi r^2 + 2\pi r \left(\frac{300}{\pi r^2} \right).$$

For what values of r is $A(r)$ defined? The radius of the can must be positive and therefore $r > 0$. How large can r be? In effect, it can be as large as desired. The height of the cylinder can always be chosen sufficiently small when r is large to satisfy the volume condition $\pi r^2 h = 300$. The can may not be aesthetically pleasing for large r and small h, but no mathematical difficulties occur in this situation. To solve the problem we must therefore minimize

$$A(r) = 2\pi r^2 + \frac{600}{r}, \quad r > 0.$$

FIGURE 4.57   Area function for pop can

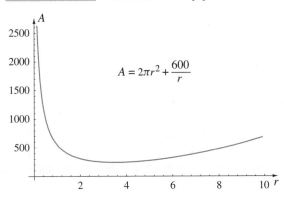

The graph of the function in Figure 4.57, together with the facts that

$$\lim_{r \to 0^+} A(r) = \infty \quad \text{and} \quad \lim_{r \to \infty} A(r) = \infty,$$

allow us to ballpark the answers. Area is minimized at the critical point near $r = 3.5$. To find this critical point, we solve

$$0 = A'(r) = 4\pi r - \frac{600}{r^2} \quad \text{or} \quad 4\pi r^3 - 600 = 0.$$

The only solution is $r = (150/\pi)^{1/3}$. Thus, minimum surface area occurs when the radius of the can is $(150/\pi)^{1/3}$ cm and its height, obtained from the fact that $h = 300/(\pi r^2)$, is $2(150/\pi)^{1/3}$ cm.

Without the graph in Figure 4.57, we would arrive at the same conclusion by noting that $A(r)$ is always positive and

$$\lim_{r \to 0^+} A(r) = \infty, \qquad A\big((150)/\pi)^{1/3}\big) = \text{a finite number}, \qquad \lim_{r \to \infty} A(r) = \infty.$$

EXAMPLE 4.24

FIGURE 4.58   Water flowing in a circular pipe

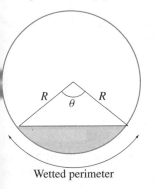

Wetted perimeter

The rate of discharge Q from a circular pipe is

$$Q = K(2R)^{5/2}\sqrt{\frac{(\theta - \sin \theta)^3}{\theta}},$$

where $K > 0$ is a constant, R is the radius of the pipe, and θ is the angle subtended at the centre of the pipe by the wetted perimeter (Figure 4.58). Find the angle for which Q is a maximum accurate to three decimal places.

SOLUTION   We first note that Q is maximized when the function $f(\theta) = (\theta - \sin \theta)^3/\theta$ under the radical is maximized. The plot of $f(\theta)$ in Figure 4.59 indicates that the maximum occurs for the critical point near $\theta = 5$ radians. To find this value, we set the derivative of $f(\theta)$ equal to zero,

$$0 = \frac{\theta(3)(\theta - \sin \theta)^2(1 - \cos \theta) - (\theta - \sin \theta)^3(1)}{\theta^2}.$$

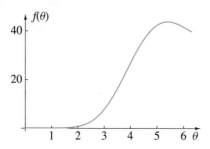

FIGURE 4.59 Discharge in a circular pipe

This implies that

$$0 = 3\theta(1 - \cos\theta) - (\theta - \sin\theta) = 2\theta - 3\theta\cos\theta + \sin\theta = g(\theta).$$

Newton's iterative procedure for finding approximate solutions defines the sequence

$$\theta_1 = 5, \qquad \theta_{n+1} = \theta_n - \frac{g(\theta_n)}{g'(\theta_n)} = \theta_n - \frac{2\theta_n - 3\theta_n\cos\theta_n + \sin\theta_n}{2 + 3\theta_n\sin\theta_n - 2\cos\theta_n}.$$

Iteration gives $\theta_2 = 5.369\,55$, $\theta_3 = 5.378\,49$, and $\theta_4 = 5.378\,51$. Since $g(5.3785) = 1.1 \times 10^{-4}$ and $g(5.3795) = -1.1 \times 10^{-2}$, we can say that to three decimal places, $\theta = 5.379$ The plot in Figure 4.59 is essential in this problem. Without it, we would not have known that $f(\theta)$ had only one critical point, and an approximation to it.

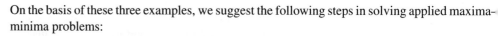

On the basis of these three examples, we suggest the following steps in solving applied maxima-minima problems:

1. Sketch a diagram illustrating the situation, if appropriate.

2. Identify the quantity that is to be maximized or minimized, choose a letter to represent it, and find an expression for this quantity.

3. If necessary, use information in the problem to rewrite the expression in 2 as a function of only one variable.

4. Determine the domain of the function in 3.

5. Maximize or minimize the function in 3 on the domain in 4.

6. Interpret the maximum or minimum values in terms of the original problem.

Step 2 is crucial. Do not consider subsidiary information in the problem until the quantity to be maximized or minimized is clearly identified, labelled, and an expression found for it. Only then should other information be considered. For instance, in Example 4.22, the restriction $2w + l = 1000$ for the length of fencing available was not introduced until the expression $A = lw$ had been identified. Likewise, in Example 4.23, volume condition $\pi r^2 h = 300$ was introduced after surface area $A = 2\pi r^2 + 2\pi rh$.

We now take a look at three additional examples.

EXAMPLE 4.25

A lighthouse is 6 km offshore and a cabin on the straight shoreline is 9 km from the point on the shore nearest the lighthouse. If a man rows at a rate of 3 km/h, and walks at a rate of 5 km/h, where should he beach his boat in order to get from the lighthouse to the cabin as quickly as possible? Repeat the problem in the case that the cabin is 4 km along the shoreline.

FIGURE 4.60   Fastest path from lighthouse to cabin

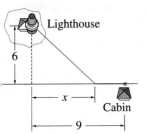

SOLUTION   Figure 4.60 illustrates the path followed by the man when he beaches the boat x kilometres from the point on land closest to the lighthouse. His travel time t for this complete journey is the sum of his time t_1 in the water and his time t_2 on land. Since speeds are constant, each of these times may be calculated by dividing distance by speed; that is, $t_1 = \sqrt{x^2 + 36}/3$ and $t_2 = (9 - x)/5$. To minimize t for some value of x between 0 and 9, we minimize the function

$$t(x) = t_1 + t_2 = \frac{\sqrt{x^2 + 36}}{3} + \frac{9 - x}{5}, \quad 0 \le x \le 9.$$

The graph in Figure 4.61 indicates that $t(x)$ has one critical point and the function has its absolute minimum at this critical point. To find it, we solve

$$0 = t'(x) = \frac{x}{3\sqrt{x^2 + 36}} - \frac{1}{5} \quad \text{or} \quad 5x = 3\sqrt{x^2 + 36}.$$

Thus,

$$25x^2 = 9(x^2 + 36),$$

from which we accept only the positive solution $x = 9/2$. The boat should therefore be beached 4.5 km from the cabin. Without Figure 4.61, we would evaluate

$$t(0) = 2 + \frac{9}{5} = 3.8, \quad t(4.5) = \frac{\sqrt{4.5^2 + 36}}{3} + \frac{9 - 4.5}{5} = 3.4, \quad t(9) = \frac{\sqrt{81 + 36}}{3} \approx 3.6.$$

FIGURE 4.61   Time function when cabin is 9 km down the shoreline

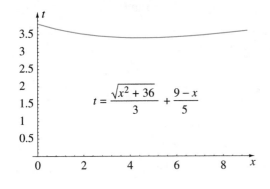

FIGURE 4.62   Time function when cabin is 4 km down the shoreline

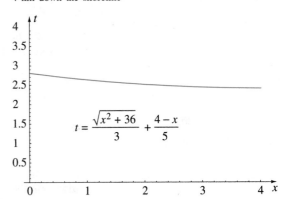

When the cabin is 4 km along the shoreline,

$$t(x) = \frac{\sqrt{x^2 + 36}}{3} + \frac{4 - x}{5}, \quad 0 \le x \le 4.$$

The graph in Figure 4.62 indicates that $f(x)$ does not have any critical points in the interval $0 \le x \le 4$. This is confirmed by the fact that critical points are again given by the equation

$$\frac{x}{3\sqrt{x^2 + 36}} - \frac{1}{5} = 0,$$

but the solution $x = 9/2$ must be rejected since it does not fall in the interval $0 \le x \le 4$. Travel time is therefore minimized by heading directly to shore or directly to the cabin, and Figure 4.62 suggests the cabin. Verification is provided by

$$t(0) = \frac{\sqrt{36}}{3} + \frac{4}{5} = \frac{14}{5}, \quad t(4) = \frac{\sqrt{16 + 36}}{3} = \frac{2\sqrt{13}}{3} < \frac{14}{5}.$$

EXAMPLE 4.26

A rectangular beam is to be cut from a circular log. Naturally, it is desirable for the beam to be as strong as possible. From our experience we know that the beam should be deeper than it is wide. For instance, place a standard $2'' \times 4'' \times 8'$ piece of lumber between two supports 2 m apart, and sit on it. It will probably support you if you sit on the narrow edge, but not if you sit on the wide edge. In other words, the 2×4 is much stronger when its depth is more than its width. Experimental evidence in structural engineering has shown that the strength of a beam is proportional to the product of its width and the square of its depth. Find the dimension of the strongest beam that can be cut from a circular log of radius R.

FIGURE 4.63a   Rectangular beam cut from circular log   FIGURE 4.63b   Width and depth related to radius of log

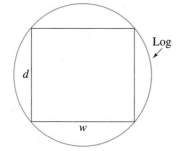

SOLUTION   For a beam with width w and depth d as shown in Figure 4.63a, its strength is given by

$$S = kwd^2,$$

where $k > 0$ is a constant. If we join the centre of the log to one of the corners of the beam (Figure 4.63b), then from the right-angled triangle

$$\left(\frac{w}{2}\right)^2 + \left(\frac{d}{2}\right)^2 = R^2 \qquad \Longrightarrow \qquad d^2 = 4R^2 - w^2.$$

Thus,

$$S = kw(4R^2 - w^2).$$

This function must be maximized on the interval $0 \leq w \leq 2R$. We begin by finding its critical point(s),

$$0 = \frac{dS}{dw} = k(4R^2 - 3w^2) \qquad \Longrightarrow \qquad w = \frac{2R}{\sqrt{3}}.$$

We now calculate

$$S(0) = 0, \qquad S\left(\frac{2R}{\sqrt{3}}\right) > 0, \qquad S(2R) = 0.$$

Consequently, the beam is strongest when it is $2R/\sqrt{3}$ centimetres wide and $\sqrt{4R^2 - 4R^2/3} = 2\sqrt{2}R/\sqrt{3}$ centimetres deep.

EXAMPLE 4.27

A wall of a building is to be braced by a square beam 10 cm by 10 cm that must pass over a parallel wall 5 m high and 2 m from the building. The beam is to sit flush against the building and on the floor (Figure 4.64). The top of the 5-m wall can be cut at any angle to accommodate flush contact with the beam, but the wall must not be shortened. Find the length of the shortest such beam.

SOLUTION The length of beam from A to B is

$$D = \|EF\| + \|AC\| + \|BG\| = x \sec\theta + \frac{1}{10}\cot\theta + \frac{1}{10}\tan\theta.$$

Since $x - 2 = 5\cot\theta$,

$$D = f(\theta) = (2 + 5\cot\theta)\sec\theta + \frac{1}{10}(\cot\theta + \tan\theta), \quad 0 < \theta < \frac{\pi}{2}.$$

The graph of this function in Figure 4.65 shows that the absolute minimum occurs at the critical point near $\theta = 1$. To find it we solve

$$0 = f'(\theta) = \sec\theta\,\tan\theta(2 + 5\cot\theta) + \sec\theta(-5\csc^2\theta) + \frac{1}{10}(-\csc^2\theta + \sec^2\theta)$$

$$= \frac{\sin\theta}{\cos^2\theta}\left(2 + \frac{5\cos\theta}{\sin\theta}\right) - \frac{5}{\sin^2\theta\,\cos\theta} + \frac{1}{10}\left(\frac{1}{\cos^2\theta} - \frac{1}{\sin^2\theta}\right).$$

Multiplication by $10\sin^2\theta\,\cos^2\theta$ gives

$$0 = 10\sin^3\theta\left(2 + \frac{5\cos\theta}{\sin\theta}\right) - 50\cos\theta + \sin^2\theta - \cos^2\theta$$

$$= 20\sin^3\theta + 50\cos\theta\,\sin^2\theta - 50\cos\theta + \sin^2\theta - (1 - \sin^2\theta)$$

$$= 20\sin^3\theta + 50\cos\theta(1 - \cos^2\theta) - 50\cos\theta + 2\sin^2\theta - 1$$

$$= 20\sin^3\theta - 50\cos^3\theta + 2\sin^2\theta - 1.$$

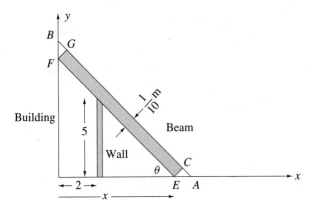

FIGURE 4.64 Beam passing over intervening wall

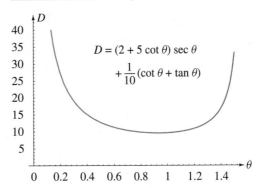

FIGURE 4.65 Length of beam

To solve this equation we use Newton's iterative procedure. Approximations are defined by

$$\theta_{n+1} = \theta_n - \frac{20\sin^3\theta_n - 50\cos^3\theta_n + 2\sin^2\theta_n - 1}{60\sin^2\theta_n\cos\theta_n + 150\cos^2\theta_n\sin\theta_n + 4\sin\theta_n\cos\theta_n}.$$

If we use $\theta_1 = 1$ as the first approximation, the next three approximations are $\theta_2 = 0.9278$, $\theta_3 = 0.9314$, and $\theta_4 = 0.9314$. Consequently, the shortest beam is $f(0.9314) = 9.8$ m.

EXAMPLE 4.28

Application Preview Revisited

In the Application Preview, we posed the problem of finding the offset that maximizes the stroke in the crank shown in Figure 4.66 if the offset must not exceed the radius of the circle.

SOLUTION   To find a formula for the length s of the stroke of the offset crank, we first draw the crank (Figure 4.67a) in what is called its *outer dead position*. Point C coincides with D and A, B, and C are collinear. The length of FD is $\sqrt{(L+r)^2 - e^2}$.

Figure 4.67b shows the crank in its *inner dead position*. The length of FE is $\sqrt{(L-r)^2 - e^2}$. Consequently, the length of the stroke is

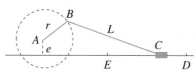

FIGURE 4.66   Maximizing stroke of offset crank

$$s = \sqrt{(L+r)^2 - e^2} - \sqrt{(L-r)^2 - e^2}.$$

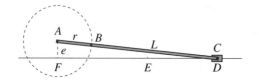

FIGURE 4.67a   Outer dead position of offset crank

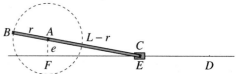

FIGURE 4.67b   Inner dead position of offset crank

We must maximize the function $s(e)$ on the interval $0 \le e \le r$. For critical points of the function, we solve

$$0 = s'(e) = \frac{-e}{\sqrt{(L+r)^2 - e^2}} - \frac{-e}{\sqrt{(L-r)^2 - e^2}}$$

$$= e\left(\frac{1}{\sqrt{(L-r)^2 - e^2}} - \frac{1}{\sqrt{(L+r)^2 - e^2}}\right).$$

This implies that $\sqrt{(L-r)^2 - e^2} = \sqrt{(L+r)^2 - e^2}$, which, upon squaring, gives

$$L^2 - 2Lr + r^2 - e^2 = L^2 + 2Lr + r^2 - e^2 \implies 4Lr = 0,$$

an impossibility. Thus, $s(e)$ has no critical points; its maximum must occur at either $e = 0$ or $e = r$. We therefore calculate

$$s(0) = (L+r) - (L-r) = 2r,$$
$$s(r) = \sqrt{(L+r)^2 - r^2} - \sqrt{(L-r)^2 - r^2} = \sqrt{L^2 + 2Lr} - \sqrt{L^2 - 2Lr}.$$

It is not apparent which of the two of these is larger, but it can be shown algebraically that $2r$ is the smaller (try it). Alternatively, we note that since $\sqrt{(L-r)^2 - e^2} < \sqrt{(L+r)^2 - e^2}$, it follows that

$$\frac{1}{\sqrt{(L-r)^2 - e^2}} > \frac{1}{\sqrt{(L+r)^2 - e^2}},$$

and therefore $s'(e) > 0$. The function $s(e)$ must therefore be increasing and its maximum value must be $s(r)$ when $e = r$.

Consulting Project 5

Figure 4.68 shows soil to depth D covering a horizontal layer of rock. A geological engineer sets up a source at A to emit a signal into the soil. The signal penetrates the soil to the rock below. Some of this signal is reflected back into the soil, some penetrates the rock, and some travels in the surface of the rock. Each part of the rock acts as an emitter sending the signal back into the soil, some of which reaches the receiver at B. This happens in many ways and the receiver records an accumulation of many signals. Our problem as consultants is to use the signals to determine depth D.

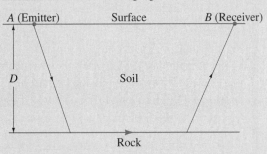

FIGURE 4.68   Sending signals into the earth and receiving them

SOLUTION   The first question to ask is: Which of the massive accumulation of interfering signals received at B from various points on the rock can be used? It seems that the only signal distinguishable from the rest is the one that arrives at B first. Let us concentrate on it and see what we can determine.

We let v_1 and v_2 represent how fast the signal travels in the soil and in the rock. It is known that $v_2 > v_1$, that v_1 is measurable, but v_2 is unknown. Next question: What is the path followed by the signal that arrives at B first? In Figure 4.69a, we have shown a signal following path $ACEB$, making acute angles θ and ϕ with the surface at emitter and receiver. This cannot be the fastest path. If E is moved to the right, the signal travels less distance in the soil where it is slow and further in the rock where it is fast. In other words, angle ϕ should be made larger. But it should not be made larger than θ, else we would make the same argument about θ. We conclude that the fastest path must occur when angles θ and ϕ are equal as shown in Figure 4.69b. The problem is now to determine angle θ for which travel time is smallest.

FIGURE 4.69a   Determination of depth of soil above bedrock

FIGURE 4.69b   Paths for signals emitted at A and received at B

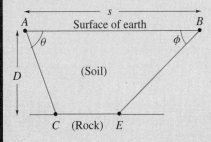

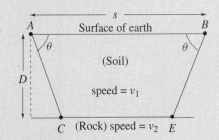

The time for the signal to travel from A to B in Figure 4.69b is

$$t = \frac{2D}{v_1} \csc \theta + \frac{s - 2D \cot \theta}{v_2}.$$

The smallest possible value for θ occurs when C and E become coincident, in which case $\theta = \text{Tan}^{-1}(2D/s)$. Thus, the domain of function $t(\theta)$ is $\text{Tan}^{-1}(2D/s) \leq \theta \leq \pi/2$. Critical points of $t(\theta)$ are given by

$$0 = \frac{dt}{d\theta} = -\frac{2D}{v_1}\csc\theta\,\cot\theta + \frac{2D}{v_2}\csc^2\theta$$

$$= \frac{-2D\cos\theta}{v_1\sin^2\theta} + \frac{2D}{v_2\sin^2\theta}$$

$$= \frac{-2D}{v_1 v_2 \sin^2\theta}(v_2\cos\theta - v_1).$$

Thus, the only critical point is $\theta = \text{Cos}^{-1}(v_1/v_2)$. It is physically clear that end points of the domain $\text{Tan}^{-1}(2D/s) \leq \theta \leq \pi/2$ cannot minimize $t(\theta)$. The critical point must do so. (It would be a good exercise for you to prove this mathematically.) When $\cos\theta = v_1/v_2$, it follows that $\sin\theta = \sqrt{1 - (v_1/v_2)^2}$, in which case minimum time is

$$\bar{t} = \frac{2D}{v_1}\frac{1}{\sqrt{1-(v_1/v_2)^2}} + \frac{1}{v_2}\left[s - \frac{2Dv_1/v_2}{\sqrt{1-(v_1/v_2)^2}}\right]$$

$$= \frac{2D}{v_1\sqrt{1-(v_1/v_2)^2}}\left[1 - \frac{v_1^2}{v_2^2}\right] + \frac{s}{v_2}$$

$$= \frac{2D}{v_1}\sqrt{1-\left(\frac{v_1}{v_2}\right)^2} + \frac{s}{v_2}.$$

Now that we have a formula for the time that the fastest signal takes to arrive at B, what do we do with it? Quantities s, v_1, and $\bar{t}$ in this formula are measurable, and therefore known. What are not known are D and v_2. If the receiver is moved to another location, however, a new distance and minimum time, say S and $\overline{T}$, are obtained. They satisfy

$$\overline{T} = \frac{2D}{v_1}\sqrt{1-\left(\frac{v_1}{v_2}\right)^2} + \frac{S}{v_2},$$

and these last two equations could be solved for D and v_2. To reduce error due to experimental measurements, we might suggest to the geological engineer that he take n measurements resulting in n pairs of values, say s_i and $\bar{t}_i$, and plot them on a graph of $\bar{t}$ against s (Figure 4.70). They look almost collinear, as they should. If we set

FIGURE 4.70   Determination of v_2 and D from line fitting experimental data

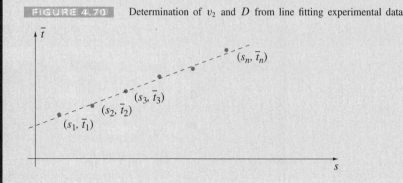

$b = (2D/v_1)\sqrt{1 - (v_1/v_2)^2}$ and $m = 1/v_2$, then

$$\bar{t} = ms + b,$$

the equation of a straight line. By measuring the slope and $\bar{t}$-intercept of the best-fitting line to the points, it is then possible to solve the equations

$$b = \frac{2D}{v_1}\sqrt{1 - \left(\frac{v_1}{v_2}\right)^2}, \qquad m = \frac{1}{v_2},$$

for v_2 and D. A mathematical way of finding the best-fitting line (instead of just eyeballing it) is discussed in Section 12.13.

EXERCISES 4.7

In Exercises 1–6 find absolute extrema, if they exist, for the function on the interval. Do so without plotting a graph of the function.

1. $f(x) = x^3 - x^2 - 5x + 4, \quad -2 \le x \le 3$

2. $f(x) = \dfrac{x-4}{x+1}, \quad 0 \le x \le 10$

3. $f(x) = x + \dfrac{1}{x}, \quad \dfrac{1}{2} \le x \le 5$

4. $f(x) = x - 2\sin x, \quad 0 \le x \le 4\pi$

5. $f(x) = x\sqrt{x+1}, \quad -1 \le x \le 1$

6. $f(x) = \dfrac{12}{x^2 + 2x + 2}, \quad x < 0$

In Exercises 7–9 find absolute extrema, if they exist, for the function on the interval. Use a plot of the graph of the function as an aid.

7. $f(x) = \dfrac{x+1}{x-1}, \quad x > 1$ 🖩 **8.** $f(x) = \dfrac{x}{x^2+3}, \quad x > 0$

9. $f(x) = \dfrac{(2x-1)(x-8)}{(x-1)(x-4)}, \quad x \le -2$

10. Use Theorem 4.2 to prove Theorem 3.17 in Section 3.14.

11. One end of a uniform beam of length L is built into a wall, and the other end is simply supported (figure below). If the beam has constant mass per unit length m, its deflection y from the horizontal at a distance x from the built-in end is given by

$$(48EI)y = -mg(2x^4 - 5Lx^3 + 3L^2x^2),$$

where E and I are constants depending on the material and cross-section of the beam, and $g > 0$ is a constant. How far from the built-in end does maximum deflection occur?

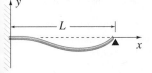

* **12.** An agronomist wishes to fence eight rectangular plots for experimentation as shown in the figure below. If each plot must contain 9000 m^2, find the minimum amount of fencing that can be used.

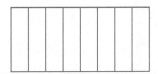

* **13.** A closed box is to have length equal to three times its width and total surface area of 30 m^2. Find the dimensions that produce maximum volume.

* **14.** A square box with open top is to have a volume of 6000 L. Find the dimensions of the box that minimize the amount of material used.

* **15.** A cherry orchard has 255 trees, each of which produces on the average 25 baskets of cherries. For each additional tree planted in the same area, the yield per tree decreases by one-twelfth of a basket. How many more trees will produce a maximum crop?

* **16.** A high school football field is being designed to accommodate a 400-m track. The track runs along the length of the field and has semicircles beyond the end zones with diameters equal to the width of the field. If the end zones must be 10 m deep, find the dimensions of the field that maximize playing area (not including the end zones).

* **17.** When a manufacturing company sells x objects per month, it sets the price r of each object at

$$r(x) = 100 - \frac{x^2}{10\,000}.$$

The total cost C of producing the x objects per month is

$$C(x) = \frac{x^2}{10} + 2x + 20.$$

Find the number of objects the company should sell per month in order to realize maximum profits.

* **18.** The base of an isosceles triangle, which is not one of the equal sides, has length b, and its altitude has length a. Find the area of the largest rectangle that can be placed inside the triangle if one of the sides of the rectangle must lie on the base of the triangle.

19. A manufacturer builds cylindrical metal cans that hold 1000 cm³. There is no waste involved in cutting material for the curved surface of the can. However, each circular end piece is cut from a square piece of metal, leaving four waste pieces. Find the height and radius of the can that uses the least amount of metal, including all waste materials.

20. Sides AB and AC of an isosceles triangle have equal length. The base BC has length $2a$, as does the altitude AD from A to BC. Find the height of a point P on AD at which the sum of the distances AP, BP, and CP is a minimum.

21. Two poles are driven into the ground 3 m apart. One pole protrudes 2 m above the ground and the other pole 1 m above the ground. A single piece of rope is attached to the top of one pole, passed through a loop on the ground, pulled taut, and attached to the top of the other pole. Where should the loop be placed in order that the rope be as short as possible?

22. In designing pages for a book, a publisher decides that the rectangular printed region on each page must have area 150 cm². If the page must have 2.5-cm margins on each side and 3.75-cm margins at top and bottom, find the dimensions of the page of smallest possible area.

23. Find the points on the hyperbola $y^2 - x^2 = 9$ closest to $(4, 0)$.

24. Find the point on the parabola $y = x^2$ closest to $(-2, 5)$.

25. A light source is to be placed directly above the centre of a circular area of radius r (figure below). The illumination at any point on the edge of the circle is directly proportional to the cosine of the angle θ and inversely proportional to the square of the distance d from the source. Find the height h above the circle at which illumination on the edge of the table is maximized.

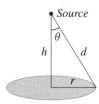

26. Among all line segments that stretch from points on the positive x-axis to points on the positive y-axis and pass through the point $(2, 5)$, find that one that makes with the positive x- and y-axes the triangle with least possible area.

27. A long piece of metal 1 m wide is to be bent in two places to form a spillway so that its cross-section is an isosceles trapezoid (figure below). Find the angle θ at which the bends should be formed in order to obtain maximum possible flow along the spillway if the lengths of AB, BC, and CD are all 1/3 m.

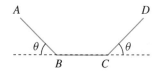

28. An automobile manufacturer has a Cobb–Douglas production function $q = x^{2/5}y^{3/5}$, where q is the number of automobiles it produces per year, x is the number of employees, and y is the daily operating budget. Annual operating costs amount to $20 000 per employee plus the operating budget of $365y$. If the manufacturer wishes to produce 1000 automobiles per year at minimum cost, how many employees should it hire?

29. The mass flow rate of gas through a nozzle is given by the function

$$Q(p) = A\sqrt{\frac{2\gamma}{\gamma - 1}p_0\rho_0\left[\left(\frac{p}{p_0}\right)^{2/\gamma} - \left(\frac{p}{p_0}\right)^{(\gamma+1)/\gamma}\right]},$$

$$0 \le p \le p_0,$$

where p is the discharge pressure of the gas from the nozzle. The discharge cross-sectional area of the nozzle A, the stagnation pressure p_0, the stagnation density ρ_0, and $\gamma > 1$ are all constants. The discharge pressure for which Q is a maximum is called the critical pressure p_c. Find p_c in terms of p_0 and γ.

30. A printer contracts to print 200 000 copies of a membership card. It costs $10 per hour to run the press, and the press produces 1000 impressions per hour. The printer is free to choose the number of set types per impression to a limit of 40; each set type costs $2. If x set types are chosen, each impression yields x cards. How many set types should be used?

31. At noon a ship S_1 is 20 km north of ship S_2. If S_1 sails south at 6 km/h, and S_2 east at 8 km/h, find when the two ships are closest together.

32. A military courier is located on a desert 6 km from a point P, which is the point on a long, straight road nearest him. He is ordered to report to a point Q on the road. If we assume that he can travel at a rate of 14 km/h on the desert and a rate of 50 km/h on the road, find the point where he should reach the road in order to get to Q in the least possible time when (a) Q is 3 km from P and (b) Q is 1 km from P.

33. Find the area of the largest rectangle that can be inscribed inside a circle of radius r.

34. Among all rectangles that can be inscribed inside the ellipse $b^2x^2 + a^2y^2 = a^2b^2$ and have sides parallel to the axes, find the one with largest area.

35. Among all rectangles that can be inscribed inside the ellipse $b^2x^2 + a^2y^2 = a^2b^2$ and have sides parallel to the axes, find the one with largest perimeter.

36. Find the area of the largest rectangle that has one side on the x-axis and two vertices on the curve $y = e^{-x^2}$.

37. Find the volume of the largest right circular cylinder that can be inscribed in a sphere of radius r.

38. Find the volume of the largest right circular cylinder that can be inscribed inside a right circular cone of radius r and height h.

39. Two beams are to be cut from the larger pieces of wood left over when the strongest beam is cut from the log in Example 4.26. What are the dimensions of the strongest such beams?

40. When a shotputter projects a shot from height h above the ground (figure below), at speed v, its range R is given by the formula

$$R = \frac{v^2 \cos \theta}{g} \left(\sin \theta + \sqrt{\sin^2 \theta + \frac{2gh}{v^2}} \right),$$

where θ is the angle of projection with the horizontal. Find the angle that maximizes R. What is the angle when $v = 13.7$ m/s and $h = 2.25$ m?

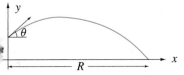

41. Two corridors, one 3 m wide and the other 6 m wide, meet at right angles. Find the length of the longest beam that can be transported horizontally around the corner. Ignore the dimensions of the beam.

42. Repeat Exercise 41 taking into account that the beam has square cross-section 1/3 m on each side.

43. A bee's cell is always constructed in the shape of a regular hexagonal cylinder open on one end and a trihedral apex at the other (figure below). It can be shown that the total area of the nine faces is given by

$$A = 6xy + \frac{3}{2}x^2(\sqrt{3}\csc\theta - \cot\theta).$$

Find the angle that minimizes A.

44. Find the absolute extrema of the function $f(x) = x/(x^2 + c)$ on the interval $0 \le x \le c$, if they exist. Treat $c > 1$ as a given constant.

45. (a) A rental company buys a new machine for p dollars, which it then rents to customers. If the company keeps the machine for t years (before replacing it), the average replacement cost per year for the t years is p/t. During these t years, the company must make repairs on the machine, the number n depending on t as given by $n(t) = t^\alpha/\beta$, where $\alpha > 1$ and $\beta > 0$ are constants. If r is the average cost per repair, then the average maintenance cost per year over the life of the machine is nr/t. The total yearly expense associated with the machine if it is kept for t years is therefore

$$C(t) = \frac{p}{t} + \frac{nr}{t}.$$

Find the optimum time at which to replace the machine.

(b) Discuss the cases where $\alpha = 1$ and $0 < \alpha < 1$.

* 46. The illumination at a point is inversely proportional to the square of the distance from the light source and directly proportional to the intensity of the source. If two light sources of intensities I_1 and I_2 are a distance d apart, at what point on the line segment joining the sources is the sum of their illuminations a minimum — relative to all other points on the line segment?

* 47. A window is in the form of a rectangle surmounted by a semicircle with diameter equal to the width of the window. If the rectangle is of clear glass while the semicircle is of coloured glass that transmits only half as much light per unit area as the clear glass, and if the total perimeter is fixed, find the proportions of the rectangular and semicircular part of the window that admit the most light.

* 48. A company wishes to construct a storage tank in the form of a rectangular parallelepiped with a square horizontal cross-section. The volume of the tank must be 100 m^3.

(a) If material for the sides and top costs $1.25/m^2, and material for the bottom costs $4.75/m^2, find the dimensions that minimize material costs.

(b) Repeat part (a) if the 12 edges must be welded at a cost of $7.50/m of weld.

* 49. When an unloaded die is thrown, there is a probability of 1/6 that it will come up "two." If the die is loaded, on the other hand, the probability that a "two" will appear is not 1/6, but is some number p between zero and one. To find p we could roll the die a large number of times, say n, and count the number of times "two" appears, say, m. It seems reasonable that an estimate for p is m/n. Mathematicians define a likelihood function, which for the present situation turns out to be

$$f(x) = \frac{n!}{m!\,(n-m)!}x^m(1-x)^{n-m}.$$

The value of x that maximizes $f(x)$ on the interval $0 \le x \le 1$ is called the *maximum likelihood estimate* of p. Show that this estimate is m/n.

* 50. An underground pipeline is to be constructed between two cities, A and B (figure below). An analysis of the substructure indicates that construction costs per kilometre in regions I ($y > 0$) and II ($y < 0$) are c_1 and c_2, respectively. Show that the total construction cost is minimized when x is chosen so that $c_1 \sin \theta_1 = c_2 \sin \theta_2$.

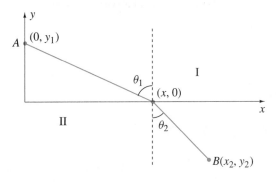

* 51. A submarine is sailing on the surface due east at a rate of s kilometres per hour. It is to pass 1 km north of a point of land on an island at midnight. Soldiers on the island wishing to escape the enemy plan to intercept the submarine by rowing a rubber raft in a straight-line course at a rate of v kilometres per hour ($v < s$). What is the last instant that they can leave the island and expect to make contact with the submarine?

* **52.** A packing company wishes to form the 1 m by 2 m piece of card-board in figure (a) below into a box as shown in figure (b). Cuts are to be made along solid lines and folds along dotted lines, and two sides are to be taped together as shown. If the outer flaps on top and bottom must meet in the centre but the inner flaps need not, find the dimensions of the box holding the most volume. How far apart will the inner flaps be?

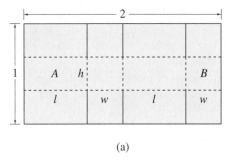

(a)

Bottom is same as top

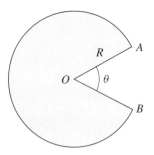

(b)

* **53.** The cost of fuel per hour for running a ship varies directly as the cube of the speed, and is B dollars per hour when the speed is b kilometres per hour. There are also fixed costs of A dollars per hour. Find the most economical speed at which to make a trip.

* **54.** A paper drinking cup in the form of a right circular cone can be made from a circular piece of paper by removing a sector and joining edges OA and OB, as shown in the figure below. If the radius of the circle is R, what choice of θ yields a cup of maximum volume?

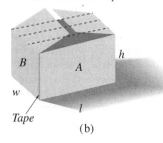

* **55.** The road information sign in the following figure specifies distance to the next city with markings 1 m high. Bottoms of markings are 5 m above road level. If the average motorist's eye level is 1.5 m above ground level, at what distance from the sign do the letters appear tallest?

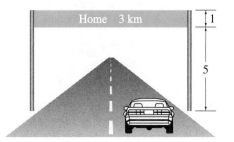

* **56.** A Boeing 727-200 jet transport has wing platform area $A = 1600$ ft^2 and gross weight $w = 150\,000$ lbs of force. At cruising speed v (in miles per hour), the thrust F of the engines is given in pounds force by

$$F = \frac{1}{2}\rho A v^2 \left(0.000182 + \frac{4w^2}{6.5\pi\rho^2 A^2 v^4} \right),$$

where ρ is the density of the atmosphere. Optimum speed occurs when the ratio of thrust to speed is a minimum.

   (a) Find the optimum speed of the jet at sea level where $\rho = 0.0238$ slugs per cubic foot.

   (b) Find the optimum speed at 30 000 ft when density of the atmosphere is about 0.375 that at sea level.

* **57.** When a force F is applied to the object of mass m in the figure below, three other forces act on m: the force of gravity mg ($g > 0$) directly downward, a reactional force of the supporting surface, and a horizontal, frictional force opposing F. The least force F that will overcome friction and produce motion is given by

$$F = \frac{\mu m g}{\cos\theta + \mu\sin\theta},$$

where μ is a constant called the *coefficient of static friction*. Find the angle θ for which F is minimal.

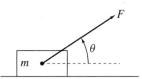

* **58.** Repeat Exercise 20 if the length of the base of the triangle is $2b$ instead of $2a$ (but the altitude is still $2a$).

* **59.** A trucking company wants to determine the highway speed to recommend to its drivers in order that company costs are kept to a minimum. Taken into account will be only hourly wage w (in dollars and assumed constant) of drivers, and gas consumption. The company has data to support the hypothesis that for speeds v between 80 km/h and 100 km/h, the number of kilometres per litre used by the trucks is a linear function $f(v) = a - bv$ (where $b > 0$). Find a formula for the recommended speed in terms of a, b, and w, and p, the price per litre for gas.

* **60.** Verify formula 1.16 by minimizing the distance function from a point to a line.

1. A window is in the form of a rectangle surmounted by a semicircle with diameter equal to the width of the rectangle. The rectangle is of clear glass costing a dollars per unit area, while the semicircle is of coloured glass costing b dollars per unit area. The coloured glass transmits only a fraction p ($0 < p < 1$) as much light per unit area as the clear glass. In addition, the curved portion of the window is surmounted by a special frame at a cost of c dollars per unit length. If the total cost of the window must not exceed A dollars, find the dimensions of the window that admit the most light.

2. Find the point on the ellipse $4x^2 + 9y^2 = 36$ closest to $(4, 13\sqrt{5}/6)$.

3. The frame for a kite is to be made from six pieces of wood as shown in the following figure. The four outside pieces have predetermined lengths a and b ($b > a$). Yet to be cut are the two diagonal pieces. How long should they be in order to make the area of the kite as large as possible?

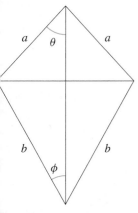

4. Suppose that the line $Ax + By + C = 0$ intersects the parabola $y = ax^2 + bx + c$ at points P and Q (figure below). Find the point R on the parabola between P and Q that maximizes the area of triangle PQR.

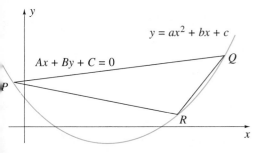

5. Tangent lines to the curve $y = \sqrt{1-x}$ make triangles with the positive x- and y-axes. Find the area of the smallest such triangle.

6. A farmer has a square plot of unirrigated land s metres by s metres. His profit is a dollars per square metre. He wishes to install an irrigation system that consists of a rotating arm that pivots at the centre of the field. The cost for irrigation is c dollars per square metre. Profit on irrigated land is b dollars per square metre. What should be the length of the rotating arm for maximum profit? Assume that $b > a + c$, otherwise it makes no sense to irrigate.

7. Four rods of lengths 1 m, 4 m, 2 m, and 3 m are hinged together as shown below to form a quadrilateral. Find the area of the largest quadrilateral that can be so formed.

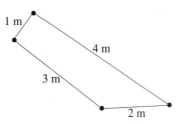

** **68.** The lake in the figure below is basically circular with radius r. It has a narrow strip of beach all round and beyond the beach is bush. You are walking from point P to point Q, both of which are on the extension of a diameter of the lake each distance s from the centre of the lake. You can walk twice as fast on the beach as in the bush. Design your travel path to get from P to Q as quickly as possible.

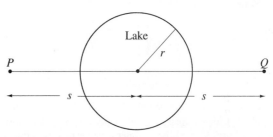

** **69.** A rectangle has width w and length L. Find the area of the largest of all rectangles that have sides passing through the corners of the given rectangle.

** **70.** There are n red stakes securely driven into the ground in a straight line. A blue stake is to be added to the line, and each red stake is to be joined to the blue one by a string. Where should the blue stake be placed in order that the total length of all strings be as small as possible?

** **71.** When blood flows through a vein or artery, it encounters resistance due to friction with the walls of the blood vessel and the viscosity of the blood itself. *Poiseuille's law* for laminar blood flow states that for a circular vessel, resistance R to blood flow is proportional to the length L of the blood vessel and inversely proportional to the fourth power of its radius r:

$$R = k\frac{L}{r^4},$$

where k is a constant. The figure below shows a blood vessel of radius r_1 from A to B and a branching vessel from D to C of radius $r_2 < r_1$.

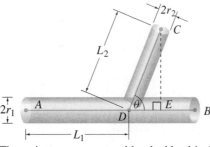

The resistance encountered by the blood in flowing from A to C is given by

$$R = k\frac{L_1}{r_1^4} + k\frac{L_2}{r_2^4},$$

where L_1 and L_2 are the lengths of AD and DC, respectively.

(a) If B is assumed to the right of E, show that R can be expressed in terms of θ as

$$R = f(\theta) = \frac{k}{r_1^4}(X - Y \cot \theta) + \frac{k}{r_2^4} Y \csc \theta,$$

where X and Y are the lengths of AE and CE, respectively.

(b) Show that $f(\theta)$ has only one critical point $\bar{\theta}$ in the range $0 < \theta < \pi/2$, and $\bar{\theta}$ is defined by

$$\cos \bar{\theta} = \frac{r_2^4}{r_1^4}.$$

(c) Verify that $\bar{\theta}$ yields a relative minimum for $f(\theta)$.

(d) Show that

$$f(\bar{\theta}) = \frac{kX}{r_1^4} + \frac{kY}{r_2^4}\sqrt{1 - \left(\frac{r_2}{r_1}\right)^8} < f(\pi/2).$$

(e) Does $\bar{\theta}$ provide an absolute minimum for $f(\theta)$?

** **72.** A right circular cone has radius R and height H. A right circular cylinder is inscribed inside the cone so that its upper edge is on the cone (figure below). Find the radius of the cylinder in order that its surface area (including top, bottom, and side) be as large as possible.

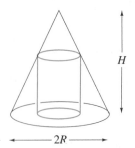

** **73.** The triangle in the following figure represents cross-sections of a prism (the length of the prism being perpendicular to the page). A ray of light (in the plane of the page) is incident on the prism at angle i relative to the normal to the prism. The light is refracted at the two faces of the prism and leaves at angle of deviation ψ relative to the incident direction. *Snell's law* relates angles of incident and refracted light at each of the faces. It states that

$$n \sin \alpha = \sin i \quad \text{and} \quad n \sin \beta = \sin \phi,$$

where n is the index of refraction of the material of the prism.

(a) Find ψ as an explicit function of i.

(b) Show that the angle of deviation ψ is a minimum when $i = \phi$. Prove that if ψ_m is the minimum angle of deviation, then

$$n = \frac{\sin[(\psi_m + \gamma)/2]}{\sin(\gamma/2)}.$$

(c) This equation can be used to determine n experimentally. Angle of incidence i is varied until ψ_m is achieved and measured. With γ also known, n can be calculated.

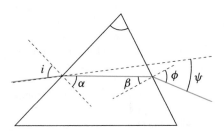

** **74.** A rope with a ring at one end is passed through two fixed rings at the same level (figure below). The end of the rope without the ring is then passed through the ring at the other end, and a mass m is attached to it. If the rope moves so as to maximize the distance from m to the line through the fixed rings, find angle θ.

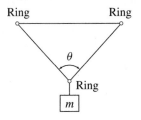

** **75.** If the fencing in Example 4.22 is to form the arc of a circle, what is the maximum possible area?

** **76.** The upper left corner of a piece of paper a units wide and b units long $(b > a)$ is folded to the right edge as shown in the following figure. Calculate length x in order that the length y of the fold be a minimum.

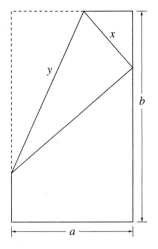

** **77.** The line $y = mx + c$ intersects the ellipse $b^2 x^2 + a^2 y^2 = a^2 b^2$ in two points P and Q. Find the point R on the ellipse in order that triangle PQR has maximum possible area.

8. A sheet of metal a metres wide and L metres long is to be bent into a trough as shown in the following figure. End pieces are also to be attached. If edge AB must be the arc of a circle, determine the radius of the circle in order that the trough hold the biggest possible volume.

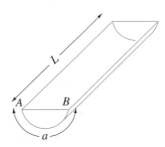

4.8 Velocity and Acceleration

Kinematics is the study of motion — relationships among the displacement, velocity, and acceleration of a body, particularly as they pertain to forces acting on the body. In this section we bring together for a final discussion everything we have learned about displacement, velocity, and acceleration, but do so mainly in a one-dimensional setting, and only from a differentiation point of view; given the position of a particle moving along a straight line, find its velocity and acceleration and use these quantities to describe the motion of the particle. In Section 5.2 we reverse these operations; beginning with the acceleration, we find velocity and position. This process is essential to physics and engineering, where acceleration of a particle is determined by the forces acting on it.

When the position of a particle moving along the x-axis is known as a function of time t, say, $x(t)$, its **instantaneous velocity** is the derivative of $x(t)$ with respect to t,

$$v(t) = \frac{dx}{dt}. \tag{4.9}$$

The **instantaneous acceleration** of the particle is defined as the rate of change of velocity with respect to time:

$$a(t) = \frac{dv}{dt} = \frac{d^2x}{dt^2}. \tag{4.10}$$

FIGURE 4.71   Displacement function

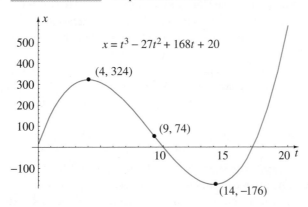

In Section 3.6 we discussed the motion of a particle moving along the x-axis with position function

$$x(t) = t^3 - 27t^2 + 168t + 20, \quad t \geq 0,$$

where x is measured in metres and t in seconds. We now add a final touch to the discussion. The best way to describe the motion of the particle is with a graph of the displacement function (Figure 4.71). With calculus we can show how geometric properties of the graph reflect important features about the velocity and acceleration of the particle. The velocity and acceleration of the particle are

$$v(t) = \frac{dx}{dt} = 3t^2 - 54t + 168 = 3(t-4)(t-14) \text{ m/s},$$

$$a(t) = \frac{d^2x}{dt^2} = 6t - 54 = 6(t-9) \text{ m/s}^2.$$

Ignoring the physical interpretations of dx/dt and d^2x/dt^2 as velocity and acceleration for the moment, and concentrating only on the fact that they are the first and second derivatives of the function $x(t)$, we immediately find that $x(t)$ has a relative maximum of $x(4) = 324$ and a relative minimum of $x(14) = -176$ at the critical points $t = 4$ and $t = 14$. The graph has a point of inflection at $(9, 74)$.

Let us now discuss what the graph tells us about the motion of the particle. Ordinates represent horizontal distances of the particle from the origin $x = 0$. When an ordinate is positive, the particle is that distance to the right of the origin; when an ordinate is negative, the particle is that distance to the left of the origin. For instance, at time $t = 0$, we calculate $x = 20$, and therefore the particle begins 20 m to the right of $x = 0$. At time $t = 4$, it is 324 m to the right of $x = 0$, and at $t = 14$, it is 176 m to the left of the origin.

The slope of the graph represents velocity of the particle. When slope is positive, namely in the intervals $0 < t < 4$ and $t > 14$, the particle is moving to the right along the x-axis, for $4 < t < 14$, velocity is negative, indicating that the particle is moving to the left. At times $t = 4$ and $t = 14$, the particle is instantaneously at rest.

The concavity of the graph reflects the sign of the acceleration. For $0 < t < 9$, the graph is concave downward, its slope is decreasing. Physically, this means that it has a negative acceleration; that is, its velocity is decreasing. For $t > 9$, the graph is concave upward, its slope is increasing. Physically, acceleration is positive; that is, velocity is increasing. At the point of inflection $(9, 74)$, the acceleration changes sign. Notice that the acceleration is not zero at $t = 4$ and $t = 14$. In spite of the fact that the velocity is zero at these times, the acceleration does not vanish. You might ask yourself what feature of the graph would reflect coincident zeros for velocity and acceleration (see Exercise 16).

 Speed is the magnitude of velocity. It represents how fast the particle is moving without regard for direction. For instance, at time $t = 0$, the velocity and speed are both 168 m/s, whereas at time $t = 10$, velocity is -72 m/s and speed is 72 m/s. Geometrically, speed is represented by the slope of the graph without regard for sign.

With the ideas above in mind, let us detail the history of the particle's motion. At time $t = 0$, it begins 20 m to the right of the origin, moving to the right with velocity 168 m/s. Since the acceleration is negative (to the left), the particle is slowing down (both velocity and speed are decreasing), until at time $t = 4$ s, it comes to an instantaneous stop 324 m to the right of the origin. Because the acceleration continues to be negative, the particle moves to the left, its velocity decreasing, but its speed increasing. At time $t = 9$ s, when the particle is 74 m to the right of the origin, the acceleration changes sign. At this instant, the velocity has attained a (relative) minimum value, but speed is a (relative) maximum. With acceleration to the right for $t > 9$ s, the particle continues to move left, but slows down, its velocity increasing until at time $t = 14$ s, when it once again comes to a stop 176 m to the left of the origin. For time $t > 14$ s, it moves to the right, picking up speed, and passes through the origin just before $t = 17$ s.

Further analysis of the interdependences of displacement, velocity, and acceleration is contained in the following example.

EXAMPLE 4.29

The position of a particle moving along the x-axis is given by the function

$$x(t) = 3t^4 - 32t^3 + 114t^2 - 144t + 40, \quad 0 \le t \le 5,$$

where x is measured in metres and t in seconds. Answer the following questions concerning its motion:

(a) What are its velocity and speed at $t = 1/2$ s?

(b) When is its acceleration increasing?

(c) Is the velocity increasing or decreasing at $t = 2$ s?

(d) Is the particle speeding up or slowing down at $t = 2$ s?

(e) What is the maximum velocity of the particle in the time interval $0 \leq t \leq 2$?

(f) What is the maximum distance the particle ever attains from the origin?

SOLUTION   We use the graph of the displacement function in Figure 4.72 to suggest answers when possible, and calculus to confirm them, when necessary. The velocity and acceleration are

$$v(t) = 12t^3 - 96t^2 + 228t - 144 = 12(t-1)(t-3)(t-4) \text{ m/s},$$

$$a(t) = 12(3t^2 - 16t + 19) \text{ m/s}^2.$$

FIGURE 4.72   Displacement function

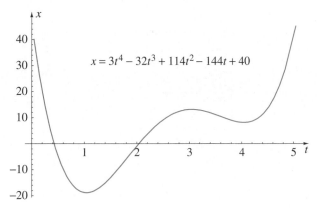

$x = 3t^4 - 32t^3 + 114t^2 - 144t + 40$

(a) The velocity is $v(1/2) = -105/2$ m/s. The speed is $105/2$ m/s.

(b) The graph does not tell us when acceleration is increasing. Algebraically, acceleration is increasing when its derivative is nonnegative. Since $da/dt = 12(6t - 16) = 24(3t - 8)$, acceleration is increasing for $8/3 \leq t \leq 5$.

(c) The graph appears to be concave downward at $t = 2$, implying that velocity is decreasing. Confirmation is provided by the fact that $a(2) = -12$ m/s^2.

(d) The graph seems to be becoming less steep around $t = 2$, indicating that the particle is slowing down. Since $v(2) = 24$ m/s, the particle is moving to the right. But according to part (c), its acceleration is to the left at this time. With velocity and acceleration in opposite directions, the particle is slowing down.

(e) Maximum velocity on $0 \leq t \leq 2$ occurs at the point of inflection to the left of $t = 2$. To find it, we set

$$0 = a(t) = 12(3t^2 - 16t + 19).$$

Of the two solutions $(8 \pm \sqrt{7})/3$, only $(8 - \sqrt{7})/3$ is in the interval $0 \leq t \leq 2$. Maximum velocity is therefore $v\big((8 - \sqrt{7})/3\big) = 25.35$ m/s.

(f) Maximum distance from the origin is represented by the point on the graph farthest from the t-axis. This is at $t = 5$ s, for which $x(5) = 45$ m.

EXAMPLE 4.30

The mass M in Figure 4.73 is pulled x_0 metres to the right of its position where the spring is unstretched, and then it is released. If at time $t = t_0$, during the subsequent oscillations, it is struck with a force of F newtons to the right, its position $x(t)$ must satisfy

$$M\frac{d^2x}{dt^2} + kx = F\delta(t - t_0), \quad x(0) = x_0, \quad x'(0) = 0,$$

where $k > 0$ is the spring constant, and $\delta(t - t_0)$ is the Dirac-delta function of Section 2.5. This assumes that air resistance due to motion is negligible, as is friction with the surface on which oscillations take place.

(a) Verify that

$$x(t) = x_0 \cos\sqrt{\frac{k}{M}}\,t + \frac{F}{\sqrt{kM}} \sin\sqrt{\frac{k}{M}}(t - t_0)\,h(t - t_0),$$

where $h(t - t_0)$ is the Heaviside unit step function, satisfies these conditions at every $t \neq t_0$.

(b) Show that the velocity of the mass changes by F/M metres per second as a result of being struck at time t_0.

FIGURE 4.73   Mass vibrating on end of spring

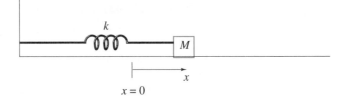

SOLUTION

(a) Since the derivative of $h(t - t_0) = 0$ for any $t \neq t_0$, we find that for $t \neq t_0$,

$$x'(t) = -\sqrt{\frac{k}{M}}x_0 \sin\sqrt{\frac{k}{M}}\,t + \frac{F}{M}\cos\sqrt{\frac{k}{M}}(t - t_0)\,h(t - t_0)$$

and

$$x''(t) = -\frac{kx_0}{M}\cos\sqrt{\frac{k}{M}}\,t - \frac{F\sqrt{k}}{M^{3/2}}\sin\sqrt{\frac{k}{M}}(t - t_0)\,h(t - t_0).$$

Thus, for $t \neq t_0$,

$$M\frac{d^2x}{dt^2} + kx = M\left[-\frac{kx_0}{M}\cos\sqrt{\frac{k}{M}}\,t - \frac{F\sqrt{k}}{M^{3/2}}\sin\sqrt{\frac{k}{M}}(t - t_0)\,h(t - t_0)\right]$$

$$+ k\left[x_0 \cos\sqrt{\frac{k}{M}}\,t + \frac{F}{\sqrt{kM}}\sin\sqrt{\frac{k}{M}}(t - t_0)\,h(t - t_0)\right]$$

$$= 0;$$

that is, $x(t)$ satisfies the differential equation $Mx'' + kx = F\delta(t - t_0)$ for $t \neq t_0$. In addition, $x(0) = x_0$, and

$$x'(0) = \left[-\sqrt{\frac{k}{M}}x_0 \sin\sqrt{\frac{k}{M}}\,t + \frac{F}{M}\cos\sqrt{\frac{k}{M}}(t - t_0)\,h(t - t_0)\right]_{t=0} = 0.$$

(b) The change in velocity at $t = t_0$ is

$$\lim_{t \to t_0^+} x'(t) - \lim_{t \to t_0^-} x'(t) = \lim_{t \to t_0^+} \left[-\sqrt{\frac{k}{M}} x_0 \sin \sqrt{\frac{k}{M}} t + \frac{F}{M} \cos \sqrt{\frac{k}{M}}(t - t_0) \right]$$

$$- \lim_{t \to t_0^-} \left[-\sqrt{\frac{k}{M}} x_0 \sin \sqrt{\frac{k}{M}} t \right]$$

$$= -\sqrt{\frac{k}{M}} x_0 \sin \sqrt{\frac{k}{M}} t_0 + \frac{F}{M} + \sqrt{\frac{k}{M}} x_0 \sin \sqrt{\frac{k}{M}} t_0$$

$$= \frac{F}{M}.$$

EXAMPLE 4.31

Rod AB in the offset slider-crank of Figure 4.74 rotates counterclockwise with constant angular speed ω about A. End C of the follower BC is confined to straight-line motion along a horizontal line between D and E. Find expressions for the velocity and acceleration of slider C.

FIGURE 4.74   Offset slider crank

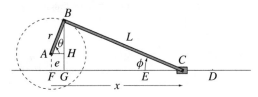

SOLUTION   Suppose we let x be the distance from F to C. Then,

$$x = \|FG\| + \|GC\| = \|AH\| + \|GC\| = r \cos \theta + L \cos \phi.$$

Now angles θ and ϕ are not independent; they are related by the offset equation

$$e = \|BG\| - \|BH\| = L \sin \phi - r \sin \theta.$$

Although these equations have been developed on the basis of Figure 4.74, which shows θ as an acute angle, it can be shown that they are valid for any value of θ whatsoever. We could use the offset equation to express x completely in terms of θ (see Exercise 24), but it is simpler to work with both θ and ϕ. Differentiation of the expression for x with respect to time t gives the velocity of C,

$$v = \frac{dx}{dt} = -r \sin \theta \frac{d\theta}{dt} - L \sin \phi \frac{d\phi}{dt} = -\omega r \sin \theta - L \sin \phi \frac{d\phi}{dt}.$$

Differentiation of the offset equation relates $d\phi/dt$ and ω,

$$0 = L \cos \phi \frac{d\phi}{dt} - r \cos \theta \frac{d\theta}{dt} = L \cos \phi \frac{d\phi}{dt} - \omega r \cos \theta \implies \frac{d\phi}{dt} = \frac{\omega r \cos \theta}{L \cos \phi}.$$

Thus,

$$v = -\omega r \sin\theta - L\sin\phi\left(\frac{\omega r \cos\theta}{L\cos\phi}\right)$$

$$= -\omega r\left(\frac{\sin\theta\cos\phi + \cos\theta\sin\phi}{\cos\phi}\right) = \frac{-\omega r \sin(\theta + \phi)}{\cos\phi}.$$

A second differentiation gives the acceleration of the slider,

$$a = \frac{dv}{dt} = -\omega r\left[\frac{\cos\phi\cos(\theta + \phi)\left(\dfrac{d\theta}{dt} + \dfrac{d\phi}{dt}\right) - \sin(\theta + \phi)(-\sin\phi)\dfrac{d\phi}{dt}}{\cos^2\phi}\right]$$

$$= \frac{-\omega r}{\cos^2\phi}\left[\cos\phi\cos(\theta + \phi)\left(\omega + \frac{\omega r \cos\theta}{L\cos\phi}\right) + \sin(\theta + \phi)\sin\phi\left(\frac{\omega r \cos\theta}{L\cos\phi}\right)\right]$$

$$= \frac{-\omega^2 r}{L\cos^3\phi}\left[\cos\phi\cos(\theta + \phi)(L\cos\phi + r\cos\theta) + r\sin(\theta + \phi)\sin\phi\cos\theta\right]$$

$$= \frac{-\omega^2 r}{L\cos^3\phi}\left[L\cos^2\phi\cos(\theta + \phi) + r\cos^2\theta\right].$$

EXERCISES 4.8

In Exercises 1–10 find the velocity and acceleration of an object that moves along the x-axis with the given position function. In each exercise, discuss the motion, including in your discussion a graph of the function $x(t)$. Assume that x is measured in metres and t in seconds.

* **1.** $x(t) = 2t + 5$, $t \geq 5$

* **2.** $x(t) = t^2 - 7t + 6$, $t \geq 0$

* **3.** $x(t) = t^2 + 5t + 10$, $t \geq 1$

* **4.** $x(t) = -2t^3 + 2t^2 + 16t - 1$, $t \geq 0$

* **5.** $x(t) = t^3 - 9t^2 + 15t + 3$, $t \geq 2$

* **6.** $x(t) = 3\cos 4t$, $t \geq 0$

* **7.** $x(t) = 1/t$, $t \geq 1$

* **8.** $x(t) = t + 4/t$, $t \geq 1$

* **9.** $x(t) = (t - 4)/t^2$, $t \geq 2$

* **10.** $x(t) = (t - 1)^2\sqrt{t}$, $t \geq 1$

* **11.** An object moving along the x-axis has position given by

$$x(t) = t^3 - 9t^2 + 24t + 1, \quad 0 \leq t \leq 6,$$

where x is measured in metres and $t \geq 0$ is time in seconds. Determine (a) whether speed is increasing or decreasing at $t = 1$ s, (b) maximum and minimum velocity, (c) maximum and minimum speed, (d) maximum and minimum acceleration, and (e) when acceleration is increasing.

* **12.** Repeat Exercise 11 if

$$x(t) = 2 - 15t + 9t^2 - t^3, \quad 0 \leq t \leq 6.$$

* **13.** A particle moves along the x-axis in such a way that its position as a function of time t is given by

$$x(t) = t\sin t, \quad 0 \leq t \leq 9.$$

(a) Plot a graph of $x(t)$.

(b) Use Newton's iterative procedure to find when velocity is zero.

(c) Use Newton's iterative procedure to find when acceleration is equal to 1.

* **14.** An object moving along the x-axis has position function given by

$$x(t) = 3t^4 - 16t^3 + 18t^2 + 2, \quad 0 \leq t \leq 4,$$

where x is measured in metres and t in seconds. Determine (a) on what intervals the velocity is increasing and decreasing, (b) on what intervals the speed is increasing and decreasing, (c) when the velocity is a maximum and a minimum, (d) when the speed is a maximum and a minimum, (e) the maximum distance from the origin achieved by the particle, and (f) the maximum distance from the point $x = 5$ achieved by the particle.

* **15.** Repeat Exercise 14 for the position function

$$x(t) = \frac{14t^3}{45} - \frac{101t^2}{45} + \frac{132t}{45} + 2, \quad 0 \leq t \leq 6.$$

* **16.** What feature on the displacement graph would indicate a time when the velocity and acceleration are simultaneously zero?

17. Are critical points for the velocity function the same as those for the speed function? Explain, using graphs.

18. What is the maximum speed of the particle in Example 4.29 over the time interval $0 \leq t \leq 2$?

19. In many velocity and acceleration problems it is more convenient to express acceleration in terms of a derivative with respect to position, as opposed to a derivative with respect to time. Show that acceleration can be written in the form

$$a = v\frac{dv}{dx}.$$

In Exercises 20–23 assume that a particle moves along the x-axis in such a way that its position, velocity, and acceleration are continuous functions on the interval $a \leq t \leq b$. Discuss the validity of each statement.

20. When position has a relative maximum, so does the absolute value of the distance from the origin to the particle.

21. When position has a relative minimum, so does the absolute value of the distance from the origin to the particle.

22. When velocity has a relative minimum, so does speed.

23. When velocity has a relative maximum, so does speed.

24. (a) Use the offset equation $e = L \sin \phi - r \sin \theta$ to show that position x of the slider in Example 4.31 can be expressed in the form

$$x = r \cos \theta + \sqrt{L^2 - (e + r \sin \theta)^2}.$$

(b) Plot this function for $0 \leq \theta \leq 2\pi$ when $L = 9$ cm, $r = 2$ cm, and $e = 1$ cm.

(c) Estimate the length of the stroke from the graph in part (b). Check this against the formula in Example 4.28.

(d) Assuming that AB rotates with constant angular speed ω, differentiate the function in part (a) to find the velocity of the slider in terms of θ. Verify that your result is consistent with the velocity formula in Example 4.31.

(e) Plot the velocity function in part (d) for $0 \leq \theta \leq 2\pi$ using the values of r, L, and e in part (b) if AB rotates one revolution each second. Does the velocity appear to be zero when the position graph in part (b) is at its highest and lowest points? Estimate maximum and minimum velocities from the graph.

** **25.** A landing approach is to be shaped generally as shown in the figure below. The following conditions are imposed on the approach pattern:

(a) Altitude must be h metres when descent commences.

(b) Smooth touchdown must occur at $x = 0$.

(c) Constant horizontal speed U metres per second must be maintained throughout.

(d) At no time must vertical acceleration in absolute value exceed some fixed positive constant k.

(e) The glide path must be a cubic polynomial.

Find when descent should commence.

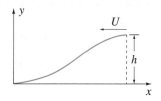

4.9 Related Rates

Many interesting and practical problems involving rates of change are commonly referred to as **related rate problems**. In these problems, two or more quantities are related to each other and rates at which some of them change are known. It is required to find rates at which the others change. Related rate problems deal almost exclusively with rates of change of quantities with respect to time. To solve these problems we first consider three examples. These will suggest the general procedures by which all related rate problems can be analyzed. We shall then discuss two somewhat more complicated problems.

EXAMPLE 4.32

FIGURE 4.75 Shadow of man walking away from lightpost

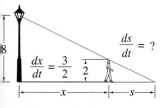

A man 2 m tall walks directly away from a streetlight that is 8 m high at the rate of $3/2$ m/s. How fast is the length of his shadow changing?

SOLUTION When x denotes the distance between the man and the lightpost (Figure 4.75), the fact that he walks directly away from the light at $3/2$ m/s means that x is changing at a rate of $3/2$ m/s; that is, $dx/dt = 3/2$ m/s. If s represents the length of the man's shadow, then we are searching for ds/dt. Similar triangles in Figure 4.75 enable us to relate s and x,

$$\frac{x + s}{s} = \frac{8}{2},$$

and this equation can be solved for s in terms of x:

$$s = \frac{x}{3}.$$

Now, s and x are each functions of time t,

$$s = f(t) \qquad \text{and} \qquad x = g(t),$$

although we have not calculated the exact form of these functions. Indeed, the essence of the related rate problem is to find ds/dt without ever knowing $f(t)$ explicitly. To do this we note that since the equation $s = x/3$ is valid at any time t when the man is walking away from the light, we may differentiate with respect to t to obtain

$$\frac{ds}{dt} = \frac{1}{3}\frac{dx}{dt}.$$

This equation relates the known rate $dx/dt = 3/2$ with the unknown rate ds/dt. It follow that

$$\frac{ds}{dt} = \left(\frac{1}{3}\right)\left(\frac{3}{2}\right) = \frac{1}{2},$$

and the man's shadow is therefore getting longer at the rate of $1/2$ m/s.

Knowing dx/dt in this example, we have calculated ds/dt, and have done so without finding s explicitly as a function of time t. This is the essence of a related rate problem. Since dx/d is a constant value, it is quite easy to find s as a function of t, and hence ds/dt. Indeed, i we choose time $t = 0$ when the man starts to walk away from the streetlight, then his distance from the light at any given time is $x = 3t/2$ m. Combine this with the fact that $s = x/3$ and we may write

$$s = \frac{1}{3}\left(\frac{3t}{2}\right) = \frac{t}{2}\text{ m.}$$

With this explicit formula for s, it is clear that $ds/dt = 1/2$. What is important to realiz is that the solution in this paragraph is possible only because the man walks at a constant rate Were his speed not constant, it might be impossible to find s explicitly in terms of t. The nex example illustrates this point in that the given rate is known only at one instant in time.

EXAMPLE 4.33

A ladder leaning against a house (Figure 4.76) is prevented from moving by a young child Suddenly, something distracts the child and she releases the ladder. The ladder begins slipping down the wall of the house, picking up speed as it falls. If the top end of the ladder is moving at 1 m/s when the lower end is 15 m from the house, how fast is the foot of the ladder moving away from the house at this instant?

FIGURE 4.76   Ladder sliding down a wall

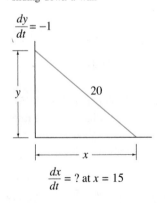

$$\frac{dy}{dt} = -1$$

20

y

x

$$\frac{dx}{dt} = ? \text{ at } x = 15$$

SOLUTION   Figure 4.76 indicates that when y denotes the height of the top of the ladde above the ground, then $dy/dt = -1$ m/s when $x = 15$ m (the negative sign because y i decreasing). We emphasize here that $dy/dt = -1$ *only* when $x = 15$. What is required i dx/dt when $x = 15$ m. Because the triangle in the figure is right-angled, we may write

$$x^2 + y^2 = 20^2,$$

and this equation is valid at any time during which the ladder is slipping. If we differentiat with respect to time t, using extended power rule 3.21,

$$2x\frac{dx}{dt} + 2y\frac{dy}{dt} = 0.$$

When $x = 15$, we calculate that $y = \sqrt{400 - 225} = 5\sqrt{7}$, and therefore at this instant

$$15\frac{dx}{dt} + 5\sqrt{7}(-1) = 0.$$

This yields $dx/dt = \sqrt{7}/3$, and we can say that when the foot of the ladder is 15 m from the wall, it is moving away from the wall at $\sqrt{7}/3$ m/s.

EXAMPLE 4.34

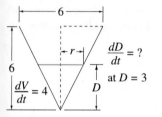

FIGURE 4.77 Tank being filled with water

A tank in the form of a right circular cone with altitude 6 m and base radius 3 m (Figure 4.77) is being filled with water at a rate of 4000 L/min. How fast is the surface of the water rising when the depth is 3 m?

SOLUTION   Figure 4.77, which illustrates a cross-section of the tank, indicates that when the depth of water in the tank is D, the volume V of water is

$$V = \frac{1}{3}\pi r^2 D.$$

Of the three variables V, r, and D in this equation, we are concerned only with V and D, since dV/dt is given and dD/dt is what we want. This suggests that we eliminate r using similar triangles. Since $r/D = 3/6$, we have $r = D/2$; therefore,

$$V = \frac{1}{3}\pi \left(\frac{D}{2}\right)^2 D = \frac{1}{12}\pi D^3.$$

Because this result is valid for all time t during the filling process, we can differentiate with respect to t, once again using extended power rule 3.21:

$$\frac{dV}{dt} = \frac{1}{4}\pi D^2 \frac{dD}{dt}.$$

Since $dV/dt = 4$ m³/min (converted from the rate of 4000 L/min since the litre is not an acceptable unit of measure for volume), we find that when $D = 3$,

$$4 = \frac{1}{4}\pi (3)^2 \frac{dD}{dt},$$

from which $dD/dt = 16/(9\pi)$. The surface is therefore rising at a rate of $16/(9\pi)$ m/min.

These examples illustrate the following general procedure for solving related rate problems:

1. Sketch a diagram illustrating all given information, especially given rates of change and desired rates of change. Do not draw the diagram at the instant in question; draw it slightly before or slightly after.

2. Find an equation valid for all time (in some interval about the instant in question) that involves only variables whose rates of change are given or required.

3. Differentiate the equation in step 2 and solve for the required rate.

Steps 1 and 3 are usually quite straightforward; step 2, on the other hand, may tax your ingenuity. To find the equation in the appropriate variables, it may be necessary to introduce and substitute for additional variables. Finding these substitutions requires you to analyze the problem very closely.

Be careful not to substitute numerical data that represent the instant at which the derivative is required before differentiation has taken place. Numerical data must be substituted after differentiation. For instance, in Example 4.34, radius r of the surface of the water when $D = 3$ is $3/2$. If we substitute this into $V = \pi r^2 D/3$, we obtain a function $V = \pi(3/2)^2 D/3 = 3\pi D/4$, which is valid only when $D = 3$. It cannot therefore be differentiated; only equations that are valid for a range of values of t can be differentiated with respect to t.

We now apply this procedure to two further examples. The first is an extension of Example 4.32.

EXAMPLE 4.35

▼

A man 2 m tall walks along the edge of a straight road 10 m wide. On the other edge of the road stands a streetlight 8 m high. If the man walks at $3/2$ m/s, how fast is his shadow lengthening when he is 10 m from the point directly opposite the light?

FIGURE 4.78   Shadow of man walking away from streetlight

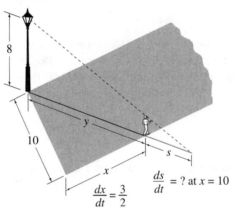

SOLUTION   First we draw Figure 4.78, wherein the man's speed is represented as the time rate of change of his distance x from the point on his side of the road directly opposite the light. What is required is the rate of change ds/dt of the length of his shadow when $x = 10$. To find an equation relating x and s, we first use similar (vertical) triangles to write

$$\frac{y + s}{s} = \frac{8}{2} = 4,$$

from which $y = 3s$, an equation that relates s to y, rather than s to x. However, since

$$y^2 = x^2 + 100,$$

we substitute to obtain

$$9s^2 = x^2 + 100.$$

The derivative of this equation with respect to time t gives

$$18s\frac{ds}{dt} = 2x\frac{dx}{dt}.$$

When $x = 10$, we obtain $s = \sqrt{100 + 100}/3 = 10\sqrt{2}/3$, and at this instant,

$$18\left(\frac{10\sqrt{2}}{3}\right)\frac{ds}{dt} = 2(10)\left(\frac{3}{2}\right).$$

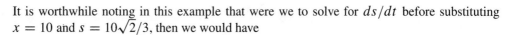

Thus,

$$\frac{ds}{dt} = \frac{30}{60\sqrt{2}} = \frac{\sqrt{2}}{4},$$

and the man's shadow is therefore lengthening at the rate of $\sqrt{2}/4$ m/s.

It is worthwhile noting in this example that were we to solve for ds/dt before substituting $x = 10$ and $s = 10\sqrt{2}/3$, then we would have

$$\frac{ds}{dt} = \frac{x}{9s}\frac{dx}{dt}.$$

Since dx/dt is always equal to $3/2$, and $s = \sqrt{x^2 + 100}/3$, we can write

$$\frac{ds}{dt} = \frac{x}{3\sqrt{x^2 + 100}}\left(\frac{3}{2}\right) = \frac{x}{2\sqrt{x^2 + 100}},$$

a general formula for ds/dt. The limit of this rate as x becomes very large is

$$\lim_{x \to \infty} \frac{ds}{dt} = \lim_{x \to \infty} \frac{x}{2\sqrt{x^2 + 100}} = \frac{1}{2}.$$

But for very large x, the man essentially walks directly away from the light, and this answer, as we might expect, is identical to that in Example 4.32.

EXAMPLE 4.36

One end of a rope is tied to a box. The other end is passed over a pulley 5 m above the floor and tied at a level 1 m above the floor to the back of a truck. If the rope is taut and the truck moves at $1/2$ m/s, how fast is the box rising when the truck is 3 m from the plumbline through the pulley?

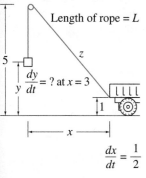

FIGURE 4.79   Mass attached to a truck passed over a pulley

SOLUTION   In Figure 4.79, we have represented the speed of the truck as the rate of change of length x, $dx/dt = 1/2$ m/s. What is required is dy/dt when $x = 3$. To find an equation relating x and y, we first use the fact that the length z of rope between pulley and truck is the hypotenuse of a right-angled triangle with sides of lengths x and 4,

$$z^2 = x^2 + 16.$$

For an equation relating y and z, we note that the length of the rope, call it L, remains constant, and is equal to the sum of z and $5 - y$,

$$L = z + (5 - y).$$

These two equations can be combined into

$$(L - 5 + y)^2 = x^2 + 16,$$

and differentiation with respect to time t now gives

$$2(L - 5 + y)\frac{dy}{dt} = 2x\frac{dx}{dt}.$$

When $x = 3$, we may write that

$$(L - 5 + y)^2 = 9 + 16 = 25.$$

We could solve this equation for y (in terms of L), but it is really not y that is needed to obtain dy/dt from the preceding equation. It is $L - 5 + y$, and this is clearly equal to 5. Thus, when $x = 3$, we have

$$2(5)\frac{dy}{dt} = 2(3)\left(\frac{1}{2}\right);$$

that is, $dy/dt = 3/10$, and the box is rising at a rate of $3/10$ m/s.

The reader should compare this example with the problem in Exercise 2. They may appear similar, but are really quite different.

Consulting Project 6

Mechanical engineers have a question concerning the mechanism in Figure 4.80. Rod OB, of length l, rotates counterclockwise in the xy-plane around the origin at ω revolutions per second. Rod AB, attached to OB, is such that A is confined to sliding horizontally along the x-axis. For unrestricted motion, the length L of AB is greater than twice l. The engineers wish to know the maximum speed attained by slider A.

FIGURE 4.80   Speed of slider in a two-bar mechanism

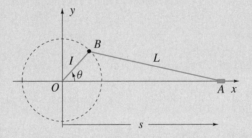

SOLUTION   Slider A moves back and forth along the x-axis, repeating its motion for each revolution of B. To determine the maximum speed of A, we need only consider its motion as B moves from $(l, 0)$ to $(-l, 0)$ along the upper semicircle. We therefore take angle θ in the interval $0 \le \theta \le \pi$. The cosine law applied to triangle OAB gives $L^2 = l^2 + x^2 - 2lx \cos\theta$. If this is differentiated with respect to time,

$$0 = 0 + 2x\frac{dx}{dt} - 2l \cos\theta \frac{dx}{dt} + 2lx \sin\theta \frac{d\theta}{dt}.$$

When we set $d\theta/dt = 2\pi\omega$, and solve for dx/dt,

$$\frac{dx}{dt} = \frac{2lx \sin\theta (2\pi\omega)}{2l \cos\theta - 2x} = \frac{2\pi\omega lx \sin\theta}{l \cos\theta - x} \text{ m/s.}$$

Two things are worth noticing. Velocity is zero when $\sin\theta = 0$, and this is when $\theta = 0$ and $\theta = \pi$, when B is on the x-axis (as we would expect). The denominator $l \cos\theta - x$ can never vanish since x/l is always greater than unity.

To find maximum speed of the follower, we should determine minimum velocity since A is moving left when $0 \le \theta \le \pi$. Since this occurs when acceleration is zero, we set

$$0 = \frac{d^2x}{dt^2} = \frac{2\pi\omega l \sin\theta}{l\cos\theta - x}\frac{dx}{dt} + \frac{2\pi\omega l x \cos\theta}{l\cos\theta - x}\frac{d\theta}{dt}$$

$$- \frac{2\pi\omega l x \sin\theta}{(l\cos\theta - x)^2}\left(-l\sin\theta\frac{d\theta}{dt} - \frac{dx}{dt}\right)$$

$$= \frac{2\pi\omega l \sin\theta}{(l\cos\theta - x)^2}(l\cos\theta - x + x)\frac{dx}{dt}$$

$$+ \frac{2\pi\omega l x}{(l\cos\theta - x)^2}[\cos\theta(l\cos\theta - x) + l\sin^2\theta]\frac{d\theta}{dt}$$

$$= \frac{2\pi\omega l^2\sin\theta\cos\theta}{(l\cos\theta - x)^2}\left(\frac{2\pi\omega l x \sin\theta}{l\cos\theta - x}\right) + \frac{4\pi^2\omega^2 l x(l - x\cos\theta)}{(l\cos\theta - x)^2}$$

$$= \frac{4\pi^2\omega^2 l^3 x\sin^2\theta\cos\theta + 4\pi^2\omega^2 l x(l - x\cos\theta)(l\cos\theta - x)}{(l\cos\theta - x)^3}\ \text{m/s}^2.$$

We now set the numerator equal to zero, at the same time removing the factor $4\pi^2\omega^2 l x$,

$$0 = l^2\sin^2\theta\cos\theta + (l - x\cos\theta)(l\cos\theta - x)$$

$$= l^2(1 - \cos^2\theta)\cos\theta + l^2\cos\theta - lx - lx\cos^2\theta + x^2\cos\theta$$

$$= -l^2\cos^3\theta - lx\cos^2\theta + (2l^2 + x^2)\cos\theta - lx.$$

This equation must be combined with $L^2 = l^2 + x^2 - 2lx\cos\theta$ to yield x and θ. If we set $y = \cos\theta$, then we must solve the following nonlinear equations for x and y,

$$l^2 y^3 + lxy^2 - (2l^2 + x^2)y + lx = 0, \qquad L^2 = l^2 + x^2 - 2lxy.$$

Normal procedure would be to solve one of these equations for x in terms of y, or y in terms of x, substitute into the other equation, and thereby obtain one equation in one unknown. Unfortunately, none of these possibilities seems appealing. If we set $y = ax$ in each of the equations, we obtain

$$l^2(a^3x^3) + lx(a^2x^2) - (2l^2 + x^2)(ax) + lx = 0, \qquad L^2 = l^2 + x^2 - 2lx(ax).$$

When we cancel an x from the first equation, then both contain only x^2's,

$$(l^2a^3 + la^2 - a)x^2 = 2l^2a - l, \qquad (1 - 2la)x^2 = L^2 - l^2.$$

When we solve each of these for x^2, and equate results, we obtain

$$\frac{2l^2a - l}{l^2a^3 + la^2 - a} = \frac{L^2 - l^2}{1 - 2la}.$$

When we cross multiply,

$$0 = L^2(l^2a^3 + la^2 - a) - l^2(l^2a^3 + la^2 - a) + l - 4l^2a + 4l^3a^2$$

$$= (l^2L^2 - l^4)a^3 + (lL^2 + 3l^3)a^2 - (L^2 + 3l^2)a + l.$$

This cubic equation must be solved for a (once l and L are specified). When this is done, $x^2 = (L^2 - l^2)/(1 - 2la)$ gives the position of maximum speed, and angle θ is given by $\theta = \text{Cos}^{-1}(ax)$. For example, if $L = 0.6$ m and $l = 0.2$ m, the equation for a reduces to

$$0.0128a^3 + 0.096a^2 - 0.48a + 0.2 = 0.$$

Of the three solutions -11.0287, $0.461\,987$, and 3.0667 of this equation, only $a = 0.461\,987$ is acceptable (a cannot be negative and the largest root leads to a negative value for x^2). With this value of a, we find $x = 0.626\,529$ and $\theta = 1.277\,15$. Maximum speed of the slider is therefore

$$\left|\frac{dx}{dt}\right| = \left|\frac{2\pi\omega(0.2)(0.626\,529)\sin 1.277\,15}{0.2\cos 1.277\,15 - 0.626\,529}\right| = 1.3253\omega.$$

EXERCISES 4.9

* **1.** A convertible is travelling along a straight highway at 100 km/h. A child in the car accidentally releases a helium-filled balloon, which then rises vertically at 10 m/s. How fast are the child and balloon separating 4 s after the balloon is released?

* **2.** A rope passes over a pulley and one end is attached to a cart as shown in the figure below. If the rope is pulled vertically downward at 2 m/s, how fast is the cart moving when $s = 6$ m?

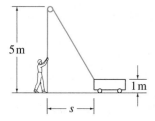

* **3.** A light is on the ground 20 m from a building. A man 2 m tall walks from the light directly toward the building at 3 m/s. How fast is the length of his shadow on the building changing when he is 8 m from the building?

* **4.** A funnel in the shape of a right circular cone is 15 cm across the top and 30 cm deep. A liquid is flowing in at the rate of 80 mL/s and flowing out at 15 mL/s. At what rate is the surface of the liquid rising when the liquid fills the funnel to a depth of 20 cm?

* **5.** A water tank is in the form of a right circular cylinder of diameter 3 m and height 3 m on top of a right circular cone of diameter 3 m and height 1 m. If water is being drawn from the bottom at the rate of 1 L/min, how fast is the water level falling when (a) it is 1 m from the top of the tank and (b) it is 3.5 m from the top of the tank?

* **6.** A point P moves along the curve $y = x^2 + x + 4$, where x and y are measured in metres. Its x-coordinate decreases at 2 m/s. If the perpendicular from P to the x-axis intersects this axis at point Q, how fast is the area of the triangle with vertices P, Q, and the origin changing when the x-coordinate of P is 2 m?

* **7.** Water is being pumped into a swimming pool which is 10 m wide, 20 m long, 1 m deep at the shallow end, and 3 m deep at the deep end.

If the water level is rising at 1 cm/min when the depth is 1 m at the deep end, at what rate is water being pumped into the pool?

* **8.** Boyle's law for a perfect gas states that the pressure exerted by the gas on its containing vessel is inversely proportional to the volume occupied by the gas. If when the volume is 10 L and the pressure is 50 N/m^2, the volume is increasing at 1/2 L/s, find the rate of change of the pressure of the gas.

* **9.** A woman driving 100 km/h along a straight highway notes that the shadow of a cloud is keeping pace with her. What can she conclude about the speed of the cloud?

* **10.** A fisherman is trolling at a rate of 2 m/s with his lure 100 m behind the boat and on the surface. Suddenly a fish strikes and dives vertically at a rate of 3 m/s. If the fisherman permits the line to run freely and it always remains straight, how fast is the line being played out when the reel is 50 m from its position at the time of the strike?

* **11.** Air expands adiabatically in accordance with the law $PV^{7/5} = $ constant. If at a given time, the volume V is 100 L and the pressure P is 40 N/cm^2, at what rate is the pressure changing when the volume is decreasing at 1 L/s?

* **12.** Sand is poured into a right circular cylinder of radius 1/2 m along its axis (figure below). Once sand completely covers the bottom, a right circular cone is formed on the top.

   (a) If 0.02 m^3 of sand enters the container every minute, how fast is the top of the sand pile rising?

   (b) How fast is the sand rising along the side of the cylinder?

13. A balloon has the shape of a right circular cylinder of radius r and length l with a hemisphere at each end of radius r. The balloon is being filled at a rate of 10 mL/s in such a way that l increases twice as fast as r. Find the rate of change of r when $r = 8$ cm and $l = 20$ cm.

14. An oval racetrack has a straight stretch 100 m long and two semi-circles, each of radius 50 m (figure below). Car 1, on the infield, moves along the x-axis from O to B. It accelerates from rest at O, attains a speed of 10 m/s at C, and maintains this speed along CB. Car 2 travels along the quarter oval $ADEB$. It is at D when Car 1 is at C. Between D and B, Car 2 maintains the same rate of change of its x-coordinate as does Car 1.

    (a) Find a formula for the rate of change of the y-coordinate of Car 2 between D and B.

    (b) How fast is the y-coordinate of Car 2 changing when it is at point E?

    (c) If the cars collide at B, which car suffers the most damage?

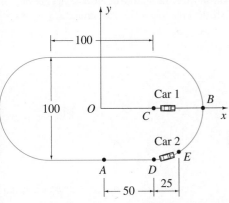

15. A ship is 1 km north of a pier and is travelling N30°E at 3 km/h. A second ship is 3/4 km east of the pier and is travelling east at 7 km/h. How fast are the ships separating?

16. The circle in the figure below represents a long-playing record which is rotating clockwise at 100/3 rpm. A bug is walking away from the centre of the record directly toward point P on the rim of the record at 1 cm/s. When the bug is at position R, 10 cm from O, angle θ is $\pi/4$ radians. Find the rate at which the distance from the bug to the fixed point Q is changing when the bug is at R.

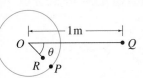

17. Eight skaters form a "whip." Show that the seventh person on the whip travels twice as fast as the fourth person.

18. A particle moves counterclockwise around a circle of radius 5 cm centred at the origin making 4 revolutions each second. How fast is the particle moving away from the point with coordinates $(5, 6)$ when it is at position $(-3, 4)$?

19. Two people (A and B in the following figure) walk along opposite sides of a road 10 m wide. A walks to the right at 1 m/s, and B walks to the left at 2 m/s. A third person, C, walks along a sidewalk 5 m from the road in such a way that B is always on the line joining A and C. Find the speed of C.

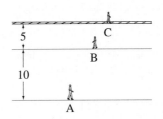

∗ 20. Let $P(x, y)$ be a point on the first-quadrant portion of the hyperbola $x^2 - y^2 = 1$. Let R be the foot of the perpendicular from P to the x-axis, and $Q(x^*, 0)$ be the x-intercept of the normal line to the hyperbola at P.

    (a) Show that $x^* = 2x$.

    (b) If P moves along the hyperbola so that its x-coordinate is decreasing at 3 units per unit time, how fast is the area of triangle QPR changing when $x = 4$?

∗ 21. A solution passes from a conical filter 24 cm deep and 16 cm across the top into a cylindrical container of diameter 12 cm. When the depth of solution in the filter is 12 cm, its level is falling at the rate of 1 cm/min. How fast is the level of solution rising in the cylinder at this instant?

∗ 22. A light is at the top of a pole 25 m high and a ball is dropped at the same height from a point 10 m from the light. How fast is the shadow of the ball moving along the ground 1 s later? The distance fallen by the ball t seconds after it has been dropped is $d = 4.905t^2$ metres.

∗ 23. A point moves along the parabola $y = x^2 - 3x$ (x and y measured in metres) in such a way that its x-coordinate changes at the rate of 2 m/s. How fast is its distance from the point $(1, 2)$ changing when it is at $(4, 4)$?

∗ 24. Repeat Exercise 23 given that the parabola is replaced by the curve $(x + y)^2 = 16x$.

∗ 25. The volume of wood in the trunk of a tree is sometimes calculated by considering it as a frustrum of a right circular cone (figure below).

    (a) Verify that the volume of the trunk is

$$V = \frac{1}{3}\pi h(R^2 + rR + r^2).$$

    (b) Suppose that at the present time the radii of the top and bottom are $r = 10$ and $R = 50$ cm, and the height is $h = 30$ m. If the tree continues to grow so that ratios r/R and r/h always remain the same as they are now, and R increases at a rate of 1/2 cm/year, how fast will the volume be changing in 2 years?

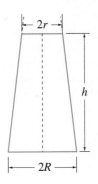

* **26.** Sand is poured into a right circular cone of radius 2 m and height 3 m along its axis (figure below). The sand forms two cones of equal height h, one inverted on top of the other.

  (a) If 0.02 m^3 of sand enters the container every minute, how fast is the top of the pile rising when it is just level with the top of the container?

  (b) How fast is the sand rising along the side of the container at this instant?

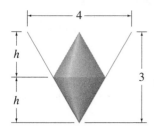

* **27.** If to the mechanism in Figure 4.80 we add a rod AC (figure below), where C is confined to sliding vertically, find the velocity of C in terms of x, θ, and y.

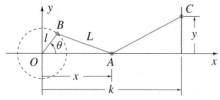

* **28.** In the figure below, a plane flies due north at 200 km/h at constant altitude 1 km. A car travels due east on a straight highway at 100 km/h. At the moment the plane crosses over the highway, the car is 2 km east of the point on the road directly below the plane. How fast are the plane and car separating 1 min after this?

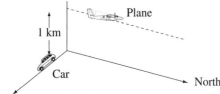

* **29.** The infield of a baseball diamond is a square with distances between bases being 27 m (approximately). The hitter hits a ground ball to the third baseman, accelerates quickly, and attains a speed of 6 m/s as she runs to first base. The third baseman catches the ball at a point 2 m from the bag on the line betwen second and third base, and throws the ball to the first baseman at 35 m/s. If the ball is halfway to first base when the batter is three-quarters of the way to first base, how fast is the distance between them changing at this instant?

* **30.** In the following figure the boy's feet make 1 revolution per second around a sprocket of radius R metres. The chain travels around a sprocket of radius r metres on the back wheel, which itself has radius $\overline{R}$ metres. If a stone embedded in the tire becomes dislodged, how fast is it travelling when it leaves the tire? Assume that the rear wheel has been placed on a stand so that the bicycle is stationary.

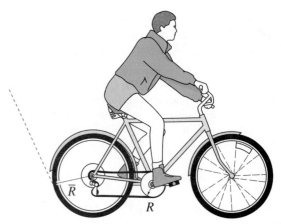

* **31.** A hemispherical tank of radius 3 m has a light on its upper edge (figure below). A stone falls vertically along the axis of symmetry of the tank, and when it is 1 m from the bottom of the tank, it is falling at 2 m/s. How fast is its shadow moving along the surface of the tank at this instant?

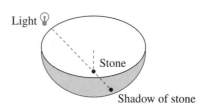

* **32.** If the sides of a triangle have lengths a, b, and c, its area is given by

$$A = \sqrt{s(s-a)(s-b)(s-c)},$$

where $s = (a+b+c)/2$ is one-half its perimeter. If the length of each side increases at a rate of 1 cm/min, how fast is A changing when $a = 3$ cm, $b = 4$ cm, and $c = 5$ cm?

* **33.** (a) If θ is the angle formed by the minute and hour hands of a clock, what is the time rate of change of θ (in radians per minute)?

  (b) If the lengths of the hands on the clock are 10 cm and 7.5 cm, find the rate at which their tips approach each other at 3:00 (i.e., find dz/dt in the figure below).

  (c) Repeat part (b) but replace time 3:00 with 8:05.

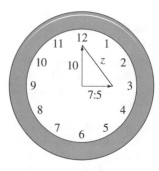

34. A runner moves counterclockwise around the track in the figure below at a rate of 4 m/s. A camera at the centre of the track is placed on a swivel so that it can follow the runner. Find the rate at which the camera turns when (a) the runner is at A and (b) the runner is at B.

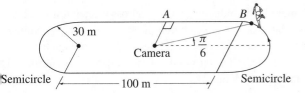

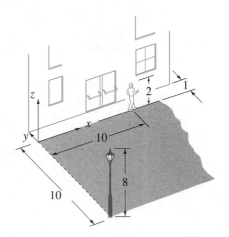

35. A man 2 m tall walks along the edge of a straight road 10 m wide (the figure to the right). On the other edge of the road stands a streetlight 8 m high. A building runs parallel to the road and 1 m from it. If the man walks away from the light at 2 m/s, how fast is the height of the shadow on the wall changing when he is 10 m from the point on the road directly opposite the light?

4.10 *LCR*-Circuits

FIGURE 4.81   Circuit containing capacitor and voltmeter

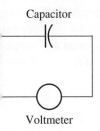

Capacitor

Voltmeter

Modern electronic equipment contains a vast array of devices; many find their origin in three fundamental elements — capacitors, resistors, and inductors. How these elements relate to one another and how they affect voltages, charges, and currents in electric circuits can be fully understood with calculus. We begin with capacitors.

A **capacitor** is a device that stores equal amounts of positive charge and negative charge in such a way that the charges cannot neutralize one another. Suppose $Q > 0$ is the amount of positive charge, and therefore $-Q$ is the negative charge. Separation of these charges creates a potential difference V at the terminals of the capacitor. It can be measured by placing a voltmeter across the terminals as shown in Figure 4.81. The size of V depends on how charges are stored in the capacitor; different configurations lead to different potential differences. When we divide Q by the potential difference V that it produces, we obtain what is called the **capacitance C** of the capacitor,

FIGURE 4.82   Circuit containing capacitor and battery

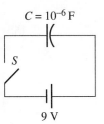

$C = 10^{-6}$ F

S

9 V

$$C = \frac{Q}{V}. \tag{4.11}$$

It has units of coulombs per volt, called *farads* (F). The higher the capacitance, the more charge that can be stored per volt of potential difference.

If a 9 V battery is connected to a 10^{-6} F capacitor (Figure 4.82) and the switch is closed, the battery creates a flow of charge in the circuit until the capacitor is charged with $Q = 9(10^{-6})$ coulombs. The rate at which charge flows is called **current**, denoted by the letter i. It is measured in amperes (A); 1 A is a flow of 1 C of charge per second. For a simple circuit like that in Figure 4.82, we can think of i as the rate of change of charge Q on the capacitor, and therefore

$$i = \frac{dQ}{dt}. \tag{4.12}$$

FIGURE 4.83   Direction of current flow related to potential difference

Potential
difference
= V

$+$ ——— ——— $-$

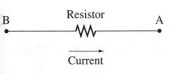

Resistor

B ————$\wedge\!\wedge\!\wedge$———— A

Current

When there is no capacitor in a circuit, or even when there is, we can think of charge Q flowing past any specific point in the circuit. How much charge flows past this point per unit time is represented by i.

A **resistor** is an electronic device that retards the flow of charge in a circuit. When a potential difference V is maintained between the terminals A and B of the resistor in Figure 4.83, where the $+$ and $-$ signs indicate that B is at higher potential than A, positive charge flows from B to A. If the rate of flow (current) is i amperes, the ratio

$$R = \frac{V}{i} \tag{4.13}$$

is called the **resistance** of the resistor. It is measured in volts per ampere, called *ohms* (Ω). The higher the resistance, the smaller the current generated by a given voltage, or the larger the voltage required to produce a given current. For example, to maintain a current of 2 A through a 3 Ω resistor requires 6 V, but to maintain the same current through a 30 Ω resistor requires 60 V.

The third fundamental circuit element is the inductor. Voltage across a capacitor is related to charge; voltage across a resistor is related to current (the rate of change of charge); voltage across an **inductor** is related to the rate of change of current. In other words, an inductor reacts to changes in current. If V is the voltage across the terminals of an inductor and current i is changing, then the inductance of the inductor is defined as

FIGURE 4.84   Circuit for Kirchhoff's loop rule

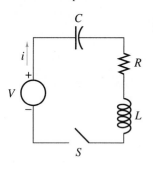

$$L = \frac{V}{di/dt} = \frac{V}{d^2Q/dt^2}. \tag{4.14}$$

It has units of volts per ampere per second, called *henries* (H). When inductance is large, a small rate of change of current produces a large voltage across the terminals of the inductor.

Charge flows through a resistor or any other electric device when a potential difference is created between its terminals. A device that creates and maintains potential difference is an **emf device** (*emf* is short for *electromotive force*). Examples are batteries, electric generators, solar cells, and thermopiles. An emf device is shown in Figure 4.84 in a circuit including a capacitor with capacitance C, a resistor with resistance R, an inductor with inductance L, and a switch S. This is called an *LCR-circuit*. When the switch S is closed, potential difference across the terminals of the emf device causes charge to flow in the circuit creating a current i. **Kirchhoff's loop rule** for electric circuits implies that the sum of the potential differences across the capacitor, resistor, and inductor must be equal to output potential of the emf device. Using equations 4.11, 4.13, and 4.14, we obtain

$$L\frac{di}{dt} + Ri + \frac{Q}{C} = V. \tag{4.15a}$$

If i is replaced by dQ/dt, we have

$$L\frac{d^2Q}{dt^2} + R\frac{dQ}{dt} + \frac{Q}{C} = V, \tag{4.15b}$$

and if this equation is differentiated with respect to t, we also have

$$L\frac{d^2i}{dt^2} + R\frac{di}{dt} + \frac{i}{C} = \frac{dV}{dt}. \tag{4.15c}$$

These are differential equations that must be solved for Q (4.15b) or i (4.15c) when V is given as a function of t.

EXAMPLE 4.37

At time $t = 0$, a 6 Ω resistor, a 1 H inductor, and a 0.04 F capacitor are connected with a generator producing a voltage of $10 \sin 5t$, where $t \geq 0$ is time in seconds, by closing the switch (Figure 4.85).

FIGURE 4.85   Current in an *LCR*-circuit

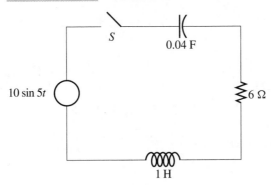

(a) What differential equation must current in the circuit satisfy for $t > 0$?

(b) Verify that

$$i(t) = \frac{5}{3}\sin 5t - \frac{1}{3}e^{-3t}(5\cos 4t + 10\sin 4t)$$

satisfies the equation in part (a).

(c) Plot a graph of $i(t)$ and explain the significance of each term.

SOLUTION

(a) With the particular values for L, C, R, and V in Figure 4.85, equation 4.15c becomes

$$\frac{d^2i}{dt^2} + 6\frac{di}{dt} + 25i = 50\cos 5t.$$

(b) Since

$$\frac{di}{dt} = \frac{25}{3}\cos 5t + e^{-3t}(5\cos 4t + 10\sin 4t)$$

$$- \frac{1}{3}e^{-3t}(-20\sin 4t + 40\cos 4t)$$

$$= \frac{25}{3}\cos 5t + \frac{1}{3}e^{-3t}(-25\cos 4t + 50\sin 4t)$$

and

$$\frac{d^2i}{dt^2} = -\frac{125}{3}\sin 5t - e^{-3t}(-25\cos 4t + 50\sin 4t)$$

$$+ \frac{1}{3}e^{-3t}(100\sin 4t + 200\cos 4t)$$

$$= -\frac{125}{3}\sin 5t + \frac{1}{3}e^{-3t}(275\cos 4t - 50\sin 4t),$$

we find that

$$\frac{d^2i}{dt^2} + 6\frac{di}{dt} + 25i = -\frac{125}{3}\sin 5t + \frac{1}{3}e^{-3t}(275\cos 4t - 50\sin 4t)$$

$$+ 50\cos 5t + 2e^{-3t}(-25\cos 4t + 50\sin 4t)$$

$$+ \frac{125}{3}\sin 5t - \frac{25}{3}e^{-3t}(5\cos 4t + 10\sin 4t)$$

$$= 50\cos 5t.$$

Thus, $i(t)$ does indeed satisfy $i'' + 6i' + 25i = 50\cos 5t$.

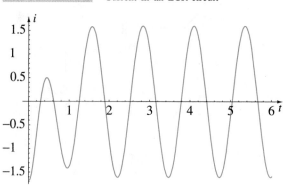

FIGURE 4.86 Current in an LCR-circuit

(c) The plot of $i(t)$ in Figure 4.86 is composed of two functions, $(5/3)\sin 5t$ and $-(1/3)e^{-3t}(5\cos 4t + 10\sin 4t)$. Just after the switch is closed, both parts contribute significantly to $i(t)$. Within a few seconds, however, the exponential factor e^{-3t} causes the second term of $i(t)$ to become negligible. This is called the *transient* part of the current; it persists for a very short time interval. The term $(5/3)\sin 5t$ remains for all time, and once the transient part of the current becomes insignificant, $i(t)$ is essentially $(5/3)\sin 5t$. This is called the *steady-state* part of the current.

EXAMPLE 4.38

When an inductor and capacitor are connected to an emf device (Figure 4.87), the charge on the capacitor must satisfy equation 4.15b with $R = 0$,

$$L\frac{d^2Q}{dt^2} + \frac{Q}{C} = V.$$

(a) Verify that if the emf device is a battery producing constant voltage V, beginning at time $t = 0$, and the capacitor has no initial charge, then the function

$$Q(t) = CV\left[1 - \cos\left(\frac{t}{\sqrt{LC}}\right)\right]$$

satisfies this equation, and the conditions $Q(0) = 0$ and $i(0) = 0$.

(b) Draw a graph of $Q(t)$ and interpret it in terms of charge on the capacitor and current in the circuit.

FIGURE 4.87 Charge on capacitor in LC-circuit

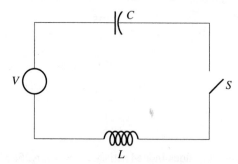

SOLUTION

(a) Since $\dfrac{dQ}{dt} = CV\left[\dfrac{1}{\sqrt{LC}}\sin\left(\dfrac{t}{\sqrt{LC}}\right)\right]$, it follows that

$$L\frac{d^2Q}{dt^2} + \frac{Q}{C} = L\left[\frac{CV}{LC}\cos\left(\frac{t}{\sqrt{LC}}\right)\right] + \frac{CV}{C}\left[1 - \cos\left(\frac{t}{\sqrt{LC}}\right)\right] = V.$$

Clearly, $Q(0) = 0$ and $i(0) = Q'(0) = 0$ also.

(b) To graph $Q(t)$, we first draw $-\cos(t/\sqrt{LC})$ in Figure 4.88a. Shifting this curve upward 1 unit and changing the scale on the vertical axis gives the graph of $Q(t)$ in Figure 4.88b.

When the circuit is closed at $t = 0$, there is no charge on the capacitor or current in the circuit. The battery immediately begins charging the capacitor, but because there is little charge on the capacitor, the voltage across its terminals is small. The remainder of the voltage (making up V which is constant for all time) is across the inductor. This is consistent with the fact that concavity is relatively large here and voltage across the inductor is proportional to the second derivative of Q, and the second derivative is positive. As time approaches $\pi\sqrt{LC}/2$, charge on the capacitor approaches CV, what would normally be its capacity if the inductor were not a part of the circuit. At time $t = \pi\sqrt{LC}/2$, voltage across the capacitor is V, and that across the inductor vanishes. There is a point of inflection at which $d^2Q/dt^2 = 0$. Current is now at a maximum in the circuit and the capacitor continues to accumulate charge. The voltage across the capacitor exceeds V, and therefore that across the inductor is negative. This agrees with the fact that the curve is concave downward and therefore $d^2Q/dt^2 < 0$. At $t = \pi\sqrt{LC}$, charge has reached a maximum value of $2CV$, voltage across the capacitor is $2V$, and that across the inductor is $-V$. Charge now begins to flow in the reverse direction and current (slope) is negative. The capacitor discharges completely at $t = 2\pi\sqrt{LC}$, and the cycle repeats.

FIGURE 4.88a   Graph of $-\cos(t/\sqrt{LC})$

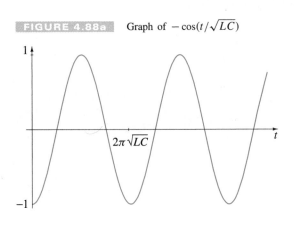

FIGURE 4.88b   Graph of charge on capacitor

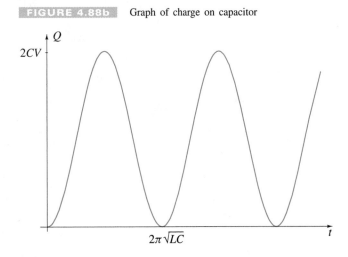

EXAMPLE 4.39

The emf device in the RC-circuit of Figure 4.89 produces a constant voltage of V volts. If the switch is closed at time $t = 0$ and then opened again at $t = t_0$, charge on the capacitor must satisfy

$$R\frac{dQ}{dt} + \frac{Q}{C} = V[1 - h(t - t_0)], \quad t > 0,$$

where $h(t - t_0)$ is the Heaviside unit step function of Section 2.5.

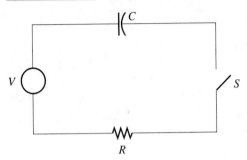

FIGURE 4.89 Charge on capacitor in RC-circuit

(a) Verify that

$$Q(t) = CV[1 - e^{-t/(RC)}] - CV[1 - e^{-(t-t_0)/(RC)}]h(t - t_0)$$

satisfies this equation for all $t \neq t_0$.

(b) Draw a graph of $Q(t)$. What is the initial charge on the capacitor?

(c) Are charge on the capacitor and current in the circuit continuous?

SOLUTION

(a) Since the derivative of the Heaviside function is zero for every $t \neq t_0$,

$$\frac{dQ}{dt} = CV\left(\frac{1}{RC}\right)e^{-t/(RC)} - CV\left(\frac{1}{RC}\right)e^{-(t-t_0)/(RC)}h(t - t_0)$$

$$= \frac{V}{R}\left[e^{-t/(RC)} - e^{-(t-t_0)/(RC)}h(t - t_0)\right].$$

Hence,

$$R\frac{dQ}{dt} + \frac{Q}{C} = V\left[e^{-t/(RC)} - e^{-(t-t_0)/(RC)}h(t - t_0)\right] + V\left[1 - e^{-t/(RC)}\right]$$

$$- V\left[1 - e^{-(t-t_0)/(RC)}\right]h(t - t_0)$$

$$= V[1 - h(t - t_0)].$$

(b) To graph $Q(t)$, we write it in the form

$$Q(t) = CV \begin{cases} 1 - e^{-t/(RC)}, & 0 \leq t < t_0 \\ e^{-(t-t_0)/(RC)} - e^{-t/(RC)}, & t > t_0 \end{cases}$$

$$= CV \begin{cases} 1 - e^{-t/(RC)}, & 0 \leq t < t_0 \\ -\left[1 - e^{t_0/(RC)}\right]e^{-t/(RC)}, & t > t_0. \end{cases}$$

For $0 \leq t < t_0$, we first draw $e^{-t/(RC)}$ as in Figure 4.90a, turn it upside down, shift it vertically one unit, and change the scale on the vertical axis (Figure 4.90b). As $t \to t_0^-$, the graph approaches $CV[1 - e^{-t_0/(RC)}]$.

For $t > t_0$, the graph declines exponentially. As $t \to t_0^+$, it approaches $CV[1 - e^{-t_0/(RC)}]$, and it is asymptotic to the t-axis (Figure 4.90c). Combining Figures 4.90b and c gives the final graph in Figure 4.90d. The capacitor has no initial charge.

FIGURE 4.90a $e^{-t/(RC)}$ for $0 \le t \le t_0$

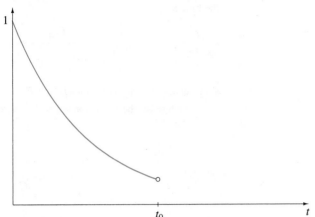

FIGURE 4.90b Charge on capacitor for $0 \le t < t_0$

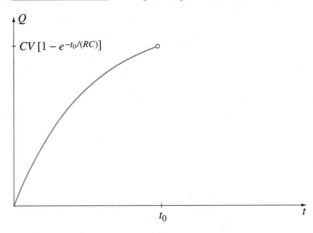

FIGURE 4.90c Charge on capacitor for $t > t_0$

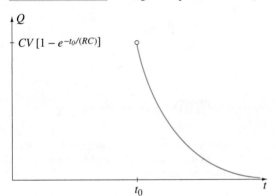

FIGURE 4.90d Charge on capacitor

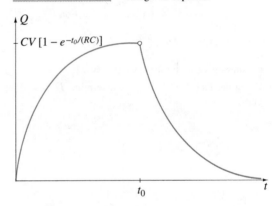

(c) If we define the value of the charge on the capacitor at t_0 to be $CV[1 - e^{-t_0/(RC)}]$, it is continuous. Since current in the circuit is the slope of the curve, it is undefined at t_0 and therefore current is discontinuous at time t_0. It suddenly reverses direction when the emf device is disconnected.

EXERCISES 4.10

1. If the switch in the RC-circuit shown to the right is closed at time $t = 0$, equation 4.15b for charge on the capacitor becomes

$$R\frac{dQ}{dt} + \frac{Q}{C} = V, \quad t > 0.$$

(a) Verify that when V is constant, the function

$$Q(t) = De^{-t/(RC)} + CV$$

satisfies the equation for any constant D.

(b) If Q_0 is the charge on the capacitor when the switch is closed, show that

$$Q(t) = CV[1 - e^{-t/(RC)}] + Q_0 e^{-t/(RC)}.$$

(c) Draw a graph of the function in part (b) when $Q_0 = 0$.

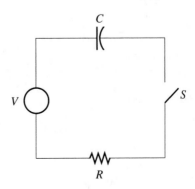

* **2.** The current i in the RC-circuit of Exercise 1 must satisfy the differential equation

$$R\frac{di}{dt} + \frac{i}{C} = \frac{dV}{dt}.$$

If $V = V_0 \sin \omega t$, where V_0 and ω are constants, verify that a solution is

$$i = f(t) = Ae^{-t/(RC)} + \frac{V_0}{Z} \sin(\omega t - \phi),$$

where A is any constant whatsoever, and

$$Z = \sqrt{R^2 + \frac{1}{\omega^2 C^2}}, \qquad \tan\phi = -\frac{1}{\omega C R}.$$

* **3.** If the switch in the LR-circuit below is closed at time $t = 0$, equation 4.15a for current in the circuit becomes

$$L\frac{di}{dt} + Ri = V, \quad t > 0.$$

(a) Verify that when V is constant, the function

$$i(t) = De^{-Rt/L} + V/R$$

satisfies the equation for any constant D.

(b) Using the fact that $i(0) = 0$, determine D, and draw a graph of $i(t)$.

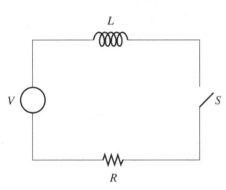

* **4.** The current i in the LR-circuit in the circuit of Exercise 3 must satisfy the differential equation

$$L\frac{di}{dt} + Ri = V.$$

If V is as in Exercise 2, verify that a solution is

$$i = f(t) = Ae^{-Rt/L} + \frac{V_0}{Z} \sin(\omega t - \phi),$$

where A is any constant, and

$$Z = \sqrt{R^2 + \omega^2 L^2}, \qquad \tan\phi = \frac{\omega L}{R}.$$

* **5.** If the switch in the following LC-circuit is closed at time $t = 0$, equation 4.15b for charge on the capacitor becomes

$$L\frac{d^2 Q}{dt^2} + \frac{Q}{C} = V, \quad t > 0.$$

(a) Verify that when $V = A \sin \omega t$, where A and ω are constants, the function

$$Q(t) = D\cos\frac{t}{\sqrt{LC}} + E\sin\frac{t}{\sqrt{LC}} - \frac{A/\omega}{\omega L - \dfrac{1}{\omega C}}\sin\omega t$$

satisfies the equation for any constants D and E.

(b) If Q_0 is the charge on the capacitor when the switch is closed, show that

$$Q(t) = Q_0\cos\frac{t}{\sqrt{LC}} + \frac{A\sqrt{LC}}{\omega L - \dfrac{1}{\omega C}}\sin\frac{t}{\sqrt{LC}} - \frac{A/\omega}{\omega L - \dfrac{1}{\omega C}}\sin\omega t.$$

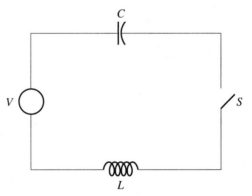

* **6.** If the voltage source labelled V in the circuit of Exercise 5 is suddenly short-circuited by the switch labelled S, the current i in the circuit thereafter must satisfy the equation

$$L\frac{d^2 i}{dt^2} + \frac{1}{C}i = 0,$$

where L and C are the sizes of the inductor and capacitor, respectively, and t is time. Verify that

$$i = f(t) = A\cos\left(\frac{t}{\sqrt{LC}}\right) + B\sin\left(\frac{t}{\sqrt{LC}}\right)$$

satisfies this equation for any constants A and B whatsoever.

* **7.** (a) If $V = A \sin \omega t$, where A and ω are constants, in Exercise 1, and there is no charge on the capacitor at time $t = 0$, verify that

$$Q(t) = \frac{CA}{1 + \omega^2 R^2 C^2}\sin\omega t + \frac{\omega RC^2 A}{1 + \omega^2 R^2 C^2}[e^{-t/(RC)} - \cos\omega t]$$

satisfies $RQ' + Q/C = V$.

(b) Show that $Q(t)$ can be expressed in the form

$$Q(t) = \frac{\omega RC^2 A}{1 + \omega^2 R^2 C^2}e^{-t/(RC)} + \frac{A/\omega}{Z}\cos(\omega t - \phi),$$

where

$$Z = \sqrt{R^2 + \frac{1}{\omega^2 C^2}} \qquad \text{and} \qquad \tan\phi = -\frac{1}{\omega C R}.$$

8. (a) If $V = A \sin \omega t$, where A and ω are constants, in Exercise 3, verify that

$$i(t) = \frac{\omega L A}{R^2 + \omega^2 L^2}(e^{-Rt/L} - \cos \omega t) + \frac{RA}{R^2 + \omega^2 L^2} \sin \omega t$$

satisfies $Li' + Ri = V$.

(b) Show that $i(t)$ can be expressed in the form

$$i(t) = \frac{\omega L A}{R^2 + \omega^2 L^2}e^{-Rt/L} + \frac{A}{Z} \sin(\omega t - \phi),$$

where

$$Z = \sqrt{R^2 + \omega^2 L^2} \quad \text{and} \quad \tan \phi = \frac{\omega L}{R}.$$

9. The current in the RCL-circuit below must satisfy equation 4.15c. If $V = A \sin \omega t$, where A and ω are constants, verify that a solution is

$$i(t) = e^{-Rt/(2L)}(D \cos vt + E \sin vt) + \frac{A}{Z} \sin(\omega t - \phi),$$

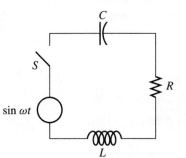

where D and E are any constants whatsoever, and

$$v = \sqrt{\frac{1}{LC} - \frac{R^2}{4L^2}},$$

$$Z = \sqrt{R^2 + \left(\omega L - \frac{1}{\omega C}\right)^2},$$

$$\tan \phi = \frac{\omega L - \dfrac{1}{\omega C}}{R}.$$

* **10.** An inductor L, a resistor R, and a capacitor C are connected with a generator, producing an oscillatory voltage $V = V_0 \cos \omega t$, for $t \geq 0$ (figure below). If L, C, R, V_0, and ω are all constants, steady-state current $i(t)$ in the circuit is given by

$$i(t) = \frac{V_0}{R^2 + \left(\omega L - \frac{1}{\omega C}\right)^2}\left[R \cos \omega t + \left(\omega L - \frac{1}{\omega C}\right) \sin \omega t\right].$$

Find the value of ω that makes the amplitude of the current a maximum. Do this in two ways:

(a) Express $i(t)$ in the form

$$i(t) = \frac{V_0}{Z} \cos(\omega t - \phi),$$

in which case amplitude V_0/Z must be maximized.

(b) Find critical points for $i(t)$ as given.

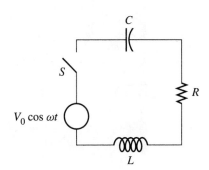

4.11 Indeterminate Forms and L'Hôpital's Rule

Derivatives are instantaneous rates of change, defined as limits of average rates of change. As a result, it was necessary to discuss limits in Chapter 2 prior to the introduction of derivatives in Section 3.1. Now that we have derivatives, it may seem quite surprising that they can be used to evaluate many limits.

The Indeterminate Form 0/0

If we let x approach zero in numerator and denominator of the limit

$$\lim_{x \to 0^+} \frac{\sqrt{1+x} - 1}{\sqrt{x}}, \quad (4.16)$$

we find that

$$\lim_{x \to 0^+} \left(\sqrt{1 + x} - 1 \right) = 0 \quad \text{and} \quad \lim_{x \to 0^+} \sqrt{x} = 0.$$

We say that limit 4.16 is of the **indeterminate form** 0/0. Similarly, the limit

$$\lim_{x \to 2} \frac{x^3 - x^2 - 8x + 12}{x^2 - 4x + 4} \tag{4.17}$$

is of the indeterminate form $0/0$ since both numerator and denominator approach zero as x approaches 2.

In Section 2.1 we evaluated limit 4.16 by rationalizing the numerator (see Example 2.6). Factoring numerator and denominator in limit 4.17 gives

$$\lim_{x \to 2} \frac{x^3 - x^2 - 8x + 12}{x^2 - 4x + 4} = \lim_{x \to 2} \frac{(x - 2)^2(x + 3)}{(x - 2)^2} = \lim_{x \to 2} (x + 3) = 5.$$

These two examples illustrate the "trickery" to which we resorted in Chapter 2 in order to evaluate limits. With Cauchy's generalized mean value theorem from Section 3.14, however, we can prove a result called L'Hôpital's rule, which makes evaluation of many limits of the indeterminate form 0/0 quite simple.

THEOREM 4.4   (L'Hôpital's rule)

Suppose functions $f(x)$ and $g(x)$ satisfy the following conditions:

1. $f(x)$ and $g(x)$ are differentiable in an open interval I except possibly at the point $x = a$ in I;
2. $g'(x) \neq 0$ in I except possibly at $x = a$;
3. $\lim\limits_{x \to a} f(x) = 0 = \lim\limits_{x \to a} g(x)$;
4. $\lim\limits_{x \to a} \dfrac{f'(x)}{g'(x)} = L$.

Then,

$$\lim_{x \to a} \frac{f(x)}{g(x)} = L.$$

PROOF   We define two functions $F(x)$ and $G(x)$ that are identical to $f(x)$ and $g(x)$ on I but have value zero at $x = a$:

$$F(x) = \begin{cases} f(x), & x \neq a \\ 0, & x = a, \end{cases} \qquad G(x) = \begin{cases} g(x), & x \neq a \\ 0, & x = a. \end{cases}$$

Since $F'(x) = f'(x)$ and $G'(x) = g'(x)$ for x in I except possibly at $x = a$, $F(x)$ and $G(x)$ are therefore differentiable on I except possibly at $x = a$. In addition, differentiability of a function implies continuity of the function (Theorem 3.6), so that $F(x)$ and $G(x)$ must certainly be continuous on I except possibly at $x = a$. But

$$\lim_{x \to a} F(x) = \lim_{x \to a} f(x) = 0 = F(a),$$

and the same is true for $G(x)$; hence, $F(x)$ and $G(x)$ are continuous for all x in I. Consequently, $F(x)$ and $G(x)$ are identical to $f(x)$ and $g(x)$ in every respect, except that they have been assigned a value at $x = a$ to guarantee their continuity there. This extra condition permits us to apply Cauchy's generalized mean value theorem to $F(x)$ and $G(x)$ on the interval between a and x, as long as x is in I. There exists a number c between a and x such that

$$\frac{F(x) - F(a)}{G(x) - G(a)} = \frac{F'(c)}{G'(c)},$$

or since $F(a) = G(a) = 0$,

$$\frac{F(x)}{G(x)} = \frac{F'(c)}{G'(c)}.$$

Since x and c are points in I, we can also write that $F(x) = f(x)$, $F'(c) = f'(c)$, $G(x) = g(x)$, $G'(c) = g'(c)$, and therefore

$$\frac{f(x)}{g(x)} = \frac{F(x)}{G(x)} = \frac{F'(c)}{G'(c)} = \frac{f'(c)}{g'(c)}.$$

If we now let x approach a, then c must also approach a since it is always between a and x. Consequently,

$$\lim_{x \to a} \frac{f(x)}{g(x)} = \lim_{c \to a} \frac{f'(c)}{g'(c)} = \lim_{x \to a} \frac{f'(x)}{g'(x)},$$

and if

$$L = \lim_{x \to a} \frac{f'(x)}{g'(x)},$$

it follows that

$$\lim_{x \to a} \frac{f(x)}{g(x)} = L. \qquad\blacksquare$$

This theorem is also valid if L is replaced by ∞ or $-\infty$. The only difference in the proof is to make the same change in the last sentence.

Theorem 4.4 is also valid if $x \to a$ is replaced by either a right-hand limit, $x \to a^+$, or a left-hand limit, $x \to a^-$. The only difference in these cases is that interval I is replaced by open intervals $a < x < b$ and $b < x < a$, respectively, and the proofs are almost identical. In addition, the following theorem indicates that $x \to a$ can be replaced by $x \to \infty$ (or $x \to -\infty$).

THEOREM 4.5 (L'Hôpital's rule)

Suppose functions $f(x)$ and $g(x)$ satisfy the following conditions:

1. $f(x)$ and $g(x)$ are differentiable for some interval $x > b > 0$;
2. $g'(x) \neq 0$ for $x > b > 0$;
3. $\lim\limits_{x \to \infty} f(x) = 0 = \lim\limits_{x \to \infty} g(x)$;
4. $\lim\limits_{x \to \infty} \dfrac{f'(x)}{g'(x)} = L \qquad (\text{or } \pm\infty)$.

Then,

$$\lim_{x \to \infty} \frac{f(x)}{g(x)} = L \qquad (\text{or } \pm\infty).$$

In other words, L'Hôpital's rule applies to any type of limit that yields the indeterminate form $0/0$ (be it $x \to a$, $x \to a^+$, $x \to a^-$, $x \to \infty$, or $x \to -\infty$). A common error when using L'Hôpital's rule is to differentiate $f(x)/g(x)$ with the quotient rule and then take the limit of the resulting derivative. L'Hôpital's rule calls for the limit of $f'(x)/g'(x)$; $f(x)$ and $g(x)$ are differentiated separately.

If we use L'Hôpital's rule on limit 4.16, we find that

$$\lim_{x \to 0^+} \frac{\sqrt{1+x} - 1}{\sqrt{x}} = \lim_{x \to 0^+} \frac{\frac{1}{2\sqrt{1+x}}}{\frac{1}{2\sqrt{x}}} = \lim_{x \to 0^+} \frac{\sqrt{x}}{\sqrt{1+x}} = 0.$$

For limit 4.17, we have

$$\lim_{x \to 2} \frac{x^3 - x^2 - 8x + 12}{x^2 - 4x + 4} = \lim_{x \to 2} \frac{3x^2 - 2x - 8}{2x - 4},$$

which is still a limit of the indeterminate form $0/0$. Note that this is a conditional equation; that is, it says that the limit on the left is equal to the limit on the right, provided that the limit on the right exists. If we apply L'Hôpital's rule a second time, to the limit on the right, we obtain

$$\lim_{x \to 2} \frac{x^3 - x^2 - 8x + 12}{x^2 - 4x + 4} = \lim_{x \to 2} \frac{6x - 2}{2} = 5.$$

EXAMPLE 4.40

Evaluate the following limits:

(a) $\displaystyle\lim_{x \to -3} \frac{3 + x}{\sqrt{3} - \sqrt{-x}}$ 　　(b) $\displaystyle\lim_{x \to 0} \frac{\tan x}{x}$

(c) $\displaystyle\lim_{x \to 4} \frac{x - 4}{x^2 - 8x + 16}$ 　　(d) $\displaystyle\lim_{x \to 4} \frac{x^3 - 4x^2 + 9x - 36}{x^2 + 5}$

SOLUTION

(a) Since we have the indeterminate form $0/0$, we use L'Hôpital's rule to write

$$\lim_{x \to -3} \frac{3 + x}{\sqrt{3} - \sqrt{-x}} = \lim_{x \to -3} \frac{1}{\frac{1}{2\sqrt{-x}}} = \lim_{x \to -3} 2\sqrt{-x} = 2\sqrt{3}.$$

(b) If we use L'Hôpital's rule, we have

$$\lim_{x \to 0} \frac{\tan x}{x} = \lim_{x \to 0} \frac{\sec^2 x}{1} = 1.$$

(c) By L'Hôpital's rule,

$$\lim_{x \to 4} \frac{x - 4}{x^2 - 8x + 16} = \lim_{x \to 4} \frac{1}{2x - 8}.$$

Since

$$\lim_{x \to 4^+} \frac{1}{2x - 8} = \infty \qquad \text{and} \qquad \lim_{x \to 4^-} \frac{1}{2x - 8} = -\infty,$$

we conclude that

$$\lim_{x \to 4^+} \frac{x - 4}{x^2 - 8x + 16} = \infty \qquad \text{and} \qquad \lim_{x \to 4^-} \frac{x - 4}{x^2 - 8x + 16} = -\infty.$$

(d) This limit is not of the indeterminate form $0/0$ since $\lim_{x \to 4} (x^2 + 5) = 21$; thus we cannot use L'Hôpital's rule. Since $\lim_{x \to 4} (x^3 - 4x^2 + 9x - 36) = 0$,

$$\lim_{x \to 4} \frac{x^3 - 4x^2 + 9x - 36}{x^2 + 5} = 0.$$

Had we used L'Hôpital's rule in part (d) of this example, we would have obtained an incorrect answer:

$$\lim_{x \to 4} \frac{x^3 - 4x^2 + 9x - 36}{x^2 + 5} = \lim_{x \to 4} \frac{3x^2 - 8x + 9}{2x} = \frac{25}{8}.$$

 In other words, L'Hôpital's rule is not to be used indiscriminately; *it must be used only on the indeterminate forms for which it is designed.*

The Indeterminate Form ∞/∞

The limit

$$\lim_{x \to \infty} \frac{1 + \sqrt{x - 1}}{2x + 5}$$

is said to be of the **indeterminate form** ∞/∞ since numerator and denominator become increasingly large as $x \to \infty$. Theorems 4.4 and 4.5 for L'Hôpital's rule can be adapted to this indeterminate form also; hence we calculate that

$$\lim_{x \to \infty} \frac{1 + \sqrt{x - 1}}{2x + 5} = \lim_{x \to \infty} \frac{\dfrac{1}{2\sqrt{x - 1}}}{2} = \lim_{x \to \infty} \frac{1}{4\sqrt{x - 1}} = 0.$$

EXAMPLE 4.41

Evaluate the following limits, if they exist:

(a) $\quad \lim_{x \to \infty} \dfrac{x^2}{e^x}$ (b) $\quad \lim_{x \to -\infty} \dfrac{\sqrt{2x^2 + 3x + 2}}{1 - x}$

SOLUTION

(a) Since this limit exhibits the indeterminate form ∞/∞, we use L'Hôpital's rule to write

$$\lim_{x \to \infty} \frac{x^2}{e^x} = \lim_{x \to \infty} \frac{2x}{e^x}.$$

Since this limit is still of the form ∞/∞, we use L'Hôpital's rule again:

$$\lim_{x \to \infty} \frac{x^2}{e^x} = \lim_{x \to \infty} \frac{2}{e^x} = 0.$$

The same result would occur for any positive power n on x; that is, $\lim_{x \to \infty} \dfrac{x^n}{e^x} = 0$. What this shows is that exponential functions grow more rapidly for large x than power functions.

(b) By L'Hôpital's rule,

$$\lim_{x \to -\infty} \frac{\sqrt{2x^2 + 3x + 2}}{1 - x} = \lim_{x \to -\infty} \frac{\dfrac{4x + 3}{2\sqrt{2x^2 + 3x + 2}}}{-1} = \lim_{x \to -\infty} \frac{-(4x + 3)}{2\sqrt{2x^2 + 3x + 2}}.$$

This limit is also of the indeterminate form ∞/∞. Further applications of L'Hôpital's rule do not lead to a simpler form for the limit. Thus, L'Hôpital's rule does not prove advantageous on this limit. It is better to divide numerator and denominator by x:

$$\lim_{x \to -\infty} \frac{\sqrt{2x^2 + 3x + 2}}{1 - x} = \lim_{x \to -\infty} \frac{\dfrac{\sqrt{2x^2 + 3x + 2}}{x}}{\dfrac{1}{x} - 1} = \lim_{x \to -\infty} \frac{-\sqrt{2 + \dfrac{3}{x} + \dfrac{2}{x^2}}}{\dfrac{1}{x} - 1} = \sqrt{2}.$$

EXAMPLE 4.42

Show that L'Hôpital's rule cannot be used to evaluate

$$\lim_{x \to \infty} \frac{x - \cos x}{x}.$$

What is the value of the limit?

SOLUTION   The limit is of the indeterminate form ∞/∞. If we apply L'Hôpital's rule we obtain

$$\lim_{x \to \infty} \frac{x - \cos x}{x} = \lim_{x \to \infty} \frac{1 + \sin x}{1}.$$

But this limit does not exist, and therefore L'Hôpital's rule has failed. But we do not need the rule since division of numerator and denominator by x gives

$$\lim_{x \to \infty} \frac{x - \cos x}{x} = \lim_{x \to \infty} \left(1 - \frac{\cos x}{x}\right) = 1.$$

The Indeterminate Form $0 \cdot \infty$

The limits

$$\lim_{x \to \infty} xe^{-2x} \qquad \text{and} \qquad \lim_{x \to 0^+} x^2 \ln x$$

are said to be of the **indeterminate form** $0 \cdot \infty$. L'Hôpital's rule can again be used if we first rearrange the limits into one of the forms $0/0$ or ∞/∞:

$$\lim_{x \to \infty} xe^{-2x} = \lim_{x \to \infty} \frac{x}{e^{2x}} = \lim_{x \to \infty} \frac{1}{2e^{2x}} = 0;$$

$$\lim_{x \to 0^+} x^2 \ln x = \lim_{x \to 0^+} \frac{\ln x}{\dfrac{1}{x^2}} = \lim_{x \to 0^+} \frac{\dfrac{1}{x}}{\dfrac{-2}{x^3}} = \lim_{x \to 0^+} \left(-\frac{x^2}{2}\right) = 0.$$

Note that had we converted the second limit into the $0/0$ form, we would have had

$$\lim_{x \to 0^+} x^2 \ln x = \lim_{x \to 0^+} \frac{x^2}{\dfrac{1}{\ln x}} = \lim_{x \to 0^+} \frac{2x}{\dfrac{-1}{x(\ln x)^2}} = \lim_{x \to 0^+} -2x^2(\ln x)^2.$$

Although this is correct, the limit on the right is more difficult to evaluate than the original. In other words, we must be judicious in converting a limit from the $0 \cdot \infty$ indeterminate form to either $0/0$ or ∞/∞.

EXAMPLE 4.43

Evaluate the following limits if they exist:

$$\text{(a)} \quad \lim_{x \to \pi/2} (x - \pi/2) \sec x \qquad \text{(b)} \quad \lim_{x \to 0^+} x e^{1/x}$$

SOLUTION

(a) $\displaystyle \lim_{x \to \pi/2} (x - \pi/2) \sec x = \lim_{x \to \pi/2} \frac{x - \pi/2}{\cos x} = \lim_{x \to \pi/2} \frac{1}{-\sin x} = -1$

(b) $\displaystyle \lim_{x \to 0^+} x e^{1/x} = \lim_{x \to 0^+} \frac{e^{1/x}}{1/x} = \lim_{x \to 0^+} \frac{e^{1/x}(-1/x^2)}{-1/x^2} = \lim_{x \to 0^+} e^{1/x} = \infty$

The Indeterminate Forms 0^0, 1^∞, ∞^0, and $\infty - \infty$

Various other indeterminate forms arise in the evaluation of limits, and many of these can be reduced to the $0/0$ and ∞/∞ forms by introducing logarithms. In particular, the limits

$$\lim_{x \to 0^+} x^x, \qquad \lim_{x \to \infty} \left(1 + \frac{1}{x}\right)^{x^2}, \qquad \lim_{x \to \pi/2^-} (\sec x)^{\cos x}, \qquad \text{and} \quad \lim_{x \to \pi/2} (\sec x - \tan x) \qquad (4.18)$$

are said to display the indeterminate forms 0^0, 1^∞, ∞^0, and $\infty - \infty$, respectively. To evaluate $\lim_{x \to 0^+} x^x$, we set

$$L = \lim_{x \to 0^+} x^x$$

and take natural logarithms of both sides,

$$\ln L = \ln \left(\lim_{x \to 0^+} x^x \right).$$

As the logarithm function is continuous, we may interchange the limit and logarithm operations (see Theorem 2.5),

$$\ln L = \lim_{x \to 0^+} (\ln x^x) = \lim_{x \to 0^+} x \ln x = \lim_{x \to 0^+} \frac{\ln x}{\dfrac{1}{x}}.$$

We are now in a position to use L'Hôpital's rule:

$$\ln L = \lim_{x \to 0^+} \frac{\dfrac{1}{x}}{\dfrac{-1}{x^2}} = \lim_{x \to 0^+} (-x) = 0.$$

Exponentiation of both sides of $\ln L = 0$ now gives $L = e^0 = 1$; that is,

$$\lim_{x \to 0^+} x^x = 1.$$

For the second limit in 4.18 we again set

$$L = \lim_{x \to \infty} \left(1 + \frac{1}{x}\right)^{x^2}$$

and take natural logarithms:

$$\ln L = \ln\left[\lim_{x\to\infty}\left(1+\frac{1}{x}\right)^{x^2}\right] = \lim_{x\to\infty}\left[\ln\left(1+\frac{1}{x}\right)^{x^2}\right]$$

$$= \lim_{x\to\infty}\left[x^2\ln\left(1+\frac{1}{x}\right)\right] = \lim_{x\to\infty}\left[\frac{\ln\left(\dfrac{x+1}{x}\right)}{\dfrac{1}{x^2}}\right].$$

By L'Hôpital's rule, we have

$$\ln L = \lim_{x\to\infty}\left[\frac{\dfrac{x}{x+1}\left(\dfrac{-1}{x^2}\right)}{\dfrac{-2}{x^3}}\right] = \lim_{x\to\infty}\frac{x^2}{2(x+1)} = \infty.$$

Consequently,

$$L = \lim_{x\to\infty}\left(1+\frac{1}{x}\right)^{x^2} = \infty.$$

In the third limit of 4.18, we set

$$L = \lim_{x\to\pi/2^-}(\sec x)^{\cos x},$$

in which case

$$\ln L = \ln\left[\lim_{x\to\pi/2^-}(\sec x)^{\cos x}\right] = \lim_{x\to\pi/2^-}[\cos x\ln(\sec x)]$$

$$= \lim_{x\to\pi/2^-}\left[\frac{\ln(\sec x)}{\sec x}\right] = \lim_{x\to\pi/2^-}\left[\frac{\dfrac{1}{\sec x}\sec x\tan x}{\sec x\tan x}\right]$$

$$= \lim_{x\to\pi/2^-}\cos x = 0.$$

Thus,

$$L = \lim_{x\to\pi/2^-}(\sec x)^{\cos x} = e^0 = 1.$$

Finally, the last limit in 4.18 is evaluated by rewriting it in the 0/0 form,

$$\lim_{x\to\pi/2}(\sec x - \tan x) = \lim_{x\to\pi/2}\left(\frac{1-\sin x}{\cos x}\right) = \lim_{x\to\pi/2}\frac{-\cos x}{-\sin x} = 0.$$

EXAMPLE 4.44

Plot a graph of the function $f(x) = x^2\ln x$. Find limits of $f(x)$ and $f'(x)$ as $x\to 0^+$. Where is the point of inflection on the graph?

FIGURE 4.91   Plot of $x^2 \ln x$

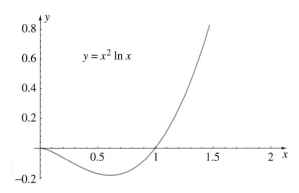

$y = x^2 \ln x$

SOLUTION   The plot in Figure 4.91 suggests that $f(x)$ and $f'(x)$ both approach 0 as $x \to 0^+$. To confirm this we use L'Hôpital's rule to calculate

$$\lim_{x \to 0^+} f(x) = \lim_{x \to 0^+} x^2 \ln x = \lim_{x \to 0^+} \frac{\ln x}{\dfrac{1}{x^2}} = \lim_{x \to 0^+} \frac{\dfrac{1}{x}}{-\dfrac{2}{x^3}} = \lim_{x \to 0^+} (-2x^2) = 0^-$$

and

$$\lim_{x \to 0^+} f'(x) = \lim_{x \to 0^+} \left(2x \ln x + \frac{x^2}{x} \right) = 2 \lim_{x \to 0^+} \frac{\ln x}{\dfrac{1}{x}} = 2 \lim_{x \to 0^+} \frac{\dfrac{1}{x}}{-\dfrac{1}{x^2}} = 2 \lim_{x \to 0^+} (-x) = 0^-.$$

For the point of inflection between $x = 0$ and $x = 1/2$, we solve

$$0 = f''(x) = 2 \ln x + \frac{2x}{x} + 1 = 2 \ln x + 3.$$

The only solution is $x = e^{-3/2}$. Since $f''(x)$ changes sign as x passes through $e^{-3/2}$, there is a point of inflection at $(e^{-3/2}, -3e^{-3}/2)$.

EXERCISES 4.11

In Exercises 1–42 evaluate the limit, if it exists.

1. $\displaystyle \lim_{x \to 0} \frac{x^2 + 3x}{x^3 + 5x^2}$

2. $\displaystyle \lim_{x \to 3} \frac{x^2 - 9}{x - 3}$

3. $\displaystyle \lim_{x \to -\infty} \frac{x^3 + 3x - 2}{x^2 + 5x + 1}$

4. $\displaystyle \lim_{x \to \infty} \frac{2x^2 + 3x}{5x^3 + 4}$

5. $\displaystyle \lim_{x \to 5} \frac{x^2 - 10x + 25}{x^3 - 125}$

6. $\displaystyle \lim_{x \to 1} \frac{1}{(x - 1)^2}$

7. $\displaystyle \lim_{x \to \infty} \frac{\sqrt{x^2 + 1}}{2x + 5}$

8. $\displaystyle \lim_{x \to -\infty} \frac{\sin x}{x}$

9. $\displaystyle \lim_{x \to \infty} \frac{\sin (2/x)}{\sin (1/x)}$

10. $\displaystyle \lim_{x \to \pi/2} \frac{\cos x}{(x - \pi/2)^2}$

11. $\displaystyle \lim_{x \to 1^+} \frac{(1 - 1/x)^3}{\sqrt{x - 1}}$

12. $\displaystyle \lim_{x \to \infty} \frac{\sin (1/x)}{1/x^2}$

13. $\displaystyle \lim_{x \to 9^-} \frac{\sqrt{x} - 3}{\sqrt{9 - x}}$

14. $\displaystyle \lim_{x \to 0} \frac{\sqrt{5 + x} - \sqrt{5 - x}}{x}$

15. $\displaystyle \lim_{x \to 0} \frac{x - \sin x}{x^3}$

16. $\displaystyle \lim_{x \to a} \frac{x^n - a^n}{x - a}$

17. $\displaystyle \lim_{x \to 0} \frac{(1 - \cos x)^2}{3x^2}$

18. $\displaystyle \lim_{x \to 0} \frac{\tan x}{x}$

19. $\displaystyle \lim_{x \to 0} \frac{\sin 3x}{\tan 2x}$

20. $\displaystyle \lim_{x \to 1} \frac{(1 - \sqrt{2 - x})^{3/2}}{x - 1}$

∗ 21. $\displaystyle \lim_{x \to 0} \frac{\sqrt{x + 1} - \sqrt{2x + 1}}{\sqrt{3x + 4} - \sqrt{2x + 4}}$

* **22.** $\lim_{x \to 0} \dfrac{(1 - \cos x)^2}{3x^4}$

* **23.** $\lim_{x \to \infty} x \sin\left(\dfrac{1}{x}\right)$

* **24.** $\lim_{x \to 2} \dfrac{(x - 2)^{10}}{\left(\sqrt{x} - \sqrt{2}\right)^{10}}$

* **25.** $\lim_{x \to 0} \left(\dfrac{4}{x^2} - \dfrac{2}{1 - \cos x}\right)$

* **26.** $\lim_{x \to \infty} xe^x$

* **27.** $\lim_{x \to \infty} x^2 e^{-4x}$

* **28.** $\lim_{x \to -\infty} x \sin\left(\dfrac{4}{x}\right)$

* **29.** $\lim_{x \to 0} x \cot x$

* **30.** $\lim_{x \to 0} \csc x(1 - \cos x)$

* **31.** $\lim_{x \to 0^+} (\sin x)^x$

* **32.** $\lim_{x \to 0^+} x^{\sin x}$

* **33.** $\lim_{x \to \infty} \left(\dfrac{x + 5}{x + 3}\right)^x$

* **34.** $\lim_{x \to 0} (1 + x)^{\cot x}$

* **35.** $\lim_{x \to \infty} x^{1/x}$

* **36.** $\lim_{x \to 0^+} |\ln x|^{\sin x}$

* **37.** $\lim_{x \to 0^+} xe^{1/x}$

* **38.** $\lim_{x \to 0} (\tan x - \csc x)$

* **39.** $\lim_{x \to 0} (\csc x - \cot x)$

* **40.** $\lim_{x \to 1} \left(\dfrac{x}{\ln x} - \dfrac{1}{x \ln x}\right)$

* **41.** $\lim_{x \to 1} \left(\dfrac{x}{x - 1} - \dfrac{1}{\ln x}\right)$

* **42.** $\lim_{x \to 0} \left(\dfrac{1}{x^2} - \dfrac{1}{\sin^2 x}\right)$

In Exercises 43–54 draw a graph of the function.

* **43.** $f(x) = xe^{-2x}$

* **44.** $f(x) = x^2 e^{3x}$

* **45.** $f(x) = xe^{-x^2}$

* **46.** $f(x) = e^{1/x}$

* **47.** $f(x) = \dfrac{\ln x}{x}$

* **48.** $f(x) = x^2 \ln x$

* **49.** $f(x) = xe^{1/x}$

* **50.** $f(x) = \dfrac{x^2}{\ln x}$

* **51.** $f(x) = x^x, \quad x > 0$

* **52.** $f(x) = x^{10} e^{-x}$

* **53.** $f(x) = e^{-x} \ln x, \quad x > 0$

* **54.** $f(x) = 2 \csc x - \cot x, \quad 0 < x < \pi/2$

* **55.** Evaluate $\lim_{x \to \infty} \left(\dfrac{x + a}{x + b}\right)^{cx}$ for any constants $a, b,$ and c.

* **56.** The indeterminate forms 0^0, 1^∞, and ∞^0 are often evaluated by introducing logarithms. Show that the limit $\lim_{x \to \infty} (x - \ln x)$ can be evaluated by introducing exponentials.

* **57.** When an electrostatic field E is applied to a gaseous or liquid polar dielectric, a net dipole moment P per unit volume is set up, where

$$P(E) = \dfrac{e^E + e^{-E}}{e^E - e^{-E}} - \dfrac{1}{E}.$$

Show that $\lim_{E \to 0^+} P(E) = 0$.

* **58.** *Planck's law* for the energy density ψ of blackbody radiation states that

$$\psi = \psi(\lambda) = \dfrac{k\lambda^{-5}}{e^{c/\lambda} - 1},$$

where k and c are positive constants and λ is the wavelength of the radiation.

   (a) Show that $\lim_{\lambda \to 0^+} \psi(\lambda) = 0$ and $\lim_{\lambda \to \infty} \psi(\lambda) = 0$.

   (b) Show that $\psi(\lambda)$ has one critical point that must satisfy the equation $(5\lambda - c)e^{c/\lambda} = 5\lambda$. Find the critical point accurate to seven decimal places when $c = 0.000\,143\,86$.

   (c) Draw a graph of the function $\psi(\lambda)$ when $c = 0.000\,143\,86$ and $k = 1$.

* **59.** The following limit arises in the calculation of the electric field intensity for a half-wave antenna: $\lim_{\theta \to 0} f(\theta)$, where

$$f(\theta) = \sin\theta \left\{ \dfrac{\sin\left[\pi/2(\cos\theta - 1)\right]}{\cos\theta - 1} + \dfrac{\sin\left[\pi/2(\cos\theta + 1)\right]}{\cos\theta + 1} \right\}.$$

Evaluate this limit by first showing that $f(\theta)$ can be written in the form

$$f(\theta) = \dfrac{2\cos\left(\dfrac{\pi}{2}\cos\theta\right)}{\sin\theta},$$

and then using L'Hôpital's rule.

* **60.** Find all values of $a, b,$ and c for which

$$\lim_{x \to 0} \dfrac{e^{ax} - bx - \cos(x + cx^2)}{2x^3 + 5x^2} = 5.$$

* **61.** The maximum flow rate of gas through a nozzle is governed by the function

$$f(x) = x\left(\dfrac{2}{x + 1}\right)^{(x+1)/(x-1)}.$$

   (a) Plot a graph of $f(x)$ on the interval $0 \leq x \leq 20$. Did you get any error messages? Should you have? Show that the function is discontinuous at $x = 1$, but $\lim_{x \to 1} f(x) = 1/e$.

   (b) Plot the function on the interval $0 \leq x \leq 200$. Does the graph appear to have a horizontal asymptote? Show that $\lim_{x \to \infty} f(x) = 2$.

** **62.**   (a) Sketch a graph of the function

$$f(x) = \begin{cases} e^{-1/x^2}, & x \neq 0 \\ 0, & x = 0. \end{cases}$$

   (b) Show that for every positive integer n,

$$\lim_{x \to 0} \dfrac{e^{-1/x^2}}{x^n} = 0.$$

   (c) Prove by mathematical induction that $f^{(n)}(0) = 0$, where $f^{(n)}(0)$ is the nth derivative of $f(x)$ evaluated at $x = 0$.

4.12 Differentials

In Section 3.1 we pointed out that the notation dy/dx for the derivative of a function $y = f(x)$ should not be considered a quotient. Beginning in Chapter 5, however, it is essential that we be able to do this, and therefore in this section we define "differentials" dx and dy so that dy/dx can be regarded as a quotient.

When we use the notation

$$\frac{dy}{dx} = \lim_{\Delta x \to 0} \frac{f(x + \Delta x) - f(x)}{\Delta x}$$

for the derivative of a function $y = f(x)$, we call Δx an *increment* in x. It represents a change in the value of the independent variable from some value x to another value $x + \Delta x$. This change can be positive or negative depending on whether we want $x + \Delta x$ to be larger or smaller than x. When the independent variable changes from x to $x + \Delta x$, the dependent variable changes by an amount Δy, where

$$\Delta y = f(x + \Delta x) - f(x). \tag{4.19}$$

In other words, Δy is the change in y resulting from the change Δx in x. For the function in Figure 4.92, Δy is positive when Δx is positive, and Δy is negative when Δx is negative.

FIGURE 4.92a dy and Δy for $dx > 0$

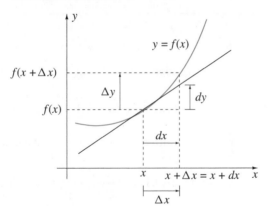

FIGURE 4.92b dy and Δy for $dx < 0$

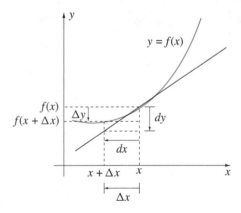

For example, when $y = x^2 - 2x$, the change Δy in y when x is changed from 3 to 3.2 is

$$\Delta y = [(3.2)^2 - 2(3.2)] - [3^2 - 2(3)] = 0.84.$$

The function increases by 0.84 when x increases from 3 to 3.2.

For purposes of integration, a topic that begins in Chapter 5 and continues in every chapter thereafter, an alternative notation for an increment in the independent variable x is more suggestive.

DEFINITION 4.7

An increment Δx in the independent variable x is denoted by

$$dx = \Delta x, \tag{4.20}$$

and when written as dx, it is called the **differential** of x.

▽
⦰ The differential dx is synonymous with the increment Δx; it represents a change in x (in most applications a very small change). The differential of the dependent variable is *not* synonymous with Δy.

DEFINITION 4.8

The differential of $y = f(x)$, corresponding to the differential dx in x, is denoted by dy and is defined by

$$dy = f'(x)\,dx. \qquad (4.21)$$

The difference between Δy and dy is most easily seen in Figures 4.92. We know that Δy is the exact change in the function $y = f(x)$ when x is changed by an amount Δx or dx. It is the difference in the height of the curve at x and $x + \Delta x = x + dx$. Now the slope of the tangent line to the graph at the point (x, y) is $f'(x)$. Definition 4.8 indicates that dy can be interpreted as the difference in the height of this tangent line at x and at $x + dx$. In other words, dy is the change in y corresponding to the change dx in x if we follow the tangent line to $y = f(x)$ at (x, y) rather than the curve itself.

Figures 4.92 also suggest that when dx is very small (close to zero), dy is approximately equal to Δy; that is,

$$dy \approx \Delta y \qquad \text{when } dx \approx 0.$$

This is illustrated in the following numerical example.

EXAMPLE 4.45

▼

Find Δy and dy for the function $y = f(x) = \sqrt{x^2 + 1}$ when $x = 2$ and $dx = 0.1$.

SOLUTION   According to equation 4.19, the change in y as x increases from 2 to 2.1 is

$$\Delta y = f(2.1) - f(2) = \sqrt{(2.1)^2 + 1} - \sqrt{2^2 + 1} = 0.089\,87.$$

Since $f'(x) = x/\sqrt{x^2 + 1}$, the differential of y for $x = 2$ and $dx = 0.1$ is

$$dy = f'(2)(0.1) = \frac{2}{\sqrt{2^2 + 1}}(0.1) = 0.089\,44.$$

The difference between dy and Δy is, therefore, $0.000\,43$, a difference of 43 parts in 8987.

Before the invention of electronic calculators, differentials were used to approximate a function near points at which it was easily evaluated. For example, imagine trying to evaluate the function $f(x) = x^{1/3}$ at $x = 126$ without a calculator. We could use differentials to approximate $126^{1/3}$ as follows:

$$126^{1/3} = f(126) = f(125) + \Delta y \approx f(125) + dy$$

$$= 5 + f'(5)(1) = 5 + \frac{1}{3(125)^{2/3}} = 5 + \frac{1}{75} = \frac{376}{75}.$$

With the advent of the electronic calculator, problems of this type are archaic. On the other hand, differentials are indispensable when we examine changes in a function without specifying values for the independent variable. Very prominent in this context are relative and percentage changes.

When a quantity y undergoes a change Δy, then its **relative change** is defined as

$$\frac{\Delta y}{y}, \tag{4.22}$$

and its **percentage change** is given by

$$100\frac{\Delta y}{y}. \tag{4.23}$$

Sometimes relative and percentage changes are more important than actual changes. To illustrate this, consider the function $V = 4\pi r^3/3$, which represents the volume of a sphere. If the radius of the sphere is increased from 0.10 m to 0.11 m, then the change in the volume of the sphere is

$$\Delta V = \frac{4}{3}\pi(0.11)^3 - \frac{4}{3}\pi(0.10)^3 = 4.4\pi \times 10^{-4} \text{ m}^3.$$

This is not a very large quantity, but in relation to the original size of the sphere, we have a relative change of

$$\frac{\Delta V}{V} = \frac{4.4\pi \times 10^{-4}}{4\pi(0.10)^3/3} = 0.33$$

and a percentage change of

$$100\frac{\Delta V}{V} = 33\%.$$

Suppose the same increase of 0.01 m is applied to a sphere with radius 100 m. The change in the volume is

$$\Delta V = \frac{4}{3}\pi(100.01)^3 - \frac{4}{3}\pi(100)^3 = 4.0\pi \times 10^2 \text{ m}^3.$$

This is quite a large change in volume, but the relative change is

$$\frac{\Delta V}{V} = \frac{4.0\pi \times 10^2}{4\pi(100)^3/3} = 3.0 \times 10^{-4},$$

and the percentage change is

$$100\frac{\Delta V}{V} = 0.03\%.$$

Although the change 400π in V when $r = 100$ is much larger than the change 0.00044π when $r = 0.1$, the relative and percentage changes are much smaller when $r = 100$. We see, then, that in certain cases it may be relative and percentage changes that are significant rather than actual changes.

In the example above, when $r = 100$ m, the change $dr = 0.01$ is certainly small compared to r. Therefore, we should be able to use the differential $dV = V'(r)\,dr = 4\pi r^2\,dr$ to approximate ΔV. With $r = 100$ and $dr = 0.01$, we have

$$dV = 4\pi(100)^2(0.01) = 4.0\pi \times 10^2 \text{ m}^3.$$

(To two significant figures, dV is equal to ΔV.) Thus, relative change in V is approximately equal to dV/V and percentage change $100dV/V$. So, for small changes in an independent variable, the differential of the dependent variable may be used in place of its increment in the calculation of relative and percentage changes; that is, equations 4.22 and 4.23 can be replaced by

$$\frac{dy}{y} \quad \text{and} \quad 100\frac{dy}{y}.$$

We do this in the following example.

EXAMPLE 4.46

When a pendulum swings, the frequency (number of cycles per second) of its oscillations is given by

$$f = h(l) = 2\pi \sqrt{\frac{g}{l}},$$

where l is the length of the pendulum and $g > 0$ is the acceleration due to gravity. If the length of the pendulum is increased by $\frac{1}{4}\%$, calculate the approximate percentage change in f.

SOLUTION   The approximate change in f is given by

$$df = h'(l)\,dl = 2\pi\sqrt{g}\left(-\frac{1}{2}\right)l^{-3/2}\,dl = -\pi\sqrt{g}\frac{dl}{l^{3/2}};$$

hence the approximate percentage change in f is

$$100\frac{df}{f} = 100\left(\frac{-\pi\sqrt{g}\,dl}{l^{3/2}}\right)\left(\frac{l^{1/2}}{2\pi\sqrt{g}}\right) = -\frac{1}{2}\left(100\frac{dl}{l}\right).$$

But because l increases by $\frac{1}{4}\%$, it follows that $100(dl/l) = 1/4$, and

$$100\frac{df}{f} = -\frac{1}{8}.$$

Therefore, the frequency changes by $-\frac{1}{8}\%$, the negative sign indicating that because l increases, f decreases.

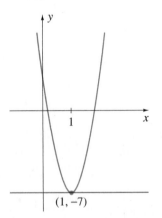

FIGURE 4.93   Differentials do not approximate changes at critical points

(1, −7)

The differential dy cannot always be used as an approximation for the actual change Δy in a function $y = f(x)$. Sometimes it cannot be used even when dx is very close to zero. For example, if $f(x) = 2x^3 + 9x^2 - 24x + 6$, then

$$dy = f'(x)\,dx = (6x^2 + 18x - 24)\,dx.$$

If x is changed from 1 to 1.01, then the approximate change in y as predicted by the differential is

$$dy = (6 + 18 - 24)(0.01) = 0.$$

In fact, for any dx whatsoever, we find that $dy = 0$. Geometrically speaking, we can *see* why. Since $f'(1) = 0$, $x = 1$ is a critical point of $f(x)$ (Figure 4.93); therefore, dy, which is the tangent line approximation to Δy, will always be zero. We cannot use differentials to approximate function changes at critical points.

The differential for the function $y = f(x) = x^{100}$ is

$$dy = 100x^{99}\,dx.$$

If x is changed by 1% from $x = 1$ to $x = 1.01$, then $dx = 0.01$ and

$$dy = 100(1)^{99}(0.01) = 1.0.$$

The actual change in y is

$$\Delta y = (1.01)^{100} - 1^{100} = 1.7.$$

We would hardly regard this dy as a very good approximation for Δy even though the change in x is only 1%.

The latter example raises the question "How small, in general, must dx be in order that dy be a reasonable approximation for Δy?" This is not a simple question to answer. We will discuss approximations in more detail in Chapter 10, and then be able to answer a more important question: How good an approximation to Δy is dy? After all, it is not much use to say that $dy \approx \Delta y$ if we cannot say to how many decimal places the approximation is accurate. Suffice it to say now that use of the differential dy to approximate Δy is to be regarded with some reservation. This is not to say that differentials are useless. We will see in Chapters 5–7 that differentials are indispensable to the topic of integration.

We make one last comment before leaving this section. If equation 4.21 is divided by differential dx, then

$$\frac{dy}{dx} = f'(x).$$

Now, the left side of this equation is the differential of y divided by the differential of x. The quotient of differentials dy and dx is equal to the derivative $f'(x)$. The entity dy/dx can henceforth be regarded either as "the derivative of y with respect to x" or as "dy divided by dx," whichever is appropriate for the discussion at hand.

EXERCISES 4.12

In Exercises 1–10 find dy in terms of x and dx.

1. $y = x^2 + 3x - 2$

2. $y = \dfrac{x+1}{x-1}$

3. $y = \sqrt{x^2 - 2x}$

4. $y = \sin(x^2 + 2) - \cos x$

5. $y = x^{1/3} - x^{5/3}$

6. $y = x^3\sqrt{3 - 4x^2}$

7. $y = x^2 \sin x$

8. $y = \dfrac{x^3 - 3x^2 + 3x + 5}{x^2 - 2x + 1}$

9. $y = \sqrt{1 + \sqrt{1-x}}$

10. $y = \dfrac{x^2(x-2)}{x^3 + 5x}$

11. The momentum M and kinetic energy K of a mass m moving with speed v are given by $M = mv$ and $K = \dfrac{1}{2}mv^2$. If v is changed by 1%, what are approximate percentage changes in M and K?

12. The magnitude of the gravitational force of attraction between two point masses m and M is given by

$$F = \frac{GmM}{r^2},$$

where $G > 0$ is a constant and r is the distance between the masses. If r changes by 2%, by how much does F change approximately?

13. According to Example 1.9, the range of a shell fired from an artillery gun with velocity v at angle θ is given by $R = (v^2 \sin 2\theta)/9.81$. Use differentials to find the approximate percentage change in R if θ is increased by 1% from an angle of $\pi/3$ radians.

14. According to Example 1.9, the maximum height attained by a shell fired from an artillery gun with velocity v at angle θ is given by $H = (v^2 \sin^2 \theta)/19.62$. Use differentials to find the approximate percentage change in H if θ is increased by 2% from an angle of $\pi/3$ radians.

15. Under adiabatic expansion, a gas obeys the law $PV^{7/5} = $ a constant, where P is pressure and V is volume. If the pressure is increased by 2%, find the approximate percentage change in the volume.

16. The magnitude of the gravitational force of attraction between two point masses m and M is defined in Exercise 12. If the earth is con-

sidered a perfect sphere (radius 6.37×10^6 m), then this law predicts a gravitational attraction of $9.81m$ newtons on a mass m on its surface. Use differentials to determine the height above the surface of the earth at which the gravitational attraction decreases to $9.80m$ newtons.

$*$ **17.** When a force F is applied to the object of mass m in the figure below, three other forces act on m: the force of gravity directly downward, a reactional force of the supporting surface, and a horizontal

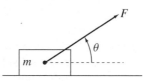

frictional force opposing motion. The least force that will overcome friction and produce motion is given by

$$F = \frac{9.81\mu m}{\cos\theta + \mu\sin\theta},$$

where μ is a constant called the *coefficient of static friction*. Use differentials to calculate the approximate percentage change in F if θ is increased by 2% from an angle of $\pi/4$ radians.

$*$ **18.** The volume of a right circular cylinder is $V = \pi r^2 h$, where r is the radius and h is the height. Use differentials to show the following:

(a) If h can be measured exactly, but r is subject to an error of $a\%$, the error in V is $2a\%$.

(b) If r can be measured exactly, but h is subject to an error of $b\%$, the error in V is $b\%$.

(c) What is the maximum percentage error in V if r is subject to an error of $a\%$ and h is subject to an error of $b\%$?

$*$ **19.** Use differentials to show that if $y = x^n$, where n is a nonzero constant, and x is subject to an error of $a\%$, then the resulting error in y is $na\%$.

* **20.** Suppose that $z = x^n y^m$, where n and m are nonzero constants.

     (a) If x is subject to an error of $a\%$, but y is subject to no error, what is the resulting error in z?

     (b) If y is subject to an error of $b\%$, but x is subject to no error, what is the resulting error in z?

     (c) What is the maximum percentage error in z if x and y are subject to errors of $a\%$ and $b\%$, respectively?

* **21.** Repeat Exercise 20 if $z = x^n / y^m$.

* **22.** A prism can be used to measure the index of refraction n of the material in the prism. According to Exercise 73 in Section 4.7, n is given by the formula

$$n = \frac{\sin\left[(\psi_m + \gamma)/2\right]}{\sin(\gamma/2)}.$$

Use differentials to find the approximate percentage error in n if the measurement of ψ_m can be out by 1% when $\psi_m = \pi/6$ and $\gamma = \pi/3$. Assume that γ is known exactly.

* **23.** Repeat Exercise 22 if ψ_m is known precisely but the measurement of γ can be out by 1%.

SUMMARY

In this chapter we discussed a number of applications of differentiation, the first of which was Newton's iterative procedure for approximating the roots of equations. It is perhaps the most popular of all approximation methods, because of its speed, simplicity, and accuracy.

     In Section 4.3 we defined a critical point of a function as a point in its domain where its first derivative either vanishes or does not exist. Geometrically, this corresponds to a point where the graph of the function has a horizontal tangent line, a vertical tangent line, or no tangent line at all. The first derivative test indicates if critical points yield relative maxima or relative minima. A function is increasing (or decreasing) on an interval if its graph slopes upward to the right (respectively, left), and this is characterized by a nonnegative (respectively, nonpositive) derivative. It is concave upward (or downward) if its slope is increasing (respectively, decreasing), and consequently if its second derivative is nonnegative (respectively, nonpositive). Points that separate intervals of opposite concavity are called points of inflection.

     In Section 4.7 we illustrated that many applied extrema problems require absolute extrema rather than relative extrema. Absolute extrema of a continuous function on a closed interval must occur at either critical points or the ends of the interval. This fact implies that to find the absolute extrema of a continuous function $f(x)$ on a closed interval $a \leq x \leq b$, we evaluate $f(x)$ at its critical points between a and b and at a and b. The largest and smallest of these values are the absolute extrema of $f(x)$ on $a \leq x \leq b$. Plotting $f(x)$ can also prove valuable.

     When an object moves along a straight line, its velocity and acceleration are the first and second derivatives, respectively, of its displacement with respect to time. In other words, if we observe straight-line motion of an object, and record its position as a function of time, then we can calculate the velocity and acceleration of that object at any instant.

     Changes in a number of interrelated quantities usually produce changes in the others — sometimes small, sometimes large. How the rates of change of these variables relate to each other was the subject of Section 4.9. Related rate problems made us acutely aware of the importance of differentiating an equation with respect to a variable only if the equation is valid for a continuous range of values of that variable.

     Potentials across resistors and inductors are expressed in terms of derivatives. Potentials across capacitors, resistors, and inductors are respectively,

$$V = \frac{Q}{C}, \qquad V = iR = R\frac{dQ}{dt}, \qquad V = L\frac{di}{dt} = L\frac{d^2Q}{dt^2}.$$

     Cauchy's generalized mean value theorem enabled us to develop L'Hôpital's rule in Section 4.11 for evaluation of various indeterminate forms such as $0/0$, ∞/∞, $0 \cdot \infty$, 0^0, 1^∞, ∞^0 and $\infty - \infty$.

     When variables change by small amounts, corresponding changes in related variables can often be approximated by differentials. Particularly important in error analyses are relative and percentage changes.

KEY TERMS

In reviewing this chapter, you should be able to define or discuss the following key terms:

Newton's iterative approach
Decreasing function
Relative (or local) maximum
First-derivative test
Concave downward
Horizontal point of inflection
Second-derivative test
Absolute (or global) minimum
Instantaneous acceleration
Related rate problems
Capacitance
Resistor
Inductor
Kirchhoff's loop rule
L'Hôpital's rule
Relative change

Increasing function
Critical point
Relative (or local) minimum
Concave upward
Points of inflection
Vertical point of inflection
Absolute (or global) maximum
Instantaneous velocity
Speed
Capacitor
Current
Resistance
Emf force
Indeterminate forms
Differential
Percentage change

REVIEW EXERCISES

1. (a) Prove that the area of the isosceles triangle in the figure below is

$$A = \frac{l^2 \sin \theta}{2}.$$

(b) If the angle θ is increasing at $1/2$ radian per minute, but l remains constant, how fast is the area of the triangle changing? What does your answer predict when $\theta = 0$, $\pi/2$, and π?

(c) When does A change most rapidly and most slowly; that is, when is $|dA/dt|$ largest and smallest?

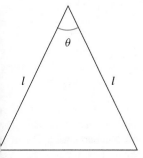

2. Draw a graph of each of the following functions, indicating all relative maxima and minima and points of inflection.

(a) $f(x) = 4x^3 + x^2 - 2x + 1$ (b) $f(x) = \dfrac{x^2 - 2x + 4}{x^2 - 2x + 1}$

3. Of all pairs of positive numbers that add to some given constant $c > 0$, find that pair which has the largest product.

4. Solve Exercise 3 for the smallest product.

* **5.** Of all pairs of positive numbers that multiply to some given constant $c > 0$, find that pair which has the smallest sum.

* **6.** Solve Exercise 5 for the largest sum.

* **7.** Two sides of the triangle in the figure below maintain constant lengths of 3 cm and 4 cm, but the length l of the third side decreases at the rate of 1 cm/min. How fast is angle θ changing when l is 4 cm?

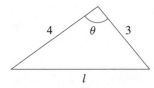

In Exercises 8–19 evaluate the limit, if it exists.

8. $\displaystyle\lim_{x \to 0} \frac{3x^2 + 2x^3}{3x^3 - 2x^2}$

9. $\displaystyle\lim_{x \to \infty} \frac{\sin 3x}{2x}$

10. $\displaystyle\lim_{x \to 4} \frac{x^2 - 16}{x - 4}$

11. $\displaystyle\lim_{x \to 0} \frac{\sin 3x}{2x}$

12. $\displaystyle\lim_{x \to -\infty} \frac{\sin x^2}{2x}$

* **13.** $\displaystyle\lim_{x \to 2^+} \frac{\sqrt{x - 2}}{\sqrt{x} - \sqrt{2}}$

* **14.** $\displaystyle\lim_{x \to \infty} x^2 e^{-3x}$

* **15.** $\displaystyle\lim_{x \to 0^+} x^{2x}$

* **16.** $\displaystyle\lim_{x \to 0^+} x^4 \ln x$

* **17.** $\displaystyle\lim_{x \to 0} \frac{\sin 2x}{\tan 3x}$

18. $\displaystyle\lim_{x \to \infty} \left(\frac{x + 1}{x - 1} \right)^x$

* **19.** $\displaystyle\lim_{x \to -\infty} x e^x$

20. Use Newton's method to find all critical points for the following functions accurate to six decimal places:

(a) $f(x) = x^4 + 3x^2 - 2x + 5$ (b) $f(x) = \dfrac{x^3 + 1}{3x^3 + 5x + 1}$

21. An object moves along the x-axis with its position defined as a function of time t by

$$x = x(t) = t^4 - \frac{44}{3}t^3 + 62t^2 - 84t, \quad t \geq 0.$$

Plot a graph of this function, indicating times when the velocity and acceleration of the object are equal to zero.

22. If the graph in the figure below represents the position $x(t)$ of an object moving along the x-axis, what could physically cause the corner at time t_0?

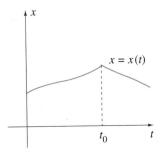

23. If an object moves along the x-axis with constant acceleration, is it possible for a graph of its position function $x(t)$ to have a point of inflection?

24. An open box is formed from a square piece of cardboard (l units long on each side) by cutting out a square at each corner and folding up the sides. What is the maximum possible volume for the box?

25. (a) If the average speed of a car for a trip is 80 km/h, must it at some time have had an instantaneous speed of 80 km/h? Explain.

(b) Must the car at some instant have had a speed of 83 km/h?

26. Draw graphs of the following functions, indicating all relative maxima and minima and points of inflection:

(a) $f(x) = \dfrac{x^3}{x^2 - 1}$ (b) $f(x) = x^2 + \sin^2 x$

27. If at some instant of time sides a and b of the triangle in the following figure form a right angle, and if these sides are increasing at equal rates, does it remain a right-angled triangle?

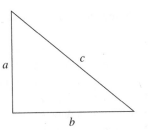

28. A football team presently sells tickets at prices of $8, $9, and $10 per seat, depending on the position of the seat. At these prices it averages sales of 10 000 at $10, 20 000 at $9, and 30 000 at $8. The team wishes to raise the price of each ticket by the same amount, but feels that for every dollar the price is raised, 10% fewer tickets of each type will be sold. What price increase per ticket will maximize revenue?

29. Repeat Exercise 28 given that the team takes into account the fact that profit from concession sales for each person at the game is 50 cents.

30. If a particle moves away from the origin along the positive x-axis with a constant speed of 10 m/s, how fast is its distance from the curve $y = x^2$ changing when it is at $x = 3$ m?

31. Each evening a cow in a pasture returns to its barn at point B (figure below). But it always does so by first walking to the river for a drink. If the cow walks at 2 km/h and stops to drink for 2 min, what is the minimum time it takes for the cow to get from the pasture to the barn? What is the minimum time if the cow walks twice as fast?

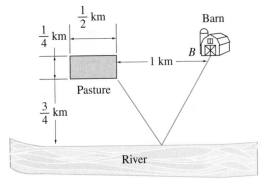

32. A farmer has 100 ha to plant in corn and potatoes. Undamaged corn yields p dollars per hectare and potatoes q dollars per hectare. For each crop, the loss due to disease and pests per unit hectare is directly proportional to the area planted. If the farmer plants x hectares of corn, the loss due to disease and pests is equal to ax per hectare. The total loss of corn is therefore ax^2 hectares. Similarly, the total loss of potatoes is by^2 if the area planted in potatoes is y hectares. Find the areas that should be planted in corn and potatoes in order to minimize monetary loss. Substitute sample values for a, b, p, and q to see whether your results look reasonable.

5 | The Indefinite Integral and the Antiderivative

Suppose all vehicles in a single lane of traffic on a highway have the same speed v. Let l be the average length of the vehicles (figure below), and let d be the distance between vehicles (assumed uniform).

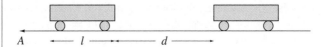

The rate r at which traffic flows is defined to be the number of vehicles passing a fixed point A per unit time. This is equal to the inverse of the time taken for one vehicle to pass A, including the distance between vehicles. At speed v, this time is $(l + d)/v$, so that $r = v/(l + d)$. Naturally a traffic engineer would like to move traffic along as quickly as possible, and this could be accomplished by increasing v and decreasing d. But safety is an important consideration, and d is not, or should not be, independent of v; it should be increased as v is increased. Experimental measurements have suggested that the shortest stopping distance is 52 m when a vehicle is travelling 22 m/s, and 96 m for 31 m/s.

THE PROBLEM   Find a reasonable dependence for d as a function of v and use it to find a suggested speed v for maximum flow rate r if cars maintain a safe driving distance. Take $l = 4$ m for the average length of a vehicle. (See Example 5.7 on page 346 for the solution.)

In Chapter 3 we introduced the derivative and ways to differentiate functions defined explicitly and implicitly. We then discussed various applications of calculus, including velocity and acceleration, related rates, maxima and minima, Newton's iterative procedure, and L'Hôpital's rule. In this chapter, we reverse the differentiation process. Instead of giving you a function and asking for its derivative, we give you the derivative and ask you to find the function. This process of backwards differentiation or antidifferentiation has such diverse applications that antidifferentiation is as important to calculus as differentiation. In many problems we find ourselves differentiating at one stage and antidifferentiating at another.

Antidifferentiation is a much more difficult process than differentiation. All but one of the rules for differentiation were developed in Chapter 3; the one remaining rule is in Section 9.1. In contrast to this, the list of formulas and techniques for finding antiderivatives is endless. In this chapter we introduce the three simplest but most important techniques; in Chapter 8 we discuss many others.

335

5.1 The Reverse Operation of Differentiation

In our discussions on velocity and acceleration in Section 4.8, we showed that when an object moves along the x-axis with its position described by the function $x(t)$, its velocity and acceleration are the first and second derivatives of $x(t)$ with respect to time t:

$$v(t) = \frac{dx}{dt} \quad \text{and} \quad a(t) = \frac{d^2x}{dt^2}.$$

For example, if $x(t) = t^3 + 3t^2$, then

$$v(t) = 3t^2 + 6t \quad \text{and} \quad a(t) = 6t + 6.$$

When engineers and physicists study the motions of objects, a more common type of problem is to determine the position of an object; that is, the position is not given. What they might know, however, is the acceleration of the object (perhaps through Newton's second law, which states that acceleration is proportional to the resultant force on the object). So the question we must now ask is: If we know the acceleration $a(t)$ of an object as a function of time, can we obtain its velocity and position by reversing the differentiations? In the example above, if we know that

$$a(t) = \frac{dv}{dt} = 6t + 6,$$

can we find the function $v(t)$ that differentiates to give $6t + 6$? We know that to arrive at the terms $6t$ and 6 after differentiation, $v(t)$ might have contained the terms $3t^2$ and $6t$. In other words, one possible velocity function that differentiates to give $a(t) = 6t + 6$ is $v(t) = 3t^2 + 6t$. It is not, however, the only one; $v(t) = 3t^2 + 6t + 10$ and $v(t) = 3t^2 + 6t - 22$ also have derivative $6t + 6$. In fact, for any constant C whatsoever, the derivative of

$$v(t) = 3t^2 + 6t + C$$

is $a(t) = 6t + 6$. In Theorem 5.1 we shall show that this velocity function represents all functions that have $6t + 6$ as their derivative. We shall also demonstrate how to evaluate the constant C.

For the present, let us set $C = 0$ so that

$$v(t) = \frac{dx}{dt} = 3t^2 + 6t.$$

We now ask what position function $x(t)$ differentiates to give $3t^2 + 6t$. One possibility is $x(t) = t^3 + 3t^2$, and for the same reason as above, so is

$$x(t) = t^3 + 3t^2 + D$$

for any constant D whatsoever.

Thus, by reversing the differentiation operation in this example we have proceeded from the acceleration $a(t) = 6t + 6$ to possible velocity functions $v(t)$, and then to possible position functions $x(t)$. This process of antidifferentiation has applications far beyond velocity and acceleration problems, and we shall see many of them as we progress through this book. We begin our formal study of antidifferentiation with the following definition.

DEFINITION 5.1

A function $F(x)$ is called an **antiderivative** of $f(x)$ on an interval I if on I,

$$F'(x) = f(x). \tag{5.1}$$

For example, since

$$\frac{d}{dx}x^4 = 4x^3,$$

we say that x^4 is an antiderivative of $4x^3$ for all x. But for any constant C, the function $x^4 + C$ is also an antiderivative of $4x^3$. The following theorem indicates that these are the only antiderivatives of $4x^3$.

THEOREM 5.1

If $F(x)$ is an antiderivative of $f(x)$ on an interval I, then every antiderivative of $f(x)$ on I is of the form

$$F(x) + C, \qquad \text{where } C \text{ is a constant.}$$

PROOF Suppose that $F(x)$ and $G(x)$ are two antiderivatives of $f(x)$ on I. If we define a function $D(x) = G(x) - F(x)$, then on I

$$D'(x) = G'(x) - F'(x) = f(x) - f(x) = 0.$$

If x_1 and x_2 are any two points in the interval I, then certainly $D'(x) = 0$ on the interval $x_1 \le x \le x_2$. But differentiability of $D(x)$ on $x_1 \le x \le x_2$ implies continuity of $D(x)$ thereon also (Theorem 3.6). We may therefore apply the mean value theorem (Theorem 3.19) to $D(x)$ on the interval $x_1 \le x \le x_2$, and conclude that there exists a number c between x_1 and x_2 such that

$$D'(c) = \frac{D(x_2) - D(x_1)}{x_2 - x_1}.$$

But because $D'(c) = 0$, it follows that $D(x_1) = D(x_2)$. Since x_1 and x_2 are any two points in I, we conclude that $D(x)$ must have the same value at every point in I; that is, on I,

$$D(x) = G(x) - F(x) = C,$$

where C is a constant. Consequently,

$$G(x) = F(x) + C. \qquad \blacksquare$$

Because of this theorem, if we find one antiderivative $F(x)$ of $f(x)$ by any means whatsoever, then we have found every antiderivative of a function $f(x)$, since every antiderivative can be written as $F(x)$ plus a constant C. Thus, every antiderivative of a function $f(x)$ is of the form $F(x) + C$, where $F(x)$ is any one antiderivative. We call $F(x) + C$ the **indefinite integral** of $f(x)$. The operation of taking the indefinite integral is denoted by

$$\int f(x)\, dx = F(x) + C, \tag{5.2}$$

where the differential dx indicates that integration is with respect to x. We call $f(x)$ the **integrand** of the indefinite integral. For example, we write

$$\int x^2\, dx = \frac{x^3}{3} + C \qquad \text{and} \qquad \int \frac{1}{x^3}\, dx = -\frac{1}{2x^2} + C.$$

Distinguish between an antiderivative of a function $f(x)$ and the indefinite integral of $f(x)$. Both reverse differentiation. An antiderivative of $f(x)$ is a function that differentiates to $f(x)$; the indefinite integral of $f(x)$ is all functions that differentate to $f(x)$, it adds an arbitrary constant to any antiderivative.

The following theorem is fundamental to the calculation of antiderivatives and indefinite integrals. Its proof is a straightforward exercise in differentiation and the use of Definition 5.1.

THEOREM 5.2

If $f(x)$ and $g(x)$ have antiderivatives on an interval I, then on I:

(i) $$\int [f(x) + g(x)]\, dx = \int f(x)\, dx + \int g(x)\, dx; \tag{5.3a}$$

(ii) $$\int kf(x)\, dx = k \int f(x)\, dx, \qquad k \text{ a constant.} \tag{5.3b}$$

For example, to evaluate the indefinite integral of $2x^3 - 4x$, we write

$$\int (2x^3 - 4x)\, dx = \int 2x^3\, dx + \int -4x\, dx \qquad \text{[by part (i) of Theorem 5.2]}$$

$$= 2 \int x^3\, dx - 4 \int x\, dx \qquad \text{[by part (ii) of Theorem 5.2)]}$$

$$= 2 \left(\frac{x^4}{4} \right) - 4 \left(\frac{x^2}{2} \right) + C$$

$$= \frac{x^4}{2} - 2x^2 + C.$$

Every differentiation formula developed in Chapter 3 can be expressed as an integration formula. In fact, equations 5.3 are integral counterparts of equations 3.9 and 3.8. Some of the differentiation formulas for trigonometric, inverse trigonometric, exponential, logarithm, and hyperbolic functions are restated below as integration formulas.

$$\int \sin x\, dx = -\cos x + C, \tag{5.4a}$$

$$\int \cos x\, dx = \sin x + C, \tag{5.4b}$$

$$\int \sec^2 x\, dx = \tan x + C, \tag{5.4c}$$

$$\int \sec x \tan x\, dx = \sec x + C, \tag{5.4d}$$

$$\int \csc^2 x\, dx = -\cot x + C, \tag{5.4e}$$

$$\int \csc x \cot x\, dx = -\csc x + C, \tag{5.4f}$$

$$\int \frac{1}{\sqrt{1 - x^2}}\, dx = \operatorname{Sin}^{-1} x + C, \tag{5.4g}$$

$$\int \frac{1}{1 + x^2}\, dx = \operatorname{Tan}^{-1} x + C, \tag{5.4h}$$

$$\int \frac{1}{x\sqrt{x^2 - 1}}\, dx = \operatorname{Sec}^{-1} x + C, \tag{5.4i}$$

$$\int e^x\, dx = e^x + C, \tag{5.4j}$$

$$\int a^x \, dx = a^x \log_a e + C, \qquad (5.4\text{k})$$

$$\int \frac{1}{x} \, dx = \ln |x| + C, \qquad (5.4\text{l})$$

$$\int \cosh x \, dx = \sinh x + C, \qquad (5.4\text{m})$$

$$\int \sinh x \, dx = \cosh x + C, \qquad (5.4\text{n})$$

$$\int \operatorname{sech}^2 x \, dx = \tanh x + C, \qquad (5.4\text{o})$$

$$\int \operatorname{csch}^2 x \, dx = -\coth x + C, \qquad (5.4\text{p})$$

$$\int \operatorname{sech} x \tanh x \, dx = -\operatorname{sech} x + C, \qquad (5.4\text{q})$$

$$\int \operatorname{csch} x \coth x \, dx = -\operatorname{csch} x + C. \qquad (5.4\text{r})$$

Equation 5.4l follows immediately from equation 3.46 with $f(x) = x$.

Perhaps the most important integration formula is the counterpart of power rule 3.7. Since

$$\frac{d}{dx} x^{n+1} = (n+1)x^n,$$

it follows that

$$\int (n+1)x^n \, dx = x^{n+1} + \overline{C}.$$

With property 5.3b, the $n+1$ can be removed from the integral and taken to the other side of the equation:

$$\int x^n \, dx = \frac{1}{n+1} x^{n+1} + C, \qquad n \neq -1, \qquad (5.5)$$

where $C = \overline{C}/(n+1)$. As indicated, this result is valid provided that $n \neq -1$, but formula 5.4l takes care of this exceptional case.

EXAMPLE 5.1

Evaluate

$$\int \left(2x + \frac{1}{x^2} \right) dx.$$

SOLUTION Using Theorem 5.2 and formula 5.5, we find that

$$\int \left(2x + \frac{1}{x^2} \right) dx = 2 \int x \, dx + \int \frac{1}{x^2} \, dx = 2 \left(\frac{x^2}{2} \right) - \frac{1}{x} + C = x^2 - \frac{1}{x} + C.$$

Where is $x^2 - 1/x + C$, the indefinite integral of $2x + 1/x^2$? It is valid on the intervals $x < 0$ and $x > 0$, but not for all x because the function is not defined at $x = 0$. To make this clear, we should write

$$\int \left(2x + \frac{1}{x^2} \right) dx = \begin{cases} x^2 - \dfrac{1}{x} + C_1, & x < 0 \\ x^2 - \dfrac{1}{x} + C_2, & x > 0 \end{cases}$$

where the constants C_1 and C_2 need not be the same. For brevity we often write

$$\int \left(2x + \frac{1}{x^2} \right) dx = x^2 - \frac{1}{x} + C,$$

thereby suppressing the complete description of the indefinite integral. When the context demands that we distinguish various intervals on which the indefinite integral is defined, we shall be careful to give the extended version.

EXAMPLE 5.2

Find a curve that passes through the point $(1, 5)$ and whose tangent line at each point (x, y) has slope $5x^4 - 3x^2 + 2$.

SOLUTION If $y = f(x)$ is the equation of the curve, then

$$\frac{dy}{dx} = 5x^4 - 3x^2 + 2.$$

If we take indefinite integrals of both sides of this equation with respect to x, we obtain

$$\int \frac{dy}{dx}\, dx = \int (5x^4 - 3x^2 + 2)\, dx \qquad \text{or}$$

$$y = x^5 - x^3 + 2x + C.$$

Since $(1, 5)$ is a point on the curve, its coordinates must satisfy the equation of the curve:

$$5 = 1^5 - 1^3 + 2(1) + C.$$

Thus $C = 3$, and the required curve is $y = x^5 - x^3 + 2x + 3$.

In taking the indefinite integral of each side of the equation

$$\frac{dy}{dx} = 5x^4 - 3x^2 + 2$$

in the example above, we added an arbitrary constant C to the right-hand side. You might question why we did not add a constant to the left-hand side. Had we done so, the result would have been

$$y + D = x^5 - x^3 + 2x + E.$$

If we had then written

$$y = x^5 - x^3 + 2x + (E - D)$$

and defined $C = E - D$, we would have obtained exactly the same result. Hence, nothing is gained by adding an arbitrary constant to both sides; a constant on one side is sufficient.

The problem in Example 5.2 was geometric: Find the equation of a curve satisfying certain properties. We quickly recast it as the problem of finding the function $y = f(x)$ that satisfies the equation

$$\frac{dy}{dx} = 5x^4 - 3x^2 + 2,$$

subject to the additional condition that $f(1) = 5$. Once again this is a differential equation. We discuss differential equations briefly in Section 5.5 and study them in detail in Chapter 15. When we solve a differential equation with no subsidiary conditions, say,

$$\frac{dy}{dx} = 4x^2 + 7x,$$

we do not get a function, but rather a one-parameter family of functions. For this differential equation, we obtain

$$y = \frac{4}{3}x^3 + \frac{7}{2}x^2 + C,$$

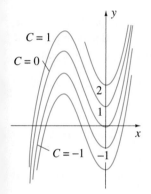

FIGURE 5.1 One-parameter family of solutions of a differential equation

a one-parameter family of cubic polynomials, C being the parameter. Geometrically, we have the one-parameter family of curves in Figure 5.1. Parameter C represents a vertical shift of one curve relative to another. Note that if a vertical line is drawn at any position x to intersect these curves, then at the points of intersection, every curve has exactly the same slope, namely $4x^2 + 7x$. For example, the slope of each cubic at $x = 0$ is zero. If an extra condition is added to the differential equation, such as to demand that y be equal to 4 when $x = 2$, then

$$4 = \frac{4}{3}(2)^3 + \frac{7}{2}(2)^2 + C,$$

or, $C = -62/3$. This condition singles out one particular function from the family, namely $y = 4x^3/3 + 7x^2/2 - 62/3$. Geometrically, it determines that curve in the family which passes through the point $(2, 4)$.

In this chapter, we discuss three basic ways to find antiderivatives. First, some antiderivatives are obvious, and the better you are at differentiation, the more obvious they will be. For example, you should have no trouble recognizing that

$$\int 7x^5 \, dx = \frac{7x^6}{6} + C \quad \text{and} \quad \int \frac{1}{x^3} \, dx = \frac{-1}{2x^2} + C.$$

Our second method results from the answer to the following question: How do we check that a function $F(x)$ is an antiderivative of $f(x)$? We differentiate $F(x)$, of course. This simple fact suggests an approach to slightly more complex problems, say,

$$\int (2x + 3)^5 \, dx.$$

We might reason that in order to have $2x + 3$ raised to power 5 after differentiation, we had $2x + 3$ to power 6 before differentiation; that is, a reasonable proposal for an antiderivative is $(2x + 3)^6$. Differentiation of this function gives

$$\frac{d}{dx}(2x + 3)^6 = 6(2x + 3)^5(2) = 12(2x + 3)^5,$$

and we see that $(2x + 3)^6$ is not a correct antiderivative. It has produced $(2x + 3)^5$, as required, but it has also given an undesirable factor of 12. We therefore adjust our original proposal by multiplying it by $1/12$; that is, the correct indefinite integral is

$$\int (2x + 3)^5 \, dx = \frac{1}{12}(2x + 3)^6 + C.$$

This is what we call *adjusting constants*: we propose an antiderivative that is within a multiplicative constant of being correct. Adjusting the original proposal by the inverse of this constant gives the correct antiderivative. But remember, our initial proposal must be within a *multiplicative constant*. We cannot adjust x's. Let us illustrate with two very similar problems,

$$\int \frac{1}{(5x + 2)^5} \, dx \quad \text{and} \quad \int \frac{1}{(5x^2 + 2)^5} \, dx.$$

For the first problem we propose as an antiderivative $(5x + 2)^{-4}$. Differentiation gives

$$\frac{d}{dx}\left[\frac{1}{(5x + 2)^4}\right] = \frac{-4}{(5x + 2)^5}(5) = \frac{-20}{(5x + 2)^5}.$$

Since we are out by a factor of -20, we adjust our original proposal with $-1/20$,

$$\int \frac{1}{(5x + 2)^5}\,dx = \frac{-1}{20(5x + 2)^4} + C.$$

It might seem as logical to propose $(5x^2 + 2)^{-4}$ as an antiderivative for the second problem, but differentiation yields

$$\frac{d}{dx}\left[\frac{1}{(5x^2 + 2)^4}\right] = \frac{-4}{(5x^2 + 2)^5}(10x) = \frac{-40x}{(5x^2 + 2)^5}.$$

This time the discrepancy is $-40x$. We cannot adjust x's. The original proposal must be abandoned. This indefinite integral is quite difficult; it will have to wait for the more powerful techniques of Chapter 8. To emphasize once again, do not try to adjust x's, only constants.

EXAMPLE 5.3

Evaluate the following indefinite integrals:

$$\text{(a)} \quad \int \cos 3x\,dx \qquad \text{(b)} \quad \int e^{-2x}\,dx \qquad \text{(c)} \quad \int \frac{x}{3x^2 - 4}\,dx$$

SOLUTION

(a) To obtain $\cos 3x$ after differentiation, we propose $\sin 3x$ as an antiderivative. Since

$$\frac{d}{dx}\sin 3x = 3\cos 3x,$$

it is necessary to adjust with $1/3$,

$$\int \cos 3x\,dx = \frac{1}{3}\sin 3x + C.$$

(b) With an initial proposal of e^{-2x} for an antiderivative based on the fact that the derivative of an exponential function always returns the same exponential, we calculate

$$\frac{d}{dx}e^{-2x} = -2e^{-2x}.$$

Consequently, we adjust with $-1/2$,

$$\int e^{-2x}\,dx = -\frac{1}{2}e^{-2x} + C.$$

(c) Since

$$\frac{d}{dx}\ln(3x^2 - 4) = \frac{1}{3x^2 - 4}(6x),$$

the required indefinite integral is

$$\int \frac{x}{3x^2 - 4}\,dx = \frac{1}{6}\ln|3x^2 - 4| + C.$$

We have inserted absolute values as suggested by formula 5.4l.

The third technique for finding antiderivatives is discussed in Section 5.3.

EXAMPLE 5.4

Find continuous indefinite integrals for the Heaviside unit step function $h(x - a)$ introduced in Section 2.5.

SOLUTION Because the function has two values (see Figure 2.35), we subdivide the integration into two parts. For $x < a$, $h(x - a) = 0$, and the indefinite integral is a constant. For $x > a$, $h(x - a) = 1$, and the indefinite integral is x plus a constant. Thus,

$$\int h(x - a)\, dx = \begin{cases} C, & x < a \\ x + D, & x > a. \end{cases}$$

For continuity at $x = a$, we require $C = a + D \implies D = C - a$. Consequently, continuous indefinite integrals are

$$\int h(x - a)\, dx = \begin{cases} C, & x < a \\ (x - a) + C, & x > a. \end{cases}$$

For future applications, it is convenient to express these in the form

$$\int h(x - a)\, dx = (x - a)h(x - a) + C,$$

where, for continuity at $x = a$, we understand that the value of the indefinite integral at $x = a$ is C. A graph of this function is shown in Figure 5.2. It is called a **ramp function**.

FIGURE 5.2   Ramp function

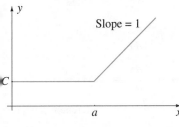

Slope = 1

EXERCISES 5.1

In Exercises 1–20 evaluate the indefinite integral.

1. $\displaystyle\int (x^3 - 2x)\, dx$

2. $\displaystyle\int (x^4 + 3x^2 + 5x)\, dx$

3. $\displaystyle\int (2x^3 - 3x^2 + 6x + 6)\, dx$

4. $\displaystyle\int \sin x\, dx$

5. $\displaystyle\int 3\cos x\, dx$

6. $\displaystyle\int \sqrt{x}\, dx$

7. $\displaystyle\int \left(x^{10} - \frac{1}{x^3}\right) dx$

8. $\displaystyle\int \left(\frac{1}{x^2} - \frac{2}{x^4}\right) dx$

9. $\displaystyle\int \left(x^{3/2} - x^{2/7}\right) dx$

10. $\displaystyle\int \left(\frac{1}{x^2} + \frac{1}{2\sqrt{x}}\right) dx$

11. $\displaystyle\int \left(\frac{4}{x^{3/2}} + 2x^{1/3}\right) dx$

12. $\displaystyle\int \left(-\frac{1}{2x^2} + 3x^3\right) dx$

13. $\displaystyle\int \frac{1}{x^\pi}\, dx$

14. $\displaystyle\int \left(2\sqrt{x} + 3x^{3/2} - 5x^{5/2}\right) dx$

15. $\displaystyle\int x^2(x^2 - 3)\, dx$

16. $\displaystyle\int \sqrt{x}(x + 1)\, dx$

17. $\displaystyle\int \left(\frac{x - 2}{x^3}\right) dx$

18. $\displaystyle\int x^2(1 + x^2)^2\, dx$

19. $\displaystyle\int (x^2 + 1)^3\, dx$

20. $\displaystyle\int \frac{(x - 1)^2}{\sqrt{x}}\, dx$

In Exercises 21–24 find the curve $y = f(x)$ that passes through the given point and whose slope at each point (x, y) is defined by the derivative indicated.

21. $dy/dx = x^2 - 3x + 2$, $(2, 1)$

22. $dy/dx = 2x^3 + 4x$, $(0, 5)$

23. $dy/dx = -2x^4 + 3x^2 + 6$, $(1, 0)$

24. $dy/dx = 2 - 4x + 8x^7$, $(1, 1)$

* **25.** Find the equation of the curve that has a second derivative equal to $6x^2$ and passes through the points $(0, 2)$ and $(-1, 3)$.

* **26.** Find a function $f(x)$ that has a relative maximum $f(2) = 3$ and has a second derivative equal to $-5x$.

* **27.** Is it possible to find a function $f(x)$ that has a relative minimum $f(2) = 3$ and has a second derivative equal to $-5x$?

In Exercises 28–67 use *adjusting constants* to evaluate the indefinite integral.

28. $\displaystyle\int \sqrt{x + 2}\, dx$

29. $\displaystyle\int (x + 5)^{3/2}\, dx$

30. $\displaystyle\int \sqrt{2 - x}\, dx$

31. $\displaystyle\int \frac{1}{\sqrt{4x + 3}}\, dx$

32. $\displaystyle\int (2x - 3)^{3/2}\, dx$

33. $\displaystyle\int (3x + 1)^5\, dx$

34. $\int (1 - 2x)^7 \, dx$

35. $\int \frac{1}{(x + 4)^2} \, dx$

36. $\int \frac{1}{(1 + 3x)^6} \, dx$

37. $\int x(x^2 + 1)^3 \, dx$

38. $\int x^2 (2 + 3x^3)^7 \, dx$

39. $\int \frac{x}{(2 + x^2)^2} \, dx$

40. $\int \cos 2x \, dx$

∗ 41. $\int \cos^2 x \sin x \, dx$

∗ 42. $\int 3 \sin 2x \cos 2x \, dx$

∗ 43. $\int \sec 12x \tan 12x \, dx$

∗ 44. $\int \csc^2 4x \, dx$

∗ 45. $\int e^{4x} \, dx$

∗ 46. $\int xe^{-x^2} \, dx$

∗ 47. $\int \frac{e^{3/x}}{x^2} \, dx$

∗ 48. $\int e^{4x-3} \, dx$

∗ 49. $\int \frac{1}{3x + 2} \, dx$

∗ 50. $\int \frac{2}{7 - 5x} \, dx$

∗ 51. $\int \frac{x}{1 - x^2} \, dx$

∗ 52. $\int \frac{3x^2}{1 - 4x^3} \, dx$

∗ 53. $\int 2^x \, dx$

∗ 54. $\int 3^{2x} \, dx$

∗ 55. $\int \frac{e^x}{e^x + 1} \, dx$

∗ 56. $\int \sin x (1 + \cos x)^4 \, dx$

∗ 57. $\int \frac{\cos x}{\sin^3 x} \, dx$

∗ 58. $\int e^{2x} (1 + e^{2x})^3 \, dx$

∗ 59. $\int \frac{\sec^2 x}{\tan^2 x} \, dx$

∗ 60. $\int \frac{1}{\sqrt{1 - 4x^2}} \, dx$

∗ 61. $\int \frac{1}{1 + 9x^2} \, dx$

∗ 62. $\int \frac{1}{x\sqrt{3x^2 - 1}} \, dx$

∗ 63. $\int \frac{3x}{1 + 5x^2} \, dx$

∗ 64. $\int \cosh 4x \, dx$

∗ 65. $\int x \sinh 3x^2 \, dx$

∗ 66. $\int \operatorname{sech} 2x \tanh 2x \, dx$

∗ 67. $\int x^2 \operatorname{csch}^2 4x^3 \, dx$

In Exercises 68–73 find a one-parameter family of functions satisfying the differential equation.

∗ 68. $\dfrac{dy}{dx} = x^3 - \dfrac{1}{x^2}$

∗ 69. $\dfrac{dy}{dx} = \sqrt{3 - 4x}$

∗ 70. $\dfrac{dy}{dx} = \dfrac{1}{(3x + 5)^{3/2}}$

∗ 71. $\dfrac{dy}{dx} = x^2 (2x^3 + 4)^4$

∗ 72. $\dfrac{dy}{dx} = \dfrac{x^3}{(2 + 3x^4)^2}$

∗ 73. $\dfrac{dy}{dx} = \sin x (1 + \cos^2 x)$

∗ 74. Find a function $y = f(x)$ that satisfies the differential equation

$$\frac{dy}{dx} = \frac{1}{x^2}$$

and passes through the two points $(1, 1)$ and $(-1, -2)$.

∗ 75. (a) Find the indefinite integral of the signum function of Exercise 47 in Section 2.4.

(b) Prove that sgn x cannot have an antiderivative on any interval containing $x = 0$.

5.2 Integrating Velocity and Acceleration

In Section 4.8 we discussed relationships among position, velocity, and acceleration from the viewpoint of derivatives. We now consider these relationships through antiderivatives, an approach providing a far more practical viewpoint when it comes to applications.

For motion along the x-axis, velocity is the derivative of position with respect to time: $v(t) = dx/dt$. We can say, therefore, that the indefinite integral of velocity represents every possible position function with this velocity:

$$x(t) = \int v(t) \, dt. \tag{5.6}$$

Similarly, as acceleration is the derivative of velocity, the indefinite integral of acceleration represents every possible velocity function,

$$v(t) = \int a(t) \, dt. \tag{5.7}$$

Thus, given the acceleration of an object moving along a straight line, we can antidifferentiate to find its velocity and antidifferentiate again for its position. Since each antidifferentiation introduces an arbitrary constant, additional information must be specified in order to evaluate these constants.

EXAMPLE 5.5

FIGURE 5.3 Car accelerating from rest

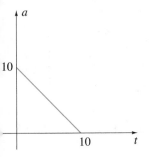

The car in Figure 5.3 accelerates from rest when the light turns green. Initially, the acceleration is 10 m/s^2, but it decreases linearly, reaching zero after 10 s. Find the velocity and position of the car during this time interval.

SOLUTION   Let us set up the coordinate system in Figure 5.3 and choose time $t = 0$ when the car pulls away from the light. Given this time convention, the acceleration of the car as a function of time t (see Figure 5.4) is

$$a(t) = 10 - t, \quad 0 \le t \le 10.$$

If $v(t)$ denotes the velocity of the car during the time interval, then

$$\frac{dv}{dt} = 10 - t,$$

and integration gives

$$v(t) = 10t - \frac{t^2}{2} + C.$$

By our time convention, the velocity of the car is zero at time $t = 0$; that is, $v(0) = 0$, and from this we obtain

$$0 = 10(0) - \frac{(0)^2}{2} + C.$$

Consequently, $C = 0$, and

$$v(t) = 10t - \frac{t^2}{2}.$$

This is the velocity of the car during the time interval $0 \le t \le 10$, measured in metres per second.

Since the position of the car with respect to its original position at time $t = 0$ is denoted by x,

$$v = \frac{dx}{dt} = 10t - \frac{t^2}{2}.$$

Thus,

$$x(t) = 5t^2 - \frac{t^3}{6} + D.$$

Since $x(0) = 0$, the constant D must also be zero, and the position of the car indicated by its distance in metres from the stoplight, for $0 \le t \le 10$, is

$$x(t) = 5t^2 - \frac{t^3}{6}.$$

EXAMPLE 5.6

A stone is thrown vertically upward over the edge of a cliff at 25 m/s. When does it hit the base of the cliff if the cliff is 100 m high?

SOLUTION   Let us measure y as positive upward, taking $y = 100$ and $t = 0$ at the point and instant of projection (Figure 5.5). A law of physics states that when an object near the earth's surface is acted on by gravity alone, it experiences an acceleration whose magnitude is 9.81 m/s^2. If a denotes the acceleration of the stone and v its velocity, then

$$a = \frac{dv}{dt} = -9.81$$

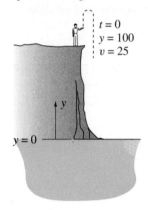

FIGURE 5.5 Stone thrown upward over edge of cliff

t = 0
y = 100
v = 25

y

y = 0

(*a* is negative since it is in the negative *y*-direction). To obtain *v* we integrate with respect to *t*:

$$v(t) = -9.81t + C.$$

By our time convention, $v(0) = 25$, so that $25 = -9.81(0) + C$. Thus, $C = 25$, and

$$v(t) = -9.81t + 25.$$

We now have the velocity of the stone at any given instant. To find the position of the stone we set $v = dy/dt$,

$$\frac{dy}{dt} = -9.81t + 25,$$

and integrate once again:

$$y(t) = -4.905t^2 + 25t + D.$$

Since $y = 100$ when $t = 0$, it follows that $100 = -4.905(0)^2 + 25(0) + D$. Hence, $D = 100$, and

$$y(t) = -4.905t^2 + 25t + 100.$$

We have found the equation that tells us exactly where the stone is at any given time. To determine when the stone strikes the base of the cliff, we set $y(t) = 0$; that is,

$$0 = -4.905t^2 + 25t + 100,$$

a quadratic equation with solutions

$$t = \frac{-25 \pm \sqrt{25^2 - 4(-4.905)(100)}}{-9.81} = \frac{25 \pm \sqrt{2587}}{9.81} = 7.7 \quad \text{or} \quad -2.6.$$

Since the negative root must be rejected, we find that the stone strikes the base of the cliff after 7.7 s.

An important point to note in Examples 5.5 and 5.6 is that the coordinate system and time convention were specified immediately; that is, we decided, and did so at the start of the problem, where to place the origin of our coordinate system, which direction to choose as positive, and when to choose $t = 0$. Only then were we able to specify the correct sign for acceleration. Furthermore, we determined constants from antidifferentiations using initial conditions expressed in terms of our coordinate system and time convention. Throughout the solutions, we were careful to refer everything to our choice of coordinates and time. Remember, then, to specify clearly the coordinate system and time convention at the beginning of a problem.

We should also note that in each of the examples we integrated with respect to time *only those equations that were valid for a range of values of time*. It is a common error to integrate equations that are only valid at one instant. For instance, students reason in Example 5.6 that the initial velocity is 25, $v = dy/dt = 25$, and hence $y(t) = 25t + C$. This is incorrect, because the equation $dy/dt = 25$ is valid only at time $t = 0$, and cannot be integrated.

EXAMPLE 5.7

Application Preview Revisited

Find the speed that maximizes flow rate in the Application Preview and draw any conclusions that you feel are justified.

SOLUTION Flow rate is, at the moment, a function of speed v and distance d between vehicles, $r = v/(l + d)$. To maximize r, we must express it as a function of one variable. As was suggested, v and d are related, d should increase when v increases. We use the fact that d should be the stopping distance for cars travelling at speed v to find a functional relationship between v and d. The stopping distance of a vehicle is composed of two parts, distance d_T that the vehicle travels during the time it takes the driver to get his foot from the accelerator to the brake, and d_B, the distance travelled while the brake is applied. If T represents the reaction time for the driver to apply the brake, then $d_T = vT$, where v is the speed of the vehicle before braking. If a is the acceleration of the vehicle during the braking period, its velocity is $V = at + C$. If we choose $t = 0$ when the brake is applied, then $V = v$ at $t = 0$, and this implies that $V = at + v$. Distance travelled during the braking period is $x = at^2/2 + vt + D$. If we choose $x = 0$ at $t = 0$, then $D = 0$, and $x = at^2/2 + vt$. The vehicle stops when $0 = V = at + v \implies t = -v/a$, and therefore $d_B = a(-v/a)^2/2 + v(-v/a) = -v^2/(2a)$. The stopping distance for a vehicle when reaction time is T and acceleration during braking is a is

$$d = d_T + d_B = vT - \frac{v^2}{2a}.$$

We can evaluate T and a using the fact that $d = 52$ m when $v = 22$ m/s, and $d = 96$ m when $v = 31$ m/s. These give

$$52 = 22T - \frac{22^2}{2a}, \quad 96 = 31T - \frac{31^2}{2a},$$

the solution of which is $a = -6.138$ m/s^2 and $T = 0.5715$ s. The flow function is

$$r(v) = \frac{v}{l + d} = \frac{v}{l + vT - v^2/(2a)}.$$

Its graph is shown in Figure 5.6 for $l = 4$. We see that it has one critical point, which is given by

$$0 = r'(v) = \frac{\left(l + vT - \dfrac{v^2}{2a}\right)(1) - v\left(T - \dfrac{v}{a}\right)}{\left(l + vT - \dfrac{v^2}{2a}\right)^2}.$$

When we set the numerator to zero, and multiply by $2a$,

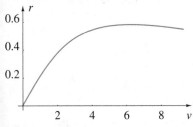
FIGURE 5.6 Flow function for cars on highway

$$0 = 2al + 2avT - v^2 - 2avT + 2v^2 = 2al + v^2 \implies v = \sqrt{-2al}.$$

The speed for maximum flow rate is $\sqrt{-2(-6.138)(4)} = 7.0$ m/s, which is about 25 km/h. Obviously highway speed limits are not set with safety in mind. On the other hand, safe distances between cars at typical highway speeds of say 100 km/h, or 27.8 m/s, would be $27.8(0.5715) + (27.8)^2/12.276 = 78.8$ m, and we know that drivers do not maintain any such distance.

In Exercises 1–8 we have defined acceleration $a(t)$ of an object moving along the x-axis during some time interval and specified the initial conditions $x(0)$ and $v(0)$. Find the velocity $v(t)$ and position $x(t)$ of the object as functions of time.

1. $a(t) = t + 2$, $\quad 0 \leq t \leq 3$; $\quad v(0) = 0, x(0) = 0$

2. $a(t) = 6 - 2t$, $\quad 0 \leq t \leq 3$; $\quad v(0) = 5, x(0) = 0$

3. $a(t) = 6 - 2t$ $\quad 0 \leq t \leq 4$; $\quad v(0) = 5, x(0) = 0$

4. $a(t) = 120t - 12t^2$, $\quad 0 \leq t \leq 10$; $\quad v(0) = 0, x(0) = 4$

5. $a(t) = t^2 + 1$, $\quad 0 \leq t \leq 5$; $\quad v(0) = -1, x(0) = 1$

6. $a(t) = t^2 + 5t + 4$, $\quad 0 \leq t \leq 15$; $\quad v(0) = -2, x(0) = -3$

7. $a(t) = \cos t$, $\quad t \geq 0$; $\quad v(0) = 0, x(0) = 0$

8. $a(t) = 3 \sin t$, $\quad t \geq 0$; $\quad v(0) = 1, x(0) = 4$

∗ 9. The velocity of an object moving along the x-axis is given in metres per second by

$$v(t) = 3t^2 - 9t + 6$$

where t is time in seconds. If the object starts from position $x = 1$ m at time $t = 0$, answer each of the following questions:

(a) What is the acceleration of the object at $t = 5$ s?

(b) What is the position of the object at $t = 2$ s?

(c) Is the object speeding up or slowing down at $t = 5/4$ s?

(d) What is the closest the object ever comes to the origin?

∗ 10. A particle moving along the x-axis has acceleration

$$a(t) = 6t - 2$$

in metres per second per second for time $t \geq 0$.

(a) If the particle starts at the point $x = 1$ moving to the left with speed 3 m/s, find its position as a function of time t.

(b) At what time does the particle have zero velocity (if any)?

∗ 11. The acceleration of a particle moving along the x-axis is given in metres per second per second by

$$a(t) = 6t - 15,$$

where $t \geq 0$ is time in seconds.

(a) If the velocity of the particle at $t = 2$ s is 6 m/s, what is its velocity at $t = 1$ s?

(b) If the particle is 10 m to the right of the origin at time $t = 0$, what is its position as a function of time t?

(c) What is the closest the particle ever comes to the origin?

∗ 12. A car is sitting at rest at a stoplight. When the light turns green at time $t = 0$, the driver immediately presses the accelerator, imparting an acceleration of $a(t) = (3 - t/5)$ m/s^2 to the car for 10 s.

(a) Where is the car after the 10 s?

(b) If the driver applies the brakes at $t = 10$ s, and the car experiences a constant deceleration of 2 m/s^2, where and when does the car come to a stop?

∗ 13. Find how far a plane will move when landing if in t seconds after touching the ground, its speed in metres per second is given by $180 - 18t$.

∗ 14. A stone is thrown directly upward with an initial speed of 10 m/s. How high will it rise?

∗ 15. You are standing on a bridge 25 m above a river. If you wish to drop a stone onto a piece of floating wood, how soon before the wood reaches the appropriate spot should you drop the stone?

∗ 16. You are standing at the base of a building and wish to throw a ball to a friend on the roof 20 m above you. With what minimum speed must you throw the ball?

∗ 17. A car is travelling at 20 m/s when the brakes are applied. What constant deceleration must the car experience if it is to stop before striking a tree that is 50 m from the car at the instant the brakes are applied? Assume that the car travels in a straight line.

∗ 18. The position of an object moving along the x-axis is given by

$$x(t) = t^3 - 6t^2 + 9t - 20, \quad t \geq 0.$$

Draw graphs of the position, velocity, and acceleration functions for this motion. Pay special attention to the fact that $v(t)$ represents the slope of $x = x(t)$ and $a(t)$ is the slope of $v = v(t)$.

∗ 19. Repeat Exercise 18 given that the acceleration of the object is

$$a(t) = 6t - 30, \quad t \geq 0,$$

and $v(0) = -33$ and $x(0) = 400$.

∗ 20. You are called on as an expert to testify at a traffic hearing. The question concerns the speed of a car that made an emergency stop with brakes locked and wheels sliding. The skid mark on the road measured 9 m. Assuming that the deceleration of the car was constant and could not exceed the acceleration due to gravity of a freely falling body (and this is indeed a reasonable assumption), what can you say about the speed of the car before the brakes were applied? Are you testifying for the prosecution or the defence?

∗ 21. When the brakes of an automobile are applied, they produce a constant deceleration of 5 m/s^2.

(a) What is the distance, from the point of application of brakes, required to stop a car travelling at 100 km/h?

(b) Repeat part (a) for 50 km/h.

(c) What is the ratio of these distances?

(d) Repeat parts (a), (b), and (c) given that the reaction time of the driver to get her foot from accelerator to brake is 3/4 s, and distances are calculated taking this reaction time into account.

∗ 22. A stone is dropped into a well and the sound of the stone striking the water is heard 3.1 s later. If the speed of sound is 340 m/s, how deep is the surface of the water in the well?

∗ 23. Two trains, one travelling at 100 km/h and the other at 60 km/h, are headed toward each other along a straight, level track. When they are 2 km apart, each engineer sees the other's train and locks his wheels.

(a) If the deceleration of each train has magnitude 1/4 m/s^2, determine whether a collision occurs.

(b) Repeat part (a) given that the deceleration is caused by the wheels being reversed rather than locked.

(c) Illustrate graphically the difference between the situations in parts (a) and (b).

24. A steel bearing is dropped from the roof of a building. An observer standing in front of a window 1 m high notes that the bearing takes $1/8$ s to fall from the top to the bottom of the window. The bearing continues to fall, makes a completely elastic collision with a horizontal sidewalk, and reappears at the bottom of the window 2 s after passing it on the way down. After a completely elastic collision, the bearing will have the same speed at a point going up as it had going down. How tall is the building?

25. A construction elevator without a ceiling is ascending with constant speed 10 m/s. A girl on the elevator throws a ball directly upward from a height of 2 m above the elevator floor just as the elevator floor is 28 m above the ground. The initial speed of the ball with respect to the elevator is 20 m/s.

   (a) What is the maximum height attained by the ball?

   (b) How long does it take for the ball to return to the elevator floor?

26. Two stones are thrown vertically upward one second apart over the edge of the cliff in Example 5.6. The first is thrown at 25 m/s, the second at 20 m/s. Determine if and when they ever pass each other.

27. In the theory of special relativity, Newton's second law ($F = ma$) is replaced by

$$F = m_0 \frac{d}{dt} \left[\frac{v}{\sqrt{1 - (v^2/c^2)}} \right],$$

where F is the applied force, m_0 the mass of the particle measured at rest, v its speed, and c the speed of light — a constant. Show that if we set $a = dv/dt$, then

$$F = \frac{m_0 a}{[1 - (v^2/c^2)]^{3/2}}.$$

Explain the difference between this law and Newton's second law.

* **28.** What speed maximizes the flow rate in Example 5.7 if cars are required to maintain only a fraction k ($0 < k < 1$) of the safe distance?

* **29.** Two stones are thrown vertically upward over the edge of a bottomless abyss, the second stone t_0 units of time after the first. The first stone has an initial speed of v_0', and the second an initial speed of v_0''.

   (a) Show that if the stones are ever to pass each other during their motions, two conditions must be satisfied:

$$gt_0 > v_0' - v_0'' \quad \text{and} \quad v_0' > gt_0/2,$$

   where $g > 0$ is the acceleration due to gravity.

   (b) Show that the first condition is equivalent to the requirement that stone 1 must begin its downward trajectory before stone 2.

   (c) Show that the second condition is equivalent to the requirement that stone 1 must not pass its original projection point before the projection of stone 2.

* **30.** Speed bumps are to be placed on a straight stretch of road in order to ensure that traffic speed does not exceed 10 m/s. The question concerns their spacing. Suppose vehicle speed is reduced to 2.5 m/s at bumps. Vehicles accelerate uniformly away from a bump at 3 m/s^2, and then decelerate uniformly toward the next bump at 7 m/s^2. Find the distance between bumps.

** **31.** It takes time T to drive your car a distance D along a straight highway. You do so by accelerating uniformly from rest, attaining maximum speed V, which you maintain for some length of time, and then decelerating uniformly to a stop. How long do you maintain speed V if the magnitudes of the acceleration and deceleration are the same?

5.3 Change of Variable in the Indefinite Integral

In Section 5.1 we suggested two methods for evaluating indefinite integrals — recognition and adjusting constants. In this section we show how a **change of variable** can often replace a complex integration problem with a simpler one.

Consider the indefinite integral

$$\int x\sqrt{2x + 1}\, dx.$$

What is annoying about this integrand is the sum of two terms $2x + 1$ under a square root. This can be changed by setting $u = 2x + 1$. As a result, $\sqrt{2x + 1} = \sqrt{u}$, and the x in front of the square root is equal to $(u - 1)/2$. Now the differential dx is used to indicate integration in the problem with respect to x. But surely there must be another reason why we have chosen the differential to denote this. If we regard $x\sqrt{2x + 1}\, dx$ as a product of $x\sqrt{2x + 1}$ and dx, then perhaps we should obtain an expression for dx in terms of du. To do this we note that since $u = 2x + 1$, then

$$\frac{du}{dx} = 2.$$

As derivatives can be regarded as quotients of differentials, we can rewrite this equation in the form

$$dx = \frac{du}{2}.$$

If we make all these substitutions into the indefinite integral

$$\int x\sqrt{2x+1}\,dx,$$

the result is an integration problem in the variable u, which is easy to evaluate:

$$\int \left(\frac{u-1}{2}\right)\sqrt{u}\,\frac{du}{2} = \frac{1}{4}\int (u^{3/2} - u^{1/2})\,du = \frac{1}{4}\left(\frac{2}{5}u^{5/2} - \frac{2}{3}u^{3/2}\right) + C.$$

If u is now replaced by $2x+1$, the result is

$$\frac{1}{10}(2x+1)^{5/2} - \frac{1}{6}(2x+1)^{3/2} + C.$$

Differentiation of this function quickly indicates that its derivative is indeed $x\sqrt{2x+1}$ and, therefore,

$$\int x\sqrt{2x+1}\,dx = \frac{1}{10}(2x+1)^{5/2} - \frac{1}{6}(2x+1)^{3/2} + C.$$

In this example, the substitution $u = 2x+1$ replaces a complex integration in x with a simple one in u. Once the problem in u is solved, replacement of u's with x's gives the solution to the original indefinite integral. This method is generally applicable and is justified in the following theorem.

THEOREM 5.3

Suppose the change of variable $u = h(x) \iff x = g(u)$ replaces the indefinite integral

$$\int f(x)\,dx \qquad \text{with} \qquad \int f\big(g(u)\big)\,g'(u)\,du,$$

where $g(u)$ is differentiable on some interval, and $f(x)$ is continuous on the range of $g(u)$. If $F(u)$ is an antiderivative of $f\big(g(u)\big)\,g'(u)$, then

$$\int f(x)\,dx = F\big(h(x)\big) + C.$$

EXAMPLE 5.8

Evaluate the following indefinite integrals:

(a) $\displaystyle\int \sqrt[5]{2x+4}\,dx$ (b) $\displaystyle\int \frac{x}{\sqrt{x+1}}\,dx$ (c) $\displaystyle\int \sin^3 x \cos^2 x\,dx$

SOLUTION

(a) We could adjust constants in this case. With an initial guess of $(2x + 4)^{6/5}$, we calculate

$$\frac{d}{dx}(2x + 4)^{6/5} = \frac{6}{5}(2x + 4)^{1/5}(2).$$

Thus,

$$\int (2x + 4)^{1/5}\, dx = \frac{5}{12}(2x + 4)^{6/5} + C.$$

Alternatively, if we set $u = 2x + 4$, then $du = 2dx$, and

$$\int (2x + 4)^{1/5}\, dx = \int u^{1/5}\,\frac{du}{2} = \frac{1}{2}\left(\frac{5}{6}u^{6/5}\right) + C = \frac{5}{12}(2x + 4)^{6/5} + C.$$

(b) If we set $u = x + 1$, then $du = dx$, and

$$\int \frac{x}{\sqrt{x + 1}}\, dx = \int \frac{u - 1}{\sqrt{u}}\, du = \int (u^{1/2} - u^{-1/2})\, du = \frac{2}{3}u^{3/2} - 2u^{1/2} + C$$

$$= \frac{2}{3}(x + 1)^{3/2} - 2(x + 1)^{1/2} + C.$$

A different substitution is also possible. If we set $u = \sqrt{x + 1}$, then

$$du = \frac{1}{2\sqrt{x + 1}}\, dx \qquad \text{and}$$

$$\int \frac{x}{\sqrt{x + 1}}\, dx = \int (u^2 - 1)(2du) = 2\left(\frac{u^3}{3} - u\right) + C$$

$$= \frac{2}{3}(x + 1)^{3/2} - 2(x + 1)^{1/2} + C.$$

(c) If we set $u = \cos x$, then $du = -\sin x\, dx$ and

$$\int \sin^3 x \cos^2 x\, dx = \int \sin^2 x \cos^2 x \sin x\, dx$$

$$= \int (1 - \cos^2 x)\cos^2 x \sin x\, dx$$

$$= \int (1 - u^2)u^2(-du) = \int (u^4 - u^2)\, du$$

$$= \frac{u^5}{5} - \frac{u^3}{3} + C = \frac{1}{5}\cos^5 x - \frac{1}{3}\cos^3 x + C.$$

EXERCISES 5.3

In Exercises 1–30 evaluate the indefinite integral.

1. $\displaystyle\int (5x + 14)^9 \, dx$

2. $\displaystyle\int \sqrt{1 - 2x} \, dx$

3. $\displaystyle\int \frac{1}{(3y - 12)^{1/4}} \, dy$

4. $\displaystyle\int \frac{5}{(5 - 42x)^{1/4}} \, dx$

5. $\displaystyle\int x^2 (3x^3 + 10)^4 \, dx$

6. $\displaystyle\int \frac{x}{(x^2 + 4)^2} \, dx$

7. $\displaystyle\int \sin^4 x \cos x \, dx$

8. $\displaystyle\int \frac{x^2}{(x - 2)^4} \, dx$

9. $\displaystyle\int z\sqrt{1 - 3z} \, dz$

10. $\displaystyle\int \frac{x}{\sqrt{2x + 3}} \, dx$

11. $\displaystyle\int \frac{1 + \sqrt{x}}{\sqrt{x}} \, dx$

12. $\displaystyle\int s^3 \sqrt{s^2 + 5} \, ds$

13. $\displaystyle\int \sin^2 x \cos^3 x \, dx$

14. $\displaystyle\int \sqrt{1 - \cos x} \sin x \, dx$

15. $\displaystyle\int \frac{x^3}{(3 - x^2)^3} \, dx$

16. $\displaystyle\int y^2 \sqrt{y - 4} \, dy$

17. $\displaystyle\int \frac{(1 + \sqrt{u})^{1/2}}{\sqrt{u}} \, du$

18. $\displaystyle\int x^8 (3x^3 - 5)^6 \, dx$

19. $\displaystyle\int \frac{1 + z^{1/4}}{\sqrt{z}} \, dz$

20. $\displaystyle\int \frac{x + 1}{(x^2 + 2x + 2)^{1/3}} \, dx$

21. $\displaystyle\int \frac{(x - 1)(x + 2)}{\sqrt{x}} \, dx$

* **22.** $\displaystyle\int \frac{\cos^3 x}{(3 - 4\sin x)^4} \, dx$

* **23.** $\displaystyle\int \sqrt{1 + \sin 4t} \cos^3 4t \, dt$

* **24.** $\displaystyle\int \sqrt{1 + \sqrt{x}} \, dx$

* **25.** $\displaystyle\int \tan^2 x \sec^2 x \, dx$

* **26.** $\displaystyle\int \tan x \sec^2 x \, dx$

* **27.** $\displaystyle\int \frac{e^{2x}}{e^{2x} + 1} \, dx$

* **28.** $\displaystyle\int \frac{\ln x}{x} \, dx$

* **29.** $\displaystyle\int \frac{1}{x \ln x} \, dx$

* **30.** $\displaystyle\int \frac{x}{(x^2 + 1)[\ln (x^2 + 1)]^2} \, dx$

* **31.** Consider the integral $\displaystyle\int \sin^3 x \cos^3 x \, dx$.

    (a) Evaluate it by making the substitution $u = \sin x$.

    (b) Evaluate it by making the substitution $u = \cos x$.

    (c) Verify that these answers are the same.

* **32.** Evaluate $\displaystyle\int \sqrt{\frac{x^2}{1 + x}} \, dx$.

In Exercises 33–36 use the suggested change of variable to evaluate the indefinite integral.

* **33.** $\displaystyle\int \frac{\sqrt{4x - x^2}}{x^3} \, dx$;   set $u = 2/x$

* **34.** $\displaystyle\int \frac{\sqrt{x - x^2}}{x^4} \, dx$;   set $u = 1/x$

* **35.** $\displaystyle\int \frac{x}{(5 - 4x - x^2)^{3/2}} \, dx$;   set $u^2 = (5 + x)/(1 - x)$

* **36.** $\displaystyle\int \frac{1}{3(1 - x^2) - (5 + 4x)\sqrt{1 - x^2}} \, dx$;

    set $u^2 = (1 - x)/(1 + x)$

* **37.** Show that the substitution $u - x = \sqrt{x^2 + x + 4}$ replaces the integral

$$\int \sqrt{x^2 + x + 4} \, dx$$

with the integral of a rational function of u.

* **38.** Show that the substitution $(x + 1)u = \sqrt{4 + 3x - x^2}$ replaces the integral

$$\int \frac{1}{\sqrt{4 + 3x - x^2}} \, dx$$

with the integral of a rational function of u.

5.4 Deflection of Beams

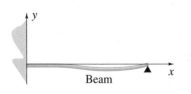

When a beam that might otherwise be horizontal is subjected to loads, it bends. By analyzing internal forces and moments, it can be shown that the shape $y(x)$ of a uniform beam with constant cross-section (Figure 5.7) is governed by the equation

$$\frac{d^4 y}{dx^4} = \frac{F(x)}{EI}, \tag{5.8}$$

where E is a constant called *Young's modulus of elasticity* (depending on the material of the beam), and I is also a constant (the moment of inertia of the cross-section of the beam).

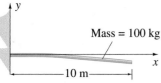

FIGURE 5.8 Beam bending under its own weight

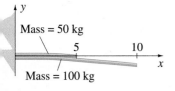

FIGURE 5.9 Beam bending under its own weight and an additional load

FIGURE 5.10 Simple support at $x = 0$

FIGURE 5.11 Built-in support at $x = 0$

FIGURE 5.12 Free end or no support at $x = 0$

Quantity $F(x)$ is the load placed on the beam; it is the vertical force per unit length in the x-direction, placed at position x, including the weight of the beam itself. For example, if a beam has mass 100 kg and length 10 m (Figure 5.8), then the load due to its weight is a constant $F(x) = -9.81(100/10) = -98.1$ N/m at every point of the beam.

Suppose a block with mass 50 kg, uniform in cross-section, and length 5 m is placed on the left half of the beam in Figure 5.8 (see Figure 5.9). It adds an additional load of $9.81(10) = 98.1$ N/m over the interval $0 < x < 5$. The total load can be represented in terms of Heaviside unit step functions as

$$F(x) = -98.1 - 98.1[h(x) - h(x - 5)].$$

[Recall that the function $h(x) - h(x - 5)$ turns the function $(-98.1$ in this case) in front of it on at $x = 0$ and off at $x = 5$.] Because the beam extends from $x = 0$ to $x = 10$, and $h(x) = 1$ thereon, we can write that

$$F(x) = -196.2 + 98.1\, h(x - 5), \qquad 0 < x < 10.$$

Accompanying equation 5.8 will be four **boundary conditions** defining the type of support (if any) at each end of the beam. Three types of supports are common. We discuss them at the left end of the beam, but they also occur at the right end.

1. Simple Support

The end of a beam is **simply supported** when it cannot move vertically but is free to rotate. Visualize a horizontal pin perpendicular to the xy-plane passing through a hole in the end of the beam at $x = 0$ (Figure 5.10). The pin is fixed, but the end of the beam can rotate on the pin. In this case, $y(x)$ must satisfy the boundary conditions

$$y(0) = y''(0) = 0. \tag{5.9a}$$

2. Built-in End

If the end $x = 0$ of the beam is permanently fixed in a horizontal position (Figure 5.11), $y(x)$ satisfies

$$y(0) = y'(0) = 0. \tag{5.9b}$$

3. Free Support

If the end $x = 0$ of the beam is not supported (Figure 5.12), $y(x)$ satisfies

$$y''(0) = y'''(0) = 0. \tag{5.9c}$$

When two boundary conditions at each end of a beam accompany differential equation 5.8, we have what is called a **boundary-value problem**.

EXAMPLE 5.9

A uniform beam with mass 100 kg and length 10 m has both ends built in horizontally. Find the deflection curve for the beam.

SOLUTION The boundary-value problem for deflection $y(x)$ is

$$\frac{d^4y}{dx^4} = -\frac{98.1}{EI}, \qquad 0 < x < 10,$$

$$y(0) = y'(0) = 0 = y(10) = y'(10).$$

Four integrations of the differential equation give

$$y(x) = \frac{1}{EI}\left(-\frac{98.1x^4}{24} + Ax^3 + Bx^2 + Cx + D\right),$$

where A, B, C, and D are constants. The boundary conditions require these constants to satisfy

$$0 = EIy(0) = D,$$

$$0 = EIy'(0) = C,$$

$$0 = EIy(10) = -\frac{98.1(10)^4}{24} + A(10)^3 + B(10)^2 + C(10) + D,$$

$$0 = EIy'(10) = -\frac{98.1(10)^3}{6} + 3A(10)^2 + 2B(10) + C.$$

These yield $A = 327/4$ and $B = -1635/4$, and therefore the curve of deflection for the beam is

$$y(x) = \frac{1}{EI}\left(-\frac{327x^4}{80} + \frac{327x^3}{4} - \frac{1635x^2}{4}\right) = \frac{327}{80EI}(-x^4 + 20x^3 - 100x^2).$$

Maximum deflection of the beam should occur at its midpoint. To confirm this, we find critical points,

$$0 = y'(x) = \frac{327}{80EI}(-4x^3 + 60x^2 - 200x) = -\frac{327}{20EI}x(x-10)(x-5).$$

Solutions are $x = 0$ and $x = 10$ (because each end is fixed horizontally) and $x = 5$. Maximum deflection is $y(5) = -40\,875/(16EI)$. For a beam such that the product $EI = 10^6$, the deflection at $x = 5$ is $y(5) = -2.55 \times 10^{-3}$ m, that is, 2.55 mm.

This problem was relatively straightforward due to the fact that the load function $F(x) = -98.1$ is continuous. Discontinuous load functions lead to more complicated calculations.

EXAMPLE 5.10

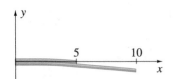

The end $x = 0$ of the beam in Figure 5.9 is horizontally built-in, and the right end is free, just like a diving board (Figure 5.13). Find the curve of deflection.

SOLUTION   The boundary-value problem for deflections is

$$\frac{d^4y}{dx^4} = \frac{1}{EI}[-196.2 + 98.1\,h(x-5)], \quad 0 < x < 10,$$

$$y(0) = y'(0) = 0, \quad y''(10) = y'''(10) = 0.$$

Integration of the differential equation four times on the intervals $0 < x < 5$ and $5 < x < 10$ gives

$$y(x) = \frac{1}{EI} \begin{cases} -8.175x^4 + Ax^3 + Bx^2 + Cx + D, & 0 < x < 5 \\ -4.0875x^4 + Px^3 + Qx^2 + Rx + S, & 5 < x < 10. \end{cases}$$

To evaluate the eight constants, we impose the four boundary conditions, and also demand that $y(x)$ and its first three derivatives be continuous at $x = 5$. This means that left- and right-hand limits of $EIy(x)$, $EIy'(x)$, $EIy''(x)$, and $EIy'''(x)$ are the same at $x = 5$.

$$\begin{aligned} EIy(0) &= 0 &&\Longrightarrow& D &= 0, \\ EIy'(0) &= 0 &&\Longrightarrow& C &= 0, \\ EIy''(10) &= 0 &&\Longrightarrow& -49.05(10)^2 + 6P(10) + 2Q &= 0, \\ EIy'''(10) &= 0 &&\Longrightarrow& -98.1(10) + 6P &= 0. \end{aligned}$$

$$\lim_{x \to 5^-} EIy(x) = \lim_{x \to 5^+} EIy(x) \quad \Longrightarrow$$

$$-8.175(5)^4 + 125A + 25B + 5C + D = -4.0875(5)^4 + 125P + 25Q + 5R + S,$$

$$\lim_{x \to 5^-} EIy'(x) = \lim_{x \to 5^+} EIy'(x) \quad \Longrightarrow$$

$$-32.7(5)^3 + 75A + 10B + C = -16.35(5)^3 + 75P + 10Q + R,$$

$$\lim_{x \to 5^-} EIy''(x) = \lim_{x \to 5^+} EIy''(x) \quad \Longrightarrow$$

$$-98.1(5)^2 + 30A + 2B = -49.05(5)^2 + 30P + 2Q,$$

$$\lim_{x \to 5^-} EIy'''(x) = \lim_{x \to 5^+} EIy'''(x) \quad \Longrightarrow$$

$$-196.2(5) + 6A = -98.1(5) + 6P.$$

Solutions of these equations are

$$A = 245.25, \quad B = -3065.625, \quad C = 0, \quad D = 0,$$

$$P = 163.5, \quad Q = -2452.5, \quad R = -2043.75, \quad S = 2554.6875.$$

The function describing deflections of the beam is, therefore,

$$y(x) = \frac{1}{EI} \begin{cases} -8.175x^4 + 245.25x^3 - 3065.625x^2, & 0 \le x \le 5 \\ -4.0875x^4 + 163.5x^3 - 2452.5x^2 - 2043.75x + 2554.6875, & 5 < x \le 10. \end{cases}$$

A graph of this function for $EI = 10^6$ is shown in Figure 5.14.

FIGURE 5.14 Deflections of a diving board

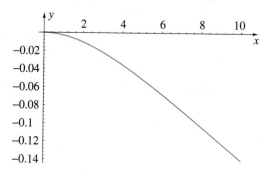

Fortunately there are easier ways to solve this problem. One such is to use what are called *Laplace transforms*. You will learn them in advanced calculus courses. Even at this stage we can simplify calculations considerably. Difficulties arose when we integrated the differential equation separately on the intervals $0 < x < 5$ and $5 < x < 10$. This introduced four additional constants of integration that were evaluated by demanding that $y(x)$, $y'(x)$, $y''(x)$, and $y'''(x)$ be continuous at $x = 5$. Suppose we demand from the beginning that $y(x)$ and its first three derivatives be continuous for the length of the beam. What this means is that we need continuous antiderivatives of $h(x - a)$. We did this in Example 5.4. The continuous indefinite integral of $h(x - a)$ is

$$\int h(x - a)\, dx = (x - a)h(x - a) + C, \tag{5.10a}$$

provided we understand that the value is C at $x = a$. The following indefinite integrals are also continuous:

$$\int (x - a)h(x - a)\, dx = \frac{1}{2}(x - a)^2 h(x - a) + C, \tag{5.10b}$$

$$\int (x - a)^2 h(x - a)\, dx = \frac{1}{3}(x - a)^3 h(x - a) + C, \tag{5.10c}$$

$$\int (x - a)^3 h(x - a)\, dx = \frac{1}{4}(x - a)^4 h(x - a) + C, \tag{5.10d}$$

provided again that each indefinite integral is given value C at $x = a$. In general,

$$\int (x - a)^n h(x - a)\, dx = \frac{1}{n + 1}(x - a)^{n+1} h(x - a) + C. \tag{5.11}$$

With these, the solution to Example 5.10 is far easier. Four integrations of

$$\frac{d^4 y}{dx^4} = \frac{1}{EI}[-196.2 + 98.1 h(x - 5)]$$

using formulas 5.10 give

$$y(x) = \frac{1}{EI}\left[-\frac{196.2 x^4}{24} + \frac{98.1}{24}(x - 5)^4 h(x - 5) + Ax^3 + Bx^2 + Cx + D\right].$$

The boundary conditions require that

$$0 = EIy(0) = D,$$

$$0 = EIy'(0) = C,$$

$$0 = EIy''(10) = -\frac{196.2(10)^2}{2} + \frac{98.1(5)^2}{2} + 6A(10) + 2B,$$

$$0 = EIy'''(10) = -196.2(10) + 98.1(5) + 6A.$$

These can be solved for $A = 245.25$ and $B = -3065.625$. The deflected curve is

$$y(x) = -\frac{196.2 x^4}{24} + \frac{98.1}{24}(x - 5)^4 h(x - 5) + 245.25 x^3 - 3065.625 x^2.$$

This is equivalent to the function $y(x)$ found in Example 5.10, but its derivation and final form are unmistakably simpler.

In Section 2.5 we suggested that the Dirac-delta function is used to model point sources in engineering and physics. We use it here to model a point force applied to a beam. The load due to a point force of magnitude F newtons applied vertically downward at a point x_0 on a beam is represented mathematically by $-F\delta(x - x_0)$. Suppose it is applied to a beam of length L, and that F is so large that the weight of the beam is negligible by comparison. In this case, the differential equation describing deflections is

$$\frac{d^4y}{dx^4} = -\frac{F}{EI}\delta(x - x_0).$$

If the beam is horizontally fixed at $x = 0$ and simply supported at $x = L$, deflections must also satisfy $y(0) = y'(0) = 0 = y(L) = y''(L)$. According to Example 3.11 in Section 3.3, $h'(x - a) = \delta(x - a)$. Consequently, we can write that

$$\int \delta(x - a)\,dx = h(x - a) + C. \tag{5.12}$$

Using this result, integration of the differential equation gives

$$\frac{d^3y}{dx^3} = \frac{1}{EI}[-Fh(x - x_0) + C].$$

Three additional integrations using formulas 5.10 yield

$$y(x) = \frac{1}{EI}\left[-\frac{F}{6}(x - x_0)^3 h(x - x_0) + Cx^3 + Ax^2 + Bx + D\right],$$

where we have absorbed a factor of 6 into C. The boundary conditions require that

$$0 = EIy(0) = D,$$

$$0 = EIy'(0) = B,$$

$$0 = EIy(L) = -\frac{F(L - x_0)^3}{6} + CL^3 + AL^2 + BL + D,$$

$$0 = EIy''(L) = -F(L - x_0) + 6CL + 2A.$$

Values of A and C are

$$A = \frac{Fx_0(L - x_0)(x_0 - 2L)}{4L^2}, \qquad C = \frac{F(L - x_0)(2L^2 + 2Lx_0 - x_0^2)}{12L^3}.$$

Thus,

$$y(x) = \frac{1}{EI}\left[-\frac{F}{6}(x - x_0)^3 h(x - x_0) + \frac{F(L - x_0)(2L^2 + 2Lx_0 - x_0^2)x^3}{12L^3}\right.$$

$$\left. + \frac{Fx_0(L - x_0)(x_0 - 2L)x^2}{4L^2}\right].$$

Consulting Project 7

A beam in a building is horizontally fixed at both ends. The size and distribution of load on the beam determine its deflection at various points; the more load at a particular point on the beam, the greater the deflection at that point. We are asked to determine the difference between deflections at the midpoint of the beam when a given total load is evenly distributed along the beam as opposed to when it is concentrated at the midpoint of the beam.

SOLUTION   Let us suppose that the length of the beam is L and the total load is $F > 0$. When this load is evenly distributed along the beam, differential equation 5.8 becomes

$$\frac{d^4 y}{dx^4} = -\frac{F}{EIL},$$

subject to the boundary conditions

$$y(0) = y'(0) = 0 = y(L) = y'(L).$$

Four integrations of the differential equation give

$$y(x) = \frac{1}{EI}\left(-\frac{Fx^4}{24L} + Ax^3 + Bx^2 + Cx + D\right).$$

The boundary conditions require

$$0 = EIy(0) = D,$$

$$0 = EIy'(0) = C,$$

$$0 = EIy(L) = -\frac{FL^3}{24} + AL^3 + BL^2 + CL + D,$$

$$0 = EIy'(L) = -\frac{FL^2}{6} + 3AL^2 + 2BL + C.$$

These give $A = F/12$ and $B = -FL/24$, so that

$$y(x) = \frac{1}{EI}\left(-\frac{Fx^4}{24L} + \frac{Fx^3}{12} - \frac{FLx^2}{24}\right).$$

Deflection at the midpoint of the beam is $y(L/2) = -FL^3/(384EI)$. When F is concentrated at the midpoint of the beam differential equation 5.8 becomes

$$\frac{d^4 y}{dx^4} = -\frac{F}{EI}\delta(x - L/2).$$

Four integrations give

$$y(x) = \frac{1}{EI}\left[-\frac{F}{6}(x - L/2)^3 h(x - L/2) + Ax^3 + Bx^2 + Cx + D\right].$$

The boundary conditions require

$$0 = EIy(0) = D,$$

$$0 = EIy'(0) = C,$$

$$0 = EIy(L) = -\frac{F(L - L/2)^3}{6} + AL^3 + BL^2 + CL + D,$$

$$0 = EIy'(L) = -\frac{F(L - L/2)^2}{2} + 3AL^2 + 2BL + C.$$

These give $A = F/12$ and $B = -FL/16$, so that

$$y(x) = \frac{1}{EI}\left[-\frac{F}{6}(x - L/2)^3 h(x - L/2) + \frac{Fx^3}{12} - \frac{FLx^2}{16}\right].$$

Deflection at the midpoint is $y(L/2) = -FL^3/(192EI)$. The ratio of concentrated deflection to distributed deflection at the midpoint is

$$\frac{-FL^3/(192EI)}{-FL^3/(384EI)} = 2.$$

EXERCISES 5.4

1. Find deflections of a uniform beam with mass m and length L when both ends are simply supported.

2. Repeat Exercise 1 if both ends are fixed horizontally.

3. Repeat Exercise 1 if the left end $x = 0$ is fixed horizontally and the right end is free.

4. Repeat Exercise 1 if the left end is fixed horizontally and the right end is simply supported.

5. A concentrated force of F newtons is applied vertically downward at the midpoint of a uniform beam of length L. Both ends of the beam are built in horizontally. If F is so large that the weight of the beam is negligible in comparison, find deflections of the beam.

6. Repeat Exercise 5 if the left end of the beam is built in horizontally and the right end is free. Is the beam straight for $x > L/2$?

7. Repeat Exercise 5 if both ends of the beam are simply supported.

8. Repeat Exercise 5 if the mass m of the beam is taken into account.

9. Repeat Exercise 6 if the mass m of the beam is taken into account.

10. Repeat Exercise 7 if the mass m of the beam is taken into account.

* **11.**   (a) A simply-supported beam of length 4 m is to carry a total uniform load of 1000 N (including its weight), and a concentrated load of 1500 N at its centre. Assuming $EI = 10^6$, use Exercise 10 to find deflections of the beam.

(b) If safety codes do not permit deflections to exceed 1/360 of span, is the beam acceptable?

* **12.** Find deflections of the beam in Figure 5.9 if the 50 kg block is on the right half of the beam. Is the deflection at the right end larger or smaller than in Figure 5.9?

* **13.** Find deflections of the beam in Figure 5.9 if the 50 kg block is centred on the beam. How does the deflection at the right end compare to that in Figure 5.9 and Exercise 12?

* **14.** A concentrated force F is applied vertically downward at the right end of a uniform beam of length L. The left end of the beam is built in horizontally and the right end is free. If the weight of the beam is negligible in comparison to F, find deflections of the beam. *Hint:* Place F at a point x_0 to the left of $x = L$, solve for deflections, and take the limit as $x_0 \to L^-$.

5.5 An Introduction to Separable Differential Equations

A **differential equation** is an equation that contains derivatives of some unknown function; the equation must be solved for all functions that satisfy it. Because differential equations arise in so many engineering problems, both elementary and advanced, we discuss what are called *separable* differential equations here. A full treatment of differential equations is taken up in Chapter 15.

When a differential equation for y as a function of x contains dy/dx, but no higher order derivatives, it is called a **first-order differential equation**. The vast majority of such differential

equations can be expressed in the form

$$\frac{dy}{dx} = F(x, y); \tag{5.13}$$

that is, they can be solved for dy/dx in terms of x and y. Examples are

$$(1 - y)\frac{dy}{dx} = 3x^2 \quad \text{and} \quad \frac{dy}{dx} + 2xy = 4x$$

Each of these can be solved for dy/dx,

$$\frac{dy}{dx} = \frac{3x^2}{1 - y} \quad \text{and} \quad \frac{dy}{dx} = 4x - 2xy.$$

Differential equation 5.13 is said to be **separable** if it can be expressed in the form

$$\frac{dy}{dx} = \frac{M(x)}{N(y)}; \tag{5.14}$$

that is, dy/dx is a function of x divided by a function of y. Both of the above examples are separable. The first is already in this form; the second can be put so,

$$\frac{dy}{dx} = 4x - 2xy = 2x(2 - y) = \frac{2x}{\dfrac{1}{2 - y}}.$$

What is equivalent to equation 5.14 is to say that 5.13 is separable if it can be expressed in the form

$$N(y)\, dy = M(x)\, dx. \tag{5.15}$$

When a differential equation is written in this way, it is said to be **separated** — separated in the sense that x- and y-variables appear on opposite sides of the equation. For a separated equation we can write therefore that

$$N(y)\frac{dy}{dx} = M(x), \tag{5.16}$$

and if we integrate both sides with respect to x, we have

$$\int N(y)\frac{dy}{dx}\, dx = \int M(x)\, dx + C. \tag{5.17}$$

Cancellation of differentials on the left leads to the solutions

$$\int N(y)\, dy = \int M(x)\, dx + C. \tag{5.18}$$

 What we mean by saying that 5.18 represents solutions for 5.14 is that any function defined *implicitly* by 5.18 is a solution of 5.14. A word of warning is appropriate here. Looking at equation 5.18 in isolation, it might appear that we have integrated the left side of equation 5.15 with respect to y and the right side with respect to x. This is not true. We do not differentiate one side of an equation with respect to one variable and the other side with respect to a different variable. Why then would we expect to be able to integrate different sides of an equation with respect to different variables? What we did was rewrite 5.15 in form 5.16 and integrate both sides of this equation with respect to x. Cancellations of differentials led to 5.18. Thus, although we did not integrate the left side of equation 5.15 with respect to y and the right side with respect to x to get equation 5.18, we can now interpret 5.18 in this way.

For instance, in the first example above, we write

$$(1 - y)dy = 3x^2\, dx.$$

According to equation 5.18, solutions are defined implicitly by

$$\int (1 - y)\, dy = \int 3x^2\, dx \qquad \Longrightarrow \qquad y - \frac{y^2}{2} = x^3 + C.$$

We can find explicit solutions by solving the equation for y in terms of x. Multiplying by -2 expresses the equation as a quadratic in y:

$$y^2 - 2y + 2(x^3 + C) = 0.$$

Therefore

$$y = \frac{2 \pm \sqrt{4 - 8(x^3 + C)}}{2} = 1 \pm \sqrt{1 - 2(x^3 + C)}.$$

Explicit solutions of the differential equation are therefore

$$y(x) = 1 + \sqrt{1 - 2(x^3 + C)} \quad \text{and} \quad y(x) = 1 - \sqrt{1 - 2(x^3 + C)},$$

provided expressions on the right are indeed functions of x. Once C is determined, this will be true only for certain values of x. For example, suppose we require the solution of the differential equation that satisfies the extra condition $y(0) = 3$. The second function $y(x) = 1 - \sqrt{1 - 2(x^3 + C)}$ cannot satisfy this condition because y cannot be greater than 1. If we substitute $x = 0$ and $y = 3$ into the other function,

$$3 = 1 + \sqrt{1 - 2C},$$

and this requires $C = -3/2$. Thus, the solution of the differential equation for which $y(0) = 3$ is

$$y(x) = 1 + \sqrt{1 - 2(x^3 - 3/2)} = 1 + \sqrt{4 - 2x^3}.$$

Since $4 - 2x^3$ must be nonnegative for this function to be defined, the solution is valid only on the interval $x \le 2^{1/3}$. In fact, because the derivative of this function is not defined at $x = 2^{1/3}$, we should consider only $x < 2^{1/3}$.

We now consider four problems that give rise to separable differential equations.

EXAMPLE 5.11

An ore sample contains, along with various impurities, an amount A_0 of radioactive material, say, uranium. Disintegrations gradually reduce this amount of uranium. Experiments have led to the following *law of radioactive disintegration*: The time rate of change of the amount of radioactive material is proportional at any instant to the amount of radioactive material present at that time. Find the amount of uranium in the sample as a function of time.

SOLUTION   If we let $A(t)$ be the amount of uranium in the sample at any time t, then the law of radioactive disintegration states that

$$\frac{dA}{dt} = kA, \tag{5.19}$$

where k is a constant. Since A is decreasing, dA/dt and hence k must be negative. If we choose $t = 0$ when the amount of uranium is A_0, then the differential equation must be solved for $A(t)$ subject to the initial condition $A(0) = A_0$. The differential equation is separable:

$$\frac{1}{A}\, dA = k\, dt,$$

and solutions are therefore defined implicitly by

$$\int \frac{1}{A}\, dA = \int k\, dt \qquad \Longrightarrow \qquad \ln |A| = kt + C.$$

We can omit the absolute values since A is always positive, in which case the initial condition requires $\ln A_0 = 0 + C$. Thus,

$$\ln A = kt + \ln A_0.$$

To find $A(t)$ explicitly, we take exponentials on both sides of the equation:

$$A = e^{kt + \ln A_0} = e^{kt} e^{\ln A_0} = A_0 e^{kt}.$$

The amount of uranium therefore decreases exponentially in time. To find k we need to know A at one additional time. For example, if we know that one ten-millionth of 1% of the original amount of uranium decays in 6.5 years, then

$$0.999\,999\,999\, A_0 = A_0 e^{13k/2}.$$

If we solve this for k, we obtain

$$k = \frac{2}{13} \ln (0.999\,999\,999) = -1.54 \times 10^{-10},$$

and therefore

$$A(t) = A_0 e^{-1.54 \times 10^{-10} t}.$$

The law of radioactive disintegration has an important application in the dating of once-living plants and animals. All living tissue contains two isotopes of carbon: C^{14} (carbon-14), which is radioactive, and C^{12} (carbon-12), which is stable. In living tissue, the ratio of the amount of C^{14} to that of C^{12} is $1/10\,000$ for all fragments of the tissue. When the tissue dies, however, the ratio changes due to the fact that no more carbon is produced, and the original C^{14} present decays radioactively into an element other than C^{12}. Thus, as the dead tissue ages, the ratio of C^{14} to C^{12} decreases, and by measuring this ratio, it is possible to predict how long ago the tissue was alive. Suppose, for example, the present ratio of C^{14} to C^{12} in a specimen is $1/100\,000$; that is, one-tenth that for a living tissue. Then 90% of the original amount of C^{14} in the specimen has disintegrated.

If we let $A(t)$ be the amount of C^{14} in the specimen at time t, taking $A = A_0$ to be the amount present in the living specimen at its death ($t = 0$), then $A = A_0 e^{kt}$. To determine k, we use the fact that the half-life of C^{14} is approximately 5550 years. (The *half-life* of a radioactive element is the time required for one-half an original sample of the material to disintegrate.) For carbon-14, this means that A is equal to $A_0/2$ when $t = 5550$; that is,

$$\frac{A_0}{2} = A_0 e^{5550k}.$$

When we divide by A_0 and take natural logarithms,

$$k = -\frac{1}{5550} \ln 2 = -0.000\,125.$$

Consequently, the amount of C^{14} in the specimen of dead tissue at any time t is given by

$$A = A_0 e^{-0.000\,125 t}.$$

If T is the present time, when the amount of C^{14} is known to be 10% of its original amount, then at this time

$$0.1A_0 = A_0 e^{-0.000\,125T}.$$

The solution of this equation is

$$T = \frac{\ln 10}{0.000\,125} = 18\,400,$$

and we conclude that the tissue died about 18 400 years ago.

EXAMPLE 5.12

When a hot (or cold) object is placed in an environment that has a different temperature, the object cools down (or heats up). For example, when a hot cup of coffee is placed on a table it cools down due to colder room temperature. Suppose, for example, that the cup of coffee is initially at temperature 95°C and the room stays at constant temperature 20°C. *Newton's law of cooling* states that the rate of change of the temperature of the coffee is proportional to the difference between temperature of the coffee and that of the room. Find the temperature of the coffee as a function of time.

SOLUTION   If $T(t)$ denotes temperature of the coffee, then Newton's law of cooling can be stated algebraically as

$$\frac{dT}{dt} = k(T - 20), \tag{5.20}$$

where $k < 0$ is a constant. It is negative because $T - 20 > 0$ and $dT/dt < 0$. According to equation 5.20, the coffee cools quickly at first because $T - 20$ is large, but more and more slowly as T approaches 20. Differential equation 5.20 for $T(t)$ is separable:

$$\frac{1}{T - 20}\, dT = k\, dt.$$

Consequently, solutions are defined implicitly by

$$\int \frac{1}{T - 20}\, dT = \int k\, dt \quad \Longrightarrow \quad \ln |T - 20| = kt + C.$$

We can drop absolute values since temperature of the coffee is never less than 20°C. Exponentiating both sides of this equation gives

$$T - 20 = e^{kt+C} = e^C e^{kt} \quad \Longrightarrow \quad T = 20 + De^{kt},$$

where $D = e^C$. If we choose $t = 0$ when temperature of the coffee is 95°C, then $95 = 20 + D \Longrightarrow D = 75$. Thus, temperature of the coffee is

$$T(t) = 20 + 75e^{kt}.$$

To find k we need to know temperature of the coffee at one other time. For example, if we knew that temperature dropped to 50°C in 5 min, then

$$50 = 20 + 75e^{5k} \quad \Longrightarrow \quad k = \frac{1}{5}\ln(2/5) = -0.183.$$

Coffee temperature is

$$T(t) = 20 + 75e^{-0.183t}.$$

We can also express $T(t)$ in the form

$$T(t) = 20 + 75e^{(t/5)\ln(2/5)} = 20 + 75e^{\ln[(2/5)^{t/5}]} = 20 + 75\left(\frac{2}{5}\right)^{t/5}.$$

The plot of these functions in Figure 5.15 indicates that coffee temperature never reaches 20°C; the graph is asymptotic to the line $T = 20$. However, temperature of the coffee is within 5° of 20°C in about 15 min, and within 1° in about 24 min.

FIGURE 5.15   Temperature of coffee

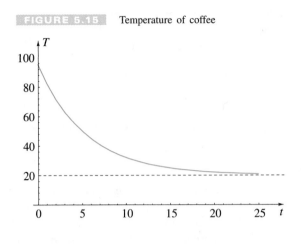

Admittedly, Newton's law of cooling is an approximation to what really happens physically. It assumes that there is no temperature variation within the cup of coffee, and that room temperature remains constant. Idealistic as this is, it is a good starting position for further discussions.

EXAMPLE 5.13

Figure 5.16 shows a liquid container with a hole in its bottom. It is interesting, and perhaps surprising, to find that the speed at which liquid exits through the hole is independent of the shape of the container; it depends only on the depth of liquid. We show this here, and then discover a technique for finding the depth of liquid in the container as a function of time.

FIGURE 5.16   Container with hole through which liquid escapes

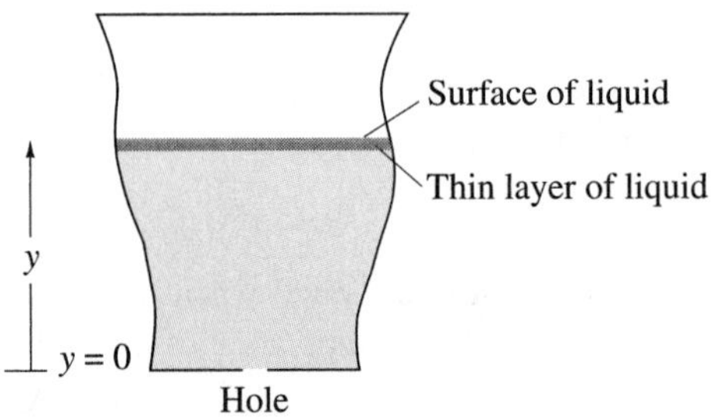

Surface of liquid

Thin layer of liquid

y

$y = 0$

Hole

If we consider a thin surface layer of the liquid, then during a small interval of time, the depth of liquid in the container drops by an amount equal to the thickness of the layer. Simultaneously a volume of liquid equal to that in the layer exits through the hole and, in so doing, causes the gravitational potential energy of the layer, due to its elevated position relative to the hole, to be converted into kinetic energy of the liquid passing through the hole. Suppose we let v be the speed at which the liquid leaves the container when the depth of liquid is y. The potential energy of the layer relative to the bottom of the container is

$$(\text{mass of layer})(g)(y),$$

where $g = 9.81$. The kinetic energy of an equal amount of liquid as it leaves the container is

$$\frac{1}{2}(\text{mass of layer})v^2.$$

When we equate these energies, and cancel the mass of the layer, we obtain

$$\frac{1}{2}v^2 = gy.$$

In other words, the speed at which liquid exits through the hole is

$$v = \sqrt{2gy}. \tag{5.21}$$

This result is known as *Torricelli's law*. It results from an idealized situation in which all potential energy is converted into kinetic energy. Experience suggests that exit speed depends on other factors as well — the size of the hole, for one. Water leaves more slowly through a small hole than through a large one. It is often assumed that v is somewhat less than $\sqrt{2gy}$, and equation 5.21 is replaced by

$$v = c\sqrt{2gy}, \tag{5.22}$$

where $0 < c < 1$ is a constant called the *discharge coefficient*. We call 5.22 the *modified Torricelli law*. We use it to find a differential equation satisfied by depth of liquid in the container. Suppose the area of the surface of the liquid is a function of depth y denoted by $A(y)$, and $V(t)$ is the volume of liquid in the container at any time t. Since depth of liquid changes at rate dy/dt, the rate at which the volume of liquid in the container changes is

$$\frac{dV}{dt} = A(y)\frac{dy}{dt}.$$

Both dV/dt and dy/dt are negative since V and y are decreasing. On the other hand, the rate at which liquid exits through the hole is the product av of the area a of the hole and exit speed v. It follows then that

$$A(y)\frac{dy}{dt} = -av = -ac\sqrt{2gy}. \tag{5.23}$$

Once the shape of the container is specified, then $A(y)$ is known, and this equation becomes a differential equation for $y(t)$. Solve for depth when the container is a right circular cylinder of radius r and height h with vertical axis. Determine how long it takes a full tank to empty.

SOLUTION   When the container is a right circular cylinder with radius r, then $A(y) = \pi r^2$, in which case differential equation 5.23 becomes

$$\pi r^2 \frac{dy}{dt} = -ac\sqrt{2gy} \qquad \Longrightarrow \qquad \frac{1}{\sqrt{y}}\, dy = -\frac{\sqrt{2g}\, ac}{\pi r^2}\, dt,$$

a separated differential equation. Solutions are defined implicitly by

$$\int \frac{1}{\sqrt{y}}\, dy = \int -\frac{\sqrt{2g}\, ac}{\pi r^2}\, dt \qquad \Longrightarrow \qquad 2\sqrt{y} = -\frac{\sqrt{2g}\, ac}{\pi r^2} t + D.$$

If the cylinder is originally full and liquid exits starting at time $t = 0$, then $2\sqrt{h} = D$, and

$$2\sqrt{y} = -\frac{\sqrt{2g}\, ac}{\pi r^2} t + 2\sqrt{h} \qquad \Longrightarrow \qquad y = \left(\sqrt{h} - \frac{\sqrt{g}\, act}{\sqrt{2}\pi r^2}\right)^2.$$

It is now a simple matter to determine how long the cylinder takes to empty. Setting $y(t) = 0$ and solving for t gives

$$t = \frac{\pi r^2}{ac}\sqrt{\frac{2h}{g}}.$$

This was a particularly simple example in that the cross-sectional area $A(y)$ of the container was constant. Containers with variable cross-sections are discussed in the exercises.

EXAMPLE 5.14

A tank originally contains 1000 L of water, in which 5 kg of salt has been dissolved.

(a) If a brine mixture containing 2 kg of salt for each 100 L of solution is poured into the tank at 10 mL/s, find the amount of salt in the tank as a function of time.

(b) If at the same time brine is being added, the mixture in the tank is being drawn off at 10 mL/s, find the amount of salt in the tank as a function of time. Assume that the mixture is stirred constantly.

SOLUTION

(a) Suppose we let $S(t)$ represent the number of grams of salt in the tank at time t. If we choose time $t = 0$ at the instant the brine mixture begins entering the original solution, then $S(0) = 5000$. Since 10 mL of mixture enters the tank each second and each millilitre contains 0.02 g of salt, it follows that 0.2 g of salt enters the tank each second. Consequently, after t seconds, $0.2t$ grams of salt has been added to the tank, and the total amount of salt in the solution is, in grams,

$$S(t) = 5000 + 0.2t.$$

(b) Once again we let $S(t)$ represent the number of grams of salt in the tank at time t. Its derivative dS/dt, the rate of change of $S(t)$, is the difference between how fast salt is being added to the tank in the brine and how fast salt is being removed as the mixture is removed,

$$\frac{dS}{dt} = \left\{ \text{rate salt added} \right\} - \left\{ \text{rate salt leaves} \right\}.$$

As in part (a), salt is being added at the constant rate of 0.2 g/s. The rate at which salt leaves the tank, on the other hand, is not constant; it depends on the concentration of salt in the tank. Since the tank always contains 10^6 mL of solution, the concentration of salt in the solution at time t is, in grams per millilitre, $S/10^6$, where $S = S(t)$ is the amount of salt in the tank at that time. As solution is being drawn off at the rate of 10 mL/s, the rate at which salt leaves the tank is, in grams per second,

$$\frac{S}{10^6}(10) = \frac{S}{10^5}.$$

Consequently,

$$\frac{dS}{dt} = \frac{1}{5} - \frac{S}{10^5} = \frac{20\,000 - S}{100\,000}.$$

To find $S(t)$ we must solve this differential equation subject to the condition that $S(0) = 5000$. The differential equation is separable:

$$\frac{1}{20\,000 - S}\, dS = \frac{1}{100\,000}\, dt.$$

Solutions are defined implicitly by

$$-\ln |20\,000 - S| = \frac{t}{100\,000} + C.$$

Let us solve for S before evaluating the constant of integration. Multiplication by -1 and exponentiation give

$$|20\,000 - S| = e^{-C - t/100\,000} \qquad \Longrightarrow \qquad 20\,000 - S = \pm e^{-C} e^{-t/100\,000}.$$

Because $\pm e^{-C}$ is an unknown constant, we simplify matters by setting $D = \pm e^{-C}$, in which case

$$S(t) = 20\,000 - De^{-t/100\,000}.$$

The initial condition $S(0) = 5000$ requires $5000 = 20\,000 - D \Longrightarrow D = 15\,000$, and therefore the number of grams of salt in the tank is

$$S(t) = 20\,000 - 15\,000e^{-t/100\,000}.$$

A graph of this function is shown in Figure 5.17. It is asymptotic to the line $S = 20\,000$. After a very long time, the amount of salt in the tank levels off so that its concentration is $20\,000/1\,000\,000 = 0.02$ g/mL. This, as might be expected, is the concentration of the incoming solution.

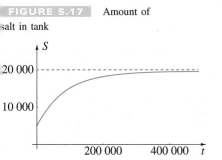

FIGURE 5.17 Amount of salt in tank

Consulting Project 8

This project concerns the rate of evaporation of water from a hemispherical tank, open on the top. All vertical cross sections of the tank are semicircles with radius r metres, one of which is shown in Figure 5.18. The tank is originally full of water and the problem is to determine how long it takes for the water to completely evaporate.

FIGURE 5.18 Evaporation of water from a hemispherical tank

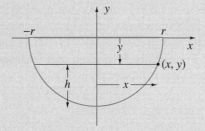

SOLUTION   We must make some kind of assumption about the rate at which water evaporates. Since evaporation takes place at the surface of the water, it seems reasonable to assume that evaporation is proportional to the surface area of the water at any given time. If $V(t)$ denotes the volume of water in the tank at time t, evaporation is represented by the derivative dV/dt, the time rate of change of the volume of water in the tank. If $A(t)$ is the surface area of the water at time t, then we have

$$\frac{dV}{dt} = kA,$$

where $k < 0$ is the constant of proportionality. It is negative because A is positive and dV/dt is negative. Essentially our problem is to solve this differential equation for $V(t)$ and determine when $V = 0$. An immediate difficulty is that the equation contains three variables, t, A, and V; one of which must be eliminated.

Certainly, t must remain, so either A or V must go. It would seem that we need a functional relationship between V and A. It is easy to see that the area of the surface of the water at any instant is πx^2, where x is the radius of the surface. What can we say about the volume V of a spherical segment? In many references we find the following formula for such a volume,

$$V = \frac{\pi h^2}{3}(3r - h),$$

where r is the radius of the sphere and h is the depth of water. (You may recall that we used this formula in Section 4.1. We will verify it in Section 7.2.) With $h = r + y$, we can write that

$$V = \frac{\pi}{3}(r + y)^2[3r - (r + y)] = \frac{\pi}{3}(2r^3 + 3r^2 - y^3).$$

Since the equation of the semicircle is $x^2 + y^2 = r^2$, we may also express A in terms of y,

$$A = \pi(r^2 - y^2).$$

These two equations determine V in terms of A, but to find exactly how would be difficult. Instead, notice that if we substitute both of them into the differential equation, we obtain

$$\frac{\pi}{3}\left(3r^2\frac{dy}{dt} - 3y^2\frac{dy}{dt}\right) = k\pi(r^2 - y^2).$$

When we simplify this equation, the result is

$$\frac{dy}{dt} = k.$$

How simple. We now have a differential equation for y as a function of t. Integration gives $y = kt + C$. If we assume that evaporation begins at time $t = 0$ when $y = 0$, we must set $C = 0$, and therefore $y = kt$. Water has completely evaporated when $y = -r$, and this occurs at time $t = -r/k$. This is the required time. It is known explicitly when k is known. For instance, if measurements indicate that the water level in the tank drops by 1% of the radius in 4 days, then

$$-\frac{1}{100}r = 4k \qquad \Longrightarrow \qquad k = -\frac{r}{400}.$$

It follows then that the tank empties in

$$-r\left(-\frac{400}{r}\right) = 400 \text{ days}.$$

EXERCISES 5.5

1. Bacteria in a culture increase at a rate proportional to the number present. If the original number increases by 25% in 2 h, when will it double?

2. If the number of bacteria in a culture doubles in 3 h, when will it triple?

3. If one-half of a sample of radioactive substance decays in 15 days, how long does it take for 90% of the sample to decay?

4. If 10% of a sample of radioactive material decays in 3 s, what is its half-life?

5. After 4 half-lives of a radioactive substance, what percentage of the original amount remains?

6. Suppose the amount of a drug injected into the body decreases at a rate proportional to the amount still present. If a dose decreases by 5% in the first hour, when will it decrease to one-half its original amount?

7. A sugar cube 1 cm on each side is dropped into a cup of coffee. If the sugar dissolves in such a way that the cube always remains a cube, compare the times for the cube to completely dissolve under the following conditions:

  (a) Dissolving occurs at a rate proportional to the surface area of the remaining cube; and

  (b) Dissolving occurs at a rate proportional to the amount of sugar remaining.

8. Solve Exercise 7 if the sugar is not in the form of a cube, but rather in free form from the sugar bowl. Assume that the sugar consists of n spherical particles each of radius r_0 cm.

9. An analysis of a sample of fossil remains shows that it contains only 1.51% of the original C^{14} in the living creature. When did the creature die?

10. If a fossilized creature died 100 000 years ago, what percentage of the original C^{14} remains?

11. The amount of a drug such as penicillin injected into the body is used up at a rate proportional to the amount still present. If a dose decreases by 5% in the first hour, when does it decrease to one-half its original amount?

12. Glucose is administered intravenously to the bloodstream at a constant rate of R units per unit time. As the glucose is added, it is converted by the body into other substances at a rate proportional to the amount of glucose in the blood at that time. Show that the amount of glucose in the blood as a function of time t is given by

$$C(t) = \frac{R}{k}(1 - e^{-kt}) + C_0 e^{-kt},$$

where k is a constant and C_0 is the amount at time $t = 0$ when the intravenous feeding is initiated. Draw a graph of this function for $C_0 < R/k$ and $C_0 > R/k$.

13. Prove that if a quantity decreases at a rate proportional to its present amount, and if its percentage decrease in some interval of time is $i\%$, then its percentage decrease in any interval of time of the same length is $i\%$.

14. Water at temperature 90°C is placed in a room at constant temperature 20°C. Newton's law of cooling states that the time rate of change of the temperature T of the water is proportional to the difference between T and the temperature of the environment:

$$\frac{dT}{dt} = k(T - 20),$$

where k is a constant. If the water cools to 60°C in 40 min, find T as a function of t.

15. A thermometer reading 25°C is taken outside where the temperature is −20°C. If the reading drops to 0°C in 4 min, when will it read −19°C?

16. A boy lives 6 km from school. He decides to walk to school at a speed that is always proportional to the square of his distance from the school. If he is half-way to school after one hour, find his distance from school at any time. How long does it take him to reach school?

17. A tank has 100 L of solution containing 4 kg of sugar. A mixture with 10 g of sugar per litre of solution is added at 200 mL/min. At the same time, 200 mL of well-stirred mixture is removed each minute. Find the amount of sugar in the tank as a function of time.

18. A tank in the form of an inverted right-circular cone of height H and radius R is filled with water. Water escapes through a hole of cross-sectional area a at the vertex. Use the modified Torricelli law 5.22 to find a formula for the time the tank takes to empty.

19. A container in the form of an inverted right-circular cone of radius 4 cm and height 10 cm is full of water. Water evaporates from the surface at a rate proportional to the area of the surface. If the water level drops 1 cm in the first 5 days, how long will it take for the water to evaporate completely?

* **20.** The water trough in the figure below is 4 m long. Its cross-section is an isosceles triangle with a half-metre base and a half-metre altitude. Water leaks out through a hole of area 1 cm² in the bottom with speed in metres per second given by $v = \sqrt{gD/2}$, where D (in metres) is the depth of water in the trough, and $g > 0$ is the acceleration due to gravity. This is the modified Torricelli law 5.22 with $c = 1/2$. Find how long a full trough takes to empty.

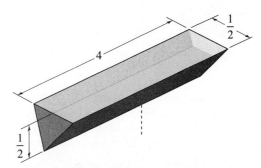

* **21.** A spring of negligible mass and elasticity constant $k > 0$ is attached to a wall at one end and a mass M at the other (figure below). The mass is free to slide horizontally along a frictionless surface. If $x = 0$ is taken as the position of M when the spring is unstretched and M is set into motion, the differential equation describing the position $x(t)$ of M is

$$\frac{d^2x}{dt^2} = -\frac{k}{M}x.$$

If motion is initiated by imparting a speed v_0 in the positive x-direction to M at position $x = 0$, find the velocity of M as a function of position. *Hint*: Express d^2x/dt^2 as follows, and then separate the differential equation

$$\frac{d^2x}{dt^2} = \frac{dv}{dt} = \frac{dv}{dx}\frac{dx}{dt} = v\frac{dv}{dx}.$$

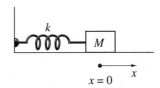

$x = 0$

* **22.** When a mass m falls under the influence of gravity alone, it experiences an acceleration d^2r/dt^2 described by

$$m\frac{d^2r}{dt^2} = -\frac{GmM}{r^2},$$

where M is the mass of the earth, G is a positive constant, and r is the distance from m to the centre of the earth (figure below). If m is dropped from a height h above the surface of the earth, find its velocity when it strikes the earth. What is the maximum attainable speed of m?

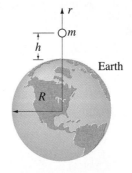

Earth

* **23.** Pieces of ice produced by a refrigerator are in the form of half disks (figure below with dimensions in centimetres). As the ice melts, the ratio of its radius to its thickness remains constant. If the rate of change of the volume of the piece is proportional to its surface area, and its radius is 1/2 cm after 10 min, when will it completely melt?

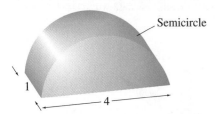

Semicircle

* **24.** When a mothball is exposed to the air, it slowly evaporates. The mothball remains spherical at all times, and evaporation is proportional to the surface area of the mothball at any given time. If its radius after one year is half its original radius, when will it disappear completely?

* **25.** As a spherical raindrop falls through a cloud, its mass increases at a rate proportional to the product of its surface area and its velocity. Assuming that the raindrop begins with zero radius, find its radius as a function of distance fallen through the cloud.

* **26.** The figure below shows an open cylinder that is always kept full by the tap. A hole is to be drilled in the side of the cylinder at a point where the stream of water will hit the ground as far from the cylinder as possible. The higher the hole, the more time the water has to reach the ground, and therefore the farther from the cylinder it will reach. On the other hand, the lower the hole, the greater the speed that the water exits through the hole. Use the modified Torricelli law to find the optimum position taking both factors into account.

FIGURE 5.19   Water escaping from hole in cylinder

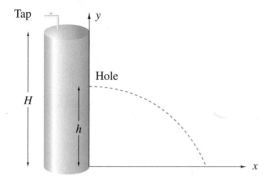

* **27.** Three boys of the same height carry a board of length L and uniform mass per unit length m horizontally as shown in the figure that follows. The weight of the board causes it to bend, and its shape is the same on either side of the middle boy. Between the first two boys, for $0 \le x \le L/2$, the displacement of the board from the horizontal $y = f(x)$ must satisfy the differential equation

$$(EI)\frac{d^2y}{dx^2} = Ax - \frac{mg}{2}x^2,$$

where E and I are constants depending on the cross-section and material of the board, A is the weight supported by the boy at $x = 0$, and $g > 0$ is the acceleration due to gravity.

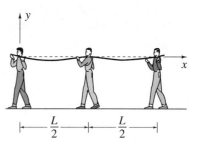

(a) Solve this equation along with the conditions $f(0) = f(L/2) = 0$ to find $f(x)$ as a function of x, E, I, m, g, and A.

(b) Find a condition that permits determination of A.

* **28.** The steady-state temperature T in a region bounded by two concentric spheres of radii 1 m and 2 m must satisfy the differential equation

$$\frac{d^2T}{dr^2} + \frac{2}{r}\frac{dT}{dr} = 0,$$

where r is the radial distance from the common centre of the spheres. If temperatures on the inner and outer spheres are maintained at $10°C$ and $20°C$ respectively, find the temperature distribution $T(r)$ between the spheres.

* **29.** A room with volume 100 m³ initially contains 0.1% carbon dioxide. Beginning at time $t = 0$, fresher air containing 0.05% carbon dioxide flows into the room at 5 m³/min. The well-mixed air in the room flows out at the same rate. Find the amount of carbon dioxide in the room as a function of time. What is the limit of the function as $t \to \infty$?

* **30.** It is sometimes assumed that the density ρ of the atmosphere is related to height h above sea level according to the differential equation

$$\frac{d\rho}{dh} = -\frac{\rho^{2-\delta}}{k\delta},$$

where $\delta > 1$ and $k > 0$ are constants.

(a) Show that if ρ_0 is density at sea level, then

$$\rho^{\delta-1} = -\frac{h}{k}\left(\frac{\delta-1}{\delta}\right) + \rho_0^{\delta-1}.$$

(b) If air pressure P and density are related by the equation $P = k\rho^\delta$, prove that

$$P^{1-1/\delta} = P_0^{1-1/\delta} - [(1-1/\delta)\rho_0 P_0^{-1/\delta}]h,$$

where P_0 is air pressure at sea level.

(c) Show that the effective height of the atmosphere is

$$\frac{\delta P_0}{(\delta-1)\rho_0}.$$

31. A certain chemical dissolves in water at a rate proportional to the product of the amount of undissolved chemical and the difference between concentration in a saturated solution and the existing concentration in the solution. A saturated solution contains 25 g of chemical in 100 mL of solution. If 50 g of chemical is added to 200 mL of water, find a formula for the amount of chemical dissolved as a function of time. Draw its graph.

* **32.** Two substances A and B react to form a third substance C in such a way that 2 g of A react with 1 g of B to produce 3 g of C. The rate at which C is formed is proportional to the amounts of A and B still present in the mixture. Find the amount of C present in the mixture as a function of time when the original amounts of A and B brought together at time $t = 0$ are 20 g and 30 g, respectively.

SUMMARY

The indefinite integral of a function $f(x)$ is a family of functions, each of which has $f(x)$ as its first derivative. If we can find one antiderivative of $f(x)$, then the indefinite integral is that antiderivative plus an arbitrary constant. According to Theorem 5.2, the indefinite integral of a sum of two functions is the sum of their indefinite integrals, and multiplicative constants may be bypassed when finding indefinite integrals:

$$\int [f(x) + g(x)]\,dx = \int f(x)\,dx + \int g(x)\,dx; \qquad \int cf(x)\,dx = c\int f(x)\,dx.$$

The most important integration formula is for powers of x,

$$\int x^n\,dx = \begin{cases} \dfrac{1}{n+1}x^{n+1} + C, & n \neq -1 \\ \ln|x| + C, & n = -1. \end{cases}$$

Other integration formulas that arise from differentiations of trigonometric, inverse trigonometric, exponential, logarithmic, and hyperbolic functions are listed in equations 5.4.

In this chapter we have studied three ways to evaluate indefinite integrals. (Others will follow in Chapter 8.) First, due to our expertise in differentiation, some antiderivatives are immediately recognizable. Second, sometimes an antiderivative can be guessed to within a multiplicative constant, and this constant can then be adjusted. Third, a change of variable can often replace a complex integration problem with a simpler one.

Integration plays a fundamental role in kinematics. Since velocity is the derivative of position, and acceleration is the derivative of velocity, it follows that position is the indefinite integral of velocity, and velocity is the indefinite integral of acceleration,

$$v(t) = \int a(t)\,dt \qquad \text{and} \qquad x(t) = \int v(t)\,dt.$$

Many physical systems are modelled by differential equations, and the solution of a differential equation usually involves one or more integrations. The differential equation for the deflection of a beam requires four integrations, resulting in four arbitrary constants that are determined by two boundary conditions at each end of the beam. The Dirac-delta function is most effective in representing point forces on beams.

First order differential equations are separable if they can be expressed in the form $N(y)\,dy = M(x)\,dx$. Solutions are then defined implicitly by

$$\int N(y)\,dy = \int M(x)\,dx.$$

In reviewing this chapter, you should be able to define or discuss the following key terms:

Antiderivative and indefinite integral	Integrand
Ramp function	Change of variable
Boundary conditions	Simply supported
Boundary-value problem	Differential equation
First-order differential equation	Separable differential equation

In Exercises 1–28 evaluate the indefinite integral.

1. $\int (3x^3 - 4x^2 + 5)\, dx$ **2.** $\int \left(\frac{1}{x^5} + 2x - \frac{1}{x^3} \right) dx$

3. $\int (2x^2 - 3x + 7x^6)\, dx$ **4.** $\int \left(\frac{1}{x^2} - 2\sqrt{x} \right) dx$

5. $\int \sqrt{x-2}\, dx$ **6.** $\int x(1 + 3x^2)^4\, dx$

7. $\int \left(\sqrt{x} - \frac{1}{\sqrt{x}} \right) dx$ **8.** $\int \left(\frac{x^2 + 5}{\sqrt{x}} \right) dx$

9. $\int \frac{1}{(x+5)^4}\, dx$ **10.** $\int \left(\frac{\sqrt{x}}{x^2} - \frac{15}{\sqrt{x}} \right) dx$

11. $\int \sin 3x\, dx$ **12.** $\int x\sqrt{1-x^2}\, dx$

13. $\int x \cos x^2\, dx$ **14.** $\int x^2 (1 - 2x^2)^2\, dx$

15. $\int x\sqrt{1+x}\, dx$ **16.** $\int \frac{x}{\sqrt{2-x}}\, dx$

17. $\int \frac{1}{(1+x)^2}\, dx$ **18.** $\int (2 + \sqrt{x})^2\, dx$

19. $\int \frac{1}{\sqrt{x}(2 + \sqrt{x})^2}\, dx$ **20.** $\int \sin^4 x \cos x\, dx$

21. $\int e^{3-5x}\, dx$ **22.** $\int x e^{-4x^2}\, dx$

23. $\int \frac{e^x - 1}{e^{2x}}\, dx$ **24.** $\int \frac{1}{5x \ln x}\, dx$

25. $\int \frac{x}{\sqrt{1-4x^4}}\, dx$ **26.** $\int \frac{3}{1+7x^2}\, dx$

27. $\int x \cosh 5x^2\, dx$ **28.** $\int \operatorname{sech}^2 5x\, dx$

29. If the number of bacteria in a culture triples in 3 days, when will it quadruple its original number?

30. Water at temperature $70°C$ is placed outside where temperature is $-20°C$. Newton's law of cooling states that the time rate of change of the temperature T of the water is proportional to the difference between

T and the temperature of the environment. If the water cools to $50°C$ in 10 min, find T as a function of t.

* **31.** Find the curve $y = f(x)$ for which $f''(x) = x^2 + 1$, and that passes through the point $(1, 1)$ with slope 4.

* **32.** Find the curve $y = f(x)$ for which $f''(x) = 12x^2$, and that passes through the two points $(1, 4)$ and $(-1, -3)$.

* **33.** Find the curve $y = f(x)$ for which $f''(x) = 24x^2 + 6x$, and that is tangent to the line $y = 4x + 4$ at $(1, 8)$.

* **34.** A boy lives 6 km from school. He decides to walk to school at a speed that is always proportional to the square root of his distance from the school. If he is halfway to school after 1 h, find his distance from school at any time. How long does it take him to reach school?

* **35.** If a ball is thrown vertically upward with a speed of 30 m/s, how high will it rise?

* **36.** A stone is thrown vertically downward over the edge of a bridge 50 m above a river. If the stone strikes the water in 2.2 s, what was its initial speed?

In Exercises 37–44 evaluate the indefinite integral.

* **37.** $\int \frac{1}{\sqrt{1 + \sqrt{x}}}\, dx$ * **38.** $\int \frac{x}{\sqrt{1+x} + 1}\, dx$

* **39.** $\int \frac{\sin x}{\sqrt{4 + 3\cos x}}\, dx$ * **40.** $\int x^8 (3 - 2x^3)^6\, dx$

* **41.** $\int \frac{(2+x)^4}{x^6}\, dx$ * **42.** $\int \sin^3 x \cos^3 x\, dx$

* **43.** $\int \frac{1}{x\sqrt{1 + 3\ln x}}\, dx$ * **44.** $\int \tan x\, dx$

* **45.** A graph of the acceleration $a(t)$ of an object is shown in the figure below. Find its velocity $v(t)$ and position $x(t)$ in the time interval $0 \le t \le 15$ if $v(0) = 0 = x(0)$, and draw graphs of each.

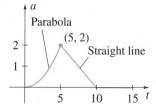

46. Find the equation of the curve that passes through the point $(1, 1)$ such that the slope of the tangent line at any point (x, y) is half the square of the slope of the line from the origin to (x, y).

47. The equation $y = x^3 + C$ describes a family of cubics where C represents the distance of its horizontal point of inflection above the x-axis. Find the equation of the curve that passes through the point $(1, 1)$ and intersects each of these curves at right angles.

48. A container in the form of an inverted right-circular cone of radius 6 cm and height 15 cm is full of water. Water evaporates from the surface at a rate proportional to the area of the surface. If the water level drops 1 cm in the first 6 days, how long will it take for half the water to evaporate?

∗ **49.** Find deflections of the beam in Figure 5.9 if the mass of the block is M kg and M is so large that the mass of the beam can be neglected in comparison. Is the beam straight for $x > 5$? Would you expect it to be?

∗ **50.** Repeat Exercise 49 if the mass is on the right half of the beam. Is the beam straight for $x < 5$? Would you expect it to be?

∗ **51.** Repeat Exercise 49 if the mass is on the middle half of the beam. Are the portions of the beam $0 < x < 5/2$ and $15/2 < x < 10$ straight? Would you expect them to be?

6 | The Definite Integral

Currents in circuits are often monitored by a controller. The controller takes action whenever the current strays significantly from its expected value. The current i in the left figure below differs from its steady-state value significantly at time t_0, but does so for a very short time interval, so short perhaps, that it might be deemed acceptable.

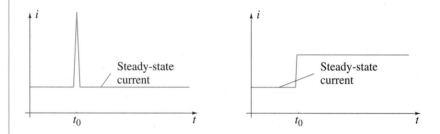

On the other hand, the current in the right figure moves only half as far from the steady-state value as in the left figure, but does so for an extended period of time. This might be deemed unacceptable.

THE PROBLEM   Devise a way to distinguish mathematically, for purposes of the controller, between very short, but very abnormal behaviour of a function, and long-term, but less dramatic changes from the norm. (See Example 6.15 on page page 401 for a solution.)

There are two aspects to calculus: differentiation and integration. We dealt with differentiation and its applications in Chapters 3 and 4. Integration in Chapter 5 was synonymous with antidifferentiation and the indefinite integral. In this chapter we investigate a new type of integral called the *definite integral*. Before doing that, we introduce sigma notation, a compact notation for sums of terms, particularly useful for definite integrals. In Section 6.2 we discuss four problems that motivate the concept of.the definite integral. In subsequent sections we develop the definition for the integral and discuss various ways to evaluate it. Section 6.6 presents an application of the definite integral — finding the average value of a function — and Chapter 7 develops a multitude of physical applications. The definite integral is, by definition, very different from the indefinite integral, yet the two are intimately related through the fundamental theorems of integral calculus (Sections 6.4 and 6.5).

⌐6.1 **Sigma Notation**

One of the most important notations in calculus, **sigma notation**, is used to represent a sum of terms, all of which are similar in form. For example, the six terms in the sum

$$\frac{1}{1+2^2} + \frac{2}{1+3^2} + \frac{3}{1+4^2} + \frac{4}{1+5^2} + \frac{5}{1+6^2} + \frac{6}{1+7^2}$$

are all formed in the same way: Each is an integer divided by 1 plus the square of the next integer. If k represents an integer, we can say that every term has the form $k/[1 + (k + 1)^2]$; the first term is obtained by setting $k = 1$, the second by setting $k = 2$, and so on, until $k = 6$. As $k/[1 + (k + 1)^2]$ represents each and every term in the sum, we can describe the sum in words by saying, "Assign k in $k/[1 + (k + 1)^2]$ the integer values between 1 and 6, inclusively, and add the resulting numbers together." The notation used to represent this statement is

$$\sum_{k=1}^{6} \frac{k}{1 + (k + 1)^2}.$$

The Σ symbol is the Greek capital letter sigma, which in this case means "sum." Summed are expressions of the form $k/[1 + (k + 1)^2]$, and the "$k = 1$" and "6" indicate that every integer from 1 to 6 is substituted into $k/[1 + (k + 1)^2]$. We call $k/[1 + (k + 1)^2]$ the **general term** of the sum, since it represents each and every term therein. The letter k is called the **index of summation** or **variable of summation**, and 1 and 6 are called the **limits of summation**. Any letter may be used to represent the index of summation; most commonly used are i, j, k, l, m, and n. Here is another example:

$$\sum_{n=5}^{15} n^3 = 5^3 + 6^3 + 7^3 + 8^3 + 9^3 + 10^3 + 11^3 + 12^3 + 13^3 + 14^3 + 15^3.$$

In summing a large number of terms, it is quite cumbersome to write them all down. One way around this difficulty is to write the first few terms to indicate the pattern by which the terms are formed, three dots, and the last term. For example, to indicate the sum of the cubes of the positive integers less than or equal to 100, we write

$$1^3 + 2^3 + 3^3 + 4^3 + 5^3 + \cdots + 100^3,$$

where the dots indicate that all numbers between 5^3 and 100^3 are to be filled in according to the same pattern suggested. Obviously, it would be preferable to express the sum in sigma notation,

$$\sum_{k=1}^{100} k^3,$$

which is compact and leaves no doubt as to the pattern by which terms are formed.

EXAMPLE 6.1

Write each of the following sums in sigma notation:

(a) $\dfrac{1}{2 \cdot 3} + \dfrac{4}{3 \cdot 4} + \dfrac{9}{4 \cdot 5} + \dfrac{16}{5 \cdot 6} + \cdots + \dfrac{169}{14 \cdot 15}$

(b) $\dfrac{16}{\sqrt{2}} + \dfrac{32}{\sqrt{3}} + \dfrac{64}{\sqrt{4}} + \dfrac{128}{\sqrt{5}} + \cdots + \dfrac{4096}{\sqrt{10}}$

SOLUTION

(a) To determine the pattern by which terms are formed, it is often advantageous to write values of the variable of summation above the terms. If we use i as the variable with $i = 1$ corresponding to the first term, we write

$$\overset{i=1}{\dfrac{1}{2 \cdot 3}} + \overset{i=2}{\dfrac{4}{3 \cdot 4}} + \overset{i=3}{\dfrac{9}{4 \cdot 5}} + \overset{i=4}{\dfrac{16}{5 \cdot 6}} + \cdots + \overset{i=?}{\dfrac{169}{14 \cdot 15}}.$$

The general term is $i^2/[(i + 1)(i + 2)]$, and

$$\frac{1}{2 \cdot 3} + \frac{4}{3 \cdot 4} + \frac{9}{4 \cdot 5} + \frac{16}{5 \cdot 6} + \cdots + \frac{169}{14 \cdot 15} = \sum_{i=1}^{13} \frac{i^2}{(i + 1)(i + 2)}.$$

(b) If n is chosen as the index of summation and $n = 2$ to correspond to the first term,

$$\overset{n=2}{\frac{16}{\sqrt{2}}} + \overset{n=3}{\frac{32}{\sqrt{3}}} + \overset{n=4}{\frac{64}{\sqrt{4}}} + \overset{n=5}{\frac{128}{\sqrt{5}}} + \cdots + \overset{n=?}{\frac{4096}{\sqrt{10}}} = \sum_{n=2}^{10} \frac{2^{n+2}}{\sqrt{n}}.$$

The representations for the sums in Example 6.1 in terms of sigma notation are not unique. In fact, there is an infinite number of representations for each sum. Consider, for example, the sum represented by

$$\sum_{i=1}^{9} \frac{2^{i+3}}{\sqrt{i + 1}}.$$

If we write out some of the terms in the summation, we find that

$$\sum_{i=1}^{9} \frac{2^{i+3}}{\sqrt{i + 1}} = \frac{16}{\sqrt{2}} + \frac{32}{\sqrt{3}} + \frac{64}{\sqrt{4}} + \frac{128}{\sqrt{5}} + \cdots + \frac{4096}{\sqrt{10}},$$

the same sum as that in Example 6.1(b). This sum can also be represented by

$$\sum_{j=-2}^{6} \frac{2^{j+6}}{\sqrt{j + 4}} \qquad \text{and} \qquad \sum_{m=15}^{23} \frac{2^{m-11}}{\sqrt{m - 13}}.$$

We can transform any one of these representations into any other by making a change of variable of summation. For example, if in the summation of Example 6.1(b) we set $i = n - 1$, then $n = i + 1$, and

$$\frac{2^{n+2}}{\sqrt{n}} = \frac{2^{i+3}}{\sqrt{i + 1}}.$$

For the limits, we find that $i = 1$ when $n = 2$, and $i = 9$ when $n = 10$. It follows that

$$\sum_{n=2}^{10} \frac{2^{n+2}}{\sqrt{n}} = \sum_{i=1}^{9} \frac{2^{i+3}}{\sqrt{i + 1}}.$$

Similarly, the changes $j = n - 4$ and $m = n + 9$ transform

$$\sum_{n=2}^{10} \frac{2^{n+2}}{\sqrt{n}}$$

into

$$\sum_{j=-2}^{6} \frac{2^{j+6}}{\sqrt{j + 4}} \qquad \text{and} \qquad \sum_{m=15}^{23} \frac{2^{m-11}}{\sqrt{m - 13}}.$$

EXAMPLE 6.2

Change each of the following summations into representations that are initiated with the integer 1:

$$\text{(a)} \quad \sum_{i=4}^{26} \frac{i^{2/3}}{i^2 + i + 1} \qquad \text{(b)} \quad \sum_{j=-3}^{102} \frac{j^2 + 2j + 5}{\sin(j + 5)}$$

SOLUTION

(a) To initiate the summation at 1, we want $n = 1$ when $i = 4$, so we set $n = i - 3$. Then $i = n + 3$, and by substitution we have

$$\sum_{i=4}^{26} \frac{i^{2/3}}{i^2 + i + 1} = \sum_{n=1}^{23} \frac{(n + 3)^{2/3}}{(n + 3)^2 + (n + 3) + 1} = \sum_{n=1}^{23} \frac{(n + 3)^{2/3}}{n^2 + 7n + 13}.$$

(b) In this case, we set $n = j + 4$. Then $j = n - 4$, and

$$\sum_{j=-3}^{102} \frac{j^2 + 2j + 5}{\sin(j + 5)} = \sum_{n=1}^{106} \frac{(n - 4)^2 + 2(n - 4) + 5}{\sin(n + 1)} = \sum_{n=1}^{106} \frac{n^2 - 6n + 13}{\sin(n + 1)}.$$

If we examine the results of Example 6.2 and the summations immediately preceding this example, we soon come to realize that there is a very simple way to change variables. To illustrate, consider once again

$$\sum_{n=2}^{10} \frac{2^{n+2}}{\sqrt{n}}.$$

Should we wish to initiate the summation with 1 rather than 2, we lower both limits by 1. To compensate, we replace each n in the general term by $n + 1$; the result is

$$\sum_{n=1}^{9} \frac{2^{(n+1)+2}}{\sqrt{n + 1}} = \sum_{n=1}^{9} \frac{2^{n+3}}{\sqrt{n + 1}}.$$

Similarly, for simplicity in the summation

$$\sum_{n=1}^{10} \frac{(n + 4)^2}{e^{n+4}},$$

it would be advisable to lower each n in the general term by 4. This can be done provided that we raise each limit by 4:

$$\sum_{n=1}^{10} \frac{(n + 4)^2}{e^{n+4}} = \sum_{n=5}^{14} \frac{n^2}{e^n}.$$

Every summation represented in sigma notation is of the form

$$\sum_{i=m}^{n} f(i), \tag{6.1}$$

where m and n are integers ($n > m$), and $f(i)$ is some function of the index of summation i. In Example 6.2(a), $f(i) = i^{2/3}/(i^2 + i + 1)$, $m = 4$, and $n = 26$; in Example 6.1(a), $f(i) = i^2/[(i + 1)(i + 2)]$, $m = 1$, and $n = 13$. The following properties of sigma notation are easily proved by writing out each summation.

THEOREM 6.1

If $f(i)$ and $g(i)$ are functions of i, and m and n are positive integers such that $n > m$, then

$$\sum_{i=m}^{n} [f(i) + g(i)] = \sum_{i=m}^{n} f(i) + \sum_{i=m}^{n} g(i); \tag{6.2a}$$

$$\sum_{i=m}^{n} cf(i) = c \sum_{i=m}^{n} f(i) \tag{6.2b}$$

if c is a constant independent of i.

Compare Theorem 6.1 with Theorem 5.2 in Section 5.1; notice the similarities between properties of summations in sigma notation and those of indefinite integrals.

We emphasize that sigma notation is simply a concise symbolism used to represent a sum of terms; it does not evaluate the sum. In the following discussion, we develop formulas for sums that prove useful in future work.

Summation $\sum_{i=1}^{n} i$ represents the sum of the first n positive integers:

$$\sum_{i=1}^{n} i = 1 + 2 + 3 + 4 + \cdots + (n - 1) + n.$$

If we write the terms on the right in reverse order, we have

$$\sum_{i=1}^{n} i = n + (n - 1) + (n - 2) + \cdots + 4 + 3 + 2 + 1.$$

Addition of these two equations gives us

$$2 \sum_{i=1}^{n} i = (n + 1) + (n + 1) + (n + 1) + \cdots + (n + 1) + (n + 1) = n(n + 1).$$

Consequently,

$$\sum_{i=1}^{n} i = \frac{n(n + 1)}{2}. \tag{6.3}$$

This result can be used to develop formulas for the sums of the squares, cubes, and so on, of the positive integers. To find the sum of the squares of the first n positive integers, we note that by expansion and simplification

$$i^3 - (i - 1)^3 = 3i^2 - 3i + 1$$

for any integer i whatsoever. It follows that

$$\sum_{i=1}^{n} [i^3 - (i - 1)^3] = \sum_{i=1}^{n} (3i^2 - 3i + 1).$$

But if we write the left-hand side in full, we find

$$\sum_{i=1}^{n} [i^3 - (i - 1)^3] = [1^3 - 0^3] + [2^3 - 1^3] + [3^3 - 2^3] + \cdots + [n^3 - (n - 1)^3].$$

Most of these terms cancel one another, leaving only n^3; that is,

$$\sum_{i=1}^{n} [i^3 - (i-1)^3] = n^3.$$

Thus,

$$n^3 = \sum_{i=1}^{n} (3i^2 - 3i + 1)$$

$$= 3\sum_{i=1}^{n} i^2 - 3\sum_{i=1}^{n} i + \sum_{i=1}^{n} 1 \quad \text{(using Theorem 6.1)}$$

$$= 3\sum_{i=1}^{n} i^2 - 3\frac{n(n+1)}{2} + n \quad \text{(using formula 6.3)}.$$

We can solve this equation for $\sum_{i=1}^{n} i^2$:

$$\sum_{i=1}^{n} i^2 = \frac{1}{3}\left[n^3 + \frac{3n(n+1)}{2} - n\right] = \frac{n(n+1)(2n+1)}{6}. \tag{6.4}$$

A similar procedure beginning with the identity $i^4 - (i-1)^4 = 4i^3 - 6i^2 + 4i - 1$ yields

$$\sum_{i=1}^{n} i^3 = \frac{n^2(n+1)^2}{4}. \tag{6.5}$$

These results can also be established independently of one another by mathematical induction. (See Appendix A for proofs of formulas 6.3 and 6.4 using this technique.)

EXERCISES 6.1

In Exercises 1–10 express the sum in sigma notation. Initiate the summation with the integer 1.

1. $2 \cdot 3 + 3 \cdot 4 + 4 \cdot 5 + 5 \cdot 6 + \cdots + 99 \cdot 100$

2. $\dfrac{1}{2} + \dfrac{2}{4} + \dfrac{3}{8} + \dfrac{4}{16} + \dfrac{5}{32} + \cdots + \dfrac{10}{1024}$

3. $\dfrac{16}{14+15} + \dfrac{17}{15+16} + \dfrac{18}{16+17} + \cdots + \dfrac{199}{197+198}$

4. $1 + \sqrt{2} + \sqrt{3} + 2 + \sqrt{5} + \sqrt{6} + \sqrt{7} + \sqrt{8} + 3 + \cdots + 121$

5. $1 + \dfrac{1}{2} + \dfrac{1}{2\cdot3} + \dfrac{1}{2\cdot3\cdot4} + \cdots + \dfrac{1}{2\cdot3\cdot4\cdot5\cdots16}$

6. $-2 + 3 - 4 + 5 - 6 + 7 - 8 + \cdots - 1020$

7. $\dfrac{2\cdot3}{1\cdot4} + \dfrac{6\cdot7}{5\cdot8} + \dfrac{10\cdot11}{9\cdot12} + \dfrac{14\cdot15}{13\cdot16} + \cdots + \dfrac{414\cdot415}{413\cdot416}$

8. $\dfrac{\tan 1}{2} + \dfrac{\tan 2}{1+2^2} + \dfrac{\tan 3}{1+3^2} + \dfrac{\tan 4}{1+4^2} + \cdots + \dfrac{\tan 225}{1+225^2}$

9. $4^3 + 5^2 + 6 + 1 + \dfrac{1}{8} + \dfrac{1}{9^2} + \cdots + \dfrac{1}{25^{18}}$

10. $0.9 + 0.99 + 0.999 + \cdots + 0.999\,999\,999$

In Exercises 11–15 verify by a change of variable of summation that the two summations are identical.

11. $\displaystyle\sum_{n=1}^{24} \dfrac{n^2}{2n+1} \qquad \sum_{i=4}^{27} \dfrac{i^2 - 6i + 9}{2i - 5}$

12. $\displaystyle\sum_{k=2}^{101} \dfrac{3k - k^2}{\sqrt{k+5}} \qquad \sum_{m=0}^{99} \dfrac{2 - m - m^2}{\sqrt{7+m}}$

13. $\displaystyle\sum_{n=5}^{20} (-1)^n \dfrac{2^n}{n^2+1} \qquad \sum_{j=1}^{16} 16(-1)^j \dfrac{2^j}{j^2+8j+17}$

14. $\displaystyle\sum_{i=0}^{37} \dfrac{3^{3i}}{i!} \qquad \sum_{m=2}^{39} \dfrac{3^{3m}}{729(m-2)!}$

15. $\displaystyle\sum_{r=15}^{225} \frac{1}{r^2 - 10r}$ $\displaystyle\sum_{n=10}^{220} \frac{1}{n^2 - 25}$

In Exercises 16–25 use Theorem 6.1 and formulas 6.3–6.5 to evaluate the sum.

16. $\displaystyle\sum_{n=1}^{12} (3n + 2)$ **17.** $\displaystyle\sum_{j=1}^{21} (2j^2 + 3j)$

18. $\displaystyle\sum_{m=1}^{n} (4m - 2)^2$ **19.** $\displaystyle\sum_{k=2}^{29} (k^3 - 3k^2)$

20. $\displaystyle\sum_{n=1}^{25} (n + 5)(n - 4)$ **21.** $\displaystyle\sum_{i=1}^{n} i(i - 3)^2$

* **22.** $\displaystyle\sum_{n=10}^{24} (n^2 - 5)$ * **23.** $\displaystyle\sum_{i=7}^{17} (i^3 - 3i^2)$

* **24.** $\displaystyle\sum_{k=5}^{n} (k + 3)(k + 4)$ * **25.** $\displaystyle\sum_{i=n}^{2n} (i^2 + 2i - 3)$

* **26.** Verify Theorem 6.1.

* **27.** Find a formula for $\displaystyle\sum_{k=1}^{n} \frac{1}{k(k + 1)}$. *Hint:* $\dfrac{1}{k(k+1)} = \dfrac{1}{k} - \dfrac{1}{k+1}$.

* **28.** If $f(x)$ is a function of x, defined for all x, simplify the sum

$$\sum_{i=1}^{n} [f(i) - f(i - 1)].$$

* **29.** Prove formula 6.5.

* **30.** Is $\displaystyle\sum_{i=1}^{n} [f(i)g(i)]$ equal to $\left[\displaystyle\sum_{i=1}^{n} f(i)\right]\left[\displaystyle\sum_{i=1}^{n} g(i)\right]$?

* **31.** A **finite geometric series** is a sum of terms of the form

$$a + ar + ar^2 + ar^3 + \cdots + ar^{n-1}.$$

There is a first term a, and every term thereafter is obtained by multiplying the preceding term by r (called the *common ratio*).

     (a) If S_n represents this sum, express S_n in sigma notation.

     (b) Prove that $S_n = \dfrac{a(1 - r^n)}{1 - r}$.

Use the formula in Exercise 31 to sum the finite geometric series in Exercises 32–35.

* **32.** $\dfrac{1}{8} + \dfrac{1}{16} + \dfrac{1}{32} + \dfrac{1}{64} + \cdots + \dfrac{1}{1\,048\,576}$

* **33.** $1 - \dfrac{1}{3} + \dfrac{1}{9} - \dfrac{1}{27} + \dfrac{1}{81} - \cdots - \dfrac{1}{19\,683}$

* **34.** $40(0.99) + 40(0.99)^2 + 40(0.99)^3 + \cdots + 40(0.99)^{15}$

* **35.** $\sqrt{0.99} + 0.99 + (0.99)^{3/2} + (0.99)^2 + \cdots + (0.99)^{10}$

* **36.** Prove that $\left|\displaystyle\sum_{i=1}^{n} f(i)\right| \leq \displaystyle\sum_{i=1}^{n} |f(i)|$.

** **37.** Express the following summation in sigma notation:

$$1 + \dfrac{1}{2} - \dfrac{1}{4} - \dfrac{1}{8} + \dfrac{1}{16} + \dfrac{1}{32} - \dfrac{1}{64} - \dfrac{1}{128} + \cdots + \dfrac{1}{4096}.$$

6.2 The Need for the Definite Integral

In this section we consider four inherently different problems: one on area, one on volume, one on blood flow, and one on work; but we shall see that a common method of solution exists for all four of them. This common theme leads to the definition of the definite integral in Section 6.3.

Problem 1

We first consider the problem of finding the area A in Figure 6.1a. At present, we have formulas for areas of very few geometric shapes — rectangles, triangles, polygons, and circles, and the

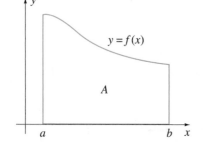

FIGURE 6.1a   Area under a curve $y = f(x)$

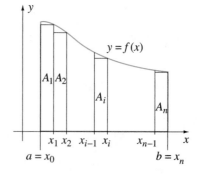

FIGURE 6.1b   Approximation of area by rectangles

shape in Figure 6.1a is not one of them. We can, however, find an approximation to the area by constructing rectangles in the following way. Between a and b pick $n-1$ points $x_1, x_2, x_3, \ldots, x_{n-1}$ on the x-axis such that

$$a = x_0 < x_1 < x_2 < \cdots < x_{n-1} < x_n = b.$$

Draw vertical lines through each of these $n-1$ points to intersect the curve $y = f(x)$ and form rectangles, as shown in Figure 6.1b.

If A_i denotes the area of the ith such rectangle $(i = 1, \ldots, n)$, then

$$A_i = \text{(height of rectangle)(width of rectangle)}$$

$$= f(x_i)(x_i - x_{i-1}).$$

The sum of these n rectangular areas is an approximation to the required area; that is, A is approximately equal to

$$\sum_{i=1}^{n} A_i = \sum_{i=1}^{n} f(x_i)(x_i - x_{i-1}).$$

If we let the number of these rectangles get larger and larger (and at the same time require each to have smaller and smaller width that eventually approaches zero), the approximation appears to get better and better. In fact, we expect that

$$A = \lim_{n \to \infty} \sum_{i=1}^{n} A_i = \lim_{n \to \infty} \sum_{i=1}^{n} f(x_i)(x_i - x_{i-1}). \tag{6.6}$$

Problem 2

If the area in Figure 6.1a is rotated around the x-axis, it traces out a volume V. This is certainly not a standard shape for which we have a volume formula, and we therefore consider finding an approximation to V. We take the rectangles in Figure 6.1b (which approximate the area) and rotate them around the x-axis (Figure 6.2). Each A_i traces out a disc of volume V_i $(i = 1, \ldots, n)$, where

$$V_i = \text{(surface area of disc)(thickness of disc)}$$

$$= \pi (\text{radius of disc})^2 (\text{thickness of disc})$$

$$= \pi [f(x_i)]^2 (x_i - x_{i-1}).$$

FIGURE 6.2   Approximation of volume by discs

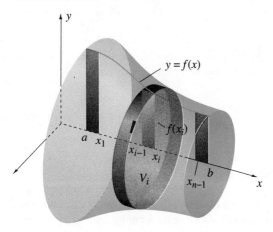

An approximation to the required volume is then

$$\sum_{i=1}^{n} V_i = \sum_{i=1}^{n} \pi [f(x_i)]^2 (x_i - x_{i-1}).$$

Again we feel intuitively that as the number of discs becomes larger and larger (and the width of each approaches zero), the approximation becomes better and better, and

$$V = \lim_{n \to \infty} \sum_{i=1}^{n} V_i = \lim_{n \to \infty} \sum_{i=1}^{n} \pi [f(x_i)]^2 (x_i - x_{i-1}). \tag{6.7}$$

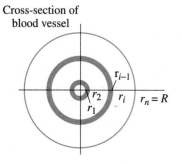

FIGURE 6.3 Blood flow through an artery

Cross-section of blood vessel

Problem 3

When blood flows through a vein or artery, it encounters resistance due to friction with the walls of the blood vessel and due to the viscosity of the blood itself. As a result, the velocity of the blood is not constant across a cross-section of the vessel; blood flows more quickly near the centre of the vessel than near its walls. It has been shown that for laminar blood flow in a vessel of circular cross-section (Figure 6.3), the velocity of blood is given by

$$v = v(r) = c(R^2 - r^2), \quad 0 \le r \le R,$$

where $c > 0$ is a constant, R is the radius of the blood vessel, and r is radial distance measured from the centre of the vessel. We wish to find the rate of blood flow through the vessel, that is, the volume of blood flowing through the cross-section per unit time.

If v were constant over the cross-section, then flow per unit time would be the product of v and the cross-sectional area. Unfortunately, this is not the case, but we can still use the idea that flow is velocity multiplied by area. We divide the cross-section into rings with radii

$$0 = r_0 < r_1 < r_2 < \cdots < r_{n-1} < r_n = R.$$

Over the ith ring, the variation in v is small and v can be approximated by $v(r_i)$. The flow through the ith ring can therefore be approximated by

$$F_i = (\text{area of ring})(\text{velocity at outer radius of ring})$$

$$= (\pi r_i^2 - \pi r_{i-1}^2) v(r_i).$$

An approximation to the required flow F is the sum of these F_i:

$$\sum_{i=1}^{n} F_i = \sum_{i=1}^{n} (\pi r_i^2 - \pi r_{i-1}^2) v(r_i),$$

and it seems reasonable that as the number of rings increases, so does the accuracy of the approximation; that is, we anticipate that

$$F = \lim_{n \to \infty} \sum_{i=1}^{n} F_i = \lim_{n \to \infty} \sum_{i=1}^{n} (\pi r_i^2 - \pi r_{i-1}^2) v(r_i). \tag{6.8}$$

Problem 4

A spring is fixed horizontally into a wall at one end, and the other end is free. Consider finding the work to stretch the spring 3 cm by pulling on its free end (Figure 6.4a).

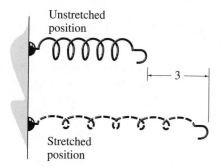

FIGURE 6.4a   Initial and final positions of stretched spring

Unstretched position

3

Stretched position

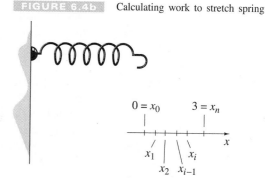

FIGURE 6.4b   Calculating work to stretch spring

$0 = x_0$ $3 = x_n$

x

x_1 x_i

x_2 x_{i-1}

Let us choose an x-axis positive to the right with $x = 0$ at the position of the free end of the spring when it is in the unstretched position (Figure 6.4b). In order to calculate the work to stretch the spring, we must know something about the forces involved. It has been shown experimentally that the force F that must be exerted on the free end of the spring in order to maintain a stretch x in the spring is proportional to x:

$$F = F(x) = kx,$$

where $k > 0$ is a constant. This then is the force that will perform the work. The basic definition of work W done by a constant force F acting along a straight-line segment of length d is $W = Fd$. Unfortunately, our force $F(x)$ is not constant; it depends on x, and we cannot therefore simply multiply force by distance. What we can find, however, is an approximation to the required work by dividing the length between $x = 0$ and $x = 3$ into n subintervals by $n - 1$ points $x_1, x_2, \ldots, x_{n-1}$ such that

$$0 = x_0 < x_1 < x_2 < \cdots < x_{n-1} < x_n = 3.$$

When the spring is stretched between x_{i-1} and x_i, the force necessary to maintain this stretch does not vary greatly and can be approximated by $F(x_i)$. It follows then that the work necessary to stretch the spring from x_{i-1} to x_i is approximately equal to

$$W_i = F(x_i)(x_i - x_{i-1}).$$

As a result, an approximation to the total work required to pull the free end of the spring from $x = 0$ to $x = 3$ is

$$\sum_{i=1}^{n} W_i = \sum_{i=1}^{n} F(x_i)(x_i - x_{i-1}).$$

Once again we expect that as n becomes indefinitely large, this approximation approaches the required work W, and

$$W = \lim_{n \to \infty} \sum_{i=1}^{n} W_i = \lim_{n \to \infty} \sum_{i=1}^{n} F(x_i)(x_i - x_{i-1}). \tag{6.9}$$

Each of these four problems on area, volume, blood flow, and work has been tackled in the same way, and the method can be described qualitatively as follows.

The quantity to be calculated, say W, cannot be obtained for the object G given because no formula exists. As a result, n smaller objects, say G_i, are constructed. The G_i are chosen in such a way that the quantity W can be calculated, exactly or approximately, for each G_i, say W_i. Then an approximation for W is

$$\sum_{i=1}^{n} W_i.$$

If the number of G_i is increased indefinitely, this approximation becomes more and more accurate and

$$W = \lim_{n\to\infty} \sum_{i=1}^{n} W_i.$$

It is this *limit-summation process* that we discuss throughout the remainder of the chapter. We begin in Section 6.3 with a mathematical description of the process, and by doing so, we obtain a unified approach to the whole idea. We then discover that there is a very simple way to calculate these limits. At that point we will be ready to use the technique in a multitude of applications, including the four problems in this section.

6.3 The Definite Integral

The four problems of Section 6.2 have a common theme: the limit of a summation. By means of a summation we approximated some quantity (area, volume, blood flow, work), and the limit led, at least intuitively, to an exact value for the quantity. In this section, we investigate the mathematics of the limit summation — but only its mathematics. We concentrate here on what a definite integral is, and how to evaluate it; interpretation of the definite integral as area, volume, work, and so on, is made in Chapter 7.

To define the definite integral of a function $f(x)$ on an interval $a \le x \le b$ (Figure 6.5), we divide the interval into n subintervals by any $n - 1$ points:

$$a = x_0 < x_1 < x_2 < \cdots < x_i < \cdots < x_{n-1} < x_n = b.$$

FIGURE 6.5   Defining the definite integral of $f(x)$ from $x = a$ to $x = b$

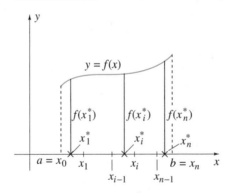

Next we choose in each subinterval $x_{i-1} \le x \le x_i$ any point x_i^* whatsoever, and evaluate $f(x_i^*)$. We now form the sum

$$f(x_1^*)(x_1 - x_0) + f(x_2^*)(x_2 - x_1) + \cdots + f(x_n^*)(x_n - x_{n-1}) = \sum_{i=1}^{n} f(x_i^*)(x_i - x_{i-1})$$

$$= \sum_{i=1}^{n} f(x_i^*)\,\Delta x_i,$$

where we have set $\Delta x_i = x_i - x_{i-1}$. We denote by $\|\Delta x_i\|$ the length of the longest of the n subintervals,

$$\|\Delta x_i\| = \max_{i=1,\dots,n} |\Delta x_i|.$$

It is often called the **norm** of the particular partition of $a \le x \le b$ into the subintervals Δx_i. With this notation, we are ready to define the definite integral of $f(x)$. It is the limit of the summation above as the number of subintervals becomes increasingly large and every subinterval shrinks to a point. An easier way to say this is to take the limit as the norm of the partition approaches zero. In other words, we define the **definite integral** of $f(x)$ with respect to x from $x = a$ to $x = b$ as

$$\int_a^b f(x)\, dx = \lim_{\|\Delta x_i\| \to 0} \sum_{i=1}^n f(x_i^*)\, \Delta x_i, \qquad (6.10)$$

provided that the limit exists. If the limit exists, but is dependent on the choice of subdivision Δx_i or star points x_i^*, then the definite integral is of little use. We stipulate, therefore, that in order for the definite integral to exist, the limit of the sum in equation 6.10 must be independent of the manner of subdivision of the interval $a \le x \le b$ and choice of star points in the subintervals. At first sight this requirement might seem rather severe, since we must now check that all subdivisions and all choices of star points lead to the same limit before concluding that the definite integral exists. Fortunately, however, the following theorem indicates that for continuous functions, this is unnecessary. A proof of this theorem can be found in advanced books on mathematical analysis.

> ## THEOREM 6.2
> If a function $f(x)$ is continuous on a finite interval $a \le x \le b$, then the definite integral of $f(x)$ with respect to x from $x = a$ to $x = b$ exists.

For a continuous function, the definite integral exists, and any choice of subdivision and star points leads to its correct value through the limiting process. We call $f(x)$ on the left-hand side of equation 6.10 the integrand, and a and b the **lower** and **upper limits of integration**, respectively. The sum $\sum_{i=1}^n f(x_i^*)\, \Delta x_i$ is called a **Riemann sum**, and because of this, definite integral 6.10 is also called the **Riemann integral**. The integral was named after German mathematician G. F. B. Riemann (1826–1866), who introduced the notion of the definite integral as a sum.

EXAMPLE 6.3

Evaluate the definite integral

$$\int_0^1 x^2\, dx.$$

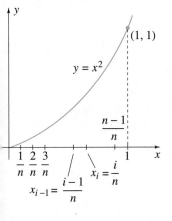

FIGURE 6.6 Definite integral of $f(x) = x^2$ from $x = 0$ to $x = 1$

SOLUTION Since $f(x) = x^2$ is continuous on the interval $0 \le x \le 1$, the definite integral exists, and we may choose any subdivision and star points in its evaluation. The simplest partition is into n equal subintervals of length $1/n$ by the points (Figure 6.6)

$$x_i = \frac{i}{n}, \qquad i = 0, \ldots, n.$$

We choose for star points the right end of each subinterval; that is, in $x_{i-1} \le x \le x_i$, we choose $x_i^* = x_i = i/n$. Then, by equation 6.10, we have

$$\int_0^1 x^2\, dx = \lim_{\|\Delta x_i\| \to 0} \sum_{i=1}^n f(x_i^*)\, \Delta x_i = \lim_{\|\Delta x_i\| \to 0} \sum_{i=1}^n (x_i^*)^2\, \Delta x_i.$$

Since all subintervals have equal length $\Delta x_i = 1/n$, the norm of the partition is $\|\Delta x_i\| = 1/n$, and taking the limit as $\|\Delta x_i\| \to 0$ is tantamount to letting $n \to \infty$. Thus,

$$\int_0^1 x^2\, dx = \lim_{n \to \infty} \sum_{i=1}^n \left(\frac{i}{n}\right)^2 \left(\frac{1}{n}\right) = \lim_{n \to \infty} \frac{1}{n^3} \sum_{i=1}^n i^2.$$

If we now use formula 6.4 for the sum of the squares of the first n positive integers, we obtain

$$\int_0^1 x^2 \, dx = \lim_{n \to \infty} \frac{1}{n^3} \frac{n(n+1)(2n+1)}{6} = \frac{1}{3}.$$

This example illustrates that even for an elementary function such as $f(x) = x^2$, evaluation of the definite integral by equation 6.10 is quite laborious. In fact, had we not known formula 6.4 for the sum of the squares of the integers, we would not have been able to complete the calculation. Imagine the magnitude of the problem were the integrand equal to $f(x) = x(x+1)^{-2/3}$. In other words, if definite integrals are to be at all useful, we must find a simpler way to evaluate them. This we do in Section 6.4, but in order to stress the definite integral as a limit summation, we consider one more example.

EXAMPLE 6.4

Evaluate the definite integral

$$\int_{-1}^1 (5x - 2) \, dx.$$

FIGURE 6.7   Definite integral of $f(x) = 5x - 2$ from $x = -1$ to $x = 1$

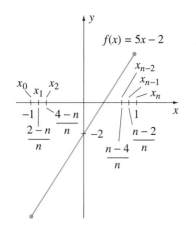

SOLUTION   Since $f(x) = 5x - 2$ is continuous on the interval $-1 \le x \le 1$, the definite integral exists, and we may choose any partition and star points in its evaluation. For n equal subdivisions of length $2/n$, we use points (Figure 6.7)

$$x_i = -1 + \frac{2i}{n}, \qquad i = 0, \dots, n.$$

If we choose the right end of each subinterval as a star point, that is, $x_i^* = x_i = -1 + 2i/n$, then equation 6.10 gives

$$\int_{-1}^1 (5x - 2) \, dx = \lim_{\|\Delta x_i\| \to 0} \sum_{i=1}^n f(x_i^*) \, \Delta x_i = \lim_{\|\Delta x_i\| \to 0} \sum_{i=1}^n (5x_i^* - 2) \, \Delta x_i.$$

Once again all subintervals have equal length $\Delta x_i = 2/n$, and therefore we may replace $\|\Delta x_i\| \to 0$ with $n \to \infty$,

$$\int_{-1}^1 (5x - 2) \, dx = \lim_{n \to \infty} \sum_{i=1}^n \left[5\left(-1 + \frac{2i}{n}\right) - 2 \right]\left(\frac{2}{n}\right) = \lim_{n \to \infty} \sum_{i=1}^n \left[\frac{20i - 14n}{n^2} \right].$$

We can break the summation into two parts, and take constants outside each summation to obtain

$$\int_{-1}^1 (5x - 2) \, dx = \lim_{n \to \infty} \left[\sum_{i=1}^n \frac{20i}{n^2} - \sum_{i=1}^n \frac{14}{n} \right] = \lim_{n \to \infty} \left[\frac{20}{n^2} \sum_{i=1}^n i - \frac{14}{n} \sum_{i=1}^n 1 \right].$$

With formula 6.3 for the sum of the first n positive integers,

$$\int_{-1}^1 (5x - 2) \, dx = \lim_{n \to \infty} \left[\frac{20}{n^2} \frac{n(n+1)}{2} - \frac{14}{n}n \right] = \lim_{n \to \infty} \left[\frac{10 - 4n}{n} \right] = -4.$$

Note that in Example 6.3 the value of the definite integral is positive, and in Example 6.4 it is negative. The value of the definite integral in Exercise 8 is zero. In other words, the value of a definite integral can be positive, negative, or zero, depending on the limits and the integrand.

In Exercises 1–8 use equation 6.10 to evaluate the definite integral.

1. $\displaystyle\int_0^1 x\,dx$

$*$ **2.** $\displaystyle\int_0^2 3x\,dx$

3. $\displaystyle\int_0^1 (3x+2)\,dx$

$*$ **4.** $\displaystyle\int_0^2 x^3\,dx$

5. $\displaystyle\int_1^2 (x^2+2x)\,dx$

$*$ **6.** $\displaystyle\int_{-1}^0 (-x+1)\,dx$

7. $\displaystyle\int_{-1}^1 x^2\,dx$

$*$ **8.** $\displaystyle\int_{-1}^1 x^3\,dx$

9. Evaluate the definite integral $\displaystyle\int_{-1}^1 x^{15}\,dx$.

10.

  (a) Consider the definite integral $\displaystyle\int_0^1 2^x\,dx$. Show that when the interval $0 \le x \le 1$ is subdivided into n equal subintervals, and star points are chosen as right-hand endpoints in each subinterval, equation 6.10 leads to

$$\int_0^1 2^x\,dx = \lim_{n\to\infty} \frac{1}{n}\sum_{i=1}^n 2^{i/n}.$$

  (b) Use the formula in Exercise 31(b) of Section 6.1 to express the summation in closed form,

$$\int_0^1 2^x\,dx = \lim_{n\to\infty} \frac{2^{1/n}}{n(2^{1/n}-1)}.$$

  (c) Use L'Hôpital's rule to evaluate this limit, and hence find the value of the definite integral.

11. Use the technique of Exercise 10 to evaluate $\displaystyle\int_1^3 e^x\,dx$.

12. Use the formula

$$\sum_{i=1}^n \sin i\theta = \frac{\sin\dfrac{(n+1)\theta}{2}\,\sin\dfrac{n\theta}{2}}{\sin\dfrac{\theta}{2}}$$

to evaluate $\displaystyle\int_0^\pi \sin x\,dx$.

$**$ **13.** Use the formula

$$\sum_{i=1}^n \cos i\theta = \frac{\cos\dfrac{(n+1)\theta}{2}\,\sin\dfrac{n\theta}{2}}{\sin\dfrac{\theta}{2}}$$

to evaluate $\displaystyle\int_0^{\pi/2} \cos x\,dx$.

$**$ **14.** In this exercise we evaluate $\displaystyle\int_a^b x^k\,dx$ for $b > a > 0$ and any $k \ne -1$.

  (a) Let $h = (b/a)^{1/n}$ and subdivide the interval $a \le x \le b$ into n subintervals by the points

$$x_0 = a, \quad x_1 = ah, \quad x_2 = ah^2, \quad \ldots,$$
$$x_i = ah^i, \quad \ldots, \quad x_n = ah^n = b.$$

Show that with the choice of $x_i^* = x_i$ in the ith subinterval $x_{i-1} \le x \le x_i$, equation 6.10 gives

$$\int_a^b x^k\,dx = a^{k+1}\lim_{n\to\infty}\left[\left(\frac{h-1}{h}\right)\sum_{i=1}^n (h^{k+1})^i\right].$$

  (b) Use the result of Exercise 31 in Section 6.1 to write the summation in closed form, and hence show that

$$\int_a^b x^k\,dx = (b^{k+1}-a^{k+1})\lim_{n\to\infty}\frac{\left(\dfrac{b}{a}\right)^{k/n}\left[\left(\dfrac{b}{a}\right)^{1/n}-1\right]}{\left(\dfrac{b}{a}\right)^{(k+1)/n}-1}.$$

  (c) Use L'Hôpital's rule to evaluate this limit, and hence obtain

$$\int_a^b x^k\,dx = \frac{b^{k+1}-a^{k+1}}{k+1}.$$

$**$ **15.** Show that the definite integral of the function in Exercise 52 of Section 2.4 does not exist on any interval $a \le x \le b$ whatsoever.

6.4 The First Fundamental Theorem of Integral Calculus

In Section 6.3 we demonstrated how to evaluate definite integrals using definition 6.10. Integrands x^2 and $5x-2$ in Examples 6.3 and 6.4 are very simple polynomials, as are the integrands in questions 1–9 of Exercises 6.3, but in spite of this, calculations were frequently laborious, and invariably required summation formulas from Section 6.1. We promised a very simple technique that would replace these calculations, and this is the substance of the *first fundamental theorem of integral calculus*.

THEOREM 6.3   (First Fundamental Theorem of Integral Calculus)

If $f(x)$ is continuous on the interval $a \leq x \leq b$, and $F(x)$ is an antiderivative of $f(x)$ thereon, then

$$\int_a^b f(x)\,dx = F(b) - F(a). \tag{6.11}$$

PROOF   Since $f(x)$ is continuous on $a \leq x \leq b$, the definite integral of $f(x)$ from $x = a$ to $x = b$ exists and is defined by equation 6.10, where we are at liberty to choose the Δx_i and x_i^* in any way whatsoever. For any choice of Δx_i, a convenient choice for the x_i^* can be found by applying the mean value theorem n times to $F(x)$, once on each subinterval $x_{i-1} \leq x \leq x_i$. This is possible since $F'(x) = f(x)$ is continuous for $a \leq x \leq b$. The mean value theorem states that for each subinterval, there exists at least one point c_i between x_{i-1} and x_i such that

$$\frac{F(x_i) - F(x_{i-1})}{x_i - x_{i-1}} = F'(c_i), \qquad i = 1, \ldots, n.$$

But $F'(c_i) = f(c_i)$, so that

$$F(x_i) - F(x_{i-1}) = f(c_i)\,\Delta x_i, \qquad i = 1, \ldots, n.$$

If we now choose $x_i^* = c_i$, then

$$f(x_i^*)\Delta x_i = F(x_i) - F(x_{i-1}),$$

and equation 6.10 gives

$$\int_a^b f(x)\,dx = \lim_{\|\Delta x_i\| \to 0} \sum_{i=1}^n f(x_i^*)\,\Delta x_i = \lim_{\|\Delta x_i\| \to 0} \sum_{i=1}^n [F(x_i) - F(x_{i-1})].$$

When we write out all terms in the summation, we find that many cancellations take place:

$$\int_a^b f(x)\,dx = \lim_{\|\Delta x_i\| \to 0} \{[F(x_1) - F(x_0)] + [F(x_2) - F(x_1)] + \cdots + [F(x_n) - F(x_{n-1})]\}$$

$$= \lim_{\|\Delta x_i\| \to 0} \{F(x_n) - F(x_0)\}$$

$$= \lim_{\|\Delta x_i\| \to 0} \{F(b) - F(a)\}$$

$$= F(b) - F(a).$$

If we introduce the notation

$$\{F(x)\}_a^b$$

to represent the difference $F(b) - F(a)$, then Theorem 6.3 can be expressed in the form

$$\int_a^b f(x)\,dx = \left\{ \int f(x)\,dx \right\}_a^b. \tag{6.12}$$

This is a fantastic result. No longer is it necessary to consider limits of summations in order to evaluate definite integrals. We simply find an antiderivative of the integrand, substitute $x = b$ and $x = a$, and subtract. For instance, to evaluate the definite integral in Example 6.3, we easily write

$$\int_0^1 x^2 \, dx = \left\{ \frac{x^3}{3} \right\}_0^1 = \frac{1}{3} - 0 = \frac{1}{3}.$$

Note that had we used *the* indefinite integral $x^3/3 + C$ for x^2, we would have had

$$\int_0^1 x^2 \, dx = \left\{ \frac{x^3}{3} + C \right\}_0^1 = \left\{ \frac{1}{3} + C \right\} - C = \frac{1}{3}.$$

Because the arbitrary constant always vanishes in the evaluation of definite integrals, we need not use the indefinite integral in this context; any antiderivative will do.

Similarly, for Example 6.4, we obtain

$$\int_{-1}^1 (5x - 2) \, dx = \left\{ \frac{5x^2}{2} - 2x \right\}_{-1}^1 = \left\{ \frac{5}{2} - 2 \right\} - \left\{ \frac{5}{2} + 2 \right\} = -4.$$

EXAMPLE 6.5 ▼

Evaluate the following definite integrals:

$$\text{(a)} \quad \int_1^2 (3x^2 - x + 4) \, dx \qquad \text{(b)} \quad \int_{-2}^4 \sqrt{x + 4} \, dx$$

SOLUTION

(a) $\displaystyle \int_1^2 (3x^2 - x + 4) \, dx = \left\{ x^3 - \frac{x^2}{2} + 4x \right\}_1^2 = \{8 - 2 + 8\} - \left\{ 1 - \frac{1}{2} + 4 \right\} = \frac{19}{2}$

(b) $\displaystyle \int_{-2}^4 \sqrt{x + 4} \, dx = \left\{ \frac{2}{3}(x + 4)^{3/2} \right\}_{-2}^4 = \frac{2}{3}(8)^{3/2} - \frac{2}{3}(2)^{3/2} = \frac{28\sqrt{2}}{3}$

▲

EXAMPLE 6.6 ▼

Evaluate

$$\int_1^3 \left(\frac{1}{x^2} + 3x^3 \right) dx.$$

SOLUTION   Since $-1/x + 3x^4/4$ is an antiderivative for $1/x^2 + 3x^3$ for $1 \leq x \leq 3$,

$$\int_1^3 \left(\frac{1}{x^2} + 3x^3 \right) dx = \left\{ -\frac{1}{x} + \frac{3x^4}{4} \right\}_1^3 = \left\{ -\frac{1}{3} + \frac{243}{4} \right\} - \left\{ -1 + \frac{3}{4} \right\} = \frac{182}{3}.$$

▲

EXAMPLE 6.7

Can the definite integral $\int_{-1}^{1} \dfrac{1}{x^2}\,dx$ be evaluated with Theorem 6.3?

SOLUTION   No! Theorem 6.3 requires the integrand $1/x^2$ to be continuous on the interval $-1 \le x \le 1$, and this is not the case. The function is discontinuous at $x = 0$.

There is a difficulty with Theorem 6.3. It is subtle, but important. The theorem states that to evaluate definite integrals of continuous functions, we use antiderivatives. But how do we know that continuous functions have antiderivatives? We don't yet. This fact will be established in Section 6.5 when we verify the second fundamental theorem. You might ask why the second fundamental theorem is not proved first. Would it not be more logical first to establish existence of antiderivatives, and then use this fact to prove Theorem 6.3? From a logic point of view, the answer is yes. However, from a practical point of view, Theorem 6.3 is so useful we want to give it every possible emphasis. To prove the second fundamental theorem first would detract from the importance and simplicity of Theorem 6.3.

Before we move on to the second fundamental theorem, we present the following theorems, which describe properties of the definite integral.

THEOREM 6.4

If $f(x)$ is continuous on $a \le x \le b$, then:

(i) $$\int_{b}^{a} f(x)\,dx = -\int_{a}^{b} f(x)\,dx;$$ (6.13a)

(ii) $$\int_{a}^{b} f(x)\,dx = \int_{a}^{c} f(x)\,dx + \int_{c}^{b} f(x)\,dx.$$ (6.13b)

THEOREM 6.5

If $f(x)$ and $g(x)$ are continuous on $a \le x \le b$, then:

(i) $$\int_{a}^{b} [f(x) + g(x)]\,dx = \int_{a}^{b} f(x)\,dx + \int_{a}^{b} g(x)\,dx;$$ (6.14a)

(ii) $$\int_{a}^{b} kf(x)\,dx = k\int_{a}^{b} f(x)\,dx,$$ (6.14b)

when k is a constant.

Properties 6.14 are analogous to 5.3 for indefinite integrals. Theorems 6.4 and 6.5 can be proved using either Theorem 6.3 or equation 6.10.

We can also establish the following property.

THEOREM 6.6

When $f(x)$ is continuous on $a \le x \le b$ and $m \le f(x) \le M$ on this interval,

$$m(b - a) \le \int_{a}^{b} f(x)\,dx \le M(b - a).$$ (6.15)

PROOF By equation 6.10, we can write

$$\int_a^b f(x)\,dx = \lim_{\|\Delta x_i\|\to 0} \sum_{i=1}^n f(x_i^*)\,\Delta x_i$$

$$\leq \lim_{\|\Delta x_i\|\to 0} \sum_{i=1}^n M\,\Delta x_i$$

$$= M \lim_{\|\Delta x_i\|\to 0} \sum_{i=1}^n \Delta x_i$$

$$= M \lim_{\|\Delta x_i\|\to 0} [(x_1 - a) + (x_2 - x_1) + (x_3 - x_2) + \cdots + (b - x_{n-1})]$$

$$= M \lim_{\|\Delta x_i\|\to 0} (b - a)$$

$$= M(b - a).$$

A similar proof establishes the inequality involving m. ∎

EXAMPLE 6.8

Use Theorem 6.6 to find a maximum possible value for

$$\int_1^4 \frac{\sin x^2}{1 + x^2}\,dx.$$

SOLUTION Clearly, $\sin x^2 \leq 1$ for all x, and on the interval $1 \leq x \leq 4$. The largest value of $1/(1+x^2)$ is $1/2$. Consequently, $(\sin x^2)/(1+x^2) \leq 1/2$ for $1 \leq x \leq 4$ and, by Theorem 6.6,

$$\int_1^4 \frac{\sin x^2}{1 + x^2}\,dx \leq \frac{1}{2}(4 - 1) = \frac{3}{2}.$$

Velocity and Speed Revisited Once Again

Suppose that $v(t) = 3t^2 - 6t - 105$ represents the velocity (in metres per second) of a particle moving along the x-axis beginning at time $t = 0$. We can easily calculate the definite integral of $v(t)$ between any two times, say $t = 0$ and $t = 12$. By doing so, we get our first glimpse of definite integrals at work in applied problems.

$$\int_0^{12} v(t)\,dt = \int_0^{12} (3t^2 - 6t - 105)\,dt = \left\{ t^3 - 3t^2 - 105t \right\}_0^{12} = 36.$$

Realizing that integration is a limit summation, and what is being added are products of velocities $v(t)$ multiplied by small time increments dt, we interpret 36 as the displacement of the particle at time $t = 12$ s relative to its displacement at $t = 0$ s. Although we do not have enough information to determine where the particle is at any given time, we can say that at $t = 12$ s, it is 36 m to the right of where it is at $t = 0$ s. In general, when $v(t)$ is the velocity of a particle moving along the x-axis,

$$\int_a^b v(t)\,dt \qquad (6.16)$$

is the displacement of the particle at time $t = b$ relative to its displacement at time $t = a$. If the definite integral is positive, then at time $t = b$ the particle is to the right of its position at $t = a$; and if the integral is negative, then at $t = b$, the particle is to the left of its position at $t = a$.

Speed is the magnitude of velocity. If we integrate speed between the same limits, we get a different result,

$$\int_0^{12} |v(t)| \, dt = \int_0^{12} |3t^2 - 6t - 105| \, dt.$$

Since $3t^2 - 6t - 105 < 0$ for $0 \le t < 7$, and $3t^2 - 6t - 105 > 0$ for $7 < t \le 12$, we divide the integration into two parts,

$$\int_0^{12} |v(t)| \, dt = \int_0^7 -(3t^2 - 6t - 105) \, dt + \int_7^{12} (3t^2 - 6t - 105) \, dt$$

$$= \left\{ -t^3 + 3t^2 + 105t \right\}_0^7 + \left\{ t^3 - 3t^2 - 105t \right\}_7^{12}$$

$$= 1114.$$

Since this integral adds products of speed $|v(t)|$ and time increments dt, which we interpret as distance travelled, 1114 m must be the distance travelled by the particle between $t = 0$ s and $t = 12$ s. In general,

$$\int_a^b |v(t)| \, dt \tag{6.17}$$

is the distance travelled between times $t = a$ and $t = b$.

These ideas are reinforced by graphs of velocity and speed in Figures 6.8. Because velocity is negative for $0 < t < 7$, the particle is moving to the left at these times. Products $v(t) \, dt$ are negative during this time interval, and therefore contribute negatively to the definite integral of $v(t)$. The particle stops at $t = 7$ s, and then moves to the right from $t = 7$ s to $t = 12$ s. Products $v(t) \, dt$ then contribute positively to the definite integral, with the ultimate result being 36 m. At time $t = 12$ s, the particle is 36 m to the right of its position at $t = 0$ s. On the other hand, speed is always positive, and therefore products $|v(t)| \, dt$ always contribute positively to the integral of speed.

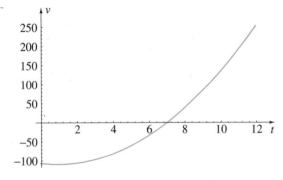

FIGURE 6.8a Velocity function $v(t) = 3t^2 - 6t - 105$

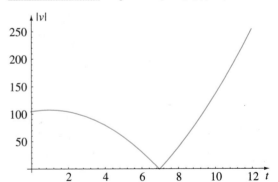

FIGURE 6.8b Speed function $|v(t)| = |3t^2 - 6t - 105|$

EXERCISES 6.4

In Exercises 1–40 evaluate the definite integral.

1. $\int_3^4 (x^3 + 3) \, dx$

2. $\int_1^3 (x^2 - 2x + 3) \, dx$

3. $\int_{-1}^1 (4x^3 + 2x) \, dx$

4. $\int_{-3}^{-1} \frac{1}{x^2} \, dx$

5. $\int_4^2 \left(x^2 + \frac{3}{x^3} \right) dx$

6. $\int_0^{\pi/2} \sin x \, dx$

7. $\displaystyle\int_{-1}^{1} (x^2 - 1 - x^4)\, dx$

8. $\displaystyle\int_{-1}^{-2} \left(\frac{1}{x^2} - 2x\right) dx$

9. $\displaystyle\int_{1}^{2} (x^4 + 3x^2 + 2)\, dx$

10. $\displaystyle\int_{0}^{1} x(x^2 + 1)\, dx$

11. $\displaystyle\int_{0}^{1} x^2(x^2 + 1)^2\, dx$

12. $\displaystyle\int_{0}^{2\pi} \cos 2x\, dx$

13. $\displaystyle\int_{1}^{3} \frac{x^2 + 3}{x^2}\, dx$

14. $\displaystyle\int_{0}^{1} (x^{2.2} - x^{\pi})\, dx$

15. $\displaystyle\int_{-1}^{1} x^2(x^3 - x)\, dx$

16. $\displaystyle\int_{3}^{4} \frac{(x^2 - 1)^2}{x^2}\, dx$

17. $\displaystyle\int_{1}^{2} \left(\sqrt{x} - \frac{1}{\sqrt{x}}\right) dx$

18. $\displaystyle\int_{-2}^{3} (x - 1)^3\, dx$

19. $\displaystyle\int_{2}^{4} \frac{(x^2 - 1)(x^2 + 1)}{x^2}\, dx$

20. $\displaystyle\int_{0}^{\pi/4} 3\cos x\, dx$

21. $\displaystyle\int_{0}^{\pi/4} \sec^2 x\, dx$

22. $\displaystyle\int_{\pi/2}^{\pi} \sin x \cos x\, dx$

23. $\displaystyle\int_{\pi/6}^{\pi/3} \csc^2 3x\, dx$

24. $\displaystyle\int_{-\pi/4}^{\pi/4} \sec x \tan x\, dx$

25. $\displaystyle\int_{0}^{2} 2^x\, dx$

26. $\displaystyle\int_{-1}^{2} e^x\, dx$

27. $\displaystyle\int_{0}^{1} e^{3x}\, dx$

28. $\displaystyle\int_{-3}^{-2} \frac{1}{x}\, dx$

29. $\displaystyle\int_{1}^{3} \frac{(x + 1)^2}{x}\, dx$

30. $\displaystyle\int_{0}^{1} 3^{4x}\, dx$

31. $\displaystyle\int_{0}^{5} |x|\, dx$

32. $\displaystyle\int_{0}^{4} x|x + 1|\, dx$

* 33. $\displaystyle\int_{-5}^{5} |x|\, dx$

* 34. $\displaystyle\int_{-2}^{1} x|x + 1|\, dx$

* 35. $\displaystyle\int_{-1/2}^{1/2} \frac{1}{\sqrt{1 - x^2}}\, dx$

* 36. $\displaystyle\int_{-1}^{1} \frac{1}{1 + x^2}\, dx$

* 37. $\displaystyle\int_{2}^{3} \frac{1}{x\sqrt{x^2 - 1}}\, dx$

* 38. $\displaystyle\int_{0}^{1} \cosh 2x\, dx$

* 39. $\displaystyle\int_{-1}^{1} \operatorname{csch}^2 x\, dx$

* 40. $\displaystyle\int_{0}^{1/2} \frac{1}{1 + 4x^2}\, dx$

In Exercises 41–46 $v(t)$ represents the velocity of a particle moving along the x-axis. Calculate the definite integrals of $v(t)$ and $|v(t)|$ between the two times shown. Interpret each number and plot (or draw) a graph of the velocity function to corroborate results.

* 41. $v(t) = 3t^2 - 6t - 105$, $t = 0$ to $t = 9$

* 42. $v(t) = 3t^2 - 6t - 105$, $t = 0$ to $t = 5$

* 43. $v(t) = -t^2 + 3t - 2$, $t = 0$ to $t = 2$

* 44. $v(t) = -t^2 + 3t - 2$, $t = 1$ to $t = 2$

* 45. $v(t) = t^3 - 3t^2 + 2t$, $t = 0$ to $t = 2$

* 46. $v(t) = t^3 - 3t^2 + 2t$, $t = 0$ to $t = 3$

In Exercises 47–52 use Theorem 6.6 to find maximum and minimum values for the integral.

* 47. $\displaystyle\int_{0}^{\pi/4} \frac{\sin x}{1 + x^2}\, dx$

* 48. $\displaystyle\int_{0}^{\pi/2} \frac{\sin x}{1 + x}\, dx$

* 49. $\displaystyle\int_{0}^{\pi} \frac{\sin x}{2 + x^2}\, dx$

* 50. $\displaystyle\int_{\pi/4}^{\pi/2} \frac{\sin 2x}{10 + x^2}\, dx$

* 51. $\displaystyle\int_{0}^{1} (1 + 4x^4) \cos(x^2)\, dx$ *** 52.** $\displaystyle\int_{1}^{3} \sqrt{4 + x^3}\, dx$

6.5 The Second Fundamental Theorem of Integral Calculus

The first fundamental theorem of integral calculus in Section 6.4 allows us to use antiderivatives to evaluate definite integrals of continuous functions. We now show that continuous functions always have antiderivatives.

When $f(t)$ is continuous for $a \le t \le b$, its definite integral

$$\int_{a}^{b} f(t)\, dt$$

is a number. If b is changed but a is kept fixed, the value of the definite integral changes; for each value of b, there is a new value for the definite integral. In other words, the value of the definite integral is a function of its upper limit. Suppose we replace b by x, and denote the resulting function by $F(x)$,

$$F(x) = \int_a^x f(t)\,dt.$$

We now show that the derivative of $F(x)$ is $f(x)$; that is, $F(x)$ is an antiderivative of $f(x)$.

THEOREM 6.7   (The Second Fundamental Theorem of Integral Calculus)

When $f(x)$ is continuous for $a \le x \le b$, the function

$$F(x) = \int_a^x f(t)\,dt \qquad\qquad (6.18)$$

is differentiable for $a \le x \le b$, and $F'(x) = f(x)$.

PROOF   If x is any point in the open interval $a < x < b$, then h can always be chosen sufficiently small that $x + h$ is also in the interval $a < x < b$. By equation 3.3, the derivative of $F(x)$ at this x is defined as

$$F'(x) = \lim_{h \to 0} \frac{F(x+h) - F(x)}{h}$$

$$= \lim_{h \to 0} \frac{1}{h} \left[\int_a^{x+h} f(t)\,dt - \int_a^x f(t)\,dt \right]$$

$$= \lim_{h \to 0} \frac{1}{h} \left[\int_a^{x+h} f(t)\,dt + \int_x^a f(t)\,dt \right] \qquad \text{(by property 6.13a)}$$

$$= \lim_{h \to 0} \frac{1}{h} \int_x^{x+h} f(t)\,dt \qquad \text{(by property 6.13b)}.$$

According to property 6.15,

$$mh \le \int_x^{x+h} f(t)\,dt \le Mh,$$

where m and M are the minimum and maximum of $f(x)$ on the interval between x and $x + h$. Division by h gives

$$m \le \frac{1}{h} \int_x^{x+h} f(t)\,dt \le M.$$

Consider what happens as we let $h \to 0$. The limit of the middle term is $F'(x)$. Furthermore, the numbers m and M must approach one another, and in the limit must both be equal to $f(x)$; they are minimum and maximum values of $f(x)$ on the interval between x and $x + h$, and h is approaching zero. We conclude therefore that

$$F'(x) = \lim_{h \to 0} \int_x^{x+h} f(t)\,dt = f(x).$$

This argument can also be used to establish that $F(x)$ has a right-hand derivative $f(a)$ at $x = a$, and a left-hand derivative $f(b)$ at $x = b$.

According to Theorem 6.3, we evaluate the definite integral of a continuous function over the interval $a \leq x \leq b$ by calculating the difference between values of any antiderivative of $f(x)$ at $x = b$ and at $x = a$. Theorem 6.7 establishes the fact that continuous functions have antiderivatives. It does not, however, yield an antiderivative of $f(x)$ in a form useful for evaluation of a definite integral of $f(x)$. If we were to use the antiderivative of equation 6.18 in equation 6.11, we would obtain

$$\int_a^b f(x)\,dx = \int_a^b f(t)\,dt - \int_a^a f(t)\,dt = \int_a^b f(t)\,dt.$$

This is certainly true, but not very helpful. To use equation 6.11 to evaluate the definite integral of a function $f(x)$, we need an antiderivative of $f(x)$ written in terms of functions that we already know. Some integrands have easily computed antiderivatives such as those in Chapter 5; others have antiderivatives that are somewhat more complicated and require the integration techniques of Chapter 8 to express them in terms of well-known functions. There are some functions, however, that do not have antiderivatives that can be expressed as the sum of a finite number of well-known functions. A simple example is e^{-x^2}. Theorem 6.7 guarantees that this function has an antiderivative; it just cannot be expressed as a finite sum of well-known functions.

Symbolically, we may write the result of Theorem 6.7 in the form

$$\frac{d}{dx}\int_a^x f(t)\,dt = f(x). \qquad (6.19)$$

We use this in the following four examples.

EXAMPLE 6.9

▼

Evaluate $\dfrac{d}{dx}\displaystyle\int_0^x \sqrt{1 - t^2}\,dt$.

SOLUTION   According to 6.19,

$$\frac{d}{dx}\int_0^x \sqrt{1 - t^2}\,dt = \sqrt{1 - x^2}.$$

This is valid for $-1 \leq x \leq 1$.

▲

EXAMPLE 6.10

▼

Evaluate $\dfrac{d}{dx}\displaystyle\int_1^{2x^2} \frac{\sin t}{1 + t^2}\,dt$.

SOLUTION   For this problem we set $u = 2x^2$, and invoke the chain rule,

$$\frac{d}{dx}\int_1^{2x^2} \frac{\sin t}{1 + t^2}\,dt = \frac{d}{dx}\int_1^u \frac{\sin t}{1 + t^2}\,dt = \left[\frac{d}{du}\int_1^u \frac{\sin t}{1 + t^2}\,dt\right]\frac{du}{dx}$$

$$= \frac{\sin u}{1 + u^2}(4x) = \frac{4x \sin(2x^2)}{1 + 4x^4}.$$

▲

EXAMPLE 6.11

Evaluate $\dfrac{d}{dx} \displaystyle\int_x^5 (1+t^3)^{2/3}\, dt$.

SOLUTION   We can solve this problem by reversing the limits on the integral, which according to equation 6.13a, introduces a negative sign. Thus,

$$\frac{d}{dx} \int_x^5 (1+t^3)^{2/3}\, dt = -\frac{d}{dx} \int_5^x (1+t^3)^{2/3}\, dt = -(1+x^3)^{2/3}.$$

In the following example, the variable x appears in both limits.

EXAMPLE 6.12

Evaluate $\dfrac{d}{dx} \displaystyle\int_{x^2}^{2x} \cos(2t^3 + 1)\, dt$.

SOLUTION   Since the integrand is continuous for all real numbers, property 6.13b permits us to write

$$\int_{x^2}^{2x} \cos(2t^3 + 1)\, dt = \int_{x^2}^{a} \cos(2t^3 + 1)\, dt + \int_{a}^{2x} \cos(2t^3 + 1)\, dt$$

for any real number a whatsoever. To find the derivative of the second integral on the right we set $v = 2x$ and use the chain rule (as in Example 6.10), and for the derivative of the first integral on the right, we reverse the limits (as in Example 6.11) and then use the chain rule with $u = x^2$:

$$\frac{d}{dx} \int_{x^2}^{2x} \cos(2t^3 + 1)\, dt = -\frac{d}{dx} \int_{a}^{x^2} \cos(2t^3 + 1)\, dt + \frac{d}{dx} \int_{a}^{2x} \cos(2t^3 + 1)\, dt$$

$$= \left[-\frac{d}{du} \int_{a}^{u} \cos(2t^3 + 1)\, dt \right] \frac{du}{dx} + \left[\frac{d}{dv} \int_{a}^{v} \cos(2t^3 + 1)\, dt \right] \frac{dv}{dx}$$

$$= -2x \cos(2u^3 + 1) + 2\cos(2v^3 + 1)$$

$$= -2x \cos(2x^6 + 1) + 2\cos(16x^3 + 1).$$

The Natural Logarithm Function

In Section 1.9 we reviewed properties of exponential and logarithm functions. First came the exponential function a^x, and the logarithm function $\log_a x$ is its inverse function. The logarithm of x to base a, $\log_a x$, is a power, the power to raise a in order to get x; that is,

$$y = \log_a x \quad \text{if} \quad x = a^y.$$

Based on this definition, it was straightforward to derive properties of the logarithm function (equations 1.66) based on corresponding properties (equations 1.63) for the exponential function.

The natural logarithm function $\ln x$ uses base e, where e is the limit of $(1 + 1/n)^n$ as n approaches infinity. We saw the advantage of $\ln x$, as opposed to $\log_a x$, for differentiation in Section 3.11; $\ln x$ avoids an extra constant.

The natural logarithm function can be introduced independently of the exponential function using a definite integral. For $x > 0$ we define

$$\ln x = \int_1^x \frac{1}{t}\,dt, \tag{6.20}$$

and with this definition we can derive all properties of the logarithm function. We begin with its graph. Since $1/t$ is positive for $t > 0$, it follows that $\ln x > 0$ for $x > 1$, and property 6.13a implies that it is negative for $0 < x < 1$. According to equation 6.19, the derivative of $\ln x$ is

$$\frac{d}{dx}\ln x = \frac{d}{dx}\int_1^x \frac{1}{t}\,dt = \frac{1}{x}. \tag{6.21}$$

Because $1/x > 0$ for $x > 0$, it follows that the derivative of $\ln x$ is positive, and therefore it is an increasing function. The second derivative is negative,

$$\frac{d^2}{dx^2}\ln x = -\frac{1}{x^2},$$

so that the graph is concave downward. The graph must therefore look like that in Figure 6.9. What is not clear at this point is that the graph is asymptotic to the negative y-axis, although this might seem reasonable on the basis that $1/t$ becomes infinite as $t \to 0^+$.

FIGURE 6.9   Graph of $\ln x$

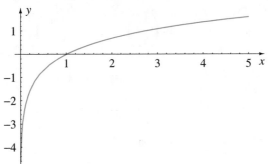

When $u(x)$ is a differentiable function of x, equation 6.21 and the chain rule give

$$\frac{d}{dx}\ln u = \frac{1}{u}\frac{du}{dx}. \tag{6.22}$$

By making specific choices for $u(x)$, we can derive properties 1.71 for the natural logarithm function. First, for $u = x_1 x$, where $x_1 > 0$ is a fixed number, 6.22 gives

$$\frac{d}{dx}\ln(x_1 x) = \frac{1}{x_1 x}x_1 = \frac{1}{x}. \tag{6.23}$$

Equations 6.21 and 6.23 show that $\ln x$ and $\ln(x_1 x)$ have the same derivative for all $x > 0$. Theorem 5.1 implies that these functions can differ by at most a constant; that is,

$$\ln(x_1 x) = \ln x + C.$$

If we set $x = 1$, and note that 6.20 implies that $\ln 1 = 0$, we obtain $\ln x_1 = 0 + C$. Consequently,

$$\ln(x_1 x) = \ln x + \ln x_1,$$

and for $x = x_2$, we obtain identity 1.71a,

$$\ln(x_1 x_2) = \ln x_1 + \ln x_2. \tag{6.24}$$

Differentiation formula 6.22 implies that for $x > 0$,

$$\frac{d}{dx} \ln\left(\frac{1}{x}\right) = \frac{1}{1/x}\left(-\frac{1}{x^2}\right) = -\frac{1}{x}.$$

When this is added to 6.21,

$$\frac{d}{dx}\left[\ln x + \ln\left(\frac{1}{x}\right)\right] = \frac{1}{x} - \frac{1}{x} = 0.$$

Because this is valid for $x > 0$, Theorem 5.1 implies that

$$\ln x + \ln\left(\frac{1}{x}\right) = C.$$

Substitution of $x = 1$ gives $C = 0$, and therefore

$$\ln\left(\frac{1}{x}\right) = -\ln x. \tag{6.25}$$

We can now verify property 1.71b,

$$\ln\left(\frac{x_1}{x_2}\right) = \ln\left[x_1\left(\frac{1}{x_2}\right)\right]$$

$$= \ln x_1 + \ln\left(\frac{1}{x_2}\right) \qquad \text{(by 6.24)}$$

$$= \ln x_1 - \ln x_2 \qquad \text{(by 6.25)}.$$

For fixed $x_2 > 0$, equation 6.22 gives

$$\frac{d}{dx}\ln(x^{x_2}) = \frac{1}{x^{x_2}}x_2\,x^{x_2-1} = \frac{x_2}{x} = x_2\frac{d}{dx}\ln x = \frac{d}{dx}(x_2 \ln x).$$

Since this is valid for all $x > 0$, Theorem 5.1 once again gives

$$\ln(x^{x_2}) = x_2 \ln x + C.$$

If we set $x = 1$, we obtain $C = 0$, and for $x = x_1$,

$$\ln(x_1^{x_2}) = x_2 \ln x_1. \tag{6.26}$$

With the logarithm function defined by 6.20, there is a natural way to introduce the number e. We define e as the number whose natural logarithm is 1:

$$\ln e = 1 \qquad \Longrightarrow \qquad \int_1^e \frac{1}{t}dt = 1. \tag{6.27}$$

The resulting exponential function e^x would satisfy properties 1.69 since all exponential functions satisfy properties of this type. Furthermore, it follows from 6.26 and 6.27 that

$$\ln(e^x) = x \ln e = x. \tag{6.28}$$

This is property 1.72b. We can also establish 1.72a by setting $y = e^{\ln x}$. When we take logarithms of both sides, and use 6.27 and 6.28,

$$\ln y = \ln (e^{\ln x}) = \ln x (\ln e) = \ln x. \tag{6.29}$$

The graph in Figure 6.9 indicates that for any given y-value on the curve, there is only one value of x that gives that y. In other words, if $\ln y = \ln x$ as in 6.29, it follows that $y = x$. Since $y = e^{\ln x}$, we have $x = e^{\ln x}$.

EXERCISES 6.5

In Exercises 1–20 differentiate the definite integral with respect to x.

1. $\displaystyle\int_0^x (3t^2 + t)\, dt$

2. $\displaystyle\int_1^x \frac{1}{\sqrt{t^2 + 1}}\, dt$

3. $\displaystyle\int_x^2 \sin (t^2)\, dt$

4. $\displaystyle\int_x^{-1} t^3 \cos t\, dt$

5. $\displaystyle\int_0^{3x} (2t - t^4)^2\, dt$

6. $\displaystyle\int_1^{2x} \sqrt{t + 1}\, dt$

7. $\displaystyle\int_4^{3x^2} \sin (3t + 4)\, dt$

8. $\displaystyle\int_{-2}^{5x+4} \sqrt{t^3 + 1}\, dt$

9. $\displaystyle\int_x^{2x} (3\sqrt{t} - 2t)\, dt$

10. $\displaystyle\int_{4x}^{4x+4} \left(t^3 - \frac{1}{\sqrt{t}} \right) dt$

11. $\displaystyle\int_{-2x}^x \tan (3t + 1)\, dt$

12. $\displaystyle\int_{-x^2}^{-2x^2} \sec (1 - t)\, dt$

13. $\displaystyle\int_0^{\sin x} \cos (t^2)\, dt$

14. $\displaystyle\int_{\cos x}^{\sin x} \frac{1}{\sqrt{t + 1}}\, dt$

15. $\displaystyle\int_0^{2\sqrt{x}} \sqrt{t}\, dt$

16. $\displaystyle\int_{\sqrt{x}}^{2\sqrt{x}} \sqrt{t}\, dt$

17. $\displaystyle\int_1^{x^2} t^2 e^{4t}\, dt$

18. $\displaystyle\int_x^2 \ln (t^2 + 1)\, dt$

19. $\displaystyle\int_x^{2x} t \ln t\, dt$

20. $\displaystyle\int_{-2x}^{3x} e^{-4t^2}\, dt$

∗ **21.** Verify that when $a(x)$ and $b(x)$ are differentiable functions of x,

$$\frac{d}{dx} \int_{a(x)}^{b(x)} f(t)\, dt = f[b(x)]\frac{db}{dx} - f[a(x)]\frac{da}{dx}.$$

Is equation 6.19 a special case of this result?

22.–28. Use the result of Exercise 21 to redo Exercises 8, 10, 12, 14, 16, 18, and 20.

6.6 Average Values

The **average value** of two numbers c and d is defined as $(c + d)/2$. The average value of a set of n numbers $y_1, y_2, \ldots, y_n$ is $(y_1 + y_2 + \cdots + y_n)/n$. We would like to extend this idea to define the average value of a function $f(x)$ over an interval $a \le x \le b$. By beginning with some simple functions we can see how to do this.

The function $f(x)$ in Figure 6.10a is equal to 1 for the first third of the interval shown, 2 for the second third, and 3 for the last third. We would expect its average value over the interval $0 \le x \le 6$ to be 2. The function in Figure 6.10b takes on the same function values, namely, 1, 2, and 3, but not on the same subintervals. The fact that it has value 2 for $2 \le x \le 5$ and value 3 for $5 \le x \le 6$ suggests that its average value should be somewhat less than the average value of 2 for the function in Figure 6.10a. The average value of the function in Figure 6.10c should be even less than that in Figure 6.10b. These three functions indicate that two factors are important when considering average values of functions: values that the function takes on, and lengths of the intervals on which they take these values. Perhaps what should be done to calculate average values for these functions is to add together the products obtained by multiplying each of the function values by the length of the interval in which it has this value, and then divide this sum by the length of the overall interval.

For the function in Figure 6.10a, this yields an average value of

$$\frac{1}{6 - 0}[1(2 - 0) + 2(4 - 2) + 3(6 - 4)] = 2;$$

FIGURE 6.10a   FIGURE 6.10b   FIGURE 6.10c

Average values of three piecewise constant functions

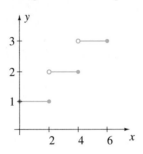
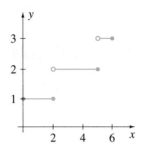
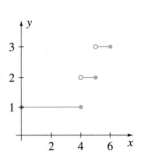

for the function in Figure 6.10b,

$$\frac{1}{6-0}[1(2-0)+2(5-2)+3(6-5)] = \frac{11}{6};$$

and for the function in Figure 6.10c,

$$\frac{1}{6-0}[1(4-0)+2(5-4)+3(6-5)] = \frac{3}{2}.$$

This procedure is applicable only to a function whose domain can be subdivided into a finite number of subintervals inside each of which the function has a constant value. Such a function is said to be **piecewise constant**. What shall we do for functions that are not piecewise constant? By rephrasing the procedure above, it will become obvious. The same three average values are obtained if we adopt the following approach [illustrated for the function $f(x)$ in Figure 6.10c]. Divide the interval $0 \le x \le 6$ into three subintervals $0 < x \le 4$, $4 < x \le 5$, and $5 < x \le 6$. Pick a point in each subinterval; call the points x_1^*, x_2^*, and x_3^*. Evaluate $f(x)$ at each point and multiply by the length of the subinterval in which the point is found. Add these results, and divide by the length of the interval to obtain the average value,

$$\frac{1}{6}[f(x_1^*)(4-0)+f(x_2^*)(5-4)+f(x_3^*)(6-5)] = \frac{1}{6}[1(4)+2(1)+3(1)] = \frac{3}{2}.$$

But this is the procedure used to define the definite integral of a function; it lacks the limit because the function is piecewise constant. In other words, for a function $f(x)$ that has a value at every point in the interval $a \le x \le b$ (but is not necessarily piecewise constant), we define its average value as

$$\text{average value} = \frac{1}{b-a}\int_a^b f(x)\,dx = \frac{1}{b-a}\lim_{\|\Delta x_i\|\to 0}\sum_{i=1}^n f(x_i^*)\,\Delta x_i. \qquad (6.30)$$

EXAMPLE 6.13

What is the average value of the function $f(x) = x^2$ on the interval $0 \le x \le 2$?

SOLUTION   By equation 6.30,

$$\text{average value} = \frac{1}{2}\int_0^2 x^2\,dx = \frac{1}{2}\left\{\frac{x^3}{3}\right\}_0^2 = \frac{1}{2}\left(\frac{8}{3}\right) = \frac{4}{3}.$$

EXAMPLE 6.14

Find the average value of $f(x) = \sin x$ on the intervals (a) $0 \le x \le \pi/2$, (b) $0 \le x \le \pi$, and (c) $0 \le x \le 2\pi$.

SOLUTION   We calculate average values on these intervals as:

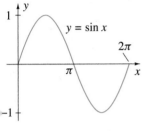

FIGURE 6.11   Average value of $\sin x$

(a) $\dfrac{1}{\pi/2} \displaystyle\int_0^{\pi/2} \sin x \, dx = \dfrac{2}{\pi} \{-\cos x\}_0^{\pi/2} = \dfrac{2}{\pi}(0 + 1) = \dfrac{2}{\pi}$

(b) $\dfrac{1}{\pi} \displaystyle\int_0^{\pi} \sin x \, dx = \dfrac{1}{\pi} \{-\cos x\}_0^{\pi} = \dfrac{1}{\pi}(1 + 1) = \dfrac{2}{\pi}$

(c) $\dfrac{1}{2\pi} \displaystyle\int_0^{2\pi} \sin x \, dx = \dfrac{1}{2\pi} \{-\cos x\}_0^{2\pi} = \dfrac{1}{2\pi}(-1 + 1) = 0$

The graph of $\sin x$ in Figure 6.11 also suggests that the average values in parts (a) and (b) should be the same, and that the average value in part (c) should be zero.

EXAMPLE 6.15

Application Preview Revisited

In the Application Preview we questioned how a controller might monitor current in a circuit. One way would be for it to constantly measure the average value of the current over some fixed interval of time. Suppose the current in the circuit is as shown in Figure 6.12, and the controller constantly measures the average value of $i(t)$ over the previous 0.1 s. If the average value ever reaches 150% of steady-state current, the controller takes corrective action. At what time, if any, does the controller react?

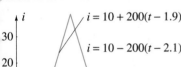
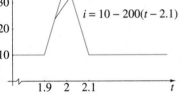

FIGURE 6.12   Abnormal current in electric circuit

SOLUTION   If we denote the average value of $i(t)$ over the 0.1 s before time t by $\bar{i}(t)$, then

$$\bar{i}(t) = \frac{1}{0.1} \int_{t-0.1}^{t} i(t) \, dt.$$

For $t \le 1.9$, $\bar{i}(t)$ is always equal to 10 and the controller takes no action. For $1.9 < t \le 2$, $i(t)$ is increasing and so also is $\bar{i}(t)$:

$$\bar{i}(t) = 10 \left\{ \int_{t-0.1}^{1.9} 10 \, dt + \int_{1.9}^{t} [10 + 200(t - 1.9)] \, dt \right\}$$

$$= 10 \left[\{10t\}_{t-0.1}^{1.9} + \{100t^2 - 370t\}_{1.9}^{t} \right]$$

$$= 20(50t^2 - 190t + 181).$$

This reaches 150% of steady-state current in the interval $1.9 < t \le 2$ if

$$20(50t^2 - 190t + 181) = 15.$$

Solutions of this quadratic equation are $t = 1.83$ and $t = 1.97$. The first is unacceptable, and therefore the controller reacts at 1.97 s.

Theorem 6.7 can be used to establish the next theorem.

THEOREM 6.8   (Mean Value Theorem for Definite Integrals)

If $f(x)$ is continuous for $a \leq x \leq b$, then there exists at least one number c between a and b such that

$$\int_a^b f(x)\, dx = (b - a)f(c). \tag{6.31}$$

PROOF   By Theorem 6.7, the function

$$F(x) = \int_a^x f(t)\, dt$$

is continuous for $a \leq x \leq b$, and has derivative $f(x)$ for $a < x < b$. Mean value theorem 3.19 applied to $F(x)$ guarantees at least one number c between a and b such that

$$F(b) - F(a) = (b - a)F'(c).$$

By substitution,

$$\int_a^b f(t)\, dt - \int_a^a f(t)\, dt = (b - a)f(c).$$

Since the second integral vanishes, we have

$$\int_a^b f(x)\, dx = (b - a)f(c).$$

By writing equation 6.31 in the form

$$f(c) = \frac{1}{b - a} \int_a^b f(x)\, dx,$$

Theorem 6.8 states that the function must take on its average value at least once in the interval.

EXAMPLE 6.16

Find all values of c satisfying Theorem 6.8 for the function $f(x) = x^2$ on the interval $0 \leq x \leq 1$.

SOLUTION   Substituting $a = 0$ and $b = 1$ in equation 6.31 gives

$$(1 - 0)c^2 = \int_0^1 x^2\, dx = \left\{ \frac{x^3}{3} \right\}_0^1 = \frac{1}{3} - 0 = \frac{1}{3}.$$

Of the two solutions $c = \pm 1/\sqrt{3}$ for this equation, only the positive one is between 0 and 1. Thus, only $c = 1/\sqrt{3}$ satisfies Theorem 6.8 for the function $f(x) = x^2$ on the interval $0 \leq x \leq 1$.

EXAMPLE 6.17

Find all values of c satisfying Theorem 6.8 for the function $f(x) = \sin x$ on the interval $\pi/4 \le x \le 3\pi/4$.

SOLUTION   Substituting $a = \pi/4$ and $b = 3\pi/4$ in equation 6.31 gives

$$\left(\frac{3\pi}{4} - \frac{\pi}{4} \right) \sin c = \int_{\pi/4}^{3\pi/4} \sin x \, dx = \{-\cos x\}_{\pi/4}^{3\pi/4} = \frac{1}{\sqrt{2}} + \frac{1}{\sqrt{2}} = \sqrt{2}.$$

Thus,

$$\sin c = \frac{2\sqrt{2}}{\pi}.$$

There are two angles between $\pi/4$ and $3\pi/4$ with a sine equal to $2\sqrt{2}/\pi$, namely $c = \text{Sin}^{-1}(2\sqrt{2}/\pi) = 1.12$ and $c = \pi - \text{Sin}^{-1}(2\sqrt{2}/\pi) = 2.02$.

EXERCISES 6.6

In Exercises 1–20 find the average value of the function over the interval.

1. $f(x) = x^2 - 2x$, $0 \le x \le 2$

2. $f(x) = x^3 - x$, $-1 \le x \le 1$

3. $f(x) = x^3 - x$, $0 \le x \le 1$

4. $f(x) = x^4$, $1 \le x \le 2$

5. $f(x) = \sqrt{x+1}$, $0 \le x \le 1$

6. $f(x) = \sqrt{x+1}$, $-1 \le x \le 1$

7. $f(x) = x^4 - 1$, $0 \le x \le 1$

8. $f(x) = x^4 - 1$, $0 \le x \le 2$

9. $f(x) = \cos x$, $-\pi/2 \le x \le \pi/2$

10. $f(x) = \cos x$, $0 \le x \le \pi/2$

11. $f(x) = |x|$, $-2 \le x \le 2$

12. $f(x) = |x|$, $0 \le x \le 2$

13. $f(x) = |x^2 - 4|$, $0 \le x \le 3$

14. $f(x) = |x^2 - 4|$, $-3 \le x \le 3$

15. $f(x) = \text{sgn}\, x$, $-1 \le x \le 1$ (See Exercise 47 in Section 2.4.)

16. $f(x) = \text{sgn}\, x$, $-1 \le x \le 3$ (See Exercise 47 in Section 2.4.)

17. $f(x) = h(x - 1)$, $0 \le x \le 2$ (See Section 2.5.)

18. $f(x) = h(x - 4)$, $0 \le x \le 2$ (See Section 2.5.)

19. $f(x) = \lfloor x \rfloor$, $0 \le x \le 3$ (See Exercise 68 in Section 1.5.)

* **20.** $f(x) = \lfloor x \rfloor$, $0 \le x \le 3.5$ (See Exercise 68 in Section 1.5.)

* **21.** If a particle moving along the x-axis is at position x_1 at time t_1 and at position x_2 at time t_2, its average velocity over this time interval is

$$\frac{x_2 - x_1}{t_2 - t_1}.$$

Verify that this is the same as the average of the velocity function over the time interval $t_1 \le t \le t_2$.

* **22.** The velocity v of blood flowing through a circular vein or artery of radius R at a distance r from the centre of the blood vessel is

$$v(r) = c(R^2 - r^2).$$

(See Problem 3 in Section 6.2.) What is the average value of $v(r)$ with respect to r?

In Exercises 23–30 find all values of c satisfying equation 6.31 for the function $f(x)$ on the specified interval.

23. $f(x) = 2x - x^2$, $0 \le x \le 2$

24. $f(x) = x^3 - 8x$, $-2 \le x \le 2$

25. $f(x) = \cos x$, $0 \le x \le \pi/2$

26. $f(x) = \cos x$, $0 \le x \le \pi$

27. $f(x) = \sqrt{x+1}$, $1 \le x \le 3$

* **28.** $f(x) = x^2(x + 1)$, $0 \le x \le 1$

* **29.** $f(x) = x\sqrt{x^2 + 1}$, $0 \le x \le 2$

* **30.** $f(x) = 1/x^2 + 1/x^3$, $1 \le x \le 2$

In many applications involving erratic functions, such as the daily price of gold, or the values of stocks and bonds, it is advantageous to define a moving average. It is the average of a function over an interval of fixed length L, but of variable position. The length L moving average of a function $f(x)$ at x is

$$\overline{f}(x) = \frac{1}{L} \int_{x-L}^{x} f(t)\,dt.$$

In Exercises 31–38 calculate the moving average of $f(x)$ for the given length L. Plot, or draw, $f(x)$ and $\overline{f}(x)$ to illustrate how $\overline{f}(x)$ aver-

ages $f(x)$ and lags behind it.

* **31.** $f(x) = \sin x$, $L = \pi$ * **32.** $f(x) = \sin x$, $L = 2\pi$
* **33.** $f(x) = x^2$, $L = 1$ * **34.** $f(x) = x^3$, $L = 2$
* **35.** $f(x) = |x|$, $L = 1$
* **36.** $f(x) = \begin{cases} 1 - |x|, & |x| \le 1 \\ 0, & |x| > 1 \end{cases}$, $L = 1$
* **37.** $f(x) = h(x - 1)$, $L = 1$, where $h(x - 1)$ is the Heaviside function
* **38.** $f(x) = h(x - a) - h(x - b)$, $L = b - a$, where $b > a > 0$ and $h(x - a)$ is the Heaviside function

6.7 Change of Variable in the Definite Integral

The first fundamental theorem of integral calculus indicates that to evaluate the definite integral

$$\int_{-4}^{-1} \frac{x}{\sqrt{x + 5}}\,dx$$

we should first find an antiderivative for $x/\sqrt{x + 5}$. To do this we set $u = x + 5$, in which case $du = dx$, and

$$\int \frac{x}{\sqrt{x + 5}}\,dx = \int \frac{u - 5}{\sqrt{u}}\,du = \int (u^{1/2} - 5u^{-1/2})\,du$$

$$= \frac{2}{3}u^{3/2} - 10u^{1/2} + C = \frac{2}{3}(x + 5)^{3/2} - 10(x + 5)^{1/2} + C.$$

Consequently,

$$\int_{-4}^{-1} \frac{x}{\sqrt{x + 5}}\,dx = \left\{ \frac{2}{3}(x + 5)^{3/2} - 10(x + 5)^{1/2} \right\}_{-4}^{-1}$$

$$= \left[\frac{2}{3}(4)^{3/2} - 10(4)^{1/2} \right] - \left[\frac{2}{3}(1)^{3/2} - 10(1)^{1/2} \right]$$

$$= -\frac{16}{3}.$$

An alternative approach, which usually turns out to be less work, is to make the **change of variable** $u = x + 5$ directly in the definite integral. In this case we again replace

$$\frac{x}{\sqrt{x + 5}}\,dx \qquad \text{by} \qquad \frac{u - 5}{\sqrt{u}}\,du.$$

In addition, we replace the limits $x = -4$ and $x = -1$ by those values of u that correspond to these values of x, namely, $u = 1$ and $u = 4$, respectively. We then obtain

$$\int_{-4}^{-1} \frac{x}{\sqrt{x + 5}}\,dx = \int_{1}^{4} \frac{u - 5}{\sqrt{u}}\,du = \int_{1}^{4} (u^{1/2} - 5u^{-1/2})\,du$$

$$= \left\{ \frac{2}{3}u^{3/2} - 10u^{1/2} \right\}_{1}^{4}$$

$$= \left[\frac{2}{3}(4)^{3/2} - 10(4)^{1/2} \right] - \left[\frac{2}{3}(1)^{3/2} - 10(1)^{1/2} \right]$$

$$= -\frac{16}{3}.$$

That this method is generally acceptable is stated in the following theorem.

> ### THEOREM 6.9
>
> Suppose $f(x)$ is continuous on $a \le x \le b$, and we set $x = g(u)$, where $a = g(\alpha)$ and $b = g(\beta)$. Then
>
> $$\int_a^b f(x)\,dx = \int_\alpha^\beta f(g(u))\,g'(u)\,du, \qquad (6.32)$$
>
> if $g'(u)$ is continuous on $\alpha \le u \le \beta$, and if when u is between α and β, $g(u)$ is between a and b.

EXAMPLE 6.18

Evaluate $\displaystyle\int_2^4 \frac{\sqrt{x}}{(\sqrt{x}-1)^4}\,dx.$

SOLUTION   If we set $u = \sqrt{x} - 1$, then $du = [1/(2\sqrt{x})]\,dx$, and

$$\int_2^4 \frac{\sqrt{x}}{(\sqrt{x}-1)^4}\,dx = \int_{\sqrt{2}-1}^1 \frac{u+1}{u^4} 2(u+1)\,du = 2\int_{\sqrt{2}-1}^1 \left(\frac{1}{u^2} + \frac{2}{u^3} + \frac{1}{u^4} \right) du$$

$$= 2\left\{ -\frac{1}{u} - \frac{1}{u^2} - \frac{1}{3u^3} \right\}_{\sqrt{2}-1}^1$$

$$= -2\left[1 + 1 + \frac{1}{3} \right] + 2\left[\frac{1}{\sqrt{2}-1} + \frac{1}{(\sqrt{2}-1)^2} + \frac{1}{3(\sqrt{2}-1)^3} \right] = 21.2.$$

What follows is a discussion leading to the well-known formula for the area of a circle. Although you should be able to follow the calculations, it is highly likely that you will not understand the source of some of the ideas. Do not be alarmed; in time, we will deal with all aspects of the derivation in detail. For now, simply take the example as an illustration of the power of integral calculus, and an indication of things to come.

When we study applications of definite integrals in Chapter 7, we shall see that the area of the circle in Figure 6.13 is given by the definite integral

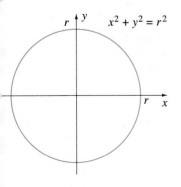

FIGURE 6.13   Area of a circle

$$A = 4\int_0^r \sqrt{r^2 - x^2}\,dx.$$

The integral itself actually gives the first quadrant area; the 4 provides a quadrupling factor for the other three quadrants. It is not possible to guess an antiderivative for $\sqrt{r^2 - x^2}$. In Section 8.4 we learn that an appropriate change of variable is to set $x = r\sin\theta$. The integrand becomes

$$\sqrt{r^2 - x^2} = \sqrt{r^2 - r^2\sin^2\theta} = r\sqrt{1 - \sin^2\theta} = r\sqrt{\cos^2\theta} = r|\cos\theta|,$$

absolute values being necessary to ensure positivity for $\sqrt{\cos^2\theta}$. Since $dx = r\cos\theta\,d\theta$, we have

$$A = 4\int_{\theta_1}^{\theta_2} r|\cos\theta|\,r\cos\theta\,d\theta,$$

where θ_1 and θ_2 are values of θ corresponding to $x = 0$ and $x = r$. When $x = 0$, the equation $x = r\sin\theta$ requires that $0 = r\sin\theta$. We choose $\theta_1 = 0$ as the solution of this equation. (There are others, and in Section 8.4 we shall see why this choice is made.) Similarly, when $x = r$, we choose the solution $\theta_2 = \pi/2$ of $r = r\sin\theta$. Then

$$A = 4\int_0^{\pi/2} r|\cos\theta|\,r\cos\theta\,d\theta.$$

For θ in the interval $0 \le \theta \le \pi/2$, $\cos\theta$ is positive, and absolute values may be dropped,

$$A = 4r^2\int_0^{\pi/2} \cos^2\theta\,d\theta.$$

To find an antiderivative for $\cos^2\theta$, we solve double-angle formula 1.46b for $\cos^2\theta$,

$$\cos 2\theta = 2\cos^2\theta - 1 \qquad \Longrightarrow \qquad \cos^2\theta = \frac{1 + \cos 2\theta}{2}.$$

Thus,

$$A = 4r^2\int_0^{\pi/2}\left(\frac{1 + \cos 2\theta}{2}\right)d\theta = 2r^2\int_0^{\pi/2}(1 + \cos 2\theta)\,d\theta$$

$$= 2r^2\left\{\theta + \frac{1}{2}\sin 2\theta\right\}_0^{\pi/2} = \pi r^2,$$

the formula for the area of a circle. Be reminded that you are not expected to solve other problems like this yet. In time, yes, but not now. What we hope is that the example gives you a glimpse of the power of integral calculus and the importance of the tools that we have developed in Chapters 5 and 6.

EXERCISES 6.7

In Exercises 1–22 evaluate the definite integral.

1. $\displaystyle\int_1^2 x(3x^2 - 2)^4\,dx$

2. $\displaystyle\int_0^1 z\sqrt{1 - z}\,dz$

3. $\displaystyle\int_{-1}^0 \frac{x}{\sqrt{x + 3}}\,dx$

4. $\displaystyle\int_{\pi/4}^{\pi/3} \cos^5 x\sin x\,dx$

5. $\displaystyle\int_1^3 x^3\sqrt{9 - x^2}\,dx$

6. $\displaystyle\int_{-5}^6 \frac{x}{\sqrt{x^2 - 12}}\,dx$

7. $\displaystyle\int_4^5 y^2\sqrt{y - 4}\,dy$

8. $\displaystyle\int_{1/2}^1 \sqrt{\frac{x^2}{1 + x}}\,dx$

9. $\displaystyle\int_1^4 \frac{\sqrt{1 + \sqrt{u}}}{\sqrt{u}}\,du$

10. $\displaystyle\int_{-2}^1 \frac{x + 1}{(x^2 + 2x + 2)^{1/3}}\,dx$

11. $\displaystyle\int_3^4 \frac{x^2}{(x - 2)^4}\,dx$

12. $\displaystyle\int_0^{\pi/6} \sqrt{2 + 3\sin x}\cos x\,dx$

13. $\displaystyle\int_{\pi/4}^{\pi/2} \frac{\sin^3 x}{(1 + \cos x)^4}\,dx$

14. $\displaystyle\int_1^4 \frac{(x + 1)(x - 1)}{\sqrt{x}}\,dx$

$*$ **15.** $\displaystyle\int_4^9 \sqrt{1 + \sqrt{x}}\,dx$

$*$ **16.** $\displaystyle\int_{-1/2}^1 \sqrt{\frac{x^2}{1 + x}}\,dx$

$*$ **17.** $\displaystyle\int_{-1}^1 \frac{|x|}{(x + 2)^3}\,dx$

$*$ **18.** $\displaystyle\int_{-1}^1 \left|\frac{x}{(x + 2)^3}\right|\,dx$

$*$ **19.** $\displaystyle\int_0^1 x^2 e^{x^3}\,dx$

$*$ **20.** $\displaystyle\int_1^2 \frac{(\ln x)^2}{x}\,dx$

$*$ **21.** $\displaystyle\int_2^4 \frac{1}{x\ln x}\,dx$

$*$ **22.** $\displaystyle\int_{-\pi/4}^{\pi/4} \frac{\sec^2 x}{\sqrt{4 + 3\tan x}}\,dx$

$*$ **23.** Show that if $f(x)$ is an odd function, then

$$\int_{-a}^a f(x)\,dx = 0;$$

and that if $f(x)$ is an even function,

$$\int_{-a}^a f(x)\,dx = 2\int_0^a f(x)\,dx.$$

24. Show algebraically that if $f(x)$ is a continuous function with period p, then

$$\int_a^{a+p} f(x)\,dx = \int_0^p f(x)\,dx.$$

In Exercises 25–27 use the suggested substitution to evaluate the definite integral.

25. $\displaystyle\int_1^3 \frac{1}{x^{3/2}\sqrt{4-x}}\,dx, \qquad u = \frac{1}{\sqrt{x}}$

* **26.** $\displaystyle\int_{-6}^{-1} \frac{\sqrt{x^2 - 6x}}{x^4}\,dx, \qquad u = \frac{1}{x}$

* **27.** $\displaystyle\int_{-4}^{0} \frac{x}{(5 - 4x - x^2)^{3/2}}\,dx, \qquad u^2 = \frac{1-x}{5+x}$

** **28.** Show that

$$\int_0^\pi \frac{\cos^2 [(\pi/2)\cos\theta]}{\sin\theta}\,d\theta = \frac{\pi}{2}\int_0^{2\pi} \frac{1 - \cos\phi}{\phi(2\pi - \phi)}\,d\phi.$$

This integral is used in calculating radiated power from a half-wave antenna.

SUMMARY

The definite integral of a continuous function $f(x)$ from $x = a$ to $x = b$ is a number, one that depends on the function $f(x)$ and the limits a and b. We have defined the definite integral by subdividing the interval $a \le x \le b$ into n parts by $n + 1$ points $a = x_0 < x_1 < \cdots < x_{n-1} < x_n = b$, and choosing a point x_i^* in each subinterval $x_{i-1} \le x \le x_i$. The definite integral is then the limit of the summation

$$\int_a^b f(x)\,dx = \lim_{\|\Delta x_i\| \to 0} \sum_{i=1}^n f(x_i^*)\,\Delta x_i,$$

where $\Delta x_i = x_i - x_{i-1}$. Since all calculations in this limit take place on the x-axis, we regard the definite integral as an integration along the x-axis from $x = a$ to $x = b$.

The first fundamental theorem of integral calculus allows us to calculate definite integrals of continuous functions using antiderivatives,

$$\int_a^b f(x)\,dx = \left\{ \int f(x)\,dx \right\}_a^b.$$

This presupposes that a continuous function has an antiderivative, as verified in Section 6.5. It was shown that when the definite integral is given a variable upper limit x, the resulting function

$$F(x) = \int_a^x f(t)\,dt$$

is an antiderivative of $f(x)$; that is, $F'(x) = f(x)$.

The average value of a function $f(x)$ over an interval $a \le x \le b$ is defined as

$$\frac{1}{b - a}\int_a^b f(x)\,dx.$$

The mean value theorem for definite integrals guarantees the existence of at least one number c between a and b, at which $f(x)$ takes on its average value,

$$f(c) = \frac{1}{b - a}\int_a^b f(x)\,dx.$$

Evaluation of a definite integral with a complex integrand can sometimes be simplified with an appropriate change of variable.

KEY TERMS

In reviewing this chapter, you should be able to define or discuss the following key terms:

Sigma notation

Index of summation or variable of summation

Finite geometric series

Definite integral

Upper limit of integration

Riemann integral

Second fundamental theorem
   of integral calculus

Piecewise constant

Change of variable

General term

Limits of summation

Norm

Lower limit of integration

Riemann sum

First fundamental theorem of integral calculus

Average value

Mean value theorem for definite integrals

REVIEW EXERCISES

In Exercises 1–20 evaluate the definite integral.

1. $\int_0^3 (x^2 + 3x - 2)\, dx$

2. $\int_{-1}^1 (x^2 - x^4)\, dx$

3. $\int_{-1}^1 (x^3 - 3x)\, dx$

4. $\int_0^2 (x^2 - 2x)\, dx$

5. $\int_1^2 (x + 1)^2\, dx$

6. $\int_{-3}^{-2} \frac{1}{x^2}\, dx$

7. $\int_4^9 \left(\frac{1}{\sqrt{x}} - \sqrt{x} \right) dx$

8. $\int_0^\pi \cos x\, dx$

9. $\int_{-1}^1 x(x + 1)^2\, dx$

10. $\int_1^2 x^2(x^2 + 3)\, dx$

11. $\int_0^3 \sqrt{x + 1}\, dx$

12. $\int_1^5 x\sqrt{x^2 - 1}\, dx$

13. $\int_1^4 \left(\frac{\sqrt{x} + 1}{\sqrt{x}} \right) dx$

14. $\int_{-1}^0 x\sqrt{x + 1}\, dx$

15. $\int_1^2 \frac{x^2 + 1}{(x + 1)^4}\, dx$

16. $\int_{-4}^{-2} x^2\sqrt{2 - x}\, dx$

17. $\int_0^{\pi/4} \frac{\cos x}{(1 + \sin x)^2}\, dx$

18. $\int_2^3 x(1 + 2x^2)^4\, dx$

* 19. $\int_1^8 \frac{(1 + x^{1/3})^2}{x^{2/3}}\, dx$

* 20. $\int_{-4}^4 |x + 2|\, dx$

* 21. Use equation 6.10 to evaluate the following definite integrals:

(a) $\int_0^2 (x - 5)\, dx$

(b) $\int_0^3 (x^2 + 3)\, dx$

* 22. Prove that each of the following answers is incorrect, but do so without evaluating the definite integral. Think about what the definite integral represents.

(a) $\int_0^4 (x^2 + 3x)\, dx = -3$

(b) $\int_{-3}^{-2} \frac{1}{x}\, dx = 5$

In Exercises 23–26 find the average value of the function on the interval.

23. $f(x) = \sqrt{x + 4}, \quad 0 \le x \le 1$

24. $f(x) = 1/x^2 - x, \quad -2 \le x \le -1$

* 25. $f(x) = x\sqrt{x + 1}, \quad 0 \le x \le 1$

* 26. $f(x) = \cos^3 x \sin^2 x, \quad 0 \le x \le \pi/2$

In Exercises 27–32 differentiate the integral with respect to x.

27. $\int_1^x t\sqrt{t^3 + 1}\, dt$

28. $\int_x^{-3} t^2(t + 1)^3\, dt$

29. $\int_1^{x^2} \sqrt{t^2 + 1}\, dt$

30. $\int_{2x}^4 t \cos t\, dt$

31. $\int_{2x+3}^{1-x} \frac{1}{t^2 + 1}\, dt$

32. $\int_{-x^2}^{x^2} \sin^2 t\, dt$

In Exercises 33–34 $v(t)$ represents the velocity of a particle moving along the x-axis. Calculate the definite integrals of $v(t)$ and $|v(t)|$ between the two times shown. Interpret each number physically and draw a graph of $v(t)$ to corroborate results.

* 33. $v(t) = t^3 - 6t^2 + 11t - 6, t = 0$ to $t = 1$

* 34. $v(t) = t^3 - 6t^2 + 11t - 6, t = 1$ to $t = 3$

n Exercises 35–42 evaluate the definite integral.

35. $\displaystyle\int_0^1 \frac{x^3}{(x^2+1)^{3/2}}\, dx$

* **36.** $\displaystyle\int_0^{\pi/6} \frac{\cos^3 x}{\sqrt{1+\sin x}}\, dx$

37. $\displaystyle\int_{-1}^1 x^3\sqrt{1-x^2}\, dx$

* **38.** $\displaystyle\int_{-1}^2 \left|\frac{x}{\sqrt{3+x}}\right|\, dx$

* **39.** $\displaystyle\int_{-1}^2 x^2(4-x^3)^5\, dx$

* **40.** $\displaystyle\int_1^5 \frac{6x^2+8x+2}{\sqrt{x^3+2x^2+x}}\, dx$

* **41.** $\displaystyle\int_1^2 \frac{x-25}{\sqrt{x}-5}\, dx$

* **42.** $\displaystyle\int_0^1 \frac{1}{\sqrt{2+x}+\sqrt{x}}\, dx$

7 | Applications of the Definite Integral

The figure on the left below shows gas in the cylinder of a steam engine, diesel engine, or internal combustion engine. It requires work to move the piston to the left and compress the gas in the cylinder. On the other hand, if the gas expands, it does work in moving the piston to the right.

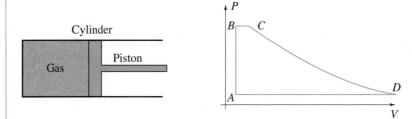

The figure on the right above is the *Rankine cycle* for an idealized steam engine; it represents the relationship between pressure P and volume V of gas in the cylinder during one cycle. Water at low temperature and pressure (point A) is heated at constant volume (along path AB). Along BC the water is converted to steam and expands slightly, and the expansion continues along CD. To complete the cycle, the steam is cooled and condensed to water along DA.

THE PROBLEM   Determine the output of the steam engine during one cycle. (See page 440 for the solution.)

In Chapter 6 we defined the definite integral of a function $f(x)$ as the limit of a summation

$$\int_a^b f(x)\,dx = \lim_{\|\Delta x_i\| \to 0} \sum_{i=1}^n f(x_i^*)\,\Delta x_i.$$

The limit summation on the right defines the value of the definite integral on the left. In this chapter we think of this equation in the reverse direction in order to evaluate geometric and physical quantities. Each quantity is expressed as a limit summation of the form in this equation. The limit summation can immediately be interpreted as a definite integral. The definite integral can then be evaluated by means of an antiderivative of the integrand.

7.1 Area

We have formulas for the **area** of very few shapes — squares, rectangles, triangles, and polygons of any shape, since they can be divided into rectangles and triangles. Consider finding the area in Figure 7.1a. Each of us has intuitive ideas about this area. In this section we take these intuitive ideas of what the area *ought to be*, and make a precise mathematical definition of what the area *is*.

FIGURE 7.1a   Area under curve $y = f(x)$

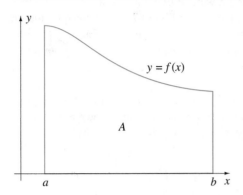

FIGURE 7.1b   Approximation of area under curve $y = f(x)$

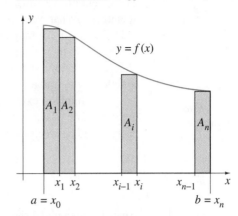

Recall from Problem 1 of Section 6.2 that the area in Figure 7.1a can be approximated by rectangles. Specifically, we partition the interval $a \leq x \leq b$ into n parts by points

$$a = x_0 < x_1 < x_2 < \cdots < x_{n-1} < x_n = b,$$

and construct rectangles as shown in Figure 7.1b. The area A_i of the ith rectangle is

$$A_i = f(x_i)(x_i - x_{i-1}) = f(x_i)\,\Delta x_i,$$

and an approximation to the required area A is therefore

$$\sum_{i=1}^{n} A_i = \sum_{i=1}^{n} f(x_i)\,\Delta x_i.$$

If we let the number of rectangles get larger and larger, and at the same time require each rectangle to have smaller and smaller width that eventually approaches zero, we feel that better and better approximations can be obtained. In fact, if we take the limit as the norm $\|\Delta x_i\|$ of the partition approaches zero, we should get A. Since we have no formal definition for the area of odd-shaped figures, we take this opportunity to make our own. We define the area A in Figure 7.1a as

$$A = \lim_{\|\Delta x_i\| \to 0} \sum_{i=1}^{n} f(x_i)\,\Delta x_i. \tag{7.1}$$

Area A then has been defined as the limit of a sum of rectangular areas.

The right side of equation 7.1 is strikingly similar to the definition of the definite integral of $f(x)$ from $x = a$ to $x = b$ (equation 6.10):

$$\int_a^b f(x)\,dx = \lim_{\|\Delta x_i\| \to 0} \sum_{i=1}^{n} f(x_i^*)\,\Delta x_i.$$

The only difference is the absence of $*$'s in equation 7.1. But when we recall that in equation 6.10, x_i^* may be chosen as any point in the subinterval $x_{i-1} \leq x \leq x_i$, we see that by choosing $x_i^* = x_i$,

$$\int_a^b f(x)\,dx = \lim_{\|\Delta x_i\| \to 0} \sum_{i=1}^{n} f(x_i)\Delta x_i.$$

It follows that area A of Figure 7.1a may be calculated by means of the definite integral

$$A = \int_a^b f(x)\,dx. \tag{7.2}$$

It is important to realize that equation 7.2 does not imply that a definite integral should always be thought of as an area; on the contrary, we will find that definite integrals can represent many quantities. What we have said is that the area in Figure 7.1a is defined by the limit in equation 7.1. But this limit may also be interpreted as the definite integral of $f(x)$ with respect to x from $x = a$ to $x = b$; hence, the area may be calculated by the definite integral in equation 7.2. Since definite integrals can be evaluated using antiderivatives, it seems that we have a very simple way to find areas.

It is simple to extend this result to the problem of finding the area in Figure 7.2. With our interpretation of equation 7.2, we can state that the area under the curve $y = f(x)$, above the x-axis, and between the vertical lines $x = a$ and $x = b$ is given by

$$A_2 = \int_a^b f(x)\,dx.$$

Since the area under the curve $y = g(x)$ is given similarly by

$$A_1 = \int_a^b g(x)\,dx,$$

it follows that the required area is

$$A = A_2 - A_1 = \int_a^b f(x)\,dx - \int_a^b g(x)\,dx = \int_a^b [f(x) - g(x)]\,dx. \qquad (7.3)$$

We now have two formulas for finding areas: equation 7.2 for the area under a curve, above the x-axis, and between two vertical lines; and equation 7.3 for the area between two curves and two vertical lines. Note that 7.2 is a special case of 7.3. At this point we could solve a number of area problems using these two results, but to do so would strongly suggest that integration should be approached from a "formula" point of view; and if there is any point of view that we wish to adopt, it is completely the opposite. By the end of this chapter we hope to have developed a sufficiently clear understanding of the limit-summation process that use of integration in situations other than those discussed here will be straightforward.

To illustrate how we arrive at the correct definite integral for area problems without memorizing the formulas in either equation 7.2 or 7.3, consider again finding the area in Figure 7.1a. We draw a rectangle of width dx at position x as shown in Figure 7.3. The area of this rectangle is

$$f(x)\,dx.$$

We visualize this rectangle as a representative for a large number of such rectangles between a and b. To find the required area we add together all such rectangular areas and take the limit as their widths approach zero. But this is the concept of the definite integral, so that we write for the limit-summation process

$$A = \int_a^b f(x)\,dx.$$

Limits a and b identify x-positions of first and last rectangles, respectively.

Similarly, for the area in Figure 7.4 we draw a rectangle of width dx and length $f(x) - g(x)$ and, therefore, of area

$$[f(x) - g(x)]\,dx.$$

To add areas of all such rectangles between a and b and, at the same time, to take the limit as their widths approach zero, we once again use the definite integral,

$$A = \int_a^b [f(x) - g(x)]\,dx.$$

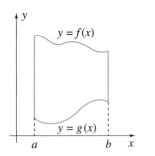

FIGURE 7.2 Area between two curves and two vertical lines

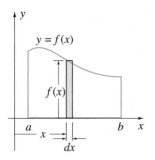

FIGURE 7.3 Integral for area under a curve

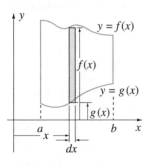

FIGURE 7.4 Integral for area between two curves and two vertical lines

For area problems, then, we start with the area of a **representative rectangle** and proceed to the required area by summation with the definite integral. We express this symbolically as follows:

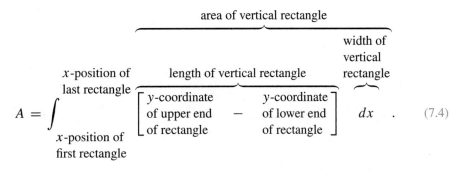

area of vertical rectangle

$$A = \int_{\substack{x\text{-position of} \\ \text{first rectangle}}}^{\substack{x\text{-position of} \\ \text{last rectangle}}} \left[\begin{array}{c} y\text{-coordinate} \\ \text{of upper end} \\ \text{of rectangle} \end{array} - \begin{array}{c} y\text{-coordinate} \\ \text{of lower end} \\ \text{of rectangle} \end{array} \right] dx . \qquad (7.4)$$

EXAMPLE 7.1

Find the area enclosed by the curves $y = x^2$ and $y = x^3$.

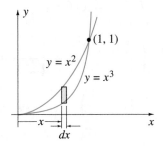

FIGURE 7.5a Area bounded by $y = x^2$ and $y = x^3$

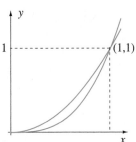

FIGURE 7.5b Estimate of area bounded by $y = x^2$ and $y = x^3$

SOLUTION   The area of the representative rectangle in Figure 7.5a is

$$(x^2 - x^3)\, dx;$$

hence,

$$A = \int_0^1 (x^2 - x^3)\, dx = \left\{ \frac{x^3}{3} - \frac{x^4}{4} \right\}_0^1 = \frac{1}{12}.$$

It is easy to check whether $1/12$ is a reasonable answer. This area is contained inside the square of area one in Figure 7.5b. It is reasonable that the area bounded by the curves is one-twelfth that of the square. We could have used this argument to ballpark the answer before making any calculations.

In the next two examples, expressions for areas of representative rectangles vary within the region specified. In such cases, we set up different integrals corresponding to different representative rectangles.

EXAMPLE 7.2

Find the area bounded by the x-axis and the curve $y = x^3 - x$.

SOLUTION   Areas of rectangles between $x = -1$ and $x = 0$ (Figure 7.6) are

$$[(x^3 - x) - 0]\, dx,$$

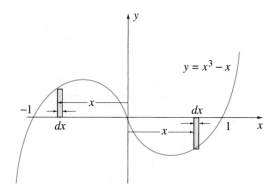

FIGURE 7.6   Area bounded by the curves $y = x^3 - x$ and $y = 0$

whereas between $x = 0$ and $x = 1$ areas are

$$[0 - (x^3 - x)] \, dx.$$

Consequently,

$$A = \int_{-1}^{0} (x^3 - x) \, dx + \int_{0}^{1} (-x^3 + x) \, dx$$

$$= \left\{ \frac{x^4}{4} - \frac{x^2}{2} \right\}_{-1}^{0} + \left\{ -\frac{x^4}{4} + \frac{x^2}{2} \right\}_{0}^{1}$$

$$= -\left(\frac{1}{4} - \frac{1}{2} \right) + \left(-\frac{1}{4} + \frac{1}{2} \right)$$

$$= \frac{1}{2}.$$

We could have saved ourselves some calculations in this example by noting that because of the symmetry of the diagram ($x^3 - x$ is an odd function), the two areas are identical. Hence, we could find the left (or right) area and double it,

$$A = 2 \int_{-1}^{0} (x^3 - x) \, dx = 2 \left\{ \frac{x^4}{4} - \frac{x^2}{2} \right\}_{-1}^{0} = -2 \left(\frac{1}{4} - \frac{1}{2} \right) = \frac{1}{2}.$$

EXAMPLE 7.3

Find the area of the triangle with edges $y = x$, $y = -x/2$, and $y = 5x - 44$.

SOLUTION   Areas of representative rectangles to the left and right of $x = 8$ (Figure 7.7a) are, respectively,

$$[x - (-x/2)] \, dx \qquad \text{and} \qquad [x - (5x - 44)] \, dx;$$

therefore,

$$A = \int_{0}^{8} [x - (-x/2)] \, dx + \int_{8}^{11} [x - (5x - 44)] \, dx$$

$$= \frac{3}{2} \int_{0}^{8} x \, dx + \int_{8}^{11} (44 - 4x) \, dx$$

$$= \frac{3}{2} \left\{ \frac{x^2}{2} \right\}_0^8 + \left\{ 44x - 2x^2 \right\}_8^{11}$$

$$= 48 + (484 - 242) - (352 - 128)$$

$$= 66.$$

Figure 7.7b provides a quick check. The required area would seem to be somewhere between a third and half that of the 15×11 rectangle with area 165 square units.

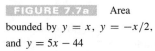

FIGURE 7.7a   Area bounded by $y = x$, $y = -x/2$, and $y = 5x - 44$

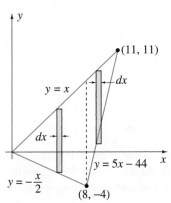

FIGURE 7.7b   Estimate of area of triangle

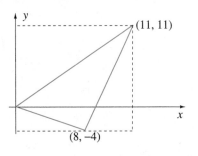

We see from the examples above that the length of a representative rectangle in equation 7.4 as "upper y minus lower y" is valid whether the rectangle is in the first quadrant (Figure 7.5a), the second and fourth quadrants (Figure 7.6), or partially in the first and partially in the fourth (Figure 7.7a). In fact, it is valid for rectangles in all quadrants. Remember this; we use it in many applications.

FIGURE 7.8   Using horizontal rectangles to find area

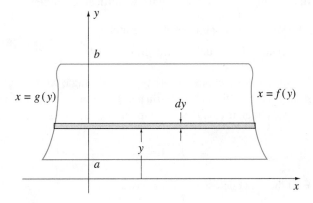

There is nothing special about vertical rectangles. Sometimes it is more convenient to subdivide an area into horizontal rectangles. For example, to find the area in Figure 7.8, we draw a representative rectangle at position y of width dy. Its length is $f(y) - g(y)$, and therefore its area is

$$[f(y) - g(y)]\, dy.$$

Adding over all rectangles gives

$$A = \int_a^b [f(y) - g(y)]\,dy. \qquad (7.5)$$

Corresponding to equation 7.4, we could write that for horizontal rectangles,

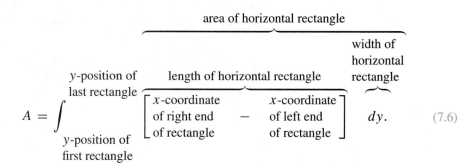

$$A = \int_{\substack{y\text{-position of} \\ \text{first rectangle}}}^{\substack{y\text{-position of} \\ \text{last rectangle}}} \left[\begin{array}{c} x\text{-coordinate} \\ \text{of right end} \\ \text{of rectangle} \end{array} - \begin{array}{c} x\text{-coordinate} \\ \text{of left end} \\ \text{of rectangle} \end{array} \right] dy. \qquad (7.6)$$

EXAMPLE 7.4

Find the area bounded by the curves $y = \sqrt{x + 14}$, $x = \sqrt{y}$, and $y = 0$.

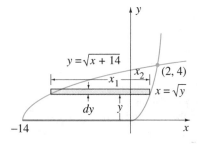

FIGURE 7.9   Area bounded by $y = \sqrt{x + 14}$, $x = \sqrt{y}$, and $y = 0$

SOLUTION   Subdivision of the region (Figure 7.9) into vertical rectangles results in two integrations: one to the left and the other to the right of the y-axis. On the other hand, throughout the required region, the area of a horizontal rectangle is

$$[\sqrt{y} - (y^2 - 14)]\,dy,$$

and therefore

$$A = \int_0^4 \left(\sqrt{y} - y^2 + 14 \right) dy = \left\{ \frac{2}{3}y^{3/2} - \frac{1}{3}y^3 + 14y \right\}_0^4 = \frac{16}{3} - \frac{64}{3} + 56 = 40.$$

In choosing between horizontal and vertical rectangles, consider two objectives:

1. Minimize the number of integrations.
2. Obtain simple definite integrals.

For instance, by choosing one type of rectangle we may obtain only one definite integral, but it may be very difficult to evaluate. If the other type of rectangle leads to two simple definite integrals, then it would be wise to choose the two simple integrals.

EXAMPLE 7.5

Find the area enclosed by the curves

$$y = \frac{x}{\sqrt{x^2 - 16}}, \qquad y = \frac{x^2}{15}, \qquad y = -x^2, \qquad x = 6, \qquad x \geq 0.$$

SOLUTION   Examination of Figure 7.10 indicates that horizontal rectangles necessitate three integrals. In addition, we would have to solve equations $y = x^2/15$ and $y = x/\sqrt{x^2 - 16}$

for x in terms of y. We therefore opt for vertical rectangles.

$$A = \int_0^5 \left[\frac{x^2}{15} - (-x^2) \right] dx + \int_5^6 \left[\frac{x}{\sqrt{x^2 - 16}} - (-x^2) \right] dx$$

$$= \frac{16}{15} \int_0^5 x^2 \, dx + \int_5^6 \left[\frac{x}{\sqrt{x^2 - 16}} + x^2 \right] dx$$

$$= \frac{16}{15} \left\{ \frac{x^3}{3} \right\}_0^5 + \left\{ \sqrt{x^2 - 16} + \frac{x^3}{3} \right\}_5^6$$

$$= \frac{16}{15} \left(\frac{125}{3} \right) + \left(\sqrt{20} + 72 \right) - \left(3 + \frac{125}{3} \right)$$

$$= 76.2.$$

FIGURE 7.10 Area bounded by four curves

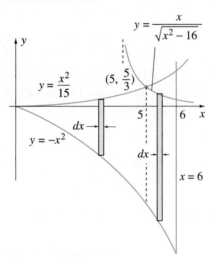

The mean value theorem for definite integrals (Theorem 6.8) states that there is a number c between a and b such that

$$\int_a^b f(x) \, dx = (b - a) f(c).$$

It has a very simple interpretation in terms of area when $f(x) \geq 0$ for $a \leq x \leq b$. Since $f(x) \geq 0$ for $a \leq x \leq b$, the definite integral may be interpreted as the area under the curve $y = f(x)$, above the x-axis, and between vertical lines at $x = a$ and $x = b$ (Figure 7.11). The right side is the area of the rectangle of width $b - a$ and height $f(c)$ shaded in Figure 7.11. The mean value theorem guarantees at least one point c between a and b for which the rectangular area is equal to the area under the curve. For the curve in Figure 7.11, there is exactly one such point c; for the curve in Figure 7.12, there are three choices for c.

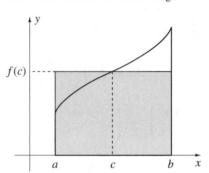

FIGURE 7.11   Area interpretation of mean value theorem for definite integrals

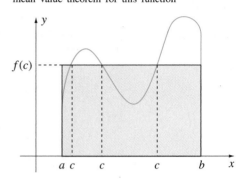

FIGURE 7.12   Three points satisfy the mean value theorem for this function

EXERCISES 7.1

In Exercises 1–16 find the area of the region bounded by the curves.

1. $y^2 = 4x$, $x^2 = 4y$

2. $y = x^3 + 8$, $y = 4x + 8$

3. $yx^2 = 4$, $y = 5 - x^2$

4. $x = y(y - 2)$, $x + y = 12$

5. $x = 4y - 4y^2$, $y = x - 3$, $y = 1$, $y = 0$

6. $y = e^{3x}$, $x = 1$, $x = 2$, $y = -x$

7. $12y = 7 - x^2$, $y = 1/(2x)$

8. $y = \sqrt{x + 4}$, $y = (x + 4)^2/8$

9. $y = (x - 1)^5$, $x = 0$, $y = 0$

10. $y = \sin x$ $(0 \le x \le \pi)$, $y = 0$

11. $y = x^5 - x$, $y = 0$

12. $x + y = 1$, $x + y = 5$, $y = 2x + 1$, $y = 2x + 6$

13. $x = \sec^2 y$, $y = 0$, $y = \pi/4$, $x = 0$

14. $xy = e$, $y = x^2$, $y = 2$ (smaller area)

15. $y = e^2(2 - x^2)$, $y = e^{2x}$, $y = e^{-2x}$, $y \ge 1$

16. $x = |y| + 1$, $x + (y - 1)^2 = 4$, $y = 0$ (above the x-axis)

∗ **17.**   (a) Find the area A of the region bounded by the curves $y = 16 - x^2$, $y = 0$, $x = 1$, and $x = 3$. In the remainder of this problem we approximate A by rectangles and show that the accuracy of the approximation increases as the number of rectangles increases. To do this you will need a large graph of the function $f(x) = 16 - x^2$ on the interval $0 \le x \le 3$.

   (b) On the graph draw two rectangles of equal width (1 unit) and with heights determined in the same way as in Figure 7.1b. If A_2 denotes the sum of the areas of these two rectangles, what is A_2? What is the error in the approximation of A by A_2? Illustrate this error on the graph.

   (c) Repeat part (b) with four rectangles all of width one-half, denoting the sum of the areas of the four rectangles by A_4. Show on the graph the extra precision of A_4 over A_2.

   (d) Repeat part (b) with eight rectangles all of equal width, denoting the sum of the areas of the eight rectangles by A_8. Show on the graph the extra precision of A_8 over A_4.

This discussion is continued in Exercise 48.

∗ **18.** Repeat Exercise 17 with the area bounded by the curves $y = x^3 + 1$, $x = 1$, $x = 3$, and $y = 0$.

In Exercises 19–22 set up (but do not evaluate) definite integral(s) for the area of the region bounded by the curves.

19. $x = 1/\sqrt{4 - y^2}$, $4x = -y^2$, $y = -1$, $y = 1$

20. $x^2 + y^2 = 4$, $x^2 + y^2 = 4x$ (interior to both)

21. $x^2 + y^2 = 4$, $x^2 + y^2 = 6x$ (interior to both)

22. $x^2 + y^2 = 16$, $x = y^2$ (smaller area)

∗ **23.** Find the area of the region bounded by $y^2 = 4ax$ and $x^2 = 4ay$, where $a > 0$ is a constant.

In Exercises 24–35 find the area of the region bounded by the curves.

∗ **24.** $y = x/\sqrt{x + 3}$, $x = 1$, $x = 6$, $y = -x^2$

∗ **25.** $x = y^2 + 2$, $x = -(y - 4)^2$, $y = -x + 4$, $y = 0$

∗ **26.** $y = x^3 - x$, $x + y + 1 = 0$, $x = \sqrt{y + 1}$

∗ **27.** $y = \left| \dfrac{x}{(x - 2)^3} \right|$, $y = 0$, $x = -1$, $x = 1$

∗ **28.** $x = 2ye^{-y^2}$, $y = x$

∗ **29.** $y = \sin^3 x$, $y = 1/8$, $0 \le x \le 2\pi$

∗ **30.** $y = \ln x^2$, $y = 1 - x^2$, $y = 1$

∗ **31.** $|x|^{1/2} + |y|^{1/2} = 1$

∗ **32.** $y^2 = x^2(4 - x^2)$

33. $y^2 = x^4(9 + x)$

34. $y^2 = x^2(x^2 - 4)$, $x = 5$

35. $(2x - y)^2 = x^3$, $x = 4$

36. The tangent line at a point on the first quadrant part of the parabola $y = 2 - x^2$ makes a triangle with the positive x- and y-axes. Find the point for which the area of the triangle is smallest.

37. Repeat Exercise 36 for the curve $y = 2 - x^4$.

In Exercises 38–45 it is necessary to use a calculator or computer to find points of intersection of the curves. Find the area of the region bounded by the curves (to three decimal places).

38. $y = x^3 + 3x^2 + 2x + 1$, $x = 0$, $y = 0$

39. $y = x^3 - 4x$, $y = 2 - x - x^2$

40. $y = x^4 - 5x^2 + 5$, $y = 0$

41. $y = e^x$, $y = 2 - x^2$

42. $y = \cos x$, $4y = x + 2$

43. $y = \dfrac{2}{x + 2}$, $y = x^3 + 3x - 1$, $x = 0$

44. $y = x^3 - 3x^2 + 4x - 2$, $x = 4 - y^2$

45. $x = y^3 - y^2 - 2y$, $x = \sqrt{2y + 1}$

46. For what values of m do the curves

$$y = \frac{x}{3x^2 + 1} \quad \text{and} \quad y = mx$$

bound a region with finite area? Find the area.

47. Find a point (a, b) on the curve $y = x/\sqrt{x^2 + 1}$ such that the region bounded by this curve, the x-axis, and the line $x = a$ has area equal to twice that of the region bounded by the curve, the y-axis, and the line $y = b$.

48. If 2^n ($n = 1, 2, \ldots$) rectangles (all of equal width) are drawn in Exercise 17 to approximate A, and A_{2^n} denotes the sum of the areas of these rectangles, show that

$$A_{2^n} = A - \left[\frac{1}{2^{n-2}} + \frac{1}{6} \left(\frac{3}{2^{n-3}} + \frac{1}{2^{2n-3}} \right) \right],$$

and hence that $\displaystyle \lim_{n \to \infty} A_{2^n} = A$.

※ 49. Show that the curves

$$y = \frac{x^3}{x^4 + 16} \quad \text{and} \quad 204y = 13x^2 - 1$$

bound three regions. Find the area of the largest region.

∗ 50. Let P be a point on the cubic curve $y = f(x) = ax^3$. Let the tangent line at P intersect $y = f(x)$ again at Q, and let A be the area of the region bounded by $y = f(x)$ and the line PQ. Let B be the area of the region defined in the same way by starting with Q instead of P. Show that B is 16 times as large as A.

∗ 51. The parabola $y = ax^2$ and the circle $x^2 + (y - r)^2 = r^2$ intersect in two points as shown in the figure below. Find the value of a that makes the area inside the parabola and below the horizontal line through the points of intersection as large as possible.

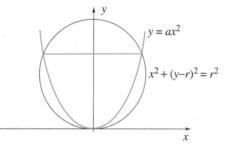

∗∗ 52. Prove that the result in Exercise 50 is valid for any cubic $y = f(x) = ax^3 + bx^2 + cx + d$.

∗∗ 53. Show that the area of the region in the figure below is

$$\frac{1}{1 + m^2} \int_{x_P}^{x_Q} [f(x) - mx - b][1 + mf'(x)] \, dx,$$

where x_P and x_Q are x-coordinates of P and Q.

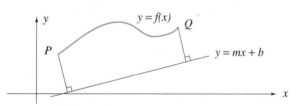

∗∗ 54. Suppose that the horizontal line $y = h$ intersects the parabola $y = ax^2 + bx + c$ ($a > 0$) in two points P and Q. Show that the area of the region so bounded is two-thirds of the length of PQ multiplied by the distance from the vertex of the parabola to the horizontal line.

∗∗ 55. A circular pasture has radius R. A cow is tied to a stake at the edge of the pasture. What length of rope permits the cow to graze on half the pasture?

7.2 Volumes of Solids of Revolution

In Section 6.2 we discussed the idea of rotating flat surfaces around coplanar lines to produce **volumes of solids of revolution**. To find the volume generated when the region in Figure 7.1a is revolved about the x-axis, we again approximate the region by n rectangles as in Figure 7.1b. If each of these rectangles is rotated around the x-axis, then n discs are formed. Since the radius

FIGURE 7.13   Approximating volume of solid of revolution by discs

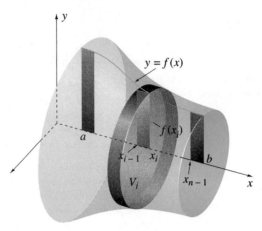

of the ith disc is $f(x_i)$ (Figure 7.13), its volume is given by

$$V_i = \pi [f(x_i)]^2 (x_i - x_{i-1}) = \pi [f(x_i)]^2 \, \Delta x_i.$$

An approximation to the required volume V is therefore

$$\sum_{i=1}^{n} V_i = \sum_{i=1}^{n} \pi [f(x_i)]^2 \, \Delta x_i.$$

If we let the number of rectangles get larger and larger, and at the same time require the widths to get smaller and smaller, it seems reasonable that we will obtain better and better approximations. Furthermore, if we take the limit as the norm of the partition approaches zero, we should get what we think is V. Since we have no formal definition for such volumes, we make our own. We define

$$V = \lim_{\|\Delta x_i\| \to 0} \sum_{i=1}^{n} \pi [f(x_i)]^2 \, \Delta x_i. \tag{7.7}$$

But we can interpret the right side of this definition as the definite integral of $\pi [f(x)]^2$ with respect to x from $x = a$ to $x = b$; that is,

$$\int_a^b \pi [f(x)]^2 \, dx = \lim_{\|\Delta x_i\| \to 0} \sum_{i=1}^{n} \pi [f(x_i)]^2 \, \Delta x_i.$$

Consequently, we can calculate the volume of the solid of revolution generated by rotating the region in Figure 7.1a around the x-axis, by means of the definite integral

$$V = \int_a^b \pi [f(x)]^2 \, dx. \tag{7.8}$$

The volume of the solid of revolution has been defined by the limit in 7.7, but for evaluation of this limit we use the definite integral in 7.8.

FIGURE 7.14 Integral for volume when area under a curve is rotated around x-axis

FIGURE 7.15 Integral for volume when area between two curves is rotated around x-axis

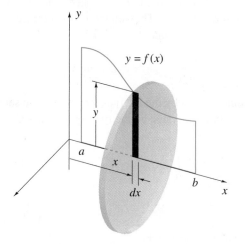

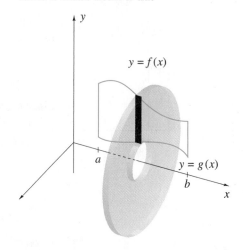

To avoid memorizing 7.8 as a formula, we use the technique introduced in Section 7.1. For the volume obtained by rotating the region in Figure 7.1a about the x-axis, we construct a rectangle of width dx at position x, as shown in either Figure 7.3 or Figure 7.14. When this rectangle is rotated around the x-axis, the volume of the disc generated is

$$\pi[f(x)]^2 \, dx,$$

where dx is the thickness of the disc and $\pi[f(x)]^2$ is the area of its flat surface. This disc is pictured as representing a large number of such discs between a and b. We find the required volume by adding volumes of all such discs and taking the limit as their widths approach zero. But this is the concept of the definite integral, so that we write for the limit-summation process

$$V = \int_a^b \pi[f(x)]^2 \, dx,$$

where limits $x = a$ and $x = b$ identify x-positions of first and last discs, respectively. This is called the **disc method** for finding the volume of a solid of revolution.

A slightly more general problem is that of finding the volume of the solid of revolution generated by rotating a region bounded by two curves and two vertical lines as shown in Figure 7.15 around the x-axis. If a representative rectangle of width dx at position x is rotated around the x-axis, the volume formed is a **washer**. Since the outer and inner radii of this washer are $f(x)$ and $g(x)$, respectively, its volume is

$$\{\pi[f(x)]^2 - \pi[g(x)]^2\} \, dx.$$

To add volumes of all such washers between a and b, and at the same time take limits as their widths approach zero, we use the definite integral

$$V = \int_a^b \{\pi[f(x)]^2 - \pi[g(x)]^2\} \, dx. \tag{7.9}$$

EXAMPLE 7.6

Volume of a sphere

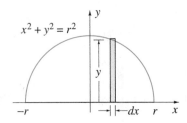

Prove that the volume of a sphere of radius r is $4\pi r^3/3$.

SOLUTION   A sphere of radius r is formed when the semicircle $x^2 + y^2 \le r^2$ $(y \ge 0)$ is rotated around the x-axis. The volume of the representative disc generated by rotating the rectangle in Figure 7.16 around the x-axis is

$$\pi y^2\, dx = \pi (r^2 - x^2)\, dx.$$

Since the volume formed by the left quarter circle is the same as for the right quarter circle, we calculate the volume generated by the right quarter and double the result:

$$V = 2 \int_0^r \pi (r^2 - x^2)\, dx = 2\pi \left\{ r^2 x - \frac{x^3}{3} \right\}_0^r = \frac{4}{3}\pi r^3.$$

EXAMPLE 7.7

Volume of a right circular cone

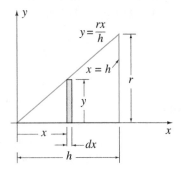

Prove that the volume of a right circular cone of base radius r and height h is $\pi r^2 h/3$.

SOLUTION   The cone can be generated by rotating the triangle in Figure 7.17 around the x-axis. Volume of the representative disc formed by rotating the rectangle shown is

$$\pi y^2\, dx = \pi \left(\frac{rx}{h} \right)^2 dx;$$

hence,

$$V = \int_0^h \frac{\pi r^2 x^2}{h^2}\, dx = \frac{\pi r^2}{h^2} \left\{ \frac{x^3}{3} \right\}_0^h = \frac{1}{3}\pi r^2 h.$$

EXAMPLE 7.8

Find the volume of the solid of revolution obtained by rotating the region bounded by the curves $y = 1 - x^2$ and $y = 4 - 4x^2$ around (a) the x-axis and (b) the line $y = -1$.

SOLUTION

(a) Let us ballpark the answer before we begin. If the rectangle in Figure 7.18a is rotated around the x-axis, it produces a cylinder with radius 4 and height 2, and therefore of

Estimation of volume of solid of revolution

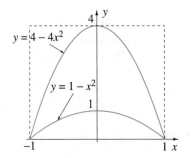

Volume when area is rotated around the x-axis

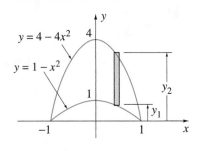

volume $\pi(4)^2(2) = 32\pi$ cubic units. We might estimate that the required volume would be about half of this. Let us find out. When the rectangle in Figure 7.18b is rotated around the x-axis, the volume of the washer formed is

$$(\pi y_2^2 - \pi y_1^2)\, dx = \pi[(4 - 4x^2)^2 - (1 - x^2)^2]\, dx = 15\pi(1 - 2x^2 + x^4)\, dx.$$

Because of the symmetry of the region, we rotate only the right half and double the result:

$$V = 2\int_0^1 15\pi(1 - 2x^2 + x^4)\, dx = 30\pi \left\{ x - \frac{2x^3}{3} + \frac{x^5}{5} \right\}_0^1 = 16\pi,$$

exactly our estimate.

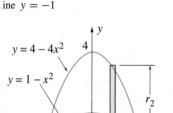

FIGURE 7.19   Volume when area is rotated around the line $y = -1$

(b) When the rectangle in Figure 7.19 is rotated around the line $y = -1$, the inner and outer radii of the washer are r_1 and r_2, respectively. Now r_1 is a length in the y-direction, and in Section 7.1 we learned that to calculate lengths in the y-direction, we take upper y minus lower y. Hence, $r_1 = (1 - x^2) - (-1) = 2 - x^2$. Similarly, $r_2 = (4 - 4x^2) - (-1) = 5 - 4x^2$. The volume of the washer is therefore

$$(\pi r_2^2 - \pi r_1^2)\, dx = [\pi(5 - 4x^2)^2 - \pi(2 - x^2)^2]\, dx = 3\pi(7 - 12x^2 + 5x^4)\, dx.$$

Consequently,

$$V = 2\int_0^1 3\pi(7 - 12x^2 + 5x^4)\, dx = 6\pi \left\{ 7x - 4x^3 + x^5 \right\}_0^1 = 24\pi.$$

We can also rotate horizontal rectangles around vertical lines to produce discs and washers as illustrated in the next two examples.

EXAMPLE 7.9

Find the volume of the solid of revolution when the region enclosed by the curves $y = \ln x$, $y = 0$, $y = 1$, and $x = 0$ is revolved around the y-axis.

SOLUTION   When the rectangle in Figure 7.20 is rotated around the y-axis, the volume of the disc formed is

$$\pi x^2\, dy = \pi(e^y)^2\, dy = \pi e^{2y}\, dy.$$

The required volume is therefore

$$V = \int_0^1 \pi e^{2y}\, dy = \pi \left\{ \frac{1}{2} e^{2y} \right\}_0^1 = \frac{\pi}{2}(e^2 - 1).$$

FIGURE 7.20   Volume when area is rotated around the y-axis

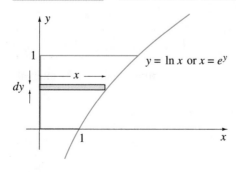

EXAMPLE 7.10

Find the volume of the solid of revolution obtained by rotating the region enclosed by the curves $y = x^2 - 1$ and $y = 0$ around the line $x = 5$.

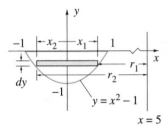

SOLUTION   When the horizontal rectangle in Figure 7.21 is rotated around the line $x = 5$, the volume of the washer formed is

$$(\pi r_2^2 - \pi r_1^2)\, dy = \pi[(5 - x_2)^2 - (5 - x_1)^2]\, dy.$$

Since x_1 and x_2 are x-coordinates of points on the curve $y = x^2 - 1$, we solve this equation for $x = \pm\sqrt{y + 1}$. Since $x_1 > 0$ and $x_2 < 0$, we set $x_1 = \sqrt{y + 1}$ and $x_2 = -\sqrt{y + 1}$. Thus, the volume of the washer can be expressed as

$$\pi\left[\left(5 + \sqrt{y + 1}\right)^2 - \left(5 - \sqrt{y + 1}\right)^2\right] dy = 20\pi\sqrt{y + 1}\, dy.$$

The required volume is therefore

$$V = \int_{-1}^{0} 20\pi\sqrt{y + 1}\, dy = 20\pi\left\{\frac{2}{3}(y + 1)^{3/2}\right\}_{-1}^{0} = \frac{40\pi}{3}.$$

Washers in Example 7.10 are not as straightforward as in previous examples. For some problems, washers are totally inappropriate. Consider, for example, rotating the region in Figure 7.22 around the x-axis. Each of the rectangles shown yields a washer with a volume formula different from the others. As a result, use of washers requires six definite integrals and only one of these is easy to set up. Determination of this volume seems to lend itself to the use of horizontal rather than vertical rectangles.

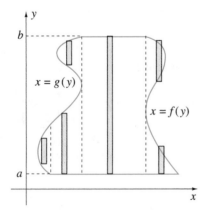

To see that this is indeed true, we divide the interval $a \le y \le b$ into n parts by the points

$$a = y_0 < y_1 < y_2 < \cdots < y_{n-1} < y_n = b.$$

In each subinterval $y_{i-1} \le y \le y_i$, we find the midpoint

$$y_i^* = \frac{y_{i-1} + y_i}{2}$$

FIGURE 7.23   Approximation of volume of
solid of revolution by cylindrical shells

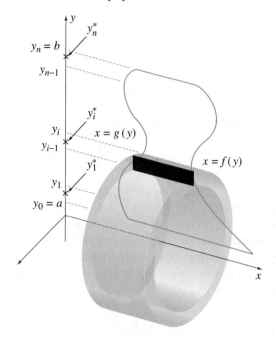

and construct a rectangle of length $f(y_i^*) - g(y_i^*)$ and width $y_i - y_{i-1}$, as shown in Figure 7.23. When this i^{th} rectangle is rotated around the x-axis, a cylindrical shell is formed. Since the length of the shell is $f(y_i^*) - g(y_i^*)$, and its inner and outer radii are y_{i-1} and y_i, respectively, its volume is

$$(\pi y_i^2 - \pi y_{i-1}^2)[f(y_i^*) - g(y_i^*)] = \pi(y_i + y_{i-1})(y_i - y_{i-1})[f(y_i^*) - g(y_i^*)]$$

$$= 2\pi y_i^*[f(y_i^*) - g(y_i^*)]\,\Delta y_i,$$

where $\Delta y_i = y_i - y_{i-1}$. When we add the volumes of all such shells, we obtain an approximation to the required volume V:

$$\sum_{i=1}^{n} 2\pi y_i^*[f(y_i^*) - g(y_i^*)]\,\Delta y_i.$$

If we let the number of rectangles get larger and larger, and at the same time require the widths to get smaller and smaller, it seems reasonable to expect that approximations will get better and better. As the norm $\|\Delta y_i\|$ approaches zero, the limit should yield what we think is V. We therefore define the required volume as

$$V = \lim_{\|\Delta y_i\| \to 0} \sum_{i=1}^{n} 2\pi y_i^*[f(y_i^*) - g(y_i^*)]\Delta y_i. \qquad (7.10)$$

But the right side of this definition is the definite integral of $2\pi y[f(y) - g(y)]$ with respect to y from $y = a$ to $y = b$; that is,

$$\int_a^b 2\pi y[f(y) - g(y)]\,dy = \lim_{\|\Delta y_i\| \to 0} \sum_{i=1}^{n} 2\pi y_i^*[f(y_i^*) - g(y_i^*)]\Delta y_i.$$

Consequently, the volume of the solid of revolution generated by rotating the region in Figure 7.22 around the x-axis is given by the definite integral

$$V = \int_a^b 2\pi y[f(y) - g(y)]\,dy. \tag{7.11}$$

In practice we develop the integral in 7.11 by drawing a rectangle of width dy at position y as shown in Figure 7.24a. When this rectangle is rotated around the x-axis, the volume of the cylindrical shell generated is approximately

$$2\pi y[f(y) - g(y)]\,dy.$$

We obtain this by picturing the shell as being cut along the rectangle and opened up into a slab with dimensions dy, $f(y) - g(y)$, and $2\pi y$ as in Figure 7.24b. Thickness dy of the shell corresponds to the thickness of the slab; length $f(y) - g(y)$ of the shell corresponds to that of the slab; and inner circumference $2\pi y$ of the shell corresponds to the width of the slab. If we now add the volumes of all such cylindrical shells, and at the same time take the limit as their widths approach zero, we obtain the required volume. But this limit summation defines the definite integral

$$\int_a^b 2\pi y[f(y) - g(y)]\,dy.$$

It follows that

$$V = \int_a^b 2\pi y[f(y) - g(y)]\,dy.$$

The method described is called the **cylindrical shell method** for finding the volume of a solid of revolution.

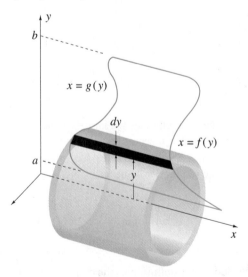

FIGURE 7.24a   Using differentials to set up integral for cylindrical shells

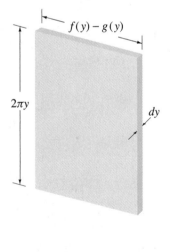

FIGURE 7.24b   Volume of cylindrical shell of thickness dy

EXAMPLE 7.11

Find the volume of the solid of revolution when the region enclosed by $x = 2y - y^2$ and the y-axis is revolved around the x-axis.

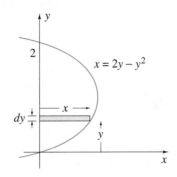

FIGURE 7.25a   Volume when area is rotated around the x-axis

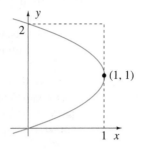

FIGURE 7.25b   Estimate of volume of solid of revolution

SOLUTION   The volume of the cylindrical shell formed by rotating the rectangle in Figure 7.25a around the x-axis is approximately

$$(2\pi y)(x)\,dy = 2\pi y(2y - y^2)\,dy = 2\pi(2y^2 - y^3)\,dy.$$

Hence,

$$V = \int_0^2 2\pi(2y^2 - y^3)\,dy = 2\pi\left\{\frac{2y^3}{3} - \frac{y^4}{4}\right\}_0^2 = 2\pi\left(\frac{16}{3} - 4\right) = \frac{8\pi}{3}.$$

We can check that this is reasonable by rotating the rectangle in Figure 7.25b around the x-axis. The cylinder so generated has volume $\pi(2)^2(1) = 4\pi$. That the above volume is two-thirds of this is acceptable.

EXAMPLE 7.12

Use cylindrical shells to calculate the volume in Example 7.10.

SOLUTION   When we rotate the rectangle in Figure 7.26 around the line $x = 5$, the volume of the cylindrical shell is approximately

$$2\pi(5 - x)(0 - y)\,dx = 2\pi(5 - x)(1 - x^2)\,dx = 2\pi(x^3 - 5x^2 - x + 5)\,dx.$$

Total volume is therefore

$$V = \int_{-1}^1 2\pi(x^3 - 5x^2 - x + 5)\,dx = 2\pi\left\{\frac{x^4}{4} - \frac{5x^3}{3} - \frac{x^2}{2} + 5x\right\}_{-1}^1$$

$$= 2\pi\left(\frac{1}{4} - \frac{5}{3} - \frac{1}{2} + 5\right) - 2\pi\left(\frac{1}{4} + \frac{5}{3} - \frac{1}{2} - 5\right) = \frac{40\pi}{3}.$$

This is the result obtained by the washer method in Example 7.10.

FIGURE 7.26   Volume when area is rotated around the line $x = 5$

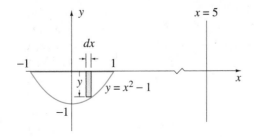

Even though the region in this example is symmetric about the y-axis, the volume generated by the right half is less than that generated by the left. For this reason we cannot integrate from $x = 0$ to $x = 1$ and double the result. If rotation were performed about the x-axis, we could indeed integrate over either half of the interval and double. Finally, for rotation about the y-axis, we would integrate over only one-half of the interval and neglect the other half in order to eliminate duplications.

The volume in Example 7.12 was also calculated with washers in Example 7.10. The volume in Example 7.11 could also be done with washers, but not so easily. Try it. Washers cannot be used in the following example.

EXAMPLE 7.13

Figure 7.27 shows a plot of the function $f(x) = x^{-1} \sin x$ on the interval $\pi/2 \le x \le \pi$. Find the volume of the solid of revolution when the area bounded by this curve and the lines $x = \pi/2$ and $y = 0$ is rotated around the y-axis.

SOLUTION   When we rotate the rectangle shown around the y-axis, the volume of the representative cylindrical shell is approximately

$$2\pi (x) \left(\frac{\sin x}{x} \right) dx = 2\pi \sin x \, dx.$$

The total volume is therefore

$$V = \int_{\pi/2}^{\pi} 2\pi \sin x \, dx = 2\pi \left\{ -\cos x \right\}_{\pi/2}^{\pi} = 2\pi.$$

FIGURE 7.27   Volume when area is rotated around the y-axis

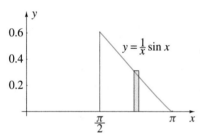

$y = \frac{1}{x} \sin x$

Equations 7.9 and 7.11 provide two methods for calculating volumes of solids of revolution: washers and shells. Where applicable, both methods give the same results; as they should. We illustrate this for the region bounded by the curves in Figure 7.28. As indicated in the figure, each curve defines y as a function of x, and x as a function of y. When this region is rotated around the x-axis, cylindrical shells can be used to calculate the volume of the solid of revolution,

$$V = \int_c^d 2\pi y[q(y) - p(y)] \, dy = 2\pi \int_c^d yq(y) \, dy - 2\pi \int_c^d yp(y) \, dy. \qquad (7.12)$$

FIGURE 7.28   Demonstration that washers and cylindrical shells give the same volume

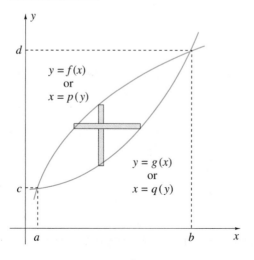

$y = f(x)$
or
$x = p(y)$

$y = g(x)$
or
$x = q(y)$

Suppose we make the change of variable $x = q(y)$ in the first integral. When this equation is solved for y in terms of x, the result is $y = g(x)$, and therefore $dy = g'(x)\,dx$. Because $x = a$ when $y = c$ and $x = b$ when $y = d$, we obtain

$$\int_c^d yq(y)\,dy = \int_a^b g(x)xg'(x)\,dx.$$

Now the product rule for differentiation gives

$$\frac{d}{dx}\left\{x[g(x)]^2\right\} = 2xg(x)g'(x) + [g(x)]^2,$$

and therefore

$$g(x)xg'(x) = \frac{1}{2}\frac{d}{dx}\left\{x[g(x)]^2\right\} - \frac{1}{2}[g(x)]^2.$$

So

$$\begin{aligned}
\int_c^d yq(y)\,dy &= \int_a^b \left\{\frac{1}{2}\frac{d}{dx}\left\{x[g(x)]^2\right\} - \frac{1}{2}[g(x)]^2\right\}dx\\
&= \left\{\frac{1}{2}x[g(x)]^2\right\}_a^b - \frac{1}{2}\int_a^b [g(x)]^2\,dx\\
&= \frac{1}{2}b[g(b)]^2 - \frac{1}{2}a[g(a)]^2 - \frac{1}{2}\int_a^b [g(x)]^2\,dx\\
&= \frac{1}{2}bd^2 - \frac{1}{2}ac^2 - \frac{1}{2}\int_a^b [g(x)]^2\,dx.
\end{aligned}$$

Similarly, the change of variable $x = p(y)$ on the second integral in 7.12 leads to

$$\int_c^d yp(y)\,dy = \frac{1}{2}bd^2 - \frac{1}{2}ac^2 - \frac{1}{2}\int_a^b [f(x)]^2\,dx.$$

Substitution of these into 7.12 gives

$$\begin{aligned}
V &= 2\pi\left\{\frac{1}{2}bd^2 - \frac{1}{2}ac^2 - \frac{1}{2}\int_a^b [g(x)]^2\,dx\right\} - 2\pi\left\{\frac{1}{2}bd^2 - \frac{1}{2}ac^2 - \frac{1}{2}\int_a^b [f(x)]^2\,dx\right\}\\
&= -\pi\int_a^b [g(x)]^2\,dx + \pi\int_a^b [f(x)]^2\,dx\\
&= \int_a^b \left\{\pi[f(x)]^2 - \pi[g(x)]^2\right\}dx.
\end{aligned}$$

But this is the integral obtained when the washer method is used to find the volume.

In Exercises 1–12 use the disc or washer method to find the volume of the solid of revolution obtained by rotating the region bounded by the curves about the line.

1. $x^2 + y^2 = 36$,   about $y = 0$

2. $y^2 = 5 - x$, $x = 0$,   about $x = 0$

3. $y = x^2 + 4$, $y = 2x^2$,   about $y = 0$

4. $x - y^2 = 16$, $x = 20$,   about $x = 0$

5. $x - 1 = y^2$, $x = 5$,   about $x = 1$

6. $x + y + 1 = 0$, $2y = x - 2$, $y = 0$,   about $y = 0$

7. $y = 4x^2 - 4x$, $y = x^3$,   about $y = -2$

8. $x = 2y - y^2 - 2$, $x = -5$,   about $x = 0$

9. $y^2 = 5 - x$, $x = 0$,   about $x = 6$

10. $y = x^2 - 2x$, $y = 2x - x^2$,   about $y = 2$

11. $y = \csc x$, $y = 0$, $x = \pi/4$, $x = 3\pi/4$,   about $y = 0$

12. $y = \ln(x + 1)$, $y = 1$, $x = 0$,   about $x = 0$

In Exercises 13–24 use the cylindrical shell method to find the volume of the solid of revolution obtained by rotating the region bounded by the curves about the line.

13. $y = 1 - x^3$, $x = 0$, $y = 0$,   about $x = 0$

14. $y = -\sqrt{4 - x}$, $x = 0$, $y = 0$,   about $y = 0$

15. $y = (x - 1)^2$, $y = 1$,   about $x = 0$

16. $x + y = 4$, $y = 2\sqrt{x - 1}$, $y = 0$,   about $y = 0$

17. $y = 3x - x^2$, $y = x^2 - 3x$,   about $x = 4$

18. $y = 2 - |x|$, $y = 0$,   about $y = -1$

19. $x = y^3$, $y = \sqrt{2 - x}$, $y = 0$,   about $y = 1$

20. $y = x^2$, $y = -x^2$, $x = -1$,   about $x = -1$

21. $y = -\sqrt{9 - x}$, $x = 0$, $y = 0$,   about $y = 0$

22. $y = x$, $xy = 9$, $x + y = 10$, $(x \geq y)$,   about $x = 0$

23. $y = 0$, $(x + 1)y = \sin x$, $0 \leq x \leq 2\pi$,   about $x = -1$

24. $y = 10 - x^2$, $x^2 y = 9$,   about $y = 0$

In Exercises 25–32 use the most appropriate method to find the volume of the solid of revolution obtained by rotating the region bounded by the curves about the line.

* **25.** $(x^2 + 1)^2 y = 4$, $y = 1$,   about $x = 0$

* **26.** $y = (x - 1)^2 - 4$, $5y = 12x$, $x = 0$, $(x \geq 0)$,   about $x = 0$

* **27.** $x^2 - y^2 = 5$, $9y = x^2 + 9$, $9y + x^2 + 9 = 0$, $(-3 \leq x \leq 3)$,   about $x = 0$

* **28.** $y = |x^2 - 1|$, $x = -2$, $x = 2$, $y = -1$,   about $y = -1$

* **29.** $y = x^2 - 2$, $y = 0$,   about $y = -1$

* **30.** $x = \sqrt{4 + 12y^2}$, $x - 20y = 24$, $y = 0$,   about $y = 0$

* **31.** $y = (x + 1)^{1/4}$, $y = -(x + 1)^2$, $x = 0$,   about $x = 0$

* **32.** $y = x^4 - 3$, $y = 0$,   about $y = -1$

In Exercises 33–36 a calculator or computer is needed to find the points of intersection of the curves. Find the volume of the solid of revolution when the region bounded by the curves is rotated about the line, correct to three decimal places.

* **33.** $y = x^3 - x$, $y = \sqrt{x}$,   about $x = 0$

* **34.** $y = e^{-2x}$, $y = 4 - x^2$,   about $y = -1$

* **35.** $y = \dfrac{1}{\sqrt{x - 1}}$, $y = 16 - x^2$,   about $y = -1$

* **36.** $y = \sqrt{4 - x}$, $y = x^3 + 1$, $y = 0$,   about $y = 0$

* **37.** A tapered rod of length L has circular cross-sections. If the radii of its ends are a and b, what is the volume of the rod?

* **38.** During one revolution an airplane propeller displaces an amount of air that can be calculated as a volume of a solid of revolution. If the region yielding the volume is that bounded by the curves $64x = y(y - 4)$ and $64x = y(4 - y)$, and is rotated about the x-axis, calculate the volume of air displaced.

* **39.**   (a) If a sphere of radius r is sliced a distance h from its centre, show that the volume of the smaller piece is

$$V = \frac{\pi}{3}(r - h)^2(2r + h).$$

   (b) Use the result in part (a) to find the ratio of h to r in order that the smaller piece will have volume equal to one-third of the sphere. Give your answer to four decimal places.

* **40.** An embankment is to be built around the circular wading pool in figure (a) below. Figure (b) shows a cross-section of the embankment.

   (a) Find a cubic polynomial $y = ax^3 + bx^2 + cx + d$ to fit the three points $(0, 0)$, $(4, 2)$, and $(6, 0)$.

   (b) Determine the amount of fill required to build the embankment.

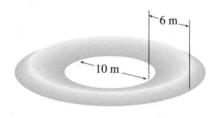

(a)

(b)

1. Find the volume of the donut obtained by rotating the circle in the figure below about the y-axis.

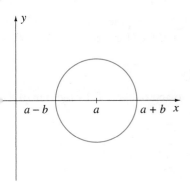

2. A cylindrical hole is bored through the centre of a sphere, the length of the hole being L. Show that no matter what the radius of the sphere, the volume of the sphere that remains is always the same and equal to the volume of a sphere of diameter L.

3. Water half fills a cylindrical pail of radius a and height L. When the pail is rotated about its axis of symmetry with angular speed ω (figure following), the surface of the water assumes a parabolic shape, the cross-section of which is given by

$$y = H + \frac{\omega^2 x^2}{2g},$$

where $g > 0$ is the acceleration due to gravity and H is a constant. Find the speed ω, in terms of L, a, and g, at which water spills over the top.

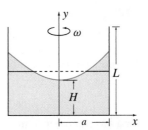

** **44.** A right circular cone of height H and base radius r has its vertex at the centre of a sphere of radius R ($R < H$). Find that part of the volume of the sphere inside the cone.

** **45.** Devise a way to calculate the volume of the solid of revolution when the region in the first quadrant bounded by

$$y = 1 - x^2, \quad x = 0, \quad y = 0$$

is rotated about the line $y = x + 1$. *Hint:* Distance formula 1.16.

** **46.** A sphere of ice cream is to be placed in a cone of height 1 unit (figure below). What radius of the sphere gives the most volume of ice cream inside the cone (as opposed to above the cone) for a cone with base angle 2θ?

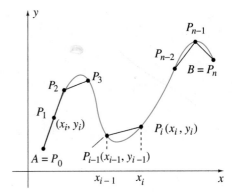

7.3 Lengths of Curves

Formula 1.10 defines the length of the straight-line segment joining two points (x_1, y_1) and (x_2, y_2). The formula $s = r\theta$ gives the length of the arc of a circle of radius r subtended by an angle θ at the centre of the circle. In this section we derive a result that will theoretically enable us to find the length of any curve.

FIGURE 7.29a Finding the length of a curve

FIGURE 7.29b Approximating the length of a curve by straight line segments

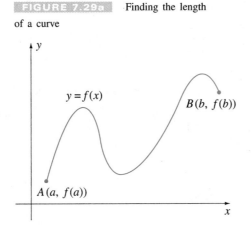

Consider finding the **length of the curve** $C : y = f(x)$ in Figure 7.29a joining points $A\big(a, f(a)\big)$ and $B\big(b, f(b)\big)$. To find its length L we begin by approximating C with a series

of straight-line segments. Specifically, we choose $n - 1$ consecutive points on C between A and B,

$$A = P_0, P_1, P_2, \ldots, P_{n-1}, P_n = B,$$

and join each P_{i-1} to P_i $(i = 1, \ldots, n)$ by means of a straight-line segment, as in Figure 7.29b. If coordinates of P_i are denoted by (x_i, y_i), then the length of the line segment joining P_{i-1} and P_i is

$$\|P_{i-1}P_i\| = \sqrt{(x_i - x_{i-1})^2 + (y_i - y_{i-1})^2}.$$

For a large number of these line segments, it is reasonable to approximate L by the sum of the lengths of the segments:

$$\sum_{i=1}^{n} \|P_{i-1}P_i\| = \sum_{i=1}^{n} \sqrt{(x_i - x_{i-1})^2 + (y_i - y_{i-1})^2}.$$

In fact, as we increase n and at the same time decrease the length of each segment, we expect the approximation to become more and more accurate. We therefore define

$$L = \lim_{\|\Delta x_i\| \to 0} \sum_{i=1}^{n} \|P_{i-1}P_i\| = \lim_{\|\Delta x_i\| \to 0} \sum_{i=1}^{n} \sqrt{(x_i - x_{i-1})^2 + (y_i - y_{i-1})^2}, \quad (7.13a)$$

where $\Delta x_i = x_i - x_{i-1}$. When we note that $y_{i-1} = f(x_{i-1})$ and $y_i = f(x_i)$, we have

$$L = \lim_{\|\Delta x_i\| \to 0} \sum_{i=1}^{n} \sqrt{(\Delta x_i)^2 + [f(x_i) - f(x_{i-1})]^2}$$

$$= \lim_{\|\Delta x_i\| \to 0} \sum_{i=1}^{n} \sqrt{1 + \left[\frac{f(x_i) - f(x_{i-1})}{\Delta x_i}\right]^2} \, \Delta x_i. \quad (7.13b)$$

In order for 7.13b to be a useful definition from the point of view of calculation, we must find a convenient way to evaluate the limit summation. To do this we assume that $f'(x)$ exists at every point in the interval $a \leq x \leq b$, and apply the mean value theorem to $f(x)$ on each subinterval $x_{i-1} \leq x \leq x_i$ (Figure 7.30). The theorem guarantees the existence of at least one point x_i^* between x_{i-1} and x_i such that

$$f'(x_i^*) = \frac{f(x_i) - f(x_{i-1})}{x_i - x_{i-1}} = \frac{f(x_i) - f(x_{i-1})}{\Delta x_i}.$$

Consequently, using these points x_i^*, we can express the length of C in the form

$$L = \lim_{\|\Delta x_i\| \to 0} \sum_{i=1}^{n} \sqrt{1 + [f'(x_i^*)]^2} \, \Delta x_i. \quad (7.14)$$

FIGURE 7.30   Mean value theorem applied to i^{th} line segment

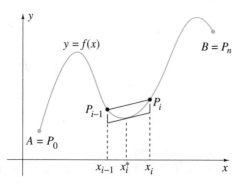

But the right side of this equation is the definition of the definite integral of the function $\sqrt{1 + [f'(x)]^2}$ with respect to x from $x = a$ to $x = b$:

$$\int_a^b \sqrt{1 + [f'(x)]^2}\, dx = \lim_{\|\Delta x_i\| \to 0} \sum_{i=1}^n \sqrt{1 + [f'(x_i^*)]^2}\, \Delta x_i.$$

Thus we can calculate the length of C by the definite integral

$$L = \int_a^b \sqrt{1 + \left(\frac{dy}{dx}\right)^2}\, dx. \tag{7.15}$$

In this derivation we assumed that $f'(x)$ was defined at each point in the interval $a \le x \le b$, necessary for the mean value theorem to apply. But in order to guarantee existence of the definite integral in 7.15, Theorem 6.2 requires continuity of the integrand $\sqrt{1 + (dy/dx)^2}$. Consequently, to ensure that the length of a curve can be calculated by means of 7.15, we assume that $f(x)$ has a continuous first derivative on the interval $a \le x \le b$. When we study *improper integrals* in Section 7.10, we shall be able to weaken the continuity requirement.

EXAMPLE 7.14

Find the length of the curve $y = x^{3/2}$ from $(1, 1)$ to $(2, 2\sqrt{2})$ (Figure 7.31).

FIGURE 7.31   Length of curve $y = x^{3/2}$ from $x = 0$ to $x = 2$

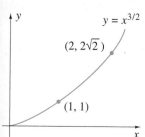

$y = x^{3/2}$

$(2, 2\sqrt{2})$

$(1, 1)$

SOLUTION   Since $dy/dx = (3/2)x^{1/2}$, equation 7.15 gives

$$L = \int_1^2 \sqrt{1 + \left(\frac{3}{2}x^{1/2}\right)^2}\, dx = \int_1^2 \sqrt{1 + \frac{9}{4}x}\, dx$$

$$= \left\{ \frac{8}{27}\left(1 + \frac{9}{4}x\right)^{3/2} \right\}_1^2 = \frac{8}{27}\left[\left(\frac{11}{2}\right)^{3/2} - \left(\frac{13}{4}\right)^{3/2}\right] = \frac{1}{27}(22^{3/2} - 13^{3/2}).$$

The result in equation 7.15 is clearly useful when the equation of the curve is expressed in the form $y = f(x)$. If the curve is defined in the form $x = g(y)$ (Figure 7.32), a similar analysis gives

$$L = \int_c^d \sqrt{1 + \left(\frac{dx}{dy}\right)^2}\, dy. \tag{7.16}$$

FIGURE 7.32   Length of curve with equation in form $x = g(y)$

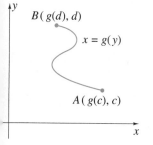

$B(g(d), d)$

$x = g(y)$

$A(g(c), c)$

For example, we may write the equation of the curve in Example 7.14 in the form $x = y^{2/3}$, in which case $dx/dy = (2/3)y^{-1/3}$, and

$$L = \int_1^{2\sqrt{2}} \sqrt{1 + \left(\frac{2}{3}y^{-1/3}\right)^2}\, dy = \int_1^{2\sqrt{2}} \sqrt{1 + \frac{4}{9}y^{-2/3}}\, dy$$

$$= \int_1^{2\sqrt{2}} \sqrt{\frac{4 + 9y^{2/3}}{9y^{2/3}}}\, dy = \int_1^{2\sqrt{2}} \frac{\sqrt{4 + 9y^{2/3}}}{3y^{1/3}}\, dy.$$

If we set $u = 4 + 9y^{2/3}$, then $du = 6y^{-1/3}dy$, and

$$L = \frac{1}{3}\int_{13}^{22} \sqrt{u}\, \frac{du}{6} = \frac{1}{18}\left\{\frac{2}{3}u^{3/2}\right\}_{13}^{22} = \frac{1}{27}(22^{3/2} - 13^{3/2}).$$

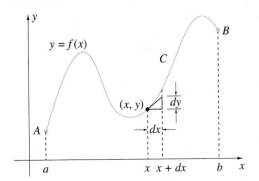

FIGURE 7.33   Using differentials to set up integral for length of a curve

Integrals 7.15 and 7.16 can both be interpreted geometrically. This interpretation will be useful when we study parametric equations of curves in Section 9.1 and line integrals in Chapter 14. Figure 7.33 shows the curve C of Figure 7.29a. At a point (x, y) on C, we draw the tangent line to the curve. If dx is a short length along the x-axis at position x, the length of that part of the tangent line between vertical lines at x and $x + dx$ is given by

$$\sqrt{(dx)^2 + (dy)^2}.$$

This length along the tangent line closely approximates the length of the curve between the same two vertical lines, the approximation being more accurate the shorter the length dx. If we picture a large number of these tangential line segments between a and b, we can find the total length of C by adding together all such lengths and taking the limit as each approaches zero. But this process is represented by the definite integral, and we therefore write

$$L = \int_{x=a}^{x=b} \sqrt{(dx)^2 + (dy)^2} \quad \text{or} \quad L = \int_{y=c}^{y=d} \sqrt{(dx)^2 + (dy)^2}. \qquad (7.17)$$

The first integral is chosen when it is more convenient to integrate with respect to x, in which case dx is taken outside the square root. We get

$$L = \int_{a}^{b} \sqrt{1 + \left(\frac{dy}{dx}\right)^2}\, dx,$$

which is equation 7.15. The second corresponds to a more convenient integral with respect to y, in which case dy is taken outside the square root. We obtain

$$L = \int_{c}^{d} \sqrt{1 + \left(\frac{dx}{dy}\right)^2}\, dy,$$

which is equation 7.16.

EXAMPLE 7.15

Find the length of the curve $24xy = x^4 + 48$ between $(2, 4/3)$ and $(3, 43/24)$.

SOLUTION   It is easier to express y in terms of x (rather than x in terms of y),

$$y = \frac{x^4 + 48}{24x} = \frac{x^3}{24} + \frac{2}{x}.$$

We now use the fact that small lengths along the curve are approximated by

$$\sqrt{(dx)^2 + (dy)^2} = \sqrt{1 + \left(\frac{dy}{dx}\right)^2}\, dx = \sqrt{1 + \left(\frac{x^2}{8} - \frac{2}{x^2}\right)^2}\, dx$$

$$= \sqrt{1 + \frac{x^4}{64} - \frac{1}{2} + \frac{4}{x^4}}\, dx = \sqrt{\frac{1}{64x^4}(x^8 + 32x^4 + 256)}\, dx$$

$$= \frac{1}{8x^2}\sqrt{(x^4 + 16)^2}\, dx = \frac{x^4 + 16}{8x^2}\, dx.$$

The total length of the curve is therefore (Figure 7.34)

$$L = \int_2^3 \frac{x^4 + 16}{8x^2}\, dx = \frac{1}{8}\int_2^3 \left(x^2 + \frac{16}{x^2}\right) dx = \frac{1}{8}\left\{\frac{x^3}{3} - \frac{16}{x}\right\}_2^3 = \frac{9}{8}.$$

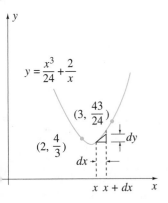

FIGURE 7.34   Length of curve $24xy = x^4 + 48$ from $x = 2$ to $x = 3$

$y = \dfrac{x^3}{24} + \dfrac{2}{x}$

$\left(3, \dfrac{43}{24}\right)$

$\left(2, \dfrac{4}{3}\right)$

dx dy

$x\ \ x + dx$ x

In Exercises 1–10 find the length of the curve.

1. $8x^2 = y^3$ from $(1, 2)$ to $(2\sqrt{2}, 4)$

2. $3y = 2(x^2 + 1)^{3/2}$ from $(-2, 10\sqrt{5}/3)$ to $(-1, 4\sqrt{2}/3)$

3. $x = 2(y - 2)^{3/2}$ from $(0, 2)$ to $(2, 3)$

4. $y = (x - 1)^{3/2}$ from $(2, 1)$ to $(10, 27)$

5. $y = \dfrac{x^3}{4} + \dfrac{1}{3x}$ from $(1, 7/12)$ to $(2, 13/6)$

6. $y = (e^x + e^{-x})/2$ between the lines $x = 0$ and $x = 1$

7. $y = \dfrac{x^4}{4} + \dfrac{1}{8x^2}$ from $(2, 129/32)$ to $(1/2, 33/64)$

8. $36xy = x^4 + 108$ from $(3, 7/4)$ to $(4, 91/36)$

9. $x = \dfrac{y^7}{20} + \dfrac{1}{7y^5}$ between the lines $y = -1$ and $y = -2$

10. $y = x^5/5 + 1/(12x^3)$ from $(1, 17/60)$ to $(2, 3077/480)$

In Exercises 11–20 set up (but do not evaluate) a definite integral for the length of the curve.

11. $y = x^2$ from $(0, 0)$ to $(1, 1)$

12. $y = 3x^2 - 4x$ from $(1, -1)$ to $(2, 4)$

13. $x^2 - y^2 = 1$ from $(1, 0)$ to $(2, \sqrt{3})$

14. $x^2 - y^2 = 1$ from $(2, -\sqrt{3})$ to $(3, 2\sqrt{2})$

15. $y = \sin x$ from $(0, 0)$ to $(\pi, 0)$

16. $y = \ln(\cos x)$ between the lines $x = 0$ and $x = \pi/4$

17. $y = \ln x$ from $(1, 0)$ to any other point on the curve

18. $8y^2 = x^2(1 - x^2)$ (complete length)

19. $x = y^2 - 2y$ from $(0, 0)$ to $(0, 2)$

20. $x^2/4 + y^2/9 = 1$ (complete length)

21. Find the length of the curve $3x = \sqrt{y}(y - 3)$ from $(-2/3, 1)$ to $(2/3, 4)$.

22. Find the length of the curve $x^{2/3} + y^{2/3} = 1$.

23. If n is any number other than 1 or -1, and $0 < a < b$, find the length of the curve

$$y = \frac{x^{n+1}}{n + 1} + \frac{1}{4(n - 1)x^{n-1}}$$

between $x = a$ and $x = b$.

24. If n is any number greater than $1/2$, and $0 < a < b$, find the length of the curve

$$y = \frac{x^{2n+1}}{4(2n - 1)} + \frac{1}{(2n + 1)x^{2n-1}}$$

between $x = a$ and $x = b$.

7.4 Work

When a body moves a distance d along a straight line l under the action of a constant force F which acts in the same direction as the motion (Figure 7.35), the **work** done on the body by F is defined as

$$W = Fd. \tag{7.18}$$

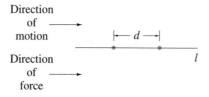

FIGURE 7.35 Definition of work done by a constant force along a straight line

FIGURE 7.36 Approximating work done by a variable force

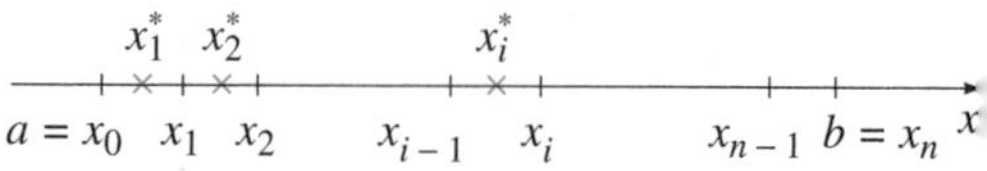

Forces often vary either in magnitude or direction, or both, and when such forces act on the body, calculation of the work done is not as simple as using equation 7.18. In this section we use definite integrals to calculate work done by forces that always act along the direction of motion, but do not have constant magnitude. Specifically, let us consider a particle that moves along the x-axis from $x = a$ to $x = b$, where $b > a$, under the action of some force. Suppose the force always acts in the positive x-direction, but its size, which we denote by $F(x)$, is not constant along the x-axis. To find the work done by the force as the particle moves from $x = a$ to $x = b$, we cannot simply multiply force by distance. [Where would we evaluate $F(x)$?] Instead, we divide the interval $a \le x \le b$ into n subintervals by points (Figure 7.36):

$$a = x_0 < x_1 < x_2 < \cdots < x_{n-1} < x_n = b.$$

In each subinterval $x_{i-1} \le x \le x_i$, we choose a point x_i^*. If lengths of subintervals are small and $F(x)$ is continuous, then $F(x)$ does not vary greatly over any given subinterval. In such a situation, we may approximate $F(x)$ by a constant force $F(x_i^*)$ on each subinterval $x_{i-1} \le x \le x_i$. It follows that work done by the force over the ith subinterval is approximately

$$F(x_i^*)(x_i - x_{i-1}) = F(x_i^*)\, \Delta x_i.$$

Furthermore, an approximation to the total work done by the force as the particle moves from $x = a$ to $x = b$ is

$$\sum_{i=1}^{n} F(x_i^*)\, \Delta x_i.$$

As n becomes larger and each Δx_i approaches zero, this approximation becomes better, and we therefore define

$$W = \lim_{\|\Delta x_i\| \to 0} \sum_{i=1}^{n} F(x_i^*)\, \Delta x_i. \tag{7.19}$$

But this limit may also be interpreted as the definite integral of $F(x)$ with respect to x from $x = a$ to $x = b$. Consequently, W may be calculated with the definite integral

$$W = \int_a^b F(x)\, dx. \tag{7.20}$$

It is simple enough to interpret parts of the integral in equation 7.20, and in so doing we begin to *feel* the definite integral at work. The integrand $F(x)$ is the force at position x, and dx is a small distance along the x-axis. The product $F(x)\, dx$ is therefore interpreted as the (approximate) work done by the force along dx. The definite integral then adds over all dx's, beginning at $x = a$ and ending at $x = b$, to give total work.

What is important about equation 7.20 is not the particular form of the definite integral, but the fact that work done by a force can be evaluated by means of a definite integral. As we solve work problems in this section, we find that the form of the definite integral varies considerably from problem to problem, but the underlying fact remains that each problem is solved with a definite integral.

With the exception of work problems that involve emptying tanks (Example 7.18), we recommend that you always make a diagram illustrating the physical setup at some intermediate stage between start and finish. Determine forces at this position in order to set up the work integral.

In the derivation of 7.20 we assumed continuity of $F(x)$, and this guarantees existence of the definite integral.

EXAMPLE 7.16

Find the work necessary to expand a spring from a stretch of 5 cm to a stretch of 15 cm if a force of 200 N stretches it 10 cm.

SOLUTION   Let the spring be stretched in the positive x-direction, and let $x = 0$ correspond to the free end of the spring in the unstretched position in the top half of Figure 7.37. In the bottom half of this figure we have shown the spring stretched to an intermediate position.

FIGURE 7.37   Work to stretch a spring

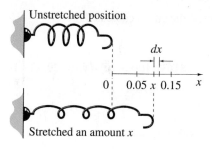

According to Hooke's law, when the spring is stretched an amount x, the restoring force in the spring is proportional to x:

$$F_s = -kx,$$

where $k > 0$ is a constant. The negative sign indicates that the force is in the negative x-direction. Since $F_s = -200$ N when $x = 0.10$ m,

$$-200 = -k(0.10),$$

and it follows that $k = 2000$ N/m. The force required to counteract the restoring force of the spring when it is stretched an amount x is therefore

$$F(x) = 2000x.$$

Work done by this force in stretching the spring from position x a further amount dx is approximately

$$2000x \, dx,$$

and hence total work to increase the stretch from 5 cm to 15 cm is

$$W = \int_{0.05}^{0.15} 2000x \, dx = 2000 \left\{ \frac{x^2}{2} \right\}_{0.05}^{0.15} = 20 \text{ J}.$$

We could check on whether this answer is reasonable as follows. The minimum force during stretching is at the beginning when the stretch is 5 cm where $F = 2000(0.05) = 100$ N. Maximum force is at 15 cm where $F = 2000(0.15) = 300$ N. Work should be between $100(0.1) = 10$ J and $300(0.1) = 30$ J. In essence, we are applying Theorem 6.6 from Section 6.4.

EXAMPLE 7.17

A cable hangs vertically from the top of a building so that a length of 100 m, having a mass of 300 kg, is hanging from the edge of the roof. What work is required to lift the entire cable to the top of the building?

SOLUTION   When the cable has been lifted to an intermediate point where its lower end is y metres above its original position (Figure 7.38), the length of cable still hanging is $100 - y$ m. Since each metre of cable has mass 3 kg, the mass of $100 - y$ m of cable is $3(100 - y)$ kg. It follows that the force that must be exerted to overcome gravity on this much cable is $9.81(3)(100 - y)$ N. The work that this force does in raising the end of the cable an additional amount dy is approximately

$$9.81(3)(100 - y)\, dy,$$

and total work to raise the entire cable is therefore

$$W = \int_0^{100} 9.81(3)(100 - y)\, dy = 29.43\left\{100y - \frac{y^2}{2}\right\}_0^{100} = 147\,150 \text{ J}.$$

FIGURE 7.38   Work to raise a cable

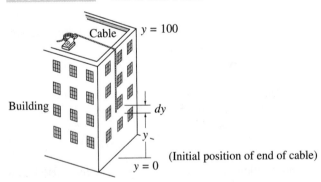

Building dy

(Initial position of end of cable)

EXAMPLE 7.18

A tank in the form of an inverted right circular cone of depth 10 m and radius 4 m is full of water. Find the work required to pump the water to a level 1 m above the top of the tank.

SOLUTION   A cross-section of the tank is shown in Figure 7.39. Suppose we approximate water in the tank with circular discs formed by rotating the rectangles shown around the y-axis

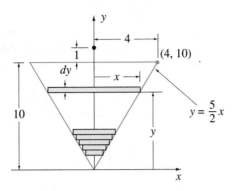

FIGURE 7.39   Work to empty a tank

The force of gravity on the representative disc at position y is its volume $\pi x^2 \, dy$, multiplied by the density of water (1000 kg/m^3), multiplied by the acceleration due to gravity (-9.81),

$$-9.81(1000)\pi x^2 \, dy.$$

It is negative because gravity acts in the negative y-direction. The work done by an equal and opposite force in lifting the disc to a level 1 m above the top of the tank (a distance $11 - y$) is

$$(11 - y)9.81(1000)\pi x^2 \, dy.$$

The total work to empty the tank is therefore

$$W = \int_0^{10} (11 - y)9810\pi x^2 \, dy.$$

To express x in terms of y, we note that x and y are coordinates of points on the straight line through the origin and the point $(4, 10)$. Since the equation of this line is $y = 5x/2$, we obtain

$$W = 9810\pi \int_0^{10} (11 - y)\left(\frac{2y}{5}\right)^2 dy = 1569.6\pi \int_0^{10} (11y^2 - y^3) \, dy$$

$$= 1569.6\pi \left\{\frac{11y^3}{3} - \frac{y^4}{4}\right\}_0^{10} = 5.75 \times 10^6 \text{ J}.$$

Two observations are noteworthy:

1. In Examples 7.16 and 7.17, the differential represents distance moved and the integrand represents force, a direct application of the discussion leading to equation 7.20. In Example 7.18, the differential is part of the force, and the distance moved is part of the integrand. This is why we suggested earlier that it is not advisable to use equation 7.20 as a formula. The integrand is not always force and the differential is not always distance moved.

2. In each of these examples we set up the coordinate system; it was not given. We can use any coordinate system whatsoever; but once we have chosen our coordinates, we must refer everything to that system. For instance, were we to use the coordinate system in Figure 7.40 for Example 7.18, the definite integral would be

$$W = \int_{-10}^{0} (1 - y)9810\pi \left[\frac{2}{5}(10 + y)\right]^2 dy.$$

Evaluation of this definite integral once again leads to $W = 5.75 \times 10^6$ J.

FIGURE 7.40   Work to empty a tank

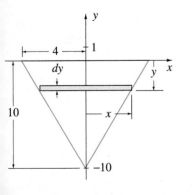

Heat Engines

In the Application Preview, we introduced the problem of converting heat to mechanical energy. With the laws of thermodynamics, we can study this conversion in heat engines. Figure 7.41 shows gas in a cylinder closed on one end with a piston on the other. The force on the piston face due to the pressure of the gas is $F = PA$, where P is the pressure of the gas in the cylinder and A is the area of the face of the piston. If the piston is moved to the right a small amount dx by the pressure of the gas in the cylinder, the work done is $F\,dx$. As the piston face moves from position x_1 to position x_2, the total work is

$$W = \int_{x_1}^{x_2} F\,dx = \int_{x_1}^{x_2} PA\,dx.$$

FIGURE 7.41   Work done by gas in moving piston

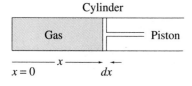

But $A\,dx$ is the change in the volume of gas in the cylinder due to the change dx in x. If we denote this change by dV, and let V_1 and V_2 represent the volumes of gas in the cylinder at positions x_1 and x_2, respectively, then

$$W = \int_{V_1}^{V_2} P\,dV. \tag{7.21}$$

Positive values of this integral correspond to the situation described above when the gas is expanding; negative values correspond to decreasing volume. Although we developed formula 7.21 on the basis of the cylinder and piston in Figure 7.41, it represents the work done by expanding or compressing gases for any shape. In order to evaluate the definite integral, it is necessary to specify P as a function of V.

FIGURE 7.42   Work done by heat engine for one cycle

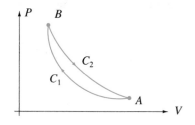

Figure 7.42 shows the states through which the gas in a heat engine might pass during one complete cycle. Beginning at point B and proceeding along C_2 to A, the volume of the gas in the engine is increasing while its pressure is decreasing. The gas is doing positive work, the amount given by the integral in 7.21, where P is defined in terms of V by the equation for C_2. As the gas returns to state B along C_1, integral 7.21 defines the work done on the gas. The difference of these integrals represents the net work done by the gas during one cycle of the engine. In Section 14.3 we interpret the integrations as line integrals, and continue discussions from that point of view. For now, notice that subtraction of the integrals represents the area enclosed by C_1 and C_2. In other words, we can calculate the output of the engine by calculating the area enclosed by the curves.

The *Rankine cycle* represents an idealized steam engine; the pressure and volume of steam during a complete cycle might be those in Figure 7.43. Water at low temperature and pressure (point A) is heated at constant volume (along path AB). Along BC the water is converted to steam and expands slightly, and the expansion continues along CD. To complete the cycle, the steam is cooled and condensed to water along DA. If the gas is expanded adiabatically along CD (temperature is held constant), then P and V are related by $PV^\gamma = k$, where $\gamma > 0$ and $k > 0$ are constants ($\gamma = 1.4$ for air). Using point C in Figure 7.43, $k = 10^5(0.02)^{1.4} = 4.2 \times 10^2$. To determine the output of this particular engine, we calculate the area enclosed by

FIGURE 7.43   Rankine cycle for steam engine

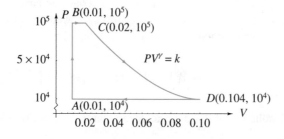

the curves:

$$W = (0.02 - 0.01)(10^5 - 10^4) + \int_{0.02}^{0.104} (kV^{-1.4} - 10^4)\, dV$$

$$= 900 + \left\{ \frac{kV^{-0.4}}{-0.4} - 10^4 V \right\}_{0.02}^{0.104} = 2.5 \times 10^3 \text{ J}.$$

EXERCISES 7.4

1. A spring requires a 10 N force to stretch it 3 cm. Find the work to increase the stretch of the spring from 5 cm to 7 cm.

2. Find the work to increase the stretch of the spring in Exercise 1 from 7 cm to 9 cm.

3. A cage of mass M kilograms is to be lifted from the bottom of a mine shaft h metres deep. If the mass of the cable used to hoist the cage is m kilograms per metre, find the work done.

4. A uniform cable of length 50 m and mass 100 kg hangs vertically from the top of a building 100 m high. How much work is required to get 10 m of the cable on top of the building?

5. A 2 m chain of mass 20 kg lies on the floor. If friction between floor and chain is ignored, how much work is required to lift one end of the chain 2 m straight up? Assume that the suspended portion of the chain makes a right angle with the portion on the floor.

6. How much work is required to lift the end of the chain in Exercise 5 only 1 m?

7. How much work is required to lift the end of the chain in Exercise 5 a distance of 4 m?

8. A 5 m chain of mass 15 kg hangs vertically. It is required to lift the lower end of the chain 5 m so that it is level with the upper end. Calculate the work done using each of the coordinate systems in the figure below.

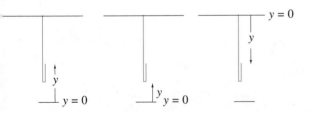

9. A tank filled with water has the form of a paraboloid of revolution with vertical axis (figure below). If the depth of the tank is 12 m and the diameter of the top is 8 m, find the work in pumping the water to the top of the tank.

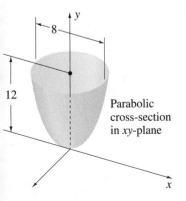

Parabolic cross-section in xy-plane

10. In Exercise 9, find the work to empty the tank through an outlet 2 m above the top of the tank.

∗ **11.** The ends of a trough are isosceles triangles with width 2 m and depth 1 m. The trough is 5 m long and it is full of water. How much work is required to lift all of the water to the top of the trough?

∗ **12.** How much work is required to lift the water in the trough of Exercise 11 to a height of 2 m above the top of the trough?

∗ **13.** A rectangular swimming pool full of water is 25 m long and 10 m wide. The depth is 3 m for the first 10 m of length, then decreases linearly to 1 m at the shallow end. How much work is required to lower the level of the water in the pool by 1/2 m?

∗ **14.** How much work is required to empty the pool of Exercise 13 over its edge?

∗ **15.** The force of repulsion between two point charges of like sign, one of size q and the other of size Q, has magnitude

$$F = \frac{qQ}{4\pi \epsilon_0 r^2},$$

where ϵ_0 is a constant and r is the distance between the charges. If Q is placed at the origin, and q is moved along the x-axis from $x = 2$ to $x = 5$, find the work done by the electrostatic force.

∗ **16.** Two positive charges q_1 and q_2 are placed at positions $x = 5$ and $x = -2$ on the x-axis. A third positive charge q_3 is moved along the x-axis from $x = 1$ to $x = -1$. Find the work done by the electrostatic forces of q_1 and q_2 on q_3 (see Exercise 15).

∗ **17.** If the chain in Exercise 5 is stretched out straight on the floor, and if the coefficient of friction between floor and chain is 0.01, what work is required to raise one end of the chain 2 m? The force of friction on that part of the chain on the floor is 0.01 times the weight of the chain on the floor. Assume, unrealistically, that the suspended portion of the chain makes a right angle with the portion on the floor.

∗ **18.** Two similar springs, each 1 m long in an unstretched position, have spring constant k newtons per metre. The springs are joined together at P (figure below), and their free ends are fastened to two posts 4 m apart. What work is done in moving the midpoint P a distance b metres to the right?

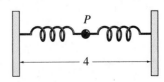

19. If the force of a crossbow (in newtons) is proportional to the draw (in metres), and it is 200 N at a full draw of 50 cm, what work is required to fully draw the crossbow?

20. A bucket of water with mass 100 kg is on the ground attached to one end of a cable with mass per unit length 5 kg/m. The other end of the cable is attached to a windlass 100 m above the bucket. If the bucket is raised at a constant speed, water runs out through a hole in the bottom at a constant rate to the extent that the bucket would have mass 80 kg when it reaches the top. To further complicate matters, a pigeon of mass 2 kg lands on the bucket when it is 50 m above the ground. He immediately begins taking a bath, splashing water over the side of the bucket at the rate of 1 kg/m. Find the work done by the windlass in raising the bucket 100 m.

21. A hemispherical tank with diametral plane at the top has radius 5 m.

    (a) Show that if the depth of oil in the tank is h metres when the tank is one-half full by volume, then h must satisfy the equation

$$h^3 - 15h^2 + 125 = 0.$$

    Find h to four decimal places.

    (b) Find the work to empty the half-full tank if the oil has density 750 kg/m^3, and the outlet is at the top of the tank.

22. *Newton's universal law of gravitation* states that the force of attraction between two point masses m and M has magnitude

$$F = \frac{GmM}{r^2},$$

where r is the distance between the masses and $G = 6.67 \times 10^{-11}$ N·m^2/kg^2 is a constant.

    (a) If M represents the mass of the earth and we regard it as a point mass concentrated at its centre, show that Newton's universal law of gravitation at the earth's surface reduces to $F = mg$, where $g = 9.82$ m/s^2. Assume for the calculation of M that the earth is a sphere with radius 6370 km and mean density 5.52×10^3 kg/m^3.

    (b) Use the original law $F = GmM/r^2$ with the earth regarded as a point mass to calculate the work required to lift a mass of 10 kg from the earth's surface to a height of 10 km.

    (c) Calculate the work in part (b) using the constant gravitational force $F = mg$ in part (a). Is there a significant difference?

23. A right circular cylinder with horizontal axis has radius r metres and length h metres. If it is full of oil with density ρ kilograms per cubic metre, how much work is required to empty the tank through an outlet at its top?

24. A gas is confined in a cylinder, closed on one end with a piston on the other. If the temperature is held constant, then $PV = C$, where P is the pressure of the gas in the cylinder, V is its volume, and C is

a constant. The force needed to move the piston is $F = PA$, where A is the area of the face of the piston. If the radius of the cylinder is 3 units, find the work done in moving the piston from a point 10 units to a point 5 units from the closed end.

25. A town's water is supplied from the water tower shown below. The tower is a cylinder of length 5 m and radius 3 m capped on top and bottom by hemispheres of radius 3 m. The water is pumped from a well 50 m below ground. If the diameter of the pipe from pump to tank is 10 cm, how much work is required to fill the tank initially?

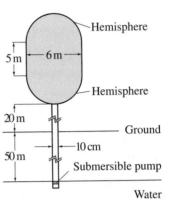

In Exercises 26–30 calculate the work done by a gas taken through the cycle shown in the figure.

26. The *Stirling cycle* in the figure below consists of two isothermal processes and two constant-volume processes.

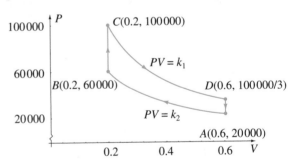

27. The *Ericsson cycle* in the figure below consists of two isothermal processes and two constant-pressure processes.

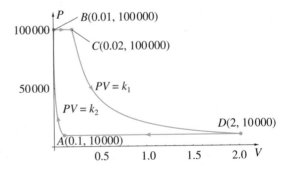

28. The *Otto cycle* for the internal combustion engine in the figure below consists of two adiabatic processes and two constant-volume processes.

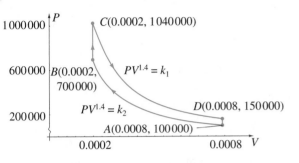

29. The *Diesel cycle* in the figure below consists of two adiabatic processes, a constant-pressure process and a constant-volume process.

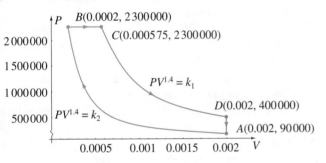

30. The *Brayton cycle* for a gas turbine in the figure below consists of two adiabatic processes and two constant-pressure processes.

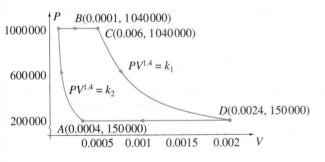

* **31.** A gas is confined in a cylinder, closed on one end with a piston on the other. If the gas expands adiabatically, then the pressure P and volume V of the gas obey the law

$$PV^{7/5} = C,$$

where C is a constant. The force on the piston face due to the pressure of the gas is given by $F = PA$, where A is the area of the face of the piston. Show that if the piston is moved so as to reduce the volume occupied by the gas from V_0 to $V_0/2$, then the work done by the piston is

$$\frac{5}{2}\left(2^{2/5} - 1\right)CV_0^{-2/5}.$$

* **32.** A drop of liquid with initial mass M kilograms falls from rest under gravity and evaporates uniformly, losing mass m kilograms each second. Neglecting air resistance, what work is done by gravity on the drop before it completely evaporates?

** **33.** A diving bell of mass 10 000 kg is attached to a chain of mass 5 kg/m, and the bell sits on the bottom of the ocean 100 m below the surface (figure below). How much work is required to lift the bell to deck level 5 m above the surface? Take into account the fact that when the bell and chain are below the surface, they weigh less than when above. The apparent loss in weight is equal to the weight of water displaced by the bell and chain. Assume that the bell is a perfect cube, 2 m on each side; therefore, when completely submerged, it displaces 8 m³ of water. Assume also that the chain displaces 1 L of water per metre of length.

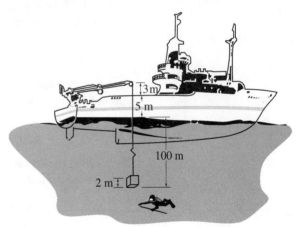

7.5 Energy

Many engineering systems are analyzed on the basis of energy, and there are many kinds of **energy** — potential, kinetic, thermal, and electrical, to name a few. Often one kind of energy is transformed into another, or others, as a system evolves. We consider a number of examples of this here.

In lifting a mass vertically, work is done against gravity. We say that the mass gains gravitational potential energy as a result of the lift, the amount being equal to the work done. For instance, when a mass m is lifted an amount $h > 0$ against gravity in Figure 7.44a, the force is constant (provided that h is relatively small), and the work done is $W = mgh$, where $g = 9.81$. We say that the mass has gained gravitational potential energy as a result of the lift,

$$\text{GPE} = mgh. \tag{7.22}$$

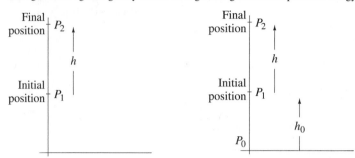

FIGURE 7.44a FIGURE 7.44b

Lifting a mass against gravity results in a gain in gravitational potential energy

We also say that the difference in gravitational potential energies between initial and final positions P_1 and P_2 is mgh. In Figure 7.44b, we have shown the same situation, but initial position P_1 is at height h_0 above some other reference point denoted by P_0. At position P_1, the mass has GPE mgh_0 relative to P_0; at position P_2, it has GPE $mg(h + h_0)$ relative to P_0. The difference in these GPEs is still $mg(h + h_0) - mgh_0 = mgh$.

We have mentioned gains in GPE, differences in GPE, and GPE at one point relative to another, and this is what is important. Often, however, it is convenient to talk about GPE at a point, rather than the difference in GPE between points. This can be accomplished by choosing some specific point as having zero GPE and saying that the GPE at any other point is the difference in GPE between the two points. For instance, if GPE is set equal to zero at P_0 in Figure 7.44b, then GPE is mgh_0 at P_1 and $mg(h + h_0)$ at P_2. The GPE at any point below P_0 is negative.

If m is allowed to drop from position P_2, it loses GPE as it falls. It picks up speed, and energy is associated with the speed. This is called **kinetic energy**; it is defined by

$$\text{KE} = \frac{1}{2}mv^2, \tag{7.23}$$

where v is the speed of m. Provided that air drag on m is ignored, whatever GPE m loses, it gains in KE. To put it another way, the sum of its GPE and KE is always the same:

$$\text{GPE} + \text{KE} = C = mgh + \frac{1}{2}mv^2. \tag{7.24}$$

This is called **conservation of energy**. Energy simply changes from gravitational potential to kinetic during the fall. We apply conservation of energy in the following example.

EXAMPLE 7.19

A stone is thrown vertically upward with speed 20 m/s at the edge of a 50-m cliff. How fast is it travelling when it strikes the base of the cliff?

SOLUTION   Let us take the base of the cliff (Figure 7.45) as our place of zero gravitational potential energy. When the stone is initially thrown upward, its GPE is $mg(50)$ and its KE is $m(20)^2/2$. If we substitute these into 7.24, we obtain

$$C = 50mg + 200m.$$

Substituting this back into 7.24 and replacing h by y gives

$$mgy + \frac{1}{2}mv^2 = 50mg + 200m \qquad \Longrightarrow \qquad 2gy + v^2 = 100g + 400.$$

This equation relates speed and height. At the bottom of the cliff $y = 0$, in which case

$$v^2 = 100g + 400 \qquad \Longrightarrow \qquad v = \sqrt{100g + 400} = 37.2.$$

The stone therefore strikes the bottom of the cliff at 37.2 m/s.

FIGURE 7.45   Schematic for stone thrown upward over the edge of a cliff

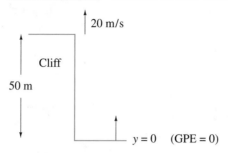

We can also solve the problem using the technique from Section 5.2. Since acceleration of the stone is $a = dv/dt = -9.81$,

$$v = -9.81t + C.$$

(Here v is velocity rather than speed.) If we choose $t = 0$ when the stone is thrown upward, then $v(0) = 20$, and this implies that $C = 20$. Hence,

$$\frac{dy}{dt} = v = -9.81t + 20.$$

A second integration gives

$$y = -4.905t^2 + 20t + D.$$

Since $y(0) = 50$, it follows that $D = 50$, and

$$y = -4.905t^2 + 20t + 50.$$

The stone strikes the bottom of the cliff when

$$0 = y(t) = -4.905t^2 + 20t + 50 \qquad \Longrightarrow \qquad t = \frac{-20 \pm \sqrt{400 - 4(-4.905)(50)}}{-9.81}.$$

The positive solution is $t = 5.827$ s, and at this time

$$v = -9.81(5.827) + 20 = -37.2 \text{ m/s}.$$

The energy solution related speed v to position y; the solution using integration relates velocity to time t. Which is preferable depends on the problem.

EXAMPLE 7.20

Show that when a spring is stretched (or compressed) an amount X, the potential energy stored in the spring is $kX^2/2$. If a mass m is attached to the end of the spring, the spring is stretched an amount X, and the mass is then released, what is its speed at the instant the spring is unstretched?

SOLUTION   The force necessary to maintain stretch x in the spring of Figure 7.46a is kx, where k is the spring constant. The work to stretch the spring X is

$$W = \int_0^X kx \, dx = \left\{ \frac{kx^2}{2} \right\}_0^X = \frac{1}{2}kX^2.$$

This is potential energy stored in the spring.

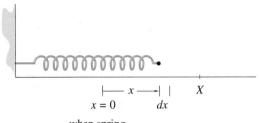

FIGURE 7.46a   Calculation of potential energy stored in a spring

when spring
unstretched

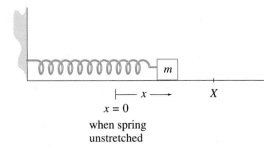

FIGURE 7.46b   Conservation of energy for vibrating mass-spring system

when spring
unstretched

Suppose now that the mass is attached to the spring, the spring is stretched X, and the mass is released. During the subsequent motion to the left (Figure 7.46b), potential energy stored in the spring turns into kinetic energy of the mass (provided there is no friction between m and the surface along which it slides, and there is no air drag on m). Conservation of energy equation 7.24 still holds, but mgh is replaced by $kx^2/2$:

$$C = \frac{1}{2}kx^2 + \frac{1}{2}mv^2.$$

Initially, the stretch is X and the mass is not moving, so that

$$C = \frac{1}{2}kX^2.$$

Hence,

$$\frac{1}{2}kX^2 = \frac{1}{2}kx^2 + \frac{1}{2}mv^2 \quad \Longrightarrow \quad v^2 = \frac{k(X^2 - x^2)}{m}.$$

When the spring is unstretched, $x = 0$, in which case

$$v = \sqrt{\frac{k}{m}}X.$$

This is the speed of m each time it passes through $x = 0$.

EXERCISES 7.5

1. If the stretch of a spring is doubled, by what factor does the potential energy in the spring change?

* **2.**   (a) Mass m in the figure below is pulled a distance x_0 to the right of its equilibrium position. It is then given speed v_0 to the right. Find an equation relating speed of the mass and stretch (or compression) in the spring thereafter.

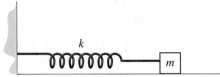

   (b) Use the equation in part (a) to find the maximum stretch the spring experiences.

   (c) Use the equation in part (a) to find the maximum speed experienced by m.

* **3.** When the crossbow of Exercise 19 in Section 7.4 is fired, the potential energy stored in the draw is transformed into kinetic energy of the arrow. If the mass of the arrow is 20 g, what is the speed of the arrow as it leaves the crossbow?

* **4.** The force of repulsion between point charges is defined in Exercise 15 of Section 7.4. If Q is placed at the origin, and q is moved along the x-axis from $x = r$ to $x = r/2$, what is the gain in electrostatic energy?

* **5.**   (a) A 50-m chain with mass 100 kg hangs from the top of a building 50 m high so that its lower end just touches the ground. What gravitational potential energy is stored in the chain? (Take ground level as zero potential.)

   (b) Use energy considerations to determine the work to lift the chain to the top of the building.

6. (a) If the mass m in the figure below is slowly allowed to compress the spring, what is the ultimate compression?

(b) If the mass is dropped, and strikes the spring with speed v_0, what is the maximum compression experienced by the spring?

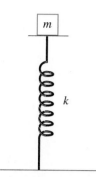

7. A particle is moved along the x-axis by a force $F(x)$ in the x-direction. Use Newton's second law to show that if $F(x)$ is the total force on the particle, then the work done by $F(x)$ is equal to the change in kinetic energy of the particle.

8. (a) Newton's universal law of gravitation is stated in Exercise 22 of Section 7.4. Suppose M represents the mass of the earth, which we regard as a sphere with radius 6370 km and mean density 5.52×10^3 kg/m^3. Calculate the gravitational potential energy gained by a 10-kg mass lifted from the earth's surface to a height of 100 km. Treat the earth as a point mass concentrated at its centre.

(b) If the mass is dropped from this height, with what speed does it strike the earth?

9. (a) Mass m in the figure for Exercise 2 is pulled a distance x_0 to the right of its equilibrium position and then released. As it moves to the left, friction between m and the surface on which it slides retards the motion. The force of friction is μmg, where $0 < \mu < 1$ is a constant and $g = 9.81$. Use energy considerations to show that when the mass is at position x, its speed is given by the equation

$$kx_0^2 = kx^2 + mv^2 + 2\mu mg(x_0 - x).$$

(b) Use the equation in part (a) to determine where the mass comes to an instantaneous stop for the first time.

(c) What is the limit of the expression in part (b) as $\mu \to 0$? Is this to be expected?

(d) What is the limit of the expression in part (b) as $\mu \to 1$? Is this to be expected?

* 10. Two springs with constants k_1 and k_2 are joined together, and then one end is fastened to a wall (figure below). It is shown in physics that when a horizontal force is applied to the free end, the ratio of the stretches s_1 and s_2 in the two springs is inversely proportional to the ratio of their spring constants:

$$\frac{s_1}{s_2} = \frac{k_2}{k_1}.$$

Show that if the free end is moved so as to produce a total stretch in the springs of L, the potential energy stored in the springs is

$$\frac{k_1 k_2 L^2}{2(k_1 + k_2)}.$$

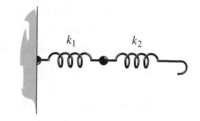

7.6 Fluid Pressure

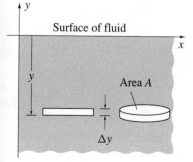

FIGURE 7.47 Determination of pressure in a fluid

When an object is immersed in a fluid, it is acted on by fluid forces. These forces are independent of the object, and are therefore a property of the fluid itself. They always act perpendicularly to the surface of the submerged object. We use the concept of pressure to describe these fluid forces. We define **pressure** at a point in a fluid as the *magnitude of the force per unit area* that would act on a surface at that point in the fluid. Because pressure is the magnitude of the fluid force at a point, it is therefore a positive quantity. Experience suggests that pressure depends on two factors: depth below the surface of the fluid and the type of fluid itself. In order to discover the precise dependence, we consider a small horizontal disc of the fluid (Figure 7.47).

Suppose we denote by $P(y)$ the functional dependence of pressure P on depth y. Then pressure at the bottom of the disc is $P(y)$ and pressure at the top of the disc is $P(y + \Delta y)$. If the fluid is stationary, then the sum of all vertical forces on the disc must be zero. There are three vertical forces acting on the disc: fluid forces on its top and bottom faces, and gravity. Since pressure $P(y)$, which is force per unit area, is the same at all points on the bottom of the disc, it follows that the force on the bottom of the disc must be $AP(y)$, that is, the product of the area A of the bottom of the disc and pressure at points on the bottom of the disc. Similarly, the fluid force on the top of the disc is $-AP(y + \Delta y)$; it is negative because it is in the negative y-direction. Finally, if ρ is the density of the fluid (mass per unit volume), the force of gravity on the disc is $-9.81\rho(A\,\Delta y)$. Since the sum of these three forces must be zero, we set

$$AP(y) - AP(y + \Delta y) - 9.81\rho(A\,\Delta y) = 0.$$

Rearrangement of this equation yields

$$\frac{P(y + \Delta y) - P(y)}{\Delta y} = -9.81\rho,$$

and if we take limits of both sides as $\Delta y \to 0$, we obtain

$$\frac{dP}{dy} = -9.81\rho.$$

This differential equation for $P(y)$ is immediately integrable:

$$P(y) = -9.81\rho y + C.$$

Since fluid pressure at the surface of the fluid is equal to zero $[P(0) = 0]$, C must be equal to zero; hence,

$$P = -9.81\rho y. \tag{7.25}$$

Since $-y$ is a measure of depth d below the surface of the fluid, we have shown that

$$P = 9.81\rho d, \tag{7.26}$$

where d is always taken as positive.

An illuminating interpretation of formula 7.26 is suggested by Figure 7.48. Above a point at depth d below the surface of a fluid, we consider a column of fluid of unit cross-sectional area. The weight of this column of fluid is its volume multiplied by 9.81ρ:

$$W = 9.81\rho V.$$

But V is the product of the length of the column, d, and the (unit) cross-sectional area; that is, $V = (1)d = d$, and hence

$$W = 9.81\rho d. \tag{7.27}$$

A comparison of equations 7.26 and 7.27 suggests that pressure at any point is precisely the weight of a column of fluid of unit cross-sectional area above that point.

We now consider the problem of determining total force on one side of a flat plate (Figure 7.49), immersed vertically in a fluid of density ρ. If the area of the plate is subdivided into horizontal rectangles of width dy, then pressure at each point of this rectangle is

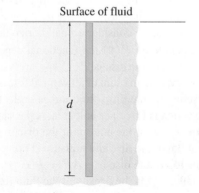

FIGURE 7.48   Pressure is weight of a column of fluid

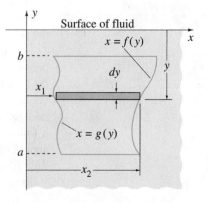

FIGURE 7.49   Force on flat surface submerged in a fluid

approximately $P = -9.81\rho y$. This is an approximation because slight variations in pressure do occur over vertical displacements within the rectangle. It follows that force on the representative rectangle is approximately equal to its area multiplied by $-9.81\rho y$,

$$-9.81\rho y(x_2 - x_1)\,dy = -9.81\rho y[f(y) - g(y)]\,dy.$$

Total force on the plate is found by adding forces on all such rectangles and taking the limit as their widths approach zero. We obtain the required force, therefore, as

$$F = \int_a^b -9.81\rho y[f(y) - g(y)]\,dy. \tag{7.28}$$

Once again we do not suggest that equation 7.28 be memorized because it is associated with the choice of coordinates in Figure 7.49. For a different coordinate system, the definite integral would be correspondingly different (see Example 7.21). What is important is the procedure: Subdivide the surface of the plate into horizontal rectangles, find the force on a representative rectangle, and finally, add over all rectangles with a definite integral.

Note that vertical rectangles cannot be used without further discussion since it is not evident how to calculate the force on such a rectangle. Consideration of vertical rectangles is given in Exercise 8.

EXAMPLE 7.21

The vertical face of a dam is parabolic with breadth 100 m and height 50 m. Find the total force due to fluid pressure on the face.

SOLUTION   If we set up the coordinate system in Figure 7.50, we see that the edge of the dam has an equation of the form $y = kx^2$. Since $(50, 50)$ is a point on this curve, it follows that $k = 1/50$, and $y = x^2/50$. Area of the representative rectangle is $2x\,dy$, and it is at depth $50 - y$ below the surface of the water. Since density of water is 1000 kg/m^3, force on the representative rectangle is approximately equal to

$$(9.81)(1000)(50 - y)2x\,dy = 19\,620(50 - y)\sqrt{50y}\,dy.$$

Total force on the dam must therefore be

$$F = \int_0^{50} 19\,620(50 - y)5\sqrt{2}y^{1/2}\,dy = 98\,100\sqrt{2}\int_0^{50}(50y^{1/2} - y^{3/2})\,dy$$

$$= 98\,100\sqrt{2}\left\{\frac{100y^{3/2}}{3} - \frac{2y^{5/2}}{5}\right\}_0^{50} = 6.54 \times 10^8\ \text{N}.$$

FIGURE 7.50   Fluid force on a dam

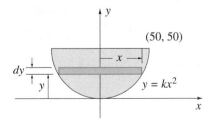

y

$(50, 50)$

dy

x

y

$y = kx^2$

x

EXAMPLE 7.22

A tank in the form of a right circular cylinder of radius 2 m and length 10 m lies on its side. If it is half-filled with oil of density ρ kilograms per cubic metre, find the force on each end of the tank.

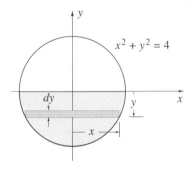

FIGURE 7.51   Fluid force on end of half-filled cylindrical tank

SOLUTION   Force on the representative rectangle in Figure 7.51 is approximately equal to the pressure $9.81\rho(-y)$ multiplied by area $2x\,dy$ of the rectangle

$$9.81\rho(-y)2x\,dy = -19.62\rho y\sqrt{4 - y^2}\,dy.$$

Total force on each end of the tank is therefore

$$F = \int_{-2}^{0} -19.62\rho y\sqrt{4 - y^2}\,dy = 19.62\rho\left\{\frac{1}{3}(4 - y^2)^{3/2}\right\}_{-2}^{0} = 52.32\rho \quad \text{N.}$$

Consulting Project 9

Marine engineers are building buoys made of wood and concrete. Both are cylinders with diameter 25 cm. The length of the concrete cylinder is 90 cm. Densities of the wood and concrete cylinders are 580 kg/m^3 and 2900 kg/m^3, respectively. They have asked us to determine the length of the wood part of the buoy if there is to be 1 m above water.

SOLUTION   To solve this problem, we need Archimedes' principle. It is discussed in Exercise 18 and states that the buoyant force on an object when immersed or partially immersed in a fluid is equal to the weight of the fluid displaced by the object. Suppose we denote the length of the wood cylinder by L (Figure 7.52). For the buoy to float in this position, the magnitude of the force of gravity on the buoy must be equal to the buoyant force. The force of gravity on the buoy is

$$580g\left[\pi\left(\frac{25}{200}\right)^2(L)\right] + 2900g\left[\pi\left(\frac{25}{200}\right)^2\left(\frac{9}{10}\right)\right].$$

The buoyant force is equal to the weight of the water displaced by the buoy

$$1000g\left[\pi\left(\frac{25}{200}\right)^2(L - 1 + 0.9)\right].$$

When we equate these two expressions and cancel $\pi g(25/200)^2$ from each term, we obtain

$$580L + 2900\left(\frac{9}{10}\right) = 1000\left(L - \frac{1}{10}\right) \qquad \Longrightarrow \qquad L = 6.45.$$

The wood cylinder should be 6.45 m long.

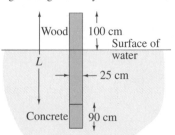

FIGURE 7.52   Determining the length of buoy

Consulting Project 10

Figure 7.53a shows an inclined-tube reservoir manometer, used to measure pressure variations on the surface of the gauge liquid in the reservoir. The reservoir is a cylinder with diameter D and the tube is a cylinder with diameter d. When pressures on liquid surfaces are the same, the surfaces are at the same level as shown. When extra pressure is exerted on surface A, the surface falls an amount H (Figure 7.53b), and liquid is pushed into the tube causing surface B to rise an amount h (both relative to their positions in Figure 7.53a). Our problem is to find an expression for the extra pressure in terms of length L and to determine design parameters that can be modified to make the manometer more effective.

FIGURE 7.53a   Analysis of parameters in an inclined-tube reservoir manometer

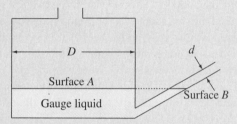

FIGURE 7.53b

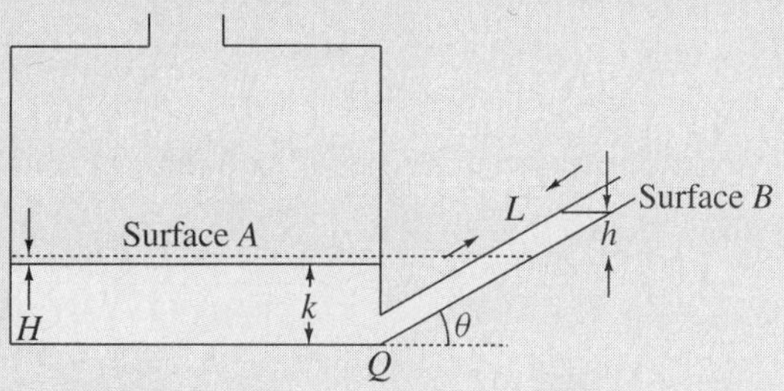

SOLUTION   Suppose we let the extra pressure on surface A be denoted by ΔP. To find a formula for ΔP in terms of h and H, we consider pressure at point Q. From the point of view of the reservoir, the pressure at Q is $\Delta P + \rho g k$, where k is the depth of liquid in the reservoir and ρ is its density. From the point of view of the tube, the pressure at Q is $\rho g(k + H + h)$. Since these must be the same

$$\Delta P + \rho g k = \rho g(k + H + h) \qquad \Longrightarrow \qquad \Delta P = \rho g(H + h).$$

Because the volume of liquid in the manometer has not changed, we can say that

$$\pi \left(\frac{D}{2}\right)^2 H = \pi \left(\frac{d}{2}\right)^2 L \qquad \Longrightarrow \qquad H = \left(\frac{d}{D}\right)^2 L.$$

In addition, $h = L \sin\theta$, so that

$$\Delta P = \rho g \left[\left(\frac{d}{D}\right)^2 L + L \sin\theta\right] = \rho g L \left[\left(\frac{d}{D}\right)^2 + \sin\theta\right].$$

The effectiveness of the manometer is in using L to predict ΔP. Most desirable is for L to be large for small changes in pressure. The larger L, the less likely errors in its measurement will affect the calculation of ΔP. When we write

$$L = \frac{\Delta P}{\rho g \left[\left(\dfrac{d}{D} \right)^2 + \sin \theta \right]},$$

we can see that L increases as ρ, d/D, and $\sin \theta$ all decrease. In other words, the gauge liquid should have small density, the tube diameter should be small relative to the reservoir diameter, and angle θ should be as small as possible. This is theoretically speaking of course.

Let us discuss each of these factors briefly. The manometer will be measuring the pressure of a fluid (gas) on top of surface A. It follows that the liquid should be immiscible with the fluid. It should also develop a reasonable meniscus so that length L can be measured satisfactorily, and it should suffer minimal loss due to evaporation. Hydrocarbon liquids turn out to be most suitable with lowest densities around 80% that of water. Use of such liquids increases effectiveness of the manometer by 25% compared to water.

The ratio d/D should be minimized mathematically. There are limits to how small d can be and how large D should be. Tube diameter, in practice, should exceed 6 mm in order to avoid excessive capillary effects. The situation $d = D$ and $\theta = \pi/2$ corresponds to a U-tube manometer. For given ΔP,

$$L = \frac{\Delta P}{\rho g (1 + 1)} = 0.5 \left(\frac{\Delta P}{\rho g} \right).$$

Consider the situation when $d = 6$ mm, $D = 60$ mm say, and $\theta = \pi/4$. Then

$$L = \frac{\Delta P}{\rho g (1/100 + 1/\sqrt{2})} = 1.39 \left(\frac{\Delta P}{\rho g} \right).$$

Finally, angle θ should be minimized. A practical lower limit is $\pi/18$ radians; below this angle, the meniscus becomes indistinct and it is difficult to get an accurate reading for L.

EXERCISES 7.6

1. A tropical fish tank has length 1 m, width 0.5 m, and depth 0.5 m. Find the force due to water pressure on each of the sides and bottom when the tank is full.

2. The vertical surface of a dam exposed to the water of a lake has the shape shown below. Find the force of the water on the face of the dam.

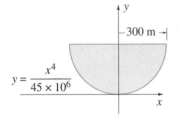

* 3. The vertical end of a water trough is an isosceles triangle with width 2 m and depth 1 m. Find the force of the water on each end when the trough is half full by volume.

* 4. A square plate, 2 m on each side, has one diagonal vertical. If it is one-half submerged in water, what is the force due to water pressure on each side of the plate?

* 5. A cylindrical oil tank of radius r and height h has its axis vertical. If the density of the oil is ρ, find the force on the bottom of the tank when it is full.

* 6. A rectangular swimming pool full of water is 25 m long and 10 m wide. The depth is 3 m for the first 10 m at the deep end, decreasing linearly to 1 m at the shallow end. Find the force due to the weight of the water on each of the sides and ends of the pool.

7. The vertical face of a dam across a river has the shape of a parabola 36 m across the top and 9 m deep at the centre. What is the force that the river exerts on the dam if the water is 0.5 m from the top?

8. Show that the force due to fluid pressure on the vertical rectangle in the figure below is

$$F = \frac{9.81\rho}{2}h(y_1^2 - y_2^2),$$

where ρ is the density of the fluid.

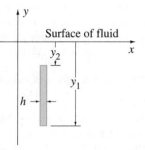

9. A flat plate in the shape of a trapezoid is submerged vertically in a fluid with density ρ. The plate has two parallel vertical sides of lengths 6 and 8 and a third side of length 5 that is perpendicular to the parallel sides and at a depth of 1 below the surface of the fluid (figure below). Find the force due to fluid pressure on each side of the plate, using both horizontal and vertical rectangles (see Exercise 8 for vertical rectangles).

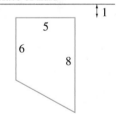

10. The base of a triangular plate, of length a, lies in the surface of a fluid of density ρ. The third vertex of the triangle is at depth b below the surface (figure below). Show that the force due to fluid pressure on each side of the plate is $9.81\rho ab^2/6$, no matter what the shape of the triangle.

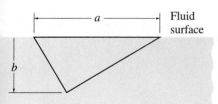

11. Set up (but do not evaluate) a definite integral(s) to find the force due to water pressure on each side of the flat vertical plate in the following figure.

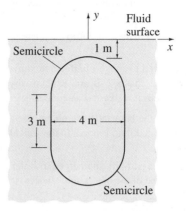

* **12.** In Exercise 6, find the force due to water pressure on each part of the bottom of the pool.

* **13.** The bow of a landing barge (figure below) consists of a rectangular flat plate A metres wide and B metres long. When the barge is stationary, this plate makes an angle of $\pi/6$ radians with the surface of the water. Find the maximum force of the water on the bow.

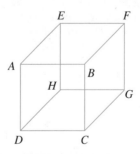

* **14.** A water tank is to be built in the form of a rectangular box with all six sides welded along their joins (figure below). Sides $ABCD$ and $EFGH$ are to be square, and all sides except $ABCD$ are supported from the outside. If the maximum force that $ABCD$ can withstand is 20 000 N, what is the largest cross-section that can be built, assuming that at some stage the tank will be full?

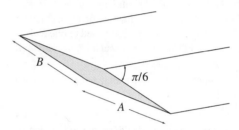

* **15.** A cylindrical oil tank of radius r and length h has its axis horizontal. If the density of the oil is ρ, find the force on each end of the tank when it is half full.

* **16.** A right circular cylinder of radius r and height h is immersed in a fluid of density ρ with its axis vertical. Show that the buoyant force on the cylinder due to the pressure of the fluid is equal to the weight of the fluid displaced. This is known as *Archimedes' principle*, and is valid for an object of any shape.

* **17.** The lower half of a cubical tank 2 m on each side is occupied by water, and the upper half by oil (density 0.90 g/cm³).

    (a) What is the force on each side of the tank due to the pressure of the water and the oil?

    (b) If the oil and water are stirred to create a uniform mixture, does the force on each side change? If not, explain why not. If so, by how much does it increase or decrease?

* **18.** *Archimedes' principle* states that the buoyant force on an object when immersed or partially immersed in a fluid is equal to the weight of the fluid displaced by the object.

    (a) Show that if an object floats partially submerged in water, the percentage of the volume of the object above water is

$$100\frac{\rho_w - \rho_o}{\rho_w},$$

    where ρ_o and ρ_w are the densities of the object and water, respectively.

    (b) If the densities of ice and water are 915 kg/m³ and 1000 kg/m³, respectively, show that only 8.5% of the volume of an iceberg is above water.

* **19.** A wood pole with radius 10 cm and length 3 m is to be used as a buoy. It has density 500 kg/m³. To make it float vertically a concrete cylinder with radius 10 cm and length 30 cm is attached to one end. If the density of the concrete is 3000 kg/m³, how much of the buoy will be above water?

* **20.** What should be the length of the concrete attachment in Exercise 19 if exactly 1 m of buoy is to be above water?

* **21.** A square log 20 cm by 20 cm and length 2 m has been floating in water for a number of years. Water gradually permeated the log so that the density of the log is no longer constant. It varies linearly with depth beginning with the density of water at the edge deepest in water to 500 kg/m³ at its top edge. Determine the height of log protruding from the water.

* **22.** If a full tube of mercury is inverted in a large container of mercury, the level of mercury in the tube will fall, but it will stabilize at a point higher than that in the container (following figure). This is due to the fact that air pressure acts on the surface of the mercury in the container but not on the surface of the mercury in the tube. The extra column of mercury, of height h, creates a force at A that counteracts the atmospheric pressure transmitted through the mercury in the container to the tube so that the total pressure at A is equal to the total pressure at B.

    (a) Show that if the density of mercury is 13.6 g/mL, then the atmospheric pressure at the surface of the mercury in the container is $1.33 \times 10^5 h$ N/m², provided that h is measured in metres.

    (b) If h is measured as 761 mm, find the atmospheric pressure.

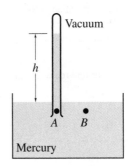

* **23.** (a) A block of wood (density 0.40 g/cm³) is cubical (0.25 m on each side). If it floats in water, how deep is its lowest point below the surface? Refer to Archimedes' principle in Exercise 18.

    (b) Repeat part (a) given that the block is a sphere of radius 0.25 m.

* **24.** A tank is to be built in the form of a right circular cylinder with horizontal axis. The ends are to be joined to the cylindrical side by a continuous weld. One end of the tank and the cylindrical side are supported from the outside. The remaining unsupported end can withstand a total force of 40 000 N less 1000 N for each metre of weld on that end. What is the maximum radius for the tank if it is to hold a fluid with density 1.019×10^3 kg/m³?

* **25.** Find the ratio L/R such that the forces due to fluid pressure on the rectangular and semicircular parts of the plate in the figure below are equal.

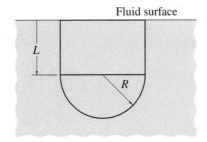

* **26.** (a) A spherical shell with inner radius 1 m and outer radius 2 m has density 2 kg/m³. It is cut in half by a plane through its centre. The flat edge of one of the halves is placed carefully on the surface of a large container of water. How far does the shell sink?

    (b) Repeat part (a) if a tiny hole is drilled through the shell at its upper most point.

7.7 Centres of Mass and Centroids

Everyone is acquainted with the action of a teeter-totter or seesaw. Two children of unequal masses can pass many hours rocking, provided that the child with greater mass sits closer to the fulcrum. In this section we discuss the mathematics of the seesaw. This requires a definition of moments of masses, and moments lead to the idea of the centre of mass of distributions of masses, lumped or continuous.

To discuss the mathematics of a seesaw, we consider in Figure 7.54 a uniform seesaw of length $2L$ balanced at its centre, with a child of mass m at one end. If a second child of equal mass is placed at the other end, the ideal seesaw situation is created. If, however, the mass of the second child is $M > m$, then this child must be placed somewhat closer to the fulcrum. To find the exact position, we must determine what might be called the *rocking power* of a mass. A little experimentation shows that when $M = 2m$, M must be placed halfway between the end and the fulcrum; when $M = 3m$, M must be placed a distance $L/3$ from the fulcrum; and in general, when $M = am$, M must be placed L/a from the fulcrum. Now the rocking power of the child of mass m is constant, and for each mass $M = am$ we have found an equal and opposite rocking power if M is placed at L/a.

FIGURE 7.54   Moments of children on a seesaw

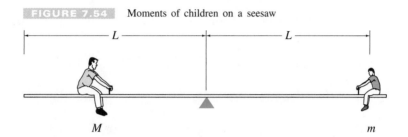

Clearly, then, rocking power depends on both mass and distance from the fulcrum. A little thought shows that the mathematical quantity that remains constant for the various masses $M = am$ is the product of M and distance to the fulcrum; in each case, this product is $(am)(L/a) = mL$, the same product as for the child of mass m. It would appear, then, that rocking power should be defined as the product of mass and distance. We do this in the following definition, and at the same time give rocking power a new name.

DEFINITION 7.1

The **first moment** of a point **mass** m about a point P is the product md, where d is the directed distance from P to m.

If directed distances to the right of point P in Figure 7.55 are chosen as positive and distances to the left are negative, then d_1 is positive and d_2 is negative. Mass m_1 has a positive first moment $m_1 d_1$ about P, and m_2 has a negative first moment $m_2 d_2$. In Figure 7.56, we have placed five children of masses m_1, m_2, m_3, m_4, and m_5 on the same seesaw. A sixth child of mass m_6 is to be placed somewhere on the seesaw so that all six children form the ideal seesaw.

FIGURE 7.55   Definition of moment of mass about a point on a line

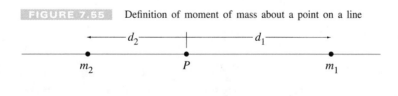

FIGURE 7.56   Determination of positions of children for ideal seesaw

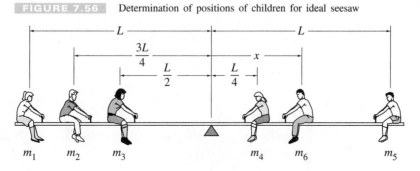

To find the appropriate position for the sixth child, we let x be the directed distance from the fulcrum to the point where this child should be placed. The total first moment of all six children about the fulcrum, choosing distances to the right as positive and to the left as negative, is

$$m_1(-L) + m_2(-3L/4) + m_3(-L/2) + m_4(L/4) + m_5(L) + m_6(x).$$

We regard this as the resultant first moment of all six children attempting to turn the seesaw — clockwise if the moment is positive, counterclockwise if the moment is negative. Balance occurs if this resultant first moment is zero:

$$0 = -m_1 L - \frac{3}{4}m_2 L - \frac{1}{2}m_3 L + \frac{1}{4}m_4 L + m_5 L + m_6 x.$$

We may solve this equation for the position of m_6:

$$x = \frac{L}{4m_6}(4m_1 + 3m_2 + 2m_3 - m_4 - 4m_5).$$

FIGURE 7.57   Placing the fulcrum for an ideal seesaw

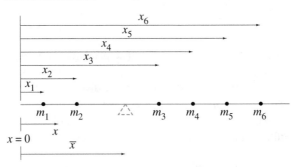

We now turn this problem around and place the six children at distances from the left end as shown in Figure 7.57. If the mass of the seesaw itself is neglected, where should the fulcrum be placed in order to create the ideal seesaw? (See Exercise 16 for the case when the mass of the seesaw is not neglected.) To solve this problem, we let the distance that the fulcrum should be placed from the left end be represented by $\overline{x}$. In order for balance to occur, the total first moment of all six children about the fulcrum must vanish; hence,

$$0 = m_1(x_1 - \overline{x}) + m_2(x_2 - \overline{x}) + m_3(x_3 - \overline{x}) + m_4(x_4 - \overline{x}) + m_5(x_5 - \overline{x}) + m_6(x_6 - \overline{x})$$

The solution of this equation is

FIGURE 7.58   Moment of a mass about a line

$$\overline{x} = \frac{m_1 x_1 + m_2 x_2 + m_3 x_3 + m_4 x_4 + m_5 x_5 + m_6 x_6}{m_1 + m_2 + m_3 + m_4 + m_5 + m_6} = \frac{1}{M}\sum_{i=1}^{6} m_i x_i, \qquad (7.29)$$

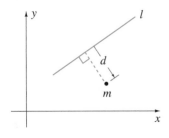

where $M = \sum_{i=1}^{6} m_i$ is the total mass of all six children. This point $\overline{x}$ at which the fulcrum creates a balancing position is called the **centre of mass** for the six children. It is a point where masses to the right are balanced by masses to the left.

In the remainder of this section we extend the idea of a centre of mass of point masses along a line (the seesaw) to the centre of mass of a distribution of point masses in a plane, and then to the centre of mass of a continuous distribution of mass. Our first step is to define the first moment of a point mass in a plane about a line in the plane (Figure 7.58).

> **DEFINITION 7.2**
>
> The first moment of a mass m about a line l is md, where d is the directed distance from l to m.

Once again directed distances are used in calculating first moments, and therefore distances on one side of the line must be chosen as positive and distances on the other side as negative. For vertical and horizontal lines, there is a natural convention for doing this. Distances to the right of a vertical line are chosen as positive, and distances to the left are negative. Distances upward from a horizontal line are positive, and distances downward are negative. In particular, when a mass m is located at position (x, y) in the xy-plane, its first moments about the x- and y-axes are my and mx, respectively.

First moments of a system of n point masses $m_1, m_2, \ldots, m_n$ located at points (x_1, y_1), $(x_2, y_2), \ldots, (x_n, y_n)$, respectively (Figure 7.59a) about the x- and y-axes are defined as the sums of the first moments of the individual masses about these lines:

$$\text{first moment of system about } x\text{-axis} = \sum_{i=1}^{n} m_i y_i, \qquad (7.30a)$$

$$\text{first moment of system about } y\text{-axis} = \sum_{i=1}^{n} m_i x_i. \qquad (7.30b)$$

FIGURE 7.59a System of n point masses embedded in a plate

FIGURE 7.59b Centre of mass of a system of n point masses

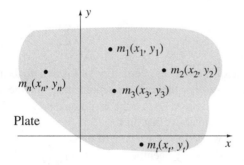

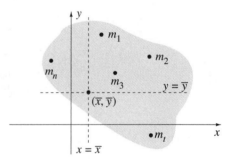

What is the physical meaning of these first moments? Do they, for instance, play the same role that first moments did for the seesaw? To see this we imagine that each point mass is embedded in a thin plastic plate in the xy-plane. The plate itself is massless and extends to include all n masses m_i. Picture now that the plate is horizontal, and a sharp edge is placed along the y-axis. Does the plate rotate about this edge, or does it balance? It is clear that the plate balances if the first moment of the system about the y-axis is equal to zero, and rotates otherwise, the direction depending on whether the first moment is positive or negative. Similarly, the plate balances on a sharp edge placed along the x-axis if the first moment of the system about the x-axis vanishes. In general, the plate balances along any straight edge if the first moment of the system about that edge vanishes.

We defined the centre of mass for a distribution of children on the seesaw as the point at which to place the fulcrum in order to obtain balance. Analogously, we define the **centre of mass** $(\bar{x}, \bar{y})$ of the distribution of point masses in Figure 7.59a as the position to place a sharp point in order to obtain balance. Remember that the plastic itself is massless and only the point masses m_i can create moments. Balance occurs at a point $(\bar{x}, \bar{y})$ if balance occurs about every straight line through $(\bar{x}, \bar{y})$. In particular, balance must occur about the lines $x = \bar{x}$ and $y = \bar{y}$ parallel to the y- and x-axes as in Figure 7.59b. Since balance occurs about $x = \bar{x}$ if the total first moment of the system about this line vanishes, we obtain the condition that

$$0 = m_1(x_1 - \bar{x}) + m_2(x_2 - \bar{x}) + \cdots + m_n(x_n - \bar{x}),$$

which can be solved for $\bar{x}$,

$$\bar{x} = \frac{1}{M} \sum_{i=1}^{n} m_i x_i, \qquad (7.31)$$

where $M = \sum_{i=1}^{n} m_i$ is the total mass of the system. Note that this equation is identical to 7.29. Similarly, for balance about $y = \overline{y}$, we find that $\overline{y}$ must be

$$\overline{y} = \frac{1}{M} \sum_{i=1}^{n} m_i y_i. \tag{7.32}$$

We have obtained a unique point $(\overline{x}, \overline{y})$ based on conditions of balance about the lines $x = \overline{x}$ and $y = \overline{y}$. Does this necessarily imply that balance occurs about every straight line through $(\overline{x}, \overline{y})$? The answer is yes (see Exercise 39).

Every planar point mass distribution has a centre of mass $(\overline{x}, \overline{y})$ defined by equations 7.31 and 7.32. Our derivation has shown that the first moment of the system about any line through $(\overline{x}, \overline{y})$ must be equal to zero. This point is significant in another way. If a particle of mass M (the total mass of the system) is located at the centre of mass $(\overline{x}, \overline{y})$, its first moment about the y-axis is $M\overline{x}$. But from equation 7.31, we have

$$M\overline{x} = \sum_{i=1}^{n} m_i x_i, \tag{7.33}$$

and we conclude that the first moment of this fictitious particle M about the y-axis is exactly the same as the first moment of the system about the y-axis. Similarly, the first moment of M about the x-axis is $M\overline{y}$, and from 7.32,

$$M\overline{y} = \sum_{i=1}^{n} m_i y_i. \tag{7.34}$$

Thus, the centre of mass of a system of point masses m_i is a point at which a single particle of mass $M = \sum_{i=1}^{n} m_i$ has the same first moments about the x- and y-axes as the system. It can be shown further (Exercise 39) that the first moment of M about any line is the same as the first moment of the system about that line.

In summary, we defined the centre of mass of a system of point masses as a balance point. We found as a result that the centre of mass is a point at which a single particle of mass equal to the total mass of the system has the same first moment about any line as the system itself. This argument is reversible. Were we to define the centre of mass as a point to place the mass of the system for equivalent first moments, it would be a balance point. In other words, we have two equivalent definitions of the centre of mass of a system of point masses — a balance point or a equivalent point for first moments.

We now make the transition from a discrete system of particles to a continuous distribution of mass in the form of a thin plate of constant mass per unit area ρ (Figure 7.60). In order to find the mass of the plate, we proceed in exactly the same way that we did for areas. We divide the plate into vertical rectangles, the mass in a representative rectangle of width dx at position x being

$$\rho[f(x) - g(x)]\,dx.$$

To find the total mass of the plate, we add over all such rectangles of ever-diminishing widths

$$M = \int_{a}^{b} \rho[f(x) - g(x)]\,dx. \tag{7.35}$$

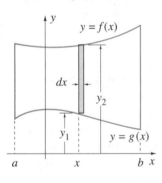

FIGURE 7.60 Centre of mass of a thin plate

Based on our discussion for systems of point masses, we define the centre of mass of a continuous distribution of mass as that point $(\overline{x}, \overline{y})$ where a particle of mass M has the same first moments about the x- and y-axes as the distribution. In algebraic terms, we note that $M\overline{x}$ is the first moment about the y-axis of a particle of mass M at $(\overline{x}, \overline{y})$. To this we must equate the first moment of the original distribution about the y-axis. Now each point in the representative rectangle in Figure 7.60 is approximately the same distance x from the y-axis — approximately, because the rectangle does have finite, though very small, width. The first moment, then, of this rectangle about the y-axis is approximately

$$x\rho[f(x) - g(x)]\,dx.$$

We find the first moment of the entire plate about the y-axis by adding first moments of all such rectangles and taking the limit as their widths approach zero. But once again this is the process defined by the definite integral, and we obtain, therefore, for the first moment of the plate about the y-axis,

$$\int_a^b x\rho[f(x) - g(x)]\,dx.$$

Consequently,

$$M\overline{x} = \int_a^b x\rho[f(x) - g(x)]\,dx, \qquad (7.36)$$

and this equation can be solved for $\overline{x}$ once M and the integral on the right have been evaluated. Equation 7.36 represents for continuous distributions what equation 7.33 does for discrete distributions.

To find $\overline{y}$ we must equate the product $M\overline{y}$ to the first moment of the plate about the x-axis. If we consider the representative rectangle in Figure 7.60 we see that not all points therein are the same distance from the x-axis. To circumvent this problem, we consider all of the mass of the rectangle to be concentrated at its centre of mass. Since the centre of mass is the midpoint of the rectangle — a point distant $[f(x) + g(x)]/2$ from the x-axis — it follows that the first moment of this rectangle about the x-axis is

$$\frac{1}{2}[f(x) + g(x)]\rho[f(x) - g(x)]\,dx.$$

The total first moment of the plate about the x-axis is the definite integral of this expression, and we set

$$M\overline{y} = \int_a^b \frac{1}{2}[f(x) + g(x)]\rho[f(x) - g(x)]\,dx. \qquad (7.37)$$

This equation is used to evaluate $\overline{y}$.

Equations 7.36 and 7.37 can be memorized as formulas for $\overline{x}$ and $\overline{y}$, but it is easier to perform the foregoing operations mentally and arrive at these equations. Besides, for various shapes of plates, we might use horizontal rectangles or combinations of horizontal and vertical rectangles, and in such cases 7.36 and 7.37 would have to be modified.

On the basis of definitions 7.36 and 7.37 for $(\overline{x}, \overline{y})$, we can show that the first moment of M at $(\overline{x}, \overline{y})$ about any line is the same as the first moment of the plate about that same line. This implies that the plate balances along any line through $(\overline{x}, \overline{y})$ and therefore at $(\overline{x}, \overline{y})$.

EXAMPLE 7.23

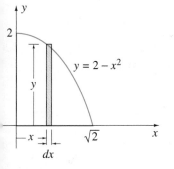

FIGURE 7.61 Centre of mass of plate bounded by $y = -x^2$, $y = 0$, and $x = 0$

Find the centre of mass of a thin plate of constant mass per unit area ρ if its edges are defined by the curves

$$y = 2 - x^2, \qquad y = 0, \qquad x = 0, \qquad x \geq 0.$$

SOLUTION   Using vertical rectangles (Figure 7.61), we find that

$$M = \int_0^{\sqrt{2}} \rho y\,dx = \rho \int_0^{\sqrt{2}} (2 - x^2)\,dx = \rho \left\{ 2x - \frac{x^3}{3} \right\}_0^{\sqrt{2}} = \frac{4\sqrt{2}\rho}{3}.$$

If $(\overline{x}, \overline{y})$ is the centre of mass of the plate, then $M\overline{x}$ is the first moment of the single particle of mass M about the y-axis. This must be equated to the first moment of the plate about the y-axis. Since $x\rho(2 - x^2)\,dx$ is approximately the first moment of the rectangle in Figure 7.61 about the y-axis, the following integral gives the first moment of the plate about the y-axis:

$$\int_0^{\sqrt{2}} x\rho(2 - x^2)\,dx = \rho \int_0^{\sqrt{2}} (2x - x^3)\,dx = \rho \left\{ x^2 - \frac{x^4}{4} \right\}_0^{\sqrt{2}} = \rho.$$

Hence we set $M\bar{x} = \rho$, and solve this equation for $\bar{x}$:

$$\bar{x} = \frac{\rho}{M} = \rho\frac{3}{4\sqrt{2}\rho} = \frac{3}{4\sqrt{2}}.$$

To find $\bar{y}$, we calculate the first moment of the plate about the x-axis. Since the centre of mass of the rectangle in Figure 7.61 is $y/2$ units above the x-axis, it follows that the first moment of this rectangle about the x-axis is $(y/2)\rho y\,dx$. When we integrate this to find the first moment of the plate about the x-axis, and equate it to $M\bar{y}$, the result is

$$M\bar{y} = \int_0^{\sqrt{2}} \frac{y}{2}\rho y\,dx = \frac{\rho}{2}\int_0^{\sqrt{2}} (2 - x^2)^2\,dx = \frac{\rho}{2}\int_0^{\sqrt{2}} (4 - 4x^2 + x^4)\,dx$$

$$= \frac{\rho}{2}\left\{4x - \frac{4x^3}{3} + \frac{x^5}{5}\right\}_0^{\sqrt{2}} = \frac{16\sqrt{2}\rho}{15}.$$

Thus,

$$\bar{y} = \frac{16\sqrt{2}\rho}{15}\frac{3}{4\sqrt{2}\rho} = \frac{4}{5}.$$

EXAMPLE 7.24

Find the centre of mass of a thin plate of constant mass per unit area ρ if its edges are defined by the curves

$$y = 2x - x^2, \qquad y = x^2 - 4.$$

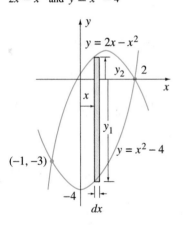

SOLUTION Using vertical rectangles (Figure 7.62) yields

$$M = \int_{-1}^{2} \rho(y_2 - y_1)\,dx = \rho\int_{-1}^{2} [(2x - x^2) - (x^2 - 4)]\,dx$$

$$= \rho\int_{-1}^{2} (4 + 2x - 2x^2)\,dx = \rho\left\{4x + x^2 - \frac{2x^3}{3}\right\}_{-1}^{2} = 9\rho.$$

If $(\bar{x}, \bar{y})$ is the centre of mass of the plate, then first moments of the plate and M about the y-axis give

$$M\bar{x} = \int_{-1}^{2} x\rho(y_2 - y_1)\,dx = \rho\int_{-1}^{2} (4x + 2x^2 - 2x^3)\,dx$$

$$= \rho\left\{2x^2 + \frac{2x^3}{3} - \frac{x^4}{2}\right\}_{-1}^{2} = \frac{9\rho}{2}.$$

Thus,

$$\bar{x} = \frac{9\rho}{2}\frac{1}{9\rho} = \frac{1}{2}.$$

To find $\bar{y}$, we use moments about the x-axis to write

$$M\bar{y} = \int_{-1}^{2} \frac{1}{2}(y_1 + y_2)\rho(y_2 - y_1)\,dx = \frac{\rho}{2}\int_{-1}^{2} (y_2^2 - y_1^2)\,dx$$

$$= \frac{\rho}{2}\int_{-1}^{2} [(2x - x^2)^2 - (x^2 - 4)^2]\,dx = 2\rho\int_{-1}^{2} (-x^3 + 3x^2 - 4)\,dx$$

$$= 2\rho\left\{-\frac{x^4}{4} + x^3 - 4x\right\}_{-1}^{2} = -\frac{27\rho}{2}.$$

Consequently,

$$\bar{y} = -\frac{27\rho}{2}\frac{1}{9\rho} = -\frac{3}{2}.$$

Looking at Figure 7.62, it would appear to balance at the point $(1/2, -3/2)$.

EXAMPLE 7.25

Find first moments of a thin plate of constant mass per unit area ρ about the lines (a) $y = 0$, (b) $y = -2$, and (c) $x = -2$, if its edges are defined by the curves

$$x = |y|^3, \qquad x = 2 - y^2.$$

FIGURE 7.63   First mo-
ments of plate about lines

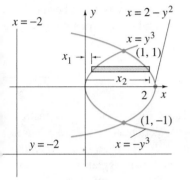

SOLUTION

(a) Since the mass is distributed symmetrically about the x-axis (Figure 7.63), the first moment about $y = 0$ is zero.

(b) Since the centre of mass of the plate is on the x-axis, its first moment about the line $y = -2$ is

$$(2)(\text{mass of plate}) = 2(2)(\text{mass of plate above the } x\text{-axis})$$

$$= 4\int_0^1 \rho(x_2 - x_1)\,dy = 4\rho\int_0^1 (2 - y^2 - y^3)\,dy$$

$$= 4\rho\left\{2y - \frac{y^3}{3} - \frac{y^4}{4}\right\}_0^1$$

$$= \frac{17\rho}{3}.$$

(c) The first moment of the plate about the line $x = -2$ is twice the first moment of its upper half. Since the x-coordinate of the centre of mass of the representative rectangle in Figure 7.63 is $(x_1 + x_2)/2$, the distance from the line $x = -2$ to the centre of mass of the rectangle is $2 + (x_1 + x_2)/2$. The first moment of the horizontal rectangle about $x = -2$ is therefore

$$\left(2 + \frac{x_1 + x_2}{2}\right)\rho(x_2 - x_1)\,dy,$$

and the first moment of the plate is

$$2\int_0^1 \left(2 + \frac{x_1 + x_2}{2}\right)\rho(x_2 - x_1)\,dy = \rho\int_0^1 [4(x_2 - x_1) + (x_2^2 - x_1^2)]\,dy$$

$$= 4\rho\int_0^1 (x_2 - x_1)\,dy + \rho\int_0^1 [(2 - y^2)^2 - (y^3)^2]\,dy$$

$$= \frac{17\rho}{3} + \rho\int_0^1 (4 - 4y^2 + y^4 - y^6)\,dy$$

$$= \frac{17\rho}{3} + \rho\left\{4y - \frac{4y^3}{3} + \frac{y^5}{5} - \frac{y^7}{7}\right\}_0^1$$

$$= \frac{881\rho}{105}.$$

It has become apparent through our discussions and examples that in calculating the centre of mass of a thin plate with constant mass per unit area ρ, ρ is really unnecessary. As a constant it is taken out of each integration and cancels in the final division. The location of the centre of mass depends only on the geometric shape of the plate, and for this reason we could replace all references to mass by area. In particular, the mass of the plate M can be replaced by its area A first moments (of mass) $M\overline{x}$ and $M\overline{y}$ can be replaced by first moments (of area) $A\overline{x}$ and $A\overline{y}$ and equations 7.36 and 7.37 then take the form

$$A\overline{x} = \int_a^b x[f(x) - g(x)]\,dx, \tag{7.38}$$

$$A\overline{y} = \int_a^b \frac{1}{2}[f(x) + g(x)][f(x) - g(x)]\,dx. \tag{7.39}$$

It is customary when using first moments of area to call $(\overline{x}, \overline{y})$ the centroid of the area rather than the centre of mass of the plate, simply because all references to mass have been deleted. We emphasize, however, that the statements in this paragraph apply only when mass per unit area is constant.

Consulting Project 11

Figure 7.64a shows an automatic valve consisting of a plate L metres wide and H metres high that pivots about a horizontal axis through point A. Water creates pressure on parts of the valve above and below A. If the force on that part of the valve below A is greater than on that part above A, the valve remains closed. If the force is greater on that part above A, the valve opens to release water from left to right. Design specifications require the valve to open when the depth of water is D. Our problem is to locate the position of A for this to happen.

FIGURE 7.64a   **FIGURE 7.64b**

Determination of pivot position in an automatic valve

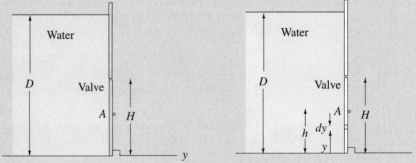

SOLUTION   First we should point out that it is not the forces of the water on the top and bottom parts of the valve that determine whether the valve opens or closes; it is moments of these forces about the pivotal axis through A. With this in mind let us consider a small horizontal strip of width dy and length L on the face of the valve (Figure 7.64b). The force due to water pressure on it is $\rho g(D - y)L\,dy$ when water depth is D. If the required position of the pivotal point A is h metres above the bottom of the valve, then the moment of this force about A is $(h - y)\rho g(D - y)L\,dy$. Moments will be positive for areas below A and negative for areas above A. The total moment on the valve is

$$\int_0^H (h - y)\rho g(D - y)L\,dy = \rho g L \int_0^H [hD - (h + D)y + y^2]\,dy$$

$$= \rho g L \left\{ hDy - (h + D)\frac{y^2}{2} + \frac{y^3}{3} \right\}_0^H$$

$$= \rho g L \left[hDH - (h + D)\frac{H^2}{2} + \frac{H^3}{3} \right].$$

The valve is on the verge of opening when this moment vanishes,

$$0 = \rho g L \left[hDH - (h + D)\frac{H^2}{2} + \frac{H^3}{3} \right].$$

The solution of this equation is $h = (3DH - 2H^2)/(6D - 3H)$.

EXERCISES 7.7

In Exercises 1–5 find the centre of mass of the thin plate with constant mass per unit area.

1.

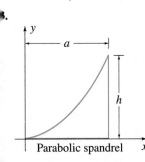

Parabolic $\longleftarrow a \longrightarrow$ h

2.

Semiparabolic $\longleftarrow a \longrightarrow$ h

3.

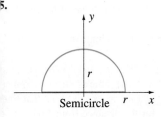

$\longleftarrow a \longrightarrow$ h

Parabolic spandrel

4.

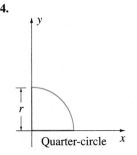

r

Quarter-circle

5.

r

Semicircle $\quad r$

In Exercises 6–13 find the centroid of the region bounded by the curves.

6. $y = x^2 - 1, \quad y = -x^2 - 2x - 1$

7. $y = \sqrt{|x|}, \quad y = 2 - x^2$

8. $y = x^3, \quad x = y^3$

9. $y = x, \quad y = 2x, \quad 2y = x + 3$

10. $x = y^2 - 2y, \quad x + y = 12$

11. $y = \sqrt{2 - x}, \quad x + y = 2$

12. $x = 4y - 4y^2, \quad y = x - 3, \quad y = 1, \quad y = 0$

13. $x^3 y = 8, \quad y = 9 - x^3$

* **14.** The edges of a thin plate with constant mass per unit area ρ are defined by the curves $y = |x|^{1/2}, y = x + 2$, and $y = 2 - x$. Find its first moment about the line $x = -5$.

* **15.** Show that the centroids of regions A and B in the figure below have coordinates

$$\bar{x}_A = \left(\frac{n + 1}{n + 2}\right)a, \quad \bar{y}_A = \left(\frac{n + 1}{4n + 2}\right)b,$$

$$\bar{x}_B = \left(\frac{n + 1}{2n + 4}\right)a, \quad \bar{y}_B = \left(\frac{n + 1}{2n + 1}\right)b.$$

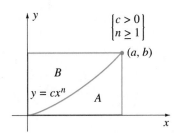

* **16.** Find the centre of mass of the seesaw in Figure 7.57 if the mass of the seesaw is not neglected. Assume that it has uniform mass per unit length ρ and length $2L$.

In Exercises 17–19 find the centroid of the region bounded by the curves.

* **17.** $x = \sqrt{y+2}$, $y = x$, $y = 0$
* **18.** $y + x^2 = 0$, $x = y + 2$, $x + y + 2 = 0$, $y = 2$ (above $y + x^2 = 0$)
* **19.** $y = \sqrt{2-x}$, $15y = x^2 - 4$
* **20.** Find the centre of mass of the thin plate in the figure below if it has constant mass per unit area.

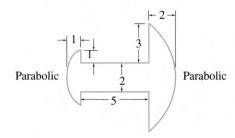

Parabolic   Parabolic

In Exercises 21–25, a plate with constant mass per unit area ρ is bounded by the given curves. Find its first moment about the suggested line.

* **21.** $x + 2 = y^2$, $y = x$ about $x + y = 1$
* **22.** $y = 2x$, $y = -2x$, $y = 4 - 2x^2$, $y \geq 0$ about $y = x$
* **23.** $y = x^2 - 2x$, $x + y = 12$ about $3x + y = 1$
* **24.** $x = y^3$, $x + y = 2$, $y = 0$ about $x + y + 1 = 0$
* **25.** $x = y^2 - 2y$, $x = 2y - y^2$ about $x + 2y = 4$
* **26.** If a region A can be subdivided into n subregions A_i ($i = 1, \ldots, n$) such that the centroid of each A_i is $(\overline{x}_i, \overline{y}_i)$, show that the centroid $(\overline{x}, \overline{y})$ of A is given by

$$\overline{x} = \frac{1}{A} \sum_{i=1}^{n} A_i \overline{x}_i, \qquad \overline{y} = \frac{1}{A} \sum_{i=1}^{n} A_i \overline{y}_i.$$

In Exercises 27–31 use the technique suggested in Exercise 26 to find the centroid of the region.

* **27.**

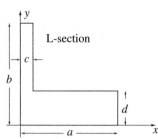

L-section

* **28.**

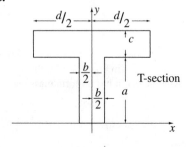

T-section

* **29.**

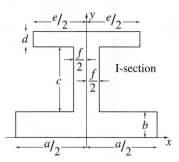

I-section

* **30.**

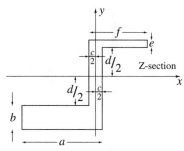

Z-section

* **31.**

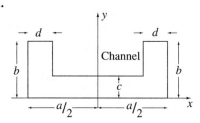

Channel

In Exercises 32–35 find coordinates of the centroid of the region accurate to three decimal places.

* **32.** $y = x^3 + 3x^2 + 2x + 1$, $x = 0$, $y = 0$
* **33.** $y = x^4 - 5x^2 + 5$, $y = 0$ (above the x-axis)
* **34.** $x = y^3 - y^2 - 2y$, $x = \sqrt{2y+1}$
* **35.** $y = x^3 - x$, $y = \sqrt{x}$

* **36.** Prove the following *theorem of Pappus*: If a plane region is revolved about a coplanar axis not crossing the region, the volume generated is equal to the product of the area of the region and the circumference of the circle described by the centroid of the region.
* **37.** Use the result of Exercise 36 to find the volume of the donut in Exercise 41 of Section 7.2.
* **38.** Use the result of Exercise 36 to find the volume in Exercise 42 of Section 7.2.
* **39.**   (a) Show that the first moment of the system of point masses in Figure 7.59b about any line is the same as the first moment of a point mass $M = \sum_{i=1}^{n} m_i$ at $(\overline{x}, \overline{y})$ about that line. *Hint:* Use formula 1.16 for the distance from a point to a line.

    (b) Does it follow that the first moment of the system about any line through $(\overline{x}, \overline{y})$ is zero?

* **40.** A thin flat plate of area A is immersed vertically in a fluid of density ρ. Show that the total force due to fluid pressure on each side of the plate is equal to the product of 9.81ρ, A, and the depth of the centroid of the plate below the surface of the fluid.

41. The dam in the figure below is 4 m high, 10 m wide, and a metres thick. It is made of concrete with density 2400 kg/m^3. Determine the minimum value of a if the dam is not to overturn about point A when $d = 4$ m.

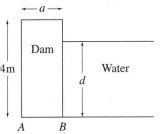

(a) Assume that a seal exists at B, so that no water pressure is present under the dam.

(b) Assume that no seal exists at B, so that full hydrostatic pressure is present under the dam from A to B.

** **42.** The gate AB in the figure below is 525 mm wide and is held in its closed position by a vertical cable and by a 525 mm hinge located along its top edge B. For a depth $d = 1.8$ m of water, determine the minimum tension in the cable required to prevent the gate from opening.

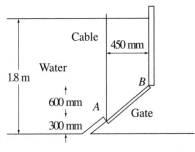

** **43.** Show that the centroid of a triangle with vertices (x_1, y_1), (x_2, y_2), and (x_3, y_3) is

$$\bar{x} = \frac{x_1 + x_2 + x_3}{3}, \qquad \bar{y} = \frac{y_1 + y_2 + y_3}{3}.$$

7.8 Moments of Inertia

Newton's second law $F = ma$ is fundamental to the study of translational motion of bodies. For rotational motion of bodies, its counterpart is $\tau = I\alpha$, where τ is torque, α is angular acceleration, and I is the *moment of inertia* of the body. The kinetic energy of a body of mass m moving with velocity v is $mv^2/2$. The kinetic energy of a body rotating with angular velocity ω is $I\omega^2/2$. Thus, for rotational motion, there is a quantity called the *moment of inertia* of a body that is analogous to mass in translational equations. In this section we define and calculate moments of inertia.

To define moments of inertia of bodies, we begin with the moment of inertia of a point mass.

> **DEFINITION 7.3**
>
> The **moment of inertia** or **second moment** of a point mass m about a line l (Figure 7.58) is the product md^2, where d is the directed distance from l to m.

In particular, if m is at position (x, y) in the xy-plane, its moments of inertia about the x- and y-axes are my^2 and mx^2, respectively. For a system of n particles of masses $m_1, m_2, \ldots, m_n$ located at points $(x_1, y_1), (x_2, y_2), \ldots, (x_n, y_n)$ as in Figure 7.59a, moments of inertia of the system about the x- and y-axes are sums of the moments of inertia of the particles about the x- and y-axes:

$$\text{moment of inertia about } x\text{-axis} = \sum_{i=1}^{n} m_i y_i^2, \tag{7.40a}$$

$$\text{moment of inertia about } y\text{-axis} = \sum_{i=1}^{n} m_i x_i^2. \tag{7.40b}$$

The transition from the discrete case to a continuous distribution in the form of a thin plate with constant mass per unit area ρ is not always so simple as for first moments. First consider the moment of inertia of the plate in Figure 7.60 in Section 7.7 about the y-axis. The mass of the representative rectangle is

$$\rho[f(x) - g(x)] \, dx,$$

and each point of the rectangle is approximately the same distance x from the y-axis. The moment of inertia, then, of this rectangle about the y-axis is approximately

$$x^2 \rho[f(x) - g(x)] \, dx.$$

The moment of inertia of the plate about the y-axis is found by adding moments of inertia of all such rectangles and taking the limit as their widths approach zero. But again this process defines a definite integral, and therefore the moment of inertia of the plate in Figure 7.60 about the y-axis is

$$I = \int_a^b x^2 \rho[f(x) - g(x)] \, dx. \tag{7.41}$$

EXAMPLE 7.26

Find the moment of inertia about the y-axis of a thin plate with constant mass per unit area ρ if its edges are defined by the curves

$$y = x^3, \qquad y = \sqrt{2 - x}, \qquad x = 0.$$

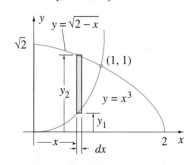

FIGURE 7.65   Moment of inertia of plate about y-axis

SOLUTION   Since the moment of inertia of the vertical rectangle in Figure 7.65 is

$$x^2 \rho(y_2 - y_1) \, dx = \rho x^2 \left(\sqrt{2 - x} - x^3 \right) dx,$$

the moment of inertia of the plate is

$$I = \int_0^1 \rho x^2 \left(\sqrt{2 - x} - x^3 \right) dx = \rho \int_0^1 x^2 \sqrt{2 - x} \, dx - \rho \int_0^1 x^5 \, dx.$$

In the first integral we set $u = 2 - x$, in which case $du = -dx$, and

$$I = \rho \int_2^1 (2 - u)^2 u^{1/2}(-du) - \rho \left\{ \frac{x^6}{6} \right\}_0^1 = \rho \int_1^2 (4u^{1/2} - 4u^{3/2} + u^{5/2}) \, du - \frac{\rho}{6}$$

$$= \rho \left\{ \frac{8u^{3/2}}{3} - \frac{8u^{5/2}}{5} + \frac{2u^{7/2}}{7} \right\}_1^2 - \frac{\rho}{6} = \frac{256\sqrt{2} - 319}{210} \rho.$$

EXAMPLE 7.27

Find the moment of inertia about the line $y = -1$ of a thin plate of constant mass per unit area ρ if its edges are defined by the curves

$$x = y^2, \qquad x = 2y.$$

FIGURE 7.66 Moment of inertia of plate about the line $y = -1$

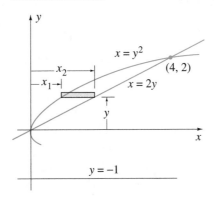

SOLUTION   Since the directed distance from the line $y = -1$ to all points in the horizontal rectangle in Figure 7.66 is approximately $y + 1$, the moment of inertia of the rectangle about the line $y = -1$ is approximately $(y + 1)^2 \rho (x_2 - x_1) \, dy$. It follows that the moment of inertia of the plate is

$$I = \int_0^2 (y + 1)^2 \rho (x_2 - x_1) \, dy = \rho \int_0^2 (y + 1)^2 (2y - y^2) \, dy$$

$$= \rho \int_0^2 (-y^4 + 3y^2 + 2y) \, dy = \rho \left\{ -\frac{y^5}{5} + y^3 + y^2 \right\}_0^2 = \frac{28\rho}{5}.$$

FIGURE 7.67   Moment of inertia of rectangle about x-axis

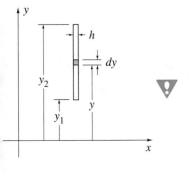

In Examples 7.26 and 7.27, and in the discussion leading to equation 7.41, we chose rectangles that had lengths parallel to the line about which we required the moment of inertia. This is not just coincidence; the use of perpendicular rectangles is more complicated. Consider, for instance, finding the moment of inertia about the x-axis of the plate in Example 7.26. To use the vertical rectangles in Figure 7.65, we first require the moment of inertia of such a rectangle about the x-axis. The mass of the rectangle, as in Example 7.26, must be multiplied by the square of the distance from the x-axis to the rectangle. Unfortunately, different points in the rectangle are at different distances. One suggestion might be to concentrate all of the mass of the rectangle at its centre of mass and use the distance from the x-axis to the centre of mass. This is *incorrect*. The centre of mass is a point at which mass can be concentrated if we are discussing first moments. We are discussing second moments. (In Exercise 12 we show that the moment of inertia of the rectangle in Figure 7.67 cannot be obtained by concentrating its mass at its centre of mass.) What are we to do then? To use this type of rectangle, we must first develop a formula for its moment of inertia. To obtain this formula, let us consider the moment of inertia about the x-axis of the rectangle of width h and length $y_2 - y_1$ in Figure 7.67. If we subdivide this rectangle into smaller rectangles of width dy, the moment of inertia of the tiny rectangle about the x-axis is approximately

$$y^2 \rho h \, dy.$$

The moment of inertia of the long, vertical rectangle can be obtained by adding over all the tiny rectangles as their widths dy approach zero,

$$\int_{y_1}^{y_2} y^2 \rho h \, dy = \rho h \int_{y_1}^{y_2} y^2 \, dy = \frac{\rho h}{3} (y_2^3 - y_1^3). \tag{7.42}$$

We can use this formula to state that the moment of inertia about the x-axis of the vertical rectangle in Figure 7.65 is

$$\frac{\rho}{3}(y_2^3 - y_1^3)\, dx = \frac{\rho}{3}[(2-x)^{3/2} - x^9]\, dx.$$

The moment of inertia of the plate about the x-axis is therefore

$$I = \int_0^1 \frac{\rho}{3}[(2-x)^{3/2} - x^9]\, dx = \frac{\rho}{3}\left\{ -\frac{2}{5}(2-x)^{5/2} - \frac{x^{10}}{10} \right\}_0^1 = \frac{(16\sqrt{2} - 5)\rho}{30}.$$

The alternative procedure for this problem is to use horizontal rectangles that are parallel to the x-axis and obtain two definite integrals:

$$I = \int_0^1 y^2 \rho y^{1/3}\, dy + \int_1^{\sqrt{2}} y^2 \rho (2 - y^2)\, dy.$$

In summary, we have two methods for determining moments of inertia of thin plates:

1. Choose rectangles parallel to the line about which the moment of inertia is required, in which case only the basic idea of mass times distance squared is needed.

2. Choose rectangles perpendicular to the line about which the moment of inertia is required, in which case formula 7.42, or a similar formula, is needed.

As for finding centres of mass, we could, in the special case of uniform mass distributions, drop all references to mass and talk about second moments of area about a line.

EXAMPLE 7.28

FIGURE 7.68   Moment of inertia of plate about line $x = 2$

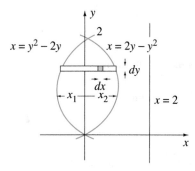

Find the moment of inertia about the line $x = 2$ of a plate with mass per unit area ρ if its edges are defined by the curves

$$x = 2y - y^2, \qquad x = y^2 - 2y.$$

SOLUTION   We first divide the plate into horizontal rectangles of width dy, and then subdivide this rectangle into smaller rectangles of width dx (Figure 7.68). Since the directed distance from the line $x = 2$ to the tiny rectangle is $x - 2$, the moment of inertia of the tiny rectangle about $x = 2$ is

$$(x - 2)^2 \rho\, dy\, dx.$$

It follows that the moment of inertia of the long horizontal rectangle about $x = 2$ is

$$\int_{x_1}^{x_2} (x - 2)^2 \rho\, dy\, dx = \rho\, dy \left\{ \frac{1}{3}(x-2)^3 \right\}_{x_1}^{x_2} = \frac{\rho}{3}[(x_2 - 2)^3 - (x_1 - 2)^3]\, dy.$$

The moment of inertia of the entire plate is now

$$I = \int_0^2 \frac{\rho}{3}[(x_2 - 2)^3 - (x_1 - 2)^3]\, dy$$

$$= \frac{\rho}{3} \int_0^2 [(2y - y^2 - 2)^3 - (y^2 - 2y - 2)^3]\, dy$$

$$= \frac{2\rho}{3} \int_0^2 (24y - 12y^2 + 8y^3 - 12y^4 + 6y^5 - y^6)\, dy$$

$$= \frac{2\rho}{3} \left\{ 12y^2 - 4y^3 + 2y^4 - \frac{12y^5}{5} + y^6 - \frac{y^7}{7} \right\}_0^2$$

$$= \frac{1184\rho}{105}.$$

EXERCISES 7.8

In Exercises 1–10 the curves define a thin plate with constant mass per unit area ρ. Find its moment of inertia about the line.

1. $y = x^2$, $y = x^3$, about the y-axis

2. $y = x$, $y = 2x + 4$, $y = 0$, about the x-axis

3. $y = x^2$, $2y = x^2 + 4$, about $y = 0$

4. $y = x^2 - 4$, $y = 2x - x^2$, about $x = -2$

5. $xy\sqrt{y^2 + 12} = 1$, $x = 0$, $y = 1$, $y = 1/2$, about $y = 0$

6. $y = |x|^{1/3}$, $y = 2 - |x|^{1/3}$, about $x = 0$

7. $x + y = 2$, $y - x = 2$, $y = 0$, about $x = -2$

8. $x = 1 - y^2$, $x = y^2 - 1$, about $x = -1$

9. $x = 1 - y^2$, $x = y^2 - 1$, about $y = 1$

10. $x = y^2$, $x + y = 2$, about $y = 3$

11. (a) If I_{Ax} and I_{Bx} represent the second moments of area about the x-axis of the regions in the figure of Exercise 15 in Section 7.7, show that $I_{Bx} = 3n I_{Ax}$.

    (b) Show that if I_{Ay} and I_{By} represent the second moments of area about the y-axis, then $n I_{Ay} = 3 I_{By}$.

12. What is the product of the mass of the rectangle in Figure 7.67 and the square of the distance from the x-axis to the centre of mass of the rectangle? Is it equal to the expression in equation 7.42?

In Exercises 13–17 find second moments of area of the section about (a) the x-axis and (b) the y-axis.

13.

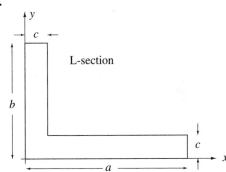

L-section

14.

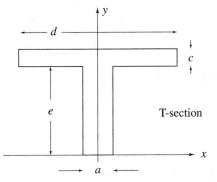

T-section

* 15.

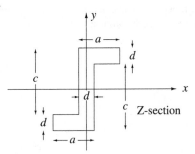

Z-section

* 16.

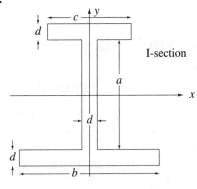

I-section

* 17.

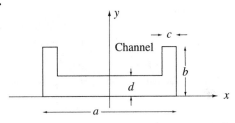

Channel

* 18. The radius of gyration r of a thin plate with constant mass per unit area about a line is defined by $I = Mr^2$, where M is the mass of the plate and I is its moment of inertia about that line. Find radii of gyration about the x- and y-axes for the plate with edges $y = 2x^3$, $y + x^3 = 0$, and $2y = x + 3$. Explain the physical significance of r.

* 19. Show that the kinetic energy of a long-playing record is equal to one-half the product of the moment of inertia of the record about a line through its centre and perpendicular to its face, and the square of its angular speed.

* 20. Prove the *parallel axis theorem*: The moment of inertia of a thin plate (with constant mass per unit area) with respect to any coplanar line is equal to the moment of inertia with respect to the parallel line through the centre of mass plus the mass multiplied by the square of the distance between the lines.

* 21. If a line $x = \tilde{x}$ is drawn through the region in Figure 7.60, what integral represents the second moment of area about this line? What should be the value of $\tilde{x}$ for the smallest possible second moment?

* **22.** The polar moment of inertia of a point mass m at (x, y) is defined as the product of m and the square of its distance from the origin, $J_0 = m(x^2 + y^2)$. For the thin plate (with constant mass per unit area) in the figure below, let I_x and I_y be its moments of inertia about the x- and y-axes. Show that $J_0 = I_x + I_y$.

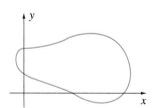

In Exercises 23–25 the curves define a plate with mass per unit area equal to 2. Find its moment of inertia about the line accurate to three decimal places.

* **23.** $y = 1 - x^2$, $x = y^2$, about $y = 0$

* **24.** $y = x^3 - x$, $y = \sqrt{x}$, about $x = 0$

* **25.** $y = x^3 - x$, $y = \sqrt{x}$, about $y = 0$

** **26.** Find the second moment of area of a rectangle about its diagonal.

7.9 Additional Applications

Volumes by Slicing

If we can represent the area of parallel cross-sections of a volume as a function of one variable, we can use a definite integral to calculate the volume. In particular, when we use the disc or washer method to determine the volume of a solid of revolution, parallel cross-sections are circles. In the following example, parallel cross-sections are squares.

EXAMPLE 7.29

A uniformly tapered rod of length 2 m has square cross-sections. If the areas of its ends are 4 cm^2 and 16 cm^2 as in Figure 7.69a, what is the volume of the rod?

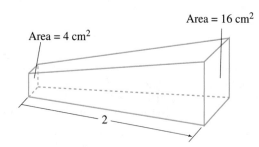

FIGURE 7.69a   Tapered rod with square cross-sections

Area = 16 cm^2

Area = 4 cm^2

2

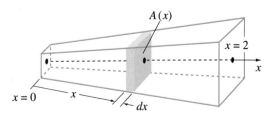

FIGURE 7.69b   Volume of tapered rod with square cross-sections

$A(x)$

$x = 2$

$x = 0$ x dx

SOLUTION   If we define an x-coordinate perpendicular to the square cross-sections as in Figure 7.69b, then the side lengths of the cross-sections at $x = 0$ and $x = 2$ are 0.02 m and 0.04 m, respectively. Since the rod is uniformly tapered, the side length of the cross-section at x is

$$0.02 + x \left(\frac{0.04 - 0.02}{2} \right) = 0.02 + 0.01x = \frac{2 + x}{100}.$$

The area of the cross section at x is therefore

$$A(x) = 10^{-4}(2 + x)^2.$$

If we construct at x a slab of cross-sectional area $A(x)$ and width dx, the volume of the slab is

$$A(x)\, dx = 10^{-4}(2 + x)^2\, dx.$$

To obtain the volume of the rod, we add volumes of all such slabs between $x = 0$ and $x = 2$, and take the limit as their widths approach zero:

$$V = \int_0^2 A(x)\, dx = \int_0^2 10^{-4}(2 + x)^2\, dx$$

$$= 10^{-4} \left\{ \frac{(2 + x)^3}{3} \right\}_0^2 = \frac{10^{-4}}{3}(64 - 8) = \frac{56}{3} \times 10^{-4}\ \text{m}^3.$$

Area of a Surface of Revolution

If a curve in the xy-plane is rotated about the x- or y-axis (or a line parallel to the x- or y-axis) in order to produce a surface, we can calculate the area of this surface using lengths along curves, discussed in Section 7.3.

EXAMPLE 7.30

If that part of the parabola $y = x^2$ between $x = 0$ and $x = 1$ is rotated around the y-axis, find the area of the surface of revolution traced out by the curve (Figure 7.70).

FIGURE 7.70 Area of surface of revolution

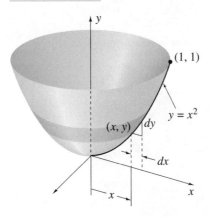

SOLUTION   We approximate length along the parabola corresponding to a change dx in x by the tangential straight-line length:

$$\sqrt{(dx)^2 + (dy)^2} = \sqrt{1 + \left(\frac{dy}{dx}\right)^2}\, dx = \sqrt{1 + (2x)^2}\, dx = \sqrt{1 + 4x^2}\, dx.$$

If this straight-line segment is rotated around the y-axis, each point follows a circular path of radius approximately equal to x; therefore, the area traced out by the line segment is approximately equal to

$$(2\pi x)\left(\sqrt{1 + 4x^2}\, dx\right).$$

We find total surface area by adding all such areas, and taking the limit as widths dx approach zero:

$$A = \int_0^1 2\pi x \sqrt{1 + 4x^2}\, dx = 2\pi \left\{ \frac{(1 + 4x^2)^{3/2}}{12} \right\}_0^1 = \frac{(5\sqrt{5} - 1)\pi}{6}.$$

EXAMPLE 7.31

Find the area of the surface of revolution traced out by rotating that part of the curve $y = x^3$ between $x = 1$ and $x = 2$ about the x-axis.

SOLUTION   We approximate length along the cubic corresponding to a change dx in x by the tangential straight-line length:

$$\sqrt{(dx)^2 + (dy)^2} = \sqrt{1 + \left(\frac{dy}{dx}\right)^2}\, dx = \sqrt{1 + (3x^2)^2}\, dx = \sqrt{1 + 9x^4}\, dx.$$

If this straight-line segment is rotated about the x-axis (Figure 7.71), each point follows a circular path of radius approximately equal to y; therefore, the area traced out by the line segment is approximately equal to

$$(2\pi y)\left(\sqrt{1 + 9x^4}\, dx\right).$$

By adding over all such areas, we obtain the area of the surface,

$$A = \int_1^2 2\pi y\sqrt{1 + 9x^4}\, dx = 2\pi \int_1^2 x^3\sqrt{1 + 9x^4}\, dx$$

$$= 2\pi \left\{\frac{(1 + 9x^4)^{3/2}}{54}\right\}_1^2 = \frac{(145^{3/2} - 10^{3/2})\pi}{27}.$$

FIGURE 7.71   Area of surface of revolution

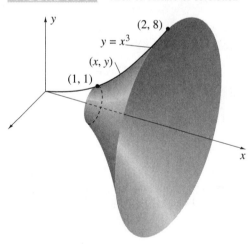

Rates of Flow

In Problem 3 of Section 6.2 we considered laminar blood flow in a circular vessel. Specifically, velocity of blood through the cross section in Figure 7.72a is a function of radial distance r from the centre of the vessel:

$$v(r) = c(R^2 - r^2), \quad 0 \le r \le R,$$

where $c > 0$ is a constant and R is the radius of the vessel. If we construct, at radius r, a thin ring of width dr as in Figure 7.72b, then the area of this ring is approximately $(2\pi r)\, dr$. Since

v does not vary greatly over this ring, the amount of blood flowing through the ring per unit time is approximately $v(r)$ multiplied by the area of the ring:

$$v(r)(2\pi r \, dr).$$

We can find the total flow through the blood vessel by adding flows through all rings and taking the limit as widths dr of the rings approach zero:

$$F = \int_0^R v(r)2\pi r \, dr = 2\pi \int_0^R rc(R^2 - r^2) \, dr = 2\pi c \left\{ \frac{R^2 r^2}{2} - \frac{r^4}{4} \right\}_0^R = \frac{\pi c R^4}{2}.$$

FIGURE 7.72a Circular blood vessel of radius R

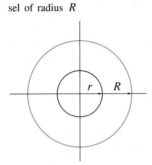

FIGURE 7.72b Rate of blood flow through a circular blood vessel

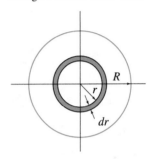

Elongation of Rods

FIGURE 7.73 Hooke's law for force in stretched spring

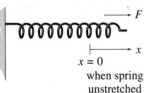

when spring unstretched

Hooke's law states that forces exerted by springs are proportional to stretch and compression. In particular, suppose x measures distance to the right in Figure 7.73 and $x = 0$ corresponds to the right end of the spring when the spring is unstretched and uncompressed. Hooke's law then states that a force F applied to the spring as shown causes stretch

$$x = \frac{F}{k},$$

where k is the spring constant.

Figure 7.74 shows a rod of natural length L. When a force F is applied to the right end of the rod, the rod acts like a spring in that the rod stretches. It stretches only minutely even when F is large; but it does stretch. It is shown in the area of strength of materials that the amount of stretch is related to F by the equation

$$x = \frac{FL}{AE}, \tag{7.43}$$

where A is the (constant) cross-sectional area of the rod and E is Young's modulus of elasticity, a constant that depends on the material of the rod. In effect, AE/L plays the role of spring constant k.

FIGURE 7.74 Stretch in rod when force is applied to one end

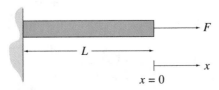

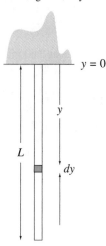

FIGURE 7.75 Stretch in rod hung vertically

Now suppose the rod is turned vertically and hung from its top end (Figure 7.75). Due to its weight, the rod stretches, and using 7.43, we can calculate how much. If we consider a small length dy at position y, the force on each cross section in this element is approximately the same, and equal to the weight of that part of the rod below it,

$$\rho g(L - y)A,$$

where ρ is the density of the material in the rod. According to 7.43, the element dy stretches by

$$\frac{\rho g(L - y)A\,dy}{AE} = \frac{\rho g(L - y)\,dy}{E}.$$

Total stretch in the rod is therefore

$$\int_0^L \frac{\rho g(L - y)}{E}dy = \frac{\rho g}{E}\left\{Ly - \frac{y^2}{2}\right\}_0^L = \frac{\rho g L^2}{2E}.$$

This may be somewhat surprising in that stretch does not depend on the cross-sectional area of the rod. This is explained by the fact that the weight of rod below any cross-section is proportional to cross-sectional area A, but stretch in 7.43 is inversely proportional to A. The two factors compensate. To get an idea of the magnitude of this stretch, suppose that $L = 2$ m, $\rho = 8000$ kg/m^3, and $E = 2 \times 10^{11}$ N/m^2. Then

$$\frac{\rho g L^2}{2E} = \frac{8000(9.81)(2)^2}{2(2 \times 10^{11})} = 7.85 \times 10^{-7} \text{ m}.$$

Suppose now that cross-sectional area A of the rod is not constant; the rod is tapered, say with circular cross-sections (Figure 7.76). Radius of the large end is r and the rod tapers to a point at $y = L$. The force on cross-sections of the element of width dy at position y is again the weight of rod below it, namely,

$$\frac{1}{3}\pi x^2(L - y)\rho g,$$

FIGURE 7.76 Elongation of tapered rod hung vertically

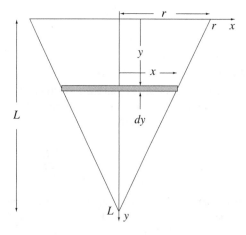

where x is the radius of the rod at position y. Since the equation of the side of the rod is

$$y = \frac{-L}{r}(x - r) \qquad \Longrightarrow \qquad x = r\left(1 - \frac{y}{L}\right),$$

the weight below y is

$$\frac{1}{3}\pi r^2\left(1 - \frac{y}{L}\right)^2 (L - y)\rho g = \frac{\pi r^2 \rho g}{3L^2}(L - y)^3.$$

The element dy therefore stretches:

$$\frac{\pi r^2 \rho g}{3L^2}(L-y)^3 \frac{dy}{\pi x^2 E} = \frac{r^2 \rho g(L-y)^3 \, dy}{3L^2 Er^2(1-y/L)^2} = \frac{\rho g(L-y)}{3E} \, dy.$$

This is one-third the stretch for the nontapered rod, and hence the total stretch of the tapered rod is $\rho g L^2/(6E)$.

Green's Functions

Green's functions are widely used to solve engineering problems, especially in the presence of quantities represented by Dirac-delta functions (see Section 2.5). We illustrate with the following problem for static deflections of a taut string of negligible mass, constant tension τ, and length L, with ends fixed at $x = 0$ and $x = L$ on the x-axis:

$$-\tau \frac{d^2 y}{dx^2} = F(x), \quad 0 < x < L, \tag{7.44a}$$

$$y(0) = y(L) = 0. \tag{7.44b}$$

Quantity $F(x)$ is the load per unit x-length on the string. To find $y(x)$ we would integrate the differential equation twice and use the end conditions to evaluate the constants of integration. For instance, if $F(x) = kx(x-L)$, where $k < 0$ is a constant, integration gives

$$y(x) = \frac{k}{\tau}\left(\frac{Lx^3}{6} - \frac{x^4}{12}\right) + Cx + D.$$

The end conditions require

$$0 = y(0) = D, \qquad 0 = y(L) = \frac{k}{\tau}\left(\frac{L^4}{6} - \frac{L^4}{12}\right) + CL + D.$$

These give $D = 0$ and $C = -kL^3(12\tau)$, and therefore

$$y(x) = \frac{k}{\tau}\left(\frac{Lx^3}{6} - \frac{x^4}{12}\right) - \frac{kL^3 x}{12\tau} = \frac{kx}{12\tau}(2Lx^2 - x^3 - L^3).$$

Green's functions provide an alternative way to solve such problems. The Green's function for this problem is

$$G(x; X) = \frac{1}{L\tau}[x(L-X)\,h(X-x) + X(L-x)\,h(x-X)], \tag{7.45}$$

where $h(x-X)$ is the Heaviside function (Section 2.5). Think of X as a parameter that can take on any value between 0 and L. Given x and X, $G(x; X)$ represents the deflection in the string at position x if a unit force in the positive y-direction is applied at position X. Given that the load on the string is $F(x)$, the deflection at any point x is given by the definite integral

$$y(x) = \int_0^L G(x; X)F(X)\,dX. \tag{7.46}$$

We reason as follows: If $G(x; X)$ is the deflection at x due to a unit force at X, then $G(x; X)F(X)$ is the deflection at x due to load $F(X)\,dX$ on that part of the string dX and X. Integration from 0 to L gives the deflection at x due to the entire load on the string. When $F(x) = kx(x - L)$, as above,

$$y(x) = \int_0^L \frac{1}{L\tau}[x(L - X)\,h(X - x) + X(L - x)\,h(x - X)]kX(X - L)\,dX.$$

It is always necessary to subdivide the integration into two parts, one from 0 to x, and the other from x to L. Since $h(X - x) = 0$ when $0 < X < x$ and $h(x - X) = 0$ when $x < X < L$, we obtain

$$y(x) = \frac{k}{L\tau}\int_0^x X(L - x)X(X - L)\,dX + \frac{k}{L\tau}\int_x^L x(L - X)X(X - L)\,dX$$

$$= \frac{k(L - x)}{L\tau}\left\{\frac{X^4}{4} - \frac{LX^3}{3}\right\}_0^x + \frac{kx}{L\tau}\left\{\frac{2LX^3}{3} - \frac{X^4}{4} - \frac{L^2X^2}{2}\right\}_x^L$$

$$= \frac{kx}{12\tau}(2Lx^2 - x^3 - L^3).$$

This is the same solution as was obtained by solving the differential equation and evaluating constants. The differential equation and end conditions are built into the Green's function. The definite integral in 7.46 takes care of the loading $F(x)$.

If the string is subjected to a concentrated load of F newtons attached at $x = L/3$, then $F(x) = -F\delta(x - L/3)$, where $\delta(x - L/3)$ is the Dirac-delta function. It is a very complicated process to solve the differential equation for such a load. Use of the Green's function is particularly simple. The deflection is again given by the integral in equation 7.46, where we use the property in Exercise 34 of Section 7.10:

$$y(x) = \int_0^L \frac{1}{L\tau}[x(L - X)\,h(X - x) + X(L - x)\,h(x - X)](-F)\delta(X - L/3)\,dX$$

$$= \frac{-F}{L\tau}\left[x\left(L - \frac{L}{3}\right)h\left(\frac{L}{3} - x\right) + \frac{L}{3}(L - x)\,h\left(x - \frac{L}{3}\right)\right]$$

$$= \frac{-F}{3\tau}\left[2x\,h\left(\frac{L}{3} - x\right) + (L - x)\,h\left(x - \frac{L}{3}\right)\right]$$

$$= \begin{cases} -\dfrac{2Fx}{3\tau}, & 0 < x < L/3 \\[2mm] -\dfrac{F}{3\tau}(L - x), & L/3 < x < L. \end{cases}$$

The graph is shown in Figure 7.77 where we have filled in the removable discontinuity at $x = L/3$.

FIGURE 7.77   Deflection of string due to a point load at $x = L/3$

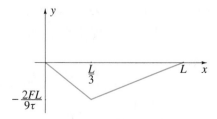

EXERCISES 7.9

1. Verify that the surface area of a sphere of radius r is $4\pi r^2$.

2. Find the area of the curved surface of a right circular cone of radius r and height h.

3. Calculate the rate of flow of blood through a circular vessel of radius R if the velocity profile is (a) $v = f(r) = cR\sqrt{R^2 - r^2}$ and (b) $v = f(r) = (c/R^2)(R^2 - r^2)^2$.

4. Find the volume of the pyramid in the figure below.

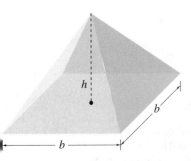

5. The amount of water consumed by a community varies throughout the day, peaking, naturally, around meal hours. During the 6-h period between 12:00 noon ($t = 0$) and 6:00 p.m. ($t = 6$), we find that the number of cubic metres of water consumed per hour at time t is given by the function

$$f(t) = 5000 + 21.65t^2 - 249.7t^3 + 97.52t^4 - 9.680t^5.$$

Find the total consumption during this 6-h period.

6. The number of bees per unit area at a distance x from a hive is given by

$$\rho(x) = \frac{600\,000}{31\pi R^5}(R^3 + 2R^2x - Rx^2 - 2x^3), \quad 0 \le x \le R,$$

where R is the maximum distance travelled by the bees.
  (a) What is the number of bees in the colony?
  (b) How many bees are within a distance $R/2$ of the hive?

7. If the radius of the blood vessel in Figure 7.72 is reduced to $R/2$ because of arteriosclerosis, the velocity profile is

$$v(r) = c(R^2 - 4r^2), \quad 0 \le r \le R/2.$$

What percentage of the normal flow ($\pi cR^4/2$) gets through the hardened vessel?

8. A tree trunk of diameter 50 cm (figure below) has a wedge cut from it by two planes. The lower plane is perpendicular to the axis of the trunk, and together the planes make an angle $\pi/3$ radians, meeting along a diameter of the circular cross-section of the trunk. Find the volume of the wedge.

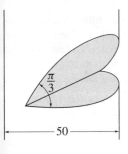

∗ 9. Find the area of the surface of revolution formed by rotating about the x-axis that part of the curve $24xy = x^4 + 48$ between $x = 1$ and $x = 2$.

∗ 10. Find the area of the surface of revolution generated by rotating the curve $8y^2 = x^2(1 - x^2)$ about the x-axis.

∗ 11. Find the area of the surface of revolution generated by rotating the loop of the curve $9y^2 = x(3 - x)^2$ about (a) the y-axis and (b) the x-axis.

∗ 12. The base of a solid is the circle $x^2 + y^2 = r^2$, and every plane section perpendicular to the x-axis is an isosceles triangle (figure below). Find the volume of the solid.

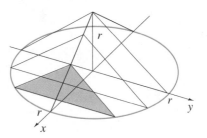

∗ 13. The end of the rod at $x = 0$ in the figure below is rigidly fixed. If a force with magnitude F is applied to the right end, how long is the rod?

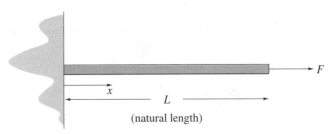

(natural length)

∗ 14. In the figure below, a mass M is placed on a vertical rod. If the length of the compressed rod is L, what is its length if M is removed and the rod is turned horizontally? Assume that M is so large that the weight of the rod can be neglected in comparison.

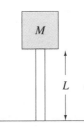

∗ 15. What is the answer to Exercise 14 if the weight of the rod is taken into account?

∗ 16. Suppose the rod in Exercise 13 is turned vertically so that its top end is fixed, and the force F pulls vertically downward on the lower end. How long is the rod if the weight of the rod is also taken into account?

∗ 17. What happens when you attempt to find the stretch of the rod in Figure 7.76 if the pointed end is at $y = 0$ and the larger end is at $y = L$?

* **18.** Suppose a mass M is attached to the lower end of the tapered rod in Figure 7.76. It is so large that the mass of the rod is negligible by comparison. What happens when you attempt to find the stretch in the rod?

* **19.** A tapered rod of length L has square cross-sections. The squares on the ends have dimensions a and b $(b > a)$. The rod is placed in a vertical position with the larger end below the smaller end, and the smaller end fixed in position (figure below). What is the length of the rod in this position?

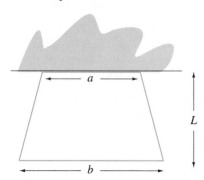

* **20.** What is the length of the rod in Exercise 19 if the rod is turned upside down?

* **21.** Repeat Exercise 20 if a mass M is distributed over the bottom of the rod.

* **22.** Repeat Exercise 19 if a mass M is distributed over the bottom of the rod.

* **23.** Electrons are fired from an electron gun at a target (figure below). The probability that an electron strikes the target in a ring of unit area at a distance x from the centre of the target is given by

$$p = f(x) = \frac{5}{3\pi R^5}(R^3 - x^3), \quad 0 \le x \le R.$$

What percentage of a cascade of electrons hits within a distance r from the centre of the target?

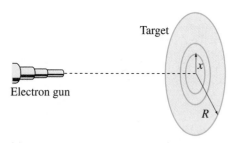

* **24.**   (a) Find the solution of formula 7.44 for deflections of the string if $F(x) = k$, where $k < 0$ is a constant.
   (b) Draw a graph of the deflected string. Is it symmetric about $x = L/2$ with minimum at $x = L/2$? Is this to be expected?

* **25.** Find the solution for formula 7.44 if the string is subjected to the constant loading of Exercise 24 and a concentrated load of F newtons at the centre of the string. Use the property in Exercise 34 of Section 7.10.

* **26.** Static deflections $y(x)$ of a diving board of length L and fixed end $x = 0$ (figure below) must satisfy

$$EI\frac{d^4y}{dx^4} = F(x), \quad 0 < x < L,$$

$$y(0) = y'(0) = 0 = y''(L) = y'''(L),$$

where E and I are material constants, and $F(x)$ is the loading. The Green's function for this problem is

$$G(x; X) = \frac{1}{6EI}(x - X)^3h(x - X) - \frac{x^3}{6EI} + \frac{Xx^2}{2EI}.$$

   (a) Use formula 7.46 to find deflections of the beam when $F(x) = k$, where $k < 0$ is a constant (perhaps the weight of the beam itself).
   (b) Where is $y(x)$ a minimum?

* **27.**   (a) Repeat Exercise 26 if the loading $F(x) = -F\delta(x - L/2)$ is due to a concentrated load F at $x = L/2$, and the weight of the board is negligible in comparison. Use the property in Exercise 34 of Section 7.10.
   (b) Draw a graph of the diving board. Is any part of it straight?

* **28.** When both ends of a beam are clamped horizontally, deflections must satisfy

$$EI\frac{d^4y}{dx^4} = F(x),$$

$$y(0) = y'(0) = 0 = y(L) = y'(L).$$

The Green's function for this problem is

$$G(x; X) = \frac{1}{6EI}(x - X)^3h(x - X)$$

$$+ \frac{x^3}{6EIL^3}(-L^3 + 3LX^2 - 2X^3)$$

$$+ \frac{x^2}{2EIL^2}(X^3 - 2LX^2 + L^2X).$$

   (a) Use formula 7.46 to find deflections of the beam when $F(x) = k$, where $k < 0$ is a constant (perhaps the weight of the beam itself).
   (b) Verify that $y(x)$ has a minimum where it should be expected.

* **29.** Repeat Exercise 28 if the loading $F(x) = -F\delta(x - L/2)$ is due to a concentrated load F at $x = L/2$ and the weight of the beam is negligible in comparison. Use the property in Exercise 34 of Section 7.10.

* **30.** The vertical wall of a rectangular container filled with water has a vertical rectangular slit with height h and width w. The upper edge of the slit is H units below the surface of the water. Use the modified Torricelli law $v = c\sqrt{2gh}$ from Section 5.5 to show that the volume rate at which water runs out of the slit, assuming that the container is kept full, is

$$\frac{2\sqrt{2g}cw}{3}[(H + h)^{3/2} - H^{3/2}].$$

31. Two right circular cylinders, each of radius r, have axes that intersect at right angles. Find the volume common to the two cylinders.

32. What is the volume of air trapped in the attic of a house if the roof has shape shown in the figure below? The peak of the roof is 2 m above the base.

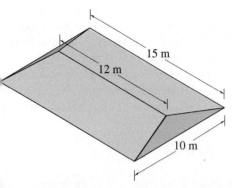

33. A certain population is being started at time $t = 0$. Once born, the probability that an individual lives to an age t is $p(t)$. For time $t > 0$, the birth rate is $r(t)$ individuals per unit time. Show that at time T the number of individuals in the population is

$$N(T) = \int_0^T p(T - t) r(t)\, dt.$$

34.
(a) An orange is spherical with radius r. Suppose it is cut into slices of equal width. If the thickness of each slice is t, find the area of peel on each slice if thickness of peel is h.

(b) Find the volume of peel in each slice.

35. The depth of water in the container in the figure below is originally 10 cm. Water evaporates from the surface at a rate proportional to the area of the surface. If the water level drops 1 cm in 5 days, how long does it take for the water to evaporate completely? Note that Exercise 19 in Section 5.5 is this same problem given a container that is a right circular cone. Now we ask you to repeat the problem with no knowledge of the shape of the container.

** **36.** An appliance retailer must be concerned with her inventory costs. For example, given that she sells N refrigerators per year, she must decide whether to order the year's supply at one time, in which case a large storage area would be necessary, or to make periodic orders throughout the year, perhaps of $N/12$ at the first of each month. In the latter case she would incur costs due to paperwork, delivery charges, and so on. Let us suppose that the retailer decides to order in equal-lot sizes, x, at equally spaced intervals, N/x times per year.

(a) If each order has fixed costs of F dollars, plus f dollars for each refrigerator, what are the total yearly ordering costs?

(b) Between successive deliveries, the retailer's stock dwindles from x to 0. If she assumes that the number of refrigerators in stock decreases linearly in time, and the yearly stocking cost per refrigerator is p dollars, what are the yearly stocking costs?

(c) If the retailer's total yearly inventory costs are her ordering costs in part (a) plus her stocking costs in part (b), what value of x minimizes inventory costs?

7.10 Improper Integrals

According to the first fundamental theorem of integral calculus (Theorem 6.3), definite integrals are evaluated with antiderivatives,

$$\int_a^b f(x)\, dx = \left\{ \int f(x)\, dx \right\}_a^b,$$

provided that $f(x)$ is continuous on $a \le x \le b$. In this section we investigate what to do when $f(x)$ is not continuous on $a \le x \le b$, or when either a or b is infinite.

Consider whether a reasonable meaning can be given to the integrals

$$\int_1^\infty \frac{1}{x^2}\, dx \quad \text{and} \quad \int_1^\infty \frac{1}{\sqrt{x}}\, dx.$$

Both integrals are *improper* in the sense that their upper limits are not finite, and we therefore call them **improper integrals**. If $b > 1$, there is no difficulty with evaluation and interpretation of

$$\int_1^b \frac{1}{x^2}\, dx \quad \text{and} \quad \int_1^b \frac{1}{\sqrt{x}}\, dx.$$

Clearly,

$$\int_1^b \frac{1}{x^2}\, dx = \left\{-\frac{1}{x}\right\}_1^b = 1 - \frac{1}{b} \quad \text{and}$$

$$\int_1^b \frac{1}{\sqrt{x}}\, dx = \left\{2\sqrt{x}\right\}_1^b = 2\sqrt{b} - 2.$$

We can interpret the first integral as the area under the curve $y = 1/x^2$, above the x-axis, and between the vertical lines $x = 1$ and $x = b$ (Figure 7.78). The second integral has exactly the same interpretation but uses the curve $y = 1/\sqrt{x}$ in place of $y = 1/x^2$ (Figure 7.79).

In Table 7.1 we have listed values of these definite integrals corresponding to various values of b. It is clear that the two integrals display completely different characteristics. As b is made very large, the integral of $1/x^2$ is always less than 1, but gets closer to 1 as b increases. In other words,

$$\lim_{b \to \infty} \int_1^b \frac{1}{x^2}\, dx = 1.$$

Geometrically, the area in Figure 7.78 is always less than 1 but approaches 1 as b approaches infinity.

Contrast this with the integral of $1/\sqrt{x}$, which becomes indefinitely large as b increases:

$$\lim_{b \to \infty} \int_1^b \frac{1}{\sqrt{x}}\, dx = \infty;$$

that is, the area in Figure 7.79 can be made as large as desired by choosing b sufficiently large.

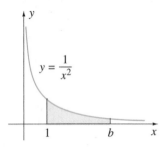

FIGURE 7.78   Area under curve $y = 1/x^2$

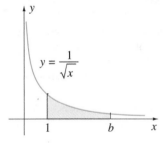

FIGURE 7.79   Area under curve $y = 1/\sqrt{x}$

TABLE 7.1

b	$\int_1^b \frac{1}{x^2}\, dx$	$\int_1^b \frac{1}{\sqrt{x}}\, dx$
100	0.99	18
10 000	0.999 9	198
1 000 000	0.999 999	1 998
100 000 000	0.999 999 99	19 998

On the basis of these calculations, we would like to say that the improper integral of $1/x^2$ has value 1, whereas the improper integral of $1/\sqrt{x}$ has no value. In other words, if these improper integrals are to have values, they should be defined by the limits

$$\int_1^\infty \frac{1}{x^2}\, dx = \lim_{b \to \infty} \int_1^b \frac{1}{x^2}\, dx \quad \text{and} \quad \int_1^\infty \frac{1}{\sqrt{x}}\, dx = \lim_{b \to \infty} \int_1^b \frac{1}{\sqrt{x}}\, dx.$$

In the second case the limit does not exist, and we interpret this to mean that the improper integral does not exist or has no value.

In general, then, an improper integral with an infinite upper limit is defined in terms of limits as follows.

DEFINITION 7.4

If $f(x)$ is continuous for $x \geq a$, we define

$$\int_a^\infty f(x)\, dx = \lim_{b \to \infty} \int_a^b f(x)\, dx, \qquad (7.47)$$

provided that the limit exists.

If the limit exists, we say that the improper integral is equal to the limit, or that the improper integral *converges* to the limit. If the limit does not exist, we say that the improper integral has no value, or that it *diverges*.

EXAMPLE 7.32

Determine whether the following improper integrals converge or diverge:

(a) $\displaystyle\int_3^\infty \frac{1}{(x-2)^3}\,dx$ (b) $\displaystyle\int_0^\infty \frac{x^2}{\sqrt{1+x^3}}\,dx$ (c) $\displaystyle\int_1^\infty \frac{2x^2+6}{3x^2+5}\,dx$

SOLUTION

(a) Using equation 7.47 gives us

$$\int_3^\infty \frac{1}{(x-2)^3}\,dx = \lim_{b\to\infty}\int_3^b \frac{1}{(x-2)^3}\,dx = \lim_{b\to\infty}\left\{\frac{-1}{2(x-2)^2}\right\}_3^b$$

$$= \lim_{b\to\infty}\left[\frac{1}{2} - \frac{1}{2(b-2)^2}\right] = \frac{1}{2},$$

and the improper integral therefore converges to $1/2$.

(b) Once again equation 7.47 gives

$$\int_0^\infty \frac{x^2}{\sqrt{1+x^3}}\,dx = \lim_{b\to\infty}\int_0^b \frac{x^2}{\sqrt{1+x^3}}\,dx = \lim_{b\to\infty}\left\{\frac{2}{3}\sqrt{1+x^3}\right\}_0^b$$

$$= \lim_{b\to\infty}\left[\frac{2}{3}\sqrt{1+b^3} - \frac{2}{3}\right] = \infty,$$

and the improper integral therefore diverges.

(c) Though we cannot at present find an antiderivative for $f(x) = (2x^2+6)/(3x^2+5)$, we can solve this problem by interpreting the improper integral as area. The graph of $f(x)$ in Figure 7.80 indicates that the curve is asymptotic to the line $y = 2/3$. Clearly, then, the improper integral

$$\int_1^\infty \frac{2x^2+6}{3x^2+5}\,dx$$

can have no value since the area under the curve is larger than the area of a rectangle of width $2/3$ and infinite length.

FIGURE 7.80   Area under curve $y = (2x^2+6)/(3x^2+5)$

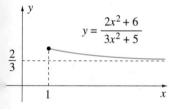

We can define improper integrals with infinite lower limits in exactly the same way as for improper integrals with infinite upper limits. Specifically, if $f(x)$ is continuous for $x \le b$, we define

$$\int_{-\infty}^b f(x)\,dx = \lim_{a\to-\infty}\int_a^b f(x)\,dx, \qquad (7.48)$$

provided that the limit exists. For example,

$$\int_{-\infty}^5 \frac{1}{(x-6)^5}\,dx = \lim_{a\to-\infty}\int_a^5 \frac{1}{(x-6)^5}\,dx = \lim_{a\to-\infty}\left\{\frac{-1}{4(x-6)^4}\right\}_a^5$$

$$= \lim_{a\to-\infty}\left[-\frac{1}{4} + \frac{1}{4(a-6)^4}\right] = -\frac{1}{4}.$$

When $f(x)$ is continuous for all x, we define

$$\int_{-\infty}^{\infty} f(x)\,dx = \lim_{a \to -\infty} \int_a^c f(x)\,dx + \lim_{b \to \infty} \int_c^b f(x)\,dx, \tag{7.49}$$

provided that both limits exist. The number c is arbitrary; existence of the limits is independent of c (see Exercise 33). For example, if we choose $c = 0$, then

$$\int_{-\infty}^{\infty} \frac{x}{(1 + x^2)^2}\,dx = \lim_{a \to -\infty} \int_a^0 \frac{x}{(1 + x^2)^2}\,dx + \lim_{b \to \infty} \int_0^b \frac{x}{(1 + x^2)^2}\,dx$$

$$= \lim_{a \to -\infty} \left\{ \frac{-1}{2(1 + x^2)} \right\}_a^0 + \lim_{b \to \infty} \left\{ \frac{-1}{2(1 + x^2)} \right\}_0^b$$

$$= \lim_{a \to -\infty} \left[\frac{1}{2(1 + a^2)} - \frac{1}{2} \right] + \lim_{b \to \infty} \left[\frac{1}{2} - \frac{1}{2(1 + b^2)} \right]$$

$$= -\frac{1}{2} + \frac{1}{2} = 0.$$

EXAMPLE 7.33

(a) Is it possible to paint the area bounded by the curves $y = 1/x$, $x = 1$, and $y = 0$?

(b) Rotate the area around the x-axis to form what is sometimes called Gabriel's horn. Does it have finite volume?

SOLUTION

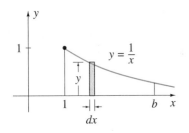

FIGURE 7.81   Painting area under $y = 1/x$

(a) Painting is possible if the area of the region bounded by the curves is finite (Figure 7.81). Since

$$\int_1^{\infty} \frac{1}{x}\,dx = \lim_{b \to \infty} \int_1^b \frac{1}{x}\,dx = \lim_{b \to \infty} \{\ln |x|\}_1^b = \lim_{b \to \infty} \ln b = \infty,$$

the area is not finite, and cannot therefore be painted.

(b) We use discs to determine whether Gabriel's horn has finite volume:

$$\int_1^{\infty} \pi y^2\,dx = \lim_{b \to \infty} \int_1^b \frac{\pi}{x^2}\,dx = \lim_{b \to \infty} \left\{ \frac{-\pi}{x} \right\}_1^b = \lim_{b \to \infty} \left(\pi - \frac{\pi}{b} \right) = \pi.$$

The volume is finite. Consider this now. The horn can be filled with paint, but the flat cross-sectional area inside the horn cannot be painted. How can this be?

EXAMPLE 7.34

Find the escape velocity of a projectile from the earth's surface.

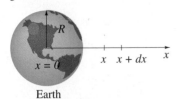

FIGURE 7.82   Escape velocity of projectile from earth's gravitational field

SOLUTION   Suppose the projectile is fired from the earth's surface with speed v in the x-direction (Figure 7.82). If the projectile is to escape the earth's gravitational pull, its initial kinetic energy must be greater than or equal to the work done against gravity as the projectile travels from the earth's surface ($x = R$) to a point where it is free of gravity ($x = \infty$). When the projectile is a distance x from the centre of the earth, the force of attraction on it is

$$F(x) = -\frac{GMm}{x^2},$$

where $G > 0$ is a constant, m is the mass of the projectile, and M is the mass of the earth. This is known as *Newton's universal law of gravitation*. The work done against gravity in travelling from $x = R$ to $x = \infty$ is

$$W = \int_R^\infty \frac{GmM}{x^2} \, dx = \lim_{b \to \infty} \int_R^b \frac{GmM}{x^2} \, dx$$

$$= \lim_{b \to \infty} \left\{ -\frac{GmM}{x} \right\}_R^b = \lim_{b \to \infty} \left[\frac{GmM}{R} - \frac{GmM}{b} \right] = \frac{GmM}{R}.$$

Since the initial kinetic energy of the projectile is $mv^2/2$, it escapes the gravitational pull of the earth if

$$\frac{1}{2}mv^2 \geq \frac{GmM}{R},$$

that is, if

$$v \geq \sqrt{\frac{2GM}{R}}.$$

Hence $\sqrt{2GM/R}$ is the escape velocity (notice that it is independent of the mass of the projectile). If we take the mean radius of the earth as 6370 km, its mean density as $\rho = 5.52 \times 10^3$ kg/m^3, and $G = 6.67 \times 10^{-11}$, we obtain $v = 11.2$ km/s. (Compare this with a 308 Winchester that fires a 150-grain bullet with a muzzle velocity of only 0.81 km/s.)

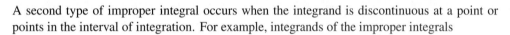

A second type of improper integral occurs when the integrand is discontinuous at a point or points in the interval of integration. For example, integrands of the improper integrals

$$\int_1^5 \frac{1}{\sqrt{x-1}} \, dx \qquad \text{and} \qquad \int_1^5 \frac{1}{(x-1)^4} \, dx$$

each have infinite discontinuities at $x = 1$. If c is a number between 1 and 5, it is easy to calculate

$$\int_c^5 \frac{1}{\sqrt{x-1}} \, dx = \left\{ 2\sqrt{x-1} \right\}_c^5 = 4 - 2\sqrt{c-1} \qquad \text{and}$$

$$\int_c^5 \frac{1}{(x-1)^4} \, dx = \left\{ \frac{-1}{3(x-1)^3} \right\}_c^5 = \frac{1}{3(c-1)^3} - \frac{1}{192},$$

and one possible interpretation of these definite integrals is the areas in Figures 7.83 and 7.84. If we let c approach 1 from the right in the first integral, we find that

$$\lim_{c \to 1^+} \int_c^5 \frac{1}{\sqrt{x-1}} \, dx = \lim_{c \to 1^+} \left(4 - 2\sqrt{c-1} \right) = 4;$$

FIGURE 7.83 Area under curve $y = 1/\sqrt{x-1}$

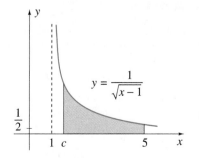

FIGURE 7.84 Area under curve $y = 1/(x-1)^4$

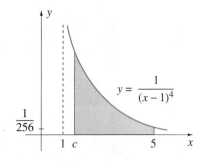

and in the second integral,

$$\lim_{c \to 1^+} \int_c^5 \frac{1}{(x-1)^4} \, dx = \lim_{c \to 1^+} \left[\frac{1}{3(c-1)^3} - \frac{1}{192} \right] = \infty.$$

It would seem reasonable then to define the area under the curve $y = 1/\sqrt{x-1}$, above the x-axis, and between the vertical lines $x = 1$ and $x = 5$ as 4 square units. On the other hand, the area bounded by $y = 1/(x-1)^4$, $y = 0$, $x = 5$, and $x = c$ becomes increasingly large as $x = c$ moves closer to $x = 1$, and we cannot therefore assign a value to this area when $c = 1$. In other words, if we define

$$\int_1^5 \frac{1}{\sqrt{x-1}} \, dx = \lim_{c \to 1^+} \int_c^5 \frac{1}{\sqrt{x-1}} \, dx \qquad \text{and}$$

$$\int_1^5 \frac{1}{(x-1)^4} \, dx = \lim_{c \to 1^+} \int_c^5 \frac{1}{(x-1)^4} \, dx,$$

then the first improper integral has a value of 4, whereas the second has no value.

These improper integrals are examples of the general situation described in the following definition and illustrated in Figures 7.85.

FIGURE 7.85a	FIGURE 7.85b	FIGURE 7.85c

Improper integrals for functions with infinite discontinuities

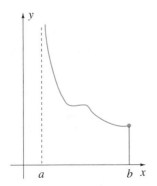
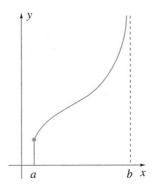
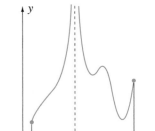

DEFINITION 7.5

If $f(x)$ is continuous at every point in the interval $a \leq x \leq b$ except at $x = a$, then

$$\int_a^b f(x) \, dx = \lim_{c \to a^+} \int_c^b f(x) \, dx, \qquad (7.50)$$

provided that the limit exists (Figure 7.85a). If $f(x)$ is continuous at every point in the interval $a \leq x \leq b$ except $x = b$, then

$$\int_a^b f(x) \, dx = \lim_{c \to b^-} \int_a^c f(x) \, dx, \qquad (7.51)$$

provided that the limit exists (Figure 7.85b). If $f(x)$ is continuous at every point in the interval $a \leq x \leq b$ except $x = d$ where $a < d < b$, then

$$\int_a^b f(x) \, dx = \lim_{c \to d^-} \int_a^c f(x) \, dx + \lim_{c \to d^+} \int_c^b f(x) \, dx, \qquad (7.52)$$

provided that both limits exist (Figure 7.85c).

EXAMPLE 7.35

Determine whether the following improper integrals converge or diverge:

(a) $\displaystyle\int_{-2}^{2} \frac{1}{(x-2)^3}\, dx$ (b) $\displaystyle\int_{-1}^{1} \frac{1}{x^4}\, dx$ (c) $\displaystyle\int_{1}^{5} \frac{x}{\sqrt{x-1}}\, dx$

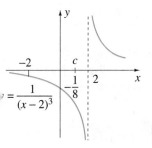

FIGURE 7.86 Improper integral of $f(x) = 1/(x-2)^3$

SOLUTION

(a) The graph of the integrand (Figure 7.86) shows a discontinuity at $x = 2$. Hence,

$$\int_{-2}^{2} \frac{1}{(x-2)^3}\, dx = \lim_{c \to 2^-} \int_{-2}^{c} \frac{1}{(x-2)^3}\, dx = \lim_{c \to 2^-} \left\{ \frac{-1}{2(x-2)^2} \right\}_{-2}^{c}$$

$$= \lim_{c \to 2^-} \left[\frac{-1}{2(c-2)^2} + \frac{1}{32} \right] = -\infty.$$

The improper integral therefore diverges.

(b) Since the discontinuity is interior to the interval of integration (Figure 7.87), we use equation 7.52:

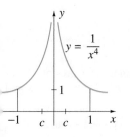

FIGURE 7.87 Improper integral of $f(x) = 1/x^4$

$$\int_{-1}^{1} \frac{1}{x^4}\, dx = \lim_{c \to 0^-} \int_{-1}^{c} \frac{1}{x^4}\, dx + \lim_{c \to 0^+} \int_{c}^{1} \frac{1}{x^4}\, dx$$

$$= \lim_{c \to 0^-} \left\{ \frac{-1}{3x^3} \right\}_{-1}^{c} + \lim_{c \to 0^+} \left\{ \frac{-1}{3x^3} \right\}_{c}^{1}$$

$$= \lim_{c \to 0^-} \left[-\frac{1}{3} - \frac{1}{3c^3} \right] + \lim_{c \to 0^+} \left[\frac{1}{3c^3} - \frac{1}{3} \right].$$

Since neither of these limits exists, the improper integral diverges.

(c) Since the integrand is discontinuous only at $x = 1$ (Figure 7.88),

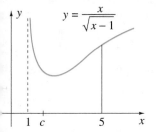

FIGURE 7.88 Improper integral of $f(x) = x/\sqrt{x-1}$

$$\int_{1}^{5} \frac{x}{\sqrt{x-1}}\, dx = \lim_{c \to 1^+} \int_{c}^{5} \frac{x}{\sqrt{x-1}}\, dx.$$

If we set $u = x - 1$, then $du = dx$, and

$$\int \frac{x}{\sqrt{x-1}}\, dx = \int \frac{u+1}{\sqrt{u}}\, du = \int (u^{1/2} + u^{-1/2})\, du$$

$$= \frac{2}{3}u^{3/2} + 2u^{1/2} + C = \frac{2}{3}(x-1)^{3/2} + 2\sqrt{x-1} + C.$$

Thus,

$$\int_{1}^{5} \frac{x}{\sqrt{x-1}}\, dx = \lim_{c \to 1^+} \left\{ \frac{2}{3}(x-1)^{3/2} + 2\sqrt{x-1} \right\}_{c}^{5}$$

$$= \lim_{c \to 1^+} \left[\frac{28}{3} - \frac{2}{3}(c-1)^{3/2} - 2\sqrt{c-1} \right] = \frac{28}{3},$$

and the improper integral converges.

FIGURE 7.89   Improper
integral of $f(x) = 1/(x^2 - 1)$

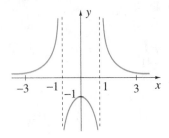

Various types of improper integrals may occur in the same problem. For example, Figure 7.89 shows that the integral of $f(x) = 1/(x^2 - 1)$ from $x = -3$ to $x = \infty$ involves the use of five limits:

$$\int_{-3}^{\infty} \frac{1}{x^2 - 1} dx = \lim_{c \to -1^-} \int_{-3}^{c} \frac{1}{x^2 - 1} dx + \lim_{c \to -1^+} \int_{c}^{0} \frac{1}{x^2 - 1} dx + \lim_{c \to 1^-} \int_{0}^{c} \frac{1}{x^2 - 1} dx$$

$$+ \lim_{c \to 1^+} \int_{c}^{10} \frac{1}{x^2 - 1} dx + \lim_{b \to \infty} \int_{10}^{b} \frac{1}{x^2 - 1} dx.$$

FIGURE 7.90   Improper
integral of function with jump
discontinuity

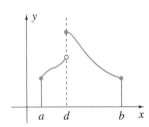

If any one of these limits fails to exist, then the improper integral diverges.

If an integrand has a jump discontinuity, such as the function $f(x)$ in Figure 7.90, then the improper integral of $f(x)$ from $x = a$ to $x = b$ is defined in terms of two limits,

$$\int_{a}^{b} f(x) \, dx = \lim_{c \to d^-} \int_{a}^{c} f(x) \, dx + \lim_{c \to d^+} \int_{c}^{b} f(x) \, dx,$$

but there is no question in this case that the improper integral converges.

EXERCISES 7.10

In Exercises 1–18 determine whether the improper integral converges or diverges. Find the value for each convergent integral.

1. $\int_{3}^{\infty} \frac{1}{(x+4)^2} dx$

2. $\int_{3}^{\infty} \frac{1}{(x+4)^{1/3}} dx$

3. $\int_{-\infty}^{-4} \frac{x}{\sqrt{x^2 - 2}} dx$

4. $\int_{-\infty}^{-4} \frac{x}{(x^2 - 2)^4} dx$

5. $\int_{-\infty}^{\infty} \frac{10^{10} x^3}{(x^4 + 5)^2} dx$

6. $\int_{-\infty}^{\infty} \frac{x^3}{(x^4 + 5)^{1/4}} dx$

7. $\int_{0}^{1} \frac{1}{(1-x)^{5/3}} dx$

8. $\int_{0}^{1} \frac{1}{\sqrt{1-x}} dx$

9. $\int_{1}^{\infty} x\sqrt{x^2 - 1} \, dx$

10. $\int_{2}^{5} \frac{x}{\sqrt{x^2 - 4}} dx$

11. $\int_{-1}^{1} \frac{x}{(1-x^2)^2} dx$

12. $\int_{-\infty}^{\infty} \frac{1}{x^2} dx$

13. $\int_{0}^{\infty} \frac{1}{\sqrt{x}} dx$

14. $\int_{-\infty}^{\pi/2} \frac{x}{(x^2 - 4)^2} dx$

15. $\int_{4}^{\infty} \cos x \, dx$

16. $\int_{-\infty}^{\infty} \sin x \, dx$

* 17. $\int_{0}^{\infty} \frac{x}{\sqrt{x+3}} dx$

* 18. $\int_{2}^{3} \frac{x^3}{\sqrt{x^2 - 4}} dx$

* 19. If $f(x)$ is a continuous odd function, is it necessarily true that

$$\int_{-\infty}^{\infty} f(x) \, dx = 0?$$

* 20.   (a) Is it possible to assign a number to represent the area bounded by the curves $y = 1 - x^{-1/4}$, $y = 1$, and $x = 1$

   (b) If the region in part (a) is rotated about the line $y = 1$, is it possible to assign a number to represent its volume?

* 21. Repeat Exercise 20 if $y = 1 - x^{-1/4}$ is replaced by $y = 1 - x^{-2/3}$

* 22. Repeat Exercise 20 if $y = 1 - x^{-1/4}$ is replaced by $y = 1 - x^{-3}$

* 23. A function $f(x)$ qualifies as a *probability density function* (pdf) on the interval $x \geq 0$ if it satisfies two conditions: $f(x) \geq 0$ for $x \geq 0$ and $\int_{0}^{\infty} f(x) \, dx = 1$.

   (a) Show that each of the following functions qualifies as a pdf

$$f(x) = \frac{6x}{(1 + 3x^2)^2}; \qquad f(x) = \frac{2x}{(1 + x)^3}$$

   (b) If a variable x has pdf $f(x)$ defined for $x \geq 0$, then the probability that x lies in an interval I is the definite integral of $f(x)$ over I. In particular, the probability that x is greater than or equal to a is

$$P(x \geq a) = \int_{a}^{\infty} f(x) \, dx.$$

Calculate $P(x \geq 3)$ for each pdf in part (a).

* 24. Verify that $\int_{1}^{\infty} \frac{1}{x^p} dx$ converges if $p > 1$ and diverges if $p \leq 1$

25. The force of repulsion between two point charges of like sign, one of size q and the other of size unity, has magnitude

$$F = \frac{q}{4\pi \epsilon_0 r^2},$$

where ϵ_0 is a constant and r is the distance between the charges. The potential V at any point P due to charge q is defined as the work required to bring the unit charge to P from infinity along the straight line joining q and P. Find a formula for V.

26. Verify that if $f(x)$ is continuous on $a \leq x \leq b$ except for a finite discontinuity at d (Figure 7.90), then the improper integral of $f(x)$ from $x = a$ to $x = b$ must converge.

27. Find the length of the loop of the curve $9y^2 = x(3 - x)^2$.

28. Is equation 7.49 equivalent to the following equation?

$$\int_{-\infty}^{\infty} f(x)\, dx = \lim_{a \to \infty} \int_{-a}^{a} f(x)\, dx$$

In Exercises 29–32 determine whether the improper integral converges or diverges. *Hint:* Compare each integral to a known convergent or divergent integral.

29. $\displaystyle \int_{2}^{\infty} \frac{x^2}{\sqrt{x^2 - 1}}\, dx$ * **30.** $\displaystyle \int_{1}^{3} \frac{x^3}{(27 - x^3)^2}\, dx$

31. $\displaystyle \int_{0}^{1} \frac{x^2}{\sqrt{1 - x^2}}\, dx$ * **32.** $\displaystyle \int_{-\infty}^{-2} \frac{\sqrt{-x}}{(x^2 + 5)^2}\, dx$

33. Show that existence of the limits in equation 7.49 is independent of the choice of c.

* **34.** One of the most important functions in physics and engineering is the *Dirac-delta function* $\delta(x - a)$ (introduced in Section 2.5 as the limit of the unit pulse function shown below):

$$\delta(x - a) = \lim_{\epsilon \to 0} P_\epsilon(x - a).$$

For a continuous function $f(x)$,

$$\int_{-\infty}^{\infty} f(x)\delta(x - a)\, dx = \int_{-\infty}^{\infty} f(x)\left[\lim_{\epsilon \to 0} P_\epsilon(x - a)\right] dx.$$

Assuming that the order of integration and the process of taking the limit as ϵ approaches zero can be interchanged, that is,

$$\int_{-\infty}^{\infty} f(x)\left[\lim_{\epsilon \to 0} P_\epsilon(x - a)\right] dx = \lim_{\epsilon \to 0} \int_{-\infty}^{\infty} f(x)P_\epsilon(x - a)\, dx,$$

show that

$$\int_{-\infty}^{\infty} f(x)\delta(x - a)\, dx = f(a).$$

This is the fundamental operational property that makes the Dirac-delta function so valuable.

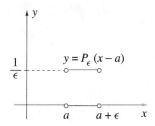

SUMMARY

In this chapter we used definite integrals in a wide variety of applications. With differentials and representative elements we were able to avoid memorization of formulas. To summarize the use of representative elements in various applications, consider the region R in Figure 7.91. To find the area of R, we draw rectangles of length $f(x) - g(x)$ and width dx. Areas $[f(x) - g(x)]\, dx$ of all such rectangles are then added to find the total area:

$$\int_{a}^{b} [f(x) - g(x)]\, dx.$$

FIGURE 7.91

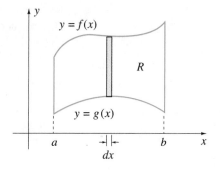

If R is rotated around the x-axis to form a solid of revolution, the rectangle generates a washer with volume $\{\pi[f(x)]^2 - \pi[g(x)]^2\}\,dx$, and therefore the total volume is

$$\int_a^b \{\pi[f(x)]^2 - \pi[g(x)]^2\}\,dx.$$

If R is rotated about the y-axis, the rectangle traces out a cylindrical shell of volume $2\pi x[f(x) - g(x)]\,dx$, and the total volume is

$$\int_a^b 2\pi x[f(x) - g(x)]\,dx.$$

First moments of this rectangle about the x-and y-axes are

$$\frac{1}{2}[f(x) + g(x)][f(x) - g(x)]\,dx \qquad \text{and} \qquad x[f(x) - g(x)]\,dx,$$

and these lead to the definite integrals

$$\int_a^b \frac{1}{2}\{[f(x)]^2 - [g(x)]^2\}\,dx \qquad \text{and} \qquad \int_a^b x[f(x) - g(x)]\,dx$$

for first moments of area of R about the x- and y-axes. Second moments of area of region R about the x- and y-axes are

$$\int_a^b \frac{1}{3}\{[f(x)]^3 - [g(x)]^3\}\,dx \qquad \text{and} \qquad \int_a^b x^2[f(x) - g(x)]\,dx,$$

where in the first integral it was necessary to use formula 7.42 for the second moment of the rectangle.

The key, then, to use of the definite integral is to divide the region into smaller elements, calculate the quantity required for a representative element, and then use a definite integral to add over all elements.

An important point to keep in mind is that for lengths in the y-direction, we always take "upper y minus lower y," and for lengths in the x-direction, we use "larger x minus smaller x." This way, many sign errors can be avoided when calculating *lengths*. For instance, note that this rule is used to find the length of a rectangle (in areas, volumes, fluid pressure), the radius of a circle (in volumes, surface area), depth (fluid pressure), and some distances (work).

An improper integral is a definite integral with an infinite limit and/or a point of discontinuity in the interval of integration. The value of an improper integral is defined by first calculating it over a finite interval in which the integrand is continuous, and then letting the limit on the integral approach either infinity or the point of discontinuity.

KEY TERMS

In reviewing this chapter, you should be able to define or discuss the following key terms:

Area	Representative rectangle
Volumes of solids of revolution	Disc method
Washer method	Cylindrical shell method
Length of a curve	Work
Energy	Kinetic energy
Conservation of energy	Fluid pressure
First moment of mass	Centre of mass
Moment of inertia or second moment	Volumes by slicing
Area of a surface of revolution	Green's function
Improper integrals	

REVIEW EXERCISES

In Exercises 1–5 calculate for the region bounded by the curves: (a) its area; (b) the volumes of the solids of revolution obtained by rotating the region about the x- and y-axes; (c) its centroid; and (d) its second moments of area about the x- and y-axes.

1. $y = 9 - x^2$, $y = 0$ **2.** $y = x^3$, $y = 2 - x$, $x = 0$

3. $x = y^2 - 1$, $x = 4 - 4y^2$ **4.** $2y = x$, $y + 1 = x$, $y = 0$

5. $2x + y = 2$, $y - 2x = 2$, $x = y + 1$, $y + x + 1 = 0$

In Exercises 6–10 find the volume of the solid of revolution obtained by rotating the region bounded by the curves about the line.

6. $y = |x| + 2$, $y = 3$,   about $y = 1$

7. $x = y^4$, $x = 4$,   about $x = 4$

8. $y = x^3 - x$, $y = 0$,   about $y = 0$

9. $yx^2 = 1$, $x = 1$, $x = 2$, $y = 0$,   about $y = -1$

10. $y = x + 2$, $y = \sqrt{4 - 2x}$, $y = 0$,   about $y = 1$

In Exercises 11–15 find the first and second moments of area of the region bounded by the curves about the line.

11. $y = 4x^2$, $y = 2 + 2x^2$,   about $y = -1$

12. $2x = \sqrt{y}$, $\sqrt{2}x = \sqrt{y - 2}$, $x = 0$,   about $x = 1$

13. $3y = 2x + 6$, $2y + x = 4$, $y = 0$,   about $y = -2$

14. $3y = 2x + 6$, $2y + x = 4$, $y = 0$,   about $y = 2$

15. $y = x^2 - 2x$ ($y \leq 0$), $y = 2x$, $y = 4 - 2x$,   about $x = 1$

16. A water tank in the form of a right circular cylinder with radius 1 m and height 3 m has its axis vertical. If it is half full, how much work is required to empty the tank through an outlet at the top of the tank?

17. A car runs off a bridge into a river and submerges. If the tops of its front and rear side windows are 2 m below the surface of the water (see figure), what is the force due to water pressure on each window?

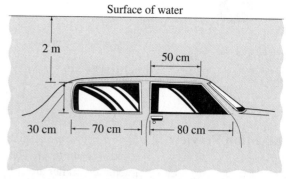

Surface of water

2 m   50 cm

30 cm   70 cm   80 cm

18. Find the surface area of the volume of the solid of revolution obtained by rotating the region bounded by the curves $2x + y = 2$, $y - 2x = 2$, and $y = 0$ about the x-axis.

* **19.** A tank in the form of an inverted right circular cone of depth 10 m and radius 4 m is full of water. How much work is required to lower the water level in the tank by 2 m by pumping water out through an outlet 1 m above the top of the tank?

* **20.** Show that if it takes W units of work to stretch a spring from equilibrium to a certain stretch, then it requires $3W$ more units of work to double that stretch.

In Exercises 21–28 determine whether the improper integral converges or diverges. Evaluate each convergent integral.

* **21.** $\displaystyle\int_{1}^{\infty} \frac{1}{x^{3/2}}\, dx$ * **22.** $\displaystyle\int_{0}^{3} \frac{1}{\sqrt{3 - x}}\, dx$

* **23.** $\displaystyle\int_{-1}^{0} \frac{1}{(x + 1)^2}\, dx$ * **24.** $\displaystyle\int_{-2}^{2} \frac{x}{\sqrt{4 - x^2}}\, dx$

* **25.** $\displaystyle\int_{-\infty}^{\infty} \frac{1}{(x + 3)^3}\, dx$ * **26.** $\displaystyle\int_{-\infty}^{-3} \frac{1}{\sqrt{-x}}\, dx$

* **27.** $\displaystyle\int_{-6}^{\infty} x\sqrt{x^2 + 4}\, dx$ * **28.** $\displaystyle\int_{-\infty}^{\infty} \frac{x}{(x^2 - 1)^2}\, dx$

* **29.** A uniformly tapered rod of length 1 m has circular cross-sections, the radii of its ends being 1 cm and 2 cm. Find the volume of the rod by (a) using the fact that every cross-section is circular and (b) considering the rod as a volume of a solid of revolution.

* **30.** In Exercise 16, how much work is required to empty the tank if its axis is horizontal and the outlet is at the top of the tank?

CHAPTER **8** | # Techniques of Integration

Application Preview

Shown below are a tractor and trailer in two positions. Both tractor and trailer begin on the x-axis (position A). The tractor then turns to the right and moves so that the pivot point where tractor and trailer are coupled together follows a curve whose equation is $Y = g(X)$. Tractor and trailer are shown at some time later in position B.

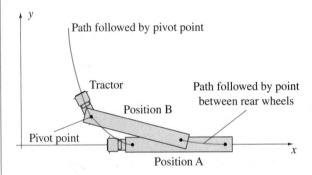

THE PROBLEM   Given that the pivot point moves along the curve $Y = g(X)$, find the equation of the curve followed by the point midway between the rear wheels. (See Example 8.13 on page 511 for the solution.)

In Chapter 5, we discussed three techniques for evaluating antiderivatives. First, since every differentiation formula can be restated as an antidifferentiation formula, it follows that the more competent we are at differentiation, the more likely we are to recognize an antiderivative. Our second technique was called "adjusting constants." It is applicable to sufficiently simple integrands that we can immediately guess the antiderivative to within a multiplicative constant. It is then a matter of adjusting the constant. The third technique was to change the variable of integration. By a suitable transformation, replace a complicated antiderivative with a simpler one.

In this chapter we develop additional techniques applicable to more complex integration problems. At the same time we take the opportunity to review the applications of definite integrals in Chapter 7.

8.1 Integration Formulas and Substitutions

Every differentiation formula can be restated as an antidifferentiation formula. A partial list of frequently encountered formulas is as follows:

$$\int u^n \, du = \frac{u^{n+1}}{n+1} + C, \quad n \neq -1; \tag{8.1a}$$

$$\int \frac{1}{u} \, du = \ln |u| + C; \tag{8.1b}$$

$$\int e^u \, du = e^u + C; \tag{8.1c}$$

$$\int a^u \, du = a^u \log_a e + C; \tag{8.1d}$$

$$\int \cos u \, du = \sin u + C; \tag{8.1e}$$

$$\int \sin u \, du = -\cos u + C; \tag{8.1f}$$

$$\int \sec^2 u \, du = \tan u + C; \tag{8.1g}$$

$$\int \csc^2 u \, du = -\cot u + C; \tag{8.1h}$$

$$\int \sec u \tan u \, du = \sec u + C; \tag{8.1i}$$

$$\int \csc u \cot u \, du = -\csc u + C; \tag{8.1j}$$

$$\int \tan u \, du = \ln |\sec u| + C = -\ln |\cos u| + C; \tag{8.1k}$$

$$\int \cot u \, du = -\ln |\csc u| + C = \ln |\sin u| + C; \tag{8.1l}$$

$$\int \sec u \, du = \ln |\sec u + \tan u| + C; \tag{8.1m}$$

$$\int \csc u \, du = \ln |\csc u - \cot u| + C; \tag{8.1n}$$

$$\int \frac{1}{\sqrt{1 - u^2}} \, du = \text{Sin}^{-1} u + C; \tag{8.1o}$$

$$\int \frac{1}{1 + u^2} \, du = \text{Tan}^{-1} u + C; \tag{8.1p}$$

$$\int \frac{1}{u\sqrt{u^2 - 1}} \, du = \text{Sec}^{-1} u + C; \tag{8.1q}$$

$$\int \cosh u \, du = \sinh u + C; \tag{8.1r}$$

$$\int \sinh u \, du = \cosh u + C; \tag{8.1s}$$

$$\int \text{sech}^2 u \, du = \tanh u + C; \tag{8.1t}$$

$$\int \text{csch}^2 u \, du = -\coth u + C; \tag{8.1u}$$

$$\int \text{sech} \, u \tanh u \, du = -\text{sech} \, u + C; \tag{8.1v}$$

$$\int \text{csch} \, u \coth u \, du = -\text{csch} \, u + C. \tag{8.1w}$$

All of these except Formulas 8.1k, l, m, n are restatements of differentiation formulas from Chapter 3. Formulas 8.1k and 8.1m are verified in Example 8.1 below. Formulas 8.1l and 8.1n are similar.

EXAMPLE 8.1

Verify formulas 8.1k and 8.1m.

SOLUTION   To verify 8.1k we express $\tan u$ in terms of $\sin u$ and $\cos u$:

$$\int \tan u \, du = \int \frac{\sin u}{\cos u} \, du = -\ln|\cos u| + C = \ln|\sec u| + C.$$

If integration to $\ln|\cos u|$ is not obvious, try the substitution $v = \cos u$.

To obtain the antiderivative of $\sec u$ requires a subtle trick. We first multiply numerator and denominator by $\sec u + \tan u$:

$$\int \sec u \, du = \int \sec u \frac{\sec u + \tan u}{\sec u + \tan u} \, du = \int \frac{\sec^2 u + \sec u \tan u}{\sec u + \tan u} \, du.$$

If we now set $v = \sec u + \tan u$, then $dv = (\sec u \tan u + \sec^2 u) \, du$, and

$$\int \sec u \, du = \int \frac{1}{v} \, dv = \ln|v| + C = \ln|\sec u + \tan u| + C.$$

We could list many other integration formulas, but one of the purposes of this chapter is to develop integration techniques that eliminate the need for excessive memorization of formulas. Our emphasis is on development of antiderivatives rather than memorization of formulas, and we prefer to keep the list of formulas short.

In Chapter 5 we demonstrated how a change of variable could sometimes replace a complex integration problem with a simpler one. When the integrand contains $x^{1/n}$, where n is a positive integer, it is often useful to set $u = x^{1/n}$, or, equivalently, $x = u^n$. We illustrate in the following example.

EXAMPLE 8.2

Evaluate the following indefinite integrals:

(a) $\displaystyle\int \frac{2 - \sqrt{x}}{2 + \sqrt{x}} \, dx$ (b) $\displaystyle\int \frac{x^{1/3}}{1 + x^{2/3}} \, dx$ (c) $\displaystyle\int \frac{\sqrt{x}}{1 + x^{1/3}} \, dx$

SOLUTION

(a) If we set $u = \sqrt{x}$ or $x = u^2$, then $dx = 2u \, du$, and

$$\int \frac{2 - \sqrt{x}}{2 + \sqrt{x}} \, dx = \int \frac{2 - u}{2 + u} (2u \, du) = 2 \int \frac{2u - u^2}{2 + u} \, du.$$

Long division of $-u^2 + 2u$ by $u + 2$ immediately gives

$$\int \frac{2 - \sqrt{x}}{2 + \sqrt{x}} \, dx = 2 \int \left(-u + 4 - \frac{8}{u + 2} \right) du$$

$$= 2 \left(-\frac{u^2}{2} + 4u - 8 \ln|u + 2| \right) + C$$

$$= -x + 8\sqrt{x} - 16 \ln \left(2 + \sqrt{x} \right) + C.$$

(b) If we set $u = x^{1/3}$ or $x = u^3$, then $dx = 3u^2\,du$, and

$$\int \frac{x^{1/3}}{1 + x^{2/3}}\,dx = \int \frac{u}{1 + u^2}(3u^2\,du) = 3\int \frac{u^3}{1 + u^2}\,du.$$

Division of u^3 by $u^2 + 1$ leads to

$$\int \frac{x^{1/3}}{1 + x^{2/3}}\,dx = 3\int \left(u - \frac{u}{u^2 + 1}\right)du$$

$$= 3\left(\frac{u^2}{2} - \frac{1}{2}\ln|u^2 + 1|\right) + C$$

$$= \frac{3}{2}x^{2/3} - \frac{3}{2}\ln(x^{2/3} + 1) + C.$$

(c) The purpose of the substitution $u = x^{1/n}$ is to rid the integrand of fractional powers. In the presence of both $x^{1/2}$ and $x^{1/3}$, we set $u = x^{1/6}$ or $x = u^6$. Then $dx = 6u^5\,du$, and

$$\int \frac{\sqrt{x}}{1 + x^{1/3}}\,dx = \int \frac{u^3}{1 + u^2}(6u^5\,du) \qquad \text{(and by long division)}$$

$$= 6\int \left(u^6 - u^4 + u^2 - 1 + \frac{1}{1 + u^2}\right)du$$

$$= 6\left(\frac{u^7}{7} - \frac{u^5}{5} + \frac{u^3}{3} - u + \mathrm{Tan}^{-1}u\right) + C$$

$$= \frac{6}{7}x^{7/6} - \frac{6}{5}x^{5/6} + 2x^{1/2} - 6x^{1/6} + 6\,\mathrm{Tan}^{-1}(x^{1/6}) + C.$$

EXERCISES 8.1

In Exercises 1–22 evaluate the indefinite integral.

1. $\displaystyle\int \frac{x^2}{5 - 3x^3}\,dx$

2. $\displaystyle\int xe^{-2x^2}\,dx$

3. $\displaystyle\int \frac{x}{(x^2 + 2)^{1/3}}\,dx$

4. $\displaystyle\int \frac{e^x}{1 + e^x}\,dx$

5. $\displaystyle\int \frac{4t + 8}{t^2 + 4t + 5}\,dt$

6. $\displaystyle\int x^2\sqrt{1 - 3x^3}\,dx$

7. $\displaystyle\int (x + 1)(x^2 + 2x)^{1/3}\,dx$

8. $\displaystyle\int \frac{x^2}{(1 + x^3)^3}\,dx$

9. $\displaystyle\int \frac{\sqrt{x}}{1 - \sqrt{x}}\,dx$

10. $\displaystyle\int \frac{1 - \sqrt{x}}{\sqrt{x}}\,dx$

11. $\displaystyle\int \frac{x + 2}{x + 1}\,dx$

12. $\displaystyle\int \frac{x^2 + 2}{x^2 + 1}\,dx$

13. $\displaystyle\int \frac{\sin\theta}{\cos\theta - 1}\,d\theta$

14. $\displaystyle\int \frac{x + 3}{\sqrt{2x + 4}}\,dx$

15. $\displaystyle\int \frac{e^x}{1 + e^{2x}}\,dx$

16. $\displaystyle\int \sin^3 2x\cos 2x\,dx$

17. $\displaystyle\int x^5(2x^2 - 5)^4\,dx$

18. $\displaystyle\int \frac{x^3}{(x + 5)^2}\,dx$

19. $\displaystyle\int z^2\sqrt{3 - z}\,dz$

20. $\displaystyle\int \tan 3x\,dx$

21. $\displaystyle\int \frac{(x - 3)^{2/3}}{(x - 3)^{2/3} + 1}\,dx$

22. $\displaystyle\int \frac{\sqrt{x}}{1 + x^{1/4}}\,dx$

* 23. Find the length of the curve $y = \ln(\cos x)$ from $(0, 0)$ to $(\pi/4, -(\ln 2)/2)$.

* 24. Find the area of the region bounded by the curves

$$y = \frac{x^2 + 1}{x + 1}, \qquad x + 3y = 7.$$

* **25.**   (a) Find the centroid of the region bounded by the curves $y = (3 + x)^{3/2}$, $x = 1$, and $y = 0$.

    (b) What is its second moment of area about the line $x = 1$?

* **26.** If $f(x)$ is a continuous even function, prove that

$$\int_{-a}^{a} f(x)\, dx = 2\int_{0}^{a} f(x)\, dx;$$

and if $f(x)$ is a continuous odd function,

$$\int_{-a}^{a} f(x)\, dx = 0.$$

* **27.**   (a) Show that the function $f(x) = \lambda e^{-\lambda x}$, where $\lambda > 0$ is a constant, qualifies as a probability density function on the interval $x \geq 0$. See Exercise 23 in Section 7.10 for the definition of a pdf.

    (b) If the probability that $x \geq 3$ is 0.5, what is λ?

* **28.** A paratrooper and his parachute fall from rest with a combined mass of 100 kg. At any instant during descent, the parachute has an air resistance force acting on it that in newtons is equal to one-half its speed (in metres per second). Assuming that the paratrooper falls vertically downward and that the parachute is already open when the jump takes place, the differential equation describing his motion is

$$200\frac{dv}{dt} = 1962 - v,$$

where v is velocity and t is time.

    (a) Find v as a function of t.

    (b) If x measures vertical distance fallen, then $v = dx/dt$. Use this to find the position of the paratrooper as a function of time.

* **29.** Evaluate $\int_{0}^{4\pi} \sqrt{1 + \cos x}\, dx$. *Hint:* $\cos x = 2\cos^2 (x/2) - 1$.

* **30.** Evaluate

$$\int \frac{1}{x(3 + 2x^n)}\, dx$$

for any $n > 0$. *Hint:* Find A and B in order that

$$\frac{1}{x(3 + 2x^n)} = \frac{A}{x} + \frac{Bx^{n-1}}{3 + 2x^n}.$$

** **31.** Use the substitution $u - x = \sqrt{x^2 + 3x + 4}$ to evaluate

$$\int \frac{1}{(x^2 + 3x + 4)^{3/2}}\, dx.$$

** **32.** Show that the substitution $u - x = \sqrt{x^2 + bx + c}$ replaces the integral $\int \sqrt{x^2 + bx + c}\, dx$ with the integral of a rational function of u.

** **33.** Show that when the quadratic $c + bx - x^2$ factors as $c + bx - x^2 = (p + x)(q - x)$, then either of the substitutions $(p + x)u = \sqrt{c + bx - x^2}$ or $(q - x)u = \sqrt{c + bx - x^2}$ replaces the integral

$$\int \frac{1}{\sqrt{c + bx - x^2}}\, dx$$

with the integral of a rational function of u.

** **34.** If a, b, and n are constants, evaluate $\int x(a + bx)^n\, dx$.

** **35.** The function

$$f(x) = \frac{1}{\sqrt{2\pi}\,\sigma} e^{-(x-\mu)^2/(2\sigma^2)}, \quad -\infty < x < \infty,$$

where μ and σ are constants, is called the *normal probability density function*. Consequently,

$$\int_{-\infty}^{\infty} f(x)\, dx = 1.$$

If the expected value of x is defined as

$$\int_{-\infty}^{\infty} xf(x)\, dx,$$

show that μ is the expected value.

** **36.** Evaluate, if possible, $\int_{-\infty}^{\infty} e^{-|a-x|} e^{-|x|}\, dx$, where a is a constant.

8.2 Integration by Parts

One of the most powerful techniques for finding antiderivatives is that called **integration by parts**. It results from the product rule for differentiation:

$$\frac{d}{dx}(uv) = u\frac{dv}{dx} + v\frac{du}{dx}.$$

If we take antiderivatives of both sides of this equation with respect to x, we obtain

$$\int \frac{d}{dx}(uv)\, dx = \int u\frac{dv}{dx}\, dx + \int v\frac{du}{dx}\, dx.$$

Since the antiderivative on the left is uv, this equation may be rewritten in the form

$$\int u\,dv = uv - \int v\,du. \tag{8.2}$$

Equation 8.2 is called the **integration-by-parts formula**. It says that given an unknown integral (the left side of 8.2), if we arbitrarily divide the function after the integral sign into two parts — one called u and the other called dv — then we can set the unknown integral equal to the right side of 8.2. What this does is replace the integral

$$\int u\,dv \qquad \text{with} \qquad \int v\,du.$$

The problem is to choose u and dv so that the new integral is simpler than the original. For many examples, it is advantageous to let dv be the *most complicated part of the integrand (plus differential) that we can integrate mentally*. The example

$$\int xe^x\,dx$$

will clarify these ideas. To use integration by parts on this problem we must define u and dv in such a way that $u\,dv = xe^x\,dx$. There are four possibilities:

$$
\begin{aligned}
u &= 1, & dv &= xe^x\,dx; \\
u &= x, & dv &= e^x\,dx; \\
u &= e^x, & dv &= x\,dx; \\
u &= xe^x, & dv &= dx.
\end{aligned}
$$

Our rule suggests the choice

$$u = x, \qquad dv = e^x\,dx,$$

in which case

$$du = dx, \qquad v = e^x.$$

Formula 8.2 then gives

$$\int xe^x\,dx = xe^x - \int e^x\,dx.$$

We have therefore replaced integration of xe^x with integration of e^x, a definite simplification. The final solution is therefore

$$\int xe^x\,dx = xe^x - e^x + C.$$

This example illustrates that integration by parts replaces an integral with two others: one mental integration (from dv to v) and a second integration that is hopefully simpler than the original. We should also note that in the evaluation of the mental integration we did not include a constant of integration. For example, in the integration from $dv = e^x\,dx$ to $v = e^x$, we did not write $v = e^x + D$. Had we done so, the solution would have proceeded as follows:

$$\int xe^x\,dx = x(e^x + D) - \int (e^x + D)\,dx$$

$$= x(e^x + D) - e^x - Dx + C$$

$$= xe^x - e^x + C.$$

Constant D has therefore disappeared from the eventual solution. This always occurs, and consequently, we do not include an arbitrary constant in the mental integration for v.

EXAMPLE 8.3 ▼

Evaluate the following integrals:

$$\text{(a) } \int \ln x \, dx \qquad \text{(b) } \int x^2 \cos x \, dx \qquad \text{(c) } \int e^x \sin x \, dx$$

SOLUTION

(a) Integration by parts leaves no choice for u and dv in this integral; they must be

$$u = \ln x, \qquad dv = dx,$$

in which case

$$du = \frac{1}{x} \, dx, \qquad v = x.$$

Then we have

$$\int \ln x \, dx = x \ln x - \int x \frac{1}{x} \, dx = x \ln x - x + C.$$

(b) If we set

$$u = x^2, \qquad dv = \cos x \, dx,$$

then

$$du = 2x \, dx, \qquad v = \sin x, \qquad \text{and}$$

$$\int x^2 \cos x \, dx = x^2 \sin x - \int 2x \sin x \, dx.$$

Integration by parts has therefore reduced the power on x from 2 to 1. To eliminate the x in front of the trigonometric function completely, we perform integration by parts once again, this time with

$$u = x, \qquad dv = \sin x \, dx.$$

Then

$$du = dx, \qquad v = -\cos x, \qquad \text{and}$$

$$\int x^2 \cos x \, dx = x^2 \sin x - 2 \left(-x \cos x - \int -\cos x \, dx \right)$$

$$= x^2 \sin x + 2x \cos x - 2 \sin x + C.$$

(c) If we set

$$u = e^x, \qquad dv = \sin x \, dx,$$

then

$$du = e^x \, dx, \qquad v = -\cos x, \qquad \text{and}$$

$$\int e^x \sin x \, dx = -e^x \cos x - \int -e^x \cos x \, dx.$$

Integration by parts appears to have led nowhere; the integration of $e^x \cos x$ is essentially as difficult as that of $e^x \sin x$. If, however, we persevere and integrate by parts once again with

$$u = e^x, \qquad dv = \cos x \, dx,$$

then

$$du = e^x \, dx, \qquad v = \sin x, \qquad \text{and}$$

$$\int e^x \sin x \, dx = -e^x \cos x + e^x \sin x - \int e^x \sin x \, dx.$$

This equation can now be solved for the unknown integral. By transposing the last term on the right to the left, we find that

$$2 \int e^x \sin x \, dx = e^x (\sin x - \cos x),$$

and therefore,

$$\int e^x \sin x \, dx = \frac{1}{2} e^x (\sin x - \cos x) + C,$$

to which we have added the arbitrary constant C.

The technique in part (c) of this example is often called **integration by reproduction**. Two applications of integration by parts enabled us to reproduce the unknown integral and then solve the equation for it.

EXAMPLE 8.4

Evaluate $\displaystyle\int \frac{x^3}{\sqrt{9 + x^2}} \, dx$.

Solution Methods

(i) If we set $u = 9 + x^2$, then $du = 2x \, dx$, and

$$\int \frac{x^3}{\sqrt{9 + x^2}} \, dx = \int \frac{x^2}{\sqrt{9 + x^2}} (x \, dx) = \int \frac{u - 9}{\sqrt{u}} \frac{du}{2} = \frac{1}{2} \int (u^{1/2} - 9u^{-1/2}) \, du$$

$$= \frac{1}{2} \left(\frac{2}{3} u^{3/2} - 18u^{1/2} \right) + C = \frac{1}{3}(9 + x^2)^{3/2} - 9\sqrt{9 + x^2} + C.$$

(ii) If we set $u = \sqrt{9 + x^2}$, then $du = \left(x/\sqrt{9 + x^2} \right) dx$, and

$$\int \frac{x^3}{\sqrt{9 + x^2}} \, dx = \int x^2 \left(\frac{x}{\sqrt{9 + x^2}} \right) dx = \int (u^2 - 9) \, du = \frac{u^3}{3} - 9u + C$$

$$= \frac{1}{3}(9 + x^2)^{3/2} - 9\sqrt{9 + x^2} + C.$$

(iii) If we set

$$u = x^2, \qquad dv = \frac{x}{\sqrt{9 + x^2}} \, dx,$$

then

$$du = 2x \, dx, \qquad v = \sqrt{9 + x^2},$$

and

$$\int \frac{x^3}{\sqrt{9 + x^2}} \, dx = x^2 \sqrt{9 + x^2} - \int 2x\sqrt{9 + x^2} \, dx$$

$$= x^2 \sqrt{9 + x^2} - 2\left[\frac{1}{3}(9 + x^2)^{3/2} \right] + C$$

$$= x^2 \sqrt{9 + x^2} - \frac{2}{3}(9 + x^2)^{3/2} + C.$$

This example illustrates that more than one technique may be effective on an indefinite integral, and the answer may appear different depending on the technique used. According to Theorem 5.1, these solutions can only differ by an additive constant. To illustrate this in Example 8.4, note that the solution obtained by method (iii) can be written as

$$\sqrt{9+x^2}\left[x^2 - \frac{2}{3}(9+x^2)\right] + C = \frac{1}{3}\sqrt{9+x^2}\left[-18+x^2\right] + C$$

$$= \frac{1}{3}\sqrt{9+x^2}\left[-27+(9+x^2)\right] + C$$

$$= -9\sqrt{9+x^2} + \frac{1}{3}(9+x^2)^{3/2} + C,$$

which is the solution obtained by the other two methods. It is also worth noting that the three methods work very well provided that the power on x in the numerator of the integrand is odd. When it is even, they fail miserably. In such cases, trigonometric substitutions (Section 8.4) come to the rescue (see Example 8.12).

Integration-by-parts formula 8.2 applies to indefinite integrals. For definite integrals, it is replaced by

$$\int_a^b u\,dv = \left\{uv\right\}_a^b - \int_a^b v\,du. \tag{8.3}$$

EXAMPLE 8.5

Evaluate $\displaystyle\int_0^4 \frac{x}{\sqrt{x+8}}\,dx$.

SOLUTION   If we set

$$u = x, \qquad dv = \frac{1}{\sqrt{x+8}}\,dx,$$

then

$$du = dx, \qquad v = 2\sqrt{x+8}, \qquad \text{and}$$

$$\int_0^4 \frac{x}{\sqrt{x+8}}\,dx = \left\{2x\sqrt{x+8}\right\}_0^4 - \int_0^4 2\sqrt{x+8}\,dx$$

$$= 8\sqrt{12} - 2\left\{\frac{2}{3}(x+8)^{3/2}\right\}_0^4$$

$$= \frac{16}{3}\left(4\sqrt{2} - 3\sqrt{3}\right).$$

EXERCISES 8.2

In Exercises 1–18 evaluate the indefinite integral.

1. $\displaystyle\int x\sin x\,dx$

2. $\displaystyle\int x^2 e^{2x}\,dx$

3. $\displaystyle\int x^4\ln x\,dx$

4. $\displaystyle\int \sqrt{x}\ln(2x)\,dx$

5. $\displaystyle\int z\sec^2(z/3)\,dz$

6. $\displaystyle\int x\sqrt{3-x}\,dx$

7. $\displaystyle\int \operatorname{Sin}^{-1} x\,dx$

8. $\displaystyle\int x^2\sqrt{x+5}\,dx$

9. $\displaystyle\int \frac{x}{\sqrt{2+x}}\,dx$

10. $\displaystyle\int \frac{x^2}{\sqrt{2+x}}\,dx$

11. $\displaystyle\int \frac{x}{\sqrt{2+x^2}}\,dx$

12. $\displaystyle\int (x-1)^2\ln x\,dx$

13. $\displaystyle\int e^x\cos x\,dx$

14. $\displaystyle\int \operatorname{Tan}^{-1} x\,dx$

15. $\displaystyle\int \cos(\ln x)\,dx$

16. $\displaystyle\int e^{2x}\cos 3x\,dx$

17. $\displaystyle\int \frac{x^3}{\sqrt{5+3x^2}}\,dx$

18. $\displaystyle\int \ln(x^2+4)\,dx$

19. Consider the evaluation of $\int x^5 e^x \, dx$. Were we to use integration by parts, the answer would eventually be of the form

$$\int x^5 e^x \, dx = Ax^5 e^x + Bx^4 e^x + Cx^3 e^x + Dx^2 e^x + Exe^x + Fe^x + G.$$

Differentiate this equation, and thereby obtain A, B, C, D, E, and F with no integration).

20. The face of a dam is shown in the figure below, where the curve has equation

$$y = e^{k|x|} - 1, \qquad k = \frac{1}{100} \ln 201.$$

Find the total force on the dam due to water pressure when the water level on the dam is 100 m.

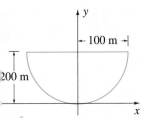

21. If we set $u = x^{-1}$ and $dv = dx$, then $du = -x^{-2} \, dx$ and $v = x$, and

$$\int \frac{1}{x} \, dx = \left(\frac{1}{x}\right) x - \int x \left(-\frac{1}{x^2} \, dx\right) = 1 + \int \frac{1}{x} \, dx.$$

Subtraction of $\int x^{-1} \, dx$ from each side of this equation now gives $0 = 1$. What is wrong with the argument?

The *Fourier series* is a topic of mathematics fundamental to all branches of engineering. We are constantly evaluating integrals of the form

$$\int_{-L}^{L} f(x) \sin \frac{n\pi x}{L} \, dx \quad \text{and} \quad \int_{-L}^{L} f(x) \cos \frac{n\pi x}{L} \, dx,$$

where $L > 0$ is a fixed constant and n is a positive integer. In Exercises 22–25 evaluate these integrals for the given function $f(x)$.

22. $f(x) = x$

23. $f(x) = x^2$

24. $f(x) = 1 - 2x$

25. $f(x) = 2x^2 - 3x$

26. The *gamma function* $\Gamma(n)$ for $n > 0$ is defined by the improper integral

$$\Gamma(n) = \int_0^\infty x^{n-1} e^{-x} \, dx.$$

Show that when n is a positive integer, $\Gamma(n) = (n-1)!$.

The *Laplace transform* of a function $f(t)$ is the function $F(s)$ defined by the improper integral

$$F(s) = \int_0^\infty e^{-st} f(t) \, dt.$$

In Exercises 27–30 find the Laplace transform of the function.

27. $f(t) = e^{3t}$

28. $f(t) = t^2$

29. $f(t) = \sin t$

30. $f(t) = te^{-t}$

31. Show that the function

$$f(x) = \frac{x^{\alpha-1} e^{-x/\beta}}{\Gamma(\alpha)\beta^\alpha}$$

for positive constants α and β qualifies as a probability density function on the interval $x \geq 0$. See Exercise 23 in Section 7.10 for the definition of a pdf and Exercise 26 for the definition of $\Gamma(\alpha)$.

In the study of *signals and communications* in electrical engineering, the *Fourier transform* of a function $f(t)$ is

$$F(\omega) = \int_{-\infty}^\infty f(t) e^{-i\omega t} \, dt,$$

where i is the complex number whose square is -1; that is, $i^2 = -1$. In Exercises 32–35 find the *Fourier transform* of the function $f(t)$.

32. $f(t) = \begin{cases} 1, & |t| < L/2 \\ 0, & |t| > L/2 \end{cases} \qquad L > 0$ a constant

33. $f(t) = \begin{cases} 0, & |t| > T \\ 1 + t/T, & -T \leq t \leq 0 \\ 1 - t/T, & 0 \leq t \leq T \end{cases} \qquad T > 0$ a constant

34. $f(t) = e^{-a|t|}$, $a > 0$ a constant

35. $f(t) = \begin{cases} 0, & t < a \\ 1, & a < t < b \\ 0, & t > b \end{cases} \qquad a$ and b constants

In Exercises 36–38 evaluate the indefinite integral.

36. $\int \operatorname{Tan}^{-1} \sqrt{x} \, dx$

37. $\int x^2 \cos^2 x \, dx$

38. $\int xe^x \sin x \, dx$

39. Use the substitution $x = \sin^2 \theta$ and integration by parts to find a formula for $\int_0^1 x^n (1-x)^m \, dx$.

40. Verify that for $n \geq 1$ an integer,

$$\int x^n \cos x \, dx = \sin x \sum_{r=0}^{\lfloor n/2 \rfloor} \frac{(-1)^r n!}{(n-2r)!} x^{n-2r}$$

$$+ \cos x \sum_{r=0}^{\lfloor (n-1)/2 \rfloor} \frac{(-1)^r n!}{(n-2r-1)!} x^{n-2r-1} + C,$$

where $\lfloor n/2 \rfloor$ indicates the floor function of Example 2.22 in Section 2.4.

8.3 Trigonometric Integrals

Differentiation formulas for trigonometric functions suggest three pairings of these functions: sine and cosine, tangent and secant, and cotangent and cosecant. The derivative of either function in a given pair involves at most functions of that pair. Note that integration formulas 8.1e–n also substantiate this pairing. Each indefinite integral gives functions in the same pairing. Because of this, a first step in every integral involving trigonometric functions is to rewrite the integrand so that each term involves only one pair. For example, to evaluate

$$\int \frac{\tan x + \sec x \cot^2 x}{\sin^3 x} \, dx,$$

we would first rewrite it in the form

$$\int \left(\frac{\csc^3 x}{\cot x} + \frac{\cos x}{\sin^5 x} \right) dx.$$

The first term contains cosecants and cotangents, and the second has sines and cosines. We now consider integrals involving each pair.

Integrals Involving Sine and Cosine

We frequently encounter integrals of the form

$$\int \cos^n x \sin^m x \, dx,$$

where m and n may or may not be integers. The key to evaluation of this type of integral is recognition of the fact that when either $n = 1$ or $m = 1$, the integrations become straightforward. To illustrate this, consider differentiation of $\sin^{n+1} x$ by power rule 3.21,

$$\frac{d}{dx} \sin^{n+1} x = (n + 1) \sin^n x \cos x.$$

This means that differentiation of a power of $\sin x$ leads to a power of $\sin x$ one lower than the original, multiplied by $\cos x$. In order to reverse the procedure and antidifferentiate a power of $\sin x$, we need a factor of $\cos x$. It then becomes a matter of raising the power on $\sin x$ by one and dividing by the new power; that is, we can say

$$\int \sin^n x \cos x \, dx = \frac{1}{n + 1} \sin^{n+1} x + C, \quad n \neq -1. \tag{8.4a}$$

Similarly,

$$\int \cos^n x \sin x \, dx = -\frac{1}{n + 1} \cos^{n+1} x + C, \quad n \neq -1. \tag{8.4b}$$

Formulas 8.1l and 8.1k contain the $n = -1$ cases for these integrals. Of course, a substitution can always be made to evaluate integrals 8.4a and b. For example, if in 8.4a we set $u = \sin x$, then $du = \cos x \, dx$, and

$$\int \sin^n x \cos x \, dx = \int u^n \, du = \frac{u^{n+1}}{n + 1} + C = \frac{\sin^{n+1} x}{n + 1} + C.$$

Also useful are the formulas

$$\int \sin nx \, dx = -\frac{1}{n} \cos nx + C \quad \text{and} \tag{8.4c}$$

$$\int \cos nx \, dx = \frac{1}{n} \sin nx + C. \tag{8.4d}$$

Trigonometric identities are used to write integrands in the forms contained in equations 8.4. Particularly helpful are

$$\sin^2 x + \cos^2 x = 1, \tag{8.5a}$$

$$\sin 2x = 2 \sin x \cos x, \tag{8.5b}$$

$$\cos 2x = 2 \cos^2 x - 1, \tag{8.5c}$$

$$\cos 2x = 1 - 2 \sin^2 x. \tag{8.5d}$$

EXAMPLE 8.6

▼

Evaluate the following integrals:

(a) $\displaystyle\int \sin^3 x \, dx$

(b) $\displaystyle\int \sin^3 x \cos^5 x \, dx$

(c) $\displaystyle\int \sin 2x \cos^6 2x \, dx$

(d) $\displaystyle\int \sqrt{\sin x} \cos^3 x \, dx$

(e) $\displaystyle\int \frac{\cos^3 3x}{\sin^2 3x} \, dx$

(f) $\displaystyle\int \sin^2 x \, dx$

(g) $\displaystyle\int \cos^4 x \, dx$

(h) $\displaystyle\int \sin^2 x \cos^2 x \, dx$

SOLUTION

(a) With identity 8.5a, we may write

$$\int \sin^3 x \, dx = \int \sin x (1 - \cos^2 x) \, dx = \int (\sin x - \cos^2 x \sin x) \, dx,$$

and the second term is in form 8.4b. Thus,

$$\int \sin^3 x \, dx = -\cos x + \frac{1}{3} \cos^3 x + C.$$

(b) Once again we use 8.5a to rewrite the integrand as two easily integrated terms,

$$\int \sin^3 x \cos^5 x \, dx = \int \sin x (1 - \cos^2 x) \cos^5 x \, dx$$

$$= \int (\cos^5 x \sin x - \cos^7 x \sin x) \, dx$$

$$= -\frac{1}{6} \cos^6 x + \frac{1}{8} \cos^8 x + C.$$

(c) This integrand is already in a convenient form for integration. Adjusting constants leads to

$$\int \sin 2x \cos^6 2x \, dx = -\frac{1}{14} \cos^7 2x + C.$$

(d) Using 8.5a, we obtain

$$\int \sqrt{\sin x} \cos^3 x \, dx = \int \sin^{1/2} x \cos x \, (1 - \sin^2 x) \, dx$$

$$= \int (\sin^{1/2} x \cos x - \sin^{5/2} x \cos x) \, dx$$

$$= \frac{2}{3} \sin^{3/2} x - \frac{2}{7} \sin^{7/2} x + C.$$

(e) Again we use 8.5a,

$$\int \frac{\cos^3 3x}{\sin^2 3x}\, dx = \int \frac{\cos 3x\,(1 - \sin^2 3x)}{\sin^2 3x}\, dx = \int \left(\frac{\cos 3x}{\sin^2 3x} - \cos 3x \right) dx$$

$$= \frac{-1}{3 \sin 3x} - \frac{1}{3} \sin 3x + C.$$

In these five examples, at least one of the powers on the sine and cosine functions was odd, and in each case we used identity 8.5a. In the next three examples, where all powers are even, identities 8.5b–d are useful.

(f) Double-angle formula 8.5d can be rearranged as $\sin^2 x = (1 - \cos 2x)/2$, and therefore

$$\int \sin^2 x\, dx = \frac{1}{2} \int (1 - \cos 2x)\, dx.$$

Formula 8.4d can now be used on the second term,

$$\int \sin^2 x\, dx = \frac{1}{2} \left(x - \frac{1}{2} \sin 2x \right) + C.$$

(g) With double-angle formula 8.5c rewritten in the form $\cos^2 x = (1 + \cos 2x)/2$, we obtain

$$\int \cos^4 x\, dx = \frac{1}{4} \int (1 + \cos 2x)^2\, dx = \frac{1}{4} \int (1 + 2 \cos 2x + \cos^2 2x)\, dx.$$

When x is replaced by $2x$ in 8.5c, the result is $\cos 4x = 2 \cos^2 2x - 1$. This can be rearranged as $\cos^2 2x = (1 + \cos 4x)/2$, and therefore

$$\int \cos^4 x\, dx = \frac{1}{4} \int \left(1 + 2 \cos 2x + \frac{1 + \cos 4x}{2} \right) dx.$$

We now use 8.4d on the two cosine terms,

$$\int \cos^4 x\, dx = \frac{1}{4} \left(\frac{3x}{2} + \sin 2x + \frac{1}{8} \sin 4x \right) + C.$$

(h) In this integral we use identities 8.5b and d,

$$\int \sin^2 x \cos^2 x\, dx = \int (\sin x \cos x)^2\, dx = \frac{1}{4} \int \sin^2 2x\, dx$$

$$= \frac{1}{4} \int \left(\frac{1 - \cos 4x}{2} \right) dx = \frac{1}{8} \left(x - \frac{1}{4} \sin 4x \right) + C.$$

EXAMPLE 8.7

Evaluate $\displaystyle\int \sin 5x \cos 2x\, dx$.

SOLUTION   Unlike Example 8.6, in which all trigonometric functions had the same argument, this integrand is the product of trigonometric functions with different arguments. Our first step is to rewrite the integrand so that we do not have such a product. Note that a sum is obviously integrable, but not a product. The product formula

$$\sin A \cos B = \frac{1}{2}[\sin(A+B) + \sin(A-B)]$$

can be used to advantage here. We can rewrite the integrand in the form

$$\int \sin 5x \cos 2x \, dx = \int \frac{1}{2}(\sin 7x + \sin 3x)\, dx,$$

and now integration is easily handled by 8.4c,

$$\int \sin 5x \cos 2x \, dx = \frac{1}{2}\left(-\frac{1}{7}\cos 7x - \frac{1}{3}\cos 3x\right) + C = -\frac{1}{14}\cos 7x - \frac{1}{6}\cos 3x + C.$$

Two-integration by parts (and reproduction) can also be used to evaluate this integral, but the above method is much simpler.

Integrals Involving Tangent and Secant

For integrals involving tangent and secant, we have two alternatives: Rewrite the integrand in terms of sines and cosines, or express the integrand in terms of easily integrated combinations of tangent and secant. It is usually more fruitful to investigate the second possibility before expressing the integrand completely in terms of sines and cosines. In this regard we note that combinations of tangent and secant that can be integrated immediately are

$$\int \tan^n x \sec^2 x \, dx = \frac{\tan^{n+1} x}{n+1} + C, \quad n \neq -1; \qquad (8.6a)$$

$$\int \sec^n x \tan x \, dx = \frac{\sec^n x}{n} + C, \quad n \neq 0. \qquad (8.6b)$$

Once again these results are suggested by differentiations of powers of $\tan x$ and $\sec x$. Alternatively, these integrals can be evaluated with substitutions: $u = \tan x$ for 8.6a and $u = \sec x$ for 8.6b.

To rearrange integrands into these integrable combinations, we use the identity

$$1 + \tan^2 x = \sec^2 x. \qquad (8.7)$$

EXAMPLE 8.8

Evaluate the following integrals:

(a) $\displaystyle\int \tan^2 x \, dx$ (b) $\displaystyle\int \tan^3 x \, dx$ (c) $\displaystyle\int \sec^4 x \, dx$

(d) $\displaystyle\int \tan^4 3x \sec^2 3x \, dx$ (e) $\displaystyle\int \tan^3 x \sec^3 x \, dx$ (f) $\displaystyle\int \frac{\sec^4 x}{\tan^2 x} \, dx$

SOLUTION

(a) $$\int \tan^2 x \, dx = \int (\sec^2 x - 1) \, dx = \tan x - x + C$$

(b) $$\int \tan^3 x \, dx = \int \tan x \, (\sec^2 x - 1) \, dx$$

$$= \int (\sec^2 x \tan x - \tan x) \, dx$$

$$= \frac{1}{2} \sec^2 x + \ln |\cos x| + C, \text{ or } \frac{1}{2} \tan^2 x + \ln |\cos x| + C$$

(c) $$\int \sec^4 x \, dx = \int \sec^2 x \, (1 + \tan^2 x) \, dx$$

$$= \int (\sec^2 x + \tan^2 x \sec^2 x) \, dx$$

$$= \tan x + \frac{1}{3} \tan^3 x + C$$

(d) $$\int \tan^4 3x \sec^2 3x \, dx = \frac{1}{15} \tan^5 3x + C$$

(e) $$\int \tan^3 x \sec^3 x \, dx = \int \tan x \, (\sec^2 x - 1) \sec^3 x \, dx$$

$$= \int (\sec^5 x \tan x - \sec^3 x \tan x) \, dx$$

$$= \frac{1}{5} \sec^5 x - \frac{1}{3} \sec^3 x + C$$

(f) $$\int \frac{\sec^4 x}{\tan^2 x} \, dx = \int \frac{\sec^2 x \, (1 + \tan^2 x)}{\tan^2 x} \, dx$$

$$= \int \left(\frac{\sec^2 x}{\tan^2 x} + \sec^2 x \right) dx$$

$$= -\frac{1}{\tan x} + \tan x + C$$

EXAMPLE 8.9

Evaluate $\int \sec^3 x \, dx$.

SOLUTION   We use integration by parts and reproduction. By setting

$$u = \sec x, \qquad dv = \sec^2 \, dx,$$

then

$$du = \sec x \tan x \, dx, \qquad v = \tan x,$$

and formula 8.2 gives

$$\int \sec^3 x \, dx = \sec x \tan x - \int \sec x \tan^2 x \, dx.$$

Since $\tan^2 x = \sec^2 x - 1$, we can write

$$\int \sec^3 x \, dx = \sec x \tan x - \int \sec x \, (\sec^2 x - 1) \, dx$$

$$= \sec x \tan x - \int \sec^3 x \, dx + \ln |\sec x + \tan x|.$$

We now solve for

$$\int \sec^3 x \, dx = \frac{1}{2} (\sec x \tan x + \ln |\sec x + \tan x|) + C.$$

Integrands in the following example cannot be transformed by means of identity 8.7 into forms 8.6, and we therefore express them in terms of sines and cosines.

EXAMPLE 8.10

Evaluate the following integrals:

$$\text{(a)} \int \frac{\tan^2 2x}{\sec^3 2x} \, dx \qquad \text{(b)} \int \frac{1}{\sec x \tan^2 x} \, dx$$

SOLUTION

(a)
$$\int \frac{\tan^2 2x}{\sec^3 2x} \, dx = \int \sin^2 2x \cos 2x \, dx = \frac{1}{6} \sin^3 2x + C$$

(b)
$$\int \frac{1}{\sec x \tan^2 x} \, dx = \int \frac{\cos^3 x}{\sin^2 x} \, dx = \int \frac{\cos x \, (1 - \sin^2 x)}{\sin^2 x} \, dx$$

$$= \int \left(\frac{\cos x}{\sin^2 x} - \cos x \right) dx = -\frac{1}{\sin x} - \sin x + C$$

Integrals Involving Cotangent and Cosecant

Since derivatives of the cotangent and cosecant pair

$$\frac{d}{dx} \cot x = -\csc^2 x, \qquad \frac{d}{dx} \csc x = -\csc x \cot x$$

are analogous to those of the tangent and secant pair

$$\frac{d}{dx} \tan x = \sec^2 x, \qquad \frac{d}{dx} \sec x = \sec x \tan x,$$

a discussion parallel to that concerning the tangent and secant could be made here. Aside from the fact that cotangent replaces tangent and cosecant replaces secant, there are also sign changes. Examples are given in the exercises.

In Exercises 1–20 evaluate the indefinite integral.

1. $\int \cos^3 x \sin x \, dx$

2. $\int \dfrac{\cos x}{\sin^3 x} \, dx$

3. $\int \tan^5 x \sec^2 x \, dx$

4. $\int \csc^3 x \cot x \, dx$

5. $\int \cos^3 (x+2) \, dx$

6. $\int \sqrt{\tan x} \sec^4 x \, dx$

7. $\int \dfrac{1}{\sin^4 t} \, dt$

8. $\int \sec^6 3x \tan 3x \, dx$

9. $\int \cos^2 x \, dx$

10. $\int \dfrac{\tan^3 x \sec^2 x}{\sin^2 x} \, dx$

11. $\int \sin^3 y \cos^2 y \, dy$

12. $\int \dfrac{\csc^2 \theta}{\cot^2 \theta} \, d\theta$

13. $\int \dfrac{\sin \theta}{1 + \cos \theta} \, d\theta$

14. $\int \dfrac{\sec^2 x}{\sqrt{1 + \tan x}} \, dx$

15. $\int \cos \theta \sin 2\theta \, d\theta$

16. $\int \dfrac{3 + 4 \csc^2 x}{\cot^2 x} \, dx$

17. $\int \sin^5 x \cos^5 x \, dx$

18. $\int \sin^4 x \, dx$

19. $\int \dfrac{\tan^3 x}{\sec^4 x} \, dx$

20. $\int \dfrac{\csc^4 x}{\cot^3 x} \, dx$

21. Find the area of the region between the x-axis and $y = \sin x$ for $0 \le x \le \pi$.

22. Find the volume of the solid of revolution obtained by rotating the region bounded by $y = \tan x$, $y = 0$, and $x = \pi/4$ about the line $y = -1$.

23. Evaluate $\displaystyle\int_0^\pi \sqrt{1 - \sin^2 x} \, dx$.

In Exercises 24–31 evaluate the indefinite integral.

24. $\int \cot^4 z \, dz$

25. $\int \dfrac{\cos^3 \theta}{3 + \sin \theta} \, d\theta$

26. $\int \dfrac{\cos^4 \theta}{1 + \sin \theta} \, d\theta$

27. $\int \sin^4 x \cos^2 x \, dx$

28. $\int \cos 6x \cos 2x \, dx$

29. $\int \cos^2 2x \sin 3x \, dx$

30. $\int \dfrac{1}{\sin x \cos^2 x} \, dx$

31. $\int \sec^5 x \, dx$

32. The power required to maintain current $i(t) = i_m \cos(\omega t + \phi_1)$, where $i_m > 0$ and ϕ_1 are constants, through a resistor that has voltage across its terminals $V(t) = V_m \cos(\omega t + \phi_2)$, where $V_m > 0$ and

ϕ_2 are constants, is $P = Vi$. Show that the average power supplied to the resistor is $P_{av} = (V_m i_m /2) \cos(\phi_1 - \phi_2)$.

33. The alternating current in a power line is given by

$$I = f(t) = A \cos \omega t + B \sin \omega t,$$

where A, B, and ω are constants, and t is time. The root-mean-square (rms) current I_{rms} is defined by

$$(I_{rms})^2 = \frac{1}{T} \int_0^T I^2 \, dt,$$

where the interval $0 \le t \le T$ represents any number of complete oscillations of the current. Show that I_{rms} is $1/\sqrt{2}$ times the amplitude of the current.

The dc value of a periodic waveform $f(t)$ with period p in the theory of signals is defined as

$$F_{dc} = \frac{1}{p} \int_{-p/2}^{p/2} f(t) \, dt.$$

In Exercises 34–37 find the dc value for the waveform.

34. $f(t)$ of Exercise 33

35. $f(t) = A + B \cos \omega t$, A, B, ω constants

36. $f(t) = \sin^2 \omega t$, ω constant

37. $f(t) = t$, $0 < t < 1$, $f(t+1) = f(t)$

38. Evaluate $\displaystyle\int (1 + \sin \theta + \sin^2 \theta + \sin^3 \theta + \cdots) \, d\theta$.

39. If n is a positive even integer, show that

$$\int \sec^n x \, dx = \sum_{r=0}^{n/2-1} \binom{n/2 - 1}{r} \frac{\tan^{2r+1} x}{2r + 1} + C.$$

40. A set of functions $f_1(x)$, $f_2(x)$, ... is said to be orthonormal over an interval $a \le x \le b$ if

$$\int_a^b f_n(x) f_m(x) \, dx = \begin{cases} 0, & m \ne n \\ 1, & m = n. \end{cases}$$

Show that the set of functions

$$\frac{1}{\sqrt{2\pi}}, \quad \frac{1}{\sqrt{\pi}} \sin x, \quad \frac{1}{\sqrt{\pi}} \cos x,$$

$$\frac{1}{\sqrt{\pi}} \sin 2x, \quad \frac{1}{\sqrt{\pi}} \cos 2x, \quad \frac{1}{\sqrt{\pi}} \sin 3x, \quad \cdots$$

is orthonormal over the interval $0 \le x \le 2\pi$. Orthonormality of these functions is the basis for *Fourier series*, a topic of fundamental importance in all branches of engineering.

8.4 Trigonometric Substitutions

Physical and geometric problems arising from circles, ellipses, and hyperbolas often give rise to integrals involving the square roots $\sqrt{a^2 - b^2x^2}$, $\sqrt{a^2 + b^2x^2}$, and $\sqrt{b^2x^2 - a^2}$. **Trigonometric substitutions** replace them with integrals involving trigonometric functions.

Consider first two integrals that involve square roots of the form $\sqrt{a^2 - b^2x^2}$:

$$(i) \quad \int \frac{1}{\sqrt{1 - x^2}}\, dx \quad \text{and} \quad (ii) \quad \int \frac{1}{\sqrt{4 - 9x^2}}\, dx.$$

Integral (i) is evident from differentiation of the inverse trigonometric functions in Section 3.10,

$$\int \frac{1}{\sqrt{1 - x^2}}\, dx = \operatorname{Sin}^{-1} x + C.$$

(See also formula 8.1o) We now show that a substitution can be made on this integral that is applicable to much more difficult problems involving the square root $\sqrt{a^2 - b^2x^2}$, including (ii). If instead of the square root $\sqrt{1 - x^2}$ in (i), we had the square root $\sqrt{1 - \sin^2 \theta}$, we could immediately simplify the latter root to $\sqrt{\cos^2 \theta} = |\cos \theta|$. This prompts us to make the substitution $x = \sin \theta$, from which we have $dx = \cos \theta\, d\theta$. Integral (i) then becomes

$$\int \frac{1}{\sqrt{1 - x^2}}\, dx = \int \frac{1}{|\cos \theta|} \cos \theta\, d\theta.$$

To eliminate the absolute values, we need to know whether $\cos \theta$ is positive or negative. But this means that we must know the possible values of θ. There is a problem. The equation $x = \sin \theta$ does not really define θ; for any given x, we do not have a unique θ, but an infinite number of possibilities. We must therefore restrict the values of θ, and we do this by specifying that $\theta = \operatorname{Sin}^{-1} x$. In other words, although we have used the equation $x = \sin \theta$ to change variables, and will continue to do so, it is really the equation $\theta = \operatorname{Sin}^{-1} x$ that properly defines the change. Since θ so defined must lie in the interval $-\pi/2 \le \theta \le \pi/2$ (the principal values of the inverse sine function), it follows that $\cos \theta \ge 0$, and absolute values may be dropped. Consequently,

$$\int \frac{1}{\sqrt{1 - x^2}}\, dx = \int \frac{\cos \theta}{|\cos \theta|}\, d\theta = \int d\theta = \theta + C = \operatorname{Sin}^{-1} x + C.$$

For integral (ii), the substitution $x = \sin \theta$ is of little use since the square root $\sqrt{4 - 9\sin^2 \theta}$ does not simplify. If, however, the square root were $\sqrt{4 - 4\sin^2 \theta}$, it would immediately reduce to $\sqrt{4 - 4\sin^2 \theta} = \sqrt{4\cos^2 \theta} = 2|\cos \theta|$. We can obtain this result if the substitution is modified to $x = (2/3)\sin \theta$ [or, more properly, $\theta = \operatorname{Sin}^{-1}(3x/2)$]. Since $dx = (2/3)\cos \theta\, d\theta$, we have

$$\int \frac{1}{\sqrt{4 - 9x^2}}\, dx = \int \frac{(2/3)\cos \theta}{\sqrt{4 - 4\sin^2 \theta}}\, d\theta = \frac{2}{3} \int \frac{\cos \theta}{2|\cos \theta|}\, d\theta.$$

Once again absolute values may be dropped because $-\pi/2 \le \theta \le \pi/2$,

$$\int \frac{1}{\sqrt{4 - 9x^2}}\, dx = \frac{1}{3} \int d\theta = \frac{\theta}{3} + C = \frac{1}{3}\operatorname{Sin}^{-1}\left(\frac{3x}{2}\right) + C.$$

For integrals containing square roots of the form $\sqrt{a^2 + b^2x^2}$, consider the pair of integrals

$$(iii) \quad \int \frac{1}{1 + x^2}\, dx \quad (iv) \quad \int \frac{1}{(3 + 5x^2)^{3/2}}\, dx.$$

Integral (iii) obviously has answer $\operatorname{Tan}^{-1} x + C$. Alternatively, if we had $1 + \tan^2 \theta$ instead of $1 + x^2$, the two terms in the denominator would simplify to the one term $\sec^2 \theta$. We therefore substitute $x = \tan\theta$ or, more properly, $\theta = \text{Tan}^{-1} x$, from which we have $dx = \sec^2\theta\, d\theta$. With this substitution, we obtain

$$\int \frac{1}{1+x^2}\, dx = \int \frac{\sec^2\theta}{\sec^2\theta}\, d\theta = \int d\theta = \theta + C = \text{Tan}^{-1} x + C.$$

The substitution $x = \tan\theta$ in (iv) yields $(3 + 5x^2)^{3/2} = (3 + 5\tan^2\theta)^{3/2}$. For simplification, we need a 3 in front of the $\tan^2\theta$ and not a 5. This can be accomplished by modification of our substitution to

$$x = \sqrt{\frac{3}{5}}\tan\theta,$$

from which

$$dx = \sqrt{\frac{3}{5}}\sec^2\theta\, d\theta.$$

Then

$$\int \frac{1}{(3+5x^2)^{3/2}}\, dx = \int \frac{1}{(3 + 3\tan^2\theta)^{3/2}}\sqrt{\frac{3}{5}}\sec^2\theta\, d\theta = \frac{1}{3\sqrt{5}}\int \frac{\sec^2\theta}{(\sec^2\theta)^{3/2}}\, d\theta$$

$$= \frac{1}{3\sqrt{5}}\int \frac{1}{\sec\theta}\, d\theta = \frac{1}{3\sqrt{5}}\int \cos\theta\, d\theta = \frac{1}{3\sqrt{5}}\sin\theta + C.$$

FIGURE 8.1   Triangle to fit
$x = \sqrt{3/5}\tan\theta$

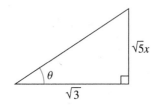

To express $\sin\theta$ in terms of x, we draw the triangle in Figure 8.1 to fit the change of variable $\sqrt{5}\,x/\sqrt{3} = \tan\theta$. Since the hypotenuse of the triangle is $\sqrt{3 + 5x^2}$, we obtain

$$\int \frac{1}{(3 + 5x^2)^{3/2}}\, dx = \frac{1}{3\sqrt{5}}\frac{\sqrt{5}\,x}{\sqrt{3 + 5x^2}} + C = \frac{x}{3\sqrt{3 + 5x^2}} + C.$$

For integrals containing square roots of the form $\sqrt{b^2x^2 - a^2}$, consider

$$\int \frac{1}{\sqrt{x^2 - 4}}\, dx.$$

If we had $\sqrt{4\sec^2\theta - 4}$ instead of $\sqrt{x^2 - 4}$, an immediate simplification would occur. We therefore substitute $x = 2\sec\theta$, from which $dx = 2\sec\theta\tan\theta\, d\theta$. Once again the real substitution is $\theta = \text{Sec}^{-1}(x/2)$, in which case θ is an angle in either the first or third quadrant. With this change,

$$\sqrt{x^2 - 4} = \sqrt{4\sec^2\theta - 4} = \sqrt{4\tan^2\theta} = 2|\tan\theta| = 2\tan\theta$$

and

$$\int \frac{1}{\sqrt{x^2 - 4}}\, dx = \int \frac{2\sec\theta\tan\theta}{2\tan\theta}\, d\theta = \int \sec\theta\, d\theta = \ln|\sec\theta + \tan\theta| + C.$$

FIGURE 8.2   Triangle to fit
$x = 2\sec\theta$

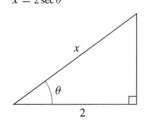

To express $\tan\theta$ in terms of x, we draw the triangle in Figure 8.2 to fit the change of variable $x/2 = \sec\theta$. Since the third side is $\sqrt{x^2 - 4}$, we have

$$\int \frac{1}{\sqrt{x^2 - 4}}\, dx = \ln\left|\frac{x}{2} + \frac{\sqrt{x^2 - 4}}{2}\right| + C = \ln\left|x + \sqrt{x^2 - 4}\right| + D,$$

where $D = C - \ln 2$.

We have illustrated by examples that trigonometric substitutions can be useful in the evaluation of integrals involving the square roots $\sqrt{a^2 - b^2x^2}$, $\sqrt{a^2 + b^2x^2}$, and $\sqrt{b^2x^2 - a^2}$.

Essentially, the method replaces terms under the square root by a perfect square and thereby rids the integrand of the square root. To obtain the trigonometric substitution appropriate to each square root, we suggest the following procedure. We have been working with the trigonometric identities

$$1 - \sin^2 \theta = \cos^2 \theta,$$

$$1 + \tan^2 \theta = \sec^2 \theta,$$

$$\sec^2 \theta - 1 = \tan^2 \theta.$$

To determine the trigonometric substitution appropriate to, say, $\sqrt{a^2 - b^2 x^2}$, we mentally set $a = b = 1$ and obtain $\sqrt{1 - x^2}$. We note that $1 - x^2$ resembles $1 - \sin^2 \theta$, and immediately set $x = \sin \theta$. To simplify the original square root $\sqrt{a^2 - b^2 x^2}$, we then modify this substitution to

$$x = \frac{a}{b} \sin \theta \qquad \left[\text{or } \theta = \text{Sin}^{-1}\left(\frac{bx}{a}\right) \right].$$

Similarly, for the square root $\sqrt{a^2 + b^2 x^2}$, we initially set $x = \tan \theta$, having mentally set $a = b = 1$, and noted that $1 + x^2$ resembles $1 + \tan^2 \theta$. We then modify the substitution to

$$x = \frac{a}{b} \tan \theta \qquad \left[\text{or } \theta = \text{Tan}^{-1}\left(\frac{bx}{a}\right) \right].$$

Finally, for the square root $\sqrt{b^2 x^2 - a^2}$, we initially set $x = \sec \theta$ and then modify to

$$x = \frac{a}{b} \sec \theta \qquad \left[\text{or } \theta = \text{Sec}^{-1}\left(\frac{bx}{a}\right) \right].$$

The square roots $\sqrt{a^2 - b^2 x^2}$, $\sqrt{a^2 + b^2 x^2}$, and $\sqrt{b^2 x^2 - a^2}$ are replaced, respectively, by

$$\sqrt{a^2 - a^2 \sin^2 \theta} \quad \sqrt{a^2 + a^2 \tan^2 \theta} \quad \sqrt{a^2 \sec^2 \theta - a^2}$$

$$= \sqrt{a^2 \cos^2 \theta} \qquad = \sqrt{a^2 \sec^2 \theta} \qquad = \sqrt{a^2 \tan^2 \theta}$$

$$= a|\cos \theta| \qquad\quad = a|\sec \theta| \qquad\quad = a|\tan \theta|$$

$$= a \cos \theta; \qquad\quad = a \sec \theta; \qquad\quad = a \tan \theta.$$

In each case it is our choice of principal values for the inverse trigonometric functions that enables us to neglect absolute values.

We now consider more complex examples.

EXAMPLE 8.11 ▼

Evaluate the following integrals:

(a) $\displaystyle\int \frac{1}{x^2(9x^2 + 2)}\, dx$ (b) $\displaystyle\int x^2\sqrt{3 - x^2}\, dx$

SOLUTION

(a) If we set $x = (\sqrt{2}/3)\tan\theta$, then $dx = (\sqrt{2}/3)\sec^2\theta\,d\theta$, and

$$\int \frac{1}{x^2(9x^2+2)}\,dx = \int \frac{1}{\left(\dfrac{2}{9}\tan^2\theta\right)(2\tan^2\theta+2)}\left(\frac{\sqrt{2}}{3}\right)\sec^2\theta\,d\theta$$

$$= \frac{3}{2\sqrt{2}}\int \frac{\sec^2\theta}{\tan^2\theta\,\sec^2\theta}\,d\theta$$

$$= \frac{3}{2\sqrt{2}}\int \cot^2\theta\,d\theta = \frac{3}{2\sqrt{2}}\int (\csc^2\theta - 1)\,d\theta$$

$$= \frac{3}{2\sqrt{2}}(-\cot\theta - \theta) + C$$

$$= \frac{-3}{2\sqrt{2}}\left[\frac{\sqrt{2}}{3x} + \text{Tan}^{-1}\left(\frac{3x}{\sqrt{2}}\right)\right] + C \qquad \text{(Figure 8.3)}$$

$$= \frac{-1}{2x} - \frac{3}{2\sqrt{2}}\text{Tan}^{-1}\left(\frac{3x}{\sqrt{2}}\right) + C.$$

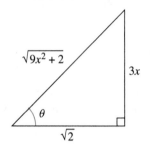

FIGURE 8.3   Triangle to fit
$x = (\sqrt{2}/3)\tan\theta$

(b) If we set $x = \sqrt{3}\sin\theta$, then $dx = \sqrt{3}\cos\theta\,d\theta$, and

$$\int x^2\sqrt{3-x^2}\,dx = \int (3\sin^2\theta)\sqrt{3 - 3\sin^2\theta}\,(\sqrt{3}\cos\theta\,d\theta)$$

$$= 9\int \sin^2\theta\sqrt{1 - \sin^2\theta}\,\cos\theta\,d\theta$$

$$= 9\int \sin^2\theta\,\cos^2\theta\,d\theta$$

$$= \frac{9}{4}\int \sin^2 2\theta\,d\theta \qquad \text{(by 8.5b)}$$

$$= \frac{9}{4}\int\left(\frac{1 - \cos 4\theta}{2}\right)d\theta \qquad \text{(by 8.5d)}$$

$$= \frac{9}{8}\left(\theta - \frac{1}{4}\sin 4\theta\right) + C \qquad \text{(by 8.4d)}$$

$$= \frac{9}{8}\text{Sin}^{-1}\left(\frac{x}{\sqrt{3}}\right) - \frac{9}{32}(2\sin 2\theta\cos 2\theta) + C \qquad \text{(by 8.5b)}$$

$$= \frac{9}{8}\text{Sin}^{-1}\left(\frac{x}{\sqrt{3}}\right) - \frac{9}{16}(2\sin\theta\cos\theta)(1 - 2\sin^2\theta) + C \qquad \text{(by 8.5b,d)}$$

$$= \frac{9}{8}\text{Sin}^{-1}\left(\frac{x}{\sqrt{3}}\right) - \frac{9}{8}\left(\frac{x}{\sqrt{3}}\right)\frac{\sqrt{3-x^2}}{\sqrt{3}}\left(1 - \frac{2x^2}{3}\right) + C \qquad \text{(Figure 8.}$$

$$= \frac{9}{8}\text{Sin}^{-1}\left(\frac{x}{\sqrt{3}}\right) - \frac{x}{8}\sqrt{3-x^2}(3 - 2x^2) + C.$$

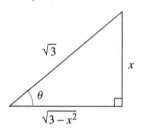

FIGURE 8.4   Triangle to fit
$x = \sqrt{3}\sin\theta$

EXAMPLE 8.12

Use a trigonometric substitution to evaluate the integral in Example 8.4.

SOLUTION   If we set $x = 3\tan\theta$, then $dx = 3\sec^2\theta\,d\theta$, and

$$\int \frac{x^3}{\sqrt{9 + x^2}}\,dx = \int \frac{27\tan^3\theta}{\sqrt{9 + 9\tan^2\theta}}3\sec^2\theta\,d\theta$$

$$= 27 \int \frac{\tan^3\theta\,\sec^2\theta}{\sec\theta}\,d\theta$$

$$= 27 \int \tan\theta\,(\sec^2\theta - 1)\sec\theta\,d\theta \qquad \text{(by 8.7)}$$

$$= 27 \int (\sec^3\theta\tan\theta - \sec\theta\tan\theta)\,d\theta$$

$$= 27\left(\frac{1}{3}\sec^3\theta - \sec\theta\right) + C \qquad \text{(by 8.6b)}$$

$$= 9\left(\frac{\sqrt{9 + x^2}}{3}\right)^3 - 27\frac{\sqrt{9 + x^2}}{3} + C \qquad \text{(Figure 8.5)}$$

$$= \frac{1}{3}(9 + x^2)^{3/2} - 9\sqrt{9 + x^2} + C.$$

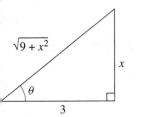

FIGURE 8.5   Triangle to fit $x = 3\tan\theta$

EXAMPLE 8.13

Application Preview Revisited

The Application Preview posed the problem of finding the path $y = f(x)$ followed by the point midway between the rear wheels of a trailer given that the pivot point of the tractor and trailer follows the curve $Y = g(X)$ (Figure 8.6).

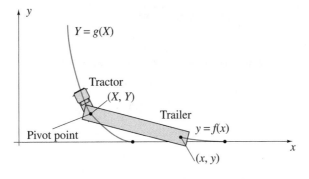

FIGURE 8.6   Path followed by point midway between rear wheels of tractor-trailer

SOLUTION   The key is to notice that the line joining (x, y) to (X, Y) is always tangent to $y = f(x)$. Consequently,

$$\frac{dy}{dx} = \frac{Y - y}{X - x} = \frac{g(X) - y}{X - x}.$$

This is a differential equation that must be solved for $y = f(x)$ subject to the initial position of the midpoint between the rear wheels on the x-axis. To solve this equation we must eliminate X. This can be done by letting L be the fixed distance between (x, y) and (X, Y), in which case

$$L^2 = (X - x)^2 + (Y - y)^2 = (X - x)^2 + [g(X) - y]^2.$$

Given $g(X)$, we can theoretically solve this equation for X in terms of x, y, and L, and substitute into the differential equation to express dy/dx in terms of x and y. To do so for given $g(X)$ could be a formidable task if $g(X)$ is at all complicated. We shall consider only the simplest possible situation in which the pivot point begins at the origin and moves up the y-axis. In this case, X is always zero, and $L^2 = x^2 + (Y - y)^2$. From this equation, $Y - y = \sqrt{L^2 - x^2}$, and the differential equation for $y = f(x)$ is

$$\frac{dy}{dx} = \frac{Y - y}{X - x} = \frac{\sqrt{L^2 - x^2}}{-x}.$$

Integration with respect to x gives

$$y = -\int \frac{\sqrt{L^2 - x^2}}{x} dx.$$

We now set $x = L \sin \theta$ and $dx = L \cos \theta \, d\theta$.

$$y = -\int \frac{L \cos \theta}{L \sin \theta} L \cos \theta \, d\theta = -L \int \frac{1 - \sin^2 \theta}{\sin \theta} d\theta = L \int (\sin \theta - \csc \theta) \, d\theta$$

$$= L[-\cos \theta - \ln |\csc \theta - \cot \theta|] + C$$

$$= -L \left(\frac{\sqrt{L^2 - x^2}}{L} + \ln \left| \frac{L}{x} - \frac{\sqrt{L^2 - x^2}}{x} \right| \right) + C$$

$$= -\sqrt{L^2 - x^2} - L \ln \left| \frac{L - \sqrt{L^2 - x^2}}{x} \right| + C$$

$$= -\sqrt{L^2 - x^2} - L \ln \left| \frac{L - \sqrt{L^2 - x^2}}{x} \frac{L + \sqrt{L^2 - x^2}}{L + \sqrt{L^2 - x^2}} \right| + C$$

$$= -\sqrt{L^2 - x^2} - L \ln \left| \frac{x}{L + \sqrt{L^2 - x^2}} \right| + C$$

$$= L \ln \left| \frac{L + \sqrt{L^2 - x^2}}{x} \right| - \sqrt{L^2 - x^2} + C.$$

Since $y = 0$ when $x = L$, it follows that $C = 0$, and

$$y = L \ln \left(\frac{L + \sqrt{L^2 - x^2}}{x} \right) - \sqrt{L^2 - x^2}.$$

EXERCISES 8.4

In Exercises 1–20 evaluate the indefinite integral.

1. $\int \dfrac{1}{x\sqrt{2x^2 - 4}} dx$

2. $\int \dfrac{1}{\sqrt{9 - 5x^2}} dx$

3. $\int \dfrac{1}{10 + x^2} dx$

4. $\int \dfrac{1}{x^2\sqrt{4 - x^2}} dx$

5. $\int \sqrt{7 - x^2} \, dx$

6. $\int x\sqrt{5x^2 + 3} \, dx$

7. $\int x^3\sqrt{4 + x^2} \, dx$

8. $\int \dfrac{1}{1 - x^2} dx$

9. $\int \dfrac{1}{\sqrt{x^2 - 5}} dx$

10. $\int \dfrac{x + 5}{10x^2 + 2} dx$

11. $\int \dfrac{1}{x\sqrt{x^2 + 3}} dx$

12. $\int \dfrac{\sqrt{4 - x^2}}{x} dx$

13. $\int \dfrac{x^2}{(2 - 9x^2)^{3/2}} dx$

14. $\int \dfrac{\sqrt{x^2 - 16}}{x^2} dx$

15. $\int \dfrac{1}{x^2\sqrt{2x^2 + 7}} dx$

16. $\int \dfrac{1}{x^3\sqrt{x^2 - 4}} dx$

17. $\displaystyle\int \frac{\sqrt{9-z^2}}{z^4}\,dz$

18. $\displaystyle\int \frac{y^3}{\sqrt{y^2+4}}\,dy$

19. $\displaystyle\int \frac{1}{(4x^2-9)^{3/2}}\,dx$

20. $\displaystyle\int \frac{x^2+2}{x^3+x}\,dx$

21. Show that if $a > 0$, then

$$\int \frac{1}{a^2-x^2}\,dx = \frac{1}{2a}\ln\left|\frac{a+x}{a-x}\right| + C.$$

22. Verify that the area inside the ellipse $b^2x^2+a^2y^2 = a^2b^2$ is πab.

23. Find the area of the region common to the circles $x^2 + y^2 = 4$ and $x^2 + y^2 = 4x$.

24. Given that the equation $2x^2 + y^2 - 2xy + 4x - 4y + 3 = 0$ describes an ellipse, find its area.

25. A boy initially at O (figure below) walks along the edge of a pier (the y-axis) towing his sailboat by a string of length L.

   (a) If the boat starts at Q and the string always remains straight, show that the equation of the curved path $y = f(x)$ followed by the boat must satisfy the differential equation

$$\frac{dy}{dx} = -\frac{\sqrt{L^2-x^2}}{x}.$$

   (b) Find $y = f(x)$.

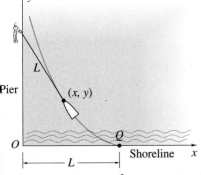

26. The parabola $x = y^2$ divides the circle $x^2 + y^2 = 4$ into two parts. Find the second moment of area of the smaller part about the x-axis.

27. Find the centroid of the region in Exercise 26.

28. Find the horizontal line that divides the ellipse $b^2x^2 + a^2y^2 = a^2b^2$ into two parts so that the area of the lower part is twice the area of the upper part.

29. If a thin circular plate has radius 2 units and constant mass per unit area ρ, find its moment of inertia about any tangent line to its edge.

30. A cylindrical oil can with horizontal axis has radius r and length h. If the density of the oil is ρ, find the force on each end of the can when it is full.

In Exercises 31–37 evaluate the indefinite integral.

31. $\displaystyle\int \frac{2x^4-x^2}{2x^2+1}\,dx$

∗ 32. $\displaystyle\int (7-x^2)^{3/2}\,dx$

∗ 33. $\displaystyle\int \frac{1}{x-x^3}\,dx$

∗ 34. $\displaystyle\int \frac{1}{x^3(4x^2-1)^{3/2}}\,dx$

∗ 35. $\displaystyle\int \sqrt{x^2-4}\,dx$

∗ 36. $\displaystyle\int \sqrt{1+3x^2}\,dx$

∗ 37. $\displaystyle\int \frac{x^2}{\sqrt{x^2-5}}\,dx$

∗ 38. Find the length of the curve $8y^2 = x^2(1-x^2)$.

∗ 39. Find the length of the parabola $y = x^2$ from $(0, 0)$ to $(1, 1)$.

∗ 40. At what distance from the centre of a circle should a line be drawn in order that the second moment of area of the circle about that line will be equal to twice the second moment of the circle about a line through its centre?

∗ 41. When water in the soil moves to the surface of the soil, depth z below the surface is related to suction head x by the equation

$$z(x) = -\int \frac{1}{1+V/K}\,dx,$$

where $V > 0$ is a constant, and $K = k/(cx^2 + 1)$ with $k > 0$ and $c > 0$ being constants. Show that if $z(0) = H$, then

$$z(x) = H - \frac{k}{\sqrt{Vc(k+V)}}\operatorname{Tan}^{-1}\sqrt{\frac{Vc}{k+V}}\,x,$$

whereas if $z(H_w - L) = L$, then

$$z(x) = L + \frac{k}{\sqrt{Vc(k+V)}}\left[\operatorname{Tan}^{-1}\sqrt{\frac{Vc}{k+V}}(H_w - L)\right.$$

$$\left. - \operatorname{Tan}^{-1}\sqrt{\frac{Vc}{k+V}}\,x\right].$$

∗ 42. When a flexible cable of constant mass per unit length ρ hangs between two fixed points A and B (figure below), the shape $y = f(x)$ of the cable must satisfy the differential equation

$$\frac{d^2y}{dx^2} = \frac{\rho g}{H}\sqrt{1+\left(\frac{dy}{dx}\right)^2},$$

where $g > 0$ and $H > 0$ are constants. If we set $k = \rho g/H$ and $p = dy/dx$, then

$$\frac{dp}{dx} = k\sqrt{1+p^2} \qquad \text{or} \qquad \frac{1}{\sqrt{1+p^2}}\frac{dp}{dx} = k.$$

It follows that

$$\int \frac{1}{\sqrt{1+p^2}}\,dp = kx + C.$$

   (a) Evaluate the integral shown to find $p = dy/dx$.

   (b) Integrate once more to find the shape of the cable (see also Example 3.39 in Section 3.13.)

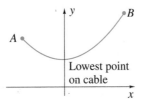

43. Repeat Exercise 21 of Section 7.6 if the log is circular with radius 10 cm.

44. Find the area inside the loop of the strophoid $y^2(a + x) = x^2(a - x)$.

45. (a) By rationalizing the denominator, show that

$$\frac{1}{x + 1 + \sqrt{x^2 + 1}} = \frac{x + 1 - \sqrt{x^2 + 1}}{2x}.$$

(b) Use part (a) to prove that

$$\int_0^1 \frac{1}{x + 1 + \sqrt{x^2 + 1}} \, dx = 1 - \frac{\sqrt{2}}{2} + \frac{1}{2} \ln \left(\frac{1 + \sqrt{2}}{2} \right).$$

46. Show that

$$\int_{-1}^1 \sqrt{\frac{1 + x}{1 - x}} \, dx = \pi.$$

8.5 Completing the Square and Trigonometric Substitutions

Trigonometric substitutions reduce integrals containing square roots of the form $\sqrt{a^2 - b^2x^2}$, $\sqrt{a^2 + b^2x^2}$, and $\sqrt{b^2x^2 - a^2}$ to trigonometric integrals. For integrals containing square roots of the form $\sqrt{ax^2 + bx + c}$ we can again reduce the integral to a trigonometric integral by a trigonometric substitution, if we first complete the square.

Consider the integral

$$\int \frac{1}{(x^2 + 2x + 5)^{3/2}} \, dx.$$

If we complete the square in the denominator, we obtain

$$\int \frac{1}{(x^2 + 2x + 5)^{3/2}} \, dx = \int \frac{1}{[(x + 1)^2 + 4]^{3/2}} \, dx.$$

Had the denominator been $(x^2 + 4)^{3/2}$, we would have set $x = 2 \tan \theta$. It is natural, then, for the denominator $[(x + 1)^2 + 4]^{3/2}$ to set

$$x + 1 = 2 \tan \theta,$$

in which case $dx = 2 \sec^2 \theta \, d\theta$, and

FIGURE 8.7 Triangle to fit $x + 1 = 2 \tan \theta$

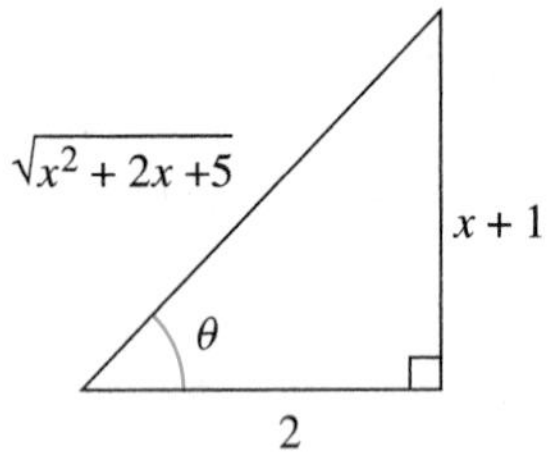

$$\int \frac{1}{(x^2 + 2x + 5)^{3/2}} \, dx = \int \frac{1}{(4 \tan^2 \theta + 4)^{3/2}} 2 \sec^2 \theta \, d\theta = \int \frac{2 \sec^2 \theta}{8 \sec^3 \theta} \, d\theta$$

$$= \frac{1}{4} \int \cos \theta \, d\theta = \frac{1}{4} \sin \theta + C$$

$$= \frac{x + 1}{4 \sqrt{x^2 + 2x + 5}} + C. \qquad \text{(Figure 8.7)}$$

Consider another example,

$$\int \sqrt{4 + x - x^2} \, dx.$$

Completion of the square leads to

$$\int \sqrt{4 + x - x^2} \, dx = \int \sqrt{-(x - 1/2)^2 + 17/4} \, dx.$$

If we now set

$$x - \frac{1}{2} = \frac{\sqrt{17}}{2} \sin \theta,$$

then

$$dx = \frac{\sqrt{17}}{2} \cos \theta \, d\theta,$$

and

$$\int \sqrt{4 + x - x^2} \, dx = \int \sqrt{-\frac{17}{4} \sin^2 \theta + \frac{17}{4}} \frac{\sqrt{17}}{2} \cos \theta \, d\theta = \int \frac{\sqrt{17}}{2} \cos \theta \frac{\sqrt{17}}{2} \cos \theta \, d\theta$$

$$= \frac{17}{4} \int \cos^2 \theta \, d\theta = \frac{17}{4} \int \left(\frac{1 + \cos 2\theta}{2} \right) d\theta$$

$$= \frac{17}{8} \left(\theta + \frac{1}{2} \sin 2\theta \right) + C = \frac{17}{8} (\theta + \sin \theta \cos \theta) + C$$

FIGURE 8.8   Triangle to fit
$x - 1 = \sqrt{17} \sin \theta$

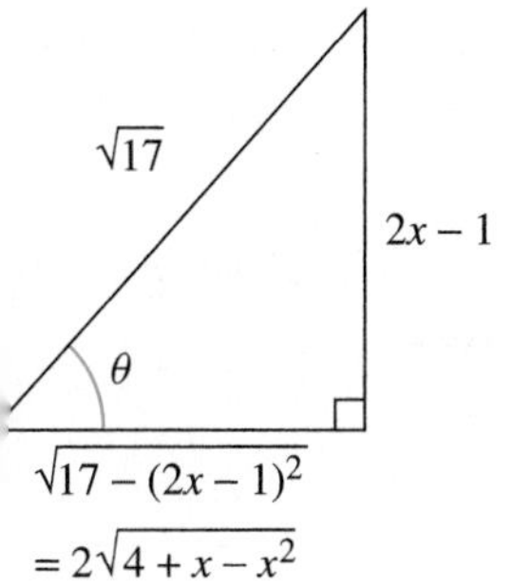

$$= \frac{17}{8} \text{Sin}^{-1}\left(\frac{2x - 1}{\sqrt{17}} \right) + \frac{17}{8} \left[\frac{2x - 1}{\sqrt{17}} \right] \frac{2\sqrt{4 + x - x^2}}{\sqrt{17}} + C \quad \text{(Figure 8.8)}$$

$$= \frac{17}{8} \text{Sin}^{-1}\left(\frac{2x - 1}{\sqrt{17}} \right) + \frac{1}{4} (2x - 1)\sqrt{4 + x - x^2} + C.$$

As a final illustrative example, we consider

$$\int \frac{x}{\sqrt{2x^2 + 3x - 6}} \, dx = \frac{1}{\sqrt{2}} \int \frac{x}{\sqrt{x^2 + \frac{3x}{2} - 3}} \, dx = \frac{1}{\sqrt{2}} \int \frac{x}{\sqrt{\left(x + \frac{3}{4} \right)^2 - \frac{57}{16}}} \, dx.$$

If we set

$$x + \frac{3}{4} = \frac{\sqrt{57}}{4} \sec \theta,$$

then

$$dx = \frac{\sqrt{57}}{4} \sec \theta \tan \theta \, d\theta$$

and

$$\int \frac{x}{\sqrt{2x^2 + 3x - 6}} \, dx = \frac{1}{\sqrt{2}} \int \frac{(\sqrt{57}/4) \sec \theta - 3/4}{\sqrt{(57/16) \sec^2 \theta - 57/16}} \frac{\sqrt{57}}{4} \sec \theta \tan \theta \, d\theta$$

$$= \frac{1}{\sqrt{2}} \int \frac{(\sqrt{57}/4) \sec \theta - 3/4}{(\sqrt{57}/4) \tan \theta} \frac{\sqrt{57}}{4} \sec \theta \tan \theta \, d\theta$$

$$= \frac{1}{\sqrt{2}} \int \left(\frac{\sqrt{57}}{4} \sec^2 \theta - \frac{3}{4} \sec \theta \right) d\theta$$

FIGURE 8.9   Triangle to fit
$x + 3 = \sqrt{57} \sec \theta$

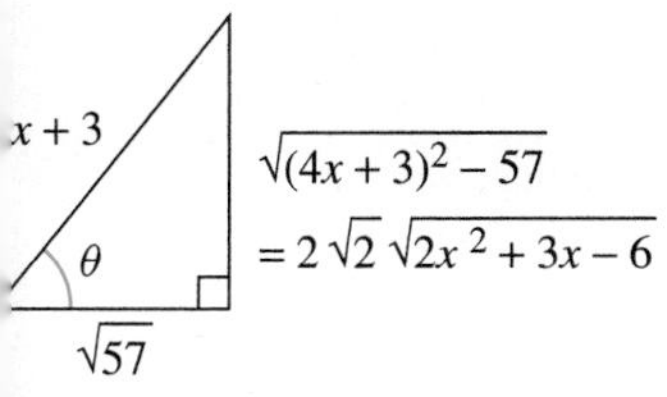

$$= \frac{\sqrt{57}}{4\sqrt{2}} \tan \theta - \frac{3}{4\sqrt{2}} \ln | \sec \theta + \tan \theta | + C$$

$$= \frac{\sqrt{57}}{4\sqrt{2}} \left[\frac{2\sqrt{2}\sqrt{2x^2 + 3x - 6}}{\sqrt{57}} \right]$$

$$- \frac{3}{4\sqrt{2}} \ln \left| \frac{4x + 3}{\sqrt{57}} + \frac{2\sqrt{2}\sqrt{2x^2 + 3x - 6}}{\sqrt{57}} \right| + C \quad \text{(Figure 8.9)}$$

$$= \frac{\sqrt{2x^2 + 3x - 6}}{2} - \frac{3}{4\sqrt{2}} \ln \left| 4x + 3 + 2\sqrt{2}\sqrt{2x^2 + 3x - 6} \right| + D,$$

where

$$D = C + \frac{3}{8\sqrt{2}} \ln 57.$$

We have shown that completing the square is the natural generalization of the technique of trigonometric substitutions. It reduces integrals containing square roots of the form $\sqrt{ax^2 + bx + c}$ to trigonometric integrals.

EXAMPLE 8.14

Find the volume of a donut.

SOLUTION   A donut is generated (as the volume of the solid of revolution) when the circle

$$(x - a)^2 + y^2 = b^2 \qquad (a > b)$$

is rotated around the y-axis (Figure 8.10). Cylindrical shells for vertical rectangles yield the volume generated by the upper semicircle, which can be doubled to give

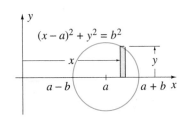

FIGURE 8.10   Donut volume by shells

$$V = 2 \int_{a-b}^{a+b} 2\pi xy \, dx.$$

If we solve the equation of the circle for $y = \pm\sqrt{b^2 - (x - a)^2}$, then

$$V = 2 \int_{a-b}^{a+b} 2\pi x \sqrt{b^2 - (x - a)^2} \, dx.$$

We now set $x - a = b \sin\theta$, in which case $dx = b \cos\theta \, d\theta$, and

$$V = 4\pi \int_{-\pi/2}^{\pi/2} (a + b\sin\theta)b\cos\theta \, b\cos\theta \, d\theta$$

$$= 4\pi b^2 \int_{-\pi/2}^{\pi/2} (a\cos^2\theta + b\cos^2\theta \sin\theta) \, d\theta$$

$$= 4\pi b^2 \int_{-\pi/2}^{\pi/2} \left[a \left(\frac{1 + \cos 2\theta}{2} \right) + b\cos^2\theta \sin\theta \right] d\theta$$

$$= 4\pi b^2 \left\{ \frac{a}{2} \left(\theta + \frac{1}{2}\sin 2\theta \right) - \frac{b}{3}\cos^3\theta \right\}_{-\pi/2}^{\pi/2}$$

$$= 4\pi b^2 \left[\frac{a}{2} \left(\frac{\pi}{2} + \frac{\pi}{2} \right) \right] = 2\pi^2 ab^2.$$

With horizontal rectangles (Figure 8.11), the washer method yields

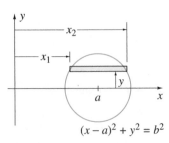

FIGURE 8.11   Volume of donut by washers

$$V = 2 \int_0^b (\pi x_2^2 - \pi x_1^2) \, dy.$$

We can solve the equation of the circle for $x = a \pm \sqrt{b^2 - y^2}$; hence,

$$x_1 = a - \sqrt{b^2 - y^2}, \qquad x_2 = a + \sqrt{b^2 - y^2}.$$

Thus

$$V = 2\pi \int_0^b \left[\left(a + \sqrt{b^2 - y^2} \right)^2 - \left(a - \sqrt{b^2 - y^2} \right)^2 \right] dy,$$

and this simplifies to

$$V = 2\pi \int_0^b 4a\sqrt{b^2 - y^2}\, dy.$$

If we set $y = b \sin\theta$, then $dy = b \cos\theta\, d\theta$, and

$$V = 8\pi a \int_0^{\pi/2} b \cos\theta\, b \cos\theta\, d\theta$$

$$= 8\pi ab^2 \int_0^{\pi/2} \left(\frac{1 + \cos 2\theta}{2}\right) d\theta$$

$$= 4\pi ab^2 \left\{\theta + \frac{1}{2}\sin 2\theta\right\}_0^{\pi/2}$$

$$= 4\pi ab^2 (\pi/2) = 2\pi^2 ab^2.$$

EXERCISES 8.5

In Exercises 1–10 evaluate the indefinite integral.

1. $\displaystyle\int \frac{x}{\sqrt{27 + 6x - x^2}}\, dx$

2. $\displaystyle\int \frac{1}{\sqrt{x^2 + 2x + 2}}\, dx$

3. $\displaystyle\int \frac{1}{(y^2 + 4y)^{3/2}}\, dy$

4. $\displaystyle\int \frac{1}{3x - x^2 - 4}\, dx$

5. $\displaystyle\int \frac{\sqrt{x^2 + 2x - 3}}{x + 1}\, dx$

6. $\displaystyle\int \frac{x}{(4x - x^2)^{3/2}}\, dx$

7. $\displaystyle\int \sqrt{-y^2 + 6y}\, dy$

8. $\displaystyle\int \frac{2x - 3}{x^2 + 6x + 13}\, dx$

9. $\displaystyle\int \frac{5 - 4x}{\sqrt{12x - 4x^2 - 8}}\, dx$

* **10.** $\displaystyle\int \frac{1}{x\sqrt{6 + 4\ln x + (\ln x)^2}}\, dx$

11. Use the substitution $z = 1/x$ to evaluate

$$\int \frac{1}{x\sqrt{x^2 + 6x + 3}}\, dx.$$

* **12.** One of the gates in a dam is circular with radius 1 m. If the gate is closed and the surface of the water is 3 m above the top of the gate, find the force due to water pressure on the gate.

* **13.** Evaluate

$$\int \frac{1}{3x - x^2}\, dx$$

(a) by completing the square, and setting $x - 3/2 = (3/2)\sin\theta$ and (b) by multiplying numerator and denominator by -1, completing the square, and setting $x - 3/2 = (3/2)\sec\theta$. (c) Explain the difference in the two answers.

In Exercises 14–16 evaluate the integral.

* **14.** $\displaystyle\int \sqrt{x^2 - 2x - 3}\, dx$

** **15.** $\displaystyle\int \frac{1}{x\sqrt{2x - x^2}}\, dx$

** **16.** $\displaystyle\int \frac{1}{(2x + 5)\sqrt{2x - 3} + 8x - 12}\, dx$

8.6 Partial Fractions

Partial fractions is a method that we apply to rational functions, integrals of the form

$$\int \frac{N(x)}{D(x)}\, dx, \tag{8.8}$$

where $N(x)$ and $D(x)$ are polynomials in x, and the degree of $N(x)$ is less than the degree of $D(x)$. When the degree of $N(x)$ is greater than or equal to that of $D(x)$, we divide $D(x)$ into $N(x)$. For example, in the integral

$$\int \frac{x^4 + 4x^3 + 2x + 4}{x^3 + 1}\, dx,$$

the numerator has degree 4 and the denominator degree 3. By long division, we obtain

$$\int \frac{x^4 + 4x^3 + 2x + 4}{x^3 + 1}\, dx = \int \left(x + 4 + \frac{x}{x^3 + 1} \right) dx.$$

The first two terms on the right, namely, $x + 4$, can be integrated immediately, and partial fractions can be applied to the remaining term, $x/(x^3 + 1)$.

As a general rule then: If the degree of $N(x)$ is greater than or equal to the degree of $D(x)$, divide $D(x)$ into $N(x)$ to produce a quotient polynomial $Q(x)$ and a remainder polynomial $R(x)$ of degree less than that of $D(x)$; that is,

$$\frac{N(x)}{D(x)} = Q(x) + \frac{R(x)}{D(x)},$$

where deg $R <$ deg D. Integral 8.8 can then be written in the form

$$\int \frac{N(x)}{D(x)}\, dx = \int Q(x)\, dx + \int \frac{R(x)}{D(x)}\, dx.$$

The first integral on the right is trivial. This leaves integration of the rational function

$$\int \frac{R(x)}{D(x)}\, dx, \qquad\qquad\qquad (8.9)$$

where deg $R <$ deg D.

To use partial fractions on integral 8.9 we must factor the denominator $D(x)$. In Section 1.2, we stated that every real polynomial, and specifically $D(x)$, can be factored into real linear factors $ax + b$ and irreducible real quadratic factors $ax^2 + bx + c$. A quadratic factor is irreducible if $b^2 - 4ac < 0$. In the complete factorization of $D(x)$, these factors may or may not be repeated; that is, the factorization of $D(x)$ contains terms of the form

$$(ax + b)^n \qquad \text{and} \qquad (ax^2 + bx + c)^n,$$

where $n \geq 1$ is an integer. When $n = 1$ the factor is nonrepeated, and when $n > 1$, it is repeated (it has multiplicity n). For example, in the factorization

$$D(x) = (x - 1)(2x + 1)^3(3x^2 + 4x + 5)(x^2 + 1)^2,$$

$x - 1$ and $3x^2 + 4x + 5$ are nonrepeated, and $2x + 1$ and $x^2 + 1$ are repeated.

It is worthwhile noting here that $D(x) = (x - 1)(2x - 2)$ does not have two distinct linear factors since we can write $D(x) = 2(x - 1)^2$. Likewise, $D(x) = (x^2 + 1)(3x^2 + 3)$ does not have distinct quadratic factors; it has a repeated quadratic factor $D(x) = 3(x^2 + 1)^2$. Distinct factors must have different zeros.

Having factored $D(x)$, we can separate the rational function in 8.9 into fractional components. We call this the **partial fraction decomposition** of the integrand. We illustrate with the rational function $(x^2 + 2)/[(x - 1)(2x + 1)^3(3x^2 + 4x + 5)(x^2 + 1)^2]$, and then state general rules for all decompositions. The partial fraction decomposition of this particular rational function is

$$\frac{x^2 + 2}{(x - 1)(2x + 1)^3(3x^2 + 4x + 5)(x^2 + 1)^2} = \frac{A}{x - 1} + \frac{B}{2x + 1} + \frac{C}{(2x + 1)^2} + \frac{D}{(2x + 1)^3}$$

$$+ \frac{Ex + F}{3x^2 + 4x + 5} + \frac{Gx + H}{x^2 + 1} + \frac{Ix + J}{(x^2 + 1)}$$

where $A, B, \ldots, J$ are constants. The first term corresponds to the nonrepeated linear factor $x - 1$, the next three correspond to the repeated linear factor $2x + 1$, the fifth term corresponds to the nonrepeated quadratic factor $3x^2 + 4x + 5$, and the last two terms result from the repeated quadratic $x^2 + 1$.

Let us now state general rules for the partial fraction decomposition of rational function $R(x)/D(x)$. There are three rules:

1. For each repeated or nonrepeated linear factor $(ax + b)^n$ in $D(x)$, include the following terms in the decomposition:

$$\frac{A_1}{ax + b} + \frac{A_2}{(ax + b)^2} + \cdots + \frac{A_n}{(ax + b)^n}. \qquad (8.10a)$$

The number of terms corresponds to the power n on $(ax + b)^n$.

2. For each repeated or nonrepeated irreducible quadratic factor $(ax^2 + bx + c)^n$ in $D(x)$, include the following terms in the decomposition:

$$\frac{B_1 x + C_1}{ax^2 + bx + c} + \frac{B_2 x + C_2}{(ax^2 + bx + c)^2} + \cdots + \frac{B_n x + C_n}{(ax^2 + bx + c)^n}. \qquad (8.10b)$$

Again the number of terms corresponds to the power n on $(ax^2 + bx + c)^n$.

3. The complete decomposition is the sum of all terms in 8.10a and all terms in 8.10b.

What is important here is to realize that all terms in expression 8.10a are immediately integrable. Terms in expression 8.10b are unlikely to be mental integrations, but if $b = 0$, they can be integrated with a trigonometric substitution. If $b \neq 0$, they can be integrated by completing the square and using a trigonometric substitution.

The following examples will clarify the above rules.

EXAMPLE 8.15

What form do partial fraction decompositions for the following rational functions take?

(a) $\dfrac{x^2 + 2x + 3}{3x^3 - x^2 - 3x + 1}$

(b) $\dfrac{x^2 + 3x - 1}{x^4 + x^3 + x^2 + x}$

(c) $\dfrac{x^3 + 3x^2 - x}{x^5 + x^4 + 2x^3 + 2x^2 + x + 1}$

(d) $\dfrac{3x^5 - 1}{(3x^2 + 5)(x^2 + 2x + 3)(2x - 1)^2}$

SOLUTION

(a) Since $3x^3 - x^2 - 3x + 1 = (3x - 1)(x - 1)(x + 1)$, we have nonrepeated linear factors. Rules 1 and 3 give the partial fraction decomposition

$$\frac{x^2 + 2x + 3}{3x^3 - x^2 - 3x + 1} = \frac{A}{3x - 1} + \frac{B}{x - 1} + \frac{C}{x + 1}.$$

(b) The factorization $x^4 + x^3 + x^2 + x = x(x + 1)(x^2 + 1)$ has two nonrepeated linear factors and a nonrepeated quadratic factor. The three rules lead to

$$\frac{x^2 + 3x - 1}{x^4 + x^3 + x^2 + x} = \frac{A}{x} + \frac{B}{x + 1} + \frac{Cx + D}{x^2 + 1}.$$

(c) Since $x^5 + x^4 + 2x^3 + 2x^2 + x + 1 = (x + 1)(x^2 + 1)^2$, the partial fraction decomposition is

$$\frac{x^3 + 3x^2 - x}{x^5 + x^4 + 2x^3 + 2x^2 + x + 1} = \frac{A}{x + 1} + \frac{Cx + D}{x^2 + 1} + \frac{Ex + F}{(x^2 + 1)^2}.$$

(d) Since $x^2 + 2x + 3$ can be factored no further, the decomposition is

$$\frac{3x^5 - 1}{(3x^2 + 5)(x^2 + 2x + 3)(2x - 1)^2} = \frac{Ax + B}{3x^2 + 5} + \frac{Cx + D}{x^2 + 2x + 3} + \frac{E}{2x - 1} + \frac{F}{(2x - 1)^2}.$$

To illustrate how to calculate coefficients in partial fraction decompositions, we use part (b) of Example 8.15. We bring the right side of the decomposition to a common denominator,

$$\frac{x^2 + 3x - 1}{x^4 + x^3 + x^2 + x} = \frac{A}{x} + \frac{B}{x+1} + \frac{Cx + D}{x^2 + 1}$$

$$= \frac{A(x+1)(x^2+1) + Bx(x^2+1) + x(x+1)(Cx+D)}{x^4 + x^3 + x^2 + x},$$

and equate numerators,

$$x^2 + 3x - 1 = A(x+1)(x^2+1) + Bx(x^2+1) + x(x+1)(Cx+D).$$

There are two methods for finding the constants A, B, C, and D.

Method 1.   First, we gather together terms in the various powers of x on the right side of the equation:

$$x^2 + 3x - 1 = (A + B + C)x^3 + (A + C + D)x^2 + (A + B + D)x + A.$$

Now, Exercise 35 in Section 3.8 states that two polynomials of the same degree can be equal for all values of x if and only if coefficients of corresponding powers of x are identical. Since we have equal cubic polynomials in the equation above, we equate coefficients of x^3, x^2, x, and x^0 (meaning terms with no x's):

$$x^3 : \qquad\qquad 0 = A + B + C,$$
$$x^2 : \qquad\qquad 1 = A + C + D,$$
$$x\ \ : \qquad\qquad 3 = A + B + D,$$
$$x^0 : \qquad\qquad -1 = A.$$

The solution of these four linear equations in four unknowns is $A = -1$, $B = 3/2$, $C = -1/2$, and $D = 5/2$. Hence, the partial fraction decomposition of $(x^2 + 3x - 1)/(x^4 + x^3 + x^2 + x)$ is

$$\frac{x^2 + 3x - 1}{x^4 + x^3 + x^2 + x} = -\frac{1}{x} + \frac{3/2}{x+1} + \frac{-x/2 + 5/2}{x^2 + 1}.$$

Method 2.   In this method we substitute convenient values of x into the equation

$$x^2 + 3x - 1 = A(x+1)(x^2+1) + Bx(x^2+1) + x(x+1)(Cx+D).$$

Clearly, $x = 0$ is most convenient, since it yields the value of A:

$$-1 = A(1)(1);$$

that is, $A = -1$. Convenient also is $x = -1$:

$$(-1)^2 + 3(-1) - 1 = B(-1)(2);$$

it gives $B = 3/2$. The values $x = 1$ and $x = 2$ yield the equations

$$1 + 3(1) - 1 = A(2)(2) + B(1)(2) + 1(2)(C + D),$$

$$4 + 3(2) - 1 = A(3)(5) + B(2)(5) + 2(3)(2C + D).$$

When $A = -1$ and $B = 3/2$ are substituted into these, the resulting equations are

$$C + D = 2,$$

$$4C + 2D = 3.$$

The solution of these is $C = -1/2$ and $D = 5/2$.

It is also possible to use a combination of methods 1 and 2 for finding coefficients in a partial fraction decomposition; that is, substitute some values of x and equate some coefficients. The total number must be equal to the number of unknown coefficients in the decomposition.

Once we have completed the partial fraction decomposition of a rational function, we can integrate the function by finding antiderivatives of the component fractions. For example, with the partial fraction decomposition of $(x^2 + 3x - 1)/(x^4 + x^3 + x^2 + x)$, it is very simple to find its indefinite integral,

$$\int \frac{x^2 + 3x - 1}{x^4 + x^3 + x^2 + x} \, dx = \int \left(-\frac{1}{x} + \frac{3/2}{x + 1} + \frac{-x/2 + 5/2}{x^2 + 1} \right) dx$$

$$= -\int \frac{1}{x} \, dx + \frac{3}{2} \int \frac{1}{x + 1} \, dx - \frac{1}{2} \int \frac{x}{x^2 + 1} \, dx + \frac{5}{2} \int \frac{1}{x^2 + 1} \, dx$$

$$= -\ln |x| + \frac{3}{2} \ln |x + 1| - \frac{1}{4} \ln (x^2 + 1) + \frac{5}{2} \text{Tan}^{-1} x + C.$$

Other integrations by partial fraction decompositions are illustrated in the following example.

EXAMPLE 8.16

Evaluate the following indefinite integrals:

(a) $\displaystyle \int \frac{x}{x^4 + 6x^2 + 5} \, dx$ (b) $\displaystyle \int \frac{x^2 + 1}{2x^3 - 5x^2 + 4x - 1} \, dx$

(c) $\displaystyle \int \frac{1}{x^3 + 8} \, dx$ (d) $\displaystyle \int \frac{x^2}{x^3 + 8} \, dx$

SOLUTION

(a) Since $x^4 + 6x^2 + 5 = (x^2 + 1)(x^2 + 5)$, the partial fraction decomposition of $x/(x^4 + 6x^2 + 5)$ has the form

$$\frac{x}{x^4 + 6x^2 + 5} = \frac{Ax + B}{x^2 + 1} + \frac{Cx + D}{x^2 + 5}.$$

When we bring the right side to a common denominator and equate numerators, the result is

$$x = (Ax + B)(x^2 + 5) + (Cx + D)(x^2 + 1).$$

We now multiply out the right side and equate coefficients of like powers of x,

$$x^3 : \qquad 0 = A + C,$$
$$x^2 : \qquad 0 = B + D,$$
$$x \; : \qquad 1 = 5A + C,$$
$$x^0 : \qquad 0 = 5B + D.$$

The solution of these equations is $A = 1/4$, $B = 0$, $C = -1/4$, and $D = 0$, and therefore

$$\int \frac{x}{x^4 + 6x^2 + 5} \, dx = \int \left(\frac{x/4}{x^2 + 1} + \frac{-x/4}{x^2 + 5} \right) dx$$

$$= \frac{1}{8} \ln (x^2 + 1) - \frac{1}{8} \ln (x^2 + 5) + C.$$

(b) Since $2x^3 - 5x^2 + 4x - 1 = (2x - 1)(x - 1)^2$, the partial fraction decomposition takes the form

$$\frac{x^2 + 1}{2x^3 - 5x^2 + 4x - 1} = \frac{A}{2x - 1} + \frac{B}{x - 1} + \frac{C}{(x - 1)^2}$$

$$= \frac{A(x - 1)^2 + B(x - 1)(2x - 1) + C(2x - 1)}{(2x - 1)(x - 1)^2}.$$

When we equate numerators,

$$x^2 + 1 = A(x - 1)^2 + B(x - 1)(2x - 1) + C(2x - 1).$$

We now set $x = 0$, $x = 1$, and $x = 1/2$:

$$x = 0 : \qquad\qquad 1 = A + B - C,$$
$$x = 1 : \qquad\qquad 2 = C(1),$$
$$x = 1/2 : \qquad\qquad 5/4 = A(1/4).$$

These give $A = 5$, $B = -2$, and $C = 2$, and therefore

$$\int \frac{x^2 + 1}{2x^3 - 5x^2 + 4x - 1}\, dx = \int \left[\frac{5}{2x - 1} - \frac{2}{x - 1} + \frac{2}{(x - 1)^2} \right] dx$$

$$= \frac{5}{2} \ln |2x - 1| - 2 \ln |x - 1| - \frac{2}{x - 1} + C.$$

(c) Since $x^3 + 8 = (x + 2)(x^2 - 2x + 4)$, we set

$$\frac{1}{x^3 + 8} = \frac{1}{(x + 2)(x^2 - 2x + 4)} = \frac{A}{x + 2} + \frac{Bx + C}{x^2 - 2x + 4}$$

$$= \frac{A(x^2 - 2x + 4) + (Bx + C)(x + 2)}{(x + 2)(x^2 - 2x + 4)},$$

and now equate numerators:

$$1 = A(x^2 - 2x + 4) + (Bx + C)(x + 2).$$

We set $x = -2$ and equate coefficients of x^2 and 1:

$$x = -2 : \qquad\qquad 1 = 12A,$$
$$x^2 : \qquad\qquad 0 = A + B,$$
$$1 : \qquad\qquad 1 = 4A + 2C.$$

These give $A = 1/12$, $B = -1/12$, and $C = 1/3$, and therefore

$$\int \frac{1}{x^3 + 8}\, dx = \int \left(\frac{1/12}{x + 2} + \frac{-x/12 + 1/3}{x^2 - 2x + 4} \right) dx$$

$$= \frac{1}{12} \ln |x + 2| + \frac{1}{12} \int \frac{4 - x}{(x - 1)^2 + 3}\, dx.$$

For the integral on the right, we set $x - 1 = \sqrt{3}\tan\theta$, in which case $dx = \sqrt{3}\sec^2\theta\,d\theta$, and

$$\int \frac{1}{x^3 + 8}\,dx = \frac{1}{12}\ln|x + 2| + \frac{1}{12}\int \frac{4 - (1 + \sqrt{3}\tan\theta)}{3\sec^2\theta}\sqrt{3}\sec^2\theta\,d\theta$$

$$= \frac{1}{12}\ln|x + 2| + \frac{1}{12\sqrt{3}}\int \left(3 - \sqrt{3}\tan\theta\right)d\theta$$

$$= \frac{1}{12}\ln|x + 2| + \frac{1}{12\sqrt{3}}\left(3\theta + \sqrt{3}\ln|\cos\theta|\right) + C$$

$$= \frac{1}{12}\ln|x + 2| + \frac{\sqrt{3}}{12}\mathrm{Tan}^{-1}\left(\frac{x - 1}{\sqrt{3}}\right) + \frac{1}{12}\ln\left|\frac{\sqrt{3}}{\sqrt{x^2 - 2x + 4}}\right| + C$$

(Figure 8.12)

$$= \frac{1}{12}\ln|x + 2| + \frac{\sqrt{3}}{12}\mathrm{Tan}^{-1}\left(\frac{x - 1}{\sqrt{3}}\right) - \frac{1}{24}\ln(x^2 - 2x + 4) + D,$$

where $D = C + (1/24)\ln 3$.

(d) Do not be misled into partial fractions in this example; the rational function is immediately integrable:

$$\int \frac{x^2}{x^3 + 8}\,dx = \frac{1}{3}\ln|x^3 + 8| + C.$$

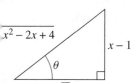

FIGURE 8.12   Triangle to $x - 1 = \sqrt{3}\tan\theta$

There are a number of useful devices that can sometimes simplify calculating coefficients in partial fraction decompositions. We indicate two of them here. First, if $ax + b$ is a nonrepeated, linear factor of the denominator of a partial fraction decomposition, then its coefficient in the decomposition can be obtained by "covering up" this term and substituting $x = -b/a$ into what remains. For example, to find the coefficient A in

$$\frac{x}{(x - 1)(2x + 3)} = \frac{A}{x - 1} + \frac{B}{2x + 3},$$

we cover up the $x - 1$ and substitute $x = 1$ into $x/(2x + 3)$. We obtain $1/(2 + 3) = 1/5$. This is A. To find B, we cover up $2x + 3$ and substitute $x = -3/2$ into $x/(x - 1)$. The result is $B = -(3/2)/(-3/2 - 1) = 3/5$. Thus,

$$\frac{x}{(x - 1)(2x + 3)} = \frac{1/5}{x - 1} + \frac{3/5}{2x + 3}.$$

Remember, however, that cover up can be used only on nonrepeated, linear factors. It could be used to find A, but not B or C, in

$$\frac{x^2 + 3}{(3x - 1)(x^2 + 2)} = \frac{A}{3x - 1} + \frac{Bx + C}{x^2 + 2}.$$

Second, when the denominator consists of only one repeated linear factor, say $(x + 3)^4$, rewriting the numerator in powers of $x + 3$ can sometimes give the partial fraction decomposition very quickly, especially when the numerator is a linear or quadratic polynomial. For example, to find the partial fraction decomposition of $(x - 4)/(x + 3)^4$, we write

$$\frac{x - 4}{(x + 3)^4} = \frac{(x + 3) - 7}{(x + 3)^4} = \frac{1}{(x + 3)^3} - \frac{7}{(x + 3)^4}.$$

Likewise,

$$\frac{x^2 + 2}{(x+4)^6} = \frac{(x+4)^2 - 8x - 14}{(x+4)^6} = \frac{(x+4)^2 - 8(x+4) + 18}{(x+4)^6}$$

$$= \frac{1}{(x+4)^4} - \frac{8}{(x+4)^5} + \frac{18}{(x+4)^6}.$$

Consulting Project 12

We have a chemical problem to solve. Two substances A and B react to form a third substance C in such a way that 2 grams of A react with 3 grams of B to produce 5 grams of C. When 20 grams of A and 40 grams of B are originally brought together, 3 grams of C are formed in the first hour. Our problem is to find the amount of C present in the mixture at any time.

SOLUTION   We need to assume something about the rate at which A and B combine to give C. Consultation with chemical engineers suggests that for many chemical reactions, the rate at which chemicals react is proportional to the amounts that are present in the mixture; the more chemicals, the faster the reaction. To express this algebraically, we let x be the number of grams of C at time t in the mixture. Some of this came from A and some from B. Specifically, $2x/5$ grams came from A and $3x/5$ grams came from B. This means that there are $20 - 2x/5$ grams of A and $40 - 3x/5$ grams of B remaining. The rate at which C is formed is represented by its derivative dx/dt and because it is proportional to the amounts of A and B present in the mixture, we write

$$\frac{dx}{dt} = K\left(20 - \frac{2x}{5}\right)\left(40 - \frac{3x}{5}\right)$$

$$= \frac{2K}{25}(50 - x)(200 - 3x) = k(50 - x)(200 - 3x),$$

where we have set $k = 2K/25$. Notice that we multiplied the amounts of A and B in the mixture rather than add them. By adding them, we would have the unacceptable situation of dx/dt being greater than zero even when one of the reactants vanishes. Since $t = 0$ when A and B are brought together, $x(t)$ must also satisfy $x(0) = 0$. In addition, $x(1) = 3$. We use partial fractions to write the differential equation in the form

$$k\,dt = \frac{1}{(50 - x)(200 - 3x)}\,dx = \left(\frac{1/50}{50 - x} - \frac{3/50}{200 - 3x}\right)dx.$$

It is separated, and solutions are defined implicitly by

$$kt + C = \frac{1}{50}\left[-\ln(50 - x) + \ln(200 - 3x)\right].$$

Absolute values are unnecessary because x cannot exceed 50. We now solve this equation for x by writing

$$50(kt + C) = \ln\left(\frac{200 - 3x}{50 - x}\right),$$

and exponentiating,

$$\frac{200 - 3x}{50 - x} = De^{50kt},$$

where $D = e^{50C}$. Cross multiplying gives $(50 - x)De^{50kt} = 200 - 3x$, and this can be solved for $x = \dfrac{200 - 50De^{50kt}}{3 - De^{50kt}}$. The initial condition $x(0) = 0$ requires $D = 4$, in which case

$$x = \frac{200 - 200e^{50kt}}{3 - 4e^{50kt}} = \frac{200\left(1' - e^{50kt}\right)}{3 - 4e^{50kt}} \text{ grams .}$$

Since $x(1) = 3$,

$$3 = \frac{200(1 - e^{50k})}{3 - 4e^{50k}} \qquad \Longrightarrow \qquad k = \frac{1}{50}\ln\left(\frac{191}{188}\right).$$

Thus, the number of grams of C in the mixture after t hours is

$$x(t) = \frac{200\left[1 - e^{t\ln(191/188)}\right]}{3 - 4e^{t\ln(191/188)}}.$$

We could plot this function or draw it using techniques from Chapter 4. What is interesting to note is that with only the differential equation and some physical reasoning, we can get a very good idea of the shape of the graph. The graph begins at $(0, 0)$, and has horizontal asymptote $x = 50$. The differential equation shows us that dx/dt, the slope of the graph is a maximum at $t = 0$ when x is smallest, and decreases as x increases. The graph can therefore have no critical points or points of inflection. It must appear as in Figure 8.13.

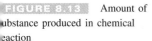

FIGURE 8.13 Amount of substance produced in chemical reaction

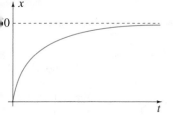

Consulting Project 13

Nuclear scientists are approaching us this time. The volume (in cubic metres) of a plug-flow-reactor to operate for 90% fractional conversion is given by the integral

$$V = \frac{1}{393} \int_0^{0.022} \frac{(0.096 - x/2)^3}{(0.024 - x)(0.024 - x/2)^2}dx.$$

Evaluation is often done numerically because it is claimed that the integration is too complex. We are asked to evaluate the integral exactly so as to reduce errors in numerical integration.

SOLUTION   First, we consider the indefinite integral

$$I = \int \frac{(0.096 - x/2)^3}{(0.024 - x)(0.024 - x/2)^2}dx = \frac{1}{2}\int \frac{(0.192 - x)^3}{(0.024 - x)(0.048 - x)^2}dx.$$

Let us set $a = 0.024$ in which case partial fractions lead to

$$I = \frac{1}{2}\int \frac{(8a - x)^3}{(a - x)(2a - x)^2}dx = \frac{1}{2}\int \left[1 + \frac{343a}{a - x} - \frac{216a^2}{(2a - x)^2} - \frac{324a}{2a - x}\right]dx$$

$$= \frac{1}{2}\left[x - 343a\ln|a - x| - \frac{216a^2}{2a - x} + 324a\ln|2a - x|\right] + C.$$

Consequently, the volume of the reactor should be

$$V = \frac{1}{2(393)} \left[0.022 - 343(0.024) \ln |0.024 - 0.022| \right.$$

$$\left. - \frac{216(0.024)^2}{0.048 - 0.022} + 324(0.024) \ln |0.048 - 0.022| \right]$$

$$= 0.023 \text{ m}^3.$$

EXERCISES 8.6

In Exercises 1–16 evaluate the indefinite integral.

1. $\displaystyle\int \frac{x+2}{x^2 - 2x + 1} \, dx$

2. $\displaystyle\int \frac{1}{y^3 + 3y^2 + 3y + 1} \, dy$

3. $\displaystyle\int \frac{1}{z^3 + z} \, dz$

4. $\displaystyle\int \frac{x^2 + 2x - 4}{x^2 - 2x - 8} \, dx$

5. $\displaystyle\int \frac{x}{(x-4)^2} \, dx$

6. $\displaystyle\int \frac{y+1}{y^3 + y^2 - 6y} \, dy$

7. $\displaystyle\int \frac{3x+5}{x^3 - x^2 - x + 1} \, dx$

8. $\displaystyle\int \frac{x^3}{(x^2 + 2)^2} \, dx$

9. $\displaystyle\int \frac{1}{x^2 - 3} \, dx$

10. $\displaystyle\int \frac{y^2}{y^2 + 3y + 2} \, dy$

11. $\displaystyle\int \frac{z^2 + 3z - 2}{z^3 + 5z} \, dz$

12. $\displaystyle\int \frac{y^2 + 6y + 4}{y^4 + 5y^2 + 4} \, dy$

13. $\displaystyle\int \frac{x}{x^4 + 7x^2 + 6} \, dx$

14. $\displaystyle\int \frac{x^2 + 3}{x^4 + x^2 - 2} \, dx$

15. $\displaystyle\int \frac{3t+4}{t^4 - 3t^3 + 3t^2 - t} \, dt$

16. $\displaystyle\int \frac{x^3 + 6}{x^4 + 2x^3 - 3x^2 - 4x + 4} \, dx$

* 17. Find the length of the curve $y = \ln(1 - x^2)$ from $x = 0$ to $x = 1/2$.

* 18. A car of mass 1500 kg starts from rest at an intersection and moves in the positive x-direction. The engine exerts a constant force of magnitude 2500 N, and air friction causes a resistive force whose magnitude in newtons is equal to the square of the speed of the car in metres per second. Newton's second law gives the following differential equation for the velocity of the car:

$$1500 \frac{dv}{dt} = 2500 - v^2.$$

(a) By separating the differential equation show that $v(t) = 50(1 - e^{-t/15})/(1 + e^{-t/15})$.

(b) Find the position of the car as a function of time.

* 19. When a raindrop with mass m falls in air, it is acted on by gravity and also by a force due to air resistance that is proportional to the square of its instantaneous speed. According to Newton's second law the differential equation describing the velocity of the raindrop is

$$m \frac{dv}{dt} = mg - kv^2,$$

where $g = 9.81$, and $k > 0$ is a constant. Separate the differential equation to show that if the raindrop exits vertically downward from the cloud with velocity v_0, then

$$v(t) = \frac{V \left[1 - \left(\dfrac{V - v_0}{V + v_0} \right) e^{-2kVt/m} \right]}{1 + \left(\dfrac{V - v_0}{V + v_0} \right) e^{-2kVt/m}},$$

where $V = \sqrt{mg/k}$. Can you interpret V physically?

* 20. If we wish to know the velocity of the raindrop in Exercise 19 as it strikes the earth, it is preferable to find velocity as a function of distance fallen (instead of time). By expressing the differential equation in the form

$$mv \frac{dv}{dy} = mg - kv^2,$$

where y is distance fallen by the raindrop, find the velocity of the raindrop when it strikes the earth if it falls from height h.

* 21. The exponential growth of bacteria in the two exercises of Section 5.5 is unrealistic in the long term when there is a limited food supply. The *logistic model* introduces a quantity C called the *carrying capacity* for the environment in which the bacteria are living. As the number $N(t)$ of bacteria approaches C, its growth rate must slow down. The logistic model to describe this is the differential equation

$$\frac{dN}{dt} = kN \left(1 - \frac{N}{C} \right).$$

Notice that when N is small, dN/dt is approximately equal to kN, thus preserving early exponential growth. The factor $1 - N/C$ causes $dN/dt \to 0$ as $N \to C$. Solve this differential equation for $N(t)$ when $k = 1$, $C = 10^6$, and $N(0) = 100$.

22. Show that the solution of the logistic model in Exercise 21 for an initial population $N(0) = N_0$ is

$$N = \frac{C}{1 + \left(\dfrac{C - N_0}{N_0}\right) e^{-kt}}.$$

23. Find the centroid of the region bounded by the curves $(x+2)^2 y = 1 - x$, $x = 0$, $y = 0$, and $(x, y \geq 0)$.

24. Suppose that N represents the number of people in a population, and $x(t)$ the number that are infected by some disease at any given time t. It is often assumed that the rate of infection is proportional to the product of the number of infected and not infected,

$$\frac{dx}{dt} = kx(N - x),$$

where $k > 0$ is a constant. Thus, $x(t)$ must satisfy this differential equation subject to an initial condition such as, perhaps, $x(0) = 1$. Therefore, one infected person is introduced into the population at time $t = 0$. Find $x(t)$.

25. Chemical reactions such as that in Project 12 are called second-order reactions (because of the x^2 term on the right). In general, they take the form

$$\frac{dx}{dt} = k(a - x)(b - x).$$

Solve this differential equation in the cases that (a) $a = b$, (b) $a \neq b$.

26. The velocity v of water, flowing from a tap that is suddenly turned on, varies initially according to

$$a \frac{dv}{dt} = v_0^2 - v^2,$$

where $a > 0$ is a constant and v_0 is the steady-state velocity. Find $v(t)$ using the initial condition $v(0) = 0$.

In Exercises 27–32 evaluate the indefinite integral.

27. $\displaystyle \int \frac{x^3 + x + 2}{x^5 + 2x^3 + x} \, dx$

28. $\displaystyle \int \frac{1}{x^5 + x^4 + 2x^3 + 2x^2 + x + 1} \, dx$

29. $\displaystyle \int \frac{1}{(x^2 + 5)(x^2 + 2x + 3)} \, dx$

30. $\displaystyle \int \frac{1}{1 + x^3} \, dx$

31. $\displaystyle \int \frac{\sin x}{\cos x (1 + \cos^2 x)} \, dx$

* **32.** $\displaystyle \int \frac{x^4 + 8x^3 - x^2 + 2x + 1}{x^5 + x^4 + x^2 + x} \, dx$

* **33.** During the initial stages of flow in a pipeline of length L, the velocity of the flow v must satisfy the differential equation

$$\frac{dv}{dt} = \frac{gH_e}{L},$$

where $g = 9.81$ and H_e is the effective head of the line, given by

$$H_e = H\left(1 - \frac{v^2}{v_f^2}\right),$$

where H is the constant head and v_f is the final velocity in the pipeline.

(a) Using the initial condition $v(0) = 0$, show that

$$t = \frac{Lv_f}{2gH} \ln\left(\frac{v_f + v}{v_f - v}\right).$$

(b) Solve the equation in part (a) to obtain v as a function of t.

* **34.** In the study of frictional fluid flow in a duct, the following indefinite integral is encountered:

$$\int \frac{M(1 - M^2)}{M^4 \left(1 + \dfrac{k - 1}{2} M^2\right)} \, dM,$$

where k is a constant. Evaluate this integral.

* **35.** If an integrand is a rational function of $\sin x$ and $\cos x$, it can be reduced to a rational function of t by the substitution

$$t = \tan\left(\frac{x}{2}\right).$$

This is often called the Weierstrass substitution. Show that with this substitution

$$dx = \frac{2}{1 + t^2} \, dt, \qquad \sin x = \frac{2t}{1 + t^2}, \qquad \cos x = \frac{1 - t^2}{1 + t^2}.$$

In Exercises 36–39 use the substitution of Exercise 35 to evaluate the integral.

* **36.** $\displaystyle \int \sec x \, dx$

* **37.** $\displaystyle \int \frac{1}{3 + 5\sin x} \, dx$

* **38.** $\displaystyle \int \frac{1}{1 - 2\cos x} \, dx$

** **39.** $\displaystyle \int \frac{1}{\sin x + \cos x} \, dx$

* **40.** Show that the answer to Exercise 36 can be expressed in the usual form, $\ln|\sec x + \tan x| + C$.

** **41.** (a) Use the change of variable in Exercise 35 to show that

$$\int \frac{1}{5 - 4 \cos x} dx = \frac{2}{3} \text{Tan}^{-1}\left(3 \tan \frac{x}{2}\right) + C.$$

(b) What happens when the antiderivative in part (a) is used to evaluate

$$\int_0^{2\pi} \frac{1}{5 - 4 \cos x} dx?$$

(c) Show that

$$\frac{x}{3} + \frac{2}{3} \text{Tan}^{-1}\left(\frac{\sin x}{2 - \cos x}\right)$$

is also an antiderivative of $(5 - 4 \cos x)^{-1}$. Use it to evaluate the definite integral in part (b).

** **42.** Evaluate $\displaystyle\int \frac{x^2 + x + 3}{x^4 + x^3 + 2x^2 + 11x - 5} dx.$

** **43.** Evaluate $\displaystyle\int \frac{2x^3 + 8x^2 - 3x + 5}{x^4 + 3x^3 + x^2 + 2x - 12} dx.$

8.7 Numerical Integration

To evaluate the definite integral of a continuous function $f(x)$ with respect to x from $x = a$ to $x = b$, we have used the first fundamental theorem of integral calculus: Find an antiderivative for $f(x)$, substitute the limits $x = b$ and $x = a$, and subtract. The evaluation procedure depends on our ability to produce an antiderivative for $f(x)$. When a function $f(x)$ is complicated, it may be difficult or even impossible to find its antiderivative. In such a case, it may be necessary to approximate the definite integral of $f(x)$ on some interval $a \le x \le b$, rather than evaluate it analytically. We discuss three methods for doing this: the rectangular, trapezoidal, and Simpson's rules. Each method divides the interval $a \le x \le b$ into subintervals and approximates $f(x)$ with an easily integrated function on each subinterval.

If $f(x) \ge 0$ on $a \le x \le b$ (Figure 8.14), the definite integral of $f(x)$ with respect to x can be interpreted as the area bounded by $y = f(x)$, $y = 0$, $x = a$, and $x = b$. We have carefully pointed out that it is not always wise to think of a definite integral as area, but for our discussion here it is convenient to do so. Our problem is to approximate the area in Figure 8.14 when it is difficult or impossible to find an antiderivative for $f(x)$.

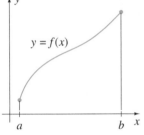
FIGURE 8.14   Definite integral as area when $f(x) \ge 0$

Rectangular Rule

The first method is to return to definition 7.1 for area in terms of approximating rectangles. We subdivide the interval $a \le x \le b$ into n subintervals by points $a = x_0 < x_1 < \cdots < x_{n-1} < x_n = b$. For simplicity, we choose n equal subdivisions, in which case

$$x_i = a + i\left(\frac{b - a}{n}\right),$$

and denote the width of each subinterval by $h = (b - a)/n$. Area under $y = f(x)$, above $y = 0$, and between $x = a$ and $x = b$, which is given by the definite integral of $f(x)$ from a to b, is approximated by the rectangles in Figure 8.15,

$$\int_a^b f(x)\, dx \approx \sum_{i=1}^n f(x_i)\, h = h \sum_{i=1}^n f(x_i). \tag{8.11}$$

This is the **rectangular rule** for approximating the definite integral; we have replaced the area under $y = f(x)$ with n rectangles. Another way of looking at it is to say that we have replaced the original function $f(x)$ by a function that is constant on each subinterval (Figure 8.16), but the constant value varies from subinterval to subinterval. Such a function is called a *step function*. The definite integral of this function from $x = a$ to $x = b$ is the right side of 8.11. It approximates the definite integral of $f(x)$.

Graphically, it is reasonable to expect that an increase in the number of subdivisions results in an increase in the accuracy of the approximation.

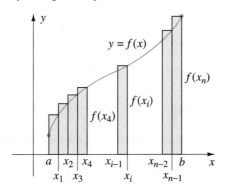

FIGURE 8.15   Approximation of area by rectangles of equal widths

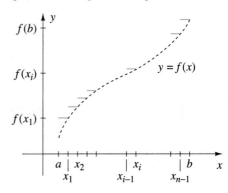

FIGURE 8.16   Interpretation of rectangular rule as integration of step function

EXAMPLE 8.17

Use a subdivision of the interval $1 \leq x \leq 3$ into 5, 10, and 20 equal parts to approximate the definite integral

$$\int_1^3 \sin x \, dx$$

with the rectangular rule.

SOLUTION   With 5 equal parts, $h = 2/5$ and $x_i = 1 + 2i/5$. Consequently,

$$\int_1^3 \sin x \, dx \approx \frac{2}{5} \sum_{i=1}^{5} \sin\left(1 + \frac{2i}{5}\right)$$

$$= \frac{2}{5}\left(\sin\frac{7}{5} + \sin\frac{9}{5} + \sin\frac{11}{5} + \sin\frac{13}{5} + \sin 3\right)$$

$$= 1.370.$$

With 10 equal parts, $h = 2/10 = 1/5$ and $x_i = 1 + i/5$, so that

$$\int_1^3 \sin x \, dx \approx \frac{1}{5} \sum_{i=1}^{10} \sin\left(1 + \frac{i}{5}\right)$$

$$= \frac{1}{5}\left(\sin\frac{6}{5} + \sin\frac{7}{5} + \sin\frac{8}{5} + \sin\frac{9}{5} + \sin 2 + \sin\frac{11}{5}\right.$$

$$\left. + \sin\frac{12}{5} + \sin\frac{13}{5} + \sin\frac{14}{5} + \sin 3\right)$$

$$= 1.455.$$

With 20 equal parts we have

$$\int_1^3 \sin x \, dx \approx \frac{1}{10} \sum_{i=1}^{20} \sin\left(1 + \frac{i}{10}\right)$$

$$= \frac{1}{10}\left(\sin\frac{11}{10} + \sin\frac{6}{5} + \sin\frac{13}{10} + \cdots + \sin\frac{29}{10} + \sin 3\right)$$

$$= 1.494.$$

When we compare these results with the correct answer,

$$\int_1^3 \sin x \, dx = \left\{ -\cos x \right\}_1^3 = -\cos 3 + \cos 1 = 1.530\,294\,8,$$

we see that as n increases from 5 to 10 to 20, the approximation improves, but it must be increased even further to give a reasonable approximation to this integral.

Trapezoidal Rule

Regarding area, it is clear in Figure 8.15 that were we to join successive points $(x_i, f(x_i))$ on the curve $y = f(x)$ with straight-line segments as in Figure 8.17a, the area under this broken straight line would be a much better approximation to the area under $y = f(x)$ than that provided by the rectangular rule. Effectively, we now approximate the area by n trapezoids.

FIGURE 8.17a   Approximation of area
by trapezoids of equal widths

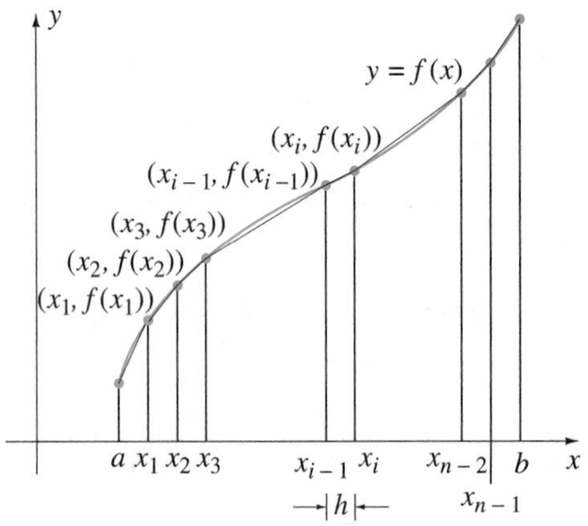

FIGURE 8.17b   Enlargement of ith
subinterval

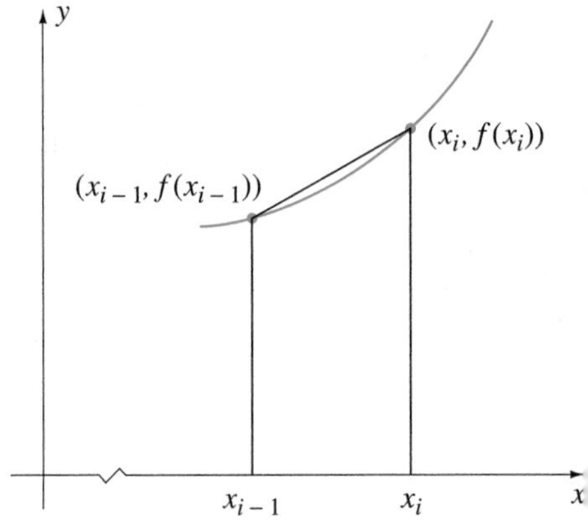

Since the area of a trapezoid is its width multiplied by the average of its parallel lengths, it follows that the area of the ith trapezoid in Figure 8.17b is given by

$$h \left[\frac{f(x_i) + f(x_{i-1})}{2} \right],$$

where again $h = x_i - x_{i-1}$. As a result, the area under $y = f(x)$ can be approximated by the sum

$$h \left[\frac{f(x_1) + f(x_0)}{2} + \frac{f(x_2) + f(x_1)}{2} + \cdots + \frac{f(x_n) + f(x_{n-1})}{2} \right]$$

$$= \frac{h}{2} [f(a) + 2f(x_1) + 2f(x_2) + \cdots + 2f(x_i) + \cdots + 2f(x_{n-1}) + f(b)]$$

$$= \frac{h}{2} \left[f(a) + 2 \sum_{i=1}^{n-1} f(x_i) + f(b) \right].$$

We write therefore that

$$\int_a^b f(x) \, dx \approx \frac{h}{2} \left[f(a) + 2 \sum_{i=1}^{n-1} f(x_i) + f(b) \right], \tag{8.12a}$$

where $x_i = a + ih = a + i(b-a)/n$, and call this the **trapezoidal rule** for approximating a definite integral. Note that if 8.12a is written in the form

$$\int_a^b f(x)\, dx \approx h \left[\frac{f(a) - f(b)}{2} \right] + h \sum_{i=1}^{n} f(x_i), \tag{8.12b}$$

the summation on the right, except for the first two terms, is the rectangular rule. In other words, the extra numerical calculation involved in using the trapezoidal rule rather than the rectangular rule is minimal, but it would appear that the accuracy is increased significantly. For this reason, the trapezoidal rule supplants the rectangular rule in most applications.

EXAMPLE 8.18

▼

Use the trapezoidal rule to approximate the definite integral in Example 8.17.

SOLUTION   With 5 equal partitions,

$$\int_1^3 \sin x\, dx \approx \frac{2/5}{2} \left[\sin 1 + 2 \sum_{i=1}^{4} \sin \left(1 + \frac{2i}{5} \right) + \sin 3 \right] = 1.5098.$$

With 10 equal parts,

$$\int_1^3 \sin x\, dx \approx \frac{1/5}{2} \left[\sin 1 + 2 \sum_{i=1}^{9} \sin \left(1 + \frac{i}{5} \right) + \sin 3 \right] = 1.5252.$$

Finally, with $n = 20$, we have

$$\int_1^3 \sin x\, dx \approx \frac{1/10}{2} \left[\sin 1 + 2 \sum_{i=1}^{19} \sin \left(1 + \frac{i}{10} \right) + \sin 3 \right] = 1.5290.$$

As expected, these approximations are significantly better than corresponding results using the rectangular rule.

▲

Simpson's Rule

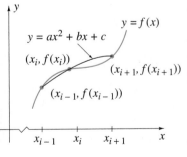

The rectangular rule replaces a function $f(x)$ with a step function; the trapezoidal rule replaces $f(x)$ with a succession of linear functions — geometrically, a broken straight line. So far as ease of integration is concerned, the next simplest function is a quadratic function. Consider, then, replacing the curve $y = f(x)$ by a succession of parabolas on the subintervals $x_{i-1} \leq x \leq x_i$. Now the equation of a parabola with a vertical axis of symmetry is of the form $y = ax^2 + bx + c$ with three constants a, b, and c to be determined. If this parabola is to approximate $y = f(x)$ on $x_{i-1} \leq x \leq x_i$, we should have the parabola pass through the end points $(x_{i-1}, f(x_{i-1}))$ and $(x_i, f(x_i))$. But this imposes only two conditions on a, b, and c, not three. To take advantage of this flexibility, we demand that the parabola also pass through the point $(x_{i+1}, f(x_{i+1}))$ (Figure 8.18).

In other words, instead of replacing $y = f(x)$ with n parabolas, one on each subinterval $x_{i-1} \leq x \leq x_i$, we replace it with $n/2$ parabolas, one on each pair of subintervals $x_{i-1} \leq x \leq x_{i+1}$. Note that this requires n to be an even integer. These three conditions imply that a, b, and c must satisfy the equations

$$ax_{i-1}^2 + bx_{i-1} + c = f(x_{i-1}), \tag{8.13a}$$

$$ax_i^2 + bx_i + c = f(x_i), \tag{8.13b}$$

$$ax_{i+1}^2 + bx_{i+1} + c = f(x_{i+1}). \tag{8.13c}$$

These equations determine the values for a, b, and c, but it will not be necessary to actually solve them. Suppose for the moment that we have solved equations 8.17 for a, b, and c, and we continue our main discussion. With the parabola $y = ax^2 + bx + c$ replacing the curve $y = f(x)$ on the interval $x_{i-1} \leq x \leq x_{i+1}$, we approximate the area under $y = f(x)$ with that under the parabola, namely,

$$\int_{x_{i-1}}^{x_{i+1}} (ax^2 + bx + c)\, dx = \frac{a}{3}(x_{i+1}^3 - x_{i-1}^3) + \frac{b}{2}(x_{i+1}^2 - x_{i-1}^2) + c(x_{i+1} - x_{i-1})$$

$$= \frac{x_{i+1} - x_{i-1}}{6}[2a(x_{i+1}^2 + x_{i+1}x_{i-1} + x_{i-1}^2)$$

$$+ 3b(x_{i+1} + x_{i-1}) + 6c].$$

Now, if we use $x_{i+1} - x_{i-1} = 2h$, $x_{i+1} = x_i + h$, and $x_{i-1} = x_i - h$ to write everything in terms of h and x_i, we obtain

$$\int_{x_{i-1}}^{x_{i+1}} (ax^2 + bx + c)\, dx = \frac{2h}{6}\left\{2a[(x_i + h)^2 + (x_i + h)(x_i - h) + (x_i - h)^2]\right.$$

$$+ 3b[(x_i + h) + (x_i - h)] + 6c\Big\}$$

$$= \frac{h}{3}\left\{a(6x_i^2 + 2h^2) + 6bx_i + 6c\right\}.$$

But if we add equation 8.13a, equation 8.13c, and four times equation 8.13b, we find that

$$f(x_{i-1}) + 4f(x_i) + f(x_{i+1}) = a(x_{i-1}^2 + 4x_i^2 + x_{i+1}^2) + b(x_{i-1} + 4x_i + x_{i+1}) + 6c$$

$$= a\left[(x_i - h)^2 + 4x_i^2 + (x_i + h)^2\right]$$

$$+ b\left[(x_i - h) + 4x_i + (x_i + h)\right] + 6c$$

$$= a(6x_i^2 + 2h^2) + 6bx_i + 6c.$$

Thus, the area under the parabola may be written in the form

$$\int_{x_{i-1}}^{x_{i+1}} (ax^2 + bx + c)\, dx = \frac{h}{3}\left[f(x_{i-1}) + 4f(x_i) + f(x_{i+1})\right],$$

and the right side is free of a, b, and c. The same expression would have resulted had we solved equations 8.13 for a, b, and c and then evaluated the integral of $ax^2 + bx + c$ from x_{i-1} to x_{i+1}. The derivation above, however, is much simpler.

When we add all such integrals over the $n/2$ subintervals $x_{i-1} \leq x \leq x_{i+1}$ between $x = a$ and $x = b$, we obtain

$$\frac{h}{3}\left[f(x_0) + 4f(x_1) + f(x_2)\right] + \frac{h}{3}\left[f(x_2) + 4f(x_3) + f(x_4)\right] + \cdots$$

$$+ \frac{h}{3}\left[f(x_{n-2}) + 4f(x_{n-1}) + f(x_n)\right]$$

$$= \frac{h}{3}\left[f(x_0) + 4f(x_1) + 2f(x_2) + 4f(x_3) + \cdots\right.$$

$$+ 2f(x_{n-2}) + 4f(x_{n-1}) + f(x_n)\Big].$$

In other words, the definite integral of $f(x)$ from $x = a$ to $x = b$ can be approximated by

$$\int_a^b f(x)\, dx \approx \frac{h}{3}\left[f(a) + 4f(x_1) + 2f(x_2) + 4f(x_3) + \cdots\right.$$

$$+ 2f(x_{n-2}) + 4f(x_{n-1}) + f(b)\Big], \qquad (8.14)$$

where $x_i = a + ih = a + i(b-a)/n$. This result is called **Simpson's rule** for approximating a definite integral. Although the formula does not display it explicitly (except by counting terms), *do not forget that n must be an even integer.*

EXAMPLE 8.19

Approximate the definite integral in Example 8.17 using Simpson's rule with $n = 10$ and $n = 20$.

SOLUTION   With $n = 10$,

$$\int_1^3 \sin x \, dx \approx \frac{1/5}{3}\left[\sin 1 + 4\sin\frac{6}{5} + 2\sin\frac{7}{5} + 4\sin\frac{8}{5} + \cdots + 2\sin\frac{13}{5} + 4\sin\frac{14}{5} + \sin 3\right]$$

$$= 1.530\,308\,5.$$

With $n = 20$,

$$\int_1^3 \sin x \, dx \approx \frac{1/10}{3}\left[\sin 1 + 4\sin\frac{11}{10} + 2\sin\frac{6}{5} + 4\sin\frac{13}{10} + \cdots + 2\sin\frac{14}{5} + 4\sin\frac{29}{10} + \sin 3\right]$$

$$= 1.530\,295\,7.$$

Table 8.1 lists the approximations in Examples 8.17–8.19. The correct answer for the integral is 1.530 294 8 (to seven decimal places). It is clear that each method gives a better approximation as the value of n increases, and that Simpson's rule is by far the most accurate.

TABLE 8.1

	Rectangular rule	Trapezoidal rule	Simpson's rule
$n = 5$	1.370	1.5098	
$n = 10$	1.455	1.5252	1.530 308 5
$n = 20$	1.494	1.5290	1.530 295 7

In practice, we use the rectangular, trapezoidal, and Simpson's rules to approximate definite integrals that cannot be handled analytically, and we will not therefore have the correct answer with which to compare the approximation. We would still like to make some statement about the accuracy of the approximation, however, since what good is the approximation otherwise? The following two theorems give error estimates for the trapezoidal rule and for Simpson's rule.

THEOREM 8.1

If $f''(x)$ exists on $a \le x \le b$, and T_n is the error in approximating the definite integral of $f(x)$ from $x = a$ to $x = b$ using the trapezoidal rule with n equal subdivisions,

$$T_n = \int_a^b f(x)\,dx - \frac{h}{2}\left[f(a) + 2\sum_{i=1}^{n-1} f(x_i) + f(b)\right],$$

then

$$|T_n| \le \frac{M(b-a)^3}{12n^2}, \tag{8.15}$$

where M is the maximum value of $|f''(x)|$ on $a \le x \le b$.

THEOREM 8.2

If $f''''(x)$ exists on $a \leq x \leq b$, and S_n is the error in approximating the definite integral of $f(x)$ from $x = a$ to $x = b$ using Simpson's rule with n equal subdivisions,

$$S_n = \int_a^b f(x)\,dx - \frac{h}{3}\left[f(a) + 4\sum_{i=1}^{n/2} f(x_{2i-1}) + 2\sum_{i=1}^{n/2-1} f(x_{2i}) + f(b) \right],$$

then

$$|S_n| \leq \frac{M(b-a)^5}{180n^4}, \tag{8.16}$$

where M is the maximum value of $|f''''(x)|$ on $a \leq x \leq b$.

Proofs of these theorems can be found in books on numerical analysis. Note that because of the n^4 factor in the denominator of 8.16, the accuracy of Simpson's rule increases much more rapidly than does that of the trapezoidal rule.

For the function $f(x) = \sin x$ in Examples 8.17–8.19,

$$f''(x) = -\sin x \quad \text{and} \quad f''''(x) = \sin x.$$

Consequently, in both cases we can state that $M = 1$, and therefore

$$|T_n| \leq \frac{(3-1)^3}{12n^2} = \frac{2}{3n^2} \quad \text{and} \quad |S_n| \leq \frac{(3-1)^5}{180n^4} = \frac{8}{45n^4}.$$

For $n = 10$ and $n = 20$, we find that

$$|T_{10}| \leq \frac{2}{3(10)^2} < 0.0067 \quad \text{and} \quad |S_{10}| \leq \frac{8}{45(10)^4} < 0.000\,018;$$

$$|T_{20}| \leq \frac{2}{3(20)^2} < 0.0017 \quad \text{and} \quad |S_{20}| \leq \frac{8}{45(20)^4} < 0.000\,001\,2.$$

Differences between the correct value for the integral and the approximations listed in Table 8.1 corroborate these predictions.

EXAMPLE 8.20

What is the maximum possible error in using the trapezoidal rule with 100 equal subdivisions to approximate

$$\int_1^3 \frac{1}{\sqrt{1+x^3}}\,dx?$$

SOLUTION   According to formula 8.15, if T_{100} is the maximum possible error, then

$$|T_{100}| \leq \frac{M(3-1)^3}{12(100)^2} = \frac{2M}{3 \times 10^4},$$

where M is the maximum of (the absolute value of) the second derivative of the integrand $1/\sqrt{1+x^3}$ on $1 \leq x \leq 3$. Now

$$\frac{d^2}{dx^2}\frac{1}{\sqrt{1+x^3}} = \frac{d}{dx}\left[\frac{-3x^2}{2(1+x^3)^{3/2}}\right] = \frac{3x(5x^3-4)}{4(1+x^3)^{5/2}}.$$

Instead of maximizing the absolute value of this function on the interval $1 \leq x \leq 3$, which would require another derivative, we note that the maximum value of the numerator is obtained for $x = 3$, and the minimum value of the denominator occurs at $x = 1$. It follows therefore that the second derivative cannot possibly be larger than

$$\frac{3(3)[5(3)^3 - 4]}{4(1 + 1)^{5/2}} = \frac{1179}{16\sqrt{2}}.$$

Thus, M must be less than or equal to $1179/(16\sqrt{2})$, and we can state that

$$|T_{100}| \leq \frac{2}{3 \times 10^4} \left(\frac{1179}{16\sqrt{2}}\right) \leq 0.0035;$$

that is, the error in using the trapezoidal rule with 100 equal subdivisions to approximate the definite integral cannot be any larger than 0.0035.

EXAMPLE 8.21

How many equal subdivisions of the interval $0 \leq x \leq 2$ guarantee an error of less than 10^{-5} in the approximation of the definite integral

$$\int_0^2 e^{-x^2} \, dx$$

using Simpson's rule?

SOLUTION   According to formula 8.16, the error in using Simpson's rule with n equal subdivisions to approximate this definite integral is

$$|S_n| \leq \frac{M(2 - 0)^5}{180n^4},$$

where M is the maximum of the (absolute value of the) fourth derivative of e^{-x^2} on the interval $0 \leq x \leq 2$. It is a short calculation to find

$$\frac{d^4}{dx^4}\left(e^{-x^2}\right) = 4(3 - 12x^2 + 4x^4)e^{-x^2}.$$

Instead of maximizing the absolute value of this function, we note that e^{-x^2} has a maximum value of 1 (when $x = 0$). Furthermore, because $|3 - 12x^2 + 4x^4| \leq 3 + 12x^2 + 4x^4$, which has a maximum at $x = 2$, it follows that

$$M \leq 4[3 + 12(2)^2 + 4(2)^4](1) = 460.$$

Consequently,

$$|S_n| \leq \frac{460(2)^5}{180n^4} = \frac{3680}{45n^4}.$$

The error is less than 10^{-5} if n is chosen sufficiently large that

$$\frac{3680}{45n^4} < 10^{-5},$$

that is, if

$$n > \left(\frac{3680}{45 \times 10^{-5}}\right)^{1/4} = 53.5.$$

Since n must be an even integer, the required accuracy is guaranteed if n is chosen greater than or equal to 54.

FIGURE 8.19 Approxi-
mating work to deflect truss from
tabulated values of deflection

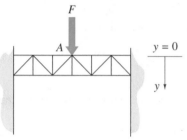

Numerical techniques are indispensable in situations where the function to be integrated is not
known, but what is available is a set of tabulated values for the function, perhaps experimental
data. For example, when the truss in Figure 8.19 is subjected to a force F at its centre, point A
deflects an amount y from its equilibrium position. Forces (in kilonewtons) required to produce
deflections from 0 to 5 cm at intervals of 0.5 cm are listed in Table 8.2.

TABLE 8.2

y	0	0.5	1.0	1.5	2.0	2.5	3.0	3.5	4.0	4.5	5.0
F	0	1.45	2.90	4.40	5.90	7.43	9.05	10.7	13.2	15.3	18.0

The work done by F in deflecting A by 5 cm is the definite integral of F with respect to y
from $y = 0$ to $y = 5$. The trapezoidal rule and Simpson's rule can be used to approximate
this definite integral even though we do not have a formula for F as a function of y. Indeed,
formulas 8.12 and 8.14 do not use the form of the function to be integrated, only its values at the
subdivision points. We have these points in Table 8.2. Since there are 11 points and, therefore,
10 subdivisions on the interval $0 \le y \le 5$, the trapezoidal rule gives

$$\int_1^5 F(y)\,dy \approx \frac{1/2}{2}[F(0) + 2F(0.5) + 2F(1.0) + \cdots + 2F(4.0) + 2F(4.5) + F(5.0)]$$

$$= \frac{1}{4}[0 + 2(1.45) + 2(2.90) + \cdots + 2(13.2) + 2(15.3) + 18.0]$$

$$= 39.67 \text{ kN} \cdot \text{cm} = 396.7 \text{ J}.$$

Simpson's rule gives

$$\int_1^5 F(y)\,dy \approx \frac{1/2}{3}[F(0) + 4F(0.5) + 2F(1.0) + \cdots + 2F(4.0) + 4F(4.5) + F(5.0)]$$

$$= \frac{1}{6}[0 + 4(1.45) + 2(2.90) + \cdots + 2(13.2) + 4(15.3) + 18.0]$$

$$= 39.54 \text{ kN} \cdot \text{cm} = 395.4 \text{ J}.$$

What we lose in this type of application is the ability to predict a maximum possible error in the
approximation since we cannot find M for formulas 8.15 and 8.16.

The error bounds in formulas 8.15 and 8.16 are somewhat idealistic in the sense that they
are error predictions based on the use of exact numbers. For instance, they predict that by
increasing n indefinitely, any degree of accuracy is attainable. Theoretically this is true, but
practically it is not. No matter how we choose to evaluate the summations in 8.12 and 8.14, be
it by hand, by an electronic hand calculator, or by a high-speed computer, each calculation is
rounded off to a certain number of decimals. The final sum takes into account many, many of
these "approximate numbers," and must therefore be inherently inaccurate. We call this *round-
off error* and it is very difficult to predict how extensive it is. It depends on both the number and
nature of the operations involved in 8.12 and 8.14. In the approximation of a definite integral
by the trapezoidal rule or Simpson's rule, there are two sources of error. Formulas 8.15 and
8.16 predict errors due to the methods themselves; round-off errors may also be appreciable for
large n.

We should emphasize once again that although we have used area as a convenient vehicle
by which to explain the approximation of definite integrals by the rectangular, trapezoidal, and
Simpson's rules, it is not necessary for $f(x)$ to be nonnegative. All three methods can be used
to approximate the definite integral of a function $f(x)$, be it positive, negative, or sometimes
positive and sometimes negative on the interval of integration. The only condition that we have
imposed is that $f(x)$ be continuous. In view of our discussion of improper integrals in Section
7.10, even this is not always necessary.

EXERCISES 8.7

In Exercises 1–10 use the trapezoidal rule and Simpson's rule with 10 equal subdivisions to approximate the definite integral. In each case, evaluate the integral analytically to get an idea of the accuracy of the approximation.

 1. $\int_1^2 \frac{1}{x}\,dx$

2. $\int_2^3 \frac{1}{\sqrt{x+2}}\,dx$

3. $\int_0^1 \tan x\,dx$

4. $\int_0^{1/2} e^x\,dx$

5. $\int_{-1}^1 \sqrt{x+1}\,dx$

6. $\int_{-3}^{-2} \frac{1}{x^3}\,dx$

7. $\int_{1/2}^1 \cos x\,dx$

8. $\int_0^1 \frac{1}{3+x^2}\,dx$

9. $\int_1^3 \frac{1}{x^2+x}\,dx$

10. $\int_0^{1/2} xe^{x^2}\,dx$

In Exercises 11–14 use the trapezoidal rule and Simpson's rule with 10 equal subdivisions to approximate the definite integral.

11. $\int_0^2 \frac{1}{1+x^3}\,dx$

12. $\int_0^1 e^{x^2}\,dx$

13. $\int_1^2 \sqrt{1+x^4}\,dx$

14. $\int_{-1}^0 \sin(x^2)\,dx$

15. Show graphically that if $y = f(x)$ is concave downward on the interval $a \le x \le b$, then the trapezoidal rule underestimates the definite integral of $f(x)$ on this interval.

16. What happens to the errors in 8.15 and 8.16 when the number of partitions is doubled?

17. The definite integral

$$\int_a^b e^{-x^2}\,dx$$

is very important in mathematical statistics. Use Simpson's rule with $n = 16$ to evaluate the integral for $a = 0$ and $b = 1$.

18. Use Simpson's rule with 10 equal intervals to approximate the definite integral for the length of the parabola $y = x^2$ between $x = 0$ and $x = 1$. Compare the answer to that of Exercise 39 in Section 8.4.

19. Use the trapezoidal rule and Simpson's rule with 10 equal subdivisions to approximate the definite integral for the length of the curve $y = \sin x$ from $x = 0$ to $x = \pi/2$.

20. The swimming pool that follows has an average depth of 1.8 m. It is to be drained and filled with dirt. To estimate the volume of dirt required, measurements across the pool are taken at 1-m intervals. Use Simpson's rule to find the estimate.

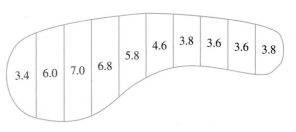

21. An aerial photograph of an oil spill shows the pattern in the figure below. Assuming that the oil slick has a uniform depth of 1 cm, estimate the number of cubic metres of oil in the spill.

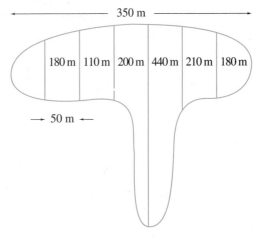

22. The numerical techniques of this section can also be used to approximate many improper integrals. Consider

$$\int_0^4 \frac{e^x}{\sqrt{x}}\,dx.$$

(a) Why can the trapezoidal rule and Simpson's rule not be used directly to approximate this integral?

(b) Show that the change of variable $u = \sqrt{x}$ replaces this improper integral with an integral that is not improper and use Simpson's rule with 20 equal subdivisions to approximate its value.

(c) Could you use the rectangular rule on the improper integral?

23. (a) Show that the definite integral

$$L = \frac{4}{3}\int_0^3 \sqrt{\frac{81-5x^2}{9-x^2}}\,dx$$

represents the length L of the ellipse $4x^2 + 9y^2 = 36$.

(b) Show that when we set $x = 3\sin\theta$, the θ-integral is no longer improper.

(c) Use the trapezoidal rule and Simpson's rule with eight equal subdivisions to approximate the θ-integral in part (b).

24. To approximate the improper integral

$$\int_1^\infty \frac{1}{1+x^4}\,dx$$

set $x = 1/t$, and use the trapezoidal rule and Simpson's rule with 10 equal subdivisions on the resulting integral.

25. Use the technique of Exercise 24 to approximate

$$\int_1^\infty \frac{x^2}{x^4+x^2+1}\,dx.$$

26. Use the trapezoidal rule and Simpson's rule to approximate the definite integral from $x = -1$ to $x = 4$ for the function tabulated in the following table.

x	−1.0	−0.5	0	0.5	1.0	1.5	2.0	2.5	3.0	3.5	4.0
y	2.287	0.395	0	0.145	0.310	0.334	1.819	0.123	0.021	−0.037	−0.055

27. Repeat Exercise 26 for the integral from $x = 1$ to $x = 3$ for the function tabulated below.

x	1.0	1.2	1.4	1.6	1.8	2.0	2.2	2.4	2.6	2.8	3.0
y	0.84	1.12	1.40	1.60	1.75	1.82	1.79	1.62	1.34	0.94	0.42

28. Show that when $f(x)$ is a cubic polynomial, evaluation of the definite integral of $f(x)$ from $x = a$ to $x = b$ by Simpson's rule always gives the *exact* answer. Illustrate with an example.

In Exercises 29–32 how many equal subdivisions of the interval of integration guarantee an error of less than 10^{-4} in the approximation of the definite integral using (a) the trapezoidal rule and (b) Simpson's rule?

29. $\displaystyle\int_1^4 \frac{1}{x}\,dx$

30. $\displaystyle\int_0^{\pi/4} \cos x\,dx$

31. $\displaystyle\int_0^{1/3} e^{2x}\,dx$

32. $\displaystyle\int_4^5 \frac{1}{\sqrt{x+2}}\,dx$

SUMMARY

Antidifferentiation is a far more complicated process than differentiation; it is not possible to state a set of rules and formulas that will suffice for most functions. Certainly, there are integration formulas that we must learn, but most of those given in this chapter have been differentiation formulas listed in Chapter 3 written in terms of integrals rather than derivatives. We have stressed the importance of knowing these obvious integration formulas, but beyond this, it becomes an organizational problem — organizing a difficult integral into a form that utilizes these simple formulas. The three most important techniques for doing this are substitutions, integration by parts, and partial fractions.

There is a variety of substitutions for the evaluation of indefinite integrals, many suggested by the form of the integrand. For example, a term in $x^{1/n}$ (n an integer) suggests the substitution $u = x^{1/n}$. Trigonometric substitutions are most important. They eliminate square roots of the form $\sqrt{a^2 \pm b^2x^2}$ and $\sqrt{b^2x^2 - a^2}$, and the general root $\sqrt{ax^2 + bx + c}$.

Integration by parts is a powerful integration technique. It is used to evaluate antiderivatives of transcendental functions that are multiplied by powers of x, it leads to the method of integration by reproduction, and it is used to develop reduction formulas.

The method of partial fractions decomposes complicated rational functions into simple fractions that are either immediately integrable or amenable to other methods, such as trigonometric substitutions.

Even with the techniques that we have studied and tables, there are many functions that either cannot be antidifferentiated at all or can be antidifferentiated only with extreme difficulty. To approximate definite integrals of such functions, numerical techniques such as the rectangular rule, the trapezoidal rule, and Simpson's rule are essential.

KEY TERMS

In reviewing this chapter, you should be able to define or discuss the following key terms:

Integration by parts
Integration by reproduction
Partial fraction decomposition
Trapezoidal rule

Integration-by-parts formula
Trigonometric substitutions
Rectangular rule
Simpson's rule

REVIEW EXERCISES

In Exercises 1–50 evaluate the indefinite integral.

1. $\int \sqrt{2-x}\, dx$

2. $\int \frac{1}{(x+3)^2}\, dx$

3. $\int \frac{x^2+3}{x}\, dx$

4. $\int \frac{x^2+3}{x+1}\, dx$

5. $\int \frac{x^2+3}{x^2+1}\, dx$

6. $\int \frac{x}{\sqrt{x+3}}\, dx$

7. $\int \sin^2 x \cos^3 x\, dx$

8. $\int x \sin x\, dx$

9. $\int \tan^2(2x)\, dx$

10. $\int \frac{x}{x^2+2x-3}\, dx$

11. $\int \frac{1}{\sqrt{4-3x^2}}\, dx$

12. $\int \frac{2-\sqrt{x}}{\sqrt{x}+5}\, dx$

13. $\int \frac{x}{3x^2+4}\, dx$

14. $\int \frac{e^x}{\sqrt{1-e^{2x}}}\, dx$

15. $\int x^2 \ln x\, dx$

16. $\int \frac{x}{(x^2+1)^2}\, dx$

17. $\int \frac{x^2}{(x^2+1)^2}\, dx$

18. $\int \frac{x^3}{(x^2+1)^2}\, dx$

19. $\int \frac{x+1}{x^3-4x}\, dx$

20. $\int \left(\frac{x+1}{x-1}\right)^2 dx$

21. $\int \frac{x^2}{(1+3x^3)^4}\, dx$

22. $\int \text{Cos}^{-1} x\, dx$

23. $\int \sin x \cos 2x\, dx$

24. $\int \sin x \cos 5x\, dx$

25. $\int e^{3x} \cos 2x\, dx$

26. $\int \frac{1}{\sqrt{x^2+4x-5}}\, dx$

27. $\int \frac{1}{x^2+4x-5}\, dx$

28. $\int x^3\sqrt{4-x^2}\, dx$

29. $\int \frac{\cos 2x}{1-\sin 2x}\, dx$

30. $\int \frac{6x}{4-x^2}\, dx$

31. $\int \frac{1}{x\sqrt{\ln x}}\, dx$

32. $\int \frac{1}{x^2+4x-4}\, dx$

33. $\int \frac{\sin x}{1+\cos^2 x}\, dx$

34. $\int \frac{1}{x^4+x^3}\, dx$

35. $\int x \sec^2(3x)\, dx$

36. $\int \frac{1}{\sqrt{16-3x+x^2}}\, dx$

37. $\int \frac{\sqrt{x^2-4}}{x^2}\, dx$

38. $\int x^2 \text{Tan}^{-1} x\, dx$

39. $\int \frac{x^2}{x^3+3x^2+3x+1}\, dx$
40. $\int \frac{\ln x}{x}\, dx$

41. $\int \frac{x^2}{1+4x^6}\, dx$

42. $\int \frac{1}{x(9+x^2)^2}\, dx$

43. $\int \frac{x^2+2}{x^3+5x^2+4x}\, dx$

44. $\int \frac{x^2+2}{x^3+4x^2+4x}\, dx$

45. $\int \frac{x^2+2}{x^3+x^2+4x}\, dx$

46. $\int \frac{3x^2+2x+4}{x^3+x^2+4x}\, dx$

47. $\int x \, \text{Sin}^{-1} x\, dx$

48. $\int \sqrt{\cot x}\, \csc^4 x\, dx$

49. $\int \ln(\sqrt{x}+1)\, dx$

50. $\int \frac{1}{(4x-x^2)^{3/2}}\, dx$

In Exercises 51–55 use the trapezoidal rule and Simpson's rule with 10 equal partitions to approximate the definite integral.

51. $\int_1^2 \frac{\sin x}{x}\, dx$

52. $\int_0^1 \sqrt{\sin x}\, dx$

53. $\int_2^4 \frac{1}{\ln x}\, dx$

54. $\int_{-1}^3 \frac{1}{1+e^x}\, dx$

55. $\int_0^1 \frac{1}{(1+x^4)^2}\, dx$

In Exercises 56–71 evaluate the indefinite integral.

* **56.** $\displaystyle\int \frac{1}{x^{1/3} - \sqrt{x}}\,dx$

* **57.** $\displaystyle\int \ln\left(1 + x^2\right) dx$

* **58.** $\displaystyle\int \frac{x}{x^4 + 16}\,dx$

* **59.** $\displaystyle\int \csc^3 x\,dx$

* **60.** $\displaystyle\int \frac{1}{(3x - x^2)^{3/2}}\,dx$

* **61.** $\displaystyle\int \frac{1}{x^3\sqrt{x^2 - 9}}\,dx$

* **62.** $\displaystyle\int \sin\sqrt{x}\,dx$

* **63.** $\displaystyle\int \sin\left(\ln x\right) dx$

* **64.** $\displaystyle\int x \cos x \sin 3x\,dx$

* **65.** $\displaystyle\int \frac{x^4 + 3x^2 + 1}{x(x^2 + 1)^2}\,dx$

* **66.** $\displaystyle\int \frac{1}{1 + \cos 2x}\,dx$

* **67.** $\displaystyle\int \frac{x^4 + 3x^2 - 2x + 5}{x^2 - 3x + 7}\,dx$

* **68.** $\displaystyle\int \sin^2 x \cos 3x\,dx$

* **69.** $\displaystyle\int \frac{1}{x^3(4 - x^2)^{3/2}}\,dx$

* **70.** $\displaystyle\int \sqrt{1 - x^2}\,\mathrm{Sin}^{-1} x\,dx$

* **71.** $\displaystyle\int \frac{1}{x + \sqrt{x^2 + 4}}\,dx$

* **72.** (a) Use the substitution $z^2 = (1 + x)/(1 - x)$ to show that

$$\int \sqrt{\frac{1 + x}{1 - x}}\,dx = 2\,\mathrm{Tan}^{-1}\sqrt{\frac{1 + x}{1 - x}} - \sqrt{1 - x^2} + C.$$

(b) Evaluate the integral in part (a) by multiplying numerator and denominator of the integrand by $\sqrt{1 + x}$. Verify that this answer is the same as that in part (a).

CHAPTER 9 | Parametric Equations and Polar Coordinates

Application Preview

The figure below shows a mechanism in which rod AB, pinned at A, rotates counterclockwise about A. Slider B moves along rod CD, causing it to rotate also. End E of rod CE is confined to slide along a horizontal line. Lengths of the members AB, CD, and CE are as shown.

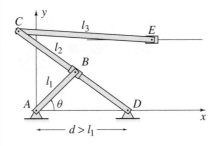

THE PROBLEM   Determine the path followed by joint C. (See Example 9.6 on page 548 for the solution.)

In this chapter we introduce parametric representations for curves and polar coordinates. Parametric representations of curves offer an alternative to the explicit and implicit forms used in Chapters 1–8. Polar coordinates provide an alternative way to identify points in a plane. They are more efficient than Cartesian coordinates in many applications. In particular, we shall see how they provide a unified approach to conic sections.

9.1 Parametric Equations

A curve is defined explicitly by equations of the form $y = f(x)$ or $x = g(y)$, and implicitly by an equation $F(x, y) = 0$. A third method is described in the following definition.

DEFINITION 9.1

A curve is said to be defined parametrically if it is given in the form

$$x = x(t), \qquad y = y(t), \qquad \alpha \le t \le \beta. \tag{9.1}$$

Variable t is called a *parameter*; it is a connecting link between x and y. Each value of t in the interval $\alpha \le t \le \beta$ is substituted into the **parametric equations** $x = x(t)$ and $y = y(t)$ in 9.1, and the pair $(x, y) = \big(x(t), y(t)\big)$ represents a point on the curve. When the interval for t is unspecified, we assume that it consists of all values for which both $x(t)$ and $y(t)$ are defined.

Graphing calculators and computers are adept at plotting curves defined parametrically. Given the functions $x(t)$ and $y(t)$, and the interval $\alpha \le t \le \beta$, they plot sufficiently many points $\big(x(t), y(t)\big)$ and join them with straight lines to give an excellent rendering of the curve.

EXAMPLE 9.1

Plot the curve

$$x = 2 - t, \qquad y = t + 5, \qquad -6 \le t \le 5.$$

What does it look like? Verify this analytically.

SOLUTION The plot in Figure 9.1 appears to be a straight line joining the points $(8, -1)$ and $(-3, 10)$. This is easily verified by solving $x = 2 - t$ for $t = 2 - x$ and substituting into $y = t + 5$:

$$y = (2 - x) + 5 = -x + 7.$$

FIGURE 9.1   Parametric plot of a straight line

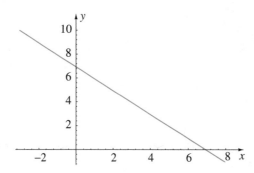

When $x(t)$ and $y(t)$ in equations 9.1 are linear functions, the curve is always a straight line or line segment. If α and β are finite, they define ends of the line segment; if they are not finite, the whole line results. Line segments can also be defined by nonlinear functions (see Exercise 31).

EXAMPLE 9.2

Plot the curve

$$x = t^2 - 2t + 4, \qquad y = 3 - 2t, \qquad -2 \le t \le 2.$$

What does it look like? Verify this analytically.

SOLUTION The plot in Figure 9.2 appears to be a parabola joining the points $(12, 7)$ and $(4, -1)$. To verify this, we solve the second equation for $t = (3 - y)/2$, and substitute into the first:

$$x = \left(\frac{3 - y}{2}\right)^2 - 2\left(\frac{3 - y}{2}\right) + 4 = \frac{y^2}{4} - \frac{y}{2} + \frac{13}{4}.$$

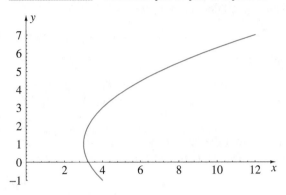

FIGURE 9.2   Parametric plot of part of a parabola

EXAMPLE 9.3

Discuss the curve defined parametrically by

$$x = 2\cos\theta, \quad y = 2\sin\theta, \quad 0 \le \theta < 2\pi.$$

SOLUTION   If the given equations are squared and added, the result is

$$x^2 + y^2 = 4\cos^2\theta + 4\sin^2\theta = 4,$$

an implicit definition of the curve. The given equations therefore define the circle parametrically. Seldom is it possible to give a geometric interpretation for parameter t in equations 9.1. Parameter θ for a circle is an exception; it can be interpreted as the angle in Figure 9.3. Values $0 \le \theta < 2\pi$ describe the complete circle in a counterclockwise direction beginning at $(2, 0)$. Additional values of θ duplicate existing points. For example, using $0 \le \theta < 4\pi$ traces the circle twice. Specifying an interval of length less than 2π describes part of the circle. For instance, $0 \le \theta \le \pi$ gives the upper semicircle $y = \sqrt{4 - x^2}$.

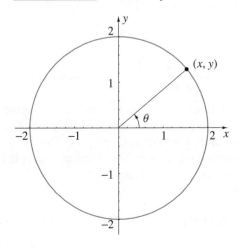

FIGURE 9.3   Parametric plot of a circle

EXAMPLE 9.4

Find an implicit definition for the curve

$$x = t - t^3, \qquad y = t + t^3,$$

and plot the curve.

SOLUTION   By adding and subtracting the given equations, we have

$$x + y = 2t \qquad \text{and} \qquad y - x = 2t^3.$$

It follows from the first of these that $t = (x + y)/2$, and when this is substituted into the second, we obtain an implicit definition of the curve

$$y - x = \frac{(x + y)^3}{4}.$$

Unfortunately, as is always the case when an electronic device is used to plot parametric equations, Figure 9.4a does not indicate which values of t give which points on the curve. We can remedy this by relating graphs of $x(t)$ and $y(t)$ in Figures 9.4b and c to the curve in Figure 9.4a. For example, beginning with $t = 0$, Figures 9.4b and c give $x = 0$ and $y = 0$, and therefore the point $(0, 0)$ in Figure 9.4a. As t increases from $t = 0$ to $t = 1$, values of x increase from 0 to a maximum value $2/(3\sqrt{3})$ at $t = 1/\sqrt{3}$, and then decrease to 0. Simultaneously, values of y increase steadily from 0 to 2. This gives the first quadrant part of the curve in Figure 9.4a. As t increases beyond 1, values of x decrease through negative numbers while y continues to increase. This is reflected in that part of the graph in Figure 9.4a in the second quadrant. Because $x = t - t^3$ and $y = t + t^3$ are odd functions, replacing t by $-t$ reverses the signs of x and y. This means that corresponding to each point (x, y) on the graph for which $t > 0$, we must have the point $(-x, -y)$ corresponding to $-t$. This gives third- and fourth-quadrant parts of the curve in Figure 9.4a.

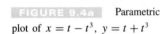

FIGURE 9.4a   Parametric plot of $x = t - t^3$, $y = t + t^3$

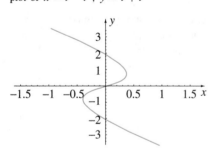

FIGURE 9.4b   Plot of $x = t - t^3$

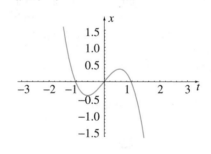

FIGURE 9.4c   Plot of $y = t + t^3$

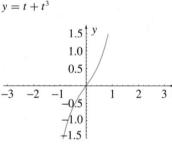

It is worthwhile noting that were calculators and/or computers not programmed to do parametric plots, we would draw parametric curves precisely as outlined in this example. Piece the curve together from separate graphs of $x(t)$ and $y(t)$. This would be the case if $x(t)$ or $y(t)$, or both, contained unspecified parameters. The following example is an illustration.

EXAMPLE 9.5

Draw the curve defined parametrically by

$$x = a \cos t, \qquad y = b \sin 2t,$$

where $a > 0$ and $b > 0$ are constants.

FIGURE 9.5a   Parametric plot of $x = 2\cos t$, $y = \sin 2t$

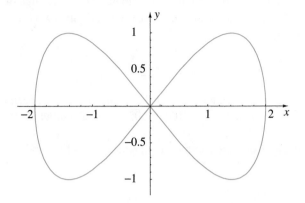

FIGURE 9.5b   Graph of $x = a\cos t$, $y = b\sin 2t$

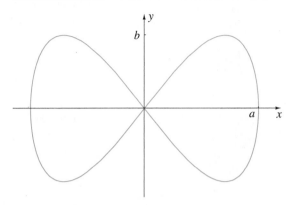

SOLUTION   Because $a\cos t$ and $b\sin 2t$ are periodic with periods 2π and π, respectively, it follows that all values of x and y are obtained for $0 \le t \le 2\pi$. We begin by plotting the curve using specific values for a and b, say $a = 2$ and $b = 1$. The result is shown in Figure 9.5a. The required curve can be obtained by changing scales on the x- and y-axes (Figure 9.5b).

To confirm this, we draw the curve using the technique of Example 9.4. We draw graphs of $x = a\cos t$ and $y = b\sin 2t$ (Figures 9.6a and b). Value $t = 0$ gives the point $(a, 0)$ in Figure 9.6c. As t increases from $t = 0$ to $t = \pi/2$, values of x decrease from a to 0 and y increases from 0 to b, and then decreases from b to 0. This gives the first-quadrant part of the curve in Figure 9.6c. As t increases from $\pi/2$ to π, values of x decrease from 0 to $-a$ and y decreases from 0 to $-b$, and then increases from $-b$ to 0. This adds the third-quadrant part of the curve in Figure 9.6d. Continuation leads to the full curve in Figure 9.5b.

FIGURE 9.6a   Graph of $x = a\cos t$

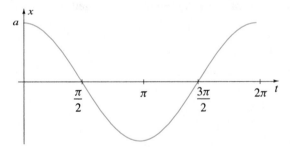

FIGURE 9.6b   Graph of $y = b\sin 2t$

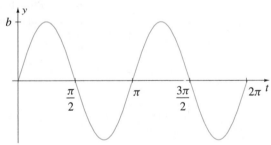

FIGURE 9.6c   First-quadrant graph of $x = a\cos t$, $y = b\sin 2t$

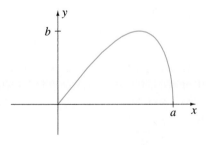

FIGURE 9.6d   First- and third-quadrant graph of $x = a\cos t$, $y = b\sin 2t$

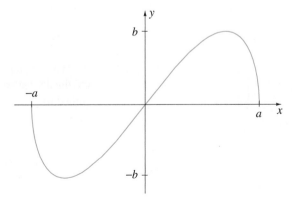

FIGURE 9.7 Path of a projectile is a parabola

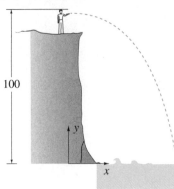

Parametric equations are frequently used in problems concerning the motion of objects in a plane or in space. For example, suppose a stone is thrown horizontally over the edge of a cliff 100 m above a river (Figure 9.7). If the initial speed of the stone is 30 m/s, the path followed by the stone is described parametrically by

$$x = 30t, \qquad y = 100 - 4.905t^2, \quad 0 \le t \le \sqrt{\frac{100}{4.905}},$$

where t is time in seconds and x and y are in metres. For these equations, t has been chosen equal to zero at the instant of projection, and $t = \sqrt{100/4.905}$ is the time at which the stone strikes the river.

When parameter t is eliminated from this parametric representation of the trajectory, the resulting equation is

$$y = 100 - 4.905 \left(\frac{x}{30}\right)^2 = 100 - 0.005\,45x^2.$$

It clearly indicates that the stone follows a parabolic path. However, if we were to discuss the motion of the stone as regards, say, velocity and acceleration, we would find it necessary to return to the parametric representation that describes the motion in the horizontal and vertical directions.

We say that equations 9.1 **define** y **parametrically** as a function of x if the curve so defined represents a function; that is, if every vertical line that intersects the curve does so exactly once. The parametric equations in Example 9.1 therefore define a function parametrically, but in Examples 9.2–9.5 they do not. However, each of the curves in these latter examples can be divided into subcurves that do represent functions. For instance, in Example 9.4, each of the subcurves

$$\begin{aligned} x = t - t^3, \quad & y = t + t^3, & t \le -1/\sqrt{3}; \\ x = t - t^3, \quad & y = t + t^3, & -1/\sqrt{3} \le t \le 1/\sqrt{3}; \\ x = t - t^3, \quad & y = t + t^3, & t \ge 1/\sqrt{3}; \end{aligned}$$

defines a function, and does so parametrically (Figures 9.8); together all three curves make up the original curve.

FIGURE 9.8a   **FIGURE 9.8b**   **FIGURE 9.8c**

Division of curve in Figure 9.4a into pieces each of which represents a function

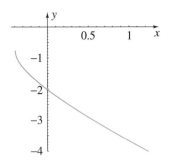 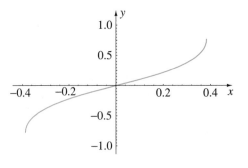

When equations 9.1 do define y parametrically as a function of x, it is straightforward to find the derivative of the function. If we denote this function by $y = f(x)$, then the chain rule applied to $y = f(x)$, $x = x(t)$ gives

$$\frac{dy}{dt} = \frac{dy}{dx}\frac{dx}{dt}.$$

Provided then that $dx/dt \ne 0$, we may solve for

$$\frac{dy}{dx} = \frac{\dfrac{dy}{dt}}{\dfrac{dx}{dt}}. \tag{9.2}$$

This is called the **parametric rule** for differentiation of a function defined parametrically by equations 9.1; it defines the derivative of y with respect to x in terms of derivatives of the given functions $x(t)$ and $y(t)$ with respect to t.

For example, the curve in Figure 9.9 is defined parametrically by

$$x(t) = \frac{1}{t^2 + 1}, \qquad y(t) = \frac{2}{t(t^2 + 1)}.$$

It can be subdivided into two parts, one corresponding to values of $t > 0$ and the other to values of $t < 0$, and each defines y as a function of x. The derivative of either function is

$$\frac{dy}{dx} = \frac{\dfrac{dy}{dt}}{\dfrac{dx}{dt}} = \frac{\dfrac{-2(3t^2 + 1)}{(t^3 + t)^2}}{\dfrac{-2t}{(t^2 + 1)^2}} = \frac{3t^2 + 1}{t^3}.$$

Neither function has a relative maximum or minimum, and this is consistent with the fact that dy/dx never vanishes. The derivative is undefined for $t = 0$, as is y. We can also calculate d^2y/dx^2 in spite of the fact that dy/dx is in terms of t rather than x. Once again it is the chain rule that comes to the rescue:

$$\frac{d^2y}{dx^2} = \frac{d}{dx}\left(\frac{dy}{dx}\right) = \frac{d}{dt}\left(\frac{dy}{dx}\right)\frac{dt}{dx} = \frac{\dfrac{d}{dt}\left(\dfrac{dy}{dx}\right)}{\dfrac{dx}{dt}}$$

$$= \frac{\dfrac{t^3(6t) - (3t^2 + 1)(3t^2)}{t^6}}{\dfrac{-2t}{(t^2 + 1)^2}} = \frac{-3t^4 - 3t^2}{t^6} \cdot \frac{(t^2 + 1)^2}{-2t} = \frac{3(t^2 + 1)^3}{2t^5}.$$

This derivative is positive for $t > 0$ and negative for $t < 0$, agreeing with the fact that the curve in Figure 9.9 is concave upward when $t > 0$ and concave downward when $t < 0$.

FIGURE 9.9   Plot of $x = (t^2 + 1)^{-1}$, $y = 2[t(t^2 + 1)]^{-1}$

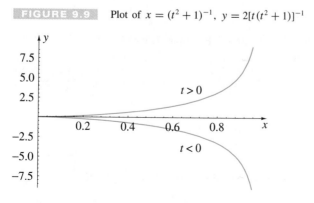

EXAMPLE 9.6

Application Preview
Revisited

Find parametric equations for the path followed by joint C of the mechanism in the Application Preview. Plot the path for $l_1 = 1/2$ m, $l_2 = 2$ m, and $d = 1$ m.

FIGURE 9.10   Determination of path followed by point C in mechanism

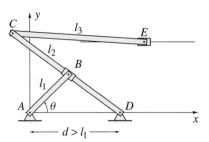

SOLUTION   Coordinates of B in terms of the angle of rotation θ from the positive x-axis (Figure 9.10) are

$$x_B = l_1 \cos \theta, \quad y_B = l_1 \sin \theta.$$

The equation of line BD is $y = m(x - d)$, and since B is on the line,

$$l_1 \sin \theta = m(l_1 \cos \theta - d) \implies m = \frac{l_1 \sin \theta}{l_1 \cos \theta - d}.$$

Because point $C(x_c, y_c)$ lies on this line, it follows that

$$y_c = \frac{l_1 \sin \theta}{l_1 \cos \theta - d}(x_c - d).$$

Furthermore, the length of CD is l_2 so that

$$(x_c - d)^2 + y_c^2 = l_2^2.$$

Consequently,

$$(x_c - d)^2 + \frac{l_1^2 \sin^2 \theta}{(l_1 \cos \theta - d)^2}(x_c - d)^2 = l_2^2.$$

This equation can be solved for x_c:

$$x_c = d \pm \sqrt{\dfrac{l_2^2}{1 + \dfrac{l_1^2 \sin^2 \theta}{(l_1 \cos \theta - d)^2}}} = d \pm \frac{l_2(d - l_1 \cos \theta)}{\sqrt{d^2 + l_1^2 - 2dl_1 \cos \theta}}.$$

Since x_c is always less than d, we choose

$$x_c = d - \frac{l_2(d - l_1 \cos \theta)}{\sqrt{d^2 + l_1^2 - 2dl_1 \cos \theta}}.$$

The y-coordinate of C is now

$$y_c = \pm\sqrt{l_2^2 - (x_c - d)^2} = \pm\sqrt{l_2^2 - \frac{l_2^2(d - l_1 \cos \theta)^2}{d^2 + l_1^2 - 2dl_1 \cos \theta}} = \frac{\pm l_1 l_2 |\sin \theta|}{\sqrt{d^2 + l_1^2 - 2dl_1 \cos \theta}}.$$

We can see that y_c is positive when θ is an angle in the first or second quadrant, and y_c is negative when θ is in the third or fourth quadrant. Hence

$$y_c = \frac{l_1 l_2 \sin \theta}{\sqrt{d^2 + l_1^2 - 2 d l_1 \cos \theta}}.$$

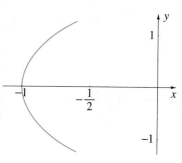

FIGURE 9.11   Motion of C

When $l_1 = 1/2$, $l_2 = 2$, and $d = 1$, parametric equations for the path followed by C are

$$x_c = 1 - \frac{2[1 - (1/2)\cos\theta]}{\sqrt{1 + 1/4 - 2(1/2)\cos\theta}} = 1 - \frac{2(2 - \cos\theta)}{\sqrt{5 - 4\cos\theta}},$$

$$y_c = \frac{(1/2)(2)\sin\theta}{\sqrt{1 + 1/4 - 2(1/2)\cos\theta}} = \frac{2\sin\theta}{\sqrt{5 - 4\cos\theta}}.$$

A plot of the curve for $0 \leq \theta \leq 2\pi$ is shown in Figure 9.11.

EXAMPLE 9.7

The tire of a car rolling along the x-axis without slipping picks up a stone at the origin. The path followed by the stone is called a *cycloid*. Show that parametric equations for the cycloid in terms of the angle θ through which the tire has rotated since picking up the stone are

$$x = R(\theta - \sin\theta), \qquad y = R(1 - \cos\theta),$$

where R is the radius of the tire. Plot the cycloid for $0 \leq \theta \leq 4\pi$. In what direction is the stone travelling when it meets the road?

SOLUTION   The x-coordinate of P in Figure 9.12a is equal to $\|OB\|$ minus $\|PC\|$. Since length $\|PB\|$ along the circle is equal to $\|OB\|$ and $\|PC\| = \|PA\|\sin\theta = R\sin\theta$, it follows that

$$x = R\theta - R\sin\theta = R(\theta - \sin\theta).$$

Furthermore,

$$y = \|AB\| - \|AC\| = R - R\cos\theta = R(1 - \cos\theta).$$

FIGURE 9.12a   Development of path followed by stone caught in tread of a tire

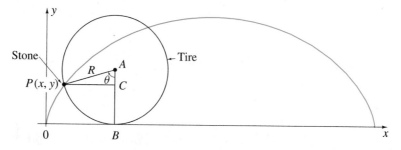

FIGURE 9.12b   Path followed by stone for two revolutions of tire

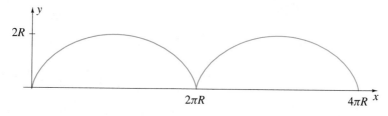

The plot in Figure 9.12b for $0 \le \theta \le 4\pi$ shows the path of the stone for two revolutions of the tire. (It was plotted for $R = 1$ and scales were then changed from 1 to R.)

The slope of the cycloid is given by

$$\frac{dy}{dx} = \frac{dy/d\theta}{dx/d\theta} = \frac{R(\sin\theta)}{R(1 - \cos\theta)} = \frac{\sin\theta}{1 - \cos\theta}.$$

It is undefined when $\theta = 2n\pi$, values of θ at which the stone meets the road. Using L'Hôpital's rule, we calculate

$$\lim_{\theta \to 2n\pi^-} \frac{\sin\theta}{1 - \cos\theta} = \lim_{\theta \to 2n\pi^-} \frac{\cos\theta}{\sin\theta} = -\infty \quad \text{and} \quad \lim_{\theta \to 2n\pi^+} \frac{\sin\theta}{1 - \cos\theta} = \infty.$$

These show that the stone is moving vertically downward as it meets the road, and then vertically upward as it leaves the road.

Many of the applications of integration in Chapter 7 can be adapted to curves defined parametrically. We illustrate in the next example and the following consultation project.

EXAMPLE 9.8

Find the area bounded by the curve in Example 9.5.

SOLUTION   Because of the symmetry of the curve, the area is four times that in the first quadrant (Figure 9.13). If we use vertical rectangles of area $y\,dx$, then the required area is

FIGURE 9.13   Area bounded by parametrically defined curve

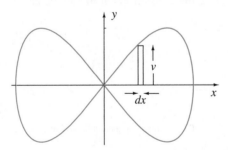

$$A = 4\int_0^a y\,dx = 4\int_{\pi/2}^0 b\sin 2t(-a\sin t)\,dt = 4ab\int_0^{\pi/2} \sin t \sin 2t\,dt.$$

With identity 1.48a,

$$A = 4ab\int_0^{\pi/2} \frac{1}{2}(-\cos 3t + \cos t)\,dt = 2ab\left\{-\frac{1}{3}\sin 3t + \sin t\right\}_0^{\pi/2} = \frac{8}{3}ab.$$

Integrals 7.17 in Section 7.3 define the length of a curve. We express them in a slightly different form,

$$L = \int_A^B \sqrt{(dx)^2 + (dy)^2},$$

FIGURE 9.14 Length of a
urve defined parametrically

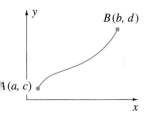

where A and B are the initial and final points on the curve (Figure 9.14). When the curve is defined parametrically by 9.1, this integral becomes

$$L = \int_{\alpha}^{\beta} \sqrt{\left(\frac{dx}{dt}\right)^2 + \left(\frac{dy}{dt}\right)^2} \, dt. \qquad (9.3)$$

Verification of this formula is left to Exercise 49. We use it in the following project.

Consulting Project 14

In a rolling mill, the right circle in Figure 9.15a represents a cylinder of radius R that rolls around a second cylinder of the same radius. Attached to point P on the end of the right cylinder is part of a linkage PZ. Many questions are being asked about the motion of point P, one of which is the distance that it travels as the right cylinder rolls once around the left cylinder. These questions can be answered if the equation for the curve followed by P can be found. Our task is to find it.

FIGURE 9.15a   FIGURE 9.15b

Distance travelled by a point on one circle as it rolls around another circle

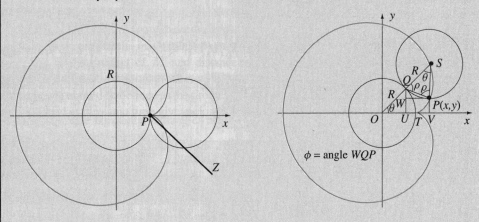

ϕ = angle WQP

SOLUTION   Suppose we let θ be the angle though which the centre of the right cylinder rotates as shown in Figure 9.15b. At this position, we can say that the x-coordinate of P is

$$x = \|OU\| + \|UV\| = R\cos\theta + \|PQ\|\sin\phi.$$

Angles θ, ϕ, and ρ are related by the equations

$$(\pi/2 - \theta) + \phi + \rho = \pi, \quad \text{and} \quad \theta + 2\rho = \pi.$$

When these are solved for ρ and results are equated, we obtain $\phi = 3\theta/2$. Furthermore, if angle θ is bisected at S to divide triangle PQS into two congruent right-angled triangles, we see that $\|PQ\|/2 = R\sin(\theta/2)$. Hence,

$$x = R\cos\theta + 2R\sin(\theta/2)\sin(3\theta/2)$$
$$= R(\cos\theta - \cos 2\theta + \cos\theta) = R(2\cos\theta - \cos 2\theta).$$

The y-coordinate of P is

$$y = \|UQ\| - \|QW\| = R\sin\theta - \|PQ\|\cos\phi$$
$$= R\sin\theta - 2R\sin(\theta/2)\cos(3\theta/2)$$
$$= R(\sin\theta + \sin\theta - \sin 2\theta) = R(2\sin\theta - \sin 2\theta).$$

We have therefore found parametric equations for the path of P,

$$x = R(2\cos\theta - \cos 2\theta), \quad y = R(2\sin\theta - \sin 2\theta).$$

According to formula 9.3, the distance that P travels for one complete rotation of the cylinder is

$$L = 2\int_0^\pi \sqrt{R^2(-2\sin\theta + 2\sin 2\theta)^2 + R^2(2\cos\theta - 2\cos 2\theta)^2}\, d\theta$$

$$= 4\sqrt{2}R\int_0^\pi \sqrt{1 - (\cos\theta\cos 2\theta + \sin\theta\sin 2\theta)}\, d\theta = 4\sqrt{2}R\int_0^\pi \sqrt{1 - \cos\theta}\, d\theta$$

$$= 4\sqrt{2}R\int_0^\pi \sqrt{1 - [1 - 2\sin^2(\theta/2)]}\, d\theta = 8R\int_0^\pi \sin(\theta/2)\, d\theta = 16R.$$

EXERCISES 9.1

In Exercises 1–12 first draw the curve and then plot it.

1. $x = 2 + t, \ y = 3t - 1$

2. $x = t^2 + 3t + 4, \ y = 1 - t$

3. $x = 1 + 2\cos t, \ y = 2 + 2\sin t, \quad 0 \le t \le 2\pi$

4. $x = -2 + 4\cos t, \ y = 3 + 4\sin t, \quad 0 \le t \le \pi$

5. $x = 1 + \cos t, \ y = -1 - \sin t, \quad 0 \le t \le \pi/4$

6. $x = t + 1/t, \ y = t - 1/t$

7. $x = 2\cos t, \ y = 4\sin t, \quad 0 \le t < 2\pi$

8. $x = 1 + 3\cos t, \ y = -2 + 2\sin t, \quad 0 \le t \le \pi$

9. $x = -1 + \sin t, \ y = -1 - 3\cos t, \ 0 \le t \le \pi/2$

10. $x = t - t^2, \ y = t + t^2$

11. $x = t^2 + 1, \ y = t^3 + 3$

12. $x = 2\cot\theta, \ y = 2\sin^2\theta, \quad -\pi/2 \le \theta \le \pi/2, \ \theta \ne 0$

In Exercises 13–20 assume that y is defined parametrically as a function of x, and find dy/dx.

13. $x = t^3 + 3t - 2, \ y = t^2 - 1$

14. $x = \dfrac{u}{u-1}, \ y = \dfrac{u^2}{u^2-1}$

15. $x = (v^3 + 2)\sqrt{v - 1}, \ y = 2v^3 + 3$

16. $x = \sqrt{\dfrac{2+t}{2-t}}, \ y = \sqrt{\dfrac{2-t}{2+t}}$

17. $x = s^{3/2} - s^{2/3}, \ y = s^2 + 2s$

18. $x = (2t + 3)^4, \ y = \dfrac{t}{t+6}$

19. $x = \left(\dfrac{1+u}{1-u}\right)^{1/3}, \ y = \left(\dfrac{1-u}{1+u}\right)^4$

20. $x = \sqrt{-t^2 + 3t + 5}, \ y = \dfrac{1}{t^2 + 2t - 5}$

21. Find equations for the tangent and normal lines to the curve

$$x = t + \frac{1}{t}, \quad y = t - \frac{1}{t}$$

at the point corresponding to $t = 4$.

22. Find point(s) on the curve

$$x = \frac{t^3}{3} - 3t, \quad y = \frac{3t^2}{2} + t$$

where the slope of the tangent line to the curve is equal to 1.

In Exercises 23–26 assume that y is defined parametrically as a function of x, and find dy/dx and d^2y/dx^2.

∗ 23. $x = t^2 + \dfrac{1}{t}, \ y = t^2 - \dfrac{1}{t}$

∗ 24. $x = \sqrt{t - 1}, \ y = \sqrt{t + 1}$

∗ 25. $x = 2u + 5, \ y = 7 - 14u$

∗ 26. $x = v^2 + 2v + 3, \ y = 2v - 4$

∗ 27. Is there a difference between the two curves

$$y = 2x^2 - 1 \quad \text{and} \quad x = \cos t, \quad y = 2\cos^2 t - 1?$$

∗ 28. What curve is described by the parametric equations

$$x = h + a\cos\theta, \quad y = k + b\sin\theta, \quad 0 \le \theta < 2\pi?$$

29. Find parametric equations for a circle with centre (h, k) and radius r.

30. Show that the straight line through two points $P_1(x_1, y_1)$ and $P_2(x_2, y_2)$ has parametric equations

$$x = x_1 + (x_2 - x_1)t, \quad y = y_1 + (y_2 - y_1)t.$$

31. Show that the equations $x = 2\sin^2 t$, $y = 4\cos^2 t$, which are not linear, define a straight-line segment.

32. Draw the following curves and determine whether they are related:

 (a) $x = \sec\theta, \quad y = \tan\theta, \quad -\pi/2 < \theta < \pi/2$

 (b) $x = \cosh\phi, \quad y = \sinh\phi$

 (c) $x = \dfrac{1}{2}\left(t + \dfrac{1}{t}\right), \quad y = \dfrac{1}{2}\left(t - \dfrac{1}{t}\right), \quad t \geq 0$

In Exercises 33–36 find parametric equations for the curve.

33. $y = \dfrac{x+1}{x-2}$

34. $x + y^3 + xy = 5y^2$

35. $x^2 + y^2 + 2x - 4y = 0$ **36.** $4 - x^2 + 2y^2 = 0$

37. Two particles move along straight lines ℓ_1 and ℓ_2 defined parametrically by

$$\ell_1: \quad x = 1 - t, \quad y = t, \quad t \geq 0;$$

$$\ell_2: \quad x = 4t - 5, \quad y = 2t - 1, \quad t \geq 0;$$

where t is time. When are they closest together?

38. If $x = x(t)$ and $y = y(t)$ define y as a function of x, show that

$$\frac{d^2y}{dx^2} = \frac{\dfrac{dx}{dt}\dfrac{d^2y}{dt^2} - \dfrac{dy}{dt}\dfrac{d^2x}{dt^2}}{\left(\dfrac{dx}{dt}\right)^3}.$$

In Exercises 39–42 find the area bounded by the curve.

39. The ellipse $x = a\cos t$, $y = b\sin t$, $0 \leq t \leq 2\pi$

40. The astroid $x = \cos^3 t$, $y = \sin^3 t$, $0 \leq t \leq 2\pi$

41. The deltoid $x = 2\cos t + \cos 2t$, $y = 2\sin t - \sin 2t$, $0 \leq t \leq 2\pi$

42. The droplet $x = 2\cos t - \sin 2t$, $y = \sin t$, $0 \leq t \leq 2\pi$

In Exercises 43–44 find the volume of the solid of revolution when the area bounded by the curve(s) is rotated about the x-axis.

43. The curve $x = a\cos t$, $y = b\sin 2t$ of Example 9.5

44. The cycloid $x = R(\theta - \sin\theta)$, $y = R(1 - \cos\theta)$, $0 \leq \theta \leq 2\pi$ of Example 9.7 and the x-axis

In Exercises 45–47 find the length of the curve.

45. $x = 3 + 4\cos t$, $y = -2 + 4\sin t$, $0 \leq t < 2\pi$

46. $x = e^{-t}\sin t$, $y = e^{-t}\cos t$, $0 \leq t \leq 1$

47. $x = t + \ln t$, $y = t - \ln t$, $1 \leq t \leq 2$

48. Set up, but do not evaluate, a definite integral representing the length of the ellipse

$$x = a\cos\theta, \quad y = b\sin\theta, \quad 0 \leq \theta < 2\pi.$$

49. Verify formula 9.3 for the length of a curve.

50. The equations $x = t^2 + 2t - 1$, $y = t + 5$, $1 \leq t \leq 4$ define a curve parametrically. Find parametric equations that describe this curve but have values of the parameter in the intervals (a) $0 \leq t \leq 3$ and (b) $0 \leq t \leq 1$:

51. Suppose $x(t)$ and $y(t)$ in equations 9.1 are continuous on $\alpha \leq t \leq \beta$ and have derivatives on $\alpha < t < \beta$, and that $x'(t) \neq 0$ on $\alpha < t < \beta$. Show that Cauchy's generalized mean value theorem (Theorem 3.18) implies that there exists a number c between α and β such that

$$\frac{y(\beta) - y(\alpha)}{x(\beta) - x(\alpha)} = \frac{y'(c)}{x'(c)}.$$

Interpret this result geometrically.

52. A particle travels around the circle $x^2 + y^2 = 4$ counterclockwise at constant speed, making 2 revolutions each second. If the particle starts at point $(2, 0)$ at time $t = 0$, find parametric equations for its position in terms of t.

53. (a) Find the area under one arch of the cycloid in Example 9.7.

  (b) Find the length of one arch of the cycloid. What does it represent physically?

54. Draw or plot the strophoid

$$x = \frac{1 - t^2}{1 + t^2}, \quad y = \frac{t(1 - t^2)}{1 + t^2},$$

and find points at which its tangent line is horizontal.

55. Plot the path followed by joint C in the mechanism of Example 9.6 when $l_1 = 1$, $l_2 = 3$, and $d = 1/2$.

56. (a) Find the x-coordinate of slider E in Example 9.6.

  (b) Plot x_E as a function of θ when $l_1 = 1/2$ m, $l_2 = 2$ m, $l_3 = 4$ m, $d = 1$ m, and $y_E = 2$ m. From the graph estimate the length of the stroke of E (the length of the line segment along which E moves).

■* 57. End A of shaft AB with length 4 cm (figure below), moves around the circle of radius 1 cm counterclockwise at 60 rpm. As it does so, a slider at end B moves back and forth along the line $y = -3$ cm. Assume that there is no binding when A is at positions $(0, \pm 1)$ so that the slider moves between quadrants three and four at these times.

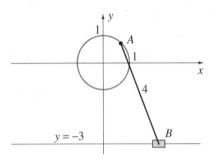

(a) Given that A starts at position $(1, 0)$ at time $t = 0$, find and plot a formula for the x-coordinate of B in terms of t for two revolutions of A. *Hint*: It is necessary to piece a function together.

(b) Estimate maximum left and right positions of the slider from the graph in part (a), and then find these positions exactly by using the fact that velocity will be zero there.

(c) Is the velocity function continuous as the slider passes between quadrants three and four?

(d) Use a graph of the velocity function to estimate maximum speed of the slider.

(e) Use a graph of the acceleration function to estimate maximum $|a(t)|$.

∗ 58. If the stone in Example 9.7 is embedded in the side of the tire rather than the tread, its path is called a *trochoid* (figure following). Show that if the distance from the centre of the tire to the stone is b, parametric equations for the trochoid are

$$x = R\theta - b\sin\theta, \quad y = R - b\cos\theta.$$

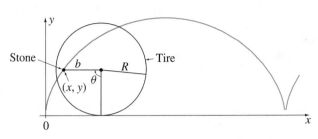

∗ 59. A string is wound around the circle $x^2 + y^2 = r^2$ in the figure below with one end at $(r, 0)$. If the string is unwound while being held taut, the curve that the end traces is called an *involute* of the circle. Show that parametric equations for the involute in terms of the angle θ shown are

$$x = r\cos\theta + r\theta\sin\theta, \quad y = r\sin\theta - r\theta\cos\theta.$$

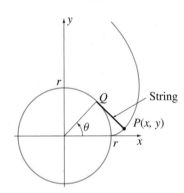

In Exercises 60–62 try to draw the curve and then plot it.

∗ 60. $x = \cos\theta, \quad y = \sin 3\theta$ (curve of Lissajous)

∗ 61. $x = \dfrac{3t}{1 + t^3}, \quad y = \dfrac{3t^2}{1 + t^3}$ (folium of Descartes)

∗ 62. $x = \cos^3\theta, \quad y = \sin^3\theta$ (astroid or hypocycloid of four cusps)

∗∗ 63. A cow is attached to the side of a silo of radius 5 m with a rope of 10 m. Determine the grazing area of the cow.

9.2 Polar Coordinates

In this section we introduce polar coordinates, an alternative coordinate system for the plane. Many problems that have complex solutions using Cartesian coordinates become much simpler in polar coordinates.

Polar coordinates are defined by choosing a point O in the plane called the **pole** and a half-line originating at O called the **polar axis** (Figure 9.16). If P is a point in the plane, we join O and P. The first polar coordinate of P, denoted by r, is the length of line segment OP. The other polar coordinate is the angle θ through which the polar axis must be rotated to coincide with line segment OP. Counterclockwise rotations are regarded as positive, and clockwise rotations as negative. In Figure 9.17, position OQ is reached through a positive rotation of $\pi/6$ radians; therefore, for point Q, $\theta = \pi/6$. But clearly we could arrive at this position in many other ways. We could, for instance, rotate the polar axis counterclockwise

FIGURE 9.16 Polar coordinates of a point

P

r

θ

O
Pole   Polar axis

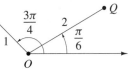

FIGURE 9.17 Polar coordinates of two specific points

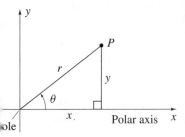

FIGURE 9.18 Relationships between polar and Cartesian coordinates

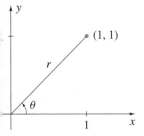

FIGURE 9.19 Polar coordinates of point (1, 1)

through any number of complete revolutions, bringing it back to its original position, and then rotate a further $\pi/6$ radians. Alternatively, we could rotate in a clockwise direction any number of complete revolutions, and then a further $-11\pi/6$ radians. In other words, polar coordinate θ for Q could be any of the values $\pi/6 + 2n\pi$, where n is an integer. Possible values of θ for point R in Figure 9.17 are $3\pi/4 + 2n\pi$. For point Q, $r = 2$, and for R, $r = 1$. Polar coordinates r and θ for a point are written in the form (r, θ) so that possible polar coordinates for Q and R are $(2, \pi/6 + 2n\pi)$ and $(1, 3\pi/4 + 2n\pi)$.

This situation is not like that for Cartesian coordinates, where each point has only one set of coordinates (x, y), and every ordered pair of real numbers specifies one point. With polar coordinates, every ordered pair of real numbers (r, θ), where r must be nonnegative, represents one and only one point, but every point has an infinity of possible representations. We should point out that in some applications this is not a desirable situation. For instance, in the branch of mathematics called *tensor analysis*, it is necessary that polar coordinates assign exactly one pair of coordinates to each point. This can be accomplished in any region that does not contain the pole by demanding, for instance, that $-\pi < \theta \leq \pi$. When we use polar coordinates to find areas in Section 9.4, we must also be particular about our choice of θ. For now, however, no advantage is gained by imposing restrictions on θ, and we therefore accept the fact that if (r, θ) are polar coordinates of a point, so are $(r, \theta + 2n\pi)$ for any integer n. Note also that polar coordinates for the pole are $(0, \theta)$ for any θ whatsoever.

If we introduce into a plane both a system of Cartesian coordinates (x, y) and a system of polar coordinates (r, θ), then relations exist between the two. Suppose the pole of polar coordinates and the origin of Cartesian coordinates are chosen as the same point, and that the polar axis is chosen as the positive x-axis (Figure 9.18). In this case, Cartesian and polar coordinates of any point P are related by the equations

$$x = r\cos\theta, \quad y = r\sin\theta. \tag{9.4}$$

These equations define Cartesian coordinates of a point in terms of its polar coordinates; that is, given its polar coordinates (r, θ), we can calculate its Cartesian coordinates (x, y) by means of 9.4. For example, if polar coordinates of a point are $(3, 2)$, then its Cartesian coordinates are

$$x = 3\cos 2 = -1.25, \quad y = 3\sin 2 = 2.73.$$

Equations 9.4 implicitly define polar coordinates of a point in terms of its Cartesian coordinates. For instance, if Cartesian coordinates of a point are $(1, 1)$ (Figure 9.19), its polar coordinates must satisfy

$$1 = r\cos\theta, \quad 1 = r\sin\theta.$$

If we square and add these equations, we have

$$1 + 1 = r^2\cos^2\theta + r^2\sin^2\theta = r^2,$$

and therefore $r = \sqrt{2}$. It follows that $\cos\theta = \sin\theta = 1/\sqrt{2}$, from which we get $\theta = \pi/4 + 2n\pi$. Thus, polar coordinates of the point are $(\sqrt{2}, \pi/4 + 2n\pi)$.

Equations 9.4 define r and θ implicitly in terms of x and y, but obviously it would be preferable to have explicit definitions. There is no problem expressing r explicitly in terms of x and y,

$$r = \sqrt{x^2 + y^2}, \tag{9.5}$$

but the case for θ is not so simple. If we substitute expression 9.5 for r into equations 9.4, we obtain

$$\cos\theta = \frac{x}{\sqrt{x^2 + y^2}}, \quad \sin\theta = \frac{y}{\sqrt{x^2 + y^2}}. \tag{9.6}$$

Except for the pole, these two equations determine all possible values of θ for given x and y. What we would like to do is obtain one equation, if possible, that defines θ. If we divide the second of these equations by the first, we have

$$\tan \theta = \frac{y}{x}, \tag{9.7}$$

and this equation suggests that we set

$$\theta = \text{Tan}^{-1}\left(\frac{y}{x}\right). \tag{9.8}$$

FIGURE 9.20 Polar coordinates of $(-1, 1)$

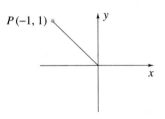

Unfortunately, neither equation 9.7 nor 9.8 is satisfactory. For instance, given the point P with Cartesian coordinates $(-1, 1)$ (Figure 9.20), equation 9.7 yields $\tan \theta = -1$, the solutions of which are $\theta = -\pi/4 + n\pi$. These are angles in the second and fourth quadrants, so only half of them are acceptable polar angles for P. Equation 9.8 gives $\theta = -\pi/4$, which is not a possible polar angle for P.

We suggest that all angles satisfying 9.7 be found, and then a diagram be used to determine those angles that are acceptable values for θ. We illustrate in the following example.

EXAMPLE 9.9

Find all polar coordinates for points with the following Cartesian coordinates:

(a) $(1, 2)$ (b) $(-2, 3)$ (c) $(3, -1)$ (d) $(-2, -4)$

FIGURE 9.21 Polar coordinates of four points

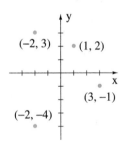

SOLUTION

(a) $r = \sqrt{1^2 + 2^2} = \sqrt{5}$. Angles that satisfy $\tan \theta = 2/1$ are $1.11 + n\pi$. Since the point is in the first quadrant (Figure 9.21), acceptable values for θ are $1.11 + 2n\pi$. Polar coordinates of the point are therefore $(\sqrt{5}, 1.11 + 2n\pi)$.

(b) $r = \sqrt{(-2)^2 + 3^2} = \sqrt{13}$. Angles satisfying $\tan \theta = -3/2$ are $-0.98 + n\pi$. Since the point is in the second quadrant, possible values for θ are $(\pi - 0.98) + 2n\pi = 2.16 + 2n\pi$. Polar coordinates are therefore $(\sqrt{13}, 2.16 + 2n\pi)$.

(c) $r = \sqrt{3^2 + (-1)^2} = \sqrt{10}$. Since values of θ satisfying $\tan \theta = -1/3$ are $-0.32 + n\pi$, and the point is in the fourth quadrant, it follows that polar coordinates are $(\sqrt{10}, -0.32 + 2n\pi)$.

(d) $r = \sqrt{(-2)^2 + (-4)^2} = 2\sqrt{5}$. Since angles satisfying $\tan \theta = 2$ are $1.11 + n\pi$ and the point is in the third quadrant, polar coordinates are $(2\sqrt{5}, 4.25 + 2n\pi)$.

The results in equations 9.4–9.8 are valid only when the pole and origin coincide and the polar axis and positive x-axis are identical. For a different arrangement, these relations must be changed accordingly. For example, if the pole is at the point with Cartesian coordinates (h, k) and the polar axis is as shown in Figure 9.22, equations 9.4 are replaced by

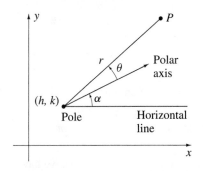

FIGURE 9.22 Polar coordinates with pole at (h, k)

$$x = h + r \cos (\theta + \alpha), \quad y = k + r \sin (\theta + \alpha). \qquad (9.9)$$

You will verify these equations in Exercise 13.

The usual choice of pole and polar axis is that in Figure 9.18, and unless otherwise stipulated, we assume this to be the case.

EXERCISES 9.2

In Exercises 1–8 plot the point having the given set of Cartesian coordinates, and find all possible polar coordinates.

1. $(1, -1)$

2. $(-1, \sqrt{3})$

3. $(4, 3)$

4. $(-2\sqrt{3}, 2)$

5. $(2, 6)$

6. $(-1, -4)$

7. $(7, -5)$

8. $(-5, 2)$

In Exercises 9–12 plot the point having the given set of polar coordinates, and find its Cartesian coordinates.

9. $(2, \pi/4)$

10. $(6, -\pi/6)$

11. $(7, 1)$

12. $(3, -2.4)$

13. Verify that polar and Cartesian coordinates as shown in Figure 9.22 are related by equations 9.9.

9.3 Curves in Polar Coordinates

A curve is defined explicitly in Cartesian coordinates by equations $y = f(x)$ or $x = g(y)$, and implicitly by $F(x, y) = 0$. A point is on a curve if and only if its Cartesian coordinates (x, y) satisfy the equation of the curve.

Analogously, a curve is defined explicitly in polar coordinates if its equation is expressed in either of the forms

$$r = f(\theta) \quad \text{or} \quad \theta = g(r), \qquad (9.10)$$

and is defined implicitly when its equation is given in the form

$$F(r, \theta) = 0. \qquad (9.11)$$

A point is on a curve if at least one of its sets of polar coordinates (r, θ) satisfies the equation of the curve. All sets of polar coordinates for a point need not satisfy the equation. For example, the origin or pole has polar coordinates $(0, \theta)$ for any θ whatsoever. But only those coordinates of the form $\left(0, (2n+1)\pi\right)$ satisfy the equation $r = 1 + \cos \theta$. Likewise, the polar coordinates $(1, \pi)$ satisfy the equation $r = \sin (\theta/2)$, but the coordinates $(1, 3\pi)$ of the same point do not satisfy this equation.

Often we are required to transform the equation of a curve from Cartesian coordinates to polar coordinates, and vice versa. To transform from Cartesian to polar is straightforward: Replace each x with $r \cos \theta$ and each y with $r \sin \theta$. For example, the equation $x^2 + y^2 = 9$ describes a circle centred at the origin with radius 3. In polar coordinates, its equation is

$$9 = (r \cos \theta)^2 + (r \sin \theta)^2 = r^2 \cos^2 \theta + r^2 \sin^2 \theta = r^2.$$

Consequently, $r = 3$ is the equation of this circle in polar coordinates, a much simpler equation than $x^2 + y^2 = 9$.

EXAMPLE 9.10

Find equations in polar coordinates for the following curves:

(a) $2x + 3y = 3$ (b) $x^2 - 2x + y^2 = 0$ (c) $x^2 + y^2 = \sqrt{x^2 + y^2} - 4x$

SOLUTION

(a) For $2x + 3y = 3$, we obtain

$$3 = 2r \cos \theta + 3r \sin \theta \qquad \text{or} \qquad r = \frac{3}{2 \cos \theta + 3 \sin \theta}.$$

(b) For the equation $x^2 - 2x + y^2 = 0$, we have

$$0 = -2x + (x^2 + y^2) = -2r \cos \theta + r^2 = r(r - 2 \cos \theta).$$

Thus,

$$r = 0 \qquad \text{or} \qquad r = 2 \cos \theta.$$

Since $r = 0$ defines the pole, and this point also satisfies $r = 2 \cos \theta$ (for $\theta = \pi/2$)
it follows that we need only write $r = 2 \cos \theta$.

(c) For the curve with equation $x^2 + y^2 = \sqrt{x^2 + y^2} - 4x$, we obtain

$$r^2 = r - 4r \cos \theta = r(1 - 4 \cos \theta).$$

Thus,

$$0 = r^2 - r(1 - 4 \cos \theta) = r(r - 1 + 4 \cos \theta),$$

from which we have

$$r = 0 \qquad \text{or} \qquad r = 1 - 4 \cos \theta.$$

Again the pole satisfies the second of these equations, and therefore the equation of
the curve in polar coordinates is $r = 1 - 4 \cos \theta$.

To transform equations of curves from polar to Cartesian coordinates can sometimes be more
difficult, principally because we have no substitution for θ. If, however, the equation involve
$\cos \theta$ and/or $\sin \theta$, we use equations 9.6.

EXAMPLE 9.11

Find equations in Cartesian coordinates for the following curves:

(a) $r = 1 + \cos \theta$ (b) $r^2 \cos 2\theta = 1$ (c) $r^2 = 9 \sin 2\theta$

SOLUTION

(a) We use equations 9.5 and 9.6 to write

$$\sqrt{x^2 + y^2} = 1 + \frac{x}{\sqrt{x^2 + y^2}},$$

and multiplication by $\sqrt{x^2 + y^2}$ gives

$$x^2 + y^2 = x + \sqrt{x^2 + y^2}.$$

(b) For $r^2 \cos 2\theta = 1$, we use double-angle formula 1.46b to write the equation in terms of $\cos \theta$ rather than $\cos 2\theta$, and then use equations 9.5 and 9.6:

$$1 = r^2(2\cos^2\theta - 1) = (x^2 + y^2)\left(2\frac{x^2}{x^2 + y^2} - 1\right) = (x^2 + y^2)\frac{x^2 - y^2}{x^2 + y^2} = x^2 - y^2.$$

(c) This time we use double-angle formula 1.45 on $\sin 2\theta$:

$$x^2 + y^2 = 18\sin\theta\cos\theta = 18\frac{y}{\sqrt{x^2 + y^2}}\frac{x}{\sqrt{x^2 + y^2}} \qquad \text{or}$$

$$(x^2 + y^2)^2 = 18xy.$$

Examples 9.10 and 9.11 illustrate that equations for some curves are simpler when expressed in polar coordinates. These polar representations can prove very efficient in producing graphs, whether we are plotting with an electronic device or drawing by hand. We illustrate with the curve $r = 1 + \cos\theta$ of Example 9.11. To plot this curve by graphing calculator or computer, we supply the function $r = 1 + \cos\theta$ and the range of values of θ. Because $1 + \cos\theta$ is 2π-periodic, the range $-\pi \le \theta \le \pi$ suffices. Other values of θ create duplications. The result is shown in Figure 9.23; it is called a *cardioid*.

To draw the cardioid by hand we first create a Cartesian coordinate system consisting of a horizontal θ-axis and a vertical r-axis. On this set of axes we graph the *function $r = f(\theta) = 1 + \cos\theta$* (Figure 9.24). This is *not* the required curve; it is a graph of the function $f(\theta)$, illustrating values of r for various values of θ. It represents an "infinite table of values" for r as a function of θ.

FIGURE 9.23 Plot of cardioid in polar coordinates

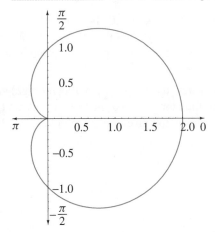

FIGURE 9.24 Cartesian graph of $r = 1 + \cos\theta$

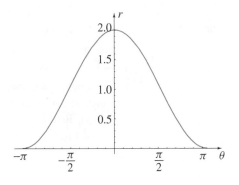

To draw the cardioid $r = 1 + \cos\theta$, we now read pairs of polar coordinates (r, θ) from Figure 9.24, interpreting r as radial distance and θ as rotation. Suppose we begin with the two points $(2, 0)$ and $(1, \pi/2)$. Figure 9.24 indicates that as θ increases from 0 to $\pi/2$, values of r decrease from 2 to 1. This means that as we rotate from the $\theta = 0$ line to the $\theta = \pi/2$ line in the first quadrant, radial distances from the origin become smaller. This is shown in Figure 9.25a. As rotation is increased from $\pi/2$ to π, Figure 9.24 shows that radial distances continue to decrease, eventually reaching 0 at an angle of π radians as in Figure 9.25b. Notice that the line $\theta = \pi$ is tangent to the curve at the pole, reflecting the fact that the pole is attained for an angle of π radians. Consideration of the graph in Figure 9.24 to the left of $\theta = 0$ leads to that part of the cardioid below the $\theta = 0$ and $\theta = \pi$ lines in Figure 9.25c. The symmetry of Figure 9.24 about the r-axis is reflected in the symmetry of Figure 9.25c about the $\theta = 0$ and $\theta = \pi$ lines. Since the function $r = f(\theta) = 1 + \cos\theta$ is 2π-periodic, only values of θ in the interval $-\pi < \theta \leq \pi$ need be considered. Values outside this interval retrace previous points.

FIGURE 9.25a Plot of $r = 1 + \cos\theta$ from 0 to $\frac{\pi}{2}$

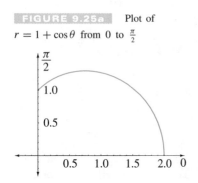

FIGURE 9.25b Plot of $r = 1 + \cos\theta$ from 0 to π

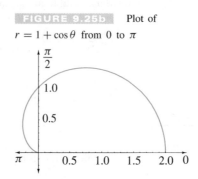

FIGURE 9.25c Plot of $r = 1 + \cos\theta$ from $-\pi$ to π

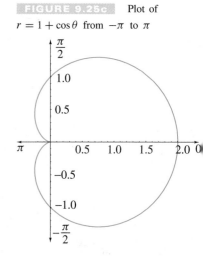

EXAMPLE 9.12

Plot and draw the curve $r^2 = 9\sin 2\theta$.

SOLUTION The explicit definition of the curve is $r = 3\sqrt{\sin 2\theta}$. For $\sin 2\theta$ to be nonnegative, θ must be restricted to the intervals $-\pi \leq \theta \leq -\pi/2$ and $0 \leq \theta \leq \pi/2$. When these are submitted to whatever electronic device you use to make polar plots, the result is as shown in Figure 9.26.

Lacking a device that does polar plots we can draw the curve by first drawing a graph of the function $\sin 2\theta$ in Figure 9.27a. A graph of the function $r = 3\sqrt{\sin 2\theta}$ then follows (Figure 9.27b). Reading pairs of polar coordinates from this graph gives the curve in Figure 9.26. It is called a *lemniscate*.

FIGURE 9.26 Polar plot of lemniscate

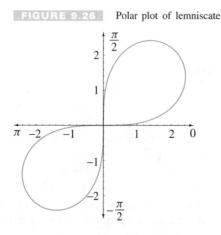

FIGURE 9.27a Cartesian graph of $\sin 2\theta$

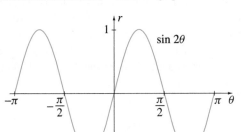

FIGURE 9.27b Cartesian graph of $r = 3\sqrt{\sin 2\theta}$

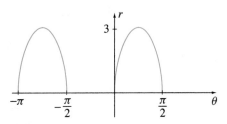

EXAMPLE 9.13

Draw the curve $r = |a - b\cos\theta|$, where $b > a > 0$ are constants.

SOLUTION   We begin by drawing the function $-b\cos\theta$ in Figure 9.28a. A shift of a units vertically gives the graph in Figure 9.28b. Absolute values lead to Figure 9.28c. Interpreting r and θ as distance and rotation leads to the curve in Figure 9.28d. Angles at which $r = 0$ are $\theta = \pm\text{Cos}^{-1}(a/b)$.

FIGURE 9.28a   Cartesian graph of $-b\cos\theta$

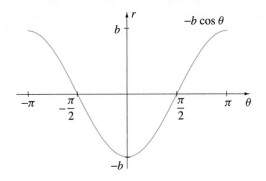

FIGURE 9.28b   Cartesian graph of $a - b\cos\theta$

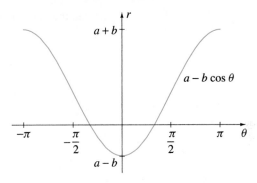

FIGURE 9.28c   Cartesian graph of $r = |a - b\cos\theta|$

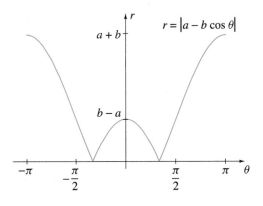

FIGURE 9.28d   Polar graph of $r = |a - b\cos\theta|$

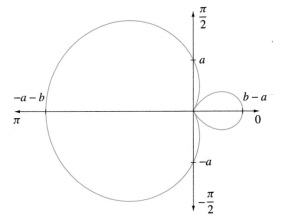

To find points of intersection of two curves whose equations are given in Cartesian coordinates, we solve the equations simultaneously for all (real) solutions. Each solution represents a distinct point of intersection. For curves whose equations are given in polar coordinates, the situation is somewhat more complex because we have multiple names for points. To find points of intersection, we again solve the equations simultaneously for all solutions. Each solution represents a point of intersection; but as points have many sets of polar coordinates, some of these solutions may represent the same point. In addition, it may also happen that one set of polar coordinates for a point of intersection satisfies one equation, whereas a different set satisfies the other equation. Particularly troublesome in this respect is the pole, which has so many sets of polar coordinates. The best way to handle these difficulties is to graph the curves.

EXAMPLE 9.14

Find points of intersection for the curves $r = \sin\theta$ and $r = 1 - \sin\theta$.

SOLUTION   If we set $\sin\theta = 1 - \sin\theta$, then $\sin\theta = 1/2$. All solutions of this equation are defined by

$$\theta = \begin{cases} \dfrac{\pi}{6} + 2n\pi \\ \dfrac{5\pi}{6} + 2n\pi \end{cases}$$

where n is an integer. Graphs of the curves in Figure 9.29 indicate that these values of θ give the points of intersection $(1/2, \pi/6)$ and $(1/2, 5\pi/6)$. The figure also indicates that the origin is a point of intersection of the curves. We did not obtain this point by solving $r = \sin\theta$ and $r = 1 - \sin\theta$ because different values of θ yield $r = 0$ in $r = \sin\theta$ and $r = 1 - \sin\theta$. To obtain $r = 0$ from $r = \sin\theta$, θ must be one of the values $n\pi$, whereas to obtain $r = 0$ from $r = 1 - \sin\theta$, θ must be one of the values $\pi/2 + 2n\pi$. Thus, both curves pass through the pole, but the pole cannot be obtained by solving the equations of the curves simultaneously.

FIGURE 9.29   Intersection points of polar curves $r = \sin\theta$ and $r = 1 - \sin\theta$

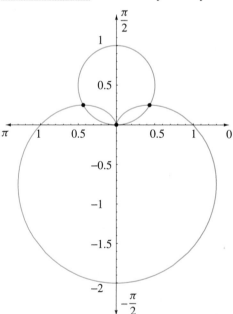

Slopes of Curves in Polar Coordinates

When a curve has polar equation $r = f(\theta), \alpha \le \theta \le \beta$, substitution into equations 9.4 gives

$$x = f(\theta)\cos\theta, \quad y = f(\theta)\sin\theta, \quad \alpha \le \theta \le \beta. \tag{9.12}$$

These are parametric equations for the curve, where the parameter is the polar angle θ. Equation 9.2 gives

$$\frac{dy}{dx} = \frac{\dfrac{dy}{d\theta}}{\dfrac{dx}{d\theta}} = \frac{f'(\theta)\sin\theta + f(\theta)\cos\theta}{f'(\theta)\cos\theta - f(\theta)\sin\theta}. \tag{9.13}$$

This formula defines the *slope of the tangent line to a curve*, which has polar equation $r = f(\theta)$.

EXAMPLE 9.15

Find points on the cardioid $r = 1 + \sin\theta$ at which the tangent line is horizontal.

SOLUTION   The graph of the cardioid in Figure 9.30 indicates three points at which the tangent line is horizontal. To find them we use equation 9.13 to write

$$0 = \frac{dy}{dx} = \frac{(\cos\theta)\sin\theta + (1 + \sin\theta)\cos\theta}{(\cos\theta)\cos\theta - (1 + \sin\theta)\sin\theta}.$$

Since the numerator must vanish, we set

$$0 = \cos\theta\sin\theta + \cos\theta + \sin\theta\cos\theta = \cos\theta(1 + 2\sin\theta).$$

From $\cos\theta = 0$, we choose $\theta = \pi/2$ and from $1 + 2\sin\theta = 0$, we take $\theta = -\pi/6$ and $-5\pi/6$. Thus, points at which the cardioid has a horizontal tangent line have Cartesian coordinates $(0, 2)$ and $(\pm\sqrt{3}/4, -1/4)$.

FIGURE 9.30   Points at which tangent line to $r = 1 + \sin\theta$ is horizontal

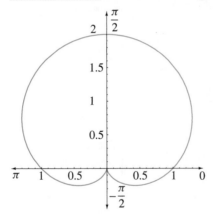

Lengths of Curves in Polar Coordinates

The *length of a curve* defined parametrically by $x = x(t)$ and $y = y(t)$ is given by formula 9.3:

$$L = \int_\alpha^\beta \sqrt{\left(\frac{dx}{dt}\right)^2 + \left(\frac{dy}{dt}\right)^2}\, dt.$$

If we substitute from equations 9.12 into this formula with t replaced by θ, we obtain

$$L = \int_\alpha^\beta \sqrt{[f'(\theta)\cos\theta - f(\theta)\sin\theta]^2 + [f'(\theta)\sin\theta + f(\theta)\cos\theta]^2}\, d\theta$$

$$= \int_\alpha^\beta \sqrt{[f'(\theta)]^2 + [f(\theta)]^2}\, d\theta.$$

Thus, we may write for the **length of a curve** $r = f(\theta)$, $\alpha \le \theta \le \beta$,

$$L = \int_{\alpha}^{\beta} \sqrt{r^2 + \left(\frac{dr}{d\theta}\right)^2}\, d\theta.$$

(9.14)

EXAMPLE 9.16

Find the length of the cardioid $r = 1 - \cos\theta$.

SOLUTION   According to equation 9.14 (see Figure 9.31),

$$L = \int_0^{2\pi} \sqrt{(1 - \cos\theta)^2 + (\sin\theta)^2}\, d\theta$$

$$= \int_0^{2\pi} \sqrt{2}\sqrt{1 - \cos\theta}\, d\theta$$

$$= \sqrt{2} \int_0^{2\pi} \sqrt{1 - \left[1 - 2\sin^2\left(\frac{\theta}{2}\right)\right]}\, d\theta$$

$$= 2 \int_0^{2\pi} \sin\left(\frac{\theta}{2}\right) d\theta = 2\left\{-2\cos\left(\frac{\theta}{2}\right)\right\}_0^{2\pi} = 8.$$

FIGURE 9.31   Length of polar curve $r = 1 - \cos\theta$

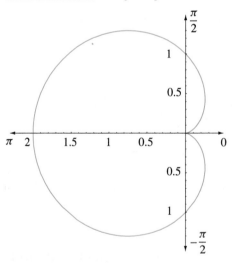

EXAMPLE 9.17

The plate cam in Figure 9.32 rotates about an axis through the origin and perpendicular to the plate. The follower moves back and forth along the x-axis as point A on its end remains in contact with the cam. Suppose $r = a + b\cos\theta$, where $a > b > 0$ are constants, is the polar equation of the edge of the cam, and that the cam rotates at ω revolutions per second. Show that the follower exhibits simple harmonic motion (called a *harmonic cam*) and find a formula for its velocity.

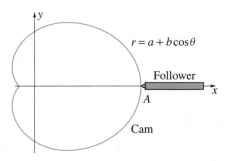

FIGURE 9.32   Velocity of follower in a plate cam

SOLUTION   As the cam rotates, the value of r represents the x-coordinate of A; that is, $x = r = a + b\cos\theta$. If we choose time $t = 0$ when the cam is in the position shown, then $\theta = 2\pi\omega t$, and $x(t) = a + b\cos(2\pi\omega t)$. This represents simple harmonic motion for the motion of A. The velocity is

$$v(t) = \frac{dx}{dt} = -2\pi b\omega \sin(2\pi\omega t).$$

EXERCISES 9.3

In Exercises 1–10 find an equation for the curve in polar coordinates. Draw, and then plot, the curve.

1. $x + 2y = 5$

2. $y = -x$

3. $x^2 + y^2 = 3$

4. $x^2 - 2x + y^2 - 2y + 1 = 0$

5. $y = 4x^2$

6. $x^2 + 2y^2 = 3$

7. $x^2 + y^2 = x$

8. $x^2 + y^2 = \sqrt{x^2 + y^2} - x$

9. $(x^2 + y^2)^2 = x$

10. $y = 1/x^2$

In Exercises 11–20 find the equation of the curve in Cartesian coordinates. Draw, and then plot, the curve.

11. $r = 5$

12. $\theta = 1$

13. $r = 3\sin\theta$

14. $r^2 = 4\sin 2\theta$

15. $r = 3 + 3\sin\theta$

16. $r = 2\sin 2\theta$

17. $r^2 = -4\cos 2\theta$

18. $r = 3 - 4\cos\theta$

19. $r = 5\csc\theta$

20. $r = \cot^2\theta\csc\theta$

21. Draw and plot the curves (a) $r = 2 + 2\sin\theta$, (b) $r = 2 + 4\sin\theta$, and (c) $r = 4 + 2\sin\theta$.

In Exercises 22–25 find all points of intersection for the curves.

22. $r = 2$, $r^2 = 8\cos 2\theta$

23. $r = \cos\theta$, $r = 1 + \cos\theta$

24. $r = 1 + \cos\theta$, $r = 2 - 2\cos\theta$

25. $r = 1$, $r = 2\cos 2\theta$

In Exercises 26–29 find the slope of the curve at the given value of θ.

26. $r = 9\cos 2\theta$ at $\theta = \pi/6$

27. $r^2 = 9\sin 2\theta$ at $\theta = -5\pi/6$

28. $r = 3 - 5\cos\theta$ at $\theta = 3\pi/4$

29. $r = 2\cos(\theta/2)$ at $\theta = \pi/2$

30. Find the slope of the tangent line to the curve $r = 3/(1 - \sin\theta)$ at the point with polar coordinates $(6, \pi/6)$ in two ways: (a) by using 9.13; (b) by finding the equation of the curve in Cartesian coordinates, and calculating dy/dx.

* **31.** Show that if $f(\theta)$ is an even function, then the curve $r = f(\theta)$ is symmetric about the lines $\theta = 0$ and $\theta = \pi$ (or x-axis). Illustrate with two examples.

In Exercises 32–39 draw, and then plot, the curve.

* **32.** $r = \sin 3\theta$

* **33.** $r = \cos 2\theta$

* **34.** $r = \sin 4\theta$

* **35.** $r^2 = \theta$

* **36.** $r = e^\theta$

* **37.** $r = 2\sin(\theta/2)$

* **38.** $r = -2\cos(\theta/2)$

* **39.** $r = 1 + \cos(\theta + \pi/6)$

* **40.** At what times and positions is the speed of the follower in Example 9.17 maximum and minimum?

* **41.** Find the length of the cardioid $r = a(1 + \sin\theta)$. (a is a constant.)

* **42.** (a) The electrostatic charge distribution consisting of a charge $q > 0$ at the point with polar coordinates $(s, 0)$ and a charge $-q$ at (s, π) is called a *dipole*. When s is very small, the lines of force for the dipole are defined by the equation $r = A \sin^2\theta$, where each value of the constant $A > 0$ defines a particular line of force. Plot or draw lines of force for $A = 1, 2$, and 3.

  (b) The equipotential lines for the dipole are defined by $r^2 = B\cos\theta$, where B is a constant. Plot or draw equipotential lines for $B = \pm 1, \pm 2$, and ± 3.

* **43.** Draw, and then plot, the *bifolium* $(x^2 + y^2)^2 = x^2 y$.

* **44.** Curves with equations of the form $r = a(1 \pm \cos\theta)$ or $r = a(1 \pm \sin\theta)$ ($a > 0$ a constant) are called *cardioids*.

  (a) Draw all such curves.

  (b) Find equations for the cardioids in Cartesian coordinates.

* **45.** Curves with equations of the form $r^2 = a^2 \cos 2\theta$ or $r^2 = a^2 \sin 2\theta$ ($a > 0$ a constant) are called *lemniscates*.

  (a) Draw all such curves.

  (b) Find equations for the lemniscates in Cartesian coordinates.

* **46.** (a) Draw the curves $r = b \pm a\cos\theta$ and $r = b \pm a\sin\theta$, where a and b are positive constants, in the three cases $a < b$, $a = b$, and $a > b$.

  (b) Find equations for the curves in Cartesian coordinates.

  (c) Compare these curves with the cardioids of Exercise 44 when $a = b$.

* **47.** A curve with equation of the form $r = a \sin n\theta$ or $r = a \cos n\theta$, where $a > 0$ is a constant and $n > 0$ is an integer, is called a *rose*. Show that the rose has n petals.

* **48.** Show that the roses $r = |a \sin n\theta|$ and $r = |a \cos n\theta|$, where $a > 0$ is a constant and $n > 0$ is an integer, have $2n$ petals.

* **49.** (a) Show that the polar equivalent for the equation of the circle $(x - a)^2 + y^2 = R^2$ is

  $$r = a\cos\theta \pm \sqrt{R^2 - a^2 \sin^2\theta}.$$

  (b) Show that when $a = R$ the equation reduces to $r = 2a\cos\theta$. Does this represent the entire circle?

  (c) Do you need both equations in part (a) to describe the entire circle when $a > R$? If so, which part of the circle corresponds to which equation?

  (d) Repeat part (c) when $a < R$.

** **50.** (a) A patrol boat at point A in the figure below spots a submarine submerging at point B at a time that we call $t = 0$. The submarine, unaware of the patrol boat, follows a straight-line path at constant speed v along some angle ϕ relative to BA (unknown to the patrol boat). The patrol boat heads directly toward point B at speed $V > v$ for $k/(v + V)$ units of time arriving at point C. Show that the submarine and patrol boat are equidistant from B at $t = k/(v + V)$.

  (b) We set up a system of polar coordinates with B as pole and BA as polar axis. Let the distance $\|BC\|$ be denoted by r_0. Suppose that the patrol boat now follows the logarithmic spiral $r = r_0 e^{\theta/\alpha}$ still at speed V, where $\alpha = \sqrt{V^2/v^2 - 1}$. Show that the patrol boat must intercept the submarine.

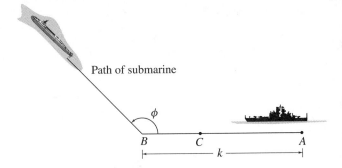

Path of submarine

9.4 Areas in Polar Coordinates

In Section 7.1 we used definite integrals to find areas bounded by curves whose equations are conveniently expressed in Cartesian coordinates. In this section we indicate how to find areas bounded by curves whose equations are expressed in **polar coordinates**. We require the formula

$$\frac{1}{2} r^2 (\theta_2 - \theta_1) \tag{9.15}$$

FIGURE 9.33 Area of sector of a circle

for the area of the shaded sector of the circle in Figure 9.33. This formula results from the fact that the area of the sector is the fractional part $(\theta_2 - \theta_1)/(2\pi)$ of the area πr^2 of the circle.

Consider finding the area of the region in Figure 9.34a bounded by the radial lines $\theta = \alpha$ and $\theta = \beta$ and the curve $r = f(\theta)$. We divide the region into subregions by means of $n + 1$ radial lines $\theta = \theta_i$, where

$$\alpha = \theta_0 < \theta_1 < \theta_2 < \cdots < \theta_{n-1} < \theta_n = \beta.$$

FIGURE 9.34a Area bounded by curves using polar coordinates

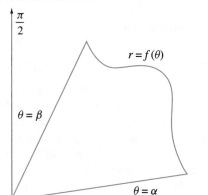

FIGURE 9.34b Approximating area with sectors of circles

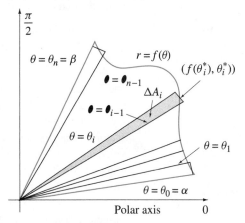

On that part of the curve $r = f(\theta)$ between $\theta = \theta_{i-1}$ and $\theta = \theta_i$, we pick any point with polar coordinates $\left(f(\theta_i^*), \theta_i^*\right)$ as in Figure 9.34b. If between the lines $\theta = \theta_{i-1}$ and $\theta = \theta_i$ we draw the arc of a circle with centre at the pole and radius $f(\theta_i^*)$, a sector is formed with area

$$\Delta A_i = \frac{1}{2}[f(\theta_i^*)]^2 \, \Delta\theta_i,$$

where $\Delta\theta_i = \theta_i - \theta_{i-1}$. Since this sector approximates that part of the required area between the radial lines $\theta = \theta_{i-1}$ and $\theta = \theta_i$, we can say that an approximation to the required area is

$$\sum_{i=1}^{n} \Delta A_i = \sum_{i=1}^{n} \frac{1}{2}[f(\theta_i^*)]^2 \, \Delta\theta_i.$$

By increasing the number of sectors indefinitely, and at the same time requiring each of the $\Delta\theta_i$ to approach zero, we obtain a better and better approximation, and in the limit

$$\text{area} = \lim_{\|\Delta\theta_i\| \to 0} \sum_{i=1}^{n} \frac{1}{2}[f(\theta_i^*)]^2 \, \Delta\theta_i.$$

But this limit is the definition of the definite integral of the function $(1/2)[f(\theta)]^2$ with respect to θ from $\theta = \alpha$ to $\theta = \beta$, and we therefore write

FIGURE 9.35 Sector area
or polar coordinates

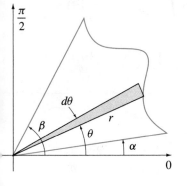

$$\text{area} = \int_{\alpha}^{\beta} \frac{1}{2}[f(\theta)]^2 \, d\theta. \tag{9.16}$$

In order to arrive at this integral in any given problem, without memorizing it, we use the procedure adopted for definite integrals in Cartesian coordinates discussed in Chapter 7. We draw at angle θ a representative sector of angular width $d\theta$ and radius r (Figure 9.35). The area of this sector is

$$\frac{1}{2}r^2 \, d\theta = \frac{1}{2}[f(\theta)]^2 \, d\theta.$$

If areas of all such sectors from angle α to angle β are added together, and the limit is taken as their widths approach zero, the required area is obtained. But this is the process defined by the definite integral, and we therefore write equation 9.16 for the area. Definite integral 9.16 exists when $f(\theta)$ is continuous for $\alpha \le \theta \le \beta$.

EXAMPLE 9.18

Find the area inside the cardioid $r = 1 + \sin\theta$.

FIGURE 9.36 Area inside cardioid $r = 1 + \sin\theta$

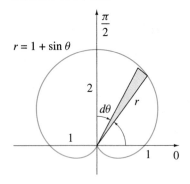

SOLUTION The area of the representative sector in Figure 9.36 is

$$\frac{1}{2}r^2\,d\theta = \frac{1}{2}(1+\sin\theta)^2\,d\theta,$$

and we must add over all sectors interior to the cardioid. Since areas on either side of the $\theta = \pi/2$ line are identical, we calculate the area to the right and double the result. To find the area to the right of the line $\theta = \pi/2$, we must identify angular positions of the first and last sectors. The first sector is at the pole, and the equation of the cardioid indicates that $r = 0$ when $\sin\theta = -1$, that is, when $\theta = -\pi/2 + 2n\pi$. But which of these values of θ shall we choose? Similarly, the last sector occurs when $r = 2$, in which case $\sin\theta = 1$, and θ could be any of the values $\pi/2 + 2n\pi$. Again, which shall we choose? If we choose $\alpha = -\pi/2$ and $\beta = \pi/2$, then all values of θ in the interval $-\pi/2 \le \theta \le \pi/2$ yield points on the right half of the cardioid with no duplications. Consequently,

$$\text{area} = 2\int_{-\pi/2}^{\pi/2}\frac{1}{2}(1+\sin\theta)^2\,d\theta = \int_{-\pi/2}^{\pi/2}(1+2\sin\theta+\sin^2\theta)\,d\theta$$

$$= \int_{-\pi/2}^{\pi/2}\left(1+2\sin\theta+\frac{1-\cos2\theta}{2}\right)d\theta$$

$$= \left\{\frac{3\theta}{2}-2\cos\theta-\frac{\sin2\theta}{4}\right\}_{-\pi/2}^{\pi/2} = \frac{3\pi}{2}.$$

EXAMPLE 9.19

Find the area common to the circles $x^2 + y^2 = 4$ and $x^2 + y^2 = 4x$.

FIGURE 9.37 Area common to circles $x^2 + y^2 = 4$ and $x^2 + y^2 = 4x$

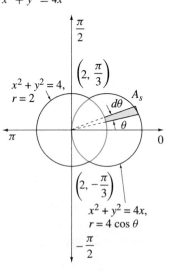

SOLUTION Equations for the circles in polar coordinates are $r = 2$ and $r = 4\cos\theta$, and they intersect in the points with polar coordinates $(2, \pm\pi/3)$ (Figure 9.37). If A_s is the area above the x-axis, outside $x^2 + y^2 = 4$ and inside $x^2 + y^2 = 4x$, then the area common to the circles is the area of either circle less twice A_s:

$$\text{area} = \pi(2)^2 - 2A_s.$$

The area of the representative element is the difference in the areas of two sectors:

$$\frac{1}{2}(4\cos\theta)^2\,d\theta - \frac{1}{2}(2)^2\,d\theta = 2(4\cos^2\theta - 1)\,d\theta.$$

Since all sectors in A_s can be identified by values of θ between 0 and $\pi/3$, the required area is

$$\text{area} = 4\pi - 2\int_0^{\pi/3}2(4\cos^2\theta - 1)\,d\theta = 4\pi - 4\int_0^{\pi/3}[2(1+\cos2\theta) - 1]\,d\theta$$

$$= 4\pi - 4\{\theta + \sin2\theta\}_0^{\pi/3} = 4\pi - 4\left(\frac{\pi}{3} + \frac{\sqrt{3}}{2}\right) = \frac{8\pi}{3} - 2\sqrt{3}.$$

EXERCISES 9.4

In Exercises 1–10 find the area of the region enclosed by the curve.

1. $r = 3\sin\theta$

2. $r = -6\cos\theta$

3. $r = 2\sin2\theta$

4. $r^2 = 2\sin2\theta$

5. $r^2 = -\cos\theta$

6. $r = 2 - 2\cos\theta$

7. $r = 4 - 4\cos\theta$

8. $r = 4 - 2\cos\theta$

9. $r = \sin3\theta$

10. $r = 2(\cos\theta + \sin\theta)$

n Exercises 11–21 find the area of the indicated region.

1. Outside $r = 3$ but inside $r = 6 \sin \theta$

2. Inside both $r = 1$ and $r = 1 - \sin \theta$

3. Inside $r = 2 \sin 2\theta$ but outside $r = 1$

4. Inside both $r = 2 + 2 \cos \theta$ and $r = 2 - 2 \cos \theta$

5. Inside both $r = \sin \theta$ and $r = \cos \theta$

6. Inside both $r = \cos \theta$ and $r = 1 - \cos \theta$

7. Inside $r = 1 - 4 \cos \theta$

18. Inside $r = 4 + 3 \sin \theta$ but outside $r = 2$

19. Inside $r = |1 - 4 \cos \theta|$

20. Inside the bifolium $r = \sin \theta \cos^2 \theta$

21. Bounded by $\theta = \pi$ and $r = \theta, 0 \le \theta \le \pi$

** **22.**   (a) Show that in polar coordinates the strophoid

$$y^2 = x^2 \frac{a - x}{a + x},$$

where $a > 0$ is a constant, takes the form $r = a \cos 2\theta \sec \theta$.

(b) Draw or plot the curve and find the area inside its loop.

9.5 Definitions of Conic Sections

In Section 1.4 we used conic sections to illustrate the algebraic-geometric interplay of plane analytic geometry. They can be visualized as curves of intersection of a plane with a pair of right circular cones (Figure 9.38). Certainly, this suggests why the conic sections are so named,

FIGURE 9.38

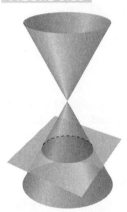

(a) Circle: Plane perpendicular to axis of cones

(b) Ellipse: Plane cuts completely across one cone but not perpendicular to axis

(c) Parabola: Plane cuts only one cone but not completely across

(d) Hyperbola: Plane cuts both cones

(e) Pair of straight lines: Plane passes through vertex and cuts through both cones

(f) One straight line: Plane passes through vertex and touches both cones

but because of the three-dimensional nature of the cone, an analysis of conic sections from this point of view is not yet possible. In this section we use plane analytic geometry to develop definitions for parabolas, ellipses, and hyperbolas.

The Parabola

> **DEFINITION 9.2**
>
> A **parabola** is the curve traced out by a point that moves in a plane so that its distances from a fixed point called the **focus** and a fixed line called the **directrix** are always the same.

Suppose the focus of a parabola is the point (p, q) and the directrix is a line $y = r$ parallel to the x-axis as in Figure 9.39a. If $P(x, y)$ is any point on the parabola, then the fact that its distance from (p, q) must be equal to its distance from $y = r$ is expressed as

$$\sqrt{(x - p)^2 + (y - q)^2} = |y - r|. \tag{9.17}$$

With the absolute values, this equation includes the case of a directrix above the focus, as in Figure 9.39b. If we square both sides of the equation and rearrange terms, we obtain

$$(x - p)^2 = (y - r)^2 - (y - q)^2 = 2y(q - r) + r^2 - q^2.$$

We can solve this equation for y in terms of x; the result is

$$y = \frac{1}{2(q - r)}[(x - p)^2 + (q^2 - r^2)]. \tag{9.18}$$

We could rewrite this equation in our accustomed form $y = ax^2 + bx + c$ for a parabola, but the present form is more informative. First, the line $x = p$ through the focus and perpendicular to the directrix is the line of symmetry for the parabola. Second, the parabola opens upward if $q > r$, in which case the focus is above the directrix, and opens downward if $q < r$. Finally, the vertex of the parabola is found by setting $x = p$, in which case $y = (q + r)/2$, halfway between the focus and directrix.

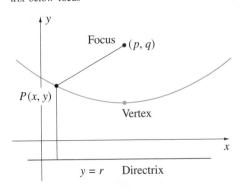

FIGURE 9.39a Parabola with directrix below focus

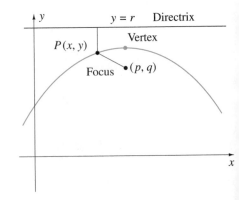

FIGURE 9.39b Parabola with directrix above focus

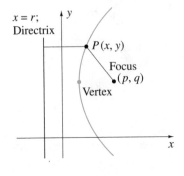

FIGURE 9.40 Parabola with vertical directrix

A similar analysis shows that when the directrix is parallel to the y-axis (Figure 9.40), the equation of the parabola is of the form

$$x = \frac{1}{2(p - r)}[(y - q)^2 + (p^2 - r^2)]. \tag{9.19}$$

Given the focus and directrix (parallel to a coordinate axis) of a parabola, we can easily find its equation: use formulas 9.18 or 9.19, or follow the algebraic steps leading from 9.17 to 9.18. Conversely, given the equation of a parabola in the form $y = ax^2 + bx + c$ or $x = ay^2 + by + c$, we can identify its focus and directrix (see Exercises 52 and 53).

EXAMPLE 9.20

FIGURE 9.41 Parabola with vertical directrix

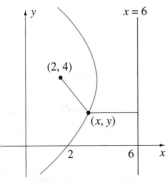

Find the equation of the parabola that has focus $(2, 4)$ and directrix $x = 6$.

SOLUTION If (x, y) is any point on the parabola (Figure 9.41), the fact that its distance from $(2, 4)$ is equal to its distance from $x = 6$ is expressed as

$$\sqrt{(x - 2)^2 + (y - 4)^2} = 6 - x.$$

If we square both sides and simplify, the result is $x = (16 + 8y - y^2)/8$.

The Ellipse

DEFINITION 9.3

An **ellipse** is the curve traced out by a point that moves in a plane so that the sum of its distances from two fixed points called *foci* remains constant.

FIGURE 9.42 Ellipse in terms of two foci

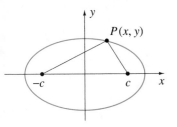

The equation of an ellipse is simplest when the foci lie on either the x- or y-axis and are equidistant from the origin. Suppose the foci are $(c, 0)$ and $(-c, 0)$ (Figure 9.42), and the sum of the distances from these foci to a point on the ellipse is $2a$, where $a > c \geq 0$.

If $P(x, y)$ is any point on the ellipse, Definition 9.3 implies that

$$\sqrt{(x + c)^2 + y^2} + \sqrt{(x - c)^2 + y^2} = 2a. \qquad (9.20)$$

If we transpose the second term to the right-hand side and square both sides, we obtain

$$(x + c)^2 + y^2 = 4a^2 - 4a\sqrt{(x - c)^2 + y^2} + (x - c)^2 + y^2,$$

and this equation simplifies to

$$a^2 - cx = a\sqrt{(x - c)^2 + y^2}.$$

Squaring once again leads to

$$a^4 - 2a^2cx + c^2x^2 = a^2(x^2 - 2cx + c^2 + y^2) \qquad \text{or}$$

$$x^2(a^2 - c^2) + a^2y^2 = a^4 - a^2c^2.$$

Division by $a^2(a^2 - c^2)$ gives

$$\frac{x^2}{a^2} + \frac{y^2}{a^2 - c^2} = 1. \qquad (9.21)$$

It is customary to denote y-intercepts of an ellipse by $\pm b$ ($b > 0$) (Figure 9.43a), in which case $b^2 = a^2 - c^2$, and the equation of the ellipse becomes

$$\frac{x^2}{a^2} + \frac{y^2}{b^2} = 1 \qquad \text{or} \qquad b^2x^2 + a^2y^2 = a^2b^2. \qquad (9.22)$$

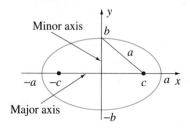

FIGURE 9.43a   Ellipse
with horizontal major axis and
vertical minor axis

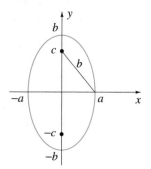

FIGURE 9.43b   Ellipse
with vertical major axis and horizontal minor axis

The line segment across the ellipse and through the foci is called the **major axis** of the ellipse; it has length $2a$ (see Figure 9.43a). The midpoint of the major axis is called the **centre of the ellipse**. The line segment across the ellipse, through its centre, and perpendicular to the major axis is called the **minor axis**; it has length $2b$. Note that the line segment joining either end of the minor axis to a focus (Figure 9.43a) has length a, and the triangle formed specifies the relationship among a, b, and c, namely $a^2 = b^2 + c^2$.

A similar analysis shows that when the foci of the ellipse are on the y-axis, equidistant from the origin, the equation of the ellipse is again in form 9.22. In this case, $2b$ is the length of the major axis, $2a$ is the length of the minor axis, and $b^2 = a^2 + c^2$ (Figure 9.43b).

What we should remember is that an equation of form 9.22 always specifies an ellipse. Foci are on the longer axis and can be located using $c^2 = |a^2 - b^2|$. The length of the major axis represents the sum of the distances from any point on the ellipse to the foci.

EXAMPLE 9.21

Draw the ellipse $16x^2 + 9y^2 = 144$, indicating its foci.

FIGURE 9.44   Ellipse $16x^2 + 9y^2 = 144$ showing foci

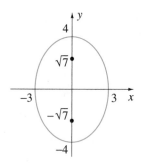

SOLUTION   If we write the ellipse in the form $x^2/9 + y^2/16 = 1$, its x- and y-intercepts are ±3 and ±4. A sketch of the ellipse is therefore as shown in Figure 9.44. The foci must lie on the y-axis at distances $\pm c = \pm\sqrt{4^2 - 3^2} = \pm\sqrt{7}$ from the origin.

If $a = b$ in equation 9.22, then $x^2 + y^2 = a^2$, and this is the equation for a circle with radius a and centre at the origin. But if $a = b$, the distance from the origin to each focus of the ellipse must be $c = 0$. In other words, a circle may be regarded as a degenerate ellipse whose foci are at one and the same point. It is also true that when c is very small compared to half the length of

FIGURE 9.45a Ellipse
when c is much less than a

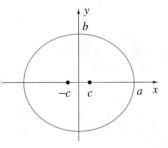

FIGURE 9.45b Ellipse
when c is approximately equal to
a

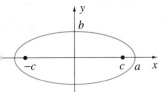

the major axis (Figure 9.45a), the ellipse is shaped very much like a circle. On the other hand, when these lengths are almost equal (Figure 9.45b), the ellipse is long and narrow.

When the centre of an ellipse is at point (h, k) and the foci lie on either the line $x = h$ or $y = k$ (Figures 9.46), the equation for the ellipse is somewhat more complex than 9.22. If $2a$ and $2b$ are again the lengths of the axes of the ellipse, a calculation similar to that leading from 9.20 to 9.22 gives (see Exercise 55)

$$\frac{(x - h)^2}{a^2} + \frac{(y - k)^2}{b^2} = 1. \tag{9.23}$$

Alternatively, the curves in Figures 9.46 are those in Figures 9.42 and 9.43b translated h units in the x-direction and k units in the y-direction. According to Section 1.5, equations for the translated ellipses can be obtained by replacing x and y in $x^2/a^2 + y^2/b^2 = 1$ by $x - h$ and $y - k$, respectively.

FIGURE 9.46a

Ellipse with centre at (h, k)

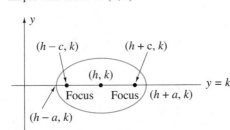

FIGURE 9.46b

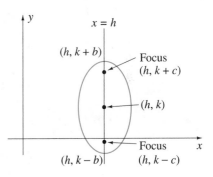

EXAMPLE 9.22

Draw the ellipse $16x^2 + 25y^2 - 160x + 50y = 1175$.

SOLUTION   If we complete the squares on the x- and y-terms, we obtain

$$16(x - 5)^2 + 25(y + 1)^2 = 1600 \quad \text{or}$$

$$\frac{(x - 5)^2}{100} + \frac{(y + 1)^2}{64} = 1.$$

The centre of the ellipse is $(5, -1)$, and lengths of its major and minor axes are 20 and 16, respectively (Figure 9.47).

FIGURE 9.47   Ellipse $16x^2 + 25y^2 - 160x + 50y = 1175$ drawn by completing the squares

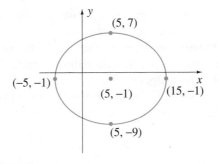

The Hyperbola

> ### DEFINITION 9.4
> A **hyperbola** is the path traced out by a point that moves in a plane so that the difference between its distances from two fixed points called *foci* remains constant.

Like the ellipse, the simplest hyperbolas have foci on either the x- or y-axis, equidistant from the origin. Suppose the foci are $(\pm c, 0)$ (Figure 9.48) and the difference in the distances from $P(x, y)$ to these foci is $2a$. Then Definition 9.4 implies that

$$\left|\sqrt{(x + c)^2 + y^2} - \sqrt{(x - c)^2 + y^2}\right| = 2a. \tag{9.24}$$

This equation can be simplified by a calculation similar to that leading to 9.22; the result is

$$\frac{x^2}{a^2} - \frac{y^2}{b^2} = 1, \tag{9.25}$$

where $b^2 = c^2 - a^2$. The hyperbola has x-intercepts equal to $\pm a$.

FIGURE 9.48　Hyperbola with foci on x-axis

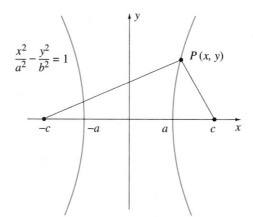

FIGURE 9.49　Hyperbola with foci on y-axis

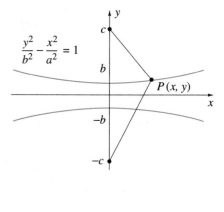

When the foci are on the y-axis (Figure 9.49), the equation of the hyperbola becomes

$$\frac{y^2}{b^2} - \frac{x^2}{a^2} = 1, \tag{9.26}$$

where $2b$ is the constant difference in the distances from a point (x, y) to the foci, and $a^2 = c^2 - b^2$. This hyperbola intersects the y-axis at $\pm b$.

That part of the line segment joining the foci of a hyperbola that is between the two branches of the curve is called the **transverse axis** of the hyperbola; it has length $2a$ in Figure 9.50a and $2b$ in Figure 9.50b. The midpoint of the transverse axis is called the *centre* of the hyperbola. The line segment perpendicular to the transverse axis, through its centre, and of length $2b$ in Figure 9.50a and $2a$ in Figure 9.50b is called the **conjugate axis**. Asymptotes of both hyperbolas are the lines $y = \pm bx/a$.

What we should remember is that an equation of form 9.25 or 9.26 specifies a hyperbola. The foci lie on the extension of the transverse axis and can be located using $c^2 = a^2 + b^2$. The length of the transverse axis represents the difference of the distances from any point on the hyperbola to the foci.

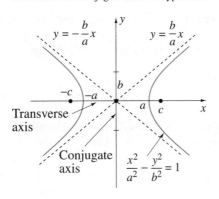

FIGURE 9.50a

Transverse and conjugate axes of hyperbolas

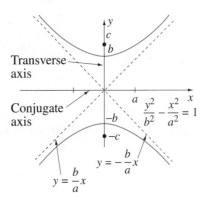

FIGURE 9.50b

EXAMPLE 9.23

FIGURE 9.51   Hyperbola $6x^2 - 9y^2 = 144$ showing foci

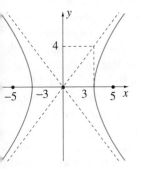

Draw the hyperbola $16x^2 - 9y^2 = 144$, indicating its foci.

SOLUTION   If we express the hyperbola in the form $x^2/9 - y^2/16 = 1$, its x-intercepts are ± 3. With asymptotes $y = \pm 4x/3$, we obtain Figure 9.51. The foci lie on the x-axis at distances $\pm c = \pm\sqrt{4^2 + 3^2} = \pm 5$ from the origin.

When the centre of a hyperbola is at point (h, k) and its foci are on the lines $x = h$ or $y = k$, equations 9.25 and 9.26 are modified in exactly the same way as equation 9.22 was modified for an ellipse. We replace each x by $x - h$ and each y by $y - k$ (see also Exercise 56). Consequently, equations for the hyperbolas in Figures 9.52 are

$$\frac{(x - h)^2}{a^2} - \frac{(y - k)^2}{b^2} = 1 \quad \text{and} \quad \frac{(y - k)^2}{b^2} - \frac{(x - h)^2}{a^2} = 1. \qquad (9.27)$$

FIGURE 9.52a

Hyperbolas with centres at (h, k)

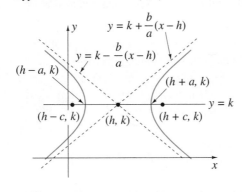

FIGURE 9.52b

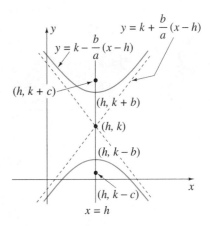

EXAMPLE 9.24

Draw the hyperbola $x^2 - y^2 + 4x + 10y = 5$.

SOLUTION If we complete squares on x- and y-terms, we find

$$(x + 2)^2 - (y - 5)^2 = -16 \quad \text{or}$$

$$\frac{(y - 5)^2}{16} - \frac{(x + 2)^2}{16} = 1.$$

The centre of the hyperbola is $(-2, 5)$, and the length of its transverse axis (along $x = -2$) is 8. If we solve the equation for y, we obtain

$$y = 5 \pm \sqrt{(x + 2)^2 + 16};$$

asymptotes of the hyperbola are then $y = 5 \pm (x + 2)$. The hyperbola can now be sketched as in Figure 9.53. Its foci are at the points $(-2, 5 + 4\sqrt{2})$ and $(-2, 5 - 4\sqrt{2})$.

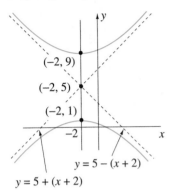

FIGURE 9.53 Hyperbola $x^2 - y^2 + 4x + 10y = 5$ drawn by completing the squares

If P is a point on a conic section, the *focal radii* at P are the lines joining P to the foci (Figure 9.54). As a result, a parabola has one focal radius at each point, and an ellipse and hyperbola each have two. One of the properties of conics that makes them so useful is the fact that the normal line to the conic at any point bisects the angle between the focal radii. For the parabola, the normal bisects the angle between the focal radius and the line through P parallel to the axis of symmetry of the parabola. We will verify these facts in Exercises 59 and 60. To obtain one physical significance of these results, suppose each conic in Figure 9.54 is rotated about the x-axis to form a surface of revolution, which we regard as a mirror. It is a law of optics that when a ray of light strikes a reflecting surface, the angle between incident light and the normal to the surface is always equal to the angle between reflected light and the normal. Consequently, if a beam of light travels in the negative x-direction and strikes the parabolic mirror in Figure 9.54a, all light is reflected toward the focus. Conversely, if F is a source of light, all light striking the mirror is reflected parallel to the x-axis. If either focus of the ellipse in Figure 9.54b is a light source, all light striking the elliptic mirror is reflected toward the other focus. Similarly, if either focus of the hyperbola in Figure 9.54c is a source, all light striking the mirror is reflected in a direction that would make it seem to originate at the other focus. Conversely, if light that is directed at one focus first strikes the mirror, it is reflected toward the other focus. This is precisely why we have parabolic reflectors in automobile headlights and searchlights, parabolic and hyperbolic reflectors in telescopes, and elliptic ceilings in whispering rooms.

Thus far, we have defined parabolas, ellipses, and hyperbolas in terms of distances; for the parabola we use a focus and a directrix, and for the ellipse and hyperbola two foci. In Section 9.6 we show that ellipses and hyperbolas can also be defined in terms of a focus and directrix.

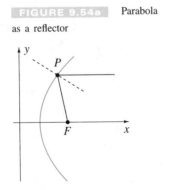

FIGURE 9.54a Parabola as a reflector

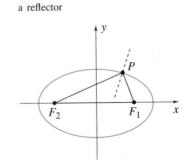

FIGURE 9.54b Ellipse as a reflector

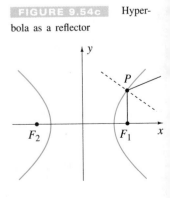

FIGURE 9.54c Hyperbola as a reflector

EXERCISES 9.5

In Exercises 1–14 identify the equation as representing a straight line, circle, a parabola, an ellipse, a hyperbola, or none of these.

1. $2x + 3y = y^2$

2. $x^2 + y^2 - 3x + 2y = 25$

3. $2x - y = 3$

4. $x^2 + y^3 = 3x + 2$

5. $5x^2 = 11 - 2y^2$

6. $2x^2 - 3y^2 + 5 = 0$

7. $y^2 - x + 3y = 14 - x^2$

8. $x^2 + 2x = 3y + 4$

9. $y^2 + x^2 - 2x + 6y + 15 = 0$

10. $x^2 + 2y^2 + 24 = 0$

11. $5 + y^2 = 3x^2$

12. $x^2 + 2y^2 = 24$

13. $y^3 = 3x + 4$

14. $3 - x = 4y$

In Exercises 15–36 draw the curve. Identify foci for each ellipse and hyperbola.

15. $y = 2x^2 - 1$

16. $\dfrac{x^2}{25} + \dfrac{y^2}{36} = 1$

17. $x^2 - \dfrac{y^2}{16} = 1$

18. $3x = 4y^2 - 1$

19. $\dfrac{y^2}{4} - \dfrac{x^2}{25} = 1$

20. $7x^2 + 3y^2 = 16$

21. $x + y^2 = 1$

22. $2y^2 + x = 3y + 5$

23. $9x^2 + 289y^2 = 2601$

24. $y^2 = 10(2 - x^2)$

25. $3x^2 - 4y^2 = 25$

26. $y^2 - x^2 = 5$

27. $y = -x^2 + 6x - 9$

28. $2x^2 - 3y^2 = 5$

29. $3x^2 + 6y^2 = 21$

30. $x^2 + 16y^2 = 2$

31. $y^2 - 3x^2 = 1$

32. $x = -(4 + y)^2$

33. $x^2 + 2x + 4y^2 - 16y + 13 = 0$

34. $x^2 - 6x - 4y^2 - 24y = 11$

35. $9x^2 + y^2 - 18x - 6y = 0$

36. $9x^2 - 16y^2 - 18x - 64y = 91$

37. Find the equation of a hyperbola that passes through the point $(1, 2)$ and has asymptotes $y = \pm 4x$.

38. Find the equation of an ellipse through the points $(-2, 4)$ and $(3, 1)$.

39. Find the width of the elliptic arch in the figure below.

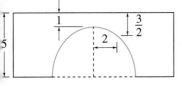

40. Find the height of the parabolic arch in the figure below.

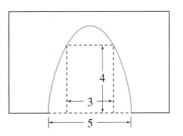

* **41.** Explain how an ellipse can be drawn with a piece of string, two tacks, and a pencil.

* **42.** Find the equation of the ellipse traced out by a point that moves so that the sum of its distances from $(\pm 4, 0)$ is always equal to 10 (a) by using equation 9.22 with suitable values for a and b and (b) by establishing and simplifying an equation similar to 9.20.

* **43.** Find the equation of the hyperbola traced out by a point that moves so that the difference between its distances from $(0, \pm 3)$ is always equal to 1 (a) by using equation 9.26 with suitable values for a and b and (b) by establishing and simplifying an equation similar to 9.24.

* **44.** Show that the equation of every straight line, every circle, and every conic section discussed in this section can be obtained by appropriate choices of constants A, C, D, E, and F in the equation

$$Ax^2 + Cy^2 + Dx + Ey + F = 0.$$

* **45.** Show that the equation of the tangent line to the ellipse $b^2x^2 + a^2y^2 = a^2b^2$ at a point (x_0, y_0) is $b^2xx_0 + a^2yy_0 = a^2b^2$.

* **46.** Show that the equation of the tangent line to the hyperbola $b^2x^2 - a^2y^2 = a^2b^2$ at a point (x_0, y_0) is $b^2xx_0 - a^2yy_0 = a^2b^2$.

* **47.** Find the point P on that part of the ellipse $2x^2 + 3y^2 = 14$ in the first quadrant where the tangent line at P is perpendicular to the line joining P and $(2, 5)$.

* **48.** Find the area inside the ellipse $b^2x^2 + a^2y^2 = a^2b^2$.

* **49.** Among all rectangles that can be inscribed inside the ellipse $b^2x^2 + a^2y^2 = a^2b^2$ and have sides parallel to the axes, find the one with largest possible area.

* **50.** A *prolate spheroid* is the solid of revolution obtained by rotating an ellipse about its major axis. An *oblate spheroid* is obtained by rotating the ellipse about its minor axis. Find volumes for the prolate and oblate spheroids generated by the ellipse $b^2x^2 + a^2y^2 = a^2b^2$ if $a > b$.

* **51.** A sharp noise originating at one focus F_1 of an ellipse is reflected by the ellipse toward the other focus F_2. Explain why all reflected noise arrives at F_2 at exactly the same time.

* **52.** Use equation 9.18 to show that when a parabola is written in the form $y = ax^2 + bx + c$, the following formulas identify its focus (p, q) and directrix $y = r$:

$$p = -\frac{b}{2a}, \quad q = \frac{1}{4a}(1 + 4ac - b^2), \quad r = \frac{1}{4a}(-1 + 4ac - b^2).$$

* **53.** What are formulas for the focus (p, q) and directrix $x = r$ for a parabola of the type $x = ay^2 + by + c$?

* **54.** Use the formulas in Exercises 52 and 53 to identify the focus and directrix for any parabolas in Exercises 15–36.

* **55.** Show that when the centre of an ellipse is at point (h, k) and its foci are on the line $x = h$ or $y = k$, Definition 9.3 leads to equation 9.23.

* **56.** Show that when the centre of a hyperbola is at point (h, k) and its foci are on the line $x = h$ or $y = k$, Definition 9.4 leads to equation 9.27.

* **57.** Prove that the normal line to the parabola $x = ay^2$ in the figure below bisects the angle between the focal radius FP and the line PQ

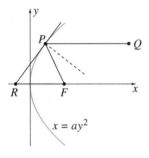

parallel to the x-axis. *Hint:* Draw the tangent line PR at P and show that $\|PF\| = \|RF\|$.

** **58.** When a beam of light travelling in the negative x-direction (figure below) strikes a parabolic mirror with cross section represented by $x = ay^2 + c$, all light rays are reflected to the focus F of the mirror. Show that all photons that pass simultaneously through $x = d$ arrive at F at the same time.

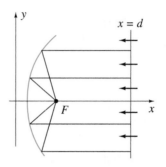

** **59.** Prove that the normal line to an ellipse or hyperbola bisects the angle between the focal radii.

** **60.** Prove that the normal line to the parabola $y = ax^2 + bx + c$ at any point P bisects the angle between the focal radius and the line through P parallel to the y-axis.

** **61.** A line segment through the focus of a parabola with ends on the parabola is called a *focal chord*. It is an established result that tangent to a parabola at the ends of a focal chord are perpendicular to each other and intersect on the directrix. Prove this for the parabola $y = ax^2$.

9.6 Conic Sections in Polar Coordinates

In Section 9.5 we defined parabolas using a focus and directrix and ellipses and hyperbolas using two foci. In this section we show that all three conics can be defined using a focus and directrix and that, in polar coordinates, one equation represents all three conics.

Let F be a fixed point (the *focus*), and l be a fixed line (the *directrix*) that does not pass through F, as in Figure 9.55a. We propose to find the equation of the curve traced out by a point that moves so that its undirected distances from F and l always remain in a constant ratio ϵ called the **eccentricity**. To do this we set up polar coordinates with F as pole and polar axis directed away from l and perpendicular to l as in Figure 9.55b.

FIGURE 9.55a

FIGURE 9.55b

Development of conic sections from focus and directrix

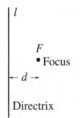

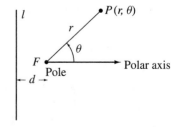

If (r, θ) are polar coordinates for any point P on the required curve, the fact that the ratio of the distances from F and l to P is equal to ϵ is expressed as

$$\frac{r}{d + r \cos \theta} = \epsilon. \tag{9.28}$$

When this equation is solved for r, we have

$$r = \frac{\epsilon d}{1 - \epsilon \cos \theta}, \tag{9.29}$$

and this is the polar equation of the curve traced out by the point. Certainly, this curve should be a parabola when $\epsilon = 1$. We now verify this, and show that the curve is an ellipse when $\epsilon < 1$ and a hyperbola when $\epsilon > 1$. To do this we transform the equation into the usual Cartesian coordinates $x = r \cos \theta$ and $y = r \sin \theta$. From equations 9.5 and 9.6, we obtain

$$\sqrt{x^2 + y^2} = \frac{\epsilon d}{1 - \dfrac{\epsilon x}{\sqrt{x^2 + y^2}}},$$

and this equation simplifies to

$$\sqrt{x^2 + y^2} = \epsilon(d + x), \tag{9.30a}$$

or when squared,

$$x^2 + y^2 = \epsilon^2(d + x)^2. \tag{9.30b}$$

When $\epsilon = 1$, the x^2-terms in 9.30b cancel, and the equation reduces to that for a parabola:

$$x = \frac{1}{2d}(y^2 - d^2), \tag{9.31}$$

as in Figure 9.56a. When $\epsilon \neq 1$, we write

$$x^2 + y^2 = \epsilon^2(d^2 + 2dx + x^2) \qquad \text{or}$$

$$(1 - \epsilon^2)x^2 - 2d\epsilon^2 x + y^2 = \epsilon^2 d^2.$$

When $\epsilon < 1$, we divide by $1 - \epsilon^2$:

$$x^2 - \frac{2d\epsilon^2}{1 - \epsilon^2}x + \frac{y^2}{1 - \epsilon^2} = \frac{\epsilon^2 d^2}{1 - \epsilon^2},$$

and complete the square on the x-terms:

$$\left(x - \frac{d\epsilon^2}{1 - \epsilon^2}\right)^2 + \frac{y^2}{1 - \epsilon^2} = \frac{\epsilon^2 d^2}{1 - \epsilon^2} + \frac{d^2 \epsilon^4}{(1 - \epsilon^2)^2} = \left(\frac{\epsilon d}{1 - \epsilon^2}\right)^2. \tag{9.32}$$

Comparing equation 9.32 with 9.23, we conclude that 9.32 is the equation for an ellipse with centre at position $x = d\epsilon^2/(1 - \epsilon^2)$ on the x-axis. Since one focus is at the origin, it follows that the other must also be on the x-axis at $x = 2d\epsilon^2/(1 - \epsilon^2)$ as in Figure 9.56b.

FIGURE 9.56a Eccentric-
ity $\epsilon = 1$ leads to parabola

FIGURE 9.56b Eccentric-
ity $\epsilon < 1$ leads to ellipse

FIGURE 9.56c Eccentric-
ity $\epsilon > 1$ leads to hyperbola

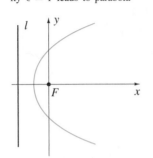

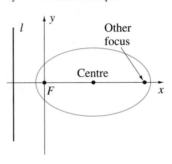

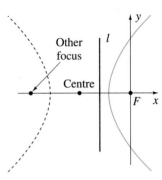

A similar calculation shows that when $\epsilon > 1$, points on the right half of the hyperbola

$$\left(x + \frac{d\epsilon^2}{\epsilon^2 - 1} \right)^2 - \frac{y^2}{\epsilon^2 - 1} = \left(\frac{\epsilon d}{\epsilon^2 - 1} \right)^2 \tag{9.33}$$

are obtained as in Figure 9.56c. Equation 9.30a is not satisfied by points with x-coordinates less than $-d$, and therefore points on the left half of the hyperbola, shown dotted, do not satisfy 9.29.

We have shown that equation 9.29 defines an ellipse when $0 < \epsilon < 1$, a parabola when $\epsilon = 1$, and a hyperbola when $\epsilon > 1$, and this provides a unifying approach to conic sections. All three conics can be studied using a focus and a directrix.

It is clear that equation 9.29 can yield only a parabola that opens to the right and has its focus on the x-axis, and an ellipse and hyperbola with foci on the x-axis, one at the origin. To obtain parabolas that open to the left, or up, or down, and ellipses and hyperbolas with foci on the y-axis, we must change the position of the directrix. The conic sections in Figures 9.57 have directrix to the right of the focus. They have equations

$$r = \frac{\epsilon d}{1 + \epsilon \cos \theta}. \tag{9.34a}$$

The conic sections in Figures 9.58 and 9.59 have equations of the form, respectively,

$$r = \frac{\epsilon d}{1 - \epsilon \sin \theta} \quad \text{and} \quad r = \frac{\epsilon d}{1 + \epsilon \sin \theta}. \tag{9.34b}$$

FIGURE 9.57a

Conic sections in the form of $r = \epsilon d / (1 + \epsilon \cos \theta)$

FIGURE 9.57b

FIGURE 9.57c

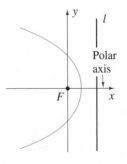

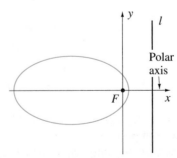

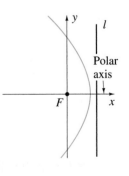

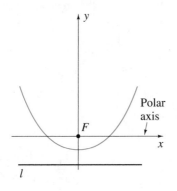

FIGURE 9.58a

Conic sections in the form of $r = \epsilon d/(1 - \epsilon \sin\theta)$

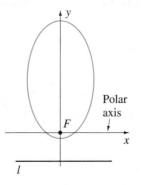

FIGURE 9.58b

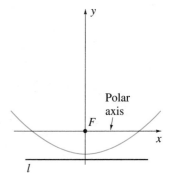

FIGURE 9.58c

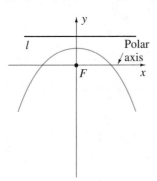

FIGURE 9.59a

Conic sections in the form of $r = \epsilon d/(1 + \epsilon \sin\theta)$

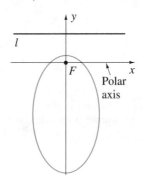

FIGURE 9.59b

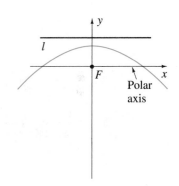

FIGURE 9.59c

In Figures 9.56–9.59, one focus of the conic is chosen as the pole. In other words, the simplicity of equations 9.29 and 9.34 to describe conic sections is a direct consequence of the fact that the pole is at a focus, and the directrix is either parallel or perpendicular to the polar axis.

EXAMPLE 9.25

Draw the curve $r = 15/(3 + 2\cos\theta)$.

SOLUTION   If we write the equation in the form 9.34a,

FIGURE 9.60   Ellipse $r = 5/(3 + 2\cos\theta)$

$$r = \frac{5}{1 + \left(\frac{2}{3}\right)\cos\theta},$$

the eccentricity $\epsilon = 2/3$ indicates that the curve is an ellipse. Both foci lie on the x-axis, and one is at the origin. The ends of the major axis occur when $\theta = 0$ and $\theta = \pi$, and for these values $r = 3$ and $r = 15$ (Figure 9.60). It now follows that the centre of the ellipse is at $x = -6$, and its other focus is at $x = -12$. If $b > 0$ denotes half the length of the minor axis (it is also the maximum y-value on the ellipse, occurring when $x = -6$), then $b^2 = a^2 - c^2 = 9^2 - 6^2 = 45$. Consequently, $b = 3\sqrt{5}$, and the ellipse is as shown in Figure 9.60. This information now permits us to write the equation of the ellipse in Cartesian coordinates:

$$\frac{(x + 6)^2}{81} + \frac{y^2}{45} = 1.$$

An alternative approach in Example 9.25 would be to substitute $r = \sqrt{x^2 + y^2}$ and $\cos\theta = x/\sqrt{x^2 + y^2}$ into the polar equation for the conic, simplify it to $(x+6)^2/81 + y^2/45 = 1$, and then draw the ellipse from this equation. We illustrate this method in the following example.

EXAMPLE 9.26

Draw the curve $r = 2/(3 - 4\sin\theta)$.

SOLUTION   If we set $r = \sqrt{x^2 + y^2}$ and $\sin\theta = y/\sqrt{x^2 + y^2}$, then

$$\sqrt{x^2 + y^2} = \frac{2}{3 - \dfrac{4y}{\sqrt{x^2 + y^2}}} = \frac{2\sqrt{x^2 + y^2}}{3\sqrt{x^2 + y^2} - 4y}.$$

Division by $\sqrt{x^2 + y^2}$ leads to

$$3\sqrt{x^2 + y^2} - 4y = 2 \qquad \text{or}$$

$$3\sqrt{x^2 + y^2} = 4y + 2.$$

If we now square both sides, we obtain

$$9(x^2 + y^2) = 4 + 16y + 16y^2 \qquad \text{or}$$

$$9x^2 - 7y^2 - 16y = 4.$$

If we complete the square on the y-terms, we obtain

$$9x^2 - 7\left(y + \frac{8}{7}\right)^2 = 4 - \frac{64}{7} = -\frac{36}{7}.$$

Division by $-36/7$ yields the equation

$$\frac{\left(y + \dfrac{8}{7}\right)^2}{\dfrac{36}{49}} - \frac{x^2}{\dfrac{4}{7}} = 1.$$

FIGURE 9.61   Hyperbola
$r = 2/(3 - 4\sin\theta)$

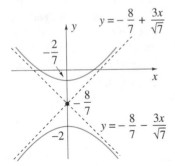

This equation describes a hyperbola with centre $(0, -8/7)$ and y-intercepts equal to $-8/7 \pm 6/7 = -2/7, -2$. Its asymptotes are

$$y = -\frac{8}{7} \pm \frac{6}{7}\sqrt{\frac{7x^2}{4}} = -\frac{8}{7} \pm \frac{3x}{\sqrt{7}}.$$

The hyperbola is shown in Figure 9.61, but only the top half is described by $r = 2/(3 - 4\sin\theta)$. The equation $3\sqrt{x^2 + y^2} = 4y + 2$ does not permit $y \leq -2$.

EXAMPLE 9.27

Find a polar representation for the ellipse

$$\frac{(x-1)^2}{4} + \frac{y^2}{9} = 1.$$

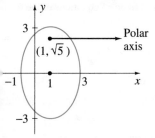

FIGURE 9.62 Ellipse $(x-1)^2/4 + y^2/9 = 1$

SOLUTION To find a polar representation for the ellipse (Figure 9.62), we could use the usual polar coordinates defined by $x = r\cos\theta$ and $y = r\sin\theta$, but the resulting equation would not be simple. Try it. We know that a simple polar representation must result if the pole is chosen at a focus of the ellipse and polar axis either parallel or perpendicular to the directrix. For this ellipse, foci are on the line $x = 1$ at distances $\pm c = \pm\sqrt{9-4} = \pm\sqrt{5}$ from the x-axis. Let us choose the pole at position $(1, \sqrt{5})$ and the polar axis parallel to the x-axis, and therefore parallel to the directrix. According to equation 9.9, polar and Cartesian coordinates are related by $x = 1 + r\cos\theta$ and $y = \sqrt{5} + r\sin\theta$. If we substitute these into the equation for the ellipse, we obtain

$$\frac{r^2\cos^2\theta}{4} + \frac{(\sqrt{5}+r\sin\theta)^2}{9} = 1$$

or

$$9r^2\cos^2\theta + 4(5 + 2\sqrt{5}r\sin\theta + r^2\sin^2\theta) = 36.$$

This equation can be expressed as a quadratic equation in r:

$$r^2(9\cos^2\theta + 4\sin^2\theta) + r(8\sqrt{5}\sin\theta) - 16 = 0.$$

Solutions for r are

$$r = \frac{-8\sqrt{5}\sin\theta \pm \sqrt{320\sin^2\theta + 64(9\cos^2\theta + 4\sin^2\theta)}}{2(9\cos^2\theta + 4\sin^2\theta)}$$

$$= \frac{-8\sqrt{5}\sin\theta \pm \sqrt{576(\cos^2\theta + \sin^2\theta)}}{2(9\cos^2\theta + 4\sin^2\theta)}$$

$$= \frac{4(\pm 3 - \sqrt{5}\sin\theta)}{9\cos^2\theta + 4\sin^2\theta}.$$

Since r must be nonnegative, we must choose $+3$, and not -3, and therefore

$$r = \frac{4(3 - \sqrt{5}\sin\theta)}{9(1 - \sin^2\theta) + 4\sin^2\theta} = \frac{4(3 - \sqrt{5}\sin\theta)}{9 - 5\sin^2\theta}$$

$$= \frac{4(3 - \sqrt{5}\sin\theta)}{(3 - \sqrt{5}\sin\theta)(3 + \sqrt{5}\sin\theta)} = \frac{4}{3 + \sqrt{5}\sin\theta}.$$

EXERCISES 9.6

In Exercises 1–10 draw the conic section.

1. $r = \dfrac{3}{1 + \cos\theta}$

2. $r = \dfrac{16}{3 + 5\cos\theta}$

3. $r = \dfrac{4}{3 - 3\sin\theta}$

4. $r = \dfrac{16}{5 + 3\cos\theta}$

5. $r = \dfrac{4}{3 - 4\sin\theta}$

6. $r = \dfrac{4}{4 - 3\sin\theta}$

7. $r = \dfrac{1}{2 - 2\cos\theta}$

8. $r = \dfrac{1}{2 + \sin\theta}$

9. $r = \dfrac{\sec\theta}{3 + 6\sec\theta}$

10. $r = \dfrac{4\csc\theta}{7\csc\theta - 2}$

In Exercises 11–16 find the Cartesian equation for the curve.

11. $r = \dfrac{3}{1 - \sin\theta}$

12. $r = \dfrac{1}{3 + \cos\theta}$

13. $r = \dfrac{1}{1 + 2\cos\theta}$

14. $r = \dfrac{2}{1 - 3\cos\theta}$

15. $r = \dfrac{4}{6 - 3\sin\theta}$

16. $r = \dfrac{4}{5 + 5\cos\theta}$

In Exercises 17–20 find a polar equation in one of the four forms of 9.29 or 9.34 for the conic.

* **17.** $x^2 - 16y^2 = 16$

* **18.** $4x^2 + 9y^2 = 36$

* **19.** $(x-1)^2 + \dfrac{(y+1)^2}{4} = 1$

* **20.** $x^2 - 9(y - 2)^2 = 9$

* **21.** (a) Show that the eccentricity ϵ of an ellipse or hyperbola is always equal to the distance from its centre to either focus, divided by half the length of the major or transverse axis.

　　(b) Discuss the eccentricities of the ellipses in Figures 9.45a and 9.45b.

* **22.** A circle has been described as a degenerate ellipse in the sense that its foci coincide. What happens to the eccentricity of an ellipse if the distance between its foci approaches zero? *Hint:* See Exercise 21.

** **23.** Paths of celestial objects can be described by equations of the form $r = a/(1 + \epsilon\cos\theta)$, where $a > 0$ is a constant. To find the time to travel from one point on the path to another, it is necessary to evaluate the integral

$$I = \int \frac{1}{(1 + \epsilon\cos\theta)^2}\,d\theta.$$

Do so in the cases that (a) $\epsilon = 0$, (b) $0 < \epsilon < 1$, (c) $\epsilon = 1$, and (d) $\epsilon > 1$.

Website

Information on **Translation and Rotation of Axes** has been placed on the Text Enrichment Site: *www.pearsoned.ca/text/trim.*

SUMMARY

In Chapter 1 we defined curves explicitly and implicitly. In this chapter we added a third description — the parametric definition

$$x = x(t), \quad y = y(t), \quad \alpha \leq t \leq \beta.$$

When such a curve defines a function, the derivative of the function can be calculated using the parametric rule:

$$\frac{dy}{dx} = \frac{\dfrac{dy}{dt}}{\dfrac{dx}{dt}};$$

second- and higher-order derivatives can be calculated using the chain rule. The length of such a curve is defined by the definite integral

$$\int_\alpha^\beta \sqrt{\left(\frac{dx}{dt}\right)^2 + \left(\frac{dy}{dt}\right)^2}\,dt.$$

Polar coordinates provide an alternative way to identify the positions of points in a plane. They use distance from a point, called the pole, and rotation of a half line, called the polar axis. Many curves that have complex equations in Cartesian coordinates can be represented very simply using polar coordinates. Particularly simple are multileaved roses, cardioids, lemniscates and some circles. Polar coordinates also provide a unified approach to parabolas, ellipses, and hyperbolas. Each curve can be described as the path traced out by a point that moves so that the ratio ϵ of its distances from a fixed point and a fixed line remain constant. The curve is an ellipse, a parabola, or a hyperbola depending on whether $0 < \epsilon < 1$, $\epsilon = 1$, or $\epsilon > 1$, respectively. Ellipses and hyperbolas can also be defined as curves traced out by a point that moves so that the sum and difference of its distances from two fixed points remain constant.

When a curve $r = f(\theta)$ in polar coordinates encloses a region R, the definite integral

$$\int_\alpha^\beta \frac{1}{2}[f(\theta)]^2\,d\theta$$

with appropriate choices of α and β can be used to find the area of R.

Parametric equations for a polar curve $r = f(\theta)$, $\alpha \le \theta \le \beta$ are

$$x(\theta) = f(\theta)\cos\theta, \quad y(\theta) = f(\theta)\sin\theta, \quad \alpha \le \theta \le \beta.$$

Its slope is given by

$$\frac{dy}{dx} = \frac{f'(\theta)\sin\theta + f(\theta)\cos\theta}{f'(\theta)\cos\theta - f(\theta)\sin\theta},$$

and its length can be calculated with the definite integral

$$\int_\alpha^\beta \sqrt{r^2 + \left(\frac{dr}{d\theta}\right)^2}\, d\theta.$$

In reviewing this chapter, you should be able to define or discuss the following key terms:

Parametric equations	Parametrically defined function
Parametric rule	Polar coordinates
Pole	Polar axis
Slopes of curves in polar coordinates	Lengths of curves in polar coordinates
Areas in polar coordinates	Parabola
Focus	Directrix
Ellipse	Major axis
Centre of an ellipse	Minor axis
Hyperbola	Transverse axis
Conjugate axis	Eccentricity

REVIEW EXERCISES

In Exercises 1–30 draw, and then plot, the curve.

1. $r = \cos\theta$

2. $r = -\sin\theta$

3. $r = \dfrac{3}{1 + 2\sin\theta}$

4. $r = \dfrac{3}{2 + \sin\theta}$

5. $r = \dfrac{1}{1 + \sin\theta}$

6. $r = \dfrac{3}{2 - 2\cos\theta}$

7. $r + 1 = 2\sin\theta$

8. $r^2 = 4\cos\theta$

9. $x = 2\cos t$, $y = 3\sin t$, $0 \le t < 2\pi$

10. $x = 4 + t$, $y = 5 - 3t^2$

11. $x = \sin^2 t$, $y = \cos^2 t$, $0 \le t \le 2\pi$

12. $x = 1 + 4\sin t$, $y = -2 + 4\cos t$, $0 \le t \le \pi$

13. $(x^2 + y^2)^3 = x$

14. $x^2 + y^2 = 2\sqrt{x^2 + y^2}$

15. $r^2 = 4\cos^2\theta$

16. $r = 3\cos 2\theta$

17. $x + y = \sqrt{x^2 + y^2}$

18. $x^2 + y^2 = x + y$

19. $r = 4\cos 3\theta$

20. $r = \sin^2\theta - \cos^2\theta$

21. $x^2 + y^2 - 3x + 2y = 1$

22. $x^2 - 3y^2 + 4 = 0$

23. $2x^2 + 3y^2 - 6y = 0$

24. $y^2 + x + 2y = 3$

25. $r(2\cos\theta - \sin\theta) = 3$

26. $r = \cos(\theta/2)$

27. $r = \sin^2\theta$

28. $x^2 = y^2 + y$

29. $x = e^{-t}$, $y = \ln t$, $t > 0$

30. $x = \sin\theta$, $y = \sin 2\theta$

In Exercises 31–35 find the area of the region indicated.

31. Inside $r = 2 + 2\cos\theta$

32. Inside $r = 4$ but outside $r = 4\sin 2\theta$

33. Common to $r = 2$ and $r^2 = 9\cos 2\theta$

34. Inside $r = \sin^2\theta$

35. Common to $r = 1 + \sin\theta$ and $r = 2 - 2\sin\theta$

In Exercises 36–37 assume that y is defined as a function of x, and find dy/dx and d^2y/dx^2.

* **36.** $x = t^3 + 2t$, $y = 3t - t^3$

* **37.** $x = 2 \sin u$, $y = 3 \cos u$

In Exercises 38–40 find the length of the curve.

* **38.** $r = 2 + 2 \cos \theta$

* **39.** $x = t^2$, $y = t^3$, $0 \le t \le 1$

* **40.** $x = e^t \cos t$, $y = e^t \sin t$, $0 \le t \le \pi/2$

* **41.** Find the equation of the tangent line to the curve $r = 2 - 2 \sin \theta$ at the point with polar coordinates $(1, \pi/6)$.

10 | Infinite Sequences and Series

Application Preview

The amount A of a certain chemical in a reactor decreases exponentially in time

$$A(t) = A_0 e^{kt},$$

where A_0 is the amount at time $t = 0$, and $k < 0$ is a constant. After time T, the amount in the reactor is $A_0 e^{kT}$, at which time an additional amount A_0 is added, resulting in an amount $A_0 + A_0 e^{kT} = A_0(1 + e^{kT})$ at that time. This amount then decreases exponentially until an additional amount A_0 is added at time $2T$, and so on, and so on. Once the amount of chemical reaches a critical level L in the reactor, the process must be terminated.

THE PROBLEM   Find an equation that determines the time at which the process must be terminated. (For a solution to the problem, see Example 10.38 on page 662.)

Sequences and series play an important role in many areas of applied mathematics. *Sequences of numbers* were first encountered in Section 4.1, although we did not use the term *sequences* at the time. Newton's iterative procedure produces a set of numbers $x_1, x_2, x_3, \ldots$ each of which approximates a root of an equation $f(x) = 0$. The first number is chosen as some initial approximation to the solution of the equation, and subsequent numbers defined by the formula

$$x_{n+1} = x_n - \frac{f(x_n)}{f'(x_n)}$$

are better and better approximations. This ordered set of numbers is called a **sequence**. Each number in the set corresponds to a positive integer, and each is calculated according to a stated formula.

A series of numbers is the sum of the numbers in a sequence. If the numbers are $x_1, x_2, x_3, \ldots$, the corresponding **series** is denoted symbolically by

$$\sum_{n=1}^{\infty} x_n = x_1 + x_2 + x_3 + \cdots,$$

where the three dots indicate that the addition is never-ending. It is all very well to write an expression like this, but it does not have meaning. No matter how fast we add, or how fast a calculator adds, or even how fast a supercomputer adds, an infinity of numbers can never be added together in a finite amount of time. We shall give meaning to such expressions, and show that they are really the only sensible way to define many of the more common transcendental functions, such as trigonometric, exponential, and hyperbolic.

The first two sections of this chapter and Section 10.8 are devoted to sequences and the remaining ten to series. This is not to say that series are more important than sequences; they are not. Discussions on series invariably become discussions on sequences associated with series. We have found that difficulties with this chapter can usually be traced back to a failure to distinguish between the two concepts. Special attention to the material in Sections 10.1, 10.2, and 10.8 will be rewarded; a cursory treatment leads to confusion in other sections.

10.1 Infinite Sequences of Numbers

Sequences of numbers are defined as follows.

> **DEFINITION 10.1**
>
> An **infinite sequence of numbers** is a function f whose domain is the set of positive integers.

For example, when $f(n) = 1/n$, the following numbers are associated with the positive integers:

$$1, \quad \frac{1}{2}, \quad \frac{1}{3}, \quad \frac{1}{4}, \quad \dots$$

The word *infinite* simply indicates that an infinity of numbers is defined by the sequence, as there is an infinity of positive integers, but it indicates nothing about the nature of the numbers. Often, we write the numbers $f(n)$ in a line separated by commas,

$$f(1), \ f(2), \ \dots, \ f(n), \ \dots \tag{10.1a}$$

and refer to this array as the sequence rather than the rule by which it is formed. Since this notation is somewhat cumbersome, we adopt a notation similar to that used for the sequence defined by Newton's iterative procedure in Section 4.1 We set $c_1 = f(1)$, $c_2 = f(2)$, ..., $c_n = f(n)$, ..., and write for 10.1a

$$c_1, c_2, c_3, \dots, c_n, \dots \tag{10.1b}$$

The first number c_1 is called the first **term** of the sequence, c_2 the second term, and for general n, c_n is called the nth term (or general term) of the sequence. For the example above, we have

$$c_1 = 1, \quad c_2 = \frac{1}{2}, \quad c_3 = \frac{1}{3}, \quad \text{etc.}$$

In some applications, it is more convenient to define a sequence as a function whose domain is the set of integers larger than or equal to some fixed integer N, and N can be positive, negative, or zero. Later in this chapter, we find it convenient to initiate the assignment with $N = 0$. For now we prefer to use Definition 10.1 where $N = 1$, in which case we have the natural situation where the first term of the sequence corresponds to $n = 1$, the second term to $n = 2$, and so on.

EXAMPLE 10.1

The general terms of four sequences are

$$\text{(a)} \ \frac{1}{2^{n-1}} \qquad \text{(b)} \ \frac{n}{n+1} \qquad \text{(c)} \ (-1)^n |n - 3| \qquad \text{(d)} \ (-1)^{n+1}.$$

Write out the first six terms of each sequence.

SOLUTION   The first six terms of these sequences are

(a) $1, \dfrac{1}{2}, \dfrac{1}{4}, \dfrac{1}{8}, \dfrac{1}{16}, \dfrac{1}{32}$;

(b) $\dfrac{1}{2}, \dfrac{2}{3}, \dfrac{3}{4}, \dfrac{4}{5}, \dfrac{5}{6}, \dfrac{6}{7}$;

(c) $-2, 1, 0, 1, -2, 3$;

(d) $1, -1, 1, -1, 1, -1$.

The sequences in Example 10.1 are said to be defined **explicitly**; we have an explicit formula for the nth term of the sequence in terms of n. This allows easy determination of any term in the sequence. For instance, to find the one-hundredth term, we simply replace n by 100 and perform the resulting arithmetic. Contrast this with the sequence in the following example.

EXAMPLE 10.2

The first term of a sequence is $c_1 = 1$ and every other term is to be obtained from the formula

$$c_{n+1} = 5 + \sqrt{2 + c_n}, \quad n \geq 1.$$

Calculate c_2, c_3, c_4, and c_5.

SOLUTION   To obtain c_2 we set $n = 1$ in the formula:

$$c_{1+1} = c_2 = 5 + \sqrt{2 + c_1} = 5 + \sqrt{2 + 1} = 5 + \sqrt{3} \approx 6.732.$$

To find c_3, we set $n = 2$:

$$c_3 = 5 + \sqrt{2 + c_2} = 5 + \sqrt{2 + \left(5 + \sqrt{3}\right)} = 5 + \sqrt{7 + \sqrt{3}} \approx 7.955.$$

Similarly,

$$c_4 = 5 + \sqrt{2 + c_3} = 5 + \sqrt{7 + \sqrt{7 + \sqrt{3}}} \approx 8.155 \quad \text{and}$$

$$c_5 = 5 + \sqrt{2 + c_4} = 5 + \sqrt{7 + \sqrt{7 + \sqrt{7 + \sqrt{3}}}} \approx 8.187.$$

When the terms of a sequence are defined by a formula such as the one in Example 10.2, the sequence is said to be defined **recursively**. The terms for a sequence obtained from Newton's iterative procedure are so defined. To find the 100th term of a recursively defined sequence, we must know the 99th; to find the 99th, we must know the 98th; to find the 98th, we need the 97th; and so on down the line. In other words, to find a term in the sequence, we must first find every term that precedes it. Obviously, it is much more convenient to have an explicit definition for c_n in terms of n, but this is not always possible. It can be very difficult to find an explicit formula for the nth term of a sequence that is defined recursively.

Sometimes it is impossible to give an algebraic formula for the terms of a sequence. This is illustrated in Exercise 31.

When the general term of a sequence is known explicitly, any term in the sequence is obtained by substituting the appropriate value of n. In other words, the general term specifies every term in the sequence. We therefore use the general term to abbreviate the notation for a sequence by writing the general term in braces. Specifically, for the sequence in Example 10.1a, we write

$$\left\{ \frac{1}{2^{n-1}} \right\}_1^\infty = 1, \frac{1}{2}, \frac{1}{4}, \frac{1}{8}, \dots, \frac{1}{2^{n-1}}, \dots,$$

where 1 and ∞ indicate that the first term corresponds to the integer $n = 1$, and that there is an infinite number of terms in the sequence. In general, we write

$$\{c_n\}_1^\infty = c_1, c_2, c_3, \dots, c_n, \dots. \tag{10.2}$$

If, as is the case in this section, the first term of a sequence corresponds to the integer $n = 1$, we abbreviate the notation further and simply write $\{c_n\}$ in place of $\{c_n\}_1^\infty$.

Since a sequence $\{c_n\}$ is a function whose domain is the set of positive integers, we can represent $\{c_n\}$ graphically. The sequences of Example 10.1 are shown in Figures 10.1.

FIGURE 10.1a
FIGURE 10.1a

Plots of terms of sequences of Example 10.1

FIGURE 10.1b

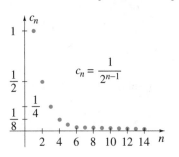

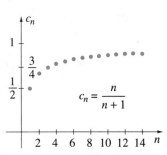

FIGURE 10.1c

FIGURE 10.1d

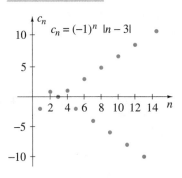

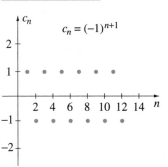

In most applications of sequences we are interested in a number called the **limit of the sequence**. Intuitively, a number L is called the limit of a sequence $\{c_n\}$ if as we go farther and farther out in the sequence, the terms get arbitrarily close to L and stay close to L. If such a number L exists, we write

$$L = \lim_{n \to \infty} c_n, \qquad (10.3)$$

and say that the sequence $\{c_n\}$ **converges** to L. If no such number exists, we say that the sequence does not have a limit, or that the sequence **diverges**.

For the sequences of Example 10.1, it is evident that:

(a) $\quad \lim_{n \to \infty} \dfrac{1}{2^{n-1}} = 0.$

(b) $\quad \lim_{n \to \infty} \dfrac{n}{n+1} = 1.$

(c) $\quad \lim_{n \to \infty} (-1)^n |n - 3|$ does not exist.

(d) $\quad \lim_{n \to \infty} (-1)^{n+1}$ does not exist.

Note how the points on the graphs in Figures 10.1a and b cluster around the limits 0 and 1 as n gets larger and larger. No such clustering occurs in the remaining two figures.

It is usually, but not always, easy to determine whether an explicitly defined sequence has a limit, and what that limit is. It is like finding the limit of a function $f(x)$ as $x \to \infty$. For example, if we divide numerator and denominator of the sequence $\{(n^2+n-3)/(2n^2+n+4)\}$ by n^2,

$$\lim_{n \to \infty} \frac{n^2 + n - 3}{2n^2 + n + 4} = \lim_{n \to \infty} \frac{1 + \dfrac{1}{n} - \dfrac{3}{n^2}}{2 + \dfrac{1}{n} + \dfrac{4}{n^2}} = \frac{1}{2}.$$

Seldom is it obvious whether a recursively defined sequence has a limit. For instance, it is not at all clear whether the recursive sequence of Example 10.2 has a limit. In spite of the fact that differences between successive terms are approaching zero, and terms of the sequence are therefore getting closer together, the sequence might not have a limit. For instance, the sequence $\{\sqrt{n}\}$ does not have a limit, yet it is easy to show that differences between successive terms get smaller and smaller as n increases. (See Exercise 30 for another example.) There are ways to verify that a sequence has a limit, and some of these will be discussed in Section 10.8.

When a sequence arises in applications, it may be perfectly clear that the sequence is convergent. Such is often the case in the field of numerical analysis, where recursive sequences commonly arise in the form of iterative procedures. We have already encountered Newton's iterative procedure as one example. When this method is applied to the equation

$$f(x) = x^3 - 3x + 1 = 0$$

with initial approximation $x_1 = 0.7$ to find the root between 0 and 1, the sequence obtained is

$$x_1 = 0.7, \qquad x_{n+1} = x_n - \frac{f(x_n)}{f'(x_n)} = x_n - \frac{x_n^3 - 3x_n + 1}{3x_n^2 - 3}, \qquad n \geq 1.$$

The first three terms of this sequence are illustrated in Figure 10.2, and the tangent line construction by which they are obtained makes it clear that the sequence must converge to the solution of the equation between 0 and 1. In other words, it is not necessary for us to verify convergence of the sequence algebraically; it is obvious geometrically.

FIGURE 10.2   Graphical illustration of sequence from Newton's iterative procedure

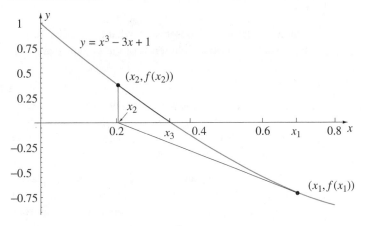

To find the solution of the equation we evaluate terms of the sequence algebraically until they repeat:

$$x_1 = 0.7, \qquad x_2 = 0.205, \qquad x_3 = 0.342, \qquad x_4 = 0.347\,285,$$

$$x_5 = 0.347\,296\,355, \qquad x_6 = 0.347\,296\,355.$$

Obviously, $x_6 = 0.347\,296\,355$ must be close to the root of $x^3 - 3x + 1 = 0$ between 0 and 1. How close can be verified with the zero intermediate value theorem from Section 1.11. For instance, to verify that $0.347\,296$ is an approximation to the solution correctly rounded to six decimal places, we calculate

$$f(0.347\,295\,5) = 2.3 \times 10^{-6} \qquad \text{and} \qquad f(0.347\,296\,5) = -3.8 \times 10^{-7}.$$

The fact that function values are opposite in sign verifies that $0.347\,296$ is correct to six decimal places.

Because Figure 10.2 illustrated that the sequence defined by Newton's iterative procedure must converge to a root of $f(x) = x^3 - 3x + 1 = 0$, it was unnecessary to verify existence of a limit of the sequence algebraically. Are we proposing that a graph of the function $f(x)$ should always be drawn when using Newton's method to solve equations? Not really, although we are of the philosophy that pictures should be drawn whenever they are helpful. Numerical analysts have proved that under very mild restrictions, the sequence defined by Newton's method always converges to a root of the equation provided that the initial approximation is sufficiently close to that root. See Exercise 68 for further discussion of this point. In practice, we approximate the solution of an equation $f(x) = 0$ using Newton's method exactly as we did in Section 4.1, and as illustrated above. We set up the appropriate sequence, choose an initial approximation, and iterate to find further approximations. We do not verify, algebraically or geometrically, that the sequence has a limit. If terms get closer and closer together, we suspect that the sequence is convergent, that is, that terms are getting closer and closer to the limit. We do not know this for sure, nor do we try to verify it. The real problem is to solve the equation $f(x) = 0$. We have what we believe is an approximation to the solution, and we verify this directly with the zero intermediate value theorem.

Another way to solve for the same root of $x^3 - 3x + 1 = 0$ is to rewrite the equation in the form

$$x = \frac{1}{3}(x^3 + 1),$$

and use this to define the following recursive sequence:

$$x_1 = \frac{1}{3}, \qquad x_{n+1} = \frac{1}{3}(x_n^3 + 1), \quad n \geq 1.$$

The initial term was chosen somewhat arbitrarily. In practice, it should be as close to the root as possible. The first eight terms of the sequence are

$x_1 = 1/3$,	$x_5 = 0.347\,293\,53$,
$x_2 = 0.345\,679$,	$x_6 = 0.347\,296\,01$,
$x_3 = 0.347\,102\,19$,	$x_7 = 0.347\,296\,31$,
$x_4 = 0.347\,272\,95$,	$x_8 = 0.347\,296\,35$.

The fact that x_7 and x_8 agree to six decimal places leads us to believe that the sequence converges and that its limit is approximately equal to 0.347 296. We cannot be certain of this, nor do we attempt to verify it. What we want is an approximation to the root of $x^3 - 3x + 1 = 0$ between 0 and 1. We can verify directly that $x = 0.347\,296$ is an approximation accurate to six decimal places, as above.

This method of finding the root of an equation is often called the **method of successive approximations** or **fixed-point iteration**. What we do is rearrange the equation $f(x) = 0$ into the form $x = g(x)$, and define a recursive sequence,

$$x_1 = A, \qquad x_{n+1} = g(x_n), \qquad n \geq 1,$$

where A is some initial approximation to the root (the closer the better). We iterate hoping that terms of the sequence get closer together. If they do, we terminate iterations when we suspect that we have an approximation to the solution of the equation with the required accuracy, and verify this directly with the zero intermediate value theorem. If terms do not seem to converge, we may try a different rearrangement of the equation into the form $x = g(x)$, or another method.

We can visualize the method of successive approximations geometrically. In terms of the example above, we are finding where the graph of the function $f(x) = x^3 - 3x + 1$ crosses the x-axis (Figure 10.3). There are three points, and we have been concentrating on the intercept between $x = 0$ and $x = 1$. When we rewrite the equation in the form $x = (x^3 + 1)/3$, an alternative interpretation is possible. We are searching for x-coordinates of points of intersection of the curves $y = x$ and $y = (x^3 + 1)/3$ (Figure 10.4).

FIGURE 10.3 Solutions of equation $x^3 - 3x + 1 = 0$

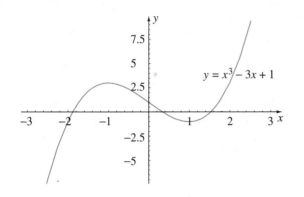

FIGURE 10.4   Solutions of $x^3 - 3x + 1 = 0$ as x-coordinates of points of intersection of curves $y = x$, $y = (x^3 + 1)/3$

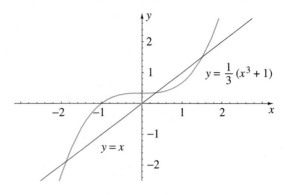

To show how the method of successive approximations works, we have expanded that part of Figure 10.4 between $x = 0.33$ and $x = 0.35$ (Figure 10.5). To find x_2 we substitute $x_1 = 1/3$ into $(x^3 + 1)/3$. It is the y-coordinate of the point A on the curve $y = (x^3 + 1)/3$. If we proceed horizontally from A to line $y = x$, then point B has coordinates (x_2, x_2). To find x_3 algebraically, we evaluate $(x^3 + 1)/3$ at x_2. Geometrically, x_3 is the y-coordinate of point C. If we move horizontally to the line $y = x$, coordinates of D are (x_3, x_3). Continuation leads to the sequence converging to the required root.

We illustrate the method again in the following example.

FIGURE 10.5   Sequence from successive approximations applied to $x = (x^3 + 1)/3$

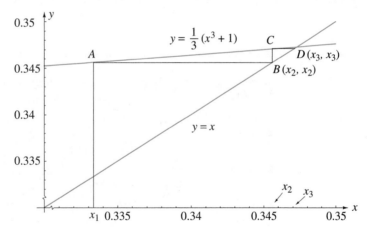

EXAMPLE 10.3

Use successive approximations to approximate the solution of $f(x) = x^3 + 25x - 50 = 0$ between $x = 1$ and $x = 2$ with error less than 10^{-6}. Illustrate the sequence graphically.

SOLUTION   The equation can be written in the form $x = (50 - x^3)/25$, leading to the sequence

$$x_1 = 1, \qquad x_{n+1} = \frac{50 - x_n^3}{25}, \qquad n \geq 1.$$

The next 15 terms are

$$
\begin{array}{lll}
x_2 = 1.96, & x_3 = 1.70, & x_4 = 1.80, \\
x_5 = 1.77, & x_6 = 1.778\,2, & x_7 = 1.775\,1, \\
x_8 = 1.776\,3, & x_9 = 1.775\,81, & x_{10} = 1.776\,00, \\
x_{11} = 1.775\,93, & x_{12} = 1.775\,954, & x_{13} = 1.775\,945, \\
x_{14} = 1.775\,948\,2, & x_{15} = 1.775\,946\,9, & x_{16} = 1.775\,947\,4.
\end{array}
$$

The terms certainly appear to be getting closer together (agreeing to six decimal places at this stage), therefore suggesting that the sequence has a limit. We do not verify this. We have what we feel is an approximation to the required solution of $x^3 + 25x - 50 = 0$ with sufficient accuracy. To verify that $1.775\,947$ is a solution with error less than 10^{-6}, we use the zero intermediate value theorem with the calculations

$$f(1.775\,946) = -4.4 \times 10^{-5}, \qquad f(1.775\,948) = 2.5 \times 10^{-5}.$$

The terms of the sequence are shown in Figure 10.6.

FIGURE 10.6   Sequence from successive approximations applied to $x = (50 - x^3)/25$

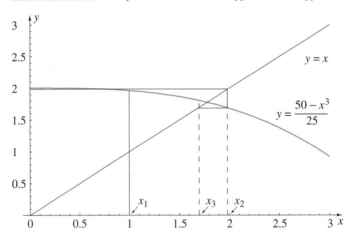

Consulting Project 15

We are being presented with a chemical problem. A precipitate rests at the bottom of a beaker that holds volume V of a mother liquid. The mother liquid is to be removed in the following way. Volume $\tilde{V}$ of pure water is added to the beaker and the liquids are mixed. Then $\tilde{V}$ of the mixture is removed, leaving volume V of a mixture that contains less mother liquid than originally held by the beaker. This is called a decantation. Our problem is to determine how many decantations are necessary before the amount of mother liquid is less than 1% of its original amount.

SOLUTION   What we need is a formula C_n for the amount of mother liquid in the beaker after the nth decantation. After the first addition of pure water, the concentration of mother liquid is $V/(V + \tilde{V})$. After mixture is removed, the amount of mother liquid remaining is

$$C_1 = V - \frac{V\tilde{V}}{V + \tilde{V}} = \frac{V^2}{V + \tilde{V}}.$$

After the addition of pure water for the $(n + 1)^{\text{st}}$ time, the concentration of mother liquid is $C_n/(V + \tilde{V})$. After mixture is removed, the amount of mother liquid remaining is

$$C_{n+1} = C_n - \frac{C_n\tilde{V}}{V + \tilde{V}} = \left(\frac{V}{V + \tilde{V}}\right) C_n.$$

Iteration of this recursive formula gives

$$C_2 = \left(\frac{V}{V + \tilde{V}}\right) C_1 = \frac{V^3}{(V + \tilde{V})^2}, \qquad C_3 = \left(\frac{V}{V + \tilde{V}}\right) C_2 = \frac{V^4}{(V + \tilde{V})^3},$$

and so on. The pattern emerging is

$$C_n = \frac{V^{n+1}}{(V + \tilde{V})^n}.$$

This will be less than 1% of V when

$$\frac{V^{n+1}}{(V + \tilde{V})^n} < \frac{V}{100} \qquad \Longrightarrow \qquad \frac{V^n}{(V + \tilde{V})^n} < \frac{1}{100}.$$

When we take logarithms,

$$n \ln \left(\frac{V}{V + \tilde{V}}\right) < \ln \left(\frac{1}{100}\right) \qquad \Longrightarrow \qquad n > \frac{- \ln 100}{\ln [V/(V + \tilde{V})]}.$$

The smallest such integer n satisfying this inequality is the minimum number of decantations required. For instance, if $V = 100$ cm^3 and $\tilde{V} = 200$ cm^3, then

$$n > - \frac{\ln 100}{\ln [100/300]} = 4.2.$$

Five decantations would be required.

EXERCISES 10.1

In Exercises 1–20 determine whether the sequence is convergent or divergent. Find limits for convergent sequences.

1. $\left\{ \dfrac{1}{n} \right\}$

2. $\left\{ 3^n + 1 \right\}$

3. $\{3\}$

4. $\left\{ \left(\dfrac{3}{4} \right)^{n+1} \right\}$

5. $\left\{ \left(\dfrac{4}{3} \right)^{n+1} \right\}$

6. $\left\{ \left(-\dfrac{15}{16} \right)^{n+5} \right\}$

7. $\left\{ \sin \left(\dfrac{n\pi}{2} \right) \right\}$

8. $\left\{ \dfrac{n}{n^2 + n + 2} \right\}$

9. $\left\{ \dfrac{(-1)^n}{n} \right\}$

10. $\left\{ \mathrm{Tan}^{-1} n \right\}$

11. $\left\{ \dfrac{\ln n}{n^2 + 1} \right\}$

12. $\left\{ (-1)^n \sqrt{n^2 + 1} \right\}$

13. $c_1 = 2, \quad c_{n+1} = \dfrac{2}{c_n - 1}, \quad n \geq 1$

14. $c_1 = 4, \quad c_{n+1} = -\dfrac{c_n}{n^2}, \quad n \geq 1$

15. $\left\{ \dfrac{n + 1}{2n + 3} \right\}$

16. $\left\{ \dfrac{2n + 3}{n^2 - 5} \right\}$

17. $\left\{ \dfrac{n^2 + 5n - 4}{n^2 + 2n - 2} \right\}$

18. $\left\{ n e^{-n} \right\}$

19. $\left\{ \dfrac{\ln n}{n} \right\}$

20. $\left\{ \dfrac{n}{n + 1} \mathrm{Tan}^{-1} n \right\}$

In Exercises 21–25 find an explicit formula for the general term of the sequence. In each case assume that the remaining terms follow the pattern suggested by the given terms.

21. $\dfrac{1}{2}, \ \dfrac{3}{4}, \ \dfrac{7}{8}, \ \dfrac{15}{16}, \ \dfrac{31}{32}, \ldots$

22. $4, \ \dfrac{7}{4}, \ \dfrac{10}{9}, \ \dfrac{13}{16}, \ \dfrac{16}{25}, \ \dfrac{19}{36}, \ldots$

23. $\dfrac{\ln 2}{\sqrt{2}}, \ -\dfrac{\ln 3}{\sqrt{3}}, \ \dfrac{\ln 4}{\sqrt{4}}, \ \dfrac{-\ln 5}{\sqrt{5}}, \ \dots$

* 24. 1, 0, 1, 0, 1, 0, ...

* 25. 1, 1, −1, −1, 1, 1, −1, −1, ...

In Exercises 26–29 show how L'Hôpital's rule can be used to evaluate the limit for the sequence.

26. $\left\{\dfrac{\ln n}{\sqrt{n}}\right\}$

27. $\left\{\dfrac{n^3 + 1}{e^n}\right\}$

* 28. $\left\{n \sin\left(\dfrac{4}{n}\right)\right\}$

* 29. $\left\{\left(\dfrac{n+5}{n+3}\right)^n\right\}$

* 30. If a sequence converges, then differences between successive terms in the sequence must approach zero. The converse is not always true. The following example is an illustration. Show that differences between successive terms of the sequence $\{\ln n\}$ approach zero, but the sequence itself diverges.

* 31. (a) The nth term of a sequence is the nth prime integer (greater than 1) when all such primes are listed in ascending size. List its first 10 terms.

  (b) Can you give a formula for the nth term?

In Exercises 32–43 use Newton's iterative procedure with the given initial approximation x_1 to define a sequence of approximations to a solution of the equation. Determine graphically whether the sequence has a limit. Approximate any limit that exists to seven decimal places.

32. $x_1 = 1, \quad x^2 + 3x + 1 = 0$

33. $x_1 = -1, \quad x^2 + 3x + 1 = 0$

34. $x_1 = -1.5, \quad x^2 + 3x + 1 = 0$

35. $x_1 = -3, \quad x^2 + 3x + 1 = 0$

36. $x_1 = 4, \quad x^3 - x^2 + x - 22 = 0$

37. $x_1 = 2, \quad x^3 - x^2 + x - 22 = 0$

38. $x_1 = 2, \quad x^5 - 3x + 1 = 0$

39. $x_1 = 1, \quad x^5 - 3x + 1 = 0$

40. $x_1 = 0, \quad x^5 - 3x + 1 = 0$

41. $x_1 = 4/5, \quad x^5 - 3x + 1 = 0$

42. $x_1 = 0.85, \quad x^5 - 3x + 1 = 0$

43. $x_1 = -2, \quad x^5 - 3x + 1 = 0$

In Exercises 44–48 illustrate that the method of successive approximations with the suggested rearrangement of the equation $f(x) = 0$, along with the initial approximation x_1, leads to a sequence that converges to a root of the equation. Find the root accurate to four decimal places.

44. $f(x) = x^2 - 2x - 1; \quad x_1 = 2; \quad$ use $x = 2 + 1/x$

45. $f(x) = x^3 + 6x + 3; \quad x_1 = -1; \quad$ use $x = -(1/6)(x^3 + 3)$

46. $f(x) = x^4 - 120x + 20; \quad x_1 = 0; \quad$ use $x = (x^4 + 20)/120$

47. $f(x) = x^3 - 2x^2 - 3x + 1; \quad x_1 = 3; \quad$ use $x = (2x^2 + 3x - 1)/x^2$

48. $f(x) = 8x^3 - x^2 - 1; \quad x_1 = 0; \quad$ use $x = (1/2)(1 + x^2)^{1/3}$

In Exercises 49–54 find a rearrangement of the equation that leads, through the method of successive approximations, to a four-decimal-place approximation to the root of the equation.

* 49. $x^3 - 6x^2 + 11x - 7 = 0 \quad$ between $x = 3$ and $x = 4$

* 50. $x^4 - 3x^2 - 3x + 1 = 0 \quad$ between $x = 0$ and $x = 1$

* 51. $x^4 + 4x^3 - 50x^2 + 100x - 50 = 0 \quad$ between $x = 0$ and $x = 1$

* 52. $\sin^2 x = 1 - x^2 \quad$ between $x = 0$ and $x = 1$

* 53. $\sec x = 2/(1 + x^4) \quad$ between $x = 0$ and $x = 1$

* 54. $e^x + e^{-x} - 10x = 0 \quad$ between $x = 0$ and $x = 1$

* 55. (a) Use Newton's iterative procedure with $x_1 = 1$ to approximate the root, between 0 and 1, of $x^4 - 15x + 2 = 0$, accurate to six decimal places.

  (b) Use the method of successive approximations with $x_1 = 1$ and $x_{n+1} = (x_n^4 + 2)/15$ to approximate the root in part (a), accurate to six decimal places.

  (c) Use Newton's method to approximate the root between 2 and 3.

  (d) What happens if the sequence in part (b) is used to approximate the root between 2 and 3 with $x_1 = 2$ and $x_1 = 3$?

* 56. A superball is dropped from the top of a building 20 m high. Each time it strikes the ground, it rebounds to 99% of the height from which it fell.

  (a) If d_n denotes the distance travelled by the ball between the nth and $(n + 1)^{\text{th}}$ bounces, find a formula for d_n.

  (b) If t_n denotes the time between the nth and $(n + 1)^{\text{th}}$ bounces, find a formula for t_n.

* 57. A dog sits at a farmhouse patiently watching for his master to return from the fields. When the farmer is 1 km from the farmhouse, the dog immediately takes off for the farmer. When he reaches the farmer, he turns and runs back to the farmhouse, whereupon he again turns and runs to the farmer. The dog continues this frantic action until the farmer reaches the farmhouse. If the dog runs twice as fast as the farmer, find the distance d_n run by the dog from the point when he reaches the farmer for the nth time to the point when he reaches the farmer for the $(n + 1)^{\text{th}}$ time. Ignore any accelerations of the dog in the turns.

* 58. The equilateral triangle in the left figure below has perimeter P. If each side of the triangle is divided into three equal parts, an equilateral triangle is drawn on the middle segment of each side, and the figure transformed into the middle figure, what is the perimeter P_1 of this figure? If each side of this figure is now subdivided into three equal

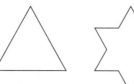
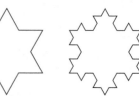

portions and equilateral triangles are similarly constructed to result in the right figure, what is the perimeter P_2 of this figure? If this subdivision process is continued indefinitely, what is the perimeter P_n after the nth subdivision? What is $\lim_{n \to \infty} P_n$?

59. A stone of mass 100 g is thrown vertically upward with speed 20 m/s. Air exerts a resistive force on the stone proportional to its speed, and has magnitude 1/10 N when the speed of the stone is 10 m/s. It can be shown that the height y (in metres) above the projection point attained by the stone is given by

$$ y = -98.1t + 1181 \left(1 - e^{-t/10}\right), $$

where t is time (measured in seconds with $t = 0$ at the instant of projection).

(a) The time taken for the stone to return to its projection point can be obtained by setting $y = 0$ and solving the equation for t. Do so (correct to two decimal places).

(b) Find the time for the stone to return if air resistance is ignored.

60. When the beam in the figure below vibrates vertically, there are certain frequencies of vibration, called *natural frequencies of the system*. They are solutions of the equation

$$ \tan x = \frac{e^x - e^{-x}}{e^x + e^{-x}} $$

divided by 20π. Find the two smallest natural frequencies.

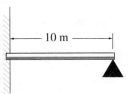

61. If A_n is the area of the figure with perimeter P_n in Exercise 58, find a formula for A_n.

62. What are the next two terms in the sequence 1, 11, 21, 1211, 111221, 312211, 13112221?

63. (a) Plot the first 20 terms of the sequence $\{c_n\}$, where $c_n = (1.02)^n + 0.5 \cos(\pi n/4 + \pi/4)$.

(b) In digital signal processing, the sequence $\{A_n\}_3^\infty$ defined in terms of c_n by $A_n = (c_n + c_{n-1} + c_{n-2})/3$ is called a *causal three-point running averager*. Plot its first 18 terms.

64. Repeat part (b) of Exercise 63 with a seven-point running averager.

65. Suppose that terms of a sequence $\{c_n\}$ represent a discrete-time signal. A FIR (finite impulse response) filter is a sequence $\{F_n\}$ whose terms are linear combinations of the terms of $\{c_n\}$. It is causal if F_n is of the form

$$ F_n = b_0 c_n + b_1 c_{n-1} + \cdots + b_{M-1} c_{n-M+1} + b_M c_{n-M} $$

$$ = \sum_{k=0}^{M} b_k c_{n-k}; $$

that is, F_n is a linear combination of c_n and the M terms immediately before c_n. The $b_0, b_1, \ldots, b_M$ are fixed constants. (The running averagers in Exercises 63 and 64 are causal FIR filters.) If $c_n = \{n/(n+1)\}$, $M = 2$ with $b_0 = 1$, $b_1 = 2$, and $b_2 = -1$, calculate the first 10 terms of $\{F_n\}_3^\infty$.

* **66.** Repeat Exercise 65 with $c_n = \{(1/n^2) \sin(n/3)\}$, $M = 3$, $b_0 = 1$, $b_1 = -2$, $b_2 = 3$, $b_3 = -4$. List terms rounded to four decimal places.

* **67.** (a) Show that if α is the only root of the equation $x = g(x)$ and the left figure below is a graph of $g(x)$, then the right figure exhibits geometrically the sequence of approximations of α determined by the method of successive approximations.

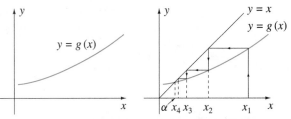

(b) Illustrate graphically how the sequence defined by the method of successive approximations converges to the root $x = \alpha$ for the equation $x = g(x)$ if $g(x)$ is as shown in the left figure below.

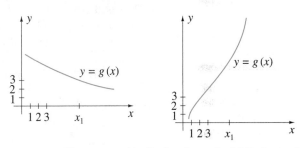

(c) Illustrate graphically that the method fails for the function $g(x)$ in the right figure above.

(d) Based on the results of parts (a)–(c), what determines success or failure of the method of successive approximations? Part (e) provides a proof of the correct answer.

(e) Prove that when $x = g(x)$ has a root $x = \alpha$, the method of successive approximations with an initial approximation of x_1 always converges to α if $|g'(x)| \leq a < 1$ on the interval $|x - \alpha| \leq |x_1 - \alpha|$. In other words, when an equation $f(x) = 0$ is rearranged into the form $x = g(x)$ for the method of successive approximations, success or failure depends on whether the derivative of $g(x)$ is between -1 and 1 near the required root.

** **68.** Suppose that $f''(x)$ exists on an open interval containing a root $x = \alpha$ of the equation $f(x) = 0$. Use the result of Exercises 67 to prove that if $f'(\alpha) \neq 0$, Newton's iterative sequence always converges to α provided the initial approximation x_1 is chosen sufficiently close to α. *Hint:* First use Exercise 67(e) to show that Newton's sequence converges to α if on the interval $|x - \alpha| \leq |x_1 - \alpha|$, $|ff''/(f')^2| \leq a < 1$. [In actual fact, Newton's method often works even when $f'(\alpha) = 0$.]

10.2 Sequences of Functions

In many applications, we encounter sequences of functions as opposed to sequences of numbers. A sequence of functions is the assignment of functions to positive integers. For instance, the first five functions in the sequence $\{x^2 + 10xe^{-nx}\}$ on the interval $0 \le x \le 1$ are

$$x^2 + 10xe^{-x}, \quad x^2 + 10xe^{-2x}, \quad x^2 + 10xe^{-3x}, \quad x^2 + 10xe^{-4x}, \quad x^2 + 10xe^{-5x}.$$

FIGURE 10.7a   Sequence $\{x^2 + 10xe^{-nx}\}$ for $0 \le x \le 1$

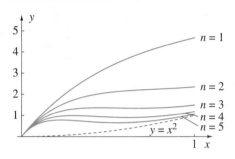

FIGURE 10.7b   Sequence $\{x^2 + 10xe^{-nx}\}$ for $-1 \le x \le 0$

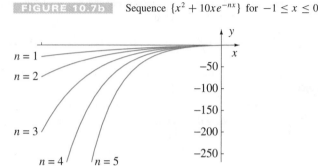

They are plotted in Figure 10.7a. As n gets larger and larger, values of $x^2 + 10xe^{-nx}$ get closer and closer to x^2 for all x in the interval $0 \le x \le 1$, and we say that the limit function for this sequence of functions is $f(x) = x^2$. We write that

$$\lim_{n \to \infty} (x^2 + 10xe^{-nx}) = x^2, \quad 0 \le x \le 1.$$

This is also true for $x > 1$, but not for $x < 0$. Figure 10.7b shows that for $x < 0$, the functions take on increasingly large negative values as n increases.

It is important to realize that a sequence of functions contains many sequences of numbers; simply substitute a value of x into the functions. For example, if we set $x = 1$ in the sequence $\{x^2 + 10xe^{-nx}\}$, we obtain the sequence of numbers $\{1 + 10e^{-n}\}$. They are the y-coordinates of the points at the ends of the curves in Figure 10.7a, and the sequence clearly converges to 1. In fact, for any nonnegative value x_0, the sequence of numbers $\{x_0^2 + 10x_0e^{-nx_0}\}$ is visualized as heights of points of intersection of the curves in Figure 10.7a with the line $x = x_0$ (Figure 10.8). Each such sequence has limit x_0^2.

FIGURE 10.8   Sequence of points obtained from a sequence of functions

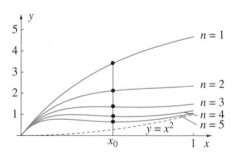

EXAMPLE 10.4

Plot the first five functions of the sequence $\{(x - 1)/[n + n^2(x - 1)^2]\}$ on the interval $-5 \le x \le 5$. What is the limit function for the sequence?

SOLUTION   The first five functions are plotted in Figure 10.9. Geometrically, and algebraically, it is clear that

$$\lim_{n \to \infty} \frac{x - 1}{n + n^2(x - 1)^2} = 0 \quad \text{for all } x.$$

FIGURE 10.9   Sequence of functions $\left\{ \frac{x-1}{n+n^2(x-1)^2} \right\}$

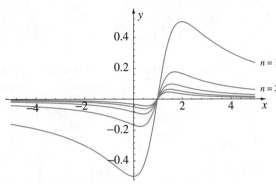

The sequence in the next example is indispensable to future discussions.

EXAMPLE 10.5

Show that the limit of the sequence of functions $\{|x|^n/n!\}$ is zero for all x; that is, verify that

$$\lim_{n \to \infty} \frac{|x|^n}{n!} = 0, \quad -\infty < x < \infty.$$

SOLUTION   We can *see* this intuitively. As n increases by 1, a fixed, extra $|x|$-factor appears in the numerator, but an ever increasing factor n occurs in the denominator. When n surpasses x, the fraction will get smaller and smaller. Let us prove this analytically.

Suppose that x is any fixed value, and let $m = \lfloor |x| \rfloor$, where $\lfloor x \rfloor$ is the floor function of Exercise 68 in Section 1.5. It is the largest integer that does not exceed x. (If $x = 3.4$, then $\lfloor |3.4| \rfloor = 3$; if $x = -22.6$, then $\lfloor |-22.6| \rfloor = \lfloor 22.6 \rfloor = 22$; and if $x = -4$, then $\lfloor |-4| \rfloor = \lfloor 4 \rfloor = 4$.) Since $|x|^n$ is $|x|$ multiplied by itself n times, and $n!$ has n multiplications, we may write that

$$\frac{|x|^n}{n!} = \left(\frac{|x|}{1} \right) \left(\frac{|x|}{2} \right) \left(\frac{|x|}{3} \right) \cdots \left(\frac{|x|}{n} \right).$$

If n is chosen greater than $m = \lfloor |x| \rfloor$, then

$$\frac{|x|^n}{n!} = \left(\frac{|x|}{1} \right) \left(\frac{|x|}{2} \right) \left(\frac{|x|}{3} \right) \cdots \left(\frac{|x|}{m} \right) \left(\frac{|x|}{m+1} \right) \cdots \left(\frac{|x|}{n} \right).$$

Suppose we let

$$M = \left(\frac{|x|}{1} \right) \left(\frac{|x|}{2} \right) \left(\frac{|x|}{3} \right) \cdots \left(\frac{|x|}{m} \right),$$

a fixed constant. Then

$$\frac{|x|^n}{n!} = M \left(\frac{|x|}{m+1} \right) \left(\frac{|x|}{m+2} \right) \cdots \left(\frac{|x|}{n} \right)$$

$$< M \left(\frac{|x|}{m+1} \right) \left(\frac{|x|}{m+1} \right) \cdots \left(\frac{|x|}{m+1} \right),$$

where we have replaced $m+2, m+3, \ldots, n$ in the denominators by the smaller integer $m+1$. Since there are $n - m$ bracketed terms on the right, all equal to each other, we have

$$\frac{|x|^n}{n!} < M\left(\frac{|x|}{m+1}\right)^{n-m} = M\left(\frac{|x|}{m+1}\right)^{-m}\left(\frac{|x|}{m+1}\right)^n.$$

Now, $M[|x|/(m+1)]^{-m}$ is a fixed number as far as the limit on n is concerned. Furthermore, $|x|/(m+1)$ is also constant and is less than 1. Consequently,

$$\lim_{n\to\infty}\frac{|x|^n}{n!} \le \lim_{n\to\infty} M\left(\frac{|x|}{m+1}\right)^{-m}\left(\frac{|x|}{m+1}\right)^n$$

$$= M\left(\frac{|x|}{m+1}\right)^{-m}\lim_{n\to\infty}\left(\frac{|x|}{m+1}\right)^n$$

$$= M\left(\frac{|x|}{m+1}\right)^{-m}(0)$$

$$= 0.$$

Since $|x|^n/n! \ge 0$, it follows that $\lim_{n\to\infty} |x|^n/n! = 0$.

EXERCISES 10.2

In Exercises 1–14 $f_n(x)$ is the nth term in a sequence of functions $\{f_n(x)\}$. Plot graphs of the first five functions in the sequence. Determine whether the sequence has a limit.

1. $f_n(x) = \dfrac{nx}{1+n^2x^2}, 0 \le x \le 1$

2. $f_n(x) = \dfrac{n^2x}{1+n^3x^2}, 0 \le x \le 1$

3. $f_n(x) = \dfrac{nx^2}{1+nx}, 0 \le x \le 1$

4. $f_n(x) = \dfrac{1}{x} + \dfrac{1}{n}\sin\left(\dfrac{1}{nx}\right), 1 \le x \le 2$

5. $f_n(x) = nx^n(1-x), 0 \le x \le 1$

6. $f_n(x) = nx^n(1-x), 0 \le x \le 2$

7. $f_n(x) = n^2x^n(2-x), 0 \le x \le 2$

8. $f_n(x) = n^2x^n(1-x^2), -1 \le x \le 1$

9. $f_n(x) = \dfrac{2+nx^2}{1+nx}, 0 \le x \le 2$

10. $f_n(x) = (\sin x)^{1/n}, 0 < x < \pi$

11. $f_n(x) = (\sin x)^{1/n}, 0 \le x \le \pi$

12. $f_n(x) = \left(\dfrac{\sin x}{x}\right)^{1/n}, 0 < x < \pi$

13. $f_n(x) = \left(\dfrac{\sin x}{x}\right)^{1/n}, 0 < x \le \pi$

14. $f_n(x) = n^2xe^{-nx}, 0 \le x < \infty$

* **15.** For what values of x does the sequence of functions $\{(1-x^n)/(1-x)\}$ have a limit?

10.3 Taylor Polynomials, Remainders, and Series

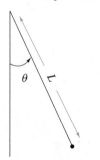

FIGURE 10.10 Oscillations of a pendulum

Figure 10.10 shows a pendulum consisting of a mass on the end of a string of length L. If the mass is pulled slightly to the side and released, it swings back and forth for some time to come. The position of the mass can be described by the angle θ that the string makes with the vertical. By analyzing the forces acting on the mass, it can be shown that when air resistance is neglected, θ, as a function of time t, must satisfy the differential equation

$$\frac{d^2\theta}{dt^2} + \frac{g}{L}\sin\theta = 0, \tag{10.4}$$

where $g = 9.81$ is the acceleration due to gravity. Thus, $\theta(t)$ is a function whose second derivative is $-g/L$ times the sine of itself. This is a very difficult differential equation to solve;

in fact, no combination of the simple functions with which we are familiar satisfies equation 10.4. What is sometimes done in examples like this is replace $\sin\theta$ with a simpler function $f(\theta)$ that simultaneously approximates $\sin\theta$ and for which the differential equation

$$\frac{d^2\theta}{dt^2} + \frac{g}{L}f(\theta) = 0 \tag{10.5}$$

is solvable. Although the solution of equation 10.5 only approximates the solution of 10.4, if $f(\theta)$ approximates $\sin\theta$ very closely, the solution of 10.5 may be sufficiently close to the solution of 10.4 to give real insight into the motion of the pendulum. In particular, as we shall show shortly, when θ is very small, $\sin\theta$ can be approximated by θ itself. In other words, for small oscillations of the pendulum, equation 10.4 can be replaced by

$$\frac{d^2\theta}{dt^2} + \frac{g}{L}\theta = 0. \tag{10.6}$$

It is straightforward to verify that

$$\theta(t) = A\cos\sqrt{\frac{g}{L}}t + B\sin\sqrt{\frac{g}{L}}t$$

is a solution of equation 10.6 for any constants A and B whatsoever. If motion is initiated at time $t = 0$ by pulling the mass an amount θ_0 to the right and then releasing it, then $\theta(t)$ must satisfy the additional conditions

$$\theta(0) = \theta_0, \qquad \theta'(0) = 0.$$

These imply that $A = \theta_0$ and $B = 0$, and hence

$$\theta(t) = \theta_0\cos\sqrt{\frac{g}{L}}t.$$

This function yields simple harmonic motion for the pendulum, as we expect.

In this section we show how to approximate complicated functions by polynomials. The interval on which the approximation is required and the accuracy of the approximation dictate the degree of the polynomial. To begin with, we recall the Mean Value Theorem from Section 3.14. It states that when $f(x)$ is continuous on the closed interval between c and some given value of x, and $f'(x)$ exists on the open interval between c and x, there exists a number between c and x, call it z_0, such that

$$\frac{f(x) - f(c)}{x - c} = f'(z_0),$$

or,

$$f(x) = f(c) + f'(z_0)(x - c). \tag{10.7}$$

This theorem was a corollary to Cauchy's generalized mean value theorem in Section 3.14. We give a direct proof of it here because a similar but more complicated argument leads to an extension called *Taylor's Remainder Formula*, a result that is fundamental to our studies in this chapter.

Consider the following function of y for fixed c and x,

$$F(y) = f(x) - f(y) - \left(\frac{x - y}{x - c}\right)[f(x) - f(c)].$$

Because f is continuous on the closed interval between c and x, $F(y)$ is continuous for y in this interval. Furthermore,

$$F'(y) = -f'(y) + \frac{1}{x - c}[f(x) - f(c)].$$

Since $f'(x)$ exists on the open interval between c and x, so also does $F'(y)$. Finally, $F(c) = 0 = F(x)$. Rolle's Theorem (Theorem 3.17 in Section 3.14) implies that there exists a number z_0 between c and x at which

$$0 = F'(z_0) = -f'(z_0) + \frac{1}{x-c}[f(x) - f(c)],$$

and this gives equation 10.7.

Suppose now that $f(x)$ and $f'(x)$ are continuous on the closed interval between c and x, and $f''(x)$ exists on the open interval between c and x. Consider the following function of y,

$$F(y) = f(x) - f(y) - f'(y)(x-y) - \left(\frac{x-y}{x-c}\right)^2 [f(x) - f(c) - f'(c)(x-c)].$$

Because $f(x)$ and $f'(x)$ are continuous on the closed interval between c and x, $F(y)$ is continuous for y in this interval. Furthermore,

$$F'(y) = -f'(y) - f''(y)(x-y) + f'(y) + \frac{2(x-y)}{(x-c)^2}[f(x) - f(c) - f'(c)(x-c)].$$

Since $f'(x)$ is continuous on the closed interval between c and x, and $f''(x)$ exists on the open interval between c and x, it follows that $F'(y)$ exists on the open interval. Finally, $F(c) = 0 = F(x)$, and therefore Rolle's Theorem implies the existence of a number z_1 between c and x at which

$$0 = F'(z_1) = -f'(z_1) - f''(z_1)(x - z_1) + f'(z_1) + \frac{2(x - z_1)}{(x-c)^2}[f(x) - f(c) - f'(c)(x - c].$$

This can be rearranged into the form

$$f(x) = f(c) + f'(c)(x - c) + \frac{f''(z_1)}{2}(x - c)^2. \tag{10.8}$$

By assuming that $f(x)$, $f'(x)$, and $f''(x)$ are continuous on the closed interval between c and x, and $f'''(x)$ exists on the open interval, it can be shown in a similar way that there exists a number z_2 between c and x such that

$$f(x) = f(c) + f'(c)(x - c) + \frac{f''(c)}{2}(x - c)^2 + \frac{f'''(z_2)}{3!}(x - c)^3. \tag{10.9}$$

These results can be extended indefinitely if we assume that $f(x)$ has a sufficient number of derivatives on the closed interval between c and x. The complete result is contained in the following theorem.

THEOREM 10.1   (Taylor's Remainder Formula)

If $f(x)$ and its first n derivatives are continuous on the closed interval between c and x, and if $f(x)$ has an $(n+1)^{\text{th}}$ derivative on the open interval between c and x, then there exists a point z_n between c and x such that

$$f(x) = f(c) + f'(c)(x - c) + \frac{f''(c)}{2!}(x - c)^2 + \frac{f'''(c)}{3!}(x - c)^3 + \cdots$$

$$+ \frac{f^{(n)}(c)}{n!}(x - c)^n + \frac{f^{(n+1)}(z_n)}{(n+1)!}(x - c)^{n+1}. \tag{10.10}$$

The notation $f^{(n)}(c)$ represents the nth derivative of $f(x)$ evaluated at $x = c$.

In the remainder of this section, we assume that $f(x)$ has derivatives of all orders on the closed interval between c and some value of x. Taylor's remainder formula can then be written down for all values of n. For $n = 0, 1, 2,$ and 3, we obtain

$$f(x) = f(c) + f'(z_0)(x - c), \tag{10.11a}$$

$$f(x) = f(c) + f'(c)(x - c) + \frac{f''(z_1)}{2!}(x - c)^2, \tag{10.11b}$$

$$f(x) = f(c) + f'(c)(x - c) + \frac{f''(c)}{2!}(x - c)^2 + \frac{f'''(z_2)}{3!}(x - c)^3, \tag{10.11c}$$

$$f(x) = f(c) + f'(c)(x - c) + \frac{f''(c)}{2!}(x - c)^2 + \frac{f'''(c)}{3!}(x - c)^3$$

$$+ \frac{f''''(z_3)}{4!}(x - c)^4. \tag{10.11d}$$

The last terms in these equations are called **Taylor remainders**, and when they are denoted by R_0, R_1, R_2, and R_3, respectively,

$$f(x) = f(c) + R_0, \tag{10.12a}$$

$$f(x) = f(c) + f'(c)(x - c) + R_1, \tag{10.12b}$$

$$f(x) = f(c) + f'(c)(x - c) + \frac{f''(c)}{2!}(x - c)^2 + R_2, \tag{10.12c}$$

$$f(x) = f(c) + f'(c)(x - c) + \frac{f''(c)}{2!}(x - c)^2 + \frac{f'''(c)}{3!}(x - c)^3 + R_3, \tag{10.12d}$$

and in general,

$$f(x) = f(c) + f'(c)(x - c) + \frac{f''(c)}{2!}(x - c)^2 + \cdots$$

$$+ \frac{f^{(n)}(c)}{n!}(x - c)^n + R_n. \tag{10.13}$$

We call R_n the remainder for the simple reason that, if we drop the remainders from equations 10.12, then what remains is a sequence of polynomial approximations to $f(x)$, called the **Taylor polynomials** of $f(x)$ about c:

$$f(x) \approx P_0(x) = f(c), \tag{10.14a}$$

$$f(x) \approx P_1(x) = f(c) + f'(c)(x - c), \tag{10.14b}$$

$$f(x) \approx P_2(x) = f(c) + f'(c)(x - c) + \frac{f''(c)}{2!}(x - c)^2, \tag{10.14c}$$

$$f(x) \approx P_3(x) = f(c) + f'(c)(x - c) + \frac{f''(c)}{2!}(x - c)^2 + \frac{f'''(c)}{3!}(x - c)^3. \tag{10.14d}$$

The nth Taylor polynomial $P_n(x)$ is a polynomial of degree n. What happens in practice is that as n increases, the polynomials $P_n(x)$ approximate $f(x)$ more and more closely on some finite or infinite interval containing c. The following examples illustrate this.

EXAMPLE 10.6 ▼

Find the first six Taylor polynomials for $\sin x$ about $x = 0$. Plot the polynomials and $\sin x$ to show how the approximations improve as more terms are included.

SOLUTION   The value of $f(x) = \sin x$ and its first five derivatives at $x = 0$ are

$$f(0) = 0, \quad f'(0) = \cos 0 = 1, \quad f''(0) = -\sin 0 = 0, \quad f'''(0) = -\cos 0 = -1,$$

$$f^{(4)}(0) = \sin 0 = 0, \quad f^{(5)}(0) = \cos 0 = 1.$$

The first six Taylor polynomials are

$$P_0(x) = f(0) = 0,$$

$$P_1(x) = f(0) + f'(0)(x - 0) = x,$$

$$P_2(x) = f(0) + f'(0)(x - 0) + \frac{f''(0)}{2!}(x - 0)^2 = x,$$

$$P_3(x) = f(0) + f'(0)(x - 0) + \frac{f''(0)}{2!}(x - 0)^2 + \frac{f'''(0)}{3!}(x - 0)^3 = x - \frac{x^3}{3!},$$

$$P_4(x) = x - \frac{x^3}{3!},$$

$$P_5(x) = x - \frac{x^3}{3!} + \frac{x^5}{5!}.$$

Because even-ordered derivatives of $\sin x$ vanish, even-numbered polynomials are the same as their odd predecessors. We have shown polynomials $P_1(x)$, $P_3(x)$, and $P_5(x)$ along with $\sin x$ for positive values of x in Figure 10.11. We have included $P_7(x)$ and $P_9(x)$ to further emphasize the following facts. The higher the degree of the polynomial, the more closely it approximates $\sin x$ for small values of x, and the larger the interval on which the approximation is reasonable.

FIGURE 10.11   Taylor polynomials of $\sin x$

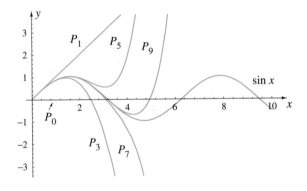

EXAMPLE 10.7

Find the first five Taylor polynomials of e^x about $x = 0$. Plot the polynomials and e^x.

SOLUTION   Since all derivatives of e^x at $x = 0$ are equal to 1, the first five polynomials are

$$P_0(x) = 1,$$

$$P_1(x) = 1 + x,$$

$$P_2(x) = 1 + x + \frac{x^2}{2!},$$

$$P_3(x) = 1 + x + \frac{x^2}{2!} + \frac{x^3}{3!},$$

$$P_4(x) = 1 + x + \frac{x^2}{2!} + \frac{x^3}{3!} + \frac{x^4}{4!}.$$

They are illustrated in Figure 10.12. When $x > 0$, the polynomials are always less than e^x and approach e^x as more terms are included. For $x < 0$, the polynomials are alternately higher and lower than e^x, gradually getting closer and closer to e^x.

FIGURE 10.12   Taylor polynomials of e^x

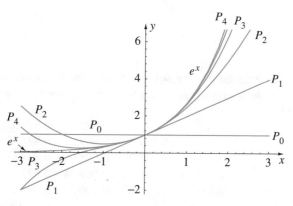

EXAMPLE 10.8

Find the first five Taylor polynomials for $\ln x$ about $x = 1$. Plot the polynomials and $\ln x$.

SOLUTION   Since

$$\ln x_{|x=1} = 0, \quad \frac{d}{dx}\ln x_{|x=1} = \frac{1}{x}\Big|_{x=1} = 1, \quad \frac{d^2}{dx^2}\ln x_{|x=1} = -\frac{1}{x^2}\Big|_{x=1} = -1,$$

$$\frac{d^3}{dx^3}\ln x_{|x=1} = \frac{2}{x^3}\Big|_{x=1} = 2, \quad \frac{d^4}{dx^4}\ln x_{|x=1} = \frac{-3!}{x^3}\Big|_{x=1} = -3!,$$

the first five polynomials are

$$P_0(x) = 0,$$

$$P_1(x) = x - 1,$$

$$P_2(x) = (x - 1) - \frac{1}{2}(x - 1)^2,$$

$$P_3(x) = (x - 1) - \frac{1}{2}(x - 1)^2 + \frac{1}{3}(x - 1)^3,$$

$$P_4(x) = (x - 1) - \frac{1}{2}(x - 1)^2 + \frac{1}{3}(x - 1)^3 - \frac{1}{4}(x - 1)^4.$$

They are plotted in Figure 10.13. For $0 < x < 1$, the polynomials are greater than $\ln x$ and approach $\ln x$ as more terms are included. For $1 < x \leq 2$, the polynomials are alternately higher and lower than $\ln x$, gradually getting closer and closer to $\ln x$. The plot of the Taylor polynomials on the interval $1 \leq x \leq 3$ in Figure 10.14 indicates that they do not approach $\ln x$ for $x > 2$.

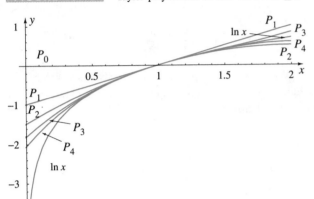

FIGURE 10.13 Taylor polynomials of $\ln x$ on $0 < x \le 2$

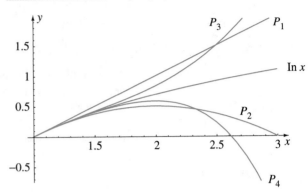

FIGURE 10.14 Taylor polynomials of $\ln x$ on $1 < x \le 3$

To make this clearer, below is a list of the first 30 Taylor polynomials evaluated at $x = 2.1$. At first these numbers get closer together, but then they separate. We have shown them in Figure 10.15.

0.0	1.1	0.495	0.938 67	0.572 64	0.894 74
0.599 48	0.877 87	0.609 92	0.871 92	0.612 54	0.871 92
0.610 38	0.875 94	0.605 69	0.883 17	0.595 99	0.893 31
0.584 43	0.906 32	0.569 94	0.922 33	0.552 32	0.941 64
0.531 23	0.964 62	0.506 23	0.991 79	0.476 75	1.023 75

FIGURE 10.15 Taylor polynomials of $\ln x$ evaluated at $x = 2.1$

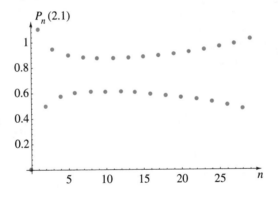

The plots in Examples 10.6–10.8 support our earlier contention that as n increases, the Taylor polynomials of a function approximate the function more and more closely on some finite or infinite interval. We would like to be able to verify analytically what we see geometrically. To do this we return to the Taylor remainders. The nth remainder is

$$R_n = R_n(c, x) = \frac{f^{(n+1)}(z_n)}{(n+1)!}(x - c)^{n+1}, \tag{10.15}$$

where z_n (which, as the notation suggests, varies with n) is always between c and x. We have written $R_n(c, x)$ to emphasize the fact that remainders depend on the point of expansion c, the value of x being considered, and the choice of n. The sequence of remainders $\{R_n(c, x)\}$ is a sequence of functions. It is these remainders that we drop from Taylor's remainder formula in equations 10.12 to form Taylor polynomials in equations 10.14. The polynomials in 10.14 will approximate $f(x)$ more and more closely as n increases if the neglected remainders $R_n(c, x)$

get smaller and smaller. In other words, to verify algebraically that the sequence of Taylor polynomials approximates a function more and more closely with increasing n on an interval I, it is sufficient to show that for each x in I,

$$\lim_{n \to \infty} R_n(c, x) = 0. \tag{10.16}$$

Interval I is called the **interval of convergence**. It is tedious to write groups of equations like 10.14 and say that, the more and more terms that are added, the better and better is the approximation. We invent a notation to stand for this. We write

$$f(x) = f(c) + f'(c)(x - c) + \frac{f''(c)}{2!}(x - c)^2 + \frac{f'''(c)}{3!}(x - c)^3 + \cdots. \tag{10.17}$$

The $\cdots$ indicate that additional terms of the same form are to be added indefinitely. This is called the **Taylor series** of $f(x)$ about the point c. Be clear on what we mean by 10.17, because it cannot be taken literally. There is an infinity of terms on the right side of the equation, and substitution of a value for x (and c) gives an infinity of numbers to add. No matter how fast you add, your calculator adds, or a super computer adds, an infinity of numbers cannot be added in a finite amount of time. Equation 10.17 does not therefore mean that adding the terms on the right gives $f(x)$. It means any one of the following three equivalent statements:

1. The more and more terms that are added on the right, the closer and closer the sum gets to $f(x)$.

2. By adding sufficiently many terms, we can make the sum on the right as close to $f(x)$ as desired.

3. If the Taylor series is truncated at any number of terms, the resulting Taylor polynomial approximates $f(x)$; the larger n, the better the approximation.

To use the terminology from Section 10.2, we say that the Taylor polynomials $P_n(x)$ converge to $f(x)$, meaning again that they get closer and closer to $f(x)$, the larger the value of n. We also say that the Taylor series converges to $f(x)$.

When $c = 0$, the Taylor series becomes

$$f(x) = f(0) + f'(0)x + \frac{f''(0)}{2!}x^2 + \frac{f'''(0)}{3!}x^3 + \cdots. \tag{10.18}$$

We call this the **Maclaurin series** for $f(x)$. It consists of powers of x rather than $x - c$.

To obtain the Taylor (or Maclaurin) series for a function $f(x)$, we evaluate all derivatives of $f(x)$ at $x = c$ (or $x = 0$), and use formula 10.17 (or 10.18). To show that the series converges to $f(x)$, we show that Taylor remainders approach 0, and we must find an interval I on which this happens. This can be quite difficult because $R_n(c, x)$ is defined in terms of z_n, an unknown value except that it must be between c and x. This problem is usually circumvented by finding a maximum value for $|R_n(c, x)|$, and showing that this maximum value approaches zero. We illustrate in the next three examples.

EXAMPLE 10.9

Find the Maclaurin series for $\sin x$ and show that it converges to $\sin x$ for all x.

SOLUTION   Using the derivatives of $\sin x$ in Example 10.6, we can write Taylor's remainder formula for $\sin x$ and $c = 0$,

$$\sin x = x - \frac{x^3}{3!} + \frac{x^5}{5!} - \frac{x^7}{7!} + \cdots + \frac{d^n}{dx^n}(\sin x)\Big|_{x=0}\frac{x^n}{n!} + R_n(0, x),$$

where

$$R_n(0, x) = \frac{d^{n+1}}{dx^{n+1}}(\sin x)\Big|_{x=z_n}\frac{x^{n+1}}{(n + 1)!}.$$

But the $(n+1)^{\text{th}}$ derivative of $\sin x$ is $\pm \sin x$ or $\pm \cos x$, so that

$$\left| \frac{d^{n+1}}{dx^{n+1}}(\sin x) \right|_{x=z_n} \leq 1.$$

Hence,

$$|R_n(0, x)| \leq \frac{|x|^{n+1}}{(n+1)!}.$$

But according to Example 10.5, $\lim_{n \to \infty} |x|^n / n! = 0$ for any x whatsoever, and therefore

$$\lim_{n \to \infty} |R_n(0, x)| = 0 \quad \Longrightarrow \quad \lim_{n \to \infty} R_n(0, x) = 0.$$

The Maclaurin series for $\sin x$ therefore converges to $\sin x$ for all x, and we may write

$$\sin x = x - \frac{x^3}{3!} + \frac{x^5}{5!} - \frac{x^7}{7!} + \cdots$$

$$= \sum_{n=0}^{\infty} \frac{(-1)^n}{(2n+1)!} x^{2n+1},$$

and this is true for all x; that is, the interval of convergence is $-\infty < x < \infty$. Let us not forget what we mean by writing the Maclaurin series for $\sin x$. If the series is truncated at any value of n, a Taylor polynomial is obtained. These polynomials approximate $\sin x$ more and more accurately for small x, and the interval on which the polynomials approximate $\sin x$ with any degree of accuracy gets larger and larger. We illustrated this graphically in Figure 10.11.

In a similar way it can be shown that the Maclaurin series for $\cos x$ is

$$\cos x = \sum_{n=0}^{\infty} \frac{(-1)^n}{(2n)!} x^{2n} = 1 - \frac{x^2}{2!} + \frac{x^4}{4!} - \frac{x^6}{6!} + \cdots, \quad -\infty < x < \infty.$$

Sometimes it is necessary to consider points on either side of the point of expansion separately. This is illustrated in the next example.

EXAMPLE 10.10

Find the Maclaurin series for e^x and show that it converges to e^x for all x.

SOLUTION   Since

$$\frac{d^n}{dx^n}(e^x) \bigg|_{x=0} = e^x \big|_{x=0} = 1,$$

Taylor's remainder formula for e^x and $c = 0$ gives

$$e^x = 1 + x + \frac{x^2}{2!} + \frac{x^3}{3!} + \cdots + \frac{x^n}{n!} + R_n(0, x),$$

where

$$R_n(0, x) = \frac{d^{n+1}}{dx^{n+1}}(e^x) \bigg|_{x=z_n} \frac{x^{n+1}}{(n+1)!} = e^{z_n} \frac{x^{n+1}}{(n+1)!}.$$

Now, if $x < 0$, then $x < z_n < 0$, and

$$|R_n(0, x)| < e^0 \frac{|x|^{n+1}}{(n+1)!},$$

which approaches zero as n becomes infinite (see Example 10.5). If $x > 0$, then $0 < z_n < x$, and

$$|R_n(0, x)| < e^x \frac{|x|^{n+1}}{(n+1)!},$$

which again has limit zero as n approaches infinity. Thus, for any x whatsoever, we can say that $\lim_{n \to \infty} R_n(0, x) = 0$, and the Maclaurin series for e^x converges to e^x:

$$e^x = 1 + x + \frac{x^2}{2!} + \frac{x^3}{3!} + \cdots = \sum_{n=0}^{\infty} \frac{1}{n!} x^n, \qquad -\infty < x < \infty.$$

Separate discussions for $x < 0$ and $x > 0$ are reflected in how the Taylor polynomials converge to e^x in Figure 10.12. For $x > 0$, polynomials are always less than e^x and increase toward e^x. For $x < 0$, polynomials are alternately larger and smaller than e^x, gradually approaching e^x.

The following example is more difficult. Remainders do not approach zero for all x so that the Taylor series does not converge to the function for all x. The problem would have been even more difficult had we not suggested values of x to consider.

EXAMPLE 10.11

Find the Taylor series for $\ln x$ about the point 1 and show that, for $1/2 \le x \le 2$, it converges to $\ln x$.

SOLUTION Using the derivatives of $\ln x$ at $c = 1$ in Example 10.8, we obtain Taylor's remainder formula for $\ln x$ and $c = 1$,

$$\ln x = (x - 1) - \frac{1}{2!}(x - 1)^2 + \frac{2!}{3!}(x - 1)^3 + \cdots + \frac{(-1)^{n+1}(n-1)!}{n!}(x - 1)^n + R_n(1, x),$$

where

$$R_n(1, x) = \frac{d^{n+1}}{dx^{n+1}}(\ln x)\Big|_{x=z_n} \frac{(x - 1)^{n+1}}{(n+1)!} = \frac{(-1)^n n!}{(z_n)^{n+1}} \frac{(x - 1)^{n+1}}{(n+1)!}$$

$$= \frac{(-1)^n}{n+1}\left(\frac{x-1}{z_n}\right)^{n+1},$$

and z_n is between 1 and x.

If $1 < x \le 2$, then the largest value of $x - 1$ is 1. Furthermore, z_n must be larger than 1. It follows that

$$|R_n(1, x)| < \frac{1}{n+1}\left(\frac{1}{1}\right)^{n+1} = \frac{1}{n+1}$$

and therefore

$$\lim_{n \to \infty} R_n(1, x) = 0.$$

If $1/2 \le x < 1$, then $-1/2 \le x - 1 < 0$. Combine this with $x < z_n < 1$, and we can state that $-1 < (x - 1)/z_n < 0$. Then,

$$|R_n(1, x)| < \frac{1}{n+1} \qquad \text{and} \qquad \lim_{n \to \infty} R_n(1, x) = 0.$$

Thus, for $1/2 \le x \le 2$, the sequence of remainders $\{R_n(1, x)\}$ approaches zero, and the Taylor series converges to $\ln x$ for those values of x:

$$\ln x = (x - 1) - \frac{1}{2}(x - 1)^2 + \frac{1}{3}(x - 1)^3 + \cdots$$

$$= \sum_{n=1}^{\infty} \frac{(-1)^{n+1}}{n}(x - 1)^n, \qquad \frac{1}{2} \leq x \leq 2.$$

This series actually converges to $\ln x$ on the larger interval $0 < x \leq 2$ as we saw graphically in Figure 10.13. We will show this algebraically in Example 10.22.

The above examples were chosen for their simplicity. It is simple to find n^{th} derivatives of $\sin x$, e^x, and $\ln x$; that remainders approach zero follows immediately. You may disagree that the examples were simple, but this only serves to emphasize the following point. If these are the simplest possible examples, and they are not simple, imagine the difficulty in finding and showing that Taylor series of other functions converge to the function. The problem is so great that we do our utmost to avoid calculating derivatives and considering remainders. Based on discussions in Section 10.4, we will find alternatives in Section 10.5.

EXERCISES 10.3

In Exercises 1–5 use Taylor's remainder formula to show that the Taylor series for the function $f(x)$ about the point indicated converges to $f(x)$ for all x. In each case, plot the first six Taylor polynomials to show how they approximate $f(x)$ more closely as the degree of the polynomial increases.

1. $f(x) = \cos x$ about $x = 0$

2. $f(x) = e^{5x}$ about $x = 0$

3. $f(x) = \sin(10x)$ about $x = 0$

4. $f(x) = \sin x$ about $x = \pi/4$

5. $f(x) = e^{2x}$ about $x = 1$

In Exercises 6–15 find the Taylor series for the function $f(x)$ about the point c. Plot enough polynomials and the function to determine the interval on which the series converges to the function.

6. $f(x) = e^{2x}$, $c = 0$

7. $f(x) = \cos 3x$, $c = 0$

8. $f(x) = \sin x$, $c = \pi/2$

9. $f(x) = 1/(1 - x)$, $c = 0$

10. $f(x) = 1/(2 - x)$, $c = 1$

11. $f(x) = x/(1 + 2x)$, $c = 0$

12. $f(x) = 1/(1 + 3x)^2$, $c = 0$

13. $f(x) = \ln x$, $c = 2$

14. $f(x) = \sqrt{1 + 3x}$, $c = 0$

15. $f(x) = 1/(4 + x)^{1/3}$, $c = 2$

* **16.** In Section 4.4 we stated the second-derivative test for determining whether a critical point x_0 at which $f'(x_0) = 0$ yields a relative maximum or a relative minimum. Use Taylor's remainder formula to verify this result when $f'(x)$ and $f''(x)$ are continuous on an open interval containing x_0.

* **17.** Extend the result of Exercise 16 to verify the extrema test of Exercise 35 in Section 4.4.

* **18.** There is an integral form for the remainder $R_n(c, x)$ in Taylor's remainder formula that is sometimes more useful than the derivative form in Theorem 10.1: If $f(x)$ and its first n derivatives are continuous on the closed interval between c and x, then

$$f(x) = f(c) + f'(c)(x - c) + \frac{f''(c)}{2!}(x - c)^2$$

$$+ \cdots + \frac{f^n(c)}{n!}(x - c)^n + R_n(c, x),$$

where

$$R_n(c, x) = \frac{1}{n!} \int_c^x (x - t)^n f^{(n+1)}(t)\, dt.$$

Use the following outline to prove this result.

(a) Show that

$$f(x) = f(c) + \int_c^x f'(t)\, dt.$$

(b) Use integration by parts with $u = f'(t)$, $du = f''(t)\, dt$, $dv = dt$, and $v = t - x$ on the integral in part (a) to obtain

$$f(x) = f(c) + f'(c)(x - c) + \int_c^x (x - t) f''(t)\, dt.$$

(c) Use integration by parts with $u = f''(t), du = f'''(t) dt$, $dv = (x - t) dt$, and $v = -(1/2)(x - t)^2$ to obtain

$$f(x) = f(c) + f'(c)(x - c) + \frac{f''(c)}{2!}(x - c)^2$$

$$+ \frac{1}{2!} \int_c^x (x - t)^2 f'''(t) \, dt.$$

(d) Continue this process to obtain the integral form for Taylor's remainder formula.

9. (a) Draw a graph of the function

$$f(x) = \begin{cases} e^{-1/x^2}, & x \neq 0, \\ 0, & x = 0. \end{cases}$$

(b) Use L'Hôpital's rule to show that for every positive integer n,

$$\lim_{x \to 0} \frac{e^{-1/x^2}}{x^n} = 0.$$

(c) Prove, by mathematical induction, that $f^{(n)}(0) = 0$ for $n \geq 1$.

(d) What is the Maclaurin series for $f(x)$?

(e) For what values of x does the Maclaurin series of $f(x)$ converge to $f(x)$?

10.4 Power Series

Our work on Taylor series in Section 10.3 began with a function and developed the Taylor (or Maclaurin) series for the function, a sequence of polynomials that approximate the function more and more closely as more and more terms are included. In this section we begin with a series called a power series and ask two questions: For what values of x does the power series converge, and to what function does it converge?

A **power series** in x is an (infinite) series of the form

$$\sum_{n=0}^{\infty} a_n x^n = a_0 + a_1 x + a_2 x^2 + \cdots + a_n x^n + \cdots, \tag{10.19}$$

where coefficients a_n are constants. For example,

$$\sum_{n=0}^{\infty} x^n = 1 + x + x^2 + x^3 + \cdots + x^n + \cdots$$

and

$$\sum_{n=0}^{\infty} \frac{(-1)^n}{2^n \, n!} x^n = 1 - \frac{x}{2 \, (1!)} + \frac{x^2}{2^2 \, (2!)} - \frac{x^3}{2^3 \, (3!)} + \cdots$$

are power series in x.

To say that the Taylor series

$$f(x) = f(c) + f'(c)(x - c) + \frac{f''(c)}{2!}(x - c)^2 + \cdots$$

of a function $f(x)$ converges to $f(x)$ is to say that the sequence of Taylor polynomials $\{P_n(x)\}$, where

$$P_n(x) = f(c) + f'(c)(x - c) + \cdots + \frac{f^{(n)}(c)}{n!}(x - c)^n,$$

converges to $f(x)$; that is, as more and more terms of the series are included, the closer and closer the sum gets to $f(x)$. We use the same idea to define what we mean by convergence of power series 10.19. Power series 10.19 is said to converge to a function $f(x)$, or have sum $f(x)$, on an interval I if the limit of the sequence $\{P_n(x)\}$ of polynomials where

$$P_n(x) = a_0 + a_1 x + \cdots + a_n x^n$$

is $f(x)$ for each x in I. In such a case we write

$$f(x) = \sum_{n=0}^{\infty} a_n x^n = a_0 + a_1 x + \cdots + a_n x^n + \cdots, \qquad x \text{ in } I, \qquad (10.20)$$

and say that the power series converges to $f(x)$, or has sum $f(x)$. Once again what we mean is that as more and more terms on the right are included, the closer and closer the sum gets to $f(x)$ on the interval I. We call I the **interval of convergence** for the power series.

There are two aspects to convergence for power series, interval I and sum $f(x)$. It is relatively simple to find the interval of convergence for a power series, but much more difficult to find its sum.

One of the easiest, and perhaps the most important, power series is

$$\sum_{n=0}^{\infty} a x^n = a + ax + ax^2 + ax^3 + \cdots, \qquad (10.21)$$

where a is some nonzero constant. It is so important in applications that it is given a special name; it is called a geometric series. A **geometric series** is a series in which every term after the first is obtained by multiplying the previous term by the same amount, called the **common ratio**. For geometric series 10.21, the common ratio is x; each term is x times the previous term. To show that the series converges we must show that the sequence of polynomials

$$P_n(x) = a + ax + ax^2 + \cdots + ax^n$$

has a limit. If we multiply $P_n(x)$ by x, then

$$x\, P_n(x) = ax + ax^2 + ax^3 + \cdots + ax^{n+1}.$$

When these are subtracted,

$$P_n(x) - x\, P_n(x) = a - ax^{n+1},$$

from which

$$P_n(x) = \frac{a(1 - x^{n+1})}{1 - x}. \qquad (10.22)$$

This is a simplified form for $P_n(x)$. Since

$$\lim_{n \to \infty} x^{n+1} = \begin{cases} 0, & -1 < x < 1 \\ 1, & x = 1 \\ \text{does not exist,} & \text{otherwise} \end{cases}$$

it follows that when the sequence $\{P_n(x)\}$ is restricted to the interval $|x| < 1$, it has a limit,

$$\lim_{n \to \infty} P_n(x) = \frac{a}{1 - x}, \qquad |x| < 1.$$

We can therefore write that

$$\sum_{n=0}^{\infty} x^n = a + ax + ax^2 + \cdots = \frac{a}{1 - x}, \qquad |x| < 1. \qquad (10.23)$$

The geometric series has sum $a/(1 - x)$ on the interval (of convergence) $|x| < 1$. It does not have a sum for $|x| \geq 1$.

Figures 10.16 show some of the polynomials $P_n(x)$ and $a/(1 - x)$ when $a = 1$. They illustrate how polynomials approximate $1/(1 - x)$ more closely as n increases. Polynomials are defined for all x and $1/(1 - x)$ is defined for all $x \neq 1$, but curves $y = P_n(x)$ approach $y = 1/(1 - x)$ only for $|x| < 1$. For $x > 0$, polynomials approach $1/(1 - x)$ from below; for $x < 0$, they oscillate about $1/(1 - x)$, but gradually approach $1/(1 - x)$.

FIGURE 10.16a

Partial sums of geometric series $1 + x + x^2 + \cdots$

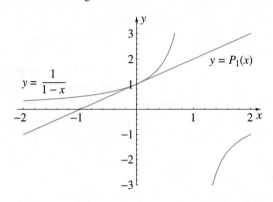

FIGURE 10.16b

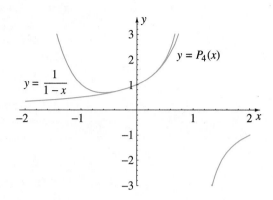

FIGURE 10.16c

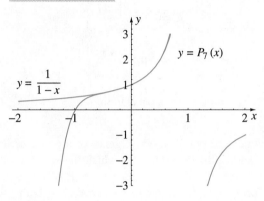

FIGURE 10.16d

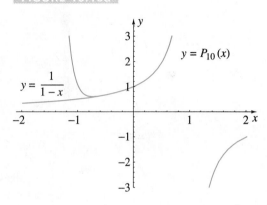

Obviously, the interval of convergence for a power series $\sum_{n=0}^{\infty} a_n x^n$ always includes the value $x = 0$, but what other possibilities are there? For geometric series 10.23, the interval of convergence is the interval $-1 < x < 1$. The terminology itself, *interval of convergence*, suggests that the values of x for which a power series converges form some kind of interval. This is indeed true, as we shall soon see.

Every Maclaurin series is a power series in x. Examples 10.9 and 10.10 developed Maclaurin series for $\sin x$ and e^x,

$$\sin x = x - \frac{x^3}{3!} + \frac{x^5}{5!} + \cdots, \qquad e^x = 1 + x + \frac{x^2}{2!} + \frac{x^3}{3!} + \cdots.$$

They are power series that add to the functions with intervals of convergence $-\infty < x < \infty$. The power series

$$\sum_{n=0}^{\infty} n!\, x^n = 1 + x + 2!\, x^2 + 3!\, x^3 + \cdots$$

converges only for $x = 0$. This is a direct result of the fact that

$$\lim_{n \to \infty} n!\, |x|^n = \infty$$

for any given nonzero value of x. How, then, could the addition of more and more terms ever converge? These examples have illustrated that there are at least three possible types of intervals of convergence for power series $\sum_{n=0}^{\infty} a_n x^n$:

1. The power series converges only for $x = 0$;

2. The power series converges for all x;

3. There exists a number $R > 0$ such that the power series converges for $|x| < R$, diverges for $|x| > R$, and may or may not converge for $x = \pm R$.

These are in fact the only possibilities for an interval of convergence. In 3 we call R the **radius of convergence** of the power series. It is half the length of the interval of convergence, or the distance we may proceed in either direction along the x-axis from $x = 0$ and expect convergence of the power series, with the possible exceptions of $x = \pm R$. In order to have a radius of convergence associated with every power series, we say in 1 and 2 above that $R = 0$ and $R = \infty$, respectively.

Every power series $\sum_{n=0}^{\infty} a_n x^n$ now has a radius of convergence R. If $R = 0$, the power series converges only for $x = 0$; if $R = \infty$, the power series converges for all x; and if $0 < R < \infty$, the power series converges for $|x| < R$, diverges for $|x| > R$, and may or may not converge for $x = \pm R$. For many power series the radius of convergence can be calculated according to the following theorem (see Section 10.12 for a proof).

THEOREM 10.2

The radius of convergence of a power series $\sum_{n=0}^{\infty} a_n x^n$ is given by

$$R = \lim_{n \to \infty} \left| \frac{a_n}{a_{n+1}} \right| \qquad \text{or} \qquad (10.24a)$$

$$R = \lim_{n \to \infty} \frac{1}{\sqrt[n]{|a_n|}}, \qquad (10.24b)$$

provided that either limit exists or is equal to infinity.

When we know that the radius of convergence of a power series $\sum_{n=0}^{\infty} a_n x^n$ is R, and $0 < R < \infty$, we know all values of x for which the series converges with the exception of two values of x. The series converges for $|x| < R$, diverges for $|x| > R$, and may or may not converge for $x = \pm R$. Some series converge at both endpoints, some at neither, and some at one end but not the other. There is no simple test that can distinguish among these situations for all power series. Every power series must be checked individually as to whether it converges at the endpoints of its interval of convergence. At $x = \pm R$, a power series reduces to a series of numbers. Sections 10.9–10.12 contain tests that determine whether series of numbers converge or diverge, and therefore testing whether endpoints of power series with finite radii of convergence can be included in the intervals of convergence will have to wait until we have covered this material. In the meantime, we will call the interval of convergence the **open interval of convergence** whenever we have not tested for inclusion of endpoints.

When Theorem 10.2 determines the radius of convergence of a power series, it also determines the open interval of convergence. We illustrate in the following examples.

EXAMPLE 10.12

Find the open interval of convergence for the power series $\displaystyle\sum_{n=1}^{\infty} \frac{(-1)^n}{n \, 5^{2n}} x^n$.

SOLUTION   Since

$$R = \lim_{n \to \infty} \left| \frac{a_n}{a_{n+1}} \right| = \lim_{n \to \infty} \left| \frac{\dfrac{(-1)^n}{n \, 5^{2n}}}{\dfrac{(-1)^{n+1}}{(n+1)5^{2n+2}}} \right| = \lim_{n \to \infty} 25 \left(\frac{n+1}{n} \right) = 25,$$

the open interval of convergence is $-25 < x < 25$. The series converges for these values of x, diverges for $x < -25$ and $x > 25$, but we do not know whether it converges for $x = -25$ or $x = 25$.

EXAMPLE 10.13

Find the open interval of convergence for the power series $\sum_{n=0}^{\infty} \frac{(n!)^2}{(2n)!} x^n$.

SOLUTION Since

$$
R = \lim_{n\to\infty} \left| \frac{a_n}{a_{n+1}} \right| = \lim_{n\to\infty} \frac{\frac{(n!)^2}{(2n)!}}{\frac{[(n+1)!]^2}{(2n+2)!}} = \lim_{n\to\infty} \left[\frac{(n!)^2}{(2n)!} \frac{(2n+2)(2n+1)(2n)!}{(n+1)^2(n!)^2} \right]
$$

$$
= \lim_{n\to\infty} \frac{(2n+2)(2n+1)}{(n+1)^2} = 4,
$$

the open interval of convergence is $-4 < x < 4$.

EXAMPLE 10.14

Find the open interval of convergence for the power series $\sum_{n=1}^{\infty} \frac{1}{n^n} x^n$.

SOLUTION Since

$$
R = \lim_{n\to\infty} \frac{1}{\sqrt[n]{|a_n|}} = \lim_{n\to\infty} \frac{1}{\sqrt[n]{1/n^n}} = \lim_{n\to\infty} n = \infty,
$$

the series converges for all x. Its interval of convergence is $-\infty < x < \infty$, which is not just its open interval.

EXAMPLE 10.15

Find the open interval of convergence for the power series $\sum_{n=1}^{\infty} \frac{1}{n^2 2^n} x^{2n+1}$.

SOLUTION Since coefficients of even powers of x are 0, the sequence $\{a_n/a_{n+1}\}$ is not defined. We cannot therefore find its radius of convergence directly using Theorem 10.2. Instead, we write

$$
\sum_{n=1}^{\infty} \frac{1}{n^2 2^n} x^{2n+1} = x \sum_{n=1}^{\infty} \frac{1}{n^2 2^n} (x^2)^n
$$

and set $y = x^2$ in the series:

$$
\sum_{n=1}^{\infty} \frac{1}{n^2 2^n} (x^2)^n = \sum_{n=1}^{\infty} \frac{1}{n^2 2^n} y^n.
$$

According to equation 10.24a, the radius of convergence of this series in y is

$$
R_y = \lim_{n\to\infty} \frac{\frac{1}{n^2 2^n}}{\frac{1}{(n+1)^2 2^{n+1}}} = \lim_{n\to\infty} 2 \left(\frac{n+1}{n} \right)^2 = 2.
$$

Since $x = \pm\sqrt{y}$, it follows that the radius of convergence of the power series in x is $R_x = \sqrt{2}$. The open interval of convergence is therefore $-\sqrt{2} < x < \sqrt{2}$.

Sums of Power Series

Finding the interval of convergence of a power series is only half the convergence problem. The other half is finding the sum of the power series. What function does the power series approximate as more and more terms are included? Sometimes it is possible to relate a given series to a series with known sum. Four very important series with known sums are

$$\frac{a}{1 - x} = a + ax + ax^2 + ax^3 + \cdots, \quad -1 < x < 1, \quad (10.25a)$$

$$e^x = 1 + x + \frac{x^2}{2!} + \frac{x^3}{3!} + \cdots, \quad -\infty < x < \infty, \quad (10.25b)$$

$$\sin x = x - \frac{x^3}{3!} + \frac{x^5}{5!} - \frac{x^7}{7!} + \cdots, \quad -\infty < x < \infty, \quad (10.25c)$$

$$\cos x = 1 - \frac{x^2}{2!} + \frac{x^4}{4!} - \frac{x^6}{6!} + \cdots, \quad -\infty < x < \infty. \quad (10.25d)$$

EXAMPLE 10.16   ▼

Find the sum of the power series $\displaystyle\sum_{n=0}^{\infty} \frac{(-1)^{n+1}}{2^n} x^n$.

SOLUTION We write

$$\sum_{n=0}^{\infty} \frac{(-1)^{n+1}}{2^n} x^n = \sum_{n=0}^{\infty} (-1) \left(\frac{-x}{2} \right)^n$$

$$= -1 + \frac{1}{2}x - \frac{1}{2^2}x^2 + \frac{1}{2^3}x^3 - \cdots,$$

and note that this is a geometric series with first term $a = -1$ and common ratio $-x/2$. If we therefore replace x by $-x/2$ in equation 10.25a, we have

$$\sum_{n=0}^{\infty} \frac{(-1)^{n+1}}{2^n} x^n = \frac{-1}{1 - (-x/2)} = \frac{-2}{x + 2}.$$

This is valid for $-1 < -x/2 < 1$, or, $-2 < x < 2$.

▲

EXAMPLE 10.17   ▼

Find the sum of the power series $\displaystyle\sum_{n=0}^{\infty} \frac{(-1)^n 2^n}{(2n)!} x^{2n+2}$.

SOLUTION The series can be expressed in the form

$$\sum_{n=0}^{\infty} \frac{(-1)^n 2^n}{(2n)!} x^{2n+2} = x^2 \sum_{n=0}^{\infty} \frac{(-1)^n}{(2n)!} \left(\sqrt{2}x \right)^{2n}$$

$$= x^2 \left[1 - \frac{(\sqrt{2}x)^2}{2!} + \frac{(\sqrt{2}x)^4}{4!} - \frac{(\sqrt{2}x)^6}{6!} + \cdots \right].$$

It is x^2 times the Maclaurin series for $\cos x$ with x replaced by $\sqrt{2}\,x$. In other words,

$$\sum_{n=0}^{\infty} \frac{(-1)^n 2^n}{(2n)!} x^{2n+2} = x^2 \cos \sqrt{2}\,x,$$

valid for $-\infty < \sqrt{2}\,x < \infty$, or $-\infty < x < \infty$.

Other methods for finding sums of power series are discussed in Section 10.6.

For many power series there is no known function to which the power series converges. In such cases we write

$$f(x) = \sum_{n=0}^{\infty} a_n x^n \tag{10.26}$$

and say that the power series defines the value of $f(x)$ at each x in the interval of convergence. For instance, a very important series in engineering and physics is

$$\sum_{n=0}^{\infty} \frac{(-1)^n}{2^{2n}(n!)^2} x^{2n},$$

which converges for all x. It arises so often in applications that it is given a special name, the *Bessel function of the first kind of order zero*, and is denoted by $J_0(x)$. In other words, we write

$$J_0(x) = \sum_{n=0}^{\infty} \frac{(-1)^n}{2^{2n}(n!)^2} x^{2n},$$

and say that the power series defines the value of $J_0(x)$ for each x. We get an idea of what $J_0(x)$ looks like by plotting the polynomial approximations obtained by adding more and more terms of the series. Figure 10.17 shows the first five polynomials for $x \geq 0$:

$$P_1(x) = 1,$$

$$P_2(x) = 1 - \frac{x^2}{2^2},$$

$$P_3(x) = 1 - \frac{x^2}{2^2} + \frac{x^4}{2^4(2!)^2},$$

$$P_4(x) = 1 - \frac{x^2}{2^2} + \frac{x^4}{2^4(2!)^2} - \frac{x^6}{2^6(3!)^2},$$

$$P_5(x) = 1 - \frac{x^2}{2^2} + \frac{x^4}{2^4(2!)^2} - \frac{x^6}{2^6(3!)^2} + \frac{x^8}{2^8(4!)^2}.$$

They are even functions, so their graphs would be symmetric about the y-axis. As the number of terms is increased, the polynomials approximate $J_0(x)$ in Figure 10.18 more and more closely.

We have called 10.19 a power series in x. It is also said to be a *power series about* 0 (meaning the point 0 on the x-axis), where we note that for any interval of convergence whatsoever, 0 is always at its centre. This suggests that power series about other points might be considered, and this is indeed the case. The general power series about a point c on the x-axis is

$$\sum_{n=0}^{\infty} a_n (x - c)^n = a_0 + a_1(x - c) + a_2(x - c)^2 + \cdots \tag{10.27}$$

and is said to be a power series in $x - c$.

FIGURE 10.17   Polynomial approximations for $J_0(x)$

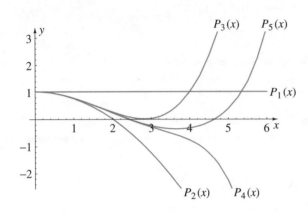

FIGURE 10.18   Bessel's function $J_0(x)$ of the first kind of order zero

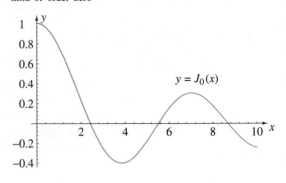

A power series in $x-c$ has an interval of convergence and a radius of convergence analogous to a power series in x. In particular, every power series in $x - c$ has a radius of convergence R such that if $R = 0$, the power series converges only for $x = c$; if $R = \infty$, the power series converges for all x; and if $0 < R < \infty$, the series converges for $|x - c| < R$, diverges for $|x - c| > R$, and may or may not converge for $x = c \pm R$. The radius of convergence is again given by equations 10.24, provided that the limits exist or are equal to infinity. For power series in $x - c$, then, the point c is the centre of the interval of convergence.

EXAMPLE 10.18

Find the open interval of convergence of the power series

$$2^2(x + 2) + 2^3(x + 2)^2 + 2^4(x + 2)^3 + \cdots + 2^{n+1}(x + 2)^n + \cdots.$$

SOLUTION   By writing the power series in sigma notation,

$$\sum_{n=1}^{\infty} 2^{n+1}(x + 2)^n,$$

we can use equation 10.24a (or 10.24b) to calculate its radius of convergence:

$$R = \lim_{n \to \infty} \left| \frac{a_n}{a_{n+1}} \right| = \lim_{n \to \infty} \frac{2^{n+1}}{2^{n+2}} = \frac{1}{2}.$$

Since this is a power series about -2, the open interval of convergence is $-5/2 < x < -3/2$. We can do better. This is a geometric series with first term $4(x+2)$ and common ratio $2(x+2)$. According to equation 10.25a, its sum is

$$\sum_{n=1}^{\infty} 2^{n+1}(x + 2)^n = \frac{4(x + 2)}{1 - 2(x + 2)} = \frac{-4(x + 2)}{2x + 3},$$

valid for $-1 < 2(x + 2) < 1 \implies -5/2 < x < -3/2$. This is now known to be the interval of convergence, not just the open interval of convergence.

EXERCISES 10.4

In Exercises 1–25 find the open interval of convergence for the power series.

1. $\displaystyle\sum_{n=1}^{\infty}\frac{1}{n}x^n$

2. $\displaystyle\sum_{n=1}^{\infty}n^2x^n$

3. $\displaystyle\sum_{n=0}^{\infty}\frac{1}{(n+1)^3}x^n$

4. $\displaystyle\sum_{n=0}^{\infty}n^23^nx^n$

5. $\displaystyle\sum_{n=0}^{\infty}\frac{1}{2^n}(x-1)^n$

6. $\displaystyle\sum_{n=4}^{\infty}(-1)^nn^3(x+3)^n$

7. $\displaystyle\sum_{n=1}^{\infty}\frac{1}{\sqrt{n}}(x+2)^n$

8. $\displaystyle\sum_{n=2}^{\infty}2^n\left(\frac{n-1}{n+2}\right)^2(x-4)^n$

9. $\displaystyle\sum_{n=1}^{\infty}\frac{1}{n^2}x^{2n}$

∗ 10. $\displaystyle\sum_{n=0}^{\infty}(-1)^nx^{3n}$

11. $\displaystyle\sum_{n=0}^{\infty}\frac{n-1}{n+1}(2x)^n$

∗ 12. $\displaystyle\sum_{n=0}^{\infty}\frac{1}{\sqrt{n+1}}x^{3n+1}$

13. $\displaystyle\sum_{n=0}^{\infty}\frac{(-1)^n}{3^n}x^{2n+1}$

∗ 14. $\displaystyle\sum_{n=1}^{\infty}\frac{(-e)^n}{n^2}x^n$

15. $\displaystyle\frac{x}{9}+\frac{4}{3^4}x^2+\frac{9}{3^6}x^3+\cdots+\frac{n^2}{3^{2n}}x^n+\cdots$

16. $x+2^2x^2+3^3x^3+\cdots+n^nx^n+\cdots$

17. $\displaystyle\frac{1}{36}(x+10)^6+\frac{1}{49}(x+10)^7+\frac{1}{64}(x+10)^8$
$\cdots+\frac{1}{n^2}(x+10)^n+\cdots$

18. $3x+8(3x)^2+27(3x)^3+\cdots+n^3(3x)^n+\cdots$

19. $\displaystyle\frac{3}{4}x^2+x^4+\frac{27}{16}x^6+\cdots+\frac{3^n}{(n+1)^2}x^{2n}+\cdots$

20. $\displaystyle 1+\frac{1}{5}x^3+\frac{1}{25}x^6+\cdots+\frac{1}{5^n}x^{3n}+\cdots$

21. $\displaystyle\sum_{n=2}^{\infty}\frac{1}{\ln n}x^n$

22. $\displaystyle\sum_{n=2}^{\infty}\frac{1}{n^2\ln n}x^n$

23. $\displaystyle\sum_{n=1}^{\infty}\frac{(n!)^3}{(3n)!}x^n$

24. $\displaystyle\sum_{n=1}^{\infty}\frac{2\cdot4\cdot6\cdot\ \ldots\ \cdot(2n)}{3\cdot5\cdot7\cdot\ \ldots\ \cdot(2n+1)}x^n$

25. $\displaystyle\sum_{n=1}^{\infty}\frac{[1\cdot3\cdot5\cdot\ \ldots\ \cdot(2n+1)]^2}{2^{2n}(2n)!}x^n$

In Exercises 26–37 find the sum of the power series.

26. $\displaystyle\sum_{n=0}^{\infty}\frac{1}{4^n}x^{3n}$

27. $\displaystyle\sum_{n=1}^{\infty}(-e)^nx^n$

28. $\displaystyle\sum_{n=1}^{\infty}\frac{1}{3^{2n}}(x-1)^n$

29. $\displaystyle\sum_{n=2}^{\infty}(x+5)^{2n}$

∗ 30. $\displaystyle\sum_{n=0}^{\infty}\frac{(-1)^n}{(2n)!}x^{4n}$

∗ 31. $\displaystyle\sum_{n=0}^{\infty}\frac{5^n}{n!}x^n$

∗ 32. $\displaystyle\sum_{n=0}^{\infty}\frac{(-1)^n}{3^{2n+1}(2n+1)!}x^{2n+2}$

∗ 33. $\displaystyle\sum_{n=0}^{\infty}\frac{(-3)^n}{n!}(x+1)^n$

∗ 34. $\displaystyle\sum_{n=1}^{\infty}\frac{(-1)^n}{n!}x^n$

∗ 35. $\displaystyle\sum_{n=0}^{\infty}\frac{(-1)^{n+1}}{(2n+1)!}(x+1)^{2n+3}$

∗ 36. $\displaystyle\sum_{n=0}^{\infty}\frac{2^n}{n!}(x-1/2)^n$

∗ 37. $\displaystyle\sum_{n=0}^{\infty}\frac{(-1)^n}{2^{2n}(2n)!}x^{4n+4}$

∗ 38. If m is a nonnegative integer, the Bessel function of order m of the first kind is defined by the power series

$$J_m(x)=\sum_{n=0}^{\infty}\frac{(-1)^n}{2^{2n+m}n!\,(n+m)!}x^{2n+m}.$$

(a) Write out the first five terms of $J_0(x)$, $J_1(x)$, and $J_m(x)$.

(b) Find the interval of convergence for each $J_m(x)$.

∗ 39. The hypergeometric series is

$$1+\frac{\alpha\beta}{\gamma}x+\frac{\alpha(\alpha+1)\beta(\beta+1)}{2!\,\gamma(\gamma+1)}x^2$$
$$+\frac{\alpha(\alpha+1)(\alpha+2)\beta(\beta+1)(\beta+2)}{3!\,\gamma(\gamma+1)(\gamma+2)}x^3+\cdots,$$

where α, β, and γ are all constants.

(a) Write this series in sigma notation.

(b) What is the radius of convergence of the hypergeometric series if γ is not zero or a negative integer?

10.5 Taylor Series Expansions of Functions

Let us summarize what we have seen in Sections 10.3 and 10.4. Given a function $f(x)$ and a point c, the Taylor series of $f(x)$ about c is

$$f(x) = \sum_{n=0}^{\infty} \frac{f^{(n)}(c)}{n!}(x-c)^n,$$

and the series converges to $f(x)$ at all values of x for which Taylor remainders approach zero. Given a power series $\sum_{n=0}^{\infty} a_n(x-c)^n$, there is an interval of convergence inside of which the series has a sum. If this sum is $f(x)$, we write

$$f(x) = \sum_{n=0}^{\infty} a_n(x-c)^n, \quad x \text{ in the interval of convergence.}$$

Clearly, Taylor series are power series. A power series would be a Taylor series if a_n were equal to $f^{(n)}(c)/n!$. If this were the case, then power series and Taylor series would be one and the same. The following theorem allows us to prove this.

THEOREM 10.3

If $f(x) = \sum_{n=0}^{\infty} a_n(x-c)^n$, and the radius of convergence R is greater than zero, then each of the following series has radius of convergence R:

$$f'(x) = \sum_{n=0}^{\infty} na_n(x-c)^{n-1}, \tag{10.28a}$$

$$\int f(x)\,dx = \sum_{n=0}^{\infty} \frac{a_n}{n+1}(x-c)^{n+1} + C. \tag{10.28b}$$

Due to the difficulty in proving this theorem, and in order to preserve the continuity of our discussion, we omit a proof. Note that the theorem is stated in terms of radii of convergence rather than intervals of convergence. This is due to the fact that in differentiating a power series, we may lose the endpoints of the original interval of convergence, and in integrating we may pick them up. It could be stated in terms of open intervals of convergence, however: term-by-term differentiation and integration of power series preserve open intervals of convergence.

The next theorem implies that power series and Taylor series are one and the same, that every power series is a Taylor series, the Taylor series of its sum.

THEOREM 10.4

If $f(x)$ is the sum of the power series $\sum_{n=0}^{\infty} a_n(x-c)^n$ with $R > 0$, then the series is the Taylor series of $f(x)$.

PROOF   When we set $x = c$ in

$$f(x) = \sum_{n=0}^{\infty} a_n(x-c)^n = a_0 + a_1(x-c) + a_2(x-c)^2 + \cdots,$$

we obtain $f(c) = a_0$. If we differentiate the power series according to Theorem 10.3, we obtain

$$f'(x) = a_1 + 2a_2(x-c) + 3a_3(x-c)^2 + \cdots.$$

When we substitute $x = c$, the result is

$$f'(c) = a_1.$$

If we differentiate the power series for $f'(x)$, we obtain

$$f''(x) = 2a_2 + 3 \cdot 2a_3(x - c) + 4 \cdot 3a_4(x - c)^2 + \cdots,$$

and substitute $x = c$,

$$f''(c) = 2a_2 \quad \text{or} \quad a_2 = \frac{f''(c)}{2!}.$$

Continued differentiation and substitution leads to the result that for all n,

$$a_n = \frac{f^{(n)}(c)}{n!}.$$

The power series $f(x) = \sum_{n=0}^{\infty} a_n(x - c)^n$ is therefore the Taylor series of $f(x)$. ∎

The following corollary is an immediate consequence of this theorem.

COROLLARY 10.4.1

If two power series $\sum_{n=0}^{\infty} a_n(x - c)^n$ and $\sum_{n=0}^{\infty} b_n(x - c)^n$ with positive radii of convergence have identical sums,

$$\sum_{n=0}^{\infty} a_n(x - c)^n = \sum_{n=0}^{\infty} b_n(x - c)^n,$$

then $a_n = b_n$ for all n.

Theorem 10.4 shows that Sections 10.3 and 10.4 were dealing with the same problem but coming at it from different directions. In Section 10.3, $f(x)$ and $x = c$ were given, and we developed the Taylor series for $f(x)$ about $x = c$. Theorem 10.4 shows that this is the only power series for $f(x)$ about $x = c$. In Section 10.4, a power series, $\sum_{n=0}^{\infty} a_n(x - c)^n$, was given, and we determined its interval of convergence and sum $f(x)$. Theorem 10.4 indicates that the power series is actually the Taylor series of $f(x)$ about $x = c$.

This equivalence of power series and Taylor series simplifies the problem of finding the Taylor series for a function $f(x)$ about a point $x = c$ immeasurably. Instead of finding $f^{(n)}(c)$ and showing that Taylor remainders approach zero (as we did is Section 10.3), we can proceed as follows. If, by any method whatsoever, we can find a power series $\sum_{n=0}^{\infty} a_n(x - c)^n$ that has sum $f(x)$, then it must be the Taylor series of $f(x)$ about $x = c$. In the remainder of this section we show how easy it is to do this. In essence, we take series with known sums, such as 10.25, and construct other series from them.

EXAMPLE 10.19

▼

Find (a) the Maclaurin series for $1/(4 + 5x)$ and (b) the Taylor series about 5 for $1/(13 - 2x)$.

SOLUTION

(a) We write

$$\frac{1}{4 + 5x} = \frac{1}{4\left(1 + \dfrac{5x}{4}\right)} = \frac{1/4}{1 + \dfrac{5x}{4}}$$

and interpret the right side as the sum of a geometric series with first term $1/4$ and common ratio $-5x/4$. Equation 10.25a then gives

$$\frac{1}{4+5x} = \sum_{n=0}^{\infty} \left(\frac{1}{4}\right)\left(-\frac{5x}{4}\right)^n, \qquad \left|-\frac{5x}{4}\right| < 1$$

$$= \sum_{n=0}^{\infty} \frac{(-1)^n 5^n}{4^{n+1}} x^n, \qquad |x| < \frac{4}{5}.$$

(b) By a similar procedure, we have

$$\frac{1}{13-2x} = \frac{1}{3-2(x-5)} = \frac{1}{3\left[1-\frac{2}{3}(x-5)\right]} = \frac{1/3}{1-\frac{2}{3}(x-5)}$$

$$= \sum_{n=0}^{\infty} (1/3)\left[\frac{2}{3}(x-5)\right]^n, \qquad \left|\frac{2}{3}(x-5)\right| < 1,$$

$$= \sum_{n=0}^{\infty} \frac{2^n}{3^{n+1}}(x-5)^n, \qquad |x-5| < \frac{3}{2}.$$

In both examples, properties of geometric series gave not only the required series, but also their intervals of convergence. To appreciate the simplicity of these solutions, we suggest using Taylor remainders in an attempt to obtain the series with the same intervals of convergence. You will quickly abort.

Addition and Subtraction of Power Series

According to the following theorem, convergent power series can be added and subtracted in their common interval of convergence.

THEOREM 10.5

If $f(x) = \sum_{n=0}^{\infty} a_n(x-c)^n$ and $g(x) = \sum_{n=0}^{\infty} b_n(x-c)^n$ have positive radii of convergence, then

$$f(x) \pm g(x) = \sum_{n=0}^{\infty} (a_n \pm b_n)(x-c)^n, \qquad (10.29)$$

valid for every x that is common to the intervals of convergence of the two series.

We use this result in the following example.

EXAMPLE 10.20

Find the Maclaurin series for $f(x) = 5x/(x^2 - 3x - 4)$.

SOLUTION   We decompose $f(x)$ into its partial fractions,

$$f(x) = \frac{5x}{x^2 - 3x - 4} = \frac{4}{x - 4} + \frac{1}{x + 1},$$

and expand each of these terms in a Maclaurin series,

$$\frac{4}{x - 4} = \frac{-1}{1 - \dfrac{x}{4}} = -\left(1 + \frac{x}{4} + \frac{x^2}{4^2} + \cdots\right), \qquad |x| < 4, \qquad \text{and}$$

$$\frac{1}{1 + x} = 1 - x + x^2 - x^3 + \cdots, \qquad |x| < 1.$$

Addition of these series within their common interval of convergence gives the Maclaurin series for $f(x)$:

$$\frac{5x}{x^2 - 3x - 4} = \left(-1 - \frac{x}{4} - \frac{x^2}{4^2} - \frac{x^3}{4^3} - \cdots\right) + \left(1 - x + x^2 - x^3 + \cdots\right)$$

$$= \left(-1 - \frac{1}{4}\right)x + \left(1 - \frac{1}{4^2}\right)x^2 + \left(-1 - \frac{1}{4^3}\right)x^3 + \cdots$$

$$= \sum_{n=1}^{\infty}\left[(-1)^n - \frac{1}{4^n}\right]x^n, \qquad |x| < 1.$$

Differentiation and Integration of Power Series

Perhaps the most powerful technique for generating Taylor series is to differentiate or integrate known expansions according to Theorem 10.3.

EXAMPLE 10.21

Find Maclaurin series for the following functions:

$$\text{(a)} \ \cos x \qquad \qquad \text{(b)} \ \frac{x}{(2 - x)^3}$$

SOLUTION

(a) We derived the Maclaurin series for $\sin x$ in Example 10.9,

$$\sin x = x - \frac{x^3}{3!} + \frac{x^5}{5!} - \frac{x^7}{7!} + \cdots.$$

We then stated that the Maclaurin series for $\cos x$ could be derived in a similar way. Term-by-term differentiation of the sine series is faster,

$$\cos x = \frac{d}{dx}(\sin x) = \frac{d}{dx}\left(x - \frac{x^3}{3!} + \frac{x^5}{5!} - \frac{x^7}{7!} + \cdots\right)$$

$$= 1 - \frac{x^2}{2!} + \frac{x^4}{4!} - \frac{x^6}{6!} + \cdots$$

$$= \sum_{n=0}^{\infty} \frac{(-1)^n}{(2n)!} x^{2n}, \qquad -\infty < x < \infty.$$

(b) We begin with the Maclaurin series for $1/(2-x)$,

$$\frac{1}{2-x} = \frac{1}{2\left(1-\frac{x}{2}\right)} = \frac{1}{2}\left[1 + \left(\frac{x}{2}\right) + \left(\frac{x}{2}\right)^2 + \cdots\right]$$

$$= \frac{1}{2} + \frac{x}{2^2} + \frac{x^2}{2^3} + \frac{x^3}{2^4} + \cdots, \qquad |x| < 2.$$

Term-by-term differentiation of this series gives

$$\frac{1}{(2-x)^2} = \frac{1}{2^2} + \frac{2x}{2^3} + \frac{3x^2}{2^4} + \frac{4x^3}{2^5} + \cdots,$$

with open interval of convergence $|x| < 2$. Another differentiation yields

$$\frac{2}{(2-x)^3} = \frac{2}{2^3} + \frac{3 \cdot 2x}{2^4} + \frac{4 \cdot 3x^2}{2^5} + \cdots, \qquad |x| < 2.$$

Multiplication by $x/2$ now gives

$$\frac{x}{(2-x)^3} = \frac{2x}{2^4} + \frac{3 \cdot 2x^2}{2^5} + \frac{4 \cdot 3x^3}{2^6} + \cdots$$

$$= \sum_{n=1}^{\infty} \frac{n(n+1)}{2^{n+3}}x^n, \qquad |x| < 2.$$

EXAMPLE 10.22

Find the Taylor series about 1 for $\ln x$.

SOLUTION   Noting that $\ln x$ is an antiderivative of $1/x$, we first expand $1/x$ in a Taylor series about 1:

$$\frac{1}{x} = \frac{1}{(x-1)+1} = 1 - (x-1) + (x-1)^2 - (x-1)^3 + \cdots, \qquad |x-1| < 1.$$

If we integrate this series term by term, we have

$$\ln|x| = \left[x - \frac{1}{2}(x-1)^2 + \frac{1}{3}(x-1)^3 - \frac{1}{4}(x-1)^4 + \cdots\right] + C.$$

Substitution of $x = 1$ implies that $0 = 1 + C$; that is, $C = -1$, and hence,

$$\ln|x| = (x-1) - \frac{1}{2}(x-1)^2 + \frac{1}{3}(x-1)^3 - \cdots$$

$$= \sum_{n=1}^{\infty} \frac{(-1)^{n+1}}{n}(x-1)^n.$$

According to Theorem 10.3, the radius of convergence of this series is also $R = 1$; that is, the open interval of convergence is $0 < x < 2$. We can therefore delete the absolute values around x.

Comparison of the solutions in Examples 10.11 and 10.22 indicates once again the advantage of avoiding the use of Taylor's remainder formula.

Multiplication and Division of Power Series

In Example 10.20 we added the Maclaurin series for $4/(x - 4)$ and $1/(1 + x)$ to obtain the Maclaurin series for $5x/(x^2 - 3x - 4)$. An alternative procedure might be to multiply the two series since

$$\frac{5x}{x^2 - 3x - 4} = 5x \left(\frac{1}{x - 4} \right) \left(\frac{1}{x + 1} \right)$$

$$= \frac{5x}{-4} \left(1 + \frac{x}{4} + \frac{x^2}{4^2} + \frac{x^3}{4^3} + \cdots \right) \left(1 - x + x^2 - x^3 + \cdots \right).$$

The rules of algebra demand that we multiply every term of the first series by every term of the second. If we do this and group all products with like powers of x, we obtain

$$\frac{5x}{x^2 - 3x - 4} = \frac{5x}{-4} \left[1 + \left(-1 + \frac{1}{4} \right) x + \left(1 - \frac{1}{4} + \frac{1}{4^2} \right) x^2 \right.$$

$$\left. + \left(-1 + \frac{1}{4} - \frac{1}{4^2} + \frac{1}{4^3} \right) x^3 + \cdots \right].$$

It is clear that the coefficient of x^n is a finite geometric series to which we can apply formula 10.22:

$$(-1)^n \left[1 - \frac{1}{4} + \frac{1}{4^2} - \cdots + \frac{(-1)^n}{4^n} \right] = (-1)^n \left[\frac{1 - \left(-\frac{1}{4} \right)^{n+1}}{1 + \frac{1}{4}} \right]$$

$$= (-1)^n \frac{4}{5} \left[1 - \left(-\frac{1}{4} \right)^{n+1} \right].$$

Consequently,

$$\frac{5x}{x^2 - 3x - 4} = \frac{5x}{-4} \sum_{n=0}^{\infty} (-1)^n \frac{4}{5} \left[1 - \left(-\frac{1}{4} \right)^{n+1} \right] x^n$$

$$= \sum_{n=0}^{\infty} (-1)^{n+1} \left[1 - \frac{(-1)^{n+1}}{4^{n+1}} \right] x^{n+1}$$

$$= \sum_{n=0}^{\infty} \left[(-1)^{n+1} - \frac{1}{4^{n+1}} \right] x^{n+1}$$

$$= \sum_{n=1}^{\infty} \left[(-1)^n - \frac{1}{4^n} \right] x^n.$$

For this example, then, multiplication as well as addition of power series leads to the Maclaurin series. Clearly, addition of power series is much simpler for this example, but we have at least demonstrated that power series can be multiplied together. That this is generally possible is stated in the following theorem.

THEOREM 10.6

If $f(x) = \sum_{n=0}^{\infty} a_n(x - c)^n$ and $g(x) = \sum_{n=0}^{\infty} b_n(x - c)^n$ have positive radii of convergence R_1 and R_2, respectively, then

$$f(x)g(x) = \sum_{n=0}^{\infty} d_n(x - c)^n, \tag{10.30a}$$

where

$$d_n = \sum_{i=0}^{n} a_i b_{n-i} = a_0 b_n + a_1 b_{n-1} + \cdots + a_{n-1} b_1 + a_n b_0, \tag{10.30b}$$

and the radius of convergence is the smaller of R_1 and R_2.

EXAMPLE 10.23

Find the Maclaurin series for $f(x) = [1/(x - 1)] \ln(1 - x)$.

SOLUTION   If we integrate the Maclaurin series

$$\frac{1}{1 - x} = 1 + x + x^2 + x^3 + \cdots, \qquad |x| < 1,$$

we find that

$$-\ln|1 - x| = \left(x + \frac{x^2}{2} + \frac{x^3}{3} + \frac{x^4}{4} + \cdots \right) + C.$$

By setting $x = 0$, we obtain $C = 0$, and

$$\ln(1 - x) = -x - \frac{x^2}{2} - \frac{x^3}{3} - \frac{x^4}{4} - \cdots.$$

We have dropped absolute value signs since the radius of convergence of the series is 1. We now form the Maclaurin series for $f(x)$:

$$\frac{1}{x - 1} \ln(1 - x) = \frac{-1}{1 - x} \ln(1 - x)$$

$$= \left(1 + x + x^2 + x^3 + \cdots \right) \left(x + \frac{x^2}{2} + \frac{x^3}{3} + \cdots \right)$$

$$= x + \left(1 + \frac{1}{2} \right) x^2 + \left(1 + \frac{1}{2} + \frac{1}{3} \right) x^3 + \cdots$$

$$= \sum_{n=1}^{\infty} \left(1 + \frac{1}{2} + \frac{1}{3} + \cdots + \frac{1}{n} \right) x^n.$$

Since both of the multiplied series have radius of convergence 1, so also does the Maclaurin series for $(x - 1)^{-1} \ln(1 - x)$. In other words, its open interval of convergence is $-1 < x < 1$.

EXAMPLE 10.24

Find the first three nonzero terms in the Maclaurin series for $\tan x$.

SOLUTION   If $\tan x = \sum_{n=0}^{\infty} a_n x^n$, then by setting $\tan x = \sin x / \cos x$, we have

$$\sin x = \cos x \sum_{n=0}^{\infty} a_n x^n.$$

We now substitute Maclaurin series for $\sin x$ and $\cos x$:

$$x - \frac{x^3}{3!} + \frac{x^5}{5!} - \frac{x^7}{7!} + \cdots = \left(1 - \frac{x^2}{2!} + \frac{x^4}{4!} - \frac{x^6}{6!} + \cdots\right)\left(a_0 + a_1 x + a_2 x^2 + \cdots\right).$$

According to the corollary to Theorem 10.4, two power series can be identical only if corresponding coefficients are equal. We therefore multiply the right side and equate coefficients of like powers of x:

$$x^0 : \quad 0 = a_0;$$

$$x : \quad 1 = a_1;$$

$$x^2 : \quad 0 = a_2 - \frac{a_0}{2!}, \quad \text{which implies } a_2 = 0;$$

$$x^3 : \quad -\frac{1}{3!} = a_3 - \frac{a_1}{2!}, \quad \text{which implies } a_3 = \frac{1}{2!} - \frac{1}{3!} = \frac{1}{3};$$

$$x^4 : \quad 0 = a_4 - \frac{a_2}{2!} + \frac{a_0}{4!}, \quad \text{from which } a_4 = 0;$$

$$x^5 : \quad \frac{1}{5!} = a_5 - \frac{a_3}{2!} + \frac{a_1}{4!}, \quad \text{from which } a_5 = \frac{1}{5!} + \frac{1}{6} - \frac{1}{4!} = \frac{2}{15}.$$

The first three nonzero terms in the Maclaurin series for $\tan x$ are therefore

$$\tan x = x + \frac{1}{3}x^3 + \frac{2}{15}x^5 + \cdots.$$

We could obtain the same result by long division of the Maclaurin series for $\sin x$ by that of $\cos x$ shown below. Long division can produce a few terms of a Maclaurin series, but seldom does it suggest a pattern for all terms in the series.

$$
\begin{array}{r}
x + \dfrac{x^3}{3} + \dfrac{2x^5}{15} + \cdots \\[2mm]
1 - \dfrac{x^2}{2} + \dfrac{x^4}{24} - \cdots \overline{\Big)\; x - \dfrac{x^3}{6} + \dfrac{x^5}{120} - \cdots} \\[2mm]
\underline{x - \dfrac{x^3}{2} + \dfrac{x^5}{24} - \cdots} \\[2mm]
\dfrac{x^3}{3} - \dfrac{x^5}{30} + \cdots \\[2mm]
\underline{\dfrac{x^3}{3} - \dfrac{x^5}{6} + \cdots} \\[2mm]
\dfrac{2x^5}{15} + \cdots \\[2mm]
\underline{\dfrac{2x^5}{15} + \cdots}
\end{array}
$$

Binomial Expansion

One of the most widely used power series is the binomial expansion. We are well acquainted with the binomial theorem, which predicts the product $(a + b)^m$ for any positive integer m:

$$(a + b)^m = \sum_{n=0}^{m} \binom{m}{n} a^n b^{m-n}. \qquad (10.31)$$

With the usual definition of the binomial coefficients,

$$\binom{m}{n} = \frac{m!}{(m-n)!\,n!} = \frac{m(m-1)(m-2)\cdots(m-n+1)}{n!},$$

the binomial theorem becomes

$$(a + b)^m = a^m + ma^{m-1}b + \frac{m(m-1)}{2!}a^{m-2}b^2 + \cdots + mab^{m-1} + b^m. \qquad (10.32)$$

Even when m is not a positive integer, this form for the binomial theorem remains almost intact. To show this, we consider the power series

$$1 + \sum_{n=1}^{\infty} \frac{m(m-1)(m-2)\cdots(m-n+1)}{n!} x^n$$

for any real number m except a nonnegative integer. The radius of convergence of this power series is

$$R = \lim_{n \to \infty} \left| \frac{m(m-1)(m-2)\cdots(m-n+1)}{n!} \frac{(n+1)!}{m(m-1)(m-2)\cdots(m-n)} \right|$$

$$= \lim_{n \to \infty} \left| \frac{n+1}{m-n} \right| = 1.$$

The open interval of convergence is therefore $|x| < 1$. Whether the series converges at the end-points $x = \pm 1$ depends on the value of m. For the time being, we will work on the interval $|x| < 1$, and at the end of the discussion, we will state the complete result. Let us denote the sum of the series by

$$f(x) = 1 + \sum_{n=1}^{\infty} \frac{m(m-1)(m-2)\cdots(m-n+1)}{n!} x^n, \qquad |x| < 1.$$

If we differentiate this series term by term according to Theorem 10.3,

$$f'(x) = \sum_{n=1}^{\infty} \frac{m(m-1)\cdots(m-n+1)}{(n-1)!} x^{n-1}, \qquad |x| < 1,$$

and then multiply both sides by x, we have

$$xf'(x) = \sum_{n=1}^{\infty} \frac{m(m-1)\cdots(m-n+1)}{(n-1)!} x^n, \qquad |x| < 1.$$

If we add these results, we obtain

$$f'(x) + xf'(x) = \sum_{n=1}^{\infty} \frac{m(m-1)\cdots(m-n+1)}{(n-1)!} x^{n-1} + \sum_{n=1}^{\infty} \frac{m(m-1)\cdots(m-n+1)}{(n-1)!} x^n$$

We now change the variable of summation in the first sum:

$$(1 + x) f'(x) = \sum_{n=0}^{\infty} \frac{m(m-1) \cdots (m-n)}{n!} x^n + \sum_{n=1}^{\infty} \frac{m(m-1) \cdots (m-n+1)}{(n-1)!} x^n.$$

When these summations are added over their common range, beginning at $n = 1$, and the $n = 0$ term in the first summation is written out separately, the result is

$$(1 + x) f'(x) = m + \sum_{n=1}^{\infty} \frac{m(m-1) \cdots (m-n+1)}{(n-1)!} \left(\frac{m-n}{n} + 1 \right) x^n$$

$$= m \left[1 + \sum_{n=1}^{\infty} \frac{m(m-1) \cdots (m-n+1)}{n!} x^n \right]$$

$$= m f(x).$$

Consequently, the function $f(x)$ must satisfy the differential equation

$$\frac{f'(x)}{f(x)} = \frac{m}{1 + x}.$$

Integration immediately gives

$$\ln |f(x)| = m \ln |1 + x| + C \qquad \text{or} \qquad f(x) = D(1 + x)^m.$$

To evaluate D, we note that from the original definition of $f(x)$ as the sum of the power series, $f(0) = 1$, and this implies that $D = 1$. Thus,

$$f(x) = (1 + x)^m,$$

and we may write finally that

$$(1 + x)^m = 1 + \sum_{n=1}^{\infty} \frac{m(m-1)(m-2) \cdots (m-n+1)}{n!} x^n \qquad (10.33a)$$

$$= 1 + mx + \frac{m(m-1)}{2!} x^2 + \frac{m(m-1)(m-2)}{3!} x^3 + \cdots, \qquad (10.33b)$$

valid for $|x| < 1$. This is called the **binomial expansion** of $(1 + x)^m$; it is the Maclaurin series for $(1 + x)^m$. We have verified the binomial expansion for m any real number except a nonnegative integer, but in the case of a nonnegative integer, the series terminates after $m + 1$ terms and is therefore valid for these values of m also. We mentioned earlier that the binomial expansion may also converge at the endpoints $x = \pm 1$, depending on the value of m. The complete result states that 10.33 is valid for

$$-\infty < x < \infty \quad \text{if } m \text{ is a nonnegative integer,}$$

$$-1 < x < 1 \quad \text{if } m \leq -1,$$

$$-1 < x \leq 1 \quad \text{if } -1 < m < 0,$$

$$-1 \leq x \leq 1 \quad \text{if } m > 0 \text{ but not an integer.}$$

It is not difficult to generalize this result to expand $(a + b)^m$ for real m. If $|b| < |a|$, we write

$$(a + b)^m = a^m \left(1 + \frac{b}{a} \right)^m$$

and now expand the bracketed term by means of 10.33:

$$(a + b)^m = a^m \left[1 + m \left(\frac{b}{a} \right) + \frac{m(m-1)}{2!} \left(\frac{b}{a} \right)^2 + \cdots \right] \qquad |b| < |a|,$$

$$= a^m + ma^{m-1}b + \frac{m(m-1)}{2!}a^{m-2}b^2 + \cdots \qquad |b| < |a|, \qquad (10.34)$$

which, as we predicted, is equation 10.32 except that the series does not terminate. We recommend use of either of equations 10.33 over 10.34; this necessitates creation of the 1. We illustrate in the following example.

EXAMPLE 10.25

▼

Use the binomial expansion to find the Maclaurin series for $x/(2-x)^3$.

SOLUTION   We demonstrate how to use both of equations 10.33a and b. By 10.33a,

$$\frac{x}{(2-x)^3} = \frac{x}{2^3 \left(1 - \frac{x}{2} \right)} = \frac{x}{2^3} \left(1 - \frac{x}{2} \right)^{-3}$$

$$= \frac{x}{2^3} \left[1 + \sum_{n=1}^{\infty} \frac{(-3)(-4)(-5) \cdots (-3 - n + 1)}{n!} \left(-\frac{x}{2} \right)^n \right]$$

$$= \frac{x}{2^3} \left[1 + \sum_{n=1}^{\infty} \frac{(-1)^n (3)(4)(5) \cdots (n+2)}{n!} \frac{(-1)^n x^n}{2^n} \right]$$

$$= \frac{x}{2^3} \left[1 + \sum_{n=1}^{\infty} \frac{(2)(3)(4) \cdots (n+2)}{2^{n+1} n!} x^n \right]$$

$$= \frac{x}{2^3} \left[1 + \sum_{n=1}^{\infty} \frac{(n+1)(n+2)}{2^{n+1}} x^n \right]$$

$$= \frac{x}{2^3} + \sum_{n=1}^{\infty} \frac{(n+1)(n+2)}{2^{n+4}} x^{n+1}$$

$$= \frac{x}{2^3} + \sum_{n=2}^{\infty} \frac{n(n+1)}{2^{n+3}} x^n$$

$$= \sum_{n=1}^{\infty} \frac{n(n+1)}{2^{n+3}} x^n, \qquad |x| < 2.$$

With Equation 10.33b, we write

$$\frac{x}{(2-x)^3} = \frac{x}{2^3 \left(1 - \frac{x}{2} \right)} = \frac{x}{2^3} \left(1 - \frac{x}{2} \right)^{-3}$$

$$= \frac{x}{2^3} \left[1 + (-3) \left(-\frac{x}{2} \right) + \frac{(-3)(-4)}{2!} \left(-\frac{x}{2} \right)^2 + \frac{(-3)(-4)(-5)}{3!} \left(-\frac{x}{2} \right)^3 + \cdots \right]$$

$$= \frac{x}{2^3} \left[1 + \frac{3x}{2} + \frac{3 \cdot 4}{2^2 \, 2!} x^2 + \frac{3 \cdot 4 \cdot 5}{2^3 \, 3!} x^3 + \frac{3 \cdot 4 \cdot 5 \cdot 6}{2^4 \, 4!} x^4 + \cdots \right]$$

$$= \frac{x}{2^3} \left[1 + \frac{3x}{2} + \frac{2 \cdot 3 \cdot 4}{2^3 \, 2!} x^2 + \frac{2 \cdot 3 \cdot 4 \cdot 5}{2^4 \, 3!} x^3 + \frac{2 \cdot 3 \cdot 4 \cdot 5 \cdot 6}{2^5 \, 4!} x^4 + \cdots \right]$$

$$= \frac{x}{2^3} \left[1 + \frac{3x}{2} + \frac{3 \cdot 4}{2^3} x^2 + \frac{4 \cdot 5}{2^4} x^3 + \frac{5 \cdot 6}{2^5} x^4 + \cdots \right]$$

$$= \frac{x}{2^3} \sum_{n=0}^{\infty} \frac{(n+1)(n+2)}{2^{n+1}} x^n$$

$$= \sum_{n=0}^{\infty} \frac{(n+1)(n+2)}{2^{n+4}} x^{n+1}$$

$$= \sum_{n=1}^{\infty} \frac{n(n+1)}{2^{n+3}} x^n, \qquad |x| < 2.$$

This result was also obtained in Example 10.21 by differentiation of the Maclaurin series for $1/(2 - x)$.

EXAMPLE 10.26

Find the Maclaurin series for $\mathrm{Sin}^{-1} x$.

SOLUTION   By the binomial expansion, we have

$$\frac{1}{\sqrt{1 - x^2}} = 1 + \left(-\frac{1}{2} \right)(-x^2) + \frac{\left(-\frac{1}{2} \right)\left(-\frac{3}{2} \right)}{2!}(-x^2)^2 + \frac{\left(-\frac{1}{2} \right)\left(-\frac{3}{2} \right)\left(-\frac{5}{2} \right)}{3!}(-x^2)^3 + \cdots$$

$$= 1 + \frac{1}{2}x^2 + \frac{3}{2^2 2!}x^4 + \frac{3 \cdot 5}{2^3 3!}x^6 + \frac{3 \cdot 5 \cdot 7}{2^4 4!}x^8 + \cdots, \qquad |x| < 1.$$

Integration of this series gives

$$\mathrm{Sin}^{-1} x = \left(x + \frac{1}{2 \cdot 3}x^3 + \frac{3}{2^2 2! \, 5}x^5 + \frac{3 \cdot 5}{2^3 3! \, 7}x^7 + \frac{3 \cdot 5 \cdot 7}{2^4 4! \, 9}x^9 + \cdots \right) + C.$$

Evaluation of both sides of this equation at $x = 0$ gives $C = 0$. According to Theorem 10.3, the radius of convergence of this series must be 1, and we can write

$$\mathrm{Sin}^{-1} x = x + \sum_{n=1}^{\infty} \frac{1 \cdot 3 \cdot 5 \cdot \ldots \cdot (2n - 1)}{2^n n! \, (2n + 1)} x^{2n+1}$$

$$= x + \sum_{n=1}^{\infty} \frac{1 \cdot 2 \cdot 3 \cdot 4 \cdot 5 \cdot \ldots \cdot (2n - 2)(2n - 1)(2n)}{2 \cdot 4 \cdot \ldots \cdot (2n) 2^n n! \, (2n + 1)} x^{2n+1}$$

$$= x + \sum_{n=1}^{\infty} \frac{(2n)!}{(2n + 1)2^{2n}(n!)^2} x^{2n+1}$$

$$= \sum_{n=0}^{\infty} \frac{(2n)!}{(2n + 1)2^{2n}(n!)^2} x^{2n+1}, \qquad |x| < 1.$$

EXAMPLE 10.27

When measuring the velocity v of gas flow using a Pitot tube, the following equation for pressure P of the gas is encountered:

$$P(v) = P_0 \left[1 + \left(\frac{k-1}{2} \right) \left(\frac{v}{c} \right)^2 \right]^{k/(k-1)},$$

where P_0, c, and k are constants. The first two terms in the binomial expansion of P represent the situation for an incompressible gas. The third term in the expansion is sometimes regarded as the error in using the tube to determine velocity for compressible flow. Find the first three terms in the binomial expansion.

SOLUTION   Using equation 10.33,

$$P(v) = P_0 \left[1 + \left(\frac{k}{k-1} \right) \left(\frac{k-1}{2} \right) \left(\frac{v}{c} \right)^2 + \frac{1}{2} \left(\frac{k}{k-1} \right) \left(\frac{k}{k-1} - 1 \right) \left(\frac{k-1}{2} \right)^2 \left(\frac{v}{c} \right)^4 + \cdots \right]$$

$$= P_0 \left(1 + \frac{k}{2c^2} v^2 + \frac{k}{8c^4} v^4 + \cdots \right).$$

EXERCISES 10.5

In Exercises 1–30 find the Maclaurin or Taylor series of the function about the indicated point.

1. $f(x) = \dfrac{1}{3x+2}$ about $x = 0$

2. $f(x) = \dfrac{1}{4+x^2}$ about $x = 0$

3. $f(x) = \cos(x^2)$ about $x = 0$

4. $f(x) = e^{5x}$ about $x = 0$

5. $f(x) = e^x$ about $x = 3$

6. $f(x) = e^{1-2x}$ about $x = 0$

7. $f(x) = e^{1-2x}$ about $x = -1$

8. $f(x) = \cosh x$ about $x = 0$

9. $f(x) = \sinh x$ about $x = 0$

10. $f(x) = x^4 + 3x^2 - 2x + 1$ about $x = 0$

11. $f(x) = x^4 + 3x^2 - 2x + 1$ about $x = -2$

* 12. $f(x) = \dfrac{1}{x+3}$ about $x = 2$

* 13. $f(x) = \dfrac{x}{2x+5}$ about $x = 1$

* 14. $f(x) = \dfrac{x^2}{3-4x}$ about $x = 2$

* 15. $f(x) = \dfrac{1}{\sqrt{1+x}}$ about $x = 0$

* 16. $f(x) = \ln(1+2x)$ about $x = 0$

* 17. $f(x) = (1+3x)^{3/2}$ about $x = 0$

* 18. $f(x) = \ln x$ about $x = 2$

* 19. $f(x) = \ln(x+3)$ about $x = -1$

* 20. $f(x) = 1/x$ about $x = 4$

* 21. $f(x) = \dfrac{1}{(x+2)^3}$ about $x = 0$

* 22. $f(x) = \dfrac{1}{(2-x)^2}$ about $x = 3$

* 23. $f(x) = \dfrac{1}{(x+3)^2}$ about $x = 1$

* 24. $f(x) = \dfrac{1}{x^2+8x+15}$ about $x = 0$

* 25. $f(x) = \text{Tan}^{-1} x$ about $x = 0$

* 26. $f(x) = \sqrt{x+3}$ about $x = 0$

* 27. $f(x) = \sqrt{x+3}$ about $x = 2$

* 28. $f(x) = (1-2x)^{1/3}$ about $x = 1$

* 29. $f(x) = \dfrac{x^2}{(1+x^2)^2}$ about $x = 0$

* 30. $f(x) = x(1-x)^{1/3}$ about $x = 0$

In Exercises 31–33 find the first four nonzero terms in the Maclaurin series for the function.

31. $f(x) = \tan 2x$

* **32.** $f(x) = \sec x$

33. $f(x) = e^x \sin x$

34. Find the Maclaurin series for $\cos^2 x$.

In Exercises 35–38 find the Maclaurin series for the function.

35. $f(x) = \dfrac{1}{x^6 - 3x^3 - 4}$

* **36.** $f(x) = \operatorname{Sin}^{-1}(x^2)$

37. $f(x) = \dfrac{2x^2 + 4}{x^2 + 4x + 3}$

* **38.** $f(x) = \ln\left[\dfrac{1 + x/\sqrt{2}}{1 - x/\sqrt{2}}\right]$

39. Prove the corollary to Theorem 10.4.

40. Prove that if a power series with positive radius of convergence has sum zero, $\sum_{n=0}^{\infty} a_n(x - c)^n = 0$, then $a_n = 0$ for all n.

41. If, during a working day, one person drinks from a fountain every 30 s (on the average), then the probability that exactly n people drink in a time interval of length t seconds is given by the *Poisson distribution*:

$$P_n(t) = \frac{1}{n!}\left(\frac{t}{30}\right)^n e^{-t/30}.$$

Calculate $\sum_{n=0}^{\infty} P_n(t)$ and interpret the result.

42. A certain experiment is to be performed until it is successful. The probability that it will be successful in any given attempt is p ($0 < p < 1$), and therefore the probability that it will fail is $q = 1 - p$. The expected number of times that the experiment must be performed in order to be successful can be shown to be represented by the infinite series

$$\sum_{n=1}^{\infty} npq^{n-1} = \sum_{n=1}^{\infty} np(1 - p)^{n-1}.$$

(a) What is the sum of this series?

(b) If p is the probability that a single die will come up 6, is the answer in part (a) what you would expect?

43. Find the Maclaurin series for the *error function* erf(x) defined by

$$\operatorname{erf}(x) = \frac{2}{\sqrt{\pi}} \int_0^x e^{-t^2}\, dt.$$

44. Find Maclaurin series for the *Fresnel integrals* $C(x)$ and $S(x)$ defined by

$$C(x) = \int_0^x \cos(\pi t^2/2)\, dt, \qquad S(x) = \int_0^x \sin(\pi t^2/2)\, dt.$$

In Exercises 45–46 use Maclaurin series to find a formula for the nth derivative of the function at $x = 0$.

* **45.** $f(x) = \dfrac{x}{(4 + 3x)^2}$

* **46.** $f(x) = xe^{-2x}$

In Exercises 47–48 use Taylor series to find a formula for the nth derivative of the function at $x = 2$.

* **47.** $f(x) = \dfrac{1}{3 + x}$

* **48.** $f(x) = xe^{-x}$

* **49.** Show that even-order derivatives of $x^2 \sin 2x$ at $x = 0$ are equal to zero.

* **50.** Show that odd-order derivatives of e^{-x^2} at $x = 0$ vanish.

Show that Bessel functions of the first kind (defined in Exercise 38 of Section 10.4) satisfy the properties in Exercises 51–52 .

* **51.** $2m J_m(x) - x J_{m-1}(x) = x J_{m+1}(x)$

* **52.** $J_{m-1}(x) - J_{m+1}(x) = 2J_m'(x)$

* **53.** If the function $(1 - 2\mu x + x^2)^{-1/2}$ is expanded in a Maclaurin series in x,

$$\frac{1}{\sqrt{1 - 2\mu x + x^2}} = \sum_{n=0}^{\infty} P_n(\mu)x^n,$$

the coefficients $P_n(\mu)$ are called the *Legendre polynomials*. Find $P_0(\mu)$, $P_1(\mu)$, $P_2(\mu)$, and $P_3(\mu)$.

* **54.**   (a) If we define $f(x) = x/(e^x - 1)$ at $x = 0$ as $f(0) = 1$, it turns out that $f(x)$ has a Maclaurin series expansion with positive radius of convergence. When this expansion is expressed in the form

$$\frac{x}{e^x - 1} = 1 + B_1 x + \frac{B_2}{2!}x^2 + \frac{B_3}{3!}x^3 + \cdots,$$

the coefficients B_1, B_2, B_3, ... are called the *Bernoulli numbers*. Write this equation in the form

$$x = (e^x - 1)\left(1 + B_1 x + \frac{B_2}{2!}x^2 + \cdots\right),$$

and substitute the Maclaurin series for e^x to find the first five Bernoulli numbers.

(b) Show that the odd Bernoulli numbers all vanish for $n \geq 3$.

* **55.** Show that

$$e^{x(t-1/t)/2} = \sum_{n=0}^{\infty} J_n(x)t^n.$$

For a definition of $J_m(x)$, see Exercise 38 in Section 10.4.

10.6 **Sums of Power Series**

Theorem 10.3 provides an important technique for finding **sums of power series**. If a series with unknown sum can be reduced to a series with known sum by differentiations or integrations, then the unknown sum can be obtained when these operations are reversed.

EXAMPLE 10.28

Find the sum of the series $\displaystyle\sum_{n=0}^{\infty} (n + 1)x^n$.

SOLUTION   Without the factor $n+1$, the series would be geometric. Integration of $(n+1)x^n$ removes the factor. This is the idea; now let us formulate it mathematically. The radius of convergence of the series is

$$R = \lim_{n\to\infty} \left| \frac{n+1}{n+2} \right| = 1.$$

If we denote the sum of the series by $S(x)$,

$$S(x) = \sum_{n=0}^{\infty} (n+1)x^n,$$

and use Theorem 10.3 to integrate the series term by term, we obtain

$$\int S(x)\,dx = \sum_{n=0}^{\infty} x^{n+1} + C.$$

But the series on the right is a geometric series with sum $x/(1 - x)$, provided that $|x| < 1$, and we may therefore write

$$\int S(x)\,dx = \frac{x}{1 - x} + C, \qquad |x| < 1.$$

If we now differentiate this equation, we obtain

$$S(x) = \frac{(1 - x)(1) - x(-1)}{(1 - x)^2} = \frac{1}{(1 - x)^2},$$

and therefore

$$\sum_{n=0}^{\infty} (n+1)x^n = \frac{1}{(1 - x)^2}.$$

The open interval of convergence is $-1 < x < 1$. If we set $x = 1$, the series reduces to $\sum_{n=0}^{\infty} (n+1) = 1 + 2 + 3 + \cdots$, which clearly does not have a sum. At $x = -1$, we have $1 - 2 + 3 - 4 + \cdots$, which diverges also. The interval of convergence is therefore $-1 < x < 1$.

EXAMPLE 10.29

Find the sum of the series $\displaystyle\sum_{n=1}^{\infty} \frac{1}{n} x^n$.

SOLUTION   We can remove the factor $1/n$ by differentiation, and thereby produce a geometric series. We proceed as follows. The radius of convergence of the series is

$$R = \lim_{n \to \infty} \left| \frac{1/n}{1/(n+1)} \right| = 1.$$

If we denote the sum of the series by

$$S(x) = \sum_{n=1}^{\infty} \frac{1}{n} x^n,$$

and differentiate the series term by term, we have

$$S'(x) = \sum_{n=1}^{\infty} x^{n-1}.$$

This is a geometric series with sum $1/(1-x)$, so that

$$S'(x) = \frac{1}{1-x}.$$

Integration now gives

$$S(x) = -\ln(1-x) + C.$$

Since $S(0) = 0$, it follows that

$$0 = -\ln(1) + C.$$

Hence, $C = 0$, and $S(x) = -\ln(1-x)$. We have shown therefore that

$$\sum_{n=1}^{\infty} \frac{1}{n} x^n = -\ln(1-x).$$

The open interval of convergence is $-1 < x < 1$.

EXAMPLE 10.30

Find the sum of the series $\displaystyle\sum_{n=0}^{\infty} \frac{(-1)^n (2n+1)}{(2n)!} x^{2n}$.

SOLUTION   Without the factor $2n + 1$, we see a cosine series (note the x^{2n} and the $(2n)!$). The factor can be eliminated by integration. We now know how to proceed. The radius of convergence of the series is

$$R = \lim_{n \to \infty} \left| \frac{\dfrac{(-1)^n (2n+1)}{(2n)!}}{\dfrac{(-1)^{n+1}(2n+3)}{(2n+2)!}} \right| = \lim_{n \to \infty} \left[\frac{(2n+1)}{(2n)!} \cdot \frac{(2n+2)(2n+1)(2n)!}{2n+3} \right]$$

$$= \lim_{n \to \infty} \frac{(2n+1)^2 (2n+2)}{2n+3} = \infty.$$

If we denote the sum of the series by $S(x)$,

$$S(x) = \sum_{n=0}^{\infty} \frac{(-1)^n (2n+1)}{(2n)!} x^{2n},$$

and use Theorem 10.3 to integrate the series term by term, we obtain

$$\int S(x)\, dx = \sum_{n=0}^{\infty} \frac{(-1)^n}{(2n)!} x^{2n+1} + C = x \sum_{n=0}^{\infty} \frac{(-1)^n}{(2n)!} x^{2n} + C = x \cos x + C.$$

Differentiation now gives

$$S(x) = -x \sin x + \cos x.$$

EXERCISES 10.6

In Exercises 1–15 find the sum of the power series.

* **1.** $\displaystyle\sum_{n=1}^{\infty} n x^{n-1}$

* **2.** $\displaystyle\sum_{n=2}^{\infty} n(n-1) x^{n-2}$

* **3.** $\displaystyle\sum_{n=1}^{\infty} (n+1) x^{n-1}$

* **4.** $\displaystyle\sum_{n=1}^{\infty} n^2 x^{n-1}$

* **5.** $\displaystyle\sum_{n=1}^{\infty} (n^2 + 2n) x^{n}$

* **6.** $\displaystyle\sum_{n=0}^{\infty} \frac{1}{n+1} x^{n}$

* **7.** $\displaystyle\sum_{n=0}^{\infty} \frac{(-1)^n}{2n+1} x^{2n+1}$

* **8.** $\displaystyle\sum_{n=1}^{\infty} \frac{(-1)^n}{n} x^{2n}$

* **9.** $\displaystyle\sum_{n=2}^{\infty} n 3^n x^{2n}$

* **10.** $\displaystyle\sum_{n=0}^{\infty} \left(\frac{n+1}{n+2} \right) x^{n}$

* **11.** $\displaystyle\sum_{n=1}^{\infty} \left(\frac{n+1}{n!} \right) x^{n}$

* **12.** $\displaystyle\sum_{n=0}^{\infty} \frac{(-1)^n (2n+1)}{(2n+1)!} x^{2n+1}$

* **13.** $\displaystyle\sum_{n=0}^{\infty} \frac{(-1)^n (n+2)}{(2n)!} x^{2n}$

* **14.** $\displaystyle\sum_{n=1}^{\infty} \frac{(2n+3) 2^n}{n!} x^{2n}$

* **15.** $\displaystyle\sum_{n=0}^{\infty} \frac{(-1)^{n+1} (2n-1)}{(2n)!} x^{2n+1}$

10.7 Applications of Taylor Series and Taylor's Remainder Formula

When a function is approximated by its nth-degree Taylor polynomial,

$$f(x) \approx f(c) + f'(c)(x-c) + \frac{f''(c)}{2!}(x-c)^2 + \cdots + \frac{f^{(n)}(c)}{n!}(x-c)^n,$$

the error in doing so is the Taylor remainder,

$$R_n(c, x) = \frac{f^{(n+1)}(z_n)}{(n+1)!}(x-c)^{n+1},$$

where z_n is between c and x. The smaller R_n is, the better the approximation. Because z_n is unknown, we cannot find $R_n(c, x)$. What we do is replace $f^{(n+1)}(z_n)$ by some larger value, thereby obtaining a maximum value for the error. For instance, consider using the first three terms of the Maclaurin series for e^x to approximate e^x on the interval $0 \le x \le 1/2$. Taylor's remainder formula, with $c = 0$, states that

$$e^x = 1 + x + \frac{x^2}{2} + R_2(0, x),$$

where

$$R_2(0, x) = \frac{d^3}{dx^3}(e^x)\bigg|_{x=z} \frac{x^3}{3!} = \frac{e^z x^3}{6},$$

and z is between 0 and x. Although z is unknown — except that it is between 0 and x, we can say that because only the values $0 \leq x \leq \frac{1}{2}$ are under consideration, z must be less than $\frac{1}{2}$. It follows that

$$R_2(0, x) < \frac{e^{1/2} x^3}{6} \leq \frac{\sqrt{e}(1/2)^3}{6} < 0.035.$$

Thus the quadratic function $1 + x + x^2/2$ approximates e^x on the interval $0 \leq x \leq 1/2$ with error no greater than 0.035.

In the following example, we determine the number of terms of a Maclaurin series required to guarantee a certain accuracy.

EXAMPLE 10.31

How many terms in the Maclaurin series for $\ln(1 + x)$ guarantee a truncation error of less than 10^{-6} for any x in the interval $0 \leq x \leq 1/2$?

SOLUTION   The nth derivative of $\ln(1 + x)$ is

$$\frac{d^n}{dx^n}\ln(1 + x) = \frac{(-1)^{n+1}(n - 1)!}{(x + 1)^n}, \qquad n \geq 1,$$

and therefore Taylor's remainder formula with $c = 0$ states that

$$\ln(1 + x) = x - \frac{x^2}{2} + \frac{x^3}{3} - \frac{x^4}{4} + \cdots + \frac{(-1)^{n+1}}{n}x^n + R_n(0, x),$$

where

$$R_n(0, x) = \frac{d^{n+1}}{dx^{n+1}}\ln(1 + x)\bigg|_{x=z_n} \frac{x^{n+1}}{(n + 1)!} = \frac{(-1)^n n!}{(z_n + 1)^{n+1}} \frac{x^{n+1}}{(n + 1)!}$$

$$= \frac{(-1)^n}{(n + 1)(z_n + 1)^{n+1}}x^{n+1}.$$

Since z_n is between 0 and x and $0 \leq x \leq 1/2$, we can state that x must be less than or equal to $1/2$, and z_n must be greater than 0. Hence,

$$|R_n(0, x)| < \frac{1}{(n + 1)(1)^{n+1}}(1/2)^{n+1} = \frac{1}{(n + 1)2^{n+1}}.$$

This is less than 10^{-6} if

$$\frac{1}{(n + 1)\,2^{n+1}} < 10^{-6} \qquad \text{or} \qquad (n + 1)2^{n+1} > 10^6.$$

A calculator quickly indicates that the smallest value of n for which this is true is $n = 15$. Consequently, if $\ln(1 + x)$ is approximated by the 15th-degree polynomial

$$\ln(1 + x) \approx x - \frac{x^2}{2} + \frac{x^3}{3} - \frac{x^4}{4} + \cdots + \frac{x^{15}}{15}$$

on the interval $0 \leq x \leq 1/2$, the truncation error is less than 10^{-6}.

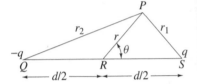

FIGURE 10.19 Potential
due to a dipole

Consulting Project 16

Two young electrical engineers are having a disagreement and we must settle the argument. Figure 10.19 shows two charges of equal size q, but of opposite signs, distance d apart. When d is small, the configuration is called a dipole. One engineer argues that when point P is very far away from the dipole, so that r is very much larger than d, the charges effectively cancel one another, and the potential due to them is zero. The other engineer disagrees. We must decide which of them is correct.

SOLUTION   The potential at P due to charge q is given by the formula $V = q/(4\pi\epsilon_0 r_1)$, where ϵ_0 is a constant. Similarly, the potential at P due to charge $-q$ is $V = -q/(4\pi\epsilon_0 r_2)$. The potential at P due to both charges is

$$V = \frac{1}{4\pi\epsilon_0}\left(\frac{q}{r_1} - \frac{q}{r_2}\right).$$

The cosine law applied to triangles PRS and PQR gives

$$r_1^2 = r^2 + \frac{d^2}{4} - rd\cos\theta,$$

$$r_2^2 = r^2 + \frac{d^2}{4} - rd\cos(\pi - \theta) = r^2 + \frac{d^2}{4} + rd\cos\theta.$$

Hence,

$$V = \frac{q}{4\pi\epsilon_0}\left(\frac{1}{\sqrt{r^2 + d^2/4 - rd\cos\theta}} - \frac{1}{\sqrt{r^2 + d^2/4 + rd\cos\theta}}\right).$$

The binomial expansion can be used to write

$$\frac{1}{\sqrt{r^2 + d^2/4 + rd\cos\theta}} = \frac{1}{r\sqrt{1 + \dfrac{d^2}{4r^2} + \dfrac{d}{r}\cos\theta}}$$

$$= \frac{1}{r}\left[1 + \left(\frac{d^2}{4r^2} + \frac{d}{r}\cos\theta\right)\right]^{-1/2}$$

$$= \frac{1}{r}\left[1 - \frac{1}{2}\left(\frac{d^2}{4r^2} + \frac{d}{r}\cos\theta\right)\right.$$

$$\left. + \frac{(-1/2)(-3/2)}{2!}\left(\frac{d^2}{4r^2} + \frac{d}{r}\cos\theta\right)^2 + \cdots\right].$$

When d is very much less than r, so that terms in d^2/r^2, d^3/r^3, etc., are negligible compared to d/r,

$$\frac{1}{\sqrt{r^2 + d^2/4 + rd\cos\theta}} \approx \frac{1}{r}\left(1 - \frac{d}{2r}\cos\theta\right).$$

Similarly, $\dfrac{1}{\sqrt{r^2 + d^2/4 - rd\cos\theta}} \approx \dfrac{1}{r}\left(1 + \dfrac{d}{2r}\cos\theta\right)$. Hence,

$$V \approx \frac{q}{4\pi\epsilon_0}\left[\frac{1}{r}\left(1+\frac{d}{2r}\cos\theta\right)-\frac{1}{r}\left(1-\frac{d}{2r}\cos\theta\right)\right]=\frac{qd\cos\theta}{4\pi\epsilon_0 r^2}.$$

This shows that the potential due to the dipole does not vanish at large distances from it. It is small because d is small and r is large, but it is not zero. Only when P is on the perpendicular bisector of the line joining the charges (so that $\theta = \pi/2$) is the potential zero.

Limits

We have customarily used L'Hôpital's rule to evaluate limits of the indeterminate form $0/0$. Maclaurin and Taylor series can sometimes be used to advantage. Consider

$$\lim_{x\to 0}\frac{x-\sin x}{x^3}.$$

Three applications of L'Hôpital's rule give a limit of $1/6$. Alternatively, if we substitute the Maclaurin series for $\sin x$,

$$\lim_{x\to 0}\frac{x-\sin x}{x^3}=\lim_{x\to 0}\frac{1}{x^3}\left[x-\left(x-\frac{x^3}{3!}+\frac{x^5}{5!}-\cdots\right)\right]$$

$$=\lim_{x\to 0}\left[\frac{1}{6}-\frac{x^2}{5!}+\cdots\right]=\frac{1}{6}.$$

Here is another example.

EXAMPLE 10.32

Evaluate $\displaystyle\lim_{\lambda\to 0^+}\frac{\lambda^{-5}}{e^{c/\lambda}-1}$, where $c > 0$ is a constant (see also Exercise 58 in Section 4.11).

SOLUTION   We begin by making the change of variable $v = 1/\lambda$ in the limit:

$$\lim_{\lambda\to 0^+}\frac{\lambda^{-5}}{e^{c/\lambda}-1}=\lim_{v\to\infty}\frac{v^5}{e^{cv}-1}.$$

We now expand e^{cv} into its Maclaurin series,

$$\lim_{\lambda\to 0^+}\frac{\lambda^{-5}}{e^{c/\lambda}-1}=\lim_{v\to\infty}\frac{v^5}{\left(1+cv+\frac{c^2v^2}{2!}+\cdots\right)-1}$$

$$=\lim_{v\to\infty}\frac{v^5}{cv+\frac{c^2v^2}{2!}+\cdots}.$$

If we now divide numerator and denominator by v^5, we obtain

$$\lim_{\lambda\to 0^+}\frac{\lambda^{-5}}{e^{c/\lambda}-1}=\lim_{v\to\infty}\frac{1}{\frac{c}{v^4}+\frac{c^2}{2v^3}+\frac{c^3}{3!\,v^2}+\frac{c^4}{4!\,v}+\frac{c^5}{5!}+\frac{c^6v}{6!}+\cdots}=0.$$

Evaluation of Definite Integrals

In Section 8.8 we developed three numerical techniques for approximating definite integrals of functions $f(x)$ that have no obvious antiderivatives: the rectangular rule, the trapezoidal rule, and Simpson's rule. Each method divides the interval of integration into a number of subintervals and approximates $f(x)$ by a more elementary function on each subinterval. The rectangular rule replaces $f(x)$ by a step function, the trapezoidal rule uses a succession of linear functions, and Simpson's rule uses quadratic functions.

Another possibility is to replace $f(x)$ by a truncated power series (a polynomial) over the entire interval of integration. For instance, consider the definite integral

$$\int_0^{1/2} \frac{\sin x}{x}\, dx,$$

where $(\sin x)/x$ is defined as 1 at $x = 0$. The integral defines the area in Figure 10.20.

FIGURE 10.20 Area under the curve
$y = x^{-1}\sin x$

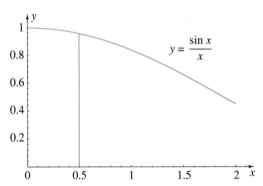

Taylor's remainder formula for $\sin x$ gives

$$\sin x = x - \frac{x^3}{3!} + \frac{x^5}{5!} - \cdots + \frac{d^n}{dx^n}(\sin x)\Big|_{x=0} \frac{x^n}{n!} + R_n(0, x)$$

where

$$R_n(0, x) = \frac{d^{n+1}(\sin x)}{dx^{n+1}}\Big|_{x=z_n} \frac{x^{n+1}}{(n+1)!}$$

and z_n is between 0 and x. Therefore,

$$\frac{\sin x}{x} = 1 - \frac{x^2}{3!} + \frac{x^4}{5!} - \cdots + \frac{1}{x}R_n(0, x),$$

where the term before $x^{-1}R_n(0, x)$ is $\dfrac{d^n}{dx^n}(\sin x)\Big|_{x=0} \dfrac{x^{n-1}}{n!}$. When we take definite integrals,

$$\int_0^{1/2} \frac{\sin x}{x}\,dx = \int_0^{1/2}\left[1 - \frac{x^2}{3!} + \frac{x^4}{5!} - \cdots + \frac{1}{x}R_n(0, x)\right]dx$$

$$= \left\{x - \frac{x^3}{3\cdot 3!} + \frac{x^5}{5\cdot 5!} - \cdots\right\}\Big|_0^{1/2} + \int_0^{1/2}\frac{1}{x}R_n(0, x)\,dx,$$

$$= \frac{1}{2} - \frac{(1/2)^3}{3\cdot 3!} + \frac{(1/2)^5}{5\cdot 5!} - \cdots + \int_0^{1/2}\frac{1}{x}R_n(0, x)\,dx.$$

Now

$$\left| \int_0^{1/2} \frac{1}{x} R_n(0, x)\, dx \right| \le \int_0^{1/2} \left| \frac{1}{x} \frac{d^{n+1}(\sin x)}{dx^{n+1}} \right|_{x=z_n} \frac{x^{n+1}}{(n+1)!} \right|\, dx.$$

Since $\left| \dfrac{d^{n+1}(\sin x)}{dx^{n+1}} \right|_{x=z_n} \le 1$, it follows that

$$\left| \int_0^{1/2} \frac{1}{x} R_n(0, x)\, dx \right| \le \int_0^{1/2} \frac{x^n}{(n+1)!}\, dx = \left\{ \frac{x^{n+1}}{(n+1)(n+1)!} \right\}_0^{1/2} = \frac{(1/2)^{n+1}}{(n+1)(n+1)!}.$$

Thus, if we write

$$\int_0^{1/2} \frac{\sin x}{x}\, dx \approx \frac{1}{2} - \frac{(1/2)^3}{3 \cdot 3!} + \frac{(1/2)^5}{5 \cdot 5!} = 0.493\,107\,639,$$

then the maximum error is

$$\frac{(1/2)^6}{6 \cdot 6!} \quad \text{or} \quad \frac{(1/2)^7}{7 \cdot 7!},$$

depending on whether we regard $x - x^3/3! + x^5/5!$ as a fifth- or sixth-degree approximation for $\sin x$. Since $(1/2)^7/(7 \cdot 7!) < 0.000\,000\,222$ gives a smaller error, we can say that

$$0.493\,107\,639 - 0.000\,000\,222 < \int_0^{1/2} \frac{\sin x}{x}\, dx < 0.493\,107\,639 + 0.000\,000\,222;$$

that is,

$$0.493\,107\,417 < \int_0^{1/2} \frac{\sin x}{x}\, dx < 0.493\,107\,861.$$

Consequently, using only three terms of the Maclaurin series for $(\sin x)/x$, we can say that to five decimal places

$$\int_0^{1/2} \frac{\sin x}{x}\, dx = 0.493\,11.$$

An easier analysis is given in Example 10.54 of Section 10.13. It uses *alternating series* instead of Taylor's remainder formula.

Differential Equations

Many differential equations arising in physics and engineering have solutions that can be expressed only in terms of infinite series. One such equation is Bessel's differential equation of order zero for a function $y = f(x)$:

$$xy'' + y' + xy = 0.$$

Before considering this somewhat difficult differential equation, we introduce the ideas through an easier example.

EXAMPLE 10.33

Determine whether the differential equation

$$\frac{dy}{dx} - 2y = x$$

has a solution that can be expressed as a power series $y = \sum_{n=0}^{\infty} a_n x^n$ with positive radius of convergence.

SOLUTION   If

$$y = f(x) = \sum_{n=0}^{\infty} a_n x^n = a_0 + a_1 x + a_2 x^2 + \cdots$$

is to be a solution of the differential equation, we may substitute the power series into the differential equation:

$$\left(a_1 + 2a_2 x + 3a_3 x^2 + 4a_4 x^3 + \cdots\right) - 2\left(a_0 + a_1 x + a_2 x^2 + \cdots\right) = x.$$

We now gather together like terms in the various powers of x:

$$0 = (a_1 - 2a_0) + (2a_2 - 2a_1 - 1)x + (3a_3 - 2a_2)x^2 + (4a_4 - 2a_3)x^3 + \cdots.$$

Since the power series on the right has sum zero, its coefficients must all vanish (see Exercise 40 in Section 10.5), and therefore we must set

$$a_1 - 2a_0 = 0,$$

$$2a_2 - 2a_1 - 1 = 0,$$

$$3a_3 - 2a_2 = 0,$$

$$4a_4 - 2a_3 = 0,$$

and so on. These equations imply that

$$a_1 = 2a_0;$$

$$a_2 = \frac{1}{2}(1 + 2a_1) = \frac{1}{2}(1 + 4a_0);$$

$$a_3 = \frac{2}{3}a_2 = \frac{2}{3!}(1 + 4a_0);$$

$$a_4 = \frac{2}{4}a_3 = \frac{2^2}{4!}(1 + 4a_0).$$

The pattern emerging is

$$a_n = \frac{2^{n-2}}{n!}(1 + 4a_0), \qquad n \geq 2.$$

Thus,

$$f(x) = a_0 + 2a_0 x + \frac{1}{2}(1 + 4a_0)x^2 + \cdots + \frac{2^{n-2}}{n!}(1 + 4a_0)x^n + \cdots$$

$$= a_0 + 2a_0 x + \frac{1}{4}(1 + 4a_0)\left(\frac{2^2}{2!}x^2 + \frac{2^3}{3!}x^3 + \cdots + \frac{2^n}{n!}x^n + \cdots\right).$$

We can find the sum of the series in parentheses by noting that the Maclaurin series for e^{2x} is

$$e^{2x} = 1 + (2x) + \frac{(2x)^2}{2!} + \frac{(2x)^3}{3!} + \cdots.$$

Therefore, the solution of the differential equation is

$$y = f(x) = a_0 + 2a_0 x + \frac{1}{4}(1 + 4a_0)\left[e^{2x} - 1 - 2x\right]$$

$$= -\frac{1}{4} - \frac{x}{2} + \frac{1}{4}(1 + 4a_0)e^{2x} = Ce^{2x} - \frac{1}{4} - \frac{x}{2}.$$

Using power series to solve the differential equation in Example 10.33 is certainly not the most expedient method. A far simpler method will be discussed in Section 15.3. But the example clearly illustrated the procedure by which power series are used to solve differential equations. We now apply the procedure to Bessel's differential equation of order zero.

EXAMPLE 10.34

Find a power series solution $y = \sum_{n=0}^{\infty} a_n x^n$, with positive radius of convergence, for Bessel's differential equation of order zero,

$$xy'' + y' + xy = 0.$$

SOLUTION   In this example we abandon the $\cdots$ notation of Example 10.33, and maintain sigma notation throughout. When we substitute $y = \sum_{n=0}^{\infty} a_n x^n$ into the differential equation, we obtain

$$0 = x \sum_{n=2}^{\infty} n(n-1)a_n x^{n-2} + \sum_{n=1}^{\infty} na_n x^{n-1} + x \sum_{n=0}^{\infty} a_n x^n$$

$$= \sum_{n=2}^{\infty} n(n-1)a_n x^{n-1} + \sum_{n=1}^{\infty} na_n x^{n-1} + \sum_{n=0}^{\infty} a_n x^{n+1}.$$

In order to bring these three summations together as one, and combine terms in like powers of x, we lower the index of summation in the last term by 2:

$$0 = \sum_{n=2}^{\infty} n(n-1)a_n x^{n-1} + \sum_{n=1}^{\infty} na_n x^{n-1} + \sum_{n=2}^{\infty} a_{n-2} x^{n-1}.$$

We now combine the three summations over their common interval, beginning at $n = 2$, and write separately the $n = 1$ term in the second summation,

$$0 = a_1 + \sum_{n=2}^{\infty} [n(n-1)a_n + na_n + a_{n-2}]x^{n-1}.$$

But the only way a power series can be equal to zero is for all of its coefficients to be equal to zero; that is,

$$a_1 = 0; \qquad n(n-1)a_n + na_n + a_{n-2} = 0, \quad n \geq 2.$$

Thus,

$$a_n = -\frac{a_{n-2}}{n^2}, \qquad n \geq 2,$$

a recursive relation defining the unknown coefficient a_n of x^n in terms of the coefficient a_{n-2} of x^{n-2}. Since $a_1 = 0$, it follows that

$$0 = a_1 = a_3 = a_5 = \cdots.$$

For $n = 2$, $$a_2 = -\frac{a_0}{2^2}.$$

For $n = 4$, $$a_4 = -\frac{a_2}{4^2} = \frac{a_0}{2^2 4^2} = \frac{a_0}{2^4 (2!)^2}.$$

For $n = 6$, $$a_6 = -\frac{a_4}{6^2} = \frac{-a_0}{2^4 (2!)^2 6^2} = -\frac{a_0}{2^6 (3!)^2}.$$

The solution is therefore

$$y = a_0 - \frac{a_0}{2^2}x^2 + \frac{a_0}{2^4(2!)^2}x^4 - \frac{a_0}{2^6(3!)^2}x^6 + \cdots$$

$$= a_0 \sum_{n=0}^{\infty} \frac{(-1)^n}{2^{2n}(n!)^2}x^{2n}.$$

The function defined by the infinite series

$$J_0(x) = \sum_{n=0}^{\infty} \frac{(-1)^n}{2^{2n}(n!)^2}x^{2n}, \qquad -\infty < x < \infty,$$

is called the zero-order Bessel function of the first kind.

EXERCISES 10.7

In Exercises 1–10 find a maximum possible error in using the given terms of the Taylor series to approximate the function on the interval specified.

* **1.** $e^x \approx 1 + x + \frac{x^2}{2} + \frac{x^3}{6}$ for $0 \le x \le 0.01$

* **2.** $e^x \approx 1 + x + \frac{x^2}{2} + \frac{x^3}{6}$ for $0 \le x < 0.01$

* **3.** $e^x \approx 1 + x + \frac{x^2}{2} + \frac{x^3}{6}$ for $-0.01 \le x \le 0$

* **4.** $e^x \approx 1 + x + \frac{x^2}{2} + \frac{x^3}{6}$ for $|x| \le 0.01$

* **5.** $\sin x \approx x - \frac{x^3}{3!}$ for $0 \le x \le 1$

* **6.** $\cos x \approx 1 - \frac{x^2}{2!} + \frac{x^4}{4!}$ for $|x| \le 0.1$

* **7.** $\ln(1 - x) \approx -x - \frac{x^2}{2} - \frac{x^3}{3}$ $0 \le x \le 0.01$

* **8.** $\frac{1}{(1-x)^3} \approx 1 + 3x + 6x^2 + 10x^3$ for $|x| < 0.2$

* **9.** $\sin 3x \approx 3x - \frac{9x^3}{2} + \frac{81x^5}{40}$ for $|x| < \pi/100$

* **10.** $\ln x \approx (x-1) - \frac{1}{2}(x-1)^2 + \frac{1}{3}(x-1)^3 - \frac{1}{4}(x-1)^4$
for $|x - 1| \le 1/2$

In Exercises 11–14 evaluate the integral correct to three decimal places.

▦* **11.** $\int_0^1 \frac{\sin x}{x}\, dx$

▦* **12.** $\int_0^{1/2} \cos(x^2)\, dx$

▦* **13.** $\int_{-1}^{1} x^{11} \sin x\, dx$

▦* **14.** $\int_0^{0.3} e^{-x^2}\, dx$

In Exercises 15–20 use series to evaluate the limit.

* **15.** $\lim\limits_{x \to 0} \dfrac{\tan x}{x}$

* **16.** $\lim\limits_{x \to 0} \dfrac{1 - \cos x}{x^2}$

* **17.** $\lim\limits_{x \to 0} \dfrac{(1 - \cos x)^2}{3x^4}$

* **18.** $\lim\limits_{x \to 0} \dfrac{\sqrt{1+x} - 1}{x}$

* **19.** $\lim\limits_{x \to \infty} x \sin\left(\dfrac{1}{x}\right)$

* **20.** $\lim\limits_{x \to 0} \left(\dfrac{e^x + e^{-x}}{e^x - e^{-x}} - \dfrac{1}{x}\right)$

In Exercises 21–24 determine where the Maclaurin series for the function may be truncated in order to guarantee the accuracy indicated.

* **21.** $\sin(x/3)$ on $|x| \le 4$ with error less than 10^{-3}

* **22.** $1/\sqrt{1+x^3}$ on $0 < x < 1/2$ with error less than 10^{-4}

* **23.** $\ln(1 - x)$ on $|x| < 1/3$ with error less than 10^{-2}

* **24.** $\cos^2 x$ on $|x| < 0.1$ with error less than 10^{-3}

In Exercises 25–30 find a series solution in powers of x for the differential equation.

* **25.** $y' + 3y = 4$

* **26.** $y'' + y' = 0$

* **27.** $xy' - 4y = 3x$

* **28.** $4xy'' + 2y' + y = 0$

* **29.** $y'' + y = 0$

* **30.** $xy'' + y = 0$

1. Find the natural logarithm of 0.999 999 999 9 accurate to 15 decimal places.

2. In special relativity theory, the kinetic energy K of an object moving with speed v is defined by

$$K = c^2(m - m_0),$$

where c is a constant (the speed of light), m_0 is the rest mass of the object, and m is its mass when moving with speed v. The masses m and m_0 are related by

$$m = \frac{m_0}{\sqrt{1 - v^2/c^2}}.$$

Use the binomial expansion to show that

$$K = \frac{1}{2}m_0 v^2 + m_0 c^2 \left(\frac{3}{8}\frac{v^4}{c^4} + \frac{5}{16}\frac{v^6}{c^6} + \cdots\right),$$

and hence, to a first approximation, kinetic energy is defined by the classical expression $m_0 v^2/2$.

3. Stagnation pressure P_0 and pressure P are related by the Mach number M in incompressible flow of an ideal gas by the equation

$$\frac{P_0}{P} = 1 + \frac{kM^2}{2},$$

where k is a constant. For compressible flow, the relation is

$$\frac{P_0}{P} = \left[1 + \left(\frac{k-1}{2}\right)M^2\right]^{k/(k-1)}.$$

Assuming that

$$\left(\frac{k-1}{2}\right)M^2 < 1,$$

expand the latter relation to show that for small M, P_0/P in compressible flow can be approximated by P_0/P in incompressible flow.

4. The figure below shows uniform, two-dimensional, compressible, adiabatic flow of a frictionless fluid around a circular object. The pressure is P_0 and the velocity is V_0 in the undisturbed flow to the left of the object.

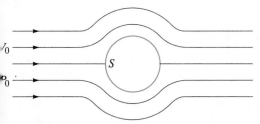

At the stagnation point S, the velocity of the fluid is zero, and we let P_s be the stagnation pressure. It is known that

$$\frac{P_s}{P_0} = \left[1 + \left(\frac{k-1}{2}\right)M_0^2\right]^{k/(k-1)},$$

where $M_0 = V_0/c_0$ is the Mach number of the flow (c_0 is the velocity of pressure propagation in the undisturbed flow), and $k > 1$ is a constant.

Show that when $(k-1)M_0^2/2 < 1$, P_s can be expressed in the form

$$P_s = P_0 + \frac{1}{2}\rho_0 V_0^2 \left[1 + \frac{M_0^2}{4} + \left(\frac{2-k}{24}\right)M_0^4 + \cdots\right],$$

where $\rho_0 = kP_0/c_0^2$.

$*$ 35. The ellipse $b^2 x^2 + a^2 y^2 = a^2 b^2$ can be represented parametrically by

$$x = a\cos t, \quad y = b\sin t, \quad 0 \le t < 2\pi.$$

(a) Show that the length of the circumference of the ellipse is defined by the definite integral

$$L = 4b \int_0^{\pi/2} \sqrt{1 - k^2 \sin^2 t}\, dt, \quad k^2 = 1 - \frac{a^2}{b^2}.$$

(b) Use the binomial expansion to show that

$$L = 2\pi b \left(1 - \frac{k^2}{4} - \frac{3k^4}{64} - \cdots\right)$$

so that to a first approximation, L is the circumference of a circle of radius b.

$*$ 36. The well function for leaky aquifers is defined by the convergent improper integral

$$W(\alpha, \beta) = \int_1^\infty \frac{1}{x} e^{-[\alpha x + \beta^2/(4\alpha x)]}\, dx.$$

(a) Show that if $e^{-\beta^2/(4\alpha x)}$ is replaced by a Maclaurin series in $\beta^2/(4\alpha x)$, and integration is done term-by-term,

$$W(\alpha, \beta) = \sum_{n=0}^\infty \frac{(-1)^n \beta^{2n}}{4^n \alpha^n n!} E_{n+1}(\alpha),$$

where $E_n(\alpha)$ is the *exponential integral*

$$E_n(\alpha) = \int_1^\infty \frac{e^{-\alpha x}}{x^n}\, dx.$$

(b) Show that the exponential integrals satisfy the recursion relation

$$E_{n+1}(\alpha) = \frac{1}{n}[e^{-\alpha} - \alpha E_n(\alpha)].$$

$*$ 37. *Planck's law* for the energy density Ψ of blackbody radiation of wavelength λ states that

$$\Psi(\lambda) = \frac{8\pi ch\lambda^{-5}}{e^{ch/(\lambda kT)} - 1},$$

where $h > 0$ is Planck's constant, c is the (constant) speed of light, and T is temperature, also assumed constant. Show that for long wavelengths, Planck's law reduces to the *Rayleigh–Jeans law*:

$$\Psi(\lambda) = \frac{8\pi kT}{\lambda^4}.$$

∗ **38.** In the figure below, charges of $q > 0$ and $-q$ coulombs are a distance d apart. The configuration is called an *electric dipole*.

(a) The electric field at point P due to these charges is

$$E = \frac{q}{4\pi\epsilon_0(x - d/2)^2} - \frac{q}{4\pi\epsilon_0(x + d/2)^2},$$

where ϵ_0 is a constant. Verify that E can be expressed in the form

$$E = \frac{q}{4\pi\epsilon_0 x^2}\left[\left(1 - \frac{d}{2x}\right)^{-2} - \left(1 + \frac{d}{2x}\right)^{-2}\right].$$

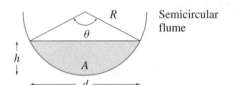

(b) Use the binomial expansion to show that when d is very much less than x, E can be approximated by

$$E = \frac{qd}{2\pi\epsilon_0 x^3}.$$

∗ **39.** Liquid flows in the semicircular flume in the figure below.

(a) Find an expression for the ratio of the cross-sectional area A of the flow to the product of h and d.

(b) Use the result in part (a) to find an approximation for the ratio that includes terms of order θ^2.

Semicircular flume

10.8 Convergence of Sequences of Numbers

In Section 10.1 we introduced the basic ideas of convergence for sequences of numbers. Examples were so simple that we had no difficulty determining whether sequences converged or diverged. As this is not always the case, we discuss two important ways to show that a sequence converges. Our concern is particularly with recursively defined sequences, since it is usually straightforward to determine whether an explicit sequence is convergent or divergent. For instance, consider the sequence

$$c_1 = 1, \qquad c_{n+1} = 5 + \sqrt{2 + c_n}, \quad n \geq 1,$$

of Example 10.2 in Section 10.1. Its next four terms are

$$c_2 = 6.732, \qquad c_3 = 7.955, \qquad c_4 = 8.155, \qquad c_5 = 8.187.$$

We suspect that the sequence has a limit. Why? Because terms are rapidly getting closer and closer to each other, or to put it another way, differences between terms are rapidly approaching zero. But this does not guarantee convergence. Differences between terms of the sequences $\{\sqrt{n}\}$ and $\{\ln n\}$ get smaller and smaller as n increases, but neither sequence has a limit. The following two definitions lead to Theorem 10.7, which can be very useful in verifying convergence of sequences like this.

DEFINITION 10.2

A sequence $\{c_n\}$ is said to be

$$(i)\ \textbf{increasing if}\qquad c_{n+1} > c_n \quad \text{for all } n \geq 1; \qquad (10.35a)$$

$$(ii)\ \textbf{nondecreasing if}\quad c_{n+1} \geq c_n \quad \text{for all } n \geq 1; \qquad (10.35b)$$

$$(iii)\ \textbf{decreasing if}\qquad c_{n+1} < c_n \quad \text{for all } n \geq 1; \qquad (10.35c)$$

$$(iv)\ \textbf{nonincreasing if}\quad c_{n+1} \leq c_n \quad \text{for all } n \geq 1. \qquad (10.35d)$$

If a sequence satisfies any one of these four properties, it is said to be **monotonic**.

DEFINITION 10.3

A sequence $\{c_n\}$ is said to have an **upper bound** U (be bounded above by U) if

$$c_n \leq U \qquad (10.36a)$$

for all $n \geq 1$. It has a **lower bound** V (is bounded below by V) if

$$c_n \geq V \qquad (10.36b)$$

for all $n \geq 1$. If a sequence has both an upper bound and a lower bound, it is said to be a **bounded sequence**.

Note that if U is an upper bound for a sequence, then any number greater than U is also an upper bound. If V is a lower bound, so too is any number smaller than V.

Let us illustrate these definitions with some simple explicit sequences before stating our first convergence theorem and applying it to the above recursive sequence. The sequence

$$\left\{ \frac{1}{2^{n-1}} \right\} = 1,\ \frac{1}{2},\ \frac{1}{4},\ \frac{1}{8},\ \cdots$$

is decreasing, has an upper bound $U = 1$, and a lower bound $V = 0$. The sequence

$$\left\{ \frac{n}{n+1} \right\} = \frac{1}{2},\ \frac{2}{3},\ \frac{3}{4},\ \frac{4}{5},\ \cdots$$

is increasing, has an upper bound $U = 5$, and a lower bound $V = -2$. The sequence

$$\{(-1)^{n+1}\} = 1,\ -1,\ 1,\ -1,\ \cdots$$

is not monotonic, has an upper bound $U = 1$, and a lower bound $V = -3$.

The above recursive sequence $c_1 = 1$, $c_{n+1} = 5 + \sqrt{2 + c_n}$ appears to be increasing, $V = 0$ is obviously a lower bound, and $U = 10$ appears to be an upper bound. The reason why we would verify these conjectures is contained in the following theorem.

THEOREM 10.7

A bounded, monotonic sequence has a limit.

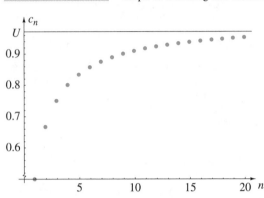

FIGURE 10.21a Graph of increasing and bounded sequence

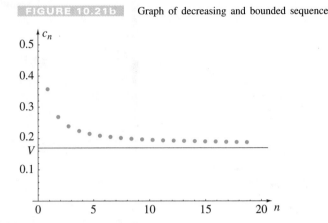

FIGURE 10.21b Graph of decreasing and bounded sequence

To expand on this statement somewhat, consider a sequence $\{c_n\}$ whose terms are illustrated graphically in Figure 10.21a. Suppose that the sequence is increasing and therefore monotonic, and that U is an upper bound for the sequence. We have shown the upper bound as a horizontal line in the figure; c_1 is a lower bound. Our intuition suggests that because the terms in the sequence always increase, and they never exceed U, the sequence must have a limit. Theorem 10.7 confirms this. The theorem does not suggest the value of the limit, but obviously it must be less than or equal to U.

Similarly, when a sequence is decreasing or nonincreasing and has a lower bound V (Figure 10.21b), it must approach a limit that is greater than or equal to V.

Another way of stating Theorem 10.7 is as follows.

COROLLARY 10.7.1

A monotonic sequence has a limit if and only if it is bounded.

We are now prepared to give a complete and typical discussion for the recursive sequence $c_1 = 1$, $c_{n+1} = 5 + \sqrt{2 + c_n}$.

EXAMPLE 10.35

Verify that the sequence

$$c_1 = 1, \qquad c_{n+1} = 5 + \sqrt{2 + c_n}, \quad n \geq 1,$$

has a limit, and find it.

SOLUTION   The first five terms of the sequence are

$$c_1 = 1, \quad c_2 = 6.732, \quad c_3 = 7.955, \quad c_4 = 8.155, \quad c_5 = 8.187.$$

They suggest that the sequence is increasing; that is, $c_{n+1} > c_n$. To prove this we use mathematical induction (see Appendix A). Certainly, the inequality is valid for $n = 1$ since $c_2 > c_1$. Suppose that k is an integer for which $c_{k+1} > c_k$. Then

$$2 + c_{k+1} > 2 + c_k,$$

from which

$$\sqrt{2 + c_{k+1}} > \sqrt{2 + c_k}.$$

It follows that

$$5 + \sqrt{2 + c_{k+1}} > 5 + \sqrt{2 + c_k}.$$

The left side is c_{k+2} and the right side is c_{k+1}. Therefore, we have proved that $c_{k+2} > c_{k+1}$. Hence, by mathematical induction $c_{n+1} > c_n$ for all $n \geq 1$. Since the sequence is increasing, its first term $c_1 = 1$ must be a lower bound. Certainly, any upper bound, if one exists, must be at least 8.187 (c_5). We can take any number greater than 8.187 and use mathematical induction to test whether it is indeed an upper bound. It appears that $U = 10$ might be a reasonable guess for an upper bound for this sequence, and we verify this by induction as follows. Clearly, $c_1 < 10$. We suppose that k is some integer for which $c_k < 10$. Then

$$c_{k+1} = 5 + \sqrt{2 + c_k} < 5 + \sqrt{2 + 10} = 5 + \sqrt{12} < 10.$$

By mathematical induction, $c_n < 10$ for $n \geq 1$.

Since the sequence is monotonic and bounded, Theorem 10.7 guarantees that it has a limit, call it L. To evaluate L, we take limits on each side of the equation defining the sequence recursively:

$$\lim_{n \to \infty} c_{n+1} = \lim_{n \to \infty} \left(5 + \sqrt{2 + c_n}\right).$$

It is important to note that this cannot be done until the conditions of Theorem 10.7 have been checked. Since terms c_n of the sequence approach L as $n \to \infty$, it follows that $5 + \sqrt{2 + c_n}$ approaches $5 + \sqrt{2 + L}$. Furthermore, as $n \to \infty$, c_{n+1} must also approach L. Do not make the mistake of saying that c_{n+1} approaches $L + 1$ as $n \to \infty$. Think about what $\lim_{n \to \infty} c_{n+1}$ means. We conclude therefore that

$$L = 5 + \sqrt{2 + L}.$$

If we transpose the 5 and square both sides of the equation, we obtain the quadratic equation

$$L^2 - 11L + 23 = 0,$$

with solutions

$$L = \frac{11 \pm \sqrt{29}}{2}.$$

Only the positive square root satisfies the original equation $L = 5 + \sqrt{2 + L}$ defining L, so that $L = (11 + \sqrt{29})/2$. The other root, $(11 - \sqrt{29})/2 \approx 2.8$, can also be eliminated on the grounds that all terms beyond the first are greater than 6.

Sequences do not have to be monotonic to be convergent. Convergence can occur for other reasons. The following example illustrates a second common way for sequences to converge. Again we have chosen a recursive sequence as illustration because for an explicit sequence, the limit is usually obvious. Consider the recursive sequence

$$c_1 = 2, \qquad c_{n+1} = 2 + \frac{1}{c_n}, \qquad n \geq 1.$$

The first six terms of the sequence are

$$c_1 = 2, \quad c_2 = 2.5, \quad c_3 = 2.4, \quad c_4 = 2.417, \quad c_5 = 2.4138, \quad c_6 = 2.414\,29.$$

The terms seem to be clustering around a number close to 2.414, one larger, one smaller, one larger, and so on (Figure 10.22). Unfortunately, lack of monotony precludes the possibility of using Theorem 10.7 to discuss convergence. In Theorem 10.8, we discuss properties that imply convergence for sequences of this type, but it is helpful first to illustrate these properties with a specific example such as the one above. What are the properties that lead us to believe that this

FIGURE 10.22 Graph of an oscillating sequence

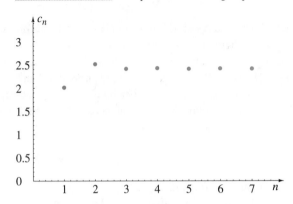

sequence converges? First, the terms are not monotonic; they are "up-down-up-down-up-down." How do we say this mathematically? Differences between successive terms in the sequence are

$$c_2 - c_1 = 2.5 - 2 = 0.5,$$
$$c_3 - c_2 = 2.4 - 2.5 = -0.1,$$
$$c_4 - c_3 = 2.417 - 2.4 = 0.017,$$
$$c_5 - c_4 = 2.4138 - 2.417 = -0.0032,$$
$$c_6 - c_5 = 2.414\,29 - 2.4138 = 0.000\,49.$$

The fact that these differences are alternately positive and negative implies that the terms in the sequence $\{c_n\}$ are up-down-up-down-up-down. This is not enough to guarantee convergence, however. For example, the sequence $1, -1, 1, -1, 1, -1, \ldots$ is up-down-up-down-up-down, but it does not converge. The added feature of the sequence above is that absolute values of the differences

$$|c_2 - c_1| = 0.5, \quad |c_3 - c_2| = 0.1, \quad |c_4 - c_3| = 0.017,$$
$$|c_5 - c_4| = 0.0032, \quad |c_6 - c_5| = 0.000\,49$$

seem to form a decreasing sequence with limit zero. We have said "seem" because we have not proved that these properties hold for all differences, only the first five. We shall provide a general verification after stating the next theorem. However, it is the up-down-up-down-up-down nature of the sequence together with the fact that absolute values of the differences decrease and approach zero that lead us to believe that the sequence $\{c_n\}$ has a limit. This is formalized in the following theorem.

THEOREM 10.8

Suppose a sequence $\{c_n\}$ has the following properties:

1. Differences $c_{n+1} - c_n$ alternate in sign.
2. Absolute values of these differences $|c_{n+1} - c_n|$ are decreasing.
3. Absolute values of the differences $|c_{n+1} - c_n|$ approach 0.

Then the sequence $\{c_n\}$ converges, and its limit lies between any two successive terms in the sequence.

A sequence that satisfies condition 1 of this theorem is said to be an **oscillating sequence**. Its terms are up-down-up-down-up-down. When terms also satisfy conditions 2 and 3, they oscillate about a limit — one below, one above, one below, one above, and so on, but gradually they get closer and closer to the limit.

We now verify that the sequence $c_1 = 2$, $c_{n+1} = 2 + 1/c_n$ on the previous page is oscillating and convergent. The difference between the $(n+1)^{th}$ and n^{th} terms is

$$c_{n+1} - c_n = \left(2 + \frac{1}{c_n}\right) - \left(2 + \frac{1}{c_{n-1}}\right) = \frac{-(c_n - c_{n-1})}{c_n c_{n-1}}.$$

Since all terms in the sequence are clearly positive, the denominator of this expression is positive. It follows that the difference $c_{n+1} - c_n$ must have the sign opposite to the previous difference $c_n - c_{n-1}$; that is, the differences $c_{n+1} - c_n$ alternate in sign. Furthermore, since 2 is a lower bound for all terms in the sequence, it follows that for $n \geq 2$,

$$|c_{n+1} - c_n| = \frac{|c_n - c_{n-1}|}{c_n c_{n-1}} < \frac{|c_n - c_{n-1}|}{(2)(2)} = \frac{|c_n - c_{n-1}|}{4}.$$

But if each difference is less than one-quarter the previous difference, the differences must be decreasing and have limit zero. By Theorem 10.8 this sequence has a limit L that we obtain by taking limits in the recursive definition:

$$\lim_{n \to \infty} c_{n+1} = \lim_{n \to \infty} \left(2 + \frac{1}{c_n}\right) \quad \Longrightarrow \quad L = 2 + \frac{1}{L}.$$

Of the two solutions $1 \pm \sqrt{2}$ of this equation, only $L = 1 + \sqrt{2}$ lies between c_1 and c_2.

Notice that we did not use mathematical induction to verify that the sequence above is oscillating and convergent. This is characteristic of oscillating sequences; mathematical induction is not required to verify the three properties of Theorem 10.8. It may be necessary to use induction to prove ancilliary results (such as a lower bound for the sequence), but induction is not needed to verify the properties of Theorem 10.8.

An oscillating sequence results when the method of **successive approximations** is applied to the equation $f(x) = x^3 + 25x - 50 = 0$ for the root between $x = 1$ and $x = 2$. With

$$x_1 = 1, \qquad x_{n+1} = \frac{50 - x_n^3}{25}, \qquad n \geq 1,$$

the next 15 terms are

$$
\begin{array}{lll}
x_2 = 1.96, & x_3 = 1.70, & x_4 = 1.80, \\
x_5 = 1.77, & x_6 = 1.7782, & x_7 = 1.7751, \\
x_8 = 1.7763, & x_9 = 1.775\,81, & x_{10} = 1.776\,00, \\
x_{11} = 1.775\,93, & x_{12} = 1.775\,954, & x_{13} = 1.775\,945, \\
x_{14} = 1.775\,948\,2, & x_{15} = 1.775\,946\,9, & x_{16} = 1.775\,947\,4.
\end{array}
$$

The terms are indeed oscillating but appear to be approaching a limit. Were we given this recursive sequence without reference to the equation $x^3 + 25x - 50 = 0$, and asked to discuss its convergence, we would verify the conditions of Theorem 10.8. In the context of solving the equation, however, we would not do this. We have what we feel is a potential candidate for the solution of the equation $x^3 + 25x - 50 = 0$ between $x = 1$ and $x = 2$. To verify that $1.775\,947$ is the solution correctly rounded to six decimal places, we use the zero intermediate value theorem with the calculations

$$f(1.775\,946\,5) = -2.7 \times 10^{-5}, \qquad f(1.775\,947\,5) = 7.3 \times 10^{-6}.$$

Now that we know what it means for a sequence to have a limit, and how to find limits, we can be more precise. To give a mathematical definition for the limit of a sequence, we start with our intuitive description and make a succession of paraphrases, each of which is one step closer to a precise definition:

A sequence $\{c_n\}$ has limit L if terms get arbitrarily close to L, and stay close to L, as n gets larger and larger.

A sequence $\{c_n\}$ has limit L if terms can be made arbitrarily close to L by choosing n sufficiently large.

A sequence $\{c_n\}$ has limit L if differences $|c_n - L|$ can be made arbitrarily close to 0 by choosing n sufficiently large.

A sequence $\{c_n\}$ has limit L if given any real number $\epsilon > 0$, no matter how small, we can make differences $|c_n - L|$ less than ϵ by choosing n sufficiently large.

Finally, we arrive at the following definition.

DEFINITION 10.4

A sequence $\{c_n\}$ has **limit** L if for any given $\epsilon > 0$, there exists an integer N such that for all $n > N$,

$$|c_n - L| < \epsilon.$$

This definition puts our intuitive idea of a limit in precise terms. For those who have studied Section 2.6, note the similarity between Definition 10.4 and Definition 2.2. For a better understanding of Definition 10.4, it is helpful to consider its geometric interpretation. The inequality $|c_n - L| < \epsilon$, when written in the form $L - \epsilon < c_n < L + \epsilon$, is interpreted as a horizontal band of width 2ϵ around L (Figure 10.23a). Definition 10.4 requires that no matter how small ϵ, we can find a stage, denoted by N, beyond which all terms in the sequence are contained in the horizontal band. For the sequence and ϵ in Figure 10.23a, N must be chosen as shown. For the same sequence, but a smaller ϵ, N must be chosen correspondingly larger (Figure 10.23b). Proofs of some results in the rest of this chapter require a working knowledge of this definition. As an example, we use it to verify that a sequence cannot have two limits, a fact that we have implicitly assumed throughout our discussions.

FIGURE 10.23a

FIGURE 10.23b

Illustration of a limit of a sequence

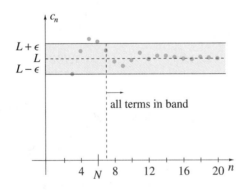

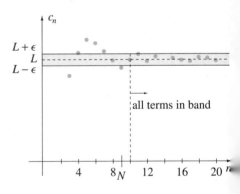

THEOREM 10.9

A sequence can have at most one limit.

PROOF We prove this by showing that a sequence cannot have two distinct limits. Suppose to the contrary that a sequence $\{c_n\}$ has two distinct limits L_1 and L_2, where $L_2 > L_1$, and let $L_2 - L_1 = \delta$. If we set $\epsilon = \delta/3$, then according to Definition 10.4, there exists an integer N_1 such that for all $n > N_1$,

$$|c_n - L_1| < \epsilon = \delta/3;$$

that is, for $n > N_1$, all terms in the sequence are within a distance $\delta/3$ of L_1.

But since L_2 is also supposed to be a limit, there exists an N_2 such that for $n > N_2$, all terms in the sequence are within a distance $\epsilon = \delta/3$ of L_2:

$$|c_n - L_2| < \epsilon = \delta/3.$$

FIGURE 10.24 Illustration of proof that a sequence cannot have two limits

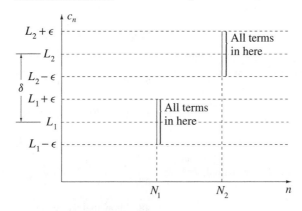

But this is impossible (Figure 10.24) if L_1 and L_2 are a distance δ apart. This contradiction therefore implies that L_1 and L_2 are the same; that is, $\{c_n\}$ cannot have two limits. ∎

The following theorem, which states some of the properties of convergent sequences, can also be proved using Definition 10.4. Only the first two parts, however, are straightforward (see Exercises 58, 66, and 67).

THEOREM 10.10

If sequences $\{c_n\}$ and $\{d_n\}$ have limits C and D, then:

(i) $\{kc_n\}$ has limit kC if k is a constant.

(ii) $\{c_n \pm d_n\}$ has limit $C \pm D$.

(iii) $\{c_n d_n\}$ has limit CD.

(iv) $\{c_n/d_n\}$ has limit C/D provided that $D \neq 0$, and none of the $d_n = 0$.

EXERCISES 10.8

In Exercises 1–24 determine whether the statement is true or false. Verify any statement that is true, and give a counterexample to any statement that is false.

1. A sequence can be increasing and nondecreasing.

2. A monotonic sequence must be bounded.

3. An increasing sequence must have a lower bound.

4. A decreasing sequence must have a lower bound.

5. An increasing sequence with a lower bound must have a limit.

6. An increasing sequence with an upper bound must have a limit.

7. If a sequence diverges, then either $\lim_{n\to\infty} c_n = \infty$ or $\lim_{n\to\infty} c_n = -\infty$.

8. A sequence cannot be both increasing and decreasing.

9. A sequence can be both nonincreasing and nondecreasing.

10. If a sequence is monotonic and has a limit, it must be bounded.

11. If a sequence is bounded and has a limit, it must be monotonic.

12. If a sequence is bounded, but not monotonic, it cannot have a limit.

13. If a sequence has a limit, it must be bounded.

14. If a sequence is not monotonic, it cannot have a limit.

15. If all terms of a sequence $\{c_n\}$ are less than U, and $L = \lim_{n\to\infty} c_n$ exists, then L must be less than U.

16. A sequence $\{c_n\}$ has a limit if and only if $\{c_n^2\}$ has a limit.

17. An oscillating sequence must converge.

18. If terms of a sequence are alternately positive and negative, the sequence is oscillating.

* **19.** A sequence $\{c_n\}$ of positive numbers converges if $c_{n+1} < c_n/2$.

* **20.** A sequence $\{c_n\}$ of numbers converges if $c_{n+1} < c_n/2$.

* **21.** If an increasing sequence has a limit L, then L must be equal to the smallest upper bound for the sequence.

* **22.** If an infinite number of terms of a sequence all have the same value a, then either a is the limit of the sequence or the sequence has no limit.

* **23.** If absolute values of differences between terms of an oscillating sequence decrease, the sequence converges.

* **24.** If terms of an oscillating sequence approach zero, then terms must be alternately positive and negative.

In Exercises 25–41 discuss, with proofs, whether the sequence is monotonic, and whether it has an upper bound, a lower bound, and a limit.

* **25.** $c_1 = 1, \quad c_{n+1} = \dfrac{1}{10}\left(c_n^3 + 12\right), \quad n \geq 1$

* **26.** $c_1 = 0, \quad c_{n+1} = \dfrac{1}{12}\left(c_n^4 + 5\right), \quad n \geq 1$

* **27.** $c_1 = 3, \quad c_{n+1} = \sqrt{5 + c_n}, \quad n \geq 1$

* **28.** $c_1 = 1, \quad c_{n+1} = \sqrt{5 + c_n}, \quad n \geq 1$

* **29.** $c_1 = 5, \quad c_{n+1} = 1 + \sqrt{6 + c_n}, \quad n \geq 1$

* **30.** $c_1 = 3, \quad c_{n+1} = 1 + \sqrt{6 + c_n}, \quad n \geq 1$

* **31.** $c_1 = 1, \quad c_{n+1} = 4 - \sqrt{5 - c_n}, \quad n \geq 1$

* **32.** $c_1 = 4, \quad c_{n+1} = 4 - \sqrt{5 - c_n}, \quad n \geq 1$

* **33.** $c_1 = 2, \quad c_{n+1} = \dfrac{1}{3 - c_n}, \quad n \geq 1$

* **34.** $c_1 = 1, \quad c_{n+1} = \dfrac{1}{4 - 2c_n}, \quad n \geq 1$

* **35.** $c_1 = 1, \quad c_{n+1} = \dfrac{7}{16 - 8c_n^2}, \quad n \geq 1$

* **36.** $c_1 = 0, \quad c_{n+1} = \dfrac{7}{16 - 8c_n^2}, \quad n \geq 1$

* **37.** $c_1 = 1, \quad c_{n+1} = \dfrac{4c_n}{4 + c_n}, \quad n \geq 1$

* **38.** $c_1 = 4, \quad c_{n+1} = \dfrac{3c_n}{2 + c_n}, \quad n \geq 1$

* **39.** $c_1 = 2, \quad c_{n+1} = \dfrac{2c_n^2}{3 + c_n}, \quad n \geq 1$

* **40.** $c_1 = \dfrac{3}{2}, \quad c_{n+1} = \dfrac{c_n + 2}{4 - c_n}, \quad n \geq 1$

* **41.** $c_1 = 0, \quad c_{n+1} = \dfrac{3 - c_n}{5 - 2c_n}, \quad n \geq 1$

* **42.** Show that the sequence

$$c_1 = 1, \quad c_{n+1} = \frac{1}{4 - c_n - c_n^2}, \quad n \geq 1$$

is monotonic and bounded. Find an approximation to its limit accurate to five decimal places.

* **43.** (a) Prove that if a sequence converges, then differences between successive terms in the sequence must approach zero.

(b) The following example illustrates that the converse is not true. Show that differences between successive terms of the sequence $\{\ln n\}$ approach zero, but the sequence itself diverges.

In Exercises 44–47 show that the sequence is convergent and find its limit.

* **44.** $c_1 = 2, \quad c_{n+1} = \dfrac{8 - c_n}{5}, \quad n \geq 1$

* **45.** $c_1 = 20, \quad c_{n+1} = 12 + \dfrac{3}{c_n}, \quad n \geq 1$

* **46.** $c_1 = 1, \quad c_{n+1} = \dfrac{1}{2 + c_n}, \quad n \geq 1$

* **47.** $c_1 = 10, \quad c_{n+1} = \dfrac{1}{3 + 2c_n}, \quad n \geq 1$

* **48.** Show that the sequence

$$c_1 = 1, \quad c_{n+1} = \frac{3}{1 + c_n}, \quad n \geq 1,$$

is convergent and find its limit. *Hint:* You may find it helpful to prove that $1 \leq c_n \leq 2$ for $n \geq 1$. Do this by mathematical induction.

* **49.** Repeat Exercise 48 if $c_1 = 0$. In this case you might want to show that $1 \leq c_n \leq 2$ for $n \geq 4$.

* **50.** Show that the sequence

$$c_1 = 1, \quad c_{n+1} = \frac{4}{2 + 5c_n}, \quad n \geq 1,$$

is convergent and find its limit. *Hint:* You may find it helpful to prove that $1/2 \leq c_n \leq 1$ for $n \geq 1$. Do this by mathematical induction.

* **51.** Show that the sequence

$$c_1 = 1, \quad c_{n+1} = \sqrt{26 - c_n}, \quad n \geq 1,$$

is convergent and find its limit. *Hint:* Show that $4 \leq c_n \leq 5$ for $n \geq 2$, and then consider $(c_{n+1})^2 - (c_n)^2$.

* **52.** Show that the sequence

$$c_1 = 4, \quad c_{n+1} = \sqrt{20 - 3c_n}, \quad n \geq 1,$$

is convergent and find its limit.

* **53.** Show that the sequence

$$c_1 = 2, \quad c_{n+1} = \frac{c_n}{4c_n - 1}, \quad n \geq 1,$$

does not converge.

* **54.** Show that the sequence in Exercise 53 diverges for any first term except $c_1 = 0$ and $c_1 = 1/2$.

* **55.** Show that the sequence

$$c_1 = 1, \quad c_{n+1} = \frac{5c_n}{2c_n - 1}, \quad n \geq 1,$$

is convergent and find its limit. *Hint:* Express c_{n+1} in the form $c_{n+1} = 5/(2 - 1/c_n)$ and show that $2 \leq c_n \leq 4$ for $n \geq 3$.

* **56.** Show that the sequence

$$c_1 = 3, \quad c_{n+1} = \sqrt[3]{10 - c_n}, \quad n \geq 1,$$

is convergent and find its limit.

7. Show that the sequence

$$c_1 = a, \qquad c_{n+1} = \frac{c_n}{bc_n - 1}, \qquad n \geq 1,$$

where a and b are nonzero constants diverges unless $ab = 2$.

8. Prove parts (i) and (ii) of Theorem 10.10.

9. Find bounds for the sequence

$$c_1 = 1, \quad c_{n+1} = \frac{1 + c_n}{1 + 2c_n}, \quad n \geq 1.$$

0. Determine whether the following sequence is monotonic, has bounds, and has a limit:

$$c_1 = -30, \quad c_2 = -20, \quad c_{n+1} = 5 + \frac{c_n}{2} + \frac{c_{n-1}}{3}, \quad n \geq 2.$$

1. A sequence $\{c_n\}$ has only positive terms. Prove that:

(a) if $\lim_{n \to \infty} c_n = L < 1$, there exists an integer N such that for all $n \geq N$,

$$c_n < \frac{L + 1}{2};$$

(b) if $\lim_{n \to \infty} c_n = L > 1$, there exists an integer N such that for all $n \geq N$,

$$c_n > \frac{L + 1}{2}.$$

These results are used in Theorem 10.16, Section 10.11.

2. Prove that if $\lim_{n \to \infty} c_n = L$, there exists an integer N such that for all $n \geq N$,
$$c_n < L + 1.$$

This result is used in Theorem 10.15, Section 10.10.

3. The *Fibonacci sequence* found in many areas of applied mathematics is defined by

$$c_1 = 1, \quad c_2 = 1, \quad c_{n+1} = c_n + c_{n-1}, \quad n \geq 2.$$

(a) Evaluate the first 10 terms of this sequence.

(b) Is the sequence monotonic, is it bounded, and does it have a limit?

(c) Prove that

$$c_n^2 - c_{n-1}c_{n+1} = (-1)^{n+1}, \quad n \geq 2.$$

(d) Verify that an explicit formula for c_n is

$$c_n = \frac{1}{\sqrt{5}} \left[\left(\frac{1 + \sqrt{5}}{2} \right)^n - \left(\frac{1 - \sqrt{5}}{2} \right)^n \right].$$

(e) Define a sequence $\{b_n\}$ as the ratio of terms in the Fibonacci sequence

$$b_n = \frac{c_{n+1}}{c_n}.$$

Is this sequence monotonic? Does it have a limit?

* **64.** A particular breed of rabbits grows and reproduces according to the following schedule. Each female rabbit becomes an adult and produces a pair of babies (one male and one female) at the age of 2 months and produces exactly one pair every month thereafter. Suppose we begin with one pair of adult rabbits who produce a pair of babies at the end of the first month. If no rabbits die, find a sequence $\{R_n\}$ representing the number of adult female rabbits after n months. Do you recognize it?

* **65.** Prove that the sequence

$$c_1 = d, \quad c_{n+1} = \sqrt{a + 2bc_n}, \quad n \geq 1$$

(a, b, and d all positive constants) is increasing if and only if $d < b + \sqrt{a + b^2}$. What happens when $d = b + \sqrt{a + b^2}$?

** **66.** In this exercise we outline a proof of part (iii) of Theorem 10.10.

(a) Verify that

$$|c_n d_n - CD| \leq |c_n||d_n - D| + |D||c_n - C|.$$

(b) Show that given any $\epsilon > 0$, there exist positive integers N_1, N_2, and N_3 such that

$$|c_n| < |C| + 1 \qquad \text{whenever } n > N_1,$$

$$|d_n - D| < \frac{\epsilon}{2(|C| + 1)} \qquad \text{whenever } n > N_2,$$

$$|c_n - C| < \frac{\epsilon}{2|D| + 1} \qquad \text{whenever } n > N_3.$$

(c) Use these results to prove part (iii) of Theorem 10.10.

** **67.** In this exercise we outline a proof for part (iv) of Theorem 10.10.

(a) Verify that when $d_n \neq 0$ and $D \neq 0$,

$$\left| \frac{c_n}{d_n} - \frac{C}{D} \right| \leq \frac{|c_n - C|}{|d_n|} + \frac{|C||d_n - D|}{|D||d_n|}.$$

(b) Show that given any $\epsilon > 0$, there exist positive integers N_1, N_2, and N_3 such that

$$|d_n| > \frac{|D|}{2} \qquad \text{whenever } n > N_1,$$

$$|c_n - C| < \frac{\epsilon|D|}{4} \qquad \text{whenever } n > N_2,$$

$$|d_n - D| < \frac{\epsilon|D|^2}{4|C| + 1} \qquad \text{whenever } n > N_3.$$

(c) Now prove part (iv) of Theorem 10.10.

** **68.** Find an explicit formula for the recursive sequence

$$c_1 = 1, \quad c_2 = 2, \quad c_{n+1} = \frac{c_n + c_{n-1}}{2}, \quad n \geq 2.$$

10.9 Infinite Series of Numbers

Infinite series of numbers arise in two ways. When $\{c_n\}$ is an infinite sequence of numbers, the expression

$$\sum_{n=1}^{\infty} c_n = c_1 + c_2 + c_3 + \cdots + c_n + \cdots \qquad (10.37)$$

is called an **infinite series of numbers**, or simply a *series*. From the sequences of Example 10.1, we may form the following series of numbers.

EXAMPLE 10.36

Write out the first six terms of the following series:

(a) $\displaystyle\sum_{n=1}^{\infty} \frac{1}{2^{n-1}}$ (b) $\displaystyle\sum_{n=1}^{\infty} \frac{n}{n+1}$ (c) $\displaystyle\sum_{n=1}^{\infty} (-1)^n |n-3|$ (d) $\displaystyle\sum_{n=1}^{\infty} (-1)^{n+1}$

SOLUTION The first six terms of these series are shown below.

(a) $\displaystyle\sum_{n=1}^{\infty} \frac{1}{2^{n-1}} = 1 + \frac{1}{2} + \frac{1}{4} + \frac{1}{8} + \frac{1}{16} + \frac{1}{32} + \cdots$

(b) $\displaystyle\sum_{n=1}^{\infty} \frac{n}{n+1} = \frac{1}{2} + \frac{2}{3} + \frac{3}{4} + \frac{4}{5} + \frac{5}{6} + \frac{6}{7} + \cdots$

(c) $\displaystyle\sum_{n=1}^{\infty} (-1)^n |n-3| = -2 + 1 - 0 + 1 - 2 + 3 - \cdots$

(d) $\displaystyle\sum_{n=1}^{\infty} (-1)^{n+1} = 1 - 1 + 1 - 1 + 1 - 1 + \cdots$

Infinite series of numbers also arise when specific values of x are substituted into power or Taylor series. For example, if we substitute $x = 1$ into the power series $\sum_{n=1}^{\infty} x^n/(n2^n)$, we obtain

$$\sum_{n=1}^{\infty} \frac{1}{n\,2^n} = \frac{1}{2} + \frac{1}{2 \cdot 2^2} + \frac{1}{3 \cdot 2^3} + \cdots + \frac{1}{n\,2^n} + \cdots;$$

if we set $x = -2$, we obtain

$$\sum_{n=1}^{\infty} \frac{(-2)^n}{n\,2^n} = -1 + \frac{1}{2} - \frac{1}{3} + \frac{1}{4} - \cdots + \frac{(-1)^n}{n} + \cdots.$$

As mentioned earlier in this chapter, it is not possible to add an infinity of numbers in a finite amount of time, and therefore expression 10.37 is as yet meaningless. To attach a meaning we take the same approach as we did for Taylor series and power series. We illustrate with a simple example, and then give a formal definition. Consider the infinite series of Example 10.36(a),

$$\sum_{n=1}^{\infty} \frac{1}{2^{n-1}} = 1 + \frac{1}{2} + \frac{1}{4} + \frac{1}{8} + \cdots.$$

If we start adding terms, a pattern soon emerges. Indeed, if we denote by S_n the sum of the first n terms of this series, we find that

$$S_1 = 1,$$

$$S_2 = 1 + \frac{1}{2} = \frac{3}{2}, \qquad \text{(sum of first two terms)},$$

$$S_3 = 1 + \frac{1}{2} + \frac{1}{4} = \frac{7}{4}, \qquad \text{(sum of first three terms)},$$

$$S_4 = 1 + \frac{1}{2} + \frac{1}{4} + \frac{1}{8} = \frac{15}{8}, \qquad \text{(sum of first four terms)},$$

$$S_5 = 1 + \frac{1}{2} + \frac{1}{4} + \frac{1}{8} + \frac{1}{16} = \frac{31}{16}, \qquad \text{(sum of first five terms)},$$

and so on. We see that the sum of the first n terms of the series is given by the formula

$$S_n = \frac{2^n - 1}{2^{n-1}} = 2 - \frac{1}{2^{n-1}}.$$

As we add more and more terms of this series together, the sum S_n gets closer and closer to 2. It is always less than 2, but S_n can be made arbitrarily close to 2 by choosing n sufficiently large. If this series is to have a sum, the only reasonable sum is 2. In practice, this is precisely what we do; we define the sum of the series $\sum_{n=1}^{\infty} 1/2^{n-1}$ to be 2.

Let us now take this idea and define sums for general infinite series. We begin by defining a sequence $\{S_n\}$ as follows:

$$S_1 = c_1,$$

$$S_2 = c_1 + c_2,$$

$$S_3 = c_1 + c_2 + c_3,$$

$$\vdots \qquad \vdots$$

$$S_n = c_1 + c_2 + c_3 + \cdots + c_n,$$

$$\vdots \qquad \vdots$$

called the **sequence of partial sums** of the series $\sum_{n=1}^{\infty} c_n$. The nth term of the sequence $\{S_n\}$ represents the sum of the first n terms of the series. If this sequence has a limit, say S, then the more terms of the series that we add together, the closer the sum gets to S. It seems reasonable, then, to call S the sum of the series. We therefore make the following definition.

DEFINITION 10.5

Let $S_n = \sum_{k=1}^{n} c_k$ be the nth partial sum of a series $\sum_{n=1}^{\infty} c_n$. If the sequence of partial sums $\{S_n\}$ has limit S,

$$S = \lim_{n \to \infty} S_n,$$

we call S the sum of the series and write

$$\sum_{n=1}^{\infty} c_n = S;$$

if $\{S_n\}$ does not have a limit, we say that the series does not have a sum.

If a series has sum S, we say that the series **converges** to S, which means that its sequence of partial sums converges to S. If a series does not have a sum, we say that the series **diverges**, which means that its sequence of partial sums diverges. Partial sums are to infinite series of numbers what Taylor polynomials are to Taylor series. Before proceeding with examples we need to make one comment about terminology and one about notation. In the next few sections we will be concentrating on series of numbers as opposed to power series or series of functions. We shall usually write *series* instead of *series of numbers*; the context will always make it clear when we are discussing series of numbers. Whenever the first term of a series corresponds to $n = 1$, we shall drop the limits $n = 1$ and ∞ on the sigma notation when the notation appears in text. For instance, we shall write $\sum c_n$ in place of $\sum_{n=1}^{\infty} c_n$. Limits will always be retained when sigma notation appears in a displayed equation, or when the lower limit is not $n = 1$.

According to Definition 10.5, the series $\sum 1/2^{n-1}$ of Example 10.36(a) has sum 2. Since every term of the series $\sum n/(n+1)$ in Example 10.36(b) is greater than or equal to $1/2$, it follows that the sum of the first n terms is $S_n \geq n(1/2) = n/2$. As the sequence of partial sums is therefore unbounded, it cannot possibly have a limit (see the corollary to Theorem 10.7). The series does not therefore have a sum; it diverges. Examination of the first few partial sums of the series $\sum (-1)^n |n - 3|$ in Example 10.36(c) leads to the result that for $n > 1$,

$$
S_n = \begin{cases} \dfrac{n-4}{2}, & \text{if } n \text{ is even,} \\[2mm] \dfrac{1-n}{2}, & \text{if } n \text{ is odd.} \end{cases}
$$

Since the sequence $\{S_n\}$ does not have a limit, the series diverges. The partial sums of the series $\sum (-1)^{n+1}$ are

$$
S_1 = 1, \quad S_2 = 0, \quad S_3 = 1, \quad S_4 = 0, \quad \dots.
$$

Since this sequence does not have a limit, the series does not have a sum.

Perhaps the most important series of numbers, and certainly the series that occurs most frequently in applications, is the *geometric series*. We discussed it in the context of power series in Section 10.4, but it is worthwhile repeating the ideas here, unencumbered by other considerations. Geometric series are of the form

$$
\sum_{n=1}^{\infty} ar^{n-1} = a + ar + ar^2 + ar^3 + \cdots. \tag{10.38}
$$

Each term is obtained by multiplying the preceding term by the same constant r, the *common ratio*. If $\{S_n\}$ is the sequence of partial sums for this series, then

$$
S_n = a + ar + ar^2 + \cdots + ar^{n-1}.
$$

If we multiply this equation by r,

$$
r S_n = ar + ar^2 + ar^3 + \cdots + ar^n.
$$

When we subtract these equations, the result is

$$
S_n - r S_n = a - ar^n.
$$

Hence, for $r \neq 1$, we obtain

$$
S_n = \frac{a(1 - r^n)}{1 - r}.
$$

Furthermore, when $r = 1$,

$$
S_n = na,
$$

and therefore the sum of the first n terms of a geometric series is

$$S_n = \begin{cases} \dfrac{a(1 - r^n)}{1 - r}, & r \neq 1 \\ na, & r = 1. \end{cases}$$ (10.39a)

To determine whether a geometric series has a sum, we consider the limit of this sequence. Certainly, $\lim_{n\to\infty} na$ does not exist, unless trivially $a = 0$. In addition, $\lim_{n\to\infty} r^n = 0$ when $|r| < 1$, and does not exist when $|r| > 1$. Nor does it exist when $r = -1$. Thus, we may state that

$$\lim_{n\to\infty} S_n = \begin{cases} \dfrac{a}{1 - r}, & |r| < 1 \\ \text{does not exist,} & |r| \geq 1. \end{cases}$$

The geometric series therefore has the sum

$$\sum_{n=1}^{\infty} ar^{n-1} = \frac{a}{1 - r}$$ (10.39b)

if $|r| < 1$, but otherwise diverges.

The series in parts (a) and (d) of Example 10.36 are geometric. The first has common ratio $1/2$ and therefore converges to $1/(1 - 1/2) = 2$; the second has common ratio -1 and therefore diverges.

We encounter the **harmonic series**

$$\sum_{n=1}^{\infty} \frac{1}{n} = 1 + \frac{1}{2} + \frac{1}{3} + \frac{1}{4} + \cdots$$ (10.40)

quite often in our work. It does not have a sum, and we can show this by considering the following partial sums of the series:

$$S_1 = 1,$$

$$S_2 = 1 + \frac{1}{2} = \frac{3}{2},$$

$$S_4 = S_2 + \frac{1}{3} + \frac{1}{4} > \frac{3}{2} + \frac{1}{4} + \frac{1}{4} = \frac{4}{2},$$

$$S_8 = S_4 + \frac{1}{5} + \frac{1}{6} + \frac{1}{7} + \frac{1}{8} > \frac{4}{2} + \frac{1}{8} + \frac{1}{8} + \frac{1}{8} + \frac{1}{8} = \frac{5}{2},$$

$$S_{16} = S_8 + \frac{1}{9} + \frac{1}{10} + \cdots + \frac{1}{16} > \frac{5}{2} + \frac{1}{16} + \frac{1}{16} + \cdots + \frac{1}{16} = \frac{6}{2}.$$

This procedure can be continued indefinitely and shows that the sequence of partial sums is unbounded. The harmonic series therefore diverges (see once again the corollary to Theorem 10.7).

In each of the examples above, we used Definition 10.5 for the sum of a series to determine whether the series converges or diverges; that is, we formed the sequence of partial sums $\{S_n\}$ in order to consider its limit. For most examples, it is either too difficult or impossible to evaluate S_n in a simple form, and in such cases consideration of the limit of the sequence $\{S_n\}$ is impractical. Consequently, we must develop alternative ways to decide on the convergence of a series. We do this in Sections 10.10–10.12.

We now discuss some fairly simple but important results on convergence of series. If a series $\sum c_n$ has a finite number of its terms altered in any fashion whatsoever, the new series converges if and only if the original series converges. The new series may converge to a different sum, but it converges if the original series converges. For example, if we double the first three terms of the geometric series in Example 10.36(a), but do not change the remaining terms, the new series is

$$2 + 1 + \frac{1}{2} + \frac{1}{8} + \frac{1}{16} + \cdots.$$

It is not geometric. But its nth partial sum, call it S_n, is very closely related to that of the geometric series, call it T_n. In fact, for $n \geq 4$, we can say that $S_n = T_n + 7/4$ (7/4 is the total change in the first three terms). Since $\lim_{n \to \infty} T_n = 2$, it follows that $\lim_{n \to \infty} S_n = 2 + 7/4 = 15/4$; that is, the new series converges, but its sum is $7/4$ greater than that of the geometric series.

On the other hand, suppose we change the first 100 terms of the harmonic series, which diverges, to 0, but leave the remaining terms unaltered. The new series is

$$\overbrace{0 + 0 + \cdots + 0}^{100 \text{ terms}} + \frac{1}{101} + \frac{1}{102} + \cdots.$$

We know that the sequence of partial sums $\{T_n\}$ of the harmonic series is increasing and unbounded. The nth partial sum S_n of the new series, for $n > 100$, is equal to $T_n - k$, where k is the sum of the first 100 terms of the harmonic series. But because $\lim_{n \to \infty} T_n = \infty$, so must $\lim_{n \to \infty} S_n = \infty$; that is, the new series diverges.

The following theorem indicates that convergent series can be added and subtracted, and multiplied by constants. These properties can be verified using Definition 10.5.

THEOREM 10.11

If series $\sum_{n=1}^{\infty} c_n$ and $\sum_{n=1}^{\infty} d_n$ have sums C and D, then:

(i) $\displaystyle\sum_{n=1}^{\infty} kc_n = kC$ (when k is a constant).   (10.41a)

(ii) $\displaystyle\sum_{n=1}^{\infty} (c_n \pm d_n) = C \pm D.$ (10.41b)

Our first convergence test is a corollary to the following theorem.

THEOREM 10.12

If a series $\sum_{n=1}^{\infty} c_n$ converges, then $\lim_{n \to \infty} c_n = 0$.

PROOF   If series $\sum c_n$ has a sum S, its sequence of partial sums

$$S_1, \ S_2, \ \ldots, \ S_n, \ \ldots$$

has limit S. The sequence

$$0, \ S_1, \ S_2, \ \ldots, \ S_{n-1}, \ \ldots$$

must also have limit S; it is the sequence of partial sums with an additional term equal to 0 at the beginning. According to Theorem 10.10, if we subtract these two sequences, the resulting sequence must have limit $S - S = 0$; that is,

$$S_1 - 0, \ S_2 - S_1, \ \cdots, \ S_n - S_{n-1}, \ \cdots \to 0.$$

But $S_1 - 0 = S_1 = c_1$, $S_2 - S_1 = c_2$, and so on; that is, we have shown that $\lim_{n \to \infty} c_n = 0$.

Theorem 10.12 states that a necessary condition for a series $\sum c_n$ to converge is that the sequence $\{c_n\}$ of its terms must approach zero. What we really want are sufficient conditions to guarantee convergence or divergence of a series. We can take the contrapositive of the theorem and obtain the following.

> ### COROLLARY 10.12.1 (n^{th}-Term Test)
> If $\lim_{n \to \infty} c_n \neq 0$, or does not exist, then the series $\sum_{n=1}^{\infty} c_n$ diverges.

This is our first convergence test, the nth-term test. It is, in fact, a test for divergence rather than convergence, stating that if $\lim_{n \to \infty} c_n$ exists and is equal to anything but zero, or the limit does not exist, then the series $\sum c_n$ diverges. Note well that the nth-term test never indicates that a series converges. Even if $\lim_{n \to \infty} c_n = 0$, we can conclude nothing about the convergence or divergence of $\sum c_n$; it may converge or it may diverge. For example, the harmonic series $\sum 1/n$ and the geometric series $\sum 1/2^{n-1}$ both satisfy the condition $\lim_{n \to \infty} c_n = 0$, yet one series diverges and the other converges. The nth-term test therefore may indicate that a series diverges, but it never indicates that a series converges.

In particular, both series $\sum (-1)^n |n - 3|$ and $\sum (-1)^{n+1}$ of Example 10.36 diverge by the nth-term test.

To understand series, it is crucial to distinguish clearly among three entities: the series itself, its sequence of partial sums, and its sequence of terms. For any series $\sum c_n$, we can form its sequence of terms $\{c_n\}$ and its sequence of partial sums $\{S_n\}$. Each of these sequences may give information about the sum of the series $\sum c_n$, but in very different ways.

The sum of the series is defined by its sequence of partial sums in that $\sum c_n$ has a sum only if $\{S_n\}$ has a limit. The sequence of partial sums therefore tells us definitely whether the series converges or diverges, provided that we can evaluate S_n in a simple form.

The sequence of terms $\{c_n\}$, on the other hand, may or may not tell us whether the series diverges. If $\{c_n\}$ has no limit or has a limit other than zero, we know that the series does not have a sum. If $\{c_n\}$ has limit zero, we obtain no information about convergence of the series, and must continue our investigation.

The interval of convergence of a power series (or Taylor series) is all values of x for which the series converges. It is determined, except possibly for endpoints, by the radius of convergence. As we uncover convergence tests for series of numbers in this and the next three sections, we will be able to discuss endpoints of intervals of convergence, and consequently, be able to convert open intervals of convergence to intervals of convergence. As a first example, we show that the open interval of convergence for the series $\sum_{n=1}^{\infty} x^{2n}/(n\, 4^n)$ is its interval of convergence; that is, endpoints are not included.

EXAMPLE 10.37

Find the interval of convergence for the power series $\displaystyle\sum_{n=1}^{\infty} \frac{1}{n\, 4^n} x^{2n}$.

SOLUTION   If we set $y = x^2$, then

$$\sum_{n=1}^{\infty} \frac{1}{n\, 4^n} x^{2n} = \sum_{n=1}^{\infty} \frac{1}{n\, 4^n} y^n.$$

According to equation 10.24a, the radius of convergence of this series is

$$R_y = \lim_{n \to \infty} \left| \frac{\dfrac{1}{n\, 4^n}}{\dfrac{1}{(n+1)4^{n+1}}} \right| = \lim_{n \to \infty} 4\left(\frac{n+1}{n} \right) = 4.$$

Consequently, the radius of convergence of the power series in x is $R_x = 2$, and the open interval of convergence is $-2 < x < 2$. At $x = \pm 2$, the series becomes

$$\sum_{n=1}^{\infty} \frac{1}{n \, 4^n}(\pm 2)^{2n} = \sum_{n=1}^{\infty} \frac{1}{n},$$

the harmonic series, which diverges. The interval of convergence is therefore $-2 < x < 2$.

EXAMPLE 10.38

Application Preview Revisited

Solve the problem in the Application Preview.

SOLUTION Were we to plot the amount of chemical in the reactor, the result would be somewhat like that in Figure 10.25.

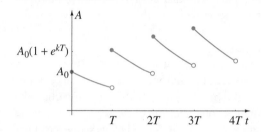

FIGURE 10.25 Amount of chemical in a reactor

We need to find a formula for the amount of chemical in the reactor at times $t = nT$ taking for granted that the amount A_0 has been injected at this time. We already know that $A(0) = A_0$, and $A(T) = A_0(1 + e^{kT})$. During the time interval $T < t < 2T$, the amount $A_0(1 + e^{kT})$ decreases exponentially so that during this time interval

$$A(t) = A_0(1 + e^{kT})e^{k(t-T)}.$$

As $t \to 2T^-$, this approaches

$$\lim_{t \to 2T^-} A(t) = A_0(1 + e^{kT})e^{k(2T-T)} = A_0(1 + e^{kT})e^{kT}.$$

Once the injection of A_0 at $t = 2T$ is made,

$$A(2T) = A_0 + A_0(1 + e^{kT})e^{kT} = A_0(1 + e^{kT} + e^{2kT}).$$

There is a pattern emerging here that we could verify by mathematical induction, but suppose we accept that

$$A(nT) = A_0(1 + e^{kT} + e^{2kT} + \cdots + e^{nkT}).$$

This is a geometric sum, and if we use formula 10.39a,

$$A(nT) = A_0 \left[\frac{1 - (e^{kT})^{n+1}}{1 - e^{kT}} \right] = A_0 \left[\frac{1 - e^{(n+1)kT}}{1 - e^{kT}} \right].$$

We can now say that the process should be terminated for the smallest integer n such that

$$A_0 \left[\frac{1 - e^{(n+1)kT}}{1 - e^{kT}} \right] > L.$$

This can be solved for

$$n > \frac{1}{kT} \ln \left[1 - \frac{L}{A_0}(1 - e^{kT}) \right] - 1.$$

EXERCISES 10.9

In Exercises 1–10 determine whether the series converges or diverges. Find the sum of each convergent series. To get a *feeling* for a series, it is helpful to write out its first few terms. Try it.

1. $\displaystyle\sum_{n=1}^{\infty} \frac{n+1}{2n}$

2. $\displaystyle\sum_{n=1}^{\infty} \frac{2^n}{5^{n+1}}$

3. $\displaystyle\sum_{n=1}^{\infty} \cos\left(\frac{n\pi}{2}\right)$

4. $\displaystyle\sum_{n=1}^{\infty} \left(\frac{n}{n+1}\right)^n$

5. $\displaystyle\sum_{n=1}^{\infty} \frac{7^{2n+3}}{3^{2n-2}}$

6. $\displaystyle\sum_{n=1}^{\infty} \frac{7^{n+3}}{3^{2n-2}}$

7. $\displaystyle\sum_{n=1}^{\infty} \sqrt{\frac{n^2-1}{n^2+1}}$

8. $\displaystyle\sum_{n=1}^{\infty} \frac{\cos(n\pi)}{2^n}$

9. $\displaystyle\sum_{n=1}^{\infty} \frac{4^n+3^n}{3^n}$

10. $\displaystyle\sum_{n=1}^{\infty} \text{Tan}^{-1} n$

In Exercises 11–14 we have given a repeating decimal. Express the decimal as a geometric series and use formula 10.39b to express it as a rational number.

11. $0.666\,666\ldots$

12. $0.131\,313\,131\,3\ldots$

13. $1.347\,346\,346\,346\ldots$

14. $43.020\,502\,050\,205\ldots$

In Exercises 15–17 complete the statement and give a short proof to substantiate your claim.

15. If $\sum c_n$ and $\sum d_n$ converge, then $\sum (c_n + d_n)$ ...

16. If $\sum c_n$ converges and $\sum d_n$ diverges, then $\sum (c_n + d_n)$ ...

17. If $\sum c_n$ and $\sum d_n$ diverge, then $\sum (c_n + d_n)$ ...

In Exercises 18–21 determine whether the series converges or diverges. Find the sum of each convergent series.

18. $\displaystyle\sum_{n=1}^{\infty} \frac{2^n+3^n}{4^n}$

19. $\displaystyle\sum_{n=1}^{\infty} \frac{3^n-1}{2^n}$

20. $\displaystyle\sum_{n=1}^{\infty} \frac{n^2+2^{2n}}{4^n}$

21. $\displaystyle\sum_{n=1}^{\infty} \frac{2^n+4^n-8^n}{2^{3n}}$

22. Find the sum of the series

$$\sum_{n=1}^{\infty} \frac{1}{n(n+1)}.$$

Hint: Use partial fractions on the nth term and find the sequence of partial sums.

* 23. Find the total distance travelled by the superball in Exercise 56 of Section 10.1 before it comes to rest.

* 24. Find the time taken for the superball in Exercise 56 of Section 10.1 to come to rest.

* 25. What distance does the dog run from the time when it sees the farmer until the farmer reaches the farmhouse in Exercise 57 of Section 10.1?

* 26. Find a simplified formula for the area A_n in Exercise 61 of Section 10.1. What is $\lim_{n\to\infty} A_n$?

* 27. Find all values of x satisfying the inequality

$$1 + x + x^2 + \cdots + x^n \geq 0,$$

where $n \geq 1$ is an integer.

* 28. (a) According to equation 10.39a, the nth partial sum of the geometric series $1 + r + r^2 + r^3 + \cdots$ is $S_n = (1 - r^n)/(1-r)$. If T_n denotes the nth partial sum of the series

$$1 + 2r + 3r^2 + 4r^3 + \cdots,$$

show that

$$T_n - S_n = r(T_n - nr^{n-1}).$$

Solve this equation for T_n and take the limit as $n \to \infty$ to show that

$$\sum_{n=1}^{\infty} nr^{n-1} = \frac{1}{(1-r)^2}, \quad |r| < 1.$$

(b) Verify this result by setting $S(r) = \sum_{n=1}^{\infty} nr^{n-1}$ and integrating.

In Exercises 29–32 use the result of Exercise 28 to find the sum of the series.

29. $\dfrac{1}{2} + \dfrac{2}{2^2} + \dfrac{3}{2^3} + \dfrac{4}{2^4} + \cdots$

30. $\dfrac{2}{5} + \dfrac{4}{25} + \dfrac{6}{125} + \dfrac{8}{625} + \cdots$

* 31. $\dfrac{2}{3} + \dfrac{3}{27} + \dfrac{4}{243} + \dfrac{5}{2187} + \cdots$

* 32. $\dfrac{12}{5} + \dfrac{48}{25} + \dfrac{192}{125} + \dfrac{768}{625} + \cdots$

* 33. Two people flip a single coin to see who can first flip a head. Show that the probability that the first person to flip wins the game is represented by the series

$$\frac{1}{2} + \frac{1}{8} + \frac{1}{32} + \cdots + \frac{1}{2^{2n-1}} + \cdots.$$

What is the sum of this series?

* **34.** Two people throw a die to see who can first throw a six. Find the probability that the person who throws first wins the game.

In Exercises 35–38 show that the interval of convergence of the power series is the same as the open interval of convergence.

* **35.** $\displaystyle\sum_{n=0}^{\infty} \frac{1}{2^n} x^n$

* **36.** $\displaystyle\sum_{n=1}^{\infty} n^2 3^n x^n$

* **37.** $\displaystyle\sum_{n=2}^{\infty} 2^n \left(\frac{n-1}{n+1}\right)^2 (x-4)^n$

* **38.** $\displaystyle\sum_{n=0}^{\infty} (-1)^n x^{3n}$

* **39.** One of *Zeno's paradoxes* describes a race between Achilles and a tortoise. Zeno claims that if the tortoise is given a head start, then no matter how fast Achilles runs, he can never catch the tortoise. He reasons as follows: In order to catch the tortoise, Achilles must first make up the length of the head start. But while he is running this distance, the tortoise "runs" a further distance. While Achilles makes up this distance, the tortoise covers a further distance, and so on. It follows that Achilles is always making up distance covered by the tortoise, and therefore can never catch the tortoise. If the tortoise is given a head start of length L and Achilles runs $c > 1$ times as fast as the tortoise, use infinite series to show that Achilles does in fact catch the tortoise and that the distance he covers in doing so is $cL(c-1)^{-1}$.

* **40.** Find the time between 1:05 and 1:10 when the minute and hour hands of a clock point in the same direction (a) by reasoning with infinite series as in Exercise 39 and (b) by finding expressions for the angular displacements of the hands as functions of time.

* **41.** Repeat Exercise 40 for the instant between 10:50 and 10:55 when the hands coincide.

* **42.** A child has a large number of identical cubical blocks, which she stacks as shown in the following figure. The top block protrudes $\frac{1}{2}$ its length over the second block, which protrudes $\frac{1}{4}$ its length over the third block, which protrudes $\frac{1}{6}$ its length over the fourth block, and so on. Assuming the centre of mass of each block is at its geometric centre, show that the centre of mass of the top n blocks lies directly over the edge of the $(n+1)^{\text{th}}$ block. Now deduce that if a sufficient number of blocks is piled, the top block can be made to protrude as far over the bottom block as desired without the stack falling.

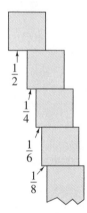

* **43.** Prove the following result: If $\sum_{n=1}^{\infty} c_n$ converges, then its terms can be grouped in any manner, and the resulting series is convergent with the same sum as the original series.

** **44.** The Laplace transform of a function $f(t)$ was defined by an im proper integral in the instructions to Exercises 27–30 in Section 8.2 Show that when $f(t)$ is a continuous function with period p, it Laplace transform is given by the ordinary integral

$$F(s) = \frac{1}{1 - e^{-ps}} \int_0^p f(t) e^{-st}\, dt.$$

** **45.** Suppose the voltage V_{in} applied to the circuit in the left figur below is as shown in the right figure. Currents and voltages in th circuit are initially zero. The rectifier passes current freely (with n resistance) in the forward direction, but prevents current in the revers direction.

(a) Show that the differential equation describing the voltag across the capacitor in the time interval $2(n-1)T < t <$ $(2n-1)T$, $n = 1, 2, \ldots$, is

$$\frac{dV}{dt} + \tau V = \alpha \overline{V}, \quad \text{where} \quad \tau = \frac{R_1 + R_2}{R_1 R_2 C} \quad \text{and} \quad \alpha = \frac{1}{R_1 C}$$

Hint: Use the fact that the current through R_1 must be the sum of the currents through R_2 and C in order to find the voltage across R_1 when the voltage across C is V.

(b) Solve the differential equation in part (a) by multiplying it by $e^{\tau t}$. By denoting the voltage across the capacitor a time $t = 2nT$ by V_n, show that for $2(n-1)T < t <$ $(2n-1)T$,

$$V = \frac{\alpha \overline{V}}{\tau} + \left(V_{n-1} - \frac{\alpha \overline{V}}{\tau}\right) - e^{-\tau[t - 2(n-1)T]}.$$

What is V at $t = (2n-1)T$?

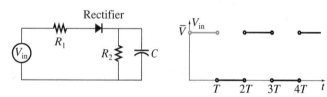

(c) Show that the differential equation describing the voltag across the capacitor in the time interval $(2n-1)T < t <$ $2nT$ is

$$\frac{dV}{dt} + \sigma V = 0, \quad \text{where} \quad \sigma = \frac{1}{R_2 C}.$$

(d) Solve the differential equation in part (c) and use the fac that $\displaystyle\lim_{t \to (2n-1)T^+} V$ should be $V\big((2n-1)T\big)$ as calculate in part (b) to show that for $(2n-1)T < t < 2nT$,

$$V = \left[\frac{\alpha \overline{V}}{\tau} + \left(V_{n-1} - \frac{\alpha \overline{V}}{\tau}\right)e^{-\tau T}\right]e^{-\sigma[t - (2n-1)T]}.$$

(e) Set $t = 2nT$ and $V = V_n$ in the function in part (d) t show that $V_n = p V_{n-1} + q$, where

$$p = e^{-T(\tau + \sigma)} \quad \text{and} \quad q = \frac{\alpha \overline{V}}{\tau}(1 - e^{-\tau T})e^{-\sigma T}.$$

Now find an explicit formula for the sequence $\{V_n\}$.

6. It is customary to assume that when a drug is administered to the human body, it will be eliminated exponentially; that is, if A represents the amount of drug in the body, then

$$A = A_0 e^{-kt},$$

where $k > 0$ is a constant and A_0 is the amount injected at time $t = 0$. Suppose n successive injections of amount A_0 are administered at equally spaced time intervals T, the first injection at time $t = 0$.

(a) Show that the amount of drug in the body at time t between the nth and $(n + 1)^{\text{th}}$ injection is given by

$$A_n(t) = A_0 e^{-kt} \left[\frac{1 - e^{knT}}{1 - e^{kT}} \right], \quad (n - 1)T < t < nT.$$

(b) Sketch graphs of these functions on one set of axes.

(c) What is the amount of drug in the body immediately after the nth injection for very large n; that is, what is

$$\lim_{n \to \infty} A_n((n - 1)T)?$$

10.10 Integral, Comparison, and Limit Comparison Tests

In Section 10.9 we derived the nth-term test, a test for divergence of a series of numbers. In order to develop further convergence and divergence tests, we consider two classes of series of numbers:

1. series with terms that are all nonnegative;

2. series with both positive and negative terms.

A series with terms that are all nonpositive is the negative of a series of type 1, and therefore any test applicable to series of type 1 is easily adapted to a series with nonpositive terms.

DEFINITION 10.6

A series $\sum_{n=1}^{\infty} c_n$ is said to be **nonnegative** if each term is nonnegative: $c_n \geq 0$.

For example, the harmonic series $\sum 1/n$ and the geometric series $\sum 1/2^{n-1}$ are both nonnegative series. We have already seen that both series have the property that $\lim_{n \to \infty} c_n = 0$, yet the harmonic series diverges and the geometric series converges. Examination of terms of these series reveals that $1/2^{n-1}$ approaches 0 much more quickly than does $1/n$. In general, whether a nonnegative series does or does not have a sum depends on how quickly its sequence of terms $\{c_n\}$ approaches zero. The study of convergence of nonnegative series is an investigation into the question: How fast must terms of a nonnegative series approach zero in order that the series have a sum? This is not a simple problem; only a partial answer is provided through the tests in this section and the next. We begin with the integral test.

THEOREM 10.13 (Integral test)

Suppose that the terms in a series $\sum_{n=1}^{\infty} c_n$ are denoted by $c_n = f(n)$, and $f(x)$ is a continuous, positive, decreasing function for $x \geq 1$. Then the series converges if and only if the improper integral $\int_1^{\infty} f(x)\, dx$ converges.

PROOF To prove this result, we return to the definition of the sum of a series as the limit of its sequence of partial sums. Suppose first that the improper integral converges to value K. We know that this value can be interpreted as the area under the curve $y = f(x)$, above the x-axis, and to the right of the line $x = 1$ in Figure 10.26a. In Figure 10.26b the area of the nth rectangle is

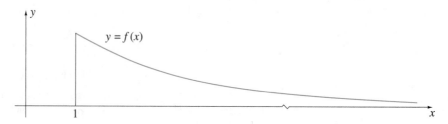

FIGURE 10.26a Comparison area for integral test

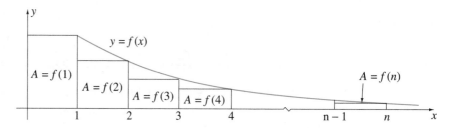

FIGURE 10.26b Rectangles to represent terms of a series to prove convergence

$f(n) = c_n$, and clearly it is less than the area under the curve from $x = n-1$ to $x = n$. Now the nth partial sum of the series $\sum c_n$ is

$$S_n = c_1 + c_2 + \cdots + c_n$$

$$= f(1) + f(2) + \cdots + f(n)$$

$$< f(1) + \int_1^2 f(x)\,dx + \int_2^3 f(x)\,dx + \cdots + \int_{n-1}^n f(x)\,dx$$

$$= f(1) + \int_1^n f(x)\,dx$$

$$< f(1) + \int_1^\infty f(x)\,dx$$

$$= f(1) + K.$$

What this shows is that the sequence $\{S_n\}$ is bounded [since $f(1) + K$ is independent of n] Because all terms of the series $\sum c_n$ are positive, the sequence $\{S_n\}$ of partial sums must be increasing. It follows by Theorem 10.7 that the sequence $\{S_n\}$ must have a limit; that is, the series $\sum c_n$ converges.

Conversely, suppose now that the improper integral diverges; the area under the curve $y = f(x)$, above the x-axis and to the right of the line $x = 1$, is "infinite." This time we draw rectangles to the right of the vertical lines at $n = 1, 2, \ldots$ (Figure 10.27). Then

$$S_n = c_1 + c_2 + \cdots + c_n$$

$$= f(1) + f(2) + \cdots + f(n)$$

$$> \int_1^2 f(x)\,dx + \int_2^3 f(x)\,dx + \cdots + \int_n^{n+1} f(x)\,dx$$

$$= \int_1^{n+1} f(x)\,dx.$$

Since $\lim_{n\to\infty} \int_1^{n+1} f(x)\,dx = \infty$, it follows that $\lim_{n\to\infty} S_n = \infty$, and the series therefore diverges.

FIGURE 10.27   Rectangles to represent terms of a series to prove divergence

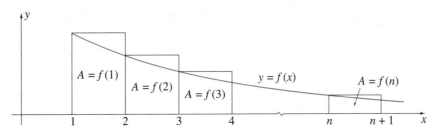

To use the integral test it is necessary to antidifferentiate the function $f(x)$, obtained by replacing n's in c_n with x's. When the antiderivative appears obvious, and other conditions of Theorem 10.13 are met, the integral test may be the easiest way to decide on convergence of the series; when the antiderivative is not obvious, it may be better to try another test.

EXAMPLE 10.39

Determine whether the following series converge or diverge:

$$\text{(a)}\ \sum_{n=1}^{\infty}\frac{1}{n^2+1} \qquad \text{(b)}\ \sum_{n=2}^{\infty}\frac{1}{n\ln n} \qquad \text{(c)}\ \sum_{n=1}^{\infty}ne^{-n}$$

SOLUTION

(a) Since $f(x)=1/(x^2+1)$ is continuous, positive, and decreasing for $x\geq 1$, and

$$\int_{1}^{\infty}\frac{1}{x^2+1}\,dx=\left\{\text{Tan}^{-1}x\right\}_{1}^{\infty}=\frac{\pi}{2}-\frac{\pi}{4}=\frac{\pi}{4},$$

it follows that the series $\sum 1/(n^2+1)$ converges.

(b) Since this series begins with $n=2$, we modify the integral test by considering the improper integral of $f(x)=1/(x\ln x)$ over the interval $x\geq 2$. Since $f(x)$ is positive, continuous, and decreasing for $x\geq 2$, and

$$\int_{2}^{\infty}\frac{1}{x\ln x}\,dx=\left\{\ln\left(\ln x\right)\right\}_{2}^{\infty}=\infty,$$

it follows that $\sum_{n=2}^{\infty}1/(n\ln n)$ diverges.

(c) Since xe^{-x} is continuous, positive, and decreasing for $x\geq 1$, and

$$\int_{1}^{\infty}xe^{-x}\,dx=\left\{-xe^{-x}-e^{-x}\right\}_{1}^{\infty}=\frac{2}{e},$$

the series converges.

We mentioned before that two very important series are geometric series and harmonic series. Convergence of geometric series was discussed in Section 10.9. The harmonic series belongs to a type of series called p-**series**, defined as follows:

$$\sum_{n=1}^{\infty}\frac{1}{n^p}=1+\frac{1}{2^p}+\frac{1}{3^p}+\cdots. \tag{10.42}$$

When $p = 1$, the p-series becomes the harmonic series, which we know diverges. The series diverges for $p < 0$ by the nth-term test. Consider the case when $p > 0$, but $p \neq 1$. The function $1/x^p$ is continuous, positive, and decreasing for $x \geq 1$, and

$$\int_1^\infty \frac{1}{x^p} \, dx = \left\{ \frac{1}{-(p-1)x^{p-1}} \right\}_1^\infty = \begin{cases} \infty, & p < 1 \\ \dfrac{1}{p-1}, & p > 1. \end{cases}$$

According to the integral test, the p-series converges when $p > 1$ and diverges when $p < 1$. Let us summarize results for geometric and p-series:

Geometric Series p**-series**

$$\sum_{n=1}^\infty ar^{n-1} = \begin{cases} \dfrac{a}{r-1}, & |r| < 1 \\ \text{diverges}, & |r| \geq 1. \end{cases} \qquad \sum_{n=1}^\infty \frac{1}{n^p} = \begin{cases} \text{converges}, & p > 1 \\ \text{diverges}, & p \leq 1. \end{cases} \quad (10.43)$$

It is unfortunate that no general formula can be given for the sum of the p-series when $p > 1$. Some interesting cases that arise frequently are cited below:

$$\sum_{n=1}^\infty \frac{1}{n^2} = \frac{\pi^2}{6}, \qquad \sum_{n=1}^\infty \frac{1}{n^3} \approx 1.202\,056\,903\,1, \qquad \sum_{n=1}^\infty \frac{1}{n^4} = \frac{\pi^4}{90}. \quad (10.44)$$

Many series can be shown to converge or diverge by comparing them to known convergent and divergent series. This is the essence of the following two tests.

THEOREM 10.14   (Comparison Test)

If $0 \leq c_n \leq a_n$ for all n and $\sum_{n=1}^\infty a_n$ converges, then $\sum_{n=1}^\infty c_n$ converges. If $c_n \geq d_n \geq 0$ for all n and $\sum_{n=1}^\infty d_n$ diverges, then $\sum_{n=1}^\infty c_n$ diverges.

PROOF   Suppose first that $\sum a_n$ converges and $0 \leq c_n \leq a_n$ for all n. Since $a_n \geq 0$ the sequence of partial sums for the series $\sum a_n$ must be nondecreasing. Since the series converges, the sequence of partial sums must also be bounded (corollary to Theorem 10.7). Because $0 \leq c_n \leq a_n$ for all n, it follows that the sequence of partial sums of $\sum c_n$ is nondecreasing and bounded also. This sequence therefore has a limit and series $\sum c_n$ converges. A similar argument can be made for the divergent case. ∎

Theorem 10.14 states that if the terms of a nonnegative series $\sum c_n$ are smaller than those of a known convergent series, then $\sum c_n$ must converge; if they are larger than a known nonnegative divergent series, then $\sum c_n$ must diverge.

In order for the comparison test to be useful, we require a catalogue of known convergent and divergent series with which we may compare other series. The geometric and p-series in equation 10.43 are extremely useful in this respect.

EXAMPLE 10.40

Determine whether the following series converge or diverge:

(a) $\displaystyle\sum_{n=2}^\infty \frac{\ln n}{n}$ (b) $\displaystyle\sum_{n=1}^\infty \frac{2n^2 - 1}{15n^4 + 14}$ (c) $\displaystyle\sum_{n=1}^\infty \frac{2n^2 + 1}{15n^4 - 14}$

SOLUTION

(a) For this series, we note that when $n \geq 3$,

$$\frac{\ln n}{n} > \frac{1}{n}.$$

Since $\sum_{n=3}^{\infty} 1/n$ diverges (harmonic series with first two terms changed to zero), so does $\sum_{n=3}^{\infty} (\ln n)/n$ by the comparison test. Thus $\sum_{n=2}^{\infty} (\ln n)/n$ diverges. This can also be verified with the integral test.

(b) For the series $\sum (2n^2 - 1)/(15n^4 + 14)$, we note that

$$\frac{2n^2 - 1}{15n^4 + 14} < \frac{2n^2}{15n^4} = \frac{2}{15n^2}.$$

Since $\sum 2/(15n^2) = (2/15) \sum 1/n^2$ converges (p-series with $p = 2$), so does the given series by the comparison test.

(c) For this series we note that the inequality

$$\frac{2n^2 + 1}{15n^4 - 14} \leq \frac{3}{n^2}$$

is valid if and only if

$$45n^4 - 42 \geq 2n^4 + n^2.$$

But this inequality is valid if and only if

$$n^2(43n^2 - 1) \geq 42,$$

which is obviously true for $n \geq 1$. Consequently,

$$\frac{2n^2 + 1}{15n^4 - 14} \leq \frac{3}{n^2},$$

and since $\sum 3/n^2 = 3 \sum 1/n^2$ converges, so does the given series.

Two observations about Example 10.40 are worthwhile:

1. To use the comparison test we must first have a suspicion as to whether the given series converges or diverges in order to discuss the correct inequality; that is, if we suspect that the given series converges, we search for a convergent series for the right side of the inequality $\leq$, and if we suspect that the given series diverges, we search for a divergent series for the right side of the opposite inequality $\geq$.

2. Recall that when the terms of a nonnegative series approach zero, whether the series has a sum depends on how fast these terms approach zero. The only difference in n^{th} terms of Examples 10.40(b) and (c) is the position of the negative sign. For very large n this difference becomes negligible since each nth term can, for large n, be closely approximated by $2/(15n^2)$. We might then expect similar analyses for these examples, yet they are quite different. In addition, it is natural to ask where we obtained the factor 3 in part (c). The answer is: "by trial and error."

Each of these observations points out weaknesses in the comparison test, but these problems can be eliminated in many examples with the following test.

THEOREM 10.15 (Limit Comparison Test)

If $0 \le c_n$ and $0 < b_n$, and

$$\lim_{n \to \infty} \frac{c_n}{b_n} = \ell, \quad 0 < \ell < \infty, \tag{10.45}$$

then series $\sum_{n=1}^{\infty} c_n$ converges if $\sum_{n=1}^{\infty} b_n$ converges, and diverges if $\sum_{n=1}^{\infty} b_n$ diverges.

PROOF   Suppose that series $\sum b_n$ converges. Since sequence $\{c_n/b_n\}$ converges to ℓ, we can use the result of Exercise 62 in Section 10.8 and say that for all n greater than or equal to some integer N,

$$\frac{c_n}{b_n} < \ell + 1 \quad \text{or} \quad c_n < (\ell + 1)b_n.$$

Since the series $\sum_{n=N}^{\infty} (\ell + 1)b_n = (\ell + 1) \sum_{n=N}^{\infty} b_n$ converges, so also must $\sum_{n=N}^{\infty} c_n$ (by the comparison test). Hence, $\sum_{n=1}^{\infty} c_n$ converges. A similar argument can be made for the divergent case.

If $\lim_{n \to \infty} c_n/b_n = \ell$, then for very large n we can say that $c_n \approx \ell b_n$. It follows that if $\{b_n\}$ approaches zero, $\{c_n\}$ approaches zero $1/\ell$ times as fast as $\{b_n\}$. Theorem 10.15 implies then that if the sequences of nth terms of two nonnegative series approach zero at proportional rates, the series converge or diverge together.

The limit comparison test avoids inequalities. To use it, we must find a series $\sum b_n$ so that the limit of the ratio c_n/b_n is finite and greater than zero. To obtain this series, it is sufficient in many examples simply to answer the question: What does the given series really look like for very large n? In both Examples 10.40(b) and (c), we see that for large n,

$$c_n \approx \frac{2n^2}{15n^4} = \frac{2}{15n^2}.$$

Consequently, we calculate in Example 10.40(c) that

$$\ell = \lim_{n \to \infty} \frac{\dfrac{2n^2 + 1}{15n^4 - 14}}{\dfrac{2}{15n^2}} = \lim_{n \to \infty} \frac{n^2 \left(2 + \dfrac{1}{n^2} \right)}{n^4 \left(15 - \dfrac{14}{n^4} \right)} \cdot \frac{15n^2}{2} = 1.$$

Since $\sum 2/(15n^2)$ converges, so does $\sum (2n^2 + 1)/(15n^4 - 14)$.

EXAMPLE 10.41

Determine whether the following series converge or diverge:

(a) $\displaystyle\sum_{n=1}^{\infty} \frac{\sqrt{n^2 + 2n - 1}}{n^{5/2} + 15n - 3}$ (b) $\displaystyle\sum_{n=1}^{\infty} \frac{2^n + 1}{3^n + 5}$

SOLUTION

(a) For very large n, the nth term of the series can be approximated by

$$\frac{\sqrt{n^2 + 2n - 1}}{n^{5/2} + 15n - 3} \approx \frac{n}{n^{5/2}} = \frac{1}{n^{3/2}}.$$

We calculate therefore that

$$\ell = \lim_{n \to \infty} \frac{\dfrac{\sqrt{n^2 + 2n - 1}}{n^{5/2} + 15n - 3}}{\dfrac{1}{n^{3/2}}} = \lim_{n \to \infty} \frac{n\sqrt{1 + \dfrac{2}{n} - \dfrac{1}{n^2}}}{n^{5/2}\left(1 + \dfrac{15}{n^{3/2}} - \dfrac{3}{n^{5/2}}\right)} \cdot n^{3/2} = 1.$$

Since $\sum 1/n^{3/2}$ converges (p-series with $p = 3/2$), so does the given series by the limit comparison test.

(b) For large n,

$$\frac{2^n + 1}{3^n + 5} \approx \left(\frac{2}{3}\right)^n.$$

We calculate therefore that

$$\ell = \lim_{n \to \infty} \frac{\dfrac{2^n + 1}{3^n + 5}}{\left(\dfrac{2}{3}\right)^n} = \lim_{n \to \infty} \frac{2^n\left(1 + \dfrac{1}{2^n}\right)}{3^n\left(1 + \dfrac{5}{3^n}\right)} \cdot \frac{3^n}{2^n} = 1.$$

Since $\sum (2/3)^n$ converges (a geometric series with $r = 2/3$), the given series converges by the limit comparison test.

EXAMPLE 10.42

Find the interval of convergence for the power series $\displaystyle\sum_{n=1}^{\infty} \frac{n}{(n + 1)^3 2^n}(x - 1)^{2n}$.

SOLUTION   If we set $y = (x - 1)^2$, then

$$\sum_{n=1}^{\infty} \frac{n}{(n + 1)^3 2^n}(x - 1)^{2n} = \sum_{n=1}^{\infty} \frac{n}{(n + 1)^3 2^n} y^n.$$

The radius of convergence of the series in y is

$$R_y = \lim_{n \to \infty} \left| \frac{\dfrac{n}{(n + 1)^3 2^n}}{\dfrac{n + 1}{(n + 2)^3 2^{n+1}}} \right| = \lim_{n \to \infty} \frac{2n(n + 2)^3}{(n + 1)^4} = 2.$$

The radius of convergence of the power series in $x - 1$ is therefore $R_x = \sqrt{2}$, and the open interval of convergence is $1 - \sqrt{2} < x < 1 + \sqrt{2}$. At endpoints $x = 1 \pm \sqrt{2}$, the power series becomes

$$\sum_{n=1}^{\infty} \frac{n}{(n + 1)^3}.$$

Since

$$\ell = \lim_{n \to \infty} \frac{\dfrac{n}{(n+1)^3}}{\dfrac{1}{n^2}} = \lim_{n \to \infty} \frac{n^3}{(n+1)^3} = 1,$$

and $\sum 1/n^2$ converges, so does $\sum n/(n+1)^3$ (by the limit comparison test). The interval of convergence is therefore $1 - \sqrt{2} \le x \le 1 + \sqrt{2}$.

EXERCISES 10.10

In Exercises 1–22 determine whether the series converges or diverges.

1. $\displaystyle\sum_{n=1}^{\infty} \frac{1}{2n+1}$

2. $\displaystyle\sum_{n=1}^{\infty} \frac{1}{4n-3}$

3. $\displaystyle\sum_{n=1}^{\infty} \frac{1}{2n^2+4}$

4. $\displaystyle\sum_{n=1}^{\infty} \frac{1}{5n^2-3n-1}$

5. $\displaystyle\sum_{n=2}^{\infty} \frac{1}{n^3-1}$

6. $\displaystyle\sum_{n=4}^{\infty} \frac{n^2}{n^4-6n^2+5}$

7. $\displaystyle\sum_{n=1}^{\infty} \frac{1}{(2n-1)(2n+1)}$

8. $\displaystyle\sum_{n=1}^{\infty} \frac{n-5}{n^2+3n-2}$

9. $\displaystyle\sum_{n=2}^{\infty} \frac{1}{\ln n}$

10. $\displaystyle\sum_{n=1}^{\infty} n^2 e^{-2n}$

11. $\displaystyle\sum_{n=2}^{\infty} \frac{\sqrt{n^2+2n-3}}{n^2+5}$

12. $\displaystyle\sum_{n=1}^{\infty} \frac{\sqrt{n+5}}{n^3+3}$

13. $\displaystyle\sum_{n=1}^{\infty} \sqrt{\frac{n^2+2n+3}{2n^4-n}}$

14. $\displaystyle\sum_{n=2}^{\infty} \frac{1}{n^2 \ln n}$

15. $\displaystyle\sum_{n=1}^{\infty} \frac{1}{2^n} \sin\left(\frac{\pi}{n}\right)$

16. $\displaystyle\sum_{n=1}^{\infty} \frac{\sqrt{n^2+1}}{n^3} \mathrm{Tan}^{-1} n$

17. $\displaystyle\sum_{n=1}^{\infty} \frac{2^n+n}{3^n+1}$

18. $\displaystyle\sum_{n=2}^{\infty} \frac{1+\ln^2 n}{n \ln^2 n}$

19. $\displaystyle\sum_{n=1}^{\infty} \frac{1+1/n}{e^n}$

20. $\displaystyle\sum_{n=1}^{\infty} \frac{\ln(n+1)}{n+1}$

21. $\displaystyle\sum_{n=1}^{\infty} n e^{-n^2}$

22. $\displaystyle\sum_{n=2}^{\infty} \frac{1}{n \sqrt[3]{\ln n}}$

In Exercises 23–26 find the interval of convergence of the power series.

∗ 23. $\displaystyle\sum_{n=1}^{\infty} \frac{1}{n} x^{2n}$

∗ 24. $\displaystyle\sum_{n=1}^{\infty} \frac{1}{n^2} x^{2n}$

∗ 25. $\displaystyle\sum_{n=1}^{\infty} \frac{1}{n 2^n} (x-1)^{4n}$

∗ 26. $\displaystyle\sum_{n=0}^{\infty} \frac{n 3^n}{(n+1)^3} (x+2)^{2n}$

In Exercises 27–29 find values of p for which the series converges.

∗ 27. $\displaystyle\sum_{n=2}^{\infty} \frac{1}{n^p \ln n}$

∗ 28. $\displaystyle\sum_{n=2}^{\infty} \frac{1}{n(\ln n)^p}$

∗∗ 29. $\displaystyle\sum_{n=2}^{\infty} \frac{1}{(\ln n)^p}$

10.11 Limit Ratio and Limit Root Tests

In this section we consider two additional tests to determine whether nonnegative series converge or diverge. The first test, the limit ratio test, indicates whether a series $\sum c_n$ resembles a geometric series for large n.

THEOREM 10.16   (Limit Ratio Test)

Suppose that $c_n > 0$ and

$$\lim_{n \to \infty} \frac{c_{n+1}}{c_n} = L. \tag{10.46}$$

Then:

(i) $\sum_{n=1}^{\infty} c_n$ converges if $L < 1$.

(ii) $\sum_{n=1}^{\infty} c_n$ diverges if $L > 1$ (or if $\lim_{n \to \infty} c_{n+1}/c_n = \infty$).

(iii) $\sum_{n=1}^{\infty} c_n$ may converge or diverge if $L = 1$.

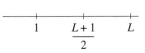

FIGURE 10.28a
FIGURE 10.28a

Schematic for proof of limit ratio test

FIGURE 10.28b

PROOF

(i) By the result of part (a) in Exercise 61 of Section 10.8 (see also Figure 10.28a), we can say that if $\lim_{n\to\infty} c_{n+1}/c_n = L < 1$, there exists an integer N such that for all $n \geq N$,

$$\frac{c_{n+1}}{c_n} < \frac{L+1}{2}.$$

Consequently,

$$c_{N+1} < \left(\frac{L+1}{2}\right) c_N;$$

$$c_{N+2} < \left(\frac{L+1}{2}\right) c_{N+1} < \left(\frac{L+1}{2}\right)^2 c_N;$$

$$c_{N+3} < \left(\frac{L+1}{2}\right) c_{N+2} < \left(\frac{L+1}{2}\right)^3 c_N;$$

and so on. Hence,

$$c_N + c_{N+1} + c_{N+2} + \cdots < c_N + \left(\frac{L+1}{2}\right) c_N + \left(\frac{L+1}{2}\right)^2 c_N + \cdots.$$

Since the right side of this inequality is a geometric series with common ratio $(L+1)/2 < 1$, it follows by the comparison test that $\sum_{n=N}^{\infty} c_n$ converges. Therefore, $\sum_{n=1}^{\infty} c_n$ converges also.

(ii) By the result of part (b) in Exercise 61 of Section 10.8 (see also Figure 10.28b), if $\lim_{n\to\infty} c_{n+1}/c_n = L > 1$, there exists an integer N such that for all $n \geq N$,

$$\frac{c_{n+1}}{c_n} > \frac{L+1}{2} > 1.$$

When $\lim_{n\to\infty} c_{n+1}/c_n = \infty$, it is also true that for n greater than or equal to some N, c_{n+1}/c_n must be greater than 1. This implies that for all $n > N$, $c_n > c_N$, and therefore

$$\lim_{n\to\infty} c_n \neq 0.$$

Hence $\sum c_n$ diverges by the nth-term test.

(iii) To show that the limit ratio test is inconclusive when $L = 1$, consider the two p-series $\sum 1/n$ and $\sum 1/n^2$. For each series, $L = 1$, yet the first series diverges and the second converges. ∎

In a nonrigorous way we can justify the limit ratio test from the following standpoint. If $\lim_{n\to\infty} c_{n+1}/c_n = L$, then for large n, each term of series $\sum c_n$ is essentially L times the term before it; that is, the series resembles a geometric series with common ratio L. We would expect convergence of the series if $L < 1$ and divergence if $L > 1$. We might also anticipate some indecision about the $L = 1$ case, depending on how this limit is reached, since this case corresponds to the common ratio that separates convergent and divergent geometric series.

EXAMPLE 10.43

Determine whether the following series converge or diverge:

$$\text{(a)} \quad \sum_{n=1}^{\infty} \frac{2^n}{n^4} \qquad \text{(b)} \quad \sum_{n=1}^{\infty} \frac{n^{100}}{1 \cdot 3 \cdot 5 \cdot \ \ldots \ \cdot (2n-1)} \qquad \text{(c)} \quad \sum_{n=1}^{\infty} \frac{n^n}{n!}$$

SOLUTION

(a) For the series $\sum 2^n / n^4$,

$$L = \lim_{n \to \infty} \frac{\dfrac{2^{n+1}}{(n+1)^4}}{\dfrac{2^n}{n^4}} = \lim_{n \to \infty} 2 \left(\frac{n}{n+1} \right)^4 = 2,$$

and the series therefore diverges by the limit ratio test.

(b) For this series,

$$L = \lim_{n \to \infty} \frac{\dfrac{(n+1)^{100}}{1 \cdot 3 \cdot 5 \cdot \ \ldots \ \cdot (2n+1)}}{\dfrac{n^{100}}{1 \cdot 3 \cdot 5 \cdot \ \ldots \ \cdot (2n-1)}} = \lim_{n \to \infty} \left(\frac{n+1}{n} \right)^{100} \frac{1}{2n+1} = 0,$$

and the series therefore converges by the limit ratio test.

(c) Since

$$L = \lim_{n \to \infty} \frac{\dfrac{(n+1)^{n+1}}{(n+1)!}}{\dfrac{n^n}{n!}} = \lim_{n \to \infty} \left(\frac{n+1}{n} \right)^n = e > 1$$

(see equation 1.68, the series $\sum n^n / n!$ diverges.

We present one last test for nonnegative series; there are many others.

THEOREM 10.17   (Limit Root Test)

Suppose that $c_n \geq 0$ and

$$\lim_{n \to \infty} \sqrt[n]{c_n} = R. \tag{10.47}$$

Then:

(i) $\sum_{n=1}^{\infty} c_n$ converges if $R < 1$.

(ii) $\sum_{n=1}^{\infty} c_n$ diverges if $R > 1$ (or if $\lim_{n \to \infty} \sqrt[n]{c_n} = \infty$).

(iii) $\sum_{n=1}^{\infty} c_n$ may converge or diverge if $R = 1$.

PROOF

(i) By the result of part (a) in Exercise 61 of Section 10.8, if $\lim_{n \to \infty} \sqrt[n]{c_n} = R < 1$, there exists an integer N such that for all $n \geq N$,

$$\sqrt[n]{c_n} < \frac{R+1}{2}.$$

Consequently,

$$\sqrt[N]{c_N} < \frac{R+1}{2} \qquad \text{or} \qquad c_N < \left(\frac{R+1}{2}\right)^N;$$

$$\sqrt[N+1]{c_{N+1}} < \frac{R+1}{2} \qquad \text{or} \qquad c_{N+1} < \left(\frac{R+1}{2}\right)^{N+1};$$

$$\sqrt[N+2]{c_{N+2}} < \frac{R+1}{2} \qquad \text{or} \qquad c_{N+2} < \left(\frac{R+1}{2}\right)^{N+2};$$

and so on. Hence,

$$c_N + c_{N+1} + c_{N+2} + \cdots < \left(\frac{R+1}{2}\right)^N + \left(\frac{R+1}{2}\right)^{N+1} + \left(\frac{R+1}{2}\right)^{N+2} + \cdots.$$

Since the right side of this inequality is a convergent geometric series, the left side must be a convergent series also by the comparison test. Thus, $\sum_{n=N}^{\infty} c_n$ converges, and so must $\sum_{n=1}^{\infty} c_n$.

(ii) When $\lim_{n \to \infty} \sqrt[n]{c_n} = R > 1$, there exists an integer N such that for all $n \geq N$,

$$\sqrt[n]{c_n} > \frac{R+1}{2} > 1 \qquad \text{or} \qquad c_n > 1,$$

as in part (b) of Exercise 61 in Section 10.8. When $\lim_{n \to \infty} \sqrt[n]{c_n} = \infty$, it is also true that for n greater than or equal to some N, $\sqrt[n]{c_n}$ must be greater than 1. But it now follows that $\lim_{n \to \infty} c_n \neq 0$, and the series diverges by the nth-term test.

(iii) To show that the test is inconclusive when $R = 1$, we show that $R = 1$ for the two p-series $\sum 1/n$ and $\sum 1/n^2$, diverging and converging, respectively. For the harmonic series, let $R = \lim_{n \to \infty} (1/n)^{1/n}$. If we take logarithms, then

$$\ln R = \ln\left[\lim_{n \to \infty} \left(\frac{1}{n}\right)^{\frac{1}{n}}\right] = \lim_{n \to \infty} \ln\left(\frac{1}{n}\right)^{1/n} = \lim_{n \to \infty} 1/n \ln\left(\frac{1}{n}\right) = -\lim_{n \to \infty} \frac{\ln n}{n}$$

$$= -\lim_{n \to \infty} \frac{1/n}{1} \qquad \text{(by L'Hôpital's rule)}$$

$$= 0.$$

Hence, $R = 1$ for this divergent series. A similar analysis gives $R = 1$ for the convergent series $\sum 1/n^2$. ∎

EXAMPLE 10.44

Determine whether the following series converge or diverge:

(a) $\displaystyle \sum_{n=1}^{\infty} \left(\frac{n+1}{n}\right)^{n^2}$ (b) $\displaystyle \sum_{n=1}^{\infty} \frac{n}{(\ln n)^n}$

SOLUTION

(a) Since

$$R = \lim_{n \to \infty} \left[\left(\frac{n+1}{n} \right)^{n^2} \right]^{1/n} = \lim_{n \to \infty} \left(\frac{n+1}{n} \right)^n = e > 1,$$

the series diverges by the limit root test.

(b) For this series,

$$R = \lim_{n \to \infty} \left[\frac{n}{(\ln n)^n} \right]^{1/n} = \lim_{n \to \infty} \frac{n^{1/n}}{\ln n}.$$

If we set $L = \lim_{n \to \infty} n^{1/n}$, then

$$\ln L = \ln \left(\lim_{n \to \infty} n^{1/n} \right) = \lim_{n \to \infty} \frac{1}{n} \ln n$$

$$= \lim_{n \to \infty} \frac{\dfrac{1}{n}}{1} \qquad \text{(by L'Hôpital's rule)}$$

$$= 0.$$

Thus, $L = 1$, and it follows that

$$R = \lim_{n \to \infty} \frac{n^{1/n}}{\ln n} = 0.$$

The series therefore converges.

We have developed six tests to determine whether series of numbers converge or diverge: the nth-term, integral, comparison, limit comparison, limit ratio, and limit root tests.

The form of the nth term of a series often suggests which test should be used. Keep the following ideas in mind when choosing a test:

1. The limit ratio test can be effective on factorials, products of the form $1 \cdot 3 \cdot 5 \cdot \ldots \cdot (2n-1)$ and constants raised to powers involving n (2^n, 3^{-n}, etc.).

2. The limit root test thrives on functions of n raised to powers involving n (see Example 10.44).

3. The limit comparison test is successful on rational functions of n, and fractional powers as well ($\sqrt{n}$, $\sqrt[3]{n/(n+1)}$, etc.).

4. The integral test can be effective when the nth term is easily integrated. Logarithms often require the integral test.

By definition, a series $\sum c_n$ converges if and only if its sequence of partial sums $\{S_n\}$ converges. The difficulty with using this definition to discuss convergence of a series is that S_n can seldom be evaluated in a simple form, and therefore consideration of $\lim_{n \to \infty} S_n$ is impossible. The tests above have the advantage of avoiding partial sums. On the other hand, they have one disadvantage. Although they may indicate that a series does indeed have a sum, the tests in no way suggest the value of the sum. The problem of calculating the sum often proves more difficult than showing that it exists in the first place. In Section 10.13 we discuss various ways to calculate and approximate sums for known convergent series.

In Exercises 1–20 determine whether the series converges or diverges.

1. $\displaystyle\sum_{n=1}^{\infty} \frac{e^n}{n^4}$

2. $\displaystyle\sum_{n=1}^{\infty} \frac{1}{n!}$

3. $\displaystyle\sum_{n=1}^{\infty} \frac{n^3}{2^n}$

4. $\displaystyle\sum_{n=1}^{\infty} \frac{1}{n^n}$

5. $\displaystyle\sum_{n=1}^{\infty} \frac{(n-1)(n-2)}{n^2 2^n}$

6. $\displaystyle\sum_{n=1}^{\infty} \frac{(2n)!}{(n!)^2}$

7. $\displaystyle\sum_{n=1}^{\infty} \frac{\sqrt{n+1}}{n^{n+1/2}}$

8. $\displaystyle\sum_{n=1}^{\infty} \frac{3^{-n} + 2^{-n}}{4^{-n} + 5^{-n}}$

9. $\displaystyle\sum_{n=1}^{\infty} \frac{e^{-n}}{\sqrt{n+\pi}}$

10. $\displaystyle\sum_{n=1}^{\infty} \frac{2 \cdot 4 \cdot \ldots \cdot (2n)}{4 \cdot 7 \cdot \ldots \cdot (3n+1)}$

11. $\displaystyle\sum_{n=1}^{\infty} \frac{n^{n-1}}{3^{n-1}(n-1)!}$

12. $\displaystyle\sum_{n=1}^{\infty} n \left(\frac{3}{4}\right)^n$

13. $\displaystyle\sum_{n=1}^{\infty} \frac{1+1/n}{e^n}$

14. $\displaystyle\sum_{n=1}^{\infty} \frac{2 \cdot 4 \cdot \ldots \cdot (2n)}{3 \cdot 5 \cdot \ldots \cdot (2n+1)} \left(\frac{1}{n^2}\right)$

* 15. $\displaystyle\sum_{n=1}^{\infty} \frac{n^n}{(n+1)^{n+1}}$

* 16. $\displaystyle\sum_{n=1}^{\infty} \frac{(n+1)^n}{n^{n+1}}$

* 17. $\displaystyle\sum_{n=1}^{\infty} \frac{n^4 + 3}{5^{n/2}}$

* 18. $\displaystyle\sum_{n=1}^{\infty} \frac{2^n + n^2 3^n}{4^n}$

* 19. $\displaystyle\sum_{n=1}^{\infty} \frac{n^2 2^n - n}{n^3 + 1}$

* 20. $\displaystyle\sum_{n=1}^{\infty} \frac{(2n)!}{(3n)!} 5^{2n}$

* 21. For what integer values of a is the series

$$\sum_{n=1}^{\infty} \frac{(n!)^2}{(an)!}$$

convergent?

10.12 Absolute and Conditional Convergence, Alternating Series

The convergence tests of Sections 10.10 and 10.11 are applicable to nonnegative series, series whose terms are all nonnegative. Series with infinitely many positive and negative terms are more complicated. It is fortunate, however, that all our tests are still useful in discussing convergence of series with positive and negative terms. What makes this possible is the following definition and Theorem 10.18.

DEFINITION 10.7

A series $\sum_{n=1}^{\infty} c_n$ is said to be **absolutely convergent** if the series of absolute values $\sum_{n=1}^{\infty} |c_n|$ converges.

At first glance it might seem that absolute convergence is a strange concept indeed. What possible good could it do to consider the series of absolute values, which is quite different from the original series? The fact is that when the series of absolute values converges, it automatically follows from the next theorem that the original series converges also. And since the series of absolute values has all nonnegative terms, we can use the comparison, limit comparison, limit ratio, limit root, or integral test to consider its convergence.

THEOREM 10.18

If a series is absolutely convergent, then it is convergent.

PROOF Let $\{S_n\}$ be the sequence of partial sums of the absolutely convergent series $\sum c_n$. Define sequences $\{P_n\}$ and $\{N_n\}$, where P_n is the sum of all positive terms in S_n, and N_n is the sum of the absolute values of all negative terms in S_n. Then

$$S_n = P_n - N_n.$$

The sequence of partial sums for the series of absolute values $\sum |c_n|$ is

$$\{P_n + N_n\},$$

and this sequence must be nondecreasing and bounded. Since each of the sequences $\{P_n\}$ and $\{N_n\}$ is nondecreasing and a part of $\{P_n + N_n\}$, it follows that each is bounded, and therefore has a limit, say, P and N, respectively. As a result, sequence $\{P_n - N_n\} = \{S_n\}$ has limit $P - N$, and series $\sum c_n$ converges to $P - N$. ∎

EXAMPLE 10.45

Show that the following series are absolutely convergent:

(a) $\displaystyle\sum_{n=1}^{\infty} \frac{(-1)^n n}{(n+1)2^n}$

(b) $1 - \dfrac{1}{2^2} - \dfrac{1}{3^3} + \dfrac{1}{4^4} + \dfrac{1}{5^5} + \dfrac{1}{6^6} - \dfrac{1}{7^7} - \dfrac{1}{8^8} - \dfrac{1}{9^9} - \dfrac{1}{10^{10}} + \cdots + \dfrac{1}{15^{15}} - \cdots$

SOLUTION

(a) The series of absolute values is $\sum n/[(n+1)2^n]$. We use the limit comparison test to show that it converges. Since

$$\ell = \lim_{n\to\infty} \frac{\dfrac{n}{(n+1)2^n}}{\dfrac{1}{2^n}} = 1,$$

and $\sum (1/2)^n$ is convergent (a geometric series with $r = 1/2$), it follows that the series of absolute values converges. The given series therefore converges absolutely.

(b) The series of absolute values is $\sum 1/n^n$. Since

$$R = \lim_{n\to\infty} \left(\frac{1}{n^n}\right)^{1/n} = \lim_{n\to\infty} \frac{1}{n} = 0,$$

the series $\sum 1/n^n$ converges by the limit root test. The given series therefore converges absolutely.

In Example 10.45, absolute convergence of the given series implies convergence of the series, but it is customary to omit such a statement. It is important to realize, however, that it is the given series that is being analyzed, and its convergence is guaranteed by Theorem 10.18.

We now ask whether series can converge without converging absolutely. If there are such series, and indeed there are, we must devise new convergence tests. We describe these series as follows.

DEFINITION 10.8

A series that converges but does not converge absolutely is said to **converge conditionally**.

The most important type of series with both positive and negative terms is an alternating series. As the name suggests, an **alternating series** has terms that are alternately positive and negative. For example,

$$\sum_{n=1}^{\infty} \frac{(-1)^{n+1}}{n} = 1 - \frac{1}{2} + \frac{1}{3} - \frac{1}{4} + \frac{1}{5} - \cdots$$

is an alternating series called the **alternating harmonic series**.

Given an alternating series to examine for convergence, we first test for absolute convergence as in Example 10.45(a). Should this fail, we check for conditional convergence with the following test.

THEOREM 10.19   (Alternating Series Test)

An alternating series $\sum_{n=1}^{\infty} c_n$ converges if the sequence of absolute values of the terms $\{|c_n|\}$ is decreasing and has limit zero.

PROOF   If $\{S_n\}$ is the sequence of partial sums of the series, then differences of partial sums are $S_n - S_{n-1} = c_n$. Because the c_n alternate in sign, so also do the differences $S_n - S_{n-1}$. This means that the partial sums $\{S_n\}$ form an oscillating sequence. Since absolute values $|S_n - S_{n-1}| = |c_n|$ are decreasing and approach zero, it follows by Theorem 10.8 of Section 10.8 that the sequence $\{S_n\}$ of partial sums has a limit. ∎

EXAMPLE 10.46

Determine whether the following series converge absolutely, converge conditionally, or diverge:

(a) $\displaystyle\sum_{n=1}^{\infty} \frac{(-1)^{n+1}}{n}$ (b) $\displaystyle\sum_{n=1}^{\infty} (-1)^n \frac{\sqrt{n^2 + 5n}}{n^{3/2}}$ (c) $\displaystyle\sum_{n=1}^{\infty} (-1)^{n+1} \frac{4^n}{n^5 3^n}$

SOLUTION

(a) The alternating harmonic series $\sum (-1)^{n+1}/n$ is not absolutely convergent because the series of absolute values $\sum 1/n$ diverges. Since the sequence of absolute values of the terms $\{1/n\}$ is decreasing with limit zero, the series $\sum (-1)^{n+1}/n$ converges conditionally.

(b) For this alternating series, we first consider the series of absolute values

$$\sum_{n=1}^{\infty} \frac{\sqrt{n^2 + 5n}}{n^{3/2}}.$$

We use the limit comparison test on this series. Since

$$\ell = \lim_{n \to \infty} \frac{\dfrac{\sqrt{n^2 + 5n}}{n^{3/2}}}{\dfrac{1}{n^{1/2}}} = \lim_{n \to \infty} \frac{n\sqrt{1 + \dfrac{5}{n}}}{n^{3/2}} \cdot n^{1/2} = 1,$$

and $\sum 1/n^{1/2}$ diverges, so does $\sum \sqrt{n^2 + 5n}/n^{3/2}$. The original series does not therefore converge absolutely. We now resort to the alternating series test. The sequence $\{\sqrt{n^2 + 5n}/n^{3/2}\}$ of absolute values of the terms of the series is decreasing if

$$\frac{\sqrt{(n + 1)^2 + 5(n + 1)}}{(n + 1)^{3/2}} < \frac{\sqrt{n^2 + 5n}}{n^{3/2}}.$$

When we square and cross multiply, the inequality becomes

$$n^3(n^2 + 7n + 6) < (n^2 + 5n)(n + 1)^3$$

$$= n^5 + 8n^4 + 18n^3 + 16n^2 + 5n;$$

that is,

$$n^4 + 12n^3 + 16n^2 + 5n > 0,$$

which is obviously valid because $n \geq 1$. Since $\lim_{n \to \infty} \sqrt{n^2 + 5n}/n^{3/2} = 0$, we conclude that the alternating series $\sum (-1)^n (\sqrt{n^2 + 5n}/n^{3/2})$ converges conditionally.

(c) If we apply the limit ratio test to the series of absolute values $\sum 4^n/(n^5 3^n)$, we have

$$L = \lim_{n \to \infty} \frac{\dfrac{4^{n+1}}{(n+1)^5 3^{n+1}}}{\dfrac{4^n}{n^5 3^n}} = \lim_{n \to \infty} \frac{4}{3} \left(\frac{n}{n+1} \right)^5 = \frac{4}{3}.$$

Since $L > 1$, the series $\sum 4^n/(n^5 3^n)$ diverges. The original alternating series does not therefore converge absolutely. But $L = 4/3$ implies that for large n, each term in the series of absolute values is approximately $4/3$ times the term that precedes it, and therefore

$$\lim_{n \to \infty} \frac{4^n}{n^5 3^n} = \infty.$$

Consequently,

$$\lim_{n \to \infty} (-1)^{n+1} \frac{4^n}{n^5 3^n}$$

cannot possibly exist, and the given series diverges by the nth-term test.

We have noted several times that the essential question for convergence of a nonnegative series is: Do the terms approach zero quickly enough to guarantee convergence of the series? With a series that has infinitely many positive and negative terms, this question is inappropriate. Such a series may converge because of a partial cancelling effect; for example, a negative term may offset the effect of a large positive term. This kind of process may produce a convergent series even though the series would be divergent if all terms were replaced by their absolute values. A specific example is the alternating harmonic series which converges (conditionally) because of this cancelling effect, whereas the harmonic series itself, which has no cancellations, diverges.

Absolute and conditional convergence are particularly important when discussing endpoints of intervals of convergence for power series since one of the endpoints often leads to an alternating series. We illustrate in the following examples.

EXAMPLE 10.47

Find the interval of convergence for the power series $\displaystyle\sum_{n=0}^{\infty} \frac{1}{(n+1)2^n} x^n$.

SOLUTION   Since the radius of convergence is

$$R = \lim_{n \to \infty} \left| \frac{\dfrac{1}{(n+1)2^n}}{\dfrac{1}{(n+2)2^{n+1}}} \right| = \lim_{n \to \infty} 2 \left(\frac{n+2}{n+1} \right) = 2,$$

the open interval of convergence is $-2 < x < 2$. At $x = 2$, the series is $\sum_{n=0}^{\infty} 1/(n+1)$, the harmonic series, which diverges. At $x = -2$, we obtain the alternating harmonic series $\sum_{n=0}^{\infty} (-1)^n/(n+1)$, which converges conditionally. The interval of convergence is therefore $-2 \leq x < 2$.

EXAMPLE 10.48

Find the interval of convergence for the power series $\sum_{n=1}^{\infty} \dfrac{n(-1)^n}{(2n+5)^3} (x-3)^n$.

SOLUTION   Since the radius of convergence is

$$R = \lim_{n \to \infty} \left| \frac{\dfrac{n(-1)^n}{(2n+5)^3}}{\dfrac{(n+1)(-1)^{n+1}}{(2n+7)^3}} \right| = \lim_{n \to \infty} \frac{n(2n+7)^3}{(n+1)(2n+5)^3} = 1,$$

the open interval of convergence is $2 < x < 4$. At $x = 2$, the series becomes

$$\sum_{n=1}^{\infty} \frac{n(-1)^n}{(2n+5)^3} (-1)^n = \sum_{n=1}^{\infty} \frac{n}{(2n+5)^3}.$$

Since

$$\ell = \lim_{n \to \infty} \left| \frac{\dfrac{n}{(2n+5)^3}}{\dfrac{1}{8n^2}} \right| = \lim_{n \to \infty} \frac{8n^3}{(2n+5)^3} = 1,$$

and $\sum 1/(8n^2) = (1/8) \sum 1/n^2$ converges ($p = 2$ series), so also does $\sum n/(2n+5)^3$ (by the limit comparison test). At $x = 4$, the power series becomes

$$\sum_{n=1}^{\infty} \frac{n(-1)^n}{(2n+5)^3}.$$

This is an alternating series that converges absolutely, as indicated in the discussion of $x = 2$. The interval of convergence is therefore $2 \leq x \leq 4$.

In Sections 10.10–10.12, we have obtained a number of tests for determining whether series of numbers converge or diverge. To test a series for convergence, we suggest the following procedure:

1. Try the nth-term test for divergence.

2. If $\{c_n\}$ has limit zero and the series is nonnegative, try the comparison, limit comparison, limit ratio, limit root, or integral test.

3. If $\{c_n\}$ has limit zero and the series contains both positive and negative terms, test for absolute convergence using the tests in 2. If this fails and the series is alternating, test for conditional convergence with the alternating series test.

Each of the comparison, limit comparison, limit ratio, limit root, integral, and alternating series tests requires conditions to be satisfied for all terms of the series. Specifically, the comparison, limit comparison, and limit root tests require that $c_n \geq 0$ for all n; the limit ratio test requires $c_n > 0$; the integral test requires $f(n)$ to be positive, continuous, and decreasing; and the alternating series test requires $\{|c_n|\}$ to be decreasing and $\{c_n\}$ to be alternately positive and negative. None of these requirements is essential for all n; in fact, so long as they are satisfied for all terms in the series beyond some point, say for n greater than or equal to some integer N, the particular test may be used on the series $\sum_{n=N}^{\infty} c_n$. The original series $\sum_{n=1}^{\infty} c_n$ then converges if and only if $\sum_{n=N}^{\infty} c_n$ converges.

Before leaving this section, we prove Theorem 10.2. We verify 10.24a using the limit ratio test; verification of 10.24b is similar, using the limit root test. If the limit ratio test is applied to the series of absolute values $\sum |a_n x^n|$,

$$L = \lim_{n\to\infty} \frac{|a_{n+1} x^{n+1}|}{|a_n x^n|} = |x| \lim_{n\to\infty} \left| \frac{a_{n+1}}{a_n} \right|.$$

Assuming that limit 10.24a exists or is equal to infinity, there are three possibilities:

(i) If $\lim_{n\to\infty} |a_n/a_{n+1}| = 0$, then $\lim_{n\to\infty} |a_{n+1}/a_n| = \infty$. Therefore, $L = \infty$, and the power series diverges for all $x \neq 0$. In other words,

$$R = 0 = \lim_{n\to\infty} \left| \frac{a_n}{a_{n+1}} \right|.$$

(ii) If $\lim_{n\to\infty} |a_n/a_{n+1}| = \infty$, then $\lim_{n\to\infty} |a_{n+1}/a_n| = 0$. Therefore, $L = 0$, and the power series converges absolutely for all x. Consequently,

$$R = \infty = \lim_{n\to\infty} \left| \frac{a_n}{a_{n+1}} \right|.$$

(iii) If $\lim_{n\to\infty} |a_n/a_{n+1}| = R$, then $\lim_{n\to\infty} |a_{n+1}/a_n| = 1/R$. In this case, $L = |x|/R$. Since the power series converges absolutely for $L < 1$ and diverges for $L > 1$, it follows that absolute convergence occurs for $|x| < R$ and divergence for $|x| > R$. This implies that R is the radius of convergence of the power series.

EXERCISES 10.12

In Exercises 1–14 determine whether the series converges absolutely, converges conditionally, or diverges.

1. $\displaystyle\sum_{n=1}^{\infty} (-1)^n \frac{n}{n^3 + 1}$

2. $\displaystyle\sum_{n=1}^{\infty} (-1)^n \frac{n}{n^2 + 1}$

3. $\displaystyle\sum_{n=1}^{\infty} \frac{\cos(n\pi/2)}{2n^2}$

4. $\displaystyle\sum_{n=1}^{\infty} (-1)^n \frac{n^3}{3^n}$

5. $\displaystyle\sum_{n=1}^{\infty} \frac{(-1)^{n+1}}{\sqrt{n}}$

6. $\displaystyle\sum_{n=1}^{\infty} (-1)^n \frac{3^n}{n^3}$

7. $\displaystyle\sum_{n=1}^{\infty} (-1)^n \frac{n}{n^2 + n + 1}$

8. $\displaystyle\sum_{n=1}^{\infty} \frac{n \sin(n\pi/4)}{2^n}$

9. $\displaystyle\sum_{n=1}^{\infty} (-1)^{n+1} \left(\frac{n}{n+1} \right)$

10. $\displaystyle\sum_{n=1}^{\infty} (-1)^{n+1} \frac{\sqrt{3n-2}}{n}$

* 11. $\displaystyle\sum_{n=1}^{\infty} (-1)^n \left(\frac{n}{n+1} \right)^n$

* 12. $\displaystyle\sum_{n=1}^{\infty} (-1)^n \frac{\sqrt{n^2 + 3}}{n^2 + 5}$

* 13. $\displaystyle\sum_{n=2}^{\infty} (-1)^{n-1} \frac{\ln n}{n}$

* 14. $\displaystyle\sum_{n=1}^{\infty} \frac{\cos(n\pi/10) \operatorname{Cot}^{-1} n}{n^3 + 5n}$

In Exercises 15–22 find the interval of convergence of the power series.

* 15. $\displaystyle\sum_{n=1}^{\infty} \frac{1}{n} x^n$

* 16. $\displaystyle\sum_{n=0}^{\infty} \frac{1}{(n+1)^2} x^n$

* 17. $\displaystyle\sum_{n=1}^{\infty} \frac{1}{n 2^n} (x-1)^n$

* 18. $\displaystyle\sum_{n=1}^{\infty} \frac{1}{\sqrt{n}} (x+2)^n$

* 19. $\displaystyle\sum_{n=0}^{\infty} \frac{n-1}{n^2 + 1} (2x)^n$

* 20. $\displaystyle\sum_{n=0}^{\infty} \frac{1}{\sqrt{n+1}} x^{3n+1}$

* 21. $\displaystyle\sum_{n=2}^{\infty} \frac{1}{\ln n} x^n$

* 22. $\displaystyle\sum_{n=2}^{\infty} \frac{1}{n^2 \ln n} (x-2)^n$

3. Discuss convergence of the series

$$\sum_{n=1}^{\infty} \frac{\sin (nx)}{n^2}.$$

4. Prove that if $\sum c_n$ converges absolutely, then $\sum c_n^p$ converges absolutely for all integers $p > 1$.

* **25.** Discuss convergence of the series

$$\sum_{n=1}^{\infty} (-1)^n \frac{n^n}{(n+1)^{n+1}}.$$

10.13 Exact and Approximate Values for Sums of Series of Numbers

In Sections 10.9–10.12 we have concentrated on whether series of numbers converge or diverge. But this is only half the problem. The comparison, limit comparison, limit ratio, limit root, integral, and alternating series tests may determine whether a series converges or diverges, but they do not determine the sum of the series in the case of a convergent series. This part of the problem, as suggested before, can sometimes be more complicated.

If the convergent series is a geometric series, no problem exists; we can use formula 10.39b to find its sum. It may also happen that the nth partial sum S_n of the series can be calculated in a simple form, in which case the sum of the series is $\lim_{n\to\infty} S_n$. Cases of the latter type are very rare. By substituting values of x into power series with known sums, we obtain formulas for sums of series of numbers. For instance, in Example 10.10, we verified that the Maclaurin series for e^x is

$$e^x = \sum_{n=0}^{\infty} \frac{1}{n!} x^n, \qquad -\infty < x < \infty.$$

By substituting $x = 1$, we obtain a series that converges to e,

$$e = \sum_{n=0}^{\infty} \frac{1}{n!} = 1 + \frac{1}{1!} + \frac{1}{2!} + \frac{1}{3!} + \cdots.$$

Another illustration is contained in the following example.

EXAMPLE 10.49

Use Example 10.22 to show that the sum of the alternating harmonic series is $\ln 2$.

SOLUTION   According to Example 10.22, the Taylor series about $x = 1$ for $\ln x$ is

$$\ln x = \sum_{n=1}^{\infty} \frac{(-1)^{n+1}}{n} (x-1)^n = (x-1) - \frac{1}{2}(x-1)^2 + \frac{1}{3}(x-1)^3 - \cdots,$$

with open interval of convergence $0 < x < 2$. At $x = 0$, the series becomes the negative of the harmonic series that diverges, and at $x = 2$, it becomes the conditionally convergent, alternating harmonic series; that is,

$$\ln 2 = \sum_{n=1}^{\infty} \frac{(-1)^{n+1}}{n} = 1 - \frac{1}{2} + \frac{1}{3} - \frac{1}{4} + \cdots.$$

Convergence of the Taylor series at $x = 2$ does not, by itself, imply convergence to $\ln 2$, as we are suggesting. It is, however, true, and this is a direct application of the following theorem.

THEOREM 10.20

If the Taylor series $\sum_{n=0}^{\infty} a_n(x - c)^n$ of a function $f(x)$ converges at the endpoint $x = c + R$ of its interval of convergence, and if $f(x)$ is continuous at $x = c + R$, then the Taylor series evaluated at $c + R$ converges to $f(c + R)$. The same result is valid at the other endpoint, $x = c - R$.

This example suggests another possibility for summing convergent series of numbers. Find a power series with known sum that reduces to the given series of numbers upon substitution of a value of x. We illustrate this in the next two examples.

EXAMPLE 10.50

Find $\sum_{n=1}^{\infty} \dfrac{n}{3^n}$.

SOLUTION   This series results if we set $x = 1/3$ in the power series

$$S(x) = \sum_{n=1}^{\infty} nx^n.$$

The radius of convergence of this series is

$$R = \lim_{n \to \infty} \left| \frac{n}{n+1} \right| = 1.$$

If we divide both sides by x,

$$\frac{1}{x} S(x) = \sum_{n=1}^{\infty} nx^{n-1},$$

and integrate according to Theorem 10.3,

$$\int \frac{1}{x} S(x)\, dx = \sum_{n=1}^{\infty} x^n.$$

This is a geometric series with sum

$$\int \frac{1}{x} S(x)\, dx = \frac{x}{1 - x}, \quad |x| < 1.$$

Differentiation now gives

$$\frac{1}{x} S(x) = \frac{(1 - x)(1) - x(-1)}{(1 - x)^2} = \frac{1}{(1 - x)^2}.$$

Thus,

$$S(x) = \sum_{n=1}^{\infty} n\, x^n = \frac{x}{(1 - x)^2}.$$

When we set $x = 1/3$,

$$\sum_{n=1}^{\infty} \frac{n}{3^n} = \frac{1/3}{(1 - 1/3)^2} = \frac{3}{4}.$$

EXAMPLE 10.51

Find the sum of the series $\displaystyle\sum_{n=0}^{\infty} \frac{(-1)^n}{(2n+1)2^n}$.

SOLUTION   There are many power series that reduce to this series upon substitution of a specific value of x. For instance, substitution of $-1/2$, 1, and $1/\sqrt{2}$ into the following power series, respectively, lead to the given series:

$$\sum_{n=0}^{\infty} \frac{1}{2n+1} x^n, \qquad \sum_{n=0}^{\infty} \frac{(-1)^n}{(2n+1)2^n} x^n, \qquad \sum_{n=0}^{\infty} \frac{\sqrt{2}(-1)^n}{2n+1} x^{2n+1}.$$

Which should we consider? Although it is not the simplest, the third series looks most promising; the fact that the power on x corresponds to the coefficient in the denominator suggests that we can find the sum of this series. We therefore set

$$S(x) = \sum_{n=0}^{\infty} \frac{\sqrt{2}(-1)^n}{2n+1} x^{2n+1}.$$

To find the radius of convergence of this series, we set $y = x^2$, in which case

$$\sum_{n=0}^{\infty} \frac{\sqrt{2}(-1)^n}{2n+1} x^{2n+1} = x \sum_{n=0}^{\infty} \frac{\sqrt{2}(-1)^n}{2n+1} y^n.$$

The radius of convergence of the y-series is

$$R_y = \lim_{n\to\infty} \left| \frac{\dfrac{\sqrt{2}(-1)^n}{2n+1}}{\dfrac{\sqrt{2}(-1)^{n+1}}{2n+3}} \right| = \lim_{n\to\infty} \left(\frac{2n+3}{2n+1} \right) = 1.$$

The radius of convergence for the power series in x is therefore $R_x = 1$ also. If we differentiate the series with respect to x,

$$S'(x) = \sum_{n=0}^{\infty} \sqrt{2}(-1)^n x^{2n} = \sum_{n=0}^{\infty} \sqrt{2}(-x^2)^n = \frac{\sqrt{2}}{1-(-x^2)} = \frac{\sqrt{2}}{1+x^2}.$$

Antidifferentiation now gives

$$S(x) = \int \frac{\sqrt{2}}{1+x^2}\, dx = \sqrt{2}\, \mathrm{Tan}^{-1} x + C.$$

Since $S(0) = 0$, it follows that $C = 0$, and

$$\sum_{n=0}^{\infty} \frac{\sqrt{2}(-1)^n}{2n+1} x^{2n+1} = \sqrt{2}\, \mathrm{Tan}^{-1} x.$$

If we now set $x = 1/\sqrt{2}$,

$$\sum_{n=0}^{\infty} \frac{\sqrt{2}(-1)^n}{2n+1} \left(\frac{1}{\sqrt{2}} \right)^{2n+1} = \sqrt{2}\, \mathrm{Tan}^{-1}\!\left(\frac{1}{\sqrt{2}} \right).$$

Consequently,

$$\sum_{n=0}^{\infty} \frac{(-1)^n}{(2n+1)2^n} = \sqrt{2}\, \mathrm{Tan}^{-1}\!\left(\frac{1}{\sqrt{2}} \right).$$

Approximating the Sum of a Series of Numbers

The techniques described above do not find sums for all convergent series of numbers. In fact there are many series for which we would find it impossible to find a sum. But in applications we might be satisfied with a reasonable approximation to the sum of a series, and we therefore turn our attention to the problem of estimating the sum of a convergent series. The easiest method for estimating the sum S of a convergent series $\sum c_n$ is simply to choose the partial sum S_N for some N as an approximation; that is, truncate the series after N terms and choose

$$S \approx S_N = c_1 + c_2 + \cdots + c_N.$$

But an approximation is of value only if we can make some definitive statement about its accuracy. In truncating the series, we have neglected the infinity of terms $\sum_{n=N+1}^{\infty} c_n$, and the accuracy of the approximation is therefore determined by the size of $\sum_{n=N+1}^{\infty} c_n$; the smaller it is, the better the approximation. The problem is that we do not know the exact value of $\sum_{n=N+1}^{\infty} c_n$; if we did, there would be no need to approximate the sum of the original series in the first place. What we must do is estimate the sum $\sum_{n=N+1}^{\infty} c_n$.

When the integral test or the alternating series test is used to prove that a series converges, simple formulas give accuracy estimates on the truncated series. Let us illustrate these first.

Truncating an Alternating Series

It is very simple to obtain the **truncation error**, an estimate of the accuracy of a truncated alternating series $\sum c_n$ provided that the sequence $\{|c_n|\}$ is decreasing with limit zero. For example, suppose that $c_1 > 0$ (a similar discussion can be made when $c_1 < 0$). If $\{S_n\}$ is the sequence of partial sums of $\sum c_n$, then even partial sums can be expressed in the form

$$S_{2n} = (c_1 + c_2) + (c_3 + c_4) + \cdots + (c_{2n-1} + c_{2n}).$$

Since $\{|c_n|\}$ is decreasing ($|c_n| > |c_{n+1}|$), each term in the parentheses is positive. Consequently, the subsequence $\{S_{2n}\}$ of even partial sums of $\{S_n\}$ is increasing and approaches the sum of the series $\sum c_n$ from below (Figure 10.29). In a similar way, we can show that the subsequence $\{S_{2n-1}\}$ of odd partial sums is decreasing and approaches the sum of the series from above. It follows that the sum $\sum c_n$ must be between any two terms of the subsequences $\{S_{2n}\}$ and $\{S_{2n-1}\}$. In particular, *the sum of the alternating series must be between any two successive partial sums*. Furthermore, *when the alternating series is truncated, the maximum possible error is the next term.*

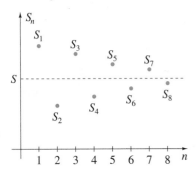

Approximating the sum of a convergent alternating series

EXAMPLE 10.52

Use the first 20 terms of the series $\sum_{n=1}^{\infty} (-1)^{n+1}/n^3$ to estimate its sum. Obtain an error estimate.

SOLUTION The sum of the first 20 terms of the series is 0.901 485. Since the series is alternating, and absolute values of terms are decreasing with limit zero, the maximum possible error in this estimate is the 21st term, $1/21^3 < 0.000\,108$. Thus,

$$0.901\,485 < \sum_{n=1}^{\infty} \frac{(-1)^{n+1}}{n^3} < 0.901\,593.$$

In practical situations, we often have to decide how many terms of a series to take in order to guarantee a certain degree of accuracy. Once again this is easy for alternating series whose terms satisfy the conditions of the alternating series test.

EXAMPLE 10.53

▼

How many terms in the series $\sum_{n=2}^{\infty} (-1)^{n+1}/(n^3 \ln n)$ ensure a truncation error of less than 10^{-5}?

SOLUTION   Because absolute values of terms are decreasing and have limit zero, the maximum error in truncating this alternating series when $n = N$ is

$$\frac{(-1)^{N+2}}{(N+1)^3 \ln (N+1)}.$$

The absolute value of this error is less than 10^{-5} when

$$\frac{1}{(N+1)^3 \ln (N+1)} < 10^{-5} \qquad \text{or}$$

$$(N+1)^3 \ln (N+1) > 10^5.$$

A calculator quickly reveals that the smallest integer for which this is valid is $N = 30$. Thus, the truncated series has the required accuracy after the 29th term (the first term corresponds to $n = 2$, not $n = 1$).

▲

In Section 10.7 we illustrated the use of Taylor's remainder formula to estimate the error when definite integrals are approximated using Taylor series. Alternating series sometimes provide an easier alternative. We redo the example of Section 10.7 to demonstrate.

EXAMPLE 10.54

▼

Approximate $\int_0^{1/2} \frac{\sin x}{x} dx$ to five decimal places using the Maclaurin series for $\sin x$.

SOLUTION   Using the Maclaurin series for $\sin x$, we obtain

$$\int_0^{1/2} \frac{\sin x}{x} dx = \int_0^{1/2} \frac{1}{x} \left(x - \frac{x^3}{3!} + \frac{x^5}{5!} - \cdots \right) dx$$

$$= \int_0^{1/2} \left(1 - \frac{x^2}{3!} + \frac{x^4}{5!} - \cdots \right) dx$$

$$= \left\{ x - \frac{x^3}{3 \cdot 3!} + \frac{x^5}{5 \cdot 5!} - \cdots \right\}_0^{1/2} = \frac{1}{2} - \frac{(1/2)^3}{3 \cdot 3!} + \frac{(1/2)^5}{5 \cdot 5!} - \cdots .$$

This is a convergent alternating series. To find a five decimal approximation, we calculate partial sums until two successive sums agree to five decimals:

$$S_1 = \frac{1}{2},$$

$$S_2 = S_1 - \frac{(1/2)^3}{3 \cdot 3!} = 0.493\,056,$$

$$S_3 = S_2 + \frac{(1/2)^5}{5 \cdot 5!} = 0.493\,108,$$

$$S_4 = S_3 - \frac{(1/2)^7}{7 \cdot 7!} = 0.493\,108.$$

Consequently, to five decimals the value of the integral is $0.493\,11$.

▲

Truncating a Series Whose Convergence Was Established with the Integral Test

Suppose now that a series $\sum_{n=1}^{\infty} c_n$ has been shown to converge with the integral test; that is, the integral

$$\int_1^{\infty} f(x)\,dx$$

converges where $f(n) = c_n$. If the series is truncated after the Nth term, the error $c_{N+1} + c_{N+2} \cdots$ is shown as the sum of the areas of the rectangles in Figure 10.30. Clearly, the sum of these areas is less than the area under $y = f(x)$ to the right of $x = N$. In other words, the error in truncating the series with the Nth term must be less than

$$\int_N^{\infty} f(x)\,dx. \qquad (10.48)$$

FIGURE 10.30 Approximating the sum of a series using the integral test

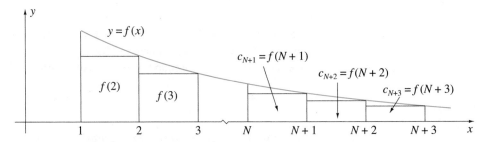

EXAMPLE 10.55

Obtain an error estimate if the series $\displaystyle\sum_{n=2}^{\infty} \frac{1}{n(\ln n)^4}$ is truncated when $n = 100$.

SOLUTION   The error cannot be larger than

$$\int_{100}^{\infty} \frac{1}{x(\ln x)^4}\,dx = \left\{\frac{-1}{3(\ln x)^3}\right\}_{100}^{\infty} = \frac{1}{3(\ln 100)^3} < 0.0035.$$

When convergence of a series is established by the comparison, limit comparison, limit ratio, or limit root tests, we often estimate the truncation error $\sum_{n=N+1}^{\infty} c_n$ by comparing it to something that is summable. We illustrate this in the following two examples.

EXAMPLE 10.56

In the first paragraph of this section we indicated that e is the sum of the series

$$\sum_{n=0}^{\infty} \frac{1}{n!}.$$

Use the first 10 terms to find an approximation for e.

SOLUTION   The sum of the first 10 terms is

$$\sum_{n=0}^{9} \frac{1}{n!} = 2.718\,281\,526.$$

The truncation error in using this as an approximation for e is

$$\sum_{n=10}^{\infty} \frac{1}{n!} = \frac{1}{10!} + \frac{1}{11!} + \frac{1}{12!} + \frac{1}{13!} + \cdots$$

$$= \frac{1}{10!}\left(1 + \frac{1}{11} + \frac{1}{11 \cdot 12} + \frac{1}{11 \cdot 12 \cdot 13} + \cdots\right)$$

$$< \frac{1}{10!}\left(1 + \frac{1}{11} + \frac{1}{11^2} + \frac{1}{11^3} + \cdots\right) \qquad \text{(a geometric series)}$$

$$= \frac{1}{10!}\frac{1}{1 - 1/11} \qquad \text{(using equation 10.39b)}$$

$$= \frac{11}{10 \cdot 10!} \qquad < 0.000\,000\,304.$$

We may write, therefore, that

$$2.718\,281\,526 < \sum_{n=0}^{\infty} \frac{1}{n!} < 2.718\,281\,830,$$

and to six decimal places, $e = 2.718\,282$.

EXAMPLE 10.57

How many terms in the convergent series $\sum_{n=1}^{\infty} n/[(n+1)3^n]$ ensure a truncation error of less than 10^{-5}?

SOLUTION   If this series is truncated after the Nth term, the error is

$$\sum_{n=N+1}^{\infty} \frac{n}{(n+1)3^n} = \frac{N+1}{(N+2)3^{N+1}} + \frac{N+2}{(N+3)3^{N+2}} + \cdots$$

$$< \frac{1}{3^{N+1}} + \frac{1}{3^{N+2}} + \cdots \qquad \text{(a geometric series)}$$

$$= \frac{\dfrac{1}{3^{N+1}}}{1 - \dfrac{1}{3}} \qquad \text{(using equation 10.39b)}$$

$$= \frac{1}{2 \cdot 3^N}.$$

Consequently, the error is guaranteed to be less than 10^{-5} if N satisfies the inequality

$$\frac{1}{2 \cdot 3^N} < 10^{-5} \qquad \text{or}$$

$$2 \cdot 3^N > 10^5 \qquad \Longrightarrow \qquad N > \frac{\ln(10^5/2)}{\ln 3} = 9.85.$$

Thus, 10 or more terms yield the required accuracy.

EXERCISES 10.13

In Exercises 1–10 verify that the sum of the series is as indicated.

1. $\displaystyle\sum_{n=0}^{\infty} \frac{2^n}{n!} = e^2$

2. $\displaystyle\sum_{n=0}^{\infty} \frac{(-1)^n}{(2n+1)!} = \sin 1$

3. $\displaystyle\sum_{n=0}^{\infty} \frac{(-1)^n 3^{2n}}{(2n)!} = \cos 3$

4. $\displaystyle\sum_{n=1}^{\infty} \frac{(-1)^n}{n!} = \frac{1}{e} - 1$

5. $\displaystyle\sum_{n=1}^{\infty} \frac{(-1)^n}{2^{2n}} = -\frac{1}{5}$

* 6. $\displaystyle\sum_{n=2}^{\infty} \frac{(-1)^{n+1} 2^{2n+3}}{(2n)!} = -8(1 + \cos 2)$

* 7. $\displaystyle\sum_{n=1}^{\infty} \frac{2^n}{n3^n} = \ln 3$

* 8. $\displaystyle\sum_{n=1}^{\infty} \frac{1}{n2^n} = \ln 2$

* 9. $\displaystyle\sum_{n=1}^{\infty} \frac{(-1)^n}{3^{2n}(2n+1)!} = 3\sin\left(\frac{1}{3}\right) - 1$

* 10. $\displaystyle\sum_{n=1}^{\infty} \frac{n}{2^n} = 2$

* 11. Find the Maclaurin series for $\text{Tan}^{-1} x$ and use it to evaluate

$$\sum_{n=1}^{\infty} \frac{(-1)^n}{2n+1}.$$

* 12. Find the Maclaurin series for $x/(1+x^2)^2$ and use it to evaluate

$$\sum_{n=1}^{\infty} \frac{n(-1)^n}{3^{2n}}.$$

In Exercises 13–14 approximate the sum of the series if it is truncated after the Nth term. Use 10.48 to find an error estimate.

13. $\displaystyle\sum_{n=2}^{\infty} \frac{n^2}{(n^3+1)^4}$, $N = 10$ **14.** $\displaystyle\sum_{n=1}^{\infty} \frac{n}{e^{3n}}$, $N = 5$

In Exercises 15–16 use 10.48 to estimate the error when the series is truncated after the Nth term.

15. $\displaystyle\sum_{n=1}^{\infty} \frac{1}{n^2+1}$, $N = 100$ **16.** $\displaystyle\sum_{n=1}^{\infty} \frac{1}{n^2}\sin\left(\frac{1}{n}\right)$, $N = 20$

In Exercises 17–24 use the number of terms indicated to find an approximation to the sum of the series. In each case, obtain an estimate of the truncation error.

17. $\displaystyle\sum_{n=2}^{\infty} \frac{(-1)^{n+1}}{n^3 3^n}$ (3 terms)   **18.** $\displaystyle\sum_{n=1}^{\infty} \frac{(-1)^n}{n^4}$ (20 terms)

* 19. $\displaystyle\sum_{n=1}^{\infty} \frac{1}{n^n}$ (5 terms)   *** 20.** $\displaystyle\sum_{n=1}^{\infty} \frac{1}{n2^n}$ (10 terms)

* 21. $\displaystyle\sum_{n=1}^{\infty} \frac{1}{2^n}\sin\left(\frac{\pi}{n}\right)$ (15 terms)

* 22. $\displaystyle\sum_{n=2}^{\infty} \frac{2^n-1}{3^n+n}$ (20 terms)   *** 23.** $\displaystyle\sum_{n=2}^{\infty} \frac{2^n+1}{3^n+n}$ (20 terms)

* 24. $\displaystyle\sum_{n=1}^{\infty} \frac{(-1)^n}{n}$ (100 terms)

In Exercises 25–27 how many terms in the series guarantee an approximation to the sum with a truncation error of less than 10^{-4}?

* 25. $\displaystyle\sum_{n=1}^{\infty} \frac{(-1)^n}{n^2}$ *** 26.** $\displaystyle\sum_{n=1}^{\infty} \frac{1}{n^2 4^n}$

* 27. $\displaystyle\sum_{n=1}^{\infty} \frac{2^n}{n!}$

* 28. Suppose the series $\displaystyle\sum_{n=1}^{\infty} e^{-n}\sin^2 n$ is truncated after the 10th term. Obtain an error estimate by (a) using 10.48 and (b) using the fact that $e^{-n}\sin^2 n < e^{-n}$. Which gives the better estimate?

In Exercises 29–36 evaluate the integral correct to three decimal places. Compare the work in Exercises 29, 30, 32, and 34 to that in Exercises 11–14 of Section 10.7.

* 29. $\displaystyle\int_0^1 \frac{\sin x}{x}\, dx$ *** 30.** $\displaystyle\int_0^{1/2} \cos(x^2)\, dx$

* 31. $\displaystyle\int_0^{2/3} \frac{1}{x^4+1}\, dx$ *** 32.** $\displaystyle\int_{-1}^{1} x^{11}\sin x\, dx$

* 33. $\displaystyle\int_0^{1/2} \frac{1}{\sqrt{1+x^3}}\, dx$ *** 34.** $\displaystyle\int_0^{0.3} e^{-x^2}\, dx$

* 35. $\displaystyle\int_{-0.1}^{0} \frac{1}{x-1}\ln(1-x)\, dx$

* 36. $\displaystyle\int_0^{1/2} \frac{1}{x^6-3x^3-4}\, dx$

* 37. In determining the radiated power from a half-wave antenna, it is necessary to evaluate

$$\int_0^{2\pi} \frac{1-\cos\theta}{\theta}\, d\theta.$$

Find a two-decimal-place approximation for this integral.

* 38. A very important function in engineering and physics is the error function $\text{erf}(x)$ defined by

$$\text{erf}(x) = \frac{2}{\sqrt{\pi}}\int_0^x e^{-t^2}\, dt.$$

Calculate $\text{erf}(1)$ correct to three decimal spaces.

39. This exercise shows that we must be very careful in predicting the accuracy of a result. Consider the series

$$S = 3.125\,100\,1 - 0.000\,090\,18\left(1 + \frac{1}{10} + \frac{1}{10^2} + \frac{1}{10^3} + \cdots\right).$$

(a) Show that the sum of this series is exactly $S = 3.124\,999\,9$. To two decimal places, then, the value of S is 3.12.

(b) Verify that the first four partial sums of the series are

$$S_1 = 3.125\,100\,1,$$

$$S_2 = 3.125\,009\,92,$$

$$S_3 = 3.125\,000\,902,$$

$$S_4 = 3.125\,000\,000\,2.$$

(c) If $E_n = S_n - S$ are the differences between the sum of the series and its first four partial sums, show that

$$E_1 = 0.000\,100\,2,$$

$$E_2 = 0.000\,010\,02,$$

$$E_3 = 0.000\,001\,002,$$

$$E_4 = 0.000\,000\,100\,2.$$

What can you say about the accuracy of S_1, S_2, S_3, and S_4 as approximations to S?

(d) If S is approximated by any of S_1, S_2, S_3, or S_4 to two decimals, the result is 3.13, not 3.12 as in part (a). Thus, in spite of the accuracy predicted in part (c), S_1, S_2, S_3, and S_4 do not predict S correctly to two decimal places. Do they predict S correctly to three or four decimal places?

SUMMARY

An infinite sequence of numbers is the assignment of numbers to positive integers. In most applications of sequences, the prime consideration is whether the sequence has a limit. If the sequence has its terms defined explicitly, then our ability to take limits of continuous functions (limits at infinity in Chapter 2 and L'Hôpital's rule in Chapter 4) can be very helpful. If the sequence is defined recursively, existence of the limit can sometimes be established by showing that the sequence is monotonic and bounded, or that it is oscillatory and convergent.

An expression of the form

$$\sum_{n=1}^{\infty} c_n = c_1 + c_2 + \cdots + c_n + \cdots$$

is called an infinite series. We define the sum of this series as the limit of its sequence of partial sums $\{S_n\}$, provided that the sequence has a limit. Unfortunately, for most series we cannot find a simple formula for S_n, and therefore analysis of the limit of the sequence $\{S_n\}$ is impossible. To remedy this, we developed various convergence tests that avoided the sequence $\{S_n\}$: n^{th} term, comparison, limit comparison, limit ratio, limit root, integral, and alternating series tests. Note the sequences that are associated with a series $\sum c_n$:

$\{S_n\}$	sequence of partial sums for the definition of a sum;
$\{c_n\}$	sequence of terms for the nth-term test;
$\{c_n/b_n\}$	sequence for the limit comparison test;
$\{c_{n+1}/c_n\}$	sequence for the limit ratio test;
$\{\sqrt[n]{c_n}\}$	sequence for the limit root test;
$\{|c_n|\}$	sequence for the alternating series test.

Depending on the limits of these sequences — if they exist — we may be able to infer something about convergence of the series.

Infinite sequences and series of functions are important in applications — in particular, power series. (As scientists, you will see other types of series: Fourier series, for example.) We considered situations where a power series was given and the sum was to be determined. We saw that every power series $\sum a_n(x - c)^n$ has a radius of convergence R and an associated interval of convergence. If $R = 0$, the interval of convergence consists of only one point $x = c$; if $R = \infty$, the power series converges for all x; and if $0 < R < \infty$, the interval of convergence must be one of four possibilities: $c - R < x < c + R$, $c - R \leq x < c + R$, $c - R < x \leq c + R$, or $c - R \leq x \leq c + R$. The radius of convergence is given by $\lim_{n \to \infty} |a_n/a_{n+1}|$ or $\lim_{n \to \infty} |a_n|^{-1/n}$ provided that the limits exist or are equal to infinity. If at each point in the interval of convergence of the power series the value of a function $f(x)$ is the same as the sum of the series, we write $f(x) = \sum a_n(x - c)^n$ and call $f(x)$ the sum of the series.

We also considered situations where a function $f(x)$ and a point c are given, and ask whether $f(x)$ has a power series expansion about c. We saw that there can be at most one power series expansion of $f(x)$ about c with a positive radius of convergence, and this series must be its Taylor series. One way to verify that $f(x)$ does indeed have a Taylor series about c and that this series converges to $f(x)$ is to show that the sequence of Taylor's remainders $\{R_n(c, x)\}$ exists and has limit zero. Often, however, it is much easier to find Taylor series by adding, multiplying, differentiating, and integrating known series.

When a Taylor series is truncated, Taylor's remainder $R_n(c, x)$ represents the truncation error and, in spite of the fact that R_n is expressed in terms of some unknown point z_n, it is often possible to calculate a maximum value for the error. Sometimes $R_n(c, x)$ can be avoided altogether. For instance, if the Taylor series is an alternating series, then the maximum possible truncation error is the value of the next term.

Power series are often used in situations that require approximations. Taylor series provide polynomial approximations to complicated functions, and they offer an alternative to the numerical techniques of Section 8.8 in the evaluation of definite integrals. Power series are also useful in situations that do not require approximations. They are sometimes helpful in evaluating limits, and they are the only way to solve many differential equations.

KEY TERMS

In reviewing this chapter, you should be able to define or discuss the following key terms:

Sequence	Series
Infinite sequence of numbers	Term
Explicit sequences	Recursive sequence
Limit of a sequence	Convergent sequence
Divergent sequence	Method of successive approximations or fixed-point iteration
Taylor Remainder Formula	Taylor remainders
Taylor polynomials	Interval of convergence
Taylor series	Maclaurin series
Power series	Geometric series
Common ratio	Radius of convergence
Open interval of convergence	Binomial expansion
Sums of power series	Increasing sequence
Nondecreasing sequence	Decreasing sequence
Nonincreasing sequence	Monotonic sequence
Upper bound	Lower bound
Bounded sequence	Oscillating sequence
Successive approximations	Infinite series of numbers
Sequence of partial sums	Convergent series
Divergent series	Harmonic series
n^{th}-term test	Nonnegative series
Integral test	p-series
Comparison test	Limit comparison test
Limit ratio test	Limit root test
Absolutely convergent series	Conditionally convergent series
Alternating series	Alternating harmonic series
Alternating series test	Truncation error

REVIEW EXERCISES

In Exercises 1–6 discuss, with all necessary proofs, whether the sequence is monotonic and has an upper bound, a lower bound, and a limit.

1. $\left\{\dfrac{n^2 - 5n + 3}{n^2 + 5n + 4}\right\}$

2. $c_1 = 1, \quad c_{n+1} = (1/2)\sqrt{c_n^2 + 1}, \quad n \geq 1$

3. $\left\{\dfrac{\operatorname{Tan}^{-1}(1/n)}{n^2 + 1}\right\}$

4. $c_1 = 7, \quad c_{n+1} = 15 + \sqrt{c_n - 2}, \quad n \geq 1$

5. $c_1 = 6, \quad c_{n+1} = 6 + \dfrac{2}{c_n}, \quad n \geq 1$

6. $c_1 = 6, \quad c_{n+1} = \dfrac{1}{5 + 4c_n}, \quad n \geq 1$

7. Use Newton's iterative procedure and the method of successive approximations to approximate the root of the equation

$$x = \left(\frac{x+5}{x+4}\right)^2$$

between $x = 1$ and $x = 2$, accurate to 5 decimal places.

8. For what values of k does the sequence

$$c_1 = k, \qquad c_{n+1} = c_n^2, \quad n \geq 1$$

converge?

9. Find an explicit definition for the sequence

$$c_1 = 1, \qquad c_{n+1} = \sqrt{1 + c_n^2}, \quad n \geq 1.$$

10. Use the derivative of the function $f(x) = (\ln x)/x$ to prove that the sequence $\{\ln n \, /n\}$ is decreasing for $n \geq 3$.

In Exercises 11–30 determine whether the series converges or diverges. In the case of a convergent series that has both positive and negative terms, indicate whether it converges absolutely or conditionally.

11. $\displaystyle\sum_{n=1}^{\infty} \dfrac{n^2 - 3n + 2}{n^3 + 4n}$

12. $\displaystyle\sum_{n=1}^{\infty} \dfrac{n^2 + 5n + 3}{n^4 - 2n + 5}$

13. $\displaystyle\sum_{n=1}^{\infty} \dfrac{5^{2n}}{n!}$

14. $\displaystyle\sum_{n=1}^{\infty} \dfrac{n^2 + 3}{n3^n}$

15. $\displaystyle\sum_{n=1}^{\infty} \dfrac{(\ln n)^2}{\sqrt{n}}$

16. $\displaystyle\sum_{n=1}^{\infty} (-1)^n \left(\dfrac{n+1}{n^2}\right)$

17. $\displaystyle\sum_{n=1}^{\infty} (-1)^n \left(\dfrac{n+1}{n^3}\right)$

18. $\displaystyle\sum_{n=1}^{\infty} \operatorname{Cos}^{-1}\left(\dfrac{1}{n}\right)$

19. $\displaystyle\sum_{n=1}^{\infty} \dfrac{1}{n} \operatorname{Cos}^{-1}\left(\dfrac{1}{n}\right)$

20. $\displaystyle\sum_{n=1}^{\infty} \dfrac{1}{n^2} \operatorname{Cos}^{-1}\left(\dfrac{1}{n}\right)$

21. $\displaystyle\sum_{n=1}^{\infty} \dfrac{2 \cdot 4 \cdot 6 \cdot \ldots \cdot (2n)}{n!}$

22. $\displaystyle\sum_{n=1}^{\infty} \dfrac{3 \cdot 6 \cdot 9 \cdot \ldots \cdot (3n)}{(2n)!}$

23. $\displaystyle\sum_{n=1}^{\infty} \sqrt{\dfrac{n^2 + 1}{n^2 + 5}}$

24. $\displaystyle\sum_{n=1}^{\infty} (-1)^{n+1} \left(1 + \dfrac{1}{n}\right)^3$

25. $\displaystyle\sum_{n=1}^{\infty} \dfrac{1}{n^2} \sin n$

26. $\displaystyle\sum_{n=1}^{\infty} \dfrac{10^n}{5^{3n+2}}$

27. $\displaystyle\sum_{n=1}^{\infty} (-1)^n \dfrac{\ln n}{n}$

28. $\displaystyle\sum_{n=1}^{\infty} \dfrac{1}{e^{n\pi}}$

29. $\displaystyle\sum_{n=1}^{\infty} \dfrac{2^n + 2^{-n}}{3^n}$

30. $\displaystyle\sum_{n=1}^{\infty} \dfrac{1}{\sqrt{n}} \cos(n\pi)$

In Exercises 31–38 find the interval of convergence for the power series.

* 31. $\displaystyle\sum_{n=0}^{\infty} \dfrac{n+1}{n^2 + 1} x^n$

* 32. $\displaystyle\sum_{n=1}^{\infty} \dfrac{1}{n^2 2^n} x^n$

* 33. $\displaystyle\sum_{n=0}^{\infty} (n+1)^3 x^n$

* 34. $\displaystyle\sum_{n=1}^{\infty} \dfrac{1}{n^n} x^n$

* 35. $\displaystyle\sum_{n=0}^{\infty} \dfrac{1}{4^n} (x-2)^n$

* 36. $\displaystyle\sum_{n=2}^{\infty} \sqrt{\dfrac{n+1}{n-1}} (x+3)^n$

* 37. $\displaystyle\sum_{n=1}^{\infty} n3^n x^{2n}$

* 38. $\displaystyle\sum_{n=1}^{\infty} \dfrac{2^n}{n} x^{3n}$

In Exercises 39–47 find the power series expansion of the function about the indicated point.

39. $f(x) = \sqrt{1 + x^2}$, about $x = 0$

40. $f(x) = e^{x+5}$, about $x = 0$

* 41. $f(x) = \cos(x + \pi/4)$, about $x = 0$

* 42. $f(x) = x \ln(2x + 1)$, about $x = 0$

* 43. $f(x) = \sin x$, about $x = \pi/4$

* 44. $f(x) = x/(x^2 + 4x + 3)$, about $x = 0$

* 45. $f(x) = e^x$, about $x = 3$

✳ **46.** $f(x) = (x+1)\ln(x+1)$, about $x = 0$

✳ **47.** $f(x) = x^3 e^{x^2}$, about $x = 0$

✳ **48.** How many terms in the Maclaurin series for $f(x) = e^{-x^2}$ guarantee a truncation error of less than 10^{-5} for all x in the interval $0 \le x \le 2$?

✳ **49.** Find a power series solution in powers of x for the differential equation

$$y'' - 4y = 0.$$

✳ **50.** Find the Maclaurin series for $f(x) = \sqrt{1 + \sin x}$ valid fo $-\pi/2 \le x \le \pi/2$ by first showing that $f(x)$ can be written i the form

$$f(x) = \sin(x/2) + \cos(x/2).$$

Why is the restriction $-\pi/2 \le x \le \pi/2$ necessary?

✳ **51.** On a calculator take the cosine of 1 (radian). Take the cosine o this result, and take it again, and again, and again, ... What happens Interpret what is going on.

CHAPTER 11

Vectors and Three-Dimensional Analytic Geometry

Application Preview

The figure below shows a boom OA carrying a mass M. The boom is supported by cables AB and AC.

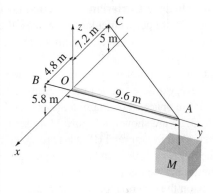

THE PROBLEM   If tensions in the cables must not exceed 20 000 N, what is the maximum mass that can be supported by the boom? (See Example 11.10 on page 719 for the solution.)

Chapters 1–10 dealt with single-variable calculus — differentiation and integration of functions $f(x)$ of one variable. In Chapters 11–14 we study multivariable calculus. Discussions of three-dimensional analytic geometry and vectors in Sections 11.1–11.5 prepare the way. In Sections 11.9–11.13 we differentiate and integrate vector functions, and apply the results to the geometry of curves in space and the motion of objects.

11.1 Rectangular Coordinates in Space

FIGURE 11.1   Coordinate axes and coordinate planes

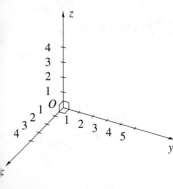

The coordinate of a point on the real line is its directed distance from the origin. Cartesian coordinates of a point in a plane are its directed distances from the coordinate axes. In space, Cartesian coordinates are directed distances from three fixed planes called the **coordinate planes**. In particular, we draw through a point O, called the **origin**, three mutually perpendicular lines called the x-, y-, and z-axes (Figure 11.1). Each of the axes is coordinatized with some unit distance (which need not be the same for all three axes). These three coordinate axes determine the three coordinate planes: The xy-coordinate plane is that plane containing the x- and y-axes, the yz-coordinate plane contains the y- and z-axes, and the xz-coordinate plane contains the x- and z-axes.

If P is any point in space, we draw lines from P perpendicular to the three coordinate planes (Figure 11.2). The directed distance from the yz-coordinate plane to P is parallel to the x-axis, and is called the x-coordinate of P. Similarly, y- and z-coordinates are defined as directed distances from the xz- and xy-coordinate planes to P. These three coordinates of P,

695

FIGURE 11.2 A point as perpendicular distances from coordinate planes

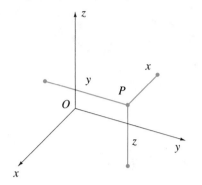

FIGURE 11.3 A point in space
as distances along coordinates axes

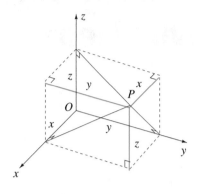

FIGURE 11.4 Coordinates of four specific points

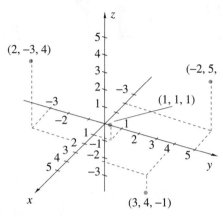

written (x, y, z), are called the **Cartesian** or **rectangular coordinates** of P. Note that if we
draw lines through P that are perpendicular to the axes, then the directed distances from O to
points of intersection of these perpendiculars with the axes are also the Cartesian coordinates of
P (Figure 11.3).

By either definition, each point in space has a unique ordered set of Cartesian coordinates
(x, y, z); conversely, every ordered triple of real numbers (x, y, z) is the set of coordinates for
one and only one point in space. For example, points with coordinates $(1, 1, 1)$, $(2, -3, 4)$,
$(3, 4, -1)$, and $(-2, 5, 3)$ are shown in Figure 11.4.

The coordinate systems in Figures 11.1–11.4 are called **right-handed coordinate systems**
because if we curl the fingers on our right hand from the positive x-direction toward the positive
y-direction, then the thumb points in the positive z-direction (Figure 11.5). The coordinate
system in Figure 11.6, on the other hand, is a **left-handed coordinate system**, since the thumb
of the left hand points in the positive z-direction when the fingers of this hand are curled from
the positive x-direction to the positive y-direction. We always use right-handed systems in this
book, as should everyone.

Suppose we construct for any two points P_1 and P_2 with coordinates (x_1, y_1, z_1) and
(x_2, y_2, z_2), respectively, a box with sides parallel to the coordinate planes and with line segment
$P_1 P_2$ as diagonal (Figure 11.7). Because triangles $P_1 A B$ and $P_1 B P_2$ are right-angled, we can
write

$$\|P_1 P_2\|^2 = \|P_1 B\|^2 + \|B P_2\|^2$$
$$= \|P_1 A\|^2 + \|A B\|^2 + \|B P_2\|^2$$
$$= (x_2 - x_1)^2 + (y_2 - y_1)^2 + (z_2 - z_1)^2.$$

FIGURE 11.5 Right-handed
coordinate system

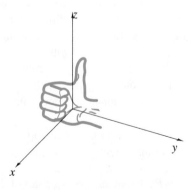

FIGURE 11.6 Left-handed coordinate system

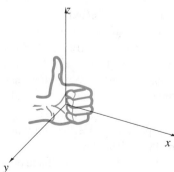

FIGURE 11.7 Distance between
two points in space

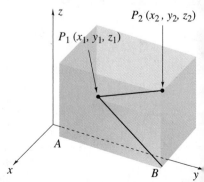

In other words, the length of the line segment joining two points $P_1(x_1, y_1, z_1)$ and $P_2(x_2, y_2, z_2)$ is

$$\|P_1 P_2\| = \sqrt{(x_2 - x_1)^2 + (y_2 - y_1)^2 + (z_2 - z_1)^2}. \tag{11.1}$$

This is the analogue of formula 1.10 for the length of a line segment joining two points in the xy-plane.

Just as the x- and y-axes divide the xy-plane into four regions called quadrants, the xy-, yz-, and xz-coordinate planes divide xyz-space into eight regions called **octants**. The region where x-, y-, and z-coordinates are all positive is called the first octant. There is no commonly accepted way to number the remaining seven octants.

EXERCISES 11.1

1. Draw a Cartesian coordinate system and show the points $(1, 2, 1)$, $(-1, 3, 2)$, $(1, -2, 4)$, $(3, 4, -5)$, $(-1, -2, -3)$, $(-2, -5, 4)$, $(8, -3, -6)$, and $(-4, 3, -5)$.

2. Find the length of the line segment joining the points $(1, -2, 5)$ and $(-3, 2, 4)$.

3. Prove that the triangle with vertices $(2, 0, 4\sqrt{2})$, $(3, -1, 5\sqrt{2})$, and $(4, -2, 4\sqrt{2})$ is right-angled and isosceles.

4. A cube has sides of length 2 units. What are coordinates of its corners if one corner is at the origin, three of its faces lie in the coordinate planes, and one corner has all three coordinates positive?

5. Show that the (undirected, perpendicular) distances from a point (x, y, z) to the x-, y-, and z-axes are, respectively, $\sqrt{y^2 + z^2}$, $\sqrt{x^2 + z^2}$, and $\sqrt{x^2 + y^2}$.

In Exercises 6–9 find the (undirected) distances from the point to (a) the origin, (b) the x-axis, (c) the y-axis, and (d) the z-axis.

6. $(2, 3, -4)$ **7.** $(1, -5, -6)$

8. $(4, 3, 0)$ **9.** $(-2, 1, -3)$

10. Prove that the three points $(1, 3, 5)$, $(-2, 0, 3)$, and $(7, 9, 9)$ are collinear.

11. Find that point in the xy-plane that is equidistant from the points $(1, 3, 2)$ and $(2, 4, 5)$ and has a y-coordinate equal to three times its x-coordinate.

12. Find an equation describing all points that are equidistant from the points $(-3, 0, 4)$ and $(2, 1, 5)$. What does this equation describe geometrically?

13. (a) If $(\sqrt{3} - 3, 2 + 2\sqrt{3}, 2\sqrt{3} - 1)$ and $(2\sqrt{3}, 4, \sqrt{3} - 2)$ are two vertices of an equilateral triangle, and if the third vertex lies on the z-axis, find the third vertex.

    (b) Can you find a third vertex on the x-axis?

14. A birdhouse is built from a box 1/2 m on each side with a roof as shown in the figure to the right. If the distance from each corner of the roof to the peak is 3/4 m, find coordinates of the nine corners of the house. (The sides of the box are parallel to the coordinate planes.)

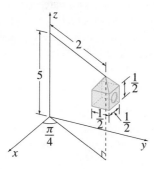

* 15. If P and Q in the figure below have coordinates (x_1, y_1, z_1) and (x_2, y_2, z_2), show that coordinates of the point R midway between P and Q are

$$\left(\frac{x_1 + x_2}{2}, \frac{y_1 + y_2}{2}, \frac{z_1 + z_2}{2} \right).$$

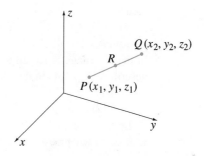

* 16. (a) Find the midpoint of the line segment joining the points $P(1, -1, -3)$ and $Q(3, 2, -4)$.

    (b) If the line segment joining P and Q is extended its own length beyond Q to a point R, find the coordinates of R.

* 17. The four-sided object in the figure below is a *tetrahedron*. If the four vertices of the tetrahedron are as shown, prove that the three lines joining the midpoints of opposite edges (one of which is PQ) meet at a point that bisects each of them.

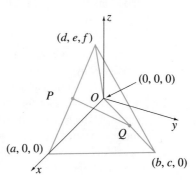

* 18. Let A, B, C, and D be the vertices of a quadrilateral in space (not necessarily planar). Show that the line segments joining midpoints of opposite sides of the quadrilateral intersect in a point that bisects each.

** **19.** Generalize the result of Exercise 15 to prove that if a point R divides the length PQ so that $\dfrac{\|PR\|}{\|RQ\|} = \dfrac{r_1}{r_2}$, where r_1 and r_2 are positive integers, then the coordinates of R are

$$x = \frac{r_1 x_2 + r_2 x_1}{r_1 + r_2}, \quad y = \frac{r_1 y_2 + r_2 y_1}{r_1 + r_2}, \quad z = \frac{r_1 z_2 + r_2 z_1}{r_1 + r_2}.$$

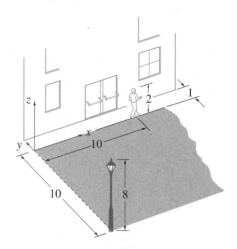

** **20.** A man 2 m tall walks along the edge of a straight road 10 m wide (figure right). On the other edge of the road stands a streetlight 8 m high. A building runs parallel to the road and 1 m from it. If Cartesian coordinates are set up as shown (with x- and y-axes in the plane of the road), find coordinates of the tip of the man's shadow when he is at the position shown.

11.2 Curves and Surfaces

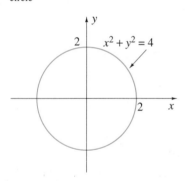

FIGURE 11.8   In the xy-plane, $x^2 + y^2 = 4$ describes a circle

An equation involving the x- and y-coordinates of points in the xy-plane usually specifies a curve. For example, the equation $x^2 + y^2 = 4$ describes a circle of radius 2 centred at the origin (Figure 11.8). We now ask what is defined by an equation involving the Cartesian coordinates (x, y, z) of points in space. For example, the equation $z = 0$ describes all points in the xy-plane since all such points have a z-coordinate equal to zero. Similarly, $y = 2$ describes all points in the plane parallel to and 2 units to the right of the xz-plane. What does the equation $x^2 + y^2 = 4$ describe? In other words, regarded as a restriction on the x-, y-, and z-coordinates of points in space, rather than a restriction on the x- and y-coordinates of points in the xy-plane, what does it represent? Because the equation says nothing about z, there is no restriction whatsoever on z. In other words, the z-coordinate can take on all possible values, but x- and y-coordinates must be restricted by $x^2 + y^2 = 4$. If we consider those points in the xy-plane $(z = 0)$ that satisfy $x^2 + y^2 = 4$, we obtain the circle in Figure 11.8. In space, each of these points has coordinates $(x, y, 0)$, where x and y still satisfy $x^2 + y^2 = 4$ (Figure 11.9). If we now take any point Q that is either directly above or directly below a point $P(x, y, 0)$ on this circle, it has exactly the same x- and y-coordinates as P; only its z-coordinate differs. Thus the x- and y-coordinates of Q also satisfy $x^2 + y^2 = 4$. Since we can do this for any point P on the circle, it follows that $x^2 + y^2 = 4$ describes the right-circular cylinder of radius 2 and infinite extent in Figure 11.9.

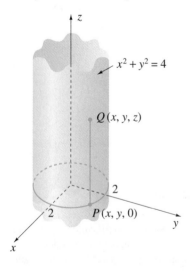

FIGURE 11.9   In space, $x^2 + y^2 = 4$ describes a cylinder

By reasoning similar to that used above, we can show that the equation $2x + y = 2$ describes the plane in Figure 11.10 parallel to the z-axis and standing on the straight line $2x + y = 2$, $z = 0$ in the xy-plane.

Finally, consider the equation $x^2 + y^2 + z^2 = 9$. Since $\sqrt{x^2 + y^2 + z^2}$ is the distance from the origin to a point with coordinates (x, y, z), this equation describes all points that are 3 units away from the origin. In other words, $x^2 + y^2 + z^2 = 9$ describes points on a sphere of radius 3 centred at the origin.

It appears that one equation in the coordinates (x, y, z) of points in space specifies a surface. The shape of the surface is determined by the form of the equation. If one equation in the coordinates (x, y, z) specifies a surface, it is easy to see what two simultaneous equations specify. For instance, suppose we ask for all points in space whose coordinates satisfy both of the equations

$$x^2 + y^2 = 4, \quad z = 1.$$

By itself, $x^2 + y^2 = 4$ describes the cylinder in Figure 11.9. The equation $z = 1$ describes all points in a plane parallel to the xy-plane and 1 unit above it. To ask for all points that satisfy $x^2 + y^2 = 4$ and $z = 1$ simultaneously is to ask for all points that lie on both surfaces.

FIGURE 11.10 Plane with equation $2x + y = 2$

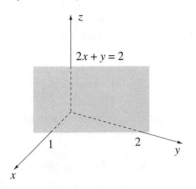

FIGURE 11.10 Plane with equation $2x + y = 2$

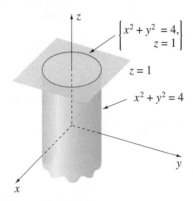

FIGURE 11.11 Curve of intersection of cylinder $x^2 + y^2 = 4$ and plane $z = 1$

Consequently, the equations $x^2 + y^2 = 4$, $z = 1$ describe the curve of intersection of the two surfaces — the circle in Figure 11.11.

The equation $x = 0$ describes the yz-plane; the equation $y = 0$ describes the xz-plane. If we put the two equations together, $x = 0$ and $y = 0$, we obtain all points that lie on both the yz-plane and the xz-plane (i.e., the z-axis). In other words, equations for the z-axis are $x = 0$, $y = 0$.

Finally, $x^2 + y^2 + z^2 = 9$ is the equation of a sphere of radius 3 centred at the origin, and $y = 2$ is the equation of a plane parallel to the xz-plane and 2 units to the right. Together, the equations $x^2 + y^2 + z^2 = 9$, $y = 2$ describe the curve of intersection of the two surfaces — the circle in Figure 11.12. Note that by substituting $y = 2$ into the equation of the sphere, we can write alternatively that $x^2 + z^2 = 5$, $y = 2$. This pair of equations is equivalent to the original pair because all points that satisfy $x^2 + y^2 + z^2 = 9$, $y = 2$ also satisfy $x^2 + z^2 = 5$, $y = 2$, and vice versa. This new pair of equations provides an alternative way of visualizing the curve. Again $y = 2$ is the plane of Figure 11.12, but $x^2 + z^2 = 5$ describes a right-circular cylinder of radius $\sqrt{5}$ and infinite extent around the y-axis (Figure 11.13). Our discussion has shown that the cylinder and plane intersect in the same curve as the sphere and plane.

In summary, we have illustrated that one equation in the coordinates (x, y, z) of a point specifies a surface; two simultaneous equations specify a curve, the curve of intersection of the two surfaces (provided, of course, that the surfaces do intersect).

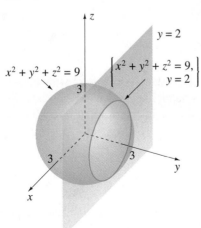

FIGURE 11.12 Curve of intersection of sphere $x^2 + y^2 + z^2 = 9$ and plane $y = 2$

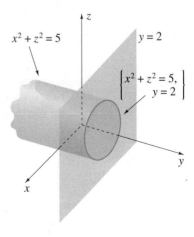

FIGURE 11.13 Curve of intersection of cylinder $x^2 + z^2 = 5$ and plane $y = 2$

In Chapters 2–9 we learned to appreciate the value of plotting and drawing curves in the xy-plane. Sometimes a plot or drawing serves as a device by which we can interpret algebraic statements geometrically (such as the mean value theorem or the interpretation of a critical point of a function as a point where the tangent line to the graph of the function is horizontal, vertical, or does not exist). Sometimes they play an integral part in the solution of a problem (such as when the definite integral is used to find areas, volumes, etc.). Sometimes a plot or drawing is a complete solution to a problem (such as to determine whether a given function has an inverse). We will find that plotting and drawing surfaces can be just as useful for multivariable calculus in Chapters 12–14.

One of the most helpful techniques for drawing a surface is to imagine the intersection of the surface with various planes — in particular, the coordinate planes. From these cross-sections of the surface, it is sometimes possible to visualize the entire surface. For example, if we intersect the surface $z = x^2 + y^2$ with the yz-plane, we obtain the parabola $z = y^2$, $x = 0$. Similarly, the parabola $z = x^2$, $y = 0$ is the intersection curve with the xz-plane. These curves, shown in Figure 11.14a, would lead us to suspect that the surface $z = x^2 + y^2$ might be shaped as shown in Figure 11.14b. To verify this, we intersect the surface with a plane $z = k$ (k a constant), giving the curve

$$z = x^2 + y^2 \qquad \text{or} \qquad x^2 + y^2 = k$$

$$z = k \qquad\qquad\qquad z = k.$$

The latter equations indicate that cross-sections of $z = x^2 + y^2$ with planes $z = k$ are circles centred on the z-axis with radii $\sqrt{k}$ that increase as k increases. This certainly confirms the sketch in Figure 11.14b.

FIGURE 11.14a   Cross-sections of surface $z = x^2 + y^2$ with xz- and yz-coordinate planes

FIGURE 11.14b   Illustration of surface $z = x^2 + y^2$

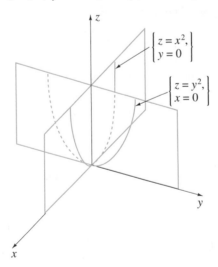

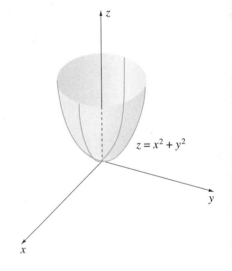

Intersections of the surface $y = z + x^2$ with the xy-, xz-, and yz-coordinate planes give two parabolas and a straight line, shown in Figure 11.15a. These really do not help us visualize the surface. If, however, we intersect the surface with planes $z = k$, we obtain the parabolas

$$y = z + x^2 \qquad \text{or} \qquad y = x^2 + k$$

$$z = k \qquad\qquad\qquad z = k.$$

These parabolas, shown in Figure 11.15b, indicate that the surface $y = z + x^2$ should be drawn as in Figure 11.15c.

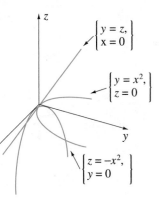

FIGURE 11.15a Cross-sections of surface $y = z + x^2$ with coordinate planes

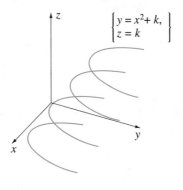

FIGURE 11.15b Cross-sections of surface $y = z + x^2$ with planes $z = k$

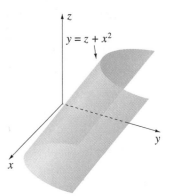

FIGURE 11.15c Illustration of surface $y = z + x^2$

We can sometimes "build" surfaces in much the same way that we "built" curves in single-variable calculus. For the surface $z = 1 - x^2 - y^2$, we first draw the surface $z = x^2 + y^2$ in Figure 11.14b. To draw $z = -(x^2 + y^2)$, we turn $z = x^2 + y^2$ upside down (Figure 11.16a), and finally we see that $z = 1 - x^2 - y^2$ is $z = -(x^2 + y^2)$ shifted upward 1 unit (Figure 11.16b).

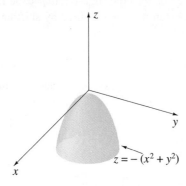

FIGURE 11.16a Illustration of surface $z = -(x^2 + y^2)$

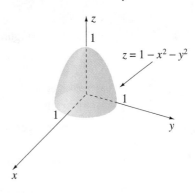

FIGURE 11.16b Illustration of surface $z = 1 - x^2 - y^2$

EXAMPLE 11.1

Draw the surface defined by each of the following equations:

$$\text{(a) } z = \sqrt{4x + 2y - x^2 - y^2 - 4} \quad \text{(b) } y = 1 + \sqrt{x^2 + z^2}$$

SOLUTION

(a) If we square the equation, and at the same time complete squares on $-x^2 + 4x$ and $-y^2 + 2y$, we have

$$z^2 = -(x - 2)^2 - (y - 1)^2 + 1,$$

or,

$$(x - 2)^2 + (y - 1)^2 + z^2 = 1.$$

Because $\sqrt{(x - 2)^2 + (y - 1)^2 + z^2}$ is the distance from a point (x, y, z) to $(2, 1, 0)$, this equation states that (x, y, z) must always be a unit distance from $(2, 1, 0)$ [i.e., the equation $(x - 2)^2 + (y - 1)^2 + z^2 = 1$ defines a sphere of radius 1 centred at $(2, 1, 0)$] (Figure 11.17a). Because the original equation requires z to be nonnegative, the required surface is the upper half of this sphere — the hemisphere in Figure 11.17b.

FIGURE 11.17a   Sphere described by $(x-2)^2 + (y-1)^2 + z^2 = 1$

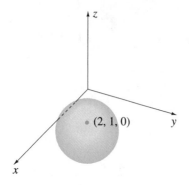

FIGURE 11.17b   Hemisphere described by $z = \sqrt{4x + 2y - x^2 - y^2 - 4}$

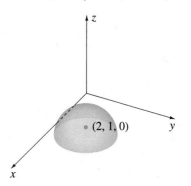

(b) If we intersect the surface $y = \sqrt{x^2 + z^2}$ with the xy-plane, we obtain the broken (i.e., bent) straight line $y = |x|$, $z = 0$ in Figure 11.18a. Intersections of the surface with planes $y = k$ (k a constant) give

$$y = \sqrt{x^2 + z^2} \qquad \text{or} \qquad x^2 + z^2 = k^2$$
$$y = k \qquad\qquad\qquad\qquad y = k.$$

These define circles of radii k in the planes $y = k$ (Figure 11.18b). Consequently, $y = \sqrt{x^2 + z^2}$ defines the right-circular cone in Figure 11.18c. The surface $y = 1 + \sqrt{x^2 + z^2}$ can now be obtained by shifting the cone 1 unit in the y-direction (Figure 11.18d).

FIGURE 11.18a   Cross-section of surface $y = \sqrt{x^2 + z^2}$ with xy-plane

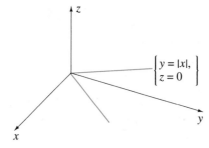

FIGURE 11.18b   Cross-sections of surface $y = \sqrt{x^2 + z^2}$ with planes $y = k$

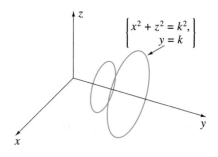

FIGURE 11.18c   Cone described by $y = \sqrt{x^2 + z^2}$

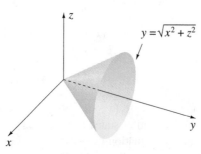

FIGURE 11.18d   Cone described by $y = 1 + \sqrt{x^2 + z^2}$

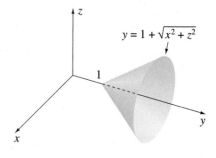

Cylinders

Suppose that l is a straight line and C is a curve that lies in some plane (the xy-plane in Figure 11.19a). A **cylinder** is the surface traced out by a line that moves along C always remaining parallel to l (Figure 11.19b). The right-circular cylinder in Figure 11.9 is generated by moving a vertical line around the circle $x^2 + y^2 = 4$, $z = 0$ in the xy-plane. Although we might not like to think of it as such, the plane in Figure 11.10 is a cylinder. The surface in Figure 11.15c is a cylinder; move lines parallel to $y = z$, $x = 0$ in Figure 11.15a along the parabola $y = x^2$, $z = 0$.

FIGURE 11.19a   Definition of a cylinder using line and curve

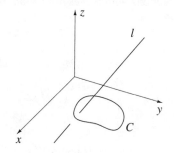

FIGURE 11.19b   Visualization of a cylinder

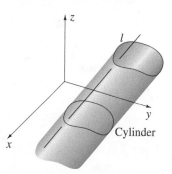

Cylinder

When one of the coordinates x, y, z is missing from the equation of a surface, a cylinder results. The right-circular cylinder in Figure 11.9 is an example (z is missing). For such cylinders, a line parallel to the axis of the missing variable (the line $x = 2$, $y = 0$, say, for $x^2 + y^2 = 4$) plays the role of l, and the cross-section of the cylinder with the plane of the remaining two variables (circle $x^2 + y^2 = 4$, $z = 0$ in the xy-plane) plays the role of C. All cross-sections of the cylinder with planes perpendicular to the axis of the missing variable are identical to C.

The equation $z = x^2$ is free of y. Each cross-section of this surface with a plane $y = k$ is the parabola $z = x^2$ in the plane $y = k$. Consequently, $z = x^2$ is the equation for the parabolic cylinder in Figure 11.20. The surface $yz = 1$, $x > 0$ is the hyperbolic cylinder in Figure 11.21. All cross-sections in planes parallel to the yz-plane are hyperbolas.

FIGURE 11.20   Parabolic cylinder $z = x^2$

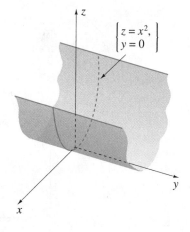

$$\begin{cases} z = x^2, \\ y = 0 \end{cases}$$

FIGURE 11.21   Hyperbolic cylinder $yz = 1$, $x > 0$

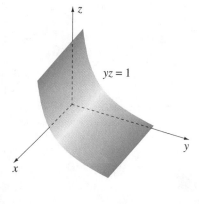

$yz = 1$

Quadric Surfaces

A **quadric surface** is a surface whose equation is quadratic in x, y, and z, the most general such equation being

$$Ax^2 + By^2 + Cz^2 + Dxy + Eyz + Fxz + Gx + Hy + Iz + J = 0. \qquad (11.2)$$

For the most part, we encounter quadric surfaces whose equations are of the form

$$Ax^2 + By^2 + Cz^2 = J \quad \text{or} \quad Ax^2 + By^2 = Iz,$$

or these equations with x, y, and z interchanged. Surfaces with these equations fall into nine major classes, depending on whether the constants are positive, negative, or zero. They are illustrated in Figures 11.22–11.30.

The names of these surfaces are derived from the fact that their cross-sections are ellipses, hyperbolas, or parabolas. For example, cross-sections of the hyperbolic paraboloid with planes $z = k$ are hyperbolas $x^2/a^2 - y^2/b^2 = k$. Cross-sections with planes $x = k$ are parabolas $z = k^2/a^2 - y^2/b^2$, as are cross-sections with planes $y = k$.

FIGURE 11.22   Elliptic cylinder

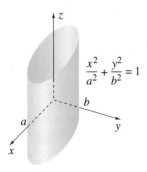

$$\frac{x^2}{a^2} + \frac{y^2}{b^2} = 1$$

FIGURE 11.23   Hyperbolic cylinder

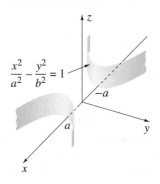

$$\frac{x^2}{a^2} - \frac{y^2}{b^2} = 1$$

FIGURE 11.24   Parabolic cylinder

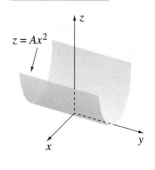

$$z = Ax^2$$

FIGURE 11.25   Ellipsoid

$$\frac{x^2}{a^2} + \frac{y^2}{b^2} + \frac{z^2}{c^2} = 1$$

FIGURE 11.26   Elliptic paraboloid

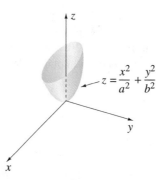

$$z = \frac{x^2}{a^2} + \frac{y^2}{b^2}$$

FIGURE 11.27   Elliptic cone

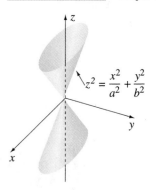

$$z^2 = \frac{x^2}{a^2} + \frac{y^2}{b^2}$$

FIGURE 11.28   Hyperbolic paraboloid

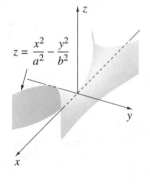

$$z = \frac{x^2}{a^2} - \frac{y^2}{b^2}$$

FIGURE 11.29   Elliptic hyperboloid of one sheet

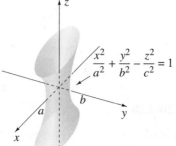

$$\frac{x^2}{a^2} + \frac{y^2}{b^2} - \frac{z^2}{c^2} = 1$$

FIGURE 11.30   Elliptic hyperboloid of two sheets

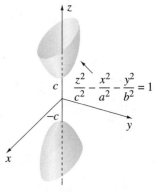

$$\frac{z^2}{c^2} - \frac{x^2}{a^2} - \frac{y^2}{b^2} = 1$$

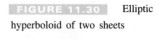
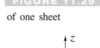

In applications of multiple integrals in Chapter 13, it is often necessary to project a space curve into one of the coordinate planes and find equations for the projection. To illustrate, consider the curve of intersection of the cylinder $x^2 + z^2 = 4$ and the plane $2y + z = 4$ (the first octant part of which is shown in Figure 11.31). Since the curve of intersection lies on the cylinder $x^2 + z^2 = 4$, its projection in the xz-plane is the circle $x^2 + z^2 = 4$, $y = 0$. To find its projection in the xy-plane, we eliminate z between the equations $2y + z = 4$ and $x^2 + z^2 = 4$. The result is $x^2 + (4 - 2y)^2 = 4$, or $x^2 + 4(y - 2)^2 = 4$. This shows that the curve of intersection lies on the elliptic cylinder $x^2 + 4(y - 2)^2 = 4$, and therefore it projects onto the ellipse $x^2 + 4(y - 2)^2 = 4$, $z = 0$ in the xy-plane. The projection of the curve in the yz-plane is that part of the line $2y + z = 4$, $x = 0$ between the points $(0,\ 1,\ 2)$ and $(0,\ 3,\ -2)$.

FIGURE 11.31   Projections of a curve in the coordinate planes

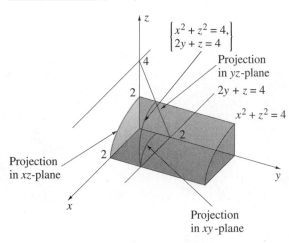

Graphing calculators and computers can plot surfaces provided that the equation of the surface is solved for z in terms of x and y. (This may not always be convenient.) We have shown some computer-generated plots in Figures 11.32 together with values for x and y specified in generating the plots. With the complexity of the expressions for z, drawing these surfaces by hand would be a formidable task.

To appreciate the shape of a plotted surface, it is often necessary to vary the point in space from which the surface is viewed. Computers usually have this ability; graphing calculators may not.

FIGURE 11.32a

Computer plots of surfaces

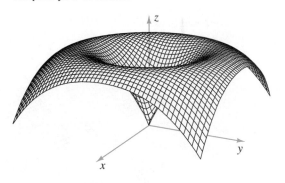

$$z = 3\sqrt{x^2 + y^2} - x^2 - y^2$$

$$-2 \le x \le 2,\, -2 \le y \le 2$$

FIGURE 11.32b

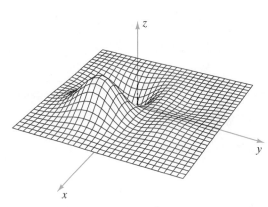

$$z = x(y - 1)^2 e^{-(x^2 + y^2)/4}$$

$$-5 \le x \le 5,\, -5 \le y \le 5$$

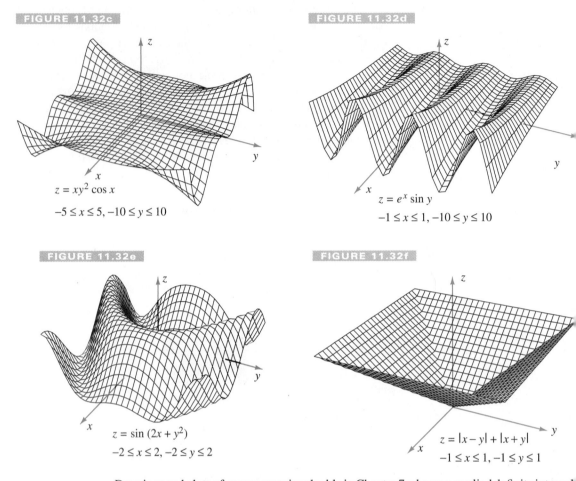

FIGURE 11.32c

$z = xy^2 \cos x$

$-5 \le x \le 5, -10 \le y \le 10$

FIGURE 11.32d

$z = e^x \sin y$

$-1 \le x \le 1, -10 \le y \le 10$

FIGURE 11.32e

$z = \sin(2x + y^2)$

$-2 \le x \le 2, -2 \le y \le 2$

FIGURE 11.32f

$z = |x - y| + |x + y|$

$-1 \le x \le 1, -1 \le y \le 1$

Drawings and plots of curves were invaluable in Chapter 7 when we applied definite integral to numerous geometric and physical problems. We encounter many of these applications in Chapter 13, but applied to volumes in space rather than areas in the plane. Our ability to visualize and draw surfaces in space proves more indispensable in these problems than plot from computers and graphing calculators. This is especially so when a picture contains a number of intersecting surfaces. For example, the surfaces $z = \sqrt{4 - x^2 - y^2}$ and $z = x^2$ bound volume in the first octant. The drawing in Figure 11.33a gives an excellent visualization of the volume; the plot in Figure 11.33b is not as satisfactory. It is often advantageous first to produce a plot of a surface and use it to render a drawing, adding whatever information is important for the application at hand.

FIGURE 11.33a   Hand drawing of two intersecting surfaces

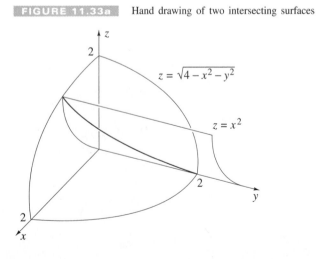

$z = \sqrt{4 - x^2 - y^2}$

$z = x^2$

FIGURE 11.33b   Computer plot of the two intersecting surface

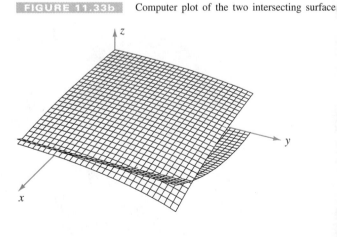

EXERCISES 11.2

In Exercises 1–35 draw the surface defined by the equation. Whenever possible, confirm your drawing with a plot generated by a computer or graphing calculator.

1. $2y + 3z = 6$

2. $2x - 3y = 0$

3. $y = x^2 + 2$

4. $z = x^3$

5. $y^2 + z^2 = 1$

6. $x^2 + y^2 + z^2 = 4$

7. $x^2 + 4y^2 = 1$

8. $y^2 - z^2 = 4$

9. $z = 2(x^2 + y^2)$

10. $x = \sqrt{y^2 + z^2}$

11. $x = \sqrt{1 - y^2}$

12. $z = 2 - x$

13. $x^2 = y^2$

14. $x = z^2 + 2$

15. $z = y + 3$

16. $4z = 3\sqrt{x^2 + y^2}$

17. $x^2 - 2x + z^2 = 0$

18. $yz = 1$

19. $x^2 + y^2 + (z - 1)^2 = 3$

20. $z + 5 = 4(x^2 + y^2)$

21. $x^2 + z^2 = y^2$

22. $x^2 + z^2 = y^2 + 1$

23. $y^2 + z^2 = x$

24. $x^2 + y^2 + 4z^2 = 1$

25. $9z^2 = x^2 + y^2 + 1$

26. $(y^2 + z^2)^2 = x + 1$

27. $z^2 + 4y^2 = 1$

28. $y - z^2 = 0$

29. $x^2 - z^2 = 4$

30. $x^2 + y^2/4 + z^2/9 = 1$

31. $z = x^2/4 + y^2/25$

32. $x^2 = z^2 + 9y^2$

33. $z = y^2/16 - x^2/4$

34. $x^2 + y^2/4 - z^2/25 = 1$

35. $z^2 - 9x^2 - 16y^2 = 1$

In Exercises 36–45 draw the curve defined by the equations.

36. $x^2 + y^2 = 2$, $z = 4$

37. $x + 2y = 6$, $y - 2z = 3$

38. $z = x^2 + y^2$, $x^2 + y^2 = 5$

39. $x^2 + y^2 = 1$, $x + z = 1$

40. $z = \sqrt{x^2 + y^2}$, $y = x$

41. $z + 2x^2 = 1$, $y = z$

42. $z = \sqrt{4 - x^2 - y^2}$, $x^2 + y^2 - 2y = 0$

43. $z = y$, $y = x^2$

44. $x^2 + z^2 = 1$, $y^2 + z^2 = 1$

45. $z = x^2$, $z = y^2$

In Exercises 46–55 find equations for projections of the curve in the xy-, yz-, and xz-coordinate planes. In each case draw the curve.

* **46.** $x + y = 3$, $2y + 3z = 4$

* **47.** $x + y + z = 4$, $2x - y + z = 6$

* **48.** $x^2 + y^2 = 4$, $z = 4$

* **49.** $x^2 + y^2 = 4$, $y = x$

* **50.** $x^2 + y^2 = 4$, $x = z$

* **51.** $x^2 + y^2 = 4$, $x + y + z = 2$

* **52.** $y^2 + z^2 = 3$, $x^2 + z^2 = 3$

* **53.** $z = x^2 + y^2$, $x + z = 1$

* **54.** $z = \sqrt{x^2 + y^2}$, $z = 6 - x^2 - y^2$

* **55.** $x^2 + y^2 + z^2 = 1$, $y = x$

In Exercises 56–61 find equations for the projection of the curve in the specified plane. Draw each curve.

* **56.** $z = x^2 - y^2$, $z = 2x + 4y$ in the xy-plane

* **57.** $x^2 + y^2 - 4z^2 = 1$, $x + y = 2$ in the xz-plane

* **58.** $y = z + x^2$, $y + z = 1$ in the xy-plane

* **59.** $x = \sqrt{1 + 2y^2 + 4z^2}$, $x^2 + 9y^2 + 4z^2 = 36$ in the yz-plane

* **60.** $z = x^2 + y^2$, $z = 4(x - 1)^2 + 4(y - 1)^2$ in the xy-plane

* **61.** $x^2 + y^2 - 2y = 0$, $z^2 = x^2 + y^2$ in the xz-plane

In Exercises 62–71 draw whatever is defined by the equation or equations.

* **62.** $(x - 2)^2 + y^2 + z^2 = 0$

* **63.** $x = 0$, $y = 5$

* **64.** $\sqrt{x} + \sqrt{y} = 1$, $z = x$

* **65.** $x + y = 15$, $y - x = 4$

* **66.** $z = 1 - (x^2 + y^2)^{1/3}$

* **67.** $z = |x|$

* **68.** $z = x^2$, $y = z^2$

* **69.** $x = \ln(y^2 + z^2)$

* **70.** $z = |x - y|$

* **71.** $x = 2$, $y = 4$, $z^2 - 1 = 0$

11.3 Vectors

Physical quantities that have associated with them only a magnitude can be represented by real numbers. Some examples are temperature, density, area, moment of inertia, speed, and pressure. They are called **scalars**. There are many quantities, however, that have associated with them both magnitude and direction, and these quantities cannot be described by a single real number. Velocity, acceleration, and force are perhaps the most notable concepts in this category. To represent such quantities mathematically, we introduce **vectors**.

DEFINITION 11.1

A vector is defined as a directed line segment.

To denote a vector we use a letter in boldface type, such as **v**. In Figures 11.34a–c we show two vectors **u** and **v** along a line, three vectors **u**, **v**, and **w** in a plane, and three vectors **u**, **v** and **w** in space, respectively. It is customary to place an arrowhead on a vector and call this end the **tip of the vector**. The other end is called the **tail of the vector**, and the direction of the vector is from tail to tip. A vector then has both *direction* and *length*.

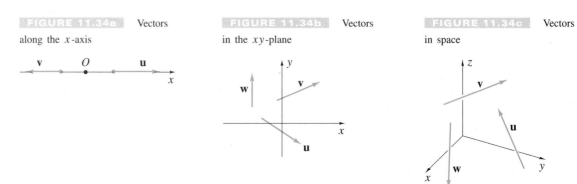

FIGURE 11.34a   Vectors along the x-axis

FIGURE 11.34b   Vectors in the xy-plane

FIGURE 11.34c   Vectors in space

Definition 11.1 for a vector says nothing about its point of application (i.e., where its tail should be placed). This means that we may place the tail anywhere we wish. This suggests the following definition for equality of vectors.

DEFINITION 11.2

Two vectors are equal if and only if they have the same length and direction. Their points of application are irrelevant.

FIGURE 11.35   Equality of vectors **u** and **v** but not of **w** and **u**

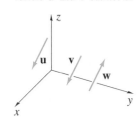

For example, vectors **u** and **v** in Figure 11.35 have exactly the same length and direction and are therefore one and the same. Although the vector **w** in the same figure is parallel to **u** and **v** and has the same length, it points in the opposite direction and is not, therefore, the same as **u** and **v**.

Components of Vectors

We realized in Chapter 1 that to solve geometric problems, it is often helpful to represent them algebraically. In fact, our entire development of single-variable calculus has hinged on our ability to represent a curve by an algebraic equation and also to draw the curve described by an equation. We now show that vectors can be represented algebraically. Suppose we denote by **PQ** the vector from point P to point Q in Figure 11.36. If P and Q have coordinates (x_1, y_1, z_1) and (x_2, y_2, z_2) in the coordinate system shown, then the length of **PQ** is

$$\sqrt{(x_2 - x_1)^2 + (y_2 - y_1)^2 + (z_2 - z_1)^2}. \tag{11.3}$$

Note also that if we start at point P, proceed $x_2 - x_1$ units in the x-direction, then $y_2 - y_1$ units in the y-direction, and finally $z_2 - z_1$ units in the z-direction, we arrive at Q. In other words, the three numbers $x_2 - x_1$, $y_2 - y_1$, and $z_2 - z_1$ characterize both the direction and the length of the vector joining P to Q. Because of this we make the following agreement.

FIGURE 11.36 Vector from point P to point Q

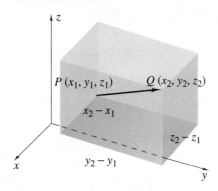

DEFINITION 11.3

If the tail of a vector $\mathbf{v}$ is at $P(x_1, y_1, z_1)$ and its tip is at $Q(x_2, y_2, z_2)$, then $\mathbf{v}$ shall be represented by the triple of numbers $x_2 - x_1$, $y_2 - y_1$, $z_2 - z_1$. In such a case we enclose the numbers in parentheses and write

$$\mathbf{v} = (x_2 - x_1, y_2 - y_1, z_2 - z_1). \tag{11.4}$$

The equal sign in 11.4 means "is represented by." The number $x_2 - x_1$ is called the x-**component** of $\mathbf{v}$, $y_2 - y_1$ the y-**component**, and $z_2 - z_1$ the z-**component**. Vectors in the xy-plane have only an x- and a y-component:

$$\mathbf{v} = (x_2 - x_1, y_2 - y_1),$$

where (x_1, y_1) and (x_2, y_2) are the coordinates of the tail and tip of $\mathbf{v}$. Vectors along the x-axis have only an x-component $x_2 - x_1$, where x_1 and x_2 are the coordinates of the tail and tip of $\mathbf{v}$.

We now have an algebraic representation for vectors. Each vector has associated with it a set of components that can be found by subtracting the coordinates of its tail from the coordinates of its tip. Conversely, given a set of real numbers (a, b, c), there is one and only one vector with these numbers as components. We can visualize this vector by placing its tail at the origin and its tip at the point with coordinates (a, b, c) (Figure 11.37). Alternatively, we can place the tail of the vector at any point (x_1, y_1, z_1) and its tip at the point $(x_1 + a, y_1 + b, z_1 + c)$.

It is worth emphasizing once again that the same components of a vector are obtained for any point of application whatsoever. For example, the two vectors in Figure 11.38 are identical, and in both cases the components $(2, 2)$ are obtained by subtracting the coordinates of the tail from those of the tip. What we are saying is that Definition 11.2 for equality of vectors can be stated algebraically as follows.

THEOREM 11.1

Two vectors are equal if and only if they have the same components.

FIGURE 11.37 Tail of a vector can be placed at any point

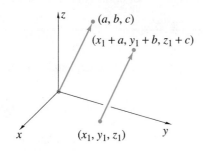

FIGURE 11.38 Two equal vectors

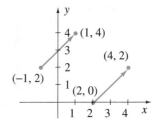

EXAMPLE 11.2

FIGURE 11.39 Components of vector with length 5 and angle $\frac{\pi}{6}$ radians with x-axis

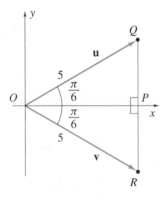

Find the components of a vector in the xy-plane that has length 5, its tail at the origin, and makes an angle of $\pi/6$ radians with the positive x-axis.

SOLUTION   Figure 11.39 illustrates that there are two such vectors, **u** and **v**. From the triangles shown, it is clear that

$$\|OP\| = 5\cos(\pi/6) = 5\sqrt{3}/2 \quad \text{and} \quad \|PQ\| = \|PR\| = 5\sin(\pi/6) = 5/2.$$

Consequently, Q and R have coordinates $Q = (5\sqrt{3}/2, 5/2)$ and $R = (5\sqrt{3}/2, -5/2)$, and

$$\mathbf{u} = \left(\frac{5\sqrt{3}}{2}, \frac{5}{2}\right), \qquad \mathbf{v} = \left(\frac{5\sqrt{3}}{2}, -\frac{5}{2}\right).$$

EXAMPLE 11.3

FIGURE 11.40 Components of a vector given its length and its direction

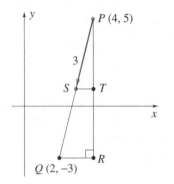

Find the components of the vector in the xy-plane that has its tail at the point $(4, 5)$, has length 3, and points directly toward the point $(2, -3)$.

SOLUTION   In Figure 11.40 $\|PQ\| = \sqrt{2^2 + 8^2} = 2\sqrt{17}$. Because of similar triangles we can write that

$$\frac{\|ST\|}{\|PS\|} = \frac{\|QR\|}{\|PQ\|} \quad \text{or} \quad \|ST\| = \frac{3(2)}{2\sqrt{17}} = \frac{3}{\sqrt{17}}.$$

Similarly,

$$\|PT\| = \|PS\|\frac{\|PR\|}{\|PQ\|} = \frac{3(8)}{2\sqrt{17}} = \frac{12}{\sqrt{17}}.$$

Since $\|ST\|$ and $\|PT\|$ represent differences in the x- and y-coordinates of P and S (except for signs), the components of **PS** are $(-3/\sqrt{17}, -12/\sqrt{17})$.

Unit Vectors and Scalar Multiplication

If the x-, y-, and z-components of a vector **v** are (v_x, v_y, v_z), often called the **Cartesian components** of **v**, then these components represent the differences in the coordinates of its tip and tail. But then, according to equation 11.3, the length of the vector, which we denote by $|\mathbf{v}|$ is

$$|\mathbf{v}| = \sqrt{v_x^2 + v_y^2 + v_z^2}. \tag{11.5}$$

In words, the length of a vector is the square root of the sum of the squares of its components.

DEFINITION 11.4

A vector **v** is said to be a **unit vector** if it has length equal to 1 unit; that is, **v** is a unit vector if

$$v_x^2 + v_y^2 + v_z^2 = 1. \tag{11.6}$$

To indicate that a vector is a unit vector, we place a circumflex $\hat{}$ above it: $\hat{\mathbf{v}}$.

EXAMPLE 11.4

What is the length of the vector from $(1, -1, 0)$ to $(2, -3, -5)$?

SOLUTION Since the components of the vector are $(1, -2, -5)$, its length is

$$\sqrt{(1)^2 + (-2)^2 + (-5)^2} = \sqrt{30}.$$

We now have vectors, which are directed line segments, and real numbers, which are scalars. We know that scalars can be added, subtracted, multiplied, and divided, but can we do the same with vectors, and can we combine vectors and scalars? In the remainder of this section we show how to add and subtract vectors and multiply vectors by scalars; in Section 11.4 we define two ways to multiply vectors. Each of these operations can be approached either algebraically or geometrically. The geometric approach uses the geometric properties of vectors, namely, length and direction; the algebraic approach uses components of vectors. Neither method is suitable for all situations. Sometimes an idea is more easily introduced with a geometric approach; sometimes an algebraic approach is more suitable. We choose whichever we feel expresses the idea more clearly. But, whenever we take a geometric approach, we are careful to follow it up with the algebraic equivalent; conversely, when an algebraic approach is taken, we always illustrate the geometric significance of the results.

To introduce multiplication of a vector by a scalar, consider the vectors **u** and **v** in Figure 11.41, both of which have their tails at the origin; **v** is in the same direction as **u** but is twice as long as **u**. In such a situation we would like to say that **v** is equal to 2**u** and write **v** = 2**u**. Vector **w** is in the opposite direction to **r** and is three times as long as **r**, and we would like to denote this vector by **w** = −3**r**. Both of these situations are realized if we adopt the following definition for multiplication of a vector by a scalar.

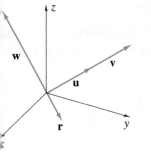

FIGURE 11.41 Geometric illustration of scalar multiplication of vectors

DEFINITION 11.5

If $\lambda > 0$ is a scalar and **v** is a vector, then λ**v** is the vector that is in the same direction as **v** and λ times as long as **v**; if $\lambda < 0$, then λ**v** is the vector that is in the opposite direction to **v** and $|\lambda|$ times as long as **v**.

This is a geometric definition of scalar multiplication; it describes the length and direction of λ**v**. We now show that the components of λ**v** are λ times the components of **v**. In Figure 11.42 we show a box with faces parallel to the coordinate planes and λ**v** as diagonal, and have given vector **v** components (v_x, v_y, v_z). From the pairs of similar triangles OAB and OCD, and OBE and ODF, we can write that

$$\frac{\|OC\|}{v_x} = \frac{\|CD\|}{v_y} = \frac{\|OD\|}{\|OB\|} = \frac{\|DF\|}{v_z} = \frac{|\lambda \mathbf{v}|}{|\mathbf{v}|} = \lambda.$$

Hence

$$\|OC\| = \lambda v_x, \quad \|CD\| = \lambda v_y, \quad \|DF\| = \lambda v_z,$$

where $\|OC\|$, $\|CD\|$, and $\|DF\|$ are the components of λ**v**. In other words, the components of λ**v** are λ times the components of **v**:

$$\lambda \mathbf{v} = \lambda(v_x, v_y, v_z) = (\lambda v_x, \lambda v_y, \lambda v_z). \tag{11.7}$$

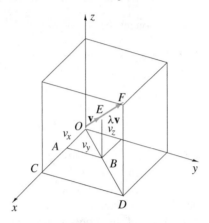

FIGURE 11.42 Components of $\lambda \mathbf{v}$ are λ times the components of $\mathbf{v}$

To multiply a vector by a scalar, then, we multiply each component by the scalar.

EXAMPLE 11.5

Find components for the unit vector in the same direction as $\mathbf{v} = (2, -2, 1)$.

SOLUTION The length of $\mathbf{v}$ is $|\mathbf{v}| = \sqrt{(2)^2 + (-2)^2 + 1^2} = 3$. According to our definition of multiplication of a vector by a scalar, the vector $\frac{1}{3}\mathbf{v}$ must have length 1 ($\frac{1}{3}$ that of $\mathbf{v}$) and the same direction as $\mathbf{v}$. Consequently, a unit vector in the same direction as $\mathbf{v}$ is

$$\hat{\mathbf{v}} = \frac{1}{3}\mathbf{v} = \frac{1}{3}(2, -2, 1) = \left(\frac{2}{3}, -\frac{2}{3}, \frac{1}{3}\right).$$

This example illustrates that a unit vector in the same direction as a given vector $\mathbf{v}$ is

$$\hat{\mathbf{v}} = \frac{\mathbf{v}}{|\mathbf{v}|}. \qquad (11.8)$$

EXAMPLE 11.6

Find components for the vector of length 4 in the direction opposite that of $\mathbf{v} = (1, 2, -3)$.

SOLUTION Since $|\mathbf{v}| = \sqrt{1 + 4 + 9} = \sqrt{14}$, a unit vector in the same direction as $\mathbf{v}$ is

$$\hat{\mathbf{v}} = \frac{1}{\sqrt{14}}\mathbf{v}.$$

The vector of length 4 in the opposite direction to $\mathbf{v}$ must therefore be

$$(-4)\hat{\mathbf{v}} = \left(\frac{-4}{\sqrt{14}}\right)\mathbf{v} = \left(\frac{-4}{\sqrt{14}}\right)(1, 2, -3) = \left(-\frac{4}{\sqrt{14}}, -\frac{8}{\sqrt{14}}, \frac{12}{\sqrt{14}}\right).$$

With the operation of scalar multiplication, we can simplify the solution of Example 11.3. The vector that points from $(4, 5)$ to $(2, -3)$ is $\mathbf{v} = (-2, -8)$, and therefore the unit vector in this direction is

$$\hat{\mathbf{v}} = \frac{1}{\sqrt{4 + 64}}(-2, -8) = \frac{1}{2\sqrt{17}}(-2, -8) = \frac{1}{\sqrt{17}}(-1, -4).$$

The required vector of length 3 is

$$3\hat{\mathbf{v}} = \frac{3}{\sqrt{17}}(-1, -4) = \left(-\frac{3}{\sqrt{17}}, -\frac{12}{\sqrt{17}}\right).$$

Addition and Subtraction of Vectors

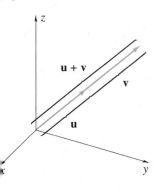

FIGURE 11.43 Addition of parallel vectors

In Figure 11.43 we show two parallel vectors $\mathbf{u}$ and $\mathbf{v}$ and have placed the tail of $\mathbf{v}$ on the tip of $\mathbf{u}$. It would seem natural to denote the vector that has its tail at the tail of $\mathbf{u}$ and its tip at the tip of $\mathbf{v}$ by $\mathbf{u} + \mathbf{v}$. For instance, if $\mathbf{u}$ and $\mathbf{v}$ were equal, then we would simply be saying that $\mathbf{u} + \mathbf{u} = 2\mathbf{u}$. We use this idea to define addition of vectors even when the vectors are not parallel.

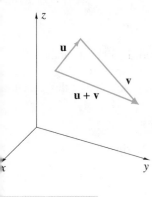

FIGURE 11.44 Triangular addition of vectors

> **DEFINITION 11.6**
>
> The sum of two vectors $\mathbf{u}$ and $\mathbf{v}$, denoted by $\mathbf{u} + \mathbf{v}$, is the vector from the tail of $\mathbf{u}$ to the tip of $\mathbf{v}$ when the tail of $\mathbf{v}$ is placed on the tip of $\mathbf{u}$.

Because the three vectors $\mathbf{u}$, $\mathbf{v}$, and $\mathbf{u} + \mathbf{v}$ then form a triangle (Figure 11.44), we call this **triangular addition of vectors**.

Note that were we to place tails of $\mathbf{u}$ and $\mathbf{v}$ both at the same point (Figure 11.45), and complete the parallelogram with $\mathbf{u}$ and $\mathbf{v}$ as sides, the diagonal of this parallelogram would also represent the vector $\mathbf{u} + \mathbf{v}$. This is an equivalent method for geometrically finding $\mathbf{u} + \mathbf{v}$, and it is called **parallelogram addition of vectors**.

Algebraically, vectors are added component by component; that is, if $\mathbf{u} = (u_x, u_y, u_z)$ and $\mathbf{v} = (v_x, v_y, v_z)$, then

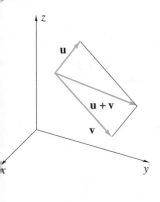

FIGURE 11.45 Parallelogram addition of vectors

$$\mathbf{u} + \mathbf{v} = (u_x + v_x, u_y + v_y, u_z + v_z). \tag{11.9}$$

To verify this we simply note that differences in the coordinates of P and Q in Figure 11.46 are (u_x, u_y, u_z), and differences in those of Q and R are (v_x, v_y, v_z). Consequently, differences in the coordinates of P and R must be $(u_x + v_x, u_y + v_y, u_z + v_z)$.

It is not difficult to show (see Exercise 26) that vector addition and scalar multiplication obey the following rules:

$$\mathbf{u} + \mathbf{v} = \mathbf{v} + \mathbf{u}; \tag{11.10a}$$

$$(\mathbf{u} + \mathbf{v}) + \mathbf{w} = \mathbf{u} + (\mathbf{v} + \mathbf{w}); \tag{11.10b}$$

$$\lambda(\mathbf{u} + \mathbf{v}) = \lambda\mathbf{u} + \lambda\mathbf{v}; \tag{11.10c}$$

$$(\lambda + \mu)\mathbf{v} = \lambda\mathbf{v} + \mu\mathbf{v}. \tag{11.10d}$$

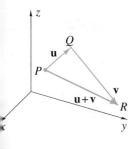

FIGURE 11.46 To add vectors, add their components

If we denote the vector $(-1)\mathbf{v}$ by $-\mathbf{v}$, then the components of $-\mathbf{v}$ are the negatives of those of $\mathbf{v}$:

$$-\mathbf{v} = (-1)\mathbf{v} = (-1)(v_x, v_y, v_z) = (-v_x, -v_y, -v_z). \tag{11.11}$$

This vector has the same length as **v**, but is opposite in direction to **v** (Figure 11.47). When **v** is added to $-\mathbf{v}$, the resultant vector has components that are all zero:

$$\mathbf{v} + (-\mathbf{v}) = (v_x, v_y, v_z) + (-v_x, -v_y, -v_z) = (0, 0, 0).$$

This vector, called the **zero vector**, is denoted by **0**, and has the property that

$$\mathbf{v} + \mathbf{0} = \mathbf{0} + \mathbf{v} = \mathbf{v} \tag{11.12}$$

for any vector **v** whatsoever.

To subtract a vector **v** from **u**, we add $-\mathbf{v}$ to **u**.

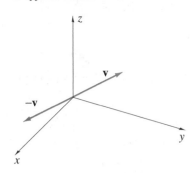

FIGURE 11.47   Vector $-\mathbf{v}$ has the same length as **v**, but is opposite in direction to **v**

DEFINITION 11.7

The difference $\mathbf{u} - \mathbf{v}$ between two vectors **u** and **v** is the vector

$$\mathbf{u} - \mathbf{v} = \mathbf{u} + (-\mathbf{v}). \tag{11.13}$$

In Figure 11.48, $\mathbf{u} - \mathbf{v}$ is determined by a triangle, and in Figure 11.49 by a parallelogram. Alternatively, if we denote by **r** the vector joining the tip of **v** to the tip of **u** in Figure 11.50, then, by triangle addition, we have $\mathbf{v} + \mathbf{r} = \mathbf{u}$. Addition of $-\mathbf{v}$ to each side of this equation gives

$$-\mathbf{v} + \mathbf{v} + \mathbf{r} = -\mathbf{v} + \mathbf{u} \quad \text{or} \quad \mathbf{0} + \mathbf{r} = -\mathbf{v} + \mathbf{u}.$$

Thus $\mathbf{r} = \mathbf{u} - \mathbf{v}$, and $\mathbf{u} - \mathbf{v}$ is the vector joining the tip of **v** to the tip of **u**. Definition 11.7 implies that vectors are subtracted component by component:

$$\mathbf{u} - \mathbf{v} = (u_x, u_y, u_z) - (v_x, v_y, v_z) = (u_x - v_x, u_y - v_y, u_z - v_z). \tag{11.14}$$

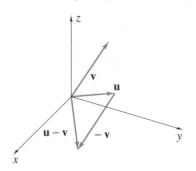

FIGURE 11.48 $\mathbf{u} - \mathbf{v}$ can be obtained by adding **u** and $-\mathbf{v}$ with triangular addition

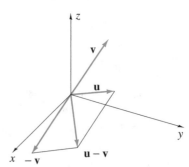

FIGURE 11.49 $\mathbf{u} - \mathbf{v}$ can be obtained by adding **u** and $-\mathbf{v}$ with parallelogram addition

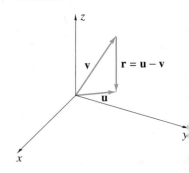

FIGURE 11.50 $\mathbf{u} - \mathbf{v}$ can be obtained directly with triangular subtraction

---

EXAMPLE 11.7

If $\mathbf{u} = (1, 1, 1)$, $\mathbf{v} = (-2, 3, 0)$, and $\mathbf{w} = (-10, 10, -2)$, find:

(a) $3\mathbf{u} + 2\mathbf{v} - \mathbf{w}$ (b) $2\mathbf{u} - 4\mathbf{v} + \mathbf{w}$ (c) $|\mathbf{u}|\mathbf{v} + \dfrac{4}{|\mathbf{v}|}\mathbf{w}$

SOLUTION

(a)

$$3\mathbf{u} + 2\mathbf{v} - \mathbf{w} = 3(1, 1, 1) + 2(-2, 3, 0) - (-10, 10, -2)$$

$$= (3, 3, 3) + (-4, 6, 0) + (10, -10, 2)$$

$$= (9, -1, 5)$$

(b)

$$2\mathbf{u} - 4\mathbf{v} + \mathbf{w} = 2(1, 1, 1) - 4(-2, 3, 0) + (-10, 10, -2)$$

$$= (0, 0, 0) = \mathbf{0}$$

(c) Since $|\mathbf{u}| = \sqrt{1^2 + 1^2 + 1^2} = \sqrt{3}$ and $|\mathbf{v}| = \sqrt{(-2)^2 + 3^2} = \sqrt{13}$,

$$|\mathbf{u}|\mathbf{v} + \frac{4}{|\mathbf{v}|}\mathbf{w} = \sqrt{3}(-2, 3, 0) + \frac{4}{\sqrt{13}}(-10, 10, -2)$$

$$= \left(-2\sqrt{3} - \frac{40}{\sqrt{13}}, 3\sqrt{3} + \frac{40}{\sqrt{13}}, -\frac{8}{\sqrt{13}}\right).$$

Forces

We have already mentioned that quantities such as temperature, area, and density have associated with them only a magnitude and are therefore represented by scalars. There are many quantities, however, that have associated with them both magnitude and direction, and these are described by vectors. The most notable of this group are forces. When we speak of a force, we mean a push or pull of some size in some specific direction. For example, when the boy in Figure 11.51a pulls his wagon, he exerts a force in the direction indicated by the handle. Suppose that he pulls with a force of 10 N and that the angle between the handle and the horizontal is $\pi/4$ radians. To represent this force as a vector $\mathbf{F}_1$, we choose the coordinate system in Figure 11.51b, and make the agreement that the *length of* $\mathbf{F}_1$ *be equal to the magnitude of the force.* Since $\mathbf{F}_1$ represents a force of 10 N, it follows that the length of $\mathbf{F}_1$ is 10 units. Furthermore, because $\mathbf{F}_1$ makes an angle of $\pi/4$ radians with the positive x- and y-axes, the difference in the x-coordinates (and the y-coordinates) of its tip and tail must be $10\cos(\pi/4) = 5\sqrt{2}$. The components of $\mathbf{F}_1$ are therefore $\mathbf{F}_1 = (5\sqrt{2}, 5\sqrt{2})$. If the boy's young sister drags her feet on the ground, then she effectively exerts a force $\mathbf{F}_2$ in the negative x-direction. If the magnitude of this force is 3 N, then its vector representation is $\mathbf{F}_2 = (-3, 0)$. Finally, if the combined weight of the wagon and the girl is 200 N, then the force $\mathbf{F}_3$ of gravity on the wagon and its load is $\mathbf{F}_3 = (0, -200)$.

FIGURE 11.51a   Boy pulling
wagon exerts a force in direction of handle

FIGURE 11.51b   Direction of forces
exerted by boy and girl dragging their feet

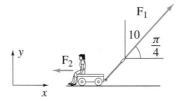

FIGURE 11.51c   Resultant
of forces of boy, girl, and gravity

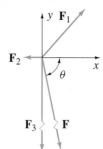

In mechanics we replace the individual forces $\mathbf{F}_1$, $\mathbf{F}_2$, and $\mathbf{F}_3$ by a single force that has the same effect on the wagon as all three forces combined. This force, called the **resultant force** of $\mathbf{F}_1$, $\mathbf{F}_2$, and $\mathbf{F}_3$, is represented by the vector $\mathbf{F}$, which is the sum of the vectors $\mathbf{F}_1$, $\mathbf{F}_2$, and $\mathbf{F}_3$:

$$\mathbf{F} = \mathbf{F}_1 + \mathbf{F}_2 + \mathbf{F}_3$$
$$= (5\sqrt{2}, 5\sqrt{2}) + (-3, 0) + (0, -200)$$
$$= (5\sqrt{2} - 3, 5\sqrt{2} - 200).$$

The magnitude of this force corresponds to the length of $\mathbf{F}$,

$$|\mathbf{F}| = \sqrt{(5\sqrt{2} - 3)^2 + (5\sqrt{2} - 200)^2} = 193.0,$$

and must therefore be 193.0 N. Its direction is shown in Figure 11.51c, where

$$\theta = \operatorname{Tan}^{-1}\left(\frac{200 - 5\sqrt{2}}{5\sqrt{2} - 3}\right) = 1.55 \text{ radians.}$$

By the x-, y-, and z-components (v_x, v_y, v_z) of a vector $\mathbf{v}$, we mean that if we start at a point P (Figure 11.52) and proceed v_x units in the x-direction, v_y units in the y-direction, and v_z units in the z-direction to a point Q, then $\mathbf{v}$ is the directed line segment joining P and Q. To

FIGURE 11.52 Geometric illustration of components of a vector

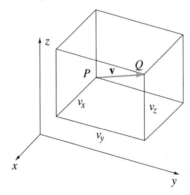

FIGURE 11.53 Unit vectors along the coordinate axes

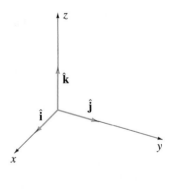

phrase this another way, we introduce three special vectors parallel to the coordinate axes. We define $\hat{\mathbf{i}}$ as a unit vector in the positive x-direction, $\hat{\mathbf{j}}$ as a unit vector in the positive y-direction, and $\hat{\mathbf{k}}$ as a unit vector in the positive z-direction. We have shown these vectors with their tail at the origin in Figure 11.53, and it is clear that their components are

$$\hat{\mathbf{i}} = (1, 0, 0), \quad \hat{\mathbf{j}} = (0, 1, 0), \quad \hat{\mathbf{k}} = (0, 0, 1). \tag{11.15}$$

But note, then, that we can write the vector $\mathbf{v} = (v_x, v_y, v_z)$ in the form

$$\mathbf{v} = (v_x, 0, 0) + (0, v_y, 0) + (0, 0, v_z)$$
$$= v_x(1, 0, 0) + v_y(0, 1, 0) + v_z(0, 0, 1)$$
$$= v_x\hat{\mathbf{i}} + v_y\hat{\mathbf{j}} + v_z\hat{\mathbf{k}}.$$

In other words, every vector in space can be written as a linear combination of the three vectors $\hat{\mathbf{i}}$, $\hat{\mathbf{j}}$, and $\hat{\mathbf{k}}$ (i.e., as a constant times $\hat{\mathbf{i}}$ plus a constant times $\hat{\mathbf{j}}$ plus a constant times $\hat{\mathbf{k}}$). Furthermore, the constants multiplying $\hat{\mathbf{i}}$, $\hat{\mathbf{j}}$, and $\hat{\mathbf{k}}$ are the Cartesian components of the vector. This result is equally clear geometrically. In Figure 11.52, we have shown the vector **v** from P to Q. If we define points A and B as shown in Figure 11.54, then

$$\mathbf{v} = \mathbf{PQ} = \mathbf{PB} + \mathbf{BQ} = \mathbf{PA} + \mathbf{AB} + \mathbf{BQ}.$$

But because **PA** is a vector in the positive x-direction and has length v_x, it follows that $\mathbf{PA} = v_x\hat{\mathbf{i}}$. Similarly, $\mathbf{AB} = v_y\hat{\mathbf{j}}$ and $\mathbf{BQ} = v_z\hat{\mathbf{k}}$, and therefore

$$\mathbf{v} = v_x\hat{\mathbf{i}} + v_y\hat{\mathbf{j}} + v_z\hat{\mathbf{k}}. \tag{11.16}$$

To say then that v_x, v_y, and v_z are the x-, y-, and z-components of a vector **v** is to say that **v** can be written in form 11.16.

Some authors refer to v_x, v_y, and v_z as the **scalar components** of the vector **v**, and the vectors $v_x\hat{\mathbf{i}}$, $v_y\hat{\mathbf{j}}$, and $v_z\hat{\mathbf{k}}$ as the **vector components** of **v**. By *component*, we always mean *scalar component*.

Vectors in the xy-plane have only an x- and a y-component, and can therefore be written in terms of $\hat{\mathbf{i}}$ and $\hat{\mathbf{j}}$. If $\mathbf{v} = (v_x, v_y)$, then we write equivalently that $\mathbf{v} = v_x\hat{\mathbf{i}} + v_y\hat{\mathbf{j}}$ (Figure 11.55). Vectors along the x-axis have only an x-component and can therefore be written in the form $\mathbf{v} = v_x\hat{\mathbf{i}}$.

We use this new notation in the following example.

FIGURE 11.54   Vectors in space can be expressed as a linear combination of $\hat{\mathbf{i}}$, $\hat{\mathbf{j}}$, and $\hat{\mathbf{k}}$

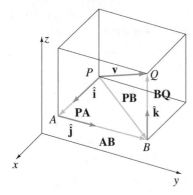

FIGURE 11.55   Vectors in the xy-plane can be expressed as a linear combination of $\hat{\mathbf{i}}$ and $\hat{\mathbf{j}}$

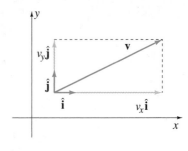

EXAMPLE 11.8

FIGURE 11.56   Electrostatic force of charge q_2 on charge q_1

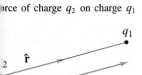

The force **F** exerted on a point charge q_1 coulombs by a charge q_2 coulombs is defined by Coulomb's law as

$$\mathbf{F} = \frac{q_1 q_2}{4\pi \epsilon_0 r^2}\hat{\mathbf{r}} \text{ N},$$

where ϵ_0 is a positive constant, r is the distance in metres between the charges, and $\hat{\mathbf{r}}$ is a unit vector in the direction from q_2 to q_1 (Figure 11.56). When q_1 and q_2 are both positive charges or both negative charges, then **F** is repulsive, and when one is positive and the other is negative, **F** is attractive. In particular, suppose that charges of 2 C and -2 C are placed at $(0, 0, 0)$ and $(3, 0, 0)$, respectively, and a third charge of 1 C is placed at $(1, 1, 1)$. According to Coulomb's law, the 2 C charge will exert a repulsive force on the 1 C charge, and the -2 C charge will exert an attractive force on the 1 C charge. Find the resultant of these two forces on the 1 C charge.

SOLUTION   If $\mathbf{F}_1$ is the force exerted on the 1 C charge by the -2 C charge (Figure 11.57), then

$$\mathbf{F}_1 = \frac{(1)(-2)}{4\pi \epsilon_0 r^2}\hat{\mathbf{r}},$$

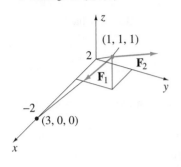

FIGURE 11.57 Force on 1-C charge at $(1, 1, 1)$ due to 2-C charge at $(0, 0, 0)$ and -2-C charge at $(3, 0, 0)$

where the distance between the charges is $r = \sqrt{(-2)^2 + 1^2 + 1^2} = \sqrt{6}$. The vector from $(3, 0, 0)$ to $(1, 1, 1)$ is $(-2, 1, 1)$, and therefore

$$\hat{\mathbf{r}} = \frac{1}{\sqrt{6}}(-2\hat{\mathbf{i}} + \hat{\mathbf{j}} + \hat{\mathbf{k}}).$$

Consequently,

$$\mathbf{F}_1 = \frac{-2}{4\pi\epsilon_0(6)} \cdot \frac{1}{\sqrt{6}}(-2\hat{\mathbf{i}} + \hat{\mathbf{j}} + \hat{\mathbf{k}}) = \frac{1}{12\sqrt{6}\,\pi\epsilon_0}(2\hat{\mathbf{i}} - \hat{\mathbf{j}} - \hat{\mathbf{k}}).$$

Similarly, the force $\mathbf{F}_2$ exerted on the 1 C charge by the charge at the origin is

$$\mathbf{F}_2 = \frac{(1)(2)}{4\pi\epsilon_0(3)} \cdot \frac{1}{\sqrt{3}}(\hat{\mathbf{i}} + \hat{\mathbf{j}} + \hat{\mathbf{k}}) = \frac{1}{6\sqrt{3}\,\pi\epsilon_0}(\hat{\mathbf{i}} + \hat{\mathbf{j}} + \hat{\mathbf{k}}).$$

The resultant of these forces is

$$\mathbf{F} = \mathbf{F}_1 + \mathbf{F}_2 = \frac{1}{12\sqrt{6}\,\pi\epsilon_0}[(2 + 2\sqrt{2})\hat{\mathbf{i}} + (2\sqrt{2} - 1)\hat{\mathbf{j}} + (2\sqrt{2} - 1)\hat{\mathbf{k}}]\text{ N}.$$

FIGURE 11.58 Resultant of n forces on a mass

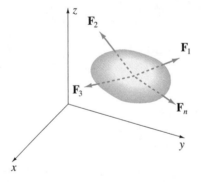

Suppose a number of forces $\mathbf{F}_1, \mathbf{F}_2, \ldots, \mathbf{F}_n$ act on the mass M in Figure 11.58. The resultant of these forces is $\mathbf{F} = \mathbf{F}_1 + \mathbf{F}_2 + \cdots + \mathbf{F}_n$. It is a principle of statics that "the mass will remain motionless if the sum of all forces on it is zero." In such circumstances, the mass is said to be in *equilibrium under the action of the forces.*

$$\mathbf{F} = \mathbf{F}_1 + \mathbf{F}_2 + \cdots + \mathbf{F}_n = \mathbf{0}. \tag{11.17a}$$

This is a vector equation. It is equivalent to three scalar equations obtained by invoking the principle that vectors are equal if and only if their components are equal; that is, if $\mathbf{F} = F_x\hat{\mathbf{i}} + F_y\hat{\mathbf{j}} + F_z\hat{\mathbf{k}}$, the equivalent to 11.17a is

$$F_x = 0, \qquad F_y = 0, \qquad F_z = 0. \tag{11.17b}$$

The following examples use this principle to advantage.

EXAMPLE 11.9

Two cables AB and AC are tied together at A and attached to a vertical wall at B and C as shown in Figure 11.59a. Determine the range of values of the magnitude of the force **P** for which both cables remain taut when a mass of 100 kg hangs at A. Force **P** acts only in the direction shown.

SOLUTION   Suppose we establish the coordinate system in Figure 11.59b. For equilibrium when both cables are taut, the x-components of all forces acting at A must sum to zero, as must the y-components. If magnitudes of the tensions in AC and BC are denoted by T_{AC} and T_{BC}, then

$$0 = -P \cos \theta + T_{AC} + T_{AB} \cos \phi, \quad 0 = P \sin \theta + T_{AB} \sin \phi - 100g.$$

FIGURE 11.59a   Force exerted on two cables attached to a wall by **P** and hanging mass

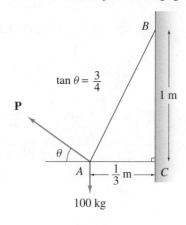

FIGURE 11.59b   Tensions in cable due to force **P** and gravity acting on mass

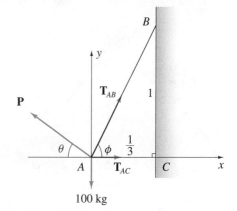

Since $\tan \theta = 3/4$, it follows that $\cos \theta = 4/5$ and $\sin \theta = 3/5$. Furthermore, $\cos \phi = 1/\sqrt{10}$ and $\sin \phi = 3/\sqrt{10}$, and therefore

$$0 = -\frac{4P}{5} + T_{AC} + \frac{T_{AB}}{\sqrt{10}}, \quad 0 = \frac{3P}{5} + \frac{3T_{AB}}{\sqrt{10}} - 100g.$$

The minimum value of P can be determined by decreasing it to the point when T_{AC} becomes zero;

$$0 = -\frac{4P}{5} + \frac{T_{AB}}{\sqrt{10}}, \quad 0 = \frac{3P}{5} + \frac{3T_{AB}}{\sqrt{10}} - 100g.$$

By eliminating T_{AB} and solving for P we obtain $P = 100g/3$ N. By setting $T_{AB} = 0$, we obtain the maximum value of P,

$$0 = -\frac{4P}{5} + T_{AC}, \quad 0 = \frac{3P}{5} - 100g.$$

The second of these gives $P = 500g/3$ N. Consequently, both cables remain taut for $100g/3 < P < 500g/3$.

EXAMPLE 11.10

Application Preview Revisited

Find the maximum mass that can be supported by the boom in the Application Preview (shown again in Figure 11.60a). Assume that the mass of the boom is negligible in comparison to the mass.

SOLUTION   We use that fact that the sum **F** of all forces acting on the end A of the boom must be zero. There is the weight **W** (Figure 11.60b), tensions $\mathbf{T}_{AB}$ and $\mathbf{T}_{AC}$, and a reaction **R** by the boom itself. They are $\mathbf{R} = (0, R, 0)$, $\mathbf{W} = (0, 0, -Mg)$, and

$$\mathbf{T}_{AB} = T_{AB}\left(\frac{\mathbf{AB}}{|\mathbf{AB}|}\right) = \frac{T_{AB}(4.8, -9.6, 5.8)}{\sqrt{4.8^2 + 9.6^2 + 5.8^2}} = \frac{T_{AB}(2.4, -4.8, 2.9)}{6.1},$$

$$\mathbf{T}_{AC} = T_{AC}\left(\frac{\mathbf{AC}}{|\mathbf{AC}|}\right) = \frac{T_{AC}(-7.2, -9.6, 5)}{\sqrt{7.2^2 + 9.6^2 + 5^2}} = \frac{T_{AC}(-7.2, -9.6, 5)}{13}.$$

Hence,

$$\mathbf{F} = (0, R, 0) + (0, 0, -Mg) + \frac{T_{AB}(2.4, -4.8, 2.9)}{6.1} + \frac{T_{AC}(-7.2, -9.6, 5)}{13}.$$

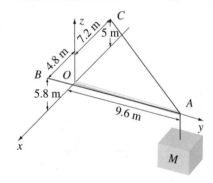

FIGURE 11.60a   Mass supported by a boom

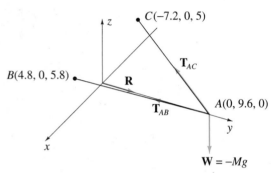

FIGURE 11.60b   Forces acting on end of boom

When we equate components of this vector to zero,

$$0 = \frac{2.4 T_{AB}}{6.1} - \frac{7.2 T_{AC}}{13}, \qquad 0 = R - \frac{4.8 T_{AB}}{6.1} - \frac{9.6 T_{AC}}{13}, \qquad 0 = -Mg + \frac{2.9 T_{AB}}{6.1} + \frac{5 T_{AC}}{13}.$$

The first and third of these can be solved for $T_{AB} = (3/2)Mg$ and $T_{AC} = (130/92)Mg$. Since the tension in cable AB is larger than that in AC, we require

$$\frac{3Mg}{2} \le 20\,000 \qquad \Longrightarrow \qquad M \le \frac{40\,000}{3g} \approx 1359.$$

Maximum mass that can be supported by the boom is 1359 kg.

EXERCISES 11.3

If $\mathbf{u} = (1, 3, 6)$, $\mathbf{v} = (-2, 0, 4)$, and $\mathbf{w} = (4, 3, -2)$, find components for the vector in Exercises 1–10.

1. $3\mathbf{u} - 2\mathbf{v}$

2. $2\mathbf{w} + 3\mathbf{v}$

3. $\mathbf{w} - 3\mathbf{u} - 3\mathbf{v}$

4. $\hat{\mathbf{v}}$

5. $2\hat{\mathbf{w}} - 3\mathbf{v}$

6. $|\mathbf{v}|\mathbf{v} - 2|\hat{\mathbf{v}}|\mathbf{w}$

7. $(15 - 2|\mathbf{w}|)(\mathbf{u} + \mathbf{v})$

8. $|3\mathbf{u}|\mathbf{v} - |-2\mathbf{v}|\mathbf{u}$

9. $|2\mathbf{u} + 3\mathbf{v} - \mathbf{w}|\hat{\mathbf{w}}$

10. $\dfrac{\mathbf{v} - \mathbf{w}}{|\mathbf{v} + \mathbf{w}|}$

If $\mathbf{u} = 2\hat{\mathbf{i}} + \hat{\mathbf{j}}$ and $\mathbf{v} = -\hat{\mathbf{i}} + 3\hat{\mathbf{j}}$, find components of the vector in Exercises 11–14 and illustrate the vector geometrically.

11. $\mathbf{u} + \mathbf{v}$

12. $\mathbf{u} - \mathbf{v}$

13. $2\hat{\mathbf{u}}$

14. $\hat{\mathbf{v}} + \hat{\mathbf{u}}$

In Exercises 15–24 find the Cartesian components for the spatial vector described. In each case, draw the vector.

15. From $(1, 3, 2)$ to $(-1, 4, 5)$

6. With length 5 in the positive x-direction

7. With length 2 in the negative z-direction

8. With tail at $(1, 1, 1)$, length 3, and pointing toward the point $(1, 3, 5)$

9. With positive y-component, length 1, and parallel to the line through $(1, 3, 6)$ and $(-2, 1, 4)$

10. In the same direction as the vector from $(1, 0, -1)$ to $(3, 2, -4)$ but only half as long

11. With positive and equal x- and y-components, length 10, and z-component equal to 4

12. Has its tail at the origin, makes angles of $\pi/3$ and $\pi/4$ radians with the positive x- and y-axes, respectively, and has length $5/2$

13. From $(1, 3, -2)$ to the midpoint of the line segment joining $(2, 4, -3)$ and $(1, 5, 6)$

14. Has its tail at the origin, makes equal angles with the positive coordinate axes, has all positive components, and has length 2

15. If P, Q, and R are the points with coordinates $(3, 2, -1)$, $(0, 1, 4)$, and $(6, 5, -2)$, respectively, find coordinates of a point S in order that $\mathbf{PQ} = \mathbf{RS}$.

16. Prove that vector addition and scalar multiplication have the properties in equations 11.10.

17. Draw all spatial vectors of length 1 that have equal x- and y-components and tails at the origin.

18. Draw all spatial vectors of length 2 that have their tails at the origin and make an angle of $\pi/4$ with the positive z-axis.

19. If $\mathbf{u} = 3\hat{\mathbf{i}} + 2\hat{\mathbf{j}} - 4\hat{\mathbf{k}}$ and $\mathbf{v} = \hat{\mathbf{i}} + 6\hat{\mathbf{j}} + 5\hat{\mathbf{k}}$, find scalars λ and ρ so that the vector $\mathbf{w} = 5\hat{\mathbf{i}} - 18\hat{\mathbf{j}} - 32\hat{\mathbf{k}}$ can be written in the form $\mathbf{w} = \lambda\mathbf{u} + \rho\mathbf{v}$.

20. Find a vector $\mathbf{T}$ of length 3 along the tangent line to the curve $y = x^2$ at the point $(2, 4)$ (figure below).

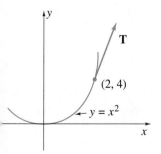

21. Use Coulomb's law (see Example 11.8) to find the force on a charge of 2 C at the origin due to equal charges of 3 C at the points $(1, 1, 2)$

and $(2, -1, -2)$.

* **32.** Newton's universal law of gravitation states that the force of attraction $\mathbf{F}$, in newtons, exerted on a mass of m kilograms by a mass of M kilograms is

$$\mathbf{F} = \frac{GmM}{r^2}\hat{\mathbf{r}},$$

where $G = 6.67 \times 10^{-11}$ is a constant, r is the distance in metres between the masses, and $\hat{\mathbf{r}}$ is a unit vector in the direction from m to M. If point masses, each of 5 kg, are situated at $(5, 1, 3)$ and $(-1, 2, 1)$, what is the resultant force on a mass of 10 kg at $(2, 2, 2)$?

* **33.** Illustrate geometrically the triangle inequality for vectors $|\mathbf{u} + \mathbf{v}| \le |\mathbf{u}| + |\mathbf{v}|$. Prove the result algebraically.

* **34.** Determine to the nearest newton the tensions in the cables AC and BC in the figure below.

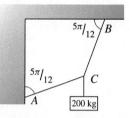

* **35.** Determine to the nearest newton the tensions in the cables AC and BC in the figure below.

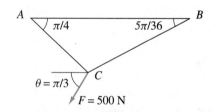

* **36.** For the cables in Exercise 35, it is known that the maximum allowable tension in cable AC is 600 N and that in BC is 750 N. Determine the maximum force $|\mathbf{F}|$ that may be applied at C, and the corresponding angle θ.

* **37.** A 200-kg crate is supported by the rope-and-pulley arrangement in the figure below. Determine the magnitude and direction of the force $\mathbf{F}$ that should be exerted on the free end of the rope.

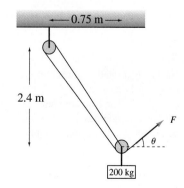

* **38.** A 16 kg, horizontal, triangular plate ABC is supported by three wires in the figure below. Determine the tension in each wire.

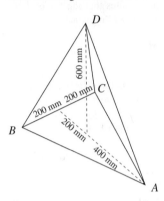

* **39.** A container with weight 360 N is supported by cables, AB and AC in the figure below. Knowing that $\mathbf{F} = 60\hat{\mathbf{i}}$ N, determine the force $\mathbf{P} = P\hat{\mathbf{j}}$ that must be applied at A to maintain the configuration. What are the tensions in the cables?

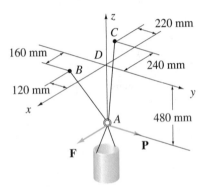

* **40.** A 200-kg mass is hung by means of two cables, AB and AC, which are attached to the top of a vertical wall (figure below). Determine the magnitude of a horizontal force $\mathbf{F}$ perpendicular to the wall that will hold the weight in place. What are the tensions in the cables?

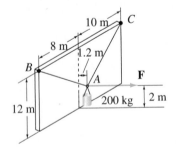

* **41.** Two bars, AB and BC, are pinned at B as well as at each of the ends A and C (figure follows). Initially each bar is of length L and point B is at a distance h above the line AC. The bars are identical, each having cross-sectional area A and Young's modulus E. A vertical force with magnitude F is applied at B. Show that the displacement y of B is related to F by the equation

$$ F = \frac{2AE}{L}(h - y)\left(\frac{L}{\sqrt{y^2 - 2hy + L^2}} - 1\right). $$

(*Hint*: See equation 7.43.)

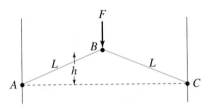

Vectors $\mathbf{u}$, $\mathbf{v}$, and $\mathbf{w}$ are said to be linearly dependent if there exist three scalars a, b, and c, not all zero, such that $a\mathbf{u} + b\mathbf{v} + c\mathbf{w} = \mathbf{0}$. If this equation can only be satisfied with $a = b = c = 0$, the vectors are said to be linearly independent. In Exercises 42–45 determine whether the vectors are linearly dependent or linearly independent.

* **42.** $\mathbf{u} = (1, 1, 1)$, $\mathbf{v} = (2, 1, 3)$, $\mathbf{w} = (4, 2, 6)$

* **43.** $\mathbf{u} = (1, 1, 1)$, $\mathbf{v} = (2, 1, 3)$, $\mathbf{w} = (1, 6, 4)$

* **44.** $\mathbf{u} = (-1, 3, -5)$, $\mathbf{v} = (2, 4, -1)$, $\mathbf{w} = (3, 11, -7)$

* **45.** $\mathbf{u} = (4, 2, 6)$, $\mathbf{v} = (1, 3, -2)$, $\mathbf{w} = (7, 1, 4)$

* **46.** Use vectors to show that the line segment joining the midpoints of two sides of a triangle is parallel to the third side and its length is one-half the length of the third side (figure below).

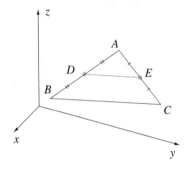

* **47.** Use vectors to show that the medians of a triangle (figure below) all meet in a point with coordinates

$$ \left(\frac{x_1 + x_2 + x_3}{3}, \frac{y_1 + y_2 + y_3}{3}, \frac{z_1 + z_2 + z_3}{3}\right). $$

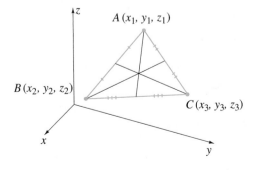

48. If n point masses m_i are located at points (x_i, y_i) in the xy-plane, equations 7.31 and 7.32 define the centre of mass $(\overline{x}, \overline{y})$ of the system. Show that these two scalar equations are represented by the one vector equation

$$M(\overline{x}, \overline{y}) = \sum_{i=1}^{n} m_i \mathbf{r}_i,$$

where $\mathbf{r}_i$ is the vector joining the origin to the point (x_i, y_i).

49. If $\mathbf{u} = u_x\hat{\mathbf{i}} + u_y\hat{\mathbf{j}}$ and $\mathbf{v} = v_x\hat{\mathbf{i}} + v_y\hat{\mathbf{j}}$, show that every vector $\mathbf{w}$ in the xy-plane can be written in the form $\mathbf{w} = \lambda\mathbf{u} + \rho\mathbf{v}$ provided that $u_xv_y - u_yv_x \neq 0$.

50. Two identical springs with constant k are joined at point C in the figure below and have their ends fixed at A and B. In this position the springs are unstretched and uncompressed.

(a) If a weight W is attached to their join, and slowly lowered until the system is in equilibrium, find an equation relating W, L, k, and the distance y from C to D.

(b) Find an approximation for W in terms of y when y is very small compared to L.

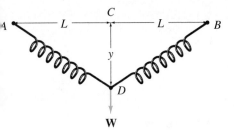

51. In the figure below, two springs (with constants k_1 and k_2 and unstretched lengths l) are fixed at points A and B. They are joined to a sleeve that slides along the x-axis. Find the resultant force of the springs on the sleeve at any point between O and C.

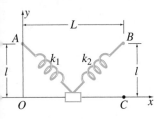

52. In the following figure, collars A and B are connected by a 440-mm wire and slide freely on frictionless rods, one along the y-axis, and the other parallel to the z-axis and passing through the x-axis 240 mm from the origin. If a force $\mathbf{F}$ of magnitude 450 N is applied to collar A, determine the force $\mathbf{P}$ required to keep collar B at the position shown. What is the tension in the wire?

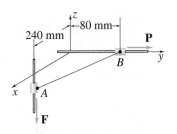

** **53.** Two wires AC and BC are attached to the top of pole CO in the figure below. The force exerted by the pole is vertical, and the

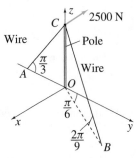

2500 N force applied at C is in the negative x-direction. Determine the tensions in the wires and the vertical force exerted by the pole.

** **54.** In the left figure below, the xy-plane is the interface between two materials that both transmit light. If a ray of light strikes the surface in a direction defined by the unit vector $\hat{\mathbf{u}} = (u_x, u_y, u_z)$, then some of the light is reflected along a vector $\hat{\mathbf{v}}$, and some is refracted along $\hat{\mathbf{w}}$. The three vectors $\hat{\mathbf{u}}, \hat{\mathbf{v}}$, and $\hat{\mathbf{w}}$ all lie in a plane that is perpendicular to the xy-plane.

(a) If the angle of incidence i in the right figure below is equal to the angle of reflection ϕ, find components for $\hat{\mathbf{v}}$ in terms of those of $\hat{\mathbf{u}}$.

(b) If the angle of refraction θ is related to the angle of incidence by $n_1 \sin i = n_2 \sin \theta$, where n_1 and n_2 are the indices of refraction of the two materials, find components of $\hat{\mathbf{w}}$.

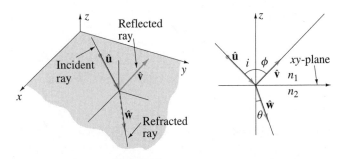

** **55.** Vectors $\mathbf{v}_1, \mathbf{v}_2, \ldots, \mathbf{v}_n$ are drawn from the centre of a regular n-sided polygon in the plane to each of its vertices. Show that the sum of these vectors is the zero vector.

11.4 Scalar and Vector Products

The Scalar Product of Vectors

In Section 11.3 we learned how to multiply a vector by a scalar and how to add and subtract vectors. The next question naturally is: Can vectors be multiplied? The answer is yes, and in fact we define two products for vectors, one of which yields a scalar and the other a vector. The first is defined algebraically as follows.

DEFINITION 11.8

The **scalar product** (**dot product** or **inner product**) of two vectors $\mathbf{u}$ and $\mathbf{v}$ with Cartesian components (u_x, u_y, u_z) and (v_x, v_y, v_z) is defined as

$$\mathbf{u} \cdot \mathbf{v} = u_x v_x + u_y v_y + u_z v_z. \tag{11.18}$$

If $\mathbf{u}$ and $\mathbf{v}$ have only x- and y-components, Definition 11.8 reduces to

$$\mathbf{u} \cdot \mathbf{v} = u_x v_x + u_y v_y, \tag{11.19}$$

and if they have only x-components, it becomes

$$\mathbf{u} \cdot \mathbf{v} = u_x v_x. \tag{11.20}$$

It is straightforward to check that

$$\hat{\mathbf{i}} \cdot \hat{\mathbf{i}} = \hat{\mathbf{j}} \cdot \hat{\mathbf{j}} = \hat{\mathbf{k}} \cdot \hat{\mathbf{k}} = 1, \tag{11.21a}$$

$$\hat{\mathbf{i}} \cdot \hat{\mathbf{j}} = \hat{\mathbf{j}} \cdot \hat{\mathbf{k}} = \hat{\mathbf{i}} \cdot \hat{\mathbf{k}} = 0, \tag{11.21b}$$

and that the scalar product is commutative and distributive:

$$\mathbf{u} \cdot \mathbf{v} = \mathbf{v} \cdot \mathbf{u}, \tag{11.22a}$$

$$\mathbf{u} \cdot (\lambda \mathbf{v} + \rho \mathbf{w}) = \lambda(\mathbf{u} \cdot \mathbf{v}) + \rho(\mathbf{u} \cdot \mathbf{w}). \tag{11.22b}$$

EXAMPLE 11.11

If $\mathbf{u} = (-2, 1, 3)$ and $\mathbf{v} = (3, -2, -1)$, evaluate each of the following:

(a) $\mathbf{u} \cdot \mathbf{v}$;   (b) $3\mathbf{u} \cdot (2\mathbf{u} - 4\mathbf{v})$.

SOLUTION

(a) $\mathbf{u} \cdot \mathbf{v} = (-2)(3) + (1)(-2) + (3)(-1) = -11$

(b) $3\mathbf{u} \cdot (2\mathbf{u} - 4\mathbf{v}) = (-6, 3, 9) \cdot (-16, 10, 10)$

$$= (-6)(-16) + (3)(10) + (9)(10)$$

$$= 216$$

By taking the scalar product of a vector $\mathbf{v} = (v_x, v_y, v_z)$ with itself, we obtain

$$\mathbf{v} \cdot \mathbf{v} = v_x^2 + v_y^2 + v_z^2 = |\mathbf{v}|^2 \iff |\mathbf{v}| = \sqrt{\mathbf{v} \cdot \mathbf{v}}. \tag{11.23}$$

Because Definition 11.8 for the scalar product of two vectors $\mathbf{u}$ and $\mathbf{v}$ is phrased in terms of the components of $\mathbf{u}$ and $\mathbf{v}$, and these components depend on the coordinate system used, it follows that this definition also depends on the fact that we have used Cartesian coordinates. Were we to use a different set of coordinates (such as polar coordinates), then the definition of $\mathbf{u} \cdot \mathbf{v}$ in terms of components in that coordinate system might be different. For this reason we now find a geometric definition for the scalar product (which is therefore independent of coordinate systems).

THEOREM 11.2

If two nonzero vectors $\mathbf{u}$ and $\mathbf{v}$ are placed tail to tail, and θ is the angle between them $(0 \leq \theta \leq \pi)$, then

$$\mathbf{u} \cdot \mathbf{v} = |\mathbf{u}||\mathbf{v}| \cos \theta. \tag{11.24}$$

FIGURE 11.61   Proof that
$\mathbf{u} \cdot \mathbf{v} = |\mathbf{u}||\mathbf{v}| \cos \theta$

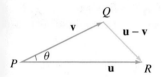

PROOF   The cosine law applied to the triangle in Figure 11.61 gives

$$|\mathbf{QR}|^2 = |\mathbf{PQ}|^2 + |\mathbf{PR}|^2 - 2|\mathbf{PQ}||\mathbf{PR}| \cos \theta$$

or

$$|\mathbf{u} - \mathbf{v}|^2 = |\mathbf{v}|^2 + |\mathbf{u}|^2 - 2|\mathbf{v}||\mathbf{u}| \cos \theta.$$

Consequently,

$$|\mathbf{u}||\mathbf{v}| \cos \theta = \frac{1}{2} \left(|\mathbf{u}|^2 + |\mathbf{v}|^2 - |\mathbf{u} - \mathbf{v}|^2 \right),$$

and if (u_x, u_y, u_z) and (v_x, v_y, v_z) are the Cartesian components of $\mathbf{u}$ and $\mathbf{v}$, then

$$\begin{aligned}
|\mathbf{u}||\mathbf{v}| \cos \theta &= \frac{1}{2} \{ (u_x^2 + u_y^2 + u_z^2) + (v_x^2 + v_y^2 + v_z^2) \\
&\quad - [(u_x - v_x)^2 + (u_y - v_y)^2 + (u_z - v_z)^2] \} \\
&= u_x v_x + u_y v_y + u_z v_z \\
&= \mathbf{u} \cdot \mathbf{v}.
\end{aligned}$$

An immediate consequence of this result is the following.

COROLLARY 11.2.1

Two nonzero vectors $\mathbf{u}$ and $\mathbf{v}$ are perpendicular if and only if

$$\mathbf{u} \cdot \mathbf{v} = 0. \tag{11.25}$$

For example, the vectors $\mathbf{u} = (1, 2, -1)$ and $\mathbf{v} = (4, 2, 8)$ are perpendicular since $\mathbf{u} \cdot \mathbf{v} = (1)(4) + (2)(2) + (-1)(8) = 0$.

Expression 11.24 doesn't just tell us whether or not the angle between two vectors is $\pi/2$ radians; it can be used to determine the angle between any two nonzero vectors $\mathbf{u}$ and $\mathbf{v}$, when they are placed tail to tail. We first solve 11.24 for $\cos \theta$:

$$\cos \theta = \frac{\mathbf{u} \cdot \mathbf{v}}{|\mathbf{u}||\mathbf{v}|}. \tag{11.26}$$

Since principal values of the inverse cosine function lie between 0 and π, precisely the range for θ, we can write that

$$\theta = \text{Cos}^{-1} \left(\frac{\mathbf{u} \cdot \mathbf{v}}{|\mathbf{u}||\mathbf{v}|} \right). \tag{11.27}$$

EXAMPLE 11.12

Find the angle between the vectors $\mathbf{u} = (2, -3, 1)$ and $\mathbf{v} = (5, 2, 4)$.

SOLUTION   According to formula 11.27,

$$\theta = \text{Cos}^{-1}\left(\frac{\mathbf{u} \cdot \mathbf{v}}{|\mathbf{u}||\mathbf{v}|}\right) = \text{Cos}^{-1}\left(\frac{10 - 6 + 4}{\sqrt{4 + 9 + 1}\sqrt{25 + 4 + 16}}\right)$$

$$= \text{Cos}^{-1}\left(\frac{8}{3\sqrt{70}}\right) = 1.25 \text{ radians.}$$

EXAMPLE 11.13

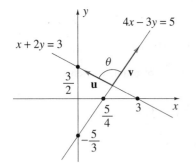

FIGURE 11.62   The angle between lines in the xy-plane using vectors along the lines

Find the angle between the lines $x + 2y = 3$ and $4x - 3y = 5$ in the xy-plane.

SOLUTION   Since the slope of $x + 2y = 3$ is $-1/2$, a vector along this line is $\mathbf{u} = (-2, 1)$. Similarly, a vector along $4x - 3y = 5$ is $\mathbf{v} = (3, 4)$. If θ is the angle between these vectors, and therefore between the lines (Figure 11.62), then

$$\theta = \text{Cos}^{-1}\left(\frac{\mathbf{u} \cdot \mathbf{v}}{|\mathbf{u}||\mathbf{v}|}\right) = \text{Cos}^{-1}\left(\frac{-6 + 4}{\sqrt{5}\sqrt{25}}\right) = \text{Cos}^{-1}\left(\frac{-2}{5\sqrt{5}}\right) = 1.75 \text{ radians.}$$

The acute angle between the lines is $\pi - 1.75 = 1.39$ radians. Formula 1.60 gives the same result.

The Vector Product of Vectors

In many applications we need to find a vector perpendicular to two given vectors. For instance, consider finding a vector $\mathbf{r} = (a, b, c)$ perpendicular to two given vectors $\mathbf{u} = (u_x, u_y, u_z)$ and $\mathbf{v} = (v_x, v_y, v_z)$. The corollary to Theorem 11.2 requires a, b, and c to satisfy

$$0 = \mathbf{r} \cdot \mathbf{u} = au_x + bu_y + cu_z,$$

$$0 = \mathbf{r} \cdot \mathbf{v} = av_x + bv_y + cv_z.$$

When we solve these equations, we find that there is an infinite number of solutions all represented by

$$a = s(u_y v_z - u_z v_y), \quad b = s(u_z v_x - u_x v_z), \quad c = s(u_x v_y - u_y v_x),$$

where s is any real number. In other words, any vector of the form

$$\mathbf{r} = s(u_y v_z - u_z v_y, u_z v_x - u_x v_z, u_x v_y - u_y v_x)$$

is perpendicular to $\mathbf{u} = (u_x, u_y, u_z)$ and $\mathbf{v} = (v_x, v_y, v_z)$. When we choose $s = 1$, the resulting vector is called the vector product of $\mathbf{u}$ and $\mathbf{v}$.

DEFINITION 11.9

The **vector product** (**cross product** or **outer product**) of two vectors $\mathbf{u}$ and $\mathbf{v}$ with Cartesian components (u_x, u_y, u_z) and (v_x, v_y, v_z) is defined as

$$\mathbf{u} \times \mathbf{v} = (u_y v_z - u_z v_y)\hat{\mathbf{i}} + (u_z v_x - u_x v_z)\hat{\mathbf{j}} + (u_x v_y - u_y v_x)\hat{\mathbf{k}}. \tag{11.28}$$

To eliminate the need for memorizing the exact placing of the six components of **u** and **v** in this definition, we borrow the notation for determinants from linear algebra. A brief discussion of determinants and their properties is given in Appendix B. We set up a 3×3 determinant with $\hat{\mathbf{i}}, \hat{\mathbf{j}}$, and $\hat{\mathbf{k}}$ across the top row and the components of **u** and **v** across the second and third rows:

$$\begin{vmatrix} \hat{\mathbf{i}} & \hat{\mathbf{j}} & \hat{\mathbf{k}} \\ u_x & u_y & u_z \\ v_x & v_y & v_z \end{vmatrix}.$$

In actual fact, this is not a determinant, since three entries are vectors and six are scalars. If we ignore this fact, and apply the rules for expansion of a 3×3 determinant along its first row (namely, $\hat{\mathbf{i}}$ times the 2×2 determinant obtained by deleting the row and column containing $\hat{\mathbf{i}}$, minus $\hat{\mathbf{j}}$ times the 2×2 determinant obtained by deleting the row and column containing $\hat{\mathbf{j}}$, plus $\hat{\mathbf{k}}$ times the 2×2 determinant obtained by deleting the row and column containing $\hat{\mathbf{k}}$), we obtain

$$\begin{vmatrix} \hat{\mathbf{i}} & \hat{\mathbf{j}} & \hat{\mathbf{k}} \\ u_x & u_y & u_z \\ v_x & v_y & v_z \end{vmatrix} = \begin{vmatrix} u_y & u_z \\ v_y & v_z \end{vmatrix} \hat{\mathbf{i}} - \begin{vmatrix} u_x & u_z \\ v_x & v_z \end{vmatrix} \hat{\mathbf{j}} + \begin{vmatrix} u_x & u_y \\ v_x & v_y \end{vmatrix} \hat{\mathbf{k}}.$$

But the value of a 2×2 determinant is

$$\begin{vmatrix} a & b \\ c & d \end{vmatrix} = ad - bc.$$

Consequently,

$$\begin{vmatrix} \hat{\mathbf{i}} & \hat{\mathbf{j}} & \hat{\mathbf{k}} \\ u_x & u_y & u_z \\ v_x & v_y & v_z \end{vmatrix} = (u_y v_z - u_z v_y)\hat{\mathbf{i}} + (u_z v_x - u_x v_z)\hat{\mathbf{j}} + (u_x v_y - u_y v_x)\hat{\mathbf{k}},$$

and this is the same as the right side of equation 11.28. We may therefore write, as a memory-saving device, that

$$\mathbf{u} \times \mathbf{v} = \begin{vmatrix} \hat{\mathbf{i}} & \hat{\mathbf{j}} & \hat{\mathbf{k}} \\ u_x & u_y & u_z \\ v_x & v_y & v_z \end{vmatrix}, \tag{11.29}$$

so long as we evaluate the right side using the general rules for expansion of a determinant along its first row.

For example, if $\mathbf{u} = (1, -1, 2)$ and $\mathbf{v} = (2, 3, -5)$, then

$$\mathbf{u} \times \mathbf{v} = \begin{vmatrix} \hat{\mathbf{i}} & \hat{\mathbf{j}} & \hat{\mathbf{k}} \\ 1 & -1 & 2 \\ 2 & 3 & -5 \end{vmatrix} = (5 - 6)\hat{\mathbf{i}} - (-5 - 4)\hat{\mathbf{j}} + (3 + 2)\hat{\mathbf{k}} = -\hat{\mathbf{i}} + 9\hat{\mathbf{j}} + 5\hat{\mathbf{k}}.$$

It is straightforward to verify that

$$\hat{\mathbf{i}} \times \hat{\mathbf{i}} = \hat{\mathbf{j}} \times \hat{\mathbf{j}} = \hat{\mathbf{k}} \times \hat{\mathbf{k}} = \mathbf{0}, \tag{11.30a}$$

$$\hat{\mathbf{i}} \times \hat{\mathbf{j}} = \hat{\mathbf{k}}, \quad \hat{\mathbf{j}} \times \hat{\mathbf{k}} = \hat{\mathbf{i}}, \quad \hat{\mathbf{k}} \times \hat{\mathbf{i}} = \hat{\mathbf{j}}, \tag{11.30b}$$

and that the cross product is anticommutative and distributive:

$$\mathbf{u} \times \mathbf{v} = -\mathbf{v} \times \mathbf{u}, \tag{11.31a}$$

$$\mathbf{u} \times (\lambda \mathbf{v} + \rho \mathbf{w}) = \lambda(\mathbf{u} \times \mathbf{v}) + \rho(\mathbf{u} \times \mathbf{w}). \tag{11.31b}$$

Our preliminary analysis indicated that $\mathbf{u} \times \mathbf{v}$ is perpendicular to both **u** and **v**. The following theorem relates the length of $\mathbf{u} \times \mathbf{v}$ to lengths of **u** and **v**.

THEOREM 11.3

If θ is the angle between two vectors $\mathbf{u}$ and $\mathbf{v}$, then

$$|\mathbf{u} \times \mathbf{v}| = |\mathbf{u}||\mathbf{v}| \sin \theta. \tag{11.32}$$

PROOF   Since θ is an angle between 0 and π, $\sin \theta$ must be positive, and we can write from equation 11.26 that

$$\sin \theta = \sqrt{1 - \cos^2 \theta} = \sqrt{1 - \left(\frac{\mathbf{u} \cdot \mathbf{v}}{|\mathbf{u}||\mathbf{v}|} \right)^2} = \frac{1}{|\mathbf{u}||\mathbf{v}|} \sqrt{|\mathbf{u}|^2 |\mathbf{v}|^2 - (\mathbf{u} \cdot \mathbf{v})^2}.$$

Consequently,

$$|\mathbf{u}||\mathbf{v}| \sin \theta = \sqrt{|\mathbf{u}|^2 |\mathbf{v}|^2 - (\mathbf{u} \cdot \mathbf{v})^2}.$$

If $\mathbf{u} = (u_x, u_y, u_z)$ and $\mathbf{v} = (v_x, v_y, v_z)$, then

$$
\begin{aligned}
|\mathbf{u}|^2 |\mathbf{v}|^2 \sin^2 \theta &= (u_x^2 + u_y^2 + u_z^2)(v_x^2 + v_y^2 + v_z^2) - (u_x v_x + u_y v_y + u_z v_z)^2 \\
&= u_x^2(v_x^2 + v_y^2 + v_z^2) + u_y^2(v_x^2 + v_y^2 + v_z^2) \\
&\quad + u_z^2(v_x^2 + v_y^2 + v_z^2) - (u_x^2 v_x^2 + u_y^2 v_y^2 + u_z^2 v_z^2 \\
&\quad + 2u_x v_x u_y v_y + 2u_x v_x u_z v_z + 2u_y v_y u_z v_z) \\
&= (u_y^2 v_z^2 - 2u_y v_y u_z v_z + u_z^2 v_y^2) + (u_x^2 v_z^2 - 2u_x v_x u_z v_z + u_z^2 v_x^2) \\
&\quad + (u_x^2 v_y^2 - 2u_x v_x u_y v_y + u_y^2 v_x^2) \\
&= (u_y v_z - u_z v_y)^2 + (u_x v_z - u_z v_x)^2 + (u_x v_y - u_y v_x)^2 \\
&= |\mathbf{u} \times \mathbf{v}|^2
\end{aligned}
$$

or

$$|\mathbf{u} \times \mathbf{v}| = |\mathbf{u}||\mathbf{v}| \sin \theta.$$

We now know that $\mathbf{u} \times \mathbf{v}$ is perpendicular to $\mathbf{u}$ and $\mathbf{v}$, and has length $|\mathbf{u}||\mathbf{v}| \sin \theta$, where θ is the angle between $\mathbf{u}$ and $\mathbf{v}$. Figure 11.63 illustrates that there are only two directions that are perpendicular to $\mathbf{u}$ and $\mathbf{v}$, and one is the negative of the other. Let us denote by $\hat{\mathbf{w}}$ the unit vector along that direction which is perpendicular to $\mathbf{u}$ and $\mathbf{v}$ and is determined by the right-hand rule (curl the fingers of the right hand from $\mathbf{u}$ toward $\mathbf{v}$ and the thumb points in direction $\hat{\mathbf{w}}$).

We now show that $\mathbf{u} \times \mathbf{v}$ always points in the direction determined by the right-hand rule (i.e., in direction $\hat{\mathbf{w}}$ rather than $-\hat{\mathbf{w}}$). To see this, we place $\mathbf{u}$ and $\mathbf{v}$ tail to tail and establish a coordinate system with this common point as origin and the positive x-axis along $\mathbf{u}$ (Figure 11.64a). Let the plane determined by $\mathbf{u}$ and $\mathbf{v}$ be the xy-plane. In this coordinate system, $\mathbf{u}$ has only an x-component, $\mathbf{u} = u_x \hat{\mathbf{i}} (u_x > 0)$, and $\mathbf{v}$ has only x- and y-components, $\mathbf{v} = v_x \hat{\mathbf{i}} + v_y \hat{\mathbf{j}}$. The cross product of $\mathbf{u}$ and $\mathbf{v}$ is therefore

$$
\mathbf{u} \times \mathbf{v} = \begin{vmatrix} \hat{\mathbf{i}} & \hat{\mathbf{j}} & \hat{\mathbf{k}} \\ u_x & 0 & 0 \\ v_x & v_y & 0 \end{vmatrix} = u_x v_y \hat{\mathbf{k}}.
$$

For $\mathbf{v}$ in Figure 11.64a, v_y is clearly positive and therefore $u_x v_y$, the component of $\mathbf{u} \times \mathbf{v}$, is also positive. But then $\mathbf{u} \times \mathbf{v}$ is indeed determined by the right-hand rule since $\hat{\mathbf{w}} = \hat{\mathbf{k}}$.

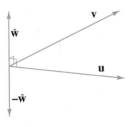

FIGURE 11.63   There are two directions perpendicular to two given vectors, one in the opposite direction to the other

FIGURE 11.64a   FIGURE 11.64b   FIGURE 11.64c

The direction of $\mathbf{u} \times \mathbf{v}$ is always determined by the right-hand rule

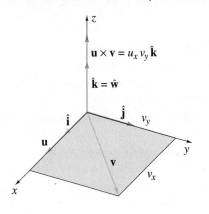   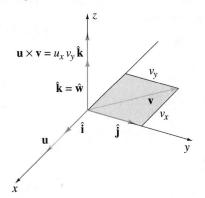   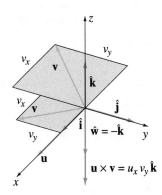

When the tip of $\mathbf{v}$ lies in the second quadrant of the xy-plane (Figure 11.64b), $\mathbf{u} \times \mathbf{v}$ is once again in the positive z-direction, the direction of $\hat{\mathbf{w}}$. When the tip of $\mathbf{v}$ is in the third or fourth quadrant (Figure 11.64c), $v_y < 0$, and $\mathbf{u} \times \mathbf{v}$ therefore has a negative z-component. But in this case $\hat{\mathbf{w}} = -\hat{\mathbf{k}}$, and the direction of $\mathbf{u} \times \mathbf{v}$ is once again determined by the right-hand rule.

What we have now established is the following coordinate-free definition for the vector product of two vectors $\mathbf{u}$ and $\mathbf{v}$:

$$\mathbf{u} \times \mathbf{v} = (|\mathbf{u}||\mathbf{v}| \sin \theta)\hat{\mathbf{w}}. \tag{11.33}$$

The unit vector $\hat{\mathbf{w}}$ defines the direction of $\mathbf{u} \times \mathbf{v}$ and the factor $|\mathbf{u}||\mathbf{v}| \sin \theta$ is its length.

The fact that the vector product $\mathbf{u} \times \mathbf{v}$ is perpendicular to both $\mathbf{u}$ and $\mathbf{v}$ makes it a powerful tool in many applications. We shall see some of them in Sections 11.5 and 11.6.

EXAMPLE 11.14

Find the cross product of the vectors $\mathbf{u} = \hat{\mathbf{i}} + 2\hat{\mathbf{j}}$ and $\mathbf{v} = 3\hat{\mathbf{i}} - 2\hat{\mathbf{j}} + \hat{\mathbf{k}}$, and show that the result is indeed perpendicular to $\mathbf{u}$ and $\mathbf{v}$.

FIGURE 11.65   Cross-product of two vectors is perpendicular to both

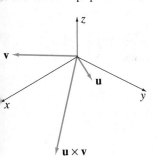

SOLUTION   Using formula 11.29,

$$\mathbf{u} \times \mathbf{v} = \begin{vmatrix} \hat{\mathbf{i}} & \hat{\mathbf{j}} & \hat{\mathbf{k}} \\ 1 & 2 & 0 \\ 3 & -2 & 1 \end{vmatrix} = 2\hat{\mathbf{i}} - \hat{\mathbf{j}} - 8\hat{\mathbf{k}}.$$

We have shown the vectors in Figure 11.65. We can check that $\mathbf{u} \times \mathbf{v}$ is perpendicular to $\mathbf{u}$ and $\mathbf{v}$ using 11.25:

$$(\mathbf{u} \times \mathbf{v}) \cdot \mathbf{u} = (2\hat{\mathbf{i}} - \hat{\mathbf{j}} - 8\hat{\mathbf{k}}) \cdot (\hat{\mathbf{i}} + 2\hat{\mathbf{j}}) = 2(1) + (-1)(2) = 0,$$

$$(\mathbf{u} \times \mathbf{v}) \cdot \mathbf{v} = (2\hat{\mathbf{i}} - \hat{\mathbf{j}} - 8\hat{\mathbf{k}}) \cdot (3\hat{\mathbf{i}} - 2\hat{\mathbf{j}} + \hat{\mathbf{k}}) = 2(3) + (-1)(-2) + (-8)(1) = 0.$$

EXERCISES 11.4

If $\mathbf{u} = 2\hat{\mathbf{i}} - 3\hat{\mathbf{j}} + \hat{\mathbf{k}}$, $\mathbf{v} = \hat{\mathbf{j}} - \hat{\mathbf{k}}$, and $\mathbf{w} = 6\hat{\mathbf{i}} - 2\hat{\mathbf{j}} + 3\hat{\mathbf{k}}$, evaluate the scalar or find the components of the vector in Exercises 1–10.

1. $\mathbf{u} \cdot \mathbf{v}$

2. $(\mathbf{v} \cdot \mathbf{w})\mathbf{u}$

3. $(2\mathbf{u} - 3\mathbf{v}) \cdot \mathbf{w}$

4. $2\hat{\mathbf{i}} \cdot \hat{\mathbf{u}}$

5. $|2\mathbf{u}|\mathbf{v} \cdot \mathbf{w}$

6. $(3\mathbf{u} - 4\mathbf{w}) \cdot (2\hat{\mathbf{i}} + 3\mathbf{u} - 2\mathbf{v})$

7. $\mathbf{w} \cdot \hat{\mathbf{w}}$

8. $\dfrac{(105\mathbf{u} + 240\mathbf{v}) \cdot (105\mathbf{u} + 240\mathbf{v})}{|105\mathbf{u} + 240\mathbf{v}|^2}$

9. $|\mathbf{u} - \mathbf{v} + \hat{\mathbf{k}}|(\hat{\mathbf{j}} + \mathbf{w}) \cdot \hat{\mathbf{k}}$

10. $\mathbf{u} \cdot \mathbf{v} + \mathbf{v} \cdot \mathbf{w} - (\mathbf{u} + \mathbf{w}) \cdot \mathbf{v}$

If $\mathbf{u} = (3, 1, 4)$, $\mathbf{v} = (-1, 2, 0)$, and $\mathbf{w} = (-2, -3, 5)$, evaluate the scalar or find the components of the vector in Exercises 11–20.

11. $\mathbf{v} \times \mathbf{w}$

12. $(-3\mathbf{u}) \times (2\mathbf{v})$

13. $\mathbf{u} \cdot (\mathbf{v} \times \mathbf{w})$

14. $\hat{\mathbf{u}} \times \hat{\mathbf{w}}$

15. $((3\mathbf{u}) \times \mathbf{w}) + (\mathbf{u} \times \mathbf{v})$

16. $\mathbf{u} \times (3\mathbf{v} - \mathbf{w})$

17. $\dfrac{\mathbf{w} \times \mathbf{u}}{|\mathbf{u} \times \mathbf{v}|}$

18. $(\mathbf{u} \times \mathbf{w}) - (\mathbf{u} \times \mathbf{v}) + (\mathbf{u} \times (2\mathbf{u} + \mathbf{v}))$

19. $(\mathbf{u} \times \mathbf{v}) \times \mathbf{w}$

20. $\mathbf{u} \times (\mathbf{v} \times \mathbf{w})$

In Exercises 21–24 determine whether the vectors are perpendicular.

21. $(1, 2)$, $(3, 5)$

22. $(2, 4)$, $(-8, 4)$

23. $(1, 3, 6)$, $(-2, 1, -4)$

24. $(2, 3, -6)$, $(-6, 6, 1)$

In Exercises 25–30 find the angle between the vectors.

25. $(3, 4)$, $(2, -5)$

26. $(1, 6)$, $(-4, 7)$

27. $(4, 2, 3)$, $(1, 5, 6)$

28. $(3, 1, -1)$, $(-2, 1, 4)$

29. $(2, 0, 5)$, $(0, 3, 0)$

30. $(1, 3, -2)$, $(-2, -6, 4)$

In Exercises 31–33 find components for the vector.

31. Perpendicular to the vectors $(1, 3, 5)$ and $(-2, 1, 4)$

32. Perpendicular to the y-axis and the vector joining the points $(2, 4, -3)$ and $(1, 5, 6)$

33. Perpendicular to the triangle with vertices $(-1, 0, 3)$, $(5, 1, 2)$, and $(-6, 2, 4)$

34. Verify the results in equations 11.22.

* 35. Use equations 11.23 to prove

$$|\mathbf{u} + \mathbf{v}|^2 + |\mathbf{u} - \mathbf{v}|^2 = 2|\mathbf{u}|^2 + 2|\mathbf{v}|^2.$$

This is often called the *parallelogram law*. Why?

* 36. The angles between a vector $\mathbf{v} = (v_x, v_y, v_z)$ and the vectors $\hat{\mathbf{i}}$, $\hat{\mathbf{j}}$, and $\hat{\mathbf{k}}$ are called *direction angles* α, β, and γ of $\mathbf{v}$. Show that

$$\alpha = \mathrm{Cos}^{-1}\left(\frac{\mathbf{v} \cdot \hat{\mathbf{i}}}{|\mathbf{v}|}\right) = \mathrm{Cos}^{-1}\left(\frac{v_x}{|\mathbf{v}|}\right),$$

$$\beta = \mathrm{Cos}^{-1}\left(\frac{\mathbf{v} \cdot \hat{\mathbf{j}}}{|\mathbf{v}|}\right) = \mathrm{Cos}^{-1}\left(\frac{v_y}{|\mathbf{v}|}\right),$$

$$\gamma = \mathrm{Cos}^{-1}\left(\frac{\mathbf{v} \cdot \hat{\mathbf{k}}}{|\mathbf{v}|}\right) = \mathrm{Cos}^{-1}\left(\frac{v_z}{|\mathbf{v}|}\right).$$

Find direction angles for the vectors in Exercises 37–40.

37. $(1, 2, -3)$

38. $(0, 1, -3)$

39. $(-1, -2, 6)$

40. $(-2, 3, 4)$

* 41. Verify the results in equations 11.31.

* 42. Show that the cross product is not associative; that is, in general $\mathbf{u} \times (\mathbf{v} \times \mathbf{w}) \neq (\mathbf{u} \times \mathbf{v}) \times \mathbf{w}$.

* 43.   (a) If $\mathbf{u} \neq 0$, show that if the conditions $\mathbf{u} \cdot \mathbf{v} = \mathbf{u} \cdot \mathbf{w}$ and $\mathbf{u} \times \mathbf{v} = \mathbf{u} \times \mathbf{w}$ are both satisfied, then $\mathbf{v} = \mathbf{w}$.

    (b) Show that if one of the conditions in part (a) is satisfied but the other is not, then $\mathbf{v}$ cannot be equal to $\mathbf{w}$.

* 44. The scalar $\mathbf{u} \cdot \mathbf{v} \times \mathbf{w}$ is called the *scalar triple product* of $\mathbf{u}$, $\mathbf{v}$ and $\mathbf{w}$.

    (a) Find $\mathbf{u} \cdot \mathbf{v} \times \mathbf{w}$ if $\mathbf{u} = (6, -1, 0)$, $\mathbf{v} = (1, 3, 4)$, and $\mathbf{w} = (-2, -1, 4)$.

    (b) Prove that $\mathbf{u} \cdot \mathbf{v} \times \mathbf{w} = \mathbf{u} \times \mathbf{v} \cdot \mathbf{w}$.

    (c) Show that $|\mathbf{u} \cdot \mathbf{v} \times \mathbf{w}|$ can be interpreted as the volume of the parallelepiped with $\mathbf{u}$, $\mathbf{v}$, and $\mathbf{w}$ as coterminal sides in the figure below.

    (d) Verify that three nonzero vectors $\mathbf{u}$, $\mathbf{v}$, and $\mathbf{w}$ all lie in the same plane if and only if $\mathbf{u} \cdot \mathbf{v} \times \mathbf{w} = 0$.

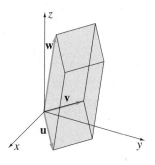

In Exercises 45–46 prove the identity.

* 45. $(\mathbf{u} \times \mathbf{v}) \cdot (\mathbf{w} \times \mathbf{r}) = (\mathbf{u} \cdot \mathbf{w})(\mathbf{v} \cdot \mathbf{r}) - (\mathbf{u} \cdot \mathbf{r})(\mathbf{v} \cdot \mathbf{w})$

* 46. $\mathbf{u} \times (\mathbf{v} \times \mathbf{w}) = (\mathbf{u} \cdot \mathbf{w})\mathbf{v} - (\mathbf{u} \cdot \mathbf{v})\mathbf{w}$

* 47. Use vectors to prove the sine law $\dfrac{\sin A}{a} = \dfrac{\sin B}{b} = \dfrac{\sin C}{c}$ for the triangle in the figure below. *Hint:* Note that $\mathbf{PQ} + \mathbf{QR} + \mathbf{RP} = 0$. Cross this equation with $\mathbf{PQ}$ and take lengths.

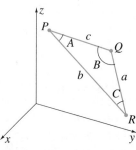

* 48. Show that the vector $(|\mathbf{v}|\mathbf{u} + |\mathbf{u}|\mathbf{v}) / ||\mathbf{u}|\mathbf{v} + |\mathbf{v}|\mathbf{u}|$ is a unit vector that bisects the angle between $\mathbf{u}$ and $\mathbf{v}$.

11.5 Planes and Lines

Planes

A plane in space can be characterized in various ways: by two intersecting lines, by a line and a point not on the line, or by three noncollinear points. For our present purposes, we use the fact that given a point $P(x_0, y_0, z_0)$ and a vector (A, B, C) (Figure 11.66), there is one and only one plane through P that is perpendicular to (A, B, C).

FIGURE 11.66   A plane is characterized by a point in it and a vector perpendicular to it

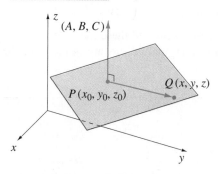

To find the equation of this plane we note that if $Q(x, y, z)$ is any other point in the plane, then vector $\mathbf{PQ} = (x - x_0, y - y_0, z - z_0)$ lies in the plane. But $\mathbf{PQ}$ must then be perpendicular to (A, B, C); hence, by the corollary to Theorem 11.2,

$$(A, B, C) \cdot (x - x_0, y - y_0, z - z_0) = 0.$$

Because this equation must be satisfied by every point (x, y, z) in the plane (and at the same time is not satisfied by any point not in the plane), it must be the equation of the plane through P and perpendicular to (A, B, C). If we expand the scalar product, we obtain the equation of the plane in the form

$$A(x - x_0) + B(y - y_0) + C(z - z_0) = 0.$$

This result is worth stating as a theorem.

THEOREM 11.4

The **equation for the plane** through the point (x_0, y_0, z_0) perpendicular to the vector (A, B, C) is

$$A(x - x_0) + B(y - y_0) + C(z - z_0) = 0. \tag{11.34}$$

Equation 11.34 can also be written in the form

$$Ax + By + Cz + D = 0, \tag{11.35}$$

where $D = -(Ax_0 + By_0 + Cz_0)$, and this equation is said to be *linear* in x, y, and z. We have shown then that every plane has a linear equation, and the coefficients A, B, and C of x, y, and z in the equation are the components of a vector (A, B, C) that is perpendicular to the plane. Instead of saying that (A, B, C) is perpendicular to the plane $Ax + By + Cz + D = 0$, we often say that (A, B, C) is **normal** to the plane or that (A, B, C) is a **normal vector to the plane**.

EXAMPLE 11.15

Find an equation for the plane through the point $(4, -3, 5)$ and normal to the vector $(4, -8, 3)$.

SOLUTION   According to 11.34, the equation of the plane is

$$4(x - 4) - 8(y + 3) + 3(z - 5) = 0 \quad \text{or} \quad 4x - 8y + 3z = 55.$$

EXAMPLE 11.16

Determine whether the planes $x + 2y - 4z = 10$ and $2x + 4y - 8z = 11$ are parallel.

SOLUTION   Normal vectors to these planes are $\mathbf{N}_1 = (1, 2, -4)$ and $\mathbf{N}_2 = (2, 4, -8)$. Since $\mathbf{N}_2 = 2\mathbf{N}_1$, the normal vectors are in the same direction, and therefore the planes are parallel.

Lines

In Section 11.2 we indicated that space curves can be described by two simultaneous equations in x, y, and z, and that such a representation describes the curve as the intersection of two surfaces. A straight line results when the surfaces are planes. We shall discuss this further, but we prefer to begin our discussion of straight lines with vectors much as we did for planes. A straight line in space is characterized by a point on it and a vector parallel to it. There is one and only one line ℓ through the point $P(x_0, y_0, z_0)$ and in the direction $\mathbf{v} = (v_x, v_y, v_z)$ in Figure 11.67.

If $Q(x, y, z)$ is any point on this line, then the vector $\mathbf{r}$ joining the origin to Q has components $\mathbf{r} = (x, y, z)$. Now this vector can be expressed as the sum of $\mathbf{r}_0$, the vector from 0 to P, and $\mathbf{PQ}$:

$$\mathbf{r} = \mathbf{r}_0 + \mathbf{PQ}.$$

FIGURE 11.67   A line is characterized by a point on it and a vector along it

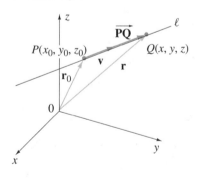

But $\mathbf{PQ}$ is in the same direction as $\mathbf{v}$; therefore, it must be some scalar multiple of $\mathbf{v}$ (i.e., $\mathbf{PQ} = t\mathbf{v}$). Consequently, the vector joining O to any point on ℓ can be written in the form

$$\mathbf{r} = \mathbf{r}_0 + t\mathbf{v}, \tag{11.36}$$

for an appropriate value of t. Because the components of $\mathbf{r} = (x, y, z)$ are also the coordinates of the point (x, y, z) on ℓ, this equation is called the **vector equation of the line** ℓ through (x_0, y_0, z_0) in direction $\mathbf{v}$. If we substitute components for $\mathbf{r}$, $\mathbf{r}_0$, and $\mathbf{v}$, then

$$(x, y, z) = (x_0, y_0, z_0) + t(v_x, v_y, v_z) = (x_0 + tv_x, y_0 + tv_y, z_0 + tv_z).$$

Since two vectors are equal if and only if corresponding components are identical, we can write that

$$x = x_0 + v_x t, \tag{11.37a}$$

$$y = y_0 + v_y t, \tag{11.37b}$$

$$z = z_0 + v_z t. \tag{11.37c}$$

These three scalar equations are equivalent to vector equation 11.36; they are called **parametric equations for line** ℓ. They illustrate once again that a line in space is characterized by a point (x_0, y_0, z_0) on it and a vector (v_x, v_y, v_z) along it. Each value of t substituted into 11.37 yields a point (x, y, z) on ℓ, and conversely, every point on ℓ is represented by some value of t. For instance, $t = 0$ yields P, and $t = 1$ gives the point at the tip of $\mathbf{v}$ in Figure 11.67.

If none of v_x, v_y, and v_z is equal to zero, we can solve equations 11.37 for t and equate the three expressions to obtain

$$\frac{x - x_0}{v_x} = \frac{y - y_0}{v_y} = \frac{z - z_0}{v_z}. \tag{11.38}$$

These are called **symmetric equations for the line** ℓ through (x_0, y_0, z_0) parallel to $\mathbf{v} = (v_x, v_y, v_z)$. There are only two independent equations in 11.38, which therefore substantiates our previous result that a curve (in this case, a line) can be described by two equations in x, y, and z. We could, for instance, write

$$v_x(y - y_0) = v_y(x - x_0) \quad \text{and} \quad v_z(y - y_0) = v_y(z - z_0)$$

or

$$v_y x - v_x y = v_y x_0 - v_x y_0, \quad v_y z - v_z y = v_y z_0 - v_z y_0.$$

Since the first of these is linear in x and y and the second is linear in y and z, each describes a plane. The line has been described as the curve of intersection of two planes.

EXAMPLE 11.17

Find, if possible, vector, parametric, and symmetric equations for the line through the points $(-1, 2, 1)$ and $(3, -2, 1)$.

SOLUTION   A vector along the line is $(3, -2, 1) - (-1, 2, 1) = (4, -4, 0)$, and so too is $(1, -1, 0)$. A vector equation for the line is

$$\mathbf{r} = (-1, 2, 1) + t(1, -1, 0) = (t - 1, -t + 2, 1).$$

Parametric equations are therefore

$$x = t - 1, \quad y = -t + 2, \quad z = 1.$$

Because the z-component of every vector along the line is zero, we cannot write full symmetric equations for the line. By eliminating t between the x- and y-equations, however, we can write

$$x + 1 = \frac{y - 2}{-1}, \quad z = 1.$$

If we set $x + 1 = 2 - y$, $z = 1$, or $x + y = 1$, $z = 1$, we represent the line as the intersection of the planes $x + y = 1$ and $z = 1$ (Figure 11.68).

FIGURE 11.68   Line represented as intersection of two planes

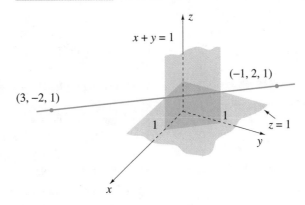

EXAMPLE 11.18

Find the equation of the plane containing the origin and the line $2x + y - z = 4$, $x + z = 5$.

SOLUTION We can easily find two more points on the plane. For instance, if we set $x = 0$, then the equations of the line require $z = 5$ and $y = 9$; and if we set $z = 0$, then $x = 5$ and $y = -6$. Thus $P(0, 9, 5)$ and $Q(5, -6, 0)$, as well as $O(0, 0, 0)$, are points on the plane. It follows then that $\mathbf{OP} = (0, 9, 5)$ and $\mathbf{OQ} = (5, -6, 0)$ are vectors in the plane, and a vector normal to the plane is

$$\mathbf{OP} \times \mathbf{OQ} = \begin{vmatrix} \hat{\mathbf{i}} & \hat{\mathbf{j}} & \hat{\mathbf{k}} \\ 0 & 9 & 5 \\ 5 & -6 & 0 \end{vmatrix} = (30, 25, -45).$$

The vector $(6, 5, -9)$ is also normal to the plane, and the equation of the plane is

$$0 = (6, 5, -9) \cdot (x - 0, y - 0, z - 0) = 6x + 5y - 9z.$$

EXAMPLE 11.19

Find symmetric equations for the line $x + y - 2z = 6$, $2x - 3y + 4z = 10$.

SOLUTION To find symmetric equations, we require a vector parallel to the line and a point on it. By setting $x = 0$ and solving $y - 2z = 6$, $-3y + 4z = 10$, we obtain $y = -22$, $z = -14$. Consequently, $(0, -22, -14)$ is a point on the line. To find a vector along the line, we could find another point on the line, say, by setting $z = 0$ and solving $x + y = 6$, $2x - 3y = 10$ for $x = 28/5$, $y = 2/5$. A vector along the line is therefore $(28/5, 2/5, 0) - (0, -22, -14) = (28/5, 112/5, 14)$, and so is $(5/14)(28/5, 112/5, 14) = (2, 8, 5)$.

Alternatively, we know that $(1, 1, -2)$ and $(2, -3, 4)$ are vectors that are normal to the planes $x + y - 2z = 6$ and $2x - 3y + 4z = 10$, and a vector along the line of intersection of the planes must be perpendicular to both of these vectors (Figure 11.69). Consequently, a vector along the line of intersection is

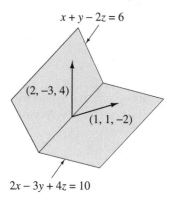

FIGURE 11.69 A vector along the line of intersection of two planes can be obtained by crossing normal vectors to the plane

$x + y - 2z = 6$

$(2, -3, 4)$

$(1, 1, -2)$

$2x - 3y + 4z = 10$

$$\begin{vmatrix} \hat{\mathbf{i}} & \hat{\mathbf{j}} & \hat{\mathbf{k}} \\ 1 & 1 & -2 \\ 2 & -3 & 4 \end{vmatrix} = (-2, -8, -5).$$

Symmetric equations for the line are therefore

$$\frac{x}{2} = \frac{y + 22}{8} = \frac{z + 14}{5}.$$

EXERCISES 11.5

In Exercises 1–10 find the equation for the plane.

1. Through the point $(1, -1, 3)$ and normal to the vector $(4, 3, -2)$

2. Through the point $(2, 1, 5)$ and normal to the vector joining $(2, 1, 5)$ and $(4, 2, 3)$

3. Containing the points $(1, 3, 2)$, $(-2, 0, -2)$, $(1, 4, 3)$

4. Containing the point $(2, -4, 3)$ and the line $(x - 1)/3 = (y + 5)/4 = z + 2$

5. Containing the lines $x = 2y = (z + 1)/4$ and $x = t$, $y = 2t$, $z = 6t - 1$

6. Containing the lines $(x - 1)/6 = y/8 = (z + 2)/2$ and $(x + 1)/3 = (y - 2)/4 = z + 5$

7. Containing the line $x - y + 2z = 4$, $2x + y + 3z = 6$ and the point $(1, -2, 4)$

8. Containing the lines $x + 2y + 4z = 21$, $x - y + 6z = 13$ and $x = 2 + 3t$, $y = 4$, $z = -3 + 5t$

9. Containing the lines $3x + 4y = -6$, $x + 2y + z = 2$ and $2y + 3z = 19$, $3x - 2y - 9z = -58$

10. Containing the line $x + y - 4z = 6$, $2x + 3y + 5z = 10$ and (a) perpendicular to the xy-plane, (b) perpendicular to the xz-plane, and (c) perpendicular to the yz-plane.

1. The acute angle between two intersecting planes is defined as the acute angle between their normals. Find the acute angle between the planes $x - 2y + 4z = 6$ and $2x + y = z + 4$.

In Exercises 12–21 find vector, parametric, and symmetric (if possible) equations for the straight line.

2. Through the point $(1, -1, 3)$ and parallel to the vector $(2, 4, -3)$
3. Through the point $(-1, 3, 6)$ and parallel to the vector $(2, -3, 0)$
4. Through the points $(2, -3, 4)$ and $(5, 2, -1)$
5. Through the points $(-2, 3, 3)$ and $(-2, -3, -3)$
6. Through the points $(1, 3, 4)$ and $(1, 3, 5)$
7. Through the point $(1, -3, 5)$ and parallel to the line

$$\frac{x}{5} = \frac{y-2}{3} = \frac{z+4}{-2}$$

8. Through the point $(2, 0, 3)$ and parallel to the line $x = 4 + t$, $y = 2$, $z = 6 - 2t$

9. Through the point of intersection of the lines

$$\frac{x-1}{2} = \frac{y+4}{-3} = \frac{z-2}{5} \quad \text{and} \quad \frac{x-1}{6} = \frac{y+4}{3} = \frac{z-2}{4},$$

and parallel to the line joining the points $(1, 3, -2)$ and $(2, -2, 1)$

10. $2x - y = 5$, $3x + 4y + z = 10$

11. Through the point $(-2, 3, 1)$ and parallel to the line $x + y = 5$, $2x - y + z = -2$

* 22. Does the line $(x - 3)/2 = y - 2 = (z + 1)/4$ lie in the plane $x - y + 2z = -1$?

* 23. Show that a vector perpendicular to the line $Ax + By + C = 0$ in the xy-plane is (A, B).

* 24. Show that if a plane has nonzero intercepts a, b, and c on the x-, y-, and z-axes, then its equation is $x/a + y/b + z/c = 1$.

* 25. Find equations for the four faces of the tetrahedron in Exercise 17 of Section 11.1.

* 26. Find equations for the nine planes forming the sides, bottom, and roof of the birdhouse in Exercise 14 of Section 11.1.

* 27. Verify that the equation of the plane passing through the three points $P_1(x_1, y_1, z_1)$, $P_2(x_2, y_2, z_2)$, and $P_3(x_3, y_3, z_3)$ can be written in the form $\mathbf{P_1P} \cdot \mathbf{P_1P_2} \times \mathbf{P_1P_3} = 0$, where $P(x, y, z)$ is any point in the plane.

* 28. The region S_{xy} in the xy-plane bounded by the straight lines $x = 0$, $x = 1$, $y = 0$, and $y = 1$ is a rectangle with unit area.

(a) Show that the region in the plane $y = z$ that projects onto S_{xy} is also a rectangle, but with area $\sqrt{2}$.

(b) Show that the region in the plane $x + y - 2z = 0$ that projects onto S_{xy} is a parallelogram with area $\sqrt{6}/2$.

(c) Generalize the results of parts (a) and (b) to show that if S is the area in a plane $Ax + By + Cz + D = 0$ $(C \neq 0)$ that projects onto S_{xy}, then the area of S is $\sec \gamma$, where γ is the acute angle between $\hat{\mathbf{k}}$ and the normal to the given plane.

11.6 Geometric Applications of Scalar and Vector Products

Components of Vectors in Arbitrary Directions

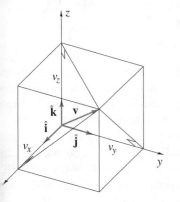

FIGURE 11.70 Components of a vector can be found by drawing perpendiculars from the tip of the vector to the coordinate axes

We have defined what is meant by Cartesian components (v_x, v_y, v_z) of a vector $\mathbf{v}$. They are scalars that multiply $\hat{\mathbf{i}}$, $\hat{\mathbf{j}}$, and $\hat{\mathbf{k}}$ so that $\mathbf{v} = v_x\hat{\mathbf{i}} + v_y\hat{\mathbf{j}} + v_z\hat{\mathbf{k}}$. They can be represented in terms of scalar products as

$$v_x = \mathbf{v} \cdot \hat{\mathbf{i}}, \quad v_y = \mathbf{v} \cdot \hat{\mathbf{j}}, \quad v_z = \mathbf{v} \cdot \hat{\mathbf{k}}. \tag{11.39}$$

Geometrically, they can be found by drawing $\mathbf{v}$, $\hat{\mathbf{i}}$, $\hat{\mathbf{j}}$, and $\hat{\mathbf{k}}$ all at the origin and dropping perpendiculars from the tip of $\mathbf{v}$ to the x-, y-, and z-axes (Figure 11.70). We now generalize our definition of a component along an axis to define the component of a vector in any direction whatsoever.

DEFINITION 11.10

To define the component of a vector $\mathbf{v}$ in a direction $\mathbf{u}$, we place $\mathbf{v}$ and $\mathbf{u}$ tail to tail at a point P and draw the perpendicular from the tip of $\mathbf{v}$ to the line containing $\mathbf{u}$ (Figure 11.71a). The directed distance PR is called the **component** of $\mathbf{v}$ in the direction $\mathbf{u}$. If R is on the same side of P as the tip of $\mathbf{u}$, PR is taken as positive; and if R is on that side of P opposite to the tip of $\mathbf{u}$, PR is negative.

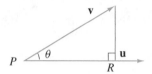

FIGURE 11.71a   Component
of vector **v** in direction **u**

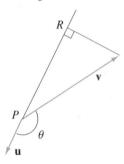

FIGURE 11.71b   Situation
where component of **v** in direction of
u is negative

In Figure 11.71a the component of **v** in direction **u** is positive, and in Figure 11.71b the component is negative. Note that the length of **u** is irrelevant; it is only the direction of **u** that determines the component of **v** in the direction **u**. If θ is the angle between **u** and **v**, then

$$||PR|| = |\mathbf{v}|\cos\theta.$$

The right side of this equation looks very much like the scalar product of **v** and a vector that makes an angle θ with **v**. It lacks only the length of this second vector. Clearly, **u** is a vector that makes an angle θ with **v**, but we cannot write $||PR|| = |\mathbf{v}||\mathbf{u}|\cos\theta$, since the length of **u** need not be 1. If, however, $\hat{\mathbf{u}}$ is the unit vector in the same direction as **u**, then we can write

$$||PR|| = |\mathbf{v}||\hat{\mathbf{u}}|\cos\theta. \tag{11.40}$$

In other words, we have the following theorem.

THEOREM 11.5

The component of a vector **v** in a direction **u** is $\mathbf{v} \cdot \hat{\mathbf{u}}$, where $\hat{\mathbf{u}}$ is the unit vector in direction **u**.

This result agrees with equations 11.39 for the x-, y-, and z-components of **v**.

EXAMPLE 11.20

Find components of $\mathbf{v} = (1, 2, -3)$ in directions specified by the following vectors:

(a) $\mathbf{u} = (-1, 3, 4)$ (b) $\mathbf{u} = (1, 3, -2)$ (c) $\mathbf{u} = (4, 4, 4)$

SOLUTION   Components of **v** in these directions are

$$(a)\ \mathbf{v} \cdot \hat{\mathbf{u}} = (1, 2, -3) \cdot \frac{(-1, 3, 4)}{\sqrt{26}} = -\frac{7}{\sqrt{26}}$$

$$(b)\ \mathbf{v} \cdot \hat{\mathbf{u}} = (1, 2, -3) \cdot \frac{(1, 3, -2)}{\sqrt{14}} = \frac{13}{\sqrt{14}}$$

$$(c)\ \mathbf{v} \cdot \hat{\mathbf{u}} = (1, 2, -3) \cdot \frac{(4, 4, 4)}{4\sqrt{3}} = 0$$

This last result means that **v** is perpendicular to $(4, 4, 4)$.

Distances Between Points, Lines, and Planes

There are three distances between geometric objects in a plane — between points, from a point to a line, and between parallel lines. They were all covered in Section 1.3. Distance formulas 1.10 and 1.16 handle the first two situations. Distance between parallel lines can be calculated by finding a point on one line and using 1.16 for the distance from this point to the other line.

In space, there are six distances of interest — point to point, point to line, point to plane, line to line, line to plane, and plane to plane. Formula 11.1 gives distance between points. We can develop a formula for the distance from a point (x_1, y_1, z_1) to a plane $Ax + By + Cz + D = 0$ very similar to formula 1.16 for distance from a point to a line in the xy-plane. Let R be the foot of the perpendicular from a point $P(x_1, y_1, z_1)$ to the plane in Figure 11.72, and $Q(x, y, z)$ be any other point in the plane. The length of vector **PR** is the distance from P to the plane, and this length is the component of **PQ** in direction **PR**. Hence, $|\mathbf{PR}| = \mathbf{PQ} \cdot \widehat{\mathbf{PR}}$. A normal vector to the plane is (A, B, C), and therefore

$$\widehat{\mathbf{PR}} = \pm\frac{(A, B, C)}{\sqrt{A^2 + B^2 + C^2}}.$$

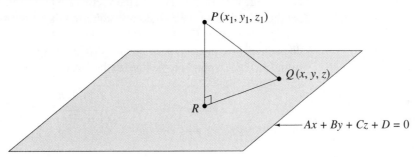

FIGURE 11.72   Distance form a point to a plane

Thus,

$$|\mathbf{PR}| = \mathbf{PQ} \cdot \left[\frac{\pm(A, B, C)}{\sqrt{A^2 + B^2 + C^2}}\right].$$

We choose the $+$ or $-$ to guarantee a positive result. This can also be accomplished by inserting absolute values,

$$|\mathbf{PR}| = \left|\mathbf{PQ} \cdot \frac{(A, B, C)}{\sqrt{A^2 + B^2 + C^2}}\right|.$$

Since $\mathbf{PQ} = (x - x_1, y - y_1, z - z_1)$,

$$|\mathbf{PR}| = \left|(x - x_1, y - y_1, z - z_1) \cdot \frac{(A, B, C)}{\sqrt{A^2 + B^2 + C^2}}\right|$$

$$= \frac{|Ax + By + Cz - (Ax_1 + By_1 + Cz_1)|}{\sqrt{A^2 + B^2 + C^2}}.$$

Because Q is in the plane, we can replace $Ax + By + Cz$ with $-D$, and the formula then becomes

$$\frac{|Ax_1 + By_1 + Cz_1 + D|}{\sqrt{A^2 + B^2 + C^2}}. \tag{11.41}$$

Notice, as we suggested, the similarity of this formula to 1.16.

EXAMPLE 11.21

Find the distance from the point $(1, 2, 5)$ to the plane $x + y + 2z = 4$.

SOLUTION   According to 11.41, the distance is

$$\frac{|1 + 2 + 2(5) - 4|}{\sqrt{1 + 1 + 4}} = \frac{9}{\sqrt{6}}.$$

With distance formula 11.41, it is straightforward to find the distance between two parallel planes, or between a plane and a line parallel to the plane.

EXAMPLE 11.22

Show that the line $x = 3 + t$, $y = 1 - 2t$, $z = 4 + 3t$ and the plane $x + 2y + z = 6$ are parallel, and find the distance between them.

SOLUTION   A vector along the line is $\mathbf{v} = (1, -2, 3)$ and a vector normal to the plane i
$\mathbf{N} = (1, 2, 1)$. Since $\mathbf{v} \cdot \mathbf{N} = 0$, the vectors are perpendicular, and this confirms that the line i
parallel to the plane. [We could also confirm this by substituting from the equations of the line
into the plane, $6 = (3 + t) + 2(1 - 2t) + (4 + 3t) = 9$, a contradiction. The line and plane
do not therefore intersect.] To find the distance between the line and plane, we find the distance
from $(3, 1, 4)$, a point on the line, to the plane:

$$\frac{|3 + 2(1) + 4 - 6|}{\sqrt{1 + 4 + 1}} = \frac{3}{\sqrt{6}}.$$

For the distance between parallel planes, choose a point on one plane and find the distance to
the other plane. There is also a formula (see Exercise 31). This leaves two distances — poin
to line and line to line. We illustrate each by example.

EXAMPLE 11.23

Find the distance from the point $(1, 3, 6)$ to the line

$$\ell : \quad x - 1 = \frac{y - 2}{2} = \frac{z - 4}{3}.$$

FIGURE 11.73   Distance
from a point to a line

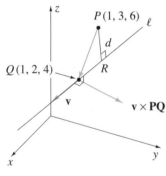

SOLUTION   Clearly, $Q(1, 2, 4)$ is a point on the line, and therefore the required distance d
is the component of $\mathbf{PQ}$ in the direction $\mathbf{PR}$ (Figure 11.73). By equation 11.40, then, $d =$
$|\mathbf{PQ} \cdot \widehat{\mathbf{PR}}|$. To find $\widehat{\mathbf{PR}}$ we need a vector in the direction $\mathbf{PR}$. Since $\mathbf{v} = (1, 2, 3)$ is a vecto
along ℓ, the vector

$$\mathbf{v} \times \mathbf{PQ} = \begin{vmatrix} \hat{\mathbf{i}} & \hat{\mathbf{j}} & \hat{\mathbf{k}} \\ 1 & 2 & 3 \\ 0 & -1 & -2 \end{vmatrix} = -\hat{\mathbf{i}} + 2\hat{\mathbf{j}} - \hat{\mathbf{k}}$$

is perpendicular to both $\mathbf{PQ}$ and $\mathbf{v}$. It now follows that a vector along $\mathbf{PR}$ is

$$(\mathbf{v} \times \mathbf{PQ}) \times \mathbf{v} = \begin{vmatrix} \hat{\mathbf{i}} & \hat{\mathbf{j}} & \hat{\mathbf{k}} \\ -1 & 2 & -1 \\ 1 & 2 & 3 \end{vmatrix} = 8\hat{\mathbf{i}} + 2\hat{\mathbf{j}} - 4\hat{\mathbf{k}}.$$

Finally, then,

$$\widehat{\mathbf{PR}} = \frac{(8, 2, -4)}{\sqrt{64 + 4 + 16}} = \frac{(4, 1, -2)}{\sqrt{21}}$$

and

$$d = \left| (0, -1, -2) \cdot \frac{(4, 1, -2)}{\sqrt{21}} \right| = \frac{\sqrt{21}}{7}.$$

Cross products can also be used to find the distance from a point to a line (see Exercises 29 and
30). The above method parallels that for other distances in this section; it uses scalar products

EXAMPLE 11.24

Find the (shortest) distance between the lines

$$\ell_1 : \quad \frac{x - 1}{2} = \frac{y + 3}{3} = z - 4 \quad \text{and} \quad \ell_2 : \quad x = -1 + t, \ y = 2t, \ z = 3 - 2t.$$

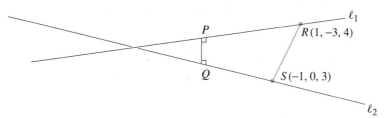

FIGURE 11.74 Shortest distance between nonintersecting lines

SOLUTION   There will be a point P on ℓ_1 and a point Q on ℓ_2 such that line segment PQ is perpendicular to ℓ_1 and ℓ_2 (Figure 11.74). The length of this line segment is the shortest distance between the lines. There is no problem finding a vector in the same direction as **PQ**: cross vectors $(2, 3, 1)$ and $(1, 2, -2)$ along ℓ_1 and ℓ_2, respectively,

$$\begin{vmatrix} \hat{\mathbf{i}} & \hat{\mathbf{j}} & \hat{\mathbf{k}} \\ 2 & 3 & 1 \\ 1 & 2 & -2 \end{vmatrix} = (-8, 5, 1).$$

Now comes the hard part to visualize. If we take any point on ℓ_1, say $R(1, -3, 4)$, and any point on ℓ_2, say $S(-1, 0, 3)$, then the component of vector **RS** along **PQ** is the length of **PQ**. (If you are not convinced, consider the following argument. It is easier to see that the component of **PS** along **PQ** is the length of **PQ** so that $|\mathbf{PQ}| = |\mathbf{PS} \cdot \widehat{\mathbf{PQ}}|$. But $\mathbf{PS} = \mathbf{PR} + \mathbf{RS}$, and therefore,

$$|\mathbf{PQ}| = |(\mathbf{PR} + \mathbf{RS}) \cdot \widehat{\mathbf{PQ}}| = |\mathbf{PR} \cdot \widehat{\mathbf{PQ}} + \mathbf{RS} \cdot \widehat{\mathbf{PQ}}|.$$

Since **PR** is perpendicular to $\widehat{\mathbf{PQ}}$, it follows that $\mathbf{PR} \cdot \widehat{\mathbf{PQ}} = 0$, and

$$|\mathbf{PQ}| = |\mathbf{RS} \cdot \widehat{\mathbf{PQ}}|.$$

The right side is the component of **RS** along $\widehat{\mathbf{PQ}}$.) Consequently,

$$|\mathbf{PQ}| = \left| (-2, 3, -1) \cdot \frac{(-8, 5, 1)}{\sqrt{90}} \right| = \sqrt{10}.$$

This procedure fails when the lines are parallel, but in this case, we can pick a point on one line and find the distance to the other line as in Example 11.23.

Areas of Triangles and Parallelograms

Vector products provide a simple way to find areas of triangles and parallelograms.

THEOREM 11.6

If A, B, and C are vertices of a triangle, then

$$\text{area of } \triangle ABC = \frac{1}{2}|\mathbf{AB} \times \mathbf{AC}|. \tag{11.42}$$

PROOF   Area of triangle ABC in Figure 11.75 is

$$\frac{1}{2}|\mathbf{AC}||\mathbf{BD}| = \frac{1}{2}|\mathbf{AC}||\mathbf{AB}| \sin\theta = \frac{1}{2}|\mathbf{AB} \times \mathbf{AC}|.$$

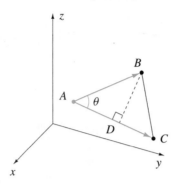

FIGURE 11.75 Area of a triangle in terms of vectors representing two sides

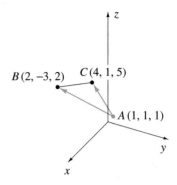

FIGURE 11.76 Area of a triangle with given vertices

For the triangle with vertices $A(1, 1, 1)$, $B(2, -3, 2)$, and $C(4, 1, 5)$ in Figure 11.76,

$$\mathbf{AB} \times \mathbf{AC} = \begin{vmatrix} \hat{\mathbf{i}} & \hat{\mathbf{j}} & \hat{\mathbf{k}} \\ 1 & -4 & 1 \\ 3 & 0 & 4 \end{vmatrix} = (-16, -1, 12).$$

The area of the triangle is, therefore,

$$\frac{1}{2}|(-16, -1, 12)| = \frac{1}{2}\sqrt{256 + 1 + 144} = \frac{1}{2}\sqrt{401}.$$

The following corollary is a direct consequence of Theorem 11.6.

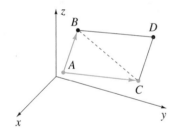

FIGURE 11.77 Area of a parallelogram

COROLLARY 11.6.1

The area of a parallelogram with coterminal sides $\mathbf{AB}$ and $\mathbf{AC}$ (Figure 11.77) is

$$\text{area} = |\mathbf{AB} \times \mathbf{AC}|. \qquad (11.43)$$

EXERCISES 11.6

In Exercises 1–4 find the area of the triangle with given vertices.

1. $(1, 0)$, $(4, 2)$, $(2, 6)$
2. $(-1, 0, 3)$, $(5, 1, 2)$, $(-6, 2, 4)$
3. $(1, 1, 1)$, $(-3, 4, -2)$, $(-1, -2, 3)$
4. $(1, 2, 3)$, $(3, 5, 10)$, $(-3, -4, -11)$

In Exercises 5–6 find the area of the parallelogram with given vertices.

5. $(1, 2, 3)$, $(4, 3, 7)$, $(-1, 3, 6)$, $(2, 4, 10)$
6. $(1, -2, 4)$, $(3, 5, 7)$, $(4, 6, 8)$, $(2, -1, 5)$

In Exercises 7–10 find the component of $\mathbf{v} = (1, -4, 3)$ in the direction specified.

7. $(1, 2, -3)$
8. In the direction from $(-1, 2, 3)$ to $(4, -3, 2)$
9. Perpendicular to the plane $x + y + 2z = 4$ in the direction of increasing z

10. Along the line $x = 1 - t$, $y = 3 + 2t$, $z = 4 - 3t$ in the direction of increasing x

In Exercises 11–28 find the distance indicated.

11. From the point $(3, 2)$ to the line $2x + 3y = 18$
12. Between the lines $x + 4 = y$ and $3x + 7 = 3y$
13. From the point $(1, 3, 4)$ to the plane $x + y - 2z = 0$
14. From the point $(-2, 3, -5)$ to the plane $2x + y + 4z = 6$
15. Between the line $x = 1 - t$, $y = 2 + 3t$, $z = 4 - 2t$ and the plane $2x + 4y + 5z = 10$
16. Between the line $x + y + z = 6$, $2x - y + z = 3$ and the plane $2x - 7y - z = 5$
17. From the line $x - 1 = 3(y + 4) = -z - 1$ to the plane $2x - 3y + z = 4$
18. From the line $x = 1 - 6t$, $y = 2 + 4t$, $z = -t$ to the plane $x + y - 2z = 1$

9. Between the planes $2x + 3y - z = 15$ and $4x + 6y - 2z = 7$

0. Between the planes $x - y + 2z = 4$ and $3x - 3y + 6z = 10$

1. From the point $(1, 2, -3)$ to the line $x = 2(y + 1) = (z - 4)/2$

2. From the point $(3, -2, 0)$ to the line $x = t$, $y = 3 - 2t$, $z = 4 + t$

3. From the point $(1, -1, 2)$ to the line $x + y - z = 2$, $2x + 3y + z = 4$

4. From the point $(1, 3, 3)$ to the line $x = 2 - t$, $y = 1 + 2t$, $z = 3$

5. Between the lines $(x - 1)/2 = (y + 3)/3 = 4 - z$ and $x = -1 + t$, $= 2t$, $z = 3 - 2t$

6. Between the lines $x = t$, $y = 3t - 1$, $z = 1 + 2t$ and $x = 2t + 1$, $= 1 - t$, $z = 4 + 2t$

7. Between the lines $x + y - z = 4$, $2x - z = 4$ and $x = y + 1)/2 = (z - 1)/3$

8. Between the lines $x + y + z = 2$, $x - y + z = 3$ and $2x - y + z = $, $y - 4z = 5$

9. Show that the distance from point $P(1, 3, 6)$ to the line $x - 1 = y - 2)/2 = (z - 4)/3$ in Figure 11.73 can also be found as follows:

$$|\mathbf{PR}| = |\mathbf{PQ}| \sin\theta = |\mathbf{PQ}||\hat{\mathbf{v}}| \sin\theta = |\mathbf{PQ} \times \hat{\mathbf{v}}|,$$

where θ is the angle between $\mathbf{PQ}$ and line ℓ.

0. Use the technique of Exercise 29 to find the distance in Exercise 4.

1. Prove that the (undirected) distance between two parallel planes $Ax + By + Cz + D_1 = 0$ and $Ax + By + Cz + D_2 = 0$ is

$$\frac{|D_1 - D_2|}{\sqrt{A^2 + B^2 + C^2}}.$$

2. Show that the lines joining the midpoints of the sides of any quadrilateral form a parallelogram.

3. Show that the three lines $(x - 4)/3 = (y - 8)/4 = (z + 7)/(-4)$; $x + 5)/3 = (y + 2)/2 = (z + 1)/2$; and $x = 1$, $y = 5 + t$,

$z = -6 - 3t$ form a triangle and find its area.

In Exercises 34–35 verify that $\hat{\mathbf{v}}$ and $\hat{\mathbf{w}}$ are perpendicular, and then find scalars λ and ρ so that $\mathbf{u} = \lambda\hat{\mathbf{v}} + \rho\hat{\mathbf{w}}$.

* **34.** $\mathbf{u} = (2, 1)$; $\hat{\mathbf{v}} = (1/\sqrt{2}, 1/\sqrt{2})$, $\hat{\mathbf{w}} = (1/\sqrt{2}, -1/\sqrt{2})$

* **35.** $\mathbf{u} = 3\hat{\mathbf{i}} - 2\hat{\mathbf{j}}$; $\hat{\mathbf{v}} = (\hat{\mathbf{i}} - 2\hat{\mathbf{j}})/\sqrt{5}$, $\hat{\mathbf{w}} = (2\hat{\mathbf{i}} + \hat{\mathbf{j}})/\sqrt{5}$

In Exercises 36–37 verify that $\hat{\mathbf{u}}$, $\hat{\mathbf{v}}$, and $\hat{\mathbf{w}}$ are mutually perpendicular, and then find scalars λ, ρ, and μ so that $\mathbf{r} = \lambda\hat{\mathbf{u}} + \rho\hat{\mathbf{v}} + \mu\hat{\mathbf{w}}$.

* **36.** $\mathbf{r} = (1, 3, -4)$; $\hat{\mathbf{u}} = (2, 1, 0)/\sqrt{5}$, $\hat{\mathbf{v}} = (-1, 2, 3)/\sqrt{14}$, $\hat{\mathbf{w}} = (3, -6, 5)/\sqrt{70}$

* **37.** $\mathbf{r} = 2\hat{\mathbf{i}} - \hat{\mathbf{k}}$; $\hat{\mathbf{u}} = (\hat{\mathbf{i}} + \hat{\mathbf{j}} + \hat{\mathbf{k}})/\sqrt{3}$, $\hat{\mathbf{v}} = (\hat{\mathbf{i}} + \hat{\mathbf{j}} - 2\hat{\mathbf{k}})/\sqrt{6}$, $\hat{\mathbf{w}} = (\hat{\mathbf{i}} - \hat{\mathbf{j}})/\sqrt{2}$

* **38.** If $\mathbf{u} = (3, 2)$, $\mathbf{v} = (1, -3)$, $\mathbf{w} = (6, 2)$, verify that $\mathbf{v}$ and $\mathbf{w}$ are perpendicular, and find scalars λ and ρ so that $\mathbf{u} = \lambda\mathbf{v} + \rho\mathbf{w}$. Be sure to recognize that $\mathbf{v}$ and $\mathbf{w}$ are not unit vectors.

* **39.** If $\mathbf{u} = (1, 0, 1)$, $\mathbf{v} = (1, 1, -1)$, $\mathbf{w} = (-1, 2, 1)$, and $\mathbf{r} = (-2, -3, 4)$, verify that $\mathbf{u}$, $\mathbf{v}$, and $\mathbf{w}$ are mutually perpendicular, and find scalars λ, ρ, and μ so that $\mathbf{r} = \lambda\mathbf{u} + \rho\mathbf{v} + \mu\mathbf{w}$. As in Exercise 38, take into account that $\mathbf{u}$, $\mathbf{v}$, and $\mathbf{w}$ are not unit vectors.

* **40.** Find the equation of a plane normal to $\hat{\mathbf{i}} - 2\hat{\mathbf{j}} + 3\hat{\mathbf{k}}$ and 2 units from the point $(1, 2, 3)$.

* **41.** If a, b, and c (all positive constants) are the intercepts of a plane with the x-, y-, and z-axes, and p is the length of the perpendicular from the origin to the plane, show that

$$\frac{1}{p^2} = \frac{1}{a^2} + \frac{1}{b^2} + \frac{1}{c^2}.$$

11.7 Physical Applications of Scalar and Vector Products

Work

FIGURE 11.78   Work one by a force along a line

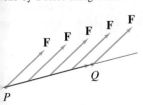

In Section 7.4, we described work as the product of force and distance. Now that we have represented forces by vectors, we can be more precise. In particular, if a particle moves along the line in Figure 11.78 from P to Q, then vector $\mathbf{PQ}$ represents its displacement. If a constant force $\mathbf{F}$ acts on the particle during this motion, then the **work** done by $\mathbf{F}$ is defined as

$$W = \mathbf{F} \cdot \mathbf{PQ}. \tag{11.44}$$

It is important to keep in mind exactly when this definition of work can be used: For constant forces acting along straight lines, and by constant $\mathbf{F}$, we mean that $\mathbf{F}$ is constant in both magnitude and direction. Note that when $\mathbf{F}$ and $\mathbf{PQ}$ are both in the same direction, then the angle between the vectors is zero. In this case,

$$W = |\mathbf{F}||\mathbf{PQ}| \cos(0) = |\mathbf{F}||\mathbf{PQ}|,$$

and this is essentially the equation dealt with in Section 7.4. When **F** and **PQ** are not in the same direction,

$$W = |\mathbf{F}||\mathbf{PQ}|\cos\theta.$$

Since $|\mathbf{F}|\cos\theta$ is the component of **F** along **PQ**, this equation simply states that when **F** and **PQ** are not in the same direction, $|\mathbf{PQ}|$ should be multiplied by the component of **F** in direction **PQ**. The component of **F** perpendicular to the displacement does no work.

EXAMPLE 11.25

FIGURE 11.79   Work
done by boy pulling wagon

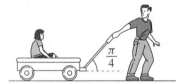

If the boy in Figure 11.79 pulls the wagon handle with a force of 10 N at an angle of $\pi/4$ radians with the horizontal, how much work does he do in walking 20 m in a straight line?

SOLUTION   The force **F** exerted by the boy has magnitude $|\mathbf{F}| = 10$, and points in a direction that makes an angle of $\pi/4$ radians with the displacement vector. If **d** is the displacement vector then

$$W = \mathbf{F}\cdot\mathbf{d} = |\mathbf{F}||\mathbf{d}|\cos\theta = 10(20)\cos\left(\frac{\pi}{4}\right) = \frac{200}{\sqrt{2}} = 100\sqrt{2}\text{ J}.$$

When motion is along a straight line, but **F** is not constant in direction or magnitude or both, we must use integration. The following example illustrates such a situation.

EXAMPLE 11.26

The spring in Figure 11.80 is fixed at A and moves the sleeve frictionlessly along the rod from B to C. If the spring is unstretched when the sleeve is at C, find the work done by the spring.

FIGURE 11.80   Work done by a spring pulling a sleeve along a line

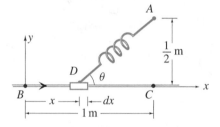

SOLUTION   If we set up a coordinate system as shown, then at position D the force **F** exerted by the spring on the sleeve has magnitude

$$|\mathbf{F}| = k\left[\sqrt{(1-x)^2 + \frac{1}{4}} - \frac{1}{2}\right],$$

where k is the spring constant. Since the spring is always stretched during motion, the direction of **F** is along the vector **DA**. Clearly, then, **F** changes in both magnitude and direction as the sleeve moves from B to C. For a small displacement dx at position D, the amount of work done by **F** is (approximately)

$$\mathbf{F}\cdot(dx\hat{\mathbf{i}}) = |\mathbf{F}|\,dx(\cos\theta) = k\left[\sqrt{(1-x)^2 + \frac{1}{4}} - \frac{1}{2}\right]dx\frac{1-x}{\sqrt{(1-x)^2 + 1/4}}$$

$$= k(1-x)\left[1 - \frac{1/2}{\sqrt{(1-x)^2 + 1/4}}\right]dx.$$

Total work done by the spring as the sleeve moves from B to C must therefore be

$$W = \int_0^1 k(1-x)\left[1 - \frac{1}{2\sqrt{(1-x)^2 + 1/4}}\right]dx$$

$$= k\int_0^1\left[1 - x - \frac{1-x}{2\sqrt{(1-x)^2 + 1/4}}\right]dx$$

$$= k\left\{x - \frac{x^2}{2} + \frac{1}{2}\sqrt{(1-x)^2 + 1/4}\right\}_0^1$$

$$= \frac{k}{4}(3 - \sqrt{5}) \text{ J.}$$

Moments

Moments play a fundamental role in structural engineering. We encountered them in Section 7.7 in the context of centres of mass, but lacking vectors we could not treat them fully. (There was a hint of vectors when we defined the first moment of an object as its mass multiplied by some *directed distance*, directed distances now being associated with vectors.) We discuss **first moments of forces** here, rather than first moments of masses. Forces were present in Section 7.7, the force of gravity on the masses, but by pushing vectors and forces into the background, we concentrated on the balance concept. Other applications of moments require a vectorial approach.

When a force is to produce a desired effect, where to apply the force may be just as important as the size of the force. For example, it is easier to insert screws using a screwdriver with a large handle than a small one; a long prybar can be more effective in removing spikes than a short one; it is easier to spin a roulette wheel by pulling on its edge rather than somewhere close to its centre. The direction to apply a force is also important. When spinning a roulette wheel, we don't pull away from the centre of the wheel (or toward the centre); we pull in a direction perpendicular to the line joining the centre of the wheel and the point of application of the force. In turning a revolving door, the best direction is perpendicular to the plane of the door. These ideas are captured in the following definition.

> **DEFINITION 11.11**
>
> The moment of a force $\mathbf{F}$, applied at a point Q, about a point P (Figure 11.81) is defined as
>
> $$\mathbf{M} = \mathbf{r} \times \mathbf{F}, \quad \text{where } \mathbf{r} = \mathbf{PQ}. \tag{11.45}$$

The magnitude of $\mathbf{M}$ is $|\mathbf{M}| = |\mathbf{r}||\mathbf{F}|\sin\theta$, making it clear that $\mathbf{M}$ depends on both the point of application (Q) and the direction of $\mathbf{F}$ (angle θ relative to $\mathbf{r}$). It is a maximum when $\mathbf{F}$ is perpendicular to $\mathbf{r}$. The direction of $\mathbf{M}$ is perpendicular to both $\mathbf{r}$ and $\mathbf{F}$. If $\mathbf{r}$ and $\mathbf{F}$ are in the plane of the page in Figure 11.81, then $\mathbf{M}$ points out of the page towards the reader.

FIGURE 11.81   Moment of a force about a point

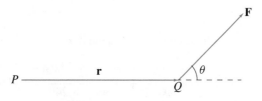

Units of moment are newtons multiplied by metres (N·m), the same as for work. For work, 1 N·m is called a *joule*. We do not associate joules with moment; moment is a vector, work is a scalar.

EXAMPLE 11.27

What is the moment of the force $\mathbf{F} = 3\hat{\mathbf{i}} - 2\hat{\mathbf{j}} + 8\hat{\mathbf{k}}$ N, applied at the point $(1, 2, -1)$ about the point $(3, 1, 4)$, all distances in metres?

SOLUTION   The moment is

$$\mathbf{M} = \begin{vmatrix} \hat{\mathbf{i}} & \hat{\mathbf{j}} & \hat{\mathbf{k}} \\ -2 & 1 & -5 \\ 3 & -2 & 8 \end{vmatrix} = -2\hat{\mathbf{i}} + \hat{\mathbf{j}} + \hat{\mathbf{k}} \text{ N·m.}$$

EXAMPLE 11.28

What is the moment of the force $\mathbf{F} = \hat{\mathbf{j}} - \hat{\mathbf{k}}$ N, applied at the point $(3, -2, 3)$ about the point $(3, 1, 0)$, all distances in metres?

SOLUTION   The moment is

$$\mathbf{M} = \begin{vmatrix} \hat{\mathbf{i}} & \hat{\mathbf{j}} & \hat{\mathbf{k}} \\ 0 & -3 & 3 \\ 0 & 1 & -1 \end{vmatrix} = \mathbf{0}.$$

It is zero because $\mathbf{F}$ is along the same line as $\mathbf{r}$. (This is like trying to close a door by pushing toward the hinges.)

EXAMPLE 11.29

The horizontal rectangular plate in Figure 11.82a is supported by brackets at A and B and by a wire CD, where D is directly above B. If the tension in the wire is 200 N, determine the moment about A of the force exerted by the wire on point C.

FIGURE 11.82a   Plate supported by hinges at A and B and wire CD

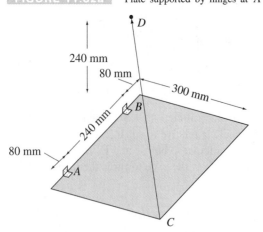

FIGURE 11.82b   Moment of tension in wire about A

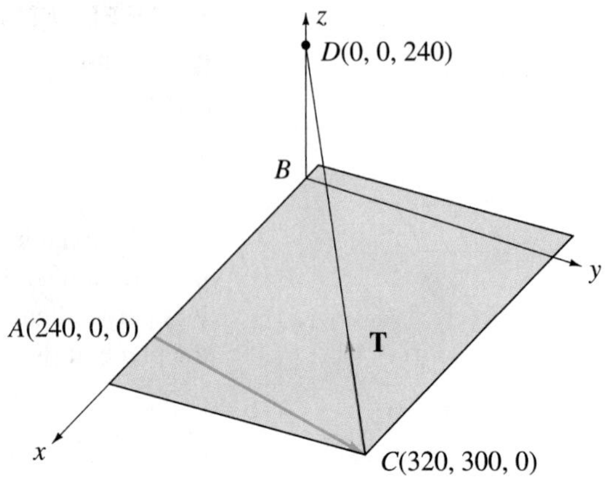

SOLUTION   With the coordinate system in Figure 11.82b, the tension in the wire acting on C is

$$\mathbf{T} = \frac{200(-320, -300, 240)}{\sqrt{320^2 + 300^2 + 240^2}} = 8(-16, -15, 12).$$

Since $\mathbf{AC} = (0.08, 0.3, 0)$, the required moment is

$$\mathbf{M} = \mathbf{AC} \times \mathbf{T} = \begin{vmatrix} \hat{\mathbf{i}} & \hat{\mathbf{j}} & \hat{\mathbf{k}} \\ 0.08 & 0.3 & 0 \\ -128 & -120 & 96 \end{vmatrix} = (28.8, -7.68, 28.8) \quad \text{N} \cdot \text{m}.$$

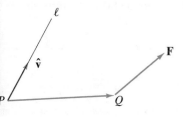

FIGURE 11.83 Moment of a force about a line

In discussing rigid bodies it is useful to define the moment of a force about a line. When a force $\mathbf{F}$ acts at a point Q, its moment about a point P (Figure 11.83) is $\mathbf{M} = \mathbf{r} \times \mathbf{F} = \mathbf{PQ} \times \mathbf{F}$. If ℓ is any line through P, we define the moment of $\mathbf{F}$ about ℓ as the component of $\mathbf{M}$ along ℓ. If $\hat{\mathbf{v}}$ is a unit vector along ℓ, then

$$\text{moment of } \mathbf{F} \text{ about } \ell = (\mathbf{PQ} \times \mathbf{F}) \cdot \hat{\mathbf{v}} = \mathbf{PQ} \times \mathbf{F} \cdot \hat{\mathbf{v}}. \tag{11.46}$$

The parentheses are unnecessary since the expression makes sense only if the vector product is performed first. This is a *scalar triple product*.

EXAMPLE 11.30

Find the moment of the force in Example 11.27 about the line through the points $P(3, 1, 4)$ and $Q(4, -1, 3)$.

SOLUTION According to the calculations in Example 11.27, the moment about P is $\mathbf{M} = -2\hat{\mathbf{i}} + \hat{\mathbf{j}} + \hat{\mathbf{k}}$. The moment about the line through P and Q is the component of $\mathbf{M}$ along PQ. Since a unit vector along PQ is $\mathbf{PQ}/|\mathbf{PQ}|$, the moment about the line is

$$(-2, 1, 1) \cdot \frac{(1, -2, -1)}{\sqrt{6}} = -\frac{5}{\sqrt{6}} \text{ N} \cdot \text{m}.$$

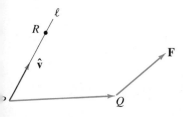

FIGURE 11.84 Moment of a force about a line is independent of point chosen on line

The moment of the force $\mathbf{F}$ about the line ℓ in Figure 11.83 is independent of the choice of point P on ℓ used in its determination. To show this, suppose that R is any other point on ℓ (Figure 11.84). Then

$$\text{moment of } \mathbf{F} \text{ about } \ell = \mathbf{PQ} \times \mathbf{F} \cdot \hat{\mathbf{v}} = (\mathbf{PR} + \mathbf{RQ}) \times \mathbf{F} \cdot \hat{\mathbf{v}} = \mathbf{PR} \times \mathbf{F} \cdot \hat{\mathbf{v}} + \mathbf{RQ} \times \mathbf{F} \cdot \hat{\mathbf{v}}.$$

Since $\mathbf{PR} \times \mathbf{F}$ is perpendicular to $\mathbf{PR}$, and $\hat{\mathbf{v}}$ is parallel to $\mathbf{PR}$, it follows that $\mathbf{PR} \times \mathbf{F}$ is perpendicular to $\hat{\mathbf{v}}$, and their scalar product is zero. Consequently,

$$\text{moment of } \mathbf{F} \text{ about } \ell = \mathbf{RQ} \times \mathbf{F} \cdot \hat{\mathbf{v}},$$

and this is formula 11.46 with P replaced by R.

The moment of a force about a line is not unique because the unit vector $\hat{\mathbf{v}}$ along the line is determined only as to sign; that is, if $\hat{\mathbf{v}}$ is a unit vector along the line, so is $-\hat{\mathbf{v}}$. In other words, moments of forces about lines are determined only as to sign. This could be remedied by specifying a direction along the line, but this is not normally done. It is agreed that when a number of moments are required about a line, the same direction will be taken along the line in all calculations.

EXAMPLE 11.31

What are the moments about the x-, y-, and z-axes due to the force in Example 11.29?

SOLUTION   Since A is on the x-axis in Figure 11.82, and the moment about A is $\mathbf{M} =$ $28.8\hat{\mathbf{i}} - 7.68\hat{\mathbf{j}} + 28.8\hat{\mathbf{k}}$ N·m, the moment about the x-axis is 28.8 N·m, the x-component of $\mathbf{M}$. To find moments about the y- and z-axes, we find the moment due to the tension $\mathbf{T}$ about B, the origin, since it lies on both axes. It is

$$\mathbf{M}_B = \mathbf{BC} \times \mathbf{T} = \begin{vmatrix} \hat{\mathbf{i}} & \hat{\mathbf{j}} & \hat{\mathbf{k}} \\ 0.32 & 0.3 & 0 \\ -128 & -120 & 96 \end{vmatrix} = (28.8, -30.72, 0) \quad \text{N} \cdot \text{m}.$$

Consequently, moments about the y- and z-axes are -30.72 N·m and 0 N·m, respectively. Also confirmed is the moment 28.8 N·m about the x-axis.

EXERCISES 11.7

In Exercises 1–5 calculate the moment of the force about the given point.

1. $\mathbf{F} = 2\hat{\mathbf{i}} + 3\hat{\mathbf{j}} - 4\hat{\mathbf{k}}$ at $(1, 3, 2)$ about $(-1, 4, 2)$

2. $\mathbf{F} = \hat{\mathbf{i}} + 2\hat{\mathbf{j}}$ at $(1, 1, 0)$ about $(2, 1, -5)$

3. $\mathbf{F} = -\hat{\mathbf{i}} + 3\hat{\mathbf{k}}$ at $(0, 0, 0)$ about $(-1, 3, 0)$

4. $\mathbf{F} = 3\hat{\mathbf{i}} - \hat{\mathbf{j}} + 4\hat{\mathbf{k}}$ at $(1, 1, 1)$ about $(2, 2, 2)$

5. $\mathbf{F} = 6\hat{\mathbf{i}}$ at $(0, 1, 3)$ about $(2, 0, 0)$

In Exercises 6–9 calculate the moment of the force about the indicated line(s).

6. $\mathbf{F} = 2\hat{\mathbf{i}} + 3\hat{\mathbf{j}} - 4\hat{\mathbf{k}}$ at $(1, 3, 2)$ about (a) the line through the points $(1, -3, 2)$ and $(-2, 4, 3)$, (b) the coordinate axes, and (c) the line $x = 2 + 3t$, $y = 4 - 2t$, $z = 1 + 5t$

7. $\mathbf{F} = 6\hat{\mathbf{i}} - 5\hat{\mathbf{j}} + \hat{\mathbf{k}}$ at $(-2, 3, 1)$ about the line $(x - 3)/2 = y + 1 = z/4$

8. $\mathbf{F} = 4\hat{\mathbf{i}} - 2\hat{\mathbf{k}}$ at $(6, -2, 1)$ about the line $x = y = z - 1$

9. $\mathbf{F} = \hat{\mathbf{i}} + \hat{\mathbf{j}} - \hat{\mathbf{k}}$ at $(-1, -1, -2)$ about the line $x - y + z = 2$, $x + 2y + 3z = 4$

10. If $\mathbf{M} = (M_x, M_y, M_z)$ are the components of the moment about the origin due to a force $\mathbf{F} = (F_x, F_y, F_z)$ acting at a point $P(x, y, z)$, what are the moments due to $\mathbf{F}$ about the x-, y-, and z-axes?

11. Suppose that a force $\mathbf{F}$ acts at a point Q, and P is a point on a line ℓ. Prove that when $\mathbf{PQ} \times \mathbf{F}$ is perpendicular to ℓ, then the moment of $\mathbf{F}$ about ℓ is zero.

* **12.** Prove that if a force $\mathbf{F}$ acts at a point P, and ℓ is a line through P, then the moment due to $\mathbf{F}$ about ℓ is zero.

* **13.** Suppose that a force $\mathbf{F} = F_x\hat{\mathbf{i}} + F_y\hat{\mathbf{j}} + F_z\hat{\mathbf{k}}$ acts at a point $P(x_1, y_1, z_1)$, and $\hat{\mathbf{v}} = v_x\hat{\mathbf{i}} + v_y\hat{\mathbf{j}} + v_z\hat{\mathbf{k}}$ are the components of a vector along a line ℓ. Show that if $Q(x_0, y_0, z_0)$ is any point on ℓ, then the moment of $\mathbf{F}$ about ℓ is the scalar triple product

$$\hat{\mathbf{v}} \cdot \mathbf{PQ} \times \mathbf{F} = \begin{vmatrix} v_x & v_y & v_z \\ x_0 - x_1 & y_0 - y_1 & z_0 - z_1 \\ F_x & F_y & F_z \end{vmatrix}.$$

* **14.** Repeat Example 11.26 if the spring has an unstretched length ℓ that is less than the length of AC.

* **15.** Two positive charges q_1 and q_2 are placed in the xy-plane at positions $(5, 5)$ and $(-2, 3)$, respectively. A third positive charge q_3 is moved along the x-axis from $x = 1$ to $x = -1$. Find the total work done by the electrostatic forces of q_1 and q_2 on q_3.

* **16.** When the rocket in the figure below passes close to the spherical asteroid, it is attracted to the asteroid by a gravitational force with magnitude GmM/r^2, where m and M are the masses of the rocket and asteroid, r is the distance from the rocket to the centre of the asteroid, and G is Newton's gravitational constant. Determine the work the rocket must do against this force in order to follow the straight-line path from A to B.

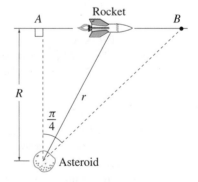

* **17.** If $\mathbf{F}$ is a constant force, show that the work done by $\mathbf{F}$ on an object moving around any closed polygon is zero.

8. Two springs with constants k and $2k$ are joined together at C and have their other ends fixed at points A and B in the figure below. When their join is at C, neither spring is compressed or stretched. If the join is pulled along the straight line CD perpendicular to AB, what work is done?

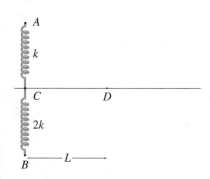

9. A small boat hangs from davits, one of which is shown in the figure below. The tension in the rope $ABAD$ is 410 N. Determine the moment about C of the resultant force of the three tensions in the rope exerted on the davit at A.

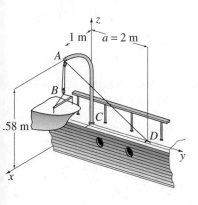

0. A 200-N force is applied to the bracket ABC in the figure below. Determine the moment of the force about A.

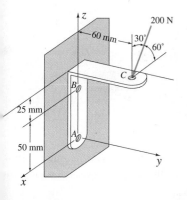

* **21.** Rod AB is held in place by cord AC in the figure below. If the tension in the cord is 1500 N, determine the magnitude of the moment about B due to the tension at A.

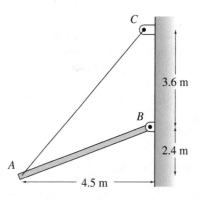

* **22.** A precast concrete wall section is temporarily held by two cables as shown in the figure below. If the tension in cable BD is 900 N, determine (a) the moment about O of the force exerted by cable BD at B, (b) the moment due to this force about the z-axis, and (c) the line BC.

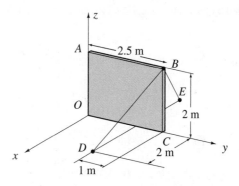

* **23.** The 6-m boom OB in the figure below has a fixed end O. A cable is stretched from the free end B to a point A in the vertical wall of the xz-plane. If the tension in the cable is 1900 N, find the moment about O of the tension at B.

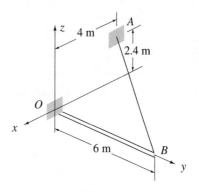

* **24.** Forces of 700 N, 1000 N, and 1200 N are applied to the bracket in the figure below. The 700-N force is directed toward point E. The 1000-N force is parallel to the xy-plane, and the 1200-N force is in the yz-plane. What is the total moment of all three forces about A?

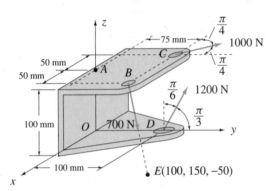

* **25.** A force $\mathbf{F}$ acts at a point A with coordinates (a, a, a). Show that the sum of the moments of $\mathbf{F}$ about the coordinate axes is zero.

* **26.** The rectangular plate in the figure below is hinged at A and B and supported by a cable that passes over a frictionless hook at E. If the tension in the cable is 1349 N, determine moments about the coordinate axes of the force exerted by the cable (a) at C and (b) at D.

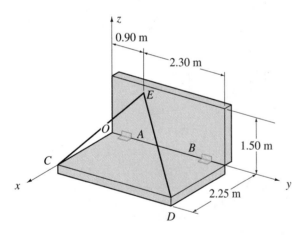

* **27.** The moment about the x-axis of the resultant force of the three tensions in the rope in Exercise 19 exerted on the davit at A must not exceed 375 N·m in absolute value. What is the largest allowable tension in the rope when $a = 2$ m?

* **28.** What is the largest allowable distance a in Exercise 27 when the tension in the rope is 300 N?

* **29.** The force $\mathbf{F}$ in the following figure has magnitude 125 N and acts in a direction perpendicular to the handle BC of the crank. Find moments of the force about the coordinate axes.

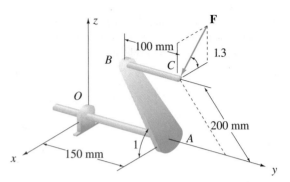

* **30.** The 575-mm vertical rod CD in the figure below is welded to the midpoint C of the 1250-mm rod AB. Determine the moments about AB of (a) the 1175-N force $\mathbf{F}_1$ and (b) the 870-N force $\mathbf{F}_2$.

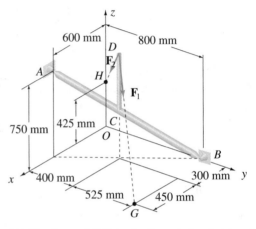

* **31.** The frame ACD in the figure below is hinged at A and D and supported by a cable that passes through a ring at B and is attached to hooks at G and H. Knowing that the tension in the cable is 1125 N, determine the moment about diagonal AD of the force exerted on the frame by (a) portion BH of the cable and (b) portion BG of the cable.

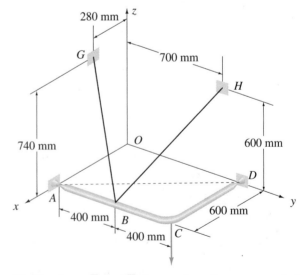

* **32.** Two forces $\mathbf{F}_1$ and $\mathbf{F}_2$ in space have the same magnitude. Prove that the moment of $\mathbf{F}_1$ about the line of action of $\mathbf{F}_2$ is equal to the moment of $\mathbf{F}_2$ about the line of action of $\mathbf{F}_1$.

11.8 Hanging Cables

Hanging cables are used in many engineering applications — suspension bridges, transmission lines, aerial gondolas, and so on. They are flexible members capable of withstanding tension forces but not shear forces. The shape of a hanging cable is determined by the loads that it supports, be they concentrated loads, distributed loads, or a combination of these.

Suppose the load per unit length in the x-direction (including the weight of the cable itself) of the cable in Figure 11.85 is $w(x)$ N/m. It is important to realize that this is not the load per unit length as measured along the cable. For simplicity, we have chosen a coordinate system where the lowest point of the cable is on the y-axis. Three forces act on that part of the cable between $x = 0$ and an arbitrary point x, a horizontal tension T_0 at $x = 0$, a tangential tension T at x, and the vertical load. They are in equilibrium so that we may equate horizontal and vertical components of their resultant to zero,

$$0 = -T_0 + T \cos\theta, \qquad 0 = T \sin\theta - \int_0^x w(t)\, dt. \qquad (11.47)$$

If we eliminate T, we obtain

$$T_0 \tan\theta = \int_0^x w(t)\, dt.$$

Since $\tan\theta$ is the slope of the cable at x, we can write

$$T_0 \frac{dy}{dx} = \int_0^x w(t)\, dt.$$

Since this equation is valid at every point on the cable we may differentiate with respect to x,

$$\frac{d^2 y}{dx^2} = \frac{w(x)}{T_0}. \qquad (11.48)$$

Two integrations of this differential equation give the shape of the cable once the load $w(x)$ is specified.

FIGURE 11.85   Differential equation describing the shape of a hanging cable

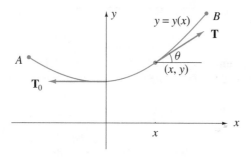

The simplest load is $w(x) = w$, a constant. Uniformly distributed loads along the horizontal would be realized in a suspension bridge where the weight of the cable would be negligible compared to that of the roadway. Two integrations of 11.48 give

$$y(x) = \frac{wx^2}{2T_0} + C_1 x + C_2.$$

Thus, uniformly loaded cables are parabolic. The fact that $y'(0) = 0$ implies that $C_1 = 0$, and therefore $y(x) = wx^2/(2T_0) + C_2$. The position of the x-axis determines C_2. If the origin is chosen as the lowest point of the cable, then $C_2 = 0$ and $y(x) = wx^2/(2T_0)$.

When the supports A and B of a cable have the same elevation, the distance L between them is called the **span of the cable**, and the vertical distance h from the supports to the lowest point of the cable is called the **sag** (Figure 11.86).

FIGURE 11.86   Span and sag of a hanging cable

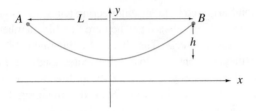

We now consider a cable carrying a load uniformly distributed along the cable itself (rather than in the horizontal direction). Such would be the case for a cable hanging under its own weight. Let w newtons per metre be the constant weight per unit length along the curve and choose a coordinate system with the y-axis through the lowest point on the cable (Figure 11.87). Since length along the cable corresponding to an increment dx along the x-axis is $\sqrt{1 + (dy/dx)^2}\, dx$, the weight of this much cable is $w\sqrt{1 + (dy/dx)^2}\, dx$. The weight per unit length in the x-direction along the cable is therefore $w\sqrt{1 + (dy/dx)^2}$. We substitute this for the function $w(x)$ in equation 11.48:

$$\frac{d^2 y}{dx^2} = \frac{w}{T_0}\sqrt{1 + \left(\frac{dy}{dx}\right)^2}. \tag{11.49}$$

FIGURE 11.87   Cable with a load uniformly distributed along the cable

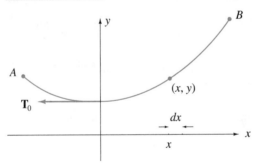

This differential equation defines the shape of the curve. To solve it, we first set $v = dy/dx$ and $dv/dx = d^2 y/dx^2$,

$$\frac{dv}{dx} = \frac{w}{T_0}\sqrt{1 + v^2} \qquad \Longrightarrow \qquad \frac{1}{\sqrt{1 + v^2}}\, dv = \frac{w}{T_0}\, dx,$$

a separated differential equation. Solutions are defined implicitly by

$$\int \frac{1}{\sqrt{1 + v^2}}\, dv = \int \frac{w}{T_0}\, dx \qquad \Longrightarrow \qquad \int \frac{1}{\sqrt{1 + v^2}}\, dv = \frac{wx}{T_0} + D.$$

In the integral we substitute $v = \tan\theta$ and $dv = \sec^2\theta\, d\theta$,

$$\frac{wx}{T_0} + D = \int \frac{1}{\sec\theta}\sec^2\theta\, d\theta = \int \sec\theta\, d\theta = \ln|\sec\theta + \tan\theta| = \ln|\sqrt{1 + v^2} + v$$

Since v is the slope of the cable and the coordinate system has been chosen so that $y'(0) = v(0) = 0$, it follows that $D = 0$. Consequently,

$$\ln\left(\sqrt{1 + v^2} + v\right) = \frac{wx}{T_0} \qquad \Longrightarrow \qquad \sqrt{1 + v^2} + v = e^{wx/T_0}.$$

When we square $\sqrt{1+v^2} = -v + e^{wx/T_0}$, the result is

$$1 + v^2 = v^2 - 2ve^{wx/T_0} + e^{2wx/T_0} \implies v = \frac{dy}{dx} = \frac{e^{2wx/T_0} - 1}{2e^{wx/T_0}} = \frac{1}{2}(e^{wx/T_0} - e^{-wx/T_0}).$$

Integration now gives

$$y = \frac{T_0}{2w}(e^{wx/T_0} + e^{-wx/T_0}) + C = \frac{T_0}{w}\cosh\left(\frac{wx}{T_0}\right) + C. \qquad (11.50)$$

This curve is called a **catenary**. If we choose the x-axis to pass through the minimum point of the cable (Figure 11.88), then $C = -T_0/w$ and

$$y = \frac{T_0}{w}\left[\cosh\left(\frac{wx}{T_0}\right) - 1\right]. \qquad (11.51)$$

FIGURE 11.88   Cable with origin chosen at lowest point

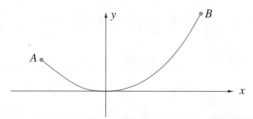

Two important properties of the catenary are derived in the next two examples.

EXAMPLE 11.32

Verify that the length of catenary 11.50 from $x = 0$ to any point with x-coordinate x is $s = (T_0/w)\sinh(wx/T_0)$.

SOLUTION   The length of the catenary from $x = 0$ to an arbitrary x is

$$s = \int_0^x \sqrt{1 + \left(\frac{dy}{dt}\right)^2}\, dt = \int_0^x \sqrt{1 + \sinh^2\left(\frac{wt}{T_0}\right)}\, dt$$

$$= \int_0^x \cosh\left(\frac{wt}{T_0}\right) dt = \left\{\frac{T_0}{w}\sinh\left(\frac{wt}{T_0}\right)\right\}_0^x = \frac{T_0}{w}\sinh\left(\frac{wx}{T_0}\right).$$

EXAMPLE 11.33

Verify that the tension T at any point $P(x, y)$ on the catenary 11.50 is related to T_0 by the equation $T = T_0 + wy$.

SOLUTION   The tension at any point in the cable is given by the first of equations 11.47,

$$T = T_0 \sec\theta = T_0\sqrt{1 + \tan^2\theta} = T_0\sqrt{1 + \left(\frac{dy}{dx}\right)^2}$$

$$= T_0\sqrt{1 + \sinh^2\left(\frac{wx}{T_0}\right)} = T_0\cosh\left(\frac{wx}{T_0}\right) = T_0\left(\frac{wy}{T_0} + 1\right) = T_0 + wy.$$

EXAMPLE 11.34

A uniform cable with mass per unit length 5 kg/m is suspended between the two points A and B in Figure 11.89a. The span is 150 m and the sag is 30 m. Determine the length of the cable and the maximum and minimum tensions in it.

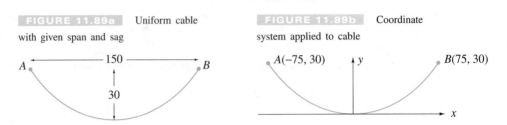

FIGURE 11.89a   Uniform cable with given span and sag

FIGURE 11.89b   Coordinate system applied to cable

SOLUTION   If we adopt the coordinate system in Figure 11.89b, the equation of the catenary takes form 11.51. With point $(75, 30)$ on the curve,

$$30 = \frac{T_0}{w}\left[\cosh\left(\frac{75w}{T_0}\right) - 1\right] \qquad \Longrightarrow \qquad \frac{30w}{T_0} = \cosh\left(\frac{75w}{T_0}\right) - 1.$$

If we set $z = w/T_0$, the equation

$$f(z) = \cosh(75z) - 30z - 1 = 0$$

can be solved numerically for z. For instance, Newton's iterative procedure with $z_1 = 0.01$ and

$$z_{n+1} = z_n - \frac{\cosh(75z_n) - 30z_n - 1}{75\sinh(75z_n) - 30}$$

gives the approximations $z_2 = 0.010\,17$, $z_3 = 0.010\,16$, and $z_4 = 0.010\,16$. Consequently, the minimum tension in the cable is $T_0 = w/z = 5(9.81)/0.010\,16 = 4828$ N. According to Example 11.32, the length of the cable is

$$L = \frac{2T_0}{w}\sinh\left(\frac{75w}{T_0}\right) = \frac{2}{0.010\,16}\sinh[75(0.010\,16)] = 164.9 \text{ m}.$$

According to Example 11.33, the maximum tension, at $x = 75$, is

$$T = T_0 + 30w = 4828 + 30(5)(9.81) = 6300 \text{ N}.$$

If a uniform cable is pulled very, very tight, it becomes almost horizontal, and it should be possible to consider its weight per unit length w along the curve as a uniform weight per unit length in the x-direction. To show that this is the case, we express equation 11.51 for the catenary in the form

$$y = \frac{T_0}{w}\left[\cosh\left(\frac{wx}{T_0}\right) - 1\right] = \frac{T_0}{w}\left[\left(1 + \frac{w^2x^2}{2T_0^2} + \frac{w^4x^4}{24T_0^4} + \cdots\right) - 1\right].$$

For a very tight cable, T_0 will be very large, so that if we retain only the first two terms of the Maclaurin series,

$$y \approx \frac{T_0}{w}\left(\frac{w^2x^2}{2T_0^2}\right) = \frac{wx^2}{2T_0}.$$

What we are saying is that as the tension in a catenary is increased, the more closely it can be approximated by a parabola.

Consulting Project 17

A cable, with uniform weight per unit length w, is to be suspended over a chasm with both ends of the cable at the same elevation. Design specification for the cable requires that tension must never exceed T_m. We are to determine the maximum allowable horizontal span for the cable.

SOLUTION   We can take the equation for the cable in form 11.51,

$$y = \frac{T_0}{w}\left[\cosh\left(\frac{wx}{T_0}\right) - 1\right],$$

where T_0 is the tension at the lowest point in the cable, provided we use the coordinate system in Figure 11.90. If h is the sag and L is the span, then

$$h = \frac{T_0}{w}\left[\cosh\left(\frac{wL}{2T_0}\right) - 1\right].$$

FIGURE 11.90   Span of hanging cable

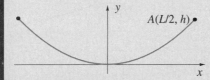

Maximum tension occurs at A and according to Example 11.33 is given by

$$T_m = T_0 + wh = T_0 + T_0\left[\cosh\left(\frac{wL}{2T_0}\right) - 1\right] = T_0\cosh\left(\frac{wL}{2T_0}\right).$$

Since T_m is specified and w is known, this equation implicitly defines L as a function of T_0, and to find the maximum value of L, we set $dL/dT_0 = 0$. Implicit differentiation gives

$$0 = \cosh\left(\frac{wL}{2T_0}\right) + T_0\sinh\left(\frac{wL}{2T_0}\right)\left(\frac{-wL}{2T_0^2} + \frac{w}{2T_0}\frac{dL}{dT_0}\right),$$

and this implies that

$$\frac{dL}{dT_0} = \frac{2}{w}\left[\frac{wL}{2T_0} - \coth\left(\frac{wL}{2T_0}\right)\right] = 0.$$

If we set $z = wL/(2T_0)$, this equation becomes

$$z - \coth z = 0 \quad \Longrightarrow \quad z\tanh z = 1.$$

This equation cannot be solved exactly, but when it is solved numerically the result is $z = 1.200$. It now follows that

$$\frac{wL}{2T_0} = 1.200 \quad \Longrightarrow \quad L = \frac{2.400T_0}{w}.$$

To replace T_0, we note that $T_m = T_0\cosh\left(\frac{wL}{2T_0}\right) = T_0\cosh 1.200$. Thus, maximum span is

$$L = \frac{2.400T_m\operatorname{sech}(1.200)}{w} = \frac{1.325T_m}{w}.$$

EXERCISES 11.8

1. A cable with ends at the same elevation has a span of 100 m and a sag of 5 m. It is subjected to a uniform horizontal load of 1000 N/m. Determine the minimum tension in the cable.

2. Cable AB in the figure below supports a uniform, horizontally distributed load of 1100 N/m. The lowest point of the cable is 3 m below support A, and the support at B is 6 m higher than the support at A. Determine the minimum and maximum tensions in the cable.

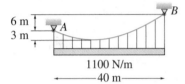

6 m
3 m

1100 N/m
—40 m—

3. Cable AB in the figure below supports a uniform, horizontally distributed load of w newtons per metre. The lowest point of the cable is 1 m below the support at A, and the support at B is 2 m higher than the support at A. Determine a formula for the minimum tension in the cable in terms of w.

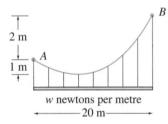

2 m

1 m

w newtons per metre
—20 m—

4. The centre span of the George Washington Bridge consists of a uniform roadway suspended from four cables. The uniform load supported by each cable is 142 kN/m along the horizontal. If the span is 1050 m and the sag is 94.8 m, determine the minimum and maximum tensions in each cable.

$*$ **5.** (a) Show that the length of a cable that supports a uniform, horizontally distributed load of w, from its minimum point to a point $P(x_P, y_P)$, is given by

$$L(P) = \int_0^{x_P} \sqrt{1 + \frac{w^2 x^2}{T_0^2}} \, dx.$$

Show that the value of this integral is

$$L(P) = \frac{x_P}{2}\sqrt{1 + \left(\frac{w x_P}{T_0}\right)^2} + \frac{T_0}{2w} \ln\left[\sqrt{1 + \left(\frac{w x_P}{T_0}\right)^2} + \frac{w x_P}{T_0}\right].$$

  (b) The expression in part (a) for $L(P)$ is unwieldy in performing calculations involving the length of a cable subject to uniform loads. In practice, it is often approximated by first expanding $\sqrt{1 + (w^2 x_P^2/T_0^2)}$ as an infinite series and integrating term by term. Show that this leads to

$$L(P) = x_P \left[1 + \frac{2}{3}\left(\frac{y_P}{x_P}\right)^2 - \frac{2}{5}\left(\frac{y_P}{x_P}\right)^4 + \cdots\right],$$

convergent for $y_P/x_P < 1/2$. In situations where y_P is much less than x_P, the series is truncated after the first two terms.

6. Use the formula in Exercise 5(a) and the two-term approximation in part (b) to calculate the length of the cable in Exercise 1.

7. Complete Exercise 6 for the cable in Exercise 4.

8. The centre span of the Verrazano-Narrows bridge consists of two uniform roadways suspended from four cables. The design of the bridge includes the effect of extreme temperature changes, which cause the sag of the centre span to vary from $h_w = 115.8$ m in winter to $h_s = 118.2$ m in summer. If the span is 1278 m, use the two-term approximation in Exercise 5(b) to determine the change in the length of the cables at these temperature extremes.

$*$ **9.** Before being fed into a printing press located to the right of D in the figure below, a continuous sheet of paper having mass per unit length 300 g/m passes over rollers at A and B. Assuming that the curve formed by the sheet is parabolic, determine (a) the location of the lowest point C and (b) the maximum tension in the sheet.

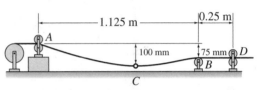

—1.125 m—   0.25 m

A

100 mm   75 mm D
          B

C

$*$ **10.** A 40-m rope is strung between the roofs of two buildings, each 14 m high. The maximum tension is 350 N and the lowest point of the cable is 6 m above the ground. Determine the horizontal distance between the buildings and the mass of the rope.

$*$ **11.** A 50-m steel measuring tape has mass 1.6 kg. If the tape is stretched between two points at the same elevation until the tension at each end is 60 N, determine the span of the tape.

$*$ **12.** An aerial tramway cable of length 150 m and with mass per unit length 4 kg/m is suspended between two points at the same elevation. If the sag is 37.5 m, find the span of the cable and the maximum tension in the cable.

$*$ **13.** A counterweight D of mass 40 kg is attached to a cable that passes over a small pulley at A and is attached to a support at B (figure below). The system is motionless. Knowing that $L = 15$ m and $h = 5$ m, determine the length of the cable from A to B and the mass per unit length of the cable. Neglect the mass of the cable from A to D.

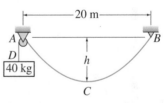

—20 m—

A   B

D
40 kg   h

C

$**$ **14.** A uniform cord 900 mm long passes over a frictionless pulley at B and is attached to a rigid support at A (figure below). If $L = 300$ mm, determine the smaller of the two sags for which the cord is in equilibrium.

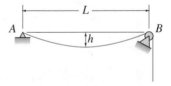

—L—

A
         h B

15. To the left of point B in the figure below, the long cable rests on a 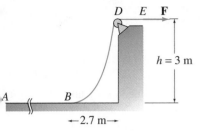 ** 16.** A cable of uniform weight per unit length w is suspended between rough horizontal surface. If the cable has mass per unit length 2.7 kg/m, two points at the same elevation a distance L apart. Determine the sag-determine the force $\mathbf{F}$ to maintain equilibrium. to-span ratio for which the maximum tension is as small as possible.

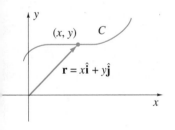

11.9 Differentiation and Integration of Vectors

In Section 11.7, vectors were used to represent forces and moments; they can also be used to describe many other physical quantities, such as position, velocity, acceleration, electric and magnetic fields, and fluid flow. In applications such as these, vectors seldom have constant components; instead, they have components that are either functions of position, or functions of some parameter, such as time, or both. For instance, the spring force in Example 11.26 varies in both magnitude and direction as the sleeve moves from B to C. Consequently, components of the vector $\mathbf{F}$ representing this force are functions of position x between B and C:

$$\mathbf{F} = F_x\hat{\mathbf{i}} + F_y\hat{\mathbf{j}} = F_x(x)\hat{\mathbf{i}} + F_y(x)\hat{\mathbf{j}}.$$

When a particle moves along a curve C in the xy-plane defined parametrically by

$$C: \quad x = x(t), \quad y = y(t), \quad \alpha \leq t \leq \beta,$$

its position (x, y) relative to the origin is represented by the vector $\mathbf{r} = x\hat{\mathbf{i}} + y\hat{\mathbf{j}}$ (Figure 11.91). This vector is called the **position vector** or **displacement vector** of the particle relative to the origin. We will have more to say about it in Section 11.13. Note, however, that if we substitute from the parametric equations for C, we have

$$\mathbf{r} = x(t)\hat{\mathbf{i}} + y(t)\hat{\mathbf{j}},$$

which indicates that the displacement vector has components that are functions of the parameter t.

In this section we consider the general situation in which the components v_x, v_y, and v_z of a vector $\mathbf{v}$ are functions of some parameter t,

$$\mathbf{v} = v_x(t)\hat{\mathbf{i}} + v_y(t)\hat{\mathbf{j}} + v_z(t)\hat{\mathbf{k}}, \tag{11.52}$$

and show how the operations of differentiation and integration can be applied to such a vector. In Sections 11.11 and 11.12 we will use these results to discuss the geometry of curves, and this will pave the way for an analysis of the motion of particles in Section 11.13.

Because $\mathbf{v}$ in 11.52 has components that are functions of t, we say that $\mathbf{v}$ itself is a vector-valued function of t, and write $\mathbf{v} = \mathbf{v}(t)$. Each of the component functions $v_x(t)$, $v_y(t)$, and $v_z(t)$ has a domain, and their common domain is called the *domain of the vector-valued function* $\mathbf{v}(t)$. Given that this domain is some interval $\alpha \leq t \leq \beta$, we express 11.52 more fully in the form

$$\mathbf{v} = \mathbf{v}(t) = v_x(t)\hat{\mathbf{i}} + v_y(t)\hat{\mathbf{j}} + v_z(t)\hat{\mathbf{k}}, \quad \alpha \leq t \leq \beta. \tag{11.53}$$

To differentiate and integrate vector-valued functions, we first require the concept of a limit.

DEFINITION 11.12

If $\mathbf{v}(t) = v_x(t)\hat{\mathbf{i}} + v_y(t)\hat{\mathbf{j}} + v_z(t)\hat{\mathbf{k}}$, then

$$\lim_{t \to t_0} \mathbf{v}(t) = \left[\lim_{t \to t_0} v_x(t)\right]\hat{\mathbf{i}} + \left[\lim_{t \to t_0} v_y(t)\right]\hat{\mathbf{j}} + \left[\lim_{t \to t_0} v_z(t)\right]\hat{\mathbf{k}}, \tag{11.54}$$

provided that each of the limits on the right exists.

This definition states that to take the limit of a vector-valued function, we take the limit of each component separately. As an illustration, consider the following example.

EXAMPLE 11.35

If $\mathbf{v} = \mathbf{v}(t) = (t^2 + 1)\hat{\mathbf{i}} + 3t\hat{\mathbf{j}} - (\sin t)\hat{\mathbf{k}}$, calculate $\lim_{t \to 5} \mathbf{v}(t)$.

SOLUTION   According to Definition 11.12,

$$\lim_{t \to 5} \mathbf{v}(t) = \left[\lim_{t \to 5}(t^2 + 1)\right]\hat{\mathbf{i}} + \left[\lim_{t \to 5}(3t)\right]\hat{\mathbf{j}} - \left[\lim_{t \to 5}\sin t\right]\hat{\mathbf{k}}$$

$$= 26\hat{\mathbf{i}} + 15\hat{\mathbf{j}} - (\sin 5)\hat{\mathbf{k}}.$$

In the remainder of this section we define continuity, derivatives, and antiderivatives for vector-valued functions. Each definition is an exact duplicate of the corresponding definition for a scalar function $y(t)$, except that $y(t)$ is replaced by $\mathbf{v}(t)$. We then show that the vector definition can be rephrased in terms of components of the vector. We begin with continuity in the following definition.

DEFINITION 11.13

A vector-valued function $\mathbf{v}(t)$ is said to be **continuous** at a point t_0 if

$$\mathbf{v}(t_0) = \lim_{t \to t_0} \mathbf{v}(t). \tag{11.55}$$

It is a simple matter to prove the next theorem.

THEOREM 11.7

A vector-valued function is continuous at a point if and only if its components are continuous at that point.

PROOF   If $\mathbf{v}(t) = v_x(t)\hat{\mathbf{i}} + v_y(t)\hat{\mathbf{j}} + v_z(t)\hat{\mathbf{k}}$, then according to 11.55, $\mathbf{v}(t)$ is continuous at t_0 if and only if

$$v_x(t_0)\hat{\mathbf{i}} + v_y(t_0)\hat{\mathbf{j}} + v_z(t_0)\hat{\mathbf{k}} = \lim_{t \to t_0}[v_x(t)\hat{\mathbf{i}} + v_y(t)\hat{\mathbf{j}} + v_z(t)\hat{\mathbf{k}}].$$

Definition 11.12 implies that we can write this condition in the form

$$v_x(t_0)\hat{\mathbf{i}} + v_y(t_0)\hat{\mathbf{j}} + v_z(t_0)\hat{\mathbf{k}} = \left[\lim_{t \to t_0} v_x(t)\right]\hat{\mathbf{i}} + \left[\lim_{t \to t_0} v_y(t)\right]\hat{\mathbf{j}} + \left[\lim_{t \to t_0} v_z(t)\right]\hat{\mathbf{k}}.$$

But because two vectors are equal if and only if their components are equal, we can say that $\mathbf{v}(t)$ is continuous at t_0 if and only if

$$v_x(t_0) = \lim_{t \to t_0} v_x(t); \quad v_y(t_0) = \lim_{t \to t_0} v_y(t); \quad v_z(t_0) = \lim_{t \to t_0} v_z(t)$$

[i.e., $\mathbf{v}(t)$ is continuous at t_0 if and only if its components are continuous at t_0].

For example, the vector-valued function

$$\mathbf{v}(t) = (t - 1)\hat{\mathbf{i}} + (1/t)\hat{\mathbf{j}} + (t^2 - 1)^{-1}\hat{\mathbf{k}}$$

is discontinuous for $t = 0$ [since $v_y(0)$ is not defined] and for $t = \pm 1$ [since $v_z(\pm 1)$ is not defined].

The derivative of a scalar function $y(t)$ is its instantaneous rate of change:

$$\frac{dy}{dt} = \lim_{h \to 0} \frac{y(t + h) - y(t)}{h}.$$

The derivative of a vector-valued function is also a rate of change defined by a similar limit.

DEFINITION 11.14

The **derivative of a vector-valued function** $\mathbf{v}(t)$ is defined as

$$\frac{d\mathbf{v}}{dt} = \lim_{h \to 0} \frac{\mathbf{v}(t + h) - \mathbf{v}(t)}{h}, \tag{11.56}$$

provided that the limit exists.

In practice, we seldom use the definition of a derivative to calculate dy/dt for a scalar function $y(t)$; formulas such as the power, product, quotient, and chain rules are more convenient. It would be helpful to have corresponding formulas for derivatives of vector-valued functions. The following theorem shows that to differentiate a vector-valued function, we simply differentiate its Cartesian components.

THEOREM 11.8

If $\mathbf{v}(t) = v_x(t)\hat{\mathbf{i}} + v_y(t)\hat{\mathbf{j}} + v_z(t)\hat{\mathbf{k}}$, then

$$\frac{d\mathbf{v}}{dt} = \frac{dv_x}{dt}\hat{\mathbf{i}} + \frac{dv_y}{dt}\hat{\mathbf{j}} + \frac{dv_z}{dt}\hat{\mathbf{k}}, \tag{11.57}$$

provided that the derivatives on the right exist.

PROOF   If we substitute the components of $\mathbf{v}(t + h)$ and $\mathbf{v}(t)$ into Definition 11.14, then we have

$$\frac{d\mathbf{v}}{dt} = \lim_{h \to 0} \left\{ \frac{[v_x(t+h)\hat{\mathbf{i}} + v_y(t+h)\hat{\mathbf{j}} + v_z(t+h)\hat{\mathbf{k}}] - [v_x(t)\hat{\mathbf{i}} + v_y(t)\hat{\mathbf{j}} + v_z(t)\hat{\mathbf{k}}]}{h} \right\}$$

$$= \lim_{h \to 0} \left\{ \left[\frac{v_x(t+h) - v_x(t)}{h} \right]\hat{\mathbf{i}} + \left[\frac{v_y(t+h) - v_y(t)}{h} \right]\hat{\mathbf{j}} + \left[\frac{v_z(t+h) - v_z(t)}{h} \right]\hat{\mathbf{k}} \right\}$$

(according to equation 11.14), and

$$\frac{d\mathbf{v}}{dt} = \left[\lim_{h\to 0}\frac{v_x(t+h)-v_x(t)}{h}\right]\hat{\mathbf{i}} + \left[\lim_{h\to 0}\frac{v_y(t+h)-v_y(t)}{h}\right]\hat{\mathbf{j}}$$
$$+ \left[\lim_{h\to 0}\frac{v_z(t+h)-v_z(t)}{h}\right]\hat{\mathbf{k}}$$

(according to equation 11.54). Since each of the limits on the right exists, we can write that

$$\frac{d\mathbf{v}}{dt} = \frac{dv_x}{dt}\hat{\mathbf{i}} + \frac{dv_y}{dt}\hat{\mathbf{j}} + \frac{dv_z}{dt}\hat{\mathbf{k}}.$$

Theorem 11.8 gives us a working rule for differentiating vector-valued functions: To differentiate a vector-valued function, we differentiate its Cartesian components.

EXAMPLE 11.36

If $\mathbf{v}(t) = t^2\hat{\mathbf{i}} + (3t^3 - 2t)\hat{\mathbf{j}} + 5\hat{\mathbf{k}}$, find $\mathbf{v}'(3)$.

SOLUTION   According to 11.57,

$$\frac{d\mathbf{v}}{dt} = 2t\hat{\mathbf{i}} + (9t^2 - 2)\hat{\mathbf{j}}.$$

Consequently, $\mathbf{v}'(3) = 6\hat{\mathbf{i}} + 79\hat{\mathbf{j}}$.

The sum rule 3.9 for differentiation of scalar functions has its counterpart in the sum rule for vector-valued functions,

$$\frac{d}{dt}(\mathbf{u} + \mathbf{v}) = \frac{d\mathbf{u}}{dt} + \frac{d\mathbf{v}}{dt} \tag{11.58}$$

(see Exercise 22). There are three types of products associated with vectors: the product of a scalar and a vector, the dot product of two vectors, and the cross product of two vectors. Corresponding to each, we have a product rule for differentiation, but all resemble the product rule for scalar functions.

THEOREM 11.9

If $f(t)$ is a differentiable function and $\mathbf{u}(t)$ and $\mathbf{v}(t)$ are differentiable vector-valued functions, then

$$\frac{d}{dt}(f\mathbf{v}) = \frac{df}{dt}\mathbf{v} + f\frac{d\mathbf{v}}{dt}, \tag{11.59a}$$

$$\frac{d}{dt}(\mathbf{u} \cdot \mathbf{v}) = \mathbf{u} \cdot \frac{d\mathbf{v}}{dt} + \frac{d\mathbf{u}}{dt} \cdot \mathbf{v}, \tag{11.59b}$$

$$\frac{d}{dt}(\mathbf{u} \times \mathbf{v}) = \mathbf{u} \times \frac{d\mathbf{v}}{dt} + \frac{d\mathbf{u}}{dt} \times \mathbf{v}. \tag{11.59c}$$

For a proof of these results, see Exercise 23.

EXAMPLE 11.37

If $f(t) = t^2 + 2t + 3$, $\mathbf{u}(t) = t\hat{\mathbf{i}} + t^2\hat{\mathbf{j}} - 3\hat{\mathbf{k}}$, and $\mathbf{v}(t) = t(\hat{\mathbf{i}} + \hat{\mathbf{j}} + \hat{\mathbf{k}})$, use 11.59 to evaluate:

$$\text{(a) } \frac{d}{dt}(f\mathbf{u}) \quad \text{(b) } \frac{d}{dt}(\mathbf{u} \cdot \mathbf{v}) \quad \text{(c) } \frac{d}{dt}(\mathbf{u} \times \mathbf{v})$$

SOLUTION

(a) With 11.59a,

$$\frac{d}{dt}(f\mathbf{u}) = \frac{df}{dt}\mathbf{u} + f\frac{d\mathbf{u}}{dt} = (2t + 2)(t\hat{\mathbf{i}} + t^2\hat{\mathbf{j}} - 3\hat{\mathbf{k}}) + (t^2 + 2t + 3)(\hat{\mathbf{i}} + 2t\hat{\mathbf{j}})$$

$$= (3t^2 + 4t + 3)\hat{\mathbf{i}} + (4t^3 + 6t^2 + 6t)\hat{\mathbf{j}} - 6(t + 1)\hat{\mathbf{k}}.$$

(b) With 11.59b,

$$\frac{d}{dt}(\mathbf{u} \cdot \mathbf{v}) = \mathbf{u} \cdot \frac{d\mathbf{v}}{dt} + \frac{d\mathbf{u}}{dt} \cdot \mathbf{v}$$

$$= (t\hat{\mathbf{i}} + t^2\hat{\mathbf{j}} - 3\hat{\mathbf{k}}) \cdot (\hat{\mathbf{i}} + \hat{\mathbf{j}} + \hat{\mathbf{k}}) + (\hat{\mathbf{i}} + 2t\hat{\mathbf{j}}) \cdot (t\hat{\mathbf{i}} + t\hat{\mathbf{j}} + t\hat{\mathbf{k}})$$

$$= (t + t^2 - 3) + (t + 2t^2) = 3t^2 + 2t - 3.$$

(c) With 11.59c,

$$\frac{d}{dt}(\mathbf{u} \times \mathbf{v}) = \mathbf{u} \times \frac{d\mathbf{v}}{dt} + \frac{d\mathbf{u}}{dt} \times \mathbf{v} = (t\hat{\mathbf{i}} + t^2\hat{\mathbf{j}} - 3\hat{\mathbf{k}}) \times (\hat{\mathbf{i}} + \hat{\mathbf{j}} + \hat{\mathbf{k}}) + (\hat{\mathbf{i}} + 2t\hat{\mathbf{j}}) \times (t\hat{\mathbf{i}} + t\hat{\mathbf{j}} + t\hat{\mathbf{k}})$$

$$= \begin{vmatrix} \hat{\mathbf{i}} & \hat{\mathbf{j}} & \hat{\mathbf{k}} \\ t & t^2 & -3 \\ 1 & 1 & 1 \end{vmatrix} + \begin{vmatrix} \hat{\mathbf{i}} & \hat{\mathbf{j}} & \hat{\mathbf{k}} \\ 1 & 2t & 0 \\ t & t & t \end{vmatrix}$$

$$= [(t^2 + 3)\hat{\mathbf{i}} - (3 + t)\hat{\mathbf{j}} + (t - t^2)\hat{\mathbf{k}}] + [2t^2\hat{\mathbf{i}} - t\hat{\mathbf{j}} + (t - 2t^2)\hat{\mathbf{k}}]$$

$$= (3t^2 + 3)\hat{\mathbf{i}} - (3 + 2t)\hat{\mathbf{j}} + (2t - 3t^2)\hat{\mathbf{k}}.$$

If vector-valued functions can be differentiated, then they can be antidifferentiated. Formally, we make the following statement.

DEFINITION 11.15

A vector-valued function $\mathbf{V}(t)$ is said to be an **antiderivative** of $\mathbf{v}(t)$ on the interval $\alpha < t < \beta$ if

$$\frac{d\mathbf{V}}{dt} = \mathbf{v}(t) \qquad \text{for } \alpha < t < \beta. \tag{11.60}$$

For example, an antiderivative of $\mathbf{v}(t) = 2t\hat{\mathbf{i}} - \hat{\mathbf{j}} + 3t^2\hat{\mathbf{k}}$ is

$$\mathbf{V}(t) = t^2\hat{\mathbf{i}} - t\hat{\mathbf{j}} + t^3\hat{\mathbf{k}}.$$

If we add to $\mathbf{V}(t)$ in 11.60 any vector with constant components, denoted by $\mathbf{C}$, then $\mathbf{V}(t) + \mathbf{C}$ is also an antiderivative of $\mathbf{v}(t)$. We call this vector the **indefinite integral** of $\mathbf{v}(t)$, and write

$$\int \mathbf{v}(t)\, dt = \mathbf{V}(t) + \mathbf{C}. \tag{11.61}$$

For our example, then,

$$\int (2t\hat{i} - \hat{j} + 3t^2\hat{k})\, dt = t^2\hat{i} - t\hat{j} + t^3\hat{k} + \mathbf{C}.$$

Because vectors can be differentiated component by component, it follows that they may also be integrated component by component; that is, if $\mathbf{v}(t) = v_x(t)\hat{i} + v_y(t)\hat{j} + v_z(t)\hat{k}$, then

$$\int \mathbf{v}(t)\, dt = \left[\int v_x(t)\, dt\right]\hat{i} + \left[\int v_y(t)\, dt\right]\hat{j} + \left[\int v_z(t)\, dt\right]\hat{k}. \qquad (11.62)$$

EXAMPLE 11.38

Find the indefinite integral of $\mathbf{v}(t) = \sqrt{t-1}\,\hat{i} + e^t\hat{j} + 6t^2\hat{k}$.

SOLUTION   According to equation 11.62,

$$\int \mathbf{v}(t)\, dt = \left[\frac{2}{3}(t-1)^{3/2} + C_1\right]\hat{i} + (e^t + C_2)\hat{j} + (2t^3 + C_3)\hat{k}$$

$$= \frac{2}{3}(t-1)^{3/2}\hat{i} + e^t\hat{j} + 2t^3\hat{k} + \mathbf{C},$$

where $\mathbf{C} = C_1\hat{i} + C_2\hat{j} + C_3\hat{k}$ is a constant vector.

EXERCISES 11.9

In Exercises 1–5 find the largest possible domain for the vector-valued function.

1. $\mathbf{v}(t) = t^2\hat{i} + \sqrt{t-1}\,\hat{j} + \hat{k}$

2. $\mathbf{v}(t) = (\sin t)\hat{i} + (\cos t)\hat{j} - t^3\hat{k}$

3. $\mathbf{v}(t) = (\text{Sin}^{-1} t)\hat{i} - t^2\hat{j} + (t+1)\hat{k}$

4. $\mathbf{v}(t) = \ln(t+4)(\hat{i} + \hat{j})$

5. $\mathbf{v}(t) = e^t\hat{i} + (\cos^2 t)\hat{j} - (e^t\cos t)\hat{k}$

If $f(t) = t^2 + 3$, $g(t) = 2t^3 - 3t$, $\mathbf{u}(t) = t\hat{i} - t^2\hat{j} + 2t\hat{k}$, and $\mathbf{v}(t) = \hat{i} - 2t\hat{j} + 3t^2\hat{k}$, find the scalar or the components of the vector in Exercises 6–21 .

6. $\dfrac{d\mathbf{u}}{dt}$

7. $\dfrac{d}{dt}[f(t)\mathbf{v}(t)]$

8. $\dfrac{d}{dt}[g(t)\mathbf{u}(t)]$

9. $\dfrac{d}{dt}(\mathbf{u} \times \mathbf{v})$

10. $\dfrac{d}{dt}(\mathbf{u} \times t\mathbf{v})$

11. $\dfrac{d}{dt}(2\mathbf{u} \cdot \mathbf{v})$

12. $\dfrac{d}{dt}(3\mathbf{u} + 4\mathbf{v})$

13. $\displaystyle\int \mathbf{u}(t)\, dt$

14. $\dfrac{d}{dt}[f(t)\mathbf{u} + g(t)\mathbf{v}]$

15. $\displaystyle\int 4\mathbf{v}(t)\, dt$

16. $\dfrac{d}{dt}[t(\mathbf{u} \times \mathbf{v})]$

17. $\displaystyle\int [f(t)\mathbf{u}(t)]\, dt$

18. $\displaystyle\int [3g(t)\mathbf{v}(t) + \mathbf{u}(t)]\, dt$

19. $\displaystyle\int [f(t)\mathbf{u} \cdot \mathbf{v}]\, dt$

20. $\mathbf{u} \times \dfrac{d\mathbf{v}}{dt} - f(t)\mathbf{u} \cdot \dfrac{d\mathbf{v}}{dt}\mathbf{v}$

21. $\mathbf{u} \cdot \dfrac{d\mathbf{v}}{dt} - \mathbf{v} \cdot \displaystyle\int \mathbf{u}(t)\, dt$

22. Prove equation 11.58.

23. Verify the results in equations 11.59.

* **24.** Prove that for differentiable functions $\mathbf{u}(t)$, $\mathbf{v}(t)$, and $\mathbf{w}(t)$,

$$\frac{d}{dt}(\mathbf{u} \cdot \mathbf{v} \times \mathbf{w}) = \frac{d\mathbf{u}}{dt} \cdot \mathbf{v} \times \mathbf{w} + \mathbf{u} \cdot \frac{d\mathbf{v}}{dt} \times \mathbf{w} + \mathbf{u} \cdot \mathbf{v} \times \frac{d\mathbf{w}}{dt}.$$

* **25.** Prove that if a differentiable function $\mathbf{v}(t)$ has constant length, then at any point at which $d\mathbf{v}/dt \neq \mathbf{0}$, the vector $d\mathbf{v}/dt$ is perpendicular to $\mathbf{v}$.

* **26.** If $\mathbf{v} = \mathbf{v}(s)$ is a differentiable vector-valued function and $s = s(t)$ is a differentiable scalar function, prove that

$$\frac{d\mathbf{v}}{dt} = \frac{d\mathbf{v}}{ds}\frac{ds}{dt}.$$

This result is called the *chain rule* for differentiation of vector-valued functions.

* **27.** Show that the following definition for the limit of a vector-valued function is equivalent to Definition 11.12: A vector-valued function $\mathbf{v}(t)$ is said to have limit $\mathbf{V}$ as t approaches t_0 if given any $\epsilon > 0$, there exists a $\delta > 0$ such that $|\mathbf{v}(t) - \mathbf{V}| < \epsilon$ whenever $0 < |t - t_0| < \delta$

11.10 Parametric and Vector Representations of Curves

In Section 11.2, we presented curves in space as the intersection of two surfaces. For example, each of the equations

$$x^2 + y^2 + z^2 = 9, \quad y = 2$$

describes a surface (the first is a sphere and the second a plane), and together they describe the curve of intersection of the surfaces — the circle in Figure 11.92.

FIGURE 11.92   Curve of intersection of a sphere and a plane is a circle

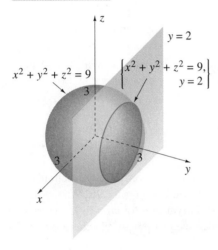

$$y = 2$$

$$x^2 + y^2 + z^2 = 9$$

$$\left\{ \begin{array}{l} x^2 + y^2 + z^2 = 9, \\ y = 2 \end{array} \right.$$

Parametric Representation of Curves

In many applications it is more convenient to have a curve defined parametrically.

> **DEFINITION 11.16**
>
> A curve in space is defined parametrically by three functions:
>
> $$C: \quad x = x(t), \quad y = y(t), \quad z = z(t), \quad \alpha \le t \le \beta. \tag{11.63}$$

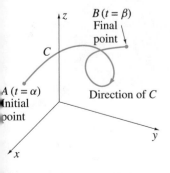

FIGURE 11.93   Curves are directed from initial point to final point

$B(t = \beta)$
Final point

C

$A(t = \alpha)$
Initial point

Direction of C

Each value of t in the interval $\alpha \le t \le \beta$ is substituted into the three functions, and the triple $(x, y, z) = (x(t), y(t), z(t))$ specifies a point on the curve. Definition 11.16 clearly corresponds to parametric Definition 9.1 for a plane curve.

It is customary to assign a direction to a curve by calling that point on C corresponding to $t = \alpha$ the initial point and that point corresponding to $t = \beta$ the final point, and the direction of C is that direction along C from initial point to final point (Figure 11.93). Note in particular that the direction of a curve always corresponds to the direction in which the parameter *increases* along the curve. Because of this, whenever we describe a curve in nonparametric form but with a specified direction, we must be careful in setting up parametric equations to ensure that the parameter increases in the appropriate direction.

When a curve is described as the curve of intersection of two surfaces, we often obtain parametric equations for the curve by specifying one of x, y, or z as a function of t, and then solving the equations for the other two as functions of t. Considerable ingenuity is sometimes required in arriving at a suitable initial function of t. We illustrate this in the following example.

EXAMPLE 11.39

Find parametric equations for each of the following curves:

(a) $z - 1 = x^2 + y^2$, $x - y = 0$ directed so that z increases when x and y are positive

(b) $x + 2y + z = 4$, $2x + y + 3z = 6$ directed so that y increases along the curve.

(c) $x^2 + (y - 1)^2 = 4$, $z = x$ directed so that y increases when x is positive.

SOLUTION

(a) The curve of intersection of the circular paraboloid $z = 1 + x^2 + y^2$ and the plane $y = x$ is shown in Figure 11.94. If we choose x as the parameter along the curve by setting $x = t$, then $y = x$ and $z = 1 + x^2 + y^2$ imply that

$$x = t, \quad y = t, \quad z = 1 + 2t^2.$$

When $t > 0$, so are x and y, and for these values of t, z increases as t increases. This means that these are acceptable parametric equations for the curve.

(b) The straight-line intersection of the two planes is shown in Figure 11.95. If we choose y as the parameter by setting $y = t$ (thus forcing y to increase as t increases), then

$$x + z = 4 - 2t, \quad 2x + 3z = 6 - t.$$

The solution of these equations for x and z in terms of t gives the parametric equations

$$x = 6 - 5t, \quad y = t, \quad z = -2 + 3t.$$

(c) The curve of intersection of the right-circular cylinder $x^2 + (y-1)^2 = 4$ and the plane $z = x$ is shown in Figure 11.96. We saw in Section 9.1 that trigonometric functions are particularly useful for circles. If we set $x = 2 \cos t$, then $y = 1 \pm 2 \sin t$. A set of parametric equations for the curve is therefore

$$x = 2 \cos t, \quad y = 1 + 2 \sin t, \quad z = 2 \cos t, \quad 0 \le t \le 2\pi.$$

Any range of values of t of length 2π traces the curve exactly once. To check that these equations specify the correct direction along the curve, we note that $t = 0$ gives the point $(2, 1, 2)$ and $t = \pi/2$ gives $(0, 3, 0)$. Since values of t between 0 and $\pi/2$ give one-quarter of the curve (rather than three-quarters), points are indeed generated in the required direction indicated by the arrowhead in Figure 11.96. Had we chosen the equations

$$x = 2 \cos t, \quad y = 1 - 2 \sin t, \quad z = 2 \cos t, \quad 0 \le t \le 2\pi,$$

we would have generated the same set of points traced in the opposite direction.

FIGURE 11.94 Curve of intersection of paraboloid and plane

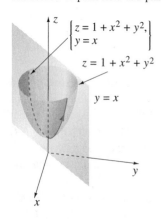

FIGURE 11.95 Line of intersection of two planes

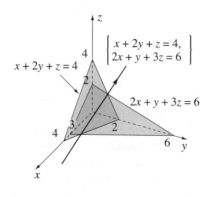

FIGURE 11.96 Curve of intersection of circular cylinder and plane

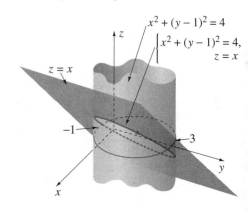

Because computers and graphing calculators produce excellent plots of parametrically defined three-dimensional curves, there is incentive to represent curves parametrically rather than as the intersection of two surfaces. Plots of the curves in Example 11.39 are shown in Figures 11.97.

FIGURE 11.97a   FIGURE 11.97b   FIGURE 11.97c

Computer plots of the curves in Example 11.39

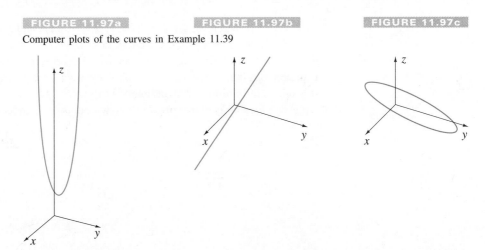

> ## DEFINITION 11.17
> A curve C: $x = x(t)$, $y = y(t)$, $z = z(t)$, $\alpha \leq t \leq \beta$, is said to be a **continuous curve** if each of the functions $x(t)$, $y(t)$, and $z(t)$ is continuous for $\alpha \leq t \leq \beta$.

Geometrically, this implies that the curve is at no point separated. Each of the curves in Example 11.39 is therefore continuous.

A curve is said to be a **closed curve** if its initial and final points are the same. Circles and ellipses are closed curves. Straight-line segments, parabolas, and hyperbolas are not closed.

Vector Representation of Curves

The position vector or displacement vector of a point $P(x, y, z)$ in space is

$$\mathbf{r} = (x, y, z) = x\hat{\mathbf{i}} + y\hat{\mathbf{j}} + z\hat{\mathbf{k}}.$$

We visualize it as the vector drawn from the origin to P (Figure 11.98). If we consider only points that lie on a curve defined parametrically by 11.63, then for these points we can write that

$$\mathbf{r} = \mathbf{r}(t) = x(t)\hat{\mathbf{i}} + y(t)\hat{\mathbf{j}} + z(t)\hat{\mathbf{k}}, \quad \alpha \leq t \leq \beta. \tag{11.64}$$

As t varies from $t = \alpha$ to $t = \beta$, the tip of this vector traces the curve C from initial point to final point. We call 11.64 the **vector representation of a curve**.

FIGURE 11.98   Vector representation of a curve

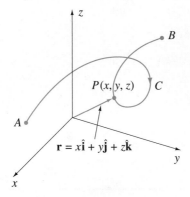

EXAMPLE 11.40

Draw and plot the curve with position vector

$$\mathbf{r} = \mathbf{r}(t) = (2\cos t)\hat{\mathbf{i}} + (3\sin t)\hat{\mathbf{j}} + (t/10)\hat{\mathbf{k}}, \quad 0 \le t \le 6\pi.$$

SOLUTION   When we set $x = 2\cos t$, $y = 3\sin t$, and $z = (t/10)$, it is clear that $x^2/4 + y^2/9 = 1$. This means that the curve lies on the elliptic cylinder $x^2/4 + y^2/9 = 1$. As t increases from 0 to 6π, values of z increase linearly from 0 to $3\pi/5$. What we have therefore is three loops that rise around the elliptic cylinder. It is part of what is called an *elliptic helix*. It is drawn in Figure 11.99a and plotted in Figure 11.99b. If the curve is given width, it could represent three coils of a spring or three windings of an inductor.

FIGURE 11.99a   Drawing an elliptic helix   FIGURE 11.99b   Computer plot of an elliptic helix

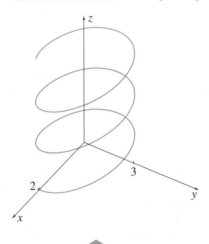

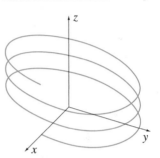

EXAMPLE 11.41

Draw and plot the curve with position vector

$$\mathbf{r} = \mathbf{r}(t) = t\hat{\mathbf{i}} + t^2\hat{\mathbf{j}} + t\hat{\mathbf{k}}, \quad t \ge 0.$$

SOLUTION   When we set $x = t$, $y = t^2$, and $z = t$, then $z = x$ and $y = x^2$. These imply that $\mathbf{r} = \mathbf{r}(t)$ describes points on the curve of intersection of the surfaces $y = x^2$ and $z = x$ (Figure 11.100a). Because $t \ge 0$, only that half of the curve of intersection in the first octant is defined by $\mathbf{r} = \mathbf{r}(t)$. The curve is plotted in Figure 11.100b.

FIGURE 11.100a   Curve $\mathbf{r} = t\hat{\mathbf{i}} + t^2\hat{\mathbf{j}} + t\hat{\mathbf{k}}$
drawn as intersection of surfaces $y = x^2$, $z = x$

FIGURE 11.100b   Computer
plot of curve $\mathbf{r} = t\hat{\mathbf{i}} + t^2\hat{\mathbf{j}} + t\hat{\mathbf{k}}$

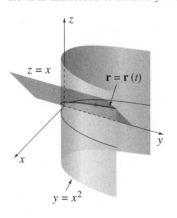

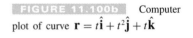

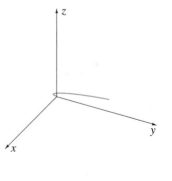

EXERCISES 11.10

In Exercises 1–10 find parametric and vector representations for the curve. Draw and plot each curve.

1. $x + 2y + 3z = 6$, $y - 2z = 3$ directed so that z increases along the curve

2. $x^2 + y^2 = 2$, $z = 4$ directed so that y increases in the first octant

3. $x^2 + y^2 = 2$, $x + y + z = 1$ directed so that y decreases when x is positive

4. $z = x^2 + y^2$, $x^2 + y^2 = 5$ directed clockwise as viewed from the origin

5. $z + 2x^2 = 1$, $y = z$ directed so that x decreases along the curve

6. $z = \sqrt{x^2 + y^2}$, $y = x$ directed so that y increases when x is positive

7. $z = x + y$, $y = x^2$ directed so that x increases along the curve

* 8. $z = \sqrt{4 - x^2 - y^2}$, $x^2 + y^2 - 2y = 0$ directed so that z decreases when x is positive

* 9. $x = \sqrt{z}$, $z = y^2$ directed away from the origin in the first octant

* 10. $z = \sqrt{x^2 + y^2}$, $y = x^2$ directed so that y decreases in the first octant.

In Exercises 11–15 draw and plot the curve with the given position vector.

11. $\mathbf{r}(t) = t\hat{\mathbf{i}} + t\hat{\mathbf{j}} + t^2\hat{\mathbf{k}}$, $t \geq 0$

12. $\mathbf{r}(t) = (2\cos t)\hat{\mathbf{i}} + (2\sin t)\hat{\mathbf{j}} + 3t\hat{\mathbf{k}}$, $0 \leq t \leq 4\pi$

13. $\mathbf{r}(t) = (t - 2)\hat{\mathbf{i}} + (2 - 3t)\hat{\mathbf{j}} + 5t\hat{\mathbf{k}}$

14. $\mathbf{r}(t) = (t^2 - t)\hat{\mathbf{i}} + t\hat{\mathbf{j}} + 5\hat{\mathbf{k}}$

15. $\mathbf{r}(t) = (\cos t)\hat{\mathbf{i}} + (\sin t)\hat{\mathbf{j}} + (\cos t)\hat{\mathbf{k}}$, $0 \leq t \leq \pi$

11.11 Tangent Vectors and Lengths of Curves

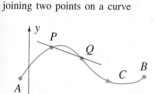

FIGURE 11.101 Line joining two points on a curve

If C is a curve in the xy-plane (Figure 11.101), then the tangent line to C at P is defined as the limiting position of the line PQ as Q moves along C toward P (see Section 3.1). We take the same approach in defining **tangent vectors to curves** in an arbitrary plane or in space. On curve C defined by 11.63, we let P and Q be the points corresponding to the parameter values t and $t + h$. Position vectors of P and Q are then

$$\mathbf{r}(t) = x(t)\hat{\mathbf{i}} + y(t)\hat{\mathbf{j}} + z(t)\hat{\mathbf{k}}$$

and

$$\mathbf{r}(t + h) = x(t + h)\hat{\mathbf{i}} + y(t + h)\hat{\mathbf{j}} + z(t + h)\hat{\mathbf{k}}$$

(Figure 11.102), and the vector joining P to Q is

$$\mathbf{PQ} = \mathbf{r}(t + h) - \mathbf{r}(t).$$

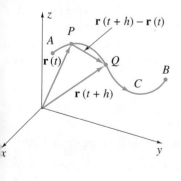

FIGURE 11.102 Limit of vector joining two points on a curve leads to tangent vector to curve

If we let h approach zero, then Q moves along C toward P, and the direction of $\mathbf{PQ}$ becomes closer to what seems to be a reasonable definition of the tangent direction to C at P. Perhaps then we should define $\lim_{h \to 0}[\mathbf{r}(t + h) - \mathbf{r}(t)]$ as a tangent vector to C at P. Unfortunately, the limit vector has length zero, and therefore

$$\lim_{h \to 0}[\mathbf{r}(t + h) - \mathbf{r}(t)] = \mathbf{0}.$$

If, however, we divide $\mathbf{r}(t + h) - \mathbf{r}(t)$ by h, then the resulting vector

$$\frac{\mathbf{r}(t + h) - \mathbf{r}(t)}{h}$$

is not equal to $\mathbf{PQ}$, but it does have the same direction as $\mathbf{PQ}$. Consider, then, taking the limit of this vector as h approaches zero:

$$\lim_{h \to 0} \frac{\mathbf{r}(t + h) - \mathbf{r}(t)}{h}.$$

FIGURE 11.103 Tangent
vector to a curve

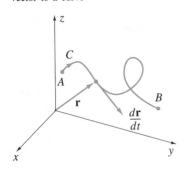

If the limit vector exists, then it will be tangent to C at P. But according to equation 11.56, this limit defines the derivative $d\mathbf{r}/dt$:

$$\frac{d\mathbf{r}}{dt} = \lim_{h \to 0} \frac{\mathbf{r}(t+h) - \mathbf{r}(t)}{h} = \frac{dx}{dt}\hat{\mathbf{i}} + \frac{dy}{dt}\hat{\mathbf{j}} + \frac{dz}{dt}\hat{\mathbf{k}}, \qquad (11.65)$$

provided that each of the derivatives dx/dt, dy/dt, and dz/dt exists. We have just established the following result.

THEOREM 11.10

If $\mathbf{r} = \mathbf{r}(t) = x(t)\hat{\mathbf{i}} + y(t)\hat{\mathbf{j}} + z(t)\hat{\mathbf{k}}$, $\alpha \le t \le \beta$, is the vector representation of a curve C, then at any point on C at which $x'(t)$, $y'(t)$, and $z'(t)$ all exist and do not vanish simultaneously,

$$\mathbf{T} = \frac{d\mathbf{r}}{dt} = \frac{dx}{dt}\hat{\mathbf{i}} + \frac{dy}{dt}\hat{\mathbf{j}} + \frac{dz}{dt}\hat{\mathbf{k}} \qquad (11.66)$$

is a tangent vector to C (Figure 11.103).

FIGURE 11.104 Two
tangent vectors exist at each point
on a curve, one in the opposite
direction to the other

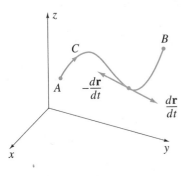

There are two tangent directions at any point on a curve. One of these has been shown to be $d\mathbf{r}/dt$; the other must be $-d\mathbf{r}/dt$ (Figure 11.104). How can we tell which one is $d\mathbf{r}/dt$? A closer analysis of the limit in 11.65 indicates the following (see Exercise 17).

COROLLARY 11.10.1

The tangent vector $d\mathbf{r}/dt$ to a curve C: $x = x(t)$, $y = y(t)$, $z = z(t)$, $\alpha \le t \le \beta$, always points in the direction in which the parameter t increases along C.

FIGURE 11.105 Smooth
curve

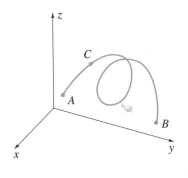

DEFINITION 11.18

A curve C: $x = x(t)$, $y = y(t)$, $z = z(t)$, $\alpha \le t \le \beta$, is said to be a **smooth curve** if the derivatives $x'(t)$, $y'(t)$, and $z'(t)$ are all continuous for $\alpha < t < \beta$ and do not vanish simultaneously for $\alpha < t < \beta$.

Since $x'(t)$, $y'(t)$, and $z'(t)$ are the components of a tangent vector to C, this definition implies that along a smooth curve, small changes in t produce small changes in the direction of the tangent vector. In other words, the tangent vector turns gradually, or "smoothly." The curve in Figure 11.105 is smooth; that in Figure 11.106 is not because abrupt changes in the direction of the curve occur at P and Q. According to the following definition, this curve is piecewise smooth.

FIGURE 11.106 Piecewise-
smooth curve

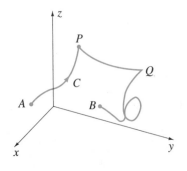

DEFINITION 11.19

A curve is said to be a **piecewise-smooth curve** if it is continuous and can be divided into a finite number of smooth subcurves.

EXAMPLE 11.42

For the curve in Example 11.41, find a tangent vector at the point $(3, 9, 3)$.

SOLUTION   A tangent vector to this curve at any point on the curve is

$$\frac{d\mathbf{r}}{dt} = \frac{dx}{dt}\hat{\mathbf{i}} + \frac{dy}{dt}\hat{\mathbf{j}} + \frac{dz}{dt}\hat{\mathbf{k}} = \hat{\mathbf{i}} + 2t\hat{\mathbf{j}} + \hat{\mathbf{k}}.$$

Since $t = 3$ yields the point $(3, 9, 3)$, a tangent vector at this point is $\mathbf{r}'(3) = \hat{\mathbf{i}} + 6\hat{\mathbf{j}} + \hat{\mathbf{k}}$.

EXAMPLE 11.43

Find a tangent vector at the point $(2, 0, 3)$ to the helix

$$x = 2\cos t, \quad y = 2\sin t, \quad z = \frac{3t}{2\pi}, \quad t \geq 0.$$

Is the helix smooth?

SOLUTION   A tangent vector to the helix at any point is

$$\frac{d\mathbf{r}}{dt} = \frac{dx}{dt}\hat{\mathbf{i}} + \frac{dy}{dt}\hat{\mathbf{j}} + \frac{dz}{dt}\hat{\mathbf{k}} = (-2\sin t)\hat{\mathbf{i}} + (2\cos t)\hat{\mathbf{j}} + \left(\frac{3}{2\pi}\right)\hat{\mathbf{k}}.$$

Since $t = 2\pi$ yields the point $(2, 0, 3)$, a tangent vector at this point is $\mathbf{r}'(2\pi) = 2\hat{\mathbf{j}} + (3/(2\pi))\hat{\mathbf{k}}$ (Figure 11.107). Since $x'(t)$, $y'(t)$, and $z'(t)$ are all continuous functions, and they are never simultaneously zero, the helix is indeed smooth.

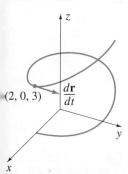

FIGURE 11.107   Tangent vector to a helix

Unit Tangent Vectors

When a curve in the xy-plane is defined parametrically by

$$C: \quad x = x(t), \quad y = y(t), \quad \alpha \leq t \leq \beta, \tag{11.67}$$

a tangent vector to C is

$$\mathbf{T} = \frac{d\mathbf{r}}{dt} = \frac{dx}{dt}\hat{\mathbf{i}} + \frac{dy}{dt}\hat{\mathbf{j}}, \tag{11.68}$$

and this tangent vector points in the direction in which t increases along C. To produce a **unit tangent vector** to C at any point, we divide $\mathbf{T}$ by its length:

$$\hat{\mathbf{T}} = \frac{\mathbf{T}}{|\mathbf{T}|} = \frac{d\mathbf{r}/dt}{|d\mathbf{r}/dt|}. \tag{11.69}$$

We now show that if length along C is used as the parameter by which to specify its points, then division by $|\mathbf{T}|$ is unnecessary.

In Section 7.3 we showed that small lengths along a plane curve C can be approximated by straight-line lengths along tangent lines to the curve, and that the total length of a smooth curve from A to B (Figure 11.108) is

$$L = \int_A^B \sqrt{(dx)^2 + (dy)^2}.$$

With parametric equations 11.67 we can write this formula as a definite integral with respect to t (see also equation 9.3):

$$L = \int_\alpha^\beta \sqrt{\left[\left(\frac{dx}{dt}\right)^2 + \left(\frac{dy}{dt}\right)^2\right](dt)^2} = \int_\alpha^\beta \sqrt{\left(\frac{dx}{dt}\right)^2 + \left(\frac{dy}{dt}\right)^2}\, dt. \tag{11.70}$$

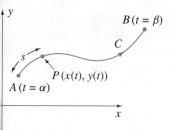

FIGURE 11.108   Length along a curve

Furthermore, if we denote by $s = s(t)$ the length of that part of C from its initial point A (where $t = \alpha$) to any point $P(x(t), y(t))$ on C (Figure 11.108), then $s(t)$ is defined by the integral

$$s(t) = \int_{\alpha}^{t} \sqrt{\left(\frac{dx}{dt}\right)^2 + \left(\frac{dy}{dt}\right)^2} \, dt. \tag{11.71}$$

It follows, then, that the derivative of $s(t)$ is

$$\frac{ds}{dt} = \sqrt{\left(\frac{dx}{dt}\right)^2 + \left(\frac{dy}{dt}\right)^2}. \tag{11.72}$$

But according to 11.68, $d\mathbf{r}/dt$ is a tangent vector to C with the same length:

$$\left|\frac{d\mathbf{r}}{dt}\right| = \sqrt{\left(\frac{dx}{dt}\right)^2 + \left(\frac{dy}{dt}\right)^2} = \frac{ds}{dt}. \tag{11.73}$$

When this equation is multiplied by dt, it gives

$$|d\mathbf{r}| = \sqrt{(dx)^2 + (dy)^2} = ds. \tag{11.74}$$

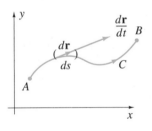

FIGURE 11.109   Small lengths along the tangent line to a curve in the xy-plane are regarded as small lengths along the curve itself

This equation states that ds is the length of the tangent vector $d\mathbf{r} = dx\hat{\mathbf{i}} + dy\hat{\mathbf{j}}$, and therefore ds is a measure of length along the tangent line to C. In spite of this we often think of ds as a measure of small lengths along C itself (Figure 11.109), and that ds is approximated by the tangential straight-line length

$$|d\mathbf{r}| = \sqrt{(dx)^2 + (dy)^2}.$$

Note too that if we use length s along C as a parameter, then the chain rule applied to $\mathbf{r} = \mathbf{r}(s)$, $s = s(t)$ gives

$$\frac{d\mathbf{r}}{dt} = \frac{d\mathbf{r}}{ds}\frac{ds}{dt}. \tag{11.75}$$

(The chain rule is proved in Exercise 26 of Section 11.9.) Consequently, equation 11.73 implies that

$$\frac{d\mathbf{r}}{ds} = \frac{d\mathbf{r}/dt}{ds/dt} = \frac{d\mathbf{r}/dt}{|d\mathbf{r}/dt|}. \tag{11.76}$$

What we have shown, then, is that if we choose length along a curve C as the parameter by which to specify points on the curve ($C: x = x(s)$, $y = y(s)$), then the vector

$$\hat{\mathbf{T}} = \frac{d\mathbf{r}}{ds} = \frac{dx}{ds}\hat{\mathbf{i}} + \frac{dy}{ds}\hat{\mathbf{j}} \tag{11.77}$$

is a unit tangent vector to C. In addition, the corollary to Theorem 11.10 implies that $d\mathbf{r}/ds$ points in the direction in which s increases along C. This suggests perhaps that we should always set up parametric equations for a curve with length along the curve as parameter. Theoretically this is quite acceptable, but practically it is impossible. For most curves we have enough difficulty just finding a set of parametric equations, let alone finding that set with length along the curve as parameter. If we then use a parameter t other than length along the curve, a unit tangent vector is calculated according to 11.69.

These results can be extended to space curves as well. When a smooth curve C has parametric equations 11.63, equation 11.69 still defines a unit tangent vector to C, but because C is a space curve, $d\mathbf{r}/dt$ is calculated according to 11.66.

Corresponding to formula 11.71 for length along a curve in the xy-plane, length along a smooth curve in space is defined by the definite integral:

$$s(t) = \int_\alpha^t \sqrt{\left(\frac{dx}{dt}\right)^2 + \left(\frac{dy}{dt}\right)^2 + \left(\frac{dz}{dt}\right)^2}\, dt. \tag{11.78}$$

These two results imply that

$$\left|\frac{d\mathbf{r}}{dt}\right| = \sqrt{\left(\frac{dx}{dt}\right)^2 + \left(\frac{dy}{dt}\right)^2 + \left(\frac{dz}{dt}\right)^2} = \frac{ds}{dt}, \tag{11.79}$$

the three-space analogue of 11.73. Once again we are led to the fact that when s is used as parameter along C, then

$$\hat{\mathbf{T}} = \frac{d\mathbf{r}}{ds} = \frac{dx}{ds}\hat{\mathbf{i}} + \frac{dy}{ds}\hat{\mathbf{j}} + \frac{dz}{ds}\hat{\mathbf{k}} \tag{11.80}$$

is a unit tangent vector to C. In addition, multiplication of 11.79 by dt yields

$$|d\mathbf{r}| = ds = \sqrt{(dx)^2 + (dy)^2 + (dz)^2}, \tag{11.81}$$

indicating that small lengths ds along C (Figure 11.110) are defined in terms of small lengths $|d\mathbf{r}|$ along the tangent line to C.

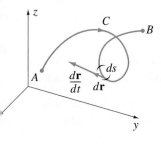

FIGURE 11.110   Small lengths along the tangent line to curve in space are regarded as small lengths along the curve itself

---

EXAMPLE 11.44

Find a unit tangent vector to the curve

$$C : \quad x = \sin t, \quad y = 2\cos t, \quad z = 2t/\pi, \quad t \geq 0$$

at the point $(0, -2, 2)$.

SOLUTION   A tangent vector to C at any point is

$$\frac{d\mathbf{r}}{dt} = \frac{dx}{dt}\hat{\mathbf{i}} + \frac{dy}{dt}\hat{\mathbf{j}} + \frac{dz}{dt}\hat{\mathbf{k}} = \cos t\,\hat{\mathbf{i}} - 2\sin t\,\hat{\mathbf{j}} + \left(\frac{2}{\pi}\right)\hat{\mathbf{k}}.$$

Since $t = \pi$ yields the point $(0, -2, 2)$, a tangent vector at this point is

$$\mathbf{r}'(\pi) = -\hat{\mathbf{i}} + (2/\pi)\hat{\mathbf{k}}.$$

A unit tangent vector is then

$$\hat{\mathbf{T}} = \frac{-\hat{\mathbf{i}} + (2/\pi)\hat{\mathbf{k}}}{\sqrt{1 + 4/\pi^2}} = \frac{-\pi\hat{\mathbf{i}} + 2\hat{\mathbf{k}}}{\sqrt{4 + \pi^2}}.$$

EXAMPLE 11.45

Find the length of that part of the curve $x = y^{2/3}$, $x = z^{2/3}$ between the points $(0, 0, 0)$ and $(4, 8, 8)$.

SOLUTION   If we use $x = t$, $y = t^{3/2}$, $z = t^{3/2}$, $0 \leq t \leq 4$, as parametric equations for the curve, then

$$L = \int_0^4 \sqrt{\left(\frac{dx}{dt}\right)^2 + \left(\frac{dy}{dt}\right)^2 + \left(\frac{dz}{dt}\right)^2}\, dt = \int_0^4 \sqrt{1 + \left(\frac{3}{2}\sqrt{t}\right)^2 + \left(\frac{3}{2}\sqrt{t}\right)^2}\, dt$$

$$= \int_0^4 \sqrt{1 + \frac{9t}{2}}\, dt = \left\{\frac{4}{27}\left(1 + \frac{9t}{2}\right)^{3/2}\right\}_0^4 = \frac{4}{27}(19\sqrt{19} - 1).$$

EXERCISES 11.11

In Exercises 1–5 express the curve in vector form and find the unit tangent vector $\hat{\mathbf{T}}$ at each point on the curve.

1. $x = \sin t, \ y = \cos t, \ z = t, \ -\infty < t < \infty$

2. $x = t, \ y = t^2, \ z = t^3, \ t \geq 1$

3. $x = (t - 1)^2, \ y = (t + 1)^2, \ z = -t, \ -3 \leq t \leq 4$

4. $x + y = 5, \ x^2 - y = z$ from $(5, 0, 25)$ to $(0, 5, -5)$

5. $x + y + z = 4, \ x^2 + y^2 = 4, \ y \geq 0$ from $(2, 0, 2)$ to $(-2, 0, 6)$

In Exercises 6–10 find $\hat{\mathbf{T}}$ at the point.

6. $x = 4\cos t, \ y = 6\sin t, \ z = 2\sin t, \ -\infty < t < \infty;$ $(2\sqrt{2}, 3\sqrt{2}, \sqrt{2})$

7. $x = 2 - 5t, \ y = 1 + t, \ z = 6 + 4t, \ -\infty < t < \infty; (7, 0, 2)$

8. $x^2 + y^2 + z^2 = 4, \ z = \sqrt{x^2 + y^2}$, directed so that x increases when y is positive; $(1, 1, \sqrt{2})$

9. $x = y^2 + 1, \ z = x + 5$, directed so that y increases along the curve; $(5, 2, 10)$

10. $x^2 + (y - 1)^2 = 4, \ z = x$, directed so that z decreases when y is negative; $(2, 1, 2)$

In Exercises 11–14 find the length of the curve. Draw each curve.

11. $x = 2\cos t, \ y = 2\sin t, \ z = 3t, \ 0 \leq t \leq 2\pi$

12. $x = 2 - 5t, \ y = 1 + t, \ z = 6 + 4t, \ -1 \leq t \leq 0$

13. $x = t^3, \ y = t^2, \ z = t^3, \ 0 \leq t \leq 1$

14. $x = t, \ y = t^{3/2}, \ z = 4t^{3/2}, \ 1 \leq t \leq 4$

* **15.** In Definition 11.18 why are the derivatives assumed not to vanish simultaneously? *Hint:* Consider the curve $x = t^3, \ y = t^2, \ z = 0$.

* **16.** Find a unit tangent vector to the curve $\mathbf{r} = (\cos t + t\sin t)\hat{\mathbf{i}} + (\sin t - t\cos t)\hat{\mathbf{j}}, \ 0 \leq t \leq 2\pi$, called an involute of a circle.

* **17.** Show that the tangent vector $d\mathbf{r}/dt$ to a curve described by equation 11.64 always points in the direction in which t increases along the curve.

* **18.**
  (a) What happens when equation 11.66 is used to determine a tangent vector to the curve $x = t^2, \ y = t^3, \ z = t^2$ $-\infty < t < \infty$, at the origin?

  (b) Can you devise a way in which to obtain a tangent vector

11.12 Normal Vectors, Curvature, and Radius of Curvature

In discussing curves we distinguish between two types of properties: intrinsic and not intrinsic. An *intrinsic property* is one that is independent of the parameter used to specify the curve; a property that is not intrinsic is parameter dependent. To illustrate, the tangent vector $\mathbf{T} = d\mathbf{r}/dt$ in 11.66 is not intrinsic; a change of parameter results in a change in the length of $\mathbf{T}$. The unit tangent vector $\hat{\mathbf{T}}$, on the other hand, is intrinsic; there is only one unit tangent vector in the direction of the curve. The length of a curve from its initial point to an arbitrary point is an intrinsic property; a change of parameter along the curve does not affect length between points.

Because length along a curve is an intrinsic property, it is customary in theoretical discussions to use it as the parameter by which to specify points on the curve. When C is a smooth curve in the xy-plane, parametric equations for C in terms of length s along C take the form

$$C: \quad x = x(s), \quad y = y(s), \quad 0 \leq s \leq L. \tag{11.82}$$

FIGURE 11.111   There is essentially one normal direction to a curve in the xy-plane

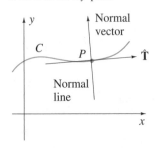

Normal Vectors to Curves

The normal line at a point P on a smooth curve C in the xy-plane is that line which is perpendicular to the tangent line to C at P (Figure 11.111). Any vector along this normal line is said to be a normal vector to the curve at P. Since the unit tangent vector to C at P is

$$\hat{\mathbf{T}} = \frac{d\mathbf{r}}{ds} = \frac{dx}{ds}\hat{\mathbf{i}} + \frac{dy}{ds}\hat{\mathbf{j}},$$

it follows that

$$\hat{\mathbf{N}} = -\frac{dy}{ds}\hat{\mathbf{i}} + \frac{dx}{ds}\hat{\mathbf{j}} \tag{11.83}$$

is a unit normal vector to C at P (note that $\hat{\mathbf{T}} \cdot \hat{\mathbf{N}} = 0$). Because there is only one direction normal to C at P, every normal vector to C at P must be some multiple $\lambda \hat{\mathbf{N}}$ of $\hat{\mathbf{N}}$.

The situation is quite different for space curves (Figure 11.112). If $\hat{\mathbf{T}}$ is the unit tangent vector to a smooth curve C, then there is an entire plane of normal vectors to C at P. In the following discussion, we single out two normal vectors called the *principal normal* and the *binormal*. Suppose that C is defined parametrically by

$$C : \quad x = x(s), \quad y = y(s), \quad z = z(s), \quad 0 \leq s \leq L, \tag{11.84}$$

and that $\hat{\mathbf{T}}$ is the unit tangent vector to C defined by 11.80. Because $\hat{\mathbf{T}}$ has unit length,

$$1 = \hat{\mathbf{T}} \cdot \hat{\mathbf{T}}.$$

If we use equation 11.59b to differentiate this equation with respect to s, we have

$$0 = \frac{d\hat{\mathbf{T}}}{ds} \cdot \hat{\mathbf{T}} + \hat{\mathbf{T}} \cdot \frac{d\hat{\mathbf{T}}}{ds} = 2\left(\hat{\mathbf{T}} \cdot \frac{d\hat{\mathbf{T}}}{ds}\right).$$

But if neither of the vectors $\hat{\mathbf{T}}$ nor $d\hat{\mathbf{T}}/ds$ is equal to zero, then the fact that their scalar product is equal to zero implies that they are perpendicular. In other words,

$$\mathbf{N} = \frac{d\hat{\mathbf{T}}}{ds} \tag{11.85}$$

is a normal vector to C at any point. The unit normal vector in this direction,

$$\hat{\mathbf{N}} = \frac{\mathbf{N}}{|\mathbf{N}|} = \frac{d\hat{\mathbf{T}}/ds}{|d\hat{\mathbf{T}}/ds|}, \tag{11.86}$$

is called the **principal normal (vector)** to C (Figure 11.113).

Because $\hat{\mathbf{N}}$ is defined in terms of intrinsic properties $\hat{\mathbf{T}}$ and s for a curve, it must also be an intrinsic property. It follows, then, that no matter what parameter is used to specify points on a curve, $\hat{\mathbf{N}}$ is always the same. But how do we find $\hat{\mathbf{N}}$ when a curve C is specified in terms of a parameter other than length along C, say, in the form

$$C : \quad x = x(t), \quad y = y(t), \quad z = z(t), \quad \alpha \leq t \leq \beta? \tag{11.87}$$

If $s = s(t)$ is length along C (measured from $t = \alpha$), then by the chain rule

$$\frac{d\hat{\mathbf{T}}}{ds} = \frac{d\hat{\mathbf{T}}}{dt}\frac{dt}{ds},$$

where dt/ds must be positive since both s and t increase along C. Consequently,

$$\hat{\mathbf{N}} = \frac{d\hat{\mathbf{T}}/ds}{|d\hat{\mathbf{T}}/ds|} = \frac{(d\hat{\mathbf{T}}/dt)(dt/ds)}{|(d\hat{\mathbf{T}}/dt)(dt/ds)|} = \frac{(d\hat{\mathbf{T}}/dt)(dt/ds)}{|d\hat{\mathbf{T}}/dt|dt/ds} = \frac{d\hat{\mathbf{T}}/dt}{|d\hat{\mathbf{T}}/dt|}. \tag{11.88}$$

In other words, for any parametrization of C whatsoever, the vector $d\hat{\mathbf{T}}/dt$ always points in the direction of the principal normal, and to find $\hat{\mathbf{N}}$, we simply find the unit vector in the direction of $d\hat{\mathbf{T}}/dt$.

In the study of space curves, a second normal vector to C, called the **binormal (vector)**, is defined by

$$\hat{\mathbf{B}} = \hat{\mathbf{T}} \times \hat{\mathbf{N}}. \tag{11.89}$$

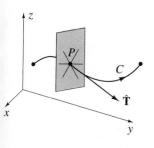

FIGURE 11.112 There is a plane of normal vectors to a curve in space

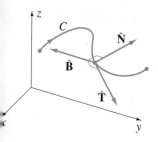

FIGURE 11.113 Principal normal and binormal to a curve

Since the cross product of two vectors is always perpendicular to each of the vectors, it follows that the binormal is perpendicular to both $\hat{\mathbf{T}}$ and $\hat{\mathbf{N}}$, and must therefore indeed be a normal vector to C (Figure 11.113).

We have singled out three vectors at each point P on a curve C: a unit tangent vector $\hat{\mathbf{T}}$ and two unit normal vectors $\hat{\mathbf{N}}$ and $\hat{\mathbf{B}}$. As P moves along C, these vectors constantly change direction but always have unit length.

EXAMPLE 11.46

Find $\hat{\mathbf{T}}$, $\hat{\mathbf{N}}$, and $\hat{\mathbf{B}}$ for the curve $x = t$, $y = t^2$, $z = t^2$, $t \geq 0$.

SOLUTION   The unit tangent vector $\hat{\mathbf{T}}$ is defined by

$$\hat{\mathbf{T}} = \frac{d\mathbf{r}/dt}{|d\mathbf{r}/dt|} = \frac{(1, 2t, 2t)}{\sqrt{1 + 4t^2 + 4t^2}} = \frac{(1, 2t, 2t)}{\sqrt{1 + 8t^2}}.$$

The principal normal $\hat{\mathbf{N}}$ lies along the vector $\mathbf{N} = d\hat{\mathbf{T}}/dt$, and according to equation 11.59a we can write that

$$\mathbf{N} = \frac{d\hat{\mathbf{T}}}{dt} = \frac{d}{dt}\left(\frac{1}{\sqrt{1 + 8t^2}}\right)(1, 2t, 2t) + \frac{1}{\sqrt{1 + 8t^2}}\frac{d}{dt}(1, 2t, 2t)$$

$$= \frac{-8t}{(1 + 8t^2)^{3/2}}(1, 2t, 2t) + \frac{1}{\sqrt{1 + 8t^2}}(0, 2, 2)$$

$$= \frac{1}{(1 + 8t^2)^{3/2}}[-8t(1, 2t, 2t) + (1 + 8t^2)(0, 2, 2)]$$

$$= \frac{(-8t, 2, 2)}{(1 + 8t^2)^{3/2}}.$$

The principal normal is therefore

$$\hat{\mathbf{N}} = \frac{\mathbf{N}}{|\mathbf{N}|} = \frac{(-8t, 2, 2)}{\sqrt{64t^2 + 4 + 4}} = \frac{(-4t, 1, 1)}{\sqrt{2 + 16t^2}}.$$

The binormal is

$$\hat{\mathbf{B}} = \hat{\mathbf{T}} \times \hat{\mathbf{N}} = \frac{(1, 2t, 2t)}{\sqrt{1 + 8t^2}} \times \frac{(-4t, 1, 1)}{\sqrt{2 + 16t^2}}$$

$$= \frac{1}{\sqrt{1 + 8t^2}\sqrt{2 + 16t^2}}\begin{vmatrix} \hat{\mathbf{i}} & \hat{\mathbf{j}} & \hat{\mathbf{k}} \\ 1 & 2t & 2t \\ -4t & 1 & 1 \end{vmatrix}$$

$$= \frac{1}{\sqrt{2}\sqrt{1 + 8t^2}\sqrt{1 + 8t^2}}(0, -1 - 8t^2, 1 + 8t^2)$$

$$= \frac{1 + 8t^2}{\sqrt{2}(1 + 8t^2)}(0, -1, 1)$$

$$= \frac{(0, -1, 1)}{\sqrt{2}}.$$

The significance of the fact that the binormal has constant direction can be seen from a drawing of the curve. Because the parametric equations imply that $y = x^2$ and $z = y$, the curve is the curve of intersection of these two surfaces (Figure 11.114). Since the curve lies in the plane $-y + z = 0$, and a normal vector to this plane is $(0, -1, 1)$, it follows that $(0, -1, 1)$ is always normal to the curve. But this is precisely the direction of $\hat{\mathbf{B}}$. In other words, constant $\hat{\mathbf{B}}$ implies that the curve lies in a plane that has $\hat{\mathbf{B}}$ as normal (see Exercise 30).

FIGURE 11.114   Tangent, principal normal, and binormal to the curve $x = t$, $y = t^2$, $z = t^2$, $t \geq 0$

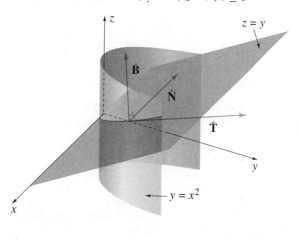

EXAMPLE 11.47

Show that for a smooth curve C: $x = x(t)$, $y = y(t)$, $\alpha \leq t \leq \beta$ in the xy-plane, the principal normal is

$$\hat{\mathbf{N}} = \operatorname{sgn}\left(\frac{dy}{dt}\frac{d^2x}{dt^2} - \frac{dx}{dt}\frac{d^2y}{dt^2}\right)\frac{(dy/dt, -dx/dt)}{\sqrt{(dx/dt)^2 + (dy/dt)^2}},$$

where the signum function $\operatorname{sgn}(u)$ is defined by

$$\operatorname{sgn}(u) = \begin{cases} 1, & \text{if } u > 0, \\ 0, & \text{if } u = 0, \\ -1, & \text{if } u < 0. \end{cases}$$

SOLUTION   The unit tangent vector to C is

$$\hat{\mathbf{T}} = \frac{(dx/dt, dy/dt)}{\sqrt{(dx/dt)^2 + (dy/dt)^2}}.$$

For simplicity in notation, we use a dot "·" above a variable to indicate that the variable is differentiated with respect to t. For example, $\dot{x} = dx/dt$ and $\ddot{x} = d^2x/dt^2$. With this notation,

$$\hat{\mathbf{T}} = \frac{(\dot{x}, \dot{y})}{\sqrt{\dot{x}^2 + \dot{y}^2}}.$$

By equation 11.88, $\hat{\mathbf{N}} = (d\hat{\mathbf{T}}/dt)/|d\hat{\mathbf{T}}/dt|$, where

$$
\begin{aligned}
\frac{d\hat{\mathbf{T}}}{dt} &= \frac{d}{dt}\frac{(\dot{x}, \dot{y})}{\sqrt{\dot{x}^2 + \dot{y}^2}} = \frac{d}{dt}\left(\frac{1}{\sqrt{\dot{x}^2 + \dot{y}^2}}\right)(\dot{x}\hat{\mathbf{i}} + \dot{y}\hat{\mathbf{j}}) + \frac{1}{\sqrt{\dot{x}^2 + \dot{y}^2}}(\ddot{x}\hat{\mathbf{i}} + \ddot{y}\hat{\mathbf{j}}) \\
&= \left[\frac{-\dot{x}\ddot{x} - \dot{y}\ddot{y}}{(\dot{x}^2 + \dot{y}^2)^{3/2}}\right](\dot{x}\hat{\mathbf{i}} + \dot{y}\hat{\mathbf{j}}) + \frac{1}{\sqrt{\dot{x}^2 + \dot{y}^2}}(\ddot{x}\hat{\mathbf{i}} + \ddot{y}\hat{\mathbf{j}}) \\
&= \frac{1}{(\dot{x}^2 + \dot{y}^2)^{3/2}}[-(\dot{x}\ddot{x} + \dot{y}\ddot{y})(\dot{x}\hat{\mathbf{i}} + \dot{y}\hat{\mathbf{j}}) + (\dot{x}^2 + \dot{y}^2)(\ddot{x}\hat{\mathbf{i}} + \ddot{y}\hat{\mathbf{j}})] \\
&= \frac{1}{(\dot{x}^2 + \dot{y}^2)^{3/2}}[(-\dot{x}^2\ddot{x} - \dot{x}\dot{y}\ddot{y} + \dot{x}^2\ddot{x} + \dot{y}^2\ddot{x})\hat{\mathbf{i}} \\
&\quad + (-\dot{x}\dot{y}\ddot{x} - \dot{y}^2\ddot{y} + \dot{x}^2\ddot{y} + \dot{y}^2\ddot{y})\hat{\mathbf{j}}] \\
&= \frac{1}{(\dot{x}^2 + \dot{y}^2)^{3/2}}[\dot{y}(\dot{y}\ddot{x} - \dot{x}\ddot{y})\hat{\mathbf{i}} + \dot{x}(\dot{x}\ddot{y} - \dot{y}\ddot{x})\hat{\mathbf{j}}] \\
&= \frac{\dot{y}\ddot{x} - \dot{x}\ddot{y}}{(\dot{x}^2 + \dot{y}^2)^{3/2}}(\dot{y}\hat{\mathbf{i}} - \dot{x}\hat{\mathbf{j}}).
\end{aligned}
$$

If $\dot{y}\ddot{x} - \dot{x}\ddot{y}$ is positive, then

$$
\hat{\mathbf{N}} = \frac{\dot{y}\hat{\mathbf{i}} - \dot{x}\hat{\mathbf{j}}}{\sqrt{\dot{x}^2 + \dot{y}^2}};
$$

whereas if $\dot{y}\ddot{x} - \dot{x}\ddot{y}$ is negative, then

$$
\hat{\mathbf{N}} = \frac{-\dot{y}\hat{\mathbf{i}} + \dot{x}\hat{\mathbf{j}}}{\sqrt{\dot{x}^2 + \dot{y}^2}}.
$$

In other words,

$$
\hat{\mathbf{N}} = \text{sgn}(\dot{y}\ddot{x} - \dot{x}\ddot{y})\left(\frac{\dot{y}\hat{\mathbf{i}} - \dot{x}\hat{\mathbf{j}}}{\sqrt{\dot{x}^2 + \dot{y}^2}}\right).
$$

Curvature and Radius of Curvature

When length s along a smooth curve C is used as the parameter by which to identify point on the curve, the vector $\hat{\mathbf{T}} = d\mathbf{r}/ds$ is a unit tangent vector to C. Suppose we differentiat $\hat{\mathbf{T}}$ with respect to s to form $d\hat{\mathbf{T}}/ds = d^2\mathbf{r}/ds^2$. Since $\hat{\mathbf{T}}$ has constant unit length, only i direction can change; therefore, the derivative $d\hat{\mathbf{T}}/ds$ must be a measure of the rate of chang of the direction of $\hat{\mathbf{T}}$. Since $\hat{\mathbf{T}}$ is really our way of specifying the direction of the curve itsel we can also say that $d\hat{\mathbf{T}}/ds$ is a measure of how fast the direction of C changes. But exactl how does a vector $d\hat{\mathbf{T}}/ds$ that has both magnitude and direction measure the rate of change c the direction of C? We illustrate by example that it cannot be the direction of $d\hat{\mathbf{T}}/ds$; it mus be its magnitude that measures the rate of change of the direction of C. In Figure 11.115 w show a number of circles in the xy-plane, all of which are tangent to the y-axis at the origi Parametric equations for the circle with centre $(R, 0)$ and radius R in terms of length s alon the circle [as measured from $(2R, 0)$] are

$$
x = R + R\cos(s/R), \quad y = R\sin(s/R), \quad 0 \le s < 2\pi R.
$$

Consequently,

$$
\hat{\mathbf{T}} = \frac{d\mathbf{r}}{ds} = -\sin\left(\frac{s}{R}\right)\hat{\mathbf{i}} + \cos\left(\frac{s}{R}\right)\hat{\mathbf{j}}
$$

and

$$\frac{d\hat{\mathbf{T}}}{ds} = -\frac{1}{R}\cos\left(\frac{s}{R}\right)\hat{\mathbf{i}} - \frac{1}{R}\sin\left(\frac{s}{R}\right)\hat{\mathbf{j}}.$$

At the origin, $s = \pi R$, and

$$\frac{d\hat{\mathbf{T}}}{ds}\bigg|_{s=\pi R} = \frac{1}{R}\hat{\mathbf{i}}.$$

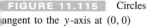

FIGURE 11.115   Circles tangent to the y-axis at $(0,0)$

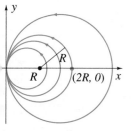

Thus, for each of the circles in Figure 11.115, the vector $d\hat{\mathbf{T}}/ds$ has exactly the same direction. Yet the rate of change of the direction of $\hat{\mathbf{T}}$ is not the same for each circle; the direction changes more rapidly as the radius of the circle decreases. We must conclude, therefore, that it cannot be the direction of $d\hat{\mathbf{T}}/ds$ that measures the rate of change of $\hat{\mathbf{T}}$. Since a vector has only length and direction, it must be the length of $d\hat{\mathbf{T}}/ds$ that measures this rate of change. The circles in Figure 11.115 certainly support this claim; the length of $d\hat{\mathbf{T}}/ds$ is $1/R$, and this quantity increases as the radii of the circles decrease. This agrees with the fact that the rate at which $\hat{\mathbf{T}}$ turns increases as R decreases. According to the following definition, we call $|d\hat{\mathbf{T}}/ds|$ *curvature* and $1/|d\hat{\mathbf{T}}/ds|$ *radius of curvature*.

DEFINITION 11.20

If $x = x(s)$, $y = y(s)$, $z = z(s)$, $0 \le s \le L$, are parametric equations for a smooth curve in terms of length s along the curve, we define the **curvature of the curve** at a point as

$$\kappa(s) = \left|\frac{d\hat{\mathbf{T}}}{ds}\right|, \tag{11.90}$$

its **radius of curvature** as

$$\rho(s) = \frac{1}{\kappa(s)}, \tag{11.91}$$

and its **circle of curvature** as that circle in the plane of $\hat{\mathbf{T}}$ and $\hat{\mathbf{N}}$ with centre at $\mathbf{r}(s)+\rho(s)\hat{\mathbf{N}}$ and radius $\rho(s)$.

The circle of curvature is illustrated in Figure 11.116.

For the circles in Figure 11.115 we have already shown that $d\hat{\mathbf{T}}/ds = R^{-1}\hat{\mathbf{i}}$. Consequently, for these circles, the curvature is always R^{-1}, and the radius of curvature is R, the radius of the circle. In other words, for a circle, the circle of curvature is the circle itself, its radius of curvature is its radius, and its curvature is the inverse of its radius. For the case when a curve is not a circle, we show in Exercise 27 that at any point on the curve the circle of curvature is in some sense the best-fitting circle to the curve at that point.

Because curvature and radius of curvature have been defined in terms of intrinsic properties $\hat{\mathbf{T}}$ and s for a curve, they must also be intrinsic properties. It follows, then, that no matter what parameter is used to specify points on a curve, curvature and radius of curvature are always the same. The following theorem shows how to calculate κ and ρ when the curve is specified in terms of a parameter other than length along the curve.

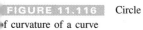

FIGURE 11.116   Circle of curvature of a curve

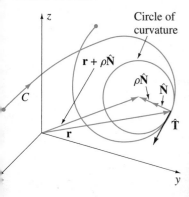

THEOREM 11.11

When a smooth curve is defined parametrically by

$$C : \quad x = x(t), \quad y = y(t), \quad z = z(t), \quad \alpha \le t \le \beta,$$

its curvature $\kappa(t)$ is given by

$$\kappa(t) = \frac{|\dot{\mathbf{r}} \times \ddot{\mathbf{r}}|}{|\dot{\mathbf{r}}|^3}, \tag{11.92}$$

where $\dot{\mathbf{r}} = d\mathbf{r}/dt$ and $\ddot{\mathbf{r}} = d^2\mathbf{r}/dt^2$.

PROOF If $s(t)$ is length along C (measured from $t = \alpha$), then by the chain rule

$$\kappa = \left|\frac{d\hat{\mathbf{T}}}{ds}\right| = \left|\frac{d\hat{\mathbf{T}}}{dt}\frac{dt}{ds}\right| = \frac{|d\hat{\mathbf{T}}/dt|}{|ds/dt|} = \frac{|d\hat{\mathbf{T}}/dt|}{ds/dt} = \frac{|d\hat{\mathbf{T}}/dt|}{|\dot{\mathbf{r}}|}$$

(see equation 11.79). Now, we can write $\dot{\mathbf{r}}$ in the form $\dot{\mathbf{r}} = |\dot{\mathbf{r}}|\hat{\mathbf{T}}$; therefore, using 11.59a, we have

$$\ddot{\mathbf{r}} = \left(\frac{d}{dt}|\dot{\mathbf{r}}|\right)\hat{\mathbf{T}} + |\dot{\mathbf{r}}|\frac{d\hat{\mathbf{T}}}{dt}$$

$$= \left(\frac{d}{dt}|\dot{\mathbf{r}}|\right)\hat{\mathbf{T}} + (|\dot{\mathbf{r}}||d\hat{\mathbf{T}}/dt|)\frac{d\hat{\mathbf{T}}/dt}{|d\hat{\mathbf{T}}/dt|}$$

$$= \left(\frac{d}{dt}|\dot{\mathbf{r}}|\right)\hat{\mathbf{T}} + \left(|\dot{\mathbf{r}}|\left|\frac{d\hat{\mathbf{T}}}{dt}\right|\right)\hat{\mathbf{N}}.$$

If we take the cross product of this vector with $\dot{\mathbf{r}}$, we get

$$\dot{\mathbf{r}} \times \ddot{\mathbf{r}} = \left(\frac{d}{dt}|\dot{\mathbf{r}}|\right)\dot{\mathbf{r}} \times \hat{\mathbf{T}} + \left(|\dot{\mathbf{r}}|\left|\frac{d\hat{\mathbf{T}}}{dt}\right|\right)\dot{\mathbf{r}} \times \hat{\mathbf{N}}$$

$$= \left(|\dot{\mathbf{r}}|\left|\frac{d\hat{\mathbf{T}}}{dt}\right|\right)\dot{\mathbf{r}} \times \hat{\mathbf{N}}. \quad \text{(since } \dot{\mathbf{r}} \text{ is parallel to } \hat{\mathbf{T}}\text{)}$$

Because $\dot{\mathbf{r}}$ is perpendicular to $\hat{\mathbf{N}}$, it follows that $|\dot{\mathbf{r}} \times \hat{\mathbf{N}}| = |\dot{\mathbf{r}}||\hat{\mathbf{N}}|\sin(\pi/2) = |\dot{\mathbf{r}}|$, and therefore

$$|\dot{\mathbf{r}} \times \ddot{\mathbf{r}}| = \left(|\dot{\mathbf{r}}|\left|\frac{d\hat{\mathbf{T}}}{dt}\right|\right)|\dot{\mathbf{r}}|.$$

Consequently,

$$\left|\frac{d\hat{\mathbf{T}}}{dt}\right| = \frac{|\dot{\mathbf{r}} \times \ddot{\mathbf{r}}|}{|\dot{\mathbf{r}}|^2}$$

and

$$\kappa = \kappa(t) = \frac{|\dot{\mathbf{r}} \times \ddot{\mathbf{r}}|}{|\dot{\mathbf{r}}|^3}.$$

For the radius of curvature, we have the following.

COROLLARY 11.11.1

When a smooth curve is defined in terms of an arbitrary parameter t,

$$\rho(t) = \frac{|\dot{\mathbf{r}}|^3}{|\dot{\mathbf{r}} \times \ddot{\mathbf{r}}|}. \tag{11.93}$$

EXAMPLE 11.48

Find curvature and radius of curvature for the curve in Example 11.46.

SOLUTION   According to 11.92,

$$\kappa(t) = \frac{|\dot{\mathbf{r}} \times \ddot{\mathbf{r}}|}{|\dot{\mathbf{r}}|^3} = \frac{|(1, 2t, 2t) \times (0, 2, 2)|}{|(1, 2t, 2t)|^3}$$

$$= \frac{1}{(1 + 4t^2 + 4t^2)^{3/2}} \left\| \begin{array}{ccc} \hat{\mathbf{i}} & \hat{\mathbf{j}} & \hat{\mathbf{k}} \\ 1 & 2t & 2t \\ 0 & 2 & 2 \end{array} \right\| = \frac{1}{(1 + 8t^2)^{3/2}} |(0, -2, 2)|$$

$$= \frac{2\sqrt{2}}{(1 + 8t^2)^{3/2}};$$

$$\rho(t) = \frac{1}{\kappa(t)} = \frac{(1 + 8t^2)^{3/2}}{2\sqrt{2}}.$$

Note in particular that as t increases, so does ρ, a fact that is certainly supported by Figure 11.114.

EXAMPLE 11.49

Show that for a smooth curve $y = y(x)$ in the xy-plane,

$$\kappa(x) = \frac{|y''|}{[1 + (y')^2]^{3/2}}.$$

SOLUTION   When we use x as parameter along the curve $y = y(x)$, parametric equations are $x = x$, $y = y(x)$. Then

$$\dot{\mathbf{r}} = (1, y'(x)), \quad \ddot{\mathbf{r}} = (0, y''(x)),$$

and

$$\dot{\mathbf{r}} \times \ddot{\mathbf{r}} = \begin{vmatrix} \hat{\mathbf{i}} & \hat{\mathbf{j}} & \hat{\mathbf{k}} \\ 1 & y' & 0 \\ 0 & y'' & 0 \end{vmatrix} = y'' \hat{\mathbf{k}}.$$

Thus,

$$\kappa(x) = \frac{|\dot{\mathbf{r}} \times \ddot{\mathbf{r}}|}{|\dot{\mathbf{r}}|^3} = \frac{|y''|}{|(1, y')|^3} = \frac{|y''|}{[1 + (y')^2]^{3/2}}.$$

According to equations 11.86 and 11.90, we can write that

$$\frac{d\hat{\mathbf{T}}}{ds} = \left| \frac{d\hat{\mathbf{T}}}{ds} \right| \hat{\mathbf{N}} = \kappa \hat{\mathbf{N}}. \tag{11.94}$$

This is called the **first Frenet–Serret formula** of differential geometry. We now derive the second of these formulas. Differentiation of $\hat{\mathbf{B}} = \hat{\mathbf{T}} \times \hat{\mathbf{N}}$ with respect to s gives

$$\frac{d\hat{\mathbf{B}}}{ds} = \frac{d\hat{\mathbf{T}}}{ds} \times \hat{\mathbf{N}} + \hat{\mathbf{T}} \times \frac{d\hat{\mathbf{N}}}{ds} = \kappa\hat{\mathbf{N}} \times \hat{\mathbf{N}} + \hat{\mathbf{T}} \times \frac{d\hat{\mathbf{N}}}{ds} = \hat{\mathbf{T}} \times \frac{d\hat{\mathbf{N}}}{ds}.$$

This equation implies that $d\hat{\mathbf{B}}/ds$ is perpendicular to $\hat{\mathbf{T}}$ (and $d\hat{\mathbf{N}}/ds$). Differentiation of $\hat{\mathbf{B}} \cdot \hat{\mathbf{B}} = 1$ with respect to s immediately implies that $d\hat{\mathbf{B}}/ds$ is perpendicular to $\hat{\mathbf{B}}$ also. It follows then that $d\hat{\mathbf{B}}/ds$ must be some multiple of $\hat{\mathbf{N}}$ ($\hat{\mathbf{N}}$ being perpendicular to $\hat{\mathbf{T}}$ and $\hat{\mathbf{B}}$), and we may therefore write that

$$\frac{d\hat{\mathbf{B}}}{ds} = -\tau\hat{\mathbf{N}}. \tag{11.95}$$

This is the **second Frenet–Serret formula**. Quantity τ is called the **torsion of the curve**. We can find formulas for τ in terms of s and in terms of an arbitrary parameter along the curve. The scalar product of equation 11.95 with $\hat{\mathbf{N}}$ gives

$$\tau = -\frac{d\hat{\mathbf{B}}}{ds} \cdot \hat{\mathbf{N}} = -\hat{\mathbf{N}} \cdot \frac{d}{ds}(\hat{\mathbf{T}} \times \hat{\mathbf{N}}) = -\hat{\mathbf{N}} \cdot \left(\hat{\mathbf{T}} \times \frac{d\hat{\mathbf{N}}}{ds} + \frac{d\hat{\mathbf{T}}}{ds} \times \hat{\mathbf{N}} \right)$$

$$= -\hat{\mathbf{N}} \cdot \left(\hat{\mathbf{T}} \times \frac{d\hat{\mathbf{N}}}{ds} + \kappa\hat{\mathbf{N}} \times \hat{\mathbf{N}} \right) = -\hat{\mathbf{N}} \cdot \hat{\mathbf{T}} \times \frac{d\hat{\mathbf{N}}}{ds}.$$

Since $\hat{\mathbf{T}} = d\mathbf{r}/ds$ and $\hat{\mathbf{N}} = \kappa^{-1}d\hat{\mathbf{T}}/ds = \kappa^{-1}d^2\mathbf{r}/ds^2$, it follows that

$$\tau = -\frac{1}{\kappa}\frac{d^2\mathbf{r}}{ds^2} \cdot \frac{d\mathbf{r}}{ds} \times \left(\frac{1}{\kappa}\frac{d^3\mathbf{r}}{ds^3} - \frac{1}{\kappa^2}\frac{d\kappa}{ds}\frac{d^2\mathbf{r}}{ds^2} \right)$$

$$= -\frac{1}{\kappa^2}\frac{d^2\mathbf{r}}{ds^2} \cdot \frac{d\mathbf{r}}{ds} \times \frac{d^3\mathbf{r}}{ds^3} + \frac{1}{\kappa^3}\frac{d\kappa}{ds}\frac{d^2\mathbf{r}}{ds^2} \cdot \frac{d\mathbf{r}}{ds} \times \frac{d^2\mathbf{r}}{ds^2}$$

$$= \frac{1}{\kappa^2}\frac{d\mathbf{r}}{ds} \cdot \frac{d^2\mathbf{r}}{ds^2} \times \frac{d^3\mathbf{r}}{ds^3}, \tag{11.96}$$

which expresses τ in terms of s. To express it in terms of an arbitrary parameter t along the curve rather than length s, we use

$$\frac{d\mathbf{r}}{ds} = \frac{dt}{ds}\frac{d\mathbf{r}}{dt} = \frac{dt}{ds}\dot{\mathbf{r}},$$

$$\frac{d^2\mathbf{r}}{ds^2} = \frac{d^2t}{ds^2}\frac{d\mathbf{r}}{dt} + \frac{dt}{ds}\frac{d}{dt}\left(\frac{d\mathbf{r}}{dt} \right)\frac{dt}{ds} = \frac{d^2t}{ds^2}\dot{\mathbf{r}} + \left(\frac{dt}{ds} \right)^2\ddot{\mathbf{r}},$$

$$\frac{d^3\mathbf{r}}{ds^3} = \frac{d^3t}{ds^3}\frac{d\mathbf{r}}{dt} + \frac{d^2t}{ds^2}\frac{d}{dt}\left(\frac{d\mathbf{r}}{dt} \right)\frac{dt}{ds} + 2\left(\frac{dt}{ds} \right)\left(\frac{d^2t}{ds^2} \right)\frac{d^2\mathbf{r}}{dt^2} + \left(\frac{dt}{ds} \right)^2\frac{d}{dt}\left(\frac{d^2\mathbf{r}}{dt^2} \right)\frac{dt}{ds}$$

$$= \frac{d^3t}{ds^3}\dot{\mathbf{r}} + 3\frac{d^2t}{ds^2}\frac{dt}{ds}\ddot{\mathbf{r}} + \left(\frac{dt}{ds} \right)^3\dddot{\mathbf{r}}.$$

With these,

$$\tau = \frac{1}{\kappa^2}\left(\frac{dt}{ds}\dot{\mathbf{r}} \right) \cdot \left[\frac{d^2t}{ds^2}\dot{\mathbf{r}} + \left(\frac{dt}{ds} \right)^2\ddot{\mathbf{r}} \right] \times \left[\frac{d^3t}{ds^3}\dot{\mathbf{r}} + 3\frac{d^2t}{ds^2}\frac{dt}{ds}\ddot{\mathbf{r}} + \left(\frac{dt}{ds} \right)^3\dddot{\mathbf{r}} \right]$$

$$= \frac{1}{\kappa^2}\left(\frac{dt}{ds} \right)^6 (\dot{\mathbf{r}} \cdot \ddot{\mathbf{r}} \times \dddot{\mathbf{r}}) = \frac{|\dot{\mathbf{r}}|^6}{|\dot{\mathbf{r}} \times \ddot{\mathbf{r}}|^2}\frac{1}{|\dot{\mathbf{r}}|^6}(\dot{\mathbf{r}} \cdot \ddot{\mathbf{r}} \times \dddot{\mathbf{r}}),$$

and hence,

$$\tau(t) = \frac{\dot{\mathbf{r}} \cdot \ddot{\mathbf{r}} \times \dddot{\mathbf{r}}}{|\dot{\mathbf{r}} \times \ddot{\mathbf{r}}|^2}. \tag{11.97}$$

The **third Frenet–Serret formula** expresses $d\hat{\mathbf{N}}/ds$ in terms of $\hat{\mathbf{T}}$ and $\hat{\mathbf{N}}$; it is developed in Exercise 29.

EXERCISES 11.12

In Exercises 1–5 find $\hat{\mathbf{N}}$ and $\hat{\mathbf{B}}$ at each point on the curve.

1. $x = \sin t$, $y = \cos t$, $z = t$, $-\infty < t < \infty$

2. $x = t$, $y = t^2$, $z = t^3$, $t \geq 1$

3. $x = (t - 1)^2$, $y = (t + 1)^2$, $z = -t$, $-3 \leq t \leq 4$

4. $x + y = 5$, $x^2 - y = z$, from $(5, 0, 25)$ to $(0, 5, -5)$

5. $z = x$, $x^2 + y^2 = 4$, $y \geq 0$, from $(2, 0, 2)$ to $(-2, 0, -2)$

In Exercises 6–10 find $\hat{\mathbf{N}}$ and $\hat{\mathbf{B}}$ at the point.

6. $x = 4 \cos t$, $y = 6 \sin t$, $z = 2 \sin t$, $-\infty < t < \infty$; $(2\sqrt{2}, 3\sqrt{2}, \sqrt{2})$

7. $x = 2 - 5t$, $y = 1 + t$, $z = 6 + 4t^3$, $-\infty < t < \infty$; $(7, 0, 2)$

8. $x^2 + y^2 + z^2 = 4$, $z = \sqrt{x^2 + y^2}$, directed so that x increases when y is positive; $(1, 1, \sqrt{2})$

9. $x = y^2 + 1$, $z = x + 5$, directed so that y increases along the curve; $(5, 2, 10)$

10. $x^2 + (y - 1)^2 = 4$, $x = z$, directed so that z decreases when y is negative; $(2, 1, 2)$

In Exercises 11–18 find the curvature and the radius of curvature of the curve (if they exist). Draw each curve.

11. $(x - h)^2 + (y - k)^2 = R^2$, $z = 0$, directed counterclockwise

12. $x = x_0 + at$, $y = y_0 + bt$, $z = z_0 + ct$, $-\infty < t < \infty$ (x_0, y_0, z_0, a, b, c all constants)

13. $x = t$, $y = t^2$, $z = 0$, $t \geq 0$

14. $x = e^t \cos t$, $y = e^t \sin t$, $z = t$, $-\infty < t < \infty$

15. $x = t$, $y = t^3$, $z = t^2$, $t \geq 0$

16. $x = 2 \cos t$, $y = 2 \sin t$, $z = 2 \sin t$, $0 \leq t < 2\pi$

17. $x = t + 1$, $y = t^2 - 1$, $z = t + 1$, $-\infty < t < \infty$

18. $x = t^2$, $y = t^4$, $z = 2t$, $-1 \leq t \leq 5$

19. At which points on the ellipse $b^2 x^2 + a^2 y^2 = a^2 b^2$ ($a > b$) is the curvature a maximum, and at which points is the curvature a minimum?

20. Show that curvature for a smooth curve $x = x(t)$, $y = y(t)$, $\alpha \leq t \leq \beta$, in the xy-plane can be expressed in the form

$$\kappa(t) = \frac{\left| \dfrac{dy}{dt} \dfrac{d^2 x}{dt^2} - \dfrac{dx}{dt} \dfrac{d^2 y}{dt^2} \right|}{\left[\left(\dfrac{dx}{dt} \right)^2 + \left(\dfrac{dy}{dt} \right)^2 \right]^{3/2}}.$$

21. Show that the only curves for which curvature is identically equal to zero are straight lines.

22. What happens to curvature at a point of inflection on the graph of function $y = f(x)$?

23. Let C be the curve $x = t$, $y = t^2$ in the xy-plane.

(a) At each point on C calculate the unit tangent vector $\hat{\mathbf{T}}$ and the principal normal $\hat{\mathbf{N}}$. What is $\hat{\mathbf{B}}$? (See Example 11.47 for $\hat{\mathbf{N}}$.)

(b) $\mathbf{F} = t^2 \hat{\mathbf{i}} + t^4 \hat{\mathbf{j}}$ is a vector that is defined at each point P on C. Denote by F_T and F_N the components of $\mathbf{F}$ in the directions $\hat{\mathbf{T}}$ and $\hat{\mathbf{N}}$ at P. Find F_T and F_N as functions of t.

(c) Express $\mathbf{F}$ in terms of $\hat{\mathbf{T}}$ and $\hat{\mathbf{N}}$.

* 24. Repeat Exercise 23 for the curve C : $x = 2 \cos t$, $y = 2 \sin t$, and the vector $\mathbf{F} = x^2 \hat{\mathbf{i}} + y^2 \hat{\mathbf{j}}$.

* 25. The vectors $\hat{\mathbf{T}}, \hat{\mathbf{N}},$ and $\hat{\mathbf{B}}$ were calculated at each point on the curve $x = t$, $y = t^2$, $z = t^2$ in Example 11.46. If $\mathbf{F} = t^2 \hat{\mathbf{i}} + 2t \hat{\mathbf{j}} - 3\hat{\mathbf{k}}$ is a vector defined along C, find the components of $\mathbf{F}$ in the directions $\hat{\mathbf{T}}, \hat{\mathbf{N}},$ and $\hat{\mathbf{B}}$. Express $\mathbf{F}$ in terms of $\hat{\mathbf{T}}, \hat{\mathbf{N}},$ and $\hat{\mathbf{B}}$.

* 26. Calculate $\hat{\mathbf{T}}, \hat{\mathbf{N}},$ and $\hat{\mathbf{B}}$ for the curve $x = \cos t$, $y = \sin t$, $z = t$. Express the vector $\mathbf{F} = x \hat{\mathbf{i}} + xy^2 \hat{\mathbf{j}} + \hat{\mathbf{k}}$ in terms of $\hat{\mathbf{T}}, \hat{\mathbf{N}},$ and $\hat{\mathbf{B}}$.

* 27. In this exercise we discuss our claim that the circle of curvature is the best-fitting circle to the curve at a point.

(a) Is it true that the circle of curvature at a point on a curve passes through that point?

(b) Show that the circle of curvature and curve share the same tangent line at their common point.

(c) Verify that the circle of curvature and curve have the same curvature at their common point.

* 28. If ϕ is the angle between $\hat{\mathbf{i}}$ and $\hat{\mathbf{T}}$ for a curve in the xy-plane (figure below), show that

$$\kappa(s) = \left| \frac{d\phi}{ds} \right|.$$

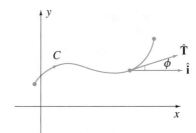

** 29. The third Frenet–Serret formula is

$$\frac{d\hat{\mathbf{N}}}{ds} = \tau \hat{\mathbf{B}} - \kappa \hat{\mathbf{T}}.$$

Verify this result by showing that $\hat{\mathbf{N}} = \hat{\mathbf{B}} \times \hat{\mathbf{T}}$ and then calculating $d\hat{\mathbf{N}}/ds$.

** 30. Show that a curve lies in a plane if and only if its torsion vanishes.

11.13 Displacement, Velocity, and Acceleration

In Sections 4.8 and 5.2 we introduced the concepts of displacement, velocity, and acceleration for moving objects, but indicated that our terminology at that time was somewhat loose. In particular, we stated that if $x = x(t)$ represents the position of a particle moving along the x-axis, then the instantaneous velocity of the particle is

$$v = \frac{dx}{dt}, \tag{11.98}$$

provided, of course, that t is time, and the acceleration of the particle is

$$a = \frac{dv}{dt} = \frac{d^2x}{dt^2}. \tag{11.99}$$

We illustrated by examples that given any one of $x(t)$, $v(t)$, or $a(t)$ and sufficient initial conditions, it is always possible to find the other two. There was nothing wrong with the calculations in the examples — they were correct — but our terminology was not quite correct. We now rectify this situation and give precise definitions of velocity and acceleration.

Suppose a particle moves along some curve C in space (under perhaps the influence of various forces), and that C is defined as a function of time t by the parametric equations

$$C: \quad x = x(t), \quad y = y(t), \quad z = z(t), \quad t \geq 0. \tag{11.100}$$

The position of the particle can then be described as a function of time by its position or displacement vector:

$$\mathbf{r} = \mathbf{r}(t) = x(t)\hat{\mathbf{i}} + y(t)\hat{\mathbf{j}} + z(t)\hat{\mathbf{k}}, \quad t \geq 0. \tag{11.101}$$

The **velocity v** of the particle at any time t is defined as the time rate of change of its displacement vector:

$$\mathbf{v} = \frac{d\mathbf{r}}{dt}. \tag{11.102}$$

Velocity, then, is a vector, and because of Theorem 11.8, the components of velocity are the derivatives of the components of displacement:

$$\mathbf{v} = \frac{d\mathbf{r}}{dt} = \frac{dx}{dt}\hat{\mathbf{i}} + \frac{dy}{dt}\hat{\mathbf{j}} + \frac{dz}{dt}\hat{\mathbf{k}}. \tag{11.103}$$

But according to Theorem 11.10, the vector $d\mathbf{r}/dt$ is tangent to the curve C (Figure 11.117). In other words, if a particle is at position P, and we draw its velocity vector with tail at P, then **v** is tangent to the trajectory.

In some applications it is the length or magnitude of velocity that is important, not its direction. This quantity, called **speed**, is therefore defined by

$$|\mathbf{v}| = \sqrt{\left(\frac{dx}{dt}\right)^2 + \left(\frac{dy}{dt}\right)^2 + \left(\frac{dz}{dt}\right)^2}. \tag{11.104}$$

Equation 11.79 implies that if $s(t)$ is length along the trajectory C [where $s(0) = 0$], then $|\mathbf{v}| = ds/dt$. In other words, speed is the time rate of change of distance travelled along C.

It is important to understand this difference between velocity and speed. Velocity is the time derivative of displacement; speed is the time derivative of distance travelled. Velocity is a vector; speed is a scalar — the magnitude of velocity.

The **acceleration** of the particle as it moves along the curve C in equations 11.100 is defined as the rate of change of velocity with respect to time:

$$\mathbf{a} = \frac{d\mathbf{v}}{dt} = \frac{d^2\mathbf{r}}{dt^2} = \frac{d^2x}{dt^2}\hat{\mathbf{i}} + \frac{d^2y}{dt^2}\hat{\mathbf{j}} + \frac{d^2z}{dt^2}\hat{\mathbf{k}}. \tag{11.105}$$

FIGURE 11.117   Velocity is always tangent to curve along which the particle travels

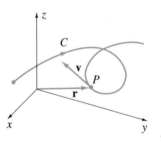

Acceleration, then, is also a vector; it is the derivative of velocity, and therefore its components are the derivatives of the components of the velocity vector. Alternatively, it is the second derivative of displacement and has components that are the second derivatives of the components of the displacement vector.

In the special case in which C is a curve in the xy-plane, definitions of displacement, velocity, speed, and acceleration become, respectively,

$$\mathbf{r} = x(t)\hat{\mathbf{i}} + y(t)\hat{\mathbf{j}}, \tag{11.106a}$$

$$\mathbf{v} = \frac{d\mathbf{r}}{dt} = \frac{dx}{dt}\hat{\mathbf{i}} + \frac{dy}{dt}\hat{\mathbf{j}}, \tag{11.106b}$$

$$|\mathbf{v}| = \sqrt{\left(\frac{dx}{dt}\right)^2 + \left(\frac{dy}{dt}\right)^2}, \tag{11.106c}$$

$$\mathbf{a} = \frac{d\mathbf{v}}{dt} = \frac{d^2\mathbf{r}}{dt^2} = \frac{d^2x}{dt^2}\hat{\mathbf{i}} + \frac{d^2y}{dt^2}\hat{\mathbf{j}}. \tag{11.106d}$$

For motion along the x-axis,

$$\mathbf{r} = x(t)\hat{\mathbf{i}}, \tag{11.107a}$$

$$\mathbf{v} = \frac{d\mathbf{r}}{dt} = \frac{dx}{dt}\hat{\mathbf{i}}, \tag{11.107b}$$

$$|\mathbf{v}| = \left|\frac{dx}{dt}\right|, \tag{11.107c}$$

$$\mathbf{a} = \frac{d\mathbf{v}}{dt} = \frac{d^2\mathbf{r}}{dt^2} = \frac{d^2x}{dt^2}\hat{\mathbf{i}}. \tag{11.107d}$$

If we compare equations 11.107b and d with equations 11.98 and 11.99, we see that for motion along the x-axis, $x(t)$, $v(t)$, and $a(t)$ are the components of the displacement, velocity, and acceleration vectors, respectively. Because these are the only components of $\mathbf{r}(t)$, $\mathbf{v}(t)$, and $\mathbf{a}(t)$, it follows that consideration of the components of the vectors is equivalent to consideration of the vectors themselves. For one-dimensional motion, then, we can drop the vector notation and work with components (and this is precisely the procedure that we followed in Sections 4.8 and 5.2).

Newton's second law describes the effects of forces on the motion of objects. It states that if an object of mass m is subjected to a force $\mathbf{F}$, then the time rate of change of its momentum $(m\mathbf{v})$ is equal to $\mathbf{F}$:

$$\mathbf{F} = \frac{d}{dt}(m\mathbf{v}). \tag{11.108}$$

In most cases, the mass of the object is constant, and this equation then yields its acceleration:

$$\mathbf{F} = m\frac{d\mathbf{v}}{dt} = m\mathbf{a}. \tag{11.109}$$

If $\mathbf{F}$ is known as a function of time t, $\mathbf{F} = \mathbf{F}(t)$, then 11.109 defines the acceleration of the object as a function of time,

$$\mathbf{a}(t) = \frac{1}{m}\mathbf{F}(t),$$

and integration of this equation leads to expressions for the velocity $\mathbf{v}(t)$ and position $\mathbf{r}(t)$ as functions of time.

EXAMPLE 11.50

A projectile is fired at angle θ to the horizontal with speed v_0 (Figure 11.118). Find the distance R from the firing place that the projectile strikes the ground (called the *range* of the projectile). What is the maximum height attained by the projectile?

FIGURE 11.118   Path of a projectile

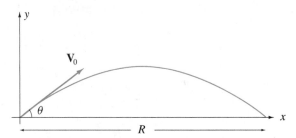

SOLUTION   Since the acceleration of the projectile is $d\mathbf{v}/dt = \mathbf{a} = -g\hat{\mathbf{j}}$, its velocity is $\mathbf{v}(t) = -gt\hat{\mathbf{j}} + \mathbf{C}$. If we choose $t = 0$ at the instant the projectile is fired, then $\mathbf{v}(0) = \mathbf{v}_0 = v_0 \cos\theta \,\hat{\mathbf{i}} + v_0 \sin\theta \,\hat{\mathbf{j}}$, and therefore $\mathbf{v}_0 = \mathbf{v}(0) = \mathbf{C}$. Integration of $\dfrac{d\mathbf{r}}{dt} = -gt\hat{\mathbf{j}} + \mathbf{v}_0$ gives

$$\mathbf{r}(t) = -\frac{1}{2}gt^2\hat{\mathbf{j}} + \mathbf{v}_0 t + \mathbf{D}.$$

Since $\mathbf{r}(0) = \mathbf{0}$, it follows that $\mathbf{D} = \mathbf{0}$, and

$$\mathbf{r}(t) = -\frac{1}{2}gt^2\hat{\mathbf{j}} + \mathbf{v}_0 t = (v_0 \cos\theta \, t)\hat{\mathbf{i}} + \left(-\frac{1}{2}gt^2 + v_0 \sin\theta \, t\right)\hat{\mathbf{j}}.$$

The projectile strikes the ground when $\mathbf{r}(t) = R\hat{\mathbf{i}}$, in which case

$$R\hat{\mathbf{i}} = (v_0 \cos\theta \, t)\hat{\mathbf{i}} + \left(-\frac{1}{2}gt^2 + v_0 \sin\theta \, t\right)\hat{\mathbf{j}}.$$

When we equate components,

$$R = v_0 \cos\theta \, t, \qquad 0 = -\frac{1}{2}gt^2 + v_0 \sin\theta \, t.$$

The second of these implies that $t = (2v_0/g)\sin\theta$, and when this is substituted into the first,

$$R = v_0 \cos\theta \left(\frac{2v_0 \sin\theta}{g}\right) = \frac{v_0^2 \sin 2\theta}{g}.$$

The projectile attains maximum height when the y-component of its velocity is zero, $0 = -gt + v_0 \sin\theta \implies t = (v_0/g)\sin\theta$. The height of the shell at this time is the y-component of its displacement,

$$-\frac{1}{2}g\left(\frac{v_0 \sin\theta}{g}\right)^2 + v_0 \sin\theta \left(\frac{v_0 \sin\theta}{g}\right) = \frac{v_0^2 \sin^2\theta}{2g}.$$

EXAMPLE 11.51

A particle starts at time $t = 0$ from position $(1, 1)$ with speed 2 m/s in the negative y-direction. It is subjected to an acceleration that is given as a function of time by

$$\mathbf{a}(t) = \frac{1}{\sqrt{t + 1}}\hat{\mathbf{i}} + 6t\hat{\mathbf{j}} \quad \text{m/s}^2.$$

Find its velocity and position as functions of time.

SOLUTION   If $\mathbf{a} = d\mathbf{v}/dt = (1/\sqrt{t + 1})\hat{\mathbf{i}} + 6t\hat{\mathbf{j}}$, then

$$\mathbf{v} = 2\sqrt{t + 1}\,\hat{\mathbf{i}} + 3t^2\hat{\mathbf{j}} + \mathbf{C},$$

where $\mathbf{C}$ is some constant vector. Because the initial velocity of the particle is 2 m/s in the negative y-direction, $\mathbf{v}(0) = -2\hat{\mathbf{j}}$. Consequently, $-2\hat{\mathbf{j}} = 2\hat{\mathbf{i}} + \mathbf{C} \implies \mathbf{C} = -2\hat{\mathbf{i}} - 2\hat{\mathbf{j}}$. The velocity, then, of the particle at any time $t \geq 0$ is

$$\mathbf{v}(t) = (2\sqrt{t + 1} - 2)\hat{\mathbf{i}} + (3t^2 - 2)\hat{\mathbf{j}} \quad \text{m/s}.$$

Because $\mathbf{v} = d\mathbf{r}/dt$, integration gives

$$\mathbf{r} = \left[\frac{4}{3}(t + 1)^{3/2} - 2t\right]\hat{\mathbf{i}} + (t^3 - 2t)\hat{\mathbf{j}} + \mathbf{D}.$$

Since the particle starts from position $(1, 1)$, $\mathbf{r}(0) = \hat{\mathbf{i}} + \hat{\mathbf{j}}$, and

$$\hat{\mathbf{i}} + \hat{\mathbf{j}} = \frac{4}{3}\hat{\mathbf{i}} + \mathbf{D}, \ \text{ or } \mathbf{D} = -\frac{1}{3}\hat{\mathbf{i}} + \hat{\mathbf{j}}.$$

The displacement of the particle is therefore

$$\mathbf{r}(t) = \left[\frac{4}{3}(t + 1)^{3/2} - 2t - \frac{1}{3}\right]\hat{\mathbf{i}} + (t^3 - 2t + 1)\hat{\mathbf{j}} \quad \text{m}.$$

EXAMPLE 11.52

The mass M in Figure 11.119a is dropped from point B. Show that if a mass m is fired from any position A directly at M at the instant M is released, m will always collide with M.

FIGURE 11.119a   Mass m is fired from A at mass M at B as M is dropped

FIGURE 11.119b   Coordinate system to analyze motions of masses m and M

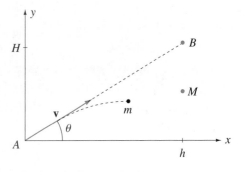

SOLUTION   We choose the coordinate system in Figure 11.119b, and take time $t = 0$ at the instant both masses begin motion. To show that the masses collide, we show that they have

the same displacement vector for some time t. The acceleration of each mass is $\mathbf{a} = -9.81\hat{\mathbf{j}}$. Integration of this gives velocities of the masses,

$$\mathbf{v}_m = -9.81t\,\hat{\mathbf{j}} + \mathbf{C}, \qquad \mathbf{v}_M = -9.81t\,\hat{\mathbf{j}} + \mathbf{D}.$$

Since M is dropped, its initial velocity is zero, $\mathbf{v}_M(0) = \mathbf{O}$. This implies that $\mathbf{D} = \mathbf{O}$. If m is fired at angle θ with speed $v > 0$, then $\mathbf{v}_m(0) = v\cos\theta\,\hat{\mathbf{i}} + v\sin\theta\,\hat{\mathbf{j}}$. This implies that $\mathbf{C} = v\cos\theta\,\hat{\mathbf{i}} + v\sin\theta\,\hat{\mathbf{j}}$. Integrations of

$$\frac{d\mathbf{r}_m}{dt} = -9.81t\,\hat{\mathbf{j}} + v\cos\theta\,\hat{\mathbf{i}} + v\sin\theta\,\hat{\mathbf{j}} \quad \text{and} \quad \frac{d\mathbf{r}_M}{dt} = -9.81t\,\hat{\mathbf{j}}$$

give

$$\mathbf{r}_m = v\cos\theta\,t\,\hat{\mathbf{i}} + (-4.905t^2 + v\sin\theta\,t)\hat{\mathbf{j}} + \mathbf{E} \quad \text{and} \quad \mathbf{r}_M = -4.905t^2\,\hat{\mathbf{j}} + \mathbf{F}.$$

Since $\mathbf{r}_m(0) = \mathbf{O}$ and $\mathbf{r}_M(0) = h\hat{\mathbf{i}} + H\hat{\mathbf{j}}$, we obtain $\mathbf{E} = \mathbf{O}$ and $\mathbf{F} = h\hat{\mathbf{i}} + H\hat{\mathbf{j}}$. Thus, $\mathbf{r}_m = v\cos\theta\,t\,\hat{\mathbf{i}} + (-4.905t^2 + v\sin\theta\,t)\hat{\mathbf{j}}$ and $\mathbf{r}_M = h\hat{\mathbf{i}} + (H - 4.905t^2)\hat{\mathbf{j}}$. The masses collide if and when

$$\mathbf{r}_m = \mathbf{r}_M \iff v\cos\theta\,t\,\hat{\mathbf{i}} + (-4.905t^2 + v\sin\theta\,t)\hat{\mathbf{j}} = h\hat{\mathbf{i}} + (H - 4.905t^2)\hat{\mathbf{j}}.$$

When we equate components,

$$v\cos\theta\,t = h, \quad -4.905t^2 + v\sin\theta\,t = H - 4.905t^2,$$

from which

$$t = \frac{h}{v\cos\theta} \quad \text{and} \quad t = \frac{H}{v\sin\theta}.$$

These are compatible since $\tan\theta = H/h$. The time for collision to occur increases as h increases, v decreases, and/or θ increases.

Tangential and Normal Components of Velocity and Acceleration

For some types of motion it is inconvenient to express velocity and acceleration of a particle in terms of Cartesian components; sometimes it is an advantage to resolve these vectors into components that are tangent and normal to the path of the particle. When the trajectory C of a particle is specified as a function of time t by 11.100, its velocity $\mathbf{v} = d\mathbf{r}/dt$ is tangent to C, and we can therefore write

$$\mathbf{v} = |\mathbf{v}|\hat{\mathbf{T}}. \tag{11.110}$$

In other words, the tangential component of velocity is speed, and $\mathbf{v}$ has no component normal to the trajectory. Differentiation of this equation gives the particle's acceleration:

$$\mathbf{a} = \frac{d\mathbf{v}}{dt} = \left(\frac{d}{dt}|\mathbf{v}|\right)\hat{\mathbf{T}} + |\mathbf{v}|\frac{d\hat{\mathbf{T}}}{dt}$$

$$= \left(\frac{d}{dt}|\mathbf{v}|\right)\hat{\mathbf{T}} + \left(|\mathbf{v}|\left|\frac{d\hat{\mathbf{T}}}{dt}\right|\right)\frac{d\hat{\mathbf{T}}/dt}{|d\hat{\mathbf{T}}/dt|}$$

$$= \left(\frac{d}{dt}|\mathbf{v}|\right)\hat{\mathbf{T}} + \left(|\mathbf{v}|\left|\frac{d\hat{\mathbf{T}}}{dt}\right|\right)\hat{\mathbf{N}}. \tag{11.111}$$

We have therefore expressed $\mathbf{a}$ in terms of the unit tangent vector $\hat{\mathbf{T}}$ to C and the principal normal $\hat{\mathbf{N}}$ (Figure 11.120). We call $d(|\mathbf{v}|)/dt$ and $|\mathbf{v}||d\hat{\mathbf{T}}/dt|$ the tangential and normal components of acceleration, respectively. If a_T and a_N denote these components, we can write that

$$\mathbf{a} = a_T\hat{\mathbf{T}} + a_N\hat{\mathbf{N}}, \tag{11.112a}$$

where

$$a_T = \mathbf{a} \cdot \hat{\mathbf{T}} = \frac{d}{dt}|\mathbf{v}|, \quad a_N = \mathbf{a} \cdot \hat{\mathbf{N}} = |\mathbf{v}|\left|\frac{d\hat{\mathbf{T}}}{dt}\right|. \tag{11.112b}$$

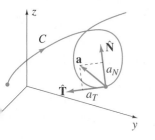

FIGURE 11.120   Tangential and normal components of acceleration

Note that the tangential component of acceleration is the time rate of change of speed. Since acceleration is the rate of change of velocity, the normal component of acceleration must determine the rate of change of the direction of $\mathbf{v}$. What is significant here is that $\mathbf{a}$ is expressed in terms of $\hat{\mathbf{T}}$ and $\hat{\mathbf{N}}$; it is not necessary to use the binormal $\hat{\mathbf{B}}$. The acceleration vector of a particle is always in the plane of $\hat{\mathbf{T}}$ and $\hat{\mathbf{N}}$.

To calculate a_N using 11.112b is often quite complicated. A far easier formula results if we take the scalar product of $\mathbf{a}$ as defined by 11.112a with itself:

$$\mathbf{a} \cdot \mathbf{a} = (a_T\hat{\mathbf{T}} + a_N\hat{\mathbf{N}}) \cdot (a_T\hat{\mathbf{T}} + a_N\hat{\mathbf{N}})$$
$$= a_T^2\hat{\mathbf{T}} \cdot \hat{\mathbf{T}} + 2a_Ta_N\hat{\mathbf{T}} \cdot \hat{\mathbf{N}} + a_N^2\hat{\mathbf{N}} \cdot \hat{\mathbf{N}}$$
$$= a_T^2 + a_N^2,$$

since $\hat{\mathbf{T}} \cdot \hat{\mathbf{N}} = 0$ and $\hat{\mathbf{T}} \cdot \hat{\mathbf{T}} = \hat{\mathbf{N}} \cdot \hat{\mathbf{N}} = 1$. Consequently,

$$a_N^2 = \mathbf{a} \cdot \mathbf{a} - a_T^2 = |\mathbf{a}|^2 - a_T^2,$$

and because a_N is always positive (see equation 11.112b),

$$a_N = \sqrt{|\mathbf{a}|^2 - a_T^2}. \tag{11.113}$$

Kepler's Laws for Planetary Motion

Based on Tycho Brahe's (1546–1601) astronomical measurements, Johannes Kepler (1571–1630) postulated three laws of planetary motion. The first states that planets move in elliptic orbits with the sun at one focus of the ellipse. This can be proved with Newton's second law and Newton's universal law of gravitation. If m is the mass of a planet, Newton's second law requires the acceleration of the planet to satisfy $\mathbf{F} = m\mathbf{a}$, where $\mathbf{F}$ is the resultant of all forces acting on the planet. If we assume that the only force acting on the planet is the force of attraction of the sun, with mass M, then $\mathbf{F} = -(GmM/r^2)\hat{\mathbf{r}}$, where $\hat{\mathbf{r}}$ is the unit vector in the direction from the sun to the planet. It follows then that

$$\mathbf{a} = -\frac{GM}{r^2}\hat{\mathbf{r}}, \tag{11.114}$$

and the acceleration of the planet always points toward the sun. Let us choose a coordinate system in space with origin at the sun (Figure 11.121).

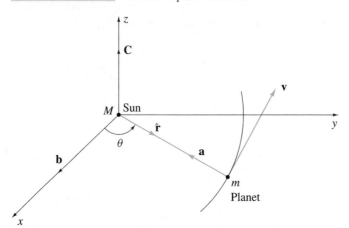

FIGURE 11.121 Motion of planet around sun

Pick any two points on the orbit of the planet and let these points and M define the xy-plane. We first show that the orbit of the planet always lies in the xy-plane. According to property 11.59c,

$$\frac{d}{dt}(\mathbf{r} \times \mathbf{v}) = \mathbf{r} \times \frac{d\mathbf{v}}{dt} + \frac{d\mathbf{r}}{dt} \times \mathbf{v} = \mathbf{r} \times \mathbf{a} + \mathbf{v} \times \mathbf{v}.$$

But $\mathbf{v} \times \mathbf{v}$ is always the zero vector, and so is $\mathbf{r} \times \mathbf{a}$ ($\mathbf{r}$ and $\mathbf{a}$ are parallel, see equation 11.114). Thus,

$$\frac{d}{dt}(\mathbf{r} \times \mathbf{v}) = \mathbf{0} \qquad \Longrightarrow \qquad \mathbf{r} \times \mathbf{v} = \mathbf{C}, \qquad (11.115)$$

where $\mathbf{C}$ is a constant vector. But this means that the displacement vector $\mathbf{r}$ and the velocity vector $\mathbf{v}$ are always perpendicular to $\mathbf{C}$. In other words, $\mathbf{C}$ is in the z-direction (or negative z-direction), and $\mathbf{r}$ and $\mathbf{v}$ are in the xy-plane; the planet moves in the xy-plane. To show that the planet follows an elliptic path we find a formula for $\mathbf{r}$ which represents the polar coordinate in the xy-plane. First we note that

$$\mathbf{C} = \mathbf{r} \times \mathbf{v} = (r\hat{\mathbf{r}}) \times \frac{d}{dt}(r\hat{\mathbf{r}}) = r\hat{\mathbf{r}} \times \left(\frac{dr}{dt}\hat{\mathbf{r}} + r\frac{d\hat{\mathbf{r}}}{dt}\right)$$

$$= r\frac{dr}{dt}(\hat{\mathbf{r}} \times \hat{\mathbf{r}}) + r^2\left(\hat{\mathbf{r}} \times \frac{d\hat{\mathbf{r}}}{dt}\right)$$

$$= r^2\left(\hat{\mathbf{r}} \times \frac{d\hat{\mathbf{r}}}{dt}\right).$$

If we cross this with $\mathbf{a} = -(GM/r^2)\hat{\mathbf{r}}$,

$$\mathbf{a} \times \mathbf{C} = -\frac{GM}{r^2}\hat{\mathbf{r}} \times \left[r^2\left(\hat{\mathbf{r}} \times \frac{d\hat{\mathbf{r}}}{dt}\right)\right] = -GM\left[\hat{\mathbf{r}} \times \left(\hat{\mathbf{r}} \times \frac{d\hat{\mathbf{r}}}{dt}\right)\right].$$

We now use Exercise 46 in Section 11.4:

$$\mathbf{a} \times \mathbf{C} = -GM\left[\left(\hat{\mathbf{r}} \cdot \frac{d\hat{\mathbf{r}}}{dt}\right)\hat{\mathbf{r}} - (\hat{\mathbf{r}} \cdot \hat{\mathbf{r}})\frac{d\hat{\mathbf{r}}}{dt}\right].$$

Differentiation of $\hat{\mathbf{r}} \cdot \hat{\mathbf{r}} = 1$ with respect to t gives

$$\hat{\mathbf{r}} \cdot \frac{d\hat{\mathbf{r}}}{dt} + \frac{d\hat{\mathbf{r}}}{dt} \cdot \hat{\mathbf{r}} = 0 \qquad \Longrightarrow \qquad \hat{\mathbf{r}} \cdot \frac{d\hat{\mathbf{r}}}{dt} = 0.$$

Thus,

$$\mathbf{a} \times \mathbf{C} = GM \frac{d\hat{\mathbf{r}}}{dt} \quad \Longrightarrow \quad GM \frac{d\hat{\mathbf{r}}}{dt} = \frac{d}{dt}(\mathbf{v} \times \mathbf{C}).$$

Integration with respect to t yields

$$\mathbf{v} \times \mathbf{C} = GM\hat{\mathbf{r}} + \mathbf{b},$$

where $\mathbf{b}$ is a constant vector. Since $\mathbf{v} \times \mathbf{C}$ and $\hat{\mathbf{r}}$ are both in the xy-plane, so is $\mathbf{b}$. Suppose the x-axis is chosen along $\mathbf{b}$ (Figure 11.121). The dot product of $\mathbf{r}$ with the equation above gives

$$\mathbf{r} \cdot (\mathbf{v} \times \mathbf{C}) = GM\hat{\mathbf{r}} \cdot \mathbf{r} + \mathbf{b} \cdot \mathbf{r} = GMr + |\mathbf{b}|r \cos\theta.$$

Thus,

$$r = \frac{\mathbf{r} \cdot (\mathbf{v} \times \mathbf{C})}{GM + |\mathbf{b}| \cos\theta}.$$

Since $\mathbf{r} \cdot (\mathbf{v} \times \mathbf{C}) = (\mathbf{r} \times \mathbf{v}) \cdot \mathbf{C} = \mathbf{C} \cdot \mathbf{C} = |\mathbf{C}|^2$, it follows that

$$r = \frac{|\mathbf{C}|^2}{GM + |\mathbf{b}| \cos\theta} = \frac{|\mathbf{C}|^2/(GM)}{1 + [|\mathbf{b}|/(GM)] \cos\theta}.$$

If we set $\epsilon = |\mathbf{b}|/(GM)$, and $d = |\mathbf{C}|^2/|\mathbf{b}|$, then

$$r = \frac{\epsilon d}{1 + \epsilon \cos\theta}. \tag{11.116}$$

According to equation 9.34a, this is a conic section with the origin as a focus. Since planets are known to follow closed paths, the conic section must be an ellipse. The second and third of Kepler's laws are discussed in Exercises 45 and 46.

EXAMPLE 11.53

A particle is confined to move in a circular path of radius R and centre (h, k) in the xy-plane if and only if its position vector is $\mathbf{r} = x\hat{\mathbf{i}} + y\hat{\mathbf{j}}$, where

$$x = h + R\cos\omega(t), \quad y = k + R\sin\omega(t),$$

and $\omega(t)$ is some function of time t. Determine the form of $\omega(t)$ if the acceleration of the particle is directed radially toward the centre of the circle.

SOLUTION The principal normal $\hat{\mathbf{N}}$ at any point on the circle is directed toward the centre of the circle. Hence, the acceleration must be along $\hat{\mathbf{N}}$ and the tangential component must vanish:

$$0 = a_T = \frac{d}{dt}|\mathbf{v}| \quad \Longrightarrow \quad |\mathbf{v}| = C = \text{constant.}$$

Since $\mathbf{v} = -R\omega'(t)\sin\omega(t)\hat{\mathbf{i}} + R\omega'(t)\cos\omega(t)\hat{\mathbf{j}}$, it follows that $C = |\mathbf{v}| = R|\omega'(t)|$. Thus,

$$\omega'(t) = \pm\frac{C}{R} \quad \Longrightarrow \quad \omega(t) = \pm\frac{Ct}{R} + D.$$

In other words, $\omega(t)$ must be a linear function of t for acceleration to be directed radially toward the centre of the circle.

Consulting Project 18

The metal roof of a structure is a hemisphere with a large radius $a = 10$ metres (Figure 11.122a); the bottom of the roof is $H = 20$ metres above the ground. During the winter, large chunks of ice that form on the roof break off, slide down the roof, and fall to the ground. An annular empty zone is to be created on the ground in order to prevent human injury or property damage. Our problem is to determine minimum width for the zone.

SOLUTION Once a chunk of ice breaks off, it picks up speed as it slides down the roof, and may leave the roof before it reaches the bottom edge. In order to determine a minimum radius for the empty zone, we shall find the farthest point at which ice can be expected to hit the ground. Chunks that strike the ground farthest from the roof are those that attain the greatest speed on the roof, and therefore leave the roof earliest. We shall assume that there is no friction between roof and ice in order to maximize speed. In addition, chunks that attain greatest speed are ones that slide from the very top of the roof.

FIGURE 11.122a   Chunk of ice sliding down a frictionless sphere

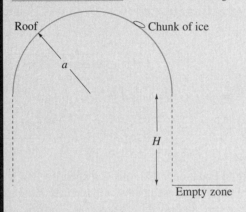

FIGURE 11.122b

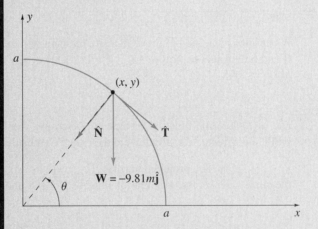

We consider, then, a mass m of ice starting from rest at the top of the roof as it slides down the circle in Figure 11.122b. It is acted on by gravity and the reaction of the sphere. As long as the mass is on the roof, the reaction of the roof on the mass is perpendicular to the sphere. It seems reasonable, then, to work with tangential and normal components of motion to the sphere (or circle). When the mass is at position (x, y), tangential and normal components of the weight $\mathbf{W} = -9.81 m \hat{\mathbf{j}}$ are

$$\mathbf{W} = 9.81m \cos\theta \,\hat{\mathbf{T}} + 9.81m \sin\theta \,\hat{\mathbf{N}}.$$

If the reaction of the sphere on m is denoted by $\mathbf{N} = -N\hat{\mathbf{N}}$, where N is therefore its magnitude, then the total force on m is

$$\mathbf{F} = 9.81m \cos\theta \,\hat{\mathbf{T}} + (-N + 9.81m \sin\theta)\,\hat{\mathbf{N}}.$$

The mass leaves the sphere when $N = 0$. To find where this happens, we use Newton's second law $\mathbf{F} = m\mathbf{a}$ with $\mathbf{F}$ as above, and $\mathbf{a}$ given by equation 11.111,

$$9.81m \cos\theta\hat{\mathbf{T}} + (-N + 9.81m \sin\theta)\hat{\mathbf{N}} = m\left(\frac{d}{dt}|\mathbf{v}|\right)\hat{\mathbf{T}} + m\left(|\mathbf{v}|\left|\frac{d\hat{\mathbf{T}}}{dt}\right|\right)\hat{\mathbf{N}}.$$

When we equate components,

$$9.81 \cos\theta = \frac{d}{dt}|\mathbf{v}|, \qquad -N + 9.81m \sin\theta = m|\mathbf{v}|\left|\frac{d\hat{\mathbf{T}}}{dt}\right|.$$

Now, $x = a\cos\theta$ and $y = a\sin\theta$, so that

$$|\mathbf{v}| = \sqrt{\left(\frac{dx}{dt}\right)^2 + \left(\frac{dy}{dt}\right)^2} = \sqrt{a^2\sin^2\theta\left(\frac{d\theta}{dt}\right)^2 + a^2\cos^2\theta\left(\frac{d\theta}{dt}\right)^2} = -a\frac{d\theta}{dt}.$$

Hence,

$$9.81\cos\theta = \frac{d}{dt}\left(-a\frac{d\theta}{dt}\right) = -a\frac{d^2\theta}{dt^2}.$$

Multiplication by $d\theta/dt$ gives

$$-9.81\cos\theta\frac{d\theta}{dt} = a\frac{d^2\theta}{dt^2}\frac{d\theta}{dt} = \frac{a}{2}\frac{d}{dt}\left(\frac{d\theta}{dt}\right)^2,$$

and we may integrate with respect to t,

$$-9.81\sin\theta = \frac{a}{2}\left(\frac{d\theta}{dt}\right)^2 + C.$$

Since $d\theta/dt = 0$ when $\theta = \pi/2$, it follows that $C = -9.81$, and

$$-9.81\sin\theta = \frac{a}{2}\left(\frac{d\theta}{dt}\right)^2 - 9.81 \qquad \Longrightarrow \qquad \left(\frac{d\theta}{dt}\right)^2 = \frac{19.62}{a}(1 - \sin\theta).$$

To tackle the normal components, we first calculate that

$$\mathbf{T} = \frac{d}{dt}(a\cos\theta, \, a\sin\theta) = a(-\sin\theta, \, \cos\theta)\frac{d\theta}{dt} \qquad \Longrightarrow \qquad \hat{\mathbf{T}} = (\sin\theta, \, -\cos\theta).$$

Thus,

$$\frac{d\hat{\mathbf{T}}}{dt} = (\cos\theta, \, \sin\theta)\frac{d\theta}{dt} \qquad \Longrightarrow \qquad \left|\frac{d\hat{\mathbf{T}}}{dt}\right| = -\frac{d\theta}{dt}.$$

Substitution into $-N + 9.81m\sin\theta = m|\mathbf{v}||d\hat{\mathbf{T}}/dt|$ gives

$$-N + 9.81m\sin\theta = m\left(-a\frac{d\theta}{dt}\right)\left(-\frac{d\theta}{dt}\right) = ma\left(\frac{d\theta}{dt}\right)^2.$$

Replacing $(d\theta/dt)^2$ by $(19.62/a)(1 - \sin\theta)$, we obtain

$$-N + 9.81m\sin\theta = 19.62m(1 - \sin\theta) \qquad \Longrightarrow \qquad N = 9.81m(3\sin\theta - 2).$$

Clearly, $N = 0$ when $\theta = \mathrm{Sin}^{-1}(2/3)$, and this is the angle at which the mass leaves the sphere. Its speed at this point is

$$|\mathbf{v}| = -a\frac{d\theta}{dt} = a\sqrt{\frac{19.62}{a}\left(1 - \frac{2}{3}\right)} = \sqrt{\frac{19.62a}{3}}.$$

Once the mass leaves the roof, the only force acting on it is gravity, and therefore its acceleration is

$$\frac{d^2\mathbf{r}}{dt^2} = -g\hat{\mathbf{j}} \qquad \Longrightarrow \qquad \frac{d\mathbf{r}}{dt} = -gt\hat{\mathbf{j}} + \mathbf{C}.$$

If we choose time $t = 0$ when the mass leaves the roof, then its velocity at this time is $\sqrt{\dfrac{19.62a}{3}}(\sin\theta\hat{\mathbf{i}} - \cos\theta\hat{\mathbf{j}})$, where angle θ is defined above. If we denote this by $\mathbf{v}_0$, then $\mathbf{C} = \mathbf{v}_0$, and integration of

$$\frac{d\mathbf{r}}{dt} = -gt\hat{\mathbf{j}} + \mathbf{v}_0 \qquad \text{gives} \qquad \mathbf{r} = -\frac{1}{2}gt^2\hat{\mathbf{j}} + \mathbf{v}_0 t + \mathbf{D}.$$

Since the initial position of the mass when it leaves the roof is $\mathbf{r}_0 = a\cos\theta\hat{\mathbf{i}} + a\sin\theta\hat{\mathbf{j}}$, it follows that $\mathbf{D} = \mathbf{r}_0$, and the position of the mass after it leaves the roof is

$$\mathbf{r}(t) = -\frac{1}{2}gt^2\hat{\mathbf{j}} + \mathbf{v}_0 t + \mathbf{r}_0.$$

The mass hits the ground when the y-component of $\mathbf{r}(t)$ is equal to $-H$,

$$-H = -\frac{1}{2}gt^2 - \sqrt{\frac{19.62a}{3}}\cos\theta\, t + a\sin\theta.$$

When we substitute $a = 10$, $H = 20$, $\sin\theta = 2/3$, and $\cos\theta = \sqrt{5}/3$, we obtain the quadratic equation

$$4.905t^2 + \sqrt{\frac{109}{3}}t - \frac{80}{3} = 0,$$

the positive solution of which is $t = 1.797$. The x-coordinate of the mass at this time is

$$\sqrt{\frac{196.2}{3}}\left(\frac{2}{3}\right)(1.797) + 10\left(\frac{\sqrt{5}}{3}\right) = 17.14 \text{ m}.$$

This is the minimum radius of the outer edge of the empty zone. In other words, it must be 7.14 metres wide.

EXERCISES 11.13

In Exercises 1–5 find the velocity, speed, and acceleration of a particle if the given equations represent its position as a function of time.

1. $x(t) = \sqrt{t^2 + 1}$, $y(t) = t\sqrt{t^2 + 1}$, $t \geq 0$

2. $x(t) = t + 1/t$, $y(t) = t - 1/t$, $t \geq 1$

3. $x(t) = \sin t$, $y(t) = 3\cos t$, $z(t) = \sin t$, $0 \leq t \leq 10\pi$

4. $x(t) = t^2 + 1$, $y(t) = 2te^t$, $z(t) = 1/t^2$, $1 \leq t \leq 5$

5. $x(t) = e^{-t^2}$, $y(t) = t \ln t$, $z(t) = 5$, $t \geq 1$

In Exercises 6–7 a particle at $(1, 2, -1)$ starts from rest at time $t = 0$. Find its position as a function of time if the given function defines its acceleration.

6. $\mathbf{a}(t) = 3t^2\hat{\mathbf{i}} + (t + 1)\hat{\mathbf{j}} - 4t^3\hat{\mathbf{k}}$, $t \geq 0$

7. $\mathbf{a}(t) = 3\hat{\mathbf{i}} + \hat{\mathbf{j}}/(t + 1)^3$, $t \geq 0$

In Exercises 8–9 find the tangential and normal components of acceleration for a particle moving with position defined by the given functions (where t is time).

8. $x(t) = t$, $y(t) = t^2 + 1$, $t \geq 0$

9. $x(t) = \cos t$, $y(t) = \sin t$, $z = t$, $t \geq 0$

10. Show that the normal component of acceleration of a particle can be expressed in the form $a_N = |\mathbf{v}|^2/\rho = \kappa|\mathbf{v}|^2$.

11. Find the kinetic energy for each particle in Exercises 1–5 if its mass is 2 g. Assume that x, y, and z are measured in metres and t in seconds.

12. A particle starts at the origin and moves along the curve $4y = x^2$ to the point $(4, 4)$.

    (a) If the y-component of its acceleration is always equal to 2 and the y-component of its velocity is initially zero, find the x-component of its acceleration.

    (b) If the x-component of its acceleration is equal to $24t^2$ (t being time) and the x-component of its velocity is initially zero, find the y-component of its acceleration.

13. A particle moves along the curve $x(t) = t$, $y(t) = t^3 - 3t^2 + t$, $0 \leq t \leq 5$ in the xy-plane (where t is time). Is there any point at which its velocity is parallel to its displacement?

14. A particle moves along the curve $y = x^3 - 2x + 3$ so that its x-component of velocity is always equal to 5. Find its acceleration.

15. If a particle starts at time $t = 0$ from rest at position $(3, 4)$ and experiences an acceleration $\mathbf{a} = -5t^4\hat{\mathbf{i}} - (2t^3 + 1)\hat{\mathbf{j}}$, find its speed at $t = 2$.

* 16. A particle travels counterclockwise around the circle $(x - h)^2 + (y - k)^2 = R^2$ in the figure below. Show that the speed of the particle at any time is $|\mathbf{v}| = \omega R$, where $\omega = d\theta/dt$ is called the *angular speed* of the particle.

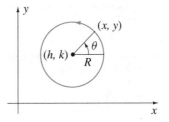

* 17. A particle travels around the circle $x^2 + y^2 = 4$ counterclockwise at constant speed, making 2 revolutions each second. If x and y are measured in metres, what is the velocity of the particle when it is at the point $(1, -\sqrt{3})$?

* 18.   (a) Show that if an object moves with constant speed in a circular path of radius R, the magnitude of its acceleration is $|\mathbf{a}| = |\mathbf{v}|^2/R$.

    (b) If a satellite moves with constant speed in a circular orbit 200 km above the earth's surface, what is its speed? *Hint:* Use Newton's universal law of gravitation (see Exercise 32 in Section 11.3) to determine the acceleration $\mathbf{a}$ of the satellite. Assume that the earth is a sphere with radius 6370 km and density 5.52×10^3 kg/m³.

* 19. Two particles move along curves C_1 and C_2 in the figure below. If at some instant of time the particles are at positions P_1 and P_2, then the vector $\mathbf{P_1P_2}$ is the displacement of P_2 with respect to P_1. Clearly, $\mathbf{OP_1} + \mathbf{P_1P_2} = \mathbf{OP_2}$. Show that when this equation is differentiated with respect to time, we have

$$\mathbf{v}_{P_1/O} + \mathbf{v}_{P_2/P_1} = \mathbf{v}_{P_2/O},$$

where $\mathbf{v}_{P_1/O}$ and $\mathbf{v}_{P_2/O}$ are velocities of P_1 and P_2 with respect to the origin, and $\mathbf{v}_{P_2/P_1}$ is the velocity of P_2 with respect to P_1. Can this equation be rewritten in the form

$$\mathbf{v}_{P_1/O} + \mathbf{v}_{O/P_2} = \mathbf{v}_{P_1/P_2}?$$

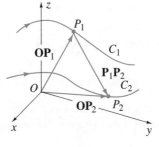

* 20. A plane flies on a course N30°E with airspeed 650 km/h (i.e., the speed of the plane relative to the air is 650). If the air is moving at 40 km/h due east, find the ground velocity and speed of the plane. *Hint:* Use Exercise 19.

* **21.** A plane flies with speed 600 km/h in still air. The plane is to fly in a straight line from city A to city B, where B is 1000 km northwest of A. What should be its bearing if the wind is blowing from the west at 50 km/h? How long will the trip take?

* **22.** A straight river is 200 m wide and the water flows at 3 km/h. If you can paddle your canoe at 4 km/h in still water, in what direction should you paddle if you wish the canoe to go straight across the river? How long will it take to cross?

* **23.** (a) In the figure below a cannon is fired up an inclined plane. If the speed at which the ball is ejected from the cannon is S, show that the range R of the ball is given by

$$R = \frac{2S^2 \cos\theta \, \sin(\theta - \alpha)}{g \cos^2\alpha},$$

where g is the acceleration due to gravity.

(b) What angle θ maximizes R?

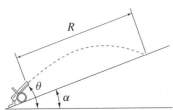

* **24.** What constant acceleration must a particle experience if it is to travel from $(1, 2, 3)$ to $(4, 5, 7)$ along the straight line joining the points, starting from rest and covering the distance in 2 units of time?

* **25.** Calculate the normal component a_N of the acceleration of a particle using equations 11.112 and 11.113 if its position is given by $x = t^2 + 1$, $y = 2t^2 - 1$, $z = t^2 + 5t$, $t \geq 0$ (t being time).

* **26.** A particle moves along the curve $x(t) = 2 + \sqrt{1 - t^2}$, $y(t) = t$, $0 \leq t \leq 1/2$, where t is time. Is there a time at which its acceleration is perpendicular to its velocity?

* **27.** (a) Show that motion along a straight line is the result in both of the following situations:

(i) The initial velocity is zero, and the acceleration is constant.

(ii) The initial velocity is nonzero, and the acceleration is constant and parallel to the initial velocity.

(b) Can we generalize the results of part (a) and state that constant acceleration produces straight-line motion? Illustrate.

* **28.** A particle starts from position $\mathbf{r}_0 = (x_0, y_0, z_0)$ at time $t = t_0$ with velocity $\mathbf{v}_0$. If it experiences constant acceleration $\mathbf{a}$, show that

$$\mathbf{r} = \mathbf{r}_0 + \mathbf{v}_0(t - t_0) + \frac{1}{2}\mathbf{a}(t - t_0)^2.$$

* **29.** A ladder 8 m long has its upper end against a vertical wall and its lower end on a horizontal floor. Suppose that the lower end slips away from the wall at constant speed 1 m/s.

(a) Find the velocity and acceleration of the middle point of the ladder when the foot of the ladder is 3 m from the wall.

(b) How fast does the middle point of the ladder strike the floor?

* **30.** The English longbow in medieval times was regarded to be accurate at 100 m or more. For an arrow to travel a horizontal distance of 100 m with maximum height 10 m, find the initial speed and angle of projection of the arrow. Ignore air friction.

* **31.** The block of mass M in the figure below slides on a thin film of oil. The film thickness is h and the area of the block in contact with the film is A. When released, mass m exerts tension in the cord, causing block M to accelerate. When the speed of M is v, the viscous force acting on it due to the film is $F = \mu A v/h$ where μ is the viscosity of the oil. Find the speed of M as a function of time t. Neglect friction in the pulley and air resistance.

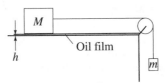

* **32.** Water issues from the nozzle of a fire hose at speed S in the figure below. Show that the maximum height attainable by the water on the building is given by $(S^4 - g^2 d^2)/(2gS^2)$, where g is the acceleration due to gravity.

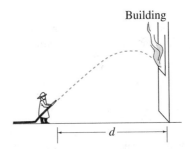

* **33.** A boy stands on a cliff 50 m high that overlooks a river 85 m wide (figure below). If he can throw a stone at 25 m/s, can he throw it across the river?

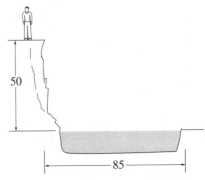

* **34.** A golfer can drive a maximum of 300 m in the air on a level fairway From the tee in the figure below, can he expect to clear the stream?

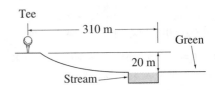

35. (a) A projectile is fired at angle θ to the horizontal from a height h above the ground with speed v (figure below). Show that the range R of the projectile is given by the formula

$$R = \frac{v^2 \cos\theta}{g}\left(\sin\theta + \sqrt{\sin^2\theta + \frac{2gh}{v^2}}\right),$$

where $g = 9.81$.

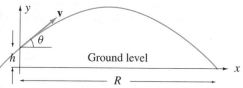

(b) What angle maximizes R for given v and h?

(c) Suppose the projectile is a shot, thrown by an Olympic athlete. What is the angle in part (b) if $v = 13.7$ m/s and $h = 2.25$ m?

(d) Prove that the maximum height attained by the projectile is

$$h + \frac{R^2 \tan^2\theta}{4(h + R\tan\theta)}.$$

36. A cannon is located on a plane inclined at angle α to the horizontal (figure below). If a projectile is fired from the cannon at angle β to the plane, prove that for the projectile to hit the plane horizontally,

$$\beta = \text{Tan}^{-1}\left(\frac{\sin 2\alpha}{3 - \cos 2\alpha}\right).$$

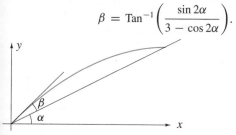

37. If $\mathbf{r}$ is the position vector of a particle with mass m moving under the action of a force $\mathbf{F}$, the torque of $\mathbf{F}$ about the origin is $\boldsymbol{\tau} = \mathbf{r} \times \mathbf{F}$. The angular momentum of m about O is defined as $\mathbf{H} = \mathbf{r} \times m\mathbf{v}$. Use Newton's second law in the form 11.108 to show that $\boldsymbol{\tau} = d\mathbf{H}/dt$.

38. When a stone is embedded in the tread of a tire and the tire rolls (without slipping) along the x-axis (figure below), the path that it traces is called a *cycloid* (see Example 9.7 in Section 9.1).

(a) Verify that if the centre of the tire moves at constant speed S, with $t = 0$ when the stone is at the origin, then $\theta = St/R$.

(b) Find velocity, speed, and acceleration of the stone at any time.

(c) What are the normal and tangential components of the stone's acceleration?

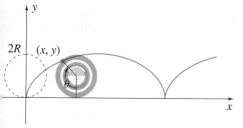

* **39.** If the stone in Exercise 38 is embedded in the side of the tire, its path is called a *trochoid* (see Exercise 58 in Section 9.1).

(a) Find velocity, speed, and acceleration of the stone if the tire rolls so that its centre has constant speed S. (Assume that $x = 0$ at time $t = 0$.)

(b) What are normal and tangential components of the stone's acceleration?

* **40.** Circles C_1 and C_2 in the figure below represent cross-sections of two cylinders. The left cylinder remains stationary while the right one rolls (without slipping) around the left one, and the cylinders always remain in contact. If the right cylinder picks up a speck of dirt at point $(R, 0)$, the path that the dirt traces out during one revolution is a cardioid.

(a) Show that parametric equations for the cardioid are

$$x = R(2\cos\theta - \cos 2\theta), \quad y = R(2\sin\theta - \sin 2\theta).$$

(b) Verify that if the point of contact moves at constant speed S, with $t = 0$ when the speck of dirt is picked up, then $\theta = St/R$.

(c) Find velocity, speed, and acceleration of the speck of dirt.

(d) What are normal and tangential components of the dirt's acceleration?

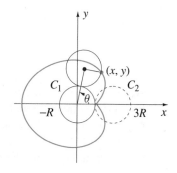

* **41.** Show that the path of a particle lies on a sphere if its displacement and velocity are always perpendicular during its motion.

* **42.** Suppose in Exercise 22 that because of an injured elbow, you can paddle only at 2 km/h. What should be your heading to travel straight to a point L kilometres downstream on the opposite shore? Are there any restrictions on L?

* **43.** Suppose that position vectors of a system of n masses m_i are denoted by $\mathbf{r}_i$ and forces acting on these masses are $\mathbf{F}_i$. Show that if $\mathbf{F} = \sum_{i=1}^{n} \mathbf{F}_i$, then the acceleration $\mathbf{a}$ of the centre of mass of the system (see Section 7.7) is given by $\mathbf{F} = M\mathbf{a}$, where $M = \sum_{i=1}^{n} m_i$.

* **44.** If the force acting on a particle is always tangent to the particle's trajectory, what can you conclude about the trajectory?

* **45.** *Kepler's second law* states that the line joining the sun to a planet sweeps out equal areas in equal time intervals. To show this, let $A(t)$ be the area swept out by the line beginning at some time t_0 in the following figure and ending at time t.

(a) Verify that

$$A(t) = \int_{\theta_0}^{\theta(t)} \frac{1}{2} r^2 d\theta,$$

and use equation 6.19 to prove that

$$\frac{dA}{dt} = \frac{1}{2} r^2 \frac{d\theta}{dt}.$$

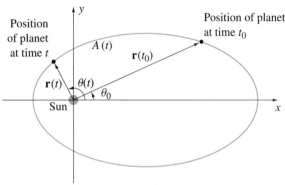

Position of planet at time t

$A(t)$

$\mathbf{r}(t_0)$

Position of planet at time t_0

$\mathbf{r}(t)$ $\theta(t)$

θ_0

Sun

y

x

(b) Show that $\hat{\mathbf{r}} = \cos\theta\hat{\mathbf{i}} + \sin\theta\hat{\mathbf{j}}$ and $d\hat{\mathbf{r}}/dt$ are perpendicular, and deduce from equation 11.115 that

$$\frac{d\theta}{dt} = \frac{|\mathbf{C}|}{r^2}.$$

(c) Use parts (a) and (b) to obtain

$$\frac{dA}{dt} = \frac{|\mathbf{C}|}{2} = \text{constant}.$$

Does this verify Kepler's second law?

* **46.** *Kepler's third law* states that the square of the period of revolution of a planet is proportional to the cube of the length of the major axis of its orbit.

(a) Use part (c) of Exercise 45 to show that if P is the time taken for one complete revolution, then

$$P = \frac{2\pi ab}{|\mathbf{C}|},$$

where $2a$ and $2b$ are lengths of the major and minor axes of the elliptic orbit.

(b) Use equation 11.116 to show that $b^2/a = |\mathbf{C}|^2/(GM)$ and therefore

$$P^2 = \frac{4\pi^2 a^3}{GM}.$$

** **47.** Suppose that in Project 18 the mass is given an initial speed v_0 at the top of the sphere. Prove that if $v_0 \le \sqrt{ag}$, the angle θ at which the mass leaves the sphere is $\text{Sin}^{-1}\left(\frac{2}{3} + \frac{v_0^2}{3ag}\right)$.

SUMMARY

We have now established the groundwork for multivariable calculus. We discussed curves and surfaces in space and introduced vectors. We described points by Cartesian coordinates (x, y, z) and then illustrated that an equation $F(x, y, z) = 0$ in these coordinates usually defines a surface. When a second equation $G(x, y, z) = 0$ also defines a surface, the pair of simultaneous equations $F(x, y, z) = 0$, $G(x, y, z) = 0$ describes the curve of intersection of the two surfaces (provided the surfaces do intersect). It is often more useful to have parametric equations for a curve, and these can be obtained by specifying one of x, y, or z as a function of a parameter t and solving the given equations for the other two in terms of t: $x = x(t)$, $y = y(t)$, $z = z(t)$.

The most common surfaces that we encountered were planes and quadric surfaces. Every plane has an equation of the form $Ax + By + Cz + D = 0$; conversely, every such equation describes a plane (provided that A, B, and C are not all zero). A plane is uniquely defined by a vector that is perpendicular to it [and (A, B, C) is one such vector] and a point on it. Quadric surfaces are surfaces whose equations are quadratic in x, y, and z, the most important of which were sketched in Figures 11.22–11.30.

Every straight line in space is characterized by a vector along it and a point on it. (Contrast this with the characterization of a plane described above.) If (a, b, c) are the components of a vector along a line and (x_0, y_0, z_0) are the coordinates of a point on it, then vector, symmetric, and parametric equations for the line are, respectively,

$$(x, y, z) = (x_0, y_0, z_0) + t(a, b, c);$$

$$\frac{x - x_0}{a} = \frac{y - y_0}{b} = \frac{z - z_0}{c};$$

$$x = x_0 + at,$$

$$y = y_0 + bt,$$

$$z = z_0 + ct.$$

Geometrically, vectors are defined as directed line segments; algebraically, they are represented by ordered sets of real numbers (v_x, v_y, v_z), called their Cartesian components. Vectors can be added or subtracted geometrically using triangles or parallelograms; algebraically, they are added and subtracted component by component. Vectors can also be multiplied by scalars to give parallel vectors of different lengths.

We defined two products of vectors: the scalar product and the vector product. The scalar product of two vectors $\mathbf{u} = (u_x, u_y, u_z)$ and $\mathbf{v} = (v_x, v_y, v_z)$ is defined as

$$\mathbf{u} \cdot \mathbf{v} = u_x v_x + u_y v_y + u_z v_z = |\mathbf{u}||\mathbf{v}| \cos \theta,$$

where $|\mathbf{u}| = \sqrt{u_x^2 + u_y^2 + u_z^2}$ is the length of $\mathbf{u}$ and θ is the angle between $\mathbf{u}$ and $\mathbf{v}$. If the components of $\mathbf{u}$ and $\mathbf{v}$ are known, this equation can be used to find the angle θ between the vectors. The scalar product has many uses: finding components of vectors in arbitrary directions, calculating distances between geometric objects, and finding mechanical work, among others.

The vector product of two vectors $\mathbf{u}$ and $\mathbf{v}$ is

$$\mathbf{u} \times \mathbf{v} = \begin{vmatrix} \hat{\mathbf{i}} & \hat{\mathbf{j}} & \hat{\mathbf{k}} \\ u_x & u_y & u_z \\ v_x & v_y & v_z \end{vmatrix} = |\mathbf{u}||\mathbf{v}| \sin \theta \hat{\mathbf{w}},$$

where $\hat{\mathbf{w}}$ is the unit vector perpendicular to $\mathbf{u}$ and $\mathbf{v}$ determined by the right-hand rule. Because of the perpendicularity property, the vector product is indispensable in finding vectors perpendicular to other vectors. We used this fact when finding a vector along the line of intersection of two planes, a vector perpendicular to the plane containing three given points, and distances between geometric objects. It can also be used to find areas of triangles and parallelograms.

If a curve is represented vectorially in the form $\mathbf{r}(t) = x(t)\hat{\mathbf{i}} + y(t)\hat{\mathbf{j}} + z(t)\hat{\mathbf{k}}$, then a unit vector tangent to the curve at any point is

$$\hat{\mathbf{T}} = \frac{d\mathbf{r}/dt}{|d\mathbf{r}/dt|}.$$

Two unit vectors normal to the curve are the principal normal $\hat{\mathbf{N}}$ and the binormal $\hat{\mathbf{B}}$:

$$\hat{\mathbf{N}} = \frac{d\hat{\mathbf{T}}/dt}{|d\hat{\mathbf{T}}/dt|}; \quad \hat{\mathbf{B}} = \hat{\mathbf{T}} \times \hat{\mathbf{N}}.$$

These three vectors form a moving triad of mutually perpendicular unit vectors along the curve. The curvature of a curve, defined by $\kappa(t) = |\dot{\mathbf{r}} \times \ddot{\mathbf{r}}|/|\dot{\mathbf{r}}|^3$, measures the rate at which the curve changes direction: The larger κ is, the faster the curve turns. The reciprocal of curvature $\rho = \kappa^{-1}$ is called radius of curvature. It is the radius of that circle which best approximates the curve at any point.

If parametric equations for a curve represent the position of a particle and t is time, then the velocity and acceleration of the particle are, respectively,

$$\mathbf{v} = \frac{d\mathbf{r}}{dt}; \quad \mathbf{a} = \frac{d\mathbf{v}}{dt} = \frac{d^2\mathbf{r}}{dt^2};$$

and its speed is the magnitude of velocity, $|\mathbf{v}|$.

Tangential and normal components of velocity and acceleration of the particle are defined by

$$\mathbf{v} = |\mathbf{v}|\hat{\mathbf{T}}; \quad \mathbf{a} = a_T\hat{\mathbf{T}} + a_N\hat{\mathbf{N}};$$

where

$$a_T = \frac{d}{dt}|\mathbf{v}| \quad \text{and} \quad a_N = |\mathbf{v}|\left|\frac{d\hat{\mathbf{T}}}{dt}\right| = \sqrt{|\mathbf{a}|^2 - a_T^2}.$$

What these results say is that velocity is always tangent to the trajectory of the particle, and its acceleration always lies in the plane of the velocity vector and the principal normal.

KEY TERMS

In reviewing this chapter, you should be able to define or discuss the following key terms:

Coordinate planes

Origin

Cartesian or rectangular coordinates

Right-handed coordinate system

Left-handed coordinate system

Octants

Cylinder

Quadric surface

Scalars

Vectors

Tip of a vector

Tail of a vector

x-, y-, and z-components of $\mathbf{v}$

Cartesian components of $\mathbf{v}$

Unit vector

Triangular addition of vectors

Parallelogram addition of vectors

Zero vector

Resultant force

Scalar components

Vector components

Scalar, dot, or inner product

Vector, cross, or outer product

Equation for a plane

Normal or normal vector to the plane

Vector equation of a line

Parametric equations for a line

Symmetric equations for a line

Component

Work

First moments of forces

Span of a cable

Sag of a cable

Catenary

Position or displacement vector

Continuous vector-valued function

Derivative of a vector-valued function

Antiderivative of a vector-valued function

Indefinite integral of a vector-valued function

Continuous curve

Parametric representation of curves

Vector representation of curve

Closed curve

Smooth curve

Tangent vectors to curves

Unit tangent vector to a curve

Piecewise-smooth curve

Binormal vector

Principal normal vector

Radius of curvature of a curve

Curvature of a curve

Frenet–Serret formulas

Circle of curvature of a curve

Velocity

Torsion of a curve

Acceleration

Speed

Normal components of velocity and acceleration

Tangential components of velocity and acceleration

Kepler's laws

REVIEW EXERCISES

In Exercises 1–10 find the value of the scalar or the components of the vector if $\mathbf{u} = (1, 3, -2)$, $\mathbf{v} = (2, 4, -1)$, $\mathbf{w} = (0, 2, 1)$, and $\mathbf{r} = (2, 0, -1)$.

1. $2\mathbf{u} - 3\mathbf{w} + \mathbf{r}$

2. $\mathbf{u} \cdot (\mathbf{v} \times \mathbf{w})$

3. $(3\mathbf{u} \times 4\mathbf{v}) - \mathbf{w}$

4. $3\mathbf{u} \times (4\mathbf{v} - \mathbf{w})$

5. $|\mathbf{u}|\mathbf{v} - |\mathbf{v}|\mathbf{r}$

6. $(\mathbf{u} + \mathbf{v}) \cdot (\mathbf{r} - \mathbf{w})$

7. $(\mathbf{u} + \mathbf{v}) \times (\mathbf{r} - \mathbf{w})$

8. $(\mathbf{u} \times \mathbf{v}) \times (\mathbf{r} \times \mathbf{w})$

9. $(\mathbf{u} \cdot \mathbf{v})\mathbf{r} - 3(\mathbf{v} \cdot \mathbf{w})\mathbf{u}$

10. $\dfrac{2\mathbf{r}}{\mathbf{v} \cdot \mathbf{w}} + 3(\mathbf{v} + \mathbf{u})$

In Exercises 11–26 draw whatever the equation, or equations, describe in space.

11. $x - y + 2z = 6$

12. $x^2 + z^2 = 1$

13. $x = \sqrt{y^2 + z^2}$

14. $x - y = 5$, $2x + y = 6$

15. $x^2 + y^2 + z^2 = 6z + 10$

16. $x^2 + y^2 + z^2 = 6z - 10$

17. $x + y = 5$, $2x - 3y + 6z = 1$, $y = z$

18. $x = t^2$, $y = t$, $z = t^3$

19. $x = t$, $y = t^3 + 1$

20. $\dfrac{x - 1}{3} = \dfrac{y + 5}{2} = z$

21. $z = 4 - x^2 - 2y^2$

22. $y^2 + z^2 = 1$, $y = z$

23. $y^2 + z^2 = 1$, $x = z$

24. $x^2 + y^2 = z^2 + 1$

25. $x = y^2$, $x = z^2$

26. $z^2 = x^2 - y^2$

In Exercises 27–30 find equations for the line.

27. Through the points $(-2, 3, 0)$ and $(1, -2, 4)$

28. Through $(6, 6, 2)$ and perpendicular to the plane $5x - 2y + z = 4$

29. Parallel to the line $x - y = 5$, $2x + 3y + 6z = 4$ and through the origin

30. Perpendicular to the line $x = t + 2$, $y = 3 - 2t$, $z = 4 + t$, intersecting this line, and through the point $(1, 3, 2)$

In Exercises 31–34 find the equation for the plane.

31. Through the points $(1, 3, 2)$, $(2, -1, 0)$, and $(6, 1, 3)$

32. Through the point $(1, 2, -1)$ and perpendicular to the line $y = z$, $x + y = 4$

33. Containing the line $x - y + z = 3$, $3x + 4y = 6$ and the point $(2, 2, 2)$

34. Containing the lines $x = 3t$, $y = 1 + 2t$, $z = 4 - t$, and $x = y = z$

In Exercises 35–39 find the distance.

35. Between the points $(1, 3, -2)$ and $(6, 4, 1)$

36. From the point $(6, 2, 1)$ to the plane $6x + 2y - z = 4$

37. From the line $x - y + z = 2$, $2x + y + z = 4$ to the plane $x - y = 5$

38. From the line $x - y + z = 2$, $2x + y + z = 4$ to the plane $3x + 6y = 4$

* **39.** From the point $(6, 2, 3)$ to the line $x - y + z = 6$, $2x + y + 4z = 1$

40. Find the area of the triangle with vertices $(1, 1, 1)$, $(-2, 1, 0)$, and $(6, 3, -2)$.

41. If the points in Exercise 40 are three vertices of a parallelogram, what are possibilities for the fourth vertex? What are areas of these parallelograms?

In Exercises 42–43 find the unit tangent vector $\hat{\mathbf{T}}$, the principal normal vector $\hat{\mathbf{N}}$, and the binormal vector $\hat{\mathbf{B}}$ for the curve.

42. $x = 2 \sin t$, $y = 2 \cos t$, $z = t$

43. $x = t^3$, $y = 2t^2$, $z = t + 4$

44. If a particle has a trajectory defined by $x = t$, $y = t^2$, $z = t^2$, where t is time, find its velocity, speed, and acceleration at any time. What are normal and tangential components of its velocity and acceleration?

* **45.** A force $\mathbf{F} = x^{-2}(2\hat{\mathbf{i}} + 3\hat{\mathbf{j}})$ acts on a particle moving from $x = 1$ to $x = 4$ along the x-axis. How much work does it do?

* **46.** A ball rolls off a table 1 m high with speed 0.5 m/s (figure below).

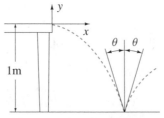

(a) With what speed does it strike the floor?

(b) What is its displacement vector relative to the point where it left the table when it strikes the floor?

(c) If it rebounds in the direction shown but loses 20% of its speed in the bounce, find the position of its second bounce.

* **47.** In the figure below, a spring (with constant k) is fixed at A and attached to a sleeve at C. The sleeve is free to slide without friction on a vertical rod, and when the spring is horizontal (at B), it is unstretched. If the sleeve is slowly lowered, there is a position at which the vertical component of the spring force on C is balanced by the force of gravity on the sleeve (ignoring the weight of the spring itself). If the mass of the sleeve is m, find an equation determining s in terms of d, m, k, and $g = 9.81$, the acceleration due to gravity.

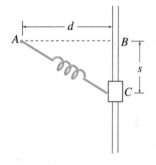

* **48.** Find Cartesian components of the spring force $\mathbf{F}$ on the sleeve in Example 11.26.

* **49.** If a toy train travels around the oval track in the figure below with constant speed, show that its acceleration at A (the point at which the circular end meets the straight section) is discontinuous.

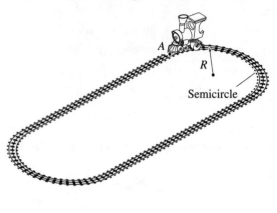

12 | Differential Calculus of Multivariable Functions

Application Preview

The figure on the left below shows gas confined in a cylinder, closed on one end, with a piston on the other. If the piston moves to the left, it compresses the gas in the cylinder. The volume V that the gas occupies decreases and the pressure P that it exerts on the piston increases. Conversely, when the piston moves to the right, the volume of the gas increases and the pressure on the piston decreases. The table below gives the pressure for various volumes of gas. They have been plotted in the figure on the right.

V	54.3	61.82	72.4	88.7	118.6	194.0
P	61.2	49.5	37.6	28.4	19.2	10.1

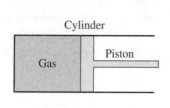

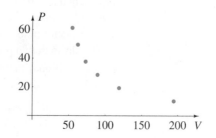

THE PROBLEM   What function represents the data in the table; or, equivalently, were the data points to be joined by a smooth curve, what would be its equation? (See the discussion on pages 882 and 883 for the solution.)

Few quantities in real life depend on only one variable; most depend on a multitude of inter-related variables. In order to understand such complicated relationships, we initiate discussions in this chapter with derivatives of functions of more than one variable. Much of the theory and many of our examples involve functions of two or three variables, because in these cases we can give geometric as well as analytic explanations. If the situation is completely analogous for functions of more variables, then it is likely that no mention of this fact will be made; on the other hand, if the situation is different for a higher number of variables, we will be careful to point out these differences.

12.1 Multivariable Functions

If a variable T depends on other variables x, y, z, and t, we write $T = f(x, y, z, t)$ and speak of T as a function of x, y, z, and t. For example, T might be temperature, x, y, and z might be the coordinates of points in some region of space, and t might be time. The stopping distance D of a car depends on many factors: the initial speed s, the reaction time t of the driver to move from the accelerator to the brake, the texture T of the road, the moisture level M on the road, and so on. We write $D = f(s, t, T, M, \ldots)$ to represent this functional dependence. The function $P = f(I, R) = I^2R$ represents the power necessary to maintain a current I through a wire with resistance R.

FIGURE 12.1   A function
$f(x, y)$ of two independent variables can be represented geometrically as a surface with equation
$z = f(x, y)$

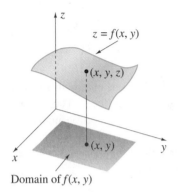

Domain of $f(x, y)$

More precisely, a variable z is said to be a function of two independent variables x and y if x and y are not related and each pair of values of x and y determines a unique value of z. We write $z = f(x, y)$ to indicate that z is a function of x and y. Each possible pair of values x and y of the independent variables can be represented geometrically as a point (x, y) in the xy-plane. The totality of all points for which $f(x, y)$ is defined forms a region in the xy-plane called the **domain** of the function. Figure 12.1, for example, illustrates a rectangular domain. If for each point (x, y) in the domain we plot a point $f(x, y)$ units above the xy-plane, we obtain a surface, such as the one in Figure 12.1. Each point on this surface has coordinates (x, y, z) that satisfy the equation

$$z = f(x, y), \tag{12.1}$$

and therefore 12.1 is the equation of the surface. This surface is a pictorial representation of the function.

It is clear that functions of more than two independent variables cannot be represented pictorially as surfaces. For example, if $u = f(x, y, z)$ is a function of three independent variables, values (x, y, z) of these independent variables can be represented geometrically as points in space. To graph $u = f(x, y, z)$ as above would require a u-axis perpendicular to the x-, y-, and z-axes, a somewhat difficult task geometrically. We can certainly think of $u = f(x, y, z)$ as defining a surface in four-dimensional $xyzu$-space, but visually we are stymied.

Although every function $f(x, y)$ of two independent variables can be represented geometrically as a surface, not every surface represents a function $f(x, y)$. A given surface does represent a function $f(x, y)$ if and only if every vertical line (in the z-direction) that intersects the surface does so in exactly one point. For example, a sphere such as $x^2 + y^2 + z^2 = 1$ does not determine z as a function of x and y. Most, but not all, vertical lines that intersect the sphere do so in two points. It is also clear algebraically that $x^2 + y^2 + z^2 = 1$ does not define z as a function of x and y. Solving for z gives $z = \pm\sqrt{1 - x^2 - y^2}$, two solutions for each x and y satisfying $x^2 + y^2 < 1$.

EXAMPLE 12.1

Draw the surface defined by the function $f(x, y) = x^2 + 4y^2$.

SOLUTION   To draw the surface $z = x^2 + 4y^2$, we note that if the surface is intersected with a plane $z = k > 0$, then the ellipse $x^2 + 4y^2 = k$, $z = k$ is obtained. As k increases, the ellipse becomes larger. In other words, cross-sections of this surface are ellipses that expand with increasing z. If we intersect the surface with the yz-plane $(x = 0)$, we obtain the parabola $z = 4y^2$, $x = 0$. Similarly, intersection of the surface with the xz-plane gives the parabola $z = x^2$, $y = 0$. These facts lead to Figure 12.2a. A computer plot of the surface for $-2 \le x \le 2$ and $-2 \le y \le 2$ is shown in Figure 12.2b.

FIGURE 12.2a   Cross-sections of surface with xz-plane, yz-plane, and planes $z = k$ lead to surface $z = x^2 + 4y^2$

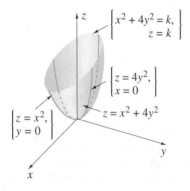

FIGURE 12.2b   Computer plot of surface $z = x^2 + 4y^2$

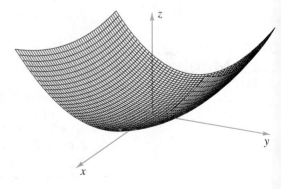

EXAMPLE 12.2

The ends of a taut string are fixed at $x = 0$ and $x = 2$ on the x-axis. At time $t = 0$, the string is given a displacement in the y-direction of $y = \sin(\pi x/2)$ (Figure 12.3a). If the string is then released, its displacement thereafter is given by

$$y = f(x, t) = \sin(\pi x/2) \cos(8\pi t).$$

Physically, this function need only be considered for $t \geq 0$ and $0 \leq x \leq 2$, and is plotted for $0 \leq t \leq 1/2$ in Figure 12.3b. Interpret physically the intersections of this surface with planes $t = t_0$ ($t_0 =$ a constant) and $x = x_0$ ($x_0 =$ a constant).

FIGURE 12.3a   Initial displacement of a string that is displaced then released

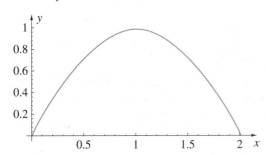

FIGURE 12.3b   Computer plot representing displacement of points in string

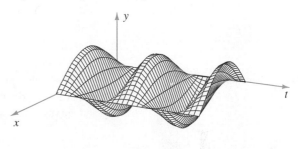

SOLUTION   Grid lines on the surface are curves of intersection of the surface with vertical planes $x = x_0$ and $t = t_0$. The curve of intersection of the surface with a plane $t = t_0$ represents the position of the string at time t_0. For $t = t_0$, the equation of the curve is $y(x, t_0) = \cos(8\pi t_0) \sin(\pi x/2)$. Thus, the string vibrates up and down always in the shape of a sine curve, its amplitude at time t_0 being $|\cos(8\pi t_0)|$.

The curve of intersection of the surface with a plane $x = x_0$, $0 < x_0 < 2$, is a graphical history of the vertical displacement of the particle in the string at position x_0. The equation of the curve is $y(x_0, t) = \sin(\pi x_0/2) \cos(8\pi t)$. The particle at x_0 undergoes simple harmonic motion, the amplitude being $|\sin(\pi x_0/2)|$.

Another way to visualize a function $f(x, y)$ of two independent variables is through level curves. Curves $f(x, y) = C$ are drawn in the xy-plane for various values of C. Effectively, the surface $z = f(x, y)$ is sliced with a plane $z = C$, and the curve of intersection is projected into the xy-plane. Each curve joins all points for which $f(x, y)$ has the same value; or it joins all points that have the same height on the surface $z = f(x, y)$. A few level curves for the surface in Figure 12.2 are shown in Figure 12.4; they are ellipses. Level curves for the function in Figure 12.5a are shown in Figure 12.5b.

FIGURE 12.4   Level curves of the function $f(x, y) = x^2 + 4y^2$

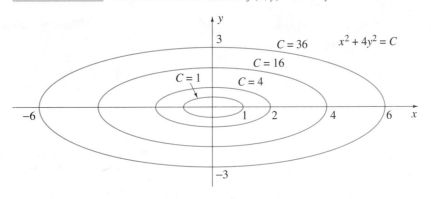

FIGURE 12.5a Computer plot of surface $z = x^2 - y^2$ defined by $f(x, y) = x^2 - y^2$

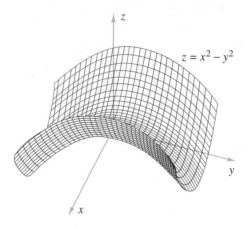

FIGURE 12.5b Level curves of $f(x, y) = x^2 - y^2$

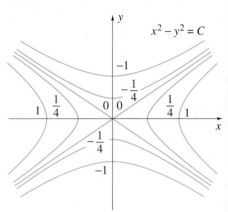

This technique is used on topographical maps to indicate land elevation, on marine charts to indicate water depth, and on climatic maps to indicate curves of constant temperature (isotherms) and curves of constant barometric pressure (isobars).

EXERCISES 12.1

1. If $f(x, y) = x^3y + x \sin y$, evaluate (a) $f(1, 2)$, (b) $f(-2, -2)$, (c) $f(x^2 + y, x - y^2)$, and (d) $f(x + h, y) - f(x, y)$.

2. If $f(x, y, z) = x^2y^2 - x^4 + 4zx^2$, show that $f(a + b, a - b, ab) = 0$.

In Exercises 3–6 find and illustrate geometrically the largest possible domain for the function.

3. $f(x, y) = \sqrt{4 - x^2 - y^2}$ **4.** $f(x, y) = \ln(1 - x^2 + y^2)$

5. $f(x, y) = \text{Sin}^{-1}(x^2y + 1)$ **6.** $f(x, y, z) = 1/(x^2 + y^2 + z^2)$

7. For what values of x and y is the function $f(x, y) = \dfrac{12xy - x^2y^2}{2(x + y)}$ equal to zero? Illustrate these values as points in the xy-plane. What is the largest domain of the function?

In Exercises 8–21 draw the surface defined by the function. Plot the surface as a check.

8. $f(x, y) = y^2$ **9.** $f(x, y) = 4 - x - 2y$

10. $f(x, y) = x^2 + y^2$ **11.** $f(x, y) = \sqrt{x^2 + y^2}$

12. $f(x, y) = y + x$ **13.** $f(x, y) = 1 - x^3$

14. $f(x, y) = 2(x^2 + y^2)$ **15.** $f(x, y) = 1 - x^2 - 4y^2$

16. $f(x, y) = xy$ * **17.** $f(x, y) = y - x^2$

* **18.** $f(x, y) = e^{-x^2-y^2}$ * **19.** $f(x, y) = |x - y|$

* **20.** $f(x, y) = x^2 - y^2$ * **21.** $f(x, y) = \sqrt{1 + x^2 - y^2}$

In Exercises 22–25 draw level curves $f(x, y) = C$ corresponding to the values $C = -2, -1, 0, 1, 2$.

* **22.** $f(x, y) = 4 - \sqrt{4x^2 + y^2}$

* **23.** $f(x, y) = y - x^2$

* **24.** $f(x, y) = \ln(x^2 + y^2)$

* **25.** $f(x, y) = x^2 - y^2$

* **26.** A closed box is to have total surface area 30 m^2. Find a formula for the volume of the box in terms of its length l and width w.

* **27.** (a) A company wishes to construct a storage tank in the form of a rectangular box. If material for sides and top costs \$1.25/m^2 and material for the bottom costs \$4.75/m^2, find the cost of building the tank as a function of its length l, width w, and height h.

    (b) If the tank must hold 1000 m^3, find the construction cost in terms of l and w.

    (c) Repeat parts (a) and (b) if the 12 edges of the tank must be welded at a cost of \$7.50/m of weld.

* **28.** A rectangular box is inscribed inside the ellipsoid $x^2/a^2 + y^2/b^2 + z^2/c^2 = 1$ with sides parallel to the coordinate planes and corners on the ellipsoid. Find a formula for the volume of the box in terms of x and y.

* **29.** (a) A silo is to be built in the shape of a right-circular cylinder surmounted by a right-circular cone. If the radius of each is 6 m, find a formula for the volume V of the silo as a function of the heights H and h of the cylinder and cone.

    (b) If the total surface area of the silo must be 200 m^2 (not including the base), find V as a function of h. (The area of the curved surface of a cone is $\pi r\sqrt{r^2 + h^2}$.)

* **30.** The figure below shows parameters taken into account in the analysis of the distance travelled by a long jumper:

θ = angle to the horizontal at which jumper leaves the ground;
T = horizontal distance from toe to centre of mass G at takeoff;
L = horizontal distance from heel to centre of mass on landing;
h = vertical distance between centre of mass on takeoff and landing;
R = horizontal distance between centre of mass on takeoff and landing

Using the formula developed in Exercise 35 of Section 11.13, the total length of the jump is

$$D(\theta, v) = T + R + L = T + L$$

$$+ \frac{v^2 \cos \theta}{g} \left(\sin \theta + \sqrt{\sin^2 \theta + \frac{2gh}{v^2}} \right),$$

where $g = 9.81$. Typical values for T, L, and h are 0.35 m, 0.9 m, and 0.5 m, respectively.

(a) Calculate $D(0.35, 9.0)$.

(b) What is the percentage change in D, from that in part (a), if takeoff speed is increased by 10%?

(c) What is the percentage change in D, from that in part (a), if takeoff angle is increased by 10%?

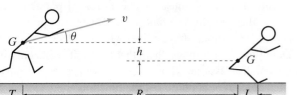

1. A long piece of metal 1 m wide is bent in two places A and B (figure below) to form a channel with three straight sides. Find a formula for the cross-sectional area of the channel in terms of x, θ, and ϕ.

2. (a) A uniform circular rod has flat ends at $x = 0$ and $x = \pi$ on the x-axis. The round side of the rod is insulated and

the faces at $x = 0$ and $x = \pi$ are both kept at temperature $0°C$ for time $t > 0$. If the initial temperature (at time $t = 0$) of the rod is given by $100 \sin x$, $0 \le x \le \pi$, then the temperature thereafter is

$$T = f(x, t) = 100e^{-kt} \sin x \quad (k > 0 \text{ constant}).$$

(b) Draw the surface $T = f(x, t)$.

(c) Interpret physically the curves of intersection of this surface with planes $x = x_0$ and $t = t_0$.

* **33.** A cow's daily diet consists of three foods: hay, grain, and supplements. The animal is always given 11 kg of hay per day, 50% of which is digestive material and 12% of which is protein. Grain is 74% digestive and 8.8% protein. Supplements are 62% digestive material and 34% protein. Hay costs $27.50 for 1000 kg, whereas grain and supplements cost $110 and $175, respectively, for 1000 kg. A healthy cow's daily diet must contain between 9.5 and 11.5 kg of digestive material and between 1.9 and 2.0 kg of protein. Find a formula for the cost per day, C, of feeding a cow in terms of the number of kilograms of grain, G, and supplements, S, fed to the cow daily. What is the domain of this function?

* **34.** The Easy University is buying computers. It has three models to choose from. Each model A computer, with 64 MB of memory and a 3 GB hard drive, costs $1300; model B, with 32 MB of memory and a 4 GB drive, costs $1200; and economy model C, with 16 MB of memory and a 1 GB drive, costs $1000. For reasons related to accreditation, the university needs at least 2000 MB of memory and 150 GB of disk space. If the computer lab must have 100 computers, set up a formula for the cost of outfitting the lab in terms of the numbers x and y of computers of models A and B. What is the domain of the function?

12.2 Limits and Continuity

The concepts of limit and continuity for multivariable functions are exactly the same as for functions of one variable; on the other hand, the work involved with the application of these concepts is more complicated for multivariable functions.

Intuitively, a function $f(x, y)$ is said to have **limit** L as x and y approach x_0 and y_0 if $f(x, y)$ gets arbitrarily close to L, and stays close to L, as x and y get arbitrarily close to x_0 and y_0. To say this in a precise mathematical way, it is convenient to represent pairs of independent variables as points (x, y) in the xy-plane. We then have the following definition.

> **DEFINITION 12.1**
>
> A function $f(x, y)$ has limit L as (x, y) approaches (x_0, y_0), written
>
> $$\lim_{(x,y) \to (x_0, y_0)} f(x, y) = L, \qquad (12.2)$$
>
> if, given any $\epsilon > 0$, we can find a $\delta > 0$ such that
>
> $$|f(x, y) - L| < \epsilon$$
>
> whenever $0 < \sqrt{(x - x_0)^2 + (y - y_0)^2} < \delta$ and (x, y) is in the domain of $f(x, y)$.

In other words, $f(x, y)$ has limit L as (x, y) approaches (x_0, y_0) if $f(x, y)$ can be made arbitrarily close to L (within ϵ) by choosing points (x, y) sufficiently close to (x_0, y_0) (within a circle of radius δ). Note the similarity of this definition to that for the limit of a function $f(x)$ of one variable in Section 2.6.

It is clear that

$$\lim_{(x,y)\to(2,1)} (x^2 + 2xy - 5) = 3,$$

but the limit

$$\lim_{(x,y)\to(0,0)} \frac{y^2 - x^2}{y^2 + x^2}$$

presents a problem, since both numerator and denominator approach zero as x and y approach zero.

To conclude that $\lim_{x\to a} f(x) = L$, the limit must be L no matter how x approaches a — be it through numbers larger than a, through numbers smaller than a, or through any other approach. For limit 12.2, the limit of $f(x, y)$ must also be L for all possible ways of approaching (x_0, y_0). But in this case there might be a multitude of ways of approaching (x_0, y_0). We might be able to approach (x_0, y_0) along straight lines with various slopes, along parabolas, along cubics, and so on. Definition 12.1 implies, then, that the limit exists only if it is independent of the manner of approach. It is assumed, however, that we approach (x_0, y_0) only through points (x, y) that lie in the domain of definition of the function.

For the second example above, suppose we approach the origin along the straight line $y = mx$. Along this line,

$$\lim_{(x,y)\to(0,0)} \frac{y^2 - x^2}{y^2 + x^2} = \lim_{x\to 0} \frac{m^2x^2 - x^2}{m^2x^2 + x^2} = \frac{m^2 - 1}{m^2 + 1}.$$

Because this result depends on m, we have shown that as (x, y) approaches $(0, 0)$ along various straight lines, the function $(y^2 - x^2)/(y^2 + x^2)$ approaches different numbers. We conclude, therefore, that the function does not have a limit as (x, y) approaches $(0, 0)$. In Figure 12.6a we show a portion of the surface $z = (y^2 - x^2)/(y^2 + x^2)$ to illustrate our conclusion. The curve of intersection of the surface with the plane $y = 2x$ is shown in Figure 12.6b. Heights of points on the curve represent values of $(y^2 - x^2)/(y^2 + x^2)$ at points on the line $y = 2x$ in the xy-plane; they are always equal to 3/5. For other vertical planes containing the z-axis, curves of intersection are at different heights.

FIGURE 12.6a   Computer plot of surface $z = \dfrac{y^2 - x^2}{y^2 + x^2}$

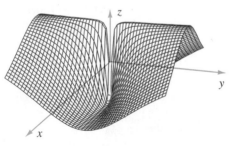

FIGURE 12.6b   Curve of intersection of surface $z = \dfrac{y^2 - x^2}{y^2 + x^2}$ and plane $y = 2x$

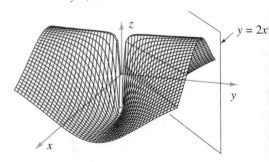

EXAMPLE 12.3

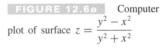

Evaluate $\displaystyle\lim_{(x,y)\to(0,0)} \frac{x^2 - y^2}{x + y}$ if it exists.

SOLUTION   Because points on the line $y = -x$ are not within the domain of definition of the function, we can write that

$$\lim_{(x,y)\to(0,0)} \frac{x^2 - y^2}{x + y} = \lim_{(x,y)\to(0,0)} \frac{(x - y)(x + y)}{x + y} = \lim_{(x,y)\to(0,0)} (x - y) = 0.$$

The concept of continuity for multivariable functions is contained in the following definition.

DEFINITION 12.2

A function $f(x, y)$ is said to be a **continuous function** at a point (x_0, y_0) if

1. $f(x, y)$ is defined at (x_0, y_0)

2. $\lim_{(x,y)\to(x_0,y_0)} f(x, y)$ exists

3. The value of the function in 1 and its limit in 2 are the same.

All three conditions can be combined by writing the single equation

$$\lim_{(x,y)\to(x_0,y_0)} f(x, y) = f(x_0, y_0). \tag{12.3}$$

The function $f(x, y) = (y^2 - x^2)/(y^2 + x^2)$ in Figure 12.6 is discontinuous at $(0, 0)$. The function is undefined at $(0, 0)$ and the limit as $(x, y) \to (0, 0)$ does not exist. Geometrically, a function $f(x, y)$ is continuous at a point (x_0, y_0) if the surface $z = f(x, y)$ is not separated at the point $(x_0, y_0, f(x_0, y_0))$. The function $f(x, y) = 1 - e^{-1/(x^2+y^2)}$ is discontinuous at $(0, 0)$ since it is undefined for $x = 0$ and $y = 0$. The surface has a hole at $(0, 0)$ (Figure 12.7). The function $f(x, y) = \text{sgn}\left[(x - 1)^2 + (y - 2)^2\right]$ (see Exercise 47 in Section 2.4) is discontinuous at $(1, 2)$; it has value 0 at $x = 1$ and $y = 2$, and value 1 for all other values of x and y (Figure 12.8). The function $h(x^2 + y^2 - 1)$ where $h(x)$ is the heaviside function of Section 2.5 has value zero everywhere inside the circle $x^2 + y^2 < 1$, does not have a value on the circle, and has value 1 outside the circle (Figure 12.9). It is discontinuous at each point on the circle. The surface $z = h(x^2 + y^2 - 1)$ is that part of the xy-plane inside the circle, and that part of the plane $z = 1$ above the outside of the circle.

FIGURE 12.7   Function $f(x, y) = 1 - e^{-1/(x^2+y^2)}$ is discontinuous at $(0, 0)$

FIGURE 12.8   Function $f(x, y) = \text{sgn}[(x - 1)^2 + (y - 2)^2]$ is discontinuous at $(1, 2)$

FIGURE 12.9   Function $f(x, y) = h(x^2 + y^2 - 1)$

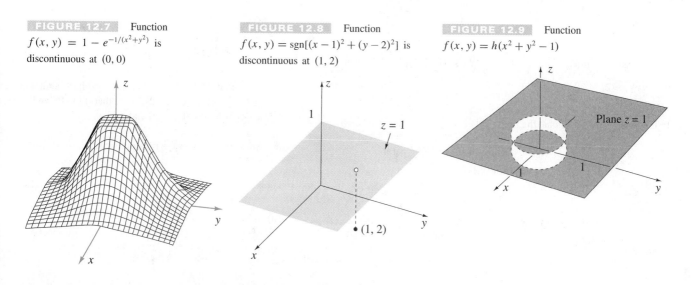

EXERCISES 12.2

In Exercises 1–20 evaluate the limit, if it exists.

1. $\lim\limits_{(x,y)\to(2,-3)} \dfrac{x^2-1}{x+y}$

2. $\lim\limits_{(x,y)\to(1,1)} \dfrac{x^3+2y^3}{x^3+4y^3}$

3. $\lim\limits_{(x,y)\to(3,2)} \dfrac{2x-3y}{x+y}$

4. $\lim\limits_{(x,y,z)\to(2,3,-1)} \dfrac{xyz}{x^2+y^2+z^2}$

5. $\lim\limits_{(x,y)\to(1,0)} \dfrac{x}{y}$

6. $\lim\limits_{(x,y,z)\to(0,\pi/2,1)} \mathrm{Tan}^{-1}\,[x/(yz)]$

7. $\lim\limits_{(x,y,z)\to(0,\pi/2,1)} \mathrm{Tan}^{-1}\,(yz/x)$

8. $\lim\limits_{(x,y,z)\to(0,\pi/2,1)} \mathrm{Tan}^{-1}\,|yz/x|$

9. $\lim\limits_{(x,y)\to(3,4)} \dfrac{|x^2-y^2|}{x^2-y^2}$

10. $\lim\limits_{(x,y)\to(3,4)} \dfrac{|x^2+y^2|}{x^2+y^2}$

11. $\lim\limits_{(x,y)\to(2,1)} \dfrac{x^2-y^2}{x-y}$

* 12. $\lim\limits_{(x,y)\to(2,2)} \dfrac{x^2-y^2}{x-y}$

* 13. $\lim\limits_{(x,y)\to(0,0)} \dfrac{x^2-y^2}{x-y}$

* 14. $\lim\limits_{(x,y)\to(0,0)} \dfrac{x-y}{x+y}$

15. $\lim\limits_{(x,y,z)\to(0,0,0)} \dfrac{x^2-y^2}{y^2+z^2+1}$

16. $\lim\limits_{(x,y)\to(2,1)} \dfrac{(x-2)^2(y+1)}{x-2}$

17. $\lim\limits_{(x,y,z)\to(1,1,1)} |2x-y-z|$

* 18. $\lim\limits_{(x,y)\to(0,0)} \dfrac{3x^3-y^3}{2x^3+4y^3}$

* 19. $\lim\limits_{(x,y)\to(0,0)} \mathrm{Sec}^{-1}\!\left(\dfrac{-1}{x^2+y^2}\right)$

* 20. $\lim\limits_{(x,y)\to(0,0)} \mathrm{Sec}^{-1}(x^2+y^2)$

In Exercises 21–26 find all points of discontinuity for the function.

21. $f(x,y)=\dfrac{x^2-1}{x+y}$

22. $f(x,y)=\dfrac{xy}{x^2+y^2}$

23. $f(x,y)=\dfrac{1}{1-x^2-y^2}$

24. $f(x,y,z)=\dfrac{1}{xyz}$

25. $f(x,y)=|x-y|$

* 26. $f(x,y)=\dfrac{x+y}{x^2y+xy^2}$

* 27. Evaluate $\lim\limits_{(x,y)\to(a,a)}\left[\cos(x+y)-\sqrt{1-\sin^2(x+y)}\right]$ where $0\le a\le \pi/2$.

In Exercises 28–35 evaluate the limit, if it exists.

* 28. $\lim\limits_{(x,y)\to(0,0)} \dfrac{x^4+y^2}{x^4-y^2}$ *Hint:* Approach $(0,0)$ along parabolas.

* 29. $\lim\limits_{(x,y)\to(0,0)} \dfrac{x^6-2y^2}{3x^6+y^2}$ *Hint:* Approach $(0,0)$ along cubic curves.

* 30. $\lim\limits_{(x,y)\to(1,0)} \dfrac{(x-1)^2+y^2}{3(x-1)^2-2y^2}$ *Hint:* Approach $(1,0)$ along straight lines.

* 31. $\lim\limits_{(x,y)\to(0,-2)} \dfrac{x^3+4(y+2)^3}{3x^3-(y+2)^2}$ *Hint:* Approach $(0,-2)$ along straight lines.

* 32. $\lim\limits_{(x,y)\to(1,1)} \dfrac{x^2-2x-y^2+2y}{x^2-2x+y^2-2y+2}$ *Hint:* Approach $(1, 1$ along straight lines.

* 33. $\lim\limits_{(x,y)\to(1,1)} \dfrac{x^2-2x+y^2+2y-2}{x^2-y^2-2x+2y}$ *Hint:* Approach $(1, 1$ along straight lines.

* 34. $\lim\limits_{(x,y)\to(1,0)} \dfrac{\sqrt{x+y}-\sqrt{x-y}}{y}$

* 35. $\lim\limits_{(x,y)\to(0,0)} \dfrac{\sin(x^2+y^2)}{x^2+y^2}$

* 36.
(a) Does the limit $\lim\limits_{(x,y)\to(1,1)} \dfrac{\sin(x-y)}{x-y}$ exist? Explain. I the function continuous at $(1,1)$?

(b) If we define the function everywhere by giving it the valu 1 along the line $y=x$, does the limit of the function exis at $(1,1)$? Is the function continuous at $(1,1)$?

* 37. Give a mathematical definition for $\lim\limits_{(x,y,z)\to(x_0,y_0,z_0)} f(x,y,z)=$

* 38. Is the following statement true or false? If a function $f(x,y)$ i undefined at every point on a curve C, then for any point (x_0,y_0) o C, $\lim\limits_{(x,y)\to(x_0,y_0)} f(x,y)$ does not exist. Explain. Give an example.

* 39. Let $f(x,y)=\begin{cases}\dfrac{x^2y^2}{x^4+y^4}, & \text{if } (x,y)\ne(0,0)\\ 0, & \text{if } (x,y)=(0,0).\end{cases}$

(a) Show that $f(x,y)$ is continuous in each variable separatel at $(0,0)$. In other words, show that $f(x,0)$ and $f(0,y$ are continuous at $x=0$ and $y=0$.

(b) Show that $f(x,y)$ is not continuous at $(0,0)$.

** 40. Prove that:

(a) $\lim_{(x,y)\to(0,0)}(xy+5)=5$

(b) $\lim_{(x,y)\to(1,1)}(x^2+2xy+5)=8$

12.3 Partial Derivatives

We now define partial derivatives of multivariable functions and interpret these derivatives algebraically and geometrically.

> **DEFINITION 12.3**
>
> The **partial derivative** of a function $f(x, y)$ with respect to x is
>
> $$\frac{\partial f}{\partial x} = \lim_{\Delta x \to 0} \frac{f(x + \Delta x, y) - f(x, y)}{\Delta x}, \qquad (12.4)$$
>
> and the partial derivative with respect to y is
>
> $$\frac{\partial f}{\partial y} = \lim_{\Delta y \to 0} \frac{f(x, y + \Delta y) - f(x, y)}{\Delta y}. \qquad (12.5)$$

It is evident from equation 12.4 that the partial derivative of $f(x, y)$ with respect to x is simply the ordinary derivative of $f(x, y)$ with respect to x, where y is considered a constant. Similarly, $\partial f / \partial y$ is the ordinary derivative of $f(x, y)$ with respect to y, holding x constant.

For the partial derivative of a function of more than two independent variables, we again permit one variable to vary, but hold all others constant. For example, the partial derivative of $f(x, y, z, t, \ldots)$ with respect to z is

$$\frac{\partial f}{\partial z} = \lim_{\Delta z \to 0} \frac{f(x, y, z + \Delta z, t, \ldots) - f(x, y, z, t, \ldots)}{\Delta z}. \qquad (12.6)$$

Hence, we differentiate with respect to z while treating $x, y, t, \ldots$ as constants.

Other notations for the partial derivative are common. In particular, for $\partial f / \partial x$ when $z = f(x, y)$, there are also

$$\frac{\partial z}{\partial x}, \quad f_x, \quad z_x, \quad \left. \frac{\partial f}{\partial x} \right)_y, \quad \text{and} \quad \left. \frac{\partial z}{\partial x} \right)_y,$$

the last two indicating that the variable y is held constant when differentiation with respect to x is performed.

EXAMPLE 12.4

Find $\partial z / \partial x$ and $\partial z / \partial y$ if $z = \sin(x^2 + y^3) + e^{xy}$.

SOLUTION   For this function,

$$\frac{\partial z}{\partial x} = 2x \cos(x^2 + y^3) + y e^{xy} \qquad \text{and} \qquad \frac{\partial z}{\partial y} = 3y^2 \cos(x^2 + y^3) + x e^{xy}.$$

EXAMPLE 12.5

Find $\partial f / \partial x$ at the point $(1, 2, 3)$ if $f(x, y, z) = x^2 / y^4 + 3xz + 4$.

SOLUTION   Since $\partial f / \partial x = 2x / y^4 + 3z$,

$$\left. \frac{\partial f}{\partial x} \right|_{(1, 2, 3)} = 2(1)/2^4 + 3(3) = \frac{73}{8}.$$

For a function $y = f(x)$ of one variable, we defined differentials dx and dy in such a way that the derivative dy/dx could be thought of as a quotient. This is *not* done for functions of more than one variable. Although we write the partial derivative $\dfrac{\partial f}{\partial x}$ in the form $\partial f/\partial x$ (for typographical reasons), we *never* consider it as a quotient.

Algebraically, the partial derivative $\partial f/\partial x$ represents the rate of change of $f(x, y, \ldots)$ with respect to x when all other variables in $f(x, y, \ldots)$ are held constant. For instance, $V = \pi r^2 h/3$ represents the volume of a right-circular cone with height h and radius r, and therefore $\partial V/\partial r = 2\pi r h/3$ represents the rate of change of the volume of the cone as the base radius changes and the height remains fixed. Similarly, $\partial V/\partial h = \pi r^2/3$ is the rate of change of the volume of the right-circular cone as the height changes and the radius is kept fixed. We shall learn later how to calculate the rate of change of the volume when the radius and height are both changing.

We can interpret the partial derivative of a function geometrically when the function can be interpreted geometrically, namely, when there are only two independent variables. Consider, then, a function $f(x, y)$ that is represented geometrically as a surface $z = f(x, y)$ in Figure 12.10. If we intersect this surface with a plane $y = y_0 =$ a constant, we obtain a curve with equations

$$y = y_0, \qquad z = f(x, y_0). \tag{12.7}$$

Because this curve lies in the plane $y = y_0$, we can talk about its tangent line at the point (x_0, y_0, z_0), where $z_0 = f(x_0, y_0)$. The slope of this tangent is the derivative of z with respect to x, but because y is being held constant at y_0, it must be the partial derivative of z with respect to x. In other words, the slope of the tangent line to the curve in Figure 12.10 at the point (x_0, y_0, z_0) is $\partial f/\partial x|_{(x_0, y_0)}$. Similarly, the partial derivative $\partial f/\partial y$ evaluated at (x_0, y_0) represents the slope of the tangent line to the curve of intersection of $z = f(x, y)$ and the plane $x = x_0$ at the point (x_0, y_0, z_0).

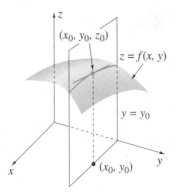

FIGURE 12.10 Geometric interpretation of partial derivatives of a function $f(x, y)$ of two variables

---

EXERCISES 12.3

In Exercises 1–20 evaluate $\partial f/\partial x$ and $\partial f/\partial y$.

1. $f(x, y) = x^3 y^2 + 2xy$

2. $f(x, y) = 3xy - 4x^4 y^4$

3. $f(x, y) = x^4/y^3$

4. $f(x, y) = x/(x + y) - x/y$

5. $f(x, y) = x/(2x^2 + y)$

6. $f(x, y) = \sin(xy)$

7. $f(x, y) = x \cos(x + y)$

8. $f(x, y) = \sqrt{x^2 + y^2}$

9. $f(x, y) = x\sqrt{x^2 - y^2}$

10. $f(x, y) = \tan(2x^2 + y^2)$

11. $f(x, y) = e^{x+y}$

12. $f(x, y) = e^{xy}$

13. $f(x, y) = xye^{xy}$

14. $f(x, y) = \ln(x^2 + y^2)$

15. $f(x, y) = (x + 1)\ln(xy)$

16. $f(x, y) = \sin(ye^x)$

17. $f(x, y) = \text{Tan}^{-1}(x/y)$

18. $f(x, y) = \sqrt[3]{1 - \cos^3(x^2 y)}$

19. $f(x, y) = \dfrac{\sin x}{\cos y}$

20. $f(x, y) = \ln\left(\sec\sqrt{x + y}\right)$

In Exercises 21–30 evaluate the derivative indicated.

21. $\partial f/\partial x$ if $f(x, y, z) = xyze^{x^2+y^2}$

22. $\partial f/\partial z$ if $f(x, z) = \text{Tan}^{-1}[1/(x^2 + z^2)]$

23. $\partial f/\partial y$ at $(1, 1, 0)$ if $f(x, y, z) = xy(x^2 + y^2 + z^2)^{1/3}$

24. $\partial f/\partial x$ at $(1, -1, 1, -1)$ if $f(x, y, z, t) = zt/(x^2 + y^2 - t^2)$

25. $\partial f/\partial t$ if $f(x, y, t) = x\sqrt{t^2 - y^2}/t^2 + (x/y)\,\text{Sec}^{-1}(t/3)$

26. $\partial f/\partial x$ if $f(x, y, z) = \text{Cot}^{-1}(1 + x + y + z)$

27. $\partial f/\partial y$ at $(1, 2, 3)$ if $f(x, y, t) = \text{Sin}^{-1}(xyt)/\text{Cos}^{-1}(xyt)$

28. $\partial f/\partial x$ if $f(x, y, z) = x^3/y + x\sin(yz/x)$

29. $\partial f/\partial t$ if $f(x, y, z, t) = xyz\ln(x^2 + y^2 + z^2)$

30. $\partial f/\partial z$ if $f(x, z) = (z^2/2)\,\text{Sin}^{-1}(x/z) + (x/2)\sqrt{z^2 - x^2}$

$*$ **31.** If $f(x, y) = x^3 y/(x - y)$, show that $x\dfrac{\partial f}{\partial x} + y\dfrac{\partial f}{\partial y} = 3f(x, y)$.

$*$ **32.** If $f(x, y, z) = (x^4 + y^4 + z^4)/(xyz)$, show that
$$x\dfrac{\partial f}{\partial x} + y\dfrac{\partial f}{\partial y} + z\dfrac{\partial f}{\partial z} = f(x, y, z).$$

$*$ **33.** If $f(x, y, z) = (x^2 + y^2)\cos[(y + z)/x]$, show that
$$x\dfrac{\partial f}{\partial x} + y\dfrac{\partial f}{\partial y} + z\dfrac{\partial f}{\partial z} = 2f(x, y, z).$$

4. To evaluate $\partial f/\partial x$ for $f(x, y)$ at the point $(1, 2)$, state which of the following are acceptable:

(a) Differentiate $f(x, y)$ with respect to x holding y constant, and then set $x = 1$ and $y = 2$.

(b) Set $x = 1$ and $y = 2$, and then differentiate with respect to x.

(c) Set $y = 2$, differentiate with respect to x, and set $x = 1$.

(d) Set $x = 1$, differentiate with respect to x, and set $y = 2$.

5. Temperature at points (x, y) in a semicircular plate defined by $x^2 + y^2 \le 4$, $y \ge 0$ is given by $T(x, y) = 16x^2 - 24xy + 40y^2$. Find, if possible, (a) $T_x(1, 1)$, (b) $T_y(1, 1)$, (c) $T_x(1, 0)$, (d) $T_y(1, 0)$, (e) $T_x(0, 2)$, and (f) $T_y(0, 2)$.

6. In the figure below, two identical bars AB and BC are pinned at B as well as at A and C. Each bar is initially of length L, and initially point B lies a distance h above line AC. When a vertical force with magnitude F is applied at B, the vertical displacement $x > 0$ of B is related to F by the equation

$$F = 2AE\left(\frac{h - x}{L}\right)\left(\frac{L}{\sqrt{L^2 - 2hx + x^2}} - 1\right),$$

where A is the cross-sectional area of the bar, and E is Young's modulus for the bars. Show that $\partial F/\partial x$ vanishes for $x = h - (L^3 - Lh^2)^{1/3}\sqrt{1 - (1 - h^2/L^2)^{1/3}}$.

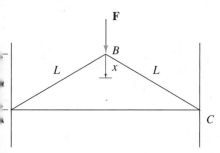

7. Can you find a function $f(x, y)$ so that $f_x(x, y) = 2x - 3y$ and $f_y(x, y) = 3x + 4y$?

8. Suppose a, b, and c are the lengths of the sides of a triangle and A, B, and C are the opposite angles. Find (a) $a_A(b, c, A)$, (b) $c_a(a, b, c)$, (c) $a_b(b, c, A)$, and (d) $A_b(a, b, c)$.

9. The equation of continuity for three-dimensional unsteady flow of a compressible fluid is

$$\frac{\partial \rho}{\partial t} + \frac{\partial}{\partial x}(\rho u) + \frac{\partial}{\partial y}(\rho v) + \frac{\partial}{\partial z}(\rho w) = 0,$$

where $\rho(x, y, z, t)$ is the density of the fluid, and $u\hat{\mathbf{i}} + v\hat{\mathbf{j}} + w\hat{\mathbf{k}}$ is the velocity of the fluid at position (x, y, z) and time t. Determine whether the continuity equation is satisfied if:

(a) $\rho = $ constant, $u = (2x^2 - xy + z^2)t$, $v = (x^2 - 4xy + y^2)t$, $w = (-2xy - yz + y^2)t$

(b) $\rho = xy + zt$, $u = x^2y + t$, $v = y^2z - 2t^2$, $w = 5x + 2z$

* **40.** A gas-filled pneumatic strut behaves like the piston-cylinder apparatus shown below. At one instant when the piston is $L = 0.15$ m away from the closed end of the cylinder, the gas density is uniform at $\rho = 18$ kg/m^3, and the piston begins to move away from the closed end at a constant rate of 12 m/s. The gas motion is one-dimensional and proportional to distance from the closed end. It varies linearly from zero at the closed end of the cylinder to 12 m/s at the piston. Gas density is always uniform throughout the cylinder, but varies in time. Use the equation of continuity in Exercise 39 to find the density of the gas as a function of time.

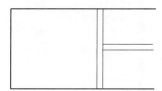

* **41.** In complex variable theory, two functions $u(x, y)$ and $v(x, y)$ are said to be *harmonic conjugates* in a region R if in R they satisfy the Cauchy–Riemann equations

$$\frac{\partial u}{\partial x} = \frac{\partial v}{\partial y}, \qquad \frac{\partial v}{\partial x} = -\frac{\partial u}{\partial y}.$$

Show that the following pairs of functions are harmonic conjugates:

(a) $u(x, y) = -3xy^2 + y + x^3$, $v(x, y) = 3x^2y - y^3 - x + 5$

(b) $u(x, y) = (x^2 + x + y^2)/(x^2 + y^2)$, $v(x, y) = -y/(x^2 + y^2)$

(c) $u(x, y) = e^x(x \cos y - y \sin y)$, $v(x, y) = e^x(x \sin y + y \cos y)$

* **42.** If r and θ are polar coordinates, then the Cauchy–Riemann equations in Exercise 41 for functions $u(r, \theta)$ and $v(r, \theta)$ take the form

$$\frac{\partial u}{\partial r} = \frac{1}{r}\frac{\partial v}{\partial \theta}, \qquad \frac{1}{r}\frac{\partial u}{\partial \theta} = -\frac{\partial v}{\partial r}, \qquad r \ne 0.$$

Show that the following pairs of functions satisfy these equations:

(a) $u(r, \theta) = (r^2 + r \cos \theta)/(1 + r^2 + 2r \cos \theta)$, $v(r, \theta) = r \sin \theta/(1 + r^2 + 2r \cos \theta)$

(b) $u(r, \theta) = \sqrt{r} \cos (\theta/2)$, $v(r, \theta) = \sqrt{r} \sin (\theta/2)$

(c) $u(r, \theta) = \ln r$, $v(r, \theta) = \theta$

12.4 Gradients

Suppose a function $f(x, y, z)$ is defined at each point in some region of space, and that at each point of the region all three partial derivatives

$$\frac{\partial f}{\partial x}, \quad \frac{\partial f}{\partial y}, \quad \frac{\partial f}{\partial z}$$

exist. For example, if $f(x, y, z)$ represents the present temperature at each point in the room in which you are working, then these derivatives represent rates of change of temperature in directions parallel to the x-, y-, and z-axes, respectively. There is a particular combination of these derivatives that proves very useful in later work. This combination is contained in the following definition.

> **DEFINITION 12.4**
>
> If a function $f(x, y, z)$ has partial derivatives $\partial f/\partial x$, $\partial f/\partial y$, and $\partial f/\partial z$ at each point in some region D of space, then at each point in D we define a vector called the **gradient** of $f(x, y, z)$, written grad f or ∇f, by
>
> $$\text{grad } f = \nabla f = \frac{\partial f}{\partial x}\hat{\mathbf{i}} + \frac{\partial f}{\partial y}\hat{\mathbf{j}} + \frac{\partial f}{\partial z}\hat{\mathbf{k}}. \qquad (12.8)$$

For a function $f(x, y)$ of only two independent variables, we have

$$\nabla f = \frac{\partial f}{\partial x}\hat{\mathbf{i}} + \frac{\partial f}{\partial y}\hat{\mathbf{j}}. \qquad (12.9)$$

EXAMPLE 12.6

If $f(x, y, z) = x^2yz - 2x/y$, find ∇f at $(1, -1, 3)$.

SOLUTION   Since $\nabla f = (2xyz - 2/y)\hat{\mathbf{i}} + (x^2z + 2x/y^2)\hat{\mathbf{j}} + (x^2y)\hat{\mathbf{k}}$, we have

$$\nabla f_{|(1,-1,3)} = -4\hat{\mathbf{i}} + 5\hat{\mathbf{j}} - \hat{\mathbf{k}}.$$

EXAMPLE 12.7

If $f(x, y, z) = \text{Tan}^{-1}(xy/z)$, what is ∇f?

SOLUTION

$$\nabla f = \left[\frac{1}{1 + (xy/z)^2}\left(\frac{y}{z}\right)\right]\hat{\mathbf{i}} + \left[\frac{1}{1 + (xy/z)^2}\left(\frac{x}{z}\right)\right]\hat{\mathbf{j}} + \left[\frac{1}{1 + (xy/z)^2}\left(\frac{-xy}{z^2}\right)\right]\hat{\mathbf{k}}$$

$$= \frac{yz}{z^2 + x^2y^2}\hat{\mathbf{i}} + \frac{xz}{z^2 + x^2y^2}\hat{\mathbf{j}} - \frac{xy}{z^2 + x^2y^2}\hat{\mathbf{k}}$$

$$= (yz\hat{\mathbf{i}} + xz\hat{\mathbf{j}} - xy\hat{\mathbf{k}})/(z^2 + x^2y^2).$$

Gradients arise in a multitude of applications in applied mathematics — heat conduction, electromagnetic theory, and fluid flow, to name a few — and two of the properties that make them so indispensable are discussed in detail in Sections 12.8 and 12.9. Examples 12.8 and 12.9 suggest these properties, but we make no attempt at a complete discussion here. For the moment we simply want you to be familiar with the definition of gradients and be able to calculate them.

EXAMPLE 12.8

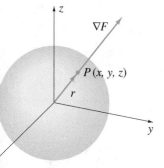

The equation $F(x, y, z) = 0$, where $F(x, y, z) = x^2 + y^2 + z^2 - 4$, defines a sphere of radius 2 centred at the origin. Show that the gradient vector ∇F at any point on the sphere is perpendicular to the sphere.

SOLUTION   If $P(x, y, z)$ is any point on the sphere (Figure 12.11), then the position vector $\mathbf{r} = x\hat{\mathbf{i}} + y\hat{\mathbf{j}} + z\hat{\mathbf{k}}$ from the origin to P is clearly perpendicular to the sphere. On the other hand,

$$\nabla F = 2x\hat{\mathbf{i}} + 2y\hat{\mathbf{j}} + 2z\hat{\mathbf{k}} = 2\mathbf{r}.$$

Consequently, at any point P on the sphere, ∇F is also perpendicular to the sphere.

This example suggests that gradients may be useful in finding perpendiculars to surfaces (and, as we will see, perpendiculars to curves).

EXAMPLE 12.9

The function $f(x, y) = 2x^2 - 4x + 3y^2 + 2y + 6$ is defined at every point in the xy-plane. If we start at the origin $(0, 0)$ and move along the positive x-axis, the rate of change of the function is $f_x(0, 0) = -4$; if we move along the y-axis, the rate of change is $f_y(0, 0) = 2$. Calculate the rate of change of $f(x, y)$ at $(0, 0)$ if we move toward the point $(1, 1)$ along the line $y = x$, and show that it is equal to the component of $\nabla f_{|(0,0)}$ in the direction $\mathbf{v} = (1, 1)$.

SOLUTION   The difference in values of $f(x, y)$ at any point (x, y) and $(0, 0)$ is $f(x, y) - f(0, 0) = (2x^2 - 4x + 3y^2 + 2y + 6) - (6) = 2x^2 - 4x + 3y^2 + 2y$. If we divide this by the length of the line joining $(0, 0)$ and (x, y), we obtain

$$\frac{f(x, y) - f(0, 0)}{\sqrt{x^2 + y^2}} = \frac{2x^2 - 4x + 3y^2 + 2y}{\sqrt{x^2 + y^2}}.$$

The limit of this quotient as (x, y) approaches $(0, 0)$ along the line $y = x$ should yield the required rate of change. We therefore set $y = x$ and take the limit as x approaches zero through positive numbers:

$$\lim_{x \to 0^+} \frac{2x^2 - 4x + 3x^2 + 2x}{\sqrt{x^2 + x^2}} = \lim_{x \to 0^+} \frac{x(5x - 2)}{\sqrt{2}x} = -\sqrt{2}.$$

This quantity, then, is the rate of change of $f(x, y)$ at $(0, 0)$ along the line $y = x$ toward the point $(1, 1)$.

On the other hand,

$$\nabla f_{|(0,0)} = (4x - 4, 6y + 2)_{|(0,0)} = (-4, 2),$$

and the component of this vector in the direction $\mathbf{v} = (1, 1)$ is

$$\nabla f_{|(0,0)} \cdot \hat{\mathbf{v}} = (-4, 2) \cdot \frac{(1, 1)}{\sqrt{2}} = \frac{-4 + 2}{\sqrt{2}} = -\sqrt{2}.$$

This example indicates that gradients may be useful in calculating rates of change of functions in directions other than those parallel to the coordinate axes.

EXAMPLE 12.10

The electrostatic potential at a point (x, y, z) in space due to a charge q fixed at the origin i given by

$$V = \frac{q}{4\pi \epsilon_0 r},$$

where $r = \sqrt{x^2 + y^2 + z^2}$. If a second charge Q is placed at (x, y, z), it experiences a force **F**, where

$$\mathbf{F} = \frac{qQ}{4\pi \epsilon_0 r^3}\mathbf{r},$$

where $\mathbf{r} = x\hat{\mathbf{i}} + y\hat{\mathbf{j}} + z\hat{\mathbf{k}}$. Show that $\mathbf{F} = -Q\nabla V$.

SOLUTION   We show that $\nabla V = -\mathbf{F}/Q$,

$$\nabla V = \nabla\left(\frac{q}{4\pi \epsilon_0 r}\right) = \frac{q}{4\pi \epsilon_0}\nabla\left(\frac{1}{\sqrt{x^2 + y^2 + z^2}}\right)$$

$$= \frac{q}{4\pi \epsilon_0}\left[\frac{-x}{(x^2 + y^2 + z^2)^{3/2}}\hat{\mathbf{i}} + \frac{-y}{(x^2 + y^2 + z^2)^{3/2}}\hat{\mathbf{j}} + \frac{-z}{(x^2 + y^2 + z^2)^{3/2}}\hat{\mathbf{k}}\right]$$

$$= \frac{-q}{4\pi \epsilon_0 r^3}(x\hat{\mathbf{i}} + y\hat{\mathbf{j}} + z\hat{\mathbf{k}}) = \frac{-q}{4\pi \epsilon_0 r^3}\mathbf{r} = -\frac{\mathbf{F}}{Q}.$$

EXERCISES 12.4

In Exercises 1–10 find the gradient of the function.

1. $f(x, y, z) = x^2 y + xz + yz^2$

2. $f(x, y, z) = x^2 yz$

3. $f(x, y, z) = x^2 y/z - 2xz^6$

4. $f(x, y) = x^2 y + xy^2$

5. $f(x, y) = \sin(x + y)$

6. $f(x, y, z) = \text{Tan}^{-1}(xyz)$

7. $f(x, y) = \text{Tan}^{-1}(y/x)$

8. $f(x, y, z) = e^{x+y+z}$

9. $f(x, y) = 1/(x^2 + y^2)$

10. $f(x, y, z) = 1/\sqrt{x^2 + y^2 + z^2}$

In Exercises 11–15 find the gradient of the function at the point.

11. $f(x, y) = xy + x + y$ at $(1, 3)$

12. $f(x, y, z) = \cos(x + y + z)$ at $(-1, 1, 1)$

13. $f(x, y, z) = (x^2 + y^2 + z^2)^2$ at $(0, 3, 6)$

14. $f(x, y) = e^{-x^2 - y^2}$ at $(2, 2)$

15. $f(x, y, z) = xy\ln(x + y)$ at $(4, -2)$

16. The equation $F(x, y, z) = Ax + By + Cz + D = 0$ defines plane in space. Show that at any point on the plane the vector ∇F i perpendicular to the plane.

17. Use the result of Exercise 16 to illustrate that a vector along the line

$$F(x, y, z) = 2x + 3y - 2z + 4 = 0,$$

$$G(x, y, z) = x - y + 3z + 6 = 0$$

is $\nabla F \times \nabla G$. Find parametric equations for the line.

18. Prove that if $f(x, y, z)$ and $g(x, y, z)$ both have gradients, the $\nabla(fg) = f\nabla g + g\nabla f$. What does this remind you of?

* **19.** Repeat Example 12.9 for the functions (a) $f(x, y) = x^2 + y$ and (b) $f(x, y) = 2x^3 - 3y$.

* **20.** The equation $F(x, y) = x^3 + xy + y^4 - 5 = 0$ implicitly define a curve in the xy-plane. Show that at any point on the curve, ∇F is normal vector to the curve.

Draw the surface defined by the equations in Exercises 21–22. At wha points on the surface is ∇F not defined?

21. $F(x, y, z) = z - \sqrt{x^2 + y^2} = 0$

22. $F(x, y, z) = z - |x - y| = 0$

3. If $f(x, y) = 1 - x^2 - y^2$, find ∇f. Find the point (x, y) at which $\nabla f = \mathbf{0}$, and illustrate graphically the nature of the surface $= f(x, y)$ at this point.

4. If the gradient of a function $f(x, y)$ is $\nabla f = (2xy - y)\hat{\mathbf{i}} + (x^2 - x)\hat{\mathbf{j}}$, what is $f(x, y)$?

5. Repeat Exercise 24 if $\nabla f = (2x/y + 1)\hat{\mathbf{i}} + (-x^2/y^2 + 2)\hat{\mathbf{j}}$.

6. If the gradient of a function $f(x, y, z)$ is $\nabla f = yz\hat{\mathbf{i}} + (xz + yz)\hat{\mathbf{j}} + (xy + y^2)\hat{\mathbf{k}}$, what is $f(x, y, z)$?

* **27.** Repeat Exercise 26 if $\nabla f = (x\hat{\mathbf{i}} + y\hat{\mathbf{j}} + z\hat{\mathbf{k}})/\sqrt{x^2 + y^2 + z^2}$.

* **28.** If $f(x, y)$ and $g(x, y)$ have first partial derivatives in a region R of the xy-plane, and if in R, $\nabla f = \nabla g$, how are $f(x, y)$ and $g(x, y)$ related?

* **29.** If $\nabla f = \mathbf{0}$ for all points in some region R of space, what can we say about $f(x, y, z)$ in R?

** **30.** Show that if the equation $F(x, y) = 0$ implicitly defines a curve C in the xy-plane, then at any point on C the vector ∇F is perpendicular to C.

12.5 Higher-Order Partial Derivatives

If $f(x, y) = x^3 y^2 + ye^x$, then

$$\frac{\partial f}{\partial x} = 3x^2 y^2 + ye^x \qquad \text{and} \qquad \frac{\partial f}{\partial y} = 2x^3 y + e^x.$$

Since each of these partial derivatives is a function of x and y, we can take further partial derivatives. The partial derivative of $\partial f/\partial x$ with respect to x is called the **second partial derivative** of $f(x, y)$ with respect to x, and is written

$$\frac{\partial}{\partial x}\left(\frac{\partial f}{\partial x}\right) = \frac{\partial^2 f}{\partial x^2} = 6xy^2 + ye^x.$$

Similarly, we have three more second partial derivatives:

$$\frac{\partial}{\partial y}\left(\frac{\partial f}{\partial x}\right) = \frac{\partial^2 f}{\partial y\, \partial x} = 6x^2 y + e^x,$$

$$\frac{\partial}{\partial x}\left(\frac{\partial f}{\partial y}\right) = \frac{\partial^2 f}{\partial x\, \partial y} = 6x^2 y + e^x,$$

$$\frac{\partial}{\partial y}\left(\frac{\partial f}{\partial y}\right) = \frac{\partial^2 f}{\partial y^2} = 2x^3.$$

Note that the second partial derivatives $\partial^2 f/\partial x\, \partial y$ and $\partial^2 f/\partial y\, \partial x$ are identical. This is not a peculiarity of this function; according to the following theorem, it is to be expected.

THEOREM 12.1

If $f(x, y)$, $\partial f/\partial x$, $\partial f/\partial y$, $\partial^2 f/\partial x\, \partial y$, and $\partial^2 f/\partial y\, \partial x$ are all defined inside a circle centred at a point P, and are continuous at P, then at P

$$\frac{\partial^2 f}{\partial x\, \partial y} = \frac{\partial^2 f}{\partial y\, \partial x}. \tag{12.10}$$

Corresponding to the subscript notation $f_x(x, y)$ for $\partial f/\partial x$, we have the following nota tions for second partial derivatives:

$$\frac{\partial^2 f}{\partial x^2} = f_{xx}, \quad \frac{\partial^2 f}{\partial x\, \partial y} = f_{yx}, \quad \frac{\partial^2 f}{\partial y\, \partial x} = f_{xy}, \quad \frac{\partial^2 f}{\partial y^2} = f_{yy}.$$

Notice the reversal in order of x and y in the middle terms. In subscript notation, derivatives ar taken in the order in which they appear (left to right, y first, x second, in f_{yx}). In $\partial^2 f/\partial x\, \partial y$ derivatives are done right to left, y first, x second. Because of Theorem 12.1, the order is usuall irrelevant anyway.

Partial derivatives of orders higher than two are also possible. For the function $f(x, y) =$ $x^3 y^2 + y e^x$ above, we have

$$f_{xxy} = \frac{\partial^3 f}{\partial y\, \partial x^2} = \frac{\partial}{\partial y}\left(\frac{\partial^2 f}{\partial x^2}\right) = \frac{\partial}{\partial y}\left(6xy^2 + ye^x\right) = 12xy + e^x$$

and

$$f_{yxyx} = \frac{\partial}{\partial x}\left[\frac{\partial}{\partial y}\left(\frac{\partial^2 f}{\partial x\, \partial y}\right)\right] = \frac{\partial}{\partial x}\left[\frac{\partial}{\partial y}\left(6x^2 y + e^x\right)\right] = \frac{\partial}{\partial x}\left[6x^2\right] = 12x.$$

For most functions with which we will be concerned, Theorem 12.1 can be extended to sa that a mixed partial derivative may be calculated in any order whatsoever. For example, if w require $\partial^{10} f/\partial x^3\, \partial y^7$, where $f(x, y) = \ln(y^y) + x^2 y^{10}$, it is advantageous to reverse th order of differentiation:

$$\frac{\partial^{10} f}{\partial x^3\, \partial y^7} = \frac{\partial^{10} f}{\partial y^7\, \partial x^3} = \frac{\partial^7}{\partial y^7}\left(\frac{\partial^3 f}{\partial x^3}\right) = 0.$$

EXAMPLE 12.11

Show that the function $f(x, y) = \operatorname{Tan}^{-1}\left(\dfrac{y}{x}\right)$ satisfies the equation

$$\frac{\partial^2 f}{\partial x^2} + \frac{\partial^2 f}{\partial y^2} = 0.$$

SOLUTION Since

$$\frac{\partial f}{\partial x} = \frac{1}{1 + \dfrac{y^2}{x^2}}\left(-\frac{y}{x^2}\right) = \frac{-y}{x^2 + y^2},$$

the second derivative with respect to x is $\dfrac{\partial^2 f}{\partial x^2} = \dfrac{2xy}{(x^2 + y^2)^2}.$ Since

$$\frac{\partial f}{\partial y} = \frac{1}{1 + \dfrac{y^2}{x^2}}\left(\frac{1}{x}\right) = \frac{x}{x^2 + y^2},$$

the second derivative with respect to y is $\dfrac{\partial^2 f}{\partial y^2} = \dfrac{-2xy}{(x^2 + y^2)^2}.$ When added, these secon derivatives cancel one another, and this completes the proof.

The equation

$$\frac{\partial^2 f}{\partial x^2} + \frac{\partial^2 f}{\partial y^2} = 0 \tag{12.11}$$

for a function $f(x, y)$ is one of the most important equations in applied mathematics. It is called **Laplace's equation** in two variables (x and y). Laplace's equation for a function $f(x, y, z)$ of three variables is

$$\frac{\partial^2 f}{\partial x^2} + \frac{\partial^2 f}{\partial y^2} + \frac{\partial^2 f}{\partial z^2} = 0. \qquad (12.12)$$

A function is said to be a **harmonic function** in a region R if it satisfies Laplace's equation in R and has continuous second partial derivatives in R. In particular, the function $f(x, y)$ in Example 12.11 is harmonic in any region that does not contain points on the y-axis. The next two examples illustrate areas of applied mathematics in which Laplace's equation is prominent. A third is contained in Exercise 30.

EXAMPLE 12.12

Show that the electrostatic potential function of Example 12.10 is harmonic in any region not containing the origin.

SOLUTION With $V = \left[q/(4\pi\epsilon_0)\right](x^2 + y^2 + z^2)^{-1/2}$,

$$\frac{\partial V}{\partial x} = \frac{q}{4\pi\epsilon_0}\left[\frac{-x}{(x^2 + y^2 + z^2)^{3/2}}\right]$$

and

$$\frac{\partial^2 V}{\partial x^2} = \frac{-q}{4\pi\epsilon_0}\left[\frac{1}{(x^2 + y^2 + z^2)^{3/2}} + \frac{-3x^2}{(x^2 + y^2 + z^2)^{5/2}}\right]$$

$$= \frac{q}{4\pi\epsilon_0}\left[\frac{2x^2 - y^2 - z^2}{(x^2 + y^2 + z^2)^{5/2}}\right].$$

Similarly,

$$\frac{\partial^2 V}{\partial y^2} = \frac{q}{4\pi\epsilon_0}\left[\frac{2y^2 - x^2 - z^2}{(x^2 + y^2 + z^2)^{5/2}}\right], \qquad \frac{\partial^2 V}{\partial z^2} = \frac{q}{4\pi\epsilon_0}\left[\frac{2z^2 - x^2 - y^2}{(x^2 + y^2 + z^2)^{5/2}}\right].$$

Addition of these shows that $V_{xx} + V_{yy} + V_{zz} = 0$. Since second partial derivatives are continuous in any region not containing $(0, 0, 0)$, $V(x, y, z)$ is harmonic therein.

EXAMPLE 12.13

Figure 12.12 shows a 1-m by 1-m metal plate that is insulated top and bottom. Temperature along the edges $x = 0$, $x = 1$, and $y = 1$ is held at $0°$C, while that along $y = 0$ is $f(x) = 4\sin \pi x$. Steady-state temperature at points inside the plate is then

$$T(x, y) = \frac{4}{e^\pi - e^{-\pi}}\left[e^{\pi(1-y)} - e^{-\pi(1-y)}\right]\sin \pi x.$$

Show that $T(x, y)$ is harmonic inside the plate.

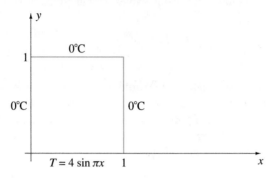

FIGURE 12.12   Temperature in a plate

SOLUTION   Partial derivatives of $T(x, y)$ are

$$\frac{\partial T}{\partial x} = \frac{4\pi}{e^{\pi} - e^{-\pi}} \left[e^{\pi(1-y)} - e^{-\pi(1-y)} \right] \cos \pi x,$$

$$\frac{\partial^2 T}{\partial x^2} = \frac{-4\pi^2}{e^{\pi} - e^{-\pi}} \left[e^{\pi(1-y)} - e^{-\pi(1-y)} \right] \sin \pi x,$$

$$\frac{\partial T}{\partial y} = \frac{4}{e^{\pi} - e^{-\pi}} \left[-\pi e^{\pi(1-y)} - \pi e^{-\pi(1-y)} \right] \sin \pi x,$$

$$\frac{\partial^2 T}{\partial y^2} = \frac{4}{e^{\pi} - e^{-\pi}} \left[\pi^2 e^{\pi(1-y)} - \pi^2 e^{-\pi(1-y)} \right] \sin \pi x.$$

Clearly, $\partial^2 T/\partial x^2 + \partial^2 T/\partial y^2 = 0$. Since second partial derivatives are continuous for $0 < x < 1$ and $0 < y < 1$, the temperature function $T(x, y)$ is harmonic in this region. In practice, we are not given $T(x, y)$; we must find it; that is, we must solve Laplace's equation

$$\frac{\partial^2 T}{\partial x^2} + \frac{\partial^2 T}{\partial y^2} = 0, \qquad 0 < x < 1, \quad 0 < y < 1,$$

subject to the conditions

$$T(0, y) = 0, \qquad 0 < y < 1,$$
$$T(1, y) = 0, \qquad 0 < y < 1,$$
$$T(x, 1) = 0, \qquad 0 < x < 1,$$
$$T(x, 0) = 4 \sin \pi x, \qquad 0 < x < 1.$$

This is called a *boundary-value problem*; it is treated in books dealing with partial differential equations. Laplace's equation is a partial differential equation, that is, a differential equation with partial derivatives.

Longitudinal Vibrations of Bars

Figure 12.13 shows a bar of length L and uniform cross-section. If a longitudinal force (in the x-direction) is applied to each cross-section, the bar stretches and/or compresses. (The bar acts like a very stiff spring.) Suppose we denote displacement of the cross-section normally at position x by $y(x, t)$, where t is time. This allows for the situation when the applied force

$F(x, t)$ varies along the bar and is also a function of time. It can be shown that $y(x, t)$ must satisfy the following partial differential equation, called the *one-dimensional wave equation*,

$$\frac{\partial^2 y}{\partial t^2} = \frac{E}{\rho} \frac{\partial^2 y}{\partial x^2} + \frac{F(x, t)}{\rho}, \qquad 0 < x < L, \quad t > 0, \qquad (12.13a)$$

where ρ is the density of the material in the bar and E is Young's modulus of elasticity. It is a constant that depends on the material of the bar; the larger the value of E, the more resistant the bar is to stretch or compression. Given $F(x, t)$, ρ, and E, the objective is to solve the wave equation for $y(x, t)$, therefore giving positions of cross-sections of the bar for all time.

FIGURE 12.13   Longitudinal vibrations of a bar

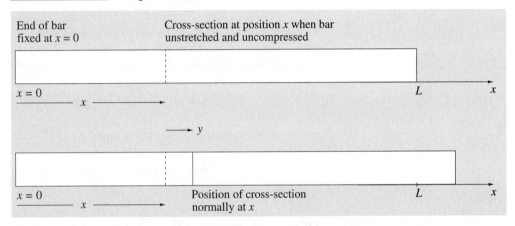

By itself, the wave equation has many solutions; other conditions must be stipulated. Newton's second law governs motion (and it was used in developing the wave equation). Our experience with particle motion suggests that we require two *initial conditions*, one specifying the initial positions of cross-sections of the bar, and a second specifying their velocities. In other words, accompanying 12.13a will be initial conditions of the form

$$y(x, 0) = f(x), \qquad 0 < x < L, \qquad (12.13b)$$

$$y_t(x, 0) = g(x), \qquad 0 < x < L. \qquad (12.13c)$$

There will also be *boundary conditions* specifying what is happening at the ends of the bar. For instance, if end $x = 0$ is clamped in position and, therefore, not allowed to move, $y(x, t)$ satisfies

$$y(0, t) = 0, \qquad t > 0. \qquad (12.13d)$$

If end $x = L$ of the bar is allowed to move freely, then the boundary condition there is

$$y_x(L, t) = 0, \qquad t > 0. \qquad (12.13e)$$

There are many types of boundary conditions that can occur at $x = 0$ and $x = L$. These are two examples.

Let us take a very simple illustration. The problem is much easier if displacement is not a function of time, in which case we solve for what are called *static displacements* of the bar. For this to occur, the applied force must be independent of time. Displacement becomes only a function of position, $y(x)$, initial conditions are dropped, time disappears from the boundary conditions, and the partial differential equation becomes an ordinary differential equation,

$$0 = E \frac{d^2 y}{dx^2} + F(x), \qquad 0 < x < L, \qquad (12.14a)$$

$$y(0) = 0, \qquad (12.14b)$$

$$y'(L) = 0. \qquad (12.14c)$$

If, for instance, all cross-sections are subjected to the same force F (a constant), then

$$\frac{d^2y}{dx^2} = -\frac{F}{E} \implies y(x) = -\frac{Fx^2}{2E} + Cx + D.$$

The boundary conditions require that

$$0 = y(0) = D, \quad 0 = y'(L) = -\frac{FL}{E} + C \implies C = \frac{FL}{E}.$$

Thus, displacement of the cross-section of the bar from its equilibrium position x is

$$y(x) = -\frac{Fx^2}{2E} + \frac{FLx}{E} = \frac{Fx}{2E}(2L - x).$$

Its graph is shown in Figure 12.14. To get an idea of the magnitude of these displacements, we find that for a 1-m steel bar with $E = 2.0 \times 10^{11}$ N/m^2, and an applied force $F = 10^5$ N, the right end has displacement

$$y(1) = \frac{10^5}{2(2 \times 10^{11})}\left[2(1) - 1\right] = 2.5 \times 10^{-5} \text{ m};$$

that is, the bar stretches by only 0.025 mm.

FIGURE 12.14   Static displacements of a bar subjected to a longitudinal force

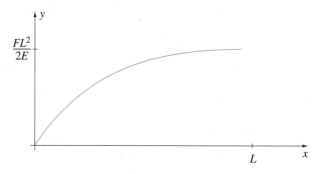

The one-dimensional wave equation 12.13a is satisfied by longitudinal vibrations of bars. It is also satisfied by transverse vibrations of strings and rotational vibrations of bars. These will be discussed in the exercises.

EXERCISES 12.5

In Exercises 1–20 find the derivative.

1. $\partial^2 f/\partial x^2$ if $f(x, y) = x^2y^2 - 2x^3y$

2. $\partial^3 f/\partial y^3$ if $f(x, y) = 2x/y + 3x^3y^4$

3. $\partial^2 f/\partial z^2$ if $f(x, y, z) = \sin(xyz)$

4. $\partial^2 f/\partial y\,\partial z$ if $f(x, y, z) = xyze^{x+y+z}$

5. $\partial^2 f/\partial y\,\partial x$ if $f(x, y) = \sqrt{x^2 + y^2}$

6. $\partial^3 f/\partial x^2\,\partial y$ if $f(x, y) = e^{x+y} - x^2/y^2$

7. $\partial^3 f/\partial y^3$ at $(1, 3)$ if $f(x, y) = 3x^3y^3 - 3x/y$

8. $\partial^3 f/\partial x\,\partial y\,\partial z$ at $(1, 0, -1)$ if $f(x, y, z) = x^2y^2 + x^2z^2 + y^2z^2$

9. $\partial^2 f/\partial x^2$ if $f(x, y) = \sqrt{1 - x^2 - y^2}$

10. $\partial^2 f/\partial z^2$ if $f(x, y, z) = \ln\sqrt{x^2 + y^2 + z^2}$

11. $\partial^3 f/\partial x^2\,\partial y$ if $f(x, y) = x^2e^y + y^2e^x$

12. $\partial^2 f/\partial x^2$ if $f(x, y) = \text{Tan}^{-1}(y/x)$

13. $\partial^3 f/\partial x\,\partial y^2$ if $f(x, y, z) = \cot(x^2 + y^2 + z^2)$

14. $\partial^2 f/\partial x\,\partial y$ at $(-2, -2)$ if $f(x, y) = \text{Sin}^{-1}(x^2 + y^2)^{-1}$

15. $\partial^{10} f/\partial x^7\,\partial y^3$ if $f(x, y) = x^7e^xy^2 + 1/y^6$

16. $\partial^8 f/\partial x^8$ if $f(x, y, z) = x^8y^9z^{10}$

17. $\partial^6 f/\partial x^2\,\partial y^2\,\partial z^2$ if $f(x, y, z) = 1/x^2 + 1/y^2 + 1/z^2$

18. $\partial^4 f/\partial x^3\,\partial y$ if $f(x, y) = \cos(x + y^3)$

19. $\partial^4 f/\partial x\,\partial y\,\partial z\,\partial t$ if $f(x, y, z, t) = \sqrt{x^2 + y^2 + z^2 - t^2}$

20. $\partial^2 f/\partial x\,\partial y$ if $f(x, y) = \text{Sec}^{-1}(xy)$

21. If $z = x^2 + xy + y^2 \sin(x/y)$, show that

$$x\frac{\partial z}{\partial x} + y\frac{\partial z}{\partial y} = 2z = x^2\frac{\partial^2 z}{\partial x^2} + 2xy\frac{\partial^2 z}{\partial x\,\partial y} + y^2\frac{\partial^2 z}{\partial y^2}.$$

22. If $u = x + y + ze^{y/x}$, show that

$$x^2\frac{\partial^2 u}{\partial x^2} + y^2\frac{\partial^2 u}{\partial y^2} + z^2\frac{\partial^2 u}{\partial z^2}$$

$$+ 2xy\frac{\partial^2 u}{\partial x\,\partial y} + 2yz\frac{\partial^2 u}{\partial y\,\partial z} + 2xz\frac{\partial^2 u}{\partial x\,\partial z} = 0.$$

In Exercises 23–28 find a region (if possible) in which the function is harmonic.

23. $f(x, y) = x^2 - y^2 + 2xy + y$

24. $f(x, y) = \ln(x^2 + y^2)$

25. $f(x, y) = x^3 y^2 - 3xy$

26. $f(x, y, z) = 3x^2 yz - y^3 z + xy$

27. $f(x, y, z) = 1/\sqrt{x^2 + y^2 + z^2}$

28. $f(x, y, z) = x^3 y^3 z^3$

29. If $V(x, y, z)$ represents the electrostatic potential at a point (x, y, z) due to a system of n point charges at points (x_i, y_i, z_i), does $V(x, y, z)$ satisfy Laplace's equation?

30. The gravitational potential at a point (x, y, z) in space due to a uniform spherical mass distribution (mass M) at the origin is defined as $V = GM/r$, where G is a constant and $r = \sqrt{x^2 + y^2 + z^2}$. Show that $V(x, y, z)$ satisfies Laplace's equation 12.12.

31. The figure below shows a plate bounded by the lines $x = 0$, $y = 0$, $x = 1$, and $y = 1$. Temperature along the first three sides is kept at $0°C$, while that along $y = 1$ varies according to $f(x) = \sin(3\pi x) - 2\sin(4\pi x)$, $0 \le x \le 1$. The temperature at any point interior to the plate is then

$$T(x, y) = C(e^{3\pi y} - e^{-3\pi y})\sin(3\pi x)$$

$$+ D(e^{4\pi y} - e^{-4\pi y})\sin(4\pi x),$$

where $C = (e^{3\pi} - e^{-3\pi})^{-1}$ and $D = -2(e^{4\pi} - e^{-4\pi})^{-1}$. Show that $T(x, y)$ is harmonic in the region $0 < x < 1$, $0 < y < 1$, and that it also satisfies the boundary conditions $T(0, y) = 0$, $T(1, y) = 0$, $T(x, 0) = 0$, and $T(x, 1) = f(x)$.

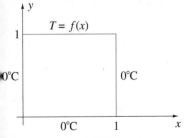

32. When the bar in Figure 12.13 is turned vertically and clamped at $x = 0$ (figure below), static deflections of the bar must satisfy the following problem:

$$0 = E\frac{d^2 y}{dx^2} + F(x), \qquad 0 < x < L,$$

$$y(0) = 0, \quad y'(L) = 0,$$

where $F(x)$ is the weight of that part of the bar below the cross-section that would be at position x if the bar were unstretched. Find the length of the bar as it stretches under its own weight.

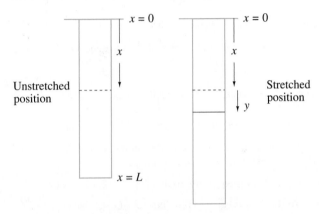

33. When the bar in Figure 12.13 is subjected to a force per unit area of magnitude F on its right end, and no other forces act on the bar, equations 12.14 for static deflections become

$$0 = E\frac{d^2 y}{dx^2}, \qquad 0 < x < L,$$

$$y(0) = 0, \quad y'(L) = F/E.$$

Find displacements for cross-sections, and the length of the bar.

34. (a) Show that when $F(x, t) \equiv 0$, wave equation 12.13a is satisfied by functions of the form

$$y(x, t) = (A\sin\lambda x + B\cos\lambda x)(C\sin c\lambda t + D\cos c\lambda t),$$

$$c^2 = E/\rho,$$

where λ, A, B, C, and D are arbitrary constants.

(b) Show that boundary conditions 12.13d and e require that $B = 0$ and $\lambda = (2n - 1)\pi/(2L)$, where n is an integer.

(c) If $g(x)$ is equal to zero [but not $f(x)$], what is C?

(d) If $f(x)$ is equal to zero [but not $g(x)$], what is D?

35. If the ends of the taut string in the figure below are fastened at $x = 0$ and $x = L$ on the x-axis, and if the string vibrates in the y-direction only, small displacements $y(x, t)$ must satisfy the one-dimensional wave equation

$$\frac{\partial^2 y}{\partial t^2} = c^2 \frac{\partial^2 y}{\partial x^2} + \frac{F(x,t)}{\rho}, \qquad 0 < x < L, \quad t > 0,$$

where $c^2 = T/\rho$, T is the (constant) tension of the string, and ρ is the (constant) mass per unit length of the string. Function $F(x,t)$ is the result of all forces per unit length acting at position x and time t in the string (except for tension in the string). Accompanying this equation will be initial conditions 12.13b and c specifying the position and velocity of the string at time $t = 0$, and the boundary conditions

$$y(0,t) = 0, \quad y(L,t) = 0, \quad t > 0,$$

since both ends of the string are attached to the x-axis.

(a) Show that when $F(x,t) \equiv 0$, the wave equation is satisfied by a function of the form

$$y(x,t) = (A \sin \lambda x + B \cos \lambda x)(C \sin c\lambda t + D \cos c\lambda t),$$

where λ, A, B, C, and D are arbitrary constants.

(b) Show that the boundary conditions require $B = 0$ and $\lambda = n\pi/L$, where n is an integer.

(c) If $g(x)$ is equal to zero [but not $f(x)$], what is C?

(d) If $f(x)$ is equal to zero [but not $g(x)$], what is D?

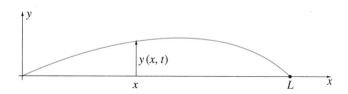

* **36.** Static deflections of the string in Exercise 35 occur when $F(x,t)$ is only a function of x, and initial conditions are ignored. The problem then is

$$0 = T \frac{d^2 y}{dx^2} + F(x), \qquad 0 < x < L,$$

$$y(0) = 0, \quad y(L) = 0.$$

When the only force acting on the string (other than internal tension) is gravity, then $F(x) = -9.81\rho$. Show that the solution for $y(x)$ is a parabola. What underlying assumption(s) make this problem and its solution different from Example 3.39 in Section 3.13?

* **37.** A string of length L is stretched tightly along the x-axis. Its right end is fixed on the x-axis, and its left end at $x = 0$ is looped around the y-axis and is free to move thereon without friction. The string is slowly lowered under the influence of gravity to take up a static position. The shape of the string is defined by the differential equation

$$0 = T \frac{d^2 y}{dx^2} - 9.81\rho, \qquad 0 < x < L,$$

$$y'(0) = 0, \quad y(L) = 0.$$

Solve this problem.

* **38.** A taut string has its ends fixed at $x = 0$ and $x = L$ on the x-axis. At time $t = 0$, the string is moved so as to take the shape of the sine curve $f(x) = 3\sin(\pi x/L)$, and then released. If the only force taken into account as acting on the string is its tension, then displacements of the string must satisfy

$$\frac{\partial^2 y}{\partial t^2} = c^2 \frac{\partial^2 y}{\partial x^2}, \qquad 0 < x < L, \quad t > 0,$$

$$y(0,t) = y(L,t) = 0, \qquad t > 0,$$

$$y(x,0) = f(x) = 3\sin(\pi x/L), \qquad 0 < x < L,$$

$$y_t(x,0) = 0, \qquad 0 < x < L.$$

(a) Show that $y(x,t) = 3\sin(\pi x/L)\cos(\pi ct/L)$ satisfies this problem.

(b) Plot the surface $y(x,t)$, interpreting cross-sections physically. (Use $L = 1$ and $c = 2$.)

(c) Plot $y(x,t)$ as a function of x for $t = 0$, $L/(8c)$, $L/(4c)$, $3L/(8c)$, and $L/(2c)$.

* **39.** Show that when $f(x) = 3\sin(\pi x/L) - 2\sin(2\pi x/L)$, in Exercise 38, the problem is satisfied by $y(x,t) = 3\sin(\pi x/L)\cos(\pi ct/L) - 2\sin(2\pi x/L)\cos(2\pi ct/L)$. Plot $y(x,t)$ as a function of x for $t = 0$, $L/(8c)$, $L/(4c)$, $3L/(8c)$, and $L/(2c)$.

* **40.** (a) If the length of the string in Exercise 38 is $L = 1$ m, and

$$f(x) = \begin{cases} x/100, & 0 \le x \le 1/2, \\ (1-x)/100, & 1/2 \le x \le 1, \end{cases}$$

displacements can be expressed in the form of an infinite series

$$y(x,t) = \frac{1}{25\pi^2} \sum_{n=1}^{\infty} \frac{(-1)^{n+1}}{(2n-1)^2} \sin(2n-1)\pi x \cos(2n-1)\pi ct.$$

Show that this function satisfies the partial differential equation, boundary conditions, and the second initial condition in Exercise 38. (Assume that differentiation and summation operations can be interchanged.)

(b) Suppose the first 20 terms of the series are used to approximate $y(x,t)$. Plot the sum of these terms as a function of x for the times $t = 0$, $1/(16c)$, $1/(8c)$, $3/(16c)$, $1/(4c)$, $5/(16c)$, $3/(8c)$, $7/(16c)$, and $1/(2c)$. Does the string appear to retain its broken-line shape?

* **41.** The uniform circular rod in the figure below has flat ends at $x = 0$ and $x = L$. If the round side of the rod is perfectly insulated, heat flows in only the x-direction. When no heat sources exist within the rod, temperature $T(x,t)$ at points in the rod must satisfy the one-dimensional heat conduction equation

$$\frac{\partial T}{\partial t} = k \frac{\partial^2 T}{\partial x^2}, \qquad 0 < x < L, \quad t > 0,$$

$x = 0$ L x

Perfect insulation

where $k > 0$ is a constant called the *thermal diffusivity* of the rod. Accompanying this equation will be two boundary conditions describing what is happening at the ends of the rod. For instance, if the left end is kept at temperature $0°C$ and the right end is insulated, $T(x, t)$ must satisfy

$$T(0, t) = 0, \quad T_x(L, t) = 0, \quad t > 0.$$

In addition, there will be an initial condition describing the temperature distribution in the rod at time $t = 0$,

$$T(x, 0) = f(x), \quad 0 < x < L.$$

(a) Show that the heat conduction equation is satisfied by functions of the form

$$T(x, t) = (A \sin \lambda x + B \cos \lambda x)e^{-k\lambda^2 t}$$

where λ, A, and B are arbitrary constants.

(b) Show that the boundary conditions require $B = 0$ and $\lambda = (2n - 1)\pi/(2L)$, where n is an integer.

(c) When the initial temperature of the rod is $f(x) = x$, so that it increases linearly from $0°C$ at its left end to $L°C$ at its right end, temperature thereafter is

$$T(x, t) = \frac{8L}{\pi^2} \sum_{n=1}^{\infty} \frac{(-1)^{n+1}}{(2n - 1)^2} e^{-(2n-1)^2\pi^2 kt/(4L^2)} \sin \frac{(2n - 1)\pi x}{2L}.$$

Show that this function satisfies the partial differential equation and boundary conditions. (Assume that differentiation and summation operations can be interchanged.)

(d) Plot the sum of the first 20 terms of the series as a function of x for the times $t = 0, 10, 100, 1000$, and $10\,000$. Use a 1-m rod and $k = 1.14 \times 10^{-4}$ m^2/s (typical for copper). Does the $t = 0$ plot approximate $f(x)$, and do the remaining plots reflect the boundary conditions?

42. If both ends of the rod in Exercise 41 are held at temperature $0°C$, the problem for $T(x, t)$ becomes

$$\frac{\partial T}{\partial t} = k\frac{\partial^2 T}{\partial x^2}, \quad 0 < x < L, \quad t > 0,$$

$$T(0, t) = T(L, t) = 0, \quad t > 0,$$

$$T(x, 0) = f(x), \quad 0 < x < L.$$

(a) Show that for the functions in part (a) of Exercise 41, the boundary conditions require that $B = 0$ and $\lambda = n\pi/L$, where n is an integer.

(b) When the initial temperature of the rod is $f(x) = x(L - x)$, temperature thereafter is

$$T(x, t) = \frac{8L^4}{\pi^3} \sum_{n=1}^{\infty} \frac{1}{(2n - 1)^3} e^{-(2n-1)^2\pi^2 kt/L^2} \sin \frac{(2n - 1)\pi x}{L}.$$

Plot the sum of the first 20 terms of this series as a function of x for the times $t = 0, 10, 100, 1000$, and $10\,000$. Use a 1-m rod and $k = 1.14 \times 10^{-4}$ m^2/s. Does the $t = 0$ plot approximate $f(x)$?

* **43.** If the ends $x = 0$ and $x = L$ of the rod in Exercise 41 are held at temperatures $T_0°C$ and $T_L°C$ for $t > 0$, temperature $T(x, t)$ must satisfy

$$\frac{\partial T}{\partial t} = k\frac{\partial^2 T}{\partial x^2}, \quad 0 < x < L, \quad t > 0,$$

$$T(0, t) = T_0, \quad T(L, t) = T_L, \quad t > 0,$$

$$T(x, 0) = f(x), \quad 0 < x < L.$$

After a very long time, the temperature at each point in the rod will remain constant, but temperature will vary from point to point. Temperature is said to have reached a steady-state situation. It can be found by removing the term $\partial T/\partial t$ from the heat conduction equation, and the initial condition. Temperature becomes a function of only x, which satisfies

$$0 = \frac{d^2 T}{dx^2}, \quad 0 < x < L,$$

$$T(0) = T_0, \quad T(L) = T_L.$$

Find $T(x)$.

* **44.** If heat is added to the rod in Exercise 41, the heat conduction equation takes the form

$$\frac{\partial T}{\partial t} = k\frac{\partial^2 T}{\partial x^2} + F(x, t), \quad 0 < x < L, \quad t > 0,$$

where $F(x, t)$ is the heat source term. If temperatures of the ends of the rod are held at $T_0°C$ and $T_L°C$, then $T(x, t)$ must also satisfy the boundary conditions

$$T(0, t) = T_0, \quad T(L, t) = T_L, \quad t > 0.$$

If $F(x, t)$ is independent of t, then the steady-state temperature of the rod (see Exercise 43) must satisfy

$$0 = k\frac{d^2 T}{dx^2} + F(x), \quad 0 < x < L,$$

$$T(0) = T_0, \quad T(L) = T_L.$$

Find $T(x)$ in the case that $F(x)$ is a constant value F.

* **45.** The figure below shows a bar of length L and uniform cross-section. When torque is applied to cross-sections, these cross-sections

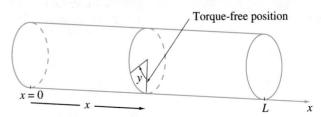

are forced to rotate. Let x represent distance from the left end of the bar, and $y(x, t)$ the angular displacement of the cross-section at position x and time t from its torque-free position. It can be shown that when $\tau(x, t)$ represents the torque per unit length at position x and time t, and ρ is the (constant) density of the bar, $y(x, t)$ must satisfy the one-dimensional wave equation

$$\frac{\partial^2 y}{\partial t^2} = c^2 \frac{\partial^2 y}{\partial x^2} + \frac{\tau(x, t)}{\rho}, \qquad 0 < x < L, \quad t > 0,$$

where $c^2 = E/\rho$ and E is Young's modulus of elasticity of the material in the bar under shear. Accompanying the partial differential equation will be initial conditions 12.13b and c specifying the initial displacements and velocities of cross-sections at time $t = 0$. In addition, there will be two boundary conditions describing end conditions. For example, if the left end is clamped so that no rotation is possible, and the right end is free to move, the boundary conditions will be 12.13d and e. The problem is therefore identical to that for longitudinal vibrations of bars. *Static rotations* occur when $\tau(x, t)$ is independent of t, the time derivative is removed from the wave equation, and initial conditions are deleted. Displacements $y(x)$ must then satisfy

$$0 = E \frac{d^2 y}{dx^2} + \tau(x), \qquad 0 < x < L,$$

$$y(0) = 0, \quad y'(L) = 0,$$

at least for the boundary conditions described above. Solve this problem when $\tau(x) = x$ so that torque increases linearly along the bar.

* **46.** Show that when $\tau(x)$ is unspecified, static rotations in Exercise 45 can be expressed in the form

$$y(x) = Cx - \frac{1}{E} \int_0^x \int_0^v \tau(u)\, du\, dv,$$

where

$$C = \frac{1}{E} \int_0^L \tau(x)\, dx.$$

* **47.** Two functions $u(x, y)$ and $v(x, y)$ are said to be harmonic conjugates if they satisfy the Cauchy–Riemann equations of Exercise 41 in Section 12.3. Show that if $u(x, y)$ and $v(x, y)$ are harmonic conjugates and have continuous second partial derivatives in a region R, then each is harmonic in R.

* **48.**   (a) Show that the function $u(x, y) = x^2 - y^2$ is harmonic in the entire xy-plane.

  (b) Use the Cauchy–Riemann equations in Exercise 41 of Section 12.3 to find a function $v(x, y)$ so that u and v are harmonic conjugates.

* **49.** Repeat Exercise 48 if $u(x, y) = e^x \cos y + x$.

* **50.** For what values of n does the function $(x^2 + y^2 + z^2)^n$ satisfy equation 12.12? In what regions are the functions harmonic?

12.6 Chain Rules for Partial Derivatives

If $y = f(u)$ and $u = g(x)$, the **chain rule** for the derivative dy/dx of the composite function $f[g(x)]$ is

$$\frac{dy}{dx} = \frac{dy}{du}\frac{du}{dx}. \tag{12.15}$$

Equation 12.15 can be extended in terms of more intermediate variables, say $y = f(u)$, $u = g(s)$, $s = h(x)$, in which case

$$\frac{dy}{dx} = \frac{dy}{du}\frac{du}{ds}\frac{ds}{dx}. \tag{12.16}$$

For multivariable functions, variations in chain rules are countless. We discuss two examples in considerable detail, and then show schematic diagrams that easily lead to chain rules for even the most complicated functional situations.

Suppose z is a function of u and v and each of u and v is a function of x and y,

$$z = f(u, v), \quad u = g(x, y), \quad v = h(x, y). \tag{12.17}$$

By the substitutions

$$z = f[g(x, y), h(x, y)], \tag{12.18}$$

we express z as a function of x and y, and can then calculate the partial derivative $\partial z/\partial x$. However, if the functions in 12.17 are at all complicated, you can imagine how difficult the composite function in 12.18 might be to differentiate. As a result, we search for an alternative procedure for calculating $\partial z/\partial x$, namely, the appropriate chain rule. It is contained in the following theorem.

THEOREM 12.2

Let $u = g(x, y)$ and $v = h(x, y)$ be continuous and have first partial derivatives with respect to x at a point (x, y), and let $z = f(u, v)$ have continuous first partial derivatives inside a circle centred at the point $(u, v) = (g(x, y), h(x, y))$. Then

$$\frac{\partial z}{\partial x} = \frac{\partial z}{\partial u}\frac{\partial u}{\partial x} + \frac{\partial z}{\partial v}\frac{\partial v}{\partial x}. \tag{12.19}$$

PROOF   This result can be proved in much the same way as chain rule 3.20a was proved in Section 3.7. By Definition 12.3,

$$\frac{\partial z}{\partial x} = \lim_{\Delta x \to 0} \frac{f[g(x + \Delta x, y), h(x + \Delta x, y)] - f[g(x, y), h(x, y)]}{\Delta x}.$$

Now the increment Δx in x produces changes in u and v, which we denote by

$$\Delta u = g(x + \Delta x, y) - g(x, y), \quad \Delta v = h(x + \Delta x, y) - h(x, y).$$

If we write u and v whenever $g(x, y)$ and $h(x, y)$ are evaluated at (x, y), and substitute for $g(x + \Delta x, y)$ and $h(x + \Delta x, y)$ in the definition for $\partial z/\partial x$, then

$$\frac{\partial z}{\partial x} = \lim_{\Delta x \to 0} \frac{f(u + \Delta u, v + \Delta v) - f(u, v)}{\Delta x}$$

$$= \lim_{\Delta x \to 0} \frac{[f(u + \Delta u, v + \Delta v) - f(u, v + \Delta v)] + [f(u, v + \Delta v) - f(u, v)]}{\Delta x}$$

$$= \lim_{\Delta x \to 0} \left[\frac{f(u + \Delta u, v + \Delta v) - f(u, v + \Delta v)}{\Delta x} + \frac{f(u, v + \Delta v) - f(u, v)}{\Delta x} \right].$$

We assumed that the derivative

$$\frac{\partial z}{\partial v} = \lim_{\Delta v \to 0} \frac{f(u, v + \Delta v) - f(u, v)}{\Delta v}$$

exists at (u, v). An equivalent way to express the fact that this limit exists is to say that

$$\frac{f(u, v + \Delta v) - f(u, v)}{\Delta v} = \frac{\partial z}{\partial v} + \epsilon_1,$$

where ϵ_1 must satisfy the condition that $\lim_{\Delta v \to 0} \epsilon_1 = 0$. We can write, therefore, that

$$f(u, v + \Delta v) - f(u, v) = [z_v(u, v) + \epsilon_1] \Delta v.$$

Similarly, we can write that

$$f(u + \Delta u, v + \Delta v) - f(u, v + \Delta v) = [z_u(u, v + \Delta v) + \epsilon_2] \Delta u,$$

where $\lim_{\Delta u \to 0} \epsilon_2 = 0$ (provided that Δv is sufficiently small). When these expressions are substituted into the limit for $\partial z/\partial x$, we have

$$\frac{\partial z}{\partial x} = \lim_{\Delta x \to 0} \left\{ [z_u(u, v + \Delta v) + \epsilon_2]\frac{\Delta u}{\Delta x} + [z_v(u, v) + \epsilon_1]\frac{\Delta v}{\Delta x} \right\}.$$

We now examine each part of this limit. Clearly,

$$\lim_{\Delta x \to 0} \frac{\Delta u}{\Delta x} = \frac{\partial u}{\partial x} \quad \text{and} \quad \lim_{\Delta x \to 0} \frac{\Delta v}{\Delta x} = \frac{\partial v}{\partial x}.$$

In addition, because $g(x, y)$ and $h(x, y)$ are continuous, $\Delta u \to 0$ and $\Delta v \to 0$ as $\Delta x \to 0$. Consequently,

$$\lim_{\Delta x \to 0} \epsilon_1 = \lim_{\Delta v \to 0} \epsilon_1 = 0 \quad \text{and} \quad \lim_{\Delta x \to 0} \epsilon_2 = \lim_{\Delta u \to 0} \epsilon_2 = 0.$$

Finally, because $\partial z / \partial u$ is continuous,

$$\lim_{\Delta x \to 0} z_u(u, v + \Delta v) = \lim_{\Delta v \to 0} z_u(u, v + \Delta v) = z_u(u, v).$$

When all these results are taken into account, we have

$$\frac{\partial z}{\partial x} = z_u(u, v)\frac{\partial u}{\partial x} + z_v(u, v)\frac{\partial v}{\partial x} = \frac{\partial z}{\partial u}\frac{\partial u}{\partial x} + \frac{\partial z}{\partial v}\frac{\partial v}{\partial x},$$

which completes the proof.

Chain rule 12.19 defines $\partial z / \partial x$ in terms of derivatives of the given functions in 12.17. We could be more explicit by indicating which variable is being held constant in each of the five derivatives:

$$\left(\frac{\partial z}{\partial x}\right)_y = \left(\frac{\partial z}{\partial u}\right)_v \left(\frac{\partial u}{\partial x}\right)_y + \left(\frac{\partial z}{\partial v}\right)_u \left(\frac{\partial v}{\partial x}\right)_y. \tag{12.20}$$

From the point of view of rates of change, this result seems quite reasonable. The left side is the rate of change of z with respect to x holding y constant. The first term $(\partial z/\partial u)(\partial u/\partial x)$ accounts for the rate of change of z with respect to those x's that affect z through u. The second term, $(\partial z/\partial v)(\partial v/\partial x)$, accounts for the rate of change of z with respect to those x's that affect z through v. The total rate of change is then the sum of the two parts.

Consider now the functional situation

$$z = f(u, v), \quad u = g(x, y, s), \quad v = h(x, y, s), \quad x = p(t), \quad y = q(t), \quad s = r(t) \tag{12.21}$$

By the substitutions

$$z = f[g(p(t), q(t), r(t)), h(p(t), q(t), r(t))], \tag{12.22}$$

we express z as a function of t alone, and can therefore pose the problem of calculating dz/dt. If we reason as in the preceding paragraph, the appropriate chain rule for dz/dt must account for all t's affecting z through u and v. We obtain, then,

$$\frac{dz}{dt} = \frac{\partial z}{\partial u}\frac{du}{dt} + \frac{\partial z}{\partial v}\frac{dv}{dt},$$

where we have written du/dt and dv/dt because u and v can be expressed entirely in terms of t:

$$u = g[p(t), q(t), r(t)], \quad v = h[p(t), q(t), r(t)].$$

Chain rules for each of du/dt and dv/dt (similar to 12.19) yield

$$\frac{du}{dt} = \frac{\partial u}{\partial x}\frac{dx}{dt} + \frac{\partial u}{\partial y}\frac{dy}{dt} + \frac{\partial u}{\partial s}\frac{ds}{dt}, \quad \frac{dv}{dt} = \frac{\partial v}{\partial x}\frac{dx}{dt} + \frac{\partial v}{\partial y}\frac{dy}{dt} + \frac{\partial v}{\partial s}\frac{ds}{dt}.$$

Finally, then,

$$\frac{dz}{dt} = \frac{\partial z}{\partial u}\left(\frac{\partial u}{\partial x}\frac{dx}{dt} + \frac{\partial u}{\partial y}\frac{dy}{dt} + \frac{\partial u}{\partial s}\frac{ds}{dt}\right) + \frac{\partial z}{\partial v}\left(\frac{\partial v}{\partial x}\frac{dx}{dt} + \frac{\partial v}{\partial y}\frac{dy}{dt} + \frac{\partial v}{\partial s}\frac{ds}{dt}\right), \tag{12.23}$$

which expresses dz/dt in terms of derivatives of the given functions in 12.21.

These two examples suggest the complexities that may be involved in finding chain rules for complicated composite functions. Fortunately, there is an amazingly simple method that gives the correct chain rule in every situation. The method is not designed to help you understand the chain rule, but to find it quickly. We suggest that you test your understanding by developing a few chain rules in the exercises with a discussion such as in the second example above, and then check your result by the quicker method.

In the first example we represent the functional situation described in 12.17 by the schematic diagram to the left. At the top of the diagram is the dependent variable z, which we wish to differentiate. In the line below z are the variables u and v in terms of which z is initially defined. In the line below u and v are x's and y's illustrating that each of u and v is defined in terms of x and y.

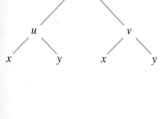

Here are the rules to obtain the partial derivative from schematic diagrams, in general, followed by $\partial z / \partial x$ for the specific example:

1. Take all possible paths in the schematic from the differentiated variable to the differentiating variable.

2. For each straight-line segment in a given path, differentiate the upper variable with respect to the lower variable and multiply together all such derivatives in that path.

3. Add the products together to form the complete chain rule.

To calculate $\partial z / \partial x$ from the schematic to the left, we note that there are two paths from z to x, one through u and one through v. For the path through u we form the product

$$\frac{\partial z}{\partial u} \frac{\partial u}{\partial x},$$

and for the path through v,

$$\frac{\partial z}{\partial v} \frac{\partial v}{\partial x}.$$

The complete chain rule is then the sum of these products,

$$\frac{\partial z}{\partial x} = \frac{\partial z}{\partial u} \frac{\partial u}{\partial x} + \frac{\partial z}{\partial v} \frac{\partial v}{\partial x},$$

and this result agrees with 12.19. The schematic diagram also indicates which variables are to be held constant in the derivatives on the right (as in 12.20). All other variables on the same level are held constant.

For the second example in equations 12.21 the schematic diagram is to the left. There are six possible paths from z to t, so that the chain rule for dz/dt must have six terms. We find

$$\frac{dz}{dt} = \frac{\partial z}{\partial u} \frac{\partial u}{\partial x} \frac{dx}{dt} + \frac{\partial z}{\partial u} \frac{\partial u}{\partial y} \frac{dy}{dt} + \frac{\partial z}{\partial u} \frac{\partial u}{\partial s} \frac{ds}{dt} + \frac{\partial z}{\partial v} \frac{\partial v}{\partial x} \frac{dx}{dt} + \frac{\partial z}{\partial v} \frac{\partial v}{\partial y} \frac{dy}{dt} + \frac{\partial z}{\partial v} \frac{\partial v}{\partial s} \frac{ds}{dt},$$

and this agrees with 12.23. Note too that if when forming a derivative from the schematic diagram, there are two or more lines emanating from a variable, then we obtain a partial derivative; if there is only one line, then we have an ordinary derivative.

EXAMPLE 12.14

Find chain rules for

$$\left. \frac{\partial z}{\partial x} \right)_y \quad \text{and} \quad \left. \frac{\partial z}{\partial y} \right)_x$$

if

$$z = f(r, s, x), \quad r = g(x, y), \quad s = h(x, y).$$

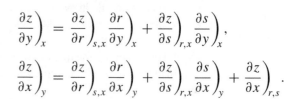

SOLUTION From the schematic diagram to the left,

$$\left.\frac{\partial z}{\partial y}\right)_x = \left.\frac{\partial z}{\partial r}\right)_{s,x}\left.\frac{\partial r}{\partial y}\right)_x + \left.\frac{\partial z}{\partial s}\right)_{r,x}\left.\frac{\partial s}{\partial y}\right)_x,$$

$$\left.\frac{\partial z}{\partial x}\right)_y = \left.\frac{\partial z}{\partial r}\right)_{s,x}\left.\frac{\partial r}{\partial x}\right)_y + \left.\frac{\partial z}{\partial s}\right)_{r,x}\left.\frac{\partial s}{\partial x}\right)_y + \left.\frac{\partial z}{\partial x}\right)_{r,s}.$$

In Example 12.14 it is essential that we indicate which variables to hold constant in the partial derivatives. If we were to omit these designations, then in the second result we would have a term $\partial z/\partial x$ on both sides of the equation but they would have different meanings. The term $\partial z/\partial x)_y$ indicates the derivative of z with respect to x holding y constant if z were expressed entirely in terms of x and y; the term $\partial z/\partial x)_{r,s}$ indicates the derivative of the given function $f(r, s, x)$ with respect to x holding r and s constant.

EXAMPLE 12.15

Find dz/dt if

$$z = x^3 y^2 + x \sin y + tx, \quad x = 2t + \frac{1}{t}, \quad y = t^2 e^t.$$

SOLUTION From the schematic diagram to the left,

$$\frac{dz}{dt} = \frac{\partial z}{\partial x}\frac{dx}{dt} + \frac{\partial z}{\partial y}\frac{dy}{dt} + \frac{\partial z}{\partial t}$$

$$= (3x^2 y^2 + \sin y + t)(2 - 1/t^2) + (2x^3 y + x \cos y)(2te^t + t^2 e^t) + x.$$

When a chain rule is used to calculate a derivative, the result usually involves all intermediate variables. For instance, the derivative dz/dt in Example 12.15 involves not only t, but the intermediate variables x and y as well. Were dz/dt required at $t = 1$, values of x and y for $t = 1$ would be calculated — $x(1) = 3$ and $y(1) = e$ — and all three values substituted to obtain

$$\left.\frac{dz}{dt}\right|_{t=1} = [3(3)^2(e)^2 + \sin(e) + 1](2 - 1)$$

$$+ [2(3)^3 e + (3)\cos(e)](2e + e) + 3$$

$$= 1378.6.$$

EXAMPLE 12.16

Find $\partial^2 z/\partial x^2$ if

$$z = s^2 t + 2 \sin t, \quad s = xy - y, \quad t = x^2 + \frac{y}{x}.$$

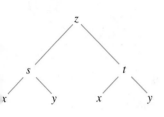

SOLUTION   From the schematic diagram to the left,

$$\frac{\partial z}{\partial x} = \frac{\partial z}{\partial s}\frac{\partial s}{\partial x} + \frac{\partial z}{\partial t}\frac{\partial t}{\partial x} = (2st)(y) + (s^2 + 2\cos t)(2x - y/x^2).$$

Now $\partial z/\partial x$ is a function of $s, t, x,$ and y, and therefore in order to find

$$\frac{\partial^2 z}{\partial x^2} = \frac{\partial}{\partial x}\left(\frac{\partial z}{\partial x}\right),$$

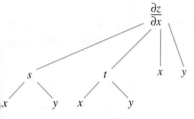

we form a schematic diagram for $\partial z/\partial x$. From this schematic diagram, we obtain

$$\frac{\partial^2 z}{\partial x^2} = \frac{\partial}{\partial s}\left(\frac{\partial z}{\partial x}\right)\frac{\partial s}{\partial x} + \frac{\partial}{\partial t}\left(\frac{\partial z}{\partial x}\right)\frac{\partial t}{\partial x} + \frac{\partial}{\partial x}\left(\frac{\partial z}{\partial x}\right)_{s,t,y}$$

$$= [2ty + 2s(2x - y/x^2)](y) + [2sy - 2\sin t(2x - y/x^2)](2x - y/x^2)$$

$$+ (s^2 + 2\cos t)(2 + 2y/x^3).$$

EXAMPLE 12.17

Temperature T at points in the atmosphere depends on both position (x, y, z) and time t: $T = T(x, y, z, t)$. When a weather balloon is released to take temperature readings, it is not free to take readings at just any point, only at those points along the path that the winds force the balloon to follow. This path is a curve in space represented parametrically by

$$C : \quad x = x(t), \quad y = y(t), \quad z = z(t), \quad t \geq 0,$$

t again being time. If we substitute from the equations for C into the temperature function, then T becomes a function of t alone,

$$T = T[x(t), y(t), z(t), t],$$

and this function of time describes the temperature at points along the path of the balloon. For the derivative of this function with respect to t, the schematic diagram yields

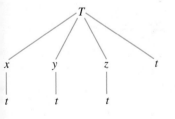

$$\frac{dT}{dt} = \frac{\partial T}{\partial x}\frac{dx}{dt} + \frac{\partial T}{\partial y}\frac{dy}{dt} + \frac{\partial T}{\partial z}\frac{dz}{dt} + \frac{\partial T}{\partial t}.$$

The question we pose is: What is the physical difference between dT/dt and $\partial T/\partial t$?

SOLUTION   Temperature at a point in space is independent of the observer measuring it; hence $T[x(t), y(t), z(t), t]$ is the temperature at points on C as measured by both the balloon and any observer fixed in the xyz-reference system. If, however, these two observers calculate the rate of change of temperature with respect to time at some point (x, y, z) on C, they calculate different results. The observer fixed in the xyz-reference system (not restricted to move along C) calculates the rate of change of T with respect to t as the derivative of the function $T(x, y, z, t)$ partially with respect to t holding x, y, and z constant (i.e., the fixed observer calculates $\partial T/\partial t$ as the rate of change of temperature in time). The balloon, on the other hand, has no alternative but to take temperature readings as it moves along C; thus its measurement of T as a function of t is

$$T[x(t), y(t), z(t), t].$$

Therefore, when the balloon calculates the time variation of temperature, it is calculating dT/dt. It follows, then, that the terms

$$\frac{\partial T}{\partial x}\frac{dx}{dt} + \frac{\partial T}{\partial y}\frac{dy}{dt} + \frac{\partial T}{\partial z}\frac{dz}{dt}$$

describe that part of dT/dt caused by the motion of the balloon through space.

Discussions like those in Example 12.17 are prominent in the study of fluid motion (gas or liquid). Sometimes rates of change from the point of view of a fixed observer are important; other times, rates of change as measured by an observer moving with the fluid are appropriate.

Many important applications of the chain rule occur in the field of partial differential equations. The following example is an illustration.

EXAMPLE 12.18

The one-dimensional wave equation

$$\frac{\partial^2 y}{\partial t^2} = c^2 \frac{\partial^2 y}{\partial x^2}, \quad c = \text{constant}$$

for functions $y(x, t)$ describes transverse vibrations of taut strings, and longitudinal and rotational vibrations of metal bars. Show that if $f(u)$ and $g(v)$ are twice-differentiable functions of u and v, then $y(x, t) = f(x + ct) + g(x - ct)$ satisfies the wave equation.

SOLUTION   The schematic diagram to the left describes the functional situation

$$y = f(u) + g(v)$$

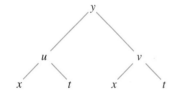

where $u = x + ct$ and $v = x - ct$. The chain rule for $\partial y/\partial t$ is

$$\frac{\partial y}{\partial t} = \frac{\partial y}{\partial u}\frac{\partial u}{\partial t} + \frac{\partial y}{\partial v}\frac{\partial v}{\partial t} = cf'(u) - cg'(v).$$

The schematic diagram for $\partial y/\partial t$ leads to

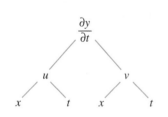

$$\frac{\partial^2 y}{\partial t^2} = \frac{\partial}{\partial u}\left(\frac{\partial y}{\partial t}\right)\frac{\partial u}{\partial t} + \frac{\partial}{\partial v}\left(\frac{\partial y}{\partial t}\right)\frac{\partial v}{\partial t}$$

$$= [cf''(u)]c + [-cg''(v)](-c)$$

$$= c^2[f''(u) + g''(v)].$$

A similar calculation gives $\dfrac{\partial^2 y}{\partial x^2} = f''(u) + g''(v)$. Hence $y(x, t)$ does indeed satisfy the wave equation.

We have suggested how important Laplace's equation is to engineering, particularly in electrostatics, heat conduction, fluid flow, and deflection of plates. The two-dimensional Laplace equation in Cartesian coordinates is 12.11. In the next example, we find its form in polar coordinates.

EXAMPLE 12.19

Find Laplace's equation in polar coordinates.

SOLUTION Cartesian coordinates are related to polar coordinates by the equations $x = r \cos \theta$, $y = r \sin \theta$. The inverse transformation is

$$r = \sqrt{x^2 + y^2}, \qquad \theta = \text{Tan}^{-1}\left(\frac{y}{x}\right).$$

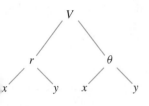

The second of these is not always correct; it may need $\pm \pi$ added to it. Since the derivation here requires only derivatives of θ, and not θ itself, the $\pm \pi$ is inconsequential. Suppose $V = f(x, y)$ is a function that satisfies Laplace's equation $\partial^2 V/\partial x^2 + \partial^2 V/\partial y^2 = 0$. We can express V in terms of r and θ by writing $V = F(r, \theta) = f(r \cos \theta, r \sin \theta)$. The schematic to the left represents the functional situation where $V = F(r, \theta)$ and r and θ are expressed in terms of x and y. From it,

$$\frac{\partial V}{\partial x} = \frac{\partial V}{\partial r}\frac{\partial r}{\partial x} + \frac{\partial V}{\partial \theta}\frac{\partial \theta}{\partial x},$$

where

$$\frac{\partial r}{\partial x} = \frac{x}{\sqrt{x^2 + y^2}} = \frac{r \cos \theta}{r} = \cos \theta,$$

$$\frac{\partial \theta}{\partial x} = \frac{1}{1 + (y/x)^2}\left(\frac{-y}{x^2}\right) = \frac{-y}{x^2 + y^2} = \frac{-\sin \theta}{r}.$$

Thus,

$$\frac{\partial V}{\partial x} = \cos \theta \frac{\partial V}{\partial r} - \frac{\sin \theta}{r}\frac{\partial V}{\partial \theta}.$$

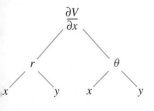

The combination of r and θ derivatives on the right of this equation must be applied to V when it is expressed in terms of r and θ to give the partial derivative with respect to x when V is expressed in terms of x and y. To find $\partial^2 V/\partial x^2$, we use the schematic to the left.

$$\frac{\partial^2 V}{\partial x^2} = \frac{\partial}{\partial r}\left(\frac{\partial V}{\partial x}\right)\frac{\partial r}{\partial x} + \frac{\partial}{\partial \theta}\left(\frac{\partial V}{\partial x}\right)\frac{\partial \theta}{\partial x}$$

$$= \left(\cos \theta \frac{\partial^2 V}{\partial r^2} + \frac{\sin \theta}{r^2}\frac{\partial V}{\partial \theta} - \frac{\sin \theta}{r}\frac{\partial^2 V}{\partial r \partial \theta}\right)\cos \theta$$

$$+ \left(-\sin \theta \frac{\partial V}{\partial r} + \cos \theta \frac{\partial^2 V}{\partial \theta \partial r} - \frac{\cos \theta}{r}\frac{\partial V}{\partial \theta} - \frac{\sin \theta}{r}\frac{\partial^2 V}{\partial \theta^2}\right)\left(-\frac{\sin \theta}{r}\right)$$

$$= \cos^2 \theta \frac{\partial^2 V}{\partial r^2} - \frac{2 \sin \theta \cos \theta}{r}\frac{\partial^2 V}{\partial r \partial \theta} + \frac{\sin^2 \theta}{r^2}\frac{\partial^2 V}{\partial \theta^2}$$

$$+ \frac{2 \sin \theta \cos \theta}{r^2}\frac{\partial V}{\partial \theta} + \frac{\sin^2 \theta}{r}\frac{\partial V}{\partial r},$$

where we have assumed that mixed partial derivatives are equal. A similar calculation gives

$$\frac{\partial^2 V}{\partial y^2} = \sin^2 \theta \frac{\partial^2 V}{\partial r^2} + \frac{2 \sin \theta \cos \theta}{r}\frac{\partial^2 V}{\partial r \partial \theta} + \frac{\cos^2 \theta}{r^2}\frac{\partial^2 V}{\partial \theta^2} - \frac{2 \sin \theta \cos \theta}{r^2}\frac{\partial V}{\partial \theta} + \frac{\cos^2 \theta}{r}\frac{\partial V}{\partial r}.$$

When these are added together, the result is Laplace's equation in polar coordinates,

$$\frac{\partial^2 V}{\partial r^2} + \frac{1}{r}\frac{\partial V}{\partial r} + \frac{1}{r^2}\frac{\partial^2 V}{\partial \theta^2} = 0. \tag{12.24}$$

Calculations in this example allow us to emphasize a point that we made in Section 12.3. We said that although the derivative dy/dx can be considered as a quotient of differentials, we never consider a partial derivative as a quotient. To do so in Example 12.19 would lead to errors. We calculated $\partial r/\partial x = \cos\theta$. Notice that $\partial x/\partial r = \cos\theta$, and therefore to regard $\partial r/\partial x$ as the reciprocal of $\partial x/\partial r$ is incorrect.

Homogeneous Functions

Homogeneous functions arise in numerous areas of applied mathematics. A function $f(x, y, z)$ is said to be a **positively homogeneous function** of degree n if for every $t > 0$,

$$f(tx, ty, tz) = t^n f(x, y, z). \tag{12.25}$$

For example, the function $f(x, y, z) = x^2 + y^2 + z^2$ is homogeneous of degree 2; the function $f(x, y) = x^3 \cos(y/x) + x^2 y + xy^2$ is homogeneous of degree 3; and $f(x, y, z, t) = \sqrt{x^2 + z^2}(x^2 y + yt^2)$ is homogeneous of degree 4. Partial derivatives of homogeneous functions satisfy many identities. In particular, their first derivatives satisfy Euler's theorem.

> **THEOREM 12.3   (Euler's Theorem)**
>
> If $f(x, y, z)$ is positively homogeneous of degree n, and has continuous first partial derivatives, then
>
> $$x\frac{\partial f}{\partial x} + y\frac{\partial f}{\partial y} + z\frac{\partial f}{\partial z} = nf(x, y, z). \tag{12.26}$$

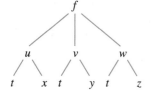

PROOF   To verify 12.26, we differentiate 12.25 with respect to t, holding x, y, and z constant. For the derivative of the left side we introduce variables $u = tx$, $v = ty$, and $w = tz$, and use the schematic to the left. The result is

$$\frac{\partial f}{\partial u}\frac{\partial u}{\partial t} + \frac{\partial f}{\partial v}\frac{\partial v}{\partial t} + \frac{\partial f}{\partial w}\frac{\partial w}{\partial t} = nt^{n-1}f(x, y, z)$$

or

$$x\frac{\partial f}{\partial u} + y\frac{\partial f}{\partial v} + z\frac{\partial f}{\partial w} = nt^{n-1}f(x, y, z).$$

When we set $t = 1$, we obtain $u = x$, $v = y$, $w = z$, and the equation above becomes 12.26. ∎

The results of Exercises 31–33 in Section 12.3 are special cases of 12.26.

In Exercises 1–10 we have defined a general functional situation and a specific example. Find the chain rule for the indicated derivative in the general situation, and then use that result to calculate the same derivative in the specific example.

1. dz/dt if $z = f(x, t)$, $x = g(t)$; $z = xt^2/(x + t)$, $x = e^{3t}$

2. $\left(\dfrac{\partial z}{\partial t}\right)_s$ if $z = f(x, y)$, $x = g(s, t)$, $y = h(s, t)$; $z = x^2 e^y +$ $y \ln x$, $x = s^2 \cos t$, $y = 4 \operatorname{Sec}^{-1}(t^2 + 2s)$

3. $\left(\dfrac{\partial u}{\partial s}\right)_t$ if $u = f(x, y, z)$, $x = g(s, t)$, $y = h(s, t)$, $z = k(s, t)$;

$u = \sqrt{x^2 + y^2 + z^2}$, $x = 2st$, $y = s^2 + t^2$, $z = st$

4. dz/du if $z = f(x, y, v)$, $x = g(u)$, $y = h(u)$, $v = k(u)$; $z = x^2 yv^3$, $x = u^3 + 2u$, $y = \ln(u^2 + 1)$, $v = ue^u$

5. $\left(\dfrac{\partial u}{\partial r}\right)_t$ if $u = f(x, y, s)$, $x = g(t)$, $y = h(r)$, $s = k(r, t)$; $u = \sqrt{x^2 + y^2 s}$, $x = t/(t + 5)$, $y = \operatorname{Sin}^{-1}(r^2 + 5)$, $s = \tan(rt)$

6. $\left(\dfrac{\partial z}{\partial t}\right)_r$ if $z = f(x)$, $x = g(y)$, $y = h(r, t)$; $z = 3^{x+2}$, $x = y^2 + 5$, $y = \csc(r^2 + t)$

7. $\left(\dfrac{\partial u}{\partial x}\right)_y$ if $u = f(x, y, z)$, $z = g(x, y)$; $u = y/\sqrt{x^2 + y^2 + z^2}$, $z = x/y$

8. $\left(\dfrac{\partial x}{\partial y}\right)_z$ if $x = f(r, s, t)$, $r = g(y)$, $s = h(y, z)$, $t = k(y, z)$; $x = s^2 r^2 t^2$, $r = y^{-5}$, $s = 1/(y^2 + z^2)$, $t = 1/y^2 + 1/z^2$

9. $\left(\dfrac{\partial z}{\partial t}\right)_s$ if $z = f(x, y)$, $x = g(r)$, $y = h(r)$, $r = k(s, t)$; $z = e^{x+y}$, $x = 2r + 5$, $y = 2r - 5$, $r = t \ln(s^2 + t^2)$

10. dz/dt if $z = f(x, y, u)$, $x = g(v)$, $u = h(x, y)$, $v = k(t)$, $y = p(t)$; $z = x^2 + y^2 + u^2$, $x = v^3 - 3v^2$, $u = 1/(x^2 - y^2)$, $v = e^t$, $y = e^{4t}$

In Exercises 11–15 find the derivative.

11. $\left(\dfrac{\partial^2 z}{\partial t^2}\right)_s$ if $z = x^2 y^2 + xe^y$, $x = s + t^2$, $y = s - t^2$

12. $d^2 x/dt^2$ if $x = y^2 + yt - t^2$, $y = t^2 e^t$

13. $\left(\dfrac{\partial^2 u}{\partial s^2}\right)_t$ if $u = x^2 + y^2 + z^2 + xyz$, $x = s^2 + t^2$, $y = s^2 - t^2$, $z = st$

14. $d^2 z/dv^2$ if $z = \sin(xy)$, $x = 3\cos v$, $y = 4\sin v$

15. $\partial^2 u/\partial x\,\partial y$ if $u = y/\sqrt{x^2 + y^2 + z^2}$, $z = x/y$

16. Suppose that u is a differentiable function of r and $r = \sqrt{x^2 + y^2 + z^2}$. Show that

$$\left(\frac{\partial u}{\partial x}\right)^2 + \left(\frac{\partial u}{\partial y}\right)^2 + \left(\frac{\partial u}{\partial z}\right)^2 = \left(\frac{du}{dr}\right)^2.$$

17. Consider a gas that is moving through some region D of space. If we follow a particular particle of the gas, it traces out some curved path (figure below)

$$C: \quad x = x(t), \quad y = y(t), \quad z = z(t), \quad t \geq 0.$$

Suppose the density of the gas at any point in the region D at time t is denoted by $\rho(x, y, z, t)$. We can write that along C, $\rho = \rho[x(t), y(t), z(t), t]$.

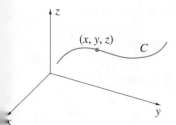

(a) Obtain the chain rule defining $d\rho/dt$ in terms of $\partial \rho/\partial t$ and derivatives of x, y, and z with respect to t.

(b) Explain the physical difference between $d\rho/dt$ and $\partial \rho/\partial t$.

* **18.** The radius and height of a right-circular cone are 10 and 20 cm, respectively. If the radius is increasing at 1 cm/min and the height is decreasing at 2 cm/min, how fast is the volume changing? Do you need multivariable calculus to solve this problem?

* **19.** If two sides of a triangle have lengths x and y and the angle between them is θ, then the area of the triangle is $A = (1/2)xy\sin\theta$. How fast is the area changing when x is 1 m, y is 2 m, and θ is 1/3 radian, if x and y are each increasing at 1/2 m/s and θ is decreasing at 1/10 radian per second?

* **20.** When a rocket rises from the earth's surface, its mass decreases because fuel is being consumed at the rate of 50 kg/s. Use Newton's universal law of gravitation (see Example 7.34 in Section 7.10) to determine how fast the force of gravity of the earth on the rocket is changing when the rocket is 100 km above the earth's surface and climbing at 2 km/s. Assume that the mass of the rocket at this height is 12×10^6 kg.

* **21.** If $z = f(u, v)$, $u = g(x, y)$, $v = h(x, y)$, find the chain rule for the second derivative $\partial^2 z/\partial x^2$.

* **22.** Determine which of the following functions are positively homogeneous:

    (a) $f(x, y) = x^2 + xy + 3y^2$

    (b) $f(x, y) = x^2 y + xy - 2xy^2$

    (c) $f(x, y, z) = x^2 \sin(y/z) + y^2 + y^3/z$

    (d) $f(x, y, z) = xe^{y/z} - xyz$

    (e) $f(x, y, z, t) = x^4 + y^4 + z^4 + t^4 - xyzt$

    (f) $f(x, y, z, t) = e^{x^2+y^2}(z^2 + t^2)$

    (g) $f(x, y, z) = \cos(xy)\sin(yz)$

    (h) $f(x, y) = \sqrt{x^2 + xy + y^2}e^{y/x}(2x^2 - 3y^2)$

* **23.** (a) Suppose that the circular plate with radius R in the figure below has its lower edge held at 0 V and its upper edge held at 1 V. Show that the electrostatic potential

$$V(r, \theta) = \frac{1}{2} + \frac{1}{\pi}\text{Tan}^{-1}\left(\frac{2Rr\sin\theta}{R^2 - r^2}\right)$$

satisfies Laplace's equation 12.24 for $r < R$.

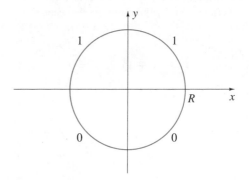

(b) The function in part (a) is not defined for $r = R$, but it can be shown that as $r \to R$, values of V approach 1 for $0 < \theta < \pi$, and approach 0 for $-\pi < \theta < 0$. Solve the expression in part (a) for r in terms of V and θ, and use the result to plot equipotential curves for $V = 1/8$, 1/4, 3/8, 1/2, 5/8, 3/4, and 7/8. Set $R = 1$ to do this.

∗ **24.** Verify that

$$V(r, \theta) = \frac{1}{2}(V_1 + V_2) + \frac{1}{\pi}(V_1 - V_2) \, \mathrm{Tan}^{-1}\left(\frac{2Rr \sin \theta}{R^2 - r^2}\right)$$

satisfies 12.24 for $r < R$. It represents potential in the circle in the figure in Exercise 23 when potential on the upper edge is V_1 and that on the lower edge is V_2.

∗ **25.** If $f(s)$ and $g(t)$ are differentiable functions, show that $\nabla f(x^2 - y^2) \cdot \nabla g(xy) = 0$.

∗ **26.** If $f(s)$ is a differentiable function, show that $f(x - y)$ satisfies the equation

$$\frac{\partial f}{\partial y} = -\frac{\partial f}{\partial x}.$$

∗ **27.** If $f(s)$ is a differentiable function, show that $u(x, y) = f(4x - 3y) + 5(y - x)$ satisfies the equation

$$3\frac{\partial u}{\partial x} + 4\frac{\partial u}{\partial y} = 5.$$

∗ **28.** If $f(s)$ and $g(t)$ are twice differentiable, show that the function $u(x, y) = xf(x + y) + yg(x + y)$ satisfies

$$\frac{\partial^2 u}{\partial x^2} - 2\frac{\partial^2 u}{\partial x \, \partial y} + \frac{\partial^2 u}{\partial y^2} = 0.$$

∗ **29.** If $f(s)$ and $g(t)$ are twice differentiable, show that $f(x - y) + g(x + y)$ satisfies

$$\frac{\partial^2 u}{\partial x^2} - \frac{\partial^2 u}{\partial y^2} = 0.$$

∗ **30.** Show that if $f(v)$ is differentiable, then $u(x, y) = x^2 f(y/x)$ satisfies

$$x\frac{\partial u}{\partial x} + y\frac{\partial u}{\partial y} = 2u.$$

In Exercises 31–33 suppose that $f(x, y)$ satisfies the first partial differential equation. Show that with the change of independent variables, function $F(u, v) = f[x(u, v), y(u, v)]$ must satisfy the second partial differential equation.

∗ **31.** $\left(\dfrac{\partial f}{\partial x}\right)^2 + \left(\dfrac{\partial f}{\partial y}\right)^2 = 0$; $u = (x + y)/2$, $v = (x - y)/2$;

$$\left(\frac{\partial F}{\partial u}\right)^2 + \left(\frac{\partial F}{\partial v}\right)^2 = 0$$

∗ **32.** $\dfrac{\partial^2 f}{\partial x^2} - \dfrac{\partial^2 f}{\partial y^2} = 0$; $u = (x+y)/2$, $v = (x-y)/2$; $\dfrac{\partial^2 F}{\partial u \, \partial v} = 0$

∗ **33.** $\left(\dfrac{\partial f}{\partial x}\right)^2 + \left(\dfrac{\partial f}{\partial y}\right)^2 = 0$; $x = u \cos v$, $y = u \sin v$;

$$\left(\frac{\partial F}{\partial u}\right)^2 + \frac{1}{u^2}\left(\frac{\partial F}{\partial v}\right)^2 = 0$$

∗ **34.** In many problems in elasticity theory, the Airy's stress function $\Phi(x, y)$ must satisfy the biharmonic equation

$$\frac{\partial^4 \Phi}{\partial x^4} + 2\frac{\partial^4 \Phi}{\partial x^2 \, \partial y^2} + \frac{\partial^4 \Phi}{\partial y^4} = 0.$$

Use Example 12.19 to show that in polar coordinates, the equation can be expressed in the form

$$\left(\frac{\partial^2}{\partial r^2} + \frac{1}{r}\frac{\partial}{\partial r} + \frac{1}{r^2}\frac{\partial^2}{\partial \theta^2}\right)\left(\frac{\partial^2 \Phi}{\partial r^2} + \frac{1}{r}\frac{\partial \Phi}{\partial r} + \frac{1}{r^2}\frac{\partial^2 \Phi}{\partial \theta^2}\right) = 0.$$

∗ **35.** An observer travels along the curve $x = t^2$, $y = 3t^3 + 1$, $z = 2t + 5$, where x, y, and z are in metres and $t \geq 0$ is in seconds. If the density ρ of a gas (in kg/m³) is given by $\rho = (3x^2 + y^2)/(z^2 + 5)$, find the time rate of change of the density of the gas as measured by the observer when $t = 2$ s.

∗ **36.** If $f(r)$ is a differentiable function and $r = \sqrt{x^2 + y^2 + z^2}$, show that

$$\nabla f = \frac{f'(r)}{r}(x\hat{\mathbf{i}} + y\hat{\mathbf{j}} + z\hat{\mathbf{k}}).$$

∗ **37.** If $f(x, y) = 0$ defines y as a function of x, show that

$$\frac{d^2 y}{dx^2} = -\frac{f_{xx}f_y^2 - 2f_{xy}f_x f_y + f_{yy}f_x^2}{f_y^3}.$$

∗ **38.** If $f(x, y)$ is a harmonic function, show that the function $F(x, y) = f(x^2 - y^2, 2xy)$ is also harmonic.

∗ **39.** (a) Show that $f(x, y) = \ln(x^2 + y^2)$ satisfies Laplace's equation 12.11.

(b) Transform $f(x, y)$ into polar coordinates and show that the function satisfies 12.24.

∗ **40.** Find an identity satisfied by the second partial derivatives of a function $f(x, y, z)$ that is positively homogeneous of degree n.

∗∗ **41.** It is postulated in one of the theories of traffic flow that the average speed u at a point x on a straight highway (along the x-axis) is related to the concentration k of traffic by the differential equation

$$u\frac{\partial u}{\partial x} + \frac{\partial u}{\partial t} = -c^2 k^n \frac{\partial k}{\partial x},$$

where t is time, and $c > 0$ and n are constants.

(a) Use chain rules for $\partial u/\partial x$ and $\partial u/\partial t$ in the functional situation $u = f(k)$ and $k = g(x, t)$ to show that

$$\frac{du}{dk}\left(u\frac{\partial k}{\partial x} + \frac{\partial k}{\partial t}\right) + c^2 k^n \frac{\partial k}{\partial x} = 0.$$

(b) The equation of continuity for traffic flow states that

$$\frac{\partial k}{\partial t} + \frac{\partial (ku)}{\partial x} = 0.$$

Use these last two equations to obtain the differential equation relating speed and concentration:

$$\frac{du}{dk} = -ck^{(n-1)/2}.$$

(c) Solve the differential equation in part (b) for $u = f(k)$.

42. A bead slides from rest at the origin on a frictionless wire in a vertical plane to the point (x_0, y_0) under the influence of gravity (figure below). As it does so, gravitational potential energy is converted into kinetic energy. At (x, y), the bead has lost potential energy mgy. If its kinetic energy is $mv^2/2$ at this point, then $mgy = mv^2/2$, so that $v = \sqrt{2gy}$. To travel a small distance $\sqrt{(dx)^2 + (dy)^2}$ along the curve at (x, y) with velocity v takes time $\sqrt{(dx)^2 + (dy)^2}/v$. Hence, the total time to traverse the entire curve is

$$t = \int_0^{x_0} \frac{\sqrt{(dx)^2 + (dy)^2}}{v}$$

$$= \int_0^{x_0} \frac{1}{\sqrt{2gy}} \sqrt{1 + \left(\frac{dy}{dx}\right)^2}\, dx = \frac{1}{\sqrt{2g}} \int_0^{x_0} \sqrt{\frac{1 + (y')^2}{y}}\, dx.$$

The problem of finding the shape of wire that makes t as small as possible is called the *brachistochrone problem*. It is shown in the calculus of variations that $y = f(x)$ must satisfy the equation

$$\frac{d}{dx}\left(\frac{\partial F}{\partial y'}\right) - \frac{\partial F}{\partial y} = 0,$$

where

$$F(y, y') = \sqrt{\frac{1 + (y')^2}{y}}.$$

(a) Show that $f(x)$ must satisfy the differential equation $1 + (y')^2 + 2yy'' = 0$.

(b) Show that the curve that satisfies the equation in part (a) is the cycloid defined parametrically by

$$x = a(\theta - \sin\theta), \quad y = a(1 - \cos\theta),$$

where a is a constant.

(c) Show that it does not matter what point on the cycloid the bead starts from, the time to get to (x_0, y_0) is always the same.

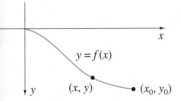

43. Two equal masses m are connected by springs having equal spring constant k so that the masses are free to slide on a frictionless table (see figure below). The walls A and B are fixed.

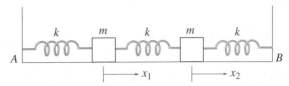

(a) Use Newton's second law to show that the differential equations for the motions of the masses are

$$m\ddot{x}_1 = k(x_2 - 2x_1), \qquad m\ddot{x}_2 = k(x_1 - 2x_2),$$

where x_1 and x_2 are the displacements of the masses from their equilibrium positions, $\ddot{x}_1 = d^2x_1/dt^2$ and $\ddot{x}_2 = d^2x_2/dt^2$.

(b) The Euler–Lagrange equations from theoretical mechanics for this system are

$$\frac{d}{dt}\left(\frac{\partial L}{\partial \dot{x}_1}\right) - \frac{\partial L}{\partial x_1} = 0, \qquad \frac{d}{dt}\left(\frac{\partial L}{\partial \dot{x}_2}\right) - \frac{\partial L}{\partial x_2} = 0,$$

where L is defined as the kinetic energy of the two masses less the energy stored in the springs. Show that

$$L(x_1, x_2, \dot{x}_1, \dot{x}_2) = \frac{m}{2}(\dot{x}_1^2 + \dot{x}_2^2) - k(x_1^2 + x_2^2 - x_1 x_2).$$

(c) Obtain the equations in part (a) from the Euler–Lagrange equations in part (b).

44. Suppose that the second-order partial differential equation

$$p\frac{\partial^2 z}{\partial x^2} + q\frac{\partial^2 z}{\partial x \partial y} + r\frac{\partial^2 z}{\partial y^2} = F\left(x, y, z, \frac{\partial z}{\partial x}, \frac{\partial z}{\partial y}\right)$$

(p, q, and r are constants) is subjected to the change of variables

$$s = ax + by, \quad t = cx + dy,$$

where a, b, c, and d are constants. Show that the partial differential equation in s and t is

$$P\frac{\partial^2 z}{\partial s^2} + Q\frac{\partial^2 z}{\partial s \partial t} + R\frac{\partial^2 z}{\partial t^2} = G\left(s, t, z, \frac{\partial z}{\partial s}, \frac{\partial z}{\partial t}\right),$$

where $Q^2 - 4PR = (q^2 - 4pr)(ad - bc)^2$.

45. Show that if a solution $u = f(x, y, z)$ of the three-dimensional Laplace equation 12.12 can be expressed in the form $u = g(r)$, where $r = \sqrt{x^2 + y^2 + z^2}$, then $f(x, y, z)$ must be of the form

$$f(x, y, z) = \frac{C}{\sqrt{x^2 + y^2 + z^2}} + D,$$

where C and D are constants.

12.7 Implicit Differentiation

In Section 3.8 we introduced the technique of **implicit differentiation** in order to obtain the derivative of a function $y = f(x)$ defined implicitly by an equation

$$F(x, y) = 0. \tag{12.27}$$

Essentially, the technique is to differentiate all terms in the equation with respect to x, considering all the while that y is a function of x. For example, if y is defined implicitly by

$$x^2 y^3 + 3xy = 3x + 2,$$

implicit differentiation gives

$$2xy^3 + 3x^2 y^2 \frac{dy}{dx} + 3y + 3x \frac{dy}{dx} = 3 \quad \Longrightarrow \quad \frac{dy}{dx} = \frac{3 - 2xy^3 - 3y}{3x^2 y^2 + 3x}.$$

With the chain rule we can actually present a formula for dy/dx. Since equation 12.27 when written in the form

$$F[x, f(x)] = 0,$$

must be valid for all x in the domain of the function $f(x)$, we can differentiate it with respect to x. From the schematic diagram to the left, the derivative of the left side of the equation is

$$\frac{dF}{dx} = \frac{\partial F}{\partial x} + \frac{\partial F}{\partial y} \frac{dy}{dx}.$$

If we equate this to the derivative of the right side of the equation, we find

$$F_x + F_y \frac{dy}{dx} = 0$$

or

$$\frac{dy}{dx} = -\frac{F_x}{F_y}. \tag{12.28}$$

For the function defined implicitly above by $x^2 y^3 + 3xy - 3x - 2 = 0$, equation 12.28 gives

$$\frac{dy}{dx} = -\frac{2xy^3 + 3y - 3}{3x^2 y^2 + 3x},$$

and this result is identical to that obtained by implicit differentiation.

Similarly, if the equation

$$F(x, y, z) = 0 \tag{12.29}$$

defines z implicitly as a function of x and y, the schematic diagram to the left immediately yields

$$\frac{\partial F}{\partial x} + \frac{\partial F}{\partial z} \frac{\partial z}{\partial x} = 0, \quad \frac{\partial F}{\partial y} + \frac{\partial F}{\partial z} \frac{\partial z}{\partial y} = 0.$$

From these we obtain the results

$$\frac{\partial z}{\partial x} = -\frac{F_x}{F_z}, \quad \frac{\partial z}{\partial y} = -\frac{F_y}{F_z}. \tag{12.30}$$

We do not suggest that formulas 12.28 and 12.30 be memorized. On the contrary, we obtain results in this section that include 12.28 and 12.30 as special cases. To develop these results we work with three equations in five variables:

$$F(x, y, u, v, w) = 0, \quad G(x, y, u, v, w) = 0, \quad H(x, y, u, v, w) = 0. \tag{12.31}$$

We assume that these equations define u, v, and w as functions of x and y for some domain of values of x and y (and do so implicitly). It might even be possible to solve the system and obtain explicit definitions of the functions

$$u = f(x, y), \quad v = g(x, y), \quad w = h(x, y). \tag{12.32}$$

We pose the problem of finding the six first-order partial derivatives of u, v, and w with respect to x and y, supposing that it is undesirable or even impossible to obtain the explicit form of the functions. To do this, we note that were results 12.32 known and substituted into 12.31, then

$$F[x, y, f(x, y), g(x, y), h(x, y)] = 0,$$

$$G[x, y, f(x, y), g(x, y), h(x, y)] = 0,$$

$$H[x, y, f(x, y), g(x, y), h(x, y)] = 0$$

would be identities in x and y. As a result we could differentiate each equation with respect to x, obtaining from the schematic diagram

$$\frac{\partial F}{\partial x} + \frac{\partial F}{\partial u}\frac{\partial u}{\partial x} + \frac{\partial F}{\partial v}\frac{\partial v}{\partial x} + \frac{\partial F}{\partial w}\frac{\partial w}{\partial x} = 0,$$

$$\frac{\partial G}{\partial x} + \frac{\partial G}{\partial u}\frac{\partial u}{\partial x} + \frac{\partial G}{\partial v}\frac{\partial v}{\partial x} + \frac{\partial G}{\partial w}\frac{\partial w}{\partial x} = 0, \tag{12.33}$$

$$\frac{\partial H}{\partial x} + \frac{\partial H}{\partial u}\frac{\partial u}{\partial x} + \frac{\partial H}{\partial v}\frac{\partial v}{\partial x} + \frac{\partial H}{\partial w}\frac{\partial w}{\partial x} = 0,$$

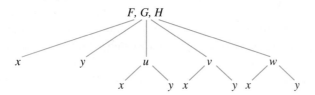

or

$$F_u\frac{\partial u}{\partial x} + F_v\frac{\partial v}{\partial x} + F_w\frac{\partial w}{\partial x} = -F_x,$$

$$G_u\frac{\partial u}{\partial x} + G_v\frac{\partial v}{\partial x} + G_w\frac{\partial w}{\partial x} = -G_x, \tag{12.34}$$

$$H_u\frac{\partial u}{\partial x} + H_v\frac{\partial v}{\partial x} + H_w\frac{\partial w}{\partial x} = -H_x.$$

We have in 12.34 three equations in the three unknowns $\partial u/\partial x$, $\partial v/\partial x$, and $\partial w/\partial x$, and because the equations are linear in the unknowns, solutions can be obtained using Cramer's rule.[†] In particular,

$$\frac{\partial u}{\partial x} = \frac{\begin{vmatrix} -F_x & F_v & F_w \\ -G_x & G_v & G_w \\ -H_x & H_v & H_w \end{vmatrix}}{\begin{vmatrix} F_u & F_v & F_w \\ G_u & G_v & G_w \\ H_u & H_v & H_w \end{vmatrix}} = -\frac{\begin{vmatrix} F_x & F_v & F_w \\ G_x & G_v & G_w \\ H_x & H_v & H_w \end{vmatrix}}{\begin{vmatrix} F_u & F_v & F_w \\ G_u & G_v & G_w \\ H_u & H_v & H_w \end{vmatrix}}. \tag{12.35}$$

[†] Cramer's rule is discussed in Appendix B.

The two determinants on the right of 12.35 involve only derivatives of the given functions F, G, and H, and we have therefore obtained a method for finding $\partial u/\partial x$ that avoids solving 12.31 for u, v, and w. We could list similar formulas for the remaining five derivatives, but first we introduce some simplifying notation.

DEFINITION 12.5

The **Jacobian determinant** of functions F, G, and H with respect to variables u, v, and w is denoted by $\dfrac{\partial(F, G, H)}{\partial(u, v, w)}$ and is defined as the determinant

$$\frac{\partial(F, G, H)}{\partial(u, v, w)} = \begin{vmatrix} F_u & F_v & F_w \\ G_u & G_v & G_w \\ H_u & H_v & H_w \end{vmatrix} = \begin{vmatrix} \dfrac{\partial F}{\partial u} & \dfrac{\partial F}{\partial v} & \dfrac{\partial F}{\partial w} \\ \dfrac{\partial G}{\partial u} & \dfrac{\partial G}{\partial v} & \dfrac{\partial G}{\partial w} \\ \dfrac{\partial H}{\partial u} & \dfrac{\partial H}{\partial v} & \dfrac{\partial H}{\partial w} \end{vmatrix}. \tag{12.36}$$

With this notation we can write 12.35 in the form

$$\frac{\partial u}{\partial x} = -\frac{\dfrac{\partial(F, G, H)}{\partial(x, v, w)}}{\dfrac{\partial(F, G, H)}{\partial(u, v, w)}}. \tag{12.37}$$

The remaining derivatives of $v = g(x, y)$ and $w = h(x, y)$ with respect to x can also be obtained from equations 12.34 by Cramer's rule:

$$\frac{\partial v}{\partial x} = -\frac{\dfrac{\partial(F, G, H)}{\partial(u, x, w)}}{\dfrac{\partial(F, G, H)}{\partial(u, v, w)}}, \quad \frac{\partial w}{\partial x} = -\frac{\dfrac{\partial(F, G, H)}{\partial(u, v, x)}}{\dfrac{\partial(F, G, H)}{\partial(u, v, w)}}. \tag{12.38}$$

A similar procedure yields

$$\frac{\partial u}{\partial y} = -\frac{\dfrac{\partial(F, G, H)}{\partial(y, v, w)}}{\dfrac{\partial(F, G, H)}{\partial(u, v, w)}}, \quad \frac{\partial v}{\partial y} = -\frac{\dfrac{\partial(F, G, H)}{\partial(u, y, w)}}{\dfrac{\partial(F, G, H)}{\partial(u, v, w)}}, \quad \frac{\partial w}{\partial y} = -\frac{\dfrac{\partial(F, G, H)}{\partial(u, v, y)}}{\dfrac{\partial(F, G, H)}{\partial(u, v, w)}}. \tag{12.39}$$

Formulas 12.37–12.39 apply only to the situation in which equations 12.31 define u, v, and w as functions of x and y. It is, however, fairly evident how to construct formulas in other situations. Here are the rules:

1. The partial derivative has a Jacobian divided by a Jacobian (and do not forget the negative sign).

2. In the denominator, it is the Jacobian of the functions defining the original equations with respect to the dependent variables.

3. The only difference in the Jacobian in the numerator is that the dependent variable that is being differentiated is replaced by the independent variable with respect to which differentiation is being performed.

The results in equations 12.37–12.39 are valid provided, of course, that the Jacobian

$$\frac{\partial(F, G, H)}{\partial(u, v, w)} \neq 0.$$

In actual fact, it is this condition that guarantees that equations 12.31 do define u, v, and w as functions of x and y in the first place.

As a second example, the equations

$$F(x, y, s, t) = x + y^2 - 2xs + t + 1 = 0, \quad G(x, y, s, t) = x^2 - y^4 - 2y^2 + y + 3s + 2t^3 + 2 = 0$$

define x and y as functions of s and t. To find $\partial x / \partial s$ when $s = 1$ and $t = 0$, we first calculate

$$\frac{\partial x}{\partial s} = -\frac{\dfrac{\partial(F, G)}{\partial(s, y)}}{\dfrac{\partial(F, G)}{\partial(x, y)}} = -\frac{\begin{vmatrix} F_s & F_y \\ G_s & G_y \end{vmatrix}}{\begin{vmatrix} F_x & F_y \\ G_x & G_y \end{vmatrix}} = -\frac{\begin{vmatrix} -2x & 2y \\ 3 & -4y^3 - 4y + 1 \end{vmatrix}}{\begin{vmatrix} 1 - 2s & 2y \\ 2x & -4y^3 - 4y + 1 \end{vmatrix}}.$$

When $s = 1$ and $t = 0$, the equations defining x and y reduce to

$$x + y^2 - 2x + 1 = 0, \qquad x^2 - y^4 - 2y^2 + y + 5 = 0.$$

The first gives $x = 1 + y^2$, which we substitute into the second:

$$0 = (1 + y^2)^2 - y^4 - 2y^2 + y + 5 = y + 6.$$

Thus, $y = -6$, and $x = 1 + 36 = 37$. With these values, the partial derivative is

$$\frac{\partial x}{\partial s} = -\frac{\begin{vmatrix} -74 & -12 \\ 3 & 889 \end{vmatrix}}{\begin{vmatrix} -1 & -12 \\ 74 & 889 \end{vmatrix}} = 65\,750.$$

EXAMPLE 12.20

If $x^2 y^2 z^3 + zx \sin y = 5$ defines z as a function of x and y, find $\partial z / \partial x$.

SOLUTION   If we set $F(x, y, z) = x^2 y^2 z^3 + zx \sin y - 5 = 0$, then

$$\frac{\partial z}{\partial x} = -\frac{\dfrac{\partial(F)}{\partial(x)}}{\dfrac{\partial(F)}{\partial(z)}} = -\frac{F_x}{F_z} = -\frac{2xy^2 z^3 + z \sin y}{3x^2 y^2 z^2 + x \sin y}.$$

EXAMPLE 12.21

The equations

$$x^2 y^3 z^3 + uvw + 1 = 0, \quad x^2 + y^2 + z^2 + u^3 + v^3 + w^2 = 6, \quad u + v + w = x + 2y,$$

define u, v, and w as functions of x, y, and z. Find $\partial v / \partial z$ when $x = 1$, $y = 0$, $z = 2$, $u = 1$, $v = -1$, and $w = 1$.

SOLUTION   If we set

$$F(x, y, z, u, v, w) = x^2 y^3 z^3 + uvw + 1, \quad G(x, y, z, u, v, w) = x^2 + y^2 + z^2 + u^3 + v^3 + w^2 - 6$$

$$H(x, y, z, u, v, w) = u + v + w - x - 2y,$$

then

$$\frac{\partial v}{\partial z} = -\frac{\dfrac{\partial(F, G, H)}{\partial(u, z, w)}}{\dfrac{\partial(F, G, H)}{\partial(u, v, w)}} = -\frac{\begin{vmatrix} F_u & F_z & F_w \\ G_u & G_z & G_w \\ H_u & H_z & H_w \end{vmatrix}}{\begin{vmatrix} F_u & F_v & F_w \\ G_u & G_v & G_w \\ H_u & H_v & H_w \end{vmatrix}} = -\frac{\begin{vmatrix} vw & 3x^2 y^3 z^2 & uv \\ 3u^2 & 2z & 2w \\ 1 & 0 & 1 \end{vmatrix}}{\begin{vmatrix} vw & uw & uv \\ 3u^2 & 3v^2 & 2w \\ 1 & 1 & 1 \end{vmatrix}}.$$

Instead of expanding these determinants, and then substituting values for the variables, we substitute first, and then expand:

$$\frac{\partial v}{\partial z} = -\frac{\begin{vmatrix} -1 & 0 & -1 \\ 3 & 4 & 2 \\ 1 & 0 & 1 \end{vmatrix}}{\begin{vmatrix} -1 & 1 & -1 \\ 3 & 3 & 2 \\ 1 & 1 & 1 \end{vmatrix}} = 0.$$

EXERCISES 12.7

In Exercises 1–4 y is defined implicitly as a function of x. Find dy/dx.

1. $x^3 y^2 - 2xy + 5 = 0$ **2.** $(x + y)^2 = 2x$

3. $x(x - y) - 4y^3 = 2e^{xy} + 6$ **4.** $\sin(x + y) + y^2 = 12x^2 + y$

In Exercises 5–8 z is defined implicitly as a function of x and y. Find $\partial z/\partial x$ and $\partial z/\partial y$.

5. $x^2 \sin z - ye^z = 2x$ **6.** $x^2 z^2 + yz + 3x = 4$

7. $z \sin^2 y + y \sin^2 x = z^3$ **8.** $\text{Tan}^{-1}(yz) = xz$

In Exercises 9–13 find the required derivative. Assume that the system of equations does define the function(s) indicated.

9. $\partial u/\partial x$ and $\partial v/\partial y$ if $x^2 - y^2 + u^2 + 2v^2 = 1$, $x^2 + y^2 = 2 + u^2 + v^2$

10. $\partial x/\partial t$ if $\sin(x + t) - \sin(x - t) = z$

11. $\partial \phi/\partial x)_{y,z}$ if $x = r \sin \phi \cos \theta$, $\quad y = r \sin \phi \sin \theta$, $\quad z = r \cos \phi$

12. dz/dx if $x^2 + y^2 - z^2 + 2xy = 1$, $\quad x^3 + y^3 - 5y = 4$

13. $\partial u/\partial y)_x$ if $xyu + vw = 4$, $\quad y^2 + u^2 - u^2 v = y$, $\quad yw + xu + v + 4 = 0$

* **14.** Given that the equations

$$x^2 - y \cos(uv) + z^2 = 0, \quad x^2 + y^2 - \sin(uv) + 2z^2 = 2,$$

$$xy - \sin u \cos v + z = 0$$

define x, y, and z as functions of u and v, find $\partial x/\partial u)_v$ at the value $x = 1$, $y = 1$, $u = \pi/2$, $v = 0$, and $z = 0$.

* **15.** If the equation $F(x, y, z) = 0$ defines each of x, y, and z as a function of the other two, show that

$$\left(\frac{\partial z}{\partial x}\right)_y \left(\frac{\partial x}{\partial y}\right)_z \left(\frac{\partial y}{\partial z}\right)_x = -1.$$

* **16.** If $z = e^x \cos y$, where x and y are functions of t defined by

$$x^3 + e^x - t^2 - t = 1, \quad yt^2 + y^2 t - t + y = 0,$$

find dz/dt.

* **17.** Find $\partial s/\partial u)_v$ if $s = x^2 + y^2$, and x and y are functions of u and v defined by

$$u = x^2 - y^2, \quad v = x^2 - y.$$

* **18.** Find $\partial z/\partial y)_x$ if $z = u^3 v + \sin(uv)$, and u and v are functions of x and y defined by

$$x = e^u \cos v, \quad y = e^u \sin v.$$

19. Given that $z^3 - xz - y = 0$ defines z as a function of x and y, show that

$$\frac{\partial^2 z}{\partial x \partial y} = -\frac{3z^2 + x}{(3z^2 - x)^3}.$$

20. If the equations $x = u^2 - v^2$, $y = 2uv$, define u and v as functions of x and y, find $\partial^2 u/\partial x^2$.

21.
(a) Given that the equation $z^4 x + y^3 z + 9x^3 = 2$ defines z as a function of x and y, and x as a function of y and z, are $\partial z/\partial x$ and $\partial x/\partial z$ reciprocals?

(b) Given that the equations $z^4 x + y^3 z + 9x^3 = 2$, $x^2 y + xz = 1$ define z as a function of x, and x as a function of z, are dz/dx and dx/dz reciprocals?

(c) Given that the equations $u^2 - v = 3x + y$, $u - 2v^2 = x - 2y$ define u and v as functions of x and y, and also define x and y as functions of u and v, are $\partial u/\partial x$ and $\partial x/\partial u$ reciprocals?

22. Given that the equations $x^2 - 2y^2 s^2 t - 2st^2 = 1$, $x^2 + 2y^2 s^2 t + 5st^2 = 1$ define s and t as functions of x and y, find $\partial^2 t/\partial y^2$.

23.
(a) Suppose the equations $F(u, v, s, t) = 0$, $G(u, v, s, t) = 0$ define u and v as functions of s and t, and the equations $H(s, t, x, y) = 0$, $I(s, t, x, y) = 0$ define s and t as functions of x and y. Show that

$$\frac{\partial(u, v)}{\partial(s, t)} \frac{\partial(s, t)}{\partial(x, y)} = \frac{\partial(u, v)}{\partial(x, y)}.$$

(b) If the equations $F(u, v, x, y) = 0$, $G(u, v, x, y) = 0$ define u and v as functions of x and y, and also define x and y as functions of u and v, show that

$$\frac{\partial(u, v)}{\partial(x, y)} = \frac{1}{\dfrac{\partial(x, y)}{\partial(u, v)}}.$$

** **24.** Suppose the system of m linear equations in n unknowns ($n > m$)

$$\sum_{j=1}^{n} a_{ij} x_j = c_i, \qquad i = 1, \ldots, m$$

defines $x_1, x_2, \ldots, x_m$ as functions of $x_{m+1}, x_{m+2}, \ldots, x_n$. Show that if $1 \le i \le m$ and $m + 1 \le j \le n$, then

$$\frac{\partial x_i}{\partial x_j} = -\frac{D_{ij}}{D},$$

where $D = |a_{ij}|_{m \times m}$, and D_{ij} is the same as determinant D except that its ith column is replaced by the jth column of $[a_{ij}]_{m \times n}$.

12.8 Directional Derivatives

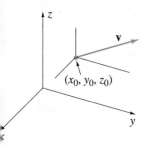

If a function $f(x, y, z)$ is defined throughout some region of space, then at any point (x_0, y_0, z_0) we can calculate its partial derivatives $\partial f/\partial x$, $\partial f/\partial y$, and $\partial f/\partial z$. These derivatives define rates of change of $f(x, y, z)$ at (x_0, y_0, z_0) in directions parallel to the x-, y-, and z-axes. But what if we want the rate of change of $f(x, y, z)$ at (x_0, y_0, z_0) in some arbitrary direction defined by a vector **v** (Figure 12.15)? By the rate of change of $f(x, y, z)$ in the direction **v**, we mean the rate of change with respect to distance as measured along a line through (x_0, y_0, z_0) in direction **v**. Let us define s as a measure of directed distance along this line, taking $s = 0$ at (x_0, y_0, z_0) and positive s in the direction of **v**. What we want, then, is the derivative of $f(x, y, z)$ with respect to s at $s = 0$. To express $f(x, y, z)$ in terms of s, we use parametric equations of the line through (x_0, y_0, z_0) along **v**. If $\hat{\mathbf{v}} = (v_x, v_y, v_z)$ is a unit vector in the direction of **v**, then parametric equations for this line (see equations 11.37) are

$$x = x_0 + v_x s, \qquad y = y_0 + v_y s, \qquad z = z_0 + v_z s. \qquad (12.40)$$

From the schematic diagram to the left, we obtain

$$\frac{df}{ds} = \frac{\partial f}{\partial x}\frac{dx}{ds} + \frac{\partial f}{\partial y}\frac{dy}{ds} + \frac{\partial f}{\partial z}\frac{dz}{ds} = \frac{\partial f}{\partial x}v_x + \frac{\partial f}{\partial y}v_y + \frac{\partial f}{\partial z}v_z,$$

where all partial derivatives of $f(x, y, z)$ are to be evaluated at (x_0, y_0, z_0). We call this a *directional derivative*. It is given an alternative notation in the following definition.

DEFINITION 12.6

The **directional derivative** of a function $f(x, y, z)$ in the direction $\hat{\mathbf{v}} = (v_x, v_y, v_z)$ at the point (x_0, y_0, z_0) is

$$D_{\mathbf{v}} f = \frac{\partial f}{\partial x} v_x + \frac{\partial f}{\partial y} v_y + \frac{\partial f}{\partial z} v_z. \qquad (12.41)$$

Now v_x, v_y, and v_z are the components of the unit vector $\hat{\mathbf{v}}$ in the direction of $\mathbf{v}$, and $\partial f/\partial x$, $\partial f/\partial y$, and $\partial f/\partial z$ are the components of the gradient of $f(x, y, z)$. We can write, therefore, that

$$D_{\mathbf{v}} f = \nabla f \cdot \hat{\mathbf{v}}. \qquad (12.42)$$

Consequently, the derivative (rate of change) of a function in any given direction is the scalar product of the gradient of the function and a unit vector in the required direction. We state this in the following theorem.

THEOREM 12.4

The directional derivative of a function in any direction is the component of the gradient of the function in that direction.

EXAMPLE 12.22

Find $D_{\mathbf{v}} f$ at $(4, 0, 16)$ if $f(x, y, z) = x^3 e^y + xz$ and $\mathbf{v}$ is the vector from $(4, 0, 16)$ to $(-2, 1, 4)$.

SOLUTION Since

$$\nabla f_{|(4,0,16)} = [(3x^2 e^y + z)\hat{\mathbf{i}} + x^3 e^y \hat{\mathbf{j}} + x\hat{\mathbf{k}}]_{|(4,0,16)}$$

$$= 64\hat{\mathbf{i}} + 64\hat{\mathbf{j}} + 4\hat{\mathbf{k}}$$

and

$$\hat{\mathbf{v}} = \frac{\mathbf{v}}{|\mathbf{v}|} = \frac{(-6, 1, -12)}{\sqrt{36 + 1 + 144}} = \frac{-1}{\sqrt{181}}(6, -1, 12),$$

we have

$$D_{\mathbf{v}} f = -(64, 64, 4) \cdot \frac{(6, -1, 12)}{\sqrt{181}} = -\frac{368}{\sqrt{181}}.$$

The fact that the derivative is negative means that $f(x, y, z)$ is decreasing in direction $\mathbf{v}$.

The directional derivative gives us insight into some of the properties of the gradient vector. In particular, we have the next theorem.

THEOREM 12.5

The gradient ∇f of a function $f(x, y, z)$ defines the direction in which the function increases most rapidly, and the maximum rate of change is $|\nabla f|$.

PROOF   Theorem 12.4 states that the directional derivative of $f(x, y, z)$ in a direction $\mathbf{v}$ is the component of ∇f in that direction. Figure 12.16, which shows components of ∇f in various directions, makes it clear that $D_{\mathbf{v}} f$ is greatest when $\mathbf{v}$ is parallel to ∇f. Alternatively, if θ is the angle between $\mathbf{v}$ and ∇f, then

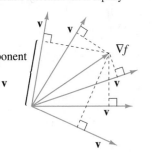

FIGURE 12.16   Gradient of a unction points in the direction in which e function increases most rapidly

$$D_{\mathbf{v}} f = \nabla f \cdot \hat{\mathbf{v}} = |\nabla f||\hat{\mathbf{v}}| \cos \theta = |\nabla f| \cos \theta.$$

Obviously $D_{\mathbf{v}} f$ is a maximum when $\cos \theta$ is a maximum (i.e., when $\cos \theta = 1$ or $\theta = 0$) and this occurs when $\mathbf{v}$ is parallel to ∇f. Finally, when $\mathbf{v}$ is parallel to ∇f, we have $D_{\mathbf{v}} f = |\nabla f|$, and this completes the proof. $\blacksquare$

Note that for any function $f(x, y, z)$,

$$D_{\hat{\mathbf{i}}} f = \frac{\partial f}{\partial x}, \quad D_{\hat{\mathbf{j}}} f = \frac{\partial f}{\partial y}, \quad D_{\hat{\mathbf{k}}} f = \frac{\partial f}{\partial z}.$$

In other words, the partial derivatives of a function are its directional derivatives along the coordinate directions.

EXAMPLE 12.23

Find the direction at the point $(1, 2, -3)$ in which the function $f(x, y, z) = x^2 y + xyz$ increases most rapidly.

SOLUTION   According to Theorem 12.5, $f(x, y, z)$ increases most rapidly in the direction

$$\nabla f_{|(1,2,-3)} = (2xy + yz, x^2 + xz, xy)_{|(1,2,-3)} = (-2, -2, 2).$$

You might feel that because the definition of the directional derivative $D_{\mathbf{v}} f$ does not involve a limit process, it is some strange new type of differentiation. To show that this is not the case, let us return to the calculation of the derivative of $f(x, y, z)$ at (x_0, y_0, z_0) in the direction $\mathbf{v}$ shown in Figure 12.15. With parametric equations 12.40 for the line through (x_0, y_0, z_0) along $\mathbf{v}$, the value of $f(x, y, z)$ at any point (x, y, z) along this line is $f(x_0 + v_x s, y_0 + v_y s, z_0 + v_z s)$. If we take the difference between this value and $f(x_0, y_0, z_0)$ and divide by the distance s between (x_0, y_0, z_0) and (x, y, z), then the limit of this expression as $s \to 0^+$ should define the derivative of $f(x, y, z)$ at (x_0, y_0, z_0) in the direction $\mathbf{v}$; that is,

$$D_{\mathbf{v}} f = \lim_{s \to 0^+} \frac{f(x_0 + v_x s, y_0 + v_y s, z_0 + v_z s) - f(x_0, y_0, z_0)}{s}. \tag{12.43}$$

It can be shown that this limit (and this is perhaps the form we might have expected the derivative to take) also leads to the result contained in 12.42 (see Exercise 35).

Consider a curve C in space that is defined parametrically by

FIGURE 12.17   Rate of hange of a function along a curve

$$C: \quad x = x(t), \quad y = y(t), \quad z = z(t), \quad \alpha \le t \le \beta$$

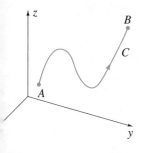

(Figure 12.17). Imagine that C is the path traced out by some particle as it moves through space under the action of some system of forces, and suppose that $f(x, y, z)$ is a function defined along C. Perhaps the particle is a weather balloon and $f(x, y, z)$ is temperature at points along its trajectory C. In such applications we are frequently asked for the rate of change of $f(x, y, z)$ with respect to distance travelled along C. If we use s as a measure of distance along C (taking $s = 0$ at A), then the required rate of change is df/ds. Since the coordinates of points (x, y, z) on C can be regarded as functions of s (although it might be difficult to find these functions explicitly), the chain rule gives

$$\frac{df}{ds} = \frac{\partial f}{\partial x}\frac{dx}{ds} + \frac{\partial f}{\partial y}\frac{dy}{ds} + \frac{\partial f}{\partial z}\frac{dz}{ds}$$

$$= \left(\frac{\partial f}{\partial x}, \frac{\partial f}{\partial y}, \frac{\partial f}{\partial z}\right) \cdot \left(\frac{dx}{ds}, \frac{dy}{ds}, \frac{dz}{ds}\right)$$

$$= \nabla f \cdot \frac{d\mathbf{r}}{ds}.$$

In Section 11.11 we saw that $d\mathbf{r}/ds$ is a unit tangent vector $\hat{\mathbf{T}}$ to C. Consequently,

$$\frac{df}{ds} = \nabla f \cdot \hat{\mathbf{T}}.$$

But this equation states that df/ds is the directional derivative of $f(x, y, z)$ along the tangent direction to the curve C. In other words, to calculate the rate of change of a function $f(x, y, z)$ with respect to distance as measured along a curve C, we calculate the directional derivative of $f(x, y, z)$ in the direction of the tangent vector to C.

EXAMPLE 12.24

Find the rate of change of the function $f(x, y, z) = x^2 y - xz$ along the curve $y = x^2$, $z = x$ in the direction of decreasing x at the point $(2, 4, 2)$.

SOLUTION   Since parametric equations for the curve are $C:\ x = -t,\ y = t^2,\ z = -t$, a tangent vector to C at any point is $\mathbf{T} = (-1, 2t, -1)$. At $(2, 4, 2)$, $t = -2$, and the tangent vector is $\mathbf{T} = (-1, -4, -1)$. A unit tangent vector to C at $(2, 4, 2)$ in the direction of decreasing x is therefore

$$\hat{\mathbf{T}} = \frac{(-1, -4, -1)}{\sqrt{18}} = \frac{-1}{3\sqrt{2}}(1, 4, 1).$$

The rate of change of $f(x, y, z)$ in this direction is

$$\nabla f \cdot \hat{\mathbf{T}} = (2xy - z, x^2, -x)_{|(2,4,2)} \cdot \frac{(1, 4, 1)}{-3\sqrt{2}}$$

$$= \frac{-1}{3\sqrt{2}}(14, 4, -2) \cdot (1, 4, 1) = -\frac{28}{3\sqrt{2}}.$$

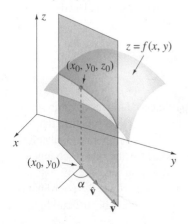

FIGURE 12.18   First and second directional derivatives of a function $f(x, y)$ of two variables

In preparation for maxima and minima of multivariable functions in Section 12.10, we now discuss directional derivatives for a function $f(x, y)$ of two independent variables. Such a function can be represented graphically as a surface $z = f(x, y)$ (Figure 12.18).

For a direction $\mathbf{v}$ at (x_0, y_0) in the xy-plane,

$$D_\mathbf{v} f = \nabla f \cdot \hat{\mathbf{v}},$$

where ∇f is evaluated at (x_0, y_0). Algebraically, this is the rate of change of $f(x, y)$ in direction $\mathbf{v}$. Geometrically, it is the rate of change of the height z of the surface as we move along the curve of intersection of the surface and a vertical plane containing the vector $\mathbf{v}$, or the slope of this curve. Each direction $\mathbf{v}$ at (x_0, y_0) defines an angle α with a line through (x_0, y_0) parallel to the positive x-axis, and for this direction

$$\hat{\mathbf{v}} = \cos\alpha\,\hat{\mathbf{i}} + \sin\alpha\,\hat{\mathbf{j}}.$$

We can write, then,

$$D_\mathbf{v} f = \nabla f \cdot \hat{\mathbf{v}} = \left(\frac{\partial f}{\partial x} \hat{\mathbf{i}} + \frac{\partial f}{\partial y} \hat{\mathbf{j}} \right) \cdot (\cos \alpha \, \hat{\mathbf{i}} + \sin \alpha \, \hat{\mathbf{j}})$$

$$= \frac{\partial f}{\partial x} \cos \alpha + \frac{\partial f}{\partial y} \sin \alpha. \qquad (12.44)$$

If $D_\mathbf{v} f$ represents the slope of the curve of intersection of the surface and the vertical plane through $\mathbf{v}$, then $D_\mathbf{v}(D_\mathbf{v} f)$ represents the rate of change of this slope. Now

$$D_\mathbf{v}(D_\mathbf{v} f) = \nabla (D_\mathbf{v} f) \cdot \hat{\mathbf{v}}$$

and

$$\nabla(D_\mathbf{v} f) = \left(\frac{\partial^2 f}{\partial x^2} \cos \alpha + \frac{\partial^2 f}{\partial x \, \partial y} \sin \alpha \right) \hat{\mathbf{i}} + \left(\frac{\partial^2 f}{\partial y \, \partial x} \cos \alpha + \frac{\partial^2 f}{\partial y^2} \sin \alpha \right) \hat{\mathbf{j}}.$$

Thus,

$$D_\mathbf{v}(D_\mathbf{v} f) = \left(\frac{\partial^2 f}{\partial x^2} \cos \alpha + \frac{\partial^2 f}{\partial x \, \partial y} \sin \alpha \right) \cos \alpha + \left(\frac{\partial^2 f}{\partial y \, \partial x} \cos \alpha + \frac{\partial^2 f}{\partial y^2} \sin \alpha \right) \sin \alpha$$

$$= \frac{\partial^2 f}{\partial x^2} \cos^2 \alpha + 2 \frac{\partial^2 f}{\partial x \, \partial y} \cos \alpha \sin \alpha + \frac{\partial^2 f}{\partial y^2} \sin^2 \alpha. \qquad (12.45)$$

We call $D_\mathbf{v}(D_\mathbf{v} f)$ the second directional derivative of $f(x, y)$ at (x_0, y_0) in direction $\mathbf{v}$. If it is positive, then the curve of intersection is concave upward, whereas if it is negative, the curve is concave downward. We will find these results useful in Section 12.10 when we discuss relative extrema of functions of two independent variables.

EXERCISES 12.8

In Exercises 1–8 calculate the directional derivative of the function at the point and in the direction indicated.

1. $f(x, y, z) = 2x^2 - y^2 + z^2$ at $(1, 2, 3)$ in the direction of the vector from $(1, 2, 3)$ to $(3, 5, 0)$

2. $f(x, y, z) = x^2 y + xz$ at $(-1, 1, -1)$ in the direction of the vector that joins $(3, 2, 1)$ to $(3, 1, -1)$

3. $f(x, y) = xe^y + y$ at $(3, 0)$ in the direction of the vector from $(3, 0)$ to $(-2, -4)$

4. $f(x, y, z) = \ln(xy + yz + xz)$ at $(1, 1, 1)$ in the direction from $(1, 1, 1)$ toward the point $(-1, -2, 3)$

5. $f(x, y) = \text{Tan}^{-1}(xy)$ at $(1, 2)$ along the line $y = 2x$ in the direction of increasing x

6. $f(x, y) = \sin(x + y)$ at $(2, -2)$ along the line $3x + 4y = -2$ in the direction of decreasing y

7. $f(x, y, z) = x^3 y \sin z$ at $(3, -1, -2)$ along the line $x = 3 + t$, $y = -1 + 4t$, $z = -2 + 2t$ in the direction of decreasing x

8. $f(x, y, z) = x^2 y + y^2 z + z^2 x$ at $(1, -1, 0)$ along the line $x + y + 1 = 0$, $x - y + 2z = 2$ in the direction of decreasing z

In Exercises 9–12 find the rate of change of the function with respect to distance travelled along the curve.

9. $f(x, y) = 2x - 3y$ at $(1, 1)$ along the curve $y = x^2$ in the direction of increasing x

10. $f(x, y) = x^2 + y$ at $(-1, 3)$ along the curve $y = -3x^3$ in the direction of decreasing x

11. $f(x, y, z) = xy + z^2$ at $(1, 0, -2)$ along the curve $y = x^2 - 1$, $z = -2x$ in the direction of increasing x

12. $f(x, y, z) = x^2 y + xy^3 z$ at $(2, -1, 2)$ along the curve $x^2 - y^2 = 3$, $z = x$ in the direction of increasing x

In Exercises 13–18 find the direction in which the function increases most rapidly at the point. What is the rate of change in that direction?

13. $f(x, y, z) = x^4 yz - xy^3 + z$ at $(1, 1, -3)$

14. $f(x, y) = 2xy + \ln(xy)$ at $(2, 1/2)$

15. $f(x, y, z) = 1/\sqrt{x^2 + y^2 + z^2}$ at $(1, -3, 2)$

16. $f(x, y, z) = -1/\sqrt{x^2 + y^2 + z^2}$ at $(1, -3, 2)$

17. $f(x, y, z) = \text{Tan}^{-1}(xyz)$ at $(3, 2, -4)$

18. $f(x, y) = xye^{xy}$ at $(1, 1)$

19. In what direction is the rate of change of $f(x, y, z) = xyz$ smallest at the point $(2, -1, 3)$?

* **20.** In what directions (if any) is the rate of change of the function $f(x, y) = x^2 y + y^3$ at the point $(1, -1)$ equal to (a) 0, (b) 1, and (c) 20?

* **21.** In what directions (if any) is the rate of change of the function $f(x, y, z) = xy + z$ at the point $(0, 1, -2)$ equal to (a) 0, (b) 1, and (c) -20?

* **22.** Must there always be a direction in which the rate of change of a function at a point is equal to (a) 0 and (b) 3?

* **23.** In the derivation of 12.41, why was it necessary to use a unit vector $\hat{\mathbf{v}}$ to determine parametric equations for the line through (x_0, y_0, z_0) along $\mathbf{v}$? In other words, why could we not use the components of $\mathbf{v}$ itself to write parametric equations for the line?

* **24.** How fast is the distance to the origin changing with respect to distance travelled along the curve $x = 2\cos t$, $y = 2\sin t$, $z = 3t$ at any point on the curve? What is the rate of change when $t = 0$? Would you expect this?

* **25.** Find points on the curve $C : x = t$, $y = 1 - 2t$, $z = t$ at which the rate of change of $f(x, y, z) = x^2 + xyz$ with respect to distance travelled along the curve vanishes.

* **26.** Repeat Exercise 25 for the curve $C : z = x$, $x = y^2$ and the function $f(x, y, z) = x^2 - y^2 + z^2$.

📱* **27.** The path of a particle is defined parametrically by $x = (\cos t + t\sin t)\hat{\mathbf{i}} + (\sin t - t\cos t)\hat{\mathbf{j}}$, where t is time. Plot the path called an *involute of a circle*. Show that the rate of change of the distance of the particle from the origin, with respect to distance travelled, is always positive.

* **28.** If we know the rate of change of a function $f(x, y, z)$ at a point P on a curve C, proceeding in one direction along C, what is the rate of change in the opposite direction along C?

* **29.** What is the rate of change of a function $f(x, y, z)$ in a direction perpendicular to ∇f?

* **30.** The rate of change of a function $f(x, y)$ at a point (x_0, y_0) in direction $\hat{\mathbf{i}} + 2\hat{\mathbf{j}}$ is 3 and the rate of change in direction $-2\hat{\mathbf{i}} - \hat{\mathbf{j}}$ is -1. Find its rate of change in direction $2\hat{\mathbf{i}} + 3\hat{\mathbf{j}}$.

* **31.** Rates of change of a function $f(x, y, z)$ at a point (x_0, y_0, z_0) in directions $\hat{\mathbf{i}} + \hat{\mathbf{j}}$, $2\hat{\mathbf{i}} - \hat{\mathbf{k}}$, and $\hat{\mathbf{i}} - \hat{\mathbf{j}} + \hat{\mathbf{k}}$ are 1, 2, and -3, respectively. What is its partial derivative with respect to z at the point?

* **32.** Find the second directional derivative of the function $f(x, y) = x^3 y^2$ at the point $(1, 1)$ in the direction of the vector $(1, -2)$.

* **33.** Find the second directional derivative of the function $f(x, y, z) = x^2 + 2y^2 + 3z^2$ at the point $(-2, -1, 3)$ in the direction $(1, 1, -1)$.

* **34.** The path followed by a stone embedded in the tread of a tire is a cycloid given parametrically by $x = R(\theta - \sin\theta)$, $y = R(1 - \cos\theta)$, $\theta \geq 0$ (see Example 9.7 in Section 9.1).
  (a) How fast is the distance from the origin changing with respect to distance travelled along the curve at the points corresponding to $\theta = \pi/2$ and $\theta = \pi$?
  (b) How fast is the y-coordinate changing at these points?
  (c) How fast is the x-coordinate changing at these points?

** **35.** Verify that expression 12.43 for $D_{\mathbf{v}} f$ leads to formula 12.42.

12.9 Tangent Lines and Tangent Planes

Tangent Lines to Curves

One equation in the coordinates x, y, and z of points in space,

$$F(x, y, z) = 0, \tag{12.46}$$

usually defines a surface. (There are exceptions. The equation $x^2 + y^2 + z^2 = 0$ defines a point, and $x^2 + y^2 + z^2 = -1$ defines nothing.) When each of the equations

$$F(x, y, z) = 0, \quad G(x, y, z) = 0 \tag{12.47}$$

defines a surface, then together they define the curve of intersection of the two surfaces (provided, of course, that the surfaces do intersect). Theoretically, we can find parametric equations for the curve by setting x equal to some function of a parameter t, say $x = x(t)$, and then solving equations 12.47 for y and z in terms of t: $y = y(t)$ and $z = z(t)$. The parametric definition, therefore, takes the form

$$x = x(t), \quad y = y(t), \quad z = z(t), \quad \alpha \leq t \leq \beta, \tag{12.48}$$

where α and β specify the endpoints of the curve. Practical difficulties arise in choosing $x(t)$ and solving for $y(t)$ and $z(t)$. For some examples, it might be more convenient to specify $y(t)$ and solve for $x(t)$ and $z(t)$ or, alternatively, to specify $z(t)$ and solve for $x(t)$ and $y(t)$. We considered examples of such conversions in Section 11.10.

In Section 11.11 we indicated that when a curve C is defined parametrically by 12.48, a tangent vector to C at any point P is

$$\frac{d\mathbf{r}}{dt} = \frac{dx}{dt}\hat{\mathbf{i}} + \frac{dy}{dt}\hat{\mathbf{j}} + \frac{dz}{dt}\hat{\mathbf{k}} \tag{12.49}$$

(Figure 12.19). The **tangent line** to C at P is defined as the line through P having direction $d\mathbf{r}/dt$. If (x_0, y_0, z_0) are the coordinates of P and t_0 is the value of t yielding P, then the vector equation for the tangent line at P is

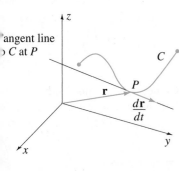

$$(x, y, z) = (x_0, y_0, z_0) + u\frac{d\mathbf{r}}{dt}\Big|_{t=t_0} \tag{12.50a}$$

(see equation 11.36). Parametric equations for the tangent line are

$$x = x_0 + x'(t_0)u, \quad y = y_0 + y'(t_0)u, \quad z = z_0 + z'(t_0)u, \tag{12.50b}$$

and in the case where none of $x'(t_0)$, $y'(t_0)$, and $z'(t_0)$ vanishes, we can also write symmetric equations for the tangent line:

$$\frac{x - x_0}{x'(t_0)} = \frac{y - y_0}{y'(t_0)} = \frac{z - z_0}{z'(t_0)}. \tag{12.50c}$$

EXAMPLE 12.25

Find equations for the tangent line to the elliptic helix

$$C: \quad x = 2\cos t, \quad y = 4\sin t, \quad z = 2t/\pi$$

at $P(\sqrt{2}, 2\sqrt{2}, 1/2)$.

SOLUTION   Since $t = \pi/4$ at P, a tangent vector to C at P is

$$\frac{d\mathbf{r}}{dt}\Big|_{t=\pi/4} = (-2\sin t, 4\cos t, 2/\pi)_{|t=\pi/4} = (-\sqrt{2}, 2\sqrt{2}, 2/\pi).$$

Symmetric equations for the tangent line are therefore

$$\frac{x - \sqrt{2}}{-\sqrt{2}} = \frac{y - 2\sqrt{2}}{2\sqrt{2}} = \frac{z - 1/2}{2/\pi}.$$

We have shown the tangent line to the helix in Figure 12.20.

FIGURE 12.20   Tangent line to an elliptic helix

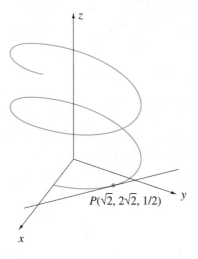

$P(\sqrt{2}, 2\sqrt{2}, 1/2)$

EXAMPLE 12.26

Find equations for the tangent line to the curve $z = 1 - x^2$, $x + y + z = 2$ at the point $P(1/2, 3/4, 3/4)$.

SOLUTION   Parametric equations for the curve are

$$x = t, \quad y = 1 - t + t^2, \quad z = 1 - t^2.$$

Since $t = 1/2$ at P, a tangent vector to the curve at P is

$$\left.\frac{d\mathbf{r}}{dt}\right|_{t=1/2} = (1, -1 + 2t, -2t)_{|t=1/2} = (1, 0, -1).$$

Because the y-component vanishes, we cannot write full symmetric equations for the tangent line, although we could write partial symmetric equations involving x and z. Alternatively, parametric equations for the tangent line are

$$x = \frac{1}{2} + u, \quad y = \frac{3}{4}, \quad z = \frac{3}{4} - u.$$

The line is shown in Figure 12.21.

FIGURE 12.21   Tangent line to curve of intersection of parabolic cylinder and plane

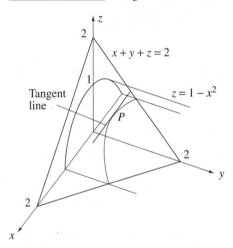

Tangent Planes to Surfaces

We now consider the problem of finding the equation for the tangent plane at a point P on a surface S (Figure 12.22). We define the **tangent plane** as that plane which contains all tangent lines at P to curves in S through P (provided, of course, that such a plane exists). Suppose that the surface is defined by the equation

$$F(x, y, z) = 0, \tag{12.51}$$

and that

$$C: \quad x = x(t), \quad y = y(t), \quad z = z(t), \quad \alpha \leq t \leq \beta$$

is any curve in S through P. Since C is in S, the equation

$$F[x(t), y(t), z(t)] = 0$$

is valid for all t in $\alpha \le t \le \beta$. If $F(x, y, z)$ has continuous first partial derivatives, and $x(t)$, $y(t)$, and $z(t)$ are all differentiable, we may differentiate this equation using the chain rule:

$$\frac{\partial F}{\partial x}\frac{dx}{dt} + \frac{\partial F}{\partial y}\frac{dy}{dt} + \frac{\partial F}{\partial z}\frac{dz}{dt} = 0.$$

This equation, which holds at all points on C, and in particular at P, can be expressed vectorially as

$$0 = \left(\frac{\partial F}{\partial x}, \frac{\partial F}{\partial y}, \frac{\partial F}{\partial z}\right) \cdot \left(\frac{dx}{dt}, \frac{dy}{dt}, \frac{dz}{dt}\right) = \nabla F \cdot \frac{d\mathbf{r}}{dt}.$$

But if the scalar product of two vectors vanishes, the vectors are perpendicular (see equation 11.25). Consequently, ∇F is perpendicular to the tangent vector $d\mathbf{r}/dt$ to C at P. Since C is an arbitrary curve in S, it follows that ∇F at P is perpendicular to the tangent line to every curve C in S at P. In other words, ∇F at P must be perpendicular to the tangent plane to S

FIGURE 12.22   Tangent plane at a point P on a surface S contains all tangent vectors at P to curves in S

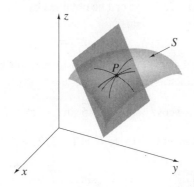

FIGURE 12.23   The gradient of the function defining a surface is perpendicular to the tangent plane to the surface

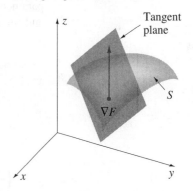

at P (Figure 12.23). If the coordinates of P are (x_0, y_0, z_0), then the equation of the tangent plane to S at P is

$$\begin{aligned}
0 &= \nabla F|_P \cdot (x - x_0, y - y_0, z - z_0) \\
&= F_x(x_0, y_0, z_0)(x - x_0) + F_y(x_0, y_0, z_0)(y - y_0) \\
&\quad + F_z(x_0, y_0, z_0)(z - z_0)
\end{aligned} \tag{12.52}$$

(see equation 11.34).

EXAMPLE 12.27

Find the equation of the tangent plane to the surface $xyz^3 + yz^2 = 4$ at the point $(1, 2, 1)$.

SOLUTION   A vector perpendicular to the tangent plane is

$$\nabla(xyz^3 + yz^2 - 4)|_{(1,2,1)} = (yz^3, xz^3 + z^2, 3xyz^2 + 2yz)|_{(1,2,1)} = (2, 2, 10).$$

But then the vector $(1, 1, 5)$ must also be perpendicular to the tangent plane, and the equation of the plane is therefore

$$0 = (1, 1, 5) \cdot (x - 1, y - 2, z - 1) = x + y + 5z - 8.$$

We have shown in this section that if the equation $F(x, y, z) = 0$ defines a surface S, and if there is a tangent plane to S at a point P, then the vector $\nabla F_{|P}$ is normal to the tangent plane (Figure 12.23). It is customary to state in this situation that $\nabla F_{|P}$ is normal to the surface itself at P, rather than to the tangent plane to the surface. This fact proves to be another of the important properties of the gradient vector, and is worth stating as a theorem.

THEOREM 12.6

If the equation $F(x, y, z) = 0$ defines a surface S, and $F(x, y, z)$ has continuous first partial derivatives, then at any point on S the vector ∇F is perpendicular to S.

A geometric application of this fact is contained in the following example.

EXAMPLE 12.28

Find equations for the tangent line at the point $(1, 2, 2)$ to the curve C: $x^2 + y^2 + z^2 = 9$, $4(x^2 + y^2) = 5z^2$.

SOLUTION   Equation 12.49 indicates that to find a tangent vector to C, we should first have parametric equations for C. These can be obtained by first solving each equation for $x^2 + y^2$ and equating the results:

$$9 - z^2 = 5z^2/4.$$

This equation implies that $z = \pm 2$, the positive result being required here. On C, then, $x^2 + y^2 = 5$, and parametric equations for C are

$$x = \sqrt{5}\cos t, \quad y = \sqrt{5}\sin t, \quad z = 2, \quad 0 \leq t < 2\pi.$$

According to 12.49, a tangent vector to C at $(1, 2, 2)$ is

$$\left(\frac{dx}{dt}, \frac{dy}{dt}, \frac{dz}{dt}\right)\Bigg|_{(1,2,2)} = (-\sqrt{5}\sin t, \sqrt{5}\cos t, 0)_{|t=\mathrm{Sin}^{-1}(2/\sqrt{5})} = (-2, 1, 0).$$

The tangent line therefore has equations

$$\frac{x - 1}{-2} = \frac{y - 2}{1}, \quad z = 2, \quad \text{or } x + 2y = 5, \quad z = 2.$$

The fact that gradients can be used to find **normals to surfaces** suggests an alternative solution. It is clear from Figure 12.24 that if we define $F(x, y, z) = x^2 + y^2 + z^2 - 9$, then

FIGURE 12.24   Cross product of gradients of sphere and cone yields a vector tangent to the curve of intersection

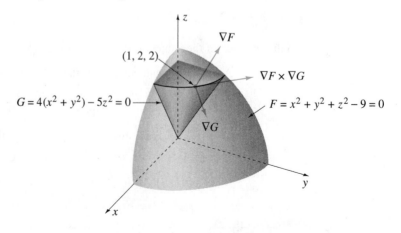

∇F evaluated at $(1, 2, 2)$ is perpendicular not only to the surface $F(x, y, z) = 0$, but also to the curve C. Similarly, if $G(x, y, z) = 4(x^2 + y^2) - 5z^2$, then ∇G at $(1, 2, 2)$ is also perpendicular to C. Since a vector along the tangent line to C at $(1, 2, 2)$ is perpendicular to both of these vectors, it follows that a vector along the tangent line is

$$(\nabla F \times \nabla G)_{|(1,2,2)} = [(2x, 2y, 2z) \times (8x, 8y, -10z)]_{|(1,2,2)}$$

$$= (2, 4, 4) \times (8, 16, -20)$$

$$= 8 \begin{vmatrix} \hat{\mathbf{i}} & \hat{\mathbf{j}} & \hat{\mathbf{k}} \\ 1 & 2 & 2 \\ 2 & 4 & -5 \end{vmatrix}$$

$$= 8(-18, 9, 0)$$

$$= 72(-2, 1, 0).$$

Once again, we have obtained $(-2, 1, 0)$ as a tangent vector to the curve, and equations for the tangent line can be written down as before.

Example 12.28 illustrates that when a curve is defined as the intersection of two surfaces $F(x, y, z) = 0$, $G(x, y, z) = 0$ (Figure 12.25), then a vector tangent to the curve is

$$\mathbf{T} = \nabla F \times \nabla G. \tag{12.53}$$

Thus to find a tangent vector to a curve we use 12.49 when the curve is defined parametrically. When the curve is defined as the intersection of two surfaces, we can either find parametric equations and use 12.49, or use 12.53. Note, too, that in order to find tangent lines to curves, it is not necessary to have a direction assigned to the curves.

FIGURE 12.25 Cross product of gradients of two surfaces yields a vector tangent to their curve of intersection   FIGURE 12.26 Unit normal vector to a surface

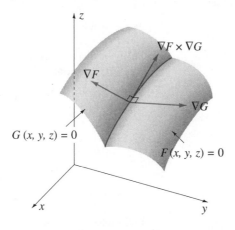

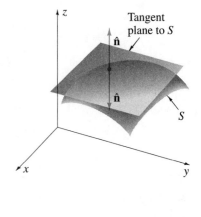

At each point on a surface S at which S has a tangent plane (Figure 12.26), we have defined a normal vector to S as a vector normal to the tangent plane to S. If we denote by $\hat{\mathbf{n}}$ a unit normal vector to S, then the direction of $\hat{\mathbf{n}}$ clearly varies as we move from point to point on S. We say that $\hat{\mathbf{n}}$ is a function of position (x, y, z) on S. Furthermore, at each point at which S has a unit normal vector, it has two such vectors, one in the opposite direction to the other. We say that a surface S is a **smooth surface** if it can be assigned a unit normal $\hat{\mathbf{n}}$ that varies continuously on S. What this means geometrically is that for small changes in position, the unit normal $\hat{\mathbf{n}}$ will undergo small changes in direction. The sphere in Figure 12.27 is smooth, as is the paraboloid in Figure 12.28.

FIGURE 12.27 A sphere
is a smooth surface

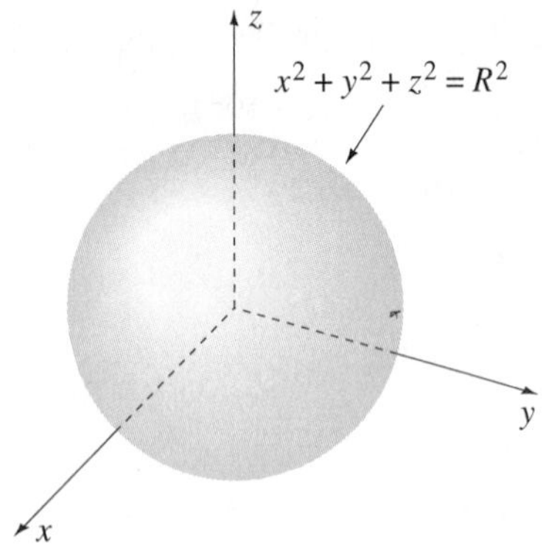

FIGURE 12.28 A paraboloid
is a smooth surface

FIGURE 12.29 The surface of the
cylinder is a piecewise-smooth surface

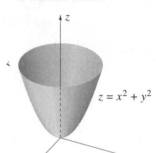

$$x^2 + y^2 + z^2 = R^2$$

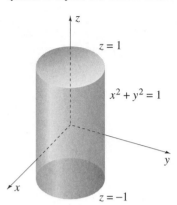

$$z = x^2 + y^2$$

$z = 1$

$$x^2 + y^2 = 1$$

$z = -1$

The surface bounding the cylindrical volume in Figure 12.29 is not smooth; a unit normal that varies continuously over the surface cannot be assigned at points on the circles $x^2 + y^2 = 1$, $z = \pm 1$. This surface can, however, be divided into a finite number of subsurfaces, each of which is smooth. In particular, we choose the three subsurfaces $S_1: z = 1,\ x^2 + y^2 \le 1$; $S_2: z = -1,\ x^2 + y^2 \le 1$; $S_3: x^2 + y^2 = 1,\ -1 < z < 1$. Such a surface is said to be a **piecewise-smooth surface**.

EXERCISES 12.9

In Exercises 1–20 find equations for the tangent line to the curve at the point.

1. $y = x^2$, $z = 0$ at $(-2, 4, 0)$

2. $x = t$, $y = t^2$, $z = t^3$ at $(1, 1, 1)$

3. $x = \cos t$, $y = \sin t$, $z = \cos t$ at $(1, 0, 1)$

4. $y = x^2$, $z = x$ at $(-2, 4, -2)$

5. $x^2 = y$, $z + x = y$ at $(1, 1, 0)$

6. $x = 2 - t^2$, $y = 3 + 2t$, $z = t$ at $(1, 5, 1)$

7. $x = 2 \cos t$, $y = 3 \sin t$, $z = 5$ at $(\sqrt{2}, -3/\sqrt{2}, 5)$

8. $x^2 y^3 + xy = 68$ at $(1, 4)$

9. $x + y + z = 4$, $x - y = 2$ at $(0, -2, 6)$

10. $x = e^{-t} \cos t$, $y = e^{-t} \sin t$, $z = t$ at $(1, 0, 0)$

11. $x = t^2 + 1$, $y = 2t - 4$, $z = t^3 + 3$ at $(2, -6, 2)$

12. $y^2 + z^2 = 6$, $x + z = 1$ at $(2, -\sqrt{5}, -1)$

13. $x^2 + y^2 + z^2 = 4$, $z^2 = x^2 + y^2$ at $(1, 1, -\sqrt{2})$

14. $x = t$, $y = 1$, $z = \sqrt{1 + t^2}$ at $(4, 1, \sqrt{17})$

15. $x = 1 + \cos t$, $y = 2 - \sin t$, $z = \sqrt{4 + t}$ at $(2, 2, 2)$

16. $x = z^2 + z^3$, $y = z - z^4$ at $(12, -14, 2)$

17. $x = y^2 + 3y^3 - 2y + 5$, $z = 0$ at $(7, 1, 0)$

18. $2x^2 + y^2 + 2y = 3$, $z = x + 1$ at $(0, 1, 1)$

19. $x = t^2$, $y = t$, $z = \sqrt{t + t^4}$ at $(1, 1, \sqrt{2})$

20. $x = t \sin t$, $y = t \cos t$, $z = 2t$ at $(0, 2\pi, 4\pi)$

In Exercises 21–26 find an equation for the tangent plane to the surface at the point.

21. $z = \sqrt{x^2 + y^2}$ at $(1, 1, \sqrt{2})$

22. $x = x^2 - y^3 z$ at $(2, -1, -2)$

23. $x^2 y + y^2 z + z^2 x + 3 = 0$ at $(2, -1, -1)$

24. $x + y + z = 4$ at $(1, 1, 2)$

25. $x = y \sin (\pi z / 2)$ at $(-1, -1, 1)$

26. $x^2 + y^2 + 2y = 1$ at $(1, 0, 3)$

* **27.** Show that the curve $x = 2(t^3 + 2)/3$, $y = 2t^2$, $z = 3t - 2$ intersects the surface $x^2 + 2y^2 + 3z^2 = 15$ at right angles at the point $(2, 2, 1)$.

* **28.** Verify that the curve $x^2 - y^2 + z^2 = 1$, $xy + xz = 2$ is tangent to the surface $xyz - x^2 - 6y + 6 = 0$ at the point $(1, 1, 1)$.

* **29.** Show that the equation of the tangent plane to a surface $S : z = f(x, y)$ at a point (x_0, y_0, z_0) on S can be written in the form

$$z - z_0 = (x - x_0) f_x(x_0, y_0) + (y - y_0) f_y(x_0, y_0).$$

In Exercises 30–32 find the indicated derivative for the function.

* **30.** $f(x, y, z) = 2x^2 + y^2 z^2$ at $(3, 1, 0)$ with respect to distance along the curve $x + y + z = 4$, $x - y + z = 2$ in the direction of increasing x

* **31.** $f(x, y, z) = xyz + xy + xz + yz$ at $(1, -2, 5)$ perpendicular to the surface $z = x^2 + y^2$

* **32.** $f(x, y, z) = x^2 + y^2 - z^2$ at $(3, 4, 5)$ with respect to distance along the curve $x^2 + y^2 - z^2 = 0$, $2x^2 + 2y^2 - z^2 = 25$ in the direction of decreasing x

33. If $F(x, y) = 0$ defines a curve implicitly in the xy-plane, prove that at any point on the curve ∇F is perpendicular to the curve.

34. Find the equation of the tangent plane to the ellipsoid $x^2/a^2 + y^2/b^2 + z^2/c^2 = 1$ at any point (x_0, y_0, z_0) on the surface.

35. Find all points on the surface $z = x^2/4 - y^2/9$ at which the tangent plane is parallel to the plane $x + y + z = 4$.

36. Find all points on the surface $z^2 = 4(x^2 + y^2)$ at which the tangent plane is parallel to the plane $x - y + 2z = 3$.

37. Suppose that the equations $F(x, y, z, t) = 0$, $G(x, y, z, t) = 0$, $H(x, y, z, t) = 0$ implicitly define parametric equations for a curve C (t being the parameter). If $P(x_0, y_0, z_0)$ is a point on C, show that

equations for the tangent line to C at P can be written in the form

$$\frac{x - x_0}{\left. \dfrac{\partial(F, G, H)}{\partial(t, y, z)} \right|_P} = \frac{y - y_0}{\left. \dfrac{\partial(F, G, H)}{\partial(x, t, z)} \right|_P} = \frac{z - z_0}{\left. \dfrac{\partial(F, G, H)}{\partial(x, y, t)} \right|_P},$$

provided that none of the Jacobians vanishes.

* **38.** Find all points on the paraboloid $z = x^2 + y^2 - 1$ at which the normal to the surface coincides with the line joining the origin to the point.

** **39.** Show that the sum of the intercepts on the x-, y-, and z-axes of the tangent plane to the surface $\sqrt{x} + \sqrt{y} + \sqrt{z} = \sqrt{a}$ at any point is a.

12.10 Relative Maxima and Minima

We now study relative extrema of functions of more than one independent variable. Most of the discussion will be confined to functions $f(x, y)$ of two independent variables because we can discuss the concepts geometrically as well as algebraically. Unfortunately, not all results are easily extended to functions of more than two independent variables, and we will therefore be careful to point out these limitations.

Before beginning the discussion, we briefly review maxima–minima results for functions $f(x)$ of one variable. We do this because maxima–minima theory for multivariable functions is essentially the same as that for single-variable functions. In fact, every definition that we make and every result that we discuss in this section has its counterpart in single-variable theory. Hence, a synopsis of single-variable results is in order. Unfortunately, proving results in the multivariable case is considerably more complicated than in the single-variable case, but if we can keep central ideas foremost in our minds and constantly make comparisons with single-variable calculus, we will find that discussions are not nearly as difficult as they might otherwise be.

Critical points of a function $f(x)$ are points at which $f'(x)$ is either equal to zero or does not exist. Geometrically, this means points at which the graph of $f(x)$ has a horizontal tangent line, a vertical tangent line, or no tangent line at all. Critical points for continuous functions can yield relative maxima, relative minima, horizontal points of inflection, vertical points of inflection, or just corners. There are two tests to determine whether a critical point x_0 yields a relative maximum or a relative minimum. The first-derivative test states that if $f'(x)$ changes from a positive quantity to a negative quantity as x increases through x_0, then x_0 gives a relative maximum; if $f'(x)$ changes from negative to positive, then a relative minimum is obtained. The second-derivative test indicates the nature of a critical point at which $f'(x_0) = 0$ whenever $f''(x_0) \neq 0$. If $f''(x_0) > 0$, then a relative minimum is obtained, and if $f''(x_0) < 0$, a relative maximum is found.

We begin our study of extrema theory for multivariable functions by defining critical points for functions of two independent variables.

DEFINITION 12.7

A point (x_0, y_0) in the domain of a function $f(x, y)$ is said to be a **critical point** of $f(x, y)$ if

$$\left. \frac{\partial f}{\partial x} \right|_{(x_0, y_0)} = 0, \quad \left. \frac{\partial f}{\partial y} \right|_{(x_0, y_0)} = 0 \tag{12.54}$$

or if one (or both) of these partial derivatives does not exist at (x_0, y_0).

There are two ways to interpret critical points of $f(x, y)$ geometrically. In Section 12.3, we interpreted $\partial f/\partial x$ at (x_0, y_0) as the slope of the tangent line to the curve of intersection of the surface $z = f(x, y)$ and the plane $y = y_0$, and $\partial f/\partial y$ as the slope of the tangent line to the curve of intersection with $x = x_0$. It follows, then, that (x_0, y_0) is critical if both curves have horizontal tangent lines or if either curve has a vertical tangent line or no tangent line at all. Alternatively, recall that the equation of the tangent plane to the surface $z = f(x, y)$ at (x_0, y_0) is

$$z - z_0 = f_x(x_0, y_0)(x - x_0) + f_y(x_0, y_0)(y - y_0)$$

(see Exercise 29 in Section 12.9). If both partial derivatives vanish, then the tangent plane is horizontal with equation $z = z_0$. For example, at each of the critical points in Figures 12.30–12.33, $\partial f/\partial x = \partial f/\partial y = 0$ and the tangent plane is horizontal. The remaining functions in Figures 12.34–12.38 have critical points at which either $\partial f/\partial x$ or $\partial f/\partial y$ or both do not exist. In Figures 12.34–12.37, the surfaces do not have tangent planes at critical points, and in Figure 12.38, the tangent plane is vertical at each critical point. Consequently, (x_0, y_0) is a critical point of a function $f(x, y)$ if at (x_0, y_0) the surface $z = f(x, y)$ has a horizontal tangent plane, a vertical tangent plane, or no tangent plane at all.

FIGURE 12.30   Tangent plane horizontal at critical point $(0, 0)$

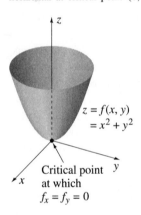

$z = f(x, y) = x^2 + y^2$

Critical point at which $f_x = f_y = 0$

FIGURE 12.31   Tangent plane horizontal at critical point $(0, 1)$

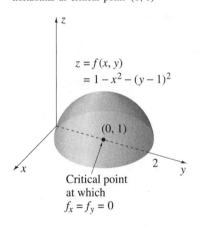

$z = f(x, y) = 1 - x^2 - (y - 1)^2$

$(0, 1)$

Critical point at which $f_x = f_y = 0$

FIGURE 12.32   Tangent plane horizontal at critical point $(0, 0)$

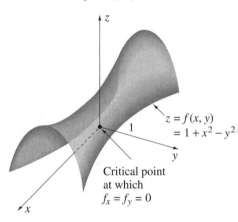

$z = f(x, y) = 1 + x^2 - y^2$

Critical point at which $f_x = f_y = 0$

FIGURE 12.33   Tangent plane horizontal at critical points $(x, 0)$

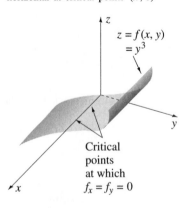

$z = f(x, y) = y^3$

Critical points at which $f_x = f_y = 0$

FIGURE 12.34   No tangent plane at critical point $(1, 0)$

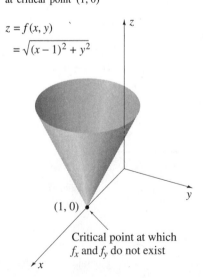

$z = f(x, y) = \sqrt{(x - 1)^2 + y^2}$

$(1, 0)$

Critical point at which f_x and f_y do not exist

FIGURE 12.35   No tangent plane at critical point (x, x)

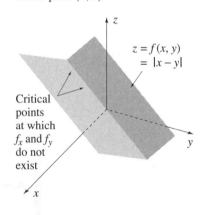

$z = f(x, y) = |x - y|$

Critical points at which f_x and f_y do not exist

FIGURE 12.36 No tangent plane at critical point $(0, 0)$

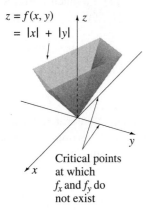

$z = f(x, y)$
$= |x| + |y|$

Critical points at which f_x and f_y do not exist

FIGURE 12.37 No tangent plane at critical point $(0, 0)$

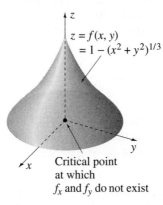

$z = f(x, y)$
$= 1 - (x^2 + y^2)^{1/3}$

Critical point at which f_x and f_y do not exist

FIGURE 12.38 Vertical tangent planes at critical points $(x, 1)$

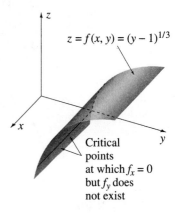

$z = f(x, y) = (y - 1)^{1/3}$

Critical points at which $f_x = 0$ but f_y does not exist

EXAMPLE 12.29

Find all critical points for the function

$$f(x, y) = x^2 y - 2xy^2 + 3xy + 4.$$

SOLUTION   For critical points, we first solve

$$0 = \frac{\partial f}{\partial x} = 2xy - 2y^2 + 3y = y(2x - 2y + 3),$$

$$0 = \frac{\partial f}{\partial y} = x^2 - 4xy + 3x = x(x - 4y + 3).$$

To satisfy these two equations simultaneously, there are four possibilities:

1. $x = 0$, $y = 0$, which gives the critical point $(0, 0)$;
2. $y = 0$, $x - 4y + 3 = 0$, which gives the critical point $(-3, 0)$;
3. $x = 0$, $2x - 2y + 3 = 0$, which gives the critical point $(0, 3/2)$;
4. $2x - 2y + 3 = 0$, $x - 4y + 3 = 0$, which gives the critical point $(-1, 1/2)$.

Since $\partial f/\partial x$ and $\partial f/\partial y$ are defined for all x and y, these are the only critical points. The plot of the surface $z = f(x, y)$ for $-10 \le x \le 10$, $-10 \le y \le 10$ in Figure 12.39 does not really illustrate the critical points. In other words, computer plots of functions of two variables are not as helpful in determining critical points of functions as were plots of functions of one variable in Chapter 4.

FIGURE 12.39   Computer plot of $z = x^2 y - 2xy^2 + 3xy + 4$

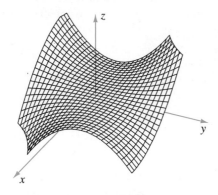

Critical points for functions of more than two independent variables can be defined algebraically, but because we have no geometric representation for such functions, there is no geometric interpretation for their critical points. For example, if $f(x, y, z, t)$ is a function of independent variables x, y, z, and t, then (x_0, y_0, z_0, t_0) is a critical point of $f(x, y, z, t)$ if all four of its first-order partial derivatives vanish at (x_0, y_0, z_0, t_0),

$$\left.\frac{\partial f}{\partial x}\right|_{(x_0,y_0,z_0,t_0)} = \left.\frac{\partial f}{\partial y}\right|_{(x_0,y_0,z_0,t_0)} = \left.\frac{\partial f}{\partial z}\right|_{(x_0,y_0,z_0,t_0)} = \left.\frac{\partial f}{\partial t}\right|_{(x_0,y_0,z_0,t_0)} = 0, \qquad (12.55)$$

or if at least one of the partial derivatives does not exist at the point. Note that because the partial derivatives of a function are the components of the gradient of the function, we can say that a critical point of a function is a point at which its gradient is either equal to zero or undefined.

EXAMPLE 12.30

Find all critical points for the function

$$f(x, y, z) = xyz\sqrt{x^2 + y^2 + z^2}.$$

SOLUTION   For critical points, we consider the equations

$$0 = \frac{\partial f}{\partial x} = yz\sqrt{x^2 + y^2 + z^2} + \frac{x^2yz}{\sqrt{x^2 + y^2 + z^2}}$$

$$= \frac{yz}{\sqrt{x^2 + y^2 + z^2}}(2x^2 + y^2 + z^2),$$

$$0 = \frac{\partial f}{\partial y} = xz\sqrt{x^2 + y^2 + z^2} + \frac{xy^2z}{\sqrt{x^2 + y^2 + z^2}}$$

$$= \frac{xz}{\sqrt{x^2 + y^2 + z^2}}(x^2 + 2y^2 + z^2),$$

$$0 = \frac{\partial f}{\partial z} = xy\sqrt{x^2 + y^2 + z^2} + \frac{xyz^2}{\sqrt{x^2 + y^2 + z^2}}$$

$$= \frac{xy}{\sqrt{x^2 + y^2 + z^2}}(x^2 + y^2 + 2z^2).$$

The partial derivatives are clearly undefined for $x = y = z = 0$, and therefore the origin $(0, 0, 0)$ is a critical point. If x, y, and z are not all zero, then the terms in parentheses cannot vanish, and we must set

$$yz = 0, \quad xz = 0, \quad xy = 0.$$

If any two of x, y, and z vanish, but the third does not, then these equations are satisfied. In other words, every point on the x-axis, every point on the y-axis, and every point on the z-axis is critical.

We now turn our attention to the classification of critical points of a function $f(x, y)$ of two independent variables. Critical points $(0, 1)$ in Figure 12.31 and $(0, 0)$ in Figure 12.37 yield "high" points on the surfaces. We describe this property in the following definition.

DEFINITION 12.8

A function $f(x, y)$ is said to have a **relative maximum** $f(x_0, y_0)$ at a point (x_0, y_0) if there exists a circle in the xy-plane centred at (x_0, y_0) such that for all points (x, y) inside this circle

$$f(x, y) \leq f(x_0, y_0). \tag{12.56}$$

The "low" points on the surfaces at $(0, 0)$ in Figure 12.30, $(1, 0)$ in Figure 12.34, and $(0, 0)$ in Figure 12.36 are relative minima according to the following.

DEFINITION 12.9

A function $f(x, y)$ is said to have a **relative minimum** $f(x_0, y_0)$ at a point (x_0, y_0) if there exists a circle in the xy-plane centred at (x_0, y_0) such that for all points (x, y) inside this circle

$$f(x, y) \geq f(x_0, y_0). \tag{12.57}$$

Every critical point in Figure 12.35 yields a relative minimum of $f(x, x) = 0$.

DEFINITION 12.10

If a critical point of a function $f(x, y)$ at which $\partial f/\partial x = \partial f/\partial y = 0$ yields neither a relative maximum nor a relative minimum, it is said to yield a **saddle point**.

The critical point $(0, 0)$ in Figure 12.32 therefore gives a saddle point, as does each of the critical points in Figure 12.33. Saddle points for surfaces $z = f(x, y)$ are clearly the analogues of horizontal points of inflection for curves $y = f(x)$. In both cases the derivative(s) of the function vanishes but there is neither a relative maximum nor a relative minimum.

The critical points in Figure 12.36 [except $(0, 0)$] are the counterparts of corners for the graph of a function $f(x)$. They are points at which one or both of the partial derivatives of $f(x, y)$ do not exist, but like corners for $f(x)$, they do not necessarily yield relative extrema. Critical points in Figure 12.38 are the analogues of vertical points of inflection for a function $f(x)$.

Our discussion has made it clear that:

(a) At a relative maximum or minimum of $f(x, y)$, either $\partial f/\partial x$ and $\partial f/\partial y$ both vanish, or one or both of the partial derivatives do not exist.

(b) Saddle points may also occur where $\partial f/\partial x = \partial f/\partial y = 0$, and points where the derivatives do not exist may fail to yield relative extrema.

In other words, every relative extremum of $f(x, y)$ occurs at a critical point, but critical points do not always give relative extrema.

Given the problem of determining all relative maxima and minima of a function $f(x, y)$, we should first find its critical points. But how do we decide whether these critical points yield relative maxima, relative minima, saddle points, or none of these? We do not have a practical test that is equivalent to the first-derivative test for functions of one variable, but we do have a test that corresponds to the second-derivative test. For functions of two independent variables the situation is more complicated, however, since there are three second-order partial derivatives, but the idea of the test is essentially the same. It determines whether certain curves are concave upward or concave downward at the critical point. The complete result is contained in the following theorem.

THEOREM 12.7

Suppose (x_0, y_0) is a critical point of $f(x, y)$ at which $\partial f/\partial x$ and $\partial f/\partial y$ both vanish. Suppose further that f_x, f_y, f_{xx}, f_{xy}, and f_{yy} are all continuous at (x_0, y_0). Define

$$A = f_{xx}(x_0, y_0), \quad B = f_{xy}(x_0, y_0), \quad C = f_{yy}(x_0, y_0).$$

If:

    (i) $B^2 - AC < 0$ and $A < 0$, then $f(x, y)$ has a relative maximum at (x_0, y_0);

    (ii) $B^2 - AC < 0$ and $A > 0$, then $f(x, y)$ has a relative minimum at (x_0, y_0);

    (iii) $B^2 - AC > 0$, then $f(x, y)$ has a saddle point at (x_0, y_0);

    (iv) $B^2 - AC = 0$, no conclusion can be made.

PROOF

(i) Suppose we intersect the surface $z = f(x, y)$ with a plane parallel to the z-axis, through the point $(x_0, y_0, 0)$, and making an angle α with the line through $(x_0, y_0, 0)$ parallel to the positive x-axis (Figure 12.40). The slope of the curve of intersection of these surfaces at the point $(x_0, y_0, f(x_0, y_0))$ is given by the directional derivative

$$D_{\mathbf{v}} f_{|(x_0, y_0)} = \left.\frac{\partial f}{\partial x}\right|_{(x_0, y_0)} \cos\alpha + \left.\frac{\partial f}{\partial y}\right|_{(x_0, y_0)} \sin\alpha$$

(see equation 12.44). Since (x_0, y_0) is a critical point at which $\nabla f = \mathbf{0}$, it follows that

$$D_{\mathbf{v}} f_{|(x_0, y_0)} = 0 \qquad \text{for all } \alpha.$$

In Figure 12.40 we have illustrated the critical point as a relative maximum. But how do we verify that this is indeed the case? If we can show that each and every curve of intersection of the surface with a vertical plane through $(x_0, y_0, 0)$ is concave downward at $(x_0, y_0, 0)$, then (x_0, y_0) must give a relative maximum. But to discuss concavity of a curve, we require the second derivative — in this case, the second directional derivative of $f(x, y)$. According to equation 12.45, the second directional derivative of $f(x, y)$ at (x_0, y_0) in the direction $\hat{\mathbf{v}} = (\cos\alpha, \sin\alpha)$ is

$$D_{\mathbf{v}}(D_{\mathbf{v}} f) = \left.\frac{\partial^2 f}{\partial x^2}\right|_{(x_0, y_0)} \cos^2\alpha + 2\left.\frac{\partial^2 f}{\partial x \partial y}\right|_{(x_0, y_0)} \cos\alpha \sin\alpha + \left.\frac{\partial^2 f}{\partial y^2}\right|_{(x_0, y_0)} \sin^2\alpha$$

$$= A \cos^2\alpha + 2B \cos\alpha \sin\alpha + C \sin^2\alpha,$$

FIGURE 12.40   Critical point yielding a relative maximum for the function

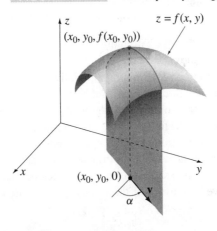

where we understand that here $D_{\mathbf{v}}(D_{\mathbf{v}}f)$ is implicitly suffixed by (x_0, y_0). In order, therefore, to verify that (x_0, y_0) gives a relative maximum, it is sufficient to show that $D_{\mathbf{v}}(D_{\mathbf{v}}f)$ is negative for each value of α in the interval $0 \leq \alpha < 2\pi$. However, because $D_{\mathbf{v}}(D_{\mathbf{v}}f)$ is unchanged if α is replaced by $\alpha + \pi$, it is sufficient to verify that $D_{\mathbf{v}}(D_{\mathbf{v}}f)$ is negative for $0 \leq \alpha < \pi$.

For any of these values of α except $\pi/2$, we can write

$$D_{\mathbf{v}}(D_{\mathbf{v}}f) = \cos^2 \alpha (A + 2B \tan \alpha + C \tan^2 \alpha),$$

and if we set $u = \tan \alpha$,

$$D_{\mathbf{v}}(D_{\mathbf{v}}f) = \cos^2 \alpha (A + 2Bu + Cu^2).$$

It is evident that $D_{\mathbf{v}}(D_{\mathbf{v}}f) < 0$ for all $\alpha \neq \pi/2$ if and only if

$$Q(u) = A + 2Bu + Cu^2 < 0 \qquad \text{for } -\infty < u < \infty.$$

Were we to draw a graph of the quadratic $Q(u)$, we would see that it crosses the u-axis where

$$u = \frac{-2B \pm \sqrt{4B^2 - 4AC}}{2C} = \frac{-B \pm \sqrt{B^2 - AC}}{C}.$$

But because $B^2 - AC < 0$, there are no real solutions of this equation, and therefore $Q(u)$ never crosses the u-axis. Since $Q(0) = A < 0$, it follows that $Q(u) < 0$ for all u. We have shown, then, that

$$D_{\mathbf{v}}(D_{\mathbf{v}}f) < 0 \qquad \text{for all } \alpha \neq \pi/2.$$

When $\alpha = \pi/2$, $D_{\mathbf{v}}(D_{\mathbf{v}}f) = C$. Since $B^2 - AC < 0$ and $A < 0$, it follows that $C < 0$ also. Consequently, if $B^2 - AC < 0$ and $A < 0$, then $D_{\mathbf{v}}(D_{\mathbf{v}}f) < 0$ for all α, and (x_0, y_0) yields a relative maximum.

(ii) If $B^2 - AC < 0$ and $A > 0$, a similar argument leads to the conclusion that (x_0, y_0) yields a relative minimum; the only difference is that inequalities are reversed.

(iii) If $B^2 - AC > 0$, then $Q(u)$ has real distinct zeros, in which case $Q(u)$ is sometimes negative and sometimes positive. This means that the curve of intersection is sometimes concave upward and sometimes concave downward, and the point (x_0, y_0) therefore gives a saddle point.

(iv) If $B^2 - AC = 0$, the classification of the point determined by (x_0, y_0) depends on which of A, B, and C vanish, if any. ∎

To illustrate that we can obtain a relative maximum, a relative minimum, or a saddle point for a critical point at which $B^2 - AC = 0$, consider the three functions $f(x, y) = -y^2$, $f(x, y) = y^2$, and $f(x, y) = y^3$ in Figures 12.41–12.43. The point $(0, 0)$ is a critical point for each function, and at this point $B^2 - AC = 0$. Yet $(0, 0)$ yields a relative maximum for $f(x, y) = -y^2$, a relative minimum for $f(x, y) = y^2$, and a saddle point for $f(x, y) = y^3$. In fact, every point on the x-axis is a relative maximum for $f(x, y) = -y^2$, a relative minimum for $f(x, y) = y^2$, and a saddle point for $f(x, y) = y^3$.

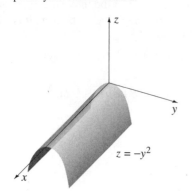

FIGURE 12.41   Critical points yield relative maxima

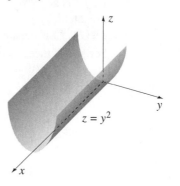

FIGURE 12.42   Critical points yield relative minima

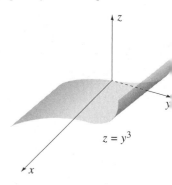

FIGURE 12.43   Critical points yield saddle points

EXAMPLE 12.31

Find and classify critical points for each of the following functions as yielding relative maxima, relative minima, saddle points, or none of these:

(a) $f(x, y) = 4xy - x^4 - y^4$

(b) $f(x, y) = x^4 y^3$

SOLUTION

(a) Critical points of $f(x, y)$ are given by

$$0 = \frac{\partial f}{\partial x} = 4y - 4x^3, \quad 0 = \frac{\partial f}{\partial y} = 4x - 4y^3.$$

Solutions of these equations are $(0, 0)$, $(1, 1)$, and $(-1, -1)$. We now calculate

$$\frac{\partial^2 f}{\partial x^2} = -12x^2, \quad \frac{\partial^2 f}{\partial x \partial y} = 4, \quad \frac{\partial^2 f}{\partial y^2} = -12y^2.$$

We could tabulate results to determine the nature of the critical points.

TABLE 12.1

Critical point	A	B	C	$B^2 - AC$	Nature
$(0, 0)$	0	4	0	16	Saddle point
$(1, 1)$	-12	4	-12	-128	Relative maximum
$(-1, -1)$	-12	4	-12	-128	Relative maximum

FIGURE 12.44   Diagram to show sign of function $f(x, y) = x^4 y^3$

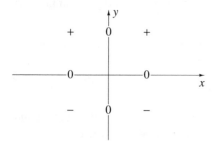

(b) For critical points we solve

$$0 = \frac{\partial f}{\partial x} = 4x^3 y^3, \quad 0 = \frac{\partial f}{\partial y} = 3x^4 y^2.$$

Every point on the x- and y-axes is critical, and at each of these points $f(x, y) = 0$. The second-derivative test fails to classify these critical points. Figure 12.44 shows a value of zero for the function on the axes and the sign of $f(x, y)$ in the four quadrants. It implies that the points $(0, y)$ for $y > 0$ yield relative minima; $(0, y)$ for $y < 0$ yield relative maxima; and $(x, 0)$ yield saddle points.

EXAMPLE 12.32

It is straightforward to verify that $(0, 0)$ is the only critical point of the function $f(x, y) = x^2 - 6xy^2 + y^4$ and that the quantity $B^2 - AC$ of Theorem 12.7 is equal to zero at this critical point. Show graphically and algebraically that the critical point gives a saddle point.

SOLUTION The value of the function at $(0, 0)$ is $f(0, 0) = 0$. There is a relative minimum at $(0, 0)$ if $f(x, y) \geq 0$ in some circle around $(0, 0)$; there is a relative maximum if $f(x, y) \leq 0$ in some such circle; and there is a saddle point if $f(x, y)$ takes on negative and positive values in every circle centred at $(0, 0)$. The plot in Figure 12.45 indicates that the last situation prevails. To show this algebraically, we first note that values of $f(x, y)$ are positive along the x-axis and the y-axis away from the origin. On the parabola $x = y^2$, values of the function are

$$(y^2)^2 - 6(y^2)y^2 + y^4 = -4y^4 \leq 0.$$

Thus, $(0, 0)$ yields a saddle point.

FIGURE 12.45   For $x^2 - 6xy^2 + y^4$, $(0, 0)$ yields a saddle point

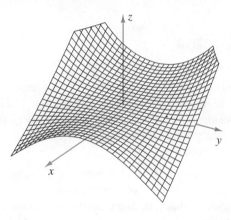

This completes our discussion of relative extrema of functions of two independent variables. Our next step should be to extend the theory to functions of more than two variables. It is a simple matter to give definitions of relative maxima and minima for such functions; they are almost identical to Definitions 12.8 and 12.9 (see Exercise 19). On the other hand, to develop a theorem for functions of more than two independent variables that is analogous to Theorem 12.7 is beyond the scope of this book. We refer the interested reader to more advanced books.

EXERCISES 12.10

In Exercises 1–14 find all critical points for the function and classify each as yielding a relative maximum, a relative minimum, a saddle point, or none of these.

1. $f(x, y) = x^2 + 2xy + 2y^2 - 6y$

2. $f(x, y) = 3xy - x^3 - y^3$

3. $f(x, y) = x^3 - 3x + y^2 + 2y$

4. $f(x, y) = x^2y^2 + 3x$

5. $f(x, y) = xy - x^2 + y^2$

6. $f(x, y) = x \sin y$

7. $f(x, y) = xye^{-(x^2+y^2)}$

8. $f(x, y) = x^2 - 2xy + y^2$

9. $f(x, y) = (x^2 + y^2)^{2/3}$

10. $f(x, y) = x^4y^3$

11. $f(x, y) = 2xy^2 + 3xy + x^2y^3$

 * **12.** $f(x, y) = |x| + y^2$

 * **13.** $f(x, y) = (1 - x)(1 - y)(x + y - 1)$

 * **14.** $f(x, y) = x^4 + y^4 - x^2 - y^2 + 1$

In Exercises 15–18 find all critical points for the function.

15. $f(x, y, z) = x^2 + y^2 - z^2 + 3x - 2y + 5$

16. $f(x, y, z, t) = x^2y^2z^2 + t^2x^2 + 3x$

17. $f(x, y, z) = xyz + x^2yz - y$

18. $f(x, y, z) = xyze^{x^2+y^2+z^2}$

* **19.** Give definitions for a relative maximum and a relative minimum for a function $f(x, y, z)$ at a point (x_0, y_0, z_0).

* **20.** Suppose that $f(x, y)$ is harmonic in the region $D : x^2 + y^2 < 1$. Show that $f(x, y)$ cannot have a relative maximum or minimum at any point in D at which either f_{xx} or f_{xy} does not vanish.

* **21.** Find and classify the critical points for the function $f(x, y) = y^2 - 4x^2y + 3x^4$.

* **22.** Find and classify the critical points of $f(x, y) = x^4 + 3xy^2 + y^2$ as yielding relative maxima, relative minima, or saddle points.

* **23.** The equation $2x^2 + 3y^2 + z^2 - 12xy + 4xz = 35$ defines function $z = f(x, y)$. Show that the point $x = 1$ and $y = 2$ is a critical point for the function with value 5 at $(1, 2)$. Does it yield a relative extremum for the function?

** **24.** (a) Show that the function $f(x, y, z) = x^2 + y^2 + z^2 - xyz$ has a critical point $(0, 0, 0)$. What are the other critical points?

(b) Use the definition in Exercise 19 to show that $f(x, y, z)$ has a relative minimum at $(0, 0, 0)$.

12.11 Absolute Maxima and Minima

Absolute maxima and minima are more important than relative maxima and minima when it comes to applications. In this section and in Section 12.12 we discuss the theory of absolute extrema and consider a number of applications. Once again we begin with functions $f(x, y)$ of two independent variables and base our discussion on the theory of absolute extrema for functions of one variable.

We learned in Section 4.7 that a function $f(x)$ that is continuous on a finite interval $a \leq x \leq b$ must have an absolute maximum and an absolute minimum on that interval. Furthermore, these absolute extrema must occur at either critical points or at the ends $x = a$ and $x = b$ of the interval. Consequently, to find the absolute extrema of a function $f(x)$, we evaluate $f(x)$ at all critical points, at $x = a$, and at $x = b$; the largest of these numbers is the absolute maximum of $f(x)$ on $a \leq x \leq b$, and the smallest is the absolute minimum.

The procedure is much the same for a function $f(x, y)$ that is continuous on a region R that is finite and includes all the points on its boundary. First, however, we define exactly what we mean by absolute extrema of $f(x, y)$ and consider a number of simple examples. We will then be able to make general statements about the nature of all absolute extrema, and proceed to the important area of applications.

> **DEFINITION 12.11**
>
> The **absolute maximum** of a function $f(x, y)$ on a region R is $f(x_0, y_0)$ if (x_0, y_0) is in R and
>
> $$f(x, y) \leq f(x_0, y_0) \qquad (12.58)$$
>
> for all (x, y) in R. The **absolute minimum** of $f(x, y)$ on R is $f(x_0, y_0)$ if (x_0, y_0) is in R and
>
> $$f(x, y) \geq f(x_0, y_0) \qquad (12.59)$$
>
> for all (x, y) in R.

In Figures 12.46–12.51, we have shown six functions defined on the circle $R : x^2 + y^2 \leq 1$. The absolute maxima and minima of these functions for this region are shown in Table 12.2.

FIGURE 12.46   Absolute maximum = 1, Absolute minimum = 0

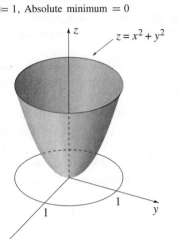

FIGURE 12.47   Absolute maximum = 2, Absolute minimum = 0

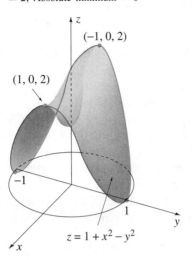

FIGURE 12.48   Absolute maximum = 2, Absolute minimum = 0

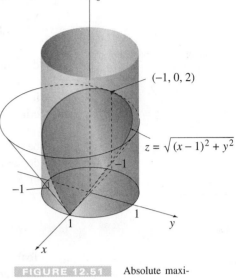

FIGURE 12.49   Absolute maximum = 4, Absolute minimum = 2

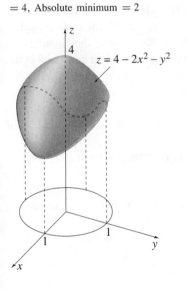

FIGURE 12.50   Absolute maximum = 1, Absolute minimum = 0

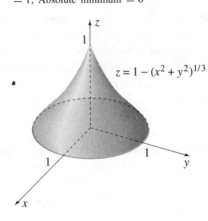

FIGURE 12.51   Absolute maximum = $\sqrt{2}$, Absolute minimum = 0

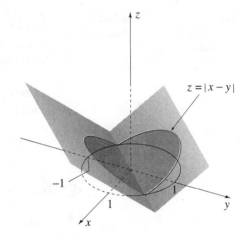

TABLE 12.2

Function $f(x, y)$	Position of absolute maximum	Value of absolute maximum	Position of absolute minimum	Value of absolute minimum
$x^2 + y^2$	Every point on $x^2 + y^2 = 1$	1	$(0, 0)$	0
$1 + x^2 - y^2$	$(\pm 1, 0)$	2	$(0, \pm 1)$	0
$\sqrt{(x-1)^2 + y^2}$	$(-1, 0)$	2	$(1, 0)$	0
$4 - 2x^2 - y^2$	$(0, 0)$	4	$(\pm 1, 0)$	2
$1 - (x^2 + y^2)^{1/3}$	$(0, 0)$	1	Every point on $x^2 + y^2 = 1$	0
$\|x - y\|$	$(\pm 1/\sqrt{2}, \mp 1/\sqrt{2})$,	$\sqrt{2}$	Every point on $y = x$, $-1/\sqrt{2} \le x \le 1/\sqrt{2}$	0

For each of the functions in these figures, *absolute extrema occur at either a critical point or a point on the boundary of R.* This result is true for any *continuous function defined on a finite region that includes all the points on its boundary.* Although this result may seem fairly obvious geometrically, to prove it analytically is very difficult; we will be content to assume its validity and carry on from there.

Sometimes a drawing or plot of the surface defined by a function makes absolute maxima and minima clear. This may not always be the case, however, and we therefore turn our attention to determining absolute extrema algebraically. Suppose, then, that a continuous function $f(x, y)$ is given and we are required to find its absolute extrema on a finite region R (which includes its boundary points). The previous discussion indicated that the extrema must occur either at critical points or on the boundary of R. Consequently, we should first determine all critical points of $f(x, y)$ in R, and evaluate $f(x, y)$ at each of these points. These values should now be compared to the maximum and minimum values of $f(x, y)$ on the boundary of R. But how do we find the maximum and minimum values of $f(x, y)$ on the boundary? If the boundary of R is denoted by C (Figure 12.52), and if C has parametric equations $x = x(t)$, $y = y(t)$, $\alpha \le t \le \beta$, then on C we can express $f(x, y)$ in terms of t, and t alone:

$$f[x(t), y(t)], \quad \alpha \le t \le \beta.$$

To find the maximum and minimum values of $f(x, y)$ on C is now an absolute extrema problem for a function of one variable. The function $f[x(t), y(t)]$ should therefore be evaluated at each of its critical points and at $t = \alpha$ and $t = \beta$. A plot of $f[x(t), y(t)]$ could be valuable here.

If the boundary of R consists of a number of curves (Figure 12.53), then this boundary procedure must be performed for each part. In other words, on each part of the boundary we express $f(x, y)$ as a function of one variable, and then evaluate this function at its critical points and at the ends of that part of the boundary to which it applies.

The absolute maximum of $f(x, y)$ on R is then the largest of all values of $f(x, y)$ evaluated at the critical points inside R, the critical points on the boundary of R, and the endpoints of each part of the boundary. The absolute minimum of $f(x, y)$ on R is the smallest of all these values.

Recall that to find the absolute extrema of a function $f(x)$, continuous on $a \le x \le b$, we evaluate $f(x)$ at all critical points and at the boundary points $x = a$ and $x = b$. The procedure that we have established here for $f(x, y)$ is much the same — the difference is that, for $f(x, y)$, the boundary consists not of two points, but of entire curves. Evaluation of $f(x, y)$ on the boundary therefore reduces to one or more extrema problems for functions of one variable. Note too that for $f(x, y)$ [or $f(x)$], it is not necessary to determine the nature of the critical points; it is necessary only to evaluate $f(x, y)$ at these points.

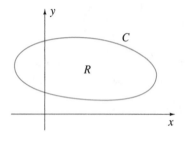

FIGURE 12.52   Absolute extrema of a function continuous on a region R and its boundary

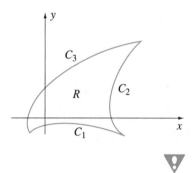

FIGURE 12.53   Boundary of a region may consist of more than one curve

EXAMPLE 12.33

Find the maximum value of the function $z = f(x, y) = 4xy - x^4 - 2y^2$ on the region R: $-2 \le x \le 2$, $-2 \le y \le 2$.

SOLUTION   A plot of the function in Figure 12.54 suggests that the maximum value occurs at a critical point in the first and third quadrants, or on the edges of the square. We confirm this with the procedure outlined above. Critical points of $f(x, y)$ are given by

$$0 = \frac{\partial f}{\partial x} = 4y - 4x^3, \quad 0 = \frac{\partial f}{\partial y} = 4x - 4y.$$

Solutions of these equations are $(0, 0)$, $(1, 1)$, and $(-1, -1)$, and the values of $f(x, y)$ at these critical points are

$$f(0, 0) = \boxed{0}, \quad f(1, 1) = \boxed{1}, \quad f(-1, -1) = \boxed{1}.$$

We denote the four parts of the boundary of R by C_1, C_2, C_3, and C_4 (Figure 12.55).

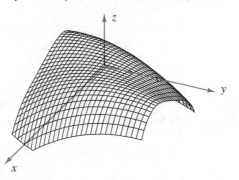

FIGURE 12.54 Computer plot of $4xy - x^4 - 2y^2$ on $-2 \le x \le 2, -2 \le y \le 2$

FIGURE 12.55 Region and its four bounding curves

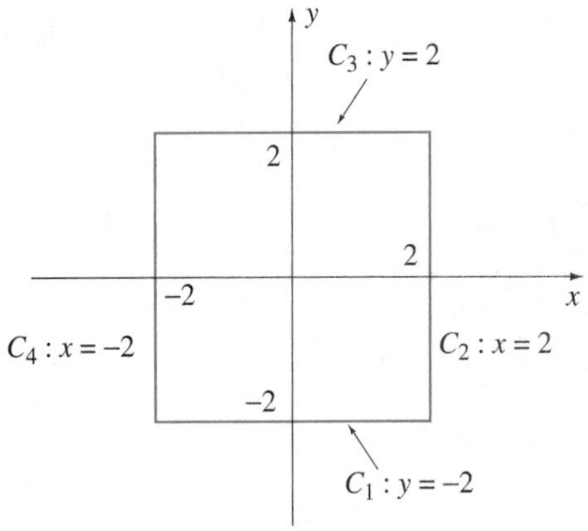

On C_1, $y = -2$, in which case

$$z = -8x - x^4 - 8, \quad -2 \le x \le 2.$$

For critical points of this function, we solve

$$0 = \frac{dz}{dx} = -8 - 4x^3.$$

The only solution is $x = -2^{1/3}$, at which the value of z is

$$z = 8 \cdot 2^{1/3} - 2^{4/3} - 8 = \boxed{-0.44}.$$

On C_2, $x = 2$, in which case

$$z = 8y - 16 - 2y^2, \quad -2 \le y \le 2.$$

Critical points are defined by

$$0 = \frac{dz}{dy} = 8 - 4y.$$

The only solution $y = 2$ defines one of the corners of the square, and at this point

$$z = \boxed{-8}.$$

On C_3, $y = 2$ and

$$z = 8x - x^4 - 8, \quad -2 \le x \le 2.$$

For critical points, we solve

$$0 = \frac{dz}{dx} = 8 - 4x^3.$$

At the single point $x = 2^{1/3}$,

$$z = 8 \cdot 2^{1/3} - 2^{4/3} - 8 = \boxed{-0.44}.$$

On the final curve C_4, $x = -2$ and

$$z = -8y - 16 - 2y^2, \quad -2 \le y \le 2.$$

Critical points are given by

$$0 = \frac{dz}{dy} = -8 - 4y.$$

The solution $y = -2$ defines another corner of the square at which

$$z = \boxed{-8}.$$

We have now evaluated $f(x, y)$ at all critical points inside R and at all critical points on the four parts of the boundary of R. It remains only to evaluate $f(x, y)$ at the corners of the square. Two corners have already been accounted for; the other two give

$$f(2, -2) = \boxed{-40}, \quad f(-2, 2) = \boxed{-40}.$$

The largest value of $f(x, y)$ is the largest of the numbers in the boxes, namely, 1, and this is therefore the maximum value of $f(x, y)$ on R.

EXAMPLE 12.34

Temperature in degrees Celsius at each point (x, y) in a semicircular plate defined by $x^2 + y^2 \le 1, y \ge 0$ is given by

$$T(x, y) = 16x^2 - 24xy + 40y^2.$$

Find the hottest and coldest points in the plate.

SOLUTION   For critical points of $T(x, y)$, we solve

$$0 = \frac{\partial T}{\partial x} = 32x - 24y, \quad 0 = \frac{\partial T}{\partial y} = -24x + 80y.$$

The only solution of these equations, $(0, 0)$, is on the boundary. On the upper edge of the plate (Figure 12.56), we set $x = \cos t$, $y = \sin t$, $0 \le t \le \pi$, in which case

$$T = 16\cos^2 t - 24\cos t \sin t + 40\sin^2 t, \quad 0 \le t \le \pi.$$

A plot of this function in Figure 12.57 shows the absolute maximum at the relative maximum and the absolute minimum at the relative minimum. To locate them we solve for critical points of this function:

$$0 = \frac{dT}{dt} = -32\cos t \sin t - 24(-\sin^2 t + \cos^2 t) + 80\sin t \cos t$$

$$= 24(\sin 2t - \cos 2t).$$

FIGURE 12.56   Maximum and minimum temperatures on a semicircular domain

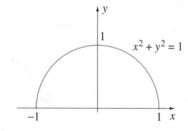

FIGURE 12.57   Plot of temperature on semicircular part of boundary

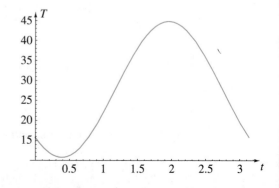

If we divide by $\cos 2t$ (since $\cos 2t = 0$ does not lead to a solution of this equation), we have

$$\tan 2t = 1.$$

The only solutions of this equation in the interval $0 \le t \le \pi$ are $t = \pi/8$ and $t = 5\pi/8$. When $t = \pi/8$, $T = \boxed{11.0}$; and when $t = 5\pi/8$, $T = \boxed{45.0}$. At the ends of this part of the boundary, $t = 0$ and $t = \pi$, and $T(1, 0) = \boxed{16}$ and $T(-1, 0) = \boxed{16}$. On the lower edge of the plate, $y = 0$, in which case

$$T = 16x^2, \quad -1 \le x \le 1.$$

The only critical point of this function is $x = 0$, at which $T = \boxed{0}$. The hottest point in the plate is therefore $(\cos(5\pi/8), \sin(5\pi/8)) = (-0.38, 0.92)$, where the temperature is $45.0°$ C, and the coldest point is $(0, 0)$ with temperature $0°$ C.

EXAMPLE 12.35

Find the point on the first octant part of the plane $6x + 3y + 4z = 6$ closest to the point $(4, 6, 7)$. Assume that lines of intersection of the plane with the coordinate planes are part of the surface.

FIGURE 12.58   Minimum distance from $(4, 6, 7)$ to plane $6x + 3y + 4z = 6$ in first octant

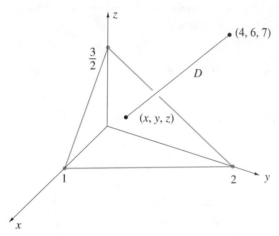

SOLUTION   The distance D from $(4, 6, 7)$ to any point (x, y, z) in space is defined by

$$D^2 = (x - 4)^2 + (y - 6)^2 + (z - 7)^2.$$

We must minimize D, but consider only points (x, y, z) that satisfy the equation of the plane and lie in the first octant (Figure 12.58). At the moment, D^2 is a function of three variables x, y, and z, but they are not all independent because of the planar restriction. If we solve the

FIGURE 12.59   Triangular domain for distance function

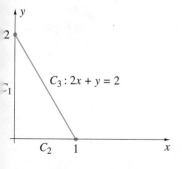

equation of the plane for z in terms of x and y, and substitute,

$$D^2 = f(x, y) = (x - 4)^2 + (y - 6)^2 + \left(\frac{6 - 6x - 3y}{4} - 7 \right)^2$$

$$= (x - 4)^2 + (y - 6)^2 + \left(\frac{6x + 3y + 22}{4} \right)^2,$$

where x and y are independent variables. Now D is minimized when D^2 is minimized, and we therefore find the point (x, y) that minimizes D^2. The values of x and y that yield points on the first octant part of the plane are those in the triangle of Figure 12.59.

For critical points of D^2, we solve

$$0 = \frac{\partial f}{\partial x} = 2(x - 4) + \frac{3}{4}(6x + 3y + 22),$$

$$0 = \frac{\partial f}{\partial y} = 2(y - 6) + \frac{3}{8}(6x + 3y + 22).$$

The solution of these equations is $(-140/61, 174/61)$, an unacceptable point since it does not lie in the triangle of Figure 12.59. The point on the triangle in Figure 12.58 closest to $(4, 6, 7)$ therefore lies along one of the edges of the triangle. We can find it by minimizing $f(x, y)$ along the edges of the triangle in Figure 12.59. On C_1, $x = 0$ in which case

$$D^2 = F(y) = f(0, y) = 16 + (y - 6)^2 + \left(\frac{3y + 22}{4}\right)^2, \quad 0 \le y \le 2.$$

For critical points we solve

$$0 = F'(y) = 2(y - 6) + \frac{3}{8}(3y + 22) \qquad \Longrightarrow \qquad y = \frac{6}{5}.$$

The value of $F(y)$ here is $F(6/5) = \boxed{80}$. We will evaluate D^2 at the corners of the triangle later. On C_2, $y = 0$ in which case

$$D^2 = G(x) = f(x, 0) = (x - 4)^2 + 36 + \left(\frac{3x + 11}{2}\right)^2, \quad 0 \le x \le 1.$$

For critical points we solve

$$0 = G'(x) = 2(x - 4) + \frac{3}{2}(3x + 11) \qquad \Longrightarrow \qquad x = -\frac{17}{13},$$

an unacceptable value. On C_3, $2x + y = 2$, in which case

$$D^2 = H(x) = (x - 4)^2 + (2 - 2x - 6)^2 + \left(\frac{6x + 6 - 6x + 22}{4}\right)^2$$

$$= (x - 4)^2 + 4(x + 2)^2 + 49, \qquad 0 \le x \le 1.$$

For critical points, we solve

$$0 = H'(x) = 2(x - 4) + 8(x + 2) \qquad \Longrightarrow \qquad x = -\frac{4}{5},$$

again an unacceptable value. We now evaluate D^2 at the corners of the triangle in Figure 12.59.

$$f(0, 0) = \boxed{329/4}, \qquad f(1, 0) = \boxed{94}, \qquad f(0, 2) = \boxed{81}.$$

The point on the first quadrant part of the plane $6x + 3y + 4z = 6$ closest to $(4, 6, 7)$ is therefore $(0, 6/5, 3/5)$, where the distance is $\sqrt{80} = 4\sqrt{5}$.

EXERCISES 12.11

In Exercises 1–8 find the maximum and minimum values of the function on the region.

1. $f(x, y) = x^2 + y^3$ on $R : x^2 + y^2 \leq 1$

2. $f(x, y) = x^2 + x + 3y^2 + y$ on the region R bounded by $y = x + 1$, $= 1 - x$, $y = x - 1$, $y = -x - 1$

3. $f(x, y) = 3x + 4y$ on region R bounded by the lines $x + y = 1$, $+ y = 4$, $y + 1 = x$, $y - 1 = x$

4. $f(x, y) = x^2 y + xy^2 + y$ on $R : -1 \leq x \leq 1, -1 \leq y \leq 1$

5. $f(x, y) = 3x^2 + 2xy - y^2 + 5$ on $R : 4x^2 + 9y^2 \leq 36$

6. $f(x, y) = x^3 - 3x + y^2 + 2y$ on the triangle bounded by $x = $, $y = 0$, $x + y = 1$

7. $f(x, y) = x^3 + y^3 - 3x - 12y + 2$ on the square $-3 \leq x \leq 3$, $-3 \leq y \leq 3$

8. $f(x, y) = x^3 + y^3 - 3x - 3y + 2$ on the circle $x^2 + y^2 \leq 1$

9. Find maximum and minimum values of the function $f(x, y, z) = y^2 z^3$ on that part of the plane $x + y + z = 6$ for which (a) $x > $, $y > 0$, $z > 0$ and (b) $x \geq 0$, $y \geq 0$, $z \geq 0$.

10. Find the point on the plane $x + y - 2z = 6$ closest to the origin.

11. Find the shortest distance from $(-1, 1, 2)$ to the plane $2x - 3y + z = 14$.

12. Find the point on the surface $z = x^2 + y^2$ closest to the point $(1, 1, 0)$.

13. Find the point on that part of the plane $x + y + 2z = 4$ in the first octant that is closest to the point $(3, 3, 1)$. For this question assume that the curves of intersection of the plane with the coordinate planes are part of the surface.

14. The electrostatic potential at each point in the region $0 \leq x \leq $, $0 \leq y \leq 1$ is given by $V(x, y) = 48xy - 32x^3 - 24y^2$. Find the maximum and minimum potentials in the region.

15. When a rectangular box is sent through the mail, the post office demands that the length of the box plus twice the sum of its height and width be no more than 250 cm. Find the dimensions of the box satisfying this requirement that encloses the largest possible volume.

16. An open tank in the form of a rectangular parallelepiped is to be built to hold 1000 L of acid. If the cost per unit area of lining the base of the tank is three times that of the sides, what dimensions minimize the cost of lining the tank?

17. Prove that minimum distance from a point (x_1, y_1, z_1) to a plane $Ax + By + Cz + D = 0$ is $|Ax_1 + By_1 + Cz_1 + D| / \sqrt{A^2 + B^2 + C^2}$.

18. Prove that for triangles the point that minimizes the sum of the squares of the distances to the vertices is the centroid.

19. Find the point on the curve $x^2 - xy + y^2 - z^2 = 1$, $x^2 + y^2 = 1$ closest to the origin.

20. Find the dimensions of the box with largest possible volume that can fit inside the ellipsoid $x^2/a^2 + y^2/b^2 + z^2/c^2 = 1$, assuming that its edges are parallel to the coordinate axes.

* 21. Find the maximum and minimum values of the function $f(x, y, z) = xyz$ on the sphere $x^2 + y^2 + z^2 = 1$.

* 22. Find the maximum and minimum values of $f(x, y) = x^2 - y^2$ on the circle $x^2 + y^2 = 1$.

* 23. Find the maximum and minimum values of $f(x, y) = |x - y|$ on the circle $x^2 + y^2 = 1$.

* 24. Find the maximum and minimum values of $f(x, y) = x^2 - y^2$ on the curve $|x| + |y| = 1$.

* 25. Find the maximum and minimum values of $f(x, y) = |x - 2y|$ on the curve $|x| + |y| = 1$.

* 26. If P is the perimeter of a triangle with sides of lengths x, y, and z, the area of the triangle is

$$A = \sqrt{\frac{P}{2}\left(\frac{P}{2} - x\right)\left(\frac{P}{2} - y\right)\left(\frac{P}{2} - z\right)},$$

where $P = x + y + z$. Show that A is maximized for fixed P when the triangle is equilateral.

* 27. Show that for any triangle with interior angles A, B, and C,

$$\sin(A/2) \sin(B/2) \sin(C/2) \leq 1/8.$$

Hint: Find the maximum value of the function $f(A, B, C) = \sin(A/2) \sin(B/2) \sin(C/2)$.

* 28. Show that $|\cos x + \cos y + \sin x \sin y| \leq 2$ for all x and y.

* 29. A silo is in the shape of a right-circular cylinder surmounted by a right-circular cone. If the radius of each is 6 m and the total surface area must be 200 m^2 (not including the base), what heights for the cone and cylinder yield maximum enclosed volume?

* 30. What values of x and y maximize the production function $P(x, y) = kx^\alpha y^\beta$, where k, α, and β are positive constants ($\alpha + \beta = 1$), when x and y must satisfy $Ax + By = C$, where A, B, and C are positive constants?

* 31. A long piece of metal 1 m wide is bent at A and B, as shown in the figure below, to form a channel with three straight sides. If the bends are equidistant from the ends, where should they be made in order to obtain maximum possible flow of fluid along the channel?

* 32. Find maximum and minimum values of the function $f(x, y, z) = xy + xz$ on the region $x^2 + y^2 + z^2 \leq 1$.

** 33. Find maximum and minimum values of the function $f(x, y, z) = x^2 yz$ on the region $x^2 + y^2 \leq 1, 0 \leq z \leq 1$.

** 34. A company produces two products, X and Y, each of which must pass through three stages of manufacture. In phase I, up to eight units per hour of X can be processed and up to four units per hour of Y. For phase II, the numbers of units per hour of X and Y are 3 and 6, respectively, whereas the total number of units that can be handled in phase III is nine per two-hour shift for each of X and Y. The profit per unit of X is \$200 and per unit of Y is \$300. How many units of each product should be manufactured for maximum profit?

** 35. A cow's daily diet consists of three foods: hay, grain, and supplements. The cow is always given 11 kg of hay per day, 50% of which is digestive material and 12% of which is protein. Grain is 74% digestive material and 8.8% protein, whereas supplements are 62% digestive material and 34% protein. The cost of hay is $27.50 for 1000 kg, and grain and supplements cost $110 and $175 for 1000 kg. A healthy cow's diet must contain between 9.5 and 11.5 kg of digestive material and between 1.9 and 2.0 kg of protein. Determine the daily amounts of grain and supplements that the cow should be fed in order that total food costs be kept to a minimum.

** 36. Find the area of the largest triangle that has vertices on the circle $x^2 + y^2 = r^2$.

** 37. A rectangle is surmounted by an isosceles triangle as shown in the figure below. Find x, y, and θ in order that the area of the figure will be as large as possible under the restriction that its perimeter must be P.

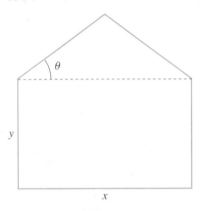

** 38. The Easy University is buying computers. It has three models to choose from. Each model A computer, with 64 MB of memory and a 3 GB hard drive, costs $1300; model B, with 32 MB of memory and a 4 GB drive, costs $1200; and economy model C, with 16 MB of memory and a 1 GB drive, costs $1000. For reasons related to accreditation, the university needs at least 2000 MB of memory and 150 GB of disk space. If the computer lab must have 100 computers, how many of each should the university purchase in order to minimize cost?

** 39. Exercises 34, 35, and 38 are examples of *linear programming problems* that abound in applications of mathematics. The general two dimensional linear programming problem is to maximize the function $f(x, y) = cx + dy$ where c and d are positive constants. Points to be considered must satisfy m inequalities of the form $A_i x + B_i y \leq C_i$ $i = 1, \ldots, m$, where A_i, B_i, and C_i are positive constants, and $x \geq 0$ and $y \geq 0$. These inequalities describe a polygon in the first quadrant of the xy-plane. Give an argument to show that $f(x, y)$ is maximized at one or more of the vertices of the polygon.

** 40. In Consulting Project 5 in Section 4.7, we considered seismic prospecting with two media. The figure below shows the situation for three media, the problem being to predict depths d_1 and d_2. Show that the time for a signal to be emitted by the source, pass through medium 1 with speed v_1, pass through medium 2 with speed v_2, travel along the interface between media 2 and 3 with speed v_3, then through medium 2 and medium 1 to the receiver is given by

$$t = \frac{2d_1 \sec \theta_1}{v_1} + \frac{2d_2 \sec \theta_2}{v_2} + \frac{s - 2d_1 \tan \theta_1 - 2d_2 \tan \theta_2}{v_3}.$$

Verify that t is a minimum when θ_1 and θ_2 are given by

$$\theta_1 = \text{Sin}^{-1}(v_1/v_3), \qquad \theta_2 = \text{Sin}^{-1}(v_2/v_3).$$

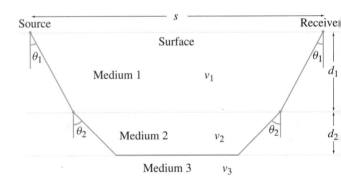

12.12 Lagrange Multipliers

Many applied maxima and minima problems result in **constraint problems**. In particular, Examples 12.34 and 12.35 contain such problems. In Example 12.34, to find extreme values of T on the edge of the plate we maximized and minimized $T(x, y) = 16x^2 - 24xy + 40y$ subject first to the constraint $x^2 + y^2 = 1$, and then to the constraint $y = 0$. Our method there was to substitute from the constraint equation into $T(x, y)$ in order to obtain a function of one variable. In Example 12.35, to find the minimum distance from $(4, 6, 7)$ to the plane $6x + 3y + 4z = 6$, we minimized $D^2 = (x - 4)^2 + (y - 6)^2 + (z - 7)^2$ subject to the constraint $6x + 3y + 4z = 6$. Again we substituted from the constraint to obtain D^2 as a function of two independent variables.

A natural question to ask is whether problems of this type can be solved without substituting from the constraint equation, for if the constraint equation is complicated, substitution may be very difficult or even impossible. To show that there is indeed an alternative, consider the situation in which a function $f(x, y, z)$ is to be maximized or minimized subject to two constraints:

$$F(x, y, z) = 0, \tag{12.60a}$$

$$G(x, y, z) = 0. \tag{12.60b}$$

Algebraically, we are to find extreme values of $f(x, y, z)$, considering only those values of x, y, and z that satisfy equations 12.60. Geometrically, we can interpret each of these conditions as specifying a surface, so that we are seeking extreme values of $f(x, y, z)$, considering only those points on the curve of intersection C of the surfaces $F(x, y, z) = 0$ and $G(x, y, z) = 0$ (Figure 12.60).

Extreme values of $f(x, y, z)$ along C will occur either at critical points of the function or at the ends of the curve. But what derivative or derivatives of $f(x, y, z)$ are we talking about when we say critical points? Since we are concerned only with values of $f(x, y, z)$ on C, we must mean the derivative of $f(x, y, z)$ along C (i.e., the directional derivative in the tangent direction to C). If $\mathbf{T}$, then, is a tangent vector to C, critical points of $f(x, y, z)$ along C are given by

$$0 = D_{\mathbf{T}} f = \nabla f \cdot \hat{\mathbf{T}},$$

or at points where the directional derivative is undefined. According to equation 12.53, a tangent vector to C is $\mathbf{T} = \nabla F \times \nabla G$, and hence a unit tangent vector is

$$\hat{\mathbf{T}} = \frac{\nabla F \times \nabla G}{|\nabla F \times \nabla G|}.$$

The directional derivative of $f(x, y, z)$ at points along C is therefore given by

$$D_{\mathbf{T}} f = \nabla f \cdot \frac{\nabla F \times \nabla G}{|\nabla F \times \nabla G|}.$$

It follows, then, that critical points of $f(x, y, z)$ are points (x, y, z) that satisfy the equation

$$\nabla f \cdot \nabla F \times \nabla G = 0,$$

or points at which the left side is not defined. Now vector $\nabla F \times \nabla G$ is perpendicular to both ∇F and ∇G. Since ∇f is perpendicular to $\nabla F \times \nabla G$ (their dot product is zero), it follows that ∇f must lie in the plane of ∇F and ∇G. Consequently, there exist scalars λ and μ such that

$$\nabla f = (-\lambda) \nabla F + (-\mu) \nabla G,$$

or

$$\nabla f + \lambda \nabla F + \mu \nabla G = \mathbf{0}. \tag{12.61}$$

This vector equation is equivalent to the three scalar equations

$$\frac{\partial f}{\partial x} + \lambda \frac{\partial F}{\partial x} + \mu \frac{\partial G}{\partial x} = 0, \tag{12.62a}$$

$$\frac{\partial f}{\partial y} + \lambda \frac{\partial F}{\partial y} + \mu \frac{\partial G}{\partial y} = 0, \tag{12.62b}$$

$$\frac{\partial f}{\partial z} + \lambda \frac{\partial F}{\partial z} + \mu \frac{\partial G}{\partial z} = 0, \tag{12.62c}$$

and these equations must be satisfied at a critical point at which the directional derivative of $f(x, y, z)$ vanishes. Note, too, that at a point at which the directional derivative of $f(x, y, z)$ does not exist, one of the partial derivatives in these equations does not exist. In other words, we have shown that critical points of $f(x, y, z)$ are points that satisfy equations 12.62 or points at which the equations are undefined. These equations, however, contain five unknowns: x, y, z, λ, and μ. To complete the system we add equations 12.60 since they must also be satisfied by a critical point. Equations 12.60 and 12.62 therefore yield a system of five

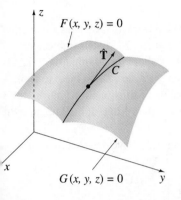

FIGURE 12.60 Extreme values of $f(x, y, z)$ along the curve of intersection of two surfaces

equations in the five unknowns x, y, z, λ, and μ; the first three unknowns (x, y, z) define a critical point of $f(x, y, z)$ along C. The advantage of this system of equations lies in the fact that differentiations in 12.62 involve only the given functions (and no substitutions from the constraint equations are necessary). What we have sacrificed is a system of three equations in the three unknowns (x, y, z) for a system of five equations in the five unknowns (x, y, z, λ, μ).

Let us not forget that the original problem was to find extreme values for the function $f(x, y, z)$ subject to constraints 12.60. What we have shown so far is that critical points at which the directional derivative of $f(x, y, z)$ vanishes can be found by solving equations 12.60 and 12.62. In addition, critical points at which the directional derivative of $f(x, y, z)$ does not exist are points at which equations 12.62 are not defined. What remains is to evaluate $f(x, y, z)$ at all critical points and at the ends of C. If C is a *closed curve* (i.e., if C rejoins itself), then $f(x, y, z)$ *needs to be evaluated only at the critical points*. This turns out to be very important in practice.

Through the directional derivative and tangent vectors to curves, we have shown that equations 12.60 and 12.62 define critical points of a function $f(x, y, z)$ that is subject to two constraints: $F(x, y, z) = 0$ and $G(x, y, z) = 0$. But what about other situations? Let us say, for example, that we require extreme values of a function $f(x, y, z, t)$ subject to a single constraint $F(x, y, z, t) = 0$. How shall we find critical points of this function? Fortunately, as we now show, there is a very simple method that yields equations 12.60 and 12.62, and this method generalizes to other situations also.

To find critical points of $f(x, y, z)$ subject to constraints 12.60, we define a function

$$L(x, y, z, \lambda, \mu) = f(x, y, z) + \lambda F(x, y, z) + \mu G(x, y, z),$$

and regard it as a function of five independent variables x, y, z, λ, and μ. To find critical points of this function, we would first solve the equations obtained by setting each of the partial derivatives of L equal to zero:

$$0 = \frac{\partial L}{\partial x} = \frac{\partial f}{\partial x} + \lambda \frac{\partial F}{\partial x} + \mu \frac{\partial G}{\partial x},$$

$$0 = \frac{\partial L}{\partial y} = \frac{\partial f}{\partial y} + \lambda \frac{\partial F}{\partial y} + \mu \frac{\partial G}{\partial y},$$

$$0 = \frac{\partial L}{\partial z} = \frac{\partial f}{\partial z} + \lambda \frac{\partial F}{\partial z} + \mu \frac{\partial G}{\partial z},$$

$$0 = \frac{\partial L}{\partial \lambda} = F(x, y, z),$$

$$0 = \frac{\partial L}{\partial \mu} = G(x, y, z).$$

In addition, we would consider points at which the partial derivatives of L do not exist. Clearly, this means points (x, y, z) at which any of the partial derivatives of $f(x, y, z)$, $F(x, y, z)$, and $G(x, y, z)$ do not exist. But these are equations 12.60 and 12.62. We have shown, then, that finding critical points (x, y, z) of $f(x, y, z)$ subject to $F(x, y, z) = 0$ and $G(x, y, z) = 0$ is equivalent to finding critical points (x, y, z, λ, μ) of $L(x, y, z, \lambda, \mu)$. The two unknowns λ and μ that accompany a critical point (x, y, z) of $f(x, y, z)$ are called **Lagrange multipliers**. They are not a part of the solution (x, y, z) to the original problem but have been introduced as a convenience by which to arrive at that solution. The function $L(x, y, z, \lambda, \mu)$ is often called the **Lagrangian** of the problem.

The method for other constraint problems should now be evident. Given a function $f(x, y, z, t, \ldots)$ of n variables to maximize or minimize subject to m constraints

$$F_1(x, y, z, t, \ldots) = 0, \quad F_2(x, y, z, t, \ldots) = 0, \ldots, \quad F_m(x, y, z, t, \ldots) = 0, \quad (12.63$$

we introduce m Lagrange multipliers $\lambda_1, \lambda_2, \ldots, \lambda_m$ into a Lagrangian of $n + m$ independent variables $x, y, z, t, \ldots, \lambda_1, \lambda_2, \ldots, \lambda_m$:

$$L(x, y, z, t, \ldots \lambda_1, \lambda_2, \ldots, \lambda_m) = f(x, y, z, t, \ldots) + \lambda_1 F_1(x, y, z, t, \ldots) + \cdots$$
$$+ \lambda_m F_m(x, y, z, t, \ldots). \tag{12.64}$$

Critical points $(x, y, z, t, \ldots)$ of $f(x, y, z, t, \ldots)$ are then determined by the equations defining critical points of $L(x, y, z, t, \ldots, \lambda_1, \lambda_2, \ldots, \lambda_m)$, namely,

$$0 = \frac{\partial L}{\partial x} = \frac{\partial f}{\partial x} + \lambda_1 \frac{\partial F_1}{\partial x} + \cdots + \lambda_m \frac{\partial F_m}{\partial x}, \tag{12.65a}$$

$$0 = \frac{\partial L}{\partial y} = \frac{\partial f}{\partial y} + \lambda_1 \frac{\partial F_1}{\partial y} + \cdots + \lambda_m \frac{\partial F_m}{\partial y}, \tag{12.65b}$$

$$\vdots \qquad \vdots \qquad \vdots$$

$$0 = \frac{\partial L}{\partial \lambda_1} = F_1(x, y, z, t, \ldots), \tag{12.65c}$$

$$0 = \frac{\partial L}{\partial \lambda_2} = F_2(x, y, z, t, \ldots), \tag{12.65d}$$

$$\vdots \qquad \vdots \qquad \vdots$$

$$0 = \frac{\partial L}{\partial \lambda_m} = F_m(x, y, z, t, \ldots). \tag{12.65e}$$

To use Lagrange multipliers in Example 12.35, where we were minimizing $D^2 = (x - 4)^2 + (y - 6)^2 + (z - 7)^2$ subject to $6x + 3y + 4z = 6$, we define the Lagrangian

$$L(x, y, z, \lambda) = D^2 + \lambda(6x + 3y + 4z - 6)$$
$$= (x - 4)^2 + (y - 6)^2 + (z - 7)^2 + \lambda(6x + 3y + 4z - 6).$$

Critical points of $L(x, y, z, \lambda)$ are defined by

$$0 = \frac{\partial L}{\partial x} = 2(x - 4) + 6\lambda,$$

$$0 = \frac{\partial L}{\partial y} = 2(y - 6) + 3\lambda,$$

$$0 = \frac{\partial L}{\partial z} = 2(z - 7) + 4\lambda,$$

$$0 = \frac{\partial L}{\partial \lambda} = 6x + 3y + 4z - 6.$$

The solution of this linear system is $(x, y, z, \lambda) = (-140/61, 174/61, 171/61, 128/61)$, yielding as before the critical point $(-140/61, 174/61, 171/61)$ of D^2.

EXAMPLE 12.36

Find the maximum and minimum values of the function $f(x, y, z) = xyz$ on the sphere $x^2 + y^2 + z^2 = 1$.

SOLUTION   If we define the Lagrangian

$$L(x, y, z, \lambda) = xyz + \lambda(x^2 + y^2 + z^2 - 1),$$

then critical points of L, and therefore of $f(x, y, z)$, are defined by the equations

$$0 = \frac{\partial L}{\partial x} = yz + 2\lambda x,$$

$$0 = \frac{\partial L}{\partial y} = xz + 2\lambda y,$$

$$0 = \frac{\partial L}{\partial z} = xy + 2\lambda z,$$

$$0 = \frac{\partial L}{\partial \lambda} = x^2 + y^2 + z^2 - 1.$$

If we multiply the first equation by y and the second by x, and equate the resulting expressions for $2\lambda xy$, we have

$$y^2 z = x^2 z.$$

Consequently, either $z = 0$ or $y = \pm x$.

Case I: $z = 0$. In this case the equations reduce to

$$\lambda x = 0, \quad \lambda y = 0, \quad xy = 0, \quad x^2 + y^2 = 1.$$

The first implies that either $x = 0$ or $\lambda = 0$. If $x = 0$, then $y = \pm 1$, and we have two critical points $(0, \pm 1, 0)$. If $\lambda = 0$, then the third equation requires $x = 0$ or $y = 0$. We therefore obtain two additional critical points $(\pm 1, 0, 0)$.

Case II: $y = x$. In this case the equations reduce to

$$xz + 2\lambda x = 0, \quad x^2 + 2\lambda z = 0, \quad 2x^2 + z^2 = 1.$$

The first implies that either $x = 0$ or $z = -2\lambda$. If $x = 0$, then $z = \pm 1$, and we have the two critical points $(0, 0, \pm 1)$. If $z = -2\lambda$, then the last two equations imply that $x = \pm 1/\sqrt{3}$, and we obtain the four critical points

$$(\pm 1/\sqrt{3}, \pm 1/\sqrt{3}, 1/\sqrt{3}) \quad \text{and} \quad (\pm 1/\sqrt{3}, \pm 1/\sqrt{3}, -1/\sqrt{3}).$$

Case III: $y = -x$. This case is similar to that for $y = x$, and leads to the additional four critical points

$$(\pm 1/\sqrt{3}, \mp 1/\sqrt{3}, 1/\sqrt{3}) \quad \text{and} \quad (\pm 1/\sqrt{3}, \mp 1/\sqrt{3}, -1/\sqrt{3}).$$

Because $x^2 + y^2 + z^2 = 1$ is a surface without a boundary, we complete the problem by evaluating $f(x, y, z)$ at each of the critical points:

$$f(\pm 1, 0, 0) = f(0, \pm 1, 0) = f(0, 0, \pm 1) = 0,$$

$$f(\pm 1/\sqrt{3}, \pm 1/\sqrt{3}, 1/\sqrt{3}) = f(\pm 1/\sqrt{3}, \mp 1/\sqrt{3}, -1/\sqrt{3}) = \sqrt{3}/9,$$

$$f(\pm 1/\sqrt{3}, \pm 1/\sqrt{3}, -1/\sqrt{3}) = f(\pm 1/\sqrt{3}, \mp 1/\sqrt{3}, 1/\sqrt{3}) = -\sqrt{3}/9.$$

The maximum and minimum values of $f(x, y, z)$ on $x^2 + y^2 + z^2 = 1$ are therefore $\sqrt{3}/9$ and $-\sqrt{3}/9$.

To compare the Lagrangian solution in this example to that without a Lagrange multiplier, see Exercise 21 in Section 12.11.

EXAMPLE 12.37

A company manufactures wheelbarrows at n of its plants. The cost of manufacturing x_i wheelbarrows at the ith plant is x_i^2/c_i, where $c_i > 0$ are known constants. The total cost of manufacturing x_1 wheelbarrows at plant 1, x_2 at plant 2, ..., x_n at plant n, is therefore

$$f(x_1, \ldots, x_n) = \frac{x_1^2}{c_1} + \frac{x_2^2}{c_2} + \cdots + \frac{x_n^2}{c_n}.$$

The production engineer wishes to schedule a total of D wheelbarrows among the plants. How many should each plant produce if costs are to be minimized, and what is minimum cost?

SOLUTION   We must minimize $f(x_1, \ldots, x_n)$ subject to the constraint $x_1 + x_2 + \cdots + x_n = D$. If we define the Lagrangian

$$L(x_1, \ldots, x_n, \lambda) = f(x_1, \ldots, x_n) + \lambda(x_1 + x_2 + \cdots + x_n - D),$$

then critical points of L (and therefore of f) are given by the $n + 1$ equations

$$0 = \frac{\partial L}{\partial x_i} = \frac{\partial f}{\partial x_i} + \lambda = \frac{2x_i}{c_i} + \lambda, \quad i = 1, \ldots, n,$$

$$0 = \frac{\partial L}{\partial \lambda} = x_1 + x_2 + \cdots + x_n - D.$$

These give $x_i = -c_i\lambda/2$ for each i, and if we substitute these into the last equation,

$$-\frac{c_1\lambda}{2} - \frac{c_2\lambda}{2} - \cdots - \frac{c_n\lambda}{2} = D \quad \Longrightarrow \quad \lambda = \frac{-2D}{c_1 + \cdots + c_n}.$$

Thus, production levels at the plants should be

$$x_i = \frac{c_i D}{c_1 + \cdots + c_n}, \quad i = 1, \ldots, n.$$

Minimum cost is

$$\frac{1}{c_1}\left(\frac{c_1 D}{c_1 + \cdots + c_n}\right)^2 + \cdots + \frac{1}{c_n}\left(\frac{c_n D}{c_1 + \cdots + c_n}\right)^2 = \frac{D^2}{c_1 + \cdots + c_n}.$$

Consulting Project 19

We are being approached by a hydraulic engineer who is fabricating an open channel from long pieces of metal joined end to end. We take the width of the metal as 1 metre, although the solution is easily adapted to any width. Each piece of metal is bent to form the channel. The engineer was once a student of this text and has solved Exercise 31 in Section 12.11, but he is not convinced that this is the optimum shape of the channel. For instance, why is it necessary that the two sides of the channel be of the same length? We are to show that the solution to Exercise 31 in Section 12.11 is indeed the best of all possible channels with a maximum of two bends.

SOLUTION   Figure 12.61a shows a channel with two bends but allows for different lengths of the sides of the channel and therefore different angels θ and ϕ. Volume of flow along the channel is maximized when its cross-sectional area is maximized, and the area for this channel is

$$A = \frac{1}{2}y^2 \sin\theta \cos\theta + xy \sin\theta + \frac{1}{2}(1 - x - y)^2 \sin\phi \cos\phi,$$

where, in order that both sides of the channel have the same height, the condition $y \sin\theta = (1 - x - y) \sin\phi$ must be satisfied. If we regard x, θ, and ϕ as independent variables, and this equation as a restriction on y, then A must be maximized for values of x, θ, and ϕ in the box of Figure 12.61b.

FIGURE 12.61a   Bending a piece of metal to form a channel

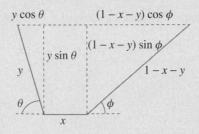

FIGURE 12.61b   Domain for optimization problem

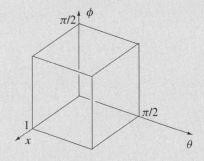

We shall find critical values of A inside the box, and then turn to the six faces of the box for maxima thereon. We shall have two difficulties. One is the complexity of the equations that must be solved, and the other is to not lose sight of exactly where we are in the solution process. For critical points of A interior to the box, we form the Lagrangian

$$L(x, y, \theta, \phi, \lambda) = \frac{1}{2}y^2 \sin\theta \cos\theta + xy \sin\theta$$

$$+ \frac{1}{2}(1 - x - y)^2 \sin\phi \cos\phi + \lambda[y \sin\theta - (1 - x - y) \sin\phi].$$

Critical points of this function are defined by the equations

$$0 = \frac{\partial L}{\partial x} = y \sin\theta - (1 - x - y) \sin\phi \cos\phi + \lambda \sin\phi, \tag{12.66a}$$

$$0 = \frac{\partial L}{\partial y} = y \sin\theta \cos\theta + x \sin\theta - (1 - x - y) \sin\phi \cos\phi$$

$$+ \lambda(\sin\theta + \sin\phi), \tag{12.66b}$$

$$0 = \frac{\partial L}{\partial \theta} = \frac{1}{2}y^2(\cos^2\theta - \sin^2\theta) + xy \cos\theta + \lambda y \cos\theta, \tag{12.66c}$$

$$0 = \frac{\partial L}{\partial \phi} = \frac{1}{2}(1 - x - y)^2(\cos^2\phi - \sin^2\phi) - \lambda(1 - x - y) \cos\phi, \tag{12.66d}$$

$$0 = \frac{\partial L}{\partial \lambda} = y \sin\theta - (1 - x - y) \sin\phi. \tag{12.66e}$$

If we subtract the first of these from the second, we obtain

$$y \sin\theta \cos\theta + x \sin\theta - y \sin\theta + \lambda \sin\theta = 0 \implies \sin\theta(y\cos\theta + x - y + \lambda) = 0.$$

One possibility is $\sin\theta = 0 \implies \theta = 0$, but this yields minimum area. The other possibility is that

$$y\cos\theta + x - y + \lambda = 0. \tag{12.66f}$$

When we solve this for λ and substitute into equation 12.66c,

$$\frac{1}{2}y^2(\cos^2\theta - \sin^2\theta) + xy\cos\theta + y\cos\theta(y - x - y\cos\theta) = 0$$

$$\implies \quad y^2\left(\cos\theta - \frac{1}{2}\right) = 0.$$

Thus, $y = 0$, which gives minimum area, or $\cos\theta = 1/2 \implies \theta = \pi/3$. If we subtract equation 12.66e from 12.66a, we find

$$-(1 - x - y)\sin\phi\cos\phi + (1 - x - y)\sin\phi + \lambda\sin\phi = 0$$

$$\implies \quad \sin\phi[(1 - x - y)(1 - \cos\phi) + \lambda] = 0.$$

Either $\sin\phi = 0 \implies \phi = 0$, giving minimum A, or $(1 - x - y)(1 - \cos\phi) + \lambda = 0$. When we substitute this into equation 12.66d,

$$0 = \frac{1}{2}(1 - x - y)^2(\cos^2\phi - \sin^2\phi) + (1 - x - y)^2(1 - \cos\phi)\cos\phi.$$

This equation implies that $x + y = 1$, giving minimum A, or,

$$\frac{1}{2}(\cos^2\phi - \sin^2\phi) - \cos^2\phi + \cos\phi = 0 \implies \cos\phi = \frac{1}{2} \implies \phi = \frac{\pi}{3}.$$

Equation 12.66e now gives $x + 2y = 1$. When this is substituted into equations 12.66a, c, two equations in y and λ result, and the solution for y is $y = 1/3$. This in turn gives $x = 1/3$, and the area of the channel with these values of x, y, θ and ϕ is

$$A = \frac{1}{2}\left(\frac{1}{3}\right)^2\left(\frac{\sqrt{3}}{2}\right)\left(\frac{1}{2}\right) + \left(\frac{1}{3}\right)\left(\frac{1}{3}\right)\left(\frac{\sqrt{3}}{2}\right)$$

$$+ \frac{1}{2}\left(1 - \frac{1}{3} - \frac{1}{3}\right)^2\left(\frac{\sqrt{3}}{2}\right)\left(\frac{1}{2}\right) = \boxed{\frac{\sqrt{3}}{12}}.$$

We now turn to the six faces of the box defining permissible values of x, θ, and ϕ. The three faces $x = 1$, $\phi = 0$, and $\theta = 0$ give minimum values 0 for area of the channel. This leaves faces $x = 0$, $\phi = \pi/2$, and $\theta = \pi/2$. We need discuss only one of the last two faces, since discussions would be identical for the other. Consider first then the face $x = 0$. In this case the metal has only one bend as shown in Figure 12.62a. Area of the cross-section of the channel is

$$A = \frac{1}{2}y^2\sin\theta\cos\theta + \frac{1}{2}(1 - y)^2\sin\phi\cos\phi,$$

subject to the restriction $y \sin \theta = (1 - y) \sin \phi$. If θ and ϕ are taken as independent variables, this function must be maximized on the square in Figure 12.62b.

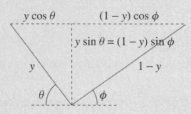

FIGURE 12.62a Triangular channel

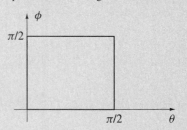

FIGURE 12.62b Domain for optimization of triangular channel

First we find critical points interior to the square, and then consider the four lines forming the boundary of the square. For critical points of A interior to the square, we form the Lagrangian

$$L(y, \theta, \phi, \lambda) = \frac{1}{2}y^2 \sin \theta \cos \theta + \frac{1}{2}(1 - y)^2 \sin \phi \cos \phi + \lambda[y \sin \theta - (1 - y) \sin \phi].$$

Critical points of this function are defined by the equations

$$0 = \frac{\partial L}{\partial y} = y \sin \theta \cos \theta - (1 - y) \sin \phi \cos \phi + \lambda(\sin \theta + \sin \phi), \quad \text{(12.67a)}$$

$$0 = \frac{\partial L}{\partial \theta} = \frac{1}{2}y^2(\cos^2 \theta - \sin^2 \theta) + \lambda y \cos \theta, \quad \text{(12.67b)}$$

$$0 = \frac{\partial L}{\partial \phi} = \frac{1}{2}(1 - y)^2(\cos^2 \phi - \sin^2 \phi) - \lambda(1 - y) \cos \phi, \quad \text{(12.67c)}$$

$$0 = \frac{\partial L}{\partial \lambda} = y \sin \theta - (1 - y) \sin \phi. \quad \text{(12.67d)}$$

The second of these implies that $y = 0$, which minimizes A, or that

$$y(\cos^2 \theta - \sin^2 \theta) + 2\lambda \cos \theta = 0. \quad \text{(12.67e)}$$

The third yields $y = 1$, which minimizes A, or

$$(1 - y)(\cos^2 \phi - \sin^2 \phi) - 2\lambda \cos \phi = 0. \quad \text{(12.67f)}$$

When we substitute from equation 12.67d into 12.67a, we obtain

$$0 = (1 - y) \sin \phi \cos \theta - (1 - y) \sin \phi \cos \phi + \lambda \left(\frac{1 - y}{y} \sin \phi + \sin \phi \right). \text{(12.67g)}$$

Either $\sin \phi = 0$, which leads to minimum A, or

$$y(1 - y) \cos \theta - y(1 - y) \cos \phi + \lambda = 0. \quad \text{(12.67h)}$$

When we solve this for $\lambda = -y(1 - y)(\cos \theta - \cos \phi)$, and substitute into equation 12.67e,

$$y(\cos^2 \theta - \sin^2 \theta) - 2y(1 - y)\cos\theta(\cos\theta - \cos\phi) = 0.$$

This requires $y = 0$, a minimizing value, or

$$\cos^2 \theta - \sin^2 \theta - 2(1 - y)\cos\theta(\cos\theta - \cos\phi) = 0. \tag{12.67i}$$

When we substitute for λ in equation 12.67f,

$$(1 - y)(\cos^2 \phi - \sin^2 \phi) + 2y(1 - y)\cos\phi(\cos\theta - \cos\phi) = 0,$$

from which $y = 1$, a minimum, or

$$\cos^2 \phi - \sin^2 \phi + 2y\cos\phi(\cos\theta - \cos\phi) = 0. \tag{12.67j}$$

We now solve equation 12.67d for $y = \sin\phi/(\sin\theta + \sin\phi)$, and substitute into 12.67i, j,

$$\cos^2 \theta - \sin^2 \theta - \frac{2\sin\theta\cos\theta(\cos\theta - \cos\phi)}{\sin\theta + \sin\phi} = 0,$$

$$\cos^2 \phi - \sin^2 \phi + \frac{2\sin\phi\cos\phi(\cos\theta - \cos\phi)}{\sin\theta + \sin\phi} = 0.$$

These imply that

$$\frac{\cos^2 \theta - \sin^2 \theta}{2\sin\theta\cos\theta} = -\frac{\cos^2 \phi - \sin^2 \phi}{2\sin\phi\cos\phi}.$$

The only way for this equation to hold is for $\theta = \phi = \pi/4$, and this implies that $y = 1/2$. The area of the channel for these values is

$$A = \frac{1}{2}\left(\frac{1}{2}\right)^2\left(\frac{1}{2}\right) + \frac{1}{2}\left(\frac{1}{2}\right)^2\left(\frac{1}{2}\right) = \boxed{\frac{1}{8}}.$$

We should now consider area on the four sides of the square. Along $\theta = 0$ and $\phi = 0$, area is zero. Consideration of $\phi = \pi/2$ is the mirror image of $\theta = \pi/2$. For $\theta = \pi/2$, we are considering channels in the shape in Figure 12.63. Area is

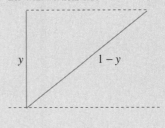

FIGURE 12.63   Triangular channel with vertical side

$$A = \frac{1}{2}y\sqrt{(1 - y)^2 - y^2} = \frac{1}{2}y\sqrt{1 - 2y},$$

where $0 \le y \le 1/2$. For critical points, we solve

$$0 = \frac{1}{2}\left(\sqrt{1 - 2y} - \frac{y}{\sqrt{1 - 2y}}\right) \implies y = \frac{1}{3}.$$

The area of the channel when $y = 1/3$ is $A = \boxed{\sqrt{3}/18}$. The area is zero when $y = 0$ and $y = 1/2$. This completes the discussion of the face $x = 0$ of the box in Figure 12.61b.

Our final consideration is face $\theta = \pi/2$ of the box. In this case, the channel has the shape in Figure 12.64a. Its area is

$$A = xy + \frac{1}{2}(1 - x - y)^2 \sin\phi\cos\phi,$$

where $y = (1 - x - y) \sin \phi$. This function must be maximized over the rectangle in Figure 12.64b.

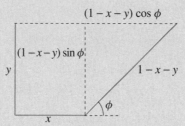

FIGURE 12.64a   Channel with one vertical side

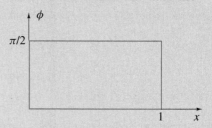

FIGURE 12.64b   Domain for optimization of channel with one vertical side

The associated Lagrangian,

$$L(x, y, \phi, \lambda) = xy + \frac{1}{2}(1 - x - y)^2 \sin \phi \cos \phi + \lambda[y - (1 - x - y) \sin \phi],$$

has critical points defined by

$$0 = \frac{\partial L}{\partial x} = y - (1 - x - y) \sin \phi \cos \phi + \lambda \sin \phi, \qquad (12.68a)$$

$$0 = \frac{\partial L}{\partial y} = x - (1 - x - y) \sin \phi \cos \phi + \lambda(1 + \sin \phi), \qquad (12.68b)$$

$$0 = \frac{\partial L}{\partial \phi} = \frac{1}{2}(1 - x - y)^2(\cos^2 \phi - \sin^2 \phi) - \lambda(1 - x - y) \cos \phi, \quad (12.68c)$$

$$0 = \frac{\partial L}{\partial \lambda} = y - (1 - x - y) \sin \phi. \qquad (12.68d)$$

The first two equations imply that $\lambda = y - x$. We substitute this into equation 12.68a, and into 12.68c, after removing the extra factor $1 - x - y$,

$$0 = y - (1 - x - y) \sin \phi \cos \phi + (y - x) \sin \phi, \qquad (12.68e)$$

$$0 = (1 - x - y)(\cos^2 \phi - \sin^2 \phi) - 2(y - x) \cos \phi. \qquad (12.68f)$$

When we multiply the first of these by $2 \cos \phi$, the second by $\sin \phi$, and add,

$$2y \cos \phi - 2(1 - x - y) \sin \phi \cos^2 \phi + (1 - x - y)(\cos^2 \phi - \sin^2 \phi) \sin \phi = 0.$$

We can use equation 12.68d to eliminate x from this equation,

$$2y \cos \phi - 2y \cos^2 \phi + (\cos^2 \phi - \sin^2 \phi)y = 0.$$

This implies that $y = 0$, which gives minimum A, or,

$$2 \cos \phi - 2 \cos^2 \phi + (\cos^2 \phi - \sin^2 \phi) = 0 \implies \cos \phi = \frac{1}{2} \implies \phi = \frac{\pi}{3}.$$

Equation 12.68f now gives

$$(1 - x - y)\left(\frac{1}{4} - \frac{3}{4}\right) - 2(y - x)\left(\frac{1}{2}\right) = 0 \qquad \Longrightarrow \qquad y = 3x - 1.$$

When we substitute this in equation 12.68d,

$$3x - 1 - (1 - x - 3x + 1)\frac{\sqrt{3}}{2} = 0 \qquad \Longrightarrow \qquad x = \frac{3 - \sqrt{3}}{3}.$$

This now gives $y = 2 - \sqrt{3}$, and area of the channel is

$$A = \left(\frac{3 - \sqrt{3}}{3}\right)(2 - \sqrt{3}) + \frac{1}{2}\left[1 - \frac{3 - \sqrt{3}}{3} - (2 - \sqrt{3})\right]^2 \left(\frac{\sqrt{3}}{2}\right)\left(\frac{1}{2}\right) = \boxed{\frac{2 - \sqrt{3}}{2}}.$$

We now consider A along the four edges of the rectangle in Figure 12.64b. Along $\phi = 0$ and $x = 1$, area is a minimum. We have already dealt with the case $x = 0$ (see Figure 12.63). This leaves the case $\phi = \pi/2$, a rectangular channel as shown in Figure 12.65. Its area is

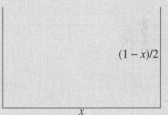

FIGURE 12.65   Channel with two vertical sides

$$A = \frac{1}{2}x(1 - x), \quad 0 \le x \le 1.$$

Critical points of this function are given by $0 = 1 - 2x$, from which $x = 1/2$. Area of the channel is $\boxed{1/8}$. Area is zero when $x = 0$ and $x = 1$.

This completes our calculations; we have considered area of the channel for values of x, θ, and ϕ interior to the box of Figure 12.61b, over each face of the box, along each edge, and at each corner. The boxed numbers are $\sqrt{3}/12$, $1/8$, $\sqrt{3}/18$ and $(2 - \sqrt{3})/2$, the largest of which is $\sqrt{3}/12$. This is the solution to Exercise 31 in Section 12.11, confirming that it provides the channel with maximum flow from all channels with a maximum of two bends. We might notice that if more bends or curved sides were allowed, the area could be increased beyond this. For instance, if the channel is semicircular, its area is $1/(2\pi)$.

In Exercises 1–8 use Lagrange multipliers to find maximum and minimum values of the function subject to the constraints. In each case, interpret the constraints geometrically.

1. $f(x, y) = x^2 + y$ subject to $x^2 + y^2 = 4$

2. $f(x, y, z) = 5x - 2y + 3z + 4$ subject to $x^2 + 2y^2 + 4z^2 = 9$

3. $f(x, y) = x + y$ subject to $(x - 1)^2 + y^2 = 1$

4. $f(x, y, z) = x^3 + y^3 + z^3$ subject to $x^2 + y^2 + z^2 = 9$

5. $f(x, y, z) = xyz$ subject to $x^2 + 2y^2 + 3z^2 = 12$

* 6. $f(x, y, z) = x^2 y + z$ subject to $x^2 + y^2 = 1$, $z = y$

* 7. $f(x, y, z) = x^2 + y^2 + z^2$ subject to $x^2 + y^2 + z^2 = 2z$, $x + y + z = 1$

* 8. $f(x, y, z) = xyz - x^2 z$ subject to $x^2 + y^2 = 1$, $z = \sqrt{x^2 + y^2}$

* 9. Use Lagrange multipliers to solve Exercise 11 in Section 12.11.

* 10. Use Lagrange multipliers to solve Exercise 12 in Section 12.11.

* 11. Use Lagrange multipliers to solve Exercise 20 in Section 12.11.

* 12. Use Lagrange multipliers to solve Exercise 26 in Section 12.11.

* 13. Use Lagrange multipliers to solve Exercise 27 in Section 12.11.

* **14.** Use Lagrange multipliers to solve Exercise 29 in Section 12.11.

* **15.** Use Lagrange multipliers to solve Exercise 22 in Section 12.11.

* **16.** Use Lagrange multipliers to solve Exercise 23 in Section 12.11.

* **17.** Use Lagrange multipliers to solve Exercise 24 in Section 12.11.

* **18.** Use Lagrange multipliers to solve Exercise 25 in Section 12.11.

* **19.** Use Lagrange multipliers to solve Exercise 17 in Section 12.11.

* **20.** Suppose that $F(x, y) = 0$ and $G(x, y) = 0$ define two curves C_1 and C_2 in the xy-plane. Let $P(x_0, y_0)$ and $Q(X_0, Y_0)$ be the points on C_1 and C_2 that minimize the distance between C_1 and C_2. If C_1 and C_2 have tangent lines at P and Q, show that the line PQ is perpendicular to these tangent lines.

* **21.** What are production levels and minimum cost in Example 12.37 for the production of 500 wheelbarrows if there are four plants with $c_1 = 26$, $c_2 = 24$, $c_3 = 23$, and $c_4 = 27$?

* **22.** When a thermonuclear reactor is built in the form of a right-circular cylinder, neutron diffusion theory requires its radius and height to satisfy the equation

$$\left(\frac{2.4048}{r}\right)^2 + \left(\frac{\pi}{h}\right)^2 = k,$$

where k is a constant. Find r and h in terms of k if the reactor is to occupy as small a volume as possible.

* **23.** Find the points on the curve $x^2 + xy + y^2 = 1$ closest to and farthest from the origin.

In Exercises 24–26 use Lagrange multipliers to find maximum and minimum values of the function.

* **24.** $f(x, y) = 3x^2 + 2xy - y^2 + 5$ for $4x^2 + 9y^2 \le 36$

* **25.** $f(x, y) = x^2y + xy^2 + y$ for $-1 \le x \le 1$, $-1 \le y \le 1$

* **26.** $f(x, y, z) = xy + xz$ for $x^2 + y^2 + z^2 \le 1$

* **27.** Find the maximum value of $f(x, y, z) = x^2yz - xzy^2$ subject to constraints $x^2 + y^2 = 1$, $z = \sqrt{x^2 + y^2}$.

* **28.** The equation $3x^2 + 4xy + 6y^2 = 140$ describes an ellipse that has its centre at the origin, but major and minor axes are not along the x- and y-axes. Find coordinates of the ends of the major and minor axes.

* **29.** Use Lagrange multipliers to find the point on the first octant part of the plane $Ax + By + Cz = D$ (A, B, C, D all positive constants) that maximizes the function $f(x, y, z) = x^p y^q z^r$, where p, q, and r are positive constants.

* **30.** The folium of Descartes has parametric equations

$$x = \frac{3at}{1 + t^3}, \qquad y = \frac{3at^2}{1 + t^3} \qquad (a > 0)$$

(see Exercise 52 in Section 3.8 and Exercise 61 in Section 9.1). Find the point in the first quadrant farthest from the origin in two ways:

(a) Express $D^2 = x^2 + y^2$ in terms of t and maximize this function of one variable.

(b) Show that an implicit equation for the curve is $x^3 + y^3 = 3axy$ and maximize $D^2 = x^2 + y^2$ subject to this constraint.

** **31.** To find the point on the curve $x^2 - xy + y^2 + z^2 = 1$, $x^2 + y^2 = 1$ closest to the origin, we must minimize the function $f(x, y, z) = x^2 + y^2 + z^2$ subject to the constraints defined by the equations of the curve. Show that this can be done by (a) using two Lagrange multipliers; (b) expressing $f(x, y, z)$ in terms of x and y alone, $u = 1 - xy$, and minimizing this function subject to $x^2 + y^2 = 1$ (with one Lagrange multiplier); (c) expressing $f(x, y, z)$ in terms of x alone, $u = 1 \pm x\sqrt{1 - x^2}$, and minimizing these functions on appropriate intervals; and (d) writing $x = \cos t$, $y = \sin t$ along the curve, expressing $f(x, y, z)$ in terms of t, $u = 1 - \sin t \cos t$, and minimizing this function on appropriate intervals.

** **32.** Find the smallest and largest distances from the origin to the curve $x^2 + y^2/4 + z^2/9 = 1$, $x + y + z = 0$.

** **33.** Find the maximum value of $f(x, y, z) = (xy + x^2)/(z^2 + 1)$ subject to the constraint $x^2(4 - x^2) = y^2$.

** **34.** Find the maximum and minimum values of the function $f(x, y, z) = 2x^2y^2 + 2y^2z^2 + 3z$ considering only values of x, y, and z satisfying the equations $z = x^2 + y^2$, $x^2 + 3y^2 = 1$. Do this with and without Lagrange multipliers.

12.13 Least Squares

The basic tool of applied mathematics is the *function*. Given a function such as $f(x) = x^2 - 2x - 3$, and any value of x, say $x = 4$, we calculate the value of the function at this x as $f(4) = 5$. In many problems, the function is not given; it must be found. What we might have is a set of experimental data suggesting that various quantities are related to one another, but the data do not specify exactly how they are related. For example, suppose a variable y is known to depend on a variable x, and an experiment is performed to measure 10 values of y corresponding to 10 values of x. The results are listed in Table 12.3.

TABLE 12.3

x	1	2	3	4	5	6	7	8	9	10
y	6.05	8.32	10.74	13.43	15.90	18.38	20.93	23.32	24.91	28.36

Consider the problem of finding that function $y = y(x)$ that best describes these data. Two considerations are important — simplicity and accuracy. We would like as simple a function as possible to describe how y depends on x. At the same time, we want the function to be as accurate as possible. For example, we could find a polynomial of degree nine to fit the data exactly; it would give the exact y-value for each value of x. But this would hardly be a simple representation. It would also be unreasonable from the following standpoint. Because y-values are determined experimentally (we assume that x-values are exact), they will be subject to errors both random and systematic. It is pointless to use a ninth-degree polynomial to reproduce the data exactly, since it therefore also reproduces the inherent errors. What we want is a simple function that best fits the *trend* of the data. To discover this *trend*, we plot the data of Table 12.3 as points (x, y) in Figure 12.66.

FIGURE 12.66   Finding the best-fitting line to data points in Table 12.3

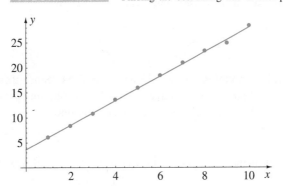

We are immediately impressed by the fact that although the points do not all lie on a straight line, to describe them by a straight line would be a simple representation, and reasonably accurate. We therefore look for a linear function

$$y = y(x) = ax + b \qquad (12.69)$$

to describe the data in Table 12.3. Many lines could be drawn to fit the points in Figure 12.66 reasonably accurately, and each line would be characterized by different values of a and b. Our problem then is to find that line (or those values of a and b) that *best* fits the points. Mathematicians have developed a method called **least squares** to arrive at a best fit.

We denote the x-values in Table 12.3 by $x_i = i$, $i = 1, \ldots, 10$, and corresponding observed values of y by $\overline{y}_i$. The linear function $y(x) = ax + b$ predicts values $y_i = y(x_i) = ax_i + b$ at the x_i. Differences between observed and predicted values of y are

$$y_i - \overline{y}_i = (ax_i + b) - \overline{y}_i.$$

We define a quantity S as the sum of the squares of these differences for all x_i:

$$S = \sum_{i=1}^{10} (y_i - \overline{y}_i)^2 = \sum_{i=1}^{10} (ax_i + b - \overline{y}_i)^2. \qquad (12.70)$$

Given any a and b, S is a measure of the degree to which the points vary from that line — the better the fit, the smaller the value of S. The method of least squares states that one way to approximate the points in Figure 12.66 by a straight line is to choose that line which minimizes the function $S = S(a, b)$. In other words, find values of a and b that minimize $S(a, b)$. For critical points of this function we solve

$$0 = \frac{\partial S}{\partial a} = \sum_{i=1}^{10} 2x_i(ax_i + b - \overline{y}_i),$$

$$0 = \frac{\partial S}{\partial b} = \sum_{i=1}^{10} 2(ax_i + b - \overline{y}_i).$$

If we rewrite these equations in the form

$$\left(\sum_{i=1}^{10} x_i^2\right) a + \left(\sum_{i=1}^{10} x_i\right) b = \sum_{i=1}^{10} x_i \overline{y}_i, \tag{12.71a}$$

$$\left(\sum_{i=1}^{10} x_i\right) a + 10b = \sum_{i=1}^{10} \overline{y}_i, \tag{12.71b}$$

we have a pair of linear equations in a and b. With the data in Table 12.3, we obtain

$$385a + 55b = 1139.27,$$

$$55a + 10b = 170.34,$$

which have solutions $a = 2.45$ and $b = 3.54$. Since there is only one critical point, and we know that $S(a, b)$ must have an absolute minimum for some a and b, it follows that these values must minimize $S(a, b)$. The straight line

$$y = y(x) = 2.45x + 3.54$$

is therefore the best straight-line fit (in the least-squares sense) to the data in Table 12.3. This is, in fact, the line in Figure 12.66. It is important to remember that least squares assumes that we have a prior knowledge of the type of function to be determined (in this case a linear function), and the method then proceeds to find the best such function.

Application Preview Revisited

Polynomials of higher order (quadratics, cubics, etcetera.) can be fitted to data points by least-squares (see Exercises 4–7). Other types of functions can also be used, often by reducing the problem to a straight-line situation. In the Application Preview we posed the problem of finding the function that represents the tabulated values below, or the equation of a curve that approximates the points plotted in Figure 12.67a.

TABLE 12.4

V	54.3	61.82	72.4	88.7	118.6	194.0
P	61.2	49.5	37.6	28.4	19.2	10.1

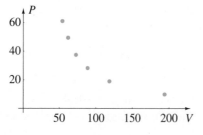

FIGURE 12.67a   Data points for pressure and volume

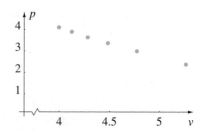

FIGURE 12.67b   Plot of logarithms of pressure and volume

   Thermodynamics suggests that when the compression and expansion of the gas is adiabatic, pressure P and volume V are related by an equation of the form $P = b/V^a$ for some constants a and b. A direct application of the above procedure leads to complicated nonlinear equations for a and b. Instead, we take logarithms of the equation $P = b/V^a$, and write

$$\ln P = \ln b - a \ln V.$$

If we define new variables, $p = \ln P$, $B = \ln b$, $v = \ln V$, and $A = -a$, then

$$p = Av + B.$$

This is the equation of a straight line in the vp-plane. We have tabulated $p = \ln P$ and $v = \ln V$ below, and plotted p against v in Figure 12.67b. The fact that the points seem reasonably collinear is confirmation of the fact that the data points in Figure 12.67a are adequately described by a function of the form $P = b/V^a$.

TABLE 12.5

$v = \ln V$	3.998	4.124	4.282	4.485	4.776	5.268
$p = \ln P$	4.114	3.902	3.627	3.346	2.955	2.313

Fitting the straight line $p = B + Av$ to the data in Table 12.5 leads to the following equations for A and B corresponding to equations 12.71:

$$\left(\sum_{i=1}^{6} v_i^2 \right) A + \left(\sum_{i=1}^{6} v_i \right) B = \sum_{i=1}^{6} v_i \overline{p}_i,$$

$$\left(\sum_{i=1}^{6} v_i \right) A + 6B = \sum_{i=1}^{6} \overline{p}_i.$$

These give

$$121.9758A + 26.9295B = 89.3605,$$

$$26.9295A + 6B = 20.2570,$$

the solution of which is $A = -1.4043$ and $B = 9.6788$. Consequently, least-squares estimates for A and B give

$$p = -1.4043v + 9.6788 \quad \Longrightarrow \quad \ln P = 9.6788 - 1.4043 \ln V.$$

When we exponentiate,

$$P = e^{9.6788 - 1.4043 \ln V} = 15975 V^{-1.4043},$$

and this function approximates the data in Table 12.4, or the points in Figure 12.67a. It is important to notice that we did not apply the method of least squares directly to the function $P = b/V^a$ and the data in Table 12.4. Were we to do so, it would be very difficult to solve the resulting equations for a and b. Try it! Were we successful in doing so, values of a and b would differ from those above, but not significantly.

   Representation of data by other types of functions is discussed in the exercises.

EXERCISES 12.13

1. The following table shows the ages M (in months) and the average test scores S (obtained on a standard intelligence test) for children between the ages of 9 and 12 years.

Age M (months)	108	112	116	120	124	128	132	136	140	144
Average score S	62	67	72	77	83	87	94	100	105	109

Find least-squares estimates for a linear function to describe $S = S(M)$.

2. The table below shows the average systolic blood pressure P of 13 children.

Age A (years)	4	5	6	7	8	9	10	11	12	13	14	15	16
Pressure P	85	87	90	92	95	98	100	105	108	110	112	115	118

Use least squares to find the equation of a straight line fitting these data.

3. The production of steel in the United States for the years 1946–1956 is shown in the table below.

Year	1946	1947	1948	1949	1950	1951	1952	1953	1954	1955	1956
Steel (tonnes $\times 10^6$)	55.2	76.9	80.5	70.9	88.0	95.6	84.7	101.5	80.3	106.4	104.7

Find the least-squares estimate for a linear function $S = at + b$ to describe the data by (a) taking $t = 0$ in year zero and (b) taking $t = 0$ in year 1946. Plot the data and line.

4. (a) Plot the 16 points in the following table.

x	3.00	3.25	3.50	3.75	4.00	4.25	4.50	4.75	5.00	5.25	5.50	5.75	6.00	6.25	6.50	6.75
y	31.5	30.4	29.2	28.1	26.9	26.4	25.3	25.2	25.1	25.2	25.4	26.3	27.0	28.2	29.3	29.9

Do they seem to follow a parabolic path?

(b) If $y = ax^2 + bx + c$ is the equation of a parabola that is to approximate the function $y = f(x)$ described by these points, then the following sum of differences between observed and predicted values is a measure of the accuracy of the fit:

$$S = S(a, b, c) = \sum_{i=1}^{16} \left(ax_i^2 + bx_i + c - \bar{y}_i\right)^2,$$

where $(x_i, \bar{y}_i)$ are the points in the table. To find the best possible fit in the least-squares sense, we choose a, b, and c to minimize S. Show that S has only one critical point (a, b, c) that is defined by the linear equations

$$\left(\sum_{i=1}^{16} x_i^4\right) a + \left(\sum_{i=1}^{16} x_i^3\right) b + \left(\sum_{i=1}^{16} x_i^2\right) c = \sum_{i=1}^{16} x_i^2 \bar{y}_i,$$

$$\left(\sum_{i=1}^{16} x_i^3\right) a + \left(\sum_{i=1}^{16} x_i^2\right) b + \left(\sum_{i=1}^{16} x_i\right) c = \sum_{i=1}^{16} x_i \bar{y}_i,$$

$$\left(\sum_{i=1}^{16} x_i^2\right) a + \left(\sum_{i=1}^{16} x_i\right) b + 16c = \sum_{i=1}^{16} \bar{y}_i.$$

(c) Solve these equations for a, b, and c.

5. (a) Fit a least-squares quadratic (as in Exercise 4) to the following data.

x	2.0	2.2	2.4	2.6	2.8	3.0	3.2	3.4	3.6	3.8	4.0	4.2
y	7.06	11.34	15.62	19.50	25.62	31.94	37.02	44.32	51.56	58.72	67.08	75.91

(b) Calculate the value of S at the critical values of a, b, and c.

6. The following table relates head H and discharge Q from a pump. Find the best-fitting parabola $H = a + bQ^2$ for the data.

Q (L/s)	0	31.5	50.4	63.0	69.3	75.6	88.2	94.5
H (m)	37.8	34.4	30.8	27.3	24.7	21.3	13.1	7.8

7. (a) Fit a least-squares cubic polynomial $y = ax^3 + bx^2 + cx + d$ to data in Exercise 5.

(b) Calculate the value of S at the critical values of $a, b, c,$ and d. How does it compare with that in Exercise 5(b)?

8. Least squares can also be used with exponential functions. Suppose, for example, that we wish to fit a curve to the data in the following table:

x	0.5	1.0	1.5	2.0	2.5	3.0	3.5	4.0	4.5	5.0	5.5
y	140	180	230	290	365	455	565	670	785	1000	1230

(a) To determine whether the data can be described by a function of the form $y(x) = be^{ax}$, we take logarithms on both sides of the equation, $\ln y = \ln b + ax$. If we define $Y = \ln y$ and $B = \ln b$, then $Y = ax + B$, and this is the equation of a straight line in the xY-plane. In other words, to test whether a set of points can be described by an exponential be^{ax}, we plot $Y = \ln y$ against x, and if these points can be approximated by a straight line, then the original data should be describable by an exponential function. Do this for the data in the table.

(b) Find least-squares estimates for a and B, and hence find the best-fitting exponential for the original data.

9. In the study of longshore sand transport on beaches, the following data were recorded at the El Moreno Beach on the Baja in California. Longshore energy flux F is in units of joules per metre per second, and immersed weight transport W is in units of newtons per second.

F	6.0	15	18	20	30	38	43	104
W	6.1	9.9	20.8	14.6	25.7	42.8	45.1	84.4

Show that W can be represented in the form $W = aF^b$ and find least-squares estimates for a and b. Plot data points and the least-squares function to show the adequacy of the fit.

10. The following table shows the number of kilometres per litre recorded by six trucks at speeds of 80, 90, and 100 km/h.

Vehicle	80	90	100
Truck 1	2.23	2.20	2.05
Truck 2	2.35	2.18	2.00
Truck 3	2.37	2.16	1.97
Truck 4	2.27	2.13	2.10
Truck 5	1.95	1.91	1.80
Truck 6	2.16	1.96	1.92

Use this information to solve Exercise 59 in Section 4.7. Use $w = \$20$ and $p = \$0.60$.

11. The following table represents census figures for the population (in millions) for the United States from 1790 to 1910.

Year	1790	1800	1810	1820	1830	1840	1850	1860	1870	1880	1890	1900	1910
Population	3.9	5.3	7.2	9.6	12.9	17.1	23.2	31.4	39.8	50.2	62.9	76.0	92.0

Show that these data can be represented by an exponential function and find its least-squares estimates.

12. (a) To fit a power function $y = bx^a$ to data, we take logarithms, obtaining $\ln y = \ln b + a \ln x$. When we set $Y = \ln y$, $B = \ln b$, and $X = \ln x$, this equation becomes $Y = B + aX$, that for a straight line. This suggests that we plot Y against X to see if data points (X, Y) are reasonably collinear. Do so for the data in the following table.

x	10	16	25	40	60
y	94	118	147	180	223

(b) Find least-squares estimates for a and b by fitting a straight line to Y as a function of X.

13. The following table represents the length of time t (in seconds) that an athlete could hold a load F (in newtons) in the position shown in the figure below. Show that t can be represented in the form $t = aF^b$, and find least-squares estimates for a and b.

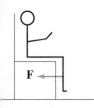

F	70	100	200	300	400
t	470	288	84	52	32

14. The number N of bacteria per unit volume in a culture after t hours is tabulated below.

t	0	1	2	3	4	5	6
N	32	47	65	92	132	190	275

    (a) Plot the data points $(t, \ln N)$ to show that it is reasonable to fit an equation of the form $N = be^{at}$ to the data.

    (b) Find least-squares estimates for a and b.

15. The following table gives experimental values of the pressure P of a given mass of gas corresponding to various volumes V. Use least squares to find estimates for constants a and b if thermodynamics suggests an equation of the form $PV^a = b$ to describe the data.

V	54.3	61.8	72.4	88.7	118.6	194.0
P	61.2	49.5	37.6	28.4	19.2	10.1

16. (a) You are given n pairs of observations $(x_i, \bar{y}_i)$ and are required to fit a curve of the form $y = b/x^2$ to them. By direct application of least squares, find a formula for b.

    (b) Apply the formula in part (a) to the data below.

x	5	10	15	20	25	30
y	0.022 43	0.005 06	0.002 97	0.001 47	0.000 98	0.000 67

    (c) Plot the data points and least-squares curve.

17. (a) Fit the curve $y = \ln(b + ax)$ to the following data by converting the given relation to linear form

$$e^y = ax + b$$

and using as data $e^{\bar{y}_i}$ and x_i.

x	2	3	4	5	6	7
y	1.952	2.156	2.413	2.549	2.670	2.821

    (b) Can you use least squares directly on the logarithm function? Explain.

18. Devise a method for using least squares to obtain a curve of the form

$$y = \frac{1}{ax + b}$$

to represent the following data.

x	5	6	7	8	9
y	1.335	1.431	1.247	1.197	1.118

19. A Cobb–Douglas production function has the form $P(x, y) = kx^q y^{1-q}$, where P is the number of items produced per unit time, x is the number of employees, and y is the operating budget for that time. The numbers $k > 0$ and $0 < q < 1$ are constants. Find least-squares estimates for k and q given the following production data.

Workers, x	100	110	90	100	95	105	110
Budget, y (dollars)	10 000	9000	9000	12 000	11 000	9500	10 000
Production, P	800	810	720	860	810	800	850

Hint: Write $P/y = k(x/y)^q$ and take logarithms.

12.14 Differentials

If $y = f(x)$ is a function of one variable, the differential of y, defined by $dy = f'(x)\,dx$, is an approximation to the increment $\Delta y = f(x + dx) - f(x)$ for small dx. In particular, dy is the change in y corresponding to the change dx in x if we follow the tangent line to the curve at (x, y) instead of the curve itself.

We take the same approach in defining differentials for multivariable functions. First consider a function $f(x, y)$ of two independent variables that can be represented geometrically as a surface with equation $z = f(x, y)$ (Figure 12.68). If we change the values of x and y by amounts $\Delta x = dx$ and $\Delta y = dy$, then the corresponding change in z is

$$\Delta z = f(x + dx, y + dy) - f(x, y).$$

Geometrically, this is the difference in the heights of the surface at the points $(x + dx, y + dy)$ and (x, y). If we draw the tangent plane to the surface at (x, y), then very near (x, y) the height of the tangent plane approximates the height of the surface (Figure 12.69). In particular, the height of the tangent plane at $(x + dx, y + dy)$ for small dx and dy approximates the height of the surface. We define the *differential* dz as the change in z corresponding to the changes dx and dy in x and y if we follow the tangent plane at (x, y) instead of the surface itself. To find dz in terms of dx and dy, we note that the vector joining the points (x, y, z) and $(x + dx, y + dy, z + dz)$ has components (dx, dy, dz), and this vector lies in the tangent plane. Since a normal vector to the tangent plane is

$$\nabla(z - f(x, y)) = (-f_x, -f_y, 1),$$

it follows that the vectors $(-f_x, -f_y, 1)$ and (dx, dy, dz) must be perpendicular. Consequently,

$$0 = (-f_x, -f_y, 1) \cdot (dx, dy, dz) = -\frac{\partial f}{\partial x}dx - \frac{\partial f}{\partial y}dy + dz,$$

and hence

$$dz = \frac{\partial f}{\partial x}dx + \frac{\partial f}{\partial y}dy. \tag{12.72}$$

Note that if y is held constant in the function $f(x, y)$, then $dy = 0$ and 12.72 for dz reduces to the definition of the differential of a function of one variable.

FIGURE 12.68 Δz is exact difference in heights of surface at (x, y) and at $(x + dx, y + dy)$

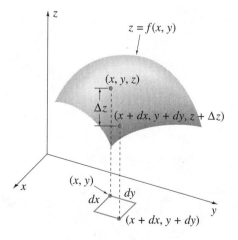

FIGURE 12.69 dz is difference in heights of surface at (x, y) and tangent plane at $(x + dx, y + dy)$

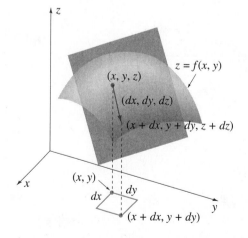

In Section 4.12 we indicated that we must be careful in using the differential dy as an approximation for the change Δy in a function $f(x)$. For the same reasons, we must be judicious in our use of dz as an approximation for Δz. Indeed, we have stated that dz is an approximation for Δz for small dx and dy, but the difficulty is deciding how small is small and how good is the approximation. In addition, note that if (x, y) is a critical point of the function $f(x, y)$, then either $dz = 0$ for all dx and dy, or dz is undefined. In other words, dz cannot be used to approximate Δz at a critical point.

EXAMPLE 12.38

If the radius of a right-circular cone is changed from 10 cm to 10.1 cm and the height is changed from 1 m to 0.99 m, use differentials to approximate the change in its volume.

SOLUTION   The volume of a cone of radius r and height h is given by the formula $V = \pi r^2 h/3$. The differential of this function is

$$dV = \frac{\partial V}{\partial r} dr + \frac{\partial V}{\partial h}\, dh = \frac{2}{3}\pi r h\, dr + \frac{1}{3}\pi r^2 dh.$$

If $r = 10$, $dr = 0.1$, $h = 100$, and $dh = -1$, then

$$dV = \frac{2}{3}\pi(10)(100)(0.1) + \frac{1}{3}\pi(10)^2(-1) = \frac{100\pi}{3}\ \text{cm}^3.$$

Equation 12.72 suggests the following definition for the differential of a function of more than two independent variables.

DEFINITION 12.12

If $u = f(x, y, z, t, \ldots, w)$, then the **differential** of $f(x, y, z, t, \ldots, w)$ is defined as

$$du = \frac{\partial f}{\partial x} dx + \frac{\partial f}{\partial y} dy + \frac{\partial f}{\partial z} dz + \frac{\partial f}{\partial t} dt + \cdots + \frac{\partial f}{\partial w} dw. \tag{12.73}$$

EXAMPLE 12.39

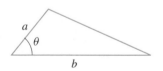

FIGURE 12.70   Using differentials to approximate change in area of a triangle

The area of the triangle in Figure 12.70 is given by the formula $A = \dfrac{1}{2}ab \sin\theta$. If when $\theta = \pi/3$, a and b are changed by $\dfrac{1}{3}\%$ and θ by $\dfrac{1}{2}\%$, use differentials to find the approximate percentage change in A.

SOLUTION   Since

$$dA = \frac{\partial A}{\partial a} da + \frac{\partial A}{\partial b} db + \frac{\partial A}{\partial \theta} d\theta$$

$$= \frac{1}{2}b\, \sin\theta\, da + \frac{1}{2}a\, \sin\theta\, db + \frac{1}{2}ab\, \cos\theta\, d\theta,$$

the approximate percentage change in A is

$$100\left(\frac{dA}{A}\right) = \frac{100}{A}\left(\frac{1}{2}b\, \sin\theta\, da + \frac{1}{2}a \sin\theta\, db + \frac{1}{2}ab\, \cos\theta\, d\theta\right)$$

$$= 100\left(\frac{da}{a} + \frac{db}{b} + \cot\theta d\theta\right).$$

Since a and b are changed by $\dfrac{1}{3}\%$ and θ by $\dfrac{1}{2}\%$,

$$100\left(\frac{da}{a}\right) = 100\left(\frac{db}{b}\right) = \frac{1}{3} \quad \text{and} \quad 100\left(\frac{d\theta}{\theta}\right) = \frac{1}{2}.$$

Thus,

$$100\left(\frac{dA}{A}\right) = \frac{1}{3} + \frac{1}{3} + \frac{\theta}{2}\cot\theta = \frac{2}{3} + \frac{\theta}{2}\cot\theta,$$

and when $\theta = \pi/3$,

$$100\left(\frac{dA}{A}\right) = \frac{2}{3} + \frac{1}{2}\left(\frac{\pi}{3}\right)\left(\frac{1}{\sqrt{3}}\right) = 0.97.$$

The approximate percentage change in A is therefore 1%.

Consulting Project 20

When n resistances R_i $(i = 1, \ldots, n)$ are connected in series, then the resultant resistance is

$$R = R_1 + R_2 + \cdots + R_n.$$

When they are connected in parallel, the effective resistance, R, is given by

$$\frac{1}{R} = \frac{1}{R_1} + \frac{1}{R_2} + \cdots + \frac{1}{R_n}.$$

We are being asked the following question. If each resistance R_i is changed by the same small percentage c, what is the percentage change in the series and parallel combinations?

SOLUTION   Let us consider the series case first. Because percentage changes in the R_i are small, we shall use differentials to calculate the percentage change in R,

$$dR = \frac{\partial R}{\partial R_1}dR_1 + \cdots + \frac{\partial R}{\partial R_n}dR_n = dR_1 + \cdots + dR_n.$$

The percentage change in R is therefore

$$100\frac{dR}{R} = \frac{100(dR_1 + \cdots + dR_n)}{R_1 + \cdots + R_n}$$

$$= \frac{100dR_1}{R_1\left(1 + \dfrac{R_2}{R_1} + \cdots + \dfrac{R_n}{R_1}\right)} + \frac{100dR_2}{R_2\left(\dfrac{R_1}{R_2} + 1 + \cdots + \dfrac{R_n}{R_2}\right)}$$

$$+ \cdots + \frac{100dR_n}{R_n\left(\dfrac{R_1}{R_n} + \cdots + \dfrac{R_{n-1}}{R_n} + 1\right)}.$$

Since the percentage change in each resistance R_i is c, it follows that $\frac{100dR_i}{R_i} = c$. Hence,

$$\frac{100dR}{R} = \frac{c}{1 + \dfrac{R_2}{R_1} + \cdots + \dfrac{R_n}{R_1}} + \frac{c}{\dfrac{R_1}{R_2} + 1 + \cdots + \dfrac{R_n}{R_2}}$$

$$+ \cdots + \frac{c}{\dfrac{R_1}{R_n} + \cdots + \dfrac{R_{n-1}}{R_n} + 1}$$

$$= \frac{cR_1}{R_1 + \cdots + R_n} + \frac{cR_2}{R_1 + \cdots + R_n} + \cdots + \frac{cR_n}{R_1 + \cdots + R_n} = c.$$

For the parallel case, we write that $R = \left(\dfrac{1}{R_1} + \cdots + \dfrac{1}{R_n} \right)^{-1}$, and take differentials,

$$dR = - \left(\frac{1}{R_1} + \cdots + \frac{1}{R_n} \right)^{-2} \left(-\frac{1}{R_1^2} dR_1 - \cdots - \frac{1}{R_n^2} dR_n \right)$$

$$= R^2 \left(\frac{1}{R_1^2} dR_1 + \cdots + \frac{1}{R_n^2} dR_n \right).$$

The percentage change in R is therefore

$$100 \frac{dR}{R} = 100 R \left(\frac{1}{R_1^2} dR_1 + \cdots + \frac{1}{R_n^2} dR_n \right)$$

$$= \frac{R}{R_1} \left(100 \frac{dR_1}{R_1} \right) + \cdots + \frac{R}{R_n} \left(100 \frac{dR_n}{R_n} \right)$$

$$= \frac{cR}{R_1} + \cdots + \frac{cR}{R_n}$$

$$= cR \left(\frac{1}{R_1} + \cdots + \frac{1}{R_n} \right) = c.$$

Thus, whether resistances are connected in series or parallel, small equal percentage changes in the individual resistances results in the same percentage change in the resultant resistance.

EXERCISES 12.14

In Exercises 1–10 find the differential of the function.

1. $f(x, y) = x^2 y - \sin y$

2. $f(x, y) = \text{Tan}^{-1}(xy)$

3. $f(x, y, z) = xyz - x^3 e^z$

4. $f(x, y, z) = \sin(xyz) - x^2 y^2 z^2$

5. $f(x, y, z) = \ln(x^2 + y^2 + z^2)$

6. $f(x, y) = \text{Sin}^{-1}(xy)$

7. $f(x, y) = \text{Sin}^{-1}(x + y) + \text{Cos}^{-1}(x + y)$

8. $f(x, y, z, t) = xy + yz + zt + xt$

9. $f(x, y, z, w) = xy \tan(zw)$

10. $f(x, y, z, t) = e^{x^2 + y^2 + z^2 - t^2}$

11. A right-circular cone has radius 10 cm and height 20 cm. If its radius increases by 0.1 cm and its height decreases by 0.3 cm, use differentials to find the approximate change in its volume. Compare this with the actual change in volume.

12. When the ellipse $b^2 x^2 + a^2 y^2 = a^2 b^2$ is rotated about the x-axis, the volume V of the spheroid is $4\pi a b^2 / 3$. If a and b are each increased by 1%, use differentials to find the approximate percentage change in V.

12.15 Taylor Series for Multivariable Functions

Taylor series for functions of one variable can be used to generate Taylor series for multivariable functions. For simplicity, we once again work with functions of two independent variables. Extensions to functions of more than two independent variables will be clear. Suppose that a function $f(x, y)$ has continuous partial derivatives of all orders in some open circle centred at the point (c, d) (Figure 12.71).

FIGURE 12.71   Finding the Taylor series of a function $f(x, y)$ about a point (c, d)

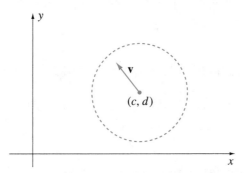

Parametric equations for the line through (c, d) in direction $\mathbf{v} = (v_x, v_y)$ are

$$x = c + v_x t, \quad y = d + v_y t.$$

If we substitute these values into $f(x, y)$, we obtain a function $F(t)$ of one variable,

$$F(t) = f(c + v_x t, d + v_y t),$$

which represents the value of $f(x, y)$ at points along the line through (c, d) in direction $\mathbf{v}$. If we expand this function into its Maclaurin series, we obtain

$$F(t) = F(0) + F'(0)t + \frac{F''(0)}{2!}t^2 + \cdots. \tag{12.74}$$

The schematic diagram to the left gives

$$F'(t) = \frac{\partial F}{\partial x}\frac{dx}{dt} + \frac{\partial F}{\partial y}\frac{dy}{dt} = f_x(x, y)v_x + f_y(x, y)v_y,$$

and therefore

$$F'(0) = f_x(c, d)v_x + f_y(c, d)v_y.$$

The schematic for $F'(t)$ gives the second derivative of $F(t)$,

$$F''(t) = \frac{\partial}{\partial x}[F'(t)]\frac{dx}{dt} + \frac{\partial}{\partial y}[F'(t)]\frac{dy}{dt}$$

$$= \frac{\partial}{\partial x}[f_x(x, y)v_x + f_y(x, y)v_y]v_x + \frac{\partial}{\partial y}[f_x(x, y)v_x + f_y(x, y)v_y]v_y$$

$$= f_{xx}(x, y)v_x^2 + 2f_{xy}(x, y)v_x v_y + f_{yy}(x, y)v_y^2,$$

and therefore

$$F''(0) = f_{xx}(c, d)v_x^2 + 2f_{xy}(c, d)v_x v_y + f_{yy}(c, d)v_y^2.$$

A similar calculation gives

$$F'''(0) = f_{xxx}(c, d)v_x^3 + 3f_{xxy}(c, d)v_x^2 v_y + 3f_{xyy}(c, d)v_x v_y^2 + f_{yyy}(c, d)v_y^3,$$

and the pattern is emerging. When these results are substituted into 12.74,

$$F(t) = f(c + v_x t, d + v_y t)$$

$$= f(c, d) + [f_x(c, d)v_x + f_y(c, d)v_y]t$$

$$+ [f_{xx}(c, d)v_x^2 + 2f_{xy}(c, d)v_x v_y + f_{yy}(c, d)v_y^2]\frac{t^2}{2!} + \cdots. \tag{12.75}$$

We now let $\mathbf{v}$ be the vector from (c, d) to point (x, y), so that $v_x = x - c$ and $v_y = y - d$, and at the same time set $t = 1$. Then $F(1) = f(c + x - c, d + y - d) = f(x, y)$, and 12.75 becomes

$$f(x, y) = f(c, d) + [f_x(c, d)(x - c) + f_y(c, d)(y - d)] + \frac{1}{2!}[f_{xx}(c, d)(x - c)^2$$

$$+ 2f_{xy}(c, d)(x - c)(y - d) + f_{yy}(c, d)(y - d)^2] + \cdots. \qquad (12.76)$$

This is the **Taylor series** for $f(x, y)$ about the point (c, d). It gives the value of the function at the point (x, y) in terms of values of the function and its derivatives at the point (c, d).

EXAMPLE 12.40

Find the first six nonzero terms in the Taylor series for $f(x, y) = \sin(2x + 3y)$ about $(0, 0)$.

SOLUTION   We calculate that

$$f(0, 0) = 0,$$

$$f_x(0, 0) = 2\cos(2x + 3y)_{|(0,0)} = 2,$$

$$f_y(0, 0) = 3\cos(2x + 3y)_{|(0,0)} = 3,$$

$$f_{xx}(0, 0) = -4\sin(2x + 3y)_{|(0,0)} = 0,$$

$$f_{xy}(0, 0) = -6\sin(2x + 3y)_{|(0,0)} = 0,$$

$$f_{yy}(0, 0) = -9\sin(2x + 3y)_{|(0,0)} = 0,$$

$$f_{xxx}(0, 0) = -8\cos(2x + 3y)_{|(0,0)} = -8,$$

$$f_{xxy}(0, 0) = -12\cos(2x + 3y)_{|(0,0)} = -12,$$

$$f_{xyy}(0, 0) = -18\cos(2x + 3y)_{|(0,0)} = -18,$$

$$f_{yyy}(0, 0) = -27\cos(2x + 3y)_{|(0,0)} = -27.$$

Formula 12.76 then gives

$$\sin(2x + 3y) = 0 + [2x + 3y] + \frac{1}{2!}[0] + \frac{1}{3!}[-8x^3 - 36x^2y - 54xy^2 - 27y^3] + \cdots$$

$$= (2x + 3y) - \frac{1}{3!}(2x + 3y)^3 + \cdots.$$

This series could also have been obtained by substituting $2x + 3y$ for x in the Maclaurin series for $\sin x$. This is not always an alternative.

EXERCISES 12.15

1. If $f(x, y) = F(x)G(y)$, is the Taylor series of $f(x, y)$ about $(0, 0)$ the product of the Maclaurin series for $F(x)$ and $G(y)$?

2. What are the cubic terms in 12.76?

In Exercises 3–8 find the Taylor series of the function about the point by using Taylor series for functions of one variable.

3. $\cos(xy)$ about $(0, 0)$

4. e^{2x-3y} about $(1, -1)$

5. $x^2 y \sqrt{1+x}$ about $(0, 0)$

6. $\ln(1 + x^2 + y^2)$ about $(0, 0)$

7. $\dfrac{1}{1+x+y}$ about $(3, -4)$

8. $\dfrac{xy^2}{1+y^2}$ about $(-1, 0)$

In Exercises 9–14 find the Taylor series of the function up to and including quadratic terms.

9. $\dfrac{xy}{x^2 + y^2}$ about $(-1, 1)$

* **10.** $\sqrt{1+xy}$ about $(2, 1)$

* **11.** $e^x \sin(3x - y)$ about $(-1, 0)$

* **12.** $(x + y)^2 \ln(x + y)$ about $(0, 1)$

* **13.** $\text{Tan}^{-1}(3x + 2y)$ about $(1, -1)$

* **14.** $x^8 y^{10}$ about $(0, 0)$

* **15.** What are the terms in the Taylor series for a function $f(x, y, z)$ about the point (c, d, e) corresponding to those in equation 12.76?

* **16.** Express 12.76 in sigma notation. *Hint:* Think about an operator,

$$\left[(x - a)\frac{\partial}{\partial x} + (y - b)\frac{\partial}{\partial y} \right]^n,$$

which is expanded as a binomial to operate on functions $f(x, y)$.

SUMMARY

We began the study of multivariable functions in this chapter, concentrating our attention on differentiation and its applications. We introduced two types of derivatives for a multivariable function: partial derivatives and directional derivatives. Partial derivatives are directional derivatives in directions parallel to the coordinate axes. The directional derivative of a function $f(x, y, z)$ in the direction $\mathbf{v}$ is given by the formula $D_{\mathbf{v}} f = \nabla f \cdot \hat{\mathbf{v}}$, where $\hat{\mathbf{v}}$ is the unit vector in the direction of $\mathbf{v}$, and the gradient ∇f is evaluated at the point at which $D_{\mathbf{v}} f$ is required. This formula leads to the fact that the gradient $\nabla f(x, y, z)$ points in the direction in which $f(x, y, z)$ increases most rapidly, and $|\nabla f|$ is the (maximum) rate of change of $f(x, y, z)$. A second property of gradient vectors (that is related to the first) is that if $F(x, y, z) = C$, C a constant, is the equation of a surface, then at any point on the surface ∇F is perpendicular to the surface. This property, along with the fact that perpendicularity to a surface is synonymous with perpendicularity to its tangent plane, enables us to find equations for tangent planes to surfaces and tangent lines to curves.

We illustrated various ways to calculate partial derivatives of a multivariable function, depending on whether the function is defined explicitly, implicitly, or as a composite function. Since partial derivatives are ordinary derivatives with other variables held constant, there is no difficulty calculating partial derivatives when the function is defined explicitly; we simply use the rules from single-variable calculus. When the partial derivative of a composite function is required, we use a schematic diagram illustrating functional dependences to develop the appropriate chain rule. Partial derivatives for functions defined implicitly are calculated using Jacobians.

Critical points of a multivariable function are points at which all of its first partial derivatives vanish or at which one or more of these partial derivatives does not exist. Critical points can yield relative maxima, relative minima, saddle points, or none of these. For functions of two independent variables, a second-derivative test exists that may determine whether a critical point at which the partial derivatives vanish yields a relative maximum, a relative minimum, or a saddle point. This test is analogous to that for functions of one variable.

A continuous function of two independent variables defined on a region that includes its boundary always takes on a maximum value and a minimum value. To find these values we evaluate the function at each of its critical points and compare these numbers to the maximum and minimum values of the function on its boundary. Finding the extreme values on the boundary involves one or more extrema problems for a function of one variable, the number of such problems depending on the complexity of the boundary.

There are two methods for finding extreme values of a function when the variables of the function are subject to constraints: solve the constraint equations for dependent variables and express the given function in terms of independent variables, or use Lagrange multipliers. Lagrange multipliers eliminate the necessity for solving the constraint equations, but they do, on the other hand, give a larger system of equations to solve for critical points.

Differentials of multivariable functions can be used to approximate changes in functions when small changes are made to its independent variables. Taylor series can also be used to approximate multivariable functions.

The method of least squares fits a function $y = f(x)$ of known form to a set of data. It minimizes the sum of the squares of the differences between measured and predicted values of y.

KEY TERMS

In reviewing this chapter, you should be able to define or discuss the following key terms:

Domain	Limit
Continuous function	Partial derivative
Gradient	Second partial derivative
Laplace's equation	Harmonic function
Chain rule	Positively homogeneous function
Euler's Theorem	Implicit differentiation
Jacobian determinant	Directional derivative
Tangent line to a curve	Tangent plane to a surface
Normals to surfaces	Smooth surface
Piecewise-smooth surface	Critical point
Relative maximum	Relative minimum
Saddle point	Absolute maximum
Absolute minimum	Constraint problems
Lagrange multipliers	Lagrangian
Least squares	Differential
Taylor series	

REVIEW EXERCISES

In Exercises 1–20 find the derivative.

1. $\partial f/\partial x$ if $f(x, y) = x^2/y^3 - \mathrm{Sin}^{-1}(xy)$

2. $\partial^2 f/\partial y^2$ if $f(x, y, z) = \ln(x^2 + y^2 + z^2)$

3. $\partial^3 f/\partial x^2 \partial y$ if $f(x, y, z, t) = x^3 e^y - xzt^2 - \sin(x + y + z + t)$

4. $\partial z/\partial x$ if $z^2 x + \mathrm{Tan}^{-1} z + y = 3x$

5. $\partial u/\partial y$ if $u \cos y + y \cos(xu) + z^2 = 5x$

6. df/dt if $f(x, y) = x^2 + y^2 - e^{xy}$, $x = t^3 + 3t$, $y = t \ln t$

7. dy/dx if $x = y^3 + 3y^2 - 2y + 4$

8. $\partial u/\partial x)_y$ if $u^2 + v^2 - xy = 5$, $3u - 2v + x^2 u = 2v^3$

9. $\partial^2 f/\partial u \partial v$ if $f(u, v) = u^2/\sqrt{v} - v/\sqrt{u}$

10. df/dt if $f(x, y) = xy - x^2 - y^2$, $x = te^t$, $y = te^{-t}$

11. $\partial z/\partial t)_u$ if $z = x^2 - y^2$, $x = 2u - 3v^2 + 3uvt$, $y = u\cos(vt)$, $v = t^2 - 2t$

12. $\partial r/\partial x)_y$ if $x = r \cos\theta$, $y = r \sin\theta$

13. $\partial\theta/\partial x)_{y,z}$ if $x = r \sin\phi \cos\theta$, $y = r \sin\phi \sin\theta$, $z = r \cos\phi$

14. $\partial u/\partial r)_\theta$ if $u = x^2 - y^2 x^3$, $x = r\cos\theta$, $y = r\sin\theta$

15. $\partial^2 u/\partial r^2)_\theta$ if $u = x^2 - y^2 x^3$, $x = r\cos\theta$, $y = r\sin\theta$

16. $d^2 u/dt^2$ if $u = x/z^2 - z/x^2$, $x = t^3 - 3$, $z = 1/t^3$

17. dz/dt if $z = y - xy^2 + x$, and $x^2 - y^2 + xt = 2t$, $xy = 4t^2$

18. $\partial^2 z/\partial x^2$ if $xz - x^2 z^3 + y^2 = 3$

19. dy/dx if $yx - x^2 z^2 + 5x = 3$, $2xz - 3x^2 y^2 = 4z^4$

20. $\partial u/\partial t)_v$ if $u = xyt^2 - 3\,\mathrm{Sin}^{-1}(xy)$, $x = v^2 t^2 - 2t$, $y = v\tan t$

＊ 21. If $u = (x^2 + y^2)[1 + \sin(x/z)]$, show that
$$x\frac{\partial u}{\partial x} + y\frac{\partial u}{\partial y} + z\frac{\partial u}{\partial z} = 2u.$$

22. If $u = 2x^2 - 3y^2 + xy$, show that

$$x^2 \frac{\partial^2 u}{\partial x^2} + 2xy \frac{\partial^2 u}{\partial x \, \partial y} + y^2 \frac{\partial^2 u}{\partial y^2} = 2u.$$

23. If $f(s)$ is a differentiable function, show that $f(3x - 2y)$ satisfies

$$2 \frac{\partial f}{\partial x} + 3 \frac{\partial f}{\partial y} = 0.$$

24. If $f(s, t)$ has continuous first partial derivatives, show that the function $f(x^2 - y^2, y^2 - x^2)$ satisfies $y \dfrac{\partial f}{\partial x} + x \dfrac{\partial f}{\partial y} = 0$.

In Exercises 25–30 find the directional derivative.

25. $f(x, y) = x^2 \sin y$ at $(3, -1)$ in the direction $\mathbf{v} = (2, 4)$

26. $f(x, y, z) = x^2 + y^2 + z^2$ at $(1, 0, 1)$ in the direction from $(1, 0, 1)$ to $(2, -1, 3)$

27. $f(x, y, z) = z \, \mathrm{Tan}^{-1}(x + y)$ at $(-1, 2, 5)$ in the direction perpendicular to the surface $z = x^2 + y^2$ with positive z-component

28. $f(x, y, z) = x^2 + y - 2z$ at $(1, -1, 2)$ along the line $x - y + z = 4$, $2x + 4y + 2 = 0$ in the direction of increasing x

29. $f(x, y) = \ln(x + y)$ at $(3, 10)$ along the curve $y = x^2 + 1$ in the direction of decreasing y

30. $f(x, y, z) = 2xyz - x^2 - z^2$ at $(0, 1, 1)$ along the curve $x^2 + y^2 + z^2 = 2$, $y = z$ in the direction of increasing x

In Exercises 31–33 find the equation of the tangent plane to the surface.

31. $z = x^2 + y^2$ at $(1, 3, 10)$ **32.** $x^2 + z^3 = y^2$ at $(-1, 3, 2)$

33. $x^2 + y^2 = z^2 + 1$ at $(1, 0, 0)$

In Exercises 34–36 find equations for the tangent line to the curve.

34. $x = t^2 + 1$, $y = t^2 - 1$, $z = t^3 + 5t$ at $(2, 0, 6)$

35. $x + y + z = 0$, $2x - 3y - 6z = 11$ at $(1, 1, -2)$

36. $z = xy$, $x^2 + y^2 = 2$ at $(1, 1, 1)$

In Exercises 37–40 find all critical points for the function and classify each as yielding a relative maximum, a relative minimum, or a saddle point.

37. $f(x, y) = x^3 + 3y^2 - 6x + 4$

38. $f(x, y) = ye^x$

39. $f(x, y) = x^2 - xy + y^2 + x - 4y$

40. $f(x, y) = (x^2 + y^2 - 1)^2$

$*$ **41.** If $f(x, y) = (x^2 + y^2) F(x, y)$ where $F(x, y) = x^3/y - y^3/x$, verify that

$$\frac{\partial^2 f}{\partial x^2} + \frac{\partial^2 f}{\partial y^2} = (x^2 + y^2) \left(\frac{\partial^2 F}{\partial x^2} + \frac{\partial^2 F}{\partial y^2} \right) + 12 F(x, y).$$

$*$ **42.** Find maximum and minimum values of the function $f(x, y) = xy$ on the circle $x^2 + y^2 \le 1$.

$*$ **43.** Find maximum and minimum values of the function $f(x, y, z) = 2x + 3y - 4z$ on the sphere $x^2 + y^2 + z^2 \le 2$.

$*$ **44.** Find the points on the curve $x^2 + x^4 + y^2 = 1$ closest to and farthest from the origin.

$*$ **45.** Find the point(s) on the surface $z^2 = 1 + xy$ closest to the origin.

$*$ **46.** Generalize Review Exercise 32 in Chapter 4 to incorporate a third crop, say sunflowers, with a yield of r dollars per hectare and a proportional loss cz.

$*$ **47.** If the equation $u = f(x - ut)$ defines u implicitly as a function of x and t, show that

$$\frac{\partial u}{\partial t} + u \frac{\partial u}{\partial x} = 0.$$

$*$ **48.** Find the first six nonzero terms in the Taylor series for $x^3 \sin(x^2 y)$ about the point $(1, \pi/4)$.

$*$ **49.** Find the best possible line, in the least-squares sense, to fit the data in the following table.

x	1	2	3	4	5	6	7	8
y	1.2	4.6	8.4	12.2	15.6	19.7	23.0	26.9

In the figure below, a freshwater marsh is drained to the ocean through an automatic tide gate that is L metres wide and 0.9 m high. The gate is held by an L-metre-long hinge at A and bears on a sill at B. The water level in the marsh is 1.8 m, and the density of ocean water is 1030 kg/m^3 compared to 1000 kg/m^3 for the fresh water in the marsh. The marsh and the ocean both create forces on the gate, and when water levels are the same, the ocean creates a greater force, thus keeping the gate closed. As the ocean level falls, the force it exerts on the gate decreases, and eventually the gate opens.

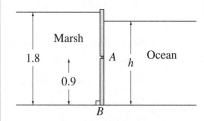

THE PROBLEM   At what depth of ocean water will the tide gate open? (See Example 13.8 on page 914 for the solution.)

The definite integral of a function $f(x)$ of one variable is defined as the limit of a sum of the form

$$f(x_1^*)\,\Delta x_1 + f(x_2^*)\,\Delta x_2 + \cdots + f(x_n^*)\,\Delta x_n, \qquad (13.1)$$

where the norm of the partition approaches zero. We have seen that definite integrals can be used to calculate area, volume, work, fluid force, and moments. In spite of the fact that some of these are two- and three-dimensional concepts, we have been careful to emphasize that a definite integral with respect to x is an integration along the x-axis, and a definite integral with respect to y is an integration along the y-axis. In other words, independent of how we interpret the result of the integration, a definite integral is a limit summation *along a line*. Generalizations of these limiting sums to functions of two and three independent variables lead to definitions of double and triple integrals.

13.1 Double Integrals and Double Iterated Integrals

Suppose a function $f(x, y)$ is defined in some region R of the xy-plane that has finite area (Figure 13.1). To define the double integral of $f(x, y)$ over R, we first divide R into n subregions of areas $\Delta A_1, \Delta A_2, \ldots, \Delta A_n$, in any manner whatsoever. In each subregion $\Delta A_i (i = 1, \ldots, n)$ we choose an arbitrary point (x_i^*, y_i^*) and form the sum

$$f(x_1^*, y_1^*)\,\Delta A_1 + f(x_2^*, y_2^*)\,\Delta A_2 + \cdots + f(x_n^*, y_n^*)\,\Delta A_n = \sum_{i=1}^{n} f(x_i^*, y_i^*)\,\Delta A_i.$$

$$(13.2)$$

The norm of the partition of R into subareas ΔA_i is the area of the largest of the subareas, denoted by $\|\Delta A_i\| = \max_{i=1,\ldots,n} \Delta A_i$.

FIGURE 13.1 Subdivide R into smaller regions to define the double integral of $f(x, y)$ over R

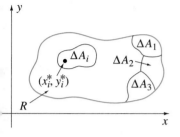

Suppose we increase the number of terms in 13.2 by increasing the number of subareas ΔA_i and decreasing the norm $\| \Delta A_i \|$. If the sum approaches a limit as the number of subareas becomes increasingly large and each subarea shrinks to a point, we call the limit the **double integral** of $f(x, y)$ over the region R, and denote it by

$$\iint_R f(x, y)\, dA = \lim_{\|\Delta A_i\| \to 0} \sum_{i=1}^{n} f(x_i^*, y_i^*)\, \Delta A_i. \tag{13.3}$$

The notation $\| \Delta A_i \| \to 0$ does not necessarily require that every ΔA_i shrink to a point. We implicitly assume, however, that this is always the case.

In some texts the R is placed below the integral signs as in

$$\iint_R f(x, y)\, dA.$$

We will use the notation in equation 13.3, but be aware of the alternative.

If the limit in 13.3 were dependent on the choice of subdivision ΔA_i or choice of star points (x_i^*, y_i^*), double integrals would be of little use. We therefore demand that the limit of the sum be independent of the manner of subdivision of R and choice of star points in the subregions. The following theorem indicates that for continuous functions this is always the case.

THEOREM 13.1

Let C be a closed, piecewise-smooth curve that encloses a region R with finite area. If $f(x, y)$ is a continuous function inside and on C, then the double integral of $f(x, y)$ over R exists.

For a continuous function, then, the double integral exists, and any choice of subdivision and star points leads to the same value through limiting process 13.3. Note that continuity was also the condition that guaranteed existence of the definite integral in Theorem 6.2.

We cannot overemphasize the fact that a double integral is simply the limit of a sum. Moreover, any limit of form 13.3 may be interpreted as the double integral of a function $f(x, y)$ over the region defined by the ΔA_i.

The following properties of double integrals are easily proved using definition 13.3:

1. If the double integral of $f(x, y)$ over R exists and c is a constant, then

$$\iint_R cf(x, y)\, dA = c \iint_R f(x, y)\, dA. \tag{13.4}$$

2. If double integrals of $f(x, y)$ and $g(x, y)$ over R exist, then

$$\iint_R [f(x, y) + g(x, y)]\, dA = \iint_R f(x, y)\, dA + \iint_R g(x, y)\, dA. \tag{13.5}$$

3. If a region R is subdivided by a piecewise-smooth curve into two parts R_1 and R_2 that have at most boundary points in common (Figures 13.2), and the double integral of $f(x, y)$ over R exists, then

$$\iint_R f(x, y)\, dA = \iint_{R_1} f(x, y)\, dA + \iint_{R_2} f(x, y)\, dA. \tag{13.6}$$

4. The area of a region R can be obtained by integrating the function $f(x, y) = 1$ over R:

$$\text{area of } R = \iint_R dA. \tag{13.7}$$

In spite of the fact that double integrals are defined as limits of sums, we do not evaluate them as such. Just as definite integrals are evaluated with antiderivatives, we evaluate double integrals with double iterated integrals.

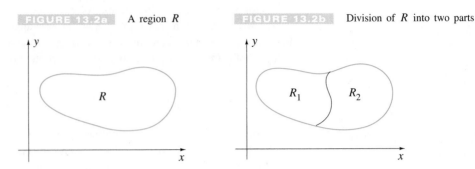

FIGURE 13.2a   A region R

FIGURE 13.2b   Division of R into two parts

Double Iterated Integrals

We have already seen that a function $f(x, y)$ of two independent variables has two first-order partial derivatives, one with respect to x holding y constant, and one with respect to y holding x constant. We now reverse this process and define "partial" integration of $f(x, y)$ with respect to x and y. Quite naturally, we define a partial antiderivative of $f(x, y)$ with respect to x as an antiderivative of $f(x, y)$ with respect to x, holding y constant. For example, since

$$\frac{\partial}{\partial x}(x^3 + x^2 y) = 3x^2 + 2xy,$$

$x^3 + x^2 y$ is an antiderivative with respect to x of $3x^2 + 2xy$. But so is $x^3 + x^2 y + y$. In fact, for any differentiable function $C(y)$ of y, $x^3 + x^2 y + C(y)$ is an antiderivative of $3x^2 + 2xy$ with respect to x. Since this expression represents all antiderivatives of $3x^2 + 2xy$, we call it the partial indefinite integral of $3x^2 + 2xy$ with respect to x, and write

$$\int (3x^2 + 2xy)\, dx = x^3 + x^2 y + C(y).$$

Similarly, the partial indefinite integral of $3x^2 + 2xy$ with respect to y is

$$\int (3x^2 + 2xy)\, dy = 3x^2 y + xy^2 + D(x),$$

where $D(x)$ is an arbitrary differentiable function of x.

In this chapter we are concerned only with partial definite integrals. Limits on a partial definite integral with respect to x must not depend on x, but may depend on y. In general, then, a partial definite integral with respect to x is of the form

$$\int_{g(y)}^{h(y)} f(x, y)\, dx;\tag{13.8}$$

and a partial definite integral with respect to y is of the form

$$\int_{g(x)}^{h(x)} f(x, y)\, dy.\tag{13.9}$$

Each of these partial definite integrals is evaluated by substituting the limits into a corresponding antiderivative. For example,

$$\int_{x^2}^{x+2} (2y + xe^y)\, dy = \{y^2 + xe^y\}_{x^2}^{x+2}$$

$$= \{(x + 2)^2 + xe^{x+2}\} - \{(x^2)^2 + xe^{x^2}\}$$

$$= (x + 2)^2 - x^4 + x(e^{x+2} - e^{x^2}).$$

Once antidifferentiation in 13.8 is completed and the limits substituted, the result is a function of y alone. It is then possible to integrate this function with respect to y between any two limits, say from $y = c$ to $y = d$:

$$\int_c^d \left\{ \int_{g(y)}^{h(y)} f(x, y) \, dx \right\} dy.$$

In practice we omit the braces and simply write

$$\int_c^d \int_{g(y)}^{h(y)} f(x, y) \, dx \, dy, \qquad (13.10)$$

understanding that in the evaluation we proceed from the inner integral to the outer. This is called a **double iterated integral** first with respect to x and then with respect to y (or, more concisely, with respect to x and y). Double iterated integrals with respect to y and x take the form

$$\int_a^b \int_{g(x)}^{h(x)} f(x, y) \, dy \, dx. \qquad (13.11)$$

EXAMPLE 13.1

Evaluate each of the following double iterated integrals:

(a) $\displaystyle\int_0^1 \int_x^{x-1} (x^2 + e^y) \, dy \, dx$ (b) $\displaystyle\int_{-1}^0 \int_y^0 \frac{x}{\sqrt{x^2 + y^2}} \, dx \, dy$

SOLUTION

(a) $\displaystyle\int_0^1 \int_x^{x-1} (x^2 + e^y) \, dy \, dx = \int_0^1 \{x^2 y + e^y\}_x^{x-1} \, dx$

$$= \int_0^1 [x^2(x - 1) + e^{x-1} - x^3 - e^x] \, dx$$

$$= \int_0^1 (-x^2 + e^{x-1} - e^x) \, dx$$

$$= \left\{ -\frac{x^3}{3} + e^{x-1} - e^x \right\}_0^1$$

$$= \left(-\frac{1}{3} + 1 - e \right) - (e^{-1} - 1)$$

$$= \frac{5}{3} - e - e^{-1}$$

(b) $\displaystyle\int_{-1}^0 \int_y^0 \frac{x}{\sqrt{x^2 + y^2}} \, dx \, dy = \int_{-1}^0 \left\{ \sqrt{x^2 + y^2} \right\}_y^0 dx = \int_{-1}^0 (\sqrt{y^2} - \sqrt{2y^2}) \, dy$

$$= \int_{-1}^0 (-y + \sqrt{2} y) \, dy = (\sqrt{2} - 1) \int_{-1}^0 y \, dy$$

$$= (\sqrt{2} - 1) \left\{ \frac{y^2}{2} \right\}_{-1}^0 = \frac{1 - \sqrt{2}}{2}$$

EXERCISES 13.1

In Exercises 1–30 evaluate the double iterated integral.

1. $\displaystyle\int_{-1}^{2}\int_{y}^{y+2}(x^2-xy)\,dx\,dy$ **2.** $\displaystyle\int_{-3}^{3}\int_{-\sqrt{18-2y^2}}^{\sqrt{18-2y^2}}x\,dx\,dy$

3. $\displaystyle\int_{0}^{1}\int_{x^2}^{x}(2xy+3y^2)\,dy\,dx$ **4.** $\displaystyle\int_{-1}^{0}\int_{y}^{2}(1+y)^2\,dx\,dy$

5. $\displaystyle\int_{3}^{4}\int_{0}^{\pi/2}x\sin y\,dy\,dx$ **6.** $\displaystyle\int_{1}^{2}\int_{1}^{y}e^{x+y}\,dx\,dy$

7. $\displaystyle\int_{-1}^{1}\int_{-x}^{5}(x^2+y^2)\,dy\,dx$ **8.** $\displaystyle\int_{-1}^{1}\int_{x}^{2x}(xy+x^3y^3)\,dy\,dx$

9. $\displaystyle\int_{0}^{1}\int_{x}^{1}(x+y)^4\,dy\,dx$ **10.** $\displaystyle\int_{1}^{2}\int_{x}^{2x}\frac{1}{(x+y)^3}\,dy\,dx$

11. $\displaystyle\int_{0}^{1}\int_{0}^{3x}\sqrt{x+y}\,dy\,dx$ **12.** $\displaystyle\int_{-1}^{1}\int_{1}^{e}\frac{y}{x}\,dx\,dy$

13. $\displaystyle\int_{1}^{4}\int_{\sqrt{x}}^{x^2}(x^2+2xy-3y^2)\,dy\,dx$

14. $\displaystyle\int_{0}^{2}\int_{x^2}^{2x^2}x\cos y\,dy\,dx$

15. $\displaystyle\int_{0}^{1}\int_{1}^{\tan x}\frac{1}{1+y^2}\,dy\,dx$ **16.** $\displaystyle\int_{0}^{1}\int_{0}^{y^3}\frac{1}{1+y^2}\,dx\,dy$

17. $\displaystyle\int_{2}^{3}\int_{0}^{1}\frac{x}{\sqrt{1-y^2}}\,dy\,dx$ **18.** $\displaystyle\int_{0}^{2}\int_{-x}^{x}(8-2x^2)^{3/2}\,dy\,dx$

* 19. $\displaystyle\int_{0}^{1}\int_{0}^{x}\frac{1}{\sqrt{1-y^2}}\,dy\,dx$ *** 20.** $\displaystyle\int_{-9}^{0}\int_{0}^{x^2\sqrt{9+x}}\,dy\,dx$

* 21. $\displaystyle\int_{0}^{2}\int_{\sqrt{4-x^2}}^{2}y^2\,dy\,dx$ *** 22.** $\displaystyle\int_{-1}^{0}\int_{y}^{0}x\sqrt{x^2+y^2}\,dx\,dy$

* 23. $\displaystyle\int_{2}^{3}\int_{1}^{2x}\frac{1}{(xy+x^2)^2}\,dy\,dx$ *** 24.** $\displaystyle\int_{0}^{1}\int_{0}^{\mathrm{Cos}^{-1}x}x\cos y\,dy\,dx$

* 25. $\displaystyle\int_{0}^{1}\int_{\sqrt{y^2+y}}^{\sqrt{2y}}x^3\sqrt{x^2-y^2}\,dx\,dy$

* 26. $\displaystyle\int_{0}^{1}\int_{\sqrt{2y}}^{\sqrt{y^2+y}}x^3\sqrt{x^2-y^2}\,dx\,dy$

* 27. $\displaystyle\int_{-2}^{0}\int_{x^4}^{4x^2}\sqrt{y-x^4}\,dy\,dx$ *** 28.** $\displaystyle\int_{-2}^{0}\int_{y}^{0}\frac{x}{\sqrt{x^2+y^2}}\,dx\,dy$

* 29. $\displaystyle\int_{-1}^{2}\int_{-1}^{y^3}\sqrt{1+y}\,dx\,dy$ **** 30.** $\displaystyle\int_{0}^{1}\int_{0}^{x}\sqrt{x^2+y^2}\,dy\,dx$

* 31. In two-dimensional, steady-state, incompressible flow, the velocit $\mathbf{v}$ of the flow has two components, $\mathbf{v}=u(x,y)\hat{\mathbf{i}}+v(x,y)\hat{\mathbf{j}}$, whic must satisfy the *continuity equation*

$$\frac{\partial u}{\partial x}+\frac{\partial v}{\partial y}=0.$$

If $u(x,y)=kx$, where k is a constant, find all possible function $v(x,y)$.

In Exercises 32–35 you are given one of the velocity components fo two-dimensional, steady-state, incompressible flow $\mathbf{v}=u(x,y)\hat{\mathbf{i}}+v(x,y)\hat{\mathbf{j}}$. Use the continuity equation of Exercise 31 to find all possibl values for the other component.

* 32. $u(x,y)=x^2+y^2$

* 33. $u(x,y)=\mathrm{Tan}^{-1}(y/x)$

* 34. $v(x,y)=x\sqrt{x^2+y^2}$

* 35. $v(x,y)=\sin x\cos y$

Stream functions $\psi(x,y)$ for two-dimensional, steady-state, incom pressible flow satisfy

$$\frac{\partial\psi}{\partial x}=-v(x,y),\qquad\frac{\partial\psi}{\partial y}=u(x,y),$$

where $\mathbf{v}=u(x,y)\hat{\mathbf{i}}+v(x,y)\hat{\mathbf{j}}$ is the velocity of the flow. In Exer cises 36–39 find all stream functions for the flow with given velocity

* 36. $\mathbf{v}=x\hat{\mathbf{i}}-y\hat{\mathbf{j}}$

* 37. $\mathbf{v}=(x^2+y^2)\hat{\mathbf{i}}-2xy\hat{\mathbf{j}}$

* 38. $\mathbf{v}=-y\sqrt{x^2+y^2}\,\hat{\mathbf{i}}+x\sqrt{x^2+y^2}\,\hat{\mathbf{j}}$

* 39. $\mathbf{v}=-\cos x\sin y\,\hat{\mathbf{i}}+(\sin x\cos y+x)\hat{\mathbf{j}}$

13.2 Evaluation of Double Integrals by Double Iterated Integrals

According to Theorem 13.1, if a function $f(x,y)$ is continuous on a finite region R with piecewise-smooth boundary, then double integral 13.3 exists, and its evaluation by means o that limit is independent of both the manner of subdivision of R into areas ΔA_i and choice o starpoints (x_i^*,y_i^*). We now show that if we make particular choices of ΔA_i, double integra can be evaluated by means of double iterated integrals in x and y.

FIGURE 13.3   Proof that double integrals can be evaluated with double iterated integrals

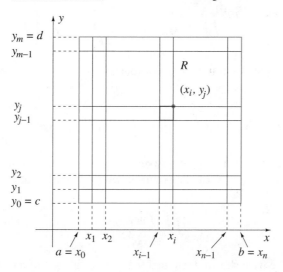

Consider first a rectangle R with edges parallel to the x- and y-axes as shown in Figure 13.3. We divide R into smaller rectangles by a network of $n+1$ vertical lines and $m+1$ horizontal lines identified by abscissae,

$$a = x_0 < x_1 < x_2 < \cdots < x_{n-1} < x_n = b,$$

and ordinates,

$$c = y_0 < y_1 < y_2 < \cdots < y_{m-1} < y_m = d.$$

If the $(i, j)^{\text{th}}$ rectangle is that rectangle bounded by the lines $x = x_{i-1}, x = x_i, y = y_{j-1}$, and $y = y_j$, then its area is $\Delta x_i \, \Delta y_j$, where $\Delta x_i = x_i - x_{i-1}$ and $\Delta y_j = y_j - y_{j-1}$. We choose as star point in the $(i, j)^{\text{th}}$ rectangle the upper right corner: $(x_i^*, y_j^*) = (x_i, y_j)$. With this rectangular subdivision of R and choice of star points, Definition 13.3 for the double integral of $f(x, y)$ over R takes the form

$$\iint_R f(x, y) \, dA = \lim_{\|\Delta x_i \Delta y_j\| \to 0} \sum_{i=1}^{n} \sum_{j=1}^{m} f(x_i, y_j) \, \Delta x_i \, \Delta y_j. \qquad (13.12a)$$

Since $\|\Delta x_i \Delta y_j\| \to 0$ if the norms $\|\Delta x_i\|$ and $\|\Delta y_j\|$ individually approach zero, we can write that

$$\iint_R f(x, y) \, dA = \lim_{\substack{\|\Delta x_i\| \to 0 \\ \|\Delta y_j\| \to 0}} \sum_{i=1}^{n} \sum_{j=1}^{m} f(x_i, y_j) \, \Delta x_i \, \Delta y_j. \qquad (13.12b)$$

Suppose we choose to first perform the limit on y and then the limit on x, and therefore write

$$\iint_R f(x, y) \, dA = \lim_{\|\Delta x_i\| \to 0} \sum_{i=1}^{n} \left\{ \lim_{\|\Delta y_j\| \to 0} \sum_{j=1}^{m} f(x_i, y_j) \Delta y_j \right\} \Delta x_i.$$

Since x_i is constant in the limit with respect to y, the y-limit is the definition of the definite integral of $f(x_i, y)$ with respect to y from $y = c$ to $y = d$; that is,

$$\lim_{\|\Delta y_j\| \to 0} \sum_{j=1}^{m} f(x_i, y_j) \, \Delta y_j = \int_c^d f(x_i, y) \, dy.$$

Consequently,

$$\iint_R f(x, y)\, dA = \lim_{\|\Delta x_i\| \to 0} \sum_{i=1}^{n} \left\{ \int_c^d f(x_i, y)\, dy \right\} \Delta x_i.$$

Because the term in braces is a function of x_i alone, we can interpret this limit as a definite integral with respect to x:

$$\iint_R f(x, y)\, dA = \int_a^b \left\{ \int_c^d f(x, y)\, dy \right\} dx = \int_a^b \int_c^d f(x, y)\, dy\, dx, \qquad (13.13)$$

a double iterated integral. By reversing the order of taking limits, we can show similarly that the double integral can be evaluated with a double iterated integral with respect to x and y:

$$\iint_R f(x, y)\, dA = \int_c^d \int_a^b f(x, y)\, dx\, dy. \qquad (13.14)$$

We have shown, then, that for the special case of a rectangle R with sides parallel to the axes, a double integral over R can be evaluated by using double iterated integrals. Conversely, every double iterated integral with constant limits represents a double integral over a rectangle. The double iterated integral simply indicates that a rectangular subdivision has been chosen to evaluate the double integral.

We have just stated that the choice of a double iterated integral to evaluate a double integral implies that the region of integration has been subdivided into small rectangles. We now show that the x- and y-integrations themselves can be interpreted geometrically. These interpretations will simplify the transition to more difficult regions of integration.

In the subdivision of R into rectangles, suppose we denote the dimensions of a representative rectangle at position (x, y) by dx and dy (Figure 13.4). In the inner integral

$$\int_c^d f(x, y)\, dy\, dx$$

of equation 13.13, x is held constant and integration is performed in the y-direction. This (partial) definite integral is therefore interpreted as summing over the rectangles in the vertical strip of width dx at position x. The limits $y = c$ and $y = d$ identify the initial and terminal positions of this vertical strip. It is important to note that we are not adding the areas of the rectangles of dimensions dx and dy in the strip. On the contrary, each rectangle of area $dy\, dx$ is multiplied by the value of $f(x, y)$ for that rectangle,

$$f(x, y)\, dy\, dx,$$

and it is these quantities that are added.

FIGURE 13.4   Addition process in a double iterated integral with respect to y and x over a rectangle

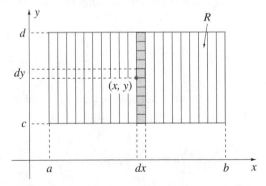

The x-integration in equation 13.13 is interpreted as adding over all strips starting at $x = a$ and ending at $x = b$. The limits on x therefore identify positions of the first and last strips. Although our diagram illustrates finite rectangles of dimensions dx and dy and finite strips of width dx, we must keep in mind that the integrations take limits as these dimensions approach zero.

Analogously, the double iterated integral in equation 13.14 is interpreted as adding over horizontal strips, as shown in Figure 13.5. Inner limits indicate where each strip starts and stops, and outer limits indicate the positions of first and last strips.

The transition now to more general regions is quite straightforward. For the double integral of $f(x, y)$ over the region in Figure 13.6, we use a double iterated integral with respect to y and x. The y-integration adds the quantities $f(x, y) \, dy \, dx$ over rectangles in a vertical strip. We write

$$\int_{g(x)}^{h(x)} f(x, y) \, dy \, dx,$$

where $g(x)$ and $h(x)$ indicate that each vertical strip starts on the curve $y = g(x)$ and ends on the curve $y = h(x)$. The x-integration now adds over all strips, beginning at $x = a$ and ending at $x = b$:

$$\iint_R f(x, y) \, dA = \int_a^b \int_{g(x)}^{h(x)} f(x, y) \, dy \, dx. \qquad (13.15)$$

A double iterated integral in the reverse order is not convenient for this region because horizontal strips neither all start on the same curve nor all end on the same curve.

For the region in Figure 13.7, we obtain

$$\iint_R f(x, y) \, dA = \int_c^d \int_{g(y)}^{h(y)} f(x, y) \, dx \, dy. \qquad (13.16)$$

The limits on double iterated integrals have been interpreted schematically as follows:

$$\int_{\substack{\text{position of first} \\ \text{horizontal strip}}}^{\substack{\text{position of last} \\ \text{horizontal strip}}} \int_{\substack{\text{where each and every} \\ \text{horizontal strip starts}}}^{\substack{\text{where each and every} \\ \text{horizontal strip stops}}} f(x, y) \, dx \, dy;$$

$$\int_{\substack{\text{position of first} \\ \text{vertical strip}}}^{\substack{\text{position of last} \\ \text{vertical strip}}} \int_{\substack{\text{where each and every} \\ \text{vertical strip starts}}}^{\substack{\text{where each and every} \\ \text{vertical strip stops}}} f(x, y) \, dy \, dx.$$

With these interpretations on the limits, you can see how important it is to have a well-labelled diagram.

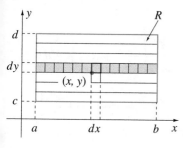

FIGURE 13.5 Addition process in a double iterated integral with respect to x and y over a rectangle

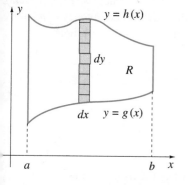

FIGURE 13.6 Integration with respect to y and x adds first inside a vertical strip, and then over all strips

FIGURE 13.7 Integration with respect to x and y adds first inside a horizontal strip, and then over all strips

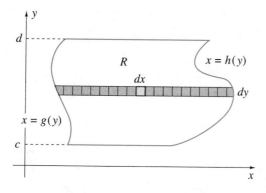

EXAMPLE 13.2

Evaluate the double integral of $f(x, y) = xy^2 + x^2$ over the region bounded by the curves $y = x^2$ and $x = y^2$.

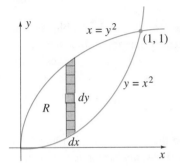

FIGURE 13.8 Integration of $xy^2 + x^2$ over R using vertical strips

SOLUTION If we use vertical strips (Figure 13.8), we have

$$\iint_R (xy^2 + x^2)\, dA = \int_0^1 \int_{x^2}^{\sqrt{x}} (xy^2 + x^2)\, dy\, dx = \int_0^1 \left\{ \frac{xy^3}{3} + x^2 y \right\}_{x^2}^{\sqrt{x}} dx$$

$$= \int_0^1 \left(\frac{4}{3} x^{5/2} - \frac{x^7}{3} - x^4 \right) dx = \left\{ \frac{8 x^{7/2}}{21} - \frac{x^8}{24} - \frac{x^5}{5} \right\}_0^1$$

$$= \frac{8}{21} - \frac{1}{24} - \frac{1}{5} = \frac{39}{280}.$$

There are two distinct parts to every double integral: first, the function $f(x, y)$ being integrated, which is the integrand; second, the region R over which integration is being performed, and this region determines the limits on the corresponding double iterated integral. Note that we do not use $f(x, y)$ to determine limits on the double iterated integral; the region determines the limits. Conversely, if we are given a double iterated integral, then we know that it represents the double integral of its integrand over some region, and the region is completely defined by the limits on the iterated integral. This point is emphasized in the following example.

EXAMPLE 13.3

Evaluate the double iterated integral $\int_0^2 \int_y^2 e^{x^2}\, dx\, dy$.

SOLUTION The function e^{x^2} does not have an elementary antiderivative with respect to x, and it is therefore impossible to evaluate the double iterated integral as it now stands. But the double iterated integral represents the double integral of e^{x^2} over some region R in the xy-plane. To find R we note that the inner integral indicates horizontal strips that all start on the line $x = y$ and stop on the line $x = 2$ (Figure 13.9a). The outer limits state that the first and last strips are at $y = 0$ and $y = 2$, respectively. This defines R as the triangle bounded by

FIGURE 13.9a Limits can be used to determine the region of integration

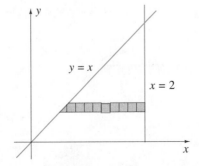

FIGURE 13.9b Reversing order of integration leads to simpler integral

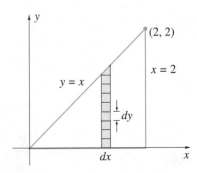

the straight lines $y = x$, $x = 2$, and $y = 0$ (Figure 13.9b). If we now reverse the order of integration and use vertical strips, we have

$$\int_0^2 \int_y^2 e^{x^2} \, dx \, dy = \iint_R e^{x^2} \, dA = \int_0^2 \int_0^x e^{x^2} \, dy \, dx = \int_0^2 \{ y e^{x^2} \}_0^x \, dx$$

$$= \int_0^2 x e^{x^2} \, dx = \left\{ \frac{1}{2} e^{x^2} \right\}_0^2 = \frac{e^4 - 1}{2}.$$

This example points out that an iterated integral in one order may be much easier to evaluate than the corresponding iterated integral in the opposite order.

EXERCISES 13.2

In Exercises 1–12 evaluate the double integral over the region.

1. $\displaystyle\iint_R (x^2 + y^2) \, dA$, where R is bounded by $y = x^2$, $x = y^2$

2. $\displaystyle\iint_R (4 - x^2 - y) \, dA$, where R is bounded by $x = \sqrt{4 - y}$, $x = 0$, $y = 0$

3. $\displaystyle\iint_R (x + y) \, dA$, where R is bounded by $x = y^3 + 2$, $x = 1$, $y = 1$

4. $\displaystyle\iint_R xy^2 \, dA$, where R is bounded by $x + y + 1 = 0$, $x + y^2 = 1$

5. $\displaystyle\iint_R x e^y \, dA$, where R is bounded by $y = x$, $y = 0$, $x = 1$

6. $\displaystyle\iint_R (x + y) \, dA$, where R is bounded by $x^2 + y^2 = 9$

7. $\displaystyle\iint_R x^2 y \, dA$, where R is bounded by $y = \sqrt{x + 4}$, $y = 0$, $x + y = 2$

8. $\displaystyle\iint_R (xy + y^2 - 3x^2) \, dA$, where R is bounded by $y = |x|$, $y = 1$, $y = 2$

9. $\displaystyle\iint_R (1 - x)^2 \, dA$, where R is bounded by $x + y = 1$, $x + y = -1$, $x - y = 1$, $y - x = 1$

10. $\displaystyle\iint_R (x + y) \, dA$, where R is bounded by $x = y^2$, $x^2 - y^2 = 12$

11. $\displaystyle\iint_R x \, dA$, where R is bounded by $y = 3x$, $y = x$, $x + y = 4$

12. $\displaystyle\iint_R y^2 \, dA$, where R is bounded by $x = 0$, $y = 1$, $y = 1/2$, $x = 1/\sqrt{y^4 + 12y^2}$

In Exercises 13–18 evaluate the double iterated integral by reversing the order of integration.

* 13. $\displaystyle\int_0^2 \int_0^{\sqrt{4 - x^2}} (4 - y^2)^{3/2} \, dy \, dx$

* 14. $\displaystyle\int_0^1 \int_y^1 \sin (x^2) \, dx \, dy$

* 15. $\displaystyle\int_{-2}^0 \int_{-y}^2 y(x^2 + y^2)^8 \, dx \, dy$

* 16. $\displaystyle\int_{-2}^0 \int_{-2}^x \frac{x}{\sqrt{x^2 + y^2}} \, dy \, dx$

* 17. $\displaystyle\int_0^2 \int_0^{x^2/2} \frac{x}{\sqrt{1 + x^2 + y^2}} \, dy \, dx$

* 18. $\displaystyle\int_0^2 \int_{-x^2/2}^0 \frac{x}{\sqrt{1 + x^2 + y^2}} \, dy \, dx$

* 19. Verify that if $m \le f(x, y) \le M$ for all (x, y) in R, then

$$m(\text{area of } R) \le \iint_R f(x, y) \, dA \le M(\text{area of } R).$$

* 20. Evaluate the double integral of $f(x, y) = 1/\sqrt{2x - x^2}$ over the region in the first quadrant bounded by $y^2 = 4 - 2x$.

In Exercises 21–28 either the integral has value 0 or it can be evaluated by doubling the double integral over half the region. By drawing the region and examining the integrand, determine which situation prevails. Do not evaluate the integral.

21. $\displaystyle\iint_R x^2 y^3 \, dA$, where R is bounded by $x = \sqrt{4 - y^2}$, $x = 0$

22. $\displaystyle\iint_R x^2 y^2 \, dA$, where R is bounded by $x = \sqrt{4 - y^2}$, $x = 0$

23. $\displaystyle\iint_R (x + y) \, dA$, where R is the square with vertices $(\pm 3, 0)$ and $(0, \pm 3)$

24. $\iint_R x^7 \cos(x^2)\,dA$, where R is bounded by $y = 4 - |x|$, $y = x^2$

25. $\iint_R e^{x^2 + y^2}\,dA$, where R is bounded by $y = 4 - 4x^2$, $y = x^2 - 1$

26. $\iint_R \cos(x^2 y)\,dA$, where R is bounded by $y = 0$, $y = x^3 - x$

27. $\iint_R \sin(x^2 y)\,dA$, where R is bounded by $y = 0$, $y = x^3 - x$

28. $\iint_R (x^2 y^3 + xy^2)\,dA$, where R is bounded by $\sqrt{|x|} + \sqrt{|y|} = 1$

The average value of a function $f(x, y)$ over a region R with area A is defined as

$$\overline{f} = \frac{1}{A} \iint_R f(x, y)\,dA.$$

In Exercises 29–32 find the average value of the function over the region.

29. $f(x, y) = xy$ over the region in the first quadrant bounded by $x = 0$, $y = \sqrt{1 - x^2}$, $y = 0$

30. $f(x, y) = x + y$ over the region bounded by $y = x$, $y = 0$, $y = \sqrt{2 - x}$

* **31.** $f(x, y) = x$ over the region between $y = \sin x$ and $y = 0$ for $0 \le x \le 2\pi$

* **32.** $f(x, y) = e^{x+y}$ over the region bounded by $y = x + 1$, $y = x - 1$, $y = 1 - x$, $y = -1 - x$

* **33.** The Cobb–Douglas production function for a widget is $P(x, y) = 10\,000x^{0.3}y^{0.7}$, where P is the number of widgets produced each month, x is the number of employees, and y is the monthly operating budget in thousands of dollars. If the company uses anywhere between 45 and 55 workers each month, and its operating budget varies from \$8000 to \$12\,000 per month, what is the average number of widgets produced each month?

* **34.** Repeat Exercise 33 using the production function $P(x, y) = 10\,000x^{0.7}y^{0.3}$.

In Exercises 35–41 evaluate the double integral over the region.

* **35.** $\iint_R x^2\,dA$, where R is bounded by $x^2 + y^2 = 4$

* **36.** $\iint_R (6 - x - 2y)\,dA$, where R is bounded by $x^2 + y^2 = 4$

* **37.** $\iint_R 6x^5\,dA$, where R is the region under $x + 5y = 16$, above $y = x - 4$, and bounded by $x = (y - 2)^2$

* **38.** $\iint_R ye^x\,dA$, where R is bounded by $y = x$, $x + y = 2$, $y = 0$

* **39.** $\iint_R \frac{\sqrt{1 + y}}{y^3}\,dA$, where R is bounded by $x = -1$, $y = 2$, $x =$

* **40.** $\iint_R y\sqrt{x^2 + y^2}\,dA$, where R is bounded by $y = x$, $x = -1$, $y = 0$

* **41.** $\iint_R (x^2 + y^2)\,dA$, where R is bounded by $x^2 + y^2 = 9$

* **42.** Evaluate the double iterated integral $\int_0^1 \int_0^1 |x - y|\,dy\,dx$.

* **43.** Evaluate the double integral $\iint_R |y - 2x^2 + 1|\,dA$, where R is the square bounded by $x = \pm 1$, $y = \pm 1$.

* **44.** For the accelerating slit system of a mass spectrometer, the number of ions within unit solid angle of the electron beam, at the plane of the first slit, is

$$n = \frac{2n_c L}{\pi} \int_0^{2d} \int_0^{\infty} \frac{x^2 (1 - a^2 y^2/c^2)}{(1 + x^2)(x^2 + a^2 y^2/c^2)}\,dx\,dy,$$

where n_c is the number of ions with initial velocity c, L is the length of the slit, d is the width of the slit, and $a > 0$ is a constant. Show that $n = 2n_c dL(1 - ad/c)$.

13.3 Areas and Volumes of Solids of Revolution

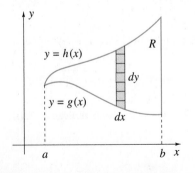

FIGURE 13.10   Double iterated integrals can be used to find areas of regions

Because equation 13.7 represents the area of a region R as a double integral, and double integrals are evaluated by means of double iterated integrals, it follows that **areas** can be calculated using double iterated integrals. In particular, to find the area of the region in Figure 13.10, we subdivide R into rectangles of dimensions dx and dy and therefore of area $dA = dy\,dx$. Areas of these rectangles are then added in the y-direction to give the area of a vertical strip

$$\int_{g(x)}^{h(x)} dy\,dx,$$

where limits indicate that every vertical strip starts on the curve $y = g(x)$ and ends on the curve $y = h(x)$. Finally, areas of the vertical strips are added together to give the total area:

$$\text{area} = \int_a^b \int_{g(x)}^{h(x)} dy \, dx,$$

where a and b indicate the x-positions of first and last strips.

EXAMPLE 13.4

Find the area of the region bounded by the curves $xy = 2$, $x = 2\sqrt{y}$, $y = 4$.

SOLUTION   If we choose horizontal strips for this region (Figure 13.11), we have

$$\text{area} = \int_1^4 \int_{2/y}^{2\sqrt{y}} dx \, dy = \int_1^4 \left(2\sqrt{y} - \frac{2}{y} \right) dy$$

$$= \left\{ \frac{4}{3} y^{3/2} - 2 \ln |y| \right\}_1^4 = \frac{28}{3} - 2 \ln 4.$$

For vertical strips (Figure 13.12), we require two iterated integrals because, to the left of the line $x = 2$, strips begin on the hyperbola $xy = 2$, whereas to the right of $x = 2$, they begin on the parabola $x = 2\sqrt{y}$. We obtain

$$\text{area} = \int_{1/2}^2 \int_{2/x}^4 dy \, dx + \int_2^4 \int_{x^2/4}^4 dy \, dx = \int_{1/2}^2 \left(4 - \frac{2}{x} \right) dx + \int_2^4 \left(4 - \frac{x^2}{4} \right) dx$$

$$= \{ 4x - 2 \ln |x| \}_{1/2}^2 + \left\{ 4x - \frac{x^3}{12} \right\}_2^4 = \frac{28}{3} - 2 \ln 4.$$

FIGURE 13.11   When horizontal strips are chosen, only one double iterated integral is needed

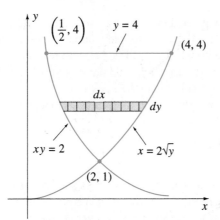

FIGURE 13.12   When vertical strips are chosen, two double iterated integrals are needed

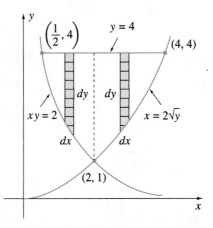

If we compare finding areas by definite integrals (Section 7.1) and finding the same areas by double integrals, it is clear that no great advantage is derived by using double integrals. In fact, it is probably more work because we must perform two, rather than one, integrations, although the first integration is trivial. The advantage of double integrals is therefore not in finding area; it is in finding volumes of solids of revolution, centres of mass, moments of inertia, and fluid forces, among other applications.

Volumes of Solids of Revolution

If the region in Figure 13.13 is rotated around the x-axis, the **volume of the** resulting **solid of revolution** can be evaluated by using the washer method introduced in Section 7.2:

$$\text{volume} = \int_a^b \{\pi [h(x)]^2 - \pi [g(x)]^2\} \, dx. \tag{13.17}$$

FIGURE 13.13 With definite integrals, volumes of washers or cylindrical shells are calculated

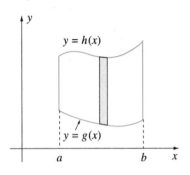

If this region is rotated around the y-axis, the volume generated is calculated by using the cylindrical shell method:

$$\text{volume} = \int_a^b 2\pi x [h(x) - g(x)] \, dx. \tag{13.18}$$

Thus, once we have chosen to use vertical rectangles, the axis of revolution determines whether we use washers or cylindrical shells. We now show that with double integrals one method works for all problems.

To rotate this region around the x-axis we subdivide R into small areas dA (Figure 13.14). If the area dA at a point (x, y) is rotated about the x-axis, it generates a "ring" with cross sectional area dA. Since (x, y) travels a distance $2\pi y$ in traversing the ring, it follows that the volume in the ring is approximately $2\pi y \, dA$. To find the total volume obtained by rotating R about the x-axis, we add the volumes of all such rings and take the limit as the areas shrink to points. But this is what we mean by the double integral of $2\pi y$ over the region R, and we therefore write

FIGURE 13.14 With double integrals, volumes of rings are calculated

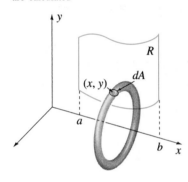

$$\text{volume} = \iint_R 2\pi y \, dA. \tag{13.19}$$

On the other hand, if dA is rotated about the y-axis, it again forms a ring, but with approximate volume $2\pi x \, dA$. The total volume, then, when R is rotated about the y-axis is

$$\text{volume} = \iint_R 2\pi x \, dA. \tag{13.20}$$

FIGURE 13.15 With double iterated integrals, volumes of rectangular rings are calculated

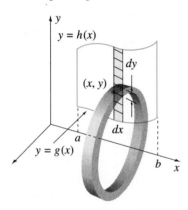

Since double iterated integrals are used to evaluate double integrals, it follows that we can set up double iterated integrals to find volumes represented by equations 13.19 and 13.20. The decision to use a double iterated integral with respect to y and x implies a subdivision of R into rectangles of dimensions dx and dy (Figure 13.15). The volume of the ring formed when this rectangle is rotated around the x-axis is $2\pi y \, dy \, dx$. If we choose to integrate first with respect to y, we are adding over all rectangles in a vertical strip

$$\int_{g(x)}^{h(x)} 2\pi y \, dy \, dx,$$

where limits indicate that all vertical strips start on the curve $y = g(x)$ and end on the curve $y = h(x)$. This integral is the volume generated by rotating the vertical strip around the x-axis. Integration now with respect to x adds over all strips to give the required volume:

$$\text{volume} = \int_a^b \int_{g(x)}^{h(x)} 2\pi y \, dy \, dx. \tag{13.21}$$

Note that when we actually do perform the inner integration, we get

$$\text{volume} = \int_a^b \{\pi y^2\}_{g(x)}^{h(x)} \, dx = \int_a^b \{\pi [h(x)]^2 - \pi [g(x)]^2\} \, dx,$$

and this is the result contained in equation 13.17.

When R is rotated around the y-axis, the rectangular area $dy\,dx$ generates a ring of volume $2\pi x\,dy\,dx$. Addition over the rectangles in a vertical strip

$$\int_{g(x)}^{h(x)} 2\pi x\,dy\,dx$$

gives the volume generated by rotating the strip about the y-axis. Finally, integration with respect to x adds over all strips to give

$$\text{volume} = \int_a^b \int_{g(x)}^{h(x)} 2\pi x\,dy\,dx. \tag{13.22}$$

This time the inner integration leads to

$$\text{volume} = \int_a^b \{2\pi xy\}_{g(x)}^{h(x)}\,dx = \int_a^b 2\pi x[h(x) - g(x)]\,dx,$$

the same result as in equation 13.18.

The advantage, then, in using double integrals to find volumes of solids of revolution is that it requires only one idea, that of rings. The first integration leads to washers or cylindrical shells, but we need never think about this.

EXAMPLE 13.5

▼

Find volumes of the solids of revolution if the region bounded by the curves $y = 2x - x^2$, $y = x^2 - 2x$ is rotated around:

(a) the y-axis   (b) $x = -3$ (c) $y = 2$ (d) $y = x + 2$.

SOLUTION

(a) If we use vertical strips (Figure 13.16), then

$$\text{volume} = \int_0^2 \int_{x^2-2x}^{2x-x^2} 2\pi x\,dy\,dx = 2\pi \int_0^2 \{xy\}_{x^2-2x}^{2x-x^2}\,dx$$

$$= 2\pi \int_0^2 x\{(2x - x^2) - (x^2 - 2x)\}\,dx$$

$$= 4\pi \int_0^2 (2x^2 - x^3)\,dx = 4\pi \left\{\frac{2x^3}{3} - \frac{x^4}{4}\right\}_0^2 = \frac{16\pi}{3}.$$

FIGURE 13.16   Area is subdivided into rectangles, volumes of rings calculated, and added over vertical strips

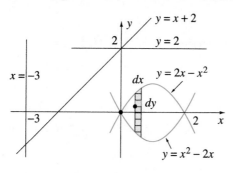

(b) In this case the radius of the ring formed by rotating the rectangle about $x = -3$ is $x + 3$, and therefore

$$\text{volume} = \int_0^2 \int_{x^2-2x}^{2x-x^2} 2\pi(x+3)\,dy\,dx = 2\pi \int_0^2 \{y(x+3)\}_{x^2-2x}^{2x-x^2}\,dx$$

$$= 2\pi \int_0^2 (x+3)(4x - 2x^2)\,dx = 4\pi \int_0^2 (6x - x^2 - x^3)\,dx$$

$$= 4\pi \left\{ 3x^2 - \frac{x^3}{3} - \frac{x^4}{4} \right\}_0^2 = \frac{64\pi}{3}.$$

(c) When the rectangle is rotated around $y = 2$, the radius of the ring is $2 - y$, and hence

$$\text{volume} = \int_0^2 \int_{x^2-2x}^{2x-x^2} 2\pi(2-y)\,dy\,dx = 2\pi \int_0^2 \left\{ -\frac{1}{2}(2-y)^2 \right\}_{x^2-2x}^{2x-x^2}\,dx$$

$$= \pi \int_0^2 [(2 - x^2 + 2x)^2 - (2 - 2x + x^2)^2]\,dx = 8\pi \int_0^2 (2x - x^2)\,dx$$

$$= 8\pi \left\{ x^2 - \frac{x^3}{3} \right\}_0^2 = \frac{32\pi}{3}.$$

(d) Using formula 1.16, the distance from the area element $dy\,dx$ at position (x, y) to the line $y = x + 2$ is $|x - y + 2|/\sqrt{2} = (x - y + 2)/\sqrt{2}$. Hence the volume of the ring obtained by rotating the rectangle around the line is $\left[2\pi(x - y + 2)/\sqrt{2} \right] dy\,dx$, and the volume of the solid of revolution is

$$\text{volume} = \int_0^2 \int_{x^2-2x}^{2x-x^2} \sqrt{2}\,\pi(x - y + 2)\,dy\,dx = \sqrt{2}\,\pi \int_0^2 \left\{ -\frac{1}{2}(x - y + 2)^2 \right\}_{x^2-2x}^{2x-x^2}\,dx$$

$$= -\frac{\pi}{\sqrt{2}} \int_0^2 [(x - 2x + x^2 + 2)^2 - (x - x^2 + 2x + 2)^2]\,dx$$

$$= \frac{\pi}{\sqrt{2}} \int_0^2 (16x - 4x^3)\,dx = \frac{\pi}{\sqrt{2}} \{8x^2 - x^4\}_0^2 = 8\sqrt{2}\,\pi.$$

Notice that double integrals allow us to calculate volumes of solids of revolution about any line (part (d) of Example 13.5). With definite integrals we were restricted to vertical and horizontal lines.

EXERCISES 13.3

In Exercises 1–10 use a double integral to find the area of the region bounded by the curves.

1. $y = 4x^2,\ x = 4y^2$

2. $y = x^2,\ y = 5x + 6$

3. $x = y^2,\ x = 3y - 2$

4. $y = x^3 + 8,\ y = 4x + 8$

5. $y = 4/x^2,\ y = 5 - x^2$

6. $y = xe^{-x},\ y = x,\ x = 2$

7. $x = 4y - 4y^2,\ y = x - 3,\ y = 1,\ y = 0$

8. $x = y(y - 2),\ x + y = 12$

9. $y = x^3 - x^2 - 2x + 2,\ y = 2$

10. $x + y = 1,\ x + y = 5,\ y = 2x + 1,\ y = 2x + 6$

In Exercises 11–20 use a double integral to find the volume of the solid of revolution obtained by rotating the region bounded by the curves around the line.

1. $y = -\sqrt{4 - x}$, $x = 0$, $y = 0$ about $y = 0$

2. $4x^2 + 9y^2 = 36$ about $y = 0$

3. $y = (x - 1)^2$, $y = 1$ about $x = 0$

4. $y = x^2 + 4$, $y = 2x^2$ about $y = 0$

5. $x - 1 = y^2$, $x = 5$ about $x = 1$

6. $x = y^3$, $y = \sqrt{2 - x}$, $y = 0$ about $y = 1$

7. $y = 4x^2 - 4x$, $y = x^3$ about $y = -2$

8. $x = 3y - y^2$, $x = y^2 - 3y$ about $y = 4$

9. $x = 2y - y^2 - 2$, $x = -5$ about $x = 1$

10. $x + y = 4$, $y = 2\sqrt{x - 1}$, $y = 0$ about $y = -1$

In Exercises 21–30 use a double integral to find the area of the region bounded by the curves.

21. $y = 2x^3$, $y = 4x + 8$, $y = 0$

22. $y = x/\sqrt{x + 3}$, $x = 1$, $x = 6$, $y = -x^2$

23. $y = \sqrt{x - 2}$, $y = 4 - \sqrt{-x}$, $y = 4 - x$, $y = 0$ ($-16 \leq x \leq$)

24. $y = x^3 - x$, $x + y + 1 = 0$, $x = \sqrt{y + 1}$

25. $y^2 = x^2(4 - x^2)$

26. $x^2 + y^2 = 4$, $x^2 + y^2 = 4x$ (interior to both)

27. $x = 1/\sqrt{4 - y^2}$, $4x + y^2 = 0$, $y + 1 = 0$, $y - 1 = 0$

* 28. $y^2 = x^4(9 + x)$

* 29. $y = (x^2 + 1)/(x + 1)$, $x + 3y = 7$

* 30. $(x + 2)^2 y = 4 - x$, $x = 0$, $y = 0$ ($x \geq 0$, $y \geq 0$)

In Exercises 31–35 use a double integral to find the volume of the solid of revolution obtained by rotating the region bounded by the curves around the line.

* 31. $y = 4/(x^2 + 1)^2$, $y = 1$ about $x = 0$

* 32. $y = x^2 - 2$, $y = 0$ about $y = -1$

* 33. $y = |x^2 - 1|$, $x = -2$, $x = 2$, $y = -1$ about $y = -2$

* 34. $x = \sqrt{4 + 12y^2}$, $x - 20y = 24$, $y = 0$ about $y = 0$

* 35. $y = (x + 1)^{1/4}$, $y = -(x + 1)^2$, $x = 0$ about $x = 0$

* 36. Find the area of the region common to the two circles $x^2 + y^2 = 4$ and $x^2 + y^2 = 6x$.

In Exercises 37–40 find the volume of the solid of revolution obtained by rotating the region bounded by the curves around the line.

* 37. $x = 1$, $y = 1$, $x = 0$, $y = 0$ about $x + y = 2$

* 38. $y = x^2$, $y = 2x + 3$ about $y = 2x + 3$

* 39. $y = \sqrt{x}$, $y = 0$, $x = 1$ about $y = 3x + 2$

* 40. $x = 2y$, $y = x - 1$, $y = 0$ about $x + y + 1 = 0$

** 41. Prove that the area above the line $y = h$ and under the circle $x^2 + y^2 = r^2$ ($r > h$) is given by

$$A = \pi r^2/2 - h\sqrt{r^2 - h^2} - r^2 \operatorname{Sin}^{-1}(h/r).$$

13.4 Fluid Pressure

In Section 7.6 we defined **pressure** at a point in a **fluid** as the magnitude of the force per unit area that would act on any surface placed at that point. We discovered that at a depth $d > 0$ below the surface of a fluid, pressure is given by

$$P = 9.81\rho d, \tag{13.23}$$

where ρ is the density of the fluid. With these ideas and the definite integral, we were able to calculate fluid forces on flat surfaces in the fluid. In particular, the magnitude of the total force on each side of the vertical surface in Figure 13.17 is given by the definite integral

$$\text{force} = \int_a^b -9.81\rho y[h(y) - g(y)]\, dy. \tag{13.24}$$

Although horizontal rectangles are convenient for this problem, it is clear that they are not reasonable for the surface in Figure 13.18.

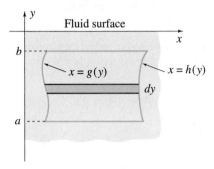

FIGURE 13.17 With definite integrals, use horizontal rectangles to calculate fluid forces

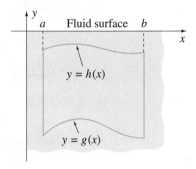

FIGURE 13.18 Use of horizontal rectangles to calculate force on this plate is inconvenient

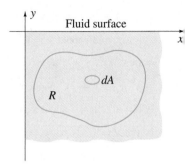

FIGURE 13.19 Double integrals to calculate forces on submerged plates are most efficient

Double integrals, on the other hand, can be applied with equal ease to both surfaces. To see this we consider, first, force on each side of the surface in Figure 13.19. If the surface is divided into small areas dA, then the force on dA is $P\,dA$, where P is pressure at that depth. Total force on R is the sum of the forces on all such areas in R as areas dA shrink to a point. But once again this is the double integral, and we therefore write

$$\text{force} = \iint_R P\,dA. \tag{13.25}$$

To set up a double iterated integral in order to evaluate this double integral, we use our interpretation of the double iterated integral as a limit of a sum in which the areas dA have been chosen as rectangles. In particular, for the surface in Figure 13.17, we draw rectangles of dimensions dx and dy, as shown in Figure 13.20. The force on this rectangle is its area $dx\,dy$ multiplied by pressure $-9.81\rho y$ at that depth, $-9.81\rho y\,dx\,dy$. Addition of these quantities over all rectangles in a horizontal strip gives the force on the strip,

$$\int_{g(y)}^{h(y)} -9.81\rho y\,dx\,dy,$$

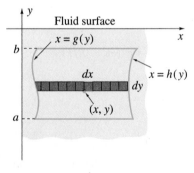

FIGURE 13.20 To find the force on this plate, use horizontal strips

where the limits indicate that all horizontal strips start on the curve $x = g(y)$ and end on the curve $x = h(y)$. Integration with respect to y now adds over all horizontal strips to give the total force on the surface:

$$\text{force} = \int_a^b \int_{g(y)}^{h(y)} -9.81\rho y\,dx\,dy. \tag{13.26}$$

When we perform the inner integration, we obtain

$$\text{force} = \int_a^b \left\{-9.81\rho y x\right\}_{g(y)}^{h(y)} dy = \int_a^b -9.81\rho y[h(y) - g(y)]\,dy,$$

and this is the result contained in equation 13.24.

For the surface in Figure 13.18 we again draw rectangles of dimensions dx and dy and calculate the force on such a rectangle, $-9.81\rho y\,dy\,dx$. In this case it is more convenient to add over rectangles in a vertical strip to give the force on the strip (Figure 13.21):

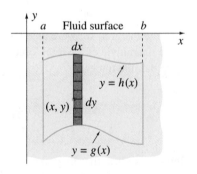

FIGURE 13.21 To find the force on this plate, use vertical strips

$$\int_{g(x)}^{h(x)} -9.81\rho y\,dy\,dx.$$

Force on the entire surface can now be found by adding over all vertical strips:

$$\text{force} = \int_a^b \int_{g(x)}^{h(x)} -9.81\rho y\,dy\,dx. \tag{13.27}$$

EXAMPLE 13.6

The face of a dam is parabolic with breadth 100 m and height 50 m. Find the magnitude of the total force due to fluid pressure on the face when the water is 1 m from the top.

FIGURE 13.22 Force of water on vertical face of a dam

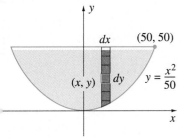

SOLUTION   If we use the coordinate system in Figure 13.22, then the edge of the dam has an equation of the form $y = kx^2$. Since $(50, 50)$ is a point on this curve, it follows that $k = \frac{1}{50}$. Because force on the left half of the dam is the same as that on the right half, we can integrate for the right half and double the result; that is,

$$\text{force} = 2 \int_0^{35\sqrt{2}} \int_{x^2/50}^{49} 9.81(1000)(49 - y)\, dy\, dx$$

$$= 19\,620 \int_0^{35\sqrt{2}} \left\{ 49y - \frac{y^2}{2} \right\}_{x^2/50}^{49} dx = 19\,620 \int_0^{35\sqrt{2}} \left(\frac{2401}{2} - \frac{49x^2}{50} + \frac{x^4}{5000} \right) dx$$

$$= 19\,620 \left\{ \frac{2401x}{2} - \frac{49x^3}{150} + \frac{x^5}{25\,000} \right\}_0^{35\sqrt{2}} = 6.22 \times 10^8 \text{ N}.$$

EXAMPLE 13.7

A tank in the form of a right-circular cylinder of radius $\dfrac{1}{2}$ m and length 10 m has its axis horizontal. If it is full of water, find the force due to water pressure on each end of the tank.

FIGURE 13.23 Force on the end of a full cylindrical tank

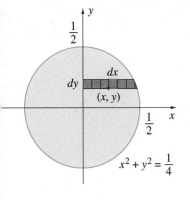

SOLUTION   Since the force on that part of the end to the left of the y-axis (Figure 13.23) is identical to the force on that part to the right, we double the force on the right half; that is,

$$\text{force} = 2 \int_{-1/2}^{1/2} \int_0^{\sqrt{1/4 - y^2}} 9.81(1000)(\tfrac{1}{2} - y)\, dx\, dy$$

$$= 9810 \int_{-1/2}^{1/2} \int_0^{\sqrt{1/4 - y^2}} dx\, dy - 19\,620 \int_{-1/2}^{1/2} \int_0^{\sqrt{1/4 - y^2}} y\, dx\, dy.$$

The first double iterated integral represents the area of one-half the end of the tank. Consequently,

$$\text{force} = 9810 \left[\frac{1}{2}\pi \left(\frac{1}{2} \right)^2 \right] - 19\,620 \int_{-1/2}^{1/2} y\sqrt{\frac{1}{4} - y^2}\, dy$$

$$= \frac{4905\pi}{4} - 19\,620 \left\{ -\frac{1}{3} \left(\frac{1}{4} - y^2 \right)^{3/2} \right\}_{-1/2}^{1/2} = \frac{4905\pi}{4} \text{ N}.$$

FIGURE 13.24 Centre of pressure on a submerged plate is a point for equivalency of moments of forces

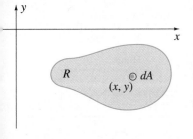

The centroid of a planar region is a point at which the area of the region can be concentrated as far as first moments of the region are concerned. It is advantageous to define a point called the **centre of pressure** for a surface submerged in a fluid; it is a point where a single force equal to that of the fluid on the surface has the same first moment about any line as does the fluid force on the surface. For example, the fluid force on the region R in Figure 13.24 is given by the double integral

$$F = \iint_R P\, dA,\tag{13.28}$$

where $P = -\rho g y$ in the coordinate system shown. It is the sum of fluid forces $P\,dA$ on elemental areas dA at points (x, y).

Each elemental force $P\,dA$ creates a moment $yP\,dA$ about the x-axis, and the total first moment of all such elemental moments is

$$\iint_R yP\,dA. \tag{13.29a}$$

Similarly, the first moment of the fluid force on the surface about the y-axis is

$$\iint_R xP\,dA. \tag{13.29b}$$

The centre of pressure of R is the point (x_c, y_c) defined by

$$F\,x_c = \iint_R xP\,dA, \qquad F\,y_c = \iint_R yP\,dA, \tag{13.30}$$

where F is given by 13.28.

For example, due to the symmetry in Example 13.6, the centre of pressure of the dam is on the y-axis. Its y-coordinate is given by

$$y_c = \frac{1}{F}\iint_R yP\,dA,$$

where R represents the dam, and $F = 6.22 \times 10^8$ N is the total force on the dam. We integrate over the right half and double the result:

$$y_c = \frac{2}{F}\int_0^{35\sqrt{2}}\int_{x^2/50}^{49} y(9810)(49 - y)\,dy\,dx = \frac{19\,620}{F}\int_0^{35\sqrt{2}}\left\{\frac{49y^2}{2} - \frac{y^3}{3}\right\}_{x^2/50}^{49}dx$$

$$= \frac{19\,620}{F}\int_0^{35\sqrt{2}}\left(\frac{117\,649}{6} - \frac{49x^4}{5000} + \frac{x^6}{375\,000}\right)dx$$

$$= \frac{19\,620}{F}\left\{\frac{117\,649x}{6} - \frac{49x^5}{25\,000} + \frac{x^7}{7(375\,000)}\right\}_0^{35\sqrt{2}} = 21.0 \text{ m}.$$

EXAMPLE 13.8

Application Preview Revisited

Find the depth h of ocean water in the Application Preview at which the tide gate opens (Figure 13.25a).

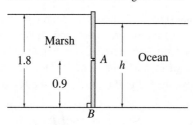

FIGURE 13.25a Tide gate is on the verge of opening when the sum of the moments of the marsh and the ocean on the gate is zero

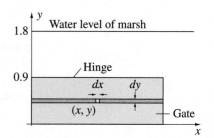

FIGURE 13.25b Calculation of moment of marsh water on gate about hinge

SOLUTION   The forces of the marsh and ocean on the gate create moments about the hinge. If we take counterclockwise moments about the hinge (marsh) as positive, and clockwise moments (ocean) as negative, the gate will be on the verge of opening when the sum of these moments is zero. Since the force of the marsh on a rectangle of dimensions dx and dy on the gate (Figure 13.25b) is $1000g(1.8 - y)\,dx\,dy$, the moment of this force about the hinge is $(0.9 - y)1000g(1.8 - y)\,dx\,dy$. The total moment about the hinge of the force of the marsh on the gate is

$$M_m = \int_0^{0.9} \int_0^L 1000g(0.9 - y)(1.8 - y)\,dx\,dy = 1000gL \int_0^{0.9} \left(\frac{81}{50} - \frac{27y}{10} + y^2\right) dy$$

$$= 1000gL \left\{\frac{81y}{50} - \frac{27y^2}{20} + \frac{y^3}{3}\right\}_0^{0.9} = \frac{1215gL}{2}.$$

Similarly, the moment about the hinge of the force of the ocean water on the gate, when depth of the ocean water is h, is

$$M_o = -\int_0^{0.9} \int_0^L 1030g(0.9 - y)(h - y)\,dx\,dy$$

$$= -1030gL \int_0^{0.9} \left[\frac{9h}{10} - \left(h + \frac{9}{10}\right)y + y^2\right] dy$$

$$= -1030gL \left\{\frac{9hy}{10} - \left(h + \frac{9}{10}\right)\frac{y^2}{2} + \frac{y^3}{3}\right\}_0^{0.9}$$

$$= -1030gL \left[\frac{81h}{100} - \frac{81}{200}\left(h + \frac{9}{10}\right) + \frac{243}{1000}\right].$$

The sum of these moments is zero when

$$0 = \frac{1215gL}{2} - 1030gL \left[\frac{81h}{100} - \frac{81}{200}\left(h + \frac{9}{10}\right) + \frac{243}{1000}\right] \quad \Longrightarrow \quad h = 1.756.$$

The gate is on the verge of opening when the ocean water is 1.756 m deep.

EXERCISES 13.4

In Exercises 1–8 the surface is submerged vertically in a fluid with density ρ. Find the force due to fluid pressure on one side of the surface.

1. An equilateral triangle of side length 2 with one edge in the surface

2. A parabolic segment of base 12 and height 4 with the base in the surface

3. A square of side length 3 with one diagonal vertical and the uppermost vertex in the surface

4. A triangle of side lengths 5, 5, and 8, with the longest side uppermost, horizontal, and 3 units below the surface

5. A triangle of side lengths 5, 5, and 8, with the longest side below the opposite vertex, horizontal, and 6 units below the surface

6. A trapezoid with vertical parallel sides of lengths 6 and 8, and a third side perpendicular to the parallel sides, of length 5, and in the surface

7. A triangle of side lengths 3, 3, and 4, with the longest side vertical, and the uppermost vertex 1 unit below the surface

8. A semicircle of radius 5 with the (diameter) base in the surface

* 9. The vertical end of a water trough is an isosceles triangle with width 2 m and depth 1 m. Find the force of the water on each end when the trough is one-half filled (by volume) with water.

* 10. A dam across a river has the shape of a parabola 36 m across the top and 9 m deep at the centre. Find the maximum force due to water pressure on the dam.

In Exercises 11–15 the surface is submerged vertically in a fluid with density ρ. Find the force due to fluid pressure on one side of the surface.

* 11. A circle of radius 2 with centre 3 units below the surface

* 12. A rectangle of side lengths 2 and 5, with one diagonal vertical and the uppermost vertex in the surface

* **13.** An ellipse with major and minor axes of lengths 8 and 6, and with the major axis horizontal and 5 units below the surface

* **14.** A parallelogram of side lengths 4 and 5, with one of the longer sides horizontal and in the surface, and two sides making an angle of $\pi/6$ radians with the surface

* **15.** A triangle of side lengths 2, 3, and 4 with the longest side vertical, the side of length 2 above the side of length 3, and the uppermost vertex 1 unit below the surface

* **16.** An oil can is in the form of a right-circular cylinder of radius r and height h. If the axis of the can is horizontal, and the can is full of oil with density ρ, find the force due to fluid pressure on each end.

* **17.** Find the force due to water pressure on each side of the flat vertical plate in the figure below.

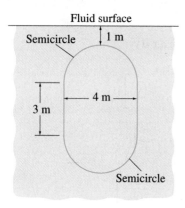

* **19.** A semicircle with radius r when the diameter is horizontal, above the rest of the semicircle, and h units below the surface

* **20.** An equilateral triangle of side length L when one edge is horizontal and in the surface

* **21.** A square with sides of length L when one diagonal is vertical and the uppermost vertex is in the surface

* **22.** The nonhypotenuse sides of a right-angled triangle have lengths L and ℓ, where $\ell < L$. The vertex containing the right angle is at the origin in the surface of the fluid and the shortest side is along the positive x-axis. Find the centre of pressure of the triangle.

* **23.** The centroid of a plane figure is a fixed point. The centre of pressure, on the other hand, changes depending on depth below the surface. Use the result of Exercise 18 to verify this.

* **24.** A square plate of side length 2 m has one side on the bottom of a swimming pool 3 m deep. The plate is inclined at an angle of $\pi/4$ radians with the bottom of the pool so that its horizontal upper edge is $3 - \sqrt{2}$ m below the surface. Find the force due to water pressure on each side of the plate.

** **25.** A thin triangular piece of wood with sides of lengths 2 m, 2 m, and 3 m floats in a pond. A piece of rope is tied to the vertex opposite the longest side. A rock is then attached to the other end of the rope and lowered into the water. When the rock sits on the bottom (and the rope is taut), the longest side of the wood still floats in the surface of the water, but the opposite vertex is 1 m below the surface. Find the force due to water pressure on each side of the piece of wood.

** **26.** Show that the centre of pressure of a plane surface is always below its centroid.

In Exercises 18–21 find the centre of pressure of the surface submerged vertically in a fluid.

* **18.** A circle with radius r when its centre is $h > r$ units below the surface

13.5 Centres of Mass and Moments of Inertia

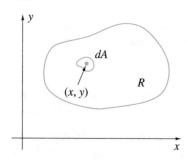

FIGURE 13.26   Double integrals are very efficient in calculating moments and centres of mass

We now show how double integrals can be used to replace definite integrals in calculating first moments, centres of mass, and moments of inertia of thin plates. Consider a thin plate with mass per unit area ρ such as that in Figure 13.26. Note that, unlike our discussion in Section 7.7 where we assumed ρ constant, we have made no such assumption here. In other words, density could be a function of position, $\rho = \rho(x, y)$.

The **centre of mass** $(\overline{x}, \overline{y})$ of the plate is a point at which a particle of mass M (equal to the total mass of the plate) has the same first moments about the x- and y-axes as the plate itself. If we divide the plate into small areas dA, then the mass in dA is $\rho\, dA$. Addition over all such areas in R as each dA shrinks to a point gives the mass of the plate

$$ M = \iint_R \rho\, dA. \tag{13.31} $$

Since the first moment of the mass in dA about the y-axis is $x\rho\, dA$, it follows that the first moment of the entire plate about the y-axis is

$$ \iint_R x\rho\, dA. $$

But this must be equal to the first moment of the particle of mass M at $(\bar{x}, \bar{y})$ about the y-axis, and hence

$$M\bar{x} = \iint_R x\rho \, dA. \tag{13.32}$$

This equation can be solved for $\bar{x}$ once the double integral on the right and M have been calculated.

Similarly, $\bar{y}$ is determined by the equation

$$M\bar{y} = \iint_R y\rho \, dA, \tag{13.33}$$

where the double integral on the right is the first moment of the plate about the x-axis.

In any given problem, the double integrals in 13.31–13.33 are evaluated by means of double iterated integrals. For example, if we divide the plate in Figure 13.27 into rectangles of dimensions dx and dy and use vertical strips, then we have

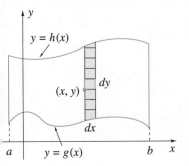

FIGURE 13.27 Vertical strips are chosen for this plate

$$M = \int_a^b \int_{g(x)}^{h(x)} \rho \, dy \, dx, \tag{13.34a}$$

$$M\bar{x} = \int_a^b \int_{g(x)}^{h(x)} x\rho \, dy \, dx, \tag{13.34b}$$

$$M\bar{y} = \int_a^b \int_{g(x)}^{h(x)} y\rho \, dy \, dx. \tag{13.34c}$$

Once again we point out that equations 13.34 should not be memorized as formulas. Indeed, each can be derived as needed. For instance, to obtain equation 13.34c, we reason that the first moment of the mass in a rectangle of dimensions dx and dy at position (x, y) about the x-axis is $y\rho \, dy \, dx$. Addition over the rectangles in a vertical strip gives the first moment of the strip about the x-axis,

$$\int_{g(x)}^{h(x)} y\rho \, dy \, dx,$$

and integration with respect to x now adds over all strips to give the first moment of the entire plate about the x-axis. Note that if ρ is constant and inner integrations are performed in each of equations 13.34, then

$$M = \int_a^b \{\rho y\}_{g(x)}^{h(x)} \, dx = \int_a^b \rho[h(x) - g(x)] \, dx,$$

$$M\bar{x} = \int_a^b \{\rho xy\}_{g(x)}^{h(x)} \, dx = \int_a^b \rho x[h(x) - g(x)] \, dx,$$

$$M\bar{y} = \int_a^b \left\{ \rho \frac{y^2}{2} \right\}_{g(x)}^{h(x)} \, dx = \int_a^b \frac{\rho}{2} \{[h(x)]^2 - [g(x)]^2\} \, dx.$$

These are equations 7.35–7.37 (with different names for the curves), but the simplicity of the discussion leading to the double iterated integrals certainly demonstrates its advantage over use of the definite integral described in Section 7.7.

EXAMPLE 13.9

Find the centre of mass of a thin plate with constant mass per unit area ρ if its edges are defined by the curves $y = 2x - x^2$ and $y = x^2 - 4$.

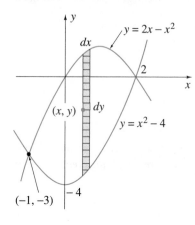

SOLUTION For vertical strips as shown in Figure 13.28,

$$M = \int_{-1}^{2} \int_{x^2-4}^{2x-x^2} \rho \, dy \, dx = \rho \int_{-1}^{2} [(2x - x^2) - (x^2 - 4)] \, dx$$

$$= \rho \left\{ x^2 - \frac{2x^3}{3} + 4x \right\}_{-1}^{2} = 9\rho.$$

If the centre of mass of the plate is $(\bar{x}, \bar{y})$, then

$$M\bar{x} = \int_{-1}^{2} \int_{x^2-4}^{2x-x^2} x\rho \, dy \, dx = \rho \int_{-1}^{2} x[(2x - x^2) - (x^2 - 4)] \, dx$$

$$= \rho \left\{ \frac{2x^3}{3} - \frac{x^4}{2} + 2x^2 \right\}_{-1}^{2} = \frac{9\rho}{2}.$$

Thus, $\bar{x} = \dfrac{9\rho}{2} \cdot \dfrac{1}{9\rho} = \dfrac{1}{2}$. Since

$$M\bar{y} = \int_{-1}^{2} \int_{x^2-4}^{2x-x^2} y\rho \, dy \, dx = \rho \int_{-1}^{2} \left\{ \frac{y^2}{2} \right\}_{x^2-4}^{2x-x^2} dx$$

$$= \frac{\rho}{2} \int_{-1}^{2} (-4x^3 + 12x^2 - 16) dx = \frac{\rho}{2} \{-x^4 + 4x^3 - 16x\}_{-1}^{2} = -\frac{27\rho}{2},$$

we find $\bar{y} = -\dfrac{27\rho}{2} \cdot \dfrac{1}{9\rho} = -\dfrac{3}{2}$.

EXAMPLE 13.10

Find the first moment of area about the line $y = -2$ for the region bounded by the curves $x = |y|^3$ and $x = 2 - y^2$.

SOLUTION Method 1 The first moment about $y = -2$ of a rectangle of dimensions dx and dy at position (x, y) is $(y + 2) \, dx \, dy$ (Figure 13.29). For the entire plate, then, the required first moment is

$$\int_{-1}^{0} \int_{-y^3}^{2-y^2} (y + 2) \, dx \, dy + \int_{0}^{1} \int_{y^3}^{2-y^2} (y + 2) \, dx \, dy$$

$$= \int_{-1}^{0} \{x(y + 2)\}_{-y^3}^{2-y^2} dy + \int_{0}^{1} \{x(y + 2)\}_{y^3}^{2-y^2} dy$$

$$= \int_{-1}^{0} (y^4 + y^3 - 2y^2 + 2y + 4) \, dy + \int_{0}^{1} (-y^4 - 3y^3 - 2y^2 + 2y + 4) \, dy$$

$$= \left\{ \frac{y^5}{5} + \frac{y^4}{4} - \frac{2y^3}{3} + y^2 + 4y \right\}_{-1}^{0} + \left\{ -\frac{y^5}{5} - \frac{3y^4}{4} - \frac{2y^3}{3} + y^2 + 4y \right\}_{0}^{1}$$

$$= \frac{17}{3}.$$

Method 2   By symmetry, the centroid of the region is somewhere along the x-axis. Hence, the required first moment is $2A$, where A is the area of the region and 2 is the distance from $y = -2$ to the centroid. Since the area of the region is equally distributed about the x-axis, we obtain the required first moment as

$$2(2)(\text{area of plate above } x\text{-axis}) = 4 \int_0^1 \int_{y^3}^{2-y^2} dx\,dy = 4 \int_0^1 (2 - y^2 - y^3)\,dy$$

$$= 4 \left\{ 2y - \frac{y^3}{3} - \frac{y^4}{4} \right\}_0^1 = \frac{17}{3}.$$

EXAMPLE 13.11

Find the first moment about the line $2x + y = 1$ of a thin plate with constant mass per unit area ρ if its edges are defined by the curves $y = x + 2$ and $y = x^2$.

SOLUTION   We could concentrate the mass M of the plate at its centre of mass $(\bar{x}, \bar{y})$, and multiply M by the distance from $2x + y = 1$ to $(\bar{x}, \bar{y})$. Because calculation of $(\bar{x}, \bar{y})$ requires three integrations, we prefer a more direct approach. The distance from $2x + y = 1$ to a rectangle of dimensions dx and dy at position (x, y) (Figure 13.30) is given by formula 1.16,

$$\frac{|2x + y - 1|}{\sqrt{4 + 1}} = \frac{|2x + y - 1|}{\sqrt{5}}.$$

If we take distances to the right of the line as positive and those to the left as negative, the directed distance is $(2x + y - 1)/\sqrt{5}$. It now follows that the first moment of the plate around the line $2x + y = 1$ is

$$\int_{-1}^2 \int_{x^2}^{x+2} \left(\frac{2x + y - 1}{\sqrt{5}} \right) \rho\,dy\,dx = \frac{\rho}{\sqrt{5}} \int_{-1}^2 \left\{ \frac{1}{2}(2x + y - 1)^2 \right\}_{x^2}^{x+2} dx$$

$$= \frac{\rho}{2\sqrt{5}} \int_{-1}^2 [(3x + 1)^2 - x^4 - 4x^3 - 2x^2 + 4x - 1]\,dx$$

$$= \frac{\rho}{2\sqrt{5}} \left\{ \frac{1}{9}(3x + 1)^3 - \frac{x^5}{5} - x^4 - \frac{2x^3}{3} + 2x^2 - x \right\}_{-1}^2$$

$$= \frac{36\sqrt{5}\,\rho}{25}.$$

FIGURE 13.30   Vertical strips are chosen to find the first moment of this plate

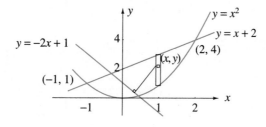

This example illustrates that double integrals allow us to calculate first moments about arbitrary lines, unlike definite integrals, which restrict first moments to horizontal and vertical lines.

Calculating **moments of inertia** (second moments) of thin plates is as easy as calculating first moments if we use double integrals. In particular, the mass in area dA in Figure 13.26 is $\rho\, dA$; thus its moments of inertia about the x- and y-axes are, respectively, $y^2 \rho\, dA$ and $x^2 \rho\, dA$. Moments of inertia of the entire plate about the x- and y-axes are therefore given by the double integrals:

$$I_x = \iint_R y^2 \rho\, dA \quad \text{and} \quad I_y = \iint_R x^2 \rho\, dA. \tag{13.35}$$

The product moment of inertia of this plate with respect to the x- and y-axes is defined as

$$I_{xy} = \iint_R xy\rho\, dA. \tag{13.36}$$

Unlike I_x and I_y, which are always positive, I_{xy} can be positive, negative, or zero.

For a plate such as that shown in Figure 13.31, we evaluate these double integrals by means of double iterated integrals with respect to x and y:

$$I_x = \int_a^b \int_{g(y)}^{h(y)} y^2 \rho\, dx\, dy, \tag{13.37a}$$

$$I_y = \int_a^b \int_{g(y)}^{h(y)} x^2 \rho\, dx\, dy, \tag{13.37b}$$

$$I_{xy} = \int_a^b \int_{g(y)}^{h(y)} xy\rho\, dx\, dy. \tag{13.37c}$$

FIGURE 13.31   Double integrals are advantageous when calculating moments of inertia of plates

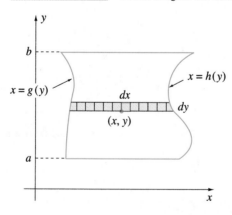

EXAMPLE 13.12

Find the moments of inertia about the x- and y-axes and the product moment of inertia of a thin plate with constant mass per unit area ρ if its edges are defined by the curves $y = x^3$, $y = \sqrt{2 - x}$, and $x = 0$.

FIGURE 13.32 Vertical strips are chosen to calculate moments of inertia of this plate

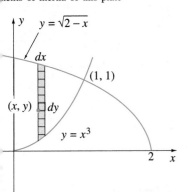

SOLUTION With vertical strips as shown in Figure 13.32, the moments of inertia are

$$
I_x = \int_0^1 \int_{x^3}^{\sqrt{2-x}} y^2 \rho \, dy \, dx = \rho \int_0^1 \left\{ \frac{y^3}{3} \right\}_{x^3}^{\sqrt{2-x}} dx = \frac{\rho}{3} \int_0^1 \left[(2-x)^{3/2} - x^9 \right] dx
$$

$$
= \frac{\rho}{3} \left\{ -\frac{2}{5}(2-x)^{5/2} - \frac{x^{10}}{10} \right\}_0^1 = \frac{16\sqrt{2} - 5}{30}\rho,
$$

$$
I_y = \int_0^1 \int_{x^3}^{\sqrt{2-x}} x^2 \rho \, dy \, dx = \rho \int_0^1 \left\{ x^2 y \right\}_{x^3}^{\sqrt{2-x}} dx = \rho \int_0^1 (x^2\sqrt{2-x} - x^5) \, dx.
$$

If we set $u = 2 - x$ in the first term,

$$
I_y = \rho \int_2^1 (2-u)^2 \sqrt{u}(-du) - \rho \left\{ \frac{x^6}{6} \right\}_0^1 = \rho \int_1^2 (4\sqrt{u} - 4u^{3/2} + u^{5/2}) \, du - \frac{\rho}{6}
$$

$$
= \rho \left\{ \frac{8u^{3/2}}{3} - \frac{8u^{5/2}}{5} + \frac{2u^{7/2}}{7} \right\}_1^2 - \frac{\rho}{6} = \frac{256\sqrt{2} - 319}{210}\rho.
$$

$$
I_{xy} = \int_0^1 \int_{x^3}^{\sqrt{2-x}} xy \rho \, dy \, dx = \rho \int_0^1 \left\{ \frac{xy^2}{2} \right\}_{x^3}^{\sqrt{2-x}} dx = \frac{\rho}{2} \int_0^1 (2x - x^2 - x^7) \, dx
$$

$$
= \frac{\rho}{2} \left\{ x^2 - \frac{x^3}{3} - \frac{x^8}{8} \right\}_0^1 = \frac{13\rho}{48}.
$$

EXAMPLE 13.13

Find the second moments of area about the lines $y = -1$ and $x + y = 1$ of the region bounded by the curves $x = y^2$ and $x = 2y$.

SOLUTION The second moment of area about the line $y = -1$ (Figure 13.33) is

$$
\int_0^2 \int_{y^2}^{2y} (y+1)^2 \, dx \, dy = \int_0^2 \{ x(y+1)^2 \}_{y^2}^{2y} \, dy = \int_0^2 (y+1)^2 (2y - y^2) \, dy
$$

$$
= \int_0^2 (-y^4 + 3y^2 + 2y) \, dy = \left\{ -\frac{y^5}{5} + y^3 + y^2 \right\}_0^2 = \frac{28}{5}.
$$

FIGURE 13.33 Horizontal strips are chosen for moments of inertia of this plate

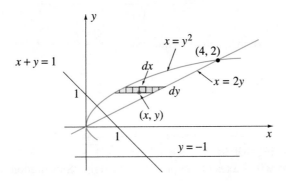

Since the undirected distance from $x + y = 1$ to (x, y) is $|x + y - 1|/\sqrt{2}$, the second moment about this line is

$$\int_0^2 \int_{y^2}^{2y} \left(\frac{x + y - 1}{\sqrt{2}}\right)^2 dx\, dy = \frac{1}{2} \int_0^2 \left\{\frac{1}{3}(x + y - 1)^3\right\}_{y^2}^{2y} dy$$

$$= \frac{1}{6} \int_0^2 [(3y - 1)^3 - y^6 - 3y^5 + 5y^3 - 3y + 1]\, dy$$

$$= \frac{1}{6} \left\{\frac{1}{12}(3y - 1)^4 - \frac{y^7}{7} - \frac{y^6}{2} + \frac{5y^4}{4} - \frac{3y^2}{2} + y\right\}_0^2 = \frac{62}{21}.$$

Principal Axes and Principal Moments of Inertia

Consider finding the lines through the origin about which the moments of inertia of the plate in Figure 13.34 are largest and smallest relative to all other lines through the origin. Using distance formula 1.16, we can say that the moment of inertia about any line $y = mx$ is

$$I(m) = \iint_R \left(\frac{|y - mx|}{\sqrt{m^2 + 1}}\right)^2 \rho\, dA = \frac{1}{m^2 + 1} \iint_R (y^2 - 2mxy + m^2 x^2)\rho\, dA$$

$$= \frac{1}{m^2 + 1}(I_x - 2mI_{xy} + m^2 I_y). \tag{13.38}$$

FIGURE 13.34 Moments of inertia have maximum and minimum values about principal axes

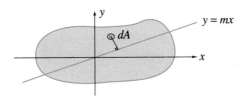

Critical points of $I(m)$ are given by

$$0 = \frac{dI}{dm} = \frac{-2m}{(m^2 + 1)^2}(I_x - 2mI_{xy} + m^2 I_y) + \frac{1}{m^2 + 1}(-2I_{xy} + 2mI_y)$$

$$= \frac{-2m(I_x - 2mI_{xy} + m^2 I_y) + (m^2 + 1)(-2I_{xy} + 2mI_y)}{(m^2 + 1)^2}$$

$$= \frac{2[(m^2 - 1)I_{xy} - mI_x + mI_y]}{(m^2 + 1)^2}.$$

Solutions of the quadratic equation $I_{xy}m^2 + (I_y - I_x)m - I_{xy} = 0$ are

$$m = \frac{I_x - I_y \pm \sqrt{(I_y - I_x)^2 + 4(I_{xy})^2}}{2I_{xy}} = \frac{I_x - I_y}{2I_{xy}} \pm \sqrt{1 + \left(\frac{I_x - I_y}{2I_{xy}}\right)^2}. \tag{13.39}$$

Lines with these slopes are perpendicular, as is easily seen by setting $a = (I_x - I_y)/(2I_{xy})$, and noting that $(a + \sqrt{1 + a^2})(a - \sqrt{1 + a^2}) = a^2 - (1 + a^2) = -1$. They are called the **principal axes** of the plate at the origin. Substitution of 13.39 into 13.38 leads to messy algebra. Instead, we note that $I_{xy}m^2 + (I_y - I_x)m - I_{xy} = 0$ can be expressed in the form $(m^2 + 1)I_{xy} + m(I_y - I_x) = 2I_{xy}$. Substituting this into 13.38 gives the moments of inertia around the principal axes:

$$I = \frac{1}{m^2 + 1}[I_x - m(m^2 + 1)I_{xy} - m^2(I_y - I_x) + m^2 I_y]$$

$$= \frac{1}{m^2 + 1}[(m^2 + 1)I_x - m(m^2 + 1)I_{xy}]$$

$$= I_x - mI_{xy} = I_x - \left[\frac{I_x - I_y \pm \sqrt{(I_x - I_y)^2 + 4(I_{xy})^2}}{2}\right]$$

$$= \frac{I_x + I_y}{2} \mp \sqrt{\left(\frac{I_x - I_y}{2}\right)^2 + (I_{xy})^2}. \tag{13.40}$$

That these yield (absolute) maximum and minimum values for $I(m)$ is shown in Exercise 42. They are called the **principal moments of inertia** of the plate about the origin; they are moments of inertia about the principal axes.

EXAMPLE 13.14

▼

Show that principal moments of inertia about the origin for the uniform rectangular plate in Figure 13.35 are I_x and I_y.

SOLUTION   Because of the symmetry of the plate about the axes, $I_{xy} = 0$, in which case equation 13.39 does not define principal directions. If we return to function 13.38, we find that

$$I(m) = \frac{1}{m^2 + 1}(I_x + m^2 I_y) = I_y + \frac{I_x - I_y}{m^2 + 1}.$$

If $I_x > I_y$ (as is the case in Figure 13.35), then this is an even function with respect to m, decreasing from $I(0) = I_x$ to $\lim_{m \to \infty} I = I_y$; that is, principal moments of inertia are I_x and I_y. If $I_x < I_y$, then this even function increases from $I(0) = I_x$ to $\lim_{m \to \infty} I = I_y$, and once again I_x and I_y are principal moments of inertia.

FIGURE 13.35   Principal moments of inertia of a rectangular plate about its centre

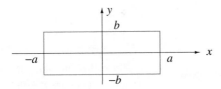

Notice that if the rectangle is a square ($a = b$), then $I_x = I_y$, and the moment of inertia of the plate about every line through the origin has the same value. In this case we say that every pair of perpendicular lines through the origin constitutes a pair of principal axes.

EXERCISES 13.5

In Exercises 1–10 find the centroid of the region bounded by the curves.

1. $x = y + 2$, $x = y^2$

2. $y = 8 - 2x^2$, $y + x^2 = 4$

3. $y = x^2 - 1$, $y + (x + 1)^2 = 0$

4. $x + y = 5$, $xy = 4$

5. $y = e^x$, $y = 0$, $x = 0$, $x = 1$

6. $y = \sqrt{4 - x^2}$, $y = x$, $x = 0$

7. $y = 1/(x - 1)$, $y = 1$, $y = 2$, $x = 0$

8. $x = 4y - 4y^2$, $x = y + 3$, $y = 1$, $y = 0$

9. $y = |x^2 - 1|$, $y = 2$

10. $y = x$, $y = 2x$, $2y = x + 3$

In Exercises 11–15 find the second moment of area of the region bounded by the curves about the line.

11. $y = x^2$, $y = x^3$ about the y-axis

12. $y = x$, $y = 2x + 4$, $y = 0$ about the x-axis

13. $y = x^2$, $2y = x^2 + 4$ about $y = 0$

14. $y = x^2 - 4$, $y = 2x - x^2$ about $x = -2$

15. $x = 1/\sqrt{y^4 + 12y^2}$, $x = 0$, $y = 1/2$, $y = 1$ about $y = 0$

16. Find the first moment about the line $y = -2$ of a thin plate of constant mass per unit area ρ if its edges are defined by the curves $y = 2 - 2x^2$ and $y = x^2 - 1$.

In Exercises 17–20 find the product moment of inertia with respect to the x- and y-axes for the plate defined by the curves if it has constant mass per unit area ρ.

17. $x^2 + y^2 = r^2$

18. $y = x^2$, $y = x^3$

* **19.** $x = -y^2$, $x + y + 2 = 0$

* **20.** $x = -2y$, $y = -x$, $x + 3y + 2 = 0$

In Exercises 21–27 find the centroid of the region bounded by the curves.

* **21.** $x = \sqrt{y + 2}$, $y = x$, $y = 0$

* **22.** $y + x^2 = 0$, $x = y + 2$, $x + y + 2 = 0$, $y = 2$ (above $y + x^2 = 0$)

* **23.** $y^2 = x^4(1 - x^2)$ (right loop)

* **24.** $3x^2 + 4y^2 = 48$, $(x - 2)^2 + y^2 = 1$

* **25.** $y = \ln x$, $y + \sqrt{x - 1} = 0$, $x = 2$

* **26.** $y = \sqrt{2 - x}$, $15y = x^2 - 4$

* **27.** $y = x\sqrt{1 - x^2}$, $x \geq 0$ and the x-axis

* **28.** Find the moment of inertia of a uniform rectangular plate a units long and b units wide about a line through the centre of the plate and perpendicular to the plate.

In Exercises 29–31 find the second moment of area of the region bounded by the curves about the line.

* **29.** $4x^2 + 9y^2 = 36$ about $y = -2$

* **30.** $x = y^2$, $x + y = 2$ about $x = -1$

* **31.** $y = \sqrt{a^2 - x^2}$, $y = a$, $x = a$ $(a > 0)$ about the x-axis

In Exercises 32–35 find the first and second moments of area of the region bounded by the curves about the line.

* **32.** $x = y^2 - 2$, $y = x$ about $x + y = 1$

* **33.** $x = y^2$, $x + y = 2$ about $y - x = 2$

* **34.** $y = x^3$, $x = y^2$ about $2x + y = 3$

* **35.** $y = 2 - x^2$, $y = |x|$ about $y = x$

* **36.** A triangular plate with constant mass per unit area ρ is bounded by the coordinate axes and the line $hx + by = hb$, where h and b are positive constants. Find its product moments of inertia about (a) the x- and y-axes and (b) the axes through the centre of mass parallel to the x- and y-axes.

* **37.** Show that the absolute value of the product moment of inertia with respect to the x- and y-axes of a plate with constant mass per unit area is always less than one-half the sum of the moments of inertia about the x- and y-axes.

* **38.** Show that for a plate with constant mass per unit area, the product moment of inertia with respect to the principal axes through a point is always zero.

* **39.** Suppose that I_x and I_y are moments of inertia of a thin plate with constant mass per unit area about the x- and y-axes. Let I'_x and I'_y be moments of inertia of the plate about any other pair of perpendicular lines through the origin. Show that $I'_x + I'_y = I_x + I_y$.

* **40.** Find the principal axes and principal moments of inertia about the origin for the uniform square plate bounded by the lines $x = 0$, $y = 0$, $x = a$, $y = a$.

* **41.** Find the principal axes and principal moments of inertia about the origin for the uniform rectangular plate bounded by the lines $x = 0$, $y = 0$, $x = a$, $y = b$, where $a > b > 0$.

* **42.** Show that the values of I in equation 13.40 are indeed maximum and minimum values of $I(m)$ as defined by 13.38 for $-\infty < m < \infty$.

* **43.** Suppose that a thin plate with constant mass per unit area is symmetric about a line ℓ and P is a point on ℓ. Show that the product moment of inertia of the plate about ℓ and a line through P perpendicular to ℓ vanishes.

* **44.** Show that if a thin plate with constant mass per unit area has an axis of symmetry, then the axis of symmetry must be a principal axis about any point on the line.

45. Show that if θ is the angle of inclination of a principal axis (about the origin), then $\tan 2\theta = \dfrac{2I_{xy}}{I_y - I_x}$.

46. Prove the *theorem of Pappus*: If a plane area is revolved about a coplanar axis not crossing the area, the volume generated is equal to the product of the area and the circumference of the circle described by the centroid of the area.

47. A thin flat plate of area A is immersed vertically in a fluid with density ρ. Show that the total force (due to fluid pressure) on each side of the plate is equal to the product of 9.81, A, ρ, and the depth of the centroid of the plate below the surface of the fluid. Use this result to find the forces in some of the problems in Exercises 13.4, say 1, 3, 4, 5, 7, 11, 12, 13, 16, and 17. For those problems involving triangles recall the result of Exercise 43 in Section 7.7 or Exercise 47 in Section 11.3.

48. Prove the *parallel axis theorem* for thin plates: The moment of inertia of a thin plate (with constant mass per unit area) with respect to any coplanar line is equal to the moment of inertia with respect to the parallel line through the centre of mass plus the mass multiplied by the square of the distance between the lines.

* **49.** Suppose that a thin plate with constant mass per unit area occupies a region R of the xy-plane. Let $x = x_1$ and $x = x_2$ be any two vertical lines and I_{x_1} and I_{x_2} be moments of inertia of the plate about these lines. Show that

$$I_{x_2} = I_{x_1} + M[x_2^2 - x_1^2 + 2\overline{x}(x_1 - x_2)],$$

where M is the mass of the plate and $\overline{x}$ is the x-coordinate of its centre of mass. Does this result reduce to the parallel axis theorem of Exercise 48 when one of the lines passes through the centre of mass?

* **50.** Suppose that a thin plate with constant mass per unit area occupies a region R of the xy-plane, and $(\overline{x}, \overline{y})$ is its centre of mass. Let I_{xy} be the product moment of inertia of the plate about the x- and y-axes, and $I_{\overline{xy}}$ be the product moment of inertia of the plate about the lines through $(\overline{x}, \overline{y})$ parallel to the x- and y-axes. Verify that $I_{\overline{xy}} = I_{xy} - M\overline{x}\,\overline{y}$, where M is the mass of the plate. (This is called the *parallel axis theorem* for product moments of inertia.)

13.6 Surface Area

To find the length of a curve in Section 7.3 we approximated the curve by tangent line segments. To find the **area of a surface** we follow a similar procedure by approximating the surface with tangential planes. In particular, consider finding the area of a smooth surface S given that every vertical line that intersects the surface does so in exactly one point (Figure 13.36). If S_{xy} is the region in the xy-plane onto which S projects, we divide S_{xy} into n subregions with areas ΔA_i in any fashion whatsoever, and choose a point (x_i, y_i) in each ΔA_i. At the point (x_i, y_i, z_i) on the surface S that projects onto (x_i, y_i), we draw the tangent plane to S. Suppose we now project ΔA_i upward onto S and onto the tangent plane at (x_i, y_i, z_i) and denote these projected areas by ΔS_i and ΔS_{Ti}, respectively.

FIGURE 13.36 Curved surface area is defined in terms of flat areas on tangent planes to the surface

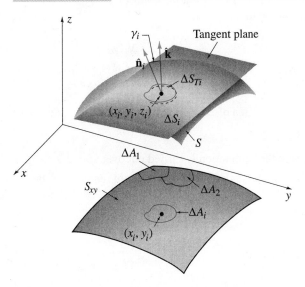

FIGURE 13.37   Area on
a slanted plane is related to its
projection in the xy-plane through
the angle between the planes

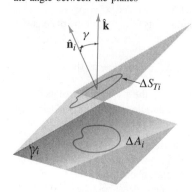

Now ΔS_{Ti} is an approximation to ΔS_i and as long as ΔA_i is small, a reasonably good approximation. In fact, the smaller ΔA_i, the better the approximation. We therefore define the area of S as follows:

$$\text{area of } S = \lim_{n \to \infty} \sum_{i=1}^{n} \Delta S_{Ti}, \qquad (13.41)$$

where in taking the limit we demand that each ΔA_i shrink to a point. We have therefore defined area on a curved surface in terms of flat areas on tangent planes to the surface. The advantage of this definition is that we can calculate ΔS_{Ti} in terms of ΔA_i. To see how, we denote by $\hat{\mathbf{n}}_i$ the unit normal vector to S at (x_i, y_i, z_i) with positive z-component, and by γ_i the acute angle between $\hat{\mathbf{n}}_i$ and $\hat{\mathbf{k}}$. Now ΔS_{Ti} projects onto ΔA_i, and γ_i is the acute angle between the planes containing ΔA_i and ΔS_{Ti} (Figure 13.37). It follows that ΔA_i and ΔS_{Ti} are related by

$$\Delta A_i = \cos \gamma_i \, \Delta S_{Ti} \qquad (13.42)$$

(see Exercise 28 in Section 11.5). Note that if ΔS_{Ti} is horizontal, then $\gamma_i = 0$ and $\Delta S_{Ti} = \Delta A_i$; and if ΔS_{Ti} tends toward the vertical ($\gamma_i \to \pi/2$), then ΔS_{Ti} becomes very large for fixed ΔA_i.

Because the surface projects in a one-to-one fashion onto the area S_{xy} in the xy-plane, we can take the equation for S in the form $z = f(x, y)$. A vector normal to S at any point is therefore

$$\nabla(z - f(x, y)) = \left(-\frac{\partial f}{\partial x}, -\frac{\partial f}{\partial y}, 1 \right);$$

hence

$$\hat{\mathbf{n}} = \frac{\left(-\dfrac{\partial f}{\partial x}, -\dfrac{\partial f}{\partial y}, 1 \right)}{\sqrt{1 + \left(\dfrac{\partial f}{\partial x} \right)^2 + \left(\dfrac{\partial f}{\partial y} \right)^2}} = \frac{\left(-\dfrac{\partial z}{\partial x}, -\dfrac{\partial z}{\partial y}, 1 \right)}{\sqrt{1 + \left(\dfrac{\partial z}{\partial x} \right)^2 + \left(\dfrac{\partial z}{\partial y} \right)^2}}.$$

Since $\hat{\mathbf{n}}_i \cdot \hat{\mathbf{k}} = |\hat{\mathbf{n}}_i||\hat{\mathbf{k}}| \cos \gamma_i = \cos \gamma_i$, it follows that

$$\cos \gamma_i = \hat{\mathbf{n}}_i \cdot \hat{\mathbf{k}} = \frac{1}{\sqrt{1 + z_x^2(x_i, y_i) + z_y^2(x_i, y_i)}}.$$

When we substitute this expression into 13.42, we obtain the result that area ΔS_{Ti} on the tangent plane to $z = f(x, y)$ at the point (x_i, y_i, z_i) is related to its projection ΔA_i in the xy-plane according to

$$\Delta S_{Ti} = \sqrt{1 + z_x^2(x_i, y_i) + z_y^2(x_i, y_i)} \, \Delta A_i. \qquad (13.43)$$

Formula 13.41 for the area of S can now be written in the form

$$\text{area of } S = \lim_{\|\Delta A_i\| \to 0} \sum_{i=1}^{n} \sqrt{1 + z_x^2(x_i, y_i) + z_y^2(x_i, y_i)} \, \Delta A_i, \qquad (13.44)$$

where the summation is carried out over all areas ΔA_i in S_{xy} as each ΔA_i shrinks to a point. But this is the definition of the double integral of the function $\sqrt{1 + (\partial z/\partial x)^2 + (\partial z/\partial y)^2}$ over the region S_{xy}. In other words, on the basis of formula 13.41, areas on surfaces can be calculated according to

$$\text{area of } S = \iint_{S_{xy}} \sqrt{1 + \left(\frac{\partial z}{\partial x} \right)^2 + \left(\frac{\partial z}{\partial y} \right)^2} \, dA. \qquad (13.45)$$

Note the analogy between equations 13.45 and 7.15. In equation 7.15 we think of $\sqrt{1 + (dy/dx)^2}\, dx$ as a small length along a curve $C : y = f(x)$ that projects onto the length dx along the x-axis. In fact, $\sqrt{1 + (dy/dx)^2}\, dx$ is along the tangent line to C, but we think of it as along C itself (see Section 11.11). The total length of C is then found by adding over all projections of C from $x = a$ to $x = b$. Similarly, in equation 13.45 we think of $\sqrt{1 + (\partial z/\partial x)^2 + (\partial z/\partial y)^2}\, dA$ as a small area on a surface $S : z = f(x, y)$ that projects onto the area dA in the xy-plane. It is, in fact, a small area on the tangent plane to S, but it is usually easier to think of it as being on S itself. The total area of S is then the addition over the projection S_{xy} of S.

FIGURE 13.38 Divide the area of this surface into two parts S_1 and S_2 with curve C and find both areas

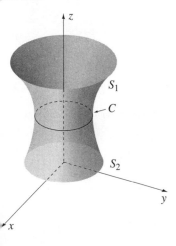

Note, too, that when S is smooth, $\partial z/\partial x$ and $\partial z/\partial y$ are continuous, thus guaranteeing existence of double integral 13.45. If S is piecewise smooth, we divide it into smooth subsurfaces and integrate over each piece separately.

This discussion has been based on the assumption that S projects one-to-one onto some region S_{xy} in the xy-plane. If this condition is not met, one possibility is to subdivide S into parts, each of which projects one-to-one onto the xy-plane. The total area of S is then the sum of the areas of its parts. For instance, to find the area of the surface S in Figure 13.38, we could subdivide S into S_1 and S_2 along the curve C. The area of S is then the sum of the areas of S_1 and S_2, each of which projects one-to-one onto the xy-plane.

Alternatively, we could note that there is nothing sacred about projecting S onto the xy-plane. We could develop similar results if S projects one-to-one onto either the yz- or the xz-plane. If S_{yz} and S_{xz} represent these projections, then

$$\text{area of } S = \iint_{S_{yz}} \sqrt{1 + \left(\frac{\partial x}{\partial y}\right)^2 + \left(\frac{\partial x}{\partial z}\right)^2}\, dA, \qquad (13.46a)$$

$$\text{area of } S = \iint_{S_{xz}} \sqrt{1 + \left(\frac{\partial y}{\partial x}\right)^2 + \left(\frac{\partial y}{\partial z}\right)^2}\, dA. \qquad (13.46b)$$

EXAMPLE 13.15

FIGURE 13.39 To find the area of $x + 2y + 3z = 6$ in the first octant, project it onto the xy-plane

Find the area of that part of the plane $x + 2y + 3z = 6$ in the first octant.

SOLUTION This area projects one-to-one onto the triangular area S_{xy} in the xy-plane bounded by the lines $x = 0$, $y = 0$, $x + 2y = 6$ (Figure 13.39). Since $z = 2 - x/3 - 2y/3$,

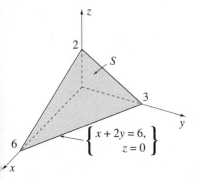

$$\text{area} = \iint_{S_{xy}} \sqrt{1 + \left(\frac{\partial z}{\partial x}\right)^2 + \left(\frac{\partial z}{\partial y}\right)^2}\, dA = \iint_{S_{xy}} \sqrt{1 + \left(-\frac{1}{3}\right)^2 + \left(-\frac{2}{3}\right)^2}\, dA$$

$$= \frac{\sqrt{14}}{3} \iint_{S_{xy}} dA = \frac{\sqrt{14}}{3} (\text{area of } S_{xy}) = \frac{\sqrt{14}}{3}\left[\frac{1}{2}(3)(6)\right] = 3\sqrt{14}.$$

We could also use formula 11.42.

EXAMPLE 13.16

Find the area of the surface $z = x^{3/2}$ that projects onto the rectangle in the xy-plane bounded by the straight lines $x = 0$, $x = 2$, $y = 1$, and $y = 3$.

SOLUTION   Since the surface projects one-to-one onto the rectangle (Figure 13.40), we find that

$$\text{area} = \iint_{S_{xy}} \sqrt{1 + \left(\frac{\partial z}{\partial x}\right)^2 + \left(\frac{\partial z}{\partial y}\right)^2} \, dA = \iint_{S_{xy}} \sqrt{1 + \left(\frac{3}{2}x^{1/2}\right)^2} \, dA$$

$$= \frac{1}{2}\int_0^2 \int_1^3 \sqrt{4 + 9x} \, dy \, dx = \frac{1}{2}\int_0^2 \{y\sqrt{4 + 9x}\}_1^3 \, dx = \int_0^2 \sqrt{4 + 9x} \, dx$$

$$= \left\{\frac{2}{27}(4 + 9x)^{3/2}\right\}_0^2 = \frac{2}{27}(22\sqrt{22} - 8).$$

FIGURE 13.40   To find the area of $z = x^{3/2}$ above the rectangle, project it onto the rectangle

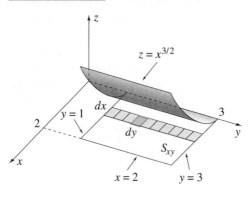

EXAMPLE 13.17

Find the area of the cone $y = \sqrt{x^2 + z^2}$ to the left of the plane $y = 1$.

SOLUTION   **Method 1**   The surface projects one-to-one onto the interior of the circle $x^2 + z^2 = 1$ in the xz-plane (Figure 13.41a). Since the area of the surface is four times that in the first octant, if we let S_{xz} be the quarter-circle $x^2 + z^2 \le 1$, $x \ge 0$, $z \ge 0$, then

$$\text{area} = 4 \iint_{S_{xz}} \sqrt{1 + \left(\frac{\partial y}{\partial x}\right)^2 + \left(\frac{\partial y}{\partial z}\right)^2} \, dA$$

$$= 4 \iint_{S_{xz}} \sqrt{1 + \left(\frac{x}{\sqrt{x^2 + z^2}}\right)^2 + \left(\frac{z}{\sqrt{x^2 + z^2}}\right)^2} \, dA$$

$$= 4 \iint_{S_{xz}} \sqrt{1 + \frac{x^2}{x^2 + z^2} + \frac{z^2}{x^2 + z^2}} \, dA = 4 \iint_{S_{xz}} \sqrt{2} \, dA$$

$$= 4\sqrt{2}(\text{area of } S_{xz}) = 4\sqrt{2}[\tfrac{1}{4}\pi(1)^2] = \sqrt{2}\,\pi.$$

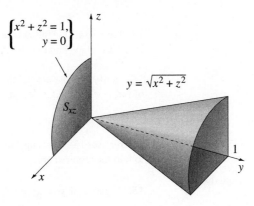

FIGURE 13.41a To
find the surface area of the cone,
project it onto the xz-plane

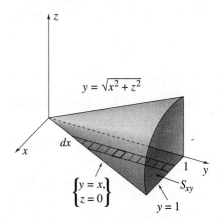

FIGURE 13.41b To
find the surface area of the cone,
project it onto the xy-plane

Method 2   Suppose instead that we project that part of the surface in the first octant onto
the triangle S_{xy} in Figure 13.41b. We write the equation of this part of the surface in the form
$z = \sqrt{y^2 - x^2}$ and calculate

$$
\begin{aligned}
\text{area} &= 4 \iint_{S_{xy}} \sqrt{1 + \left(\frac{\partial z}{\partial x}\right)^2 + \left(\frac{\partial z}{\partial y}\right)^2} \, dA \\
&= 4 \iint_{S_{xy}} \sqrt{1 + \left(\frac{-x}{\sqrt{y^2 - x^2}}\right)^2 + \left(\frac{y}{\sqrt{y^2 - x^2}}\right)^2} \, dA \\
&= 4 \iint_{S_{xy}} \sqrt{1 + \frac{x^2}{y^2 - x^2} + \frac{y^2}{y^2 - x^2}} \, dA = 4\sqrt{2} \iint_{S_{xy}} \frac{y}{\sqrt{y^2 - x^2}} \, dA.
\end{aligned}
$$

To evaluate this double integral, it is advantageous to integrate first with respect to y:

$$
\begin{aligned}
\text{area} &= 4\sqrt{2} \int_0^1 \int_x^1 \frac{y}{\sqrt{y^2 - x^2}} \, dy \, dx = 4\sqrt{2} \int_0^1 \{\sqrt{y^2 - x^2}\}_x^1 \, dx \\
&= 4\sqrt{2} \int_0^1 \sqrt{1 - x^2} \, dx.
\end{aligned}
$$

If we now set $x = \sin\theta$, then $dx = \cos\theta \, d\theta$, and

$$
\begin{aligned}
\text{area} &= 4\sqrt{2} \int_0^{\pi/2} \cos\theta \cos\theta \, d\theta = 4\sqrt{2} \int_0^{\pi/2} \left(\frac{1 + \cos 2\theta}{2}\right) d\theta \\
&= 2\sqrt{2} \left\{\theta + \frac{1}{2} \sin 2\theta\right\}_0^{\pi/2} = \sqrt{2}\,\pi.
\end{aligned}
$$

EXERCISES 13.6

In Exercises 1–6 find the area required.

1. The area of $2x + 3y + 6z = 1$ in the first octant

2. The area of $x + 2y - 3z + 4 = 0$ for which $x \le 0$, $y \le 0$ and $z \ge 0$

3. The area of $z = 1 - 4\sqrt{x^2 + y^2}$ above the xy-plane

4. The area of $z = \sqrt{2xy}$ cut out by the planes $x = 1$, $x = 2$, $y = 1$, $y = 3$

* **5.** The area in the first octant cut out from the surface $z = x + y$ by the plane $x + 2y = 4$

* **6.** The area of $z = x^{3/2} + y^{3/2}$ in the first octant cut off by the plane $x + y = 1$

In Exercises 7–12 set up, but do not evaluate, double iterated integrals to find the required area.

* **7.** The area of $x^2 + y^2 + z^2 = 2$ inside the cone $z = \sqrt{x^2 + y^2}$

* **8.** The area of $4x = y^2 + z^2$ cut off by $x = 4$

* **9.** The area in the first octant cut from $y = xz$ by the cylinder $x^2 + z^2 = 1$

* **10.** The area of $z = (x^2 + y^2)^2$ below $z = 4$

* **11.** The area of $y = 1 - x^2 - 3z^2$ to the right of the xz-plane

* **12.** The area of $z = \ln(1 + x + y)$ in the first octant cut off by $y = 1 - x^2$

* **13.** Find the area of the surface $z = \ln x$ that projects onto the rectangle in the xy-plane bounded by the lines $x = 1$, $x = 2$, $y = 0$, $y = 2$.

* **14.** Verify that the area of the curved portion of a right-circular cone of radius r and height h is $\pi r\sqrt{r^2 + h^2}$.

In Exercises 15–19 set up, but do not evaluate, double iterated integrals to find the required area.

* **15.** The area of $y = x^2 + z^2$ cut off by $y + z = 1$

* **16.** The area of $y = z^2 + x$ inside $x^2 + y^2 = 1$

* **17.** The area of $y^2 = z + x^2$ inside $x^2 + y^2 = 4$

* **18.** The area of $z = x^3 + y^3$ that is in the first octant and between the planes $x + y = 1$ and $x + y = 2$

* **19.** The area of $x^2 + y^2 = z^2 + 1$ between the planes $z = 1$ and $z = 4$

* **20.** Find the area of that part of the surface $z = 2x^2 + 3y$ bounded by the planes $x = 2$, $y = 0$, and $y = x$.

13.7 Double Iterated Integrals in Polar Coordinates

So far we have used only double iterated integrals in x and y to evaluate double integrals. But for some problems this is not convenient. For instance, the double integral of a continuous function $f(x, y)$ over the region R in Figure 13.42 requires three double iterated integrals in x and y. In other words, a subdivision of R into rectangles by coordinate lines $x = $ constant and $y = $ constant is simply not convenient for this region. For such an area, polar coordinates are more suitable.

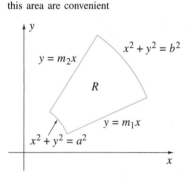

FIGURE 13.42 Double integrals in polar coordinates over this area are convenient

Polar coordinates with the origin as pole and the positive x-axis as polar axis are defined by

$$x = r\cos\theta, \qquad y = r\sin\theta$$

(see Section 9.2). We wish to obtain **double iterated integrals in polar coordinates** that represent the double integral of $f(x, y)$ over the region R in Figure 13.42. To do this we return to definition 13.3 for a double integral, and choose a subdivision of R into subareas convenient to polar coordinates. When using Cartesian coordinates we drew coordinate lines $x = $ constant and $y = $ constant. When using polar coordinates we draw coordinate curves $r = $ constant and $\theta = $ constant. In particular, we subdivide R by a network of $n + 1$ circles $r = r_i$, where

$$a = r_0 < r_1 < r_2 < \cdots < r_{n-1} < r_n = b,$$

and $m + 1$ radial lines $\theta = \theta_j$, where

$$\alpha = \theta_0 < \theta_1 < \theta_2 < \cdots < \theta_{m-1} < \theta_m = \beta$$

(Figure 13.43a). If ΔA_{ij} represents the area bounded by the circles $r = r_{i-1}$ and $r = r_i$ and the radial lines $\theta = \theta_{j-1}$ and $\theta = \theta_j$ (Figure 13.43b), then it is straightforward to show that

$$\Delta A_{ij} = \frac{1}{2}(r_i^2 - r_{i-1}^2)(\theta_j - \theta_{j-1}).$$

If we set $\Delta r_i = r_i - r_{i-1}$ and $\Delta\theta_j = \theta_j - \theta_{j-1}$, then

$$\Delta A_{ij} = \frac{1}{2}(r_i + r_{i-1})(r_i - r_{i-1})(\theta_j - \theta_{j-1}) = \left(\frac{r_i + r_{i-1}}{2}\right)\Delta r_i \, \Delta\theta_j. \qquad (13.47)$$

Our next task in using equation 13.3 for the double integral of $f(x, y)$ over R is to choose a star point in each ΔA_{ij}. If we select

$$(r_i^*, \theta_j^*) = \left(\frac{r_i + r_{i-1}}{2}, \theta_j\right),$$

FIGURE 13.43a Division of region into subregions using polar coordinates

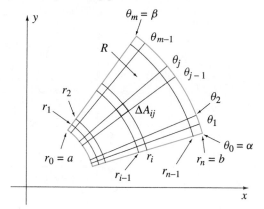

FIGURE 13.43b Area of small element using polar coordinates

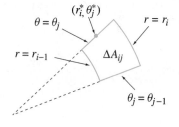

then

$$\Delta A_{ij} = r_i^* \, \Delta r_i \, \Delta\theta_j,$$

and by equation 13.3,

$$\iint_R f(x, y)\, dA = \lim_{\|\Delta A_{ij}\| \to 0} \sum_{j=1}^{m} \sum_{i=1}^{n} f(r_i^* \cos\theta_j^*, r_i^* \sin\theta_j^*)\, \Delta A_{ij}$$

$$= \lim_{\substack{\|\Delta r_i\| \to 0 \\ \|\Delta\theta_j\| \to 0}} \sum_{j=1}^{m} \sum_{i=1}^{n} f(r_i^* \cos\theta_j^*, r_i^* \sin\theta_j^*)r_i^* \, \Delta r_i \, \Delta\theta_j.$$

If we take the limit first as $\|\Delta r_i\| \to 0$ and then as $\|\Delta\theta_j\| \to 0$, we obtain the double iterated integral

$$\iint_R f(x, y)\, dA = \int_\alpha^\beta \int_a^b f(r\cos\theta, r\sin\theta)r\, dr\, d\theta. \qquad (13.48)$$

Reversing the order of taking limits reverses the order of the iterated integrals:

$$\iint_R f(x, y)\, dA = \int_a^b \int_\alpha^\beta f(r\cos\theta, r\sin\theta)r\, d\theta\, dr. \qquad (13.49)$$

For the region R of Figure 13.43a, then, there are two double iterated integrals in polar coordinates representing the double integral of $f(x, y)$ over R.

We have interpreted double iterated integrals in Cartesian coordinates as integrations over horizontal or vertical strips. Double iterated integrals in polar coordinates can also be interpreted geometrically. Take, for instance, equation 13.48. A double iterated integral in polar coordinates implies a subdivision of the region R into areas as shown in Figure 13.44. Let us denote small variations in r and θ for a representative piece of area at position (r, θ) by dr and $d\theta$. If dr and $d\theta$ are very small (as is implied in the definition of the double integral), then this piece of area is almost rectangular with an approximate area of $(r\, d\theta)\, dr$. In polar coordinates, then, we think of dA in equation 13.48 as being replaced by

$$dA = r\, dr\, d\theta. \tag{13.50}$$

Each such area at (r, θ) is multiplied by the value of $f(x, y)$ at (r, θ) to give the product

$$f(r\cos\theta, r\sin\theta)r\, dr\, d\theta.$$

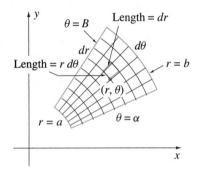

FIGURE 13.44   Area element in polar coordinates expressed in terms of differentials is $dA = r\, dr d\theta$

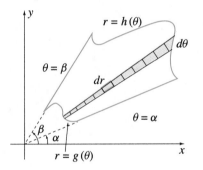

FIGURE 13.45   Double iterated integral with respect to r and θ adds first inside a wedge, and then over all wedges

The inner integral

$$\int_a^b f(r\cos\theta, r\sin\theta)r\, dr\, d\theta$$

with respect to r holds θ constant and is therefore interpreted as a summation over the small areas in a wedge $d\theta$ from $r = a$ to $r = b$. The θ-integration then adds over all wedges starting at $\theta = \alpha$ and ending at $\theta = \beta$. Limits on θ therefore identify positions of first and last wedges.

If the order of integration is reversed (equation 13.49), then the inner integral

$$\int_c^d f(r\cos\theta, r\sin\theta)r\, d\theta\, dr$$

holds r constant. We interpret this as an addition over the small areas in a ring dr, where the limits indicate that each and every ring starts on the curve $\theta = \alpha$ and ends on the curve $\theta = \beta$. The outer r-integration is an addition over all rings with the first ring at $r = a$ and the last at $r = b$.

Double iterated integrals in polar coordinates for more general regions are now quite simple. For the region R of Figure 13.45,

$$\iint_R f(x, y)\, dA = \int_\alpha^\beta \int_{g(\theta)}^{h(\theta)} f(r\cos\theta, r\sin\theta)r\, dr\, d\theta.$$

EXAMPLE 13.18

Evaluate the double iterated integral

$$\int_0^1 \int_0^{\sqrt{-x^2+x}} y^2 \, dy \, dx.$$

SOLUTION   The limits identify the region of integration as the interior of the semicircle in Figure 13.46a. The integrand suggests an interpretation of the integral as the second moment of area of this semicircle about the x-axis. Since the semicircle R in Figure 13.46b has exactly the same second moment about the x-axis, we can state that

$$\int_0^1 \int_0^{\sqrt{-x^2+x}} y^2 \, dy \, dx = \iint_R y^2 \, dA = \int_0^\pi \int_0^{1/2} (r^2 \sin^2 \theta) r \, dr \, d\theta$$

$$= \int_0^\pi \left\{ \frac{r^4}{4} \sin^2 \theta \right\}_0^{1/2} d\theta = \frac{1}{64} \int_0^\pi \sin^2 \theta \, d\theta$$

$$= \frac{1}{128} \int_0^\pi (1 - \cos 2\theta) \, d\theta = \frac{1}{128} \left\{ \theta - \frac{\sin 2\theta}{2} \right\}_0^\pi = \frac{\pi}{128}.$$

FIGURE 13.46a Limits indicate that the region of integration is the semicircle	FIGURE 13.46b Integration of x^2 over this semicircle gives the same result

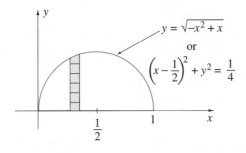

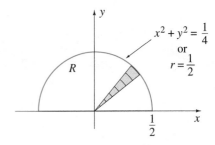

EXAMPLE 13.19

FIGURE 13.47   Centroid of the semicircular annulus

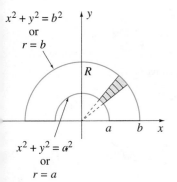

Find the centroid of the region R in Figure 13.47.

SOLUTION   Evidently, $\bar{x} = 0$, and the area of the region is $A = (\pi b^2 - \pi a^2)/2$. Since

$$A\bar{y} = \iint_R y \, dA = \int_0^\pi \int_a^b (r \sin \theta) r \, dr \, d\theta = \int_0^\pi \left\{ \frac{r^3}{3} \sin \theta \right\}_a^b d\theta$$

$$= \frac{1}{3}(b^3 - a^3) \int_0^\pi \sin \theta \, d\theta = \frac{1}{3}(b^3 - a^3)\{-\cos \theta\}_0^\pi = \frac{2}{3}(b^3 - a^3),$$

it follows that $\bar{y} = \frac{2}{3}(b^3 - a^3)\dfrac{2}{\pi(b^2 - a^2)} = \dfrac{4}{3\pi} \dfrac{b^2 + ab + a^2}{a + b}.$

EXAMPLE 13.20

Find the area of that portion of the sphere $x^2 + y^2 + z^2 = 2$ inside the cone $z = \sqrt{x^2 + y^2}$.

SOLUTION   If S is that portion of the sphere that is inside the cone and also in the first octant (Figure 13.48), then the required area is four times that of S; that is,

$$\text{area} = 4 \iint_{S_{xy}} \sqrt{1 + \left(\frac{\partial z}{\partial x}\right)^2 + \left(\frac{\partial z}{\partial y}\right)^2} \, dA,$$

where S_{xy} is the projection of S on the xy-plane. The curve of intersection of the cone and the sphere has equations

$$x^2 + y^2 + z^2 = 2, \qquad \text{or equivalently,} \qquad x^2 + y^2 = 1,$$

$$z = \sqrt{x^2 + y^2}, \qquad\qquad\qquad\qquad\qquad z = 1.$$

FIGURE 13.48   Polar coordinates are advantageous in calculating the area of that part of a sphere inside a cone

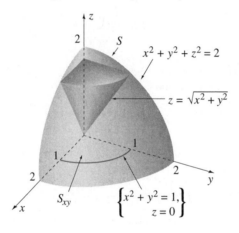

Consequently, S_{xy} is the interior of the quarter-circle $x^2 + y^2 \leq 1$, $x \geq 0$, $y \geq 0$. On S,

$$\frac{\partial z}{\partial x} = -\frac{x}{z} \qquad \text{and} \qquad \frac{\partial z}{\partial y} = -\frac{y}{z},$$

so that

$$\text{area} = 4 \iint_{S_{xy}} \sqrt{1 + \frac{x^2}{z^2} + \frac{y^2}{z^2}} \, dA = 4 \iint_{S_{xy}} \sqrt{\frac{x^2 + y^2 + z^2}{z^2}} \, dA$$

$$= 4 \iint_{S_{xy}} \sqrt{\frac{2}{z^2}} \, dA = 4\sqrt{2} \iint_{S_{xy}} \frac{1}{\sqrt{2 - x^2 - y^2}} \, dA.$$

If we now use polar coordinates to evaluate this double integral, we have

$$\text{area} = 4\sqrt{2} \int_0^{\pi/2} \int_0^1 \frac{1}{\sqrt{2 - r^2}} r \, dr \, d\theta = 4\sqrt{2} \int_0^{\pi/2} \{-\sqrt{2 - r^2}\}_0^1 \, d\theta$$

$$= 4\sqrt{2}(\sqrt{2} - 1) \int_0^{\pi/2} d\theta = 4\sqrt{2}(\sqrt{2} - 1)\frac{\pi}{2} = 2\sqrt{2}\pi(\sqrt{2} - 1).$$

Consulting Project 21

Here is perhaps our trickiest consultation project. Liquid flow through a pipe of radius R is controlled by a circular valve of the same radius that moves right-to-left across the pipe (Figure 13.49). In order to calibrate the amount of flow through the pipe, we are being asked to find a formula for the area of flow through the pipe as a function of the position a of the centre of the valve; that is, when the centre of the valve is at position $x = a$, what is the area inside the pipe not covered by the valve?

FIGURE 13.49   Valve to restrict flow in a circular pipe

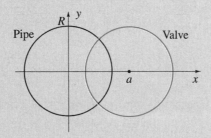

SOLUTION   Why do we claim that this project is so tricky? It would seem to be a simple matter to use a double integral in polar coordinates to find the area common to the pipe and the valve and subtract this from πR^2, the area of the pipe. In principle this is correct, but setting up the double iterated integral for the common area has hidden difficulties. We shall see this as we proceed. In polar coordinates, the equation for the pipe is $r = R$; in Cartesian coordinates, the equation for the valve is $(x - a)^2 + y^2 = R^2$. When we change this to polar coordinates, the result is

$$(r \cos \theta - a)^2 + (r \sin \theta)^2 = R^2 \quad \Longrightarrow \quad r^2 - 2ar \cos \theta + (a^2 - R^2) = 0.$$

When we use the quadratic formula to find an explicit definition of the curve, we obtain

$$r = a \cos \theta \pm \sqrt{R^2 - a^2 \sin^2 \theta}.$$

To find the point of intersection of the pipe and valve in the first quadrant, we set

$$R = a \cos \theta - \sqrt{R^2 - a^2 \sin^2 \theta}, \quad \text{which simplifies to} \quad \cos \theta = \frac{a}{2R}.$$

Let us denote the solution by $\overline{\theta} = \text{Cos}^{-1} \left(\dfrac{a}{2R} \right)$. Figure 13.50 shows the situation when the valve is only slightly closed.

FIGURE 13.50   Flow when valve almost fully open

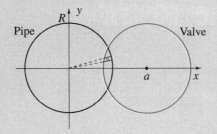

To find the area of the flow through the pipe, we subtract the area common to pipe and valve from πR^2. The common area is

$$\overline{A} = 2 \int_0^{\overline{\theta}} \int_{a\cos\theta - \sqrt{R^2 - a^2\sin^2\theta}}^{R} r \, dr \, d\theta$$

$$= \int_0^{\overline{\theta}} [R^2 - (a\cos\theta - \sqrt{R^2 - a^2\sin^2\theta})^2] \, d\theta$$

$$= \int_0^{\overline{\theta}} [R^2 - a^2\cos^2\theta + 2a\cos\theta\sqrt{R^2 - a^2\sin^2\theta} - R^2 + a^2\sin^2\theta] \, d\theta$$

$$= a \int_0^{\overline{\theta}} [a(-\cos 2\theta) + 2\cos\theta\sqrt{R^2 - a^2\sin^2\theta}] \, d\theta.$$

On the second term, we make the change of variable $R\sin\phi = a\sin\theta$, in which case $R\cos\phi \, d\phi = a\cos\theta \, d\theta$. If $R\sin\overline{\phi} = a\sin\overline{\theta}$, then

$$\overline{A} = a^2 \left\{ \frac{-\sin 2\theta}{2} \right\}_0^{\overline{\theta}} + 2a \int_0^{\overline{\phi}} \frac{R}{a}\cos\phi \, R\cos\phi \, d\phi$$

$$= -\frac{a^2}{2}\sin 2\overline{\theta} + 2R^2 \int_0^{\overline{\phi}} \left(\frac{1+\cos 2\phi}{2} \right) d\phi = -\frac{a^2}{2}\sin 2\overline{\theta} + R^2 \left\{ \phi + \frac{\sin 2\phi}{2} \right\}_0^{\overline{\phi}}$$

$$= -\frac{a^2}{2}\sin 2\overline{\theta} + \frac{R^2}{2}\left(2\overline{\phi} + \sin 2\overline{\phi} \right) = -\frac{a^2}{2}\sin 2\overline{\theta} + R^2\left(\overline{\phi} + \sin\overline{\phi}\cos\overline{\phi} \right)$$

$$= -a^2\sin\overline{\theta}\cos\overline{\theta} + R^2\left[\operatorname{Sin}^{-1}\left(\frac{a}{R}\sin\overline{\theta} \right) + \frac{a}{R}\sin\overline{\theta}\sqrt{1 - \frac{a^2}{R^2}\sin^2\overline{\theta}} \right]$$

$$= -a^2\left(\frac{a}{2R} \right)\sqrt{1 - \frac{a^2}{4R^2}} + R^2\operatorname{Sin}^{-1}\left(\frac{a}{R}\sqrt{1 - \frac{a^2}{4R^2}} \right)$$

$$\qquad + aR\sqrt{1 - \frac{a^2}{4R^2}}\sqrt{1 - \frac{a^2}{R^2}\left(1 - \frac{a^2}{4R^2} \right)}$$

$$= R^2\operatorname{Sin}^{-1}\left(\frac{a\sqrt{4R^2 - a^2}}{2R^2} \right) - \frac{a(a^2 - R^2)\sqrt{4R^2 - a^2}}{2R^2}.$$

Area of the flow is therefore

$$A = \pi R^2 - R^2\operatorname{Sin}^{-1}\left(\frac{a\sqrt{4R^2 - a^2}}{2R^2} \right) + \frac{a(a^2 - R^2)\sqrt{4R^2 - a^2}}{2R^2}.$$

It is worth noting that when $a = 2R$, the largest value for a, this reduces to πR^2, as it should.

As the valve continues to close, it reaches the position in Figure 13.51a where the line joining the origin to the point of intersection of the curves is tangent to the valve. When the valve is to the left of this position, the previous calculation of $\overline{A}$ is no longer valid.

To find the value of a at this stage, we calculate the slope of the valve by differentiating $(x - a)^2 + y^2 = R^2$, to get $2(x - a) + 2yy' = 0$, and evaluate y' at the point of intersection $(a/2, \sqrt{R^2 - a^2/4})$,

$$y' = -\frac{a/2 - a}{\sqrt{R^2 - a^2/4}} = \frac{a}{\sqrt{4R^2 - a^2}}.$$

When we equate this to the slope of the line joining the origin to $(a/2, \sqrt{R^2 - a^2/4})$, we obtain

$$\frac{a}{\sqrt{4R^2 - a^2}} = \frac{\sqrt{R^2 - a^2/4}}{a/2},$$

and when this equation is solved for a, the result is $a = \sqrt{2}R$. Thus, the above calculation for A is valid only in the interval $R \le a \le \sqrt{2}R$.

Flow when valve is at intermediate position

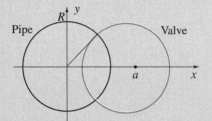

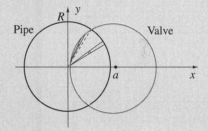

When the centre of the valve is just to the left of this position (Figure 13.51b), the area common to the pipe and valve is the sum of $\overline{A}$ as calculated above and twice the small area $\overline{\overline{A}}$ between the dashed line and the valve,

$$\overline{\overline{A}} = 2 \int_{\overline{\theta}}^{\overline{\overline{\theta}}} \int_{a\cos\theta - \sqrt{R^2 - a^2\sin^2\theta}}^{a\cos\theta + \sqrt{R^2 - a^2\sin^2\theta}} r \, dr \, d\theta,$$

where $\overline{\theta} = \mathrm{Cos}^{-1}[a/(2R)]$, as above. Angle $\overline{\overline{\theta}}$ is defined by the equation

$$a\cos\theta - \sqrt{R^2 - a^2\sin^2\theta} = a\cos\theta + \sqrt{R^2 - a^2\sin^2\theta} \implies \overline{\overline{\theta}} = \mathrm{Sin}^{-1}\left(\frac{R}{a}\right).$$

Thus,

$$\overline{\overline{A}} = \int_{\overline{\theta}}^{\overline{\overline{\theta}}} [(a\cos\theta + \sqrt{R^2 - a^2\sin^2\theta})^2 - (a\cos\theta - \sqrt{R^2 - a^2\sin^2\theta})^2] \, d\theta$$

$$= 4a \int_{\overline{\theta}}^{\overline{\overline{\theta}}} \cos\theta \sqrt{R^2 - a^2\sin^2\theta} \, d\theta.$$

When we use the above substitution $R\sin\phi = a\sin\theta$ to evaluate this integral, the result is

$$\overline{\overline{A}} = R^2 \left[\pi - 2\mathrm{Sin}^{-1}\left(\frac{a\sqrt{4R^2 - a^2}}{2R^2} \right) - \frac{a(2R^2 - a^2)\sqrt{4R^2 - a^2}}{2R^4} \right].$$

Area of the flow is therefore

$$A = \pi R^2 - \overline{A} - \overline{\overline{A}} = \pi R^2 - R^2 \mathrm{Sin}^{-1}\left(\frac{a\sqrt{4R^2 - a^2}}{2R^2}\right)$$

$$+ \frac{a(a^2 - R^2)\sqrt{4R^2 - a^2}}{2R^2}$$

$$- R^2\left[\pi - 2\mathrm{Sin}^{-1}\left(\frac{a\sqrt{4R^2 - a^2}}{2R^2}\right) - \frac{a(2R^2 - a^2)\sqrt{4R^2 - a^2}}{2R^4}\right]$$

$$= R^2 \mathrm{Sin}^{-1}\left(\frac{a\sqrt{4R^2 - a^2}}{2R^2}\right) + \frac{a\sqrt{4R^2 - a^2}}{2}.$$

This is valid when the position of the centre of the valve satisfies $R \le a \le \sqrt{2}R$. Note that this calculation agrees with that for A when $a = \sqrt{2}R$; both give $A = (\pi + 2)R^2/2$. Finally, we calculate that for $0 \le a \le R$ (Figure 13.51c),

$$A = 2\int_{\overline{\theta}}^{\pi}\int_{a\cos\theta + \sqrt{R^2 - a^2\sin^2\theta}}^{R} r\,dr\,d\theta$$

$$= \int_{\overline{\theta}}^{\pi}[R^2 - (a\cos\theta + \sqrt{R^2 - a^2\sin^2\theta})^2]\,d\theta$$

$$= \int_{\overline{\theta}}^{\pi}[-a^2\cos 2\theta - 2a\cos\theta\sqrt{R^2 - a^2\sin^2\theta}]\,d\theta.$$

Once again we use the substitution $R\sin\phi = a\sin\theta$ to evaluate the second term, the final result being

$$A = R^2 \mathrm{Sin}^{-1}\left(\frac{a\sqrt{4R^2 - a^2}}{2R^2}\right) + \frac{a\sqrt{4R^2 - a^2}}{2}.$$

Putting all these results together, the area A of flow as a function of position a of the centre of the valve is

$$A = \begin{cases} R^2 \mathrm{Sin}^{-1}\left(\dfrac{a\sqrt{4R^2 - a^2}}{2R^2}\right) + \dfrac{a\sqrt{4R^2 - a^2}}{2}, & 0 \le a \le \sqrt{2}R \\[4mm] \pi R^2 - R^2 \mathrm{Sin}^{-1}\left(\dfrac{a\sqrt{4R^2 - a^2}}{2R^2}\right) + \dfrac{a(a^2 - R^2)\sqrt{4R^2 - a^2}}{2R^2}, & \sqrt{2}R < a \le 2R. \end{cases}$$

FIGURE 13.51c   Flow when valve almost closed

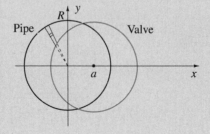

EXERCISES 13.7

In Exercises 1–5 evaluate the double integral of the function over the region R.

1. $f(x, y) = e^{x^2+y^2}$, where R is bounded by $x^2 + y^2 = a^2$

2. $f(x, y) = x$, where R is bounded by $x = \sqrt{2y - y^2}$, $x = 0$

3. $f(x, y) = \sqrt{x^2 + y^2}$, where R is bounded by $y = \sqrt{9 - x^2}$, $y = x$, $x = 0$

4. $f(x, y) = 1/\sqrt{x^2 + y^2}$, where R is the region outside $x^2 + y^2 = 4$ and inside $x^2 + y^2 = 4x$

5. $f(x, y) = \sqrt{1 + 2x^2 + 2y^2}$, where R is bounded by $x^2 + y^2 = 4$, $x^2 + y^2 = 4$

In Exercises 6–7 evaluate the double iterated integral.

6. $\int_0^1 \int_0^{\sqrt{1-x^2}} \sqrt{x^2 + y^2}\, dy\, dx$ **7.** $\int_{-\sqrt{2}}^0 \int_{-y}^{\sqrt{4-y^2}} x^2\, dx\, dy$

In Exercises 8–12 find the area of the region bounded by the curves.

8. Outside $x^2 + y^2 = 9$ and inside $x^2 + y^2 = 2\sqrt{3}\, y$

9. $r = 9(1 + \cos\theta)$

10. $r = \cos 3\theta$

11. Common to $r = 2$ and $r^2 = 9\cos 2\theta$

12. Common to $r = 1 + \sin\theta$ and $r = 2 - 2\sin\theta$

13. Find the centroid of the region bounded by the curves $y = x$, $y = -x$, $x = \sqrt{2 - y^2}$.

14. Find the second moment of area for a circular plate of radius R about any diameter.

15. A water tank in the form of a right-circular cylinder with radius R and length h has its axis horizontal. If it is full, what is the force due to water pressure on each end?

In Exercises 16–18 find the area of the surface.

16. The area of $z = x^2 + y^2$ below $z = 4$

17. The area of $x^2 + y^2 + z^2 = 4$ inside $x^2 + y^2 = 1$

18. The area of $z = xy$ inside $x^2 + y^2 = 9$

19. Prove that the area of a sphere of radius R is $4\pi R^2$.

20. Find the area of the hyperbolic paraboloid $z = x^2 - y^2$ between the cylinders $x^2 + y^2 = 1$ and $x^2 + y^2 = 4$.

In Exercises 21–22 find the volume of the solid of revolution obtained by rotating the region bounded by the curve about the line.

21. $r = \cos^2\theta$ about the x-axis

22. $r = 1 + \sin\theta$ about the y-axis

23. Find the area of the region bounded by the curve $(x^2 + y^2)^3 = a^2 x^2 y^2$.

∗ 24. The roof of an exhibition hall consists of cylindrical concrete shells as shown below. The length of each shell (not shown) is 50 m. The underside of each shell is half of a circular cylinder with radius 7 m. The thickness of the shell (measured radially) is 10 cm at the base and decreases linearly with angle θ to 5 cm at the top. Find the volume of each shell.

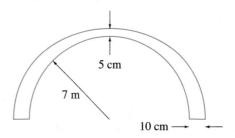

∗ 25. Find the area inside the circle $x^2 + y^2 = 4x$ and outside the circle $x^2 + y^2 = 1$.

∗ 26. Find the volume of the solid of revolution when a circle of radius R is rotated about a tangent line.

∗ 27. A circular plate of radius R (figure below) has a uniform charge distribution of ρ coulombs per square metre. If P is a point directly above the centre of the plate and dA is a small area on the plate, then the potential at P due to dA is given by

$$\frac{1}{4\pi\epsilon_0}\frac{\rho\, dA}{s},$$

where s is the distance from P to dA.

(a) Show that in terms of polar coordinates, the potential V at P due to the entire plate is

$$V = \frac{\rho}{4\pi\epsilon_0}\int_{-\pi}^{\pi}\int_0^R \frac{r}{\sqrt{r^2 + d^2}}dr\, d\theta,$$

where d is the distance from P to the centre of the plate.

(b) Evaluate the double iterated integral to find V.

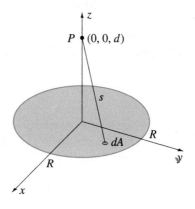

∗ 28. Use Coulomb's law (see Example 11.8 in Section 11.3) to find the force on a charge q at point P due to the charge on the plate in Exercise 27. What happens to this force as the radius of the plate gets very large?

In Exercises 29–31 find the area of the region bounded by the curves.

* **29.** $(x^2 + y^2)^2 = 2xy$
* **30.** Inside both $r = 6\cos\theta$ and $r = 4 - 2\cos\theta$
* **31.** $r = \cos^2\theta\sin\theta$

* **32.** Find the centroid of the region bounded by the cardioid $r = 1 + \cos\theta$.
* **33.** Find the second moment of area about the x-axis for the region bounded by $r^2 = 9\cos 2\theta$.
* **34.** Evaluate the double integral of

$$f(x, y) = \sqrt{\frac{1 - x^2 - y^2}{1 + x^2 + y^2}}$$

over the region inside the circle $x^2 + y^2 = 1$.

* **35.** The figure below illustrates a piece of an artery or vein with circular cross-section (radius R). The speed of blood flowing through the blood vessel is not uniform because of the viscosity of the blood and friction at the walls. *Poiseuille's law* states that for laminar blood flow, the speed v of blood at a distance r from the centre of the vessel is given by

$$v = \frac{P}{4nL}(R^2 - r^2),$$

where P is the pressure difference between the ends of the vessel, L is the length of the vessel, and n is the viscosity of the blood. Find the amount of blood flowing over a cross-section of the blood vessel per unit time.

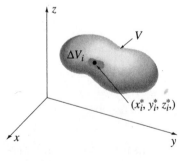

* **36.** Suppose we denote $PR^2/(4nL)$ in Exercise 35 by V_{max}, the maximum velocity at the centre of the blood vessel, $v = V_{max}[1 - (r/R)^2]$. Suppose the vessel converges to a smaller vessel of radius $R_1 = \alpha R$, $0 < \alpha < 1$. If blood flow in the smaller vessel also has a parabolic profile $v = U_{max}[1 - (r/R_1)^2]$, find U_{max} in terms of V_{max} and α. *Hint:* Assume that the volume flow rates in the two vessels must be the same.

* **37.** Repeat Exercise 36 for flow in a square pipe that reduces from a width of L to one of width αL ($0 < \alpha < 1$). Assume a velocity profile in the larger pipe of the form $v = V_{max}(1 - 4x^2/L^2)(1 - 4y^2/L^2)$, $-L/2 \le x \le L/2$, $-L/2 \le y \le L/2$, and a similar profile in the smaller pipe.

** **38.** Find the area of that part of the sphere $x^2 + y^2 + z^2 = a^2$ inside $(x^2 + y^2)^2 = a^2(x^2 - y^2)$.

** **39.** Find the area of that portion of the surface $x^2 + z^2 = a^2$ cut out by $x^2 + y^2 = a^2$.

** **40.** A very important integral in statistics is

$$I = \int_0^\infty e^{-x^2}\, dx.$$

To evaluate the integral we set

$$I = \int_0^\infty e^{-y^2}\, dy,$$

and then multiply these two equations. Do this to prove that $I = \sqrt{\pi}/2$.

** **41.** Use the result of Exercise 40 to evaluate the gamma function

$$\Gamma(n) = \int_0^\infty x^{n-1} e^{-x}\, dx$$

at $n = 1/2$.

13.8 Triple Integrals and Triple Iterated Integrals

Triple integrals are defined in much the same way as double integrals. Suppose $f(x, y, z)$ is a function defined in some region V of space that has finite volume (Figure 13.52). We divide V into n subregions of volumes $\Delta V_1, \Delta V_2, \ldots, \Delta V_n$ in any manner whatsoever, and in each subregion ΔV_i ($i = 1, \ldots, n$) we choose an arbitrary point (x_i^*, y_i^*, z_i^*). We then form the sum

$$f(x_1^*, y_1^*, z_1^*)\,\Delta V_1 + f(x_2^*, y_2^*, z_2^*)\,\Delta V_2 + \cdots + f(x_n^*, y_n^*, z_n^*)\,\Delta V_n$$

$$= \sum_{i=1}^n f(x_i^*, y_i^*, z_i^*)\,\Delta V_i. \qquad (13.51)$$

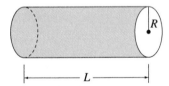 **FIGURE 13.52** Subdivide a region V into smaller volumes to define the triple integral of $f(x, y, z)$

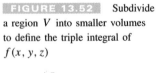

If this sum approaches a limit as the number of subregions becomes increasingly large and every subregion shrinks to a point, we call the limit the **triple integral** of $f(x, y, z)$ over the region V and denote it by

$$\iiint_V f(x, y, z)\, dV = \lim_{\|\Delta V_i\| \to 0} \sum_{i=1}^n f(x_i^*, y_i^*, z_i^*)\,\Delta V_i. \qquad (13.52)$$

As in the case of double integrals, we require that this limit be independent of the manner of subdivision of V and the choice of star points in the subregions. This is guaranteed for continuous functions by the following theorem.

THEOREM 13.2

Let S be a piecewise-smooth surface that encloses a region V with finite volume. If $f(x, y, z)$ is continuous inside and on S, then the triple integral of $f(x, y, z)$ over V exists.

Properties analogous to those in equations 13.4–13.7 hold for triple integrals, although we will not list the first three here. Corresponding to equation 13.7, the volume of a region V is given by the triple integral

$$\text{volume of } V = \iiint_V dV. \tag{13.53}$$

We evaluate triple integrals with **triple iterated integrals**. If we use Cartesian coordinates there are six possible triple iterated integrals of a function $f(x, y, z)$, corresponding to the six permutations of the product of the differentials $dx, dy,$ and dz:

$$dz\,dy\,dx, \quad dz\,dx\,dy, \quad dx\,dz\,dy, \quad dx\,dy\,dz, \quad dy\,dx\,dz, \quad dy\,dz\,dx.$$

The general triple iterated integral of $f(x, y, z)$ with respect to z, y, and x is of the form

$$\int_a^b \int_{g_1(x)}^{g_2(x)} \int_{h_1(x,y)}^{h_2(x,y)} f(x, y, z)\, dz\, dy\, dx. \tag{13.54}$$

Because the first integration with respect to z holds x and y constant, the limits on z may therefore depend on x and y. Similarly, the second integration with respect to y holds x constant, and the limits may be functions of x.

EXAMPLE 13.21

Evaluate the triple iterated integral $\displaystyle \int_0^1 \int_0^{x^2} \int_{xy}^{x+y} xyz\, dz\, dy\, dx.$

SOLUTION

$$\int_0^1 \int_0^{x^2} \int_{xy}^{x+y} xyz\, dz\, dy\, dx = \int_0^1 \int_0^{x^2} \left\{ \frac{xyz^2}{2} \right\}_{xy}^{x+y} dy\, dx$$

$$= \frac{1}{2} \int_0^1 \int_0^{x^2} [xy(x+y)^2 - xy(xy)^2]\, dy\, dx$$

$$= \frac{1}{2} \int_0^1 \int_0^{x^2} (x^3 y + 2x^2 y^2 + xy^3 - x^3 y^3)\, dy\, dx$$

$$= \frac{1}{2} \int_0^1 \left\{ \frac{x^3 y^2}{2} + \frac{2x^2 y^3}{3} + \frac{xy^4}{4} - \frac{x^3 y^4}{4} \right\}_0^{x^2} dx$$

$$= \frac{1}{24} \int_0^1 (6x^7 + 8x^8 + 3x^9 - 3x^{11})\, dx$$

$$= \frac{1}{24} \left\{ \frac{3x^8}{4} + \frac{8x^9}{9} + \frac{3x^{10}}{10} - \frac{x^{12}}{4} \right\}_0^1 = \frac{19}{270}.$$

Because of the analogy between double and triple integrals, we accept without proof that triple integrals can be evaluated with triple iterated integrals. We must, however, examine how triple iterated integrals bring about the summations represented by triple integrals, for it is only by understanding this process thoroughly that we can obtain limits for triple iterated integrals.

In Section 13.2 we discussed in considerable detail evaluation of double integrals by means of double iterated integrals. In particular, we showed that double iterated integrals in Cartesian coordinates represent the subdivision of an area into small rectangles by coordinate lines $x =$ constant and $y =$ constant. The first integration creates a summation over rectangles in a strip, and the second integration adds over all strips. It is fairly straightforward to generalize these ideas to triple integrals. Consider evaluating the triple integral

$$\iiint_V f(x, y, z)\, dV$$

over the region V in Figure 13.53a bounded above by the surface $z = h(x, y)$, below by the region R in the xy-plane, and on the sides by a cylindrical wall standing on the curve bounding R.

The choice of a triple iterated integral in Cartesian coordinates to evaluate this triple integral implies a subdivision of V into small rectangular parallelepipeds (boxes, for short) by means of coordinate planes $x = $ constant, $y = $ constant, and $z = $ constant. The dimensions of a representative box at position (x, y, z) in V are denoted by dx, dy, and dz, with resulting volume $dx\, dy\, dz$. If we decide on a triple iterated integral with respect to z, y, and x, then the first integration on z holds x and y constant. This integration therefore adds the quantities $f(x, y, z)\, dz\, dy\, dx$ over boxes in a vertical column of cross-sectional dimensions dx and dy. Lower and upper limits on z identify where each and every column starts and stops, and must consequently be 0 and $h(x, y)$:

$$\int_0^{h(x,y)} f(x, y, z)\, dz\, dy\, dx.$$

Since this integration produces a function of x and y alone, the remaining integration with respect to y and x is essentially a double iterated integral in the xy-plane. These integrations must account for all columns in V and therefore the region in the xy-plane over which this double iterated integral is performed is the region R upon which all columns in V stand. Since the y-integration adds inside a strip in the y-direction and limits identify where all strips start and stop, they must therefore be $g_1(x)$ and $g_2(x)$. We now have

$$\int_{g_1(x)}^{g_2(x)} \int_0^{h(x,y)} f(x, y, z)\, dz\, dy\, dx.$$

Finally, the x-integration adds over all strips and the limits are $x = a$ and $x = b$:

$$\iiint_V f(x, y, z)\, dV = \int_a^b \int_{g_1(x)}^{g_2(x)} \int_0^{h(x,y)} f(x, y, z)\, dz\, dy\, dx.$$

Suppose now that V is the region bounded above by the surface $z = h_2(x, y)$ and below by $z = h_1(x, y)$ (Figure 13.53b). In this case, the limits on the first integration with respect to z are $h_1(x, y)$ and $h_2(x, y)$ since every column starts on the surface $z = h_1(x, y)$ and ends on the surface $z = h_2(x, y)$:

$$\int_{h_1(x,y)}^{h_2(x,y)} f(x, y, z)\, dz\, dy\, dx.$$

For the volume of Figure 13.53a we interpreted the final two integrations as a double iterated integral in the xy-plane over the region R from which all columns emanated. For the present volume we interpret R as the region in the xy-plane onto which all columns project. We then obtain

$$\iiint_V f(x, y, z)\, dV = \int_a^b \int_{g_1(x)}^{g_2(x)} \int_{h_1(x,y)}^{h_2(x,y)} f(x, y, z)\, dz\, dy\, dx.$$

FIGURE 13.53a Columns start on the xy-plane $z = 0$ and end on surface $z = h(x, y)$. R is the area in the xy-plane on which all columns stand

FIGURE 13.53b Columns start and stop on surfaces $z = h_1(x, y)$ and $z = h_2(x, y)$. R is the area in the xy-plane onto which all columns project

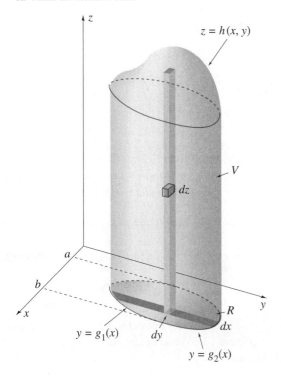

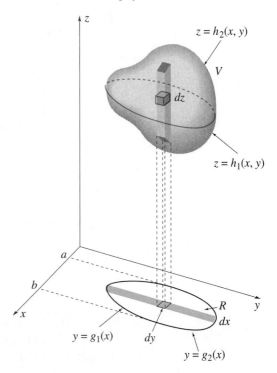

Schematically, we have obtained the following interpretation for the limits of triple iterated integrals in Cartesian coordinates:

$$\int_{\substack{\text{position of}\\ \text{first strip}}}^{\substack{\text{position of}\\ \text{last strip}}} \int_{\substack{\text{where every}\\ \text{strip starts}}}^{\substack{\text{where every}\\ \text{strip stops}}} \int_{\substack{\text{where every}\\ \text{column starts}}}^{\substack{\text{where every}\\ \text{column stops}}} f(x, y, z) \begin{Bmatrix} dz\,dy\,dx \\ dz\,dx\,dy \\ dx\,dz\,dy \\ dx\,dy\,dz \\ dy\,dx\,dz \\ dy\,dz\,dx \end{Bmatrix}.$$

EXAMPLE 13.22

Set up the six triple iterated integrals in Cartesian coordinates for the triple integral of a function $f(x, y, z)$ over the region V in the first octant bounded by the surfaces

$$y^2 + z^2 = 1, \quad y = x, \quad z = 0, \quad x = 0.$$

SOLUTION   The triple integral of $f(x, y, z)$ over V is given by each of the following triple iterated integrals (see Figure 13.54):

$$\int_0^1 \int_x^1 \int_0^{\sqrt{1-y^2}} f(x, y, z)\,dz\,dy\,dx, \qquad \int_0^1 \int_0^y \int_0^{\sqrt{1-y^2}} f(x, y, z)\,dz\,dx\,dy,$$

$$\int_0^1 \int_0^{\sqrt{1-z^2}} \int_0^y f(x, y, z)\,dx\,dy\,dz, \qquad \int_0^1 \int_0^{\sqrt{1-y^2}} \int_0^y f(x, y, z)\,dx\,dz\,dy,$$

$$\int_0^1 \int_0^{\sqrt{1-x^2}} \int_x^{\sqrt{1-z^2}} f(x, y, z)\,dy\,dz\,dx, \qquad \int_0^1 \int_0^{\sqrt{1-z^2}} \int_x^{\sqrt{1-z^2}} f(x, y, z)\,dy\,dx\,dz.$$

The six triple iterated integrals for the volume bounded by $y = x$, $y^2 + z^2 = 1$, $z = 0$, $x = 0$

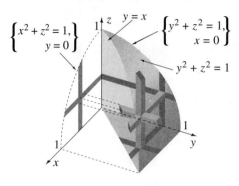

$$\left\{ \begin{matrix} x^2 + z^2 = 1, \\ y = 0 \end{matrix} \right\}$$

$y = x$

$$\left\{ \begin{matrix} y^2 + z^2 = 1, \\ x = 0 \end{matrix} \right\}$$

$y^2 + z^2 = 1$

EXAMPLE 13.23

Evaluate the triple integral of $f(x, y, z) = xyz$ over the region V bounded by the surfaces

$$z = \sqrt{y}, \quad y + z = 2, \quad x = 0, \quad z = 0, \quad x = 2.$$

SOLUTION   If we choose a triple iterated integral with respect to z, y, and x, some columns end on the parabolic cylinder $z = \sqrt{y}$ (Figure 13.55) and others end on the plane $y + z = 2$. We therefore require two such iterated integrals,

Columns in the z-direction necessitate two triple iterated integrals

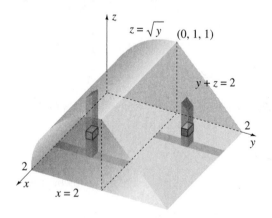

$z = \sqrt{y}$ $(0, 1, 1)$

$y + z = 2$

$x = 2$

$$\iiint_V xyz \, dz \, dy \, dx = \int_0^2 \int_0^1 \int_0^{\sqrt{y}} xyz \, dz \, dy \, dx + \int_0^2 \int_1^2 \int_0^{2-y} xyz \, dz \, dy \, dx$$

$$= \int_0^2 \int_0^1 \left\{ \frac{xyz^2}{2} \right\}_0^{\sqrt{y}} dy \, dx + \int_0^2 \int_1^2 \left\{ \frac{xyz^2}{2} \right\}_0^{2-y} dy \, dx$$

$$= \frac{1}{2} \int_0^2 \int_0^1 xy^2 \, dy \, dx + \frac{1}{2} \int_0^2 \int_1^2 x(4y - 4y^2 + y^3) \, dy \, dx$$

$$= \frac{1}{2} \int_0^2 \left\{ \frac{xy^3}{3} \right\}_0^1 dx + \frac{1}{2} \int_0^2 \left\{ x \left(2y^2 - \frac{4y^3}{3} + \frac{y^4}{4} \right) \right\}_1^2 dx$$

$$= \frac{1}{6} \int_0^2 x \, dx + \frac{5}{24} \int_0^2 x \, dx$$

$$= \frac{3}{8} \left\{ \frac{x^2}{2} \right\}_0^2$$

$$= \frac{3}{4}.$$

Only one iterated integral is required if integration is first performed with respect to y, namely

$$\int_0^1 \int_0^2 \int_{z^2}^{2-z} xyz \, dy \, dx \, dz \quad \text{or} \quad \int_0^2 \int_0^1 \int_{z^2}^{2-z} xyz \, dy \, dz \, dx.$$

EXERCISES 13.8

In Exercises 1–12 evaluate the triple integral over the region.

1. $\iiint_V (x^2z + ye^x)\,dV$, where V is bounded by $x = 0$, $x = 1$, $y = 1$, $y = 2$, $z = 0$, $z = 1$

2. $\iiint_V x\,dV$, where V is bounded by $x = 0$, $y = 0$, $z = 0$, $x + y + z = 4$

3. $\iiint_V \sin(y + z)\,dV$, where V is bounded by $z = 0$, $y = 2x$, $y = 0$, $x = 1$, $z = x + 2y$

4. $\iiint_V xy\,dV$, where V is enclosed by $z = \sqrt{1 - x^2 - y^2}$, $z = 0$

5. $\iiint_V dV$, where V is bounded by $x^2 + y^2 = 1$, $x^2 + z^2 = 1$

6. $\iiint_V (x^2 + 2z)\,dV$, where V is bounded by $z = 0$, $y + z = 4$, $y = x^2$

7. $\iiint_V x^2y^2z^2\,dV$, where V is bounded by $z = 1 + y$, $y + z = 1$, $x = 1$, $x = 0$, $z = 0$

8. $\iiint_V xyz\,dV$, where V is the first octant region cut out by $z = x^2 + y^2$, $z = \sqrt{x^2 + y^2}$

9. $\iiint_V dV$, where V is bounded by $z = x^2$, $y + z = 4$, $y = 0$

10. $\iiint_V (x + y + z)\,dV$, where V is bounded by $x = 0$, $x = 1$, $z = 0$, $y + z = 2$, $y = z$

11. $\iiint_V xyz\,dV$, where V is bounded by $z = 1$, $z = x^2/4 + y^2/9$

12. $\iiint_V x^2y\,dV$, where V is the first octant region bounded by $z = 1$, $z = x^2/4 + y^2/9$

13. Set up the six triple iterated integrals in Cartesian coordinates for the triple integral of a function $f(x, y, z)$ over the region enclosed by the surfaces $y = 1 - x^2$, $z = 0$, and $y = z$.

In Exercises 14–17 set up, but do not evaluate, a triple iterated integral for the triple integral.

∗ **14.** $\iiint_V (x^2 + y^2 + z^2)\,dV$, where V is bounded by $z = \sqrt{1 - x^2 - y^2}$, $z = x^2$

∗ **15.** $\iiint_V xz \sin(x + y)\,dV$, where V is bounded by $y^2 = 1 + 4x^2 + 4z^2$, $y = \sqrt{4 + x^2}$

∗ **16.** $\iiint_V xyz\,dV$, where V is bounded by $z = x^2 + 4y^2$, $2x + 8y + z = 4$

∗ **17.** $\iiint_V x^2y^2z^2\,dV$, where V is bounded by $x = y^2 + z^2$, $x + 1 = (y^2 + z^2)^2$

In Exercises 18–23 evaluate the triple integral over the region.

∗ **18.** $\iiint_V (y + x^2)\,dV$, where V is bounded by $x + z^2 = 1$, $z = x + 1$, $y = 1$, $y = -1$

∗ **19.** $\iiint_V (xy + z)\,dV$, where V is bounded by $y + z = 1$, $z = 2y$, $z = y$, $x = 0$, $x = 3$

∗ **20.** $\iiint_V dV$, where V is bounded by $z = 0$, $x^2 + y^2 = 1$, $x + y + z = 2$

∗ **21.** $\iiint_V dV$, where V is bounded by $z = x^2 + y^2$, $z = 4 - x^2 - y^2$

∗ **22.** $\iiint_V (x + y + z)\,dV$, where V is bounded by $2z = y^2 - x^2$, $z = 1 - x^2$

∗ **23.** $\iiint_V |yz|\,dV$, where V is bounded by $z^2 = 1 + x^2 + y^2$, $z = \sqrt{4 - x^2 - y^2}$

∗ **24.** Set up, but do not evaluate, triple iterated integrals to evaluate the triple integral of the function $f(x, y, z) = x^2 + y^2 + z^2$ over the region bounded by the surfaces $x^2 + y^2 = z^2 + 1$, $2z = \sqrt{x^2 + y^2}$, $z = 0$.

13.9 Volumes

Because the volume of a region V is represented by triple integral 13.53 and triple integrals are evaluated by means of triple iterated integrals, it follows that volumes can be evaluated with triple iterated integrals. For example, to evaluate the volume of the region in Figure 13.53b using a triple iterated integral in x, y, and z, we subdivide V into boxes of dimensions dx, dy, and dz and therefore of volume $dz\,dy\,dx$. Integration with respect to z adds these volumes in the z-direction to give the volume of a vertical column (Figure 13.56):

$$\int_{h_1(x,y)}^{h_2(x,y)} dz\,dy\,dx.$$

Limits indicate that all columns start on the surface $z = h_1(x, y)$ and end on the surface $z = h_2(x, y)$. Integration with respect to y now adds volumes of columns that project onto a strip in the y-direction:

$$\int_{g_1(x)}^{g_2(x)} \int_{h_1(x,y)}^{h_2(x,y)} dz \, dy \, dx,$$

where the limits indicate that all strips start on the curve $y = g_1(x)$ and end on the curve $y = g_2(x)$. Evidently, this integration yields the volume of a slab as shown in Figure 13.56. Finally, integration with respect to x adds the volumes of all such slabs in V:

$$\int_a^b \int_{g_1(x)}^{g_2(x)} \int_{h_1(x,y)}^{h_2(x,y)} dz \, dy \, dx,$$

where a and b designate positions of first and last strips.

FIGURE 13.56 The integrations when a triple iterated integral is used to find volume

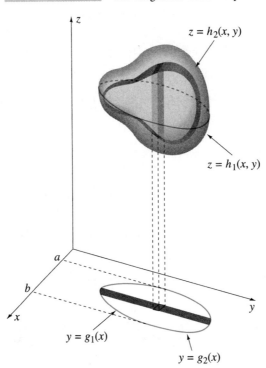

EXAMPLE 13.24

Find the volume bounded by the planes $z = x + y$, $y = 2x$, $z = 0$, $x = 0$, $y = 2$.

SOLUTION   If we use vertical columns (Figure 13.57), we have

$$\text{volume} = \int_0^1 \int_{2x}^2 \int_0^{x+y} dz \, dy \, dx = \int_0^1 \int_{2x}^2 (x + y) \, dy \, dx = \int_0^1 \left\{ xy + \frac{y^2}{2} \right\}_{2x}^2 dx$$

$$= 2 \int_0^1 (1 + x - 2x^2) \, dx = 2 \left\{ x + \frac{x^2}{2} - \frac{2x^3}{3} \right\}_0^1 = \frac{5}{3}.$$

FIGURE 13.57   The volume bounded by the planes $z = x + y$, $y = 2x$, $z = 0$, $x = 0$, $y = 2$

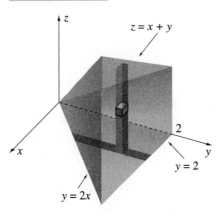

EXAMPLE 13.25

Find the volume in the first octant cut from the cyclinder $x^2 + z^2 = 4$ by the plane $y + z = 6$.

SOLUTION   With columns in the y-direction (Figure 13.58), we have

$$
\text{volume} = \int_0^2 \int_0^{\sqrt{4-x^2}} \int_0^{6-z} dy \, dz \, dx = \int_0^2 \int_0^{\sqrt{4-x^2}} (6 - z) \, dz \, dx
$$

$$
= \int_0^2 \left\{ 6z - \frac{z^2}{2} \right\}_0^{\sqrt{4-x^2}} dx
$$

$$
= \int_0^2 \left\{ 6\sqrt{4 - x^2} - \frac{1}{2}(4 - x^2) \right\} dx.
$$

In the first term we set $x = 2 \sin \theta$, from which we get $dx = 2 \cos \theta \, d\theta$, and

$$
\text{volume} = 6 \int_0^{\pi/2} (2 \cos \theta) \cdot 2 \cos \theta \, d\theta + \left\{ -2x + \frac{x^3}{6} \right\}_0^2
$$

$$
= 12 \int_0^{\pi/2} (1 + \cos 2\theta) \, d\theta - \frac{8}{3}
$$

$$
= 12 \left\{ \theta + \frac{1}{2} \sin 2\theta \right\}_0^{\pi/2} - \frac{8}{3} = 6\pi - \frac{8}{3}.
$$

FIGURE 13.58   Volume in first octant cut from cylinder $x^2 + z^2 = 4$ by plane $y + z = 6$

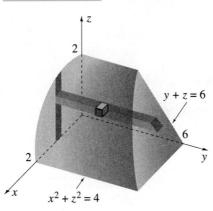

Had we used iterated integrals with respect to z, y, and x in this example, we would have had two integrals,

$$\text{volume} = \int_0^2 \int_0^{6-\sqrt{4-x^2}} \int_0^{\sqrt{4-x^2}} dz\,dy\,dx + \int_0^2 \int_{6-\sqrt{4-x^2}}^6 \int_0^{6-y} dz\,dy\,dx.$$

Consulting Project 22

We are being consulted by an architect who wants to know the volume of air contained in a certain type of structure. It has a horizontal, polygonal base (Figure 13.59a) on which stand vertical walls all of the same height H. Above this is a roof formed in the following way. There is a peak, height h above the top of the walls, that is joined to the tops of the walls by planes.

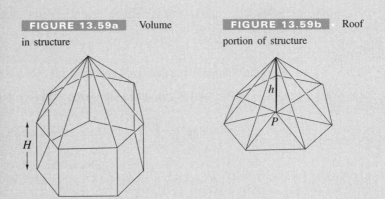

FIGURE 13.59a   Volume in structure

FIGURE 13.59b   Roof portion of structure

SOLUTION   There is no problem with the lower part of the structure; the volume of air contained in it is the area of the base multiplied by H. To calculate the volume of the upper part, we take it aside as in Figure 13.59b and drop a perpendicular from the top to the polygonal base. Now join the foot of the perpendicular P to each vertex of the polygon. Finally draw vertical planes each containing one of these lines, the perpendicular from top to base, and a slanted edge of the roof. What this does is divide the volume into tetrahedrons (four-sided figures) each with one horizontal and two vertical planes. The remaining face of the tetrahedron is a slanted face of the roof. If we find the volume in one such tetrahedron, we can add to find the total volume in the upper portion of the structure.

To find the volume of any one of the tetrahedrons, let us place its triangular base in the xy-plane with one vertex at the origin, a second vertex on the x-axis, and the third vertex in the first quadrant (Figure 13.60). The top of the tetrahedron will be on the z-axis. Denote coordinates of the tetrahedron by $O(0,0,0)$, $A(a,0,0)$, $B(b,c,0)$, and $D(0,0,h)$. Equations of lines AB and OB are $y = c(x-a)/(b-a)$ and $y = cx/b$, respectively. A normal vector to plane ABD is

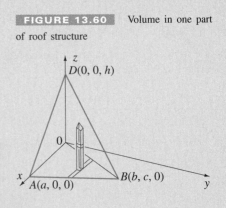

FIGURE 13.60   Volume in one part of roof structure

$$\mathbf{AB} \times \mathbf{AD} = \begin{vmatrix} \hat{\mathbf{i}} & \hat{\mathbf{j}} & \hat{\mathbf{k}} \\ b-a & c & 0 \\ -a & 0 & h \end{vmatrix} = (ch, ah - bh, ac).$$

The equation of plane ABD is therefore $ch(x-a) + (ah - bh)y + acz = 0$. The volume of the tetrahedron can now be calculated with a triple integral,

$$V = \int_0^c \int_{by/c}^{a+(b-a)y/c} \int_0^{[(bh-ah)y-ch(x-a)]/(ac)} dz\, dx\, dy$$

$$= \int_0^c \int_{by/c}^{a+(b-a)y/c} \left[\frac{(bh-ah)y - ch(x-a)}{ac} \right] dx\, dy$$

$$= \frac{1}{ac} \int_0^c \left\{ h(b-a)xy - \frac{ch}{2}(x-a)^2 \right\}_{by/c}^{a+(b-a)y/c} dy$$

$$= \frac{1}{ac} \int_0^c \left\{ h(b-a)y \left[a + \frac{(b-a)y}{c} \right] - \frac{ch}{2} \left[a + \frac{(b-a)y}{c} - a \right]^2 \right.$$

$$\left. - h(b-a)y \left(\frac{by}{c} \right) + \frac{ch}{2} \left(\frac{by}{c} - a \right)^2 \right\} dy$$

$$= \frac{1}{ac} \left\{ \frac{h}{c}(b-a)^2 \frac{y^3}{3} + ah(b-a)\frac{y^2}{2} - \frac{h}{2c}(b-a)^2 \frac{y^3}{3} \right.$$

$$\left. - \frac{bh}{c}(b-a)\frac{y^3}{3} + \frac{c^2 h}{6b} \left(\frac{by}{c} - a \right)^3 \right\}_0^c$$

$$= \frac{1}{6} ach.$$

Since the area of the triangular base is $ac/2$, we have shown that the volume of the tetrahedron is the area of the triangular base multiplied by one-third of the height h. It follows that the volume of the upper structure is the area of the polygonal base multiplied by $h/3$. Finally, then, if A represents the area of the polygonal base, the volume of air in the structure is

$$V = A \left(H + \frac{h}{3} \right).$$

EXERCISES 13.9

In Exercises 1–17 find the volume of the region bounded by the surfaces.

1. $y = x^2$, $y = 1$, $z = 0$, $z = 4$

2. $x = z^2$, $z = x^2$, $y = 0$, $y = 2$

3. $x = 3z$, $z = 3x$, $y = 1$, $y = 0$, $x = 2$

4. $x + y + z = 6$, $y = 4 - x^2$, $z = 0$, $y = 0$

5. $z = x^2 + y^2$, $y = x^2$, $y = 4$, $z = 0$

6. $x + y + z = 4$, $y = 3z$, $x = 0$, $y = 0$

7. $x^2 + y^2 = 4$, $y^2 + z^2 = 4$

8. $y = x^2 - 1$, $y = 1 - x^2$, $x + z = 1$, $z = 0$

9. $z = 16 - x^2 - 4y^2$, $x + y = 1$, $z = 16$, $x \geq 0$, $y \geq 0$

10. $z = x^2 + y^2$, $x = 1$, $z = 0$, $x = y$, $x = 2y$

11. $z = 1 - x^2 - y^2$, $z = 0$

12. $x - z = 0$, $x + z = 3$, $y + z = 1$, $z = y + 1$, $z = 0$

13. $x + y + z = 2$, $x^2 + y^2 = 1$, $z = 0$

14. $y + z = 1$, $z = 2y$, $z = y$, $x = 0$, $x + y + z = 4$

15. $x^2 + 4y^2 = z$, $x^2 + 4y^2 = 12 - 2z$

* **16.** $y = 1 - z^2$, $y = z^2 - 1$, $x = 1 - z^2$, $x = z^2 - 1$

* **17.** $x + 3y + 2z = 6$, $z = 0$, $y = x$, $y = 2x$

* **18.** Find the volume in the first octant bounded by the plane $2x + y + z = 2$ and inside the cylinder $y^2 + z^2 = 1$.

* **19.** A pyramid has a square base with side length b and has height h at its centre.

   (a) Find its volume by taking cross-sections parallel to the base (see Section 7.9).

   (b) Find its volume using triple integrals.

The average value of a function $f(x, y, z)$ over a region with volume V is defined as

$$\overline{f} = \frac{1}{V} \iiint_V f(x, y, z)\, dV.$$

In Exercises 20–22 find the average value of the function over the region.

20. $f(x, y, z) = xy$ over the region bounded by the surfaces $x = 0$, $y = 0$, $z = 0$, $x + y + z = 1$

21. $f(x, y, z) = x + y + z$ over the region in the first octant bounded by the surfaces $z = 9 - x^2 - y^2$, $z = 0$ and for which $0 \le x \le 1$, $0 \le y \le 1$

22. $f(x, y, z) = x^2 + y^2 + z^2$ over the region bounded by the surfaces $x = 0$, $x = 1$, $y + z = 2$, $y = 2$, $z = 2$

* **23.** Find the volume bounded by the surfaces $z = x^2 - y^2$ and $z = 4 - 2(x^2 + y^2)$.

* **24.** Verify that the surfaces $z = x^2 - y^2$ and $z = 4 - x^2 - y^2$ do not bound a finite volume.

* **25.** Find the volume bounded by the surfaces $x + z = 2$, $z = 0$, $4y^2 = x(2 - z)$.

* **26.** Find the volume bounded by the surfaces $z = (x-1)^2 + y^2$, $2x + z = 2$.

* **27.** Find the volume inside the ellipsoid $x^2/a^2 + y^2/b^2 + z^2/c^2 = 1$.

** **28.** The bottom and sides of a boat are defined by the surface equation $x = 10(1 - y^2 - z^2)$, $0 \le x \le 10$, where all dimensions are in metres.

   (a) Find the volume of water displaced by the boat when the water level on the side of the boat is d metres below the top of the boat.

   (b) Archimedes' principle states that the buoyant force on an object when immersed or partially immersed in a fluid is equal to the weight of the fluid displaced by the object. Find the maximum weight of the boat and contents just before sinking.

** **29.** Find the volume inside all three surfaces $x^2 + y^2 = a^2$, $x^2 + z^2 = a^2$, $y^2 + z^2 = a^2$.

13.10 Centres of Mass and Moments of Inertia

In this section we discuss centres of mass and moments of inertia for three-dimensional objects of density $\rho(x, y, z)$ (mass per unit volume). If we divide the object occupying region V in Figure 13.61 into small volumes dV, then the mass in dV is $\rho\, dV$. The triple integral

$$M = \iiint_V \rho\, dV \qquad (13.55)$$

adds the masses of all such volumes (of ever-decreasing size) to produce the total mass M of the object.

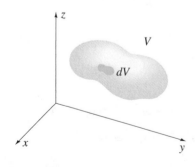
FIGURE 13.61   Centre of mass of a 3-dimensional object

Corresponding to equations 13.32 and 13.33 for first moments of planar masses about the coordinate axes are the following formulas for first moments of the object about the coordinate planes:

$$\text{first moment of object about } yz\text{-plane} = \iiint_V x\rho\, dV, \qquad (13.56a)$$

$$\text{first moment of object about } xz\text{-plane} = \iiint_V y\rho\, dV, \qquad (13.56b)$$

$$\text{first moment of object about } xy\text{-plane} = \iiint_V z\rho\, dV. \qquad (13.56c)$$

The centre of mass of the object is defined as that point $(\bar{x}, \bar{y}, \bar{z})$ at which a particle of mass M would have the same first moments about the coordinate planes as the object itself. Since the first moments of M at $(\bar{x}, \bar{y}, \bar{z})$ about the coordinate planes are $M\bar{x}$, $M\bar{y}$, and $M\bar{z}$, it follows that we can use the equations

$$M\bar{x} = \iiint_V x\rho \, dV, \qquad M\bar{y} = \iiint_V y\rho \, dV, \qquad M\bar{z} = \iiint_V z\rho \, dV, \quad (13.57)$$

to solve for $(\bar{x}, \bar{y}, \bar{z})$ once M and the integrals on the right have been evaluated.

If we use a triple iterated integral with respect to z, y, and x to evaluate the third of these, say, for the object in Figure 13.56, then

$$\iiint_V z\rho \, dV = \int_a^b \int_{g_1(x)}^{g_2(x)} \int_{h_1(x,y)}^{h_2(x,y)} z\rho \, dz \, dy \, dx.$$

Quantity $z\rho \, dz \, dy \, dx$ is the first moment about the xy-plane of the mass in an elemental box of dimensions dx, dy, and dz. The z-integration then adds these moments over boxes in the z-direction to give the first moment about the xy-plane of the mass in a vertical column:

$$\int_{h_1(x,y)}^{h_2(x,y)} z\rho \, dz \, dy \, dx.$$

The y-integration then adds the first moments of columns that project onto a strip in the y-direction:

$$\int_{g_1(x)}^{g_2(x)} \int_{h_1(x,y)}^{h_2(x,y)} z\rho \, dz \, dy \, dx.$$

This quantity therefore represents the first moment about the xy-plane of the slab in Figure 13.56. Finally, the x-integration adds first moments of all such slabs to give the total first moment of V about the xy-plane:

$$\int_a^b \int_{g_1(x)}^{g_2(x)} \int_{h_1(x,y)}^{h_2(x,y)} z\rho \, dz \, dy \, dx.$$

EXAMPLE 13.26

Find the centre of mass of an object of constant density ρ if it is bounded by the surfaces $z = 1 - y^2$, $x = 0$, $z = 0$, $x = 2$.

FIGURE 13.62 Centre of mass of uniform solid bounded by $z = 1 - y^2$, $x = 0$, $z = 0$, $x = 2$

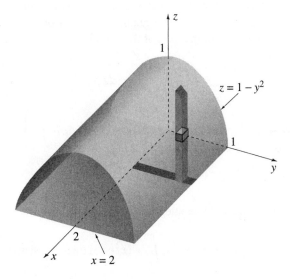

SOLUTION   From the symmetry of the object (Figure 13.62), we see $\bar{x} = 1$ and $\bar{y} = 0$. Now

$$M = 2 \int_0^2 \int_0^1 \int_0^{1-y^2} \rho \, dz \, dy \, dx = 2\rho \int_0^2 \int_0^1 (1 - y^2) \, dy \, dx$$

$$= 2\rho \int_0^2 \left\{ y - \frac{y^3}{3} \right\}_0^1 dx = \frac{4\rho}{3} \int_0^2 dx = \frac{8\rho}{3}$$

and

$$M\bar{z} = 2 \int_0^2 \int_0^1 \int_0^{1-y^2} z\rho \, dz \, dy \, dx = 2\rho \int_0^2 \int_0^1 \left\{ \frac{z^2}{2} \right\}_0^{1-y^2} dy \, dx$$

$$= \rho \int_0^2 \int_0^1 (1 - 2y^2 + y^4) \, dy \, dx$$

$$= \rho \int_0^2 \left\{ y - \frac{2y^3}{3} + \frac{y^5}{5} \right\}_0^1 dx = \frac{8\rho}{15} \int_0^2 dx = \frac{16\rho}{15}.$$

Thus, $\bar{z} = \dfrac{16\rho}{15} \cdot \dfrac{3}{8\rho} = \dfrac{2}{5}$.

In view of our discussion on first moments in this section and on moments of inertia of thin plates in Section 13.5, it should not be necessary to give a full treatment of moments of inertia of three-dimensional objects. Instead, we simply note that the distances from a point (x, y, z) to the x-, y-, and z-axes are, respectively, $\sqrt{y^2 + z^2}$, $\sqrt{x^2 + z^2}$, and $\sqrt{x^2 + y^2}$. It follows, then, that if an object of density $\rho(x, y, z)$ occupies a region V of space, its moments of inertia about the x-, y-, and z-axes are, respectively:

$$I_x = \iiint_V (y^2 + z^2)\rho \, dV, \quad I_y = \iiint_V (x^2 + z^2)\rho \, dV, \quad I_z = \iiint_V (x^2 + y^2)\rho \, dV.$$

$$(13.58)$$

EXAMPLE 13.27

Find the moment of inertia of a right-circular cylinder of constant density ρ about its axis.

SOLUTION   Let the length and radius of the cylinder be h and r and choose the coordinate system in Figure 13.63. The required moment of inertia about the z-axis is four times the moment of inertia of that part of the cylinder in the first octant. Hence,

$$I = 4 \int_0^r \int_0^{\sqrt{r^2-x^2}} \int_0^h (x^2 + y^2)\rho \, dz \, dy \, dx = 4\rho h \int_0^r \int_0^{\sqrt{r^2-x^2}} (x^2 + y^2) \, dy \, dx$$

$$= 4\rho h \int_0^r \left\{ x^2 y + \frac{y^3}{3} \right\}_0^{\sqrt{r^2-x^2}} dx = 4\rho h \int_0^r \left[x^2\sqrt{r^2 - x^2} + \frac{1}{3}(r^2 - x^2)^{3/2} \right] dx$$

$$= \frac{4\rho h}{3} \int_0^r (r^2\sqrt{r^2 - x^2} + 2x^2\sqrt{r^2 - x^2}) \, dx.$$

To evaluate this definite integral we set $x = r \sin\theta$, which implies that $dx = r \cos\theta \, d\theta$, and

$$I = \frac{4\rho h}{3} \int_0^{\pi/2} (r^2 r \cos\theta + 2r^2 \sin^2\theta r \cos\theta) r \cos\theta \, d\theta$$

$$= \frac{4\rho h r^4}{3} \int_0^{\pi/2} (\cos^2\theta + 2\sin^2\theta\cos^2\theta)\, d\theta$$

$$= \frac{4\rho h r^4}{3} \int_0^{\pi/2} \left(\frac{1+\cos 2\theta}{2} + \frac{1-\cos 4\theta}{4} \right) d\theta$$

$$= \frac{4\rho h r^4}{3} \left\{ \frac{3\theta}{4} + \frac{\sin 2\theta}{4} - \frac{\sin 4\theta}{16} \right\}_0^{\pi/2} = \frac{\rho h r^4 \pi}{2}.$$

FIGURE 13.63 Moment of inertia of a uniform right-circular cylinder about its axis

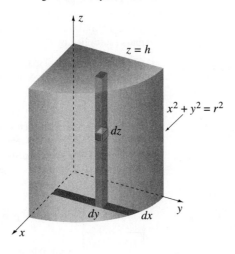

The product moments of inertia of a mass with density ρ occupying a region V about the yz- and xz-planes, the xz- and xy-planes, and the yz- and xy-planes are

$$I_{xy} = \iiint_V xy\,\rho\,dV, \qquad I_{yz} = \iiint_V yz\,\rho\,dV, \qquad I_{xz} = \iiint_V xz\,\rho\,dV. \quad (13.59)$$

They can be positive, negative, or zero. Symmetry of the object plays a key role in this determination.

EXAMPLE 13.28

Find the product moments of inertia of the object in Example 13.26.

SOLUTION   Because of the symmetry of the object about the xz-plane, product moments of inertia I_{xy} and I_{yz} are both zero.

$$I_{xz} = 2\int_0^2 \int_0^1 \int_0^{1-y^2} \rho xz\,dz\,dy\,dx = 2\rho \int_0^2 \int_0^1 \left\{ \frac{xz^2}{2} \right\}_0^{1-y^2} dy\,dx$$

$$= \rho \int_0^2 \int_0^1 x(1-y^2)^2\,dy\,dx = \rho \int_0^2 \left\{ x\left(y - \frac{2y^3}{3} + \frac{y^5}{5} \right) \right\}_0^1 dx = \frac{8\rho}{15} \left\{ \frac{x^2}{2} \right\}_0^2 = \frac{16\rho}{15}.$$

EXERCISES 13.10

In Exercises 1–5 the surfaces bound a solid object of constant density. Find its centre of mass.

1. $z = x^2 + y^2$, $z = 0$, $x = 0$, $y = 0$, $x = 1$, $y = 1$

2. $x + y + z = 1$, $x = 0$, $y = 0$, $z = 0$

3. $z = x^2$, $y + z = 4$, $y = 0$

4. $y = 4 - x^2$, $z = 0$, $y = z$

5. $x + y + z = 4$, $y = 3z$, $x = 0$, $y = 0$

In Exercises 6–10 the surfaces bound a solid object of constant density ρ. Find its moment of inertia about the line.

6. $x = 0$, $y = 0$, $z = 0$, $x = 1$, $y = 1$, $z = 1$ about the x-axis

7. $z = 2x$, $z = 0$, $y = 0$, $y = 2$, $x = 3$ about the y-axis

8. $x + y + z = 2$, $y = 0$, $x = 0$, $0 \le z \le \sqrt{1 - y}$ about the x-axis

9. $z = xy$, $x^2 + y^2 = 1$, $z = 0$ (first octant) about the z-axis

10. $y + z = 2$, $x + z = 2$, $x = 0$, $y = 0$, $z = 0$ about the z-axis

11. Find the first moment about the xy-plane of a solid of constant density ρ if it is bounded by the surfaces $x = z$, $x + z = 0$, $z = 2$, $y = 0$, $y = 2$.

In Exercises 12–14 the surfaces bound a solid object of constant density. Find its centre of mass.

* **12.** $y = x^3$, $x = y^2$, $z = 1 + x^2 + y^2$, $z = -x^2 - y^2$

* **13.** $z = x^2$, $x + z = 2$, $z = y$, $y = 0$

* **14.** $y + z = 0$, $y - z = 0$, $x + z = 0$, $x - z = 0$, $z = 2$

In Exercises 15–17 the surfaces bound a solid object of constant density ρ. Find its moment of inertia about the line.

* **15.** $z = 4 - x^2$, $x + z + 2 = 0$, $y = 0$, $y = 2$ about the y-axis

* **16.** $x^2 + z^2 = a^2$, $x^2 + y^2 = a^2$ about the x-axis

* **17.** $x + y - z = 0$, $x = 3y$, $3y = 2x$, $x = 3$, $z = 0$ about the z-axis

* **18.** Find the first moment about the plane $x + y + z = 1$ of a solid object of constant density ρ if it is bounded by the surfaces $x + 2y + 4z = 12$, $x = 0$, $y = 0$, $z = 0$.

* **19.** Find the product moments of inertia with respect to the coordinate planes of an object with constant density ρ if it is bounded by the

surfaces $x = 0$, $y = 0$, $z = 0$, and $ax + by + cz = 1$, where a, b, and c are positive constants.

* **20.** Repeat Exercise 19 for the object in the first octant bounded by the surfaces $z = 0$, $y = x$, $y = 2x$, $x + z = 2$.

* **21.** Show that the moment of inertia of an object, with constant density, about any one of the three coordinate axes is less than or equal to the sum of the moments of inertia about the other two axes.

* **22.** Prove the parallel axis theorem for solid objects: The moment of inertia of a uniform solid about a line is equal to the moment of inertia about a parallel line through the centre of mass of the solid plus the mass multiplied by the square of the distance between the lines.

* **23.** Find the centre of mass of a uniform solid in the first octant bounded by the ellipsoid $x^2/a^2 + y^2/b^2 + z^2/c^2 = 1$.

** **24.** The figure below shows a pop can (in the form of a right-circular cylinder) partially full of pop. The centre of mass of the can plus contents is between the centre of mass of an empty can and that of the pop itself. Show that the centre of mass of can plus contents is lowest when it coincides with the surface of the pop. Assume that the hole in the can is so small that the centre of mass is the same as if there were no hole.

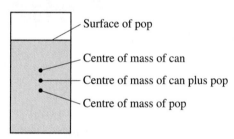

When an object floats partially submerged in a fluid, its centre of buoyancy is defined as the centre of mass of the water displaced by the object. In Exercises 25–26 find the depth of the centre of buoyancy below the surface as the object floats in water.

* **25.** A sphere with density 500 kg/m^3 and radius R

* **26.** A right-circular cone (radius R and height H) with density 800 kg/m^3 floating with its apex pointing downward

** **27.** Let $\hat{\mathbf{v}} = (v_x, v_y, v_z)$ be a unit vector with its tail at the origin. Show that the moment of inertia I of any solid object occupying a region V about the line containing $\hat{\mathbf{v}}$ can be expressed in the form

$$I = v_x^2 I_x + v_y^2 I_y + v_z^2 I_z - 2v_x v_y I_{xy} - 2v_y v_z I_{yz} - 2v_z v_x I_{xz}.$$

** **28.** Find the moment of inertia of a uniform solid sphere of radius R about any tangent line.

** **29.** A homogeneous object with mass M occupies a region V. The sum of its moments of inertia about the coordinate axes is $I_x + I_y + I_z$. A homogeneous sphere with the same mass and the same density is centred at the origin. Show that the sum of its moments of inertia about the axes is less than or equal to $I_x + I_y + I_z$.

13.11 Triple Iterated Integrals in Cylindrical Coordinates

In Section 13.7 we saw that polar coordinates are sometimes more convenient than Cartesian coordinates in evaluating double integrals. It should come as no surprise, then, that other coordinate systems can simplify the evaluation of triple integrals. Two of the most common are cylindrical and spherical coordinates. Cylindrical coordinates are useful in problems involving an axis of symmetry. They are based on a Cartesian coordinate along the axis of symmetry and polar coordinates in a plane perpendicular to the axis of symmetry. If the z-axis is the axis of symmetry and polar coordinates are defined in the xy-plane with the origin as pole and the positive x-axis as polar axis, then cylindrical coordinates and Cartesian coordinates are related by the equations

$$x = r \cos \theta, \quad y = r \sin \theta, \quad z = z \tag{13.60}$$

FIGURE 13.64 Relationships between Cartesian and cylindrical coordinates

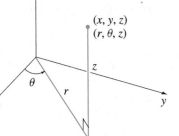

(see Figure 13.64). Recall that r can be expressed in terms of x and y by

$$r = \sqrt{x^2 + y^2}, \tag{13.61a}$$

and θ is defined implicitly by the equations

$$\cos \theta = \frac{x}{\sqrt{x^2 + y^2}}, \quad \sin \theta = \frac{y}{\sqrt{x^2 + y^2}}. \tag{13.61b}$$

To use cylindrical coordinates in the evaluation of triple integrals, we must express equations of surfaces in terms of these coordinates. But this is very simple, for if $F(x, y, z) = 0$ is the equation of a surface in Cartesian coordinates, then to express this equation in cylindrical coordinates we substitute from equations 13.60: $F(r \cos \theta, r \sin \theta, z) = 0$. For example, the right-circular cylinder $x^2 + y^2 = 9$, which has the z-axis as its axis of symmetry, has the very simple equation $r = 3$ in cylindrical coordinates. The right-circular cone $z = \sqrt{x^2 + y^2}$ also has the z-axis as its axis of symmetry, and its equation in cylindrical coordinates takes the simple form $z = r$.

Suppose that we are to evaluate the triple integral of a continuous function $f(x, y, z)$,

$$\iiint_V f(x, y, z) \, dV,$$

over some region V of space. The choice of a **triple iterated integral in cylindrical coordinates** implies a subdivision of V into small volumes dV by means of coordinate surfaces $r = $ constant, $\theta = $ constant, and $z = $ constant (Figure 13.65). Surfaces $r = $ constant are right-circular cylinders coaxial with the z-axis; surfaces $\theta = $ constant are half planes containing the z-axis and therefore perpendicular to the xy-plane; and surfaces $z = $ constant are planes parallel to the xy-plane.

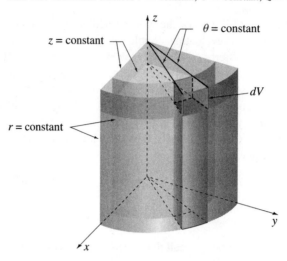

FIGURE 13.65 Elemental volume in cylindrical coordinates is created with coordinate surfaces $r = $ constant, $\theta = $ constant, $z = $ constant

FIGURE 13.66 Elemental volume in cylindrical coordinates is $dV = r\,dz\,dr\,d\theta$

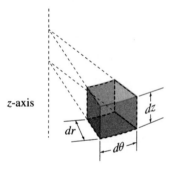

If we denote small variations in r, θ, and z for the element dV by dr, $d\theta$, and dz (Figure 13.66), then volume of the element is approximately $(r\,dr\,d\theta)\,dz$, where $r\,dr\,d\theta$ is the polar cross-sectional area parallel to the xy-plane. Hence, in cylindrical coordinates we set

$$dV = r\,dz\,dr\,d\theta. \tag{13.62}$$

The integrand $f(x, y, z)$ is expressed in cylindrical coordinates as $f(r\cos\theta, r\sin\theta, z)$. It remains only to affix appropriate limits to the triple iterated integral, and these will, of course, depend on which of the six possible iterated integrals in cylindrical coordinates we choose. The most commonly used triple iterated integral is with respect to z, r, and θ, and in this case the z-integration adds the quantities $f(r\cos\theta, r\sin\theta, z)r\,dz\,dr\,d\theta$ in a vertical column, where r and θ are constant. The limits therefore identify surfaces on which each and every column starts and stops, and generally depend on r and θ:

$$\int_{h_1(r,\theta)}^{h_2(r,\theta)} f(r\cos\theta, r\sin\theta, z)r\,dz\,dr\,d\theta.$$

The remaining integrations with respect to r and θ perform additions over the region in the xy-plane onto which all vertical columns project. Since r and θ are simply polar coordinates, the r-integration adds over small areas in a wedge and the θ-integration adds over all wedges. The triple iterated integral with respect to z, r, and θ therefore has the form

$$\int_{\alpha}^{\beta}\int_{g_1(\theta)}^{g_2(\theta)}\int_{h_1(r,\theta)}^{h_2(r,\theta)} f(r\cos\theta, r\sin\theta, z)r\,dz\,dr\,d\theta.$$

We comment on the geometric aspects of these additions more fully in the following examples.

EXAMPLE 13.29

Find the volume inside both the sphere $x^2 + y^2 + z^2 = 2$ and the cylinder $x^2 + y^2 = 1$.

SOLUTION   The required volume is eight times the first octant volume shown in Figure 13.67. If we use cylindrical coordinates, volume of an elemental piece is $r\,dz\,dr\,d\theta$. A z-integration adds these pieces to give volume in a vertical column:

$$\int_0^{\sqrt{2-r^2}} r\,dz\,dr\,d\theta,$$

where limits indicate that for volume in the first octant all columns start on the xy-plane (where $z = 0$) and end on the sphere (where $r^2 + z^2 = 2$). An r-integration now adds volumes of all columns that stand on a wedge:

$$\int_0^1 \int_0^{\sqrt{2-r^2}} r \, dz \, dr \, d\theta,$$

where limits indicate that all wedges start at the origin (where $r = 0$) and end on the curve $x^2 + y^2 = 1$ (or $r = 1$) in the xy-plane. This integration therefore yields the volume of a slice (Figure 13.67). Finally, the θ-integration adds volumes of all such slices

$$\int_0^{\pi/2} \int_0^1 \int_0^{\sqrt{2-r^2}} r \, dz \, dr \, d\theta,$$

where limits 0 and $\pi/2$ identify positions of first and last wedges, respectively, in the first quadrant. We obtain the required volume, then, as

$$8 \int_0^{\pi/2} \int_0^1 \int_0^{\sqrt{2-r^2}} r \, dz \, dr \, d\theta = 8 \int_0^{\pi/2} \int_0^1 r\sqrt{2 - r^2} \, dr \, d\theta$$

$$= 8 \int_0^{\pi/2} \left\{ -\frac{1}{3}(2 - r^2)^{3/2} \right\}_0^1 d\theta$$

$$= \frac{8}{3}(2\sqrt{2} - 1) \int_0^{\pi/2} d\theta = \frac{4\pi}{3}(2\sqrt{2} - 1).$$

FIGURE 13.67   Cylindrical coordinates to calculate the volume inside the sphere $x^2 + y^2 + z^2 = 2$ and the cylinder $x^2 + y^2 = 1$

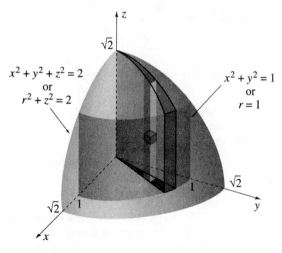

Had we used a triple iterated integral with respect to z, y, and x in Example 13.29, we would have obtained

$$\text{volume} = 8 \int_0^1 \int_0^{\sqrt{1-x^2}} \int_0^{\sqrt{2-x^2-y^2}} dz \, dy \, dx.$$

To appreciate the value of cylindrical coordinates, try to evaluate this triple iterated integral.

As further evidence of the value of cylindrical coordinates, repeat Examples 13.25 and 13.27 using cylindrical coordinates.

EXAMPLE 13.30

Find the moment of inertia of a uniform right-circular cone of radius R and height h about its axis.

SOLUTION One-quarter of such a cone is shown in Figure 13.68. We multiply its moment of inertia about the z-axis by 4:

$$
I_z = 4 \int_0^{\pi/2} \int_0^R \int_{hr/R}^h r^2 \, \rho r \, dz \, dr \, d\theta = 4\rho \int_0^{\pi/2} \int_0^R \left\{ r^3 z \right\}_{hr/R}^h dr \, d\theta
$$

$$
= \frac{4\rho h}{R} \int_0^{\pi/2} \int_0^R r^3 (R - r) \, dr \, d\theta = \frac{4\rho h}{R} \int_0^{\pi/2} \left\{ \frac{R r^4}{4} - \frac{r^5}{5} \right\}_0^R d\theta
$$

$$
= \frac{\rho h R^4}{5} \left\{ \theta \right\}_0^{\pi/2} = \frac{\rho \pi h R^4}{10}.
$$

FIGURE 13.68 Moment of inertia of a right-circular cone

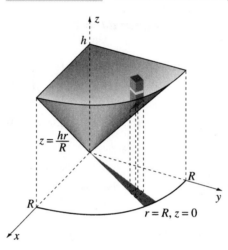

EXERCISES 13.11

In Exercises 1–10 find the equation for the surface in cylindrical coordinates. Draw each surface and indicate whether it is symmetric about the z-axis.

1. $x^2 + y^2 + z^2 = 4$ **2.** $x^2 + y^2 = 1$

3. $y^2 + z^2 = 6$ **4.** $x + y = 5$

5. $z = 2\sqrt{x^2 + y^2}$ **6.** $z = x^2$

7. $x^2 + 4y^2 = 4$ **8.** $4z = x^2 + y^2$

9. $y = x$ **10.** $x^2 + y^2 = 1 + z^2$

In Exercises 11–15 find the volume bounded by the surfaces.

11. $z = \sqrt{x^2 + y^2}$, $x^2 + y^2 = 4$, $z = 0$

12. $z = \sqrt{2 - x^2 - y^2}$, $z = x^2 + y^2$

13. $z = xy$, $x^2 + y^2 = 1$, $z = 0$

14. $z = x^2 + y^2$, $z = 4 - x^2 - y^2$

15. $x + y + z = 2$, $x^2 + y^2 = 1$, $z = 0$

∗ **16.** Find the volume inside the sphere $x^2 + y^2 + z^2 = 4$ but outside the cylinder $x^2 + y^2 = 1$.

∗ **17.** Find the centre of mass of a uniform hemispherical solid.

∗ **18.** Set up the six triple iterated integrals in cylindrical coordinates for the triple integral of a function $f(x, y, z)$ over the region bounded by surfaces $z = 1 + x^2 + y^2$, $x^2 + y^2 = 9$, $z = 0$.

∗ **19.** Find the moment of inertia of a uniform right-circular cylinder of radius R and height h (a) about its axis and (b) about a line through the centre of its base and perpendicular to its axis.

∗ **20.** Find the moment of inertia of a uniform sphere of radius R about any line through its centre.

In Exercises 21–25 evaluate the triple iterated integral.

∗ **21.** $\displaystyle \int_0^3 \int_0^{\sqrt{9-x^2}} \int_0^{\sqrt{x^2+y^2}} dz \, dy \, dx$

22. $\displaystyle\int_0^9 \int_0^{\sqrt{81-y^2}} \int_0^{\sqrt{81-x^2-y^2}} \frac{1}{\sqrt{x^2+y^2}}\, dz\, dx\, dy$

23. $\displaystyle\int_0^4 \int_0^{\sqrt{4y-y^2}} \int_0^{y+x^2} dz\, dx\, dy$

24. $\displaystyle\int_0^{\sqrt{3}/2} \int_{5-\sqrt{21-y^2}}^{\sqrt{1-y^2}} \int_0^{x^2+y^2} y\, dz\, dx\, dy$

25. $\displaystyle\int_0^1 \int_0^{\sqrt{1-y^2}} \int_0^{x^2+y^2} y^2\, dz\, dx\, dy$

26. To analyze running technique for sprinters, it is necessary to cal-culate the moment of inertia of the legs about a horizontal line through the hips H (left figure below). Suppose the upper and lower portions of the leg of an athlete (including the foot) are modelled as cylinders with lengths 0.45 m and 0.5 m and radii 0.07 m and 0.05 m, respec-tively (right figure). The athlete has mass 73 kg, of which 13.7% is in each upper leg and 6% in each lower leg. Use Exercise 27 and the parallel axis theorem from Exercise 22 in Section 13.10 to determine the moment of inertia when axes of upper and lower legs make an angle of $\pi/3$ radians.

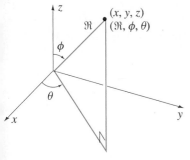

G_U = Centre of mass of upper leg

G_L = Centre of mass of lower leg

27. Show that the moment of inertia of a right-circular cylinder of radius R, length L, and constant density, about an axis through its centre and perpendicular to its length is $m(R^2/4 + L^2/12)$, where m is the mass of the cylinder.

28. Find the centre of mass for the uniform solid bounded by the sur-faces $x^2 + y^2 = 2x$, $z = \sqrt{x^2 + y^2}$, $z = 0$.

* **29.** A casting is in the form of a sphere of radius b with two cylindrical holes of radius $a < b$ such that the axes of the holes pass through the centre of the sphere and intersect at right angles. What volume of metal is required for the casting?

* **30.** Show that the volume bounded by the cylinder $x^2 + y^2 = R^2$ and the parallel planes $z = my$ and $z = my + h$, where m and h are constants, is the same as a cylinder of radius R and length h.

In Exercises 31–41 find the volume described.

* **31.** Bounded by $x^2 + y^2 - z^2 = 1$, $4z^2 = x^2 + y^2$
* **32.** Bounded by $z = x^2 + y^2$, $z = 0$, $(x^2 + y^2)^2 = x^2 - y^2$
* **33.** Bounded by $x^2+y^2+z^2 = 4$, $x^2+y^2+z^2 = 16$, $z = \sqrt{x^2 + y^2}$ (smaller piece)
* **34.** Bounded by $x^2 + y^2 + z^2 = 1$, $y = x$, $x = 2y$, $z = 0$ (in the first octant)
* **35.** Inside both $2x^2 + 2y^2 + z^2 = 8$ and $x^2 + y^2 = 1$
* **36.** Bounded by $z^2 = (x^2 + y^2)^2$, $x^2 + y^2 = 2y$
* **37.** Inside $x^2 + y^2 + z^2 = a^2$ but outside $x^2 + y^2 = ay$
* **38.** Bounded by $z = 0$, $x^2 + y^2 = 1$, $z = e^{-x^2-y^2}$
* **39.** Inside $x^2 + y^2 + z^2 = 9$ but outside $x^2 + y^2 = 1 + z^2$
* **40.** Cut off from $z = x^2 + y^2$ by $z = x + y$
* **41.** Inside $x^2 + y^2 + z^2 = 4$ and below $3z = x^2 + y^2$

* **42.** Evaluate the triple integral of $\sqrt{x^2 + y^2 + z^2}$ over the region bounded by $z = 3$ and $z = \sqrt{x^2 + y^2}$.

* **43.** Evaluate the triple integral of $f(y, z) = |yz|$ over the region bounded by $z^2 = 1 + x^2 + y^2$ and $z = \sqrt{4 - x^2 - y^2}$.

** **44.** A tumbler in the form of a right-circular cylinder of radius R and height h is full of water. As the axis of the tumbler is tilted from the vertical, water pours over the side. Find the volume of water remaining in the tumbler as a function of the angle between the vertical and the axis of the tumbler.

** **45.** Use cylindrical coordinates to find the volume of the torus $(\sqrt{x^2 + y^2} - a)^2 + z^2 = b^2$, $b < a$.

13.12 Triple Iterated Integrals in Spherical Coordinates

FIGURE 13.69 Rela-tionships between Cartesian and spherical coordinates

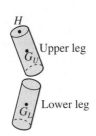

Spherical coordinates are useful in solving problems where figures are symmetric about a point. If the origin is that point, then spherical coordinates $(\Re, \phi, \theta)$[†] in Figure 13.69 are related to Cartesian coordinates (x, y, z) by the equations

$$x = \Re \sin\phi \cos\theta, \quad y = \Re \sin\phi \sin\theta, \quad z = \Re \cos\phi. \tag{13.63}$$

As is the case for polar and cylindrical coordinates, without restrictions on $\Re$, ϕ, and θ, each point in space has many sets of spherical coordinates. The positive value of its spherical coor-

---

[†] Mathematicians often use the letter $\Re$ to stand for the set of real numbers. We have avoided using it for this purpose so that no confusion would arise in this section.

dinate $\Re$ is given by

$$\Re = \sqrt{x^2 + y^2 + z^2}. \tag{13.64a}$$

The θ-coordinate in cylindrical and spherical coordinates is identical, so that no simple formula for θ in terms of x, y, and z exists. The ϕ-coordinate is the angle between the positive z-axis and the line joining the origin to the point (x, y, z), and that value of ϕ in the range $0 \leq \phi \leq \pi$ is determined by the formula

$$\phi = \text{Cos}^{-1}\left(\frac{z}{\sqrt{x^2 + y^2 + z^2}}\right). \tag{13.64b}$$

To transform equations $F(x, y, z) = 0$ of surfaces from Cartesian to spherical coordinates, we substitute from equations 13.63:

$$F(\Re \sin\phi \cos\theta, \Re \sin\phi \sin\theta, \Re \cos\phi) = 0.$$

For example, the sphere $x^2 + y^2 + z^2 = 4$ is symmetric about its centre, and its equation in spherical coordinates is simply $\Re = 2$. For the right-circular cone $z = \sqrt{x^2 + y^2}$, we write

$$\Re \cos\phi = \sqrt{\Re^2 \sin^2\phi \cos^2\theta + \Re^2 \sin^2\phi \sin^2\theta} = \Re \sin\phi.$$

Consequently, $\tan\phi = 1$ or $\phi = \pi/4$ (i.e., $\phi = \pi/4$ is the equation of the cone in spherical coordinates).

Suppose that we are to evaluate the triple integral of a function $f(x, y, z)$,

$$\iiint_V f(x, y, z)\, dV,$$

over some region V of space. The choice of a **triple iterated integral in spherical coordinates** implies a subdivision of V into small volumes by means of coordinate surfaces $\Re = $ constant, $\phi = $ constant, and $\theta = $ constant (Figure 13.70). Surfaces $\Re = $ constant are spheres centred at the origin; surfaces $\phi = $ constant are right-circular cones symmetric about the z-axis with the origin as apex; and surfaces $\theta = $ constant are half planes containing the z-axis.

FIGURE 13.70   Elemental volume in spherical coordinates is created with coordinate surfaces $\Re = $ constant, $\phi = $ constant, $\theta = $ constant

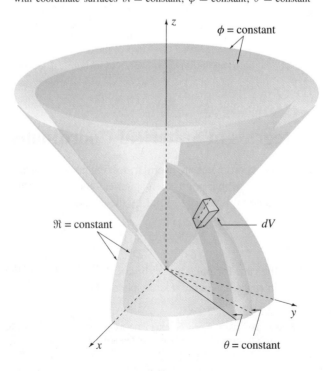

If we denote small variations in $\Re$, ϕ, and θ for the element dV by $d\Re$, $d\phi$, and $d\theta$ (Figure 13.71) and approximate dV by a rectangular parallelepiped with dimensions $d\Re$, $\Re\,d\phi$, and $\Re\sin\phi\,d\theta$, then

$$dV = (\Re\sin\phi\,d\theta)(\Re\,d\phi)\,d\Re = \Re^2\sin\phi\,d\Re\,d\theta\,d\phi. \tag{13.65}$$

FIGURE 13.71 Elemental volume in spherical coordinates is $dV = \Re^2\sin\phi\,d\Re\,d\phi\,d\theta$

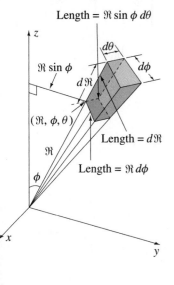

Length $= \Re\sin\phi\,d\theta$

$\Re\sin\phi$

$d\theta$

$d\phi$

$d\Re$

$(\Re, \phi, \theta)$

Length $= d\Re$

$\Re$

Length $= \Re\,d\phi$

ϕ

The integrand $f(x, y, z)$ is expressed in spherical coordinates as

$$f(\Re\sin\phi\cos\theta, \Re\sin\phi\sin\theta, \Re\cos\phi).$$

It remains only to affix appropriate limits to the triple iterated integral, and these limits depend on which of the six possible triple iterated integrals in spherical coordinates we choose. If we use a triple iterated integral with respect to $\Re$, ϕ, and θ, then it is of the form

$$\iiint_V f(x, y, z)\,dV$$
$$= \int_\alpha^\beta \int_{g_1(\theta)}^{g_2(\theta)} \int_{h_1(\phi,\theta)}^{h_2(\phi,\theta)} f(\Re\sin\phi\cos\theta, \Re\sin\phi\sin\theta, \Re\cos\phi)\Re^2\sin\phi\,d\Re\,d\phi\,d\theta.$$

The geometric interpretations of the additions represented by these integrations with respect to $\Re$, ϕ, and θ are left to the examples.

EXAMPLE 13.31

Find the volume of a sphere.

SOLUTION   The equation of a sphere of radius R centred at the origin is $x^2 + y^2 + z^2 = R^2$ or, in spherical coordinates, $\Re = R$. The volume of this sphere is eight times the first octant volume shown in Figure 13.72. If we use spherical coordinates, the volume of an elemental piece is

$$\Re^2\sin\phi\,d\Re\,d\phi\,d\theta.$$

An $\Re$-integration adds these volumes for constant ϕ and θ to give the volume in a "spike,"

$$\int_0^R \Re^2\sin\phi\,d\Re\,d\phi\,d\theta,$$

where limits indicate that all spikes start at the origin (where $\Re = 0$) and end on the sphere (where $\Re = R$). A ϕ-integration now adds the volumes of spikes for constant θ. This yields the volume of a slice

$$\int_0^{\pi/2} \int_0^R \Re^2\sin\phi\,d\Re\,d\phi\,d\theta,$$

where limits indicate that all slices in the first octant start on the z-axis (where $\phi = 0$) and end on the xy-plane (where $\phi = \pi/2$). Finally, the θ-integration adds the volumes of all such slices

$$\int_0^{\pi/2} \int_0^{\pi/2} \int_0^R \Re^2\sin\phi\,d\Re\,d\phi\,d\theta,$$

where limits 0 and $\pi/2$ identify positions of first and last slices, respectively, in the first octant. The volume of the sphere is therefore

$$
8 \int_0^{\pi/2} \int_0^{\pi/2} \int_0^R \Re^2 \sin \phi \, d\Re \, d\phi \, d\theta = 8 \int_0^{\pi/2} \int_0^{\pi/2} \left\{ \frac{\Re^3}{3} \sin \phi \right\}_0^R d\phi \, d\theta
$$

$$
= \frac{8R^3}{3} \int_0^{\pi/2} \{-\cos \phi\}_0^{\pi/2} \, d\theta
$$

$$
= \frac{8R^3}{3} \{\theta\}_0^{\pi/2} = \frac{4}{3}\pi R^3.
$$

FIGURE 13.72   Volume of a sphere calculated with spherical coordinates

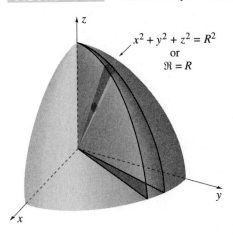

$x^2 + y^2 + z^2 = R^2$
or
$\Re = R$

EXAMPLE 13.32

Find the centre of mass of a solid object of constant density if it is in the shape of a right-circular cone.

SOLUTION   Let the cone have altitude h and base radius R. Then its mass is

$$
M = \frac{1}{3}\pi R^2 h \rho,
$$

where ρ is the density of the object. If we place axes as shown in Figure 13.73, then $\overline{x} = \overline{y} = 0$ (i.e., the centre of mass is on the axis of symmetry of the cone). To find $\overline{z}$ we offer three solutions.

Method 1   If we use Cartesian coordinates, the equation of the surface of the cone is of the form $z = k\sqrt{x^2 + y^2}$. Since $(0, R, h)$ is a point on the cone, $h = kR$, and therefore $k = h/R$. Now

$$
M\overline{z} = 4 \int_0^R \int_0^{\sqrt{R^2-x^2}} \int_{k\sqrt{x^2+y^2}}^h z\rho \, dz \, dy \, dx = 4\rho \int_0^R \int_0^{\sqrt{R^2-x^2}} \left\{ \frac{z^2}{2} \right\}_{k\sqrt{x^2+y^2}}^h dy \, dx
$$

$$
= 2\rho \int_0^R \int_0^{\sqrt{R^2-x^2}} [h^2 - k^2(x^2 + y^2)] \, dy \, dx = 2\rho \int_0^R \left\{ h^2 y - k^2 x^2 y - \frac{k^2 y^3}{3} \right\}_0^{\sqrt{R^2-x^2}} dx
$$

$$
= 2\rho \int_0^R \left[h^2\sqrt{R^2 - x^2} - k^2 x^2\sqrt{R^2 - x^2} - \frac{k^2}{3}(R^2 - x^2)^{3/2} \right] dx.
$$

FIGURE 13.73   Centre of mass of a cone using Cartesian and cylindrical coordinates

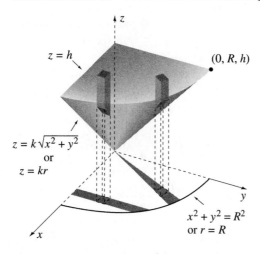

If we set $x = R \sin \theta$, then $dx = R \cos \theta \, d\theta$, and

$$M\bar{z} = 2\rho \int_0^{\pi/2} \left[h^2 R \cos \theta - k^2 (R^2 \sin^2 \theta) R \cos \theta - \frac{k^2}{3} R^3 \cos^3 \theta \right] R \cos \theta \, d\theta$$

$$= 2\rho R^2 \int_0^{\pi/2} \left[\frac{h^2}{2} (1 + \cos 2\theta) - \frac{k^2 R^2}{8} (1 - \cos 4\theta) \right.$$

$$\left. - \frac{k^2 R^2}{12} \left(1 + 2 \cos 2\theta + \frac{1 + \cos 4\theta}{2} \right) \right] d\theta$$

$$= \rho R^2 \left\{ h^2 \left(\theta + \frac{\sin 2\theta}{2} \right) - \frac{k^2 R^2}{4} \left(\theta - \frac{\sin 4\theta}{4} \right) \right.$$

$$\left. - \frac{k^2 R^2}{6} \left(\frac{3\theta}{2} + \sin 2\theta + \frac{\sin 4\theta}{8} \right) \right\}_0^{\pi/2}$$

$$= \rho R^2 \left(\frac{\pi h^2}{2} - \frac{k^2 R^2 \pi}{8} - \frac{k^2 R^2 \pi}{8} \right) = \frac{\pi R^2 h^2 \rho}{4}.$$

Thus, $\bar{z} = \dfrac{\pi R^2 h^2 \rho}{4} \cdot \dfrac{3}{\pi R^2 h \rho} = \dfrac{3h}{4}.$

Method 2   If we use cylindrical coordinates, the equation of the surface of the cone is $z = kr$. From Figure 13.73, we see that

$$M\bar{z} = 4 \int_0^{\pi/2} \int_0^R \int_{kr}^h z\rho r \, dz \, dr \, d\theta = 4\rho \int_0^{\pi/2} \int_0^R \left\{ \frac{rz^2}{2} \right\}_{kr}^h dr \, d\theta$$

$$= 2\rho \int_0^{\pi/2} \int_0^R r(h^2 - k^2 r^2) \, dr \, d\theta = 2\rho \int_0^{\pi/2} \left\{ \frac{r^2 h^2}{2} - \frac{k^2 r^4}{4} \right\}_0^R d\theta$$

$$= \rho \left(R^2 h^2 - \frac{k^2 R^4}{2} \right) \{ \theta \}_0^{\pi/2} = \rho \left(R^2 h^2 - \frac{k^2 R^4}{2} \right) \frac{\pi}{2} = \frac{\pi R^2 h^2 \rho}{4}.$$

Again, then, $\bar{z} = 3h/4$.

FIGURE 13.74 Centre
of mass of cone using spherical
coordinates

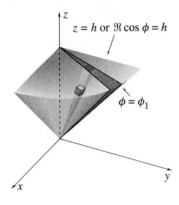

$z = h$ or $\Re \cos \phi = h$

$\phi = \phi_1$

Method 3 If we use spherical coordinates, then the equation of the surface of the cone is

$$\phi = \text{Cos}^{-1}\frac{h}{\sqrt{h^2 + R^2}} = \phi_1.$$

From Figure 13.74, we see that

$$M\bar{z} = 4 \int_0^{\pi/2} \int_0^{\phi_1} \int_0^{h\sec\phi} (\Re \cos \phi)\rho\Re^2 \sin \phi \, d\Re \, d\phi \, d\theta$$

$$= 4\rho \int_0^{\pi/2} \int_0^{\phi_1} \left\{\frac{\Re^4}{4} \sin \phi \cos \phi \right\}_0^{h\sec\phi} d\phi \, d\theta$$

$$= \rho h^4 \int_0^{\pi/2} \int_0^{\phi_1} \frac{\sin \phi}{\cos^3 \phi} d\phi \, d\theta = \rho h^4 \int_0^{\pi/2} \left\{\frac{1}{2\cos^2 \phi}\right\}_0^{\phi_1} d\theta$$

$$= \frac{\rho h^4}{2}\left(\frac{1}{\cos^2 \phi_1} - 1\right)\{\theta\}_0^{\pi/2} = \frac{\pi\rho h^4}{4}\left(\frac{h^2 + R^2}{h^2} - 1\right) = \frac{\pi h^2 R^2 \rho}{4}.$$

Again, $\bar{z} = 3h/4$.

In Exercises 1–7 find the equation of the surface in spherical coordinates. Draw the surface.

1. $x^2 + y^2 + z^2 = 4$

2. $x^2 + y^2 = 1$

3. $3z = \sqrt{x^2 + y^2}$

4. $4z = x^2 + y^2$

5. $y = x$

6. $x^2 + y^2 = 1 + z^2$

7. $z = -2\sqrt{x^2 + y^2}$

In Exercises 8–12 find the volume described.

8. Bounded by $z = \sqrt{x^2 + y^2}$, $z = \sqrt{1 - x^2 - y^2}$

9. Bounded by $z = 1$, $z = \sqrt{4 - x^2 - y^2}$

10. Bounded by $x^2 + y^2 + z^2 = 1$, $y = x$, $y = 2x$, $z = 0$ (in the first octant)

11. Inside $x^2 + y^2 + z^2 = 2$ but outside $x^2 + y^2 = 1$

12. Bounded by $z = 2\sqrt{x^2 + y^2}$, $x^2 + y^2 = 4$, $z = 0$

* **13.** Find the centre of mass of a uniform hemispherical solid.

* **14.** Find the moment of inertia of a uniform solid sphere of radius R about any line through its centre.

* **15.** Find the first moment about the yz-plane of the uniform solid in the first octant bounded by the surfaces $x^2+y^2+z^2 = 4$, $x^2+y^2+z^2 = 9$, $y = 0$, and $y = \sqrt{3}x$.

* **16.** A solid sphere of radius R and centre at the origin has a continuous charge distribution throughout. If the density of the charge is $\rho(x, y, z) = k\sqrt{x^2 + y^2 + z^2}$ coulombs per cubic metre, find the total charge in the sphere.

* **17.** Set up the six triple iterated integrals in spherical coordinates for the triple integral of a function $f(x, y, z)$ over the region in the first octant under the sphere $x^2 + y^2 + z^2 = 2$ and inside the cylinder $x^2 + y^2 = 1$.

In Exercises 18–19 evaluate the triple iterated integral.

* **18.** $\int_0^9 \int_0^{\sqrt{81-y^2}} \int_0^{\sqrt{81-x^2-y^2}} \frac{1}{x^2 + y^2 + z^2} dz \, dx \, dy$

* **19.** $\int_0^1 \int_0^{\sqrt{1-x^2}} \int_{\sqrt{x^2+y^2}}^{\sqrt{2-x^2-y^2}} dz \, dy \, dx$

* **20.** Find a formula for the volume of the smaller region bounded by $x^2 + y^2 + z^2 = R^2$ and $z = k\sqrt{x^2 + y^2}$ $(k > 0)$.

* **21.** Find the volume bounded by $(x^2 + y^2 + z^2)^2 = x$.

* **22.** (a) Use Archimedes' principle to determine the density of a spherical ball if it floats half submerged in water.

(b) What force is required to keep the ball with its centre at a depth of one-half the radius of the ball?

* **23.** Find the volume bounded by the surface $(x^2 + y^2 + z^2)^2 = 2z(x^2 + y^2)$.

24. A sphere of radius R carries a uniform charge distribution of ρ coulombs per cubic metre (figure below). If P is a point on the z-axis (distance $d > R$ from the centre of the sphere) and dV is a small element of volume of the sphere, then potential at P due to dV is given by

$$\frac{1}{4\pi\epsilon_0}\frac{\rho\,dV}{s},$$

where s is distance from P to dV.

(a) Show that in terms of spherical coordinates potential V at P due to the entire sphere is

$$V = \frac{\rho}{4\pi\epsilon_0}\int_{-\pi}^{\pi}\int_0^{\pi}\int_0^{R}\frac{\Re^2\sin\phi}{\sqrt{\Re^2 + d^2 - 2\Re d\cos\phi}}\,d\Re\,d\phi\,d\theta.$$

(b) Because this iterated integral is very difficult to evaluate, we replace ϕ with the variable $s = \sqrt{\Re^2 + d^2 - 2\Re d\cos\phi}$. Show that with this change

$$V = \frac{\rho}{4\pi\epsilon_0 d}\int_{-\pi}^{\pi}\int_0^{R}\int_{d-\Re}^{d+\Re}\Re\,ds\,d\Re\,d\theta.$$

(c) Evaluate the integral in part (b) to verify that $V = Q/(4\pi\epsilon_0 d)$, where Q is total charge on the sphere.

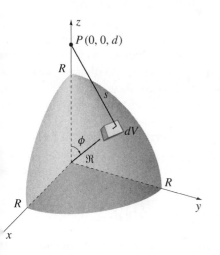

∗ 25. A sphere of constant density ρ and radius R is located at the origin (figure below). If a mass m is situated at a point P on the z-axis (distance $d > R$ from the centre of the sphere) and dV is a small element of volume of the sphere, then according to Newton's universal law of gravitation, the z-component of the force on m due to the mass in dV is given by

$$-\frac{Gm\rho\,dV\cos\psi}{s^2},$$

where G is a constant and s is distance between P and dV.

(a) Show that in spherical coordinates total force on m due to the entire sphere has z-component

$$F_z = -\frac{Gm\rho}{2d}\int_{-\pi}^{\pi}\int_0^{\pi}\int_0^{R}\left(\frac{s^2 + d^2 - \Re^2}{s^3}\right)\Re^2\sin\phi\,d\Re\,d\phi\,d\theta.$$

(b) Use the transformation in Exercise 24(b) to write F_z in the form

$$F_z = -\frac{Gm\rho}{2d^2}\int_{-\pi}^{\pi}\int_0^{R}\int_{d-\Re}^{d+\Re}\Re\left(\frac{s^2 + d^2 - \Re^2}{s^2}\right)ds\,d\Re\,d\theta,$$

and show that $F_z = -GmM/d^2$, where M is the total mass of the sphere.

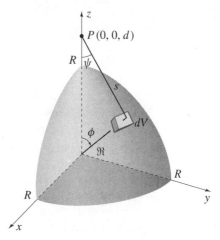

∗∗ 26. A homogeneous solid is bounded by two concentric spheres of radii a and b ($a < b$). Verify that the force that this layer exerts on a point mass at any point interior to the shell vanishes.

13.13 General Transformations in Multiple Integrals

Geometric discussions that led to the area element $dA = r\,dr\,d\theta$ in polar coordinates and the volume elements $dV = r\,dz\,dr\,d\theta$ and $dV = \Re^2\sin\phi\,d\Re\,d\phi\,d\theta$ in cylindrical and spherical coordinates were successful due to the simplicity of coordinate curves and surfaces in these coordinate systems. In coordinate systems with a less simple geometry, corresponding discussions could prove much less transparent. In this section we discover how to evaluate multiple integrals in arbitrary coordinate systems. Our approach is geometric, but the final result is stated algebraically.

Suppose that we are to evaluate the double integral of a function $f(x, y)$ over some region R in the xy-plane, using a coordinate system (u, v) related to Cartesian coordinates by equations

$$u = u(x, y), \quad v = v(x, y). \tag{13.66}$$

We assume that the functions in equations 13.66 and their first derivatives are continuous in R, and that at every point in R, the Jacobian

$$\frac{\partial(u, v)}{\partial(x, y)} \neq 0. \tag{13.67}$$

These conditions guarantee the existence of the inverse transformation

$$x = x(u, v), \quad y = y(u, v). \tag{13.68}$$

A curve in R along which u is constant, say $u = u_0$, is called a *u-coordinate curve*. A *v-coordinate curve* is a curve along which $v = v_0$ is constant. Nonvanishing of the Jacobian 13.67 guarantees that at every point P of R, there is exactly one coordinate curve of each type passing through P, say $u = u_0$ and $v = v_0$ (Figure 13.75). We regard (u_0, v_0) as the uv-coordinates of P.

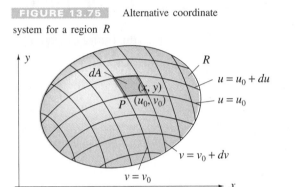

FIGURE 13.75 Alternative coordinate system for a region R

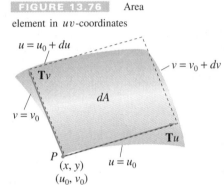

FIGURE 13.76 Area element in uv-coordinates

In using Cartesian coordinates to evaluate the double integral of $f(x, y)$ over R, we subdivide R into rectangles by means of a grid of coordinate lines $x =$ constant and $y =$ constant. The area element $dA = dx\,dy$ is then the (rectangular) area bounded by coordinate curves whose separations are dx and dy. In a similar way, to use uv-coordinates, we form a curvilinear grid with coordinate curves $u =$ constant and $v =$ constant (Figure 13.75). The area element dA is the area bounded by coordinate curves that differ by amounts du and dv. If du and dv are very small, then dA can be approximated by a parallelogram (Figure 13.76), the accuracy of the approximation increasing as du and dv decrease. To find the area of the approximating parallelogram, we need tangent vectors to the coordinate curves $u = u_0$ and $v = v_0$ at P. Since parametric equations for $u = u_0$ are $x = x(u_0, v)$, $y = y(u_0, v)$, a tangent vector is

$$\frac{\partial x}{\partial v}\hat{\mathbf{i}} + \frac{\partial y}{\partial v}\hat{\mathbf{j}}, \qquad \text{as is} \qquad \mathbf{T}_u = \left(\frac{\partial x}{\partial v}\hat{\mathbf{i}} + \frac{\partial y}{\partial v}\hat{\mathbf{j}}\right) dv.$$

The components of $\mathbf{T}_u$ represent the changes in x and y corresponding to the change dv in v between coordinate curves, and therefore the length of this tangent vector does indeed represent the length of the side of the parallelogram in Figure 13.76. Similarly, the side of the parallelogram corresponding to a change du in u along $v = v_0$ is

$$\mathbf{T}_v = \left(\frac{\partial x}{\partial u}\hat{\mathbf{i}} + \frac{\partial y}{\partial u}\hat{\mathbf{j}}\right) du.$$

According to equation 11.43, the area of the approximating parallelogram is the length of the vector

$$\begin{aligned}
\mathbf{T}_v \times \mathbf{T}_u &= \left(\frac{\partial x}{\partial u}\hat{\mathbf{i}} + \frac{\partial y}{\partial u}\hat{\mathbf{j}}\right) du \times \left(\frac{\partial x}{\partial v}\hat{\mathbf{i}} + \frac{\partial y}{\partial v}\hat{\mathbf{j}}\right) dv \\
&= \left(\frac{\partial x}{\partial u}\frac{\partial y}{\partial v} - \frac{\partial x}{\partial v}\frac{\partial y}{\partial u}\right) du\,dv\,\hat{\mathbf{k}} \\
&= \frac{\partial(x, y)}{\partial(u, v)} du\,dv\,\hat{\mathbf{k}}.
\end{aligned}$$

Consequently, if du and dv are positive, the element of area dA in uv-coordinates is

$$dA = \left| \frac{\partial(x, y)}{\partial(u, v)} \right| du \, dv. \tag{13.69}$$

To evaluate the double integral of $f(x, y)$ over R, we use 13.68 to express $f(x, y)$ in terms of u and v, and write

$$\iint_R f(x, y) \, dA = \iint_{R_{uv}} f[x(u, v), y(u, v)] \left| \frac{\partial(x, y)}{\partial(u, v)} \right| du \, dv. \tag{13.70}$$

Since the right side of this equation represents a double iterated integral in u and v, the integrals should have limits. Because this can be done only when a region R has been specified, we have used the notation R_{uv} to represent limits describing R in terms of coordinates u and v. This result can be stated in terms of double integrals also. Under a *transformation of coordinates* 13.66 or 13.68, double iterated integrals transform according to

$$\iint_{R_{xy}} f(x, y) \, dx \, dy = \iint_{R_{uv}} f[x(u, v), y(u, v)] \left| \frac{\partial(x, y)}{\partial(u, v)} \right| du \, dv. \tag{13.71}$$

The **general transformation for triple integrals** is as follows. If Cartesian coordinates (x, y, z) are related to curvilinear coordinates (u, v, w) in some region V of space by equations

$$x = x(u, v, w), \quad y = y(u, v, w), \quad z = z(u, v, w), \tag{13.72}$$

then

$$\iiint_V f(x, y, z) \, dV = \iiint_{V_{uvw}} f[x(u, v, w), y(u, v, w), z(u, v, w)] \left| \frac{\partial(x, y, z)}{\partial(u, v, w)} \right| du \, dv \, dw \tag{13.73}$$

or

$$\iiint_{V_{xyz}} f(x, y, z) \, dV = \iiint_{V_{uvw}} f[x(u, v, w), y(u, v, w), z(u, v, w)] \left| \frac{\partial(x, y, z)}{\partial(u, v, w)} \right| du \, dv \, dw. \tag{13.74}$$

EXAMPLE 13.33

Show that transformation laws 13.71 and 13.74 lead to the correct differential expressions for area and volume in polar, cylindrical, and spherical coordinates.

SOLUTION Transformation law 13.71 leads in polar coordinates to the area element $dA = \left| \frac{\partial(x, y)}{\partial(r, \theta)} \right| dr \, d\theta$ where $x = r \cos \theta$ and $y = r \sin \theta$. Since

$$\frac{\partial(x, y)}{\partial(r, \theta)} = \left| \begin{array}{cc} \partial x/\partial r & \partial x/\partial \theta \\ \partial y/\partial r & \partial y/\partial \theta \end{array} \right| = \left| \begin{array}{cc} \cos \theta & -r \sin \theta \\ \sin \theta & r \cos \theta \end{array} \right| = r,$$

we have $dA = r \, dr \, d\theta$. Transformation law 13.74 in cylindrical coordinates gives the volume element $dV = \left| \frac{\partial(x, y, z)}{\partial(r, \theta, z)} \right| dr \, d\theta \, dz$, where $x = r \cos \theta$, $y = r \sin \theta$ and $z = z$. Since

$$\frac{\partial(x, y, z)}{\partial(r, \theta, z)} = \left| \begin{array}{ccc} \partial x/\partial r & \partial x/\partial \theta & \partial x/\partial z \\ \partial y/\partial r & \partial y/\partial \theta & \partial y/\partial z \\ \partial z/\partial r & \partial z/\partial \theta & \partial z/\partial z \end{array} \right| = \left| \begin{array}{ccc} \cos \theta & -r \sin \theta & 0 \\ \sin \theta & r \cos \theta & 0 \\ 0 & 0 & 1 \end{array} \right| = r,$$

we have $dV = r \, dr \, d\theta \, dz$.

For spherical coordinates, we obtain $dV = \left| \dfrac{\partial(x, y, z)}{\partial(\mathfrak{R}, \phi, \theta)} \right| d\mathfrak{R}\, d\phi\, d\theta$. Since

$$\frac{\partial(x, y, z)}{\partial(\mathfrak{R}, \phi, \theta)} = \begin{vmatrix} \partial x/\partial \mathfrak{R} & \partial x/\partial \phi & \partial x/\partial \theta \\ \partial y/\partial \mathfrak{R} & \partial y/\partial \phi & \partial y/\partial \theta \\ \partial z/\partial \mathfrak{R} & \partial z/\partial \phi & \partial z/\partial \theta \end{vmatrix} = \begin{vmatrix} \sin\phi\cos\theta & \mathfrak{R}\cos\phi\cos\theta & -\mathfrak{R}\sin\phi\sin\theta \\ \sin\phi\sin\theta & \mathfrak{R}\cos\phi\sin\theta & \mathfrak{R}\sin\phi\cos\theta \\ \cos\phi & -\mathfrak{R}\sin\phi & 0 \end{vmatrix} = \mathfrak{R}^2\sin\phi,$$

it follows that $dV = \mathfrak{R}^2 \sin\phi\, d\mathfrak{R}\, d\phi\, d\theta$.

EXAMPLE 13.34

Evaluate the double integral of $f(x, y) = x^3 y - xy^3$ over the region R in the first quadrant bounded by the hyperbolas $xy = 2$, $xy = 4$, $x^2 - y^2 = 1$, $x^2 - y^2 = 9$.

SOLUTION   Evaluation of any double integral over R by means of Cartesian coordinates requires three double iterated integrals (Figure 13.77). To improve the situation, consider new coordinates (u, v) defined by $u = x^2 - y^2$ and $v = xy$. The integrand $f(x, y)$ is expressed in terms of u and v as $xy(x^2 - y^2) = uv$, and therefore

$$\iint_R f(x, y)\, dA = \iint_{R_{uv}} uv \left| \frac{\partial(x, y)}{\partial(u, v)} \right| du\, dv.$$

FIGURE 13.77   Evaluation of the double integral of $x^3 y - xy^3$ over a region bounded by hyperbolas using a change of coordinates

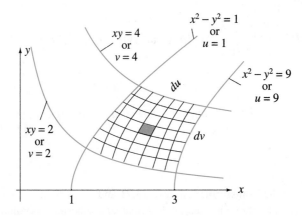

To evaluate the Jacobian we could solve for the inverse coordinate transformation defining x and y in terms of u and v. Instead, we recall from Exercise 23 in Section 12.7 that

$$\frac{\partial(x, y)}{\partial(u, v)} = \frac{1}{\dfrac{\partial(u, v)}{\partial(x, y)}} = \frac{1}{\begin{vmatrix} \partial u/\partial x & \partial u/\partial y \\ \partial v/\partial x & \partial v/\partial y \end{vmatrix}} = \frac{1}{\begin{vmatrix} 2x & -2y \\ y & x \end{vmatrix}} = \frac{1}{2(x^2 + y^2)}.$$

This can be expressed in terms of u and v by noting that the combination $x^2 + y^2$ is obtained by squaring the first equation defining the coordinate transformation and adding four times the square of the second: $u^2 + 4v^2 = (x^4 - 2x^2y^2 + y^4) + 4x^2y^2 = (x^2 + y^2)^2$. Consequently

$$\iint_R f(x, y)\, dA = \iint_{R_{uv}} \frac{uv}{2\sqrt{u^2 + 4v^2}}\, du\, dv.$$

The reason u and v were chosen as above was to simplify the limits of integration. Coordinate curves $u = $ constant and $v = $ constant define a hyperbolic grid with the hyperbolas $u = 1$

$u = 9$, $v = 2$, and $v = 4$ forming the boundary of R. Integration with respect to u holding v constant sums over the area elements in a hyperbolic strip, and the v-integration then adds over all strips. We obtain

$$\iint_R f(x, y)\, dA = \int_2^4 \int_1^9 \frac{uv}{2\sqrt{u^2 + 4v^2}}\, du\, dv = \frac{1}{2}\int_2^4 \left\{v\sqrt{u^2 + 4v^2}\right\}_1^9 dv$$

$$= \frac{1}{2}\int_2^4 (v\sqrt{81 + 4v^2} - v\sqrt{1 + 4v^2})\, dv = \frac{1}{24}\left\{(81 + 4v^2)^{3/2} - (1 + 4v^2)^{3/2}\right\}_2^4$$

$$= \frac{1}{24}(145^{3/2} - 65^{3/2} - 97^{3/2} + 17^{3/2}).$$

An alternative view is to regard the equations $u = x^2 - y^2$ and $v = xy$ as a mapping of the region R in the xy-plane to a region R_{uv} in the uv-plane. The four edges of R are mapped to a rectangle in the uv-plane (Figure 13.78).

FIGURE 13.78   A coordinate transformation can be regarded as a mapping from the xy-plane to the uv-plane

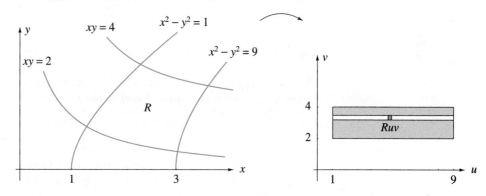

The double integral of $x^3y - xy^3$ over R is equivalent to the double integral of $(uv/2)/\sqrt{u^2 + 4v^2}$ over the rectangle R_{uv}. Horizontal strips lead to

$$\iint_{R_{uv}} \frac{uv}{2\sqrt{u^2 + 4v^2}}\, du\, dv = \int_2^4 \int_1^9 \frac{uv}{2\sqrt{u^2 + 4v^2}}\, du\, dv,$$

and evaluation proceeds as before.

The next example contains a further illustration wherein equations $x = r\cos\theta$, $y = r\sin\theta$ are regarded as a mapping from the xy-plane to the $r\theta$-plane rather than a transformation to polar coordinates.

EXAMPLE 13.35

Use the equations $x = r\cos\theta$, $y = r\sin\theta$ as a mapping to evaluate the double integral of $f(x, y) = \sqrt{x^2 + y^2}$ over the region R bounded by $y = \sqrt{a^2 - x^2}$, $y = 0$, where $a > 0$ is a constant.

SOLUTION   We first note that the Jacobian of the mapping, $\partial(x, y)/\partial(r, \theta) = r$, is equal to zero at the origin, a point on the edge of the region. We shall see the ramification of this momentarily. To map R into the $r\theta$-plane, we map its edges. The semicircle $C_1 : y = \sqrt{a^2 - x^2}$ is mapped to the vertical line segment $C_1': r = a$ from $\theta = 0$ to $\theta = \pi$ (Figure 13.79). The positive x-axis C_2 with $0 < x \le a$ is mapped to the line segment $C_2' : \theta = 0, 0 < r \le a$, and the

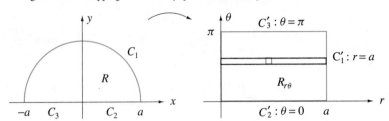

negative x-axis C_3 with $-a \leq x < 0$ is mapped to the line segment $C_3' : \theta = \pi$ with
$0 < r \leq a$. The origin in the xy-plane is mapped to the θ-axis where $r = 0$. Hence R is
mapped to the rectangle $R_{r\theta}$ bounded by $r = 0$, $r = a$, $\theta = 0$, and $\theta = \pi$ in the $r\theta$-plane.
With horizontal strips in $R_{r\theta}$,

$$\iint_R \sqrt{x^2 + y^2}\, dA = \iint_{R_{r\theta}} rr\, dr\, d\theta = \int_0^\pi \int_0^a r^2\, dr\, d\theta = \int_0^\pi \left\{ \frac{r^3}{3} \right\}_0^a d\theta = \frac{a^3}{3} \{\theta\}_0^\pi = \frac{\pi a^3}{3}.$$

EXAMPLE 13.36

Use the transformation $x = au$, $y = bv$, $z = cw$ to evaluate the triple integral of $f(x, y) = x^2 + y^2$ over the region inside the ellipsoid $x^2/a^2 + y^2/b^2 + z^2/c^2 = 1$.

SOLUTION   The Jacobian of the transformation is

$$\frac{\partial(x, y, z)}{\partial(u, v, w)} = \begin{vmatrix} \partial x/\partial u & \partial x/\partial v & \partial x/\partial w \\ \partial y/\partial u & \partial y/\partial v & \partial y/\partial w \\ \partial z/\partial u & \partial z/\partial v & \partial z/\partial w \end{vmatrix} = \begin{vmatrix} a & 0 & 0 \\ 0 & b & 0 \\ 0 & 0 & c \end{vmatrix} = abc.$$

Regarded as a mapping from xyz-space to uvw-space, the ellipsoid is mapped to the sphere $u^2 + v^2 + w^2 = 1$. If we change to spherical coordinates in uvw-space,

$$\iiint_V (x^2 + y^2)\, dV = \iiint_{V_{uvw}} (a^2 u^2 + b^2 v^2)(abc)\, du\, dv\, dw$$

$$= abc \int_0^{2\pi} \int_0^\pi \int_0^1 (a^2 \Re^2 \sin^2 \phi \cos^2 \theta + b^2 \Re^2 \sin^2 \phi \sin^2 \theta)\Re^2 \sin \phi\, d\Re\, d\phi\, d\theta$$

$$= abc \int_0^{2\pi} \int_0^\pi \left\{ \frac{\Re^5}{5}(a^2 \sin^2 \phi \cos^2 \theta + b^2 \sin^2 \phi \sin^2 \theta) \sin \phi \right\}_0^1 d\phi\, d\theta$$

$$= \frac{abc}{5} \int_0^{2\pi} \int_0^\pi (a^2 \cos^2 \theta + b^2 \sin^2 \theta)(1 - \cos^2 \phi) \sin \phi\, d\phi\, d\theta$$

$$= \frac{abc}{5} \int_0^{2\pi} \left\{ (a^2 \cos^2 \theta + b^2 \sin^2 \theta)\left(-\cos \phi + \frac{\cos^3 \phi}{3} \right) \right\}_0^\pi d\theta$$

$$= \frac{4abc}{15} \int_0^{2\pi} \left[a^2 \left(\frac{1 + \cos 2\theta}{2} \right) + b^2 \left(\frac{1 - \cos 2\theta}{2} \right) \right] d\theta$$

$$= \frac{2abc}{15} \left\{ (a^2 + b^2)\theta + \frac{(a^2 - b^2) \sin 2\theta}{2} \right\}_0^{2\pi} = \frac{4\pi abc(a^2 + b^2)}{15}.$$

In the previous examples, the transformations facilitated limits on double and triple integrals. The following example illustrates that they can be used to simplify an integrand.

EXAMPLE 13.37

Use the transformation $u = x + y$, $v = y/(x + y)$ to evaluate the double integral of $f(x, y) = e^{y/(x+y)}$ over the region R bounded by the lines $x + y = 1$, $x = 0$, $y = 0$.

SOLUTION The transformation maps the triangle R in the xy-plane to a square in the uv-plane (Figure 13.80). The Jacobian of the transformation is

$$\frac{\partial(x, y)}{\partial(u, v)} = \frac{1}{\dfrac{\partial(u, v)}{\partial(x, y)}} = \frac{1}{\begin{vmatrix} \partial u/\partial x & \partial u/\partial y \\ \partial v/\partial x & \partial v/\partial y \end{vmatrix}} = \frac{1}{\begin{vmatrix} 1 & 1 \\ -y/(x + y)^2 & x/(x + y)^2 \end{vmatrix}} = x + y = u.$$

Consequently,

$$\iint_R e^{y/(x+y)} \, dA = \iint_{R_{uv}} e^v u \, dv \, du = \int_0^1 \int_0^1 u e^v \, dv \, du$$

$$= \int_0^1 \left\{ u e^v \right\}_0^1 \, du = \int_0^1 u(e - 1) \, du = (e - 1) \left\{ \frac{u^2}{2} \right\}_0^1 = \frac{e - 1}{2}.$$

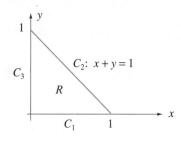

FIGURE 13.80 The transformation $u = x+y$, $v = y/(x+y)$ maps the triangle to a square

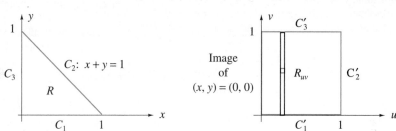

EXERCISES 13.13

In Exercises 1–4 use the suggested equations to set up a double iterated integral for the double integral. In Exercises 1 and 2 do this in two ways: (a) Consider the equations as a coordinate transformation. (b) Consider the equations as a mapping from the xy-plane to another plane. Do not evaluate the double iterated integral.

1. $\iint_R \sqrt{x^2 + y^2} \, dA$, where R is bounded by $x^2 + y^2 = 4$, $x^2 + y^2 = 9$, $x = 0$, $y = 0$, $(x, y \geq 0)$. Use $x = r \cos \theta$, $y = r \sin \theta$.

2. $\iint_R xy \, dA$, where R is bounded by $x^2 + y^2 + x = \sqrt{x^2 + y^2}$, $x = \sqrt{3}y$, $y = \sqrt{3}x$, $(x, y \geq 0)$. Use $x = r \cos \theta$, $y = r \sin \theta$.

3. $\iint_R x^2 \cos y \, dA$, where R is bounded by $y = 2x+1$, $y = 2x-1$, $x + y = 1$, $x + y = 4$. Use $u = x + y$, $v = y - 2x$.

* 4. $\iint_R (x^2 + y) \, dA$, where R is the area in the first quadrant bounded by $y = x^2$, $y = x^2+4$, $y = 5-2x^2$, $y = 6-2x^2$. Use $u = y-x^2$, $v = y + 2x^2$.

In Exercises 5–8 use the suggested transformation to set up a triple iterated integral for the triple integral. Do this in two ways: (a) Consider the equations as a coordinate transformation. (b) Consider the equations as a mapping from the xyz-space to another space. Do not evaluate the triple iterated integral.

5. $\iiint_V z e^{x^2+y^2} \, dV$, where V is the region bounded by $x^2+y^2 = 4$, $z = 0$, $z = 1$. Use $x = r \cos \theta$, $y = r \sin \theta$, $z = z$.

* 6. $\iiint_V (x^2 + y^2) \, dV$, where V is the region in the first octant bounded by $x^2 + y^2 = 9$, $z = \sqrt{x^2 + y^2}$, $x = 0$, $y = 0$, $z = 0$. Use $x = r \cos \theta$, $y = r \sin \theta$, $z = z$.

* **7.** $\iiint_V \dfrac{1}{x^2 + y^2} dV$, where V is the region bounded by $x^2 + y^2 + z^2 = 4$, $y = x$, $y = 0$, $z = 0$ ($x, y, z \geq 0$). Use $x = \Re \sin\phi \cos\theta$, $y = \Re \sin\phi \sin\theta$, $z = \Re \cos\phi$.

* **8.** $\iiint_V x^2 y^2 z \, dV$, where V is the region bounded by $z = \sqrt{x^2 + y^2}$, $z = \sqrt{4 - x^2 - y^2}$. Use $x = \Re \sin\phi \cos\theta$, $y = \Re \sin\phi \sin\theta$, $z = \Re \cos\phi$.

* **9.** Use a transformation to map the region R inside the ellipse $b^2 x^2 + a^2 y^2 = a^2 b^2$ onto a circle in order to evaluate the double integral of $f(x, y) = \sqrt{x^2/a^2 + y^2/b^2}$ over R.

* **10.** Evaluate the double integral of $f(x, y) = xy$ over the first quadrant region bounded by $x^2 - y^2 = 1$, $x^2 - y^2 = 4$, $x^2 + y^2 = 9$, $x^2 + y^2 = 16$.

* **11.** Evaluate the double integral of $f(x, y) = 2x^2 - xy - y^2$ over the region bounded by $y = 8 - 2x$, $y = 4 - 2x$, $y = x - 1$, $y = x + 3$.

* **12.** Use the transformation $u = x - y$, $v = y$ to evaluate the double integral of $f(x, y) = x + y$ over the parallelogram bounded by $y = 0$, $y = 2$, $y = x$, $y = x - 2$.

* **13.** Evaluate the triple integral of $f(x, y, z) = x^2 y^2 z^2$ over the interior of the ellipsoid $x^2/a^2 + y^2/b^2 + z^2/c^2 = 1$.

* **14.** Use the transformation $x = (v/u) \cos w$, $y = (v/u) \sin w$, $z = v^2$ ($u, v > 0$) to evaluate the triple integral of $f(x, y) = x^2 + y^2$ over the region bounded by $z = x^2 + y^2$, $z = 4(x^2 + y^2)$, $z = 1$, $z = 4$.

* **15.** Evaluate the triple integral of $f(x, y, z) = y$ where V is the region bounded by $z = 0$, $y + z = 2$, $x^2 + y^2 = 2y$ using: (a)

cylindrical coordinates $x = r\cos\theta$, $y = r\sin\theta$, $z = z$, and (b) cylindrical coordinates based at $(0, 1)$: $x = r\cos\theta$, $y = 1 + r\sin\theta$, $z = z$.

* **16.** Evaluate the double integral of $f(x, y) = \cos[(x - y)/(x + y)]$ over the triangle bounded by $x + y = 1$, $x = 0$, $y = 0$ by making the transformation $u = x - y$, $v = x + y$.

* **17.** Use the transformation $x = \sqrt{v - u}$, $y = u + v$ to evaluate the double integral of $f(x, y) = x/(x^2 + y)$ over the smaller region bounded by $y = x^2$, $y = 4 - x^2$, $x = 1$.

* **18.** Use the transformation $u = 2x/(x^2 + y^2)$, $v = 2y/(x^2 + y^2)$ to evaluate the double integral of $f(x, y) = 1/(x^2 + y^2)^2$ over the region bounded by the circles $x^2 + y^2 = 6x$, $x^2 + y^2 = 4x$, $x^2 + y^2 = 2y$ and $x^2 + y^2 = 8y$.

* **19.** Use the transformation $x = u + uv$, $y = v + uv$ to evaluate the double integral of $f(x, y) = 1/\sqrt{(x - y)^2 + 2x + 2y + 1}$ over the triangle with vertices $(0, 0)$, $(2, 0)$, and $(2, 2)$.

* **20.** Evaluate the triple integral of $f(x, y, z) = x + y + z$ over the region bounded by the planes $y = x + 1$, $y = x - 1$, $x + y = 1$, $x + y = 3$, $z = 2x + y$, $z = 2x + y + 1$ using (a) triple iterated integrals in Cartesian coordinates, and (b) the transformation $u = x - y$, $v = x + y$, $w = z - 2x - y$.

* **21.** Evaluate the double integral of $f(x, y) = x + y$ over the region in the fourth quadrant bounded by the parabolas $y = x^2 - 1$, $y = x^2 - 4$, $y = -x^2$, $y = 1 - x^2$ using (a) double iterated integrals in Cartesian coordinates, and (b) the transformation $u = x^2 - y$, $v = x^2 + y$.

13.14 Derivatives of Definite Integrals

If a function $f(x, y)$ of two independent variables is integrated with respect to y from $y = a$ to $y = b$, the result depends on x. Suppose we denote this function by $F(x)$:

$$F(x) = \int_a^b f(x, y) \, dy. \tag{13.75}$$

To calculate the derivative $F'(x)$ of this function, we should first integrate with respect to y and then differentiate with respect to x. The following theorem indicates that differentiation can be done first and integration later.

THEOREM 13.3

If the partial derivative $\partial f/\partial x$ of $f(x, y)$ is continuous on a rectangle $a \leq y \leq b$, $c \leq x \leq d$, then for $c < x < d$,

$$\frac{d}{dx} \int_a^b f(x, y) \, dy = \int_a^b \frac{\partial f(x, y)}{\partial x} \, dy. \tag{13.76}$$

PROOF   If we define

$$g(x) = \int_a^b \frac{\partial f(x, y)}{\partial x} \, dy$$

FIGURE 13.81   Proof that
the order of operations of integration with respect to one variable
and differentiation with respect to
another can be interchanged

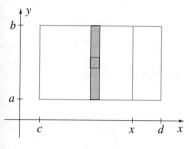

as the right side of equation 13.76, then this function is defined for $c \leq x \leq d$. In fact, because $\partial f / \partial x$ is continuous, $g(x)$ is also continuous. We can therefore integrate $g(x)$ with respect to x from $x = c$ to any value of x in the interval $c \leq x \leq d$:

$$\int_c^x g(x)\, dx = \int_c^x \int_a^b \frac{\partial f(x, y)}{\partial x}\, dy\, dx.$$

This double iterated integral represents the double integral of $\partial f / \partial x$ over the rectangle in Figure 13.81, and if we reverse the order of integration, we have

$$\int_c^x g(x)\, dx = \int_a^b \int_c^x \frac{\partial f(x, y)}{\partial x}\, dx\, dy = \int_a^b \{f(x, y)\}_c^x\, dy$$

$$= \int_a^b [f(x, y) - f(c, y)]\, dy = \int_a^b f(x, y)\, dy - \int_a^b f(c, y)\, dy.$$

Because the second integral on the right is independent of x, if we differentiate this equation with respect to x, we get

$$\frac{d}{dx} \int_c^x g(x)\, dx = \frac{d}{dx} \int_a^b f(x, y)\, dy.$$

But Theorem 6.7 gives

$$g(x) = \frac{d}{dx} \int_a^b f(x, y)\, dy.$$

This completes the proof.

The limits on the integral in 13.75 need not be numerical constants; as far as the integration with respect to y is concerned, x is constant, and therefore a and b could be functions of x:

$$F(x) = \int_{a(x)}^{b(x)} f(x, y)\, dy.$$

Indeed, this is precisely what does occur in the evaluation of double integrals by means of double iterated integrals. To differentiate $F(x)$ now is more complicated than in Theorem 13.3 because we must also account for the x's in $a(x)$ and $b(x)$. The chain rule can be used to develop a formula for $F'(x)$.

THEOREM 13.4   (Leibniz's Rule)

If the partial derivative $\partial f / \partial x$ is continuous on the area bounded by the curves $y = a(x)$, $y = b(x)$, $x = c$, and $x = d$, then

$$\frac{d}{dx} \int_{a(x)}^{b(x)} f(x, y)\, dy = \int_{a(x)}^{b(x)} \frac{\partial f(x, y)}{\partial x}\, dy + f[x, b(x)] \frac{db}{dx} - f[x, a(x)] \frac{da}{dx}.$$

(13.77)

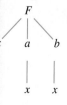

PROOF   If $F(x) = \int_{a(x)}^{b(x)} f(x, y)\, dy$, the schematic diagram gives the chain rule

$$F'(x) = \frac{\partial F}{\partial x}\bigg)_{a,b} + \frac{\partial F}{\partial b}\bigg)_{a,x} \frac{db}{dx} + \frac{\partial F}{\partial a}\bigg)_{b,x} \frac{da}{dx}.$$

The first term is the situation covered in Theorem 13.3, and therefore

$$\frac{\partial F}{\partial x}\bigg)_{a,b} = \int_a^b \frac{\partial f(x, y)}{\partial x}\, dy.$$

Since Theorem 6.7 indicates that

$$\frac{d}{db}\int_a^b f(y)\,dy = f(b),$$

it follows that

$$\frac{\partial}{\partial b}\int_a^b f(x,y)\,dy = f(x,b).$$

In other words,

$$\left(\frac{\partial F}{\partial b}\right)_{a,x} = f(x,b).$$

Furthermore,

$$\left(\frac{\partial F}{\partial a}\right)_{b,x} = \frac{\partial}{\partial a}\int_a^b f(x,y)\,dy = -\frac{\partial}{\partial a}\int_b^a f(x,y)\,dy = -f(x,a).$$

Substitution of these facts into the chain rule now gives Leibniz's rule.

Our derivation of Leibniz's rule for the differentiation of a definite integral that depends on a parameter (x in this case) shows that the first term accounts for those x's in the integrand, and the second and third terms for the x's in the upper and lower limits. The following geometric interpretation of Leibniz's rule emphasizes this same point.

Suppose the function $f(x,y)$ has only positive values so that the surface $z = f(x,y)$ lies completely above the xy-plane (Figure 13.82). Equations $y = a(x)$ and $y = b(x)$ describe cylindrical walls standing on the curves $y = a(x), z = 0$ and $y = b(x), z = 0$ in the xy plane. Were we to slice through the surface $z = f(x,y)$ with a plane $x =$ constant, an area would be defined in this plane bounded on the top by $z = f(x,y)$, on the sides by $y = a(x)$ and $y = b(x)$, and on the bottom by the xy-plane. This area is clearly defined by the definite integral

$$\int_{a(x)}^{b(x)} f(x,y)\,dy$$

in Leibniz's rule, and as x varies so too does the area. Note, in particular, that as the plane varies, area changes, not only because height of the surface $z = f(x,y)$ changes but also because width of the area varies (i.e., the two cylindrical walls are not a constant distance apart). The first term in Leibniz's rule accounts for vertical variation, whereas the remaining two terms represent variations due to fluctuating width.

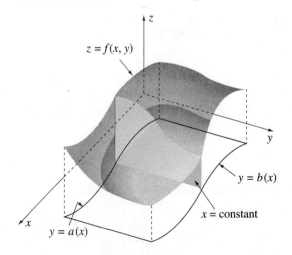

FIGURE 13.82   Geometric interpretation of Leibniz's rule

EXAMPLE 13.38

If $F(x) = \int_x^{2x} (y^3 \ln y + x^3 e^y) \, dy$, find $F'(x)$.

SOLUTION   With Leibniz's rule, we have

$$F'(x) = \int_x^{2x} (3x^2 e^y) \, dy + [(2x)^3 \ln (2x) + x^3 e^{2x}](2) - [x^3 \ln x + x^3 e^x](1)$$

$$= \{3x^2 e^y\}_x^{2x} + 16x^3 (\ln 2 + \ln x) + 2x^3 e^{2x} - x^3 \ln x - x^3 e^x$$

$$= 3x^2 e^{2x} - 3x^2 e^x + (16 \ln 2)x^3 + 15x^3 \ln x + 2x^3 e^{2x} - x^3 e^x$$

$$= 3x^2 e^x (e^x - 1) + x^3 (16 \ln 2 + 15 \ln x + 2e^{2x} - e^x).$$

In Exercise 6 you are asked to evaluate the integral defining $F(x)$ in this example and then to differentiate the resulting function with respect to x. It will be clear, then, that for this example Leibniz's rule simplifies the calculations considerably.

EXAMPLE 13.39

Evaluate

$$\int_0^1 \frac{y^x - 1}{\ln y} \, dy \qquad \text{for } x > -1.$$

SOLUTION   We use Leibniz's rule in this example to avoid finding an antiderivative for $(y^x - 1)/\ln y$. If we set

$$F(x) = \int_0^1 \frac{y^x - 1}{\ln y} \, dy$$

and use Leibniz's rule, we have

$$F'(x) = \int_0^1 \frac{\partial}{\partial x} \left\{ \frac{y^x - 1}{\ln y} \right\} \, dy = \int_0^1 \frac{y^x \ln y}{\ln y} \, dy = \int_0^1 y^x \, dy = \left\{ \frac{y^{x+1}}{x+1} \right\}_0^1 = \frac{1}{x+1}.$$

It follows, therefore, that $F(x)$ must be of the form

$$F(x) = \ln (x + 1) + C.$$

But from the definition of $F(x)$ as an integral, it is clear that $F(0) = 0$, and hence $C = 0$. Thus,

$$\int_0^1 \frac{y^x - 1}{\ln y} \, dy = \ln(x + 1).$$

Note that this problem originally had nothing whatsoever to do with Leibniz's rule. But by introducing the rule, we were able to find a simple formula for $F'(x)$, and this immediately led to $F(x)$. This can be a very useful technique for evaluating definite integrals that depend on a parameter.

EXAMPLE 13.40

When a shell is fired from the artillery gun in Figure 13.83, the barrel recoils along a well-lubricated guide and its motion is braked by a battery of heavy springs. We set up a coordinate system where $x = 0$ represents the firing position of the gun when no stretch or compression exists in the springs. Suppose that when the gun is fired, the horizontal component of the force

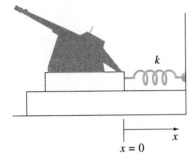

FIGURE 13.83 Displacement of an artillery gun after its shell is fired

causing recoil is $g(t)$ (t = time). If the mass of the gun is m and the effective spring constant for the battery of springs is k, then Newton's second law states that

$$m\frac{d^2x}{dt^2} = g(t) - kx \quad\text{or}\quad m\frac{d^2x}{dt^2} + kx = g(t).$$

This differential equation can be solved by a method called **variation of parameters**, and the solution is

$$x(t) = A\cos\sqrt{\frac{k}{m}}\,t + B\sin\sqrt{\frac{k}{m}}\,t + \frac{1}{\sqrt{mk}}\int_0^t g(u)\sin\left[\sqrt{\frac{k}{m}}(t-u)\right]du,$$

where A and B are arbitrary constants. We do not discuss differential equations until Chapter 15, but with Leibniz's rule it is possible to verify that this function is indeed a solution. Do so.

SOLUTION To differentiate the integral we rewrite Leibniz's rule in terms of the variables of this problem:

$$\frac{d}{dt}\int_{a(t)}^{b(t)} f(t,u)\,du = \int_{a(t)}^{b(t)} \frac{\partial f(t,u)}{\partial t}\,du + f[t,b(t)]\frac{db}{dt} - f[t,a(t)]\frac{da}{dt}.$$

If we now apply this formula to the definite integral in $x(t)$ where

$$f(t,u) = g(u)\sin\left[\sqrt{\frac{k}{m}}(t-u)\right],$$

then

$$\frac{dx}{dt} = -\sqrt{\frac{k}{m}}A\sin\sqrt{\frac{k}{m}}\,t + \sqrt{\frac{k}{m}}B\cos\sqrt{\frac{k}{m}}\,t$$

$$+ \frac{1}{\sqrt{mk}}\left\{\int_0^t g(u)\sqrt{\frac{k}{m}}\cos\left[\sqrt{\frac{k}{m}}(t-u)\right]du + g(t)\sin\left[\sqrt{\frac{k}{m}}(t-t)\right]\right\}$$

$$= \sqrt{\frac{k}{m}}\left(-A\sin\sqrt{\frac{k}{m}}\,t + B\cos\sqrt{\frac{k}{m}}\,t\right) + \frac{1}{m}\int_0^t g(u)\cos\left[\sqrt{\frac{k}{m}}(t-u)\right]du.$$

We now use Leibniz's rule once more to find d^2x/dt^2:

$$\frac{d^2x}{dt^2} = \sqrt{\frac{k}{m}}\left(-\sqrt{\frac{k}{m}}A\cos\sqrt{\frac{k}{m}}\,t - \sqrt{\frac{k}{m}}B\sin\sqrt{\frac{k}{m}}\,t\right)$$

$$+ \frac{1}{m}\left\{\int_0^t -g(u)\sqrt{\frac{k}{m}}\sin\left[\sqrt{\frac{k}{m}}(t-u)\right]du + g(t)\cos\left[\sqrt{\frac{k}{m}}(t-t)\right]\right\}$$

$$= -\frac{k}{m}\left(A\cos\sqrt{\frac{k}{m}}\,t + B\sin\sqrt{\frac{k}{m}}\,t\right)$$

$$- \sqrt{\frac{k}{m^3}}\int_0^t g(u)\sin\left[\sqrt{\frac{k}{m}}(t-u)\right]du + \frac{1}{m}g(t)$$

$$= -\frac{k}{m}\left\{A\cos\sqrt{\frac{k}{m}}\,t + B\sin\sqrt{\frac{k}{m}}\,t + \frac{1}{\sqrt{mk}}\int_0^t g(u)\sin\left[\sqrt{\frac{k}{m}}(t-u)\right]du\right\}$$

$$+ \frac{1}{m}g(t)$$

$$= -\frac{k}{m}x(t) + \frac{1}{m}g(t).$$

In other words, the function $x(t)$ satisfies the differential equation

$$m\frac{d^2x}{dt^2} = -kx(t) + g(t),$$

and the proof is complete.

This example illustrates that Leibniz's rule is essential to the manipulation of solutions of differential equations that are represented as definite integrals. Here is another example, but with a partial differential equation instead of an ordinary differential equation.

EXAMPLE 13.41

The one-dimensional heat conduction equation

$$\frac{\partial T}{\partial t} = k\frac{\partial^2 T}{\partial x^2},$$

where $k > 0$ is a constant, describes temperature in a rod (see Exercise 41 in Section 12.5). Show that the function defined by the definite integral

$$T(x, t) = \frac{2T_0}{\sqrt{\pi}} \int_0^{x/(2\sqrt{kt})} e^{-v^2}\, dv$$

satisfies the heat conduction equation.

SOLUTION   We use Leibniz's rule 13.77 to obtain $\partial T/\partial t$ and $\partial T/\partial x$:

$$\frac{\partial T}{\partial t} = \frac{2T_0}{\sqrt{\pi}}e^{-[x/(2\sqrt{kt})]^2}\left(\frac{-x}{4\sqrt{k}\, t^{3/2}}\right) = \frac{-xT_0}{2\sqrt{k\pi}\, t^{3/2}}e^{-x^2/(4kt)},$$

$$\frac{\partial T}{\partial x} = \frac{2T_0}{\sqrt{\pi}}e^{-[x/(2\sqrt{kt})]^2}\left(\frac{1}{2\sqrt{kt}}\right) = \frac{T_0}{\sqrt{k\pi t}}e^{-x^2/(4kt)}.$$

It follows that

$$\frac{\partial^2 T}{\partial x^2} = \frac{T_0}{\sqrt{k\pi}\, t}e^{-x^2/(4kt)}\left(\frac{-2x}{4kt}\right) = \frac{1}{k}\left[\frac{-xT_0}{2\sqrt{k\pi}\, t^{3/2}}e^{-x^2/(4kt)}\right] = \frac{1}{k}\frac{\partial T}{\partial t}.$$

EXERCISES 13.14

In Exercises 1–5 use Leibniz's rule to find the derivative of $F(x)$. Check your result by evaluating the integral and then differentiating.

1. $F(x) = \displaystyle\int_0^3 (x^2y^2 + 3xy)\, dy$

2. $F(x) = \displaystyle\int_1^x (x^2/y^2 + e^y)\, dy$

3. $F(x) = \displaystyle\int_{x-1}^{x^2} (x^3y + y^2 + 1)\, dy$

4. $F(x) = \displaystyle\int_{x^2}^{x^3-1} (x + y\ln y)\, dy$

5. $F(x) = \displaystyle\int_0^x \frac{y-x}{y+x}\, dy$

6. Find $F'(x)$ in Example 13.38 by first evaluating the definite integral.

7. Use the result of Example 13.39 to prove that

$$\int_0^1 \frac{x^p - x^q}{\ln x}\, dx = \ln\left(\frac{p+1}{q+1}\right),$$

provided that $p > -1$ and $q > -1$.

8. Find $F'(x)$ if $F(x) = \displaystyle\int_{\sin x}^{e^x} \sqrt{1 + y^3}\, dy$.

In Exercises 9–11 use Leibniz's rule to verify that the function $y(x)$ satisfies the differential equation.

* **9.** $y(x) = \dfrac{1}{x^2} \displaystyle\int_0^x t^2 f(t)\, dt; \quad x\dfrac{dy}{dx} + 2y = xf(x)$

* **10.** $y(x) = \dfrac{1}{2} \displaystyle\int_0^x f(t)(e^{x-t} - e^{t-x})\, dt; \quad \dfrac{d^2 y}{dx^2} - y = f(x)$

* **11.** $y(x) = \dfrac{1}{\sqrt{2}} \displaystyle\int_0^x e^{2(t-x)} \sin\left[\sqrt{2}(x-t)\right] f(t)\, dt;$

$\dfrac{d^2 y}{dx^2} + 4\dfrac{dy}{dx} + 6y = f(x)$

* **12.** Given that

$$\int_0^b \frac{1}{1+ax}\, dx = \frac{1}{a}\ln(1+ab),$$

find a formula for $\displaystyle\int_0^b \frac{x}{(1+ax)^2}\, dx$.

* **13.** Given that

$$\int \frac{1}{\sqrt{a^2 - x^2}}\, dx = \operatorname{Sin}^{-1}\left(\frac{x}{a}\right) + C,$$

find a formula for $\displaystyle\int \frac{1}{(a^2 - x^2)^{3/2}}\, dx$.

* **14.** Given that

$$\int \frac{1}{a^2 + x^2}\, dx = \frac{1}{a}\operatorname{Tan}^{-1}\left(\frac{x}{a}\right) + C,$$

find a formula for $\displaystyle\int \frac{1}{(a^2 + x^2)^3}\, dx$.

* **15.** Use the result that

$$\int_0^{\pi/2} \frac{1}{a^2 \cos^2 x + b^2 \sin^2 x}\, dx = \frac{\pi}{2|ab|}$$

to find a formula for $\displaystyle\int_0^{\pi/2} \frac{1}{(a^2 \cos^2 x + b^2 \sin^2 x)^2}\, dx$.

In Exercises 16–17 use Leibniz's rule to evaluate the integral.

* **16.** $\displaystyle\int_0^\pi \frac{\ln(1 + a\cos x)}{\cos x}\, dx$, where $|a| < 1$

* **17.** $\displaystyle\int_0^\infty \frac{\operatorname{Tan}^{-1}(ax)}{x(1+x^2)}\, dx$, where $a > 0$

* **18.** Verify that the function

$$T(x,\, t) = \int_0^{(1-x)/(2\sqrt{t})} e^{-v^2}\, dv + \int_0^{(1+x)/(2\sqrt{t})} e^{-v^2}\, dv$$

satisfies the partial differential equation

$$\frac{\partial T}{\partial t} = \frac{\partial^2 T}{\partial x^2}.$$

* **19.**   (a) What is the domain of the function

$$F(x) = \int_0^9 \ln(1 - x^2 y^2)\, dy?$$ What is $F(0)$?

   (b) Find $F'(x)$ by Leibniz's rule. What is $F'(0)$?

   (c) Show that the graph of the function $F(x)$ is concave downward for all x in its domain of definition.

** **20.** Laplace's equation for a function $u(r, \theta)$ in polar coordinates is

$$\frac{\partial^2 u}{\partial r^2} + \frac{1}{r}\frac{\partial u}{\partial r} + \frac{1}{r^2}\frac{\partial^2 u}{\partial \theta^2} = 0$$

(see Example 12.19 in Section 12.6). If the values of $u(r, \theta)$ are specified on the circle $r = R$ as $u(R, \phi)$, $-\pi < \phi \le \pi$, then Poisson's integral formula states that the value of $u(r, \theta)$ interior to this circle is defined by

$$u(r, \theta) = \frac{R^2 - r^2}{2\pi} \int_{-\pi}^\pi \frac{u(R, \phi)}{R^2 + r^2 - 2rR\cos(\theta - \phi)}\, d\phi.$$

Show that this function does indeed satisfy Laplace's equation.

** **21.** If the function $u(r, \theta)$ in Exercise 20 must satisfy the condition

$$\frac{\partial u(R, \theta)}{\partial r} = f(\theta), \quad -\pi < \theta \le \pi,$$

on the circle $r = R$ (rather than have its values prescribed on the circle), then values of $u(r, \theta)$ interior to the circle are given by

$$u(r, \theta) = C - \frac{R}{2\pi} \int_{-\pi}^\pi f(u) \ln\left[R^2 + r^2 - 2rR\cos(\theta - u)\right] du$$

where C is an arbitrary constant. This is called *Dini's integral*. Verify that it satisfies Laplace's equation in polar coordinates.

SUMMARY

The definite integral of a function $f(x)$ from $x = a$ to $x = b$ is a limit of a sum

$$\int_a^b f(x)\, dx = \lim_{\|\Delta x_i\| \to 0} \sum_{i=1}^n f(x_i^*)\, \Delta x_i.$$

In this chapter we extended this idea to define double integrals of functions $f(x, y)$ over regions in the xy-plane and triple integrals of functions $f(x, y, z)$ over regions of space. Each is once again the limit of a sum:

$$\iint_R f(x, y)\, dA = \lim_{\|\Delta A_i\| \to 0} \sum_{i=1}^{n} f(x_i^*, y_i^*)\, \Delta A_i,$$

$$\iiint_V f(x, y, z)\, dV = \lim_{\|\Delta V_i\| \to 0} \sum_{i=1}^{n} f(x_i^*, y_i^*, z_i^*)\, \Delta V_i.$$

To evaluate double integrals we use double iterated integrals in Cartesian or polar coordinates. Which is the more useful in a given problem depends on the shape of the region R and the form of the function $f(x, y)$. For instance, circles centred at the origin and straight lines through the origin are represented very simply in polar coordinates, and therefore a region R with these curves as boundaries immediately suggests the use of polar coordinates. On the other hand, curves that can be described in the form $y = f(x)$, where $f(x)$ is a polynomial, a rational function, or a transcendental function, often suggest using double iterated integrals in Cartesian coordinates. Each integration in a double iterated integral can be interpreted geometrically, and through these interpretations it is a simple matter to find appropriate limits for the integrals. In particular, for a double iterated integral in Cartesian coordinates, the inner integration is over the rectangles in a strip (horizontal or vertical), and the outer integration adds over all strips. In polar coordinates, the inner integral is inside either a wedge or a ring, and the outer integral adds over all wedges or rings.

To evaluate triple integrals we use triple iterated integrals in Cartesian, cylindrical, or spherical coordinates. Once again the limits on these integrals can be determined by interpreting the summations geometrically. For example, the first integration in a triple iterated integral in Cartesian coordinates is always over the boxes in a column, the second over the rectangles inside a strip, and the third over all strips.

We used double integrals to find plane areas, volumes of solids of revolution, centroids, moments of inertia, fluid forces, and areas of surfaces. We dealt with these same applications (with the exception of surface area) in Chapter 7 using the definite integral, but with some difficulty: Volumes required two methods, shells and washers; centroids required an averaging formula for the first moment of a rectangle that has its length perpendicular to the axis about which a moment is required; moments of inertia needed a "one-third cubed formula" for rectangles with lengths perpendicular to the axis about which the moment of inertia is required; and fluid forces required horizontal rectangles. On the other hand, double integrals eliminate these difficulties, but more important, provide a unified approach to all applications.

In Section 13.13 we showed how to transform double and triple integrals from one set of variables to another. Finally, in Section 13.14 we used double integrals to verify Leibniz's rule for differentiating definite integrals that depend on a parameter.

KEY TERMS

In reviewing this chapter, you should be able to define or discuss the following key terms:

Double integral	Double iterated integral
Areas	Volume of a solid of revolution
Fluid pressure	Centre of pressure
Centre of mass	Moments of inertia
Principal axes	Principal moments of inertia
Surface area	Double iterated integrals in polar coordinates
Triple integral	Triple iterated integral
Triple iterated integrals in cylindrical coordinates	Triple iterated integrals in spherical coordinates
General transformation for multiple integrals	Leibniz's rule

In Exercises 1–21 evaluate the integral over the region.

1. $\iint_R (2x + y) \, dA$, where R is bounded by $y = x$, $y = 0$, $x = 2$

2. $\iiint_V xyz \, dV$, where V is bounded by $y = z$, $x = 0$, $y = 3$, $x = 1$, $z = 0$

3. $\iint_R x^3 y^2 \, dA$, where R is bounded by $x = 1$, $x = -1$, $y = 2$, $y = -2$

4. $\iiint_V (x^2 - y^3) \, dV$, where V is bounded by $z = xy$, $z = 0$, $x = 1$, $y = 1$

5. $\iiint_V (x^2 - y^2) \, dV$, where V is the region in Exercise 4

6. $\iint_R y \, dA$, where R is bounded by $y = (x - 1)^2$, $y = x + 1$

7. $\iint_R xy^2 \, dA$, where R is bounded by $x = 2 - 2y^2$, $x = -y^2$

8. $\iint_R x^2 y \, dA$, where R is the region of Exercise 7

9. $\iiint_V (x^2 + y^2 + z^2) \, dV$, where V is bounded by $z = x$, $z = -x$, $y = 0$, $y = 1$, $z = 2$

10. $\iint_R (xy - x^2 y^2) \, dA$, where R is bounded by $y = 2x^2$, $y = 4 - 2x^2$

11. $\iint_R x \sin y \, dA$, where R is bounded by $x = \sqrt{1 - y}$, $x = 0$, $y = 0$

12. $\iiint_V (x + y + z) \, dV$, where V is bounded by $z = 1 - x^2 - y^2$, $z = 0$

13. $\iint_R x e^y \, dA$, where R is bounded by $x = 0$, $y = 5$, $y = 2x + 1$

14. $\iiint_V dV$, where V is bounded by $z^2 = x^2 + y^2$, $x^2 + y^2 = 4$

15. $\iint_R (x + y) \, dA$, where R is bounded by $x = -\sqrt{1 - y}$, $y = 1 - \sqrt{x}$, $y = x - 1$

16. $\iiint_V (x^2 + y^2 + z^2) \, dV$, where V is bounded by $z = \sqrt{1 - x^2 - y^2}$, $z = 0$

17. $\iint_R \dfrac{x}{x + y} \, dA$, where R is bounded by $y = x - 1$, $x = 2$, $y = 0$

18. $\iint_R (x^2 + y^2) \, dA$, where R is bounded by $x^2 + y^2 = 2x$

19. $\iiint_V \dfrac{x^2}{z^2} \, dV$, where V is bounded by $z = 1$, $z = \sqrt{4 - x^2 - y^2}$

20. $\iint_R \dfrac{1}{x^2 + y^2} \, dA$, where R is bounded by $y = x$, $x = 1$, $x = 2$, $y = 0$

21. $\iint_R (x^2 - y^2) \, dA$, where R is bounded by $y = x$, $y = x - 1$, $y = 5 - 2x$, $y = 14 - 2x$

22. If R represents the region of the xy-plane in the figure below, identify the double integrals that represent the following:
  (a) The area of R
  (b) The volumes of the solids of revolution when R is rotated about the lines $x = 2$ and $y = -4$
  (c) The first moments of area of R about the lines $x = 1$ and $y = -1$
  (d) The second moments of area of R about the lines $x = -1$ and $y = 4$
  (e) The total charge on R if it carries a charge per unit area $\sigma(x, y)$
  (f) The total mass if R is a plate with mass per unit area $\rho(x, y)$
  (g) The probability of an electron being in R if the probability of the electron being in unit area at point (x, y) is $P(x, y)$

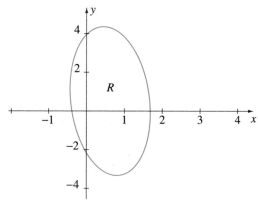

23. Find the volume bounded by the surfaces $z = 0$, $z = 1 - 2e^{-x^2 - y^2}$.

24. Find the area bounded by the curve $x^2(4 - x^2) = y^2$.

25. Find the centroid of the region bounded by the curves $4x = 4 - y^2$, $x = \sqrt{8 - 2y^2}$.

26. Find the volume bounded by the surfaces $y + z = 2$, $y - z = 2$, $y = x^2 + 1$.

27. If the viewing glass on a deep-sea diver's helmet is circular with diameter 10 cm, what is the force on the glass when the centre of the glass is 50 m below the surface?

28. Find the area of that part of $z = 1 - 4x^2 - 4y^2$ above the xy-plane.

29. Find the area bounded by the curve $x^4(4 - x^2) = y^2$.

30. Find the volumes of the solids of revolution when the region in Exercise 24 is rotated about the x- and y-axes.

31. Find the moment of inertia about the z-axis of the solid bounded by the surfaces $z = \sqrt{x^2 + y^2}$, $z = 2 - \sqrt{x^2 + y^2}$ if its density ρ is constant.

32. Find the second moment of area about the y-axis of the region in the first quadrant bounded by the curve $x^3 + y^3 = 1$.

33. Find the centre of mass of a uniform solid bounded by $z = 1 + x^2 + y^2$, $x^2 + y^2 = 1$, $z = 0$.

34. Find the average value of the function $f(x, y) = x^2 + y^2$ over the region bounded by the curve $x^2 + y^2 = 4$.

35. Find the force due to water pressure on each side of the vertical parallelogram in the figure below. All measurements are in metres.

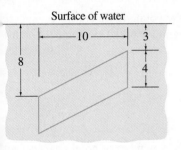

Surface of water

36. Find the moment of inertia about the z-axis of a uniform solid bounded by the surfaces $z = 0$, $z = 1$, $x^2 + y^2 = 1 + z^2$.

37. Find the average value of the function $f(x, y, z) = x + y + z$ over the region bounded by the surfaces $-x + 2y + 2z = 4$, $x + y + z = 2$, $y = 0$, $z = 0$.

38. Find the centroid of the bifolium $r = \sin\theta \, \cos^2\theta$.

39. Verify that

$$y(x) = e^{-3x}(C_1 \cos x + C_2 \sin x)$$

$$+ \int_0^x f(t)e^{3(t-x)} \sin(x - t) \, dt$$

is a solution of the differential equation $\dfrac{d^2y}{dx^2} + 6\dfrac{dy}{dx} + 10y = f(x)$ for any constants C_1 and C_2.

40. Find the centre of mass of the uniform solid in the first octant common to the cylinders $x^2 + z^2 = 1$, $y^2 + z^2 = 1$.

41. Find the area of that part of $z = \ln(x^2 + y^2)$ between the cylinders $x^2 + y^2 = 1$ and $x^2 + y^2 = 4$.

42. Find the volume bounded by the surface $\sqrt{x^2 + z^2} = y(2 - y)$.

14 | Vector Calculus

Application Preview

It is well known that the electric field $\mathbf{E}$ at a point (x, y, z) due to a charge q at the origin is

$$\mathbf{E} = \frac{q}{4\pi\epsilon_0|\mathbf{r}|^3}\mathbf{r},$$

where $\mathbf{r} = x\hat{\mathbf{i}} + y\hat{\mathbf{j}} + z\hat{\mathbf{k}}$ is the vector from the origin to the point. If we consider a sphere of radius R centred at the origin, then the product of the magnitude of the electric field on the sphere and the area of the sphere is

$$\frac{q}{4\pi\epsilon_0 R^2}(4\pi R^2) = \frac{q}{\epsilon_0},$$

and this is independent of the radius of the sphere.

THE PROBLEM   If a surface (not just a sphere) contains a number of point charges (not just one at the origin), what is the relationship between values of the electric field on the surface due to the charges, and the charges themselves? (See Example 14.25 on page page 1037 for the solution.)

In Sections 11.9 and 11.10 we considered vectors whose components are functions of a single variable. In particular, if an object moves along a curve C defined parametrically by $x = x(t)$, $y = y(t)$, $z = z(t)$, $t \geq 0$, where t is time, then the components of the position, velocity, and acceleration vectors are functions of time:

$$\mathbf{r} = \mathbf{r}(t) = x(t)\hat{\mathbf{i}} + y(t)\hat{\mathbf{j}} + z(t)\hat{\mathbf{k}},$$

$$\mathbf{v} = \frac{d\mathbf{r}}{dt} = \frac{dx}{dt}\hat{\mathbf{i}} + \frac{dy}{dt}\hat{\mathbf{j}} + \frac{dz}{dt}\hat{\mathbf{k}},$$

$$\mathbf{a} = \frac{d\mathbf{v}}{dt} = \frac{d^2x}{dt^2}\hat{\mathbf{i}} + \frac{d^2y}{dt^2}\hat{\mathbf{j}} + \frac{d^2z}{dt^2}\hat{\mathbf{k}}.$$

FIGURE 14.1   Force on charge Q at position (x, y, z) due to charge q at the origin

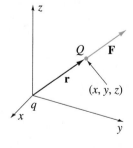

If the object has constant mass m and is subjected to a force $\mathbf{F}$, which is given as a function of time t, $\mathbf{F} = \mathbf{F}(t)$, then Newton's second law expresses the acceleration of the object as $\mathbf{a} = \mathbf{F}(t)/m$. This equation can then be integrated to yield the velocity and position of the object as functions of time. Unfortunately, what often happens is that we do not know $\mathbf{F}$ as a function of time. Instead, we know that if the object were at such and such a position, then the force on it would be such and such (i.e., we know $\mathbf{F}$ as a function of position). For example, suppose a positive charge q is placed at the origin in space (Figure 14.1), and a second positive charge Q is placed at position (x, y, z). According to Coulomb's law, the force on Q due to q is

$$\mathbf{F} = \frac{qQ}{4\pi\epsilon_0|\mathbf{r}|^3}\mathbf{r} = \frac{qQ}{4\pi\epsilon_0(x^2 + y^2 + z^2)^{3/2}}(x\hat{\mathbf{i}} + y\hat{\mathbf{j}} + z\hat{\mathbf{k}}).$$

If we allow Q to move under the influence of this force, we will not know $\mathbf{F}$ as a function of time, but rather, as a function of position. This makes Newton's second law much more difficult to deal with, but it is in fact the normal situation. Most forces are represented as a function of position rather than time. Besides electrostatic forces, consider, for instance, spring forces, gravitational forces, and fluid forces — all of these are functions of position. Forces that are functions of position are examples of vectors that are functions of position. In this chapter we study vectors that are functions of position. In particular, we differentiate them and integrate them along curves and over surfaces.

14.1 Vector Fields

FIGURE 14.2 Interior, exterior, and boundary points for an open set

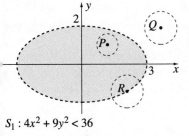

$S_1 : 4x^2 + 9y^2 < 36$

FIGURE 14.3 Interior, exterior, and boundary points for a closed set

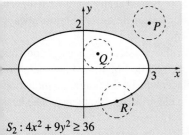

$S_2 : 4x^2 + 9y^2 \geq 36$

Domains for functions $f(x)$ of one variable are open, closed, half-open, or half-closed intervals on the x-axis. Domains for functions $f(x, y)$ of two variables are sets of points in the xy-plane, and domains for functions $f(x, y, z)$ are sets of points in xyz-space. In order to state definitions and theorems for multivariable functions in this chapter, we require corresponding definitions of open and closed sets of points. We define them for the xy-plane; analogous definitions for three-dimensional space are given in Exercise 10.

Consider a set S of points in the xy-plane. A point P in the plane is called an **interior point** of S if there exists a circle centred at P that contains only points of S. A point Q is called an **exterior point** of S if there exists a circle centred at Q that contains no point of S. A point R is called a **boundary point** of S if every circle with centre R contains at least one point in S and at least one point not in S. For example, consider the set of points $S_1 : 4x^2 + 9y^2 < 36$ (Figure 14.2). The fact that we have dotted the ellipse indicates that these points are not in S_1. Every point inside the ellipse is an interior point of S_1, every point outside the ellipse is an exterior point, and every point on the ellipse is a boundary point. For the set $S_2 : 4x^2 + 9y^2 \geq 36$ (Figure 14.3), every point outside the ellipse is interior to S_2, every point inside the ellipse is exterior to S_2, and every point on the ellipse is a boundary point.

A set of points S is said to be an **open set** if all points in S are interior points. Alternatively, a set is open if it contains none of its boundary points. A set is said to be a **closed set** if it contains all of its boundary points. Set S_1 above is open; set S_2 is closed. If to S_1 we add the points on the upper half of the ellipse (Figure 14.4), this set, call it S_3, is neither open nor closed. It contains some of its boundary points but not all of them.

A set S is said to be a **connected set** if every pair of points in S can be joined by a piecewise-smooth curve lying entirely within S. Sets S_1, S_2, and S_3 are all connected. Set S_4 in Figure 14.5 is also connected. Set S_5 in Figure 14.6 is not connected; it consists of two disjoint pieces.

A **domain** is an open, connected set. *Domain* is perhaps a poor choice of word, as it might be confused with *domain of a function*, but it has become accepted terminology. Context always makes it clear which interpretation is intended. Sets S_1 and S_4 are domains; sets S_2, S_3, and S_5 are not.

A domain is said to be a **simply connected domain** if every closed curve in the domain contains in its interior only points of the domain. In essence, a simply connected domain has

FIGURE 14.4 A set that is neither open nor closed

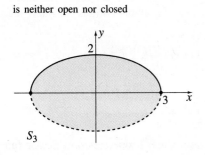

S_3

FIGURE 14.5 A connected set

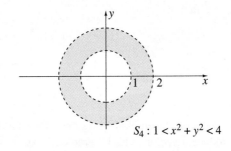

$S_4 : 1 < x^2 + y^2 < 4$

FIGURE 14.6 A set that is not connected

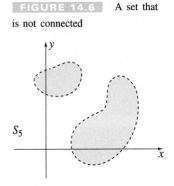

S_5

FIGURE 14.7   A domain that is not simply connected

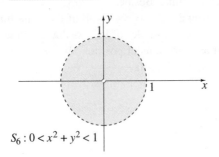

$S_6 : 0 < x^2 + y^2 < 1$

no holes. Domain S_1 is simply connected, S_4 is not. Set $S_6 : 0 < x^2 + y^2 < 1$ (Figure 14.7) is a domain, but it is not simply connected.

Analogous definitions for sets of points in space are given in Exercise 10. One must be somewhat more careful here as there are different kinds of holes. See Exercises 13 and 15.

When it is not necessary to indicate the particular characteristics of a set of points, we often use the word *region*.

Many vectors are functions of position. We call such vectors **vector fields**. To be precise, we say that **F** is a vector field in a region D if **F** assigns a vector to each point in D. If D is a region of space, then **F** assigns a vector $\mathbf{F}(x, y, z)$ to each point in D. If P, Q, and R are the components of $\mathbf{F}(x, y, z)$, then each of these is a function of x, y, and z:

$$\mathbf{F} = \mathbf{F}(x, y, z) = P(x, y, z)\hat{\mathbf{i}} + Q(x, y, z)\hat{\mathbf{j}} + R(x, y, z)\hat{\mathbf{k}}. \tag{14.1}$$

If D is a region of the xy-plane, then

$$\mathbf{F} = \mathbf{F}(x, y) = P(x, y)\hat{\mathbf{i}} + Q(x, y)\hat{\mathbf{j}}. \tag{14.2}$$

In Chapter 11 we stressed the fact that the tail of a vector could be placed at any point whatsoever. What was important was relative positions of tip and tail. For vector fields, we almost always place the tail of the vector associated with a point at that point.

Vector fields are essential to the study of most areas in the physical sciences. Moments $(\mathbf{M} = \mathbf{r} \times \mathbf{F})$ in mechanics (see Section 11.7) depend on the position of **F**, and are therefore vector fields. The electric field intensity **E**, the electric displacement **D**, the magnetic induction **B**, and the current density **J** are all important vector fields in electromagnetic theory. The heat flux vector **q** is the basis for the study of heat conduction.

We have also encountered vector fields that are of geometric importance. For example, the gradient of a scalar function $f(x, y, z)$ is a vector field,

$$\text{grad}\, f = \nabla f = \frac{\partial f}{\partial x}\hat{\mathbf{i}} + \frac{\partial f}{\partial y}\hat{\mathbf{j}} + \frac{\partial f}{\partial z}\hat{\mathbf{k}}. \tag{14.3}$$

It assigns the vector ∇f to each point (x, y, z) in some region of space. We have seen that ∇f points in the direction in which $f(x, y, z)$ increases most rapidly, and its magnitude $|\nabla f|$ is the rate of increase in that direction. In addition, we know that if $f(x, y, z) = c$ is the equation of a surface that passes through a point (x, y, z), then ∇f at that point is normal to the surface.

Often, we write

$$\nabla f = \left(\frac{\partial}{\partial x}\hat{\mathbf{i}} + \frac{\partial}{\partial y}\hat{\mathbf{j}} + \frac{\partial}{\partial z}\hat{\mathbf{k}} \right) f$$

and regard

$$\nabla = \frac{\partial}{\partial x}\hat{\mathbf{i}} + \frac{\partial}{\partial y}\hat{\mathbf{j}} + \frac{\partial}{\partial z}\hat{\mathbf{k}}$$

as a vector differential operator, called the **del operator**. It operates on a scalar function to produce a vector field, its gradient. As an operator, ∇ should never stand alone, but should always be followed by something on which to operate. Because of this, ∇ should not itself be considered a vector, in spite of the fact that it has the form of a vector. It is a differential operator, and must therefore operate on something. In the remainder of this section we use the del operator to define two extremely useful operations on vector fields: the *divergence* and the *curl*.

DEFINITION 14.1

If $\mathbf{F}(x, y, z) = P(x, y, z)\hat{\mathbf{i}} + Q(x, y, z)\hat{\mathbf{j}} + R(x, y, z)\hat{\mathbf{k}}$ is a vector field in a region D, then the **divergence** of $\mathbf{F}$ is a scalar field in D defined by

$$\text{div } \mathbf{F} = \nabla \cdot \mathbf{F} = \frac{\partial P}{\partial x} + \frac{\partial Q}{\partial y} + \frac{\partial R}{\partial z}, \qquad (14.4)$$

provided that the partial derivatives exist at each point in D.

It is clear why we use the notation $\nabla \cdot \mathbf{F}$, in spite of the fact that ∇ is not a vector in the true sense of the word.

EXAMPLE 14.1

Calculate $\nabla \cdot \mathbf{F}$ if

$$\text{(a) } \mathbf{F} = 2xy\hat{\mathbf{i}} + z\hat{\mathbf{j}} + x^2 \cos(yz)\hat{\mathbf{k}} \quad \text{(b) } \mathbf{F} = \frac{qQ}{4\pi\epsilon_0|\mathbf{r}|^3}\mathbf{r},$$

where $\mathbf{r} = x\hat{\mathbf{i}} + y\hat{\mathbf{j}} + z\hat{\mathbf{k}}$.

SOLUTION

(a) $\nabla \cdot \mathbf{F} = \dfrac{\partial}{\partial x}(2xy) + \dfrac{\partial}{\partial y}(z) + \dfrac{\partial}{\partial z}[x^2 \cos(yz)] = 2y - x^2 y \sin(yz).$

(b) For the derivative of the x-component of $\mathbf{F}$ with respect to x, we calculate

$$\frac{\partial}{\partial x}\left[\frac{qQx}{4\pi\epsilon_0(x^2 + y^2 + z^2)^{3/2}}\right] = \frac{qQ}{4\pi\epsilon_0}\left[\frac{1}{(x^2 + y^2 + z^2)^{3/2}} - \frac{3x^2}{(x^2 + y^2 + z^2)^{5/2}}\right]$$

$$= \frac{qQ}{4\pi\epsilon_0}\left[\frac{-2x^2 + y^2 + z^2}{(x^2 + y^2 + z^2)^{5/2}}\right].$$

With similar results for the remaining two derivatives, we obtain

$$\nabla \cdot \mathbf{F} = \frac{qQ}{4\pi\epsilon_0}\left[\frac{-2x^2 + y^2 + z^2}{(x^2 + y^2 + z^2)^{5/2}} + \frac{x^2 - 2y^2 + z^2}{(x^2 + y^2 + z^2)^{5/2}} + \frac{x^2 + y^2 - 2z^2}{(x^2 + y^2 + z^2)^{5/2}}\right]$$

$$= 0.$$

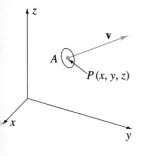

FIGURE 14.8 Vector $\mathbf{v}$ represents direction of gas flow at point P. A is a unit area at P perpendicular to $\mathbf{v}$

Any physical interpretation of $\nabla \cdot \mathbf{F}$ depends on the interpretation of $\mathbf{F}$. The following discussion describes the interpretation of the divergence of a certain vector field in the theory of fluid flow. If a gas flows through a region D of space, then it flows with some velocity $\mathbf{v}$ past point $P(x, y, z)$ in D at time t (Figure 14.8). Consider a unit area A around P perpendicular to $\mathbf{v}$. If at time t the density of gas at P is ρ, then the vector $\rho\mathbf{v}$ represents the mass of gas flowing through A per unit time. At each point P in D, the direction of $\rho\mathbf{v}$ tells us the direction of gas flow, and its length indicates the mass of gas flowing in that direction. For changing conditions, each of ρ and $\mathbf{v}$ depend not only on position (x, y, z) but also on time t:

$$\rho\mathbf{v} = \rho(x, y, z, t)\mathbf{v}(x, y, z, t).$$

The vector $\rho\mathbf{v}$, then, is a vector field that also depends on time.

In fluid dynamics it is shown that at each point in D, $\rho\mathbf{v}$ must satisfy

$$\nabla \cdot (\rho\mathbf{v}) = -\frac{\partial\rho}{\partial t},$$

called the **equation of continuity**. It is this equation that gives us an interpretation of the divergence of $\rho\mathbf{v}$. Density is the mass per unit volume of gas. If $\partial\rho/\partial t$ is positive, then density is increasing. This means that more mass must be entering unit volume than leaving it. Similarly, if $\partial\rho/\partial t$ is negative, more mass is leaving than entering. Since $\nabla \cdot (\rho\mathbf{v})$ is the negative of $\partial\rho/\partial t$, it follows that $\nabla \cdot (\rho\mathbf{v})$ must be a measure of how much more gas is leaving unit volume than entering. We can see, then, that the word *divergence* is appropriately chosen for this application.

The curl of a vector field is defined as follows.

DEFINITION 14.2

If $\mathbf{F}(x, y, z) = P(x, y, z)\hat{\mathbf{i}} + Q(x, y, z)\hat{\mathbf{j}} + R(x, y, z)\hat{\mathbf{k}}$ is a vector field in a region D, then the **curl** of $\mathbf{F}$ is a vector function in D defined by

$$\text{curl } \mathbf{F} = \nabla \times \mathbf{F} = \begin{vmatrix} \hat{\mathbf{i}} & \hat{\mathbf{j}} & \hat{\mathbf{k}} \\ \dfrac{\partial}{\partial x} & \dfrac{\partial}{\partial y} & \dfrac{\partial}{\partial z} \\ P & Q & R \end{vmatrix}$$

$$= \left(\frac{\partial R}{\partial y} - \frac{\partial Q}{\partial z}\right)\hat{\mathbf{i}} + \left(\frac{\partial P}{\partial z} - \frac{\partial R}{\partial x}\right)\hat{\mathbf{j}} + \left(\frac{\partial Q}{\partial x} - \frac{\partial P}{\partial y}\right)\hat{\mathbf{k}}, \quad (14.5)$$

provided that the partial derivatives exist at each point in D.

In representing curl $\mathbf{F}$ in the form of a determinant we agree to expand the determinant along the first row.

EXAMPLE 14.2

Calculate the curls of the vector fields in Example 14.1.

SOLUTION

(a)
$$\nabla \times \mathbf{F} = \begin{vmatrix} \hat{\mathbf{i}} & \hat{\mathbf{j}} & \hat{\mathbf{k}} \\ \dfrac{\partial}{\partial x} & \dfrac{\partial}{\partial y} & \dfrac{\partial}{\partial z} \\ 2xy & z & x^2\cos(yz) \end{vmatrix}$$

$$= [-x^2 z \sin(yz) - 1]\hat{\mathbf{i}} + [-2x\cos(yz)]\hat{\mathbf{j}} + (-2x)\hat{\mathbf{k}}$$

(b)
$$\nabla \times \mathbf{F} = \frac{qQ}{4\pi\epsilon_0} \begin{vmatrix} \hat{\mathbf{i}} & \hat{\mathbf{j}} & \hat{\mathbf{k}} \\ \dfrac{\partial}{\partial x} & \dfrac{\partial}{\partial y} & \dfrac{\partial}{\partial z} \\ \dfrac{x}{|\mathbf{r}|^3} & \dfrac{y}{|\mathbf{r}|^3} & \dfrac{z}{|\mathbf{r}|^3} \end{vmatrix}$$

The x-component of $\nabla \times \mathbf{F}$ is $qQ/(4\pi\epsilon_0)$ multiplied by

$$\frac{\partial}{\partial y}\left(\frac{z}{|\mathbf{r}|^3}\right) - \frac{\partial}{\partial z}\left(\frac{y}{|\mathbf{r}|^3}\right) = \frac{\partial}{\partial y}\left[\frac{z}{(x^2 + y^2 + z^2)^{3/2}}\right]$$

$$- \frac{\partial}{\partial z}\left[\frac{y}{(x^2 + y^2 + z^2)^{3/2}}\right]$$

$$= \frac{-3yz}{(x^2 + y^2 + z^2)^{5/2}} + \frac{3yz}{(x^2 + y^2 + z^2)^{5/2}} = 0.$$

Similar results for the y- and z-components give $\nabla \times \mathbf{F} = \mathbf{0}$.

If a vector $\rho\mathbf{v}$ is defined as above for gas flow through a region D, then the curl of $\rho\mathbf{v}$ describes the tendency for the motion of the gas to be circular rather than flowing in a straight line. This suggests why the term *curl* is used. With this interpretation, the following definition seems reasonable.

DEFINITION 14.3

A vector field $\mathbf{F}$ is said to be **irrotational** in a region D if in D

$$\nabla \times \mathbf{F} = \mathbf{0}. \tag{14.6}$$

Applications of divergence and curl extend far beyond the topic of fluid dynamics. Both concepts are indispensable in many areas of applied mathematics, such as electromagnetism, continuum mechanics, and heat conduction, to name a few.

The del operator ∇ operates on a scalar field to produce the gradient of the scalar field and on a vector field to give the divergence and the curl of the vector field. The following list of properties for the del operator is straightforward to verify (see Exercise 43).

If f and g are scalar fields that have first partial derivatives in a region D, and if $\mathbf{F}$ and $\mathbf{G}$ are vector fields in D with components that have first partial derivatives, then

$$\nabla(f + g) = \nabla f + \nabla g, \tag{14.7}$$

$$\nabla \cdot (\mathbf{F} + \mathbf{G}) = \nabla \cdot \mathbf{F} + \nabla \cdot \mathbf{G}, \tag{14.8}$$

$$\nabla \times (\mathbf{F} + \mathbf{G}) = \nabla \times \mathbf{F} + \nabla \times \mathbf{G}, \tag{14.9}$$

$$\nabla(fg) = f\nabla g + g\nabla f, \tag{14.10}$$

$$\nabla \cdot (f\mathbf{F}) = \nabla f \cdot \mathbf{F} + f\nabla \cdot \mathbf{F}, \tag{14.11}$$

$$\nabla \times (f\mathbf{F}) = \nabla f \times \mathbf{F} + f\nabla \times \mathbf{F}, \tag{14.12}$$

$$\nabla \cdot (\mathbf{F} \times \mathbf{G}) = \mathbf{G} \cdot (\nabla \times \mathbf{F}) - \mathbf{F} \cdot (\nabla \times \mathbf{G}), \tag{14.13}$$

$$\nabla \times (\nabla f) = \mathbf{0}, \tag{14.14}$$

$$\nabla \cdot (\nabla \times \mathbf{F}) = 0. \tag{14.15}$$

For properties 14.14 and 14.15, we assume that f and $\mathbf{F}$ have continuous second-order partial derivatives in D. A typical way to verify these identities is to reduce each side of the identity to the same quantity. For example, to verify 14.13, we set $\mathbf{F} = P\hat{\mathbf{i}} + Q\hat{\mathbf{j}} + R\hat{\mathbf{k}}$ and $\mathbf{G} = L\hat{\mathbf{i}} + M\hat{\mathbf{j}} + N\hat{\mathbf{k}}$. Then

$$\mathbf{F} \times \mathbf{G} = (QN - RM)\hat{\mathbf{i}} + (RL - PN)\hat{\mathbf{j}} + (PM - QL)\hat{\mathbf{k}};$$

thus,

$$\nabla \cdot (\mathbf{F} \times \mathbf{G}) = \frac{\partial}{\partial x}(QN - RM) + \frac{\partial}{\partial y}(RL - PN) + \frac{\partial}{\partial z}(PM - QL)$$

$$= Q\frac{\partial N}{\partial x} + N\frac{\partial Q}{\partial x} - R\frac{\partial M}{\partial x} - M\frac{\partial R}{\partial x} + R\frac{\partial L}{\partial y} + L\frac{\partial R}{\partial y} - P\frac{\partial N}{\partial y}$$

$$- N\frac{\partial P}{\partial y} + P\frac{\partial M}{\partial z} + M\frac{\partial P}{\partial z} - Q\frac{\partial L}{\partial z} - L\frac{\partial Q}{\partial z}$$

$$= L\left(\frac{\partial R}{\partial y} - \frac{\partial Q}{\partial z}\right) + M\left(\frac{\partial P}{\partial z} - \frac{\partial R}{\partial x}\right) + N\left(\frac{\partial Q}{\partial x} - \frac{\partial P}{\partial y}\right)$$

$$+ P\left(\frac{\partial M}{\partial z} - \frac{\partial N}{\partial y}\right) + Q\left(\frac{\partial N}{\partial x} - \frac{\partial L}{\partial z}\right) + R\left(\frac{\partial L}{\partial y} - \frac{\partial M}{\partial x}\right).$$

On the other hand,

$$\mathbf{G}\cdot(\nabla \times \mathbf{F}) - \mathbf{F} \cdot (\nabla \times \mathbf{G})$$

$$= \mathbf{G} \cdot \begin{vmatrix} \hat{\mathbf{i}} & \hat{\mathbf{j}} & \hat{\mathbf{k}} \\ \partial/\partial x & \partial/\partial y & \partial/\partial z \\ P & Q & R \end{vmatrix} - \mathbf{F} \cdot \begin{vmatrix} \hat{\mathbf{i}} & \hat{\mathbf{j}} & \hat{\mathbf{k}} \\ \partial/\partial x & \partial/\partial y & \partial/\partial z \\ L & M & N \end{vmatrix}$$

$$= (L, M, N) \cdot \left(\frac{\partial R}{\partial y} - \frac{\partial Q}{\partial z}, \frac{\partial P}{\partial z} - \frac{\partial R}{\partial x}, \frac{\partial Q}{\partial x} - \frac{\partial P}{\partial y}\right)$$

$$- (P, Q, R) \cdot \left(\frac{\partial N}{\partial y} - \frac{\partial M}{\partial z}, \frac{\partial L}{\partial z} - \frac{\partial N}{\partial x}, \frac{\partial M}{\partial x} - \frac{\partial L}{\partial y}\right)$$

$$= L\left(\frac{\partial R}{\partial y} - \frac{\partial Q}{\partial z}\right) + M\left(\frac{\partial P}{\partial z} - \frac{\partial R}{\partial x}\right) + N\left(\frac{\partial Q}{\partial x} - \frac{\partial P}{\partial y}\right)$$

$$- P\left(\frac{\partial N}{\partial y} - \frac{\partial M}{\partial z}\right) - Q\left(\frac{\partial L}{\partial z} - \frac{\partial N}{\partial x}\right) - R\left(\frac{\partial M}{\partial x} - \frac{\partial L}{\partial y}\right),$$

and this is the same expression as for $\nabla \cdot (\mathbf{F} \times \mathbf{G})$.

Given a scalar function $f(x, y, z)$ it is straightforward to calculate its gradient ∇f. Conversely, given the gradient of a function ∇f, it is possible to find the function $f(x, y, z)$. For example, if the vector field

$$\mathbf{F} = (3x^2yz + z^2)\hat{\mathbf{i}} + (x^3z + 2y)\hat{\mathbf{j}} + (x^3y + 2xz + 1)\hat{\mathbf{k}}$$

is known to be the gradient of some function $f(x, y, z)$, then

$$\frac{\partial f}{\partial x}\hat{\mathbf{i}} + \frac{\partial f}{\partial y}\hat{\mathbf{j}} + \frac{\partial f}{\partial z}\hat{\mathbf{k}} = (3x^2yz + z^2)\hat{\mathbf{i}} + (x^3z + 2y)\hat{\mathbf{j}} + (x^3y + 2xz + 1)\hat{\mathbf{k}}.$$

Since two vectors are equal if and only if they have identical components, we can say that

$$\frac{\partial f}{\partial x} = 3x^2yz + z^2, \quad \frac{\partial f}{\partial y} = x^3z + 2y, \quad \frac{\partial f}{\partial z} = x^3y + 2xz + 1.$$

Integration of the first of these with respect to x, holding y and z constant, implies that $f(x, y, z)$ must be of the form

$$f(x, y, z) = x^3yz + xz^2 + v(y, z)$$

for some function $v(y, z)$. To determine $v(y, z)$, we substitute this $f(x, y, z)$ into the second equation,

$$x^3z + \frac{\partial v}{\partial y} = x^3z + 2y \quad \Longrightarrow \quad \frac{\partial v}{\partial y} = 2y.$$

Consequently, $v(y, z) = y^2 + w(z)$ for some $w(z)$, and therefore $f(x, y, z) = x^3yz + xz^2 + y^2 + w(z)$. We now know both the x- and y-dependence of $f(x, y, z)$. To find $w(z)$ we substitute into the equation for $\partial f/\partial z$:

$$x^3y + 2xz + \frac{dw}{dz} = x^3y + 2xz + 1 \quad \implies \quad \frac{dw}{dz} = 1,$$

from which we have $w(z) = z + C$, where C is a constant. Thus,

$$f(x, y, z) = x^3yz + xz^2 + y^2 + z + C,$$

and this represents all functions that have a gradient equal to the given vector $\mathbf{F}$.

A more difficult question is to determine whether a given vector field $\mathbf{F}(x, y, z)$ is the gradient of some scalar function $f(x, y, z)$. In many examples, we can say that if the procedure described above fails, then the answer is no. For instance, if $\mathbf{F} = x^2y\hat{\mathbf{i}} + xy\hat{\mathbf{j}} + z\hat{\mathbf{k}}$, and we attempt to find a function $f(x, y, z)$ so that $\nabla f = \mathbf{F}$, then

$$\frac{\partial f}{\partial x} = x^2y, \quad \frac{\partial f}{\partial y} = xy, \quad \frac{\partial f}{\partial z} = z.$$

The first implies that

$$f(x, y, z) = \frac{x^3y}{3} + v(y, z),$$

and when this is substituted into the second, we get

$$\frac{x^3}{3} + \frac{\partial v}{\partial y} = xy \quad \implies \quad \frac{\partial v}{\partial y} = xy - \frac{x^3}{3}.$$

But this is an impossible situation since v is to be a function of y and z only. How then could its derivative depend on x? Although this type of argument will suffice in most examples, it is really not a satisfactory mathematical answer. The following theorem gives a test by which to determine whether a given vector function is the gradient of some scalar function.

THEOREM 14.1

Suppose the components $P(x, y, z)$, $Q(x, y, z)$, and $R(x, y, z)$ of $\mathbf{F} = P\hat{\mathbf{i}} + Q\hat{\mathbf{j}} + R\hat{\mathbf{k}}$ have continuous first partial derivatives in a domain D. If there exists a function $f(x, y, z)$ defined in D such that $\nabla f = \mathbf{F}$, then $\nabla \times \mathbf{F} = \mathbf{0}$. Conversely, if D is simply connected, and $\nabla \times \mathbf{F} = \mathbf{0}$ in D, then there exists a function $f(x, y, z)$ such that $\nabla f = \mathbf{F}$ in D.

It is obvious that if $\mathbf{F} = \nabla f$, then $\nabla \times \mathbf{F} = \mathbf{0}$, for this is the result of equation 14.14. To prove the converse result requires Stokes's theorem from Section 14.10, and a proof is therefore delayed until that time. Notice that in the converse result the domain (open, connected set) must be simply connected. This is our first encounter with a situation where the nature of a region is important to the result.

In the special case in which $\mathbf{F}$ is a vector field in the xy-plane, the equation $\nabla \times \mathbf{F} = \mathbf{0}$ is still the condition for existence of a function $f(x, y)$ such that $\nabla f = \mathbf{F} = P\hat{\mathbf{i}} + Q\hat{\mathbf{j}}$, but the condition reduces to

$$\mathbf{0} = \begin{vmatrix} \hat{\mathbf{i}} & \hat{\mathbf{j}} & \hat{\mathbf{k}} \\ \partial/\partial x & \partial/\partial y & \partial/\partial z \\ P & Q & 0 \end{vmatrix} = \left(\frac{\partial Q}{\partial x} - \frac{\partial P}{\partial y} \right) \hat{\mathbf{k}}$$

or

$$\frac{\partial Q}{\partial x} = \frac{\partial P}{\partial y}. \tag{14.16}$$

EXAMPLE 14.3

Find, if possible, a function $f(x, y)$ such that

$$\nabla f = \left(\frac{x^3 - 2y^2}{x^3 y} \right) \hat{\mathbf{i}} + \left(\frac{y^2 - x^3}{x^2 y^2} \right) \hat{\mathbf{j}}.$$

SOLUTION   We first note that

$$\frac{\partial}{\partial x} \left(\frac{y^2 - x^3}{x^2 y^2} \right) = \frac{\partial}{\partial x} \left(\frac{1}{x^2} - \frac{x}{y^2} \right) = \frac{-2}{x^3} - \frac{1}{y^2}$$

and

$$\frac{\partial}{\partial y} \left(\frac{x^3 - 2y^2}{x^3 y} \right) = \frac{\partial}{\partial y} \left(\frac{1}{y} - \frac{2y}{x^3} \right) = \frac{-1}{y^2} + \frac{-2}{x^3}.$$

Since the components of **F** are undefined whenever $x = 0$ or $y = 0$, we can state that in any simply connected domain that does not contain points on either of the axes, there is a function $f(x, y)$ such that $\nabla f = \mathbf{F}$. To find $f(x, y)$, we set

$$\frac{\partial f}{\partial x} = \frac{1}{y} - \frac{2y}{x^3}, \quad \frac{\partial f}{\partial y} = \frac{1}{x^2} - \frac{x}{y^2}.$$

From the first equation, we have

$$f(x, y) = \frac{x}{y} + \frac{y}{x^2} + v(y),$$

which, substituted into the second equation, gives us

$$-\frac{x}{y^2} + \frac{1}{x^2} + \frac{dv}{dy} = \frac{1}{x^2} - \frac{x}{y^2} \quad \Longrightarrow \quad \frac{dv}{dy} = 0.$$

Thus, $v(y) = C$, and $f(x, y) = \dfrac{x}{y} + \dfrac{y}{x^2} + C$.

EXERCISES 14.1

In Exercises 1–9 determine whether the set of points in the xy-plane is open, closed, connected, a domain, and/or a simply connected domain.

1. $x^2 + (y + 1)^2 < 4$

2. $x^2 + (y - 3)^2 \leq 4$

3. $0 < x^2 + (y - 1)^2 < 16$

4. $1 < (x - 4)^2 + (y + 1)^2 \leq 9$

5. $x > 3$

6. $y \leq -2$

7. $2(x - 1)^2 - (y + 2)^2 < 16$

8. All points satisfying $x^2 + y^2 < 1$ or $(x - 2)^2 + y^2 < 1$

9. $4(x + 1)^2 + 9(y - 2)^2 > 20$

∗ **10.** Give definitions of the following for sets of points in xyz-space: interior, exterior, and boundary points; open, closed, and connected sets; domain and simply connected domain.

In Exercises 11–19 determine whether the set of points in space is open, closed, connected, a domain, and/or a simply connected domain.

11. $x^2 + y^2/4 + z^2/9 < 1$

12. $z \geq x^2 + y^2$

13. $x^2 + y^2 + z^2 > 0$

14. $1 < x^2 + y^2 + z^2 < 4,\ x \geq 0,\ y \geq 0,\ z \geq 0$

15. $x^2 + y^2 > 0$

16. $|z| > 0$

17. $z^2 > x^2 + y^2$

18. $z^2 > x^2 + y^2 - 1$

19. $z^2 < x^2 + y^2 - 1$

∗ **20.** Prove that the only nonempty set in the xy-plane that is both open and closed is the whole plane.

In Exercises 21–40 calculate the required quantity.

21. ∇f if $f(x, y, z) = 3x^2 y - y^3 z^2$

22. ∇f if $f(x, y, z) = (x^2 + y^2 + z^2)^{-1/2}$

23. ∇f if $f(x, y) = \text{Tan}^{-1}(y/x)$

24. ∇f at $(1, 2)$ if $f(x, y) = x^3y - 2x \cos y$

25. ∇f at $(1, -1, 4)$ if $f(x, y, z) = e^{xyz}$

26. $\nabla \cdot \mathbf{F}$ if $\mathbf{F}(x, y, z) = 2xe^y\hat{\mathbf{i}} + 3x^2z\hat{\mathbf{j}} - 2x^2yz\hat{\mathbf{k}}$

27. $\nabla \cdot \mathbf{F}$ if $\mathbf{F}(x, y) = x \ln y\hat{\mathbf{i}} - y^3e^x\hat{\mathbf{j}}$

28. $\nabla \cdot \mathbf{F}$ if $\mathbf{F}(x, y, z) = \sin(x^2 + y^2 + z^2)\hat{\mathbf{i}} + \cos(y + z)\hat{\mathbf{j}}$

29. $\nabla \cdot \mathbf{F}$ if $\mathbf{F}(x, y) = e^x\hat{\mathbf{i}} + e^y\hat{\mathbf{j}}$

30. $\nabla \cdot \mathbf{F}$ at $(1, 1, 1)$ if $\mathbf{F}(x, y, z) = x^2y^3\hat{\mathbf{i}} - 3xy\hat{\mathbf{j}} + z^2\hat{\mathbf{k}}$

31. $\nabla \cdot \mathbf{F}$ at $(-1, 3)$ if $\mathbf{F}(x, y) = (x + y)^2(\hat{\mathbf{i}} + \hat{\mathbf{j}})$

32. $\nabla \cdot \mathbf{F}$ if $\mathbf{F}(x, y, z) = (x\hat{\mathbf{i}} + y\hat{\mathbf{j}} + z\hat{\mathbf{k}})/\sqrt{x^2 + y^2 + z^2}$

33. $\nabla \cdot \mathbf{F}$ if $\mathbf{F}(x, y) = \text{Cot}^{-1}(xy)\hat{\mathbf{i}} + \text{Tan}^{-1}(xy)\hat{\mathbf{j}}$

34. $\nabla \times \mathbf{F}$ if $\mathbf{F}(x, y, z) = x^2z\hat{\mathbf{i}} + 12xyz\hat{\mathbf{j}} + 32y^2z^4\hat{\mathbf{k}}$

35. $\nabla \times \mathbf{F}$ if $\mathbf{F}(x, y) = xe^y\hat{\mathbf{i}} - 2xy^2\hat{\mathbf{j}}$

36. $\nabla \times \mathbf{F}$ if $\mathbf{F}(x, y, z) = x^2\hat{\mathbf{i}} + y^2\hat{\mathbf{j}} + z^2\hat{\mathbf{k}}$

37. $\nabla \times \mathbf{F}$ at $(1, -1, 1)$ if $\mathbf{F}(x, y, z) = xz^3\hat{\mathbf{i}} - 2x^2yz\hat{\mathbf{j}} + 2yz^4\hat{\mathbf{k}}$

38. $\nabla \times \mathbf{F}$ at $(2, 0)$ if $\mathbf{F}(x, y) = y\hat{\mathbf{i}} - x\hat{\mathbf{j}}$

39. $\nabla \times \mathbf{F}$ if $\mathbf{F}(x, y, z) = \ln(x + y + z)(\hat{\mathbf{i}} + \hat{\mathbf{j}} + \hat{\mathbf{k}})$

40. $\nabla \times \mathbf{F}$ if $\mathbf{F}(x, y) = \text{Sec}^{-1}(x + y)\hat{\mathbf{i}} + \text{Csc}^{-1}(y + x)\hat{\mathbf{j}}$

41. If $\mathbf{F} = x^2y\hat{\mathbf{i}} - 2xz\hat{\mathbf{j}} + 2yz\hat{\mathbf{k}}$, find $\nabla \times (\nabla \times \mathbf{F})$.

42. (a) Verify that Laplace's equation 12.12 can be expressed in the form $\nabla \cdot \nabla f = 0$.

(b) Show that $f(x, y, z) = (x^2 + y^2 + z^2)^{-1/2}$ satisfies Laplace's equation.

43. Prove properties 14.7–14.12, 14.14, and 14.15.

44. A gas is moving through some region D of space. If we follow a particular particle of the gas, it traces out some curved path $C : x = x(t)$, $y = y(t)$, $z = z(t)$, where t is time. If $\rho(x, y, z, t)$ is the density of gas at any point in D at time t, then along C we can express density in terms of t only, $\rho = \rho[x(t), y(t), z(t), t]$. Show that along C,

$$\frac{d\rho}{dt} = \frac{\partial\rho}{\partial t} + \nabla\rho \cdot \frac{d\mathbf{r}}{dt},$$

where $\mathbf{r} = x\hat{\mathbf{i}} + y\hat{\mathbf{j}} + z\hat{\mathbf{k}}$.

In Exercises 45–49 find all functions $f(x, y)$ such that $\nabla f = \mathbf{F}$.

45. $\mathbf{F}(x, y) = 2xy\hat{\mathbf{i}} + x^2\hat{\mathbf{j}}$

46. $\mathbf{F}(x, y) = (3x^2y^2 + 3)\hat{\mathbf{i}} + (2x^3y + 2)\hat{\mathbf{j}}$

47. $\mathbf{F}(x, y) = e^y\hat{\mathbf{i}} + (xe^y + 4y^2)\hat{\mathbf{j}}$

48. $\mathbf{F}(x, y) = (x + y)^{-1}(\hat{\mathbf{i}} + \hat{\mathbf{j}})$

∗ 49. $\mathbf{F}(x, y) = -xy(1 - x^2y^2)^{-1/2}(y\hat{\mathbf{i}} + x\hat{\mathbf{j}})$

In Exercises 50–55 find all functions $f(x, y, z)$ such that $\nabla f = \mathbf{F}$.

50. $\mathbf{F}(x, y, z) = x\hat{\mathbf{i}} + y\hat{\mathbf{j}} + z\hat{\mathbf{k}}$

51. $\mathbf{F}(x, y, z) = yz\hat{\mathbf{i}} + xz\hat{\mathbf{j}} + (yx - 3)\hat{\mathbf{k}}$

52. $\mathbf{F}(x, y, z) = (1 + x + y + z)^{-1}(\hat{\mathbf{i}} + \hat{\mathbf{j}} + \hat{\mathbf{k}})$

53. $\mathbf{F}(x, y, z) = (2x/y^2 + 1)\hat{\mathbf{i}} - (2x^2/y^3)\hat{\mathbf{j}} - 2z\hat{\mathbf{k}}$

54. $\mathbf{F}(x, y, z) = (1 + x^2y^2)^{-1}(y\hat{\mathbf{i}} + x\hat{\mathbf{j}}) + z\hat{\mathbf{k}}$

55. $\mathbf{F}(x, y, z) = (3x^2y + yz + 2xz^2)\hat{\mathbf{i}} + (xz + x^3 + 3z^2 - 6y^2z)\hat{\mathbf{j}} + (2x^2z + 6yz - 2y^3 + xy)\hat{\mathbf{k}}$

∗ 56. (a) Find constants a, b, and c in order that the vector field

$$\mathbf{F} = (x^2 + 2y + az)\hat{\mathbf{i}} + (bx - 3y - z)\hat{\mathbf{j}} + (4x + cy + 2z)\hat{\mathbf{k}}$$

will be irrotational.

(b) If $\mathbf{F}$ is irrotational, find a scalar function $f(x, y, z)$ such that $\nabla f = \mathbf{F}$.

∗ 57. A vector field $\mathbf{F}(x, y, z)$ is said to be *solenoidal* if $\nabla \cdot \mathbf{F} = 0$.

(a) Is either of $\mathbf{F} = (2x^2 + 8xy^2z)\hat{\mathbf{i}} + (3x^3y - 3xy)\hat{\mathbf{j}} - (4y^2z^2 + 2x^3z)\hat{\mathbf{k}}$ or $xyz^2\mathbf{F}$ solenoidal?

(b) Show that $\nabla f \times \nabla g$ is solenoidal for arbitrary functions $f(x, y, z)$ and $g(x, y, z)$ that have continuous second partial derivatives.

∗ 58. Associated with every electric field is a scalar function $V(x, y, z)$ called *potential*. It is defined by $\mathbf{E} = -\nabla V$, where $\mathbf{E}$ is a vector field called the *electric field intensity*. In addition, if a point charge Q is placed at a point (x, y, z) in the electric field, then the force $\mathbf{F}$ on Q is $\mathbf{F} = Q\mathbf{E}$.

(a) If the force on Q due to a charge q at the origin is

$$\mathbf{F} = \frac{qQ}{4\pi\epsilon_0|\mathbf{r}|^3}\mathbf{r},$$

where $\mathbf{r} = x\hat{\mathbf{i}} + y\hat{\mathbf{j}} + z\hat{\mathbf{k}}$ and ϵ_0 is a constant, find $V(x, y, z)$ for the field due to q.

(b) If the entire xy-plane is given a uniform charge density σ units of charge per unit area, it is found that the force on a charge Q placed z units above the plane is $\mathbf{F} = [Q\sigma/(2\epsilon_0)]\hat{\mathbf{k}}$. Find the potential V for the electric field due to this charge distribution.

∗ 59. Show that if $\mathbf{v} = \boldsymbol{\omega} \times \mathbf{r}$, where $\boldsymbol{\omega}$ is a constant vector, and $\mathbf{r} = x\hat{\mathbf{i}} + y\hat{\mathbf{j}} + z\hat{\mathbf{k}}$, then $\boldsymbol{\omega} = (1/2)(\nabla \times \mathbf{v})$.

∗ 60. Show that if a function $f(x, y, z)$ satisfies Laplace's equation 12.12, then its gradient is both irrotational and solenoidal.

∗ **61.** Theorem 14.1 indicates that a vector field $\mathbf{F}$ is the gradient of some scalar field if $\nabla \times \mathbf{F} = \mathbf{0}$. Sometimes a given vector field $\mathbf{F}$ is the curl of another field $\mathbf{v}$; that is, $\mathbf{F} = \nabla \times \mathbf{v}$. The following theorem indicates when this is the case: Let D be the interior of a sphere in which the components of a vector field $\mathbf{F}$ have continuous first partial derivatives. Then there exists a vector field $\mathbf{v}$ defined in D such that $\mathbf{F} = \nabla \times \mathbf{v}$ if and only if $\nabla \cdot \mathbf{F} = 0$ in D. In other words, $\mathbf{F}$ is the curl of a vector field if and only if $\mathbf{F}$ is solenoidal.

(a) Show that if $\mathbf{F} = P\hat{\mathbf{i}} + Q\hat{\mathbf{j}} + R\hat{\mathbf{k}}$ is solenoidal, then the components of $\mathbf{v} = L\hat{\mathbf{i}} + M\hat{\mathbf{j}} + N\hat{\mathbf{k}}$ would have to satisfy the equations

$$P = \frac{\partial N}{\partial y} - \frac{\partial M}{\partial z}, \quad Q = \frac{\partial L}{\partial z} - \frac{\partial N}{\partial x}, \quad R = \frac{\partial M}{\partial x} - \frac{\partial L}{\partial y}.$$

(b) Show that the vector field $\mathbf{v}(x, y, z)$ defined by

$$\mathbf{v}(x, y, z) = \int_0^1 t\mathbf{F}(tx, ty, tz) \times (x, y, z)\, dt$$

satisfies these equations. In other words, this formula defines a possible vector $\mathbf{v}$. Is it unique?

(c) Show that if $\mathbf{F}$ satisfies the property that $\mathbf{F}(tx, ty, tz) = t^n \mathbf{F}(x, y, z)$, then

$$\mathbf{v}(x, y, z) = \frac{1}{n + 2}\mathbf{F} \times \mathbf{r}, \quad \mathbf{r} = x\hat{\mathbf{i}} + y\hat{\mathbf{j}} + z\hat{\mathbf{k}}.$$

In Exercises 62–64 verify that the vector field is solenoidal, and then use the formulas in Exercise 61 to find a vector field $\mathbf{v}$ such that $\mathbf{F} = \nabla \times \mathbf{v}$.

∗ **62.** $\mathbf{F} = x\hat{\mathbf{i}} + y\hat{\mathbf{j}} - 2z\hat{\mathbf{k}}$

∗ **63.** $\mathbf{F} = (1 + x)\hat{\mathbf{i}} - (x + z)\hat{\mathbf{k}}$

∗ **64.** $\mathbf{F} = 2x^2\hat{\mathbf{i}} - y^2\hat{\mathbf{j}} + (2yz - 4xz)\hat{\mathbf{k}}$

14.2 Line Integrals

Just as definite integrals, double integrals, and triple integrals are defined as limits of sums, so are *line* and *surface* integrals. The only difference is that line integrals are applied to functions defined along curves, and surface integrals involve functions defined on surfaces.

A curve C in space is defined parametrically by three functions,

$$C: \quad x = x(t), \quad y = y(t), \quad z = z(t), \quad \alpha \le t \le \beta, \tag{14.17}$$

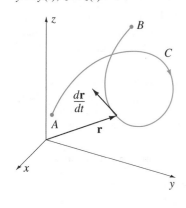

FIGURE 14.9 $\dfrac{d\mathbf{r}}{dt}$ is a tangent vector to a curve C: $x = x(t)$, $y = y(t)$, $z = z(t)$

where α and β specify initial and final points A and B of the curve, respectively (Figure 14.9). The direction of a curve is from initial to final point, and in Section 11.9 we agreed to parametrize a curve using parameters that increase in the direction of the curve.

The curve is said to be *continuous* if each of the functions $x(t)$, $y(t)$, and $z(t)$ is continuous (implying that C is at no point separated). It is said to be *smooth* if each of these functions has a continuous first derivative; geometrically, this means that the tangent vector $d\mathbf{r}/dt = x'(t)\hat{\mathbf{i}} + y'(t)\hat{\mathbf{j}} + z'(t)\hat{\mathbf{k}}$ turns gradually or smoothly along C. A continuous curve that is not smooth but can be divided into a finite number of smooth subcurves is said to be *piecewise smooth*.

Suppose a function $f(x, y, z)$ is defined along a curve C joining A to B (Figure 14.10). We divide C into n subcurves of lengths $\Delta s_1, \Delta s_2, \ldots, \Delta s_n$ by any $n - 1$ consecutive points $A = P_0, P_1, P_2, \ldots, P_{n-1}, P_n = B$, whatsoever. On each subcurve of length Δs_i ($i = 1, \ldots, n$) we choose an arbitrary point $P_i^*(x_i^*, y_i^*, z_i^*)$. We then form the sum

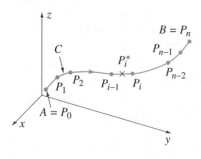

FIGURE 14.10   Divide C into shorter curves to define the line integral of $f(x, y, z)$ along C

$$f(x_1^*, y_1^*, z_1^*)\, \Delta s_1 + f(x_2^*, y_2^*, z_2^*)\, \Delta s_2 + \cdots + f(x_n^*, y_n^*, z_n^*)\, \Delta s_n$$

$$= \sum_{i=1}^{n} f(x_i^*, y_i^*, z_i^*)\, \Delta s_i. \tag{14.18}$$

If this sum approaches a limit as the number of subcurves becomes increasingly large and the length of every subcurve approaches zero, we call the limit the **line integral** of $f(x, y, z)$ along the curve C, and denote it by

$$\int_C f(x, y, z)\, ds = \lim_{\|\Delta s_i\| \to 0} \sum_{i=1}^{n} f(x_i^*, y_i^*, z_i^*)\, \Delta s_i. \tag{14.19}$$

A more appropriate name might be *curvilinear integral* rather than *line integral*, but *line integral* is the accepted terminology. We regard the word *line* as meaning *curved line* rather than *straight line*.

For definition 14.19 to be useful, we demand that the limit be independent of the manner of subdivision of C and choice of star points on the subcurves. Theorem 14.2 indicates that for continuous functions defined on smooth curves, this requirement is always satisfied.

THEOREM 14.2

Let $f(x, y, z)$ be continuous on a smooth curve C of finite length, $C : x = x(t)$, $y = y(t), z = z(t), \alpha \leq t \leq \beta$. Then the line integral of $f(x, y, z)$ along C exists and can be evaluated by means of the following definite integral:

$$\int_C f(x, y, z)\, ds = \int_\alpha^\beta f[x(t), y(t), z(t)] \sqrt{\left(\frac{dx}{dt}\right)^2 + \left(\frac{dy}{dt}\right)^2 + \left(\frac{dz}{dt}\right)^2}\, dt.$$

(14.20)

It is not necessary to memorize 14.20 as a formula. Simply put, the right side is obtained by expressing x, y, z, and ds in terms of t and interpreting the result as a definite integral with respect to t. To be more explicit, recall from equation 11.81 that when length along a curve is measured from its initial point, then an infinitesimal length ds along C corresponding to an increment dt in t is given by

$$ds = \sqrt{(dx)^2 + (dy)^2 + (dz)^2} = \sqrt{\left[\left(\frac{dx}{dt}\right)^2 + \left(\frac{dy}{dt}\right)^2 + \left(\frac{dz}{dt}\right)^2\right](dt)^2}$$

$$= \sqrt{\left(\frac{dx}{dt}\right)^2 + \left(\frac{dy}{dt}\right)^2 + \left(\frac{dz}{dt}\right)^2}\, dt.$$

If we substitute this into the left side of 14.20 and at the same time use the equations for C to express $f(x, y, z)$ in terms of t, then

$$\int_C f(x, y, z)\, ds = \int_C f[x(t), y(t), z(t)] \sqrt{\left(\frac{dx}{dt}\right)^2 + \left(\frac{dy}{dt}\right)^2 + \left(\frac{dz}{dt}\right)^2}\, dt.$$

But if we now interpret the right side of this equation as the definite integral of the function $f[x(t), y(t), z(t)] \sqrt{(dx/dt)^2 + (dy/dt)^2 + (dz/dt)^2}$ with respect to t and affix limits $t = \alpha$ and $t = \beta$ that identify endpoints of C, we obtain 14.20.

To evaluate a line integral, then, we express $f(x, y, z)$ and ds in terms of some parameter along C and evaluate the resulting definite integral. If equations for C are given in the form $C : y = y(x), z = z(x), x_A \leq x \leq x_B$, then x is a convenient parameter, and equation 14.20 takes the form

$$\int_C f(x, y, z)\, ds = \int_{x_A}^{x_B} f[x, y(x), z(x)] \sqrt{1 + \left(\frac{dy}{dx}\right)^2 + \left(\frac{dz}{dx}\right)^2}\, dx.$$

(14.21)

Similar expressions exist if either y or z is a convenient parameter.

When C is piecewise smooth rather than smooth, we evaluate line integrals along each smooth subcurve and add results.

EXAMPLE 14.4

Evaluate the line integral of $f(x, y, z) = 8x + 6xy + 30z$ from $A(0, 0, 0)$ to $B(1, 1, 1)$

(a) Along the straight line joining A to B with parametrization:

$$C_1 : x = t, \quad y = t, \quad z = t, \quad 0 \leq t \leq 1.$$

FIGURE 14.11 The line integral along a curve usually depends on the curve between the points

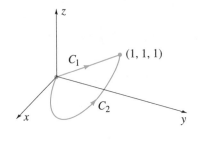

(b) Along the straight line in part (a) with parametrization:

$$C_1 : x = -1 + \frac{t}{2}, \quad y = -1 + \frac{t}{2}, \quad z = -1 + \frac{t}{2}, \quad 2 \leq t \leq 4.$$

(c) Along the curve (Figure 14.11)

$$C_2 : x = t, \quad y = t^2, \quad z = t^3, \quad 0 \leq t \leq 1.$$

SOLUTION

(a) $\displaystyle\int_{C_1} (8x + 6xy + 30z)\, ds = \int_0^1 (8t + 6t^2 + 30t)\sqrt{(1)^2 + (1)^2 + (1)^2}\, dt$

$$= \sqrt{3} \int_0^1 (38t + 6t^2)\, dt = \sqrt{3}\{19t^2 + 2t^3\}_0^1$$

$$= 21\sqrt{3}.$$

(b) $\displaystyle\int_{C_1} (8x + 6xy + 30z)\, ds = \int_2^4 \{8(-1 + t/2) + 6(-1 + t/2)^2 + 30(-1 + t/2)\}$

$$\times \sqrt{(1/2)^2 + (1/2)^2 + (1/2)^2}\, dt$$

$$= \frac{\sqrt{3}}{2} \int_2^4 [38(-1 + t/2) + 6(-1 + t/2)^2]\, dt$$

$$= \frac{\sqrt{3}}{2}\{38(-1 + t/2)^2 + 4(-1 + t/2)^3\}_2^4$$

$$= \frac{\sqrt{3}}{2}\{38 + 4\} = 21\sqrt{3}.$$

(c) $\displaystyle\int_{C_2} (8x + 6xy + 30z)\, ds = \int_0^1 (8t + 6t^3 + 30t^3)\sqrt{(1)^2 + (2t)^2 + (3t^2)^2}\, dt$

$$= \int_0^1 (8t + 36t^3)\sqrt{1 + 4t^2 + 9t^4}\, dt$$

$$= \{\tfrac{2}{3}(1 + 4t^2 + 9t^4)^{3/2}\}_0^1 = \tfrac{2}{3}(14\sqrt{14} - 1).$$

Parts (a) and (b) of this example suggest that the value of a line integral does not depend on the particular parametrization of the curve used in its evaluation. This is indeed true, and should perhaps be expected since definition 14.20 makes no reference whatsoever to parametrization of the curve. For a proof of this fact, see Exercise 37. Different parameters normally lead to different definite integrals, but they all give the same value for the line integral. Parts (a) and (c) illustrate that the line integral does depend on the curve joining points A and B (i.e., the value of the line integral changed when the curve C joining A and B changed).

EXAMPLE 14.5

Evaluate the line integral of $f(x, y) = x^2 + y^2$ once clockwise around the circle $x^2 + y^2 = 4$, $z = 0$.

SOLUTION   If we use the parametrization

$$C : x = 2\cos t, \quad y = -2\sin t, \quad z = 0, \quad 0 \le t \le 2\pi$$

(Figure 14.12), then

$$\int_C (x^2 + y^2) \, ds = \int_0^{2\pi} (4)\sqrt{\left(\frac{dx}{dt}\right)^2 + \left(\frac{dy}{dt}\right)^2} \, dt$$

$$= 4\int_0^{2\pi} \sqrt{(-2\sin t)^2 + (-2\cos t)^2} \, dt = 8\int_0^{2\pi} dt = 16\pi.$$

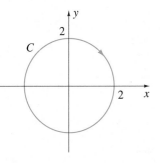

FIGURE 14.12   Evaluation of line integral of $x^2 + y^2$ around $x^2 + y^2 = 4$, $z = 0$

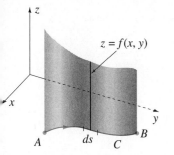

FIGURE 14.13   Planar line integrals of nonnegative functions can be interpreted as areas of walls

The value of this line integral, as well as many other line integrals in the xy-plane, can be given a geometric interpretation. Suppose a function $f(x, y)$ is positive along a curve C in the xy-plane. If at each point of C we draw a vertical line of height $z = f(x, y)$, then a vertical wall is constructed as shown in Figure 14.13. Since ds is an elemental piece of length along C, the quantity $f(x, y) \, ds$ can be interpreted as approximately the area of the vertical wall projecting onto ds. Because the line integral

$$\int_C f(x, y) \, ds,$$

like all integrals, is a limit-summation process, we interpret the value of this line integral as the total area of the vertical wall. Correct as this interpretation is, it really is of little use in the evaluation of line integrals and, in addition, the interpretation is valid only if the curve C along which the line integral is performed is contained in a plane.

Because a line integral is a limit summation, it should be obvious that the line integral

$$\int_C ds \tag{14.22}$$

represents the length of the curve C. If C is a curve in the xy-plane, we substitute $ds = \sqrt{(dx)^2 + (dy)^2}$, and if C is a curve in space, then $ds = \sqrt{(dx)^2 + (dy)^2 + (dz)^2}$. This agrees with the results of equations 11.74 and 11.81.

We make one last point about notation. To indicate that a line integral is being evaluated around a closed curve, we usually draw a circle on the integral sign, as follows:

$$\oint_C f(x, y, z) \, ds.$$

Such would be the case for the curve of intersection of the cylinder $x^2 + y^2 = 1$ and the plane $x + z = 1$ (Figure 14.14).

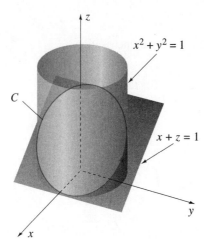

FIGURE 14.14 Curve of intersection of the cylinder $x^2 + y^2 = 1$ and the plane $x + z = 1$ is closed

When C is a closed curve in the xy-plane that does not cross itself, we indicate the direction along C by an arrowhead on the circle. For the curves shown in Figures 14.15 we write

$$\oint_{C_1} f(x, y)\, ds \qquad \text{and} \qquad \oint_{C_2} f(x, y)\, ds.$$

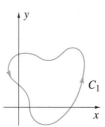

FIGURE 14.15a

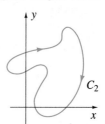

FIGURE 14.15b

Arrowheads are placed on circles representing line integrals around closed curves in a plane

EXERCISES 14.2

In Exercises 1–6 evaluate the line integral.

1. $\displaystyle\int_C x\, ds$, where C is the curve $y = x^2$, $z = 0$ from $(0, 0, 0)$ to $(1, 1, 0)$

2. $\displaystyle\oint_C (x^2 + y^2)\, ds$ once around the square C in the xy-plane with vertices $(\pm 1, 1)$ and $(\pm 1, -1)$

3. $\displaystyle\oint_C (2 + x - 2xy)\, ds$ once around the circle $x^2 + y^2 = 4$, $z = 0$

4. $\displaystyle\int_C (x^2 + yz)\, ds$ along the straight line from $(1, 2, -1)$ to $(3, 2, 5)$

5. $\displaystyle\int_C xy\, ds$, where C is the first octant part of $x^2 + y^2 = 1$, $x^2 + z^2 = 1$ from $(1, 0, 0)$ to $(0, 1, 1)$

6. $\displaystyle\int_C x^2 yz\, ds$, where C is the curve $z = x + y$, $x + y + z = 1$ from $(1, -1/2, 1/2)$ to $(-3, 7/2, 1/2)$

7. Prove that the length of the circumference of a circle is 2π multiplied by the radius.

8. A spring has six coils in the form of the helix

$$x = 3\cos t, \quad y = 3\sin t, \quad z = 3t/(4\pi), \quad 0 \le t \le 12\pi,$$

where all dimensions are in centimetres. Find the length of the spring.

9. Use parametric equations $x = \cos^3 \theta$, $y = \sin^3 \theta$, $0 \le \theta < 2\pi$, to draw or plot the astroid $x^{2/3} + y^{2/3} = 1$ in the xy-plane. At each point (x, y) on the astroid, a vertical line is drawn with height $z = x^2 + y^2$, thus forming a cylindrical wall. Find the area of the wall.

In Exercises 10–16 evaluate the line integral.

10. $\int_C xz\,ds$ along the first octant part of $y = x^2$, $z + y = 1$ from $(0, 0, 1)$ to $(1, 1, 0)$

11. $\int_C (x + y)^5\,ds$ along C : $x = t + 1/t$, $y = t - 1/t$ from $(2, 0)$ to $(17/4, 15/4)$

12. $\int_C x\sqrt{y + z}\,ds$, where C is that part of the curve $3x + 2y + 3z = 5$, $x - 2y + 4z = 5$ from $(1, 0, 1)$ to $(0, 9/14, 11/7)$

13. $\int_C xy\,ds$, where C is the curve $x = 1 - y^2$, $z = 0$ from $(1, 0, 0)$ to $(0, 1, 0)$

14. $\int_C (x + y)z\,ds$, where C is the curve $y = x$, $z = 1 + y^4$ from $(-1, -1, 2)$ to $(1, 1, 2)$

15. $\int_C \dfrac{1}{y + z}\,ds$, where C is the curve $y = x^2$, $z = x^2$ from $(1, 1, 1)$ to $(2, 4, 4)$

16. $\int_C (2y + 9z)\,ds$, where C is the curve $z = xy$, $x = y^2$ from $(0, 0, 0)$ to $(4, -2, -8)$

17. (a) If the curve C below is rotated around the y-axis, show that the area of the surface that it traces out is represented by the line integral $\int_C 2\pi x\,ds$.

(b) If C is rotated around the x-axis, what line integral represents the area of the surface traced out?

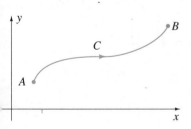

In Exercises 18–20 use the method in Exercise 17 to find the area of the surface traced out when the curve is rotated around the line.

18. $y = x^3$, $1 \le x \le 2$, around $y = 0$

19. $24xy = x^4 + 48$, $1 \le x \le 2$, around $x = 0$

20. $8y^2 = x^2(1 - x^2)$, around $y = 0$

21. Find the length of the parabola $y = x^2$ from $(0, 0)$ to $(1, 1)$.

In Exercises 22–23 find a definite integral that can be used to evaluate the line integral. Use power series to approximate the definite integral accurate to three decimal places.

22. $\int_C xy\,ds$, where C is the curve $y = x^3$, $z = 0$ from $(0, 0, 0)$ to $(1/2, 1/8, 0)$

23. $\int_C e^{-(x+y-2)^2}\,ds$, where C is the curve $x + y + z = 2$, $y + 2z = 3$ from $(-1, 3, 0)$ to $(0, 1, 1)$

The average value of a function $f(x, y, z)$ along a curve C is defined as the value of the line integral of the function along the curve divided by the length of the curve. In Exercises 24–27 find the average value of the function along the curve.

* **24.** $f(x, y) = x^2y^2$ along C : $x^2 + y^2 = 4$, $z = 0$
* **25.** $f(x, y, z) = x^2 + y^2 + z^2$ along C : $x = \cos t$, $y = \sin t$, $z = t$, $0 \le t \le \pi$
* **26.** $f(x, y, z) = xyz$ along C : $z = x^2$, $y = x^2$ from $(0, 0, 0)$ to $(1, 1, 1)$
* **27.** $f(x, y) = y$ along C : $y = x^3/4 + 1/(3x)$ from $(1, 7/12)$ to $(2, 13/6)$

* **28.** At each point on the curve $(x^2 + y^2)^2 = x^2 - y^2$ a vertical line is drawn with height equal to the distance from the point to the origin. Find the area of the vertical wall so formed.

* **29.** During a sleet storm, a power line between two poles at positions $x = \pm 20$ hangs in the shape $y = 40\cosh(x/40) - 10$, where all distances are measured in metres. Ice accumulates more heavily on the middle part of the line than at the ends. In fact, the combined mass of ice and line per unit length in the x-direction at position x is given in kilograms per metre by the formula $\rho(x) = 1 - |x|/40$. Find the total mass of the line.

In Exercises 30–33 use the fact that in polar coordinates small lengths along a curve can be expressed in the form $ds = \sqrt{r^2 + (dr/d\theta)^2}\,d\theta$ (see formula 9.14) to evaluate the line integral.

* **30.** $\int_C \dfrac{x}{\sqrt{x^2 + y^2}}\,ds$, where C is the first quadrant part of the limaçon $r = 2 - \sin\theta$ starting from the point on the x-axis

* **31.** $\oint_C (x^2 + y^2)\,ds$, where C is the cardioid $r = 1 + \cos\theta$

* **32.** $\int_C xy\,ds$, where C is the spiral $r = e^\theta$ from $\theta = 0$ to $\theta = 2\pi$

* **33.** $\oint_C \cos^3 2\theta\,ds$ around the first quadrant loop of the lemniscate $r^2 = \sin 2\theta$

In Exercises 34–35 find a definite integral which can be used to evaluate the line integral. Use Simpson's rule with 10 equal subdivisions to approximate the definite integral.

* **34.** $\int_C (x^2y + z)\,ds$, where C is the curve $z = x^2 + y^2$, $y + x = 1$ from $(-1, 2, 5)$ to $(1, 0, 1)$

* **35.** $\oint_C x^2y^2\,ds$, where C is the ellipse $4x^2 + 9y^2 = 36$, $z = 0$

** **36.** Find the surface area of the torus obtained by rotating, around the y-axis, the circle $(x - a)^2 + y^2 = b^2$ $(a > b)$.

** **37.** Show that the value of a line integral is independent of the parameter used to specify the curve.

14.3 Line Integrals Involving Vector Functions

There are many ways in which $f(x, y, z)$ in the line integral

$$\int_C f(x, y, z)\, ds \qquad (14.23)$$

can arise. According to equation 14.22, we choose $f(x, y, z) = 1$ in order to find the length of the curve C; Exercise 17 in Section 14.2 indicates that for areas of the surfaces traced out when a curve in the xy-plane is rotated around the y- and x-axes, we choose $f(x, y)$ equal to $2\pi x$ and $2\pi y$, respectively.

The most important and common type of line integral occurs when $f(x, y, z)$ is specified as the tangential component of some given vector field $\mathbf{F}(x, y, z)$ defined along C [i.e., $f(x, y, z)$ itself is not given, but $\mathbf{F}$ is, and to find $f(x, y, z)$ we must calculate the tangential component of $\mathbf{F}$ along C]. By the tangential component of $\mathbf{F}(x, y, z)$ along C we mean the component of $\mathbf{F}$ along that tangent vector to C that points in the same direction as C (Figure 14.16).

In Section 11.11 we saw that if s is a measure of length along a curve C from A to B (Figure 14.17), and if s is chosen equal to zero at A, then a unit tangent vector pointing in the direction of motion along C is

$$\hat{\mathbf{T}} = \frac{d\mathbf{r}}{ds}. \qquad (14.24)$$

Consequently, if $f(x, y, z)$ is the tangential component of $\mathbf{F}(x, y, z)$ along C, then

$$f(x, y, z) = \mathbf{F} \cdot \hat{\mathbf{T}} = \mathbf{F} \cdot \frac{d\mathbf{r}}{ds}, \qquad (14.25)$$

and we can write that

$$\int_C f(x, y, z)\, ds = \int_C \mathbf{F} \cdot \hat{\mathbf{T}}\, ds = \int_C \mathbf{F} \cdot \frac{d\mathbf{r}}{ds}\, ds = \int_C \mathbf{F} \cdot d\mathbf{r}. \qquad (14.26)$$

If the components of the vector field $\mathbf{F}(x, y, z)$ are

$$\mathbf{F}(x, y, z) = P(x, y, z)\hat{\mathbf{i}} + Q(x, y, z)\hat{\mathbf{j}} + R(x, y, z)\hat{\mathbf{k}}, \qquad (14.27)$$

then

$$\int_C \mathbf{F} \cdot d\mathbf{r} = \int_C (P\, dx + Q\, dy + R\, dz),$$

and if parentheses are omitted, we have

$$\int_C \mathbf{F} \cdot d\mathbf{r} = \int_C P\, dx + Q\, dy + R\, dz. \qquad (14.28)$$

FIGURE 14.16   Of the two tangent directions to a curve, always use the one that points in the direction of the curve

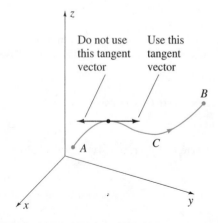

FIGURE 14.17   The integrand of a line integral is often the tangential component of a vector field defined along the curve

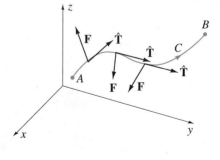

This discussion has shown that when the integrand $f(x, y, z)$ of a line integral is specified as the tangential component of $\mathbf{F} = P\hat{\mathbf{i}} + Q\hat{\mathbf{j}} + R\hat{\mathbf{k}}$ along C, the product $f(x, y, z)\,ds$ can be replaced by the sum of products $P\,dx + Q\,dy + R\,dz$:

$$\int_C f(x, y, z)\,ds = \int_C \mathbf{F} \cdot d\mathbf{r} = \int_C P\,dx + Q\,dy + R\,dz. \qquad (14.29)$$

According to the results of Section 14.2, evaluation of this line integral can be accomplished by expressing $P\,dx + Q\,dy + R\,dz$ in terms of any parametric representation of C and evaluating the resulting definite integral.

EXAMPLE 14.6

Evaluate

$$\int_C \frac{z}{y}\,dx + (x^2 + y^2 + z^2)\,dz,$$

where C is the first octant intersection of $x^2 + y^2 = 1$ and $z = 2x + 4$ joining $(1, 0, 6)$ to $(0, 1, 4)$.

SOLUTION   If we choose the parametrization

$$x = \cos t, \quad y = \sin t, \quad z = 2\cos t + 4, \quad 0 \le t \le \pi/2,$$

for C (Figure 14.18), then

$$\int_C \frac{z}{y}\,dx + (x^2 + y^2 + z^2)\,dz$$

$$= \int_0^{\pi/2} \left[\left(\frac{2\cos t + 4}{\sin t} \right)(-\sin t\,dt) \right.$$

$$\left. + (\cos^2 t + \sin^2 t + 4\cos^2 t + 16\cos t + 16)(-2\sin t\,dt) \right]$$

$$= -2 \int_0^{\pi/2} (\cos t + 2 + 17\sin t + 4\cos^2 t\,\sin t + 16\cos t\,\sin t)\,dt$$

$$= -2 \left\{ \sin t + 2t - 17\cos t - \frac{4\cos^3 t}{3} + 8\sin^2 t \right\}_0^{\pi/2} = -2\pi - \frac{164}{3}.$$

FIGURE 14.18   Line integral along the first octant intersection of $x^2 + y^2 = 1$ and $z = 2x + 4$

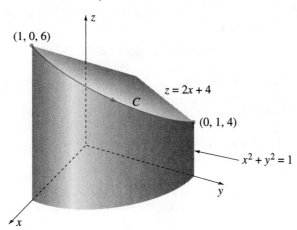

EXAMPLE 14.7

Evaluate

$$\oint_C y^2 \, dx + x^2 \, dy,$$

where C is the closed curve in Figure 14.19.

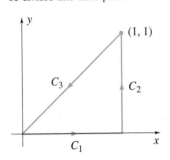

FIGURE 14.19   Line integral around the boundary of a triangle must be divided into three parts

SOLUTION   If we start at the origin and denote the three straight-line paths by

$$C_1: \quad y = 0, \quad 0 \le x \le 1,$$

$$C_2: \quad x = 1, \quad 0 \le y \le 1,$$

$$C_3: \quad y = x = 1 - t, \quad 0 \le t \le 1,$$

then

$$\oint_C y^2 dx + x^2 \, dy$$

$$= \int_{C_1} y^2 dx + x^2 \, dy + \int_{C_2} y^2 dx + x^2 \, dy + \int_{C_3} y^2 \, dx + x^2 \, dy$$

$$= \int_0^1 0 \, dx + x^2 0 + \int_0^1 y^2 0 + 1 \, dy + \int_0^1 (1 - t)^2 (-dt) + (1 - t)^2 (-dt)$$

$$= \{y\}_0^1 + \left\{ \frac{2}{3}(1 - t)^3 \right\}_0^1 = 1 - \frac{2}{3} = \frac{1}{3}.$$

If a curve C has initial and final points A and B, then the curve that traces the same points in the opposite direction, and therefore has initial and final points B and A, is denoted by $-C$. Values of an integral of type 14.21 are the same along $-C$ and C. This is so because length ds is the same along C and $-C$. This is not the case for integrals of type 14.28. In Exercise 36 it is shown that

$$\int_{-C} \mathbf{F} \cdot d\mathbf{r} = - \int_C \mathbf{F} \cdot d\mathbf{r}. \qquad (14.30)$$

This is illustrated in the following example.

EXAMPLE 14.8

Show that the line integral along $-C$ in Example 14.6 has value $2\pi + 164/3$, the negative of its value along C.

SOLUTION   Parametric equations for $-C$, with an increasing parameter along the curve, are

$$x = \cos t, \quad y = -\sin t, \quad z = 2\cos t + 4, \quad -\pi/2 \le t \le 0.$$

Using these,

$$\int_{-C} \frac{z}{y} \, dx + (x^2 + y^2 + z^2) \, dz$$

$$= \int_{-\pi/2}^0 \left[\left(\frac{2\cos t + 4}{-\sin t} \right) (-\sin t \, dt) \right.$$

$$\left. + (\cos^2 t + \sin^2 t + 4\cos^2 t + 16\cos t + 16)(-2\sin t \, dt) \right]$$

$$= 2 \int_{-\pi/2}^{0} (\cos t + 2 - 17 \sin t - 4 \cos^2 t \sin t - 16 \cos t \sin t) \, dt$$

$$= 2 \left\{ \sin t + 2t + 17 \cos t + \frac{4 \cos^3 t}{3} - 8 \sin^2 t \right\}_{-\pi/2}^{0}$$

$$= 2\pi + \frac{164}{3}.$$

According to this example, when the direction along a curve is reversed, the value of a line integral of form 14.28 along the new curve is the negative of its value along the original curve. This is because the signs of dx, dy, and dz are reversed when the direction along C is reversed. This is not the case for line integral 14.20; ds does not change sign when the direction along C is reversed.

Line integrals of form 14.28 are singled out for special consideration because they arise in so many physical problems. For example, suppose $\mathbf{F}$ represents a force, and we consider work done by this force as a particle moves along a curve C from A to B. We begin by dividing C into n subcurves of lengths Δs_i as shown in Figure 14.20.

FIGURE 14.20   To find work done by a force $\mathbf{F}$ along a curve, divide the curve into subcurves and approximate the work along subcurves

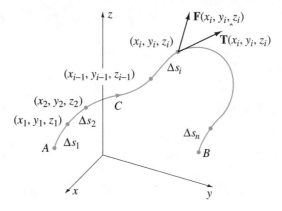

If $\mathbf{F}(x, y, z)$ is continuous along C, then along any given Δs_i, $\mathbf{F}(x, y, z)$ does not vary greatly (provided, of course, that Δs_i is small). If we approximate $\mathbf{F}(x, y, z)$ along Δs_i by its value $\mathbf{F}(x_i, y_i, z_i)$ at the final point (x_i, y_i, z_i) of Δs_i, then an approximation to the work done by $\mathbf{F}$ along Δs_i is $\mathbf{F}(x_i, y_i, z_i) \cdot \hat{\mathbf{T}}(x_i, y_i, z_i) \, \Delta s_i$. An approximation to the total work done by $\mathbf{F}$ along C is therefore

$$\sum_{i=1}^{n} \mathbf{F}(x_i, y_i, z_i) \cdot \hat{\mathbf{T}}(x_i, y_i, z_i) \, \Delta s_i.$$

To obtain the work done by $\mathbf{F}$ along C, we take the limit of this sum as the number of subdivisions becomes larger and larger and each Δs_i approaches zero. But this is the definition of the line integral of $\mathbf{F} \cdot \hat{\mathbf{T}}$, and we therefore write

$$W = \int_{C} \mathbf{F} \cdot \hat{\mathbf{T}} \, ds = \int_{C} \mathbf{F} \cdot d\mathbf{r}. \tag{14.31}$$

EXAMPLE 14.9

The force of repulsion between two positive point charges, one of size q and the other of size unity, has magnitude $q/(4\pi\epsilon_0 r^2)$, where ϵ_0 is a constant and r is the distance between the charges. The potential V at any point P due to charge q is defined as the work required to bring the unit charge to P from an infinite distance along the straight line joining q and P. Find V.

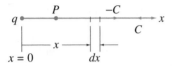 **FIGURE 14.21** Electrostatic potential at a point P due to a charge q is the work to bring a unit charge from infinity to P

SOLUTION   If that part of the line joining q and P from infinity to P is denoted by C (Figure 14.21), then

$$V = \int_C \mathbf{F} \cdot d\mathbf{r} = -\int_{-C} \mathbf{F} \cdot d\mathbf{r}$$

(see equation 14.30), where $\mathbf{F}$, the force necessary to overcome the electrostatic repulsion, is given by

$$\mathbf{F} = \frac{-q}{4\pi\epsilon_0 x^2}\hat{\mathbf{i}}.$$

Along $-C$, $d\mathbf{r} = dx\hat{\mathbf{i}}$, and therefore V can be evaluated by the (improper) definite integral

$$V = -\int_r^\infty \frac{-q}{4\pi\epsilon_0 x^2}dx = -\left\{\frac{q}{4\pi\epsilon_0 x}\right\}_r^\infty = \frac{q}{4\pi\epsilon_0 r}.$$

If the vector field $\mathbf{F}$ in equation 14.27 has an x-component that is only a function of x, $P = P(x)$, and if the curve C is a portion of the x-axis from $x = a$ to $x = b$, then

$$\int_C \mathbf{F} \cdot d\mathbf{r} = \int_a^b P(x)\,dx.$$

This equation indicates that definite integrals with respect to x can be regarded as line integrals along the x-axis.

Heat Engines

In Section 7.4 we showed that when a gas completes a cycle such as that in Figure 14.22, the work (output) of the gas corresponding to the states represented by C_2 is

$$W = \int_{V_1}^{V_2} P\,dV, \tag{14.32}$$

where the equation of C_2 is used to express P in terms of V. Output of the gas during the return cycle along C_1 is negative; it is given by integral 14.32 with limits reversed and P is expressed in terms of V using the equation of C_1. What is important is to notice that if integral 14.32 is interpreted as a line integral in the VP-plane, then it is evaluated in exactly the same way as the definite integral: substitute for P in terms of V and evaluate the definite integral. In other words, we can replace definite integral 14.32 with line integral

$$W = \int_C P\,dV, \tag{14.33}$$

where C is the curve in the VP-plane representing the succession of states through which the gas is taken. For the cycle in Figure 14.22, the work done is

$$W = \oint_{C_1+C_2} P\,dV. \tag{14.34}$$

We discussed the output of a gas for the Rankine cycle of an idealized steam engine in Figure 14.23 using area in Section 7.4. From a line integral point of view, the output is line integral

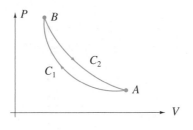

FIGURE 14.22   Gas output is the line integral of $P\,dV$ around the cycle

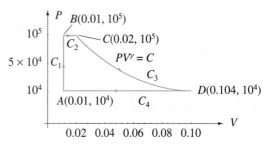

FIGURE 14.23   Rankine cycle for a steam engine

14.33, where C is made up of C_1, C_2, C_3, and C_4. If the gas is expanded adiabatically along C_3 (temperature is held constant), then P and V are related by $PV^{\gamma} = k$, where $\gamma > 0$ and $k > 0$ are constants ($\gamma = 1.4$ for air). Using point C in Figure 14.23, $k = 10^5 (0.02)^{1.4} = 4.2 \times 10^2$. Consequently,

$$
W = \oint_C P\,dV = \int_{C_1} P\,dV + \int_{C_2} P\,dV + \int_{C_3} P\,dV + \int_{C_4} P\,dV
$$

$$
= 0 + \int_{0.01}^{0.02} 10^5\,dV + \int_{0.02}^{0.104} kV^{-1.4}\,dV + \int_{0.104}^{0.01} 10^4\,dV
$$

$$
= 10^5(0.02 - 0.01) + k\left\{\frac{V^{-0.4}}{-0.4}\right\}_{0.02}^{0.104} + 10^4(0.01 - 0.104) = 2.5 \times 10^3 \text{ J.}
$$

EXERCISES 14.3

In Exercises 1–10 evaluate the line integral.

1. $\int_C x\,dx + x^2 y\,dy$, where C is the curve $y = x^3$, $z = 0$ from $(-1, -1, 0)$ to $(2, 8, 0)$

2. $\int_C x\,dx + yz\,dy + x^2\,dz$, where C is the curve $y = x$, $z = x^2$ from $(-1, -1, 1)$ to $(2, 2, 4)$

3. $\int_C x\,dx + (x + y)\,dy$, where C is the curve $x = 1 + y^2$ from $(2, 1)$ to $(2, -1)$

4. $\int_C x^2\,dx + y^2\,dy + z^2\,dz$, where C is the curve $x + y = 1$, $x + z = 1$ from $(-2, 3, 3)$ to $(1, 0, 0)$

5. $\int_C (y + 2x^2 z)\,dx$, where C is the curve $x = y^2$, $z = x^2$ from $(4, -2, 16)$ to $(1, 1, 1)$

6. $\oint_C x^2 y\,dx + (x - y)\,dy$ once counterclockwise around the curve bounding the region described by the curves $x = 1 - y^2$, $y = x + 1$

7. $\int_C y^2\,dx + x^2\,dy$, where C is the semicircle $x = \sqrt{1 - y^2}$ from $(0, 1)$ to $(0, -1)$

8. $\int_C y\,dx + x\,dy + z\,dz$, where C is the curve $z = x^2 + y^2$, $x + y = 1$ from $(1, 0, 1)$ to $(-1, 2, 5)$

9. $\oint_C x^2 y\,dy + z\,dx$, where C is the curve $x^2 + y^2 = 1$, $x + y + z = 1$ directed so that x decreases when y is positive

10. $\oint_C y^2\,dx + x^2\,dy$ once clockwise around the curve $|x| + |y| = 1$

11. Find the work done by the force $\mathbf{F} = x^2 y\hat{\mathbf{i}} + x\hat{\mathbf{j}}$ as a particle moves from $(1, 0)$ to $(6, 5)$ along the straight line joining these points.

12. Consider the line integral $\int_C xy\,dx + x^2\,dy$, where C is the quarter-circle $x^2 + y^2 = 9$ from $(3, 0)$ to $(0, 3)$. Show that for each of the following parametrizations of C the value of the line integral is the same: (a) $x = 3\cos t$, $y = 3\sin t$, $0 \le t \le \pi/2$; (b) $x = \sqrt{9 - y^2}$, $0 \le y \le 3$.

13. Evaluate the line integral $\int_C xy\,dx + x\,dy$ from $(-5, 3, 0)$ to $(4, 0, 0)$ along each of the following curves: (a) the straight line joining the points; (b) $x = 4 - y^2$, $z = 0$; (c) $3y = x^2 - 16$, $z = 0$.

14. Find the work done by a force $\mathbf{F} = x\hat{\mathbf{i}} + y\hat{\mathbf{j}}$ on a particle as it moves once counterclockwise around the ellipse $b^2 x^2 + a^2 y^2 = a^2 b^2$, $z = 0$.

In Exercises 15–19 evaluate the line integral.

* **15.** $\int_C \dfrac{1}{yz}\,dx$, where C is the curve $z = \sqrt{1 - x^2}$, $y = \sqrt{1 - x^2}$ from $(1/\sqrt{2}, 1/\sqrt{2}, 1/\sqrt{2})$ to $(-1\sqrt{2}, 1/\sqrt{2}, 1/\sqrt{2})$

* **16.** $\oint_C (x^2 + 2y^2)\,dy$ twice clockwise around the circle $(x - 2)^2 + y^2 = 1$, $z = 0$

* **17.** $\int_C y\,dx - y(x - 1)\,dy + y^2 z\,dz$, where C is the first octant intersection of $x^2 + y^2 + z^2 = 4$ and $(x - 1)^2 + y^2 = 1$ from $(2, 0, 0)$ to $(0, 0, 2)$

* **18.** $\int_C x^2 y\,dx + y\,dy + \sqrt{1 - x^2}\,dz$, where C is the curve $y - 2z^2 = 1$, $z = x + 1$ from $(0, 3, 1)$ to $(1, 9, 2)$

* **19.** $\int_C x\,dx + xy\,dy + 2\,dz$, where C is the curve $x + 2y + z = 4$, $4x + 3y + 2z = 13$ from $(2, -1, 4)$ to $(3, 1, -1)$

* **20.** Evaluate the line integral $\int_C \dfrac{x^3}{(1+x^4)^3}\,dx + y^2 e^y\,dy + \dfrac{z}{\sqrt{1+z^2}}\,dz$, where C consists of line segments joining successively the points $(0, -1, 1)$, $(1, -1, 1)$, $(1, 0, 1)$, and $(1, 0, 2)$.

* **21.** One end of a spring (with constant k) is fixed at point D in the figure below. The other end is moved along the x-axis from A to B. If the spring is stretched an amount l at A, find the work done against the spring.

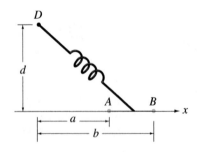

* **22.** Two positive charges q_1 and q_2 are placed at positions $(5, 5)$ and $(-2, 3)$, respectively, in the xy-plane. A third positive charge q_3 is moved along the x-axis from $x = 1$ to $x = -1$. Find the work done by the electrostatic forces of q_1 and q_2 on q_3.

23.–27. Repeat Exercises 26–30 in Section 7.4 but do so from a line integral point of view.

* **28.** Find the work done for the Rankine cycle in the figure below in terms of P_2, V_1, V_2, V_3, and γ.

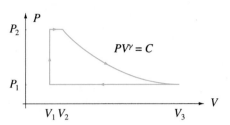

In Exercises 29–32 set up a definite integral to evaluate the line integral. Use Simpson's rule with 10 equal subdivisions to approximate the definite integral.

* **29.** $\int_C xy\,dx + xy^2\,dy$, where C is the curve $z = 0$, $y = 1/\sqrt{1 + x^3}$, from $(0, 1, 0)$ to $(2, 1/3, 0)$

* **30.** $\int_C xz\,dx + \tan x\,dy + e^{xy}\,dz$, where C is the curve $x = y^2$, $z = y^3$ from $(1, -1, -1)$ to $(1, 1, 1)$

* **31.** $\int_C \sqrt{1 + y^2}\,dz + zy\,dy$, where C is curve $y = \cos^3 t$, $z = \sin^3 t$, $x = 0$, $0 \le t \le \pi/2$

* **32.** $\int_C xyz\,dy$, where C is the curve $x = (1 - t^2)/(1 + t^2)$, $y = t(1 - t^2)/(1 + t^2)$, $z = t$, $-1 \le t \le 1$

In Exercises 33–34 evaluate the line integral along the polar coordinate curve.

* **33.** $\oint_C y\,dx$, where C is the cardioid $r = 1 - \cos\theta$

* **34.** $\int_C y\,dx + x\,dy$, where C is the curve $r = \theta$, $0 \le \theta \le \pi$

* **35.** Suppose a gas flows through a region D of space. At each point $P(x, y, z)$ in D and time t, the gas has velocity $\mathbf{v}(x, y, z, t)$. If C is a closed curve in D, the line integral

$$\Gamma = \oint_C \mathbf{v} \cdot d\mathbf{r}$$

is called the *circulation of the flow* for the curve C. If C is the circle $x^2 + y^2 = r^2$, $z = 1$ (directed clockwise as viewed from the origin), calculate Γ for the following flow vectors: (a) $\mathbf{v}(x, y, z) = (x\hat{\mathbf{i}} + y\hat{\mathbf{j}} + z\hat{\mathbf{k}})/(x^2 + y^2 + z^2)^{3/2}$; (b) $\mathbf{v}(x, y, z) = -y\hat{\mathbf{i}} + x\hat{\mathbf{j}}$.

* **36.** Verify the result in equation 14.30.

* **37.** We have shown that given a line integral 14.28, it is always possible to write it uniquely in form 14.23, where $f = \mathbf{F} \cdot \hat{\mathbf{T}}$. Show that the converse is not true; that is, given $f(x, y, z)$, there does not exist a unique $\mathbf{F}(x, y, z)$ such that $\mathbf{F} \cdot d\mathbf{r} = f(x, y, z)\,ds$.

8. The cycloid $x = R(\theta - \sin\theta)$, $y = R(1 - \cos\theta)$ is the curve traced out by a fixed point on the circumference of a circle of radius R rolling along the x-axis (see Example 9.7 in Section 9.1). Suppose the point is acted on by a force of unit magnitude directed toward the centre of the rolling circle.

  (a) Find the work done by the force as the point moves from $\theta = 0$ to $\theta = \pi$.

  (b) How much of the work in part (a) is done by the vertical component of the force?

* **39.** Explain why the line integral $\oint_C f(x)\,dx + g(y)\,dy + h(z)\,dz$ must have value zero when $f(x)$, $g(y)$, and $h(z)$ are continuous functions in some domain containing C.

14.4 Independence of Path

In Sections 14.2 and 14.3 we illustrated that the value of a line integral joining two points usually depends on the curve joining the points. In this section we show that certain line integrals have the same value for all curves joining the same two points. We formalize this idea in the following definition.

DEFINITION 14.4

A line integral $\int \mathbf{F} \cdot d\mathbf{r}$ is said to be **independent of path** in a domain D if for each pair of points A and B in D, the value of the line integral

$$\int_C \mathbf{F} \cdot d\mathbf{r}$$

is the same for all piecewise-smooth paths C in D from A to B.

The value of such a line integral for given $\mathbf{F}$ will then depend only on the endpoints A and B. Note that we speak of independence of path only for the special class of line integrals of the form $\int \mathbf{F} \cdot d\mathbf{r}$. The question we must now ask is: How do we determine whether a given line integral is independent of path? One answer is contained in the following theorem.

THEOREM 14.3

Suppose that $P(x, y, z)$, $Q(x, y, z)$, and $R(x, y, z)$ are continuous functions in some domain D. The line integral

$$\int \mathbf{F} \cdot d\mathbf{r} = \int P\,dx + Q\,dy + R\,dz$$

is independent of path in D if and only if there exists a function $\phi(x, y, z)$ defined in D such that

$$\nabla\phi = \mathbf{F} = P\hat{\mathbf{i}} + Q\hat{\mathbf{j}} + R\hat{\mathbf{k}}. \qquad (14.35)$$

Essentially, then, a line integral is independent of path if $\mathbf{F}$ is the gradient of some scalar function.

PROOF Suppose, first, that in D there exists a function $\phi(x, y, z)$ such that $\nabla\phi = P\hat{\mathbf{i}} + Q\hat{\mathbf{j}} + R\hat{\mathbf{k}}$. If

$$C: \quad x = x(t), \quad y = y(t), \quad z = z(t), \quad \alpha \le t \le \beta$$

is any smooth curve in D from A to B, then

$$\int_C \mathbf{F} \cdot d\mathbf{r} = \int_C P\,dx + Q\,dy + R\,dz = \int_\alpha^\beta \left(\frac{\partial\phi}{\partial x}\frac{dx}{dt} + \frac{\partial\phi}{\partial y}\frac{dy}{dt} + \frac{\partial\phi}{\partial z}\frac{dz}{dt} \right) dt.$$

The term in parentheses is the chain rule for the derivative of composite function $\phi[x(t), y(t), z(t)]$, and we can therefore write that

$$\int_C \mathbf{F} \cdot d\mathbf{r} = \int_\alpha^\beta \frac{d\phi}{dt} dt = \{\phi[x(t), y(t), z(t)]\}_\alpha^\beta$$

$$= \phi[x(\beta), y(\beta), z(\beta)] - \phi[x(\alpha), y(\alpha), z(\alpha)]$$

$$= \phi(x_B, y_B, z_B) - \phi(x_A, y_A, z_A).$$

(The same result is obtained even when C is piecewise smooth rather than smooth.) Because this last expression does not depend on the curve C taken from A to B, it follows that the line integral is independent of path in D.

Conversely, suppose now that the line integral

$$\int \mathbf{F} \cdot d\mathbf{r} = \int P\, dx + Q\, dy + R\, dz$$

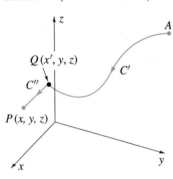

FIGURE 14.24   Proof that independence of path implies existence of ϕ such that $\mathbf{F} = \nabla\phi$

is independent of path in D, and A is chosen as some fixed point in D. If $P(x, y, z)$ is any other point in D (Figure 14.24), and C is a piecewise-smooth curve in D from A to P, then the line integral

$$\phi(x, y, z) = \int_C \mathbf{F} \cdot d\mathbf{r}$$

defines a single-valued function $\phi(x, y, z)$ in D, and the value of $\phi(x, y, z)$ is the same for all piecewise-smooth curves from A to P.

Consider a curve C composed of two parts: a straight-line portion C'' parallel to the x-axis from a fixed point $Q(x', y, z)$ to $P(x, y, z)$, and any other piecewise-smooth curve C' in D from A to Q. Then

$$\phi(x, y, z) = \int_{C'} \mathbf{F} \cdot d\mathbf{r} + \int_{C''} \mathbf{F} \cdot d\mathbf{r}.$$

Now along C'', y and z are both constant, and therefore

$$\phi(x, y, z) = \int_{C'} \mathbf{F} \cdot d\mathbf{r} + \int_{x'}^x P(t, y, z)\, dt.$$

The partial derivative of this function with respect to x is

$$\frac{\partial \phi}{\partial x} = \frac{\partial}{\partial x} \int_{C'} \mathbf{F} \cdot d\mathbf{r} + \frac{\partial}{\partial x} \int_{x'}^x P(t, y, z)\, dt,$$

but because $Q(x', y, z)$ is fixed,

$$\frac{\partial}{\partial x} \int_{C'} \mathbf{F} \cdot d\mathbf{r} = 0.$$

Consequently,

$$\frac{\partial \phi}{\partial x} = \frac{\partial}{\partial x} \int_{x'}^x P(t, y, z)\, dt = P(x, y, z).$$

By choosing other curves with straight-line portions parallel to the y- and z-axes, we can also show that $\partial \phi / \partial y = Q$ and $\partial \phi / \partial z = R$. Thus, $\mathbf{F} = \nabla\phi$, and this completes the proof. ∎

This theorem points out that it is very simple to evaluate a line integral that is independent of path. We state this in the following corollary.

COROLLARY 14.3.1

When a line integral is independent of path in a domain D, and A and B are points in D, then

$$\int_C \mathbf{F} \cdot d\mathbf{r} = \phi(x_B, y_B, z_B) - \phi(x_A, y_A, z_A),$$

where $\nabla \phi = \mathbf{F}$, for every piecewise-smooth curve C in D from A to B.

EXAMPLE 14.10

Evaluate $\int_C 2xy\, dx + x^2\, dy + 2z\, dz$, where C is the first octant intersection of $x^2 + y^2 = 1$ and $z = 2x + 4$ from $(0, 1, 4)$ to $(1, 0, 6)$.

SOLUTION   Since $\nabla(x^2 y + z^2) = 2xy\hat{\mathbf{i}} + x^2\hat{\mathbf{j}} + 2z\hat{\mathbf{k}}$, the line integral is independent of path everywhere, and

$$\int_C 2xy\, dx + x^2 dy + 2z\, dz = \left\{ x^2 y + z^2 \right\}_{(0,1,4)}^{(1,0,6)} = 36 - 16 = 20.$$

The following corollary is also an immediate consequence of Theorem 14.3.

COROLLARY 14.3.2

The line integral $\int \mathbf{F} \cdot d\mathbf{r}$ is independent of path in a domain D if and only if

$$\oint_C \mathbf{F} \cdot d\mathbf{r} = 0 \qquad\qquad (14.36)$$

for every closed path in D.

Theorem 14.3 states that a necessary and sufficient condition for line integral 14.28 to be independent of path is the existence of a function $\phi(x, y, z)$ such that $\nabla \phi = \mathbf{F}$. For most problems it is obvious whether such a function $\phi(x, y, z)$ exists; but for a few, it is not. Since much time could be wasted searching for $\phi(x, y, z)$ (when in fact it does not exist), it would be helpful to have a test that states *a priori* whether $\phi(x, y, z)$ does indeed exist. Such a test is contained in Theorem 14.1. It states that $\mathbf{F}$ is the gradient of some scalar function $\phi(x, y, z)$ if $\nabla \times \mathbf{F} = \mathbf{0}$. When this result is combined with Theorem 14.3, we obtain this important theorem.

THEOREM 14.4

Let D be a domain in which $P(x, y, z)$, $Q(x, y, z)$, and $R(x, y, z)$ have continuous first derivatives. If the line integral $\int \mathbf{F} \cdot d\mathbf{r} = \int P\, dx + Q\, dy + R\, dz$ is independent of path in D, then $\nabla \times \mathbf{F} = \mathbf{0}$ in D. Conversely, if D is simply connected, and $\nabla \times \mathbf{F} = \mathbf{0}$ in D, then the line integral is independent of path in D.

We have in Theorem 14.4 a simple test to determine whether a given line integral is independent of path: We see whether the curl of $\mathbf{F}$ is zero. Evaluation of a line integral that is independent of path still requires the function $\phi(x, y, z)$, but it is at least nice to know that ϕ exists before searching for it.

Theorems 14.1, 14.3, and 14.4 have identified an important equivalence, at least in simply connected domains:

1. $\int \mathbf{F} \cdot d\mathbf{r}$ is independent of path in D.
2. $\mathbf{F} = \nabla\phi$ for some function $\phi(x, y, z)$ defined in D.
3. $\nabla \times \mathbf{F} = \mathbf{0}$ in D.

Theorem 14.1 states that (2) and (3) are equivalent; Theorem 14.3 verifies the equivalence of (1) and (2); and these two imply the equivalence of (1) and (3) (Theorem 14.4).

For line integrals in the xy-plane, this equivalence is still valid except that $\nabla \times \mathbf{F} = \mathbf{0}$ can be stated more simply as

$$\frac{\partial Q}{\partial x} = \frac{\partial P}{\partial y} \tag{14.37}$$

(see equation 14.16).

EXAMPLE 14.11

Evaluate $\int_C 2xye^z\, dx + (x^2e^z + y)\, dy + (x^2ye^z - z)\, dz$ along the straight line C from $(0, 1, 2)$ to $(2, 1, -8)$.

SOLUTION **Method 1** Parametric equations for the straight line are

$$C: \quad x = 2t, \quad y = 1, \quad z = 2 - 10t, \quad 0 \le t \le 1.$$

If I is the value of the line integral, then

$$I = \int_0^1 \{2(2t)(1)e^{2-10t}(2\,dt) + [(2t)^2e^{2-10t} + 1](0)$$

$$+ [(2t)^2(1)e^{2-10t} - 2 + 10t](-10\,dt)\}$$

$$= \int_0^1 [8e^2(t - 5t^2)e^{-10t} + 20 - 100t]\, dt$$

$$= 8e^2\left\{(t - 5t^2)\frac{e^{-10t}}{-10}\right\}_0^1 + \frac{4e^2}{5}\int_0^1 (1 - 10t)e^{-10t}\, dt + 10\{2t - 5t^2\}_0^1$$

$$= 8e^2\left\{\frac{2}{5}e^{-10}\right\} - 30 + \frac{4}{5}e^2\left\{(1 - 10t)\frac{e^{-10t}}{-10}\right\}_0^1 + \frac{2e^2}{25}\int_0^1 -10e^{-10t}\, dt$$

$$= \frac{16}{5}e^{-8} - 30 + \frac{4}{5}e^2\left(\frac{9}{10}e^{-10} + \frac{1}{10}\right) + \frac{2e^2}{25}\left\{e^{-10t}\right\}_0^1 = 4e^{-8} - 30.$$

Method 2 It is evident that

$$\nabla\left(x^2ye^z + \frac{y^2}{2} - \frac{z^2}{2}\right) = 2xye^z\hat{\mathbf{i}} + (x^2e^z + y)\hat{\mathbf{j}} + (x^2ye^z - z)\hat{\mathbf{k}},$$

and hence the line integral is independent of path. Its value is therefore

$$I = \left\{x^2ye^z + \frac{y^2}{2} - \frac{z^2}{2}\right\}_{(0,1,2)}^{(2,1,-8)} = \left(4e^{-8} + \frac{1}{2} - 32\right) - \left(\frac{1}{2} - 2\right) = 4e^{-8} - 30.$$

Method 3   Since

$$\nabla \times (2xye^z\hat{\mathbf{i}} + (x^2e^z + y)\hat{\mathbf{j}} + (x^2ye^z - z)\hat{\mathbf{k}})$$

$$= \begin{vmatrix} \hat{\mathbf{i}} & \hat{\mathbf{j}} & \hat{\mathbf{k}} \\ \partial/\partial x & \partial/\partial y & \partial/\partial z \\ 2xye^z & x^2e^z + y & x^2ye^z - z \end{vmatrix}$$

$$= (x^2e^z - x^2e^z)\hat{\mathbf{i}} + (2xye^z - 2xye^z)\hat{\mathbf{j}} + (2xe^z - 2xe^z)\hat{\mathbf{k}} = \mathbf{0},$$

the line integral is independent of path. Thus, there exists a function $\phi(x, y, z)$ such that

$$\nabla\phi = 2xye^z\hat{\mathbf{i}} + (x^2e^z + y)\hat{\mathbf{j}} + (x^2ye^z - z)\hat{\mathbf{k}}$$

or

$$\frac{\partial\phi}{\partial x} = 2xye^z, \quad \frac{\partial\phi}{\partial y} = x^2e^z + y, \quad \frac{\partial\phi}{\partial z} = x^2ye^z - z.$$

Integration of the first of these equations yields

$$\phi(x, y, z) = x^2ye^z + K(y, z).$$

Substitution of this function into the left side of the second equation gives

$$x^2e^z + \frac{\partial K}{\partial y} = x^2e^z + y \quad \Longrightarrow \quad \frac{\partial K}{\partial y} = y.$$

Consequently, $K(y, z) = \dfrac{y^2}{2} + L(z)$, and we know both the x- and y-dependence of ϕ:

$$\phi(x, y, z) = x^2ye^z + \frac{y^2}{2} + L(z).$$

To obtain the z-dependence contained in $L(z)$, we substitute into the left side of the third equation,

$$x^2ye^z + \frac{dL}{dz} = x^2ye^z - z \quad \Longrightarrow \quad \frac{dL}{dz} = -z.$$

Hence, $L(z) = -z^2/2 + C$, where C is a constant, and

$$\phi(x, y, z) = x^2ye^z + \frac{y^2}{2} - \frac{z^2}{2} + C.$$

Finally, then, we have

$$I = \left\{ x^2ye^z + \frac{y^2}{2} - \frac{z^2}{2} \right\}_{(0,1,2)}^{(2,1,-8)} = 4e^{-8} - 30.$$

Method 1 is one of "brute force." The function $x^2ye^z + y^2/2 - z^2/2$ in method 2 was obtained by observation. Method 3 is the systematic procedure suggested in Section 14.1 for finding the function $\phi(x, y, z)$.

EXAMPLE 14.12

Evaluate

$$I = \int_C \left(\frac{x^3 - 2y^2}{x^3 y} \right) dx + \left(\frac{y^2 - x^3}{x^2 y^2} \right) dy + 2z^2 \, dz,$$

where C is the curve $y = x^2$, $z = x - 1$ from $(1, 1, 0)$ to $(2, 4, 1)$.

SOLUTION   If we set

$$\mathbf{F} = \left(\frac{x^3 - 2y^2}{x^3 y} \right) \hat{\mathbf{i}} + \left(\frac{y^2 - x^3}{x^2 y^2} \right) \hat{\mathbf{j}} + 2z^2 \hat{\mathbf{k}},$$

$$= \left(\frac{1}{y} - \frac{2y}{x^3} \right) \hat{\mathbf{i}} + \left(\frac{1}{x^2} - \frac{x}{y^2} \right) \hat{\mathbf{j}} + 2z^2 \hat{\mathbf{k}},$$

it is evident that $\mathbf{F} = \nabla \phi$ if

$$\phi(x, y, z) = \frac{x}{y} + \frac{y}{x^2} + \frac{2z^3}{3},$$

and this is valid in any domain that does not contain points on the yz-plane ($x = 0$) or xz-plane ($y = 0$). Since C does not pass through either of these planes, then

$$I = \left\{ \frac{x}{y} + \frac{y}{x^2} + \frac{2z^3}{3} \right\}_{(1,1,0)}^{(2,4,1)} = \left(\frac{1}{2} + 1 + \frac{2}{3} \right) - \left(1 + 1 \right) = \frac{1}{6}.$$

EXAMPLE 14.13

In thermodynamics the state of a gas is described by four variables — pressure P, absolute temperature T, internal energy U, and volume V. These variables are related by two equations of state,

$$F(P, T, U, V) = 0 \quad \text{and} \quad G(P, T, U, V) = 0,$$

so that two of the variables are independent and two are dependent. If U and V are chosen as independent variables, then $T = T(U, V)$ and $P = P(U, V)$. An experimental law called the *second law of thermodynamics* states that the line integral

$$\int_C \frac{1}{T} dU + \frac{P}{T} dV$$

is independent of path in the UV-plane. The first term, dU/T, is an incremental change in internal energy of the gas per degree of absolute temperature. The second term, PdV/T, is an increment of work done by the gas per degree of absolute temperature. The line integral gives the total change of these as the gas is taken from one state to another; it is independent of the states (U, V) through which the gas is taken, depending only on initial and final values of U and V. Show that the second law can be expressed in the differential form

$$T \frac{\partial P}{\partial U} - P \frac{\partial T}{\partial U} + \frac{\partial T}{\partial V} = 0.$$

SOLUTION   According to equation 14.37, the line integral is independent of path if and only if

$$\frac{\partial}{\partial U} \left(\frac{P}{T} \right) = \frac{\partial}{\partial V} \left(\frac{1}{T} \right),$$

or

$$0 = \frac{T\frac{\partial P}{\partial U} - P\frac{\partial T}{\partial U}}{T^2} + \frac{1}{T^2}\frac{\partial T}{\partial V};$$

that is,

$$0 = T\frac{\partial P}{\partial U} - P\frac{\partial T}{\partial U} + \frac{\partial T}{\partial V},$$

and the proof is complete.

Since the line integral above is independent of path, there exists a function $S(U, V)$ such that

$$\frac{\partial S}{\partial U} = \frac{1}{T}, \quad \frac{\partial S}{\partial V} = \frac{P}{T},$$

and the value of the line integral is given by

$$\int_C \frac{1}{T}\, dU + \frac{P}{T}\, dV = S(B) - S(A),$$

where C joins A and B. This function, called *entropy*, plays a key role in thermodynamics. The differential dS of $S(U, V)$ can be expressed in the form

$$dS = \frac{\partial S}{\partial U}\, dU + \frac{\partial S}{\partial V}\, dV = \frac{1}{T}dU + \frac{P}{T}\, dV.$$

The equation $T\, dS = dU + P\, dV$ is called the *first of the Gibbs equations*; it is the basis for much of the work in thermodynamics.

EXERCISES 14.4

In Exercises 1–10 show that the line integral is independent of path, and evaluate it.

1. $\int_C xy^2\, dx + x^2 y\, dy$, where C is the curve $y = x^2$, $z = 0$ from $(0, 0, 0)$ to $(1, 1, 0)$

2. $\int_C (3x^2 + y)\, dx + x\, dy$, where C is the straight line from $(2, 1, 5)$ to $(-3, 2, 4)$

3. $\int_C 2xe^y\, dx + (x^2 e^y + 3)\, dy$, where C is the curve $y = \sqrt{1 - x^2}$, $z = 0$ from $(1, 0, 0)$ to $(-1, 0, 0)$

4. $\int_C 3x^2 yz\, dx + x^3 z\, dy + (x^3 y - 4z)\, dz$, where C is the curve $x^2 + y^2 + z^2 = 3$, $y = x$ from $(-1, -1, 1)$ to $(1, 1, -1)$

5. $\int_C -\frac{y}{z}\sin x\, dx + \frac{1}{z}\cos x\, dy - \frac{y}{z^2}\cos x\, dz$, where C is the helix $x = 2\cos t$, $y = 2\sin t$, $z = t$ from $(2, 0, 2\pi)$ to $(2, 0, 4\pi)$

6. $\oint_C y\cos x\, dx + \sin x\, dy$ once clockwise around the circle $x^2 + y^2 - 2x + 4y = 7$, $z = 0$

7. $\int_C x^2\, dx + y^2\, dy + z^2\, dz$, where C is the curve $x + y = 1$, $x + z = 1$ from $(-2, 3, 3)$ to $(1, 0, 0)$

8. $\int_C y\, dx + x\, dy + z\, dz$, where C is the curve $z = x^2 + y^2$, $x + y = 1$ from $(1, 0, 1)$ to $(-1, 2, 5)$

9. $\int_C \frac{1}{y}\, dx - \frac{x}{y^2}\, dy + dz$, where C is the curve $y = x^2 + 1$, $x + y + z = 2$ from $(0, 1, 1)$ to $(3, 10, -11)$

10. $\int_C 3x^2 y^3\, dx + 3x^3 y^2\, dy$, where C is the curve $y = e^x$ from $(0, 1)$ to $(1, e)$

11. Show that if $f(x)$, $g(y)$, and $h(z)$ have continuous first derivatives, then the line integral

$$\int_C f(x)\, dx + g(y)\, dy + h(z)\, dz$$

is independent of path.

12. If $\nabla \times \mathbf{F} = \mathbf{0}$ in a domain D that is not simply connected, can you conclude that the line integral $\int_C \mathbf{F} \cdot d\mathbf{r}$ is not independent of path in D? Explain.

In Exercises 13–18 evaluate the line integral.

* **13.** $\int_C zye^{xy}\,dx + zxe^{xy}\,dy + (e^{xy} - 1)\,dz$, where C is the curve $y = x^2$, $z = x^3$ from $(1, 1, 1)$ to $(2, 4, 8)$

* **14.** $\oint_C y(\tan x + x \sec^2 x)\,dx + x \tan x\,dy + dz$ once around the circle $x^2 + y^2 = 1$, $z = 0$

* **15.** $\int_C \left(\dfrac{1 + y^2}{x^3}\right)dx - \left(\dfrac{y + x^2 y}{x^2}\right)dy + z\,dz$, where C is the broken line joining successively $(1, 0, 0)$, $(25, 2, 3)$, and $(5, 2, 1)$

* **16.** $\oint_C \dfrac{zy\,dx - xz\,dy + xy\,dz}{y^2}$, where C is the curve $x^2 + z^2 = 1$, $y + z = 2$

* **17.** $\int_C -\dfrac{1}{x^2}\operatorname{Tan}^{-1} y\,dx + \dfrac{1}{x + xy^2}\,dy$, where C is the curve $x = y^2 + 1$ from $(2, -1)$ to $(10, 3)$

* **18.** $\int_C \dfrac{1}{(x - 3)^2(y + 5)}\,dx + \dfrac{1}{(x - 3)(y + 5)^2}\,dy + \dfrac{1}{z + 4}\,dz$, where C is the curve $x = y = z$ from $(0, 0, 0)$ to $(2, 2, 2)$

* **19.** Evaluate $\oint_C \dfrac{-y\,dx + x\,dy}{x^2 + y^2}$ (a) once counterclockwise around the circle $x^2 + y^2 = 1$, $z = 0$ and (b) once counterclockwise around the circle $(x - 2)^2 + y^2 = 1$, $z = 0$.

* **20.** Evaluate $\int_C \dfrac{y}{x^2 + y^2}\,dx - \dfrac{x}{x^2 + y^2}\,dy$, where C consists of line segments joining successively the points $(1, 0)$, $(1, 1)$, $(-1, 1)$, and $(-1, 0)$.

* **21.** Is the line integral

$$\int_C \frac{x}{\sqrt{x^2 + y^2}}\,dx + \frac{y}{\sqrt{x^2 + y^2}}\,dy$$

independent of path in the domain consisting of the xy-plane with the origin removed? Is the line integral

$$\int_C \frac{x}{\sqrt{x^2 + y^2 + z^2}}\,dx + \frac{y}{\sqrt{x^2 + y^2 + z^2}}\,dy + \frac{z}{\sqrt{x^2 + y^2 + z^2}}\,dz$$

independent of path in the domain consisting of xyz-space with the origin removed?

* **22.** In which of the following domains is the line integral $\int_C \dfrac{y\,dx - x\,dy}{x^2 + y^2}$ independent of path: (a) $x > 0$, (b) $x < 0$, (c) $y > 0$, (d) $y < 0$, or (e) $x^2 + y^2 > 0$?

* **23.** The second law of thermodynamics states that the line integral $I = \int T^{-1}(dU + P\,dV)$ is independent of path in the UV-plane (see Example 14.13).

  (a) The equations of state for an ideal gas are $PV = nRT$, $U = f(T)$, where n and R are constants and $f(T)$ is some given function. Because of these, it is more convenient to choose T and V as independent variables and to express P and U in terms of T and V. If this is done, show that

  $$I = \int kT^{-1}\,dT + nRV^{-1}\,dV, \qquad \text{where } k = dU/dT.$$

  (b) Since the line integral is independent of path, there exists a function $S(T, V)$, called *entropy*, such that

  $$\int_C \frac{k}{T}\,dT + \frac{nR}{V}\,dV = S(B) - S(A),$$

  where C is any curve joining points A and B. In the case that k is constant, show that $S = k \ln T + nR \ln V + S_0$ where S_0 is a constant.

* **24.** Evaluate the line integral $\oint_C (2xye^{x^2 y} + x^2 y)\,dx + x^2 e^{x^2 y}\,dy$ once clockwise around the ellipse $x^2 + 4y^2 = 4$, $z = 0$.

* **25.** A spring has one end fixed at point P. The other end is moved along the curve $y = f(x)$ from $(x_0, f(x_0))$ to $(x_1, f(x_1))$. If initial and final stretches in the spring are a and b $(b > a)$, what work is done against the spring?

** **26.** Electrostatic forces due to point charges and gravitational forces due to point masses are examples of inverse square force fields — force fields of the form $\mathbf{F} = k\hat{\mathbf{r}}/|\mathbf{r}|^2$, where k is a constant and $\mathbf{r} = x\hat{\mathbf{i}} + y\hat{\mathbf{j}} + z\hat{\mathbf{k}}$.

  (a) Is the line integral representing work done by such a force field independent of path?

  (b) What is the work done by $\mathbf{F}$ in moving a particle from (x_1, y_1, z_1) to (x_2, y_2, z_2)?

14.5 Energy and Conservative Force Fields

In Section 7.5 we introduced the idea of potential energy associated with forces and the law of conservation of energy. We now complete this discussion by showing to which forces can be attributed potential energy.

> **DEFINITION 14.5**
>
> A force field $\mathbf{F}(x, y, z)$, defined in a domain D of space, is said to be **conservative** in D if the line integral $\int \mathbf{F} \cdot d\mathbf{r}$ is independent of path in D.

Since $\int_C \mathbf{F} \cdot d\mathbf{r}$ can be interpreted as the work done by $\mathbf{F}$ along C, a force field is conservative if work that it does is independent of path taken from one point to another. According to the results of Section 14.4, we can also state that a force field $\mathbf{F}$ is conservative if and only if there exists a function $\phi(x, y, z)$ such that $\mathbf{F} = \nabla\phi$.

It is customary to associate a potential energy function $U(x, y, z)$ with a conservative force field $\mathbf{F}$. This function assigns a potential energy to each point (x, y, z) in such a way that if a particle moves from point A to point B, then the difference in potential energy $U(A) - U(B)$ is the work done by $\mathbf{F}$; that is, if C is any curve joining A and B, then

$$U(A) - U(B) = \int_C \mathbf{F} \cdot d\mathbf{r}. \tag{14.38}$$

If $\int \mathbf{F} \cdot d\mathbf{r} > 0$, then potential energy at A is greater than potential energy at B; if $\int \mathbf{F} \cdot d\mathbf{r} < 0$, then potential energy at B is greater than that at A. To find $U(x, y, z)$, we use the fact that because $\mathbf{F}$ is conservative, there exists a function $\phi(x, y, z)$ such that $\mathbf{F} = \nabla\phi$, and

$$\int_C \mathbf{F} \cdot d\mathbf{r} = \phi(B) - \phi(A).$$

If follows, then, that $U(x, y, z)$ must satisfy the equation

$$U(A) - U(B) = \phi(B) - \phi(A) \quad\Longrightarrow\quad U(A) + \phi(A) = U(B) + \phi(B).$$

Since A and B are arbitrary points in D, this last equation states that the value of the function $U(x, y, z) + \phi(x, y, z)$ is the same at every point in the force field,

$$U(x, y, z) + \phi(x, y, z) = C,$$

where C is a constant. Thus,

$$U(x, y, z) = -\phi(x, y, z) + C. \tag{14.39}$$

Equation 14.39 shows that the force field $\mathbf{F}$ defines a potential energy function $U(x, y, z)$ up to an additive constant. (This seems reasonable in that ϕ itself is defined only to an additive constant.) Because $U = -\phi + C$ and $\mathbf{F} = \nabla\phi$, we can also regard U as being defined by the equation

$$\mathbf{F} = -\nabla U. \tag{14.40}$$

The advantage of this equation is that it defines U directly, not through the function ϕ.

For a conservative force field $\mathbf{F}$, then, we define a potential energy function $U(x, y, z)$ by equation 14.40. If a particle moves from A to B, then the work done by $\mathbf{F}$ is

$$W = U(A) - U(B); \tag{14.41}$$

in other words, work done by a conservative force field is equal to loss in potential energy.

On the other hand, if a particle moves under the action of a force $\mathbf{F}$ (and only $\mathbf{F}$), be it conservative or nonconservative, then it does so according to Newton's second law,

$$\mathbf{F} = \frac{d}{dt}(m\mathbf{v}) = m\frac{d\mathbf{v}}{dt},$$

where m is the mass of the particle (assumed constant), $\mathbf{v}$ is its velocity, and t is time. The action of $\mathbf{F}$ produces motion along some curve C, and the work done by $\mathbf{F}$ along this curve from A to B is

$$W = \int_C \mathbf{F} \cdot d\mathbf{r} = \int_\alpha^\beta m\frac{d\mathbf{v}}{dt} \cdot \frac{d\mathbf{r}}{dt} dt = \int_\alpha^\beta m\frac{d\mathbf{v}}{dt} \cdot \mathbf{v}\, dt = \int_\alpha^\beta \frac{d}{dt}\left(\frac{1}{2}m\mathbf{v} \cdot \mathbf{v}\right) dt$$

$$= \left\{\frac{1}{2}m\mathbf{v} \cdot \mathbf{v}\right\}_\alpha^\beta = \left\{\frac{1}{2}m|\mathbf{v}|^2\right\}_\alpha^\beta.$$

Thus, if $K(x, y, z) = \frac{1}{2}m|\mathbf{v}|^2$ represents kinetic energy of the particle, the work done by $\mathbf{F}$ is equal to the gain in kinetic energy of the particle,

$$W = K(B) - K(A) \qquad (14.42)$$

(and this is true for any force $\mathbf{F}$ as long as $\mathbf{F}$ is the total resultant force producing motion).

If the total force producing motion is a conservative force field $\mathbf{F}$, we have two expressions, 14.41 and 14.42, for the work done as a particle moves from one point to another under the action of $\mathbf{F}$. If we equate them, we have

$$U(A) - U(B) = K(B) - K(A)$$

or

$$U(A) + K(A) = U(B) + K(B). \qquad (14.43)$$

We have shown then that if a particle moves under the action of a conservative force field *only*, the sum of the kinetic and potential energies at B must be the same as the sum of the kinetic and potential energies at A. In other words, if E is the total energy of the particle, kinetic plus potential, then

$$E(A) = E(B). \qquad (14.44)$$

Since B can be any point along the path of the particle, it follows that when a particle moves under the action of a conservative force field, and only a conservative force field, then at every point along its trajectory

$$E = \text{a constant.} \qquad (14.45)$$

This is the **law of conservation of energy** for a conservative force field.

EXAMPLE 14.14

Show that the electrostatic force due to a point charge is conservative, and determine a potential energy function for the field.

FIGURE 14.25   Electrostatic force due to a point charge is conservative

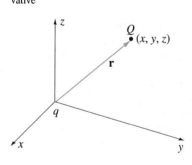

SOLUTION   The electrostatic force on a charge Q due to a charge q is

$$\mathbf{F} = \frac{qQ}{4\pi\epsilon_0|\mathbf{r}|^3}\mathbf{r},$$

where $\mathbf{r}$ is the vector from q to Q. If we choose a Cartesian coordinate system with q at the origin (Figure 14.25), then

$$\mathbf{F} = \frac{qQ}{4\pi\epsilon_0(x^2 + y^2 + z^2)^{3/2}}(x\hat{\mathbf{i}} + y\hat{\mathbf{j}} + z\hat{\mathbf{k}}).$$

Since

$$\nabla\left[\frac{-qQ}{4\pi\epsilon_0(x^2 + y^2 + z^2)^{1/2}}\right] = \mathbf{F},$$

the force field is conservative, and possible potential energy functions are

$$U(x, y, z) = \frac{qQ}{4\pi\epsilon_0(x^2 + y^2 + z^2)^{1/2}} + C = \frac{qQ}{4\pi\epsilon_0 r} + C,$$

where $r = |\mathbf{r}|$. In electrostatics it is customary to choose $U(x, y, z)$ so that $\lim_{r\to\infty} U = 0$, in which case $C = 0$, and

$$U(x, y, z) = \frac{qQ}{4\pi\epsilon_0 r}.$$

In addition, if V is defined as the potential energy per unit test charge Q, then

$$V(x, y, z) = \frac{U}{Q} = \frac{q}{4\pi \epsilon_0 r}.$$

This result agrees with that in Example 14.9.

In Exercises 1–5 determine whether the force field is conservative. Identify each conservative force field, and find a potential energy function.

1. $\mathbf{F}(x, y, z) = \dfrac{q_1 q_2}{4\pi \epsilon_0} \dfrac{x\hat{\mathbf{i}} + y\hat{\mathbf{j}} + z\hat{\mathbf{k}}}{(x^2 + y^2 + z^2)^{3/2}}$

2. $\mathbf{F}(x, y) = mx\hat{\mathbf{i}} + xy\hat{\mathbf{j}}$; m is a constant

3. $\mathbf{F}(x) = -kx\hat{\mathbf{i}}$; k is a constant

4. $\mathbf{F}(x, y, z) = -mg\hat{\mathbf{k}}$; m and g are constants

5. $\mathbf{F}(x, y, z) = GMm\dfrac{x\hat{\mathbf{i}} + y\hat{\mathbf{j}} + z\hat{\mathbf{k}}}{(x^2 + y^2 + z^2)^{3/2}}$; G, M, and m are constants

6. Suppose that $\mathbf{F}(x, y, z)$ is a conservative force field in some domain D, and $U(x, y, z)$ is a potential energy function associated with $\mathbf{F}$. The surfaces $U(x, y, z) = C$, where C is a constant, are called *equipotential surfaces*. Through each point P in D there is one and only one such equipotential surface for $\mathbf{F}$. Show that at P the force $\mathbf{F}$ is normal to the equipotential surface through P.

7. Draw equipotential surfaces for the forces in Exercises 1, 4, and 5.

8. One end of a spring with unstretched length L is fixed at the origin in space. If the other end is at point (x, y, z) (all coordinates in metres), what is the force exerted by the spring? Is this force conservative?

9. Explain why friction is not conservative.

10. (a) When students in Universityland leave their houses, a supernatural power attracts them to the university in such a way that the magnitude of the force at any point is inversely proportional to the square of the distance from the university. This force acts until they are 100 m from the university and then it disappears. Is this force conservative?

   (b) If someone diverts the power so that the force attracts students to the local donut shop, is this force conservative?

11. (a) A particle with mass m moves along the x-axis under the influence of a conservative force field with potential $U(x)$. If the particle is at position x_0 at time $t = 0$, and E is its total energy, show that

$$\int_{x_0}^{x} \frac{dx}{\sqrt{E - U(x)}}dx = \sqrt{\frac{2}{m}}t.$$

   (b) If $U(x) = kx^2/2$ (as would be the case for a mass on the end of a spring), use the result of part (a) to find x as a function of t. Is it what you would expect? Simplify the solution when the initial velocity of the mass is zero.

** **12.** A force field $\mathbf{F}(x, y, z)$ is said to be radially symmetric about the origin if it can be written in the form

$$\mathbf{F}(x, y, z) = f\left(\sqrt{x^2 + y^2 + z^2}\right)\mathbf{r}, \qquad \mathbf{r} = x\hat{\mathbf{i}} + y\hat{\mathbf{j}} + z\hat{\mathbf{k}},$$

for some function f. We often write in such a case that

$$\mathbf{F}(x, y, z) = f(r)\mathbf{r}, \quad \text{where } r = |\mathbf{r}| = \sqrt{x^2 + y^2 + z^2}.$$

   (a) Use Theorem 14.4 to show that such a force is conservative in suitably defined domains [provided that $f(r)$ has a continuous first derivative].

   (b) If A and B are the points in the figure below joined by the curve C, show that

$$\int_C \mathbf{F} \cdot d\mathbf{r} = \int_a^b rf(r)\,dr,$$

   where the limits a and b are the distances from the origin to A and B.

   (c) Have we discussed any radially symmetric force fields in this chapter?

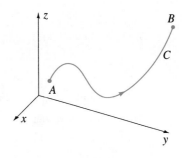

14.6 Green's Theorem

Line integrals in the xy-plane are of the form $\int_C f(x, y)\,ds$, and in the special case that $f(x, y)$ is the tangential component of some vector field $\mathbf{F}(x, y) = P(x, y)\hat{\mathbf{i}} + Q(x, y)\hat{\mathbf{j}}$ along C, they take the form

$$\int_C \mathbf{F} \cdot d\mathbf{r} = \int_C P(x, y)\,dx + Q(x, y)\,dy. \tag{14.46}$$

We now show that when C is a closed curve, line integral 14.46 can usually be replaced by a double integral. The precise result is contained in the following theorem.

THEOREM 14.5   (Green's Theorem)

Let C be a piecewise-smooth, closed curve in the xy-plane that does not intersect itself and that encloses a region R (Figure 14.26). If $P(x, y)$ and $Q(x, y)$ have continuous first partial derivatives in a domain D containing C and R, then

$$\oint_C P\,dx + Q\,dy = \iint_R \left(\frac{\partial Q}{\partial x} - \frac{\partial P}{\partial y} \right) dA. \tag{14.47}$$

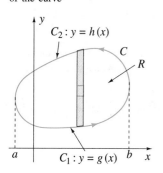

FIGURE 14.26   Green's theorem replaces a line integral around a closed curve with a double integral over the interior of the curve

PROOF   First consider a simple region R for which every line parallel to the x- and y-axes that intersects C does so in at most two points (Figure 14.26). Then C can be subdivided into an upper and a lower part,

$$C_2: \quad y = h(x) \qquad \text{and} \qquad C_1: \quad y = g(x).$$

If we consider the second term on the right of equation 14.47, we have

$$\iint_R -\frac{\partial P}{\partial y}\,dA = \int_a^b \int_{g(x)}^{h(x)} -\frac{\partial P}{\partial y}\,dy\,dx = \int_a^b \left\{ -P \right\}_{g(x)}^{h(x)} dx$$

$$= \int_a^b \{P[x, g(x)] - P[x, h(x)]\}\,dx.$$

On the other hand, the first term on the left of 14.47 is

$$\oint_C P\,dx = \int_{C_1} P\,dx + \int_{C_2} P\,dx = \int_{C_1} P\,dx - \int_{-C_2} P\,dx,$$

and if we use x as a parameter along C_1 and $-C_2$, then

$$\oint_C P\,dx = \int_a^b P[x, g(x)]\,dx - \int_a^b P[x, h(x)]\,dx$$

$$= \int_a^b \{P[x, g(x)] - P[x, h(x)]\}\,dx.$$

We have shown therefore that

$$\oint_C P\,dx = \iint_R -\frac{\partial P}{\partial y}\,dA.$$

By subdividing C into two parts of the type $x = g(y)$ and $x = h(y)$ [where $g(y) \le h(y)$], we can also show that

$$\oint_C Q\,dy = \iint_R \frac{\partial Q}{\partial x}\,dA.$$

Addition of these results gives Green's theorem for this C and R.

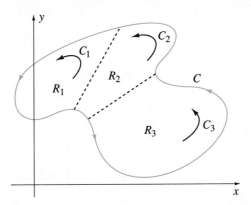

FIGURE 14.27   This region can be divided into regions
where the condition is that every line parallel to the x- and y-
axes that intersects its boundary does so at most twice

Now consider a more general region R such as that in Figure 14.27; it can be decomposed
into n subregions R_i, each of which satisfies the condition that lines parallel to the coordinate
axes intersect its boundary in at most two points. For each subregion R_i, Green's theorem gives

$$\oint_{C_i} P\,dx + Q\,dy = \iint_{R_i} \left(\frac{\partial Q}{\partial x} - \frac{\partial P}{\partial y} \right) dA.$$

If these results are added, we get

$$\sum_{i=1}^{n} \oint_{C_i} P\,dx + Q\,dy = \sum_{i=1}^{n} \iint_{R_i} \left(\frac{\partial Q}{\partial x} - \frac{\partial P}{\partial y} \right) dA.$$

Now R is composed of the subregions R_i; thus the right side of this equation is the double integral
over R. Figure 14.27 illustrates that when line integrals over the C_i are added, contributions from
ancillary (interior) curves cancel in pairs, leaving the line integral around C. This completes
the proof. ∎

We omit a proof for even more general regions that cannot be divided into a finite number
of these subregions. The interested reader should consult more advanced books.

EXAMPLE 14.15

FIGURE 14.28   Green's
theorem applied to a triangle

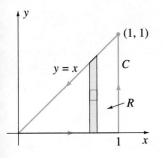

Evaluate the line integral of Example 14.7.

SOLUTION   By Green's theorem (see Figure 14.28), we have

$$\oint_C y^2\,dx + x^2\,dy = \iint_R (2x - 2y)\,dA = 2\int_0^1 \int_0^x (x - y)\,dy\,dx$$

$$= 2\int_0^1 \left\{ xy - \frac{y^2}{2} \right\}_0^x dx = \int_0^1 x^2\,dx = \left\{ \frac{x^3}{3} \right\}_0^1 = \frac{1}{3}.$$

EXAMPLE 14.16

Show that the area of a region R is defined by each of the line integrals

$$\oint_C x\,dy = \oint_C -y\,dx = \frac{1}{2}\oint_C x\,dy - y\,dx,$$

where C is the boundary of R.

SOLUTION   By Green's theorem, we have

$$\oint_C x\,dy = \iint_R 1\,dA = \text{area of } R$$

and

$$\oint_C -y\,dx = \iint_R 1\,dA = \text{area of } R.$$

The third expression for area is the average of these two equations.

The area formulas in this example are of particular value when the curve C is defined parametrically (see Exercises 15–19). We also use it in our next project.

Consulting Project 23

We all know that the area of any plane polygon can be found by dividing the polygon into rectangles and triangles. We are being asked whether we can provide a faster way to find such areas.

SOLUTION   Consider finding the area inside the polygon in Figure 14.29a. It has n vertices labelled $P_i(x_i, y_i)$. According to Example 14.16, the area is one-half the line integral of $x\,dy - y\,dx$ around the edge C of the polygon. Suppose we denote the individual edges of the polygon by C_i so that $C = C_1 + \cdots + C_n$. Then the area of the polygon is

$$A = \frac{1}{2}\oint_C x\,dy - y\,dx = \frac{1}{2}\sum_{i=1}^{n}\int_{C_i} x\,dy - y\,dx.$$

Evaluation of the line integral along each of the C_i will be similar. If we evaluate the line integral along C_1, we will see how to evaluate the line integral along each of the C_i. With parametric equations $x = x_1 + (x_2 - x_1)t$, $y = y_1 + (y_2 - y_1)t$, $0 \le t \le 1$, for C_1, we find

$$\int_{C_1} x\,dy - y\,dx = \int_0^1 [x_1 + (x_2 - x_1)t](y_2 - y_1)\,dt$$

$$- [y_1 + (y_2 - y_1)t](x_2 - x_1)\,dt$$

$$= \int_0^1 (x_1 y_2 - x_2 y_1)\,dt = x_1 y_2 - x_2 y_1.$$

With similar results along the other line segments, we write that

$$A = \frac{1}{2}(x_1 y_2 - x_2 y_1) + \frac{1}{2}(x_2 y_3 - x_3 y_2) + \cdots + \frac{1}{2}(x_n y_1 - x_1 y_n).$$

We can take advantage of the fact that each term in parentheses has the form of a 2×2 determinant. Suppose we arrange the vertices in a vertical column as shown in Figure 14.29b, and take

$$\frac{1}{2}[(\text{sum of downward products to right}) - (\text{sum of downward products to left})].$$

We obtain

$$\frac{1}{2}[(x_1 y_2 + x_2 y_3 + x_3 y_4 + \cdots + x_n y_1) - (x_2 y_1 + x_3 y_2 + x_4 y_3 + \cdots + x_1 y_n)]$$

$$= \frac{1}{2}[(x_1 y_2 - x_2 y_1) + (x_2 y_3 - x_3 y_2) + \cdots + (x_n y_1 - x_1 y_n)].$$

This is A. Here then is what we have discovered. To find the area of a polygon, label its vertices $(x_1, y_1), \cdots, (x_n, y_n)$ consecutively, proceeding counterclockwise around the polygon. Arrange the vertices in a column repeating the first vertex. The area of the polygon is then one-half the sum of downward products to the right less the sum of downward products to the left.

FIGURE 14.29a

Fast way to calculate area inside a polygon

FIGURE 14.29b

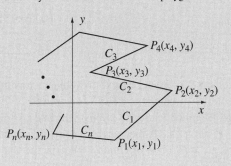

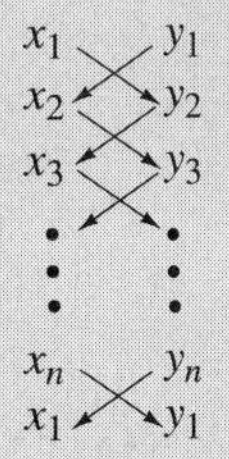

Green's theorem is applicable to line integrals of type 14.46 only when curve C is closed. However, when C is not closed, it can be made so with the addition of another curve. In this way, Green's theorem may be brought to bear even on line integrals when C is not closed. The following example is an illustration.

EXAMPLE 14.17

Use Green's theorem to evaluate the line integral

$$\int_C (x - y^3)\, dx + y^4 \sin y^2\, dy,$$

where C is the parabola $y = x^2$ from $(1, 1)$ to $(-1, 1)$.

FIGURE 14.30 Green's
theorem applied to a parabola

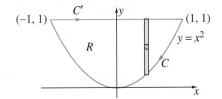

SOLUTION If to C we add the straight-line segment C' from $(-1, 1)$ to $(1, 1)$ (Figure 14.30), then the combined curve $C + C'$ is closed. Green's theorem then gives

$$\oint_{C+C'} (x - y^3) \, dx + y^4 \sin y^2 \, dy = -\iint_R [0 - (-3y^2)] \, dA = -\int_{-1}^{1} \int_{x^2}^{1} 3y^2 \, dy \, dx$$

$$= -\int_{-1}^{1} \left\{ y^3 \right\}_{x^2}^{1} dx = -\int_{-1}^{1} (1 - x^6) \, dx$$

$$= -\left\{ x - \frac{x^7}{7} \right\}_{-1}^{1} = -\frac{12}{7}.$$

But, because

$$\oint_{C+C'} (x - y^3) \, dx + y^4 \sin y^2 \, dy$$

$$= \int_C (x - y^3) \, dx + y^4 \sin y^2 \, dy + \int_{C'} (x - y^3) \, dx + y^4 \sin y^2 \, dy,$$

we can write that

$$\int_C (x - y^3) \, dx + y^4 \sin y^2 \, dy = -\frac{12}{7} - \int_{C'} (x - y^3) \, dx + y^4 \sin y^2 \, dy$$

$$= -\frac{12}{7} - \int_{-1}^{1} (x - 1) \, dx = -\frac{12}{7} - \left\{ \frac{x^2}{2} - x \right\}_{-1}^{1} = \frac{2}{7}.$$

Green's theorem cannot be used to evaluate a line integral around a closed curve that contains a point at which either $P(x, y)$ or $Q(x, y)$ fails to have continuous first partial derivatives. For example, it cannot be used to evaluate the line integral in Exercise 19(a) of Section 14.4 because neither P nor Q is defined at the origin. Try it and compare the result to the correct answer (2π). A generalization of Green's theorem that can be useful in such situations is discussed in Exercise 30.

EXERCISES 14.6

In Exercises 1–11 use Green's theorem (if possible) to evaluate the line integral.

1. $\oint_C y^2 \, dx + x^2 \, dy$, where C is the circle $x^2 + y^2 = 1$

2. $\oint_C (x^2 + 2y^2) \, dy$, where C is the curve $(x - 2)^2 + y^2 = 1$

3. $\oint_C x^2 e^y \, dx + (x + y) \, dy$, where C is the square with vertices $(\pm 1, 1)$ and $(\pm 1, -1)$

4. $\oint_C xy^3 \, dx + x^2 \, dy$, where C is the curve enclosing the region bounded by $x = \sqrt{1 + y^2}$, $x = 2$

5. $\oint_C (x^3 + y^3) \, dx + (x^3 - y^3) \, dy$, where C is the curve enclosing the region bounded by $x = y^2 - 1$, $x = 1 - y^2$

6. $\oint_C 2 \operatorname{Tan}^{-1}(y/x) \, dx + \ln(x^2 + y^2) \, dy$, where C is the circle $(x - 4)^2 + (y - 1)^2 = 2$

7. $\oint_C (3x^2 y^3 + y) \, dx + (3x^3 y^2 + 2x) \, dy$, where C is the boundary of the region enclosed by $x + y = 1$, $x = -1$, $y = -1$

8. $\oint_C (x^3 + y^3) \, dx + (x^3 - y^3) \, dy$, where C is the curve $2|x| + |y| = 1$

9. $\oint_C (x^2 y^2 + 3x) \, dx + (2xy - y) \, dy$, where C is the boundary of the region enclosed by $x = 1 - y^2$ ($x \geq 0$), $y = x + 1$, $y + x + 1 = 0$

10. $\oint_C (xy^2 + 2x) \, dx + (x^2 y + y + x^2) \, dy$, where C is the boundary of the region enclosed by $y^2 - x^2 = 4$, $x = 0$, $x = 3$

11. $\oint_C \frac{-y \, dx + x \, dy}{x^2 + y^2}$, where C is the circle $x^2 + y^2 = 1$

12. Show that Green's theorem can be expressed vectorially in the form

$$\oint_C \mathbf{F} \cdot d\mathbf{r} = \iint_R (\nabla \times \mathbf{F}) \cdot \hat{\mathbf{k}} \, dA.$$

13. If a curve C is traced out in the direction defined by Green's theorem, it can be shown that a normal vector to C that always points to the outside of C is $\mathbf{n} = (dy, -dx)$. Show that Green's theorem can be written vectorially in the form

$$\oint_C \mathbf{F} \cdot \hat{\mathbf{n}} \, ds = \iint_R \nabla \cdot \mathbf{F} \, dA.$$

In Exercises 14–19 use the results of Example 14.16 to find the area enclosed by the curve.

14. $x^2/a^2 + y^2/b^2 = 1$

15. The strophoid $x = (1 - t^2)/(1 + t^2)$, $y = (t - t^3)/(1 + t^2)$ (see Exercise 54 in Section 9.1)

16. The astroid $x = \cos^3\theta$, $y = \sin^3\theta$ (see Exercise 62 in Section 9.1)

17. The right loop of the curve of Lissajous (see Exercise 60 in Section 9.1)

18. The deltoid $x = 2\cos t + \cos 2t$, $y = 2\sin t - \sin 2t$

19. The droplet $x = 2\cos t - \sin 2t$, $y = \sin t$

In Exercises 20–24 use the result of Project 23 to find the area of the polygon with the points as successive vertices.

20. $(1, 0)$, $(0, 1)$, $(-1, 0)$, $(0, -1)$

21. $(1, 2)$, $(-3, 2)$, $(4, 1)$

22. $(2, -2)$, $(1, -3)$, $(-2, 1)$, $(5, 6)$

23. $(3, 0)$, $(1, 1)$, $(2, 5)$, $(-4, -4)$

24. $(0, 4)$, $(-1, 0)$, $(-2, 0)$, $(-3, -4)$, $(0, -5)$, $(6, -2)$, $(3, 0)$, $(2, 2)$

In Exercises 25–29 evaluate the line integral.

25. $\oint_C (2xye^{x^2y} + 3x^2y) \, dx + x^2e^{x^2y} \, dy$, where C is the ellipse $x^2 + 4y^2 = 4$

26. $\oint_C (3x^2y^3 - x^2y) \, dx + (xy^2 + 3x^3y^2) \, dy$, where C is the circle $x^2 + y^2 = 9$

27. $\oint_C -x^3y^2 \, dx + x^2y^3 \, dy$, where C is the right loop of $(x^2 + y^2)^{3/2} = x^2 - y^2$

∗ 28. $\int_C (x - y)(dx + dy)$, where C is the semicircular part of $x^2 + y^2 = 4$ above $y = x$ from $(-\sqrt{2}, -\sqrt{2})$ to $(\sqrt{2}, \sqrt{2})$

∗ 29. $\int_C (e^y - y\sin x) \, dx + (\cos x + xe^y) \, dy$, where C is the curve $x = 1 - y^2$ from $(0, -1)$ to $(0, 1)$

∗ 30. The result of this exercise is useful when the curve C in Green's theorem contains a point (or points) at which either P or Q fails to have continuous first partial derivatives (see Exercises 31–35).

(a) Suppose a piecewise-smooth curve C (figure below) contains in its interior another piecewise-smooth curve C', and $P(x, y)$ and $Q(x, y)$ have continuous first partial derivatives in a domain containing C and C' and the region R between them. Prove that

$$\oint_C P \, dx + Q \, dy + \oint_{C'} P \, dx + Q \, dy = \iint_R \left(\frac{\partial Q}{\partial x} - \frac{\partial P}{\partial y} \right) dA.$$

Hint: Join C and C' by two curves such as those in the figure.

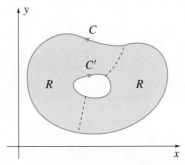

(b) Extend this result to show that when C' is replaced by n distinct curves C_i (figure below), and P and Q have continuous first partial derivatives in a domain containing C and the C_i and the region R between them,

$$\oint_C P \, dx + Q \, dy + \oint_{C_1} P \, dx + Q \, dy + \cdots$$

$$+ \oint_{C_n} P \, dx + Q \, dy = \iint_R \left(\frac{\partial Q}{\partial x} - \frac{\partial P}{\partial y} \right) dA.$$

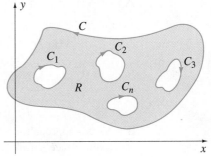

(c) What can we conclude in parts (a) and (b) if $\partial Q/\partial x = \partial P/\partial y$ in R?

In Exercises 31–33 use the result of Exercise 30 to evaluate the line integral.

* **31.** $\oint_C \dfrac{y\,dx - (x-1)\,dy}{(x-1)^2 + y^2}$, where C is the circle $x^2 + y^2 = 4$

* **32.** $\oint_C \dfrac{-x^2 y\,dx + x^3\,dy}{(x^2 + y^2)^2}$, where C is the ellipse $4x^2 + y^2 = 1$

* **33.** $\oint_C \dfrac{-y\,dx + x\,dy}{x^2 + y^2}$, where C is the square with vertices $(\pm 2, 0)$ and $(0, \pm 2)$

* **34.** Show that the line integral of Exercise 19 in Section 14.4 has value $\pm 2\pi$ for every piecewise-smooth, closed curve enclosing the origin that does not intersect itself.

* **35.**
  (a) In what domains is the line integral $\displaystyle\int_C \dfrac{x\,dx + y\,dy}{x^2 + y^2}$ independent of path?

  (b) Evaluate the integral clockwise around the curve $x^2 + y^2 - 2y = 1$.

In Exercises 36–38 assume that $P(x, y)$ and $Q(x, y)$ have continuous second partial derivatives in a domain containing R and C (figure follows). Let $\hat{\mathbf{n}} = (dy/ds, -dx/ds)$ be the outward-pointing normal to C, and let

$$\frac{\partial P}{\partial n} = \nabla P \cdot \hat{\mathbf{n}}, \qquad \frac{\partial Q}{\partial n} = \nabla Q \cdot \hat{\mathbf{n}}$$

be the directional derivatives of P and Q in the direction $\hat{\mathbf{n}}$.

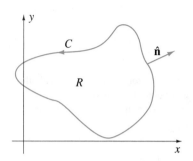

* **36.** Show that

$$\oint_C \frac{\partial P}{\partial n}\,ds = \iint_R \nabla^2 P\,dA,$$

where $\nabla^2 P = \dfrac{\partial^2 P}{\partial x^2} + \dfrac{\partial^2 P}{\partial y^2}$. *Hint:* See Exercise 13. What can we conclude if $P(x, y)$ satisfies Laplace's equation in R?

* **37.** Show that

$$\oint_C P\frac{\partial Q}{\partial n}\,ds = \iint_R P\nabla^2 Q\,dA + \iint_R \nabla P \cdot \nabla Q\,dA.$$

Hint: Use identity 14.11. This result is often called *Green's first identity* (in the plane).

* **38.** Prove that

$$\oint_C \left(P\frac{\partial Q}{\partial n} - Q\frac{\partial P}{\partial n}\right)ds = \iint_R (P\,\nabla^2 Q - Q\,\nabla^2 P)\,dA.$$

This is often called *Green's second identity* (in the plane).

** **39.** Find all possible values for the line integral $\displaystyle\oint_C \dfrac{-y\,dx + x\,dy}{x^2 + y^2}$ for curves in the xy-plane not passing through the origin.

14.7 Surface Integrals

FIGURE 14.31 Divide S into smaller surfaces to define the surface integral of $f(x, y, z)$ over S

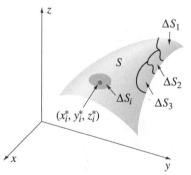

Consider a function $f(x, y, z)$ defined on some surface S (Figure 14.31). We divide S into n subsurfaces of areas $\Delta S_1, \Delta S_2, \ldots, \Delta S_n$ in any manner whatsoever. On each subsurface ΔS_i ($i = 1, \ldots, n$) we choose an arbitrary point (x_i^*, y_i^*, z_i^*), and form the sum

$$f(x_1^*, y_1^*, z_1^*)\,\Delta S_1 + f(x_2^*, y_2^*, z_2^*)\,\Delta S_2 + \cdots + f(x_n^*, y_n^*, z_n^*)\,\Delta S_n$$

$$= \sum_{i=1}^{n} f(x_i^*, y_i^*, z_i^*)\,\Delta S_i. \qquad (14.48)$$

If this sum approaches a limit as the number of subsurfaces becomes increasingly large and every subsurface shrinks to a point, we call the limit the **surface integral** of $f(x, y, z)$ over the surface S, and denote it by

$$\iint_S f(x, y, z)\,dS = \lim_{\|\Delta S_i\| \to 0} \sum_{i=1}^{n} f(x_i^*, y_i^*, z_i^*)\,\Delta S_i. \qquad (14.49)$$

Surface integrals, like all integrals, are limit summations. We think of dS as a small piece of area on S, and each dS is multiplied by the value of $f(x, y, z)$ for that area. All such products are then added together and the limit taken as the pieces of area shrink to points.

The following theorem guarantees existence of surface integrals of continuous functions over smooth surfaces.

> ### THEOREM 14.6
>
> If a function $f(x, y, z)$ is continuous on a smooth surface S of finite area, then the surface integral of $f(x, y, z)$ over S exists. If S projects in a one-to-one fashion onto a region S_{xy} in the xy-plane, then the surface integral of $f(x, y, z)$ over S can be evaluated by means of the following double integral:
>
> $$\iint_S f(x, y, z)\, dS = \iint_{S_{xy}} f[x, y, g(x, y)]\sqrt{1 + \left(\frac{\partial z}{\partial x}\right)^2 + \left(\frac{\partial z}{\partial y}\right)^2}\, dA, \quad (14.50)$$
>
> where $z = g(x, y)$ is the equation of S.

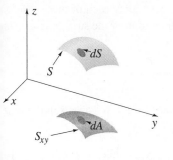

It is not necessary to memorize 14.50 as a formula. The right side is obtained by expressing z and dS in terms of x and y and interpreting the result as a double integral over the projection of S in the xy-plane. To be more explicit, recall from Section 13.6 that when a surface S can be represented in the form $z = g(x, y)$, a small area dS on S is related to its projection dA in the xy-plane according to the formula

$$dS = \sqrt{1 + \left(\frac{\partial z}{\partial x}\right)^2 + \left(\frac{\partial z}{\partial y}\right)^2}\, dA \quad (14.51)$$

(Figure 14.32). If we substitute this into the left side of 14.50 and at the same time use $z = g(x, y)$ to express $f(x, y, z)$ in terms of x and y, then

$$\iint_S f(x, y, z)\, dS = \iint_S f[x, y, g(x, y)]\sqrt{1 + \left(\frac{\partial z}{\partial x}\right)^2 + \left(\frac{\partial z}{\partial y}\right)^2}\, dA.$$

But if we now interpret the right side of this equation as the double integral of the function $f[x, y, g(x, y)]\sqrt{1 + (\partial z/\partial x)^2 + (\partial z/\partial y)^2}$ over the projection S_{xy} of S onto the xy-plane, we obtain 14.50.

Note the analogy between equations 14.50 and 14.20. Equation 14.20 states that the line integral on the left can be evaluated by means of the definite integral on the right. Equation 14.50 states that the surface integral on the left can be evaluated by means of the double integral on the right.

If a surface does not project one-to-one onto an area in the xy-plane, then one possibility is to divide it into parts, each of which projects one-to-one onto the xy-plane. The total surface integral over the surface is then the sum of the surface integrals over the parts. For example, if we require the surface integral of a function $f(x, y, z)$ over the sphere $S : x^2 + y^2 + z^2 = 1$ (Figure 14.33), we could divide S into two hemispheres,

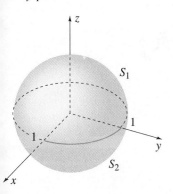

$$S_1 : \quad z = \sqrt{1 - x^2 - y^2}, \qquad S_2 : \quad z = -\sqrt{1 - x^2 - y^2},$$

each of which projects onto $S_{xy} : x^2 + y^2 \leq 1$. Then

$$\iint_S f(x, y, z)\, dS = \iint_{S_1} f(x, y, z)\, dS + \iint_{S_2} f(x, y, z)\, dS.$$

A second possibility is to project surfaces onto either the xz-plane or the yz-plane. If a surface projects one-to-one onto regions S_{xz} and S_{yz} in these planes, then

$$\iint_S f(x, y, z)\, dS = \iint_{S_{xz}} f[x, g(x, z), z]\sqrt{1 + \left(\frac{\partial y}{\partial x}\right)^2 + \left(\frac{\partial y}{\partial z}\right)^2}\, dA \qquad (14.52)$$

and

$$\iint_S f(x, y, z)\, dS = \iint_{S_{yz}} f[g(y, z), y, z]\sqrt{1 + \left(\frac{\partial x}{\partial y}\right)^2 + \left(\frac{\partial x}{\partial z}\right)^2}\, dA. \qquad (14.53)$$

We should also note that the area of a surface S is defined by the surface integral

$$\iint_S dS, \qquad (14.54)$$

and if S projects one-to-one onto S_{xy}, then

$$\text{area of } S = \iint_{S_{xy}} \sqrt{1 + \left(\frac{\partial z}{\partial x}\right)^2 + \left(\frac{\partial z}{\partial y}\right)^2}\, dA. \qquad (14.55)$$

This is formula 13.45.

The results in equations 14.50–14.55 were based on a smooth surface S. If S is piecewise smooth, rather than smooth, we subdivide S into smooth parts and apply each of these results to the smooth parts. The surface integral over S is then the summation of the surface integrals over its parts.

EXAMPLE 14.18

Evaluate $\displaystyle\iint_S (x + y + z)\, dS$, where S is that part of the plane $x + 2y + 4z = 8$ in the first octant.

SOLUTION   The surface S projects one-to-one onto the triangle S_{xy} in the xy-plane in Figure 14.34. Since $z = (8 - x - 2y)/4$ on S,

$$\iint_S (x + y + z)\, dS = \iint_{S_{xy}} \left(x + y + 2 - \frac{x}{4} - \frac{y}{2}\right)\sqrt{1 + \left(\frac{\partial z}{\partial x}\right)^2 + \left(\frac{\partial z}{\partial y}\right)^2}\, dA$$

$$= \frac{1}{4}\iint_{S_{xy}} (3x + 2y + 8)\sqrt{1 + \left(-\frac{1}{4}\right)^2 + \left(-\frac{1}{2}\right)^2}\, dA$$

$$= \frac{\sqrt{21}}{16}\int_0^4 \int_0^{8-2y} (3x + 2y + 8)\, dx\, dy$$

$$= \frac{\sqrt{21}}{16}\int_0^4 \left\{\frac{3x^2}{2} + (2y + 8)x\right\}_0^{8-2y}\, dy$$

$$= \frac{\sqrt{21}}{8}\int_0^4 (80 - 24y + y^2)\, dy$$

$$= \frac{\sqrt{21}}{8}\left\{80y - 12y^2 + \frac{y^3}{3}\right\}_0^4 = \frac{56\sqrt{21}}{3}.$$

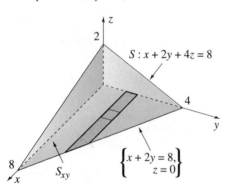

FIGURE 14.34   Surface integral of $x + y + z$ over the first octant part of $x + 2y + 4z = 8$

When a surface S, such as a sphere, encloses a volume, it is said to be a **closed surface**. When a function $f(x, y, z)$ is integrated over a closed surface S, the notation used is

$$\oiint_S f(x, y, z) \, dS.$$

This is similar to the notation for a line integral around a closed curve.

EXAMPLE 14.19

Evaluate $\oiint_S z^2 \, dS$, where S is the sphere $x^2 + y^2 + z^2 = 4$.

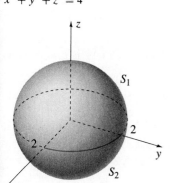

FIGURE 14.35 Surface integral of z^2 over the sphere $x^2 + y^2 + z^2 = 4$

SOLUTION We divide S into two hemispheres (Figure 14.35),

$$S_1 : z = \sqrt{4 - x^2 - y^2}, \qquad S_2 : z = -\sqrt{4 - x^2 - y^2}$$

each of which projects one-to-one onto the circle $S_{xy} : x^2 + y^2 \leq 4, \, z = 0$ in the xy-plane. For each hemisphere,

$$dS = \sqrt{1 + \left(\frac{\partial z}{\partial x}\right)^2 + \left(\frac{\partial z}{\partial y}\right)^2} \, dA$$

$$= \sqrt{1 + \frac{x^2}{4 - x^2 - y^2} + \frac{y^2}{4 - x^2 - y^2}} \, dA = \frac{2}{\sqrt{4 - x^2 - y^2}} \, dA,$$

and therefore

$$\oiint_S z^2 \, dS = \iint_{S_1} z^2 \, dS + \iint_{S_2} z^2 \, dS$$

$$= \iint_{S_{xy}} (4 - x^2 - y^2) \frac{2}{\sqrt{4 - x^2 - y^2}} \, dA$$

$$+ \iint_{S_{xy}} (4 - x^2 - y^2) \frac{2}{\sqrt{4 - x^2 - y^2}} \, dA$$

$$= 4 \iint_{S_{xy}} \sqrt{4 - x^2 - y^2} \, dA.$$

If we use polar coordinates to evaluate this integral over S_{xy}, then

$$\oiint_S z^2 \, dS = 4 \int_0^{2\pi} \int_0^2 \sqrt{4 - r^2} \, r \, dr \, d\theta$$

$$= 4 \int_0^{2\pi} \left\{ -\frac{1}{3}(4 - r^2)^{3/2} \right\}_0^2 d\theta = \frac{32}{3} \{\theta\}_0^{2\pi} = \frac{64\pi}{3}.$$

If parameters can be found to describe a surface, it may not be necessary to project the surface into one of the coordinate planes. Such is the case for a sphere centred at the origin. Formula 13.65 for the volume element in spherical coordinates indicates that an area element on the surface of a sphere of radius R can be expressed in terms of angles θ and ϕ (Figure 14.36) as

$$dS = R^2 \sin \phi \, d\phi \, d\theta. \tag{14.56}$$

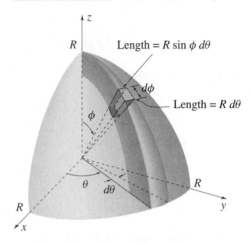

With this choice of area element, the surface integral in Example 14.19 is evaluated as follows:

$$\iint_S z^2 \, dS = \int_0^{2\pi} \int_0^{\pi} (2\cos\phi)^2 \cdot 4 \sin\phi \, d\phi \, d\theta = 16 \int_0^{2\pi} \int_0^{\pi} \cos^2\phi \sin\phi \, d\phi \, d\theta$$

$$= 16 \int_0^{2\pi} \left\{ -\frac{1}{3} \cos^3\phi \right\}_0^{\pi} d\theta = \frac{32}{3} \{\theta\}_0^{2\pi} = \frac{64\pi}{3}.$$

EXERCISES 14.7

In Exercises 1–8 evaluate the surface integral.

1. $\iint_S (x^2 y + z) \, dS$, where S is the first octant part of $2x + 3y + z = 6$

2. $\iint_S (x^2 + y^2)z \, dS$, where S is that part of $z = x + y$ cut out by $x = 0$, $y = 0$, $x + y = 1$

3. $\oiint_S xyz \, dS$, where S is the surface of the cube $0 \le x \le 1$, $0 \le y \le 1$, $0 \le z \le 1$

4. $\iint_S xy \, dS$, where S is the first octant part of $z = \sqrt{x^2 + y^2}$ cut out by $x^2 + y^2 = 1$

5. $\iint_S \frac{1}{\sqrt{z - y + 1}} \, dS$, where S is the surface defined by $2z = x^2 + 2y$, $0 \le x \le 1$, $0 \le y \le 1$

6. $\iint_S \sqrt{4y + 1} \, dS$, where S is the first octant part of $y = x^2$ cut out by $2x + y + z = 1$

7. $\iint_S x^2 z \, dS$, where S is the surface $x^2 + y^2 = 1$ for $0 \le z \le 1$

8. $\oiint_S (x + y) \, dS$, where S is the surface bounding the volume enclosed by $x = 0$, $y = 0$, $z = 0$, $6x - 3y + 2z = 6$

* **9.** Set up double iterated integrals for the surface integral of a function $f(x, y, z)$ over the surface defined by $z = 4 - x^2 - 4y^2$ ($x, y, z \ge 0$) if the surface is projected onto the xy-, the xz-, and the yz-planes.

* **10.** Use a surface integral to find the area of the curved portion of a right-circular cone of radius R and height h.

In Exercises 11–17 evaluate the surface integral.

* **11.** $\iint_S xyz^3 \, dS$, where S is the surface defined by $x = y^2$, $0 \le x \le 4$, $0 \le z \le 1$

* **12.** $\iint_S xyz \, dS$, where S is the surface defined by $2y = \sqrt{9 - x}$, $x \ge 0$, $0 \le z \le 3$

* **13.** $\iint_S \frac{1}{\sqrt{2az - z^2}} \, dS$, where S is that part of $x^2 + y^2 + (z - a)^2 = a^2$ inside the cylinder $x^2 + y^2 = ay$, underneath the plane $z = a$, and in the first octant

* **14.** $\iint_S z \, dS$, where S is that part of the surface $x^2 + y^2 - z^2 = 1$ between the planes $z = 0$ and $z = 1$

* **15.** $\iint_S x^2 y^2 \, dS$, where S is that part of $z = x^2 + y^2$ inside $x^2 + y^2 + z^2 = 2$

16. $\displaystyle\iint_S x^2\,dS$, where S is that part of $z = xy$ inside $x^2 + y^2 = 4$

17. $\displaystyle\iint_S z(y + x^2)\,dS$, where S is that part of $y = 1 - x^2$ bounded by $z = 0$, $z = 2$, and $y = 0$

In Exercises 18–22 evaluate the surface integral by projecting the surface into one of the coordinate planes and also by using area element 14.56.

18. $\displaystyle\oiint_S dS$, where S is the sphere $x^2 + y^2 + z^2 = R^2$. Is this the formula for the area of a sphere?

19. $\displaystyle\oiint_S x^2 z^2\,dS$, where S is the sphere $x^2 + y^2 + z^2 = 1$

20. $\displaystyle\iint_S (x^2 - y^2)\,dS$, where S is the hemisphere $z = \sqrt{9 - x^2 - y^2}$

21. $\displaystyle\oiint_S (x^2 + y^2)\,dS$, where S is the sphere $x^2 + y^2 + z^2 = R^2$

* **22.** $\displaystyle\iint_S \frac{1}{x^2 + y^2}\,dS$, where S is that part of the sphere $x^2 + y^2 + z^2 = 4R^2$ between the planes $z = 0$ and $z = R$

* **23.** A viscous material is allowed to drip onto the sphere $x^2 + y^2 + z^2 = 1$ at the point $(0, 0, 1)$ (all dimensions in metres). The material spreads out evenly in all directions and runs down the sphere, becoming more and more viscous as it does so. The thickness of the material increases linearly with respect to angle ϕ (Figure 14.36) from 0.001 m at $(0, 0, 1)$ to 0.005 m at $(0, 0, -1)$. Find the volume of material on the sphere.

* **24.** Show that if a surface S defined implicitly by the equation $F(x, y, z) = 0$ projects one-to-one onto the region S_{xy} in the xy-plane, then

$$\iint_S f(x, y, z)\,dS = \iint_{S_{xy}} f[x, y, g(x, y)]\frac{|\nabla F|}{|\partial F/\partial z|}\,dA.$$

* **25.** (a) Find the area cut from the cones $z^2 = x^2 + y^2$ by the cylinder $x^2 + y^2 = 2x$.
  (b) Find the area cut from the cylinder $x^2 + y^2 = 2x$ by the cones $z^2 = x^2 + y^2$.

14.8 Surface Integrals Involving Vector Fields

The most important and common type of surface integral occurs when $f(x, y, z)$ in 14.49 is specified as the normal component of some given vector field $\mathbf{F}(x, y, z)$ defined on S. In other words, $f(x, y, z)$ itself is not given, but $\mathbf{F}$ is, and to find $f(x, y, z)$ we must calculate the component of $\mathbf{F}$ normal to S.

This presupposes that surfaces are two-sided and that a normal vector to a surface can be assigned in an unambiguous way. When this is possible, the surface is said to be **orientable**. All surfaces in this book are orientable, with the exception of the Möbius strip mentioned below.

Take a thin rectangular strip of paper and label its corners A, B, C, and D (Figure 14.37a). Give the strip a half twist and join A and C, and B and D (Figure 14.37b). This surface, called a *Möbius strip*, cannot be assigned a unique normal vector that varies continuously over the surface. To illustrate, suppose that at point P in Figure 14.37b, we assign a unit normal vector $\hat{\mathbf{n}}$ as shown. By moving once around the strip, we can vary the direction of $\hat{\mathbf{n}}$ continuously and arrive back at P with $\hat{\mathbf{n}}$ pointing in the opposite direction. This surface is said to have only one side, or to be nonorientable.

We consider only surfaces that are orientable, or have two sides, and can therefore be assigned a unit normal vector in an unambiguous way.

FIGURE 14.37a　Thin, rectangular strip of paper

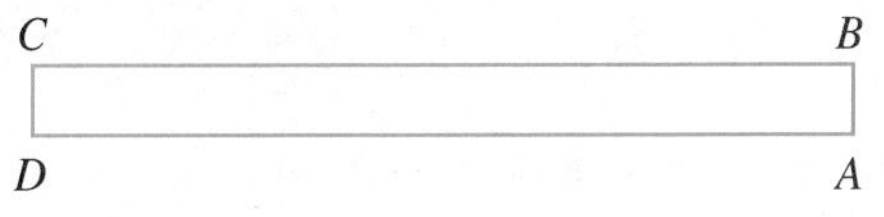

FIGURE 14.37b　Example of a nonorientable surface

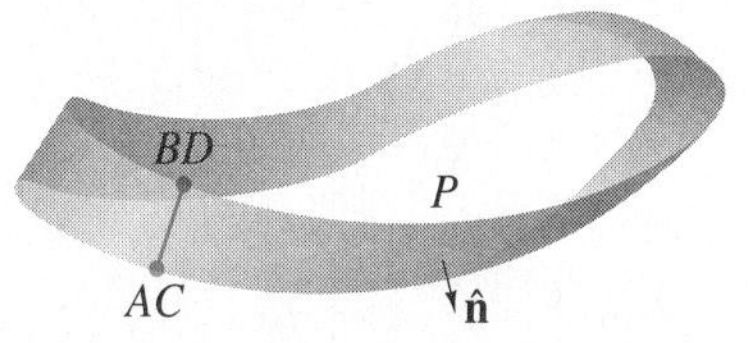

Suppose again that $\mathbf{F}(x, y, z)$ is a vector field defined on an (orientable) surface S and $\hat{\mathbf{n}}$ is the unit normal on one side of S. If $f(x, y, z)$ is the component of $\mathbf{F}$ in the direction $\hat{\mathbf{n}}$, then $f(x, y, z) = \mathbf{F} \cdot \hat{\mathbf{n}}$, and the surface integral of $f(x, y, z)$ over S can be expressed in the form

$$\iint_S f(x, y, z)\, dS = \iint_S \mathbf{F} \cdot \hat{\mathbf{n}}\, dS. \tag{14.57}$$

EXAMPLE 14.20

Evaluate $\displaystyle\iint_S \mathbf{F} \cdot \hat{\mathbf{n}}\, dS$, where $\mathbf{F} = x^2 y\hat{\mathbf{i}} + xz\hat{\mathbf{j}}$ and $\hat{\mathbf{n}}$ is the upper normal to the surface $S : z = 4 - x^2 - y^2,\ z \geq 0$.

SOLUTION   Since a normal vector to S is $\nabla(z - 4 + x^2 + y^2) = (2x, 2y, 1)$, it follows that

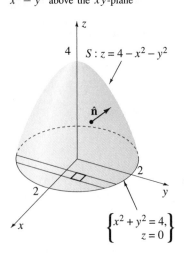

FIGURE 14.38   Surface integral over that part of $z = 4 - x^2 - y^2$ above the xy-plane

$$\hat{\mathbf{n}} = \frac{(2x, 2y, 1)}{\sqrt{4x^2 + 4y^2 + 1}}.$$

Thus,

$$\mathbf{F} \cdot \hat{\mathbf{n}} = \frac{2x^3 y + 2xyz}{\sqrt{4x^2 + 4y^2 + 1}}.$$

If we project S onto $S_{xy} : x^2 + y^2 \leq 4$ in the xy-plane (Figure 14.38), then

$$\iint_S \mathbf{F} \cdot \hat{\mathbf{n}}\, dS = \iint_{S_{xy}} \frac{2x^3 y + 2xyz}{\sqrt{4x^2 + 4y^2 + 1}} \sqrt{1 + \left(\frac{\partial z}{\partial x}\right)^2 + \left(\frac{\partial z}{\partial y}\right)^2}\, dA$$

$$= \iint_{S_{xy}} \frac{2x^3 y + 2xy(4 - x^2 - y^2)}{\sqrt{4x^2 + 4y^2 + 1}} \sqrt{1 + (-2x)^2 + (-2y)^2}\, dA$$

$$= 2\iint_{S_{xy}} (4xy - xy^3)\, dA = 2 \int_{-2}^{2} \int_{-\sqrt{4-x^2}}^{\sqrt{4-x^2}} (4xy - xy^3)\, dy\, dx$$

$$= 2 \int_{-2}^{2} \left\{ 2xy^2 - \frac{xy^4}{4} \right\}_{-\sqrt{4-x^2}}^{\sqrt{4-x^2}}\, dx = 0.$$

EXAMPLE 14.21

Evaluate $\displaystyle\oiint_S \mathbf{F} \cdot \hat{\mathbf{n}}\, dS$, where $\mathbf{F} = x\hat{\mathbf{i}} + y\hat{\mathbf{j}} + z\hat{\mathbf{k}}$ and $\hat{\mathbf{n}}$ is the unit outward-pointing normal to the surface S enclosing the volume bounded by $x^2 + y^2 = 4,\ z = 0,\ z = 2$.

SOLUTION   We divide S into four parts (Figure 14.39):

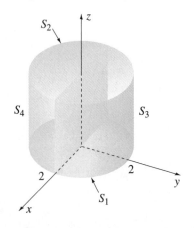

FIGURE 14.39   Surface integral over a closed surface

$$S_1 : \quad z = 0, \quad x^2 + y^2 \leq 4;$$

$$S_2 : \quad z = 2, \quad x^2 + y^2 \leq 4;$$

$$S_3 : \quad y = \sqrt{4 - x^2}, \quad 0 \leq z \leq 2;$$

$$S_4 : \quad y = -\sqrt{4 - x^2}, \quad 0 \leq z \leq 2.$$

On S_1, $\hat{\mathbf{n}} = -\hat{\mathbf{k}}$; on S_2, $\hat{\mathbf{n}} = \hat{\mathbf{k}}$; and on S_3 and S_4,

$$\hat{\mathbf{n}} = \frac{\nabla(x^2 + y^2 - 4)}{|\nabla(x^2 + y^2 - 4)|} = \frac{(2x, 2y, 0)}{\sqrt{4x^2 + 4y^2}} = \frac{(x, y, 0)}{2}.$$

We know that S_1 and S_2 project onto $S_{xy} : x^2 + y^2 \le 4$ in the xy-plane, and S_3 and S_4 project onto the rectangle $S_{xz} : -2 \le x \le 2, 0 \le z \le 2$ in the xz-plane. Consequently,

$$\oiint_S \mathbf{F} \cdot \hat{\mathbf{n}}\, dS = \iint_{S_1} -z\, dS + \iint_{S_2} z\, dS + \iint_{S_3} \left(\frac{x^2 + y^2}{2}\right) dS$$

$$+ \iint_{S_4} \left(\frac{x^2 + y^2}{2}\right) dS$$

$$= 0 + \iint_{S_{xy}} 2\sqrt{1}\, dA$$

$$+ \frac{1}{2} \iint_{S_{xz}} (x^2 + 4 - x^2)\sqrt{1 + \left(\frac{-x}{\sqrt{4 - x^2}}\right)^2}\, dA$$

$$+ \frac{1}{2} \iint_{S_{xz}} (x^2 + 4 - x^2)\sqrt{1 + \left(\frac{x}{\sqrt{4 - x^2}}\right)^2}\, dA$$

$$= 2 \iint_{S_{xy}} dA + 4 \iint_{S_{xz}} \frac{2}{\sqrt{4 - x^2}}\, dA$$

$$= 2(\text{area of } S_{xy}) + 8 \int_{-2}^{2} \int_{0}^{2} \frac{1}{\sqrt{4 - x^2}}\, dz\, dx$$

$$= 2(4\pi) + 16 \int_{-2}^{2} \frac{1}{\sqrt{4 - x^2}}\, dx.$$

If we set $x = 2\sin\theta$, then $dx = 2\cos\theta\, d\theta$ and

$$\oiint_S \mathbf{F} \cdot \hat{\mathbf{n}}\, dS = 8\pi + 16 \int_{-\pi/2}^{\pi/2} \frac{1}{2\cos\theta} \cdot 2\cos\theta\, d\theta = 8\pi + 16\{\theta\}_{-\pi/2}^{\pi/2} = 24\pi.$$

EXAMPLE 14.22

A viewing window on the side of a submersible vehicle in a marine theme park is a hemisphere of radius 1/2 m. Find the force due to water pressure on the window when its centre is h metres below the surface.

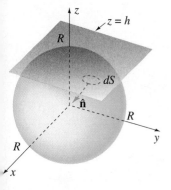

FIGURE 14.40   Force of water on a hemispherical window of a submersible vehicle

SOLUTION   Let us choose a coordinate system with the plane $z = h$ in the surface of the water, and take the equation of the window as $S : x^2 + y^2 + z^2 = 1/4, y \ge 0$ (Figure 14.40). If dS is a small area on S, the force due to water pressure on dS has magnitude $P\, dS$, where P is pressure, and this force acts normal to dS. If $\hat{\mathbf{n}}$ is the unit normal to the hemisphere with negative y-component, then the force on dS is $(P\, dS)\hat{\mathbf{n}}$. Because of the symmetry of the hemisphere, the x-component of the resultant force will be zero. Since the y-component of the force on dS is $(P\, dS)\hat{\mathbf{n}} \cdot \hat{\mathbf{j}}$, the y-component of the resultant force on the window is

$$\iint_S P\hat{\mathbf{j}} \cdot \hat{\mathbf{n}}\, dS.$$

A normal to the surface is $\nabla(x^2 + y^2 + z^2 - 1/4) = (2x, 2y, 2z)$, and therefore the unit normal in the negative y-direction is

$$\hat{\mathbf{n}} = \frac{-(x, y, z)}{\sqrt{x^2 + y^2 + z^2}} = -2(x, y, z).$$

With $P = 9810(h - z)$ and formula 14.56 for an area element on the hemisphere (where $R = 1/2$),

$$\iint_S P\hat{\mathbf{j}} \cdot \hat{\mathbf{n}} \, dS = \iint_S 9810(h - z)(-2y) \, dS$$

$$= -19\,620 \int_0^\pi \int_0^\pi \left(h - \frac{1}{2} \cos \phi \right) \left(\frac{1}{2} \sin \phi \sin \theta \right) \left(\frac{1}{4} \sin \phi \right) d\phi \, d\theta$$

$$= -\frac{4905}{4} \int_0^\pi \int_0^\pi (2h - \cos \phi) \sin^2 \phi \sin \theta \, d\phi \, d\theta$$

$$= -\frac{4905}{4} \int_0^\pi \int_0^\pi [h(1 - \cos 2\phi) - \sin^2 \phi \cos \phi] \sin \theta \, d\phi \, d\theta$$

$$= -\frac{4905}{4} \int_0^\pi \left\{ h \left(\phi - \frac{\sin 2\phi}{2} \right) - \frac{\sin^3 \phi}{3} \right\}_0^\pi \sin \theta \, d\theta$$

$$= -\frac{4905\pi h}{4} \{-\cos \theta\}_0^\pi = -\frac{4905\pi h}{2} \quad \text{N.}$$

The z-component of the resultant force on the window is

$$\iint_S P\hat{\mathbf{k}} \cdot \hat{\mathbf{n}} \, dS = \iint_S 9810(h - z)(-2z) \, dS$$

$$= -19\,620 \int_0^\pi \int_0^\pi \left(h - \frac{1}{2} \cos \phi \right) \left(\frac{1}{2} \cos \phi \right) \left(\frac{1}{4} \sin \phi \right) d\phi \, d\theta$$

$$= -\frac{4905}{4} \int_0^\pi \int_0^\pi (2h - \cos \phi) \cos \phi \sin \phi \, d\phi \, d\theta$$

$$= -\frac{4905}{4} \int_0^\pi \left\{ -h \cos^2 \phi + \frac{1}{3} \cos^3 \phi \right\}_0^\pi d\theta$$

$$= \frac{1635}{2} \{\theta\}_0^\pi = \frac{1635\pi}{2} \quad \text{N.}$$

The resultant force on the window is therefore $\pi(-4905h\hat{\mathbf{j}} + 1635\hat{\mathbf{k}})/2$ N.

Consulting Project 24

The gate in Figure 14.41a has constant width $w = 5$ m (into the page). Its cross-section is in the shape of the parabola $4x = y^2$ and it is hinged along a line through O into the page. Water with depth 4 m pushes on the concave side. We are asked to determine the magnitude of a vertical force G at P that will maintain the gate in an equilibrium state. We are to ignore the weight of the gate itself in our calculations.

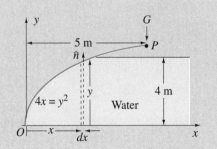

FIGURE 14.41a

Forces on a hinged gate

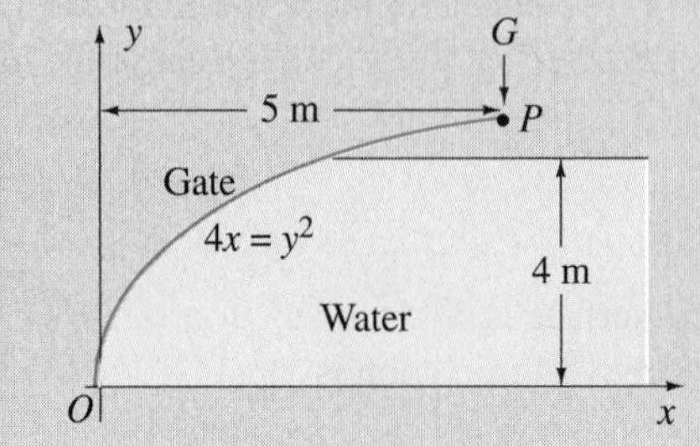

FIGURE 14.41b

SOLUTION The force of the water against the gate creates a moment that attempts to turn the gate counterclockwise around O. The force G at P creates a moment in the opposite direction. They must combine to prevent motion. If we take the moment of G about O as negative, then it is $M_G = -5G$. The moment of the water on the gate is more difficult to calculate. We divide the gate into horizontal strips of length 5 m and width corresponding to length dx along the x-axis (Figure 14.41b). The magnitude of the force of water on the strip is

$$5\rho g(4-y)\sqrt{1+\left(\frac{dy}{dx}\right)^2}\,dx = 5\rho g(4-y)\sqrt{1+\left(\frac{2}{y}\right)^2}\,dx$$

$$= \frac{5\rho g(4-y)\sqrt{y^2+4}}{y}\,dx.$$

This force acts perpendicular to the gate. Since the unit normal to the gate is $\hat{\mathbf{n}} = (-4, 2y)/\sqrt{16+4y^2} = (-2, y)/\sqrt{4+y^2}$, the force of water on the strip is

$$\left[\frac{5\rho g(4-y)\sqrt{y^2+4}}{y}\,dx\right]\frac{(-2, y)}{\sqrt{4+y^2}} = \frac{5\rho g(4-y)\,dx}{y}(-2, y).$$

The horizontal component of the force of water on the gate is therefore

$$F_x = \int_0^4 -\frac{10\rho g(4-y)}{y}\,dx = -10\rho g\int_0^4 \frac{4-y}{y}\left(\frac{y}{2}dy\right)$$

$$= -5\rho g\int_0^4 (4-y)\,dy = -5\rho g\left\{4y - \frac{y^2}{2}\right\}_0^4 = -40\rho g \text{ N}.$$

The vertical component is

$$F_y = \int_0^4 5\rho g(4-y)\,dx = 5\rho g\int_0^4 (4-y)\left(\frac{y}{2}dy\right)$$

$$= \frac{5\rho g}{2}\left\{2y^2 - \frac{y^3}{3}\right\}_0^4 = \frac{80\rho g}{3} \text{ N}.$$

To determine the moment of $\mathbf{F} = F_x\hat{\mathbf{i}} + F_y\hat{\mathbf{j}}$ about O, we must know its point of application, the centre of pressure (x_c, y_c). Using moments of the components of $\mathbf{F}$ about O (Figure 14.42), we calculate that

$$\frac{80\rho g}{3}x_c = \int_0^4 x[5\rho g(4-y)]\,dx = 5\rho g\int_0^4 \frac{y^2}{4}(4-y)\left(\frac{y}{2}dy\right)$$

$$= \frac{5\rho g}{8}\left\{y^4 - \frac{y^5}{5}\right\}_0^4 = 32\rho g,$$

$$-40\rho g y_c = \int_0^4 y\left[\frac{-10\rho g(4-y)}{y}\right]\left(\frac{y}{2}dy\right)$$

$$= -5\rho g\left\{2y^2 - \frac{y^3}{3}\right\}_0^4 = -\frac{160\rho g}{3}.$$

Consequently,

$$x_c = 32\rho g \cdot \frac{3}{80\rho g} = \frac{6}{5} \quad \text{and} \quad y_c = -\frac{160\rho g}{3} \cdot \frac{1}{-40\rho g} = \frac{4}{3}.$$

We now take moments of G, F_x, and F_y about O in order to calculate G,

$$0 = -5G + 40\rho g \left(\frac{4}{3}\right) + \frac{80\rho g}{3}\left(\frac{6}{5}\right) \quad \Longrightarrow \quad G = 167\,000 \text{ N}.$$

FIGURE 14.42 Moments of forces on hinged gate

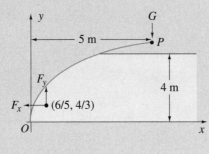

EXERCISES 14.8

In Exercises 1–16 evaluate the surface integral.

1. $\displaystyle\iint_S (x\hat{\mathbf{i}} + y\hat{\mathbf{k}}) \cdot \hat{\mathbf{n}}\, dS$, where S is the first octant part of $x + y + z = 3$, and $\hat{\mathbf{n}}$ is the unit normal to S with positive z-component

2. $\displaystyle\iint_S (yz^2\hat{\mathbf{i}} + ye^x\hat{\mathbf{j}} + x\hat{\mathbf{k}}) \cdot \hat{\mathbf{n}}\, dS$, where S is defined by $y = x^2$, $0 \leq y \leq 4$, $0 \leq z \leq 1$, and $\hat{\mathbf{n}}$ is the unit normal to S with positive y-component

3. $\displaystyle\iint_S (x\hat{\mathbf{i}} + y\hat{\mathbf{j}} + z\hat{\mathbf{k}}) \cdot \hat{\mathbf{n}}\, dS$, where S is the hemisphere $z = \sqrt{1 - x^2 - y^2}$, and $\hat{\mathbf{n}}$ is its upper normal

4. $\displaystyle\iint_S (yz\hat{\mathbf{i}} + zx\hat{\mathbf{j}} + xy\hat{\mathbf{k}}) \cdot \hat{\mathbf{n}}\, dS$, where S is that part of the surface $z = x^2 + y^2$ cut out by the planes $x = 1$, $x = -1$, $y = 1$, $y = -1$, and $\hat{\mathbf{n}}$ is the unit lower normal to S

5. $\displaystyle\oiint_S (z\hat{\mathbf{i}} - x\hat{\mathbf{j}} + y\hat{\mathbf{k}}) \cdot \hat{\mathbf{n}}\, dS$, where S is the surface that encloses the volume defined by $z = \sqrt{4 - x^2 - y^2}$, $z = 0$, and $\hat{\mathbf{n}}$ is the unit outer normal to S

* **6.** $\displaystyle\iint_S (x\hat{\mathbf{i}} + y\hat{\mathbf{j}}) \cdot \hat{\mathbf{n}}\, dS$, where S is that part of the surface $z = \sqrt{x^2 + y^2}$ below $z = 1$, and $\hat{\mathbf{n}}$ is the unit normal to S with negative z-component

* **7.** $\displaystyle\iint_S (xyz\hat{\mathbf{i}} - x\hat{\mathbf{j}} + z\hat{\mathbf{k}}) \cdot \hat{\mathbf{n}}\, dS$, where S is the smaller part of $x^2 + y^2 = 9$ cut out by $z = 0$, $z = 2$, $y = |x|$, and $\hat{\mathbf{n}}$ is the unit normal to S with positive y-component

* **8.** $\displaystyle\iint_S (x^2y\hat{\mathbf{i}} + xy\hat{\mathbf{j}} + z\hat{\mathbf{k}}) \cdot \hat{\mathbf{n}}\, dS$, where S is defined by $z = 2 - x^2 - y^2$, $z \geq 0$, and $\hat{\mathbf{n}}$ is the unit normal to S with negative z-component

* **9.** $\displaystyle\oiint_S (yz\hat{\mathbf{i}} + xz\hat{\mathbf{j}} + xy\hat{\mathbf{k}}) \cdot \hat{\mathbf{n}}\, dS$, where S is the surface enclosing the volume defined by $x = 0$, $x = 2$, $z = 0$, $z = y$, $y + z = 2$, and $\hat{\mathbf{n}}$ is the unit outer normal to S

* **10.** $\displaystyle\iint_S (x\hat{\mathbf{i}} + y\hat{\mathbf{j}}) \cdot \hat{\mathbf{n}}\, dS$, where S is the surface $x^2 + y^2 + z^2 = 4$, $z \geq 1$, and $\hat{\mathbf{n}}$ is the unit upper normal to S

* **11.** $\displaystyle\oiint_S (x^2\hat{\mathbf{i}} + y^2\hat{\mathbf{j}} + z^2\hat{\mathbf{k}}) \cdot \hat{\mathbf{n}}\, dS$, where S is the sphere $x^2 + y^2 + z^2 = a^2$, and $\hat{\mathbf{n}}$ is the unit outer normal to S

* **12.** $\displaystyle\iint_S (y\hat{\mathbf{i}} - x\hat{\mathbf{j}} + \hat{\mathbf{k}}) \cdot \hat{\mathbf{n}}\, dS$, where S is the smaller surface cut, by the plane $y + z = 1$, from the sphere $x^2 + y^2 + z^2 = 1$, and $\hat{\mathbf{n}}$ is the unit upper normal to S

3. $\oiint_S \mathbf{F} \cdot \hat{\mathbf{n}} \, dS$, where $\mathbf{F} = (z^2 - x)\hat{\mathbf{i}} - xy\hat{\mathbf{j}} + 3z\hat{\mathbf{k}}$, S is the surface enclosing the volume defined by $z = 4 - y^2$, $x = 0$, $x = 3$, $z = 0$, and $\hat{\mathbf{n}}$ is the unit outer normal to S

4. $\iint_S (x^2\hat{\mathbf{i}} + xy\hat{\mathbf{j}} + xz\hat{\mathbf{k}}) \cdot \hat{\mathbf{n}} \, dS$, where S is that part of the surface $z = \sqrt{4 + y^2 - x^2}$ in the first octant cut out by the planes $y = 0$, $y = 1$, $x = 0$, $z = 0$, and $\hat{\mathbf{n}}$ is the unit normal to S with positive z-component

5. $\iint_S (x^2\hat{\mathbf{i}} + yz\hat{\mathbf{j}} - x\hat{\mathbf{k}}) \cdot \hat{\mathbf{n}} \, dS$, where S is that part of the surface $z = yz$ in the first octant cut out by $y^2 + z^2 = 1$, and $\hat{\mathbf{n}}$ is the unit normal to S with positive x-component

6. $\oiint_S (yx\hat{\mathbf{i}} + y^2\hat{\mathbf{j}} + yz\hat{\mathbf{k}}) \cdot \hat{\mathbf{n}} \, dS$, where S is the ellipsoid $x^2 + y^2/4 + z^2 = 1$, and $\hat{\mathbf{n}}$ is the unit outer normal to S

7. Show that if a surface S projects one-to-one onto a region S_{xy} in the xy-plane, then

$$\iint_S (P\hat{\mathbf{i}} + Q\hat{\mathbf{j}} + R\hat{\mathbf{k}}) \cdot \hat{\mathbf{n}} \, dS$$

$$= \pm \iint_{S_{xy}} \left(-P\frac{\partial z}{\partial x} - Q\frac{\partial z}{\partial y} + R \right) dA,$$

the $\pm$ depending on whether $\hat{\mathbf{n}}$ is the upper or lower normal to S. What are corresponding formulas when S projects one-to-one onto regions S_{yz} and S_{xz} in the yz- and xz-coordinate planes?

8. Evaluate $\iint_S (y\hat{\mathbf{i}} - x\hat{\mathbf{j}} + z\hat{\mathbf{k}}) \cdot \hat{\mathbf{n}} \, dS$, where (a) S is that part of $z = 9 - x^2 - y^2$ cut out by $z = 2y$ and (b) S is that part of $z = 2y$ cut out by $z = 9 - x^2 - y^2$, and $\hat{\mathbf{n}}$ is the unit upper normal to S in each case. *Hint:* Use polar coordinates with pole at $(0, -1)$.

9. Show that if a surface S, defined implicitly by the equation $G(x, y, z) = 0$, projects one-to-one onto the region S_{xy} in the xy-plane, then

$$\iint_S \mathbf{F} \cdot \hat{\mathbf{n}} \, dS = \pm \iint_{S_{xy}} \frac{\mathbf{F} \cdot \nabla G}{|\partial G/\partial z|} \, dA.$$

10. A water trough is bounded on the sides by the surfaces $x = 0$, $x = 2$, $y + z = 0$, and $z = y^2$, all dimensions in metres, and is 1 m deep.

 (a) Find the force on each of the four sides when the trough is full.

 (b) Add all four forces in part (a). Is the result equal to the weight of the water in the trough?

* **21.** A very long water channel extends in the x-direction and has constant cross-section that, in the yz-plane, is defined by

$$z = \begin{cases} -(y + 1)^3, & -2 \leq y < -1 \\ 0, & -1 \leq y \leq 1 \\ (y - 1)^3, & 1 < y \leq 2. \end{cases}$$

If the channel is full, find the force per unit length in the x-direction on (a) the bottom of the channel ($-1 \leq y \leq 1$), (b) the right wall of the channel ($1 \leq y \leq 2$), and (c) the left wall of the channel ($-2 \leq y \leq -1$). If all three forces are added, is the result equal to the weight per unit length of the water in the channel?

* **22.** Suppose the surface of a fluid is in the xy-plane and a flat surface S is submerged in the fluid. Let S be a part of the plane $Ax + By + Cz + D = 0$, and let the projection of S in the xy-plane be denoted by R_{xy}. Show that the magnitude of the z-component of the fluid force on S is equal to the weight of the column of fluid above S and below R_{xy}.

* **23.** Extend the result of Exercise 22 to any surface S that projects one-to-one onto R_{xy}.

* **24.** A circular tube $S : x^2 + z^2 = 1$, $0 \leq y \leq 2$ is a model for a part of an artery. Blood flows through the artery and the force per unit area at any point on the arterial wall is given by

$$\mathbf{F} = e^{-y}\hat{\mathbf{n}} + \frac{1}{y^2 + 1}\hat{\mathbf{j}},$$

where $\hat{\mathbf{n}}$ is the unit outer normal to the arterial wall. Blood diffuses through the wall in such a way that if dS is a small area on S, the amount of diffusion through dS in 1 s is $\mathbf{F} \cdot \hat{\mathbf{n}} \, dS$. Find the total amount of blood leaving the entire wall per second.

* **25.** A beam of light travelling in the positive y-direction has circular cross-section $x^2 + z^2 \leq a^2$. It strikes a surface $S : x^2 + y^2 + z^2 = a^2$, $y \geq a/2$. The intensity of the beam is given by

$$\mathbf{I} = \frac{e^{-t}}{y^2}\hat{\mathbf{j}},$$

where t is time. The absorption of light by a small area dS on S in time dt is $\mathbf{I} \cdot \hat{\mathbf{n}} \, dS \, dt$, where $\hat{\mathbf{n}}$ is the unit normal to S at dS.

 (a) Find the total absorption over S in time dt.

 (b) Find the total absorption over S from time $t = 0$ to time $t = 5$.

14.9 The Divergence Theorem

In this section and in Section 14.10 we show that relationships may exist between line integrals and surface integrals and between surface integrals and triple integrals.

The divergence theorem relates certain surface integrals over surfaces that enclose volumes to triple integrals over the enclosed volume. More precisely, we have the following.

THEOREM 14.7   (Divergence Theorem)

Let S be a piecewise-smooth surface enclosing a region V (Figure 14.43a). Let $\mathbf{F}(x, y, z) = L(x, y, z)\hat{\mathbf{i}} + M(x, y, z)\hat{\mathbf{j}} + N(x, y, z)\hat{\mathbf{k}}$ be a vector field whose components L, M, and N have continuous first partial derivatives in a domain containing S and V. If $\hat{\mathbf{n}}$ is the unit outer normal to S, then

$$\oiint_S \mathbf{F} \cdot \hat{\mathbf{n}}\, dS = \iiint_V \nabla \cdot \mathbf{F}\, dV \qquad (14.58a)$$

or

$$\oiint_S (L\hat{\mathbf{i}} + M\hat{\mathbf{j}} + N\hat{\mathbf{k}}) \cdot \hat{\mathbf{n}}\, dS = \iiint_V \left(\frac{\partial L}{\partial x} + \frac{\partial M}{\partial y} + \frac{\partial N}{\partial z} \right) dV. \qquad (14.58b)$$

FIGURE 14.43a   Elementary surfaces enclosing volume V

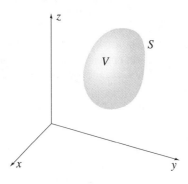

FIGURE 14.43b   Division of surface into two parts to verify the divergence thereom

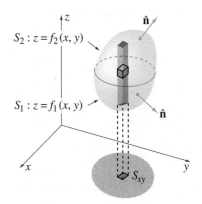

PROOF   We consider first of all a surface S for which any line parallel to any coordinate axis intersects S in at most two points (Figure 14.43b). We can then divide S into an upper and a lower portion, $S_2 : z = f_2(x, y)$ and $S_1 : z = f_1(x, y)$, both of which have the same projection S_{xy} in the xy-plane. We consider the third term in the surface integral on the left side of equation 14.58b:

$$\oiint_S N\hat{\mathbf{k}} \cdot \hat{\mathbf{n}}\, dS = \iint_{S_1} N\hat{\mathbf{k}} \cdot \hat{\mathbf{n}}\, dS + \iint_{S_2} N\hat{\mathbf{k}} \cdot \hat{\mathbf{n}}\, dS.$$

On S_1,

$$\hat{\mathbf{n}} = \frac{\left(\dfrac{\partial f_1}{\partial x}, \dfrac{\partial f_1}{\partial y}, -1 \right)}{\sqrt{1 + \left(\dfrac{\partial f_1}{\partial x} \right)^2 + \left(\dfrac{\partial f_1}{\partial y} \right)^2}},$$

and on S_2,

$$\hat{\mathbf{n}} = \frac{-\left(\dfrac{\partial f_2}{\partial x}, \dfrac{\partial f_2}{\partial y}, -1 \right)}{\sqrt{1 + \left(\dfrac{\partial f_2}{\partial x} \right)^2 + \left(\dfrac{\partial f_2}{\partial y} \right)^2}}.$$

Consequently,

$$\oiint_S N\hat{\mathbf{k}} \cdot \hat{\mathbf{n}}\, dS = \iint_{S_1} \frac{-N}{\sqrt{1 + \left(\dfrac{\partial f_1}{\partial x}\right)^2 + \left(\dfrac{\partial f_1}{\partial y}\right)^2}}\, dS$$

$$+ \iint_{S_2} \frac{N}{\sqrt{1 + \left(\dfrac{\partial f_2}{\partial x}\right)^2 + \left(\dfrac{\partial f_2}{\partial y}\right)^2}}\, dS$$

$$= \iint_{S_{xy}} \frac{-N[x, y, f_1(x, y)]}{\sqrt{1 + \left(\dfrac{\partial f_1}{\partial x}\right)^2 + \left(\dfrac{\partial f_1}{\partial y}\right)^2}} \sqrt{1 + \left(\frac{\partial f_1}{\partial x}\right)^2 + \left(\frac{\partial f_1}{\partial y}\right)^2}\, dA$$

$$+ \iint_{S_{xy}} \frac{N[x, y, f_2(x, y)]}{\sqrt{1 + \left(\dfrac{\partial f_2}{\partial x}\right)^2 + \left(\dfrac{\partial f_2}{\partial y}\right)^2}} \sqrt{1 + \left(\frac{\partial f_2}{\partial x}\right)^2 + \left(\frac{\partial f_2}{\partial y}\right)^2}\, dA$$

$$= \iint_{S_{xy}} \{N[x, y, f_2(x, y)] - N[x, y, f_1(x, y)]\}\, dA.$$

On the other hand,

$$\iiint_V \frac{\partial N}{\partial z}\, dV = \iint_{S_{xy}} \left\{ \int_{f_1(x,y)}^{f_2(x,y)} \frac{\partial N}{\partial z}\, dz \right\} dA = \iint_{S_{xy}} \left\{ N \right\}_{f_1(x,y)}^{f_2(x,y)} dA$$

$$= \iint_{S_{xy}} \{N[x, y, f_2(x, y)] - N[x, y, f_1(x, y)]\}\, dA.$$

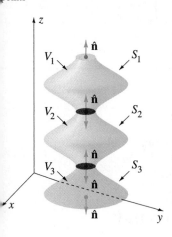

FIGURE 14.44 Division of a surface into pieces such that lines parallel to the coordinate axes intersect the surface in at most two points

We have shown then that

$$\oiint_S N\hat{\mathbf{k}} \cdot \hat{\mathbf{n}}\, dS = \iiint_V \frac{\partial N}{\partial z}\, dV.$$

Projections of S onto the xz- and yz-planes lead in a similar way to

$$\oiint_S M\hat{\mathbf{j}} \cdot \hat{\mathbf{n}}\, dS = \iiint_V \frac{\partial M}{\partial y}\, dV \quad \text{and} \quad \oiint_S L\hat{\mathbf{i}} \cdot \hat{\mathbf{n}}\, dS = \iiint_V \frac{\partial L}{\partial x}\, dV.$$

By adding these three results, we obtain the divergence theorem for $\mathbf{F}$ and S.

The proof can be extended to more general surfaces for which lines parallel to the coordinate axes intersect the surfaces in more than two points. Indeed, most volumes V bounded by surfaces S can be divided into n subvolumes V_i whose bounding surfaces S_i do satisfy this condition (Figure 14.44). For each such subvolume the divergence theorem is now known to apply:

$$\oiint_{S_i} \mathbf{F} \cdot \hat{\mathbf{n}}\, dS = \iiint_{V_i} \nabla \cdot \mathbf{F}\, dV, \quad i = 1, \ldots, n.$$

If these n equations are then added together, we have

$$\sum_{i=1}^{n} \oiint_{S_i} \mathbf{F} \cdot \hat{\mathbf{n}}\, dS = \sum_{i=1}^{n} \iiint_{V_i} \nabla \cdot \mathbf{F}\, dV.$$

The right side is the triple integral of $\nabla \cdot \mathbf{F}$ over V since the V_i constitute V. Figure 14.44 illustrates that when surface integrals over the S_i are added, contributions from auxiliary (interior) surfaces cancel in pairs, and the remaining surface integrals add to give the surface integral of $\mathbf{F} \cdot \hat{\mathbf{n}}$ over S. We omit a proof for even more general surfaces that cannot be divided into a finite number of subsurfaces of this type. The interested reader should consult more advanced books. ■

EXAMPLE 14.23

Use the divergence theorem to evaluate the surface integral of the normal component of $\mathbf{F} = x^2\hat{\mathbf{i}} + yz\hat{\mathbf{j}} + x\hat{\mathbf{k}}$ over the surface S enclosing the volume V bounded by the surfaces $x + y + z = 1$, $x = 0$, $y = 0$, and $z = 0$.

SOLUTION   The divergence theorem (see Figure 14.45) gives

FIGURE 14.45   Divergence theorem applied to a tetrahedron

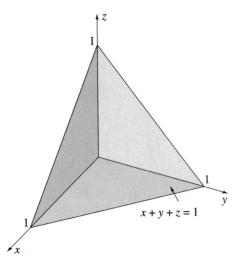

$$\oiint_S \mathbf{F} \cdot \hat{\mathbf{n}}\, dS = \iiint_V \nabla \cdot \mathbf{F}\, dV = \iiint_V (2x + z)\, dV$$

$$= \int_0^1 \int_0^{1-x} \int_0^{1-x-y} (2x + z)\, dz\, dy\, dx$$

$$= \int_0^1 \int_0^{1-x} \left\{ 2xz + \frac{z^2}{2} \right\}_0^{1-x-y} dy\, dx$$

$$= \frac{1}{2} \int_0^1 \int_0^{1-x} [4x(1 - x - y) + (1 - x - y)^2]\, dy\, dx$$

$$= \frac{1}{2} \int_0^1 \left\{ 4x\left(y - xy - \frac{y^2}{2} \right) - \frac{1}{3}(1 - x - y)^3 \right\}_0^{1-x} dx$$

$$= \frac{1}{2} \int_0^1 \left[2x(1 - x)^2 + \frac{1}{3}(1 - x)^3 \right] dx$$

$$= \frac{1}{2} \int_0^1 \left[2x - 4x^2 + 2x^3 + \frac{1}{3}(1 - x)^3 \right] dx$$

$$= \frac{1}{2} \left\{ x^2 - \frac{4x^3}{3} + \frac{x^4}{2} - \frac{1}{12}(1 - x)^4 \right\}_0^1 = \frac{1}{8}.$$

EXAMPLE 14.24

Use the divergence theorem to evaluate the surface integral of Example 14.21.

SOLUTION   By the divergence theorem (Figure 14.39), we have

$$\oiint_S \mathbf{F} \cdot \hat{\mathbf{n}}\, dS = \iiint_V \nabla \cdot \mathbf{F}\, dV = \iiint_V (1 + 1 + 1)\, dV$$

$$= 3 \iiint_V dV = 3(\text{volume of } V) = 3(4\pi)(2) = 24\pi.$$

EXAMPLE 14.25

Application Preview
Revisited

Suppose that a surface S contains n point charges q_i ($i = 1, \ldots, n$) at points $\mathbf{r}_i = (x_i, y_i, z_i)$ in its interior. According to Coulomb's law, the electric field $\mathbf{E}$ at a point $\mathbf{r} = (x, y, z)$ due to the charges is defined by

$$\mathbf{E} = \sum_{i=1}^{n} \frac{q_i(\mathbf{r} - \mathbf{r}_i)}{4\pi\epsilon_0|\mathbf{r} - \mathbf{r}_i|^3}.$$

Use the result of Exercise 28 in this section to verify Gauss's law

$$\oiint_S \mathbf{E} \cdot \hat{\mathbf{n}}\,dS = \frac{Q}{\epsilon_0},$$

where Q is the total charge inside S. This is the problem of the Application Preview.

SOLUTION According to Exercise 28,

$$\oiint_S \mathbf{E} \cdot \hat{\mathbf{n}}\,dS = \oiint_S \left[\sum_{i=1}^{n} \frac{q_i(\mathbf{r} - \mathbf{r}_i)}{4\pi\epsilon_0|\mathbf{r} - \mathbf{r}_i|^3}\right] \cdot \hat{\mathbf{n}}\,dS$$

$$= \sum_{i=1}^{n} \frac{q_i}{4\pi\epsilon_0}\left(\oiint_S \frac{\mathbf{r} - \mathbf{r}_i}{|\mathbf{r} - \mathbf{r}_i|^3} \cdot \hat{\mathbf{n}}\,dS\right)$$

$$= \sum_{i=1}^{n} \frac{q_i}{4\pi\epsilon_0}(4\pi) \qquad \text{(because each } q_i \text{ is inside } S)$$

$$= \sum_{i=1}^{n} \frac{q_i}{\epsilon_0} = \frac{Q}{\epsilon_0}.$$

EXAMPLE 14.26

Prove Archimedes' principle, which states that when an object is submerged in a fluid, it experiences a buoyant force equal to the weight of the fluid displaced.

SOLUTION Suppose the surface of the fluid is taken as the xy-plane ($z = 0$), and the object occupies a region V with bounding surface S (Figure 14.46). The force due to fluid pressure P on a small area dS on S is $(P\,dS)\hat{\mathbf{n}}$, where $\hat{\mathbf{n}}$ is the unit inner normal to S at dS. If ρ is the density of the fluid, then $P = -9.81\rho z$, and the force on dS is $(-9.81\rho z\,dS)\hat{\mathbf{n}}$. The resultant buoyant force is in the positive z-direction (the x- and y-components cancelling), so that we require only the z-component of this force, $(-9.81\rho z\,dS)\hat{\mathbf{n}} \cdot \hat{\mathbf{k}}$. The total buoyant force must therefore have z-component

$$\oiint_S (-9.81\rho z)\hat{\mathbf{n}} \cdot \hat{\mathbf{k}}\,dS = \oiint_S (-9.81\rho z\hat{\mathbf{k}}) \cdot \hat{\mathbf{n}}\,dS = \oiint_S (9.81\rho z\hat{\mathbf{k}}) \cdot (-\hat{\mathbf{n}})\,dS,$$

where $-\hat{\mathbf{n}}$ is the unit outer normal to S. If we now use the divergence theorem, we have

$$\oiint_S (-9.81\rho z)\hat{\mathbf{n}} \cdot \hat{\mathbf{k}}\,dS = \iiint_V \nabla \cdot (9.81\rho z\hat{\mathbf{k}})\,dV$$

$$= \iiint_V 9.81\rho\,dV = 9.81\rho \text{ (volume of } V),$$

and this is the weight of the fluid displaced by the object.

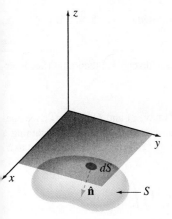
FIGURE 14.46   Verification of Archimedes' principle for any object

In Exercises 1–12 use the divergence theorem to evaluate the surface integral.

1. $\iint_S (x\hat{\mathbf{i}} + y\hat{\mathbf{j}} - 2z\hat{\mathbf{k}}) \cdot \hat{\mathbf{n}}\, dS$, where S is the surface bounding the volume defined by the surfaces $z = 2x^2 + y^2$, $x^2 + y^2 = 3$, $z = 0$, and $\hat{\mathbf{n}}$ is the unit outer normal to S

2. $\iint_S (x^2\hat{\mathbf{i}} + y^2\hat{\mathbf{j}} + z^2\hat{\mathbf{k}}) \cdot \hat{\mathbf{n}}\, dS$, where S is the sphere $x^2 + y^2 + z^2 = a^2$, and $\hat{\mathbf{n}}$ is the unit outer normal to S

3. $\iint_S (yz\hat{\mathbf{i}} + xz\hat{\mathbf{j}} + xy\hat{\mathbf{k}}) \cdot \hat{\mathbf{n}}\, dS$, where S is the surface enclosing the volume defined by $x = 0$, $x = 2$, $z = 0$, $z = y$, $y + z = 2$, and $\hat{\mathbf{n}}$ is the unit outer normal to S

4. $\iint_S [(z^2 - x)\hat{\mathbf{i}} - xy\hat{\mathbf{j}} + 3z\hat{\mathbf{k}}] \cdot \hat{\mathbf{n}}\, dS$, where S is the surface enclosing the volume defined by $z = 4 - y^2$, $x = 0$, $x = 3$, $z = 0$, and $\hat{\mathbf{n}}$ is the unit outer normal to S

5. $\iint_S \mathbf{F} \cdot \hat{\mathbf{n}}\, dS$, where $\mathbf{F} = (x^2 y\hat{\mathbf{i}} + y^2 z\hat{\mathbf{j}} + z^2 x\hat{\mathbf{k}})/2$, S is the surface bounding the volume in the first octant defined by $x = 0$, $y = 0$, $z = 1$, $z = 0$, $x^2 + y^2 = 1$, and $\hat{\mathbf{n}}$ is the unit inner normal to S

* **6.** $\iint_S (x\hat{\mathbf{i}} + y\hat{\mathbf{j}} + 2z\hat{\mathbf{k}}) \cdot \hat{\mathbf{n}}\, dS$, where S is the surface bounding the volume defined by the surfaces $z = 2x^2 + y^2$, $x^2 + y^2 = 3$, $z = 0$, and $\hat{\mathbf{n}}$ is the unit outer normal to S

* **7.** $\iint_S (z\hat{\mathbf{i}} - x\hat{\mathbf{j}} + y\hat{\mathbf{k}}) \cdot \hat{\mathbf{n}}\, dS$, where S is the surface enclosing the volume defined by the surfaces $z = \sqrt{4 - x^2 - y^2}$, $z = 0$, and $\hat{\mathbf{n}}$ is the unit outer normal to S

* **8.** $\iint_S (2x^2 y\hat{\mathbf{i}} - y^2\hat{\mathbf{j}} + 4xz^2\hat{\mathbf{k}}) \cdot \hat{\mathbf{n}}\, dS$, where S is the surface enclosing the volume in the first octant defined by $y^2 + z^2 = 9$, $x = 2$, and $\hat{\mathbf{n}}$ is the unit outer normal to S

* **9.** $\iint_S (yx\hat{\mathbf{i}} + y^2\hat{\mathbf{j}} + yz\hat{\mathbf{k}}) \cdot \hat{\mathbf{n}}\, dS$, where S is the ellipsoid $x^2 + y^2/4 + z^2 = 1$, and $\hat{\mathbf{n}}$ is the unit outer normal to S

* **10.** $\iint_S (x^3\hat{\mathbf{i}} + y^3\hat{\mathbf{j}} - z^3\hat{\mathbf{k}}) \cdot \hat{\mathbf{n}}\, dS$, where S is the surface enclosing the volume defined by $z = 6 - x^2 - y^2$, $z = \sqrt{x^2 + y^2}$, and $\hat{\mathbf{n}}$ is the unit outer normal to S

* **11.** $\iint_S (y\hat{\mathbf{i}} - xy\hat{\mathbf{j}} + zy^2\hat{\mathbf{k}}) \cdot \hat{\mathbf{n}}\, dS$, where S is the surface enclosing the volume defined by $y^2 - x^2 - z^2 = 4$, $y = 4$, and $\hat{\mathbf{n}}$ is the unit inner normal to S

* **12.** $\iint_S (xy\hat{\mathbf{i}} + z^2\hat{\mathbf{k}}) \cdot \hat{\mathbf{n}}\, dS$, where S is the surface enclosing the volume in the first octant bounded by the planes $z = 0$, $y = x$, $y = 2x$, $x + y + z = 6$, and $\hat{\mathbf{n}}$ is the unit outer normal to S

In Exercises 13–15 use the divergence theorem to evaluate the surface integral. In each case an additional surface must be introduced in order to enclose a volume.

* **13.** $\iint_S (x\hat{\mathbf{i}} + y\hat{\mathbf{j}} + z\hat{\mathbf{k}}) \cdot \hat{\mathbf{n}}\, dS$, where S is the top half of the ellipsoid $x^2 + 4y^2 + 9z^2 = 36$, and $\hat{\mathbf{n}}$ is the unit upper normal to S

* **14.** $\iint_S (xy\hat{\mathbf{i}} - yz\hat{\mathbf{j}} + x^2 z\hat{\mathbf{k}}) \cdot \hat{\mathbf{n}}\, dS$, where S is that part of the cone $z = \sqrt{x^2 + y^2}$ below $z = 2$, and $\hat{\mathbf{n}}$ is the unit normal to S with positive z-component

* **15.** $\iint_S (y^2 e^z\hat{\mathbf{i}} - xy\hat{\mathbf{j}} + z\hat{\mathbf{k}}) \cdot \hat{\mathbf{n}}\, dS$, where S is that part of $z = 4 - x^2 - y^2$ cut out by $z = 2y$, and $\hat{\mathbf{n}}$ is the unit upper normal to S

* **16.** Show that if $\hat{\mathbf{n}}$ is the unit outer normal to a surface S, then the region enclosed by S has volume

$$V = \frac{1}{3}\iint_S \mathbf{r} \cdot \hat{\mathbf{n}}\, dS, \qquad \mathbf{r} = x\hat{\mathbf{i}} + y\hat{\mathbf{j}} + z\hat{\mathbf{k}}.$$

* **17.** If $\hat{\mathbf{n}}$ is the unit outer normal to a surface S that encloses a region V, show that the area of S can be expressed in the form

$$\text{area}(S) = \iiint_V \nabla \cdot \hat{\mathbf{n}}\, dV.$$

* **18.** How would you prove Archimedes' principle in the case that an object is only partially submerged? (See Example 14.26.)

In Exercises 19–21 evaluate the surface integral.

* **19.** $\iint_S [(x + y)\hat{\mathbf{i}} + y^3\hat{\mathbf{j}} + x^2 z\hat{\mathbf{k}}] \cdot \hat{\mathbf{n}}\, dS$, where S is the surface enclosing the volume defined by $x^2 + y^2 - z^2 = 1$, $2z^2 = x^2 + y^2$, and $\hat{\mathbf{n}}$ is the unit outer normal to S

* **20.** $\iint_S [(x + y)^2\hat{\mathbf{i}} + x^2 y\hat{\mathbf{j}} - x^2 z\hat{\mathbf{k}}] \cdot \hat{\mathbf{n}}\, dS$, where $\hat{\mathbf{n}}$ is the unit inner normal to the surface S enclosing the volume defined by $z^2 = (1 - x^2 - 2y^2)^2$

* **21.** $\iint_S [(y^3 + x^2 y)\hat{\mathbf{i}} + (x^3 - xy^2)\hat{\mathbf{j}} + z\hat{\mathbf{k}}] \cdot \hat{\mathbf{n}}\, dS$, where $\hat{\mathbf{n}}$ is the unit upper normal to the surface $S : z = \sqrt{1 - x^2 - y^2}$

* **22.** If V is a region bounded by a closed surface S, and $\mathbf{B} = \nabla \times \mathbf{A}$, show that $\iint_S \mathbf{B} \cdot \hat{\mathbf{n}}\, dS = 0$.

* **23.** Is Green's theorem related to the divergence theorem? (See Exercise 13 in Section 14.6.)

In Exercises 24–26 assume that $P(x, y, z)$ and $Q(x, y, z)$ have continuous first and second partial derivatives in a domain containing a closed surface S and its interior V. Let $\hat{\mathbf{n}}$ be the unit outer normal to S.

* **24.** Show that

$$\iint_S \nabla P \cdot \hat{\mathbf{n}}\, dS = \iiint_V \nabla^2 P\, dV,$$

where $\nabla^2 P = \dfrac{\partial^2 P}{\partial x^2} + \dfrac{\partial^2 P}{\partial y^2} + \dfrac{\partial^2 P}{\partial z^2}$. What can we conclude if $P(x, y, z)$ satisfies Laplace's equation in V?

25. Show that

$$\oiint_S P\nabla Q \cdot \hat{\mathbf{n}}\,dS = \iiint_V (P\,\nabla^2 Q + \nabla P \cdot \nabla Q)\,dV.$$

This result is called *Green's first identity*.

26. Prove that

$$\oiint_S (P\,\nabla Q - Q\,\nabla P) \cdot \hat{\mathbf{n}}\,dS = \iiint_V (P\,\nabla^2 Q - Q\,\nabla^2 P)\,dV.$$

This result is called *Green's second identity*.

* **27.** Compare Exercises 24–26 with Exercises 35–38 in Section 14.6.

** **28.** Let S be a closed surface, and let $\hat{\mathbf{n}}$ be the unit outer normal to S. If $\mathbf{r}_0 = x_0\hat{\mathbf{i}} + y_0\hat{\mathbf{j}} + z_0\hat{\mathbf{k}}$ is the position vector of some fixed point P_0, show that

$$\oiint_S \frac{\mathbf{r} - \mathbf{r}_0}{|\mathbf{r} - \mathbf{r}_0|^3} \cdot \hat{\mathbf{n}}\,dS = \begin{cases} 0, & \text{if } S \text{ does not enclose } P_0, \\ 4\pi, & \text{if } S \text{ does enclose } P_0. \end{cases}$$

14.10 Stokes's Theorem

Stokes's theorem relates certain line integrals around closed curves to surface integrals over surfaces that have the curves as boundaries.

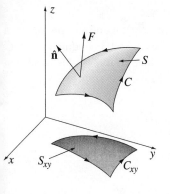

FIGURE 14.47 Elementary surface for proof of Stokes's theorem

> ### THEOREM 14.8 (Stokes's Theorem)
>
> Let C be a closed, piecewise-smooth curve that does not intersect itself and let S be a piecewise-smooth, orientable surface with C as boundary (Figure 14.47). Let $\mathbf{F}(x, y, z) = P(x, y, z)\hat{\mathbf{i}} + Q(x, y, z)\hat{\mathbf{j}} + R(x, y, z)\hat{\mathbf{k}}$ be a vector field whose components P, Q, and R have continuous first partial derivatives in a domain that contains S and C. Then
>
> $$\oint_C \mathbf{F} \cdot d\mathbf{r} = \iint_S (\nabla \times \mathbf{F}) \cdot \hat{\mathbf{n}}\,dS, \qquad (14.59a)$$
>
> or
>
> $$\oint_C P\,dx + Q\,dy + R\,dz$$
> $$= \iint_S \left[\left(\frac{\partial R}{\partial y} - \frac{\partial Q}{\partial z}\right)\hat{\mathbf{i}} + \left(\frac{\partial P}{\partial z} - \frac{\partial R}{\partial x}\right)\hat{\mathbf{j}} + \left(\frac{\partial Q}{\partial x} - \frac{\partial P}{\partial y}\right)\hat{\mathbf{k}} \right] \cdot \hat{\mathbf{n}}\,dS, \quad (14.59b)$$
>
> where $\hat{\mathbf{n}}$ is the unit normal to S chosen in the following way: If when moving along C the surface S is on the left, then $\hat{\mathbf{n}}$ must be chosen as the unit normal on that side of S. On the other hand, if when moving along C, the surface is on the right, then $\hat{\mathbf{n}}$ must be chosen on the opposite side of S.

PROOF We first consider a surface S that projects in a one-to-one fashion onto each of the three coordinate planes. Because S projects one-to-one onto some region S_{xy} in the xy-plane, we can take the equation for S in the form $z = f(x, y)$. If the direction along C is as indicated in Figure 14.47, and C_{xy} is the projection of C on the xy-plane, then

$$\oint_C P\,dx = \oint_{C_{xy}} P[x, y, f(x, y)]\,dx.$$

If we use Green's theorem on this line integral around C_{xy}, we have

$$\oint_C P\,dx = \iint_{S_{xy}} -\left(\frac{\partial P}{\partial y} + \frac{\partial P}{\partial z}\frac{\partial z}{\partial y}\right) dA.$$

On the other hand, since a unit normal to S is

$$\hat{\mathbf{n}} = \frac{\left(-\dfrac{\partial z}{\partial x}, -\dfrac{\partial z}{\partial y}, 1\right)}{\sqrt{1 + \left(\dfrac{\partial z}{\partial x}\right)^2 + \left(\dfrac{\partial z}{\partial y}\right)^2}},$$

it also follows that

$$\iint_S \left(\frac{\partial P}{\partial z}\hat{\mathbf{j}} - \frac{\partial P}{\partial y}\hat{\mathbf{k}} \right) \cdot \hat{\mathbf{n}}\, dS = \iint_S \frac{-\dfrac{\partial P}{\partial z}\dfrac{\partial z}{\partial y} - \dfrac{\partial P}{\partial y}}{\sqrt{1 + \left(\dfrac{\partial z}{\partial x}\right)^2 + \left(\dfrac{\partial z}{\partial y}\right)^2}}\, dS$$

$$= \iint_{S_{xy}} \frac{-\dfrac{\partial P}{\partial z}\dfrac{\partial z}{\partial y} - \dfrac{\partial P}{\partial y}}{\sqrt{1 + \left(\dfrac{\partial z}{\partial x}\right)^2 + \left(\dfrac{\partial z}{\partial y}\right)^2}} \sqrt{1 + \left(\dfrac{\partial z}{\partial x}\right)^2 + \left(\dfrac{\partial z}{\partial y}\right)^2}\, dA$$

$$= \iint_{S_{xy}} -\left(\frac{\partial P}{\partial y} + \frac{\partial P}{\partial z}\frac{\partial z}{\partial y} \right) dA.$$

We have shown, then, that

$$\oint_C P\, dx = \iint_S \left(\frac{\partial P}{\partial z}\hat{\mathbf{j}} - \frac{\partial P}{\partial y}\hat{\mathbf{k}} \right) \cdot \hat{\mathbf{n}}\, dS.$$

By projecting C and S onto the xz- and yz-planes, we can show similarly that

$$\oint_C R\, dz = \iint_S \left(\frac{\partial R}{\partial y}\hat{\mathbf{i}} - \frac{\partial R}{\partial x}\hat{\mathbf{j}} \right) \cdot \hat{\mathbf{n}}\, dS$$

and

$$\oint_C Q\, dy = \iint_S \left(\frac{\partial Q}{\partial x}\hat{\mathbf{k}} - \frac{\partial Q}{\partial z}\hat{\mathbf{i}} \right) \cdot \hat{\mathbf{n}}\, dS.$$

Addition of these three results gives Stokes's theorem for $\mathbf{F}$ and S.

The proof can be extended to more general curves and surfaces that do not project in a one-to-one fashion onto all three coordinate planes. Most surfaces S with bounding curves C can be divided into n subsurfaces S_i with bounding curves C_i that do satisfy this condition (Figure 14.48). For each such subsurface, Stokes's theorem applies:

$$\oint_{C_i} \mathbf{F} \cdot d\mathbf{r} = \iint_{S_i} (\nabla \times \mathbf{F}) \cdot \hat{\mathbf{n}}\, dS, \quad i = 1, \ldots, n.$$

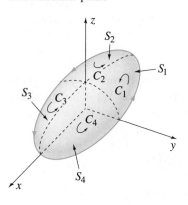

FIGURE 14.48   Division of a surface into pieces, each of which projects one-to-one onto all three coordinate planes

If these n equations are now added together, we have

$$\sum_{i=1}^{n} \oint_{C_i} \mathbf{F} \cdot d\mathbf{r} = \sum_{i=1}^{n} \iint_{S_i} (\nabla \times \mathbf{F}) \cdot \hat{\mathbf{n}}\, dS.$$

Since the S_i constitute S, the right side of this equation is the surface integral of $(\nabla \times \mathbf{F}) \cdot \hat{\mathbf{n}}$ over S. Figure 14.48 illustrates that when line integrals over the C_i are added, contributions from auxiliary (interior) curves cancel in pairs, and the remaining line integrals give the line integral of $\mathbf{F} \cdot d\mathbf{r}$ along C. For general surfaces that cannot be divided into a finite number of subsurfaces of this type, the reader should consult a more advanced book.

Green's theorem is a special case of Stokes's theorem. For if $\mathbf{F} = P(x, y)\hat{\mathbf{i}} + Q(x, y)\hat{\mathbf{j}}$, and C is a closed curve in the xy-plane, then by Stokes's theorem,

$$\oint_C P\, dx + Q\, dy = \iint_S \left(\frac{\partial Q}{\partial x} - \frac{\partial P}{\partial y} \right) \hat{\mathbf{k}} \cdot \hat{\mathbf{n}}\, dS,$$

where S is any surface for which C is the boundary. If we choose S as that part of the xy-plane bounded by C, then $\hat{\mathbf{n}} = \hat{\mathbf{k}}$ and

$$\oint_C P\, dx + Q\, dy = \iint_S \left(\frac{\partial Q}{\partial x} - \frac{\partial P}{\partial y} \right) dA.$$

With Stokes's theorem, it is straightforward to verify the sufficiency half of Theorem 14.1. Suppose that the curl of a vector field $\mathbf{F}$ vanishes in a simply connected domain D. If C is any piecewise-smooth, closed curve in D, then there exists a piecewise-smooth surface S in D with C as boundary. By Stokes's theorem,

$$\oint_C \mathbf{F} \cdot d\mathbf{r} = \iint_S (\nabla \times \mathbf{F}) \cdot \hat{\mathbf{n}} \, dS = 0.$$

According to Corollary 2 of Theorem 14.3, the line integral is independent of path in D, and the theorem itself implies the existence of a function $f(x, y, z)$ such that $\nabla f = \mathbf{F}$.

EXAMPLE 14.27

▼

Verify Stokes's theorem if $\mathbf{F} = x^2\hat{\mathbf{i}} + x\hat{\mathbf{j}} + xyz\hat{\mathbf{k}}$, and S is that part of the sphere $x^2 + y^2 + z^2 = 4$ above the plane $z = 1$.

FIGURE 14.49   Verification of Stokes's theorem for the part of $x^2 + y^2 + z^2 = 4$ above the plane $z = 1$

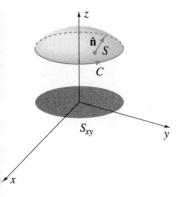

SOLUTION   If we choose $\hat{\mathbf{n}}$ as the upper normal to S, then C, the boundary of S, must be traversed in the direction shown in Figure 14.49. (If $\hat{\mathbf{n}}$ is chosen as the lower normal, then C must be traversed in the opposite direction.) Since parametric equations for C are

$$x = \sqrt{3}\cos t, \quad y = \sqrt{3}\sin t, \quad z = 1, \quad 0 \le t \le 2\pi,$$

$$\begin{aligned}
\oint_C \mathbf{F} \cdot d\mathbf{r} &= \oint_C x^2 \, dx + x \, dy + xyz \, dz \\
&= \int_0^{2\pi} [3\cos^2 t(-\sqrt{3}\sin t \, dt) + \sqrt{3}\cos t(\sqrt{3}\cos t \, dt)] \\
&= \int_0^{2\pi} \left[-3\sqrt{3}\cos^2 t \sin t + \frac{3}{2}(1 + \cos 2t) \right] dt \\
&= \left\{ \sqrt{3}\cos^3 t + \frac{3t}{2} + \frac{3\sin 2t}{4} \right\}_0^{2\pi} = 3\pi.
\end{aligned}$$

On the other hand,

$$\begin{aligned}
\iint_S (\nabla \times \mathbf{F}) \cdot \hat{\mathbf{n}} \, dS &= \iint_S (xz\hat{\mathbf{i}} - yz\hat{\mathbf{j}} + \hat{\mathbf{k}}) \cdot \frac{(2x, 2y, 2z)}{\sqrt{4x^2 + 4y^2 + 4z^2}} \, dS \\
&= \iint_S \frac{x^2 z - y^2 z + z}{\sqrt{x^2 + y^2 + z^2}} \, dS \\
&= \iint_{S_{xy}} \frac{z(x^2 - y^2 + 1)}{2} \sqrt{1 + \left(\frac{\partial z}{\partial x}\right)^2 + \left(\frac{\partial z}{\partial y}\right)^2} \, dA \\
&= \frac{1}{2} \iint_{S_{xy}} z(x^2 - y^2 + 1)\sqrt{1 + (-x/z)^2 + (-y/z)^2} \, dA \\
&= \frac{1}{2} \iint_{S_{xy}} z(x^2 - y^2 + 1)\frac{\sqrt{x^2 + y^2 + z^2}}{z} \, dA \\
&= \iint_{S_{xy}} (x^2 - y^2 + 1) \, dA.
\end{aligned}$$

If we use polar coordinates to evaluate this double integral over $S_{xy} : x^2 + y^2 \leq 3$, we have

$$\iint_S (\nabla \times \mathbf{F}) \cdot \hat{\mathbf{n}} \, dS = \int_{-\pi}^{\pi} \int_0^{\sqrt{3}} (r^2 \cos^2 \theta - r^2 \sin^2 \theta + 1) \, r \, dr \, d\theta$$

$$= \int_{-\pi}^{\pi} \left\{ \frac{r^4}{4} (\cos^2 \theta - \sin^2 \theta) + \frac{r^2}{2} \right\}_0^{\sqrt{3}} d\theta$$

$$= \int_{-\pi}^{\pi} \left(\frac{9}{4} \cos 2\theta + \frac{3}{2} \right) d\theta$$

$$= \left\{ \frac{9}{8} \sin 2\theta + \frac{3\theta}{2} \right\}_{-\pi}^{\pi} = 3\pi.$$

EXAMPLE 14.28

Evaluate

$$\oint_C 2xy^3 \, dx + 3x^2 y^2 \, dy + (2z + x) \, dz,$$

where C consists of line segments joining $A(2, 0, 0)$ to $B(0, 1, 0)$ to $D(0, 0, 1)$ to A.

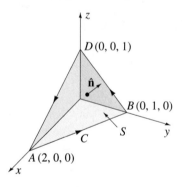

SOLUTION By Stokes's theorem,

$$\oint_C 2xy^3 \, dx + 3x^2 y^2 \, dy + (2z + x) \, dz = \iint_S \nabla \times (2xy^3, 3x^2 y^2, 2z + x) \cdot \hat{\mathbf{n}} \, dS,$$

where S is any surface with C as boundary. If we choose S as the flat triangle bounded by C (Figure 14.50), then a normal vector to S is

$$\mathbf{BD} \times \mathbf{BA} = \begin{vmatrix} \hat{\mathbf{i}} & \hat{\mathbf{j}} & \hat{\mathbf{k}} \\ 0 & -1 & 1 \\ 2 & -1 & 0 \end{vmatrix} = (1, 2, 2),$$

and therefore, $\hat{\mathbf{n}} = \dfrac{(1, 2, 2)}{3}$. Since

$$\nabla \times (2xy^3, 3x^2 y^2, 2z + x) = \begin{vmatrix} \hat{\mathbf{i}} & \hat{\mathbf{j}} & \hat{\mathbf{k}} \\ \partial/\partial x & \partial/\partial y & \partial/\partial z \\ 2xy^3 & 3x^2 y^2 & 2z + x \end{vmatrix} = -\hat{\mathbf{j}},$$

it follows that $\nabla \times (2xy^3, 3x^2 y^2, 2z + x) \cdot \hat{\mathbf{n}} = -\hat{\mathbf{j}} \cdot \dfrac{(1, 2, 2)}{3} = -\dfrac{2}{3}$, and

$$\oint_C 2xy^3 \, dx + 3x^2 y^2 \, dy + (2z + x) \, dz = \iint_S -\frac{2}{3} dS = -\frac{2}{3} (\text{area of } S).$$

But from equation 11.42, the area of triangle S is

$$\frac{1}{2} |\mathbf{BD} \times \mathbf{BA}| = \frac{1}{2} |(1, 2, 2)| = \frac{3}{2}.$$

Finally, then,

$$\oint_C 2xy^3 \, dx + 3x^2 y^2 \, dy + (2z + x) \, dz = -\frac{2}{3} \left(\frac{3}{2} \right) = -1.$$

In Exercises 1–14 use Stokes's theorem to evaluate the line integral.

1. $\oint_C x^2 y\, dx + y^2 z\, dy + z^2 x\, dz$, where C is the curve $z = x^2 + y^2$, $x^2 + y^2 = 4$, directed counterclockwise as viewed from the origin

2. $\oint_C y^2\, dx + xy\, dy + xz\, dz$, where C is the curve $x^2 + y^2 = 2y$, $y = z$, directed so that y increases when x is positive

3. $\oint_C (xyz + 2yz)\, dx + xz\, dy + 2xy\, dz$, where C is the curve $z = 1$, $x^2 + y^2 + z^2 = 4$, directed clockwise as viewed from the origin

4. $\oint_C (2xy + y)\, dx + (x^2 + xy - 3y)\, dy + 2xz\, dz$, where C is the curve $z = \sqrt{x^2 + y^2}$, $z = 4$

5. $\oint_C x^2\, dx + y^2\, dy + (x^2 + y^2)\, dz$, where C is the boundary of the first octant part of the plane $x + y + z = 1$, directed counterclockwise as viewed from the origin

6. $\oint_C y\, dx + x\, dy + (x^2 + y^2 + z^2)\, dz$, where C is the curve $x^2 + y^2 = 1$, $z = xy$, directed clockwise as viewed from the point $(0, 0, 1)$

7. $\oint_C zy^2\, dx + xy\, dy + (y^2 + z^2)\, dz$, where C is the curve $x^2 + z^2 = 9$, $y = \sqrt{x^2 + z^2}$, directed counterclockwise as viewed from the origin

8. $\oint_C y\, dx + z\, dy + x\, dz$, where C is the curve $x + y = 2b$, $x^2 + y^2 + z^2 = 2b(x + y)$, directed clockwise as viewed from the origin

9. $\oint_C y^2\, dx + (x + y)\, dy + yz\, dz$, where C is the curve $x^2 + y^2 = 2$, $x + y + z = 2$, directed clockwise as viewed from the origin

10. $\oint_C (x + y)^2\, dx + (x + y)^2\, dy + yz^3\, dz$, where C is the curve with equations $z = \sqrt{x^2 + y^2}$, $(x - 1)^2 + y^2 = 1$

* **11.** $\oint_C xy\, dx - zx\, dy + yz\, dz$, where C is the boundary of that part of $z = x + y$ in the first octant cut off by $x + y = 1$, directed counterclockwise as viewed from the point $(0, 0, 1)$

* **12.** $\oint_C y^3\, dx - x^3\, dy + xyz\, dz$, where C is the curve $x^2 + y^2 = z^2 + 3$, $z = 3 - \sqrt{x^2 + y^2}$, directed clockwise as viewed from the origin

* **13.** $\oint_C z(x + y)^2\, dx + (y - x)^2\, dy + z^2\, dz$, where C is the smooth curve of intersection of the surfaces $x^2 + z^2 = a^2$, $y^2 + z^2 = a^2$ which has a portion in the first octant, directed so that z decreases in the first octant

* **14.** $\oint_C -2y^3 x^2\, dx + x^3 y^2\, dy + z\, dz$, where C is the curve $x^2 + y^2 + z^2 = 4$, $x^2 + 4y^2 = 4$, directed so that x decreases along that part of the curve in the first octant

* **15.** Evaluate the line integral $\oint_C 2x^2 y\, dx - yz\, dy + xz\, dz$, where C is the curve $x^2 + y^2 + z^2 = 4$, $z = \sqrt{3}(x^2 + y^2)$, directed clockwise as viewed from the origin, in four ways: (a) directly as a line integral, (b) using Stokes's theorem with S as that part of $z = \sqrt{4 - x^2 - y^2}$ bounded by C, (c) using Stokes's theorem with S as that part of $z = \sqrt{3}(x^2 + y^2)$ bounded by C, and (d) using Stokes's theorem with S as that part of $z = \sqrt{3}$ bounded by C.

* **16.** Let S_1 be that part of $x^2 + y^2 + z^2 = 1$ above the xy-plane and S_2 be that part of $z = 1 - x^2 - y^2$ above the xy-plane. Show that if $\hat{\mathbf{n}}_1$ and $\hat{\mathbf{n}}_2$ are the unit upper normals to these surfaces, and $\mathbf{F}$ is a vector field defined on both S_1 and S_2, then

$$\iint_{S_1} (\nabla \times \mathbf{F}) \cdot \hat{\mathbf{n}}_1\, dS = \iint_{S_2} (\nabla \times \mathbf{F}) \cdot \hat{\mathbf{n}}_2\, dS.$$

SUMMARY

When a vector is a function of position, it becomes susceptible to the operations of differentiation and integration. In this chapter we developed various ways of differentiating and integrating vector fields beginning with the operations of divergence and curl. The divergence of a vector field is a scalar field, and the curl of a vector field is another vector field. Both are extremely useful in applied mathematics. Mathematically, the curl appeared in our discussion of independence of path for line integrals and in Stokes's theorem; we saw its physical importance in our study of fluid flow and electromagnetic theory. We introduced the divergence of a vector field in our discussion of the divergence theorem and in the same applications as those for the curl.

The line integral of a function $f(x, y, z)$ along a curve C is defined in the same way as a definite integral, a double integral, or a triple integral — that is, the limit of a sum,

$$\int_C f(x, y, z)\, ds = \lim_{\|\Delta s_i\| \to 0} \sum_{i=1}^n f(x_i^*, y_i^*, z_i^*)\, \Delta s_i.$$

The most important type of line integral occurs when $f(x, y, z)$ is the tangential component of a vector field $\mathbf{F} = P\hat{\mathbf{i}} + Q\hat{\mathbf{j}} + R\hat{\mathbf{k}}$ defined along C, and in this case we write

$$\int_C f(x, y, z)\, ds = \int_C \mathbf{F} \cdot d\mathbf{r} = \int_C P\, dx + Q\, dy + R\, dz.$$

We developed three methods for evaluating line integrals:

1. Express all parts of the line integral in terms of any parameter along C and evaluate the resulting definite integral. All line integrals can be evaluated in this way, but often methods 2 and 3 lead to much simpler calculations.

2. If a line integral is independent of path, then we can evaluate it by taking the difference in values of a function ϕ (where $\nabla \phi = \mathbf{F}$) at the ends of the curve.

3. If C is a closed curve, we can sometimes use Stokes's theorem to replace a line integral with a simpler surface integral. In this regard, Green's theorem is a special case of Stokes's theorem.

If the line integral of a force field $\mathbf{F}$ is independent of path, the force field is said to be conservative. Associated with every conservative force field is a potential function U such that the work done by $\mathbf{F}$ along a curve C from A to B is equal to the difference in U at A and B. In addition, motion of an object in a conservative force field is always characterized by an exchange of potential energy for kinetic energy in such a way that the sum of the two energies is always a constant value.

Surface integrals are also limits of sums,

$$\iint_S f(x, y, z)\, dS = \lim_{\|\Delta S_i\| \to 0} \sum_{i=1}^n f(x_i^*, y_i^*, z_i^*)\, \Delta S_i,$$

and the most important type of surface integral occurs when $f(x, y, z)$ is the normal component of a vector field $\mathbf{F}$ on S:

$$\iint_S f(x, y, z)\, dS = \iint_S \mathbf{F} \cdot \hat{\mathbf{n}}\, dS.$$

We suggested two methods for the evaluation of surface integrals:

1. Project S onto some region R in one of the coordinate planes, express all parts of the integral in terms of coordinates in that plane, and evaluate the resulting double integral over R.

2. If S is closed, it could be advantageous to replace a surface integral with the triple integral of $\nabla \cdot \mathbf{F}$ over the volume bounded by S (the divergence theorem).

In the final section of the chapter (see www.pearsoned.ca/text/trim), we learned how to express scalar and vector functions in orthogonal, curvilinear coordinates, including polar, cylindrical, and spherical coordinates. We also developed formulas for the gradient, divergence, and curl in these coordinate systems.

In reviewing this chapter, you should be able to define or discuss the following key terms:

Interior point	Exterior point
Boundary point	Open set
Closed set	Connected set
Domain	Simply connected domain
Vector fields	Del operator
Divergence of a vector field	Equation of continuity
Curl of a vector field	Irrotational vector field
Line integrals	Path independence of a line integral
Conservative force field	Law of conservation of energy
Green's theorem	Surface integral
Closed surface	Orientable surface
Divergence theorem	Stokes's theorem

In Exercises 1–10 calculate the quantity.

1. ∇f if $f(x, y, z) = x^2 y^3 - xy + z$

2. $\nabla \cdot \mathbf{F}$ if $\mathbf{F}(x, y) = x^3 y \hat{\mathbf{i}} - (x^2/y) \hat{\mathbf{j}}$

3. $\nabla \times \mathbf{F}$ if $\mathbf{F}(x, y) = \sin(xy) \hat{\mathbf{i}} + \cos(xy) \hat{\mathbf{j}} + xy \hat{\mathbf{k}}$

4. $\nabla \times \mathbf{F}$ if $\mathbf{F}(x, y, z) = (x + y + z)(\hat{\mathbf{i}} + \hat{\mathbf{j}} + \hat{\mathbf{k}})$

5. ∇f if $f(x, y, z) = \ln(x^2 + y^2 + z^2)$

6. $\nabla \cdot \mathbf{F}$ if $\mathbf{F}(x, y, z) = ye^x \hat{\mathbf{i}} + ze^y \hat{\mathbf{j}} + xe^z \hat{\mathbf{k}}$

7. $\nabla \times \mathbf{F}$ if $\mathbf{F}(x, y, z) = xyz \hat{\mathbf{j}}$

8. ∇f if $f(x, y) = \mathrm{Sin}^{-1}(x + y)$

9. $\nabla \cdot \mathbf{F}$ if $\mathbf{F}(y, z) = yz \hat{\mathbf{i}} - (y^2 + z^2) \hat{\mathbf{j}} + y^2 z^2 \hat{\mathbf{k}}$

10. $\nabla \times \mathbf{F}$ if $\mathbf{F}(x, y, z) = \mathrm{Cot}^{-1}(xyz) \hat{\mathbf{i}}$

In Exercises 11–30 evaluate the integral.

11. $\displaystyle\int_C y \, ds$, where C is the curve $y = x^3$ from $(-1, -1)$ to $(2, 8)$

12. $\displaystyle\iint_S (x^2 + yz) \, dS$, where S is that part of $x + y + z = 2$ in the first octant

13. $\displaystyle\iint_S (x \hat{\mathbf{i}} + y \hat{\mathbf{j}}) \cdot \hat{\mathbf{n}} \, dS$, where S is that part of $z = x^2 + y^2$ bounded by the surfaces $x = \pm 1$, $y = \pm 1$, and $\hat{\mathbf{n}}$ is the lower normal

14. $\displaystyle\iint_S (x \hat{\mathbf{i}} + y \hat{\mathbf{j}}) \cdot \hat{\mathbf{n}} \, dS$, where S is that part of $z = x^2 + y^2$ below $z = 1$, and $\hat{\mathbf{n}}$ is the lower normal

15. $\displaystyle\oint_C x \, dx + y \, dy - z^2 \, dz$, where C is the curve $x^2 + y^2 = 1$, $y = z$

16. $\displaystyle\int_C xy \, dx + xz \, dz$, where C is the curve $y = \sqrt{1 + x^2}$, $z = \sqrt{2 - x^2 - y^2}$, from $(1/\sqrt{2}, \sqrt{3/2}, 0)$ to $(-1/\sqrt{2}, \sqrt{3/2}, 0)$

17. $\displaystyle\oint_C 2xy^3 \, dx + (3x^2 y^2 + 2xy) \, dy$, where C is the curve $(x - 1)^2 + y^2 = 1$

18. $\displaystyle\oint_C 2xy^3 \, dx + (3x^2 y^2 + x^2) \, dy$, where C is the curve $(x-1)^2 + y^2 = 1$

19. $\displaystyle\iint_S (x^2 \hat{\mathbf{i}} + y^2 \hat{\mathbf{j}} + z^2 \hat{\mathbf{k}}) \cdot \hat{\mathbf{n}} \, dS$, where S is the surface bounding the volume enclosed by $y = z$, $y + z = 2$, $x = 0$, $x = 1$, $z = 0$, and $\hat{\mathbf{n}}$ is the outer normal to S

★ 20. $\displaystyle\iint_S (x^2 + y^2) \, dS$, where S is that part of $x^2 + y^2 + z^2 = 6$ inside $z = x^2 + y^2$

★ 21. $\displaystyle\iint_S (x^2 + y^2) \hat{\mathbf{i}} \cdot \hat{\mathbf{n}} \, dS$, where S is that part of $x^2 + y^2 + z^2 = 6$ inside $z = x^2 + y^2$, and $\hat{\mathbf{n}}$ is the upper normal to S

★ 22. $\displaystyle\oint_C (x^2 \hat{\mathbf{i}} + y \hat{\mathbf{j}} - xz \hat{\mathbf{k}}) \cdot d\mathbf{r}$, where C is the curve $x^2 + y^2 = 1$, $z = x + 1$, directed clockwise as viewed from the origin

★ 23. $\displaystyle\oint_C (xy \hat{\mathbf{i}} + z \hat{\mathbf{j}} - x^2 \hat{\mathbf{k}}) \cdot d\mathbf{r}$, where C is the curve in Exercise 22

★ 24. $\displaystyle\oint_C y \, dx + 2x \, dy - 3z^2 \, dz$, where C is the curve $y = \sqrt{1 + z^2 - x^2}$, $x^2 + z^2 = 1$, directed counterclockwise as viewed from the origin

* **25.** $\oint_C (xy + 4x^3y^2)\,dx + (z + 2x^4y)\,dy + (z^5 + x^2z^2)\,dz$, where C is the curve with equations $x^2 + z^2 = 4$, $x^2 + y^2 = 4$, $y = z$, directed counterclockwise as viewed from a point far up the positive z-axis

* **26.** $\iint_S (x^2yz\hat{\mathbf{i}} - x^2yz\hat{\mathbf{j}} - xyz^2\hat{\mathbf{k}}) \cdot \hat{\mathbf{n}}\,dS$, where S is that part of $z = 1 - \sqrt{x^2 + y^2}$ above the xy-plane, and $\hat{\mathbf{n}}$ is the upper normal

* **27.** $\iint_S dS$, where S is that part of $z = x^2 - y^2$ inside $x^2 + y^2 = 4$

* **28.** $\iint_S y\,dS$, where S is that part of $x = y^2 + 1$ in the first octant which is under $x + z = 2$

* **29.** $\oint_C (ye^{xy} + xy^2e^{xy})\,dx + (xe^{xy} + x^2ye^{xy} + x^3y)\,dy$, where C is the curve with equations $x^2 + y^2 = 2y$, $z = 0$

* **30.** $\iint_S (x\hat{\mathbf{i}} + y\hat{\mathbf{j}}) \cdot \hat{\mathbf{n}}\,dS$, where S is that part of $z^2 - x^2 - y^2 = 1$ between the planes $z = 0$ and $z = 2$, and $\hat{\mathbf{n}}$ is the lower normal

* **31.** Let S be that part of the sphere $x^2 + y^2 + z^2 = 1$ that lies above the parabolic cylinder $2z = x^2$. Set up, but do not evaluate, double iterated integrals to calculate the surface integral

$$\iint_S x^2y^2z^2\,dS$$

by projecting S onto (a) the xy-coordinate plane, (b) the yz-coordinate plane, and (c) the xz-coordinate plane.

* **32.** If $\mathbf{r} = x\hat{\mathbf{i}} + y\hat{\mathbf{j}} + z\hat{\mathbf{k}}$, show that $\nabla(|\mathbf{r}|^n) = n|\mathbf{r}|^{n-2}\mathbf{r}$.

* **33.** Verify that $\nabla \times (\nabla \times \mathbf{F}) = \nabla(\nabla \cdot \mathbf{F}) - \nabla^2\mathbf{F}$, where $\nabla^2 = \partial^2/\partial x^2 + \partial^2/\partial y^2 + \partial^2/\partial z^2$.

CHAPTER 15 | Differential Equations

Application Preview

The figure on the left below shows a car moving along a road represented by the x-axis. Unfortunately, many roads are not flat; they have undulations that cause the car to oscillate vertically. To reduce or eliminate these oscillations, there is a spring and a shock absorber (also known as a dashpot). These are shown schematically in the figure on the right.

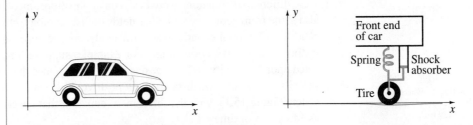

THE PROBLEM   Given the equation of the road and the speed of the car, determine an equation defining the oscillations of the front end of the car. (For the solution, see Example 15.29 on page 1109.)

Differential equations serve as models for many problems in engineering and physics. In this chapter we discuss some of the methods for solving first-order and simple second-order equations. We also give a fairly thorough treatment of linear differential equations. We include a wide variety of applications to illustrate the relevance of differential equations in applied mathematics.

15.1 Introduction

A **differential equation** is an equation that must be solved for an unknown function. What distinguishes a differential equation from other equations is the fact that it contains at least one derivative of the unknown function. For example, each of the following equations is a differential equation in y as a function of x:

$$\frac{dy}{dx} + \frac{k}{m}y^2 = 9.81, \tag{15.1}$$

$$\frac{d^2y}{dx^2} = k\sqrt{1 + \left(\frac{dy}{dx}\right)^2}, \tag{15.2}$$

$$x\frac{d^2y}{dx^2} + \frac{dy}{dx} + xy = 0, \tag{15.3}$$

$$\frac{d^4y}{dx^4} - k^4y = 0. \tag{15.4}$$

Equation 15.1 can be used to determine the position of a skydiver who falls under the influences of gravity and of air resistance, which is proportional to the square of velocity (see Example 15.9); equation 15.2 describes the shape of a hanging cable (equation 11.49 in Section 11.8); equation 15.3, one of Bessel's differential equations, is found in heat flow and vibration problems; and equation 15.4 is used to determine the deflection of beams.

DEFINITION 15.1

The **order of a differential equation** is the order of the highest derivative in the equation.

Of the four differential equations 15.1–15.4, the first is first order, the second and third are second order, and the last is fourth order.

We have considered quite a number of differential equations in Chapters 3, 5, and 8. In Section 5.5 we dealt with separable differential equations; in Chapter 3 we verified that particular combinations of transcendental functions satisfied certain differential equations; and in Chapter 8 we used our integration techniques to solve many separable equations. Almost all of these differential equations were based on applications, most from physics and engineering, but also some from geometry and other fields such as ecology, chemistry, and psychology. In applications, differential equations are almost always accompanied by subsidiary conditions called **initial** or **boundary conditions**. For example, suppose a mass m, while sinking in water, is acted upon by gravity and a force due to water resistance that is proportional to its instantaneous velocity. If we choose distance y as positive downward, taking $y = 0$ at the surface of the water (Figure 15.1), then the differential equation that describes the velocity $v(t)$ of the mass as a function of time t is

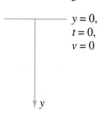

FIGURE 15.1 Schematic for stone sinking in water

$$m\frac{dv}{dt} = -kv + mg, \quad \text{where } k > 0 \text{ is a constant and } g = 9.81. \tag{15.5}$$

This is simply a statement of *Newton's second law*, where dv/dt is the vertical component of the acceleration of m, and $-kv + mg$ is the vertical component of the total force on m due to gravity (mg) and water resistance ($-kv$). If the mass is released from rest (in the surface) at time $t = 0$, the condition $v(0) = 0$ must be added to the differential equation. In other words, the real problem is to find the solution of the differential equation that also satisfies the initial condition:

$$m\frac{dv}{dt} = -kv + mg, \quad v(0) = 0. \tag{15.6}$$

This is the form in which applied mathematicians use differential equations — the differential equation is accompanied by subsidiary conditions that express extra requirements of the solution. It is not difficult to show that the solution of equation 15.6 is

$$v(t) = \frac{mg}{k} - \frac{mg}{k}e^{-kt/m}. \tag{15.7}$$

(All we need do to verify this is substitute the function into the differential equation to see that it does indeed satisfy the equation. It is clear that it does satisfy the initial condition.)

If we change the initial condition to $v(0) = v_0$, so that the initial velocity of m has vertical component v_0 as it enters the water, then the solution becomes

$$v(t) = \frac{mg}{k} - \left(\frac{mg}{k} - v_0\right)e^{-kt/m}. \tag{15.8}$$

In other words, every solution of differential equation 15.5 can be written in the form

$$v(t) = \frac{mg}{k} + Ce^{-kt/m}, \tag{15.9}$$

and when we impose the initial condition $v(0) = v_0$, then $C = v_0 - mg/k$.

In a similar way, equation 15.4 for beam deflections is normally accompanied by four boundary conditions that specify the types of supports at the ends of the beam (see Section 5.4). For simple supports at both $x = 0$ and $x = L$, $y(x)$ must satisfy[†]

$$y(0) = y(L) = 0;$$

$$y''(0) = y''(L) = 0.$$

It can be shown that every solution of equation 15.4 can be expressed in the form

$$y(x) = C_1 e^{kx} + C_2 e^{-kx} + C_3 \sin(kx) + C_4 \cos(kx), \qquad (15.10)$$

where C_1, C_2, C_3, and C_4 are arbitrary constants, and when the boundary conditions are applied, these constants must satisfy the four equations

$$0 = C_1 + C_2 + C_4,$$
$$0 = C_1 e^{kL} + C_2 e^{-kL} + C_3 \sin(kL) + C_4 \cos(kL),$$
$$0 = C_1 + C_2 - C_4,$$
$$0 = C_1 e^{kL} + C_2 e^{-kL} - C_3 \sin(kL) - C_4 \cos(kL).$$

We have stated that every solution of 15.5 can be written in form 15.9, and every solution of 15.4 can be expressed as 15.10. Note that 15.9 contains one arbitrary constant whereas 15.10 has four, but in both cases the number of arbitrary constants is the same as the order of the differential equation. We might suspect that every solution of an nth-order differential equation can be expressed as a function involving n arbitrary constants. For many differential equations this is indeed true, but unfortunately it is not true for all equations. As an illustration, consider the equation

$$\frac{d^2 y}{dx^2} = \left(\frac{dy}{dx} \right)^2. \qquad (15.11)$$

In Example 15.6 we apply standard techniques for solving differential equations to obtain the solution $y(x) = C_1 - \ln(C_2 + x)$, which contains two arbitrary constants C_1 and C_2. This two-parameter family of solutions does not, however, contain all solutions of the differential equation, for no choice of C_1 and C_2 will give the perfectly acceptable solution $y(x) \equiv 1$. This solution is not particularly interesting, but it is nonetheless a solution that is not contained within the two-parameter family. Such a solution is called a **singular solution** for the two-parameter family.

We have illustrated that a solution that contains the same number of arbitrary constants as the order of the differential equation may or may not contain all solutions of the differential equation. In spite of this unfortunate circumstance, there do exist large classes of differential equations for which a solution with the same number of arbitrary constants as the order of the equation does indeed represent all possible solutions. Because of this we make the following definition.

> **DEFINITION 15.2**
>
> An n-parameter family of solutions of an nth-order differential equation is said to be a **general solution** if it contains all solutions of the differential equation.[‡]

[†] In this chapter it is frequently convenient to use the notation $y', y'', y''', \ldots$ to represent $dy/dx, d^2y/dx^2, d^3y/dx^3$, and so on. In this notation $y''(a)$ is the second derivative of y evaluated at $x = a$. In addition, we denote the solution of a differential equation in y as a function of x by $y(x)$.

[‡] Readers should be aware that not all authors agree on this definition of a general solution of a differential equation. Some do not require a general solution to contain all solutions of the differential equation.

Consequently, in order for a function to be a general solution of a differential equation, it must, first, be a solution; second, contain the requisite number of arbitrary constants; and third, contain all solutions of the differential equation.

EXAMPLE 15.1

Find a general solution for the differential equation $\dfrac{d^2y}{dx^2} = xe^{-x}$.

SOLUTION   Integration of both sides of the differential equation, with integration by parts on the right, gives

$$\frac{dy}{dx} = -xe^{-x} - e^{-x} + C_1.$$

A second integration yields

$$y(x) = xe^{-x} + 2e^{-x} + C_1x + C_2.$$

Inclusion of constants of integration assures us that all solutions of the differential equation are included in this two-parameter family, and hence we have a general solution.

There is no procedure that always determines whether an n-parameter family of solutions of an nth-order differential equation is a general solution. It may happen, as in Example 15.1, that the method of arriving at the n-parameter family of solutions guarantees that all solutions are captured. Although this is the exception rather than the rule, there are classes of differential equations for which an n-parameter family of solutions is automatically a general solution. We shall certainly point these out.

It is straightforward to illustrate that $y = (C_1 + C_2x)e^{-2x} + 1/4$ is a two-parameter family of solutions of the differential equation $y'' + 4y' + 4y = 1$. At this time we cannot be sure that all solutions of this differential equation can be obtained by specifying values for C_1 and C_2, and hence we cannot claim to have a general solution. We shall be able to do so in Section 15.8.

> **DEFINITION 15.3**
>
> A **particular solution** of a differential equation is a solution that contains no arbitrary constants.

It follows, therefore, that particular solutions can be obtained by assigning specific values to the arbitrary constants in a family of solutions. For example, $y(x) = 5 - \ln(3 + x)$ is a particular solution of differential equation $d^2y/dx^2 = (dy/dx)^2$, as is $y(x) = -\ln x$, both being obtained from the two-parameter family of solutions $y(x) = C_1 - \ln(C_2 + x)$ by specifying values for C_1 and C_2. On the other hand, the solution $y(x) = 10$ is also a particular solution, but it cannot be obtained from the two-parameter family.

EXAMPLE 15.2

Find a particular solution of the differential equation

$$5\frac{d^3y}{dx^3} + 3\frac{d^2y}{dx^2} + 2y = 4.$$

SOLUTION   In Section 15.9 we develop systematic techniques for finding general and particular solutions for differential equations such as this. But clearly those techniques are not needed here; a simple glance tells us that $y(x) = 2$ is a solution.

Some differential equations are immediately solvable (or, as we often say, immediately integrable). For example, to solve a differential equation of the form

$$\frac{dy}{dx} = M(x),\qquad(15.12)$$

where $M(x)$ is given, we integrate both sides of the equation with respect to x:

$$y(x) = \int M(x)\,dx + C.\qquad(15.13)$$

Because the right side represents all antiderivatives of $M(x)$, this is a general solution of 15.12. For the nth-order equation

$$\frac{d^n y}{dx^n} = M(x),\quad n\text{ a positive integer},\qquad(15.14)$$

we integrate successively n times to obtain a general solution

$$y(x) = \int \cdots \int M(x)\,dx \cdots dx + C_1 + C_2 x + \cdots + C_n x^{n-1}.\qquad(15.15)$$

EXERCISES 15.1

In Exercises 1–10 show that each function in the family satisfies the differential equation.

1. $y(x) = 2 + Ce^{-x^2};\ \dfrac{dy}{dx} + 2xy = 4x$

2. $y(x) = \dfrac{x}{1+Cx};\ \dfrac{dy}{dx} = \dfrac{y^2}{x^2}$

3. $y(x) = \dfrac{x^3}{2} + Cx^3 e^{1/x^2};\ x^3\dfrac{dy}{dx} + (2 - 3x^2)y = x^3$

4. $y(x) = C_1 \sin 3x + C_2 \cos 3x;\ \dfrac{d^2 y}{dx^2} + 9y = 0$

5. $y(x) = \dfrac{C_1^2 e^{2x} + 1}{2C_1 e^x} + C_2;\ \left(\dfrac{d^2 y}{dx^2}\right)^2 = 1 + \left(\dfrac{dy}{dx}\right)^2$

6. $y(x) = C_1 e^{2x} \cos\left(\dfrac{x}{\sqrt{2}}\right) + C_2 e^{2x} \sin\left(\dfrac{x}{\sqrt{2}}\right);$
$2\dfrac{d^2 y}{dx^2} - 8\dfrac{dy}{dx} + 9y = 0$

7. $y(x) = C_1 \cos 2x + C_2 \sin 2x + C_3 \cos x + C_4 \sin x;$
$\dfrac{d^4 y}{dx^4} + 5\dfrac{d^2 y}{dx^2} + 4y = 0$

8. $y(x) = \left(C_1 + C_2 x - \dfrac{x^2}{4}\right)e^{4x};$
$2\dfrac{d^2 y}{dx^2} - 16\dfrac{dy}{dx} + 32y = -e^{4x}$

9. $y(x) = C_1 \cos(2\ln x) + C_2 \sin(2\ln x) + \dfrac{1}{4};$
$x^2\dfrac{d^2 y}{dx^2} + x\dfrac{dy}{dx} + 4y = 1$

10. $y(x) = C_1\dfrac{\sin x}{\sqrt{x}} + C_2\dfrac{\cos x}{\sqrt{x}};$
$x^2\dfrac{d^2 y}{dx^2} + x\dfrac{dy}{dx} + \left(x^2 - \dfrac{1}{4}\right)y = 0$

In Exercises 11–14 find a particular solution of the differential equation in Exercise 4 that satisfies the conditions.

11. $y(0) = 1,\ y'(0) = 6$

12. $y(0) = 2,\ y(\pi/2) = 3$

13. $y(\pi/12) = 0,\ y'(\pi/12) = 1$

14. $y(1) = 1,\ y(2) = 2$

In Exercises 15–19 find a general solution for the differential equation.

15. $\dfrac{dy}{dx} = 6x^2 + 2x$

16. $\dfrac{dy}{dx} = \dfrac{1}{9 + x^2}$

17. $\dfrac{d^2 y}{dx^2} = 2x + e^x$

18. $\dfrac{d^2 y}{dx^2} = x \ln x$

19. $\dfrac{d^3 y}{dx^3} = \dfrac{1}{3x^5}$

* **20.** (a) A boy initially at O in the figure below walks along the edge of a swimming pool (the y-axis) towing his sailboat by a string of length L. If the boat starts at Q and the string always remains straight, show that the equation of the curved path $y = y(x)$ followed by the boat must satisfy the differential equation

$$\frac{dy}{dx} = -\frac{\sqrt{L^2 - x^2}}{x}.$$

(b) Solve this differential equation for $y(x)$.

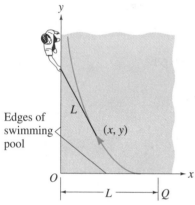

Edges of swimming pool

* **21.** Show that $y(x) = -(x^2 + C)^{-1}$ is a one-parameter family of solutions for the differential equation

$$\frac{dy}{dx} = 2xy^2.$$

Find a singular solution.

* **22.** Show that $y(x) = 1 - (x^3 + C)^{-1}$ is a one-parameter family of solutions for the differential equation

$$\frac{dy}{dx} = 3x^2(y - 1)^2.$$

Find a singular solution.

* **23.** (a) Verify that $y(x) = Ce^{2x}$ is a one-parameter family of solutions for the differential equation

$$\frac{dy}{dx} = 2y.$$

(b) Draw the one-parameter family of curves defined by this solution.

(c) Show that there is a particular solution that passes through any given point (x_0, y_0), and that this solution can be obtained by choosing C appropriately.

* **24.** (a) Verify that a one-parameter family of solutions for the differential equation

$$2x \frac{dy}{dx} = y$$

is defined implicitly by the equation $y^2 = Cx$.

(b) Draw the one-parameter family of curves defined by this equation.

(c) Show that with the exception of points on the y-axis, there is a particular solution that passes through any given point (x_0, y_0), and that this solution can be obtained by specifying C appropriately.

* **25.** (a) Draw the one-parameter family of curves defined by the solution in Exercise 21.

(b) Show that with the exception of points on the x-axis, there is a solution passing through any given point in the xy-plane.

* **26.** Consider the differential equation

$$\frac{dy}{dx} = \frac{1}{x^2}.$$

(a) Find a solution that satisfies the condition $y(1) = 1$.

(b) Find a solution that satisfies the condition $y(-1) = 2$.

(c) Find a solution that satisfies the conditions in both parts (a) and (b).

** **27.** As a VCR plays a movie, its counter (assumed set equal to zero when the movie begins at time $t = 0$) records the number n of revolutions of the takeup reel. In this exercise we develop a formula for $n(t)$.

(a) If $\theta(t)$ represents the angle through which the takeup reel has rotated from time $t = 0$, then $n = \theta/(2\pi)$. How fast the takeup reel rotates depends on the amount of tape on the reel, the more tape on the reel, the slower the angular rotation. Suppose we let $r(t)$ be the distance from the centre of the reel to the point P where the tape joins the reel at time t (figure below). Show that if v is the constant speed at which the tape passes through the head(s) of the VCR, then $dn/dt = v/(2\pi r)$.

(b) The area of the annulus occupied by the tape at time t can be calculated in two ways. First, it is the difference in the areas of two circles $\pi(r^2 - r_0^2)$, where r_0 is the radius of the tape at $t = 0$. Second, it is the width w of the tape multiplied by the length of tape placed on the reel by time t, namely vt. Use this to show that $n(t)$ satisfies the differential equation

$$\frac{dn}{dt} = \frac{v/(2\pi r_0)}{\sqrt{\dfrac{wvt}{\pi r_0^2} + 1}}.$$

(c) Solve the differential equation in part (b) to find the dependence of n on t.

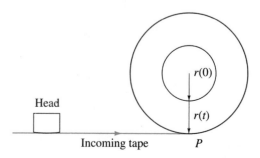

Head

Incoming tape

P

15.2 Separable Differential Equations

Separable differential equations were discussed in Section 5.5. We reproduce the discussion here, giving a slightly more exhaustive treatment and an abundance of exercises.

We consider only first-order differential equations that can be written in the form

$$\frac{dy}{dx} = F(x, y). \tag{15.16}$$

Since any function $F(x, y)$ can always be considered as the quotient of two other functions,

$$F(x, y) = \frac{M(x, y)}{N(x, y)},$$

equation 15.16 can also be written in the equivalent form

$$N(x, y) \, dy = M(x, y) \, dx. \tag{15.17}$$

Depending on the form of $F(x, y)$ [or $M(x, y)$ and $N(x, y)$] various methods can be used to obtain the unknown function $y(x)$. Two of the more important techniques are considered here and in Section 15.3; others are discussed in the exercises.

Differential equation 15.16 is said to be a **separable differential equation** if it can be expressed in the form

$$\frac{dy}{dx} = \frac{M(x)}{N(y)}, \tag{15.18}$$

that is, if dy/dx is equal to a function of x divided by a function of y. Equivalently, a differential equation is said to be separable if it can be written in the form

$$N(y) \, dy = M(x) \, dx. \tag{15.19}$$

When a differential equation is written in this way, it is said to be separated — separated in the sense that x- and y-variables appear on opposite sides of the equation. For a separated equation we can write therefore that

$$N(y)\frac{dy}{dx} = M(x), \tag{15.20}$$

and if we integrate both sides with respect to x, we have

$$\int N(y)\frac{dy}{dx} \, dx = \int M(x) \, dx + C. \tag{15.21}$$

Cancellation of differentials on the left leads to the solution

$$\int N(y) \, dy = \int M(x) \, dx + C. \tag{15.22}$$

What we mean by saying that 15.21 and 15.22 represent one-parameter families of solutions for 15.19 is that any function defined *implicitly* by 15.21 or 15.22 is a solution of 15.19. For example, the differential equation

$$\frac{dy}{dx} = \frac{3x^2}{1 - y}$$

is separable. Multiplying by $(1 - y)dx$ leads to

$$(1 - y) \, dy = 3x^2 \, dx.$$

According to 15.21, we should divide by dx and integrate both sides with respect to x:

$$\int (1 - y)\frac{dy}{dx} dx = \int 3x^2 \, dx = x^3 + C.$$

Antidifferentiation on the left must be interpreted as *implicit antidifferentiation*, asking for that function which when differentiated with respect to x gives $(1 - y)\,dy/dx$. Since an antiderivative is $y - y^2/2$, a one-parameter family of solutions for the differential equation is

$$y - \frac{y^2}{2} = x^3 + C.$$

Were we to use 15.22 after separation (instead of 15.21), we would write

$$\int (1 - y)\,dy = \int 3x^2\,dx + C,$$

and integrate for

$$y - \frac{y^2}{2} = x^3 + C.$$

By saying that $y - y^2/2 = x^3 + C$ represents a one-parameter family of solutions of the original differential equation, we mean that any function defined implicitly by this equation is a solution. We can find explicit solutions by solving the equation for y in terms of x. Multiplying by -2 expresses the equation as a quadratic in y,

$$y^2 - 2y + 2(x^3 + C) = 0,$$

and therefore,

$$y = \frac{2 \pm \sqrt{4 - 8(x^3 + C)}}{2} = 1 \pm \sqrt{1 - 2(x^3 + C)}.$$

Explicit solutions of the differential equation are therefore

$$y(x) = 1 + \sqrt{1 - 2(x^3 + C)} \qquad \text{and} \qquad y(x) = 1 - \sqrt{1 - 2(x^3 + C)},$$

provided that expressions on the right are indeed functions of x. Once C is determined, this will be true only for certain values of x. For example, suppose we require the solution of the differential equation that satisfies the initial condition $y(0) = 3$. The second function $y(x) = 1 - \sqrt{1 - 2(x^3 + C)}$ cannot satisfy this condition because y cannot be greater than 1. If we substitute $x = 0$ and $y = 3$ into the other function,

$$3 = 1 + \sqrt{1 - 2C},$$

and this requires $C = -3/2$. Thus, the solution of the differential equation for which $y(0) = 3$ is

$$y(x) = 1 + \sqrt{1 - 2(x^3 - 3/2)} = 1 + \sqrt{4 - 2x^3}.$$

Since $4 - 2x^3$ must be nonnegative for this function to be defined, x must be restricted to the interval $x \le 2^{1/3}$. Because the derivative is undefined at $x = 2^{1/3}$, the solution has domain $x < 2^{1/3}$.

EXAMPLE 15.3

Find one-parameter families of solutions for the following differential equations:

(a) $2x^3 y^2\,dx = xy^3\,dy$ (b) $\dfrac{dy}{dx} = \dfrac{y \sin x + y^3 \sin x}{(1 + y^2)^2}$

SOLUTION

(a) If we divide the differential equation by xy^2 (requiring therefore that $y \neq 0$), we obtain

$$y \, dy = 2x^2 \, dx,$$

which is separated. A one-parameter family of solutions is therefore defined implicitly by

$$\int y \, dy = \int 2x^2 \, dx + C$$

or

$$\frac{y^2}{2} = \frac{2x^3}{3} + C.$$

We note that $y(x) = 0$ is also a solution, but it cannot be obtained by specifying C. In other words, $y(x) = 0$ is a singular solution. We removed this solution when we divided the original equation by xy^2. Always be careful of this. If an equation is separated by dividing by x's and y's, say $F(x, y)$, determine whether setting $F(x, y) = 0$ leads to functions of x that satisfy the original differential equation. If it does, and the functions are included in the family of solutions obtained, nothing further need be said; if they are not in the family, then they are singular solutions for that family.

(b) Since

$$\frac{dy}{dx} = \frac{y \sin x (1 + y^2)}{(1 + y^2)^2} = \frac{y \sin x}{1 + y^2},$$

the differential equation can be separated:

$$\sin x \, dx = \frac{1 + y^2}{y} dy = \left(\frac{1}{y} + y \right) dy \quad (y \neq 0).$$

A one-parameter family of solutions is therefore defined implicitly by

$$\int \sin x \, dx + C = \int \left(\frac{1}{y} + y \right) dy$$

or

$$-\cos x + C = \ln |y| + \frac{y^2}{2}.$$

Since $y(x) = 0$ is a solution of the differential equation, and it cannot be obtained by specifying C, it is a singular solution of the one-parameter family of solutions.

EXAMPLE 15.4

Figure 15.2 shows a straight, prismatic channel carrying uniform flow. Experimentally it has been shown that the mean velocity of the flow is proportional to $(A/p)^n$, where A is the cross-sectional area of the flow, p is the wetted perimeter, and n is a constant (between $1/2$ and $2/3$). Let A_0 and p_0 be values of A and p that yield minimum flow for the given cross-section below the x-axis. The problem is to find the equation $x = x(y)$ for the (symmetric) side of the channel above the x-axis so that the ratio A/p remains A_0/p_0.

SOLUTION   When the height of the water in the channel is Y, the cross-sectional area of the flow and the wetted perimeter are

$$A = A_0 + 2 \int_0^Y x(y) \, dy, \qquad p = p_0 + 2 \int_0^Y \sqrt{1 + \left(\frac{dx}{dy} \right)^2} \, dy.$$

FIGURE 15.2   Flow in a straight, prismatic channel

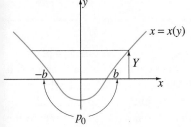

To have $A/p = A_0/p_0$ or $Ap_0 = A_0p$, we set

$$p_0 \left[A_0 + 2 \int_0^Y x(y)\,dy \right] = A_0 \left[p_0 + 2 \int_0^Y \sqrt{1 + \left(\frac{dx}{dy}\right)^2}\,dy \right].$$

If we cancel $p_0 A_0$, divide by 2, and then differentiate both sides with respect to Y, we obtain

$$p_0 x(Y) = A_0 \sqrt{1 + [x'(Y)]^2}.$$

If we now replace Y with y, we obtain a differential equation for $x(y)$,

$$p_0 x = A_0 \sqrt{1 + \left(\frac{dx}{dy}\right)^2} \quad \Longrightarrow \quad \frac{dx}{dy} = \pm \sqrt{\frac{p_0^2 x^2}{A_0^2} - 1}.$$

Since the slope is always positive for the right half of the channel, we choose the positive sign, and separate the equation

$$\frac{1}{\sqrt{p_0^2 x^2 - A_0^2}}\,dx = \frac{1}{A_0}\,dy.$$

A one-parameter family of solutions is defined by

$$\frac{y}{A_0} + C = \int \frac{1}{\sqrt{p_0^2 x^2 - A_0^2}}\,dx.$$

We set $x = (A_0/p_0) \sec\theta$, and $dx = (A_0/p_0) \sec\theta \tan\theta\,d\theta$,

$$\frac{y}{A_0} + C = \int \frac{1}{A_0 \tan\theta} \frac{A_0}{p_0} \sec\theta \tan\theta\,d\theta = \frac{1}{p_0} \int \sec\theta\,d\theta$$

$$= \frac{1}{p_0} \ln|\sec\theta + \tan\theta| = \frac{1}{p_0} \ln\left| \frac{p_0 x}{A_0} + \frac{\sqrt{p_0^2 x^2 - A_0^2}}{A_0} \right|$$

$$= \frac{1}{p_0} \ln|p_0 x + \sqrt{p_0^2 x^2 - A_0^2}| - \frac{1}{p_0} \ln A_0.$$

We absorb the last constant into C. Since $x(0) = b$, $C = \frac{1}{p_0} \ln|p_0 b + \sqrt{p_0^2 b^2 - A_0^2}|$, so that

$$\frac{y}{A_0} + \frac{1}{p_0} \ln|p_0 b + \sqrt{p_0^2 b^2 - A_0^2}| = \frac{1}{p_0} \ln|p_0 x + \sqrt{p_0^2 x^2 - A_0^2}|.$$

Thus,

$$y = \frac{A_0}{p_0} \ln\left| \frac{p_0 x + \sqrt{p_0^2 x^2 - A_0^2}}{p_0 b + \sqrt{p_0^2 b^2 - A_0^2}} \right| = \frac{A_0}{p_0} \ln\left[\frac{x + \sqrt{x^2 - (A_0/p_0)^2}}{b + \sqrt{b^2 - (A_0/p_0)^2}} \right].$$

In the 1968 Olympic Games in Mexico City, Bob Beamon of the United States made a phenomenal long jump of 8.90 m that has yet to be approached. Many have claimed that this was a result of less air drag due to the rarefied air at the high altitude of Mexico City. We refute this claim here. Experimentally, it has been shown that the air drag on a long jumper is proportional to air density ρ, the square of speed v, and the cross-sectional area A that the body presents to the air. Air drag can then be represented in the form $F = C\rho A v^2$, where $0 < C < 1$ is a constant called the *drag coefficient*. For a long jumper, typical values of C and A are $C = 0.375$ and $A = 0.75 \text{ m}^2$. Newton's second law for the acceleration $\mathbf{a}$ of the long jumper while in flight gives

$$m\mathbf{a} = -mg\hat{\mathbf{j}} - C\rho A v^2 \hat{\mathbf{v}},$$

where $\hat{\mathbf{v}}$ is a unit vector in the direction of motion. Separating this into horizontal and vertical components leads to a coupled system of differential equations for the components of velocity

$$m\frac{dv_x}{dt} = -C\rho A(v_x^2 + v_y^2)\frac{v_x}{\sqrt{v_x^2 + v_y^2}}, \qquad m\frac{dv_y}{dt} = -mg - C\rho A(v_x^2 + v_y^2)\frac{v_y}{\sqrt{v_x^2 + v_y^2}}.$$

Realizing, however, that there is very little motion in the vertical direction compared to the horizontal direction, we simplify the problem by ignoring vertical motion. Let us assume that motion is horizontal then, with air drag in the negative x-direction. Then

$$m\frac{dv}{dt} = -C\rho A v^2,$$

where v now denotes the horizontal component of velocity v_x. The differential equation is separable,

$$\frac{dv}{v^2} = -\frac{C\rho A}{m}\,dt,$$

and therefore a one-parameter family of solutions is defined implicitly by

$$-\frac{1}{v} = -\frac{C\rho A}{m}t + D.$$

If we take time $t = 0$ when the long jumper takes off, and use v_0 as takeoff speed, then $D = -1/v_0$. This gives

$$v = \frac{dx}{dt} = \frac{1}{\dfrac{C\rho A t}{m} + \dfrac{1}{v_0}}.$$

Integration now gives

$$x(t) = \frac{m}{C\rho A}\ln\left(\frac{C\rho A t}{m} + \frac{1}{v_0}\right) + E.$$

With $x(0) = 0$, we obtain $E = \dfrac{m}{C\rho A}\ln v_0$, and

$$x(t) = \frac{m}{C\rho A}\ln\left(\frac{C\rho A t}{m} + \frac{1}{v_0}\right) + \frac{m}{C\rho A}\ln v_0 = \frac{m}{C\rho A}\ln\left(1 + \frac{C\rho A v_0 t}{m}\right).$$

Typical time for the duration of a long jump is 1 s, and typical initial speeds at takeoff are around 10 m/s. With these values and $m = 80$ kg, $A = 0.75 \text{ m}^2$, and $C = 0.375$, we can calculate the difference in lengths of the long jump at sea level, where $\rho = 1.225 \text{ kg/m}^3$, and in Mexico City, where $\rho = 0.984 \text{ kg/m}^3$:

$$\frac{80}{0.375(0.984)(0.75)}\ln\left[1 + \frac{0.375(0.984)(0.75)(10)(1)}{80}\right]$$

$$-\frac{80}{0.375(1.225)(0.75)}\ln\left[1 + \frac{0.375(1.225)(0.75)(10)(1)}{80}\right]$$

$$= 0.04 \text{ m}.$$

In other words, the rarefied atmosphere in Mexico City would have made a difference of only approximately 4 cm.

EXERCISES 15.2

In Exercises 1–10 find a one-parameter family of solutions for the differential equation. Be careful to identify any singular solutions.

1. $y^2 \, dx - x^2 \, dy = 0$

2. $\dfrac{dy}{dx} + 2xy = 4x$

3. $2xy \, dx + (x^2 + 1) \, dy = 0$

4. $\dfrac{dy}{dx} = 3y + 2$

5. $3(y^2 + 2) \, dx = 4y(x - 1) \, dy$

6. $(x^2 y + x^2) \, dx + (xy^2 - y^2) \, dy = 0$

7. $\dfrac{dy}{dx} = -\dfrac{\cos y}{\sin x}$

8. $(x^2 y e^x - y) \, dx + xy^3 \, dy = 0$

9. $(x^2 y^2 \sec x \, \tan x + xy^2 \sec x) \, dx + xy^3 \, dy = 0$

10. $\dfrac{dy}{dx} = \dfrac{1 + y^2}{1 + x^2}$

In Exercises 11–15 find a solution of the differential equation that also satisfies the given condition.

11. $2y \, dx + (x + 1) \, dy = 0, \quad y(1) = 2$

12. $(xy + y) \, dx - (xy - x) \, dy = 0, \quad y(1) = 2$

13. $\dfrac{dy}{dx} = e^{x+y}, \quad y(0) = 0$

14. $\dfrac{dy}{dx} = 2x(1 + y^2), \quad y(2) = 4$

15. $\dfrac{dy}{dx} = \dfrac{\sin^2 y}{\cos^2 x}, \quad y(0) = \pi/2$

* **16.** A girl lives 6 km from school. She decides to travel to school so that her speed is always proportional to the square of her distance from the school.

    (a) Find her distance from school at any time.

    (b) When does she reach school?

* **17.** Find a one-parameter family of solutions for the differential equation
$$\dfrac{dy}{dx} = -\dfrac{1 + y^3}{xy^2 + x^3 y^2}.$$

* **18.** When a container of water at temperature $80°$C is placed in a room at temperature $20°$C, Newton's law of cooling states that the time rate of change of the temperature of the water is proportional to the difference between the temperature of the water and room temperature. If the water cools to $60°$C in 2 min, find a formula for its temperature as a function of time.

* **19.** A thermometer reading $23°$C is taken outside where the temperature is $-20°$C. If the reading drops to $0°$C in 4 min, when will it read $-19°$C?

* **20.** The amount of a drug such as penicillin injected into the body is used up at a rate proportional to the amount still present. If a dose decreases by 5% in the first hour, when does it decrease to one-half its original amount?

* **21.** When a deep-sea diver inhales air, his body tissues absorb extra amounts of nitrogen. Suppose the diver enters the water at time $t = 0$, drops very quickly to depth d, and remains at this depth for a very long time. The amount N of nitrogen in his body tissues increases as he remains at this depth until a maximum amount $\overline{N}$ is reached. The time rate of change of N is proportional to the difference $\overline{N} - N$. Show that if N_0 is the amount of nitrogen in his body tissues when he enters the water, then
$$N = N_0 e^{-kt} + \overline{N}(1 - e^{-kt}),$$
where $k > 0$ is a constant.

* **22.** When a substance such as glucose is administered intravenously into the bloodstream, it is used up by the body at a rate proportional to the amount present at that time. If it is added at a constant rate of R units per unit time, and A_0 is the amount present in the bloodstream when the intravenous feeding begins, find a formula for the amount in the bloodstream at any time.

* **23.** Find the equations for all curves that satisfy the condition that the normal at any point on the curve, and the line joining the point to the origin, form an isosceles triangle with the x-axis as base.

* **24.** When two substances A and B are brought together in one solution, they react to form a third substance C in such a way that 1 g of A reacts with 1 g of B to produce 2 g of C. The rate at which C is formed is proportional to the product of the amounts of A and B still present in the solution. If 10 g of A and 15 g of B are originally brought together, find a formula for the amount of C present in the mixture at any time.

* **25.** What is the solution to Exercise 24 when the initial amounts of A and B are both 10 g?

* **26.** Two spherical containers each of radius R are connected by a pipe of cross-sectional area a (figure below). The rate of flow from one container to the other is $(a/3)\sqrt{2gh}$ where $g = 9.81$, and h is the difference in water levels in the containers. If one container is initially empty and the other is full, how long does it take for water levels to be equal? *Hint*: See equation 5.22 in Section 5.5.

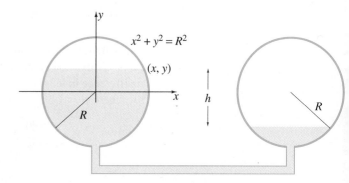

* **27.** A tank in the form of a circular cylinder (radius R and length L) with hemispherical ends has its axis horizontal. Water exits through a hole in the bottom at rate $0.6a\sqrt{2gh}$, where a is the area of the hole and h is the depth of water in the tank. Find a formula for the length of time it takes a full tank to empty. *Hint*: See equation 5.22 in Section 5.5.

28. The schematic diagram of a canal lock 8 m by 16 m is shown below. The water level in the lock is raised by closing valve B and opening valve A. The rate at which water enters the lock is $Q = 0.04\sqrt{2gh}$ m^3/s where $g = 9.81$, and h is the difference in heights of the upstream water level and water in the lock. If the upstream gate is opened when $h = 2$ cm, and the initial value of h is 2 m, how long will it be before the upstream gate is opened?

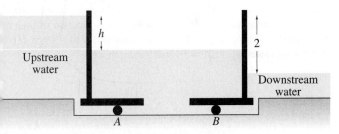

29. A cubical container, filled with water, is 1 m on each side. One of its sides has a slit 20 cm high and 1 mm wide starting at the top of the container. Use the result of Exercise 30 in Section 7.9 to determine how long it takes for the water level in the container to fall by 10 cm. Use $c = 0.6$.

30. A pipe of length L is connected to a large reservoir of depth h (figure below). If water is allowed to flow from the pipe at time $t = 0$, the velocity v of the flow from the pipe must, in the short term, satisfy the equation

$$gh = \frac{v^2}{2} + L\frac{dv}{dt},$$

where $g = 9.81$. Find v as a function of t.

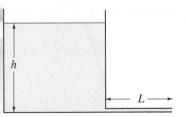

31. Chemical reactors are of third order when the amount $x(t)$ of substance being formed satisfies a differential equation of the form

$$\frac{dx}{dt} = k(a - x)(b - x)(c - x),$$

where a, b, c, and k are positive constants. Solve this differential equation in the cases: (a) $a = b = c$ (b) $a = b \neq c$ and (c) $a \neq b \neq c$.

32. When fluid flows through a tube of varying cross-section A, the Mach number M of the flow is related to A by the differential equation

$$\frac{dA}{dM} = \frac{A(M^2 - 1)}{M\left[\left(\dfrac{k-1}{2}\right)M^2 + 1\right]},$$

where $k > 1$ is a constant. If $A = A_0$ when $M = 1$, find A in terms of M.

★ 33. In transient free-surface flow through porous media, velocity $v(t)$ must satisfy the differential equation

$$\frac{dv}{dt} = \frac{K}{n}\left[\frac{v + H(t)}{v}\right],$$

where $K > 0$ is the hydraulic conductivity, $n > 0$ is the porosity of the media, both assumed constant, and $H(t)$ is the head as a function of time t. Solve this differential equation in each of the following situations, assuming that $v(0) = 0$:

(a) $H(t)$ is a constant.

(b) $H(t) = qt - nv$, where $q > 0$ is a constant. Hint: Try a solution of the form $v = At$, and find A.

★ 34. A first-order differential equation in $y(x)$ is said to be *homogeneous* if it can be written in the form

$$\frac{dy}{dx} = f\left(\frac{y}{x}\right).$$

Show that the change of dependent variable $v = y/x$ yields a differential equation in $v(x)$ that is always separable.

★ 35. State a condition on the functions M and N in equation 15.17 in order that the differential equation be homogeneous in the sense of Exercise 34.

In Exercises 36–41 show that the differential equation is homogeneous, and use the change of variable in Exercise 34 to find a one-parameter family of solutions.

★ 36. $(y^2 - x^2)\,dx + xy\,dy = 0$

★ 37. $2x\,dy - 2y\,dx = \sqrt{x^2 + 4y^2}\,dx$

★ 38. $\dfrac{dy}{dx} = \dfrac{y + x}{y - x}$

★ 39. $x\,dy - y\,dx = x\cos(y/x)\,dx$

★ 40. $\dfrac{dy}{dx} = \dfrac{x^2 e^{-y/x} + y^2}{xy}$ **★ 41.** $(x^2 y + y^3)\,dx + x^3\,dy = 0$

★ 42. If a curve passes through the point $(1, 2)$ and is such that the length of that part of the tangent line at (x, y) from (x, y) to the y-axis is equal to the y-intercept of the tangent line, find the equation of the curve.

★ 43. Find in explicit form a function that satisfies the differential equation $dy/dx = \csc y$ and the following conditions: (a) $y(0) = \pi/4$ and (b) $y(0) = 7\pi/4$.

★ 44. In a chemical reaction, one molecule of trypsinogen yields one molecule of trypsin. In order for the reaction to take place, an initial amount of trypsin must be present. Suppose that the initial amount is y_0. Thereafter, the rate at which trypsinogen is changed into trypsin is proportional to the product of the amounts of each chemical in the reaction. Find a formula for the amount of trypsin if the initial amount of trypsinogen is A.

* **45.** If a first-order differential equation can be written in the form

$$\frac{dy}{dx} = f(ax + by),$$

where a and b are constants, show that the change of dependent variable $v = ax + by$ always gives a differential equation in $v(x)$ that is separable.

Use the method of Exercise 45 to find a one-parameter family of solutions for the differential equation in Exercises 46–49.

* **46.** $\dfrac{dy}{dx} = x + y$

* **47.** $\dfrac{dy}{dx} = (x + y)^2$

* **48.** $\dfrac{dy}{dx} = \dfrac{1}{2x + 3y}$

* **49.** $\dfrac{dy}{dx} = \sin^2(x - y)$

* **50.** The *logistic model* for growth of bacteria introduces a quantity C called the *carrying capacity* for the environment in which the bacteria are living. As the number $N(t)$ of bacteria approaches C, its growth rate must slow down. The logistic model to describe this is the differential equation

$$\frac{dN}{dt} = kN\left(1 - \frac{N}{C}\right).$$

Notice that when N is small, dN/dt is approximately equal to kN, thus preserving early exponential growth. The factor $1 - N/C$ causes $dN/dt \to 0$ as $N \to C$. Solve this differential equation for $N(t)$ when $k = 1$, $C = 10^6$, and $N(0) = 100$.

* **51.** Show that the solution of the logistic model in Exercise 50 for an initial population $N(0) = N_0$ is

$$N = \frac{C}{1 + \left(\dfrac{C - N_0}{N_0}\right)e^{-kt}}.$$

* **52.** One of the models to describe the weight $w(t)$ of a fish as a function of time t is

$$\frac{dw}{dt} = aw^n - bw^m,$$

where a, b, n, and m are positive constants depending on the type of fish and its environment. Solve this differential equation subject to the initial condition $w(0) = w_0$ when $n = 2/3$ and $m = 1$.

* **53.** A certain chemical dissolves in water at a rate proportional to the product of the amount of undissolved chemical and the difference between concentrations in a saturated solution and the existing concentration in the solution. A saturated solution contains 25 g of chemical in 100 mL of solution.

(a) If 50 g of chemical is added to 200 mL of water, find a formula for the amount of chemical dissolved as a function of time. Draw its graph.

(b) Repeat part (a) if the 50 g of chemical is added to 100 mL of water.

(c) Repeat part (a) if 10 g of chemical is added to 100 mL of water.

* **54.** In transient free-surface flow through porous media with a vertical barrier, velocity $v(t)$ must satisfy the barrier flow equation

$$\frac{dv}{dt} = \frac{K}{n}\left\{\frac{v^2 - [D - \gamma - H(t)]v - D\,H(t)}{v(v - D)}\right\},$$

where $K > 0$ is the hydraulic conductivity, $n > 0$ is the porosity of the media, both assumed constant, $D > 0$ is the depth of the barrier, $\gamma > 0$ is a constant, and $H(t)$ is the head as a function of time t. Show that the solution of this differential equation for $v(t)$ when $H(t)$ is constant, assuming $v(0) = 0$, is defined implicitly by

$$v + \frac{H}{r_1 - r_2}\left\{[DH - (H + \gamma)r_1]\ln\left|\frac{v - r_1}{r_1}\right|\right.$$

$$\left. - [DH - (H + \gamma)r_2]\ln\left|\frac{v - r_2}{r_2}\right|\right\} = \frac{Kt}{n},$$

where $r_1 = [(D - \gamma - H) + \sqrt{(D - \gamma - H)^2 + 4DH}]/2$ and $r_2 = [(D - \gamma - H) - \sqrt{(D - \gamma - H)^2 + 4DH}]/2$.

* **55.** Snow has been falling for some time when a snowplow starts plowing the highway. The plow begins at 12:00 and travels 2 km during the first hour and 1 km during the second hour. Make reasonable assumptions to find out when the snow started falling.

* **56.** In order to perform a 1 h operation on a dog, a veterinarian anesthetizes the dog with sodium pentobarbital. During the operation, the dog's body breaks down the drug at a rate proportional to the amount still present, and only half an original dose remains after 5 h. If the dog has mass 20 kg, and 20 mg of sodium pentobarbital per kilogram of body mass is required to maintain surgical anesthesia, what original dose is required?

* **57.** Explain what the cancellation of differentials in proceeding from 15.21 to 15.22 really means.

* **58.** Solve the differential equation

$$(x^3y^4 + 2xy^4)\,dx + (x - xy^6)\,dy = 0, \quad y(1) = 1.$$

* **59.** Two substances A and B react to form a third substance C in such a way that 2 g of A reacts with 1 g of B to produce 3 g of C. The rate at which C is formed is proportional to the amounts of A and B still present in the mixture. Find the amount of C present in the mixture as a function of time when the original amounts of A and B brought together at time $t = 0$ are as follows: (a) 20 g and 10 g, (b) 20 g and 5 g, and (c) 20 g and 20 g. Draw graphs of all three functions on the same axes.

** **60.** A bird is due east of its nest a distance L away and at the same height above the ground as the nest. Wind is blowing due north at speed v. If the bird flies horizontally with constant speed V always pointing straight at its nest, what is the equation of the curve that it follows? Take the nest at the origin and the x- and y-directions as east and north.

** **61.** Find the equation of the curve that passes through $(1, 1)$ and is such that the tangent and normal lines at any point (x, y) make with the x-axis a triangle whose area is equal to the slope of the tangent line at (x, y).

15.3 Linear First-Order Differential Equations

A first-order differential equation that can be written in the form

$$\frac{dy}{dx} + P(x)y = Q(x) \qquad (15.23)$$

is said to be **linear**. We will explain the significance of the adjective *linear* in Section 15.6. To illustrate how to solve such differential equations, consider the equation

$$\frac{dy}{dx} + \frac{1}{x}y = 1.$$

If we multiply both sides by x, we have

$$x\frac{dy}{dx} + y = x.$$

But note now that the left side is the derivative of the product xy; that is,

$$\frac{d}{dx}(xy) = x\frac{dy}{dx} + y.$$

In other words, we can write the differential equation in the form

$$\frac{d}{dx}(xy) = x,$$

and integration immediately gives a one-parameter family of solutions

$$xy = \frac{x^2}{2} + C \qquad \Longrightarrow \qquad y(x) = \frac{x}{2} + \frac{C}{x}.$$

This is the principle behind all linear first-order equations: Multiply the equation by a function of x in order that the left side can be expressible as the derivative of a product. To show that this is always possible, we turn now to the general equation 15.23. If the equation is multiplied by a function $\mu(x)$,

$$\mu\frac{dy}{dx} + \mu P(x)y = \mu Q(x). \qquad (15.24)$$

This equation is equivalent to 15.23 in the sense that $y(x)$ is a solution of 15.23 if and only if it is a solution of 15.24. We ask whether it is possible to find μ so that the left side of 15.24 can be written as the derivative of the product μy; that is, can we find $\mu(x)$ so that

$$\mu\frac{dy}{dx} + \mu P(x)y = \frac{d}{dx}(\mu y)?$$

If we expand the right side, μ must satisfy

$$\mu\frac{dy}{dx} + \mu P(x)y = \mu\frac{dy}{dx} + y\frac{d\mu}{dx},$$

from which we get

$$\mu P(x) = \frac{d\mu}{dx} \quad \text{or} \quad \frac{d\mu}{\mu} = P(x)\,dx.$$

Thus μ must satisfy a separated differential equation, one solution of which is

$$\ln|\mu| = \int P(x)dx \quad \text{or} \quad \mu = \pm e^{\int P(x)dx}.$$

We have shown then that if 15.23 is multiplied by the factor

$$e^{\int P(x)dx} \qquad (\text{or by } -e^{\int P(x)dx}),$$

then the differential equation becomes

$$e^{\int P(x)dx}\frac{dy}{dx} + P(x)ye^{\int P(x)dx} = Q(x)e^{\int P(x)dx},$$

and the left side can be expressed as the derivative of a product:

$$\frac{d}{dx}\{ye^{\int P(x)dx}\} = Q(x)e^{\int P(x)dx}.$$

Integration now gives a one-parameter family of solutions

$$ye^{\int P(x)dx} = \int Q(x)e^{\int P(x)dx}dx + C. \qquad (15.25)$$

The quantity $e^{\int P(x)dx}$ is called an **integrating factor** for equation 15.23 because when the equation is multiplied by this factor, it becomes immediately integrable.

In summary, if linear differential equation 15.23 is multiplied by the function $e^{\int P(x)dx}$, then the left side of the equation becomes the derivative of the product of y and this function. In this form the equation can immediately be integrated. There are two things that you should remember in specific examples:

1. The differential equation must be expressed in form 15.23.

2. An integrating factor is $e^{\int P(x)\,dx}$. Multiply the differential equation by this factor and the left side is the derivative of a product. Now integrate both sides of the differential equation.

EXAMPLE 15.5

Find one-parameter families of solutions for the following differential equations:

(a) $\dfrac{dy}{dx} + xy = x$ (b) $(y - x\sin x)\,dx + x\,dy = 0$ (c) $\cos x\dfrac{dy}{dx} + y\sin x = 1$

SOLUTION

(a) An integrating factor for this linear equation is

$$e^{\int x\,dx} = e^{x^2/2}.$$

If we multiply the equation by this integrating factor, we have

$$e^{x^2/2}\frac{dy}{dx} + yxe^{x^2/2} = xe^{x^2/2} \qquad \text{or} \qquad \frac{d}{dx}\{ye^{x^2/2}\} = xe^{x^2/2}.$$

Integration yields

$$ye^{x^2/2} = \int xe^{x^2/2}dx = e^{x^2/2} + C \qquad \Longrightarrow \qquad y(x) = 1 + Ce^{-x^2/2}.$$

(b) If we write the differential equation in the form

$$\frac{dy}{dx} + \frac{y}{x} = \sin x,$$

then we see that it is linear first order. An integrating factor is therefore

$$e^{\int 1/x\,dx} = e^{\ln|x|} = |x|.$$

If we multiply the differential equation by this factor, we get

$$|x|\frac{dy}{dx} + \frac{|x|}{x}y = |x|\sin x.$$

If $x > 0$, then we write

$$x\frac{dy}{dx} + \frac{x}{x}y = x\sin x,$$

whereas if $x < 0$,

$$-x\frac{dy}{dx} - \frac{x}{x}y = -x\sin x.$$

In either case, however, the equation simplifies to

$$x\frac{dy}{dx} + y = x\sin x \qquad \text{or} \qquad \frac{d}{dx}(xy) = x\sin x.$$

Integration now gives

$$xy = \int x\sin x\,dx = -x\cos x + \sin x + C.$$

Finally, then,

$$y(x) = -\cos x + \frac{\sin x}{x} + \frac{C}{x}.$$

(c) An integrating factor for the linear equation

$$\frac{dy}{dx} + y\tan x = \frac{1}{\cos x}$$

is

$$e^{\int \tan x\,dx} = e^{\ln|\sec x|} = |\sec x|.$$

For either $\sec x < 0$ or $\sec x > 0$, we obtain

$$\frac{1}{\cos x}\frac{dy}{dx} + y\frac{\sin x}{\cos^2 x} = \frac{1}{\cos^2 x} \qquad \text{or} \qquad \frac{d}{dx}\left\{\frac{y}{\cos x}\right\} = \sec^2 x.$$

Integration now yields

$$\frac{y}{\cos x} = \tan x + C \qquad \Longrightarrow \qquad y(x) = \sin x + C\cos x.$$

A one-parameter family of solutions of a first-order differential equation is a general solution of the differential equation if and only if it contains all solutions of the differential equation. If the family has singular solutions, it cannot be a general solution. In Section 15.8, we indicate that a one-parameter family of solutions of a linear first-order differential equation cannot have singular solutions. In other words, a one-parameter family of solutions for a linear first-order differential equation is always a general solution.

In Section 3.11 we introduced differential equations that arise when energy balance is applied to a system such as that in Figure 15.3. We repeat the discussion here but in a more general way. Liquid at temperature $T_0\,^\circ$C enters the tank, the liquid is heated, and it then leaves at higher temperature $T\,^\circ$C. The tank is perfectly insulated so that no heat can escape from its sides, and therefore all heat supplied by the heater raises the temperature of the liquid. When the tank is full, and we assume that this is always the case, the mass of liquid is M kilograms. Liquid enters the tank at a rate denoted by $\dot{m}$ kilograms per second and leaves at the same rate. The heater adds energy at the rate of q joules per second. In reality, temperature of the liquid in the tank depends on both time and position in the tank. The mixer is added to remove spatial dependence. We assume that the mixer is so efficient that temperature of the liquid is the same at every point in the tank. Temperature then depends only on time t, denoted by $T(t)$, and this is also the temperature at which liquid leaves the tank.

FIGURE 15.3   A linear differential equation describes the temperature of the liquid in this tank

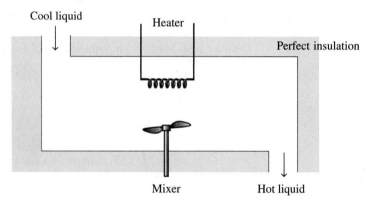

Energy balance for this system states that

$$\left\{\begin{array}{c}\text{rate at which}\\ \text{energy enters tank}\\ \text{in cool water}\end{array}\right\} + \left\{\begin{array}{c}\text{rate at which}\\ \text{energy is added}\\ \text{by heater}\end{array}\right\} = \left\{\begin{array}{c}\text{rate at which}\\ \text{energy leaves tank}\\ \text{in warmer water}\end{array}\right\} + \left\{\begin{array}{c}\text{rate at which}\\ \text{energy is used to raise}\\ \text{temperature of}\\ \text{water in tank}\end{array}\right\}$$

(For simplicity, we have ignored energy associated with the mixer.) The first three rates are known; the fourth can be expressed in terms of dT/dt, the time rate of change of temperature. The second term on the left is q. For the remaining terms, we must introduce the *specific heat* c_p of the liquid. It is the energy required to raise the temperature of 1 kg of the liquid by $1\,^\circ$C. [For example, the specific heat of water is $c_p = 4190$ J/kg$\cdot\,^\circ$C. It requires 4190 J of energy to raise 1 kg of water $1\,^\circ$C. To raise 5 kg of water $10\,^\circ$C requires $(5)(10)(4190) = 209\,500$ J.] Since $\dot{m}$ kilograms of liquid at temperature $T_0\,^\circ$C enters the tank each second, the rate at which energy enters the tank due to this liquid is $\dot{m}c_p T_0$ joules per second. (This is the energy required to raise $\dot{m}$ kilograms of liquid from $0\,^\circ$C to $T_0\,^\circ$C.) In a similar way, the rate at which energy leaves the tank in the liquid at temperature T is $\dot{m}c_p T$ joules per second. This leaves only the last term on the right side of the energy balance equation. The rate of change of temperature of the M kilograms of liquid in the tank is dT/dt; that is, the temperature changes dT/dt degrees each second. It follows that the rate at which energy is used to raise the temperature of this liquid is $Mc_p(dT/dt)$.

When these rates are substituted into the energy balance equation, the result is

$$\dot{m}c_p T_0 + q = \dot{m}c_p T + Mc_p\frac{dT}{dt}.$$

Division by Mc_p gives

$$\frac{dT}{dt} + \frac{\dot{m}}{M}T = \frac{\dot{m}}{M}T_0 + \frac{q}{Mc_p}.$$

Of the quantities $\dot{m}$, c_p, T_0, q, and M in this equation, only c_p and M must be constant; $\dot{m}$, T_0, and q could be functions of time t. In an actual problem, all five quantities are given, and the objective is to solve this linear first-order differential equation for $T(t)$. Multiplication of the differential equation by the integrating factor

$$e^{\int (\dot{m}/M)dt}$$

leads to

$$\frac{d}{dt}\left[T e^{\int(\dot{m}/M)dt} \right] = \left(\frac{\dot{m}}{M}T_0 + \frac{q}{Mc_p} \right) e^{\int(\dot{m}/M)dt}.$$

Integration gives

$$T e^{\int(\dot{m}/M)dt} = \int \left(\frac{\dot{m}}{M}T_0 + \frac{q}{Mc_p} \right) e^{\int(\dot{m}/M)dt} \, dt + F$$

or

$$T(t) = F e^{-\int(\dot{m}/M)dt} + e^{-\int(\dot{m}/M)dt} \int \left(\frac{\dot{m}}{M}T_0 + \frac{q}{Mc_p} \right) e^{\int(\dot{m}/M)dt} \, dt,$$

where F is the constant of integration. Once $\dot{m}$, c_p, T_0, q, and M are specified, integrations can be performed to yield $T(t)$. Let us consider the special case in which all five quantities are constants (independent of time). Then

$$T(t) = F e^{-\dot{m}t/M} + e^{-\dot{m}t/M} \int \left(\frac{\dot{m}}{M}T_0 + \frac{q}{Mc_p} \right) e^{\dot{m}t/M} \, dt$$

$$= F e^{-\dot{m}t/M} + e^{-\dot{m}t/M} \left(\frac{\dot{m}}{M}T_0 + \frac{q}{Mc_p} \right) \left(\frac{M}{\dot{m}} \right) e^{\dot{m}t/M}$$

$$= T_0 + \frac{q}{\dot{m}c_p} + F e^{-\dot{m}t/M}.$$

If we assume that T_0 is not only the temperature at which liquid enters the tank, but also the temperature of the water at time $t = 0$, then $F = -q/(\dot{m}c_p)$, and

$$T(t) = T_0 + \frac{q}{\dot{m}c_p}\left(1 - e^{-\dot{m}t/M} \right).$$

A graph of this function is shown in Figure 15.4.

FIGURE 15.4   Graph of temperature of the liquid in a heat tank

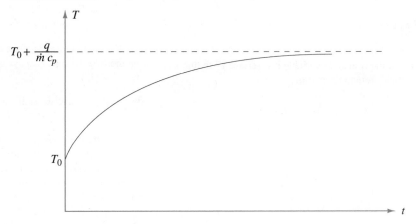

EXERCISES 15.3

In Exercises 1–12 find a general solution for the differential equation.

1. $\dfrac{dy}{dx} + 2xy = 4x$

2. $\dfrac{dy}{dx} + \dfrac{2}{x}y = 6x^3$

3. $(2y - x)\,dx + dy = 0$

4. $\dfrac{dy}{dx} + y \cot x = 5e^{\cos x}$

5. $(x^2 + 2xy)\,dx + (x^2 + 1)\,dy = 0$

6. $(x + 1)\dfrac{dy}{dx} - 2y = 2(x + 1)$

7. $\dfrac{1}{x}\dfrac{dy}{dx} - \dfrac{y}{x^2} = \dfrac{1}{x^3}$

8. $(y + e^{2x})\,dx = dy$

9. $\dfrac{dy}{dx} + y = 2\cos x$

10. $x^3\dfrac{dy}{dx} + (2 - 3x^2)y = x^3$

11. $\dfrac{dy}{dx} + \dfrac{y}{x \ln x} = x^2$

12. $(-2y \cot 2x - 1 + 2x \cot 2x + 2 \csc 2x)\,dx + dy = 0$

In Exercises 13–15 solve the differential equation.

13. $\dfrac{dy}{dx} + 3x^2 y = x^2, \quad y(1) = 2$

14. $(-e^x \sin x + y)\,dx + dy = 0, \quad y(0) = -1$

15. $\dfrac{dy}{dx} + \dfrac{x^3 y}{x^4 + 1} = x^7, \quad y(0) = 1$

∗ 16. Find a general solution for the differential equation $(y^3 - x)\,dy = y\,dx$.

∗ 17. A differential equation of the form

$$\frac{dy}{dx} + P(x)y = y^n Q(x)$$

is called a *Bernoulli equation*. Show that the change of dependent variable $z = y^{1-n}$ gives

$$\frac{dz}{dx} + (1 - n)Pz = (1 - n)Q,$$

a linear first-order equation in $z(x)$.

In Exercises 18–22 use the change of variable in Exercise 17 to find a general solution for the differential equation.

∗ 18. $\dfrac{dy}{dx} + y = y^2 e^x$

∗ 19. $\dfrac{dy}{dx} + \dfrac{y}{x} = \dfrac{y^2}{x^2}$

∗ 20. $\dfrac{dy}{dx} - y + (x^2 + 2x)y^2 = 0$

∗ 21. $x\,dy + y\,dx = x^3 y^5\,dx$

∗ 22. $\dfrac{dy}{dx} + y \tan x = y^4 \sin x$

∗ 23. Repeat Exercise 22 in Section 15.2 if the glucose is added at a rate $R(t)$ that is a function of time t.

∗ 24. A tank has 100 L of solution containing 4 kg of sugar. A mixture with 10 g of sugar per litre of solution is added at a rate of 200 mL per minute. At the same time, 100 mL of well-stirred mixture is removed each minute. Find the amount of sugar in the tank as a function of time.

∗ 25. Repeat Exercise 24 if 300 mL of mixture is removed each minute.

∗ 26. A tank originally contains 1000 L of water in which 5 kg of salt has been dissolved. A mixture containing 2 kg of salt for each 100 L of solution is added to the tank at 10 mL/s. At the same time, the well-stirred mixture in the tank is removed at the rate of 5 mL/s. Find the amount of salt in the tank as a function of time. Draw a graph of the function.

∗ 27. Repeat Exercise 26 if the mixture is removed at 10 mL/s. What is the limit of the amount of salt in the tank for long periods of time?

∗ 28. Repeat Exercise 26 if the mixture is removed at 20 mL/s.

∗ 29. A room with volume 100 m^3 initially contains 0.1% carbon dioxide. Beginning at time $t = 0$, fresher air containing 0.05% carbon dioxide flows into the room at 5 m^3/min. The well-mixed air in the room flows out at the same rate. Find the amount of carbon dioxide in the room as a function of time. What is the limit of the function as $t \to \infty$?

∗ 30. A potato at room temperature 20°C is placed in the oven at the moment the oven is set to 200°C. It takes the oven 5 min to reach 200°C, the temperature increasing at a constant rate, and the temperature of the oven remains at 200°C thereafter. Assuming that the temperature of the potato obeys Newton's law of cooling, find its temperature as a function of time.

∗ 31. Water at temperature 10°C enters the tank in Figure 15.3 at a rate of 0.03 kg/s. At time $t = 0$, the tank reaches capacity (100 kg), and water then leaves at the same rate. The heater is turned on at $t = 0$ adding energy to the water at 2000 J/s. Find the temperature of the water in the tank as a function of time t, and plot its graph.

∗ 32. Repeat Exercise 31 if the rate at which water enters the tank is $100/(t + 1)$ kilograms per second.

∗ 33. Repeat Exercise 31 if the amount of energy supplied by the heater for the first 10 min is $q = 20t$ joules per second. Solve only for $0 \le t \le 600$.

∗ 34. Repeat Exercise 31 if the temperature of the incoming water is $T_0(t) = 10e^{-t}, t \ge 0$.

∗ 35. The step response for the RC-circuit in the following figure is the voltage $V(t)$ across the capacitor when the applied voltage is $E = h(t)$, where $h(t)$ is the Heaviside unit step function and the initial voltage across the capacitor at time $t = 0$ is zero. Given that $V(t)$ must satisfy

$$RC\frac{dV}{dt} + V = E, \qquad V(0) = 0,$$

find the step response and draw its graph.

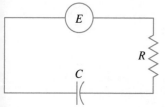

36. The pulse response $i(t)$ for the RL-circuit in the figure below is the current in the circuit when the current source is $I = h(t) - h(t-1)$ where $h(t)$ is the Heaviside unit step function, and the initial current in the circuit at time $t = 0$ is zero. Given that $i(t)$ must satisfy

$$\frac{L}{R}\frac{di}{dt} + i = I, \qquad i(0) = 0,$$

find the pulse response and draw its graph.

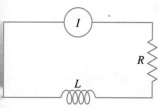

37. (a) The current I in the LR-circuit of Exercise 36 (with I replaced by E) must satisfy the differential equation

$$L\frac{dI}{dt} + RI = E.$$

If $E(t) = E_0 \sin(\omega t)$, $t \geq 0$, where E_0 and ω are constants, solve this differential equation for $I(t)$, and show that the solution can be written in the form

$$I(t) = Ae^{-Rt/L} + \frac{E_0}{Z} \sin(\omega t - \phi),$$

where A is an arbitrary constant and

$$Z = \sqrt{R^2 + \omega^2 L^2}, \qquad \phi = \text{Tan}^{-1}\left(\frac{\omega L}{R}\right).$$

(b) What is the value of A if the current in the circuit at time $t = 0$ when $E(t)$ is connected is I_0?

$*$ **38.** (a) The current I in the RC-circuit of Exercise 35 must satisfy the differential equation

$$R\frac{dI}{dt} + \frac{I}{C} = \frac{dE}{dt}.$$

If $E(t)$ is as in Exercise 37, show that the solution can be written in the form

$$I(t) = Ae^{-t/(RC)} + \frac{E_0}{Z} \sin(\omega t - \phi),$$

where A is an arbitrary constant and

$$Z = \sqrt{R^2 + \frac{1}{\omega^2 C^2}}, \qquad \phi = \text{Tan}^{-1}\left(-\frac{1}{\omega CR}\right).$$

(b) What is the value of A if the current in the circuit at time $t = 0$ is I_0 when $E(t)$ is connected?

$**$ **39.** Repeat Exercise 30 but do not assume that the temperature of the oven is constant for $t > 5$. Assume instead that it is sinusoidal with period 10 min, oscillating between $210°$C and $190°$C.

$*$ **40.** A channel is empty below a dam. The dam breaks at time $t = 0$ and releases water to the channel at a high rate at first, decreasing in time. Let the input to the channel be $I(t) = I_0 e^{-\lambda t}$, where I_0 and λ are positive constants. The storage S of the channel and the discharge Q from the channel are related to $I(t)$ by the Muskingum routing equations

$$\frac{dS}{dt} = I - Q, \qquad S = K[xI + (1-x)Q],$$

where K and $0 < x < 1$ are constants. Use these to find $Q(t)$.

$**$ **41.** Repeat Exercise 21 in Section 15.2 given that the diver descends slowly to the bottom. Assume that his descent is at a constant rate over a time interval of length T. Assume also that maximum pressure $\overline{N}$ is proportional to depth below the surface.

15.4 Second-Order Equations Reducible to Two First-Order Equations

Second-order differential equations in y as a function of x can be expressed symbolically in the form

$$F(x, y, y', y'') = 0. \tag{15.26}$$

When F is independent of x or y, the second-order equation can be replaced by two first-order equations.

Type I: Dependent Variable Missing

If a second-order differential equation in $y(x)$ is explicitly independent of y, then it is of the form

$$F(x, y', y'') = 0. \tag{15.27}$$

In such a case we set

$$v = \frac{dy}{dx} \quad \text{and} \quad \frac{dv}{dx} = \frac{d^2y}{dx^2}. \tag{15.28}$$

If we substitute these into 15.27, we obtain

$$F(x, v, v') = 0,$$

a first-order differential equation in $v(x)$. If we can solve this equation for $v(x)$, we can then integrate $dy/dx = v(x)$ for $y(x)$.

EXAMPLE 15.6

Find two-parameter families of solutions for the following differential equations:

$$\text{(a) } xy'' - y' = 0 \qquad \text{(b) } y'' = (y')^2$$

SOLUTION

(a) Since y is explicitly missing, we use substitutions 15.28:

$$x\frac{dv}{dx} - v = 0.$$

Variables are now separable,

$$\frac{dv}{v} = \frac{dx}{x}$$

(provided $v \neq 0$), and a solution for $v(x)$ is

$$\ln|v| = \ln|x| + C \quad \text{or} \quad v = Dx \quad (D = \pm e^C).$$

Because $v = dy/dx$,

$$\frac{dy}{dx} = Dx,$$

and we can integrate for

$$y(x) = \frac{D}{2}x^2 + E = Fx^2 + E \quad (F = D/2).$$

When $v = 0$, we obtain $y = $ constant, which satisfies the differential equation. Since such functions are contained in the two-parameter family $y = Fx^2 + E$, they are not singular.

(b) Since y is again missing, we substitute from 15.28 to get

$$\frac{dv}{dx} = v^2.$$

Variables are again separable,

$$\frac{dv}{v^2} = dx,$$

(provided $v \neq 0$), and a solution for $v(x)$ is

$$-\frac{1}{v} = x + C.$$

Consequently,

$$v = \frac{dy}{dx} = -\frac{1}{x + C}.$$

Integration now yields

$$y(x) = -\ln|x + C| + D.$$

When $v = 0$, we obtain $y = $ constant, which satisfies the differential equation. Since these functions cannot be obtained from the two-parameter family, they are singular solutions.

Type II: Independent Variable Missing

If a second-order differential equation in $y(x)$ is explicitly independent of x, then it is of the form

$$F(y, y', y'') = 0. \tag{15.29}$$

In this case we set

$$\frac{dy}{dx} = v \quad \text{and} \quad \frac{d^2y}{dx^2} = \frac{dv}{dx} = \frac{dv}{dy}\frac{dy}{dx} = v\frac{dv}{dy}. \tag{15.30}$$

When we substitute these into 15.29, we obtain

$$F\left(y, v, v\frac{dv}{dy}\right) = 0,$$

a first-order differential equation in $v(y)$. If we can solve this equation for $v(y)$, we can separate $dy/dx = v(y)$ and integrate for $y(x)$.

EXAMPLE 15.7

Find explicit two-parameter families of solutions for the following differential equations:

$$\text{(a) } yy'' + (y')^2 = 1 \qquad \text{(b) } y'' = (y')^2$$

SOLUTION

(a) Since x is explicitly missing, we substitute from 15.30:

$$yv\frac{dv}{dy} + v^2 = 1.$$

If variables are separated, we have

$$\frac{v\,dv}{v^2 - 1} = -\frac{dy}{y}$$

(provided $v \neq \pm 1$), and integration gives

$$\frac{1}{2} \ln |v^2 - 1| = -\ln |y| + C.$$

Thus,

$$|v^2 - 1| = \frac{e^{2C}}{y^2},$$

from which we have

$$\frac{dy}{dx} = v = \frac{\pm\sqrt{D + y^2}}{y} \quad (D = \pm e^{2C}).$$

We separate variables again to get

$$\frac{y\,dy}{\sqrt{y^2 + D}} = \pm dx,$$

and obtain an implicit definition of the solution $y(x)$,

$$\sqrt{y^2 + D} = \pm x + E.$$

Explicit solutions are

$$y = \pm\sqrt{(E \pm x)^2 - D} = \pm\sqrt{x^2 \pm 2Ex + (E^2 - D)} = \pm\sqrt{x^2 + Fx + G},$$

where $F = \pm 2E$ and $G = E^2 - D$. When $v = \pm 1$, we obtain $y = \pm x + C$, which are solutions of the differential equation. Since they can be obtained by choosing $F = 2C$ and $G = C^2$, they are not singular solutions.

(b) Since x is explicitly missing, we again substitute from 15.30 to obtain

$$v\frac{dv}{dy} = v^2.$$

If variables are separated,

$$\frac{dv}{v} = dy$$

(provided $v \neq 0$), and a solution for $v(y)$ is

$$\ln |v| = y + C.$$

Thus,

$$|v| = e^{y+C}$$

and

$$\frac{dy}{dx} = v = De^y \quad (D = \pm e^C).$$

We separate variables again,

$$e^{-y}\,dy = D\,dx,$$

and find an implicit definition of the solution $y(x)$,

$$-e^{-y} = Dx + E.$$

An explicit form for the solution is

$$y = -\ln(-Dx - E) = -\ln(Fx + G),$$

where $F = -D$ and $G = -E$. From $v = 0$, we obtain $y = $ constant, which are solutions of the differential equation. Since these solutions can be obtained by setting $F = 0$ in the two-parameter family, they are not singular solutions.

In each of Examples 15.6 and 15.7 we solved the differential equation $y'' = (y')^2$ since both the independent variable x and the dependent variable y are missing. Although the solutions appear different, each is easily derivable from the other. Notice that solutions $y = $ constant are singular for the two-parameter family $y(x) = D - \ln|x + C|$, but not for $y(x) = -\ln(Fx + G)$.

Let us summarize the results of this section. Substitutions 15.28 for differential equation 15.27 with the dependent variable missing replace the second-order differential equation with two first-order equations: a first-order equation in $v(x)$, followed by a first-order equation in $y(x)$. Contrast this with the method for equation 15.29 with the independent variable missing. Substitutions 15.30 again replace the second-order equation with two first-order equations. However, the first first-order equation is in $v(y)$, so that for this equation y is the independent variable rather than the dependent variable. The second first-order equation is again one for $y(x)$.

EXERCISES 15.4

In Exercises 1–10 find a two-parameter family of solutions for the differential equation.

1. $xy'' + y' = 4x$

2. $2yy'' = 1 + (y')^2$

3. $y'' = y' + 2x$

4. $x^2 y'' = (y')^2$

5. $y'' \sin x + y' \cos x = \sin x$ * **6.** $y'' = [1 + (y')^2]^{3/2}$

7. $y'' + 4y = 0$ * **8.** $y'' = yy'$

9. $y'' + (y')^2 = 1$·   * **10.** $(y'')^2 = 1 + (y')^2$

11. The figure below shows a long cylindrical cable. Copper wire runs down the centre of the cable and insulation covers the wire. If r measures radial distance from the centre of the cable, then steady-state temperature T in the insulation is a function of r that must satisfy the differential equation

$$r \frac{d^2 T}{dr^2} + \frac{dT}{dr} = 0.$$

Find $T(r)$ if temperatures on inner and outer edges $r = a$ and $r = b$ of the insulation are constant values T_a and T_b.

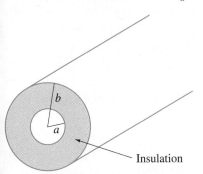

Insulation

12. If heat is generated at a constant rate in the copper wire of Exercise 11 (perhaps because of electrical current), the differential equation is replaced by

$$r \frac{d^2 T}{dr^2} + \frac{dT}{dr} = k,$$

where k is a constant. Find $T(r)$ if temperatures on inner and outer edges $r = a$ and $r = b$ of the insulation are constant values T_a and T_b.

* **13.** The well in the figure below penetrates an aquifer of depth b. The head h at a radius r from the well must satisfy the differential equation

$$r \frac{d^2 h}{dr^2} + \frac{dh}{dr} = 0,$$

subject to $h = h_w$ at the edge $r = r_w$ of the well and $h = h_i$ at some distance $r = r_i$ from the well. Find $h(r)$.

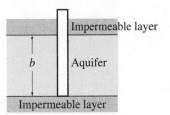

** **14.** A dog at position $(L, 0)$ in the xy-plane spots a rabbit at position $(0, 0)$ running in the positive y-direction. If the dog runs at the same speed as the rabbit and always moves directly toward the rabbit, find the equation of the path followed by the dog.

** **15.** Repeat Exercise 14 if the dog runs twice as fast as the rabbit.

** **16.** A hawk at position $(L, 0)$ in the figure below spots a pigeon at the origin flying with speed v in the positive y-direction. The hawk immediately takes off after the pigeon with speed $V > v$, always heading directly toward the pigeon. After time t, the pigeon is at position $P(0, vt)$. If the equation of the pursuit curve of the hawk is $y = y(x)$, then during time t the hawk travels distance Vt along this curve. But distance along this curve can be calculated by means of the definite integral

$$\int_x^L \sqrt{1 + \left(\frac{dy}{dx}\right)^2}\, dx.$$

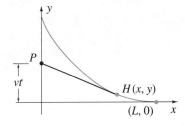

1072   Chapter 15   Differential Equations

(a) Show that $y(x)$ must satisfy the integrodifferential equation

$$x\frac{dy}{dx} - y = \frac{v}{V}\int_L^x \sqrt{1 + \left(\frac{dy}{dx}\right)^2}\, dx.$$

(b) Differentiate this equation to obtain the second-order differential equation

$$x\frac{d^2y}{dx^2} = \frac{v}{V}\sqrt{1 + \left(\frac{dy}{dx}\right)^2}.$$

(c) Solve this differential equation for the pursuit curve of the hawk.

(d) Show that the hawk catches the pigeon a distance $vVL/(V^2 - v^2)$ up the y-axis.

** **17.**

(a) An aircraft at position $C(a, b)$ is flying to the right along the line $y = b$ with speed v. A missile is fired from the origin with speed $V > v$, always heading directly toward the aircraft. Use a discussion like that in Exercise 16 to find the path followed by the missile.

(b) Plot the curve in part (a) when $a = b = 1$ and $V = 2v$. Where does the missile catch the aircraft?

** **18.**

(a) The following differential equation arises in the physics of bubble coalescence. Variable y is dimensionless film thickness (joining two touching bubbles) and T is dimensionless time:

$$\frac{d^2y}{dT^2} = \frac{5}{4y}\left(\frac{dy}{dT}\right)^2 - ay + \frac{b}{y} - \frac{c}{y^3}.$$

Show that the substitutions $v = dy/dT$ and $d^2y/dT^2 = v\,dv/dy$ lead to a linear first-order differential equation in v^2. Using the fact that y must be a decreasing function of T, and setting the constant of integration equal to zero, show that

$$\frac{dy}{dT} = -2\sqrt{ay^2 - \frac{b}{5} + \frac{c}{9y^2}}.$$

(b) Assuming that $100ac - 9b^2 > 0$ and that $y(0) = 1$, show that $y(T)$ is defined implicitly by

$$\ln\left|\sqrt{y^4 - \frac{by^2}{5a} + \frac{c}{9a}} + y^2 - \frac{b}{10a}\right|$$

$$= -4\sqrt{a}\,T + \ln\left|\sqrt{1 - \frac{b}{5a} + \frac{c}{9a}} + 1 - \frac{b}{10a}\right|.$$

15.5 Newtonian Mechanics

One of the most important applications of differential equations is in the study of moving particles and objects. In classical mechanics, motion is governed by Newton's second law 11.109, which states that when an object of constant mass m is subjected to a force $\mathbf{F}$, the resultant acceleration is described by

$$\mathbf{F} = m\mathbf{a}. \tag{15.31}$$

When $\mathbf{F}$ is given, this is an algebraic equation giving acceleration $\mathbf{a}$. If we substitute $\mathbf{a} = d\mathbf{v}/dt$, we obtain a first-order differential equation

$$\mathbf{F} = m\frac{d\mathbf{v}}{dt} \tag{15.32}$$

for velocity $\mathbf{v}$ as a function of time t. If we substitute $\mathbf{a} = d^2\mathbf{r}/dt^2$, we obtain a second-order differential equation for position $\mathbf{r}$ as a function of time:

$$\mathbf{F} = m\frac{d^2\mathbf{r}}{dt^2}. \tag{15.33}$$

In practice, it is seldom this simple. Often, $\mathbf{F}$ is not given as a function of time, but as a function of position, or velocity, or some other variable. In such cases we may have to change the dependent or independent variable, or both, in order to solve the differential equation.

When motion is in one direction only, we may dispense with vectors and consider the single component of these vectors in that direction. If $r(t)$, $v(t)$, $a(t)$, and F are the components of $\mathbf{r}(t)$, $\mathbf{v}(t)$, $\mathbf{a}(t)$, and $\mathbf{F}$ in this direction, then equations 15.31–15.33 take the forms

$$F = ma, \tag{15.34}$$

$$F = m\frac{dv}{dt}, \tag{15.35}$$

$$F = m\frac{d^2r}{dt^2},$$

(15.36)

respectively. We studied applications of the last two equations in Section 5.2, and many of these applications involved objects falling under the influence of gravity. We were not able to take air resistance into account at that time, but with the differential equation-solving techniques in Sections 15.2–15.3, we can obtain more realistic results. It has been shown experimentally that when an object moves slowly through a medium (such as for an object sinking in water), the frictional drag is proportional to velocity, whereas when it moves quickly through the medium (such as for a skydiver), the frictional drag is proportional to the square of velocity. We give examples of each.

EXAMPLE 15.8

A stone with mass m, dropped into water, has speed v_0 as it penetrates the surface. During its descent to the bottom, it is acted on by gravity and water resistance that is proportional to the speed of the stone. Ignore the apparent loss in weight of the stone due to Archimedes' principle. Find the speed and position of the stone as functions of time.

FIGURE 15.5 Schematic for mass sinking in water

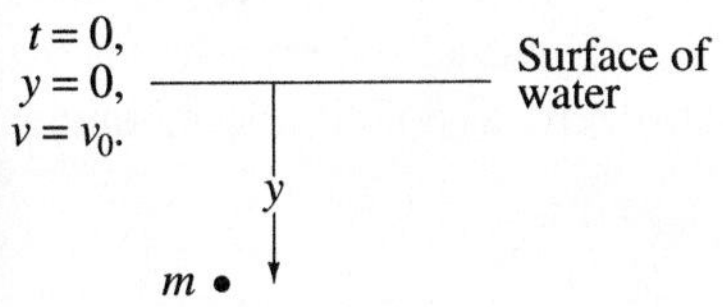

$t = 0,$
$y = 0,$ ——————— Surface of water
$v = v_0.$

y

$m \bullet$

SOLUTION Let us measure y as positive downward taking $y = 0$ at the surface of the water, and taking time $t = 0$ when the stone penetrates the surface (Figure 15.5).

If $F_{\mathbf{w}}$ is (the vertical component of) the force of water resistance on the stone, then $F_{\mathbf{w}} = -kv$, where $k > 0$ is a constant. Since the total force on the stone has vertical component $mg - kv$, Newton's second law gives

$$m\frac{dv}{dt} = mg - kv, \quad t \geq 0.$$

Note that the force on the right is not a function of t, so that the equation is not immediately integrable. It is, however, separable,

$$\frac{dv}{kv - mg} = -\frac{dt}{m},$$

and solutions are defined implicitly by

$$\frac{1}{k}\ln|kv - mg| = -\frac{t}{m} + C.$$

An explicit solution is found by solving this equation for v; the result is

$$v(t) = \frac{mg}{k} + De^{-kt/m} \quad (D = \pm e^{kC}/k).$$

Since $v = v_0$ when $t = 0$,

$$v_0 = \frac{mg}{k} + D \quad \text{or} \quad D = v_0 - \frac{mg}{k}.$$

The velocity of the stone as a function of time is therefore

$$v(t) = \frac{mg}{k} + \left(v_0 - \frac{mg}{k}\right)e^{-kt/m}.$$

Because $v(t) = dy/dt$, we set

$$\frac{dy}{dt} = \frac{mg}{k} + \left(v_0 - \frac{mg}{k}\right)e^{-kt/m},$$

and this equation is immediately integrable to

$$y(t) = \frac{mg}{k}t - \frac{m}{k}\left(v_0 - \frac{mg}{k}\right)e^{-kt/m} + E.$$

Since $y(0) = 0$,

$$0 = -\frac{m}{k}\left(v_0 - \frac{mg}{k}\right) + E \quad \text{or} \quad E = \frac{m}{k}\left(v_0 - \frac{mg}{k}\right).$$

Consequently, the distance sunk by the stone as a function of time is given by

$$y(t) = \frac{mg}{k}t - \frac{m}{k}\left(v_0 - \frac{mg}{k}\right)(1 - e^{-kt/m}).$$

Note that the velocity of the stone does not increase indefinitely. In fact, as time passes, a limiting velocity is approached:

$$\lim_{t\to\infty} v(t) = \lim_{t\to\infty}\left[\frac{mg}{k} + \left(v_0 - \frac{mg}{k}\right)e^{-kt/m}\right] = \frac{mg}{k}.$$

This is called the *terminal velocity* of the stone; it is a direct result of the assumption that water resistance is proportional to instantaneous velocity. We could have predicted it from the differential equation describing the motion of the stone. With the initial velocity of the stone being v_0, the force on it, namely $mg - kv$, is initially $mg - v_0$, gradually decreasing as the velocity increases. As v approaches mg/k, the force approaches zero, as does the acceleration, and the stone attains terminal velocity.

EXAMPLE 15.9

A skydiver and his parachute have mass m kilograms. As he plunges toward earth (because of gravity), he also experiences air resistance that is directly proportional to the square of his instantaneous velocity. Assuming that his vertical velocity is zero when he leaves the plane, find the vertical component of his velocity and his vertical position as functions of time.

FIGURE 15.6 Schematic for falling skydiver

$t = 0,$
$y = 0,$
$v = 0.$

$\downarrow y$

SOLUTION Let us measure y as positive in the downward direction, taking $y = 0$ and time $t = 0$ at the instant the skydiver leaves the plane (Figure 15.6). If F_a is the vertical component of the force of air resistance, then $F_a = -kv^2$, where $k > 0$ is a constant. Since the total force on the skydiver during the fall has vertical component $mg - kv^2$, Newton's second law gives

$$m\frac{dv}{dt} = mg - kv^2, \quad t \ge 0.$$

As in Example 15.8, we see that the skydiver has a terminal velocity defined by $mg - kv^2 = 0$. If we denote it by $V = \sqrt{mg/k}$, then the differential equation for $v(t)$ can be separated in the form

$$\frac{dv}{v^2 - V^2} = -\frac{k}{m}dt.$$

Solutions are defined implicitly by

$$-\frac{kt}{m} + C = \int \frac{1}{(v+V)(v-V)}\,dv = \frac{1}{2V}\int\left(\frac{-1}{v+V} + \frac{1}{v-V}\right)dv$$

$$= \frac{1}{2V}\left(-\ln|v+V| + \ln|v-V|\right) = \frac{1}{2V}\ln\left|\frac{v-V}{v+V}\right|.$$

To find explicit solutions, we multiply by $2V$ and exponentiate,

$$\frac{v-V}{v+V} = De^{-2kVt/m},$$

where $D = \pm e^{2VC}$. When we cross-multiply, $v - V = D(v + V)e^{-2kVt/m}$, from which

$$v = \frac{V(1 + De^{-2kVt/m})}{1 - De^{-2kVt/m}}.$$

The initial velocity $v(0) = 0$ requires that $0 = V(1 + D)/(1 - D) \implies D = -1$. If we substitute $V = \sqrt{mg/k}$ in the exponential functions, we obtain the velocity of the skydiver in the form

$$v(t) = \frac{V(1 - e^{-2\sqrt{kg/m}\,t})}{1 + e^{-2\sqrt{kg/m}\,t}}.$$

It reflects the fact that the limiting velocity is V since $\lim_{t\to\infty} v = V$. To obtain the distance fallen by the skydiver, we set the velocity equal to dy/dt and integrate with respect to t:

$$y(t) = V \int \frac{1 - e^{-2\sqrt{kg/m}\,t}}{1 + e^{-2\sqrt{kg/m}\,t}}\,dt = V \int \left(1 - \frac{2e^{-2\sqrt{kg/m}\,t}}{1 + e^{-2\sqrt{kg/m}\,t}}\right) dt$$

$$= V \left[t + \sqrt{\frac{m}{kg}}\, \ln\left(1 + e^{-2\sqrt{kg/m}\,t}\right)\right] + C.$$

The initial condition $y(0) = 0$ implies that $0 = V\sqrt{m/(kg)}\, \ln 2 + C$, and therefore

$$y(t) = Vt + \sqrt{\frac{m}{kg}}\, V \ln\left(1 + e^{-\sqrt{kg/m}\,t}\right) - \sqrt{\frac{m}{kg}}\, V \ln 2$$

$$= Vt + \frac{m}{k} \ln\left[\frac{1}{2}(1 + e^{-2\sqrt{kg/m}\,t})\right].$$

The first term is distance fallen by an object with constant speed V. Since the second term is always negative, the distance fallen by the skydiver is always less than Vt, as should be expected, since the skydiver never achieves terminal velocity.

Whenever an object such as the block in Figure 15.7 is moving over a surface, there is resistance to the motion. This resistance, called **friction**, is due to the fact that the interface between the block and the surface is not smooth; each surface is inherently rough, and this roughness retards the motion of one surface over the other. In effect, a force slowing the motion of the block is created, and this force is called the **force of friction**. Many experiments have been performed to obtain a functional representation for this force. It turns out that when the block in Figure 15.7

FIGURE 15.7   Friction exists when a block moves over a surface

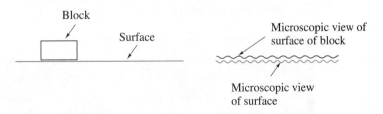

slides along a horizontal surface, the magnitude of the force of friction opposing the motion is given by

$$|\mathbf{F}| = \mu mg, \tag{15.37}$$

where m is the mass of the block, g is the acceleration due to gravity, and μ is a constant called the **coefficient of kinetic friction**. In other words, the force of friction is directly proportional to the weight mg of the block. We caution the reader that this result is valid for the situation shown in Figure 15.7, but it may not be valid for other configurations (say, perhaps, for an inclined plane). Furthermore, there is a coefficient of static friction that is used in place of μ if the block is being accelerated from rest.

EXAMPLE 15.10

A block of mass 2 kg is given initial speed 5 m/s along a horizontal surface. If the coefficient of kinetic friction between the block and surface is $\mu = 0.25$, how far does the block slide before stopping?

SOLUTION   Let us measure x as positive in the direction of motion (Figure 15.8), taking $x = 0$ and $t = 0$ at the instant the block is released. The x-component of the force of friction on the block is

$$F = -0.25(2)g = -\frac{g}{2} \quad (g = 9.81).$$

FIGURE 15.8   Schematic for block moving over a surface

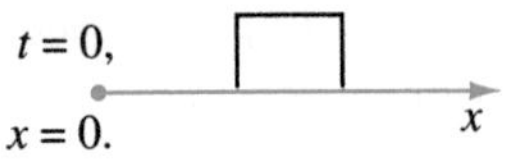

$t = 0,$

$x = 0.$

x

According to Newton's second law,

$$2\frac{dv}{dt} = -\frac{g}{2},$$

from which we get

$$v(t) = -\frac{g}{4}t + C.$$

Since $v(0) = 5$, it follows that $C = 5$, and

$$v(t) = -\frac{g}{4}t + 5.$$

But $v = dx/dt$, and hence

$$\frac{dx}{dt} = -\frac{g}{4}t + 5.$$

Integration gives

$$x(t) = -\frac{g}{8}t^2 + 5t + D.$$

Because we chose $x = 0$ at time $t = 0$, D must also be zero, and

$$x(t) = -\frac{g}{8}t^2 + 5t.$$

The block comes to rest when $v = 0$, that is, when

$$-\frac{g}{4}t + 5 = 0 \quad \text{or} \quad t = \frac{20}{g}.$$

The position of the block at this time is

$$x = -\frac{g}{8}\left(\frac{20}{g}\right)^2 + 5\left(\frac{20}{g}\right) = 5.10.$$

The block therefore slides 5.10 m before stopping.

Consulting Project 25

We are being hired as consultants by the olympic ski team to analyze race times of downhill skiers. The team knows the factors that determine the speed of a skier, and has a wealth of information for us to use, but it is unable to determine which factors have the greatest influence, and therefore which ones it should concentrate on in training.

SOLUTION   Let us begin by establishing a framework in which to analyze the speed of skiers and then identify factors that influence their speed. Certainly, downhill courses have many turns and areas where the course is very steep and quite flat. Let us begin by simplifying things and assume that the course is straight and that it makes a constant angle θ with the horizontal (Figure 15.9a). What the olympic team really wants to do is win races and to do this, its skiers must minimize their times. With our simplified set up with no curves and constant angle, this occurs when speed is maximized. We can find speed as a function of either time or distance travelled; which would be better? There would seem to be no advantage of one over the other so lets set out to find, as usual, speed as a function of time.

FIGURE 15.9a

Analysis of speed of a downhill skier

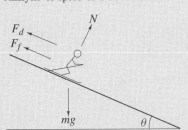

FIGURE 15.9b

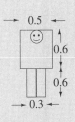

There are three forces acting on a skier, gravity down the hill, air drag, and friction between skis and snow. If the mass of the skier is m, the component of gravity down the hill is $mg \sin \theta$, where $g = 9.81$. Air drag F_d, which acts up the hill, has been shown experimentally to be directly proportional to the area A of the skier facing downhill (Figure 15.9b), the square of speed v, and the density ρ of the air; that is $F_d = \eta \rho A v^2$, where η is a constant called the drag coefficient. Finally, friction F_f, which also acts up the hill, is proportional to the normal force N exerted by the hill on the skier, $F_f = \mu mg \cos \theta$, where μ is the coefficient of kinetic friction. When we substitute these into Newton's second law for motion down the hill, we obtain

$$m\frac{dv}{dt} = mg \sin \theta - \eta \rho A v^2 - \mu mg \cos \theta = mg(\sin \theta - \mu \cos \theta) - \eta \rho A v^2.$$

This differential equation is separable, but in order to simplify calculations, let us for the moment set $a^2 = mg(\sin \theta - \mu \cos \theta)$ and $b^2 = \eta \rho A$, in which case

$$\frac{dt}{m} = \frac{dv}{a^2 - b^2 v^2} = \frac{1}{2a}\left(\frac{1}{a + bv} + \frac{1}{a - bv}\right) dv.$$

Notice that we need not worry about v being equal to a/b since a/b is the terminal speed of the skier. Solutions of the differential equation are defined implicitly by

$$\frac{t}{m} + C = \frac{1}{2ab}\left(\ln|a + bv| - \ln|a - bv|\right) = \frac{1}{2ab}\ln\left|\frac{a + bv}{a - bv}\right|.$$

To solve this for v, we exponentiate and cross multiply,

$$a + bv = D(a - bv)e^{2abt/m} \quad \Longrightarrow \quad v = \frac{a(De^{2abt/m} - 1)}{b(De^{2abt/m} + 1)},$$

where $D = \pm e^{2abC}$. Now all skiers start with zero initial speed, but in the next couple of seconds, they use their poles to accelerate quickly. Let us assume that all skiers pole for the same length of time, and we begin our analysis thereafter, which we take as $t = 0$. To differentiate between skiers, let us assume that the initial speed is an unspecified v_0. This requires

$$v_0 = \frac{a(D - 1)}{b(D + 1)} \quad \Longrightarrow \quad D = \frac{a + bv_0}{a - bv_0}.$$

Thus,

$$v(t) = \frac{a}{b} \left[\frac{(a + bv_0) - (a - bv_0)e^{-2abt/m}}{(a + bv_0) + (a - bv_0)e^{-2abt/m}} \right].$$

This is somewhat complicated; perhaps it would have been better to find v in terms of distance travelled. Let us quickly find out. If we let s be distance skied down the slope, then we can write

$$m\frac{dv}{ds}\frac{ds}{dt} = a^2 - b^2v^2 \quad \Longrightarrow \quad \frac{v\,dv}{a^2 - b^2v^2} = \frac{ds}{m},$$

integration of which gives

$$-\frac{1}{2b^2}\ln|a^2 - b^2v^2| = \frac{s}{m} + C.$$

When this is solved for v^2, the result is

$$v^2 = \frac{a^2}{b^2} - De^{-2b^2s/m}.$$

If we take $s = 0$ when poling ceases so that $v(0) = v_0$, then $v_0^2 = a^2/b^2 - D$, and therefore

$$v^2 = \frac{a^2}{b^2} - \left(\frac{a^2}{b^2} - v_0^2 \right) e^{-2b^2s/m}$$

$$= \frac{mg(\sin\theta - \mu\cos\theta)}{\eta\rho A}(1 - e^{-2\eta\rho As/m}) + v_0^2 e^{-2\eta\rho As/m}.$$

This does appear to be simpler than the expression for $v(t)$. The parameters over which we have no influence are s, θ, g, and ρ. The others m, η, A, μ, and v_0 are alterable. To determine which of these has the greatest effect on speed, we shall change each in turn by 10% and calculate the percentage change in v. Let us specify the following base values: the distance at which to compare speeds $s = 50$ m; the angle of the hill $\theta = \pi/12$; the acceleration due to gravity $g = 9.81$; the density of air $\rho = 1.25$ kg/m^3; the mass of the skier $m = 80$ kg; the coefficient of dynamic friction $\mu = 0.05$ for well-waxed skis; the drag coefficient $\eta = 0.7$ determined experimentally; and the initial speed $v_0 = 4$ m/s. For these values, the speed of the skier at 50 m is 13.0 m/s, compared to a terminal speed of 19.8 m/s. The numbers in the Table 15.1 are percentage changes in speed at the 50 m distance due to 10% changes from base values in m, η, A, μ and v_0, respectively. Perhaps surprisingly, an increase of the initial speed v_0 of the skier after poling is the least significant. Decreasing the area of the skier facing downhill is most significant, suggesting that skiers should be in a tuck position as much as possible. Decreasing the drag coefficient η is equally significant, but not much can be done to decrease it, except for making ski-suits slippery. Heavier skiers should have an advantage over lighter ones, but an increase in mass is likely to be offset by an increase in A. Decreases in μ are likely to be small unless major technological advances can be made in waxes and/or ski materials.

TABLE 15.1

m	η	A	μ	v_0
1.17	1.29	1.29	1.08	0.58

EXERCISES 15.5

1. A body with mass m is caused to move in the positive x-direction by a constant force with magnitude F. It is also acted on by a resistive force that is proportional to the square root of the speed at any instant. If the magnitude of this resistive force is F^* when the speed is v^*, find an expression for its terminal speed.

2. In Example 15.8 it was necessary to solve the equation

$$\frac{1}{k} \ln |kv - mg| = -\frac{t}{m} + C$$

for $v = v(t)$ and use the condition $v(0) = v_0$ to evaluate the constant. We chose first to solve for $v(t)$ and then to evaluate the constant. Instead, first use the condition $v(0) = v_0$ to evaluate C, and then solve the equation for $v(t)$.

3. A boat and its contents have mass 250 kg. Water exerts a resistive force on the motion of the boat that is proportional to the instantaneous speed of the boat and is 200 N when the speed is 30 km/h.

    (a) If the boat starts from rest and the engine exerts a constant force of 250 N in the direction of motion, find the speed of the boat as a function of time.

    (b) What is the limiting speed of the boat?

4. You are called on as an expert to testify in a traffic accident hearing. The question concerns the speed of a car that made an emergency stop with brakes locked and wheels sliding. The skid mark on the road measured 9 m. If you assume that the coefficient of kinetic friction between the tires and road was less than 1, what can you say about the speed of the car before the brakes were applied? Are you testifying for the prosecution or the defence?

5. A car of mass 1500 kg starts from rest at an intersection. The engine exerts a constant force of 2500 N, and air friction causes a resistive force whose magnitude in newtons is equal to the square of the speed of the car in metres per second. Find the speed of the car and its distance from the intersection after 10 s.

6. Small particles moving in a fluid experience a drag force kv proportional to their speed v.

    (a) Determine the time required for a particle with mass m to accelerate from rest to 95% of its terminal speed in terms of m, g, and k.

    (b) Determine the distance travelled by the particle in achieving 95% of its terminal speed.

7. The English longbow in medieval times was regarded to be accurate at 100 m or more. For an arrow to travel a horizontal distance of 100 m with maximum height 10 m, find the initial speed and angle of projection of the arrow. Ignore air friction.

8. A spring (with constant k) is attached on one end to a wall and on the other end to a mass M (figure follows). The mass is set into motion along the x-axis by pulling it a distance x_0 to the right of the position it would occupy were the spring unstretched and given speed v_0 to the left. During the subsequent motion, there is a frictional force between M and the horizontal surface with coefficient of kinetic friction equal to μ.

    (a) Show that the differential equation describing the motion of M is

$$M\frac{d^2x}{dt^2} = -kx + \mu Mg, \quad x(0) = x_0, \quad v(0) = -v_0,$$

if we take $t = 0$ at the instant that motion is initiated. When is this equation valid?

    (b) Since t is explicitly missing from the equation in part (a), show that it can be rewritten in the form

$$Mv\frac{dv}{dx} = -kx + \mu Mg, \quad v(x_0) = -v_0,$$

and that therefore

$$\frac{k}{2}(x_0^2 - x^2) = \frac{M}{2}(v^2 - v_0^2) + \mu Mg(x_0 - x).$$

Interpret each of the terms in this equation physically.

    (c) If x^* represents the position at which M comes to rest for the first time, use the equation in part (b) to determine x^* as a function of μ, M, g, k, x_0, and v_0. Discuss the possibilities of x^* being positive, negative, and zero.

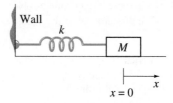

Wall

k

M

$x = 0$

x

* **9.** A 1-kg mass falls under the influence of gravity. It is also acted on by air resistance proportional to the square of its velocity and is 5 N when its velocity is 50 m/s. If the velocity of the mass has magnitude 20 m/s at time $t = 0$, find a formula for its velocity as a function of time.

* **10.** Repeat Exercise 9 if the initial velocity has magnitude 100 m/s.

* **11.** For how long does the mass in Exercise 9 rise if it is thrown upward with velocity 20 m/s?

* **12.** When a body falls in air, it is acted on by gravity and also by a force due to air resistance that is proportional to the square of its instantaneous speed. If the body is initially projected downward with velocity less than its terminal velocity, find its velocity as a function of time.

* **13.** Repeat Exercise 12 if the body is projected downward with velocity greater than its terminal velocity.

* **14.** For how long does the mass in Exercise 12 rise if it is thrown upward with velocity v_0?

* **15.** A 1-kg rock is thrown vertically upward with speed 20 m/s. Air resistance to its motion when measured in newtons has magnitude equal to one-tenth the square of its speed in metres per second. Find the maximum height attained by the rock.

∗ **16.**   (a) A mass m is thrown upward with speed v_0. It is acted on by gravity and air resistance proportional to the square of velocity. Find a formula for speed of the mass on its ascent. Is this formula also valid for its speed when it begins to fall?

     (b) Find a formula for its height. How high does it rise?

∗ **17.** An object of mass m is injected into a medium at speed v_0. The medium exerts a resistive force proportional to velocity on the mass with constant of proportionality β. In addition, a constant force with magnitude F opposes the motion of the mass. Show that the mass comes to rest after it has travelled a distance

$$\frac{mv_0}{\beta} - \frac{Fm}{\beta^2}\ln\left(1 + \frac{\beta v_0}{F}\right),$$

taking time $(m/\beta)\ln(1 + \beta v_0/F)$ to do so.

∗ **18.** A projectile with mass m is launched from the origin with initial speed v_0 at an angle θ with the positive x-axis. If it is acted on by air resistance $-\beta\mathbf{v}$ that is proportional to its velocity $\mathbf{v}$, find the velocity and position of the projectile at any time.

∗ **19.** We have solved one-dimensional displacement problems for objects acted on by resistive forces proportional to velocity and to the square of velocity (Examples 15.8 and 15.9). In Exercise 18 we solved a two-dimensional problem with resistance proportional to velocity. Use the projectile problem of Exercise 18 to show that we cannot solve two-dimensional problems with resistance proportional to the square of velocity.

∗ **20.** A mass m slides from rest down a frictionless plane inclined at angle α to the horizontal. Find a formula for the time taken to travel a distance D down the plane. What is its speed at this time?

∗ **21.** Find formulas for speed and distance travelled for the mass in Exercise 20 if air resistance proportional to velocity also acts on the mass.

∗ **22.** The coordinates of an electron moving in the xy-plane about its nucleus are given at any time t by

$$2\frac{dx}{dt} + y = 3, \qquad \frac{dy}{dt} - 2x = 4.$$

Find the path followed by the electron by eliminating t and using the condition that the electron passes through the point $(0, 3)$. Identify the path.

∗ **23.** In Example 7.34 of Section 7.10 we derived the escape velocity of a projectile from the earth's surface based on energy principles. In this exercise we obtain the same result using differential equations. When a projectile of mass m, fired from the earth's surface, is a distance r from the centre of the earth, the magnitude of the force of attraction on it is given by Newton's universal law of gravitation, $F = GMm/r^2$, where M is the mass of the earth and G is a constant. Use Newton's second law and a substitution corresponding to 15.30 to find the velocity of the projectile as a function of r. What minimum initial velocity guarantees that the projectile escapes the gravitational field of the earth?

∗∗ **24.** An object is dropped from a height of 5000 km above the earth. The only force acting on it is gravity with magnitude $F = GmM/r^2$, where $G > 0$ is a constant, m and M are the mass of the object and the earth, and r is the distance between the object and the centre of the earth.

     (a) Assuming that $g = 9.81$ on the surface of the earth, where $r = 6370$ km, show that GM, as a single constant, is equal to 3.98×10^{14}.

     (b) The differential equation for the motion of the object is $m\,dv/dt = -GmM/r^2$. Show that it can be expressed in the form $v\,dv/dr = -GM/r^2$, and solve it for

$$v = -8.37 \times 10^3\sqrt{\frac{11.37 \times 10^6 - r}{r}} \qquad \text{m/s}.$$

     (c) Substitute dr/dt for v in part (b) and solve for an implicit equation defining r as a function of t. *Hint:* Think about setting $r = 11.37 \times 10^6 \sin^2\theta$.

     (d) How long does the object take to fall to earth?

∗∗ **25.** A stone of mass 100 g is thrown vertically upward with speed 20 m/s. Air exerts a resistive force on the stone proportional to the square of its instantaneous speed, and has magnitude 1/10 N when the speed of the stone is 10 m/s. Find the time when the stone returns to its projection point. Compare the result with the time taken if air resistance is neglected.

∗∗ **26.** A huge cannon fires a projectile with initial velocity v_0 directly toward the moon (figure below). When the projectile is a distance r above the earth's surface, the force of attraction of the earth on the projectile has magnitude

$$\frac{GMm}{(r + R)^2},$$

where G is a constant, R is the radius of the earth, M is the mass of the earth, and m is the mass of the projectile. At this point the moon's gravitational attraction has magnitude

$$\frac{GM^*m}{(a + R^* - r)^2},$$

where M^* is the mass of the moon and R^* is its radius.

     (a) Show that if only the two forces above are considered to act on the projectile, then the differential equation describing its motion is

$$\frac{d^2r}{dt^2} = -\frac{gR^2}{(r + R)^2} + \frac{g^*R^{*2}}{(a + R^* - r)^2},$$

where g and g^* are gravitational accelerations on the surfaces of the earth and the moon.

     (b) Prove that the velocity of the projectile at a distance r above the surface of the earth is defined by

$$v^2 = \frac{2gR^2}{r + R} + \frac{2g^*R^{*2}}{a + R^* - r} + v_0^2 - 2gR - \frac{2g^*R^{*2}}{a + R^*}.$$

∗∗ **27.** Newton's second law states that if an object of variable mass $m(t)$ is subjected to a force $\mathbf{F}(t)$, then

$$\frac{d}{dt}(m\mathbf{v}) = \mathbf{F},$$

where t is time and $\mathbf{v}$ is the velocity of the object. A uniform chain of length 3 m and mass 6 kg is held by one end so that the other end just touches the floor. If the chain is released, find the velocity of the falling chain as a function of the length of chain still falling. How fast does the end hit the floor?

15.6 Linear Differential Equations

Throughout Chapters 5 and 8 and Sections 15.1–15.5, we have stressed the use of differential equations in solving applied problems. We have considered examples from such diverse areas as engineering, geometry, ecology, and psychology, hoping thereby to illustrate how valuable differential equations can be in modelling situations mathematically. Perhaps the most important type of differential equation is the linear differential equation.

DEFINITION 15.4

A differential equation of the form

$$a_0(x)\frac{d^n y}{dx^n} + a_1(x)\frac{d^{n-1} y}{dx^{n-1}} + a_2(x)\frac{d^{n-2} y}{dx^{n-2}} + \cdots + a_{n-1}(x)\frac{dy}{dx} + a_n(x)y = F(x)$$

(15.38)

is called a **linear differential equation**. It is n^{th} order when $a_0(x) \not\equiv 0$.

Note in particular that none of the derivatives of $y(x)$ are multiplied together, nor are they squared or cubed or taken to any other power, nor do they appear as the argument of any transcendental function. All we see is a function of x multiplying y, plus a function of x multiplying the first derivative of y, plus a function of x multiplying the second derivative of y, and so on.

If $n = 1$, equation 15.38 reduces to

$$a_0(x)\frac{dy}{dx} + a_1(x)y = F(x),$$

and at any point at which $a_0(x) \neq 0$, we can divide to obtain

$$\frac{dy}{dx} + \frac{a_1(x)}{a_0(x)}y = \frac{F(x)}{a_0(x)}.$$

If we set $P(x) = a_1(x)/a_0(x)$ and $Q(x) = F(x)/a_0(x)$, we have

$$\frac{dy}{dx} + P(x)y = Q(x);$$

that is, every linear first-order differential equation can be expressed in this form. We discussed equations of this type in Section 15.3, where it was shown that such equations have a general solution

$$y(x) = e^{-\int P(x)dx}\left\{\int Q(x)e^{\int P(x)dx}dx + C\right\}.$$

In other words, we already know how to solve linear first-order differential equations, and therefore our discussion in the next five sections is directed primarily at second- and higher-order equations. Keep in mind, however, that all results are also valid for first-order linear equations.

Because equation 15.38 is so cumbersome, we introduce notation to simplify its representation. In particular, if we use the notation $D = d/dx$, $D^2 = d^2/dx^2$, and so on, we can write

$$a_0(x)D^n y + a_1(x)D^{n-1}y + a_2(x)D^{n-2}y + \cdots + a_{n-1}(x)Dy + a_n(x)y = F(x)$$

(15.39)

or

$$\{a_0(x)D^n + a_1(x)D^{n-1} + a_2(x)D^{n-2} + \cdots + a_{n-1}(x)D + a_n(x)\}y = F(x).$$

(15.40)

The quantity in braces is called a **differential operator**; it operates on whatever follows it — in this case, y. It is a "differential" operator because it operates by taking derivatives. Because the operator involves only x's and D's, we denote it by

$$\phi(x, D) = a_0(x)D^n + a_1(x)D^{n-1} + a_2(x)D^{n-2} + \cdots + a_{n-1}(x)D + a_n(x).$$
(15.41)

The general linear n^{th}-order differential equation can then be represented very simply by

$$\phi(x, D)y = F(x).$$
(15.42)

For example, differential equation 15.3,

$$xy'' + y' + xy = 0,$$

is called *Bessel's differential equation of order zero*. In operator notation we write

$$(xD^2 + D + x)y = 0$$

or

$$\phi(x, D)y = 0,$$

where $\phi(x, D) = xD^2 + D + x$.

For the differential equation

$$y'' + 2y' - 3y = e^{-x},$$

we write

$$\phi(D)y = e^{-x},$$

where $\phi(D) = D^2 + 2D - 3$.

We now indicate the meaning of the term *linear*. Suppose that L is an operator that operates on each function $y(x)$ in some set S. For example, L might be the operation that multiplies each function by 5, or perhaps squares each function, or perhaps differentiates each function. It is said to be a linear operator if it satisfies the following definition.

DEFINITION 15.5

An operator L is said to be a **linear operator** on a set of functions S if for any two functions $y_1(x)$ and $y_2(x)$ in S, and any constant c,

$$L(y_1 + y_2) = Ly_1 + Ly_2, \qquad (15.43a)$$

$$L(cy_1) = c(Ly_1). \qquad (15.43b)$$

Many of the operations in calculus are therefore linear. For instance, taking limits is a linear operation, as is differentiation, antidifferentiation, and taking definite integrals. On the other hand, taking the square root of a positive function is not a linear operation, since

$$L(y_1 + y_2) = \sqrt{y_1 + y_2} \neq L(y_1) + L(y_2).$$

It is not difficult to show that the differential operator $\phi(x, D)$ in 15.42 is linear; that is,

$$\phi(x, D)(y_1 + y_2) = \phi(x, D)y_1 + \phi(x, D)y_2,$$

$$\phi(x, D)(cy_1) = c[\phi(x, D)y_1],$$

and because of this, differential equation 15.42 is also said to be linear. The following two differential equations are not linear. We say that they are *nonlinear*.

$$\frac{d^2y}{dx^2} + y^2 = x, \qquad \frac{d^2y}{dx^2}\frac{dy}{dx} = e^x.$$

In particular, if we substitute $y_1(x) + y_2(x)$ into the left side of the first equation, we obtain

$$\frac{d^2}{dx^2}(y_1 + y_2) + (y_1 + y_2)^2 = \frac{d^2y_1}{dx^2} + \frac{d^2y_2}{dx^2} + y_1^2 + 2y_1y_2 + y_2^2.$$

If we substitute $y_1(x)$, then $y_2(x)$, and then add the results, we find a different expression:

$$\frac{d^2y_1}{dx^2} + y_1^2 + \frac{d^2y_2}{dx^2} + y_2^2.$$

Unless otherwise indicated, we assume that coefficient functions $a_0(x), a_1(x), \ldots, a_n(x)$ and the function $F(x)$ in linear differential equation 15.38 are all continuous on some open interval I. We also assume that $a_0(x) \neq 0$ at any point in I. Solutions of the differential equation must necessarily have derivatives of orders up to and including n at each point in I. But existence of the nth derivative implies continuity of all lower-order derivatives. Furthermore, we can write 15.38 in the form

$$\frac{d^ny}{dx^n} = \frac{1}{a_0(x)}\left[F(x) - a_1(x)\frac{d^{n-1}y}{dx^{n-1}} - \cdots - a_n(x)y\right],$$

where all functions on the right are continuous and $a_0(x) \neq 0$. Hence, when derivatives of orders up to and including $n - 1$ are continuous on I, so also is the nth derivative d^ny/dx^n. Thus, solutions must have continuous derivatives of orders up to and including n.

EXERCISES 15.6

In Exercises 1–10 prove either that the operator L is linear or that it is not linear. In each case assume that the set S of functions on which L operates is the set of all functions on which L can operate. For instance, in Exercise 6, assume that S is the set of all functions $y(x)$ that have a first derivative dy/dx.

1. L multiplies functions $y(x)$ by 5

2. L multiplies functions $y(x)$ by $15x$

3. L adds the fixed function $z(x)$ to functions $y(x)$

4. L takes the limit of functions $y(x)$ as x approaches 3

5. L takes the limit of functions $y(x)$ as x approaches infinity

6. L takes the first derivative of functions $y(x)$ with respect to x

7. L takes the third derivative of functions $y(x)$ with respect to x

8. L takes the antiderivative of functions $y(x)$ with respect to x

9. L takes the definite integral of functions $y(x)$ with respect to x from $x = -1$ to $x = 4$

10. L takes the cube root of functions $y(x)$

In Exercises 11–20 determine whether the differential equation is linear or nonlinear. Write those equations that are linear in operator notation 15.42.

11. $2x\dfrac{d^2y}{dx^2} + x^3y = x^2 + 5$

12. $2x\dfrac{d^2y}{dx^2} + x^3y = x^2 + 5y$

13. $2x\dfrac{d^2y}{dx^2} + x^3y = x^2 + 5y^2$

14. $x\dfrac{d^3y}{dx^3} + 3x\dfrac{d^2y}{dx^2} - 2\dfrac{dy}{dx} + y = 10\sin x$

15. $x\dfrac{d^3y}{dx^3} + 3x\dfrac{d^2y}{dx^2} - 2\dfrac{dy}{dx} + y^2 = 10\sin x$

16. $y\dfrac{d^3y}{dx^3} + 3x\dfrac{d^2y}{dx^2} - 2\dfrac{dy}{dx} + y = 10\sin x$

17. $y'' - 3y' - 2y = 9\sec^2 x$ **18.** $yy'' + 3y' - 2y = e^x$

19. $\sqrt{1 + y'} + x^2 = 4$ **20.** $y'''' + y'' - y = \ln x$

* **21.** The Laplace transform of a function $y(t)$ is defined as the function $L(y) = \displaystyle\int_0^\infty e^{-st}y(t)\,dt$, provided that the improper integral converges. Show that if S is the set of all functions that have a Laplace transform, then the operation of taking the Laplace transform is linear on S.

* **22.** The finite Fourier cosine transform of a function $y(x)$ is defined as $L(y) = \displaystyle\int_0^{2\pi} y(x)\cos nx\,dx$, where n is a nonnegative integer, provided that the definite integral exists. Show that if S is the set of all functions that have a finite Fourier cosine transform, then the operation of taking the transform is linear on S.

15.7 Homogeneous Linear Differential Equations

Two classes of linear differential equations present themselves: those for which $F(x) \equiv 0$, and those for which $F(x) \not\equiv 0$. In this section and the next we consider equations for which $F(x) \equiv 0$; we discuss the more difficult class, in which $F(x) \not\equiv 0$, in Section 15.9. First let us name each of these classes of linear differential equations.

> ### DEFINITION 15.6
>
> A linear differential equation $\phi(x, D)y = F(x)$ is said to be **homogeneous** if $F(x) \equiv 0$, and **nonhomogeneous** otherwise.

The meaning of *homogeneous* to describe a property of linear differential equations in this definition is totally different from the meaning in Exercise 34 of Section 15.2.

The fundamental idea behind the solution of all linear differential equations is the following theorem.

> ### THEOREM 15.1 (Superposition principle)
>
> If $y_1(x), y_2(x), \ldots, y_m(x)$ are solutions of a homogeneous linear differential equation
>
> $$\phi(x, D)y = 0,$$
>
> on an interval I, then so too is any linear combination of them,
>
> $$C_1 y_1(x) + C_2 y_2(x) + \cdots + C_m y_m(x)$$
>
> (for arbitrary constants $C_1, C_2, \ldots, C_m$).

PROOF The proof requires only linearity of the operator $\phi(x, D)$, for

$$\phi(x, D)[C_1 y_1(x) + \cdots + C_m y_m(x)] = C_1[\phi(x, D)y_1] + \cdots + C_m[\phi(x, D)y_m]$$

$$= 0 + 0 + \cdots + 0 = 0. \qquad \blacksquare$$

Solutions of a linear differential equation that are linearly combined to produce other solutions are said to be *superposed* — consequently, the name *superposition principle* for Theorem 15.1. For example, it is straightforward to verify that $y_1(x) = e^{4x}$ and $y_2(x) = e^{-3x}$ are solutions of the homogeneous equation $y'' - y' - 12y = 0$. The superposition principle then states that for any constants C_1 and C_2, the function $y(x) = C_1 y_1(x) + C_2 y_2(x) = C_1 e^{4x} + C_2 e^{-3x}$ must also be a solution. It is a two-parameter family of solutions.

Similarly, superposition of the three solutions e^x, xe^x, and $x^2 e^x$ of the linear differential equation $y''' - 3y'' + 3y' - y = 0$ gives a three-parameter family of solutions $y(x) = C_1 e^x + C_2 x e^x + C_3 x^2 e^x$.

Now we begin to see the importance of the superposition principle. If we can find n solutions $y_1(x), y_2(x), \ldots, y_n(x)$ of an nth-order homogeneous linear differential equation, then an n-parameter family of solutions is

$$y(x) = C_1 y_1(x) + C_2 y_2(x) + \cdots + C_n y_n(x).$$

In other words, all that we need do is find n solutions; the superposition principle will do the rest. There is a problem, however, if we take things a little too literally. For instance, $y_1(x) = e^{4x}$ and $y_2(x) = 10e^{4x}$ are both solutions of $y'' - y' - 12y = 0$. By superposition, so, too, then is $y(x) = C_1 y_1 + C_2 y_2 = C_1 e^{4x} + 10C_2 e^{4x}$. But is it a two-parameter family of solutions? The answer is no, because we could write $y(x) = (C_1 + 10C_2)e^{4x}$, and by setting

$C_3 = C_1 + 10C_2$, we have $y(x) = C_3 e^{4x}$. Superposition of the solutions e^{4x} and $10e^{4x}$ has not therefore led to a two-parameter family of solutions, and the reason is that they are essentially the same solution: $y_2(x)$ is $y_1(x)$ multiplied by a constant. Superposition does not therefore lead to a solution with two arbitrary constants.

In a similar way, $y_1(x) = e^x$, $y_2(x) = xe^x$, and $y_3(x) = 2e^x - 3xe^x$ are all solutions of $y''' - 3y'' + 3y' - y = 0$, and therefore so is $y(x) = C_1 y_1 + C_2 y_2 + C_3 y_3$. But because we can write

$$y(x) = C_1 e^x + C_2 xe^x + C_3(2e^x - 3xe^x)$$
$$= (C_1 + 2C_3)e^x + (C_2 - 3C_3)xe^x$$
$$= C_4 e^x + C_5 xe^x,$$

$y(x)$ is not a three-parameter family of solutions; it is only a two-parameter family. This is a direct result of the fact that $y_3(x)$ is a linear combination of the solutions $y_1(x)$ and $y_2(x)$; it is twice $y_1(x)$ minus three times $y_2(x)$.

Our problem seems to come down to this: If we have n solutions of an nth-order homogeneous linear differential equation, how can we determine whether superposition leads to a solution that contains n arbitrary constants? Our examples have suggested that if any one of the solutions is a linear combination of the others, then an n-parameter family of solutions is not obtained, and this is indeed true. If one of the solutions is a linear combination of the others, we say that the n solutions are **linearly dependent**; if no solution is a linear combination of the others, we say that the n solutions are **linearly independent**. We summarize these results in the following theorem.

THEOREM 15.2

If $y_1(x)$, $y_2(x)$, ..., $y_n(x)$ are n linearly independent solutions of an nth-order homogeneous linear differential equation on an interval I, then $y(x) = C_1 y_1(x) + C_2 y_2(x) + \cdots + C_n y_n(x)$ is an n-parameter family of solutions of the differential equation on I.

What we should now do is devise a test to determine whether a set of n solutions is linearly independent or linearly dependent. For most examples, no test is really necessary; it is obvious whether one of the solutions can be written as a linear combination of the others. For those rare occasions when it is not obvious, Exercise 10 describes a test that can be used to determine whether functions are linearly independent.

We pointed out in Section 15.1 that an n-parameter family of solutions for an nth-order differential equation might not contain all solutions of the equation, and might not therefore be a general solution of the differential equation. This is not the case for linear n^{th}-order differential equations. It can be shown that if $y(x)$ is an n-parameter family of solutions for a linear n^{th}-order differential equation on an interval I, then every solution of the differential equation on I can be obtained by specifying particular values for the arbitrary constants in $y(x)$. Here then is a very important class of differential equations for which an n-parameter family of solutions is always a general solution. Let us state this as a corollary to Theorem 15.2.

COROLLARY 15.2.1

If $y_1(x)$, $y_2(x)$, ..., $y_n(x)$ are n linearly independent solutions of an nth-order homogeneous linear differential equation on an interval I, then $y(x) = C_1 y_1(x) + C_2 y_2(x) + \cdots + C_n y_n(x)$ is a general solution of the differential equation on I.

In summary, the superposition principle states that solutions of a homogeneous linear differential equation can be superposed to produce other solutions. If n linearly independent solutions of an nth-order equation are superposed, a general solution is obtained. This is the importance of the superposition principle. We need not devise a method that takes us directly to a general solution; we need a method for finding n linearly independent solutions — superposition does the rest. Unfortunately, for completely general coefficients $a_i(x)$ in $\phi(x, D)$ (see equation 15.41), it is impossible to give a method that always yields n linearly independent solutions of a homogeneous equation. There is, however, one special case of great practical importance in which it is always possible to produce n linearly independent solutions in a very simple way. This special case occurs when the coefficients $a_i(x)$ are all constants a_i, and this is the subject of Section 15.8.

EXAMPLE 15.11

If $y_1(x) = \cos 3x$ and $y_2(x) = \sin 3x$ are solutions of the differential equation $y'' + 9y = 0$, find a general solution.

SOLUTION   Since $y_1(x)$ and $y_2(x)$ are linearly independent solutions (one is not a constant times the other), a general solution can be obtained by superposition:

$$y(x) = C_1 \cos 3x + C_2 \sin 3x.$$

EXAMPLE 15.12

Given that $y_1(x) = e^{2x} \cos x$ and $y_2(x) = e^{2x} \sin x$ are solutions of the homogeneous linear differential equation $y'' - 4y' + 5y = 0$, find that solution that satisfies the conditions $y(\pi/4) = 1$, $y(\pi/3) = 2$.

SOLUTION   By superposition, a general solution of the differential equation is

$$y(x) = C_1 e^{2x} \cos x + C_2 e^{2x} \sin x = e^{2x}(C_1 \cos x + C_2 \sin x).$$

To satisfy the conditions $y(\pi/4) = 1$ and $y(\pi/3) = 2$, we have

$$1 = e^{\pi/2}(C_1/\sqrt{2} + C_2/\sqrt{2}), \quad 2 = e^{2\pi/3}(C_1/2 + \sqrt{3}\,C_2/2).$$

Thus C_1 and C_2 are defined by the pair of equations

$$C_1 + C_2 = \sqrt{2}\,e^{-\pi/2}, \quad C_1 + \sqrt{3}\,C_2 = 4e^{-2\pi/3},$$

the solution of which is

$$C_2 = \frac{4e^{-2\pi/3} - \sqrt{2}\,e^{-\pi/2}}{\sqrt{3} - 1}, \quad C_1 = \sqrt{2}\,e^{-\pi/2} - C_2.$$

The required solution is therefore $y(x) = e^{2x}(C_1 \cos x + C_2 \sin x)$ with these values for C_1 and C_2.

EXERCISES 15.7

In Exercises 1–8 show that the functions are solutions of the differential equation. Check that the differential equation is linear and homogeneous, and then find a general solution.

1. $y'' + y' - 6y = 0$; $y_1(x) = e^{2x}$, $y_2(x) = e^{-3x}$

2. $y' + y \tan x = 0$; $y_1(x) = \cos x$

3. $y'''' + 5y'' + 4y = 0$; $y_1(x) = \cos 2x$, $y_2(x) = \sin 2x$, $y_3(x) = \cos x$, $y_4(x) = \sin x$

4. $2y'' - 16y' + 32y = 0$; $y_1(x) = 3e^{4x}$, $y_2(x) = -2xe^{4x}$

5. $y''' - 3y'' + 2y' = 0$; $y_1(x) = 10$, $y_2(x) = 3e^x$, $y_3(x) = 4e^{2x}$

6. $2y'' - 8y' + 9y = 0$; $y_1(x) = e^{2x}\cos(x/\sqrt{2})$, $y_2(x) = e^{2x}\sin(x/\sqrt{2})$

7. $x^2 y'' + xy' + (x^2 - 1/4)y = 0$; $y_1(x) = (\sin x)/\sqrt{x}$, $y_2(x) = (\cos x)/\sqrt{x}$

8. $x^2 y'' + xy' + 4y = 0$; $y_1(x) = \cos(2\ln x)$, $y_2(x) = \sin(2\ln x)$

9. Show that $y_1(x) = -2/(x+1)$ and $y_2(x) = -2/(x+2)$ are both solutions of the differential equation $y'' = yy'$. Is $y(x) = y_1(x) + y_2(x)$ a solution? Explain.

10. We stated in this section that n functions $y_1(x), \ldots, y_n(x)$ are linearly dependent if at least one of the functions can be expressed as a linear combination of the others; they are linearly independent if none of the functions is a linear combination of the others. Another way of saying this is as follows: Functions $y_1(x), \ldots, y_n(x)$ are linearly dependent on an interval I if there exist constants $C_1, \ldots, C_n$, not all zero, such that on I,

$$C_1 y_1(x) + \cdots + C_n y_n(x) \equiv 0.$$

If this equation can be satisfied only with $C_1 = C_2 = \cdots = C_n = 0$, the functions are linearly independent. In this exercise we give a test to determine whether functions are linearly independent or dependent. If $y_1(x), \ldots, y_n(x)$ have derivatives up to and including order $n-1$ on the interval I, we define the Wronskian of the functions as the $n \times n$ determinant:

$$W(y_1, \ldots, y_n) = \begin{vmatrix} y_1 & y_2 & \cdots & y_n \\ y_1' & y_2' & \cdots & y_n' \\ y_1'' & y_2'' & \cdots & y_n'' \\ \vdots & \vdots & \ddots & \vdots \\ y_1^{(n-1)} & y_2^{(n-1)} & \cdots & y_n^{(n-1)} \end{vmatrix}.$$

Show that if $y_1, \ldots, y_n$ are linearly dependent on I, then $W(y_1, \ldots, y_n) \equiv 0$ on I. It follows then that if there exists at least one point in I at which $W(y_1, \ldots, y_n) \neq 0$, the functions $y_1, \ldots, y_n$ are linearly independent on I.

In Exercises 11–15 use the method of Exercise 10 to determine whether the functions are linearly dependent or independent on the interval.

11. $\{1, x, x^2\}$ on $-\infty < x < \infty$

12. $\{x, 2x - 3x^2, x^2\}$ on $-\infty < x < \infty$

13. $\{\sin x, \cos x\}$ on $0 \le x \le 2\pi$

14. $\{x, xe^x, x^2 e^x\}$ on $0 \le x \le 1$

15. $\{x \sin x, e^{2x}\}$ on $-\infty < x < \infty$

15.8 Homogeneous Linear Differential Equations with Constant Coefficients

We now consider homogeneous linear differential equations

$$a_0 D^n y + a_1 D^{n-1} y + a_2 D^{n-2} y + \cdots + a_{n-1} D y + a_n y = 0, \qquad (15.44)$$

where the coefficients $a_0, a_1, \ldots, a_n$ are all constants. In operator notation we write

$$\phi(D) y = 0, \qquad (15.45\text{a})$$

where

$$\phi(D) = a_0 D^n + a_1 D^{n-1} + \cdots + a_{n-1} D + a_n. \qquad (15.45\text{b})$$

The superposition principle states that a general solution of equation 15.44 is $y(x) = C_1 y_1(x) + \cdots + C_n y_n(x)$, provided that $y_1(x), \ldots, y_n(x)$ are any n linearly independent solutions of the equation. Our problem, then, is to devise a technique for finding n linearly independent solutions; to illustrate a possible procedure, we first consider three second-order equations. The first is

$$y'' + 2y' - 3y = 0.$$

It is not unreasonable to expect that for some value of m, the function $y(x) = e^{mx}$ might be a solution of this equation. After all, the equation says that the second derivative of the function must be equal to three times the function minus twice its first derivative. Since the exponential function reproduces itself when differentiated, perhaps m can be chosen to produce this combination. To see whether this is possible, we substitute $y = e^{mx}$ into the differential equation, and find that if $y = e^{mx}$ is to be a solution, then

$$m^2 e^{mx} + 2m e^{mx} - 3 e^{mx} = 0;$$

this implies that

$$0 = m^2 + 2m - 3 = (m+3)(m-1).$$

Thus $y = e^{mx}$ is a solution if m is chosen as either 1 or -3 (i.e., $y_1 = e^x$ and $y_2 = e^{-3x}$ are solutions of the differential equation). Since they are linearly independent, a general solution is $y(x) = C_1 e^x + C_2 e^{-3x}$.

For our second example, we take

$$y'' + 2y' + y = 0.$$

Since exponentials worked in the first example, we once again try a solution of the form $y(x) = e^{mx}$. If we substitute into the differential equation, we obtain

$$m^2 e^{mx} + 2m e^{mx} + e^{mx} = 0,$$

which implies that

$$0 = m^2 + 2m + 1 = (m + 1)^2.$$

Thus $y_1 = e^{-x}$ is a solution, but unfortunately it is the only solution that we obtain as a result of our guess. We need a second linearly independent solution $y_2(x)$ in order to obtain a general solution. Clearly, no other exponential will work. Perhaps if we multiplied e^{-x} by another function, we might find a second solution; in other words, perhaps there is a solution of the form $y(x) = v(x)e^{-x}$ for some $v(x)$. To see, we again substitute into the differential equation,

$$0 = \{v''e^{-x} - 2v'e^{-x} + ve^{-x}\} + 2\{v'e^{-x} - ve^{-x}\} + ve^{-x} = v''e^{-x}.$$

Consequently, $v'' = 0$, and this implies that $v(x) = Ax + B$, for any constants A and B. In particular, if $A = 1$ and $B = 0$, $v(x) = x$, and $y_2(x) = v(x)e^{-x} = xe^{-x}$ is also a solution of the differential equation. By superposition, then, we find that a general solution is $y(x) = C_1 e^{-x} + C_2 x e^{-x} = (C_1 + C_2 x)e^{-x}$. Note that if we had set $A = 0$ and $B = 1$, then $v(x) = 1$, and the solution $y(x) = v(x)e^{-x}$ would have been $y_1(x)$. Further, if we had simply set $y(x) = v(x)e^{-x} = (Ax + B)e^{-x}$, we would have the general solution.

Our third example is

$$y'' + 2y' + 10y = 0.$$

As in the preceding two examples, if we assume a solution $y = e^{mx}$, then

$$m^2 e^{mx} + 2m e^{mx} + 10 e^{mx} = 0$$

or

$$0 = m^2 + 2m + 10.$$

The solutions of this quadratic equation are the complex numbers

$$m = \frac{-2 \pm \sqrt{4 - 40}}{2} = -1 \pm 3i.$$

(A brief introduction to complex numbers can be found in Appendix C.) Because m is complex, no real exponential $y = e^{mx}$ satisfies the differential equation. If, however, we form complex exponentials $y_1(x) = e^{(-1+3i)x}$ and $y_2(x) = e^{(-1-3i)x}$, and superpose these solutions, then

$$y(x) = A e^{(-1+3i)x} + B e^{(-1-3i)x},$$

and this must also be a solution. When we use Euler's identity for complex exponentials, $e^{i\theta} = \cos\theta + i\sin\theta$, we can write $y(x)$ in the form

$$y(x) = A e^{-x} e^{3xi} + B e^{-x} e^{-3xi}$$

$$= A e^{-x}(\cos 3x + i \sin 3x) + B e^{-x}(\cos 3x - i \sin 3x)$$

$$= e^{-x}[(A + B)\cos 3x + i(A - B)\sin 3x]$$

$$= e^{-x}(C_1 \cos 3x + C_2 \sin 3x),$$

where $C_1 = A + B$ and $C_2 = i(A - B)$. In other words, the function $y(x) = e^{-x}(C_1 \cos 3x + C_2 \sin 3x)$ is a general solution of the differential equation, and it has been derived from the complex roots $m = -1 \pm 3i$ of the equation $m^2 + 2m + 10 = 0$. Note that what multiplies x in the exponential is the real part of these complex numbers, and what multiplies x in the trigonometric functions is the imaginary part. In Exercise 17 we show that this solution can also be derived without complex numbers, but we feel that in general the use of complex numbers is the best method.

In each of these examples we guessed $y = e^{mx}$ as a possible solution. We then substituted into the differential equation to obtain an algebraic equation for m. In each case the equation for m was

$$\phi(m) = 0;$$

that is, take the operator $\phi(D)$, replace D by m, and set the polynomial equal to zero. This is not a peculiarity of these examples for it is straightforward to show that for any homogeneous linear equation 15.45, if we assume a solution of the form $y = e^{mx}$, then m must satisfy the equation $\phi(m) = 0$. We name this equation in the following definition.

DEFINITION 15.7

With every linear differential equation that has constant coefficients $\phi(D)y = F(x)$, we associate an equation

$$\phi(m) = 0 \qquad (15.46)$$

called the **auxiliary equation**.

To summarize, in each of the examples we assumed a solution $y = e^{mx}$ and found that m had to satisfy the auxiliary equation $\phi(m) = 0$. From the roots of the auxiliary equation we obtained solutions of the differential equation, and superposition then led to a general solution. This procedure works on every homogeneous linear differential equation with constant coefficients. But if the procedure is the same in every case, surely we can set down rules that eliminate the necessity of tediously repeating these steps in every example. This we do in the following theorem.

THEOREM 15.3

If $\phi(m) = 0$ is the auxiliary equation associated with the homogeneous linear differential equation $\phi(D)y = 0$, then there are two possibilities:

(i) $\phi(m) = 0$ has a real root m of multiplicity k. Then a solution of the differential equation is

$$(C_1 + C_2 x + \cdots + C_k x^{k-1})e^{mx}. \qquad (15.47a)$$

(ii) $\phi(m) = 0$ has a pair of complex conjugate roots $a \pm bi$ each of multiplicity k. Then a solution of the differential equation is

$$e^{ax}[(C_1 + C_2 x + \cdots + C_k x^{k-1})\cos bx$$
$$+ (D_1 + D_2 x + \cdots + D_k x^{k-1})\sin bx]. \qquad (15.47b)$$

A general solution of the differential equation is obtained by superposing all solutions in (i) and (ii).

For a proof of this theorem see Exercise 26. Let us now apply the theorem to our previous examples. The auxiliary equation for $y'' + 2y' - 3y = 0$ is

$$0 = m^2 + 2m - 3 = (m + 3)(m - 1)$$

with solutions $m = 1$ and $m = -3$. If we now use part (i) of Theorem 15.3 with two real roots, each of multiplicity 1, a general solution of the differential equation is

$$y(x) = C_1 e^x + C_2 e^{-3x}.$$

The auxiliary equation for $y'' + 2y' + y = 0$ is

$$0 = m^2 + 2m + 1 = (m + 1)^2$$

with solutions $m = -1$ and $m = -1$. Part (i) of Theorem 15.3 with a single real root of multiplicity 2 gives the general solution

$$y(x) = (C_1 + C_2 x)e^{-x}.$$

The auxiliary equation for $y'' + 2y' + 10y = 0$ is

$$0 = m^2 + 2m + 10$$

with solutions $m = -1 \pm 3i$. Part (ii) of Theorem 15.3 with a pair of complex conjugate roots, each of multiplicity 1, gives the general solution

$$y(x) = e^{-x}(C_1 \cos 3x + C_2 \sin 3x).$$

EXAMPLE 15.13

Find a general solution for $y''' - y = 0$.

SOLUTION   The auxiliary equation is

$$0 = m^3 - 1 = (m - 1)(m^2 + m + 1)$$

with solutions $m = 1$ and $m = -(1/2) \pm (\sqrt{3}/2)i$. A general solution of the differential equation is, therefore,

$$y(x) = C_1 e^x + e^{-x/2}[C_2 \cos(\sqrt{3}\,x/2) + C_3 \sin(\sqrt{3}\,x/2)].$$

EXAMPLE 15.14

When temperature $T(x)$ in a very long, thin wire with one end at the origin is analyzed, the following differential equation is sometimes encountered:

$$\frac{d^2 T}{dx^2} - hT = 0,$$

where $h > 0$ is a constant. Find $T(x)$ if it must also satisfy the conditions $T(0) = T_0$ and $\lim_{x \to \infty} T = 0$.

SOLUTION   The auxiliary equation is $m^2 - h = 0$ with solutions $m = \pm\sqrt{h}$. Then,

$$T(x) = C_1 e^{\sqrt{h}x} + C_2 e^{-\sqrt{h}x}.$$

Since $\lim_{x \to \infty} T = 0$, we must set $C_1 = 0$, in which case $T_0 = T(0) = C_2$. Thus, $T(x) = T_0 e^{-\sqrt{h}x}$.

EXAMPLE 15.15

If the roots of the auxiliary equation $\phi(m) = 0$ are

$$3, \ 3, \ 3, \ \pm 2i, \ -2, \ 1 \pm \sqrt{3}, \ -4 \pm i, \ -4 \pm i,$$

find a general solution of the differential equation $\phi(D)y = 0$.

SOLUTION   A general solution is

$$y(x) = (C_1 + C_2 x + C_3 x^2)e^{3x} + C_4 \cos 2x + C_5 \sin 2x + C_6 e^{-2x}$$
$$+ C_7 e^{(1+\sqrt{3})x} + C_8 e^{(1-\sqrt{3})x} + e^{-4x}[(C_9 + C_{10}x)\cos x$$
$$+ (C_{11} + C_{12}x)\sin x].$$

EXERCISES 15.8

In Exercises 1–12 find a general solution for the homogeneous differential equation.

1. $y'' + y' - 6y = 0$

2. $2y'' - 16y' + 32y = 0$

3. $2y'' + 16y' + 82y = 0$

4. $y'' + 2y' - 2y = 0$

5. $y'' - 4y' + 5y = 0$

6. $y''' - 3y'' + y' - 3y = 0$

7. $y'''' + 2y'' + y = 0$

8. $y''' - 6y'' + 12y' - 8y = 0$

9. $3y''' - 12y'' + 18y' - 12y = 0$

10. $y'''' + 5y'' + 4y = 0$

11. $y''' - 3y'' + 2y' = 0$

12. $y'''' + 16y = 0$

In Exercises 13–16 find a homogeneous linear differential equation that has the function as general solution.

13. $y(x) = C_1 e^x + (C_2 + C_3 x)e^{-4x}$

14. $y(x) = e^{-2x}(C_1 \cos 4x + C_2 \sin 4x)$

15. $y(x) = C_1 + C_2 e^{\sqrt{3}x} + C_3 e^{-\sqrt{3}x}$

16. $y(x) = e^x(C_1 + C_2 x)\cos \sqrt{2}\,x + e^x(C_3 + C_4 x)\sin \sqrt{2}\,x$

17. Show that if we assume that $y(x) = e^{ax} \sin bx$ is a solution of the differential equation $y'' + 2y' + 10y = 0$, then a and b must be equal to -1 and ± 3, respectively. Verify that, for this a and b, $y(x) = e^{ax} \cos bx$ is also a solution, and therefore a general solution is $y(x) = e^{-x}(C_1 \cos 3x + C_2 \sin 3x)$.

18. The equation $y''' + ay'' + by' + cy = 0$, where a, b, and c are constants, has solution $y(x) = C_1 e^{-x} + e^{-2x}(C_2 \sin 4x + C_3 \cos 4x)$. Find a, b, and c.

Sturm-Liouville systems play a prominent role in the study of partial differential equations. The system consists of a homogeneous, linear, second-order differential equation containing a parameter λ and two boundary conditions. Only for certain values of λ, called eigenvalues, do solutions exist, and corresponding solutions are called eigenfunctions. In Exercises 19–23, find eigenvalues and eigenfunctions for the Sturm-Liouville system.

∗ 19. $\dfrac{d^2 y}{dx^2} + \lambda y = 0, \ 0 < x < 3; \ y(0) = 0 = y(3)$

∗ 20. $\dfrac{d^2 y}{dx^2} + \lambda y = 0, \ 0 < x < 4; \ y'(0) = 0 = y'(4)$

∗ 21. $\dfrac{d^2 y}{dx^2} + \lambda y = 0, \ 0 < x < 2; \ y(0) = 0 = y'(2)$

∗ 22. $\dfrac{d^2 y}{dx^2} + \lambda y = 0, \ 0 < x < 5; \ y'(0) = 0 = y(5)$

∗ 23. $\dfrac{d^2 y}{dx^2} - \dfrac{dy}{dx} + \lambda y = 0, \ 0 < x < 1; \ y(0) = 0 = y(1)$

∗ 24. A mass M moving in the xy-plane is subjected to a force always directed toward the origin with magnitude proportional to its distance from the origin. At a certain instant, the mass passes through the point $(x_0, 0)$ with speed v in the positive y-direction. (a) Find the path of the mass. Assume no friction between the mass and the xy-plane, and no damping. (b) Find the path if the force is away from the origin instead of towards it.

∗∗ 25. Show that if p is constant and $f(x)$ is differentiable, then

$$D\{e^{px} f(x)\} = e^{px}\{(D + p)f(x)\}.$$

Now use mathematical induction to prove that if $f(x)$ is k times differentiable, then

$$D^k\{e^{px} f(x)\} = e^{px}\{(D + p)^k f(x)\}.$$

Finally, verify that

$$\phi(D)\{e^{px} f(x)\} = e^{px}\{\phi(D + p)f(x)\},$$

a result called the *operator shift theorem*.

** **26.**   (a) If m_0 is a real root of multiplicity k for the auxiliary equation $\phi(m) = 0$, show that the operator $\phi(D)$ can be expressed in the form

$$\phi(D) = (D - m_0)^k \psi(D),$$

where $\psi(D)$ is a polynomial in D. Now use the operator shift theorem of Exercise 25 to verify that $(C_1 + C_2 x + \cdots + C_k x^{k-1})e^{m_0 x}$ is a solution of $\phi(D)y = 0$.

(b) If $a \pm bi$ are complex conjugate roots each of multiplicity k for the auxiliary equation $\phi(m) = 0$, show that $\phi(D)$ can be expressed in the form

$$\phi(D) = (D - a - bi)^k (D - a + bi)^k \psi(D),$$

where $\psi(D)$ is a polynomial in D. Now use the operator shift theorem of Exercise 25 to verify that 15.47b is a solution of $\phi(D)y = 0$.

** **27.** If M, β, and k are all positive constants, find a general solution for the linear differential equation

$$M\frac{d^2 x}{dt^2} + \beta\frac{dx}{dt} + kx = 0.$$

Discuss all possibilities.

15.9 Nonhomogeneous Linear Differential Equations with Constant Coefficients

The general nonhomogeneous linear differential equation with constant coefficients is

$$\phi(D)y = F(x), \tag{15.48a}$$

where

$$\phi(D) = a_0 D^n + a_1 D^{n-1} + \cdots + a_{n-1}D + a_n. \tag{15.48b}$$

It is natural to ask whether we can use the results of Section 15.8 concerning homogeneous equations with constant coefficients to solve nonhomogeneous problems. Fortunately, the answer is yes, as shown by the following definition and theorem.

DEFINITION 15.8

With every nonhomogeneous linear differential equation with constant coefficients

$$\phi(D)y = F(x),$$

we associate a homogeneous equation

$$\phi(D)y = 0, \tag{15.49}$$

called the **homogeneous (reduced,** or **complementary) equation** associated with $\phi(D)y = F(x)$.

We now prove the following theorem.

THEOREM 15.4

A general solution of the linear differential equation $\phi(D)y = F(x)$ is $y(x) = y_h(x) + y_p(x)$, where $y_h(x)$ is a general solution of the associated homogeneous equation, and $y_p(x)$ is any particular solution of the given equation.

PROOF   Since $\phi(D)$ is a linear operator,

$$\phi(D)(y_h + y_p) = \phi(D)y_h + \phi(D)y_p = 0 + F(x) = F(x),$$

so that $y_h + y_p$ is indeed a solution of the given differential equation. Because $y_h(x)$ is a general solution of the associated homogeneous equation, it contains the requisite number of arbitrary constants for $y(x)$ to be an n-parameter family of solutions of $\phi(D)y = F(x)$. What remains is to show that any (and every) solution $y_1(x)$ of $\phi(D)y = F(x)$ can be expressed in the form $y_h(x) + y_p(x)$ for some choice of the constants in $y_h(x)$. Consider the function $y(x)$ defined as the difference between $y_1(x)$ and $y_p(x)$, that is, $y(x) = y_1(x) - y_p(x)$. Because of linearity,

$$\phi(D)y = \phi(D)(y_1 - y_p) = \phi(D)y_1 - \phi(D)y_p = F(x) - F(x) = 0.$$

But every solution of the homogeneous equation $\phi(D)y = 0$ can be expressed in the form $y_h(x)$ for some choice of the constants; that is, $y(x) = y_1(x) - y_p(x)$ can be expressed in the form

$$y_1(x) - y_p(x) = y_h(x)$$

for some choice of constants in $y_h(x)$. This completes the proof that $y_1(x)$ can be expressed in the form $y_h(x) + y_p(x)$, and therefore $y_h(x) + y_p(x)$ is a general solution of 15.48a.   ∎

We note in passing that Theorem 15.4 is also valid for linear differential equations with variable coefficients.

Theorem 15.4 indicates that discussions of nonhomogeneous differential equations can be divided into two parts. First, find a general solution $y_h(x)$ of the associated homogeneous equation 15.49, and this can be done using the results of Section 15.8. To this, add any particular solution $y_p(x)$ of 15.48. We present two methods for finding a particular solution: (1) the method of undetermined coefficients, and (2) the method of operators. Both methods apply in general only to differential equations in which $F(x)$ is a power (x^n, n a nonnegative integer), an exponential (e^{px}), a sine ($\sin px$), a cosine ($\cos px$), and/or any sums or products thereof.

Method of Undetermined Coefficients for a Particular Solution

The **method of undetermined coefficients** is to be used only when $F(x)$ in equation 15.48 is of the form x^n, e^{px}, $\sin px$, $\cos px$, and/or sums or products thereof. For example, if

$$y'' + y' - 6y = e^{4x},$$

the method essentially says that Ae^{4x} is the simplest function that could conceivably yield e^{4x} when substituted into the left side of the differential equation. Consequently, it is natural to assume that $y_p = Ae^{4x}$ and attempt to determine the unknown coefficient A. Substitution of this function into the differential equation gives

$$16Ae^{4x} + 4Ae^{4x} - 6Ae^{4x} = e^{4x}.$$

If we divide by e^{4x}, then

$$14A = 1 \quad \text{and} \quad A = \tfrac{1}{14}.$$

A particular solution is therefore $y_p = e^{4x}/14$.

Before stating a general rule, we illustrate a few more possibilities in the following example.

EXAMPLE 15.16

Find a particular solution of $y'' + y' - 6y = F(x)$ in each case:

(a) $F(x) = 6x^2 + 2x + 3$ (b) $F(x) = 2\sin 2x$ (c) $F(x) = xe^{-x} - e^{-x}$

SOLUTION

(a) Since terms in x^2, x, and constants yield terms in x^2, x, and constants when substituted into the left side of the differential equation, we attempt to find a particular solution of the form $y_p = Ax^2 + Bx + C$. Substitution into the differential equation gives

$$(2A) + (2Ax + B) - 6(Ax^2 + Bx + C) = 6x^2 + 2x + 3$$

or $(-6A)x^2 + (2A - 6B)x + (2A + B - 6C) = 6x^2 + 2x + 3.$

But this equation can hold for all values of x only if coefficients of corresponding powers of x are identical (see Exercise 35 in Section 3.8). Equating coefficients then gives

$$-6A = 6, \qquad 2A - 6B = 2, \qquad 2A + B - 6C = 3.$$

These imply that $A = -1$, $B = -2/3$, $C = -17/18$, and

$$y_p = -x^2 - \frac{2x}{3} - \frac{17}{18}.$$

(b) Since terms in $\sin 2x$ and $\cos 2x$ yield terms in $\sin 2x$ when substituted into the left side of the differential equation, we assume that $y_p = A\sin 2x + B\cos 2x$. Substitution into the differential equation gives

$$(-4A\sin 2x - 4B\cos 2x) + (2A\cos 2x - 2B\sin 2x)$$
$$- 6(A\sin 2x + B\cos 2x) = 2\sin 2x$$

or
$$(-10A - 2B)\sin 2x + (2A - 10B)\cos 2x = 2\sin 2x.$$

Equating coefficients of $\sin 2x$ and $\cos 2x$ gives

$$-10A - 2B = 2, \qquad 2A - 10B = 0.$$

These imply that $A = -5/26$, $B = -1/26$, and hence

$$y_p = -\frac{1}{26}(5\sin 2x + \cos 2x).$$

(c) Since terms in xe^{-x} and e^{-x} yield terms in xe^{-x} and e^{-x} when substituted into the left side of the differential equation, we assume that $y_p = Axe^{-x} + Be^{-x}$. Substitution into the differential equation gives

$$(Axe^{-x} - 2Ae^{-x} + Be^{-x}) + (-Axe^{-x} + Ae^{-x} - Be^{-x})$$
$$- 6(Axe^{-x} + Be^{-x}) = xe^{-x} - e^{-x}$$

or
$$(-6A)xe^{-x} + (-A - 6B)e^{-x} = xe^{-x} - e^{-x}.$$

Equating coefficients of e^{-x} and xe^{-x} yields

$$-6A = 1, \qquad -A - 6B = -1.$$

These imply that $A = -1/6$, $B = 7/36$, and hence

$$y_p = -(1/6)xe^{-x} + (7/36)e^{-x}.$$

The following rule encompasses each part of this example.

RULE 1

If a term of $F(x)$ consists of a power (x^n), an exponential (e^{px}), a sine $(\sin px)$, a cosine $(\cos px)$, or any product thereof, assume as a part of y_p a constant multiplied by that term plus a constant multiplied by any linearly independent function arising from it by differentiation.

For Example 15.16(a), since $F(x)$ contains the term $6x^2$, we assume that y_p contains Ax^2. Differentiation of Ax^2 yields a term in x and a constant so that we form $y_p = Ax^2 + Bx + C$. No new terms for y_p are obtained from the terms $2x$ and 3 in $F(x)$.

For Example 15.16(b), we assume that y_p contains $A \sin 2x$ to account for the term $2 \sin 2x$ in $F(x)$. Differentiation of $A \sin 2x$ gives a linearly independent term in $\cos 2x$ so that we form $y_p = A \sin 2x + B \cos 2x$.

For Example 15.16(c), since $F(x)$ contains the term xe^{-x}, we assume that y_p contains Axe^{-x}. Differentiation of Axe^{-x} yields a term in e^{-x}, so that we form $y_p = Axe^{-x} + Be^{-x}$. No new terms for y_p are obtained from the term $-e^{-x}$ in $F(x)$.

EXAMPLE 15.17

What is the form of the particular solution predicted by rule 1 for the differential equation

$$y'' + 15y' - 6y = x^2 e^{4x} + x + x \cos x?$$

SOLUTION   Rule 1 suggests that

$$y_p = Ax^2 e^{4x} + Bxe^{4x} + Ce^{4x} + Dx + E + Fx \cos x + Gx \sin x$$

$$+ H \cos x + I \sin x.$$

Unfortunately, exceptions to rule 1 do occur. For the differential equation $y'' + y = \cos x$, rule 1 would predict $y_p = A \cos x + B \sin x$. If we substitute this into the differential equation we obtain the absurd identity $0 = \cos x$, and certainly no equations to solve for A and B. This result could have been predicted had we first calculated $y_h(x)$. The auxiliary equation $m^2 + 1 = 0$ has solutions $m = \pm i$, so that $y_h(x) = C_1 \cos x + C_2 \sin x$. Since y_p as suggested by rule 1 is precisely y_h with different names for the constants, then certainly $y_p'' + y_p = 0$. Suppose that as an alternative we multiply this y_p by x, and assume that $y_p = Ax \cos x + Bx \sin x$. Substitution into the differential equation now gives

$$-2A \sin x + 2B \cos x = \cos x.$$

Identification of coefficients requires $A = 0$, $B = 1/2$, and $y_p = (1/2)x \sin x$.

This example suggests that if y_p predicted by rule 1 is already contained in y_h, then a modification of y_p is necessary. A precise statement of the situation is given in the following rule.

RULE 2

Suppose that a term in $F(x)$ is of the form $x^n f(x)$ (n a nonnegative integer). Suppose further that $f(x)$ can be obtained from $y_h(x)$ by specifying values for the arbitrary constants. If this term in y_h results from a root of the auxiliary equation of multiplicity k, then corresponding to $x^n f(x)$, assume as a part of y_p the term $Ax^k(x^n f(x)) = Ax^{n+k} f(x)$, plus a constant multiplied by any linearly independent function arising from it by differentiation. Do not include any terms that are already in y_h.

To use this rule we first require $y_h(x)$. Then, and only then, can we decide on the form of $y_p(x)$. As an illustration, consider the following example.

EXAMPLE 15.18

Find a general solution for $y''' - y = x^3 e^x$.

SOLUTION   In Example 15.13 we solved the auxiliary equation to obtain $m = 1$ and $m = -1/2 \pm (\sqrt{3}/2)i$, from which we formed

$$y_h(x) = C_1 e^x + e^{-x/2}[C_2 \cos(\sqrt{3}\,x/2) + C_3 \sin(\sqrt{3}\,x/2)].$$

Now $x^3 e^x$ is x^3 times e^x, and e^x can be obtained from y_h by specifying $C_1 = 1$ and $C_2 = C_3 = 0$. Since this term results from the root $m = 1$ of multiplicity 1, we assume that y_p contains $Ax(x^3 e^x) = Ax^4 e^x$. Differentiation of this function gives terms in $x^3 e^x$, $x^2 e^x$, xe^x, and e^x. We therefore take

$$y_p = Ax^4 e^x + Bx^3 e^x + Cx^2 e^x + Dxe^x.$$

(We do not include a term in e^x since it is already in y_h.) Substitution into the differential equation and simplification gives

$$(12A)x^3 e^x + (36A + 9B)x^2 e^x + (24A + 18B + 6C)xe^x$$

$$+ (6B + 6C + 3D)e^x = x^3 e^x.$$

Equating coefficients gives

$$12A = 1, \quad 36A + 9B = 0, \quad 24A + 18B + 6C = 0, \quad 6B + 6C + 3D = 0.$$

These imply that $A = 1/12$, $B = -1/3$, $C = 2/3$, $D = -2/3$, and

$$y_p = \tfrac{1}{12}x^4 e^x - \tfrac{1}{3}x^3 e^x + \tfrac{2}{3}x^2 e^x - \tfrac{2}{3}xe^x.$$

A general solution of the differential equation is therefore

$$y(x) = C_1 e^x + e^{-x/2}[C_2 \cos(\sqrt{3}\,x/2) + C_3 \sin(\sqrt{3}\,x/2)]$$

$$+ \frac{x^4 e^x}{12} - \frac{x^3 e^x}{3} + \frac{2x^2 e^x}{3} - \frac{2xe^x}{3}.$$

EXAMPLE 15.19

If the roots of the auxiliary equation $\phi(m) = 0$ for the differential equation $\phi(D)y = x^2 - 2\sin x + xe^{-2x}$ are $\pm i, -2, -2, -2, 4$, and 4, what is the form of y_p predicted by the method of undetermined coefficients?

SOLUTION   From the roots of the auxiliary equation, we can form

$$y_h(x) = C_1 \cos x + C_2 \sin x + (C_3 + C_4 x + C_5 x^2)e^{-2x} + (C_6 + C_7 x)e^{4x}.$$

Corresponding to the term x^2 in $F(x)$, rule 1 requires that y_p contain $Ax^2 + Bx + C$. Because $-2\sin x$ can be obtained from $y_h(x)$ by specifying $C_2 = -2$, $C_1 = C_3 = C_4 = C_5 = C_6 = C_7 = 0$, and this term results from the roots $m = \pm i$, each of multiplicity 1, rule 2 suggests that y_p contain $Dx \sin x + Ex \cos x$. (We do not include terms in $\sin x$ and $\cos x$ since they are already in y_h.) Finally, xe^{-2x} is x times e^{-2x}, and this function can be obtained from $y_h(x)$ by setting $C_3 = 1$, and $C_1 = C_2 = C_4 = C_5 = C_6 = C_7 = 0$. Because this term results from the root $m = -2$ of multiplicity 3, y_p must contain $Fx^4e^{-2x} + Gx^3e^{-2x}$ (but not terms in x^2e^{-2x}, xe^{-2x}, and e^{-2x} since they are in y_h). The total particular solution is therefore

$$y_p = Ax^2 + Bx + C + Dx \sin x + Ex \cos x + Fx^4e^{-2x} + Gx^3e^{-2x}.$$

Operator Method for a Particular Solution

The **operator method**, like that of undetermined coefficients, is designed only for functions $F(x)$ in equation 15.48 of the form x^n, e^{px}, $\sin px$, $\cos px$, and/or sums or products thereof. Essentially, the operator method says that if

$$\phi(D)y = F(x), \tag{15.50}$$

then

$$y = \frac{1}{\phi(D)}F(x). \tag{15.51}$$

But there is a problem. What does it mean to say that

$$\frac{1}{\phi(D)} = \frac{1}{a_0D^n + a_1D^{n-1} + \cdots + a_{n-1}D + a_n} \tag{15.52}$$

operates on $F(x)$? The operator method then depends on our explaining how $1/\phi(D)$ operates on $F(x)$. The simplest $\phi(D)$ is $\phi(D) = D$. In this case the differential equation is

$$Dy = F(x),$$

and the solution is

$$y(x) = \int F(x)\,dx.$$

If the solution is also to be represented in operator notation by

$$y(x) = \frac{1}{D}F(x),$$

then we must define the operator $1/D$ by

$$\frac{1}{D}F(x) = \int F(x)\,dx. \tag{15.53}$$

If $1/D$ means integrate, then $1/D^2$ must mean integrate twice, $1/D^3$ integrate three times, and so on.

We have stated that the method of operators is applicable in general only when $F(x)$ consists of powers, exponentials, and/or sines and cosines. Consider first the case in which $F(x)$ is a power x^n, in which case equations 15.50 and 15.51 become

$$\phi(D)y = x^n \tag{15.54}$$

and

$$y = \frac{1}{\phi(D)}x^n. \tag{15.55}$$

If we forget for the moment that D is a differential operator, and simply regard $1/\phi(D)$ as a rational function of a variable D, then we can express $1/\phi(D)$ as an infinite series of the form

$$\frac{1}{\phi(D)} = \frac{1}{D^k}\{b_0 + b_1 D + b_2 D^2 + \cdots\} \tag{15.56}$$

for some nonnegative integer k. (We will show how in a moment.) But this suggests that we write 15.55 in the form

$$y = \frac{1}{D^k}\{b_0 + b_1 D + b_2 D^2 + \cdots\}x^n. \tag{15.57}$$

If we now reinterpret D as d/dx, then the operator on the right will produce a polynomial in x. (Note that $D^m x^n = 0$ if $m > n$.) It turns out that if we ignore all arbitrary constants that result from the integrations (when $k \geq 1$), this polynomial is a solution of 15.54. In other words, when $F(x) = x^n$, a particular solution of 15.54 is

$$y_p = \frac{1}{D^k}\{b_0 + b_1 D + b_2 D^2 + \cdots\}x^n. \tag{15.58}$$

We now show how to expand $1/\phi(D)$ as a series of form 15.56. Only two situations arise and each of these can be illustrated with a simple example. First suppose that $\phi(D) = D^2 + 4D + 5$. For $1/\phi(D)$ we write

$$\frac{1}{\phi(D)} = \frac{1}{D^2 + 4D + 5} = \frac{1/5}{1 + \dfrac{4D + D^2}{5}},$$

and interpret the right side as the sum of an infinite geometric series with first term $1/5$ and common ratio $-(4D + D^2)/5$. When we write this series out, we have

$$\frac{1}{D^2 + 4D + 5} = \frac{1}{5}\left\{1 - \left(\frac{4D + D^2}{5}\right) + \left(\frac{4D + D^2}{5}\right)^2 - \cdots\right\}$$

$$= \frac{1}{5}\left\{1 - \frac{4}{5}D + \frac{11}{25}D^2 + \cdots\right\},$$

which is of the required form 15.56 with $k = 0$.

Power k in 15.56 is positive only when $\phi(D)$ has no constant term. For example, if $\phi(D) = D^4 - 2D^3 + 3D^2$, then we factor out D^2, and proceed as above:

$$\frac{1}{\phi(D)} = \frac{1}{D^2(3 - 2D + D^2)} = \frac{1}{3D^2\left(1 - \dfrac{2D - D^2}{3}\right)}$$

$$= \frac{1}{3D^2}\left\{1 + \left(\frac{2D - D^2}{3}\right) + \left(\frac{2D - D^2}{3}\right)^2 + \cdots\right\}$$

$$= \frac{1}{3D^2}\left\{1 + \frac{2}{3}D + \frac{1}{9}D^2 + \cdots\right\}.$$

EXAMPLE 15.20

Find a particular solution of the differential equation

$$y'' + 6y' + 4y = x^2 + 4.$$

SOLUTION   We write

$$y_p = \frac{1}{D^2 + 6D + 4}(x^2 + 4) = \frac{1}{4\left(1 + \dfrac{6D + D^2}{4}\right)}(x^2 + 4)$$

$$= \frac{1}{4}\left\{1 - \left(\frac{6D + D^2}{4}\right) + \left(\frac{6D + D^2}{4}\right)^2 - \cdots\right\}(x^2 + 4).$$

Since $D^n(x^2 + 4) = 0$ if $n > 2$, we require only the constant term and terms in D and D^2:

$$y_p = \frac{1}{4}\left\{1 - \frac{3}{2}D + 2D^2 + \cdots\right\}(x^2 + 4)$$

$$= \frac{1}{4}(x^2 + 4) - \frac{3}{8}(2x) + \frac{1}{2}(2) = \frac{x^2}{4} - \frac{3x}{4} + 2.$$

EXAMPLE 15.21

Find a particular solution of the differential equation

$$y''' + 2y' = x^2 - x.$$

SOLUTION   A particular solution is

$$y_p = \frac{1}{D^3 + 2D}(x^2 - x) = \frac{1}{D(2 + D^2)}(x^2 - x)$$

$$= \frac{1}{2D\left(1 + \dfrac{D^2}{2}\right)}(x^2 - x)$$

$$= \frac{1}{2D}\left\{1 - \frac{D^2}{2} + \cdots\right\}(x^2 - x)$$

$$= \frac{1}{2D}\left\{x^2 - x - \frac{1}{2}(2)\right\},$$

and since $1/D$ means to integrate with respect to x, we have

$$y_p = \frac{1}{2}\left\{\frac{x^3}{3} - \frac{x^2}{2} - x\right\} = \frac{x^3}{6} - \frac{x^2}{4} - \frac{x}{2}.$$

In summary, to evaluate 15.51 when $F(x)$ is a power x^n (or a polynomial), we expand $1/\phi(D)$ in a series of form 15.56 and perform the indicated differentiations (and integrations if $k > 0$).

For all other cases of $F(x)$, we make use of a theorem called the **inverse operator shift theorem**. This theorem states that

$$\frac{1}{\phi(D)}\{e^{px} f(x)\} = e^{px}\frac{1}{\phi(D + p)} f(x) \qquad (15.59)$$

(see Exercise 16). The theorem enables us to shift the exponential e^{px} past the operator $1/\phi(D)$, but to do so, the operator must be modified to $1/\phi(D+p)$. Equation 15.59 immediately yields $y_p(x)$ whenever $f(x)$ is a power x^n, n a nonnegative integer (or a polynomial). This is illustrated in the following example.

EXAMPLE 15.22

Find particular solutions for the following differential equations:

(a) $y'' + 3y' + 10y = x^2 e^{-x}$

(b) $y'' - 4y' + 4y = e^{2x}$

(c) $y''' + 2y'' + 3y' - y = e^{2x}$

SOLUTION

(a) Equation 15.59 gives

$$y_p = \frac{1}{D^2 + 3D + 10} x^2 e^{-x} = e^{-x} \frac{1}{(D-1)^2 + 3(D-1) + 10} x^2$$

$$= e^{-x} \frac{1}{D^2 + D + 8} x^2,$$

and we now proceed as in Example 15.20:

$$y_p = e^{-x} \frac{1}{8\left(1 + \dfrac{D + D^2}{8}\right)} x^2$$

$$= \frac{e^{-x}}{8} \left\{ 1 - \left(\frac{D + D^2}{8}\right) + \left(\frac{D + D^2}{8}\right)^2 - \cdots \right\} x^2$$

$$= \frac{e^{-x}}{8} \left\{ 1 - \frac{D}{8} - \frac{7D^2}{64} + \cdots \right\} x^2 = \frac{e^{-x}}{8} \left\{ x^2 - \frac{x}{4} - \frac{7}{32} \right\}.$$

(b) Once again we use the inverse operator shift theorem to get

$$y_p = \frac{1}{D^2 - 4D + 4} e^{2x} = e^{2x} \frac{1}{(D+2)^2 - 4(D+2) + 4}(1) \tag{1}$$

$$= e^{2x} \frac{1}{D^2}(1) = \frac{x^2}{2} e^{2x},$$

since $1/D^2$ means to integrate twice.

(c) For this differential equation,

$$y_p = \frac{1}{D^3 + 2D^2 + 3D - 1} e^{2x} = e^{2x} \frac{1}{(D+2)^3 + 2(D+2)^2 + 3(D+2) - 1}(1) \tag{1}$$

$$= e^{2x} \frac{1}{D^3 + 8D^2 + 23D + 21}(1) = e^{2x} \frac{1}{21\left(1 + \dfrac{23D + 8D^2 + D^3}{21}\right)}(1) \tag{1}$$

$$= \frac{e^{2x}}{21} \{1 + \cdots\}(1) = \frac{e^{2x}}{21}.$$

EXAMPLE 15.23

Find a general solution for $y''' - 3y'' + 3y' - y = 2x^2 e^x$.

SOLUTION   The auxiliary equation is

$$0 = m^3 - 3m^2 + 3m - 1 = (m - 1)^3$$

with solutions 1, 1, and 1. Thus,

$$y_h(x) = (C_1 + C_2 x + C_3 x^2)e^x.$$

A particular solution is

$$y_p(x) = \frac{2}{(D-1)^3}x^2 e^x = 2e^x \frac{1}{(D+1-1)^3}x^2 = 2e^x \frac{1}{D^3}x^2 = \frac{x^5}{30}e^x.$$

A general solution of the differential equation is therefore

$$y(x) = (C_1 + C_2 x + C_3 x^2)e^x + \frac{x^5}{30}e^x.$$

EXAMPLE 15.24

Find a particular solution for the differential equation

$$y''' + 3y'' - 4y = xe^{-2x} + x^2.$$

SOLUTION

$$y_p = \frac{1}{D^3 + 3D^2 - 4}(xe^{-2x} + x^2)$$

$$= e^{-2x}\frac{1}{(D-2)^3 + 3(D-2)^2 - 4}x + \frac{1}{D^3 + 3D^2 - 4}x^2$$

$$= e^{-2x}\frac{1}{D^3 - 3D^2}x - \frac{1}{4\left(1 - \dfrac{3D^2 + D^3}{4}\right)}x^2$$

$$= e^{-2x}\frac{1}{D^2}\frac{1}{D-3}x - \frac{1}{4}\left\{1 + \left(\frac{3D^2 + D^3}{4}\right) + \cdots\right\}x^2$$

$$= e^{-2x}\frac{1}{D^2}\frac{1}{-3\left(1 - \dfrac{D}{3}\right)}x - \frac{1}{4}\left(x^2 + \frac{3}{2}\right)$$

$$= \frac{e^{-2x}}{-3}\frac{1}{D^2}\left\{1 + \frac{D}{3} + \cdots\right\}x - \frac{x^2}{4} - \frac{3}{8}$$

$$= \frac{e^{-2x}}{-3}\frac{1}{D^2}\left\{x + \frac{1}{3}\right\} - \frac{x^2}{4} - \frac{3}{8}$$

$$= \frac{e^{-2x}}{-3}\left\{\frac{x^3}{6} + \frac{x^2}{6}\right\} - \frac{x^2}{4} - \frac{3}{8}.$$

When $F(x)$ in 15.51 is of the form $x^n \sin px$ or $x^n \cos px$, n a nonnegative integer and p a constant, we introduce complex exponentials. Specifically, because

$$x^n e^{ipx} = x^n (\cos px + i \sin px),$$

we can write

$$x^n \cos px = \text{real part of } x^n e^{ipx} = \text{Re}\{x^n e^{ipx}\}, \tag{15.60a}$$

$$x^n \sin px = \text{imaginary part of } x^n e^{ipx} = \text{Im}\{x^n e^{ipx}\}. \tag{15.60b}$$

To operate on either of these functions by $1/\phi(D)$, we interchange the operations of $1/\phi(D)$ and of taking real and imaginary parts:

$$\frac{1}{\phi(D)}\{x^n \cos px\} = \frac{1}{\phi(D)}\text{Re}\{x^n e^{ipx}\} = \text{Re}\left\{\frac{1}{\phi(D)}(x^n e^{ipx})\right\}, \tag{15.61a}$$

$$\frac{1}{\phi(D)}\{x^n \sin px\} = \frac{1}{\phi(D)}\text{Im}\{x^n e^{ipx}\} = \text{Im}\left\{\frac{1}{\phi(D)}(x^n e^{ipx})\right\}. \tag{15.61b}$$

We can now proceed by using inverse operator shift theorem 15.59.

EXAMPLE 15.25

Find particular solutions for the following differential equations:

(a) $y'' + y = \sin 2x$

(b) $y'' + 4y = x^2 \cos x$

(c) $y'' + 9y = \sin 3x$

(d) $y'' + 4y = x \sin 2x$

(e) $y'' + 2y' + 4y = e^{-x} \sin \sqrt{3}x$

SOLUTION

(a) Equation 15.61b gives

$$y_p = \frac{1}{D^2 + 1} \sin 2x = \frac{1}{D^2 + 1}\{\text{Im}(e^{2ix})\} = \text{Im}\left\{\frac{1}{D^2 + 1}e^{2ix}\right\}.$$

We now use inverse operator shift theorem 15.59,

$$y_p = \text{Im}\left\{e^{2ix} \frac{1}{(D + 2i)^2 + 1}(1)\right\}$$

$$= \text{Im}\left\{e^{2ix} \frac{1}{D^2 + 4iD - 3}(1)\right\} = \text{Im}\left\{e^{2ix} \frac{1}{-3\left(1 - \dfrac{4iD + D^2}{3}\right)}(1)\right\}$$

$$= \text{Im}\left\{\frac{e^{2ix}}{-3}\right\} = -\frac{1}{3}\sin 2x.$$

(b) We begin by using equation 15.61a,

$$y_p = \frac{1}{D^2 + 4}x^2 \cos x = \frac{1}{D^2 + 4}\mathrm{Re}(x^2 e^{ix}) = \mathrm{Re}\left\{\frac{1}{D^2 + 4}x^2 e^{ix}\right\}$$

$$= \mathrm{Re}\left\{e^{ix}\frac{1}{(D + i)^2 + 4}x^2\right\} = \mathrm{Re}\left\{e^{ix}\frac{1}{D^2 + 2iD + 3}x^2\right\}$$

$$= \mathrm{Re}\left\{e^{ix}\frac{1}{3\left(1 + \dfrac{2iD + D^2}{3}\right)}x^2\right\}$$

$$= \mathrm{Re}\left\{\frac{e^{ix}}{3}\left[1 - \left(\frac{2iD + D^2}{3}\right) + \left(\frac{2iD + D^2}{3}\right)^2 - \cdots\right]x^2\right\}$$

$$= \frac{1}{3}\mathrm{Re}\left\{e^{ix}\left[1 - \frac{2iD}{3} - \frac{7D^2}{9} + \cdots\right]x^2\right\} = \frac{1}{3}\mathrm{Re}\left\{e^{ix}\left[x^2 - \frac{4ix}{3} - \frac{14}{9}\right]\right\}$$

$$= \frac{1}{3}\left\{x^2 \cos x + \frac{4x}{3}\sin x - \frac{14}{9}\cos x\right\}.$$

(c) For the differential equation $y'' + 9y = \sin 3x$, we have

$$y_p = \frac{1}{D^2 + 9}\sin 3x = \frac{1}{D^2 + 9}\mathrm{Im}(e^{3ix}) = \mathrm{Im}\left\{\frac{1}{D^2 + 9}e^{3ix}\right\}$$

$$= \mathrm{Im}\left\{e^{3ix}\frac{1}{(D + 3i)^2 + 9}(1)\right\} = \mathrm{Im}\left\{e^{3ix}\frac{1}{D^2 + 6iD}(1)\right\}$$

$$= \mathrm{Im}\left\{e^{3ix}\frac{1}{D}\frac{1}{D + 6i}(1)\right\} = \mathrm{Im}\left\{\frac{e^{3ix}}{6i}\frac{1}{D}\frac{1}{1 + D/6i}(1)\right\}$$

$$= \mathrm{Im}\left\{-\frac{i}{6}e^{3ix}\frac{1}{D}\left(1 - \frac{D}{6i} + \cdots\right)(1)\right\} = \mathrm{Im}\left\{-\frac{i}{6}e^{3ix}\frac{1}{D}(1)\right\}$$

$$= \mathrm{Im}\left\{-\frac{i}{6}e^{3ix}x\right\} = -\frac{x}{6}\cos 3x.$$

(d) For the differential equation $y'' + 4y = x \sin 2x$, we get

$$y_p = \frac{1}{D^2 + 4}x \sin 2x = \frac{1}{D^2 + 4}\mathrm{Im}(xe^{2ix}) = \mathrm{Im}\left\{\frac{1}{D^2 + 4}xe^{2ix}\right\}$$

$$= \mathrm{Im}\left\{e^{2ix}\frac{1}{(D + 2i)^2 + 4}x\right\} = \mathrm{Im}\left\{e^{2ix}\frac{1}{D^2 + 4iD}x\right\}$$

$$= \mathrm{Im}\left\{e^{2ix}\frac{1}{D}\frac{1}{D + 4i}x\right\} = \mathrm{Im}\left\{\frac{1}{4i}e^{2ix}\frac{1}{D}\left(1 - \frac{D}{4i} + \cdots\right)x\right\}$$

$$= \mathrm{Im}\left\{-\frac{i}{4}e^{2ix}\frac{1}{D}\left(x - \frac{1}{4i}\right)\right\} = \mathrm{Im}\left\{-\frac{i}{4}e^{2ix}\left(\frac{x^2}{2} + \frac{ix}{4}\right)\right\}$$

$$= -\frac{x^2}{8}\cos 2x + \frac{x}{16}\sin 2x.$$

(e) For the differential equation $y'' + 2y' + 4y = e^{-x} \sin \sqrt{3}x$, we have

$$y_p = \frac{1}{D^2 + 2D + 4} e^{-x} \sin \sqrt{3}x = \frac{1}{D^2 + 2D + 4} \mathrm{Im}(e^{-x} e^{\sqrt{3}ix})$$

$$= \mathrm{Im}\left\{ \frac{1}{D^2 + 2D + 4} e^{(-1+\sqrt{3}i)x} \right\}$$

$$= \mathrm{Im}\left\{ e^{(-1+\sqrt{3}i)x} \frac{1}{(D - 1 + \sqrt{3}i)^2 + 2(D - 1 + \sqrt{3}i) + 4}(1) \right\}$$

$$= \mathrm{Im}\left\{ e^{(-1+\sqrt{3}i)x} \frac{1}{D^2 + 2\sqrt{3}i\,D}(1) \right\} = \mathrm{Im}\left\{ e^{(-1+\sqrt{3}i)x} \frac{1}{D} \frac{1}{D + 2\sqrt{3}i}(1) \right\}$$

$$= \mathrm{Im}\left\{ e^{(-1+\sqrt{3}i)x} \frac{1}{D} \frac{1}{2\sqrt{3}i} \right\} = \mathrm{Im}\left\{ -\frac{i}{2\sqrt{3}} x e^{(-1+\sqrt{3}i)x} \right\}$$

$$= -\frac{xe^{-x}}{2\sqrt{3}} \cos \sqrt{3}x.$$

EXAMPLE 15.26

Find a general solution of $y'' + 6y' + y = \sin 3x$.

SOLUTION   The auxiliary equation is $m^2 + 6m + 1 = 0$ with solutions $m = -3 \pm 2\sqrt{2}$. Consequently,

$$y_h(x) = C_1 e^{(-3+2\sqrt{2})x} + C_2 e^{(-3-2\sqrt{2})x}.$$

A particular solution is

$$y_p = \frac{1}{D^2 + 6D + 1} \mathrm{Im}(e^{3ix}) = \mathrm{Im}\left\{ \frac{1}{D^2 + 6D + 1} e^{3ix} \right\}$$

$$= \mathrm{Im}\left\{ e^{3ix} \frac{1}{(D + 3i)^2 + 6(D + 3i) + 1}(1) \right\}$$

$$= \mathrm{Im}\left\{ e^{3ix} \frac{1}{D^2 + (6 + 6i)D + (-8 + 18i)}(1) \right\} = \mathrm{Im}\left\{ e^{3ix} \frac{1}{-8 + 18i} \right\}$$

$$= \mathrm{Im}\left\{ e^{3ix} \frac{1}{-8 + 18i} \frac{-8 - 18i}{-8 - 18i} \right\} = -\frac{1}{194}(4 \sin 3x + 9 \cos 3x).$$

Finally, then, a general solution of the differential equation is

$$y(x) = C_1 e^{(-3+2\sqrt{2})x} + C_2 e^{(-3-2\sqrt{2})x} - \frac{1}{194}(4 \sin 3x + 9 \cos 3x).$$

Examples 15.20–15.26 have illustrated that with identity 15.59 and the concept of series, we can obtain a particular solution for 15.50 whenever $F(x)$ is x^n, e^{px}, $\sin px$, $\cos px$, and/or any sums or products thereof. In summary:

when $F(x) = x^n$, use series 15.56 for $1/\phi(D)$.

In any other situation, use 15.59, and then series 15.56 for $1/\phi(D + p)$.

We make one final comment. We introduced complex exponentials e^{ipx} to handle terms involving $\sin px$ and $\cos px$. This is not the only way to treat trigonometric functions, since there do exist other methods that completely avoid complex numbers. Unfortunately, these methods require memorization of somewhat involved identities. Because of this, we prefer the use of complex numbers.

EXERCISES 15.9

In Exercises 1–11 find a particular solution for the differential equation, by both the method of operators and the method of undetermined coefficients. Find a general solution of the equation.

1. $2y'' - 16y' + 32y = -e^{4x}$

2. $y'' + 2y' - 2y = x^2 e^{-x}$

3. $y''' - 3y'' + y' - 3y = 3xe^x + 2$

4. $y'''' + 2y'' + y = \cos 2x$

5. $y''' - 6y'' + 12y' - 8y = 2e^{2x}$

6. $y'''' + 5y'' + 4y = e^{-2x}$

7. $y''' - 3y'' + 2y' = x^2 + e^{-x}$

8. $2y'' + 16y' + 82y = -2e^{2x} \sin x$

9. $y'' + y' - 6y = x + \cos x$

10. $y'' - 4y' + 5y = x \cos x$

11. $3y''' - 12y'' + 18y' - 12y = x^2 + 3x - 4$

In Exercises 12–15 state the form for the particular solution predicted by the method of undetermined coefficients. Do not evaluate the coefficients.

12. $y''' + 9y'' + 27y' + 27y = xe^{3x} + 2x \cos x$

13. $y''' + 4y'' + y' + 4y = xe^x \sin x$

14. $2y''' - 6y'' - 12y' + 16y = xe^x + 2x^3 - 4\cos x$

15. $2y'' - 4y' + 10y = 5e^x \sin 2x$

16. Use the operator shift theorem of Exercise 25 in Section 15.8 to verify the inverse operator shift theorem 15.59.

In Exercises 17–18 find a general solution for the differential equation.

17. $y'' + 2y' - 4y = \cos^2 x$

18. $2y'' - 4y' + 3y = \cos x \sin 2x$

In Exercises 19–20 find a solution for the differential equation.

19. $y'' - 3y' + 2y = 8x^2 + 12e^{-x}, \ y(0) = 0, \ y'(0) = 2$

20. $y'' + 9y = x(\sin 3x + \cos 3x), \ y(0) = y'(0) = 0$

21. If J, k, and w are positive constants, find a general solution for
$$J\frac{d^4 y}{dx^4} + ky = w.$$

22. The second-order linear differential equation
$$x^2\frac{d^2 y}{dx^2} + ax\frac{dy}{dx} + by = F(x), \qquad a, b \text{ constants,}$$

is called the *Cauchy–Euler linear equation*. Because of the x^2- and x-factors, it does not have constant coefficients, and is therefore not immediately amenable to the techniques of this chapter. Show that if we make a change of independent variable $x = e^z$, then

$$x\frac{dy}{dx} = \frac{dy}{dz}, \qquad x^2\frac{d^2 y}{dx^2} = \frac{d^2 y}{dz^2} - \frac{dy}{dz},$$

and that as a result the Cauchy–Euler equation is transformed into a linear equation in $y(z)$ with constant coefficients.

In Exercises 23–24 use the technique of Exercise 22 to find a general solution for the differential equation.

∗ 23. $\dfrac{d^2 u}{dr^2} + \dfrac{1}{r}\dfrac{du}{dr} - \dfrac{u}{r^2} = 0, \ r > 0$

∗ 24. $x^2 y'' + xy' + 4y = 1, \ x > 0$

∗∗ 25. If M, β, k, A, and ω are all positive constants, find a particular solution of the linear differential equation
$$M\frac{d^2 x}{dt^2} + \beta\frac{dx}{dt} + kx = A \sin \omega t.$$

∗∗ 26. We have assumed that the nonhomogeneity in 15.48a is continuous on the interval I on which the differential equation is to be considered. It is possible to adapt our procedure to nonhomogeneities that are piecewise continuous. For example, consider the initial-value problem
$$y'' - 3y' + 2y = F(x), \qquad y(0) = 2, \quad y'(0) = -1/2,$$
where
$$F(x) = \begin{cases} x, & 0 \le x \le 1, \\ 0, & x > 1 \end{cases}$$
has a discontinuity at $x = 1$. Solve this problem using the following method.

(a) Find the solution to the initial-value problem on the interval $0 \le x \le 1$.

(b) Find a general solution for $x > 1$.

(c) Choose the constants in the solution to part (b) so that the solution and its first derivative are continuous at $x = 1$.

(d) Does the function in part (c) satisfy the differential equation for all $x > 0$?

∗∗ 27. Use the method in Exercise 26 to find the solution, with a continuous first derivative, for the initial-value problem
$$y'' + y = F(x), \qquad y(0) = 0, \quad y'(0) = 0,$$
where
$$F(x) = \begin{cases} x - 1, & 0 \le x \le \pi, \\ e^{-x}, & x > \pi. \end{cases}$$

** **28.** The biharmonic differential equation for a function $\Phi(r)$ arises in two-dimensional elasticity theory:

$$r^3 \frac{d^4\Phi}{dr^4} + 2r^2 \frac{d^3\Phi}{dr^3} - r \frac{d^2\Phi}{dr^2} + \frac{d\Phi}{dr} = 0.$$

Extend the technique of Exercise 22 to show that a general solution of this equation is

$$\Phi(r) = C_1 + C_2 \ln r + C_3 r^2 + C_4 r^2 \ln r.$$

** **29.**   (a) Show that if a function $\Phi(r) = f(r) \cos n\theta$, where $n \geq 1$ is an integer, is to satisfy the biharmonic equation of Exercise 34 in Section 12.6, then $f(r)$ must satisfy

$$\frac{d^4 f}{dr^4} + \frac{2}{r} \frac{d^3 f}{dr^3} - \frac{1 + 2n^2}{r^2} \frac{d^2 f}{dr^2} + \frac{1 + 2n^2}{r^3} \frac{df}{dr} + \frac{n^4 - 4n^2}{r^4} f = 0.$$

(b) Extend the result of Exercise 22 to find all solutions for $f(r)$. Consider the cases $n = 1$ and $n > 1$ separately.

15.10 Applications of Linear Differential Equations

Vibrating Mass-Spring Systems

In Figure 15.10 we have shown a mass M suspended vertically from a spring. If M is given an initial motion in the vertical direction (and the vertical direction only), then we expect M to oscillate up and down for some time. In this section we show how to describe these oscillations mathematically.

In order to describe the position of M as a function of time t, we must choose a vertical coordinate system. There are two natural places to choose the origin $y = 0$, one being the position of M when the spring is unstretched. Suppose we do this and choose y as positive upward. When M is a distance y away from the origin, the restoring force of the spring has y-component $-ky$ ($k > 0$). In addition, if $g = 9.81$ is the acceleration due to gravity, then the force of gravity on M has y-component $-Mg$. Finally, suppose oscillations take place in a medium that exerts a damping force proportional to the instantaneous velocity of M. This damping force must therefore have a y-component of the form $-\beta(dy/dt)$, where β is a positive constant. The total force on M therefore has y-component

$$-ky - Mg - \beta \frac{dy}{dt},$$

and Newton's second law states that the acceleration d^2y/dt^2 of M must satisfy the equation

$$-ky - Mg - \beta \frac{dy}{dt} = M \frac{d^2y}{dt^2}.$$

Consequently, the differential equation that determines the position $y(t)$ of M relative to the unstretched position of the spring is

$$M \frac{d^2y}{dt^2} + \beta \frac{dy}{dt} + ky = -Mg. \tag{15.62}$$

The alternative possibility for describing oscillations is to attach M to the spring and slowly lower M until it reaches an equilibrium position. At this position, the restoring force of the spring is exactly equal to the force of gravity on the mass, and the mass, left by itself, will remain motionless. If s is the amount of stretch in the spring at equilibrium, then at equilibrium

$$ks - Mg = 0, \quad \text{where } s > 0. \tag{15.63}$$

Suppose we take the equilibrium position as $x = 0$ and x as positive upward (Figure 15.11). When M is a distance x away from its equilibrium position, the restoring force on M has

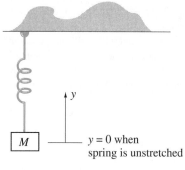

FIGURE 15.10 Displacement of vibrating mass relative to unstretched position of spring

y

M — $y = 0$ when spring is unstretched

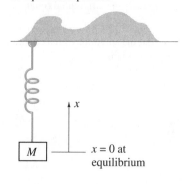

FIGURE 15.11 Displacement of vibrating mass relative to its equilibrium position

x

M — $x = 0$ at equilibrium

x-component $k(s - x)$. The x-component of the force of gravity remains as $-Mg$, and that of the damping force is $-\beta(dx/dt)$. Newton's second law therefore implies that

$$M\frac{d^2x}{dt^2} = k(s - x) - Mg - \beta\frac{dx}{dt}$$

or

$$M\frac{d^2x}{dt^2} + \beta\frac{dx}{dt} + kx = -Mg + ks.$$

But according to 15.63, $-Mg + ks = 0$, and hence

$$M\frac{d^2x}{dt^2} + \beta\frac{dx}{dt} + kx = 0. \tag{15.64}$$

This is the differential equation describing the displacement $x(t)$ of M relative to the equilibrium position of M.

Note that both equations 15.62 and 15.64 are linear second-order differential equations with constant coefficients. The advantage of 15.64 is that it is homogeneous as well, and this is simply due to a convenient choice of dependent variable (x as opposed to y). Physically, we are saying that there are two parts to the spring force $k(s - x)$: a part ks and a part $-kx$. Gravity is always acting on M, and that part ks of the spring force is counteracting it in an attempt to restore the spring to its unstretched position. Because these forces always cancel, we might just as well eliminate both of them from our discussion. This would leave us $-kx$, and we therefore interpret $-kx$ as the *spring force attempting to restore the mass to its equilibrium position*.

If we choose equation 15.64 to describe the motion of M (and this equation is usually chosen over equation 15.62), we must remember three things: x is measured from equilibrium, $-kx$ is the spring force attempting to restore M to its equilibrium position, and gravity has been taken into account.

There are three basic ways to initiate the motion. First, we can move the mass away from its equilibrium position and then release it, giving it an initial displacement but no initial velocity. Second, we can strike the mass at the equilibrium position, imparting an initial velocity but no initial displacement. And finally, we can give the mass both an initial displacement and an initial velocity. Each of these methods adds two initial conditions to the differential equation.

To be complete we note that when (in addition to the forces already mentioned) there is an externally applied force acting on the mass that is represented as a function of time by $F(t)$, then equation 15.64 is modified to

$$M\frac{d^2x}{dt^2} + \beta\frac{dx}{dt} + kx = F(t). \tag{15.65}$$

Perhaps, for example, M contains some iron, and $F(t)$ is due to a magnet directly below M that exerts a time-dependent attractive force on M.

EXAMPLE 15.27

A 2-kg mass is suspended vertically from a spring with constant 16 N/m. The mass is raised 10 cm above its equilibrium position and then released. If damping is ignored, find the amplitude, period, and frequency of the motion.

SOLUTION   If we choose $x = 0$ at the equilibrium position of the mass and x positive upward (Figure 15.11), differential equation 15.64 for motion $x(t)$ of the mass is

$$2\frac{d^2x}{dt^2} + 16x = 0$$

or

$$\frac{d^2x}{dt^2} + 8x = 0,$$

along with the initial conditions

$$x(0) = 1/10, \quad x'(0) = 0.$$

The auxiliary equation is $m^2 + 8 = 0$ with solutions $m = \pm 2\sqrt{2}\,i$. Consequently,

$$x(t) = C_1 \cos(2\sqrt{2}\,t) + C_2 \sin(2\sqrt{2}\,t).$$

The initial conditions require that

$$1/10 = C_1, \quad 0 = 2\sqrt{2}\,C_2.$$

Thus,

$$x(t) = \frac{1}{10}\cos(2\sqrt{2}\,t).$$

The amplitude of the oscillations is 1/10 m, the period is $2\pi/(2\sqrt{2}) = \pi/\sqrt{2}$ s, and the frequency is $\sqrt{2}/\pi$ s^{-1}. A graph of this function (Figure 15.12) illustrates the oscillations of the mass about its equilibrium position. This is an example of **simple harmonic motion**.

FIGURE 15.12 Simple harmonic motion of vibrating mass when damping is negligible

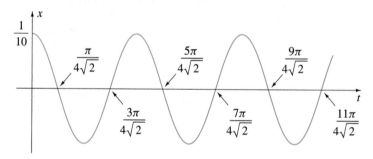

EXAMPLE 15.28

A 100-g mass is suspended vertically from a spring with constant 5 N/m. The mass is pulled 5 cm below its equilibrium position and given velocity 2 m/s upward. If during the motion the mass is acted on by a damping force in newtons numerically equal to one-twentieth the instantaneous velocity in metres per second, find the position of the mass at any time.

SOLUTION   If we choose $x = 0$ at the equilibrium position of the mass and x positive upward (Figure 15.11), then differential equation 15.64 for motion $x(t)$ of the mass is

$$\frac{1}{10}\frac{d^2x}{dt^2} + \frac{1}{20}\frac{dx}{dt} + 5x = 0$$

or

$$2\frac{d^2x}{dt^2} + \frac{dx}{dt} + 100x = 0,$$

along with the initial conditions

$$x(0) = -1/20, \quad x'(0) = 2.$$

The auxiliary equation is $2m^2 + m + 100 = 0$ with solutions

$$m = \frac{-1 \pm \sqrt{1 - 800}}{4} = \frac{-1 \pm \sqrt{799}\,i}{4}.$$

Consequently,

$$x(t) = e^{-t/4}[C_1 \cos(\sqrt{799}\,t/4) + C_2 \sin(\sqrt{799}\,t/4)].$$

The initial conditions require that

$$-1/20 = C_1, \quad 2 = -C_1/4 + \sqrt{799}\,C_2/4,$$

from which we get

$$C_2 = \frac{159\sqrt{799}}{15\,980}.$$

Finally, then,

$$x(t) = e^{-t/4}\left[-\frac{1}{20}\cos\left(\frac{\sqrt{799}\,t}{4}\right) + \frac{159\sqrt{799}}{15\,980}\sin\left(\frac{\sqrt{799}\,t}{4}\right)\right].$$

The graph of this function in Figure 15.13 clearly indicates how the amplitude of the oscillations decreases in time.

FIGURE 15.13   Displacement of mass when damping is present

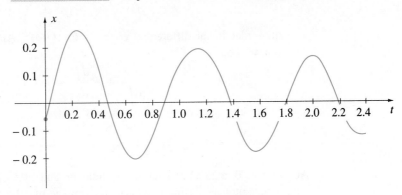

EXAMPLE 15.29

Application Preview Revisited

FIGURE 15.14   Displacement of front end of a car as wheel follows uneven road

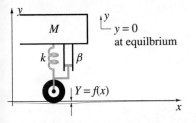

If the car in the Application Preview moves with constant speed v, and the equation of the road is $Y = 0$ for $x < 0$, and $Y = f(x)$ for $x \geq 0$, find an initial-value problem describing the vertical oscillations of the front end of the car.

SOLUTION   We assume that the four wheels act independently and that M is the mass of that part of the car supported by the wheel (Figure 15.14). Let y measure the height of the car above its equilibrium position when it is motionless. Newton's second law for the motion of M gives

$$M\frac{d^2y}{dt^2} = -\beta\frac{dy}{dt} - k(y - Y - s) - Mg,$$

where $g = 9.81$ and s is the compression in the spring when the car is at equilibrium. Because $ks - Mg = 0$ at equilibrium, it follows that

$$M\frac{d^2y}{dt^2} + \beta\frac{dy}{dt} + ky = kY.$$

Since the speed of the car is v, the x-coordinate of the car is $x = vt$ (taking $t = 0$ when it passes through the origin), and therefore

$$M\frac{d^2y}{dt^2} + \beta\frac{dy}{dt} + ky = kf(vt), \quad t \geq 0.$$

The solution would also be subject to the initial conditions $y(0) = y'(0) = 0$.

LCR **Circuits**

FIGURE 15.15   Schematic for current in LCR-circuit

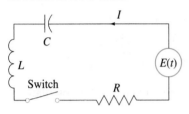

If a resistance R, an inductance L, and a capacitance C are connected in series with an electromotive force $E(t)$ (Figure 15.15) and the switch is closed, current flows in the circuit and charge builds up in the capacitor. If at any time t, Q is the charge on the capacitor and I is the current in the loop, then Kirchhoff's law states that

$$L\frac{dI}{dt} + RI + \frac{Q}{C} = E(t), \tag{15.66}$$

where $L\,dI/dt$, RI, and Q/C represent the voltage drops across the inductor, the resistor, and the capacitor, respectively. If we substitute $I = dQ/dt$, then

$$L\frac{d^2Q}{dt^2} + R\frac{dQ}{dt} + \frac{1}{C}Q = E(t), \tag{15.67}$$

a second-order linear differential equation for $Q(t)$. Alternatively, if we differentiate this equation, we obtain

$$L\frac{d^2I}{dt^2} + R\frac{dI}{dt} + \frac{1}{C}I = E'(t), \tag{15.68}$$

a second-order linear differential equation for $I(t)$.

EXAMPLE 15.30

FIGURE 15.16   Charge on the capacitor and current in an LCR-circuit

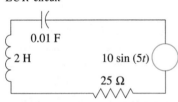

At time $t = 0$, a 25 Ω resistor, a 2 H inductor, and a 0.01 F capacitor are connected in series with a generator producing an alternating voltage of $10\sin(5t)$, $t \geq 0$ (Figure 15.16). Find the charge on the capacitor and the current in the circuit if the capacitor is uncharged when the circuit is closed.

SOLUTION   Differential equation 15.67 for the charge Q on the capacitor is

$$2\frac{d^2Q}{dt^2} + 25\frac{dQ}{dt} + 100Q = 10\sin(5t),$$

to which we add the initial conditions

$$Q(0) = 0, \quad Q'(0) = I(0) = 0.$$

The auxiliary equation is $2m^2 + 25m + 100 = 0$ with solutions

$$m = \frac{-25 \pm \sqrt{625 - 800}}{4} = \frac{-25 \pm 5\sqrt{7}i}{4}.$$

Consequently, a general solution of the homogeneous equation is

$$Q_h(t) = e^{-25t/4}[C_1\cos(5\sqrt{7}t/4) + C_2\sin(5\sqrt{7}t/4)].$$

To find a particular solution of the nonhomogeneous equation by undetermined coefficients, we set

$$Q_p(t) = A \sin(5t) + B \cos(5t).$$

Substitution of this function into the differential equation gives

$$2\{-25A \sin(5t) - 25B \cos(5t)\} + 25\{5A \cos(5t) - 5B \sin(5t)\}$$
$$+ 100\{A \sin(5t) + B \cos(5t)\} = 10 \sin(5t).$$

This equation requires A and B to satisfy

$$50A - 125B = 10, \quad 125A + 50B = 0,$$

the solution of which is $A = 4/145$, $B = -10/145$. A particular solution is therefore

$$Q_p(t) = \frac{2}{145}\left[2 \sin(5t) - 5 \cos(5t)\right],$$

and a general solution is

$$Q(t) = Q_h(t) + Q_p(t)$$
$$= e^{-25t/4}\left[C_1 \cos\left(\frac{5\sqrt{7}t}{4}\right) + C_2 \sin\left(\frac{5\sqrt{7}t}{4}\right)\right] + \frac{2}{145}[2 \sin(5t) - 5 \cos(5t)].$$

The initial conditions require that

$$0 = C_1 - \frac{10}{145}, \quad 0 = -\frac{25}{4}C_1 + \frac{5\sqrt{7}}{4}C_2 + \frac{20}{145},$$

and these imply that $C_1 = 10/145$, $C_2 = 34/(145\sqrt{7})$. Consequently,

$$Q(t) = \frac{e^{-25t/4}}{145\sqrt{7}}\left[10\sqrt{7} \cos\left(\frac{5\sqrt{7}t}{4}\right) + 34 \sin\left(\frac{5\sqrt{7}t}{4}\right)\right]$$
$$+ \frac{2}{145}[2 \sin(5t) - 5 \cos(5t)].$$

The current in the circuit is

$$I(t) = \frac{dQ}{dt} = \left(-\frac{25}{4}\right)\frac{e^{-25t/4}}{145\sqrt{7}}\left[10\sqrt{7} \cos\left(\frac{5\sqrt{7}t}{4}\right) + 34 \sin\left(\frac{5\sqrt{7}t}{4}\right)\right]$$
$$+ \frac{e^{-25t/4}}{145\sqrt{7}}\left[-\frac{175}{2} \sin\left(\frac{5\sqrt{7}t}{4}\right) + \frac{85\sqrt{7}}{2} \cos\left(\frac{5\sqrt{7}t}{4}\right)\right]$$
$$+ \frac{2}{145}[10 \cos(5t) + 25 \sin(5t)]$$
$$= -\frac{e^{-25t/4}}{29\sqrt{7}}\left[4\sqrt{7} \cos\left(\frac{5\sqrt{7}t}{4}\right) + 60 \sin\left(\frac{5\sqrt{7}t}{4}\right)\right]$$
$$+ \frac{2}{29}[2 \cos(5t) + 5 \sin(5t)].$$

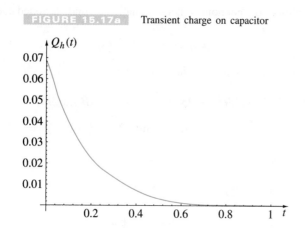

FIGURE 15.17a Transient charge on capacitor

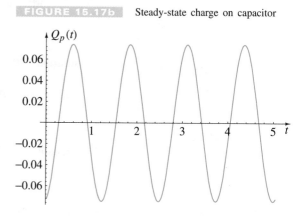

FIGURE 15.17b Steady-state charge on capacitor

The solution $Q(t)$ contains two parts. The first two terms (containing the exponential $e^{-25t/4}$) are $Q_h(t)$ with the constants C_1 and C_2 determined by the initial conditions; the last two terms are $Q_p(t)$. We point this out because the two parts display completely different characteristics. Their graphs are shown in Figures 15.17. For small t, both parts of $Q(t)$ are present and contribute significantly, but for large t, the first two terms become negligible. In other words, after a long time, the charge $Q(t)$ on the capacitor is defined essentially by $Q_p(t)$. We call $Q_p(t)$ the steady-state part of the solution, and the two other terms in $Q(t)$ are called the transient part of the solution. Similarly, the first two terms in $I(t)$ are called the transient part of the current and the last two terms the steady-state part of the current.

Finally, note that the frequency of the steady-state part of either $Q(t)$ or $I(t)$ is exactly that of the forcing voltage $E(t)$.

The similarity between differential equations 15.65 and 15.67 cannot go unmentioned:

$$M\frac{d^2x}{dt^2} + \beta\frac{dx}{dt} + kx = F(t),$$

$$L\frac{d^2Q}{dt^2} + R\frac{dQ}{dt} + \frac{1}{C}Q = E(t).$$

Each of the coefficients M, β, and k for the mechanical system has its analogue L, R, and $1/C$ in the electrical system.

Consulting Project 26

Resonance in a physical system is something that can be detrimental to the system, or it can be advantageous. Resonance in a vibrating mass-spring system is to be avoided, it may destroy the system; resonance in some electrical networks is a necessity. Resonance occurs when a periodic force is applied to an oscillating system, and the frequency of the force is chosen to maximize the amplitude of the oscillations. Sometimes the maximum amplitude can be excessively large; other times, it is less dramatic. We have been asked to analyze resonance for current in an LCR-circuit (Figure 15.15) with an applied voltage $E(t) = F \sin \omega t$. In particular, we are to determine the value of ω that yields resonance and the resulting maximum amplitude. We are also to find the frequencies at which the amplitude of the current falls to $p\%$ of its resonance value.

SOLUTION Example 15.30 showed that current in an LCR-circuit has two components, a transient part, which dies out quickly, and a steady-state part that persists forever. Furthermore, the steady-state part is the particular solution of the differential equation defining the current, namely, equation 15.68,

$$L\frac{d^2i}{dt^2} + R\frac{di}{dt} + \frac{i}{C} = \omega F \cos \omega t.$$

If we drop the "p"-subscript for particular solutions and let $i(t) = B \cos \omega t + C \sin \omega t$ be the particular (steady-state) solution of this differential equation, then substituting into the differential equation gives

$$L(-\omega^2 B \cos \omega t - \omega^2 C \sin \omega t) + R(-\omega B \sin \omega t + \omega C \cos \omega t)$$

$$+ \frac{1}{C}(B \cos \omega t + C \sin \omega t) = \omega F \cos \omega t.$$

When we equate coefficients of terms in $\cos \omega t$ and $\sin \omega t$, we obtain

$$\left(\frac{1}{C} - \omega^2 L\right) B + \omega R C = \omega F, \qquad -\omega R B + \left(\frac{1}{C} - \omega^2 L\right) C = 0.$$

Solutions are

$$B = \frac{\left(\dfrac{1}{\omega C} - \omega L\right) F}{\left(\dfrac{1}{\omega C} - \omega L\right)^2 + R^2}, \qquad C = \frac{-RF}{\left(\dfrac{1}{\omega C} - \omega L\right)^2 + R^2}.$$

Consequently,

$$i(t) = \frac{\left(\dfrac{1}{\omega C} - \omega L\right) F \cos \omega t - RF \sin \omega t}{\left(\dfrac{1}{\omega C} - \omega L\right)^2 + R^2}.$$

According to formula 1.52 in Section 1.7, the amplitude of this oscillating current is

$$A = \sqrt{\left[\frac{\left(\dfrac{1}{\omega C} - \omega L\right) F}{\left(\dfrac{1}{\omega C} - \omega L\right)^2 + R^2}\right]^2 + \left[\frac{-RF}{\left(\dfrac{1}{\omega C} - \omega L\right)^2 + R^2}\right]^2}$$

$$= \frac{F}{\sqrt{\left(\dfrac{1}{\omega C} - \omega L\right)^2 + R^2}}.$$

The amplitude is a maximum when the denominator is a minimum, and this occurs when

$$\frac{1}{\omega C} - \omega L = 0 \qquad \Longrightarrow \qquad \omega = \frac{1}{\sqrt{LC}}.$$

This is the frequency giving resonance, and the resonance amplitude is F/R. Notice that the resonance frequency is independent of the resistance in the circuit; it depends on the inductance and capacitance. On the other hand, the resonance amplitude depends only on resistance. The smaller the resistance, the larger the resonance amplitude.

Amplitude of the current will fall to $p\%$ of peak value when

$$\frac{F}{\sqrt{\left(\dfrac{1}{\omega C} - \omega L\right)^2 + R^2}} = \frac{p}{100}\left(\frac{F}{R}\right) \implies R^2 = \frac{p^2}{10^4}\left[\left(\frac{1}{\omega C} - \omega L\right)^2 + R^2\right].$$

This equation reduces to

$$LC\omega^2 \pm \sqrt{1 - p^2/10^4}\,RC\omega - 1 = 0,$$

a quadratic in ω. Solutions are

$$\omega = \frac{\pm\sqrt{1 - p^2/10^4}\,RC + \sqrt{(1 - p^2/10^4)R^2C^2 + 4LC}}{2LC}.$$

Resonance for mechanical systems is discussed in the exercises; it is somewhat different than that for the electrical circuit.

Buckling Loads for Beams

Figure 15.18a shows a uniform beam of length L that is pinned at each end. A compressive force of P newtons is applied to each end of the beam along the axis of the beam. We assume that gravity can be ignored so that for small P, the beam will be horizontal. If P is allowed to increase, a stage will be reached at which the column buckles slightly upward (or downward) (Figure 15.18b). We can find the smallest force that causes buckling and the shape of the deflected curve.

FIGURE 15.18a   Compressive force P applied to ends of a beam

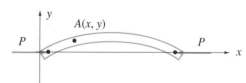

FIGURE 15.18b   Buckling resulting from compressive load

It is shown in the branch of engineering known as strength of materials that for small deflections, the concavity of the deflected curve at a point (x, y) is proportional to the moment of P about A,

$$\frac{d^2y}{dx^2} = \frac{M}{EI} = -\frac{Py}{EI},$$

where E is a constant called *Young's modulus of elasticity*. It depends on the material of the beam. Constant I is the moment of inertia of cross-sections of the beam. If we set $k^2 = P/(EI)$, then

$$\frac{d^2y}{dx^2} + k^2y = 0,$$

a second-order, homogeneous, linear differential equation with constant coefficients. With roots $\pm ki$ of the auxiliary equation, a general solution is

$$y(x) = C_1 \cos kx + C_2 \sin kx.$$

Since deflection at the left end $x = 0$ of the beam is $y = 0$, it follows that $C_1 = 0$. Similarly, because $y(L) = 0$, we must also have $0 = C_2 \sin kL$. We cannot set $C_2 = 0$ [else $y(x) \equiv 0$], so that $\sin kL = 0 \Longrightarrow kL = n\pi$, where $n \neq 0$ is an integer. Consequently,

$$k^2 = \frac{n^2\pi^2}{L^2} = \frac{P}{EI} \quad \Longrightarrow \quad P = \frac{n^2\pi^2 EI}{L^2}.$$

The smallest value of the load for which deflection occurs is $n = 1$. This load $P = \pi^2 EI/L^2$ is called **Euler's buckling load** for a pin-ended column. The deflected shape for this load is a sine curve

$$y(x) = C_2 \sin \frac{\pi x}{L}.$$

It is not possible to find the value of C_2.

Euler's buckling load depends on the boundary conditions at the ends of the beam. Other boundary conditions are discussed in the exercises.

EXERCISES 15.10

1. A 1-kg mass is suspended vertically from a spring with constant 16 N/m. The mass is pulled 10 cm below its equilibrium position, and then released. Find the position of the mass, relative to its equilibrium position, at any time if (a) damping is ignored, (b) a damping force in newtons equal to one-tenth the instantaneous velocity in metres per second acts on the mass, and (c) a damping force in newtons equal to ten times the instantaneous velocity in metres per second acts on the mass.

2. A 200-g mass suspended vertically from a spring with constant 10 N/m is set into vibration by an external force in newtons given by $4 \sin 10t$, $t \geq 0$. During the motion a damping force in newtons equal to 3/2 the velocity of the mass in metres per second acts on the mass. Find the position of the mass as a function of time t.

3. A 0.001 F capacitor and a 2 H inductor are connected in series with a 20 V battery. If there is no charge on the capacitor before the battery is connected, find the current in the circuit as a function of time.

4. At time $t = 0$, a 0.02 F capacitor, a 100 Ω resistor, and 1 H inductor are connected in series. If the charge on the capacitor is initially 5 C, find its charge as a function of time.

5. A 5 H inductor and 20 Ω resistor are connected in series with a generator supplying an oscillating voltage of $10 \sin 2t$, $t \geq 0$. What are the transient and steady-state currents in the circuit?

6. At time $t = 0$, a mass M is attached to the end of a hanging spring with constant k, and then released. Assuming that damping is negligible, find the subsequent displacement of the mass as a function of time.

7. A 0.5-kg mass sits on a table attached to a spring with constant 18 N/m (figure follows). The mass is pulled so as to stretch the spring 5 cm and is then released.

(a) If friction between the mass and the table creates a force of 0.5 N that opposes motion, show that the differential equation determining motion is

$$\frac{d^2x}{dt^2} + 36x = 1, \quad x(0) = 0.05, \quad x'(0) = 0.$$

(b) Find where the mass comes to rest for the first time. Will it move from this position?

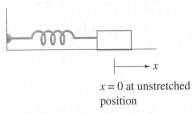

$x = 0$ at unstretched
position

* 8. Repeat Exercise 7 given that the mass is pulled 25 cm to the right.

* 9. At time $t = 0$ an uncharged 0.1 F capacitor is connected in series with a 0.5 H inductor and a 3 Ω resistor. If the current in the circuit at this instant is 1 A, find the maximum charge that the capacitor stores.

* 10. Differential equation 15.64 describes the motion of a mass M at the end of a spring, taking damping proportional to velocity into account. Show each of the following:

(a) If $\beta = 0$,

$$x(t) = C_1 \cos \left(\sqrt{k/M}\, t\right) + C_2 \sin \left(\sqrt{k/M}\, t\right),$$

called *simple harmonic motion*.

(b) If $\beta \neq 0$ and $\beta^2 - 4kM < 0$,

$$x(t) = e^{-\beta t/(2M)}(C_1 \cos \omega t + C_2 \sin \omega t), \quad \text{where } \omega = \frac{\sqrt{4kM - \beta^2}}{2M},$$

called *damped oscillatory motion*.

(c) If $\beta \neq 0$ and $\beta^2 - 4kM > 0$,

$$x(t) = e^{-\beta t/(2M)}(C_1 e^{\omega t} + C_2 e^{-\omega t}), \quad \text{where } \omega = \frac{\sqrt{\beta^2 - 4kM}}{2M},$$

called *overdamped motion*.

(d) If $\beta \neq 0$ and $\beta^2 - 4kM = 0$,

$$x(t) = (C_1 + C_2 t)e^{-\beta t/(2M)},$$

called *critically damped motion*.

* **11.** In the figure below, the piston is caused to move by changing the pressure P of the gas by way of the valve. The spring restrains the motion of the piston, as does the dashpot, the latter exerting a damping force proportional to the velocity of the piston. If the mass of the piston is m, and A is its surface area exposed to the gas, the differential equation for motion of the piston is

$$m\frac{d^2x}{dt^2} + \beta\frac{dx}{dt} + kx = AP(t).$$

Find the steady-state solution when $P(t) = P_0 \sin \omega t$, where $P_0 > 0$ and $\omega > 0$ are constants.

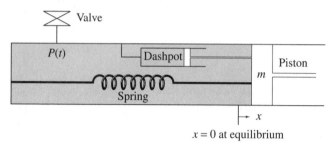

$x = 0$ at equilibrium

* **12.** A 100-g mass is suspended from a spring with constant 4000 N/m. At its equilibrium position, it is suddenly (time $t = 0$) given an upward velocity of 10 m/s. If an external force $3 \cos 200t$, $t \geq 0$, acts on the mass, find its displacement as a function of time. Does resonance occur?

* **13.** A vertical spring having constant 64 N/m has a 1-kg mass attached to it. An external force $F(t) = 2 \sin 8t$, $t \geq 0$ is applied to the mass. If the mass is at rest at its equilibrium position at time $t = 0$, and damping is negligible, find the position of the mass as a function of time. What happens to the oscillations as time progresses?

* **14.** A mass M is suspended from a vertical spring with constant k. If an external force $F(t) = A \cos \omega t$, $t \geq 0$, is applied to the mass, find the value of ω that causes resonance.

* **15.** A 25/9 H inductor, a 0.04 F capacitor, and a generator with voltage $15 \cos 3t$ are connected in series at time $t = 0$. Find the current in the circuit as a function of time. Does resonance occur?

* **16.** A mass M is suspended from a vertical spring with constant k. When an external force $F(t) = A \cos \omega t$, $t \geq 0$ is applied to the mass, and a damping force proportional to velocity acts on the mass during its subsequent oscillations, the differential equation governing the motion of M is

$$M\frac{d^2x}{dt^2} + \beta\frac{dx}{dt} + kx = A \cos \omega t.$$

(a) Show that steady-state oscillations of M are defined by

$$x(t) = \frac{A}{(k - M\omega^2)^2 + \beta^2\omega^2}\left[(k - M\omega^2)\cos \omega t + \beta\omega \sin \omega t\right].$$

(b) Verify that this function can be expressed in the form

$$x(t) = \frac{A}{\sqrt{(k - M\omega^2)^2 + \beta^2\omega^2}}\sin (\omega t + \phi),$$

where

$$\sin \phi = \frac{k - M\omega^2}{\sqrt{(k - M\omega^2)^2 + \beta^2\omega^2}},$$

$$\cos \phi = \frac{\beta\omega}{\sqrt{(k - M\omega^2)^2 + \beta^2\omega^2}}.$$

(c) If resonance is said to occur when the amplitude of the steady-state oscillations is a maximum, what value of ω yields resonance? What is the maximum amplitude?

* **17.** (a) A cube L metres on each side and with mass M kilograms floats half submerged in water. If it is pushed down slightly and then released, oscillations take place. Use Archimedes' principle to find the differential equation governing these oscillations. Assume no damping forces due to the viscosity of the water.

(b) What is the frequency of the oscillations?

* **18.** A weighing platform has weight W and is supported by springs with combined spring constant k. A package with weight w is dropped on the platform so that the two move together. Find a formula for the maximum value of w so that oscillations do not occur. Assume that there is damping in the motion with constant β.

* **19.** A spring-mounted two-wheel trailer of mass 400 kg is towed at speed v over an undulating road surface whose contour is $Y = A \cos (\pi x/5)$ m. Assume no damping to the vertical oscillations of the trailer, and a spring constant of 40 000 N/m. Use the differential equation of Example 15.29 to determine the speed at which resonance occurs for vibrations of the trailer.

* **20.** Suppose in Example 15.29 that $M = 200$ kg, $k = 50 000$ N/m, $\beta = 3000$ kg/s, and the equation of the road is sinusoidal $Y = 0.1 \sin (\pi x/40)$ m.

(a) Find $y(t)$ for $v = 10$ m/s. What is the amplitude of the steady-state part of the solution?

(b) Repeat part (a) for $v = 20$ m/s.

* **21.** In a reversible reaction from substance A to substance B, the amount $x(t)$ of A at time t satisfies the differential equation

$$\frac{d^2x}{dt^2} + 3\frac{dx}{dt} + 2x = a,$$

where $a > 0$ is a constant. If $x(0) = 0$ and $x'(0) = 2a/3$, find $x(t)$ in terms of a and t.

* **22.** A particle of mass 3 units moves in the xy-plane under the influence of a conservative force field with potential function $U(x, y) = 12x(3y - 4x)$. The particle starts at time $t = 0$ from rest at position $(10, -10)$.

(a) Show that differential equations defining the position of the particle are

$$\frac{d^2x}{dt^2} = -12y + 32x, \qquad \frac{d^2y}{dt^2} = -12x.$$

(b) Use these to show that $y(t)$ must satisfy the fourth-order equation

$$\frac{d^4y}{dt^2} - 32\frac{d^2y}{dt^2} - 144y = 0.$$

Solve this differential equation to find $x(t)$ and $y(t)$.

(c) Plot the path of the particle and make any conclusions that seem justified.

23. When a particle with charge q and mass m moves in an electromagnetic field $\mathbf{E}$ and magnetic field $\mathbf{B}$, the force acting on it is called the *Lorentz force* $\mathbf{F} = q(\mathbf{E} + \mathbf{v} \times \mathbf{B})$, where $\mathbf{v}$ is its velocity. Suppose that $\mathbf{B}$ and $\mathbf{E}$ are constants, $\mathbf{B}$ in the negative x-direction, and $\mathbf{E}$ in the positive y-direction. Find the path followed by the particle if it starts from rest at the origin.

24. The piezometric head distribution $h(x)$ in the leaky aquifer below must satisfy the boundary-value problem

$$\frac{d^2h}{dx^2} + \frac{K_v(H_0 - h)}{KbB} = 0, \quad h(0) = H_0, \quad h(L) = b,$$

where K and K_v are hydraulic conductivities of the semipervious layer of thickness B and the aquifer of thickness b, and H_0 is a fixed external head.

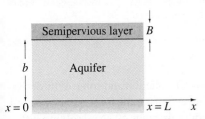

(a) Find $h(x)$.

(b) Plot a graph of $h(x)$ to show that the head remains relatively flat for $0 \le x \le 600$ when $K_v = 10^{-8}$ m/s, $K = 10^{-6}$ m/s, $b = 100$ m, $B = 1$ m, $H_0 = 125$ m, and $L = 1000$ m.

25. A cylindrical buoy 20 cm in diameter floats partially submerged with its axis vertical. When it is depressed slightly and released, its oscillations have a period equal to 4 s. What is the mass of the buoy?

26. A uniform chain of length a has a portion $0 < b < a$ hanging over the edge of a smooth table (figure below). Prove that the time taken for the chain to slide off the table if it starts from rest is $\sqrt{a/g} \ln[(a + \sqrt{a^2 - b^2})/b]$.

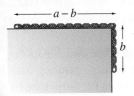

27. (a) When axially compressive forces P are applied to the ends of the beam in the following figure, clamped horizontally at $x = 0$ and free at $x = L$, small deflections must satisfy the differential equation

$$\frac{d^4y}{dx^4} + \frac{P}{EI}\frac{d^2y}{dx^2} = 0,$$

where E and I are positive constants. Show that a general solution is

$$y(x) = C_1 \cos\sqrt{\frac{P}{EI}}x + C_2 \sin\sqrt{\frac{P}{EI}}x + C_3x + C_4.$$

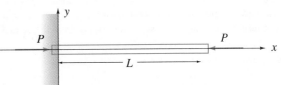

(b) Show that the boundary conditions $y(0) = y'(0) = 0 = y''(L) = y'''(L) + Py'(L)/(EI)$ require that

$$y(x) = C_4\left[1 - \cos\frac{(2n+1)\pi x}{2L}\right],$$

where n is an integer and $P = (2n+1)^2\pi^2 EI/(4L^2)$.

(c) Find Euler's buckling load for this configuration.

* **28.** (a) When axially compressive forces P are applied to the ends of the beam in the figure below, clamped horizontally at $x = 0$ and $x = L$, small deflections must satisfy the differential equation

$$\frac{d^4y}{dx^4} + \frac{P}{EI}\frac{d^2y}{dx^2} = 0,$$

where E and I are positive constants. Show that a general solution is

$$y(x) = C_1 \cos\sqrt{\frac{P}{EI}}x + C_2 \sin\sqrt{\frac{P}{EI}}x + C_3x + C_4.$$

(b) Show that the boundary conditions $y(0) = y'(0) = 0 = y(L) = y'(L)$ require that

$$C_2 \sin\frac{\mu L}{2}\left(2\sin\frac{\mu L}{2} - \mu L \cos\frac{\mu L}{2}\right) = 0, \quad \mu = \sqrt{P/(EI)},$$

where n is an integer and $P = 4n^2\pi^2 EI/L^2$.

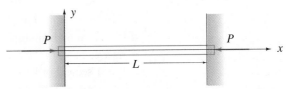

(c) What is Euler's buckling load for this configuration?

* **29.** Reversible reactions in a batch reactor under constant volume and temperature are represented as follows.

$$A \underset{k_2}{\overset{k_1}{\rightleftarrows}} B \underset{k_4}{\overset{k_3}{\rightleftarrows}} C$$

Values of the k's represent proportionality constants for the rates of reaction. If N_A, N_B, and N_C denote the numbers of moles of A, B, and C at any given time, then

$$\frac{dN_A}{dt} = -k_1N_A + k_2N_B, \quad \frac{dN_B}{dt} = -(k_2 + k_3)N_B + k_1N_A + k_4N_C.$$

If one mole of A is present initially, but no B or C, then $N_A(0) = 1$, and N_A, N_B, and N_C must satisfy the condition $N_A + N_B + N_C = 1$ for all t. Because of this condition, we do not formulate a differential equation for N_C.

(a) Show that $N_A(t)$ must satisfy

$$\frac{d^2 N_A}{dt^2} + (k_1 + k_2 + k_3 + k_4)\frac{dN_A}{dt} + (k_1 k_3 + k_2 k_4 + k_1 k_4)N_A = k_2 k_4.$$

(b) Solve the differential equation in part (a) subject to $N_A(0) = 1$ and $N'_A(0) = 0$ when $k_1 = 2$, $k_2 = 1$, $k_3 = 4$, and $k_4 = 3$. What is the limit of $N_A(t)$ as $t \to \infty$?

(c) Find $N_B(t)$. What is its limit as $t \to \infty$?

(d) Find $N_C(t)$. What is its limit as $t \to \infty$?

** **30.** Repeat parts (b), (c), and (d) of Exercise 29 for arbitrary k_1, k_2, k_3, and k_4.

** **31.** Repeat Exercise 26 if the coefficient of kinetic friction between the chain and top of the table is μ.

** **32.** A sphere of radius R floats half submerged in water. It is set into vibration by pushing it down slightly and then releasing it. If y denotes the instantaneous distance of its centre below the surface, show that

$$\frac{d^2 y}{dt^2} = \frac{-3g}{2R^3}\left(R^2 y - \frac{y^3}{3}\right),$$

where g is the acceleration due to gravity.

** **33.** A cable hangs over a peg, 10 m on one side and 15 m on the other. Find the time for it to slide off the peg (a) if friction at the peg is negligible and (b) if friction at the peg is equal to the weight of 1 m of cable.

SUMMARY

A differential equation is an equation that contains an unknown function and some of its derivatives, and the equation must be solved for this function. Depending on the form of the equation, various techniques may be used to find the solution. From this point of view, solving differential equations is much like evaluating antiderivatives: We must first recognize the technique appropriate to the particular problem at hand, and then proceed through the mechanics of the technique. It is important, then, to recognize immediately the type of differential equation under consideration.

Broadly speaking, we divided the differential equations we considered into two main groups: first-order equations together with simple second-order equations, and linear differential equations. A first-order differential equation in $y(x)$ is said to be separable if it can be written in the form $N(y)dy = M(x)\,dx$, and a one-parameter family of solution for such an equation is defined implicitly by

$$\int N(y)\,dy = \int M(x)\,dx + C.$$

A differential equation of the form $dy/dx + P(x)y = Q(x)$ is said to be linear first order. If this equation is multiplied by the integrating factor $e^{\int P(x)dx}$, then the left side becomes the derivative of $ye^{\int P(x)dx}$, and the equation is immediately integrable.

If a second-order differential equation in $y(x)$ has either the dependent variable or the independent variable explicitly missing, it can be reduced to a pair of first-order equations. This is accomplished in the former case by setting $v = y'$ and $v' = y''$, and in the latter case by setting $v = y'$ and $y'' = v\,dv/dy$.

A differential equation in $y(x)$ is said to be linear if it is of the form

$$a_0\frac{d^n y}{dx^n} + a_1\frac{d^{n-1}y}{dx^{n-1}} + \cdots + a_{n-1}\frac{dy}{dx} + a_n y = F(x).$$

The general solution of such an equation is composed of two parts: $y(x) = y_h(x) + y_p(x)$. The function $y_h(x)$ is a general solution of the associated homogeneous equation obtained by replacing $F(x)$ by 0; $y_p(x)$ is any particular solution of the given equation whatsoever. In the special case that the a_i are constants, it is always possible to find $y_h(x)$. This is done by calculating all solutions of the auxiliary equation

$$a_0 m^n + a_1 m^{n-1} + \cdots + a_{n-1}m + a_n = 0,$$

a polynomial equation in m, and then using the rules of Theorem 15.3.

When the a_i are constants and $F(x)$ is a polynomial, an exponential, a sine, a cosine, or any sums or products thereof, we can find $y_p(x)$ either by undetermined coefficients or by operators. The method of undetermined coefficients is simply an intelligent way of guessing at $y_p(x)$ on the basis of $F(x)$ and $y_h(x)$. Note once again that $y_h(x)$ must be calculated before using the method of undetermined coefficients. The method of operators, on the other hand, is based on formal algebraic manipulations and the inverse operator shift theorem. It does not require prior calculation of $y_h(x)$ and, in its simplest form, uses complex numbers.

KEY TERMS

In reviewing this chapter, you should be able to define or discuss the following key terms:

Differential equation

Initial or boundary conditions

General solution

Separable differential equation

Integrating factor

Force of friction

Linear differential equation

Linear operator

Nonhomogeneous linear differential equation

Linearly dependent solutions

Auxiliary equation

Method of undetermined coefficients

Operator method

Simple harmonic motion

Order of a differential equation

Singular solution

Particular solution

Linear first-order differential equation

Friction

Coefficient of kinetic friction

Differential operator

Homogeneous linear differential equation

Superposition principle

Linearly independent solutions

Homogeneous, reduced, or complementary equation

Inverse operator shift theorem

Euler's buckling load

REVIEW EXERCISES

In Exercises 1–20 find a general solution for the differential equation.

1. $x^2\,dy - y\,dx = 0$

2. $(x+1)\,dx - xy\,dy = 0$

3. $\dfrac{dy}{dx} + 3xy = 2x$

4. $\dfrac{dy}{dx} + 4y = x^2$

5. $\dfrac{d^2y}{dx^2} + 4\dfrac{dy}{dx} + 3y = 2$

6. $\dfrac{d^2y}{dx^2} + 3\dfrac{dy}{dx} + 4y = 2$

7. $yy' = \sqrt{1+y^2}$

8. $\dfrac{d^2y}{dx^2} + \dfrac{1}{x}\dfrac{dy}{dx} = x$

9. $y'' + 6y' + 3y = xe^x$

10. $\dfrac{dy}{dx} + 2xy = 2x^3$

11. $y^2y'' = y'$

12. $y'' - 4y' + 4y = \sin x$

13. $y'' - 4y' + 4y = x^2e^{2x}$

14. $y'' + 4y = \sin 2x$

15. $y'' + 4y' = x^2$

16. $2xy^2\dfrac{dy}{dx} + (x+1)^2y^3 = 0$

17. $\dfrac{d^3y}{dx^3} + 3\dfrac{d^2y}{dx^2} + 3\dfrac{dy}{dx} + y = 2e^{-x}$

18. $y'' + 2y' + 4y = e^{-x}\cos\sqrt{3}x$

19. $\dfrac{dy}{dx} = y\tan x + \cos x$

20. $(2y^2 + 3x)\,dy + dx = 0$

In Exercises 21–24 find the solution of the differential equation satisfying the given condition(s).

21. $y^2\,dx + (x+1)\,dy = 0$, $y(0) = 3$

22. $y'' - 8y' - 9y = 2x + 4$, $y(0) = 3$, $y'(0) = 7$

23. $y'' + 9y = e^x$, $y(0) = 0$, $y(\pi/2) = 4$

* **24.** $y' + \dfrac{2}{x}y = \sin x$, $y(1) = 1$

* **25.** The quantity of radioactive material present in a sample decays at a rate proportional to the amount of the material in the sample (see Section 5.5). If one-fourth of a sample decays in 5 years, how long does it take for 90% of the sample to decay?

* **26.** (a) A piece of wood rises from the bottom of a container of oil 1 m deep. If the wood has mass 0.5 g and volume 1 cm³, and the density of the oil is 0.9 g/cm³, show that Archimedes' principle predicts a buoyant force due to fluid pressure of 8.829×10^{-3} N. What is the force on the piece of wood due to gravity and fluid pressure?

(b) The viscosity of the oil opposes motion by exerting a force equal (in newtons) to twice its velocity (in metres per second). Find the distance travelled by the wood as a function of time, assuming that it starts from rest on the bottom.

∗ **27.** A 100-g mass is suspended vertically from a spring with constant 1 N/m. The mass is pulled 4 cm above its equilibrium position and then released. Find the position of the mass at any time if (a) damping is ignored, (b) a damping force in newtons equal to one-fifth the instantaneous velocity of the mass in metres per second acts on the mass, and (c) a damping force in newtons equal to $\sqrt{2}/5$ the instantaneous velocity of the mass in metres per second acts on the mass.

∗ **28.** A 10-g stone is dropped over the side of a bridge 50 m above a river 10 m deep. Air resistance is negligible, but as the stone sinks in the water, its motion is retarded by a force in newtons equal to one-fifth the speed of the stone in metres per second. If the stone loses 10% of its speed in penetrating the surface of the water, find (a) its velocity as a function of time, (b) its position relative to the drop point as a function of time, and (c) the time it takes to reach the bottom of the river from the point at which it was dropped. Assume that the water is stationary and that buoyancy due to Archimedes' principle may be neglected.

∗ **29.** Repeat Exercise 28 but assume that for the purpose of Archimedes' principle, the volume of the stone is 3 cm^3.

APPENDIX A

Mathematical Induction

Mathematical induction is a type of proof used to verify propositions that involve integers. It is best introduced through a simple example.

Suppose we are required to verify that the sum of the first n positive integers is

$$1 + 2 + 3 + \cdots + n = \frac{n(n + 1)}{2}. \tag{A.1}$$

It is a simple matter to check that this result is valid for the first few integers:

For $n = 1$, the sum is 1, and the formula gives $1(1 + 1)/2 = 1$.
For $n = 2$, the sum is $1 + 2 = 3$; the formula gives $2(2 + 1)/2 = 3$.
For $n = 3$, the sum is $1 + 2 + 3 = 6$; the formula gives $3(3 + 1)/2 = 6$.

We shall see later that to use induction, it is necessary to verify the result at this stage only for $n = 1$.

We now suppose that k is some integer for which the result is valid; that is, we suppose that

$$1 + 2 + 3 + \cdots + k = \frac{k(k + 1)}{2}. \tag{A.2}$$

On the basis of this supposition, the sum of the first $k + 1$ integers is

$$
\begin{aligned}
1 + 2 + \cdots + (k + 1) &= (1 + 2 + \cdots + k) + (k + 1) \\
&= \frac{k(k + 1)}{2} + (k + 1) \qquad \text{(by assumption A.2)} \\
&= \frac{1}{2}[k(k + 1) + 2(k + 1)] \\
&= \frac{(k + 1)(k + 2)}{2}.
\end{aligned}
$$

But this is equation A.1 with n replaced by $k + 1$. We have shown that if A.1 is valid for some integer k, then it must also be valid for the next integer $k + 1$.

We now put the above facts together to verify that A.1 is valid for all $n \geq 1$. We have proved that it is valid for $n = 1$, 2, and 3. But if it is true for 3, then it must be true for 4. If it is true for 4, then it is true for 5. If it is true for 5, it is true for 6; and so on, and so on.

Consequently, A.1 must be valid for every positive integer n.

This example contains the two essential parts of every proof by what we call **mathematical induction**:

1. Verification of the result for the smallest integer in question

2. Verification of the fact that if the result is valid for some integer k, then it must be valid for the next integer $k + 1$

Once these two facts are established, it follows by the same argument as above that the result is valid for the given set of integers. This is the principle of mathematical induction, which we state formally as follows.

Principle of Mathematical Induction

Suppose that with each integer n greater than or equal to some fixed integer N, there is associated a proposition P_n. Then P_n is true for all $n \geq N$ provided that:

1. P_N is valid.

2. The validity of P_k implies the validity of P_{k+1}.

All inductive arguments must contain these two essential parts. Part 1 is usually quite simple to establish; but part 2 can sometimes be difficult. Be sure you understand what we are doing in part 2. We are proving that under the assumption that P_k is true, then P_{k+1} must also be true. It is the principle of mathematical induction that puts parts 1 and 2 together to state that P_n is true for all $n \geq N$.

EXAMPLE A.1

Verify that

$$1^2 + 2^2 + 3^2 + \cdots + n^2 = \frac{n(n+1)(2n+1)}{6}, \quad n \geq 1. \qquad (A.3)$$

SOLUTION   When $n = 1$ the left side of this equation has value 1, and the right side is $1(2)(3)/6 = 1$. The required result is therefore true for $n = 1$. Next we suppose that k is some integer for which A.3 is valid; that is, we suppose that

$$1^2 + 2^2 + 3^2 + \cdots + k^2 = \frac{k(k+1)(2k+1)}{6}, \qquad (A.4)$$

and we must write this down because we are going to need it later. Our objective now is to verify that the result is valid for $k + 1$; that is, we must verify that

$$1^2 + 2^2 + 3^2 + \cdots + (k+1)^2 = \frac{(k+1)(k+2)[2(k+1)+1]}{6}. \qquad (A.5)$$

If we begin on the left of A.5, we have

$$1^2 + 2^2 + \cdots + (k+1)^2 = (1^2 + 2^2 + \cdots + k^2) + (k+1)^2$$

$$= \frac{k(k+1)(2k+1)}{6} + (k+1)^2 \qquad \text{(by A.4)}$$

$$= \frac{(k+1)}{6}[k(2k+1) + 6(k+1)]$$

$$= \frac{(k+1)}{6}(2k^2 + 7k + 6)$$

$$= \frac{(k+1)(k+2)(2k+3)}{6}$$

$$= \frac{(k+1)(k+2)[2(k+1)+1]}{6},$$

which is the right side of A.5. We have therefore verified A.5. Consequently, by mathematical induction, formula A.3 is valid for all $n \geq 1$.

EXAMPLE A.2

Prove that 7 divides $23^{3n} - 1$ for all positive integers n.

SOLUTION When $n = 1, 23^{3n} - 1 = 23^3 - 1 = 12\ 166$, and this is divisible by 7. The result is therefore true for $n = 1$. Next we suppose that k is some integer for which the result is valid; that is, we suppose that 7 divides $23^{3k} - 1$. We must now verify that 7 divides $23^{3(k+1)} - 1$. We do this by expressing $23^{3(k+1)} - 1$ in the form

$$23^{3(k+1)} - 1 = 23^{3k+3} - 1 = (23^{3k})(23^3) - 1$$
$$= [(23^{3k})(23^3) - 23^3] + (23^3 - 1)$$
$$= 23^3(23^{3k} - 1) + (23^3 - 1).$$

Since both $23^{3k} - 1$ and $23^3 - 1$ are divisible by 7, it follows that $23^{3(k+1)} - 1$ must also be divisible by 7. The result is therefore true for $k + 1$, and by mathematical induction, it is true for all $n \geq 1$.

EXAMPLE A.3

Verify that if n lines are drawn in the plane, no two of which are identical, then the maximum number of points of intersection of these lines is $n(n - 1)/2$.

SOLUTION If we draw two lines in the plane, there can be at most one point of intersection. This is precisely what is predicted by $n(n - 1)/2$ when $n = 2$. The result is therefore correct for $n = 2$. Suppose the proposition is true for some integer k; that is, suppose when k lines are drawn in the plane, the maximum number of points of intersection is $k(k - 1)/2$. We must now verify that when any $k + 1$ lines are drawn, the maximum number of points of intersection is $(k + 1)k/2$. To do this, we remove one of the $k + 1$ lines, leaving k lines. But for these k lines the maximum number of points of intersection is $k(k - 1)/2$. If the line that was removed is now replaced, it can add at most k more points of intersection, one with each of the k lines. The maximum number of points of intersection of the $k + 1$ lines is therefore

$$\frac{k(k - 1)}{2} + k = \frac{k}{2}(k - 1 + 2) = \frac{k(k + 1)}{2},$$

and this is the result for $n = k + 1$. By mathematical induction, then, the result is valid for all $n \geq 2$.

EXERCISES

In Exercises 1–10 use mathematical induction to establish the formula for positive integer n.

1. $1^3 + 2^3 + 3^3 + \cdots + n^3 = \dfrac{n^2(n + 1)^2}{4}$

2. $1 + 4 + 7 + 10 + \cdots + (3n - 2) = \dfrac{n(3n - 1)}{2}$

3. $1 + 3 + 6 + 10 + \cdots + \dfrac{n(n + 1)}{2} = \dfrac{n(n + 1)(n + 2)}{6}$

4. $1^2 + 3^2 + 5^2 + \cdots + (2n - 1)^2 = \dfrac{n(2n - 1)(2n + 1)}{3}$

5. $\dfrac{1}{1 \cdot 2} + \dfrac{1}{2 \cdot 3} + \dfrac{1}{3 \cdot 4} + \cdots + \dfrac{1}{n(n + 1)} = \dfrac{n}{n + 1}$

6. $\dfrac{1}{1 \cdot 2 \cdot 3} + \dfrac{1}{2 \cdot 3 \cdot 4} + \dfrac{1}{3 \cdot 4 \cdot 5} + \cdots + \dfrac{1}{n(n + 1)(n + 2)}$
$= \dfrac{n(n + 3)}{4(n + 1)(n + 2)}$

7. $2 + 2^3 + 2^5 + \cdots + 2^{2n-1} = \dfrac{2(2^{2n} - 1)}{3}$

8. $\dfrac{1}{5^2} + \dfrac{1}{5^4} + \dfrac{1}{5^6} + \cdots + \dfrac{1}{5^{2n}} = \dfrac{1}{24}\left(1 - \dfrac{1}{25^n}\right)$

9. $1(1!) + 2(2!) + 3(3!) + \cdots + n(n!) = (n+1)! - 1$

10. $1(2^{-1}) + 2(2^{-2}) + 3(2^{-3}) + \cdots + n(2^{-n})$
    $= 2 - (n+1)2^{-n}$

11. Prove that 15 divides $4^{2n} - 1$ for $n \geq 1$.

12. Prove that 7 divides $8^n - 1$ for $n \geq 1$.

13. Prove that 4 divides $3(7^{2n}) - 3$ for $n \geq 1$.

14. Prove that $x - y$ divides $x^n - y^n$ for $n \geq 1$.

15. Verify that $x + y$ divides $x^{2n+1} + y^{2n+1}$ for $n \geq 0$.

16. Prove that 576 divides $5^{2n+2} - 24n - 25$ for $n \geq 1$.

17. Prove that if the proposition

$$1 + 3 + 5 + \cdots + (2n - 1) = n^2 + 4$$

is valid for some integer k, then it is also valid for $k + 1$. Is the result valid for all $n \geq 1$?

* 18. Verify that the sum of the interior angles of a polygon with n sides is $(n - 2)\pi$ radians.

* 19. Show that if n is a positive integer, so is $(n^3 + 6n^2 + 2n)/3$.

* 20. Verify that $2 + 4 + 6 + \cdots + (2n) < n^2 + 2n$ for $n \geq 1$.

* 21. Verify that $1 + 3 + 5 + \cdots + (2n - 1) < n^2 + n$ for $n \geq 1$.

* 22. Prove that $9(n!) > 2^{2n}$ for $n \geq 5$.

* 23. Verify that when $a > 0$ and $n \geq 2$ is an integer, then $(1 + a)^n \geq 1 + na$.

* 24. Verify that:

(a) $\left(1 + \dfrac{1}{1}\right)\left(1 + \dfrac{1}{2}\right)\left(1 + \dfrac{1}{3}\right)\cdots\left(1 + \dfrac{1}{n}\right) = n + 1, \quad n \geq 1$

(b) $\left(1 - \dfrac{1}{4}\right)\left(1 - \dfrac{1}{9}\right)\left(1 - \dfrac{1}{16}\right)\cdots\left(1 - \dfrac{1}{n^2}\right) = \dfrac{n+1}{2n}, \quad n \geq 2$

* 25. Suppose n points are drawn in the plane, no three of which are collinear ($n \geq 3$). If a line is drawn through each pair of points, find and verify a formula for the number of such lines.

In Exercises 26–33 use mathematical induction to establish the result for positive integer n.

* 26. $1^2 + 2^2 + 3^2 + \cdots + (2n)^2 = \dfrac{n(2n+1)(4n+1)}{3}$

* 27. $2^3 + 4^3 + 6^3 + \cdots + (4n)^3 = 8n^2(2n+1)^2$

* 28. $1 + 4 + 7 + \cdots + (6n - 2) = n(6n - 1)$

* 29. $n + (n+1) + (n+2) + \cdots + (2n) = \dfrac{3n(n+1)}{2}$

* 30. $3^n + 3^{n+1} + 3^{n+2} + \cdots + 3^{2n} = \dfrac{3^n(3^{n+1} - 1)}{2}$

* 31. $2^{2n} + 2^{2n+2} + 2^{2n+4} + \cdots + 2^{6n} = \dfrac{4^n(4^{2n+1} - 1)}{3}$

* 32. $2n + (2n+1) + (2n+2) + \cdots + (5n) = \dfrac{7n(3n+1)}{2}$

* 33. $(3n+1) + (3n+4) + (3n+7) + \cdots + (6n - 2)$
    $= \dfrac{n(9n-1)}{2}$

* 34. Verify that $1 + \dfrac{1}{2} + \dfrac{1}{3} + \cdots + \dfrac{1}{2^n} \leq n + 1$ for $n \geq 1$.

* 35. Verify that $1 + \dfrac{1}{1!} + \dfrac{1}{2!} + \dfrac{1}{3!} + \cdots + \dfrac{1}{n!} \leq 3 - \dfrac{1}{n}$ for $n \geq 1$.

* 36. Verify that any integer $n \geq 14$ can always be expressed in the form $n = 3p + 8q$, where p and q are nonnegative integers.

* 37. Verify that when $r \neq 1$,

$$1 + 2r + 3r^2 + 4r^3 + \cdots + nr^{n-1} = \dfrac{1 - (n+1)r^n + nr^{n+1}}{(1-r)^2},$$

$n \geq 1$.

* 38. The *tower of Hanoi* problem consists in moving the n rings on peg 1 in the figure below to peg 2 in as few moves as possible. There are two rules to be followed:

(i) Only one ring may be moved at a time.

(ii) A ring may never be placed on top of a smaller ring.

Solve the problem for $n = 2$ and $n = 3$, conjecture a solution for arbitrary n, and prove it by mathematical induction.

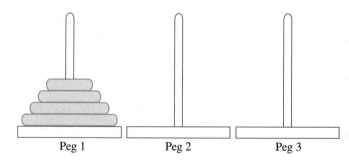

Peg 1   Peg 2   Peg 3

* 39. Prove that:

(a) $1 + \cos\theta + \cos 2\theta + \cdots + \cos n\theta$
    $= \dfrac{1 - \cos\theta + \cos n\theta - \cos(n+1)\theta}{2 - 2\cos\theta}$

(b) $\sin\theta + \sin 3\theta + \cdots + \sin(2n-1)\theta = \dfrac{\sin^2 n\theta}{\sin\theta}$

** 40. Use mathematical induction to show that for $n \geq 2$, there exist constants a_r and b_r, $r = 0, 1, 2, \ldots, n$ such that

$$\sin^n x = \sum_{r=0}^{n}(a_r \cos rx + b_r \sin rx).$$

Determinants

A **determinant of order** n is n^2 numbers arranged in n rows and n columns enclosed by two vertical lines. Thus,

$$\begin{vmatrix} 2 & 1 \\ 3 & 4 \end{vmatrix}, \qquad \begin{vmatrix} -1 & 0 & 3 \\ 2 & 1 & 6 \\ 7 & -1 & 4 \end{vmatrix}, \qquad \begin{vmatrix} 0 & 0 & 1 & 5 \\ 2 & -1 & 3 & 6 \\ -1 & -2 & -3 & 4 \\ 6 & 7 & 8 & 10 \end{vmatrix}$$

are determinants of orders 2, 3, and 4, respectively. The general determinant of order n is written in the form

$$D = \begin{vmatrix} a_{11} & a_{12} & a_{13} & \cdots & a_{1n} \\ a_{21} & a_{22} & a_{23} & \cdots & a_{2n} \\ \vdots & \vdots & \vdots & \ddots & \vdots \\ a_{n1} & a_{n2} & a_{n3} & \cdots & a_{nn} \end{vmatrix}. \tag{B.1}$$

The element in the ith row and j^{th} column is called the $(i, j)^{\text{th}}$ element and is denoted by a_{ij}. For brevity we write

$$D = \big|a_{ij}\big|_{n \times n} \tag{B.2}$$

to identify the general determinant of order n.

We wish to assign a value to every determinant, and to do this we first define what is meant by a minor and a cofactor of an element in a determinant.

DEFINITION B.1

The **minor** M_{ij} of a_{ij} in $D = \big|a_{ij}\big|_{n \times n}$ is the determinant of order $n - 1$ obtained by deleting the ith row and j^{th} column of D.

For example, if

$$D = \begin{vmatrix} -1 & 0 & 3 \\ 2 & 1 & 6 \\ 7 & -1 & 4 \end{vmatrix},$$

then

$$M_{11} = \begin{vmatrix} 1 & 6 \\ -1 & 4 \end{vmatrix}, \qquad M_{23} = \begin{vmatrix} -1 & 0 \\ 7 & -1 \end{vmatrix}, \qquad M_{31} = \begin{vmatrix} 0 & 3 \\ 1 & 6 \end{vmatrix}.$$

DEFINITION B.2

The **cofactor** A_{ij} of a_{ij} in $D = \big|a_{ij}\big|_{n \times n}$ is $(-1)^{i+j} M_{ij}$.

If D is as in the preceding paragraph, then

$$A_{11} = (-1)^{1+1} M_{11} = \begin{vmatrix} 1 & 6 \\ -1 & 4 \end{vmatrix}, \qquad A_{23} = (-1)^{2+3} M_{23} = -\begin{vmatrix} -1 & 0 \\ 7 & -1 \end{vmatrix}.$$

The following two rules now specify how to find the value of every determinant:

1. The value of a determinant $D = |a_{11}|$ of order 1 is a_{11}.

2. The value of a determinant $D = \left|a_{ij}\right|_{n \times n}$ of order n is obtained by choosing any line (row or column) and adding elements in that line each multiplied by its cofactor.

If we select row i, then

$$D = a_{i1}A_{i1} + a_{i2}A_{i2} + \cdots + a_{in}A_{in} = \sum_{j=1}^{n} a_{ij}A_{ij}, \tag{B.3a}$$

and if we select column j,

$$D = a_{1j}A_{1j} + a_{2j}A_{2j} + \cdots + a_{nj}A_{nj} = \sum_{i=1}^{n} a_{ij}A_{ij}. \tag{B.3b}$$

Rule 2 defines a determinant of order n in terms of n determinants of order $n-1$ (the cofactors); each of these determinants is defined in terms of determinants of order $n-2$, and the process is continued until only determinants of order 1 are involved.

It is not clear that the *value of a determinant is independent of the line chosen in its evaluation*, but this is indeed the case. In other words, rules 1 and 2 define a unique value for every determinant.

The following result is essential to the speedy evaluation of determinants.

THEOREM B.1

The value of a determinant of order 2 is

$$D = \begin{vmatrix} a_{11} & a_{12} \\ a_{21} & a_{22} \end{vmatrix} = a_{11}a_{22} - a_{12}a_{21}. \tag{B.4}$$

PROOF   If we expand D along its first row, we have

$$D = a_{11}A_{11} + a_{12}A_{12} = a_{11}(-1)^{1+1}M_{11} + a_{12}(-1)^{1+2}M_{12} = a_{11}(a_{22}) - a_{12}(a_{21}). \quad \blacksquare$$

EXAMPLE B.1

Evaluate $\begin{vmatrix} 2 & 3 \\ -4 & 6 \end{vmatrix}$.

SOLUTION   By Theorem B.1,

$$\begin{vmatrix} 2 & 3 \\ -4 & 6 \end{vmatrix} = (2)(6) - (3)(-4) = 24.$$

EXAMPLE B.2

Evaluate $\begin{vmatrix} 3 & -2 & 6 \\ 1 & 3 & 4 \\ 2 & -1 & 2 \end{vmatrix}$.

SOLUTION   If we expand along the first column, we obtain

$$D = 3(-1)^2 \begin{vmatrix} 3 & 4 \\ -1 & 2 \end{vmatrix} + 1(-1)^3 \begin{vmatrix} -2 & 6 \\ -1 & 2 \end{vmatrix} + 2(-1)^4 \begin{vmatrix} -2 & 6 \\ 3 & 4 \end{vmatrix}$$

$$= 3(6 + 4) - (-4 + 6) + 2(-8 - 18) = -24.$$

Using equations B.3 to evaluate determinants is not always particularly easy. For instance, even for a determinant of order 5, it is necessary to evaluate 60 determinants of order 2. As a result, we now prove two theorems that are used to simplify determinants.

THEOREM B.2

If any line of a determinant with value D has its elements multiplied by c, the new determinant has value cD.

PROOF If the ith row of $D = \left|a_{ij}\right|_{n \times n}$ is multiplied by c, and the resulting determinant is expanded along this row, its value is

$$\sum_{j=1}^{n} ca_{ij}A_{ij} = c\sum_{j=1}^{n} a_{ij}A_{ij} = cD.$$

THEOREM B.3

If a multiple of one line of a determinant with value D is added to a parallel line, the resulting determinant also has value D.

PROOF Suppose that c times the ith row is added to the kth row of $D = \left|a_{ij}\right|_{n \times n}$ to form

$$E = \begin{vmatrix} a_{11} & a_{12} & \cdots & a_{1n} \\ a_{21} & a_{22} & \cdots & a_{2n} \\ \vdots & \vdots & \ddots & \vdots \\ a_{i1} & a_{i2} & \cdots & a_{in} \\ \vdots & \vdots & \ddots & \vdots \\ a_{k1}+ca_{i1} & a_{k2}+ca_{i2} & \cdots & a_{kn}+ca_{in} \\ \vdots & \vdots & \ddots & \vdots \\ a_{n1} & a_{n2} & \cdots & a_{nn} \end{vmatrix}.$$

If we expand E along the kth row,

$$E = \sum_{j=1}^{n}(a_{kj}+ca_{ij})A_{kj} = \sum_{j=1}^{n} a_{kj}A_{kj} + c\sum_{j=1}^{n} a_{ij}A_{kj}.$$

Now $\sum_{j=1}^{n} a_{kj}A_{kj} = D$, and $\sum_{j=1}^{n} a_{ij}A_{kj}$ is the value of the following determinant, which has identical i^{th} and k^{th} rows:

$$\begin{vmatrix} a_{11} & a_{12} & \cdots & a_{1n} \\ \vdots & \vdots & \ddots & \vdots \\ a_{i1} & a_{i2} & \cdots & a_{in} \\ \vdots & \vdots & \ddots & \vdots \\ a_{i1} & a_{i2} & \cdots & a_{in} \\ \vdots & \vdots & \ddots & \vdots \\ a_{n1} & a_{n2} & \cdots & a_{nn} \end{vmatrix} \begin{matrix} \\ \\ \leftarrow i^{\text{th}} \text{ row} \\ \\ \leftarrow k^{\text{th}} \text{ row} \\ \\ \end{matrix}.$$

In Exercise 16 we show that a determinant with two identical parallel lines has value zero, and therefore $E = D$, which completes the proof.

This theorem is the key to evaluation of determinants. We use it to replace a determinant with an equivalent determinant that has many zeros in a line. This makes evaluation by cofactors much simpler.

EXAMPLE B.3

Evaluate the determinant $\quad D = \begin{vmatrix} 1 & 2 & -3 & 4 \\ 2 & 4 & 0 & -1 \\ 3 & 6 & 1 & 2 \\ 4 & 0 & 1 & 5 \end{vmatrix}.$

SOLUTION   Instead of expanding the determinant immediately according to equation B.3, we use Theorem B.3 to create zeros in column 3. We do this by first adding 3 times row 3 to row 1, and second, adding -1 times row 3 to row 4. The result is

$$D = \begin{vmatrix} 10 & 20 & 0 & 10 \\ 2 & 4 & 0 & -1 \\ 3 & 6 & 1 & 2 \\ 1 & -6 & 0 & 3 \end{vmatrix}.$$

We now expand D along the third column:

$$D = (1)(-1)^6 \begin{vmatrix} 10 & 20 & 10 \\ 2 & 4 & -1 \\ 1 & -6 & 3 \end{vmatrix}.$$

When 10 is factored from the first row and 2 from the second column,

$$D = 20 \begin{vmatrix} 1 & 1 & 1 \\ 2 & 2 & -1 \\ 1 & -3 & 3 \end{vmatrix}.$$

Finally, we expand D along the first row to obtain

$$D = 20[1(3) - 1(7) + 1(-8)] = -240.$$

EXAMPLE B.4

Evaluate $D = \begin{vmatrix} 3 & 0 & 2 & -1 & 4 \\ 6 & 3 & 7 & -2 & 8 \\ 10 & 3 & 1 & 0 & 6 \\ 2 & 1 & 3 & -4 & 1 \\ 6 & -2 & -3 & -5 & 2 \end{vmatrix}.$

SOLUTION   By adding multiples of column 4 to columns 1, 3, and 5, we create zeros in the first row:

$$D = \begin{vmatrix} 0 & 0 & 0 & -1 & 0 \\ 0 & 3 & 3 & -2 & 0 \\ 10 & 3 & 1 & 0 & 6 \\ -10 & 1 & -5 & -4 & -15 \\ -9 & -2 & -13 & -5 & -18 \end{vmatrix}.$$

Expansion along the first row gives

$$D = \begin{vmatrix} 0 & 3 & 3 & 0 \\ 10 & 3 & 1 & 6 \\ -10 & 1 & -5 & -15 \\ -9 & -2 & -13 & -18 \end{vmatrix}.$$

We now add -1 times column 2 to column 3:

$$D = \begin{vmatrix} 0 & 3 & 0 & 0 \\ 10 & 3 & -2 & 6 \\ -10 & 1 & -6 & -15 \\ -9 & -2 & -11 & -18 \end{vmatrix}.$$

Expansion along row 1 gives

$$D = 3(-1) \begin{vmatrix} 10 & -2 & 6 \\ -10 & -6 & -15 \\ -9 & -11 & -18 \end{vmatrix}.$$

We now factor -1 from rows 2 and 3, 2 from row 1, and 3 from column 3, and then expand along row 1:

$$D = -18 \begin{vmatrix} 5 & -1 & 1 \\ 10 & 6 & 5 \\ 9 & 11 & 6 \end{vmatrix} = -18[5(-19) + 1(15) + 1(56)] = 432.$$

Solution of Linear Equations by Cramer's Rule

To solve a pair of linear equations in two unknowns, such as

$$2x + 3y = 6, \qquad x - 4y = -1,$$

we can eliminate one of the unknowns, say y, solve the resulting equation for x, and then substitute this value of x into either of the original equations to obtain y. The result for the pair above is $x = 21/11$, $y = 8/11$. A similar procedure can be followed for three linear equations in three unknowns:

$$2x - 3y + 4z = 6, \qquad x + 4y - 2z = 7, \qquad 3x - 2y + z = -2.$$

First one variable is eliminated, say x, to obtain two equations in the two unknowns y and z. These are then solved for y and z and substituted into one of the original equations to find x. The solution is $x = 3/7$, $y = 128/35$, $z = 141/35$.

A formula can be derived for the solution of linear equations. It can be stated simply using determinants; the result is called **Cramer's rule**. We illustrate it for the second example above and then demonstrate its general validity. The coefficients of the unknowns are arranged in a determinant called the *determinant of the system of equations*. For the system of three equations above, it is

$$D = \begin{vmatrix} 2 & -3 & 4 \\ 1 & 4 & -2 \\ 3 & -2 & 1 \end{vmatrix}.$$

We now define three other determinants. They are obtained by replacing the first, second, and third columns in D by the coefficients on the right sides of the equations:

$$D_x = \begin{vmatrix} 6 & -3 & 4 \\ 7 & 4 & -2 \\ -2 & -2 & 1 \end{vmatrix}, \qquad D_y = \begin{vmatrix} 2 & 6 & 4 \\ 1 & 7 & -2 \\ 3 & -2 & 1 \end{vmatrix}, \qquad D_z = \begin{vmatrix} 2 & -3 & 6 \\ 1 & 4 & 7 \\ 3 & -2 & -2 \end{vmatrix}.$$

Cramer's rule states that

$$x = \frac{D_x}{D}, \qquad y = \frac{D_y}{D}, \qquad z = \frac{D_z}{D}.$$

We check the first of these and leave it to the reader to verify the other two. If we expand D_x and D along their first rows, then

$$x = \frac{D_x}{D} = \frac{6(0) + 3(3) + 4(-6)}{2(0) + 3(7) + 4(-14)} = \frac{-15}{-35} = \frac{3}{7}.$$

For our first example of two linear equations in x and y, Cramer's rule gives

$$x = \frac{D_x}{D} = \frac{\begin{vmatrix} 6 & 3 \\ -1 & -4 \end{vmatrix}}{\begin{vmatrix} 2 & 3 \\ 1 & -4 \end{vmatrix}} = \frac{-21}{-11} = \frac{21}{11}, \qquad y = \frac{D_y}{D} = \frac{\begin{vmatrix} 2 & 6 \\ 1 & -1 \end{vmatrix}}{-11} = \frac{-8}{-11} = \frac{8}{11}.$$

To prove Cramer's rule we require the following theorem.

THEOREM B.4

If the elements in any line of a determinant are multiplied by the cofactors of corresponding elements of a distinct parallel line, the resulting sum is zero.

PROOF Suppose we take $D = \left| a_{ij} \right|_{n \times n}$, and construct a determinant E from D by replacing the kth row of D by its ith row:

$$
\begin{vmatrix}
a_{11} & a_{12} & \cdots & a_{1n} \\
\vdots & \vdots & \ddots & \vdots \\
a_{i1} & a_{i2} & \cdots & a_{in} \\
\vdots & \vdots & \ddots & \vdots \\
a_{i1} & a_{i2} & \cdots & a_{in} \\
\vdots & \vdots & \ddots & \vdots \\
a_{n1} & a_{n2} & \cdots & a_{nn}
\end{vmatrix}
\quad
\begin{array}{l} \\ \\ i^{\text{th}} \text{ row} \\ \\ k^{\text{th}} \text{ row} \\ \\ \end{array}
.
$$

If we expand this determinant along its kth row, the net effect is to multiply the elements of row i by the cofactors of row k:

$$
E = \sum_{j=1}^{n} a_{ij} A_{kj}.
$$

However, because E has two identical parallel lines, its value must be zero (see Exercise 16). Thus,

$$
\sum_{j=1}^{n} a_{ij} A_{kj} = 0 \qquad \text{whenever } i \neq k. \tag{B.5a}
$$

A similar proof for columns gives

$$
\sum_{i=1}^{n} a_{ij} A_{ik} = 0 \qquad \text{whenever } j \neq k. \tag{B.5b}
$$

We now prove the following theorem.

THEOREM B.5   (Cramer's Rule)

A system of n linear equations in n unknowns, $x_1, x_2, \ldots, x_n$,

$$
\begin{aligned}
a_{11}x_1 + a_{12}x_2 + \cdots + a_{1n}x_n &= c_1, \\
a_{21}x_1 + a_{22}x_2 + \cdots + a_{2n}x_n &= c_2, \\
&\vdots \\
a_{n1}x_1 + a_{n2}x_2 + \cdots + a_{nn}x_n &= c_n,
\end{aligned} \tag{B.6a}
$$

can be represented compactly in the form

$$
\sum_{j=1}^{n} a_{ij}x_j = c_i, \qquad i = 1, \ldots, n. \tag{B.6b}
$$

From the determinant $D = \left| a_{ij} \right|_{n \times n}$ of the system, we define n other determinants D_k by replacing the kth column of D by the column of constants $c_1, c_2, \ldots, c_n$. The solution of equations B.6 is then

$$
x_k = \frac{D_k}{D}, \qquad k = 1, \ldots, n, \tag{B.7}
$$

provided that $D \neq 0$.

PROOF We multiply the first equation in B.6 by the cofactor A_{1k}, the second equation by A_{2k}, and so on, until the last equation is multiplied by A_{nk}. Symbolically, this is represented by multiplying the ith equation by A_{ik} to get

$$
\sum_{j=1}^{n} a_{ij} A_{ik} x_j = c_i A_{ik}, \qquad i = 1, \ldots, n.
$$

We now add all these equations together:

$$\sum_{i=1}^{n}\sum_{j=1}^{n}a_{ij}A_{ik}x_j = \sum_{i=1}^{n}c_iA_{ik}.$$

The right side of this equation is the expansion of D_k along its kth column. Consequently,

$$D_k = \sum_{j=1}^{n}\left(\sum_{i=1}^{n}a_{ij}A_{ik}\right)x_j.$$

But according to Theorem B.4, the summation in parentheses is zero unless $j = k$; and when $j = k$, the result is

$$D_k = \left(\sum_{i=1}^{n}a_{ik}A_{ik}\right)x_k = D\,x_k \qquad \text{or} \qquad x_k = \frac{D_k}{D}. \qquad \blacksquare$$

EXAMPLE B.5

Use Cramer's rule to solve

$$3x - 2y = -1,$$
$$x + 4y - 2z = 6,$$
$$3y + 4z = 7.$$

SOLUTION   We have

$$D = \begin{vmatrix} 3 & -2 & 0 \\ 1 & 4 & -2 \\ 0 & 3 & 4 \end{vmatrix} = 3(22) + 2(4) = 74; \qquad D_x = \begin{vmatrix} -1 & -2 & 0 \\ 6 & 4 & -2 \\ 7 & 3 & 4 \end{vmatrix} = -1(22) + 2(38) = 54;$$

$$D_y = \begin{vmatrix} 3 & -1 & 0 \\ 1 & 6 & -2 \\ 0 & 7 & 4 \end{vmatrix} = 3(38) + 1(4) = 118; \qquad D_z = \begin{vmatrix} 3 & -2 & -1 \\ 1 & 4 & 6 \\ 0 & 3 & 7 \end{vmatrix} = 3(10) - 1(-11) = 41.$$

Hence, $x = 54/74 = 27/37$, $y = 118/74 = 59/37$, and $z = 41/74$.

EXERCISES

In Exercises 1–9 evaluate the determinant.

1. $\begin{vmatrix} 3 & 2 \\ -1 & 4 \end{vmatrix}$

2. $\begin{vmatrix} 1 & 0 \\ 0 & 1 \end{vmatrix}$

3. $\begin{vmatrix} -2 & -4 \\ -6 & -8 \end{vmatrix}$

4. $\begin{vmatrix} 1 & 2 & 3 \\ 2 & 4 & 6 \\ -1 & 3 & 0 \end{vmatrix}$

5. $\begin{vmatrix} -2 & 0 & 5 \\ 1 & 3 & 6 \\ -7 & 8 & 10 \end{vmatrix}$

6. $\begin{vmatrix} 10 & 20 & 30 \\ 16 & 32 & 64 \\ -1 & 2 & -3 \end{vmatrix}$

7. $\begin{vmatrix} 1 & 1 & 1 & -3 \\ 2 & 1 & 3 & 6 \\ 7 & -8 & 9 & 10 \\ 3 & 4 & -2 & 1 \end{vmatrix}$

* 8. $\begin{vmatrix} 3 & -2 & 1 & 6 \\ 4 & 5 & 2 & -1 \\ 0 & 0 & 3 & 2 \\ -1 & 3 & 4 & -1 \end{vmatrix}$

* 9. $\begin{vmatrix} 0 & 1 & 2 & 3 & 4 \\ -1 & 0 & 5 & 6 & 7 \\ -2 & -5 & 0 & 8 & 9 \\ -3 & -6 & -8 & 0 & 10 \\ -4 & -7 & -9 & -10 & 0 \end{vmatrix}$

In Exercises 10–15 use Cramer's rule to solve the system of equations.

10. $-3x + 4y = 2,\ x - 2y = 6$

11. $2x - 3y = -10,\ x + y = 0$

12. $4r - 2s = 6,\ 3r - s = -1$

13. $2x - 3y + z = 2,\ 6x - y + 2z = 4,\ x - y = 1$

14. $3z - 2y + x = 6,\ z + y + 4x = 2,\ 2y - x - z = -1$

* 15. $2x - 3y + 4z + w = 1,\ x - 3y + 2w = 6,\ 3y + 4z - w = 2,\ 3x - y + z = 0$

* **16.** (a) Use mathematical induction to verify that when two parallel lines of a determinant with value D are interchanged, the value of the new determinant is $-D$.

   (b) Use part (a) to prove that a determinant with two identical parallel lines has value zero.

* **17.** A determinant is said to be *skew-symmetric* if its elements satisfy the property $a_{ij} + a_{ji} = 0$. (The determinant in Exercise 9 is skew-symmetric.) Show that a skew-symmetric determinant of odd order has value zero.

* **18.** Solve the system of equations

$$2x + 3y - 4z + w = 0, \quad x + y - 2z + 3w = 0,$$

$$2x - 3y + z - 2w = 0, \quad x + y - z + w = 0.$$

* **19.** Can we use Cramer's rule to solve the system

$$2x + 3y - 4z + w = 0, \quad x + y - 2z + 3w = 0,$$

$$2x - 3y + z - 2w = 0, \quad 5x + y - 5z + 2w = 0?$$

* **20.** (a) Show that the equation of the straight line in the xy-plane through the two points (x_1, y_1) and (x_2, y_2) can be expressed in the form

$$\begin{vmatrix} x & y & 1 \\ x_1 & y_1 & 1 \\ x_2 & y_2 & 1 \end{vmatrix} = 0.$$

   (b) Use the result in part (a) to find the equation of the line through $(1, 1)$ and $(-2, 3)$.

* **21.** (a) Show that the equation of the circle in the xy-plane through the three points (x_1, y_1), (x_2, y_2), and (x_3, y_3) can be expressed in the form

$$\begin{vmatrix} x^2 + y^2 & x & y & 1 \\ x_1^2 + y_1^2 & x_1 & y_1 & 1 \\ x_2^2 + y_2^2 & x_2 & y_2 & 1 \\ x_3^2 + y_3^2 & x_3 & y_3 & 1 \end{vmatrix} = 0.$$

   (b) Use the result in part (a) to find the circle through $(2, 1)$, $(-3, -3)$, and $(7, -5)$.

* **22.** A parabola of the form $y = ax^2 + bx + c$ is to pass through the three points $(1, 0)$, $(2, 11)$, and $(-2, 1)$. Find a determinant that defines its equation implicitly.

* **23.** Find a condition in the form of a determinant that serves as a test to determine whether a circle can be drawn through four given points, no three of which are collinear. *Hint:* See Exercise 21.

* **24.** Evaluate

$$\begin{vmatrix} a+b & a & a & \cdots & a \\ a & a+b & a & \cdots & a \\ a & a & a+b & \cdots & a \\ \vdots & \vdots & \vdots & \ddots & \vdots \\ a & a & a & \cdots & a+b \end{vmatrix}_{n \times n}.$$

* **25.** (a) Evaluate

$$\begin{vmatrix} a & b & b & b & \cdots & b & b \\ b & a & b & b & \cdots & b & b \\ b & b & a & b & \cdots & b & b \\ \vdots & \vdots & \vdots & \vdots & \ddots & \vdots & \vdots \\ b & b & b & b & \cdots & a & b \\ b & b & b & b & \cdots & b & a \end{vmatrix}_{n \times n}.$$

   (b) Use the result in part (a) to solve the system of equations

$$ax_1 + bx_2 + bx_3 + \cdots + bx_n = 1,$$

$$bx_1 + ax_2 + bx_3 + \cdots + bx_n = 1,$$

$$bx_1 + bx_2 + ax_3 + \cdots + bx_n = 1,$$

$$\vdots \qquad\qquad \vdots \quad \vdots$$

$$bx_1 + bx_2 + bx_3 + \cdots + ax_n = 1.$$

Complex Numbers

The fundamental complex number is i, a number whose square is -1; that is, i is defined as a number satisfying $i^2 = -1$. The **complex number system** is all numbers of the form

$$z = x + yi, \tag{C.1}$$

where x and y are real. The number x is called the **real part** of z, and y is called the **imaginary part** of z. For example, real and imaginary parts of $6 - 2i$ are 6 and -2. Both real and imaginary parts of a complex number are themselves real numbers. The real number system is a subset of the complex number system obtained when $y = 0$. We call $x + yi$ the *Cartesian form* for a complex number.

Complex numbers can be visualized geometrically as points in the complex (Argand) plane. Some fixed point O is chosen to represent the complex number $0 + 0i$. Through O are drawn two mutually perpendicular axes (Figure C.1), one called the *real axis* and the other called the *imaginary axis*. The complex number $x + yi$ is then represented by the point x units in the real direction and y units in the imaginary direction. For example, the complex numbers $1 + 2i$, $-1 - i$, $4 - 3i$, and $-2 + 2i$ are shown in Figure C.2. The real number system is represented by points on the real axis.

FIGURE C.1

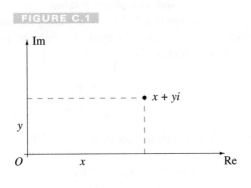

FIGURE C.2

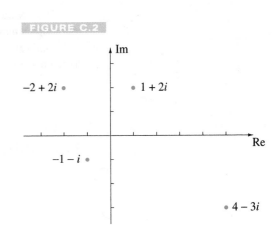

Two complex numbers $x + yi$ and $a + bi$ are said to be *equal* if their real and imaginary parts are identical; that is,

$$x + yi = a + bi \quad \Longleftrightarrow \quad x = a \quad \text{and} \quad y = b. \tag{C.2}$$

Geometrically, two complex numbers are equal if they correspond to the same point in the complex plane.

We add and subtract complex numbers $z_1 = x + yi$ and $z_2 = a + bi$ as follows:

$$z_1 + z_2 = (x + a) + (y + b)i, \tag{C.3a}$$

$$z_1 - z_2 = (x - a) + (y - b)i. \tag{C.3b}$$

In words, complex numbers are added and subtracted by adding and subtracting their real and imaginary parts. For example,

$$(3 - 2i) + (6 + i) = (3 + 6) + (-2 + 1)i = 9 - i,$$

$$(3 - 2i) - (6 + i) = (3 - 6) + (-2 - 1)i = -3 - 3i.$$

Complex numbers are multiplied according to the following definition. If $z_1 = x + yi$ and $z_2 = a + bi$, then

$$z_1 z_2 = (x + yi)(a + bi) = (xa - yb) + (xb + ya)i. \tag{C.4}$$

For example,

$$(3 - 2i)(6 + i) = [(3)(6) - (-2)(1)] + [(3)(1) + (-2)(6)]i = 20 - 9i.$$

It is not necessary to memorize C.4 when we note that this definition is precisely what we would expect if the usual laws for multiplying binomials were applied, together with the fact that $i^2 = -1$:

$$(3 - 2i)(6 + i) = (3)(6) + (3)(i) + (-2i)(6) + (-2i)(i)$$

$$= 18 + 3i - 12i - 2i^2$$

$$= 18 - 9i - 2(-1)$$

$$= 20 - 9i.$$

With addition, subtraction, and multiplication taken care of, it is natural to turn to division of complex numbers. If we accept that division of any complex number by itself should be equal to 1, and ordinary rules of algebra should prevail, a definition of division of complex numbers is not necessary; it follows from C.4. When $z_1 = x + yi$ and $z_2 = a + bi$, we calculate

$$\frac{z_1}{z_2} = \frac{x + yi}{a + bi}$$

by multiplying numerator and denominator by $a - bi$. This results in

$$\frac{z_1}{z_2} = \frac{x + yi}{a + bi} = \frac{(x + yi)(a - bi)}{(a + bi)(a - bi)}$$

$$= \frac{(xa + yb) + (-xb + ya)i}{a^2 + b^2} \qquad \text{(using C.4)}$$

$$= \left(\frac{xa + yb}{a^2 + b^2}\right) + \left(\frac{ya - xb}{a^2 + b^2}\right)i.$$

For example,

$$\frac{3 - 2i}{6 + i} = \frac{(3 - 2i)(6 - i)}{(6 + i)(6 - i)} = \frac{16 - 15i}{37} = \frac{16}{37} - \frac{15}{37}i.$$

In summary, addition, subtraction, multiplication, and division of complex numbers are performed using ordinary rules of algebra, with the extra condition that i^2 is always replaced by -1.

EXAMPLE C.1

Write the following complex numbers in Cartesian form:

$$\text{(a)} \quad (3+i)(2-i)^2 - i \qquad \text{(b)} \quad \frac{i^3}{2+i} \qquad \text{(c)} \quad \frac{4 - 3i^2 + 2i}{(2 - 2i^3)^2}$$

SOLUTION

(a) $(3+i)(2-i)^2 - i = (3+i)(3-4i) - i = (13 - 9i) - i = 13 - 10i$

(b) $\dfrac{i^3}{2+i} = \dfrac{-i(2-i)}{(2+i)(2-i)} = \dfrac{-1-2i}{5} = -\dfrac{1}{5} - \dfrac{2}{5}i$

(c) $\dfrac{4 - 3i^2 + 2i}{(2 - 2i^3)^2} = \dfrac{4+3+2i}{(2+2i)^2} = \dfrac{7+2i}{8i} = \dfrac{(7+2i)(-i)}{(8i)(-i)} = \dfrac{2-7i}{8} = \dfrac{1}{4} - \dfrac{7}{8}i$

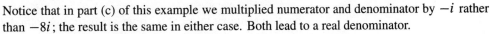

Notice that in part (c) of this example we multiplied numerator and denominator by $-i$ rather than $-8i$; the result is the same in either case. Both lead to a real denominator.

The complex conjugate $\bar{z}$ of a complex number $z = x + yi$ is

$$\bar{z} = x - yi. \tag{C.5}$$

Geometrically, $\bar{z}$ is the reflection of z in the real axis (Figure C.3).

FIGURE C.3

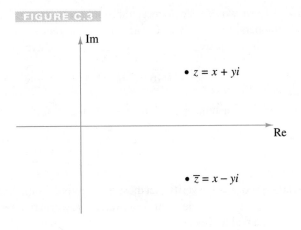

The procedure by which two complex numbers are divided can be stated as follows. To divide z_1 by z_2, multiply z_1 and z_2 by $\bar{z}_2$,

$$\frac{z_1}{z_2} = \frac{z_1 \bar{z}_2}{z_2 \bar{z}_2}.$$

The denominator will be real, and the Cartesian form is immediate.

With complex numbers in place, quadratic equations can be discussed fully. When the discriminant of a quadratic equation is positive, the equation has two real solutions. For example, the discriminant of

$$x^2 + 4x - 2 = 0$$

is $16 + 8 = 24$, and solutions of the equation are

$$x = \frac{-4 \pm \sqrt{16 + 8}}{2} = -2 \pm \sqrt{6}.$$

When the discriminant is zero, we regard the quadratic as having two real solutions that are identical. For example, the discriminant of

$$x^2 + 4x + 4 = 0$$

is zero. The left side may be factored in the form

$$(x + 2)^2 = 0.$$

We say that -2 is a double root of the equation or a root of multiplicity 2.

For quadratics with negative discriminants, we first consider the equation

$$x^2 + 1 = 0.$$

The complex number i is a solution, but so is $-i$ since $(-i)^2 + 1 = -1 + 1 = 0$. The quadratic equation

$$x^2 + 16 = 0$$

has two solutions, $x = \pm 4i$. If we apply the quadratic formula to the equation

$$x^2 + 2x + 5 = 0,$$

the result is

$$x = \frac{-2 \pm \sqrt{4 - 20}}{2} = \frac{-2 \pm \sqrt{-16}}{2}.$$

By $\sqrt{-16}$ we would seem to mean the number that multiplied by itself is -16. But there are two such numbers, namely $\pm 4i$. Let us make the agreement that $\sqrt{-16}$ shall denote that complex number whose square is -16, and which has a positive imaginary part. By this agreement,

$$\sqrt{-16} = 4i \qquad \text{and} \qquad -\sqrt{-16} = -4i.$$

The quadratic formula applied to $x^2 + 2x + 5 = 0$ therefore gives two complex numbers,

$$x = \frac{-2 \pm 4i}{2} = -1 \pm 2i.$$

It is straightforward to verify that these two complex conjugates actually satisfy $x^2 + 2x + 5 = 0$.

The agreement made in this last example is worth reiterating as a general principle: When $a > 0$ (is a real number),

$$\sqrt{-a} = \sqrt{a}\, i. \tag{C.6}$$

We call $\sqrt{a}\, i$ the **principal square root** of $-a$; the other square root is $-\sqrt{a}\, i$.

The examples above lead us to state that every real quadratic equation

$$ax^2 + bx + c = 0 \tag{C.7a}$$

has two solutions,

$$x = \frac{-b \pm \sqrt{b^2 - 4ac}}{2a}. \tag{C.7b}$$

When $b^2 - 4ac > 0$, roots are real and distinct; when $b^2 - 4ac = 0$, roots are real and equal; and when $b^2 - 4ac < 0$, roots are complex conjugates. Verification of this is a matter of substituting C.7b into C.7a.

The only other aspect of complex numbers that we use in the context of differential equations in Chapter 15 is Euler's identity. This is usually developed in courses in complex function theory. We can get to it in a formal way using Maclaurin series from Chapter 10. The real exponential function e^x can be defined by its Maclaurin series (see Example 10.10 in Section 10.3). Consider whether meaning can be given to e raised to a purely imaginary number θi, where θ is real; that is, can we give meaning to $e^{\theta i}$? One way would be to replace x in the Maclaurin series for e^x by θi, and call the result $e^{\theta i}$:

$$e^{\theta i} = \sum_{n=0}^{\infty} \frac{(\theta i)^n}{n!} = 1 + \theta i + \frac{(\theta i)^2}{2!} + \frac{(\theta i)^3}{3!} + \frac{(\theta i)^4}{4!} + \cdots.$$

If we gather even and odd terms in separate groups,

$$e^{\theta i} = \left(1 - \frac{\theta^2}{2!} + \frac{\theta^4}{4!} - \frac{\theta^6}{6!} + \cdots \right) + \left(\theta - \frac{\theta^3}{3!} + \frac{\theta^5}{5!} - \cdots \right) i.$$

But the two series on the right are Maclaurin series for $\cos \theta$ and $\sin \theta$; that is, we may write that

$$e^{\theta i} = \cos \theta + \sin \theta\, i.$$

In other words, if $e^{\theta i}$ is to have a value and this value is to be consistent with our theory on infinite series, then it must be defined as

$$e^{\theta i} = \cos \theta + \sin \theta\, i. \tag{C.8}$$

This is called *Euler's identity*. For instance, e^{2i} must be the complex number

$$e^{2i} = \cos 2 + \sin 2\, i.$$

It follows immediately that

$$e^{\pi i/2} = i, \quad e^{\pi i} = -1, \quad e^{3\pi i/2} = -i, \quad e^{2\pi i} = 1.$$

Euler's identity C.8 implies that

$$e^{-\theta i} = \cos (-\theta) + \sin (-\theta)\, i = \cos \theta - \sin \theta\, i. \tag{C.9}$$

When C.8 is added to this equation, we obtain

$$e^{\theta i} + e^{-\theta i} = 2 \cos \theta \quad \text{or} \quad \cos \theta = \frac{e^{\theta i} + e^{-\theta i}}{2}.$$

This expresses the real cosine function in terms of two complex exponentials. By subtracting rather than adding, we can express $\sin \theta$ in terms of complex exponentials,

$$\sin \theta = \frac{e^{\theta i} - e^{-\theta i}}{2i}.$$

EXERCISES

1. Show each of the following complex numbers in the complex plane: $2 - i, 3 + 4i, -1 - 5i, -3 + 2i, 5i, 2(1 + i)$.

In Exercises 2–26 write the complex expression in Cartesian form.

2. $(2 + 4i) - (3 - 2i)$

3. $(1 + 2i)^2$

4. $(-2 + i)(3 - 4i)$

5. $3i(4i - 1)^2$

6. $i^3 - 3i^2 + 2i + 4$

7. $(1 + i)^6$

8. $\dfrac{1 - i}{3 + 2i}$

9. $\dfrac{(3 + i)^2}{2 - i}$

10. $i^{24} - 3i^{13} + 4$

11. $(i - 2)[(2 + i)(1 - i) + 3i - 2]$

12. $6i\left(\dfrac{1 + i}{2 - i}\right) + 3\left(\dfrac{i - 4}{2i + 1}\right)$

13. $\overline{2 + i} - \overline{(3 + 4i)}$

14. $\overline{1 + i}^2 + \overline{(1 + i)^2}$

15. $\left(\dfrac{1}{2} - \dfrac{\sqrt{3}}{2}i\right)^3$

16. $(1 + \overline{2 - i})^2$

17. $\dfrac{(2i + 3)(4 - i)}{(3 + i)(-6 + 2i)}$

18. $\dfrac{(1 + i)^2(2 - i)}{(3 + 2i)^2}$

19. $(1 - i)^{12}(2i + 3)$

* **20.** $\overline{(4 - i)^2}$

* **21.** $\dfrac{1}{2i}\left(\dfrac{1}{\sqrt{2}} + \dfrac{i}{\sqrt{2}}\right)^{18}$

* **22.** $\dfrac{1}{1 + \dfrac{1}{1 + 2i}}$

* **23.** $\dfrac{i}{\dfrac{3 + i}{(2 - i)^2}}$

* **24.** $\dfrac{(1 + i^3)^2(2 - i)}{4 - 5i}$

* **25.** $\left(\dfrac{1}{\sqrt{2}} + \dfrac{i}{\sqrt{2}}\right)^4\left(\dfrac{1}{\sqrt{2}} - \dfrac{i}{\sqrt{2}}\right)^4$

* **26.** $\left(\dfrac{2i}{1 + i}\dfrac{3 - 4i}{3 + 4i}\right)^2$

In Exercises 27–36 find all solutions of the equation.

27. $x^2 + 5x + 3 = 0$

28. $x^2 + 3x + 5 = 0$

29. $x^2 + 8x + 16 = 0$

30. $x^2 + 2x - 7 = 0$

31. $x^2 + 2x + 7 = 0$

32. $4x^2 - 2x + 5 = 0$

* **33.** $\sqrt{3}\,x^2 + 5x + \sqrt{15} = 0$ * **34.** $x^4 + 4x^2 - 5 = 0$

* **35.** $x^4 + 4x^2 + 3 = 0$ * **36.** $x^4 + 6x^2 + 3 = 0$

* **37.** Verify the following properties for the complex conjugation operation:

(a) $\overline{z_1 + z_2} = \overline{z_1} + \overline{z_2}$

(b) $\overline{z_1 - z_2} = \overline{z_1} - \overline{z_2}$

(c) $\overline{z_1 z_2} = \overline{z_1}\,\overline{z_2}$

(d) $\overline{\left(\dfrac{z_1}{z_2}\right)} = \dfrac{\overline{z_1}}{\overline{z_2}}$

(e) $\overline{z^n} = \overline{z}^n$, n a positive integer

* **38.** Verify that all complex numbers z satisfying the equation $z\overline{z} = r^2$, $r > 0$ a real constant, lie on a circle. What are its centre and radius?

* **39.** Prove that if $z_1 z_2 = 0$, then at least one of z_1 and z_2 must be zero.

* **40.** We have made the agreement that when $a > 0$ is a real number, $\sqrt{-a}$ denotes a complex number with positive imaginary part. Show that with this agreement, $\sqrt{z_1 z_2}$ is not always equal to $\sqrt{z_1}\,\sqrt{z_2}$.

* **41.** Explain the fallacy in

$$-1 = \sqrt{-1}\,\sqrt{-1} = \sqrt{(-1)(-1)} = \sqrt{1} = 1.$$

* **42.** Find two numbers whose sum is 6 and whose product is 10.

* **43.** Verify that the values of x in C.7b satisfy C.7a.

* **44.** To find the square roots of a complex number, say i, we could set $(x + yi)^2 = i$, and solve the equation for x and y. Do this by using C.2 for equality of complex numbers.

* **45.** Use the technique of Exercise 44 to find square roots for (a) $-7 - 24i$ and (b) $2 + i$.

Answers to Even-Numbered Exercises

Chapter 1

Exercises 1.2

2. $-5/14$

4. $-5/4, 1/3$

6. $(5 \pm \sqrt{61})/4$

8. $9/2$ multiplicity 2

10. No real solutions

12. $-1/2$ multiplicity 3

14. -1

16. $-3, -4, \pm\sqrt{3}$

18. $-5, 3/2$

20. No real solutions

22. $-4, 6$ both of multiplicity 2

24. $5/6$

26. $1, 2, 3, 4, 5$

28. $2 \pm \sqrt{3}, 2/5$ multiplicity 2

30. ± 3

32. $2[x - (3 + \sqrt{65})/4][x - (3 - \sqrt{65})/4]$

34. $24(x + 5/3)(x - 1/4)(x - 1/2)$

36. $16(x + 1/2)^2(x - 1/2)^2$

Exercises 1.3

2. $3\sqrt{5}$

4. $\sqrt{58}$

6. $(1, -1/2)$

8. $(-1/2, 1/2)$

10. $y = -6$

12. $x + 2y + 2 = 0$

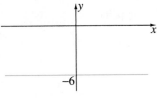

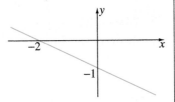

14. $y = 0$ is the x-axis

16. $y = 6x + 4$

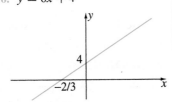

18. $y = -3x$

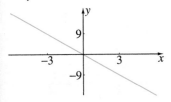

20. Parallel

22. Perpendicular

24. Neither

26. Neither

28. $(1, 2)$

30. $(-10, -14)$

32. $(37/73, 153/146)$

34. $8/\sqrt{5}$

36. $2\sqrt{2}$

38. $67/\sqrt{229}$

40. $x + y = 2$

42. $3y = 2x$

44. $5x + 3y = 30$

46. $y = -1$

48. (a) $T_K = 5(T_F - 32)/9 + 273.16$

(b) $T_F = 9(T_K - 273.16)/5 + 32$

(c)

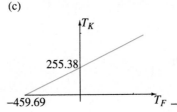

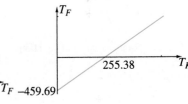

(d) 574.61

50. $(4/3, 10/3)$

52. $2x = 3y + 5$

54. No

Exercises 1.4A

2.

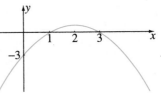

4.

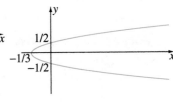

6.

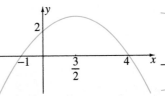

8.

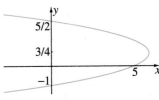

10.

12.

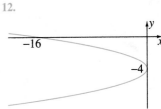

14. $y = x^2/2 + 1$

16. $y = (x^2 - 2x - 3)/3$

18. $(0, 1)$, $(-1, 0)$

20. No points

22. $(0, 1)$, $(-3, -2)$

24. $\pi/4$ radians

26. $(1, 1)$, $(4, 4)$

28. $y = (x^2 - 2x + 5)/2$

30. $(1, 0)$

Exercises 1.4B

2.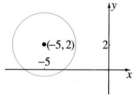

4.

6.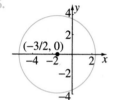

8.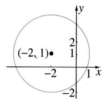

10. No points satisfy the equation

12. $(x - 1)^2 + y^2 = 1$

14. $(x - 3/2)^2 + (y + 3/2)^2 = 9/2$

16. $3x^2 + 3y^2 - 14x - 4y = 4$

18. (a) $(x + 5)^2 + (y + 2)^2 = 100$ (b) $(x + 7k + 19)^2 + (y - k)^2 = 50k^2 + 300k + 500$

20. $\big((-2 \pm \sqrt{14})/5, (9 \mp 2\sqrt{14})/5\big)$

22. $(0, \pm 4)$

26. $x^2 + (y + 140/3)^2 = 28\,000/9$

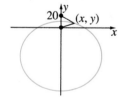

28. $(6/7, 12/7)$

Exercises 1.4C

2.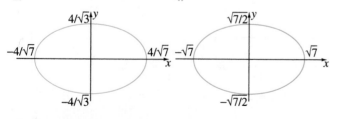

4.

6.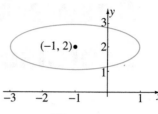

8.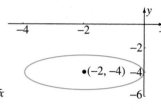

10. $32/\sqrt{15}$

12. $(-3, 0)$, $(21/25, 96/25)$

14. $(2, 3\sqrt{3}/2)$

16. $(\pm 2, 0)$, $(\pm\sqrt{15}/2, -1/4)$

Exercises 1.4D

2.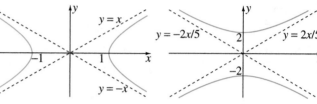

4.

6.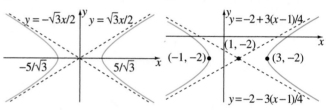

8.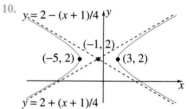

10.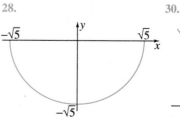

12. $\pm(\sqrt{2}, 1/\sqrt{2})$

14. No points

16. $(0, 0)$, $(3, \pm\sqrt{3})$

18. $(5, 1 \pm 3\sqrt{3})$

Exercises 1.5

2. $x \neq 2$

4. $|x| > 2$

6. $2 \leq |x| < 3$

8. $x = 0$ and $x \geq 1$

10. Odd

12. Odd

14. Neither

16. Neither

18. Even

20. $(x^2 + 10)/(x^2 - 25)$, $-7x/(x^2 - 25)$

22. 0, $x^3/(x^2 + 3)$

24. No even and odd parts

28.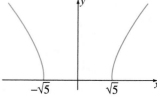

30.

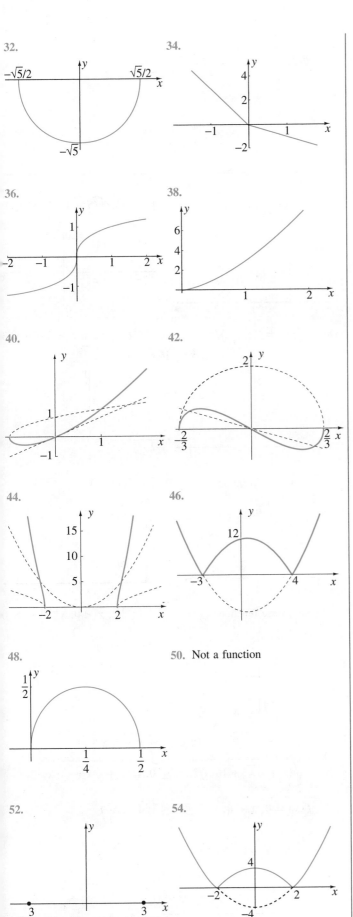

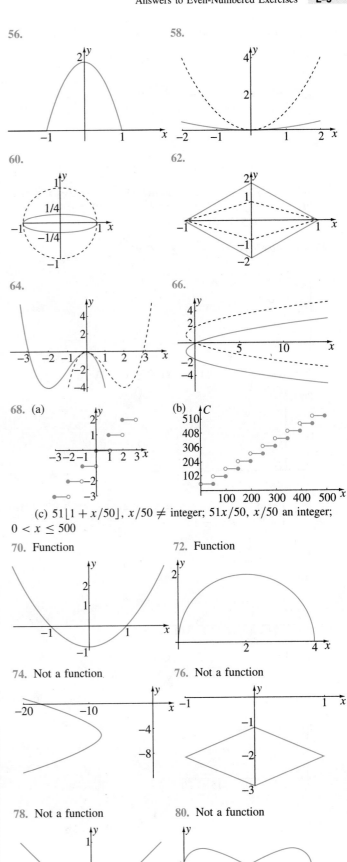

(c) $51\lfloor 1 + x/50 \rfloor$, $x/50 \neq$ integer; $51x/50$, $x/50$ an integer; $0 < x \leq 500$

70. Function

72. Function

74. Not a function

76. Not a function

78. Not a function

80. Not a function

82. (a) $s(t) = 0$, $t < 0$; $4t^2$, $0 \le t \le 1$; $2(3-t)$, $1 < t \le 3$; 0, $t > 3$

(b)

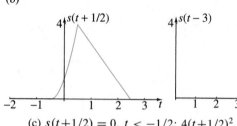

(c) $s(t+1/2) = 0$, $t < -1/2$; $4(t+1/2)^2$, $-1/2 \le t \le 1/2$; $5 - 2t$, $1/2 < t \le 5/2$; 0, $t > 5/2$ $s(t-3) = 0$, $t < 3$; $4(t-3)^2$, $3 \le t \le 4$; $2(6-t)$, $4 < t \le 6$; 0, $t > 6$

84. $A = (4bx/a)\sqrt{a^2 - x^2}$, $0 \le x \le a$

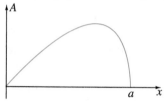

86. $R = (2k/9)(30 - x)(120 - x)$, $0 \le x \le 30$

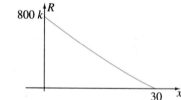

88. $A = \begin{cases} t^2/4, & 0 \le t \le 4\sqrt{2} \\ 4\sqrt{2}t - t^2/4 - 16, & 4\sqrt{2} < t \le 8\sqrt{2} \end{cases}$

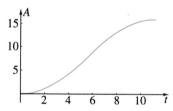

90. $y = \begin{cases} 0, & 5 \le x \le 5\sqrt{7} \\ -16 + \sqrt{x^2 + 81}, & 5\sqrt{7} < x \le 4\sqrt{34} \end{cases}$

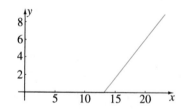

92.
94.

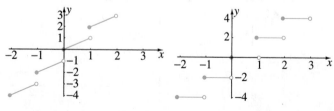

96.

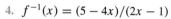
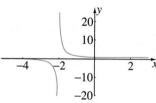
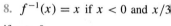

2. $f^{-1}(x) = x^2 - 1$, $x \ge 0$

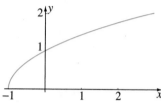

4. $f^{-1}(x) = (5 - 4x)/(2x - 1)$

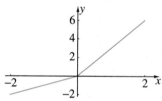

6. $f^{-1}(x) = [(x-2)/3]^{1/3}$

8. $f^{-1}(x) = x$ if $x < 0$ and $x/3$ if $x \ge 0$

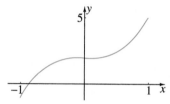 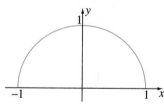

10. No inverse function

12. $f^{-1}(x) = 1 + \sqrt{x - 3}$

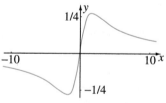 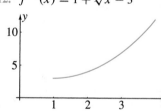

14. No inverse function

16. $-x^{-1/4}$, $x < 0$; $x^{-1/4}$, $x > 0$

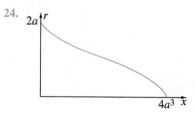

18. $-\sqrt{\sqrt{2 + x} - 2}$, $x \le 0$; $\sqrt{\sqrt{2 + x} - 2}$, $x \ge 0$

20. $-\sqrt{(x + \sqrt{x^2 + 16x})/2}$, $x \le 0$; $\sqrt{(x + \sqrt{x^2 + 16x})/2}$, $x \ge 0$

22. $f(x) = 0$, $x < 0$; x, $0 \le x \le 1$; 1, $x > 1$

24.

Exercises 1.7

2. $\pi/3$ rad   4. $-\pi/2$ rad   6. $17\pi/4$ rad

8. $-32\pi/45$ rad   10. $-213\pi/180$ rad   12. $-225°$

14. $1440°$ 16. $180/\pi°$ 18. $450/\pi°$

20. $1980/\pi°$ 22. 1.10×10^2 m   24. 20.6 m

26. 0.927 rad, 0.644 rad, $\pi/2$ rad   28. 2.57, 2.19, $19\pi/30$ rad

34. 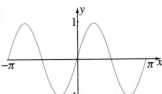   36.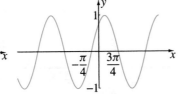

38. 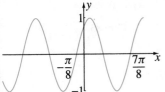   40.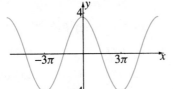

42. 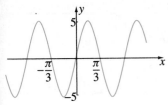   44.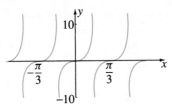

46. 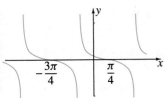   48.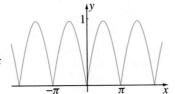

50. 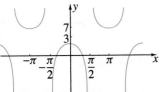   52.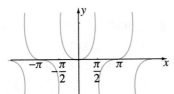

54.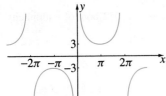

56. $(4n+1)\pi/2 \pm \pi/6$, n an integer

58. $\pm 2\pi/3 + 2n\pi$, n an integer

60. $(4n+1)\pi/4$, n an integer

62. $(8n \pm 3)\pi/8$, n an integer

64. $(4n+1)\pi/6 \pm \pi/9$, n an integer

66. $n\pi$, $(6n \pm 1)\pi/3$, n an integer

68. $(3n \pm 1)\pi/3$, n an integer

70. (b)

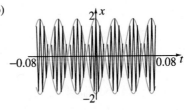

72. $3\sqrt{2} \sin(3x + \pi/4)$; Amplitude $= 3\sqrt{2}$, Period $= 2\pi/3$, Phase shift $= -\pi/12$

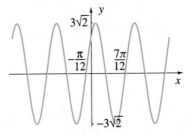

74. $4\sin(x + 2\pi/3)$; Amplitude $= 4$, Period $= 2\pi$, Phase shift $= -2\pi/3$

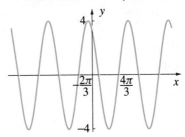

76. $(1/2)\sin 2x$; Amplitude $= 1/2$, Period $= \pi$, Phase shift $= 0$

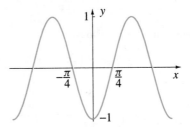

84. No   86. $\pi/4$, $\pi/2$ 88. None

Exercises 1.8

2. 0.253   4. $-2\pi/3$

6. Does not exist   8. $-\pi/4$

10. Does not exist   12. 0.164

14. $\pi/4$ 16. $-\pi/4$

18. $-0.876 + n\pi$, n an integer

20. $-0.107 + n\pi/4$, n an integer

22. $-0.273 + n\pi/3$, n an integer

24. No solutions   **26.** No solutions

28.

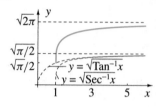

30.

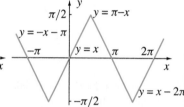

32. 2.68 rad   **34.** 1.25 rad

36. 0 rad   **38.** Parallel

40. Perpendicular   **42.** 1.37 rad

44. 1.11 rad   **46.** $(3n \pm 1)\pi/3$, n an integer

48. No solutions

50. $\text{Tan}^{-1}(2n\pi \pm 3\pi/4) + m\pi$, m and n integers

52. (a)   (b)

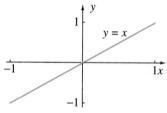

54. $\sqrt{17}\sin(2x + 0.245)$ **56.** $2\sqrt{5}\sin(2x + 2.03)$

58. $\sqrt{25 - 12\sqrt{3}}\sin(\omega t + 2.846)$

60. $\sqrt{13 + 12\cos 3}\cos(\omega t - 0.301)$

62. $\sqrt{45 + 14\sqrt{2} - 20\sqrt{3} - 4\sqrt{6}}\sin(\omega t + 0.858)$

64. $\text{Tan}^{-1}\left(\dfrac{R\sin\phi}{L + R\cos\phi}\right)$

72. $A = 10\cos 1$, $\phi = 2 + (4n - 1)\pi/2$, n an integer

Exercises 1.9

2. $(1/3)\log_{10} 5$ **4.** No solution   **6.** No solution

8. No solution   **10.** 5   **12.** $-1 \pm \sqrt{1 + a^2}$

14.   **16.**

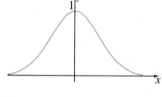

18.   **20.**

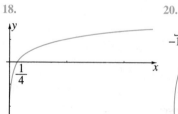

22.

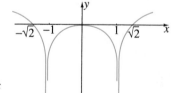

24.

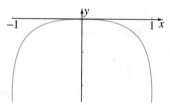

26.

28. $x = 11$

30. Yes. The domain of $f(x)$ is $x \neq 0$, but that of $g(x)$ is $x > 0$.

32. $20\,000(3/4)^t$ **36.** $x_1 > 0$ and $x_2 > 0$

38. (b) 9 years (c) $Pe^{it/100}$ (d) 1822.12, 1790.85

40. (a) L/R **42.** 0.694   **44.** $\sqrt{2}$

46. $\{20\,000[1 - (3/4)^t] + 250[(6/5)^t - 1]\}/t$

48. $(1/2)\ln(y + \sqrt{1 + y^2})$ **50.** $(1/2)\ln[(1 + y)/(1 - y)]$

Exercises 1.10

2. 2.30   **4.** 9.08×10^{-5} **6.** 1.00

8. 0.266   **10.** 5.07

Exercises 1.11

2. $-1.561\,553$, $2.561\,553$ **4.** $3.044\,723$

6. $0.754\,878$ **8.** $2.485\,584$

10. $\pm 0.795\,324$

12. $-2.931\,137$, $-2.467\,518$, $-1.555\,365$, $-0.787\,653$, $0.056\,258$, $0.642\,851$

14. $4.188\,760$ **16.** $0.852\,606$

18. 0, -1.2483 **20.** -1.892, -0.172, 3.064

22. ± 1.0986 **24.** 0.3212

26. $(-0.1725, 0.6848)$, $(-1.8920, 0.7956)$, $(3.0644, 16.1596)$

28. $(-1.3532, 3.8312)$

30. (a) 3.83 s, (b) 4.08 s   **32.** 0.000 0290 0

Review Exercises

2. $-3/2$, 3 multiplicity 2   **4.** $-5/3$, $3/2$ both of multiplicity 2

6. $5\sqrt{2}$, $(-1/2, -3/2)$ **8.** $y = 2x + 4$ **10.** $x + y = 2$

12. $|x| \geq \sqrt{5}$ **14.** $x \neq 0$, -1 **16.** $x \geq 0$

18. $|x + 1| > \sqrt{14}/2$ **20.** $-1 \leq x < 0$, $x \geq 1$ **22.** Parabola

24. Ellipse   **26.** None of these   **28.** Parabola

30. Circle   **32.** Hyperbola   **34.** $21/\sqrt{10}$

36.

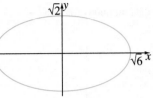

38.

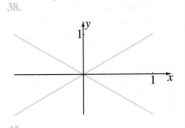

40.

42.

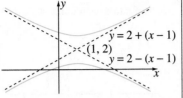

44.

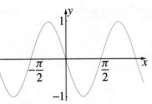

46.

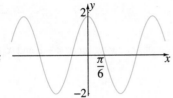

48.

50.

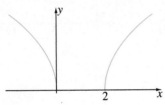

52.

54.

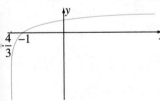

56.

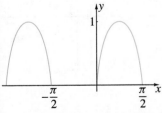

58.

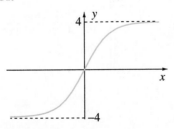

60. 0.833 rad

62. $\sqrt{(x+1)(2-x)}$

64. $1 + \sqrt{-x}$

66. $-\sqrt{4 + \sqrt{16+x}},\ x \le -2;\ -\sqrt{4 - \sqrt{16-x}},\ -2 \le x \le 0;$
$\sqrt{4 - \sqrt{16-x}},\ 0 \le x \le 2;\ \sqrt{4 + \sqrt{16+x}},\ x \ge 2$

68. $(x^2 - \sqrt{x^4 + 4x^2})/2,\ -1 \le x \le 0;\ (x^2 + \sqrt{x^4 + 4x^2})/2,\ x \ge 0$

70. 2π

72. $(4n+1)\pi/4 \pm 0.659$, n an integer

74. -1.79

76. $(4n+1)\pi/2 \pm 0.604$, n an integer

78. $\ln\left[(4n+1)\pi/2 \pm 0.84\right] - 2$, $n \ge 0$ an integer

80. 2.09 **82.** 0.760 s **84.** 1.526

86. -11.6187, -0.8738, 0.4925

Chapter 2

Exercises 2.1

2. 0 **4.** $-3/2$ **6.** 0

8. -12 **10.** 6 **12.** -4

14. 0 **16.** -1 **18.** 0

20. $-7/4$ **22.** Does not exist **24.** $1/\sqrt{2}$

26. 2 **28.** $2\sqrt{2}$ **30.** 1

32. Does not exist **34.** $-1/4$ **36.** 1

38. 4 **40.** -2 **42.** $2a$

44. $-1/(a+1)$ **46.** $1/(2\sqrt{a})$ **48.** $1/\sqrt{a}$

50. 1 **52.** 0 **54.** False

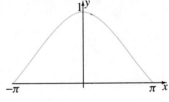

56. $1/(2\sqrt{x})$ **60.** L

62. Cannot find **64.** $-L$

66. 0 if $F = 0$; does not exist if $F \ne 0$

Exercises 2.2

2. $-\infty$ **4.** ∞ **6.** ∞

8. $13/2$ **10.** Does not exist **12.** 1

14. ∞ **16.** ∞ **18.** Does not exist

20. ∞ **22.** 0 **24.** ∞

26. ∞ **28.** Does not exist **30.** ∞

Exercises 2.3

2. $-1/2$ **4.** ∞ **6.** $1/3$

8. $-\infty$ **10.** $3/2$ **12.** 0

14. 1 **16.** ∞ **18.** ∞

20. $3/4^{1/3}$ **22.** 0 **24.** 0

26. Does not exist **28.** 0 **30.** ∞

32. 1 **34.** -1 **36.** 0

38. ∞ **40.** $1/\sqrt{a}$ **42.** $-\sqrt{a}$

44. Vertical $x = 5/2$; horizontal $y = 1/2$ approached from above as $x \to \infty$, and from below as $x \to -\infty$

46. Vertical $x = -3/2$; horizontal $y = \sqrt{5}/2$ approached from below as $x \to \infty$, and $y = -\sqrt{5}/2$ approached from below as $x \to -\infty$

48. Vertical $x = (3 \pm \sqrt{5})/2$; oblique $y = 3x + 9$ approached from above as $x \to \infty$, and from below as $x \to -\infty$

Exercises 2.4

2. Removable discontinuity at $x = -4$

4. No discontinuities

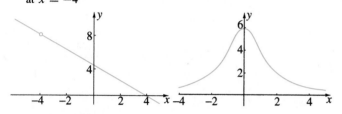

6. No discontinuities

8. Removable discontinuities at $x = 0, 1$

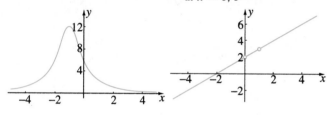

10. Infinite discontinuities at $x = (2n + 1)\pi/2$

12. Discontinuous at $x = 0$

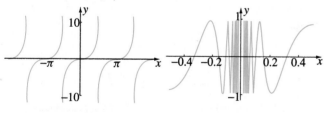

14. Infinite discontinuities at $x = \pm 3$

16. Jump discontinuity at $x = 3$

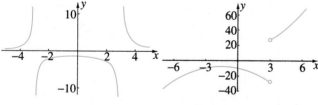

18. Infinite discontinuities at $x = -1, 2$

20. Infinite discontinuities at $x = -1, 7$

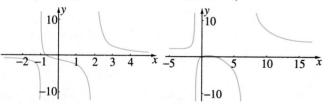

22. Infinite discontinuity at $x = 1$

24. Infinite discontinuities at $x = -5, 0$

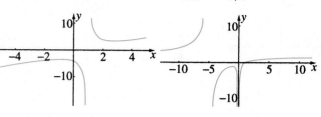

26. Infinite discontinuity at $x = -5$

28. Continuous for $x > -5$

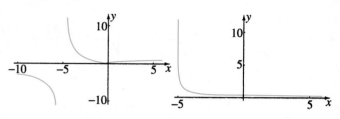

30. Continuous for $x < -2$ and $x \geq 3$

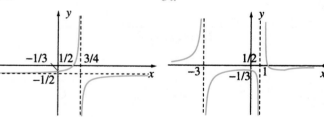

32.   34.

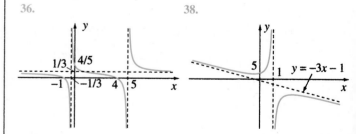

36.   38.

40.   42.

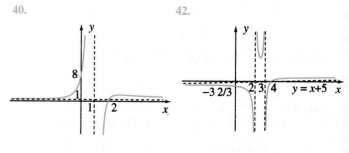

44. No

48. (a) Discontinuous at
$x = n/10$

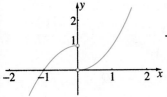

50. (a) Discontinuous at
$x = n + 1/2$

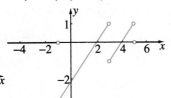

Exercises 2.5

2. $1 - x^2 + (2x^2 - 1)h(x)$

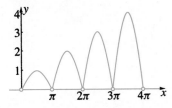

4. $(x - 2)h(x + 1) - 2h(x - 3) - (x - 4)h(x - 5)$

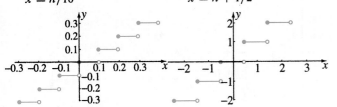

6. $\sin x[h(x) - 3h(x - \pi) + 5h(x - 2\pi) - 7h(x - 3\pi) + 4h(x - 4\pi)]$

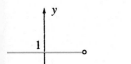

8. $10 \sin 4t[h(t - 1) - h(t - 1 - \pi)]$

10. $(2t + 80)[h(t - 10) - h(t - 60)]$

12. $-(2mg/L)[h(x) - h(x - L/2)]$

14. $F_1\delta(x - x_1) - F_2\delta(x - x_2)$

16. $h(x - a) - h(x - b) + h(x - c)$

18. Yes, except at $x = a$ where $h(x - a)h(x - b)$ is undefined

20.

22.

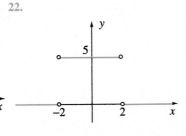

24.

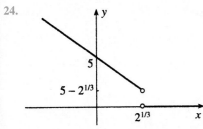

Exercises 2.6

10. $\lim_{x \to a^+} f(x) = L$ if given any $\epsilon > 0$, there exists a $\delta > 0$ such that $|f(x) - L| < \epsilon$ whenever $0 < x - a < \delta$.

12. $\lim_{x \to \infty} f(x) = L$ if given any $\epsilon > 0$, there exists an $X > 0$ such that $|f(x) - L| < \epsilon$ whenever $x > X$.

14. $\lim_{x \to a} f(x) = \infty$ if given any $M > 0$, there exists a $\delta > 0$ such that $f(x) > M$ whenever $0 < |x - a| < \delta$.

16. $\lim_{x \to \infty} f(x) = \infty$ if given any $M > 0$, there exists an $X > 0$ such that $f(x) > M$ whenever $x > X$.

18. $\lim_{x \to -\infty} f(x) = \infty$ if given any $M > 0$, there exists an $X < 0$ such that $f(x) > M$ whenever $x < X$.

30. No

Review Exercises

2. -2	4. 1	6. $-1/4$
8. 0	10. 0	12. ∞
14. $1/2$	16. 0	18. $-\sqrt{3}/2$

20. $-\infty$

22. Infinite discontinuity
at $x = 2$

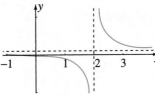

24. Removable discontinuity
at $x = 6$

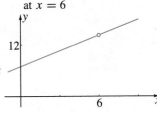

26. Infinite discontinuity
at $x = 1$

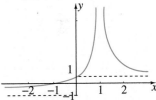

28. Infinite discontinuity
at $x = 1$

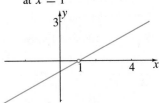

30. Infinite discontinuities
at $x = -1, 4$

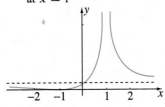

32. Removable discontinuity
at $x = 1$

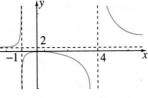

34. $3 + x^3 + (-x^3 + x^2 - 1)h(x + 1) + (2 - x^2)h(x - 2)$

Chapter 3

Exercises 3.1

2. $6x$	4. $3x^2 + 4x$	6. $-9/(x - 5)^2$
8. $3x^2 + 4x$	10. $(2x^2 - 2)/(x^2 + x + 1)^2$	
12. $2\pi r$	14. $4\pi r^2$	16. m
18. 0	20. $y = 2x + 2$	22. $x + 4y = 3$

24. 1.107 rad   **26.** 2.897 rad   **28.** $8x^7$

30. $-4/(x-2)^5$ **32.** $(3x+2)/(2\sqrt{x+1})$

34. $\sqrt{A}/(4\sqrt{\pi})$ **36.** $|x|/x$

38.

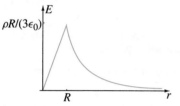

Exercises 3.2

2. $9x^2 + 4$ **4.** $20x^4 - 30x^2 + 3$

6. $-6/x^4$ **8.** $1/x^3 - 12/x^5$

10. $20x^3 - 5/(4x^6)$ **12.** $1/(2\sqrt{x})$

14. $-(3/2)/x^{5/2} + (3/2)\sqrt{x}$ **16.** $\pi^2 x^{\pi-1}$

18. $4 + 3/x^4$ **20.** $6(2x+5)^2$

22. $4y = x + 24,\ 4x + y = 23$ **24.** $y = 6x - 10,\ x + 6y = 51$

28. (a)   (b) 50 Hz, 150 Hz

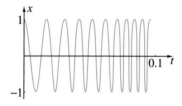

30. $(2, -12)$ **34.** $x = 15(4 + \sqrt{194})$ m

36. $((1+\sqrt{3})/2, (2+\sqrt{3})/2),\ ((1-\sqrt{3})/2, (-4-\sqrt{3})/2)$ and $((1-\sqrt{3})/2, (2-\sqrt{3})/2),\ ((1+\sqrt{3})/2, (-4+\sqrt{3})/2)$

38. $n|x|^{n-1}\mathrm{sgn}(x)$

Exercises 3.3

2. Right; no left; no derivative   **4.** No right; no left; no derivative

6. Right; no left; no derivative   **8.** Right; no left; no derivative

10. True   **12.** True

14.   **16.**

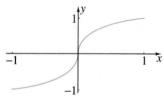

18. 1.231 rad at $(\pm 1/\sqrt{2}, 1/2)$

20. Orthogonal at $(1, 2)$ and $(-1, 2)$

24. 2/3   **26.** Yes   **28.** Not necessarily

30. (a)   (b)

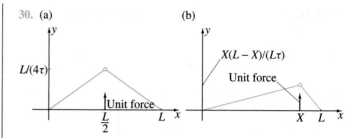

32. No

Exercises 3.4

2. $4(2 - 3x^2 - x^3)$ **4.** $-10x/(4x^2 - 5)^2$

6. $x^2(4x^2 + 3)/(4x^2 + 1)^2$ **8.** $(2 - 3x)/[2\sqrt{x}(3x + 2)^2]$

10. $-(2x^2 + 20x + 1)/(2x^2 - 1)^2$

12. $(17 - 4x - 7x^2)/(x^2 - 5x + 1)^2$

14. $-27/(x + 1)^4$ **16.** $(x - 8\sqrt{x} - 2)/[\sqrt{x}(\sqrt{x} - 4)^2]$

18. (a)   (b) $x(x-4)/(x^2 + x - 2)^2$

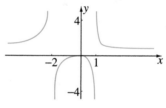

20. $f'(x)g(x)h(x) + f(x)g'(x)h(x) + f(x)g(x)h'(x)$

22. 0.668 at $(\sqrt{2}, 2\sqrt{2} + 2)$; 0.415 at $(-\sqrt{2}, 2 - 2\sqrt{2})$

24. $(5 + \sqrt{55}, -\sqrt{55}/(11 + \sqrt{55})),\ (5 - \sqrt{55}, \sqrt{55}/(11 - \sqrt{55}))$

26. (a) $(2x + 1)h(x + 1)$

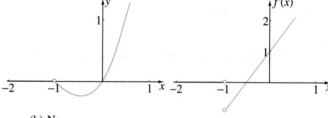

(b) No

Exercises 3.5

2. 6   **4.** 18   **6.** $6 + 60/t^6$

8. $(3u^2 - 6u - 1)/[4u^{3/2}(u + 1)^3]$

10. $-(\sqrt{x} + 3)/[4\sqrt{x}(\sqrt{x} + 1)^3]$

12. (b) $c = \dfrac{ab}{b-a}\left[(T_a - T_b) + \frac{k}{6}(b^2 - a^2)\right],\ d = \dfrac{bT_b - aT_a}{b-a} - \frac{k}{6}(a^2 + ab + b^2)$

14. (a) $f'''g + 3f''g' + 3f'g'' + fg'''$
(b) $f''''g + 4f'''g' + 6f''g'' + 4f'g''' + fg''''$

16. (a) Yes

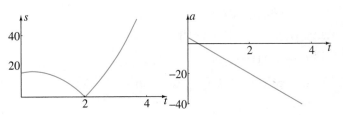

Exercises 3.6

2. (a) Left, right (b) Left, right (c) Three (d) 7/2 (e) 9/2

4. (a) Right, left (b) Right, left (c) Two (d) 9/2 (e) 9/2

6. 28/15

8. $v = 2t - 7$ m/s, $a = 2$ m/s^2

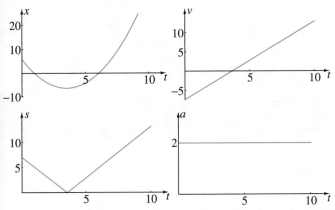

10. $v = 4 - 3t^2$ m/s, $a = -6t$ m/s^2

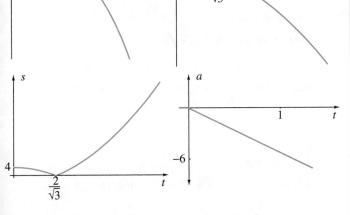

12. $v = -6t^2 + 4t + 16$ m/s, $a = -12t + 4$ m/s^2

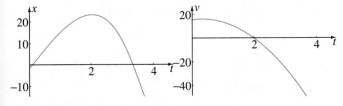

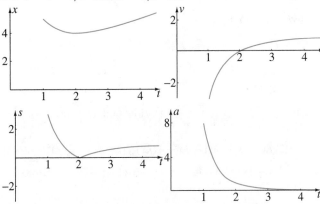

14. $v = 1 - 4/t^2$ m/s, $a = 8/t^3$ m/s^2

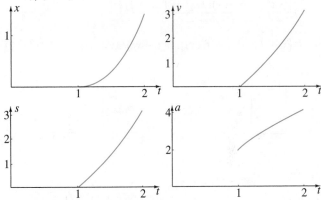

16. $v = (5t^2 - 6t + 1)/(2\sqrt{t})$ m/s, $a = (15t^2 - 6t - 1)/(4t^{3/2})$ m/s^2

18. (a) -11 m, -12 m/s, 12 m/s, 0 m/s^2 (b) $t = 1$ s, $t = 5$ s
(c) $t = 3$ s (d) $0 \le t < 1$, $t > 5$; $1 < t < 5$ (e) $(9 \pm \sqrt{39})/3$
s (f) $(9 \pm \sqrt{39})/3$ s, $(9 \pm \sqrt{33})/3$ s (g) $(9 + 4\sqrt{6})/3$ s
(h) $(9 + 4\sqrt{6})/3$ s

20. (a) $a(t_1 + t_2)/2 + b$ (b) $(t_1 + t_2)/2$, closer to x_1

22. $a = (y_1 - y_2 + 2y_3)^2/(4y_1\theta_3^2)$, $m = (y_1 - y_2 + 2y_3)/\theta_3$,
$b = -y_1$, $A = (y_1 - y_2 + 2y_3)^2/[4\theta_3^2(y_2 - y_3)]$

Exercises 3.7

2. $1/[2\sqrt{x}(u + 1)^2]$ **4.** $2(1 - x)(s^2 + 2)/(s^2 - 2)^2$

6. $-21/[(t - 4)^2(x + 1)^2]$

8. $(4u + 5u^{3/2})(x^2 + 2x - 1)/[2(x - x^2)^2]$

10. $(3x + 2)/[2\sqrt{x + 1}]$ **12.** $(x + 1)/(2x + 1)^{3/2}$

14. $(2x - 1)(6x + 23)/(3x + 5)^2$

16. $x^2(18 - 55x^2)/[3(2 - 5x^2)^{2/3}]$

18. $(x + 1)(15x + 11)(3x + 1)^2$

20. $(9x - 8)/[2x^3\sqrt{2 - 3x}]$

22. $-1/[(2 + x)^{5/4}(2 - x)^{3/4}]$

24. $(x + 5)^3(11x^3 + 15x^2 + 8)/[2\sqrt{1 + x^3}]$

26. $(x + 5)^3(13x^4 + 25x^3 + 10x + 10)/[2\sqrt{1 + x^3}]$

28. $(7x^2 + 6x + 4\sqrt{1 + x})/[4\sqrt{1 + x}\sqrt{1 + x\sqrt{1 + x}}]$

30. $(4/3)x(s - 1)/[(x^2 + 5)^2(2s - s^2)^{2/3}]$

32. $(5/2)(2 - x)/[x^2\sqrt{x - 1}(u + 5)^2]$

34. $[1 + (2\sqrt{t} + 1)/(4\sqrt{t}\sqrt{t + \sqrt{t}})][-4x/(x^2 - 1)^2]$

36. $(1 - k - 3k^2)(x^2 + 5)^4(11x^2 + 5)/[2\sqrt{k}(1 + k + k^2)^2]$

38. -0.067 N/s

44. $(6u^4 + 6u^3 - 2)(2\sqrt{x + 1} + 1)^2/[4u^3(x + 1)] - [3(u^2 + u)^2 + 1]/[4u^2(x + 1)^{3/2}]$

46. $-3[2\sqrt{x} + (s + 6)(1 + 3x + 4\sqrt{x})]/[2x^{3/2}(1 + \sqrt{x})^4(s + 6)^3]$

48. $-8f(3 - 4x)f'(3 - 4x); 8(4x - 3)(16x^2 - 24x + 7)(48x^2 - 72x + 25)$

50. $(1 - 1/x^2)f'(x + 1/x); (x^2 - 1)(3x^4 + 4x^2 + 3)/x^4$

52. $12f(1 - 3x)f'(1 - 3x)/\sqrt{3 - 4[f(1 - 3x)]^2}; 12(1 - 3x)(9x^2 - 6x - 1)(27x^2 - 18x + 1)/\sqrt{3 - 4(1 - 3x)^2(9x^2 - 6x - 1)^2}$

54. $f'(x - f(x))[1 - f'(x)]; 3(1 - x^2)(3x^6 - 18x^4 + 27x^2 - 2)$

56.

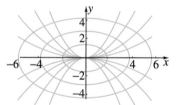

58.

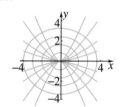

60. 444

62. $-x(x + 1)^2/\sqrt{1 - x^2}$

64. $((-8 \pm 24\sqrt{39})/35, (-12 \pm \sqrt{39})/35)$

66. $(-4^{1/3}, \pm(4^{1/3} - 1)^{3/2})$

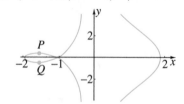

Exercises 3.8

2. $-4x^3/(2y + 3y^2)$

4. $(6x^2 - 3y^4 + 5y)/(12xy^3 - 5x)$

6. $(1 - x - y)/(x + y)$

8. $(48x\sqrt{x + y} - 1)/[1 + (4y - 2)\sqrt{x + y}]$

10. $-(6x + 5y)/(5x + 2y)$

12. $-[2(3y^2 + 1)^2 + 24x^2y]/(3y^2 + 1)^3$

14. $-(25/4)/(y + 1)^3$

16. -1

18. $-(2xy^3 + 2)/(3x^2y^2 + 4), [-2y^3(3x^2y^2 + 4)^2 + 12xy^2(2xy^3 + 2)(3x^2y^2 + 4) - 6x^2y(2xy^3 + 2)^2]/(3x^2y^2 + 4)^3$

20. $\sqrt{1 - y^2}/(1 - 2y^2), (3y - 2y^3)/(1 - 2y^2)^3$

22. $-3/4$

24. $0, -1/2$

26. $(1, \pm 1), (-1, \pm 1)$

28. (a) $(x^2 + 4x + 2)/(x^2 + 3x + 2)$ (b) $(2y + 1)(3 - y)/(7y)$

30. $(1, 2)$

34. $(3r^3 - 6ar^2 + x)/(3r^2 - 6ar)$

36. (a) -1

42. (a) $-\sqrt{y/x}$ (b)

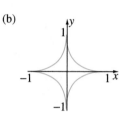

44. $2xu/[(u^2 - 1)(1 + 2x^2\sqrt{u^2 - 1})]$

52. $(3a/2, 3a/2)$

Exercises 3.9

2. $-\sin x - 20\cos 5x$

4. $-9\sec^2 3x/\tan^4 3x$

6. $2\csc(4 - 2x)\cot(4 - 2x)$

8. $\cot x^2 - 2x^2\csc^2 x^2$

10. $[(x^2 + x)\cos x + \sin x]/(x + 1)^2$

12. $2\cos 4x$

14. $(3/4)\tan^2 x\sec^2 x/(1 + \tan^3 x)^{3/4}$

16. $(\cos y + y\sin x)/(\cos x + x\sin y)$

18. $\sec^2(x + y)/[1 - \sec^2(x + y)]$

20. $3(1 - x^2y)/(x^3 + 2\tan y\sec^2 y)$

22. $-\sec^2 x\sin(\tan x)$

24. $4x\sin(2x^2)$

26. $-\sec v\tan v\sec^2\sqrt{x}/[4\sqrt{x}\sqrt{3 - \sec v}]$

28. $-x\sin x^2\sec^3 u\tan u/[\sqrt{1 + \cos x^2}(1 + \sec^3 u)^{2/3}]$

30. $[18x^2\tan^2(3x^2 - 4)\sec^2(3x^2 - 4) - (2 + x\cot x)(1 + \tan^3(3x^2 - 4))]/(x^3\sin x)$

32. $2(y + 1)[\sec^2 y - x - (y + 1)\sec^2 y\tan y]/(\sec^2 y - x)^3$

34. 0

36. 2

38. Does not exist

40. No

44. $\pm 1, (n\pi \pm \sqrt{n^2\pi^2 - 4})/2, n > 0$ an integer

Exercises 3.10

2. $-2x/[1 + (x^2 + 2)^2]$

4. $-2x/[1 + (2 - x^2)^2]$

6. $\text{Csc}^{-1}(x^2 + 5) - 2x^2/[(x^2 + 5)\sqrt{(x^2 + 5)^2 - 1}]$

8. $1/[2(x + 3)\sqrt{x + 2}]$

10. $-1/(x\sqrt{x^2 - 1})$

12. $2x\,\text{Sec}^{-1} x + x/\sqrt{x^2 - 1}$

14. $-1/(1+x^2)$ 16. $-1/[\sqrt{x}(x+1)]$

18. $x(t - \sqrt{1-t^2}\,\mathrm{Cos}^{-1}t)/[\sqrt{1-x^2}\sqrt{1-t^2}]$

20. $(5\sqrt{1-x^2y^2} - y)/(x - 2\sqrt{1-x^2y^2})$

22. $-18/(x^3\sqrt{x^2-9}$

24. $-(1/x^2)\,\mathrm{Csc}^{-1}(3x) - 2/(x^2\sqrt{9x^2-1})$

26. $(\mathrm{Cos}^{-1}x)^2 - 2 + 2x(1 - \mathrm{Cos}^{-1}x)/\sqrt{1-x^2}$

28. $2\sqrt{4x - x^2}$

30. $\sqrt{2}(1-x^2)/[(1+x^2)\sqrt{1+x^4}]$

32. $-1/(|x|\sqrt{x^2-1})$

Exercises 3.11

2. $6x/(3x^2+1)$ 4. $-2e^{1-2x}$ 6. $\ln x + 1$

8. $-4\log_{10} e/(3 - 4x)$ 10. $-\tan x$

12. $2x + (4x^3 + 3x^2)e^{4x}$ 14. $2e^{2x}\cos(e^{2x})$

16. $e^{-2x}(3\cos 3x - 2\sin 3x)$ 18. $4(e^x + e^{-x})^{-2}$

20. $2\sin(\ln x)$ 22. $-\tan v \sin 2x$

24. $-(y\cos x + 2x\ln y + e^y)/(xe^y + \sin x + x^2/y)$

26. $2x^2/\sqrt{x^2+1}$ 28. $(1 + 5e^{4x})^{-1}$

30. $(y^2/x^2)(e^{1/x} - 1)/(1 - e^{1/y})$

32. (b) $[T_b - T_a + k(a - b)]/\ln(b/a),\ [T_a\ln b - T_b\ln a + k(b\ln a - a\ln b)]/\ln(b/a)$

34. (b) $\lim_{t\to\infty} x(t) = 30$ 36. (a) $dT/dt + 3T/10\,000 = (20t + 1257)/419\,000$

(b)

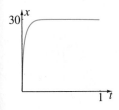

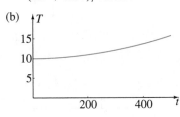

40. (b) Asymptotic to $S = 500 + t/10$

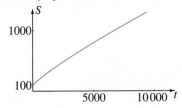

42. (b)   (c) $1000\ln(4/3)$ s

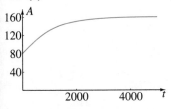

44.

$$\frac{\mu_0 i}{2\pi}\ln\left[\frac{R(r+w)}{r(R+w)}\right]$$

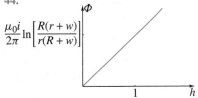

$$\frac{\mu_0 h i}{2\pi}\ln\left(\frac{R}{r}\right)$$

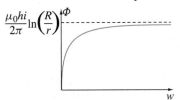

46. $1/\sqrt{x^2+4}$ 48. $2(x^2+y^2)/(x-y)^3$

Exercises 3.12

2. $4x^{4\cos x}(\cos x/x - \sin x\ln x)$

4. $(\sin x)^x[x\cot x + \ln(\sin x)]$

6. $x(1 + 1/x)^{x^2}[2\ln(1 + 1/x) - 1/(1+x)]$

8. $(3/x^2)(2/x)^{3/x}(\ln x - \ln 2 - 1)$

10. $(1/x)(\ln x)^{\ln x}[\ln(\ln x) + 1]$

12. $(1 + 9x^2 + 6x^4)/[2\sqrt{x}(1+x^2)^{3/2}]$

14. $(x^2 + 3x)^2(x^2+5)^3(14x^3 + 33x^2 + 30x + 45)$

16. $\sqrt{x}e^{-2x}(3 - 4x)/2$

18. $e^x[(x-1)\ln(x-1) - 1]/\{(x-1)[\ln(x-1)]^2\}$

20. $(1 - 6x^2 - 3x^4)/[2\sqrt{x}(1+x^2)^{3/2}]$

22. $x^3(13x^4 - 44x^3 + 10x - 32)/(2\sqrt{1+x^3})$

24. $\sin 2x\sec 5x[2\cot 2x + 5\tan 5x - 6\csc^2 x/(1 - 2\cot x)]/(1 - 2\cot x)^3$

26. (b) $(a - br)^{-1}$

Exercises 3.13

2. $\sinh(x/2) + (x/2)\cosh(x/2)$

4. $(1/x)\,\mathrm{sech}^2(\ln x)$

6. $2\sqrt{1+y}\,\mathrm{csch}^2 x/(2\sqrt{1+y} - 1)$

8. $-\sinh t\,\mathrm{sech}^2 x\sec^2(\cosh t)\sin(\tanh x)$

10. $2\,\mathrm{csch}\,4x$

Exercises 3.14

2. 2

4. Cannot be applied since $f'(0)$ does not exist

6. $(-2 + \sqrt{19})/3$ 8. $1 + \sqrt{3}$ 10. $-3 + \sqrt{6}$

12. $(4 - \ln 5)/(2\ln 5)$

14. Cannot be applied since $f(\pi/2)$ is not defined

16. Cannot be applied since $g'(0) = 0$

18. $(1 + \sqrt{6})/(1 - \sqrt{6})$

Review Exercises

2. $6x + 2 - 1/x^2$

4. $(1/3)x^{-2/3} - (10/9)x^{2/3}$

6. $x(x^2 + 2)(x^3 - 3)^2(13x^3 + 18x - 12)$

8. $17/(x + 5)^2$

10. $-4(x^2 + 2)/(x^2 + 5x - 2)^2$

12. $2x(y - 1)/(2y - x^2)$

14. $-(4xy\sqrt{1 + x} + y)/[2\sqrt{1 + x}(x^2 + \sqrt{1 + x})]$

16. $-8\sec^2(1 - 4x)\tan(1 - 4x)$

18. $2\sec^2 2x\sec(\tan 2x)\tan(\tan 2x)$

20. $(1/2)\sin 4x$

22. $(1 - 2\sin 2t)(1 + 2\sin 2x)$

24. $(-x/v)(\cos^2 v - v\sin 2v)$ 26. $(1/2)e^{-2y}$

28. $2x(x^2 + 1)/(x^2 - 1) + 2x\ln(x^2 - 1)$

30. 1

32. $6x\cosh(x^2)$

34. $(x^2 - 1)/(x^4 + 3x^2 + 1)$ 36. $\cos x\operatorname{sech} y$

38. $\operatorname{Csc}^{-1}(1/x^2) + 2x^2/\sqrt{1 - x^4}$

40. -1

42. $16x/(x^2 + 1)^2$

44. $x(20 - 23x - 3x^2)/[2\sqrt{1 - x}(x + 5)^2]$

46. $(x - y)/(x + y)$

48. $(y^3 + 4y + 6)^2/(-y^4 + 4y^3 + 4y^2 + 12y - 12)$

50. $1/(2\sqrt{x})$

52. $(\cos x)^x[\ln(\cos x) - x\tan x]$

54. $(\log_{10} e)^2/(x\log_{10} x)$ 56. $1/(e^y - e^{-y})$

58. $[(x + y)(1 - 2xy) - 1]/[x^2(x + y) + 1]$

60. $x + 25y = 5$, $125x - 5y + 1 = 0$

62. $23x + 9y = 11$, $27x - 69y = 119$

64. $-42(x^2 - 5xy + y^2)/(5x - 2y)^3$

66. No

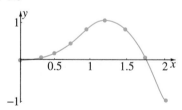

68. $(-3, 1)$; finding the shortest distance from $(-6, 7)$ to the curve

70. $\sqrt{3}L/2$

72. $(\sqrt{6}/4, \pm\sqrt{2}/4)$, $(-\sqrt{6}/4, \pm\sqrt{2}/4)$

74. $(\sqrt{93} - 9)/6$

Chapter 4

Exercises 4.1

2. $-1.561\,553$, $2.561\,553$

4. $3.044\,723$

6. $0.754\,878$

8. $2.485\,584$

10. $\pm 0.795\,324$

12. $-2.931\,137$, $-2.467\,518$, $-1.555\,365$, $-0.787\,653$, $0.056\,258$, $0.642\,851$

14. $4.188\,760$

16. $0.852\,606$

18. 0, -1.2483

20. -1.892, -0.172, 3.064

22. ± 1.0986

24. 0.3212

26. $(-0.1725, 0.6848)$, $(-1.8920, 0.7956)$, $(3.0644, 16.5196)$

28. $(-1.3532, 3.8312)$

32. 0.0625, 0.1125

34. $0.000\,029\,0$

Exercises 4.2

2. Decreasing for all x

4. Increasing for $x \leq 5/4$; decreasing for $x \geq 5/4$

6. Increasing for $x \leq 1/4$; decreasing for $x \geq 1/4$

8. Increasing for all x

10. Decreasing for all x

12. Decreasing for $x \leq 1$; increasing for $x \geq 1$

14. Decreasing for $-1 \leq x \leq 1$; increasing for $x \leq -1$ and $x \geq 1$

16. Decreasing for $x \leq -1$ and $0 < x \leq 1$; increasing for $-1 \leq x < 0$ and $x \geq 1$

18. Decreasing for $0 \leq x < 1$ and $x > 1$; increasing for $x < -1$ and $-1 < x \leq 0$

20. Decreasing for $x \leq -1$ and $0 \leq x \leq 1$; increasing for $-1 \leq x \leq 0$ and $x \geq 1$

22. Decreasing for $x \leq 0$ and $x \geq 2$; increasing for $0 \leq x \leq 2$

24. Decreasing for $0 < x \leq 1/e$; increasing for $x \geq 1/e$

26. Decreasing for $x < -2$, $-2 < x < 1$, $x > 1$

28.

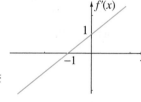

30.

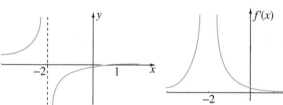

32.

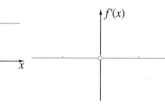

34.

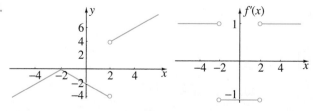

36. Decreasing for $x \leq 0.8612$; increasing for $x \geq 0.8612$

38. Decreasing for $-\pi \leq x \leq -2.2889$ and $2.2889 \leq x \leq \pi$; increasing for $-2.2889 \leq x \leq 2.2889$

50. Yes

52. One at most

Exercises 4.3

2. Relative minimum at $x = -1$; relative maximum at $x = -4$

4. $x = 1$ critical, but does not give a relative extremum

6. Relative maximum at $x = 1 - \sqrt{2}$; relative minimum at $x = 1 + \sqrt{2}$

8. $x = 2$ critical, but does not give a relative extremum

10. $x = 0$ critical, but does not give a relative extremum

12. Relative minimum at $x = -3^{1/4}$; relative maximum at $x = 3^{1/4}$; $x = 0$ critical, but does not give a relative extremum

14. Relative maximum at $x = -1$; relative minimum at $x = 1$

16. Relative minimum at $x = 0$, every point on negative x-axis gives a relative maximum and a relative minimum

18. Relative maximum at $x = 7$; $x = 1$ critical, but does not give a relative extremum

20. Relative minima at $x = 2(3n + 2)\pi/3$; relative maxima at $x = 2(3n + 1)\pi/3$

22. Relative minima at $x = 0, 2 + 50^{1/3}$

24. Relative maximum at $x = 1$

26. Relative maximum at $x = 0$

28. Relative maximum at $x = -2$; relative minimum at $x = 2$

30. Relative maxima at $x = (8n + 1)\pi/4$; relative minima at $x = (8n + 5)\pi/4$

32. Relative minimum at $x = 0.464$

34. $x = 3\pi/4$ and $7\pi/4$ are critical, but do not give relative extrema

36. Relative minimum at $x = 1/e$

38. Relative maximum at $x = 1/2$

40. Relative maximum at $x = -2/3$; relative minimum at $x = 0$

42. Relative minimum at $x = -1/\sqrt{5}$; relative maximum at $x = -\sqrt{5}/10$

44. Every integer gives a relative maximum, every other value of x gives a relative maximum and a relative minimum

46. Relative maximum at $x = 0$ **48.** None

50. Relative maximum at $x = 0$ **52.** $\pm 1/3^{1/16}$

54. $x = 0$

56. (a) $x = 0$ critical; $x = 1$ not critical (b) No

58. False **60.** True

62. False **64.** False

68. (a) (b) $x = \pi/4 + n\pi$

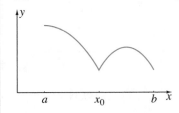

70. Relative minima at $x = 2.3301$ and $x = -2.1284$; relative maximum at $x = -0.2016$

72. Relative maximum at $x = -4.1072$

74. $3\sqrt{3}/4$

76. (c) No

78. (c) Relative maximum

Exercises 4.4

2. Concave upward on $x \leq -2/3$ and $x \geq 0$; concave downward on $-2/3 \leq x \leq 0$; points of inflection at $(0, 2)$ and $(-2/3, 470/27)$

4. Concave upward on $x < -1$ and $x > 1$; concave downward on $-1 < x < 1$

6. Concave upward for all x

8. Concave upward on $-2\pi < x \leq -5\pi/6, -\pi/6 \leq x \leq 7\pi/6, 11\pi/6 \leq x < 2\pi$; concave downward on $-5\pi/6 \leq x \leq -\pi/6, 7\pi/6 \leq x \leq 11\pi/6$; points of inflection at $(-5\pi/6, 2 + 25\pi^2/36), (-\pi/6, 2 + \pi^2/36), (7\pi/6, 2 + 49\pi^2/36), (11\pi/6, 2 + 121\pi^2/36)$

10. Concave downward on $0 < x \leq e^{-3/2}$; concave upward on $x \geq e^{-3/2}$; point of inflection at $(e^{-3/2}, -3/(2e^3))$

12. Concave downward on $x \leq 1$; concave upward on $x \geq 1$; point of inflection at $(1, e^{-2})$

14. Concave downward on $x \leq -\ln 2$; concave upward on $x \geq -\ln 2$; point of inflection at $(-\ln 2, (\ln 2)^2 - 2)$

16. Relative minimum at $x = 1$; relative maximum at $x = -1$

18. Relative minumum at $x = 1/5$

20. Relative minimum at $x = 1/\sqrt{e}$

22. Relative minimum at $x = 0$; relative maximum at $x = 1$

26.

28.

30.

32. Not necessarily

Exercises 4.5

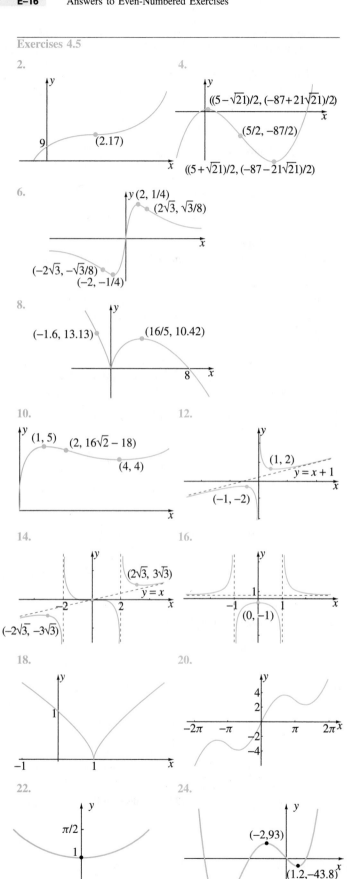

2.

4. $((5-\sqrt{21})/2, (-87+21\sqrt{21})/2)$

$(5/2, -87/2)$

$((5+\sqrt{21})/2, (-87-21\sqrt{21})/2)$

9 (2.17)

6. $y\,(2, 1/4)$

$(2\sqrt{3}, \sqrt{3}/8)$

$(-2\sqrt{3}, -\sqrt{3}/8)$

$(-2, -1/4)$

8. $(-1.6, 13.13)$ $(16/5, 10.42)$

8

10. $(1, 5)$ $(2, 16\sqrt{2}-18)$

$(4, 4)$

12. $(1, 2)$

$y = x + 1$

$(-1, -2)$

14. $(2\sqrt{3}, 3\sqrt{3})$

$y = x$

$(-2\sqrt{3}, -3\sqrt{3})$

16. $(0, -1)$

18.

20.

22. $\pi/2$

24. $(-2, 93)$

$(1.2, -43.8)$

$(-6.7, -289.4)$

26.

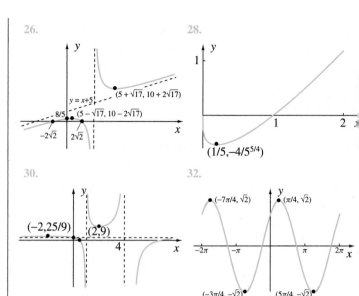

$y = x+5$

$8/5$ $(5 - \sqrt{17}, 10 - 2\sqrt{17})$

$(5 + \sqrt{17}, 10 + 2\sqrt{17})$

$-2\sqrt{2}$ $2\sqrt{2}$

28.

$(1/5, -4/5^{5/4})$

30.

$(-2, 25/9)$ $(2, 9)$

4

32.

$(-7\pi/4, \sqrt{2})$ $(\pi/4, \sqrt{2})$

-2π $-\pi$ π 2π

$(-3\pi/4, -\sqrt{2})$ $(5\pi/4, -\sqrt{2})$

34.

$(800^{1/3}, 25.4)$

-8

36.

$M - R$

$RK/(M - R)$

$-R$

38. $0, (2/a)(9 \pm 4\sqrt{5})e^{-3\mp\sqrt{5}}$ **40.** Point of inflection at $(E_f, 1/2)$

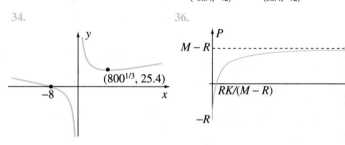

$2a$

$(3-\sqrt{5})a$ $(3+\sqrt{5})a$

$1/2$

E_f

42. (a) $\sqrt{2RT/M}, \sqrt{(5 \pm \sqrt{17})RT/(2M)}$

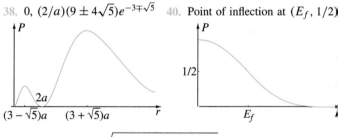

0.005 $T = 50$

0.0025 $T = 300$

200 400 600 800 v

44.

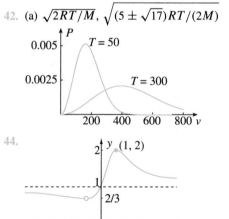

2 y $(1, 2)$

1

$2/3$

-4 -2 -1 2 4 x

Exercises 4.6

2. No relative extrema; horizontal point of inflection $(2, 17)$

4. Relative maximum $(-87 + 21\sqrt{21})/2$ at $x = (5 - \sqrt{21})/2$; relative minimum $(-87 - 21\sqrt{21})/2$ at $x = (5 + \sqrt{21})/2$; point of inflection $(5/2, -87/2)$

6. Relative maximum $1/4$ at $x = 2$; relative minimum $-1/4$ at $x = -2$; points of inflection $(0, 0)$, $(\pm 2\sqrt{3}, \pm\sqrt{3}/8)$

8. Relative maximum 10.42 at $x = 16/5$; relative minimum 0 at $x = 0$; point of inflection $(-8/5, 13.13)$

10. Relative maximum 5 at $x = 1$; relative minimum 4 at $x = 4$; point of inflection $(2, 16\sqrt{2} - 18)$

12. Relative maximum -1 at $x = -1$; relative minimum 3 at $x = 1$; no points of inflection; vertical asymptote $x = 0$; oblique asymptote $y = x + 1$

14. Relative maximum $-3\sqrt{3}$ at $x = -2\sqrt{3}$; relative minimum $3\sqrt{3}$ at $x = 2\sqrt{3}$; horizontal point of inflection $(0, 0)$; vertical asymptotes $x = \pm 2$; oblique asymptote $y = x$

16. Relative maximum -1 at $x = 0$; vertical asymptotes $x = \pm 1$; horizontal asymptote $y = 1$

18. Relative minimum 0 at $x = 0$

20. Relative maxima $\sqrt{3} + 2\pi/3 + 2n\pi$ at $x = 2\pi/3 + 2n\pi$; relative minima $-\sqrt{3} - 2\pi/3 + 2n\pi$ at $x = -2\pi/3 + 2n\pi$; points of inflection $(n\pi, n\pi)$

22. Relative maximum 93; relative minima -289.4 and -43.8

24. Relative maximum $10 - 2\sqrt{17}$; relative minimum $10 + 2\sqrt{17}$

26. Relative minimum $-4/5^{5/4}$

28. Relative maximum $25/9$; relative minimum 9

30. Relative maxima $\sqrt{2}$; relative minima $-\sqrt{2}$

32. Relative minimum 25.4

34. (a)   (b)

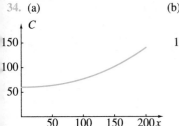 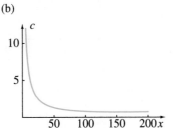

Exercises 4.7

2. $-4, 6/11$ 4. $\pi/3 - \sqrt{3}, 11\pi/3 + \sqrt{3}$

6. No absolute minimum, 12   8. No absolute minimum, $\sqrt{3}/6$

12. $720\sqrt{10}$ m   14. $12^{1/3} \times 12^{1/3} \times 12^{1/3}/2$ m

16. $90 \times 180/\pi$ m   18. $ab/4$

20. $a/\sqrt{3}$ 22. 15 cm $\times$ 22.5 cm

24. $(-1 - \sqrt{6}/2, 5/2 + \sqrt{6})$ 26. Smallest area is 20

28. 71   30. 32

32. (a) $7/4$ km from P (b) Directly to Q

34. Lengths in x- and y-directions are $\sqrt{2}a$ and $\sqrt{2}b$

36. $\sqrt{2/e}$ 38. $4\pi hr^2/27$

40. $(1/2) \cos^{-1}[gh/(v^2 + gh)]$, 0.733 rad

42. 11.8 m   44. $0, 1/(2\sqrt{c})$

46. $dI_1^{1/3}/(I_1^{1/3} + I_2^{1/3})$ from I_1 source

48. (a) Width $= 5/3^{1/3}$ m, height $= 4(3^{2/3})$ m (b) Width $= 4.19$ m, height $= 5.70$ m

52. $l = 1/3$ m, $w = h = 2/3$ m; $1/3$ m apart

54. $2\pi(1 - \sqrt{6}/3)$ 56. (a) 323 mph   (b) 527 mph

58. $b/\sqrt{3}$ 62. $(2, 2\sqrt{5}/3)$

64. $x = -(A + bB)/(2aB)$

Exercises 4.8

2. $v = 2t - 7$ m/s, $a = 2$ m/s^2

4. $v = -6t^2 + 4t + 16$ m/s, $a = -12t + 4$ m/s^2

6. $v = -12\sin 4t$ m/s, $a = -48\cos 4t$ m/s^2

8. $v = 1 - 4/t^2$ m/s, $a = 8/t^3$ m/s^2

10. $v = (5t^2 - 6t + 1)/(2\sqrt{t})$ m/s, $a = (15t^2 - 6t - 1)/(4t^{3/2})$ m/s^2

12. (a) Neither (b) 12 m/s, -15 m/s (c) 15 m/s, 0 m/s (d) 18 m/s^2, -18 m/s^2 (e) Never

14. (a) Increasing on $0 \le t \le (4 - \sqrt{7})/3$, $(4 + \sqrt{7})/3 \le t \le 4$; decreasing on $(4 - \sqrt{7})/3 \le t \le (4 + \sqrt{7})/3$ (b) Increasing on $0 \le t \le (4 - \sqrt{7})/3$, $1 \le t \le (4 + \sqrt{7})/3$, $3 \le t \le 4$; decreasing on $(4 - \sqrt{7})/3 \le t \le 1$, $(4 + \sqrt{7})/3 \le t \le 3$ (c) $t = 4$ s, $t = (4 + \sqrt{7})/3$ s (d) $t = 4$ s, $t = 0, 1, 3$ s (e) 34 m (f) 30 m

16. Horizontal point of inflection

18. 144 m/s   20. Not always   22. Not always

24. (b)

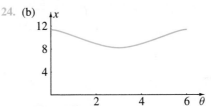

(c) 4.08 cm
(d) $-\omega r[\sin\theta + \cos\theta(e + r\sin\theta)/\sqrt{L^2 - (e + r\sin\theta)^2}]$
(e) ± 13 cm/s

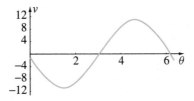

Exercises 4.9

2. $2\sqrt{13}/3$ m/s   4. $13/(5\pi)$ cm/s

6. Decreasing at 20 m^2/s   8. -2.5 N/m^2/s

10. $7/\sqrt{5}$ m/s

12. (a) $2/(25\pi)$ m/min (b) $2/(25\pi)$ m/min

14. (a) $10(50 - x)/y$ m/s (b) $10/\sqrt{3}$ m/s (c) Car 1

16. 25.83 cm/s   18. 115.8 cm/s

20. $-31\sqrt{15}/10$ 22. 102 m/s

24. $6/\sqrt{13}$ m/s

26. (a) $1/(50\pi)$ m/min (b) $\sqrt{13}/(300\pi)$ m/min

28. $3100/\sqrt{230}$ km/h **30.** $2\pi \overline{R} R/r$ m/s

32. $7/2$ cm^2/min

Exercises 4.10

10. (b) $1/\sqrt{LC}$

Exercises 4.11

2. 6 **4.** 0 **6.** ∞

8. 0 **10.** Does not exist **12.** ∞

14. $1/\sqrt{5}$ **16.** na^{n-1} **18.** 1

20. 0 **22.** $1/12$ **24.** 2^{15}

26. ∞ **28.** 4 **30.** 0

32. 1 **34.** e **36.** 1

38. Does not exist **40.** 2 **42.** $-1/3$

44. **46.**

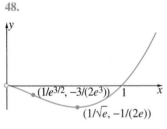
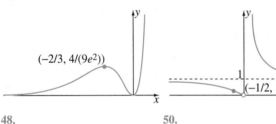

48. **50.**

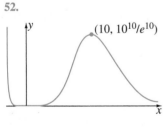
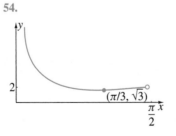

52. **54.**

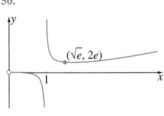

56. ∞

58. (b) 0.000 029 0 (c)

60. $a = b = \pm 7$, c arbitrary

Exercises 4.12

2. $-2dx/(x-1)^2$ **4.** $[2x \cos(x^2+2) + \sin x]dx$

6. $x^2(9 - 16x^2)dx/\sqrt{3 - 4x^2}$ **8.** $[1 - 12/(x-1)^3]dx$

10. $2(x^2 + 5x - 5)dx/(x^2+5)^2$

12. -4 **14.** $4\sqrt{3}\pi/9$

16. 3250 m **18.** (c) $2a + b$

20. (a) na (b) mb (c) $na + mb$ **22.** $\pi/12$

Review Exercises

2. (a) (b)

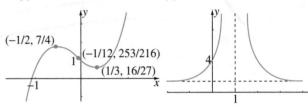

4. No solution **6.** No solution **8.** $-3/2$

10. 8 **12.** 0 **14.** 0

16. 0 **18.** e^2

20. (a) 0.312 908 (b) 1.051 888

24. $2l^3/27$

26. (a)

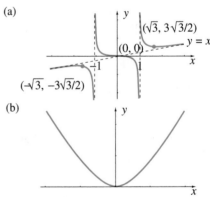

(b)

28. 67 ¢ **30.** $4\sqrt{5}$ m/s

32. $x = 100bq/(ap + bq)$ hectares, $y = 100ap/(ap + bq)$ hectares

Chapter 5

Exercises 5.1

2. $x^5/5 + x^3 + 5x^2/2 + C$ **4.** $-\cos x + C$

6. $(2/3)x^{3/2} + C$ **8.** $-1/x + 2/(3x^3) + C$

10. $-1/x + \sqrt{x} + C$ **12.** $1/(2x) + (3/4)x^4 + C$

14. $(4/3)x^{3/2} + (6/5)x^{5/2} - (10/7)x^{7/2} + C$

16. $(2/5)x^{5/2} + (2/3)x^{3/2} + C$ **18.** $x^3/3 + 2x^5/5 + x^7/7 + C$

20. $(2/5)x^{5/2} - (4/3)x^{3/2} + 2\sqrt{x} + C$

22. $y = x^4/2 + 2x^2 + 5$ **24.** $y = 2x - 2x^2 + x^8$

26. $f(x) = -5x^3/6 + 10x - 31/3$

28. $(2/3)(x + 2)^{3/2} + C$

30. $-(2/3)(2 - x)^{3/2} + C$

32. $(1/5)(2x - 3)^{5/2} + C$

34. $-(1/16)(1 - 2x)^8 + C$

36. $-(1/15)/(1 + 3x)^5 + C$

38. $(1/72)(2 + 3x^3)^8 + C$

40. $(1/2)\sin 2x + C$

42. $(3/4)\sin^2 2x + C$

44. $-(1/4)\cot 4x + C$

46. $-(1/2)e^{-x^2} + C$

48. $(1/4)e^{4x-3} + C$

50. $-(2/5)\ln |7 - 5x| + C$

52. $-(1/4)\ln |1 - 4x^3| + C$

54. $(1/2)3^{2x}\log_3 e + C$

56. $-(1/5)(1 + \cos x)^5 + C$

58. $(1/8)(1 + e^{2x})^4 + C$

60. $(1/2)\,\mathrm{Sin}^{-1}(2x) + C$

62. $\mathrm{Sec}^{-1}(\sqrt{3}x) + C$

64. $(1/4)\sinh 4x + C$

66. $-(1/2)\,\mathrm{sech}\, 2x + C$

68. $y = x^4/4 + 1/x + C$

70. $-(2/3)/(3x + 5)^{1/2} + C$

72. $-(1/12)/(2 + 3x^4) + C$

74. $y = -3 - 1/x,\ x < 0;\ y = 2 - 1/x,\ x > 0$

Exercises 5.2

2. $6t - t^2 + 5,\ 3t^2 - t^3/3 + 5t$ **4.** $60t^2 - 4t^3,\ 20t^3 - t^4 + 4$

6. $t^3/3 + 5t^2/2 + 4t - 2,\ t^4/12 + 5t^3/6 + 2t^2 - 2t - 3$

8. $4 - 3\cos t,\ 4 + 4t - 3\sin t$

10. (a) $t^3 - t^2 - 3t + 1$ (b) $(\sqrt{10} + 1)/3$

12. (a) 350/3 m (b) 650/3 m, 20 s

14. 5.1 m

16. $|v| \geq 19.8$ m/s

18.

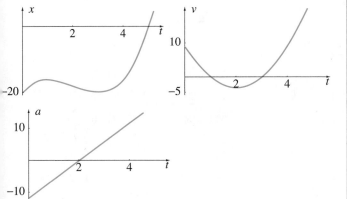

20. Less than or equal to 47.8 km/h; defence

22. 43.3 m

24. 17.3 m

26. 5.2 s

28. $\sqrt{-2al/k}$

30. 22.3 m

Exercises 5.3

2. $-(1/3)(1 - 2x)^{3/2} + C$ **4.** $-(10/63)(5 - 42x)^{3/4} + C$

6. $-(1/2)/(x^2 + 4) + C$

8. $-1/(x - 2) - 2/(x - 2)^2 - (4/3)/(x - 2)^3 + C$

10. $(1/6)(2x + 3)^{3/2} - (3/2)\sqrt{2x + 3} + C$

---

12. $(1/5)(s^2 + 5)^{5/2} - (5/3)(s^2 + 5)^{3/2} + C$

14. $(2/3)(1 - \cos x)^{3/2} + C$

16. $(2/7)(y - 4)^{7/2} + (16/5)(y - 4)^{5/2} + (32/3)(y - 4)^{3/2} + C$

18. $(1/729)(3x^3 - 5)^9 + (5/324)(3x^3 - 5)^8 + (25/567)(3x^3 - 5)^7 + C$

20. $(3/4)(x^2 + 2x + 2)^{2/3} + C$

22. $(7/192)/(3 - 4\sin x)^3 + (3/64)/(3 - 4\sin x)^2 - (1/64)/(3 - 4\sin x) + C$

24. $(4/5)(1 + \sqrt{x})^{5/2} - (4/3)(1 + \sqrt{x})^{3/2} + C$

26. $(1/2)\tan^2 x + C$ **28.** $(\ln x)^2/2 + C$

30. $-(1/2)/\ln(x^2 + 1) + C$

32. $(2/3)(1 + x)^{3/2} - 2\sqrt{1 + x} + C$ if $x \geq 0$; $-(2/3)(1 + x)^{3/2} + 2\sqrt{1 + x} + C$ if $-1 < x < 0$

34. $-(2/5)(1/x - 1)^{5/2} - (2/3)(1/x - 1)^{3/2} + C$

36. $-2/[\sqrt{(1 - x)/(1 + x)} - 3] + C$

Exercises 5.4

2. $y = -9.81m(x^4 - 2Lx^3 + L^2x^2)/(24EIL)$

4. $y = -9.81m(2x^4 - 5Lx^3 + 3L^2x^2)/(48EIL)$

6. $y = [-(F/6)(x - L/2)^3 h(x - L/2) + (Fx^3/6) - (FLx^2/4)]/(EI)$

8. $y = [-(F/6)(x - L/2)^3 h(x - L/2) - mgx^4/(24L) + (F + mg)x^3/12 - L(3F + 2mg)x^2/48]/(EI)$

10. $y = [-(F/6)(x - L/2)^3 h(x - L/2) - mgx^4/(24L) + (F + mg)x^3/12 - L^2(3F + 2mg)x/48]/(EI)$

12. $y(x) = -4.0875[x^4 + (x - 5)^4 h(x - 5) - 60x^3 + 1050x^2]/(EI)$ greater

14. $Fx^2(x - 3L)/(6EI)$

Exercises 5.5

2. 4.75 h **4.** 19.74 s

6. 13.51 h **8.** (a) In finite time (b) Infinitely long

10. $3.8 \times 10^{-4}\%$

12. $C(t) = (R/k)(1 - e^{-kt}) + C_0 e^{-kt}$

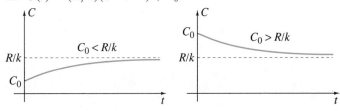

14. $T = 20 + 70e^{-0.01399t}$ **16.** Infinitely long

18. $\pi R^2 \sqrt{2H/g}/(5cA)$ **20.** 70.95 min

22. $\sqrt{2GM/R}$ **24.** 2 years

26. $H/2$ **28.** $30 - 20/r$

32. $600kt/(20kt + 3)$ g

Review Exercises

2. $-(1/4)/x^4 + x^2 + (1/2)/x^2 + C$

4. $-1/x - (4/3)x^{3/2} + C$

6. $(1/30)(1 + 3x^2)^5 + C$ **8.** $(2/5)x^{5/2} + 10\sqrt{x} + C$

10. $-2/\sqrt{x} - 30\sqrt{x} + C$ **12.** $-(1/3)(1 - x^2)^{3/2} + C$

14. $x^3/3 - (4/5)x^5 + (4/7)x^7 + C$

16. $(2/3)(2 - x)^{3/2} - 4\sqrt{2 - x} + C$

18. $4x + (8/3)x^{3/2} + x^2/2 + C$ **20.** $(1/5)\sin^5 x + C$

22. $-(1/8)e^{-4x^2} + C$ **24.** $(1/5)\ln|\ln x| + C$

26. $(3/\sqrt{7})\text{Tan}^{-1}\sqrt{7}x + C$ **28.** $(1/5)\tanh 5x + C$

30. $-20 + 90e^{\ln(7/9)t/4}$ **32.** $y = x^4 + 7x/2 - 1/2$

34. $[(\sqrt{3} - \sqrt{6})t + \sqrt{6}]^2$ km, $\sqrt{6}/(\sqrt{6} - \sqrt{3})$ h

36. 11.9 m/s **38.** $(2/3)(1 + x)^{3/2} - x + C$

40. $-(3/56)(3-2x^3)^7+(1/32)(3-2x^3)^8-(1/216)(3-2x^3)^9+ C$

42. $(1/4)\sin^4 x - (1/6)\sin^6 x + C$

44. $-\ln|\cos x| + C$ **46.** $y = 2x/(1 + x)$

48. $15(6 - 108^{1/3})$ days

50. $y = -9.81M[(x-5)^4h(x-5)/24 - 5x^3/6 + 75x^2/4]/(5EI)$; no

Chapter 6

Exercises 6.1

2. $\sum_{k=1}^{10} k/2^k$ **4.** $\sum_{k=1}^{14\,641} \sqrt{k}$

6. $\sum_{k=1}^{1019} (-1)^k(k + 1)$ **8.** $\sum_{k=1}^{225} (\tan k)/(1 + k^2)$

10. $\sum_{k=1}^{9} (10^k - 1)/10^k$ **16.** 258

18. $4n(4n^2 - 1)/3$ **20.** 5350

22. 4540 **24.** $(n^3 + 12n^2 + 47n - 444)/3$

28. $f(n) - f(0)$ **30.** No

32. $(1 - 2^{-18})/4$ **34.** $3960[1 - (0.99)^{15}]$

Exercises 6.3

2. 6 **4.** 4 **6.** 3/2

8. 0 **10.** (c) $\log_2 e$

Exercises 6.4

2. 20/3 **4.** 2/3 **6.** 1

8. $-7/2$ **10.** 3/4 **12.** 0

14. $5/16 - 1/(\pi + 1)$ **16.** 125/12 **18.** $-65/4$

20. $3/\sqrt{2}$ **22.** $-1/2$ **24.** 0

26. $(e^3 - 1)/e$ **28.** $\ln(2/3)$ **30.** $20\log_3 e$

32. 88/3 **34.** $-1/6$ **36.** $\pi/2$

38. $(1/2)\sinh 2$ **40.** $\pi/8$

42. $-475, 475$ **44.** $1/6, 1/6$

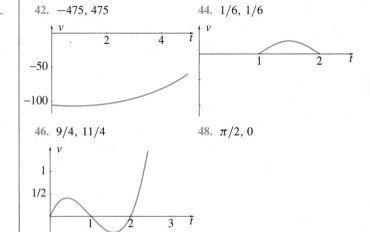

46. 9/4, 11/4 **48.** $\pi/2, 0$

50. $4\pi/(\pi^2 + 160), 0$ **52.** $2\sqrt{31}, 2\sqrt{5}$

Exercises 6.5

2. $1/\sqrt{x^2 + 1}$ **4.** $-x^3\cos x$

6. $2\sqrt{2x + 1}$ **8.** $5\sqrt{(5x + 4)^3 + 1}$

10. $256(3x^2 + 3x + 1) + 2/\sqrt{x} - 2/\sqrt{x + 1}$

12. $2x[-2\sec(1 + 2x^2) + \sec(1 + x^2)]$

14. $\cos x/\sqrt{\sin x + 1} + \sin x/\sqrt{\cos x + 1}$

16. $(2\sqrt{2} - 1)/(2x^{1/4})$

18. $-\ln(x^2 + 1)$ **20.** $3e^{-36x^2} + 2e^{-16x^2}$

Exercises 6.6

2. 0 **4.** 31/5 **6.** $2\sqrt{2}/3$

8. 11/5 **10.** $2/\pi$ **12.** 1

14. 23/9 **16.** 1/2 **18.** 0

20. 9/7 **22.** $2cR^2/3$ **24.** 0

26. $\pi/2$ **28.** 0.6032 **30.** 1.400

32. $\overline{f}(x) = 0$

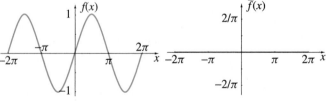

34. $\overline{f}(x) = x^3 - 3x^2 + 4x - 2$

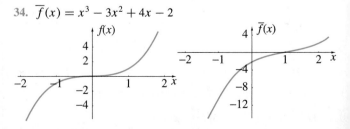

36. $\overline{f}(x) = 0$ for $x \le -1$, $x^2/2 + x + 1/2$ for $-1 < x < 0$, $-x^2 + x + 1/2$ for $0 \le x < 1$, $x^2/2 - 2x + 2$ for $1 \le x < 2$, 0 for $x \ge 2$

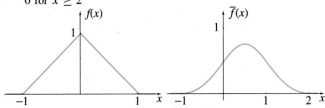

38. $\overline{f}(x) = 0$ for $x < a$, $(x - a)/(b - a)$ for $a \le x < b$, $(2b - x - a)/(b - a)$ for $b \le x < 2b - a$, 0 for $x \ge 2b - a$

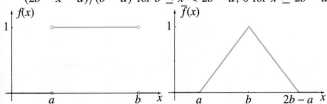

Exercises 6.7

2. 4/15   **4.** 7/384   **6.** Does not exist

8. $\sqrt{3/2} - 2\sqrt{2}/3$ **10.** $3(5^{2/3} - 2^{2/3})/4$ **12.** $\sqrt{2}(7\sqrt{7} - 8)/18$

14. 52/5   **16.** $(16 - 9\sqrt{2})/6$ **18.** 5/18

20. $(1/3)(\ln 2)^3$ **22.** $2(\sqrt{7} - 1)/3$ **26.** $(56\sqrt{7} - \sqrt{2})/135$

Review Exercises

2. 4/15   **4.** $-4/3$ **6.** 1/6

8. 0   **10.** 66/5   **12.** $16\sqrt{6}$

14. $-4/15$ **16.** $(2112\sqrt{6} - 704)/105$

18. 120 852.5   **20.** 20   **24.** 2

26. $4/(15\pi)$ **28.** $-x^2(x + 1)^3$ **30.** $-4x \cos 2x$

32. $4x \sin^2 x^2$

34. 0, 1/2

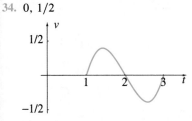

36. $(33\sqrt{6} - 56)/60$ **38.** $(24\sqrt{3} - 14\sqrt{2} - 8\sqrt{5})/3$

40. $24\sqrt{5} - 8$ **42.** $(3\sqrt{3} - 2\sqrt{2} - 1)/3$

Chapter 7

Exercises 7.1

2. 8   **4.** 343/6   **6.** $(e^6 - e^3)/3 + 3/2$

8. 8/3   **10.** 2   **12.** 20/3

14. $4\sqrt{2}/3 - e \ln 2$ **16.** 10/3

18. (a) 22 (b) 37, 15 (c) 29, 7 (d) 203/8, 27/8

20. $4\int_1^2 \sqrt{4 - x^2}\, dx$ **22.** $2\int_0^{\sqrt{(\sqrt{65}-1)/2}} (\sqrt{16 - y^2} - y^2)\, dy$

24. 235/3   **26.** 7/6   **28.** $1 - \ln 2$

30. $4\sqrt{e} - 16/3$ **32.** 32/3   **34.** $14\sqrt{21}$

36. $(\sqrt{2/3}, 4/3)$ **38.** 2.182   **40.** 8.436

42. 2.067   **44.** 7.177

46. $0 < m < 1$, $(m - 1 - \ln m)/3$

Exercises 7.2

2. $80\sqrt{5}\pi/3$ **4.** $5888\pi/15$ **6.** π

8. $1088\pi/15$ **10.** $32\pi/3$ **12.** $\pi(e^2 - 4e + 5)/2$

14. 8π **16.** $14\pi/3$ **18.** $40\pi/3$

20. $\pi/3$ **22.** $344\pi/3$ **24.** $1472\pi/15$

26. $775\pi/6$ **28.** $272\pi/15$ **30.** $68\pi/9$

32. $16\pi(13 + 2^{17/4} - 3^{9/4})/45$

34. 111.303   **36.** 21.186   **38.** $4\pi/3$

40. (a) $x^2(6 - x)/16$ (b) $1161\pi/10$ m^3

Exercises 7.3

2. 17/3   **4.** $(85^{3/2} - 13^{3/2})/27$

6. $(e - e^{-1})/2$ **8.** 23/18

10. 3011/480   **12.** $\int_1^2 \sqrt{36x^2 - 48x + 17}\, dx$

14. $\int_{-\sqrt{3}}^{2\sqrt{2}} \sqrt{(1 + 2y^2)/(1 + y^2)}\, dy$

16. $\int_0^{\pi/4} \sec x\, dx$ **18.** $\sqrt{2}\int_0^1 (3 - 2x^2)/\sqrt{1 - x^2}\, dx$

20. $2\int_0^2 \sqrt{(16 + 5x^2)/(4 - x^2)}\, dx$

22. 6

24. $(b^{2n+1} - a^{2n+1})/[4(2n - 1)] + (1/a^{2n-1} - 1/b^{2n-1})/(2n + 1)$

Exercises 7.4

2. 8/15 J   **4.** 8829 J   **6.** 49.05 J

8. 184 J   **10.** 1.78×10^7 J   **12.** 114 450 J

14. 7.60×10^6 J   **16.** $q_3(q_1 - 8q_2)/(48\pi\epsilon_0)$

18. kb^2 J   **20.** 3.22×10^5 J

22. (b) 9.8087×10^5 J (c) 9.82×10^5 J

24. $C \ln 2$ **26.** 8.8×10^3 J   **28.** 72 J

30. 7.8×10^2 J   **32.** $g^2 M^3/(6m^2)$ J

Exercises 7.5

2. (a) $kx^2 + mv^2 = kx_0^2 + mv_0^2$ (b) $\sqrt{x_0^2 + mv_0^2/k}$ (c) $\sqrt{v_0^2 + kx_0^2/m}$

4. $qQ/(4\pi\epsilon_0 r)$

6. (a) $19.62m/k$ (b) $(9.81m + \sqrt{9.81^2m^2 + kmv_0^2})/k$

8. (a) 9.67×10^6 J (b) 1.4×10^3 m/s

Exercises 7.6

2. 6.78×10^{10} N
4. 9.25×10^3 N

6. 7.60×10^5 N, 4.41×10^5 N, 4.91×10^4 N

12. 2.943×10^6 N, 2.969×10^6 N

14. 1.60 m × 1.60 m
20. 1/4 m

22. (b) 1.01×10^5 N/m^2
24. 1.02 m

26. (a) 2.33 mm (b) 3.11 mm

Exercises 7.7

2. $(3a/8, 3h/5)$
4. $(4r/(3\pi), 4r/(3\pi))$

6. $(-1/2, -1/2)$
8. $(0, 0)$

10. $(33/5, 1/2)$
12. $(177/85, 9/17)$

14. $25\rho/3$

16. $(M + 2\rho L)^{-1}(\sum_{i=1}^{6} m_i x_i + 2\rho L^2)$

18. $(0, 192/205)$

20. On the axis of symmetry, 1637/350 from smaller end

22. $76\sqrt{2}\rho/15$
24. $1181\sqrt{2}\rho/840$

28. $(0, (a^2 b + 2acd + c^2 d)/(2ab + 2cd))$

30. $((abc - a^2 b + ef^2 - cef)/(2ab + 2cd + 2ef),$
    $(def + e^2 f - ab^2 - abd)/(2ab + 2cd + 2ef))$

32. $(-1.215, 0.523)$
34. $(0.219, 1.134)$

38. $13\sqrt{2}\pi/20$

Exercises 7.8

2. $32\rho/3$
4. $603\rho/10$
6. $2\rho/15$

8. $344\rho/105$
10. $1143\rho/20$

12. $\rho h(y_2^3 + y_2^2 y_1 - y_1^2 y_2 - y_1^3)/4$; no

14. $[ae^3 + cd(c^2 + 3ce + 3e^2)]/3$, $(cd^3 + ea^3)/12$

16. $[2a^3 d + (b+c)(6a^2 d + 12ad^2 + 8d^3)]/24$, $d(ad^2 + b^3 + c^3)/12$

18. $\sqrt{44/45}$, $\sqrt{2}/3$
24. 1.278

Exercises 7.9

2. $\pi r \sqrt{r^2 + h^2}$
4. $b^2 h/3$

6. (a) 20 000 (b) 7000
8. $31\,250/\sqrt{3}$ cm^3

10. $\pi/2$
12. $r^3(3\pi - 4)/6$

14. $AEL/(AE - 9.81M)$

16. $L + FL/(AE) + 9.81\rho L^2/(2E)$

18. Discontinuity at $y = L$

20. $L + 9.81\rho L^2(2a^3 - 3a^2 b + b^3)/[6Eb(a - b)^2]$

22. $L + 9.81\rho L^2(2b^3 - 3ab^2 + a^3)/[6Eb(a - b)^2] + 9.81ML/(abE)$

24. (a) $kx(L - x)/(2\tau)$ (b) Yes, Yes

26. (a) $kx^2(x^2 - 4Lx + 6L^2)/(24EI)$ (b) $x = L$

28. (a) $x^2(L - x)^2/(24EI)$

32. 140 m^3
34. (a) $2\pi rt$ (b) $\pi(2rh - h^2)t$

Exercises 7.10

2. Diverges
4. $-1/16\,464$
6. Diverges

8. 2
10. $\sqrt{21}$
12. Diverges

14. Diverges
16. Diverges
18. $17\sqrt{5}/3$

20. (a) No (b) No
22. (a) 1/2 (b) $\pi/5$
28. No

30. Diverges
32. Converges

Review Exercises

2. (a) 5/4 (b) $46\pi/21$, $14\pi/15$ (c) $(28/75, 92/105)$ (d) 73/60, 1/4

4. (a) 1/2 (b) $\pi/3$, π (c) $(1, 1/3)$ (d) 1/12, 7/12

6. $10\pi/3$
8. $16\pi/105$
10. $43\pi/12$

12. $-5/6$, 3/5
14. $-28/3$, 14
16. 1.04×10^5 J

18. $4\sqrt{5}\pi$
22. $2\sqrt{3}$
24. 0

26. Diverges
28. Diverges
30. 6.58×10^4 J

Chapter 8

Exercises 8.1

2. $-(1/4)e^{-2x^2} + C$
4. $\ln(1 + e^x) + C$

6. $-(2/27)(1 - 3x^3)^{3/2} + C$
8. $-(1/6)(1 + x^3)^{-2} + C$

10. $2\sqrt{x} - x + C$
12. $\text{Tan}^{-1} x + x + C$

14. $(1/6)(2x + 4)^{3/2} + \sqrt{2x + 4} + C$

16. $(1/8)\sin^4 2x + C$

18. $(1/2)(x + 5)^2 - 15x + 75\ln|x + 5| + 125/(x + 5) + C$

20. $(1/3)\ln|\sec 3x| + C$

22. $(4/5)x^{5/4} - x + (4/3)x^{3/4} - 2\sqrt{x} + 4x^{1/4} - 4\ln(x^{1/4} + 1) + C$

24. $(35 - 12\ln 6)/6$

28. (a) $1962(1 - e^{-t/200})$ m/s (b) $1962t + 392\,400(e^{-t/200} - 1)$ m

30. $(1/(3n))\ln|x^n/(3 + 2x^n)| + C$

Exercises 8.2

2. $(x^2/2 - x/2 + 1/4)e^{2x} + C$

4. $(2/3)x^{3/2} \ln(2x) - (4/9)x^{3/2} + C$

6. $-(2x/3)(3-x)^{3/2} - (4/15)(3-x)^{5/2} + C$

8. $(2x^2/3)(x+5)^{3/2} - (8x/15)(x+5)^{5/2} + (16/105)(x+5)^{7/2} + C$

10. $2x^2\sqrt{x+2} - (8x/3)(x+2)^{3/2} + (16/15)(x+2)^{5/2} + C$

12. $(1/3)(x-1)^3 \ln x - x^3/9 + x^2/2 - x + (1/3)\ln x + C$

14. $x\,\mathrm{Tan}^{-1}\,x - (1/2)\ln(1+x^2) + C$

16. $e^{2x}(2\cos 3x + 3\sin 3x)/13 + C$

18. $x\ln(x^2+4) - 2x + 4\mathrm{Tan}^{-1}(x/2) + C$

20. 5.92×10^9 N   22. $2(-1)^{n+1}L^2/(n\pi), 0$

24. $4(-1)^n L^2/(n\pi), 0$ 28. $2/s^3$

30. $1/(s+1)^2$ 32. $i(e^{-i\omega L/2} - e^{i\omega L/2})/\omega$

34. $2a/(\omega^2 + a^2)$

36. $x\,\mathrm{Tan}^{-1}\sqrt{x} - \sqrt{x} + \mathrm{Tan}^{-1}\sqrt{x} + C$

38. $(x/2)e^x(\sin x - \cos x) + (1/2)e^x \cos x + C$

Exercises 8.3

2. $-(1/2)\sin^{-2} x + C$ 4. $-(1/3)\csc^3 x + C$

6. $(2/3)\tan^{3/2} x + (2/7)\tan^{7/2} x + C$

8. $(1/18)\sec^6 3x + C$ 10. $(1/4)\sec^4 x + C$

12. $\tan\theta + C$ 14. $2\sqrt{1+\tan x} + C$

16. $7\tan x - 3x + C$

18. $(1/32)(12x - 8\sin 2x + \sin 4x) + C$

20. $(1/2)\tan^2 x + \ln|\tan x| + C$

22. $\pi(1 - \pi/4 + \ln 2)$ 24. $-(1/3)\cot^3 z + \cot z + z + C$

26. $\theta/2 + (1/4)\sin 2\theta + (1/3)\cos^3\theta + C$

28. $(1/8)\sin 4x + (1/16)\sin 8x + C$

30. $\sec x + \ln|\csc x - \cot x| + C$

34. 0 36. $1/2$

Exercises 8.4

2. $(1/\sqrt{5})\,\mathrm{Sin}^{-1}(\sqrt{5}x/3) + C$ 4. $-\sqrt{4-x^2}/(4x) + C$

6. $(1/15)(5x^2 + 3)^{3/2} + C$

8. $(1/2)\ln|(1+x)/(1-x)| + C$

10. $(1/20)\ln(5x^2 + 1) + (\sqrt{5}/2)\,\mathrm{Tan}^{-1}(\sqrt{5}x) + C$

12. $\sqrt{4-x^2} + 2\ln|(2 - \sqrt{4-x^2})/x| + C$

14. $\ln|x + \sqrt{x^2 - 16}| - \sqrt{x^2 - 16}/x + C$

16. $(1/16)\,\mathrm{Sec}^{-1}(x/2) + \sqrt{x^2 - 4}/(8x^2) + C$

18. $(1/3)(y^2 + 4)^{3/2} - 4\sqrt{y^2 + 4} + C$

20. $2\ln|x| - (1/2)\ln(1+x^2) + C$

24. π

26. 1.053 28. $0.265b$

30. $9.81\pi\rho r^3$

32. $(147/8)\,\mathrm{Sin}^{-1}(x/\sqrt{7}) + (x/8)(35 - 2x^2)\sqrt{7 - x^2} + C$

34. $-6\,\mathrm{Sec}^{-1}(2x) + (1 - 12x^2)/(2x^2\sqrt{4x^2 - 1}) + C$

36. $(1/2)[x\sqrt{1+3x^2} + (1/\sqrt{3})\ln|\sqrt{3}x + \sqrt{1+3x^2}|] + C$

38. $\sqrt{2}\pi$ 40. $r/2$

42. (a) $(1/2)(e^{kx} - e^{-kx})$ (b) $(e^{kx} + e^{-kx})/(2k) + C$

Exercises 8.5

2. $\ln|\sqrt{x^2 + 2x + 2} + x + 1| + C$

4. $-(2/\sqrt{7})\,\mathrm{Tan}^{-1}[(2x-3)/\sqrt{7}] + C$

6. $(x/2)/\sqrt{4x - x^2} + C$

8. $\ln(x^2 + 6x + 13) - (9/2)\,\mathrm{Tan}^{-1}[(x+3)/2] + C$

10. $\ln|\sqrt{6 + 4\ln x + (\ln x)^2} + \ln x + 2| + C$

12. $39\,240\pi$ N

14. $(1/2)(x-1)\sqrt{x^2 - 2x - 3} - 2\ln|x - 1 + \sqrt{x^2 - 2x - 3}| + C$

Exercises 8.6

2. $-(1/2)(y+1)^{-2} + C$

4. $x + (10/3)\ln|x - 4| + (2/3)\ln|x + 2| + C$

6. $-(1/6)\ln|y| - (2/15)\ln|y + 3| + (3/10)\ln|y - 2| + C$

8. $(1/2)\ln(x^2 + 2) + (x^2 + 2)^{-1} + C$

10. $y - 4\ln|y + 2| + \ln|y + 1| + C$

12. $\ln(y^2 + 1) + \mathrm{Tan}^{-1} y - \ln(y^2 + 4) + C$

14. $-[1/(3\sqrt{2})]\,\mathrm{Tan}^{-1}(x/\sqrt{2}) + (2/3)\ln|(x-1)/(x+1)| + C$

16. $-(5/27)\ln|x - 1| - (7/9)(x - 1)^{-1} + (32/27)\ln|x + 2| + (2/9)(x + 2)^{-1} + C$

18. (b) $750[t/15 + 2\ln((1 + e^{-t/15})/2)]$

20. $\sqrt{mg/k - (mg/k - v_0^2)e^{-2kh/m}}$

24. $N/[1 + (N-1)e^{-Ft}]$ 26. $v_0(e^{2v_0t/a} - 1)/(e^{2v_0t/a} + 1)$

28. $(1/4)\ln|x + 1| - (1/8)\ln(x^2 + 1) + (1/2)\,\mathrm{Tan}^{-1} x + (1/4)(x + 1)/(x^2 + 1) + C$

30. $(1/3)\ln|x + 1| + (1/\sqrt{3})\,\mathrm{Tan}^{-1}[(2x - 1)/\sqrt{3}] - (1/6)\ln(x^2 - x + 1) + C$

32. $\ln|x| + \ln(x^2 - x + 1) - 2\ln|x + 1| - 3/(x + 1) + (2/\sqrt{3})\,\mathrm{Tan}^{-1}[(2x - 1)/\sqrt{3}] + C$

34. $-1/(2M^2) + [(k + 1)/4]\ln\{[1 + (k - 1)M^2/2]/M^2\} + C$

36. $\ln|[1 + \tan(x/2)]/[1 - \tan(x/2)]| + C$

38. $(1/\sqrt{3})\ln|[\sqrt{3}\tan(x/2) - 1]/[\sqrt{3}\tan(x/2) + 1]| + C$

Exercises 8.7

2. 0.472 15, 0.472 14   **4.** 0.648 86, 0.648 72

6. −0.069 570, −0.069 445   **8.** 0.302 20, 0.302 30

10. 0.142 21, 0.142 01   **12.** 1.4672, 1.4627

14. 0.311 17, 0.310 26   **16.** 1/4 previous; 1/16 previous

18. 1.4789   **20.** 83.76

22. 32.91   **24.** 0.2437, 0.2438

26. 2.113, 1.729   **30.** (a) 21 (b) 4

32. (a) 3 (b) 2

Review Exercises

2. $-1/(x + 3) + C$ **4.** $x^2/2 - x + 4\ln|x + 1| + C$

6. $(2/3)(x + 3)^{3/2} - 6\sqrt{x + 3} + C$

8. $-x\cos x + \sin x + C$

10. $(3/4)\ln|x + 3| + (1/4)\ln|x - 1| + C$

12. $14\sqrt{x} - x - 70\ln(5 + \sqrt{x}) + C$

14. $\text{Sin}^{-1}(e^x) + C$ **16.** $-(1/2)/(x^2 + 1) + C$

18. $(1/2)\ln(x^2 + 1) + (1/2)/(x^2 + 1) + C$

20. $x + 4\ln|x - 1| - 4/(x - 1) + C$

22. $x\,\text{Cos}^{-1}x - \sqrt{1 - x^2} + C$

24. $-(1/12)\cos 6x + (1/8)\cos 4x + C$

26. $\ln|x + 2 + \sqrt{x^2 + 4x - 5}| + C$

28. $(1/5)(4 - x^2)^{5/2} - (4/3)(4 - x^2)^{3/2} + C$

30. $-3\ln|4 - x^2| + C$

32. $[1/(2\sqrt{2})]\ln|(x + 2 - 2\sqrt{2})/\sqrt{x^2 + 4x - 4}| + C$

34. $\ln|x| + 1/x - 1/(2x^2) - \ln|x + 1| + C$

36. $\ln|2\sqrt{x^2 - 3x - 16} + 2x - 3| + C$

38. $-x^2/6 + (x^3/3)\,\text{Tan}^{-1}x + (1/6)\ln(x^2 + 1) + C$

40. $(\ln x)^2/2 + C$

42. $(1/81)\ln|x| - (1/162)\ln(x^2 + 9) + (1/18)/(x^2 + 9) + C$

44. $(1/2)\ln|x| + (1/2)\ln|x + 2| + 3/(x + 2) + C$

46. $\ln|x| + \ln(x^2 + x + 4) + C$

48. $-(2/3)\cot^{3/2}x - (2/7)\cot^{7/2}x + C$

50. $(x - 2)/(4\sqrt{4x - x^2}) + C$

52. 0.636 65, 0.640 41   **54.** 1.2667, 1.2647

56. $-2\sqrt{x} - 3x^{1/3} - 6x^{1/6} - 6\ln|1 - x^{1/6}| + C$

58. $(1/8)\,\text{Tan}^{-1}(x^2/4) + C$ **60.** $(4x - 6)/(9\sqrt{3x - x^2}) + C$

62. $2\sin\sqrt{x} - 2\sqrt{x}\cos\sqrt{x} + C$

64. $(1/32)\sin 4x + (1/8)\sin 2x - (x/8)\cos 4x - (x/4)\cos 2x + C$

66. $(1/2)\tan x + C$

68. $(1/6)\sin 3x - (1/20)\sin 5x - (1/4)\sin x + C$

70. $(1/4)(\text{Sin}^{-1}x)^2 + (x/2)\sqrt{1 - x^2}\,\text{Sin}^{-1}x - x^2/4 + C$

72. (b) $\text{Sin}^{-1}x - \sqrt{1 - x^2} + C$

Chapter 9

Exercises 9.1

2.   **4.**

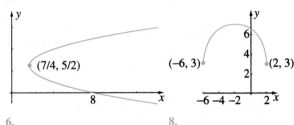

6.   **8.**

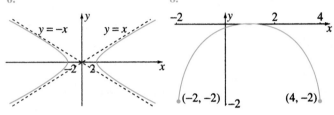

10.   **12.**

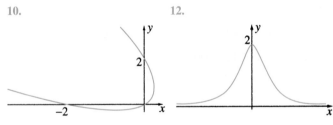

14. $2u/(u + 1)^2$ **16.** $-1/x^2$

18. $3/[4(t + 6)^2(2t + 3)^3]$

20. $4(t + 1)\sqrt{-t^2 + 3t + 5}/[(2t - 3)(t^2 + 2t - 5)^2]$

22. $(8/3, 1/2)$, $(28/3, 28)$ **24.** x/y, $2/y^3$

26. $1/(v + 1)$, $-(1/2)/(v + 1)^3$

28. Ellipse $b^2(x - h)^2 + a^2(y - k)^2 = a^2b^2$

32.

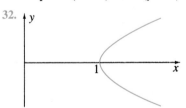

34. $x = (5t^2 - t^3)/(t + 1)$, $y = t$

36. $x = 2\sec t$, $y = \sqrt{2}\tan t$, $-\pi \le t \le \pi$, $t \ne \pm\pi/2$

40. $3\pi/8$ **42.** 2π **44.** $5\pi^2 R^3$

46. $\sqrt{2}(e - 1)/e$

48. $4\int_0^{\pi/2}\sqrt{a^2\sin^2\theta + b^2\cos^2\theta}\,d\theta$

50. (a) $x = t^2 + 4t + 2$, $y = t + 6$ (b) $x = 9t^2 + 12t + 2$, $y = 3t + 6$

52. $x = 2\cos 4\pi t$, $y = 2\sin 4\pi t$, $t \geq 0$

54. $(0.62, \pm 0.30)$

56. (a) $x_E = d - \dfrac{l_2(d - l_1 \cos\theta)}{\sqrt{d^2 + l_1^2 - 2dl_1\cos\theta}}$

$+ \sqrt{l_3^2 - \left[y_E - \dfrac{l_1 l_2 \sin\theta}{\sqrt{d^2 + l_1^2 - 2dl_1\cos\theta}} \right]^2}$

(b)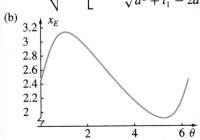

60. 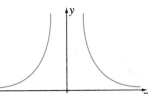 62.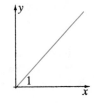

Exercises 9.2

2. $(2, 2\pi/3 + 2n\pi)$

4. $(4, 5\pi/6 + 2n\pi)$

6. $(\sqrt{17}, -1.82 + 2n\pi)$

8. $(\sqrt{29}, 2.76 + 2n\pi)$

10. $(3\sqrt{3}, -3)$

12. $(-2.21, -2.03)$

Exercises 9.3

2. $\theta = 3\pi/4$, $\theta = -\pi/4$

4. $r^2 - 2r(\cos\theta + \sin\theta) + 1 = 0$

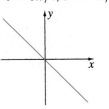 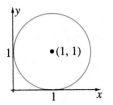

6. $r^2 = 3/(1 + \sin^2\theta)$

8. $r = 1 - \cos\theta$

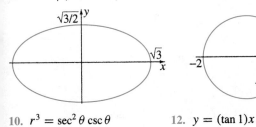 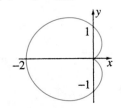

10. $r^3 = \sec^2\theta \csc\theta$

12. $y = (\tan 1)x$

14. $(x^2 + y^2)^2 = 8xy$

16. $(x^2 + y^2)^{3/2} = 4xy$

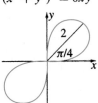 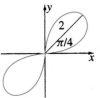

18. $x^2 + y^2 = 3\sqrt{x^2 + y^2} - 4x$

20. $y^3 = x^2$

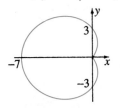 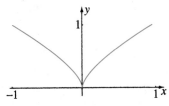

22. $(2, \pm\pi/6)$, $(2, \pm 5\pi/6)$

24. $(0, \theta)$, $(4/3, \pm 1.23)$

26. $\sqrt{3}/7$

28. $3/(5\sqrt{2} + 3)$

30. $\sqrt{3}$

32. 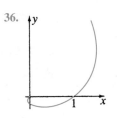 34.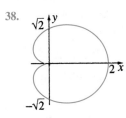

36. 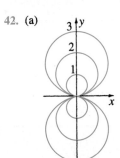 38.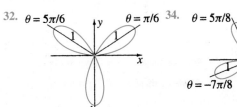

40. Maximum for $t = (2n \pm 1)/(4\omega)$ and $x = a$; minimum for $t = n/(2\omega)$ and $x = a \pm b$

42. (a) 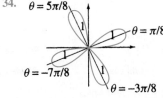 (b)

44. $x^2 + y^2 = a(\sqrt{x^2 + y^2} \pm x)$, $x^2 + y^2 = a(\sqrt{x^2 + y^2} \pm y)$

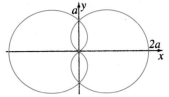 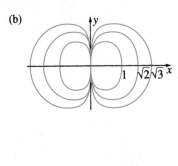

46. (a) Turn figures below sideways for $r = b \pm a\sin\theta$

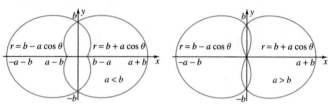

(b) $x^2 + y^2 = b\sqrt{x^2 + y^2} \pm ax$, $x^2 + y^2 = b\sqrt{x^2 + y^2} \pm ay$.

(c) For the case $a = b$, curves are the cardioids in Exercise 44.

Exercises 9.4

2. 9π **4.** 2 **6.** 6π

8. 18π **10.** 2π **12.** $(5\pi - 8)/4$

14. $6\pi - 16$ **16.** $(7\pi - 12\sqrt{3})/12$

18. $33[\pi + 2\,\mathrm{Sin}^{-1}(2/3)]/4 + 7\sqrt{5}$ **20.** $\pi/32$

Exercises 9.5

2. Circle   **4.** None of these   **6.** Hyperbola

8. Parabola   **10.** None of these   **12.** Ellipse

14. Straight line

16. 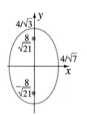   **18.**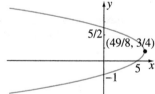

20. 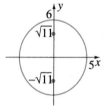   **22.**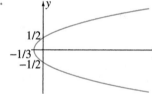

24.   **26.**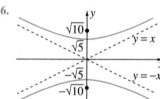

28. 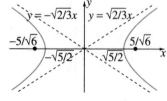   **30.**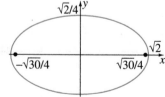

---

32. 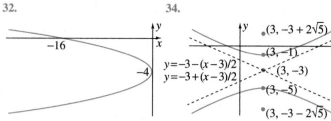   **34.**

36.
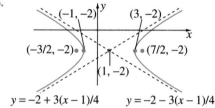

$y = -2 + 3(x-1)/4$ $y = -2 - 3(x-1)/4$

38. $3x^2 + y^2 = 28$

40. $25/4$

42. $9x^2 + 25y^2 = 225$

48. πab

50. $4\pi ab^2/3$, $4\pi a^2 b/3$

54. In Exercise 15, focus is $(0, -7/8)$, directrix is $y = -9/8$; in Exercise 18, focus is $(-7/48, 0)$, directrix is $x = -25/48$; in Exercise 21, focus is $(3/4, 0)$, directrix is $x = 5/4$; in Exercise 22, focus is $(6, 3/4)$, directrix is $x = 25/4$; in Exercise 27, focus is $(3, -1/4)$, directrix is $y = 1/4$; in Exercise 32, focus is $(-1/4, -4)$, directrix is $x = 1/4$.

Exercises 9.6

2. 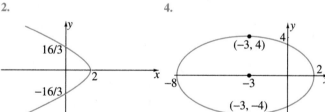   **4.**

6.   **8.** 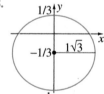   **10.**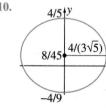

12. $8x^2 + 2x + 9y^2 = 1$

14. $8x^2 + 12x - y^2 + 4 = 0$ **16.** $40x + 25y^2 = 16$

18. With $x = \sqrt{5} + r\cos\theta$, $y = r\sin\theta$, the equation becomes $r = 4/(3 + \sqrt{5}\cos\theta)$.

20. With $x = \sqrt{10} + r\cos\theta$, $y = 2 + r\sin\theta$, the equation becomes $r = \pm 1/(3 \mp \sqrt{10}\cos\theta)$.

22. $\epsilon \to 0$

Review Exercises

2.
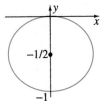

4.
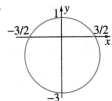

6.
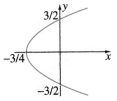

8.
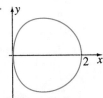

10.
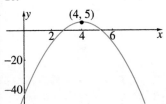

12.

14.
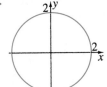

16.
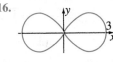

18.
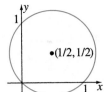

20.

22.
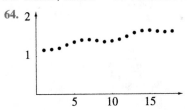

24.
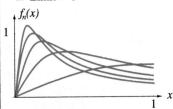

26.
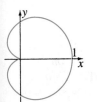

28.
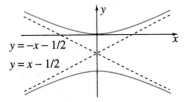

30.
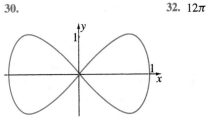

32. 12π

34. $3\pi/8$

36. $(3 - 3t^2)/(2 + 3t^2), \; -30t/(2 + 3t^2)^3$

38. 16

40. $\sqrt{2}(e^{\pi/2} - 1)$

Chapter 10

Exercises 10.1

2. Divergent   **4.** Convergent, 0   **6.** Convergent, 0

8. Convergent, 0   **10.** Convergent, $\pi/2$ **12.** Divergent

14. Convergent, 0   **16.** Convergent, 0   **18.** Convergent, 0

20. Convergent, $\pi/2$ **22.** $(3n + 1)/n^2$ **24.** $[1 + (-1)^{n+1}]/2$

26. 0   **28.** 4   **32.** $-0.381\,966\,0$

34. x_2 is not defined   **36.** $3.044\,723\,1$ **38.** $1.214\,648\,0$

40. $0.334\,734\,1$ **42.** $-1.388\,792\,0$ **44.** 2.4142

46. 0.1667 **48.** 0.5453 **50.** 0.2648

52. 0.7391 **54.** 0.2042

56. (a) $40(0.99)^n$ m (b) $4(0.981)^{-1/2}(0.99)^{n/2}$ s

58. $4P/3, \; 16P/9, \; (4/3)^n P$; does not exist

60. $0.0625, 0.1125$ **62.** $1113213211, 31131211131221$

64.

66. $-0.9712, \; -0.4196, \; -0.2461, \; -0.1593, \; -0.1059, \; -0.0693,$
$-0.0430, \; -0.0237, \; -0.0096, \; 0.0002$

Exercises 10.2

2. Limit = 0
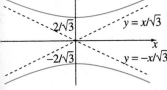

4. Limit = $1/x$
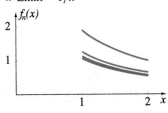

6. No limit
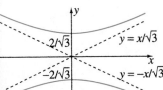

8. Limit = 0
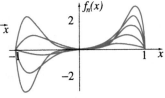

10. Limit = 1

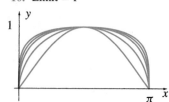

12. Limit = 1

14. Limit = 0

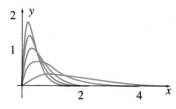

Exercises 10.3

2. $\sum_{n=0}^{\infty} (5^n/n!)x^n$, $-\infty < x < \infty$

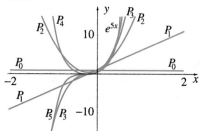

4. $(1/\sqrt{2})[1 + (x - \pi/4) - (x - \pi/4)^2/2! - (x - \pi/4)^3/3! + (x - \pi/4)^4/4! + \cdots]$, $-\infty < x < \infty$

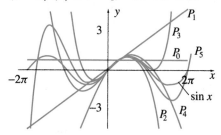

6. $\sum_{n=0}^{\infty} (2^n/n!)x^n$, $-\infty < x < \infty$

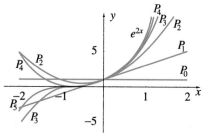

8. $\sum_{n=0}^{\infty} [(-1)^n/(2n)!](x - \pi/2)^{2n}$

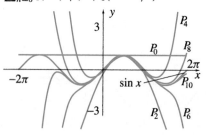

10. $\sum_{n=0}^{\infty} (x - 1)^n$, $0 < x < 2$

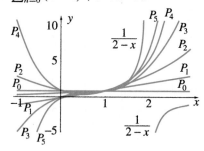

12. $\sum_{n=0}^{\infty} (n + 1)(-1)^n 3^n x^n$, $-1/3 < x < 1/3$

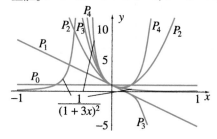

14. $1 + 3x/2 + \sum_{n=2}^{\infty} \{[(-1)^{n+1} 3^n (2n - 3)!]/[2^{2n-2} n! (n - 2)!]\}x^n$, $-1/3 \le x \le 1/3$

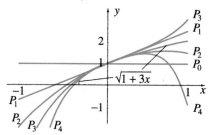

Exercises 10.4

2. $-1 < x < 1$ **4.** $-1/3 < x < 1/3$

6. $-4 < x < -2$ **8.** $7/2 < x < 9/2$

10. $-1 < x < 1$ **12.** $-1 < x < 1$

14. $-1/e < x < 1/e$ **16.** $x = 0$

18. $-1/3 < x < 1/3$ **20.** $-5^{1/3} < x < 5^{1/3}$

22. $-1 < x < 1$ **24.** $-1 < x < 1$

26. $4/(4 - x^3)$, $-4^{1/3} < x < 4^{1/3}$

28. $(x - 1)/(10 - x)$, $-8 < x < 10$

30. $\cos(x^2)$, $-\infty < x < \infty$

32. $x \sin(x/3)$, $-\infty < x < \infty$

34. $e^{-x} - 1$, $-\infty < x < \infty$

36. e^{2x-1}, $-\infty < x < \infty$

38. (a) $1 - x^2/2^2 + x^4/[2^4(2!)^2] - x^6/[2^6(3!)^2] + x^8/[2^8(4!)^2] - \cdots$; $x/2 - x^3/(2^3 2!) + x^5/(2^5 2! 3!) - x^7/(2^7 3! 4!) + x^9/(2^9 4! 5!) + \cdots$; $x^m/(2^m m!) - x^{m+2}/[2^{m+2}(m + 1)!] + x^{m+4}/[2^{m+4} 2!(m + 2)!] - x^{m+6}/[2^{m+6} 3!(m + 3)!] + x^{m+8}/[2^{m+8} 4!(m + 4)!] - \cdots$ (b) $-\infty < x < \infty$

Exercises 10.5

2. $\sum_{n=0}^{\infty} [(-1)^n/4^{n+1}]x^{2n}$, $-2 < x < 2$

4. $\sum_{n=0}^{\infty} (5^n/n!)x^n$, $-\infty < x < \infty$

6. $\sum_{n=0}^{\infty} [e(-2)^n/n!]x^n$, $-\infty < x < \infty$

8. $\sum_{n=0}^{\infty} [1/(2n)!]x^{2n}$, $-\infty < x < \infty$

10. $x^4 + 3x^2 - 2x + 1$, $-\infty < x < \infty$

12. $\sum_{n=0}^{\infty} [(-1)^n/5^{n+1}](x-2)^n$, $-3 < x < 7$

14. $-4/5 - 4(x-2)/25 + \sum_{n=2}^{\infty} \{9(-1)^{n+1}4^{n-2}/[(5^{n+1})]\}(x-2)^n$,
$3/4 < x < 13/4$

16. $\sum_{n=1}^{\infty} [(-1)^{n+1}2^n/n]x^n$, $-1/2 < x < 1/2$

18. $\ln 2 + \sum_{n=1}^{\infty} [(-1)^{n+1}/(n2^n)](x-2)^n$, $0 < x < 4$

20. $\sum_{n=0}^{\infty} [(-1)^n/4^{n+1}](x-4)^n$, $0 < x < 8$

22. $\sum_{n=0}^{\infty} (-1)^n(n+1)(x-3)^n$, $2 < x < 4$

24. $\sum_{n=0}^{\infty} [(-1)^n(1/3^{n+1} - 1/5^{n+1})/2]x^n$, $-3 < x < 3$

26. $\sqrt{3} + \sum_{n=1}^{\infty} \{2\sqrt{3}(-1)^{n+1}(2n-2)!/[12^n n!\,(n-1)!]\}\,x^n$,
$-3 \leq x \leq 3$

28. $-1 + \sum_{n=1}^{\infty} \{(-1)^{n-1}2^n[2\cdot5\cdot8\cdots(3n-4)]/(3^n n!)\}\,(x-1)^n$, $1/2 \leq x \leq 3/2$

30. $x + \sum_{n=2}^{\infty} \{(-1)\cdot2\cdot5\cdot8\cdots(3n-7)/[3^{n-1}(n-1)!]\}x^n$,
$-1 \leq x \leq 1$

32. $1 + x^2/2 + 5x^4/24 + 61x^6/720$

34. $1 + \sum_{n=1}^{\infty} [(-1)^n 2^{2n-1}/(2n)!]x^{2n}$, $-\infty < x < \infty$

36. $\sum_{n=0}^{\infty} \{(2n)!/[(2n+1)2^{2n}(n!)^2]\}x^{4n+2}$, $-1 < x < 1$

38. $\sqrt{2}\sum_{n=0}^{\infty} \{1/[2^n(2n+1)]\}x^{2n+1}$, $-\sqrt{2} < x < \sqrt{2}$

42. (a) $1/p$ (b) 6

44. $\sum_{n=0}^{\infty} \{(-1)^n\pi^{2n}/[(4n+1)2^{2n}(2n)!]\}x^{4n+1}$, $-\infty < x < \infty$; $\sum_{n=0}^{\infty} \{(-1)^n\pi^{2n+1}/[(4n+3)2^{2n+1}(2n+1)!]\}x^{4n+3}$,
$-\infty < x < \infty$

46. $n(-2)^{n-1}$

48. $(-1)^{n+1}(n-2)/e^2$

54. (a) $-1/2$, $1/6$, 0, $-1/30$, 0

Exercises 10.6

2. $2/(1-x)^3$, $-1 < x < 1$ 4. $(x+1)/(1-x)^3$, $-1 < x < 1$

6. $-(1/x)\ln(1-x)$, $-1 < x < 1$

8. $-\ln(1+x^2)$, $-1 < x < 1$

10. $1/(x-x^2) + (1/x^2)\ln(1-x)$, $-1 < x < 1$

12. $x\cos x$ 14. $(4x^2+3)e^{2x^2} - 3$

Exercises 10.7

2. 4.2×10^{-10} 4. 4.2×10^{-10} 6. 1.4×10^{-9}

8. 0.115   10. 0.2   12. 0.497

14. 0.291   16. 1/2   18. 1/2

20. 0   22. $1 - x^3/2 + 3x^6/8 - 5x^9/16$

24. $1 - x^2$

26. $a_0 + a_1\sum_{n=1}^{\infty} [(-1)^{n+1}/n!]x^n = C + De^{-x}$

28. $a_0\sum_{n=0}^{\infty} [(-1)^n/(2n)!]x^n$

30. $a_1\sum_{n=1}^{\infty} \{(-1)^{n+1}/[n!\,(n-1)!]\}x^n$

Exercises 10.8

2. False, $\{n\}$ 4. False, $\{-n\}$ 6. True

8. True   10. True   12. False, $\{(-1)^n/n\}$

14. False, $\{(-1)^n/n\}$ 16. False, $\{(-1)^n\}$ 18. True

20. False, $\{-3^n\}$ 22. True

24. False, $\{[1 + (-1)^{n+1}]/(2n)\}$

26. Increasing, $U = 1$, $V = 0$, $L = 0.419\,241$

28. Increasing, $U = 5$, $V = 0$, $L = (\sqrt{21}+1)/2$

30. Increasing, $U = 10$, $V = 3$, $L = (3+\sqrt{29})/2$

32. Decreasing, $U = 4$, $V = 0$, $L = (7-\sqrt{5})/2$

34. Decreasing, $U = 1$, $V = 0$, $L = (2-\sqrt{2})/2$

36. Increasing, $U = 1$, $V = 0$, $L = 1/2$

38. Decreasing, $U = 4$, $V = 0$, $L = 0$

40. Decreasing, $U = 2$, $V = 1$, $L = 1$

42. 0.273 89   44. 4/3   46. $\sqrt{2} - 1$

48. $(\sqrt{13}-1)/2$ 50. $(\sqrt{21}-1)/5$ 52. $(\sqrt{89}-3)/2$

56. 2

60. Increasing, $U = 30$, $V = -30$, $L = 30$

64. Fibonacci sequence

Exercises 10.9

2. 2/15   4. Diverges   6. 10 804.5

8. $-1/3$ 10. Diverges   12. 13/99

14. 430 162/9999   16. Diverges   18. 4

20. Diverges   22. 1   24. 804 s

26. $(\sqrt{3}P^2/180)[8 - 3(4/9)^n]$, $2\sqrt{3}P^2/45$

30. 5/8   32. 12   34. 6/11

36. $-1/3 < x < 1/3$

38. $-1 < x < 1$ 40. 60/11 minutes after 1:00

Exercises 10.10

2. Diverges   **4.** Converges   **6.** Converges

8. Diverges   **10.** Converges   **12.** Converges

14. Converges   **16.** Converges   **18.** Diverges

20. Diverges   **22.** Diverges   **24.** $-1 \leq x \leq 1$

26. $-2 - 1/\sqrt{3} \leq x \leq -2 + 1/\sqrt{3}$

28. Converges for $p > 1$, diverges for $p \leq 1$

Exercises 10.11

2. Converges   **4.** Converges   **6.** Diverges

8. Diverges   **10.** Converges   **12.** Converges

14. Converges   **16.** Diverges   **18.** Converges

20. Converges

Exercises 10.12

2. Converges conditionally   **4.** Converges absolutely

6. Diverges   **8.** Converges absolutely

10. Converges conditionally   **12.** Converges conditionally

14. Converges absolutely   **16.** $-1 \leq x \leq 1$

18. $-3 \leq x < -1$ **20.** $-1 \leq x < 1$

22. $1 \leq x \leq 3$

Exercises 10.13

12. $-9/100$

14. $0.055\,140\,9$, error $< 6 \times 10^{-7}$

16. error $< 0.001\,25$

18. $-0.947\,030$, error $< 5.2 \times 10^{-6}$

20. $0.693\,065$, error $< 1/(11 \cdot 2^{10})$

22. $1.067\,49$, error $< 4.01 \times 10^{-4}$

24. $-0.688\,172\,2$, error $< 1/101$

26. 4

28. (a) Error $< 2.92 \times 10^{-5}$ (b) Error $< 2.65 \times 10^{-5}$

30. 0.497   **32.** 0.133   **34.** 0.291

36. -0.122 **38.** 0.843

Review Exercises

2. Decreasing, $U = 1$, $V = 0$, $L = 1/\sqrt{3}$

4. Increasing, $U = 100$, $V = 7$, $L = (31 + \sqrt{53})/2$

6. Oscillatory, $L = (\sqrt{41} - 5)/8$

8. $|k| \leq 1$

12. Converges   **14.** Converges

16. Converges conditionally   **18.** Diverges

20. Converges   **22.** Converges

24. Diverges   **26.** Converges

28. Converges   **30.** Converges conditionally

32. $-2 \leq x \leq 2$ **34.** $-\infty < x < \infty$

36. $-4 < x < -2$ **38.** $-2^{-1/3} \leq x < 2^{-1/3}$

40. $e^5 \sum_{n=0}^{\infty} (1/n!)x^n$, $-\infty < x < \infty$

42. $\sum_{n=2}^{\infty} [(-1)^n 2^{n-1}/(n-1)]x^n$, $-1/2 < x \leq 1/2$

44. $(1/2) \sum_{n=0}^{\infty} (-1)^{n+1}(1 - 1/3^n)x^n$, $|x| < 1$

46. $x + \sum_{n=2}^{\infty} \{(-1)^n/[n(n-1)]\}x^n$, $-1 < x \leq 1$

48. $\sum_{n=0}^{18} [(-1)^n/n!]x^{2n}$

50. $\sum_{n=0}^{\infty} \{(-1)^n/[2^{2n}(2n)!]\}x^{2n} + \sum_{n=0}^{\infty} \{(-1)^n/[2^{2n+1}(2n + 1)!]\}x^{2n+1}$

Chapter 11

Exercises 11.1

2. $\sqrt{33}$

4. $(0, 0, 0)$, $(2, 0, 0)$, $(0, 2, 0)$, $(0, 0, 2)$, $(2, 2, 0)$, $(2, 0, 2)$, $(0, 2, 2)$, $(2, 2, 2)$

6. $\sqrt{29}$, 5, $2\sqrt{5}$, $\sqrt{13}$ **8.** 5, 3, 4, 5

12. $10x + 2y + 2z = 5$, plane   **14.** $(\sqrt{2}, \sqrt{2}, 5)$, $(\sqrt{2} \pm 1/4, \sqrt{2} \pm 1/4, 5 - \sqrt{7}/4)$, $(\sqrt{2} \pm 1/4, \sqrt{2} \pm 1/4, 9/2 - \sqrt{7}/4)$

16. (a) $(2, 1/2, -7/2)$ (b) $(5, 5, -5)$

Exercises 11.2

2.

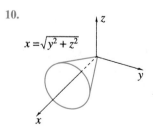

4.

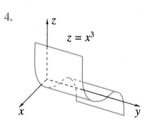

6.

8.

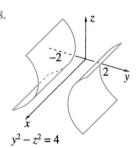

10.

12.

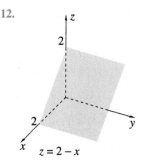

14.
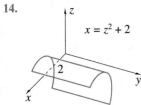
$x = z^2 + 2$

16.
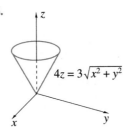
$4z = 3\sqrt{x^2 + y^2}$

18.
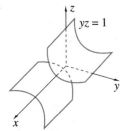
$yz = 1$

20.
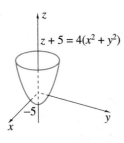
$z + 5 = 4(x^2 + y^2)$
-5

22.
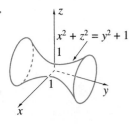
$x^2 + z^2 = y^2 + 1$

24.
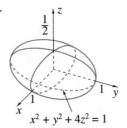
$\frac{1}{2}$
$x^2 + y^2 + 4z^2 = 1$

26.
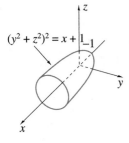
$(y^2 + z^2)^2 = x + 1$

28.
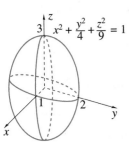
$y - z^2 = 0$

30.
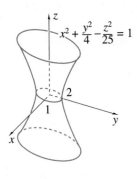
$x^2 + \frac{y^2}{4} + \frac{z^2}{9} = 1$

32.
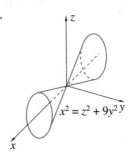
$x^2 = z^2 + 9y^2$

34.
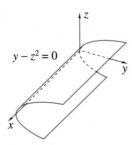
$x^2 + \frac{y^2}{4} - \frac{z^2}{25} = 1$

36.
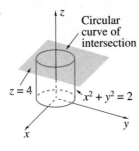
Circular curve of intersection
$z = 4$
$x^2 + y^2 = 2$

38.
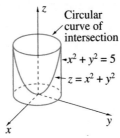
Circular curve of intersection
$x^2 + y^2 = 5$
$z = x^2 + y^2$

40.
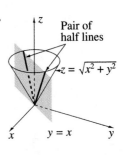
Pair of half lines
$z = \sqrt{x^2 + y^2}$
$y = x$

42.
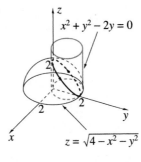
$x^2 + y^2 - 2y = 0$
$z = \sqrt{4 - x^2 - y^2}$

44.
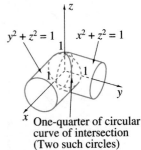
$y^2 + z^2 = 1$
$x^2 + z^2 = 1$
One-quarter of circular curve of intersection
(Two such circles)

46. $x + y = 3$, $z = 0$; $2y + 3z = 4$, $x = 0$; $2x - 3z = 2$, $y = 0$

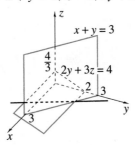
$x + y = 3$
$\frac{4}{3}$
$2y + 3z = 4$

48. $x^2 + y^2 = 4$, $z = 0$; $z = 4$, $x = 0$, $|y| \le 2$; $z = 4$, $y = 0$, $|x| \le 2$

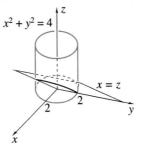
$z = 4$
$x^2 + y^2 = 4$

50. $y^2 + z^2 = 4$, $x = 0$; $x^2 + y^2 = 4$, $z = 0$; $z = x$, $y = 0$, $|x| \le 2$

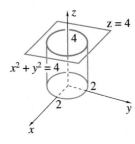
$x^2 + y^2 = 4$
$x = z$

52. $y = \pm x, z = 0, |x| \le \sqrt{3}; y^2 + z^2 = 3, x = 0; x^2 + z^2 = 3,$
$y = 0$

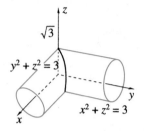

54. $x^2 + y^2 = 4, z = 0; z = 2, x = 0, |y| \le 2; z = 2, y = 0,$
$|x| \le 2$

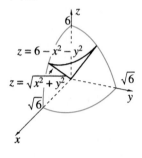

56. $(y + 2)^2 - (x - 1)^2 = 3, z = 0$

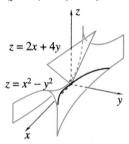

58. $2y = 1 + x^2, z = 0$

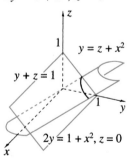

60. $(x - 4/3)^2 + (y - 4/3)^2 = 8/9, z = 0$

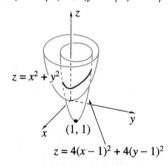

62.

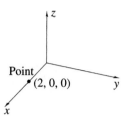

64.

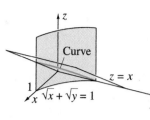

66.

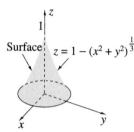

68.

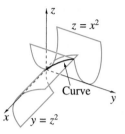

70.

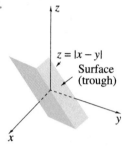

Exercises 11.3

2. $(2, 6, 8)$ 4. $(-1/\sqrt{5}, 0, 2/\sqrt{5})$

6. $(-4\sqrt{5} - 8, -6, 8\sqrt{5} + 4)$

8. $(-6\sqrt{46} - 4\sqrt{5}, -12\sqrt{5}, 12\sqrt{46} - 24\sqrt{5})$

10. $(-6/\sqrt{17}, -3/\sqrt{17}, 6/\sqrt{17})$

12. $3\hat{\mathbf{i}} - 2\hat{\mathbf{j}}$

14. $(2/\sqrt{5} - 1/\sqrt{10})\hat{\mathbf{i}} + (1/\sqrt{5} + 3/\sqrt{10})\hat{\mathbf{j}}$

16. $(5, 0, 0)$ 18. $(0, 3/\sqrt{5}, 6/\sqrt{5})$

20. $(1, 1, -3/2)$ 22. $(5/4, 5\sqrt{2}/4, 5/4)$

24. $(2/\sqrt{3}, 2/\sqrt{3}, 2/\sqrt{3})$ 30. $(3/\sqrt{17}, 12/\sqrt{17})$

32. $5G(-3/\sqrt{10} + 30\sqrt{11}/121, -10\sqrt{11}/121, -1/\sqrt{10} + 10\sqrt{11}/121)$ N

34. 586 N, 2188 N   36. 784 N

38. $16\sqrt{13}g/9$ N, $16\sqrt{11}g/9$ N, $16\sqrt{11}g/9$ N

40. $24g$ N, $40\sqrt{1034}g/9$ N, $32\sqrt{1259}g/9$ N

42. Linearly dependent   44. Linearly dependent

50. (a) $W = 2ky(1 - L/\sqrt{L^2 + y^2})$ (b) ky^3/L^2

Exercises 11.4

2. $(-10, 15, -5)$

4. $4/\sqrt{14}$

6. -178

8. 1

10. 0

12. $(48, 24, -42)$

14. $(2\sqrt{247})^{-1}(17, -23, -7)$

16. $(-41, 11, 28)$

18. $(17, -23, -7)$

20. $(-13, 19, 5)$

22. Yes

24. Yes

26. 0.684 rad

28. 2.20 rad

30. π rad

32. $\lambda(9, 0, 1)$

38. $\pi/2$, 1.25, 2.82

40. 1.95, 0.980, 0.734

44. (a) 108

Exercises 11.5

2. $2x + y - 2z + 5 = 0$

4. $19x - 14y - z = 91$

6. $2x - y - 2z = 6$

8. $10x + 89y - 6z = 394$

10. (a) $13x + 17y = 70$ (b) $x - 17z = 8$ (c) $y + 13z = -2$

12. $(x, y, z) = (1, -1, 3) + t(2, 4, -3)$; $x = 1 + 2t$, $y = -1 + 4t$, $z = 3 - 3t$; $(x-1)/2 = (y+1)/4 = (z-3)/(-3)$

14. $(x, y, z) = (2, -3, 4) + t(3, 5, -5)$; $x = 2 + 3t$, $y = -3 + 5t$, $z = 4 - 5t$; $(x-2)/3 = (y+3)/5 = (z-4)/(-5)$

16. $(x, y, z) = (1, 3, 4) + t(0, 0, 1)$; $x = 1$, $y = 3$, $z = 4 + t$

18. $(x, y, z) = (2, 0, 3) + t(1, 0, -2)$; $x = 2 + t$, $y = 0$, $z = 3 - 2t$; $x - 2 = (z - 3)/(-2)$, $y = 0$

20. $(x, y, z) = (0, -5, 30) + t(1, 2, -11)$; $x = t$, $y = -5 + 2t$, $z = 30 - 11t$; $x = (y+5)/2 = (z-30)/(-11)$

22. No

26. $x = \sqrt{2} \pm 1/4$, $y = \sqrt{2} \pm 1/4$, $z = 9/2 - \sqrt{7}/4$, $\sqrt{7}x - z = \sqrt{14} - 5$, $\sqrt{7}x + z = \sqrt{14} + 5$, $\sqrt{7}y - z = \sqrt{14} - 5$, $\sqrt{7}y + z = \sqrt{14} + 5$

Exercises 11.6

2. $\sqrt{299}/2$

4. 0

6. $\sqrt{42}$

8. $22/\sqrt{51}$

10. $18/\sqrt{14}$

12. $5/(3\sqrt{2})$

14. $27/\sqrt{21}$

16. $20/(3\sqrt{6})$

18. $2/\sqrt{6}$

20. $2/(3\sqrt{6})$

22. $\sqrt{146}/2$

24. 0

26. $9/\sqrt{117}$

28. $31/(2\sqrt{153})$

34. $3/\sqrt{2}$, $1/\sqrt{2}$

36. $\sqrt{5}$, $-\sqrt{14}/2$, $-\sqrt{70}/2$

38. $-3/10$, $11/20$

40. $x - 2y + 3z = 6 \pm 2\sqrt{14}$

Exercises 11.7

2. $(-10, 5, -2)$

4. $(-5, 1, 4)$

6. (a) $\pm 60/\sqrt{59}$ (b) $-18, 8, -3$ (c) $\pm 2/\sqrt{38}$

8. $\pm 8\sqrt{3}$

10. M_x, M_y, M_z

14. $k[1 + l(1 - \sqrt{5})]/2$ J

16. $GMm(\sqrt{2} - 1)/(\sqrt{2}R)$

18. $3k(L^2/2 - l\sqrt{L^2 + l^2} + l^2)$

20. $(-6\sqrt{3}, 15/2, -6)$ N·m

22. (a) $300(-3, 4, -5)$ N·m (b) -1500 N·m (c) 0

24. $(60\sqrt{3} + 15, 30, 25/\sqrt{2})$ N·m

26. (a) 0, $-3195/2$ N·m, $1917/2$ N·m (b) 1824 N·m, $-2565/2$ N·m, $1539/2$ N·m

28. 1.532 m

30. (a) 310.5 N·m (b) -264.96 N·m

Exercises 11.8

2. 3.93×10^4 N, 4.82×10^4 N

4. 2.06×10^8 N, 2.19×10^8 N

6. 100.663 m, 100.667 m

8. 1.17 m

10. 35.6 m, 49.2 kg

12. 123.6 m, 3679 N

Exercises 11.9

2. $-\infty < t < \infty$

4. $t > -4$

6. $\hat{\mathbf{i}} - 2t\hat{\mathbf{j}} + 2\hat{\mathbf{k}}$

8. $2t(4t^2 - 3)\hat{\mathbf{i}} + t^2(9 - 10t^2)\hat{\mathbf{j}} + 4t(4t^2 - 3)\hat{\mathbf{k}}$

10. $3t^2(4 - 5t^2)\hat{\mathbf{i}} + 4t(1 - 3t^2)\hat{\mathbf{j}} - 3t^2\hat{\mathbf{k}}$

12. $3\hat{\mathbf{i}} - 2(3t + 4)\hat{\mathbf{j}} + 6(1 + 4t)\hat{\mathbf{k}}$

14. $9t^2\hat{\mathbf{i}} + (6t - 20t^3)\hat{\mathbf{j}} + (30t^4 - 21t^2 + 6)\hat{\mathbf{k}}$

16. $(-15t^4 + 12t^2)\hat{\mathbf{i}} + (4t - 12t^3)\hat{\mathbf{j}} - 3t^2\hat{\mathbf{k}}$

18. $(3t^4/2 - 4t^2)\hat{\mathbf{i}} + (17t^3/3 - 12t^5/5)\hat{\mathbf{j}} + (3t^6 - 27t^4/4 + t^2)\hat{\mathbf{k}} + \mathbf{C}$

20. $(-14t^4 - 6t^3 - 42t^2 + 4t)\hat{\mathbf{i}} + (28t^5 + 84t^3 - 6t^2)\hat{\mathbf{j}} - (126t^4 + 42t^6 + 2t)\hat{\mathbf{k}}$

Exercises 11.10

2. $x = \sqrt{2}\cos t$, $y = \sqrt{2}\sin t$, $z = 4$; $0 \le t < 2\pi$ $\mathbf{r} = \sqrt{2}\cos t\,\hat{\mathbf{i}} + \sqrt{2}\sin t\,\hat{\mathbf{j}} + 4\hat{\mathbf{k}}$

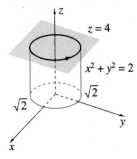

4. $x = \sqrt{5}\cos t$, $y = \sqrt{5}\sin t$, $z = 5$; $0 \le t < 2\pi$ $\mathbf{r} = \sqrt{5}\cos t\,\hat{\mathbf{i}} + \sqrt{5}\sin t\,\hat{\mathbf{j}} + 5\hat{\mathbf{k}}$

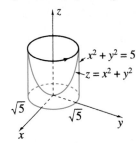

6. $x = t$, $y = t$, $z = \sqrt{2}|t|$ $\mathbf{r} = t\hat{\mathbf{i}} + t\hat{\mathbf{j}} + \sqrt{2}|t|\hat{\mathbf{k}}$

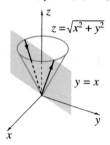

8. $x = \cos t$, $y = 1 + \sin t$, $z = \sqrt{2 - 2\sin t}$; $0 \le t < 2\pi$
$\mathbf{r} = \cos t\,\hat{\mathbf{i}} + (1 + \sin t)\hat{\mathbf{j}} + \sqrt{2 - 2\sin t}\,\hat{\mathbf{k}}$

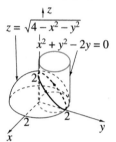

10. $x = -t$, $y = t^2$, $z = \sqrt{t^2 + t^4}$
$\mathbf{r} = -t\hat{\mathbf{i}} + t^2\hat{\mathbf{j}} + \sqrt{t^2 + t^4}\,\hat{\mathbf{k}}$

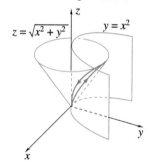

12.

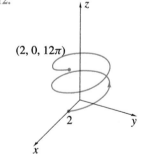

14.

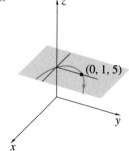

Exercises 11.11

2. $\mathbf{r} = t\hat{\mathbf{i}} + t^2\hat{\mathbf{j}} + t^3\hat{\mathbf{k}}$; $\hat{\mathbf{T}} = (\hat{\mathbf{i}} + 2t\hat{\mathbf{j}} + 3t^2\hat{\mathbf{k}})/\sqrt{1 + 4t^2 + 9t^4}$

4. $\mathbf{r} = -t\hat{\mathbf{i}} + (5 + t)\hat{\mathbf{j}} + (t^2 - t - 5)\hat{\mathbf{k}}$, $-5 \le t \le 0$; $\hat{\mathbf{T}} = [-\hat{\mathbf{i}} + \hat{\mathbf{j}} + (2t - 1)\hat{\mathbf{k}}]/\sqrt{4t^2 - 4t + 3}$

6. $(-2\hat{\mathbf{i}} + 3\hat{\mathbf{j}} + \hat{\mathbf{k}})/\sqrt{14}$ 8. $(\hat{\mathbf{i}} - \hat{\mathbf{j}})/\sqrt{2}$

10. $-\hat{\mathbf{j}}$ 12. $\sqrt{42}$

14. $(616\sqrt{616} - 157\sqrt{157})/459$

16. $\cos t\,\hat{\mathbf{i}} + \sin t\,\hat{\mathbf{j}}$

18. (a) $(0, 0, 0)$ (b) $2\hat{\mathbf{i}} + 2\hat{\mathbf{k}}$

Exercises 11.12

2. $[-(2t + 9t^3)\hat{\mathbf{i}} + (1 - 9t^4)\hat{\mathbf{j}} + (3t + 6t^3)\hat{\mathbf{k}}]/$
$\sqrt{1 + 13t^2 + 54t^4 + 117t^6 + 81t^8}$;
$(3t^2\hat{\mathbf{i}} - 3t\hat{\mathbf{j}} + \hat{\mathbf{k}})/\sqrt{1 + 9t^2 + 9t^4}$

4. $[(2t - 1)\hat{\mathbf{i}} + (1 - 2t)\hat{\mathbf{j}} + 2\hat{\mathbf{k}}]/\sqrt{8t^2 - 8t + 6}$; $(\hat{\mathbf{i}} + \hat{\mathbf{j}})/\sqrt{2}$

6. $-(5\hat{\mathbf{i}} + 3\hat{\mathbf{j}} + \hat{\mathbf{k}})/\sqrt{35}$; $(-\hat{\mathbf{j}} + 3\hat{\mathbf{k}})/\sqrt{10}$

8. $-(\hat{\mathbf{i}} + \hat{\mathbf{j}})/\sqrt{2}$; $-\hat{\mathbf{k}}$

10. $-(\hat{\mathbf{i}} + \hat{\mathbf{k}})/\sqrt{2}$; $(\hat{\mathbf{i}} - \hat{\mathbf{k}})/\sqrt{2}$

12. $\kappa = 0$, ρ undefined

14. $\kappa = 2e^t\sqrt{1 + e^{2t}}/(1 + 2e^{2t})^{3/2}$, $\rho = (1 + 2e^{2t})^{3/2}/(2e^t\sqrt{1 + e^{2t}})$

16. $\kappa = 1/[\sqrt{2}(1 + \cos^2 t)^{3/2}]$, $\rho = \sqrt{2}(1 + \cos^2 t)^{3/2}$

18. $\kappa = \sqrt{1 + 36t^4 + 16t^6}/[2(1 + t^2 + 4t^6)^{3/2}]$, $\rho = 2(1 + t^2 + 4t^6)^{3/2}/\sqrt{1 + 36t^4 + 16t^6}$

22. 0

24. (a) $-\sin t\,\hat{\mathbf{i}} + \cos t\,\hat{\mathbf{j}}$; $-(\cos t\,\hat{\mathbf{i}} + \sin t\,\hat{\mathbf{j}})$; $\hat{\mathbf{k}}$ (b) $2\sin 2t(\sin t - \cos t)$, $-4(\cos^3 t + \sin^3 t)$; (c) $2\sin 2t(\sin t - \cos t)\hat{\mathbf{T}} - 4(\cos^3 t + \sin^3 t)\hat{\mathbf{N}}$

26. $(-\sin t\,\hat{\mathbf{i}} + \cos t\,\hat{\mathbf{j}} + \hat{\mathbf{k}})/\sqrt{2}$; $-(\cos t\,\hat{\mathbf{i}} + \sin t\,\hat{\mathbf{j}})$; $(\sin t\,\hat{\mathbf{i}} - \cos t\,\hat{\mathbf{j}} + \hat{\mathbf{k}})/\sqrt{2}$; $(1 - \cos t\,\sin t + \cos^2 t\,\sin^2 t)\hat{\mathbf{T}}/\sqrt{2} - \cos t(\cos t + \sin^3 t)\hat{\mathbf{N}} + (-\cos^2 t\,\sin^2 t + \cos\,\sin t + 1)\hat{\mathbf{B}}/\sqrt{2}$

Exercises 11.13

2. $[(t^2 - 1)\hat{\mathbf{i}} + (t^2 + 1)\hat{\mathbf{j}}]/t^2$; $\sqrt{2 + 2t^4}/t^2$; $2(\hat{\mathbf{i}} - \hat{\mathbf{j}})/t^3$

4. $2t\hat{\mathbf{i}} + 2e^t(t + 1)\hat{\mathbf{j}} - 2\hat{\mathbf{k}}/t^3$; $2\sqrt{t^2 + e^{2t}(t + 1)^2 + 1/t^6}$; $2\hat{\mathbf{i}} + 2e^t(t + 2)\hat{\mathbf{j}} + 6\hat{\mathbf{k}}/t^4$

6. $(t^4/4 + 1)\hat{\mathbf{i}} + (t^3 + 3t^2 + 12)\hat{\mathbf{j}}/6 - (t^5/5 + 1)\hat{\mathbf{k}}$

8. $4t/\sqrt{1 + 4t^2}$; $2/\sqrt{1 + 4t^2}$

12. (a) 0 (b) $56t^6$ 14. $150x\hat{\mathbf{j}}$

18. (b) 7.79 km/s

20. $365\hat{\mathbf{i}} + 325\sqrt{3}\hat{\mathbf{j}}$ km/h; 670.9 km/h

22. $\mathrm{Sin}^{-1}(3/4)$ radians upstream, $1/(5\sqrt{7})$ h

24. $(3/2, 3/2, 2)$ 26. $t = 0$

30. 37.7 m/s, 0.381 rad   34. Yes

38. (b) $S[(1 - \cos\theta)\hat{\mathbf{i}} + \sin\theta\,\hat{\mathbf{j}}]$, $S\sqrt{2 - 2\cos\theta}$, $(S^2/R)(\sin\theta\,\hat{\mathbf{i}} + \cos\theta\,\hat{\mathbf{j}})$ (c) $(S^2/R)\sqrt{(1 - \cos\theta)/2}$, $(S^2/R)\sin\theta/\sqrt{2 - 2\cos\theta}$

40. (c) $2S[(\sin 2\theta - \sin\theta)\hat{\mathbf{i}} + (\cos\theta - \cos 2\theta)\hat{\mathbf{j}}]$, $2\sqrt{2}S\sqrt{1 - \cos\theta}$, $(2S^2/R)[(2\cos 2\theta - \cos\theta)\hat{\mathbf{i}} + (2\sin 2\theta - \sin\theta)\hat{\mathbf{j}}]$ (d) $(3\sqrt{2}S^2/R)\sqrt{1 - \cos\theta}$, $(\sqrt{2}S^2/R)\sin\theta/\sqrt{1 - \cos\theta}$

42. At an angle $\mathrm{Sin}^{-1}[(15L + \sqrt{100L^2 - 5})/(50L^2 + 2)]$ downstream. $L \ge \frac{\sqrt{5}}{10}$

44. Straight line path

Review Exercises

2. -8

4. $(39, -33, -30)$

6. -2

8. $(-16, -24, -4)$

10. $(67/7, 21, -65/7)$

12.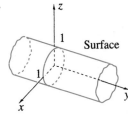

14.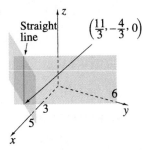

16. No points

18.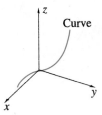

20.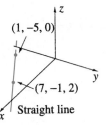

22.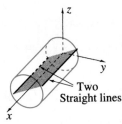

24.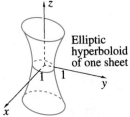

26.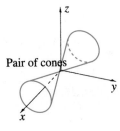

28. $x = 6 + 5t$, $y = 6 - 2t$, $z = 2 + t$

30. $x = 1 + u$, $y = 3 + 2u$, $z = 2 + 3u$

32. $x = y + z$ 34. $3x - 4y + z = 0$

36. $35/\sqrt{41}$ 38. $2\sqrt{5}/15$ 40. $\sqrt{59}$

42. $(2\cos t\,\hat{\mathbf{i}} - 2\sin t\,\hat{\mathbf{j}} + \hat{\mathbf{k}})/\sqrt{5}$; $-\sin t\,\hat{\mathbf{i}} - \cos t\,\hat{\mathbf{j}}$; $(\cos t\,\hat{\mathbf{i}} - \sin t\,\hat{\mathbf{j}} - 2\hat{\mathbf{k}})/\sqrt{5}$

44. $\hat{\mathbf{i}} + 2t\hat{\mathbf{j}} + 2t\hat{\mathbf{k}}$; $\sqrt{1 + 8t^2}$; $2(\hat{\mathbf{j}} + \hat{\mathbf{k}})$; 0, $\sqrt{1 + 8t^2}$; $2\sqrt{2}/\sqrt{1 + 8t^2}$, $8t/\sqrt{1 + 8t^2}$

46. (a) 4.46 m/s (b) $0.226\hat{\mathbf{i}} - \hat{\mathbf{j}}$ m (c) 0.515 m from point on floor directly below point it left table

48. $k(1 - x)[1 - 1/\sqrt{1 + 4(1 - x)^2}]\hat{\mathbf{i}} + (k/2)[1 - 1/\sqrt{1 + 4(1 - x)^2}]\hat{\mathbf{j}}$ N

Chapter 12

Exercises 12.1

4. All points between the branches of the hyperbola $x^2 - y^2 = 1$

6. All points in space except $(0, 0, 0)$

8. 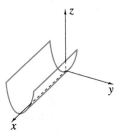   10.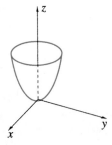

12. 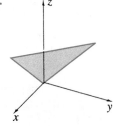   14.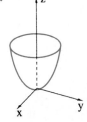

16. 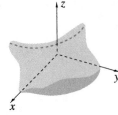   18.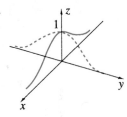

20. 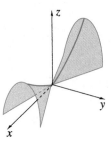   22.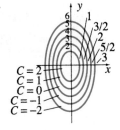

24.

26. $lw(15 - lw)/(w + l)$ 28. $8c|xy|\sqrt{1 - x^2/a^2 - y^2/b^2}$

30. (a) 7.70 m (b) 14.9% (c) 4.4%

32.

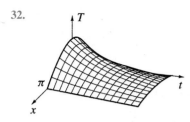

34. $100\,000 + 300x + 200y$; $3x + y \geq 25$, $2x + 3y \geq 50$, $x \geq 0$, $y \geq 0$

Exercises 12.2

2. $3/5$ **4.** $-3/7$ **6.** 0

8. $\pi/2$ **10.** 1 **12.** 4

14. Does not exist   **16.** 0 **18.** Does not exist

20. Does not exist   **22.** $(0, 0)$

24. $x = 0$; $y = 0$; $z = 0$ **26.** $x = 0$; $y = 0$; $y = -x$

28. Does not exist   **30.** Does not exist

32. Does not exist   **34.** 1

36. (a) 1, no (b) 1, yes   **38.** False

Exercises 12.3

2. $3y - 16x^3y^4$, $3x - 16x^4y^3$

4. $-x(x + 2y)/[y(x + y)^2]$, $x^2(x + 2y)/[y^2(x + y)^2]$

6. $y \cos(xy)$, $x \cos(xy)$

8. $x/\sqrt{x^2 + y^2}$, $y/\sqrt{x^2 + y^2}$

10. $4x \sec^2(2x^2 + y^2)$, $2y \sec^2(2x^2 + y^2)$

12. ye^{xy}, xe^{xy}

14. $2x/(x^2 + y^2)$, $2y/(x^2 + y^2)$

16. $ye^x \cos(ye^x)$, $e^x \cos(ye^x)$

18. $2xy \cos^2(x^2y) \sin(x^2y)/[1 - \cos^3(x^2y)]^{2/3}$, $x^2 \cos^2(x^2y) \sin(x^2y)/[1 - \cos^3(x^2y)]^{2/3}$

20. $\tan\sqrt{x + y}/(2\sqrt{x + y})$, $\tan\sqrt{x + y}/(2\sqrt{x + y})$

22. $-2z/[1 + (x^2 + z^2)^2]$

24. 2

26. $-1/[1 + (1 + x + y + z)^2]$

28. $3x^2/y + \sin(yz/x) - (yz/x)\cos(yz/x)$

30. $z \operatorname{Sin}^{-1}(x/z)$ if $z > 0$; $z \operatorname{Sin}^{-1}(x/z) + xz/\sqrt{z^2 - x^2}$ if $z < 0$

34. (a) Yes (b) No (c) Yes (d) No

38. (a) $bc \sin A/\sqrt{b^2 + c^2 - 2bc \cos A}$
(b) $2a/\sqrt{4b^2c^2 - (b^2 + c^2 - a^2)^2}$
(c) $(b - c \cos A)/\sqrt{b^2 + c^2 - 2bc \cos A}$
(d) $(c^2 - a^2 - b^2)/[b\sqrt{4b^2c^2 - (b^2 + c^2 - a^2)^2}]$

40. $2.7/(0.15 + 12t)$ kg/m^3

Exercises 12.4

2. $2xyz\hat{\mathbf{i}} + x^2z\hat{\mathbf{j}} + x^2y\hat{\mathbf{k}}$ **4.** $(2xy + y^2)\hat{\mathbf{i}} + (x^2 + 2xy)\hat{\mathbf{j}}$

6. $(1 + x^2y^2z^2)^{-1}(yz\hat{\mathbf{i}} + xz\hat{\mathbf{j}} + xy\hat{\mathbf{k}})$

8. $e^{x+y+z}(\hat{\mathbf{i}} + \hat{\mathbf{j}} + \hat{\mathbf{k}})$

10. $-(x^2 + y^2 + z^2)^{-3/2}(x\hat{\mathbf{i}} + y\hat{\mathbf{j}} + z\hat{\mathbf{k}})$

12. $-\sin 1(\hat{\mathbf{i}} + \hat{\mathbf{j}} + \hat{\mathbf{k}})$

14. $-4e^{-8}(\hat{\mathbf{i}} + \hat{\mathbf{j}})$

22. ∇F not defined along the line $y = x$

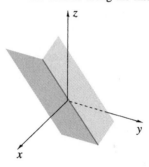

24. $x^2y - xy + C$ **26.** $xyz + y^2z + C$

28. $f(x, y) = g(x, y) + C$

Exercises 12.5

2. $-12x/y^4 + 72x^3y$ **4.** $x(1 + z + y + yz)e^{x+y+z}$

6. $e^{x+y} + 4/y^3$ **8.** 0

10. $(x^2 + y^2 - z^2)/(x^2 + y^2 + z^2)^2$

12. $2xy/(x^2 + y^2)^2$

14. $127/(756\sqrt{7})$ **16.** $8!y^9z^{10}$

18. $3y^2 \cos(x + y^3)$ **20.** $-xy/(x^2y^2 - 1)^{3/2}$

24. Entire plane with $(0, 0)$ deleted

26. All space   **28.** Not harmonic

32. $L + 9.81\rho AL^3/(6E)$ **34.** (c) 0 (d) 0

38. (a)

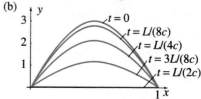

(b)

40. (b)

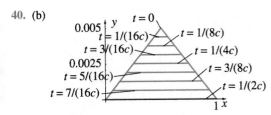

42. (b)

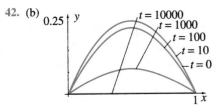

44. $T_0 + (T_L - T_0)x/L + Fx(L - x)/(2k)$

48. (b) $2xy + C$

50. $n = 0$, all space; $n = -1/2$, $x^2 + y^2 + z^2 > 0$

Exercises 12.6

2. $(2xe^y + y/x)(-s^2 \sin t) + (x^2e^y + \ln x)\left[8t/\left((t^2 + 2s)\sqrt{(t^2 + 2s)^2 - 1}\right)\right]$

4. $xv^2[2yv(3u^2 + 2) + 2xuv/(u^2 + 1) + 3xye^u(u + 1)]$

6. $-2(\ln 3)y3^{x+2} \csc(r^2 + t) \cot(r^2 + t)$

8. $-2rst[5st/y^6 + 2yrt/(y^2 + z^2)^2 + 2rs/y^3]$

10. $6xve^t(v - 2)[1 - 2u/(x^2 - y^2)^2] + 8ye^{4t}[1 + 2u/(x^2 - y^2)^2]$

12. $2(t^2 + 2t)^2e^{2t} + (t^3 + 6t^2 + 6t + 2yt^2 + 4y + 8yt)e^t - 2$

14. $-(3y \sin v - 4x \cos v)^2 \sin(xy) - (24 \sin v \cos v + 3y \cos v + 4x \sin v) \cos(xy)$

18. $200\pi/3$ cc/min; no

20. -7.11×10^4 N/s

22. (a) Yes (b) No (c) Yes (d) No (e) Yes (f) No (g) No (h) Yes

40. $n(n-1)f(x, y, z) = x^2\partial^2 f/\partial x^2 + y^2\partial^2 f/\partial y^2 + z^2\partial^2 f/\partial z^2 + 2xy\partial^2 f/\partial x\partial y + 2yz\partial^2 f/\partial y\partial z + 2zx\partial^2 f/\partial z\partial x$

Exercises 12.7

2. $1/(x + y) - 1$

4. $[24x - \cos(x + y)]/[2y - 1 + \cos(x + y)]$

6. $-(2xz^2 + 3)/(2x^2z + y)$, $-z/(2x^2z + y)$

8. $z(1 + y^2z^2)/(y - x - xy^2z^2)$, $z/(x + xy^2z^2 - y)$

10. $[\cos(x + t) + \cos(x - t)]/[\cos(x - t) - \cos(x + t)]$

12. $(x + y)(3y^2 - 3x^2 - 5)/(3y^2z - 5z)$

14. 0

16. $e^x[(2t + 1)\cos y/(3x^2 + e^x) + (y^2 + 2yt - 1)\sin y/(t^2 + 2yt + 1)]$

18. $[u^3 + u\cos(uv)]e^{-u}\cos v + [3u^2v + v\cos(uv)]e^{-u}\sin v$

20. $u(3v^2 - u^2)/[4(u^2 + v^2)^3]$

22. $-2t/(9y^2)$

Exercises 12.8

2. $1/\sqrt{5}$

4. $-2/\sqrt{17}$

6. $1/5$

8. $-13/\sqrt{29}$

10. $11/\sqrt{82}$

12. $-40/\sqrt{6}$

14. $(1, 4)$, $\sqrt{153}/2$

16. $(1, -3, 2)$, $1/14$

18. $(1, 1)$, $2\sqrt{2}e$

20. (a) $\pm(2, 1)$ (b) $(-1 \pm 2\sqrt{19}, 2 \pm \sqrt{19})$ (c) No direction

22. (a) Yes (b) No

24. $9t/[\sqrt{13}\sqrt{4 + 9t^2}]$, 0

26. $(0, 0, 0)$, $(1/4, \pm 1/2, 1/4)$

28. The negative thereof

30. $\sqrt{65}/3$

32. -2

34. (a) $\pi/(\sqrt{2}\sqrt{8 - 4\pi + \pi^2})$, $\pi/\sqrt{4 + \pi^2}$ (b) $1/\sqrt{2}$, 0 (c) $1/\sqrt{2}$, 1

Exercises 12.9

2. $x = 1 + u$, $y = 1 + 2u$, $z = 1 + 3u$

4. $x = -2 + t$, $y = 4 - 4t$, $z = -2 + t$

6. $x = 1 - 2u$, $y = 5 + 2u$, $z = 1 + u$

8. $132x + 49y = 328$, $z = 0$

10. $x = 1 - u$, $y = u$, $z = u$

12. $x = 2 + t$, $y = -\sqrt{5} + t/\sqrt{5}$, $z = -1 - t$

14. $x = 4 + \sqrt{17}u$, $y = 1$, $z = \sqrt{17} + 4u$

16. $x = 12 + 16t$, $y = -14 - 31t$, $z = 2 + t$

18. $x = t$, $y = 1$, $z = 1 + t$

20. $x = 2\pi u$, $y = 2\pi + u$, $z = 4\pi + 2u$

22. $3x + 6y + z + 2 = 0$

24. $x + y + z = 4$

26. $x + y = 1$

30. $6\sqrt{2}$

32. 0

34. $x_0x/a^2 + y_0y/b^2 + z_0z/c^2 = 1$

36. No points

38. $(0, 0, -1)$ and $x^2 + y^2 = 1/2$, $z = -1/2$

Exercises 12.10

2. $(0, 0)$ saddle point; $(1, 1)$ relative maximum

4. None

6. $(0, n\pi)$ saddle points

8. (x, x) relative minima

10. $(x, 0)$ saddle points; $(0, y)$ $y > 0$ relative minima; $(0, y)$ $y < 0$ relative maxima

12. $(0, 0)$ relative minimum; $(0, y)$ $y \neq 0$ none of these

14. $(0, 0)$ relative maximum; $(0, \pm 1/\sqrt{2})$, $(\pm 1/\sqrt{2}, 0)$ saddle points; $(1/\sqrt{2}, \pm 1/\sqrt{2})$, $(-1/\sqrt{2}, \pm 1/\sqrt{2})$ relative minima

16. None

18. All points on the coordinate axes

22. $(0, 0)$ relative minimum; $(-1/3, \pm 2/9)$ saddle points

Exercises 12.11

2. $4, -1/3$

4. $3, -3$

6. $3, -2$

8. $(2\sqrt{2}+5)/\sqrt{2}, (2\sqrt{2}-5)/\sqrt{2}$

10. $(1, 1, -2)$

12. $(1/2, 1/2, 1/2)$

14. $2, -32$

16. $100(2/3)^{1/3} \times 100(2/3)^{1/3} \times 100(3/2)^{2/3}$ cm

20. $2a/\sqrt{3} \times 2b/\sqrt{3} \times 2c/\sqrt{3}$ 22. $1, -1$

24. $1, -1$ 30. $C\alpha/A, C\beta/B$ 32. $\pm 1/\sqrt{2}$

Exercises 12.12

2. $(8 \pm 9\sqrt{13})/2$

4. ± 27

6. $\pm\sqrt{32/27}$

8. $(\pm\sqrt{2}-1)/2$

16. $\sqrt{2}, 0$

18. $2, 0$

22. $2.4048\sqrt{3/(2k)}, \pi\sqrt{3/k}$ 24. $33.12, -0.12$

26. $\pm 1/\sqrt{2}$

28. $(\pm 2\sqrt{14}, \mp\sqrt{14}), (\pm 2, \pm 4)$

30. $(3a/2, 3a/2)$

Exercises 12.13

2. $P = 2.8187A + 72.967$

4. (c) $y = 1.6653x^2 - 16.642x + 66.802$

6. $H = 38.82 - 0.003265Q^2$ 8. $y = 120.3e^{0.43x}$

10. 101.3 km/h

12. $y = 31.35x^{0.4776}$

14. $N = 32.1476e^{0.35554t}$

16. (a) $\left(\sum_{i=1}^{n} \bar{y}_i/x_i^2\right) / \left(\sum_{i=1}^{n} 1/x_i^4\right)$ (b) 0.5590

18. $y = 1/(0.04274x + 0.4968)$

Exercises 12.14

2. $y(1+x^2y^2)^{-1}dx + x(1+x^2y^2)^{-1}dy$

4. $[yz\cos(xyz) - 2xy^2z^2]dx + [xz\cos(xyz) - 2x^2yz^2]dy + [xy\cos(xyz) - 2x^2y^2z]dz$

6. $(1 - x^2y^2)^{-1/2}(y\,dx + x\,dy)$

8. $(y+t)\,dx + (x+z)\,dy + (y+t)\,dz + (z+x)\,dt$

10. $2e^{x^2+y^2+z^2-t^2}(x\,dx + y\,dy + z\,dz - t\,dt)$

12. 3%

Exercises 12.15

2. $\frac{1}{3!}[f_{xxx}(c,d)(x-c)^3 + 3f_{xxy}(c,d)(x-c)^2(y-d) + 3f_{xyy}(c,d)(x-c)(y-d)^2 + f_{yyy}(c,d)(y-d)^3]$

4. $\sum_{n=0}^{\infty}\sum_{r=0}^{n} \frac{e^5(-1)^{n-r}2^r3^{n-r}}{(n-r)!\,r!}(x-1)^r(y+1)^{n-r}$

6. $\sum_{n=1}^{\infty}\sum_{r=0}^{n} \frac{(-1)^{n+1}(n-1)!}{(n-r)!\,r!}x^{2r}y^{2n-2r}$

8. $[(x+1)-1]\sum_{n=0}^{\infty}(-1)^ny^{2n+2}$

10. $[72 + 12(x-2) + 24(y-1) - (x-2)^2 + 8(x-2)(y-1) - 4(y-1)^2]/(24\sqrt{3})$

12. $[2x + 2(y-1) + 3x^2 + 6x(y-1) + 3(y-1)^2]/2$

14. 0

16. $\sum_{n=0}^{\infty} \frac{1}{n!}\left[(x-c)\frac{\partial}{\partial x} + (y-d)\frac{\partial}{\partial y}\right]^n f(c,d)$

Review Exercises

2. $2(x^2 - y^2 + z^2)/(x^2 + y^2 + z^2)^2$

4. $(3 - z^2)(1 + z^2)/(1 + 2xz + 2xz^3)$

6. $3(2x - ye^{xy})(t^2 + 1) + (2y - xe^{xy})(\ln t + 1)$

8. $(y + 3yv^2 - 2xuv)/(2u + 6uv^2 + 3v + vx^2)$

10. $e^t(t+1)(y - 2x) + e^{-t}(1 - t)(x - 2y)$

12. $\cos\theta$

14. $(2x - 3x^2y^2)\cos\theta - 2yx^3\sin\theta$

16. $6(6x^4t^6z + 9x^5 + x^4t^9z^2 + 2xt^9z^5 - 9t^{12}z^5 - 6xt^6z^4 - 4x^5t^3z - 2x^2t^3z^4)/(x^4t^8z^4)$

18. $xz[(1 - 3xz^2)(1 - 6xz^2 + 6x^2z^4) + (1 - 2xz^2)(1 - 3xz^2 + 6x^2z^4)]/(x - 3x^2z^2)^3$

20. $2y(v^2t - 1)(t^2 - 3/\sqrt{1 - x^2y^2}) + xv\sec^2 t\,(t^2 - 3/\sqrt{1 - x^2y^2}) + 2xyt$

26. $\sqrt{6}$ 28. $9/\sqrt{14}$ 30. 2

32. $x + 3y - 6z + 4 = 0$

34. $x = 2 + u, y = u, z = 6 + 4u$

36. $x = 1 + t, y = 1 - t, z = 1$

38. None

40. $(0, 0)$ relative maximum; all points on $x^2 + y^2 = 1$ relative minima

42. $1/2, -1/2$ 44. $\left(\pm\sqrt{(\sqrt{5}-1)/2}, 0\right), (0, \pm 1)$

46. $x = 100bcqr/(acpr + abpq + bcqr), y = 100acpr/(acpr + abpq + bcqr), z = 100abpq/(acpr + abpq + bcqr)$

48. $1/\sqrt{2} + \sqrt{2}(\pi + 6)(x-1)/4 + (y - \pi/4)/\sqrt{2} + \sqrt{2}(24 + 14\pi - \pi^2)(x-1)^2/16 + \sqrt{2}(10 - \pi)(x-1)(y - \pi/4)/4 - \sqrt{2}(y - \pi/4)^2/4 + \cdots$

Chapter 13

Exercises 13.1

2. 0   4. 3/4   6. $e^2(1-e)^2/2$

8. 0   10. 5/144   12. 0

14. -0.54 16. $(1 - \ln 2)/2$ 18. $128\sqrt{2}/5$

20. $11664/35$ 22. $(1 - 2\sqrt{2})/12$ 24. 1/3

26. 8/189   28. $2(1 - \sqrt{2})$ 32. $v = -2xy + f(x)$

34. $u = -y\sqrt{x^2 + y^2} + f(y)$ 36. $xy + C$

38. $-(1/3)(x^2 + y^2)^{3/2} + C$

Exercises 13.2

2. $128/15$
4. $-621/140$
6. 0

8. 0
10. $304/15$
12. $\sqrt{13} - 7/2$

14. $(1 - \cos 1)/2$
16. $2(1 - \sqrt{2})$
18. $(5/4)\ln 5 - 1$

20. 4
22. Double
24. 0

26. Double
28. 0
30. $101/70$

32. $(e^2 - 1)/(2e)$
34. $307\,973$
36. 24π

38. $e^2 - 2e - 1$
40. $(1 - 2\sqrt{2})/12$
42. $1/3$

Exercises 13.3

2. $343/6$
4. 8
6. $1 + 3e^{-2}$

8. $343/6$
10. $20/3$
12. 16π

14. $1024\pi/15$
16. $26\pi/15$
18. 45π

20. $34\pi/3$
22. $235/3$
24. $7/6$

26. $8\pi/3 - 2\sqrt{3}$
28. $23\,328/35$
30. $2 - \ln 3$

32. $16\pi(1 + \sqrt{2})/15$ 34. $68\pi/9$
36. 5.38

38. $512\pi/(15\sqrt{5})$
40. $7\sqrt{2}\pi/6$

Exercises 13.4

2. $256\rho g/5$
4. $48\rho g$
6. $370\rho g/3$

8. $250\rho g/3$
10. 7.63×10^6 N
12. $5\sqrt{29}\rho g$

14. $10\rho g$
16. $\pi\rho g r^3$

18. On vertical diameter, $r^2/(4h)$ units below centre of circle

20. On vertical altitude, $\sqrt{3}L/4$ units above lowest point

22. $(l/4, -L/2)$
24. 9.00×10^4 N

Exercises 13.5

2. $(0, 24/5)$
4. $\bar{x} = \bar{y} = 9/(15 - 16\ln 2)$

6. $\big(8(2 - \sqrt{2})/(3\pi), 8\sqrt{2}/(3\pi)\big)$

8. $(177/85, 9/17)$

10. $(4/3, 5/3)$
12. $32/3$

14. $603/10$
16. $48\rho/5$

18. $\rho/48$
20. -2ρ

22. $(0, 192/205)$
24. $(-2/(8\sqrt{3} - 1), 0)$

26. $(-61/28, 807/700)$
28. $\rho ab(a^2 + b^2)/12$

30. $4761/140$
32. $-81\sqrt{2}/40, 1863/280$

34. $-47\sqrt{5}/350, 89/350$
36. (a) $\rho b^2 h^2/24$ (b) $-\rho b^2 h^2/72$

40. $y = \pm x, 7\rho a^4/12, \rho a^4/12$

Exercises 13.6

2. $4\sqrt{14}/3$
4. $4(5\sqrt{3} - 2\sqrt{6} - 3 + \sqrt{2})/3$

6. $(247\sqrt{13} + 64)/1215$

8. $2\int_0^4 \int_0^{\sqrt{16-y^2}} \sqrt{4 + y^2 + z^2}\, dz\, dy$

10. $4\int_0^{\sqrt{2}} \int_0^{\sqrt{2-x^2}} \sqrt{1 + 16(x^2 + y^2)^3}\, dy\, dx$

12. $\int_0^1 \int_0^{1-x^2} \sqrt{1 + 2/(1 + x + y)^2}\, dy\, dx$

16. $\sqrt{2}\int_{-1/\sqrt{2}}^{1/\sqrt{2}} \int_x^{\sqrt{1-x^2}} \sqrt{2 + 1/(y - x)}\, dy\, dx + \sqrt{2}\int_{-1}^{-1/\sqrt{2}}$
$\int_{-\sqrt{1-x^2}}^{\sqrt{1-x^2}} \sqrt{2 + 1/(y - x)}\, dy\, dx$

18. $\int_0^1 \int_{1-x}^{2-x} \sqrt{1 + 9x^4 + 9y^4}\, dy\, dx + \int_1^2 \int_0^{2-x} \sqrt{1 + 9x^4 + 9y^4}$
$dy\, dx$

20. $(37\sqrt{74} - 5\sqrt{10})/24$

Exercises 13.7

2. $2/3$
4. $4(3\sqrt{3} - \pi)/3$

6. $\pi/6$
8. $(3\sqrt{3} - \pi)/2$

10. $\pi/4$

12. $15\pi/4 - 19\sqrt{2}/3 - (9/2)\,\text{Sin}^{-1}(1/3)$

14. $\pi R^4/4$
16. $\pi(17\sqrt{17} - 1)/6$

18. $2\pi(10\sqrt{10} - 1)/3$
20. $(17\sqrt{17} - 5\sqrt{5})\pi/6$

22. $8\pi/3$
24. 8.29×10^7 cc

26. $2\pi^2 R^3$

28. $[q\rho/(2\epsilon_0)](1 - d/\sqrt{R^2 + d^2}), q\rho/(2\epsilon_0)$

30. $9\pi - 12\sqrt{3}$
32. $(5/6, 0)$

34. $\pi(\pi - 2)/2$
36. $U_{\max} = V_{\max}/\alpha^2$

Exercises 13.8

2. $32/3$
4. 0
6. $1024/21$

8. $1/96$
10. $11/6$
12. $48/35$

14. $4\int_0^{\sqrt{(\sqrt{5}-1)/2}} \int_{x^2}^{\sqrt{1-x^2}} \int_0^{\sqrt{1-x^2-z^2}} (x^2 + y^2 + z^2)\, dy\, dz\, dx$

16. $\int_{-5/2}^{1/2} \int_{-1-\sqrt{9-4(y+1)^2}}^{-1+\sqrt{9-4(y+1)^2}} \int_{x^2+4y^2}^{4-2x-8y} xyz\, dz\, dx\, dy$

18. $729/70$
20. 2π
22. $\pi/3$

24. $4\int_0^1 \int_0^{\sqrt{1-x^2}} \int_0^{\sqrt{x^2+y^2}/2} (x^2 + y^2 + z^2)\, dz\, dy\, dx\ +$
$4\int_0^1 \int_{\sqrt{1-x^2}}^{\sqrt{4/3-x^2}} \int_{\sqrt{x^2+y^2-1}}^{\sqrt{x^2+y^2}/2} (x^2 + y^2 + z^2)\, dz\, dy\, dx +$
$4\int_1^{2/\sqrt{3}} \int_0^{\sqrt{4/3-x^2}} \int_{\sqrt{x^2+y^2-1}}^{\sqrt{x^2+y^2}/2} (x^2 + y^2 + z^2)\, dz\, dy\, dx$

Exercises 13.9

2. 2/3   4. 704/15   6. 8

8. 8/3   10. 19/96   12. 7/3

14. 5/18   16. 64/15   18. $\pi/4 - 1/3$

20. 1/20   22. 13/3   26. $\pi/2$

Exercises 13.10

2. $(1/4, 1/4, 1/4)$ 4. $(0, 16/7, 8/7)$

6. $2\rho/3$ 8. $773\rho/2520$

10. $64\rho/15$ 12. $(6772/11\,847, 7300/14\,001, 1/2)$

14. $(0, 0, 3/2)$ 16. $128\rho a^5/45$

18. $51\sqrt{3}\rho$ 20. $12\rho/5, 4\rho/5, 8\rho/15$

26. $(2/5)(4/5)^{1/3} H$ below surface

Exercises 13.11

2. $r = 1$ 4. $r = 5/(\cos\theta + \sin\theta)$

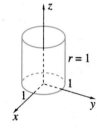   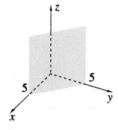

6. $z = r^2 \cos^2\theta$ 8. $4z = r^2$

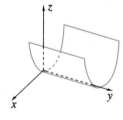   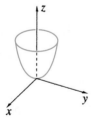

10. $r^2 = 1 + z^2$ 12. $(8\sqrt{2} - 7)\pi/6$

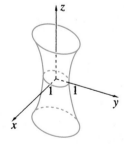

14. 4π 16. $4\sqrt{3}\pi$

18. $\int_{-\pi}^{\pi} \int_0^3 \int_0^{1+r^2} f(r\cos\theta, r\sin\theta, z)\, r\, dz\, dr\, d\theta,$

$\int_0^3 \int_{-\pi}^{\pi} \int_0^{1+r^2} f(r\cos\theta, r\sin\theta, z)\, r\, dz\, d\theta\, dr,$

$\int_0^3 \int_0^{1+r^2} \int_{-\pi}^{\pi} f(r\cos\theta, r\sin\theta, z)\, r\, d\theta\, dz\, dr,$

$\int_0^1 \int_0^3 \int_{-\pi}^{\pi} f(r\cos\theta, r\sin\theta, z)\, r\, d\theta\, dr\, dz +$

$\int_1^{10} \int_{\sqrt{z-1}}^3 \int_{-\pi}^{\pi} f(r\cos\theta, r\sin\theta, z)\, r\, d\theta\, dr\, dz,$

$\int_{-\pi}^{\pi} \int_0^1 \int_0^3 f(r\cos\theta, r\sin\theta, z)\, r\, dr\, dz\, d\theta +$

$\int_{-\pi}^{\pi} \int_1^{10} \int_{\sqrt{z-1}}^3 f(r\cos\theta, r\sin\theta, z)\, r\, dr\, dz\, d\theta,$

$\int_0^1 \int_{-\pi}^{\pi} \int_0^3 f(r\cos\theta, r\sin\theta, z)\, r\, dr\, d\theta\, dz +$

$\int_1^{10} \int_{-\pi}^{\pi} \int_{\sqrt{z-1}}^3 f(r\cos\theta, r\sin\theta, z)\, r\, dr\, d\theta\, dz$

20. $8\pi\rho R^5/15$ 22. $81\pi^2/8$

24. 0.084   26. 1.45 kg·m^2

28. $(6/5, 0, 27\pi/128)$ 32. $\pi/8$

34. $[\pi - 4\,\mathrm{Tan}^{-1}(1/2)]/12$ 36. 3π

38. $\pi(1 - 1/e)$ 40. $\pi/8$

42. $27\pi(2\sqrt{2} - 1)/2$

Exercises 13.12

2. $\Re \sin\phi = 1$ (see figure for exercise 13.11–2)

4. $\Re = 4\cot\phi\,\csc\phi$ (see figure for exercise 13.11–8)

6. $\Re^2 = -\sec 2\phi$ (see figure for exercise 13.11–10)

8. $(2 - \sqrt{2})\pi/3$ 10. $[4\,\mathrm{Tan}^{-1} 2 - \pi]/12$

12. $32\pi/3$ 14. $8\pi\rho R^5/15$

16. $k\pi R^4$ coulombs   18. $9\pi/2$

20. $(2\pi R^3/3)(1 - k/\sqrt{1 + k^2})$

22. (a) $\rho_b = \rho_w/2$ (b) $11\pi\rho_w g R^3/24$

Exercises 13.13

2. $\int_{\pi/6}^{\pi/3} \int_0^{1-\cos\theta} r^3 \cos\theta \sin\theta\, dr\, d\theta$

4. $(1/(6\sqrt{3})) \int_0^4 \int_5^6 [(2v + u)/(\sqrt{v - u}]\, dv\, du$

6. $\int_0^{\pi/2} \int_0^3 \int_0^r r^3\, dz\, dr\, d\theta$

8. $\int_{-\pi}^{\pi} \int_0^{\pi/4} \int_0^2 \Re^7 \sin^5\phi \cos\phi \sin^2\theta \cos^2\theta\, d\Re\, d\phi\, d\theta$

10. 21/8   12. 12   14. $315\pi/32$

16. $(\sin 1)/2$ 18. 1/32   20. 11

Exercises 13.14

2. $2x - 1 + e^x$

4. $4x^3 - 3x^2 - 1 + 3x^2(x^3 - 1)\ln(x^3 - 1) - 2x^3 \ln(x^2)$

8. $e^x\sqrt{1 + e^{3x}} - \cos x\sqrt{1 + \sin^3 x}$

12. $(1/a^2)\ln(1 + ab) - b/(a + a^2 b)$

14. $[3/(8a^5)]\,\mathrm{Tan}^{-1}(x/a) + x(3x^2 + 5a^2)/[8a^4(a^2 + x^2)^2] + C$

16. $\pi\,\mathrm{Sin}^{-1} a$

Review Exercises

2. $81/16$ **4.** $1/40$ **6.** $36/5$

8. 0 **10.** $-4544/945$ **12.** $\pi/6$

14. $32\pi/3$ **16.** $2\pi/5$ **18.** $3\pi/2$

20. $(\pi \ln 2)/4$

22. (a) $\iint_R dA$ (b) $\iint_R 2\pi(2-x)\, dA$, $\iint_R 2\pi(y+4)\, dA$ (c) $\iint_R (x-1)\, dA$, $\iint_R (y+1)\, dA$ (d) $\iint_R (x+1)^2\, dA$, $\iint_R (y-4)^2\, dA$ (e) $\iint_R \sigma(x,y)\, dA$ (f) $\iint_R \rho(x,y)\, dA$ (g) $\iint_R P(x,y)\, dA$

24. $32/3$ **26.** $16/15$ **28.** $\pi(17\sqrt{17}-1)/96$

30. $128\pi/15$, $4\pi^2$ **32.** $1/4$ **34.** 2

36. $14\rho\pi/15$ **38.** $(0, 1/8)$

40. $(9\pi/64, 9\pi/64, 3/8)$ **42.** $16\pi/15$

Chapter 14

Exercises 14.1

2. Closed, connected   **4.** Connected

6. Closed, connected   **8.** Open

10. For interior, exterior, and boundary points replace circle with sphere in planar definitions. Open, closed, connected, and domain definitions are identical. A domain is simply-connected if every closed curve in the domain is the boundary of a surface that contains only points of the domain.

12. Closed, connected   **14.** Connected

16. Open

18. Open, connected, simply-connected domain

22. $-(x^2+y^2+z^2)^{-3/2}(x\hat{\mathbf{i}} + y\hat{\mathbf{j}} + z\hat{\mathbf{k}})$

24. $(6-2\cos 2)\hat{\mathbf{i}} + (1+2\sin 2)\hat{\mathbf{j}}$

26. $2(e^y - x^2 y)$

28. $2x\cos(x^2+y^2+z^2) - \sin(y+z)$

30. 1 **32.** $2/\sqrt{x^2+y^2+z^2}$

34. $4y(16z^4 - 3x)\hat{\mathbf{i}} + x^2\hat{\mathbf{j}} + 12yz\hat{\mathbf{k}}$

36. 0 **38.** $-2\hat{\mathbf{k}}$

40. $-2\hat{\mathbf{k}}/[(x+y)\sqrt{(x+y)^2 - 1}]$

46. $x^3 y^2 + 3x + 2y + C$

48. $\ln|x+y| + C$ **50.** $(x^2+y^2+z^2)/2 + C$

52. $\ln|1+x+y+z| + C$ **54.** $\text{Tan}^{-1}(xy) + z^2/2 + C$

56. (a) $4, 2, -1$ (b) $x^3/3 + 2xy + 4xz - 3y^2/2 - yz + z^2 + C$

58. (a) $q/(4\pi\epsilon_0|\mathbf{r}|) + C$ (b) $-\sigma z/(2\epsilon_0) + C$

62. $yz\hat{\mathbf{i}} - xz\hat{\mathbf{j}}$

64. $(1/4)[(4xyz - 3y^2 z)\hat{\mathbf{i}} + (2xyz - 6x^2 z)\hat{\mathbf{j}} + (2x^2 y + xy^2)\hat{\mathbf{k}}]$

Exercises 14.2

2. $32/3$ **4.** $50\sqrt{10}/3$

6. $37\sqrt{2}/3$ **8.** $9\sqrt{1+16\pi^2}$ cm

10. $37/80$ **12.** 0.78

14. 0 **16.** $(1 - 161\sqrt{161})/6$

18. $\pi(145\sqrt{145} - 10\sqrt{10})/27$ **20.** $\pi/2$

22. 0.007 **24.** 2

26. 0.242 **28.** π

30. $(5\sqrt{5} - 1)/6$ **32.** $\sqrt{2}(1 - e^{6\pi})/13$

34. 17.08 **36.** $4\pi^2 ab$

Exercises 14.3

2. $51/4$ **4.** -15 **6.** $-99/140$

8. 10 **10.** 0 **12.** 9

14. 0 **16.** -8π **18.** $(768 + 5\pi)/20$

20. $67/32 + \sqrt{5} - \sqrt{2} - 5/e$

22. $(4\pi\epsilon_0)^{-1}[q_1 q_3(1/\sqrt{41} - 1/\sqrt{61}) + q_2 q_3(1/\sqrt{18} - 1/\sqrt{10})]$

24. 4.4×10^4 J   **26.** 1.5×10^3 J

28. $P_2(V_2 - V_1) + P_2 V_2^\gamma (V_3^{1-\gamma} - V_2^{1-\gamma})/(1-\gamma) + P_2(V_2/V_3)^\gamma(V_2 - V_3)$

30. 3.719 **32.** -4.26×10^{-4} **34.** 0

38. (a) $2R$ (b) 0

Exercises 14.4

2. -43 **4.** -2 **6.** 0

8. 10 **10.** e^3 **12.** No

14. 0 **16.** 0 **18.** $8/105 + \ln(3/2)$

20. $-\pi$ **22.** (a) Yes (b) Yes (c) Yes (d) Yes (e) No

24. 2π

Exercises 14.5

2. Not conservative   **4.** mgz

8. $-k(\sqrt{x^2+y^2+z^2} - L)(x\hat{\mathbf{i}} + y\hat{\mathbf{j}} + z\hat{\mathbf{k}})/\sqrt{x^2+y^2+z^2}$, yes

10. (a) Yes (b) Yes

Exercises 14.6

2. 4π **4.** $2\sqrt{3}/5$ **6.** 0

8. $-3/8$ **10.** $4(13\sqrt{13} - 8)/3$

12. πab **14.** $3\pi/8$ **16.** 2π

20. 2 **22.** 24 **24.** $77/2$

26. $81\pi/2$ **28.** -4π **32.** π

Exercises 14.7

2. $2\sqrt{3}/15$ 4. $\sqrt{2}/8$ 6. $(-61 + 44\sqrt{2})/5$

8. -3 10. $\pi R\sqrt{R^2 + h^2}$

12. $3(145^{5/2} - 361)/5120$

14. $\pi(3\sqrt{3} - 1)/3$ 16. $(50\sqrt{5} + 2)\pi/15$

18. $4\pi R^2$ 20. 0 22. $\pi \ln 3$

Exercises 14.8

2. $2e^2 - 10e^{-2}$ 4. 0 6. $2\pi/3$

8. -2π 10. $10\pi/3$ 12. $\sqrt{2}\pi/4$

14. $2\sqrt{5} + 8\ln[(\sqrt{5} + 1)/2]$ 16. 0

18. (a) 30π (b) -20π

20. (a) $-4251\hat{\mathbf{i}}$ N, $4251\hat{\mathbf{i}}$ N, $-9810(\hat{\mathbf{j}} + \hat{\mathbf{k}})$ N, $9810\hat{\mathbf{j}} - 13080\hat{\mathbf{k}}$ N
    (b) $-22890\hat{\mathbf{k}}$

24. $2\pi(1 - e^{-2})$

Exercises 14.9

2. 0 4. 16 6. 27π

8. 180 10. $-1328\pi/5$ 12. $57/2$

14. $52\pi/5$ 20. 0

Exercises 14.10

2. 0 4. $\pm 16\pi$ 6. 0

8. $-2\sqrt{2}\pi b^2$ 10. $\pm 2\pi$ 12. -24π

14. 3π

Review Exercises

2. $3x^2y + x^2/y^2$ 4. $\mathbf{0}$

6. $ye^x + ze^y + xe^z$ 8. $(\hat{\mathbf{i}} + \hat{\mathbf{j}})/\sqrt{1 - (x + y)^2}$

10. $x(-y\hat{\mathbf{j}} + z\hat{\mathbf{k}})/(1 + x^2y^2z^2)$ 12. $2\sqrt{3}$

14. π 16. $\sqrt{2}/3$

18. 2π 20. $8\pi(18 - 7\sqrt{6})/3$

22. 0 24. 0

26. $-\pi/120$ 28. $(25\sqrt{5} - 11)/120$

30. $8\pi/3$

Chapter 15

Exercises 15.1

12. $-3\sin 3x + 2\cos 3x$

14. $[(2\cos 3 - \cos 6)\sin 3x + (\sin 6 - 2\sin 3)\cos 3x]/\sin 3$

16. $(1/3)\,\text{Tan}^{-1}(x/3) + C$

18. $(x^3/6)\ln x - 5x^3/36 + C_1x + C_2$

20. (b) $-L\ln x - \sqrt{L^2 - x^2} + L\ln(L + \sqrt{L^2 - x^2})$

22. $y = 1$

24. (b)

$y^2 = Cx\ (C < 0)$ | $y^2 = Cx\ (C > 0)$

26. (a) $y(x) = C - 1/x$ for $x < 0$ and $2 - 1/x$ for $x > 0$
    (b) $y(x) = 1 - 1/x$ for $x < 0$ and $D - 1/x$ for $x > 0$
    (c) $y(x) = 1 - 1/x$ for $x < 0$ and $2 - 1/x$ for $x > 0$

Exercises 15.2

2. $y(x) = 2 + Ce^{-x^2}$ 4. $y(x) = Ce^{3x} - 2/3$

6. $y^2 + x^2 + 2(x - y) + 2\ln|y + 1| + 2\ln|x - 1| = C;\ y = -1$

8. $y(x) = [C + 3\ln|x| + 3e^x(1 - x)]^{1/3}$

10. $y(x) = (x + C)/(1 - Cx)$ 12. $xy = 2e^{y-x-1}$

14. $y(x) = [\tan(x^2 - 4) + 4]/[1 - 4\tan(x^2 - 4)]$

16. (a) $6(1 - 6kt)^{-1}$ km (b) Never

18. $20 + 60e^{-0.203t}$ 20. 13.51 h

22. $(R/k)(1 - e^{-kt}) + A_0e^{-kt}$ 24. $60(1 - e^{kt})/(2 - 3e^{kt})$ g

26. $12\pi R^{5/2}/(5a\sqrt{g})$ 28. 1839 s

30. $\sqrt{2gh}(e^{\sqrt{2gh}t/L} - 1)/[e^{\sqrt{2gh}t/L} + 1]$

32. $(A_o/M)\{[(k - 1)M^2 + 2]/)(k + 1)\}^{(k+1)/(2k-2)}$

36. $x^2(x^2 - 2y^2) = C$ 38. $x^2 + 2xy - y^2 = C$

40. $(y - x)e^{y/x} = x\ln|x| + Cx$

42. $y = \sqrt{5x - x^2}$ 44. $y_0(A + y_0)/[y_0 + Ae^{-k(A+y_0)t}]$

46. $y = Ce^x - x - 1$

48. $6y - 3\ln|4x + 6y + 3| = C$

50. $N = 10^6/(1 + 9999e^{-t})$

52. $w = [(a/b)(1 - e^{-bt/3}) + w_0^{1/3}e^{-bt/3}]^3$

56. 459.5 mg

58. $y^3(x^3 + 6x - 5) = 1 + y^6$

Exercises 15.3

2. $y(x) = x^4 + C/x^2$ 4. $y\sin x = C - 5e^{\cos x}$

6. $y(x) = C(x + 1)^2 - 2x - 2$ 8. $y(x) = Ce^x + e^{2x}$

10. $y(x) = x^3/2 + Cx^3e^{1/x^2}$ 12. $y(x) = x + \cos 2x + C\sin 2x$

14. $y(x) = e^x(2\sin x - \cos x)/5 - 4e^{-x}/5$

16. $xy = C + y^4/4$ 18. $y(x) = e^{-x}/(C - x)$

20. $x^2y + Cye^{-x} = 1$

22. $y(x) = [(3/4)\cos x + C\sec^3 x]^{-1/3}$

24. $1000 + t + 3 \times 10^6/(1000 + t)$ g

26. $(10^6 + 5t)/50 - 15 \times 10^9/(10^6 + 5t)$ g

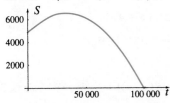

28. $20\,000 - t/5 - 15(10^5 - t)^2/10^7$ g

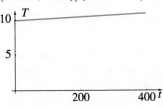

30. $20 + 36t + (36/k)(e^{-5k} - 1), t \le 5; 200 + (36/k)(1 - e^{5k})e^{-kt}, t > 5$

32. $(4190t + 4189)/(419t + 419) + (t + 1)/419$

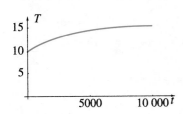

34. $20\,000/1257 - (30/9997)e^{-t} - (74\,240\,000/12\,566\,229)e^{-3t/10\,000}$

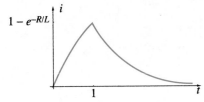

36. $i = 1 - e^{-Rt/L}, 0 \le t \le 1; (e^{R/L} - 1)e^{-Rt/L}, t > 1$

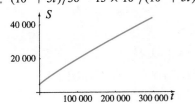

38. (b) $I_0 - E_0/(\omega C Z^2)$

40. $I_0 \left[\frac{1}{(1-x)(1-\lambda b)} - \frac{x}{1-x} \right](e^{-\lambda t} - e^{-t/b})$

Exercises 15.4

2. $2\sqrt{Cy - 1} = \pm Cx + D$

4. $y(x) = Cx - C^2 \ln |x + C| + D$

6. $(x + C)^2 + (y + D)^2 = 1$

8. $D + x/2 = \left\{-1/y; \text{ or } C\text{Tan}^{-1}(Cy); \text{ or } [1/(2C)] \ln |(y - C)/(y + C)|\right\}$

10. $y(x) = (1/2)(Ce^x + C^{-1}e^{-x}) + D$

12. $kr + \ln r[(T_b - T_a) + k(a - b)]/ \ln (b/a) + [T_a \ln b - T_b \ln a + k(b \ln a - a \ln b)]/ \ln (b/a)$

Exercises 15.5

4. Less than 47.8 km/h; defence

6. (a) $(m/k) \ln 20$ (b) $(m^2 g/k^2)(\ln 20 - 19/20)$

8. (c) $x^* = (\mu Mg \pm \sqrt{\mu^2 M^2 g^2 + Mkv_0^2 + k^2 x_0^2 - 2k\mu Mgx_0})/k$

10. $70.0(1 + 0.176e^{-0.280t})/(1 - 0.176e^{-0.280t})$ m/s

12. $V[v_0 + V + (v_0 - V)e^{-2kVt/m}]/[v_0 + V - (v_0 - V)e^{-2kVt/m}]$, where $V = \sqrt{mg/k}$

14. $\sqrt{m/(kg)}\text{Tan}^{-1}(\sqrt{k}v_0/\sqrt{mg})$

16. (a) $V \tan [\text{Tan}^{-1}(v_0/V) - kVt/m]$, where $V = \sqrt{mg/k}$; No
(b) $(m/k) \ln (\sqrt{v_0^2 + V^2}/V)$

18. $-(mg/\beta)\hat{\mathbf{j}} + [\mathbf{v}_0 + (mg/\beta)\hat{\mathbf{j}}]e^{-\beta t/m}$,
$-(mgt/\beta)\hat{\mathbf{j}} + (m/\beta)[\mathbf{v}_0 + (mg/\beta)\hat{\mathbf{j}}](1 - e^{-\beta t/m})$

20. $\sqrt{2gD \sin \alpha}$

22. $4x^2 + y^2 + 16x - 6y + 9 = 0$

Exercises 15.6

2. Linear   4. Linear   6. Linear
8. Linear   10. Not linear   12. Linear
14. Linear   16. Not linear   18. Not linear
20. Linear

Exercises 15.7

2. $C \cos x$ 4. $C_1 e^{4x} + C_2 x e^{4x}$

6. $e^{2x}[C_1 \cos (x/\sqrt{2}) + C_2 \sin (x/\sqrt{2})]$

8. $C_1 \cos (2 \ln x) + C_2 \sin (2 \ln x)$

12. Dependent   14. Independent

Exercises 15.8

2. $(C_1 + C_2 x)e^{4x}$

4. $C_1 e^{(-1+\sqrt{3})x} + C_2 e^{(-1-\sqrt{3})x}$

6. $C_1 e^{3x} + C_2 \cos x + C_3 \sin x$

8. $(C_1 + C_2 x + C_3 x^2)e^{2x}$

10. $C_1 \cos x + C_2 \sin x + C_3 \cos 2x + C_4 \sin 2x$

12. $e^{\sqrt{2}x}[C_1 \cos (\sqrt{2}x) + C_2 \sin (\sqrt{2}x)] + e^{-\sqrt{2}x}[C_3 \cos (\sqrt{2}x) + C_4 \sin (\sqrt{2}x)]$

14. $y'' + 4y' + 20y = 0$

16. $y'''' - 4y''' + 10y'' - 12y' + 9y = 0$

18. 5, 24, 20

20. $\lambda_0 = 0, y_0(x) = C_1; \lambda_n = n^2\pi^2/16, y_n(x) = C_1 \cos(n\pi x/4)$

22. $\lambda_n = (2n - 1)^2\pi^2/100, y_n(x) = C_1 \cos[(2n - 1)\pi x/10]$

24. (a) $x^2/x_0^2 + ky^2/(Mv^2) = 1$ (b) $x^2/x_0^2 - ky^2/(Mv^2) = 1$

Exercises 15.9

2. $-(x^2/3 + 2/9)e^{-x}$, $C_1 e^{(-1+\sqrt{3})x} + C_2 e^{-(1+\sqrt{3})x} - (x^2/3 + 2/9)e^{-x}$

4. $(1/9)\cos 2x$, $(C_1+C_2 x)\cos x+(C_3+C_4 x)\sin x+(1/9)\cos 2x$

6. $e^{-2x}/40$, $C_1 \cos x + C_2 \sin x + C_3 \cos 2x + C_4 \sin 2x + e^{-2x}/40$

8. $e^{2x}(\cos x - 5\sin x)/312$, $e^{-4x}(C_1 \cos 5x + C_2 \sin 5x) + e^{2x}(\cos x - 5\sin x)/312$

10. $[(2x+1)\cos x - 2(x+1)\sin x]/16$, $e^{2x}(C_1 \cos x + C_2 \sin x) + [(2x+1)\cos x - 2(x+1)\sin x]/16$

12. $Axe^{3x} + Be^{3x} + Cx\cos x + Dx\sin x + E\cos x + F\sin x$

14. $Ax^2 e^x + Bxe^x + Cx^3 + Dx^2 + Ex + F + G\cos x + H\sin x$

18. $e^x[C_1 \cos(x/\sqrt{2}) + C_2 \sin(x/\sqrt{2})] + (4\cos 3x - 5\sin 3x)/246 + (\sin x + 4\cos x)/34$

20. $-(1/108)\sin 3x + (x^2/12)(\sin 3x - \cos 3x) + (x/36)(\sin 3x + \cos 3x)$

24. $C_1 \cos(2\ln x) + C_2 \sin(2\ln x) + 1/4$

Exercises 15.10

2. $e^{-15t/4}[(12/65)\cos(5\sqrt{23}t/4)+(20/(13\sqrt{23}))\sin(5\sqrt{23}t/4)]-(4/65)[3\cos(10t) + 2\sin(10t)]$ m

4. $-0.0253e^{-99.50t} + 5.03e^{-0.50t}$ C

6. $(Mg/k)\cos(\sqrt{k/M}t)$ from equilibrium

8. (b) $x = -7/36$ m, yes

12. $(1/20 + 3t/40)\sin 200t$ m

14. $\sqrt{k/M}$

16. (c) $\sqrt{k/M - \beta^2/(2M^2)}$, $2AM/[\beta\sqrt{4kM - \beta^2}]$

18. $\beta^2 g/(4k) - W$

20. (a) $e^{-15t/2}[0.00473\cos(5\sqrt{31}t/2) - 0.00228\sin(5\sqrt{31}t/2)] - 0.00473\cos(\pi t/4) + 0.100\sin(\pi t/4)$, 0.100
(b) $e^{-15t/2}[0.00953\cos(5\sqrt{31}t/2) + 0.00406\sin(5\sqrt{31}t/2)] - 0.00953\cos(\pi t/2) + 0.100\sin(\pi t/2)$, 0.100

22. (b) $x = 6(e^{6t}+e^{-6t})-2\cos 2t$, $y = -2(e^{6t}+e^{-6t})-6\cos 2t$
(c) Graph appears to be a straight line, but in actuality it is not.

24. (a) $(b - H_0)(e^{mx} - e^{-mx})/(e^{mL} - e^{-mL}) + H_0$
(b)
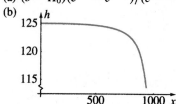

28. $4\pi^2 EI/L^2$

Review Exercises

2. $y = \pm\sqrt{2(x + \ln|x|)} + C$

4. $y = Ce^{-4x} + (8x^2 - 4x + 1)/32$

6. $y = e^{-3x/2}[C_1 \cos(\sqrt{7}x/2) + C_2 \sin(\sqrt{7}x/2)] + 1/2$

8. $y = x^3/9 + C_1 \ln|x| + C_2$

10. $y = Ce^{-x^2} + x^2 - 1$

12. $y = (C_1 + C_2 x)e^{2x} + (3\sin x + 4\cos x)/25$

14. $y = C_1 \cos 2x + C_2 \sin 2x - (x/4)\cos 2x$

16. $xy^2 = Ce^{-2x-x^2/2}$

18. $e^{-x}[C_1 \cos(\sqrt{3}x) + C_2 \sin(\sqrt{3}x)] + (\sqrt{3}x/6)e^{-x}\sin(\sqrt{3}x)$

20. $x = Ce^{-3y} + (2/27)(-9y^2 + 6y - 2)$

22. $y = (424/405)e^{9x} + (11/5)e^{-x} - 2x/9 - 20/81$

24. $y = -\cos x + (2\sin x)/x + (2\cos x + 1 - \cos 1 - 2\sin 1)/x^2$

26. (a) 3.924×10^{-3} N (b) $1.962 \times 10^{-3}t + 4.905 \times 10^{-7}(e^{-4000t} - 1)$ m

28. (a) In air, $v = 9.81t$; in water $v = 0.4905 + 1.494 \times 10^{29}e^{-20t}$
(b) In air, $y = 4.905t^2$; in water, $y = 49.82 + 0.4905t - 7.47 \times 10^{27}e^{-20t}$ (c) 20.75 s

Appendix B

Exercises

2. 1   **4.** 0   **6.** -640

8. 68   **10.** $x = -14$, $y = -10$

12. $r = -4$, $s = -11$

14. $x = -5/18$, $y = 11/18$, $z = 5/2$

18. $x = y = z = w = 0$

20. (b) $2x + 3y = 5$

22. $\begin{vmatrix} y & x^2 & x & 1 \\ 0 & 1 & 1 & 1 \\ 11 & 4 & 2 & 1 \\ 1 & 4 & -2 & 1 \end{vmatrix} = 0$

24. $b^{n-1}(an + b)$

Appendix C

Exercises

2. $-1 + 6i$ **4.** $-2 + 11i$ **6.** $7 + i$

8. $1/13 - 5i/13$ **10.** $5 - 3i$ **12.** $-24/5 + 33i/5$

14. $-4i$ **16.** $8 + 6i$ **18.** $58/169 - 4i/169$

20. $15 - 8i$ **22.** $3/4 + i/4$ **24.** $-28/41 - 6i/41$

26. $-672/625 - 1054i/625$ **28.** $-3/2 \pm \sqrt{11}i/2$

30. $-1 \pm 2\sqrt{2}$ **32.** $1/4 \pm \sqrt{19}i/4$ **34.** ± 1, $\pm\sqrt{5}i$

36. $\pm\sqrt{3 + \sqrt{6}}i$, $\pm\sqrt{3 - \sqrt{6}}i$ **38.** $0 + 0i$, r

42. $3 \pm i$ **44.** $\pm(1/\sqrt{2} + i/\sqrt{2})$

Index

Derivative Formulas

$$\frac{d}{du}(u + v) = \frac{du}{dx} + \frac{dv}{dx}$$

$$\frac{d}{dx}(u^n) = nu^{n-1}\frac{du}{dx}$$

$$\frac{d}{dx}(uv) = u\frac{dv}{dx} + v\frac{du}{dx}$$

$$\frac{d}{dx}\left(\frac{u}{v}\right) = \frac{v\dfrac{du}{dx} - u\dfrac{dv}{dx}}{v^2}$$

$$\frac{d}{dx}\sin u = \cos u\frac{du}{dx}$$

$$\frac{d}{dx}\cos u = -\sin u\frac{du}{dx}$$

$$\frac{d}{dx}\tan u = \sec^2 u\frac{du}{dx}$$

$$\frac{d}{dx}\cot u = -\csc^2 u\frac{du}{dx}$$

$$\frac{d}{dx}\sec u = \sec u\,\tan u\frac{du}{dx}$$

$$\frac{d}{dx}\csc u = -\csc u\,\cot u\frac{du}{dx}$$

$$\frac{d}{dx}\text{Sin}^{-1}u = \frac{1}{\sqrt{1 - u^2}}\frac{du}{dx}$$

$$\frac{d}{dx}\text{Cos}^{-1}u = \frac{-1}{\sqrt{1 - u^2}}\frac{du}{dx}$$

$$\frac{d}{dx}\text{Tan}^{-1}u = \frac{1}{1 + u^2}\frac{du}{dx}$$

$$\frac{d}{dx}\text{Cot}^{-1}u = \frac{-1}{1 + u^2}\frac{du}{dx}$$

$$\frac{d}{dx}\text{Sec}^{-1}u = \frac{1}{u\sqrt{u^2 - 1}}\frac{du}{dx}$$

$$\frac{d}{dx}\text{Csc}^{-1}u = \frac{-1}{u\sqrt{u^2 - 1}}\frac{du}{dx}$$

$$\frac{d}{dx}a^u = a^u \ln a\frac{du}{dx}$$

$$\frac{d}{dx}e^u = e^u\frac{du}{dx}$$

$$\frac{d}{dx}\log_a u = \frac{1}{u}\log_a e\frac{du}{dx}$$

$$\frac{d}{dx}\ln u = \frac{1}{u}\frac{du}{dx}$$

$$\frac{d}{dx}\sinh u = \cosh u\frac{du}{dx}$$

$$\frac{d}{dx}\cosh u = \sinh u\frac{du}{dx}$$

$$\frac{d}{dx}\tanh u = \text{sech}^2 u\frac{du}{dx}$$

$$\frac{d}{dx}\coth u = -\text{csch}^2 u\frac{du}{dx}$$

$$\frac{d}{dx}\text{sech } u = -\text{sech } u\,\tanh u\frac{du}{dx}$$

$$\frac{d}{dx}\text{csch } u = -\text{csch } u\,\coth u\frac{du}{dx}$$

$$\frac{d}{dx}\text{Sinh}^{-1}u = \frac{1}{\sqrt{u^2 + 1}}\frac{du}{dx}$$

$$\frac{d}{dx}\text{Cosh}^{-1}u = \frac{1}{\sqrt{u^2 - 1}}\frac{du}{dx}$$

$$\frac{d}{dx}\text{Tanh}^{-1}u = \frac{1}{1 - u^2}\frac{du}{dx}$$

$$\frac{d}{dx}\text{Coth}^{-1}u = \frac{1}{1 - u^2}\frac{du}{dx}$$

$$\frac{d}{dx}\text{Sech}^{-1}u = \frac{-1}{u\sqrt{1 - u^2}}\frac{du}{dx}$$

$$\frac{d}{dx}\text{Csch}^{-1}u = \frac{-1}{|u|\sqrt{1 + u^2}}\frac{du}{dx}$$